SPOKANE COUNTY LIBRARY DISTRICT

38th YEAR OF PUBLICATION

CHASE'S 1995
CALENDAR OF EVENTS

For Reference

Not to be taken from this room

CB
CONTEMPORARY
BOOKS

A TRIBUNE NEW MEDIA COMPANY

★ SPECIAL BONUS SECTION ★
SPOTLIGHT ON 1995 BANNER EVENTS

☆ Chase's 1995 Calendar of Events ☆

COPYRIGHT © 1994 BY CONTEMPORARY BOOKS, INC.

— WARNING —

THE INFORMATION CONTAINED IN *CHASE'S* HAS BEEN DEVELOPED, MAINTAINED AND UPDATED AT GREAT EXPENSE OF TIME AND MONEY AND IS PROPRIETARY TO CONTEMPORARY BOOKS, INC. NO PART OF THIS WORK MAY BE REPRODUCED OR COPIED IN ANY FORM OR BY ANY MEANS—INCLUDING GRAPHIC, ELECTRONIC, MECHANICAL, PHOTOCOPYING, OR INFORMATION STORAGE AND RETRIEVAL SYSTEMS—WITHOUT WRITTEN PERMISSION OF THE PUBLISHER. ANY SUCH REPRODUCTION OR COPYING WILL BE PROSECUTED TO THE FULLEST EXTENT UNDER COPYRIGHT LAW AND UNDER COMMON LAW.

FOR PERMISSION TO REPRODUCE OR COPY ANY PORTION OF THE WORK, CONTACT THE PUBLISHER:

> CONTEMPORARY BOOKS, INC.
> A TRIBUNE NEW MEDIA COMPANY
> 2 PRUDENTIAL PLAZA, SUITE 1200
> CHICAGO, ILLINOIS 60601
> FAX: (312) 540-4687
> PHONE: (312) 540-4500

Printed in USA

— NOTICE —

Events listed herein are not necessarily endorsed by the editors or publisher. Every effort has been made to assure the correctness of all entries, but neither the authors nor the publisher can warrant their accuracy. IT IS IMPERATIVE, IF FINANCIAL PLANS ARE TO BE MADE IN CONNECTION WITH DATES OR EVENTS LISTED HEREIN, THAT PRINCIPALS BE CONSULTED FOR FINAL INFORMATION.

ILLUSTRATIONS

Woodcuts at the beginning of each month are from *The Shepheards Calendar*, by Edmund Spenser, London, 1579. The woodcuts of people at various types of work are from a collection of drawings by Swiss painter and print maker Jost Amman, that was published in Leipzig in the 16th century. Many are from *The Book of Days*, by R. Chambers, London, 1864. Other engravings used in this book have been selected to represent almanac, advertising and book illustrations of the 15th-19th and early 20th centuries.

Chase's 1995 calendar of events,
Chicago, Contemporary Books, Inc. (c. 1994)

592 p., ill.
Includes index and tables.
Published annually since 1958.
Previous title: (1992–1994): Chase's Annual Events: The Day-by-Day Directory
(1984–1991): Chase's Annual Events
(1958–1983): Chases' Calendar of Annual Events
1. Calendars, 2. Almanacs, 3. Holidays, 4. Festivals, 5. Chronology, 6. Anniversaries, 7. Manners and customs, 8. Year-books. I. Title.

D11.5C48
LC 57-14540
ISSN: 0740-5286

ISBN: 0-8092-3634-6
529.3 (calendars)
394.26 (holidays)

☆ *Chase's 1995 Calendar of Events* ☆

TABLE OF CONTENTS

How to Use This Book	Inside front cover
Copyright and Cataloging Information	2
Welcome to *Chase's 1995*	4
Spotlight on 1995 Banner Events	5–48
Chronological Calendar of Events: Jan 1–Dec 31	49–487
Presidential Proclamations	488–490
A Word About Presidential Proclamations	490
The Congressional Process for Declaring Special Observances	491
Decade of the Brain Proclamation	491
National Education Goals	492
National Days of the World for 1995	493
Calendar Information for 1995	494
Calendar Information for 1996	495
Astronomical Phenomena 1995, 1996, 1997	496
Looking Forward	497
Selected Special Years	497
Space Objects Box Score	497
Universal, Standard and Daylight Times	498
Leap Seconds	498
World Map of Time Zones	499
Some Facts About the States	500–501
State & Territory Abbreviations: United States	501
Province & Territory Abbreviations: Canada	501
Some Facts About Canada	502
Some Facts About the Presidents	502–503
US Supreme Court Justices	504
State Governors and United States Senators	505
The Naming of Hurricanes	506–507
National Film Registry	508
ATAS Television Hall of Fame	509
Major Awards Presented in 1994	510–515

    Academy Awards (Oscars), Tony Awards, American Music Awards, Grammy Awards, Black Achievement Awards, Dove Awards, ACE Awards, PEN/Faulkner Award, Golden Globe Awards, George Polk Awards, American Library Association Awards for Children's Books, National Endowment for the Humanities Awards, NAB Crystal Awards, Pulitzer Prizes, Daytime Emmy Awards

Alphabetical Index	516–589
Anniversary Gifts	590
Order Form for *Chase's* Additional Copies	591
How to Submit a New Entry	592
Calendars 1994, 1995, 1996, 1997	Inside back cover

★ in text indicates Presidential Proclamations
◈ in text indicates that entry is related to material found in the Spotlight section

☆ Chase's 1995 Calendar of Events ☆

WELCOME TO CHASE'S 1995

Welcome to the 38th edition of *Chase's*. This edition includes more than 10,000 entries for events and observances in 1995. We've again included many entries concerning World War II battles and events of 50 years ago, as well as devoting a large portion of our Spotlight section to the war. By 1945 it seemed inevitable that the Allies would be successful; the troops would be coming home soon and life would return to normal. Few Americans imagined the Soviet dominance of Eastern Europe or the full unspeakable truth of Hitler's Final Solution and none but a few scientists could imagine the horrible new weapon that would be premiered in August. So while, yes, the war would be ended in 1945, there would be some surprises with far-reaching consequences—the cold war, the nuclear threat, the Middle-East conflict.

Other Spotlight-related entries have to do with important women, the UN's 50th anniversary, Earth Day, Frederick Douglass, the Montgomery Bus Boycott, Charlie Parker and the Great Bambino. These are marked with the symbol ◈.

As usual we have also included interesting local events from all across the country. The selection of these items reflects our editorial opinion as to the important, interesting and/or unusual events coming up in the new year. Choices are often difficult and at times involve lengthy discussion and/or research. Sometimes the name of a town is so unusual we'll include an otherwise pretty ordinary event. Sometimes we include very ordinary events because as Bill Chase (our originator) says, every event is important to the people holding it.

We have many new special observances, while some old ones have fallen away. We try to track down the significant ones, but the deadline eventually comes around. *Chase's* does not create observances of that nature but selects and reports them.

What's New in Chase's 1995?

We spent a lot of time and energy this year to add some identifying information on the people listed under "Birthdays Today." This was (and is since it's still ongoing) an enormous task, much more time-consuming than we anticipated, but it's something many of you have requested. If you have suggestions on this, please write or fax us. (Be aware our current database program allows only limited space in the pertinent field—and we don't intend to be a volume of biographies.) In addition, you may have noticed we've changed our title again.

Types of Events in Chase's

PRESIDENTIAL PROCLAMATIONS: In addition to the complete list of proclamations issued January 1, 1993–June 30, 1994, we have included in the day-by-day directory special observance proclamations that have continuing authority and a clear formula for calculating dates of observance. We consult the White House on this information each year.

NATIONAL DAYS AND STATE DAYS: National Days and public holidays of other nations are gleaned from United Nations documents and from information we obtain through letters and phone calls to embassies or other agencies of the individual countries. For state days and statutory observances we consult governors' offices or other official state agencies.

SPONSORED EVENTS: Events for which there is individual or organizational sponsorship are listed with the name of the event, inclusive dates and place of observance (if local or regional), brief description, approximate attendance when provided and the sponsor's name and address, plus a phone number if sponsor agrees. We obtain information for these from a variety of sources, often involving a lot of correspondence and/or numerous phone calls (and the persistent and incredible tracking skills of Assistant Editor Beth Johnson).

ASTRONOMICAL PHENOMENA: Information about eclipses, equinoxes and solstices, moon phases, and other astronomically related information is calculated largely from data prepared by the US Naval Observatory's Nautical Almanac Office and Her Majesty's Nautical Almanac Office (Royal Greenwich Observatory), as well as some other sources.

HISTORIC ANNIVERSARIES, FOLKLORIC EVENTS AND BIRTHDAYS: Compiled from a wide variety of materials dating from the 15th century to the present, and covering events in virtually every part of the world, these entries are supported whenever possible by two or more independent sources. Usually, living persons are listed under "Birthdays Today."

RELIGIOUS OBSERVANCES: Principal observances of the Christian, Jewish, Hindu and Baha'i faiths are presented with available background information from their respective calendars. Where known, the special observances of the Orient, religious and secular, including traditional Chinese and Japanese events, also are listed. We include dates for Muslim holidays based on what appears to us to be the most widely used method of determining them. Different groups have differing views on calculating them—including whether they should be calculated at all—and we have not found a sensible way to reflect them all, while we do wish to include them so that concerned individuals will at least be alerted when they are coming up.

Omissions/"Errors"

The omission of an event usually means that the dates were set too late for inclusion. Errors in dates are most often the result of inaccurate or (unknown to us) tentative information which is later changed by the sponsoring organization.

We welcome the submission of information for our consideration. Full instructions for this are provided at the end of the book on page 592. Final selection and format of information included in *Chase's* is, of course, the decision of the editors.

Acknowledgments

Our thanks to the many people who help us, particularly the many library personnel; Bob Sheets at the National Hurricane Center; Stanley Elswick at the National Oceanic and Atmospheric Administration Library; Dr Brad Schaefer; the people involved with the major awards; the helpful people at the Pentagon and the White House Clerk's Office.

And thanks also to those enthusiastic persons—Dawn Barker, Craig Bolt, Cheri Chenoweth, Elena Delaney, Louise El, Gigi Grajdura, Gerilee Hundt, Grant Keiser, Audrey Sails, Kathy Willhoite—who helped in a variety of ways with the work involved in compiling *Chase's*. Also, special thanks to computer ace Martha Best, artist Jan Geist, expert typesetter Terry Stone and the rest of Contemporary's production team—Ellen Kollmon, Dana Draxten, Sue Springston, Fran Westbrook and Mary Viglione—for their fine work. And, finally, to Jon and David Eley.

And now we invite our readers to join with us in the celebration of the coming year.

September 1994

The Editorial Staff
Mary M. Eley, Editor
Beth Johnson, Assistant Editor
Warner Crocker, Contributing Editor

☆ Chase's 1995 Calendar of Events ☆

SPOTLIGHT
BANNER 1995 EVENTS

The following section focuses on some of the major events and milestone anniversary observances taking place in 1995. It provides background information and lists addresses for further information where appropriate. Related commemorative events, exhibits and historical entries have been included in the Chronological Calendar of Events section of *Chase's* and are indicated there by the symbol ◈. The subjects we elaborate on in this Spotlight section were chosen and developed on the basis of what we felt would be of importance and special interest to our readers. Not every subject could be included and deciding which subjects to present and which material to include within those subjects chosen was a difficult process, with a little fun and education thrown in for good measure. (We each added a number of subjects to our "sometime lists" for further reading, and we hope our readers also will be prompted to visit their local libraries to find further reading material on subjects that pique their interest.)

SPOTLIGHT CONTENTS

Mayflower to America • 375 Years 6
Crispus Attucks and the Boston Massacre • 225 Years 7
William Blackstone • Birth • 400 Years 7
Ludwig van Beethoven • Birth • 225 Years 8
Béla Bartók • Death • 50 Years 9
Paul Hindemith • Birth • 100 Years 9
James Knox Polk • Birth • 200 Years 10
State Birthdays: Maine • Florida • Texas 11–13
"America the Beautiful" • 100 Years 14
Florence Nightingale • Birth • 175 Years 15
Robert E. Lee • Death • 125 Years 16–17
Willam T. Sherman • Birth • 175 Years 17
15th Amendment • 125 Years 18
1965 Voting Rights Act • 30 Years 18
Dept of Justice • 125 Years 19
Frederick Douglass • Death • 100 Years 19
Women's Suffrage • 75th Anniversary 20–21
Susan B. Anthony • Birth • 175 Years 21
League of Women Voters • Founded • 75 Years 22
Women's Rights National Historical Park • Established • 15 Years 22
Elizabeth's List • Sisters of Distinction 23
Jonathan Swift • Death • 250 Years 24
Charles Dickens • Death • 125 Years 25
John Keats • Birth • 200 Years 26
Wilhelm Roentgen: Birth • 150 Years • Discovery of X-Rays • 100 Years 27
Rudolph Valentino • Birth • 100 Years 28
New York Public Library • Established • 100 Years 28
Busby Berkeley • Birth • 100 Years 29
John Ford • Birth • 100 Years 29
Babe Ruth • Birth • 100 Years 30
Jack Dempsey • Birth • 100 Years 31
Anthracite Miner's Memorial • Unveiling 31
Americans at War 1945 • World War II • 50 Years 32–41
United Nations: Special Activities • 50th Anniversary Celebration 42–43
World War II • 50th Anniversary Commemoration 44
Charlie Parker • Birth • 75 Years 45
Rosa Parks Says No • 40 Years 45
1970: Year to Remember • 25 Years 46–47
Other 1995 Milestones 47
Environmental Milestones 48

Mayflower to America
SEPT 16 – DEC 21 • 375 YEARS

Faced with religious intolerance, a group of separatists from the Church of England left England in search of a home where they could live, meet and practice their beliefs as they saw fit. To escape persecution, they first sought refuge in the Netherlands, in the town of Leiden, in 1608. Because as aliens they could not join the Dutch guilds, they were relegated to poorly paid, unskilled jobs; this fact, along with a fear that their children would lose their heritage and language, led some of the immigrants to determine to move again. This time they would start a new life in the New World.

PLANS FOR THE NEW WORLD

After three years of planning and preparation, they obtained a patent from the Virginia Company of London to settle as an independent community on its land. In exchange for a share of the colonists' crops during their first six years in America, the English merchants provided financing for the venture. When the separatists requested from King James I a guarantee of religious freedom, he refused but put forth "that he would . . . not molest them, provided they carried themselves peaceably"—a historic concession that opened English America to settlement by dissenting Protestants.

A SHAKY START

The separatists sailed from Leiden to England in a small ship called the *Speedwell* and joined the *Mayflower*, a merchant vessel, in Southampton harbor. The expedition twice set sail and twice was forced to return because the *Speedwell* sprang leaks. Finally, those in the Leiden group who wished to continue abandoned the smaller ship and joined other separatists and employees of the merchant sponsors on the *Mayflower*. Under the command of Capt Christopher Jones, the ship set sail for Virginia from Plymouth on Sept 16, 1620—just as the North Atlantic's stormy season was approaching.

A STORMY VOYAGE

The stormy conditions and crowded quarters made for an unpleasant voyage. With 102 passengers, privacy and sanitary conditions were nonexistent. Cold food was the norm and scurvy was a common affliction. The rough waters sent sea spray through seams where caulking had loosened, and the main beam buckled and had to be repaired midcourse. Because of harsh winds and navigational errors, when the immigrants finally landed it was at a site more than 500 miles northeast of their original destination. Rounding Cape Cod on Nov 21, the *Mayflower* dropped anchor off the coast of what today is Provincetown, MA.

THE MAYFLOWER COMPACT

We now know the *Mayflower* passengers as the Pilgrims, but it was a disparate group that found itself far from where it had been authorized to settle, and some of them wanted to go their own way and to settle on their own. To prevent this, 41 passengers, including John Alden, Myles Standish (both nonseparatists), William Bradford and William Brewster, convened in the ship's cabin on Nov 21 and drew up a document called the Mayflower Compact. It formed the colonists into a civil body politic for the purpose of establishing a majority-rule form of government, and it bound the signers to abide by the ordinances that would be enacted later. Every adult male was required to sign the covenant, which was the first constitution written in America.

PLYMOUTH COLONY

The immigrants explored the land for about a month—during which the first European child born in New England, Peregrine White, was born aboard the ship. They finally settled on a site for their colony and disembarked on Dec 21, 1620, and founded Plymouth, the first permanent settlement in New England. Plymouth Colony remained a virtually independent republic governed by the Mayflower Compact for over 70 years, until it was annexed to the much larger Massachusetts Bay Colony in 1691.

The *Mayflower* itself remained with the colonists throughout the Pilgrims' long and difficult first winter and then returned to England in the spring of 1621. There is no record of the ship's exploits after its historic voyage to Plymouth.

FOR INFO

The Plymouth County Development council is planning several events to mark the historic landing. The target date for most of these events, including a reenactment of the landing of the *Mayflower*, is July 1995. For further information, including specific dates and times, contact the Plymouth Development Council at (800) 231-1620 or (617) 826-3136.

CRISPUS ATTUCKS
AND THE BOSTON MASSACRE
MAR 5 • 225 YEARS

The cyclical history of revolutions and political uprisings usually begins with the bold acts and arrogant attitudes of individuals placed in the path of change. The political foment that led to the American Revolution of 1776 featured such a beginning with the Boston Massacre on Mar 5, 1770. Boston citizens had been demonstrating against the imposition of duties on imports to New World colonies prescribed by the Townshend Acts. To quell the demonstrations, British authorities had quartered troops in the city. The troops were constantly harassed by the citizenry.

TANGLIN' WITH THE TROOPS

Toward evening on Mar 5, 1770, Crispus Attucks led a large group, consisting mostly of sailors, from Dock Square to the British garrison on King Street. The crowd hurled taunts and insults as well as rocks and other objects at a group of British soldiers there. The violent actions that occurred next were disputed, but the outcome was not. Attucks allegedly grabbed the bayonet of a British soldier after trying to hit British captain John Preston with a long stick. (This is the account that John Adams, who later became a US president, gave in his defense of the British soldiers.) The soldiers fired into the crowd, killing five people. Attucks was the first to fall and died almost instantly.

The eight soldiers and Preston were arrested and brought to trial for murder. Defended by Adams and Josia Quincy, two of the accused were convicted of manslaughter and the others were acquitted. The two convicted soldiers were punished by being branded on the thumb.

PUTTIN' ON THE SPIN

The event did not end with the trial and its verdict. Seizing the moment with oratorical zeal, Samuel Adams, an eloquent and passionate opponent of British tyranny, disallowed the suggestion that Attucks had provoked the attack. Through pamphlets, newspapers and public pronouncements he decried the "massacre." The Boston Massacre was the first of many actions to fuel the flames of political discontent that eventually led to the American Revolution.

MYSTERY MARTYR

Crispus Attucks's body lay in state for three days at Faneuil Hall. He and the others slain were then buried in a common grave.

Although Attucks became widely known posthumously as the first to fall in the revolutionary cause, very little is known about his life. One account described him as a very large man "whose looks was enough to terrify." Most historians agree that Crispus Attucks was a black man, but some believe that he was an American Indian or that his heritage included both races. Records exist of a 1750 advertisement seeking the return of a runaway slave named Crispus, believed to be the same man. Historians speculate that he spent the 20 years between his escape and the massacre serving aboard whaling ships sailing out of the New England area.

Crispus Attucks's ultimate sacrifice to the cause of freedom did not go unnoticed. In 1888 a monument to his martyrdom by sculptor Augustus Saint-Gaudens was erected in Boston Common. May 5 is remembered both as the anniversary of the Boston Massacre and as Crispus Attucks Day.

FOR INFO

Bostonian Society
Phone: (617) 720-1713

WILLIAM BLACKSTONE • BIRTH • MAR 5 • 400 YEARS

Born at Gibside, England, on Mar 5, 1595, William Blackstone, an ordained minister in the Church of England, came to New England with the Capt Robert Gorges expedition in 1623. The colony failed and most returned, but Blackstone stayed and settled in what is now the Beacon Hill section of Boston, MA. He sold most of that property in 1634 and moved on to further expore the unspoiled wilderness, eventually settling on the shores of the (now) Blackstone River in (now) Rhode Island. In 1635 Roger Williams was banished from the Massachusetts Bay Colony and with a group of exiled Puritans settled at Providence and gradually others settled in the area. Blackstone preached the first Anglican sermon here in (now) Wickford, RI. He died on May 26, 1675, at what is now Cumberland, RI. The Blackstone River and the Blackstone Valley are named for him.

LUDWIG VAN BEETHOVEN
BIRTH • DEC 16 • 225 YEARS

If J. S. Bach was the culmination of the Baroque musical tradition and Mozart the sublime exponent of the classical period that followed, then Ludwig van Beethoven was the great bridge from the ordered world of classicism to the emotive power of Romanticism. Beethoven's influence reached beyond music to dominate the world of the arts as a whole.

EARLY LIFE

Ludwig van Beethoven was born at Bonn, Germany (though of Dutch descent), on Dec 16, 1770. His father was a court musician, though not a particularly good one. Growing up in a musical environment in which he learned to play piano, violin and French horn and became a court organist before the age of 12, Ludwig showed early evidence of the musical gifts that would win the praise of both Mozart and Haydn. His first teacher, C. G. Neefe, a man of modest abilities, did give him an excellent grounding in music. But Beethoven's father, besides being a mediocre musician, also proved to be deficient as a parent, and the young Beethoven found himself having to take over the responsibilities of maintaining his family when his father lost his job.

TO VIENNA

In 1792 a 22-year-old Beethoven moved to Vienna, Austria, where he was able to study with the renowned Haydn. While he had great respect and affection for "Papa Haydn," he found he needed more demanding instruction and so moved on to study with J. G. Albrechtsberger and Antonio Salieri (whose portrayal as a jealous schemer in the musical *Amadeus* has no historical basis).

Beethoven first gained fame in Vienna as a concert pianist and made the acquaintance of many important figures in Austrian society. Yet due to the growth of the music publishing business, he became one of the first composers to be able to support himself working on a freelance basis rather than being in the employ of a particular noble or church official.

HEARING LOSS

In 1796, when he was still under 30, Beethoven experienced the first signs of the hearing loss that would become complete deafness. In 1802, having become convinced the loss was worsening and irreversible, he expressed to his brothers the depth of his despair and embarrassment in a letter that has come to be called the "Heiligenstadt Testament." In the last 10 years of his life the deafness was total, and people had to communicate with him by writing in so-called "conversation books," yet during this time he composed the extraordinary and expressive *Ninth Symphony* and *Missa Solemnis* as well as many sonatas and chamber pieces.

THE MUSIC

Beethoven composed music in all the major musical forms of his time. Many musicologists break his work into three periods. The early music (1794-1800) is clearly influenced by Mozart and Haydn, as is evidenced by his first two piano concertos and chamber music of that era. During his middle period (1801-14) he expanded the musical forms he was working in, incorporating more improvisation and much greater emotive power. And in his last years (1814-27) he composed his greatest works, even though he was almost totally deaf and had to rely almost completely on his own inner resources. The music of this period is astounding in its depth and originality. For example, the *Ninth Symphony* is marked by its long developmental sections brimming with new musical ideas, any one of which most lesser composers would have based whole symphonies on. The symphony ends with a climactic choral piece, the famous "Ode to Joy," which seems to lift the listener out of his ordinary self and transport him to a higher plane of existence.

Beethoven's works for voice were equally innovative. He was one of the first to write *Lieder*—songs in which music and poetry were melded. Beethoven's one opera, *Fidelio*, combined the musical and dramatic elements in a powerful whole that foreshadowed the works of Wagner.

The influence of Beethoven on European music was all-encompassing. During his lifetime and afterward composers were influenced by his work. For example, before Beethoven composers wrote symphonies by the dozens; Mozart wrote over 30 and Haydn more than 100. Beethoven wrote only nine, and most composers following him wrote fewer still. The reason for this is that Beethoven redefined the very nature of the symphony. One could say that the symphony of the classical era was more like a short story while the symphonies of Beethoven were like novels filled with the strong emotions of life and death. Beethoven's music drove the Romantic era that followed him by giving voice to the mystical and emotive forces that were at work not only in music but also in art and literature. Even today Beethoven's music has great power and influence, for our modern world is struggling with the forces of order and chaos, rationality and emotion, which are at the very core of Beethoven's work.

DEATH AND LEGACY

Ludwig van Beethoven died at Vienna on Mar 26, 1827. Although his music had not been in public popularity for some time, some 20,000 people attended his funeral and tens of thousands of people lined the route of the funeral procession. Besides Beethoven's actual compositions, he left another priceless legacy: more than 7,000 pages of his notes and drafts for compositions that he made over the years, a treasury of his genius.

BÉLA BARTÓK
DEATH • SEPT 26 • 50 YEARS

Composer and ethnomusicologist Béla Bartók was born at Nagyszentmiklós, Hungary (now Sinnicolau Mare, Romania), on Mar 25, 1881. During a childhood spent in a number of provincial towns he became familiar with the peasant music that would have such a profound influence on his studies and his musical compositions. In 1899 Bartók entered the Royal Hungarian Academy of Music in Budapest. He was appointed to the faculty of the academy as professor of pianoforte in 1907 and remained there until 1934.

In 1903 Bartók began researching authentic Hungarian peasant music when he discovered that the source of the Hungarian folk music upon which he had based many of his compositions was actually the less-than-authentic efforts of city-dwelling Gypsies. He and fellow composer Zoltán Kodály spent their vacations traveling to remote areas collecting and transcribing authentic Hungarian folk music. These elements increasingly permeated their compositions.

Bartók conducted two concert tours in the US and emigrated there in 1940 when it appeared that Hungary would soon form an alliance with Nazi Germany. He was appointed to a research assistant position in music at Columbia University, where he transcribed and edited a collection of Serbo-Croatian women's songs. He died at New York City on Sept 26, 1945.

Bartók's compositions—including several stage works, a cantata, orchestral works, string quartets, piano solos and many folk songs for piano and voice—are considered to be among the classics of Western music. Bartók's most famous compositions include the sonata for two pianos and percussion (1937); the opera, *Bluebeard's Castle* (1911); *The Wooden Prince* (1914–1916); *The Miraculous Mandarin* (1919); *Music for Strings, Percussion and Celeste* (1936); *Divertissement for Strings* (1939) and *Concerto for Orchestra* (1943). His work in the field of music ethnology (the study of the music of a specific race or culture) was as important as his accomplishments as a composer. He published several invaluable studies of Hungarian and Romanian folk music during his lifetime. In addition, a first volume on Slovakian folk music (1959) and a three-volume study of Romanian folk music (1967) were published after his death.

PAUL HINDEMITH
BIRTH • NOV 16, 1895 • 100 YEARS

Paul Hindemith, the leading German composer of the first half of the 20th century, was born at Hanau, Germany, on Nov 16, 1895. As a young man Hindemith earned his living by playing the violin in dance bands and in theaters and cafes. He studied music in Frankfurt and at age 20 became the leader of the Frankfurt Opera Orchestra. Undoubtedly influenced by a functional trend in postwar German culture, he came to regard himself as a craftsman rather than an artist, turning out music to meet social needs. He was a pioneer of *Gebrauchsmusik*, or "utility music" and of *Gemeinschaftmusik* or "music for amateurs." He believed that the composition of music was a craft that could be applied in a utilitarian manner for social purposes either through education or diversion. He rebelled against the notion that musical composition should dwell only in the realm of the knowledgable elite. His works included pieces for youth groups, brass bands, children's games and radio plays. He made his mark as a composer in the early 1920s when the debut of his first opera, *Sancta Suzanna*, resulted in the police being summoned to disperse the crowd reacting to his "scandalous" approach to composition.

In 1934 his greatest work, the opera *Mathis der Maler*, which was about painter Mathias Grünewald's struggles with society, was banned by the Nazi cultural authorities. Joseph Goebbels denounced Hindemith as a "cultural Bolshevist" and "spiritual non-Aryan." He left his native land for Turkey, where he taught at the conservatory in Ankara from 1935 to 1937. He also taught at Yale University (1940-53) and at the University of Zurich (1951-58). He died on Dec 28, 1963, at Frankfurt am Main, Germany.

Hindemith's other well-known compositions include the operas *Cardillac* (1926) and *The Harmony of the World* (1957); the ballet *Noblissima Visione* (1938); *Konzertmusik* (1939); *Symphonic Metamorphoses on themes by C. M. von Weber* (1943) and *The Four Temperaments* (1940).

JAMES KNOX POLK
EXPANSIONIST PRESIDENT
BIRTH • NOV 2 • 200 YEARS

In today's environment of under-the-microscope examination of political candidates, it is difficult to imagine an earlier political era when candidates were not as well known to the general public. The Democratic convention of 1844 found the party split over the choice of a presidential nominee between prominent politicians Martin Van Buren, Lewis Cass and James Buchanan. Seeking a compromise candidate, the Democrats made James Knox Polk the first "dark horse" candidate. In a three-party race that saw the Liberty Party pull votes away from Whig candidate Henry Clay, Polk won by a margin of approximately 38,000 votes. Who was this new president? Even today many know little of the 11th President of the United States.

ROOTS IN THE HOUSE

James Knox Polk was born Nov 2, 1795, at Mecklenburg, NC. The eldest of 10 children, he spent his youth devoted to education, graduating with honors from the University of North Carolina in 1818. Practicing law in Columbia, TN, he quickly became a leading advocate known for his oratorical skills. His first foray into politics came in 1823 when he was elected to the Tennessee legislature. Then in 1825 he ran for the US House of Representatives and served there for 14 years. As an ally and supporter of Andrew Jackson, he was pitted against the politics of then-president John Quincy Adams and his eventual presidential opponent Henry Clay. When Jackson won the presidency, Polk shepherded Jackson's policies through the often contentious House. He served as the Speaker of the House during the 24th and 25th Congresses, from December 1835 through March 1839.

During his career as a legislator, Polk acquired a reputation as a diligent and fair public servant. Guided by his integrity, he negotiated many divisive procedural issues, with the result usually being supported by the leaders of both parties. At the end of each legislative session, he and his wife returned to Tennessee, where he continued his law practice until the next session. He left the House in 1839 to run for governor of Tennessee and served one term, being defeated twice (in 1841 and 1843) for a second term.

COMPROMISE CANDIDATE

Although the Democrats had planned to nominate Polk for vice president in 1844, his reputation took him one step further when the party could not settle a dispute over who should be the presidential nominee. As a compromise candidate he surprised everyone by taking aggressive positions on two of the burning political issues of the day. He advocated the annexation of Texas into the Union and insisted that there be no joint occupancy of the Oregon territory with England. These two issues would become centerpieces of his tenure in office. To those two campaign planks he added three more goals that would occupy his administration: the acquisition of California, the reduction of tariffs and the establishment of an independent treasury. Polk also announced at the outset of his term of office that he would not seek a second term, which allowed him a measure of independence in his attempts to achieve his goals.

PUSHING TO THE PACIFIC

Achieve he did. In 1846, the Walker Tariff Act lowered tariffs and provided the still young country with its first efforts at a free trade policy. In that same year he established the national treasury. That financial system, sparked by his efforts and vision, remains largely untouched today. His goals for territorial expansion reshaped the map of the US into the continental entity it is today. But his efforts resulted in considerable diplomatic wrangling and even war. His successful Texas annexation policy led to an unpopular two-year war with Mexico. The US's victory brought (both by conquest and by purchase) more than 522,000 square miles under its domain in the Southwest and far West and fulfilled Polk's goal of acquiring California. During his campaign, Polk had stood out front in the dispute between Britain and the US over the boundary of the Oregon territory. His party's slogan "Fifty-four Forty or Fight!" seemed to beg for another war with England. Polk offered a compromise that dropped the border south from a latitude of 54° 40' to 49°. It was rejected by the British. He then instructed his Secretary of State, James Buchanan, to claim the entire territory. Howls of protest from within and without his administration followed the bold move as even Buchanan predicted war would result. Polk held fast, and the British capitulated and accepted his initial offer.

This agenda of expansion led to another of Polk's achievements, the creation of the Department of the Interior to manage the new acquisitions. His administration also created the US Naval Academy and authorized the creation of the Smithsonian Institution.

BROAD VISION

Polk's achievements lay largely in his ability to work with Congress and convince them that national interests superseded parochial ones. His strength of character brought dignity to the office that saw its antechambers filled each day with office seekers. He addressed his work each day with a diligent and tireless approach. The burdens of guiding the still fledgling country exhausted him, however, and three months after his term of office ended he died at his home in Nashville, on June 15, 1849.

STATE BIRTHDAYS
MAINE • FLORIDA • TEXAS

1995 marks milestone anniversaries of statehood for three states, each situated along an extreme border of the continental US. Florida and Texas celebrate their 150th and Maine celebrates its 175th. The history and development of these states are as different as their geographical locations.

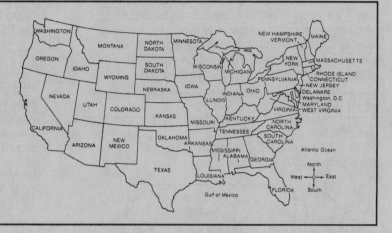

MAINE
MAR 15, 1820 • 175 YEARS

Buffeted by the waves of the North Atlantic, Maine's rocky shore lies precipitously along the northeasternmost corner of the US. Famous for its coastline, its lobster industry and its large coniferous forests that gave the state its nickname as the Pine Tree State, Maine will celebrate its 175th anniversary of statehood on Mar 15, 1995.

The state's early years under colonial rule found it more a haven for fur trapping and the timber industry than a target for large settlements. Prior to the arrival of the Europeans the land was inhabited by more than 20 related Algonquian tribes that were united in a loose confederation known as the Abnaki or Wabanaki. Early attempts by the English and the French to establish settlements failed, and in 1658 Massachusetts claimed jurisdiction over the area. Although the population remained small (numbering about one fourth the size of Massachusetts), the separatist movement began as early as 1785 and gained popularity in 1816 with the Brunswick Convention.

THE 23RD STATE

A population boom spurred by the lumber and shipbuilding industries increased the populace to a size more than half that of its mother state. At that point the Massachusetts General Court agreed to a separation, after which a constitutional convention was held in Portland and a petition for statehood was submitted in December 1819. Under the Missouri Compromise, statehood was granted, and Maine entered the Union the following year.

Maine first entered the political and social consciousness of the country soon after achieving statehood. It was a leader in the temperance movement in the 1820s, later passing an alcohol prohibition law in 1851.

Maine was also known for an ongoing dispute with Canada over the border, which escalated in the years after Maine became a state. Various lumber interests coveted the rich harvest from the Maine forests, and the dispute grew into the so-called Aroostook War. Interestingly enough, without one shot being fired, the Aroostook War ended in 1842 with the Webster-Ashburton Treaty establishing the border by dividing the disputed area almost equally between the two complainants.

Maine's economy grew on the strength of the lumber, shipbuilding and fishing industries and also as a provider of ice for food packing and as a burgeoning center for textile manufacturing. But following the Civil War, shipbuilding shifted to the more durable steel-hulled craft, and the textile industry began its gradual move out of New England, causing a shift in Maine's industrial base to the paper and pulp industries. Tourism, which is now one of the leading industries in the state, first began to make its mark on the economy in the 1880s.

The geography of Maine has had a major impact on the state's history and economic development. Maine is the only state in the US that shares a border with only

one other state—New Hampshire. In addition to its Atlantic coastline, the state is crisscrossed by more than 5,100 rivers and streams and dotted with more than 2,200 lakes and ponds. Forests blanket more than 80% of the state and contain many species of pine as well as hardwoods such as black cherry, white ash, oak, maple and birch. The access to water and the vast amounts of lumber led to development of fishing, shipbuilding, lumbering and paper manufacturing industries, and those same natural resources attract the many tourists who journey northeastward to visit the "Down-Easters" of the most sparsely populated state east of the Mississippi.

FOR INFO

Maine Tourism Information
Out-of-state phone: (800) 533-9595
Maine phone: (207) 623-0363

FLORIDA
MAR 3, 1845 • 150 YEARS

The image of the Sunshine State, with its beaches, citrus groves and semitropical climate, has long made Florida a haven for tourists, and tourism is one of its prime industries. In the 1920s fewer than 1 million people lived on the peninsula, but in the latter half of this century Florida has become one of the fastest-growing states, with an average of more than 5,000 people moving there each week. Transplanted northerners, many of whom are senior citizens, account for a large percentage of those seeking a warm-weather lifestyle, and their migration has led to the rapid expansion of Miami, Jacksonville, Orlando, Tampa Bay and St. Petersburg among others.

The home of the Everglades, the Kennedy Space Center and Mickey Mouse and his compatriots at Disney World has a fascinating history as well. Prior to being claimed for Spain in 1513 by Juan Ponce de León, the area was populated by various native tribes, including the Timucua, the Apalachee and the Calusa. These tribes later came under the influence of the Seminoles. Ponce de León's search for the mythical fountain of youth was followed by Pánfilo de Narváez, who first sailed into Tampa Bay in 1528. After he and his men were lost at sea attempting to reach Mexico, King Charles V (as Charles I of Spain) then appointed Hernando de Soto governor of Spain's new province. The Spanish were not the only Old World settlers to land on the beaches of Florida. Fleeing religious persecution, a group of French Huguenots, led by Jean Ribaut, landed at the mouth of the St. Johns River. Later a settlement named Fort Caroline was built near the river's mouth. In 1565 it was conquered by Pedro Menéndez de Avilés, who renamed it San Mateo. De Avilés had earlier founded the settlement of St. Augustine, the oldest continually inhabited city in the US.

CHANGING HANDS

Following, if not setting, the pattern of disputing ownership over land between Old World empires, the Spanish and the English feuded over the territory until the end of the Seven Years' War in 1763. By treaty, Florida was given to England in exchange for Havana. British dominion lasted until the American Revolution, when Spain declared war on England. The 1783 Treaty of Paris returned Florida to the Spanish. The newly formed United States claimed western Florida as a part of the Louisiana Purchase of 1803, but Spain ignored this claim and boundaries remained in dispute through the War of 1812. After the war US troops invaded, attempting to capture escaped slaves, and met fierce resistance from the Seminoles. The fighting continued until US forces led by Andrew Jackson defeated the Seminoles and captured Pensacola in 1818. Spain relinquished rights to Florida in 1819 with the Adams-Onís Treaty, and in exchange the US gave up its claims to Texas. Florida was organized as a territory in 1822 though the remaining Seminoles continued to resist. Seven years of warfare, from 1835 to 1842, ended only when the Seminoles were forced to move to the Oklahoma territory.

THE 27TH STATE

Florida was admitted to the Union on March 3, 1845, as the 27th state. It seceded from the Union and joined the Confederacy in 1861. Later, during Reconstruction, the state was under federal military control. Florida was readmitted to the union in 1868, after adopting a new constitution that affirmed the abolition of slavery and the right of black suffrage.

FOR INFO

Florida Division of Tourism
(904) 487-1462

TEXAS
DEC 29, 1845 • 150 YEARS

The Lone Star State has flown the flags of six nations during its history. Spain, France, Spain again, Mexico, the Republic of Texas, the United States, the Confederate States and the United States again all have declared dominion over the vast territory. Texas's tumultuous history is filled with a rough frontier spirit that is richly colored by the Alamo, cattle drives, Texas Rangers, big oil, rowdy politicians and bold businessmen. Covering more area than Illinois, Indiana, Michigan, Ohio and Wisconsin combined, Texas sprawls across the center of the continental US's southern border, second only to Alaska in size and, along with Hawaii, one of two states that were once recognized as independent countries. Its big size is matched only by the big dreams and bold talk of its people. The name Texas comes from a Caddo Indian word meaning friends, and the Texas state motto is "Friendship." The social, cultural and political life is influenced by a large Spanish-speaking population.

Early residents included Indians from the Coahuiltecan, Karankawa, Caddo, Jumano, Apache and Tonkawa tribes. In later years migrating Cherokee, Comanche and Wichita tribes were stirred into the mix. As in Florida, the first old-world explorers into Texas were the Spanish. Álvar Núñez Cabeza de Vaca, Francisco Vásquez de Coronado and Luís Moscoso all led expeditions through the Texas terrain. None of the expeditions produced gold or treasure, and Spain focused its exploring efforts elsewhere until the French made a claim on the land in 1684. On the shores of Matagorda Bay, Robert Cavelier, sieur de La Salle founded Fort Saint Louis, prompting the Spanish to found missions in east Texas beginning in 1690. Although both countries laid claim to areas of Texas, the area remained only sparsely settled by Europeans for almost three centuries.

The Americans first began taking an interest in Texas at the turn of the 19th century. After the Louisiana Purchase in 1803, the US recognized east Texas as its territory, though naturally the Spanish did not validate that claim. The US relinquished its interest by swapping Texas for Florida in the Treaty of Adams-Onís in 1819. With Mexico's independence from Spain, Texas became a Mexican state, and for a brief time immigration was encouraged. Stephen Austin, among others, began establishing settlements in Texas in the 1820s on land granted by Mexico. Austin and those like him were called *empresarios*, Spanish for colonizers. Around this time the US began to take a greater interest and offered to purchase the territory but was rebuffed. Tension between the new American immigrants and the Mexican government increased in 1826 when a frustrated empresario tried to set up the independent Republic of Fredonia in east Texas. This led to a crackdown on immigration policies that also prohibited the importation of African slaves. When Mexican President Antonio López de Santa Anna set aside the Mexican constitution and took power in 1835, the Texans rebelled against the new dictator and the Texas revolution was on.

THE ALAMO AND INDEPENDENCE

The Texan leaders proclaimed their loyalty to the Mexican constitution and demanded Santa Anna step down. They faired well in the early battles driving Santa Anna's army south of the Rio Grande. But later, Santa Anna marched northward with a larger army and defeated the Texans at the battle of the Alamo in San Antonio, as well as in other battles across south Texas. The Alamo and the massacre of 300 Texas prisoners at Goliad became rallying cries for Texans. While the fighting waged in the south,

the politicians convened in the north at Washington on the Brazos. On Mar 2, 1836, they adopted a declaration of independence and a constitution based on the US model and declared themselves the Republic of Texas. Led by General Sam Houston, the Texans won their independence on Apr 21, 1836, when they defeated Santa Anna on the banks of the San Jacinto River near Houston. Houston was elected the republic's first president.

THE 28TH STATE

The Republic of Texas existed for the next nine years but not without conflict. Houston worked toward a policy of annexation by the US, but his successor, Mirabeau Lamar, tilted policy in the opposite direction, wanting to remain independent. Although Texas won recognition from the US, Great Britain and France, the young republic still reeled from the problems of debt and insecure borders. Lamar found himself in the minority, and although most Texans favored annexation, many in the US did not favor the admission of another slave state. Finally, under President Polk's guidance, Texas was admitted to the Union on Dec 29, 1845, and the final transfer of authority took place on Feb 19, 1846. As predicted by several of the annexation foes, war soon broke out between the US and Mexico and lasted for two years. Its conclusion established the Rio Grande as the border. The US later purchased more than 100,000 square miles from Texas, an area which became parts of the states of Colorado, Kansas, New Mexico, Oklahoma and Wyoming.

With the coming of the Civil War, Texas sided with the South against the wishes of then Governor Sam Houston, who was deposed for his efforts to keep Texas in the Union. Cotton was king. And it remained so in the years following Reconstruction, although cattle also became a major source of profit for the state's growing economy. The population increased significantly prior to 1900 and soared again after the discovery of oil (especially in the East Texas oil field in 1930) changed the economy of Texas dramatically. Oil led to industrialization, urbanization and rapid population growth.

FOR INFO

Texas Department of Transportation
Texas Travel Information
Phone: (800) 452-9292

For Patriot Dream

"America the Beautiful"
July 4, 1895 • 100 Years

Patriotic music has always stirred the emotions. It becomes so ingrained in the fabric of society and the individual that the first few familiar strains strike chords of recognition that resonate with feelings of pride and belonging. A stirring example of that quality, "America the Beautiful," the unofficial but indisputable national hymn of the United States, was first published 100 years ago on July 4, 1895.

THE WORDS

Katharine Lee Bates was born on Aug 12, 1859, at Falmouth, MA. She was educated at Wellesley College and returned there in 1885 to teach English; she was made a professor in 1891. While on a tour of the western United States in 1893, Bates climbed Pike's Peak. After viewing the magnificent, awesome landscape below her, she expressed her feelings in a poem entitled "America the Beautiful." The poem was first published in The Congregationalist, a church publication, on July 4, 1895. It was published again in a revised edition in 1904 and in its final form in 1911 in her collection America the Beautiful and Other Poems.

THE MUSIC

There were several attempts to set the lines of Professor Bates's poem to music, but the now familiar melody was composed by one Samuel Augustus Ward. Ward had written a song entitled "O Mother, Dear Jerusalem," or "Materna," that was first performed by a parish choir in Boston on July 12, 1888. In 1895, the music of this song was married to the poetry of Bates's "America the Beautiful" and was first published by the Bryan and Taylor Company in a volume entitled Famous Songs and Those Who Made Them, compiled by Silas G. Pratt. No record exists of the first time that the words were sung to this music.

Ms. Bates taught English Literature at Wellesley College for 40 years, retiring in 1925. She wrote many books including travel, poetry, short stories, children's literature and scholarly works. She died at Wellesley, MA, Mar 28, 1929; her expression of her love for the physical beauty and political promise of her country lives on as a legacy to inspire our own patriot dreams.

America the Beautiful

O beautiful for spacious skies,
For amber waves of grain,
For purple mountain majesties
Above the fruited plain!
America! America!
God shed his grace on thee,
And crown thy good with brotherhood
From sea to shining sea.

O beautiful for pilgrim feet,
Whose stern, impassioned stress
A thoroughfare for freedom beat
Across the wilderness!
America! America!
God mend thine every flaw,
Confirm thy soul in self-control,
Thy liberty in law.

O beautiful for heroes proved
In liberating strife,
Who more than self their country loved,
And mercy more than life!
America! America!
May God thy gold refine,
Till all success be nobleness,
And every gain divine.

O beautiful for patriot dream
That sees beyond the years
Thine alabaster cities gleam,
Un-dimmed by human tears!
America! America!
God shed his grace on thee.
And crown thy good with brotherhood
From sea to shining sea.

Lady of the Lamp
FLORENCE NIGHTINGALE
BIRTH • MAY 12 • 175 YEARS

A well-educated woman of English society (born May 12, 1820, at Florence, Italy), Florence Nightingale was expected to conform to the feminine roles of her class—raising a family, entertaining and pursuing culture. But Nightingale believed that when she was 16 the voice of God had charged her to be of service to humankind. She decided that the field of nursing was the path to fulfilling this mission. Nightingale's requests to study nursing were repeatedly rebuffed by her mother and sister. Instead she studied parliamentary reports on the subject and in three years she became an acknowledged authority on health and hospitals.

Visits to European hospitals furthered her education; finally, in 1850 she received formal nursing training in Germany. In 1853 she became administrator of the Institution for the Care of Sick Gentlewomen in London and improved its nursing care and patient services. It was here that her remarkable administrative abilities began to emerge, but it was in wartime that they were honed.

THE CRIMEAN WAR

In early spring of 1854 Great Britain and France went to war against Russia in the Crimea. When news accounts of the horrifying conditions endured by sick and wounded British soldiers threatened to bring down the government, Nightingale was among the early volunteers. She was appointed head of a group of 38 nurses to combat the appalling conditions at Barrack Hospital at Scutari, Turkey.

Rejected by the surgeons (who resented the authority of a woman) upon her arrival in Turkey, Nightingale bided her time, initially taking over the kitchen administration and food distribution. Before long she had quietly and deliberately taken on every aspect of caring for the wounded and managing the hospital. After grueling days of work she would walk through the wards with a lamp, offering aid and comfort to her charges—and as a result she became known as the "Lady of the Lamp." Nightingale contracted Crimean fever and almost died but refused to return to England. She remained and the hospital's death rate declined from 45 percent to 2 percent—a staggering accomplishment.

HOME WITH HONOR

While at Scutari, Nightingale became an internationally known and respected figure. People in high places, including Queen Victoria, viewed her as a font of knowledge in the health field. Back in England, Nightingale avoided the many fetes held to honor her, instead retiring to her family home in Hampshire and later to London. She stayed out of the public eye the rest of her life but used the respect she had earned to persuade others to help her accomplish an enormous amount of work.

Her *Notes on Matters Affecting the Health, Efficiency, and Hospital Administration of the British Army* (1858) led to improved conditions for the average British soldier. She demonstrated that even in peacetime the military's death rate was twice that of civilians. As a direct result of her writings, the Army Medical School was founded.

Notes on Nursing (1859), Nightingale's most popular work, taught housewives how to care for their families' health. Using £45,000 given to her by a grateful public, in 1860 she established the Nightingale School for Nurses at St. Thomas Hospital. She implemented midwife training and improved health care in workhouses, cofounded the first Visiting Nurse Association and wrote *Suggestions for Improving the Nursing Service for the Sick Poor* (1874).

During the course of most of this work Nightingale was largely forgotten by the general public. She spent her last 40 years in her room as a semi-invalid. She received visitors from around the world, taking in and disseminating information while reclining on a couch. It is uncertain whether Nightingale actually had any physical ailments or if the affliction that kept her at home was largely psychological. Whatever the case, it didn't prevent her from leaving a lasting legacy in the field of health care: she was an orginator of the Red Cross movement and the first woman to receive the British Order of Merit.

When Florence Nightingale died at London on Aug 13, 1910, true to form, she left wishes rejecting a state funeral and burial at Westminster Abbey. Her final resting place was the family plot in East Wellow, Hampshire. Her tombstone reads "F.N. Born 1820. Died 1910."

FOR INFO

The American Nurses Association is planning a celebration in honor of Florence Nightingale.

American Nurses Association
600 Maryland Avenue SW
Suite 100W
Washington, DC 20024
Phone: (202) 651-7000

TWO WHO SERVED
GENERALS GRAY AND BLUE

Two of America's most controversial and colorful military figures will be celebrated with coming and going anniversaries in 1995. Jan 19 marks the 125th death anniversary of Confederate general Robert E. Lee, and Feb 8 marks the 175th birth anniversary of Union general William Tecumseh Sherman. Praised for their military genius, both were despised and vilified by their opponents during and after the Civil War.

ROBERT E. LEE
DEATH • OCT 12 • 125 YEARS

Robert E. Lee was born to a distinguished military and political family on Jan 19, 1807, at Stratford, VA. The son of Revolutionary War hero Light-Horse Harry Lee, Robert followed his father's footsteps and chose a military career, graduating from West Point in 1829. Like many of the men who would lead Union and Confederate forces in the Civil War, Lee served during the Mexican War in the 1840s. His distinguished service led to a rapid rise in rank. He served as superintendent at the US Military Academy at West Point and later as lieutenant colonel of the Department of Texas. In 1859, while on leave at home in Virginia, he was recalled to duty to stop the insurrection of abolitionist John Brown at Harpers Ferry. After capturing Brown, Lee returned to Texas. With the secession movement in full swing, he was summoned back to Washington in 1861 by President Abraham Lincoln, who offered him field command of the Union forces. Like many men of his profession, Lee was torn between loyalty to the Union and loyalty to his home state. After deliberation, he resigned his commission in the army and returned to Virginia to offer his services there.

REBEL GENERAL

Virginia had seceded from the Union but had not joined the Confederacy when Lee arrived in Richmond and was appointed commander of the Virginia military forces. After the state formally joined the Confederacy, however, Lee became a leader with no followers, for the direction of military forces was handled by the new Confederate government. He was eventually sent to northwestern Virginia to repel a Union advance, but he failed in this mission. He was then stationed along the South Carolina and Georgia coastline to construct fortifications.

Lee was recalled to the Confederate capital of Richmond in 1862 and appointed general-in-chief of all Confederate forces, but he was still serving under the rigid authority of Confederate president Jefferson Davis. Lee's first major achievement in the war came in his 1862 defense of Richmond. He was able to force Union general George B. McClellan's forces away from Richmond, which prompted the US government to recall the Union forces to northern Virginia to merge with a smaller army commanded by General John Pope. One of Lee's most daring and successful military moves followed, as he tried to intercept and destroy Pope's forces before McClellan could reach them. Going against traditional military strategy, Lee split his forces, sending a portion of his army into Pope's rear. The attack, known as the Second Battle of Bull Run or Manassas, defeated Pope and sent his troops fleeing back to Washington.

Lee then pushed the advantage and invaded Maryland. The climactic battle of that campaign occurred at Antietam, when McClellan moved against the Southern forces but could not breach their positions. Although the battle could be called a draw, Lee was forced to retreat to Virginia to preserve his supply line. There he defeated the Union forces, which were this time under the command of General Ambrose E. Burnside, at the battle of Fredericksburg.

GETTYSBURG

The year 1863 began as 1862 had ended, with Lee's daring strategic gambles leaving the Federals grasping for a victory. At Chancellorsville, he once again split his forces and defeated the Union troops, then under General Joseph Hooker. Lee then launched what some consider the deciding campaign in the long and bloody Civil War, again invading the North. Union general George G. Meade's forces met Lee's at Gettysburg, PA, on July 1 in the deadliest battle ever fought on American ground. About 53,000 men were killed in three days of fighting. Lee has been criticized for trying to attack a well-fortified position with a smaller force. Ironically, he lost Gettysburg while using tactics employed by the Union forces in their earlier defeat at Fredericksburg. On the climactic day of the battle, Lee launched a historic but futile frontal assault on the Union lines. Known as Pickett's Charge, the attack was repulsed and Lee was forced to withdraw again to the safety of Virginia.

After Gettysburg, Lee's strategic genius soon became no match for the overwhelming numbers of Union troops.

With smaller and smaller armies and a dearth of supplies, he was forced into a defensive posture. Although he was still able to strike victorious blows at the battles of the Wilderness, Spotsylvania and Cold Harbor, the war in Virginia settled into a siege of Richmond in 1864. When Union forces under the command of Ulysses S. Grant finally seized the railroads leading into the town in the spring of 1865, Lee was forced to withdraw, and he eventually decided that the war was lost. He surrendered to Grant at Appomattox Court House on Apr 9, 1865, effectively ending the war.

POSTWAR LIFE

While Lee was a hero during wartime, his stature only increased after the war. He accepted the defeat and urged fellow Southerners to strive for national unity. He took the job of president of Washington College in Lexington, VA, which was later renamed Washington and Lee University. He died on Oct 12, 1870, at Lexington. In 1975 the US Congress posthumously restored his citizenship. Although he is remembered as the "Rebel General," Robert E. Lee did not support slavery or secession. Ruthless in battle, he also was a firm believer that war should not be used to resolve political conflicts. In his earlier writings Lee stated, "[if] strife and civil war are to take the place of brotherly love and kindness, I shall mourn for my country and for the welfare and progress of mankind."

WILLIAM T. SHERMAN
BIRTH • FEB 8 • 175 YEARS

William Tecumseh Sherman was born on Feb 8, 1820, at Lancaster, OH. He was named for the famed Shawnee Indian chief Tecumseh. After his father's death when he was nine, Sherman was adopted by a family friend. His foster mother added William to his name, afraid that he could not be baptized with a heathen name. He had a wild streak and a roughness about him that seemed well suited to his upbringing in the then "far west." He was appointed to West Point at the age of 16, and he excelled both in his studies and in collecting demerits. Unlike his future colleagues and opponents in the Civil War, his early military career was an undistinguished one. He sat out the Mexican War in California as an administrative officer. He resigned his commission in 1853 to enter banking in San Francisco but lost money in the collapse of the gold rush. His attempts at a legal career also met with misfortune when he lost his first case. Frustrated at his failures, he sought reinstatement in the army, but he was refused.

UNION GENERAL

Old military compatriots came to Sherman's aid and secured him an appointment as superintendent of the Louisiana State Seminary of Learning and Military Academy. Although he enjoyed the work, he left the position when Louisiana seceded from the Union; he again offered his services to the US Army. Through the efforts of his brother, Senator John Sherman, he was reinstated. As a brigadier general in charge of volunteers he led a division in the Union victory at the battle of Shiloh in 1862 and received a promotion to major general for his efforts. Sherman pleaded with President Lincoln not to trust him with an independent command, but he was placed in charge of troops in Kentucky. Sherman refused to mount a decisive campaign, complaining that he was vastly outnumbered and needed more recruits. His complaints and enemy assessments were viewed by many as hallucinations and madness.

MARCH TO THE SEA

When he came under the command of Ulysses S. Grant in 1863, Sherman's military fortunes changed. Together they conquered the city of Vicksburg, MS, opening the Mississippi River to Union commerce again. When Grant became commander of all US forces in 1864, Sherman was promoted to supreme commander of the western armies, and there he began the campaign that brought him both fame and infamy. Driving southward from Chattanooga, TN, it took him almost four months of constant fighting against Confederate General Joseph Johnston to reach Atlanta, GA, on Sept 2, 1864. He cleared the city, destroying anything of military value, and then, against the advice of Grant and Lincoln, he began his 400-mile march to the sea. For 32 days Sherman's troops marched toward Savannah, GA, without communication to his superiors and cut off from their supply line. They foraged the land and ruthlessly destroyed anything of military value in their 60-mile-wide path. On Dec 20, 1864, he reached Savannah. After resting his troops he turned north to link up with Grant, now laying siege to Richmond, but Confederate General Robert E. Lee surrendered before Sherman reached Virginia, so he accepted the surrender of General Johnston's troops in North Carolina and marched his troops victoriously to Washington.

POST-WAR LIFE

Sherman remained in the army after the war and was appointed commander of the entire army when Grant became president in 1869. Upon his retirement in 1884, Sherman was offered the opportunity to run for president. His famous response was, "If nominated, I will not run. If elected, I will not serve." Sherman was famous for two other quotations that showed the character of what many claimed was a brutal soldier. On the eve of the war he described the growing war fever to one of his daughters by saying, "Men are blind and crazy." Although he despised the notion of war, once engaged he fought ruthlessly and sought to inflict as much damage on his foe as possible. To him, this was the only humane way of bringing a war to an end. In describing those efforts in a speech at Columbus, OH, in 1880, he uttered perhaps his most famous quote, "War is hell." Sherman died at New York City on Feb 14, 1891.

VOTING RIGHTS
OPENING THE POLLS

Voting is a privilege that allows any citizen to participate in the political life of his or her country. The power of the polls has brought down governments and changed the face of nations. Each step in extending that right from the privileged few to others has been fought vigorously by those already enfranchised. The world has recently witnessed one of the most dramatic extensions of voting rights to a disenfranchised people ever. Although the first all-race election in South Africa's history was marred by violence, the inauguration of Nelson Mandela as the first black president of South Africa on May 10, 1994, brought an end to the long struggle in South Africa for voting rights and political inclusion. The issue of voting rights in the US has existed since the country's founding. This year is the 125th anniversary of the 15th Amendment, which granted the right to vote to all men, regardless of race or prior condition of servitude. On Aug 18, 1920, 75 years ago, the 19th Amendment was ratified, extending that same right to women (see page 20). On Aug 6, 1965, 30 years ago, the Voting Rights Act was signed into law by Lyndon Johnson, further clarifying the rights of minorities and removing restrictions placed on their participation in the electoral process.

15TH AMENDMENT
RATIFIED • FEB 3 • 125 YEARS

Article XV
Section 1: The right of citizens of the United States to vote shall not be denied or abridged by the United States or by any State on account of race, color, or previous condition of servitude.
Section 2: The Congress shall have power to enforce this article by appropriate legislation.

Following the US Civil War and the freeing of the slaves, great efforts were made to integrate African-Americans into the political process. During the period of Reconstruction three amendments to the Constitution were passed. These "Civil War amendments"—the 13th, 14th and 15th—established the foundation for civil rights legislation that would be enacted in later years. With the passing of the 15th Amendment, African-Americans were swiftly integrated into the political process in the Reconstruction South, participating not only as voters but as elected representatives to local and national office. Hiram R. Revels of Mississippi became the first black senator and Joeseph H. Rainey became the first black elected to the US House of Representatives. But, as the aggressive northern stewardship over the southern states came to an end, the political clock rolled back. Various restrictions such as poll taxes and literacy and other voter-qualification tests were put in place to inhibit minority participation in the electoral process.

1965 VOTING RIGHTS ACT
SIGNED • AUG 6 • 30 YEARS

As the struggle for civil rights intensified during the middle decades of the 20th century, much attention was given to the enfranchisement issue. The first federal civil rights law enacted by Congress since the Reconstruction created the Civil Rights Commission in 1957 and authorized the US attorney general to enforce voting rights. The 1964 Civil Rights Act swept away more barriers; the 24th Amendment to the constitution, also ratified in 1964, did away with poll taxes in federal elections. Resistance still existed, and in 1965 President Lyndon Johnson called for a sweeping reform of electoral procedures. The Voting Rights Act of 1965, signed into law on Aug 6, suspended literacy tests and other disqualifying tests, authorized the appointment of federal voting examiners and provided for judicial relief on the federal level to bar discriminatory poll taxes. The act was signed into law on Aug 6, 1965, and extended again in 1975, 1984 and 1991.

The removal of existing barriers certainly spurred minority participation in the electoral process. But the battle for total access to the voting booth and the issues of minority representation continue. Proponents of increased minority representation have argued for models other than the one-man-one-vote model. Attempts at redrawing voting district lines to ensure more equitable minority representation have sometimes succeeded, but sometimes such gerrymandering has been challenged and overturned after lengthy court disputes. The latest effort to remove barriers to voter registration resulted in the passage of the Motor Voter Bill, signed into law by President William Clinton on May 20, 1993. This bill requires states to allow voters to register by mail or upon applying for or renewing a driver's license.

DEPT OF JUSTICE
JUNE 22 • 125 YEARS

It should come as no surprise that the US Justice Department is the largest law office in the world: its chief function is to represent the US government in court. Legislation creating the position of attorney general was enacted on Sept 24, 1789, and Edmund Randolph was the first person to fill it. On June 22, 1870, acknowledging the fact that the attorney general's office had grown to become a de facto executive department of the government, Congress made it official by passing legislation entitled "An Act to Establish the Department of Justice."

The attorney general (at present, Janet Reno) supervises the administration of the department; provides the president, the cabinet, the heads of the executive departments and the agencies of the government with legal advice; and proposes appointees to the federal judiciary and to her own department. Of the approximately 96,800 people employed by the Justice Department, half of these hold positions in the 93 US attorney's offices across the country.

FOR INFO

US Department of Justice
Pennsylvania Avenue at 10th Street NW
Washington, DC 20530

FREDERICK DOUGLASS
DEATH • FEB 20 • 100 YEARS

Frederick Douglass, originally named Frederick Augustus Washington Bailey, was born into slavery Feb 7, 1817, at Tuckahoe, MD. He escaped from slavery Sept 3, 1838, fully 25 years before Abraham Lincoln issued the Emancipation Proclamation. Using a free black seaman's identification, he traveled from Baltimore to New York City, where he changed his name to Frederick Douglass.

At an antislavery convention in 1841 Douglass gave a speech describing his life as a slave—so moving that he was urged to help promote the cause of abolition by speaking before antislavery gatherings.

In 1845 Douglass published his autobiography, *The Narrative of the Life of Frederick Douglass: an American Slave* (revised as *Life and Times of Frederick Douglass* in 1882), as a response to those who believed that he was too intelligent and articulate to have been raised in slavery. Ironically, the book's release put his freedom in jeopardy, because in it he revealed his former owner's identity. To avoid re-enslavement, Douglass fled to Great Britain, where he attempted to enlist British support for the antislavery movement. He remained abroad until 1847, when English Quakers raised money for him to buy his freedom.

Returning to the US, Douglass settled for a time in Rochester, NY, where he founded the abolitionist newspaper *The North Star* (later renamed *Frederick Douglass's Paper*). In it he wrote against the existing segregation of Rochester schools, encouraged the employment of blacks in positions other than those of servants and laborers and proposed the formation of schools to train blacks in skilled crafts. During his years in Rochester he led the city's branch of the Underground Railroad, helping to smuggle escaped slaves to Canada.

Increasingly Douglass saw political activity rather than moral persuasion as the most productive road to freedom for American slaves. He didn't actively encourage the use of violence as a means to freedom for blacks, yet he did not categorically reject it as a tool to that end: "It can never be wrong for the imbruted and whip-scarred slaves . . . or their friends . . . to hunt, harass, and even strike down the traffickers in human flesh." Accusations, however, that Douglass aided the white abolitionist John Brown in his 1859 attack on the federal arsenal at Harpers Ferry, WV, were false. Douglass had refused to join the revolt, but, unable to stem the accusations, he was again forced to flee to Great Britain.

Douglass was able to return to the US in 1860. He then worked for the presidential campaign of abolitionist Gerrit Smith and later supported Abraham Lincoln for president. Douglass was disappointed when the Emancipation Proclamation, issued by Lincoln in 1863, guaranteed freedom for slaves only in the Rebel states.

Anxious for blacks to actively participate in the Civil War, Douglass urged the Union Army to accept them as soldiers. He helped to organize two black regiments and urged Lincoln to grant blacks the same pay and opportunities for promotion that whites were provided.

After the war Douglass served in several federal offices and participated in the struggle for the 15th Amendment, which was finally ratified in 1870, granting all male citizens the right to vote. He also was an active participant in the long battle for women's rights and a delegate at the first women's-rights convention, at Seneca Falls, NY, in 1848.

Frederick Douglass died at Washington, DC, on Feb 20, 1895.

WOMEN'S SUFFRAGE
75TH ANNIVERSARY • AUG 26

"We hold these truths to be self-evident, that all men are created equal, that they are endowed by their Creator with certain unalienable Rights, that among these are Life, Liberty and the pursuit of Happiness...."
—Declaration of Independence, July 4, 1776

By the middle of the 19th century increasing numbers of American women were no longer willing to accept a second-class citizenship that denied them property rights, provided them little opportunity for higher education, allowed them no right to guardianship of their own children in the event of divorce and granted them no vote to redress the wrongs they were forced to endure. Women were beginning to demand their "unalienable" rights.

SENECA FALLS CONVENTION

On July 14, 1848, the *Seneca Falls Courier* carried an announcement that a women's rights convention would be held in the Wesleyan Chapel at Seneca Falls, NY, on July 19 and 20. That convention is generally regarded as the beginning of the women's movement in America.

Elizabeth Cady Stanton, who would become a forceful feminist speaker, presented her first speech at that convention. Her Declaration of Sentiments, patterned after the Declaration of Independence, included a call for equal opportunities for women in education and employment as well as a demand for the right to vote. Stanton pointed out that women had never been permitted to "exercise [their] inalienable right to the elective franchise." During the debate on the "Declaration of Sentiments," only the declaration's ninth resolution, which advocated women's suffrage, caused frenzied argument and dissension. A strong speech by Frederick Douglass in favor of the resolution eventually carried the day, and the resolution narrowly passed. What this pioneering group of 300 couldn't know at the time was the magnitude of the task ahead.

THE LONG STRUGGLE

Over the next 72 years myriad tactics were employed to further the cause of women's suffrage. Millions of women barraged state legislatures with petitions demanding the vote. Demonstrations and parades were held. Literature was distributed at polling places. Some women actually attempted to vote, citing their right to do so under the 14th Amendment to the US Constitution, which says, "no State shall make or enforce any law which shall abridge the privileges or immunities of citizens of the United States." Most of the few who succeeded in casting ballots were rewarded with time in jail.

A MOVEMENT DIVIDED

In the years leading up to the Civil War many suffragists selflessly dedicated their time and efforts to the abolition of slavery. Susan B. Anthony, for example, organized the collection of 400,000 signatures (an enormous figure for the time) urging the passage of the 13th Amendment. Once slavery had been officially outlawed, however, issues of race and gender led to tactical disagreements among leaders of the suffrage movement, resulting in an unfortunate split. The National Woman Suffrage Association (NWSA), formed by Stanton and Anthony, refused to separate the struggle for women's suffrage from efforts to secure the vote for the newly freed slaves. The NWSA wanted to work for an amendment to the Constitution that would enfranchise blacks as well as women, a proposal made moot the next year, when the passage of the 15th Amendment gave the vote to all men. On the other hand, the American Woman Suffrage Association (AWSA), founded by Lucy Stone, was willing to separate the struggle for the black vote from that for the women's vote and focused on municipal and state women's suffrage rather than a Constitutional change.

As each side looked for allies, Anthony wrote to abolitionist and former slave Sojourner Truth asking for her support. Always a champion for women's suffrage, Truth responded: "There is a great stir about colored men getting their rights, but not a word about the colored women; and if colored men get their rights, and not colored women theirs, you see the colored men will be masters over the women, and it will be just as bad as it was before. So I am for keeping the thing going while things are stirring; because if we wait till it is still, it will take a great while to get it going again."

And it did take a while to get things going again. Not until 21 years later, in 1890, did the two groups opt for strength in numbers and again combine their efforts. Under the banner of the National American Woman Suffrage Association (NAWSA), they worked for both state suffrage and a Constitutional Amendment.

UNITED FRONT

In spite of the movement's united front, by 1915 only 13 states had amended their constitutions or enacted legislation to allow women the vote. Twice brought before the US Congress, the issue had failed to pass there as well. Agitated by this lack of progress, NAWSA called upon the indomitable Carrie Chapman Catt to take the reigns as their president. Catt had relinquished that position in 1904 to nurse her husband. Fifteen years later, NAWSA saw Catt as the only leader with the personal strength and fortitude to get the job done and prevailed upon her to resume control. Catt's tactics, which came to be known as the Winning Plan, were kept secret because past experience showed that powerful antisuffrage forces—liquor interests, political machines, clergy and big business—would put formidable obstacles in the way of the suffragists. Catt predicted the vote would be theirs in six years. It would actually take only four years, but in the interim the US entered World War I, on Apr 16, 1917. Most NAWSA members gave full support to the American role in the war, folding bandages and working in government offices and hospitals. They hoped their loyal service would be rewarded with speedy passage of legislation giving them the vote. Indeed, their efforts did weaken some opposition. But in the war for the vote, several battles remained to be fought.

STATE CONSIDERATION

For women to gain the vote through a constitutional amendment, that amendment had to pass both houses of Congress by a two-thirds majority and be ratified by the states, again by a two-thirds majority. An amendment supporting women's suffrage had been introduced into Congress in 1878, when it was overwhelmingly defeated, and then again in 1914, when it was narrowly rejected. On May 20, 1919, the House of Representatives finally passed the proposed amendment to the Constitution, and on June 4, 1919, the Senate followed suit. The constitutional right of women to vote finally had become a matter for consideration by the states.

Once the amendment was submitted to the states for ratification, the process went relatively fast. On Mar 22, 1920, Washington became the 35th state to ratify. Affirmation by only one more state was needed. Suffragists saw Tennessee as the state where they had their best chance, and so focused their efforts there. Antisuffragists moved a sizable lobby into the state and made it a bitter battle to the end.

SHALL NOT BE DENIED

On Aug 13, 1920, the Tennessee senate passed the Amendment 25-14. But the antisuffragists had concentrated their efforts on Tennessee's house of representatives. Time and again they prevented the issue from coming to a vote, stalling for more time to influence legislators. Finally, after repeated postponements, on Aug 18, 1920, the Tennessee house of representatives voted to ratify. On Aug 26, 1920, US Secretary of State Bainbridge Colby signed the proclamation finalizing the adoption of the amendment.

The Susan B. Anthony Amendment (as it came to be called) enfranchised 25 million American women. It contained the original, simple and precise language with which it had been submitted to Congress 42 years earlier:
Article XIX
Section 1: The right of citizens of the United States to vote shall not be denied or abridged by the United States or by any State on account of sex.
Section 2: Congress shall have power to enforce this article by appropriate legislation.

SUSAN B. ANTHONY
BIRTH • FEB 15 • 175 YEARS

"Failure is impossible." —Susan B. Anthony

Susan B. Anthony was born Feb 15, 1820, at Adams, MA. Her Quaker parents raised her in the belief that women should enjoy the same civil rights as men. Anthony began her career as a reformer working for the temperance movement. A rebuff by male temperance workers at an 1852 convention led her and others to form the Woman's State Temperance Society of New York, and also contributed to her decision to dedicate more than 50 years of her life to the women's movement.

Anthony's role in the movement required considerable sacrifice and left little time for a personal life. She traveled all over the country to organize meetings and to lecture on the rights of women—particularly the right to vote. She also developed the tactic of inundating government officials with petitions to demonstrate support for women's issues.

Also active in the antislavery movement, Anthony dedicated considerable time to that cause before she began her efforts on behalf of women's suffrage. She was unsuccessful, however, in her fight to include women in the 14th Amendment, which discussed voting in terms of "male inhabitants," including former slaves.

Anthony attempted to challenge the extent of the 15th Amendment by registering and voting in Rochester, NY, in 1872. She was arrested, tried and fined for violating the law. Remaining firm in her position that the wording of the 15th Amendment—"The right of *citizens* of the United States to vote shall not be denied"—should be interpreted to include women, Anthony refused to pay the fine. The court, unwilling to draw additional attention to the cause of suffrage, never pursued the issue.

Anthony served as president of the National American Woman Suffrage Association from 1892 to 1900. What spare time she had in the 1880s, and '90s she spent with her good friends Elizabeth Cady Stanton and Matilda Gage, compiling the first four volumes of the *History of Woman Suffrage*.

Susan B. Anthony, reformer and advocate of women's rights, died at Rochester, NY, on Mar 13, 1906, 14 years before the Anthony Amendment became law.

LEAGUE OF WOMEN VOTERS
FOUNDED • FEB 14 • 75 YEARS

On Feb 14, 1920, representatives of the National American Woman Suffrage Association (NAWSA) met in Chicago to celebrate their impending victory in the 72-year battle for the vote. At this gathering they formed the League of Women Voters to assist American women in responsibly exercising their newly granted right. The main objective of the League was, and remains, nonpartisan participation in the electoral process. Through education and advocacy its members hope to affect public policy.

Today the League functions on the local, state and national levels of government. Its official positions are a result of a thorough process of research and study. Drawing on material presenting both sides of an issue, members discuss each view, debate the issue in convention and then take action based on the general consensus of members.

To influence government policy toward its views, the League lobbies elected officials, monitors government activity and litigates when legislation is not fully or fairly implemented or is deemed unjust.

In 1957 the League of Women Voters Education Fund (LWVEF) was founded to help inform citizens about current issues through workshops, conferences and the distribution of publications. Nonpartisan services are provided to voters throughout the year and especially at election time.

The most readily recognized service provided by the League of Women Voters is the sponsorship of presidential debates. These debates, now expected and eagerly anticipated by a large segment of the voting public, provide viewers with an opportunity to hear leading presidential candidates discuss the issues. Every two years the League of Women Voters chooses new issues for debate in the areas of government, international relations and social policy. Subjects debated in the past include national security, equal opportunities for women and minorities, voting rights, hazardous waste, reproductive choice and support of the United Nations.

To obtain further information on the League of Women Voters call your local chapter.

FOR INFO

League of Women Voters of the United States
1730 M Street, NW
Washington, DC 20036
Phone: (202) 429-1965

WOMEN'S RIGHTS
NATIONAL HISTORICAL PARK
ESTABLISHED • DEC 28 • 15 YEARS

Seneca Falls, NY, is rightly known as the birthplace of the women's movement in the United States. Thus it is fitting that this historically significant town was chosen in 1980 as the location for the Women's Rights National Historical Park.

The homes of three of the five women who called for and organized the first women's rights convention in the US are in the area: the Elizabeth Cady Stanton House is in Seneca Falls, and the homes of Mary Ann McClintock and Jane Hunt are in the nearby town of Waterloo. Wesleyan Chapel, the site of the 1848 convention that initiated the long struggle for the enfranchisement of women, is in Seneca Falls. In the words of the legislation that helped to establish the park, the convention was "an event of major importance in the history of the United States because it marked the formal beginning of the struggle of women for their equal rights." The purpose of the park is to preserve these nationally significant historical and cultural sites and structures associated with the 72-year fight for women's suffrage.

The National Women's Hall of Fame, a soon-to-be-named university and park officials are forging a cooperative agreement for the development of the Women's Education and Cultural Center.

Women's Rights National Historical Park is located on 5.5 acres of land about 15 miles south of the New York State Thruway. It can be reached by taking Exit 41 via Route 414 and Route 5/20.

FOR INFO:

Women's Rights National Historical Park
136 Fall Street
Seneca Falls, NY 13148
Phone: (315) 568-2991
Fax: (315) 568-5563

ELIZABETH'S LIST
SISTERS OF DISTINCTION

> The following is a partial list of birth dates of American women who have made significant contributions in myriad endeavors. You can find them in the chronology section under the dates mentioned. The list was selected by *Chase's* Assistant Editor, Beth Johnson. We know it's a tiny sampling of our talented and usually unsung foremothers, so please do not call us to quibble about the choices. Do feel free, however, to write us and suggest other names for inclusion in future editions. Please furnish biographical info for the names you submit. Mail to **Chase's Calendar**, Attn: Elizabeth's List, at the address on page 2.

Jane Addams, Sept 6, 1860, social reformer
Susan B. Anthony, Feb 15, 1820, women's-rights advocate
Clara Barton, Dec 25, 1821, founder American Red Cross
Catherine E. Beecher, Sept 6, 1800, educator
Mary McLeon Bethune, July 10, 1875, educator
Elizabeth Blackwell, Feb 3, 1821, physician
Amelia J. Bloomer, May 27, 1818, women's-rights advocate
Jane M. Bolin, Apr 11, 1908, federal judge
Gwendolyn Brooks, June 7, 1917, poet
Pearl S. Buck, June 26, 1892, author
Hattie Caraway, Feb 1, 1878, US senator
Rachel Carson, May 27, 1907, scientist
Mary Cassatt, May 22, 1844 (or 1845), artist
Carrie Catt, Jan 9, 1859, women's-rights advocate
Shirley Chisholm, Nov 30, 1924, US congresswoman
Lydia Darragh (no date available), 1729, American Revolution spy
Dorothea Dix, Apr 4, 1802, social reformer
Isadora Duncan, May 27, 1878, dancer
Amelia Earhart, July 24, 1898, aviatrix
Edna Ferber, Aug 15, 1887, author
Ella Fitzgerald, Apr 25, 1918, singer
Betty Friedan, Feb 4, 1921, women's-rights advocate
Margaret Fuller, May 23, 1810, author, critic and social reformer
Katharine Graham, June 16, 1917, newspaper executive
Helen Hayes, Oct 10, 1900, actress
Grace Hopper, Dec 9, 1906, COBOL computer language developer
Mother Jones, May 1, 1830, labor leader
Helen Keller, June 27, 1880, author
Julie C. Lathrop, June 29, 1858, social reformer
Belva Lockwood, Oct 24, 1830, lawyer
Juliette G. Low, Oct 31, 1860, founder Girl Scouts of USA
Claire Boothe Luce, Mar 10, 1903, ambassador
Mary Lyon, Feb 28, 1797, educator
Arabella Mansfield, May 23, 1846, lawyer
Maria Goeppert Mayer, June 28, 1906, physicist
Mildred (Horton) McAfee, May 12, 1900, director US Navy WAVES
Margaret Mead, Dec 16, 1901, anthropologist
Maria Mitchell, Aug 1, 1818, astronomer
Esther H. M. Morris, Aug 8, 1814, women's-rights advocate
Constance Baker Motley, Sept 14, 1921, lawyer
Lucretia Mott, Jan 3, 1793, women's-rights advocate
Alice Freeman Palmer, Feb 21, 1855, educator
Elizabeth Palmer Peabody, Jan 3, 1894, educator
Annie S. Peck, Oct 19, 1850, mountain climber
Frances Perkins, Apr 10, 1880, member US presidential cabinet
Pocahontas, Mar 21, 1617,* mediator and unofficial ambassador
Jeannette Rankin, June 11, 1880, US congresswoman
Libby Riddles, Apr 1, 1956, dogsledder
Sally Ride, May 26, 1951, astronaut
Edith Nourse Rogers, Mar 19, 1881, US congresswoman
Eleanor Roosevelt, Oct 11, 1884, social reformer
Nellie Tayloe Ross, Nov 29, 1880, governor
Sacajawea, Dec 20, 1812,* interpreter
Deborah Sampson, Dec 17, 1760, Revolutionary Army soldier
Margaret Sanger, Sept 14, 1879 (or 1883), women's-rights advocate and nurse
Rose Schneiderman, Apr 6, 1884, labor leader
Margaret Chase Smith, Dec 14, 1897, US congresswoman and senator
Elizabeth Cady Stanton, Nov 12, 1815, women's-rights advocate
Lucy Stone, Aug 13, 1818, women's-rights advocate
Dorothy C. Stratton, Mar 24, 1898, commander SPARS, US Coast Guard Women's Reserve
Dorothy Chaney Streeter, Oct 2, 1895, director US Marine Corps Women's Reserve
Anne (Macy) Sullivan, Apr 14, 1866, educator
Ida Tarbell, Nov 5, 1857, author and historian
Helen B. Taussig, May 24, 1898, physician
(Martha) Carey Thomas, Jan 2, 1857, educator and social reformer
Sojourner Truth, Nov 26, 1883,* abolitionist
Harriett Tubman, Mar 10, 1913,* abolitionist
Barbara Tuchman, Jan 30, 1912, historian
Lillian Wald, Mar 10, 1867, public health reformer
Ida B. Wells, July 16, 1862, journalist
Phillis Wheatley, Dec 5, 1784,* poet
Margaret Bourke-White, June 14, 1906, photojournalist
Victoria Woodhull, Sept 23, 1838, social reformer
Rosalyn Yalow, July 19, 1921, medical physicist
Babe Didrikson Zaharias, June 26, 1914, athlete

*death anniversary

JONATHAN SWIFT
DEATH • OCT 19 • 250 YEARS

Jonathan Swift penned his own epitaph. Translated from Latin, it reads, "The body of Jonathan Swift, Doctor of Sacred Theology, dean of this cathedral church, is buried here, where fierce indignation can no more lacerate his heart. Go, traveler, and imitate, if you can, one who strove with all his strength to champion liberty." Of his most famous work, *Gulliver's Travels*, he declared in a letter to his friend Alexander Pope that he intended to "vex the world rather than divert it." His writings did indeed vex the world, although they succeeded in diverting it also, as he channeled his indignation at man's treatment of his fellows into some of the wittiest and most biting satirical prose ever written in English.

Swift was born on Nov 30, 1667, at Dublin, Ireland, during a tumultuous period of Anglo-Irish history. His grandfather had supported the Royalists during the English Civil War and lost everything. His father had died before his birth, so Jonathan was raised by an uncle he considered ungenerous. With a degree in hand, he left for England after the Revolution of 1688 and was introduced to Sir William Temple, beginning what today is still a mysterious personal and familial life and a tempestuous relationship with Temple, whom he served as secretary. After being advised by his cousin, the poet Dryden, to abandon his attempts at poetry, Swift returned in frustration to Ireland in 1694 and entered the Church of England. Soon bored with his rural parish near Belfast, he returned to the Temple household.

His chief attachment to the household was a young child, Esther Johnson, the daughter of Temple's housekeeper. Swift taught the child (14 years his junior) to read and write, and over time she became his closest friend. During this period he turned to prose and wrote in 1696 and 1697 his first important satires, *A Tale of a Tub* and "The Battle of the Books," both published in 1704. After Temple died in 1699, Swift, followed by Esther, returned to Ireland and the church. His new assignment allowed him to leave the small country parish in the curate's care, so he spent much of his time in Dublin and London.

SATIRE AND POLITICS

In London, Swift's career as a satirist and political lobbyist bloomed. While he was there to gain benefits for the Irish church, he also sought personal advancement in the church hierarchy. *A Tale of a Tub* had brought him instant fame. A Whig by political heritage, in 1710 he joined the Tories, primarily because he felt they showed more concern for the church. Writing anonymously as editor of the *Examiner* in that publication and in a number of pamphlets, he defended the Tory positions and scathingly attacked his opponents. In *The Conduct of the Allies* he charged that the Whigs had prolonged the war of the Spanish Succession out of self-interest, bringing about the dismissal of John Churchill, the first Duke of Marlborough, who was waging the campaign. Swift was rewarded for his partisan participation in 1713 with an appointment as dean of Saint Patrick's Cathedral in Dublin. The next year his political influence ended with the electoral defeat of the Tory party.

WHO WAS STELLA?

Swift returned to Ireland and his personal life there remains a mystery. Intimate letters to Esther Johnson, called *Journal to Stella*, caused speculation that Swift and "Stella" were secretly married and that perhaps they were both the children of Sir William Temple, this fact being made known to them only after they were married. Then there was Esther Vanhomrigh, "Vanessa" to Swift, who followed him to Ireland. Though Swift apparently did not return her devotion, one story suggests she bore him a child later cared for by Stella. Still another legend has Swift and Temple being sired by the same father, thus making Stella Swift's niece.

CHAMPION OF JUSTICE

Back in Ireland, Swift's satirical jibes focused on, among other things, the mistreatment of the Irish by the British. *Drapier's Letters* and *A Modest Proposal* made Swift a hero to his countrymen. *A Modest Proposal* is perhaps his most outrageous satire; in it he suggested that the Irish sell their children to be eaten by the wealthy, turning an economic burden into a profitable situation.

Swift then turned to his most memorable work, *Travels into Several Remote Nations of the World*, better known as *Gulliver's Travels*. Published in 1726, it was an immediate success. The allegory was intended as an arrow launched at the heart of hypocrisy that Swift found in the political, judicial and social institutions of his day. Although bitter in tone, Gulliver's travels are so comically described and the fantastical characters so engaging that this masterpiece of satire was assumed to be—and still endures as—a popular children's book.

FAILING HEALTH

In Swift's last years his temper and fear of mental decay led some to label him insane. But his mental acuity remained intact and he was still regarded as an Irish hero. Spells of dizziness, deafness and nausea, in recent years described as vertigo or Ménière's syndrome, eventually were followed by a stroke, and Swift was declared incapable of caring for himself in 1742. He was tended by a guardian until his death on Oct 19, 1745, at Dublin and was interred next to his Stella, who had died in 1728.

Satire is a sort of glass, wherein beholders do generally discover everybody's face but their own.
—from "The Battle of the Books"

CHARLES DICKENS
DEATH • JUNE 9 • 125 YEARS

Tiny Tim from *A Christmas Carol*.

After seeing their fortunes dwindle and the large family forced to move about London from one poorhouse to the next, he, the eldest son, had to end his education and take up manual work in a blacking warehouse. His only respite from his long hours of work and living in a garret were his Sunday visits to the debtor's prison where his parents were held.

It sounds like a particularly Dickensian beginning to a life, and in fact it is. It was the beginning of the life of Charles Dickens himself, and the impact of those times and conditions are depicted in what many critics regard as his best work, the semi-autobiographical novel *David Copperfield* (1849).

Charles John Huffam Dickens was born Feb 7, 1812, at Portsmouth, England, to a middle-class family and spent much of his youth in Kent and London. His father, on whom Dickens later modeled the character of Mr. Micawber, was indeed extravagant, and his financial failings did lead to the difficulties described. A timely inheritance rescued the family from its financial difficulties and Dickens was able to continue his private education for a few years. His formal schooling, however, was less important to his later success than his remarkable powers of observation and imagination and his ability to educate himself.

After serving as a legal clerk, Dickens taught himself shorthand and worked as a court reporter. He later reported on the House of Commons and traveled the countryside covering election speeches. In his fiction Dickens later parlayed his observations and experiences into sometimes comic, sometimes bleak depictions of the legal and political professions. Throughout his career his observations were sharpened into scalpellike satire that indelibly portrayed Victorian England.

Although his love for the stage almost led him to a theater career, Dickens pursued writing. After an anonymously submitted manuscript was published, he began writing *Sketches by Boz* (1836), after the nickname of a younger brother, which led to a commission of similar sketches that became *The Pickwick Papers*. Bursting with creative energy, he honed his craft in early novels such as *Oliver Twist*, *Nicholas Nickleby*, *The Old Curiosity Shop* and *Barnaby Rudge*—all populated with a multitude of unforgettable characters.

In April 1836, during the same week the first monthly installment of *The Pickwick Papers* appeared, Dickens married Catherine Hogarth, daughter of Scottish journalist George Hogarth.

MEETING AMERICA

When Dickens and his wife journeyed to America in 1842, the novelist received a welcome second only to that of Lafayette. Soon, however, Dickens became disillusioned with the US, and his arguments in favor of an international copyright agreement and against slavery led newspapers to accuse him of abusing American hospitality. Dickens wrote satirically, sometimes bitterly, about the US in *American Notes* (1842) and the novel *Martin Chuzzlewit* (1844).

Back in London Dickens's serialized novels brought him financial security and immense popularity, but soon he sought another outlet for his message of social reform. He cofounded the *Daily News* in 1846 but, frustrated with the business of producing a newspaper, abruptly left 19 days later. His journalistic talents later manifested themselves in the weeklies *Household Words* and *All the Year Round*, in which *Child's History of England*, *Christmas Stories*, *Hard Times*, *A Tale of Two Cities* and *Great Expectations* were serialized.

"FEARFUL LOCOMOTIVE"

Described by Ralph Waldo Emerson as a man who "has too much talent for his genius; it is a fearful locomotive to which he is bound and can never be free from it nor set to rest," the prolific novelist continued to seek other avenues of expression. He directed a reformatory home for young women, spoke and raised funds for social causes, managed a small theatrical company, gave public lectures and presented dramatic readings of his works.

LEGACY

Charles Dickens died on June 9, 1870. He is remembered as an incomparable chronicler of his time and a champion of social justice as well as a literary genius. His empathy with and sympathy for the downtrodden were often reflected in his work, never more so than in the novella *A Christmas Carol* (1843), a myth of redemption and goodness that has forever tied Charles Dickens to his favorite day of the year.

JOHN KEATS
BIRTH • OCT 31 • 200 YEARS

A thing of beauty is a joy for ever:
Its loveliness increases; it will never
Pass into nothingness; but still will keep
A bower quiet for us, and a sleep
Full of sweet dreams, and health, and quiet
breathing.

From Endymion

Beauty is truth, truth beauty,—that is all
Ye know on earth, and all ye need to know.

From "Ode on a Grecian Urn"

A seminal influence in the romantic movement, John Keats was passionate about beauty, searching for it and composing his feelings into some of the most enduring poetry of the period. He was born at London on Oct 31, 1795, and unlike many poets and artists of the romantic period he did not spend the majority of his youth in the countryside. His father was a livery-stable owner who died when Keats was nine years old. After his mother's second marriage failed, he, his sister and two brothers lived with their widowed grandmother at Edmonton, Middlesex. Keats attended the Clarke School at Enfield and after his mother's death was apprenticed at the age of 15 to a surgeon. He left for London in 1814 and studied medicine for two years while working as a junior house surgeon at Guy's and St. Thomas's hospitals. In 1816 he became a licensed apothecary, but his literary interests replaced his desire for a career in medicine and he never practiced the profession. By the next year he was devoting his life to the pursuit of poetry.

Keats's early career was influenced by the poet and journalist Leigh Hunt, to whom he was introduced by Charles Cowden Clarke, the son of his former schoolmaster. His first published poems, "To Solitude" and "On First Looking into Chapman's Homer," appeared in *The Examiner*, which was edited by Hunt. Hunt introduced Keats to his circle of artistic and literary friends, which included the painter Benjamin Haydon and the poets John Hamilton Reynolds and Percy Bysshe Shelley. Keats's first volume of verse, *Poems*, was published in 1817 and included "Sleep and Poetry," a defense of the school of romanticism as espoused by Hunt and an attack on the romantic principles as practiced by George Gordon, Lord Byron. Keats then began working on his first long poem, *Endymion*, which was published in 1818. This effort came under severe critical attack with *Blackwood's Magazine* labeling *Endymion* "nonsense" and urging Keats to give up poetry.

Around this same time, from 1818 through 1820, Keats's passionate love life took center stage and battled for attention with his brother's illness and with his own suffering from tuberculosis. His brother, Tom, was ill with tuberculosis and Keats spent much time caring for him until Tom's death. Also at this time, Keats met Fanny Brawne and began a relationship that would have a profound effect on his development as a poet. They became engaged in 1819, and that year featured his greatest writing. *Lamia*, *The Eve of St. Agnes*, "Ode on Indolence," "Ode on a Grecian Urn," "Ode to Psyche," "Ode to a Nightingale," "Ode on Melancholy," "To Autumn" and two versions of *Hyperion* were written that year and published in 1820. Also in 1820, under doctor's orders, Keats left for the warmer climate of Rome. His correspondence with Fanny and his brother and sister during this time has been praised as some of the finest literary letters written in the English language. He died at Rome on Feb 23, 1821.

John Keats's short life included a writing career of only seven years, but his body of work and his power for observing beauty and truth in the most minute details of nature had a profound effect on the romantic poets and artists of his day and on those who followed. Three of his odes ("On a Grecian Urn," "On Melancholy," "To a Nightingale") are considered some of the finest poetry in the English language. Keats suffered the fate of many artists whose work becomes greatly appreciated only after their death. His popularity and the acknowledgment of his contributions have increased with the passage of time, with the 20th century producing the greatest appreciation of his efforts. Not aware of his legacy, and knowing that tuberculosis would kill him, Keats composed his own epitaph: "Here lies one whose name was writ in water."

ROENTGEN AND X-RAYS
BIRTH • MAR 27 • 150 YEARS
DISCOVERY • NOV 8 • 100 YEARS

Wilhelm Conrad Roentgen (also spelled Röntgen) was born on Mar 27, 1845, at Lennep, Prussia, which is now a part of Remscheid, Germany. He studied at the Polytechnic in Zurich, Switzerland, where he received his doctorate in physics in 1869. He then taught and conducted research at the universities of Strasbourg, Giessen, Würzburg and Munich. He began experiments that led to the discovery of x-rays in 1895 at the University of Würzburg. In 1901 he received the first Nobel Prize for physics for his discovery of the x-ray, also called the Roentgen ray. Roentgen continued research in a variety of fields, including elasticity, fluids and crystals, until his death at Munich on Feb 10, 1923.

Some of the most important discoveries in science have been accidental. One such discovery was the x-ray. On Nov 8, 1895, German physicist Wilhelm Conrad Roentgen was conducting experiments with the flow of electricity in a cathode-ray tube encased in a black cardboard box. Roentgen noticed that when the tube was operating, a nearby barium-platinocyanide screen glowed with fluorescent light. Intrigued, he theorized that a form of radiation was escaping from the box and altering the chemical properties of the screen across the room. Finding that many materials were transparent to the radiation but that it left an image on a photographic plate, he took the first photos of the interiors of metal objects and of the bones in his wife's hand. Roentgen also noted that the radiation passed through materials without reflection or refraction and mistakenly concluded that these rays were not related to light. He labeled his discovery x-radiation (x represents the unknown in mathematics), later also known as Roentgen rays. After Roentgen revealed his discovery to the Physico-Medical Society of Würzburg, he sent photographs and reprints to his colleagues and friends. The news spread, and he was summoned to appear before the kaiser, where he demonstrated his findings on Jan 13, 1896.

WHAT ARE X-RAYS?

X-rays are a type of electromagnetic radiation. Their short wavelengths and high energy enable them to penetrate opaque objects, whereas longer-wavelength, lower-energy light waves are reflected or absorbed. Similar to light rays, x-rays are oscillating electric and magnetic fields that travel at the speed of light. Both rays are produced by the transitions of electrons that orbit atoms—light rays by electrons orbiting on an outer path, and x-rays by electrons orbiting on an inner path.

USE IN MEDICINE

Modern medicine in particular benefited from the development of Roentgen's discovery. First used to detect foreign objects and abnormalities inside the human body, x-rays were soon used to provide diagnostic assistance, becoming the single most important method of diagnosing tuberculosis. It wasn't until later that scientists discovered the energy of x-radiation can ionize atoms within a cell and cause damage. This discovery led to guidelines protecting medical staff and patients from harm and also to radiotherapy, which exposes the rapidly dividing cells of cancerous tumors to x-radiation.

Computerized axial tomography (CAT) has taken the technology to a point where three-dimensional pictures of an area can be displayed with approximately 100 times more clarity than with traditional fluoroscopes. Dental x-rays are now as routine as regular checkups and cleanings.

IN INDUSTRY AND SCIENCE

Industry uses x-rays to detect flaws inside metallic castings without destroying the castings. Ultrasoft x-rays are used to determine authenticity of works of art and for art restoration. X-ray inspection of luggage is common at airports.

X-rays also played a leading role in decreasing the size of computer microchips, replacing longer-wavelength light in exposing the photosensitive material on silicon wafers used in the smaller circuitry.

Scientific research also has benefited. The use of x-rays made developments possible in theoretical physics, quantum mechanics, astronomy and crystallography.

THE GREAT LOVER
RUDOLPH VALENTINO
BIRTH • MAY 6 • 100 YEARS

Rodolfo Pietro Filiberto Raffaello Guglielmi, the man who would come to be idolized as the "Great Lover," was born on May 6, 1895, at Castellaneta, Italy. He came to the US in 1913 and, like most immigrants, spent his first years in menial labor, as a gardener and a dishwasher. His rise out of New York's Little Italy began, however, on the dance floor. With a talent for dancing and a way with the ladies, Valentino became a favored "taxi-dancer": young women would come to nightclubs and for a small fee would be partnered with young men for the evening's dancing. He came to the attention of dancer Bonnie Glass, who was looking for a new partner. Seeking to incorporate the tango into her act, she found that the moves and romantic Latin looks of the renamed Rodolpho di Valentino filled the bill.

After a scrape with the law Valentino feared deportation and left New York as a cast member of a touring musical, finally settling in Los Angeles in 1918. While working as a dancer, he sought bit parts in films. A role in *The Eyes of Youth* landed him the role of Julio in *The Four Horsemen of the Apocalypse*, which would change his life and career. Originally Julio was conceived as a much smaller part, but Valentino's onscreen personality quickly convinced the filmmakers to enlarge the role's function to add a romantic element to the film. With his dramatic entrance and swashbuckling tango (a scene lifted from an earlier film by the same director, Rex Ingram), Valentino danced into the hearts of American filmgoers and never left. The films that followed in his short seven-year career featured more of the same romantic characters with seductive eyes and bewitching ways. *The Shiek*, *Blood and Sand*, *The Young Rajah*, *A Sainted Devil* and *Son of the Sheik*, among others, all attracted throngs to the box office.

On Aug 15, 1926, Valentino was rushed to a hospital in New York. Suffering from a gastric ulcer and a ruptured appendix, he underwent surgery. Reports of his steadily declining health following the operation were broadcast on the new phenomenon, radio. His deathwatch took on the surreal quality of a national catastrophe that occupied the public's consciousness on a scale equal only to the death of a president or royalty. On Aug 23, 1926, Valentino died and pandemonium broke out among his legions of fans. Suicides were reported and riots erupted as his body lay in state. Wishing to pay their final respects, a crowd stretched for 11 blocks. Mounted police had to beat back the crowds. More than 100 mourners were injured. In both real life and his larger-than-life screen roles, Valentino was a romantic enigma who enchanted millions with his exploits. His death just contributed to the legend. For many years afterward, a "woman in black," and sometimes several "women in black," appeared at his tomb to mark the anniversary of his passing.

JOURNEYS OF THE SPIRIT
NEW YORK PUBLIC LIBRARY
FOUNDED • MAY 23 • 100 YEARS

One hundred years ago New York City was coming of age and experiencing an explosion in prominence, but while New York had two fine private libraries, the premier city in the land of opportunity had no public library. That was a source of great disappointment to then-Governor Samuel J. Tilden, who became the driving force behind the creation of The New York Public Library. As a result, on May 23, 1895, the Astor Library (a private reference library founded in 1849 with money donated by John Jacob Astor) and the Lenox Library (the private collection of rare books, manuscripts and Americana of distinguished book collector James Lenox, founded in 1870) were combined with a $2 million endowment and 15,000 volumes from the Tilden Trust to become the New York Public Library.

The Croton Reservoir, a popular strolling place that occupied a two-block section of Fifth Avenue between 40th and 42nd Streets, was chosen as the location for the new building. Today the library also includes four special research libraries and 83 branch locations.

The New York Public Library will celebrate its centennial by hosting a series of special events beginning in May 1995 and concluding in April 1996, including special exhibitions, educational programs, a birthday party, gala and more, plus the opening of the Library's new Science, Industry and Business Library. See Library in index for two events scheduled as of press time.

For info
Nancy Donner, PR Manager
The New York Public Library
5th Ave and 42nd St
New York, NY 10018-2788   Phone: (212) 768-7439

BUSBY BERKELEY
BIRTH • NOV 29 • 100 YEARS

William Berkeley Enos was born into a theatrical family in Los Angeles, CA, on November 29, 1895. He made an inauspicious stage debut at the age of five when his brother George, playing an Arab in the melodrama *Under Two Flags*, smuggled him onstage hidden in the folds of his costume. Hoping for a better life for their children, his parents sent him to military school in New York, and upon graduation he began work as a management trainee in a Massachusetts shoe factory. Dissatisfied with his employment and with the adventure of World War I beckoning, Berkeley enlisted in the army. While serving in France he had his first experience choreographing movements for large numbers of people. He devised a drill scheme for the 1,200 men in his battalion that involved complex patterns and that was executed without any verbal instructions. The military brass were impressed, and a star was born.

After his military service, Berkeley had limited success as an actor mixed with bouts of unemployment before landing a job as dance director of *A Connecticut Yankee* in 1927. Even though he admitted knowing little about music, Berkeley's career as a choreographer/director took off, and by 1930 he was in great demand on the Great White Way. But Hollywood beckoned. In his first film assignment, *Whoopee* (1930), his reputation as an innovator was solidified when he moved the previously stationary camera aloft to film the dancers' geometric patterns from above. His films featured lavish spectacle and intricate patterns of choreography, usually executed by a gaggle of beautiful women known as the "Berkeley Girls," whom he retained under personal contract. Among the few Berkeley Girls who were able to leap from his chorus lines into stardom were Lucille Ball and Betty Grable.

Berkeley's extravaganzas include *42nd Street*, *Footlight Parade*, *Gold Diggers of 1933* (all 1933); *Gold Diggers of 1935* (1935), the first movie he directed as well as choreographed; *Stage Struck* (1936), *Babes in Arms* (1939), *Strike Up the Band* (1940) and *Take Me Out to the Ball Game* (1949), the last movie he directed. He continued to choreograph, however, working on films such as *Two Tickets to Broadway* (1951), *Rose Marie* (1954), and *Billy Rose's Jumbo* (1962).

He died on March 14, 1976, at Palm Desert, CA.

JOHN FORD
BIRTH • FEB 1 • 100 YEARS

Born Sean Aloysius O'Fienne at Cape Elizabeth, ME, John Ford became one of America's preeminent film directors, famous for his cinematic depictions of the old West. Ford cut his teeth in Hollywood in the property department of Universal Studios and later as an assistant director. He also did some acting and had a small role in *The Birth of a Nation*, where he came under the influence of D. W. Griffith. As an assistant director Ford worked on many short subjects and early westerns and by 1921 had made approximately 30 films. His first big success came in 1924 with *The Iron Horse*, but his career as a full-scale director took off in the 1930s.

Ford had a deserved reputation for human dramas and literary adaptations, including *The Informer* (1935) and *The Grapes of Wrath* (1940). His legacy is best exemplified, however, in his robust and sentimental films of the American West made with John Wayne. Movies such as *Stagecoach* (1939), *My Darling Clementine* (1946), *The Searchers* (1956) and the highly acclaimed trilogy of *Fort Apache* (1948), *She Wore a Yellow Ribbon* (1949) and *Rio Grande* (1950) firmly placed both Ford's and Wayne's stars in the Hollywood firmament. Ford's achievements were honored with Best Director Oscars for *The Informer*, *The Grapes of Wrath*, *How Green Was My Valley* (1941) and *The Quiet Man* (1952). He died on August 31, 1973, at Palm Desert, CA.

WHAT A BABE!
GEORGE HERMAN RUTH, JR
BIRTH • FEB 6 • 100 YEARS

George Herman Ruth, Jr.—called "Jidge" by his teammates on the New York Yankees and "the Babe" by millions of fans all over the world—was born on Feb 6, 1895, in his grandparents' house at Baltimore, MD. Ruth was a rough kid who grew up in a tough world. He lived above the family tavern but spent most of his time roaming Baltimore's streets. By the time he was seven, his parents felt he was uncontrollable and sent him to St. Mary's Industrial School for Boys. Ruth was in and out of St. Mary's over the succeeding years, spending between seven and ten years there and learning to play baseball.

In 1914 Brother Gilbert, one of the Xaverian Brothers who ran St. Mary's, brought Ruth's pitching and hitting talents to the attention of Jack Dunn, owner and manager of the Baltimore Orioles, at that time a minor league team. Ruth left for spring training with the Orioles on Feb 27, 1914. In less than five months his contract had been sold to the Boston Red Sox and he was headed for the "bigs" (baseball parlance for the major leagues).

A lefty, Ruth started in the majors as a pitcher. In 1915, his first full year with the Red Sox, he achieved an 18-8 record, and over the next two years he won 47 games. In the 1916 and 1918 World Series, Ruth pitched a record 29 2/3 consecutive scoreless innings, which broke the Series standard set by the legendary Christy Mathewson in 1905. (This record stood for four decades; it was finally surpassed by Whitey Ford in the 1961 World Series.) The Babe was arguably the best left-handed pitcher of his era.

During Ruth's last two years with the Red Sox, his slugging prowess had become somewhat of a problem: his towering home runs had become the talk of the league, and fans were packing the ballparks to see the Babe hit. (Home runs were an extremely rare event in those days.) But pitchers played only every third or fourth day, and Red Sox officials were reluctant to convert their ace pitcher into an outfielder. So Ruth continued to pitch, although less frequently, during the 1918 and 1919 seasons, but he got more and more playing time in the outfield and occasionally at first base. His at-bats more than doubled from 1917 to 1918.

On Dec 26, 1919, the financially strapped owner of the Red Sox, Harry Frazee, sold Ruth's contract to the New York franchise for $125,000—a trade that to this day Red Sox fans are not allowed to forget—and thus he began his long, glorious and tumultuous career with the Yankees.

Ruth quickly assumed his rightful place among the league leaders in almost every offensive power category. In 1919 he led the American League in slugging average, home runs, runs scored, runs batted in and bases on balls—a feat he would repeat with astonishing consistency throughout the 1920s. The Yankees wasted no time switching "the Sultan of Swat" to the outfield to get his bat into the lineup every day. The rest, as they say, is baseball history.

Babe Ruth chalked up some pretty hefty contracts in his career. He made $80,000 in 1930 and 1931—phenomenal numbers for those days. Asked in March of 1930 (five months after the infamous stock market crash of Oct 1929) if he deserved such a huge salary, which was greater than the $75,000 Herbert Hoover was paid as President of the United States, he replied, "Why not? I had a better year than he did."

Notorious for curfew breaking and general carousing, Ruth was a bit of a discipline problem for his managers. In 1925, after he ignored curfew for two consecutive nights during a series in St. Louis, Yankee manager Miller Huggins fined him $5,000 and suspended him. Ultimately, Ruth's history of irresponsibility cost him what he wanted most after his playing days ended—a major league coaching job.

To say that Babe Ruth was baseball's greatest hitter is to understate his importance to the game. Perhaps no other individual has had such an impact on the sport. Before the Babe, baseball was dominated by singles hitters, low-scoring games were the rule, and the home run was considered a wasteful extravagance. In only his third season as an everyday player, Babe Ruth passed George Connor's lifetime total of 136 homers and became baseball's most prolific home-run hitter; each of his next 577 round-trippers was frosting on the cake. He hit at least 40 home runs in 11 of his 22 seasons of professional baseball; in four of those seasons he hit at least 50. Over a 14-year span, he won 12 American League home-run titles and led the league in slugging average 13 times. His single-season record of 60 homers, set in 1927, was not broken until 1961, by Roger Maris. And his lifetime total of 714 regular-season homers, long considered unreachable, stood for 34 years until it was finally topped by the great Henry Aaron in 1974.

In 1936 Babe Ruth became one of the five original inductees into the Baseball Hall of Fame. He died Aug 16, 1948, at New York City.

MANASSA MAULER
JACK DEMPSEY • BIRTH
JUNE 24 • 100 YEARS

William Harrison (Jack) Dempsey answered the first bell of his life when he was born June 24, 1895, at Manassa, CO. Using the nickname Kid Blackie, he began his boxing career on Aug 17, 1914, against Young Herman, and the match ended in a draw. Taking the name Jack he continued boxing, and by the age of 24 "Jack the Giant Killer" had battled enough opponents and had rung up enough knockouts to earn a shot at the title. On July 4, 1919, he squared off against the older title holder, Jess Willard, in Toledo, OH. The younger Dempsey threw such devastating punches that Willard went to the floor seven times in the first round. When Willard failed to answer the bell at the beginning of the fourth round, the world had a new heavyweight champion, by then known as the Manassa Mauler in recognition of his brutal style and home town.

Dempsey went on to compile an impressive list of title defenses, defeating the likes of Billy Miske, Bill Brennan, Georges Carpentier, Tom Gibbons and Luis Firpo. During his 1923 bout with the Argentinean Firpo, he was knocked out of the ring in the first round, climbed back in and pummeled Firpo, knocking him out in the second round. After this fifth successful title defense, Dempsey fought only exhibition matches for the next three years. Then in Philadelphia, on Sept 23, 1926, he lost his title to Gene Tunney in a 10-round decision.

Amid much hype and hoopla, a rematch was scheduled for Sept 22, 1927, in Chicago. The fight would live up to its hype in both ferocity and controversy. In the seventh round Dempsey knocked Tunney to the mat. Instead of retreating immediately to a neutral corner, as dictated by the rules, Dempsey—somewhat dazed himself—towered over his fallen foe, delaying the referee's beginning the ten count. Those few seconds allowed Tunney to regain his composure and make it to his feet on the count of nine. The match continued and Tunney was once again the victor by decision after ten rounds. The controversial "Battle of the Long Count" entered boxing history.

After this fight Dempsey retired, although he re-entered the ring for a series of exhibitions in 1931 and 1932 and again in 1940. After working as a boxing and wrestling referee and serving in the Coast Guard during World War II, Jack Dempsey became a successful restaurant owner and businessman in New York City, where he died on May 31, 1983, at the age of 87.

From the beginning of his professional career through 1940 Jack Dempsey fought 78 bouts, excluding exhibitions. Defeated only six times, he won 49 of his 62 victories by knockout, and 10 of his matches ended in a draw. He was inducted into boxing's first Hall of Fame in 1954 and into the International Boxing Hall of Fame in Canastota, NY, in June of 1990.

TO THEIR LABOR
SEPT 4
ANTHRACITE MINER'S MEMORIAL

In the heart of Pennsylvania's Anthracite Region, on Labor Day, Sept 4, 1995, the Shenandoah Area Chamber of Commerce will unveil the Anthracite Miner's Memorial in tribute to the people and industry that characterize the area's heritage and history. Anthracite coal has the greatest heat value and so is the most valuable, but is the least plentiful, found mostly in the eastern US. Its mining was one of the largest industries in the area, providing the fuel to drive the industrial revolution. In 1917, at the region's economic height, 329,000 miners brought up 278 million tons worth $705 million. The same year, 581 miners died in the deep underground.

The men who worked in the mines suffered through hazardous working conditions, risking cave-ins and explosions, to make a small living for their families. Children worked in the mines to supplement family income. Miners often spent 10 to 12 hours a day crouched or on their stomachs in narrow seams digging coal with picks. Many who survived the grueling days in the mines died years later of black lung disease.

Aside from the terrible conditions of the work itself, there were additional grievances that affected entire families and towns. The mines were in isolated areas with homes and stores completely controlled by the mining companies. Money was taken right out of the miners' wages to pay for rent, bills from the company store and all of the miners' equipment. Often there was little or nothing left of the pay. In slow seasons miners were forced to borrow from the mining companies; it often took years to pay off the debts.

To combat the unsafe working conditions and low pay, the first coal miners' union in the US was established in 1848 in the Pennsylvania anthracite region.

FOR INFO

Shenandoah Area Chamber of Commerce
PO Box 606
Shenandoah, PA 17976
Phone: (800) 755-1942

AMERICANS AT WAR 1945
WORLD WAR II • 50 YEARS

Nineteen forty-four had been a watershed year. An almost unbroken series of Allied conquests had left Germany and Japan surging inevitably toward defeat. But the triumphant course that began in Leningrad and ended in Leyte would branch and fork in uncharted directions before the war was concluded.

AS 1945 DAWNS

The Soviets had driven out the Germans, defeated Romania, Bulgaria and Finland, and then linked up with Tito in Yugoslavia and forced Germany out of the Balkans. On the western front the Allies had successfully staged the long-planned invasion of France; D-Day became the launch of a relentless press to drive the Germans back from there and from Italy. With the Red Army poised to sweep through Eastern Europe and Allied forces positioned to cross the Rhine, Germany was now being squeezed from all fronts. Clearly, the Third Reich was nearing its end.

The Japanese too were watching the enemy converge on their homeland. The Solomons, the Carolines, the Marshalls, New Guinea and most of the Philippines—all had been lost in the Allies' inexorable sea march toward Tokyo. In the wake of scores of hard-won battles, Japanese military strength was left at a fraction of its original magnitude.

It was, in fact, the technological and industrial might of the Allies that would help turn the final tide. Germany's touted secret weapons were proving to have no appreciable effect, while war production in the USSR was in high gear in 1944 and US efforts were producing military machinery at a stupendous pace—134,000 aircraft and 148,000 tanks in that year alone. Meanwhile, the US was on the brink of introducing a "new and most cruel bomb," as Japanese Emperor Hirohito later called it, and as the war entered another year, the Allied leaders decided the time had come to take this last resort. It was a weapon that would cast a shadow over major world events for the next 50 years, and the decision to use it was perhaps the most momentous of many by which the Big Three shaped the world that emerged from the war.

JAN 17—WARSAW LIBERATION

Late in 1944 Poland's struggle to free itself from military and political domination had come to a head. With five years of German occupation bearing down on them and the sound of approaching Russian bombs in their ears, the partisans had decided it was time to rise up and reestablish their rightful government before a Russian occupation supplanted the German occupation. The two-month ordeal ultimately failed and left Poland with nearly a quarter-million dead. In its aftermath the Polish military establishment, which largely supported the Polish government in exile, had been virtually annihilated by its German captors, and the balance of Warsaw's population had been deported, leaving the Russians a clear path to dictate the political makeup of the "liberated" Poland. In August 1944 the communist-controlled National Council in Poland had declared the city of Lublin the temporary capital, and on Jan 1, 1945, 16 days before the liberation of Warsaw, the Soviets declared the Lublin Committee the provisional government of Poland, preempting the Polish government in exile. After liberating Warsaw on Jan 17, the Russians restored the capital, but the city was in ruin and the country was under the control of the USSR.

FEB 4—YALTA CONFERENCE

Germany's fall was so certain by the end of 1944 that the need to plan for anticipated postwar problems—such as the disposition of the vanquished—became urgent. In response, the leaders of the Big Three—Franklin D. Roosevelt, Josef Stalin, and Winston Churchill—convened at the Livadia Palace in Yalta on Feb 4, 1945, to negotiate the fate of Europe.

Churchill and Stalin

Churchill and Stalin arrived at Yalta already at odds over postwar Germany. The Soviet premier insisted that Germany be deprived of 80 percent of its heavy industry and be required to pay $20 million in reparations, half of that to go to the USSR because of its heavy losses suffered at Germany's hands. Further, Germany should be partitioned and occupied by Russia, England and the United States.

Churchill objected to leaving Germany as crippled economically as it had been after World War I. Its strength might be needed should England and France have to defend the English Channel against the now formidable Russian Bear. To bolster France's position as well, Churchill demanded a fourth German zone and a seat for France on the Allied Control Commission scheduled to administer Germany's affairs after the war.

At the end of the conference Germany's fate remained largely undecided. Partitioning would be mentioned in Germany's surrender terms, but no binding power would be dictated. Likewise, reparations would be a jumping-off point for discussion, but final decisions were left for another time. Churchill's only success was in gaining a position for France on the Allied Control Commission.

Roosevelt and Stalin

At the top of Roosevelt's agenda for the Yalta conference was gaining Stalin's commitment to enter the war in the Far East as quickly as possible. Despite the assurances of Pacific Commanders MacArthur and Nimitz that Japan could be vanquished with a naval blockade and naval and aerial bombardment, FDR relied on warnings delivered by the Joint Chiefs—who, having underestimated the Japanese at the onset of the war, were now inclined to overestimate their shrinking powers—that without Russian assistance conquering Japan might cost a million American lives.

In a December 1944 meeting with US Ambassador W. Averell Harriman, Stalin stated that for its involvement in the Far East the USSR wanted in return the Kurile Islands, lower Sakhalin, leases to Port Arthur and the Manchurian railroads and acknowledgment from China of Outer Mongolia's independence. The US State Department advised against outright transfer of the lower Sakhalin and the Kuriles to the Soviets, but written materials stating that position were inexplicably missing from the briefing book provided Roosevelt when he departed for Yalta. Roosevelt accepted the position of the Joint Chiefs and in a private meeting with Stalin granted him his entire list of demands. In return the Soviet premier promised to enter the war against Japan two to three months after a German surrender.

The United Nations

At Yalta the proposed United Nations was also discussed. The participants ensured the autonomy of their separate nations as well as their world power by unanimously agreeing that each permanent member of a UN Security Council would have veto power and that no direct decision-making powers would be granted to the United Nations.

Stalin continued to demonstrate his negotiating skills when he claimed a seat on the General Assembly for each of the 16 republics of the Soviet Union but ultimately accepted only three seats (for Russia, Byelorussia and the Ukraine). In the process he impressed Churchill and Roosevelt with his willingness to compromise while putting himself in a better position to demand what he'd really come to Yalta for: Poland.

Poland

Because Great Britain had entered World War II in direct response to Germany's invasion of Poland, Churchill considered it his duty to restore Polish sovereignty after the war. To that end he had tried to bring Stalin and the London-based exiled Polish government to agreement prior to the Yalta conference.

Stalin considered the question of Poland a matter not of honor but of national defense. Historically Poland had been used as a gateway to invade Russia, and Stalin wanted the Soviet-Polish border pushed west to the old Curzon Line proposed by the British in 1920. Since the area was populated mostly by Ukrainians and Byelorussians, Churchill and Roosevelt readily agreed.

When Stalin proposed that Poland be compensated for the lost territory by extending its western frontier to the River Neisse in Germany, however, Churchill countered that "it would be a pity to stuff the Polish goose so full of German food that he will die of indigestion" and reminded Stalin that Hitler used the presence of ethnic Germans in the Balkans as an excuse to invade there.

Continued discussion about the western border was postponed for the Peace Conference, but Russia's presence throughout Eastern Europe made Stalin's proposals a fait accompli. In fact virtually every concession Stalin made on Poland at Yalta later evaporated in the frenzy of postwar regrouping. The open elections agreed to were never supervised; the organization of the provisional government on broader and more democratic lines was circumvented by trumped-up technicalities.

In essence, the Yalta agreement gave Stalin written permission to do as he wished in the Balkans, Poland and China irrespective of the Atlantic Charter and the United Nations.

FEB 13—BUDAPEST FALLS TO THE RUSSIANS

Hungarian Regent Admiral Miklos Horthy had allied his country with Germany against their Russian nemesis in 1941. Three years later Hungary declared war on Germany, just two months before Budapest fell to the Red Army, a crushing blow to the European Axis.

When Stalin's forces entered Dubno, Poland, just 170 miles from Hungary's eastern border, in March 1944, the need for oil was already one of Germany's biggest concerns, so Hitler informed his Hungarian allies that German troops were taking control of their oil fields as well as other raw materials. The Allies responded with heavy bombing of Budapest in July, making the oil refineries and storage tanks their main targets. They also began dropping leaflets warning that anyone who carried out Germany's threatened deportation to Auschwitz of the 750,000 Jews residing in Hungary would be tried and punished. The Hungarians refused to assist in the deportations, and they ground to a halt, but not before more than 437,000 Hungarian Jews had been sent to the death camp.

Wholesale deportation resumed in October following the kidnapping of Adm Horthy, who had never cooperated in the removal of Hungarian Jews. Thousands of men, women and children were forced to dig antitank trenches against the advancing Soviets as they were marched by foot toward Germany.

The contest for Budapest began in earnest in the declining days of 1944. By Dec 4 the Russians had practically cut off the Germans' stronghold in the capital, and Hitler began ordering reinforcements from the western front. After a three-day battle the Soviets surrounded the city and severed all German supply routes.

On Dec 29 the Russians demanded that Budapest be surrendered. The pair of Red Army emissaries who approached the city under a white flag were promptly shot, and two days later Hungary declared war on Germany.

By mid-January the Red Army and Germans were fighting hand to hand on the streets of Budapest. On Feb 13 Russian soldiers reached the central city, and Hungary was transferred into the hands of the Soviets. With Budapest, left ravaged on both sides of the Danube, went three-quarters of Germany's oil supply and its last grasp on Europe.

FEB 13–14—FIRE BOMBING OF DRESDEN

While the Russians were breaching Budapest, Allied bombers were targeting a relatively wide area a few hundred miles to the northwest. The aim was to cut off the mass movement of German military divisions form the western to the eastern front. The result was the total devastation of Dresden, an accomplishment of dubious strategic benefit that became a symbol for the horror of war.

The Berlin-Dresden-Leipzig area was proposed for attack by Soviet Deputy Chief of Staff General Antonov, who as the Yalta Conference got under way on Feb 4 made a plea to the Allied Chiefs of Staff to redirect bombers form German oil reserves and supplies to enemy rail lines and roadways. The resulting decision was approved by Air Chief Marshal Arthur "Bomber" Harris, who had little patience with the limited success of precision bombing and advocated "area" bombing on the order of Germany's air attack on Coventry, England, on Nov 14, 1940. Harris argued in favor of saturating whole areas with high explosives and using incendiaries to destroy military targets. The firestorm that had engulfed 60,000 of Coventry's 75,000 buildings and killed some 568 men, women and children had demoralized British citizens to Germany's advantage. Harris hoped to accomplish the same in the Dresden area.

On Feb 13, 245 British bombers attacked the railway marshaling yards at Dresden; a second wave of 529 British bombers followed three and a half hours later. The resulting firestorm burned through 11 square miles. The next morning 450 American bombers hit Dresden's historic city center, previously untouched by the war. The flames lasted seven days and nights.

Death toll figures for Dresden range from 40,000 to almost 135,000. An exact count was impossible. Of the 39,773 bodies officially identified, most had burned to death. At least 20,000 bodies were buried under debris or incinerated beyond recognition. The inscription on a mass grave in Dresden's main cemetery reads, "How many died? Who knows the number?"

Dresden was bombed a total of six times—the last attack was on Apr 17, when 572 American bombers hit the city. In retrospect it became clear that the bombing of Dresden had little military impact.

FEB 19—US LANDINGS ON IWO JIMA

The Japanese defined the area surrounding their homeland as the Absolute National Defense Zone. Its parameters were considered sacred, and the military had vowed that the enemy would never penetrate the outer edges. Yet by November 1944 the Allies had not only penetrated most of the zone but were regularly bombing the Japanese home islands.

These long-distance bombing missions whetted the Allied appetite for an island base closer to Japan. An air base on Iwo Jima, only 760 miles south of Tokyo, would reduce the heavy losses of B-29 Superfortresses due to Iwo Jima's radar and to the prohibitive 1,500 miles the planes had to cross to return to their bases on Saipan and Tinian in the Marianas. A home base on Iwo Jima would enable American planes to carry less fuel and a greater bomb load while preventing Japanese raids on the Marianas and making a subsequent Allied invasion of Okinawa possible.

Less than five miles long and half as wide, bleak and barren Iwo Jima gave the Japanese a perfect fixed position from which to defend the homeland. Its 1,500 invisible caves made an ideal network of underground fortifications. Commander Lt Gen Tadamichi Kuribayashi installed in them blockhouses and pillboxes with nearly impregnable walls surrounded by 50 feet of sandbags and topped with machine-gun nests. Kuribayashi, a harsh perfectionist, also built two formidable defenses a mile apart and reaching from shore to shore. The only beaches accessible for landing were on either side of the 548-foot-high inactive volcano Mt Suribachi at the southern tip of the island.

Allied aerial bombardment of Iwo Jima's military installations began Feb 16, 1945, simultaneously with the departure of the American expeditionary landing forces from Saipan. Taking part in the landing on Feb 19 were the Fourth and Fifth Marine Divisions, with the Third in reserve. As the Marines came ashore, the beaches exploded with land mines but the soldiers were otherwise unopposed. As soon as the landing was completed, the Japanese began furious shelling from every small incline. Attempts to dig foxholes resulted in landslides that immediately refilled them. Tanks sank in the sand or were blown up as they came ashore.

Kuribayashi's 21,000 soldiers manning thousands of fixed positions, some large enough to hurl rockets weighing 200 to 550 pounds, gave the Marines their worst landing of the Pacific action. Among the 2,420 casualties, 600 resulting in death, were many that were rewounded while waiting to be attended or evacuated.

A week after the landing, the Fifth Division fought fiercely to take the Meatgrinder, the high ground under which lay the island's massive communications center. After seven days of concurrent attacks on the three mutually supporting strongholds capable of turning their guns on one another—the Hill, the Amphitheater and Turkey Knob—the Marines captured the Meatgrinder at a cost of 6,591 casualties.

The battle for Iwo Jima continued relentlessly until the entire island was secured Mar 21. The toll was more than 5,000 Marines killed on Iwo Jima and almost 20,000 more injured. Of 22 battalion commanders to land with their men, 19 were killed or wounded. In the words of Adm Nimitz, "On Iwo Jima, uncommon valor was a common virtue." Kuribayashi's notorious devotion to duty never wavered; after weeks of battle only a handful of his troops were alive.

MAR 3—MANILA SECURED

On Jan 2 an attack fleet of battleships, cruisers and destroyers, along with air-support carriers, left Leyte headed for Lingayen Gulf, on the western shores of Luzon, in the Philippines. Ten divisions and five regiment combat

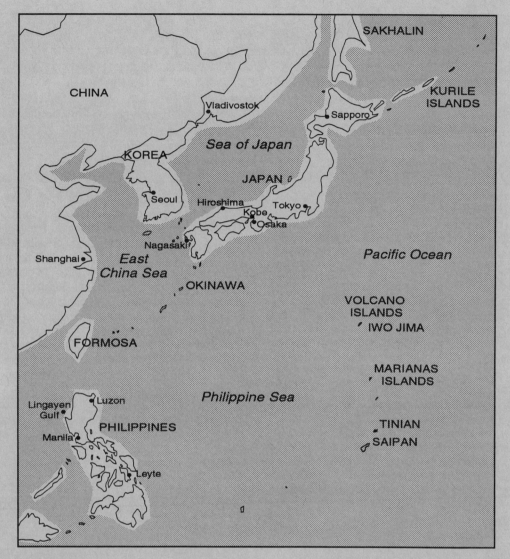

teams of the Eighth and Sixth Armies were about to take part in the largest campaign of the Pacific war, the fulfillment of Gen Douglas MacArthur's vow to recapture the northern end of the Malay Archipelago. Lined with fine beaches, a nearby railroad and good roads, the gulf made an ideal landing site. But as Allied ships arrived off the coast, kamikaze attacks began. During two days of suicide attacks the US Navy lost 24 ships, and 67 more were damaged. But by Jan 9, when American troops began going ashore, the Japanese Air Force was completely decimated.

Aware that he hadn't enough manpower to defend Manila or Bataan, Japanese commander Lt General Tomoyuki Yamashita divided his remaining forces into three groups and moved to the interior to stall the Americans and inflict as many casualties as possible. Encountering no resistance during their landing, the Sixth Army was able to form a 20-mile beachhead within several days. The XIV Corps then moved south through farmland. Heading north and east to seize an important road junction, I Corps was hit by Japanese artillery, mortar and machine guns so well placed in the mountainous terrain that it took until the end of January to take the all-weather airfield at the Clark Field complex coveted by MacArthur.

Yamashita's strategy so fragmented his defenses, however, that the American XI Corps landed unopposed southwest of Clark airfield on Jan 29 and quickly advanced to close off Bataan. MacArthur then launched a three-pronged advance by the 37th Infantry, the 1st Cavalry Division and the 11th Airborne to reach Manila with the utmost speed. With the 11th stopped on Feb 4 by the Japanese defenses south of Manila and the 37th halted by unfordable streams, ultimately it came down to a race between the 1st Calvary's two flying columns. On Feb 3 demolition expert Naval Lt James P. Sutton braved enemy fire to cut the fuse of a bomb meant to blow up the only bridge crossing the gorge of the Tuliahan River, and the 8th Regiment claimed glory when they reached the outskirts of Manila by nightfall.

From there Allied progress slowed drastically. Liberation did not come to Manila until almost a month later, after bloody combat that often came down to one-on-one struggles for individual streets. Despite MacArthur's refusal to use devastating air support, nearly 100,000 Filipino lives were lost in the heavy artillery crossfire. Japanese resistance finally ended on Mar 3, 1945, freeing the US commander of the southwest Pacific to complete his mission to return victorious to Bataan and Corregidor.

> ### *Divine Wind*
> In 1281, as a fleet of Mongol invaders approached Japan, a typhoon formed at sea, and its forceful winds dispersed the encroaching threat. That typhoon came to be called *kamikaze*, meaning "divine wind." In the waning months of World War II, Japan again looked to "Divine Wind" to aid the empire. In this case the kamikazes were young and inexperienced pilots who swooped down out of the sky in deliberately suicidal attacks on Allied targets—particularly ships. The greatest damage inflicted upon the US Navy during the war—indeed, during the entire history of the Navy—occurred during the two-month-long Battle of Okinawa as a result of kamikaze attacks. More than 1,900 kamikaze missions resulted in the sinking of 38 ships and the damaging of dozens of others. Close to 5,000 men lost their lives in these attacks.

MAR 7—US TROOPS SEIZE BRIDGE AT REMAGEN

As the last winter of the war moved into February, the British Second and Canadian First Armies were taking the first step in the Allies' plan to surround the Germans in their own land by invading from two points on the Rhine, and by mid-February the German army had been pushed back to the river. The retreating Nazi army fled across to the east bank while blowing up bridges behind them.

Hitler considered the homeland sacrosanct. Faced with the possibility that Germany would be breached by an enemy for the first time in 140 years, he threatened the firing squad for any officer who allowed a bridge to be captured intact.

Eight armies lined up along the west bank of the Rhine—the American Ninth, First, and Seventh; the Canadian First; the Allied First Airborne; the British Second; and the French First. Eisenhower planned a direct assault over a wide front and expected the retreating aggressors to inflict heavy losses. But a chance event altered these plans.

On Mar 7 the 9th Armored Division, at the Bonn-Cologne sector, split into two columns: one made for the Ahr River in search of a remaining bridge; the other aimed at capturing the west-bank city of Remagen. At Remagen the Americans were taken completely by surprise: the Ludendorff Railway Bridge was still standing. As they approached the bridge an explosive discharged, creating a crater at the western end but leaving the catwalk in place. The main charge intended to demolish the bridge had failed to go off. A company of American soldiers braved heavy enemy machine-gun fire and rushed the bridge towers. Desperate Army engineers hurriedly cut fuses and cables before the Germans had another chance at the bridge. Miraculously, the Allies had gained access to the land of the Third Reich. Hitler ordered four officers court-martialed and shot. Within 24 hours 8,000 American soldiers made their way to the east bank of the Rhine. By the time the German army finally destroyed the Ludendorff Bridge 10 days later, American engineers had built 62 spans across the Rhine. By the end of March seven armies had crossed the river.

MAR 9–10—US FIRE RAIDS ON TOKYO

The first US aerial strike at the Japanese homeland had been a success, but in June 1944 the Allies were unwilling to wait for the manufacture of the many more Superfortresses that continual precision bombing would require. So on Feb 4, 1945, they used the incendiary (area) type of bombing (which had brought a halt to the Japanese offensive in China in late 1944) to launch their full-scale attacks on Japan. The initial target was Kobe, Japan's sixth-largest city, and the Superfortress attack cut the city's shipbuilding capability in half and damaged 5 of the 12 largest factories. Incendiary bombing of Tokyo was undertaken a month later. Attacks on Mar 8 and 9 destroyed 18 percent of the industrial area of the city but also cost 83,000 civilian lives and injured more than 41,000 people, leveling some 267,000 buildings in a 16-square-mile area.

After 600 B-29s were made ready, American bombing turned from the major cities of Tokyo, Nagoya, Kobe, Osaka, Yokohama and Kawasake to 50 lesser industrial areas. As with the German air war against Great Britain, the psychological effect on the populace was incalculable.

APR 1—US LANDS ON OKINAWA

Air raids were chipping away at Japan, and by April 1945 the Allies wanted an improved air base to continue their assault on the homeland. In the greatest amphibious assault of all time, the battle for Okinawa became one of the most horrific battles of the Pacific war.

The 485-square-mile island of Okinawa is a mere 380 miles from the southernmost Japanese home island of Kyushu. Like Iwo Jima, it is honeycombed with tunnels capable of holding high-caliber weapons. Japan's plan to defend it, code-named Ten-Ishigo, featured suicide air strikes against American ships transporting and supporting the amphibious assault. These Special Attack Forces, which came to be known as *kamikazes*, consisted of old, stripped-down planes loaded with high explosives and flown by young and inexperienced pilots. Their major objective was to drive the American landing fleet away from shore and leave the beached assault forces without naval support.

As on Iwo Jima, the Japanese held back their defenses when on Apr 1 an armada of 1,300 ships landed 50,000 American troops of the 1st and 6th Marine and 7th and 96th Army Divisions on an 8-mile-long beachhead. The Allies expended 30,000 heavy-caliber shells in the war's largest prelanding bombardment, though the Japanese offered no resistance. Knowing it was futile to attempt to oppose the superior firepower, the Japanese waited for the Americans to make their way inland. When the US Army's XXIV Corps pushed directly across the island and turned south, the Japanese began some of their most ferocious fighting of the war.

On Apr 6 the Japanese air and sea attack, dubbed Operation Floating Chrysanthemum, hit the American fleet with 355 kamikazes in a single stroke, sinking three destroyers, two ammunition ships and a tank-landing ship. The next day another kamikaze attack damaged a battleship, a carrier and two destroyers. Hoping at least to hinder, and perhaps even prevent, an Allied invasion of the homeland since victory on Okinawa was virtually impossible, Gen Mitsuru Ishijima made wholesale destruction of American sea power his chief objective. Between Apr 6 and June 10, kamikaze missions involving 50 to 300 aircraft inflicted significant physical and psychological damage on American naval forces off the shore of Okinawa; but the Japanese air force quickly dwindled, and by the end of June its supply of pilots and aircraft was depleted.

Switching to a new suicide tactic, on Apr 7 the Japanese launched the 72,800-ton battleship *Yamato* with 2,500 tons of fuel—enough for a one-way trip from Japan. Its mission was to break through the amphibious screen around Okinawa and devastate the enemy with its 18-inch guns. The *Yamato* never reached its destination. After being attacked by 280 American aircraft and torpedoed six times the first day out, she rolled over and sank with 2,300 Japanese sailors still aboard.

All in all, a quarter of a million men of the US Tenth Army had landed on Okinawa, and the Japanese had failed to turn back the American invasion. Awaiting the onslaught from their underground bunkers, most of the Japanese suffocated or were burned alive in their positions or were forced out of their fortifications by explosive charges. For a time the advancing Allied tanks were stopped by a quagmire resulting from torrential rains, and the battle for Okinawa raged on into May.

APR 12—ROOSEVELT DIES

At Yalta in February, President Roosevelt had been noticeably pale. The man who once towered over Churchill and Stalin now seemed barely their equal in stature. On Apr 12, at 1:00 PM, while sitting for a portrait at his home in Warm Springs, GA, he suffered a cerebral hemorrhage and was dead by 3:35 PM.

Americans mourned Roosevelt's passing on a scale not seen since the assassination of Abraham Lincoln. Soldiers wept at the news of his death, and as the funeral train from Warm Springs to Washington traveled through small towns in the Carolinas and Virginia, mourners packed the rails at every station. A shocked Harry S. Truman later described his first months as president as "on-the-job training."

APR 25—US AND RUSSIAN TROOPS MEET AT THE ELBE

Shortly after noon on Apr 25, 1945, a lone American officer and a single Red Army soldier met on the western bank of the Elbe near the town of Lechwitz, Germany. The east and west Allied armies were drawing together at the demarcation line between the Soviet and western occupation zones—they had successfully split Germany down the middle. Ten miles northwest and four hours later, American patrols came upon more Soviet soldiers at the village of Torgau, 75 miles south of Berlin. Completely disregarding rank or nationality, the two armies embraced and toasted one another. In Moscow 324 guns fired a 24-salvo salute upon receiving the news. In the US crowds danced and sang in Times Square.

APR 28—MUSSOLINI KILLED BY PARTISANS

On Feb 25, 1945, SS Gen Karl Wolff had dispatched an emissary to Switzerland to meet with US Secret Service Chief Allan Dulles and offer a negotiated surrender of all German forces remaining in Italy. All that remained for Wolff was to convince the German commanders in the field. Time was not on his side. On Apr 2, as the Allies began their final offensive in Italy, Hitler sent Field Marshal Kesselring an order to replace all commanders who failed to stand their ground.

Hitler's commands notwithstanding, a German retreat had begun. On Apr 20, attempting to cut off all lines of escape, Allied bombers began Operation Corncob, a three-day attack on the bridges over the rivers Adige and Brenta. On Apr 24 Allied forces crossed the River Po and the Italian Committee for National Liberation ordered a general revolt in those areas still under German control. Meanwhile, Benito Mussolini had returned to Milan on Apr 18 hoping to rally his supporters with a mass demonstration. When he realized American troops were rapidly converging on the city, he fled in haste, accompanied by Fascists and the SS guard assigned to "protect" him when he had begun his 600-day rule of the collapsed republic. By Apr 25 the city of Milan had been liberated by Italian partisans. On Apr 27 partisans in Turin also took up arms, and the Fascist-Nazi partnership effectively came to an end.

On Apr 28, 1945, Mussolini's motorcade, which had been redirected toward Germany by the SS, was diverted again. This time it was the Italian Communists, and they summarily imposed a death sentence. Mussolini was shot nine times, along with his mistress, Clara Petacci, and their bodies were taken to Milan, where they were hung upside down in the Piazzale Loreto on the morning of Apr 29. The 23-year period of Fascist control of Italy had come to an end. The same day, Gen von Vietinghoff's representatives signed the unconditional surrender of German troops in Italy.

APR 30—HITLER SUICIDE AS RUSSIANS CLOSE IN

April 1945 brought not only the promise of spring but also the promise of victory in Europe at last. By previous agreement among the Allies, the Red Army was given the honor of conducting the assault on Berlin. Because both had been promised the command, Stalin made the offensive a contest between his military adviser Zhukov and Konev, who could be expected to take Berlin without significant casualties. The charge began from two fronts at 5:00 AM on Apr 16. Zhukov quickly broke German defense lines and was ready to assault Berlin by Apr 19. On Apr 20 Konev reached the outskirts of Berlin and was ordering his army into the capital as Zhukov brought up guns from the 6th Breakthrough Artillery Division and began bombarding the streets of the capital.

The Red Army chose that day, Hitler's 56th birthday, to open artillery fire on Berlin.

Hitler's Last Days

Hitler celebrated his birthday with a tea party in his Berlin bunker under the Chancellery, from which he emerged only briefly to inspect the SS unit of Frunksber Division and decorate a squad of Hitler Youth made up of boys orphaned by the Dresden bombings and given the honor of defending the capital. It was Hitler's last public appearance. Some guests were impressed by Hitler's intention to continue the fight, while others were convinced he intended to take his own life.

The next day Russian troops entered Berlin's northern suburbs. On Apr 22 Hitler called for an attack on the Red Army at Eberswalde but failed to rally even one soldier. He finally recognized that the war was lost. The next day he personally assumed command of Berlin and vowed to remain there and shoot himself when the end came.

By Apr 26 the Red Army had entered Potsdam. Generals Konev and Zhukov had encircled Berlin with 464,000 troops while gathering forces to eliminate the remaining resistance from within the capital. Food and water were running out, and Russian bombardment had cut off electrical, gas and sewage service. The residents of Berlin were reduced to living in cellars.

By Apr 27 the portion of Berlin remaining in German hands had been reduced to a 10-by-3-mile strip.

On Apr 28, 1945, Josef Goebbels officiated at the wedding of Adolf Hitler and Eva Braun. While waiting for the ceremony to commence, Hitler wrote his last will and testament. He named Karl Doenitz president of the Reich and appointed Freidrich Schoerner Commander in Chief of the armed forces. A wedding feast followed the ceremony.

By Apr 29 fighting was just a quarter of a mile from the Reich Chancellery, and beneath its garden Hitler was giving his final orders. Goering and Himmler were expelled from the Nazi Party—Goering for presuming he would replace Hitler as head of state and Himmler for making overtures of peace to the Allies. Hitler then called for the destruction of all railroads, canals, locks, docks and ships as well as the electrical, gas and water systems. He gave orders to armies that no longer existed.

That evening phials of cyanide were given to those who wanted them. Goebbels, who had moved into the bunker with his wife and six children, asked for eight.

At midnight Hitler went to bed. At 2:00 AM he arose again and passed among those in the bunker, shaking hands but saying nothing. On Apr 30, shortly after having a last meal, Hitler simultaneously took cyanide and shot himself in the mouth. Eva took cyanide only. Their bodies were incinerated in a shell crater in the Chancellery garden.

An hour earlier soldiers from Zhukov's 1st Battalion, 756th Rifle Regiment, 150th Division of the Third Shock Army, had run a Red Victory banner outside the second floor of the Reichstag.

MAY 7—GERMANY'S FORMAL SURRENDER

In the wake of Hitler's suicide, Germany's high command launched a desperate attempt to unite with the US and Great Britain in an anti-communist association against Russia. Heinrich Himmler offered to surrender all German forces to the western Allies, but President Truman promptly reiterated the Allied demand for Germany's unconditional surrender on all fronts. In fact, German soldiers had already begun to surrender, though many who had fought on the eastern front traveled west to give themselves up to the British or Americans. German civilians were traveling west as well, hoping to avoid Russian occupation.

On May 4 Grand Adm Karl Doenitz sent Adm Hans von Friedeburg to Supreme Headquarters Allied Expedition Force (SHAEF) headquarters to request authorization to surrender to the West. The Americans again demanded a surrender on all fronts. A second stipulation that German citizens not travel west of the Elbe was never enforced, and uncounted numbers of Germans continued to flee.

Finally Germany accepted the inevitable. At 1:41 AM on May 7, 1945, Generals Bedell Smith, Sir Frederick Morgan, Carl Spaatz, Francois Sevez, Ivan Susloparov and other ranking officers gathered on the upper floor of a school in Rheims. In a room barely large enough to hold them, they looked on as the chief of the German general staff, Field Marshal Alfred Gustav Jodl, signed a full and unconditional surrender for the German state. After his complete capitulation, Jodl was ushered into an adjoining

---

A Sign of the Times
American officer Colonel Donovan P. Yeuell acquired a particularly unusual souvenir at the close of World War II—a plaque that had been hung over the prison cell at Landsberg, Germany, that Hitler occupied for a time in the 1920s. It read, "Here a dishonorable system imprisoned Germany's greatest son from November 11, 1923 to December 20, 1924. During this time Adolf Hitler wrote the book of the National Socialist Revolution, *Mein Kampf* [*My Battle*]." This book was Hitler's blueprint of his political and military plans. The plaque can be seen at the Kentucky Military History Museum, in Frankfort, KY.

room where the Allied Supreme Commander, Dwight D. Eisenhower, asked if he understood what he had signed and informed him that he would be held personally responsible for upholding the terms of the surrender.

In Berlin the following day, May 8, 1945, at half an hour before midnight, the Germans signed the document of capitulation demanded by the Soviets. The war in Europe had ended.

JUNE 21—OKINAWA SECURED

The war in the Pacific raged on. Eighty-two days of fighting on Okinawa cost the US Army 4,000 men, the Marines 2,938 and the Navy more than 5,000 offshore, most from kamikaze attacks. Between 70,000 and 160,000 of Okinawa's civilian population of 450,000 were killed. The Japanese lost 110,000—including servicemen, clerks, cooks and Okinawa labor conscripts. Large numbers of them were suicides. Only 7,400 Japanese prisoners were taken. When the Marines reached the Japanese command cave at Mabuni on the evening of June 21, they found the remains of Lt Gen Mitsuru Ushijima and Lt Gen Isanu Cho, who had committed hara-kiri (ritual suicide) that morning.

JULY 16—FIRST SUCCESSFUL ATOMIC BOMB TEST

At Alamogordo Air Base in the New Mexico desert at 5:30 AM on July 16, 1945, a plutonium bomb, dubbed "Fat Boy," had been set off from atop a 100-foot-tall steel tower at the Manhattan Project's Trinity headquarters, and the blast had set the sky aglow as the resulting cloud grew to a height of 8,000 feet within a fraction of a second. A ball of fire more than a mile in diameter had then spread into the shape of a mushroom and climbed to a height of 41,000 feet. Manhattan Project leader Dr Robert Oppenheimer, one of the witnesses of this awesome event, quoted from the *Bhagavad-Gita*, the sacred epic of Hinduism: "I am become Death, the shatterer of worlds."

The weapon had been in the works since March 1941, when plutonium (named after the planet discovered only 11 years earlier) was discovered. Now armed with the necessary components to accomplish nuclear fission, the United States made the decision on Dec 6, 1941, to proceed with the production of an atomic bomb under a unified program. The bombing of Pearl Harbor the very next day convinced many previously reluctant scientists to join the Manhattan Project.

Less than a year later, on Dec 2, 1942, under the stands of the unused football stadium at the University of Chicago, Italian emigre scientist Enrico Fermi and colleagues achieved the first man-made and self-sustaining nuclear chain reaction. When ground was broken two months later at Oak Ridge, TN, for a huge plant to manufacture uranium-235, the stage was set for the making of an atomic bomb.

President Truman had first learned of the Manhattan Project upon his inauguration. On Apr 25, 1945, barely two weeks later, Secretary of War Henry Stimson told Truman of the rapid progress being made and said, "Within four months we shall in all probability have completed the most terrible weapon ever known in human history, one bomb of which could destroy a whole city."

JULY 17–AUG 2
POTSDAM CONFERENCE

At Potsdam, Josef Stalin got what he had gone after in Yalta three months before Germany's surrender. The decisions made at the Potsdam Conference ruthlessly simplified the postwar map and resulted in Soviet domination of central and eastern Europe for years to come. For example, Soviet territorial claims on the Baltic eventually led to Estonia, Latvia and Lithuania's becoming Soviet republics. Russia got the reparations from Germany that Stalin had demanded at Yalta, and in effect the Stalin-dictated Polish frontiers gave Soviet-dominated Poland an occupied zone equivalent to those assigned to the US, Great Britain, France and the USSR.

The communique issued from the conference on Aug 1 detailed Germany's fate: the Nazi Party was abolished, and Nazi leaders charged with war crimes were to be tried.

Meanwhile, on July 21, the day before receiving news at the conference of the first successful atomic test, Truman and Churchill had agreed in principle that the atomic bomb should be used when ready. Three days later, knowing for certain the destructive capability of the new weapon, Churchill and Truman sent Japan a message that ended, "We call upon the Government of Japan to proclaim now the unconditional surrender of all the Japanese armed forces, and to provide proper and adequate assurances of their good faith in such action. The alternative for Japan is complete and utter destruction."

AUG 6—ATOMIC BOMB
DROPPED ON HIROSHIMA

Secretary of War Stimson's earlier prediction had been pessimistic. The goal was reached in less than three months, and four preliminary targets were already chosen. At the end of July 1945 Truman was notified that of the four cities being considered only Hiroshima did not have Allied prisoner-of-war camps. Thus the fate of Japan's seventh largest city, with a population of 250,000 plus 150,000 soldiers of the Second General Army who were headquartered there, was sealed.

On the morning of Aug 6, 1945, a plane named the *Enola Gay* took off from Tinian Island in the Marianas with its historic payload. Its crew of six was chosen from a group of 75 who had trained for this mission for months. Because of fears of a crash during takeoff, Capt William S. Parson would assemble the atomic bomb after the B-29 was airborne.

It was a clear day over Hiroshima, with visibility extending from 10 to 12 miles. At 7:07 AM, an air-raid siren signaled residents to rush for shelters. A mere 24 minutes later sirens prematurely alerted people that all was clear.

As the bomb hit Hiroshima an indescribable white-pinkish light permeated the city and a blast louder than 100 thunderclaps saturated the air. Suffocating heat and howling winds consumed the center of the city and swept away everything in their path. Eighty thousand people caught outdoors were killed instantly, and 35,000 thousand more lay mortally wounded and writhing in pain as cars and buses were flung about among them. Nothing within three-quarters of a mile was left erect or alive.

At 9:20 AM, Parsons sent two words from the *Enola Gay* across the Pacific to its home base on Tinian Island: "Mission successful!"

Beyond the center of the blast thousands more were killed or wounded as structures collapsed like dominos. So intense was the heat that people swarmed into rivers and streams, only to be boiled to death. A half hour after the bombing, cries for water were answered by a gentle rain that stopped abruptly five minutes later—the result of humidity rising from the intensely hot earth. Of 90,000 buildings in the city about 62,000, or 60 percent, were destroyed by the blast and ensuing fires. Before the day ended, as many as 150,000 people died or received wounds from which they would later die. Thousands more perished over the next two or three weeks from the effects of gamma rays. Among those killed were several American pilots who had been shot down only eight days earlier. Only 20 of Hiroshima's 200 doctors and 150 of the city's 1,780 nurses were able to attend to the wounded. Illnesses resulting from radiation are still being diagnosed a half century after the bombing.

Several hours after the atomic bomb was dropped on Hiroshima, the US government issued a statement calling on the Japanese to surrender or to expect complete destruction.

AUG 9—ATOMIC BOMB
DROPPED ON NAGASAKI

On Aug 9, after three days had passed with no response from the Japanese, a B-29 named *Bock's Car* departed Tinian Island for the city of Kokura, Japan. Kokura was obscured by clouds, though, so the city of Nagasaki became the target of the second atomic bomb. More than 35,000 were killed in this raid, and an additional 5,000 died from related injuries before year's end.

Just hours before the bomb was dropped on Nagasaki the Japanese Supreme Council for the Direction

of the War met to discuss the demands of the Potsdam Declaration, but the meeting had broken down in bitter argument.

The following day Emperor Hirohito communicated through the Swiss that Japan was willing to accept all Allied terms except elimination of the imperial institution. A compromise was accepted by the Allies: "The authority of the Emperor and the Japanese Government to rule the State shall be subject to the Supreme Commander of the Allied Powers."

AUG 14/15—JAPAN SURRENDERS AND THE WAR IS ENDED

On Aug 15 (Aug 14 in the US due to the time difference), Emperor Hirohito, who normally never spoke in public, announced by radio to the army and people of Japan that the Japanese government would negotiate with the enemy. A few hardline militarists at first refused to accept the terms and tried to continue the fight; a few others committed hara-kiri. In the US President Truman announced the surrender. The official ratification would not be signed until Sept 2 aboard the USS *Missouri* in Tokyo Bay, but the war was over.

On Aug 28, 1945, MacArthur arrived at Yokohama to initiate the American occupation of Japan.

World War II Estimated Military Casualties

Country	Dead	Wounded	POW/MIA
Allied Powers			
Belgium	7,760	*	70,000
British Empire	353,652	475,070	417,303
China	1,310,224(d)	1,752,591(d)	115,248(d)(e)
France	166,195	408,895	1,500,000(b)
Greece	77,200	*	*
The Netherlands	6,238	20,000(b)	25,000(a)
Norway	1,000(a)	*	*
Poland	125,000(a)	141,000(a)	542,000(a)
Soviet Union	6,750,000(a)	*	*
United States	293,986	670,846	*(c)
Total	9,186,111	3,468,402	2,669,551
Axis Powers			
Finland	52,609	125,000	*
Germany	3,250,000	4,000,000	11,741,000
Hungary	75,000(a)	*	*
Italy	60,000(f)	*(f)	700,000(f)
Japan	1,862,499	4,616,000	635,464(b)
Romania	80,000	*	594,000(b)
Yugoslavia	75,000(a)	*	125,000(b)
Total	5,380,000	8,741,000	13,795,464
Grand total	14,566,111	12,209,402	16,465,015

* No reliable figures available
(a) Broadly estimated
(b) Prisoners only
(c) All such personnel now officially accounted for
(d) Chinese casualties are for 1937–45
(e) Missing only
(f) Italian casualties on Allied side were 69,774 dead, 76,610 wounded, and 17,647 POW/MIA

Table from *Compton's Encyclopedia & Fact-Index*

Estimated civilian casualty figures ranged from 35,000,000 to 60,000,000. Of this figure were 5,700,000 Jews who died in Nazi concentration and death camps. As many as 21,000,000 people became refugees. Of this number more than half were "displaced persons" who had been driven from their homes to perform slave labor. Countless millions were made homeless.

UNITED NATIONS
1995 SPECIAL ACTIVITIES

UNITED NATIONS YEAR FOR TOLERANCE

The General Assembly on Dec. 20, 1993, proclaimed 1995 the United Nations Year for Tolerance (resolution 48/126). The United Nations Educational, Scientific and Cultural Organization (UNESCO) was invited to assume the role of the lead organization for the Year and to prepare a declaration on tolerance. Member states were invited to cooperate with UNESCO in preparing national and international programs and to participate in implementing the activities to be organized. Interested intergovernmental and nongovernmental organizations were invited to contribute to the preparation of programs for the year. The Assembly also recommended that the specialized agencies, regional commissions and other organizations of the UN system consider the contributions they could make to the success of the Year.

CONFERENCES

WORLD SUMMIT FOR SOCIAL DEVELOPMENT

The World Summit for Social Development will be held Mar 6–12, 1995, at Copenhagen, Denmark.

The summit's objectives are to express a shared worldwide commitment to put the needs of people at the center of development and international cooperation; and to address, in creative ways, the need to attain a balance between economic efficiency and social justice in a growth-oriented, equitable and sustainable development environment in accordance with nationally defined priorities. The core issues affecting all societies to be discussed are enhancing social integration, particularly that of the more disadvantaged and marginalized groups; alleviating and reducing poverty; and expanding productive employment.

FOR INFO

World Summit for Social Development Secretariat
Two United Nations Plaza
Rm DC2—1370
United Nations
New York, NY 10017
Phone: (212) 963-5855
Fax: (212) 963-3062

CONFERENCE ON CLIMATE CHANGE

The first session of the conference of the United Nations Framework Convention on Climate Change will be held at Berlin, Germany March 28–Apr 7, 1995.

FOURTH WORLD CONFERENCE ON WOMEN

The Fourth World Conference on Women will be held Sept 4–15, 1995, at Beijing, China.

The conference was called by the UN General Assembly to further review the Nairobi Forward-Looking Strategies for the Advancement of Women and determine the measures necessary to ensure that the objectives of the Nairobi Strategies will be achieved. A platform will be provided for national reviews and appraisals of policies for the advancement of women and to aid in setting targets and mobilizing interest in this issue. The General Assembly called upon member states to establish specific targets to increase the participation of women in professional, management and decision-making positions in their countries.

FOR INFO

Gertrude Mongella
Two United Nations Plaza
Fourth World Conference on Women Secretariat
Rm DC2—12th Floor
New York, NY 10017
Phone: (212) 963-5086
Fax: (212) 963-5086

THIRD UNITED NATIONS CONFERENCE ON EQUITABLE PRINCIPLES AND RULES FOR THE CONTROL OF RESTRICTIVE BUSINESS PRACTICES

The Second United Nations Conference to Review All Aspects of the Set of Multilaterally Agreed Equitable Principles and Rules for the Control of Restrictive Business Practices recommended that a third conference be held in 1995. The location and date for this conference had not been determined at press time.

UN 50TH CELEBRATION
APR 24 • OCT 24 • 50 YEARS

The United Nations, the world body established to work for world peace, was born out of the consuming flames of World War II, the deadliest war in the history of humankind. On Jan 1, 1942, 26 nations that were allied against Germany, Italy and Japan signed a document stating their aims in the war. These countries became known as the United Nations. This name was later passed on to the organization formulated to make these goals a reality.

Between Aug 21 and Oct 7, 1944, representatives of the Big Four—the US, the USSR, Great Britain and China—met at the private estate of Dumbarton Oaks, in Washington, DC, to hammer out the structure for a permanent organization. Although the meeting was for the most part successful, disagreement over the "veto problem" of the proposed Security Council plagued the agency for years to come.

DUMBARTON CONFERENCE

The Dumbarton Conference results provided the elements for negotiation when delegates from 50 nations gathered in San Francisco on Apr 25, 1945, for the United Nations Conference on International Organization. Their mission was to modify and complete the Charter for the United Nations. By Oct 24, 1945, the results of their efforts had been ratified by the required number of nations, and the United Nations officially came into existence. Oct 24 has since been known as United Nations Day.

UN CELEBRATION PLANS

"The 50th anniversary of the United Nations will be a milestone for the international community. It is a time to rekindle the ideals of the Charter, to spread the spirit of cooperation among the peoples of the world, to celebrate and to share a vision of the great potential of this unique organization. In this way, the United Nations will be both strengthened and renewed."
—Secretary-General Boutros Boutros-Ghali

As the 21st century rapidly approaches, hundreds of local, national and international activities will be held to celebrate the 50th anniversary of the organization established to work for world peace. Films, TV documentaries and series, exhibits, artistic events and a world tour by the United Kingdom's Royal Philharmonic Orchestra are among the projects planned. At the same time grass-roots initiatives are greatly encouraged.

The official UN50 emblem, used with permission

See United Nations in the index for a small sampling of events. Organizations are urged to convey their plans to the UN for inclusion in the UN50 master calendar.

FOR INFO

United Nations Headquarters
Fiftieth Anniversary Secretariat
S-3194, United Nations
New York, NY 10017
Phone: (212) 963-1995
Fax: (212) 963-9545

New York
UNA-USA
485 5th Avenue
New York, NY 10017-6104
Phone: (212) 697-3232
Fax: (212) 682-9185

San Francisco
UN50 Committee
312 Sutter Street, Suite 610
San Francisco, CA 94108
Phone: (415) 989-1996
Fax: (415) 989-1996

LONDON

London played a singular role in UN history by serving as host to the first session of the United Nations General Assembly from Jan 10 to Feb 14, 1946. To mark this event, the City of London will hold a special commemorative ceremony with the UN Secretary-General in attendance.

FOR INFO

United Nations Association of Great Britain
  and Northern Ireland
3 Whitehall Court
London SW1 A22L
England

COMMEMORATION WWII
50TH ANNIVERSARY EVENTS

THE WALL OF LIBERTY

Fifty years ago, millions of American soldiers, sailors and airmen left their homes and their families to join the war to restore freedom to the continent of Europe. These brave Americans marched into battle from Normandy to Sicily and the Bulge in the name of liberty. Many of them never made it home. As the 50th Anniversary of V-E Day approaches on May 8, 1995, The Battle of Normandy Foundation, a non-profit organization, is developing projects to honor the men and women who served in the ETO and to educate younger generations of Americans about the importance of remembering WWII.

One of the Foundation's projects is the erection of a wall in Normandy honoring all Americans, men and women, living or deceased, who served in World War II's European Theater of Operations (ETO) during the period from Dec 7, 1941, to May 8, 1945. (ETO includes Europe, North Africa, the Mediterranean and Italy.) The names of all of them are to be inscribed on the wall. Pierre Salinger, former Kennedy White House press secretary and WWII veteran, is serving as Chairman of the Foundation's Board of Directors and volunteer campaign.

The Wall of Liberty is being built in Caen, France, the capital of Normandy, near the D-Day invasion beaches. Investiture ceremonies for the first panel of the wall took place on June 6, 1994, to coincide with the 50th anniversary of D-Day. The first phase of the Wall is to be completed by V-E Day, May 8, 1995.

To register for The Wall of Liberty, all ETO veterans, relatives or friends of ETO veterans can call 1-800-WW2-VETS to obtain a registration form. A donation of $40 is requested for each veteran's name to be inscribed. Individuals, corporations and other foundations are also being asked to join in this nationwide campaign by providing financial support to ensure that every ETO veteran, even any who may be unable or cannot afford to contribute the $40, will be included on the Wall of Liberty.

FOR INFO

The Battle of Normandy Foundation
1730 Rhode Island Ave NW
Washington, DC 20036

COMMEMORATIVE COMMUNITY PROGRAM

The period 1991 through Veterans Day 1995 marks the 50th anniversary of World War II. The Department of Defense has formed the 50th Anniversary of World War II Commemorative Committee to thank and honor the veterans of World War II, their families and all those on the home front, and to develop programs and materials that provide the American people with a clearer understanding and appreciation of the lessons and history of World War II.

The flagship program of the committee is the World War II Commemorative Community Program, consisting of military and civilian communities coming together to develop activities and events that honor veterans and educate Americans on the role of the military and those on the home front. It also offers an opportunity for Americans of all generations to say to our veterans, "a grateful Nation remembers."

FOR INFO

Contact your nearest military public affairs office, including Reserve and National Guard units, or one of the following:

(For Upcoming Commemorative Events)
Lt Colonel Roger King
Deputy Director of Operations
1213 Jefferson Davis Hwy
Crystal Gateway Four, Ste 702
Crystal City, VA 22202
Phone: (703) 604-0820 or 0824

(For Commemorative Communities)
Master Sargeant Richard Jung
1213 Jefferson Davis Hwy
Crystal Gateway Four, Ste 702
Crystal City, VA 22202
Phone: (703) 604-0826 or 0818

(For written materials relating to WWII 50th Anniversary)
Lt Colonel Alfred Lott
Public Affairs Officer
1213 Jefferson Davis Hwy
Crystal Gateway Four, Ste 702
Crystal City, VA 22202
Phone: (703) 604-0819 or 0822

CHARLIE PARKER
BIRTH • AUG 29 • 75 YEARS

Charlie "Bird" (or "Yardbird") Parker, perhaps the greatest improviser in jazz history, was born Charles Parker, Jr, in Kansas City, KS, Aug 29, 1920. He grew up in the rich musical atmosphere across the river in Kansas City, MO, where his family had moved in 1927. Ignoring his mother's dictate, the young Parker frequented the "Combat Zone" (Twelfth Street), where he lied about his age to get into nightclubs and listen to some of the country's greatest jazz musicians and big bands. Before he was out of his teens, the self-taught Parker was trying out his saxophone in some of these Kansas City hot spots.

Parker's style, originally patterned after the advanced swing-era music he heard in Kansas City, eventually assimilated a variety of influences, including African-American folk forms and 20th-century concert music. In his lifetime Charlie Parker became an accomplished alto saxophone player, composer and band leader. With jazz greats trumpet player Dizzy Gillespie and pianist Thelonious Monk, he originated and developed the style known as bebop, which changed jazz forever.

Parker was totally devoted to his music: once when asked about his religious beliefs, he replied that he was "a devout musician." Parker's artistry was little appreciated by the American public, although he did enjoy a certain amount of recognition in New York City, where he had moved in 1939. His greatest acceptance was in Europe, where jazz enjoyed more widespread popularity.

His fellow artists viewed him differently. He not only influenced other saxophonists but effected a profound change in the entire genre. Jazz musicians copied Parker so routinely that his phrasings became standard fare for modern jazz improvisers. John Coltrane, Sonny Rollins and Miles Davis are only a few jazz stars heavily influenced by Parker. Many of Parker's songs, including "Confirmation," "Now's the Time," and "Ornithology," are considered classics today.

Parker began using drugs, including heroin, before the age of 20. Years of hard living undoubtedly played a part in the illness that developed into lobar pneumonia and took his life in New York City on Mar 12, 1955. He was only 35. His last public appearance was at Birdland, the New York jazz club named for him.

Soon after his death the graffiti began to appear: "Bird Lives!"

ONCE UPON A BUS
ROSA SAYS NO • DEC 1 • 40 YEARS

Rosa Louise McCauley Parks was born at Tuskegee, AL, on Feb 4, 1913, and grew up in Pine Level, AL. As a child and young woman Parks experienced firsthand the indignities of American segregation. She saw a post-World War I resurgence of the Ku Klux Klan, which was determined to put "back in their place" the "uppity" African-American soldiers who returned from the trenches of France talking about being treated more respectfully overseas. Because black children went to segregated schools for only six months and worked the fields the rest of the year, Parks's mother, a schoolteacher determined to see her daughter receive a good education, sent Rosa to Montgomery to a school with well-trained teachers for nine months of classes—Alabama State College—as there were no area high schools for blacks.

Even with a good education Parks was limited to menial jobs. Only her volunteer services to the National Association for the Advancement of Colored People (NAACP) allowed her to use her full abilities. She served as secretary to E. D. Nixon, president of the local branch, and joined the Montgomery Voters' League, which taught blacks how to respond to the 24-question test devised by election officials to prevent them from voting.

On Dec 1, 1955, returning from a tiring day of stitching and ironing shirts at the Montgomery Fair Department Store, Parks was ordered by a white bus driver to relinquish her seat to a white male passenger. Thoroughly fed up, she refused, and the bus driver had her arrested. Instead of pleading guilty and paying a fine, Parks chose to challenge the segregation law. Her subsequent trial and appeals went all the way to the US Supreme Court. A ruling by the high court on Nov 13, 1956, confirmed an earlier federal court decision that segregation on public transportation was illegal.

Between Dec 5, 1955, and Dec 20, 1956, when the Court's decision was implemented, the African-American community of Montgomery conducted a boycott against the city's public bus system. Walking, carpooling, even riding in horse-drawn wagons, blacks denied the bus company and local merchants hundreds of thousands of dollars. Rosa's act also inspired lunch-counter sit-ins, marches and additional boycotts—activities that developed into the civil rights movement. Today Rosa Parks is viewed as the mother of that movement.

YEAR TO REMEMBER · 1970
25 YEARS · A QUICK LOOK

Richard Nixon's Secretary of the Interior, Walter J. Hickel, sent a letter (later leaked to the press) warning the president that the administration's lack of response was a leading cause of the alienation of America's young people and was contributing to anarchy and revolt. Hickel was later fired.

Between May 1 and June 30, 1970, campus unrest erupted 508 times. Nixon ordered the invasion of Cambodia on Apr 29, and on May 4 National Guardsmen killed four and wounded nine students while breaking up a protest demonstration at Kent State University in Ohio; two more were killed by police in Jackson, MS, on May 15. The "plumbers" went to work for the Nixon White House. The 19th US census on Apr 1 reported the smallest ratio of men to women for the first time in US history (94.8 to 100). Congress began the good fight over equality between the sexes, with the Senate holding the first hearings on an Equal Rights Amendment since 1956 and the House of Representatives beginning subcommittee hearings on sex discrimination in education. Nationwide, women answered feminist Betty Friedan's call for a national strike for equality on the 50th anniversary of women's suffrage. The military commissioned the first women generals. *Newsweek* reported that 7 out of 10 college students believed there was too little emphasis on family life. The census reported that there was a 700 percent increase since 1960 in unmarried couples living together. Vassar College went co-ed. On Mar 18 the Protestants' *New English Bible* was published, followed by the Catholics' *New American Bible* on Sept 30.

Campus activism moved into another realm, emphasizing ecology. Earth Day teach-ins sought to combat pollution with mock funerals for cars and "Dishonor Rolls" for recalcitrant polluters. The rate of garbage collected in urban areas reached 5 pounds per person, compared to 2.75 per person 50 years earlier. The Environmental Protection Agency went into effect on Dec 2, and the Air Quality Control Act was signed by Nixon on Dec 31.

For the first time white-collar workers outnumbered blue-collar workers. The nation's unemployment rate reached its highest peak since 1965. Uniformed military personnel reached a high of 3,066,000. Postal workers were on strike Mar 18 to 25. Congress passed a law that would reduce the voting age to 18 in 1971.

The Supreme Court on June 15 ruled in *Welsh v. US* that an individual could qualify for a draft exemption as a conscientious objector on moral as well as religious grounds. The Supreme Court also moved up the deadline set for the desegregation of all public schools in six southern states (Alabama, Florida, Georgia, Louisiana, Mississippi, Tennessee) to Feb 1, 1970. The Chicago Seven were acquitted in Chicago on Feb 18, Angela Davis was arrested on Oct 13, and the court-martial of Lt William Colby for the My Lai massacre began Nov 12. The US and the USSR began the second round of Strategic Arms Limitation talks (SALT) in Vienna on Apr 16.

Blood, Sweat and Tears became the first US rock group to tour behind the Iron Curtain. Twenty-four-year-old Tammy Terrell succumbed to a brain tumor on Mar 16. Jimi Hendrix died Sept 18 at the age of 24 and Janis Joplin died Oct 4 at the age of 25, and both deaths were drug-related. Jerry Garcia was ruled obscene by the Federal Communications Commission for an FM radio interview and the station was fined. *Jesus Christ Superstar* made its mark as a record album, signaling a trend in rock with a religious theme. The Who launched the *Tommy* tour at the Metropolitan Opera House in New York City. The second Atlanta Pop Festival drew over 40,000, but only 18 of 48 major rock festivals planned for 1970 were actually held as local government officials found ways to block them. On Dec 31 Paul McCartney filed suit seeking legal dissolution of the Beatles' association. The Fab Four as a group were no more.

The Controlled Substances Act of 1970 reduced possession of marijuana from a felony to a misdemeanor and decriminalization movements followed in a few states. Health food sales topped $3 billion. The world Barbie doll population reached 12 million. $7.5 million worth of Mickey Mouse watches and clocks sold in six months and over $400,000 a month was spent on Mickey Mouse sweatshirts. Tie-dyed clothes were in.

Airport, *Five Easy Pieces*, *Love Story*, *M*A*S*H* and *Patton* dominated cinema box offices and were nominated for the Best Picture Oscar, with *Patton* later claiming the prize. Television said good-bye to old favorites, presenting the final seasons of "The Ed Sullivan Show," "Hogan's Heroes," "The Lawrence Welk Show" and "Mayberry RFD"; and hello to the new with the debuts of "NFL Monday Night Football," "The Flip Wilson Show," "The Mary Tyler Moore Show" and "The Partridge Family."

HOT FILMS

Airport
Catch-22
Fellini Satyricon
Five Easy Pieces
The Graduate
The Great White Hope
Little Big Man
Love Story
M*A*S*H
Patton
Ryan's Daughter
Tora! Tora! Tora!

HIT SINGLES

"After Midnight," Eric Clapton
"Ain't No Mountain High Enough," Diana Ross
"Ball of Confusion," Temptations
"Bridge Over Troubled Water," Simon & Garfunkel
"Do the Funky Chicken," Rufus Thomas
"Evil Ways," Santana
"Fire and Rain," James Taylor
"Get Ready," Rare Earth

"I Think I Love You," the Partridge Family
"I'll Be There," Jackson Five
"In the Summertime," Mungo Jerry
"Lay Down (Candles in the Rain)," Melanie
"Let It Be," the Beatles
"The Letter," Joe Cocker
"The Long and Winding Road," the Beatles
"My Sweet Lord," George Harrison

"Signed, Sealed, Delivered I'm Yours," Stevie Wonder
"Something's Burning," Kenny Rogers & the First Edition
"Tears of a Clown," Smokey Robinson & the Miracles
"(They Long to Be) Close to You," the Carpenters
"The Thrill Is Gone," B. B. King
"25 or 6 to 4," Chicago
"War," Edwin Starr
"The Wonder of You," Elvis Presley

OTHER 1995 MILESTONES

RAPHAEL (RAFAELLO SANZIO) DEATH
APR 6, 1520 • 475 YEARS
Italian Renaissance master painter and architect, Raphael died at Rome, Italy.

HENRY PURCELL DEATH
NOV 21, 1695 • 300 YEARS
Composer Henry Purcell was the most influential English composer of his day, the early Baroque period. He crossed genres to compose for the church, the stage, the court and popular entertainment.

WILLIAM WORDSWORTH BIRTH
APR 7, 1770 • 225 YEARS
The poet of nature William Wordsworth was born at Cockermouth, Cumberland, England. His work marked the beginning of the romantic movement in English poetry.

THOMAS CARLYLE BIRTH
DEC 4, 1795 • 200 YEARS
British historian and essayist Thomas Carlyle was born at Ecclefechan, Dumfriesshire, Scotland.

JOHN BARTLETT BIRTH
JUNE 14, 1820 • 175 YEARS
The author of *Bartlett's Familiar Quotations* was born at Plymouth, MA.

ANDREW JACKSON DEATH
JUNE 8, 1845 • 150 YEARS
Andrew Jackson, the seventh president of the US, died at at Nashville, TN.

(RICHARD) BUCKMINSTER FULLER BIRTH
JULY 12, 1895 • 100 YEARS
American inventor and futurist Buckminster Fuller was born at Milton, MA.

GOODWILL INDUSTRIES CENTENNIAL CELEBRATION
1895 • 100 YEARS
Rev Dr Edgar James Helms began employing the poor in the repair and resale unwanted clothes and goods. For info: Morgen Memorial Goodwill Industries, Inc, Nancy Cicco, Pulic Affairs Associate, 1010 Harrison Ave, Boston, MA 02119-2540. Phone: (617) 445-1010.

LEO SOWERBY BIRTH
MAY 1, 1895 • 100 YEARS
Pulitzer Prize-winning composer Leo Sowerby was born at Grand Rapids, MI.

LAST MAJOR LEAGUE TRIPLEHEADER
Oct 2, 1920 • 75 YEARS
The Cincinnati Reds defeated the Pittsburgh Pirates two games out of three ending the season in third place. The fourth-place Pirates needed all three games to overtake the Reds but lost the first game and their bid to rise in the standings.

ELVIS PRESLEY BIRTH
JAN 8, 1935 • 60 YEARS
The King of Rock 'n' Roll was born at Tupelo, MS.

POLIO VACCINE RELEASED FOR GENERAL USE IN US
APR 12, 1955 • 40 YEARS
After much research and testing, Jonas Salk's vaccine to prevent poliomyelitis was approved for use.

APOLLO 11 MOON LANDING SCRAPPED
APR 11, 1970 • 25 YEARS
After an oxygen tank ruptured, the *Apollo 11*'s moon landing was canceled and the team of astronauts returned safely to Earth.

APOLLO-SOYUZ SPACE LINKUP
JULY 15, 1975 • 20 YEARS
After years of planning, the first US-USSR joint space project came to fruition when the *Apollo 18* and *Soyuz 19* linked up for 47 hours while orbiting the earth.

GERMANY REUNIFIED
OCT 3, 1990 • 5 YEARS
After years of separation by the Iron Curtain and the Berlin wall, East and West Germany were reunited as one country.

HUBBEL SPACE TELESCOPE DEPLOYED
APR 25, 1990 • 5 YEARS
The Hubbel space telescope was deployed by the space shuttle *Discovery*.

"SAVE THE EARTH"
ENVIRONMENTAL MILESTONES

The debate over the environment never seems to end—and it probably shouldn't. Activists espousing "green" policy changes continually battle industries and governments reluctant to increase regulations that might hinder profits and competitiveness. Both sides enlist the scientific community to support their respective positions, proving only that at this point the science of determining the hard facts about some environmental issues is imprecise and still developing. While the debate rages, some businesses are offering an increasing number of product lines that cater to the environmentally conscious. Computers are now available with energy-saving options that shut down or suspend system activity when it is not required. Environmentally safe labels appear on consumer products, and the recycling industry grows. In the midst of the controversies and technological innovations, 1995 marks the anniversaries of four major green events.

EARTH DAY
APR 20 • 25 YEARS

Earth Day began in 1970. A surprisingly ambitious response to a remark made by Wisconsin Senator Gaylord Nelson, the first Earth Day included more than 1,000 cities and towns rallying around a "Save the Earth" theme on Apr 22. Nelson had casually suggested that the day be set aside for serious discussion of environmental issues. From this modest beginning Earth Day has grown into an annual celebration and educational exploration of green issues in more than 140 countries worldwide. For a small sampling of events that celebrate the 25th anniversary of Earth Day, see Environment in the Index.
For info
Earth Day USA
Box 9827
San Diego, CA 92169-9827
Phone: (619) 272-7370
Fax: (619) 272-2933

THE EPA
DEC 2 • 25 YEARS

In July 1970 the Environmental Protection Agency (EPA) was established by executive order from President Richard Nixon. The agency was activated on Dec 2, 1970, and William D. Ruckelshaus was confirmed as its first director on the same day. While seeking to guard against environmental problems and disasters, the EPA has become a lightning rod for those who oppose government regulation. The EPA's activism has always been colored by the leanings of the president and administration it serves, and this on-again/off-again approach has led to even greater controversy.
For info
US Environmental Protection Agency
401 M Street, SW
Washington, DC 20460   Phone: (202) 260-2090

RAINBOW WARRIOR SINKING
JULY 10 • 10 YEARS

The sinking of Greenpeace's *Rainbow Warrior* in the harbor at Aukland, Australia, on July 10, 1985, by French intelligence agents, exemplifies the high stakes involved in the environmental debate. The flagship was to be used in a campaign against French nuclear testing in the South Pacific. Two explosions ripped through its hull, sinking the ship and killing Greenpeace photographer Fernando Pereira. The resulting scandal in France led to the resignation of several political leaders when France eventually admitted its guilt in the incident.
For info
Greenpeace International
1436 U Street, NW
Washington, DC 20009
Phone: (202) 462-1177
Fax: (202) 462-4507

ENVIRONMENTAL EDUCATION ACT
NOV 16 • 5 YEARS

On Nov 16, 1990, President George Bush signed into law the National Environmental Education Act. The bill is intended to increase public understanding of the environment and to facilitate environmental education and training. The newly formed Environmental Education Division (EED) of the EPA was called upon to take a leadership position in implementing the law by encouraging mutual effort on the part of federal and state government agencies, local education institutions, not-for-profit organizations and the private sector.

To carry out their goal of educating citizens about environmental issues, the EED and EPA award grants to support environmental education projects; administer the national Network for Environmental Management Studies, which encourages students to pursue professional environmental careers; and have developed Students Watching Over Our Planet Earth (SWOOPE), a laboratory program that teaches science through hands-on environmental research. Along with the University of Michigan and other education institutions and agencies, the EED is developing an interactive, computer-based environmental education resource library.

The EED publishes a periodical entitled *Earth Notes*, in which educators share firsthand experience with teaching environmental issues. The EED also is responsible for administering the President's Environmental Youth Awards program, which honors young people who have shown extraordinary commitment to the environment.
For info
Office of Environmental Education
US Environmental Protection Agency
401 M Street, SW (1707)
Washington, DC 20460
Phone: (202) 260-4965
Fax: (202) 260-4095

☆ Chase's 1995 Calendar of Events ☆ Jan 1

Januarye.

JANUARY 1 — SUNDAY
1st Day — Remaining, 364

SUNDAY, JANUARY ONE, 1995. First day of the first month of the Gregorian calendar year, Anno Domini 1995, being third after Leap Year, and (until July 4th) 219th year of American Independence. 1995 will be year 6708 of the Julian Period, a time frame consisting of 7,980 years, which began at noon, universal time, Jan 1, 4713 BC. Astronomers will note that Julian Day number 2,449,719 begins at noon, universal time, Jan 1, 1995 (representing the number of days since Julian calendar date Jan 1, 4713 BC). New Year's Day is a public holiday in the US and in many other countries. Traditionally, it is a time for personal stocktaking, for making resolutions for the coming year, and sometimes for recovering from the festivities of New Year's Eve. Financial accounting begins anew for businesses and individuals whose fiscal year is the calendar year. Jan 1 has been observed as the beginning of the year in most English-speaking countries since the British Calendar Act of 1751, prior to which the New Year began on Mar 25 (approximating the vernal equinox). Earth begins another orbit of the sun, during which it, and we, will travel some 583,416,000 miles in 365.2422 days. New Year's Day has been called "Everyman's Birthday," and in some countries a year is added to everyone's age on Jan 1 rather than on the anniversary of each person's birth.

ANNOUNCEMENT OF TEN BEST PUNS OF THE YEAR. Jan 1. A salute to the best punsters of the year, thus encouraging wordplay and combatting illiteracy. Annually, Jan 1. Phone/Fax: (416) 223-2236. Sponsor: Intl Save the Pun Foundation, Norman Gilbert, Box 5040, Station A, Toronto, Ont, Canada M5W 1N4. Phone: (416) 223-3351.

BONZA BOTTLER DAY™. Jan 1. (Also Feb 2, Mar 3, Apr 4, May 5, June 6, July 7, Aug 8, Sept 9, Oct 10, Nov 11 and Dec 12.) To celebrate when the number of the day is the same as the number of the month. Bonza Bottler Day™ is an excuse to have a party at least once a month. T-shirts available. For info: Elaine Fremont, 203 Waddell Rd, Taylors, SC 29687. Phone: (803) 244-2023.

CANADA: CN TOWER: CELEBRATION 1995. Jan 1. Toronto, Ontario. CN Tower celebrates 1995 with the world's longest ball drop and the 1st party of 1995. Outdoor concert, dramatic laser effects and fireworks all combine to create a bunch of family fun. For info: CN Tower, 301 Front St W, Toronto, Ont, Canada M5V 2T6. Phone: (416) 360-8500.

January 1995

S	M	T	W	T	F	S
1	2	3	4	5	6	7
8	9	10	11	12	13	14
15	16	17	18	19	20	21
22	23	24	25	26	27	28
29	30	31				

CHIMNEY ROCK PARK PHOTO CONTEST. Jan 1–Dec 15. Chimney Rock Park, Chimney Rock, NC. Capture the beauty of the park in color prints, slides or black-and-white photos and win cash prizes. Open to amateur and professional photographers. Winning entries to be announced Mar 1, 1996. For info: Mary Jaeger-Gale, Mktg Mgr, PO Box 39, Chimney Rock, NC 28720. Phone: (800) 277-9611.

CIRCUMCISION OF CHRIST. Jan 1. Holy day in many Christian churches. Celebrates Jesus's submission to Jewish law; on the octave day of Christmas. See also: "Solemnity of Mary, Mother of God" (Jan 1) for Roman Catholic observance since 1969 calendar reorganization.

CUBA: LIBERATION DAY. Jan 1. A national holiday that celebrates the end of Spanish rule—Jan 1, 1899. Cuba, largest island of the West Indies, was a Spanish possession from its discovery by Columbus (Oct 27, 1492) until 1899. Under US military control 1899–1902 and 1906–1909, a republican government took over on Jan 28, 1909, and controlled the island until it was overthrown Jan 1, 1959, by Fidel Castro's revolutionary movement (which had begun on July 26, 1953).

DECADE OF THE BRAIN: YEAR SIX. Jan 1–Dec 31. H.J. Res. 174, Public Law 101-58 of July 25, 1989, established the period from Jan 1, 1990, through Dec 31, 1999, as the Decade of the Brain. Presidential Proclamation 6158 was issued July 17, 1990. The purpose of this decade is to enhance public awareness of the benefits to be derived from brain research. (To find the full text of the proclamation, see Contents.)

DEFENSE DEPOT TRACY: ANNIVERSARY. Jan 1. Anniversary of the establishment of Defense Depot Tracy, CA, as the first depot in the Defense Logistics Agency distribution system.

ELLIS ISLAND OPENED: ANNIVERSARY. Jan 1. Ellis Island was opened on New Year's Day, Jan 1, 1892. Over the years more than 20 million individuals were processed through the stations. The island was used as a point of deportation as well: in 1932 alone, 20,000 people were deported from Ellis Island. When the US entered WW II in 1941, Ellis Island became a Coast Guard Station. It closed Nov 12, 1954, and was declared a national park in 1956. After years of disuse it was restored and in 1990 reopened as a museum.

☆ ☆ ☆

EMANCIPATION PROCLAMATION: ANNIVERSARY. Jan 1. Two of the most important presidential proclamations of American history are those of Sept 22, 1862, and Jan 1, 1863, in which Abraham Lincoln, by executive proclamation, freed the slaves in the rebelling states. "That on . . . [Jan 1, 1863] . . . all persons held as slaves within any state or designated part of a state, the people whereof shall then be in rebellion against the United States, shall be then, thenceforward, and forever, free. . . ." See also: "13th Amendment Anniversary" (Dec 18) for abolition of slavery in all states.

ENGLAND: A ROYAL PALETTE. Jan 1–Apr 2. (Began Oct 22.) The Gallery, Windsor Castle, Windsor. Drawings, watercolour paintings and etchings by members of the Royal Family from the seventeenth century to the present day. Est attendance: 50,000. For info: Exhibition Dept, Windsor Castle, The Royal Library, Windsor, Berkshire, England SL4 1NJ.

ENGLAND: LONDON PARADE. Jan 1. London. Begins at noon at Westminster Abbey and progresses via Trafalgar Square to Berkeley Square. For info: New Year's Day Parade, Research House, Fraser Road, Perivale, Middlesex, England UB6 7AQ.

ENGLAND: UK YEAR OF LITERATURE AND WRITING. Jan 1–Dec 31. Various venues, Swansea, West Glamorgan. A year of literature festivities which will include music, dance, sculpture and drama from all around the world. For info: UK Year of Literature and Writing, Somerset Place, Swansea, England SA1 1SE.

Jan 1 ☆ *Chase's 1995 Calendar of Events* ☆

FAMILY HISTORY BEGINS WITH ME: WRITING MY OWN LIFE STORY. Jan 1–Dec 31. Make 1995 the year that you give your unique gift (and it won't cost a thing!): your memories. If writing your memoirs or autobiography seems an overwhelming task, you might break it into smaller steps or monthly "homework assignments." Each month write a chapter or topic. Sponsor: Family History Month, Monmouth Co Genealogy Club, Monmouth Co Historical Assn, 70 Court St, Freehold, NJ 07728.

FIRST MONDAY TRADE DAYS. Jan 1–2. Jackson County Courthouse Square, Scottsboro AL. One of the Deep South's oldest and largest trade days where bartering, haggling and swapping of goods has not passed on with time; blend of antique shows, craft fairs, rummage sales. Believed to have begun in the mid-1850s when Jackson County Circuit Court used to meet on the first Monday of each month, thus attracting tradesmen. For info: Rick Roden, Tourism Dir, PO Box 973, Scottsboro, AL 35768. Phone: (205) 574-3333.

FORSTER, E.M.: BIRTH ANNIVERSARY. Jan 1. Edward Morgan Forster, English author born at London, England, Jan 1, 1879. Especially remembered for his six novels: *Where Angels Fear to Tread* (1905), *The Longest Journey* (1907), *A Room with a View* (1908), *Howard's End* (1910), *A Passage to India* (1924) and the posthumously published *Maurice* (1971), which Forster had written about 1913–1914. He also achieved eminence for his short stories and essays, and he collaborated on the libretto for one opera, Benjamin Britten's *Billy Budd* (1951). Forster died at Coventry, England, June 7, 1970.

FROM GEORGE TO GEORGE: THE PRESIDENTIAL PORTRAITS OF MORGAN MONCEAUX. Jan 1–Jan 16. (Began Nov 1, 1994.) Gerald R. Ford Museum, Grand Rapids, MI. Morgan Monceaux's brightly colored oil pastel portraits are embellished with braid, campaign buttons and other flea market finds, as well as with text describing the subject and his career. Est attendance: 21,000. For info: Gerald R. Ford Museum, 303 Pearl St NW, Grand Rapids, MI 49504-5353. Phone: (616) 451-9263.

GARDEN OF LIGHTS. Jan 1–2. (Began Nov 25.) Honor Heights Park, Muskogee, OK. Park sparkles and shimmers with more than 900,000 lights. Three graceful doves with 12-ft wingspans; four lighted nature exhibits; lighted waterfall with sound effects is 200 ft long with a family of deer resting nearby. Lights all over! Est attendance: 295,000. For info: Greater Muskogee Chamber of Commerce, Ervalene Jenkins, PO Box 797, Muskogee, OK 74402-0797. Phone: (918) 682-2401.

HAITI: INDEPENDENCE DAY. Jan 1. A national holiday commemorating the proclamation of independence on Jan 1, 1804. Haiti, occupying the western third of the island Hispaniola (second largest of the West Indies), was a Spanish colony from the time of its discovery by Columbus in 1492 until 1697, then a French colony until the proclamation of independence in 1804.

HUMAN RESOURCES MONTH. Jan 1–31. Recognizes importance of human resources executives in corporate management and honors professionals in that area with the Personnel Journal Optimas Award for Excellence. For info: Stephanie Peck, Promo Coord, Personnel Journal, 245 Fischer Ave B-2, Costa Mesa, CA 92626. Phone: (714) 751-1883.

JANUARY DIET MONTH. Jan 1–31. To promote the importance of getting back on a healthy eating plan after the winter holidays, and to remind dieters that nutritious popcorn can be a part of that plan. Annually, the month of January. Sponsor: The Popcorn Institute, 401 N Michigan Ave, Chicago, IL 60611-4267. Phone: (312) 644-6610.

January 1995

S	M	T	W	T	F	S
1	2	3	4	5	6	7
8	9	10	11	12	13	14
15	16	17	18	19	20	21
22	23	24	25	26	27	28
29	30	31				

JAPANESE ERA NEW YEAR. Jan 1–3. Celebration of the beginning of the year 2655 of the Japanese era.

MARCH OF DIMES BIRTH DEFECTS PREVENTION MONTH. Jan 1–31. To heighten awareness of birth defects and how they may be prevented, to inform the public about the work of the March of Dimes and to offer opportunities to new volunteers for service in the prevention of birth defects. Sponsor: March of Dimes Birth Defects Foundation, 1275 Mamaroneck Ave, White Plains, NY 10605. Phone: (914) 428-7100.

MARKET ABILITY MONTH. Jan 1–31. To draw attention to the growing need for workers and entrepreneurs to increase their "market ability" to gain professional and personal success. Tips will be released throughout the month. Annually, the month of January. For info: Nancy Michaels, Impression Impact, 113 Hill St, Concord, MA 01742. Phone: (508) 287-0718.

MOON PHASE: NEW MOON. Jan 1. Moon enters New Moon phase at 5:56 AM, EST.

NATIONAL BE ON-PURPOSE MONTH. Jan 1–31. An observance to encourage us to start the new year by putting our good intentions into action, personally and professionally, and to trade confusion for clarity as we balance our lives with more meaning and purpose. Sponsor: US Partners, Kevin W. McCarthy, 120 University Park Dr, Ste 200, Winter Park, FL 32792. Phone: (407) 657-6000.

NATIONAL BOOK BLITZ MONTH. Jan 1–31. Santa Barbara, CA. Focuses attention on improving authors' relationships with the media in order to create a bestselling book. Free book PR evaluation available. Annually, January. May Call Fax-On-Demand: (805)96FAX-IT. For info: Barbara Gaughen, 226 E Canon Perdido #B, Santa Barbara, CA 93101. Phone: (805) 965-8482.

NATIONAL ENVIRONMENTAL POLICY ACT: ANNIVERSARY. Jan 1. The National Environmental Policy Act of 1969, establishing the Council on Environmental Quality and making it federal government policy to protect the environment, took effect Jan 1, 1970.

NATIONAL EYE CARE MONTH. Jan 1–31. To educate the public to seek medical eye care at the earliest sign of eye trouble to prevent blindness. Bob Hope, Honorary Chairman. Sponsor: Optic Foundation, Rosemary Rushka, PO Box 429098, San Francisco, CA 94142-9098. Phone: (415) 561-8500.

NATIONAL HOT TEA MONTH. Jan 1–31. To celebrate one of nature's most popular, soothing and relaxing beverages, the only beverage in America commonly served hot or iced, anytime, anywhere, for any occasion. Sponsor: The Tea Council of the USA, 230 Park Ave, New York, NY 10169. Phone: (212) 968-6998.

NATIONAL PRUNE BREAKFAST MONTH. Jan 1–31. To encourage Americans to cut the fat from their diets by using prune puree (a simple mixture of prunes & water) as a substitute for butter and oil in baked goods; to communicate the many benefits of prunes, their good taste, high fiber and versatility; and to dispel misconceptions about prunes. More information, recipes and telephone interviews are available. Sponsor: California Prune Board. For info: Ketchum PR, Daphne Scofield, 55 Union St, San Francisco, CA 94111. Phone: (415) 984-6307.

NATIONAL RETAIL BAKERS MONTH. Jan 1–31. To highlight retail bakers' contribution to their communities and the economy and to promote their dedication to producing fresh, nutritious bakery foods. Fax: (301) 725-2187. Sponsor: Retail Bakers of America, 14239 Park Center Dr, Laurel, MD 20707. Phone: (301) 725-2149.

NATIONAL SOUP MONTH. Jan 1–31. To celebrate one of the world's favorite foods, and its relevance to lifestyles today. Sponsor: Campbell Soup Co, Public Affairs, Campbell Pl, Camden, NJ 08103-1799. Phone: (609) 342-8588.

★ Chase's 1995 Calendar of Events ★ Jan 1

NATIONAL YOURS, MINE AND OURS MONTH. Jan 1–31. Blending families and creating positive step-relationships can be one of the most challenging aspects of a couple's remarriage. Each member of the family is affected in different ways. This observance focuses on what parents and children can expect as a step- or blended-family and offers tips for a smooth transition and enhanced long-term relationships. Annually, the month of January. For info: Teresa Langston, Parenting Without Pressure, 1330 Boyer St, Longwood, FL 32750. Phone: (407) 767-2524.

NEW YEAR'S DAY. Jan 1. Legal holiday in all states and territories of the US and in most other countries. The world's most widely celebrated holiday.

NEW YEAR'S DISHONOR LIST. Jan 1. Since 1976, America's dishonor list of words banished from Queen's English. Overworked words and phrases (e.g., *uniquely unique, first time ever, safe sex*). Send nominations to following address. For info: PR Office, Lake Superior State University, Sault Ste. Marie, MI 49783. Phone: (906) 635-2315.

NEWPORT YACHT CLUB FROSTBITE FLEET RACE. Jan 1. Newport Yacht Club, Long Wharf, Newport, RI. Annual New Year's Day race. Sponsor: Newport Yacht Club. Est attendance: 75. For info: Rhode Island Tourism, 7 Jackson Walkway, Providence, RI 02903. Phone: (800) 556-2484.

OATMEAL MONTH. Jan 1–31. "Celebrate oatmeal, a low-fat, sodium-free whole grain which is a good source of fiber. Include versatile oatmeal in menus all day long. Enjoy hot oatmeal for breakfast, oatmeal cookies for snacks, and oatmeal in breads, muffins and other baked goods." For info: The Quaker Oats Co, c/o Oatmeal Month, PO Box 049003, Chicago, IL 60604-9003. Phone: (312) 222-7843.

OREGON SHAKESPEARE FESTIVAL. Jan 1–Apr 15. (Began Oct 36, 1994.) Portland, OR. A 6-month season of 5 plays by Shakespeare, classical and contemporary playwrights in the Intermediate Theatre of the Portland Center for the Performing Arts. Est attendance: 90,000. For info: Oregon Shakespeare Festival, PO Box 9008, Portland, OR 97207. Phone: (503) 274-6588.

PENGUIN PLUNGE. Jan 1. Mackeral Cove, Jamestown, RI. Annual plunge into the icy waters of Narragansett Bay to benefit Rhode Island Special Olympics. Annually, Jan 1. Est attendance: 2,000. For info: Rhode Island Special Olympics, 33 College Hill Rd, Bldg 31, Warwick, RI 02886. Phone: (401) 823-7411.

PHILIPPINES: BLACK NAZARENE FIESTA. Jan 1–9. Manila. A traditional nine-day fiesta honors Quiapo district's patron saint. Cultural events, fireworks and parades culminate in a procession with the life-size statue of the Black Nazarene. Procession begins at the historic Quiapo church.

POLAR BEAR SWIM. Jan 1. Sheboygan Armory, Sheboygan, WI. Each New Year's Day at 1 PM, more than 450 daring swimmers brave Lake Michigan's ice floes. Most are costumed, all are crazy. Refreshments and free live entertainment from 10–6. Sponsor: Sheboygan Polar Bear Club. Est attendance: 450. For info: Sheboygan Conv and Visitors Bureau, 712 Riverfront Drive, Ste 101, Sheboygan, WI 53081. Phone: (414) 457-9495.

REELFOOT EAGLE TOURS. Jan 1–Mar 15. Reelfoot Lake State Resort Park, Tiptonville, TN. A prime winter nesting area for the American bald eagle, Reelfoot Lake annually hosts 150 to 200 eagles. For info: Tennessee Tourist Dvmt, Box 23170, Nashville, TN 37202. Phone: (615) 741-9021.

REVERE, PAUL: BIRTH ANNIVERSARY. Jan 1. American patriot, silversmith and engraver, maker of false teeth, eyeglasses, picture frames and surgical instruments. Best remembered for his famous ride on Apr 18, 1775, celebrated in Longfellow's poem, "The Midnight Ride of Paul Revere." Born at Boston, MA, on Jan 1, 1735. Died there on May 10, 1818. See also: "Paul Revere's Ride Anniversary" (Apr 18).

ROSS, BETSY: BIRTH ANNIVERSARY. Jan 1. According to legend based largely on her grandson's revelations in 1870, needleworker Betsy Ross created the first stars-and-stripes flag in 1775, under instructions from George Washington. Her sewing and her making of flags were well known, but there is little corroborative evidence of her role in making the first stars-and-stripes. The account is generally accepted, however, in the absence of any documented claims to the contrary. She was born Elizabeth Griscom at Philadelphia, PA, on Jan 1, 1752, and died there on Jan 30, 1836.

RUSSIA: NEW YEAR'S DAY OBSERVANCE. Jan 1. National holiday. Modern tradition calls for setting up New Year's trees in homes, halls, clubs, palaces of culture and the hall of the Kremlin Palace. Children's parties with Granddad Frost and his granddaughter, Snow Girl. Games, songs, dancing, special foods, family gatherings and exchanges of gifts and New Year's cards.

ST. BASIL'S DAY. Jan 1. St. Basil's or St. Vasily's feast day observed on Jan 1 by Eastern Orthodox churches. Special traditions for the day include serving St. Basil cakes, which each contain a coin. Feast day observed on Jan 14 by those churches using Julian calendar.

SHANGHAI PARADE. Jan 1. Lewisburg, WV. A bit of Americana. Originally a spontaneous promenade that included everyone who happened by. Est attendance: 1,750. Sponsor: Lewisburg Visitors Ctr, 105 Church St, Lewisburg, WV 24901. Phone: (304) 645-1000.

SILENT RECORD WEEK. Jan 1–7. To commemorate the 33rd anniversary of the invention of the silent record, which was played on Detroit jukeboxes. The following year a Silent Record Concert and Recording Session featured emcee Henry Morgan, Soupy Sales and the 120-piece Hush Symphonic Band. [Originated by the late W.T. Rabe of Sault Ste. Marie, MI.]

SNO'FLY: THE FIRST KITE FLY OF THE YEAR. Jan 1. Prairie View Park, Kalamazoo, MI. Keep the New Year's celebration in full flight at this high-flying alternative to (seemingly) endless football games. An official event of Kalamazoo's New Year's Fest, Sno'Fly takes off from the frozen lake at Prairie View Park. Est attendance: 500. For info: John Cosby, Kalamazoo County Parks Dept, 2900 Lake St, Kalamazoo, MI 49001. Phone: (616) 383-8778.

SOCIETY FOR THE PREVENTION OF CRUELTY TO YOUR MONEY MONTH OR SECOND HAND BUT SIMPLY GRAND. Jan 1–31. Kankakee, IL. "Not new should never mean having to say you're sorry" is the theme of style seminars that will be started this month sharing the secrets of living well: spending time, money and energy effectively and graciously with constructive care and sharing communications. The seminars are done as community service projects. In the area of fashion the idea is "Second hand, but simply grand," based on the concept "We start with what we have and add by adjusting, accessorizing and altering." Sponsor: Patricia J. Reynolds, Dir, Now Is Your Time, Midtown Towers #223, 340 N Dearborn, Kankakee, IL 60901. Phone: (815) 933-6109.

SOLEMNITY OF MARY, MOTHER OF GOD. Jan 1. Holy Day of Obligation in Roman Catholic Church since calendar reorganization of 1969, replacing the Feast of the Circumcision, which had been recognized for more than 14 centuries. See also: "Circumcision of Christ" (Jan 1).

SUDAN: INDEPENDENCE DAY. Jan 1. National holiday. Sudan was proclaimed a sovereign independent republic on Jan 1, 1956, ending its status as an Anglo-Egyptian condominium (since 1899).

Jan 1 ☆ Chase's 1995 Calendar of Events ☆

THYROID DISEASE AWARENESS MONTH. Jan 1-31. A special public health outreach to emphasize the need for better detection and diagnosis of conditions that are often easily treated, but undetected until serious complications result. Thyroid disease affects 6-8 million Americans—primarily women—but as many as 4 million remain undiagnosed and untreated because of the condition's slow onset and diverse symptoms. This observance is a national initiative to educate people at risk for hypo- and hyperthyroidism about warning signs and the need for screening and to educate the general public about the thyroid, its function and the diseases affecting it. Sponsor: American Assn of Clinical Endocrinologists (AACE). For info: Interscience Communications Ltd, 1120 Avenue of the Americas, New York, NY 10036.

TOURNAMENT OF ROSES ASSOCIATION FOOTBALL GAME: ANNIVERSARY. Jan 1. Michigan defeated Stanford 49-0 in the first post-season football game, played on Jan 1, 1902. Called the Rose Bowl since 1923, it is preceded each year by the Tournament of Roses Parade in Pasadena, CA.

UNITED NATIONS: ASIAN AND PACIFIC DECADE OF DISABLED PERSONS: YEAR THREE. Jan 1-Dec 31. The decade 1993-2002 was proclaimed on Apr 23, 1992, by the Economic and Social Commission for Asia and the Pacific (ESCAP) and was endorsed by the General Assembly on Dec 16, 1992 (Res 47/88), to give further impetus to the implementation of the World Programme of Action concerning Disabled Persons in the ESCAP region and to strengthen regional cooperation in achieving the goals. Info from: United Nations, Dept of Public Info, New York, NY 10017.

UNITED NATIONS: DECADE AGAINST DRUG ABUSE: YEAR FIVE. Jan 1-Dec 31. United Nations program for the decade 1991-2000 to strengthen international, regional and national efforts in the fight against drug abuse. Proclaimed by General Assembly on Feb 23, 1990 (Res S-17/2). Info from: United Nations, Dept of Public Info, New York, NY 10017.

UNITED NATIONS: DECADE OF INTERNATIONAL LAW: YEAR SIX. Jan 1-Dec 31. The General Assembly declared this decade (1990-1999) for the purpose of promoting acceptance of and respect for international law principles; promoting peaceful settlement of disputes between states, including full respect for the International Court of Justice; encouraging development and codification of international law; and seeking to expand the teaching, study, dissemination and appreciation of international law. Info from: United Nations, Dept of Public Info, New York, NY 10017.

UNITED NATIONS: FOURTH UNITED NATIONS DEVELOPMENT DECADE: YEAR FIVE. Jan 1-Dec 31. Major United Nations program (1991-2000) to promote world efforts to bridge enormous gap between advanced and developing countries, where two-thirds of the world's people live. (Res 45/199, Dec 21, 1990.) Info from: United Nations, Dept of Public Info, New York, NY 10017.

UNITED NATIONS: INTERNATIONAL DECADE FOR NATURAL DISASTER REDUCTION: YEAR SIX. Jan 1-Dec 31. A decade (1990-2000) in which the international community, under UN auspices, will pay special attention to fostering international cooperation in natural disaster reduction in order to reduce loss of life, property damage and social and economic disruption caused by natural hazards such as earthquakes, floods and hurricanes. Info from: United Nations, Dept of Public Info, New York, NY 10017.

January 1995

S	M	T	W	T	F	S
1	2	3	4	5	6	7
8	9	10	11	12	13	14
15	16	17	18	19	20	21
22	23	24	25	26	27	28
29	30	31				

UNITED NATIONS: INTERNATIONAL DECADE FOR THE ERADICATION OF COLONIALISM: YEAR SIX. Jan 1-Dec 31. On Nov 22, 1988, the General Assembly proclaimed 1990-2000 the International Decade for Eradication of Colonialism. 1994 marks the 34th anniversary of the adoption of the Declaration of the Granting of Independence to Colonial Countries and Peoples. Info from: United Nations, Dept of Public Info, New York, NY 10017.

UNITED NATIONS: INTERNATIONAL DECADE FOR THE WORLD'S INDIGENOUS PEOPLE: YEAR ONE. Jan 1-Dec 9. (Began Dec 10, 1994.) Proclaimed by the General Assembly on Dec 21, 1993 (Res 48/163), this decade is for the purpose of focusing international attention and cooperation on the problems of indigenous people in a range of areas, such as human rights, health, education, development and environment. Governments are encouraged to include representatives of these people in planning and executing goals and activities for the decade.

UNITED NATIONS: SECOND INDUSTRIAL DEVELOPMENT DECADE FOR AFRICA: YEAR FIVE. Jan 1-Dec 31. On Dec 22, 1989, the General Assembly proclaimed the decade for the years 1991-2000 for the purpose of mobilizing increased political commitment to and financial and technical support for the industrialization of Africa. On Dec 22, 1992, (Res 47/177), the Assembly changed the time period for the decade to 1993-2002. Info from: United Nations, Dept of Public Info, New York, NY 10017.

UNITED NATIONS: SECOND TRANSPORT AND COMMUNICATIONS DECADE IN AFRICA: YEAR FIVE. Jan 1-Dec 31. On Dec 20, 1988 (Res 43/179), the General Assembly proclaimed the decade 1991-2000 for the purpose of continuing the progress made in the UN's first decade (1978-1988) in solving the problems of the continent in this field and mobilizing the needed technical and financial resources. Info from: United Nations, Dept of Public Info, New York, NY 10017.

UNITED NATIONS: THIRD DECADE TO COMBAT RACISM AND RACIAL DISCRIMINATION: YEAR TWO. Jan 1-Dec 9. (Began Dec 10, 1994.) In 1973 the United Nations General Assembly proclaimed the years 1973-1983, beginning on Dec 10, UN Human Rights Day, as the Decade to Combat Racism and Racial Discrimination. Renewing its efforts, the UN designated the years 1983-1993 as the Second Decade to Combat Racism and Racial Discrimination. The adopted Program of Action states the decade's goals and outlines the measures to be taken at the regional, national and international levels to achieve them. Due to a lack of resources, activities planned for the last year of the second decade were not implemented, and in Res 47/77 of Dec 16, 1992, the Assembly called upon the international community to provide resources for the program to be carried out during a third decade, particularly for the monitoring of the transition from apartheid scheduled to take place in South Africa. Info from: United Nations, Dept of Public Info, New York, NY 10017.

UNITED NATIONS: THIRD DISARMAMENT DECADE: YEAR SIX. Jan 1-Dec 31. United Nations General Assembly, in 1969, proclaimed the 1970s as a Disarmament Decade. Governments are urged to intensify efforts to stop the nuclear arms race, to promote nuclear disarmament and elimination of other weapons of mass destruction, and to develop a treaty on complete disarmament under strict and effective international control. The observance was extended into a second decade, the 1980s, and a third, the 1990s. Info from: United Nations, Dept of Public Info, New York, NY 10017.

☆ Chase's 1995 Calendar of Events ☆   Jan 1-2

UNITED NATIONS: TRANSPORT AND COMMUNICATIONS DECADE FOR ASIA AND THE PACIFIC: YEAR NINE (FOURTH YEAR OF PHASE II). Jan 1–Dec 31. On Dec 18, 1984, the United Nations General Assembly proclaimed the Transport and Communications Decade for Asia and the Pacific for the years 1985–1994. Subsequently, the Assembly decreed the second half of the decade to be the years 1992–1996. Among the objectives: to raise the transport and communications infrastructures of the developing Economic and Social Commission for Asia and the Pacific (ESCAP) member countries to a level capable of serving their development objectives and priorities. Info from: United Nations, Dept of Public Info, New York, NY 10017.

UNITED NATIONS: WORLD DECADE FOR CULTURAL DEVELOPMENT: YEAR EIGHT. Jan 1–Dec 31. On Dec 8, 1986, the United Nations General Assembly proclaimed 1988–1997 the World Decade for Cultural Development—to be observed under the auspices of UNESCO. The UN has four main objectives for voluntary efforts by states, organizations and individuals: acknowledgment of the cultural dimension of development; recognition and support of cultural identities; increased participation in culture; and increased international cultural cooperation. A global mid-term review of the decade is to be presented to the assembly. Info from: Dept of Public Info, New York, NY 10017.

UNITED NATIONS: YEAR FOR TOLERANCE. Jan 1–Dec 31. Proclaimed by the General Assembly (res 48/126) on Dec 20, 1993. See Spotlight section for full description.

UNIVERSAL LETTER-WRITING WEEK. Jan 1–7. The purpose of this week is to get people all over the world to get the new year off to a good start by sending letters and cards to friends and acquaintances not only in their own country but to people throughout the world. For complete information and suggestions about writing good letters, send $1 to cover expense of printing, handling and postage. Annually, the first seven days of January. Sponsor: Intl Soc of Friendship and Goodwill, Dr. Stanley Drake, Pres, 9538 Summerfield St, Spring Valley, San Diego County, CA 91977-2852. Phone: (619) 466-8882.

WESTERN PACIFIC HURRICANE SEASON. Jan 1–Dec 31. Most hurricanes occur from June 1 through Oct 1, though the season lasts all year. (Western Pacific: West of International Dateline.) Info from: US Dept of Commerce, Natl Oceanic and Atmospheric Admin, Rockville, MD 20852.

YOSEMITE CHEFS' HOLIDAYS. Jan 1–26. (Sundays through Thursdays.) The Ahwahnee Hotel, Yosemite National Park, CA. Receptions, seminars, tastings and lectures presented by culinary experts and influential guest chefs. Each session is highlighted by a chefs' dinner. Est attendance: 250. Sponsor: Yosemite Park and Curry Co, Yosemite Reservations/Chefs' Holidays, 5410 E Home Ave, Fresno, CA 93727. Phone: (209) 252-4848.

Z DAY. Jan 1. To give recognition on the first day of the year to all persons and places whose names begin with the letter "Z" and are always listed or thought of last in any alphabetized list. Sponsor: Tom Zager, Box 875, Sterling Heights, MI 48310.

ZWINGLI, ULRICH: BIRTH ANNIVERSARY. Jan 1. Swiss humanist, author, preacher, politician and reformer. Born on Jan 1, 1484, at Wildhaus, St. Gall, Switzerland. While serving as a military chaplain in the Second War of Kappel, Zwingli was killed on Oct 11, 1531. A monument marks the place where he fell during the battle.

BIRTHDAYS TODAY

Ernest F. Hollings, 73, US Senator (D, South Carolina), born at Charleston, SC, Jan 1, 1922.
LaMarr Hoyt (Dewey LaMarr Hoyt), 40, former baseball player, born at Columbia, SC, Jan 1, 1955.
Milt Jackson, 72, musician, born at Detroit, MI, Jan 1, 1923.
Helmut Jahn, 55, architect, born at Nuremberg, Germany, Jan 1, 1940.

Elliot Janeway, 82, economist, author, born at New York, NY, Jan 1, 1913.
Frank Langella, 55, actor (*The Twelve Chairs*), born at Bayonne, NJ, Jan 1, 1940.
Don Novello (Father Guido Sarducci), 52, actor, comedian ("The Smothers Brother Show," "Saturday Night Live"), born at Ashtabula, OH, Jan 1, 1943.
J.D. Salinger, 76, author (*Franny & Zooey, Raise High the Roof Beam*), born at New York, NY, Jan 1, 1919.

JANUARY 2 — MONDAY
2nd Day — Remaining, 363

BACON, NATHANIEL: BIRTH ANNIVERSARY. Jan 2. Leader of Bacon's Rebellion in 1676 (Virginia). Born, Suffolk, England, Jan 2, 1647. Died, Gloucester County, VA, Oct 1676 (exact date unknown).

CARQUEST AUTO PARTS BOWL. Jan 2. Joe Robbie Stadium, Miami, FL. College football bowl game. For info: Greater Ft. Lauderdale Conv/Visitors Bureau, 200 E Las Olas Blvd, Ste 1500, Fort Lauderdale, FL 33301. Phone: (305) 564-5000.

FEDERAL EXPRESS ORANGE BOWL FOOTBALL GAME. Jan 2. Miami, FL. The champion of the Big-8 Conference meets a nationally ranked opponent in the Orange Bowl Stadium. Host of seven national championship games in last 13 years. Est attendance: 80,000. For info: Orange Bowl Committee, Lisa Franson, 601 Brickell Key Dr, Ste 206, Miami, FL 33131. Phone: (305) 371-4600.

55mph SPEED LIMIT: ANNIVERSARY. Jan 2. President Richard Nixon signed a bill on Jan 2, 1974, requiring states to limit highway speeds to a maximum of 55 mph. This measure was meant to conserve energy during the crisis precipitated by the embargo imposed by the Arab oil-producing countries on the United States. A plan, used by some states, limited sale of gasoline on odd-numbered days for cars whose plates ended in odd numbers and even-numbered days for even-numbered plates. Some states limited purchases to $2–$3 per auto, and lines as long as six miles resulted in some locations. See also: "Arab Oil Embargo Lifted" (Mar 13).

FLORIDA CITRUS BOWL. Jan 2. Orlando, FL. Post-season college football game matching two teams selected from the Big Ten and Southeastern Conferences. Sponsors: The Florida Citrus Assn, Inc. Est attendance: 65,000. For info: Dylan Thomas, Florida Citrus Sports Assn, Inc, One Citrus Bowl Place, Orlando, FL 32805. Phone: (407) 423-2476.

GEORGIA: RATIFICATION DAY. Jan 2. By unanimous vote on Jan 2, 1788, Georgia became the fourth state to ratify the Constitution.

HAITI: ANCESTORS' DAY. Jan 2. Commemoration of the ancestors. Also known as Hero's Day. Public holiday.

Jan 2–3 ☆ *Chase's 1995 Calendar of Events* ☆

HALL OF FAME BOWL GAME. Jan 2. Tampa Stadium, Tampa, FL. The New Year's Day Hall of Fame Bowl Game brings together top college football teams for post-season play. In addition, the bowl is highlighted by a variety of special events, sports activities, concerts and private functions. Est attendance: 60,000. For info: Mike Schulze, Hall of Fame Bowl Assn, 4511 N Himes Ave, Ste 260, Tampa, FL 33614. Phone: (813) 874-2695.

JAPAN: KAKIZOME. Jan 2. Traditional Japanese festival gets underway when the first strokes of the year are made on paper with the traditional brushes.

MILLER, ROGER: BIRTH ANNIVERSARY. Jan 2, 1936. Country and western singer, songwriter and musician "King of the Road" Roger Miller was born at Ft Worth, TX. Miller won 11 Grammy Awards and a Tony (1986 for the score to the Broadway play *Big River*). Miller died Oct 25, 1992, at Los Angeles, CA.

MOBIL COTTON BOWL CLASSIC. Jan 2. Dallas, TX. Post-season football game matching the Southwest Conference champion team and top-ranked at-large team. Sponsor: Mobil Corp. Est attendance: 71,456. For info: Cotton Bowl Athletic Assn, Box 569420, Dallas, TX 75356-9420. Phone: (800) 638-2695.

NATIONAL GEOGRAPHY BEE, SCHOOL LEVEL. Jan 2–13. (Began Dec 5.) Principals must register their schools by Oct 15, 1994. Nationwide contest involving millions of students at the school level. The Bee is designed to encourage the teaching and study of geography. There are three levels of competition. A student must win school-level Bee in order to win the right to take a written exam. The written test determines the top 100 students in each state who are eligible to go on to the state level. National Geographic brings the state winner and his/her teacher to Washington for the national level in May. Alex Trebek moderates the national level. For info: Natl Geography Bee, Natl Geographic Society, 1145 17th St NW, Washington, DC 20036. Phone: (202) 828-6635.

OGLEBAY INSTITUTE: WINTER LODGE PROGRAM. Jan 2–31. Nature Center, Oglebay Park, Wheeling, WV. Nature and history education. Est attendance: 100. Sponsor: Oglebay Institute, Jeff Donahue, Oglebay Park, Wheeling, WV 26003. Phone: (304) 242-6855.

ROSE BOWL GAME. Jan 2. Pasadena, CA. Football conference champions from Big Ten and Pacific-10 meet in the 81st Rose Bowl game. Tournament of Roses has been annual New Year's Day event since 1890; Rose Bowl football game since 1902. Est attendance: 100,000. Sponsor: Pasadena Tournament of Roses Assn, 391 S Orange Grove Blvd, Pasadena, CA 91184. Phone: (818) 449-4100.

RUSSIA: PASSPORT PRESENTATION. Jan 2. A ceremony for 16-year-olds, who are recognized as citizens of the country on this day. Always on the first working day of the New Year.

SPACE MILESTONE: *LUNA 1* (USSR). Jan 2. First moon shot missed and became first spacecraft from Earth to orbit sun. Jan 2, 1959.

STAR OF WONDER PLANETARIUM SHOW. Jan 2–8. (Began Nov 19.) Newport News, VA. A Christmas tradition in Virginia's Hampton Roads region. This multi-media planetarium show explores the combination of scientific knowledge and religious history relating to the famous "star" that has intrigued astronomers for nearly 2,000 years. 30th annual show. Est attendance: 4,000. For info: Virginia Living Museum, Dave Maness, 524 J. Clyde Morris Blvd, Newport News, VA 23601. Phone: (804) 595-1900.

January 1995

S	M	T	W	T	F	S
1	2	3	4	5	6	7
8	9	10	11	12	13	14
15	16	17	18	19	20	21
22	23	24	25	26	27	28
29	30	31				

STOCK EXCHANGE HOLIDAY (NEW YEAR'S DAY OBSERVED). Jan 2. The holiday schedules for the various exchanges are subject to change if relevant rules, regulations or exchange policies are revised. If you have questions, phone: American Stock Exchange (212) 306-1212; Chicago Board of Options Exchange (312) 786-7760; Chicago Board of Trade (312) 435-3500; New York Stock Exchange (212) 656-2065; Pacific Stock Exchange (415) 393-4000; Philadelphia Stock Exchange (215) 496-5000.

TAFT, HELEN HERRON: BIRTH ANNIVERSARY. Jan 2. Wife of William Howard Taft, 27th president of the US, born at Cincinnati, OH, Jan 2, 1861. Died May 22, 1943.

THOMAS, MARTHA CAREY: BIRTH ANNIVERSARY. Jan 2, 1857. The second president of Bryn Mawr College, Martha Carey Thomas gained a reputation for her insistence that the education of women should be as rigorous as that of men. A zealous suffragist, she served as the first president of the National College Women's Equal Suffrage League. Thomas promoted Bryn Mawr's Summer School for Women in Industry (opened in 1921) to provide a liberal education for working women. Born at Baltimore, MD, she died at Philadelphia, Dec 2, 1935.

TOURNAMENT OF ROSES PARADE. Jan 2. Pasadena, CA. 106th annual parade. Theme: "Sports—Quest for Excellence." Rose Parade starting at 8:05 AM, PST, includes floats, bands and equestrians. Est attendance: 1,000,000. Sponsor: Pasadena Tournament of Roses Assn, 391 S Orange Grove Blvd, Pasadena, CA 91184. Phone: (818) 449-4100.

UNITED KINGDOM AND REPUBLIC OF IRELAND: NEW YEAR'S HOLIDAY. Jan 2.

USF&G SUGAR BOWL. Jan 2. Louisiana Superdome, New Orleans, LA. Annual football classic. Est attendance: 77,000. For info: USF&G Sugar Bowl, Louisiana Superdome, Mezzanine Level, 1500 Sugar Bowl Dr, New Orleans, LA 70112. Phone: (504) 525-8573.

WOLFE, JAMES: BIRTH ANNIVERSARY. Jan 2. English general who commanded the British army's victory over Montcalm's French forces on the Plains of Abraham at Quebec City in 1759. As a result, France surrendered Canada to England. Wolfe was born on Jan 2, 1727 (New Style), at Westerham, Kent, England. He died at the Plains of Abraham, Quebec, Canada, of battle wounds on Sept 13, 1759.

BIRTHDAYS TODAY

Jim Bakker (James Orsen), 56, former TV evangelist, born at Muskegon, MI, Jan 2, 1939.

Christopher Durang, 46, playwright, actor, born at Montclair, NJ, Jan 2, 1949.

Kay A. Orr, 56, former Governor of Nebraska (R), born at Burlington, IA, Jan 2, 1939.

Wendy Philips, 43, actress (Stacey Walling on "Executive Suite"), born at Brooklyn, NY, Jan 2, 1952.

Richard Riley, 62, US Secretary of Education in Clinton administration, born at Greenville, SC, Jan 2, 1933.

Renata Tebaldi, 73, singer, born at Pesaro, Italy, Jan 2, 1922.

Michael Tippett, 90, composer, born at London, England, Jan 2, 1905.

JANUARY 3 — TUESDAY
3rd Day — Remaining, 362

ALASKA ADMISSION DAY: ANNIVERSARY. Jan 3. Alaska, which had been purchased from Russia in 1867, became the 49th state on Jan 3, 1959. The area of Alaska is nearly one-fifth the size of the rest of the United States.

☆ Chase's 1995 Calendar of Events ☆ Jan 3–4

CONGRESS ASSEMBLES. Jan 3. The Constitution provides that "the Congress shall assemble at least once in every year..." and the 20th Amendment specifies "and such meeting shall begin at noon on the 3rd day of January, unless they shall by law appoint a different day."

COOLIDGE, GRACE ANNA GOODHUE: BIRTH ANNIVERSARY. Jan 3. Wife of Calvin Coolidge, 30th president of the US, born at Burlington, VT, Jan 3, 1879. Died July 8, 1957.

LENNON-ONO ALBUM CONFISCATION: ANNIVERSARY. Jan 3. John Lennon and Yoko Ono posed nude for the cover of their album *Two Virgins*. On Jan 3, 1969, a shipment of 30,000 of the albums was confiscated by police in Newark, NJ, as a violation of pornography statutes.

◈ **MOTT, LUCRETIA (COFFIN): BIRTH ANNIVERSARY.** Jan 3. American teacher, minister, antislavery leader and (with Elizabeth Cady Stanton) one of the founders of the women's rights movement in the US. Born at Nantucket, MA, Jan 3, 1793. Died near Philadelphia, PA, Nov 11, 1880.

NORTHERN ILLINOIS FARM SHOW. Jan 3–5. Rockford, IL. One of the largest full-line agriculture trade shows in the Midwest. Est attendance: 25,000. For info: Rockford Area Conv and Visitors Bureau, Memorial Hall, 211 N Main St, Rockford, IL 61101. Phone: (800) 521-0849.

PAVLOVA, ANNA: BIRTH ANNIVERSARY. Jan 3. Russian ballerina Anna Pavlova, thought by some to have been the greatest dancer of all time, was born at St. Petersburg, Russia, on Jan 3, 1885. After performing with great success with the Ballet Russe and other companies, she formed her own company in 1910 and performed on tour for enthusiastic audiences in nearly every country in the world. Pavlova died at The Hague, Netherlands, on Jan 23, 1931.

SCOTLAND: NEW YEAR'S BANK HOLIDAY. Jan 3.

TOM SAWYER'S CAT'S BIRTHDAY. Jan 3. To bring into focus Tom Sawyer's mania for doing things the hard way—e.g., he wanted to dig a tunnel to free Jim instead of just announcing that Jim had been freed. In that spirit, Tom Sawyerists worldwide may show how "burrow-crats" complicate life by finding difficult and expensive ways of doing things, like pretending that Tom's cat taught Tom, by telepathy, how to become a Congressman and then a general in the Civil War, and what to do then. The pretense is that the cat is super intelligent and is the mastermind behind the confusion in Congress and the Army, to make sure that no one will challenge the waste and cushy jobs set up to get simple jobs done. Sponsor: The Puns Corps, c/o Bob Birch, Box 2364, Falls Church, VA 22042-0364. Phone: (703) 533-3668.

BIRTHDAYS TODAY

Joan Walsh Anglund, 69, author, illustrator of children's books (*Crocus in the Snow, Bedtime Book*), born at Hinsdale, IL, Jan 3, 1926.

Victor Borge (Borge Rosenbaum), 86, comedian, pianist, born at Copenhagen, Denmark, Jan 3, 1909.

Dabney Coleman, 63, actor ("Buffalo Bill," *Nine to Five, Tootsie*), born at Austin, TX, Jan 3, 1932.

Jim (James Samuel) Everett III, 32, football player, born at Emporia, KS, Jan 3, 1963.

Mel Gibson, 44, actor (*The Road Warrior, Lethal Weapon*), born at New York, NY, Jan 3, 1951.

Carla Anderson Hills, 61, Former US Trade Representative in Bush administration, born at Los Angeles, CA, Jan 3, 1934.

Bobby Hull, 56, former hockey player, born at Point Anne, Ontario, Canada, Jan 3, 1939.

Robert Loggia, 65, actor, born at Staten Island, NY, Jan 3, 1930.

Victoria Principal, 45, actress (Pam Ewing on "Dallas"), born at Fukuoka, Japan, Jan 3, 1950.

Stephen Stills, 50, musician, songwriter, born at Dallas, TX, Jan 3, 1945.

Vernon A. Walters, 78, former US Representative to the United Nations, born at New York, NY, Jan 3, 1917.

JANUARY 4 — WEDNESDAY
4th Day — Remaining, 361

BRAILLE, LOUIS: BIRTH ANNIVERSARY. Jan 4. The inventor of a widely used touch system of reading and writing for the blind was born at Coupvray, France, Jan 4, 1809. Permanently blinded at the age of three (by a leatherworking awl in his father's saddlemaking shop), Braille developed a system of writing that used, ironically, an awl-like stylus to punch marks in paper that could be felt and interpreted by the blind. The system was largely ignored until after Braille died in poverty, suffering from tuberculosis, on Mar 28, 1852.

BURMA: INDEPENDENCE DAY. Jan 4. National holiday. Became independent nation on this day in 1948 by virtue of treaty with Great Britain.

EARTH AT PERIHELION. Jan 4. At approximately 6 AM, EST, planet Earth will reach Perihelion, that point in its orbit when it is closest to the sun (about 91,400,000 miles). The Earth's mean distance from the sun (mean radius of its orbit) is reached early in the months of April and October. Note that Earth is closest to the sun during Northern Hemisphere winter. See also: "Earth at Aphelion" (July 3).

GENERAL TOM THUMB: BIRTH ANNIVERSARY. Jan 4. Charles Sherwood Stratton, perhaps the most famous midget in history, was born at Bridgeport, CT, on Jan 4, 1838. His growth almost stopped during his first year, but he eventually reached a height of three feet, four inches and a weight of 70 pounds. "Discovered" by P.T. Barnum in 1842, Stratton, as "General Tom Thumb" became an internationally known entertainer and on tour performed before Queen Victoria and other heads of state. On Feb 10, 1863, he married another midget, Lavinia Warren. Stratton died at Middleborough, MA, on July 15, 1883.

GRIMM, JACOB: BIRTH ANNIVERSARY. Jan 4. Librarian, mythologist and philologist, born at Hanau, Germany, Jan 4, 1785. Best remembered for *Grimm's Fairy Tales* (in collaboration with his brother Willhelm). Died at Berlin, Germany, Sept 20, 1863.

HOLLOWAY, STERLING: 90th BIRTH ANNIVERSARY. Jan 4. Actor Sterling Holloway prospered in films and television, but he is probably best remembered as the voice of Winnie the Pooh. Holloway, who was born Jan 4, 1905, provided the voices for a number of characters in several full-length animated features, including *Alice in Wonderland* (the Cheshire Cat), *The Aristocats* and *The Jungle Book*. Holloway died Nov 22, 1992, in Los Angeles, CA.

Jan 4–5 ☆ *Chase's 1995 Calendar of Events* ☆

MYANMAR: INDEPENDENCE DAY. Jan 4. National Day.

SETON, ELIZABETH ANN BAYLEY: FEAST DAY. Jan 4. First American-born saint (beatified Mar 17, 1963, canonized Sept 14, 1975). Born at New York, NY, Aug 28, 1774. She was founder of American Sisters of Charity, the first American order of Roman Catholic nuns. Died at Baltimore, MD, Jan 4, 1821.

TRIVIA DAY. Jan 4. In celebration of those who know all sorts of facts and/or have doctorates in uselessology. Annually, Jan 4. Sponsor: Puns Corps, Robert L. Birch, Box 2364, Falls Church, VA 22042-0364. Phone: (703) 533-3668.

UTAH: ADMISSION DAY. Jan 4. Became 45th state on this day in 1896.

WYMAN, JANE: BIRTHDAY. Jan 4. Award-winning American actress, first wife of Ronald Wilson Reagan, 40th president of the US. Born at St. Joseph, MO, Jan 4, 1914. Married Reagan Jan 24, 1940. They were divorced in 1948.

BIRTHDAYS TODAY

Norman H. Bangerter, 62, former Governor of Utah (R), born at Granger, UT, Jan 4, 1933.

Dyan Cannon, 58, actress (Oscar nominition for *Heaven Can Wait, Bob and Carol and Ted and Alice*), born at Tacoma, WA, Jan 4, 1937.

Lauro F. Cavazos, 68, Secretary of Education, born at King Ranch, TX, Jan 4, 1927.

Clifford Eugene Levingston, 34, basketball player, born at San Diego, CA, Jan 4, 1961.

Floyd Patterson, 60, former boxer, born at Waco, NC, Jan 4, 1935.

Don Shula, 65, football coach, born at Grand River, OH, Jan 4, 1930.

JANUARY 5 — THURSDAY
5th Day — Remaining, 360

AILEY, ALVIN: BIRTH ANNIVERSARY. Jan 5. Dancer and choreographer. Born at Rogers, TX, on Jan 5, 1931, Alvin Ailey began his noted career as a choreographer in the late 1950s after a successful career as a dancer. He founded the Alvin Ailey American Dance Theater, drawing from classical ballet, jazz, Afro-Caribbean and modern dance idioms to create the 79 ballets of the company's repertoire. He and his work played a central part in establishing a role for blacks in the world of modern dance. Ailey died on Dec 1, 1989, at New York, NY.

AMERICAN HISTORICAL ASSOCIATION: ANNUAL MEETING. Jan 5–8. Chicago, IL. Approximately 140 sessions covering a wide range of scholarly, professional and pedagogical topics dealing with world history. Est attendance: 3,800. Sponsor: American Historical Assn, 400 A St SE, Washington, DC 20003. Phone: (202) 544-2422.

CANADA: SUNRISE FESTIVAL. Jan 5. Inuvik, Northwest Territories. Fireworks, bonfire and refreshments on evening preceding return of sun above horizon, following the 6-week dark period. Annually, the night preceding first sunrise. Phone: (403) 979-2607 or (403) 979-2185. Est attendance: 2,500. For info: Everett Giles, Firefighter, PO Box 1160, Inuvik, NWT, Canada, X0E 0T0.

CARVER, GEORGE WASHINGTON: DEATH ANNIVERSARY. Jan 5. Black American agricultural scientist, author, inventor and teacher. Born into slavery at Diamond Grove, MO, probably in 1864. His research led to the creation of synthetic products made from peanuts, potatoes and wood. Carver died at Tuskegee, AL, on Jan 5, 1943. His birthplace became a national monument in 1953.

DECATUR, STEPHEN: BIRTH ANNIVERSARY. Jan 5. American naval officer (whose father and grandfather, both named Stephen Decatur, were also seafaring men). Born at Sinepuxent, MD, Jan 5, 1779. In a toast at a dinner in Norfolk in 1815, Decatur spoke his most famous words: "Our country! In her intercourse with foreign nations may she always be in the right; but our country, right or wrong." Mortally wounded in a duel with Commodore James Barron, at Bladensburg, MD, on the morning of Mar 22, 1820. Carried to his home in Washington, he died a few hours later.

ENGLAND: LONDON INTERNATIONAL BOAT SHOW. Jan 5–15. Earls Court Exhibition Centre, London. One of the largest international boat shows in Europe, displaying more than 600 craft plus accessories for the marine enthusiast. Est attendance: 200,000. For info: British Marine Industries Federation/Natl Boat Shows Ltd, Meadlake Place, Thorpe Lea Rd, Egham, Surrey, England TW20 8HE.

FIVE-DOLLAR-A-DAY MINIMUM WAGE: ANNIVERSARY. Jan 5. Henry Ford announced on Jan 5, 1914, that all worthy Ford Motor Company employees would receive a minimum wage of $5.00 a day. Ford explained the policy as "profit sharing and efficiency engineering." The more cynical attributed it to an attempt to prevent unionization and to obtain a docile work force that would accept job speedups. To obtain this minimum wage an employee had to be of "good personal habits." Whether an individual fit these criteria was determined by a new office created by Ford Motor Company—the Sociological Department.

ITALY: EPIPHANY FAIR. Jan 5. Piazza Navona, Rome, Italy. On the eve of Epiphany a fair of toys, sweets and presents takes place among the beautiful Bernini Fountains.

PICCARD, JEANNETTE RIDLON: 100th BIRTH ANNIVERSARY. Jan 5. First American woman to qualify as free balloon pilot (1934). One of first women to be ordained as Episcopal priest (1976). Pilot for record-setting balloon ascent into stratosphere (from Dearborn, MI, Oct 23, 1934) (57,579 ft) with her husband, Jean Felix Piccard. See also: "Piccard, Jean Felix: Birth Anniversary" (Jan 28). Identical twin married to identical twin. Born at Chicago, IL, Jan 5, 1895. Died at Minneapolis, MN, May 17, 1981.

RUFFIN, EDMUND: BIRTH ANNIVERSARY. Jan 5. Born in Prince George County, VA, on Jan 5, 1794, Edmund Ruffin was an early American agriculturist whose discoveries about crop rotation and fertilizer were influential in the early agrarian culture of the US. He published the *Farmer's Register* from 1833 to 1842, a journal which promoted scientific agriculture. A noted politician as well as a farmer, he was an early advocate of Southern secession whose views were widely circulated in pamphlets. As a member of the Palmetto Guards of Charleston, he was given the honor of firing the first shot on Fort Sumter on Apr 12, 1861. According to legend, after the South's defeat he became despondent and, wrapping himself in the Confederate flag, took his own life on June 18, 1865, at Amelia County, Virginia.

SAN DIEGO BOAT AND SPORT FISHING SHOW. Jan 5–8. San Diego Convention Center and Marriott Marina, San Diego, CA. Sponsor: Natl Marine Manufacturers Assn. Sponsor: Jeff Hancock, NMMA, 4901 Morena Blvd, Ste 901, San Diego, CA 92117. Phone: (619) 274-9924.

January 1995

S	M	T	W	T	F	S
1	2	3	4	5	6	7
8	9	10	11	12	13	14
15	16	17	18	19	20	21
22	23	24	25	26	27	28
29	30	31				

☆ Chase's 1995 Calendar of Events ☆   Jan 5-6

TWELFTH NIGHT. Jan 5. Evening before Epiphany. Twelfth Night marks the end of medieval Christmas festivities and the end of Twelfthtide (the 12-day season after Christmas and ending with Epiphany). Also called Twelfth Day Eve.

WYOMING INAUGURATES FIRST WOMAN GOVERNOR IN US. Jan 5. Mrs William B. Ross became the first woman to serve as governor upon her inauguration as governor of Wyoming on Jan 5, 1925.

BIRTHDAYS TODAY

Robert Bernstein, 72, president of Random House, human rights activist, born at New York, NY, Jan 5, 1923.
Jeanne Dixon, 77, clairvoyant, astrologer, born at Medford, WI, Jan 5, 1918.
Robert Duvall, 64, actor ("Naked City," *The Godfather*), born at San Diego, CA, Jan 5, 1931.
Alex English, 41, former basketball player, born at Columbia, SC, Jan 5, 1954.
Charles Oliver Hough, 47, baseball player, born at Honolulu, HI, Jan 5, 1948.
Diane Keaton (Diane Hall), 49, actress (Oscar for *Annie Hall*; *Sleeper, Love and Death, Hair*), born at Los Angeles, CA, Jan 5, 1946.
Ron Kittle, 37, former baseball player, born at Gary, IN, Jan 5, 1958.
Pamela Sue Martin, 41, actress (*The Poseidon Adventure*, "The Nancy Drew Mysteries," "Dynasty"), born at Westport, CT, Jan 5, 1954.
Walter Frederick "Fritz" Mondale, 67, former US vice president and senator, born at Ceylon, MN, Jan 5, 1928.
W.D. Snodgrass, 69, poet, born at Wilkinsburg, PA, Jan 5, 1926.

JANUARY 6 — FRIDAY
6th Day — Remaining, 359

ALL-COLLEGE BASKETBALL TOURNAMENT. Jan 6-7. Myriad Convention Center, Oklahoma City, OK. The oldest college basketball tournament in the country. Usually held in December, due to unusual circumstances the 59th annual tournament scheduled for Dec 1994 had to be postponed to Jan 1995. The 60th will be held Dec 29-30, 1995. Est attendance: 25,000. For info: Stanley Draper, Jr, OK City All Sports Assn, 100 W Main, Ste 287, Oklahoma City, OK 73102. Phone: (405) 236-5000.

ARMENIAN CHRISTMAS. Jan 6. Christmas is observed on this day in the Armenian Church, the oldest Christian national church.

AUGUSTA FUTURITY. Jan 6-14. Augusta, GA. Brings together the top cutting horses and riders in the world to compete for purse and awards of more than $600,000. Sponsors: Wrangler, Jeep Eagle, John Deere, Budweiser, WRDW-TV12, Manna Pro Feed Co, Augusta Regional Medical Center, Justin Boots, Delta Air Lines, Palmer & Cay/Carswell Insurance, Tex Tan Western Leather Co, CellularOne, Bailey Hats, Augusta Richmond County Conv & Visitors Bureau and The Augusta Chronicle. Est attendance: 40,000. For info: Skip Peterson, Dir of Mktg, Augusta Futurity, PO Box 936, Augusta, GA 30903. Phone: (706) 823-3370.

BAKER STREET IRREGULARS' DINNER AND COCKTAIL PARTY. Jan 6-7. New York, NY. Annual dinner, meeting and cocktail party. Est attendance: 200. For info: Thomas L. Stix, Jr, 34 Pierson Ave, Norwood, NJ 07648.

CANADA: TORONTO INTERNATIONAL BOAT SHOW. Jan 6-15. Coliseum and Automotive Buildings, Exhibition Place, Toronto, Ontario. Largest indoor display of boats, marine products, services and accessories for the consumer in North America. Grand Opening to general public on Fri, Jan 6. For info: NMMA Boat Shows, 401 N Michigan Ave, Ste 1150, Chicago, IL 60611. Phone: (312) 836-4740.

CARNIVAL SEASON. Jan 6-Feb 28. A secular festival preceding Lent. A time of merrymaking and feasting before the austere days of Lenten fasting and penitence (40 weekdays between Ash Wednesday and Easter Sunday). The word *carnival* probably is derived from the Latin *carnem levare*, meaning "to remove meat." Depending on local custom, the carnival season may start any time between Nov 11 and Shrove Tuesday. Conclusion of the season is much less variable, being the close of Shrove Tuesday in most places. Celebrations vary considerably, but the festival often includes many theatrical aspects (masks, costumes and songs) and has given its name (in the US) to traveling amusement shows that may be seen throughout the year. Observed traditionally in Roman Catholic countries from Epiphany through Shrove Tuesday.

CHARLOTTE OBSERVER MARATHON, 10K, AND RUNNER'S EXPO. Jan 6-7. Charlotte, NC. A 26.2-mile marathon (Jan 7) plus 10K and 1-mile races. Runners Expo at Charlotte Convention Center. Est attendance: 32,000. For info: Observer Marathon, Box 30294, Charlotte, NC 28230. Phone: (704) 358-5425.

EPIPHANY. Jan 6. See "Twelfth Day" this date.

FIRST FRIDAY. Jan 6. (Also Feb 3, Mar 3, Apr 7, May 5, Sept 8, Oct 6, Nov 3, Dec 1.) Milwaukee Art Museum, Milwaukee, WI. Milwaukee's most unique happy hour featuring live jazz, tours, hors d'oeuvres and cash bar. Begins at 5:30 in the east entrance of the museum. Admission $5/members, $7/non-members and free to those who buy Art Museum membership at the door. Est attendance: 250. For info: Milwaukee Art Museum, 750 N Lincoln Memorial Dr, Milwaukee, WI 53202. Phone: (414) 224-3200.

GRAND AMERICAN COON HUNT. Jan 6-7. County Fairgrounds, Orangeburg, SC. Coon hunters and sportsmen from all over the United States and Canada bring their dogs to compete for Grand American Champion. An A.C.H.A. qualifying hunt. Est attendance: 25,000. For info: Orangeburg County Chamber of Commerce, Carol P. Whisenhunt, PO Box 328, Orangeburg, SC 29116-0328. Phone: (803) 534-6821.

ITALY: LA BEFANA. Jan 6. Epiphany festival in which the "Befana," a kindly witch, bestows gifts on children—toys and candy for those who have been good, but a lump of coal or a pebble for those who have been naughty. The festival begins on the night of Jan 5 with much noise and merrymaking (when the Befana is supposed to come down the chimneys on her broom, leaving gifts in children's stockings) and continues with joyous fairs, parades and other activities throughout Jan 6.

JAMAICA: MAROON FESTIVAL. Jan 6. Commemorates the 18th-century Treaty of Cudjoe. While Jamaica was a Spanish colony, its native inhabitants (Arawaks) were exterminated. The Spanish then imported African slaves to work their plantations. When the Spanish were driven out (1655), the black slaves fled to the mountains. The "Maroons" (fugitive slaves) were permitted to settle in the north of the island in 1738.

MIX, TOM: BIRTH ANNIVERSARY. Jan 6. American motion picture actor, especially remembered for western cowboy films. Born at Driftwood, PA, Jan 6, 1880. Died near Florence, AZ, Oct 12, 1940.

NEW MEXICO: ADMISSION DAY: ANNIVERSARY. Jan 6. Became 47th state on this day in 1912.

57

Jan 6–7 ☆ *Chase's 1995 Calendar of Events* ☆

NEW YORK NATIONAL BOAT SHOW. Jan 6–15. Jacob Javits Convention Center, New York, NY. Sponsor: Natl Marine Manufacturers Assn, Michael Duffy, 600 Third Ave, New York, NY 10016. Phone: (212) 922-1212.

SALOMON, HAYM: DEATH ANNIVERSARY. Jan 6. American Revolutionary War patriot and financier was born at Lissa, Poland, in 1740 (exact date unknown). Salomon died at Philadelphia, PA, Jan 6, 1785.

SANDBURG, CARL: BIRTH ANNIVERSARY. Jan 6. American writer—poet, biographer, historian and folklorist—born Jan 6, 1878, at Galesburg, IL. Died at Flat Rock, NC, July 22, 1967.

STAMP EXPO. Jan 6–8. Travelodge Hotel, Long Beach, CA. Fax (818) 988-4337 or (818) 988-4337. Est attendance: 7,500. Sponsor: Intl Stamp Collectors Soc, PO Box 854, Van Nuys, CA 91408. Phone: (818) 997-6496.

THERMOPOLIS DAY. Jan 6. Thermopolis, WY. Curbside musicians, roving sing-alongs, an impromptu local and tourist parade and much more. Annually, the first Friday in January. Est attendance: 1,000. Sponsor: KTHE Radio, Neal A. Behnke, PO Box 591, Thermopolis, WY 82443. Phone: (307) 864-2119.

THREE KINGS DAY. Jan 6. Major festival of Christian Church observed in many parts of the world with gifts, feasting, last lighting of Christmas lights and burning of Christmas greens. Twelfth and last day of Feast of the Nativity. Commemorates visit of the Three Wise Men (Kings or Magi) to Bethlehem.

TWELFTH DAY or EPIPHANY. Jan 6. Known also as Old Christmas Day and Twelfthtide. On the twelfth day after Christmas, Christians celebrate the visit of the Magi, the first Gentile recognition of Christ. Epiphany of Our Lord, one of the oldest Christian feasts, is observed in Roman Catholic churches in the US on a Sunday between Jan 2 and 8. Theophany of the Eastern Orthodox Church is observed on this day in churches using the Gregorian calendar and on Jan 19 in those churches using the Julian calendar and celebrates the manifestation of the divinity of Jesus at the time of his baptism in the Jordan River by John the Baptist.

BIRTHDAYS TODAY

John DeLorean, 70, auto executive, author, born at Detroit, MI, Jan 6, 1925.
E.L. Doctorow, 64, writer (*Ragtime, Welcome to Hard Times*), born at New York, NY, Jan 6, 1931.
Bonnie Franklin, 51, actress (Ann Romano on "One Day at a Time"; *Applause*), born at Santa Monica, CA, Jan 6, 1944.
Charles Lewis Haley, 31, football player, born at Gladys, VA, Jan 6, 1964.
Lou Harris, 74, public opinion analyst, author, born at New Haven, CT, Jan 6, 1921.
Lou Holtz, 58, football coach, born at Follansbee, WV, Jan 6, 1937.
Howie Long, 35, former football player, born at Somerville, MA, Jan 6, 1960.
Nancy Lopez, 38, golfer, born at Torrance, CA, Jan 6, 1957.

January 1995

S	M	T	W	T	F	S
1	2	3	4	5	6	7
8	9	10	11	12	13	14
15	16	17	18	19	20	21
22	23	24	25	26	27	28
29	30	31				

Earl Scruggs, 71, musician, born at Cleveland County, NC, Jan 6, 1924.
Loretta Young, 82, actress (Oscar for *The Farmer's Daughter*; "The Loretta Young Show"), born at Salt Lake City, UT, Jan 6, 1913.

JANUARY 7 — SATURDAY
7th Day — Remaining, 358

BEYOND CATEGORY: THE MUSICAL GENIUS OF DUKE ELLINGTON. Jan 7–Mar 19. California Afro-American Museum, Los Angeles, CA. Exhibition celebrating 50 years of Duke Ellington's musical innovation and accomplishment. The jazz musician's unpublished manuscripts serve as the basis for this innovative and original audio-visual presentation. Call or write the Smithsonian for other dates and venues. For info: Smithsonian Institution Traveling Exhibition Service, 1100 Jefferson Dr SW, Ste 3146, Washington, DC 20560. Phone: (202) 357-2700.

CHAMPIONSHIP CAT SHOW. Jan 7–8. (Also July 1–2.) Fairgrounds, Indianapolis, IN. Exhibition and judging of longhair and shorthair purebred and household pet cats and kittens. Est attendance: 1,500. For info: American Cat Fanciers Assn, Circle City Cat Club, Maribeth Echard, 8507 N Illinois St, Indianapolis, IN 46260. Phone: (317) 251-4486.

CHICAGO'S WINDY CITY JITTERBUG CLUB DANCE. Jan 7. (Also Jan 21.) American Legion Hall, Franklin Park, IL. Get that old feeling dancing the jitterbug/swing. These are the first of twice monthly dances (first and third Saturday of each month—with occasional Fridays) intended to promote and exchange steps and styling among dancers throughout the country. Sponsor: CWCJC, PO Box 713, Franklin Park, IL 60131. Phone: (312) 725-7230.

FILLMORE, MILLARD: BIRTH ANNIVERSARY. Jan 7. Thirteenth president of the US (July 10, 1850–Mar 3, 1853). Fillmore succeeded to the presidency upon the death of Zachary Taylor, but was unsuccessful in getting hoped-for nomination from his party in 1852. He ran for president in 1856 as candidate of the "Know-Nothing Party," whose platform demanded, among other things, that every government employee (federal, state and local) should be a native-born citizen. Fillmore was born at Summerhill, NY, Jan 7, 1800, and died at Buffalo, NY, Mar 8, 1874. Now his birthday is often occasion for parties for which there is no other reason.

FIRST BALLOON FLIGHT ACROSS ENGLISH CHANNEL: ANNIVERSARY. Jan 7. Dr. John Jeffries, a Boston physician, and Jean-Pierre Blanchard, French aeronaut, on Jan 7, 1785, crossed the English Channel from Dover, England, to Calais, France, landing in a forest after being forced to throw overboard all ballast, equipment and even most of their clothing to avoid forced landing in the icy waters of the English Channel. Blanchard's trousers are said to have been the last article thrown overboard.

FIRST CROSSING OF PANAMA CANAL: ANNIVERSARY. Jan 7. An interoceanic waterway connecting the Atlantic and the Pacific oceans became a focus of explorers and countries seeking quicker trade routes as early as the 16th century. Planning for such a waterway reached a peak in the 19th century as the French began a plan to engineer and finance a sea-level canal project in 1876. From an engineering standpoint, this plan proved impractical and plans were laid to construct a lock canal, but the French company went bankrupt in 1889, plagued by heat, disease and financial difficulties. US interest peaked again during the Spanish-American War. After a volcanic eruption in Nicaragua cooled interest in a canal in that country, Congress authorized President Theodore Roosevelt to acquire rights and property of the New Panama Canal Zone from the French, providing that Colombia, of which the Isthmus of Panama was a province, agreed to grant perpetual control of the required land. On Jan 22, 1903, a treaty was signed leasing the US a 10-mile wide zone for 100 years. Colombia delayed ratification of the treaty, and the citizens of Panama revolted, encouraged by the presence of a US warship offshore. Panama declared its independence on Nov 3, 1903, and on Nov 18, two weeks later, Panama signed the Hay-Buana-Varilla Treaty, giving the US the right to build a canal in return for a payment of $10 million. The treaty gave the US all rights, in perpetuity, to the 10- by 50-mile strip of land. Construction began in 1906, and on Jan 7, 1914, a self-propelled crane boat, *Alex La Valley*, made the first passage through the canal. The first ocean steamer, the SS *Ancon*, passed through on Aug 3, 1914, and the canal officially opened on Aug 15, 1914.

GARDENIA, VINCENT (VINCENT SCOGNAMIGLIO): BIRTH ANNIVERSARY. Jan 7. Stage, screen and television performer Vincent Gardenia was born Jan 7, 1922, at Naples, Italy. Gardenia began performing at age 5 and once estimated he had played 500 parts in his lifetime. He received two Oscar nominations, one for playing a baseball manager in *Bang the Drum Slowly* and again for the role of patriarch of a goofy Brooklyn family in *Moonstruck*. He won a Tony for his part in *The Prisoner of Second Avenue* and an Emmy for his portrayal in *Age Old Friends*. Vincent Gardenia died Dec 9, 1992, at Philadelphia, PA.

GERMANY: MUNICH FASCHING CARNIVAL. Jan 7–Feb 28. Munich. From Jan 7 until Shrove Tuesday is Munich's famous carnival season. Costume balls are popular throughout carnival. "High points on Fasching Sunday (Feb 26) and Shrove Tuesday (Feb 28) with great carnival doings outside at the Viktualienmarkt and on Pedestrian Mall."

JAPAN: NANAKUSA. Jan 7. Festival dates back to the 7th century and recalls the seven plants served to the emperor that are believed to have great medicinal value—shepherd's purse, chickweed, parsley, cottonweed, radish, hotoke-no-za and aona.

JAPAN: USOKAE (BULLFINCH EXCHANGE FESTIVAL). Jan 7. Dazaifu, Fukuoka Prefecture. "Good Luck" gilded wood bullfinches, mixed among many plain ones, are sought after by the throngs as priests of the Dazaifu Shrine pass them out in the dim light of a small bonfire.

MONTGOLFIER, JACQUES ETIENNE: 250th BIRTH ANNIVERSARY. Jan 7. Merchant and inventor born at Vidalon-lez Annonay, Ardèche, France, Jan 7, 1745. With his older brother, Joseph Michel, in Nov 1782, conducted experiments with paper and fabric bags filled with smoke and hot air, which led to invention of the hot air balloon and human's first flight. Died at Serrieres, France, Aug 2, 1799. See also: "First Balloon Flight: Anniversary" (June 5); "Aviation History Month" (Nov 1).

MORE THAN MEETS THE EYE. Jan 7–Mar 19. New Mexico Museum of Natural History, Albuquerque, NM. With an original approach to the topics of sight, visual impairment and blindness, this exhibit offers visitors a new awareness of how they "see" their world. Through hands-on, low-tech activities, visitors explore communications, mobility and orientation to the environment and come to understand how the visually impaired or blind train their other senses. For info: Smithsonian Institution Traveling Exhibition Service, 1100 Jefferson Dr SW, Ste 3146, Washington, DC 20560. Phone: (202) 357-2700.

OLD CALENDAR ORTHODOX CHRISTMAS. Jan 7. Some Orthodox Churches celebrate Christmas on this day, which is the "Old" (Julian) calendar date.

STRENGTH AND DIVERSITY: JAPANESE-AMERICAN WOMEN 1885–1990. Jan 7–Mar 19. Field Museum of Natural History, Chicago, IL. Exhibition chronicles the lives of Issei, Nisei and Sansei women as they adjusted to life in the US, faced wartime relocation and reconciled two very different cultures. Historical photographs, artifacts, literary and artistic works and oral histories document the blending of Japanese heritage with daily life in the US. Call or write the Smithsonian for other dates and venues. For info: Smithsonian Institution Traveling Exhibition Service, 1100 Jefferson Dr SW, Ste 3146, Washington, DC 20560. Phone: (202) 357-2700.

TRY THIS ON—A HISTORY OF CLOTHING, GENDER AND POWER. Jan 7–Feb 19. Frances Tavern Museum, New York, NY. A social history about clothing and appearance in the US over the past 200 years, this exhibit examines links between rules for appearance and different roles, opportunities and rewards that are open to men and women. Artifact and graphics drawn from the Smithsonian Institution's National Museum of American History examine corsets and bustles, blue jeans and bandannas. Low-tech interactive stations allow visitors to identify the social rules and values reflected. For info: Smithsonian Institution Traveling Exhibition Service, 1100 Jefferson Dr SW, Ste 3146, Washington, DC 20560. Phone: (202) 357-2700.

BIRTHDAYS TODAY

William Blatty, 67, screenwriter, born at New York, NY, Jan 7, 1928.
Erin Gray, 43, actress (Col. Wilma Deering in "Buck Rogers in the 25th Century"), born at Honolulu, HI, Jan 7, 1952.
Douglas Kiker, 65, journalist, author ("NBC Magazine with David Brinkley"; *Death Below Deck*), born at Griffin, GA, Jan 7, 1930.
Kenny Loggins, 47, singer, songwriter, born at Everett, WA, Jan 7, 1948.
Jean-Pierre Rampal, 73, musician, born at Marseilles, France, Jan 7, 1922.
Jann Wenner, 48, journalist, publisher, born at New York, NY, Jan 7, 1947.

JANUARY 8 — SUNDAY
8th Day — Remaining, 357

AT&T DIVESTITURE: ANNIVERSARY. Jan 8. In the most significant antitrust suit since the breakup of Standard Oil in 1911, American Telephone and Telegraph agreed Jan 8, 1982, to give up its 22 local Bell System companies ("Baby Bells"). These companies represented 80% of AT&T's assets. This ended the corporation's virtual monopoly on US telephone service.

BATTLE OF NEW ORLEANS: ANNIVERSARY. Jan 8. British forces suffered crushing losses (more than 2,000 casualties) in an attack on New Orleans, LA, Jan 8, 1815. Defending US troops were led by General Andrew Jackson, who became a popular hero as a result of the victory. Neither side knew that the war had ended two weeks previously with the signing of the Treaty of Ghent on Dec 24, 1814. Battle of New Orleans Day is observed in Louisiana and Massachusetts.

ELVIS PRESLEY BIRTHDAY PARTY AT BLUEBERRY HILL. Jan 8. St. Louis, MO. Join the fun at the biggest annual celebration outside Memphis, TN. Includes a live show plus impersonation contest. Est attendance: 900. For info: Joe Edwards, Blueberry Hill, 6504 Delmar, St. Louis, MO 63130. Phone: (314) 727-0880.

Jan 8-9 ☆ *Chase's 1995 Calendar of Events* ☆

FERRER, JOSE: BIRTH ANNIVERSARY. Jan 8. Award-winning actor, producer, writer and director Jose Ferrer was born Jan 8, 1912, at Santurce, Puerto Rico. A promising pianist in his youth, he was expected to become a concert performer. Nominated three times for an Academy Award, he won best actor for his role in *Cyrano de Bergerac*. Additionally, Ferrer was awarded Tonys and Critics' Circle prizes during half a century in the entertainment world. Jose Ferrer died Jan 26, 1992, at Coral Gables, FL.

GREECE: MIDWIFE'S DAY or WOMEN'S DAY. Jan 8. Midwife's Day or Women's Day is celebrated on Jan 8 each year to honor midwives and all women. "On this day women stop their housework and spend their time in cafés, while the men do all the housework chores and look after the children." In some villages, men caught outside "will be stripped ... and drenched with cold water."

LAS VEGAS OPEN HORSESHOE TOURNAMENT. Jan 8-12. Hacienda Hotel, Las Vegas, NV. Attracts horseshoe pitchers from all over the world. Est attendance: 4,500. For info: Natl Horseshoe Pitching Assn, Donnie Roberts, Box 7927, Columbus, OH 43207. Phone: (614) 444-8510.

MOON PHASE: FIRST QUARTER. Jan 8. Moon enters First Quarter phase at 10:46 AM, EST.

NATIONAL JOYGERM DAY. Jan 8. A celebration of jubilation as all parts of the nation employ tons of tenderness; smatterings of smiles; cartons of kindness; gleanings of glee; graspings of grins as they win over gruff and grumpy grouches to the Joygerm Generation. No force, but, of course, Joygerm Joan's Sulk Vaccine will be on the scene. For cynics: Smile Check-Up Clinics. For info: Joygerm Joan E. White, Founder, Joygerms Unlimited, PO Box 219, Eastwood Station, Syracuse, NY 13206. Phone: (315) 472-2779.

NATIONAL WORD PROCESSING TRANSCRIPTIONIST WEEK. Jan 8-14. Norfolk, NE. To recognize the contribution of word processing personnel in all areas of business and healthcare. To honor the professionalism in making a positive difference in achieving quality workmanship. Annually, the second full week in January. Est attendance: 74. For info: Norma Rohlff, Word Processing Supv, Norfolk Regional Ctr, PO Box 1209, Norfolk, NE 68701. Phone: (402) 370-3203.

PRESLEY, ELVIS AARON: 60th BIRTH ANNIVERSARY. Jan 8. Popular American rock singer. Born at Tupelo, MS, Jan 8, 1935. Died at Memphis, TN, Aug 16, 1977. Although his middle name was spelled incorrectly as "Aron" on his birth certificate, Elvis had it legally changed to "Aaron," which is how it is spelled on his gravestone.

ST. GUDULA'S FEAST DAY. Jan 8. Virgin, patron saint of the city of Brussels. Died Jan 8, probably in the year 712. Her relics were transferred to the church of St. Michael in Brussels.

"SOMEDAY WE'll LAUGH ABOUT THIS" WEEK. Jan 8-14. We've all used the expression, "Someday we'll laugh about this!" Why wait? It usually takes less than seven days for people to violate 90% of their New Year's resolutions. This week helps us to remember the art of laughing at ourselves. This week tickles the yoke and joke of perfectionism while encouraging people to strive for excellence at the same time. This week is a great way to start the new year—laughing at the humorous human condition. For a free information packet on the positive power of humor send a stamped (98¢) self-addressed envelope. Sponsor: The Humor Project, Inc, Dept C, 110 Spring St, Saratoga Springs, NY 12866. Phone: (515) 587-8770.

January 1995

S	M	T	W	T	F	S
1	2	3	4	5	6	7
8	9	10	11	12	13	14
15	16	17	18	19	20	21
22	23	24	25	26	27	28
29	30	31				

SPACE MILESTONE: *LUNA 21* (USSR). Jan 8. Unmanned vehicle, launched Jan 8, 1973, landed on the moon Jan 16, carrying Lunakho, a radio-controlled vehicle that explored a 37-km distance over a four-month period.

SWITZERLAND: MEITLISUNNTIG. Jan 8. On Meitlisunntig, the second Sunday in January, the girls of Meisterschwanden and Fahrwangen, in the Seetal district of Aargau, Switzerland, stage a procession in historical uniforms and a military parade before a female General Staff. According to tradition, the custom dates from the Villmergen War of 1712, when the women of both communes gave vital help that led to victory. Popular festival follows the processions.

WAR ON POVERTY: ANNIVERSARY. Jan 8. President Lyndon Johnson declared a War on Poverty in his State of the Union message Jan 8, 1964. He stressed improved education as one of the cornerstones of the program. The following Aug 20, he signed a $947.5 million anti-poverty bill designed to assist 30-35 million citizens.

BIRTHDAYS TODAY

David Bowie (David Robert Jones), 48, musician, actor, born at London, England, Jan 8, 1947.
Vladimir Feltsman, 43, Russian pianist, born at Moscow, USSR, Jan 8, 1952.
Slade Gorton, 67, US Senator (R, Washington) up for reelection in Nov '94, born at Chicago, IL, Jan 8, 1928.
Bill Graham (Wolfgang Grajonca), 64, concert promoter, born at Berlin, Germany, Jan 8, 1931.
Yvette Mimieux, 54, actress (*The Light in the Piazza, The Time Machine*, "The Most Deadly Game"), born Los Angeles, CA, Jan 8, 1941.
Soupy Sales (Morton Supman), 65, comedian, author ("The Soupy Sales Show"; may hold record for pies in the face), born at Wake Forest, NC, Jan 8, 1930.

JANUARY 9 — MONDAY
9th Day — Remaining, 356

AVIATION IN AMERICA: ANNIVERSARY. Jan 9. A Frenchman, Jean Pierre Blanchard, made the first manned free-balloon flight in America's history on Jan 9, 1793, at Philadelphia, PA. The event was watched by President George Washington and many other high government officials. The hydrogen-filled balloon rose to a height of about 5,800 feet, traveled some 15 miles and landed 46 minutes later. Reportedly Blanchard had one passenger on the flight—a little black dog.

BALANCHINE, GEORGE: BIRTH ANNIVERSARY. Jan 9. Born Georgi Militonovitch Balanchivadze on Jan 9, 1904, in St. Petersburg, Russia, George Balanchine became one of the leading influences in 20th-century ballet. He choreographed more than 200 ballets including *Concerto Barocco, Apollo, Orpheus, Serande, Firebird, Swan Lake, The Triumph of Bacchus, Waltz Academy* and *The Nutcracker*. In 1933 he was invited to the US by Boston philanthropist Lincoln Kirstein to establish a school for American dancers. Together they founded the School of American Ballet in 1934 and then formed several ballet companies, including the New York City Ballet which was led by Balanchine. Died at New York, NY, Apr 30, 1983.

Chase's 1995 Calendar of Events — Jan 9–10

◆ **CATT, CARRIE LANE CHAPMAN: BIRTH ANNIVERSARY.** Jan 9. American women's rights leader, founder (in 1919) of National League of Women Voters, born at Ripon, WI, Jan 9, 1859. Died at New Rochelle, NY, Mar 9, 1947.

CONNECTICUT RATIFIES CONSTITUTION: ANNIVERSARY. Jan 9. By a vote of 128 to 40, on Jan 9, 1788, Connecticut became the fifth state to ratify the Constitution.

ENGLAND: PLOUGH MONDAY. Jan 9. Always the Monday after Twelfth Day. Work on the farm is resumed after the festivities of the 12 days of Christmas. On preceding Sunday ploughs may be blessed in churches. Celebrated with dances and plays.

NATIONAL CLEAN-OFF-YOUR-DESK DAY. Jan 9. To provide one day early each year for every desk worker to see the top of the desk and prepare for the following year's paperwork. Annually, the second Monday in January. Sponsor: A.C. Moeller, Box 71, Clio, MI 48420.

NIXON, RICHARD MILHOUS: BIRTHDAY. Jan 9, 1913. Richard Nixon served as 36th vice president of the US (under President Dwight D. Eisenhower) Jan 20, 1953, to Jan 20, 1961. He was the 37th president of the US, serving Jan 20, 1969, to Aug 9, 1974, when he resigned the presidency while under threat of impeachment. First US president to resign that office. He was born at Yorba Linda, CA, and died at New York City, NY, Apr 22, 1994.

PANAMA: MARTYRS' DAY. Jan 9. Public holiday.

PHILIPPINES: FEAST OF THE BLACK NAZARENE. Jan 9. Culmination of a nine-day fiesta. Manila's largest procession takes place in the afternoon of Jan 9, in honor of the Black Nazarene, whose shrine is at the Quiapo Church.

SHOW AND TELL DAY AT WORK. Jan 9. Since students have show and tell at school, adults should get to do the same. Phone: (212) 388-8673. Or (717) 274-8451. Sponsor: Wellness Permission League, Tom and Ruth Roy, 2105 Water St, Lebanon, PA 17046.

ULLR FEST. Jan 9–15. Breckenridge, CO. The festival celebrates Ullr, the mythical god of winter, with ice sculpture, parade, fireworks, ski torchlight, world freestyle skiing championships, Ullr Ball and Mr and Mrs Ullr Beauty Pageant. Est attendance: 130,000. Sponsor: Breckenridge Resort Chamber, John Hendryson, PO Box 1909, Breckenridge, CO 80424. Phone: (303) 453-2913.

◆ **US LANDING ON LUZON: 50th ANNIVERSARY.** Jan 9, 1945. The US forces began the final push to retake the Philippines by attacking at the same location where the Japanese had begun their invasion nearly four years earlier. General Douglas MacArthur landed 67,000 troops in the Gulf of Lingayen on the western coast of the big island of Luzon. The Japanese offered little opposition to the landing itself but fought fiercely against Allied advancement, particularly around Clarke Field, the major air base in the islands.

YORBA LINDA, CA: NIXON BIRTHDAY HOLIDAY. Jan 9. Yorba Linda, CA. Yorba Linda, the birthplace in 1913 of former president Richard M. Nixon, became the first community officially to declare his birth anniversary a public holiday. In announcing the declaration on Sept 20, 1989, Mayor Henry Wedaa said, "We're not here to judge history—we're here to recognize it." The first observance by Yorba Linda's municipal employees took place in 1990.

BIRTHDAYS TODAY

Joan Baez, 54, folksinger, born at Staten Island, NY, Jan 9, 1941.
Muggsy (Tyrone) Bogues, 30, basketball player, born at Baltimore, MD, Jan 9, 1965.
Robert P. Casey, 63, Governor of Pennsylvania (D), born at Jackson Heights, NY, Jan 9, 1932.
Bob Denver, 60, actor ("The Many Loves of Dobie Gillis," "Gilligan's Island"), born at New Rochelle, NY, Jan 9, 1935.
Dick Enberg, 60, sportscaster, host, born at Mt Clemens, MI, Jan 9, 1935.

Crystal Gayle (Brenda Gayle Webb), 44, singer ("Don't It Make My Brown Eyes Blue"), born at Paintsville, KY, Jan 9, 1951.
Judith Krantz (Judith Tarcher), 67, author (*Dazzle, Scruples*), born at New York, NY, Jan 9, 1928.
Bart Starr, 61, Pro Football Hall of Famer and coach, born at Montgomery, AL, Jan 9, 1934.

JANUARY 10 — TUESDAY
10th Day — Remaining, 355

ALLEN, ETHAN: BIRTH ANNIVERSARY. Jan 10. Revolutionary War hero and leader of the Vermont "Green Mountain Boys." Born at Litchfield, CT, Jan 10, 1738. Died at Burlington, VT, Feb 12, 1789.

JEFFERS, ROBINSON: BIRTH ANNIVERSARY. Jan 10. American poet and playwright. Born at Pittsburgh, PA, Jan 10, 1887. Died at Carmel, CA, Jan 20, 1962.

LEAGUE OF NATIONS: 75th ANNIVERSARY. Jan 10. Through the Treaty of Versailles, the League of Nations came into existence on Jan 10, 1920. Fifty nations entered into a covenant designed to avoid war. The US never joined the League of Nations, which was dissolved on Apr 18, 1946.

NATIONAL WESTERN STOCK SHOW AND RODEO. Jan 10–22. Denver, CO. One of the nation's largest livestock shows with over 30 breeds of animals, 23 rodeo performances, cutting horse contest and sheep shearing contest. 89th annual. Est attendance: 525,000. For info: Natl Western Stock Show and Rodeo, 4655 Humboldt St, Denver, CO 80216. Phone: (303) 297-1166.

SPACE MILESTONE: *SOYUZ 17* (USSR). Jan 10. Launched Jan 10, 1975, Cosmonauts A. Gubarev and G. Grechko completed 30-day space flight, landing Feb 9. Cosmonauts spent 28 days aboard *Salyut 4*, orbiting space station.

SPACE MILESTONE: *SOYUZ 27* (USSR). Jan 10. Launched Jan 10, 1978, Cosmonauts Vladimir Dzhanibekov and Oleg Makarov linked with *Salyut 6* space station, which was already occupied by crew of *Soyuz 26*. Returned to Earth Jan 16 in *Soyuz 26*.

UNITED NATIONS GENERAL ASSEMBLY: ANNIVERSARY. Jan 10. On the 26th anniversary of the establishment of the unsuccessful League of Nations, delegates from 51 nations met in London, England, Jan 10, 1946, for the first meeting of the UN General Assembly.

◆ **WOMEN'S SUFFRAGE AMENDMENT INTRODUCED IN CONGRESS: ANNIVERSARY.** Jan 10, 1878. Senator A. A. Sargent of California, a close friend of Susan B. Anthony, introduced into the US Senate a women's suffrage amendment known as the Susan B. Anthony Amendment. It wasn't until Aug 26, 1920, 42 years later, that the amendment was signed into law.

BIRTHDAYS TODAY

Pat Benatar (Patricia Andrejewski), 42, singer, born at Brooklyn, NY, Jan 10, 1953.
George Foreman, 46, former boxer, born at Marshall, TX, Jan 10, 1949.

Gisele Mackenzie, 68, singer ("Hard To Get"; "Your Hit Parade"), born at Winnipeg, Manitoba, Canada, Jan 10, 1927.
Rod Stewart, 50, singer, musician ("Maggie May," "Da Ya Think I'm Sexy?"), born at Glasgow, Scotland, Jan 10, 1945.

JANUARY 11 — WEDNESDAY
11th Day — Remaining, 354

ASPEN/SNOWMASS WINTERSKOL. Jan 11-15. Snowmass Village, CO. A winter carnival—time to enjoy winter and have fun. Mad Hatter's Ball, ski and snowshoe races, ice sculpture contest and torchlight parade. Est attendance: 5,000. For info: Snowmass Resort Assn, Box 5566, Snowmass Village, CO 81615. Phone: (303) 923-2000.

CUCKOO DANCING WEEK. Jan 11-17. To honor the memory of Laurel and Hardy, whose theme, "The Dancing Cuckoos," shall be heard throughout the land as their movies are seen and their antics greeted by laughter by old and new fans of these unique masters of comedy. [Originated by the late William T. Rabe of Sault Ste. Marie, MI.]

DE HOSTOS, EUGENIO MARIA: BIRTH ANNIVERSARY. Jan 11. Puerto Rican patriot, scholar and author of more than 50 books. Born at Rio Canas, Puerto Rico, on Jan 11, 1839. Died at Santo Domingo, Dominican Republic, on Aug 11, 1903. The anniversary of his birth is observed as a public holiday in Puerto Rico.

HAMILTON, ALEXANDER: 240th [?] BIRTH ANNIVERSARY. Jan 11. American statesman born at British West Indies, Jan 11, 1755 or 1757. Engaged in a duel with Aaron Burr the morning of July 11, 1804, at Weehawken, NJ. Mortally wounded there and died July 12, 1804.

HUSKER FEED GRAINS AND SOYBEAN CONFERENCE. Jan 11-13. Lincoln, NE. Nebraska's premier agricultural conference highlighting crop production, research, industrial use and worldwide marketing activities. Sponsors: Nebraska Soybean Check-off and Growers Assn Boards, Nebraska Grain Sorghum Check-off and Growers Assn Boards, Nebraska Corn Check-off and Growers Assn Boards. Est attendance: 400. For info: Nebraska Soybean Program, Box 95144, 301 Centennial Mall South, Lincoln, NE 68509. Phone: (402) 471-4894.

MacDONALD, JOHN A.: BIRTH ANNIVERSARY. Jan 11. Canadian statesman, first prime minister of Canada. Born at Glasgow, Scotland, Jan 11, 1815. Died June 6, 1891. His birth anniversary is observed in Canada.

NATIONAL THANK YOU DAY. Jan 11. A day to thank someone from your past or present who did something nice for you. Write, call or fax him or her and say thank you. Sponsor: Thanks Alot, Adrienne Sioux Koopersmith, 1437 W Rosemont, #1W, Chicago, IL 60660. Phone: (312) 743-5341.

January 1995

S	M	T	W	T	F	S
1	2	3	4	5	6	7
8	9	10	11	12	13	14
15	16	17	18	19	20	21
22	23	24	25	26	27	28
29	30	31				

NEPAL: NATIONAL UNITY DAY. Jan 11. Celebration paying homage to King Prithvinarayan Shah (1723-1775), founder of present house of rulers of Nepal and creator of the unified Nepal of today.

PAUL, ALICE: 110th BIRTH ANNIVERSARY. Jan 11. Women's rights leader and founder of the National Woman's Party in 1913, advocate of an equal rights amendment to the US Constitution. Born at Moorestown, NJ, Jan 11, 1885. Died there July 10, 1977.

BIRTHDAYS TODAY

Clarence Clemons, 53, musician, singer, born at Norfolk, VA, Jan 11, 1942.
Ben Crenshaw, 43, golfer, born at Austin, TX, Jan 11, 1952.
Darryl Dawkins, 38, former basketball player, born at Orlando, FL, Jan 11, 1957.
Juanita Kreps, 74, government, university administrator, born at Harlan County, KY, Jan 11, 1921.
Rod Taylor, 65, actor (*The Birds*; "Masquerade"), born at Sydney, Australia, Jan 11, 1930.
Grant Tinker, 69, TV executive, born at Stamford, CT, Jan 11, 1926.

JANUARY 12 — THURSDAY
12th Day — Remaining, 353

BURKE, EDMUND: BIRTH ANNIVERSARY. Jan 12. British orator, politician and philosopher. Born at Dublin, Ireland, Jan 12, 1729. "Superstition is the religion of feeble minds," he wrote in 1790, but best remembered is "The only thing necessary for the triumph of evil is for good men to do nothing," not found in his writings but almost universally attributed to Burke. Died at Beaconsfield, England, July 9, 1797.

CONGRESS AUTHORIZES USE OF FORCE AGAINST IRAQ: ANNIVERSARY. Jan 12. On Jan 12, 1991, the US Congress passed a resolution authorizing the President of the US to use force to expel Iraq from Kuwait. This was the sixth congressional vote in US history declaring war or authorizing force on another nation.

ENGLAND: WEST LONDON ANTIQUES FAIR. Jan 12-15. (Also Aug 17-20.) Kensington Town Hall, Hornton St, London. Twice-yearly antiques fair with wide range of quality antiques, most pre-1890. Est attendance: 6,000. For info: Penman Antiques Fair, PO Box 114, Haywards Heath, West Sussex, England RH16 2YU.

HAYES, IRA HAMILTON: BIRTH ANNIVERSARY. Jan 12. Ira Hayes was one of six US Marines who raised the American flag on Iwo Jima's Mt Suribachi, on Feb 23, 1945, following a US assault on the Japanese stronghold. The event was immortalized by AP photographer Joe Rosenthal's famous photo (and later by a Marine War Memorial monument at Arlington, VA). Hayes was born on a Pima Indian Reservation in Arizona, Jan 12, 1922. He returned home after World War II a much celebrated hero. A postage stamp depicting the flag-raising by the six Marines was issued in 1945. Hayes, a hero to everyone except himself, was unable to cope with fame. He became an alcoholic, was arrested more than 50 times in 13 years and was found dead of "exposure to freezing weather and over-consumption of alcohol" on the Sacaton Indian Reservation in Arizona, on Jan 24, 1955.

LONDON, JACK: BIRTH ANNIVERSARY. Jan 12. American author of more than 50 books: short stories, novels, travel, stories of the sea and of the far north, many marked by brutal realism. He was born at San Francisco, CA, on Jan 12, 1876, and died near Santa Rosa, CA, Nov 22, 1916.

MISSION SANTA CLARA DE ASIS: ANNIVERSARY. Jan 12. California mission to the Indians founded Jan 12, 1777.

PHARMACISTS DAY. Jan 12. A day to honor pharmacists who are a very important part of the health care team. Annually, the second Thursday in January. For info: Connie Heidebrecht, Activity Dir, Mary Mashall Manor, 810 N 18th St, Marysville, KS 66508.

TANZANIA: ZANZIBAR REVOLUTION DAY. Jan 12. National day. Zanzibar became independent in Dec 1963, under a sultan.

BIRTHDAYS TODAY

Kirstie Alley, 40, actress (Emmy for "Cheers"; *Star Trek: The Motion Picture, Look Who's Talking*), Wichita, KS, Jan 12, 1955.

James Farmer, 75, civil rights leader, born at Marshall, TX, Jan 12, 1920.

Joe Frazier, 51, former boxer, born at Beaufort, SC, Jan 12, 1944.

HAL, 3, computer, "born" at Urbana, IL, Jan 12, 1992.

Ray Price, 69, country singer, born at Perryville, TX, Jan 12, 1926.

Luise Rainer, 83, actress (Oscars for *The Great Ziegfield* and *The Good Earth*), born at Vienna, Austria, Jan 12, 1912.

Dominique (Jacques) Wilkins, 35, basketball player, born at Paris, France, Jan 12, 1960.

JANUARY 13 — FRIDAY
13th Day — Remaining, 352

ALGER, HORATIO, JR: BIRTH ANNIVERSARY. Jan 13. American clergyman and author of more than 100 popular books for boys (some 20 million copies sold). Born at Revere, MA, Jan 13, 1834. Died at Natick, MA, July 18, 1899. Honesty, frugality and hard work assured that the heroes of his books would find success, wealth and fame.

BLAME SOMEONE ELSE DAY. Jan 13. To share the responsibility and the guilt for the mess we're in. Blame someone else! Annually, the first Friday the 13th of the year. Sponsor: A.C. Moeller, Box 71, Clio, MI 48420.

CANADA: JASPER IN JANUARY. Jan 13-29. Jasper, Alberta. Fun-filled days of winter activities and events including skiing, canyon crawling, sleigh rides, skating, fun ski races, a chili bake-off, interpretive programs, dining, dances, parties, a parade, fireworks, shopping, plus fantastic savings at Marmot Basin and participating Jasper hotels. Est attendance: 30,000. For info: Jasper Tourism & Commerce, Box 98, Jasper, AB, Canada T0E 1E0. Phone: (403) 852-3858.

FRIDAY THE THIRTEENTH. Jan 13. Variously believed to be a lucky or unlucky day. Every year has at least one Friday the 13th, but never more than three. Two Fridays in 1995 fall on the 13th day, this one in January and one in October. Fear of the number 13 is known as triskaidekaphobia.

FULLER, ALFRED CARL: BIRTH ANNIVERSARY. Jan 13. Founder of the Fuller Brush Company. Born at Kings County, Nova Scotia, Canada, on Jan 13, 1885. In 1906 the young brush salesman went into business on his own, making brushes at a bench between the furnace and the coal bin in his sister's basement. Died at Hartford, CT, on Dec 4, 1973.

HOUSTON-TENNECO MARATHON'S RUNNING WORLD EXPOSITION. Jan 13-14. Hyatt Regency Hotel, Houston, TX. Sports and fitness consumer show provides education and enjoyment to a broad spectrum of runners, sports and fitness fans, while promoting interest in and demand for all related products and services. Annually, in conjunction with the Houston-Tenneco Marathon (Jan 15). Est attendance: 10,000. For info: Kenneth D. Knezick, Genl Mgr, 8582 Katy, Ste 118, Houston, TX 77024. Phone: (713) 973-8590.

JAZZ FESTIVAL. Jan 13-15. Ramada London Bridge Resort, Lake Havasu, AZ. For info: Visitors and Conv Bureau, 1930 Mesquite Ave, Ste 3, Lake Havasu City, AZ 86403. Phone: (800) 242-8278.

NC RV AND TRAVEL SHOW. Jan 13-15. Charlie Rose Expo Center, Fayetteville, NC. A display of the latest in recreation vehicles and accessories by various dealers. Est attendance: 5,000. For info: Apple Rock Advertising & Promotion, 1200 Eastchester Dr, High Point, NC 27265. Phone: (910) 883-7199.

POETRY BREAK. Jan 13. Celebrate poetry by announcing a "poetry break and reading a poem aloud—at home, at school, in the office, in the market . . . anywhere!" Sponsor: Dr. Caroline Feller Bauer, 10175 Collins, Ave, #201, Miami Beach, FL 33154.

RADIO BROADCASTING: 85th ANNIVERSARY. Jan 13. Radio pioneer and electron tube inventor, Lee Deforest arranged the world's first radio broadcast to the public—in New York, NY, on Jan 13, 1910. He succeeded in broadcasting the voice of Enrico Caruso along with other stars of the Metropolitan Opera to several receiving locations in the city where listeners with earphones marveled at wireless music from the air. Though only a few were able to listen to that performance, it was the first broadcast to reach the public and the beginning of a new era in which wireless radio communication became almost universal.

SOUTHWEST ANTIQUE SHOW AND SALE. Jan 13-15. Yuma Civic and Convention Center, Yuma, AZ. A fine collection of antiques and collectibles. Sponsor: Jack Black Enterprises. For info: Yuma Civic and Conv Ctr, 1440 Desert Hills Dr, Yuma, AZ 85364. Phone: (602) 344-3800.

STAMP EXPO '95: WEST. Jan 13-15. Grand Hotel, Anaheim, CA. Fax (818) 988-4337. Est attendance: 4,000. Sponsor: Intl Stamp Collectors Soc, PO Box 854, Van Nuys, CA 91408. Phone: (818) 997-6496.

★ **STEPHEN FOSTER MEMORIAL DAY.** Jan 13. Presidential Proclamation 2957 of Dec 13, 1951 (designating Jan 13, 1952), covers all succeeding years. (PL82-225 of Oct 27, 1951.) Observed on the anniversary of Foster's death, Jan 13, 1864, at New York, NY. See also: "Foster, Stephen: Birth Anniversary" (July 4).

WINTER FEST. Jan 13-15. New Glarus, WI. Sleigh ride, skiing and parade. Est attendance: 1,700. Sponsor: New Glarus Chamber of Commerce, PO Box 713, New Glarus, WI 53574. Phone: (608) 527-2095.

BIRTHDAYS TODAY

Frank Gallo, 62, artist, sculptor, born at Toledo, OH, Jan 13, 1933.

Edward Madigan, 59, former US Secretary of Agriculture in Bush administration, born in Illinois, Jan 13, 1936.

Jay McInerney, 40, writer (*Bright Lights, Big City*), born at Hartford, CT, Jan 13, 1955.

Charles Nelson Reilly, 64, actor ("The Ghost and Mrs. Muir," "Match Game P.M."), born at New York, NY, Jan 13, 1931.

Robert Stack, 76, actor (Emmy for "The Untouchables"), born at Los Angeles, CA, Jan 13, 1919.

Brandon Tartikoff, 46, broadcast executive, born at Long Island, NY, Jan 13, 1949.

Gwen Verdon, 69, actress (*Damn Yankees, Cocoon*; stage: *Damn Yankees, High Button Shoes*), born at Los Angeles, CA, Jan 13, 1926.

JANUARY 14 — SATURDAY
14th Day — Remaining, 351

AFRICAN AMERICAN ARTS FESTIVAL. Jan 14–Mar 19. Greensboro, NC. A series of arts events held in celebration of the cultural achievements of local, regional and national black artists in all disciplines of the arts (music, dance, visual arts, literature and theatre). Annually, during the months of January, February and March. Sponsors: Miller Brewing Co, News and Record and WQMG FM. Est attendance: 20,000. For info: United Arts Council of Greensboro, PO Box 877, Greensboro, NC 27402. Phone: (910) 333-7440.

ARNOLD, BENEDICT: BIRTH ANNIVERSARY. Jan 14. American officer who deserted to the British during the Revolutionary War and whose name has since become synonymous with treachery. Born on Jan 14, 1741, at Norwich, CT. Died June 14, 1801.

BULGARIA: VINEGROWER'S DAY. Jan 14. Ancient holiday rite called Trifon Zarezan is inherited from the Thracians. Early in morning vines are pruned. Vinegrowers then sprinkle the pruned vine shoots with wine from a decorated wooden vessel, with wishes for fertility. Vine king chosen, followed by a feast in the meadow, with music, dancing and horseracing.

CHOCOLATE FESTIVAL. Jan 14. Hyatt Regency Hotel, Knoxville, TN. Benefits the National Kidney Foundation. For info: Tennessee Tourist Development, PO Box 23170, Nashville, TN 37202. Phone: (615) 741-2158.

FAMILY FOLKLORE. Jan 14–Apr 2. Lexington Children's Museum, Lexington, KY. To remind viewers of their own family's folklore, this exhibition features the tales and histories of such people as cabdrivers, steelworkers, urban school children, Oregon lumberjacks and the people of Appalachia. Call or write the Smithsonian for other dates and venues. For info: Smithsonian Institution Traveling Exhibition Service, 1100 Jefferson Dr SW, Ste 3146, Washington, DC 20560. Phone: (202) 357-2700.

FRED E. MILLER: PHOTOGRAPHER OF THE CROWS. Jan 14–Feb 25. University of Montana, Missoula, MT. Exhibition highlights 104 photographs taken by Miller between 1898 and 1912, the years he lived on the Crow Reservation in Montana. His photos offer an intimate view of the Crow Indians during the last years of their traditional life on the Great Plains. Call or write the Smithsonian for other dates and venues. For info: Smithsonian Institution Traveling Exhibition Service, 1100 Jefferson Dr SW, Ste 3146, Washington, DC 20560. Phone: (202) 357-2700.

JULIAN CALENDAR NEW YEAR'S DAY. Jan 14. Begins year 6708 of the Julian Period.

NAUTICAL AND WILDLIFE ART FESTIVAL AND NORTH AMERICAN ARTS AND CRAFTS SHOW. Jan 14–15. Ocean City, MD. Nature lovers and art lovers alike will want to attend these two wonderful exhibitions featuring original and limited edition works from the nation's leading artists, skilled artisans and crafters. Dual shows, single admission. Phone: (800) 626-2326 or (410) 524-9177. Est attendance: 10,000. For info: Ocean City Convention Center, PO Box 158, Ocean City, MD 21842.

PRODUCE FOR VICTORY: POSTERS ON THE AMERICAN HOME FRONT, 1941–1945. Jan 14–Feb 26. Dane G. Hansen Memorial Museum, Logan, KS. The posters that populated the home front during WWII proclaimed that every American was a "production soldier," as essential to the war effort as the soldier on the front line. This form of wartime propaganda showed up in factory lunchrooms, offices, grocery store windows and street billboards, encouraging citizens to boost production at home. The exhibition traces the evolution of these posters, a key to mobilizing and maintaining stateside support for the war effort. Call or write the Smithsonian for other dates and venues. For info: Smithsonian Institution Traveling Exhibition Service, 1100 Jefferson Dr SW, Ste 3146, Washington, DC 20560. Phone: (202) 357-2700.

RATIFICATION DAY. Jan 14. Anniversary of the act that officially ended the American Revolution and established the US as a sovereign power. On Jan 14, 1784, the Continental Congress, meeting at Annapolis, MD, ratified the Treaty of Paris, thus fulfilling the Declaration of Independence of July 4, 1776.

SCHWEITZER, ALBERT: 120th BIRTH ANNIVERSARY. Jan 14. Alsatian philosopher, musician, physician and winner of the 1952 Nobel Peace Prize, born at Kayserberg, Upper Alsace, Jan 14, 1875. Died at Lambarene, Gabon, Sept 4, 1965.

SECRET PAL DAY. Jan 14. A day for secret pals to remember and do something special for each other. On this day, the ladies of Eagles Lodge #4080 reveal identities of the past year's secret sisters and new secret sisters are chosen for the ensuing year. Annually, the second Sunday in January. Sponsor: Eagles Lodge #4080, Donna Coontz or Claire Price, PO Box 1319, Hayden Lake, ID 83835. Phone: (208) 772-4901.

SIMPSONS PREMIERE: 5th ANNIVERSARY. Jan 14, 1990. TV's hottest animated family, "The Simpsons," premiered as a half-hour weekly sitcom. The originator of Homer, Marge, Bart, Lisa and Maggie is cartoonist Matt Groening.

SNOW SHOVEL RIDING CONTEST. Jan 14. Old Economy Park, Ambridge, PA. Contest begins at 1 PM. Participants ride snow shovels downhill, handles extended, instead of sleds. Best time is the winner. For info: April Koehler, Exec Dir, Beaver Co Tourist Promo Agency, 215B Ninth St, Monaca, PA 15061. Phone: (412) 728-0212.

SNOWBIRD BREAKFAST. Jan 14. Bucklin Park, El Centro, CA. To welcome winter visitors to the Imperial Valley. Est attendance: 5,000. Sponsor: El Centro Chamber of Commerce, Box 3006, El Centro, CA 92244. Phone: (619) 352-3681.

SPACE MILESTONE: *Soyuz 4* (USSR). Jan 14. First docking of two manned spacecraft (with *Soyuz 5*) and first interchange of spaceship personnel in orbit. Launch date was Jan 14, 1969.

WHALE OF A WINE FESTIVAL/BREWING UP A STORM. Jan 14–15. Curry County Fairgrounds, Gold Beach, OR. View gray whales as they migrate south along the coast while you sample a variety of Oregon wines, gourmet munchies and coffees. Seminars, demos, art exhibit and more. For info: Gold Beach Chamber of Commerce, 1225 S Ellensburg, #3, Gold Beach, OR 97444. Phone: (800) 525-2334.

WHIPPLE, WILLIAM: BIRTH ANNIVERSARY. Jan 14. American patriot and signer of the Declaration of Independence. Born at Kittery, ME, on Jan 14, 1730. Died at Portsmouth, NH, on Nov 10, 1785.

WHO'S IN CHARGE? WORKERS AND MANAGERS IN THE UNITED STATES. Jan 14–Feb 26. Public Museum of Grand Rapids, Grand Rapids, MI. Artifacts, graphics and photographs trace 150 years of conflict and change in the relationships between workers and managers. Beginning with the pre-factory system, the exhibit explores the industrial era, scientific management, unions and automation and how they altered the roles of labor and administration. For info: Smithsonian Institution Traveling Exhibition Service, 1100 Jefferson Dr SW, Ste 3146, Washington, DC 20560. Phone: (202) 357-2700.

January 1995

S	M	T	W	T	F	S
1	2	3	4	5	6	7
8	9	10	11	12	13	14
15	16	17	18	19	20	21
22	23	24	25	26	27	28
29	30	31				

✮ Chase's 1995 Calendar of Events ✮   Jan 14–15

BIRTHDAYS TODAY

Jason Bateman, 26, actor ("Little House on the Prairie," "The Hogan Family"), born at Rye, NY, Jan 14, 1969.
Julian Bond, 55, legislator, civil rights leader, born at Nashville, TN, Jan 14, 1940.
Faye Dunaway, 54, actress (*Bonnie and Clyde, Chinatown, Network* [Oscar '76]), born Bascom, FL, Jan 14, 1941.
Lauch Faircloth, 67, US Senator (R, North Carolina), born at Sampson County, NC, Jan 14, 1928.
Marjoe Gortner, 50, ex-evangelist, actor, singer, born at Long Beach, CA, Jan 14, 1945.
Lawrence Kasdan, 46, filmmaker (*Body Heat, The Big Chill, Grand Canyon*), born at Miami Beach, FL, Jan 14, 1949.
Andy Rooney, 76, writer, columnist, ("60 Minutes," *Pieces of My Mind*), born at Albany, NY, Jan 14, 1919.

JANUARY 15 — SUNDAY
15th Day — Remaining, 350

ACE, GOODMAN: BIRTH ANNIVERSARY. Jan 15. Radio and TV writer, actor, columnist and humorist. Born at Kansas City, MO, Jan 15, 1899. With his wife, Jane, created and acted in the popular series of radio programs (1928–1945) "Easy Aces." Called "America's greatest wit" by Fred Allen. Died at New York, NY, Mar 25, 1982, soon after asking that his tombstone be inscribed "No flowers, please, I'm allergic."

BEARGREASE SLED DOG MARATHON. Jan 15–21. Duluth, MN. To commemorate John Beargrease, a Chippewa sled dog mail carrier along the North Shore of Lake Superior from 1887–1900. A 500-mile endurance race with mushers and dogs from the US and Canada. Sponsor: Grand Portage Chippewa. Est attendance: 7,000. For info: Beargrease, Box 500, Duluth, MN 55801. Phone: (218) 722-7631.

BRITISH MUSEUM: ANNIVERSARY. Jan 15. On Jan 15, 1759, the British Museum opened its doors at Montague House in London. Incorporated by an act of Parliament in 1753, following the death of British medical doctor and naturalist Sir Hans Sloane, who had bequeathed his personal collection of books, manuscripts, coins, medals and antiquities to Britain. As the national museum and the national library of the United Kingdom, the British Museum houses many of the world's most prized treasures.

HEALTHY WEIGHT WEEK. Jan 15–21. To focus attention in this traditional dieting month on the importance of striving for a healthy weight. Emphasizes healthy lifestyle habits, positive family patterns, prevention of weight problems and the acceptance of a wider range of sizes. Free report "A New Look at the Problems of Weight Loss," is available. Annually, the third week in January. Sponsor: Frances M Berg Editor/Publisher, Healthy Weight Journal, 402 S 14th St, Hettinger, ND 58639. Phone: (701) 567-2646.

HOUSTON-TENNECO MARATHON. Jan 15. Houston, TX. 23rd annual city-wide race in conjunction with the Houston-Tenneco Marathon's Running World Exposition (Jan 13–14). Est attendance: 250,000. For info: Greg D. Goss, 5900 Memorial Dr, Ste 200, Houston, TX 77007. Phone: (713) 864-9305.

HUMANITARIAN DAY. Jan 15. One of the three Kingdom Respect Days Color Commemorated by wearing White Colors to highlight human and civil rights, one of the three key principles of the Civil Rights Renaissance of the '60s. Held on the anniversary of Dr. Martin Luther King, Jr's birth to show visible unity, respect and remembrance for all unsung humanitarians, regardless of their race, who challenged white supremacy practices and changed America's unjust laws of racial segregation. It kicks off Humanitarian Week. See also: "Victims of Violence Holy Day" (Apr 4) and "Dream Day" (Aug 28). Phone: (312 RESPECT (737-7328). Sponsor: Global Committee Commemorating Kingdom Respect Days, PO Box 21050, Chicago, IL 60621.

INTERNATIONAL PRINTING WEEK. Jan 15–21. To develop public awareness of the printing/graphic arts industry. Annually, the week including Ben Franklin's birthday, Jan 17. For info: Intl Assn of Printing House Craftsmen, Kevin P. Keane, Exec Dir, 7042 Brooklyn Blvd, Minneapolis, MN 55429-1370. Phone: (612) 560-1620.

JAPAN: ADULTS' DAY. Jan 15. National holiday. Day is set apart for youth of the country who have reached adulthood during the preceding year.

KING, MARTIN LUTHER, JR: BIRTH ANNIVERSARY. Jan 15. Black civil rights leader, minister and recipient of the Nobel Peace Prize (1964). Born at Atlanta, GA, on Jan 15, 1929. He was assassinated at Memphis, TN, Apr 4, 1968. After his death many states and territories observed his birthday as a holiday. In 1983 the Congress approved HR 3706, "A bill to amend Title 5, United States Code, to make the birthday of Martin Luther King, Jr, a legal public holiday." Signed by the president on Nov 2, 1983, it became Public Law 98-144. The law sets the third Monday in January for observance of King's birthday. First observance was Jan 20, 1986. See also: "King, Martin Luther, Jr: Birthday Observed" (Jan 16).

LIVINGSTON, PHILIP: BIRTH ANNIVERSARY. Jan 15. Merchant and signer of the Declaration of Independence, born at Albany, NY, Jan 15, 1716. Died at York, PA, June 12, 1778.

MOLIERE DAY. Jan 15. Most celebrated of French authors and dramatists, Jean Baptiste Poquelin, baptized at Paris, France, Jan 15, 1622, took the stage name Molière when he was about 22 years old. While playing in a performance of his last play, *Le Malade Imaginaire* (about a hypochondriac afraid of death), Molière became ill and died within a few hours at Paris, Feb 17, 1673.

MUSICAL TRIBUTE TO DR. MARTIN LUTHER KING. Jan 15. Abraham Lincoln's Birthplace NHS, Hodgenville, KY. Local Baptist and Methodist choirs present a selection of hymns and spirituals in commemoration of Dr. King's birthday. Est attendance: 200. For info: Gary V Talley, 2995 Lincoln Farm Rd, Hodgenville, KY 42748. Phone: (502) 358-3137.

QUARTERLY ESTIMATED FEDERAL INCOME TAX PAYERS' DUE DATE. Jan 15. For those individuals whose fiscal year is the calendar year and who make quarterly estimated federal income tax payments, today is one of the due dates. (Jan 15, Apr 15, June 15 and Sept 15, 1995.)

TOP JUNK-FOOD NEWS STORIES OF 1994. Jan 15. To announce the top ten sensationalized, junk-food news stories of 1994. Fax: (707) 664-2505. For info: Carl Jensen, PhD, Dir, Project Censored or Mark Lowenthal, Asst Dir, Sonoma State Univ, Rohnert Park, CA 94928. Phone: (707) 664-2500.

◈ **TRAIN FOR PARIS: 50th ANNIVERSARY.** Jan 15, 1945. The civilian populations of England and France had their first direct contact since May 1940: a boat train left London's Victoria Station headed for the coast, crossed the English Channel by boat and then continued on to Paris.

65

Jan 15-16 ☆ Chase's 1995 Calendar of Events ☆

WORLD RELIGION DAY. Jan 15. To proclaim the oneness of religion and the belief that world religion will unify the peoples of the earth. Baha'i-sponsored observance established in 1950 by the Baha'is of the United States. Annually, the third Sunday in January. For info: Natl Spiritual Assembly of the US, 1320 Nineteenth St, NW, Ste 701, Washington, DC 20036. Phone: (202) 833-8990.

WORLDWIDE KIWANIS WEEK. Jan 15-21. To encourage the creation of permanent relationships among Kiwanis clubs in different parts of the world. Sponsor: Kiwanis Intl, Program Dvmt Dept, 3636 Woodview Trace, Indianapolis, IN 46268.

BIRTHDAYS TODAY

Lloyd Bridges, 82, actor ("Seahunt," *Roots*, *The Goddess*), born at San Leandro, CA, Jan 15, 1913.
Charo (Maria Martinez), 44, actress ("Chico and the Man"), born at Murcia, Spain, Jan 15, 1951.
Rod MacLeish, 69, journalist, born at Bryn Mawr, PA, Jan 15, 1926.
Margaret O'Brien, 58, actress (Beth in *Little Women*; Tootie in *Meet Me in St. Louis*), born at Los Angeles, CA, Jan 15, 1937.
Edward Teller, 87, physicist, born at Budapest, Hungary, Jan 15, 1908.

JANUARY 16 — MONDAY
16th Day — Remaining, 349

DEAN, DIZZY: BIRTH ANNIVERSARY. Jan 16. Jay Hanna "Dizzy" Dean, major league pitcher (St. Louis Cardinals) and Baseball Hall of Fame member. Born at Lucas, AR, on Jan 16, 1911. Following his baseball playing career, Dean established himself as a radio and TV sports announcer and commentator, becoming famous for his innovative delivery. "He slud into third," reported Dizzy, who on another occasion explained that "Me and Paul [baseball player brother Paul "Daffy" Dean]... didn't get much education." Died at Reno, NV, July 17, 1974.

ELEMENTARY SCHOOL TEACHER DAY. Jan 16. A day of recognition for these professionals who teach our children and fill their minds with knowledge and dreams. Annually, the third Monday in Jan. Sponsor: Dawn T. Brue, 399 SW 8th St, #1, Boca Raton, FL 33432-5732.

HUMAN RELATIONS DAY. Jan 16. Flint, MI. To focus attention on human relations and acquisition of skills enabling achievement of the dreams of Dr. Martin Luther King, Jr, for social justice for all mankind. Celebrated in the public schools of Flint, MI, as a tribute to Dr. Martin Luther King, Jr, on the date set aside each year as Dr. King's birthday holiday. For info: Flint Community Schools, Staff Dvmt Office, 923 E Kearsley St, Flint, MI 48502. Phone: (810) 760-1122.

JAPAN: HARU-NO-YABUIRI. Jan 16. Employees and servants who have been working over the holidays are given a day off.

KING, MARTIN LUTHER, JR: BIRTHDAY OBSERVED. Jan 16. Public Law 98-144 designates the third Monday in January as an annual legal public holiday observing the birth of Martin Luther King, Jr. First observed in 1986. See also: "King, Martin Luther, Jr: Birth Anniversary" (Jan 15).

LEE-JACKSON-KING DAY IN VIRGINIA. Jan 16. Annually, the third Monday in January.

MAN WATCHERS WEEK. Jan 16-21. A week of appreciation for men who are well worth watching. Announcement of the 10 most watchable men in the world. List of activities available. Est attendance: 600. Sponsor: Suzy Mallery's Man Watchers Inc, Suzy Mallery, 12308 Darlington Ave, Los Angeles, CA 90049. Phone: (310) 826-9101.

★ **MARTIN LUTHER KING, JR, FEDERAL HOLIDAY.** Jan 16. Presidential Proclamation has been issued without request each year for the third Monday in January since 1986.

MERMAN, ETHEL: BIRTH ANNIVERSARY. Jan 16. Musical-comedy star famous for her belting voice and brassy style. Her rendition of "I've Got Rhythm" in the 1930 show *Girl Crazy* was an overnight hit and that performance launched her to stardom. Born Ethel Agnes Zimmerman on Jan 16, 1909 (or 1912—the date changed the older she got, but most sources say 1909) at Queens, NY. Died Feb 15, 1984, at New York, NY.

MOON PHASE: FULL MOON. Jan 16. Moon enters Full Moon phase at 3:26 PM, EST.

NATIONAL NOTHING DAY. Jan 16. Anniversary of National Nothing Day, an event created by newspaperman Harold Pullman Coffin and first observed in 1973 "to provide Americans with one national day when they can just sit without celebrating, observing or honoring anything." Since 1975, though many other events have been listed on that day, lighthearted traditional observance of Coffin's idea has continued. Coffin, a native of Reno, NV, died at Capitola, CA, Sept 12, 1981, at age 76.

PROHIBITION AMENDMENT: ANNIVERSARY. Jan 16. Nebraska became 36th state to ratify the prohibition amendment on Jan 16, 1919, and the 18th Amendment became part of the US Constitution. One year later, on Jan 16, 1920, the 18th Amendment took effect and the sale of alcoholic beverages became illegal in the US with the Volstead Act providing for enforcement. This was the first time that an amendment to the Constitution dealt with a social issue. The 21st Amendment, repealing the 18th, went into effect on Dec 6, 1933.

RELIGIOUS FREEDOM DAY. Jan 16. On Jan 16, 1786, the legislature of Virginia adopted a religious freedom statute that protected Virginians against any requirement to attend or support any church and against discrimination. This statute, which had been drafted by Thomas Jefferson and introduced by James Madison, later was the model for the 1st Amendment to the US Constitution.

SERVICE, ROBERT WILLIAM: BIRTH ANNIVERSARY. Jan 16. Canadian poet. Born at Preston, England, Jan 16, 1874. Lived in the Canadian northwest for many years and perhaps is best remembered for such ballads as "The Shooting of Dan McGrew" and "The Cremation of Sam McGee" and for such books as *Songs of a Sourdough*, *Rhymes of a Rolling Stone* and *The Spell of the Yukon*. Died in France, Sept 11, 1958.

TU B'SHVAT. Jan 16. Hebrew calendar date: Shebat 15, 5755. The 15th day of the month of Shebat in the Hebrew calendar year is set aside as Hamishah Asar (New Year of the Trees or Jewish Arbor Day), a time to show respect and appreciation for trees and plants.

VIRGINIA BOAT SHOW. Jan 16-23. Richmond Convention Center, Richmond, VA. Introducing the new line of boats. Display of boats, boating equipment and related industries. Est attendance: 300,000. For info: Virginia Boat Show, 20 Park Ave, Ste 301, Baltimore, MD 21201-3423. Phone: (401) 385-1800.

WAR AGAINST IRAQ BEGINS: ANNIVERSARY. Jan 16. On the night of Jan 16, 1991, the Allied forces launched a major air offensive at Iraq to begin the War in the Gulf. The strike was designed to destroy Iraqi air defenses, command, control and communications centers. As Desert Shield became Desert Storm, the world was able to see and hear for the first time an initial engagement of war as CNN broadcasters stationed in Baghdad broadcast the attack live.

January 1995

S	M	T	W	T	F	S
1	2	3	4	5	6	7
8	9	10	11	12	13	14
15	16	17	18	19	20	21
22	23	24	25	26	27	28
29	30	31				

☆ Chase's 1995 Calendar of Events ☆ Jan 16–17

BIRTHDAYS TODAY
Debbie Allen, 45, dancer, singer, actress ("Fame"), born at Houston, TX, Jan 16, 1950.
A.J. Foyt, 60, auto racer, born at Houston, TX, Jan 16, 1935.
Marilyn Horne, 52, opera singer, born at Bradford, PA, Jan 16, 1943.
Eartha Kitt, 67, singer, born at North, SC, Jan 16, 1928.
Jack Burns McDowell, 29, baseball player, born at Van Nuys, CA, Jan 16, 1966.
Ronnie Milsap, 51, singer ("[There's] No Gettin' Over Me"), born at Robinsville, NC, Jan 16, 1944.
Francesco Scavullo, 66, photographer, born at Staten Island, NY, Jan 16, 1929.

JANUARY 17 — TUESDAY
17th Day — Remaining, 348

CHICAGO'S FREAK GAS EXPLOSION: ANNIVERSARY. Jan 17. On Jan 17, 1992, a gas explosion ripped through a six-block area of Chicago's River West neighborhood. Four people were killed and 10 fires spread through 21 buildings. The explosion was later blamed on a faulty piece of pressure regulating equipment.

FIRST NUCLEAR-POWERED SUBMARINE VOYAGE: 40th ANNIVERSARY. Jan 17. The world's first nuclear-powered submarine, the *Nautilus*, now forms part of the *Nautilus* Memorial Submarine Force Library and Museum at the Naval Submarine Base New London in Groton, CT. At 11:00 AM, EST, on Jan 17, 1955, her commanding officer, Commander Eugene P. Wilkerson, ordered all lines cast off and sent the historic message: "Under way on nuclear power." Highlights of the *Nautilus*: keel laid by President Harry S. Truman on June 14, 1952; christened and launched by Mrs Dwight D. Eisenhower on Jan 21, 1954; commissioned to the US Navy on Sept 30, 1954.

FRANKLIN, BENJAMIN: BIRTH ANNIVERSARY. Jan 17. "Elder statesman of the American Revolution," oldest signer of both the Declaration of Independence and the Constitution, scientist, diplomat, author, printer, publisher, philosopher, philanthropist and self-made, self-educated man. Author of *Poor Richard's Almanack*, 1733–1758. Born at Boston, MA, Jan 17, 1706. Died at Philadelphia, PA, Apr 17, 1790. Franklin's birthday is commemorated each year by the Poor Richard Club of Philadelphia, with graveside observance. In 1728 Franklin wrote a premature epitaph for himself. It first appeared in print in Ames's 1771 almanac:

> The Body of BENJAMIN FRANKLIN
> Printer
> Like a Covering of an old Book
> Its contents torn out
> And stript of its Lettering and Gilding,
> Lies here, Food for Worms;
> But the work shall not be lost,
> It will (as he believ'd) appear once more
> In a New and more beautiful Edition
> Corrected and amended
> By the Author.

FREEDOM FORUM ASIAN CENTER OPENING: ANNIVERSARY. Jan 17, 1994. The Asian Center opened in Hong Kong to coordinate Freedom Forum training programs and other activities in Asia. "The Freedom Forum is the largest media-oriented foundation in the USA devoted to promoting free press, free speech and free spirit around the world." For info: Freedom Forum Asian Center, 1305-06 13th Fl, Shui On Centre, 6-8 Harbour Rd, Wanchai, Hong Kong. Phone: (703) 528-0800.

MEXICO: BLESSING OF THE ANIMALS AT THE CATHEDRAL. Jan 17. Church of San Antonio in Mexico City or Xochimilco provides best sights of chickens, cows and household pets gaily decorated with flowers. (Saint's day for San Antonio Abad, patron saint of domestic animals.)

MINOW, NEWTON: BIRTHDAY. Jan 17. American lawyer and former head of the Federal Communications Commission (1961–1963) was born at Milwaukee, WI, on Jan 17, 1926. Perhaps his most repeated words are "vast wasteland," his description of American television, but more recently he was quoted as saying, "The past is not always worse than the present."

NATIONAL PRINTING INK DAY. Jan 17. To highlight the essential contribution of printing inks to the dissemination of knowledge, culture, education and entertainment through service to the graphic arts and communications industries. Annually, the Tuesday of Printing Week. Sponsor: Natl Assn of Printing Ink Mfrs, Heights Plaza, 777 Terrace Ave, Hasbrouck Hghts, NJ 07604. Phone: (201) 288-9454.

PALOMARES HYDROGEN BOMB ACCIDENT: ANNIVERSARY. Jan 17. At 10:16 AM on Jan 17, 1966, according to villagers, fire fell from the sky over Palomares, Spain. An American B-52 bomber carrying four hydrogen bombs collided with its refueling plane, spilling the bombs (two of which had "chemical explosions," scattering radioactive plutonium over the area). In a cleanup, American soldiers burned crops, slaughtered animals and removed tons of topsoil (which was sent to South Carolina for burial). More than 19 years later, in Nov 1985, the Nuclear Energy Board permitted villagers to see their medical reports for the first time.

PHILIPPINES: CONSTITUTION DAY: 50th ANNIVERSARY. Jan 17. National holiday. Simple rites to commemorate the 1935 ratification of the amended Philippine constitution.

◈ **POLAND: LIBERATION DAY: 50th ANNIVERSARY.** Jan 17. Celebration of liberation from Nazi oppression on Jan 17, 1945, of the city of Warsaw by Soviet troops. Special ceremonies at the Monument to the Unknown Soldier in Warsaw's Victory Square (which had been called Adolf Hitler Platz during the German occupation).

QUEEN LILIUOKALANI DEPOSED: ANNIVERSARY. Jan 17. Hawaiian Queen Liliuokalani, the last monarch of Hawaii, lost her throne on Jan 17, 1893, when the monarchy was abolished by the "Committee of Safety," with the foreknowledge of US minister John L. Stevens, who encouraged the revolutionaries. The Queen's supporters were intimidated by the 300 US Marines sent to protect American lives and property. Judge Sanford B. Dole became president of the republic and later was Hawaii's first governor after the US annexed it by joint resolution of Congress on July 7, 1898. Hawaii held incorporated territory status for 60 years. President Dwight D. Eisenhower signed the proclamation making Hawaii the 50th state on Aug 21, 1959.

RID THE WORLD OF FAD DIETS AND GIMMICKS DAY. Jan 17. To discourage false and misleading weight loss promotions, fads and gimmicks. The Slim Chance Awards will be announced—the worst weight loss promotions of the year, judged by Healthy Weight Journal and the National Council Against Health Fraud. Annually, Tuesday of Healthy Weight Week. Sponsor: Frances M. Berg, Editor/Publisher, Healthy Weight Journal, 402 S 14th St, Hettinger, ND 58639. Phone: (701) 567-2646.

Jan 17–18 ☆ Chase's 1995 Calendar of Events ☆

RUSH, WILLIAM: DEATH ANNIVERSARY. Jan 17. First native American sculptor. William Rush's work in wood and clay included busts of many notables, American and European alike; carved wooden female figureheads for ships; the masks of Tragedy and Comedy seen at the Actor's House outside Philadelphia, PA, and the *Spirit of Schuylkill*, in Fairmount Park in Philadelphia. He died on Jan 17, 1833.

ST. ANTHONY'S DAY. Jan 17. Feast day honoring Egyptian hermit who became the first Christian monk and who established communities of hermits; patron saint of domestic animals and patriarch of all monks. Lived about AD 251–354.

SOUTHERN CALIFORNIA EARTHQUAKE: ANNIVERSARY. Jan 17, 1994. A powerful earthquake measuring 6.6 on the Richter scale struck Southern California at 4:31 AM. Centered beneath the community of Northridge in an unidentified fault zone, the tremblor caused the most extensive damage in the San Fernando valley, collapsing freeways, apartment and garage structures. Fires and continued aftershocks added to the chaos and destruction. Although not "The Big One" that many fear will one day rock the area along the San Andrea fault, this quake shook an already jostled community still recovering from the LA riots following the Rodney King incident and recent fire storms that devasted vast areas of the sundrenched Pacific southwest.

SOUTHWEST SENIOR INVITATIONAL GOLF CHAMPIONSHIP. Jan 17–20. Yuma Golf and Country Club, Yuma, AZ. A 36-hole medal play tournament, limited to 100 players 50 and older. Est attendance: 100. Sponsor: Caballeros de Yuma, Inc, Box 5987, Yuma, AZ 85366. Phone: (602) 343-1715.

BIRTHDAYS TODAY

Muhammad Ali (Cassius Marcellus Clay, Jr), 53, former boxer, born at Louisville, KY, Jan 17, 1942.
Jim Carrey, 33, actor, comedian, born in Ontario, Canada, Jan 17, 1962.
Troy Donahue, 58, actor ("Hawaiian Eye," *Parrish*), born at New York, NY, Jan 17, 1937.
James Earl Jones, 64, actor (*The Great White Hope, Roots: The Next Generation*), born at Tate County, MS, Jan 17, 1931.
Shari Lewis (Shari Hurwitz), 61, puppeteer, born at New York, NY, Jan 17, 1934.
Harlan Mathews, 68, former US Senator (D, Tennessee-appointed), born at Sumiton, AL, Jan 17, 1927.
Sheree North (Dawn Bethel), 62, actress (*Marilyn: The Untold Story; How to Be Very Very Popular*), born at New York, NY, Jan 17, 1933.
Vidal Sassoon, 67, hair stylist, born at London, England, Jan 17, 1928.
Betty White, 71, actress ("Mary Tyler Moore," "The Golden Girls"; animal rights activist), born at Oak Park, IL, Jan 17, 1924.
Paul Young, 39, singer ("Everytime You Go Away"), born at Luton, England, Jan 17, 1956.
Don Zimmer, 64, former baseball player and manager, born at Cincinnati, OH, Jan 17, 1931.

JANUARY 18 — WEDNESDAY
18th Day — Remaining, 347

ALL STATES PICNIC (FOR SENIOR CITIZENS). Jan 18. Yuma, AZ. Seniors from all over the US and Canada gather for picnic lunch, entertainment and fun. Sponsor: City of Yuma Parks and Recreation Department. Est attendance: 1,000. For info: Yuma Civic and Conv Ctr, 1440 Desert Hills Dr, Yuma, AZ 85364. Phone: (602) 344-3800.

FIRST BLACK US CABINET MEMBER: ANNIVERSARY. Jan 18. On Jan 18, 1966, Robert Clifton Weaver was sworn in as Secretary of Housing and Urban Development, becoming the first black cabinet member in US history. He was nominated by President Lyndon Johnson.

INTERNATIONAL SNOW SCULPTING CHAMPIONSHIPS. Jan 18–22. Breckenridge, CO. More than 20 snow sculptures from huge blocks of snow (12′ × 12′ × 12′) will line the streets of Breckenridge to determine the international champion. Est attendance: 175,000. Sponsor: John Hendryson, Breckenridge Resort Chamber, PO Box 1909, Breckenridge, CO 80424. Phone: (303) 453-2913.

MAINTENANCE DAY. Jan 18. A day to honor the unsung heroes who work so hard to keep buildings, hospitals, nursing homes, business places, factories, etc, in good, safe working order. They are jacks of all trades: plumbers, carpenters, mechanics, painters, gardeners—they have to be skilled in many fields. Annually, the third Wednesday in January. Sponsor: Jim Heidebrecht, Maintenance Super, Blue Valley Nursing Home, Blue Rapids, KS 66411. Phone: (913) 363-7777.

POOH DAY: A.A. MILNE BIRTH ANNIVERSARY. Jan 18. Anniversary of the birth of A(lan) A(lexander) Milne, English author, especially remembered for his children's stories: *Winnie the Pooh* and *The House at Pooh Corner*. Also the author of *Mr. Pim Passes By, When We Were Very Young* and *Now We Are Six*. Born at London, England, Jan 18, 1882. Died at Hartfield, England, Jan 31, 1956.

ROGET, PETER MARK: BIRTH ANNIVERSARY. Jan 18, 1779. English physician, best known as author of *Roget's Thesaurus of English Words and Phrases*, first published in 1852. Also the inventor of the "log-log" slide rule. Born at London, died at West Malvern, Worcestershire, England, Sept 12, 1869.

TUNISIA: REVOLUTION DAY. Jan 18. Public holiday commemorating the Tunisian Revolution.

VERSAILLES PEACE CONFERENCE: ANNIVERSARY. Jan 18. French President Raymond Poincare formally opened the (World War I) Peace Conference at Versailles, France, Jan 18, 1919. It proceeded under the chairmanship of Georges Clemenceau. In May the conference disposed of Germany's colonies and delivered a treaty to the German delegates on May 7, 1919, fourth anniversary of the sinking of the *Lusitania*. Final treaty-signing ceremonies were completed at the palace in Versailles, on June 28, 1919.

WEBSTER, DANIEL: BIRTH ANNIVERSARY. Jan 18. American statesman and orator who said, on Apr 6, 1830, "The people's government, made for the people, made by the people, and answerable to the people." Born at Salisbury, NH, Jan 18, 1782. Died at Marshfield, MA, Oct 24, 1852.

BIRTHDAYS TODAY

John Boorman, 62, filmmaker (*Deliverance, Excalibur*), born at Shepperton, England, Jan 18, 1933.
Kevin Costner, 40, actor (*Field of Dreams, Dances with Wolves* [Oscar for directing]), born Compton, CA, Jan 18, 1955.

	S	M	T	W	T	F	S
January 1995	1	2	3	4	5	6	7
	8	9	10	11	12	13	14
	15	16	17	18	19	20	21
	22	23	24	25	26	27	28
	29	30	31				

☆ Chase's 1995 Calendar of Events ☆   Jan 18–19

Steven L. DeBerg, 41, football player, born at Oakland, CA, Jan 18, 1954.
Ray Dolby, 62, inventor, born at Portland, OR, Jan 18, 1933.
Curt Flood, 57, former baseball player, born at Houston, TX, Jan 18, 1938.
Evelyn Lear, 64, opera singer, born at New York, NY, Jan 18, 1931.

JANUARY 19 — THURSDAY
19th Day — Remaining, 346

CONFEDERATE HEROES DAY. Jan 19. Texas. Observed on Robert E. Lee's birthday.

ENGLAND: KNITTING, NEEDLECRAFT & DESIGN EXHIBITION. Jan 19–22. Sandown Park Racecourse, High Street, Esher, Surrey. Everything for the hand and machine knitter, and the needlecraft enthusiast including fashion shows and lectures. Open daily from 10 AM–6 PM. Est attendance: 30,000. For info: Robert Ewin, Nationwide Exhibitions (UK) Ltd, PO Box 20, Fishponds, Bristol, England BS16 5QU.

HEALTHY WEIGHT, HEALTHY LOOK DAY. Jan 19. A day to honor American women of all sizes and promote their good health. This day culminates a national contest held annually to support business and media who portray a healthy, realistic look for women. 1995 Healhty Weight, Healthy Look Contest winners to be announced include magazines, advertisements, TV networks and other media visuals which portray real people and confirm that beauty, health and strength come in all sizes. Annually, Thursday of Healthy Weight Week. Sponsor: Frances M Berg, Editor/Publisher, Healthy Weight Journal, 402 S 14th St, Hettinger, ND 58639. Phone: (701) 562-2646.

◆ **HELMS, EDGAR J.: BIRTH ANNIVERSARY.** Jan 19, 1863. Born near Malone, NY, Rev. Dr. Helms became a minister to a parish of poor immigrants in Boston's South End. In that capacity he developed the philosophy and organization that eventually became Goodwill Industries. Edgar died Dec 23, 1942, at Boston.

HISTORIC BIRTHDAY PARTIES IN LEXINGTON. Jan 19–21. Lexington, VA. Celebrates the birthdays of Robert E. Lee and Stonewall Jackson. Annually, Jan 19–21. For info: Lexington Visitors Bureau, 106 E Washington St, Lexington, VA 24450. Phone: (703) 463-3777.

JOPLIN, JANIS: BIRTH ANNIVERSARY. Jan 19. Possibly the most highly regarded white female blues singer of all time, Janis Joplin was born Jan 19, 1943, at Port Arthur, TX. Joplin's appearance with Big Brother and The Holding Company at the Monterey International Pop Festival in August 1967 launched her to superstar status. Among her recording hits are, "Get It While You Can," "Piece of My Heart" and "Ball and Chain." Janis Joplin died of a heroin overdose Oct 4, 1970, at Hollywood, CA.

◆ **LEE, ROBERT EDWARD: BIRTH ANNIVERSARY.** Jan 19. Greatest military leader of the Confederacy, son of Revolutionary War General Henry (Light Horse Harry) Lee. His surrender on Apr 9, 1865, to Union General Ulysses S. Grant brought an end to the Civil War. Born in Westmoreland County, VA, Jan 19, 1807. Died at Lexington, VA, Oct 12, 1870. His birthday is observed on this day in Arkansas, Georgia, Louisiana, South Carolina and Texas. Observed on third Monday in January in Alabama and Mississippi.

NORWAY: THE NORTHERN LIGHTS FESTIVAL. Jan 19–22. Tromso. Classical and contemporary music performances by notable Norwegian and international musicians. Annually. For info: Norwegian Tourist Bd, 655 Third Ave, New York, NY 10017. Phone: (212) 949-2333.

POE, EDGAR ALLAN: BIRTH ANNIVERSARY. Jan 19. American poet and story writer, called "America's most famous man of letters." Born at Boston, MA, Jan 19, 1809. He was orphaned, in dire poverty, in 1811, and was raised by Virginia merchant John Allan. In 1836 he married his 13-year-old cousin, Virginia Clemm. A magazine editor of note, he is best remembered for his poetry (especially "The Raven") and for his tales of suspense. Alcohol addiction, a problem throughout his life, was the immediate cause of his death. Died at Baltimore, MD, Oct 7, 1849.

SEA LION "HAUL OUT": FIFTH ANNIVERSARY. Jan 19. PIER 39, Beach St and the Embarcadero, San Francisco, CA. Since 1990, more than 600 California sea lions yearly have begun "hauling out" at the PIER and making PIER 39 their yearly habitat of choice. The public can view the sea lion habitat free of charge and attend talks about the sea lions' behavior, marine habitat and migratory patterns. Annually, in January. Est attendance: 10,000. For info: Alicia Vargas, PR Dir, PIER 39, PO Box 193730, San Francisco, CA 94119. Phone: (415) 705-5500.

SOUTH AFRICA: HANSA DUSI MARATHON. Jan 19–21. Pietermaritzburg to Durban. Ultra-marathon water race (the world's longest) spans three challenging days of rocks, white water rapids and cross-country portage. Large portion of race is on Umgeni River through the Valley of 1000 Hills. Sponsor: Hansa Beer. For info: John Oliver, Natal Canoe Club, Box 1966, Pietermaritzburg, 3200, Natal, South Africa. Phone: (212) 730-2929.

SOUTHWESTERN BLACK STUDENT LEADERSHIP CONFERENCE. Jan 19–22. Rudder Tower and Memorial Student Center, Texas A&M University, College Station, TX. Approximately 1,000 black students from the Southwest attend workshops and special presentations focusing on ways to accept and conquer the challenges facing them in society. Annually, during January. Sponsor: Southwestern Black Student Leadership Conference. For info: Multicultural Services Dept, Texas A&M University, College Station, TX 77843-1121. Phone: (409) 845-4565.

SUNDANCE FILM FESTIVAL. Jan 19–29. Park City, UT. "The premier US festival for American independent filmmakers." Est attendance: 5,000. For info: Utah Travel Council, Council Hall, Capitol Hill, Salt Lake City, UT 84114. Phone: (801) 538-1030.

TAMPA BOAT SHOW. Jan 19–22. Tampa Convention Center, Tampa, FL. Sponsor: Cathy Johnston, Natl Marine Mfrs Assn, 14502 N Dale Mabry, Ste 332, Tampa, FL 33618. Phone: (813) 264-0490.

THOMAS, ISAIAH: BIRTH ANNIVERSARY. Jan 19. American printer, editor, almanac publisher and historian. Born Jan 19, 1749, at Boston, MA. Died, Apr 4, 1831, at Worcester, MA.

WATT, JAMES: BIRTH ANNIVERSARY. Jan 19. Scottish engineer and inventor. Born at Greenock, Scotland, on Jan 19, 1736. The modern steam engine grew out of his efficiency-improving inventions. Died at Heathfield, England, on Aug 19, 1819.

BIRTHDAYS TODAY

Desi Arnaz, Jr, 42, singer, actor, born at Los Angeles, CA, Jan 19, 1953.
Michael Crawford, 53, actor, singer (*Phantom of the Opera*), born Salisbury, Wiltshire, England, Jan 19, 1942.
Phil Everly, 56, singer, born at Brownie, KY, Jan 19, 1939.
Roy Leonard, 64, Chicago radio personality, born at Redwood Falls, MN, Jan 19, 1931.
Robert MacNeil, 64, broadcast journalist, born at Montreal, Quebec, Canada, Jan 19, 1931.
Robert Palmer, 46, singer, born at Babley, England, Jan 19, 1949.
Dolly Parton, 49, singer ("Jolene"), actress (*Nine to Five*), born at Sevier County, TN, Jan 19, 1946.

Jan 19-20 ☆ Chase's 1995 Calendar of Events ☆

Javier Perez de Cuellar, 75, Peruvian diplomat, former United Nations Secretary General, born at Lima, Peru, Jan 19, 1920.
Simon Rattle, 40, British orchestra conductor, born at Liverpool, England, Jan 19, 1955.
Christopher Andrew Sabo, 33, baseball player, born at Detroit, MI, Jan 19, 1962.
Jean Stapleton (Jeanne Murray), 72, actress (*Klute*; Emmy for Edith Bunker [Dingbat] in "All in the Family"), born at New York, NY, Jan 19, 1923.
Fritz Weaver, 69, actor (*Holocaust, Marathon Man*), born at Philadelphia, PA, Jan 19, 1926.

JANUARY 20 — FRIDAY
20th Day — Remaining, 345

AQUARIUS, THE WATER CARRIER. Jan 20–Feb 19. In the astronomical/astrological zodiac, which divides the sun's apparent orbit into 12 segments, the period Jan 20–Feb 19 is identified, traditionally, as the sun-sign of Aquarius, the Water Carrier. The ruling planet is Uranus or Saturn.

AUSTRIA: MOZART WEEK. Jan 20–29. Salzburg. Focuses on Mozart's works and features international soloists, conductors and orchestras. Est attendance: 20,000. For info: Intl Stiftung Mozarteum, Schwarzstrasse 26, A-5024 Salzburg, Postfach 34, Austria.

BRAZIL: NOSSO SENHOR DO BONFIM FESTIVAL. Jan 20–30. Salvador, Bahia, Brazil. Our Lady of the Happy Ending Festival is one of Salvador's most colorful religious feasts. Climax comes with people carrying water to pour over church stairs and sidewalks to cleanse them of impurities.

BRAZIL: SAN SEBASTIAN'S DAY. Jan 20. Patron Saint of Rio de Janeiro.

CLOCK TOWER JAZZ FESTIVAL. Jan 20–22. Rockford, IL. One of the largest jazz festivals with over 20 different jazz bands from mainstream to big band. Est attendance: 4,000. For info: Clock Tower Resort, 7801 E State St, Rockford, IL 61125. Phone: (815) 398-6000.

FELLINI, FEDERICO: BIRTH ANNIVERSARY. Jan 20. Director and screenwriter Federico Fellini was born Jan 20, 1920, at Rimini, Italy. Four of Fellini's movies won Oscars for best foreign-language film: *La Strada* (1956), *The Nights of Cabiria* (1957), *8 1/2* (1963) and *Amarcord* (1974). He received an honorary Oscar in 1993 in recognition of his cinematic accomplishments. Federico Fellini died Oct 31, 1993, at Rome.

GALLERY NIGHT. Jan 20. (Also, Apr 21, July 21 and Oct 20.) Milwaukee Art Museum, Milwaukee, WI. A night of gallery and museum-hopping. Begins at 5:30 PM at the Milwaukee Art Museum with a mini-lecture, from there take the Gallery Night Express Bus, or walk or drive to participating galleries. Gallery Night is free of charge. The Gallery Night Express Bus has a fee of $9 per person. Bus reservations required. Est attendance: 400. For info: Milwaukee Art Museum, 750 N Lincoln Memorial Dr, Milwaukee, WI 53202. Phone: (414) 224-3200.

January 1995

S	M	T	W	T	F	S
1	2	3	4	5	6	7
8	9	10	11	12	13	14
15	16	17	18	19	20	21
22	23	24	25	26	27	28
29	30	31				

GUINEA-BISSAU: NATIONAL HEROES DAY. Jan 20. National holiday in the Republic of Guinea-Bissau.

ICEBOX DAYS. Jan 20–29. International Falls, MN. Smoosh racing, "Freeze Yer Gizzard Blizzard Run," softball on ice, ski races, mutt races, smoosh races and beach party. Est attendance: 5,000. For info: Shawn M. Mason, Chamber of Commerce, PO Box 169, 200 Fourth St, International Falls, MN 56649. Phone: (800) 325-5766.

LESOTHO: ARMY DAY. Jan 20. Lesotho, South Africa.

SEWARD: 10th ANNIVERSARY POLAR BEAR JUMP FESTIVAL. Jan 20–22. Resurrection Bay, Seward, AK. Volunteers collect pledges to jump into bay in costumes. Festivities include golf tournament, parade, Dog Weight Pull, Dog Sled Race, Polar Bear Expo/Trade Show, Ski-joring, Bachelor/Bachelorette Auction, Ice Bowling, Talent Show, Seafood Feed and much more. Benefits local non-profit organizations and the American Cancer Society. The actual "Plunge" takes place on Saturday, Jan 21. Est attendance: 2,500. For info: Seward Polar Bear Jump Fest, PO Box 386, Seward, AK 99664. Phone: (907) 224-5230.

SOUTHWESTERN EXPOSITION LIVESTOCK SHOW AND RODEO. Jan 20–Feb 5. Fort Worth, TX. Western-flavored extravaganza begun in 1896. World's first indoor rodeo added in 1918. Prize livestock displays, horse shows, midway, commercial exhibits and quality entertainment. Est attendance: 800,000. For info: Delbert Bailey, PO Box 150, Fort Worth, TX 76101-0150. Phone: (817) 877-2400.

US REVOLUTIONARY WAR: CESSATION OF HOSTILITIES: ANNIVERSARY. Jan 20. On Jan 20, 1783, the British and US Commissioners signed a preliminary "Cessation of Hostilities," which was ratified by England's King George III on Feb 14 and led to the Treaties of Paris and Versailles, Sept 3, 1783, ending the war.

WISCONSIN DELLS FLAKE OUT FESTIVAL. Jan 20–22. Tommy Bartlett Show, Wisconsin Dells, WI. Wisconsin sanctioned snow sculpting competition. Winners will compete in the National Snow Sculpting Competition. Other activities include ice carving competition, snowmobile radar run, sleigh rides, ice skating and children's snow sculpting. Est attendance: 10,000. For info: Wisc Dells Visitor and Conv Bureau, 701 Superior St, PO Box 390, Wisconsin Dells, WI 53965. Phone: (800) 223-3557.

BIRTHDAYS TODAY

Edwin "Buzz" Aldrin, 65, former astronaut, one of first three men on moon, born at Montclair, NJ, Jan 20, 1930.
George Burns (Nathan Birnbaum), 99, actor, comedian ("The George Burns and Gracie Allen Show," *Oh, God!*), born at New York, NY, Jan 20, 1896.
Paul D. Coverdell, 56, US Senator (R, Georgia) up for reelection in '94, born at Des Moines, IA, Jan 20, 1939.
Shelley Fabares, 53, actress (Mary on "The Donna Reed Show"; "Coach," sang "Johnny Angel"), born at Santa Monica, CA, Jan 20, 1942.
Arte Johnson, 61, comedian, actor (Emmy for "Rowan & Martin's Laugh-In"), born at Chicago, IL, Jan 20, 1934.
Lorenzo Lamas, 37, actor ("Falcon Crest," "Renegade"), born at Los Angeles, CA, Jan 20, 1958.
David Lynch, 49, director ("Twin Peaks," *Blue Velvet*), writer, producer, born at Missoula, MT, Jan 20, 1946.
Anatoly Scharansky, 47, expatriate Soviet dissident, born at Donetsk, USSR, Jan 20, 1948.
Bruce Sundlun, 75, Governor of Rhode Island (D), born at Providence, RI, Jan 20, 1920.
Otis Dewey "Slim" Whitman, 71, singer (first country performer to play at the London Palladium), born at Tampa, FL, Jan 20, 1924.

JANUARY 21 — SATURDAY
21st Day — Remaining, 344

BALD EAGLE APPRECIATION DAYS. Jan 21–22. Riverfront, Keokuk, IA. Features trained personnel stationed at observation points for viewing the American bald eagle. Est attendance: 10,000. For info: Keokuk Area Conv and Tourism Bureau, Marilyn Pohorsky, Dir, 401 Main, Keokuk, IA 52632. Phone: (800) 383-1219.

BALDWIN, ROGER NASH: BIRTH ANNIVERSARY. Jan 21. Founder of the American Civil Liberties Union, called the "country's unofficial agitator for, and defender of, its civil liberties." Born Jan 21, 1884, at Wellesley, MA. Died Aug 26, 1981, at Ridgewood, NJ.

BRECKINRIDGE, JOHN CABELL: BIRTH ANNIVERSARY. Jan 21. Fourteenth vice president of the US (1857–1861). Born at Lexington, KY, Jan 21, 1821. Died there May 17, 1875.

CANADA: MINDEN SLED DOG DERBY. Jan 21–22. Minden, Ontario. World's largest limited-class speed sled-dog derby. Non-toll-free phone: (705) 286-1760. Est attendance: 7,000. For info: Minden Sled Dog Derby, Box 97, Minden, Ont, Canada K0M 2K0. Phone: (800) 461-7677.

CLARK, BARNEY: BIRTH ANNIVERSARY. Jan 21. First person to receive a permanent artificial heart. Born at Provo, UT, on Jan 21, 1921. The artificial heart (made of polyurethane plastic and aluminum) was implanted Dec 2, 1982, at the University of Utah Medical Center, Salt Lake City, UT. Clark lived almost 112 days after implantation of the artificial heart. Died Mar 23, 1983.

EAGLE DAYS. Jan 21–22. Milford Lake Nature Center/Fish Hatchery, Junction City, KS. Learn more about this magnificent bird which is our national emblem. Meet both a live bald eagle and a golden eagle! Guides with spotting scopes and binoculars will be waiting to show you eagles as they roost and soar around Milford Lake. Sponsor: Kansas Wildlife and Parks; US Army Corps of Engineers. Est attendance: 400. For info: Milford Nature Center, 3115 Hatchery Dr, Junction City, KS 66441. Phone: (913) 238-2638.

FITCH, JOHN: BIRTH ANNIVERSARY. Jan 21. American inventor, clockmaker, gunsmith, surveyor and steamboat developer. Born at East Windsor, CT, Jan 21, 1743. Died at Bardstown, KY, July 2, 1798.

FUN-IN-THE-SUN POSTCARD SALE. Jan 21–22. Las Palmas Hotel, Orlando, FL. Sale of antique and modern postcards by 20 dealers. Est attendance: 200. Sponsor: Postcard Society, Inc, John H. McClintock, Dir, Box 1765, Manassas, VA 22110. Phone: (703) 368-2757.

HOT AND SPICY FOOD INTERNATIONAL DAY. Jan 21. Des Moines, IA, and many other places around the country. It is very important for the Hot and Spicy Food lovers of the world to have a day of their own to celebrate and enjoy the pleasures of appropriate culinary creation. Hot and Spicy Food is something special and it is time for the world to recognize it. Annually, the third Saturday of January. For info: Thai Network, Prasong Nurack, 215 E Walnut St, Des Moines, IA 50309. Phone: (515) 282-0044.

JACKSON, THOMAS JONATHAN "STONEWALL": BIRTH ANNIVERSARY. Jan 21. Confederate general and one of the most famous soldiers of the American Civil War, best known as "Stonewall" Jackson. Born at Clarksburg, VA (now WV) Jan 21, 1824. Died of wounds received in battle near Chancellorsville, VA, May 10, 1863.

KIWANIS INTERNATIONAL: ANNIVERSARY. Jan 21. First Kiwanis Club chartered Jan 21, 1915, at Detroit, MI.

LONGWOOD GARDENS WELCOME SPRING. Jan 21–Apr 30. Kennett Square, PA. Indoor conservatory display features thousands of colorful, fragrant spring bulbs, green lawns, palm trees, orchids and roses. Organ concerts on most Sunday afternoons. Est attendance: 110,000. For info: Colvin Randall, PR Mgr, Longwood Gardens, PO Box 501, Kennett Sq, PA 19348-0501. Phone: (610) 388-6741.

NATIONAL HUGGING DAY™. Jan 21. Since hugging is something everyone can do and since it is a healthful form of touching, this day should be spent hugging anyone who will accept a hug, especially family and friends. The most famous huggers of the year will be announced. For more information, please send S.A.S.E. Sponsor: Kevin C. Zaborney, Box 123, Clio, MI 48420. Phone: (313) 982-8436.

PHILADELPHIA BOAT SHOW. Jan 21–29. Civic Center, Philadelphia, PA. Show features more than 600 sail and power boats, engines and accessories from all major manufacturers. Est attendance: 65,000. For info: Natl Marine Mfrs Assn, Jim Ranieri, 514 Harriet Lane, Havertown, PA 19083. Phone: (610) 449-9910.

PHILIPPINES: ATI-ATIHAN FESTIVAL. Jan 21–22. Kalibo, Aklan. One of the most colorful celebrations in the Philippines, the Ati-Atihan Festival commemorates the peace pact between the Ati of Panay (pygmies) and the Malays, who were early migrants in the islands. The townspeople blacken their bodies with soot, don colorful and bizarre costumes and sing and dance in the streets. The festival also celebrates the Feast Day of Santo Niño (the infant Jesus). Annually, the third weekend in January.

SENIOR BOWL FOOTBALL GAME. Jan 21. Ladd Memorial Stadium, Mobile, AL. All-star football game featuring the nation's top collegiate seniors. Proceeds go to charities. Est attendance: 40,000. For info: Mobile Arts and Sports Assn, Vic Knight, PR Dir, 63 S Royal St, Ste 107, Mobile, AL 36602. Phone: (205) 438-2276.

STAR-ATHON '95: 17th ANNUAL TELETHON FOR CEREBRAL PALSY. Jan 21–22. Annual telethon to raise money to support research and to provide programs and services for people with cerebral palsy and other disabilities. Averages around $24 million in receipts each year. It can be seen on 60 station networks. For info: United Cerebral Palsy Assn, Inc, 1522 K St NW, Ste 1112, Washington, DC 20005. Phone: (800) 872-5827.

TIP-UP TOWN USA ICE FESTIVAL. Jan 21–22. (Also Jan 28–29.) Houghton Lake, MI. Activities on ice. Ice sculpting, fishing, games and contests. Est attendance: 45,000. For info: Chamber of Commerce, Houghton Lake, MI 48629. Phone: (800) 248-5253.

BIRTHDAYS TODAY

Robbie Benson (Robert Segal), 39, actor ("Search for Tomorrow," *Ode to Billie Joe*), born at Dallas, TX, Jan 21, 1956.

Geena Davis, 38, actress (Oscar for *The Accidental Tourist*; *Thelma and Louise*; "Buffalo Bill"), born at Ware, MA, Jan 21, 1957.

Mac Davis, 53, actor, songwriter ("The Mac Davis Show," *North Dallas Forty*), born at Lubbock, TX, Jan 21, 1942.

Placido Domingo, 54, opera singer, born at Madrid, Spain, Jan 21, 1941.

Jill Eikenberry, 48, actress (Ann Kelsey in "L.A. Law"), born at New Haven, CT, Jan 21, 1947.

Jack Nicklaus, 55, golfer, born at Columbus, OH, Jan 21, 1940.

Jan 21-22 ★ Chase's 1995 Calendar of Events ★

Billy Ocean, 45, musician, songwriter, born at London, England, Jan 21, 1950.
Hakeem Aboul Olajuwon (Akeem), 32, basketball player, born at Lagos, Nigeria, Jan 21, 1963.
Detlef Schrempf, 32, basketball player, born at Leverkusen, West Germany, Jan 21, 1963.
Wolfman Jack, 56, disc jockey, born at Brooklyn, NY, Jan 21, 1939.

JANUARY 22 — SUNDAY
22nd Day — Remaining, 343

AMPERE, ANDRE: 220th BIRTH ANNIVERSARY. Jan 22. Physicist, student of electrical and magnetic phenomena, founder of the science of electrodynamics. Born at Lyons, France, on Jan 20, 1775. From his early childhood, tragedy and depression pursued him. His father was executed during the French Revolution. His first wife died young. His second marriage was unhappy. Ampère died at Marseilles, France, on June 10, 1836. The epitaph he selected for his tombstone was *tandem felix* ("happy at last"). The ampere, a unit of electrical current, is named for him.

ANSWER YOUR CAT'S QUESTION DAY. Jan 22. If you will stop what you are doing and take a look at your cat, you will observe that the cat is looking at you with a serious question. Meditate upon it, then answer the question! Annually, Jan 22. Phone: (212) 388-8673 or (717) 274-8451. Sponsor: Wellness Permission League, Tom or Ruth Roy, 2105 Water St, Lebanon, PA 17046.

BYRON, GEORGE GORDON: BIRTH ANNIVERSARY. Jan 22. Romantic poet, born at London, England, Jan 22, 1788. Died of fever at Missolonghi, Greece, Apr 19, 1824.

CELEBRITY READ A BOOK WEEK. Jan 22–28. Homewood's Corporate Complex, Williamsport, MD. Celebrities read books to children at the Book Fair hosted by Choo Choo Child Care Center. Also, new books are donated to the local public library and books are collected for the center's library. Annually, the last full week in January. For info: Denise Feeser, Dir, Choo Choo Child Care Center, PO Box 250, Williamsport, MD 21795. Phone: (301) 582-4894.

CHRIS STRATTON DART TOURNAMENT. Jan 22. Springbrook Golf Course, Battle Creek, MI. Third annual benefit tournament dedicated to Chris Stratton, a great darter and well-liked person. Fifty percent of entry fees donated to American Cancer Society. Sponsor: Springbrook Golf Course, Bill Buckner, 1600 Ave A, Battle Creek, MI 49015. Phone: (616) 965-6512.

COCOANUT GROVE BRIDAL EXPO. Jan 22. Cocoanut Grove, Santa Cruz, CA. Features more than 60 booths with every aspect needed for that perfect wedding, reception and honeymoon. Est attendance: 2,000. For info: Ann Parker, Pub Mgr, Cocoanut Grove, 400 Beach St, Santa Cruz, CA 95060-5491. Phone: (408) 423-5590.

GRIFFITH, DAVID (LEWELYN) WARK: BIRTH ANNIVERSARY. Jan 22. D.W. Griffith, pioneer producer-director in the American motion picture industry, best remembered for his film *The Birth of a Nation* (1915). Born at LaGrange, KY, Jan 22, 1875. Died at Hollywood, CA, July 23, 1948.

HULA BOWL GAME. Jan 22. Aloha Stadium, Honolulu, Oahu, HI. College all-star football classic. Est attendance: 19,000. For info: Hula Bowl Office, University of Hawaii Fdtn, PO Box 11270, Honolulu, HI 96828. Phone: (808) 956-4852.

LEE BIRTHDAY CELEBRATIONS. Jan 22. Alexandria, VA. The birthdays of Revolutionary War Colonel "Light Horse Harry" Lee and his son Civil War General Robert E. Lee, are celebrated at the Lee-Fendall House and The Boyhood Home of Robert E. Lee. Refreshments, period music, house tours. Admission. Fourth Sunday in January. Est attendance: 125. For info: Diane Bechtol, Mgr, Media Relations, Alexandria Conv and Visitors Bureau, 221 King St, Alexandria, VA 22314. Phone: (703) 838-4200.

NATIONAL ACTIVITY PROFESSIONALS WEEK. Jan 22–28. A week-long celebration to recognize and honor activity professionals who work in the nation's long-term care facilities, adult day care centers, retirement living facilities and senior center. National Activity Professionals Day is Friday Jan 27. Info from: Natl Assn of Activity Professionals, Charles Price, Exec Dir, 1225 Eye St. NW, Ste 300, Washington, DC 20005. Phone: (202) 289-0722.

NATIONAL GLAUCOMA AWARENESS WEEK. Jan 22–28. African-Americans are 15 times more likely than the general population to go blind from glaucoma. Throughout this week, Prevent Blindness America (formerly known as National Society to Prevent Blindness) will provide valuable information about this "Sneak thief of sight." Organizations are encouraged to educate the community through screenings, forums and programs. Sponsor: Prevent Blindness America, Marita Gomez, Media Relations, 500 E Remington Rd, Schaumburg, IL 60173. Phone: (800) 331-2020.

ROE V WADE DECISION: ANNIVERSARY. Jan 22. On Jan 22, 1973, in the case of Roe v Wade, the US Supreme Court struck down state laws restricting abortions during the first six months of pregnancy. In the following two decades debate has continued to rage between those who believe a woman has a right to choose whether or not to continue a pregnancy that has begun in her body and those who believe that aborting such a pregnancy is murder of an unborn child.

ST. VINCENT'S FEAST DAY. Jan 22. Spanish deacon and martyr who died AD 304. Patron saint of wine growers. Old weather lore says if there is sun on this day, good wine crops may be expected in the ensuing season.

SIGHT-SAVING SABBATH. Jan 22. (Also Jan 28.) Religious denominations throughout the United States are invited to participate in this ecumenical observance that alerts their members about the importance of periodic eye examinations and screenings to prevent blindness stemming from eye diseases, such as glaucoma. Materials that can easily be posted or distributed to the community will be provided. Sponsor: Prevent Blindness America, Marita Gomez, Media Relations, 500 E Remington Rd, Schaumburg, IL 60173. Phone: (800) 331-2020.

SOCIETY FOR THE PRESERVATION AND ENCOURAGEMENT OF BARBER SHOP QUARTET SINGING IN AMERICA, INC (SPEBSQSA) MID-WINTER CONVENTION. Jan 22–28. Tucson, AZ. Some 3,000 members meet for administrative conferences, shows, tours and contest to select national Seniors Quartet Champion. Est attendance: 3,000. For info: SPEBSQSA, Inc, Brian Lynch, Communications Specialist, 6315 Third Ave, Kenosha, WI 53143-5199. Phone: (414) 653-8440.

TEXAS STATE CHAMPIONSHIP DOMINO TOURNAMENT. Jan 22. Knights of Columbus Hall, Halletsville, TX. Determines the "partner domino" champions in Texas. Est attendance: 1,500. For info: Knights of Columbus, Thomas Grahmann, Chair, Box 383, Hallettsville, TX 77964. Phone: (512) 798-2181.

VINSON, FRED M.: BIRTH ANNIVERSARY. Jan 22. The 13th Chief Justice of the US, born at Louisa, KY, on Jan 22, 1890. Served in the House of Representatives, appointed Director of War Mobilization during World War II, Secretary of the Treasury under Harry Truman. Nominated by Truman to succeed Harlan F. Stone as Chief Justice of the US Supreme Court. Died at Washington, DC, on Sept 8, 1953.

January 1995

S	M	T	W	T	F	S
1	2	3	4	5	6	7
8	9	10	11	12	13	14
15	16	17	18	19	20	21
22	23	24	25	26	27	28
29	30	31				

Chase's 1995 Calendar of Events — Jan 22-23

BIRTHDAYS TODAY

Linda Blair, 36, actress (*The Exorcist, Airport*), born at Westport, CT, Jan 22, 1959.
Pierre S. duPont IV, 60, former governor of Delaware, born at Wilmington, DE, Jan 22, 1935.
John Hurt, 55, actor ("And the Band Played On," *The Elephant Man*), born at Lincolnshire, England, Jan 22, 1940.
Piper Laurie (Rosetta Jacobs), 63, actress, born at Detroit, MI, Jan 22, 1932.
Steve Perry, 46, singer, born at Hanford, CA, Jan 22, 1949.
Joseph Wambaugh, 58, author, ex-police officer (*The Blooding, Fugitive Nights*), born at East Pittsburgh, PA Jan 22, 1937.

JANUARY 23 — MONDAY
23rd Day — Remaining, 342

CHINESE NEW YEAR CELEBRATION. Jan 23–Feb 26. Chinatown, Los Angeles, CA. Activities: press conference on Mon Jan 23; fashion show luncheon introducing Miss Los Angeles Chinatown contestants on Sat Jan 28; street fair on Feb 18–19; Little Miss King and Queen contest on Sat Feb 4; Chinese New Year Day—Year of the Boar (Pig) 4693—with traditional festivities on Tues Jan 31; Miss LA Chinatown Beauty Pageant on Sat Feb 11; Golden Dragon Parade on Sat Feb 18; Chinese New Year Banquet on Sat Feb 25; Firecracker 5K/10K Run on Sun Feb 26. Est attendance: 90,000. For info: Chinese Chamber of Commerce, Nancy Yee, 977 Broadway, Ste E, Los Angeles, CA 90012. Phone: (213) 617-0396.

FIRST LADIES GOWNS. Jan 23–Apr 30. Gerald R. Ford Museum, Grand Rapids, MI. Exhibition includes original gowns as well as reproductions of gowns worn by First Ladies, supplemented with photos and accessories. Organized by the Gerald R. Ford Museum. For info: Gerald R. Ford Museum, 303 Pearl St NW, Grand Rapids, MI 49504-5353. Phone: (616) 451-9263.

HANCOCK, JOHN: BIRTH ANNIVERSARY. Jan 23. American patriot and statesman, first signer of the Declaration of Independence. Born Jan 23, 1737, at Braintree, MA. Died at Quincy, MA, Oct 8, 1793. His name has become part of the American language, referring to any handwritten signature, "Put your John Hancock on that!"

HEWES, JOSEPH: BIRTH ANNIVERSARY. Jan 23. Signer of the Declaration of Independence. Born at Kingston, NJ, Jan 23, 1730. Died Nov 10, 1779.

MOON PHASE: LAST QUARTER. Jan 23. Moon enters Last Quarter phase at 11:58 PM, EST.

NATIONAL HANDWRITING ANALYSIS WEEK. Jan 23–29. To inform the public that handwriting is a form of behavior that can be analyzed for personality traits; that handwriting originates in the brain; that handwriting style, like personality, remains constant over a period of time while reflecting development; that personality traits can be changed by making changes in one's handwriting. Annually, a week in January to include National Handwriting Day (John Hancock's birthday). Sponsors: American Assn of Handwriting Analysts (AAHA), 820 W Maple St, Hinsdale, IL 60521; phone (708) 323-5647; fax (313) 645-2511; and American Handwriting Analysis Fdtn (AHAF), PO Box 6201, San Jose, CA 95150; phone (408) 377-6775; fax (408) 377-3739.

NATIONAL HANDWRITING DAY. Jan 23. Popularly observed on birthday of John Hancock to encourage more legible handwriting.

NATIONAL PIE DAY. Jan 23. To focus attention on pie as an art form and gastronomical delight and on the joy it has brought to millions. Sponsor: Charlie Papazian, Box 1825, Boulder, CO 80306. Phone: (303) 665-6489.

ONE-TOOTH RHEE LANDING DAY. Jan 23. Observed in vicinity of all government offices to celebrate beginning of "confusionist" branch of American "burrocracy." One-Tooth Rhee, mythical Korean inventor of the custom of each official wearing four hats so that contradictory sets of instructions can be given with each job title. Annually, Jan 23. Sponsor: Puns Corps, Robert L. Birch, Coord, Box 2364, Falls Church, VA 22042-0364. Phone: (703) 533-3668.

STENDHAL: BIRTH ANNIVERSARY. Jan 23. French author Marie Henri Beyle, whose best-known pseudonym was Stendhal. Born at Grenoble, France, Jan 23, 1783. Best remembered are his novels *The Red and the Black* (1831) and *The Charterhouse of Parma* (1839). Died at Paris, France, Mar 23, 1842.

STEWART, POTTER: BIRTH ANNIVERSARY. Jan 23. Associate Justice of the Supreme Court of the United States, nominated by President Eisenhower on Jan 17, 1959. (Oath of office, May 15, 1959.) Born at Jackson, MI, on Jan 23, 1915. He retired in July 1981 and died on Dec 7, 1985, five days after suffering a stroke, at Putney, VT. Buried at Arlington National Cemetery.

TWENTIETH AMENDMENT TO US CONSTITUTION RATIFIED: ANNIVERSARY. Jan 23. The 20th Amendment was ratified on Jan 23, 1933, fixing the date of the presidential inauguration at the current Jan 20 instead of the previous Mar 4. It also specified that were the president-elect to die before taking office, the vice president-elect would succeed to the presidency. In addition, it set Jan 3 as the official opening date of Congress each year.

TWENTY-FOURTH AMENDMENT TO US CONSTITUTION RATIFIED: ANNIVERSARY. Jan 23. Poll taxes and other taxes were eliminated as a prerequisite for voting in all federal elections by the 24th Amendment, ratified on Jan 23, 1964.

WINSLOW HOMER THE ILLUSTRATOR: HIS WOOD ENGRAVINGS, 1857–1888. Jan 23–July 10. LBJ Library and Museum, Austin, TX. An exhibition of 145 prints representing nearly 80% of the illustrations done by one of America's greatest artists. A visual history lesson of America in the latter 1800s, including Civil War, national holidays, factory life, the seasons and more. Engravings are from the Cornell Fine Arts Museum of Winter Park, FL. Est attendance: 1,000,000. For info: Lawrence Reed, Asst Dir, LBJ Fdtn, 2313 Red River St, Austin, TX 78705.

BIRTHDAYS TODAY

Richard Anderson, 45, actor ("General Hospital," "MacGyver"), born at Minneapolis, MN, Jan 23, 1950.
Princess Caroline, 38, born at Monte Carlo, Monaco, Jan 23, 1957.
Tom Carper, 48, Governor of Delaware (D), born at Beckley, WV, Jan 23, 1947.
David Douglas Duncan, 79, photojournalist, author, born at Kansas City, MO, Jan 23, 1916.
Gil Gerard, 52, actor ("Buck Rogers," "Sidekicks"), born at Little Rock, AR, Jan 23, 1943.

Jan 23-25 ☆ *Chase's 1995 Calendar of Events* ☆

Pat Haden, 42, former football player, born at Westbury, NY, Jan 23, 1953.
Rutger Hauer, 51, actor, born at Breukelen, Netherlands, Jan 23, 1944.
Frank R. Lautenberg, 71, US Senator (D, New Jersey) up for reelection in Nov '94, born at Paterson, NJ, Jan 23, 1924.
Jeanne Moreau, 67, actress (*Jules and Jim, Viva Maria*), born at Paris, France, Jan 23, 1928.
Chita Rivera (Conchita del Rivero), 62, actress, singer ("The New Dick Van Dyke Show"), born at Washington, DC, Jan 23, 1933.

JANUARY 24 — TUESDAY
24th Day — Remaining, 341

ACHELIS, ELISABETH: BIRTH ANNIVERSARY. Jan 24. Calendar reform advocate, editor, author. Born at Brooklyn, NY, Jan 24, 1880. Author of *The World Calendar*. Proposed calendar made every year the same, with equal quarters, each year beginning on Sunday, Jan 1, and each date falling on same day of week every year. Died at New York, NY, Feb 11, 1973.

BOLIVIA: ALACITIS FAIR. Jan 24-26. La Paz. Traditional annual celebration by Aymara Indians on this date with prayers and offerings to god of prosperity.

CALIFORNIA GOLD DISCOVERY: ANNIVERSARY. Jan 24. James W. Marshal, an employee of John Sutter, accidentally discovered gold while building a sawmill near Coloma, CA, on Jan 24, 1848. Efforts to keep the discovery secret failed, and the gold rush was quickly under way.

GOODSON, MARK: BIRTH ANNIVERSARY. Jan 24. Producer and creator of TV game shows, Mark Goodson was born in Sacramento, CA, on Jan 24, 1915. His career in entertainment began in radio where he created his first game show, "Pop the Question." He later teamed with Bill Todman and that partnership led to "What's My Line?," "I've Got a Secret," "Password," "The Price Is Right" and "Family Feud." He died on Dec 18, 1992, at New York, NY.

SAN ANTONIO SPORT, BOAT AND RV SHOW. Jan 24-29. San Antonio Convention Center, San Antonio, TX. Boating, travel, hunting, fishing, camping, RVs and recreation show. 39th annual show. Sponsor: Boating Trades Assn of San Antonio, TX. For info: Mike Coffen, Double C Productions, Inc, PO Box 1678, Huntsville, TX 77342. Phone: (409) 295-9677.

SPACE MILESTONE: *COSMOS 954* (USSR). Jan 24. Nuclear-equipped reconnaissance satellite launched Sept 18, 1977. Fell into Earth's atmosphere and burned over northern Canada. Some radioactive debris reached ground on Jan 24, 1978.

SPACE MILESTONE: *DISCOVERY* (US). Jan 24. Space shuttle *Discovery* launched from and returned to Kennedy Space Center, FL, deploying eavesdropping satellite in secret, all-military mission, Jan 24-27, 1985.

TOGO: ECONOMIC LIBERATION DAY. Jan 24. National holiday.

US INTERNATIONAL SNOW SCULPTING COMPETITION. Jan 24-29. Performing Arts Center grounds, Milwaukee, WI. 10th anniversary event. All international 3-person sculpting teams will have competed at the Olympic Winter Games International Snow Sculpting competitions in Calgary, Albertville or Lillehammer. Est attendance: 100,000. For info: Gene Kempfer, Dir, 1554 S. 74th St, Milwaukee, WI 53214. Phone: (414) 476-5573.

January 1995

S	M	T	W	T	F	S
1	2	3	4	5	6	7
8	9	10	11	12	13	14
15	16	17	18	19	20	21
22	23	24	25	26	27	28
29	30	31				

WHARTON, EDITH: BIRTH ANNIVERSARY. Jan 24. American author and Pulitzer Prize winner. Born at New York, NY, on Jan 24, 1862. Died Aug 11, 1937.

BIRTHDAYS TODAY

Ernest Borgnine, 78, actor (*Marty*, "McHale's Navy"), born at Hamden, CT, Jan 24, 1917.
Jack Brickhouse, 79, former radio and TV sports announcer, born at Peoria, IL, Jan 24, 1916.
Neil Diamond, 54, singer, composer ("Cracklin' Rosie," "Song Sung Blue"), born at Coney Island, NY, Jan 24, 1941.
Robert Keith Dibble, 31, baseball player, born at Bridgeport, CT, Jan 24, 1964.
Mary Lou Retton, 27, gymnast, Olympic medalist, born at Fairmont, WV, Jan 24, 1968.
Oral Roberts, 77, evangelist, born at Tulsa, OK, Jan 24, 1918.
Yakov Smirnoff, 44, comedian ("What a Country," "Night Court"), born at Odessa, USSR, Jan 24, 1951.
Maria Tallchief, 70, former ballet dancer, born at Fairfax, OK, Jan 24, 1925.

JANUARY 25 — WEDNESDAY
25th Day — Remaining, 340

CURTIS, CHARLES: BIRTH ANNIVERSARY. Jan 25. Thirty-first vice president of the US (1929-1933). Born at Topeka, KS, Jan 25, 1860. Died at Washington, DC, Feb 8, 1936.

ENGLAND: WORLD OF DRAWINGS AND WATERCOLOURS. Jan 25-29. Park Lane Hotel, Piccadilly, London. Original works of art are offered for sale by 50 leading galleries and dealers. The only fair of its kind in the world. Est attendance: 10,000. For info: Angela Wynn, World of Drawings and Watercolours, 10 Dukes Avenue, London, England W4 2AE.

FIRST TELEVISED PRESIDENTIAL NEWS CONFERENCE: ANNIVERSARY. Jan 25. Beginning a tradition that survives to this day, John F. Kennedy held the first televised presidential news conference five days after being inaugurated the 35th president.

GRENADA: GAME FISHING TOURNAMENT. Jan 25-30. Anglers of international repute enjoy the rich fishing grounds of sailfish, marlin, yellowfin tuna and more. Est attendance: 2,000. For info: Grenada Tourist Info Office, 820 Second Ave, Ste 900-D, New York, NY 10017. Phone: (800) 927-9554.

MAUGHAM, W. SOMERSET: BIRTH ANNIVERSARY. Jan 25. English short story writer, novelist and playwright. Born at Paris, France, Jan 25, 1874. Among his best remembered books: *Of Human Bondage, Cakes and Ale* and *The Razor's Edge*. Died at Cap Ferrat, France, Dec 16, 1965.

NATIONAL SCHOOL NURSE DAY. Jan 25. A day to honor and recognize the school nurse, School Nurse Day has been established to foster a better understanding of the role of school nurses in the educational setting. Annually, the fourth Wednesday in January. For info: Natl Assn of School Nurses, Inc, PO Box 1300, Scarborough, ME 04070. Phone: (207) 883-2117.

☆ Chase's 1995 Calendar of Events ☆ Jan 25–26

PONTIAC SILVERDOME CAMPER, TRAVEL AND RV SHOW. Jan 25-28. Pontiac Silverdome, Pontiac, MI. This event brings together buyers and sellers of RVs, motor homes, campers and camping accessories, as well as buyers and sellers of camping vacations and travel destinations. Est attendance: 28,000. For info: Mike Wilbraham, ShowSpan, Inc, 1400 28th St SW, Grand Rapids, MI 49509. Phone: (616) 530-1919.

A ROOM OF ONE'S OWN DAY. Jan 25. For anyone who knows or longs for the sheer bliss and rightness of having a private place, no matter how humble, to call one's own. Phone: (212) 388-8673 or (717) 274-8451. Sponsor: Wellness Permission League, Tom and Ruth Roy, 2105 Water St, Lebanon, PA 17046.

WOOLF, VIRGINIA: BIRTH ANNIVERSARY. Jan 25. English writer, critic and novelist. Author of *Jacob's Room* and *To the Lighthouse*. Women's rights activist. Born at London, England, on Jan 25, 1882. After completing her last novel, *Between the Acts,* she collapsed under the strain and drowned herself in the River Ouse near Rodmell, England, on Mar 28, 1941.

BIRTHDAYS TODAY

Elizabeth Allen, 61, actress, singer ("The Paul Lynde Show," "C.P.O. Sharkey"), born at Jersey City, NJ, Jan 25, 1934.
Corazon "Cory" Aquino, 72, former president of the Philippines, born at Tarlac Province, Philippine Islands, Jan 25, 1923.
Conrad Burns, 60, US Senator (R, Montana), born at Gallatin, MO, Jan 25, 1935.
Mark "Super" Duper, 36, football player, born at Pineville, LA, on Jan 25, 1959.
Ernie Harwell, 77, baseball broadcaster, born at Atlanta, GA, Jan 25, 1918.
Dean Jones, 64, actor (*Tea and Sympathy, Love Bug, Beethoven*), born at Decatur, AL, Jan 25, 1931.
Edwin Newman, 76, journalist, author ("Comet," *A Civil Tongue*), born at New York, NY, Jan 25, 1919.
Eduard Amvrosiyevich Shevardnadze, 67, former Russian minister of foreign affairs, born at Mamati, USSR, Jan 25, 1928.

JANUARY 26 — THURSDAY
26th Day — Remaining, 339

AUSTRALIA: AUSTRALIA DAY. Jan 26. Anniversary of first British settlement, Jan 26, 1788, when a shipload of convicts arrived briefly at Botany Bay (which proved to be unsuitable) and then at Port Jackson (later the site of the city of Sydney). Establishment of an Australian prison colony was to relieve crowding of British prisons. Australia Day, formerly known as Foundation Day or Anniversary Day, has been observed since about 1817, and a public holiday since 1838. Observed on Jan 26 if a Monday, otherwise on the first Monday thereafter. See also: "Australia: Australia Day Observance" (Jan 30).

DOMINICAN REPUBLIC: NATIONAL HOLIDAY. Jan 26. An official public holiday celebrates the birth anniversary of Juan Pablo Duarte, one of the fathers of the republic.

FLORIDA CITRUS FESTIVAL. Jan 26–Feb 5. Florida Citrus Showcase Fairgrounds, Winter Haven, FL. Celebrating 100 years of Florida citrus growing, packing and processing. Featuring an industry-wide fresh fruit competition, citrus fruit displays, an automated fresh fruit packing line and appearances by Miss Florida Citrus. FFA and 4-H exhibits, crafts and horticultural competitions, livestock exhibits, giant-sized midway and big-name country and western entertainers. Citrus Parade on Jan 28 at 10 AM. Est attendance: 144,000. For info: Christa Deason, Florida Citrus Showcase, PO Box 2008 CAE, Auburndale, FL 33823. Phone: (813) 967-3175.

GRANT, JULIA DENT: BIRTH ANNIVERSARY. Jan 26. Wife of Ulysses Simpson Grant, 18th president of the US. Born at St. Louis, MO, Jan 26, 1826. Died Dec 14, 1902.

INDIA: REPUBLIC DAY: 45th ANNIVERSARY. Jan 26. National holiday. Anniversary of Proclamation of the Republic, Basant Panchmi. On Jan 26, 1929, Indian National Congress resolved to work for establishment of a sovereign republic, a goal that was realized on Jan 26, 1950, when India became a democratic republic.

MacARTHUR, DOUGLAS: 115th BIRTH ANNIVERSARY. Jan 26. US general and supreme commander of Allied forces in Southwest Pacific during World War II. Born at Little Rock, AR, Jan 26, 1880. He served as commander of the Rainbow Division's 84th Infantry Brigade in World War I, leading it in the St. Mihiel, Meuse-Argonne and Sedan offensives. MacArthur died at Washington, DC, Apr 5, 1964. Remembered for his "I will return" prediction made when forced out of the Philippines by the Japanese, a promise he fulfilled. Relieved of Far Eastern command by President Harry Truman on Apr 11, 1951, during the Korean War.

MICHIGAN: ADMISSION DAY: ANNIVERSARY. Jan 26. Became 26th state on this day in 1837.

ZORA NEALE HURSTON FESTIVAL OF THE ARTS AND HUMANITIES. Jan 26-29. Eatonville, FL (10 miles east of Orlando). Festival celebrates Hurston, her work and the community she celebrated with 4 days of cultural arts programs highlighting theatre, music, folklore and literature through seminars, symposia, master classes, intellectual conversations/dialogues, art exhibitions, workshops, self-tours, concert-fare and performances. Festival attracts both "domestic and international travelers looking for 'the other Florida'—the Africa-descendant cultural matrix." Fax: (407) 647-3959. Phone: (407) 647-3307 or (407) 647-4006. Est attendance: 75,000. Sponsor: Assn to Preserve Eatonville, Inc, PO Box 2586, Eatonville, FL 32751.

BIRTHDAYS TODAY

Father George Harold Clements, 63, clergyman who adopted child, civil rights leader, born Chicago, IL, Jan 26, 1932.
Angela Davis, 51, political activist, born at Birmingham, AL, Jan 26, 1944.
Philip Jose Farmer, 77, science fiction writer, born at North Terre Haute, IN, on Jan 26, 1918.
Jules Feiffer, 66, cartoonist, writer, born at New York, NY, Jan 26, 1929.
Scott Glenn, 53, actor (*The Right Stuff, Silverado*), born at Pittsburgh, PA, Jan 26, 1942.
Wayne Gretzky, 34, hockey player, born at Brantford, Ontario, Canada, Jan 26, 1961.
Paul Newman, 70, actor (Oscar for *The Color of Money; Cool Hand Luke, Hud*), born at Cleveland, OH, Jan 26, 1925.
Andrew Ridgeley, 32, musician, born at Bushey, England, Jan 26, 1963.
Gene Siskel, 49, movie critic, born at Chicago, IL, Jan 26, 1946.
Bob Uecker, 60, sportscaster, actor, born at Milwaukee, WI, Jan 26, 1935.
Roger Vadim, 67, filmmaker (*Barbarella, And God Created Woman*), born at Paris, France, Jan 26, 1928.
C. William Verity, 78, business executive, born at Middletown, OH, Jan 26, 1917.

JANUARY 27 — FRIDAY
27th Day — Remaining, 338

AFRO-AMERICAN HISTORY MONTH KICKOFF. Jan 27–28. National kickoff workshop to launch Afro-American History Month. Annually, the last Friday and Saturday in January. Fax: (202) 328-8677 Est attendance: 500. For info: W. Leanna Miles, Managing Dir, The Associated Publishers, Inc, 1407 14th St NW, Washington, DC 20005-3704. Phone: (202) 265-1441.

APOLLO I: SPACECRAFT FIRE: ANNIVERSARY. Jan 27. Three American astronauts, Virgil I. Grissom, Edward H. White and Roger B. Chaffee, died when fire suddenly broke out at 6:31 PM in *Apollo I* during a launching simulation test, as it stood on the ground at Cape Kennedy, FL, Jan 27, 1967. First launching in the Apollo program had been scheduled for Feb 27, 1967.

BACKWARDS DAY. Jan 27. In the rush hour of life, take time to look back on your life's accomplishments. For those daring enough: wear your clothes backwards, eat backwards (dessert first), walk backwards, school children attend classes backwards (last period first). Last comes first. Sponsor: Unity Lutheran School. Annually, the last Friday of January. For info: Unity Lutheran School, Pamela Wolfe, 5401 S Calhoun, Fort Wayne, IN 46807. Phone: (219) 747-2958.

CANADA: ONTARIO WINTER CARNIVAL BON SOO. Jan 27–Feb 5. Sault Ste. Marie, Ontario. One of Canada's largest winter carnivals features more than 100 festive indoor and hearty outdoor events for all ages during a 10-day winter extravaganza. Annually, the last weekend in January-the first weekend in February. Fax: (705) 759-6950. Est attendance: 100,000. For info: Bon Soo Winter Carnival Inc, Donna Gregg, PO Box 781, Sault Ste. Marie, Ont, Canada P6A 5N3. Phone: (705) 759-3000.

CANADA: ROSSLAND WINTER CARNIVAL. Jan 27–29. Rossland, British Columbia. Celebrate winter in Canada's Alpine City with snowmobile races, winter relay races, bobsled races, the British Columbia Cup Luge, a wine festival, dinner and dance socials, casinos and more. Annually, the last weekend in January. Est attendance: 1,000. For info: Rossland Chamber of Commerce, Box 1385, Rossland, BC, Canada V0G 1Y0. Phone: (604) 362-5666.

CARTER CAVES CRAWLATHON. Jan 27–29. Carter Caves State Resort Park, Olive Hill, KY. Weekend of caving for beginners and experienced. Est attendance: 500. For info: Carter Caves State Resort Pk, John Tierney, Naturalist, Olive Hill, KY 41164.

DODGSON, CHARLES LUTWIDGE (LEWIS CARROLL): BIRTH ANNIVERSARY. Jan 27. English mathematician and author, better known by his pseudonym, Lewis Carroll, creator of *Alice's Adventures in Wonderland*, was born at Cheshire, England, on Jan 27, 1832. *Alice* was written for Alice Liddell, daughter of a friend, and first published in 1886. *Through the Looking-Glass*, a sequel, and *The Hunting of the Snark* followed. Dodgson's books for children proved equally enjoyable to adults, and they overshadowed his serious works on mathematics. Dodgson died at Guildford, Surrey, England, on Jan 14, 1898.

EAGLES, ETC. Jan 27–29. Bismarck, AR. To see bald eagles in the wild and to learn about and observe birds of prey. Est attendance: 1,000. For info: DeGray State Park, Park Naturalist, Rte 3, Box 490, Bismarck, AR 71929-8194. Phone: (800) 737-8355.

GOMPERS, SAMUEL: BIRTH ANNIVERSARY. Jan 27. Labor leader, first president of the American Federation of Labor, born at London, England, Jan 27, 1850. Died Dec 13, 1924, at San Antonio, TX.

KERN, JEROME: BIRTH ANNIVERSARY. Jan 27. American composer born at New York, NY, Jan 27, 1885. Died there Nov 11, 1945. In addition to scores for stage and screen are many memorable song hits, including "Who," "Make Believe," "Ol' Man River," "Bill," "The Night Was Made for Love," "Smoke Gets in Your Eyes," "Lovely to Look At," "I Won't Dance," "The Way You Look Tonight," "All the Things You Are," "The Last Time I Saw Paris" and "Dearly Beloved."

MOZART, WOLFGANG AMADEUS: BIRTH ANNIVERSARY. Jan 27. One of the world's greatest music-makers. Born at Salzburg, Austria, Jan 27, 1756. A member of a gifted musical family, he began performing at age three and began composing at age five. Died at Vienna, Austria, Dec 5, 1791. While his birthplace is a shrine, his burial place is unknown.

NATIONAL ACTIVITY PROFESSIONALS DAY. Jan 27. To recognize the contribution of activity directors to the well being of nursing home, adult day care center and senior center clients. Sponsor: Natl Assn of Activity Professionals, Charles Price, Exec Dir, 1225 Eye Street NW, Ste 300, Washington, DC 20005. Phone: (202) 289-0722.

RICKOVER, HYMAN GEORGE: 95th BIRTH ANNIVERSARY. Jan 27. American naval officer. Known as the "Father of the Nuclear Navy." Admiral Rickover directed development of nuclear reactor-powered submarines, the first of which was the *Nautilus*, launched in 1954. Rickover was noted for his blunt remarks: "To increase the efficiency of the Department of Defense," he said, "you must first abolish it." The four-star admiral retired (unwillingly) at the age of 81, after 63 years in the navy. Born in Russia on Jan 27, 1900, Rickover died at Arlington, VA, on July 9, 1986, and was buried at Arlington National Cemetery.

SAINT PAUL WINTER CARNIVAL. Jan 27–Feb 5. St. Paul, MN. Minnesota's largest tourist attraction, the 109-year-old festival provides 10 fun-filled days with more than 100 indoor and outdoor events celebrating the thrills and chills of wintertime fun. Est attendance: 1,500,000. Sponsor: Saint Paul Festival and Heritage Fdtn, 101 Norwest Ctr, 55 E 5th St, St. Paul, MN 55101. Phone: (612) 297-6953.

THOMAS CRAPPER DAY. Jan 27. Possibly apocryphal stories claim this date as the anniversary of the death of Thomas Crapper. Born at Thorpe, Yorkshire, England, in 1837 (exact date unknown). Died Jan 27, 1910. Said to be prime developer of flush toilet mechanism as it is known today. Founder, London, 1861, of Thomas Crapper & Co, later patentees and manufacturers of sanitary appliances; engineers by appointment to His Majesty the King, and HRH, the Prince of Wales. [Editor's Note: The date of Crapper's death has been revised based on info sent by Dr. Andy Gibbons of the Intl Thomas Crapper Society, who has viewed Crapper's gravestone, interviewed a Crapper relation and obtained a copy of Crapper's death certificate. Dr. Gibbons is a professional librarian.]

January 1995

S	M	T	W	T	F	S
1	2	3	4	5	6	7
8	9	10	11	12	13	14
15	16	17	18	19	20	21
22	23	24	25	26	27	28
29	30	31				

VIETNAM WAR ENDED: ANNIVERSARY. Jan 27. US and North Vietnam, along with South Vietnam and the Viet Cong, signed an "Agreement on ending the war and restoring peace in Vietnam." Signed at Paris, France, Jan 27, 1973, to take effect on January 28 at 8 AM Saigon time, thus ending US combat role in a war that had involved American personnel stationed in Vietnam since defeated French forces had departed under terms of the Geneva Accords in 1954. Longest war in US history. More than one million combat deaths (US: 46,079).

BIRTHDAYS TODAY

Mikhail Baryshnikov, 47, ballet dancer, born at Riga, Latvia, USSR, Jan 27, 1948.
Cris Collinsworth, 36, football player, born at Dayton, OH, Jan 27, 1959.
Mairead Corrigan, 51, pacifist, Nobel Peace Prize winner, born at Belfast, Northern Ireland, Jan 27, 1944.
Skitch Henderson, 77, orchestra leader, born at Halstad, MN, Jan 27, 1918.

JANUARY 28 — SATURDAY

28th Day — Remaining, 337

AFTER THE REVOLUTION—EVERDAY LIFE IN AMERICA, 1780–1800. Jan 28–Feb 26. University of Georgia, Tate Student Center, Athens, GA. An intimate view of rural and urban America in the late 1700s, this exhibit traces life in the five regions that made up the US and sheds light on issues that affect Americans into the 19th and 20th centuries by surveying work, social customs, entertainment, conflicts and lives of ordinary Americans. For info: Smithsonian Institution Traveling Exhibition, 1100 Jefferson Dr SW, Ste 3146, Washington, DC 20560. Phone: (202) 357-2700.

BALD EAGLE DAY. Jan 28–29. Nelson Dewey State Park, Cassville, WI. Indoor programs and exhibits. Outdoor viewing of eagles at Riverside Park (Mississippi River), Cassville and Nelson Dewey State Park with knowledgeable attendants. Bus tours around countryside. Candlelight Ski in Nelson Dewey State Park on Jan 28. For info: Mona Hudson, Tourism Coord, Cassville Tourism, PO 576, Cassville, WI 53806. Phone: (608) 725-5399.

BRIDAL EXPO. Jan 28. Adler Theatre, Davenport, IA. Exhibits for every wedding need and a fashion show featuring the latest in all wedding fashions. Est attendance: 5,500. For info: Davenport Jaycees, PO Box 3099, Davenport, IA 52808. Phone: (319) 322-6271.

BROOKFIELD ICE HARVEST. Jan 28. Brookfield, VT. Demonstrations of ice harvesting using the original equipment near the Brookfield Floating Bridge, one of only two such bridges remaining in the US today. Annually, the last Saturday in January. Est attendance: 1,000. For info: Al Wilder, PO Box 405, Brookfield, VT 05036. Phone: (802) 276-3959.

***CHALLENGER* SPACE SHUTTLE EXPLOSION: ANNIVERSARY.** Jan 28. At 11:39 AM, EST, on Jan 28, 1986, the Space Shuttle *Challenger* exploded, 74 seconds into its flight and about 10 miles above the earth. Hundreds of millions around the world watched television replays of the horrifying event that killed seven people, destroyed the billion-dollar craft, suspended all shuttle flights and halted, at least temporarily, much of the US manned space flight program. Killed were teacher Christa McAuliffe (who was to have been the first ordinary citizen in space) and six crew members: Francis R. Scobee, Michael J. Smith, Judith A. Resnik, Ellison S. Onizuka, Ronald E. McNair and Gregory B. Jarvis. See also individual names.

FANCY RAT AND MOUSE ANNUAL SHOW. Jan 28. Hacienda Heights, CA. Annual show where trophies are awarded to the winners. As urban sprawl continues to limit the necessary space needed to keep dogs and cats, rats and mice as pets are gradually emerging as an ideal substitute. They provide all the pleasure and satisfaction of a warm, cuddly, intelligent and friendly pet companion. The American Fancy Rat and Mouse Assn (AFRMA) was founded in 1983 to promote and encourage the breeding and exhibition of fancy rats and mice. For info: AFRMA (CAE), 9230 64th St, Riverside, CA 92509. Phone: (909) 685-2350.

LOUIS ARMSTRONG: CULTURAL LEGACY. Jan 28–Apr 9. Museum of African American Life & Culture, Dallas, TX. With paintings, sculptures, photographs, drawings, print artifacts, music and videos, this exhibit presents the life of one of the world's outstanding musicians and explores the importance of jazz in the history of 20th-century creativity. For info: Smithsonian Institution Traveling Exhibition Service, 1100 Jefferson Dr SW, Ste 3146, Washington, DC 20560. Phone: (202) 357-2700.

NATIONAL KAZOO DAY. Jan 28. To recognize the kazoo as a musical instrument enjoyed by young and old alike and promote its use among stroke patients and others who enjoy music. Sponsor: Leader Nursing and Rehabilitation Ctr, 1070 Stouffer Ave, Chambersburg, PA 17201. Phone: (717) 263-0436.

NATIONAL SPIELING DAY. Jan 28. On this date America celebrates its unofficial national pastime—blathering. All citizens are encouraged to give lengthy lectures to their friends, families and workmates. Speakers should take extra care to discourse in loud, grating voices and to expand upon arcane topics. Listeners should applaud no matter how ear-wrenching the spiel. Fax: (503) 254-5761. Sponsor: Donahue Brothers Productions, Bill Donahue, 7626 SE Alder, Portland, OR 97214. Phone: (503) 254-5315.

SNOW FEST. Jan 28. Pensacola Beach Fishing Pier, Pensacola Beach, FL. Brings a winter theme to the island that has sand as white as snow. Snow is blown onto the beach to create a "blizzard" and a ramp extending into the gulf is converted into a giant snow slide. Est attendance: 25,000. For info: Santa Rosa Island Authority, Dee Lucas, PO Drawer 1208, Pensacola Beach, FL 32562. Phone: (904) 932-2259.

SONORA SHOWCASE. Jan 28–29. Yuma, AZ. A fiesta to promote the state of Sonora, Mexico, features dancers, food and beverage samples, vacation information and curios. For info: Yuma Civic and Conv Ctr, 1440 Desert Hills Dr, Yuma, AZ 85364. Phone: (602) 344-3800.

STANLEY, HENRY MORTON: BIRTH ANNIVERSARY. Jan 28. Explorer. Born in Wales, Jan 28, 1841. Leader of African expedition to find the missing missionary-explorer David Livingstone, who had not been heard from for more than two years. He began his search in Africa on Mar 21, 1871. Stanley found him at Ujiji, near Lake Tanganyika, on Nov 10, 1871, and his first words are said to have been the now-famous phrase: "Dr. Livingstone, I presume?" Stanley died at London, England, May 10, 1904.

WINTER FEST. Jan 28–29. Lake City, MN. Snowmobile events, winter golf classic, arts and crafts, cross-country ski race, taste fest, sleigh and cutter parade, snowmobile torchlight parade, ice skating and fireworks. Est attendance: 5,000. For info: Lake City Area Chamber of Commerce, 212 S Washington St, Box 150, Lake City, MN 55041. Phone: (800) 369-4123.

Jan 28–30 ☆ Chase's 1995 Calendar of Events ☆

BIRTHDAYS TODAY

Alan Alda (Alphonso D'Abruzzo), 59, actor (*Paper Lion, The Four Seasons, M*A*S*H*), director, born New York, NY, Jan 28, 1936.
Susan Howard (Jeri Lynn Mooney), 52, actress (Donna Culver Krebbs on "Dallas"), born at Marshall, TX, Jan 28, 1943.
Claes Oldenburg, 66, artist, sculptor, (painted steel, *Standing Trowel*), born at Stockholm, Sweden, Jan 28, 1929.
Susan Sontag, 62, author (*Against Interpretation, The Volcano Lover: A Romance*), born at New York, NY, Jan 28, 1933.

JANUARY 29 — SUNDAY
29th Day — Remaining, 336

CHEKHOV, ANTON PAVLOVICH: 135th BIRTH ANNIVERSARY. Jan 29. Russian playwright and short story writer, especially remembered for *The Sea Gull, The Three Sisters* and *The Cherry Orchard*. Born at Taganrog, Russia, on Jan 29, 1860 (New Style). Died on July 15, 1904, at the Black Forest spa at Badenweiler, Germany.

HOBBY INDUSTRY ASSOCIATION ANNUAL TRADE SHOW. Jan 29–Feb 1. Anaheim Convention Center, Anaheim, CA. 54th annual showcase for wholesale and retail buyers of craft and hobby products. Est attendance: 16,000. For info: Hobby Industry Association, 319 E 54th St, Elmwood Park, NJ 07407. Phone: (201) 794-1133.

KANSAS: ADMISSION DAY. Jan 29. Became 34th state on this day in 1861.

McKINLEY, WILLIAM: BIRTH ANNIVERSARY. Jan 29. 25th president of the US. Born at Niles, OH, Jan 29, 1843. Died in office, at Buffalo, NY, Sept 14, 1901, as the result of a gunshot wound by an anarchist assassin on Sept 6, 1901, while he was attending Pan American Exposition.

NATIONAL PUZZLE DAY. Jan 29. To recognize different puzzles and games and their creators. Call or write for free information on the origins and creators of puzzles and games. Est attendance: 10,000. For info: Carol Handz, Coord, Jodi Jill Features, 1705 14th St, Ste 321, Boulder, CO 80301. Phone: (303) 575-1319.

PAINE, THOMAS: BIRTH ANNIVERSARY. Jan 29. American Revolutionary leader, a corset-maker by trade, author of *Common Sense, The Age of Reason* and many other influential works. Born at Thetford, England, Jan 29, 1737. Paine died at New York, NY, June 8, 1809, and his remains were moved to England by William Cobbett for reburial there in 1819. Reburial was refused, and the location of Paine's bones, said to have been distributed, is unknown. "These are the times that try men's souls" are opening words of his inspirational tract *The Crisis*.

SPACE MILESTONE: SATURN SA-5 (US). Jan 29. Orbits record 19 tons. Launched on Jan 29, 1964.

SUPER BOWL XXIX. Jan 29. Joe Robbie Stadium, Miami, FL. Est attendance: 75,000. For info: The Natl Football League, PR Dept, 410 Park Ave, New York, NY 10022. Phone: (212) 758-1500.

SWEDENBORG, EMANUEL: BIRTH ANNIVERSARY. Jan 29. Born at Stockholm, Sweden, on Jan 29, 1688, Swedenborg is remembered as a scientist, inventor, writer and religious leader. He died at London, England, on Mar 29, 1772.

January 1995

S	M	T	W	T	F	S
1	2	3	4	5	6	7
8	9	10	11	12	13	14
15	16	17	18	19	20	21
22	23	24	25	26	27	28
29	30	31				

BIRTHDAYS TODAY

John Forsythe (John Freund), 77, actor ("Bachelor Father"; Charlie's voice on "Charlie's Angels"; "Dynasty"), born Penn's Grove, NJ, Jan 29, 1918.
Germaine Greer, 56, author (*Daddy We Hardly Knew You, The Change*), born at Melbourne, Victoria, Australia, Jan 29, 1939.
Martha Wright Griffiths, 83, political leader, congresswoman, advocate for feminist issues, born at Pierce City, MO, Jan 29, 1912.
Ann Jillian, 44, actress ("It's a Living," "The Ann Jillian Story"), born at Cambridge, MA, Jan 29, 1951.
Stacey (Ronald) King, 28, basketball player, born at Lawton, OK, Jan 29, 1967.
Claudine Longet, 53, actress, born at Paris, France, Jan 29, 1942.
Greg Louganis, 35, athlete, actor, born at San Diego, CA, Jan 29, 1960.
Katharine Ross, 52, actress (Elaine Robinson in *The Graduate*), born at Los Angeles, CA, Jan 29, 1943.
Steve Sax, 35, baseball player, born at Sacramento, CA, Jan 29, 1960.
Tom Selleck, 50, actor ("Magnum, P.I."; *Three Men and a Baby, Mr. Baseball*), born at Detroit, MI Jan 29, 1945.
Oprah Winfrey, 41, TV talk show hostess, born at Kosciusko, MS, Jan 29, 1954.

JANUARY 30 — MONDAY
30th Day — Remaining, 335

AUSTRALIA: AUSTRALIA DAY OBSERVANCE. Jan 30. Public holiday. Commemorates beginning of settlement when Governor Phillip landed at Sydney Cove on Jan 26, 1788. When the date falls other than on a Monday, the holiday is held on the Monday following Jan 26, in order to give a long weekend. First proclaimed a public holiday in 1838. See also: "Australia: Australia Day" (Jan 26).

BEATLES LAST PUBLIC APPEARANCE: ANNIVERSARY. Jan 30. On this day in 1969 The Beatles performed together in public for the last time. The show took place on the roof of their Apple Studios in London, England, but it was interrupted by police after they received complaints from the neighbors about the noise.

BLOODY SUNDAY: ANNIVERSARY. Jan 30. In Derry, Northern Ireland, on Jan 30, 1972, 13 Roman Catholics were shot dead by British troops during a banned civil rights march. During 1972, which was the first year of British direct rule, 467 people were killed in the fighting.

CHARLES I EXECUTION: ANNIVERSARY. Jan 30. English king beheaded by order of Parliament under Oliver Cromwell Jan 30, 1649; considered a martyr by some.

DIXIE NATIONAL LIVESTOCK SHOW & RODEO. Jan 30–Feb 20. Jackson, MS. Largest rodeo and second largest livestock show east of the Mississippi River. Est attendance: 250,000. For info: Mississippi Fair Commission, PO Box 892, Jackson, MS 39205. Phone: (601) 961-4000.

☆ Chase's 1995 Calendar of Events ☆   Jan 30–31

FIRST BRAWL IN THE US HOUSE OF REPRESENTATIVES: ANNIVERSARY. Jan 30. The first brawl to break out on the floor of the US House of Representatives occurred in Philadelphia, PA, on Jan 30, 1798. The fight was precipitated by an argument between Matthew Lyon of Vermont and Roger Griswold of Connecticut. Lyon spat in Griswold's face. Although a resolution to expel Lyon was introduced, the measure failed and Lyon maintained his seat.

GANDHI, MOHANDAS: ASSASSINATION: ANNIVERSARY. Jan 30. Indian religious and political leader, assassinated at New Delhi, India, Jan 30, 1948. The assassin was a Hindu extremist, Ram Naturam. See also: "Gandhi, Mohandas Karamchand (Mahatma): Birth Anniversary" (Oct 2).

GRANDEUR, SIMPLICITY AND CONVENIENCE: THE US CAPITOL: 1793-1993. Jan 30–Mar 13. Gerald R. Ford Museum, Grand Rapids, MI. Organized by the American Architectural Foundation and the Architect of the Capitol, this exhibit consists of framed images and text exploring the architectural history of the Capitol building. Est attendance: 13,000. For info: Gerald R. Ford Museum, 303 Pearl St NW, Grand Rapids, MI 49504-5353. Phone: (616) 451-9263.

MOON PHASE: NEW MOON. Jan 30. Moon enters New Moon phase at 5:48 PM, EST.

QUILT COMPETITION AND DISPLAY. Jan 30–Apr 2. Peddler's Village, Lahaska, PA. Create a beautiful quilt and compete for cash prizes. Six categories: Traditional, Applique, Amish, Machine Quilted, Amateur and Quilted Clothing. Winning entries displayed in Gazebo. Open daily to public. Free admission. Est attendance: 3,000. For info: Peddler's Village, Box 218, Lahaska, PA 18931. Phone: (215) 794-4000.

ROOSEVELT, FRANKLIN DELANO: BIRTH ANNIVERSARY. Jan 30. 32nd president of the US (Mar 4, 1933–Apr 12, 1945). Only president to serve more than two terms; elected to four consecutive terms. He supported the Allies in WWII before US entered the struggle by supplying them with war materials through the Lend-Lease Act; and he became deeply involved in broad decision-making after the attack on Pearl Harbor Dec 7, 1941. Born at Hyde Park, NY, Jan 30, 1882. Died in office at Warm Springs, GA, Apr 12, 1945.

TET OFFENSIVE BEGINS: ANNIVERSARY. Jan 30. After calling for a cease-fire during the Tet holiday celebrations, North Vietnam and the National Liberation Front launched a major offensive throughout South Vietnam on Jan 30, 1968, during the Tet celebrations. The attacks erupted in thirty-six of the forty-four provincial capitals and five of the six major cities. In addition, the Vietcong attacked the US embassy in Saigon, Tan Son Nhut Air Base, the presidential palace and South Vietnam general staff headquarters. Suffering as many as 40,000 battlefield deaths, the offensive was a tactical defeat for the Vietcong and North Vietnam. The South Vietnamese held their ground and the US was able to airlift troops into the critical areas and quickly regain control. However, the offensive is credited as a strategic success in that it continued the demoralization of American public opinion. After Tet, American policy toward Vietnam shifted from winning the war to seeking an honorable way out.

TEXAS CITRUS FIESTA. Jan 30–Feb 5. Theme: "A Slice of Nostalgia." Fiesta Fun Fair: Arts, crafts, exhibits, races and music. Coronation of Queen Citrianna and King Citrus, product costume shows and Parade of Oranges. Product Style Show features participants wearing their original creations, adorned with Valley products. Est attendance: 100,000. For info: Texas Citrus Fiesta, Box 407, Mission, TX 78573. Phone: (210) 585-9724.

TUCHMAN, BARBARA W.: BIRTH ANNIVERSARY. Jan 30, 1912. Historian and journalist Barbara Tuchman's most famous works were her Pulitzer Prize-winning books *The Guns of August* (1962) and *Stilwell and the American Experience in China, 1911–45* (1971). Tuchman's books of history were known for making history live, never dry. Other well known books included *The Proud Tower* (1966) and *The First Salute* (1988). Barbara Wertheim Tuchman was born at New York City and died at Greenwich, CT, Feb 6, 1988.

BIRTHDAYS TODAY

Richard B. Cheney, 54, US Secretary of Defense in Bush administration, born at Lincoln, NE, Jan 30, 1941.
Gene Hackman, 65, actor (*French Connection, Bonnie and Clyde*), born at San Bernadino, CA, Jan 30, 1930.
Dick Martin, 73, comedian, actor (Emmy for "Rowan & Martin's Laugh-In"), born at Detroit, MI, Jan 30, 1922.
Vanessa Redgrave, 58, actress (*Mary Queen of Scots, Julia*), born at London, England, Jan 30, 1937.
Boris Spassky, 58, chess player, journalist, born at Leningrad, USSR, Jan 30, 1937.
Curtis Strange, 40, golfer, born at Norfolk, VA, Jan 30, 1955.
Jody Watley, 34, singer, born at Chicago, IL, Jan 30, 1961.

JANUARY 31 — TUESDAY
31st Day — Remaining, 334

BIRMINGHAM SPORT AND BOAT SHOW. Jan 31–Feb 5. Birmingham/Jefferson Civic Center, Birmingham, AL. For info: Double C Productions, Inc, Box 1678, Huntsville, TX 77342. Phone: (409) 295-9677.

CANADA'S COLDEST RECORDED TEMPERATURE: ANNIVERSARY. Jan 31. At Whitehorse, in Canada's Yukon Territory, a temperature of 62 degrees below zero (Fahrenheit) was recorded on Jan 31, 1947, a record low for all of Canada's provinces and territories. Next lowest temperature was −60°F at Yellowknife, in the Northwest Territories, recorded also on Jan 31, 1947.

CHINESE NEW YEAR. Jan 31. Traditional Chinese lunar year begins at sunset on day of second New Moon following the winter solstice. Begins year 4693 of the ancient Chinese calendar, designated as the Year of the Pig. Generally celebrated until the Lantern Festival fifteen days later, but merchants usually reopen their stores and places of business on the fifth day of the first lunar month. See also: "China: Lantern Fesival" (Feb 14).

CHINESE NEW YEAR FESTIVAL. Jan 31–Feb 18. San Francisco, CA. North America's largest Chinese community salutes the Year of the Pig, 4693 on the lunar calendar (Jan 31, 1995). Set in the heart of traditional Chinatown, the Chinatown street fair showcases the diversity of Chinese culture. Stages feature Chinese opera and ballet, martial arts, fashion shows of ancient dynastic costumes and traditional dance. Booths feature cooking demonstrations, calligraphy and arts and crafts. The famous Golden Dragon Parade takes place on Feb 18 at 6 PM. Est attendance: 400,000. For info: Chinese Chamber of Commerce, New Year Festival, 809 Montgomery St, San Francisco, CA 94133-5142. Phone: (415) 391-9680.

Jan 31 ☆ Chase's 1995 Calendar of Events ☆

FIRST MISS ALBANIA CROWNED: ANNIVERSARY. Jan 31. Valbona Selimllari, an unemployed 19-year-old, was crowned the first Miss Albania on Jan 31, 1992, in the first beauty pageant organized in Europe's poorest nation. Selimllari, of the capital city of Tirana, received a trip to Italy, France and Germany and $600, which is approximately equal to three years' average wages in Albania.

GREY, ZANE: 120th BIRTH ANNIVERSARY. Jan 31. American dentist who turned to writing. Author of numerous stories of the American western frontier. Translated into many languages and sold more than 10 million copies. *Riders of the Purple Sage* is one of his best remembered. Born at Zanesville, OH, on Jan 31, 1875. Died at Altadena, CA, Oct 23, 1939.

IRONCLADS BATTLE IN CHARLESTON HARBOR: ANNIVERSARY. Jan 31. On Jan 31, 1863, two Confederate ironclad ships, the *Chicora* and *Palmetto State*, severely damaged two Union ships participating in the blockade of Charleston, SC, harbor. Despite the successful battle, the harbor remained under blockade, although the South declared otherwise.

NAURU: NATIONAL HOLIDAY. Jan 31. Republic of Nauru.

ROBINSON, JACKIE: BIRTH ANNIVERSARY. Jan 31. Jack Roosevelt Robinson, athlete and business executive, first black to enter professional major league (Brooklyn Dodgers, 1947) baseball. Born at Cairo, GA, Jan 31, 1919. Played with Brooklyn Dodgers 1947–1956. Voted National League's Most Valuable Player in 1949 and elected to the Baseball Hall of Fame in 1962. Died at Stamford, CT, Oct 24, 1972. A special commemorative Jackie Robinson postage stamp was issued in the Black Heritage USA Series in 1982.

SCHUBERT, FRANZ: BIRTH ANNIVERSARY. Jan 31. Composer, born at Vienna, Austria, on Jan 31, 1797. Died there, of typhus, on Nov 19, 1828. Buried, at his request, near the grave of Beethoven. Schubert finished his work on his "Unfinished Symphony" (No 8) in 1822. On the hundredth anniversary of his death, 1928, a $10,000 prize was offered to "finish" the work. The protests were so great that the offer was withdrawn.

SCOTLAND: UP HELLY AA. Jan 31. Lerwick, Shetland. Norse galley burned in impressive ceremony symbolizing sacrifice to sun. Old Viking custom. Annually, the last Tuesday in January.

SLOVIK, EDDIE D.: 50th EXECUTION ANNIVERSARY. Jan 31. Anniversary of execution by firing squad in Jan 1945 of 24-year-old Private Eddie D. Slovik. Born at Detroit, MI, Feb 18, 1920. Member of Company G, 109th Infantry, 28th Division, US Army. His death sentence, the first for desertion since the Civil War, has been a subject of controversy since. First buried in France, Slovik's remains were exhumed in 1987 for reburial beside his wife, Antoinette, who died in 1979, after years of effort to clear Slovik's name and have his body returned to the US.

SPACE MILESTONE: *APOLLO 14* (US). Jan 31. Launched Jan 31, 1971. Astronauts Alan B. Shepard, Jr, and Edgar D. Mitchell landed on moon (Lunar Module "Antares") Feb 5. Command Module "Kitty Hawk" piloted by Stuart A. Roosa. Pacific splashdown on Feb 9.

SPACE MILESTONE: *LUNA 9* (USSR). Jan 31. First soft landing on Moon, Feb 3. Transmitted photos. Launched Jan 31, 1966.

SPACE MILESTONE: PROJECT MERCURY TEST (US). Jan 31. First US recovery of large animal from space; Ham, the chimpanzee, successfully transmitted signals. Launched Jan 31, 1961.

SPACE MILESTONE: UNITED STATES SPACE EXPLORATION: ANNIVERSARY. Jan 31. *Explorer I*, the first successful US satellite, was launched Jan 31, 1958. Weighing 31 lbs, *Explorer I* transmitted radio signals until May 23, 1958, and discovered the Van Allen Belt. Decayed Mar 31, 1970.

WAR IN EL SALVADOR OFFICIALLY ENDED: ANNIVERSARY. Jan 31. On Jan 31, 1992, the army of El Salvador officially announced the end of its 12-year war with leftist insurgents. The announcement took place in a huge ceremony as the army informed President Alfredo Cristiani of the end of hostilities that had claimed 75,000 lives. After the announcement the rebel insurgents flew home from Mexico and were met by a flag-waving, jubilant crowd.

BIRTHDAYS TODAY

Ernie Banks, 64, Hall of Fame shortstop, Chicago Cubs (Mr. Cub), born at Dallas, TX, Jan 31, 1931.

Carol Channing (Carol Channing Lowe), 72, actress (*Hello, Dolly!*; *Thoroughly Modern Millie*), born at Seattle, WA, Jan 31, 1923.

Phil Collins, 44, musician, singer, songwriter, born at Chiswick, England, Jan 31, 1951.

Norman Mailer, 72, author (*The Executioner's Song, The Naked and the Dead*), born at Long Branch, NJ, Jan 31, 1923.

Suzanne Pleshette, 58, actress (Emily Hartley in "The Bob Newhart Show"), born at New York, NY, Jan 31, 1937.

Nolan Ryan, 48, former baseball player, born at Refugio, TX, Jan 31, 1947.

Chase's 1995 Calendar of Events — Feb 1

Februarie.

FEBRUARY 1 — WEDNESDAY
32nd Day — Remaining, 333

AFRO-AMERICAN HISTORY MONTH. Feb 1-28. Traditionally the month containing Abraham Lincoln's birthday (Feb 12) and Frederick Douglass's presumed birthday (Feb 14). Observance of a special period to recognize achievements and contributions by Afro-Americans dates from Feb 1926, when it was launched by Dr. Carter G. Woodson and others. Variously designated Negro History, Black History, Afro-American History, Black Heritage and Black Expressions, the observance period was initially for a week, but since 1976 for the entire month of February. 1995 Theme: "Reflections on 1895: Douglass, Dubois, Washington." Fax (202) 328-8677. Est attendance: 500. For info: W. Leanna Miles, Managing Dir, The Associated Publishers, Inc, 1407 14th St NW, Washington, DC 20005. Phone: (202) 265-1441.

AMD AWARENESS MONTH. Feb 1-28. More than 13 million Americans age 40 and older show signs of age-related macular degeneration. AMD is now the leading cause of blindness among older Americans. During this month, Prevent Blindness America (formerly known as National Society to Prevent Blindness) steps up its efforts to inform people about AMD. Sponsor: Prevent Blindness America, Marita Gomez, Media Relations, 500 E Remington Rd, Schaumburg, IL 60173. Phone: (800) 331-2020.

★ **AMERICAN HEART MONTH.** Feb 1-28. Presidential Proclamation always issued each year for February since 1964. (PL88-254 of Dec 30, 1963.)

AMERICAN HEART MONTH. Feb 1-28. Volunteers across the country spend one to four weeks canvassing neighborhoods and providing educational information about heart disease and stroke. 1995 campaign will focus on how research saves lives. Call local AHA offices at (800) AHA-USA1.

AMERICAN HISTORY MONTH. Feb 1-28. Massachusetts.

AMERICAN HISTORY MONTH. Feb 1-28. Individual DAR chapters plan activities to promote the study of American history. Events vary but might include programs, displays, spot announcements and recognition of the annual American History Essay Contests winners. Sponsor: Natl Soc Daughters of the American Revolution, Historian-General, Admin Bldg, 1776 D St NW, Washington, DC 20006-5392. Phone: (202) 628-1776.

February 1995

S	M	T	W	T	F	S
			1	2	3	4
5	6	7	8	9	10	11
12	13	14	15	16	17	18
19	20	21	22	23	24	25
26	27	28				

BE AN ENCOURAGER® DAY. Feb 1. "An Encourager"® literally means "one who fills the heart"; hence this is a special day to share heartfelt humor and encouragement with family, friends and co-workers. Annually, on February 1. For info: Liz Curtis Higgs, "An Encourager,"® PO Box 43577, Louisville, KY 40253-0577. Phone: (502) 254-5454.

CANNED FOOD MONTH. Feb 1-28. National trade promotion to promote the nutrition, convenience and great taste of canned foods by providing information to consumers and the food industry through special events, publications and in-store promotions. Sponsor: Canned Food Info Council, 500 N Michigan Ave, Ste 200, Chicago, IL 60611. Phone: (312) 836-7279.

◆ **CARAWAY, HATTIE WYATT: BIRTH ANNIVERSARY.** Feb 1, 1878. Born in Bakersville, TN, Hattie Caraway became a US senator from Arkansas when her husband died in 1931 and she was appointed to fill out his term. The following year she ran for the seat herself and became the first woman elected to the US Senate. She served 14 years there, becoming an adept and tireless legislator (once introducing 43 bills on the same day) who worked for women's rights (once cosponsoring an equal rights amendment), supported New Deal policies as well as Prohibition and opposed the increasing influence of lobbyists. Caraway died at Falls Church, VA, Dec 21, 1950.

THE CAROLINA CAMPAIGN BEGINS: 130th ANNIVERSARY. Feb 1. Union troops under General William Tecumseh Sherman began to move into South Carolina from their positions around Savannah, GA, on Feb 1, 1865. On this march through South Carolina, Federal troops destroyed not only military targets but much private property in an attempt to punish the first state to secede from the Union.

CELEBRATION OF CHOCOLATE. Feb 1-28. Afternoon and evening reservations for private Chocolate Tea Parties. Collection of hot chocolates, imported and domestic boxed chocolates, specialty fresh-baked Pinehill chocolate desserts, gourmet fudge collection, etc. For info: Sharon Burdick, Innkeeper, Pinehill Bed & Breakfast, 400 Mix St, Oregon, IL 61061. Phone: (815) 732-2061.

CONSERVATION AWARENESS PROGRAM. Feb 1-Nov 30. Sea Life Park Hawaii, Makapuu Point, Waimanalo, HI. Ongoing environmental programs include special fairs, lectures, presentations, demonstrations and activities in the park. Community classes for the public, including family groups, pre-schoolers, elementary students and adults. Est attendance: 40,000. For info: Marilyn Lee, Curator of Education, Sea Life Park Hawaii, Makapuu Point, Waimanalo, HI 96795. Phone: (808) 259-7933.

CREATIVE ROMANCE MONTH. Feb 1-28. To encourage couples to keep the sizzle in their relationships by celebrating romance in unique ways; to remind couples that a change of location and a sprinkle of intrigue can turn an ordinary occurrence into a romantic adventure. Phone: (800) 368-7978 or (714) 459-7620. For info: Eileen Buchheim, Celebrate Romance, 5199 E Pacific Coast Hwy, Ste 303A, Long Beach, CA 90804.

FORD, JOHN: 100th BIRTH ANNIVERSARY. Feb 1. Film director John Ford was born Feb 1, 1895, at Cape Elizabeth, ME, as Sean Aloysius O'Feeney. After moving to Hollywood he took the name of John Ford. His first feature film was *Silver Wings* in 1922. He won his first Academy Award in 1935 for *The Informer*. His other films include *The Plough and the Stars, Mary of Scotland, Stagecoach, Young Mr. Lincoln, Grapes of Wrath, How Green Was My Valley, Fort Apache, Three Godfathers, She Wore a Yellow Ribbon, Rio Grande, The Quiet Man, What Price Glory?, The Searchers, Mister Roberts* and *The Man Who Shot Liberty Valance* among many others. During World War II he served as chief of the Field Photographic Branch of the OSS. Two of his documentaries made during the war earned him Academy Awards. He died on Aug 31, 1973, at Palm Desert, CA.

Feb 1 ☆ *Chase's 1995 Calendar of Events* ☆

FREEDOM DAY: 130th ANNIVERSARY. Feb 1. Anniversary of President Abraham Lincoln's approval, Feb 1, 1865, of the 13th Amendment to the US Constitution (abolishing slavery). "1. Neither slavery nor involuntary servitude, except as a punishment for crime whereof the party shall have been duly convicted, shall exist within the United States or any place subject to their jurisdiction. 2. Congress shall have power to enforce this article by appropriate legislation." The amendment had been proposed by the Congress on Jan 31, 1865; ratification was completed Dec 18, 1865.

GABLE, CLARK: BIRTH ANNIVERSARY. Feb 1. William Clark Gable's first film was *The Painted Desert* in 1931 during the era when talking films were replacing silent films. He won an Academy Award for his role in the comedy *It Happened One Night*, which established him as a romantic screen idol. Other films included *China Seas, Mutiny on the Bounty, Saratoga, Command Decision, Run Silent Run Deep*, and *Gone with the Wind*, for which his casting as Rhett Butler seemed a foregone conclusion due to his popularity as the acknowledged "King of Movies." Gable was born on Feb 1, 1901, at Cadiz, OH, and died on Nov 16, 1960, shortly after completing his last film, Arthur Miller's *The Misfits*, in which he starred with Marilyn Monroe.

GREAT AMERICAN PIES MONTH. Feb 1-28. To encourage home preparation of America's favorite dessert, emphasizing the convenient ingredients available to make pie-making contemporary and easy. Sponsor: Eagle® Brand Sweetened Condensed Milk, Borden, Inc, Veronica Petta, Mgr Product Publicity, 180 E Broad St, Columbus, OH 43215-3367. Phone: (614) 225-4037.

GREENSBORO SIT-IN: 35th ANNIVERSARY. Feb 1. Commercial discrimination against blacks and other minorities provoked a nonviolent protest, Feb 1, 1960. In Greensboro, NC, four students from the Agricultural and Technical College at Greensboro (Ezell Blair, Jr, Franklin McCain, Joseph McNeill and David Richmond) sat down at a Woolworth store lunch counter and ordered coffee. Refused service, they remained all day. The following days similar sit-ins took place at the Woolworth lunch counter. Before the week was over they were joined by a few white students. The protest spread rapidly, especially in southern states. More than 1,600 persons were arrested before the year was over for participating in them. Civil rights for all became a cause for thousands of students and activists. In response, equal accommodation regardless of race reached lunch counters, hotels and business establishments in thousands of places where it had been unknown before.

HUMAN RELATIONS MONTH. Feb 1-28. New Hanover County, NC. A month to promote interaction and understanding among people of different races, beliefs, cultures and needs—and celebrate those things that unify us as a people. Traditional events include an awards gala, theatrical productions, school contests, interfaith service and seminars. Co-sponsored by the Greater Wilmington Chamber of Commerce and New Hanover Human Relations Commission. Est attendance: 400. For info: New Hanover Human Relations Commission, 320 Chestnut St, Ste 409, Wilmington, NC 28401. Phone: (910) 341-7171.

HUMPBACK WHALE AWARENESS MONTH. Feb 1-28. Sea Life Park Hawaii, Waimanalo, HI. Lecture and slide presentations by marine mammal experts; Hawaii Marine Youth Artist Competition and Exhibit; daily informative Discover the Sea cart on humpback whales. Est attendance: 4,000. For info: Sea Life Park Hawaii, Makapuu Point, Waimanalo, HI 96795. Phone: (808) 259-7933.

INTERNATIONAL "BOOST" YOUR SELF-ESTEEM MONTH. Feb 1-28. A month to focus on the importance of building self-esteem to beat the winter blahs, to boost morale and to inspire yourself and others to seize new challenges. For info: Valla Dana Fotiades, M.Ed., PO Box 812, West Side Stn, Worcester, MA 01602-0812. Phone: (508) 799-9860.

INTERNATIONAL EMBROIDERY MONTH. Feb 1-28. A time to promote commercial monogramming and embroidery and heighten apparel consumers' awareness of the role that this industry plays in fashion, home furnishings, advertising and sports around the world. Fax: (303) 793-0454. Sponsor: Melissa J. Thompson Maher, Exec Ed, Stitches Magazine, 5660 Greenwood Plaza Blvd., Ste 350, Englewood, CO 80111. Phone: (303) 793-0448.

MARFAN SYNDROME AWARENESS MONTH. Feb 1-28. Volunteers across the country distribute heart-shaped objects and educational information about the Marfan syndrome, an inheritable disorder of connective tissue that can result in life-threatening cardiovascular problems as well as orthopedic and ophthalmologic handicaps. Annually, the month of February. Sponsor: National Marfan Foundation, 382 Main St, Port Washington, NY 11050. Phone: (800) 862-7326.

MUSLIM HOLIDAY: RAMADHAN. Feb 1-2. Beginning of month of fasting. Muslim calendar date: Ramadhan 1, 1414. Gregorian calendar date of observance may vary by one day. The month of Ramadhan traditionally begins when Muslim sheiks in Saudi Arabia sight the new moon. During this holy month, Muslims abstain from food, drink, smoking, gambling and sex from sunrise to sunset. Different methods for calculating the Gregorian date are used by different Muslim sects or groups. *Chase's* dates are based on astronomical calculations of the visibility of the new moon crescent at Mecca; EST date may vary.

NATIONAL CHERRY MONTH. Feb 1-28. To publicize the colorful red tart cherry. Recipes, posters, and table tents available. For info: Cherry Marketing Institute, 2220 University Park Drive, Okemos, MI 48864. Phone: (517) 347-0010.

NATIONAL CHILDREN'S DENTAL HEALTH MONTH. Feb 1-28. To increase dental awareness and stress the importance of regular dental care. To purchase materials, phone in US: (800) 621-8099. Sponsor: American Dental Assn, 211 E Chicago Ave, Chicago, IL 60611.

NATIONAL FIBER FOCUS MONTH. Feb 1-28. To recognize the contributions fiber makes toward maintaining good health. For info: Kathryn Newton, Fiber One Cereal, General Mills, Inc, PO Box 1113, Minneapolis, MN 55440. Phone: (612) 540-2469.

★ **NATIONAL FREEDOM DAY.** Feb 1. Presidential Proclamation 2824, Jan 25, 1949, covers all succeeding years (PL80-842 of June 30, 1948).

NATIONAL SNACK FOOD MONTH. Feb 1-28. Potato chips and other crunchy munchies are promoted as "fun foods" in radio ads, radio contests, publicity and in-store promotions. Sponsors: Snack Food Assn and The Potato Board. For info: Kimberly Testerman, Comm Assist, Snack Food Assn, 1711 King St, Alexandria, VA 22314. Phone: (703) 836-4500.

NATIONAL WEDDINGS MONTH. Feb 1-28. As the "weddings season" gets into high gear, this observance is to call attention to the fact that approximately 2.5 million weddings are celebrated in the US each year. FAX: (203) 354-1404. Sponsor: Assn of Bridal Consultants, Gerard J. Monaghan, Pres, 200 Chestnutland Rd, New Milford, CT 06776-2521. Phone: (203) 355-0464.

February 1995

S	M	T	W	T	F	S
			1	2	3	4
5	6	7	8	9	10	11
12	13	14	15	16	17	18
19	20	21	22	23	24	25
26	27	28				

☆ Chase's 1995 Calendar of Events ☆   Feb 1-2

NATIONAL WILD BIRD FEEDING MONTH. Feb 1-28. To recognize that February is one of the most difficult winter months in much of the US for birds to survive in the wild, and to encourage people to provide food, water and shelter to supplement the wild birds' natural diet of weed seeds and harmful insects. Sponsor: Sue Wells, Exec Dir, National Bird-Feeding Society, 2218 Crabtree Lane, Northbrook, IL 60062-3520. Phone: (708) 272-0135.

NEW ORLEANS BOAT & SPORTFISHING SHOW. Feb 1-5. Louisiana Superdome, New Orleans, LA. Boat and marine products, fishing equipment and resort info. Display and sales. For info: Natl Marine Manufacturers Assn, Sherron Smith, 4051 Veterans Memorial Blvd, Metairie, LA 70002. Phone: (504) 885-9709.

NO LAUGHING MATTER: POLITICAL CARTOONISTS ON THE ENVIRONMENT. Feb 1-Apr 2. Community Memorial Museum, Yuba City, CA. Political cartoonists address environmental issues in this exhibition featuring the work of editorial cartoonists from more than 20 countries in addition to the US. Cartoons transcend cultural barriers, and by using simple graphic images, metaphors and humor, they can stimulate thought and incite action on the truly global environmental issues. Call or write the Smithsonian for other dates and venues. For info: Smithsonian Institution Traveling Exhibition Service, 1100 Jefferson Dr SW, Ste 3146, Washington, DC 20560. Phone: (202) 357-2700.

RESPONSIBLE PET OWNER MONTH. Feb 1-28. To highlight the major components of proper pet care, this ASPCA-sponsored observance focuses on such important issues as health, exercise, veterinary care, grooming and nutrition. Info packets are available; send SASE. Phone: (212) 876-7700, ext 4655. Sponsor: American Soc for the Prevention of Cruelty to Animals, (ASPCA), 424 E 92nd St, New York, NY 10128.

RETURN SHOPPING CARTS TO THE SUPERMARKET MONTH. Feb 1-28. Chicago, IL. A month-long opportunity to return stolen shopping carts, milk crates, bread trays and ice cream baskets to supermarkets and avoid the increased food prices that these thefts cause. Annually, the month of February. Sponsor: Illinois Food Retailers Assn. For info: Anthony A. Dinolfo, Grocer-"Retired," 8148 S Homan Ave, Chicago, IL 60652. Phone: (312) 737-6540.

ROBINSON CRUSOE DAY. Feb 1. Anniversary of the rescue, Feb 1, 1709, of Alexander Selkirk, Scottish sailor who had been put ashore (in Sept 1704) on the uninhabited island Juan Fernandez at his own request after a quarrel with his captain. His adventures formed the basis for Daniel Defoe's book *Robinson Crusoe*. A day to be adventurous and self-reliant.

SLEEP SAFETY MONTH. Feb 1-28. Established by the Sleep Products Safety Council (SPSC) to focus attention on fire safety in the bedroom. Programs emphasize a few key messages: do not smoke in bed, keep matches and lighters out of children's reach, check that smoke alarms are operational and develop a fire escape plan. For info: Patricia Martin, Dir, Sleep Products Safety Council, 333 Commerce, Alexandria, VA 22314. Phone: (703) 683-8371.

WISCONSIN FARM WOMAN OF THE YEAR MONTH. Feb 1-28. Recognition of farm women for their great importance and role in agriculture. They milk cows, do field work, run errands, pay bills, take care of children, make meals plus other household chores. Because of her dedication, a program has been developed to honor the farm woman. The local radio stations hold a contest for county winners. Each station chooses one farm woman from the county winners to be a representative for the state-wide competition. From those station contestants a state-wide winner is chosen to be the Wisconsin State Farm Woman of the Year. Sponsor: Greg Steward, Mgr, WWIB Radio, 5558 Hallie Rd, Chippewa Falls, WI 54729. Phone: (715) 723-1037.

WOMEN'S HEART HEALTH DAY. Feb 1. As American Heart Month Begins, here is a day to promote awareness that heart disease is the number one killer of American Women. For info: Charlotte Libov, 71 Judson Lane, Bethlehem, CT 06751. Phone: (203) 266-5904.

BIRTHDAYS TODAY

T.R. Dunn, 40, former basketball player, born at Birmingham, AL, Feb 1, 1955.
Don Everly, 58, singer, musician ("Bye Bye Love," "Wake Up Little Susie" with brother Phil), born Brownie, KY, Feb 1, 1937.
Sherman Hemsley, 57, actor ("All in the Family," "The Jeffersons"), born at Philadelphia, PA, Feb 1, 1938.
Rick James (James Johnson), 43, singer (King of Funk in the '80s, *Street Songs*), born at Buffalo, NY, Feb 1, 1952.
Bob Jamieson, 52, broadcast journalist, born at Streator, IL, Feb 1, 1943.
Terry Jones, 53, actor ("MPFC"), director ("Monty Python's Flying Circus"), born at Colwyn Bay, Wales, Feb 1, 1942.
Garrett Morris, 58, comedian ("Saturday Night Live"), singer, born New Orleans, LA, Feb 1, 1937.
Stuart Whitman, 59, actor ("Cimarron Strip," *The Seekers*), born at San Francisco, CA, Feb 1, 1936.

FEBRUARY 2 — THURSDAY
33rd Day — Remaining, 332

◆ **BABE VOTED INTO BASEBALL HALL OF FAME: ANNIVERSARY.** Feb 2, 1936. The five charter members of the brand-new Baseball Hall of Fame at Cooperstown, NY, were announced. Of 226 ballots cast, Ty Cobb was named on 222, Babe Ruth on 215, Honus Wagner on 215, Christy Mathewson on 205 and Walter Johnson on 189. A total of 170 votes were necessary to be elected to the Hall of Fame.

BENET, WILLIAM ROSE: BIRTH ANNIVERSARY. Feb 2. American poet and critic. Born at Fort Hamilton, NY, on Feb 2, 1886. Died at New York, NY, on May 4, 1950.

BONZA BOTTLER DAY™. Feb 2. To celebrate when the number of the day is the same as the number of the month. Bonza Bottler Day™ is an excuse to have a party at least once a month. T-shirts available. For info: Elaine Fremont, 203 Waddell Rd, Taylors, SC 29687. Phone: (704) 646-2887.

CANDLEMAS DAY OR PRESENTATION OF THE LORD. Feb 2. Observed in Roman Catholic and Eastern Orthodox Churches. Commemorates presentation of Jesus in the Temple and the purification of Mary 40 days after his birth. Candles have been blessed on this day since the 11th century. Formerly called the Feast of Purification of the Blessed Virgin Mary. Old Scottish couplet proclaims: "If Candlemas is fair and clear/ There'll be two winters in the year."

GROUNDHOG DAY. Feb 2. Old belief that if the sun shines on Candlemas Day, or if the groundhog sees his shadow when he emerges on this day, six weeks of winter will ensue.

GROUNDHOG DAY IN PUNXSUTAWNEY, PENNSYLVANIA. Feb 2. Widely observed traditional annual Candlemas Day event at which "Punxsutawney Phil, king of the weather prophets," is the object of a search. Tradition is said to have been established by early German settlers. The official trek (which began in 1887) is followed by a weather prediction for the next six weeks.

Feb 2-3 ☆ Chase's 1995 Calendar of Events ☆

GROUNDHOG DAY IN SUN PRAIRIE, WISCONSIN. Feb 2. Sun Prairie, WI. To predict the weather for the balance of winter. Prognostication at 7:15 AM, CST, to see if Jimmy the Groundhog has seen his shadow. Persons born on this date are eligible for "official" groundhog birth certificate and/or groundhog club (for a small fee). Est attendance: 350. For info: Chamber of Commerce, 109 E Main St, Sun Prairie, WI 53590. Phone: (608) 837-4547.

IMBOLC. Feb 2. (Also called Imbolg, Candlemas, Lupercalia, Feast of Pan, Feast of Torches, Feast of Waxing Light, Brigit's Day and Oimelc.) One of the "Greater Sabbats" during the Wiccan year, Imbolc marks the recovery of the Goddess (after giving birth to the Sun, or the God, at Yule) and celebrates the anticipation of spring. Annually, Feb 2.

JOYCE, JAMES: BIRTH ANNIVERSARY. Feb 2. Irish novelist and poet, author of *Dubliners*, *A Portrait of the Artist as a Young Man*, *Ulysses*, *Finnegan's Wake* and *Chamber Music*. Born at Dublin, Ireland, on Feb 2, 1882. "A man of genius," he wrote, in *Ulysses*, "makes no mistakes. His errors are volitional and are portals of discovery." Of *Finnegan's Wake* Joyce is reported to have replied to an academic whose letter had asked for clues to its meaning, "If I can throw any obscurity on the subject, let me know." (Quote is from *The Ann Arbor News*, Sunday, Feb 7, 1988, page F-8, "1000 years of Irish Memories..." by Hugh A. Mulligan, AP.) Joyce died of peritonitis at Zurich, Switzerland, on Jan 13, 1941, and was buried there.

LUXEMBOURG: CANDLEMAS. Feb 2. Traditional observance of Candlemas. At night children sing a customary song wishing health and prosperity to their neighbors and receive sweets in return. They carry special candles called *Lichtebengel* symbolizing the coming of spring.

MAJOR LEAGUE/MINOR LEAGUE: PHOTOGRAPHS OF AMERICA'S BASEBALL STADIUMS BY JIM DOW. Feb 2-Mar 19. Art and Culture Center, Hollywood, FL. Dow's vivid color images offer an unusual glimpse of America's ballparks through a fan's-eye view. Positioned as if in the stands, a viewer can watch the people who work, play and dream on the playing field and can understand something of the civic pride that led local communities to build them. Call or write the Smithsonian for other dates and venues. For info: Smithsonian Institution Traveling Exhibition Service, 1100 Jefferson Dr SW, Ste 3146, Washington, DC 20560. Phone: (202) 357-2700.

MEXICO: DIA DE LA CANDELARIA. Feb 2. All Mexico celebrates. Dances, processions, bullfights.

MOBIUS ADVERTISING AWARDS. Feb 2. Chicago, IL. Selection and recognition of the world's most outstanding television and radio commercials, print advertising and package designs. Founded in 1971. Patricia Meyer, Exec Dir. Fax: (708) 834-5655. Est attendance: 500. For info: The Mobius Advertising Awards, J.W. Anderson, Chair, 841 N Addison Ave, Elmhurst, IL 60126-1291. Phone: (708) 834-7773.

THE RECORD OF A SNEEZE: ANNIVERSARY. Feb 2. One day after Thomas Edison's "Black Maria" studio was completed in West Orange, NJ, a studio cameraman took the first "close-up" in film history, Feb 2, 1893. *"The Record of a Sneeze"*, starring Edison's assistant Fred P. Ott, was also the first motion picture to receive a copyright (1894). See also: "Black Maria Studio: Anniversary" (Feb 1).

February 1995

S	M	T	W	T	F	S
			1	2	3	4
5	6	7	8	9	10	11
12	13	14	15	16	17	18
19	20	21	22	23	24	25
26	27	28				

TREATY OF GUADALUPE HIDALGO: ANNIVERSARY. Feb 2. The war between Mexico and the United States formally ended with the signing of the Treaty of Guadalupe Hidalgo, signed in the village for which it was named, on Feb 2, 1848. The treaty provided for Mexico's cession to the US of the territory that became the states of California, Nevada, Utah, most of Arizona, and parts of New Mexico, Colorado and Wyoming, in exchange for $15 million from the US. In addition, Mexico relinquished all rights to Texas, north of the Rio Grande. The Senate ratified the treaty on Mar 10, 1848.

WALTON, GEORGE: DEATH ANNIVERSARY. Feb 2. Signer of the Declaration of Independence. Born at Prince Edward County, VA, 1741 (exact date unknown). Died Feb 2, 1804.

BIRTHDAYS TODAY

Christie Brinkley, 42, model, born at Monroe, MI, Feb 2, 1953.
Vlade Divac, 27, basketball player, born at Belgrade, Yugoslavia, Feb 2, 1968.
Sean Michael Elliott, 27, basketball player, born at Tucson, AZ, Feb 2, 1968.
Farrah Fawcett, 48, actress, model ("Charlie's Angels," *The Burning Bed*), born at Corpus Christi, TX, Feb 2, 1947.
Dexter Manley, 36, football player, born at Houston, TX, Feb 2, 1959.
Graham Nash, 53, musician, singer, born at Blackpool, England, Feb 2, 1942.
Liz Smith, 72, journalist, author, born at Fort Worth, TX, Feb 2, 1923.
Tom Smothers, 58, comedian, folksinger (Dick's brother; "Smothers Brothers Comedy Hour"; Yo-Yo Man), born at New York, NY, Feb 2, 1937.
Elaine Stritch, 70, actress (*Company*, sang "The Ladies Who Lunch"), born Birmingham, MI, Feb 2, 1925.

FEBRUARY 3 — FRIDAY
34th Day — Remaining, 331

AFRICAN CULTURAL BALL. Feb 3. Chattanooga African-American Museum, Chattanooga, TN. Ball to celebrate African heritage at the beginning of Black History Month. In-state phone: (615) 267-1076. Est attendance: 2,000. For info: Vilma Fields, 730 M.L. King Blvd, Chattanooga, TN 37403. Phone: (800) 322-3344.

AMERICANA INDIAN AND WESTERN ART SHOW AND SALE. Feb 3-5. Yuma, AZ. Indian and Western paintings, rugs and jewelry. For info: Yuma Civic and Conv Ctr, 1440 Desert Hills Dr, Yuma, AZ 85364. Phone: (602) 344-3800.

ARBOR DAY (ARIZONA). Feb 3. In the counties of Apache, Navajo, Coconino, Mohave and Yavapai, the Friday following Apr 1, and in all other counties the Friday following Feb 1, in each year, shall be known as Arbor Day. See also: "Arbor Day (Arizona)" (Apr 7).

BADGER STATE WINTER GAMES. Feb 3-5. Wausau, WI, and nine other Central Wisconsin communities. 7th annual Olympic-style competition for Wisconsin residents all ages and abilities features nine sports: cross country skiing, curling, downhill skiing, figure skating, youth & adult ice hockey, ski jumping, snowshoe racing, snowboarding and speedskating. America's largest Olympic-style winter sports festival with more than 4,600 participants. Annually, the first weekend of Febru-

☆ Chase's 1995 Calendar of Events ☆ Feb 3

ary. Major sponsors are: AT&T, Ameritech, GTE, Wisconsin Milk Marketing Board, and WPS Insurance Corp. Presenting sponsor is Wausau Insurance Companies. Member of the National Congress of State Games. Est attendance: 21,000. For info: Otto Breitenbach, Exec Dir, or Rick Foy, PR Dir, Badger State Games, PO Box 1377, Madison, WI 53701-1377. Phone: (608) 251-3333.

BULLNANZA. Feb 3-4. Lazy E Arena, Guthrie, OK. Present and past champions compete in bullriding. Annually, the first Friday and Saturday in February. Est attendance: 14,000. For info: Lazy E Arena, Rte 5, Box 393, Guthrie, OK 73044. Phone: (800) 234-3393.

CANADA: WINTERLUDE. Feb 3-5. (Also Feb 10-12, 17-19.) Ottawa, Ontario. 17th annual celebration of Canadian winter and traditions for the whole family. Skate on the Rideau Canal, the world's longest skating rink. Enjoy spectacular snow and ice sculptures, world class figure skating, North America's largest snow playground and exciting Winterlude Triathlon. Phone: (613) 239-5000 or (800) 465-1867. Est attendance: 633,000. For info: Natl Capital Commission, 161 Laurier Ave W, Ottawa, Ont, Canada K1P 6J6.

CIVIL WAR PEACE TALKS: 130th ANNIVERSARY. Feb 3. On Feb 3, 1865, Abraham Lincoln and his Secretary of State, William Seward, met to discuss peace with Confederate Vice-President Alexander Stephens and others at Hampton Road, VA. The meeting, which took place on board the ship *River Queen*, lasted four hours and produced no positive results. The Confederates sought an armistice first and discussion of reunion later, while Lincoln was insistent that recognition of Federal authority must be the first step towards peace.

"THE DAY THE MUSIC DIED." Feb 3. The anniversary of the death of rock-and-roll pioneer and legend Charles Hardin "Buddy" Holly. "The Day the Music Died," so called in singer Don McLean's song "American Pie," is the date on which in 1959 Holly was killed in a plane crash along with J.P. Richardson (otherwise known as "The Big Bopper") and Richie Valens. Holly was born Sept 7, 1936, at Lubbock, TX, and died in a cornfield near Mason City, IA, Feb 3, 1959.

DULCIBRRR. Feb 3-4. Falls of Rough, KY. Dulcimer workshops and concerts. Est attendance: 200. For info: Tom DeHaven, Recreation Super, Rough River State Resort Pk, Falls of Rough, KY 40119. Phone: (502) 257-2311.

EDISON FESTIVAL OF LIGHT. Feb 3-19. Ft Myers, FL. Community celebration commemorating the birth of Thomas Alva Edison. Est attendance: 450,000. For info: Edison Festival of Light, Inc, Margie B. Willis, 2210 Bay Street, Ft Myers, FL 33901. Phone: (813) 334-2999.

FESTIVAL OF THE NORTH. Feb 3-5. (Also Feb 10-12, 17-19 and 24-26.) Ketchikan, AK. A cultural event encompassing performing, visual and literary arts, including a myriad of workshops. Annually, every weekend in February. Est attendance: 1,000. For info: Ketchikan Area Arts and Humanities Council, 338 Main St, Ketchikan, AK 99901. Phone: (907) 225-2211.

◈ **FIFTEENTH AMENDMENT TO US CONSTITUTION RATIFIED: 125th ANNIVERSARY.** Feb 3. The Fifteenth Amendment, ratified on Feb 3, 1870, granted that the right of citizens to vote shall not be denied on account of race, color or previous condition of servitude.

FLORIDA STATE FAIR. Feb 3-19. Florida Expo Park, Tampa, FL. The fair features the best arts, crafts, livestock, entertainment and food found in Florida, as well as one of the most spectacular midways in the country. Est attendance: 1,300,000. For info: Sandy Lee, Florida State Fair, PO Box 11766, Tampa, FL 33680. Phone: (813) 621-7821.

FLORIDA WILDLIFE EXPOSITION. Feb 3-5. Orlando Centroplex, Expo Centre, Downtown Orlando, FL. Featuring more than 200 of the nation's finest wildlife artists, wood sculptors, conservation groups and live wildlife exhibits. Est attendance: 8,000. For info: Florida Wildlife Exposition Inc, PO Box 15693, Sarasota, FL 34277. Phone: (813) 364-9453.

GREAT NORTHEAST HOME SHOW. Feb 3-5. Knickerbocker Arena and Empire State Plaza, Albany, NY. This is the largest home show in the northeast. In addition to exhibits, informative seminars are held throughout the show on topics about the home. Annually, the first full weekend in February. Sponsor: The Times Union Newspaper. Est attendance: 30,000. For info: Heather Schlachter, Show Promoter, Ed Lewi Associates, 8 Wade Rd, Latham, NY 12110. Phone: (518) 783-1333.

HOMESTEAD RODEO. Feb 3-5. Homestead, FL. Championship cowboys and prize-winning stock will be the "stars" of the rodeo. Saddle bronc, bareback riding, bull riding, calf roping, steer wrestling, team roping and barrel racing. Also Rodeo Parade and All-Star Cowboys Barn Dance. Est attendance: 45,000. For info: Susan Neuman, Homestead Rodeo Assn, 555 NE 15th St, Ste 25K, Miami, FL 33132. Phone: (305) 372-9966.

INCOME TAX BIRTHDAY: SIXTEENTH AMENDMENT TO US CONSTITUTION: RATIFICATION ANNIVERSARY. Feb 3. On Feb 3, 1913, the 16th Amendment was ratified, granting Congress the authority to levy taxes on income. (Church bells did not ring throughout the land, and no dancing in the streets was reported.)

LONGHORN WORLD CHAMPIONSHIP RODEO. Feb 3-5. Tulsa Convention Center, Tulsa, OK. More than 500 of North America's best cowboys and cowgirls compete in 7 professionally sanctioned contests ranging from bronc riding to bull riding for top prize money and world championship points. The theme of the production changes annually. Additional attractions include special animal trainers, trick ropers, trick riders and professional clowns. 5th annual. Est attendance: 15,000. For info: Longhorn World Chmpshp Rodeo Inc, PO Box 70159, Nashville, TN 37207. Phone: (615) 876-1016.

MICHIGAN SCIENCE TEACHERS ASSOCIATION CONFERENCE. Feb 3-4. Lansing Center, Lansing, MI. 250 sessions and workshops for science teachers of all grade levels. Also exhibits on the latest science teaching books and hands-on materials from approximately 130 companies. Sponsor: Michigan Science Teachers Assn. Est attendance: 3,000. For info: WMU, Office of Conferences, Attn: MSTA, Ellsworth Hall, Kalamazoo, MI 49008-5161. Phone: (616) 387-4174.

NAUVOO LEGION CHARTERED: ANNIVERSARY. Feb 3, 1841. Created by Illinois Charter and comprised of 5,000 Mormon men under the command of Lt General Joseph Smith, the Nauvoo Legion was considered the "largest trained soldiery in the US" except for the US Army. For info: Angus Belliston, Pres, Sons of Utah Pioneers, 3301 E—2920 S, Salt Lake City, UT 84109. Phone: (801) 484-4441.

NC RV AND CAMPING SHOW. Feb 3-5. Charlotte Merchandise Mart, Charlotte, NC. A display of the latest in recreation vehicles and accessories by various dealers. Est attendance: 11,000. For info: Apple Rock Advertising & Promotion, 1200 Eastchester Dr, High Point, NC 27265. Phone: (910) 883-7199.

NFL PRO BOWL BEACH CHALLENGE. Feb 3. Oahu, HI. The very best of the NFL and AFC football pros take to the beach for a series of exciting competitions, which include obstacle courses, outrigger canoe races, tug-of-war and sprint relays. Open to the public. For info: Event Marketing, Inc., 1001 Bishop St, #477, Pauahi Tower, Honolulu, HI 96813. Phone: (808) 521-4322.

NRA RODEO FINALS. Feb 3-5. MetraPark, Billings, MT. Rodeo action at its best. A fun filled rodeo weekend in cowboy country! Est attendance: 16,000. For info: Northern Rodeo Assn, PO Box 1122, Billings, MT 59103. Phone: (406) 252-1122.

Feb 3-4 ☆ Chase's 1995 Calendar of Events ☆

PERCHVILLE USA. Feb 3-5. Tawas Bay, MI. Coronation of Perchville King and Queen, parade, fishing contests, ice sculptures, perch dinners, polar bear swims, demolition derby on ice, craft show and ATV races. Est attendance: 15,000. For info: Chamber of Commerce, Box 608, Tawas City, MI 48764. Phone: (800) 558-2927.

ROCKWELL, NORMAN: BIRTH ANNIVERSARY. Feb 3. American artist and illustrator especially noted for his realistic and homey magazine cover art for the *Saturday Evening Post*. Born at New York, NY, Feb 3, 1894. Died at Stockbridge, MA, Nov 8, 1978.

SPACE MILESTONE: STS-10 (US). Feb 3. Shuttle *Challenger* launched Feb 3, 1984, from Kennedy Space Center, FL, with crew of five (Vance Brand, Robert Gibson, Ronald McNair, Bruce McCandless and Robert Stewart). On Feb 7 two astronauts became first to fly freely in space (propelled by their backpack jets), untethered to any craft. Landed at Cape Canaveral, FL, on Feb 11.

STEIN, GERTRUDE: BIRTH ANNIVERSARY. Feb 3. Avantgarde expatriate American writer, perhaps best remembered for her poetic declaration (in 1913): "A rose is a rose is a rose." Born at Allegheny, PA, Feb 3, 1874. Died at Paris, France, July 29, 1946.

WINTERFEST BALLOON RALLY. Feb 3-5. Angel Fire, NM. Two morning lift-offs and a balloon glow. Rides given for crew assistance. Est attendance: 200. For info: Angel Fire Chamber of Commerce, PO Box 547, Angel Fire, NM 87710. Phone: (800) 446-8117.

WORLD SHOVEL RACE CHAMPIONSHIPS. Feb 3-5. Angel Fire, MN. This famous event highlights thrilling competition in production and modified divisions for several age groups. "Modified" competition reaches speeds of 60 mph. Spectator competition on stock grain scoop shovels. Est attendance: 1,500. For info: Angel Fire Resort, PO Drawer B, Angel Fire, NM 87710. Phone: (800) 633-7463.

BIRTHDAYS TODAY

Shelley Berman, 69, comedian ("Mary Hartman, Mary Hartman"), born at Chicago, IL, Feb 3, 1926.
Joey Bishop (Joseph Abraham Gottlieb), 77, actor ("The Joey Bishop Show," "Liar's Club"), born at New York, NY, Feb 3, 1918.
Morgan Fairchild (Patsy McClenny), 45, actress ("Dallas," "Falcon Crest," "Flamingo Road," "North and South"), born at Dallas, TX, Feb 3, 1950.
Bob Griese, 50, former football player, born at Evansville, IN, Feb 3, 1945.
James Michener, 88, author (*Journey, The World Is My Home: Memoirs*), born at New York, NY, Feb 3, 1907.
Bibi Osterwald, 75, actress ("Bridget Loves Bernie"), born at New Brunswick, NJ, Feb 3, 1920.
Paul S. Sarbanes, 62, US Senator (D, Maryland) up for reelection in Nov '94, born at Salisbury, MD, Feb 3, 1933.

February 1995

S	M	T	W	T	F	S
			1	2	3	4
5	6	7	8	9	10	11
12	13	14	15	16	17	18
19	20	21	22	23	24	25
26	27	28				

Fran Tarkenton, 55, former football player, born at Richmond, VA, Feb 3, 1940.

FEBRUARY 4 — SATURDAY
35th Day — Remaining, 330

AFRICAN-AMERICAN HISTORY AND HERITAGE TOUR. Feb 4. (Also Feb 11, 18 and 25.) Old City Park, Dallas, TX. A bus tour of sites important in the history of the Dallas African-American community including the Knights of Pythias Buildings, Good Street Baptist Church and Juanita Craft Home. Tour leaves from Old City Park front gate. Annually, every Saturday in February, African-American Heritage Month. Est attendance: 1,500. For info: Old City Park, Dallas Co Heritage Soc, 1717 Gano, Dallas, TX 75215. Phone: (214) 421-5141.

AMERICAN BOWLING CONGRESS CHAMPIONSHIPS TOURNAMENT. Feb 4-June 30. Reno, NV. 100,000 bowlers from around the country and several foreign locations compete for titles in singles, doubles, five-man team and all events in an arena setting. Lanes are specially built into the convention center each year. Est attendance: 170,000. For info: American Bowling Congress, 5301 S 76th St, Greendale, WI 53129-0500. Phone: (414) 421-6400.

APACHE WARS BEGIN: ANNIVERSARY. Feb 4, 1861. The period of conflict known as the Apache Wars began on this date at Apache Pass, AZ, when Army Lt George Bascom arrested Apache Chief Cochise for raiding a ranch. Cochise escaped and declared war. The wars lasted 25 years under the leadership of Cochise and, later, Geronimo. For info: Edwin R. Sweeney, Cochise Historian, 323 White Pine, St. Charles, MO. Phone: (314) 441-9157.

ARTISTS' SALUTE TO BLACK HISTORY MONTH. Feb 4-12. Los Angeles, CA. This 13th annual exhibition of fine art by artists of African descent features more than 150 artists from approximately 32 states. Paintings, graphics, enameling, sculpture, ceramics and batiks will be on display along with a special exhibit, "Legends," which highlights artists of the 1930s and 1940s. Free seminars on framing, collecting and developing art will be offered. Annually, the first week in February. Est attendance: 350,000. For info: Barbara Wesson, Founder/Coord, PO Box 36B75, Los Angeles, CA 90036-1203. Phone: (213) 939-0250.

CO-ED BROOMBALL HOCKEY TOURNAMENT. Feb 4-5. Blue Line Ice Center, Sheboygan, WI. Provides action, thrills and laughter on the ice. Annually, the first weekend in February. Sponsors: Sheboygan Community Recreation Dept and Lakeside Pepsi-Cola Bottling Co. Est attendance: 200. For info: Sheboygan Area Conv and Visitors Bureau, 631 New York Ave, Sheboygan, WI 53081. Phone: (414) 457-9495.

◈ **A DAY IN THE WARSAW GHETTO: A BIRTHDAY TRIP IN HELL.** Feb 4-Mar 19. Robert I. Kahn Gallery, Houston, TX. Jews relocated by the Nazis to the Warsaw ghetto in Poland during WWII numbered more than 400,000. This exhibit features photographs of that ghetto taken by Heinz Jost, German hotelkeeper and former Wehrnacht soldier, on his day off, his birthday: Sept 18, 1941. Not until 1980 did Jost reveal the existence of this photographic portrait of ghetto life: beggars, children, burials. Also included in the exhibit is a 20-minute video from original Nazi footage taken by German soldiers. Call or write the Smithsonian for other dates and venues. For info: Smithsonian Institution Traveling Exhibition Service, 1100 Jefferson Dr SW, Ste 3146, Washington, DC 20560. Phone: (202) 357-2700.

DOLL SHOW. Feb 4. Yuma Civic and Convention Center, Yuma, AZ. Assorted antique, handmade and collector's item dolls on display and for sale. Also, miniature furniture and doll-making supplies on sale. Sponsor: Yuma Doll Club. Est attendance: 600. For info: Yuma Civic and Conv Ctr, Yuma Doll Club, 1440 Desert Hills Dr, Yuma, AZ 85364. Phone: (602) 782-5797.

Chase's 1995 Calendar of Events — Feb 4

◆ **FRIEDAN, BETTY NAOMI: BIRTHDAY.** Feb 4, 1921. Betty Goldstein was born at Peoria, IL, and grew up to go to Smith College and work in New York City until marrying and moving to the suburbs where she kept house, raised children, did freelance writing, surveyed the feelings of her college classmates and wrote a book. Published in 1963, *The Feminine Mystique* became an overnight bestseller and sparked the dialogue among American women that grew into the contemporary women's rights movement, as well as the ongoing controversy and confusion as men and women try to redefine or reaffirm their roles and their rights. In 1966 Friedan founded the National Organization for Women (NOW), serving as its president for four years.

GASPARILLA PIRATE FEST, INVASION AND PARADE. Feb 4. Tampa, FL. The celebration begins at 10 AM with the Pirate Fest. Live entertainment ranging from rock to reggae is scheduled on five stages. The world's only fully rigged pirate ship sets sail from Ballast Point Pier. Hundreds of boats accompany the pirate ship and its crew of bloodthirsty buccaneers as it heads toward downtown Tampa. The crew of several hundred costumed pirates launches a day-long celebration at the Tampa Convention Center. Sponsor: Ye Mystic Krewe of Gasparilla. Est attendance: 400,000. For info: Tampa/Hillsborough Conv and Visitors Assn, 111 Madison St, Ste 1010, Tampa, FL 33602-4706. Phone: (800) 448-2672.

HALFWAY POINT OF WINTER. Feb 4. At 9:18 AM, EST, on Feb 4, 44 days, 11 hours and 55 minutes of winter will have elapsed and the equivalent remains before 9:14 PM, EST, on Mar 20, 1995, which is the spring equinox and the beginning of spring.

HOPKINS, MARK: BIRTH ANNIVERSARY. Feb 4. American educator, author and college president (Williams College). Born at Stockbridge, MA, Feb 4, 1802. Died at Williamstown, MA, June 17, 1887.

INTERNATIONAL AUTO SHOW. Feb 4–12. Timonicium State Fairgrounds, Baltimore, MD. Introducing the new line of automobiles. Display of new cars, foreign and domestic. Est attendance: 300,000. For info: Intl Auto Shows Ltd, 20 Park Ave, Ste 301, Baltimore, MD 21201-3423. Phone: (410) 385-1800.

LAURA INGALLS WILDER GINGERBREAD SOCIABLE. Feb 4. Pomona, CA. The 28th annual event commemorates the birthday (Feb 7, 1867) of the renowned author of the "Little House" books. The library has on permanent display the handwritten manuscript of *Little Town on the Prairie* and other Wilder memorabilia. The sociable includes entertainment by fiddlers, craft displays and refreshments of apple cider and gingerbread. Annually, the first Saturday in February. Est attendance: 300. Sponsor: Friends of the Pomona Public Library, Marguerite F. Raybould, 625 S Garey Ave, Pomona, CA 91766. Phone: (909) 620-2017.

LINDBERGH, CHARLES AUGUSTUS: BIRTH ANNIVERSARY. Feb 4. American aviator Charles "Lucky Lindy" Lindbergh was the first to fly solo and nonstop over the Atlantic Ocean, New York to Paris, May 20–21, 1927. Born at Detroit, MI, Feb 4, 1902. Died at Kipahula, Maui, HI, Aug 27, 1974. See also: "Lindbergh Flight Anniversary" (May 20).

MACKINAW MUSH SLED DOG RACE. Feb 4–5. Mackinaw City, MI. Sanctioned ISDA event with purse and various classes. Also children's events and weight pull. Annually, the first weekend in February. Est attendance: 1,000. For info: Mackinaw Area Tourist Bureau, 706 S Huron, Mackinaw City, MI 49701. Phone: (800) 666-0160.

PERRY'S "BRR" (BIKE RIDE TO RIPPEY). Feb 4. Perry, IA. Winter bike riding. Twenty-two miles of frigid fun. Annually, the first Saturday in February. Est attendance: 2,000. Sponsor: Chamber of Commerce, Kathy Hoskinson, 1226 Second St, Perry, IA 50220. Phone: (515) 465-4601.

RAILROAD AND HOBBY SHOW. Feb 4–5. West Springfield, MA. Now expanded to two buildings, nearly 4½ acres with more than 30 operating layouts; displays; art; railroadiana; Shortline, Tourist and Class 1 railroads; flea market, dealers and more. Est attendance: 18,000. For info: Robert Buck, PO Box 718, Warren, MA 01083-0718. Phone: (413) 436-0242.

ROLEX 24 AT DAYTONA. Feb 4–5. Daytona International Speedway, Daytona Beach, FL. IMSA Camel Grand Prix of Endurance. For info: Daytona Intl Speedway, Glenn Barber, PR, Box 2801, Daytona Beach, FL 32120-2801. Phone: (904) 254-6782.

SPEED WEEKS. Feb 4–19. Daytona International Speedway, Daytona Beach, FL. For info: Daytona Intl Speedway, Box 2801, Daytona Beach, FL 32120-2801. Phone: (904) 254-6782.

SPORTSFEST '95. Feb 4–5. Held indoors at the Myriad Convention Center, Oklahoma City, OK. Oklahoma's amateur winter sports festival includes some 3,000 athletes competing in about 14 events, including gymnastics, basketball, tumbling, table tennis, karate and cheerleading plus a trade show, sports memorabilia show and sports celebrities. Est attendance: 6,000. For info: Sooner State Games, 100 W Main, Ste 287, Oklahoma City, OK 73102. Phone: (405) 235-4222.

SRI LANKA: INDEPENDENCE DAY. Feb 4. Democratic Socialist Republic of Sri Lanka observes National Day. Public holiday.

TOMS RIVER WILDFOWL ART AND DECOY SHOW. Feb 4–5. Toms River Intermediate School East, Hooper Ave, Toms River, NJ. 125 artists and carvers, decoy carving competitions, decorative and gunning decoys, junior competition, new seminars and free clinic; NJ Ducks Unlimited art and decoy painting competitions held at the school while the show is on. Adults pick up decoys on Saturday. They are judged Sunday at 1 PM. Kids sit at the table on Saturday and paint a duck while parents look at show. Annually, first weekend in February. Est attendance: 6,000. For info: Maureen Rohrs, Ocean County YMCA, PO Box 130, Toms River, NJ 08754. Phone: (908) 341-9622.

THE TONGASS—ALASKA'S MAGNIFICENT RAIN FOREST. Feb 4–Mar 19. Central Florida Community College, Ocala, FL. This exhibit explores the intricate harmony of the largest non-equatorial rain forest on earth. This balanced ecosystem provides a vast laboratory in which to study the complexities of interdepence, the diversity of life and the range of land use, including wilderness preservation, habitat protection and the benefits and dangers of commercial interests. For info: Smithsonian Institution Traveling Exhibition Service, 1100 Jefferson Dr SW, Ste 3146, Washington, DC 20560. Phone: (202) 357-2700.

TORTURE ABOLITION DAY: 10th ANNIVERSARY. Feb 4, 1985. Twenty countries signed a UN document entitled "Convention Against Torture and Other Cruel, Inhuman or Degrading Treatment or Punishment." Adopted Dec 10, 1984, by the UN General Assembly, it defined torture as any act "by which severe pain or suffering, whether physical or mental, is intentionally inflicted" to obtain information or a confession. Signatory countries were Afghanistan, Argentina, Belgium, Bolivia, Costa Rica, Denmark, Dominican Republic, Finland, France, Greece, Iceland, Italy, Netherlands, Norway, Portugal, Senegal, Spain, Sweden, Switzerland and Uruguay. While the US did sign the document, signing is only a preliminary stage that must be followed by ratification—which the US has never done.

TWIN CITIES' KREW OF JANUS MARDI GRAS PARADE. Feb 4. Monroe, LA. Festive parade to celebrate Mardi Gras. Annually, the second Saturday before Fat Tuesday. For info: Judy M. Lowentritt, Tourism Sales Mgr, 1333 State Farm Dr, Monroe, LA 71202. Phone: (318) 325-6402.

USO: BIRTHDAY. Feb 4. To honor the civilian agency founded in 1941 that provides support worldwide for US service people and their families. The United Service Organizations (USO) centers have served as a home away from home for hundreds of thousands of Americans.

Feb 4–5 ☆ *Chase's 1995 Calendar of Events* ☆

YUMA JAYCEE'S SILVER SPUR PARADE. Feb 4. Yuma, AZ. Parade kicks off a full week of activities, such as Senior Citizen Day, Southwest Heritage Day and a rodeo on the following weekend. Annually, the first Saturday in February. Sponsors: Budweiser, Dodge, Pepsi, Best Western Inn Suites and Jack-In-The-Box. Est attendance: 28,000. For info: Yuma Jaycees, Steve Ames, Rodeo Chair, 1798 S Arizona Ave, Yuma, AZ 85364. Phone: (602) 783-3641.

BIRTHDAYS TODAY

David Brenner, 50, comedian ("Nightlife," short lived talk show opposite "The Tonight Show"), born at Philadelphia, PA, Feb 4, 1945.
Alice Cooper (Vincent Damon Furnier), 47, singer, songwriter, born at Detroit, MI, Feb 4, 1948.
Betty Friedan, 74, author, feminist (*The Feminine Mystique, The Fountain of Age*), born at Peoria, IL, Feb 4, 1921.
Russel Hoban, 70, artist, author, born at Lansdale, PA, Feb 4, 1925.
Ida Lupino, 77, director, actress (Eve Drake in "Mr Adams and Eve"; High Sierra), born at London, England, Feb 4, 1918.
◆ *Rosa Lee Parks*, 82, civil rights leader who refused to give up her seat on the bus, born at Tuskegee, AL, Feb 4, 1913.
Dan Quayle, 48, 44th US Vice President, born at Indianapolis, IN, Feb 4, 1947.
Donald W. Riegle, Jr, 57, US Senator (D, Michigan), born at Flint, MI, Feb 4, 1938.
Denis Savard, 34, hockey player, born at Pointe Gatineau, Quebec, Canada, Feb 4, 1961.
Lawrence Taylor, 36, football player, born at Williamsburg, VA, Feb 4, 1959.

FEBRUARY 5 — SUNDAY
36th Day — Remaining, 329

BOY SCOUTS OF AMERICA ANNIVERSARY WEEK. Feb 5–11. Celebrates founding of youth organization incorporated in Washington, DC, in 1910 and chartered by Congress in 1916. For info: Boy Scouts of America, Richard Walker, Natl Spokesperson, 3131 Turtle Creek Blvd, Ste 500, Dallas, TX 75219. Phone: (214) 443-7580.

CHINESE LUNAR NEW YEAR FESTIVAL: YEAR OF THE BOAR (JEE). Feb 5. Baltimore, MD. Dragon (Mo Tze) dance, fashion show of Chinese historic gowns, Chinese culture and traditions. Phone: (410) 539-1395 or (410) 377-8143. Est attendance: 900. For info: Grace and St. Peter's Parish, Lillian Lee Kim, 524 Anneslie Rd, Baltimore, MD 21212-2009.

CIRCLE K INTERNATIONAL WEEK. Feb 5–11. To highlight the work and activities of Circle K for the general public. Circle K is a sponsored program of Kiwanis Intl for college students. Annually, the second week of February. Sponsor: Circle K Intl, 3636 Woodview Trace, Indianapolis, IN 46268-3196. Phone: (317) 875-8755.

GROUNDHOG RUN. Feb 5. Kansas City, MO. 13th annual. The only 10K underground run in the world takes place on the Sunday closest to Groundhog Day at the Hunt Midwest Enterprises SubTropolis. More than 1,400 runners from all over the country participate in this event to benefit Children's TLC. Est attendance: 2,000. Sponsor: Children's TLC, Paula Yehle, 2928 Main St, Kansas City, MO 64108. Phone: (816) 234-3392.

February 1995

S	M	T	W	T	F	S
			1	2	3	4
5	6	7	8	9	10	11
12	13	14	15	16	17	18
19	20	21	22	23	24	25
26	27	28				

LONGEST WAR IN HISTORY: 10th ENDING ANNIVERSARY. Feb 5. The Third Punic War, between Rome and Carthage, started in the year 149 BC. It culminated in the year 146 BC, when Roman soldiers led by Scipio razed Carthage to the ground. The desolated site was cursed and rebuilding forbidden. On Feb 5, 1985, 2,131 years after the war began, Ugo Vetere, mayor of Rome, and Chedli Klibi, mayor of Carthage, met at Tunis to sign a treaty of friendship officially ending the Third Punic War.

MEXICO: ANNIVERSARY OF THE CONSTITUTION. Feb 5. Present constitution, embracing major social reforms, adopted on this day, 1917.

NATIONAL CRIME PREVENTION WEEK. Feb 5–11. To alert America to the growing menace and cost of crime and stimulate public interest in year-round crime prevention activities. Sponsor: The Natl Exchange Club, 3050 Central Ave, Toledo, OH 43606-1700. Phone: (419) 535-3232.

NFL PRO BOWL '95. Feb 5. Aloha Stadium, Honolulu, HI. All-star football game involving the National and American Conferences of the National Football League. Est attendance: 50,000. For info: Aloha Stadium, PO Box 30666, Honolulu, HI 96820. Phone: (808) 486-9300.

PEEL, ROBERT: BIRTH ANNIVERSARY. Feb 5. English statesman, established the Irish constabulary (known as the "Peelers"). Later, as England's Home Secretary, he reorganized the London police, thereafter known as "Bobbies." Born at Lancashire, England, Feb 5, 1788. Died July 2, 1850, at London from injuries received in a fall from his horse.

STEVENSON, ADLAI EWING: BIRTH ANNIVERSARY. Feb 5. American statesman, governor of Illinois, Democratic candidate for president in 1952 and 1956, US representative to the UN, 1961–1965. Born at Los Angeles, CA, Feb 5, 1900. Died at London, England, July 14, 1965. (Not to be confused with his grandfather, Vice President Adlai Ewing Stevenson, born Oct 23, 1835.) See also: "Stevenson, Adlai Ewing: Birth Anniversary" (Oct 23).

SWITZERLAND: HOMSTROM. Feb 5. Scuol. Burning of straw men on poles as a symbol of winter's imminent departure. Annually, the first Sunday in February.

TANZANIA: CCM DAY. Feb 5. Chama cha Mapinduzi (CCM), Tanzania's sole political party, was born through the merger of Tanganyika African National Union (TANU) and Afro-Shirazi Party (ASP) of Zanzibar, Feb 5, 1977. Feb 5 is also the anniversary of the Feb 5, 1967, Arusha Declaration, defining Tanzania's policy of socialism and self-reliance.

WEATHERMAN'S DAY. Feb 5. Commemorates the birth of one of America's first weathermen, John Jeffries. Born on Feb 5, 1744, and died on Sept 16, 1819, a Boston physician who kept detailed records of weather conditions, 1774–1816. See also: "First Balloon Flight Across English Channel: Anniversary" (Jan 7).

WITHERSPOON, JOHN: BIRTH ANNIVERSARY. Feb 5. Clergyman, signer of the Declaration of Independence and reputed coiner of the word *Americanism* (in 1781). Born near Edinburgh, Scotland, Feb 5, 1723. Died Nov 15, 1794.

BIRTHDAYS TODAY

Hank Aaron, 61, baseball executive, coach, Hall of Famer, all-time home-run leader, born at Mobile, AL, Feb 5, 1934.
Red Buttons (Aaron Chwatt), 76, actor ("The Red Buttons Show," "The Double Life of Henry Phyfe"), born at the Bronx, NY, Feb 5, 1919.

☆ Chase's 1995 Calendar of Events ☆   Feb 5–6

Andrew Greeley, 67, author (*Happy are the Merciful, An Occasion of Sin*), born at Oak Park, IL, Feb 5, 1928.
Christopher Guest, 47, writer, comedian (Emmy for writing *Lily Tomlin*; regular "Saturday Night Live"), born New York, NY, Feb 5, 1948.
Barbara Hershey (Barbara Hertzstein), 47, actress (*From Here to Eternity, With Six You Get Eggrolls*, "The Monroes"), born at Los Angeles, CA, Feb 5, 1948.
Jennifer Jason Leigh (Marrow), 33, actress (*Miami Blues, Rush, Backdraft*), born at Los Angeles, CA, Feb 5, 1962.
Roger Staubach, 53, former football player, born at Cincinnati, OH, Feb 5, 1942.

FEBRUARY 6 — MONDAY
37th Day — Remaining, 328

ACCESSION OF QUEEN ELIZABETH II: ANNIVERSARY. Feb 6. Princess Elizabeth Alexandra Mary succeeded to the British throne (becoming Elizabeth II, Queen of the United Kingdom of Great Britain and Northern Ireland and Head of the Commonwealth) upon the death of her father, King George VI on Feb 6, 1952. Her coronation took place on June 2, 1953, in Westminster Abbey at London.

DIETARY MANAGERS' PRIDE IN FOOD SERVICE WEEK. Feb 6–10. An annual commemorative event honoring dietary managers and their employees and celebrating the important role they play in institutional food service. Sponsor: Dietary Managers Assn, Diane Everett, Dir of Communications, One Pierce Pl, Ste 1220W, Itasca, IL 60143-1277. Phone: (708) 775-9200.

GTE SUNCOAST CLASSIC. Feb 6–12. TPC of Tampa Bay, Tampa, FL. One of the premier stops on the Senior PGA Tour, attracting one one of the strongest fields of top Senior Tour pros. The TPC of Tampa Bay affords outstanding "Stadium Golf"® viewing with amphitheater-style mounds along fairways and around greens. Pro-Am event Wed and Thurs, followed by professional competitive rounds Fri–Sun. Est attendance: 151,000. Sponsor: GTE Suncoast Classic, 16002 N Dale Mabry, 2nd Fl, Tampa, FL 33618. Phone: (813) 265-4653.

LINCOLN, ABRAHAM: DELAWARE AND OREGON BIRTHDAY OBSERVANCE. Feb 6. Observed annually in Delaware and Oregon on the first Monday in February. See also: "Lincoln, Abraham: Birth Anniversary" (Feb 12).

MASSACHUSETTS RATIFIES CONSTITUTION: ANNIVERSARY. Feb 6. By a vote of 187 to 168 on Feb 6, 1788, Massachusetts became the sixth state to ratify the Constitution.

MID-WINTER'S DAY CELEBRATION. Feb 6. Ann Arbor, MI. To create euphoria by fiat in celebration that winter is half over. Sponsor: Richard Ankli, The Fifth Wheel Tavern, 639 Fifth St, Ann Arbor, MI 48103.

MILITARY AIR TRANSPORT CRASHES IN INDIANA: ANNIVERSARY. Feb 6. A Lockhead C-130 transport on training maneuvers crashed into the rear of a restaurant and a hotel in Evansville, IN, on Feb 6, 1992. Sixteen people were killed including the five crew members who were with the Kentucky National Guard based in Louisville, KY. The plane was practicing touch-and-go maneuvers at Evansville Regional Airport when the accident occurred. The crash marked the second time in 4½ years that a military plane crashed into an Indiana hotel. On Oct 20, 1987, an Air Force fighter jet crashed into the Ramada Inn-Airport in Indianapolis killing 10.

MOVE HOLLYWOOD AND BROADWAY TO LEBANON, PENNSYLVANIA DAY. Feb 6. We have lots of room, friendly folks and Amish farms! Original hometown of event's sponsor, the Wellness Permission League and former home office of "The David Letterman Show." Phone: (212) 388-8673 or (717) 274-8451. For info: Tom and Ruth Roy, Wellness Permission League, 2105 Water St, Lebanon, PA 17046.

NATIONAL SCHOOL COUNSELING WEEK. Feb 6–10. Promotes school counseling in the school and community. Phone: (703) 823-9800, ext 388. Sponsor: American School Counselor Assn, Scott R. Swirling, 5999 Stevenson Ave, Alexandria, VA 22304.

NEW ZEALAND: WAITANGI DAY. Feb 6. National Day. Commemorates signing of the Treaty of Waitangi (at Waitangi, Chatham Islands, New Zealand) on Feb 6, 1840. The treaty between the native Maori and the European peoples provided for development of New Zealand under the British Crown.

NORMAN ROCKWELL'S COLONIAL SIGN PAINTER COVER FOR *SATURDAY EVENING POST*: ANNIVERSARY. Feb 6. Norman Rockwell's Feb 6, 1926 cover for the *Post* was the first to appear in full color. It depicted a Rockwell kindred spirit—a Colonial sign painter.

REAGAN, RONALD WILSON: BIRTHDAY. Feb 6. Fortieth president of the US. Former sportscaster, motion picture actor, rancher, businessman, author (a 1965 autobiography titled *Where's the Rest of Me?* in collaboration with Richard G. Hubler), governor of California (1967–1974), he was the oldest and the first divorced person to become president. Born Feb 6, 1911, at Tampico, IL. Married Jane Wyman, Jan 25, 1940 (they were divorced in 1948); Nancy Davis, Mar 4, 1952.

❖ **RUTH, "BABE": 100th BIRTH ANNIVERSARY.** Feb 6, 1895. One of baseball's greatest heroes, George Herman "Babe" Ruth was born at Baltimore, MD. The left-handed pitcher—"the Sultan of Swat," "the Great Bambino"—hit 714 home runs in 22 major league seasons of play and played in 10 World Series. Died at New York, NY, on Aug 16, 1948.

TWENTIETH AMENDMENT TO US CONSTITUTION: ADOPTION ANNIVERSARY. Feb 6. The 20th Amendment was adopted on this date in 1933. The amendment fixed the date of the presidential inauguration as January 20th, instead of the previous date of March 4th; specified that were the president-elect to die before taking office, the vice president-elect would succeed to the presidency. In addition, it set Jan 3 as the official opening date of Congress each year.

BIRTHDAYS TODAY

Tom Brokaw, 55, journalist, born at Yankton, SD, Feb 6, 1940.
Natalie Cole, 45, singer ("This Will Be," "I've Got Love On My Mind"), born at Los Angeles, CA, Feb 6, 1950.
Fabian (Fabian Forte), 52, singer, actor, born at Philadelphia, PA, Feb 6, 1943.
Mike Farrell, 56, actor ("M*A*S*H," "The Interns"), born at St. Paul, MN, Feb 6, 1939.
Walter Edward Fauntroy, 62, Delegate to Congress (D, District of Columbia), born at Washington, DC, Feb 6, 1933.
Zsa Zsa Gabor (Sari Gabor), 76, actress ("Ninotchka," *Special Tonight*), born at Budapest, Hungary, Feb 6, 1919.
Ronald Wilson Reagan, 84, 40th US president, born at Tampico, IL, Feb 6, 1911.
Rip Torn (Elmore Torn, Jr), 64, actor (*The Blue and the Gray*), born at Temple, TX, Feb 6, 1931.
Michael Tucker, 51, actor (Stuart Markowitz on "L.A. Law"), born at Baltimore, MD, Feb 6, 1944.

FEBRUARY 7 — TUESDAY
38th Day — Remaining, 327

BALLET INTRODUCED TO THE US: ANNIVERSARY. Feb 7. Renowned French danseuse Mme Francisquy Hutin introduced ballet to the US on Feb 7, 1827, with a performance of *The Deserter*, staged at the Bowery Theater, New York, NY. A minor scandal erupted when the ladies in the lower boxes left the theatre upon viewing the light and scanty attire of Mme Hutin and her troupe.

BLAKE, EUBIE: BIRTH ANNIVERSARY. Feb 7. James Hubert "Eubie" Blake, American composer and pianist, writer of nearly 1,000 songs (including "I'm Just Wild About Harry" and "Memories of You"). Born at Baltimore, MD, on Feb 7, 1883. Recipient of the Presidential Medal of Freedom in 1981. Last professional performance was in Jan 1982. Died at Brooklyn, NY, five days after his 100th birthday, on Feb 12, 1983.

DICKENS, CHARLES: BIRTH ANNIVERSARY. Feb 7. English social critic and novelist, born at Portsmouth, England, on Feb 7, 1812. Among his most successful books: *Oliver Twist, The Posthumous Papers of the Pickwick Club, David Copperfield* and *A Christmas Carol*. Died at Gad's Hill, England, June 9, 1870, and was buried at Westminster Abbey.

ELEVENTH AMENDMENT TO US CONSTITUTION (SOVEREIGNTY OF THE STATES): 200th RATIFICATION ANNIVERSARY. Feb 7. On Feb 7, 1795, the 11th Amendment to the Constitution was ratified, curbing the powers of the federal judiciary in relation to the states. The amendment reaffirmed the sovereignty of the states by prohibiting suits against them.

THE GREAT AMERICAN PIZZA BAKE. Feb 7-14. Pizza parlors, restaurants and volunteer agencies nationwide create healthy pizza recipes to increase the public's awareness of the benefits of controlling high cholesterol levels through diet. For info: Frederick S. Mayer, Pres, Cholesterol Council of America, c/o PPSI, PO Box 1336, Sausalito, CA 94966. Phone: (415) 332-4066.

GRENADA: INDEPENDENCE DAY. Feb 7. National Day.

LEWIS, SINCLAIR: 110th BIRTH ANNIVERSARY. Feb 7. American novelist and social critic. Recipient of Nobel Prize for Literature (1930). Among his novels: *Main Street, Babbitt* and *It Can't Happen Here*. Born Harry Sinclair Lewis at Sauk Center, MN, on Feb 7, 1885. Died at Rome, Italy, Jan 10, 1951.

MOON PHASE: FIRST QUARTER. Feb 7. Moon enters First Quarter phase at 7:54 AM, EST.

MORE, SIR THOMAS: BIRTH ANNIVERSARY. Feb 7. Anniversary of birth of lawyer, scholar, author, Lord Chancellor of England, martyr and saint at London, England, on Feb 7, 1478. Refusing to recognize Henry VIII's divorce from Queen Catherine, the "Man for All Seasons" was found guilty of treason and imprisoned in the Tower of London, Apr 17, 1534. He was beheaded at Tower Hill on July 6, 1535, and his head displayed from Tower Bridge. Canonized in 1935. Memorial observed on June 22.

NATIONAL HANGOVER AWARENESS DAY. Feb 7. As this is doubtless the anniversary of some of the most memorable hangovers in a long career by the "Bambino," his fans and followers take stock of their lifetime imbibing stats, alcohol abuse in general, and how they will survive the morning, after "Babe's Night Out." Sponsor: Participants must sponsor their own hangover. Annually, the day after Babe Ruth's birthday.

Sponsor: Brian McCullough, Morning Announcer, WRNX/WTTT, PO Box 67, Amherst, MA 01004. Phone: (413) 256-6794.

SPACE MILESTONE: *SOYUZ 24* (USSR). Feb 7. Two cosmonauts docked at *Salyut 5* space lab. Returned to Earth Feb 25. Launched on Feb 7, 1977.

BIRTHDAYS TODAY

Oscar Brand, 75, singer, born at Winnipeg, Manitoba, Canada, Feb 7, 1920.
Garth Brooks, 33, singer ("Friends in Low Places"), born at Tulsa, OK, Feb 7, 1962.
Herb Kohl, 60, US Senator (D, Wisconsin) up for reelection in '94, born at Milwaukee, WI, Feb 7, 1935.
Dan Quisenberry, 42, former baseball player, born at Santa Monica, CA, Feb 7, 1953.
Gay Talese, 63, author (*The Kingdom and the Power, Unto the Sons*), born at Ocean City, NJ, Feb 7, 1932.

FEBRUARY 8 — WEDNESDAY
39th Day — Remaining, 326

***THE BIRTH OF A NATION*: 80th PREMIERE ANNIVERSARY.** Feb 8. On Feb 8, 1915, D.W. Griffith's landmark motion picture, *The Birth of a Nation*, premiered at Clune's Auditorium in Los Angeles, CA. Based on Thomas Dixon's novel *The Clansman*, the film aroused bitter protests for its sympathetic treatment of the Ku Klux Klan during the Reconstruction period.

BOY SCOUTS OF AMERICA FOUNDED: 85th ANNIVERSARY. Feb 8. On Feb 8, 1910, the Boy Scouts of America were founded in Washington, DC, by William Boyce, based on the work of Sir Robert Baden-Powell with the British Boy Scout Association.

DEAN, JAMES: BIRTH ANNIVERSARY. Feb 8. American stage, film and television actor, who achieved immense popularity during a brief career. Born Feb 8, 1931, at Fairmont, IN. Best remembered for his role in *Rebel Without a Cause*. Died in an automobile accident on Sept 30, 1955, at age 24.

JAPAN: HA-RI-KU-YO (NEEDLE MASS). Feb 8. Ha-Ri-Ku-Yo, a Needle Mass, may be observed on either Feb 8 or Dec 8. On this day girls do no needlework; instead they gather old and broken needles which they dedicate to the Awashima Shrine in Wakayama. Girls pray to Awashima Myozin (their protecting deity) that their needlework, symbolic of love and marriage, will be good. Participation in the Needle Mass hopefully leads to a happy marriage.

JAPAN: SNOW FESTIVAL. Feb 8-12. Sapporo, Hokkaido. Huge, elaborate snow and ice sculptures are erected on the Odori-Koen Promenade.

LOVE MAY MAKE THE WORLD GO 'ROUND, BUT LAUGHTER KEEPS US FROM GETTING DIZZY WEEK. Feb 8-14. This week is dedicated to Victor Borge's notion that "Laughter is the shortest distance between two people" and Joel Goodman's notion that "Seven days without laughter make one weak." This is a chance to lighten your relationships and to reinforce the connection between "heart" and "hearty laughter." Annually, the week leading up to and including Valentine's Day. To receive a free live-love-laugh info packet on the positive power of humor, send a stamped (98¢) self-addressed envelope. Sponsor: The HUMOR Project, Inc., Dept C, 110 Spring St, Saratoga Springs, NY 12866. Phone: (518) 587-8770.

February 1995

S	M	T	W	T	F	S
			1	2	3	4
5	6	7	8	9	10	11
12	13	14	15	16	17	18
19	20	21	22	23	24	25
26	27	28				

MARTHA GRIFFITHS SPEAKS OUT AGAINST SEX DISCRIMINATION: ANNIVERSARY. Feb 8. On Feb 8, 1964, during the congressional debate over the 1964 Civil Rights Act, Representative Martha Griffiths delivered a memorable speech advocating the prohibition of discrimination based on sex. Her efforts resulted in adding Civil Rights protection for women in the 1964 Act. She later successfully led the campaign for the Equal Rights Amendment in the House of Representatives.

OPERA DEBUT IN THE COLONIES: 260th ANNIVERSARY. Feb 8. The first opera produced in the colonies was performed on Feb 8, 1735, at the Courtroom, in Charleston, SC. The opera was *Flora; or the Hob in the Well*, written by Colley Cibber.

REENACTMENT OF COWTOWN'S LAST OLD WEST GUNFIGHT. Feb 8. White Elephant Saloon, Fort Worth, TX. Annual reenactment of Fort Worth's last Old West gunfight, which took place on Feb 8, 1887, between White Elephant Saloon owner Luke Short and former Marshal T. I. "Longhaired Jim" Courtright. Annually, Feb 8. Est attendance: 400. Sponsor: Jane Wade, Adm Asst, 108 E Exchange Ave, Fort Worth, TX 76106-8210. Phone: (817) 624-9712.

◈ **SHERMAN, WILLIAM TECUMSEH: 175th BIRTH ANNIVERSARY.** Feb 8. American soldier. Born at Lancaster, OH, Feb 8, 1820. General Sherman is especially remembered for his devastating march through Georgia during the Civil War and his statement "War is hell." Died at New York, NY, Feb 14, 1891.

SPACE MILESTONE: *ARABSAT-1*. Feb 8. League of Arab States communications satellite launched into geosynchronous orbit from Kourou, French Guiana, by ESA, Feb 8, 1985.

SPACE MILESTONE: *BRASILSAT-1* (BRAZIL). Feb 8. Brazilian communications satellite launched into geosynchronous orbit from Kourou, French Guiana, by ESA, Feb 8, 1985.

VERNE, JULES: BIRTH ANNIVERSARY. Feb 8. French writer, sometimes called "the father of science fiction." Born at Nantes, France, on Feb 8, 1828. Author of *Around the World in Eighty Days, Twenty Thousand Leagues Under the Sea* and many other novels. Died at Amiens, France, on Mar 24, 1905.

BIRTHDAYS TODAY

Gary Coleman, 27, actor ("Diff'rent Strokes," *The Kid from Left Field*), born at Zion, IL, Feb 8, 1968.

Robert Klein, 53, comedian ("Comedy Tonight," "TV's Bloopers & Practical Jokes"), born at New York, NY, Feb 8, 1942.

Ted Koppel, 55, journalist (anchor of "Nightline"), born at Lancashire, England, Feb 8, 1940.

Jack Lemmon (John Uhler, III), 70, actor (*Mr. Roberts, The Odd Couple*), born at Boston, MA, Feb 8, 1925.

Audrey Meadows, 71, actress (Alice on "The Honeymooners"), born at Wu Chang, China, Feb 8, 1924.

Nick Nolte, 61, actor (*48 hours, Prince of Tides*, "Rich Man, Poor Man"), born at Omaha, NE, Feb 8, 1934.

Lana Turner (Julia Jean Mildred Frances), 75, actress (*The Postman Always Rang Twice, Imitation of Life*), born at Wallace, ID, Feb 8, 1920.

John Williams, 63, musician, composer (conducted Boston Pops; scored *Jaws, Star Wars, Schindler's List*), born New York, NY, Feb 8, 1932.

FEBRUARY 9 — THURSDAY
40th Day — Remaining, 325

ADE, GEORGE: BIRTH ANNIVERSARY. Feb 9. American newspaperman, playwright and humorist. Born Feb 9, 1866, at Kentland, IN. Remembered especially for his *Fables in Slang*, published in 1900. Died at Brook, IN, May 16, 1944.

AFL-CIO MERGER: 40th ANNIVERSARY. Feb 9. A merger agreement was announced by the American Federation of Labor and the Congress of Industrial Organizations on Feb 9, 1955. Employers condemned this merger as a union monopoly. In 1968 the United Auto Workers, the largest AFL-CIO affiliate, disaffiliated. Shortly after this the UAW announced a "functional alliance" with the Teamsters union.

ANTIQUE VALENTINE EXHIBIT. Feb 9–12. Surratt House and Tavern, Clinton, MD. Display of 19th-century Valentines and memorabilia. Est attendance: 450. Sponsor: Surratt House and Tavern, 9110 Brandywine Rd, Clinton, MD 20735. Phone: (301) 868-1121.

BEHAN, BRENDAN: BIRTH ANNIVERSARY. Feb 9. Irish playwright and poet. Born at Dublin, Ireland, on Feb 9, 1923. Died there on Mar 20, 1964.

CANADA: MARDI GRAS FESTIVAL. Feb 9–13. Prince George, British Columbia. World championship snogolf, a parade and more than 30 other events. Est attendance: 70,000. For info: Mardi Gras Soc, PO Box 383, Street A, Prince George, BC, Canada V2L 4S2. Phone: (604) 564-3737.

GYPSY ROSE LEE: BIRTH ANNIVERSARY. Feb 9. American ecdysiast and author whose real name was Rose Louise Hovick. Born at Seattle, WA, on Feb 9, 1914. Her autobiography, *Gypsy*, was made into a broadway musical and a motion picture. Died at Los Angeles, CA, on Apr 26, 1970.

HARRISON, WILLIAM HENRY: BIRTH ANNIVERSARY. Feb 9. Ninth president of the US (Mar 4–Apr 4, 1841). His term of office was the shortest in our nation's history—32 days. He was the first president to die in office (of pneumonia contracted during inaugural ceremonies). Born at Berkeley, VA, Feb 9, 1773. Died at Washington, DC, Apr 4, 1841.

LOWELL, AMY: BIRTH ANNIVERSARY. Feb 9. American poet born at Brookline, MA, Feb 9, 1874. Died there May 12, 1925.

MILWAUKEE HOME IMPROVEMENT SHOW. Feb 9–12. Wisconsin State Fair Park, West Allis, WI. Co-sponsored by Milwaukee/NARI Home Improvement Council and the *Milwaukee Sentinel*, this all-member show is for remodeling contractors and is of particular interest to homeowners planning a remodeling project for their home. Est attendance: 28,500. For info: Milwaukee/NARI Home Improvement Council, Mary Fox, 13390 Watertown Plank Rd, Elm Grove, WI 53122. Phone: (414) 789-4800.

NATIONAL KRAUT AND FRANKFURTER WEEK. Feb 9–18. To celebrate the fabulous pair of sauerkraut and frankfurters as one of America's favorite taste combinations. Sponsor: Natl Kraut Packers Assn. For info: DHM Group, Inc, PO Box 767, Holmdel, NJ 07733.

RICHWOOD WINTERFEST. Feb 9–11. Richwood, WV. A romp in the West Virginia snowlands featuring a chili cook-off, snow games and cross country volksski. Est attendance: 300. Sponsor: Richwood Chamber of Commerce, 50 Oakford Ave, Richwood, WV 26261. Phone: (304) 846-6790.

TUBB, ERNEST: BIRTH ANNIVERSARY. Feb 9. Country and western singer, born at Crisp, TX, on Feb 9, 1914. Ernest Tubb was the sixth member to be elected to the Country Music Hall of Fame and the headliner on the first country music show ever to be presented at Carnegie Hall. His first major hit "Walking the Floor Over You" gained him his first appearance at the Grand Ole Opry in 1942, and he attained regular membership in 1943. He died on Sept 6, 1984, at Nashville, TN.

WASHINGTON'S BIRTHDAY CELEBRATION. Feb 9–19. Laredo, TX. Founded in 1898, the event celebrates the cultures of both the US and Mexico with festivities ranging from the Jalapeño Festival to parades, fireworks and a carnival. The largest celebration of George Washington's birthday in the nation. Est attendance: 400,000. Sponsor: Washington's Birthday Celebration Assn, PO Box 816, Laredo, TX 78042-0816. Phone: (210) 722-0589.

Feb 9-10 ☆ Chase's 1995 Calendar of Events ☆

BIRTHDAYS TODAY

Mia Farrow (Maria de Lourdes Villers), 50, actress ("Peyton Place," *Rosemary's Baby, Hannah and Her Sisters*), born at Los Angeles, CA, Feb 9, 1945.

Kathryn Grayson (Zelma Hedrick), 72, actress (*Kiss Me Kate, The Kissing Bandit*), born at Winston-Salem, NC, Feb 9, 1923.

Carole King, 53, singer, songwriter, born at Brooklyn, NY, Feb 9, 1942.

John Martin Kruk, 34, baseball player, born at Charleston, WV, Feb 9, 1961.

Judith Light, 46, actress ("One Life to Live," "Who's the Boss?"), born at Trenton, NJ, Feb 9, 1949.

Roger Mudd, 67, journalist (former anchorman "ABC Evening News"), born at Washington, DC, Feb 9, 1928.

Dean Rusk, 86, statesman, teacher, born at Cherokee County, GA, Feb 9, 1909.

Alice Walker, 51, author (*The Color Purple, Possessing the Secret of Joy*), born at Eatonton, GA, Feb 9, 1944.

James Webb, 49, former US Secretary of the Navy, born at St. Joseph, MO, Feb 9, 1946.

FEBRUARY 10 — FRIDAY
41st Day — Remaining, 324

"ALL THE NEWS THAT'S FIT TO PRINT": ANNIVERSARY. Feb 10. The familiar slogan "All the News That's Fit to Print" has appeared on page one of *The New York Times* since Feb 10, 1897. Although in October 1896 *The New York Times* offered a prize of $100 to anyone who could come up with a better slogan of 10 words or less, none was found.

ANCHORAGE FUR RENDEZVOUS. Feb 10-19. Anchorage, AK. "Alaska's largest celebration." Features four world championships in three sled dog races and dog weight-pulling contest. Carnival, native dances, Eskimo blanket toss, fur auction, snow sculpture, snow sports and much more—more than 120 events. 60th Annual Rondy! Annually, beginning the second Friday in February and running for 10 days. Est attendance: 250,000. For info: Greater Anchorage, Inc, 327 Eagle St, Anchorage, AK 99501. Phone: (907) 277-8615.

AUSTRALIA: MELBOURNE MUSIC FESTIVAL. Feb 10-19. Melbourne. Australia's largest contemporary music festival, featuring hundreds of performances by local and international artists in various venues all over Melbourne. Also incorporates the National Music Industry Conference, seminars, skill shops, Melbourne Guitar Show, master classes and Hall of Fame dinner. Annually, in February. Est attendance: 700,000. Sponsor: Dobe Newton, Exec Producer, Victorian Rock Foundation, PO Box 297, Port Melbourne, Vic 3207, Australia.

CANADA: FESTIVAL DU VOYAGEUR. Feb 10-19. Winnipeg, Manitoba. More than 400 shows, food and "joie de vivre" of the fur trade era. Sled dog races, snow sculptures and much more at Western Canada's largest winter festival. Est attendance: 150,000. For info: Festival du Voyageur, 768 Tache Ave, Winnipeg, MB, Canada R2H 2C4. Phone: (204) 237-7692.

CANADA: WINTER CARNIVAL. Feb 10-19. Bancroft, Ontario, Canada. Special attractions on Feb 18. Bancroft sled dog races. Est attendance: 3,000. For info: Bancroft and Dist Chmbr of Commerce, PO Box 539, Bancroft, Ont, Canada K0L 1C0. Phone: (613) 332-1513.

ENGLAND: JORVIK VIKING FESTIVAL. Feb 10-25. Various venues, York, North Yorkshire. Colorful festival with a viking theme which includes fireworks displays, torchlit processions, longboat races and concerts. For info: Jorvik Viking Festival Office, Piccadilly House, 55 Piccadilly, York, England YO1 1PL.

FIRST WORLD WAR II MEDAL OF HONOR: ANNIVERSARY. Feb 10. Second Lieut Alexander Ramsey ("Sandy") Nininger, Jr was posthumously awarded World War II's first Medal of Honor on Feb 10, 1942, for heroism at the Battle of Bataan. He had graduated from West Point in 1941 and was on his first assignment after being commissioned.

GOLD RUSH DAYS. Feb 10-12. Wickenburg, AZ. 47th annual communitywide celebration of the Old West: rodeo, parade, carnival, gold panning. Named one of the top 100 events in North America by the American Bus Assn. Annually, the second weekend in February. Est attendance: 43,000. Sponsor: Chamber of Commerce, Drawer CC, Wickenburg, AZ 85358. Phone: (602) 684-5479.

LAMB, CHARLES: 220th BIRTH ANNIVERSARY. Feb 10. Literary critic, poet and essayist, born at London, England, on Feb 10, 1775. "The greatest pleasure I know," he wrote in 1834, "is to do a good action by stealth, and to have it found out by accident." Died at Edmonton, England, Dec 27, 1834.

LONGHORN WORLD CHAMPIONSHIP RODEO. Feb 10-12. Riverfront Coliseum, Cincinnati, OH. More than 200 of North America's best cowboys and cowgirls compete in 6 professionally sanctioned contests ranging from bronc riding to bull riding for top money and world championship points. Additional attractions include special animal trainers, trick ropers, trick riders and professional clowns. 21st annual. Est attendance: 20,000. For info: Longhorn World Chmpshp Rodeo, Inc, PO Box 70159, Nashville, TN 37207. Phone: (615) 876-1016.

NATIONAL BASKETBALL ASSOCIATION: ALL-STAR WEEKEND. Feb 10-12. America West Arena, Phoenix, AZ. For info: Brian McIntyre, VP, PR, Natl Basketball Assn, Olympic Tower, 645 Fifth Ave, New York, NY 10022. Phone: (212) 826-7000.

PASTERNAK, BORIS LEONIDOVICH: 105th BIRTH ANNIVERSARY. Feb 10. Russian poet and novelist, born at Moscow, Russia, Feb 10, 1890. Best-known work: *Doctor Zhivago*. Died at Moscow, May 30, 1960.

PLIMSOLL DAY. Feb 10. A day to remember Samuel Plimsoll, "The Sailor's Friend," a coal merchant turned reformer and politician, who was elected to Parliament in 1868. He attacked the practice of overloading heavily insured ships, calling them "coffin ships." His persistence brought about amendment of Britain's Merchant Shipping Act. The Plimsoll Line, named for him, is a line on the side of ships marking maximum load allowed by law. Born at Bristol, England, Feb 10, 1824. Died at Folkestone, England, June 3, 1898.

PROFESSIONAL SECRETARIES INTERNATIONAL LEADERSHIP CONFERENCE. Feb 10-12. Philadelphia, PA. Est attendance: 150. For info: Professional Secretaries Intl, Dale Shuter, Conv Mgr, PO Box 20404, Kansas City, MO 64195-0404. Phone: (816) 891-6600.

STAMP EXPO/USA. Feb 10-12. Grand Hotel, Anaheim, CA. Fax: (818) 988-4337. Est attendance: 4,000. Sponsor: Intl Stamp Collectors Society, PO Box 854, Van Nuys, CA 91408. Phone: (818) 997-6496.

February 1995

S	M	T	W	T	F	S
			1	2	3	4
5	6	7	8	9	10	11
12	13	14	15	16	17	18
19	20	21	22	23	24	25
26	27	28				

Chase's 1995 Calendar of Events — Feb 10-11

TWENTY-FIFTH AMENDMENT TO US CONSTITUTION (PRESIDENTIAL SUCCESSION, DISABILITY): RATIFICATION ANNIVERSARY. Feb 10. Procedures for presidential succession were further clarified by the 25th Amendment, along with provisions for continuity of power in the event of a disability or illness of the president. The 25th Amendment was ratified on Feb 10, 1967.

WASHINGTON BIRTHDAY FESTIVAL. Feb 10-12. Pendleton Park, Eustis, FL. Juried arts and crafts show. Est attendance: 40,000. Sponsor: Chamber of Commerce, Arts and Crafts Chairperson, Box 1210, Eustis, FL 32727-1210. Phone: (904) 357-3434.

WHITE, WILLIAM ALLEN: BIRTH ANNIVERSARY. Feb 10. American newspaperman, owner and editor of the *Emporia Gazette.* Coined the phrase "tinhorn politician" and, in one obituary, wrote of the deceased that he had "the talent of a meat-packer, the morals of a money changer and the manners of an undertaker." Born at Emporia, KS, Feb 10, 1868. Died there Jan 29, 1944.

YUMA JAYCEE'S SILVER SPUR RODEO. Feb 10-12. Yuma County Fairgrounds, Yuma, AZ. Three-day rodeo features professional rodeo cowboy action, preceded by a week of activities. Annually, the second weekend in February. Sponsors: Budweiser, Dodge, Pepsi, Best Western Inn Suites and Jack-in-the-Box. Est attendance: 28,000. For info: Yuma Jaycees, Steve Ames, Rodeo Chair, 1798 S Arizona Ave, Yuma, AZ 85364. Phone: (602) 783-3641.

BIRTHDAYS TODAY

Donovan (Donovan P. Leitch), 49, singer, songwriter, born at Glasgow, Scotland, Feb 10, 1946.
Lenny (Leonard Kyle) Dykstra, 32, baseball player, born at Santa Ana, CA, Feb 10, 1963.
Roberta Flack, 56, singer, born at Black Mountain, NC, Feb 10, 1939.
Kirk Fordice, 61, Governor of Mississippi (R), born at Memphis, TN, Feb 10, 1934.
Dennis Louis Gentry, 36, former football player, born at Lubbock, TX, Feb 10, 1959.
Frances Moore Lappe, 51, author (*Diet For a Small Planet, Rediscovering America's Values*), born at Pendleton, OR, Feb 10, 1944.
Greg Norman, 40, golfer, born at Melbourne, Australia, Feb 10, 1955.
Leontyne Price, 68, opera singer, born at Laurel, MS, Feb 10, 1927.
Mark Spitz, 45, Olympic gold medalist swimmer, born at Modesto, CA, Feb 10, 1950.
Robert Wagner, 65, actor ("It Takes a Thief," "Hart to Hart"), born at Detroit, MI, Feb 10, 1930.

FEBRUARY 11 — SATURDAY
42nd Day — Remaining, 323

AFRICA'S LEGACY IN MEXICO: PHOTOGRAPHS BY TONY GLEATON. Feb 11-Mar 26. Toledo Museum of Art, Toledo, OH. Costeños, black Mexicans who live in the coastal states of Guerrero, Oaxaca and Veracruz, have held on to their African heritage through centuries of racial mixing despite the isolation and bigotry that still threaten them. Gleaton chronicles their survival on film, recognizing the Costeños for their rich contributions to the nation's history. Call or write the Smithsonian for other dates and venues. For info: Smithsonian Institution Traveling Exhibition Service, 1100 Jefferson Dr SW, Ste 3146, Washington, DC 20560. Phone: (202) 357-2700.

ARIZONA RENAISSANCE FESTIVAL. Feb 11-Mar 26. (Saturdays and Sundays only.) Apache Junction, AZ. The King and his court, duelists and faire maidens, crafts, food and games, jousting knights and all the other fun of the days of yore. The official sister event to the Robin Hood Festival in Sherwood Forest, England. For info: Arizona Renaissance Festival, 12601 E Highway 60, Apache Junction, AZ 85219. Phone: (602) 463-2700.

BATTLE OF WEST POINT AND PRAIRIE. Feb 11-12. Columbus and West Point, MS. Civil war reenactment with campsites, battles, entertainment and sutler stores taking place on Windmill Rd (off Hwy 50) between Columbus and West Point. Annually, the second weekend in February. Sponsor: Possum Tour Reenactors, Inc. Est attendance: 2,500. For info: Carolyn Denton, Dir, Columbus Conv & Visitors Bureau, PO Box 789, Columbus, MS 39703. Phone: (800) 327-2686.

BLACKPOWDER TRADE FAIR AND GUN SHOW. Feb 11-12. National Guard Armory, Albert Lea, MN. Reenactors and craftsmen from 7 states dress in full costume and sell their fur trade period items. Entertainment, food and drink, workshops and demonstrations. Est attendance: 3,000. For info: Big Island Rendezvous, Inc, Box 686, Albert Lea, MN 56007. Phone: (800) 658-2526.

CAMELLIA FESTIVAL. Feb 11-19. Massee Lane Gardens, Fort Valley, GA. 5th annual celebration of the beautiful flower, the camellia. Events include: garden tours and workshops, art show and sale, fashion show and luncheon, Candlelight and Camellia Ball and a tour of homes. Call for more details and ticket prices. Annually, the second week of February. Sponsor: The American Camellia Society. Fax: (912) 967-2083. Est attendance: 4,000. For info: Vennie Moss, Activities Dir, Massee Lane Gardens, One Massee Lane, Fort Valley, GA 31030. Phone: (912) 967-2722.

CAMEROON: YOUTH DAY. Feb 11. Public holiday.

CANADA: YUKON QUEST 1,000-MILE SLED DOG RACE. Feb 11. Whitehorse, Yukon. The 12th annual "challenge of the North" 1,000-mile sled dog race to Fairbanks, Alaska, from Whitehorse, Yukon. Top mushers from across North America compete for the $100,000 purse. Est attendance: 3,500. For info: Yukon Tourism, Box 2703, Whitehorse, Yukon, Canada Y1A 2C6. Phone: (403) 668-4711.

CHOCOLATE FESTIVAL. Feb 11. Firehouse Art Center, Norman, OK. Tasting Sessions and tantalizing displays of chocolate as an art form are a treat for the eyes as well as the palate. Annually, the weekend before Valentine's Day. Est attendance: 3,000. For info: The Firehouse Art Center, 444 S Flood, Norman, OK 73069. Phone: (405) 329-4523.

COOKBOOK FESTIVAL. Feb 11-12. West Acres Shopping Center, Fargo, ND. Open to churches and non-profit organizations to sell their cookbooks. Est attendance: 44,000. For info: Dee Lander, Promotions/Marketing, West Acres Office, West Acres Shopping Center, Fargo, ND 58103. Phone: (701) 282-2222.

CORVETTE AND HIGH PERFORMANCE WINTER MEET. Feb 11-12. West Washington Fairgrounds, Puyallup, WA. Automotive-related buying and selling of new and used cars and parts. Heavy on Corvettes and GM cars of '60s and '70s. Held in conjunction with Silver's Collector Car Auction. Est attendance: 11,000. For info: Larry or Karen Johnson, Show Organizers, PO Box 7753, Olympia, WA 98507. Phone: (206) 786-8844.

DOG SLED WEIGHT PULL CHAMPIONSHIPS AND WINTER CARNIVAL. Feb 11-12. Christmas Mountain Village, Wisconsin Dells, WI. Sled dog weight pull contest. Sleigh rides, chili cook-off and skiing, downhill ski competition and winter golf. Craft fair, refreshments, sky diving, fireworks, children's activities, live entertainment and woodsplitting contests. Est attendance: 4,000. For info: Wisconsin Dells Visitor and Conv Bureau, Box 390, Wisconsin Dells, WI 53965. Phone: (800) 223-3557.

Feb 11 ☆ *Chase's 1995 Calendar of Events* ☆

EDISON, THOMAS ALVA: BIRTH ANNIVERSARY. Feb 11. American inventive genius and holder of more than 1,200 patents (including the incandescent electric lamp, phonograph, electric dynamo and key parts of many now familiar devices such as the movie camera, telephone transmitter, etc.). Edison said, "Genius is 1 percent inspiration and 99 percent perspiration." His birthday is now widely observed as Inventor's Day. Born at Milan, OH, Feb 11, 1847. Died at Menlo Park, NJ, Oct 18, 1931.

FULLER, MELVILLE WESTON: BIRTH ANNIVERSARY. Feb 11. Eighth chief justice of the US Supreme Court. Born at Augusta, ME, Feb 11, 1833. Died at Sorrento, ME, July 4, 1910.

GALESBURG CHOCOLATE FESTIVAL. Feb 11–12. Browning Mansion, Galesburg, IL. A chocolate lover's dream! Homemade and commercially made chocolates, tortes, cakes, pies and creams—all you can eat for a small admission fee. Est attendance: 700. For info: Galesburg Area Conv and Visitors Bureau, PO Box 631, Galesburg, IL 61402-0631. Phone: (309) 343-1194.

IMPERIAL VALLEY PRODUCE CHARITY BALL. Feb 11. Imperial Valley, CA. To focus attention on the billion-dollar produce industry in Imperial Valley. Est attendance: 3,000. Sponsor: Los Vigilantes, Box 3006, El Centro, CA 92244. Phone: (619) 352-3681.

IRAN: NATIONAL DAY. Feb 11. National holiday observed in Islamic Republic of Iran.

JAPAN: NATIONAL FOUNDATION DAY: 2655th ANNIVERSARY. Feb 11. Marks the founding of the Japanese nation. In 1872 the government officially set Feb 11, 660 BC, as the date of accession to the throne of the Emperor Jimmu (said to be Japan's first emperor) and designated the day a national holiday by the name of Empire Day. The holiday was abolished after World War II, but was revived as National Foundation Day in 1966. Ceremonies are held with Their Imperial Majesties the Emperor and Empress, the Prime Minister and other dignitaries attending. National holiday.

KNIGHTS OF ST. YAGO ILLUMINATED NIGHT PARADE. Feb 11. Ybor City, Tampa, FL. Dozens of illuminated floats and parade units, the Gasparilla pirates, dignitaries and marching bands light up the night in the city's biggest night parade. Est attendance: 200,000. For info: Tampa/Hillsborough Conv and Visitors Assn, 111 Madison St, Ste 1010, Tampa, FL 33602-4706. Phone: (800) 448-2672.

MANDELA, NELSON: PRISON RELEASE 5th ANNIVERSARY. Feb 11. At 4:14 PM (locally), Feb 11, 1990, after serving more than 27½ years of a life sentence (convicted, with eight others, of sabotage and conspiracy to overthrow the government), South Africa's Nelson Mandela, 71 years old, walked away from the Victor Verster prison farm at Paarl, South Africa, a free man. He had survived the cruelty accorded to blacks by a governmental system of apartheid, the legally enforced separation and oppression of non-white persons. Mandela greeted a cheering throng of well-wishers, along with hundreds of millions of television viewers worldwide, with demands for an intensification of the struggle for equality for blacks, who make up nearly 75% of South Africa's population.

MANKIEWICZ, JOSEPH L.: BIRTH ANNIVERSARY. Feb 11. Oscar-winning American film writer, director and producer was born at Wilkes-Barre, PA, Feb 11, 1909. His career in film began in Berlin, where while working as a reporter he also translated silent-film intertitles. With the aid of his brother, Herman (Oscar winner as co-writer of *Citizen Kane*), he became a Hollywood screenwriter and received his first Oscar nomination for *Skippy* in 1931. He coined the famous W.C. Fields phrase "my little chickadee" for the 1932 film *If I Had a Million*. His producing career began with Metro-Goldwyn-Mayer in 1935 and included *The Philadelphia Story*, *Strange Cargo* and *Woman of the Year*. His directorial debut came in 1946 with *Dragonwyck* and his stature grew with such films as *The Late George Apley, The Ghost and Mrs. Muir, A Letter to Three Wives, All About Eve, Five Fingers, Guys and Dolls, Cleopatra* and *Sleuth*, among others. Mankiewicz won four Academy awards for directing and screenwriting. He died Feb 5, 1993, at Mount Kisco, NY.

RACE TO THE SKY DOG SLED RACE. Feb 11–18. Starts near Helena, MT. This 500-mile race along the Continental Divide is the longest in the lower 48 states. There is also a 300-mile race. Annually, the second weekend in February. Phone: (406) 442-2335 or (406) 442-4008. Est attendance: 10,000. For info: Montana Sled Dog, Inc., 736 N Ewing, Helena, MT 59601.

SAN FRANCISCO CHINESE NEW YEAR GOLDEN DRAGON PARADE. Feb 11. San Francisco, CA. The largest Chinese New Year Parade outside Mainland China. Est attendance: 400,000. For info: Wayne Hu, Chair, Chinese New Year Parade, PO Box 2653, San Francisco, CA 94126. Phone: (415) 391-9680.

SOCIETY OF ILLUSTRATORS ANNUAL EXHIBITION. Feb 11–Mar 11. (Also Mar 18–Apr 15.) Museum of American Illustration, New York, NY. A juried exhibition for the best in contemporary American illustration in four categories: book, editorial, advertising and institutional. Deadline for entries is Oct 1, annually. Est attendance: 7,000. Sponsor: Soc of Illustrators, Terry Brown, Dir, 128 E 63rd St, New York, NY 10021. Phone: (212) 838-2560.

SPACE MILESTONE: FIRST SOVIET COMMERCIAL SATELLITE MISSION: 5th ANNIVERSARY. Feb 11. Anatoly Solovyov and Aleksandr Balandin departed the Baikonur launching site, Feb 11, 1990, on the Soviet Union's first satellite mission designed for profit—by producing industrial crystals in the weightlessness of space. The craft arrived at the *Mir* orbital space station on Feb 13. Launching of the *Soyuz TM-9* capsule was witnessed by four American astronauts and televised live. The mission was hailed as initiating a new level of openness of information about Soviet space projects.

SPACE MILESTONE: *OSUMI* (JAPAN): 25th ANNIVERSARY. Feb 11. First Japanese satellite launched Feb 11, 1970. Japan became fourth nation to send a satellite into space.

TAIWAN: TOURISM WEEK. Feb 11–17. Tourism Week embraces Tourism Day and the three days preceding and succeeding it. See also: "Taiwan: Lantern Festival and Tourism Day" (Feb 14).

TEDDY BEAR TEA. Feb 11. Barstow Branch Library, Barstow, CA. Teddy bears and their people gather to celebrate a kid's best friend (no matter what the kid's age). Est attendance: 125. Sponsor: Barstow Branch, San Bernardino County Library, Paul Kopach or Karen Wosika, 304 E Buena Vista, Barstow, CA 92311. Phone: (619) 256-4850.

VALENTINE HEART THROB BIATHLON. Feb 11. Nome, AK. Battle the elements of an Alaskan winter to run the outdoor 5K followed by a 1K indoor swim. Est attendance: 50. For info: Nome Community Schools, PO Box 131, Nome, AK 99762. Phone: (907) 443-5161.

VATICAN CITY: INDEPENDENCE ANNIVERSARY. Feb 11. The Lateran Treaty, signed by Pietro Cardinal Gasparri and Benito Mussolini on Feb 11, 1929, guaranteed the independence of the State of Vatican City and recognized the sovereignty of the Holy See over it. Area is about 109 acres.

WATER CLOSET INCIDENT: 35th ANNIVERSARY. Feb 11. Jack Paar, then host of "Tonight," walked out of his late-night TV show on Feb 11, 1960. The incident was prompted by NBC's censoring of a slightly off-color "water closet" joke the previous night. After a meeting with network officials, Paar agreed to return to the show on Mar 7.

February 1995

S	M	T	W	T	F	S
			1	2	3	4
5	6	7	8	9	10	11
12	13	14	15	16	17	18
19	20	21	22	23	24	25
26	27	28				

☆ Chase's 1995 Calendar of Events ☆ Feb 11-12

WHITE SHIRT DAY. Feb 11. Anniversary of UAW-GM agreement following 44-day sit-down strike at General Motors' Flint, MI, factories in 1937. "Blue-collar" workers traditionally wear white shirts to work on this day, symbolic of workingman's dignity won. Has been observed by proclamation in Flint, MI.

BIRTHDAYS TODAY

Lloyd Bentsen, 74, US Secretary of Treasury in Clinton administration, born at Mission, TX, Feb 11, 1921.
Paul Bocuse, 69, chef, born at Collonges-au-Mont-d'Or, France, Feb 11, 1926.
Mel Carnahan, 61, Governor of Missouri (D), born at Birch Tree, MO, Feb 11, 1934.
Eva Gabor, 74, actress (Lisa Douglas in "Green Acres," sisters Zsa Zsa and Magda Gabor), born at Budapest, Hungary, Feb 11, 1921.
Virginia Johnson, 70, psychologist, born at Springfield, MO, Feb 11, 1925.
Mike Leavitt, 44, Governor of Utah (R), born at Cedar City, UT, Feb 11, 1951.
Tina Louise, 61, actress (Ginger Grant in "Gilligan's Island's"; *The Stepford Wives*), born at New York, NY, Feb 11, 1934.
Sergio Mendes, 54, musician, band leader, born at Niteroi, Brazil, Feb 11, 1941.
Leslie Nielsen, 73, actor (*Naked Gun* films, *Airplane!*, "Peyton Place"), born at Regina, Sask, Canada, Feb 11, 1922.
Burt Reynolds, 59, actor (*Hooper, Deliverance, Cannonball Run*, "Evening Shade"), born at Waycross, GA, Feb 11, 1936.
Sidney Sheldon, 78, author (*Bloodline, The Doomsday Conspiracy*), born at Chicago, IL, Feb 11, 1917.

FEBRUARY 12 — SUNDAY
43rd Day — Remaining, 322

ADAMS, LOUISA CATHERINE JOHNSON: BIRTH ANNIVERSARY. Feb 12. Wife of John Quincy Adams, sixth president of the US. Born at London, England, Feb 12, 1775. Died May 14, 1852.

AMERICAN COUNCIL ON EDUCATION ANNUAL MEETING. Feb 12-15. Hyatt Regency Embarcadero, San Francisco, CA. Est attendance: 1,200. For info: Amer Council on Educ, Stephanie Marshall, One Dupont Circle, Washington, DC 20036. Phone: (202) 939-9410.

ARCA 200 LATE MODEL STOCK CAR RACE. Feb 12. Daytona International Speedway, Daytona Beach, FL. For info: Daytona Intl Speedway, Glenn Barber, PR, Box 2801, Daytona Beach, FL 32120-2801. Phone: (904) 254-6782.

BRADLEY, OMAR NELSON: BIRTH ANNIVERSARY. Feb 12. A graduate with Dwight D. Eisenhower in the class of 1915 at the US Military Academy, Bradley became Ike's Protégé as his career progressed. He returned to West Point on two occasions as a math instructor and later to teach military tactics. He was also a graduate of the Command and General Staff School and the Army War College. From 1938-41 Bradley served on the Army General Staff in Washington, becoming assistant secretary in July 1939. He reached the rank of brigadier general while commandant of the Infantry School at Fort Benning, GA, from 1941 until the US entered WWII. Appointed major general in February 1943, he was sent to North Africa as field aide to Eisenhower. As successor to Gen George S. Patton as commander of the US 2nd Corps, Bradley stormed Bizerte on May 7, 1943, taking 40,000 prisoners, and took part in the attack on Sicily. Bradley then went to England to participate in the planning of the invasion of Western Europe. He led the US 1st Army in the D-Day invasion of Normandy on June 5-6, 1944, then commanded the 12th Army Group of 1.2 million combat troops, the single largest command ever under one American general. His troops crossed the Rhine in March 1945 and in April met Soviet troops on the Elbe. Bradley was a calm man with a strong grasp of tactics who inspired confidence. In 1948, he succeeded Eisenhower as chief of staff of the Army and in 1949, became the first permanent chairman of the Joint Chiefs of Staff. He published his memoirs, entitled *Soldier's Story*, after his retirement in 1951. Bradley was born on Feb 12, 1893, at Clark, MO, and died at New York City on Apr 8, 1981.

BUSCH CLASH OF '95. Feb 12. Daytona International Speedway, Daytona Beach, FL. For info: Daytona Intl Speedway, Glenn Barber, PR, Box 2801, Daytona Beach, FL 32120-2801. Phone: (904) 254-6782.

CELEBRATION OF LOVE WEEK. Feb 12-18. To stress the importance and value of love in making the world a much better place in which to live. For complete info, send $1 to cover expense of printing, handling and postage. Annually, the second full week of February. Sponsor: Intl Soc of Friendship and Good Will, Dr. Stanley Drake, Pres, 9538 Summerfield St, Spring Valley, San Diego Cty, CA 91977-2852. Phone: (619) 466-8882.

DARWIN, CHARLES ROBERT: BIRTH ANNIVERSARY. Feb 12. Author and naturalist. Born at Shrewsbury, England, Feb 12, 1809. Best remembered for his books: *On the Origin of Species by Means of Natural Selection, or the Preservation of Favoured Races in the Struggle for Life*, and *The Descent of Man and Selection in Relation to Sex*. Died at Down, Kent, England, Apr 19, 1882.

LAKE WINNEBAGO STURGEON SEASON. Feb 12-Mar 1. Fond du Lac, WI, at the foot of the lake is the Sturgeon Capital of the World. The sturgeon species have remained unchanged for more than 50 million years and the Lake Winnebago region claims one of the major populations of them today. Shanty and equipment rental available; sturgeon fishing tag required. Phone: (800) 937-9123, ext 95. For info: Fond du Lac Conv Bureau, 19 W Scott St, Fond du Lac, WI 54935.

LEWIS, JOHN LLEWELLYN: 115th BIRTH ANNIVERSARY. Feb 12. American labor leader born near Lucas, IA, Feb 12, 1880. His parents came to the US from Welsh mining towns, and Lewis left school in the seventh grade to become a miner himself. Became leader of United Mine Workers of America and champion of all miners' causes. Died at Washington, DC, June 11, 1969.

LINCOLN, ABRAHAM: BIRTH ANNIVERSARY. Feb 12. Sixteenth president of the US (Mar 4, 1861-Apr 15, 1865) and the first to be assassinated (on Good Friday, Apr 14, 1865, while watching a performance of *Our American Cousin*, at Ford's Theatre in Washington, DC). His presidency encompassed the tragic Civil War. Especially remembered are his Emancipation Proclamation (Jan 1, 1863), his Gettysburg Address (Nov 19, 1863) and his proclamation establishing the last Thursday of November as Thanksgiving Day. Lincoln's birthday is observed on this day in most states, but on the first Monday in February in Delaware and Oregon. Born in Hardin County, KY, Feb 12, 1809. Died at Washington, DC, Apr 15, 1865.

LINCOLN'S BIRTHPLACE CABIN WREATH LAYING. Feb 12. Abraham Lincoln's Birthplace NHS, Hodgenville, KY. A wreath is placed at the door of the "Birthplace Cabin" in commemoration of the birth of Abraham Lincoln. Refreshments are also served to park visitors. Est attendance: 200. For info: Gary V. Talley, 2995 Lincoln Farm Rd, Hodgenville, KY 42748. Phone: (502) 358-3137.

Feb 12-13 ☆ Chase's 1995 Calendar of Events ☆

LUXEMBOURG: BURGSONNDEG. Feb 12. Young people build a huge bonfire on a hill to celebrate the victorious sun, marking the end of winter. A tradition dating to pre-Christian times.

NATIONAL FUTURE HOMEMAKERS OF AMERICA FHA/HERO WEEK. Feb 12-19. To call the nation's attention to the activities and goals of the organization and vocational home economics education. Theme: "FHA/HERO-50 years Leading the Way." Sponsor: Future Homemakers of America, Catherine Burdette, 1910 Association Dr, Reston, VA 22091. Phone: (703) 476-4900.

OGLETHORPE DAY. Feb 12. General James Edward Oglethorpe (born at London, England, Dec 22, 1696), with some 100 other Englishmen, landed at what is now Savannah, GA, on Feb 12, 1733. Naming the new colony Georgia for England's King George II, Oglethorpe was organizer and first governor of the colony and founder of the city of Savannah. Oglethorpe Day and Georgia Day observed on this date.

SAFETYPUP'S BIRTHDAY. Feb 12. This year Safetypup, created by the National Child Safety Council, joyously celebrates his birthday of bringing safety awareness/education messages to children in a positive nonthreatening manner. Safetypup has achieved a wonderful balance of safety sense, caution and childlike enthusiasm about life and helping kids "Stay Safe and Sound." Sponsor: NCSC, Box 1368, Jackson, MI 49204-1368. Phone: (517) 764-6070.

SPACE MILESTONE: *VENERA 1* (USSR). Feb 12. Spacecraft launched from space platform in experiment projecting satellite into interplanetary space, Feb 12, 1961.

WINTER FUN FESTIVAL. Feb 12. The INN on Lake Waramaug, New Preston, CT. Snowman building, snow sculpturing and snowball throwing contests and a traditional New England ice harvest. An ice skating party all afternoon with hot stew, hot chocolate and hot toddies available for purchase. Est attendance: 400. For info: Nancy Conant, The INN on Lake Waramaug, 107 N Shore Rd, New Preston, CT 06777. Phone: (203) 868-0563.

WORLD MARRIAGE DAY. Feb 12. A special day to honor all marriages. Celebrations worldwide include: parades, dances, shopping mall events, balloon releases, dinners, religious services, renewing wedding vows, local contests for poster and essays and a national Couple Married Longest Search. Annually, the second Sunday in February. For info: Mike and Sandy Vanoudenhaegen, Natl Coords, World Marriage Day 1995, 2785 NW Cornelius, Schefflin Rd, Cornelius, OR 97113. Phone: (503) 359-9131.

WSBA and WARM 103 EASTER CRAFT SHOW. Feb 12. York, PA. Quality craft displays will be presented with everything from miniatures and ornaments to quilts and country furniture. Admission fee. Est attendance: 6,500. For info: Gina M. Koch, Promotions Dir, PO Box 910, York, PA 17405. Phone: (717) 764-1155.

BIRTHDAYS TODAY

Joe Don Baker, 59, actor (*Charlie Varrick, Cool Hand Luke*), born at Groesbeck, TX, Feb 12, 1936.
Judy Blume, 57, author (*Blubber, Superfudge*), born at Elizabeth, NJ, Feb 12, 1938.
Hank Brown, 55, US Senator (R, Colorado), born at Denver, CO, Feb 12, 1940.
Joan Finney, 70, Governor of Kansas (D), born at Topeka, KS, Feb 12, 1925.
Joe Garagiola, 69, sportscaster, former baseball player, born at St. Louis, MO, Feb 12, 1926.
Arsenio Hall, 40, comedian, actor (*Coming to America*), TV talk show host, born Cleveland, OH, Feb 12, 1955.
Joanna Kerns, 42, actress ("Growing Pains;" former national caliber gymnast), born at San Francisco, CA, Feb 12, 1953.
Eddie Robinson, 76, football coach, born at Jackson, LA, Feb 12, 1919.
Bill Russell (William Felton), 61, basketball Hall of Famer, former coach, born at Monroe, LA, Feb 12, 1934.
Arlen Specter, 65, US Senator (R, Pennsylvania), born at Wichita, KS, Feb 12, 1930.
Franco Zeffirelli, 72, film director, born at Florence, Italy, Feb 12, 1923.

FEBRUARY 13 — MONDAY
44th Day — Remaining, 321

CLEAN OUT YOUR COMPUTER DAY. Feb 13. A day dedicated to purging and organizing computer files that are devouring disk space, slowing systems and confounding users who try to locate documents. An opportunity to scrape the plaque from your information arteries and review your protocols for managing the e-mail which now inundates your electronic mailbox like paper predecessors inundated your desk. Annually, the second Monday of February. Sponsor: Institute for Business Technology, Ira Chaleff, Pres, 513 Capitol Ct NE, Washington, DC 20002. Phone: (202) 544-0097.

FIRST MAGAZINE PUBLISHED IN AMERICA: ANNIVERSARY. Feb 13. Andrew Bradford published *The American Magazine*, Feb 13, 1741, just three days ahead of Benjamin Franklin's *General Magazine*.

FIRST PUBLIC SCHOOL IN AMERICA: 360th ANNIVERSARY. Feb 13. The Boston Latin School opened on Feb 13, 1635, and is America's oldest public school.

GET A DIFFERENT NAME DAY. Feb 13. If you dislike your name, or merely find it boring, today is the day to adopt the moniker of your choice. Phone: (212) 388-8673 or (717) 274-8451. For info: Wellness Permission League, Thomas or Ruth Roy, 2105 Water St, Lebanon, PA 17046.

NATIONAL FIELD TRIAL CHAMPIONSHIP. Feb 13-24. (Monday-Friday only.) Ames Plantation, Grand Junction, TN. To select the national champion all-age bird dog. Fax: (901) 878-1068. Est attendance: 12,000. Sponsor: Natl Field Trial Champion Assn, Box 389, Grand Junction, TN 38039. Phone: (901) 878-1067.

TENNIS: INDOOR CHAMPIONSHIPS. Feb 13-19. Racquet Club of Memphis, Memphis, TN. The world's finest tennis players compete in one of the most prestigious tour events of the year. Sponsor: Volvo. For info: Tennessee Tourist Development, PO Box 23170, Nashville, TN 37202. Phone: (615) 741-2158.

TRUMAN, BESS (ELIZABETH) VIRGINIA WALLACE: BIRTH ANNIVERSARY. Feb 13. Wife of Harry S. Truman, 33rd president of the US. Born at Independence, MO, Feb 13, 1885. Died there on Oct 18, 1982.

WOOD, GRANT: BIRTH ANNIVERSARY. Feb 13. American artist especially noted for his powerful realism and satirical paintings of the American scene. Born near Anamosa, IA, on Feb 13, 1892. He was a printer, sculptor, woodworker and a high school and college teacher. Among his best-remembered works are *American Gothic, Fall Plowing* and *Stone City*. Died at Iowa City, IA, Feb 12, 1942. Buried at the Riverside Cemetery in Anamosa, IA.

February 1995

S	M	T	W	T	F	S
			1	2	3	4
5	6	7	8	9	10	11
12	13	14	15	16	17	18
19	20	21	22	23	24	25
26	27	28				

☆ Chase's 1995 Calendar of Events ☆ Feb 13-14

BIRTHDAYS TODAY

Stockard Channing (Susan Stockard), 51, actress (*Six Degrees of Separation, House of Blue Leaves*), born at New York, NY, Feb 13, 1944.

Robert J. Eaton, 55, Vice-Chairman and Chief Operating Officer at Chrysler Corp, born at Buena Vista, CO, Feb 13, 1940.

Carol Lynley, 53, actress (*Harlow, Bunny Lake Is Missing*), born at New York, NY, Feb 13, 1942.

Kim Novak (Marilyn Novak), 62, actress (*Bell, Book and Candle; Vertigo*) born at Chicago, IL, Feb 13, 1933.

Oliver Reed, 57, actor (*The Jokers, Women in Love*) born at Wimbledon, England, Feb 13, 1938.

George Segal, 61, actor (*A Touch of Class, King Rat*), born at Great Neck, NY, Feb 13, 1934.

Peter Tork, 51, singer, actor, born at Washington, DC, Feb 13, 1944.

Chuck Yeager, 72, pilot who broke sound barrier, born at Myra, WV, Feb 13, 1923.

FEBRUARY 14 — TUESDAY
45th Day — Remaining, 320

ARIZONA: ADMISSION DAY: ANNIVERSARY. Feb 14. Became 48th state on this day in 1912.

BENNY, JACK: BIRTH ANNIVERSARY. Feb 14. American comedian. Born Benjamin Kubelsky, Jack Benny entered vaudeville in Waukegan, IL, at age 17, using the violin as a comic stage prop. His radio show first aired in 1932 and continued for 20 years with little change in format. He also had a long-running television show. One of his most well-known comic gimmicks was his purported stinginess. In 1927 Benny married Mary Livingstone (Sadie Marks). He was born Feb 14, 1894, at Chicago, IL, and died Dec 26, 1974, at Beverly Hills, CA.

BULGARIA: VITICULTURISTS' DAY (TRIFON ZAREZAN). Feb 14. Celebrated since Thracian times. Festivities are based on cult of Dionysus, god of merriment and wine.

CHERISHED TREASURES: A QUILT DISPLAY. Feb 14–Mar 14. Mormon Historic Sites Vistor Center, Omaha, NE. Quilt display; no admission charged. Annually, Feb 14–Mar 14. Est attendance: 2,500. For info: Kenneth R. Barker, Dir, Mormon Historic Sites Center, 3215 State St, Omaha, NE 68104. Phone: (402) 453-9372.

CHINA: LANTERN FESTIVAL. Feb 14. Traditional Chinese festival falls on 15th day of first month of Chinese lunar calendar year. Lantern processions mark end of the Chinese New Year holiday season. See also: "Chinese New Year" (Jan 31).

KGBX VALENTINE'S DAY CARNATION GIVE AWAY. Feb 14. Springfield, MO. KGBX personnel dress in tuxedos and visit businesses and other public areas throughout Springfield, giving away single carnations to women in honor of Valentine's Day. Several thousand flowers are distributed. 6th annual give away. For info: Mark Phillips, KGBX Creative Service Dir, 840 S Glenstone, Springfield, MO 65802. Phone: (417) 869-1059.

◆ **LEAGUE OF WOMEN VOTERS 75th ANNIVERSARY CELEBRATION.** Feb 14. Local Leagues of Women Voters Chapters around the country, together with the national office, celebrate the 75th birthday of their League. For info: Public Affairs Dept League of Women, 1730 M St NW, Washington, DC 20036. Phone: (202) 429-1965.

◆ **LEAGUE OF WOMEN VOTERS FORMED: 75th ANNIVERSARY.** Feb 14. While meeting in Chicago at the Congress Hotel on Feb 14, 1920, to celebrate the imminent ratification of the 19th Amendment to the Constitution, leaders of the National American Woman Suffrage Association (NAWSA) approved the formation of a new organization—the League of Women Voters. With the vote for women just a few months away, the new organization was created to help American women exercise their new political rights and responsibilities. For info: League of Women Voters of Illinois, 332 S Michgan Ave, Ste 1142, Chicago, IL 60604. Phone: (312) 939-5935.

NATIONAL CONDOM WEEK. Feb 14–21. To educate consumers, patients, students and professionals on the prevention of sexually transmitted diseases, including AIDS and teenage pregnancies. Sponsor: Pharmacists Planning Service Inc, Frederick S. Mayer, Pres, Box 1336, Sausalito, CA 94966. Phone: (415) 332-4066.

NATIONAL HAVE-A-HEART DAY. Feb 14. To celebrate life and to create a new consumer consciousness concerning the consequences of our cuisine choices for the environment, world hunger, animal welfare and human health—especially heart health. Sponsor: Vegetarian Awareness Network/VEGANET, PO Box 321, Knoxville, TN 37901. Phone: (800) 872-8343.

OREGON: ADMISSION DAY. Feb 14. Became 33rd state on this day in 1859.

RACE RELATIONS DAY. Feb 14. A day designated by some churches to recognize the importance of interracial relations. Formerly was observed on Abraham Lincoln's birthday or on the Sunday preceding it. Since 1970 observance has generally been on Feb 14.

READ TO YOUR CHILD DAY. Feb 14. Motto: Show your kids you love them: Read to them. To encourage parents, teachers and other caregivers to engage in the wonderfully beneficial and delightfully fun practice of reading to children. For a free packet of materials, send a self-addressed, stamped envelope to the address below. Packet includes ideas for running campaigns to promote literacy plus reproducible flyers on classroom reading in schools, family reading at home and sharing books with babies. Each flyer describes the benefits of read-aloud sessions, gives tips for oral reading, and lists books people can read for more information. Annually, on Valentine's Day. For info: Dee Anderson, Children's Librarian, 1023 25 St, #1, Moline, IL 61265. Phone: (309) 755-9614.

SALMON RUSHDIE'S DEATH SENTENCE: ANNIVERSARY. Feb 14. Iranian leader Ayatollah Ruholla Khomeini, offended by *The Satanic Verses*, called on Muslims Feb 14, 1989, to kill the book's British author, Salmon Rushdie. On the following day the Ayatollah offered a $1 million reward for execution of his sentence. Rushdie, suddenly the center of international controversy and fearful for his life, went into hiding. Worldwide protests against the efforts to abridge academic and literary freedoms, countered by protests of Muslim and other religious fundamentalists, stimulated the sales of *The Satanic Verses*, but Rushdie remained virtually a prisoner unable to resume a public life because of the Ayatollah's death sentence. The reward was later increased to $2 million by an Iranian foundation.

SPACE MILESTONE: *LUNA 20* (USSR). Feb 14. Launched Feb 14, 1972. Unmanned lunar probe soft-landed on Moon, Feb 21, collected samples and returned to Earth, Feb 25.

SPACE MILESTONE: SMM (US). Feb 14. Unmanned Delta rocket, a "Solar Maximum Mission Observatory," launched Feb 14, 1980. Intended to study solar flares.

97

Feb 14-15 ☆ Chase's 1995 Calendar of Events ☆

TAIWAN: LANTERN FESTIVAL AND TOURISM DAY. Feb 14. Fifteenth day of the First Moon of the lunar calendar marks end of New Year holiday season. Lantern processions and contests.

VALENTINE'S DAY. Feb 14. St. Valentine's Day celebrates the feasts of two Christian martyrs of this name. One, a priest and physician, was beaten and beheaded on the Flaminian Way at Rome, Italy, on Feb 14, AD 269, during the reign of Emperor Claudius II (who died of the plague less than a year later). Another Valentine, the Bishop of Terni, is said to have been beheaded, also on the Flaminian Way at Rome, on Feb 14 (possibly in a later year). Both history and legend are vague and contradictory about details of the Valentines and some say that Feb 14 was selected for the celebration of Christian martyrs as a diversion from the ancient pagan observance of Lupercalia. An old legend has it that birds choose their mates on Valentine's Day. Now it is one of the most widely observed unofficial holidays. It is an occasion for the exchange of gifts (usually books, flowers or sweets) and greeting cards with affectionate or humorous messages. See also: "Lupercalia" (Feb 15).

VALENTINE'S DAY MASSACRE: ANNIVERSARY. Feb 14. Anniversary of Chicago gangland executions in 1929, when gunmen posing as police shot seven members of the George "Bugs" Moran gang.

WALLET, SKEEZIX: "BIRTHDAY." Feb 14. Comic strip character in "Gasoline Alley" by Frank King. First cartoon character to grow and age with the days, weeks and years of publication. Foundling child of Walt and Phyllis Wallet, discovered on doorstep Feb 14, 1921. Skeezix grew through childhood, marriage, military service in World War II, returning home to parenthood and business after the war. Comic strip began in *Chicago Tribune*, Aug 23, 1919.

BIRTHDAYS TODAY

Mel Allen, 82, sportscaster, born at Birmingham, AL, Feb 14, 1913.

Carl Bernstein, 51, journalist, author (investigated Watergate story with Bob Woodward), born at Washington, DC, Feb 14, 1944.

Ben Blaz, 67, Delegate to Congress (R, Guam), born at Agana, GU, Feb 14, 1928.

Hugh Downs, 74, broadcaster ("Today," "20/20"), actor, born Akron, OH, Feb 14, 1921.

Judd Gregg, 48, US Senator (R, New Hampshire), born at Nashua, NH, Feb 14, 1947.

Florence Henderson, 61, singer, actress ("The Brady Bunch"), born Dale, IN, Feb 14, 1934.

Gregory Hines, 49, dancer, actor (*Tap*), born at New York, NY, Feb 14, 1946.

Edmund George Love, 83, teacher, historian, author, born at Flushing, MI, Feb 14, 1912.

Donna Shalala, 54, US Secretary of Health and Human Services in Clinton administration, born at Cleveland, OH, Feb 14, 1941.

February 1995

S	M	T	W	T	F	S
			1	2	3	4
5	6	7	8	9	10	11
12	13	14	15	16	17	18
19	20	21	22	23	24	25
26	27	28				

FEBRUARY 15 — WEDNESDAY
46th Day — Remaining, 319

ANTHONY, SUSAN BROWNELL: 175th BIRTH ANNIVERSARY. Feb 15. American reformer and militant advocate of women's suffrage. She was arrested and fined in 1872 for voting—a criminal act if done by a woman! First American woman to have her likeness on coinage (1979, Susan B. Anthony dollar). Born at Adams, MA, Feb 15, 1820. Died at Rochester, NY, Mar 13, 1906.

ARLEN, HAROLD: 90th BIRTH ANNIVERSARY. Feb 15. American composer and songwriter. Born at Buffalo, NY, on Feb 15, 1905. Arlen wrote many popular songs, including *Over the Rainbow* (for which he won 1939 Oscar for best song), *That Old Black Magic*, *Blues in the Night* and *Stormy Weather*. Died at New York, NY, on Apr 23, 1986.

BARRYMORE, JOHN: BIRTH ANNIVERSARY. Feb 15. American actor of famous acting family. Born John Blythe at Philadelphia, PA, Feb 15, 1882. Died at Los Angeles, CA, May 29, 1942. US Postal Service stamp was issued in 1982 featuring Ethel, John and Lionel Barrymore.

BOB HOPE CHRYSLER GOLF CLASSIC. Feb 15-19. Bermuda Dunes, CA. The nation's largest sports event for charity. It features PGA pros, celebrities and amateurs. Est attendance: 100,000. For info: Connie Whelchel, Press Dir, 39000 Bob Hope Dr, Rancho Mirage, CA 92270. Phone: (619) 346-8184.

CANADA: INTERNATIONAL CURLING BONSPIEL. Feb 15-18. Dawson City, Yukon. 96th annual. Top of the World Curling Club hosts a 32-team bonspiel with prizes valued in excess of $6,000. For info: Yukon Tourism, Box 2703, Whitehorse, Yukon, Canada Y1A 2C6. Phone: (403) 993-5398.

CANADA: NORTHERN MANITOBA TRAPPERS' FESTIVAL. Feb 15-19. The Pas, Manitoba. Since 1916 this 5-day event has commemorated the annual winter gathering of the fur trappers. For info: Northern Manitoba Trappers' Fest, Sonny Lavallee, Box 475, The Pas, Manitoba, Canada R9A 1K6. Phone: (204) 623-5483.

CLARK, ABRAHAM: BIRTH ANNIVERSARY. Feb 15. Signer of the Declaration of Independence, farmer and lawyer. Born Feb 15, 1726. Died Sept 15, 1794.

DESERT FOOTHILLS MUSIC FEST. Feb 15-18. Cactus Shadows Fine Arts Center, also Feb 16 and 19 at Desert Hills Presbyterian Church, Carefree, AZ. A five-day festival of classical music performed by artists from across the US features chamber music ensembles, solos and vocalists. Admission charged. Est attendance: 1,500. For info: Desert Foothills Music Fest, PO Box 5254, Carefree, AZ 85377. Phone: (602) 488-0806.

LUPERCALIA. Feb 15. Anniversary of ancient Roman fertility festival. Thought by some to have been established by Romulus and Remus who, legend says, were suckled by a she-wolf at Lupercal (a cave in Palestine). Goats and dogs were sacrificed. Lupercalia celebration persisted until the fifth century of the Christian era. Possibly a forerunner of Valentine's Day customs.

MENENDEZ DE AVILES, PEDRO: BIRTH ANNIVERSARY. Feb 15. Spanish explorer and naval adventurer. Born Feb 15, 1519. Explored Florida coastal regions for king of Spain and established a fort at St. Augustine in September 1565. Died Sept 17, 1574, at Santander, Spain.

MILWAUKEE BOAT SHOW AT MECCA. Feb 15-19. MECCA, Milwaukee, WI. This event brings together buyers and sellers of sail and power boats, including fishing boats, pontoons and boating accessories, as well as vacation property and travel destinations. For info: Henri Boucher, ShowSpan, Inc, 1400 28th St SW, Grand Rapids, MI 49509. Phone: (616) 530-1919.

MOON PHASE: FULL MOON. Feb 15. Moon enters Full Moon phase at 7:15 AM, EST.

☆ Chase's 1995 Calendar of Events ☆ Feb 15-16

PONTIAC SILVERDOME BOAT, SPORT AND FISHING SHOW. Feb 15–19. Pontiac Silverdome, Pontiac, MI. This event brings together buyers and sellers of boating, fishing and outdoor sporting products. US and Canadian hunting and fishing trips, as well as other vacation travel destinations are featured. Est attendance: 35,000. For info: Henri Boucher, ShowSpan, Inc, 1400 28th St SW, Grand Rapids, MI 49509. Phone: (616) 530-1919.

REMEMBER THE *MAINE* DAY. Feb 15. American battleship, *Maine*, was blown up while at anchor in Havana harbor, at 9:40 PM, on this day in 1898. The ship, under the command of Captain Charles G. Sigsbee, sank quickly, and 260 members of its crew were lost. Inflamed public opinion in the US ignored the lack of evidence to establish responsibility for the explosion. "Remember the *Maine*" became the war cry, and a formal declaration of war against Spain followed on Apr 25, 1898.

SENIOR DAY. Feb 15. (Also Apr 19, June 21, Oct 18.) Milwaukee Art Museum, Milwaukee, WI. A special day for adults 55+ featuring free admission for senior citizens from 10–2, discounts at Arts Deli and the Art Museum Gift Shop and complimentary refreshments. Docent-guided exhibition tours will also be featured. Est attendance: 250. For info: Milwaukee Art Museum, 750 N Lincoln Memorial Dr, Milwaukee, WI 53202. Phone: (414) 224-3200.

SPANISH WAR MEMORIAL DAY AND MAINE MEMORIAL DAY. Feb 15. Massachusetts.

TIFFANY, CHARLES LEWIS: BIRTH ANNIVERSARY. Feb 15. American jeweler whose name became synonymous with high standards of quality. Born at Killingly, CT, Feb 15, 1812, and died at New York, NY, on Feb 18, 1902. Father of artist Louis Comfort Tiffany. See also: "Tiffany, Louis Comfort: Birth Anniversary" (Feb 18).

BIRTHDAYS TODAY

Adolfo (Adolfo F. Sardina), 62, fashion designer, born at Havana, Cuba, Feb 15, 1933.
Marisa Berenson, 47, actress (*Cabaret, Barry Lyndon*), model, born at New York, NY, Feb 15, 1948.
Claire Bloom, 64, actress (*Intimate Contact, Look Back in Anger*), born at London, England, Feb 15, 1931.
Susan Brownmiller, 60, author, feminist (*Against Our Will, Feminity*), born at Brooklyn, NY, Feb 15, 1935.
Matt Groening, 41, cartoonist ("The Simpsons"), born at Portland, OR, Feb 15, 1954.
Harvey Korman, 68, actor, comedian (Two Emmys for "The Carol Burnett Show"; *High Anxiety*), born at Chicago, IL, Feb 15, 1927.
Melissa Manchester, 44, singer ("Don't Cry Out Loud"), born at the Bronx, NY, Feb 15, 1951.
Jane Seymour, 44, actress (Emmy for "East of Eden," "Dr. Quinn: Medicine Woman"), born at Hillingdon, England, Feb 15, 1951.

FEBRUARY 16 — THURSDAY
47th Day — Remaining, 318

BERGEN, EDGAR: BIRTH ANNIVERSARY. Feb 16. Actor, radio entertainer and ventriloquist, voice of Charlie McCarthy, Mortimer Snerd and Effie Klinker. Born at Chicago, IL, Feb 16, 1903. Died at Las Vegas, NV, Sept 30, 1978.

BIG TEN WOMEN'S SWIMMING AND DIVING CHAMPIONSHIP. Feb 16–18. IUPUI. Indianapolis, IN. For info: Big Ten Conference, 1500 W Higgins Rd, Park Ridge, IL 60068-6300. Phone: (708) 696-1010.

CULTURAL DIVERSITY DAY. Feb 16. Westboro, MA. Youth raising awareness about cultural diversity through activities and education. Annually, Feb 16. Sponsor: Westboro Youth & Family Services, Explorer Post 24, Town Hall, 34 W Main St, Westboro, MA 01581. Phone: (508) 366-3090.

FLAHERTY, ROBERT JOSEPH: BIRTH ANNIVERSARY. Feb 16. American filmmaker, explorer and author, called "father of the documentary film." Born at Iron Mountain, MI, Feb 16, 1884. Died at Dunnerston, VT, July 23, 1951. Films included *Nanook of the North, Moana* and *Man of Aran*.

GATORADE TWIN 125-MILE QUALIFYING RACES (FOR DAYTONA 500). Feb 16. Daytona International Speedway, Daytona Beach, FL. Qualifying races for the Daytona 500. Sponsor: Gatorade. For info: Daytona Intl Speedway, Glenn Barber, PR, Box 2801, Daytona Beach, FL 32120-2801. Phone: (904) 254-6782.

KING, WAYNE: BIRTH ANNIVERSARY. Feb 16. American saxophonist and band leader, widely known as "the Waltz King." Born at Savannah, IL, on Feb 16, 1901. His own composition, "The Waltz You Save for Me," was his theme song. Died at age 84 at Paradise Valley, AZ, on July 16, 1985.

LITHUANIA: INDEPENDENCE DAY. Feb 16. National Day. The anniversary of Lithuania's Feb 16, 1918, declaration of independence is observed by many Lithuanians, especially those outside Lithuania, as the Baltic state's Independence Day. Dec 16 is recognized as the anniversary of statehood and of the proclamation of Lithuania as a Soviet Republic in 1918. In 1940, Lithuania became a part of the Soviet Union, under an agreement between Joseph Stalin and Adolf Hitler. On Mar 11, 1990, Lithuania declared its independence from the Soviet Union, holding free elections and severing ties with the Communist party in Moscow. Lithuania was the first of the Soviet republics to declare its independence. After demanding independence, Lithuania set up a border police force and aided young men in efforts to avoid the Soviet military draft, prompting then Soviet leader Mikhail Gorbachev to send tanks into the capital of Vilnius and impose oil and gas embargos. In the wake of the failed coup attempt in Moscow on Aug 19, 1991, Lithuanian independence finally was recognized.

MIAMI INTERNATIONAL BOAT SHOW AND SAILBOAT SHOW. Feb 16–22. Miami Beach Convention Center, Miami Beach, FL. (Miami Beach Marina and Biscayne Bay Mariott.) Sponsor: NMMA, 1770 Bay Rd, Miami Beach, FL 33139. Phone: (305) 531-8410.

RICHARDS, IVOR ARMSTRONG: BIRTH ANNIVERSARY. Feb 16. Author of many books on literary criticism, himself a "critic's critic." Born at Sandbach, Cheshire, England, Feb 16, 1893. Died at Cambridge, England, Sept 7, 1979.

WILSON, HENRY: BIRTH ANNIVERSARY. Feb 16. Eighteenth vice president of the US (1873–1875). Born at Farmington, NH, Feb 16, 1812. Died at Washington, DC, Nov 22, 1875.

BIRTHDAYS TODAY

Sonny Bono (Salvatore Bono), 60, singer ("I Got you Babe," "The Beat Goes On" [with Cher]), born Detroit, MI, Feb 16, 1935.
LeVar Burton, 38, actor ("Roots," "Star Trek: Next Generation"), born at Landsthul, Germany, Feb 16, 1957.
James Ingram, 39, singer ("Baby, Come to Me" with Patti Austin), songwriter, born at Akron, OH, Feb 16, 1956.
George Frost Kennan, 91, historian, diplomat, born at Milwaukee, WI, Feb 16, 1904.
John Patrick McEnroe, Jr, 36, tennis player, born at Wiesbaden, Germany, Feb 16, 1959.
(William) Mark Price, 31, basketball player, born at Bartlesville, GA, Feb 16, 1964.

Feb 16-17 ☆ Chase's 1995 Calendar of Events ☆

Kelly Tripucka, 36, former basketball player, born at Glen Ridge, NJ, Feb 16, 1959.
Leonard F. Woodcock, 84, labor union official, statesman, born at Providence, RI, Feb 16, 1911.

FEBRUARY 17 — FRIDAY
48th Day — Remaining, 317

ATLANTIC CITY CLASSIC CAR AUCTION, FLEA MARKET AND ANTIQUE SHOW. Feb 17-19. Convention Center, Atlantic City, NJ. Hundreds of antique and classic cars on display and on sale. Est attendance: 45,000. For info: Greater Atlantic City Conv and Visitors Bureau, 2314 Pacific Ave, Atlantic City, NJ 08401. Phone: (609) 348-7100.

BARBER, WALTER LANIER "RED": BIRTH ANNIVERSARY. Feb 17. Sports reporter who along with Mel Allen became one of the first broadcasters inducted into the Baseball Hall of Fame, "Red" Barber was born Feb 17, 1908, in Columbus, MS. Barber's first professional play-by-play experience was announcing the Cincinnati Reds opening day on radio in 1934. That game was also the first major league game he had ever seen. He broadcast baseball's first night game (in Brooklyn) on Aug 26, 1939, the game in which Jackie Robinson broke the color barrier in 1947 and while a Yankee radio announcer he called Roger Maris's 61st home run in 1961. "Red" Barber died Oct 22, 1992, at Tallahassee, FL.

BOSKONE 32. Feb 17-19. Framingham, MA. New England's regional science fiction convention. Includes program, a large professional art show and numerous other activities. Est attendance: 850. For info: New England Science Fiction Assn, PO Box 809, Framingham, MA 01701-0203. Phone: (617) 625-2311.

CANADA: CALGARY WINTER FESTIVAL. Feb 17-26. Calgary, Alberta. Sporting, cultural and community events are presented during the festival's ten days. Something for everyone to enjoy! Est attendance: 50,000. For info: Calgary Winter Festival, 710, 237-8th Ave, SE, Calgary, AB, Canada T2G 5C3. Phone: (403) 268-2688.

CANADA: FROSTBITE MUSIC FESTIVAL. Feb 17-19. Whitehorse, Yukon. Concert from Friday, 7:00 PM to Sunday at midnight. All-day workshops on Saturday and Sunday. Evening dance on Friday and Saturday night. Family dance and Kidsartfest on Sunday. Musicians and entertainers from across Canada and Yukon Territory. Est attendance: 2,500. For info: Frostbite Music Soc, PO Box 5004, Whitehorse, Yukon, Canada Y1A 4S2. Phone: (403) 668-4921.

COLUMBIA SURRENDERS TO SHERMAN: 130th ANNIVERSARY. Feb 17. On the morning of Feb 17, 1865, city officials of the capital city of South Carolina, Columbia, rode out to Union lines to surrender the city to Union General William Tecumseh Sherman. The city was occupied during the day by the Union forces, and during the night fires broke out which eventually consumed two-thirds of the city. Although Sherman blamed retreating Southern soldiers for the fires, the symbol of a burning Columbia stands in Southern memory as a testament to the cruelty of Sherman's army.

FLORIDA 200. Feb 17. Daytona Beach, FL. For info: Daytona Intl Speedway, Glenn Barber, PR, Box 2801, Daytona Beach, FL 32120-2801. Phone: (904) 254-6782.

FORT SUMTER RETURNED TO UNION CONTROL: 130th ANNIVERSARY. Feb 17. After a siege that lasted almost a year and a half, Fort Sumter in South Carolina returned to Union hands on Feb 17, 1865. The site of the first shots fired in the American Civil War, the fort had become a symbol for both sides during the war. As Union attempts to retake it by shelling diminished the fort's capacity with large bombardments, Southern forces managed to hold out with few casualties.

FRANCE: NICE CARNIVAL. Feb 17-28. Dates from the 14th century and is celebrated each year during the 12 days ending with Shrove Tuesday. Derived from ancient rites of spring, the carnival offers parades, floats, battles of flowers and confetti, a fireworks display lighting up the entire Baie des Anges. King Carnival is burnt on his pyre at the end of the event.

GERONIMO: DEATH ANNIVERSARY. Feb 17. American Indian of the Chiricahua (Apache) tribe was born about 1829 in Arizona. He was the leader of a small band of warriors whose devastating raids in Arizona, New Mexico and Mexico caused the US Army to send 5,000 men to recapture him after his first escape. He was confined at Fort Sill, OK, where he died on Feb 17, 1909, after dictating, for publication, the story of his life.

HOGGETOWNE MEDIEVAL FAIRE. Feb 17-19. Alachua County Fairgrounds, Gainesville, FL. 9th annual Faire features jousting, birds of prey, medieval arts and crafts, food and continuous entertainment on five stages. Est attendance: 18,000. For info: City of Gainesville, Division of Cultural Affairs, PO Box 490-30, Gainesville, FL 32602. Phone: (904) 334-2197.

ICESCAPE, THE FOX CITIES WINTER FESTIVAL. Feb 17-19. Houdini Plaza, Appleton, WI. Winter festival featuring the Wisconsin State Ice Carving Championship, food, the "Avenue of Ice," family activities, ice skating, flaming cherries jubilee, carriage rides and other winter fun. Est attendance: 20,000. Sponsor: Icescape Steering Committee, c/o Avenue Art, 10 College Ave, Appleton, WI 54911. Phone: (414) 954-9112.

INDIANAPOLIS BOAT, SPORT AND TRAVEL SHOW. Feb 17-26. Indiana State Fairgrounds, Indianapolis, IN. Est attendance: 200,000. For info: Kevin Renfro, 2511 E 46th St, Ste E-2, Corporate Sq East, Indianapolis, IN 46205. Phone: (317) 546-4344.

INTERNATIONAL RACE OF CHAMPIONS. Feb 17. Daytona International Speedway, Daytona Beach, FL. Sponsor: Budweiser. For info: Daytona Intl Speedway, Glenn Barber, PR, Box 2801, Daytona Beach, FL 32120-2801. Phone: (904) 254-6782.

JALAPENO FESTIVAL. Feb 17-18. Laredo, TX. Celebration of the Jalapeño Pepper. Events include the world famous Jalapeño Eating Contest, "The Some Like It Hot Recipe" Contest, the crowning of Ms. Jalapeño, Jalapeño Olympic games, spicy food, cold drinks, big name entertainment, music and dancing. Held during Laredo's annual Washington's Birthday Celebration, one of the largest celebrations honoring George Washington. Est attendance: 22,000. Sponsor: Jalapeño Festival Assn, Inc, c/o Anselmo Castro, Jr, PO Drawer 1359, Laredo, TX 78042-1359. Phone: (210) 726-6697.

LONGHORN WORLD CHAMPIONSHIP RODEO. Feb 17-19. Show Me Center, Cape Girardeau, MO. More than 220 of North America's best cowboys and cowgirls compete in six professionally sanctioned contests ranging from bronc riding to bull riding for top prize money and world championship points. The theme of the production changes annually. Additional attractions include special animal trainers, trick ropers, trick riders and professional clowns. 8th annual. Est attendance: 15,000. For info: Longhorn World Chmpshp Rodeo, Inc, PO Box 70159, Nashville, TN 37207. Phone: (615) 876-1016.

February 1995

S	M	T	W	T	F	S
			1	2	3	4
5	6	7	8	9	10	11
12	13	14	15	16	17	18
19	20	21	22	23	24	25
26	27	28				

Chase's 1995 Calendar of Events — Feb 17-18

MID-CONTINENT RAILWAY'S STEAM SNOW TRAIN. Feb 17-19. Mid-Continent Railway Museum, North Freedom, WI. Snow tours aboard authentic steam train. Annually, the third weekend in February. Est attendance: 3,000. For info: Wisc Dells Visitor and Conv Bureau, 701 Superior St, PO Box 390, Wisconsin Dells, WI 53965. Phone: (800) 223-3557.

MIDWINTER BLUEGRASS FESTIVAL. Feb 17-19. Holiday Inn, Hannibal, MO. Annually, third weekend in Feb. Est attendance: 1,500. Sponsor: Delbert Spray, Program Dir, Tri-State Bluegrass Assn, RR1, Kahoka, MO 63445. Phone: (314) 853-4344.

MSC STUDENT CONFERENCE ON NATIONAL AFFAIRS (SCONA). Feb 17-20. Memorial Student Center and Rudder Tower, Texas A&M University, College Station, TX. During a three-day conference, university students and faculty members from across the US and abroad meet to discuss political, social and world issues with experts in the field. Students have an opportunity to become involved in world affairs through discussions with world leaders who participate in SCONA. For info: MSC SCONA, Anna Olivari, Chair, PO Box J-1, College Station, TX 77844-9081.

MUSIC EDUCATORS NORTHWEST DIVISION IN-SERVICE CONFERENCE. Feb 17-19. Spokane, WA. Clinics, showcases, performances and trade exhibits. Biennially, in odd years. Est attendance: 1,000. Sponsor: Music Educators Natl Conf (MENC), Dir Mtgs and Convs, 1806 Robert Fulton Dr, Reston, VA 22091.

NATIONAL PTA FOUNDER'S DAY. Feb 17. Celebrates the PTA's founding by Phoebe Apperson Hearst and Alice McLellan Birney in 1897. Sponsor: Natl PTA, 330 N. Wabash, Ste 2100, Chicago, IL 60611. Phone: (312) 670-6782.

OREGON SHAKESPEARE FESTIVAL. Feb 17-Oct 29. An 8-month season of 11 plays by Shakespeare, classical and contemporary playwrights on 3 stages: the outdoor Elizabethan Stage, the versatile Angus Bowmer Theatre and the intimate Black Swan. Est attendance: 349,000. For info: Oregon Shakespeare Festival, Box 158, Ashland, OR 97520. Phone: (503) 482-4331.

PAY YOUR BILLS WEEK. Feb 17-25. Various. A week during which members of the American Collectors Association, Inc. put on seminars, contests and other events to educate consumers about the importance of financial responsibility and maintaining a good credit reputation. Annually, the third week in February. Sponsor: American Collectors Assn, Inc., PO Box 39106, Minneapolis, MN 55439-0106. Phone: (612) 926-6547.

PEALE, RAPHAEL: BIRTH ANNIVERSARY. Feb 17. American painter, member of famous family of early American painters, born at Annapolis, MD, Feb 17, 1774. Died Mar 4, 1825.

RECREATIONAL VEHICLE SHOW. Feb 17-19. (Also Feb 23-26.) Timonium State Fairgrounds, Timonium, MD. Mid-Atlantic's oldest, largest and best-attended RV show with exhibitors to display all the latest in motor homes, camping and RV accessories. Est attendance: 20,000. Sponsor: Maryland Recreational Vehicle Assn, 8332 Pulaski Hwy, Baltimore, MD 21237. Phone: (410) 687-7200.

SIMPLOT GAMES. Feb 17-18. Holt Arena, Idaho State University, Pocatello, ID. One of the nation's largest indoor high school track and field events, featuring the top high school athletes in the US and Canada. Est attendance: 2,000. Sponsor: J.R. Simplot Co, Carol Lish, PO Box 912, Pocatello, ID 83204. Phone: (800) 635-9444.

SOUTHEASTERN WILDLIFE EXPOSITION. Feb 17-19. Spectacular wildlife and western art exposition showcasing more than $20 million in art. Features original paintings, prints, sculpture, carvings, collectibles and crafts. Fourteen exhibition sites throughout historic Charleston. Est attendance: 42,000. For info: Southeastern Management Co, 211 Meeting Street, Charleston, SC 29401. Phone: (800) 221-5273.

STAMP EXPO. Feb 17-19. Elks Lodge, Pasadena, CA, and Palm Springs Convention Center, Palm Springs, CA. Fax: (818) 988-4337. Est attendance: 4,000. Sponsor: Intl Stamp Collectors Society, PO Box 854, Van Nuys, CA 91408. Phone: (818) 997-6496.

TODAY'S WOMAN. Feb 17-19. Benton Convention Center, Winston-Salem, NC. A Forum for Women with interest in home, health, beauty and business. Sponsor: WKII-TV. Est attendance: 6,000. For info: Apple rock Advertising & Promotion, 1200 Eastchester Dr, High Point, NC 27265.

WHISKEY FLAT DAYS. Feb 17-20. Kernville, CA. Whiskey Flat Days commemorates the old Kernville that was called Whiskey Flat in 1860 before the name was changed to Kernville. We turn back the clock as the parade, stores and residents go back to the 1860s. Frog races, craft booths, rodeo, parade, games for small children, carnival, street dances, puppet shows, costume and whiskerino contests, melodrama and much more. Annually, the Friday-Monday of President's Day weekend. Est attendance: 25,000. For info: Kernville Chamber of Commerce, Rose Ann Viton, PO Box 397, Kernville, CA 93238. Phone: (619) 376-2629.

BIRTHDAYS TODAY

Alan Bates (Arthur Bates), 61, actor (*An Unmarried Woman, Women in Love*), born at Derbyshire, England, Feb 17, 1934.
Jim Brown, 59, Pro Football Hall of Famer, actor, born at St. Simon's Island, GA, Feb 17, 1936.
Lee Hoiby, 69, composer, concert pianist, born at Madison, WI, Feb 17, 1926.
Hal Holbrook (Harold Rowe, Jr), 70, actor (*Magnum Force, All the President's Men*), born Cleveland, OH, Feb 17, 1925.
Michael Jordan, 32, former basketball player, baseball player, born at Wilmington, NC, Feb 17, 1963.

FEBRUARY 18 — SATURDAY
49th Day — Remaining, 316

CHINESE NEW YEAR GOLDEN DRAGON PARADE. Feb 18. Chinatown, Los Angeles, CA. Traditional, colorful parade in celebration of the Chinese New Year. The most popular event of the New Year celebration activities that began on Jan 23. Est attendance: 50,000. For info: Chinese Chamber of Commerce, 977 N Broadway, G/F, #E, Los Angeles, CA 90012. Phone: (213) 617-0396.

COCONUT GROVE ARTS FESTIVAL. Feb 18-20. Coconut Grove, FL. More than 300 artists exhibit and sell their work at this 3-day festival. Various ethnic foods will be offered and nationally-known jazz musicians will perform as well. Est attendance: 750,000. For info: Coconut Grove Assn, Inc, Suzanne Kores, Asst Dir, 2980 McFarlane Rd, Ste 204, Coconut Grove, FL 33133. Phone: (305) 447-0401.

CONFERENCE ON STUDENT GOVERNMENT ASSOCIATIONS (COSGA). Feb 18-21. Texas A&M University, College Station, TX. College and university students from throughout the US and abroad meet annually to learn leadership skills and to study how student governments function. Est attendance: 400. Sponsor: COSGA, Angela Blakeney, Dir, PO Box A-8, College Station, TX 77844. Phone: (409) 845-9394.

COW MILKED WHILE FLYING IN AN AIRPLANE: 65th ANNIVERSARY. Feb 18, 1930. Elm Farm Ollie became the first cow to fly in an airplane. During the flight, which was attended by reporters, she was milked and the milk was sealed in paper containers and parachuted over St. Louis, MO.

Feb 18 ☆ Chase's 1995 Calendar of Events ☆

COWBOY STATE GAMES WINTER SPORTS FESTIVAL. Feb 18–20. Casper, WY. The festival features a variety of winter sporting events for athletes of all ages. Est attendance: 1,000. For info: Cowboy State Games, Eileen Ford, PO Box 3485, Casper, WY 82602. Phone: (307) 577-1125.

ELM FARM OLLIE DAY CELEBRATION. Feb 18. Wisconsin. This day marks the first flight in an airplane by a dairy cow, Elm Farm Ollie, on Feb 18, 1930, at the St. Louis International Air Exposition. Although this brave bovine hailed from Missouri, residents of the "Dairy State" mark this festive event by consuming large amounts of cheese and ice cream and by singing the *Bovine Cantata* from Moocini's lyric opera *Madame Butterfat*. Est attendance: 200. Sponsor: Mount Horeb Mustard Museum, Barry Levenson, Curator, 109 E Main St, Mount Horeb, WI 53572. Phone: (608) 437-3986.

ESA MID-WINTER SURFING CHAMPIONSHIP. Feb 18. Narragansett Town Beach, Narragansett, RI. Competition in all age categories and specialty events with prizes and trophies. Est attendance: 300. Sponsor: Peter Pan, ESA Dir, 396 Main St, Wakefield, RI 02879. Phone: (401) 789-3399.

GAMBIA: INDEPENDENCE DAY: 30th ANNIVERSARY. Feb 18. National holiday. Independence from Britain granted Feb 18, 1965. Referendum in April 1970 established Gambia as a republic within the Commonwealth.

GASPARILLA DISTANCE CLASSIC. Feb 18. Tampa, FL. Thousands of runners compete for cash or fun in the 18th annual Gasparilla Distance Classic. Includes 5K and 15K running races and a 15K wheelchair race. Please send a self-addressed, stamped envelope for entry application to: Gasparilla '95, PO Box 1881, Tampa, FL 33601-1881. Sponsors: First Florida Bank, Nike, WUSA/W101 FM, Tampa Tribune and the city of Tampa, FL. Est attendance: 13,000. For info: Tampa/Hillsborough Conv and Visitors Assn, 111 Madison St, Ste 1010, Tampa, FL 33602-4706. Phone: (800) 448-2672.

GEORGE WASHINGTON BIRTHNIGHT BANQUET AND BALL. Feb 18. Alexandria, VA. Recreation of dinner, dancing and toasts offered to George Washington during his lifetime in Alexandria. Staged in original setting—Gadsby's Tavern. Eighteenth-century dancing, clothing. Alexandrians portray historic characters including George and Martha Washington. Annually, the Saturday of the federal holiday weekend. Sponsor: George Washington Birthday Celebration Committee. Est attendance: 275. For info: Alexandria Conv and Visitors Bureau, 221 King St, Alexandria, VA 22314. Phone: (703) 838-4200.

THE GOOD, THE BAD AND THE CUDDLY. Feb 18–Apr 16. Detroit Zoo, Royal Oak, MI. This exhibit takes a sometimes whimsical and always thoughtful look at the attitudes people hold toward wildlife, domesticated animals and environment. Myths and truths are presented. For info: Smithsonian Institution Traveling Exhibition Service, 1100 Jefferson Dr SW, Ste 3146, Washington, DC 20560. Phone: (202) 357-2700.

GOODY'S 300 (NASCAR BUSCH GRAND NATIONAL SERIES STOCK CAR RACE). Feb 18. Daytona International Speedway, Daytona Beach, FL. For info: Daytona Intl Speedway, Glenn Barber, PR, Box 2801, Daytona Beach, FL 32120-2801. Phone: (904) 254-6782.

GRANT SEAFOOD FESTIVAL. Feb 18–19. Grant, FL. 29th annual festival promotes Indian River seafood, environmental awareness and conservation. Annually, the third weekend in February. Est attendance: 50,000. For info: Grant Community Club, Inc, Margaret B. Senne, PO Box 44, Grant, FL 32949. Phone: (407) 723-6811.

HAPPY BIRTHDAY, A.C.! Feb 18. The Gilbert House Children's Museum, Salem, OR. Celebrate the birth of Erector Set inventor Alfred Carlton Gilbert (A.C.), who was born in Salem, Oregon, on Feb 15, 1884. Gilbert's namesake, The Gilbert House Children's Museum, will herald the inventor's birth with games, magic, Erector Sets, chemistry and engineering demonstrations and more. The museum will also offer week-long activities to coincide with National Engineers Week, Feb 18–24. Annually, the Saturday closest to A.C.'s birthday. Sponsor: The Gilbert House Children's Museum. Est attendance: 600. For info: Tracey Etzel, PR, 116 Marion St NE, Salem, OR 97301. Phone: (503) 371-3631.

NEW YORK DOLLHOUSE AND MINIATURES TRADE SHOW. Feb 18–20. Sheraton New York Hotel, New York City. Manufacturers, importers, wholesalers and handcrafters exhibit their wares for wholesale and retail buyers only. 13th annual show. Sponsor: Miniatures Industry Assn of America. Est attendance: 2,000. For info: Patty Parrish, Assn/Trade Show Mgr, PO Box 2188, Zanesville, OH 49702-2188,. Phone: (614) 452-4541.

PEABODY, GEORGE: 200th BIRTH ANNIVERSARY. Feb 18. American merchant and philanthropist. Born Feb 18, 1795. Died Nov 4, 1869.

RABINOWITZ, SOLOMON: BIRTH ANNIVERSARY. Feb 18. Russian-born author and humorist better known by his pen name, Sholem Aleichem. Born in the Ukraine, Feb 18, 1859. Affectionately known in the US as the "Jewish Mark Twain." Died at New York, NY, May 13, 1916.

SHRINER'S INVITATIONAL CUTTER RACES. Feb 18–19. Melody Ranch, Jackson, WY. Horse-drawn cutters race each other down a ¼-mile track, the western version of chariot racing. Annually, Presidents Day weekend. Sponsor: Jackson Hole Shrine Club. For info: Jackson Hole, Box 401, Jackson, WY 83001. Phone: (307) 733-3316.

SPEARFISH CROSS COUNTRY SKI CHALLENGE. Feb 18. Spearfish, SD. A variety of cross-country ski events through the impressive Black Hills along 26K trails. Est attendance: 70. For info: Everett or Joann Follette, Ski Cross Country, 701 Third St, Spearfish, SD 57783. Phone: (605) 642-3851.

TIFFANY, LOUIS COMFORT: BIRTH ANNIVERSARY. Feb 18. American artist, son of famed jeweler Charles L. Tiffany. Best remembered for his remarkable work with decorative iridescent "favrile" glass. Born at New York, NY, on Feb 18, 1848. Died there Jan 17, 1933. See also: "Tiffany, Charles Lewis: Birth Anniversary" (Feb 15).

WILLKIE, WENDELL LEWIS: BIRTH ANNIVERSARY. Feb 18. American lawyer, author, public utility executive and politician. Born at Elwood, IN, Feb 18, 1892. Presidential nominee of the Republican Party in 1940. Remembered too for his book, *One World*, published in 1943. Died at New York, NY, Oct 8, 1944.

BIRTHDAYS TODAY

Helen Gurley Brown, 73, author (*Sex and the Office*), publisher (*Cosmopolitan*), born Green Forest, AR, Feb 18, 1922.
Aldo Ceccato, 61, conductor, born at Milan, Italy, Feb 18, 1934.
Matt Dillon, 31, actor (*My Bodyguard, Drugstore Cowboy*), born at Westchester, NY, Feb 18, 1964.

February 1995

S	M	T	W	T	F	S
			1	2	3	4
5	6	7	8	9	10	11
12	13	14	15	16	17	18
19	20	21	22	23	24	25
26	27	28				

✯ Chase's 1995 Calendar of Events ✯ Feb 18-19

Milos Forman, 63, film director (Oscars *Amadeus, One Flew Over the Cuckoo's Nest*), born Caslaz, Czechoslovakia, Feb 18, 1932.
George Kennedy, 68, actor (Oscar for *Cool Hand Luke*; "The Blue Knight"), born at New York, NY, Feb 18, 1927.
Toni Morrison (Chloe Anthony), 64, novelist (*Beloved, Jazz, Tar Baby, Sula*), born at Lorain, OH, Feb 18, 1931.
Juice Newton (Judy Cohen), 43, singer (platinum album *Juice*, gold *Quiet Lives*), born Virginia Beach, VA, Feb 18, 1952.
Yoko Ono, 62, artist, musician, widow of John Lennon, born at Tokyo, Japan, Feb 18, 1933.
Jack Palance, 75, actor (Oscar for *City Slickers*; "Bronk," "Ripley's Believe it or Not"), born at Lattimer, PA, Feb 18, 1920.
Molly Ringwald, 27, actress (*Sixteen Candles, Breakfast Club, Pretty in Pink*), born at Roseville, CA, Feb 18, 1968.
Cybill Shepherd, 45, actress (*The Last Picture Show*; Maddie Hayes on "Moonlighting"), born at Memphis, TN, Feb 18, 1950.
John Travolta, 40, actor (*Urban Cowboy, Saturday Night Fever*, "Welcome Back Kotter"), born at Englewood, NJ, Feb 18, 1955.
John William Warner, 68, US Senator (R, Virginia), (ex-husband of Elizabeth Taylor) born Washington, DC, Feb 18, 1927.
Vanna White, 38, TV personality ("Wheel of Fortune"), born Conway, SC, Feb 18, 1957.
Gahan Wilson, 65, cartoonist, author, born at Evanston, IL, Feb 18, 1930.

FEBRUARY 19 — SUNDAY
50th Day — Remaining, 315

BOLLINGEN PRIZE AWARD: ANNIVERSARY. Feb 19. On Feb 19, 1949, the first Bollingen Prize for poetry was awarded to Ezra Pound for his collection *The Pisan Cantos*. The first award was steeped in controversy because Pound had been charged with treason after making pro-Fascist broadcasts in Italy during World War II.

BROTHERHOOD/SISTERHOOD WEEK. Feb 19-25. A kickoff period for programs emphasizing the need for year-round commitment to brotherhood/sisterhood. The National Program Office develops educational materials for use during this period that can be used year round. Sponsor: The National Conference founded as the Natl Conference of Christians and Jews. For info: The Natl Conference, 71 Fifth Ave, New York, NY 10003. Phone: (212) 206-0006.

COPERNICUS, NICOLAUS: BIRTH ANNIVERSARY. Feb 19. Polish astronomer and priest who revolutionized scientific thought with what came to be called the Copernican theory, which placed the sun instead of the earth at the center of our planetary system. Born at Torun, Poland, Feb 19, 1473. Died in East Prussia on May 24, 1543.

DAYTONA 500 BY STP NASCAR-FIA (WINSTON CUP STOCK CAR CLASSIC). Feb 19. Daytona International Speedway, Daytona Beach, FL. Daytona 500 by STP NASCAR-FIA. For info: Daytona Intl Speedway, Glenn Barber, PR, Box 2801, Daytona Beach, FL 32120-2801. Phone: (904) 254-6782.

GARRICK, DAVID: BIRTH ANNIVERSARY. Feb 19. English actor and playwright. Born Feb 19, 1717. Died Jan 20, 1779.

HEALTH EDUCATION WEEK IN NEW YORK STATE. Feb 19-25. Promotes healthful behavior and healthy lifestyles. Annually, the third full week in February. Sponsor: New York State Health Dept, Bureau of Community Relations, Corning Tower Bldg, Rm 1084, Empire State Plaza, Albany, NY 12237. Phone: (518) 474-5370.

HEDIN, SVEN: BIRTH ANNIVERSARY. Feb 19. Explorer and scientist, Sven Anders Hedin was born at Stockholm, Sweden, on Feb 19, 1865, and died there on Nov 26, 1952. His Tibetan explorations provided the first substantial knowledge of that region to the rest of the world.

HOMES FOR BIRDS WEEK. Feb 19-25. A week to encourage people to clean out, fix up and put up homes for wild birds. Annually, the third week in February. Sponsor: Wild Bird Marketplace, 710 W Main St, PO Box 1184, New Holland, PA 17557. Phone: (717) 354-2841.

INTERNATIONAL FRIENDSHIP WEEK. Feb 19-25. Promotion of international friendship and the international language Esperanto. For complete information, send $1 to cover expense of printing, handling and postage. Annually, the last full week in February. Sponsor: Intl Society of Friendship & Good Will, Dr. Stanley Drake, Pres, 9538 Summerfield St, Spring Valley, San Diego County, CA 91977-2852. Phone: (619) 466-8882.

JAPANESE INTERNMENT: ANNIVERSARY. Feb 19. As a result of President Franklin Roosevelt's Executive Order 9066, issued on Feb 19, 1942, some 110,000 Japanese Americans living in coastal Pacific areas were placed in concentration camps in remote areas of Arizona, Arkansas, inland California, Colorado, Idaho, Utah and Wyoming. The interned Japanese Americans (two-thirds were US citizens) lost an estimated $400 million in property.

KNIGHTS OF PYTHIAS: FOUNDING ANNIVERSARY. Feb 19. The social and fraternal order of the Knights of Pythias was founded at Washington, DC, Feb 19, 1864.

NATIONAL ENGINEERS WEEK. Feb 19-25. Brings visibility to professional engineers. Celebrated annually during the week of George Washington's birthday because Washington was a military engineer and land surveyor, and he was responsible for establishing the first US engineering school at Valley Forge, PA, which later became the US Military Academy at West Point, NY. For info: Natl Society of Professional Engineers, Leslie Collins, 1420 King St, Alexandria, VA 22314-2715. Phone: (703) 684-2852.

REVOLUTIONARY WAR ENCAMPMENT. Feb 19. Alexandria, VA. Sunday afternoon of Washington's Birthday Weekend. Camp life demonstrated at Fort Ward Museum and Historic Site. British and Colonial uniformed troops engage in a skirmish. Free. Est attendance: 2,000. For info: Diane Bechtol, Mgr, Media Relations, Alexandria Conv and Visitors Bureau, 221 King St, Alexandria, VA 22314. Phone: (703) 838-4200.

◆ **US LANDING ON IWO JIMA: 50th ANNIVERSARY.** Feb 19, 1945. Beginning at dawn, the landing of 30,000 American troops took place on the barren 12-square-mile island of Iwo Jima. Initially there was little resistance, but 21,500 Japanese stood ready underground to fight to the last man to protect massive strategic fortifications linked by tunnels.

BIRTHDAYS TODAY

Prince Andrew, 35, Duke of York, born at London, England, Feb 19, 1960.
Eddie Arcaro, 79, former jockey, broadcaster, born at Cincinnati, OH, Feb 19, 1916.
Justine Bateman, 29, actress ("Family Ties"), born at Rye, NY, Feb 19, 1966.
Falco, 38, musician, born at Vienna, Austria, Feb 19, 1957.
Margaux Hemingway, 40, actress, born at Portland, OR, Feb 19, 1955.
Hana Mandlikova, 32, tennis player, born at Prague, Czechoslovakia, Feb 19, 1963.

Feb 19–20 ☆ *Chase's 1995 Calendar of Events* ☆

Smokey Robinson (William Robinson, Jr), 55, singer, songwriter (lead with The Miracles; "Cruisin'," "Being With You"), born Detroit, MI, Feb 19, 1940.

FEBRUARY 20 — MONDAY
51st Day — Remaining, 314

CANADA: YUKON SOURDOUGH RENDEZVOUS. Feb 20–26. Whitehorse, Yukon. Mad trapper competitions, bike races, flour packing, beard-growing contests, Old Time Fiddle contest, sourdough pancake breakfasts, can-can girls, talent shows, etc. Also, many family-oriented activities. Visitors welcome to participate. Est attendance: 10,000. For info: Yukon Sourdough Rendezvous, Box 5108, Whitehorse, Yukon, Canada Y1A 4S3. Phone: (403) 667-2148.

COBAIN, KURT: BIRTH ANNIVERSARY. Feb 20, 1967. Lead singer, guitarist and songwriter for the rock group Nirvana. Born at Aberdeen, WA, he died at Seattle, WA, on Apr 8, 1994.

◆ **DOUGLASS, FREDERICK: 100th DEATH ANNIVERSARY.** Feb 20. American journalist, orator and antislavery leader. Born at Tuckahoe, MD, probably in February 1817. Died at Anacostia Heights, DC, Feb 20, 1895. His original name before his escape from slavery was Frederick Augustus Washington Bailey.

GEORGE WASHINGTON BIRTHDAY PARADE. Feb 20. Alexandria, VA. The nation's largest parade honoring George Washington takes place as a part of a month-long celebration in his honor staged by his home town. Annually, the third Monday in February. Est attendance: 55,000. Sponsor: George Washington Birthday Celebration Committee, Alexandria Conv & Visitors Bureau, 221 King St, Alexandria, VA 22314. Phone: (703) 838-4200.

NORTHERN HEMISPHERE HOODIE-HOO DAY. Feb 20. At high noon (local time) citizens are asked to go outdoors and yell "Hoodie-Hoo" to chase away winter and make ready for spring, one month away. Phone: (22) 388-8673 or (717) 274-8451. For info: Wellness Permission League, Tom or Ruth Roy, 2105 Water St, Lebanon, PA 17046.

PISCES, THE FISH. Feb 20–Mar 20. In the astronomical/astrological zodiac, which divides the sun's apparent orbit into 12 segments, the period Feb 20–Mar 20 is identified, traditionally, as the sun sign of Pisces, the Fish. The ruling planet is Neptune.

PRESCOTT, WILLIAM: BIRTH ANNIVERSARY. Feb 20. American Revolutionary soldier. Born at Groton, MA, Feb 20, 1726. Died at Pepperell, MA, Oct 13, 1795. Credited with the order, "Don't fire until you see the whites of their eyes," at the Battle of Bunker Hill, June 17, 1775.

PRESIDENTS' DAY. Feb 20. The third Monday in February. Presidents' Day observes the birthdays of George Washington (Feb 22) and Abraham Lincoln (Feb 12). With the adoption of the Monday Holiday Law (which moved the observance of George Washington's birthday from Feb 22 each year to the third Monday in February), some of the specific significance of the event was lost, and added impetus was given to the popular description of that holiday as Presidents' Day. Present usage often regards Presidents' Day as a day to honor all former presidents of the US. Presidents' Day has statutory authority in Hawaii, Nebraska, Ohio and the Commonwealth of the Northern Mariana Islands, and popular recognition in most states.

ROSS PEROT ANSWERS A QUESTION: ANNIVERSARY. Feb 20. Toward the end of his Feb 20, 1992, appearance on the "Larry King Live" TV talk show, Texas billionaire and businessman H. Ross Perot was asked by King whether there were any circumstances under which Perot would run for president of the US. Perot responded that if supporters put his name on the ballot in all 50 states, he would agree to run, thus becoming a footnote in history. His anti-Washington, anti-politician appeal generated an historic enthusiasm in the form of a grassroots movement organizing petition drives in all 50 states to place his name on the ballot. Perot's popularity grew early with steady rises in public opinion polls as journalists struggled to keep up with the phenomenon. See also: "Ross Perot's July Surprise" (July 16).

SPACE MILESTONE: *FRIENDSHIP 7* (US): FIRST AMERICAN TO ORBIT EARTH: ANNIVERSARY. Feb 20. On Feb 20, 1962, John Herschel Glenn, Jr, became the first American, and the third man, to orbit Earth. Aboard the capsule *Friendship-7*, he made three orbits of Earth. Spacecraft was *Mercury-Atlas 6*.

SPACE MILESTONE: *MIR* SPACE STATION (USSR). Feb 20. A "third-generation" orbiting space station, *Mir* (Peace), was launched without crew on Feb 20, 1986, from the Baikonur space center at Leninsk, Kazakhstan. Believed to be 40 feet long, weigh 47 tons and have six docking ports.

STOCK EXCHANGE HOLIDAY (WASHINGTON'S BIRTHDAY). Feb 20. The holiday schedules for the various exchanges are subject to change if relevant rules, regulations or exchange policies are revised. If you have questions, phone: American Stock Exchange (212) 306-1212; Chicago Board of Options Exchange (312) 786-7760; Chicago Board of Trade (312) 435-3500; New York Stock Exchange (212) 656-2065; Pacific Stock Exchange (415) 393-4000; Philadelphia Stock Exchange (215) 496-5000.

STUDENT VOLUNTEER DAY. Feb 20. To honor students who give of themselves and of their personal time to improve the lives of others and their communities. Annually, February 20. Sponsor: Susquehanna University, Center for Volunteer Programs, Selinsgrove, PA 17870. Phone: (717) 372-4139.

WASHINGTON, GEORGE: BIRTHDAY OBSERVANCE (LEGAL HOLIDAY). Feb 20. Legal public holiday (Public Law 90-363 sets Washington's birthday observance on the third Monday in February each year—applicable to federal employees and to the District of Columbia). Observed on this day in all states. See also: "Washington, George: Birth Anniversary" (Feb 22).

BIRTHDAYS TODAY

Robert Altman, 70, film director (*M*A*S*H*, *Nashville*), born at Kansas City, MO, Feb 20, 1925.

Charles Barkley, 32, basketball player, born at Leeds, AL, Feb 20, 1963.

Sandy Duncan, 49, actress (*Funny Face*, "The Hogan Family," *Peter Pan*), born at Henderson, TX, Feb 20, 1946.

Muriel Fay Buck Humphrey, 83, former senator, widow of Hubert Humphrey, born at Huron, SD, Feb 20, 1912.

Mitch McConnell, 53, US Senator (R, Kentucky), born in Colbert County, AL, Feb 20, 1942.

Jennifer O'Neill, 47, actress (*The Summer of '42*, "Cover-Up"), born at Rio de Janeiro, Brazil, Feb 20, 1948.

Kenneth Olsen, 69, founder of Digital Equipment Corporation, born at Bridgeport, CT, Feb 20, 1926.

February 1995

S	M	T	W	T	F	S
			1	2	3	4
5	6	7	8	9	10	11
12	13	14	15	16	17	18
19	20	21	22	23	24	25
26	27	28				

Sidney Poitier, 68, actor (*In the Heat of the Night*, Oscar for *Lilies of the Field*), born at Miami, FL, Feb 20, 1927.
Buffy Sainte-Marie (Beverly Sainte-Marie), 54, singer, born at Craven, Saskatchewan, Canada, Feb 20, 1941.
Patty Hearst Shaw, 41, newspaper heir (kidnapped by radical group Symbionese Liberation Army), born at San Francisco, CA, Feb 20, 1954.
Bobby Unser, 61, auto racer, born at Albuquerque, NM, Feb 20, 1934.
Gloria Vanderbilt, 71, fashion designer, artist, born at New York, NY, Feb 20, 1924.
Nancy Wilson, 58, singer ("Yesterday's Love Songs/Today's Blues"), born at Chillicothe, OH, Feb 20, 1937.

FEBRUARY 21 — TUESDAY
52nd Day — Remaining, 313

AUDEN, WYSTAN HUGH: BIRTH ANNIVERSARY. Feb 21. Pulitzer Prize–winning American poet. Born at York, England, on Feb 21, 1907. "Some books," he wrote in *The Dyer's Hand* (1962), "are undeservedly forgotten; none are undeservedly remembered." Died on Sept 28, 1973, at Vienna, Austria.

BANGLADESH: MARTYRS DAY. Feb 21. National mourning day in memory of martyrs of the Bengali Language Movement in 1952. Mourners gather at the Azimpur graveyard and proceed to Shaheed Minar.

FIRST WOMAN TO GRADUATE FROM DENTAL SCHOOL: ANNIVERSARY. Feb 21. On Feb 21, 1866, Lucy Hobbs became the first woman to graduate from a dental school, in Cincinnati, OH.

GRAND CENTER BOAT SHOW. Feb 21–26. Grand Center, Grand Rapids, MI. This event brings together buyers and sellers of power and sail boats, boating accessories, dock, dockominiums and vacation properties. For info: Mike Wilbraham, ShowSpan, Inc, 1400 28th St SW, Grand Rapids, MI 49509. Phone: (616) 530-1919.

MALCOLM X: 30th ASSASSINATION ANNIVERSARY. Feb 21. Malcolm X, a black leader who renounced the Black Muslim sect to form the Organization of Afro-American Unity, was shot and killed as he spoke to a rally at the Audubon Ballroom in New York, NY, on Feb 21, 1965. Three men were convicted of the murder in 1966 and sentenced to life in prison. Born Malcolm Little, the son of a Baptist preacher, on May 19, 1925, at Omaha, NE.

MUSLIM FESTIVAL: LAILAT AL-QADR. Feb 21–22. "The Night of Power" is observed on this date. Muslim calendar date: Ramadhan 21, 1415. Different methods for calculating the Gregorian date are used by different Muslim sects or groups. *Chase's* dates are based on astronomical calculations of the visibility of the new moon crescent at Mecca; EST date may vary.

***NEW YORKER* MAGAZINE: 70th BIRTHDAY.** Feb 21. First issue published on Feb 21, 1925.

◆ **PALMER, ALICE FREEMAN: BIRTH ANNIVERSARY.** Feb 21, 1855. Born at Colesville, NY, Alice Freeman Palmer became president of Wellesley College at the age of 27. Under her leadership the little-known school for girls grew into one of the leading women's colleges. She was also instrumental in bringing the women's school Radcliffe College into its association with Harvard University. One of the organizers of the American Association of University Women, she served as its president for two terms. She was named to the Massachusetts Board of Education in 1889 and was appointed the first dean of women at the University of Chicago when it opened in 1892. Palmer died Dec 6, 1902, at Paris.

RICHARD NIXON'S TRIP TO CHINA: ANNIVERSARY. Feb 21–28. Richard Nixon became the first US president to visit any country not diplomatically recognized by the US when he went to the People's Republic of China for meetings with Chairman Mao Tse-tung and Premier Chou En-lai. Nixon arrived at Peking on Feb 21, 1972, and departed China on Feb 28. The "Shanghai Communique" was issued Feb 27. See also: "Shanghai Communique Anniversary" (Feb 27).

WASHINGTON MONUMENT: 110th DEDICATION ANNIVERSARY. Feb 21. Monument to first president was dedicated on Feb 21, 1885, at Washington, DC.

BIRTHDAYS TODAY

Erma Bombeck, 68, columnist, author (*The Grass Is Always Greener over the Septic Tank*), born at Dayton, OH, Feb 21, 1927.
Gaston Caperton, 55, Governor of West Virginia (D), born at Charleston, WV, Feb 21, 1940.
Tyne Daly, 48, actress (Emmy for Lacey on "Cagney and Lacey"; *Gypsy*), born Madison, WI, Feb 21, 1947.
Hubert de Givenchy, 68, fashion designer, born at Beauvais, France, Feb 21, 1927.
Barbara Jordan, 59, former US congresswoman, lawyer, educator, born in Houston, TX, Feb 21, 1936.
Rue McClanahan, 61, actress (Vivian on "Maude," Blanche on "Golden Girls"), born Healdton, OK, Feb 21, 1934.
Nina Simone (Eunice Waymon), 62, singer ("I Loves You Porgy," "Trouble in Mind"), born Tryon, NC, Feb 21, 1933.
Alan Trammel, 37, baseball player, born at Garden Grove, CA, Feb 21, 1958.

FEBRUARY 22 — WEDNESDAY
53rd Day — Remaining, 312

BADEN-POWELL, ROBERT: BIRTH ANNIVERSARY. Feb 22. British army officer who founded the Boy Scouts and Girl Guides. Born at London, England, Feb 22, 1857. Died at Kenya, Africa, Jan 8, 1941.

LA FIESTA DE LOS VAQUEROS. Feb 22–26. Tucson, AZ. Tucson celebrates its Old West heritage with parade, PRCA rodeo and other related rodeo events. Est attendance: 55,000. For info: Tucson Rodeo Committee, Inc, PO Box 11006, Tucson, AZ 85734. Phone: (602) 741-2233.

LIONEL HAMPTON JAZZ FESTIVAL. Feb 22–25. University of Idaho, Moscow, Idaho. Each year, college, high school and junior high school vocal and instrumental jazz ensembles come from all over the US in the festival and attend concerts and clinics given by the world's greatest jazz artists. Annually, during the last full week of February. For info: Dr. Lynn J. Skinner, Ex Dir, UI Lionel Hampton Jazz Fest, Lionel Hampton School of Music, Univ of Idaho, Moscow, ID 83844-4014. Phone: (208) 885-6765.

LOWELL, JAMES RUSSELL: BIRTH ANNIVERSARY. Feb 22. American essayist, poet and diplomat. Born at Cambridge, MA, on Feb 22, 1819. Died there on Aug 12, 1891.

◆ **MONTGOMERY BOYCOTT ARRESTS: ANNIVERSARY.** Feb 22, 1956. On Feb 20 white city leaders of Montgomery, AL, issued an ultimatum to black organizers of the three-month-old Montgomery bus boycott. They said if the boycott ended immediately there would be "no retaliation whatsoever." If it did not end, it was made clear they would begin arresting black leaders. Two days later, 80 well-known boycotters, including Rosa Parks, Martin Luther King, Jr and E.D. Nixon marched to the sheriff's office in the county courthouse, where they gave themselves up for arrest. They were booked, fingerprinted and photographed. The next day the story was carried by newspapers all over the world.

MOON PHASE: LAST QUARTER. Feb 22. Moon enters Last Quarter phase at 8:04 AM, EST.

Feb 22-23 ☆ *Chase's 1995 Calendar of Events* ☆

NORTHWEST FLOWER AND GARDEN SHOW. Feb 22-26. Washington State Conv Center, Seattle, WA. The 1995 theme is "Rainbow of Color." Est attendance: 80,000. For info: Celice Eldred, 1515 NW 51st St, Seattle, WA 98107. Phone: (206) 789-5333.

PEALE, REMBRANDT: BIRTH ANNIVERSARY. Feb 22. American portrait and historical painter, born at Bucks County, PA, on Feb 22, 1778. Died at Philadelphia, PA, on Oct 3, 1860.

POPCORN INTRODUCED TO COLONISTS: ANNIVERSARY. Feb 22. As his contribution to the first Thanksgiving, celebrated Feb 22, 1630, a Native American named Quadequina (the brother of Wampanoag chief Massasoit) contributed a deerskin bag filled with several bushels of "popped" corn.

ST. LUCIA: INDEPENDENCE DAY. Feb 22. National holiday.

SCHOPENHAUER, ARTHUR: BIRTH ANNIVERSARY. Feb 22. Philosopher and author, born at Danzig, Germany, Feb 22, 1788, and died at Frankfurt am Main, Germany, Sept 21, 1860. Generally regarded as a misanthrope, the never-married Schopenhauer wrote, in 1819, "To marry is to halve your rights and double your duties."

WADLOW, ROBERT PERSHING: BIRTH ANNIVERSARY. Feb 22. Tallest man in recorded history, born on Feb 22, 1918, at Alton, IL. Though only 9 lbs at birth, by age 10 Wadlow already stood over 6 feet tall and weighed 210 lbs; Wadlow died at age 22, a remarkable 8 feet 11.1 inches tall, 490 lbs. His gentle, friendly manner in the face of constant public attention earned him the name the "Gentle Giant." Wadlow died on July 15, 1940, at Manistee, MI, of complications resulting from a foot infection.

WASHINGTON, GEORGE: BIRTH ANNIVERSARY. Feb 22. First president of the US ("First in war, first in peace and first in the hearts of his countrymen" in the words of Henry "Light-Horse Harry" Lee). Born at Westmoreland County, VA, on Feb 22 (New Style), Feb 11 (Old Style), 1732. Died at Mt Vernon, VA, Dec 14, 1799. See also: "Washington, George: Birthday Observance (Legal Holiday)" (Feb 20 in 1995).

BIRTHDAYS TODAY

Amy Strum Alcott, 39, golfer, born at Kansas City, MO, Feb 22, 1956.
Sparky Anderson (George Lee), 61, baseball manager, born at Bridgewater, SD, Feb 22, 1934.
Drew Barrymore, 20, actress (Gertie in *E.T.*; *Irreconcilable Differences*, *Bad Girls*), born at Los Angeles, CA, Feb 22, 1975.
Julius "Doctor J" Erving, 45, former basketball player, born at Roosevelt, NY, Feb 22, 1950.
Nelson Bunker Hunt, 69, business executive, born at El Dorado, TX, Feb 22, 1926.
Edward Moore Kennedy, 63, US Senator (D, Massachusetts) up for reelection in '94, born Boston, MA, Feb 22, 1932.
Niki Lauda, 46, auto racer, author, born at Vienna, Austria, Feb 22, 1949.
Robert Young, 88, actor ("Father Knows Best," "Marcus Welby, M.D."; *Claudia*), born Chicago, IL, Feb 22, 1907.

FEBRUARY 23 — THURSDAY
54th Day — Remaining, 311

ACBL TOURNAMENT. Feb 23-26. Yuma, AZ. Bridge sectional. Annually, the last weekend in February. For info: Yuma Civic and Conv Ctr, 1440 Desert Hills Dr, Yuma, AZ 85364. Phone: (602) 344-3800.

February 1995

S	M	T	W	T	F	S
			1	2	3	4
5	6	7	8	9	10	11
12	13	14	15	16	17	18
19	20	21	22	23	24	25
26	27	28				

AMERICAN BIRKEBEINER XXIII. Feb 23-25. Cable to Hayward, WI. The largest and most prestigious ski marathon and nordic festival in North America. Est attendance: 8,500. For info: American Birkebeiner Ski Foundation, Inc, Box 911, Hayward, WI 54843. Phone: (715) 634-5025.

BIG TEN MEN'S SWIMMING AND DIVING CHAMPIONSHIP. Feb 23-25. Site to be announced. For info: Big Ten Conference, 1500 W Higgins Rd, Park Ridge, IL 60068-6300. Phone: (708) 696-1010.

BRUNEI DARUSSALAM: NATIONAL DAY. Feb 23. National holiday observed in Brunei Darussalam.

CANADA: MUSKOKA WINTER CARNIVAL. Feb 23-26. Gravenhurst, Ontario. Snowmobile activities and events, arts and crafts, children's ice and snow village, dances, winter baseball, broomball, helicopter rides, wreck 'm race, snow box derby, snow sculpture, turkey bowling, pancake breakfasts. More than 50 events—"the best family winter fun under the sun." Est attendance: 10,000. For info: Muskoka Winter Carnival Assn, PO Box 129, Gravenhurst, Ont, Canada P1P 1T5. Phone: (705) 687-8432.

CHARRO DAYS. Feb 23-26. Brownsville, TX. Colorful celebration of the charro horseman of Mexico, a man of great riding skills. Dances, parades and carnival. Est attendance: 100,000. For info: Charro Days, Inc, Cookie Wimer, PO Box 3247, Brownsville, TX 78523. Phone: (210) 542-4245.

DUBOIS, W.E.B.: BIRTH ANNIVERSARY. Feb 23. William Edward Burghardt Dubois, American educator and leader of movement for black equality. Born at Great Barrington, MA, Feb 23, 1868. Died at Accra, Ghana, Aug 27, 1963. "The cost of liberty," he wrote in 1909, "is less than the price of repression."

ENGLAND: NORTHWEST KNITTING AND NEEDLECRAFT EXHIBITION. Feb 23-26. G-Mex Centre, Manchester, Greater Manchester. Everything for the hand and machine knitter and the needlecraft enthusiast, fashion shows, lectures, demonstrations. Est attendance: 30,000. For info: Robert Ewin, Dir, Nationwide Exhibitions (UK) Ltd, PO Box 20, Fishponds, Bristol, England BS16 5QU.

FESTIVAL OF NATIVE ARTS. Feb 23-25. University of Alaska, Fairbanks. Native cultural event features craft exhibits and sales, demonstrations and dances presented by native people from all over Alaska. Est attendance: 4,000. For info: Festival of Native Arts, 5th Fl Gruening, University of Alaska, Fairbanks, AK 99775. Phone: (907) 474-7181.

GROUND WAR AGAINST IRAQ BEGINS: ANNIVERSARY. Feb 23. On Feb 23, 1991, after an air campaign lasting slightly more than a month, Allied forces launched the ground offensive against Iraqi forces. The relentless air attacks had devastated troops and targets in both Iraq and Kuwait. A world that had watched and anticipated the "The Mother of All Battles" was surprised at the swiftness and ease with which Allied forces were able to subdue Iraqi forces.

GUYANA: 25th ANNIVERSARY OF REPUBLIC. Feb 23. Guyana became a republic Feb 23, 1970.

HANDEL, GEORGE FREDERICK: 310th BIRTH ANNIVERSARY. Feb 23. Born at Halle, Saxony, Germany, on Feb 23, 1685. Handel and Johann Sebastian Bach (see also: "Bach, Johann Sebastian: Birth Anniversary" [Mar 21]), born the same year, were perhaps the greatest masters of Baroque music.

Chase's 1995 Calendar of Events — Feb 23-24

Handel's most frequently performed work is the oratorio *Messiah*, which was first heard in 1742. He died at London, England, on Apr 14, 1759.

ITALY: FEAST OF THE INCAPPUCCIATI. Feb 23. Gradoli (near Viterbo). On the Thursday before Ash Wednesday the members of the Confraternity of Purgatory make the rounds of the town dressed in traditional hooded robes, bearing a banner and walking to the beat of a drum. They stop at every house to collect foodstuffs in the name of the souls in purgatory; the food is then served at the banquet on Ash Wednesday. For info: Italian Govt Travel Office, 500 N Michigan Ave, Chicago, IL 60611. Phone: (312) 644-0990.

TAYLOR, GEORGE: DEATH ANNIVERSARY. Feb 23. Signer of the Declaration of Independence. Born 1716 (exact place and date unknown). Died Feb 23, 1781.

BIRTHDAYS TODAY

Bobby (Roberto Martin Antonio) Bonilla, 32, baseball player, born at New York, NY, Feb 23, 1963.

Sylvia Chase, 57, newscaster, born at Northfield, MN, Feb 23, 1938.

Peter Fonda, 56, actor (*Easy Rider, Race with the Devil*), born at New York, NY, Feb 23, 1939.

Howard Jones, 40, singer (gold album *Dream Into Action*), born at Southampton, England, Feb 23, 1955.

Edward "Too Tall" Jones, 44, former football player, boxer, born at Jackson, TN, Feb 23, 1951.

Johnny Winter (John Dawson, III), 51, singer, musician (*Still Alive and Well, Second Winter*), born at Beaumont, TX, Feb 23, 1944.

FEBRUARY 24 — FRIDAY

55th Day — Remaining, 310

BIG TEN MEN'S INDOOR TRACK AND FIELD CHAMPIONSHIP. Feb 24-25. University of Illinois, Champaign, IL. For info: Big Ten Conference, 1500 W Higgins Rd, Park Ridge, IL 60068-6300. Phone: (708) 696-1010.

BIG TEN WOMEN'S INDOOR TRACK AND FIELD CHAMPIONSHIP. Feb 24-25. University of Michigan, Ann Arbor, MI. For info: Big Ten Conference, 1500 W Higgins Rd, Park Ridge, IL 60068-6300. Phone: (708) 696-1010.

BLUEBERRY HILL OPEN DART TOURNAMENT. Feb 24-26. St. Louis, MO. America's largest pub dart tournament open to everyone. 23rd annual tournament. Sponsor: Rock and Roll Beer. Est attendance: 400. For info: Joe Edwards, Blueberry Hill, 6504 Delmar, St. Louis, MO 63130. Phone: (314) 727-0880.

CANADA: KLONDYKE CENTENNIAL SOCIETY CASINO NIGHT. Feb 24. Dawson City, Yukon. Enjoy an evening of gambling and slot machines, blackjack and roulette. For info: Yukon Tourism, Box 2703, Whitehorse, Yukon, Canada Y1A 2C6. Phone: (403) 993-1996.

ESTONIA: INDEPENDENCE DAY. Feb 24. National holiday.

GREGORIAN CALENDAR DAY. Feb 24. Pope Gregory XIII, enlisting the expertise of distinguished astronomers and mathematicians, issued a bull, Feb 24, 1582, correcting the Julian calendar that was then 10 days in error. The new calendar named for him, the Gregorian calendar, became effective on Oct 4, 1582, in most Catholic countries, in 1752 in Britain and the American colonies, in 1917 in the Soviet Union and on various other dates in other countries. It is the most widely used calendar in the world today. See also: "Calendar Adjustment Day Anniversary" (Sept 2) and "Gregorian Calendar Adjustment Anniversary" (Oct 4).

GRIMM, WILHELM CARL: BIRTH ANNIVERSARY. Feb 24. Mythologist and author, born at Hanau, Germany, Feb 24, 1786. Best remembered for *Grimm's Fairy Tales*, in collaboration with his brother, Jacob. Died at Berlin, Germany, Dec 16, 1859. See also: "Grimm, Jacob: Birth Anniversary" (Jan 4).

HADASSAH: ANNIVERSARY. Feb 24. On Feb 24, 1912, 12 members of the Daughters of Zion Study Circle met in New York City under the leadership of Henrietta Szold. A constitution was drafted to expand the study group into a national organization called Hadassah (Hebrew for "myrtle" and the biblical name of Queen Esther) to foster Jewish education in America and to create public health nursing and nurses training in Palestine. Hadassah is now the largest and oldest Zionist women's organization in the world with 1,500 chapters throughout the US and Puerto Rico rooted in health care delivery, education and vocational training, child rescue and land reclamation in Israel. See also "Szold, Henrietta: Birth Anniversary" (Dec 21). For info: Wendy Hirschhorn, Dir Public Affairs, Hadassah, 50 W 58th St, New York, NY 10019. Phone: (212) 303-8153.

HOMER, WINSLOW: BIRTH ANNIVERSARY. Feb 24. American artist. Born at Boston, MA, Feb 24, 1836. Noted for the realism of his work, from the Civil War reportage to the highly regarded rugged outdoor scenes of hunting and fishing. Died at his home at Prout's Neck, ME, Sept 29, 1910.

LONGHORN WORLD CHAMPIONSHIP RODEO. Feb 24-26. The Palace of Auburn Hills, Auburn Hills, MI. More than 250 of North America's best cowboys and cowgirls compete in 6 professionally sanctioned contests ranging from bronc riding to bull riding for top prize money and world championship points. The theme of the production changes annually. Additional attractions include special animal trainers, trick ropers, trick riders and professional clowns. 17th annual. Est attendance: 42,000. For info: Longhorn World Chmpshp Rodeo, Inc, PO Box 70159, Nashville, TN 37207. Phone: (615) 876-1016.

LOST DUTCHMAN DAYS. Feb 24-26. Apache Junction, AZ. Top entertainment, arts and crafts show, carnival, parade, 3-day rodeo competition and dance to celebrate the legend of the Superstitions and the Old Dutchman. Fax: (602) 982-3234. Est attendance: 100,000. For info: Apache Junction Chamber of Commerce, PO Box 1747, Apache Junction, AZ 85217-1747. Phone: (602) 982-3141.

NEWPORT SEAFOOD AND WINE FESTIVAL. Feb 24-26. Newport, OR. Central coastal festival featuring seafood and wines from Oregon, Washington, California and Idaho. Est attendance: 20,000. For info: Chamber of Commerce, 555 SW Coast Hwy, Newport, OR 97365. Phone: (503) 265-8801.

NIMITZ, CHESTER: 110th BIRTH ANNIVERSARY. Feb 24. Commander of all Allied naval, land and air forces in the southwest Pacific during a portion of WWII, Admiral Chester William Nimitz was born Feb 24, 1885, at Fredericksburg, TX. During the final assault on Japan in April 1945, Nimitz resumed command of the entire naval operation in the Pacific which he had shared with MacArthur for some time. Nimitz was one of the signers of the Japanese document of surrender Sept 2, 1945, aboard the *USS Missouri* in Tokyo Bay. Nimitz died Feb 20, 1966, at Treasure Island, San Francisco Bay, CA. The USS *Nimitz* was so named in his honor.

VOTE TO IMPEACH PRESIDENT ANDREW JOHNSON: ANNIVERSARY. Feb 24. In an escalating showdown over reconstruction policy following the American Civil War, the House of Representatives voted to impeach President Andrew Johnson on Feb 24, 1867. During the two years following the end of the war, the Republican-controlled Congress had sought to severely punish the South by undoing the policies of President Johnson. Congress passed the Reconstruction Act that divided the South into five military districts headed by general officers who were to take their orders from General Grant, the head of the army, instead of from President Johnson. In addition, Congress passed the Tenure of Office Act, which required Senate approval before Johnson could remove any official whose appointment was originally approved by the Senate. Johnson vetoed this act but the veto was overridden by Congress. To test the constitutionality of the act, Johnson dismissed Secretary of War Edwin Stanton, triggering the impeachment vote.

Feb 24–25   ☆ *Chase's 1995 Calendar of Events* ☆

WAGNER, HONUS: BIRTH ANNIVERSARY. Feb 24. American baseball great born John Peter Wagner at Carnegie, PA, on Feb 24, 1874. Nicknamed the "Flying Dutchman," Wagner was among the first five players elected to the Baseball Hall of Fame in 1936. Died in the town of his birth on Dec 6, 1955.

BIRTHDAYS TODAY

James Farentino, 57, actor ("Dynasty," "Cool Million," *The Story of a Woman*), born at New York, NY, Feb 24, 1938.
Rupert Holmes, 48, musician, songwriter ("Escape"), born at Tenafly, NJ, Feb 24, 1947.
Mark Lane, 75, lawyer, feminist, former congresswoman (hats), born at New York, NY, Feb 24, 1920.
Michel Legrand, 63, composer, conductor, born at Paris, France, Feb 24, 1932.
Joseph I. Lieberman, 53, US Senator (D, Connecticut) up for reelections in Nov '94, born at Stamford, CT, Feb 24, 1942.
Donald Vincent Majkowski, 31, football player, born at Buffalo, NY, Feb 24, 1964.
Zell Miller, 63, Governor of Georgia (D), born at Young Harris, GA, Feb 24, 1932.
Eddie Murray, 39, baseball player, born at Los Angeles, CA, Feb 24, 1956.
Edward James Olmos, 48, actor (*Stand and Deliver*, "Miami Vice"), born at East Los Angeles, CA, Feb 24, 1947.
Renata Scotto, 59, soprano, born at Savona, Italy, Feb 24, 1936.
Abe Vigoda, 74, actor ("Barney Miller," "Fish"), born at New York, NY, Feb 24, 1921.

FEBRUARY 25 — SATURDAY
56th Day — Remaining, 309

BASCOM, "TEXAS ROSE": BIRTH ANNIVERSARY. Feb 25, 1922. A Cherokee-Choctaw Indian born in Covington County, MS, Rose Flynt married a rodeo cowboy and learned trick roping, becoming known as the greatest female trick roper in the world. She appeared on stage, in movies and on early TV. She toured with the USO during WWII, performing at every military base and military hospital in the US. After the war she toured the world over entertaining servicemen stationed overseas. In 1981 she was inducted into the National Cowgirl Hall of Fame. She died Sept 23, 1993. For info: Margaret Formby, Exec Dir, Natl Cowgirl Hall of Fame, PO Box 1742, Hereford, TX 79045. Phone: (806) 364-5252.

BLACK HISTORY PARADE. Feb 25. Monroe, LA. Informative parade to celebrate Black History Month. For info: Judy M. Lowentritt, Tourism Sales Mgr, Monroe-W Monroe CVB, PO Box 6054, Monroe, LA 71211. Phone: (318) 362-2662.

BRAZIL: CARNIVAL. Feb 25–28. Especially in Rio de Janeiro, this carnival is said to be one of the last great folk festivals, and the big annual event in the life of Brazilians. Begins on Saturday night before Ash Wednesday and continues through Shrove Tuesday.

CARUSO, ENRICO: BIRTH ANNIVERSARY. Feb 25. Operatic tenor of legendary voice and fame, born at Naples, Italy, on Feb 25, 1873. Died there Aug 2, 1921.

CHAMORRO, VIOLETA: 5th ELECTION ANNIVERSARY. Feb 25. On Feb 25, 1990, Violeta Barrios de Chamorro defeated communist leader Daniel Ortega in a free election bringing Marxist rule to an end in Nicaragua. Her election was the result of a broad-based coalition that sought an end to the communist policies and tensions between the US-backed Contras and Ortega's forces.

February 1995

S	M	T	W	T	F	S
			1	2	3	4
5	6	7	8	9	10	11
12	13	14	15	16	17	18
19	20	21	22	23	24	25
26	27	28				

CLAM CHOWDER COOKOFF AND GREAT CHOWDER CHASE. Feb 25–26. Santa Cruz Beach Boardwalk, Santa Cruz, CA. Who makes the world's greatest clam chowder? Up to 80 teams compete to find out. Prizes for best booth encourages wacky costumes, elaborate props! Separate categories for restaurants, media and individuals, Boston and Manhattan style chowder. Free admission. Est attendance: 15,000. For info: Ann Parker, Publicity Mgr, Santa Cruz Beach Boardwalk, 400 Beach St, Santa Cruz, CA 95060-5491. Phone: (408) 423-5590.

DAVIS, ADELLE: 90th BIRTH ANNIVERSARY. Feb 25. American nutritionist and author. Born at Lizton, IN, on Feb 25, 1905. Her message "You are what you eat" found an eager readership for her books, including *Let's Cook It Right* (1947) and *Let's Eat Right to Keep Fit* (1954). Davis died of bone cancer at Palo Verdes Estates, CA, on May 31, 1974.

DORAL-RYDER OPEN. Feb 25–Mar 5. Doral Resort and Country Club, Miami, FL. A PGA Tour golf tournament with a full week of events, including a free outdoor pops concert, a skins game, three celebrity pro-ams and the four-day tournament which features 144 of the top golfers in the world. The Doral-Ryder Open is the nation's largest sports fundraiser for the American Cancer Society. Est attendance: 120,000. Sponsor: Susan Lanier, Doral-Ryder Open, PO Box 522927, Miami, FL 33152. Phone: (305) 477-4653.

FENWICK, MILLICENT HAMMOND: BIRTH ANNIVERSARY. Feb 25. Former fashion model, author, editor, member NJ General Assembly and US Congresswoman, Millicent Fenwick was born at NYC, Feb 25, 1910. A champion of liberal causes, Fenwick pointed to her sponsorship of the resolution creating the commission to monitor the 1975 Helsinki accords on human rights as her proudest achievement. She fought for civil rights, peace in Vietnam, aid for the poor, reduction of military programs, gun control and restrictions on capital punishment. She was widely known because of her sense of humor and idiosyncratic habit of smoking a pipe. Millicent Fenwick, who was the inspiration for Garry Trudeau's Doonesbury character Lacey Davenport, died at Bernardsville, NJ, on Sept 16, 1992.

GEM AND MINERAL SHOW. Feb 25–26. Jackson, MS. Gemstones, fossils, minerals, jewelry cutting and stone-polishing demonstrations. 36th annual show. Est attendance: 5,000. For info: Mississippi Gem and Mineral Soc, Terry Barney, Show Chairman, Jackson, MS 39205. Phone: (601) 372-5525.

HATSUME FAIR. Feb 25–26. Morikami Museum, Delray Beach, FL. Celebrates the coming of spring with demonstrations and performances of Japanese artists. Bonsai, folk drumming, music, traditional and contemporary dance and martial arts. Annually, the last weekend in February. Est attendance: 14,000. Sponsor: The Morikami Museum, Larry Rosensweig, Dir, 4000 Morikami Park Rd, Delray Beach, FL 33446. Phone: (407) 495-0233.

KCCR FARM, HOME & SPORT SHOW. Feb 25–26. Ramkota Rivercentre, Pierre, SD. 23rd annual home show features more than 80 exhibitors displaying merchandise and services. More than 17,000 square feet of exhibits. Open to the public. Booth space for rent through KCCR Radio. Est attendance: 6,000. For info: Mark Pautsch, Sales Mgr, KCCR Radio, 106 W Capitol, Pierre, SD 57501. Phone: (605) 224-1240.

KUWAIT: NATIONAL DAY. Feb 25. National holiday.

Chase's 1995 Calendar of Events — Feb 25–26

MARDI GRAS IN CAJUN COUNTRY. Feb 25–28. Lafayette, LA. Celebration before the beginning of Lent. Included festivities: parades, costume contest, annual Southwest Louisiana Mardi Gras Association Pageant and Ball. The surrounding rural areas offer the traditional Courir du Mardi Gras. Second largest carnival celebration in the US. Est attendance: 300,000. For info: Lafayette Conv and Visitors Commission, Kelly Strenge, PO Box 52066, Lafayette, LA 70505. Phone: (800) 346-1958.

NENANA TRIPOD RAISING FESTIVAL. Feb 25–26. Nenana, AK. Festival centers around the guessing of the exact time of the ice break-up on the Tanana River. Highlights include the raising of the tripod (Feb 26), the Nenana Banana Eating Contest, native dancers, sled dog races, craft bazaar and much more. Est attendance: 1,000. For info: Nenana Ice Classic, Box 272, Nenana, AK 99760. Phone: (907) 832-5446.

PARKE COUNTY MAPLE FAIR. Feb 25–26. (Also Mar 4–5.) Rockville, IN. Sugar camps feature unique process of making maple syrup. Tours. Pancakes, sausage and maple syrup served, arts and crafts, farmers' market and butcher shop. Annually, the last weekend in February and the first weekend in March. Est attendance: 20,000. Sponsor: Parke County, Inc, Anne Lynk, PO Box 165, Rockville, IN 47872. Phone: (317) 569-5226.

PORTUGAL: OVAR CARNIVAL FESTIVAL. Feb 25–28. Costa de Prata. Masked processions, carnival floats and battles of flowers. Annually, the four days before Lent. For info: Portuguese Natl Tourist Office, 590 Fifth Ave, New York, NY 10036-4704. Phone: (212) 354-4403.

RENOIR, PIERRE AUGUSTE: BIRTH ANNIVERSARY. Feb 25. Impressionist painter, born at Limoges, France, on Feb 25, 1841. His paintings are known for their joy and sensuousness as well as the light techniques Renoir employed in them. In his later years he was crippled by arthritis and would paint with the brush strapped to his hand. He died at Cagnes-sur-Mer, Provence, France, Dec 17, 1919.

SOUTHERN SPRING SHOW. Feb 25–Mar 5. Charlotte, NC. The sights and scents of springtime—everything for home and garden. Dozens of professionally landscaped gardens, judged and full flower show, outdoor furniture and all the supplies and equipment for lawn and garden. Plus designer rooms, home building products and accessories, crafts, art and sculpture pavilion, travel and leisure exhibits and special show features and offerings. Est attendance: 100,000. For info: Southern Shows, Inc, PO Box 36859, Charlotte, NC 28236. Phone: (800) 849-0248.

SPACE MILESTONE: *SOYUZ 32* (USSR). Feb 25. Launched on Feb 25, 1979, from Baikonur space center in Soviet Central Asia. Cosmonauts Vladimir Lyakhov and Valery Ryumin aboard, docked at *Salyut 6* space station on Feb 26. Returned to Earth in *Soyuz 34* after record 175 days in space on Aug 19, 1979.

SURINAME: REVOLUTION DAY. Feb 25.

WORLD CHAMPIONSHIP HOG CALLING CONTEST. Feb 25. Downtown Weatherford, OK. Food vendors, arts and crafts booths, pork barbecue, Miss Pigtail Contest, hog weight-guessing, a parade, greased pig chase, crowning of the Hog Queen and the hog-calling competition. An evening dance—the Hog Call Ball. 9th annual. Annually, the last Saturday in February. Sponsor: Milton-Bradley. For info: Weatherford Daily News, PO Box 191, Weatherford, OK 73096. Phone: (405) 772-3301.

BIRTHDAYS TODAY

Anthony Burgess, 78, author (*Clockwork Orange, Any Old Iron*), born at Manchester, England, Feb 25, 1917.
Tom Courtenay, 58, actor (*The Loneliness of the Long Distance Runner, Otley*), born at Hull, England, Feb 25, 1937.
Larry Gelbart, 67, writer, producer ("M*A*S*H"), born at Chicago, IL, Feb 25, 1928.
Karen Grassle, 51, actress (Caroline Ingalls on "Little House on the Prairie"), born Berkeley, CA, Feb 25, 1944.
Philip Habib, 75, diplomat, born at Brooklyn, NY, Feb 25, 1920.
George Harrison, 52, musician, born at Liverpool, England, Feb 25, 1943.
Kurt Rambis, 37, basketball player, born at Cupertino, CA, Feb 25, 1958.
Sally Jessy Raphael, 52, broadcaster, born at Easton, PA, Feb 25, 1943.
Bobby Riggs (Robert Larimore), 77, tennis player (lost to Billie Jean King in the battle of the sexes), born at Los Angeles, CA, Feb 25, 1918.
Faron Young, 63, singer ("Hello Walls"), born at Shreveport, LA, Feb 25, 1932.

FEBRUARY 26 — SUNDAY
57th Day — Remaining, 308

CODY, WILLIAM FREDERIC "BUFFALO BILL": BIRTH ANNIVERSARY. Feb 26. American frontiersman who claimed to have killed more than 4,000 buffaloes. Born at Scott County, IA, on Feb 26, 1846. Subject of many heroic Wild West yarns, Cody became successful as a showman and exhibitionist, taking his acts across the US and to Europe. Died on Jan 10, 1917, at Denver, CO.

DAUMIER, HONORE: BIRTH ANNIVERSARY. Feb 26. French painter and caricaturist famous for his satirical and comic lithographs. Once spent six months in prison for a caricature of Louis Philippe shown as *Gargantua* consuming the heavy taxes of the citizens. Born on Feb 26, 1808, at Marseilles, France. He died on Feb 11, 1879, at Volmondois, France.

FASCHING SUNDAY. Feb 26. Germany and Austria. The last Sunday before Lent.

GLEASON, JACKIE: BIRTH ANNIVERSARY. Feb 26. American musician, comedian and actor, Herbert John "Jackie" Gleason. Born at Brooklyn, NY, Feb 26, 1916. Best known for his role as Ralph Kramden in the long-running television series "The Honeymooners." Died at Fort Lauderdale, FL, June 24, 1987.

GRAND CANYON NATIONAL PARK ESTABLISHED: ANNIVERSARY. Feb 26. By an act of Congress, Grand Canyon National Park was established Feb 26, 1919. An immense gorge cut through the high plateaus of northwest Arizona by the raging Colorado River and covering 1,218,375 acres, Grand Canyon National Park is considered one of the most spectacular natural phenomena in the world.

HORTON BAY WINTER OLYMPICS. Feb 26. Horton Bay, MI. Unique series of unusual athletic competitions including a frozen-fish-tossing contest, ice bowling, golf obstacle course, bobbing for apples (through the ice on Lake Charlevoix) and a shovel-pulling contest. For info: Boyne County Conv and Visitors Bureau, Peter Fitzsimons, PO Box 694, Petoskey, MI 49770.

HUGO, VICTOR: BIRTH ANNIVERSARY. Feb 26. French author, born Feb 26, 1802, at Besançon, France. "An invasion of armies can be resisted," he wrote in 1852, "but not an idea whose time has come." His most well-known work was the novel *Les Misérables*. Died at Paris, May 22, 1885.

ITALY: CARNIVAL WEEK. Feb 26–Mar 4. Milan. Carnival week is held according to local tradition, with shows and festive events for children on Tuesday and Thursday. Parades of floats, figures in the costume of local folk characters, Meneghin and Cecca, parties and more traditional events are held on Saturday. Annually, the Sunday–Saturday of Ash Wednesday week. For info: Italian Govt Travel Office, 500 N Michigan Ave, Chicago, IL 60611. Phone: (312) 644-0990.

JOE CAIN PROCESSION. Feb 26. Mobile, AL. Honors the man who revived the Mardi Gras in Mobile in 1866 following the Civil War. Annually, the Sunday before Shrove Tuesday. Est attendance: 100,000. For info: Wayne Dean, Sr, VP, Joe Cain Society, 1064 Palmetto St, Mobile, AL 36604-3041. Phone: (334) 432-3960.

NATIONAL PANCAKE WEEK. Feb 26–Mar 4. Traditional celebration surrounding Shrove or Pancake Tuesday to recognize the history and continuing popularity of pancakes. Fax: (612) 540-3232. For info: Pam Becker, Bisquick Baking Mix, General Mills, Inc, #1 General Mills Blvd, Minneapolis, MN 55426. Phone: (612) 540-2470.

SHROVETIDE. Feb 26–28. The three days before Ash Wednesday: Shrove Sunday, Monday and Tuesday—a time for confession and for festivity before the beginning of Lent.

STRAUSS, LEVI: BIRTH ANNIVERSARY. Feb 26. Bavarian immigrant Levi Strauss created the world's first pair of jeans—Levi's 501 jeans—for California's gold miners in the 1800s. Born at Buttenheim, Bavaria, Germany, Feb 26, 1829. Died in 1902.

VERCORS, JEAN: BIRTH ANNIVERSARY. Feb 26. Jean Vercors was the author of the first clandestine novel published during the Nazi occupation of France. Vercors, whose real name was Jean-Marcel de Bruller, was best known for his novel *Silence of the Sea*, which he published with Pierre de Lescure for their publishing house, Les Editions de Minuit, after the Nazis occupied France in 1941. Vercors was born Feb 26, 1902, at Paris, France, and died there June 10, 1991.

YELTSIN FOES FREED: ANNIVERSARY. Feb 26, 1994. Former speaker of the Russian parliament Ruslan Khasbulatov and former vice-president Alexander Rutskoi along with many of their followers walked out of prison, freed by general amnesty granted by the new Russian parliament. Their attempt in October 1993 to overthrow the government of Boris Yeltsin turned violent and ended with the military shelling the Russian parliament building forcing their capitulation. The uncertain Russian political landscape was further clouded when the amnesty also ended the treason trials of the former hardliners who sought to overthrow Mikhail Gorbachev in August 1991.

BIRTHDAYS TODAY

Mason Adams, 76, actor ("Lou Grant," "Morningstar/Evening Star"), born at New York, NY, Feb 26, 1919.

Rolando Blackman, 36, basketball player, born at Panama City, Panama, Feb 26, 1959.

Johnny Cash, 63, singer ("Guess Things Happen That Way," "Ring of Fire"), born at Kingsland, AR, Feb 26, 1932.

Fats Domino (Antoine Domino), 67, singer, songwriter ("Blueberry Hill"), born New Orleans, LA, Feb 26, 1928.

Betty Hutton (Elizabeth June Thornberg), 74, actress, singer (*Annie Get Your Gun*, "The Betty Hutton Show"), born Battle Creek, MI, Feb 26, 1921.

Tony Randall (Leonard Rosenberg), 75, actor (*Pillow Talk*, "The Odd Couple"), born Tulsa, OK, Feb 26, 1920.

	S	M	T	W	T	F	S
February 1995				1	2	3	4
	5	6	7	8	9	10	11
	12	13	14	15	16	17	18
	19	20	21	22	23	24	25
	26	27	28				

FEBRUARY 27 — MONDAY
58th Day — Remaining, 307

AMATEUR ATHLETIC UNION JAMES E. SULLIVAN MEMORIAL AWARD DINNER. Feb 27. Indiana Convention Center, Indianapolis, IN. This award is presented annually to the outstanding amateur athlete based on the qualities of leadership, character, sportsmanship and the ideals of amateurism. For info: VP of Mktg, Indiana Sports Corp, 201 S Capitol Ave, Ste 1200, Indianapolis, IN 46225. Phone: (317) 237-5000.

CARNIVAL. Feb 27–28. Period of festivities, feasts, foolishness and gaiety immediately before Lent begins on Ash Wednesday. Ordinarily Carnival includes only Fasching (the Feast of Fools), being the Monday and Tuesday immediately preceding Ash Wednesday. The period of carnival may also be extended to include the preceding three days, Friday, Saturday and Sunday, or even longer periods in some areas.

DENMARK: STREET URCHINS' CARNIVAL. Feb 27. Observed on Shrove Monday.

DOMINICAN REPUBLIC: INDEPENDENCE DAY. Feb 27. National Day. Independence gained in 1844 at withdrawal of Haitians, who had controlled the area for the past 22 years.

FARRELL, JAMES THOMAS: BIRTH ANNIVERSARY. Feb 27. American author, novelist and short-story writer, best known for his Studs Lonigan trilogy. Born at Chicago, IL, Feb 27, 1904. Died at New York, NY, Aug 22, 1979.

FASCHING. Feb 27–28. In Germany and Austria, Fasching, also called Fasnacht, Fasnet or Feast of Fools, is a Shrovetide festival with processions of masked figures, both beautiful and grotesque. Always the two days (Rose Monday and Shrove Tuesday) between Fasching Sunday and Ash Wednesday.

GRENADA: CARRIACOU CARNIVAL. Feb 27–28. Carriacou. On Shrove Tuesday ("Sweet Mas"), "Pierots" dressed in well-padded costumes recite Shakespeare and must do so correctly or be subjected to a bull whip. Est attendance: 2,000. For info: Grenada Tourist Information Office, 820 Second Ave, Ste 900-D, New York, NY 10017. Phone: (800) 927-9554.

ICELAND: BUN DAY. Feb 27. Children invade homes in the morning with colorful sticks and receive gifts of whipped cream buns (on the Monday before Shrove Tuesday).

INTERNATIONAL WEEK AT TEXAS A&M. Feb 27–Mar 3. Memorial Student Center, Texas A&M University, College Station, TX. Texas A&M's more than 2,600 international students from some 115 countries introduce their cultures to the university and community during this 16th annual International Week. Cultural displays, an international buffet, talent show and traditional dress parade are some of the week's highlights. Annually, the first week in March. Est attendance: 2,300. For info: Intl Student Services, Texas A&M University, College Station, TX 77843-1226. Phone: (409) 845-1825.

JOHN STEINBECK'S BIRTHDAY. Feb 27. Steinbeck Home, Salinas, CA. Celebrations of Steinbeck's birth at his birthplace which is now a museum and restaurant run by volunteers. Proceeds go to local charities. Est attendance: 400. For info: Steinbeck Fdtn, Amanda Piper, Promo Mgr, PO Box 2495, Salinas, CA 93902. Phone: (408) 753-6411.

KUWAIT CITY LIBERATED AND 100-HOUR WAR ENDS: ANNIVERSARY. Feb 27. On Feb 27, 1991, Allied troops liberated Kuwait City, Kuwait, four days after launching a ground offensive. President George Bush declared Kuwait to be liberated and ceased all offensive military operations. The end of military operations at midnight, eastern standard time, came 100 hours after the beginning of the land attack.

LONGFELLOW, HENRY WADSWORTH: BIRTH ANNIVERSARY. Feb 27. American poet and writer, born at Portland, ME, Feb 27, 1807. Best remembered for his classic narrative poems such as *The Song of Hiawatha, Paul Revere's Ride* and *The Wreck of the Hesperus*, he died at Cambridge, MA, Mar 24, 1882.

ST. GABRIEL POSSENTI FEAST DAY. Feb 27. St. Gabriel of the Sorrowful Mother (Francis Possenti, 1838-1862). Italian patron saint of young seminarians.

SHANGHAI COMMUNIQUE: ANNIVERSARY. Feb 27. Feb 27, 1972, President Richard Nixon and Premier Chou En-Lai released a joint communique (the Shanghai Communique) after Nixon's week-long visit to the People's Republic of China. The two nations agreed to work toward normalizing relations. Stopping short of establishing diplomatic relations, this was the first step in that direction. The two nations entered full diplomatic relations on Jan 1, 1979, during the Carter administration.

SHROVE MONDAY. Feb 27. The Monday before Ash Wednesday. In Germany and Austria, this is called Rose Monday.

TRINIDAD: CARNIVAL. Feb 27-28. Port of Spain. An amazing spiritual tradition that brings together people from all over the world in an incredible colorful setting that includes the world's most celebrated calypsonians, steel band players, costume designers and masqueraders. Annually, the two days before Ash Wednesday. For info: Trinidad and Tobago Tourist Board, 25 W 43rd St, Ste 1508, New York, NY 10036. Phone: (800) 232-0082.

TWENTY-SECOND AMENDMENT TO US CONSTITUTION (TWO-TERM LIMIT): RATIFICATION ANNIVERSARY. Feb 27. After the four successive presidential terms of Franklin Roosevelt, the 22nd Amendment, ratified on Feb 27, 1950, limited the tenure of presidential office to two terms.

BIRTHDAYS TODAY

Mary Frann, 52, actress (Joanna on "Newhart"), born at St. Louis, MO, Feb 27, 1943.
Alan Guth, 48, physicist, born at New Brunswick, NJ, Feb 27, 1947.

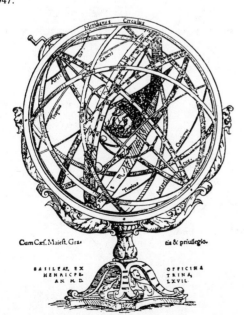

Howard Hesseman, 55, actor ("WKRP in Cincinnati," "Head of the Class"), born at Salem, OR, Feb 27, 1940.
Charlayne Hunter-Gault, 53, broadcast journalist, born at Due West, SC, Feb 27, 1942.
Ralph Nader, 61, consumer advocate, lawyer, born at Winsted, CT, Feb 27, 1934.
Irwin Shaw, 82, writer (*Rich Man, Poor Man; The Young Lions*), born New York, NY, Feb 27, 1913.
Elizabeth Taylor, 63, actress (*National Velvet, Cat on a Hot Tin Roof, Cleopatra*), AIDS activist, born London, England, Feb 27, 1932.
Malcolm Wallop, 62, former US Senator (R, Wyoming), born at New York, NY, Feb 27, 1933.
Joanne Woodward, 65, actress (Oscar for *The Three Faces of Eve; Mr & Mrs Bridge*), born at Thomasville, GA, Feb 27, 1930.
James Ager Worthy, 34, former basketball player, born at Gastonia, NC, Feb 27, 1961

FEBRUARY 28 — TUESDAY
59th Day — Remaining, 306

◆ **AMERICANS LAND AT PUERTO PRINCESA: 50th ANNIVERSARY.** Feb 28, 1945. Having retaken the big Philippine island of Luzon, American forces began "mopping-up" activities in the southern islands with the landing of troops at Puerto Princesa on Palawan Island. By mid-July at least 13,000 American lives had been lost in the series of 38 attacks, needed to overcome 450,000 Japanese remaining in the islands.

BLONDIN, CHARLES: BIRTH ANNIVERSARY. Feb 28. Daring French acrobat and aerialist (whose real name was Jean Francois Gravelet). Born at St. Omer, France, Feb 28, 1824. Especially remembered for his conquest of Niagara Falls. See also: "Charles Blondin's Conquest of Niagara Falls: Anniversary" (June 30).

DE MONTAIGNE, MICHEL: BIRTH ANNIVERSARY. Feb 28. French essayist and philosopher. Born Feb 28, 1533. Died Sept 13, 1592. "And if you have lived a day," he wrote in Book I of his *Essays*, "you have seen everything. One day is equal to all days. There is no other light, no other night. This sun, this moon, these stars, the way they are arranged, all is the very same your ancestors enjoyed and that will entertain your grandchildren...."

HENDO'S FAT TUESDAY AT BRECKENRIDGE. Feb 28. Breckenridge Ski Area, Breckenridge, CO. A Mardi Gras celebration at 10,000 feet—beads, face painting, costumes, Cajun music and fresh seafood at the Maggie Base of Peak 9. Est attendance: 25,000. For info: John Hendryson, PO Box 7868, Breckenridge, CO 80424. Phone: (303) 453-2913.

ICELAND: BURSTING DAY. Feb 28. Feasts with salted mutton and thick pea soup. (Shrove Tuesday.)

INTERNATIONAL PANCAKE DAY. Feb 28. Liberal, KS. The 1995 International Pancake Race will be the 45th annual competition between the women of Liberal, KS, and Olney, Bucks, England. The women, wearing the traditional dress, apron and scarf, run a 415-yard "S"-shaped course, carrying a pancake in a skillet. Record time is 58.5 seconds, set in 1975. Other events include a parade, talent show, eating and flipping contests and 8K and 2-mile Fun Runs. Est attendance: 5,000. For info: Rosalee Phillips, Exec Secy, PO Box 665, Liberal, KS 67905. Phone: (316) 624-6427.

◆ **LYON, MARY: BIRTH ANNIVERSARY.** Feb 28, 1797. Mary Lyon, born near Buckland, MA, became a pioneer in the field of higher education for women. Teaching since age 17, she founded Mount Holyoke Seminary (forerunner of Mount Holyoke College) in South Hadley, MA, in 1837 at a time when American women were educated primarily by ministers in classes held in their homes. Mount Holyoke was one of the first permanent women's colleges. She died Mar 5, 1849, at South Hadley.

Feb 28–29 ☆ Chase's 1995 Calendar of Events ☆

MAMOU MARDI GRAS CELEBRATION. Feb 28. Mamou, LA. Country-style Courir de Mardi Gras with a band of horsemen who leave town at 8:30 AM, travel the countryside gathering various items for their gumbo and performing for the residents and return to town at 4:00 PM. Live Cajun music is provided from 11 AM to 6 PM downtown at no charge. A dance with live Cajun music is held at the American Legion Hall from 6 PM to 12 midnight with a cover charge. Sponsor: Mamou Mardi Gras Assn. Est attendance: 15,000. For info: Gerald J. Fontenot, Chamber of Commerce, PO Box 34, Mamou, LA 70554. Phone: (318) 468-3272.

MARDI GRAS. Feb 28. Celebrated especially in New Orleans, LA, Mobile, AL, and certain Mississippi and Florida cities. Last feast before Lent. Although *Mardi Gras* (Fat Tuesday, literally) is properly limited to Shrove Tuesday, it has come to be popularly applied to the preceding two weeks of celebration.

"M*A*S*H": THE FINAL EPISODE: ANNIVERSARY. Feb 28. Concluding a run of 255 episodes, this 2½-hour finale broadcast on Feb 28, 1983, was the most-watched television show at that time—77% of the viewing public were tuned in. Cast members included Alan Alda as Capt Hawkeye Pierce, Wayne Rogers as Capt Trapper John McIntyre, McLean Stevenson as Lt Col Henry Blake, Harry Morgan as Col Sherman Potter, Loretta Swit as Maj Margaret "Hot Lips" Houlihan, Mike Farrell as Capt B. J. Hunnicut, Larry Linville as Maj Frank Burns, Gary Burghoff as Pvt Walter "Radar" O'Reilly, William Christopher as Father Francis Mulcahy, Jamie Farr as Cpl Maxwell Klinger and David Ogden Stiers as Maj Charles Emerson Winchester III. The show premiered on Sept 17, 1972.

PACZKI DAY IN HAMTRAMCK. Feb 28. Hamtramck, MI. Shrove Tuesday, the day before Lenten fasting begins, is the occasion for baking and selling the delicious pastry known as *paczki* (pronounced "panchkey"), which is described as a "distant cousin of the jelly doughnut." Especially observed among Polish citizens.

ST. OSWALD OF WORCESTER FEAST DAY. Feb 28. Bishop of Worcester, England, from 961, and Archbishop of York from 972. Oswald died on Feb 29, 992, but Feb 28 is generally celebrated as his Feast Day.

SHROVE TUESDAY. Feb 28. Always the day before Ash Wednesday. Sometimes called Pancake Tuesday. Public holiday in Florida.

SHROVETIDE PANCAKE RACE. Feb 28. Olney, Buckinghamshire, England, and Liberal, Kansas. The pancake race at Olney has been run since 1445. Competitors must be women over 16 years of age, wearing traditional housewife's costume, including apron and headcovering. With a toss and flip of the pancake on the griddle that each must carry, the women dash from marketplace to the parish church, where the winner receives a kiss from the ringer of the Pancake Bell. Shriving service follows. Starting time for the race is usually 11:45 AM. Annually, on Shrove Tuesday.

SNOWMASS MARDI GRAS. Feb 28. Snowmass Village, CO. Parade, costume contest and masquerade ball. Est attendance: 3,000. For info: Snowmass Resort Assn, PO Box 5566, Snowmass Village, CO 81615. Phone: (303) 923-2000.

TENNIEL, JOHN: 175th BIRTH ANNIVERSARY. Feb 28. Illustrator and cartoonist, born at London, England, Feb 28, 1820. Best remembered for his illustrations for Lewis Carroll's *Alice's Adventures in Wonderland*. Died at London, Feb 25, 1914.

USS *PRINCETON* EXPLOSION: ANNIVERSARY. Feb 28. On Feb 28, 1844, the newly built "war steamer," USS *Princeton*, cruising on the Potomac River with top government officials as its passengers, fired one of its guns (known, ironically, as the "Peacemaker") to demonstrate the latest in naval armament. The gun exploded, killing Abel P. Upshur, Secretary of State; Thomas W. Gilmer, Secretary of the Navy; David Gardiner, of Gardiners Island, NY; and several others. Many were injured. The president of the US, John Tyler, was on board and narrowly escaped death.

BIRTHDAYS TODAY

Svetlana Allilueva, 69, author, daughter of Joseph Stalin (*The Faraway Music*), born at Moscow, USSR, Feb 28, 1926.
Mario Andretti, 55, race car driver, born at Montona, Trieste, Italy, Feb 28, 1940.
Adrian Dantley, 39, former basketball player, born at Washington, DC, Feb 28, 1956.
Charles Durning, 72, actor (*Dog Day Afternoon*, "Evening Shade"), born at Highland Falls, NY, Feb 28, 1923.
Frank Gehry, 66, architect, born at Toronto, Ontario, Canada, Feb 28, 1929.
Gavin MacLeod, 65, actor (Murray on "The Mary Tyler Moore Show"; captain on "The Love Boat"), born at Mt Kisco, NY, Feb 28, 1930.
Bernadette Peters, 51, actress, singer (*George M.*, *Dames at Sea*), born at New York, NY, Feb 28, 1944.
Bubba Smith (Charles Aaron Smith), 50, former football player, actor, born at Orange, TX, Feb 28, 1945.
Tommy Tune, 56, actor, singer, dancer (Tony for *My One and Only*; "Dean Martin Presents . . ."), musical theater director, born at Wichita Falls, TX, Feb 28, 1939.

FEBRUARY 29 ANNIVERSARY

NEMEROV, HOWARD: 75th BIRTH ANNIVERSARY. Feb 29. Howard Nemerov was the third poet laureate of the US from 1988 to 1990. Among his works are 26 books, including five novels. He won a Pulitzer Prize and the National Book Award for his *Collected Works* in 1978. He was also a recipient of the National Medal of the Arts. As poet laureate he penned verses commemorating the 200th anniversary of the US Congress and the launch of the Space Shuttle, *Atlantis*. Nemerov was born on Feb 29, 1920, at New York City, NY and died July 5, 1991, at St. Louis, MO.

FEBRUARY 29 BIRTHDAYS

Joss Ackland, 67, actor (*The House That Dripped Blood*, *Penny Gold*), born at London, England, Feb 29, 1928.
Jack Lousma, 59, astronaut, born at Grand Rapids, MI, Feb 29, 1936.
Michelle Morgan (Simone Roussel), 75, actress (*The Fallen Idol*), born at Neuilly, France, Feb 29, 1920.

☆ Chase's 1995 Calendar of Events ☆   Mar 1

March.

MARCH 1 — WEDNESDAY
60th Day — Remaining, 305

★ **AMERICAN RED CROSS MONTH.** Mar 1–31. Presidential Proclamation for Red Cross Month issued each year for March since 1943. Issued as American Red Cross Month since 1987.

ARTICLES OF CONFEDERATION: RATIFICATION ANNIVERSARY. Mar 1. This compact made among the original 13 states had been adopted by the Congress Nov 15, 1777, and submitted to the states for ratification on Nov 17, 1777. Maryland was the last state to approve, on Feb 27, 1781, but Congress named Mar 1, 1781, as the day of formal ratification. The Articles of Confederation remained the supreme law of the nation until Mar 4, 1789.

ASH WEDNESDAY. Mar 1. Marks the beginning of Lent. Forty weekdays and six Sundays (Saturday considered a weekday) remain until Easter Sunday. Named for use of ashes in ceremonial penance.

CATARACT AWARENESS MONTH. Mar 1–31. Activities during this month are designed to educate people about one of the most common types of vision loss and the treatment options now available. Materials that can easily be posted or distributed to the community will be provided. Sponsor: Prevent Blindness America, Marita Gomez, Media Relations, 500 E Remington Rd, Schaumburg, IL 60173. Phone: (800) 331-2020.

EAST TEXAS WALKING AND RACKING HORSE SHOW. Mar 1–31. Central East Texas Fairgrounds, Marshall, TX. Call for specific dates in March. A charity horse show consisting of 50 classes of walking horses, racking horses and spotted saddle horses. Three classes held for other breeds. Stick horse riding for kids. Prizes awarded to the top five entries. Est attendance: 1,200. For info: Greater Marshall Chamber of Commerce, PO Box 520, Marshall, TX 75671. Phone: (903) 935-7868.

HELP SOMEONE SEE WEEK OBSERVANCE. Mar 1–5. Rockford, OH. Collect used eyeglasses, offer new glass cases for nursing home residents, arrange visits from guide dogs, have programs or videos about sight. For info: Shane Hill Nursing Home, Dorothy Trisel, Activities Dir, 10731 State Rte 118, Rockford, OH 45882-0159.

HUMORISTS ARE ARTISTS MONTH (HAAM). Mar 1–31. To recognize the important contributions made by various types of humorists to the high art of living. For info: Lone Star Publications of Humor, Attn: Lauren Barnett, Box 29000, Ste 103, San Antonio, TX 78229.

March 1995

S	M	T	W	T	F	S
			1	2	3	4
5	6	7	8	9	10	11
12	13	14	15	16	17	18
19	20	21	22	23	24	25
26	27	28	29	30	31	

ITALY: PURGATORY BANQUET. Mar 1. Gradoli (near Viterbo). On Ash Wednesday, gourmands are on hand for the banquet of penitence for the souls in purgatory, held on the premises of the cooperative winery. For info: Italian Govt Travel Office, 500 N Michigan Ave, Chicago, IL 60611. Phone: (312) 644-0990.

JAPAN: OMIZUTORI (WATER-DRAWING FESTIVAL). Mar 1–14. Todaiji, Nara. At midnight, a solemn rite is performed in the flickering light of pine torches. People rush for sparks from the torches, which are believed to have magic power against evil. Most spectacular on the night of Mar 12. The ceremony of drawing water is observed at 2:00 AM on Mar 13, to the accompaniment of ancient Japanese music.

KOREA: SAMILJOL OR INDEPENDENCE DAY. Mar 1. Koreans observe the anniversary of the Mar 1, 1919, independence movement against Japanese colonial rule.

LEARNING DISABILITIES ASSOCIATION OF AMERICA INTERNATIONAL CONFERENCE. Mar 1–4. Twin Towers, Orlando, FL. Est attendance: 4,000. For info: Learning Disabilities Assn of America, Jean S. Petersen, Exec Dir, 4156 Library Rd, Pittsburgh, PA 15234. Phone: (412) 341-1515.

LENT BEGINS. Mar 1–Apr 15. Most Christian churches observe period of fasting and penitence (40 weekdays and 6 Sundays—Saturday considered a weekday) beginning on Ash Wednesday and ending on the Saturday before Easter.

LINDBERGH KIDNAPPING: ANNIVERSARY. Mar 1. 20-month-old Charles A. Lindbergh, Jr, the son of Charles A. and Anne Morrow Lindbergh, was kidnapped from their home at Hopewell, NJ, on Mar 1, 1932. Even though the Lindberghs paid a $50,000 ransom, their child's body was found in a wooded area less than five miles from the family home on May 12. As a result of the kidnapping and murder of the Lindbergh baby, the Crime Control Act was passed on May 18, 1934. It authorized the death penalty for kidnappers who take their victims across state lines.

MENTAL RETARDATION AWARENESS MONTH. Mar 1–31. To educate the public about the needs of this nation's more than 7 million citizens with mental retardation and about ways to prevent retardation. The Arc is a national organization on mental retardation, formerly the Assn for Retarded Citizens. Sponsor: The Arc, Liz Moore, 500 E Border St, Ste 300, Arlington, TX 76010. Phone: (817) 261-6003.

MILLER, GLENN: BIRTH ANNIVERSARY. Mar 1. American band leader and composer (Alton) Glenn Miller. Born at Clarinda, IA, on Mar 1, 1904. Miller enjoyed great popularity, especially preceding and during the World War II era. His hit recordings included "Moonlight Serenade," "String of Pearls," "Jersey Bounce" and "Sleepy Lagoon." Major Miller, leader of the US Army Air Force band, disappeared Dec 15, 1944, over the English Channel, on a flight to Paris where he was scheduled to give a show. There were many explanations of his disappearance, but 41 years later, in December 1985, new accounts reported that crew members of an aborted RAF bombing said they believed they had seen Miller's plane, a Norseman D-64, go down, the victim of bombs being jettisoned by the RAF over the English Channel.

MOON PHASE: NEW MOON. Mar 1. Moon enters New Moon phase at 6:48 AM, EST.

MUSIC IN OUR SCHOOLS MONTH. Mar 1–31. To draw public awareness to the importance of music education as part of a balanced curriculum. The 1995 theme is "Music Means the World to Me." Additional information and awareness items are available from MENC. Phone: (800) 336-3768 or (703) 860-4000. Approx participation: 8,000,000. For info: Judy Reinhardt, Mgr Special Programs, Music Educators Natl Conference, 1806 Robert Fulton Dr, Reston, VA 22091. Phone: (800) 336-3768.

Mar 1 ☆ Chase's 1995 Calendar of Events ☆

NAIA MEN'S AND WOMEN'S SWIMMING AND DIVING CHAMPIONSHIPS. Mar 1-4. Palo Alto Natatorium, San Antonio, TX. Individuals compete for All-America honors while teams compete for the national championship. 15th annual competitions. Fax: (918) 494-8841. Est attendance: 3,000. For info: Natl Assn of Intercollegiate Athletics, 6120 S Yale Ave, Ste 1450, Tulsa, OK 74136. Phone: (918) 494-8828.

NATIONAL CHRONIC FATIGUE SYNDROME AWARENESS MONTH. Mar 1-31. To educate patients, their families, the public and the medical profession about the nature and impact of CFS, "the Thief of Vitality," and related disorders, as well as to encourage and provide research funding. Annually, the month of March. For info: Natl Chronic Fatigue Syndrome Assn, Inc, 3521 Broadway, Ste 222, Kansas City, MO 64111. Phone: (816) 931-4777.

NATIONAL CRAFT MONTH. Mar 1-31. To promote the fun and creativity of crafts and hobbies. Sponsor: Hobby Industry Assn. For info: Allan Fliss, Richartz & Fliss, 259 Baldwin Rd, Parsippany, NJ 07054. Phone: (201) 299-0070.

NATIONAL FEMININE EMPOWERMENT MONTH. Mar 1-31. This month celebrates the essence of women's spirituality and power, and acknowledges the richness of the female and male aspects of all humans, regardless of gender. It encourages growth, mental and physical wellness, and emotional and spiritual wholeness. For info: Jean Benedict Raffa, 17 South Osceola Ave, Orlando, FL 32801. Phone: (407) 426-7077.

NATIONAL FROZEN FOOD MONTH. Mar 1-31. Promotes a national awareness of the economical and nutritional benefits of frozen foods. Annually, the month of March. Sponsor: Cindi G. Rockwell, VP Comm, Natl Frozen Food Assn, 4755 Linglestown Rd, Ste 300, Harrisburg, PA 17112. Phone: (717) 657-8601.

NATIONAL NOODLE MONTH. Mar 1-31. Celebration of the misunderstood egg noodle. Honoring the egg noodle, National Noodle Month educates noodle lovers everywhere about the real nutrition facts about egg noodles. Annually, the month of March. Sponsor: National Pasta Assn. Phone: Donna Chowning Reid (703) 841-0818. For info: Christina Salcedo, Fleishman-Hillard, 1330 Ave of the Americas, New York, NY 10019.

NATIONAL NUTRITION MONTH®. Mar 1-31. To educate consumers about the importance of good nutrition by providing the latest practical information on how simple it can be to eat healthfully. The "Eat right America"™ campaign. Fax: (312) 899-1739. Sponsor: The American Dietetic Assn, Natl Ctr for Nutrition and Dietetics, 216 W Jackson Blvd, Chicago, IL 60606-6995. Phone: (312) 899-0040.

NATIONAL "ON-HOLD" MONTH. Mar 1-31. A month to recognize everyone who has been placed "on hold" after calling a place of business, and to honor those businesses who make this hold time more enjoyable by supplying informative messages and music for their callers waiting on hold. Sponsor: The Hold Co, Inc, 540 Township Line Rd, Blue Bell, PA 19422. Phone: (215) 643-0700.

NATIONAL PEANUT MONTH. Mar 1-31. To celebrate America's favorite nut—the peanut! Sponsor: National Peanut Council, Inc., 1500 King St, Ste 301, Alexandria, VA 22314. Phone: (703) 838-9500.

NATIONAL PIG DAY. Mar 1. To accord to the pig its rightful, though generally unrecognized, place as one of man's most intelligent and useful domesticated animals. Annually, Mar 1. For further information send a self-addressed, stamped envelope. Sponsor: Ellen Stanley, 7006 Miami, Lubbock, TX 79413.

March 1995

S	M	T	W	T	F	S
			1	2	3	4
5	6	7	8	9	10	11
12	13	14	15	16	17	18
19	20	21	22	23	24	25
26	27	28	29	30	31	

NATIONAL PROFESSIONAL SOCIAL WORK MONTH. Mar 1-31. To honor the social work profession and to recognize the contributions social workers and concerned citizens make within their communities. "Social Worker of the Year" and "Public Citizen of the Year" awards are announced and each winner receives a plaque. Sponsor: Natl Assn of Social Workers, Inc, 750 First St NE, Washington, DC 20002. Phone: (202) 408-8600.

NATIONAL SAUCE MONTH. Mar 1-31. Held each year in March to celebrate the diversity of sauces used in cooking. Celebrated with cooking contests, samplings and taste-offs across the country. Sponsor: Martini and Rossi Vermouth. For info: Laura Baddish, The Alden Group, 52 Vanderbilt Ave, New York, NY 10017. Phone: (212) 867-6400.

NATIONAL "TALK WITH YOUR TEEN ABOUT SEX" MONTH. Mar 1-31. The importance of frank talk about sex with teenagers is emphasized. Parents are encouraged to provide their teenage children with current, accurate information and open lines for communication, as well as support their self-esteem. The purpose of this commemoration is to reduce misinformation and guide teenagers toward making responsible decisions regarding sex. Annually, the month of March. For further info, send an SASE. Sponsor: Teresa Langston, Founder, Parenting Without Pressure (PWOP), 1330 Boyer St, Longwood, FL 32750. Phone: (407) 767-2524.

NATIONAL WOMEN'S HISTORY MONTH. Mar 1-31. A time for re-examining and celebrating the wide range of women's contributions and achievements, too often overlooked in the telling of US history. Information catalog available. For info: Natl Women's History Project, 7738 Bell Road, Windsor, CA 95492. Phone: (707) 838-6000.

NEBRASKA: ADMISSION DAY. Mar 1. Became 37th state on this day in 1867.

OHIO: ADMISSION DAY. Mar 1. Became 17th state on this day in 1803.

PEACE CORPS: FOUNDING ANNIVERSARY. Mar 1. Official establishment of the Peace Corps by President John F. Kennedy's signing of executive order. Since 1961, the Peace Corps has sent more than 140,000 volunteers to over 100 developing countries to help their people help themselves. The volunteers have been assisting these people in projects such as health, education, water sanitation, agriculture, nutrition and forestry. Annually, Mar 1. Sponsor: Peace Corps, 1990 K St, Washington, DC 20526. Phone: (202) 606-0062.

POISON PREVENTION AWARENESS MONTH. Mar 1-31. To educate parents, grandparents, school children and PTAs about accidental poisoning and how to prevent it. PPSI is a nonprofit organization. There is a $15 charge for kit materials. Annually, the month of March. Fax: (415) 332-1832. For info: Pharmacists Planning Service, Inc, 200 Gate Five Rd, PO Box 1336, Sausalito, CA 94966. Phone: (800) 523-2222.

POLO: INTERNATIONAL OPEN AND HANDICAP. Mar 1-19. Wellington, FL. Est attendance: 10,000. For info: Palm Beach Polo and Country Club, 13420 South Shore Blvd, Wellington, FL 33414. Phone: (407) 798-7634.

Chase's 1995 Calendar of Events — Mar 1–2

RED CROSS MONTH. Mar 1–31. To make the public aware of American Red Cross service in the community. There are nearly 2,000 Red Cross chapters nationwide; each local chapter plans its activities. For info on activities in your area, contact your local Red Cross chapter. Sponsor: External Communications, American Red Cross Natl HQ, 18th and D St NW, Washington, DC 20006-5399.

RETURN THE BORROWED BOOKS WEEK. Mar 1–7. To remind you to make room for those precious old volumes that will be returned to you, by cleaning out all that worthless trash that your friends are waiting for. Annually, the first seven days of March. Sponsor: Inter-Global Soc for Prevention of Cruelty to Cartoonists, Al Kaelin, Secy, 3119 Chadwick Dr, Los Angeles, CA 90032. Phone: (213) 221-7909.

ROSACEA AWARENESS MONTH. Mar 1–31. Rosacea Awareness Month has been designated by the National Rosacea Society to raise understanding of this increasingly common disease. Rosacea is a skin condition that can cause permanent physical and psychological damage if it is not diagnosed and treated. Est attendance: 25,000. For info: Sam Huff, Natl Rosacea Soc, 220 S Cook St, Ste 201, Barrington, IL 60010. Phone: (708) 382-8971.

SHORE, DINAH: BIRTH ANNIVERSARY. Mar 1, 1917. Singer and TV personality, Shore recorded approximately 75 hit songs including "Dear Hearts and Gentle People," and hosted TV's "Dinah Shore Chevy Show," winning 10 Emmy awards. She was born at Winchester, TN, and died at Beverly Hills, CA, on Feb 24, 1994.

SLAYTON, DONALD "DEKE" K.: BIRTH ANNIVERSARY. Mar 1. "Deke" Slayton, longtime chief of flight operations at the Johnson Space Center, was born Mar 1, 1924, in Sparta, WI. Slayton was a member of Mercury Seven, the original group of young military aviators chosen to inaugurate America's sojourn into space. Unfortunately for Slayton, a heart problem prevented him from participating in any of the Mercury flights. When in 1971 the heart condition mysteriously went away, Slayton flew on the last Apollo Mission. This July 1975 flight involving a docking with a Soviet Soyuz spacecraft symbolized a momentary thaw in relations between the two nations. During his years as chief of flight operations, Slayton directed astronaut training and selected the crews for nearly all missions. Donald Slayton died June 13, 1993, at League City, TX.

SWITZERLAND: CHALANDRA MARZ. Mar 1. Engadine. Springtime traditional event when costumed young people, ringing bells and cracking whips, drive away the demons of winter.

UNIVERSAL HUMAN BEINGS WEEK. Mar 1–7. The purpose of this observance is to inspire men and women to become Universal Human Beings in a world that is rapidly becoming a "global village." Annually, Mar 1–7. For complete info send $2 to cover printing, handling and postage. Sponsor: Intl Soc of Friendship and Good Will, Dr. Stanley Drake, Pres, 9538 Summerfield St, Spring Valley, CA 91977-7852. Phone: (619) 466-8882.

WALES: ST. DAVID'S DAY. Mar 1. Celebrates patron saint of Wales. Welsh tradition calls for the wearing of a leek on this day.

YOUTH ART MONTH. Mar 1–31. To emphasize the value and importance of participation in art in the development of all children and youth. Sponsor: Council for Art Education, Inc, 100 Boylston St, Ste 1050, Boston, MA 02116. Phone: (617) 426-6400.

BIRTHDAYS TODAY

Catherine Bach, 41, actress ("The Dukes of Hazzard"), born at Warren, OH, Mar 1, 1954.
Harry Belafonte, 68, singer ("Mary's Boy Child," "Island in the Sun," "Banana Boat Song"), born New York, NY, Mar 1, 1927.
Robert Heron Bork, 68, former US Court of Appeals circuit judge, born at Pittsburg, PA, Mar 1, 1927.
John B. Breaux, 51, US Senator (D, Louisiana), born at Crowley, LA, Mar 1, 1944.
Harry Caray, 76, sports announcer, born at St. Louis, MO, Mar 1, 1919.
Robert Conrad, 60, actor ("The Wild Wild West"), born at Chicago, IL, Mar 1, 1935.
Roger Daltry, 51, singer (lead singer of the Who), born at London, England, Mar 1, 1944.
Ron Howard, 41, actor (Opie on "Andy Griffith Show"; "Happy Days"); director (*Cocoon, Backdraft*), born Duncan, OK, Mar 1, 1954.
Judith Rossner, 60, novelist, born at New York, NY, Mar 1, 1935.
Pete Rozelle, 69, former commissioner of the National Football League, born at South Gate, CA, Mar 1, 1926.
Alan Thicke, 48, actor ("Thicke of the Night," "Growing Pains"), former disc jockey, born at Kirkland Lake, Ont, Canada, Mar 1, 1947.
Richard (Purdy) Wilbur, 74, former Poet Laureate of the US, born at New York, NY, Mar 1, 1921.

MARCH 2 — THURSDAY
61st Day — Remaining, 304

BATTLE OF WAYNESBOROUGH: 130th ANNIVERSARY. Mar 2, 1865. Union troops under the command of General George Armstrong Custer defeated Rebel forces at Waynesborough, VA. With the rout of Confederate General Jubal Early's forces, the Battle of Waynesborough brought an end to the Union's campaign in the Shenandoah Valley.

BELGIUM: CAT FESTIVAL. Mar 2. Ieper. Traditional cultural observance. Annually, on the second day of Lent.

CORAY, MELISSA BURTON: BIRTH ANNIVERSARY. Mar 2, 1828. Coray was born at Mersey, Ont, Canada. At the age of 18 she accompanied her Mormon Battalion soldier husband, William Coray, on a 2,000-mile military march on foot from Council Bluffs, IA, to San Diego, CA, then 1,500 more miles across the Sierra Nevada Mountains and the Nevada desert to Salt Lake City, UT, the only woman to make the entire trip. On July 30, 1994, a mountain peak near Carson Pass was named for her, the 2nd peak in California to be named for a woman. For info: Ben E. Lofgren, Historian, 4337 Figwood Way, Sacramento, CA 95864. Phone: (916) 487-1353.

THE ENDURING MR. LINCOLN. Mar 2–May 28. Clinton, MD. Special exhibit of Lincoln memorabilia. Sponsor: Surratt House & Tavern, 9110 Brandywine Rd, Clinton, MD 20735. Phone: (301) 868-1121.

FLORIDA STRAWBERRY FESTIVAL. Mar 2–12. Plant City, FL. Celebration of winter strawberry harvest. Ticket Office: (813) 754-1996. Est attendance: 761,000. For info: Florida Strawberry Fest, Patsy Brooks, Asst Mgr, PO Drawer 1869, Plant City, FL 33564-1869. Phone: (813) 752-9194.

GEISEL, THEODOR "DR. SEUSS": BIRTH ANNIVERSARY. Mar 2. Theodor Seuss Geisel, the creator of *The Cat in the Hat* and *The Grinch Who Stole Christmas*, was born on Mar 2, 1904, at Springfield, MA. Known to children and parents as Dr. Seuss, Geisel's books have sold more than 200 million copies and have been translated into 20 languages. His career began with *And to Think That I Saw it on Mulberry Street*, which was turned down by 27 publishing houses before being published by Vanguard Press. His books included many messages, from the environmentally conscious *The Lorax* to the dangers of pacifism in *Horton Hatches the Egg* and *Yertel the Turtle*'s thinly veiled references to Hitler as the title character. He was awarded a Pulitzer Prize in 1984 "for his contribution over nearly half a century to the education and enjoyment of America's children and their parents." He died on Sept 24, 1991, at La Jolla, CA.

Mar 2-3 ☆ *Chase's 1995 Calendar of Events* ☆

LOS ANGELES MARATHON QUALITY OF LIFE EXPO. Mar 2-4. Los Angeles Convention Center, Los Angeles, CA. Massive consumer show features health and fitness products and services, food sampling, seminars and demonstrations. Phone: (310) 444-5544, ext 45 or 56. Est attendance: 50,000. For info: Mike Gerlowski, Dir Sales and Mktg, City of Los Angeles Marathon, Inc, 11110 W Ohio Ave, Ste 100, Los Angeles, CA 90025.

MICHIGAN HOME AND GARDEN SHOW AT THE PONTIAC SILVERDOME. Mar 2-5. Pontiac Silverdome, Pontiac MI. This event brings together buyers and sellers of products and services for home building/remodeling, home furnishings and interior design, lawn and garden and related areas. It also features on-site constructions, theme gardens, seminars and demonstrations, plus The Standard Flower Show sponsored by The Federated Garden Clubs of Michigan, District 1. Est attendance: 45,000. For info: Mike Wilbraham, ShowSpan, Inc, 1400 28th St SW, Grand Rapids, MI 49509. Phone: (616) 530-1919.

MUSLIM HOLIDAY: ID AL-FITR. Mar 2-3. Feast marks end of month-long fast of Ramadhan. Muslim calendar date: Shawwal 1, 1415. Feast ordinarily continues for two or three days. Different methods for calculating the Gregorian date are used by different Muslim sects or groups. *Chase's* dates are based on astronomical calculations of the visibility of the new moon crescent at Mecca; EST date may vary.

NCAA MEN'S AND WOMEN'S RIFLE CHAMPIONSHIPS. Mar 2-4. Site to be determined. For info: NCAA, 6201 College Blvd, Overland Park, KS 66211-2422. Phone: (913) 339-1906.

NEW YORK FLOWER SHOW. Mar 2-6. (Gala preview to benefit the Horticultural Society of New York on Mar 1.) New York, NY. Spectacular flower and garden exposition with the goal of educating the public about plants and flowers. In addition to major landscape designs, there is a competition area open to the public and a gardener's marketplace for purchasing a variety of garden accessories and wares. Visitors will learn about new plants, how to develop gardens within specific surroundings and be inspired to realize their own gardening dreams. Est attendance: 70,000. For info: The Horticultural Soc of NY, 128 W. 58th St, New York, NY 10019. Phone: (212) 757-0915.

PEACE OVERTURES IN THE CIVIL WAR: 130th ANNIVERSARY. Mar 2, 1965. Confederate General-in-Chief Robert E. Lee sent a message through Union lines to Union commander Ulysess S. Grant proposing a "military convention" to attempt a "satisfactory adjustment of the present unhappy difficulties." The next day, Grant was instructed by Lincoln not to confer with Lee unless it was to accept surrender, and that all political matters would be settled by the President.

POPE PIUS XII: BIRTH ANNIVERSARY. Mar 2. Eugenio Maria Giovanni Pacelli, 260th pope of the Roman Catholic Church. Born Mar 2, 1876, at Rome, Italy. Elected pope Mar 2, 1939. Died at Castel Gandolfo, near Rome, Oct 9, 1958.

SCHURZ, CARL: BIRTH ANNIVERSARY. Mar 2. American journalist, political reformer and army officer in Civil War. Born near Cologne, Germany, on Mar 2, 1829. Died at New York, NY, on May 14, 1906.

SPACE MILESTONE: *PIONEER 10* (US). Mar 2. On Mar 2, 1972, this probe began a journey on which it was scheduled to pass and photograph Jupiter, 620 million miles from Earth, in December 1973, cross the orbit of Pluto in 1987, and then become the first known Earth object to leave our solar system. On Sept 22, Pioneer 10 reached another space milestone at 4:19 PM, when it reached a distance 50 times farther from the sun than the sun is from Earth.

SPACE MILESTONE: *SOYUZ 28* (USSR). Mar 2. Cosmonauts Alexi Gubarev and Vladimir Remek linked with *Salyut 6* space station Mar 3, visiting crew of *Soyuz 26*. Returned to Earth Mar 10. Remek, from Czechoslovakia, was the first person in space from a country other than the US or USSR. Launched Mar 2, 1978.

TEXAS INDEPENDENCE DAY. Mar 2. Texas adopted Declaration of Independence from Mexico on this day, 1836.

WCUZ WEST MICHIGAN HOME AND GARDEN SHOW. Mar 2-5. Grand Center, Grand Rapids, MI. Est attendance: 65,000. For info: WCUZ Radio, PR Dept, 140 Monroe Ctr, Grand Rapids, MI 49503. Phone: (616) 451-2551.

WORLD'S LARGEST CONCERT. Mar 2. To kick off Music in Our Schools Month, the World's Largest Concert will be broadcast on PBS stations nationally and over the Armed Forces Television Network overseas. Approx participation: 8,000,000. Phone: (800) 336-3768 or (703) 860-4000. For info: Judy Reinhardt, Mgr Special Programs, Music Educators Natl Conference, 1806 Robert Fulton Dr, Reston, VA 22091.

BIRTHDAYS TODAY

Jon Bon Jovi (John Bongiovi), 33, singer ("Living on a Prayer," "You Give Love a Bad Name"), musician, songwriter, born Sayreville, NJ, Mar 2, 1962.

Russell D. Feingold, 42, US Senator (D, Wisconsin), born at Janesville, WI, Mar 2, 1953.

Mikhail Sergeyvich Gorbachev, 64, former Soviet political leader, born at Privolnoye, Stavropol, Russia, Mar 2, 1931.

John Irving, 53, author (*Cider House Rules, World According to Garp*), born at Exeter, NH, Mar 2, 1942.

Jennifer Jones (Phyllis Isley), 76, actress (Oscar for *The Song of Bernadette*), born at Tulsa, OK, Mar 2, 1919.

Eddie Money, 46, musician, born at Brooklyn, NY, Mar 2, 1949.

Laraine Newman, 43, comedienne ("Saturday Night Live"), born at Los Angeles, CA, Mar 2, 1952.

Doc Watson, 72, singer, musician, born at Deep Gap, NC, Mar 2, 1923.

Tom Wolfe, 64, author, journalist (*The Bonfire of the Vanities, The Right Stuff*), born at Richmond, VA, Mar 2, 1931.

MARCH 3 — FRIDAY
62nd Day — Remaining, 303

BELL, ALEXANDER GRAHAM: BIRTH ANNIVERSARY. March 3. Inventor of the telephone, born Mar 3, 1847, at Edinburgh, Scotland, Alexander Graham Bell acquired his interest in the transmission of sound from his father, Melville Bell, a teacher of the deaf. Bell's use of visual devices to teach articulation to the deaf contributed to the theory from which he derived the principle of the vibrating membrane used in the telephone. On Mar 10, 1876, Bell spoke the first electrically transmitted sentence to his assistant in the next room, "Mr Watson, come here, I want you." Bell's other accomplishments include a refinement of Mr Edison's phonograph, the first successful phonograph record, ailerons and the audiometer, and he continued exploring the nature and causes of deafness. He died near Baddeck, Nova Scotia, Canada, on Aug 2, 1922.

March 1995

S	M	T	W	T	F	S
			1	2	3	4
5	6	7	8	9	10	11
12	13	14	15	16	17	18
19	20	21	22	23	24	25
26	27	28	29	30	31	

★ Chase's 1995 Calendar of Events ★ Mar 3

BETHUNE, NORMAN: BIRTH ANNIVERSARY. Mar 3. Canadian physician who worked in the front lines during World War I, the Spanish Civil War and the Chinese Revolution. Bethune was born at Gravenhurst, Ontario, Canada, Mar 3, 1890. He died in China while treating a soldier of Mao's Eighth Route Army, Nov 11, 1939. He is said to be the only Western man recognized as a hero of the Chinese Revolution.

BLACK HERITAGE FESTIVAL OF LOUISIANA. Mar 3-6. Civic Center, Lake Charles, LA. Cultural, educational and fun activities that showcase the achievements and talents of blacks. Features an essay contest, art exhibit, job fair, talent show, story tellers, national performers and exhibitors. Est attendance: 10,000. For info: Stella H. Miller, Exec Dir, PO Box 16365-6365, Lake Charles, LA 70616. Phone: (318) 478-2127.

BONZA BOTTLER DAY™. Mar 3. To celebrate when the number of the day is the same as the number of the month. Bonza Bottler Day™ is an excuse to have a party at least once a month. T-shirts available. For info: Elaine Fremont, 203 Waddell Rd, Taylors, SC 29687. Phone: (803) 244-2023.

BULGARIA: LIBERATION DAY. Mar 3. Grateful tribute to the Russian, Romanian and Finnish soldiers and Bulgarian volunteers who, in the Russo-Turkish War, 1877-1878, liberated Bulgaria from Ottoman rule and five centuries of oppression.

CARNAVAL MIAMI. Mar 3-12. Little Havana, Miami, FL. A Latin-flavored festival that includes a beauty pageant, variety shows, sports, parade and "the world's largest block party"—Calle Ocho: Open House Festival. Est attendance: 1,500,000. Sponsor: Kiwanis Club of Little Havana, 1312 SW 27th Ave, 3rd Fl, Miami, FL 33145. Phone: (305) 644-8888.

CAROLINA CRAFTSMEN'S SPRING CLASSIC. Mar 3-5. State Fairgrounds, Columbia, NC. Arts and crafts. Est attendance: 20,000. For info: The Carolina Craftsmen, 1240 Oakland Ave, Greensboro, NC 27403. Phone: (910) 274-5550.

DAYTONA MOTORCYCLE WEEK. Mar 3-12. Daytona International Speedway, Daytona Beach, FL. For info: Daytona Intl Speedway, Glenn Barber, PR, Box 2801, Daytona Beach, FL 32120-2801. Phone: (904) 254-6782.

◆ **FLORIDA: ADMISSION DAY: 150th ANNIVERSARY.** Mar 3. Became 27th state on this day in 1845.

FULTON OYSTERFEST. Mar 3-5. Fulton Navigation Park, Fulton, TX. This salute to the oyster industry features oyster-shucking, raw oyster-eating contests, more than 100 arts and crafts booths, food and fun, live bands and other entertainment. Annually, the first full weekend in March. Est attendance: 40,000. For info: Fulton Oysterfest, PO Box 393, Fulton, TX 78358. Phone: (800) 242-0071.

GOLD WING GETAWAY. Mar 3. This 4th Annual event will be held during Daytona's Bike Week. Activities include a free vendor expo in the mall, bike show, light show, poker runs and many other activities around town. Est attendance: 400,000. For info: Dusty Carder, Chapter Dir, GWRRA Chapter FLI-H, PO Box 250092, Holly Hill, FL 32125. Phone: (904) 428-1839.

I WANT YOU TO BE HAPPY DAY. Mar 3. A day dedicated to reminding people to be thoughtful of others even when things are not going well for themselves. For info: Harriette W. Grimes, Grandmother, PO Box 545, Winter Garden, FL 34777-0545. Phone: (407) 656-3830.

JAPAN: HINAMATSURI (DOLL FESTIVAL). Mar 3. Special festival for girls, observed throughout Japan. Annually, Mar 3.

KENTUCKY HILLS WEEKEND. Mar 3-4. Cumberland Falls State Resort Park, Corbin, KY. To pay tribute to Appalachian folk life. Folk music, crafts and demonstrations. Est attendance: 800. For info: Cumberland Falls State Resort Pk, Kim Burchett, Corbin, KY 40701. Phone: (606) 528-4121.

LEADVILLE'S CRYSTAL CARNIVAL. Mar 3-5. Leadville's Historic Winter Celebration with entertainment, Parade of Lights, Hot-Air Balloon Glow, Ski Joring, Snow Sculpting and other events in charming Victorian town. Toll Free Phone: (800) 933-3901. For info: Chamber of Commerce, PO Box 861, Leadville, CO 80461. Phone: (719) 486-3900.

LONGHORN WORLD CHAMPIONSHIP RODEO. Mar 3-5. Von Braun Civic Center, Huntsville, AL. More than 220 of North America's best cowboys and cowgirls compete in 6 professionally sanctioned contests ranging from bronc riding to bull riding for top prize money and world championship points. The theme of the production changes annually. Additional attractions include special animal trainers, trick ropers, trick riders and professional clowns. 16th annual. Est attendance: 19,000. For info: Longhorn World Chmpshp Rodeo, Inc, PO Box 70159, Nashville, TN 37207. Phone: (615) 876-1016.

MALAWI: MARTYR'S DAY. Mar 3. Public holiday in Malawi.

MOROCCO: ANNIVERSARY OF THE THRONE. Mar 3. National holiday.

NAIA MEN'S AND WOMEN'S INDOOR TRACK AND FIELD CHAMPIONSHIPS. Mar 3-4. Site to be determined. Individuals compete for All-America honors while teams compete for the national championship. 30th annual competition. Fax: (918) 494-8841. Est attendance: 2,500. For info: Natl Assn Intercollegiate Athletics, 6120 S Yale Ave, Ste 1450, Tulsa, OK 74136. Phone: (918) 494-8828.

NATIONAL ANTHEM DAY. Mar 3. On Mar 3, 1931, the bill designating "The Star-Spangled Banner" as our national anthem was adopted by the US Senate and went to President Herbert Hoover for signature. The president signed it the same day.

PULLMAN, GEORGE MORTIMER: BIRTH ANNIVERSARY. Mar 3. American inventor and cabinetmaker, originator of the railway sleeping car. The first to be called a "Pullman" was the *Pioneer*. Later he was president of the Pullman Palace Car Company. Born Mar 3, 1831, at Brocton, NY. Died at Chicago, IL, on Oct 19, 1897.

RIDGWAY, MATTHEW BUNKER: 100th BIRTH ANNIVERSARY. Mar 3. American Army officer, Matthew Bunker Ridgway was born Mar 3, 1895, at Fort Monroe, VA. As major general commanding the newly formed 82nd Airborn Division, he led it in the invasion of Sicily in July of 1943 and the invasion of the Italian mainland in 1944. Ridgway replaced MacArthur as commander of the US Eighth Army in Korea in 1951 and succeeded Eisenhower as Supreme Allied Commander of the North Atlantic Treaty Organization in 1952. He became US Army Chief of Staff in 1953. Ridgway retired in June of 1955. He died at Fox Chapel, PA, July 26, 1993.

SNOWFEST. Mar 3-12. North Lake Tahoe, and Truckee, CA. Snowfest is a fantastic vacation opportunity, showcasing America's largest concentration of skiing and outdoor recreation combined with the fun, sparkle and excitement of 10 full days of 120 special and unique events. Annually, beginning, the Friday before the first Sunday in March. Est attendance: 120,000. For info: Festivals at Tahoe, PO Box 7590, Tahoe City, CA 96145. Phone: (916) 583-7625.

TEXAS COWBOY POETRY GATHERING. Mar 3-5. Sul Ross State University, Alpine, TX. Cowboys from neighboring states gather for poetry readings and music. Includes Trappings of Texas Art and gear show. Est attendance: 4,000. Sponsor: Texas Cowboy Poetry Gathering Committee, PO Box 395, Alpine, TX 79831. Phone: (915) 837-8143.

Mar 3-4 ☆ *Chase's 1995 Calendar of Events* ☆

USA/MOBIL INDOOR MEN'S & WOMEN'S TRACK & FIELD CHAMPIONSHIPS. Mar 3-4. Georgia Dome, Atlanta, GA. National (Open) Track & Field competition. Sponsor: Mobil Corp. For info: P. Duffy Mahoney, Dir of Operations, USA Track & Field, PO Box 120, Indianapolis, IN 46206-0120. Phone: (317) 261-0500.

WINTER CARNIVAL. Mar 3-4. Red Lodge, MT. Winter Carnival features parades, snow sculptures, King and Queen contest, Snow Ball, costume contest, Cardboard Classic Race, live music and prizes. Torchlight parade and spaghetti dinner. Annually, the first weekend in March. Est attendance: 2,000. For info: Red Lodge Area Chamber of Commerce, Joan Cline, Exec Secy, PO Box 988, Red Lodge, MT 59068. Phone: (406) 446-1718.

◆ **WOMAN SUFFRAGE PARADE ATTACKED: ANNIVERSARY.** Mar 3. A parade held by the National American Woman Suffrage Association in Washington, DC, on Mar 3, 1913—the day before Woodrow Wilson's inauguration—turned into a near riot when people in the crowd began jeering and shoving the marchers. The 5,000 women and their supporters were spit upon, struck in the face and pelted with burning cigar stubs while police looked on and made no effort to intervene. Secretary of War Henry Stimson was forced to send soldiers from Fort Myer to restore order.

WORLD DAY OF PRAYER. Mar 3. Theme: "The Earth Is a House for All People," written by Christian women from Ghana. An ecumenical event that reinforces bonds between peoples of the world as they join in a global circle of prayer. Annually, the first Friday in March. Sponsor: International Committee for World Day of Prayer. Church Women United is the National World Day of Peace Committee for the US. For info: Church Women United, 475 Riverside Dr, Rm 812, New York, NY 10115. Phone: (212) 870-2347.

YUMA SQUARE AND ROUND DANCE FESTIVAL. Mar 3-5. Yuma Civic and Convention Center, Yuma, AZ. Square and round dance enthusiasts from Southern California and Arizona participate. For info: Yuma Civic and Conv Ctr, 1440 Desert Hills Dr, Yuma, AZ 85364. Phone: (602) 344-3800.

BIRTHDAYS TODAY

Jackie Joyner-Kersee, 33, heptathlon athlete, born at East St. Louis, IL, Mar 3, 1962.
Tim Kazurinsky, 45, actor, comedian, writer ("Saturday Night Live"), born at Johnstown, PA, Mar 3, 1950.
Arnold Newman, 77, photographer, born at New York, NY, Mar 3, 1918.
Princess Radziwill (Caroline Lee Bouvier), 62, sister of the late Jackie Kennedy Onassis, born New York, NY, Mar 3, 1933.

	S	M	T	W	T	F	S
March				1	2	3	4
1995	5	6	7	8	9	10	11
	12	13	14	15	16	17	18
	19	20	21	22	23	24	25
	26	27	28	29	30	31	

Herschel Walker, 33, football player, born at Wrightsville, GA, Mar 3, 1962.
Robert Whitehead, 79, producer, born at Montreal, Quebec, Canada, Mar 3, 1916.

MARCH 4 — SATURDAY
63rd Day — Remaining, 302

ADAMS, JOHN QUINCY: RETURN TO CONGRESS ANNIVERSARY. Mar 4. On Mar 4, 1830, John Quincy Adams returned to the House of Representatives to represent the district of Plymouth, MA. He was the first former president to do so and served for eight consecutive terms.

BIG TEN WRESTLING CHAMPIONSHIP. Mar 4-5. Indiana University, Bloomington, IN. For info: Big Ten Conference, 1500 W Higgins Rd, Park Ridge, IL 60068-6300. Phone: (708) 696-1010.

CONGRESS: ANNIVERSARY OF FIRST MEETING UNDER CONSTITUTION. Mar 4. The first Congress met in New York, NY, on Mar 4, 1789. A quorum was obtained in the House on Apr 1 and in the Senate on Apr 5, and the 1st Congress was formally organized on April 6. Electoral votes were counted, and George Washington was declared president (69 votes) and John Adams vice president (34 votes).

DING LING: DEATH ANNIVERSARY. Mar 4. Writer and champion of women's rights, born in Hunan Province, China, in 1904. Miss Ding was a prolific author, having written nearly 300 novels as well as plays, short stories and essays. She received the 1951 Stalin Prize for Literature for her novel *The Sun Shines Over the Sanggan River* (1949). In the 1950s she fell from favor, was exiled and, in 1970, was imprisoned. After the death of Chairman Mao she was freed and during her last years she enjoyed renewed attention and favor. Died at age 82 at Peking, China, on Mar 4, 1986.

GASPARILLA SIDEWALK ART FESTIVAL. Mar 4-5. Tampa, FL. Painters, artisans, photographers, sculptors, jewelers and artists from around the world display their work. The festival awards more than $60,000 in prize money, including $15,000 for Best of Show. Exhibits, food booths and stages of live entertainment line the banks of the Hillsborough River and the entrance of the Tampa Bay Performing Arts Center. Hours for the show are Saturday, 9-5:30, and Sunday, 10-5. Admission: Free. For applications to submit entries, send a SASE to: GSAF, PO Box 10591, Tampa, FL 33679. Fax: (813) 229-6616. Est attendance: 200,000. For info: Tampa/Hillsborough Conv and Visitors Assn, Inc, 111 Madison St, Ste 1010, Tampa, FL 33601-0519. Phone: (800) 448-2672.

GIRL SCOUT SABBATH. Mar 4. Girl Scouts worship together in the temple of their choice. Sponsor: Girl Scouts of the USA, Media Services, 420 Fifth Ave, New York, NY 10018.

HARLEM: PHOTOGRAPHS BY AARON SISKIND. Mar 4-Apr 30. California Afro-American Museum. Los Angeles, CA. Nightclubs, union rallies and children at play are a few of the vivid vignettes Siskind captures in his black and white photographs of New York City's famous black neighborhood in the 1930s. Call or write the Smithsonian for other dates and venus. For info: Smithsonian Institution Traveling Exhibition Service, 1100 Jefferson Dr SW, Ste 3146, Washington, DC 20560. Phone: (202) 357-2700.

HELP SOMEONE SEE WEEK. Mar 4-11. Save and donate your discarded eyeglasses for distribution in Third World countries by Medical Group Mission of the Christian Medical and Dental Society. For info: Dr. and Mrs Fleming Barbour, 2015 Lincoln Dr, Flint, MI 48503. Phone: (810) 235-4752.

HONG KONG FOOD FESTIVAL. Mar 4-19. Hong Kong. This annual festival engages the entire territory with focus on one of its favorite pastimes: eating. Included are special food-related tours, exhibitions, a street carnival and special events at hotels and restaurants. Organized by the Hong Kong Tourist Assn with sponsorship by local businesses. For info: Hong Kong

Tourist Assn, 590 Fifth Ave, 5th Fl, New York, NY 10036-4706. Phone: (212) 869-5008.

IRISH HERITAGE SEASON. Mar 4-31. Biloxi, MS. Irish Derby, 5K run, parties, golf tournament and Irish displays. Big event of season is St. Patrick's Parade (see Mar 11). Est attendance: 40,000. Sponsor: Hibernia Marching Society of Mississippi, PO Box 707, Biloxi, MS 39533. Phone: (601) 896-6363.

MUSEUM OF ART LAS OLAS ART FESTIVAL. Mar 4-5. Museum of Art, Fort Lauderdale, FL. Fine arts and crafts festival held annually. Est attendance: 75,000. For info: Las Olas Art Festival, PO Box 2211, Fort Lauderdale, FL 33303. Phone: (305) 525-3838.

NATCHEZ SPRING PILGRIMAGE. Mar 4-Apr 2. Corner of Canal and State Sts, Natchez, MS. 64th annual tour of 30 antebellum mansions furnished with period antiques; set in formal gardens. Exciting Confederate Pageant "Southern Exposure" and other evening entertainment. Carriage and bus sightseeing tours daily. Est attendance: 40,000. Sponsor: Natchez Pilgrimage Tours, PO Box 347, Natchez, MS 39121. Phone: (800) 647-6742.

NATIONAL EASTER SEAL TELETHON. Mar 4-5. This 20-hour television show provides a format for educating the public about Easter Seal's services as well as promoting the independence of people with disabilities. Raises funds for children and adults with disabilities. Annually, the first weekend in March. Sponsor: Natl Easter Seal Soc, Sandra Gordon, Sr VP, 230 W Monroe St, Chicago, IL 60606.

OLD INAUGURATION DAY. Mar 4. Anniversary of the date set for beginning the US presidential term of office, 1789-1933. Although the Continental Congress had set the first Wednesday of March 1789 as the date for the new government to convene, a quorum was not present to count the electoral votes until Apr 6. Though George Washington's term of office began on Mar 4, he did not take the oath of office until Apr 30, 1789. All subsequent presidential terms (except successions following the death of an incumbent) until Franklin D. Roosevelt's second term began on Mar 4. The 20th Amendment (ratified Jan 23, 1933) provided that "the terms of the President and Vice President shall end at noon on the 20th day of January ... and the terms of their successors shall then begin."

PENNSYLVANIA DEEDED TO WILLIAM PENN: ANNIVERSARY. Mar 4. To satisfy a debt of £16,000, King Charles II of England granted a royal charter, deed and governorship of Pennsylvania to William Penn on Mar 4, 1681.

PEOPLE MAGAZINE: ANNIVERSARY. Mar 4. The popular gossip magazine was officially launched with the Mar 4, 1974, issue featuring a cover photo of Mia Farrow.

PERKINS, FRANCES: CABINET APPOINTMENT ANNIVERSARY. Mar 4. Frances Perkins became the first woman appointed to the president's cabinet on Mar 4, 1933. She was appointed Secretary of Labor by President Franklin D. Roosevelt.

PIER 39's TULIPMANIA. Mar 4-18. PIER 39, San Francisco, CA. More than 35,000 multi-colored tulips in 70 different varieties of blooming colors will be on display in the 17th annual Tulipmania, a spring tradition in San Francisco. This year's festival features tulips from as far away as Holland, Europe, and Holland, Michigan. Free guided landscaping tours are presented daily at 10 AM. Est attendance: 10,000. For info: Alicia Vargas, PR Dir, Pier 39, PO Box 193730, San Francisco, CA 94119. Phone: (415) 705-5500.

PULASKI, CASIMIR: BIRTH ANNIVERSARY. Mar 4. American Revolutionary hero, General (and chief of cavalry) Kazimierz (Casimir) Pulaski, born at Winiary, Mazovia, Poland, Mar 4, 1747. The son of a count, Pulaski was a Polish patriot and military leader in the fight against Russia of 1770-71, who went into exile at the partition of Poland in 1772. He went to America in 1777 to join in the Revolution, fighting with General Washington at Brandywine and also serving at Germantown and Valley Forge. Organized the Pulaski Legion, waging guerrilla warfare against the British. He was mortally wounded in a heroic charge at the siege of Savannah, GA, on Oct 9, 1779, and died aboard the warship *Wasp* on Oct 11, 1779. Pulaski Day is celebrated on the first Monday of March in Illinois.

ST. LOUIS VARIETY CLUB TELETHON AND DINNER WITH THE STARS. Mar 4-5. Adam's Mark Hotel, St. Louis, MO. A 19-hour telethon to raise funds for disabled and disadvantaged children in the greater St. Louis area. Aired on KMOV Channel 4 (CBS). Est attendance: 1,200. For info: St. Louis Variety Club, 1000 Des Peres Rd, Ste 120, St. Louis, MO 63131. Phone: (314) 821-8184.

SCHOHARIE COLONIAL HERITAGE ASSOCIATION ANTIQUE SHOW AND SALE. Mar 4-5. (Also Sept 16-17.) Schoharie, NY. Est attendance: 2,000. For info: Schoharie Colonial Heritage Assn, PO Box 554, Schoharie, NY 12157. Phone: (518) 295-7505.

SONGS OF MY PEOPLE. Mar 4-Apr 16. Paterson Museum, Paterson, NJ. The more than 150 images in this exhibition are the work of 50 photojournalists who during the summer of 1990 traveled the country to document the lives of fellow African-Americans. From the cotton fields of Mississippi to the financial houses of Wall Street, these powerful and penetrating photographs capture the character and contradictions of black life in America. Call or write the Smithsonian for other dates and venues. For info: Smithsonian Institution Traveling Exhibition Service, 1100 Jefferson Dr SW, Ste 3146, Washington, DC 20560. Phone: (202) 357-2700.

SOUTHEAST FLORIDA SCOTTISH FESTIVAL AND GAMES. Mar 4. Key Biscayne, FL. To promote Scottish heritage. Est attendance: 8,000. Sponsor: Scottish-American Soc of Southeast Florida, 5901 NE 21 Rd, Fort Lauderdale, FL 33308. Phone: (305) 776-5675.

SPACE MILESTONE: *OGO 5* (US). Mar 4. Orbiting geophysical laboratory collected data on sun's influence on Earth. Launched Mar 4, 1968.

TELEVISION ACADEMY HALL OF FAME: FIRST INDUCTEES ANNOUNCED. Mar 4. On Mar 4, 1984, the Television Academy of Arts and Sciences announced the formation of the Television Academy Hall of Fame in Burbank, CA. The first inductees were Lucille Ball, Milton Berle, Paddy Chayefsky, Norman Lear, Edward R. Murrow, William S. Paley and David Sarnoff.

TRAIL'S END MARATHON. Mar 4. Seaside, OR. Local and regional runners from across the nation are attracted to this officially sanctioned marathon run on Oregon's northern coast. One of the west coast's oldest and finest runs. 26-mile, 385-yard marathon also includes an 8K run and walk. Est attendance: 1,000. For info: Oregon Road Runners Club, PO Box 549, Beaverton, OR 97075. Phone: (503) 646-7867.

VERMONT: ADMISSION DAY ANNIVERSARY. Mar 4. Became 14th state on this day in 1791.

YELLOWSTONE ART CENTER'S ART AUCTION: 27th ANNUAL. Mar 4. Billings, MT. Nationally advertised contemporary art auction (regional/national work). 27th annual auction. Est attendance: 500. For info: Donna M. Forbes, Dir, Yellowstone Art Ctr, 401 N 27th St, Billings, MT 59101. Phone: (406) 256-6804.

YOSEMITE NORDIC HOLIDAY RACE. Mar 4. Badger Pass Ski Area, Yosemite National Park, CA. The oldest cross-country ski race in California. 17K, 10:30 AM. For info: Yosemite Park and Curry Co, 5410 E Home Ave, Fresno, CA 93727. Phone: (209) 372-1244.

Mar 4-5 ☆ *Chase's 1995 Calendar of Events* ☆

BIRTHDAYS TODAY

Chastity Bono, 26, daughter of Sonny and Cher, born at Los Angeles, CA, Mar 4, 1969.

Jane Goodall (Baroness VanLawick-Goodall), 61, anthropologist known for study of chimpanzees, born at London, England, Mar 4, 1934.

Charles Goren, 94, bridge expert, columnist, born at Philadelphia, PA, Mar 4, 1901.

Kevin Maurice Johnson, 29, basketball player, born at Sacramento, CA, Mar 4, 1966.

Kay Lenz, 42, actress ("Rich Man, Poor Man"), born at Los Angeles, CA, Mar 4, 1953.

Miriam Makeba, 63, actress, singer (antiapartheid activist), born at Johannesburg, South Africa, Mar 4, 1932.

Barbara McNair, 61, singer, actress ("The Barbara McNair Show"), born at Racine, WI, Mar 4, 1934.

Catherine O'Hara, 41, comedienne, writer ("Second City TV," "SCTV Network 90"), actress (mom in *Home Alone*) born at Toronto, Ontario, Canada, Mar 4, 1954.

Paula Prentiss (Paula Ragusa), 56, actress ("He & She," *Man's Favorite Sport, What's New Pussycat?*), born San Antonio, TX, Mar 4, 1939.

Alice Rivlin, 64, Director, Office of Management and Budget, born at Philadelphia, PA, March 4, 1931.

Mary Wilson, 51, singer (original member of the Supremes [with Diana Ross and Florance Ballard]), born Detroit, MI, Mar 4, 1944.

MARCH 5 — SUNDAY
64th Day — Remaining, 301

AMERICAN CAMP WEEK. Mar 5-11. A national tradition, more than 5 million children attend day or resident camps. Building self-confidence, learning new skills and making memories that last a lifetime are just the beginning of what makes camp special. To find the right program, parents begin looking at summer camps now—before they are filled. For info: American Camping Assn, Attn: Shirley Boltz, 5000 State Rd 67N, Martinsville, IN 46151. Phone: (317) 342-8456.

AUSTRALIA: EIGHT HOUR DAY OR LABOR DAY. Mar 5. Western Australia and Tasmania. Parades and celebrations commemorate trade union efforts during the 19th century to limit working hours. Their slogan: "Eight hours labor, eight hours recreation, eight hours rest!" In New South Wales it is called "Six Hour Day."

◆ **BLACKSTONE, WILLIAM: 400th BIRTH ANNIVERSARY.** Mar 5, 1595. William Blackstone, born in Gibside, Whickham, Durham County, England, was the first settler in what is now Boston, MA, and also the first in what is now Rhode Island. Blackstone came to New England with the Capt Robert Gorges expedition in 1623. When the expedition failed and most returned to England, he stayed and settled on what later became Beacon Hill. In 1634, he sold most of his Boston property and moved on to the shores of the river that now bears his name. He died there in what is now Cumberland, RI, on May 26, 1675. Blackstone is thought to be the first to cultivate apples in North America, and, being an ordained minister of the Church of England, he preached the first Anglican sermon here in Wickford, RI. The Blackstone River, the Blackstone Valley, the Town of Blackstone, MA, are all named for him as well as numerous streets, parks and institutions in the Blackstone River Valley.

	S	M	T	W	T	F	S
March				1	2	3	4
1995	5	6	7	8	9	10	11
	12	13	14	15	16	17	18
	19	20	21	22	23	24	25
	26	27	28	29	30	31	

◆ **BOSTON MASSACRE: 225th ANNIVERSARY.** Mar 5. On Mar 5, 1770, a skirmish between British troops and a crowd in Boston, MA, became widely publicized and contributed to the unpopularity of the British regime in America before the American Revolution. Five men were killed and six more were injured by British troops commanded by Capt Thomas Preston.

◆ **CRISPUS ATTUCKS DAY—225th DEATH ANNIVERSARY.** Mar 5. New Jersey. Honors Crispus Attucks, possibly a runaway slave, who was a leader of the American colonists involved in the Boston Massacre and the first to die in the skirmish there on Mar 5, 1770.

◆ **GERMAN 16-YEAR-OLDS DRAFTED: 50th ANNIVERSARY.** Mar 5, 1945. The Nazis began inducting German boys of the Hilter Youth born in or before 1929 into the regular German army, thereby lowering the draft age to 16.

GIANTS RIDGE INTERNATIONAL CLASSIC MARATHON. Mar 5. Biwabik, MN. Top international men and women compete in a citizen's 30K classical cross country and a 15K classical ski race. Est attendance: 250. For info: John Filander, PO Box 190, Biwabik, MN 55708. Phone: (800) 688-7669.

GIRL SCOUT SUNDAY. Mar 5. Girl Scouts worship together in the place of worship of their choice. Sponsor: Girl Scouts of the USA, Media Services, 420 Fifth Ave, New York, NY 10018.

GIRL SCOUT WEEK. Mar 5-11. To observe the anniversary of the founding of the Girl Scouts of the United States of America, the largest voluntary organization for girls and women in the world, which began on Mar 12, 1912. Special observances include: Girl Scout Sabbath, Mar 4, and Girl Scout Sunday, Mar 5, when Girl Scouts gather to attend religious services together; Girl Scout birthday, Mar 12. Sponsor: Girl Scouts of the USA, Media Services, 420 Fifth Ave, New York, NY 10018.

HARRISON, REX: BIRTH ANNIVERSARY. Mar 5. Born Reginald Carey on Mar 5, 1908, at Huyton, England. Rex Harrison's career as an actor encompassed more than 40 films and scores of plays. He won both a Tony and an Oscar for the role of Henry Higgins in *My Fair Lady*, perhaps his most famous role. Among other films, he appeared in *Dr. Doolittle, Cleopatra, Blithe Spirit* and *Major Barbara*. He claimed he would never retire from acting, and he was appearing in a Broadway revival of Somerset Maugham's *The Circle*, three weeks before his death on June 2, 1990, at his home in New York, NY.

HEMLOCK DAY. Mar 5. To honor the legendary Taffy Hemlock, supposed inventrix of a procedure for attaching padlocks around the hem of her skirt so that it would not be blown upward by inquisitive winds that might let her ankles get freckled. Celebrated by the True-Blue Law Society with patent searches to determine when/if the idea was ever patented. Annually, Mar 5. Sponsor: Puns Corps, Robert L. Birch, Box 2364, Falls Church, VA 22042-0364. Phone: (703) 533-3668.

HONDA GOLF CLASSIC. Mar 5-12. Weston Hills, FL. PGA golf event. For info: Greater Fort Lauderdale Conv/Visitors Bureau, 200 E Las Olas Blvd, Ste 1500, Fort Lauderdale, FL 33301. Phone: (305) 384-6000.

☆ Chase's 1995 Calendar of Events ☆   Mar 5–6

INTERNATIONAL DAY OF THE SEAL. Mar 5. In 1982 Congress declared an International Day of the Seal to draw attention to the cruelty of seal hunts and the virtual inevitability of these creatures' extinction. Zoos and aquariums around the world observe this day with special programs and activities; contact your local affiliate for a schedule of activities. Est attendance: 5,000. For info: Friends of The Natl Zoo (FONZ), c/o Natl Zoological Park, 3001 Connecticut Ave NW, Washington, DC 20008. Phone: (202) 673-4956.

JOHNSON IMPEACHMENT PROCEEDINGS BEGIN: ANNIVERSARY. Mar 5. On Mar 5, 1868, the Senate convened as a court to hear the charges against President Andrew Johnson. Supreme Court Chief Justice Salmon Chase was the presiding officer and the prosecution was led by radical Republicans Benjamin Butler and Thaddeus Stevens. Johnson was defended ably, and the Senate vote of 35–19 fell one vote short of the two-thirds majority necessary for impeachment.

LOS ANGELES MARATHON AND FAMILY REUNION FESTIVAL. Mar 5. Los Angeles, CA. A multicultural athletic competition designed to foster community spirit as well as pride in one's physical well-being. The Marathon Family Reunion Festival is held at Los Angeles Sports Arena. Food, arts and crafts, sampling, general merchandise. Phone for reunion: (310) 444-5544, ext 45 or 56. Est attendance: 1,500,000. For info: Los Angeles Marathon, 11110 W Ohio Ave, Ste 100, Los Angeles, CA 90025.

NATIONAL AARDVARK WEEK. Mar 5–11. To promote and enhance the image of the aardvark, while helping to fight off the "mid-winter drearies." The original aardvark advocacy group; holds the annual Miss Aardvark Contest. Sponsor: Amer Assn of Aardvark Aficionados, The Aardvark Group, Robert L. Bogart, Pres, Box 200, Parsippany, NJ 07054-0200. Phone: (201) 729-4555.

NATIONAL LUTHERAN SCHOOLS WEEK. Mar 5–11. To recognize the contributions of Lutheran schools to the US. Music festivals, mass children's worship, kite flying and special clothing days are typical of the events planned by the more than 3,000 Lutheran schools in the United States. Sponsor: School Services Unit, 1333 S Kirkwood Rd, St. Louis, MO 63122-7295.

NATIONAL PTA DRUG AND ALCOHOL AWARENESS WEEK. Mar 5–11. Designed to prevent drug and alcohol abuse through parent awareness. "Prevention Begins At Home." Annually, the first full week in March. Sponsor: Natl PTA, Attn: Program Outreach Div, 330 N Wabash Ave, Ste 2100, Chicago, IL 60611-3690. Phone: (312) 670-6782.

NATIONAL VOLUNTEERS OF AMERICA WEEK. Mar 5–11. To celebrate the founding of Volunteers of America on Mar 8, 1896. Sponsor: Volunteers of America, 3939 N Causeway Blvd, Metairie, LA 70002. Phone: (800) 899-0089.

NEW PROSPECTS. Mar 5–May 14. (Also Oct 1–Dec 10) Sundays only. Picnic House, Prospect Park, Brooklyn, NY. Performing arts on Sunday afternoons. Showcases new artists and performers, culturally diverse. Est attendance: 2,000. For info: Public Info Office, Prospect Park, 95 Prospect Park W, Brooklyn, NY 11215. Phone: (718) 965-8999.

PHILADELPHIA FLOWER SHOW "MOMENTS IN TIME—A GALAXY OF GARDENS." Mar 5–12. Philadelphia Civic Center, Philadelphia, PA. Celebrate spring at the world's largest indoor Flower Show. Est attendance: 220,000. For info: The Pennsylvania Horticultural Soc, Lisa Stephano, Communications Mgr, 325 Walnut St, Philadelphia, PA 19106-2777. Phone: (215) 625-8253.

SANDHILL CRANE MIGRATION. Mar 5–Apr 7. Platte River, Grand Island, NE. Season to view migrating birds. For info: Platte River Whooping Crane Trust, 2550 N Diers Ave, Ste H, Grand Island, NE 68803-1214. Phone: (308) 384-4633.

★ **SAVE YOUR VISION WEEK.** Mar 5–11. Presidential Proclamation issued for the first week of March since 1964, except 1971 and 1982 when issued for the second week of March. (PL88-1942, of Dec 30, 1963.)

SAVE YOUR VISION WEEK. Mar 5–11. To remind Americans that vision is one of the most vital of all human needs and its protection is of great significance to the health and welfare of every individual. Annually, the first full week in March. Sponsor: American Optometric Assn, 243 N Lindbergh Blvd, St. Louis, MO 63141. Phone: (314) 991-4100.

SURFSIDE SALUTES CANADA WEEK. Mar 5–11. Surfside, FL. A salute to Canada in appreciation for decades of patronage each winter season. Sports competitions daily and entertainment nightly. On Mar 11—a day-long festival featuring contests, ethnic foods, arts and crafts and live music. Est attendance: 2,000. Sponsor: Surfside Tourist Bd, 9301 Collins Ave, Surfside, FL 33154. Phone: (305) 864-0722.

TV TURN-OFF. Mar 5–11. Rather than an anti-TV campaign, this week-long event encourages communities to turn off their televisions and engage in alternative activities. Lyle the Crocodile and other book characters visit the Plymouth Public Library Mar 11 for the wrap-up program, "The Storybook Breakfast." Annually, the second week in March. $1.00 fee for postage and handling for printed material. Est attendance: 400. Sponsor: Friends of the Plymouth Public Library, 132 South St, Plymouth, MA 02360. Phone: (508) 746-4669.

BIRTHDAYS TODAY

Samantha Eggar, 56, actress ("Samantha and the King," *The Collector*), born London, England, Mar 5, 1939.
Penn Jillette, 40, magician, born at Greenfield, MA, Mar 5, 1955.
Paul Sand (Paul Sanchez), 51, actor ("St. Elsewhere;" Tony for *Story Theatre*), born at Los Angeles, CA, Mar 5, 1944.
Dean Stockwell, 59, actor ("Quantum Leap"), born at Los Angeles, CA, Mar 5, 1936.
Laurence Tisch, 72, business executive, born at Brooklyn, NY, Mar 5, 1923.
Michael Warren, 49, actor ("Paris," "Hill Street Blues") born at South Bend, IN, Mar 5, 1946.
Fred Williamson, 57, actor ("Julia," "Half Nelson"), born at Gary, IN, Mar 5, 1938.

MARCH 6 — MONDAY
65th Day — Remaining, 300

BROWNING, ELIZABETH BARRETT: BIRTH ANNIVERSARY. Mar 6. English poet, author of *Sonnets from the Portuguese*, wife of poet Robert Browning and subject of the play *The Barretts of Wimpole Street*, was born near Durham, England, Mar 6, 1806. She died at Florence, Italy, June 29, 1861.

CYPRUS: GREEN MONDAY. Mar 6. Green, or Clean, Monday is the first Monday of Lent. Lunch in the fields, with bread, olives and uncooked vegetables and no meat or dairy products.

FALL OF THE ALAMO: ANNIVERSARY. Mar 6. Anniversary of the fall of the Texan fort, the Alamo, on Mar 6, 1836. The siege, led by Mexican general Santa Anna, began on Feb 23 and reached its climax on Mar 6, when the last of the defenders was slain. Texans, under General Sam Houston, rallied with the war-cry "Remember the Alamo" and, at the Battle of San Jacinto, on Apr 21, defeated and captured Santa Anna, who signed a treaty recognizing Texas's independence.

GHANA: INDEPENDENCE DAY. Mar 6. National holiday. Received independence from Great Britain on this day, 1957.

GUAM: DISCOVERY DAY or MAGELLAN DAY. Mar 6. Commemorates discovery of Guam on this day in 1521.

Mar 6–7 ☆ *Chase's 1995 Calendar of Events* ☆

MICHELANGELO: 520th BIRTH ANNIVERSARY. Mar 6. Anniversary of the birth, Mar 6, 1475, at Caprese, Italy, of Michelangelo di Lodovico Buonarroti Simoni. A prolific Renaissance painter, sculptor, architect and poet, who had a profound impact on Western art. Michelangelo's fresco painting on the ceiling of the Sistine Chapel at the Vatican in Rome, Italy, is often considered the pinnacle of his achievement in painting, as well as the highest achievement of the Renaissance. Also among his works were the sculptures *David* and *The Pieta*. Appointed architect of St. Peter's in 1542, a post he held until his death on Feb 18, 1564, at Rome.

NATIONAL PROCRASTINATION WEEK. Mar 6–12. To promote the benefits of relaxing through putting off till tomorrow everything that needn't be done today. Sponsor: Procrastinators' Club of America Inc, Les Waas, Pres, PO Box 712, Bryn Athyn, PA 19009. Phone: (215) 947-9020.

NATIONAL PROFESSIONAL PET SITTERS WEEK. Mar 6–12. A week to show appreciation for the pet sitters who work 365 days a year for customers. For info: Bill Foster, Membership Dir, Pet Sitters Intl (PSI), 418 E King St, King, NC 27021. Phone: (910) 983-9222.

NATIONAL SCHOOL BREAKFAST WEEK. Mar 6–10. To focus on the importance of a nutritious breakfast served in the schools, giving children a good start to their day. Annually, the first full week in March. Sponsor: American School Food Service Assn, 1600 Duke St, 7th Fl, Alexandria, VA 22314. Phone: (703) 739-3900.

NEWSPAPER IN EDUCATION WEEK. Mar 6–10. A weeklong celebration of using newspapers in the classroom as living textbooks. Annually, the first full week in March (weekdays). For info: Gwendolyn Kirk, Manager Edu Programs, NAA Foundation, 11600 Sunrise Valley Dr, Reston, VA 22091. Phone: (703) 648-1251.

ORTHODOX LENT. Mar 6–Apr 15. Great Lent or Easter Lent, observed by Eastern Orthodox Churches, lasts until Holy Week begins on Orthodox Palm Sunday (Apr 16).

PEALE, ANNA CLAYPOOLE: BIRTH ANNIVERSARY. Mar 6. American painter of miniatures. Born at Philadelphia, PA, Mar 6, 1791. Died on Dec 25, 1878.

BIRTHDAYS TODAY

Marion Barry, 59, former mayor of Washington, DC, born at Itta Bena, MS, Mar 6, 1936.
Christopher Samuel Bond, 56, US Senator (R, Missouri), born at St. Louis, MO, Mar 6, 1939.
Sarah Caldwell, 71, conductor, born at Maryville, MO, Mar 6, 1924.
L. Gordon Cooper, 68, astronaut, born at Shawnee, OK, Mar 6, 1927.
Gabriel Garcia-Marquez, 67, author (*A Hundred Years of Solitude, Love in the Time of Cholera*), born at Aracaracca, Columbia, Mar 6, 1928.
Dave Gilmour, 51, singer, musician (member Pink Floyd, *The Dark Side of the Moon*), born at Cambridge, England, Mar 6, 1944.
Alan Greenspan, 69, economist, Chairman of the Federal Reserve Board, born at New York City, NY, Mar 6, 1926.
Ed McMahon, 72, actor, TV host ("Big Top" clown; "The Tonight Show," "Star Search"), born Detroit, MI, Mar 6, 1923.
Rob Reiner, 50, actor ("All in the Family"), director (*When Harry Met Sally, This Is Spinal Tap*), born New York, NY, Mar 6, 1945.
Valentina Tereshkova-Nikolaeva, 58, cosmonaut, born at Maslennikovo, USSR, Mar 6, 1937.

MARCH 7 — TUESDAY
66th Day — Remaining, 299

BURBANK, LUTHER: BIRTH ANNIVERSARY. Mar 7. Anniversary of birth of American naturalist and author, creator and developer of many new varieties of flowers, fruits, vegetables and trees. Luther Burbank's birthday is observed by some as Bird and Arbor Day. Born at Lancaster, MA, Mar 7, 1849. Died at Santa Rosa, CA, Apr 11, 1926.

DISTINGUISHED SERVICE MEDAL: ANNIVERSARY. Mar 7, 1918. With US troops fighting in the trenches in France during the first world war, Pres Woodrow Wilson authorized a new bronze, beribboned medal to be given to US Army personnel who performed "exceptionally meritorious service."

ENGLAND: CHELSEA ANTIQUES FAIR. Mar 7–18. (Also Sept 12–23.) Chelsea Old Town Hall, King's Rd, London. Twice-yearly antiques fair with a wide range of pre-1830 furniture and other items with a pre-1860 dateline for sale. Est attendance: 12,000. For info: Penman Antiques Fair, PO Box 114, Haywards Heath, West Sussex, England RH16 2YU.

HOPKINS, STEPHEN: BIRTH ANNIVERSARY. Mar 7. Colonial governor (Rhode Island) and signer of the Declaration of Independence. Born at Providence, RI, Mar 7, 1707. Died July 13, 1785.

◈ **REMAGEN BRIDGE CAPTURE: 50th ANNIVERSARY.** Mar 7. On this date in 1945, a small advance force of the US 1st Army captured the Ludendorff railway bridge across the Rhine River at Remagen (between Bonn and Coblenz)—the only bridge across the Rhine that had not been blown up by the German defenders—thus acquiring the first bridgehead onto the east bank and the beginning of the Allied advance into Germany, a turning point in World War II.

SALVADOR DALI MUSEUM OPENING: ANNIVERSARY. Mar 7. The museum was opened in St. Petersburg, FL, on Mar 7, 1982, by A. Reynolds and Eleanor R. Morse as a permanent home for the Morse collection. It is located on the waterfront in a former marine warehouse. Est attendance: 500. For info: Salvador Dali Museum, 1000 Third St S, St. Petersburg, FL 33701. Phone: (813) 823-3767.

TOWN MEETING DAY. Mar 7. Vermont. The first Tuesday in March is an official state holiday in Vermont. "Nearly every town elects officers, approves budget items and deals with a multitude of other items in a day-long public meeting of the voters. This is a very significant day for Vermonters."

BIRTHDAYS TODAY

Anthony Armstrong-Jones (Lord Snowdon), 65, photographer, born at London, England, Mar 7, 1930.
Joseph Carter, 35, baseball player, born at Oklahoma City, OK, Mar 7, 1960.
Michael Eisner, 53, corporation executive, born at Mt. Kisco, NY, Mar 7, 1942.
Janet Guthrie, 57, auto racer, born at Iowa City, IA, Mar 7, 1938.

March 1995

S	M	T	W	T	F	S
			1	2	3	4
5	6	7	8	9	10	11
12	13	14	15	16	17	18
19	20	21	22	23	24	25
26	27	28	29	30	31	

Franco Harris, 45, Pro Football Hall of Famer, born at Ft Dix, NJ, Mar 7, 1950.
Ivan Lendl, 35, tennis player, born at Ostrava, Czechoslovakia, Mar 7, 1960.
Willard Herman Scott, 61, weatherman (the "Today Show"; friend of centenarians), born at Alexandria, VA, Mar 7, 1934.
Daniel J. Travanti, 55, actor ("Hill Street Blues"), born at Kenosha, WI, Mar 7, 1940.
Peter Wolf, 49, singer (lead with J. Geils Band, "Centerfold"), born at Boston, MA, Mar 7, 1946.

MARCH 8 — WEDNESDAY
67th Day — Remaining, 298

CAXTON'S "MIRROR OF THE WORLD" TRANSLATION: ANNIVERSARY. Mar 8. William Caxton, England's first printer, completed the translation from French into English of *The Mirror of the World*, a popular account of astronomy and other sciences, on Mar 8, 1481. In print soon afterward, *Mirror of the World* became the first illustrated book printed in England.

INTERNATIONAL (WORKING) WOMEN'S DAY. Mar 8. A day to honor women, especially working women. Said to commemorate an 1857 march and demonstration in New York, NY, by female garment and textile workers. Believed to have been first proclaimed for this date at an international conference of women held in Helsinki, Finland, in 1910, "that henceforth March 8 should be declared International Women's Day." The 50th anniversary observance, at Peking, China, in 1960, cited Clara Zetkin (1857–1933) as "initiator of Women's Day on March 8." This is perhaps the most widely observed holiday of recent origin and is unusual among holidays originating in the US in having been widely adopted and observed in other nations, including socialist countries. In the USSR and the People's Republic of China it is a national holiday, and flowers or gifts are presented to women workers.

MARYLAND HOME AND FLOWER SHOW. Mar 8–12. Timonium Fairgrounds, Baltimore, MD. The largest display of home building, remodeling and home decorating exhibits in the Baltimore area. Includes educational exhibits, crafts, plant marketplace and displays of landscaped gardens. Est attendance: 71,000. For info: S & L Productions, Inc, 7870 Spruce Hill Rd, PO Box 639, Severn, MD 21144. Phone: (410) 969-8585.

SYRIAN ARAB REPUBLIC REVOLUTION DAY: ANNIVERSARY. Mar 8. Official public holiday commemorating assumption of power by Revolutionary National Council, Mar 8, 1963.

UNITED NATIONS: INTERNATIONAL WOMEN'S DAY. Mar 8. An international day observed by the organizations of the United Nations system. Info from: United Nations, Dept of Public Info, New York, NY 10017.

UNIVERSAL WOMEN'S WEEK. Mar 8–14. To remind ourselves and others of the value of women of all ages and classes, of their rights and dignity, and to honor outstanding women in the fields of government, business, industry, science, health, education, social work and the cultural arts by election to Universal Hall of Fame. Please send $1 to cover expense of printing, handling and postage. Annually, Mar 8–14. Sponsor: Intl Soc of Friendship and Good Will, Dr. Stanley Drake, Pres, 9538 Summerfield St, Spring Valley, San Diego County, CA 91977-2852. Phone: (619) 446-8882.

VAN BUREN, HANNAH HOES: BIRTH ANNIVERSARY. Mar 8. Wife of Martin Van Buren, eighth president of the US. Born at Kinderhook, NY, Mar 8, 1783. Died Feb 5, 1819.

WARMEST US WINTER ON RECORD DECLARED: ANNIVERSARY. Mar 8. On Mar 8, 1992, the National Climatic Data Center declared the winter of 1991–1992 the warmest US winter in the 97 years that the Federal Government had been keeping record of climatic conditions. The average temperature of 36.87 degrees Fahrenheit eclipsed the previous record set in 1953–1954 of 36 degrees Fahrenheit.

BIRTHDAYS TODAY

George F. Allen, 43, Governor of Virginia (R), born at Whittier, CA, Mar 8, 1952.
Cyd Charisse (Tula Finklea), 72, actress (*Silk Stockings*), dancer (*Grand Hotel*), born Amarillo, TX, Mar 8, 1923.
Susan Clark, 55, actress ("Webster," "Babe"), born at Sarnia, Ontario, Canada, Mar 8, 1940.
Mickey Dolenz, 50, singer (Monkees, "I'm a Believer"), actor ("The Monkees," "Circus Boy"), born Los Angeles, CA, Mar 8, 1945.
Mike Lowry, 56, Governor of Washington (D), born at St. John, WA, Mar 8, 1939.
Charley Pride, 57, singer ("Just Between You and Me," "Wonder Could I Live There Anymore," "Kiss an Angel Good Morning," "Hope You're Feelin' Me," "Why Baby Why?"), born Sledge, MS, Mar 8, 1938.
Lynn Redgrave, 52, actress (*Georgy Girl*; Ann Anderson in "House Calls"), born at London, England, Mar 8, 1943.
Jim Rice, 42, former baseball player, born at Anderson, SC, Mar 8, 1953.
Carole Bayer Sager, 48, singer, songwriter ("That's What Friends Are For"), born at New York, NY, Mar 8, 1947.
Raynoma (Mayberry Liles) Gordy Singleton, 58, co-founder of Motown, born at Detroit, MI, Mar 8, 1937.
Claire Trevor, 86, actress (*Key Largo*, *Stagecoach*), born New York, NY, Mar 8, 1909.

MARCH 9 — THURSDAY
68th Day — Remaining, 297

ADAMS, RICHARD C.: DEATH ANNIVERSARY. Mar 9. Richard C. Adams, inspired by the shortage of paint brushes during World War II, invented the paint roller in 1940, while working in his basement. Adams died at La Mesa, CA, on Mar 9, 1988.

BELIZE: BARON BLISS DAY. Mar 9. Official public holiday in Belize. Celebrated in honor of Sir Henry Edward Ernest Victor Bliss, a great benefactor of Belize.

GAGARIN, YURI ALEXSEYEVICH: BIRTH ANNIVERSARY. Mar 9. Russian cosmonaut Yuri Gagarin, the first man to travel in space (Apr 12, 1961). Born at Gzhatsk, USSR, on Mar 9, 1934. The 27-year-old Soviet Air Force major made man's first flight in space, lasting 108 minutes and orbiting Earth in a rocket-propelled, five-ton space capsule, 187 miles above the Earth's surface. Gagarin was killed in an airplane crash near Moscow, USSR, on Mar 27, 1968. After his death the town in which he was born was renamed Gagarin, and the Gagarin Museum was established in the frame house where he spent his childhood.

MOON PHASE: FIRST QUARTER. Mar 9. Moon enters First Quarter phase at 5:14 AM, EST.

NAIA MEN'S DIVISION II BASKETBALL CHAMPIONSHIP TOURNAMENT. Mar 9–14. Northwest Nazarene College, Nampa, ID. 24-team field competes for the national championship. 4th annual. Fax: (918) 494-8841. Est attendance: 28,500. For info: Natl Assn of Intercollegiate Athletics, 6120 S Yale Ave, Ste 1450, Tulsa, OK 74136. Phone: (918) 494-8828.

Chase's 1995 Calendar of Events

Mar 9-10

NAIA WOMEN'S DIVISION II BASKETBALL CHAMPIONSHIP TOURNAMENT. Mar 9-14. Western Oregon State College, Monmouth, OR. 24-team field competes for the national championship. 4th annual. Fax: (918) 494-8841. Est attendance: 12,500. For info: Natl Assn of Intercollegiate Athletics, 6120 S Yale Ave, Ste 1450, Tulsa, OK 74136. Phone: (918) 494-8828.

PANIC DAY. Mar 9. Run around all day in a panic, telling others you can't handle it anymore. Phone: (212) 388-8673 or (717) 274-8451. Sponsor: Wellness Permission League, Tom or Ruth Roy, 2105 Water St, Lebanon, PA 17046.

ST. FRANCES OF ROME: FEAST DAY. Mar 9. Patron of motorists and model for housewives and widows (1384-1440). After 40 years of marriage she was widowed in 1436, and later joined the community of Benedictine Oblates. Canonized in 1608.

SPACE MILESTONE: *SPUTNIK 9* (USSR). Mar 9. Dog, Chernushka (Blackie), was a passenger. Mar 9, 1961.

◆ **TOKYO BLANKET BOMBING: 50th ANNIVERSARY.** Mar 9, 1945. The Japanese capital of Tokyo was bombed by 343 Superfortresses carrying all the incendiary bombs they could hold. Within the targeted areas of the city, population densities were four times greater than those of most American cities, and homes were made primarily of wood and paper. Carried by the wind, the fires leveled 16 sq miles. More than a quarter million buildings were destroyed, including 18% of the industrial area. The death toll was 83,000, 41,000 were injured. For the balance of the war American strategic bombing followed this pattern.

VESPUCCI, AMERIGO: BIRTH ANNIVERSARY. Mar 9. Italian navigator, merchant and explorer for whom the Americas were named. Born at Florence, Italy, probably on Mar 9, 1451. He participated in at least two expeditions between 1499 and 1502, which took him to the coast of South America, where he discovered the Amazon and Plate rivers. Vespucci's expeditions were of great importance because he believed that he had discovered a *new continent*, not just a new route to the Orient. Neither Vespucci nor his exploits achieved the fame of Columbus, but the New World was to be named not for Columbus, but for Amerigo Vespucci, by an obscure German geographer and mapmaker, Martin Waldseemuller. See also: "Waldseemuller, Martin: Remembrance Day" (Apr 25). Ironically, in his work as an outfitter of ships, Vespucci had been personally acquainted with Christopher Columbus. Vespucci died at Seville, Spain, on Feb 22, 1512.

BIRTHDAYS TODAY

Brian Bosworth, 30, former football player, born Oklahoma City, OK, Mar 9, 1965.
Bobby Fischer, 52, World Chess Champion ('72), born at Chicago, IL, Mar 9, 1943.
Mickey Gilley, 59, musician, born at Natchez, MS, Mar 9, 1936.
Raul Julia, 55, actor (*Eating Raoul, The Addams Family*), born at San Juan, Puerto Rico, Mar 9, 1940.
David Hume Kennerly, 48, photographer, born at Rosenburg, OR, Mar 9, 1947.
Emmanuel Lewis, 24, actor ("Webster"), born at Brooklyn, NY, Mar 9, 1971.
Jeffrey Osborne, 47, musician, songwriter, born at Providence, RI, Mar 9, 1948.
Benito Santiago, 30, baseball player, born at Ponce, Puerto Rico, Mar 9, 1965.

March 1995

S	M	T	W	T	F	S
			1	2	3	4
5	6	7	8	9	10	11
12	13	14	15	16	17	18
19	20	21	22	23	24	25
26	27	28	29	30	31	

Mickey Spillane (Frank Morrison), 77, author (*The Killing Man, Vengeance Is Mine*), born at Brooklyn, NY, Mar 9, 1918.
Trish Van Devere, 52, actress (*Where's Poppa?; One Is a Lonely Number*), born Tenafly, NJ, Mar 9, 1943.
Joyce Van Patten, 61, actress (*Monkey Shines*, "The Goodbye Guys"), born at Queens, NY, Mar 9, 1934.

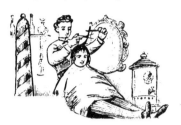

MARCH 10 — FRIDAY
69th Day — Remaining, 296

ALL-NORTHWEST BARBERSHOP BALLAD CONTEST. Mar 10-11. Forest Grove, OR. Barbershop quartets throughout the Pacific Northwest compete in a gay nineties setting. Est attendance: 3,000. For info: Chamber of Commerce, 2417 Pacific Ave, Forest Grove, OR 97116. Phone: (503) 357-3006.

CANADA: SPAGHETTI BRIDGE BUILDING CONTEST. Mar 10. Kelowna, British Columbia. To challenge the world to match our expertise in Spaghetti Bridge design. World record load 176kg (388 lbs), suspended from a 1-meter long bridge weighing less than 0.75kg (1.66 lbs). Annually, the second Friday in March. Est attendance: 500. Sponsor: Okanagan University College, Garry Gaudet, Info Officer, 1000 KLO Rd, Kelowna, BC, Canada V1Y 4X8. Phone: (604) 862-5662.

JUPITER EFFECT: ANNIVERSARY. Mar 10. The much talked-about and sometimes-feared planetary configuration of Mar 10, 1982 (a semi-alignment of the planets on the same side of the sun), occurred on that date without causing any of the disasters or unusual natural phenomena that some had predicted. See the 1982 edition of *Chase's Annual Events* for diagram and detailed description.

LONGHORN WORLD CHAMPIONSHIP RODEO. Mar 10-12. UTC Arena, Chattanooga, TN. More than 220 of North America's best cowboys and cowgirls compete in 6 professionally sanctioned contests ranging from bronc riding to bull riding for top prize money and world championship points. The theme of the production changes annually. Additional attractions include special animal trainers, trick ropers, trick riders and professional clowns. 13th annual. Est attendance: 16,000. For info: Longhorn World Championship Rodeo, Inc, PO Box 70159, Nashville, TN 37207. Phone: (615) 876-1016.

◆ **LUCE, CLAIRE BOOTHE: BIRTH ANNIVERSARY.** Mar 10, 1903. Playwright and politician Claire Boothe Luce was born at New York City. Luce wrote and edited for *Vogue* and *Vanity Fair* as well as writing plays, three of which were later adapted into motion pictures—*The Women* (1936), *Kiss the Boys Goodbye* (1938) and *Margin of Error* (1939). She served in the US House of Representatives (1943-47) and, as ambassador to Italy (1953-56)—the first woman appointed ambassador to a major country. Luce died Oct 9, 1987, at Washington, DC.

MALPIGHI, MARCELLO: BIRTH ANNIVERSARY. Mar 10. Italian physician, physiologist, author and teacher, called the "father of microscopic anatomy." Born near Bologna, Italy, on Mar 10, 1628. Malpighi was a pioneer in the use of the microscope for biological and botanical study. Died at Rome, Italy, on Nov 30, 1694.

MILWAUKEE SENTINEL SPORTS SHOW. Mar 10-19. Milwaukee Exhibition Convention Center, Milwaukee, WI. Travel and resort exhibits, boating and marine sporting goods, sport vehicles, hunting gear. Est attendance: 118,000. Sponsor: *Milwaukee Sentinel*, 333 W State St, PO Box 661, Milwaukee, WI 53201-0661. Phone: (424) 786-5600.

☆ Chase's 1995 Calendar of Events ☆ Mar 10-11

NAIA WRESTLING CHAMPIONSHIP. Mar 10-11. Jamestown Civic Center, Jamestown, N.D. Individuals compete for All-America honors in 12 weight divisions, while teams compete for the national championship. Fax: (918) 494-8841. 38th annual competition. Est attendance: 4,000. For info: Natl Assn of Intercollegiate Athletics, 6120 S Yale Ave, Ste 1450, Tulsa, OK 74136. Phone: (918) 494-8828.

NATIONAL WOMEN'S GET-AWAY WEEKEND. Mar 10-12. A weekend for women everywhere to get away from it all by giving themselves a weekend away from home to country inns, bed and breakfasts or other pleasant places. Sponsor: *Country Inns Magazine* For info: Laura Baddish, The Alden Group, 52 Vanderbilt Ave, New York, NY 10017. Phone: (212) 867-6400.

NCAA MEN'S AND WOMEN'S DIVISION I INDOOR TRACK AND FIELD CHAMPIONSHIPS. Mar 10-11. Indianapolis, IN. Organizing Host: USA Track and Field. Est attendance: 18,000. For info: P. Duffy Mahoney, Dir of Operations, USATF, One Hoosier Dome, Ste 140, Indianapolis, IN 46225. Phone: (317) 261-0500.

NC RV AND CAMPING SHOW. Mar 10-12. Raleigh, NC. A display of the latest in recreation vehicles and accessories by various dealers. Est attendance: 8,500. For info: Apple Rock Advertising & Promotion, 1200 Eastchester Dr, High Point, NC 27265. Phone: (910) 883-7199.

NORTHWEST CRIBBAGE TOURNAMENT. Mar 10-12. Baker City, OR. Est attendance: 200. For info: Baker County VCB, 490 Campbell St, Baker City, OR 97814. Phone: (800) 523-1235.

PAPER MONEY ISSUED: ANNIVERSARY. Mar 10. The first paper money was issued in the US on March 10, 1862. The denominations were $5 (Hamilton), $10 (Lincoln) and $20 (Liberty). They were not legal tender when first issued but became so by act of March 17, 1862.

SALVATION ARMY IN THE USA: ANNIVERSARY. Mar 10, 1880. On this day Commissioner George Scott Railton and seven women officers landed in New York to officially begin the work of The Salvation Army in the United States.

TELEPHONE INVENTION: ANNIVERSARY. Mar 10. Alexander Graham Bell transmitted the first telephone message to his assistant in the next room: "Mr Watson, come here, I want you," on Mar 10, 1876, at Cambridge, MA. See also: "Bell, Alexander Graham: Birth Anniversary" (Mar 3).

◆ **TUBMAN, HARRIET: DEATH ANNIVERSARY.** Mar 10. American abolitionist, Underground Railroad leader, born a slave at Bucktown, Dorchester County, MD, about 1820 or 1821. Died at Auburn, NY, Mar 10, 1913. She escaped from Maryland plantation in 1849, later helped more than 300 slaves reach freedom.

VIRGINIA SPRING SHOW. Mar 10-12. Showplace Exhibition Center, Richmond, VA. Features 400 artisans and craftspeople, food shops, spring entertainment. Holiday Cooking Theatre 7th annual show. Est attendance: 25,000. For info: Virginia Show Productions, Patricia Wagstaff, PO Box 305, Chase City, VA 23924. Phone: (804) 372-3996.

◆ **WALD, LILLIAN: BIRTH ANNIVERSARY.** Mar 10, 1867. Founder of the internationally known Henry Street Settlement, Lillian Wald was born at Cincinnati, OH, and trained in nursing and medicine in New York. Sociologist, nurse and social worker, by 1902 Wald was using the Henry Street location, not only for traditional social work activities but also headquarters for the first large public health nursing center in America. By 1913 it employed 92 nurses who made 200,000 home visits a year. In 1912 Wald helped found and became first president of the National Organization for Public Health Nursing and in response to her urging, the US Congress established the US Children's Bureau. Lillian Wald died Sept 1, 1940, at Westport, CT.

BIRTHDAYS TODAY

Heywood Hale Broun, 77, broadcaster, born at New York, NY, Mar 10, 1918.
Prince Edward, 31, son of Queen Elizabeth II, born at London, England, Mar 10, 1964.
Bob Greene, 48, journalist, born at Columbus, OH, Mar 10, 1947.
Pamela Mason, 77, actress, born at Westgate, England, Mar 10, 1918.
Chuck Norris, 55, actor (*Missing in Action*, "Chuck Norris Karate Commandos"), born Ryan, OK, Mar 10, 1940.
David Rabe, 55, playwright, born at Dubuque, IA, Mar 10, 1940.

MARCH 11 — SATURDAY
70th Day — Remaining, 295

CAMPBELL, MALCOLM: BIRTH ANNIVERSARY. Mar 11. Record-making British auto racer, the first man to travel five miles a minute (300 mph) in an automobile. Born at Chislehurst, Kent, England, on Mar 11, 1885. Died at his home in Surrey, England, Dec 31, 1948.

CANADIAN AMERICAN DAYS FESTIVAL. Mar 11-19. Myrtle Beach, SC. Concerts, square dances, beach games and sports events. Est attendance: 100,000. Sponsor: Myrtle Beach Area Chamber of Commerce, 1301 N Kings Hwy, Myrtle Beach, SC 29578. Phone: (803) 626-7444.

CHESAPEAKE CAT CLUB CFA ALL-BREED CAT SHOW. Mar 11-12. Baltimore, MD. One of the most prestigious CFA cat shows in the US. Est attendance: 5,000. Sponsor: Chesapeake Cat Club, Inc, c/o Pamela Swanson, 801 Upper Glencoe Rd, Sparks, MD 21152. Phone: (410) 771-4880.

DAYTONA SUPERCROSS BY HONDA. Mar 11. Daytona International Speedway, Daytona Beach, FL. For info: Daytona Intl Speedway, Glenn Barber, PR, Box 2801, Daytona Beach, FL 32120-2801. Phone: (904) 254-6782.

HIGHLAND COUNTY MAPLE FESTIVAL. Mar 11-12. (Also Mar 18-19.) Highland County, VA. To welcome visitors to view the process of syrup making. Large craft shows. Est attendance: 60,000. For info: Highland County Chamber of Commerce, PO Box 223, Monterey, VA 24465. Phone: (703) 468-2550.

INDIANA FLOWER AND PATIO SHOW. Mar 11-19. Indiana State Fairgrounds Event Center, Indianapolis, IN. The oldest show of its kind in the Midwest, featuring 12 gardens with products and services for home, yard, and patio. Est attendance: 95,000. For info: Patrick Buchen, Pres, HSI Show Productions, Box 20760, Indianapolis, IN 46220. Phone: (317) 255-4151.

JOHNNY APPLESEED DAY. Mar 11. Anniversary of the death on Mar 11, 1847, of John Chapman, better known as Johnny Appleseed, believed to have been born at Leominster, MA, Sept 26, 1774. The planter of orchards and friend of wild animals was regarded by the Indians as a great medicine man. He died at Allen County, IN. See also: "Johnny Appleseed: Birth Anniversary" (Sept 26).

McGUIRE'S 5K ST. PATRICK'S DAY RUN. Mar 11. Pensacola, FL. The largest, most popular 5K run in the history of Pensacola. 9:30 AM. Est attendance: 2,500. For info: Susi Lyon, McGuires Irish Pub, 600 E Gregory St, Pensacola, FL 32501. Phone: (904) 433-6789.

125

MONTEREY COUNTY HOT AIR AFFAIR. Mar 11-12. Laguna Seca Recreational Park, Monterey, CA. Ballon pilots from all over the country compete for cash and prizes in a field of approximately 40 ballons. Also skydiving, tethered ballon rides, food, arts and crafts. Benefits local Monterey County charities. Annually, in March. Est attendance: 6,000. Sponsor: Bostrom Management, 2600 Garden Rd, Ste 208, Monterey, CA 93940. Phone: (408) 649-6544.

NATIONAL SKI-JORING FINALS. Mar 11-12. Red Lodge Rodeo Grounds, Red Lodge, MT. Horsemen and skiers provide action entertainment. Derived from the Scandinavian sport of pulling a skier behind a horse, Ski-Joring has evolved from a leisure winter diversion into lively, regulated competition. Annually, the second weekend in March. Est attendance: 3,000. For info: Red Lodge Chamber of Commerce, Box 988, Red Lodge, MT 59068. Phone: (406) 446-1718.

NEW ENGLAND SLED DOG RACES. Mar 11-12. Intown Trail, Rangeley, ME. Eight-dog, six-dog, four-dog pro and junior classes race 1-9 miles, depending on the size of the team. Annually, the second weekend in March. Sponsor: New England Sled Dog Club. For info: Chamber of Commerce, PO Box 317, Rangeley, ME 04970. Phone: (207) 864-5364.

RBA—THE RETAILERS BAKERY-DELI ASSOCIATION: 77th CONVENTION EXHIBITION. Mar 11-13. St. Louis, MO. Fax: (301) 725-2187. Phone: (301) 725-2149 or (800) 638-0924. Est attendance: 10,000. Sponsor: RBA Retailers' Bakery-Deli Assn, J. Stewart Taylor, Conv Dir, 14239 Park Center Dr, Laurel, MD 20707.

ST. PATRICK'S DAY DOG FUN FAIR, PARADE AND CELEBRATION. Mar 11. Alexandria, VA. A celebration of Irish heritage and culture featuring a parade. Preceded by "dog fun fair." Parties in local restaurants with live entertainment immediately after parade. Free. Second Saturday in March. Est attendance: 30,000. For info: Diane Bechtol, Mgr, Media Relations, Alexandria Conv and Visitors Bureau, 221 King St, Alexandria, VA 22314. Phone: (703) 838-4200.

SPIDERS! Mar 11-June 4. American Museum of Natural History, New York, NY. An exhibit to "spider-wise" visitors to how spiders go about their daily lives—finding food, fending off dangers, finding a mate and producing offspring. A variety of audiovisual programs and dynamic interactives reveal how spiders manage these feats. For info: Smithsonian Institution Traveling Exhibition Service, 1100 Jefferson Dr SW, Ste 3146, Washington, DC 20560. Phone: (202) 357-2700.

SPRING CRAFT AND GIFT SHOW. Mar 11-12. Wisconsin State Fair Park, Milwaukee, WI. Show combined with commercial gift exhibitors and craftsmen. Est attendance: 23,000. For info: Dennis R. Hill, Dir, 3233 S Villa Circle, Milwaukee, WI 53227. Phone: (414) 321-2100.

March 1995

S	M	T	W	T	F	S
			1	2	3	4
5	6	7	8	9	10	11
12	13	14	15	16	17	18
19	20	21	22	23	24	25
26	27	28	29	30	31	

TASSO, TORQUATO: BIRTH ANNIVERSARY. Mar 11. Poet of the late Renaissance, born at Sorrento, Italy, on Mar 11, 1544. His violent outbursts and acute sensitivity to criticism led to his imprisonment for seven years, during which the "misunderstood genius" continued his literary creativity. Died at Rome, Italy, Apr 25, 1595.

TRY THIS ON—A HISTORY OF CLOTHING, GENDER AND POWER. Mar 11-Apr 23. Rock County Historical Society, Janesville, WI. A social history about clothing and appearance in the US over the past 200 years. See Jan 7 for full description. For info: Smithsonian Institution Traveling Exhibition Service, 1100 Jefferson Dr SW, Ste 3146, Washington, DC 20560. Phone: (202) 357-2700.

WELK, LAWRENCE: BIRTH ANNIVERSARY. Mar 11. Bandleader Lawrence Welk was born in Strasburg, ND, on Mar 11, 1903. He began his career playing the accordion and at 17 formed the Biggest Little Band in America for the inaugural of radio station WNAX in Yankton, SD. He left home at the age of 21 and, after playing ballrooms, hotels and radio stations in the Midwest, he moved to Los Angeles where in 1955 his show began its nationwide television broadcast of "Champagne Music." The longest-running program in TV history, "The Lawrence Welk Show" played each Saturday on ABC from 1955 until 1971 when it was dropped because sponsors thought its audience was too old. Welk managed to keep the show on a network of more than 250 independent stations for 11 more years and it still can be seen on public television, repackaged as "Memories with Lawrence Welk." Welk's entertainment empire expanded to include the purchase of royalty rights to songs, including the entire collection of songs by Jerome Kern, and resort and retirement complex Lawrence Welk Village in Escondido, CA. Welk died in Santa Monica, CA, on May 17, 1992.

WINTER GAMES OF OREGON. Mar 11-12. Portland and various sites, OR. 7th annual games. Alpine and Nordic competition to be held at Timberline, Mt. Hood Meadows, Ski Bowl/Multorpor, Teacup. Est attendance: 800. For info: Portland, Oregon Visitors Assn, Three World Trade Ctr, 26 SW Salmon, Portland, OR 97204-3299. Phone: (503) 520-1319.

WOOD RADIO ST. PATRICK'S DAY PARADE. Mar 11. Grand Rapids, MI. Parade focusing on Celtic cultural tradition featuring local Irish bands and floats. Followed by traditional Irish St. Patrick's Day ceremony. Est attendance: 3,000. For info: Juli Agacinski, Promo Dir, WOOD-AM 1300/EZ-105.7 FM, 180 N Division, Grand Rapids, MI 49503. Phone: (616) 459-1919.

BIRTHDAYS TODAY

Sam Donaldson, 61, journalist, born at El Paso, TX, Mar 11, 1934.
Nigel Lawson, 63, British government official, born at London, England, Mar 11, 1932.
Bobby McFerrin, 45, jazz musician, singer, songwriter, born at New York City, NY, Mar 11, 1950.
Robert Adam Mosbacher, 68, Secretary of Commerce, born at Mt Vernon, NY, Mar 11, 1927.
Antonin Scalia, 59, Associate Justice of the US Supreme Court, born at Trenton, NJ, Mar 11, 1936.
Harold Wilson, 79, statesman, born at Huddersfield, England, Mar 11, 1916.

MARCH 12 — SUNDAY
71st Day — Remaining, 294

AMERICAN SOCIETY OF ASSOCIATION EXECUTIVES MANAGEMENT AND MEETINGS FORUM. Mar 12-15. Nashville, TN. For info: Lorri Lee, American Soc of Assn Executives, 1575 Eye St NW, Washington, DC 20005-1168. Phone: (202) 626-2789.

CALLE OCHO: OPEN HOUSE. Mar 12. Little Havana, Miami, FL. Endless entertainment in this 23-block party with more than 40 musical stages, ethnic foods, dancing and events for the entire family. Billed as the largest block party in the world. Est

☆ Chase's 1995 Calendar of Events ☆ Mar 12-13

attendance: 1,000,000. Sponsor: Kiwanis Club of Little Havana, 1312 SW 27 Ave, 3rd Fl, Miami, FL 33145. Phone: (305) 644-8888.

DAYTONA 200 BY ARAI MOTORCYCLE ROAD RACE. Mar 12. Daytona International Speedway, Daytona Beach, FL. For info: Daytona Intl Speedway, Glenn Barber, PR, Box 2801, Daytona Beach, FL 32120-2801. Phone: (904) 254-6782.

FDR's FIRST FIRESIDE CHAT: ANNIVERSARY. Mar 12. President Franklin Delano Roosevelt made the first of his Sunday evening "fireside chats" to the American people on Mar 12, 1933. Speaking by radio from the White House, he reported rather informally on the economic problems of the nation and on his actions to deal with them.

GIRL SCOUTS OF THE USA FOUNDING: ANNIVERSARY. Mar 12. Juliet Low founded Girl Scouts of the USA on Mar 12, 1912, in Savannah, GA.

LESOTHO: MOSHOESHOE'S DAY. Mar 12. Public holiday.

MAURITIUS: INDEPENDENCE DAY: ANNIVERSARY. Mar 12. National holiday commemorates attainment of independent nationhood (within the British Commonwealth) on Mar 12, 1968.

◆ **PARKER, CHARLES (CHARLIE, BIRD, YARDBIRD): 40th DEATH ANNIVERSARY.** Mar 12, 1955. Charlie Parker, alto saxophonist, composer, band leader and father of the improvisational bebop style of modern jazz, died at New York City. Parker was born Aug 29, 1920, at Kansas City, KS.

PIERCE, JANE MEANS APPLETON: BIRTH ANNIVERSARY. Mar 12. Wife of Franklin Pierce, 14th president of the US. Born at Hampton, NH, Mar 12, 1806. Died Dec 2, 1863.

"SEZ WHO?" FOURPLAY! MARCH MADNESS. Mar 12-Apr 3. Hooters, Chicago, IL. 2-player teams compete against one another, trying to complete humorous, provocative and otherwise memorable quotes made by basketball personalities. Annually, from the Sunday in March when NCAA Tournament field is announced through Monday, 22 days later, when the championship game is played. Est attendance: 1,000. For info: Rich Bysina, Exec Producer, 551 Rosevelt Rd, Ste 230, Glen Ellyn, IL 60137. Phone: (708) 627-2540.

SPACE MILESTONE: SOYUZ T-4 (USSR). Mar 12. Launched Mar 12, 1981. Two cosmonauts (V. Kovalyonok and V. Savinykh) docked at Salyut 6 space station (in orbit since Sept 29, 1977) on Mar 13. Returned to Earth May 26, after 75 days in space.

SUN YAT-SEN: 70th DEATH ANNIVERSARY. Mar 12, 1925. The heroic leader of China's 1911 revolution is remembered on the anniversary of his death at Peking, China. Observed as Arbor Day in Taiwan.

WAGONS, WIRES & RAILS ALONG THE OREGON TRAIL: EXHIBITION. Mar 12-Oct 31. Various venues, Omaha, NE. Exhibit links the wagon trains of the 1840s-50s to the telegraph and railway system that sprang up across Nebraska in the 1860s-70s along the Oregon-California Trail routes and shows how the communications highway tying our nation together was thus born. For info: Historical Soc of Douglas County, PR Dept, PO Box 11398, CAE, Omaha, NE 68111-0398. Phone: (402) 455-9990.

WEBE 108 KIDS FEST. Mar 12. Trumbull Marriott, Trumbull, CT. Giant exposition celebrating kids. Crafts, demonstrations, games and fun for all ages. Est attendance: 10,000. For info: Megan O'Connell, Promo Dir, 2 Lafayette Sq, Bridgeport, CT 06604. Phone: (203) 333-9108.

YAM '95 PROGRAMS FOR YOUTH ART MONTH. Mar 12-Apr 30. Huntsville Museum of Art, Huntsville, AL. A juried selection of the best art works created during the past year by North Alabama area students in grades K-12. Annually, near or during "Youth Art Month" (March). Est attendance: 2,500. For info: Marylyn Coffey or Deborah Taylor, Huntsville Museum of Art, Von Braun Civic Ctr, 700 Monroe St, Huntsville, AL 35801. Phone: (205) 535-4350.

BIRTHDAYS TODAY

Edward Albee, 67, playwright, born at Washington, DC, Mar 12, 1928.
Kent Conrad, 47, US Senator (D, North Dakota) up for reelection in Nov '94, born at Bismark, ND, Mar 12, 1948.
Barbara Feldon, 54, actress (played 99 on "Get Smart"; Smile), born at Pittsburgh, PA, Mar 12, 1941.
Marlon Jackson, 38, singer (Jackson 5), born at Gary, IN, Mar 12, 1957.
Al (Alwin) Jarreau, 55, singer, songwriter, born at Milwaukee, WI, Mar 12, 1940.
Liza Minnelli, 49, singer, actress (Oscar for Cabaret; The Sterile Cuckoo, Come Saturday Morning) born Los Angeles, CA, Mar 12, 1946.
Dale Murphy, 39, former baseball player, born at Portland, OR, Mar 12, 1956.
Wally Schirra, 72, former astronaut, born at Hackensack, NJ, Mar 12, 1923.
Darryl Strawberry, 33, baseball player, born Los Angeles, CA, Mar 12, 1962.
James Taylor, 47, singer, musician ("You've Got a Friend," "Handy Man"), born Boston, MA, Mar 12, 1948.
John D. Waihee, III, 69, Governor of Hawaii (D), born at Honolulu, HI, Mar 12, 1926.
Andrew Young, 63, civil rights leader, former mayor of Atlanta, GA, born at New Orleans, LA, Mar 12, 1932.

MARCH 13 — MONDAY
72nd Day — Remaining, 293

AMERICAN BOWLING CONGRESS CONVENTION (WITH HALL OF FAME INDUCTION CEREMONIES). Mar 13-18. Mobile, AL. Local, state and national bowling leaders gather to decide the rules of the game in a democratic setting. The week features board of directors' meetings, special seminars, dinners honoring top leaders, Hall of Fame induction ceremonies and workshops. Est attendance: 5,000. For info: American Bowling Congress Conv, 5301 S 76th St, Greendale, WI 53129-0500. Phone: (414) 421-6400.

◆ **ANTHONY, SUSAN BROWNELL: DEATH ANNIVERSARY.** Mar 13, 1906. American reformer and militant advocate of women's suffrage Susan B. Anthony was arrested and fined in 1872 for voting—a criminal act if done by a woman! She was the first American woman to have her likeness on coinage (1979, Susan B. Anthony dollar). Born at Adams, MA, Feb 15, 1820, died at Rochester, NY.

ARAB OIL EMBARGO LIFTED: ANNIVERSARY. Mar 13. The oil-producing Arab countries agreed to lift their five-month embargo on petroleum sales to the US Mar 13, 1974. During the embargo prices went up 300 percent and a ban was imposed on Sunday gasoline sales. The embargo was in retaliation for US support of Israel during the October 1973 Middle-East War.

CAMP FIRE BOYS AND GIRLS BIRTHDAY WEEK. Mar 13-19. To celebrate the 85th anniversary of Camp Fire Boys and Girls, (founded in 1910 as Camp Fire Girls). Sponsor: Camp Fire Boys and Girls, 4601 Madison Ave, Kansas City, MO 64112. Phone: (816) 756-1950.

CANADA: COMMONWEALTH DAY. Mar 13. Second Monday in March is observed as Commonwealth Day but is not a public holiday.

CLARENCE DARROW DEATH COMMEMORATION. Mar 13. Jackson Park, Chicago, IL. Annually, on the anniversary of his death, a wreath is tossed from the Jackson Park Clarence Darrow Bridge, named in honor of the famed lawyer and civil libertarian. At 10 AM. Est attendance: 100. For info: Herb Kraus, 875 N Michigan, Ste 2250, Chicago, IL 60611. Phone: (312) 266-7800.

DELMONICO, LORENZO: BIRTH ANNIVERSARY. Mar 13. Famed restaurateur and gastronomic authority. Born Mar 13, 1813, at Marengo, Switzerland. Operated a number of restaurants in New York, NY. Died there Sept 3, 1881.

FILLMORE, ABIGAIL POWERS: BIRTH ANNIVERSARY. Mar 13. First wife of Millard Fillmore, 13th president of the US. Born at Stillwater, NY, Mar 13, 1798. Died Mar 30, 1853. It is said that the White House was without any books until Abigail Fillmore, formerly a teacher, made a room on the second floor into a library. Within a year, Congress appropriated $250 for the president to spend on library books for the White House.

GOOD SAMARITAN INVOLVEMENT DAY. Mar 13. A day to emphasize the importance of unselfish aid to those who need it. Recognized on the anniversary of the killing of Catherine (Kitty) Genovese, Mar 13, 1964, in the Kew Gardens community, Queens, NY. Reportedly no less than 38 of her neighbors, not wanting "to get involved," witnessed and watched for nearly 30 minutes as the fleeing girl was pursued and repeatedly stabbed by her 29-year-old attacker.

HUBBARD, L. RON: BIRTH ANNIVERSARY. Mar 13. Lafayette Ronald Hubbard, science fiction writer, recluse and founder of the Church of Scientology. Born at Tilden, NE, Mar 13, 1911. His best known book was *Dianetics: The Modern Science of Mental Health*. Died in San Luis Obispo County, CA, on Jan 24, 1986.

LOWELL, PERCIVAL: BIRTH ANNIVERSARY. Mar 13. American astronomer, founder of the Lowell Observatory at Flagstaff, AZ. Born at Boston, MA, Mar 13, 1855. Died at Flagstaff on Nov 12, 1916. Lowell was initiator of the search that resulted (25 years after the search began and 14 years after his death) in discovery of the planet Pluto. The discovery was announced on Lowell's birthday, Mar 13, 1930, by the Lowell Observatory.

MAPLE SUGARING FESTIVAL. Mar 13. The INN on Lake Waramaug, New Preston, CT. Maple syrup-making demonstrations, free samples of syrup over ice cream, festival king and queen selection, maple syrup and candies for sale. A waffle and maple syrup brunch is featured at the INN. For info: Nancy Conant, The INN on Lake Waramaug, 107 N Shore Rd, New Preston, CT 06777. Phone: (203) 868-0563.

MOST BORING FILM AWARDS. Mar 13. Tenth annual Awards for the most boring films of the previous year. Categories include Comedy, Action and Big Stars-Big Flops. For info: Alan Caruba, Founder, The Boring Inst, PO Box 40, Maplewood, NJ 07040. Phone: (201) 763-6392.

NATIONAL MANUFACTURING WEEK. Mar 13-16. Chicago, IL. More than 1,800 exhibits and hundreds of hours of conference seminars on what's new in manufacturing. Est attendance: 67,000. For info: Natl Assn of Manufacturers, 1331 Pennsylvania Ave NW, Ste 1500N, Washington, DC 20004. Phone: (202) 637-3047.

PLANET URANUS DISCOVERY: ANNIVERSARY. Mar 13. German-born English astronomer Sir William Herschel discovered the 7th planet from the sun, Uranus, on Mar 13, 1781.

PRIESTLY, JOSEPH: BIRTH ANNIVERSARY. Mar 13. English clergyman and scientist, discoverer of oxygen. Born at Fieldhead, England, Mar 13, 1733. He and his family narrowly escaped an angry mob attacking their home because of his religious and political views. They moved to the US in 1794. Died at Northumberland, PA, Feb 6, 1804.

ST. AUBIN, HELEN "CALLAGHAN" CANDAELE: BIRTH ANNIVERSARY. Mar 13. Helen Candaele St. Aubin, known as Helen Callaghan during her baseball days, was born Mar 13, 1929, at Vancouver, British Columbia, Canada. St. Aubin and her sister, Margaret Maxwell, were recruited for the all-American Girls Professional Baseball League, which flourished in the 1940s when many major league players were off fighting World War II. She first played at age 15 for the Minneapolis Millerettes, an expansion team that moved to Indiana and became the Ft. Wayne Daisies. The left-handed outfielder spent five years with the Daisies. For the 1945 season she led the league with a .2999 average and 24 extra base hits. In 1946 she stole 114 bases in 111 games. Her son Kelly Candaele's documentary on the women's 1940s baseball league inspired the film *A League of Their Own*. Helen St. Aubin, who was known as the "Ted Williams of women's baseball," died Dec 8, 1992, at Santa Barbara, CA.

UNITED KINGDOM: COMMONWEALTH DAY. Mar 13. Replaces Empire Day observance recognized until 1958. Observed on second Monday in March.

BIRTHDAYS TODAY

Walter Annenberg, 87, publisher, born at Milwaukee, WI, Mar 13, 1908.
Andy Bean, 42, golfer, born at Lafayette, GA, Mar 13, 1953.
Adam Clayton, 35, musician (U2), born at Dublin, Ireland, Mar 13, 1960.
Vance Johnson, 32, football player, born at Trenton, NJ, Mar 13, 1963.
Deborah Raffin, 42, actress ("Foul Play"), born at Los Angeles, CA, Mar 13, 1953.
Neil Sedaka, 56, singer, songwriter ("Breaking Up Is Hard to Do" with Howard Greenfield), born Brooklyn, NY, Mar 13, 1939.

MARCH 14 — TUESDAY
73rd Day — Remaining, 292

ANN ARBOR FILM FESTIVAL. Mar 14-19. Ann Arbor, MI. Independent 16mm film festival in its 33rd year. Genre: experimental, animation, documentary, narrative, avant-garde. No video for pre-entry. Entry deadline: Feb 15, 1995. $8,000 awarded in prizes. Est attendance: 5,000. Sponsor: Ann Arbor Film Festival, Vicki Honeyman, Festival Dir, PO Box 8232, Ann Arbor, MI 48107. Phone: (313) 995-5356.

ASSOCIATION OF AMERICAN GEOGRAPHERS ANNUAL MEETING. Mar 14-18. Chicago, IL. National meeting of members, workshops, paper and poster sessions and field trips. Est attendance: 3,000. Sponsor: Assn of American Geographers, 1710 16th St NW, Washington, DC 20009-3198. Phone: (202) 234-1450.

EINSTEIN, ALBERT: BIRTH ANNIVERSARY. Mar 14. Theoretical physicist best known for his theory of relativity. Born at Ulm, Germany, Mar 14, 1879. Nobel Prize, 1921. Died at Princeton, NJ, Apr 18, 1955.

ENGLAND: CHELTENHAM GOLD CUP MEETING. Mar 14-16. Cheltenham Racecourse, Prestbury, Cheltenham, Gloucestershire. Major national hunt meeting. Est attendance: 120,000. For info: Cheltenham Racecourse, Prestbury Park, Cheltenham, Gloucestershire, England.

JONES, CASEY: BIRTH ANNIVERSARY. Mar 14. Railroad engineer and hero of ballad, whose real name was John Luther Jones. Born near Cayce, KY, Mar 14, 1864. Died in railroad wreck Apr 30, 1900.

March 1995

S	M	T	W	T	F	S
			1	2	3	4
5	6	7	8	9	10	11
12	13	14	15	16	17	18
19	20	21	22	23	24	25
26	27	28	29	30	31	

☆ Chase's 1995 Calendar of Events ☆   Mar 14–15

MOTH-ER DAY. Mar 14. A day set aside to honor moth collectors and specialists. Celebrated in museums or libraries with moth collections. Annually, Mar 14. Sponsor: Puns Corps, c/o Bob Birch, Grand Punscorpion, Box 2364, Falls Church, VA 22042-0364. Phone: (703) 533-3668.

NAIA MEN'S DIVISION I NATIONAL BASKETBALL CHAMPIONSHIP. Mar 14–20. (Tentative.) Mabee Center, Tulsa, OK. 58th annual tournament. Fax: (918) 494-8841. Est attendance: 45,000. For info: Natl Assn of Intercollegiate Athletics, 6120 S Yale Ave, Ste 1450, Tulsa, OK 74136. Phone: (918) 494-8828.

NURSING CONFERENCE ON PEDIATRIC PRIMARY CARE. Mar 14–18. Nashville, TN. Provides continuing education offerings on clinical practice, professional development, and legislative issues relevant to nursing practice. Est attendance: 950. Sponsor: Natl Assn of Pediatric Nurse Associates and Practitioners, 1101 Kings Hwy N, Ste 206, Cherry Hill, NJ 08034. Phone: (609) 256-2300.

PIZZA EXPO. Mar 14–17. Las Vegas, NV. International trade show for pizza owners, operators and industry. Site of the *Pizzaaahlympics*, the world championship contests for making the fastest, largest and highest pizza, as well as innovative pizza-throwing routines choreographed and set to music. *Pizza Festiva* recipe winners will also be announced (see Oct 1). Sponsor: *Pizza Today* Magazine and the Natl Assn of Pizza Operators. Est attendance: 10,000. For info: Gerry Durnell, Exec Dir, PO Box 1347, New Albany, IN 47151. Phone: (812) 949-0909.

TAYLOR, LUCY HOBBS: BIRTH ANNIVERSARY. Mar 14. Lucy Beaman Hobbs, first woman in America to receive a degree in dentistry (Ohio College of Dental Surgery, 1866) or to be admitted to membership in a state dental association. Born in New York state, Mar 14, 1833. In 1867 she married James M. Taylor, a painter for a railroad, who also became a dentist (after she instructed him in the essentials). Active women's rights advocate. Died at Lawrence, KS, Oct 3, 1910.

TRIUMPH OF AGRICULTURE EXPOSITION FARM AND RANCH MACHINERY SHOW. Mar 14–15. Civic Auditorium, Omaha, NE. For info: Mid-America Expositions, Inc, 666 Farnam Bldg, Omaha, NE 68102. Phone: (402) 346-8003.

BIRTHDAYS TODAY

Frank Borman, 67, former astronaut, airline executive, born at Gary, IN, Mar 14, 1928.
Michael Caine (Maurice Joseph Micklewhite), 62, actor (*Alfie, The Ipcress File, Sleuth*), born at London, England, Mar 14, 1933.
Billy Crystal, 48, actor ("Soap," *When Harry Met Sally, City Slickers*), born Long Beach, NY, Mar 14, 1947.

Quincy Jones, 62, composer, born at Chicago, IL, Mar 14, 1933.
Hank Ketcham, 75, cartoonist ("Dennis the Menace"), born at Seattle, WA, Mar 14, 1920.
Kirby Puckett, 34, baseball player, born at Chicago, IL, Mar 14, 1961.
Rita Tushingham, 53, actress (*Dr. Zhivago, A Taste of Honey*), born at Liverpool, England, Mar 14, 1942.

MARCH 15 — WEDNESDAY
74th Day — Remaining, 291

BATTLE OF GUILFORD COURTHOUSE OBSERVANCE. Mar 15. Guilford Courthouse National Military Park, Greensboro, NC. A program to observe the 214th anniversary of the Battle of Guilford Courthouse, the largest Revolutionary War battle in North Carolina, lasting from noon until 3 PM, the hours of the original battle. Annually, Mar 15. Est attendance: 100. For info: Superintendent, Guilford Courthouse NMP, 2332 New Garden Rd, Greensboro, NC 27410-2355. Phone: (910) 288-1776.

BUZZARDS' DAY. Mar 15. Hinckley, OH. Tradition says that on this day the buzzards (also known as *turkey vultures* or *carrion crows*) return to Hinckley, OH, from their winter quarters in the Great Smoky Mountains to rear their young. Local celebration usually held on following Sunday.

C.M. RUSSELL AUCTION OF ORIGINAL WESTERN ART. Mar 15–18. Heritage Inn, Great Falls, MT. Western art event features auctions, seminars, 102 display/sale rooms with thousands of art works for sale, entertainment, elegant receptions and the best of Montana hospitality. Annually, the third Thursday–Saturday in March. Est attendance: 6,000. For info: DeAnn Andre, Exec Dir, Great Falls Adv Fed, PO Box 634, Great Falls, MT 59403-0634. Phone: (406) 761-6453.

IDES OF MARCH. Mar 15. Julius Caesar assassinated this day in 44 BC.

JACKSON, ANDREW: BIRTH ANNIVERSARY. Mar 15. Seventh president of the US (Mar 4, 1829–Mar 3, 1837). Born in a log cabin at Waxhaw, SC, Mar 15, 1767. Had reputation as a brawler, reportedly participant in countless duels (in at least one of which his opponent was mortally wounded). Married the same woman (Rachel Robards) twice (1791 and 1794)—once before and once after her divorce. Public reception for his first inauguration, attended by 20,000 persons, was a rowdy affair that left ruined rugs and furnishings and damage in the thousands of dollars at the White House. Jackson was the first president since George Washington who had not attended college. Died at Nashville, TN, June 8, 1845.

◆ **MAINE: ADMISSION DAY: 175th ANNIVERSARY.** Mar 15. Became 23rd state on this day in 1820.

MARY STRATTON'S BIRTHDAY. Mar 15. Pewabic Pottery, Detroit, MI. Open house celebration in observance of birth of Mary Chase Perry (later Stratton) on Mar 15, 1867, at Hancock, MI. Stratton founded Pewabic Pottery, now a National Historic Landmark, on Oct 8, 1903, and became renowned for her handcrafted tiles and vessels and her unique iridescent glazes. She died at Detroit, MI, in 1961, but her legacy is carried on by the Pottery, now a nonprofit production, education and exhibition center for ceramic arts. Pewabic tile installations can be seen in private residences and public buildings throughout the country. Annually, Mar 15, unless that date is a Sunday (in which case it is celebrated on the following Monday). Est attendance: 300. For info: Dir of Development and Comm, Pewabic Pottery, 10125 E Jefferson, Detroit, MI 48214. Phone: (313) 822-0954.

NAIA WOMEN'S DIVISION I BASKETBALL CHAMPIONSHIP TOURNAMENT. Mar 15–21. Oman Arena, Jackson, TN. 32-team field competes for the national championship. Fax: (918) 494-8841. 15th annual competition. Est attendance: 38,000. For info: Natl Assn of Intercollegiate Athletics, 6120 S Yale Ave, Ste 1450, Tulsa, OK 74136. Phone: (918) 494-8828.

Mar 15–16 ☆ Chase's 1995 Calendar of Events ☆

RIO GRANDE VALLEY LIVESTOCK SHOW. Mar 15–19. Mercedes, TX. For the youth in the four counties in the valley to exhibit their projects. Est attendance: 170,000. For info: Jack D. Schwarz, Rio Grande Valley Livestock Show, Inc, Box 867, Mercedes, TX 78570. Phone: (210) 565-2456.

SPRINGTIME TALLAHASSEE. Mar 15–Apr 1. Tallahassee, FL. 1995 theme: "Celebrating Florida's Sesquicentennial Anniversary (150 Years of Statehood!) This 27th annual festival celebrates Florida's capital city, Tallahassee, "the city where Spring begins!" Mar 15: A.J. Breakfast in the Park; Mar 18: Springtacular: A Children's parade and Family Festival; Apr 1: Grand Parade and Jubilee on the streets of downtown Tallahassee. Fax: (904) 224-0833. Sponsored in part by Coca-Cola, Tallahassee Democrat, AMSOUTH Bank. Est attendance: 250,000. For info: Holly Floyd, Exec Dir, Springtime Tallahassee, PO Box 1465, Tallahassee, FL 32302-1465. Phone: (904) 224-1373.

TA'ANIT ESTHER (FAST OF ESTHER). Mar 15. Hebrew calendar date: Adar 13, 5755. Commemorates Queen Esther's fast, in the 6th century BC, to save the Jews of ancient Persia. Ordinarily observed Adar 13, the Fast of Esther is observed on the previous Thursday (Adar 11) when Adar 13 is a Sabbath.

TRUE CONFESSIONS DAY. Mar 15. Confession is good for the soul. Go into work today and tell all. If you plan to stay home, make an appointment with your mirror. Phone (212) 388-8673 or (717) 274-8451. Sponsor: Wellness Permission League, Tom and Ruth Roy, 2105 Water St, Lebanon, PA 17046.

WINTER EQUESTRIAN FESTIVAL. Mar 15–Apr 1. Bob Thomas Equestrian Center, Florida State Fairgrounds, Tampa, FL. Top riders compete in this 36th annual festival. Known as the "world's richest hunter-jumper horse show circuit," with more than $1 million in prize money. Features 10 weeks of qualifying competition which begin at West Palm Beach and continue through the final three weeks at the FL Expo Park in Tampa. The tournament is divided as follows: Suncoast Internationale Mar 15–19, Tampa Bay Classic Mar 22–26, Tournament of Champions Mar 28–Apr 1. Grand finale of the festival is the $100,000 Budweiser American Invitational Apr 1, 7 PM, in Tampa Stadium, an invitational world-class jumping competition (the first jewel in the Triple Crown of Show Jumping), also featuring band performances, parades and dressage demonstrations. Fax: (813) 229-6616. For info: Tampa/Hillsborough Conv and Visitors Assn, Inc, 111 Madison St, Ste 1010, Tampa, FL 33601-0519. Phone: (800) 448-2672.

BIRTHDAYS TODAY

Harold Douglas Baines, 36, baseball player, born at Easton, MD, Mar 15, 1959.
Alan Bean, 63, former astronaut, born at Wheeler, TX, Mar 15, 1932.
Terry Cummings, 34, basketball player, born at Chicago, IL, Mar 15, 1961.
Mark Green, 50, lawyer, author, social activist (*Ronald Reagan's Reign of Error*), born Brooklyn, NY, Mar 15, 1945.
Mickey Hatcher (Michael Vaughn Hatcher, Jr), 40, baseball player, born at Cleveland, OH, Mar 15, 1955.
Judd Hirsch, 60, actor (Emmy for "Taxi"; *Ordinary People*), born at New York, NY, Mar 15, 1935.
Mike Love, 54, singer, musician (member Beach Boys), born at Los Angeles, CA, Mar 15, 1941.
Federico Pena, 48, US Secretary of Transportation in Clinton administration, born at Laredo, TX, Mar 15, 1947.
Dee Snider, 40, singer, composer, born at Massapequa, NY, Mar 15, 1955.
Sly Stone (Sylvester Stewart), 51, singer, musician (funk/rock leader Sly & The Family Stone), born at Dallas, TX, Mar 15, 1944.

MARCH 16 — THURSDAY
75th Day — Remaining, 290

BLACK PRESS DAY: ANNIVERSARY OF THE FIRST BLACK NEWSPAPER. Mar 16. Anniversary of the founding of the first Black newspaper in the US, *Freedom's Journal*, on Varick St in New York, NY, on Mar 16, 1827. Observed as "Black Press Day in New York City" by mayoral proclamation in 1977.

CLYMER, GEORGE: BIRTH ANNIVERSARY. Mar 16. Signer of the Declaration of Independence and of the US Constitution. Born Mar 16, 1739. Died at Philadelphia, PA, Jan 24, 1813.

CURLEW DAY. Mar 16. Traditional arrival date for the long-billed curlew at the Umatilla (Oregon) Natl Wildlife Refuge. More than 500 of the long-billed curlews have been reported at this location during their nesting season.

ENGLAND: CRUFTS DOG SHOW. Mar 16–19. National Exhibition Centre, Birmingham, West Midlands. Championship dog show in which more than 160 breeds compete for the Best in Show title. Est attendance: 90,000. For info: Crufts Office, The Kennel Club, 1-5 Clarges St, London, England W1 8AB.

FESTIVAL OF HOUSES AND GARDENS. Mar 16–Apr 15. Charleston, SC. Held annually since 1947. Provides a rare opportunity to explore the private dwellings and gardens of historic Charleston. Est attendance: 70,000. Sponsor: Historic Charleston Fdtn, PO Box 1120, Charleston, SC 29402. Phone: (803) 723-1623.

FREEDOM OF INFORMATION DAY. Mar 16. Observed by organizations interested in freedom of government information, freedom of the press, and President James Madison's contribution to these freedoms through his statements and introduction of the Bill of Rights in the first Congress. For info: American Library Assn, 50 E Huron St, Chicago, IL 60611. Phone: (312) 280-5041.

GODDARD DAY. Mar 16. Commemorates first liquid-fuel-powered rocket flight on this day, 1926, devised by Robert Hutchings Goddard (1882–1945) at Auburn, MA.

GRAND CENTER SPORT, FISHING AND RV SHOW. Mar 16–19. Grand Center, Grand Rapids, MI. This event brings together buyers and sellers of fishing boats and equipment, RVs, campers and their accessories, as well as other outdoor sporting goods. US and Canadian hunting and fishing trips and other vacation travel destinations are featured. All aspects of fishing, including tackle boats, seminars, demonstrations and displays are emphasized. Est attendance: 45,000. For info: Henri Boucher, ShowSpan, Inc, 1400 28th St SW, Grand Rapids, MI 49509. Phone: (616) 530-1919.

MADISON, JAMES: BIRTH ANNIVERSARY. Mar 16. Fourth president of the US (Mar 4, 1809–Mar 3, 1817). Born at Port Conway, VA, Mar 16, 1751. He was president when British forces invaded Washington, DC, requiring Madison and other

March 1995

S	M	T	W	T	F	S
			1	2	3	4
5	6	7	8	9	10	11
12	13	14	15	16	17	18
19	20	21	22	23	24	25
26	27	28	29	30	31	

high officials to flee while the British burned the Capitol, the president's residence and most other public buildings (Aug 24–25, 1814). Died at Montpelier, VA, June 28, 1836.

MOON PHASE: FULL MOON. Mar 16. Moon enters Full Moon phase at 8:26 PM, EST.

MY LAI MASSACRE: ANNIVERSARY. Mar 16. Most publicized atrocity of Vietnam War. On Mar 16, 1968, according to findings of US Army's Peers investigating team, approximately 300 noncombatant Vietnamese villagers (at My Lai and Mykhe, near the South China Sea) were killed by infantrymen of the American Division.

NCAA WOMEN'S DIVISION I SWIMMING AND DIVING CHAMPIONSHIPS. Mar 16–18. University of Texas at Austin, Austin, TX. Est attendance: 3,000. For info: NCAA, 6201 College Blvd, Overland Park, KS 66211-2422. Phone: (913) 339-1906.

NIXON, THELMA CATHERINE PATRICIA RYAN: BIRTH ANNIVERSARY. Mar 16. Wife of Richard Milhous Nixon, 37th president of the US. Born at Ely, NV, Mar 16, 1912. Died at Park Ridge, NJ, June 22, 1993.

NJCAA DIVISION II NATIONAL BASKETBALL FINALS. Mar 16–18. Danville, IL. Junior College Division II national basketball finals tournament. Est attendance: 8,000. For info: Jeanie Cooke, Ex Dir, Danville Area Conv/Visitors Bureau, PO Box 992, Danville, IL 61834. Phone: (217) 442-2096.

OYSTER FESTIVAL. Mar 16–19. Point Cadet Plaza, Biloxi, MS. The festival features an oyster shucking contest, cooking contest, musical entertainment and many games for the kids. Est attendance: 11,000. For info: Boys and Girls Clubs of the Gulf Coast, Inc, PO Box 688, Biloxi, MS 39533. Phone: (601) 374-2330.

PURIM. Mar 16. Hebrew calendar date: Adar 14, 5755. Feasts, gifts, charity and the reading of the Book of Esther mark this joyous commemoration of Queen Esther's intervention, in the 6th century BC, to save the Jews of ancient Persia. Haman's plot to exterminate the Jews was thwarted, and he was hanged on the very day he had set for execution of the Jews.

REALLY ROSIE, A MUSICAL. Mar 16–18. Koger Center, Columbia, SC. Book and lyrics written by Maurice Sendak, music by Carole King, designed and directed by Maurice Sendak, presented by the Night Kitchen Theater, a theater company devoted to the development of quality performing arts productions. Est attendance: 6,000. For info: Judy McClendon, PR Librarian, Richland County Public Library, 1431 Assembly St, Columbia, SC 29201. Phone: (803) 929-3440.

ST. PATRICK'S DAY DART TOURNAMENT. Mar 16–19. Battle Creek, MI. 11th annual tournament to celebrate the luck of the Irish by encouraging the fast-growing interest in the US in the sport of darting. Annually, beginning the Thursday on or before St. Patrick's Day. Est attendance: 300. Sponsor: Springbrook Golf Course, Bill Buckner, Owner, 1600 Ave A, Battle Creek, MI 49015. Phone: (616) 965-6512.

ST. URHO'S DAY. Mar 16. Hood River, OR. Join the parade and party to honor the tongue-in-cheek patron saint who drove the grasshoppers out of the vineyards in Finland. For info: Camille Hukari, 3009 Dethman Ridge, Hood River, OR 97031. Phone: (503) 386-5785.

SPACE MILESTONE: *GEMINI 8* (US). Mar 16. Executed (with *Agena*) first docking of orbiting spacecraft. Safe emergency landing after malfunction. Launched Mar 16, 1966.

SPIFFS INTERNATIONAL FOLK FAIR. Mar 16–19. St. Petersburg, FL. An international folk fair with arts, crafts, music, dance and foods of 55 different ethnicities. Exhibitors are members only. Colorful, exciting Parade of Nations. Est attendance: 50,000. Sponsor: The St. Petersburg Intl Folk Fair Soc (SPIFFS), 2201 First Ave N, St. Petersburg, FL 33713. Phone: (813) 327-7999.

SPRING POSTAGE STAMP MEGA EVENT. Mar 16–19. Madison Square Garden, New York, NY. 150 dealers offering stamps, postal history materials for sale. Display of private collections, seminars, lectures, free appraisal service, free admission; highlights US & UN. Est attendance: 11,000. Sponsor: Joseph B. Savarese, Exec VP, American Stamp Dealers Assn., 3 School St, Glen Cove, NY 11542. Phone: (516) 759-7000.

BIRTHDAYS TODAY

Bernardo Bertolucci, 54, filmmaker (*Once Upon a Time in the West, Last Tango in Paris*), born at Parma, Italy, Mar 16, 1941.

Erik Estrada, 46, actor ("CHIPS," *Honey Boy*), born New York, NY, Mar 16, 1949.

Jerry Lewis, 70, comedian, actor (*My Friend Irma*), director (*The Bellboy*), born at Newark, NJ, Mar 16, 1925.

Daniel Patrick Moynihan, 68, US Senator (D, New York) up for reelection in Nov '94, born at Tulsa, OK, Mar 16, 1927.

MARCH 17 — FRIDAY
76th Day — Remaining, 289

AMERICAN CROSSWORD PUZZLE TOURNAMENT. Mar 17–19. Stamford Marriott Hotel, Stamford, CT. 250 solvers from the US and Canada compete on 8 puzzles during this 18th annual event. The final puzzle is played on giant Plexiglas boards for everyone to watch, and points are awarded for accuracy and speed. Prizes are awarded in 20 categories and the grand prize is $1,000. Annually, in March. Est attendance: 300. For info: Will Shortz, Dir, Amer Crossword Puzzle Tournament, 55 Great Oak Lane, Pleasantville, NY 10570. Phone: (914) 769-9128.

CAMP FIRE BOYS AND GIRLS FOUNDERS DAY. Mar 17. To commemorate the anniversary of the founding of Camp Fire Boys and Girls, and the service given to children and youth across the nation. Sponsor: Camp Fire Boys and Girls, 4601 Madison Ave, Kansas City, MO 64112. Phone: (816) 756-1950.

CITY OF BALTIMORE'S ST. PATRICK'S DAY CELEBRATION. Mar 17. Broadway Market Square, Baltimore, MD. Held in observance of St. Patrick's Day. Program features Irish band, Irish dance group, free balloons and a Leprechaun Look-A-Like Contest. Est attendance: 250. Sponsor: Dept of Rec and Parks, Office of Adventures in Fun, Clarence "Du" Burns Arena, 1301 S Ellwood Ave, Baltimore, MD 21224. Phone: (410) 396-9177.

COLE, NAT "KING" (NATHANIEL ADAMS): BIRTH ANNIVERSARY. Mar 17. Nat "King" Cole was born Mar 17, 1919, at Montgomery, AL, and began his musical career at an early age, playing the piano at age four. His career included many highlights, among which was his role as the first black entertainer to host a national television show. His many songs included "The Christmas Song," "Nature Boy," "Mona Lisa," "Rambling Rose" and "Unforgettable." Although he was dogged by racial discrimination throughout his career, including the cancellation of his television show because opposition from southern white viewers decreased advertising revenue, Cole was criticized by prominent black newspapers for not joining other black entertainers in the civil rights struggle. Cole contributed more than $50,000 to civil rights organizations in response to the criticism. Nat "King" Cole died Feb 25, 1965, at Santa Monica, CA.

ENGLAND: ROYAL SHAKESPEARE THEATRE SEASON. Mar 17–Jan 28. Royal Shakespeare Theatre, Stratford-upon-Avon, Warwickshire. Season of plays performed by world-renowned theatre company. Est attendance: 550,000. For info: Membership Office, Royal Shakespeare Theatre, Stratford-upon-Avon, Warwickshire, England CV37 6BB.

Mar 17 ☆ Chase's 1995 Calendar of Events ☆

ENGLAND: SOUTHERN KNITTING AND NEEDLECRAFT EXHIBITION. Mar 17–19. South West Exhibition Centre, Bristol, Avon. Everything for the hand and machine knitter and the needlecraft enthusiast, fashion shows, lectures, demonstrations. Est attendance: 18,000. For info: Robert Ewin, Dir, Nationwide Exhibitions (UK) Ltd, PO Box 20, Fishponds, Bristol, England BS16 5QU.

EVACUATION DAY. Mar 17. A public holiday in Boston and Suffolk County, MA, celebrates anniversary of the evacuation from Boston of British troops on Mar 17, 1776.

IRELAND: NATIONAL DAY. Mar 17. St. Patrick's Day is observed in the Republic of Ireland as a legal national holiday.

JONES, BOBBY: BIRTH ANNIVERSARY. Mar 17. Golfing great Robert Tyre Jones, Jr, first golfer to win the grand slam (the four major British and American tournaments in one year). Born at Atlanta, GA, on Mar 17, 1902. Died there Dec 18, 1971.

MACON CHERRY BLOSSOM FESTIVAL. Mar 17–26. Macon, GA. The 13th annual Cherry Blossom Festival will feature concerts, tours, exhibits, parades, children's events, hot air balloons, arts and crafts, a street party, fireworks, food, fun and family entertainment. 197,500 Yoshino cherry trees. Est attendance: 450,000. For info: Carolyn Crayton, Macon Cherry Blossom Fest, 794 Cherry St, Macon, GA 31201. Phone: (912) 751-7429.

NORTHEAST GREAT OUTDOORS SHOW. Mar 17–19. Empire State Plaza and Knickerbocker Arena, Albany, NY. 10th annual expo with seminars by professional sportsmen, archery range, Deer Calling Contest, casting pool and lots of exhibitors. Annually, the third weekend in March. Sponsor: Albany Times Union Newspapers, Schenectady Gazette. Est attendance: 30,000. For info: Heather Schlachter, Exec Dir, 8 Wade Rd, Latham, NY 12110. Phone: (518) 783-1333.

NORTHERN IRELAND: ST. PATRICK'S DAY. Mar 17. National Holiday.

NUREYEV, RUDOLF HAMETOVICH: BIRTH ANNIVERSARY. Mar 17. Rudolf Nureyev, one of the most charismatic ballet stars of the 20th century, was born Mar 17, 1938, on a train in southeastern Siberia. Nureyev's defection from the Soviet Union on June 17, 1961, while on tour with the Kirov Ballet, made headlines worldwide. The dancer was known for his ability to combine passion with a high level of perfectionism. His long partnership with Dame Margot Fonteyn of the Royal Ballet was legendary, and he also performed frequently with the Martha Graham Dance Company. During the years Nureyev performed he also choreographed, restaged many classics and served as the Paris Opera Ballet's artistic director. Nureyev died Jan 6, 1993, at Levallois, France, a suburb of Paris.

PARKER, GEORGE: DEATH ANNIVERSARY. Mar 17. George Parker, the second Earl of Macclesfield, was born in 1697 (exact date unknown). The eminent English astronomer was president of the Royal Society from 1752 until his death on Mar 17, 1764. He was one of the principal authors of the Bill for Regulating the Commencement of the Year (British Calendar Act of 1751), which was introduced in Parliament by Lord Chesterfield. That act caused the adoption, in 1752, of the "New Style" Gregorian calendar, which is still in use today. Parker died at Shirburn Castle, England.

POLO: USPA GOLD CUP. Mar 17–Apr 2. Wellington, FL. Est attendance: 10,000. For info: Palm Beach Polo and Country Club, 13420 South Shore Blvd, Wellington, FL 33414. Phone: (407) 798-7634.

March 1995

S	M	T	W	T	F	S
			1	2	3	4
5	6	7	8	9	10	11
12	13	14	15	16	17	18
19	20	21	22	23	24	25
26	27	28	29	30	31	

RUSTIN, BAYARD: BIRTH ANNIVERSARY. Mar 17. Black pacifist and civil rights leader, Bayard Rustin was an organizer and participant in many of the great social protest marches—for jobs, freedom, nuclear disarmament. He was arrested and imprisoned more than 20 times for his civil rights and pacifist activities. Born at West Chester, PA, on Mar 17, 1910, Rustin died at New York, NY, Aug 24, 1987.

ST. PATRICK'S DAY. Mar 17. Commemorates the patron saint of Ireland, Bishop Patrick (AD 389–461) who, about AD 432, left his home in the Severn Valley, England, and introduced Christianity into Ireland. Feast Day in the Roman Catholic Church. A national holiday in Ireland and Northern Ireland, the highlight of a week of festivity there.

ST. PATRICK'S DAY CELEBRATION. Mar 17. 111th St & Third Ave, Huntington, WV. Live entertainment, music and dance appropriate for St. Patrick's Day. Food, games, beverages, concessions and parade. Est attendance: 2,000. For info: Rick Abel, Greater Huntington Parks & Recreation, PO Box 9361, Huntington, WV 25704. Phone: (304) 696-5954.

ST. PATRICK'S DAY CELEBRATION BEARD GROWING CONTEST. Mar 17–19. Shamrock, TX. All men in town must either grow a beard or purchase a permit, and they must begin growing beard after Jan 1, 1995. Celebration also includes Miss Irish Rose Pageant, chili cookoff, rodeo events, and dances. Est attendance: 6,000. For info: Nan L. Reeves, PO Box 588, 121 N Main, Shamrock, TX 79079. Phone: (806) 256-2501.

ST. PATRICK'S DAY PARADE. Mar 17. Roanoke, VA. The parade is participatory with everybody invited to join in. Annually, Mar 17. Sponsors: The Roanoke Special Events Committee and WROV AM & FM. Est attendance: 1,000. For info: E. Laban Johnson, Special Events Coord, City of Roanoke, 210 Reserve Ave SW, Roanoke, VA 24016. Phone: (703) 981-2889.

ST. PATRICK'S DAY PARADE. Mar 17. Fifth Avenue, New York, NY. Held since 1762, the parade of 125,000 begin their two-mile march at 11 AM. It lasts about six hours. Est attendance: 1,000,000. For info: NY Conv and Visitors Bureau, 2 Columbus Circle, New York, NY 10019. Phone: (212) 397-8222.

ST. PETERSBURG FESTIVAL OF STATES (WITH NATIONAL BAND CHAMPIONSHIPS). Mar 17–Apr 9. To salute civic endeavors and highlight the 50 states. National band championships, art show, parades, antique cars, three-day nationally known jazz festival, plus 5 more nights of major concert entertainment and a World War II air show. Est attendance: 600,000. Sponsor: St. Petersburg Festival of States, Box 1731, St. Petersburg, FL 33731. Phone: (813) 898-3654.

SHAMROCK SPORTSFEST & MARATHON. Mar 17–18. Pavilion Convention Center, Virginia Beach, VA. 8K, masters 8K, 5K walk, sports and fitness expo, runner's clinic and children's marathon. Friday night pasta party. Sat night party. FAX: (804) 481-2942. Est attendance: 25,000. For info: Lori Bocrie, 2308 Maple St, Virginia Beach, VA 23451. Phone: (804) 481-5090.

SHAMROCKS FOR DYSTROPHY. Mar 17. Participating retailers nationwide sell "shamrock" certificates to help the Muscular Dystrophy Assn. When a certificate is purchased, the buyer's name is printed on it's face and it is attached to the retailer's wall. Proceeds go to neuromuscular disease research, medical and support services and public health education. Phone: (602) 529-2000, ext 5317. Sponsor: Muscular Dystrophy assn, Jim Brown, Dir Public Affairs, 3300 E Sunrise Dr, Tucson, AZ 85718-3208.

SPACE MILESTONE: *DISCOVERY* (US). Mar 17. Space shuttle *Discovery* went into orbit, carrying five astronauts on mission to put into orbit a $100 million tracking and data relay satellite. Landed at Edwards Air Force Base on Mar 17, 1989.

132

Chase's 1995 Calendar of Events — Mar 17-18

SPACE MILESTONE: VANGUARD 1 (US): ANNIVERSARY. Mar 17. Established "pear shape" of Earth. Three pounds. First solar-powered satellite. Mar 17, 1958.

WINGS OVER THE PLATTE. Mar 17-19. Grand Island, NE. Celebrate the arrival of the world's largest concentration of Sandhill cranes. Each spring up to 500,000 cranes gather along the Platte River during their northward migration. Seminars, tours, films. Est attendance: 900. For info: Grand Island Hall County Conv and Visitors Bureau, Theresa Kuzelka, PO Box 1486, Grand Island, NE 68802. Phone: (800) 658-3178.

WORLD'S SHORTEST ST. PATRICK'S DAY PARADE. Mar 17. Fourth and Buchanan, Maryville, MO. An annual parade featuring floats and marchers, winding down Buchanan Street for less than one-half of a block (1994 parade was 97 ft long). The entire parade route is painted green. The parade will be shortened annually to always set what is believed to be a new world's record. Annually, Mar 17. Est attendance: 600. For info: Bruce Judd, Chair, 422 N Buchanan, St Maryville, MO 64468. Phone: (816) 562-9965.

BIRTHDAYS TODAY

Daniel Ray Ainge, 36, basketball player, born at Eugene, OR, Mar 17, 1959.
Betty Allen, 65, educator at Harlem School of the Arts, former opera and concert singer, born at Campbell, OH, Mar 17, 1930.
Frederick Brisson, 82, producer, born at Copenhagen, Denmark, Mar 17, 1913.
Patrick Duffy, 46, actor ("Man from Atlantis," "Dallas"), born at Townsend, MT, Mar 17, 1949.
Paul Horn, 65, composer, musician, born at New York, NY, Mar 17, 1930.
Rob Lowe, 31, actor (*Saint Elmo's Fire*, *About Last Night*), born Charlottesville, VA, Mar 17, 1964.
Mercedes McCambridge, 77, actress (voice of Satan in *The Exorcist*; Oscar for *All the King's Men*), born Joliet, IL, Mar 17, 1918.
Kurt Russell, 44, actor (*Backdraft*; Emmy nomination for *Elvis*) born at Springfield, MA, Mar 17, 1951.

MARCH 18 — SATURDAY
77th Day — Remaining, 288

ABCs OF GENEALOGY: GETTING STARTED. Mar 18. The Ellen Payne Odom Genealogy Library, Moultrie, GA. A seminar that will help researchers get started doing their family history and explain the "how-tos" of getting started. Registration fee. Continuing Education credit for Abraham Baldwin College, Tifton, GA. Est attendance: 100. Sponsor: Beth Gay, PR, The Ellen Payne Odom Genealogy Library, PO Box 1110, Moultrie, GA 31776-1110. Phone: (912) 985-6540.

ARUBA: FLAG DAY. Mar 18. Aruba national holiday. Display of flags, national music and folkloric events.

ATLANTIQUE CITY ANTIQUE AND COLLECTIBLES EXPOSITION. Mar 18-19. Convention Center, Atlantic City, NJ. Largest indoor antique show in the US. Est attendance: 49,000. For info: Greater Atlantic City Conv and Visitors Bureau, 2314 Pacific Ave, Atlantic City, NJ 08401. Phone: (609) 348-7100.

BERING SEA ICE GOLF CLASSIC. Mar 18. Nome, AK. A 6-hole course played on the frozen Bering Sea. The object is to land the bright-orange ricocheting golf ball into the sunken, flagged coffee cans before losing it among the built-up chunks of ice. Starts promptly at 10 AM at the Breakers Bar. Approx 60 golfers. 12th annual classic. Sponsor: Bering Sea Lions Club, Box 326, Nome, AK 99762. Phone: (907) 443-2494.

CALHOUN, JOHN CALDWELL: BIRTH ANNIVERSARY. Mar 18. American statesman. Born at Abbeville District, SC, Mar 18, 1782. First vice president of the US to resign that office (Dec 28, 1832). Died at Washington, DC, Mar 31, 1850.

CAMP FIRE BOYS AND GIRLS BIRTHDAY SABBATH. Mar 18. A day when Camp Fire Boys and Girls worship together and participate in the services in their churches or temples. Sponsor: Camp Fire Boys and Girls, 4601 Madison Ave, Kansas City, MO 64112. Phone: (816) 756-1950.

CANADA: MAPLE FESTIVAL OF NOVA SCOTIA. Mar 18-Apr 8. (Saturdays only.) Northern and central Nova Scotia. Promotion of the maple industry in Nova Scotia. Est attendance: 5,000. For info: Lorna A. Crowe, RR1, Southampton, Cumberland Co, NS, Canada B0M 1W0.. Phone: (902) 546-2844.

CLEVELAND, GROVER: BIRTH ANNIVERSARY. Mar 18. The 22nd and 24th president of the US. Born at Caldwell, NJ, Mar 18, 1837 (given name was Stephen Grover Cleveland). Terms of office as president: Mar 4, 1885-Mar 3, 1889, and Mar 4, 1893-Mar 3, 1897. He ran for president for the intervening term and received a plurality of votes cast, but failed to win electoral college victory for that term. Only president to serve two nonconsecutive terms. Also the only president to be married in the White House. He married 21-year-old Frances Folsom, his ward. Their daughter, Esther, was the first child of a president to be born in the White House. Died at Princeton, NJ, June 24, 1908.

CONFEDERATE STATES CONGRESS ADJOURNMENT. Mar 18. The Congress of the Confederate States adjourned for the last time on Mar 18, 1865.

DIESEL, RUDOLPH: BIRTH ANNIVERSARY. Mar 18. German engineer and inventor of the Diesel oil-burning internal combustion engine (about 1897). Born at Paris, France, on Mar 18, 1858, Diesel drowned in the English Channel on Sept 29, 1913.

FIRST ELECTRIC RAZOR MARKETED: ANNIVERSARY. Mar 18. The first electric razor was marketed by Shick Inc. on Mar 18, 1931.

MILITARY THROUGH THE AGES. Mar 18-19. Jamestown Settlement, Williamsburg, VA. Reenactment groups depicting soldiers and military encounters throughout history join forces with modern-day veterans and active units to demonstrate camp life, tactics and weaponry. Est attendance: 3,000. For info: Media Relations, Jamestown-Yorktown Fdtn, Box JF, Williamsburg, VA 23187. Phone: (804) 253-4838.

PRODUCE FOR VICTORY: POSTERS ON THE AMERICAN HOME FRONT, 1941-1945. Mar 18-Apr 30. Virginia Air & Space Center, Hampton, VA. See Jan 14 for full description. Call or write the Smithsonian for other dates and venues. For info: Smithsonian Institution Traveling Exhibition Service, 1100 Jefferson Dr SW, Ste 3146, Washington, DC 20560. Phone: (202) 357-2700.

ST. PATRICK'S DAY PARADE: "THE WEARIN' OF THE GREEN." Mar 18. Baton Rouge, LA. Headquarters: Zee Zee Gardens. Includes floats, marching bands, walking groups, bagpipers and more. Largest St. Patrick's Day celebration in area. Street celebration follows with live entertainment. Coordinators: The Mabyn Kean Agency. Sponsored by Southern Beverage Budweiser. Est attendance: 25,000. For info: Baton Rouge Irish Club, 573 Jefferson Hwy, Baton Rouge, LA 70806. Phone: (504) 925-8295.

Mar 18-19 ☆ *Chase's 1995 Calendar of Events* ☆

ST. URHO'S DAY. Mar 18. Finland, MN. Wacky, light-hearted fest salutes the great St. Urho, who rid Finland of grasshoppers; includes spoof beauty pageant, parade, dances, ethnic food and heritage center. Est attendance: 2,000. For info: Bonnie Tikkanen, St. Urho Committee, Box 516, Finland, MN 55603. Phone: (218) 353-7359.

SEA LION SUDS FEST. Mar 18. Curry County Fairgrounds, Gold Beach, OR. Celebration of Oregon's micro-brewed suds with a slew of beers to sample. Fun games, craft booths, food from pizza to seafood and live music and dancing. Est attendance: 500. For info: Gold Beach Chamber of Commerce, 1225 S Ellensburg, #3, Gold Beach, OR 97444. Phone: (800) 525-2334.

SPACE MILESTONE: *VOSKHOD 2* (USSR). Mar 18. Colonel Leonov steps out of capsule for 20 minutes in special space suit. Launched Mar 18, 1965.

SPRING STROLL. Mar 18-19. (Also Mar 25-26.) MainStrasse Village, Covington, KY. The Village welcomes Spring with an Easter Basket Extravaganza! Drawing for two Easter baskets, brimming with delightful gifts for adults and children. No purchase necessary, entry forms available at all participating MainStrasse Village businesses. Free refreshments. Saturday 10-7; Sunday noon-5. Est attendance: 275,000. Sponsor: MainStrasse Village, Cindy Scheidt, 616 Main St, Covington, KY 41011. Phone: (606) 491-0458.

◆ **SUICIDE WEAPON INTRODUCED: 50th ANNIVERSARY.** Mar 18, 1945. The Japanese released mechanized flying bombs piloted by young Japanese men. These suicide bombs, directed against the US aircraft carrier fleet attacking the Japanese fleet in the Kure-Kobe area, inflicted serious damage on the *Enterprise*, *Intrepid* and *Wasp*.

WACO WIND FESTIVAL. Mar 18-19. Waco, TX. A festival of "wind" events—kites, frisbees, and more. Free event open for the whole family. Annually, the third weekend of March. Est attendance: 5,000. For info: Sally Gavlik, Dept of Leisure Services, PO Box 2570, Waco, TX 76702. Phone: (817) 750-5980.

WHO'S IN CHARGE? WORKERS AND MANAGERS IN THE UNITED STATES. Mar 18-Apr 30. Center for Financial Studies, Fairfield, CT. 150 years of conflict and change in worker-manager relations. See Jan 14 for full description. For info: Smithsonian Institution Traveling Exhibition Service, 1100 Jefferson Dr SW, Ste 3146, Washington, DC 20560. Phone: (202) 357-2700.

WILL ROGERS BIRTHPLACE TEXAS LONGHORN CATTLE SALE WITH BARBECUE AND CHILI COOK-OFFS. Mar 18. Will Rogers Living History Ranch, Oologah, OK. Sale of Registered Texas Longhorns from Will Rogers Herd. CASI-sanctioned chili cookoff and IBS-sanctioned barbecue cookoff. For info: Will Rogers Memorial, Joseph Carter, Box 157, Claremore, OK 74018. Phone: (800) 828-9643.

WYATT EARP BIRTHDAY CELEBRATION. Mar 18. Monmouth, IL. Luncheons with Wyatt Earp portrayal and Earp cousins. Birthplace and Pioneer Cemetery Tours. Also music, exhibits. Phone for Chamber of Commerce: (304) 734-3181. Est attendance: 150. For info: Wyatt Earp Birthplace Historic House Museum, c/o 1020 E Detroit Ave, Monmouth, IL 61462. Phone: (309) 734-6419.

BIRTHDAYS TODAY

Bonnie Blair, 31, Olympic gold medal speed skater, born Cornwall, NY, Mar 18, 1964.
Irene Cara, 36, singer ("Fame," "The Dream"), actress (*Ain't Misbehavin'*), born at the Bronx, NY, Mar 18, 1959.
Frederik Willem de Klerk, 59, former President of South Africa, born Johannesburg, South Africa, Mar 18, 1936.
Kevin Dobson, 51, actor ("Kojak," "Knots Landing"), born at New York, NY, Mar 18, 1944.
Peter Graves, 59, actor ("Mission: Impossible," "The Winds of War"), born at Minneapolis, MN, Mar 18, 1936.
George Kander, 68, composer, born at Kansas City, MO, Mar 18, 1927.
Wilson Pickett, 54, singer, songwriter ("If You Need Me," "It's Too Late"), born at Prattville, AL, Mar 18, 1941.
George Plimpton, 68, author (*Paper Lion, Shadow Box*), TV host, editor, born New York, NY, Mar 18, 1927.
John Updike, 63, writer (*Rabbit Run, The Witches of Eastwick*), born at Shillington, PA, Mar 18, 1932.

MARCH 19 — SUNDAY
78th Day — Remaining, 287

AMERICAN CHOCOLATE WEEK. Mar 19-25. To salute and enjoy one of America's favorite flavors—chocolate. Sponsor: Chocolate Manufacturers Assn of the USA, Susan Smith, 7900 Westpark Dr, Ste A-320, McLean, VA 22102. Phone: (703) 790-5011.

AUSTRALIA: CANBERRA DAY. Mar 19. Australian Capital Territory. Public holiday.

BRADFORD, WILLIAM: BIRTH ANNIVERSARY. Mar 19. Pilgrim father, governor of Plymouth Colony. Born at Yorkshire, England, and baptized on Mar 19, 1589. Sailed from Southampton, England, on the *Mayflower* in 1620. Died at Plymouth, MA, on May 9, 1657.

BRYAN, WILLIAM JENNINGS: BIRTH ANNIVERSARY. Mar 19. American political leader, member of Congress, Democratic presidential nominee (1896), "free silver" advocate, assisted in prosecution at Scopes trial, known as "the Silver-Tongued Orator." Born at Salem, IL, Mar 19, 1860. Died at Dayton, TN, July 26, 1925.

CAMP FIRE BOYS AND GIRLS BIRTHDAY SUNDAY. Mar 19. A day when Camp Fire Boys and Girls worship together and participate in the services of their churches or temples. Sponsor: Camp Fire Boys and Girls, 4601 Madison Ave, Kansas City, MO 64112. Phone: (816) 756-1950.

CHILDREN AND HOSPITALS WEEK™. Mar 19-25. The goal of this annual event is to increase awareness among families, schools, local communities and all health care professionals of the special needs of children and their families in health care settings. Phone: (812) 949-0909, ext 301. Sponsor: Assn for the Care of Children's Health, Trish Mclean, Coord, 7910 Woodmont Ave, Ste 300, Bethesda, MD 20814.

KITE FESTIVAL. Mar 19. Gunston Hall Plantation, Lorton, VA. Children through age 15 will be admitted free, when accompanied by an adult, to fly their kites in the field. Puppet show in Ann Mason Building at 1:30 and 2:30. Hands-on activities for children, 11-4 PM. Est attendance: 1,000. For info: Special Events, Gunston Hall, 10709 Gunston Rd, Lorton, VA 22079. Phone: (703) 550-9220.

LIVINGSTONE, DAVID: BIRTH ANNIVERSARY. Mar 19. Scottish physician, missionary and explorer. Born at Blantyre, Scotland, Mar 19, 1813. Subject of a famous search by Henry

	S	M	T	W	T	F	S
March 1995				1	2	3	4
	5	6	7	8	9	10	11
	12	13	14	15	16	17	18
	19	20	21	22	23	24	25
	26	27	28	29	30	31	

M. Stanley, who found him at Ujiji, near Lake Tanganyika in Africa, on Nov 10, 1871. Dr. Livingstone died in Africa, May 1, 1873. See also: "Stanley, Henry Morton: Birth Anniversary" (Jan 28).

MARY TYLER MOORE SHOW: THE FINAL EPISODE: ANNIVERSARY. Mar 19. "Mary Tyler Moore" was the first of a new wave of sitcoms to make it big in the early '70s. It combined good writing, an effective supporting cast and contemporary attitudes. The show centered around the two most important places in Mary Richards's (Mary Tyler Moore) life—the WJM-TV newsroom and her apartment on North Waverly in Minneapolis. At home she shared the ups and downs of life with her friend Rhoda Morgenstern (Valerie Harper) and the manager of her apartment building, Phyllis Lindstrom (Cloris Leachman). At work, as the associate producer (later producer) of "The Six O'Clock News," Mary struggled to function in a man's world. Figuring in her professional life were her irascible boss Lou Grant (Ed Asner), level-headed and soft-hearted newswriter Murray Slaughter (Gavin MacLeod) and self-obsessed, narcissistic anchorman Ted Baxter (Ted Knight). In the last episode the unthinkable happened—everyone in the WJM newsroom except the inept Ted was fired. (Premiered fall of 1970 and ran for 168 episodes, the last being in 1977.)

NATIONAL FREE PAPER WEEK. Mar 19-25. To promote the contributions free circulation community newspapers make to their communities each week. Sponsor: Assn of Free Community Papers, Deirdre Flynn, Member Services Dir, 401 N Michigan, Chicago, IL 60611-4267. Phone: (312) 644-6610.

★ **NATIONAL POISON PREVENTION WEEK.** Mar 19-25. Presidential Proclamation issued each year for the third week of March since 1962. (PL87-319 of Sept 26, 1961.)

NATIONAL POISON PREVENTION WEEK. Mar 19-25. To aid in encouraging the American people to learn of the dangers of accidental poisoning and to take preventive measures against it. Sponsor: Poison Prevention Week Council, Ken Giles, Secy, Box 1543, Washington, DC 20013.

NEW JERSEY RESTAURANT AND HOSPITALITY EXPO. Mar 19-20. Garden State Convention Center, Somerset, NJ. Showcase for more than 300 booths featuring the latest in products and services for the food service industry. Est attendance: 8,000. For info: Diane DiNuzzo, Expo Mgr, NJ Restaurant Assn, 1 Executive Dr, Somerset, NJ 08873. Phone: (908) 302-1800.

PERIGEAN SPRING TIDES. Mar 19-20. Spring tides, the highest possible tides, which occur when New Moon or Full Moon takes place within 24 hours of the moment the Moon is nearest Earth (perigee) in its monthly orbit, on Mar 20, at 8 AM, EST.

◈ **ROGERS, EDITH NOURSE: BIRTH ANNIVERSARY.** Mar 19, 1881. Edith Nourse Rogers was a YMCA and Red Cross volunteer in France during World War I. In 1925 she was elected to the US Congress to fill the vacancy left by the death of her husband. An able legislator, she was reelected to the House of Representatives 17 times and became the first woman to have her name attached to major legislation. She was a major force in the legislation creating the Women's Army Auxiliary Corps (May 14, 1942) during World War II. Rogers was born at Saco, ME, and died Sept 10, 1960, at Boston.

RUSSELL, CHARLES M.: BIRTH ANNIVERSARY. Mar 19, 1864. Born in St. Louis, MO, Charles M. Russell moved to Montana at about age 16 and became a cowboy. Considered one of the greatest Western artists, Russell recorded the life of a cowboy in his art work. He died Oct 26, 1926, at Great Falls, MT. For info: Elizabeth Dear, Curator, C.M. Russell Museum, 400 13th St N, Great Falls, MT 59401. Phone: (406) 727-8787.

ST. JOSEPH'S DAY AND FEAST OF ST. JOSEPH. Mar 19. Holy day in Catholic Church, recognizing spouse of the Virgin Mary, foster father of Jesus, patriarch, patron of Catholic Church.

SIRICA, JOHN JOSEPH: BIRTH ANNIVERSARY. Mar 19. John Sirica, "the Watergate Judge," was born Mar 19, 1904, at Waterbury, CT. During two years of trials and hearings, Sirica relentlessly pushed for the names of those responsible for the June 17, 1972, burglary of the Democratic National Committee headquarters in Washington's Watergate Complex. Sirica's unwavering search for the truth ultimately resulted in the toppling of the Nixon Administration. Judge John Sirica died Aug 15, 1992, at Washington, DC.

SWALLOWS RETURN TO SAN JUAN CAPISTRANO. Mar 19. Traditional date (St. Joseph's Day) for swallows to return to old mission of San Juan Capistrano, CA, since 1776. See also: "St. John of Capistrano: Death Anniversary" (Oct 23).

TAIWAN: BIRTHDAY OF KUAN YIN, GODDESS OF MERCY. Mar 19. Nineteenth day of Second Moon of the lunar calendar, celebrated at Taipei's Lungshan (Dragon Mountain) and other temples.

US STANDARD TIME ACT: ANNIVERSARY. Mar 19. Anniversary of passage by the Congress of the Standard Time Act, Mar 19, 1918, which authorized the Interstate Commerce Commission to establish standard time zones for the US. The Act also established "Daylight-Saving Time," to save fuel and to promote other economies in a country at war. Daylight-saving time first went into operation on Easter Sunday, Mar 31, 1918. The Uniform Time Act of 1966, as amended in 1986, by Public Law 99-359, now governs standard time in the US. See also: "Daylight-Saving Time" (Apr 2).

WARREN, EARL: BIRTH ANNIVERSARY. Mar 19. American jurist, 14th Chief Justice of the US Supreme Court. Born at Los Angeles, CA, Mar 19, 1891. Died at Washington, DC, July 9, 1974.

BIRTHDAYS TODAY

Ursula Andress, 59, actress (*Dr. No, What's New Pussycat?*), born at Bern, Switzerland, Mar 19, 1936.
Glenn Close, 48, actress (*The Big Chill, Fatal Attraction, Sunset Boulevard*), born at Greenwich, CT, Mar 19, 1947.
Ornette Coleman, 65, composer, saxophonist, born at Fort Worth, TX, Mar 19, 1930.
Patrick McGoohan, 67, director, actor ("The Prisoner"), born New York, NY, Mar 19, 1928.
Philip Roth, 62, author (*The Counterlife, Letting Go*), born at Newark, NJ, Mar 19, 1933.
Brent Scowcroft, 70, business executive, consultant, born at Ogden, UT, Mar 19, 1925.
Bruce Willis, 40, actor ("Moonlighting," *In Country*), born at Penn's Grove, NJ, Mar 19, 1955.

MARCH 20 — MONDAY
79th Day — Remaining, 286

DANGEROUS DAN'S ANNUAL COFFEE CUP WASHING. Mar 20. Annual observance to give all early morning people a reason to wash out their coffee mugs at least once a year, whether they need it or not. Annually, the third Monday in March. For info: Dangerous Dan's Morning Madness, WCVL/WIMC Radio, PO Box 603, Crawfordsville, IN 47933. Phone: (317) 362-8200.

EARTH DAY. Mar 20. Day of the Vernal Equinox. In 1979 children rang the UN peace bell in New York at the exact moment of the equinox—when the sun crossed the equator. This is the beginning of spring in the Northern Hemisphere and of autumn in the Southern. Participation also in Paris, Tokyo and other cities. In 1995, the vernal equinox occurs at 9:14 PM, EST.

Mar 20-21 ☆ Chase's 1995 Calendar of Events ☆

GREAT AMERICAN MEATOUT. Mar 20. To publicize animal abuse and other destructive impacts of intensive animal agriculture on human health and natural resources by asking Americans to kick the meat habit and to consider a less violent diet. Toll-free phone: (800) MEATOUT. Fax: (301) 530-5747. Sponsor: Farm Animal Reform Movement, Box 30654, Bethesda, MD 20824. Phone: (301) 530-1737.

IBSEN, HENRIK: BIRTH ANNIVERSARY. Mar 20. Norwegian playwright. Born at Skien, Norway, Mar 20, 1828. Among his best remembered plays: *Peer Gynt, The Pillars of Society, The Wild Duck, An Enemy of the People* and *Hedda Gabler*. Died at Oslo, Norway, May 23, 1906.

NATIONAL AGRICULTURE DAY. Mar 20. A day to honor America's providers of food and fiber and to educate the general public about the US agricultural system. Week of celebration: March 20-26. Annually, the first day of Spring. For info: Agriculture Council of America, 927 15th St NW, Ste 800, Washington, DC 20005. Phone: (202) 682-9200.

NATIONAL AGRICULTURE WEEK. Mar 20-26. To honor America's providers of food and fiber and to educate the general public about the US agricultural system. Annually, the week that includes the first day of spring. For info: Agriculture Council of America, 927 15th St NW, Ste 800, Washington, DC 20005. Phone: (800) 982-4329.

OSTARA. Mar 20. (Also called Alban Eilir.) One of the "Lesser Sabbats" during the Wiccan year, Ostara is a fire and fertility festival that marks the beginning of spring. Annually, on the spring equinox.

PIGEONS RETURN TO THE CITY-COUNTY BUILDING. Mar 20. City-County Bldg, Ft Wayne, IN. Pigeons flock back to Ft Wayne's municipal building after the long winter. Umbrellas are recommended for viewing the event! Est attendance: 750. For info: Barb Richards, Asst Program Dir, WAJI Radio, 347 W Berry, Ste 600, Ft Wayne, IN 46802. Phone: (219) 423-3676.

PROPOSAL DAY!® Mar 20. (Also Sept 23.) Proposal Day honors unmarried adults everywhere who are seeking marriage. Both men and women are encouraged to propose marriage to their true love on Proposal Day. List released today of the 10 currently most eligible singles in the world, according to American Singles. Annually, on the first day of spring (Vernal Equinox). For info: John Michael O'Loughlin, 1333 W Campbell, #125, Richardson, TX 75080. Phone: (214) 721-9975.

SNOWMAN BURNING. Mar 20. Reading of poetry heralding the end of winter and the arrival of spring, followed by sacrifice in effigy, toasts and cheers. Annually, on or near the first day of spring. Est attendance: 300. Sponsor: Lake Superior State University, Sault Ste. Marie, MI 49783. Phone: (906) 635-2315.

SPRING. Mar 20-June 21. In the Northern Hemisphere spring begins today with the vernal equinox, at 9:14 PM, EST. Note that in the Southern Hemisphere today is the beginning of autumn. Sun rises due east and sets due west everywhere on Earth (except near poles) and the daylight length (interval between sunrise and sunset) is virtually the same everywhere: 12 hours, 8 minutes.

SPRING DAY CELEBRATION. Mar 20. Broadway Market Square, Baltimore, MD. To welcome the incoming spring season. Program including a variety of musical and dance entertainment, a Best Decorated Hat Contest and the distribution of free flower and vegetable seed packets to the audience. Est attendance: 250. Sponsor: Dept of Recreation and Parks, Office of Adventures in Fun, Clarence "Du" Burns Arena, 1301 S Ellwood Ave, Baltimore, MD 21224. Phone: (410) 396-9177.

TUNISIA: INDEPENDENCE DAY. Mar 20. Commemorates treaty of Mar 20, 1956, by which France recognized Tunisian autonomy.

BIRTHDAYS TODAY

Holly Hunter, 37, actress (Oscar and Best Actress at Cannes for *The Piano*; *Broadcast News*, *The Firm*), born Conyers, GA, Mar 20, 1958.
William Hurt, 45, actor (*The Accidental Tourist*, *Broadcast News*) born at Washington, DC, Mar 20 1950.
Hal Linden (Harold Lipshitz), 64, actor ("Barney Miller," "Blacke's Magic"), born at the Bronx, NY, Mar 20, 1931.
Bobby Orr, 47, former hockey player, born at Parry Sound, Ontario, Canada, Mar 20, 1948.
Jerry Reed (Jerry Hubbard), 58, singer, songwriter ("When You're Hot, You're Hot"), born Atlanta, GA, Mar 20, 1937.
Carl Reiner, 73, actor ("The Dick Van Dyke Show," "Your Show of Shows"), writer, director, born Bronx, NY, Mar 20, 1922.
Pat Riley, 50, basketball coach and former player, born at Schenectady, NY, Mar 20, 1945.
Fred Rogers, 67, producer, TV personality ("Mr Rogers' Neighborhood"), born at Latrobe, PA, Mar 20, 1928.
Paul Junger Witt, 52, producer, director, born at New York, NY, Mar 20, 1943.

MARCH 21 — TUESDAY
80th Day — Remaining, 285

ARIES, THE RAM. Mar 21-Apr 19. In the astronomical/astrological zodiac, which divides the sun's apparent orbit into 12 segments, the period Mar 21-Apr 19 is identified, traditionally, as the sun sign of Aries, the Ram. The ruling planet is Mars.

BACH, JOHANN SEBASTIAN: BIRTH ANNIVERSARY. Mar 21. Organist and composer, one of the most influential composers in musical history. Born on Mar 21, 1685, at Eisenach, Germany. Died at Leipzig, Germany, on July 28, 1750.

BIRD DAY. Mar 21. Iowa.

INTERNATIONAL ASTROLOGY DAY. Mar 21. To foster a spirit of networking among the astrological community worldwide; to build and enhance ties among individual astrologers and astrological organizations; and to educate the media and the public about the nature of serious astrology. Annually, Mar 21. Sponsor: Assn for Astrological Networking (AFAN), 8306 Wilshire Blvd, Ste 537, Beverly Hills, CA 90211.

IRANIAN NEW YEAR. Mar 21. National celebration for all Iranians, this is the traditional Persian New Year. For info: Mahvash Tafreshi, Librarian, Farmingdale Public Library, 116 Merritts Road, Farmingdale, NY 11735. Phone: (516) 249-9090.

JUAREZ, BENITO PABLO: BIRTH ANNIVERSARY. Mar 21. Anniversary of the birth of Benito Pablo Juarez (1806–1872), born to Zapotec Indian parents and orphaned at an early age. Learned Spanish at age 12, became symbol of liberation and of Mexican resistance to foreign intervention.

LESOTHO: NATIONAL TREE PLANTING DAY. Mar 21. Lesotho, South Africa.

LEWIS, FRANCIS: BIRTH ANNIVERSARY. Mar 21. Signer of the Declaration of Independence. Born in Wales, Mar 21, 1713. Died Dec 31, 1802.

March 1995

S	M	T	W	T	F	S
			1	2	3	4
5	6	7	8	9	10	11
12	13	14	15	16	17	18
19	20	21	22	23	24	25
26	27	28	29	30	31	

MASTER GARDENER DAY. Mar 21. A day to recognize the Master Gardener Program, which was begun by Dr. Dave Gibby in the state of Washington in 1972. Becoming an official Master Gardener involves completing a gardening course and serving an internship of volunteer work on various projects with a Master Gardener. Requirements vary from state to state. An international organization called MaGic was started in 1988. Annually, on the first full day of spring. For info: Jim Arnold, Master Gardener, 543 Wagner St, Ft. Wayne, IN 46805. Phone: (219) 426-9904.

MEMORY DAY. Mar 21. To encourage awareness of traditional memory system using pattern t,d = 1; n = 2; m = 3; r = 4; L = 5; j,ch = 6; k,q,g-hard = 7; f,v = 8; b,p = 9. Study historic examples of the use of the memory system in the writings of Milton, Thomas Gray, Longfellow, Lincoln and others. Annually, Mar 21. Sponsor: Puns Corps, Robert L. Birch, Coord, Box 2364, Falls Church, VA 22042-0364. Phone: (703) 533-3668.

NAMIBIA: INDEPENDENCE DAY. Mar 21. National Day.

NAW-RUZ. Mar 21. Baha'i New Year's Day. Astronomically fixed to commence the year. One of the nine days of the year when Baha'is suspend work. For info: Natl Spiritual Assembly of the US, 1320 Nineteenth St, NW, Ste 701, Washington, DC 20036. Phone: (202) 833-8990.

NORUZ: IRANIAN NEW YEAR. Mar 21. Always the first day of spring, recognizing the rebirth of nature with rituals that have been practiced for more than 3,000 years. The Noruz ceremonies begin two weeks before spring with the germinating of seeds and the traditional cleaning and decorating of homes. The festivities conclude about two weeks later when the sprouted seeds are thrown into the water to symbolize the end of one year and the rebirth of another. Many of these rituals focus on the concepts of Good and Evil, Beginning and End. For info: Yassaman Djalali, Librarian, West Valley Branch Library, 1243 San Tomas Aquino Rd, San Jose, CA 95117. Phone: (408) 244-4766.

"ORIGINAL" WESTERN MASSACHUSETTS HOME SHOW. Mar 21-26. West Springfield, MA. The largest home show in New England with more than 700 booths and a large outside area covering more than six acres of exhibits. Annually, the fourth week of March. Est attendance: 96,000. For info: Edward R. Pedersen Home Builders Assn of Greater Springfield, 260 Worthington St, Ste 203, Springfield, MA 01103. Phone: (413) 733-3126.

◈ **POCAHONTAS (REBECCA ROLFE): DEATH ANNIVERSARY.** Mar 21, 1617. Pocahontas, daughter of Powhatan, leader of the Indian union of Algonkin nations, helped to foster good will between the colonists of the Jamestown settlement and her people. Pocahontas converted to Christianity, was baptized with the name Rebecca, and married John Rolfe Apr 5, 1614. In 1616, she accompanied Rolfe on a trip to his native England, where she was regarded as an overseas "ambassador." Pocahontas's stay in England drew so much attention to the Virginia Company's Jamestown settlement that lotteries were held to help support the colony. Shortly before she was scheduled to return to Jamestown, Pocahontas died of either smallpox or pneumonia.

UNITED NATIONS: INTERNATIONAL DAY FOR THE ELIMINATION OF RACIAL DISCRIMINATION. Mar 21. Initiated by the United Nations General Assembly in 1966 to be observed annually on March 21, the anniversary of the killing of 69 African demonstrators at Sharpeville, South Africa in 1960, as a day to remember "the victims of Sharpeville and those countless others in different parts of the world who have fallen victim to racial injustice" and to promote efforts to eradicate racial discrimination worldwide. Info from: United Nations, Dept of Public Info, New York, NY 10017.

UNITED NATIONS: WEEK OF SOLIDARITY WITH THE PEOPLES STRUGGLING AGAINST RACISM AND RACIAL DISCRIMINATION. Mar 21-27. Annual observance initiated by United Nations General Assembly as part of its program of the Decade for Action to Combat Racism and Racial Discrimination. Info from: United Nations, Dept of Public Info, New York, NY 10017.

BIRTHDAYS TODAY

Ed Broadbent (John Edward Broadbent), 59, Canadian professor and political leader (New Democratic Party), born at Oshawa, Ontario, Canada, Mar 21, 1936.
Matthew Broderick, 33, actor (*War Games, The Freshman, Family Business*), born at New York, NY, Mar 21, 1962.
Peter Brook, 70, theatre director, born at London, England, Mar 21, 1925.
Timothy Dalton, 49, actor (*Centennial, Licence to Kill*), born at Colwyn Bay, Wales, Mar 21, 1946.
Al Freeman, Jr, 61, actor (*A Patch of Blue; Roots: The Next Generation*), born at San Antonio, TX, Mar 21, 1934.
Jay Walter Hilgenberg, 35, football player, born Iowa City, IA, Mar 21, 1960.
Phyllis McGinley, 90, poet, born at Ontario, Canada, Mar 21, 1905.

MARCH 22 — WEDNESDAY
81st Day — Remaining, 284

BONHEUR, ROSA: BIRTH ANNIVERSARY. Mar 22. French landscape painter. Born Mar 22, 1822. Died May 25, 1899.

EQUAL RIGHTS AMENDMENT SENT TO STATES FOR RATIFICATION: ANNIVERSARY. Mar 22. On Mar 22, 1972, the Senate passed the 27th Amendment, prohibiting discrimination on the basis of sex, sending it to the states for ratification. Hawaii led the way as the first state to ratify and by the end of the year, 22 of the required states had ratified it. On Oct 6, 1978, the deadline for ratification was extended to June 30, 1982, by Congress. The amendment still lacked 3 of the required 38 states for ratification. This was the first extension granted since Congress set seven years as the limit for ratification. The amendment failed to achieve ratification as the deadline came and passed and no additional states ratified the measure.

FIRST WOMEN'S COLLEGIATE BASKETBALL GAME: ANNIVERSARY. Mar 22. The first women's collegiate basketball game was played at Smith College in Northampton, MA, on Mar 22, 1893. Senda Berenson, then Smith's director of physical education and "mother of women's basketball," supervised the game, in which Smith's sophomore team beat the freshman team 5-4. For info: Stacey Schmeidel, Dir of Media Relations, Smith College, Office of College Relations, Northampton, MA 01063. Phone: (413) 585-2190.

FUTURE BIRTHDAY OF CAPTAIN JAMES T. KIRK. Mar 22. Riverside, IA. On Mar 22, 1985, the Riverside City Council voted unanimously to declare a spot behind what used to be the town barbershop as the "future birth place" of Captain James T. Kirk, commander of the starship *Enterprise* NCC1701. Gene Roddenberry's book, *The Making of Star Trek*, says Kirk "was born in a small town in the State of Iowa." Through city proclamation, TV, radio, magazine and newspaper interviews, and with a certificate of commendation from Roddenberry, Riverside has become known for this "Future Historical Event." Whereas Captain Kirk's future birthday is Mar 21, 2228, the mayor of the City of Riverside has proclaimed that Mar 22 will be set aside to honor and celebrate the future Captain James T. Kirk's birth in the City of Riverside. Est attendance: 150. Sponsor: Riverside Area Comm Club, Box 55, Riverside, IA 52327. Phone: (319) 648-4808.

Mar 22-23 ☆ Chase's 1995 Calendar of Events ☆

JORDAN: ARAB LEAGUE DAY. Mar 22. National holiday, the Hashemite Kingdom of Jordan.

NATIONAL GOOF-OFF DAY. Mar 22. A day of relaxation and a time to be oneself; a day for some good-humored fun and some good-natured silliness. Everyone needs one special day each year to goof off. Sponsor: Monica A. Dufour, 121 Taylor Ct, Davison, MI 48423-8535. Phone: (810) 658-3147.

NATIONAL SING-OUT DAY™. Mar 22. Break out in song today like they do in the musical. Sing out your words in conversations instead of speaking them. You can even add a few dance steps if you like. Annually, on Stephen Sondheim's birthday. Sponsor: Adrienne Sioux Koopersmith, 1437 W Rosemont, #1W, Chicago, IL 60660. Phone: (312) 743-5341.

SPACE MILESTONE: *SOYUZ 39* (USSR). Mar 22. Launched on Mar 22, 1981, two cosmonauts (V. Dzhanibekov and, from Mongolia, J. Gurragcha) docked at *Salyut 6* space station Mar 23, where they were greeted by the previous team of cosmonauts. Returned to Earth Mar 30.

SPACE MILESTONE: STS-3 (US). Mar 22. Shuttle *Columbia* launched Mar 22, 1982, on third test flight from Kennedy Space Center, FL, with astronauts Jack Lousma and Gordon Fullerton. Landed at White Sands Missile Range, NM, Mar 30.

UNITED NATIONS: WORLD DAY FOR WATER. Mar 22. The General Assembly declared this observance (Res 47/193) to promote public awareness of how water resource development contributes to economic productivity and social well-being.

BIRTHDAYS TODAY

George Benson, 52, singer, guitarist ("On Broadway," "Give Me the Night"), born Pittsburgh, PA, Mar 22, 1943.
Bob Costas (Robert Quinlan), 43, sportscaster, TV host, born at Queens, NY, Mar 22, 1952.
Orrin Grant Hatch, 61, US Senator (R, Utah) up for reelection in Nov '94, born at Pittsburgh, PA, Mar 22, 1934.
Werner Klemperer, 75, actor (Emmy for "Hogan's Heroes"; *Ship of Fools*), born at Cologne, Germany, Mar 22, 1920.
Andrew Lloyd Webber, 47, composer (*Cats, Phantom of the Opera*), born London, England, Mar 22, 1948.
Karl Malden (Mladen Sekulovich), 81, actor (*Streetcar Named Desire*, "The Streets of San Francisco"), born Gary, IN, Mar 22, 1914.
Marcel Marceau, 72, actor, pantomimist (had the only speaking part in *Silent Movie*), born at Strasbourg, France, Mar 22, 1923.
Pat (Marion Gordon) Robertson, 65, TV evangelist, born at Lexington, VA, Mar 22, 1930.
William Shatner, 64, actor (Captain Kirk in "Star Trek"; "T.J. Hooker"), author ("Tek" novels), born Montreal, Que, Canada, Mar 22, 1931.
Stephen Sondheim, 65, composer (*A Little Night Music*), born New York, NY, Mar 22, 1930.
Bill Wendell, 71, announcer ("Tonight," "Late Night with David Letterman"), born at London, England, Mar 22, 1924.

March 1995	S	M	T	W	T	F	S
				1	2	3	4
	5	6	7	8	9	10	11
	12	13	14	15	16	17	18
	19	20	21	22	23	24	25
	26	27	28	29	30	31	

MARCH 23 — THURSDAY
82nd Day — Remaining, 283

AGGIECON. Mar 23-26. Memorial Student Center and Rudder Tower, Texas A&M University, College Station, TX. This 26th annual science fiction convention organized by Texas A&M students features celebrities, artists, writers, art show and auction, movies, lectures and a masquerade ball. Annually, in the spring. Est attendance: 2,000. Sponsor: MSC Cepheid Variable Committee, Student Programs Office, Memorial Student Center, Texas A&M University, College Station, TX 77843-1237.

CLARK, BARNEY: DEATH ANNIVERSARY. Mar 23. Barney Clark died on Mar 23, 1983, after living almost 112 days with an artificial heart. The heart, made of polyurethane plastic and aluminum, was implanted in Clark at the University of Utah Medical Center, Salt Lake City, on Dec 2, 1982. Clark was the first person ever to receive a permanent artificial heart. Born at Provo, UT, on Jan 21, 1921, Clark was 61 at the time of the implantation and 62 when he died.

COLFAX, SCHUYLER: BIRTH ANNIVERSARY. Mar 23. Seventeenth vice president of the US (1869-1873). Born Mar 23, 1823, at New York, NY. Died Jan 13, 1885, at Mankato, MN.

COLLEGIATE SECRETARIES INTERNATIONAL/FUTURE SECRETARIES ASSOCIATION CONFERENCE. Mar 23-25. Bloomington, MN. For info: Professional Secretaries Intl, Dale Shuter, 10502 NW Ambassador Dr, Kansas City, MO 64195-0404. Phone: (816) 891-6600.

LIBERTY DAY: 220th ANNIVERSARY. Mar 23. Anniversary of Patrick Henry's speech for arming the Virginia militia—at St. Johns Church, Richmond, VA, on Mar 23, 1775. "I know not what course others may take, but as for me, give me liberty or give me death."

MOON PHASE: LAST QUARTER. Mar 23. Moon enters Last Quarter phase at 3:10 PM, EST.

NCAA DIVISION I WOMEN'S BASKETBALL CHAMPIONSHIP. Mar 23-Apr 2. Regionals held Mar 23-25. East: Univ of Connecticut; Mideast: Univ of Tennessee-Knoxville; Midwest: Drake Univ; West: Univ of California-LA. Finals at the Target Center in Minneapolis Apr 1-2, hosted by Univ of Minnesota-Twin Cities. For info: Natl Collegiate Athletic Assn, 6201 College Blvd, Overland Park, KS 66211. Phone: (913) 339-1906.

NCAA MEN'S DIVISION I SWIMMING AND DIVING CHAMPIONSHIPS. Mar 23-25. Indiana University Natatorium, Indianapolis, IN. For info: NCAA, 6201 College Blvd, Overland Park, KS 66211-2422. Phone: (913) 339-1906.

NEAR MISS DAY. Mar 23, 1989. A mountain-sized asteroid passed within 500,000 miles of Earth, a very close call according to NASA. Impact would have equalled strength of 40,000 hydrogen bombs, created a crater the size of the District of Columbia and devastated everything for 100 miles in all directions.

NEW ZEALAND: OTAGO AND SOUTHLAND PROVINCIAL ANNIVERSARY. Mar 23. In addition to the statutory public holidays of New Zealand, there is in each provincial district a holiday for the provincial anniversary. This is observed in Otago and Southland on Mar 23.

PAKISTAN: REPUBLIC DAY. Mar 23. National holiday. On this day in 1940 the All-India-Muslim league adopted resolution calling for a Muslim homeland. On the same day in 1956 Pakistan declared itself a republic.

RALLY FOR DECENCY: ANNIVERSARY. Mar 23. Anita Bryant, Jackie Gleason and Kate Smith rallied with 30,000 others in Miami on this day in 1969 in reaction to Jim Morrison's arrest for indecent exposure.

SCOTLAND: EDINBURGH INTERNATIONAL SCIENCE FESTIVAL. Mar 23-Apr 17. More than 450 events at more than 50 venues, Edinburgh. Includes workshops, talks, films and exhibitions. Est attendance: 250,000. For info: Edinburgh

Science Festival Ltd, 1 Broughton Market, Edinburgh, Scotland EH3 6NU.

SCRATCH ANKLE '95. Mar 23. Milton, FL. Festival celebrating Milton's history. Est attendance: 17,500. For info: Donna Adams, Box 909, Milton, FL 32572.

UNITED NATIONS: WORLD METEOROLOGICAL DAY. Mar 23. An international day observed by meteorological services throughout the world and by the organizations of the UN system. "Natural Disaster Reduction: How Meteorological and Hydrological Services Can Help." Info from: United Nations, Dept of Public Info, New York, NY 10017.

BIRTHDAYS TODAY

Roger Bannister, 66, former track athlete, physician, born at Harrow, England, Mar 23, 1929.
Teresa Ganzel, 38, actress, born at Toledo, OH, Mar 23, 1957.
Chaka Khan (Yvette Marie Stevens), 42, singer ("Tell Me Something Good," "You Got the Love"), born at Chicago, IL, Mar 23, 1953.
Akira Kurosawa, 85, filmmaker (*Rashomon, The Seven Samurai, Ran*), born at Tokyo, Japan, Mar 23, 1910.
Moses Malone, 41, former basketball player, born at Petersburg, VA, Mar 23, 1954.
Amanda Plummer, 38, actress (Tony for *Agnes of God*; *The Fisher King*), born at New York City, NY, Mar 23, 1957.

MARCH 24 — FRIDAY
83rd Day — Remaining, 282

CANADA: CANADIAN CHAMPIONSHIP DOG DERBY. Mar 24–26. Yellowknife, Northwest Territories. Three days of 50-mile circuits allow for excellent spectator viewing. Attracts mushers from all over Canada and the US. Richest event on the Canadian Dog racing circuit. Est attendance: 5,000. For info: Garth Wallbridge, Race Marshall, Box 1766, Yellowknife, NT, Canada X1A 2P3. Phone: (403) 873-5901.

EPSTEIN YOUNG ARTISTS PROGRAM: ANNIVERSARY. Mar 24. Boys & Girls Clubs of America's Epstein Young Artists Program was established in 1949 by Mr and Mrs Julius Epstein of Chicago in memory of their son, Steven David. The program provides talented young people with the type of scholarship assistance usually unavailable from conventional sources, and may include private lessons with master teachers, travel abroad to study and compete in competitions, master classes, and professional debuts or exhibitions. For info: Boys and Girls Clubs of America, 230 W Peachtree St NW, Atlanta, GA 30309.

EXXON VALDEZ OIL SPILL: ANNIVERSARY. Mar 24. On Mar 24, 1989, the tanker *Exxon Valdez* ran aground in Prince William Sound. The resultant oil spill leaked 11 million gallons of oil into one of nature's richest habitats.

HOUDINI, HARRY: BIRTH ANNIVERSARY. Mar 24. Magician and escape artist. Born at Budapest, Hungary, on Mar 24, 1874. Died at Detroit, MI, on Oct 31, 1926. Lecturer, athlete, author, expert on history of magic, exposer of fraudulent mediums and motion picture actor. Was best known for his ability to escape from locked restraints (handcuffs, straitjackets, coffins, boxes and milk cans). Anniversary of his death (Halloween) has been occasion for meetings of magicians and attempts at communication by mediums.

MAPLE SYRUP FESTIVAL. Mar 24–25. Wakarusa, IN. Maple syrup camps, quilting demonstration, antique and craft displays, wood carving, sheep shearing, horse shoeing. Est attendance: 7,500. Sponsor: Wakarusa Chamber of Commerce, Box 291, Wakarusa, IN 46573. Phone: (219) 862-4344.

MELLON, ANDREW W.: BIRTH ANNIVERSARY. Mar 24. American financier, industrialist, government official (Secretary of the Treasury), art and book collector, born Mar 24, 1855. Died Aug 27, 1937.

MORRIS, WILLIAM: BIRTH ANNIVERSARY. Mar 24. English poet and artist. Born Mar 24, 1834. Died at Hammersmith, London, England, Oct 3, 1896.

NC RV & CAMPING SHOW. Mar 24–26. Special Events Center, Greensboro Coliseum Complex, Greensboro, NC. A display of the latest in recreation vehicles and accessories by various dealers. Est attendance: 8,000. For info: Apple Rock Advertising & Promotion, 1200 Eastchester Dr, High Point, NC 27265. Phone: (910) 883-7199.

PHILIPPINE INDEPENDENCE: ANNIVERSARY. Mar 24. President Franklin Roosevelt signed a bill Mar 24, 1934, granting independence to the Philippines. The bill, which took effect July 4, 1946, brought to a close almost half a century of US control of the islands.

POWELL, JOHN WESLEY: BIRTH ANNIVERSARY. Mar 24. American geologist, explorer, ethnologist. Born at Mt Morris, NY, Mar 24, 1834. Died at Haven, ME, Sept 23, 1902.

RHODE ISLAND VOTERS REJECT CONSTITUTION: ANNIVERSARY. Mar 24. In a popular referendum on Mar 24, 1788, Rhode Island rejected the new Constitution by a vote of 2,708 to 237. The state later (May 29, 1790) ratified the Constitution and on June 7, 1790, ratified the Bill of Rights.

ST. GABRIEL FEAST DAY. Mar 24. St. Gabriel the Archangel, patron saint of postal, telephone and telegraph workers.

STAMP EXPO. Mar 24–26. Grand Hotel, Anaheim, CA. Fax: (818) 988-4337. Est attendance: 4,000. Sponsor: Intl Stamp Collectors Soc, PO Box 854, Van Nuys, CA 91408. Phone: (818) 997-6496.

◈ **STRATTON, DOROTHY CONSTANCE: BIRTHDAY.** Mar 24, 1898. Dorothy Constance Stratton, born at Brookfield, MO, was instrumental during WWII in organizing the SPARS, the women's branch of the US Coast Guard (authorized Nov 23, 1942). Under Lt Com Stratton's command some 10,000 women were trained for supportive noncombat roles in the Coast Guard. SPARS was dissolved in 1946 after the war had ended. Stratton worked with many women's organizations, including the Girl Scouts as national executive director in the '50s.

WORLD CHAMPIONSHIP SNOWMOBILE HILLCLIMB. Mar 24–26. Jackson, WY. Snowmobiles attempt to climb Snow King Mountain, 1,500 vertical feet. Sponsor: Jackson Hole Snow Devils. For info: Jackson Hole Chamber of Commerce, Box E, Jackson, WY 83001. Phone: (307) 733-3316.

BIRTHDAYS TODAY

Norman Fell, 71, actor (*The Graduate*, "Three's Company"), born Philadelphia, PA, Mar 24, 1924.
Lawrence Ferlinghetti, 76, poet, author (*Coney Island of the Mind*), born Yonkers, NY, Mar 24, 1919.
Byron Janis, 67, pianist, born at McKeesport, PA, Mar 24, 1928.
Bob Mackie, 55, costume and fashion designer, born at Monterey Park, CA, Mar 24, 1940.

MARCH 25 — SATURDAY
84th Day — Remaining, 281

ARTISTS' NATIONAL JURIED COMPETITION. Mar 25–Apr 22. Coastal Center for the Arts (formerly Island Arts Center), St. Simon's Island, GA. Artists throughout the country submit work to be selected by juror for entry and prizes. Deadline for entry is Feb 11, 1995. The exhibit opens the Saturday of the Annual Tour of Homes as a special event of the tour with official opening awards presentation. 42nd annual competition. Est attendance: 2,000. For info: Coastal Center for the Arts, 2012 Demere Rd, St Simons Island, GA 31522. Phone: (912) 634-0404.

BARTOK, BELA: BIRTH ANNIVERSARY. Mar 25. Hungarian composer. Born Mar 25, 1881, at Nagyszentmiklos (now in Yugoslavia). Died at New York, NY, Sept 26, 1945.

BIG TEN MEN'S GYMNASTICS CHAMPIONSHIP. Mar 25–26. Site to be announced. For info: Big Ten Conference, 1500 W Higgins Rd, Park Ridge, IL 60068-6300. Phone: (708) 696-1010.

BIG TEN WOMEN'S GYMNASTICS CHAMPIONSHIP. Mar 25. Site to be announced. For info: Big Ten Conference, 1500 W Higgins Rd, Park Ridge, IL 60068-6300. Phone: (708) 696-1010.

BORGLUM, GUTZON: BIRTH ANNIVERSARY. Mar 25. American sculptor who created the huge sculpture of four American presidents (Washington, Jefferson, Lincoln and Theodore Roosevelt) at Mt Rushmore National Memorial in the Black Hills of South Dakota. Borglum, born at Bear Lake, ID, on Mar 25, 1867, worked the last 14 years of his life on the Mt Rushmore sculpture. He died at Chicago, IL, on Mar 6, 1941.

CHURCHILL ENTERS GERMANY: 50th ANNIVERSARY. Mar 25, 1945. Winston Churchill briefly crossed to the eastern bank of the Rhine, the first British leader to enter Germany since Chamberlain signed the Munich Pact in September 1938. Churchill later wrote to Montgomery, "The Rhine and all its fortress lines lie behind the 21st Group of Armies. A beaten army, not long ago Master of Europe, retreats before its pursuers."

COXEY'S ARMY MARCH ON WASHINGTON: ANNIVERSARY. Mar 25. Anniversary of a march on the nation's Capitol, began Mar 25, 1894. Jacob S. Coxey, businessman, economic reformer, advocate of interest-free government bonds, left Massillon, OH, on foot with an "army" of about 100 followers. Arrived at Washington, DC, May 1. His hope to influence Congress was thwarted when he and part of his army were arrested for trespassing on government property. Fifty years later he spoke from the Capitol steps, reiterating his belief in non-interest-bearing government bonds.

ENGLAND: HEAD OF THE RIVER RACE. Mar 25. Mortlake to Putney, River Thames, London. At 11:45 AM. Processional race for 420 eight-oared crews, starting at ten-second intervals. Est attendance: 15,000. For info: Mrs. Churcher, 58 Forest Rd, Kew, Richmond, Surrey, England TW9 3BZ. Phone: (081) 940-2219.

FEAST OF ANNUNCIATION. Mar 25. Celebrated in the Roman Catholic Church in commemoration of the message of the Angel Gabriel to Mary that she was to be the Mother of Christ.

GREECE: INDEPENDENCE DAY. Mar 25. National holiday. Celebrates independence from Turkey, 1821.

HAPPY LAND SOCIAL CLUB FIRE: 5th ANNIVERSARY. Mar 25. Julio Gonzalez, described as a jilted boyfriend, returned to the Happy Land Social Club after being evicted by a bouncer and set fire to it as an act of revenge on Mar 25, 1990. Most of the 87 victims were Honduran and Dominican immigrants who attended the unlicensed dance club in the East Tremont section of the Bronx, NY. Rejected by his girlfriend, an employee of the club, Gonzalez set fire to the only entrance and exit, cutting off escape for the assembled crowd.

JAZZ AT THE PHILHARMONIC: ANNIVERSARY. Mar 25, 1946. One of the most influential solos of jazz alto saxophonist Charlie (Bird) Parker's career was his rendition of "Lady Be Good," performed at the Los Angles Philharmonic Auditorium. Every aspect of the performance became part of the language of modern jazz.

LEAN, DAVID: BIRTH ANNIVERSARY. Mar 25. Sir David Lean was born in Croydon, England, on Mar 25, 1908. As a film director he was known for his grand, larger than life epics including *Lawrence of Arabia*, *The Bridge Over the River Kwai*, *Dr. Zhivago*, *A Passage to India*, *Oliver Twist* and *Great Expectations*. He died Apr 16, 1991, at London.

MAPLE SYRUP SATURDAY. Mar 25. Gordon Bubolz Nature Preserve, Appleton, WI. Find out how to make maple syrup. Est attendance: 800. For info: Gordon Bubolz Nature Preserve, 4815 N Lynndale Dr, Appleton, WI 54915. Phone: (414) 731-6041.

MARYLAND DAY. Mar 25. Commemorates arrival of Lord Baltimore's first settlers in Maryland in 1634.

NATCHEZ POWWOW. Mar 25–26. Natchez, MS. Native American dancing and crafts. Est attendance: 4,000. For info: Jim Barnett, Grand Village of the Natchez Indians, 400 Jefferson Davis Blvd, Natchez, MS 39120. Phone: (601) 446-6502.

PECAN DAY. Mar 25. Anniversary of the planting by George Washington, on Mar 25, 1775, of pecan trees (some of which still survive) at Mount Vernon. The trees were a gift to Washington from Thomas Jefferson, who had planted a few pecan trees from the southern US at Monticello, VA. The pecan native to southern North America is sometimes called "America's own nut." First cultivated by American Indians, it has been transplanted to other continents but has failed to achieve wide use or popularity outside the US.

PLAINFIELD HOME AND PRODUCT SHOW. Mar 25–26. Plainfield High School, Plainfield, CT. More than 90 exhibits of local and area products and services, workshops, seminars, food court, raffles and door prizes. Est attendance: 6,000. For info: Cheryl Vargas, Dir, Plainfield Chamber of Commerce, 619 Norwich Rd, Plainfield, CT 06374. Phone: (203) 564-5566.

POSSUM PEDAL 100 BICYCLE RIDE/RACE. Mar 25. Courthouse Square, Graham, TX. Bicycle fun ride through the rolling hills and around Graham that annually attracts more than 3,000 participants. Sponsor: Rotary Club of Graham Scholarship Fund. Annually, the "Taste of Graham." Est attendance: 2,500. Sponsor: Graham Chamber of Commerce, PO Box 299, Graham, TX 76450. Phone: (817) 549-3355.

STATE STREET STATION COOKBOOK FAIR. Mar 25. Greensboro, NC. Recipe tastings, non-profit cookbooks are sold within the 35 shops and restaurants of a unique restored shopping village includes cookbooks from throughout the southeast! Est attendance: 3,000. For info: Betsy Seale, State Street Station, PO Box 13563, Greensboro, NC 27415-3563. Phone: (910) 230-0623.

March 1995

S	M	T	W	T	F	S
			1	2	3	4
5	6	7	8	9	10	11
12	13	14	15	16	17	18
19	20	21	22	23	24	25
26	27	28	29	30	31	

☆ Chase's 1995 Calendar of Events ☆ Mar 25–26

BIRTHDAYS TODAY

Anita Bryant, 55, singer ("The George Gobel Show"; hit song "Paper Roses"), Miss America ('58), born Barnsdall, OK, Mar 25, 1940.

Howard Cosell (Howard William Cohen), 75, former sportscaster ("Monday Night Football"), born at Winston-Salem, NC, Mar 25, 1920.

Eileen Ford, 73, model agency executive, born at New York, NY, Mar 25, 1922.

Aretha Franklin, 53, singer ("Respect," "Think," "Ain't Nothin' Like the Real Thing"), born Memphis, TN, Mar 25, 1942.

Paul Michael Glaser, 52, actor ("Starsky and Hutch"), director (*Butterflies Are Free*), born Cambridge, MA, Mar 25, 1943.

Mary Gross, 42, comedienne, actress ("NBC's Saturday Night Live"), born at Chicago, IL, Mar 25, 1953.

Elton John (Reginald Kenneth Dwight), 48, musician, singer, songwriter, born at Pinner, England, Mar 25, 1947.

Tom Monaghan, 58, founder of Domino's Pizza and owner of Detroit Tigers, born at Ann Arbor, MI, Mar 25, 1937.

Gloria Steinem, 60, feminist (original publisher of *MS.* magazine), journalist, born at Toledo, OH, Mar 25, 1935.

Debi Thomas, 28, figure skater, born at Poughkeepsie, NY, Mar 25, 1967.

MARCH 26 — SUNDAY
85th Day — Remaining, 280

BANGLADESH: INDEPENDENCE DAY. Mar 26. Celebrated with parades, youth festivals and symposia.

BELLAMY, EDWARD: BIRTH ANNIVERSARY. Mar 26. American author best remembered for his novel *Looking Backward* (1888). Born at Chicopee Falls, MA, on Mar 26, 1850. Died there on May 28, 1898.

BOWDITCH, NATHANIEL: BIRTH ANNIVERSARY. Mar 26. American mathematician and astronomer, author of the *New American Practical Navigator*. Born at Salem, MA, Mar 26, 1773. Died at Boston, MA, Mar 16, 1838.

DELANO, JANE: BIRTH ANNIVERSARY. Mar 26. Jane Arminda Delano, dedicated American nurse and teacher, superintendent of the US Army Nurse Corps, chairman of the American Red Cross Nursing Service and recipient (posthumously) of the Distinguished Service Medal of the US, was long believed to have been born Mar 12, 1862. Records in Miss Delano's handwriting reveal that she was born Mar 26, 1858, near Townsend, NY. While on an official visit to review Red Cross activities, she died Apr 15, 1919, in an army hospital at Savenay, France. Her last words: "What about my work? I must get back to my work." Buried first at Loire, France, her remains were reinterred at Arlington Cemetery in 1920.

ENGLAND: MOTHERING SUNDAY. Mar 26. Fourth Sunday of Lent, formerly occasion for attending services at Mother Church, family gatherings and visits to parents. Now popularly known as Mother's Day, and a time for visiting and taking gifts to mothers.

ENGLAND: SUMMER TIME. Mar 26–Oct 29. In England and much of Europe, "Summer Time" (one hour in advance of Standard Time), similar to "Daylight-Saving Time," is observed from 01 hours on the day after the fourth Saturday in March until 01 hours on the day after the fourth Saturday in October.

EUROPE: SUMMER DAYLIGHT-SAVING TIME. Mar 26–Sept 24. Many European countries observe daylight-saving (summer) time from 2 AM on the last Sunday in March until 3 AM on the last Sunday in September.

FROST, ROBERT LEE: BIRTH ANNIVERSARY. Mar 26. American poet who tried his hand at farming, teaching, shoe-making and editing before winning acclaim as a poet. Pulitzer Prize winner. Born at San Francisco, CA, Mar 26, 1874. Died at Boston, MA, Jan 29, 1963.

KUHIO DAY. Mar 26. Hawaii. Also known as Prince Jonah Kuhio Kalanianaole Day, honoring birthday of Hawaii's second delegate to Congress.

LUXEMBOURG: BRETZELSONNDEG. Mar 26. Fourth Sunday in Lent is occasion for boys to give pretzel-shaped cakes to sweethearts who may respond, on Easter Sunday, with a gift of decorated egg or sweet.

MAKE UP YOUR OWN HOLIDAY DAY. Mar 26. This day is a day you may name for whatever you wish. Reach for the stars! Make up a holiday! Annually, Mar 26. Phone (212) 388-8673 or (717) 274-8451. For info: Thomas and Ruth Roy, Wellness Permission League, 2105 Water St, Lebanon, PA 17046.

O'CONNOR, SANDRA DAY: 65th BIRTHDAY. Mar 26. Associate Justice of the Supreme Court of the US, nominated by President Reagan on July 7, 1981, and sworn in on Sept 25, 1981. Born at El Paso, TX, on Mar 26, 1930. First woman nominated and first appointed to US Supreme Court.

SOVIET COSMONAUT RETURNS TO NEW WORLD: ANNIVERSARY. Mar 26. After spending 313 days in space in the Soviet *Mir* space station, Cosmonaut Serge Krikalev returned to Earth and to what was for him a new world, on Mar 26, 1992. He left Earth on May 18, 1991, a citizen of the Soviet Union, but during his stay aboard the space station, the Soviet Union crumbled and quickly became the Commonwealth of Independent States. Originally scheduled to return in October 1991, Krikalev's return was delayed by five months due to his country's disintegration and the ensuing monetary problems.

WILLIAMS, TENNESSEE: BIRTH ANNIVERSARY. Mar 26. Tennessee Williams was born in Columbus, MS, on Mar 26, 1911. He became one of America's most prolific playwrights, producing such works as *The Glass Menagerie*, *A Streetcar Named Desire*, which won a Pulitzer Prize, *Cat on a Hot Tin Roof*, which won a second Pulitzer, *Night of the Iguana*, *Summer and Smoke*, *The Rose Tattoo*, *Camino Real*, *Sweet Bird of Youth* and *Small Craft Warnings*, among others. Williams died Feb 25, 1983.

BIRTHDAYS TODAY

Marcus Allen, 35, football player, born at San Diego, CA, Mar 26, 1960.

Alan Arkin, 61, actor (*Catch-22*), director (*Little Murders*), born at New York, NY, Mar 26, 1934.

Uwe Blab, 33, basketball player, born at Munich, Germany, Mar 26, 1962.

Pierre Boulez, 70, composer, conductor, born at Montbrison, France, Mar 26, 1925.

James Caan, 55, actor (*Rabbit Run*, *The Godfather*), director, born at New York, NY, Mar 26, 1940.

Jennifer Grey, 35, actress (Joel Grey's daughter; *Dirty Dancing*, *Ferris Bueller's Day Off*), born at New York, NY, Mar 26, 1960.

Erica Jong, 53, author, poet (*Becoming Light*, *How to Save Your Own Life*), born at New York, NY, Mar 26, 1942.

Vicki Lawrence, 46, actress, singer ("The Carol Burnett Show," "Mama's Family"), born at Inglewood, CA, Mar 26, 1949.

Leonard Nimoy, 64, actor (Spock, "Star Trek"), director (*Three Men and a Baby*), writer, born Boston, MA, Mar 26, 1931.

Sandra Day O'Connor, 65, Associate Justice of the US Supreme Court, born at El Paso, TX, Mar 26, 1930.

Mar 26–27 ☆ Chase's 1995 Calendar of Events ☆

Teddy Pendergrass, 45, singer, born at Philadelphia, PA, Mar 26, 1950.
Diana Ross, 51, singer (lead, The Supremes; "Love Hangover," "Upside Down"), actress (*Lady Sings the Blues, Mahogany*), born at Detroit, MI, Mar 26, 1944.
Martin Short, 45, actor, comedian ("SCTV Network 90," "Saturday Night Live," *Innerspace, Three Amigos*), born at Hamilton, Ontario, Canada, Mar 26, 1950.
John Houston Stockton, 33, basketball player, born Spokane, WA, Mar 26, 1962.
Bob Woodward, 52, journalist (investigated Watergate with Carl Bernstein), born Geneva, IL, Mar 26, 1943.

MARCH 27 — MONDAY
86th Day — Remaining, 279

ALASKA EARTHQUAKE: ANNIVERSARY. Mar 27. Severe earthquake Mar 27, 1964.

CANADA: ROYAL MANITOBA WINTER FAIR. Mar 27–Apr 1. Keystone Center, Brandon, Manitoba. Equestrian and heavy horse show, commercial exhibits, Royal Farm Yard, cattle shows and sales, entertainment stages. Only show in Canada excepting Toronto being granted Royal Patronage. Est attendance: 132,800. For info: Provincial Exhibition of Manitoba, #3–1175 18th St, Brandon, Man, Canada R7A 7C5. Phone: (204) 726-3590.

FUNKY WINKERBEAN: 23rd ANNIVERSARY. Mar 27. Anniversary of the nationally syndicated comic strip. For info: Tom Batiuk, Creator, 2750 Substation Rd, Medina, OH 44256. Phone: (216) 722-8755.

HILL, PATTY SMITH: BIRTH ANNIVERSARY. Mar 27. Patty Smith Hill, schoolteacher, author and education specialist was born at Anchorage (suburb of Louisville), KY, on Mar 27, 1868. She was author of the lyrics of the song "Good Morning to All," which later became known as "Happy Birthday to You." Her older sister, Mildred J. Hill, composed the melody for the song which was first published in 1893 as a classroom greeting, in the book *Song Stories for the Sunday School*. A stanza beginning "Happy Birthday to You" was added in 1924, and the song became arguably the most frequently sung song in the world. See also: "Hill, Mildred J.: Birth Anniversary" (June 27) and "Happy Birthday to 'Happy Birthday to You'" (June 27).

LUXEMBOURG: OSWEILER. Mar 27. Blessing of horses, tractors and cars.

NATIONAL EXCHANGE CLUB BIRTHDAY. Mar 27. Anniversary of the day in 1911 when the first Exchange Club was founded in Detroit, MI. Celebrated annually by nearly 40,000 Exchangites in the US and Puerto Rico. Sponsor: The Natl Exchange Club, 3050 Central Ave, Toledo, OH 43606-1700. Phone: (419) 535-3232.

NORTH SEA OIL RIG DISASTER: 15th ANNIVERSARY. Mar 27. The Alexander L. Keilland Oil Rig capsized Mar 27, 1980, during a heavy storm in the Norwegian sector of the North Sea. The pentagon-type, French-built oil rig had about 200 persons aboard, and 123 lives were lost.

March 1995

S	M	T	W	T	F	S
			1	2	3	4
5	6	7	8	9	10	11
12	13	14	15	16	17	18
19	20	21	22	23	24	25
26	27	28	29	30	31	

◈ ROENTGEN, WILHELM KONRAD: 150th BIRTH ANNIVERSARY. Mar 27. German scientist who discovered X-rays (1895) and won Nobel Prize in 1901. Born at Lennep, Prussia, Mar 27, 1845. Died at Munich, Germany, Feb 10, 1923.

SMITH, THORNE: BIRTH ANNIVERSARY. Mar 27. Perhaps the most critically neglected popular author of the 20th century, he was born James Thorne Smith, Jr, at Annapolis, MD, on Mar 27, 1892, and died in Florida on June 20, 1934. Author of numerous humorous supernatural fantasy novels, including *Rain in the Doorway, The Stray Lamb* and, of course, *Topper*. The "Thorne Smith" touch has inspired several motion pictures and television series, including "Bewitched." For info: The Thorne Smith Soc, George H. Scheetz, Exec Secy, 1901 Bittersweet Dr, Champaign, IL 61821-6371.

SPACE MILESTONE: *VENERA 8* (USSR). Mar 27. Soft landing on Venus July 22, 1972, with radio transmission of surface data. Launched Mar 27, 1972.

STEICHEN, EDWARD: BIRTH ANNIVERSARY. Mar 27. Celebrated American photographer. Born Mar 27, 1879. Died Mar 25, 1973.

SWANSON, GLORIA: BIRTH ANNIVERSARY. Mar 27. American film actress and businesswoman. Born Gloria May Josephine Svensson at Chicago, Mar 27, 1899. Author of an autobiography, *Swanson on Swanson*, published in 1980. Died at New York, NY, Apr 4, 1983.

VAUGHAN, SARAH: BIRTH ANNIVERSARY. Mar 27. Born at Newark, NJ, on Mar 27, 1924. Legendary jazz singer renowned for her melodic improvising, wide vocal range and extraordinary technique. She began her career by winning an amateur contest sponsored by New York's Apollo Theater in 1943, which she entered on a dare. She was spotted and hired by Earl Hines to accompany his band as his relief pianist as well as singer. Her solo career began in 1945. As her career took off, she was given the nickname "The Divine One," by Chicago disc jockey Dave Garroway, a moniker that would remain with her the rest of her life. She used her three-octave range to explore a repertoire including Duke Ellington, George Gershwin, Irving Berlin, Rodgers and Hammerstein and an album of poems written by Pope John Paul II, set to music by Tito Fontana and Sante Palumbo. Sarah Vaughan died at Los Angeles, CA, Apr 3, 1990.

BIRTHDAYS TODAY

Jesse Brown, 51, US Secretary of Veterans Affairs in Clinton administration, born at Detroit, MI, Mar 27, 1944.
Randall Cunningham, 32, football player, born at Santa Barbara, CA, Mar 27, 1963.
Anthony Lewis, 68, journalist, author (*Gideon's Trumpet; Make No Law: Sullivan Case and the First Amendment*), born New York, NY, Mar 27, 1927.
Maria Schneider, 43, actress (*Last Tango in Paris*), born at Paris, France, Mar 27, 1952.
Cale Yarborough, 55, auto racer, born at Timminsville, SC, Mar 27, 1940.
Michael York, 53, actor (*Cabaret, The Three Musketeers*), born Fulmer, England, Mar 27, 1942.

☆ Chase's 1995 Calendar of Events ☆   Mar 28–29

MARCH 28 — TUESDAY
87th Day — Remaining, 278

AMERICAN DIABETES ALERT. Mar 28. To increase public awareness of the seriousness of diabetes and its complications, and to educate the person with diabetes about managing the disease. Annually, the fourth Tuesday in March. Sponsor: American Diabetes Assn. Contact your local affiliate office. For info: American Diabetes Assn, Natl HQ, PR Dept, 1660 Duke St, Alexandria, VA 22314. Phone: (800) 342-2383.

CZECHOSLOVAKIA: TEACHERS' DAY. Mar 28. Celebrates birth on this day of Jan Amos Komensky (Comenius), Moravian educational reformer (1592–1671).

HAIR BROADWAY OPENING: ANNIVERSARY. Mar 28. The controversial rock musical Hair, produced by Michael Butler, opened at the Biltmore Theatre in New York City on Mar 28, 1968. For those who opposed the Vietnam War and the "Establishment," this was a defining piece of work—as evidenced by some of its songs, such as "Aquarius," "Hair," "Let the Sunshine In."

NATIONAL ORGANIZE YOUR HOME OFFICE DAY. Mar 28. To set aside one day each year for the home office professional to organize his or her office. Annually, the fourth Tuesday in March. For info: Lisa Kanarek, Organizational Consultant, 660 Preston Forest Ctr, Ste 120, Dallas, TX 75230. Phone: (214) 361-0556.

PERTINAX: ASSASSINATION ANNIVERSARY. Mar 28. Roman emperor Pertinax, on Mar 28, AD 193, according to Edward Gibbon, "disdaining either flight or concealment, advanced to meet his assassins and recalled to their minds his own innocence, and the sanctity of their recent oath. For a few moments they stood in silent suspense, ashamed of their own atrocious design, and awed by the venerable aspect and majestic firmness of their sovereign.... His head, separated from his body and placed on a lance, was carried in triumph to Praetorian camp, in the sight of a mournful and indignant people, who lamented the unworthy fate of that excellent prince, and the transient blessings of a reign the memory of which could only serve to aggravate their approaching misfortunes."

ST. JOHN NEPOMUCENE NEUMANN: BIRTH ANNIVERSARY. Mar 28. First male saint of the US. Born at Prachatice, Bohemia, on Mar 28, 1811. Came to the US in 1836. As Bishop of Philadelphia, he was affectionately known as the "Little Bishop." Died at Philadelphia, PA, Jan 5, 1860. Beatified Oct 13, 1963. Canonized June 19, 1977. Pope Paul VI, in proclaiming sainthood, said: "For the honor of the most Holy Trinity, for the exaltation of the Catholic faith and the increase of the Christian life, by the authority of our Lord Jesus Christ, of the Holy Apostles Peter and Paul and by our own authority, after having reflected for a long time and invoked the counsel of many of our brother bishops, we inscribe his name in the Calendar of Saints and establish that he should be devoutly honored among the saints of the Universal church."

SPRING PILGRIMAGE TO ANTEBELLUM HOMES. Mar 28–Apr 9. Columbus, MS. Tours of pre–Civil War homes for visiting public. Sponsor: Columbus Historic Fdtn, PO Box 46, Columbus, MS 39703. Est attendance: 6,000. For info: Columbus Conv and Visitor Bureau, PO Box 789, Columbus, MS 39703. Phone: (800) 327-2686.

THREE MILE ISLAND NUCLEAR POWER PLANT ACCIDENT: ANNIVERSARY. Mar 28. A series of accidents beginning at 4 AM, EST, Mar 28, 1979, at Three Mile Island on the Susquehanna River about 10 miles southeast of Harrisburg, PA, were responsible for extensive reevaluation of the safety of existing nuclear power generating operations. Equipment and other failures reportedly brought Three Mile Island close to a meltdown of the uranium core, threatening extensive radiation contamination.

VIETNAM MORATORIUM CONCERT: 25th ANNIVERSARY. Mar 28. A seven-hour concert at Madison Square Garden in New York City on Mar 28, 1970, featured many stars who had donated their services for the anti-war cause. Among them were Jimi Hendrix; Dave Brubeck; Harry Belafonte; Peter, Paul and Mary; Judy Collins; the Rascals; Blood, Sweat and Tears; and the Broadway cast of Hair.

BIRTHDAYS TODAY

Dirk Bogarde (Derek van Den Bogaerde), 74, actor (The Damned, Death in Venice), born at London, England, Mar 28, 1921.

Ken Howard, 51, actor ("The White Shadow"), born at El Centro, CA, Mar 28, 1944.

Reba McEntire, 41, singer ("For My Broken Heart"), born at Chockie, OK, Mar 28, 1954.

Frank Hughes Murkowski, 62, US Senator (R, Alaska), born at Seattle, WA, Mar 28, 1933.

Edmund S. Muskie, 81, former senator, secretary of state, presidential and vice presidential candidate, born Rumford, ME, Mar 28, 1914.

Byron Anton Scott, 34, basketball player, born at Ogden, UT, Mar 28, 1961.

Dianne Wiest, 47, actress (Oscar for Hannah and Her Sisters; Radio Days), born at Kansas City, MO, Mar 28, 1948.

MARCH 29 — WEDNESDAY
88th Day — Remaining, 277

COMMITTEE ON ASSASSINATIONS REPORT: ANNIVERSARY. Mar 29. The House Select Committee on Assassinations released the final report on its investigation into the assassinations of President John F. Kennedy, Martin Luther King, Jr, and Robert Kennedy on Mar 29, 1979. Based on available evidence, the committee concluded that President Kennedy was assassinated as a result of a conspiracy, although no trail of a conspiracy could be established. The Soviet Union, Cuba, CIA, FBI, Secret Service, Anti-Castro Cuban groups and Organized Crime were ruled out as possible sources of a conspiracy, although in the latter two groups evidence did not preclude the possibility that members of those two groups may have been involved as individuals not representing their organizations. They also concluded that on the basis of scientific acoustical evidence two gunmen fired at the President, although no second gunman could be identified, and they found that the Warren Commission failed to adequately investigate the possibility of a conspiracy primarily because the Commission did not receive all the relevant information that was in the possession of other government agencies at the time. (In Dec 1980, the FBI released a report discounting the two-gunman theory stating that the distinguishable sounds of two separate guns were not proven scientifically.) In addition the committee concluded that the possibility of conspiracy did exist in the cases of Dr. King and Robert Kennedy, although no specific individuals or organizations could be pinpointed as being involved. See also: "Warren Commission Report: Anniversary" (Sept 27).

Mar 29-30 ☆ *Chase's 1995 Calendar of Events* ☆

HENRIED, PAUL: DEATH ANNIVERSARY. Mar 29. Actor Paul Henried once estimated that he had played in or directed more than 300 feature or made-for-TV films. He was born at Trieste, Austria, on Jan 10, 1908. Otto Preminger discovered his talent, launching his career with Max Reinhardt's theater. His career as a dashing leading man often found him playing the Continental lover, but his career expanded when he played Prince Albert in a London production of *Victoria Regina*. Though a staunch anti-Nazi, his early film roles included a number of German roles, including *Goodbye Mr. Chips, Mad, Man of Europe* and *Night Train*. He eventually moved away from the German stereotype in such films as *Of Human Bondage, Rope of Sand, Deep in My Heart, The Four Horsemen of the Apocalypse, Exorcist III* and as Victor Laslo in *Casablanca*. His film career cut short by the anti-Communist blacklist in Hollywood during the 1940s, Henried found a second calling as a director, directing more than 80 episodes of TV's "Alfred Hitchcock Presents." He died on Mar 29, 1992, at Pacific Palisades, CA.

HOOVER, LOU HENRY: BIRTH ANNIVERSARY. Mar 29. Wife of Herbert Clark Hoover, 31st president of the US. Born Mar 29, 1875, at Waterloo, IA. Died Jan 7, 1944.

JOHN PARTRIDGE "DEATH" HOAX: ANNIVERSARY. Mar 29. English astrologer John Partridge (real name: John Hewson) so offended readers by his foolish predictions that he became the target of parodies and jokes, most serious of which was that of the satirist Jonathan Swift. Under the pseudonym Isaac Bickerstaff, Swift published his own almanac, for the year 1708, in which he predicted that Partridge would die at 11 PM, Mar 29, 1708, "of a raging fever." Poor Partridge made the mistake of trying to prove he was still alive, only to find writers, citizens and even the court were more amused by continuing the fiction of his death.

KEYSTONE CLUB CONFERENCE. Mar 29-31. Atlanta, GA. Established in 1964 for Boys & Girls Club members 14-18 years old, Keystone Clubs stress service to Club and community, leadership development, education and free enterprise. At each annual national conference, Keystoners undertake a new year-long, humanitarian national project. Est attendance: 1,200. For info: Boys and Girls Clubs of America, 1230 W Peachtree St NW, Atlanta, GA 30309.

KNIGHTS OF COLUMBUS FOUNDERS' DAY. Mar 29. The first Knights of Columbus charter was granted in 1882 by the state of Connecticut. This Catholic, family, fraternal, service organization has grown into a volunteer force of Knights and family members totaling nearly six million who annually donate tens of millions of dollars and volunteer hours to countless charitable projects. For info: Harvey G. Bacqueé, Asst to Supreme Knight, Knights of Columbus, One Columbus Plaza, New Haven, CT 06510-3326.

MADAGASCAR: COMMEMORATION DAY. Mar 29. Commemoration Day for the victims of the rebellion in 1947 against French colonization.

NATIONAL ASSOCIATION OF ACTIVITY PROFESSIONALS ANNUAL CONVENTION. Mar 29-Apr 1. Nashville, TN. Est attendance: 500. Sponsor: Natl Assn of Activity Professionals, Charles Price, Exec Dir, 1225 Eye St NW, Ste 300, Washington, DC 20005. Phone: (202) 289-0722.

QUINLAN, KAREN ANN: BIRTH ANNIVERSARY. Mar 29. Born at Scranton, PA, on Mar 29, 1954, Karen Ann Quinlan became the center of an international legal, medical and ethical controversy over the right to die. Reportedly she became irreversibly comatose on Apr 14, 1975. A petition filed by her adoptive parents in New Jersey's Superior Court, Sept 12, 1975, sought permission to discontinue use of a respirator, allowing her to die "with grace and dignity." Eventually (in 1976) the petition was upheld (by New Jersey's Supreme Court). Miss Quinlan lived nearly a decade without the respirator—until June 11, 1985. Described by some as unconsciously one of the world's great teachers, her life and tragic plight brought into focus the ethical dilemmas of advancing medical technology—the need for new understanding of life and death; the right to die; the role of judges, doctors and hospital committees in deciding when to prolong or when *not* to prolong life.

TAIWAN: YOUTH DAY. Mar 29. Public holiday observed annually on Mar 29.

TWENTY-THIRD AMENDMENT TO US CONSTITUTION RATIFIED: ANNIVERSARY. Mar 29. District of Columbia residents were given the right to vote in presidential elections under the 23rd Amendment, ratified Mar 29, 1961.

TYLER, JOHN: BIRTH ANNIVERSARY. Mar 29. Tenth president of the US (Apr 6, 1841-Mar 3, 1845). Born at Charles City County, VA, Mar 29, 1790, Tyler succeeded to the presidency upon the death of William Henry Harrison. Tyler's first wife died while he was president, and he remarried before the end of his term of office, becoming the first president to marry while in office. Fifteen children were born of the two marriages. In 1861 he was elected to the Congress of the Confederate States but died at Richmond, VA, Jan 18, 1862, before being seated. His death received no official tribute from the US government.

BIRTHDAYS TODAY

Earl Campbell, 40, Pro Football Hall of Famer, born Tyler, TX, Mar 29, 1955.

Bud Cort, 45, actor (*Harold and Maude, Brewster McCloud*), born at New Rochelle, NY, Mar 29, 1950.

Eric Idle, 52, actor ("Monty Python's Flying Circus"), author, born at Durham, England, Mar 29, 1943.

Eugene McCarthy, 79, former US Senator (anti-Vietnam War presidential hopeful in '68), born at Watkins, MN, Mar 29, 1916.

John Joseph McLaughlin, 68, editor, columnist, TV host, born at Providence, RI, Mar 29, 1927.

Larry Pressler, 53, US Senator (R, South Dakota), born at Humbolt, SD, Mar 29, 1942.

Norman Tebbit, 64, British political leader, born at London, England, Mar 29, 1931.

Kurt Thomas, 39, gymnast, born at Terre Haute, IN, Mar 29, 1956.

MARCH 30 — THURSDAY
89th Day — Remaining, 276

ABRAHAM BALDWIN AGRICULTURAL COLLEGE HOMECOMING CELEBRATION. Mar 30-Apr 1. ABAC Campus, Tifton, GA. Class reunions, alumni awards luncheon, dance. Est attendance: 500. For info: ABAC Alumni Assn, Nancy Coleman, Exec Secy, ABAC 13, 2802 Moore Hwy, Tifton, GA 31794-2693. Phone: (912) 386-3321.

ANESTHETIC FIRST USED IN SURGERY: ANNIVERSARY. Mar 30. Dr. Crawford W. Long, having seen the use of nitrous oxide and sulfuric ether at "laughing gas" parties, observed that individuals under their influences felt no pain. On Mar 30, 1842, he removed a tumor from the neck of a man who was under the influence of ether.

	S	M	T	W	T	F	S
March 1995				1	2	3	4
	5	6	7	8	9	10	11
	12	13	14	15	16	17	18
	19	20	21	22	23	24	25
	26	27	28	29	30	31	

✯ Chase's 1995 Calendar of Events ✯ Mar 30–31

DOCTORS' DAY. Mar 30. Traditional annual observance since 1933 to honor America's physicians on anniversary of occasion when Dr. Crawford W. Long became the first acclaimed physician to use ether as an anesthetic agent in a surgical technique, on Mar 30, 1842. The red carnation has been designated the official flower of Doctors' Day.

MOON PHASE: NEW MOON. Mar 30. Moon enters New Moon phase at 9:09 PM, EST.

NATIONAL BADMINTON DAY. Mar 30. To salute badminton on the day of Opening Ceremonies of the 1995 Yonex US National Championship at Georgia State University, Atlanta, GA The best in badminton will compete for national titles in weekend tournament, Mar 30–Apr 2. For info: Jim Hadley, Exec Dir, US Badminton Assn, One Olympic Plaza, Colorado Springs, CO 80909. Phone: (719) 574-4808.

NCAA DIVISION I MEN'S ICE HOCKEY CHAMPIONSHIP FINALS. Mar 30–Apr 1. Providence Civic Center, Providence, RI, host, Providence College. Est attendance: 80,500. For info: NCAA, 6201 College Blvd, Overland Park, KS 66211-2422. Phone: (913) 339-1906.

O'CASEY, SEAN: BIRTH ANNIVERSARY. Mar 30. Irish playwright. Born at Dublin, Ireland, Mar 30, 1880. Died at Torquay, England, on Sept 18, 1964.

PIN THE EAR ON THE VAN GOGH. Mar 30. Museum of Art, Ft Wayne, IN. In celebration of Vincent Van Gogh's birthday, contestants are blindfolded and are given an ear to pin on a life size poster of Van Gogh! Est attendance: 200. Sponsor: WAJI Radio, Barb Richards, Asst Program Dir, 347 W Berry, Ste 600, Ft Wayne, IN 46802. Phone: (219) 423-3676.

REAGAN, RONALD: ASSASSINATION ATTEMPT ANNIVERSARY. Mar 30. President Ronald Reagan was shot in the chest by a 25-year-old gunman, about 2:30 PM EST, Mar 30, 1981, in Washington, DC. Three other persons were wounded. John W. Hinckley, Jr, the accused attacker, was arrested at the scene. On June 21, 1982, a federal jury in the District of Columbia found Hinckley not guilty by reason of insanity and he was committed to St. Elizabeth's Hospital, at Washington, DC, for an indefinite time.

SCHMECKFEST. Mar 30–Apr 1. Freeman, SD. Bratwurst and sauerkraut, kuchen and pluma moos. These are just a few of the dishes served at this German "festival of tasting" where visitors can also watch cooking and craft demonstrations and an evening musical. Est attendance: 5,000. For info: Elaine Glanzer, RR 1, Box 57, Bridgewater, SD 57319. Phone: (605) 729-2649.

SEWARD'S DAY: ANNIVERSARY OF THE ACQUISITION OF ALASKA. Mar 30. Observed near anniversary of the acquisition of Alaska from Russia in 1867. Annually, the last Monday in March. The treaty of purchase was signed between the Russians and the Americans on Mar 30, 1867, and ratified by the Senate on May 28, 1867. The territory was formally transferred on Oct 18, 1867.

VAN GOGH, VINCENT: BIRTH ANNIVERSARY. Mar 30. Dutch post-Impressionist painter, especially known for his bold and powerful use of color. Born at Groot Zundert, Holland, Mar 30, 1853. Died at Auvers-sur-Oise, France, July 29, 1890.

WOMEN'S INTERNATIONAL BOWLING CONGRESS CHAMPIONSHIP TOURNAMENT. Mar 30–31. Tucson, AZ. Est attendance: 46,000. Sponsor: WIBC, 5301 S 76th St, Greendale, WI 53129. Phone: (414) 421-9000.

BIRTHDAYS TODAY

John Astin, 65, actor ("The Addams Family," *The Three Penny Opera*), director, born Baltimore, MD, Mar 30, 1930.
Warren Beatty, 57, actor (*Bonnie and Clyde*), director (*Reds, Dick Tracy*), producer, born Richmond, VA, Mar 30, 1938.
McGeorge Bundy, 76, statesman, educator, born at Boston, MA, Mar 30, 1919.
Eric Clapton, 50, singer (member Yardbirds, Cream), songwriter (Grammy for "Tears in Heaven" [with Will Jennings]; "Layla" [with Jim Gordon]), born at Ripley, England, Mar 30, 1945.

Richard Dysart, 66, actor (LeLand McKenzie on "L.A. Law"), born at Augusta, ME, Mar 30, 1929.
Frankie Laine, 82, actor, singer ("Frankie Laine Time," *Viva Las Vegas*), born Chicago, IL, Mar 30, 1913.
Peter Marshall, 68, TV host, actor, born at Huntington, WV, Mar 30, 1927.
Bob Miller, 50, Governor of Nevada (D), born at Evanston, IL, Mar 30, 1945.
Bob C. Smith, 54, US Senator (R, New Hampshire), born at Trenton, NJ, Mar 30, 1941.

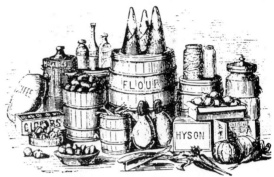

MARCH 31 — FRIDAY
90th Day — Remaining, 275

AFRO-AMERICAN CULTURAL CENTER AT YALE: 25th ANNIVERSARY CONFERENCE. Mar 31–Apr 2. New Haven, CT. A special weekend conference to commemorate the Center's 25 years as a student/community center. For info: Afro-American Cultural Center, Yale University, 3439 Yale Station, 211 Park St, New Haven, CT 06520. Phone: (203) 432-4131.

ANTEBELLUM JUBILEE. Mar 31–Apr 2. (Also Apr 7–9.) Georgia's Stone Mountain Park, Stone Mountain, GA. Set in the Park's Antebellum Plantation Complex (a multi-structure group of historic buildings), the Jubilee is a living history event spotlighting the arts, music, entertainment and lifestyles of pre-Civil War America. Annually, the first two weekends of April (if there is no conflict with Easter). Est attendance: 125,000. For info: K Thweatt, PR Office, Georgia's Stone Mountain Park, PO Box 778, Stone Mountain, GA 30086. Phone: (404) 498-5702.

BUNSEN BURNER DAY. Mar 31. A day to honor the inventor of the Bunsen burner, Robert Wilhelm Eberhard von Bunsen, who provided chemists and chemistry students with one of their most indispensable instruments. The Bunsen burner allows the user to regulate the proportions of flammable gas and air to create the most efficient flame. Bunsen was born at Gottingen, Germany, on Mar 31, 1811, and was a professor of chemistry at the universities at Kassel, Marburg, Breslau and Heidelberg. He died at Heidelberg, Germany, Aug 16, 1899.

CHAVEZ, CESAR ESTRADA: BIRTH ANNIVERSARY. Mar 31, 1927. Labor leader who organized migrant farm workers in support of better working conditions. Chavez initiated the National Farm Workers Assn in 1962, attracting attention to the migrant farm workers' plight by organizing boycotts of products including grapes and lettuce. He was born at Yuma, AZ, and died Apr 23, 1993, at San Luis, AZ.

CHRYSLER OPENS DETROIT FACTORY: ANNIVERSARY. Mar 31. In an unorthodox move for recessionary times, the Chrysler Corporation opened a new plant on Jefferson Ave in Detroit, MI, on Mar 31, 1992. The opening came at a time when American automakers were scaling back and closing plants while Japanese competitors were opening facilities in mid-South rural locations to avoid unions and urban problems. Located across the street from the site of its 85-year-old predecessor that was torn down in 1991, the new plant produces Jeep Grand Cherokee utility vehicles.

145

Mar 31 ☆ Chase's 1995 Calendar of Events ☆

DESCARTES, RENE: BIRTH ANNIVERSARY. Mar 31. French philosopher and mathematician, known as the "father of modern philosophy." Born at La Haye, Touraine, France, on Mar 31, 1596. Cartesian philosophical precepts are often remembered because of his famous proposition "I think, therefore I am" (*Cogito ergo sum*...). Died of pneumonia at Stockholm, Sweden, on Feb 11, 1650.

EIFFEL TOWER: ANNIVERSARY. Mar 31. Built for the Paris Exhibition of 1889, the tower was named for its architect, Alexandre Gustave Eiffel, and is one of the world's most well-known landmarks.

FITZGERALD, EDWARD: BIRTH ANNIVERSARY. Mar 31. English author. Born Mar 31, 1809. Perhaps best known for his translation of Omar Khayyam's *Rubaiyat*. Died June 14, 1883.

GOGOL, NIKOLAI VASILEVICH: BIRTH ANNIVERSARY. Mar 31. Russian author of plays, novels and short stories. Born at Sorochinsk, Russia, on Mar 31, 1809. Died at Moscow, Russia, Mar 4, 1852. Gogol's most famous work was the novel *Dead Souls*.

HAYDN, FRANZ JOSEPH: BIRTH ANNIVERSARY. Mar 31. "Father of the symphony." Born at Rohrau, Austria-Hungary, on Mar 31, 1732. Composed about 120 symphonies, more than a hundred works for chamber groups, a dozen operas and hundreds of other musical works. Died at Vienna, Austria, on May 31, 1809.

MARVELL, ANDREW: BIRTH ANNIVERSARY. Mar 31. English poet. Born at Winestead, Yorkshire, England, Mar 31, 1621. Died at London, England, Aug 18, 1678.

MUSIC EDUCATORS EASTERN DIVISION IN-SERVICE CONFERENCE. Mar 31-Apr 2. Rochester, NY. Clinics, showcases, performances and trade exhibits. Biennially, in odd years. Est attendance: 1,200. Sponsor: Music Educators Natl Conf (MENC), Dir Mtgs and Convs, 1806 Robert Fulton Dr, Reston, VA 22091.

PEARSE, RICHARD: FLIGHT ANNIVERSARY. Mar 31. According to claim, Richard Pearse, a farmer and inventor, flew a monoplane of his own design several hundred yards along a road near Temuka, New Zealand, and then landed it on top of a 12-foot-high hedge, on Mar 31, 1903. Pearse also built the craft, which consisted of a steerable tricycle undercarriage and an internal combustion engine. A Pearse commemorative medal was issued on Sept 19, 1971, by the Museum of Transport and Technology, Auckland, New Zealand.

ROUGH RIVER HUMOR FESTIVAL. Mar 31-Apr 1. Falls of Rough, KY. Festival includes variety shows, plus workshops on how to tell funny stories and jokes. Est attendance: 200. For info: Rough River Dam State Resort PK, Tom DeHaven, Rec Super, Falls of Rough, KY 40119. Phone: (502) 257-2311.

SKAGIT VALLEY TULIP FESTIVAL. Mar 31-Apr 16. Skagit County, Mount Vernon, WA. To celebrate and share the spectacular beauty of more than 1,500 acres of blooming daffodils and tulips that herald the arrival of spring in the Skagit Valley of Washington state. Est attendance: 500,000. For info: Audrey Smith, SVTF Coord, PO Box 1007, Mt Vernon, WA 98273. Phone: (206) 428-8547.

SOVIET GEORGIA VOTES FOR INDEPENDENCE: ANNIVERSARY. Mar 31. On Mar 31, 1991, the Soviet Republic of Georgia voted to declare their republic independent from the Soviet Union. Georgia followed the Baltic States of Lithuania, Estonia and Latvia by becoming the fourth republic to reject Mikhail Gorbachev's new vision of the Soviet Union as espoused in a new Union Treaty. Vote totals revealed that 98.9 percent of those voting favored independence from Moscow. Hours after the election, troops were dispatched from Moscow to Georgia under a state of emergency.

VIRGIN ISLANDS: TRANSFER DAY. Mar 31. Commemorates transfer resulting from purchase of the Virgin Islands by the US from Denmark, Mar 31, 1917, for $25 million.

BIRTHDAYS TODAY

Herb Alpert, 60, musician (Tijuana Brass), born Los Angeles, CA, Mar 31, 1935.

Leo Buscaglia (Felice Leonardo Buscaglia), 70, educator, author, motivational speaker, born Los Angeles, CA, Mar 31, 1925.

Richard Chamberlain, 60, actor ("Dr. Kildare," *Shogun*; an accomplished Shakespearian actor), born at Los Angeles, CA, Mar 31, 1935.

Liz Claiborne, 66, fashion designer, born at Brussels, Belgium, Mar 31, 1929.

William Daniels, 68, actor (Emmy for "St. Elsewhere"; voice of KITT on "Knight Rider"), born Brooklyn, NY, Mar 31, 1927.

Albert Gore, Jr, 47, 45th US vice president, born at Washington, DC, Mar 31, 1948.

Gordie Howe, 67, former hockey player, born at Saskatoon, Saskatchewan, Canada, Mar 31, 1928.

John Jakes, 63, author (*California Gold, In the Big Country*), born at Chicago, IL, Mar 31, 1932.

Shirley Jones, 61, singer, actress ("The Partridge Family," *Elmer Gantry, Oklahoma!*), born Smithton, PA, Mar 31, 1934.

Gabe Kaplan, 49, actor ("Welcome Back Kotter"), born at Brooklyn, NY, Mar 31, 1946.

Richard Kiley, 73, actor (Emmy for "The Thorn Birds"; *The Blackboard Jungle*), born at Chicago, IL, Mar 31, 1922

Patrick J. Leahy, 55, US Senator (D, Vermont), born at Montpelier, VT, Mar 31, 1940.

Ed Marinaro, 44, actor ("Hill Street Blues," "Sisters"), born at New York, NY, Mar 31, 1951.

Rhea Perlman, 47, actress (Karla Tortelli in "Cheers" [three Emmy Awards]; *Carwash*), born at Brooklyn, NY, Mar 31, 1948.

Christopher Walken, 52, actor (*Deerhunter, Batman Returns*), born Queens, NY, Mar 31, 1943.

☆ Chase's 1995 Calendar of Events ☆ Apr 1

Aprill.

APRIL 1 — SATURDAY
91st Day — Remaining, 274

ALCOHOL AWARENESS MONTH. Apr 1–30. To help raise awareness among community prevention leaders and citizens of underage drinking. Concentrates on community grassroots, activities. Sponsor: Rebecca Fenson, Public Info Asst, 12 W 21st St, New York, NY 10010.

APRIL FOOLS' OR ALL FOOLS' DAY. Apr 1. "The joke of the day is to deceive persons by sending them upon frivolous and nonsensical errands; to pretend they are wanted when they are not, or, in fact, any way to betray them into some supposed ludicrous situation, so as to enable you to call them 'An April Fool.'"—Brady's *Clavis Calendaria*, 1812. "The first of April, some do say, Is set apart for All Fools' Day, But why the people call it so, Nor I nor they themselves do know."—*Poor Robin's Almanack* for 1760.

ASPEN/SNOWMASS BANANA SEASON. Apr 1–8. Snowmass Village, CO. Events: Banana Bonanza Hunt; High Altitude Beach Party and volleyball games; Bartenders' Banana Brawl Drink Contest; Banana Sidewalk Sale. For the Banana Bonanza Hunt, hundreds of prize-filled plastic bananas are hidden on the ski slopes. Clues lead to one "Big Banana Prize." Est attendance: 3,000. For info: Snowmass Resort Assn, PO Box 5566, Snowmass Village, CO 81615. Phone: (303) 923-2000.

◆ **BATTLE OF OKINAWA BEGINS: 50th ANNIVERSARY.** Apr 1, 1945. On Easter Sunday, the US 10th Army began operation *Iceberg*, the invasion of the Ryukyu Islands of Okinawa. Ground troops numbering 180,000 plus 368,000 men in support services made a total of 548,000 troops involved—the biggest amphibious operation of the Pacific war.

BRIDGE OVER THE NEPONSET: ANNIVERSARY. Apr 1. The first bridge built in the US spanned the Neponset River between Milton and Dorcheser, MA. The authority to build the bridge and an adjoining mill was issued to Israel Stoughton, on Apr 1, 1634, by the Massachusetts General Court.

BULGARIA: ST. LASARUS'S DAY. Apr 1. Ancient Slav holiday of young girls, in honor of the goddess of spring and love.

CANADA: ELMIRA MAPLE SYRUP FESTIVAL. Apr 1. Elmira, Ontario. Tours of maple bush by hay wagon, sugaring-off shanty in operation, Pennsylvania Dutch cuisine, handcrafted goods, arts and crafts and antiques. Est attendance: 40,000. For info: Elmira & Woolwich Chamber of Commerce, 5 First St E, Elmira, Ont, Canada N3B 2E3. Phone: (519) 669-2605.

	S	M	T	W	T	F	S
April							1
1995	2	3	4	5	6	7	8
	9	10	11	12	13	14	15
	16	17	18	19	20	21	22
	23	24	25	26	27	28	29
	30						

CAROLINA MODEL RAILROADERS' SCALE MODEL TRAIN SHOW AND SWAP MEET. Apr 1. The Old Southern Railway Passenger Station, Greensboro, NC. Club's "HO" "N" and "HON3" scale layouts on display. The train swap will have dealer tables with brass, books, rolling stock and accessories. Annually, the first Saturday in April. Admission $3. Est attendance: 750. For info: Neil Jones, c/o Carolina Model Railroaders, PO Box 313, Jamestown, NC 27282. Phone: (910) 674-0576.

CHASE'S DEADLINE APPROACHING. Apr 1. Time to plan ahead. Schedule 1996 celebrations and observances and submit information to *Chase's 1996 Calendar of Events* by May 8, 1995. Sponsors/information suppliers of events in this book should have received confirmation/revision ("Re-Up") forms for the 1996 edition by this time. Sponsor: Chase's Editor, Chase's Calendar of Events, 2 Prudential Plaza, Ste 1200, Chicago, IL 60601. Phone: (312) 540-4500. Fax: (312) 540-4687.

COMMUNITY SERVICES MONTH IN CALIFORNIA. Apr 1–30. Focuses attention on the variety of community services provided by California cities; Child Care, Recreation Services, Housing Services, Public Information, Social Services, Health Services, Library and Art Programs. All are essential to the development and maintenance of a healthy and well-balanced community and contribute significantly to the vitality and quality of life in our cities. Annually, the month of April. Sponsor: League of California Cities, 1400 K St, Sacramento, CA 95814. Phone: (916) 444-5790. For info: INTERCOM, 10601 Magnolia Ave, Santee, CA 92071-1222. Phone: (619) 258-4100.

COOKING IN PARADISE. Apr 1–8. (2nd class, Apr 8–15.) Sea Horse Hotel, St. Barthelemy. Renowned cooking instructor and James Beard Award winner Steven Raichlen offers his 6th annual "Cooking in Paradise" class, focusing on Creole and Caribbean cuisine in sophisticated St. Barts. The course also includes visits to restaurants, a gourmet picnic and a sailing cruise. Est attendance: 15. For info: Barbara Raichlen, 1746 Espanola Dr, Coconut Grove, FL 33133. Phone: (305) 854-9550.

DAYTONA BEACH SPRING '95 SPEEDWAY SPECTACULAR. Apr 1–2. Daytona International Speedway, Daytona Beach, FL. Car show of all makes and models of collector vehicles. Many car clubs make this their largest annual event. Show includes display of antiques, classics, sportscars, muscle cars, race cars, custom and special interest vehicles on the speedway infield both days, with a large swap meet of auto parts and accessories. Auto art, collector car sales corral, crafts sale. Also, marine dealers boat show. Annually, the last or next to the last weekend in March and first weekend in April. Est attendance: 25,000. For info: Daytona Beach Racing and Recreational Facilities District, Rick D'Louhy, Exec Dir, PO Box 1958, Daytona Beach, FL 32115-1958. Phone: (904) 255-7355.

DEFEAT AT FIVE FORKS: ANNIVERSARY. Apr 1. After withdrawing to Five Forks, VA, Confederate troops under George Pickett were defeated and cut off by Union troops on Apr 1, 1865. This defeat, according to many military historians, sealed the immediate fate of Robert E. Lee's armies at Petersburg and Richmond. On Apr 2, Lee informed Confederate President Jefferson Davis that he would have to evacuate Richmond. Davis and his cabinet fled by train to Danville, VA.

EGGSIBIT '95. Apr 1–2. Firth Youth Center, Phillipsburg, NJ. 25th anniversary show to encourage the art of decorating egg shells. Annually, the weekend prior to Palm Sunday. Est attendance: 3,000. For info: Kit Stansbury, Dir, 525/419 Fisher Ave, Phillipsburg, NJ 08865. Phone: (908) 859-4496.

ENGLAND: OXFORD v CAMBRIDGE UNIVERSITY BOAT RACE. Apr 1. Putney to Mortlake, River Thames, London. Popular amateur rowing event held since 1829 with crowds watching from towpaths along the river bank. For info: Alison Moor, Press Office, Scope Communications Mgmnt, Towers House, 8-14 Southampton St, London, England WC2E 7HA.

Apr 1 ☆ Chase's 1995 Calendar of Events ☆

EXCHANGE CLUB CHILD ABUSE PREVENTION MONTH. Apr 1-30. Nationwide effort to raise awareness of child abuse and how to prevent it. Sponsor: The Natl Exchange Club, Foundation for Prevention of Child Abuse, 3050 Central Ave, Toledo, OH 43606-1700. Phone: (419) 535-3232.

GRAHAM, MARTHA: DEATH ANNIVERSARY. Apr 1. Martha Graham, renowned dancer and choreographer, died Apr 1, 1991, at the age of 96. Graham is credited with creating the first lasting alternative dance idiom to classical ballet.

GRAND STRAND FISHING RODEO. Apr 1-Oct 31. Myrtle Beach, SC. Awards for surf, inlet and deep-sea fish catches. Est attendance: 700. For info: Chamber of Commerce, Box 2115, Myrtle Beach, SC 29578. Phone: (803) 626-7444.

◆ **HITLER MOVES TO BUNKER: 50th ANNIVERSARY.** Apr 1, 1945. With the Red Army relentlessly approaching Berlin, Adolf Hitler moved his headquarters to a bunker 50 feet below the Reich Chancellery.

HOLY HUMOR MONTH. Apr 1-30. To recognize the healing power of Christian joy, humor and celebration; to be "Fools for Christ" on April Fool's Day (Apr 1); to celebrate the Easter Monday of Western Christianity (Apr 17) and the Easter Monday of Eastern Christianity (Apr 24). Churches and prayer groups nationwide participate. Sponsor: The Fellowship of Merry Christians, Inc, PO Box 895, Portage, MI 49081-0895. Phone: (616) 324-0990.

HOME IMPROVEMENT TIME. Apr 1-Sept 30. To explain the investment advantages of spending disposable income for home improvement to create better family living and improved community environment. (May is a promotion focal point.) For info: Home Improvement Time, J.A. Stewart, Sr, Program Admin, 7425 Steubenville Pike, Oakdale, PA 15071.

HOPE SLED DOG FRIENDSHIP RUN (INTERNATIONAL-INTERCONTINENTAL SLED DOG RACE). Apr 1. Nome, Alaska to Anadyr, Chukotka, Siberia, Russia. 1,000+-mile sled dog race across the Bering Straits in some of the most desolate lands in the world. Fax/msg: (907) 443-5777. For info: Leo Rasmussen, Box 2, Nome, AK 99762-0002. Phone: (907) 443-2798.

INDIAN DANCE AND CRAFTS FESTIVAL. Apr 1-2. DeSoto Caverns Park, Childersburg, AL. Artists and craftspeople from across the US. Indian dancing with dances from Oklahoma and other states. 20th annual festival. Est attendance: 11,000. For info: Rebecca Grevas, DeSoto Caverns Pkwy, Childersburg, AL 35044. Phone: (205) 378-7252.

INTERNATIONAL AMATEUR RADIO MONTH. Apr 1-30. To disseminate information about the important part amateur radio operators or "hams" throughout the world are playing in promoting friendship, peace and good will. To obtain complete information about becoming an International Good Will Ambassador as well as a list of amateur radio operators in many countries, send $1 to cover expense of printing, handling and postage. Annually, the month of April. Sponsor: Intl Soc of Friendship and Good Will, Dr. Stanley Drake, Pres, 9538 Summerfield St, Spring Valley, San Diego County, CA 91977-2852. Phone: (619) 466-8882.

INTERNATIONAL GUITAR MONTH. Apr 1-30. To celebrate world-wide popularity of all types of guitars. Annually, the month of April. Sponsor: Guitar & Accessories Mktg Assn, Jerome Hershman, Exec VP, 38 W 21st St, 5th Fl, New York, NY 10010-6906. Phone: (212) 924-9175.

April 1995

S	M	T	W	T	F	S
						1
2	3	4	5	6	7	8
9	10	11	12	13	14	15
16	17	18	19	20	21	22
23	24	25	26	27	28	29
30						

INTERNATIONAL TWIT AWARD MONTH. Apr 1-30. Any famous name (celebrity with the worst sense of humor) is eligible for most Tiresome Wit (TWIT) of 1995. For info: Lone Star Publications of Humor, Attn: Lauren Barnett, Box 29000, Ste #103, San Antonio, TX 78229.

KEEP AMERICA BEAUTIFUL MONTH. Apr 1-30. To educate Americans about their personal responsibility for litter prevention, proper solid waste disposal and environmental improvement through various community projects. Annually, the month of April. Sponsor: Keep America Beautiful, Inc, Anne-Marie Korbut, Media Relations Mgr, 9 W Broad St, Stamford, CT 06902.

KENNEDY CENTER IMAGINATION CELEBRATION. Apr 1-30. Colorado Springs, CO. A national festival program of the John F. Kennedy Center for the Performing Arts, this community-wide celebration has something for everyone: from lively performances that delight the senses to friendly, small-town happenings that warm the heart. Most events are free or at low cost and open to the public. Annually, the month of Apr. Phone (719) 531-6333, ext 1205 or 1206. Est attendance: 188,000. For info: Mary Mashburn, Coord, Pikes Peak Library Dist, 5550 N Union Blvd, Colorado Springs, CO 80918.

LISTENING AWARENESS MONTH. Apr 1-30. To help raise awareness of the need for effective listening skill building by the presentation of listening seminars, book readings, skill-building workshops, newspaper/magazine articles, radio/TV guest interviews and listening parties throughout the USA. Sponsor: Effective Listeners Assn, Christine Carpenter, Dir, PO Box 8716, Portland, OR 97207. Phone: (503) 228-6003.

LONGWOOD GARDENS' ACRES OF SPRING. Apr 1-May 26. Kennett Square, PA. Thousands of spring bulbs, flowering shrubs and trees bloom throughout 1,050 acres of formal gardens, woodlands and meadows. Sunday afternoon organ and chamber music concerts. Est attendance: 145,000. For info: Colvin Randall, PR Mgr, Longwood Gardens, Kennett Square, PA 19348-0501. Phone: (610) 388-6741.

MATHEMATICS EDUCATION MONTH. Apr 1-30. An opportunity for students, teachers, parents and the community as a whole to focus on the importance of mathematics and the changes taking place in the mathematics curriculum. Sponsor: Natl Council of Teachers of Mathematics, Communications Mgr, 1906 Association Dr, Reston, VA 22091-1593. Phone: (703) 620-9840.

MONTH OF THE YOUNG CHILD. Apr 1-30. Michigan. To promote awareness of the importance of young children and their specific needs in today's society. Many communities celebrate with special events for children and families. Phone: (800) 336-6424 or (517) 336-9700. Sponsor: Michigan Assn for Education of Young Children, Beacon Pl, Ste 1-D, 4572 S Hagadorn Road, East Lansing, MI 48823-5385.

MULTICULTURAL COMMUNICATION MONTH. Apr 1-30. To encourage all people to improve their crosscultural knowledge and to inspire all media to explore and communicate that ethnic diversity to the audiences they serve. Workshops designed to improve multicultural literacy and understanding are held for professionals in communication fields. Sponsor: Intl Center of Multicultural Communication, Wanice LaMoyne, Dir, 6100 Wilshire Blvd, 16th Fl, Los Angeles, CA 90048. Phone: (213) 954-9703.

NATIONAL ANXIETY MONTH. Apr 1-30. 5th annual sponsorship of month-long "self awareness" event to encourage a review of individuals' personal anxieties, the media's role in stimulating high levels of national anxiety, and the provision of advice on how to reduce, eliminate, and seek professional assistance to deal with anxieties. Sponsor: The Natl Anxiety Center, Box 40, Maplewood, NJ 07040. Phone: (201) 763-6392.

NATIONAL FLORIDA TOMATO MONTH. Apr 1-30. To publicize Florida tomatoes as a versatile, nutritious, flavorful food. For info: Teri Cirelli-Pedersen, Acct Services, Lewis & Neale, Inc, 928 Broadway, New York, NY 10010. Phone: (212) 420-8808.

NATIONAL GARDEN MONTH. Apr 1-30. National celebration of the joys and benefits of gardening. Annually, the month of April. For info: Vicki Bendure, Garden Council, 10210 Bald Hill Rd, Mitchellville, MD 20721. Phone: (301) 577-4073.

NATIONAL HUMOR MONTH. Apr 1-30. Celebrate the joy of laughter and how it can reduce stress, improve job performance and enrich the quality of our lives. Begins April Fool's Day. Send self-addressed 52¢-stamped envelope for info. Sponsor: Larry Wilde, Founder, Jester Press, 25470 Cañada Dr, Carmel, CA 93923.

NATIONAL KNUCKLES DOWN MONTH. Apr 1-30. To recognize and revive the American tradition of playing marbles and keep it rolling along. Please send self-addressed-stamped envelope with inquiries. Sponsor: Cathy C. Runyan, The Marble Lady, 7812 NW Hampton Rd, Kansas City, MO 64152.

NATIONAL OCCUPATIONAL THERAPY MONTH. Apr 1-30. To recognize the services and accomplishments of occupational therapy practitioners and promote awareness of the benefits of occupational therapy. Sponsor: The American Occupational Therapy Assn, Inc, Suzanne Carleton, 1383 Piccard Dr, Rockville, MD 20850. Phone: (301) 948-9626.

NATIONAL SEXUALLY TRANSMITTED DISEASES (STDs) EDUCATION AND AWARENESS MONTH. Apr 1-30. To educate consumers, patients, students and professionals about the prevention of sexually transmitted diseases. Kit of materials available for $15. Fax: (415) 332-1832. Sponsor: Pharmacists Planning Service, Inc, Frederick Mayer, PO Box 1336, Sausalito, CA 94966. Phone: (415) 332-4066.

NATIONAL WELDING MONTH. Apr 1-30. A celebration of welding and how the process has helped all Americans live better lives. Used in products ranging from toasters to spacecraft, welding touches the lives of everyone. In addition, the field of welding offers a wide variety of opportunities for personal career growth and job satisfaction. Welding Month will involve 145 AWS local sections. Phone: (800) 443-WELD. For info: Mktg Communications Dept, Amer Welding Soc, 550 NW LeJeune Rd, Miami, FL 33126.

NATIONAL WOODWORKING MONTH. Apr 1-30. To focus attention on the beauty and satisfaction of working with wood. To help motivate Americans to undertake woodworking projects to improve their home and general surroundings. To increase consumers' knowledge and skill in woodworking and wood finishing endeavors. For info: Gilbert Whitney and Johns, Inc, Minwax Natl Woodworking Month Program, 110 S Jefferson Rd, Whippany, NJ 07981.

NCAA DIVISION I WOMEN'S BASKETBALL CHAMPIONSHIP FINALS. Apr 1-2. Richmond Coliseum, Richmond, VA. Est attendance: 17,500. For info: Natl Collegiate Athletic Assn, 6201 College Blvd, Overland Park, KS 66211. Phone: (913) 339-1906.

NEW ENGLAND SPEAKERS ASSOCIATION "BUILDING PROFESSIONAL COMPETENCE SPEAKING" MONTH. Apr 1-30. Professional speakers from around the country share practical ideas and techniques for enhancing personal and professional skills, speaking, marketing and business at the NESA's annual Speakers' School held Apr 7-9 in Massachusetts. Open to professionals from any field, options are offered for every skill level, individualized learning and networking. For info: NESA, PO Box 812, West Side Stn, Worcester, MA 01602-0812. Phone: (508) 799-9860.

POLE, PEDAL, PADDLE. Apr 1. Jackson, WY. The original relay race of this kind. Alpine skiing, cross country skiing, biking, and boating from Teton Village to the Snake River. Sponsor: Jackson Hole Ski Club. For info: Jackson Hole Chamber of Commerce, Box E, Jackson, WY 83001. Phone: (307) 733-3316.

POLO: WORLD CUP. Apr 1-16. Wellington, FL. Est attendance: 15,000. For info: Palm Beach Polo and Country Club, 13420 South Shore Blvd, Wellington, FL 33414. Phone: (407) 798-7634.

PONY EXPRESS–JESSE JAMES DAYS. Apr 1-2. Patee House and Pony Express National Memorial, St. Joseph, MO. Events in two locations celebrate the beginning of the Pony Express and the ending of Jesse James. Annually, the weekend nearest Apr 3. For info: Patee House, 12 and Penn, St. Joseph, MO 64503. Phone: (816) 232-8206.

PRAIRIE DOG CHILI COOKOFF AND WORLD CHAMPIONSHIP OF PICKLED QUAIL-EGG EATING. Apr 1-2. Traders Village, Grand Prairie, TX. Tongue-in-cheek salute to the official state dish of Texas, chili con carne, or "Texas Red." World championship of pickled quail-egg eating featuring contestants devouring as many of these gourmet delights as possible in the 60-second time limit. Est attendance: 57,000. For info: Traders Village, Doug Beich, 2602 Mayfield Rd, Grand Prairie, TX 75052. Phone: (214) 647-2331.

PREVENTION OF ANIMAL CRUELTY MONTH. Apr 1-30. Animals are defenseless, feeling creatures. In most states it is against the law to abandon a pet or cause it any malicious harm. To find out how to join the ASPCA's campaign to alleviate pain, fear and suffering in the lives of all animals, send SASE. Phone: (212) 876-7700, ext 4655. Sponsor: American Soc for Prevention of Cruelty to Animals (ASPCA), 424 E 92nd St, New York, NY 10128.

PRO-AM SNIPE EXCURSION AND HUNT. Apr 1. Moultrie, GA. Celebrating the time-honored custom of snipe hunting. The Denim Snipe has come back from the brink of extinction and will be honored at the 1995 event which will include a Snipe Parade, Snipe Ball and festivities at the Denim Wing of the Snipe Museum at New Elm, GA. New Snipe-O-Rama racing oval open! Annually, Apr 1 For info: Beth Gay, PO Box 1110, Moultrie, GA 31776. Phone: (912) 985-6540.

ROBERT THE HERMIT: ANNIVERSARY. Apr 1. One of the most famous hermits in American history died in his hermitage at Seekonk, MA, on Apr 1, 1832. Robert was a bonded slave, the son of an African mother and probably an Anglo-Saxon father. After obtaining his freedom, he was swindled out of it and shipped to a foreign slave market, then later he escaped to America. He was separated from his first wife by force and rejected by his second wife after a long sea voyage, before withdrawing from society.

Apr 1-2 ☆ Chase's 1995 Calendar of Events ☆

SEA CADET MONTH. Apr 1-30. Nationwide youth program for boys and girls 11 through 17 teaching leadership and self-discipline with emphasis on Navy-oriented training without military obligation. Sponsor: US Naval Sea Cadet Corps, 2300 Wilson Blvd, Arlington, VA 22201. Phone: (703) 243-6910.

SEAPORT SENIORS MONTH. Apr 1-30. Mystic Seaport, Mystic, CT. Everyone 50 years and older receives $2 off the regular admission every day during the month of April. Discount coupons included for shopping, dining. Sponsor: Mystic Seaport, 75 Greenmanville, Box 6000, Mystic, CT 06355. Phone: (203) 572-5315.

SFCC SPRING ARTS FESTIVAL. Apr 1-2. Northeast First St, and the Thomas Center, Gainesville, FL. Artists and craftsmen from all areas of the US and the Children's Rainbow-Garden—a complete art fest for kids. 26th annual festival. Est attendance: 80,000. Sponsor: Santa Fe Community College, 3000 NW 83rd St, Gainesville, FL 32606. Phone: (904) 395-5355.

SORRY CHARLIE DAY. Apr 1. To honor Charlie the Tuna, who has been rejected for 30 years and still keeps his spunk. A day to recognize anyone who has been rejected, and lived through it. Join the "Sorry Charlie, No-Fan-Club-For-You Club." Please send SASE envelope with inquiries. Sponsor: Cathy Runyan, 7812 NW Hampton Rd, Kansas City, MO 64152.

SPORTS EYE SAFETY MONTH. Apr 1-30. During this month's observance, Prevent Blindness America (formerly known as the National Society to Prevent Blindness), will encourage young athletes to wear eye/face protection when participating in sports. Materials that can easily be posted or distributed to the community will be provided. Sponsor: Prevent Blindness America, Marita Gomez, Media Relations, 500 E Remington Rd, Schaumburg, IL 60173. Phone: (800) 331-2020.

STRESS AWARENESS MONTH. Apr 1-30. To promote public awareness of what stress is, what causes it to occur and what can be done about it. A month-long focus on the dangers of stress, successful coping strategies and the myths about stress that are prevalent in our society. Sponsor: The Health Resource Network, Morton C. Orman, MD, Dir, 2936 E Baltimore St, Baltimore, MD 21224. Phone: (410) 732-1900.

TATER DAYS. Apr 1-3. Benton, KY. Nation's oldest trade day has been held annually since 1843. Parade; horse and mule pulling contests; arts and crafts; beauty pageant; quilt show and flea market. Est attendance: 20,000. For info: Chamber of Commerce, Rte 8, Box 3, Benton, KY 42025. Phone: (502) 527-7665.

TEN BEST CENSORED STORIES OF 1994. Apr 1. To announce the top ten underreported news stories of 1994 as determined by a national panel of jurors. On Apr 1 a 250–300 page Sourcebook detailing the "Top 25 Censored Stories of 1994" will be released by Project Censored. Fax: (707) 664-2505. For info: Mark Lowenthal, Asst Dir, Project Censored, Sonoma State University, Rohnert Park, CA 94928. Phone: (707) 664-2500.

THAI HERITAGE MONTH. Apr 1-30. "To get together and to learn about the Thai Heritage, in order to promote prosperity of life and harmonious living in the community, country and the world." Month's activities include Thai New Year Day, Songkran Day and Water Festival. Annually, the month of April. For info: Office of Information, Royal Thai Embassy, Washington, DC; phone: (202) 944-3625. Sponsor: Prasong Nurack, *Thai Network*, 215 E Walnut St, Des Moines, IA 50309. Phone: (515) 282-0044.

US HOUSE OF REPRESENTATIVES ACHIEVES A QUORUM: ANNIVERSARY. Apr 1. First session of Congress was held Mar 4, 1789 but not enough representatives arrived to achieve a quorum until Apr 1.

VAN AND CONNIE INTERNATIONAL LEISURE SUIT CONVENTION. Apr 1. Val Air Ballroom, Des Moines, IA. While the main event is mingling, the convention also features disco dance lessons, white shoe throw, leisure suit fashion show and white belt toss. Awards presented for various categories of leisure suits, including "most flammable." 7th annual convention. Sponsor: Val Air Ballroom. Est attendance: 2,000. For info: WHO Radio, Van Harden, 1801 Grand Ave, Des Moines, IA 50309. Phone: (515) 242-3671.

ZOO FLING '95. Apr 1-30. (Also May 1-30.) Asheboro, NC. Daily. Free Flight-a spectacular education program featuring birds of prey in demonstrations of natural flight. Est attendance: 60,000. Sponsor: North Carolina Zoological Park, Events and Projects, 4401 Zoo Pkwy, Asheboro, NC 27203. Phone: (800) 488-0444.

BIRTHDAYS TODAY

Kevin Jerome Duckworth, 31, basketball player, born at Harvey, IL, Apr 1, 1964.
David Eisenhower, 48, author, lawyer, grandson former president Dwight Eisenhower (*Eisenhower at War*), born West Point, NY, Apr 1, 1947.
Ali MacGraw, 56, actress (*Goodbye, Columbus; Love Story*), born at Pound Ridge, NY, Apr 1, 1939.
Jane Powell (Suzanne Burce), 66, actress (*Seven Brides for Seven Brothers*), born at Portland, OR, Apr 1, 1929.
Debbie Reynolds, 63, actress (*Singin' in the Rain, Tammy and the Bachelor*), born at El Paso, TX, Apr 1, 1932.
Libby Riddles, 39, first woman to win the 1,135-mile Iditarod Alaskan dogsled race, in '85, born Madison, WI, Apr 1, 1956.
Glen Edward "Bo" Schembechler, 66, baseball executive, former football coach, born Barberton, OH, Apr 1, 1929.
Rusty Staub, 51, former baseball player, born at New Orleans, LA, Apr 1, 1944.

APRIL 2 — SUNDAY
92nd Day — Remaining, 273

ANDERSEN, HANS CHRISTIAN: 190th BIRTH ANNIVERSARY. Apr 2. Author chiefly remembered for his more than 150 fairy tales, many of which are classics of children's literature. Andersen was born at Odense, Denmark, on Apr 2, 1805, and died at Copenhagen, Denmark, on Aug 4, 1875.

ARGENTINA: NATIONAL HOLIDAY. Apr 2. Commemoration of the intent of recovery of the Malvinas Islands.

ARTHRITIS FOUNDATION NATIONAL TELETHON. Apr 2. Fundraiser for programs of research, patient services and public and professional education for the 37 million Americans affected by arthritis. Sponsor: Arthritis Foundation, Natl Office, Roy Scott, 1314 Spring St NW, Atlanta, GA 30309. Phone: (404) 872-7100.

BERTHOLDI, FREDERIC AUGUSTE: BIRTH ANNIVERSARY. Apr 2. French sculptor who created *Liberty Enlightening the World*, which stands in New York Harbor. Also remembered for the *Lion of Belfort* in Belfort, France. Born at Colman, in Alsace, France, on Apr 2, 1834. Died at Paris, France, Oct 4, 1904.

CASANOVA, GIACOMO GIROLAMO: 270th BIRTH ANNIVERSARY. Apr 2. Celebrated Italian writer-librarian and, by his own account, philanderer, adventurer, rogue, seminarian, soldier and spy. Born at Venice, Italy, Apr 2, 1725. As the Chevalier de Seingalt, he died at Dux, Bohemia, June 4, 1798, while serving as librarian and working on his lively and frank *History of My Life*, a brilliant picture of 18th-century life.

	S	M	T	W	T	F	S
April							1
	2	3	4	5	6	7	8
1995	9	10	11	12	13	14	15
	16	17	18	19	20	21	22
	23	24	25	26	27	28	29
	30						

✩ Chase's 1995 Calendar of Events ✩ Apr 2-3

CONSIDER CHRISTIANITY WEEK. Apr 2-8. A week to encourage Christians to examine the evidence and reasons for their faith, and for Non-Christians to take another look at the faith that has played such an important role in shaping the history and culture in which we live. Annually, the week beginning two Sundays before Easter. Sponsor: Aletheia Publishing, Hanna Hushbeck, PR, 1814 Commercenter W, Ste G, San Bernadino, CA 92408. Phone: (909) 794-8941.

ENGLAND: CARE SUNDAY. Apr 2. The fifth Sunday of Lent, also known as Carling Sunday and Passion Sunday. First day of Passiontide, remembering the sorrow and passion of Christ.

FALKLAND ISLANDS WAR: ANNIVERSARY. Apr 2-June 15. Argentina, claiming sovereignty over the nearby Falkland Islands (called by them the Malvinas), invaded and occupied the British Crown Colony on Apr 2, 1982. British forces defeated the Argentinians on June 15, 1982. About 250 British and 600 Argentine lives were lost in the conflict. In 1986, three military officers, including General Leopoldo Galtieri (who was president of Argentina at the time of the invasion), were convicted and sentenced for the military crime of negligence. Commemorative ceremonies are observed on Apr 2 in Argentina and on June 15 in the Falkland Islands.

FIRST WHITE HOUSE EASTER EGG ROLL: ANNIVERSARY. Apr 2. The first White House Easter Egg Roll took place on Apr 2, 1877, during the administration of Rutherford B. Hayes. The traditional event was discontinued by President Franklin D. Roosevelt in 1942 and reinstated on Apr 6, 1953, by President Dwight D. Eisenhower.

GREECE: DUMB WEEK. Apr 2-8. The week preceding Holy Week is known as Dumb Week, as no services are held in churches throughout this period except on Friday, eve of the Saturday of Lazarus.

INTERNATIONAL CHILDREN'S BOOK DAY. Apr 2. Commemorates the international aspects of children's literature and observes Hans Christian Andersen's birthday. Sponsor: Intl Bd on Books for Young People, Nonnenweg 12, Postfach, CH-4003 Basel, Switzerland. For info: USBBY Secretariat, c/o Intl Reading Assn, Box 8139, Newark, DE 19714-8139.

NATIONAL BIRTHPARENTS WEEK. Apr 2-8. A time to recognize and honor the 5 million mothers and fathers who, at great personal sacrifice, have been separated from their children by adoption and to focus attention on the rights of all family members to know each other. Annually, the first full week in April. Est attendance: 100. Sponsor: Concerned United Birthparents, Inc (CUB), 2000 Walker St, Des Moines, IA 50317. Phone: (800) 822-2777.

PASSION WEEK. Apr 2-8. The week beginning on the fifth Sunday in Lent; the week before Holy Week.

PASSIONTIDE. Apr 2-15. The last two weeks of Lent (Passion Week and Holy Week), beginning with the fifth Sunday of Lent (Passion Sunday) and continuing through the day before Easter (Holy Saturday or Easter Even).

PONCE DE LEON DISCOVERS FLORIDA. Apr 2. On Apr 2, 1513, Juan Ponce de Leon discovered Florida by landing at the site that became the city of St. Augustine. He claimed the land for the King of Spain.

STRAW HAT WEEK. Apr 2-9. A week of celebration during which the felt has is put aside in favor of the straw hat by both men and women. Local businesses are encouraged to hat-related activities. Annually, about two weeks before Easter. For info: Casey Bush, Exec Dir, Millinery Info Bureau, 302 W 12 St, PH-C, New York, NY 10014. Phone: (212) 627-8333.

US: DAYLIGHT-SAVING TIME. Apr 2-Oct 29. Daylight-Saving Time begins at 2:00 AM. The Uniform Time Act of 1966 (as amended in 1986 by Public Law 99-359), administered by the US Dept of Transportation, provides that Standard Time in each zone be advanced one hour from 2:00 AM on the first Sunday in April until 2:00 AM on the last Sunday in October (except where state legislatures provide exemption). Many use the popular rule "spring forward, fall back," to remember which way to turn their clocks. See also: "Standard Time" (Oct 29).

US MINT: ANNIVERSARY. Apr 2. The first US Mint was established at Philadelphia, PA, as authorized by act of Congress dated Apr 2, 1792.

ZOLA, EMILE: BIRTH ANNIVERSARY. Apr 2. Prolific French novelist of the Naturalist School, remembered especially for his role in the Dreyfus case (resulting in retrial and vindication of Alfred Dreyfus). Emile Edouard Charles Antoine Zola was born at Paris, France, Apr 2, 1840. Defective venting of a stove flue in his bedroom (which some believed to be the work of political enemies) resulted in his death from carbon monoxide poisoning at Paris, Sept 28, 1902.

BIRTHDAYS TODAY

Dana Carvey, 40, comedian, actor (*Wayne's World*, "Saturday Night Live"), born Missoula, MT, Apr 2, 1955.
Buddy Ebsen, 87, actor (Jed Clampett on "Beverly Hillbillies;" "Barnaby Jones"), born at Belleville, IL, Apr 2, 1908.
Emmylou Harris, 47, singer ("Amarillo," "Till I Gain Control Again"), born at Birmingham, AL, Apr 2, 1947.
Linda Hunt, 50, actress (Mother Superior on "The Flying Nun"; Mae in *Cat on a Hot Tin Roof*), born Moristown, NJ, Apr 2, 1945.
Leon Russell, 54, musician, born at Lawton, OK, Apr 2, 1941.

APRIL 3 — MONDAY
93rd Day — Remaining, 272

ARMENIAN APPRECIATION DAY. Apr 3. Lighthearted look at the contribution of legendary Armenians such as Palboonian (Paul Bunyan) and Torontonian to American folklore, with special emphasis on studies of the relationship of the Smithsonian Institution Collection to the history of Armenian-American folklore. Annually, Apr 3. Sponsor: Puns Corps, c/o Robert L. Birch, Box 2364, Falls Church, VA 22042-0364. Phone: (703) 533-3668.

BLACKS RULED ELIGIBLE TO VOTE: ANNIVERSARY. Apr 3. On Apr 3, 1944, the United States Supreme Court, in an 8-1 ruling, declared that blacks could not be barred from voting in the Texas Democratic primaries. The high court repudiated the contention that political parties are private associations and held that discrimination against blacks violated the 15th Amendment.

BURROUGHS, JOHN: BIRTH ANNIVERSARY. Apr 3. American naturalist and author. Born at Roxbury, NY, Apr 3, 1837. "Time does not become sacred to us until we have lived it," he wrote in 1877. Died Mar 29, 1921.

CHICKEN LITTLE AWARDS. Apr 3. 5th annual awards to organizations and/or individuals "who have frightened the daylights out of large numbers of people" with scientifically-dubious predictions, theories and statements. For info: Alan Caruba, Natl Anxiety Center, PO Box 40, Maplewood, NJ 07040. Phone: (201) 763-6392.

DON'T GO TO WORK UNLESS IT'S FUN DAY. Apr 3. Two-thirds of the American workforce are unhappy in their jobs, yet few know what to do about it. This day is designed to encourage people to enjoy work, not just endure it. Sponsor: Frank Sanitate, 1152 Camino Manadero, Santa Barbara, CA 93111. Phone: (805) 967-7899.

Apr 3-4 ☆ Chase's 1995 Calendar of Events ☆

FALL OF RICHMOND: ANNIVERSARY. Apr 3. After the withdrawal of Robert E. Lee's troops, the Confederate capital of Richmond and nearby Petersburg surrendered to Union forces on Apr 3, 1865. Richmond had survived four years of continuous threats from the North. On Apr 4, the city was toured by President Abraham Lincoln.

GRAHAM, CALVIN "BABY VET": BIRTH ANNIVERSARY. Apr 3. The man who became known as World War II's "baby vet," Calvin Graham was born Apr 3, 1930, at Canton, TX. Graham enlisted in the Navy at the age of 12. As a gunner on the USS *South Dakota*, he was struck by shrapnel during the battle of Guadalcanal in 1942 but still helped pull fellow crew members to safety. The navy gave Graham a dishonorable discharge, revoked his disability benefits and stripped him of his decorations, including a Purple Heart and Bronze Star, after discovering his age. Eventually, through congressional efforts, he was granted an honorable discharge and won back all but the Purple Heart. His benefits were restored in 1988. Graham died Nov 6, 1992 at Fort Worth, TX.

INAUGURATION OF PONY EXPRESS: ANNIVERSARY. Apr 3. The Pony Express began on Apr 3, 1860, when the first rider left St. Joseph, MO. The following day another rider headed east from Sacramento, CA. For $5 an ounce letters were delivered within 10 days. There were 190 way stations between 10 and 15 miles apart, and each rider had a "run" of between 75 and 100 miles. The Pony Express lasted less than two years, ceasing operation in Oct, 1861, when the overland telegraph was completed.

IRVING, WASHINGTON: BIRTH ANNIVERSARY. Apr 3. American author, attorney and one time US Minister to Spain, Irving was born at New York, NY, Apr 3, 1783. Creator of *Rip Van Winkle* and *The Legend of Sleepy Hollow*. Also author of many historical and biographical works, including *A History of the Life and Voyages of Christopher Columbus* and the *Life of Washington*. Died at Tarrytown, NY, Nov 28, 1859.

PATENT GRANTED FOR HAT BLOCKING MACHINE: ANNIVERSARY. Apr 3. On Apr 3, 1866, Rudolph Eickemeyer and G. Osterheld were granted the first patent for a hat blocking and shaping machine.

RAINEY, MA (GERTRUDE BRIDGET): BIRTH ANNIVERSARY. Apr 3. Known as the "Mother of the Blues," Gertrude "Ma" Rainey was born Apr 3, 1888, at Columbus, GA. She made her stage debut at the Columbus Opera House in 1900 in a talent show called "The Bunch of Blackberries." After touring with her husband as "Rainey and Rainey, the Assassinators of the Blues," they eventually separated and she toured on her own under the auspices of the Theater Owners Booking Association. She made her first recording in 1923 and her last on Dec 28, 1928, after being told that the rural southern blues she sang had gone out of style. She died Dec 22, 1939, at Columbus, GA.

RAND, SALLY: BIRTH ANNIVERSARY. Apr 3. American actress, ecdysiast and inventor of the fan dance (which gained fame in 1933 at the Chicago World's Fair). Born Helen Gould Beck at Hickory County, MO, on Apr 3, 1904. Died at Glendora, CA, Aug 31, 1979.

TWEED DAY. Apr 3. Day to consider the cost of political corruption. Birthday of William March Tweed, New York City political boss, whose "Tweed Ring" is said to have stolen $30 to $200 million from the city. Born at New York, NY, Apr 3, 1823. Died in his cell at New York's Ludlow Street Jail, Apr 12, 1878. Cartoonist Thomas Nast deserves much credit for Tweed's arrests and convictions.

	S	M	T	W	T	F	S
April 1995							1
	2	3	4	5	6	7	8
	9	10	11	12	13	14	15
	16	17	18	19	20	21	22
	23	24	25	26	27	28	29
	30						

BIRTHDAYS TODAY

Alec Baldwin, 37, actor ("Knots Landing," *The Getaway, Miami Blues*), born at Massapequa, NY, Apr 3, 1958.

Marlon Brando, 71, actor (Oscars *On the Waterfront, The Godfather*; Emmy *Roots: Next Generation*), born Omaha, NE, Apr 3, 1924.

Doris Day (Doris VonKappelhoff), 71, actress, singer (*Pillow Talk*, "The Doris Day Show"), born at Cincinnati, OH, Apr 3, 1924.

Max Frankel, 65, journalist, born at Gera, Germany, Apr 3, 1930.

Helmut Kohl, 65, Chancellor of Germany, born at Ludwigshafen, Germany, Apr 3, 1930.

Marsha Mason, 53, actress (Oscar nominations for *The Goodbye Girl, Cinderella Liberty*), born at St. Louis, MO, Apr 3, 1942.

Eddie Murphy, 34, comedian, actor (*48 HRS; Beverly Hills Cop* movies), born at Brooklyn, NY, Apr 3, 1961.

Wayne Newton, 53, singer ("Danke Schoen," "Daddy Don't You Walk So Fast"), born at Norfolk, VA, Apr 3, 1942.

Tony Orlando (Michael Orlando Cassivitis), 51, singer (Tony Orlando and Dawn; "Tie a Yellow Ribbon Round the Old Oak Tree"), born New York, NY, Apr 3, 1944.

APRIL 4 — TUESDAY
94th Day — Remaining, 271

BONZA BOTTLER DAY™. Apr 4. To celebrate when the number of the day is the same as the number of the month. Bonza Bottler Day™ is an excuse to have a party at least once a month. T-shirts available. For info: Elaine Fremont, 203 Waddell Rd, Taylors, SC 29687. Phone: (803) 244-2023.

◈ **DIX, DOROTHEA LYNDE: BIRTH ANNIVERSARY.** Apr 4. American social reformer and author. Born at Hampden, ME, on Apr 4, 1802. Left home at age 10, was teaching at age 14 and founded a home for girls in Boston while still in her teens. In spite of frail health, she was a vigorous crusader for humane conditions in insane asylums, jails, and almshouses and for the establishment of state-supported institutions to serve those needs. Named superintendent of women nurses during the Civil War. Died at Trenton, NJ, on July 17, 1887.

FLAG ACT OF 1818: ANNIVERSARY. Apr 4. Congress approved the first flag of the United States on this day in 1818.

GIAMATTI, ANGELO BARTLETT: BIRTH ANNIVERSARY. Apr 4. Baseball Commissioner and former president of Yale University. Born at Boston, MA, on Apr 4, 1938. A. Bartlett Giamatti received his education and taught at Yale University, and later became the youngest person to be named president of the university, at the age of 39, in 1978. Following his tenure at Yale he became the president of Major League Baseball's National League in 1986 and served in that capacity until he was appointed Commissioner of Baseball on Apr 1, 1989. An accomplished author, he moved freely between the worlds of literature and baseball, often linking the two in the many articles he wrote. One week prior to his death, he suspended Pete Rose for life for betting on baseball games. Giamatti died at Martha's Vineyard, MA, on Sept 1, 1989.

HATE WEEK. Apr 4-10. Recognizes the day on which the fictional character Winston Smith started his secret diary and wrote the words "DOWN WITH BIG BROTHER," Wednesday, Apr 4, 1984. From George Orwell's anti-Utopian novel, *1984*, portraying the end of human privacy and the destruction of the individual in a totalitarian state (first published in 1949). "Hates" varied from the daily two-minute concentrated hate to the grand culmination observed during Hate Week.

☆ Chase's 1995 Calendar of Events ☆ Apr 4–5

KING, MARTIN LUTHER, JR: ASSASSINATION ANNIVERSARY. Apr 4. The Reverend Martin Luther King, Jr, was shot at Memphis, TN, Apr 4, 1968. James Earl Ray is serving 99-year sentence for the crime. See also: "King, Martin Luther, Jr: Birth Anniversary" (Jan 15).

NATIONAL READING A ROAD MAP WEEK. Apr 4–10. To promote map reading as an enjoyable pastime and as a survival skill for present and future drivers and all arm-chair travelers. Motto: Happiness is knowing how to read a road map. For info: Rosalind Schilder, PO Box 708, Plymouth Meeting, PA 19462.

NATO FOUNDED: ANNIVERSARY. Apr 4. Delegates from 12 nations met in Washington, DC, to sign the North Atlantic Treaty, Apr 4, 1949.

SALTER ELECTED FIRST WOMAN MAYOR IN US: ANNIVERSARY. Apr 4. The first woman elected mayor in the US was Susanna Medora Salter, who was elected mayor of Argonia, KS, on Apr 4, 1887. Her name had been submitted for election without her knowledge by the Woman's Christian Temperance Union, and she did not know she was a candidate until she went to the polls to vote. She received a two-thirds majority vote and served one year for the salary of $1.00.

SENEGAL: INDEPENDENCE DAY. Apr 4. National holiday.

SMOTHERS BROTHERS' FIRING: ANNIVERSARY. Apr 4. CBS canceled this popular comedy series Apr 4, 1969. The hour-long show strongly influenced television humor during the two years it aired. Tom and Dick, however, frequently found themselves at odds with the censors over material that would be considered tame today. Guests and cast members frequently knocked the Vietnam War and the Nixon Administration. Acts featuring antiwar protestors such as Harry Belafonte were often cut.

SPACE MILESTONE: *STS-6* (US). Apr 4. Shuttle *Challenger* launched from Kennedy Space Center, FL, Apr 4, 1983, with four astronauts (Paul Weitz, Karol Bobko, Story Musgrave and Donald Peterson). Four-hour spacewalk by Musgrave and Peterson. Landed at Edwards Air Force Base, CA, Apr 9.

VICTIMS OF VIOLENCE HOLY DAY. Apr 4. One of the three Kingdom Respect Days Color Commemorated by wearing Black Colors to highlight Anti-Violence, one of the three key principles of the Civil Rights Renaissance of the '60s on the anniversary of Dr. Martin Luther King Jr's assassination to show visible unity, respect and remembrance for slavery victims, present innocent victims, survivors of rape and missing children who are potential victims of violence. See also: "Humanitarian Day" (Jan 15) and "Dream Day" (Aug 28). Phone: (312) RESPECT (737-7328). Sponsor: Global Committee Commemorating Kingdom Respect Days, PO Box 21050, Chicago, IL 60621.

YALE, LINUS: BIRTH ANNIVERSARY. Apr 4. American portrait painter and inventor of the lock that is named for him was born at Salisbury, NY, Apr 4, 1821. He was creator of the Yale Infallible Bank Lock and developer of the cylinder lock. Yale died at New York, NY, Dec 25, 1868.

YAMAMOTO, ISOROKU: BIRTH ANNIVERSARY. Apr 4. Considered Japan's greatest naval strategist, Admiral Isoroku Yamamoto who planned the attack on Pearl Harbor, was born Apr 4, 1884, at Nagaoko, Honshu. Yamamoto also devised the complex attack on Midway Island which ended in defeat for the Japanese because the Allies had the key to the Imperial fleet code and were prepared for the June 4, 1942 attack. The US intercepted reports of Yamamoto's proposed 1943 tour of the Western Solomons and shot down his plane Apr 18, while he was touring Japanese installations in the area.

BIRTHDAYS TODAY

Maya Angelou, 67, author, poet (*All God's Children Need Travelling Shoes*), born at St. Louis, MO, Apr 4, 1928.

Elmer Bernstein, 73, composer, born at New York, NY, Apr 4, 1922.

Kitty Kelley, 53, author (*Jackie Oh!, Nancy Reagan*), born at Hartford, CT, Apr 4, 1942.

Christine Lahti, 45, actress (Maggie Kavanaugh on "The Harvey Korman Show"; *Swing Shift*), born at Detroit, MI, Apr 4, 1950.

Richard G. Lugar, 63, US Senator (R, Indiana) up for reelection in '94, born at Indianapolis, IN, Apr 4, 1932.

William Manchester, 73, author (*The Last Lion, A World Lit Only by Fire*), born at Attleboro, MA, Apr 4, 1922.

Nancy McKeon, 29, actress (Jo on "Facts of Life"), born at Westbury, NY, Apr 4, 1966.

APRIL 5 — WEDNESDAY
95th Day — Remaining, 270

DAVIS, BETTE: BIRTH ANNIVERSARY. Apr 5. Born at Lowell, MA, on Apr 5, 1908, Bette Davis began her remarkable film career in 1931, performing in over 80 films. She was nominated for 10 Academy Awards, receiving two for her performances in *Dangerous* and *Jezebel*. Her films include *Now, Voyager, All About Eve, The Little Foxes, Watch on the Rhine, What Ever Happened to Baby Jane?* and *The Whales of August* among others. Bette Davis died at Paris, France, on Oct 7, 1989.

LISTER, JOSEPH: BIRTH ANNIVERSARY. Apr 5. English physician who was the founder of aseptic surgery. Born Apr 5, 1827, at Upton, Essex, England. Died at Walmer, England, on Feb 10, 1912.

RESNIK, JUDITH A.: BIRTH ANNIVERSARY. Apr 5. Dr. Judith A. Resnik, 36-year-old electrical engineer, was mission specialist on Space Shuttle *Challenger* on Jan 28, 1986. Dr. Resnik, the second American woman in space (1984), was born at Akron, OH, on Apr 5, 1949. She perished with all others aboard when the Space Shuttle *Challenger* exploded on Jan 28, 1986. See also: "*Challenger* Space Shuttle Explosion Anniversary" (Jan 28).

ROBERT PRAGER MEMORIAL DAY: A GERMAN-AMERICAN DAY OF REMEMBRANCE. Apr 5. A day to commemorate the anti-German hysteria of the two world wars, to pay tribute to German-American heritage and recognize its contribution to the American community and to remember all people who have suffered from prejudice and hate. Annually, Apr 5. For info: Dr. Don Heinrich Tolzmann, Dir, German-American Studies Program, Univ of Cincinnati, ML33, Cincinnati, OH 45221-0033. Phone: (513) 556-1859.

TAIWAN: NATIONAL TOMB-SWEEPING DAY. Apr 5. National holiday since 1972. According to Chinese custom, the tombs of ancestors are swept "clear and bright" on this day, and rites honoring ancestors are held. Tomb-Sweeping Day is observed on Apr 5, except in Leap Years, when it falls on Apr 4.

WASHINGTON, BOOKER TALIAFERRO: BIRTH ANNIVERSARY. Apr 5. Black educator and leader. Born at Franklin County, VA, Apr 5, 1856. "No race can prosper," he wrote in *Up from Slavery*, "till it learns that there is as much dignity in tilling a field as in writing a poem." Died at Tuskegee, AL, on Nov 14, 1915.

BIRTHDAYS TODAY

Eric Burdon, 54, singer (*Eric Burdon Declares War, Black Man's Burdon*), songwriter, born at Walker-on-Tyne, England, Apr 5, 1941.

153

Apr 5-6 ☆ Chase's 1995 Calendar of Events ☆

Roger Corman, 69, filmmaker (King of B Horror Movies), born at Detroit, MI, Apr 5, 1926.

Max Gail, 52, actor ("Barney Miller," *Pearl*), born at Grosse Point, MI, Apr 5, 1943.

Arthur Hailey, 75, author (*Airport, The Final Diagnosis*), born at Luton, England, Apr 5, 1920.

Michael Moriarty, 53, actor (*The Last Detail, Bang the Drum Slowly*), born at Detroit, MI, Apr 5, 1942.

Gregory Peck, 79, actor (Oscar for *To Kill a Mockingbird; Roman Holiday, Gentleman's Agreement*), born La Jolla, CA, Apr 5, 1916.

Colin Luther Powell, 58, former Chairman US Joint Chiefs of Staff, retired, born at Harlem, NY, Apr 5, 1937.

Gale Storm, 73, actress ("My Little Margie," "NBC Comedy Hour," "The Gale Storm Show"), born Bloomington, TX, Apr 5, 1922.

APRIL 6 — THURSDAY
96th Day — Remaining, 269

ASIMOV, ISAAC: DEATH ANNIVERSARY. Apr 6. Although Isaac Asimov was one of the world's best-known writers of science fiction, his almost 500 books dealt with subjects as diverse as the Bible, works for pre-schoolers, college textbooks, mysteries, chemistry, biology, limericks, Shakespeare, Gilbert and Sullivan and modern history, among others. During his prolific career he helped to elevate science fiction from pulp-magazines to a more intellectual level. Some of his works include the *Foundation Trilogy, The Robots of Dawn, Robots and Empire, Nemesis, Murder at the A.B.A.* (in which he himself was a character), *The Gods Themselves* and *I, Robot*, in which he posited the famous Three Laws of Robotics. Asimov was born Jan 2, 1920, near Smolensk, Russia, and died on Apr 6, 1992.

CHING MING FESTIVAL. Apr 6. Widely observed Chinese festival (literally: "Pure and Bright"), Ching Ming is now regarded as an All Souls' Day. Families gather at graves of ancestors, leaving flowers and food after tidying the graves. A picnic spirit prevails, rather than solemnity. Set for the 106th day after the winter solstice, the Gregorian calendar date may vary slightly.

DEMOCRACY STIFLED IN PERU: ANNIVERSARY. Apr 6. Peruvian President Alberto K. Fujimori dissolved his Congress, suspended the Constitution, imposed censorship and arrested political leaders in an attempt to retain power on Apr 6, 1992. Battered by a stifled economy and continued insurgency by the Maoist "Shining Path" movement, Fujimori claimed that his actions were an offensive against rebels and drug traffickers. His actions were condemned by most nations, and the US halted all but humanitarian aid to the country in protest.

DOGWOOD ARTS FESTIVAL. Apr 6-23. Maryville, TN. To celebrate spring. Activities and events for all ages. For info: Blount County Dogwood Arts Fest, Chamber of Commerce, 309 S Washington St, Maryville, TN 37801. Phone: (615) 983-2241.

ENGLAND: GRAND NATIONAL HORSERACING MEETING. Apr 6-8. Aintree Racecourse, Aintree, Liverpool. Britain's premier horse race and most famous steeplechase is run over 4.5 miles on this famous course. The Grand National race takes place on Apr 8, the culmination of the race meeting. Est attendance: 85,000. For info: Racecourse Mgr, Aintree Racecourse, Liverpool, Merseyside, England L9 5AS.

	S	M	T	W	T	F	S
April 1995							1
	2	3	4	5	6	7	8
	9	10	11	12	13	14	15
	16	17	18	19	20	21	22
	23	24	25	26	27	28	29
	30						

FIRST US CREDIT UNION LAW: ANNIVERSARY. Apr 6. To recognize the Apr 6, 1909, charter of the St. Canadian credit union of Manchester, NH, with the help of Alphonse Desjardins, Canadian credit union pioneer.

FOUNDING OF THE MORMON CHURCH: 165th ANNIVERSARY. Apr 6. On this day in 1830 Joseph C. Smith and Oliver Cowdery organized the Church of Jesus Christ of Latter Day Saints.

JOYCE PAPERS RELEASED: ANNIVERSARY. Apr 6. On Apr 6, 1992, after 51 years, many of the personal papers, letters, notes and bills of James Joyce were put on display at the National Library of Ireland in Dublin. The papers were gathered by Joyce's secretary Paul Léon during the early days of World War II in Paris. Léon, who was eventually arrested and killed at Auschwitz, gave the papers to the Irish envoy in Paris to send to the National Library in Dublin, with instructions not to allow their viewing until fifty years after Joyce's death. Some of the documents are being withheld from viewing until the year 2050.

◆ OPERATION FLOATING CHRYSANTHEMUM: 50th ANNIVERSARY. Apr 6, 1945. Two US destroyers, two ammunition ships and a tank-landing ship were sunk off the coast of Okinawa when the Japanese Air Force launched 355 Kamikaze (suicide) pilots against the Allied fleet in Operation Floating Chrysanthemum.

PLAN YOUR EPITAPH DAY. Apr 6. (Also, Nov 1.) Dedicated to the proposition that a forgettable gravestone is a fate worse than death, and that everyone can be in the same league with Shakespeare and W.C. Fields. Semiannually, concides with Ching Ming Festival and the Day of the Dead. For info: Lance Hardie, Dead or Alive, PO Box 4595, Arcata, CA 95521. Phone: (707) 822-6924.

RAPHAEL: BIRTH ANNIVERSARY. Apr 6. Raffaello Santi (Sanzio), Italian painter and architect. Probably born on Apr 6, 1483, at Urbino, Italy. Died on his 37th birthday, at Rome, Italy, Apr 6, 1520.

◆ SCHNEIDERMAN, ROSE: BIRTH ANNIVERSARY. Apr 6, 1882. A pioneer in the battle to increase wages and improve working conditions for women, Rose Schneiderman was born at Saven, Poland, but with her family immigrated to the US six years later. At age 16 she began factory work in NYC's garment district and quickly became a union organizer. A leading opponent of the open-shop policy, which permitted nonunion members to work in a unionized shop, Schneiderman organized a 1913 strike of 25,000 women shirtwaist makers. She worked as an organizer for the International Ladies Garment Workers Union (ILGWU) as well as for the Women's Trade Union League (WTUL, for which she served as president for more than 20 years. During the Great Depression Pres Roosevelt appointed her to his Labor Advisory Board—the only woman member. Rose Schneiderman died Aug 11, 1972, at New York City.

SCOTLAND: EDINBURGH INTERNATIONAL SCIENCE FESTIVAL. Apr 6-22. Various venues, Edinburgh, Lothian. Includes workshops, lectures and exhibitions. For info: Edinburgh Science Festival Ltd, 1 Broughton Market, Edinburgh, Scotland EH3 6NU.

SOUTH AFRICA: VAN RIEBEECK DAY. Apr 6. Republic of South Africa. Jan van Riebeeck was the first commander of the Dutch East India Company that established a halfway station at the Cape of Good Hope in April 1652. This day is regarded in South Africa as the day on which Western civilization was established on the southern tip of the continent. A holiday in South Africa. Also known as Founder's Day.

☆ Chase's 1995 Calendar of Events ☆   Apr 6–7

SPACE MILESTONE: STS-11 (US). Apr 6. Shuttle *Challenger* launched Apr 6, 1984, with five astronauts (Robert Crippen, Francis Scobee, George Nelson, Terry Hart and James Van Hoften) had mission requiring recovery and repair of damaged satellite. Landed at Edwards Air Force Base, CA, on Apr 13.

SPRING FESTIVAL. Apr 6–11. Westville Village, Lumpkin, GA. Celebrate the arrival of Spring with activities similar to those of the Spring of 1850 with emphasis on music of the period. Est attendance: 4,000. For info: Patty Cannington, Westville, Box 1850, Lumpkin, GA 31815. Phone: (912) 838-6310.

SWITZERLAND: NAFELS PILGRIMAGE. Apr 6. Canton Glarus. Commemoration of the Battle of Nafels, fought on Apr 9, 1388. Observed annually on first Thursday in April, with processions, prayers, sermon and a reading out of the names of those killed in the battle.

THAILAND: CHAKRI DAY. Apr 6. Commemorates foundation of present dynasty by King Rama I (1782–1809), who also established Bangkok as capital.

THOMAS, LOWELL: BIRTH ANNIVERSARY. Apr 6. World traveler, reporter, editor and radio newscaster, whose broadcasts spanned more than half a century, 1925–1976. Born at Woodington, OH, on Apr 6, 1892. Died at Pawling, NY, on Aug 29, 1981. His radio sign-off, "So long until tomorrow," was known to millions of listeners and he is said to have been the first to broadcast from a ship, an airplane, a submarine and a coal mine.

TRAGEDY IN RWANDA. Apr 6, 1994. A plane carrying the presidents of Rwanda and Burundi was shot down near Kigali, the Rwandan capital, exacerbating a brutal ethnic war that led to the massacre of hundreds of thousands. Presidents Juvenal Habyarimana of Rwanda and Cyprien Ntaryamira of Burundi were returning from a summit in Tanzania where they discussed ways of ending the killing in their countries sparked by ethnic rivalries between the Hutu and Tutsi tribes. Following the attack on the two leaders, Rwanda descended into unspeakable chaos as the two tribes began viciously killing each other in a genocidal battle for power, leading to a mass exodus of bereaved civilians caught in the maelstrom.

TREE-MENDOUS MONTH. Apr 6–30. Statewide, MD. Apr 6th is kickoff date for this celebration honoring Maryland's PLANT communities, tree-mendous Marylanders and others whose volunteer efforts make it possible to plant 400,000 seedlings and trees each April. For info: Wally Orlinsky, Tree-Mendous Maryland, Tawes State Office Bldg, Forestry Program, E-1, Annapolis, MD 21401. Phone: (410) 974-3776.

US SENATE ACHIEVES A QUORUM: ANNIVERSARY. Apr 6. The US Senate was formally organized after achieving a quorum on Apr 6, 1789.

WORLD BEEF EXPO. Apr 6–9. Madison, WI. International cattle and trade show featuring beef breed shows and sales, educational programs, and over 200 commercial exhibitors. The trade show consists of beef equipment, veterinary supplies, feed, computers and software, and other agricultural production inputs. Est attendance: 22,000. For info: World Beef Expo, 122 E Olin Ave, Ste 270, Madison, WI 53513. Phone: (608) 251-2606.

BIRTHDAYS TODAY

Bert (Rik Aalbert) Blyleven, 44, former baseball player, born at Zeist, The Netherlands, Apr 6, 1951.
Merle Haggard, 58, singer, songwriter ("Okie From Muskogee," "If We Make It Through December"), born at Bakersfield, CA, Apr 6, 1937.
Marilu Henner, 43, actress (Elaine Nardo on "Taxi"; Ava Newton on "Evening Shade"), born at Chicago, IL, Apr 6, 1952.
Bruce King, 71, Governor of New Mexico (D), born at Stanley, NM, Apr 6, 1924.

Andre Previn, 66, composer, born at Berlin, Germany, Apr 6, 1929.
John Ratzenberger, 48, actor (Cliff on "Cheers"), born at Bridgeport, CT Apr 6, 1947.
John Sculley, 56, business executive, born at New York, NY, Apr 6, 1939.
Roy Thinnes, 57, actor ("The Invaders," "Outer Limits"), born at Chicago, IL, Apr 6, 1938.
Billy Dee Williams, 58, actor (*Brian's Song*, *Lady Sings the Blues*), born at New York, NY, Apr 6, 1937.

APRIL 7 — FRIDAY
97th Day — Remaining, 268

ALCOHOL-FREE WEEKEND. Apr 7–9. To increase public awareness of the problems associated with drinking alcoholic beverages by asking Americans to refrain from drinking them for this weekend. Sponsor: Natl Council on Alcoholism and Drug Dependence, Inc, Rebecca Fenson, 12 W 21st St, New York, NY 10010. Phone: (212) 206-6770.

ARBOR DAY (ARIZONA). Apr 7. In the counties of Apache, Navajo, Coconino, Mohave and Yavapai, the Friday following Apr 1, and in all other counties the Friday following Feb 1, in each year, shall be known as Arbor Day. See also: "Arbor Day (Arizona)" (Feb 3).

CAROLINA CRAFTSMEN'S SPRING CLASSIC. Apr 7–9. Coliseum Exhibition Building, Greensboro, NC. Arts and crafts. Est attendance: 25,000. For info: The Carolina Craftsmen, 1240 Oakland Ave, Greensboro, NC 27403. Phone: (910) 274-5550.

DOGWOOD ARTS FESTIVAL. Apr 7–23. Knoxville, TN. A springtime celebration with sixty miles of Dogwood trails and more than 100 events highlighting arts, crafts, parades and musical performances. It's the "Best 17 Days of Spring in America." Est attendance: 250,000. Sponsor: Dogwood Arts Festival, 111 N Central Ave, Knoxville, TN 37902. Phone: (615) 522-8733.

EASTER CELEBRATED IN RED SQUARE AGAIN: ANNIVERSARY. Apr 7. In a sign of expanding religious tolerance in the former Soviet Union, more than 3,000 people gathered at St. Basil's Cathedral in Red Square for Orthodox Easter services on Sunday, Apr 7, 1991, for the first time in decades. In addition to the Red Square services, the state television news program Vremya began its Sunday broadcast with an Easter message from Russian Patriarch Alexi II, and national television broadcast Saturday services from the Cathedral of the Epiphany, the seat of Russian Orthodoxy.

FESTIVAL OF FLOWERS. Apr 7–May 7. Biltmore Estate, Asheville, NC. A Victorian celebration of spring with floral displays filling each room of Biltmore House. Call for a schedule of events. Est attendance: 86,000. For info: The Biltmore Co, 1 N Pack Sq, Asheville, NC 28801. Phone: (800) 543-2961.

FRENCH QUARTER FESTIVAL. Apr 7–9. French Quarter, New Orleans, LA. This free festival focuses on all that makes the Quarter special—art, antiques, food, music, shopping, lifestyles and the people. Free concerts on 13 stages, historic patio tours, parade, children's activities, fireworks, 5K race, and other family activities. Est attendance: 300,000. Sponsor: Sandra Dartus, Exec Dir, 1008 N Peters St, New Orleans, LA 70116. Phone: (800) 673-5725.

GLOBAL FEST. Apr 7–8. Indian Spring Park. An international festival featuring food, entertainment and educational booths. Est attendance: 3,000. For info: Sally Gavlik, Rec Super, City of Waco Dept of Leisure Services, PO Box 2570, Waco, TX 76702. Phone: (817) 750-5980.

HOOSIER HORSE FAIR EXPO. Apr 7-9. Indiana State Fairground Event Center, Indianapolis, IN. All-breed gathering and informative exposition. Est attendance: 55,000. For info: John Cloe, PO Box 728, Cloverdale, IN 46120. Phone: (317) 795-6932.

MALINOWSKI, BRONISLAW: BIRTH ANNIVERSARY. Apr 7. Leading British anthropologist, author and teacher. Born on Apr 7, 1884, at Cracow, Poland. His pioneering anthropological field work in Melanesia inspired his colleagues and students. In 1939 he became a visiting professor at Yale University. Died at New Haven, CT, on May 16, 1942.

MEDIEVAL FAIR. Apr 7-9. University of Oklahoma Brandt Park Duck Pond, Norman, OK. Arts and Crafts and living history fair. The Middle Ages come alive with dancers, music, theater, jousting, knights in combat and a human chess match. Feasts and follies include games and food "fit for a king." Meet such characters as King Arthur, Sir Lancelot and Merlin. Shop as more than 200 artists and craftsmen display their wares. Admission is free. Annually, the second weekend in April. Est attendance: 70,000. Sponsor: Linda Linn, 1700 Asp, Norman, OK 73037. Phone: (405) 321-7227.

MENNONITE RELIEF SALE. Apr 7-8. Kansas State Fair Grounds, Hutchinson, KS. More than 70 Mennonite, Brethren in Christ and Amish congregations in Kansas sponsor this annual festival and benefit auction for the worldwide hunger relief and community aid programs of the Mennonite Central Committee. Auctions of quilts, grandfather clocks, furniture, tools and crafts. Great food and lots more. No vendors. Est attendance: 35,000. For info: LaVern Stucky, Pres, Rte 1, Box 54, Peabody, KS 66866. Phone: (316) 983-2348.

NATIONAL GEOGRAPHY BEE STATE LEVEL. Apr 7. Site is different in each state, many are in state capital. Winners of school-level competitions who scored in the top 100 in their state on a written test compete in the State Geography Bees. The winner of each state Bee will go to Washington, DC, for the national level in May. Est attendance: 450. For info: Natl Geography Bee, Natl Geographic Soc, 1145 17th St, NW, Washington, DC 20036. Phone: (202) 828-6635.

NATIONAL TEACHER APPRECIATION DAY. Apr 7. A day for elementary through high school students to show appreciation to their teachers. Students are urged to thank their teachers for their care and concerned effort, to be extra cooperative with them and to bring a can of food in their honor. This food can be donated to a school family in need or a local food bank. Annually, the Friday before Palm Sunday. Sponsor: Joseph C. Kearley, 2129 Benbrook Dr, Carrollton, TX 75007. Phone: (214) 245-6192.

NO HOUSEWORK DAY. Apr 7. No trash. No dishes. No making of beds or washing of laundry. And no guilt. Give it a rest. Phone: (212) 388-8673 or (717) 274-8451. Sponsor: Wellness Permission League, Tom or Ruth Roy, 2105 Water St, Lebanon, PA 17046.

NORWAY: VOSS JAZZ FESTIVAL. Apr 7-9. Voss. Three days of jazz as well as folk music at Voss, a major ski resort, by Norwegian, European and American performers. Annually, the weekend before Easter. For info: Norwegian Tourist Bd, 655 Third Ave, New York, NY 10017. Phone: (212) 949-2333.

OZARK UFO CONFERENCE. Apr 7-9. Inn of the Ozarks Convention Center, Eureka Springs, AR. 7th annual meeting of researchers from various states and foreign countries to inform the public of the latest news concerning UFOs. Speakers include authors of books on the subject and people who have investigated UFO cases; program includes audio visual presentations of UFO evidence. Est attendance: 425. For info: Ozark UFO Conference, Lucius Farish, Rt 1, Box 220, Plumerville, AR 72127. Phone: (501) 354-2558.

POTEET STRAWBERRY FESTIVAL. Apr 7-9. Poteet, TX. One of the oldest and largest festivals in Texas established to promote Poteet's crop—strawberries. Great food and family entertainment. Est attendance: 100,000. For info: Poteet Strawberry Festival Assn, Nita Harvey, Festival Coord, PSFA, PO Box 227, Poteet, TX 78065. Phone: (210) 742-8144.

RATTLESNAKE HUNT. Apr 7-9. Waurika, OK. The annual roundup that originated to aid area ranchers against rattlesnakes now draws more than 20,000 visitors to this southwestern Oklahoma community each year. You can join in on the hunt or simply observe the action. For info: Rattlesnake Hunt Organization, 905 W Anderson, Waurika, OK 73573. Phone: (405) 228-2787.

STAMP EXPO. Apr 7-9. Wilshire Ebell Convention Complex, Los Angeles, CA. Fax: (818) 988-4337. Est attendance: 4,000. Sponsor: Intl Stamp Collectors Soc, PO Box 854, Van Nuys, CA 91408. Phone: (818) 997-6496.

STUDENT GOVERNMENT DAY IN MASSACHUSETTS. Apr 7. Annually, the first Friday of April.

SUGARLOAF CRAFT FESTIVAL. Apr 7-9. Gaithersburg, MD. More than 480 professional artists and craftspeople, demonstrations, live music, and children's entertainment. Est attendance: 40,000. Sponsor: Sugarloaf Mt Works, Inc, 200 Orchard Ridge Dr, #215, Gaithersburg, MD 20878. Phone: (301) 990-1400.

UNITED NATIONS: WORLD HEALTH DAY. Apr 7. A United Nations observance commemorating the establishment of the World Health Organization on Apr 7, 1948. Info from: United Nations, Dept of Public Info, New York, NY 10017.

WINCHELL, WALTER: BIRTH ANNIVERSARY. Apr 7. Controversial American journalist, gossip columnist and radio broadcaster, singer and actor, who roused the ire of many prominent public figures. Born at New York, NY, on Apr 7, 1879. Died at Los Angeles, CA, on Feb 20, 1972.

WORDSWORTH, WILLIAM: 225th BIRTH ANNIVERSARY. Apr 7. English Lake Poet and philosopher born on this day, 1770. "Poetry," he said, "is the spontaneous overflow of powerful feelings: it takes its origin from emotion recollected in tranquility." Wordsworth died on Apr 23, 1850.

WORLD HEALTH DAY. Apr 7. A "complete planning kit" available focusing on theme of child immunization from birth to two years of age. For info: World Health Day, American Assn for World Health, 1129 20th St NW, Ste 400, Washington, DC 20036. Phone: (202) 466-5883.

WORLD'S LARGEST TRIVIA CONTEST. Apr 7-9. Stevens Point, WI. More than 11,000 players including more than 525 teams compete to answer eight questions every hour for 54 hours straight. Prize: Oscar-like trophy. Sponsor: Jim Oliva, WWSP-Radio (89.9 FM), Rm 105 CAC, University of Wisconsin, Stevens Point, WI 54481. Phone: (715) 346-3755.

April 1995

S	M	T	W	T	F	S
						1
2	3	4	5	6	7	8
9	10	11	12	13	14	15
16	17	18	19	20	21	22
23	24	25	26	27	28	29
30						

◈ **YAMATO SUICIDE: 50th ANNIVERSARY.** Apr 7, 1945. The 72,000-ton *Yamato* (at that time the largest battleship) set sail for Okinawa carrying only enough fuel for a one-way trip— a suicide mission against the American transport fleet. But she failed to reach her target: struck 19 times by American aerial torpedoes, she sank with 2,498 of her crew aboard.

BIRTHDAYS TODAY

(William) Hodding Carter, III, 60, television and newspaper journalist, born at New Orleans, LA, Apr 7, 1935.
Francis Ford Coppola, 56, filmmaker (the Godfather movies; *Apocalypse Now*), born at Detroit, MI, Apr 7, 1939.
Tony Dorsett, 41, former football player, born at Aliquippa, PA, Apr 7, 1954.
Daniel Ellsberg, 64, author (released the "Pentagon Papers" to the *New York Times*), born at Chicago, IL, Apr 7, 1931.
David Frost, 56, entertainer ("That Was the Week That Was"), interviewer, born Tenterden, England, Apr 7, 1939.
James Garner (James Baumgardner), 67, actor (*The Americanization of Emily*, "Maverick," "The Rockford Files"), born at Norman, OK, Apr 7, 1928.
John Oates, 47, singer ("Maneater" with Hall and Oates), songwriter, born at New York, NY, Apr 7, 1948.
Wayne Rogers, 62, actor ("M*A*S*H," "House Calls"), born at Birmingham, AL, Apr 7, 1933.

APRIL 8 — SATURDAY
98th Day — Remaining, 267

ART IN THE PARK FESTIVAL. Apr 8. Oakland Park Library, Oakland Park, FL. Annual celebration of National Library Week, features a juried, fine-arts-only show with cash awards. Activities for all ages, including storytimes and sidewalk chalk art contest for children, plant sales and displays, used book sale, live music, voter registration, facepainting and more. Annually, the Saturday before National Library Week. Est attendance: 5,000. For info: Joanne B. Fischer, Art In The Park Coord, Oakland Park Library, 1298 NE 37 St, Oakland Park, FL 33334-4576. Phone: (305) 561-6289.

AZALEA FESTIVAL. Apr 8-23. Honor Heights Park, Muskogee, OK. More than 600 varieties of 70,000 azaleas in bloom. Parade (on Apr 8), rodeo, arts and crafts and chili cook-off. Est attendance: 285,000. For info: Chamber of Commerce, Box 797, Muskogee, OK 74401. Phone: (918) 682-2401.

BEYOND CATEGORY: THE MUSICAL GENIUS OF DUKE ELLINGTON. Apr 8–June 18. Strong Museum, Rochester, NY. See Jan 7 for full description. Call or write the Smithsonian for other dates and venues. For info: Smithsonian Institution Traveling Exhibition Service, 1100 Jefferson Dr SW, Ste 3146, Washington, DC 20560.

BIRTHDAY OF THE BUDDHA: BIRTH ANNIVERSARY. Apr 8. Among Buddhist holidays, April 8th is the most important as it commemorates the birthday of the Buddha. The founder of Buddhism had the given name Siddhartha, the family name Gautama, the clan name Shaka, and he is commonly called the Buddha, which in Sanskrit means "the enlightened one." He is thought to have lived in India from c. 563 BC to 483 BC.

CANADA: WORLD CURLING CHAMPIONSHIPS. Apr 8-16. Brandon, Manitoba. Curlers from around the world compete for championship. For info: Travel Manitoba, 7-155 Carlton St, Winnipeg, Manitoba, Canada R3C 3H8. Phone: (204) 945-3796.

CHERRY BLOSSOM FESTIVAL. Apr 8-9. Micke Grove Park and Zoo, Lodi, CA. Event coincides with the blooming of cherry trees in the park and includes Japanese foods, tea ceremonies, flower arrangements, Bonsai displays, traditional foods, dancing and art. Est attendance: 2,000. For info: San Joaquin Cnty Pks & Recreation, 11793 N Micke Grove Rd, Lodi, CA 95240. Phone: (209) 953-8800.

COSMONAUTS' DAY. Apr 8. Space Center, Alamogordo, NM. A celebration of Yuri Gagarin's flight, the first man in space. Annual celebration with kite flight (1-4). Est attendance: 300. For info: Public Affairs, Space Center, PO Box 533, Alamogordo, NM 88311. Phone: (800) 545-4021.

◈ **A DAY IN THE WARSAW GHETTO: A BIRTHDAY TRIP IN HELL.** Apr 8-May 28. Mizel Museum of Judaica, Denver, CO. See Feb 4 for full description. Call or write the Smithsonian for other dates and venues. For info: Smithsonian Institution Traveling Exhibition Service, 1100 Jefferson Dr SW, Ste 3146, Washington, DC 20560. Phone: (202) 357-2700.

ENGLAND: FLOWERS EXHIBITION. Apr 8–Sept 9. The Gallery, Windsor Castle, Windsor. An exhibition of drawings and watercolours from the Royal Library. Est attendance: 50,000. For info: Exhibitions Dept, The Royal Library, Windsor Castle, Windsor, Berkshire, England SL4 1NJ.

FINGER LAKES CRAFTSMEN SPRING-EASTER ARTS AND CRAFTS SHOW. Apr 8-9. Monroe County Fairgrounds Dome Arena, Rochester, NY. All media and categories including photography and prints. 24th annual show. Est attendance: 9,500. For info: Finger Lakes Craftsmen Shows, Ronald L. Johnson, Freshour Dr, Shortsville, NY 14548. Phone: (716) 289-9439.

FISCUS, KATHY: DEATH ANNIVERSARY. Apr 8. Three-year-old Kathy Fiscus of San Marino, CA, fell while playing into an abandoned well pipe 14 inches wide and 120 feet deep. Rescue workers toiled ceaselessly for two days while thousands watched and while national attention was focused on the tragedy. Her body was recovered Apr 10, 1949. An alarmed nation suddenly became attentive to other abandoned wells and similar hazards, and "Kathy Fiscus laws" were enacted in a number of places requiring new safety measures to prevent recurrence of such a tragic accident.

FORD, ELIZABETH (BETTY) BLOOMER WARREN: BIRTHDAY. Apr 8. Wife of Gerald Rudolph Ford, 38th president of the US. Born at Chicago, IL, Apr 8, 1918.

HANDEL ORATORIO SOCIETY SPRING PERFORMANCE. Apr 8. Centennial Hall, Rock Island, IL. 114th year of performance. The 350-voice ensemble, composed of the Handel Oratorio Society, The Augustana Choir and the Augustana Symphony Orchestra, offers a masterwork from the classical oratorio repertoire. For info: Augustana College, Ticket Office, Rock Island, IL 61201. Phone: (309) 794-7306.

HERITAGE FAIR/SPRING FEST. Apr 8. Poor Farm Park, Ashland, VA. Features demonstrations by Native American artists and craftspeople. Also, encampments trace US history from the arrival of Europeans to modern military. Est attendance: 12,000. For info: Hanover Co Parks and Recreation, Jeannie Chewning, 200 Berkley St, Ashland, VA 23005. Phone: (804) 798-8062.

HOLLAND: ART AND ANTIQUE FAIR. Apr 8-17. Het Turfschip, Breda. International art and antique fair. Est attendance: 19,000. For info: Netherlands Board of Tourism, 225 N. Michigan Ave, Chicago, IL 60601. Phone: (312) 819-0300.

HOME RUN RECORD SET BY HANK AARON: ANNIVERSARY. Apr 8. Henry "Hammerin' Hank" Aaron hit the 715th home run of his career on April 8, 1974, breaking the record set by Babe Ruth in 1935. Playing for the Atlanta Braves, Aaron broke the record in Atlanta in a game against the Los Angeles Dodgers. He finished his career in 1976 with a total of 755 home runs. This record remains unbroken. At the time of his retirement, Aaron also held records for first in RBIs, second in at-bats and runs scored and third in base hits.

JAPAN: FLOWER FESTIVAL (HANA MATSURI). Apr 8. Commemorates Buddha's birthday. Ceremonies in all temples.

Apr 8 ☆ *Chase's 1995 Calendar of Events* ☆

KNIGHT, O. RAYMOND: BIRTH ANNIVERSARY. Apr 8, 1872. The "Father of Canadian Rodeo," O. Raymond Knight was born at Payson, UT. His father, the Utah mining magnate, Jesse Knight, founded the town of Raymond, Alberta in 1901. In 1902, Raymond produced Canada's first rodeo, "Raymond Stampede." He also built rodeo's first grandstand and first chute in 1903. O. Raymond Knight died on Feb 7, 1947. For info: Raymond Sports Hall of Fame, Max Court, Box 511, Raymond, Alberta, Canada T0K 2S0. Phone: (403) 752-3094.

MAJOR LEAGUE/MINOR LEAGUE: PHOTOGRAPHS OF AMERICA'S BASEBALL STADIUMS BY JIM DOW. Apr 8–May 21. Philharmonic Center for the Arts, Naples, FL. See Feb 2 for full description. Call or write the Smithsonian for other dates and venues. For info: Smithsonian Institution Traveling Exhibition Service, 1100 Jefferson Dr SW, Ste 3146, Washington, DC 20560. Phone: (202) 357-2700.

MAPLE SYRUP FESTIVAL. Apr 8–9. Bradys Run County Park, Fallston, PA. Demonstrates the maple tree tapping process and builds a 19th-century arts, crafts and educational festival around it. Est attendance: 20,000. For info: Beaver County Conservation District, 1000 Third St, Ste 202, Beaver, PA 15009-2026. Phone: (412) 774-7090.

MOON PHASE: FIRST QUARTER. Apr 8. Moon enters First Quarter phase at 1:35 AM, EDT.

MORRIS, LEWIS: BIRTH ANNIVERSARY. Apr 8. Signer of the Declaration of Independence, born in Westchester County, NY, Apr 8, 1726. Died Jan 22, 1798.

NO LAUGHING MATTER: POLITICAL CARTOONISTS ON THE ENVIRONMENT. Apr 8–30. Mount San Jacinto College, San Jacinto, CA. See Feb 1 for full description. Call or write the Smithsonian for other dates and venues. For info: Smithsonian Institution Traveling Exhibition Service, 1100 Jefferson Dr SW, Ste 3146, Washington, DC 20560. Phone: (202) 357-2700.

OKEFENOKEE SPRING FLING. Apr 8–9. Okefenokee Swamp Park, Waycross, GA. Guided boat tours and special wildlife shows (featuring Doris Mager "The Eagle Lady"), musical entertainment, clogging, an outdoor melodrama, Cliff Patton & Skeeter, fun food and a fish fry. Admission charged. Food not included in admission price. Annually, the weekend before Easter. Est attendance: 2,000. For info: Okefenokee Swamp Park, US 1 S, Waycross, GA 31501. Phone: (912) 283-0583.

PEANUT-KIDS-BASEBALL DAY. Apr 8. Edwards-Freeman Nut Co, Conshohocken, PA. In a salute to the ingredients that relate to our national pastime, an annual observance is held at which MVP (Master Vendor, Peanuts) Cheryl Spielvogel throws out the first "ball" (bag of peanuts shaped like a baseball) to Philadelphia Phillies players, then to Little League players and others present. Annually, the Saturday prior to the Phillies' home opening game. Sponsors: Conshohocken Little League, Edwards-Freeman Nut Co. Est attendance: 150. For info: Abe S. Rosen, Rosen-Coren Agency, Inc, 902 Fox Pavilion, PO Box 643, Jenkintown, PA 19046. Phone: (215) 937-1017.

PHILADELPHIA ANTIQUES SHOW—A BENEFIT FOR THE HOSPITAL OF THE UNIVERSITY OF PENNSYLVANIA. Apr 8–12. 103rd Engineers Armory, Philadelphia, PA. Museum-quality antiques, symposiums, appraisals and guided tours by Philadelphia Museum of Art. To raise funds for the Hospital of the University of Pennsylvania. Est attendance: 6,000. Sponsor: Board of Women Visitors of Hospital of Univ of Pennsylvania, 3400 Spruce, Philadelphia, PA 19104. Phone: (215) 662-3941.

	S	M	T	W	T	F	S
April 1995							1
	2	3	4	5	6	7	8
	9	10	11	12	13	14	15
	16	17	18	19	20	21	22
	23	24	25	26	27	28	29
	30						

SEVENTEENTH AMENDMENT TO US CONSTITUTION RATIFIED. Apr 8. Prior to the 17th Amendment, ratified on Apr 8, 1913, members of the Senate were elected by each state's respective legislature. The advent and popularity of primary elections during the last decade of the 19th century and the early 20th century and a string of senatorial scandals, most notably a scandal involving William Lorimer, an Illinois political boss in 1909, forced the Senate to end its resistance to a constitutional amendment requiring direct popular election of senators.

SHEEP SHEARING/KIDS DAY. Apr 8. Petaluma Adobe State Historic Park, Petaluma, CA. Miss Petaluma Ambassador candidates will be there to help the visitors to see how the sheep shearing is done and feel the wool. A professional sheep shearer will be performing the modern sheep-shearing techniques, and the Rangers will be explaining the way it was done in the 1800s. 10–5. (send S.A.S.E.) Est attendance: 500. For info: Petaluma Old Adobe Assn, PO Box 631, Petaluma, CA 94953. Phone: (707) 762-4871.

TARPON SPRINGS ARTS AND CRAFTS FESTIVAL. Apr 8–9. Tarpon Springs, FL. Est attendance: 55,000. For info: Tarpon Springs Chamber of Commerce, 210 S Pinellas Ave, #120, Tarpon Springs, FL 34689. Phone: (813) 937-6109.

THE TONGASS—ALASKA'S MAGNIFICENT RAIN FOREST. Apr 8–May 21. University of Alaska Museum, Fairbanks, AK. This exhibit displays the intricate harmony of the largest non-equatorial rain forest on earth. See Feb 4 for full description. For info: Smithsonian Institution Traveling Exhibition Service, 1100 Jefferson Dr SW, Ste 3146, Washington, DC 20560. Phone: (202) 357-2700.

◆ **WHITE, RYAN: 5th DEATH ANNIVERSARY.** Apr 8. The young man from Kokomo, IN, who put the face of a child on AIDS and served as a leader for gaining greater understanding of the disease. Ryan, a hemophiliac, contracted Acquired Immune Deficiency Syndrome through a blood transfusion. He was banned from the public school system in Central Indiana in 1984 and then moved with his mother and 16-year-old sister to Cicero, IN. There he was accepted by students and faculty alike. In Cicero, Ryan helped pierce the myths of AIDS and showed how the disease affects people from many walks of life. Ryan once stated that he only wanted to be treated as a normal teenager, but that was not to be. He lived life as a celebrity in his last years, even attending the Academy Awards. He was a voracious reader of auto magazines and received a red Mustang convertible from his friend Michael Jackson. A few days after attending the Academy Awards ceremony, Ryan had trouble swallowing. Shortly thereafter the 18-year-old was hospitalized at the Riley Hospital for Children. It was his decision to use life support systems, which were removed on Palm Sunday, April 8, 1990, when Ryan lost his valiant fight. His funeral was attended by many celebrities, including Michael Jackson, First Lady Barbara Bush and Elton John, who also sang at the service.

WILLIAMS, WILLIAM: BIRTH ANNIVERSARY. Apr 8. Signer of the Declaration of Independence. Born at Lebanon, CT, Apr 8, 1731. Died there Aug 2, 1811.

YOSEMITE SPRING SKIFEST. Apr 8. Yosemite National Park, CA. A traditional ski carnival at the oldest ski area in California, Badger Pass. Est attendance: 600. For info: Yosemite Park and Curry Co, 5410 E Home Ave, Fresno, CA 93727. Phone: (209) 372-1332.

☆ Chase's 1995 Calendar of Events ☆   Apr 8-9

BIRTHDAYS TODAY

Gary Carter, 41, former baseball player, born at Culver City, CA, Apr 8, 1954.
William D. Chase, 73, librarian and chronicler of contemporary civilization as co-founder and co-editor (retired) of *Chase's Annual Events*, born at Lakeview, MI, Apr 8, 1922.
Mark Gregory Clayton, 34, football player, born at Indianapolis, IN, Apr 8, 1961.
Betty Ford, 77, former first lady, born at Chicago, IL, Apr 8, 1918.
Shecky Greene, 70, comedian, actor, born at Chicago, IL, Apr 8, 1925.
John Havlicek, 55, basketball Hall of Famer, born at Martins Ferry, OH, Apr 8, 1940.
Seymour Hersh, 58, journalist, born at Chicago, IL, Apr 8, 1937.
Julian Lennon, 32, musician, singer, son of John Lennon, born at Liverpool, England, Apr 8, 1963.
Carmen McRae, 73, singer ("The Next Time It Happens"), born at New York, NY, Apr 8, 1922.
Terry Porter, 32, basketball player, born at Milwaukee, WI, Apr 8, 1963.

APRIL 9 — SUNDAY
99th Day — Remaining, 266

CIVIL WAR ENDING: 130th ANNIVERSARY. Apr 9. At 1:30 PM, on Sunday, Apr 9, 1865, General Robert E. Lee, commander of the Army of Northern Virginia, surrendered to General Ulysses S. Grant, commander-in-chief of the Union Army, ending four years of civil war. Meeting took place in the house of Wilmer McLean at the village of Appomattox Court House, VA. Confederate soldiers were permitted to keep their horses and go free to their homes, while Confederate officers were allowed to retain their swords and side arms as well. Grant wrote the terms of surrender in his own hand. Formal surrender took place at the Courthouse on Apr 12. Death toll for the Civil War is estimated at 500,000 men.

CURTIS EASTER PAGEANT. Apr 9. Medicine Valley High School Auditorium, Curtis, NE. "Truly an unforgettable, inspirational, touching spiritual experience!" A choir, acting cast and supporting cast of more than 200 people depict the last week in the life of Christ through music and narration. Seventeen magnificent "Living Pictures." Curtis is Nebraska's official "Easter City," so-named by the Nebraska Legislature. Free will offering. Annually, Palm Sunday afternoon. Sponsor: Curtis Community. Est attendance: 2,000. For info: Joe Beals, Easter Pageant Pres, Box 112, Curtis, NE 69025. Phone: (308) 367-5225.

FIELDS, W.C.: BIRTH ANNIVERSARY. Apr 9. Claude William Dukenfield (W.C. Fields), stage and motion picture actor and expert juggler. Born at Philadelphia, PA, Apr 9, 1879. Died Dec 25, 1946, at Pasadena, CA. He wrote his own epitaph: "On the whole, I'd rather be in Philadelphia."

HOLY WEEK. Apr 9-15. Christian observance dating from the fourth century, known also as Great Week. The seven days beginning on the sixth and final Sunday in Lent (Palm Sunday), consisting of: Palm Sunday, Monday of Holy Week, Tuesday of Holy Week, Spy Wednesday (or Wednesday of Holy Week), Maundy Thursday, Good Friday and Holy Saturday (or Great Sabbath or Easter Even). A time of solemn devotion to and memorializing of the suffering (passion), death and burial of Christ. Formerly a time of strict fasting.

MARIAN ANDERSON EASTER CONCERT: ANNIVERSARY. Apr 9. On Easter Sunday, Apr 9, 1939, black American contralto Marian Anderson sang an open-air concert from the steps of the Lincoln Memorial in Washington, DC, to an audience of 75,000, after having been denied use of the Daughters of the American Revolution (DAR) Constitution Hall. The event became an American anti-discrimination *cause célèbre* and led Eleanor Roosevelt, wife of the US president, to resign from the DAR.

MULTI-CULTURAL EASTER EGG DISPLAY. Apr 9-23. Belleville, IL. A multi-cultural exhibit of Easter eggs decorated in various ethnic and modern artistic styles, and featuring daily demonstrations of individual techniques. No admission fee. Est attendance: 6,600. For info: Shrine of Our Lady of the Snows, 9500 W Illinois, Hwy 15, Belleville, IL 62223. Phone: (618) 397-6700.

MUYBRIDGE, EADWEARD: BIRTH ANNIVERSARY. Apr 9. English photographer famed for his studies of animals in motion. Born Edward James Muggeridge, at Kingston-on-Thames, England, Apr 9, 1830. Died there May 8, 1904.

NATIONAL BUILDINGS SAFETY WEEK. Apr 9-15. To make all Americans aware of the important health and life safety services available to them from their state and local professional building departments. Sponsors: Natl Conference of States on Building Codes and Standards, Inc, and Council of American Building Officials. For info: Deborah Brettner, NCSBCS, 505 Huntmar Park Dr, Ste 210, Herndon, VA 22070. Phone: (703) 437-0100.

NATIONAL GARDEN WEEK. Apr 9-15. To recognize and honor the 43 million American households that eagerly garden each year. These American gardeners enhance and improve the environment with their efforts. Annually, the second full week of April. Sponsor: Natl Garden Bureau, 1311 Butterfield Rd, Ste 310, Downers Grove, IL 60515.

NATIONAL HOME SAFETY WEEK. Apr 9-15. To better inform the public of products, tips and steps to make your home a safer and more enjoyable habitat. The week will be marked with activities reaffirming concern for the preservation and enhancement of the home and its habitants. Annually, beginning the second Sunday in April. Sponsor: Sean Gannon, Program Dir, Newton Products Co, 3874 Virginia Ave, Cincinnati, OH 45227-0175. Phone: (800) 543-9149.

NATIONAL LIBRARY WEEK. Apr 9-15. To promote use and support of all types of libraries: public, school, academic and special, and to make the public aware of the many services available at their local library. Sponsor: American Library Assn, Pamela Goodes, 50 E Huron St, Chicago, IL 60611. Phone: (312) 280-5041.

NATIONAL MEDICAL LABORATORY WEEK. Apr 9-15. To promote the vital role of pathologists, technologists and technicians in the diagnosis and treatment of illness. Special events scheduled in thousands of laboratories across the country. For information on planning a Lab Week celebration, call toll-free 800-621-4142 (in Illinois: 312-738-1336). Sponsor: American Soc of Clinical Pathologists, Suzanne Stock, 2100 W Harrison St, Chicago, IL 60612-3798. Phone: (312) 738-4886.

NORIEGA CONVICTION: ANNIVERSARY. Apr 9. Former president of Panama General Manuel Antonion Noriega was convicted in US federal court on eight counts of cocaine trafficking, racketeering and money laundering on Apr 9, 1992. His conviction marked the first time that a US jury had convicted a foreign head of state on criminal charges. Noriega had surrendered to authorities after the US invaded Panama in December 1989 in Operation Just Cause, which was intended to topple Noriega from power and bring him to justice for drug crimes.

PALM SUNDAY. Apr 9. Commemorates Christ's last entry into Jerusalem, when His way was covered with palms by the multitude. Beginning of Holy (or Great) Week in Western Christian churches.

PALM SUNDAY EGG HUNT. Apr 9. Beach areas, Seaside Heights, NJ. 1 PM. More than 10,000 eggs are buried in the sand. Children in 5 different age groups dig for these plastic eggs filled with various tickets that are redeemed for prizes. Annually, Palm Sunday. Est attendance: 25,000. For info: Seaside Business Assn, Attn: Stacy Palmieri, PO Box 98, Seaside Heights, NJ 08751. Phone: (908) 793-1510.

Apr 9-10 ☆ *Chase's 1995 Calendar of Events* ☆

★ **PAN AMERICAN WEEK.** Apr 9-15. Presidential Proclamation customarily issued as "Pan American Day and Pan American Week." Always issued for the week including Apr 14, except in 1965, from 1946 through 1948, 1955 through 1977, and 1979.

PHILIPPINES: HOLY WEEK. Apr 9-15. National observance. Flagellants in the streets, *cenaculos* (passion plays) and other colorful and solemn rituals mark the country's observance of Holy Week.

PORTUGAL: HOLY WEEK FESTIVITIES. Apr 9-16. Braga, Ovar, Povoa de Varzim and other major cities. The Holy Week celebrations attain a great splendor in these places, especially in Braga, a bulwark of Christianity from early times. The most important events take place on Monday, Thursday and Good Friday when imposing parades march through the streets. For info: Portuguese Natl Tourist Office, 590 Fifth Ave, New York, NY 10036. Phone: (212) 354-4403.

ROBESON, PAUL BUSTILL: BIRTH ANNIVERSARY. Apr 9. Paul Robeson was born on Apr 9, 1898, in Princeton, NJ. He was an All-American football player at Rutgers University and received his law degree from Columbia University in 1923. After being seen by Eugene O'Neill in an amateur stage production, he was offered a part in O'Neill's play, *The Emperor Jones*. His performance in that play with the Provincetown Players established him as an actor. Without ever having taken a voice lesson, he also became a popular singer. His stage credits include *Show Boat, Porgy, The Hairy Ape,* and *Othello* which enjoyed the longest Broadway run of a Shakespearean play. He displayed his talents during his tours of Europe and the Soviet Union, and in 1950 he was denied a passport by the US for refusing to sign an affidavit stating whether he was or ever had been a member of the Communist Party. The action was overturned by the Supreme Court in 1958. His film credits include *Emperor Jones, Showboat, King Solomon's Mines* and *Song of Freedom* among others. Robeson died at Philadelphia, PA, on Jan 23, 1976.

SPACE MILESTONE: *SOYUZ 35* (USSR). Apr 9. Two cosmonauts (Valery Ryumin and Leonid Popov) launched Apr 9, 1980, from Baikonur space center in Kazakhstan, USSR. Docked at *Salyut 6* on Apr 10. Ryumin and Popov returned to Earth Oct 11, 1980, after setting a new space endurance record of 185 days.

TUNISIA: MARTYRS' DAY. Apr 9.

WINSTON CHURCHILL DAY. Apr 9. Anniversary of enactment of legislation in 1963 that made the late British statesman an honorary citizen of the US.

BIRTHDAYS TODAY

Seve Ballesteros, 38, golfer, born at Pedrena, Spain, Apr 9, 1957.
Jean-Paul Belmondo, 62, actor (*The Man from Rio, Is Paris Burning?*), born at Neuilly-sur-Seine, France, Apr 9, 1933.
Hugh Hefner, 69, founder of *Playboy*, born at Chicago, IL, Apr 9, 1926.
Paul Krassner, 63, editor, journalist, born at Brooklyn, NY, Apr 9, 1932.
Michael Learned, 56, actress ("The Waltons," "Nurses," *The Sisters Rosenzweig*), born at Washington, DC, Apr 9, 1939.
Tom Lehrer, 67, songwriter ("Vatican Rag," "New Math"), pianist, mathematician, born at New York, NY, Apr 9, 1928.

	S	M	T	W	T	F	S
April 1995							1
	2	3	4	5	6	7	8
	9	10	11	12	13	14	15
	16	17	18	19	20	21	22
	23	24	25	26	27	28	29
	30						

Harvey Lichtenstein, 66, president of the Brooklyn Academy of Music, born at Brooklyn, NY, Apr 9, 1929.
Carl Perkins, 63, singer, songwriter ("Blue Suede Shoes"), born at Jackson, TN, Apr 9, 1932.
Keshia Knight Pulliam, 16, actress (Rudy Huxtable in "The Cosby Show"), born at Newark, NJ, Apr 9, 1979.
Dennis Quaid, 41, actor (*The Big Easy, Everybody's All American*), born at Houston, TX, Apr 9, 1954.
Harris Wofford, 69, US Senator (D, Pennsylvania) up for reelection in Nov '94, born at New York, NY, Apr 9, 1926.

APRIL 10 — MONDAY
100th Day — Remaining, 265

BATAAN DEATH MARCH: ANNIVERSARY. Apr 10. On the morning of Apr 10, 1942, American and Filipino prisoners were herded together by Japanese soldiers on Mariveles Airfield on Bataan (in the Philippine islands) and began the Death March to Camp O'Donnell, near Cabanatuan. During the six-day march they were given only one bowl of rice. More than 5,200 Americans and many more Filipinos lost their lives in the course of the march.

BOOTH, WILLIAM: BIRTH ANNIVERSARY. Apr 10. General William Booth, founder of the movement that became known, in 1878, as the Salvation Army, was born at Nottingham, England, on Apr 10, 1829. Apprenticed to a pawnbroker at the age of 13, Booth experienced first-hand the misery of poverty. He broke with conventional church religion and established a quasi-military religious organization with military uniforms and ranks. Recruiting from the poor, from converted criminals and from many other social outcasts, his organization grew rapidly and its influence spread from England to the US and to most other countries. Booth was the author of *In Darkest England* and *The Way Out*. At revivals in slum areas the itinerant evangelist offered help for the poor, homes for the homeless, sobriety for alcoholics, rescue homes for women and girls, training centers and legal aid. Booth died at London, England, Aug 20, 1912.

COMMODORE PERRY DAY. Apr 10. Matthew Calbraith Perry, commodore in the US Navy, negotiator of first treaty between US and Japan (Mar 31, 1854). Born Apr 10, 1794. Died Mar 4, 1858.

CONNORS, CHUCK (KEVIN JOSEPH): BIRTH ANNIVERSARY. Apr 10. "The Rifleman" of television fame, Chuck Connors was born Apr 10, 1921, at Brooklyn, NY. Connors played Lucas McCain, a homesteader and single father who dealt with villians with his Winchester rifle in the "Rifleman" series from 1958 to 1963. He was nominated for an Emmy Award for his portrayal of a slave owner in the mini-series *Roots* and won a Golden Globe Award in 1959. Connors acted in more than 45 films and appeared on many television series and specials. He played professional basketball and baseball before entering the field of entertainment. Connors died Nov 10, 1992, at Los Angeles, CA.

GROTIUS, HUGO: BIRTH ANNIVERSARY. Apr 10. Anniversary of the birth of Hugo Grotius, the Dutch theologian, attorney, scholar and statesman whose beliefs profoundly influenced American thinking, especially with regard to the conscience of humanity. Born at Delft, Holland, Apr 10, 1583, Grotius was long an exile from his own country. He died at Rostock, Germany, Aug 28, 1645.

Chase's 1995 Calendar of Events — Apr 10-11

◆ **LIBERATION OF BUCHENWALD CONCENTRATION CAMP: 50th ANNIVERSARY.** Apr 10, 1945. Buchenwald, north of Weiner, Germany, was entered by Allied troops on this date. It was the first of the Nazi concentration camps to be liberated. It had been established in 1937, and about 56,000 people died there.

MCI HERITAGE CLASSIC GOLF TOURNAMENT. Apr 10-16. Hilton Head Island, SC. Top PGA professionals compete for the $1.125 million purse. Pre-tournament festivities include a parade enacting all the pomp and ceremony of ancient golfing traditions, a booming cannon and bagpipe music. Annually, the second week of April. Est attendance: 125,000. For info: Classic Sports, Mike Stevens, Dir, 79 Lighthouse Rd, Ste 414, Hilton Head Is, SC 29928. Phone: (803) 671-2448.

◆ **PERKINS, FRANCES: BIRTH ANNIVERSARY.** Apr 10. First woman member of a US presidential cabinet. Born at Boston, MA, Apr 10, 1880. Appointed secretary of labor by President Franklin D. Roosevelt in 1933, a post in which she served until 1945. She was married in 1915 to Paul Caldwell Wilson, but used her maiden name in public life. Died at New York, NY, May 14, 1965.

PULITZER, JOSEPH: BIRTH ANNIVERSARY. Apr 10. American journalist, founder of the Pulitzer Prizes, born at Budapest, Hungary, on Apr 10, 1847. Died at Charleston, SC, Oct 29, 1911. Pulitzer Prizes awarded annually since 1917. Write for entry and deadline info. (Please specify Book, Journalism, Drama or Music Competition.) For info: Pulitzer Prize Bd, 702 Journalism, Columbia Univ, New York, NY 10027. Phone: (212) 854-3841.

ROBERT GRAY BECOMES FIRST AMERICAN TO CIRCUMNAVIGATE THE EARTH. Apr 10. Robert Gray docked the ship *Columbia* in Boston Harbor on April 10, 1790, becoming the first American to circumnavigate the earth. He sailed from Boston, MA, in Sept 1787 to trade with Indians of the Pacific Northwest. After doing so, he sailed for China and then continued around the world. His 42,000-mile journey opened trade between New England and the Pacific Northwest and helped the US establish claims to the Oregon Territory.

WOODWARD, ROBERT BURNS: BIRTH ANNIVERSARY. Apr 10. Nobel Prize-winning (1965) Harvard University science professor whose special field of study was molecular structure of complex organic compounds. Called "one of the most outstanding scientific minds of the century." Born at Boston, MA, Apr 10, 1917. Died July 8, 1979.

BIRTHDAYS TODAY

David Halberstam, 61, journalist, author (*The Best and the Brightest, The Summer of Forty-Nine*), born at New York, NY, Apr 10, 1934.

John Madden, 59, sportscaster, born at Austin, NM, Apr 10, 1936.

Don Meredith, 57, sportscaster, actor, born at Mount Vernon, TX, Apr 10, 1938.

Harry Morgan (Harry Bratsburg), 80, actor ("M*A*S*H," "Dragnet"), born at Detroit, MI, Apr 10, 1915.

Omar Sharif (Michael Shalhoub), 63, actor (*Lawrence of Arabia, Dr. Zhivago*), born at Alexandria, Egypt, Apr 10, 1932.

Paul Edward Theroux, 54, author (*The Mosquito Coast, Millroy the Magician*), born at Medford, MS, Apr 10, 1941.

Max Von Sydow, 66, actor (*The Seventh Seal, The Emigrants*), born at Lund, Sweden, Apr 10, 1929.

APRIL 11 — TUESDAY
101st Day — Remaining, 264

BARBERSHOP QUARTET DAY. Apr 11. Commemorates the gathering of some 26 persons in Tulsa, OK, on Apr 11, 1938, and the founding there of the Society for the Preservation and Encouragement of Barbershop Quartet Singing in America.

BLISS, LIZZIE "LILLIE": BIRTH ANNIVERSARY. Apr 11, 1864. Lizzie "Lillie" Bliss was born at Boston, MA. She was one of the three founders (all women) of the Museum of Modern Art in New York City in 1929. She died on March 12, 1931, at New York City.

◆ **BOLIN, JANE MATILDA: BIRTHDAY.** Apr 11, 1908. Jane Matilda Bolin, born at Poughkeepsie, NY, was the first black woman to graduate from the Yale School of Law (1931) and went on to become the first black woman judge in the US. She served as assistant corporation counsel for the city of New York before being appointed to the city's Domestic Relations Court and the Family Court of the State of New York.

CIVIL RIGHTS ACT OF 1968: ANNIVERSARY. Apr 11. On Apr 11, 1968, exactly one week after the assassination of Martin Luther King, Jr, the Civil Rights Act of 1968 (protecting civil rights workers, expanding the rights of Native Americans and providing anti-discrimination measures in housing) was signed into law by President Lyndon B. Johnson, who said: " . . . the proudest moments of my presidency have been times such as this when I have signed into law the promises of a century."

EVERETT, EDWARD: BIRTH ANNIVERSARY. Apr 11. American statesman and orator, born at Dorcester, MA, on Apr 11, 1794. It was Edward Everett who delivered the main address at the dedication of Gettysburg National Cemetery, Nov 19, 1863. President Abraham Lincoln also spoke at the dedication, and his brief speech (less than two minutes) has been called one of the most eloquent in the English language. Once a candidate for vice president of the US (1860), Everett died at Boston, MA, on Jan 15, 1865.

HAROLD WASHINGTON ELECTED FIRST BLACK MAYOR OF CHICAGO: ANNIVERSARY. Apr 11. Harold Washington defeated Bernard Epton on Apr 11, 1983, and became the first black mayor of Chicago.

HUGHES, CHARLES EVANS: BIRTH ANNIVERSARY. Apr 11. Eleventh chief justice of US Supreme Court. Born at Glens Falls, NY, Apr 11, 1862. Died at Osterville, MA, Aug 27, 1948.

JULIAN, PERCY: BIRTH ANNIVERSARY. Apr 11. Percy Julian, producer of a synthetic progesterone using soy beans was born Apr 11, 1899, at Montgomery, AL. He also developed a cheaper method of producing cortisone, a drug to treat glaucoma and a chemical foam to fight petroleum fires. Percy Julian died Apr 19, 1975, at Waukegan, IL.

LIBRARY LEGISLATIVE DAY. Apr 11. Washington, DC. Legislative activities are planned by the Washington, DC Library Association, American Library Association, and Special Libraries Association to express support for specific library-related issues with members of Congress. Est attendance: 600. For info: Sandy Morton-Schwalb, Dir of Govt Relations, SLA, 1700 18th St, NW, Washington, DC 20009. Phone: (202) 234-4700.

MERRIAM, EVE: DEATH ANNIVERSARY. Apr 11. A poet, playwright and author of more than 50 books for both adults and children, Eve Merriam's works, which often focused on feminism, include *It Doesn't Always Have to Rhyme, Blackberry Ink, After Nora Slammed the Door: The Women's Unfinished Revolution, Halloween ABC, Mommies at Work*, and a book of urban poems attacked by authorities as glamorizing crime, *The Urban Mother Goose*. For the stage she penned the musical *The Club*, set in an all-male private club with conversations centering on derogatory remarks about women, in which all the male characters are played by women. Her play *Out of Our Father's House* portrayed the lives of American women, and in 1978 it was presented at the White House as well as on public television's "Great Performances" series. She also wrote the first documentary on women's rights for network TV, *We the Women*. She died at New York, NY on Apr 11, 1992.

Apr 11-12 ☆ Chase's 1995 Calendar of Events ☆

SPACE MILESTONE: *APOLLO 13* (US): 25th ANNIVERSARY. Apr 11. Astronauts Lovell, Haise and Swigert endangered when oxygen tank ruptured. Planned moon landing canceled. Details of accident made public and world shared concern for crew who splashed down successfully in Pacific Apr 17. Launched Apr 11, 1970.

UGANDA: LIBERATION DAY. Apr 11. Republic of Uganda celebrates anniversary of "overthrow of Idi Amin's dictatorship," on Apr 11, 1979.

◆ **YUGOSLAVIA/USSR TREATY: 50th ANNIVERSARY.** Apr 11, 1945. The Soviet Union and Yugoslavia signed a treaty of friendship, mutual aid and post-war collaboration.

BIRTHDAYS TODAY

Nicholas F. Brady, 65, US Secretary of the Treasury in Bush administration, born at New York, NY, Apr 11, 1930.

Oleg Cassini, 82, fashion designer, born at Paris, France, Apr 11, 1913.

Joel Grey (Joe Katz), 63, actor (Oscar for *Cabaret*; *The Seven-Per-Cent Solution*), born at Cleveland, OH, Apr 11, 1932.

Bill Irwin, 45, actor, choreographer (*The Regard of Flight*), born at Santa Monica, CA, Apr 11, 1950.

Ethel Kennedy, 67, widow of Robert Kennedy, born at Greenwich, CT, Apr 11, 1928.

Louise Lasser, 56, actress ("Mary Hartman, Mary Hartman"), born at New York, NY, Apr 11, 1939.

Jean-Claude Servan-Schreiber, 77, journalist, author (*The Chosen and the Choice*), born at Paris, France, Apr 11, 1918.

APRIL 12 — WEDNESDAY
102nd Day — Remaining, 263

ANNIVERSARY OF THE BIG WIND. Apr 12. Mount Washington, NH. The highest-velocity natural wind ever recorded occurred on the morning of Apr 12, 1934, at the Mount Washington, NH Observatory. Three weather observers, Wendell Stephenson, Alexander McKenzie and Salvatore Pagliuca, observed and recorded the phenomenon in which gusts reached 231 miles per hour—"the strongest natural wind ever recorded on the earth's surface." The 50th anniversary was observed at the site in 1984, with the three original observers participating in the ceremony.

ATTACK ON FORT SUMTER: ANNIVERSARY. Apr 12. After months of escalating tension, Major Robert Anderson refused to evacuate Fort Sumter in Charleston, SC. On April 12, 1861, Confederate troops under the command of General P.T. Beauregard opened fire on the harbor fort at 4:30 AM and continued until Major Anderson surrendered on April 13. No lives were lost despite the firing of some 40,000 shells in the first major engagement of the American Civil War.

BILLINGS, JOHN SHAW: BIRTH ANNIVERSARY. Apr 12. American librarian and army physician. Born Apr 12, 1838. Died Mar 11, 1913.

	S	M	T	W	T	F	S
April							1
	2	3	4	5	6	7	8
1995	9	10	11	12	13	14	15
	16	17	18	19	20	21	22
	23	24	25	26	27	28	29
	30						

CITY OF BALTIMORE'S EASTER CELEBRATION. Apr 12. Broadway Market Square, Baltimore, MD. Musical entertainment and performances given in celebration of Easter. A best-decorated Easter egg contest and the Easter bunny also featured. Est attendance: 250. For info: Dept of Recreation & Parks, Office of Adventures in Fun, Clarence "Du" Burns Arena, 1301 S Ellwood Ave, Baltimore, MD 21224. Phone: (410) 396-9177.

HALIFAX INDEPENDENCE DAY. Apr 12. North Carolina. Anniversary of the resolution adopted by the Provincial Congress of North Carolina at Halifax, NC, Apr 12, 1776, authorizing the delegates from North Carolina to the Continental Congress to vote for a Declaration of Independence.

NIGHT OF A THOUSAND STARS. Apr 12. A "Great American" read-aloud with special guest readers at libraries nationwide. Annually, the Wednesday of Natl Library Week. Sponsor: American Library Assn, Public Info Office, 50 E Huron St, Chicago, IL 60611. Phone: (312) 280-5041.

◆ **ROOSEVELT, FRANKLIN DELANO: 50th DEATH ANNIVERSARY.** Apr 12, 1945. With the end of WWII only months away, the nation and the world were stunned by the sudden death of the president shortly into his fourth term of office. Roosevelt, 32nd president of the US (Mar 4, 1933–Apr 12, 1945), was the only president to serve more than two terms—he was elected to four consecutive terms. He died at Warm Springs, GA.

◆ **SALK VACCINE: 40th ANNIVERSARY.** Apr 12. Anniversary of announcement in 1955 that the Salk vaccine (developed by American physician Dr. Jonas E. Salk) was declared "safe, potent and effective." Incidence of the dreaded infantile paralysis, or poliomyelitis, almost miraculously declined following introduction of preventive vaccines.

SPACE MILESTONE: *DISCOVERY* (US). Apr 12. On its 16th mission (from Kennedy Space Center, FL) *Discovery* carried a US Senator (Jake Garn) as a member of its crew of seven. Apr 12, 1985.

SPACE MILESTONE: *STS 1* (US). Apr 12. First flight Apr 12, 1981, of Shuttle *Columbia*. Two astronauts (John Young and Robert Crippen) on first manned US space mission since *Apollo-Soyuz* in July 1976, spent 54 hours in space (36 orbits of Earth) before landing at Edwards Air Force Base, CA, Apr 14.

SPACE MILESTONE: *VOSTOK I*, FIRST MAN IN SPACE: ANNIVERSARY. Apr 12. Yuri Gagarin became the first man in space on Apr 12, 1961, when he made a 108-minute voyage, orbiting Earth in a 10,395-lb vehicle, *Vostok I*, launched by the USSR.

THANK YOU SCHOOL LIBRARIAN DAY. Apr 12. Recognizes the unique contribution made by school librarians who are resource people extraordinaire, supporting the myriad educational needs of faculty, staff, students and parents *all year long*! Three cheers to all the public, private and parochial school infomaniacs whose true love of reading and lifelong learning make them great role models for kids of all ages. To help celebrate, take your school librarian to lunch, donate a book in his/her honor to the library, tell your librarian what a difference he/she has made in your life. Sponsor: "Carpe Libris" (Seize the Book), a loosely knit group of underappreciated librarians. For info: Judyth Lesse, Organizer, Carpe Libris, PO Box 1285, Tucson, AZ 85702-1285. Phone: (602) 798-8912.

TRUANCY LAW: ANNIVERSARY. Apr 12. The first truancy law was enacted in New York on April 12, 1853. A $50 fine was charged against parents whose children between the ages of 5 and 15 were absent from school.

BIRTHDAYS TODAY

David Cassidy, 45, singer ("Cherish"), actor ("The Partridge Family"), born New York, NY, Apr 12, 1950.

Lionel Hampton, 82, bandleader, born at Louisville, KY, Apr 12, 1913.

Herbie Hancock, 55, musician, born at Chicago, IL, Apr 12, 1940.
David Letterman, 48, comedian, TV talk show host, born at Indianapolis, IN, Apr 12, 1947.
Ann Miller (Lucille Ann Collier), 72, actress (*Sugar Babies, You Can't Take It With You, Easter Parade*), born at Houston, TX, Apr 12, 1923.
Tiny Tim (Herbert Buckingham Khaury), 73, entertainer ("Tiptoe Through the Tulips"), was married on "The Tonight Show" in 1969, born at New York, NY, Apr 12, 1922.
Scott Turow, 46, writer (*Presumed Innocent, The Burden of Proof*), born at Chicago, IL, Apr 12, 1949.

APRIL 13 — THURSDAY
103rd Day — Remaining, 262

ENGLAND: WEST SUSSEX INTERNATIONAL YOUTH MUSIC FESTIVAL. Apr 13-18. West Sussex. Non-competitive festival brings together youth (up to age 25) in bands, choirs and orchestras from all over the world. Est attendance: 3,000. For info: Concertworld (UK) Ltd, 150 Waterloo Rd, London, England SE1 8SB.

FIRST BASEBALL STRIKE ENDS: ANNIVERSARY. Apr 13. On Apr 13, 1972, major league baseball players and owners agreed on a settlement in which owners added an additional $500,000 to the players' pension fund. This ended the first baseball strike, which had begun Apr 5 when the season opener was cancelled.

FORT LAUDERDALE SPRING BOAT SHOW. Apr 13-16. Greater Fort Lauderdale/Broward Co Conv Center. Everything from small boats to mega-yachts to boating equipment. For info: Greater Ft. Lauderdale Conv/Visitors Bureau, 200 E Las Olas Blvd, Ste 1500, Fort Lauderdale, FL 33301. Phone: (305) 765-5900.

GREAT CHICAGO FLOOD: ANNIVERSARY. Apr 13. The morning of Apr 13, 1992, Chicago, IL, awoke to one of the most unusual disasters of modern times; the Chicago River broke through a rupture in an old underground freight tunnel wall, sending millions of gallons of water flooding into the tunnel system of the downtown business district known as the Loop. The water filled the freight tunnels and began pouring into basements of buildings that previously had been connected to the tunnel system. The Loop area had to be evacuated as electricity was cut off ahead of the rising water. A hectic effort mounted to plug the leak in the river eventually succeeded, even though the exact nature of the damage could not be ascertained. Once the flow of water had stopped, the city began the slow and expensive process of draining the water from beneath its structures and trying to cope with the losses due to what for most Chicagoans was an unseen disaster.

INTERNATIONAL SPECIAL LIBRARIANS DAY. Apr 13. Recognition for the important role that special librarians play in the international sharing of information. For info: Mark Serepca, Dir of Communications, SLA, 1700 18th St, NW, Washington, DC 20009. Phone: (202) 234-4700.

ITALY: PROCESSION OF THE ADDOLORATA AND PROCESSION OF THE MYSTERIES. Apr 13-14. Taranto. Procession of the Addolorata is held on Holy Thursday, while the Procession of the Mysteries takes place on Good Friday. Both processions have in common the very slow pace of the participants and their unusual costumes. For info: Italian Govt Travel Office, 500 N Michigan Ave, Chicago, IL 60611. Phone: (312) 644-0990.

★ **JEFFERSON, THOMAS: BIRTH ANNIVERSARY.** Apr 13. Presidential Proclamation 2276, of Mar 21, 1938, covers all succeeding years. (Pub Res No. 60 of Aug 16, 1937).

JEFFERSON, THOMAS: BIRTH ANNIVERSARY. Apr 13. Third president of the US (Mar 4, 1801-Mar 3, 1809). Born at Albermarle County, VA, Apr 13, 1743. Jefferson, who died at Charlottesville, VA, on July 4, 1826, wrote his own epitaph: "Here was buried Thomas Jefferson, author of the Declaration of American Independence, of the statute of Virginia for religious freedom, and father of the University of Virginia."

MAUNDY THURSDAY OR HOLY THURSDAY. Apr 13. The Thursday before Easter, originally "dies mandate," celebrates Christ's injunction to love one another, "Mandatus novum do vobis. . . ." ("A new commandment I give to you. . . .")

NORWAY: EASTER FESTIVAL. Apr 13-17. Kautokeino and Karasjok. Traditional celebration with Sami (Lapp) weddings, traditional fair, concerts and the World Championship Reindeer Races. For info: Norwegian Tourist Board, 655 Third Ave, New York, NY 10017. Phone: (212) 949-2333.

PGA SENIORS' CHAMPIONSHIP. Apr 13-16. Palm Beach Gardens, FL. Conducted by the Professional Golfers' Association of America. For info: PGA, Box 109601, Palm Beach Grdns, FL 33410-9601. Phone: (407) 622-4653.

PHILIPPINES: MORIONE'S FESTIVAL. Apr 13-16. Marinduque Island. Provincewide masquerade, Lenten plays and celebrations. Annually, Holy Thursday through Easter Sunday.

SILENT SPRING PUBLICATION: ANNIVERSARY. Apr 13. Rachel Carson's *Silent Spring*, released Apr 12, 1962, warned humankind that for the first time in history every person is subjected to contact with dangerous chemicals from conception until death. Ms Carson painted a vivid picture of how chemicals—used in many ways but particularly in pesticides—have upset the balance of nature, undermining the survival of countless species. This enormously popular and influential book became a softspoken battle cry to protect our natural surroundings. Ms Carson's intention was to shock and stir people into action; and, indeed, the publication of this book might well have signaled the beginning of the environmental movement.

SINGAPORE: SONGKRAN FESTIVAL. Apr 13-15. Public holiday. Thai water festival. To welcome the new year the image of Buddha is bathed with holy or fragrant water, and lustral water is sprinkled on celebrants. Joyous event, especially observed at Thai Buddhist temples. (Dates of observance subject to alteration.)

UNIVERSITY OF VIRGINIA FOUNDER'S DAY. Apr 13. Charlottesville, VA. Celebration marks the 252nd anniversary of the birth of the University's founder, Thomas Jefferson. For info: University of Virginia, Univ News Office, Booker House, Box 9018, Charlottesville, VA 22906-9018.

WAYNE STATE UNIVERSITY: FUNERAL FOR WINTER. Apr 13. Detroit, MI. Events include a New Orleans-style procession with jazz band playing Dixieland music. "Miss Spring" and a local TV personality conduct an irreverent burial ceremony. Est attendance: 1,000. For info: Wayne State University, Pat Borninski, MR Officer, 3222 FAB, Detroit, MI 48202. Phone: (313) 577-2150.

BIRTHDAYS TODAY

Peabo Bryson, 44, singer, born at Greenville, SC, Apr 13, 1951.
Ben Nighthorse Campbell, 62, US Senator (D, Colorado), born at Auburn, CA, Apr 13, 1933.
Jack Casady, 51, musician, born at Washington, DC, Apr 13, 1944.
Al Green, 49, singer ("Let's Stay Together," "You Ought To Be With Me"), born at Forrest City, AR, Apr 13, 1946.
Howard Keel (Howard Leek), 76, actor (*Annie Get Your Gun, Seven Brides for Seven Brothers*), born Gillespie, IL, Apr 13, 1919.
Madalyn Murray O'Hair, 76, atheist, lawyer, author (*Why I Am an Atheist*), born at Pittsburgh, PA, Apr 13, 1919.
Bret Saberhagen, 31, baseball player, born at Chicago Heights, IL, Apr 13, 1964.

Saundra Santiago, 38, actress ("Miami Vice"), born at the Bronx, NY, Apr 13, 1957.
Paul Sorvino, 56, actor ("Law and Order"), born at Brooklyn, NY, Apr 13, 1939.
Max M. Weinberg, 44, musician (drummer with E Street Band), born at South Orange, NJ, Apr 13, 1951.
Eudora Welty, 86, writer (*Delta Wedding*, *Losing Battles*), born at Jackson, MS, Apr 13, 1909.

APRIL 14 — FRIDAY
104th Day — Remaining, 261

AMERICAN BOMBING OF LIBYA: ANNIVERSARY. Apr 14. American warplanes, on orders from President Reagan, bombed the Libyan cities of Tripoli and Benghaze at 7 PM, EST, on Apr 14, 1986 (1 AM, Apr 15, Libyan time), reportedly killing 37 persons (including the infant daughter of head-of-state Muammar el-Qaddafi) and wounding another 93, many of them civilians. The American raid (condemned in a US-vetoed UN resolution) was, according to President Reagan's broadcast on Apr 14, in retaliation for the Apr 5 bomb explosion at La Belle, a popular West Berlin discotheque, which caused the deaths of two American soldiers and a Turkish woman. The La Belle explosion was later attributed to the work of a German woman, Christine Gabriele Endrigkeit, who was arrested Jan 11, 1988. See also: "La Belle Bombing: Anniversary" (Apr 5).

BIG MUDDY FOLK MUSIC FESTIVAL. Apr 14–15. Thespian Hall, Boonville, MO. Performing folk festival with instructional workshop by artists appearing. Est attendance: 3,000. For info: Judy Shields, Admin, Friends of Historic Boonville, PO Box 1776, Boonville, MO 65233. Phone: (816) 882-7977.

CAMPBELL BECOMES FIRST AMERICAN AIR ACE: ANNIVERSARY. Apr 14. Lt Douglas Campbell became the first American pilot to achieve the designation of ACE when he shot down his fifth German aircraft on Apr 14, 1918.

CHERRY BLOSSOM FESTIVAL. Apr 14–16. (Also Apr 21–23.) Japantown, San Francisco, CA. More than 2,000 Californians of Japanese descent and performers from Japan participate in most elaborate offering of Japanese culture and customs this side of Honshu highlighted by colorful parade. Est attendance: 100,000. For info: Cherry Blossom Fest, PO Box 15147, San Francisco, CA 94115-0147. Phone: (415) 922-7171.

ENGLAND: DEVIZES TO WESTMINSTER INTERNATIONAL CANOE RACE. Apr 14–17. Starts from Wharf Car Park, Wharf St, Devizes, Wiltshire. Canoes race along 125 miles of the Kennet and Avon canal and the River Thames, ending at County Hall Steps, Westminster Bridge Rd, London. Est attendance: 6,000. For info: Paul Owen, 14 Milldown Ave, Goring on Thames, Reading, Berkshire, England RG8 0AS.

ENGLAND: HARROGATE INTERNATIONAL YOUTH MUSIC FESTIVAL. Apr 14–20. Harrogate, North Yorkshire. Non-competitive festival which brings together youth (up to 25 years of age) in bands, choirs, orchestras and dance groups from all over the world. Est attendance: 7,000. For info: Concertworld (UK) Ltd, 150 Waterloo Rd, London, England SE1 8SB.

ENGLAND: HARVEYS WINE MUSEUM: 30th ANNIVERSARY. Apr 14–Dec 24. Harveys Wine Museum, Bristol, Avon. Anniversary celebrations with various events and demonstrations leading up to the 200th anniversary of Harveys of Bristol in 1996. For info: J Hawkes, Harveys Wine Museum Anniv, John Harvey and Sons, 12 Denmark St, Bristol, England BS1 5AQ.

April 1995

S	M	T	W	T	F	S
						1
2	3	4	5	6	7	8
9	10	11	12	13	14	15
16	17	18	19	20	21	22
23	24	25	26	27	28	29
30						

FIRST AMERICAN ABOLITION SOCIETY FOUNDED: ANNIVERSARY. Apr 14. The first abolition organization formed in the US was The Society for the Relief of Free Negroes Unlawfully Held in Bondage, which was founded Apr 14, 1775 in Philadelphia, PA.

GOOD EGG TREASURE HUNT. Apr 14–15. Good Children's Zoo, Oglebay, Wheeling, WV. Treasure hunt and magic shows. Annually, the Friday and Saturday before Easter. Est attendance: 3,400. Sponsor: Good Children's Zoo, John Hargleroad, Operating Mgr, Oglebay, Wheeling, WV 26003. Phone: (304) 243-4028.

GOOD FRIDAY. Apr 14. Observed in commemoration of the crucifixion. Oldest Christian celebration. Possible corruption of "God's Friday." Observed in some manner by most Christian sects and as public holiday or part holiday in many places.

GREAT PLAINS ROWING CHAMPIONSHIPS. Apr 14–16. Lake Shawnee, Topeka, KS. 11th annual 60-event rowing regatta for high school, college, and master age oarspeople. Annually, on Income Tax weekend. Sponsors: US Rowing and many other groups. Fax: (913) 233-9952 on 9th ring. Est attendance: 10,000. For info: Don Craig, 4336 SE 25th St Terrace, Topeka, KS 66605. Phone: (913) 233-9951.

GRENADA: EASTER REGATTA. Apr 14–17. Boats race to Grenada from Trinidad and from Union Island in the Grenadines, and then in Grenada waters. Est attendance: 2,000. For info: Grenada Tourist Info Office, 820 Second Ave, Ste 900-D, New York, NY 10017. Phone: (800) 927-9554.

HONDURAS: DIA DE LAS AMERICAS. Apr 14. Honduras. Pan American Day, a national holiday.

ITALY: GOOD FRIDAY. Apr 14. Calitri (near Avellino). At dawn, members of a religious confraternity, wearing white hoods surmounted by crowns of thorns, carry crosses on their shoulders in a procession accompanied by choirs singing psalms and folk songs. For info: Italian Govt Travel Office, 500 N Michigan Ave, Chicago, IL 60611. Phone: (312) 644-0990.

LINCOLN, ABRAHAM: 130th ASSASSINATION ANNIVERSARY. Apr 14. President Abraham Lincoln was shot while watching a performance of *Our American Cousin* at Ford's Theatre, Washington, DC, on Apr 14, 1865. He died the following day. Assassin was John Wilkes Booth, a young actor.

THE MAJIC 95.1 BODACIOUS BUNNY HUNT. Apr 14. Franke Park, Ft Wayne, IN. Inflatable bunnies are hidden all around the park. Each bunny is worth a specific number of points. Contestants have one minute to hunt for bunnies. The one with the most points wins an Easter Brunch and a chocolate bunny! Annually, the Friday prior to Easter. Est attendance: 200. Sponsor: WAJI Radio, Barb Richards, Asst Program Dir, 347 W Berry, Ste 600, Ft Wayne, IN 46802. Phone: (219) 423-3676.

MARION EASTER PAGEANT. Apr 14–16. Marion, IN. Portrays the biblical events from Palm Sunday to the resurrection as told through music, pantomime and pageantry. Est attendance: 10,000. For info: Jim Hill, PO Box 1201, Marion, IN 46952. Phone: (317) 664-3947.

NATIONAL WEEK OF THE OCEAN FESTIVAL SEASON. Apr 14–May 14. Fort Lauderdale, FL. Includes Fort Lauderdale Billfish tournament, sea chanty concerts, school marine fair, Earth Day, underwater seminar and kickoff party. Concludes with Mother Ocean Day. Est attendance: 100,000. For info: Cynthia Hancock, Natl Week of the Ocean, Inc, PO Box 179, Ft Lauderdale, FL 33302. Phone: (305) 462-5573.

★ **PAN AMERICAN DAY.** Apr 14. Presidential Proclamation 1912, of May 28, 1930, had effect of covering every Apr 14 (req'd by Governing Board of Pan American Union). Proc 2386, Feb 12, 1940, observed Union's 50th Anniversary. Proclamation issued each year since 1948.

★ Chase's 1995 Calendar of Events ★ Apr 14–15

PASSOVER BEGINS AT SUNDOWN. Apr 14. See "Pesach" (Apr 15).

SPINACH FESTIVAL. Apr 14–15. Alma, AR. Spinach, beans and cornbread supper; country western and gospel music; custom and antique car show, arts and crafts; carnival and food booths; race car display and car bash; softball, baseball and tournaments. "Fifties" celebration on Friday evening, hula-hoop contest, bubble-gum blowing contest, fifties dress-up contest and Twist competition. 9th annual festival. Annually, the third weekend in April. Est attendance: 6,000. For info: Chamber of Commerce, PO Box 2607, 1101 Hwy 71 N, Alma, AR 72921. Phone: (501) 632-4127.

STOCK EXCHANGE HOLIDAY (GOOD FRIDAY). Apr 14. The holiday schedules for the various exchanges are subject to change if relevant rules, regulations or exchange policies are revised. If you have questions, phone: American Stock Exchange (212) 306-1212; Chicago Board of Options Exchange (312) 786-7760; Chicago Board of Trade (312) 435-3500; New York Stock Exchange (212) 656-2065; Pacific Stock Exchange (415) 393-4000; Philadelphia Stock Exchange (215) 496-5000.

◆ **SULLIVAN, ANNE: BIRTH ANNIVERSARY.** Apr 14, 1866. Anne Sullivan, became well known for "working miracles" with Helen Keller, who was blind and deaf. Nearly blind herself, Sullivan used a manual alphabet, communicated by the sense of touch to teach Keller to read, write and speak and then to help her to go on to higher education. Anne Sullivan died Oct 20, 1936, at Forest Hills, NY.

TAFT OPENED BASEBALL SEASON: ANNIVERSARY. Apr 14. On Apr 14, 1910, President William Howard Taft began a sports tradition by throwing out the first baseball of the season at an American League game between Washington and Philadelphia. Washington won 3–0.

TOYNBEE, ARNOLD JOSEPH: BIRTH ANNIVERSARY. Apr 14. English historian, author of monumental *Study of History*. Born at London, England, Apr 14, 1889. Died at York, England, Oct 22, 1975.

UNITED KINGDOM: GOOD FRIDAY BANK HOLIDAY. Apr 14. Bank and public holiday in England, Wales, Scotland and Northern Ireland.

BIRTHDAYS TODAY

Julie Christie, 55, actress (*Dr. Zhivago, Petulia, Shampoo*), born Chukua, India, Apr 14, 1940.
Bradford Dillman, 65, actor (*Compulsion*, "Falcon Crest"), born at San Francisco, CA, Apr 14, 1930.
Sir John Gielgud, 91, actor (*Becket, The Charge of the Light Brigade*), director, born at London, England, Apr 14, 1904.
Anthony Michael Hall, 27, actor, comedian ("Saturday Night Live," *Sixteen Candles, The Breakfast Club*), born at Boston, MA, Apr 14, 1968.
David Christopher Justice, 29, baseball player, born at Cincinnati, OH, Apr 14, 1966.
Loretta Lynn, 60, singer ("Coal Miner's Daughter," "The Pill"), actress, born at Butcher's Hollow, KY, Apr 14, 1935.
Gregory Alan Maddux, 29, baseball player, born at San Angelo, TX, Apr 14, 1966.
Pete Rose, 54, former baseball player, born at Cincinnati, OH, Apr 14, 1941.
Rod Steiger, 70, actor (*On the Water Front*; Oscar for *In The Heat of the Night*), born Westhampton, NY, Apr 14, 1925.

APRIL 15 — SATURDAY
105th Day — Remaining, 260

AFRICA'S LEGACY IN MEXICO: PHOTOGRAPHS BY TONY GLEATON. Apr 15–May 28. Ann Arbor Public Library, Ann Arbor, MI. See Feb 11 for full description. Call or write the Smithsonian for other dates and venues. For info: Smithsonian Institution Traveling Exhibition Service, 1100 Jefferson Dr SW, Ste 3146, Washington, DC 20560. Phone: (202) 357-2700.

◆ **BELSEN CONCENTRATION CAMP LIBERATED: 50th ANNIVERSARY.** Apr 15, 1945. British troops reached the concentration camp at Belsen, Germany. They counted approximately 35,000 corpses there.

BENTON, THOMAS HART: BIRTH ANNIVERSARY. Apr 15. Thomas Hart Benton was an artist whose work was indicative of the American style of painting known as Regionalism. His works of life in the Midwest and South were not always flattering to their subjects, but his style became known as a truly American style of painting. He was born on Apr 15, 1889, at Neosho, MO, and died at Kansas City, MO, Jan 19, 1975.

CHINA: CANTON SPRING TRADE FAIR. Apr 15–May 15. The Guangzhou (Canton) Spring Trade Fair is held during the same dates each year.

CHINCOTEAGUE ISLAND EASTER DECOY SHOW. Apr 15–16. Chincoteague Island, VA. Wildfowl carving and wildlife art exhibits. Annually, on Easter weekend. Est attendance: 2,500. Sponsor: Chincoteague Chamber of Commerce, Box 258, Chincoteague Island, VA 23336. Phone: (804) 336-6161.

CYPRUS: THE PROCESSION OF ICON OF ST. LAZARUS. Apr 15. Larnaca, Cyprus. Annually, the day before Easter Sunday.

EASTER BEACH RUN. Apr 15. Daytona Beach, FL. The 27th annual beach run on "the world's most famous beach" includes a 4-mile run for 26 age divisions and a 2-mile run for youth 11 years old and younger. Est attendance: 3,000. For info: Easter Beach Run, Daytona Parks and Rec Dept, PO Box 2451, Daytona Beach, FL 32115. Phone: (904) 258-3106.

EASTER EGG HUNT. Apr 15. Shane Hill Nursing Home, Rockford, OH. Residents color 60 dozen Easter Eggs to hide for children and grandchildren of nursing home employees. Est attendance: 70. For info: Dorothy Trisel, Activities Dir, Shane Hill Nursing Home, 10731 State Rte 118, Rockford, OH 45882-0159.

EASTER EVEN. Apr 15. The Saturday before Easter. Last day of Holy Week and of Lent.

FEAST OF THE RAMSON. Apr 15. Richwood, WV. A dinner dedicated to the ramp, a wild leek, which grows wild in the mountains. Arts and crafts; mountain song and dance. Est attendance: 1,500. For info: Richwood Area Chamber of Commerce, Maxine Corbet, Exec Secy, 50 Oakford Ave, Richwood, WV 26261. Phone: (304) 846-6790.

HOLY CITY EASTER PAGEANT: *THE PRINCE OF PEACE*. Apr 15. Holy City of the Wichitas in the Wichita Mountains Wildlife Refuge, Lawton, Oklahoma. The life of Jesus is retold by a cast of hundreds during an Easter Eve, nighttime performance. This is the 70th year for this longest-running passion play drama in America, an Oklahoma Easter tradition. Est attendance: 5,000. For info: Holy City Office, PO Box 2003, Lawton, OK 73502. Phone: (405) 248-4043.

HOMEFRONT: AMERICA IN THE 1940s. Apr 15–Oct 29. Herbert Hoover Library/Museum, West Branch, IA. Return to the era of swing bands and swing shifts! This exhibit will tell the story of how America changed itself and how the world changed after World War II. Est attendance: 80,000. For info: Herbert Hoover Library/Museum, PO Box 488, West Branch, IA 52358. Phone: (319) 643-5301.

HOOD RIVER VALLEY BLOSSOM FESTIVAL. Apr 15–16. Hood River, OR. Breathtaking views of the Hood River Valley's orchards in bloom. Arts and crafts, dinners, seasonal opening of the Mt Hood Railroad. 41th annual festival. For info: Hood River Co Chamber of Commerce, Port Marina Park, Hood River, OR 97031. Phone: (800) 366-3530.

INCOME TAX PAY-DAY—BUT NOT THIS YEAR. Apr 15. A day all Americans need to know—the day by which taxpayers are supposed to make their accounting of the previous year and pay their share of the cost of government. The US Internal Revenue Service provides free forms. But since Apr 15 is a Saturday in 1995, the deadline is extended to Apr 17.

INDIAN DAY. Apr 15. Russell Cave National Monument, Bridgeport, AL. Programs on Native American culture with demonstrations of how everyday things were done in the past presented by expert volunteers. Archaeological programs present information on how science can help us learn about the past. Activities include flintknapping, pipe carving, bows, arrows, fire, pottery, finger weaving, petroglyphs and much more. For info: Larry Beane, Archaeologist, 3729 Cnty Rd 98, Bridgeport, AL 35740. Phone: (205) 495-2672.

JOHN WILKES BOOTH ESCAPE ROUTE TOUR. Apr 15 and 29. (Also Sept 9 and 23.) Clinton, MD. A 12-hour bus tour over the route used by Lincoln's assassin. Est attendance: 45. Sponsor: Surratt House & Tavern, Box 427, Clinton, MD 20735. Phone: (301) 868-1121.

LONGYEAR, JOHN MUNROE: BIRTH ANNIVERSARY. Apr 15. American capitalist, president of a bank and a button factory, land owner, philanthropist, one-time mayor of Marquette, MI. Disapproving of a railway route through Marquette, he caused his home, a stone castle-like showplace, to be torn down in 1903 and moved, stone by stone and stick by stick (in more than 190 freight cars), a distance, he said, of more than 1,300 miles, by rail, and reerected at Brookline, MA. Born at Lansing, MI, Apr 15, 1850. Died May 28, 1922.

LOUIS ARMSTRONG: CULTURAL LEGACY. Apr 15–June 25. Terra Museum of American Art, Chicago, IL. See Jan 28 for full description. For info: Smithsonian Institution Traveling Exhibition Service, 1100 Jefferson Dr SW, Ste 3146, Washington, DC 20560. Phone: (202) 357-2700.

LUNAR ECLIPSE. Apr 15. Partial eclipse of the moon. The beginning of the umbral phase is at approximately 7:40 AM EDT, middle of eclipse at 8:18 AM EDT and the end at approximately 8:55 AM EDT. The eclipse will be visible over part of Antarctica, Mexico, western half of North America, Pacific Ocean, Australia and eastern part of Asia.

MONACO: PRINTEMPS DES ARTS DE MONTE-CARLO. Apr 15–May 14. Monte-Carlo. For info: Printemps des Arts de Monte-Carlo, 8 rue Louis-Notari, MC-98000, Monaco. Phone: (339) 315-8303.

MOON PHASE: FULL MOON. Apr 15. Moon enters Full Moon phase at 8:08 AM, EDT.

MORE THAN MEETS THE EYE. Apr 15–June 25. Museum of Science and History, Jackson, FL. An original approach to the topics of sight, visual impairment and blindness. See Jan 7 for full description. For info: Smithsonian Institution Traveling Exhibition Service, 1100 Jefferson Dr SW, Ste 3146, Washington, DC 20560. Phone: (202) 357-2700.

NORTHERN CALIFORNIA CHERRY BLOSSOM FESTIVAL (WITH PARADE). Apr 15–16. (Also Apr 22–23.) Japantown, San Francisco, CA. More than 2,000 Californians of Japanese descent and performers from Japan participate in this most elaborate offering of Japanese culture and customs this side of the Pacific, highlighted by colorful parade. Sponsors: Japanese American communities of Northern California and several corporations. Media contact address: Louise Hanford, Hanford Assoc, Assoc, 582 Market St, Ste 608, San Francisco, CA 94104. Est attendance: 150,000. For info: Cherry Blossom Festival, PO Box 15147, San Francisco, CA 94115-0147. Phone: (415) 563-2313.

PEALE, CHARLES WILSON: BIRTH ANNIVERSARY. Apr 15. American portrait painter (best known for his many portraits of colonial and American Revolutionary War figures) was born in Queen Anne County, MD, Apr 15, 1741. Died at Philadelphia, PA, on Feb 22, 1827.

PESACH OR PASSOVER. Apr 15–22. Hebrew calendar dates: Nisan 15–22, 5755. Apr 15, the first day of Passover, begins an eight-day celebration of the delivery of Jews from slavery in Egypt. Unleavened bread (matzoh) is eaten at this time.

QUARTERLY ESTIMATED FEDERAL INCOME TAX PAYERS' DUE DATE. Apr 15. For those individuals whose fiscal year is the calendar year and who make quarterly estimated federal income tax payments, today is one of the due dates. (Jan 15, Apr 15, June 15 and Sept 15, 1995.)

RABBIT SHOW. Apr 15. Oglebay Park, Wheeling, WV. National competitive rabbit show with awards. Educates the public on rabbit care in recognition of the day before Easter. Est attendance: 800. For info: Oglebay Institute, Nature/Environmental Ed Dept, Ford Parker, Dir, AB Brooks Nature Center, Oglebay Park, Wheeling, WV 26003. Phone: (304) 242-6855.

REMEMBRANCE DAY. Apr 15. A special day to remember a loved one who has died. A way to say "I still love you; I have not forgotten" by planting a tree, community service, saying a prayer, etc. Send an L.S.A.S.E. for info. Annually, the Saturday before Easter. Sponsor: Chaplain Mike Miller, Happy Days Ministries, 35 Hilcreek Blvd, Charleston, SC 29412.

RUBBER ERASER DAY. Apr 15. In 1770 English Chemist Joseph Priestley coined the term eraser when he found that a small cube of latex could be used to rub out pencil marks. Use Priestley's discovery to make those corrections before mailing your income tax return today. For info: Dorothy Dudley Muth, 7328 San Bartolo, Carlsbad, CA 92009.

SINKING OF THE *TITANIC*: ANNIVERSARY. Apr 15. The "unsinkable" luxury liner *Titanic*, on its maiden voyage (from Southampton, England, to New York, NY) struck an iceberg just before midnight on Apr 14, 1912, and sank at 2:27 AM, Apr 15. The *Titanic* had 2,224 persons aboard. Of these, more than 1,500 were lost. About 700 people were rescued from the icy waters off Newfoundland by the liner *Carpathia*, which reached the scene about 20 minutes after the *Titanic* went down. The sunken *Titanic* was located and photographed in Sept 1985. In July 1986, an expedition aboard the *Atlantis II*, headed by Dr. Robert D. Ballard, descended to the deck of the *Titanic* in a submersible craft, *Alvin*, and guided a robot named Jason, Jr, in a search of the ship. Two memorial bronze plaques were left on the deck of the sunken ship.

SMITH, BESSIE: BIRTH ANNIVERSARY. Apr 15. The "Empress of the Blues," Bessie Smith, was born at Chattanooga, TN, Apr 15, 1894 (year varies as late as 1900). She was assisted in her efforts to break into show business by Ma Rainey, the

	S	M	T	W	T	F	S
April 1995							1
	2	3	4	5	6	7	8
	9	10	11	12	13	14	15
	16	17	18	19	20	21	22
	23	24	25	26	27	28	29
	30						

first great blues singer. Her first recording was made in February, 1923. Smith died of injuries she sustained in an automobile accident at Clarksdale, MS, Sept 26, 1937. It has been said that had she been white she would have received earlier medical treatment. The black hospital to which Bessie was taken after the accident is now the Riverside Hotel where legendary bluesmen have stayed since 1944.

SPRING PRAIRIE FESTIVAL. Apr 15. The Prairie Center, Olathe, KS. Demonstrations of a controlled burning of the prairie, guided nature trail walks and environmental exhibits are a part of this environmental awareness festival. Annually, the third weekend in April. Est attendance: 1,000. For info: The Prairie Center, 26325 Prairie Center Rd, Olathe, KS 66061. Phone: (913) 856-8832.

STRAWBERRY HILL RACES. Apr 15. Fairgrounds on Strawberry Hill, Richmond, VA. Annual steeplechase featuring a week of festivities leading up to the event. Elegant pre-race dinner dance and tailgate competition. More than $65,000 in purses. Sponsored by Atlantic Rural Exposition, Inc. Benefits The Richmond Symphony. Est attendance: 20,000. For info: Sue Mullins, Equine Dir, Strawberry Hill Races, PO Box 26805, Richmond, VA 23261. Phone: (804) 228-3238.

STRENGTH AND DIVERSITY: JAPANESE-AMERICAN WOMEN 1885-1990. Apr 15–June 25. Dane G. Hansen Memorial Museum, Logan, KS. See Jan 7 for full description of event. Call or write the Smithsonian for other dates and venues. For info: Smithsonian Institution Traveling Exhibition Service, 1100 Jefferson Dr SW, Ste 3146, Washington, DC 20560. Phone: (202) 357-2700.

BIRTHDAYS TODAY

Evelyn Ashford, 38, track athlete, born at Shreveport, LA, Apr 15, 1957.
Claudia Cardinale, 57, actress (*The Pink Panther, Once Upon a Time in the West*), born at Tunis, Italy, Apr 15, 1938.
Roy Clark, 62, singer ("Yesterday, When I Was Young"), born at Meherrin, VA, Apr 15, 1933.
Michael Cooper, 39, former basketball player, born at Los Angeles, CA, Apr 15, 1956.
Heloise Cruse Evans, 44, newspaper columnist ("Hints from Heloise"), born at Waco, TX, Apr 15, 1951.
Elizabeth Montgomery, 62, actress (Samantha on "Bewitched"), born at Los Angeles, CA, Apr 15, 1933.

APRIL 16 — SUNDAY
106th Day — Remaining, 259

BOYS AND GIRLS CLUB WEEK. Apr 16. An annual event commemorating the founding of the first Club 130 years ago. Every president since Herbert Hoover has served as Honorary Chairman of Boys and Girls Clubs of America. Today there are 1,600 Clubs providing educational, arts and sports programs to two million young people in the US and the Virgin Islands. Annually, the third week in April. For info: Boys and Girls Clubs of Amer, 1230 W Peachtree St NW, Atlanta, GA 30309.

CHAPLIN, CHARLES SPENCER: BIRTH ANNIVERSARY. Apr 16. Celebrated film comedian. Born at London, England, Apr 16, 1889. Film debut in 1914. Knighted in 1975. Died at Vevey, Switzerland, Dec 25, 1977. In his autobiography Chaplin wrote: "There are more valid facts and details in works of art than there are in history books."

DENMARK: QUEEN MARGRETHE'S BIRTHDAY. Apr 16. Thousands of children gather to cheer the queen at Amalienborg Palace, and the Royal Guard wears scarlet gala uniforms. A national holiday.

DIEGO, JOSE DE: BIRTH ANNIVERSARY. Apr 16. Puerto Rico. Celebrates birth on this day, 1867, of Puerto Rican patriot and political leader Jose de Diego.

EASTER PARADE. Apr 16. Brighton Park, Atlantic City, NJ. Awards presented for best bonnet and best dressed: woman, couple, gentleman, senior citizen, girl and boy. Annually, Easter Sunday. Est attendance: 8,000. For info: Greater Atlantic City Conv and Visitors Bureau, 2314 Pacific Ave, Atlantic City, NJ 08401. Phone: (609) 348-7100.

EASTER SUNDAY. Apr 16. Commemorates the Resurrection of Christ. Most joyous festival of the Christian year. The date of Easter, a movable feast, is derived from the lunar calendar (as prescribed by the Council of Nicaea, AD 325): the first Sunday following the first full moon on or after the vernal equinox (Mar 20)—always between Mar 22 and Apr 25. Many other dates in the Christian year are derived from the date of Easter.

EASTER SUNDAYS THROUGH THE YEAR 2000

1995, Apr 16	1998, Apr 12
1996, Apr 7	1999, Apr 4
1997, Mar 30	2000, Apr 23

EASTER SUNRISE SERVICE. Apr 16. Chimney Rock Park, Chimney Rock, NC. Celebrate the glory of Easter at this 40th annual nondenominational community worship service overlooking beautiful Lake Lure. Gates open at 4:30 AM. No admission charge. Est attendance: 1,500. For info: Mary Jaeger-Gale, Mktg Mgr, PO Box 39, Chimney Rock, NC 28720. Phone: (800) 277-9611.

INTERNATIONAL LATHE-TURNED OBJECTS: CHALLENGE V: EXHIBITION. Apr 16–June 11. Huntsville Museum of Art, Huntsville, AL. A display of beautiful decorative and functional wood and metal objects selected by the Wood Turning Center, Philadelphia. Est attendance: 2,000. For info: Marylyn Coffey, Publicist, or Peter Baldaia, Curator, Huntsville Museum of Art, Von Braun Civic Center, 700 Monroe St SW, Huntsville, AL 35801. Phone: (205) 535-4350.

ITALY: EXPLOSION OF THE CART. Apr 16. Florence. At noon in Piazza del Duomo a cart full of fireworks is exploded, perpetuating a ceremony of ancient origin and recalling the fire that used to be kindled during the *Gloria* at Easter mass, and was then distributed to all of Florence's households. The tradition is held to date back to the time of the First Crusade, when the valorous Pazzino dei Pazzi was awarded some pieces of flint from the Holy Sepulcher. After his return to Florence the holy fire was kindled with these flints, now preserved in the church of Santi Apostoli. For info: Italian Govt Travel Office, 500 N Michigan Ave, Chicago, IL 60611. Phone: (312) 644-0990.

MARKSVILLE EASTER EGG KNOCKING CONTEST. Apr 16. Marksville, LA. Competition among owners of chicken, goose, turkey and guinea eggs after they have been boiled and dyed. Annually, on Easter Sunday. Est attendance: 300. Sponsor: Chamber of Commerce, Box 365, Marksville, LA 71351. Phone: (318) 253-9222.

MERRIE MONARCH FESTIVAL (WITH WORLD'S LARGEST HULA COMPETITION). Apr 16–22. Hilo, HI. Cultural event honoring King David Kalakaua. Festival culminates with the world's largest hula competition. Hawaii's finest hula schools compete in ancient and modern divisions. Annually, beginning on Easter Sunday. Est attendance: 6,000. For info: Dorothy Thompson, Merrie Monarch Office, 93 Banyan Dr, Hawaii Naniloa Hotel, Hilo, HI 96720. Phone: (808) 935-9168.

MORAVIAN EASTER SUNRISE SERVICE. Apr 16. Winston-Salem, NC. Outdoor religious service featuring Moravian bands playing in streets to awaken sleepers. Service begins in Salem Square, and concludes in God's Acre, the Moravian Graveyard, at daybreak. Est attendance: 12,500. For info: Salem Congregation, 459 S Church St, Winston-Salem, NC 27101-5314. Phone: (910) 722-6504.

Apr 16–17 ☆ Chase's 1995 Calendar of Events ☆

MT RUBIDOUX EASTER SUNRISE SERVICE. Apr 16. Mt. Rubidoux, Riverside, CA. 86th annual non-denominational outdoor Easter Sunrise Service, the nation's oldest. Worshippers hike to mountain top, as Boy Scouts light the trails—the oldest "Good Turn" in the history of scouting. Service starts approximately 30 minutes before sunrise. Annually, Easter Sunday. Est attendance: 2,000. For info: Lee McIntyre, Pres, KSGN Radio, 11498 Pierce St, Riverside, CA 92505. Phone: (909) 687-5746.

NATIONAL COIN WEEK. Apr 16–22. To promote the history and lore of numismatics and the hobby of coin collecting. Est attendance: 30,000. Sponsor: American Numismatic Assn, James Taylor, Dir of Educ, 818 N Cascade Ave, Colorado Springs, CO 80903. Phone: (719) 632-2646.

NATIONAL STRESS AWARENESS DAY. Apr 16. To focus public awareness on one of the leading health problems in the world today. Health-related organizations throughout the country are encouraged to sponsor stress education programs and events. Annually the first day after income taxes due! Sponsor: The Health Resource Network, Morton C. Orman, MD, Dir, 2936 E Baltimore St, Baltimore, MD 21224. Phone: (410) 732-1900.

NATIONAL WEEK OF THE OCEAN. Apr 16–22. A week focusing on humanity's interdependence with the ocean, asking each of us to appreciate, protect and use the ocean wisely. 12th annual observance. Sponsor: Natl Week of the Ocean, Inc, PO Box 179, Fort Lauderdale, FL 33302. Phone: (305) 462-5573.

NATURAL BRIDGES NATIONAL MONUMENT: ANNIVERSARY. Apr 16. Utah. Natural Bridges National Monument was established Apr 16, 1908.

ORTHODOX PALM SUNDAY. Apr 16. Celebration of Christ's entry into Jerusalem, when His way was covered with palms by the multitudes. Beginning of Holy Week in the Orthodox Church.

SLAVERY ABOLISHED IN DISTRICT OF COLUMBIA: ANNIVERSARY. Apr 16. Congress abolished slavery in the District of Columbia on Apr 16, 1862. One million dollars was appropriated to compensate owners of freed slaves, and $100,000 was set aside to pay district slaves who wished to emigrate to Haiti, Liberia or any other country outside the US.

SLOANE, HANS: BIRTH ANNIVERSARY. Apr 16. British medical doctor and naturalist whose personal collection became the nucleus of the British Museum. Born at County Down, Ireland, on Apr 16, 1660. Upon his death at Chelsea, England, on Jan 11, 1753, his collections of books, manuscripts, medals and antiquities were bequeathed to Britain and accepted by an act of Parliament that incorporated the British Museum. It was opened to the public at London, England, on Jan 15, 1759. It is the national depository, museum and library of the United Kingdom.

SPACE MILESTONE: *APOLLO 16* (US). Apr 16. On Apr 16, 1972, astronauts John W. Young, Charles M. Duke, Jr, and Thomas K. Mattingly II (command module pilot), began 11-day mission that included 71-hour exploration of moon (Apr 20–23). Landing Module (LM) named *Orion*. Splashdown in Pacific Ocean within a mile of target, Apr 27.

SYNGE, JOHN MILLINGTON: BIRTH ANNIVERSARY. Apr 16. Irish dramatist and poet, most of whose plays were written in the brief span of six years before his death at age 37, of lymphatic sarcoma. His best-known work was *The Playboy of the Western World* (1907), which caused protests and rioting at early performances. Synge (pronounced "Sing") was born near Dublin, on Apr 16, 1871, and died there on Mar 24, 1909.

	S	M	T	W	T	F	S
April 1995	2	3	4	5	6	7	1 8
	9	10	11	12	13	14	15
	16	17	18	19	20	21	22
	23	24	25	26	27	28	29
	30						

WRIGHT, WILBUR: BIRTH ANNIVERSARY. Apr 16. Aviation pioneer. Born Apr 16, 1867, at Millville, IN. Died at Dayton, OH, May 30, 1912.

BIRTHDAYS TODAY

Kareem Abdul-Jabbar (Lewis Alcindor, Jr), 48, former basketball player, born at New York, NY, Apr 16, 1947.
Edie Adams (Elizabeth Edith Enke), 64, singer, actress, born at Kingston, PA, Apr 16, 1931.
Kingsley Amis, 73, author (*The Crime of the Century, Lucky Jim*), born at London, England, Apr 16, 1922.
Ellen Barkin, 40, actress, (*Tender Mercies, Diner*), born at New York, NY, Apr 16, 1955.
Merce Cunningham, 76, dancer, choreographer, born at Centralia, WA, Apr 16, 1919.
Peter Ustinov, 74, actor (Oscars for *Spartacus, Topkapi*), born London, England, Apr 16, 1921.
Bobby Vinton, 60, singer ("Mr. Lonely," "Roses Are Red [My Love]"), born at Canonsburg, PA, Apr 16, 1935.

APRIL 17 — MONDAY
107th Day — Remaining, 258

AMERICAN SAMOA: FLAG DAY. Apr 17. Pago Pago. National Holiday to observe the Tutuila Cession to the United States in 1900. An international celebration with competitive sports competition, dance and song fests and speeches from Pacific leaders to recognize the significance of the event. American Samoa Government plays host to this affair. Annually, Apr 17. Est attendance: 2,500. For info: Emma F. Randall, Dir, Office of Tourism, American Samoa Govt, Pago Pago, Amer Samoa 96799. Phone: (684) 633-1091.

BANDARANAIKE, SIRIMAVO: BIRTHDAY. Apr 17. World's first woman prime minister (Sri Lanka). Born at Ratnapura, Sri Lanka (formerly called Ceylon), on Apr 17, 1916.

BOSTON MARATHON. Apr 17. Boston, MA. The marathon begins in the rural New England town of Hopkinton, winds through eight cities and towns and finishes near downtown Boston. 1995 will be the 95th year of this historic running event. 10,000 participants. Est attendance: 1,000,000. For info: Boston Athletic Assn, Boston Marathon, PO Box 1994, Hopkinton, MA 01748. Phone: (617) 236-1652.

CHASE, SAMUEL: BIRTH ANNIVERSARY. Apr 17. American Revolutionary leader, signer of the Declaration of Independence and justice of the US Supreme Court. Born at Somerset County, MD, Apr 17, 1741. Died June 19, 1811.

CIMARRON TERRITORY CELEBRATION. Apr 17–22. Beaver, OK. A homesteader festival highlighted by the World Champion Cow Chip Throwing Contest. Stock car racing, parade and artisans demonstrating homesteading techniques and crafts. Antiques, coins, guns and crafts show. Annually, the third week in April. Est attendance: 4,500. For info: Beaver Chamber of Commerce, Box 878, Beaver, OK 73932. Phone: (405) 625-4726.

EASTER MONDAY. Apr 17. Holiday or bank holiday in many places, including England, Northern Ireland, Wales, Canada and North Carolina in the US.

EASTER MONDAY CELEBRATION. Apr 17. Portage, MI. To resurrect the old Christian custom of celebrating Easter Monday in observance of Jesus rising from the dead, "the practical joke that God played on Satan." (Easter Monday for Eastern Christians is celebrated on Apr 24.) Churches and prayer groups nationwide participate. Sponsor: The Fellowship of Merry Christians, Inc, PO Box 895, Portage, MI 49081-0895. Phone: (616) 324-0990.

EGG SALAD WEEK. Apr 17–23. Dedicated to the many delicious uses for all of the Easter eggs that have been cooked, colored, hidden and found. Annually, the full week after Easter. For info: American Egg Bd, 1460 Renaissance Dr, Park Ridge, IL 60068.

ENGLAND: HALLATON BOTTLE KICKING. Apr 17. Hallaton, Leicestershire. Ancient custom dating back at least 600 years. Annually, on Easter Monday.

GENERAL TIRE PROFESSIONAL BOWLERS ASSOCIATION MEN'S TOURNAMENT OF CHAMPIONS. Apr 17–22. Fairlawn, OH. Est attendance: 8,000. For info: Professional Bowlers Assn, PO Box 5118, Akron, OH 44334-0118. Phone: (216) 836-5568.

GREAT LA CLEAN UP. Apr 17–May 3. Citywide, Los Angeles, CA. In 1994, 31,094 people participated in this major civic event to clean up the city of Los Angeles, collecting trash, planting plants and removing graffiti. For info: Earth Service, Inc, 2401 Colorado Ave, Ste 170, Santa Monica, CA 90404. Phone: (310) 829-9190.

HERITAGE WEEK. Apr 17–22. New Harmony, IN. Head to this historic village and watch crafters create 19th-century-style chairs, candles, baskets and pottery. Handmade souvenirs for sale. Annually, the third week in April. Est attendance: 4,000. For info: Special Projects Coord, PO Box 579, New Harmony, IN 47631. Phone: (812) 682-4488.

INCOME TAX PAY-DAY: THIS IS REALLY IT. Apr 17. Taxpayers got a temporary reprieve this year. Since Apr 15 fell on a Saturday, we all had two extra days to procrastinate. But now it's time to face the music and pay the piper—get those returns in the mail by midnight! (Don't get excited about next year—in '96 the 15th is going to Leap right over Sunday and fall on Monday.)

LUXEMBOURG: EMAISHEN. Apr 17. Luxembourg (city). Popular traditional market and festival on the "Marche-aux-Poissons." Young lovers present each other with earthenware articles, sold only on this day. Annually, on Easter Monday.

MORGAN, JOHN PIERPONT: BIRTH ANNIVERSARY. Apr 17. American financier and corporation director. Born at Hartford, CT, on Apr 17, 1837. Morgan died on Mar 31, 1913, leaving an estate valued at more than $70 million.

NEEDHAM, THERESA: BIRTH ANNIVERSARY. Apr 17. Owner of Chicago's legendary South Side blues bar, Theresa's Lounge, Theresa Needham was born Apr 17, 1912, at Meridian, MS. Especially memorable at the bar were "Blue Monday" all-day jams at which the city's top blues performers locked horns in musical battles. Needham, who was bartender, bouncer and talent agent, came to be known as "the Godmother of Chicago Blues." Died at Chicago, IL, Oct 16, 1992.

NETHERLANDS AND SCILLY ISLES PEACE: ANNIVERSARY. Apr 17. The 335-year "state of war" that had existed between the Netherlands and the Scilly Isles came to an end on Apr 17, 1986. On that date Dutch ambassador Jonkheer Huydecoper flew to the Scilly Isles to deliver a proclamation terminating the war that had started in 1651. Though hostilities had ceased three centuries earlier, a standing joke in the islands was that no one had bothered to declare an end to the war—until 1986. One of history's longer wars thus was concluded.

PATRIOT'S DAY IN MASSACHUSETTS AND MAINE. Apr 17. Commemorates Battle of Lexington and Concord, 1775. Annually, the third Monday in April.

SOLIDARITY GRANTED LEGAL STATUS: ANNIVERSARY. Apr 17. After nearly a decade of struggle and suppression under martial law, the Polish labor union Solidarity was granted legal status on Apr 17, 1989, clearing the way for the downfall of the Polish Communist Party. Solidarity and the Polish people surprised the government by winning 99 of the 100 parliamentary seats in the election. General Wojciech Jaruzelski was elected president on July 19 and nominated Czeslaw Kiszczak prime minister, a move that enraged the Lech Walesa-led Solidarity. On Aug 7 Walesa swayed the traditional allies of the Communist Party—the United Peasant and Democratic Parties—to switch sides. Kiszczak resigned as prime minister a week later after failing to form a government, forcing Jaruzelski to accept the principal of a government led by Solidarity.

SPACE MILESTONE: *SURVEYOR 3* (US). Apr 17. Lunar probe vehicle made soft landing on Moon on Apr 20 and with its digging apparatus established surface qualities. Launched Apr 17, 1967.

SWITZERLAND: EGG RACES. Apr 17. Rural northwest Swiss Easter Monday custom. Race among competitors carrying large number of eggs while running to neighboring villages.

SYRIAN ARAB REPUBLIC: INDEPENDENCE DAY. Apr 17. Official holiday.

TALKING BOOK WEEK. Apr 17–21. To recognize volunteers and descriptive organizations who provide reading and informational services for the visually impaired and physically disabled. Sponsor: Dawn L. Jordan, Outreach Dir, Audio Descriptive Network, 115 Brenton St, Richmond, VA 23222. Phone: (804) 321-2063.

UNITED KINGDOM: EASTER MONDAY BANK HOLIDAY. Apr 17. Bank and public holiday in England, Wales and Northern Ireland. (Scotland not included.)

VERRAZANO DAY. Apr 17. Celebrates discovery of New York harbor, 1524, by Giovanni Verrazano, Florentine navigator, 1480 (?)–1527.

WHITE HOUSE EASTER EGG ROLL. Apr 17. Traditionally held at executive mansion's south lawn on Easter Monday. Custom said to have started at Capitol grounds about 1810. Transferred to White House lawn in 1870s.

WILDER, THORNTON: BIRTH ANNIVERSARY. Apr 17. Pulitzer prize-winning American playwright and novelist, born at Madison, WI, Apr 17, 1897. Died at Hamden, CT, Dec 7, 1975.

WORLD COW CHIP THROWING CHAMPIONSHIP. Apr 17–22. Beaver, OK. "A highly specialized athletic event which draws dung flingers from around the world. A special division of this competition is held for politicians, who are known to be highly practiced in this area." Annually, in April. Est attendance: 4,500. For info: Beaver Chamber of Commerce, Box 878, Beaver, OK 73932. Phone: (405) 625-4726.

BIRTHDAYS TODAY

Lindsay Anderson, 72, director, born at Bangalore, India, Apr 17, 1923.

Norman Julius "Boomer" Esiason, 34, football player, born at West Islip, NY, Apr 17, 1961.

Joseph Foss, 80, former president of the National Rifle Association, born at Sioux Falls, SD, Apr 17, 1915.

Roy Gallant, 71, author of children's books (*Mirages & Sundogs, Earth's Vanishing Forests*), born at Portland, ME, Apr 17, 1924.
Anna Marie Holmes, 52, dancer, born at Mission City, British Columbia, Canada, Apr 17, 1943.
Olivia Hussey, 44, actress (Juliet in Franco Zeffirelli's *Romeo and Juliet*), born at Buenos Aires, Argentina, Apr 17, 1951.
Don Kirshner, 61, music publisher, promoter, born at the Bronx, NY, Apr 17, 1934.
Cynthia Ozick, 67, feminist, writer, born at New York, NY, Apr 17, 1928.

APRIL 18 — TUESDAY
108th Day — Remaining, 257

CANADA: CONSTITUTION ACT OF 1982: ANNIVERSARY. Apr 18. Replacing the British North America Act of 1867, the Canadian Constitution Act of 1982 provides Canada with a new set of fundamental laws and civil rights. Signed by Queen Elizabeth II, at Parliament Hill, Ottawa, Canada, on Saturday, Apr 18, 1982, it went into effect at 12:01 AM, Sunday, Apr 18, 1982.

CANADA: SHAW FESTIVAL. Apr 18–Oct 28. Niagara-on-the-Lake, Ontario. The only professional theatre company in the world featuring the works of George Bernard Shaw and his contemporaries. In North America phone: 800-267-4759. Est attendance: 280,000. For info: Shaw Festival, Box 774, Niagara-on-the-Lake, Ont, Canada L0S 1J0.

CRAWFORD, SAMUEL EARL "WAHOO SAM": BIRTH ANNIVERSARY. Apr 18. Major league baseball player with the Detroit Tigers, born Apr 18, 1880, at Wahoo, NE. Wahoo Sam played pro ball for 20 years, racking up a career batting average of .309. His record of 312 career triples still stands. He was inducted into the Baseball Hall of Fame in 1957. Crawford died on June 15, 1968, at Hollywood, CA.

DARROW, CLARENCE SEWARD: BIRTH ANNIVERSARY. Apr 18. American attorney often associated with unpopular causes, from the Pullman strike in 1894 to the Scottsboro case in 1932, born at Kinsman, OH, Apr 18, 1857. At the Scopes trial, July 13, 1925, Darrow said: "I do not consider it an insult, but rather a compliment, to be called an agnostic. I do not pretend to know where many ignorant men are sure—that is all that agnosticism means." Darrow died at Chicago, IL, on Mar 13, 1938.

EIGHTY-NINER CELEBRATION. Apr 18–22. Guthrie, OK. In celebration of its heritage, this historically restored town features Old West gunfights, chuckwagon feed, professional rodeo and Oklahoma's largest parade of bands, floats and round-up clubs from across the state. Phone: (800) 299-1889 or (405) 282-1947. Est attendance: 10,000. For info: Chamber of Commerce, Box 995, Guthrie, OK 73044.

◆ **THE HOUSE THAT RUTH BUILT: ANNIVERSARY.** Apr 18, 1923. More than 74,000 fans attended Opening Day festivities as the New York Yankees inaugurated their new stadium in the Bronx. Babe Ruth christened it with a game-winning three-run homer into the right-field bleachers. In his coverage of the game for the *New York Evening Telegram* sportswriter Fred Lieb described Yankee Stadium as "The House that Ruth Built"—the name stuck.

LOOK-ALIKE DAY. Apr 18. A day for recognizing those people who are often recognized for looking like someone famous, but enjoy no recognition for having that famous face. The news media in every town can find people who are look alikes of the rich and famous and give them a little fame of their own. Annually, the third Tuesday in April. For info: Jack Etzel, Feature Reporter, WPXI-TV, 11 Television Hill, Pittsburgh, PA 15214. Phone: (412) 237-4952.

★ **NATIONAL YOUTH SERVICE DAY.** Apr 18. Presidential Proclamation 6674 of Apr 19, 1994, covered this observance for 1994 and 1995.

NATIONAL CATHOLIC EDUCATIONAL ASSOCIATION CONVENTION AND EXPOSITION. Apr 18–21. Cincinnati Convention Center, Cincinnati, OH. Annual meeting for NCEA members and anyone working in, or interested in, the welfare of Catholic education. Est attendance: 10,000. Sponsor: Natl Catholic Educational Assn, Nancy Brewer, Conv Dir, 1077 30th St NW, Ste 100, Washington, DC 20007. Phone: (202) 337-6232.

NATIONAL CPAs' GOOF OFF-DAY. Apr 18. After working on tax returns, this is the day that CPAs throughout the US are allowed to goof off. Annually, the first business work day after the Apr 15 federal tax filing deadline. Note: Because Apr 15 is a Saturday in 1995, the deadline becomes Monday, Apr 17. For info: Joseph C. Kearley, CPA, 2129 Benbrook Dr, Carrollton, TX 75007. Phone: (214) 245-6192.

PAUL REVERE'S RIDE: 220th ANNIVERSARY. Apr 18. The "Midnight Ride" of Paul Revere and William Dawes started at about 10 PM on Apr 18, 1775, to warn American patriots between Boston, MA, and Concord, of the approaching British.

PET OWNERS INDEPENDENCE DAY. Apr 18. Dog and cat owners take day off from work and the pets go to work in their place, since most pets are jobless, sleep all day and do not even take out the trash. Phone: (212) 388-8673 or (717) 274-8451. For info: Wellness Permission League, Tom or Ruth Roy, 2105 Water St, New York, NY 10101-1506.

◆ **PYLE, ERNEST TAYLOR: 50th DEATH ANNIVERSARY.** Apr 18, 1945. Journalist Ernie Pyle was born Aug 3, 1900, at Dana, IN. As a roving reporter he wrote a column that was syndicated by nearly 200 newspapers. His reports of the bombing of London in 1940 and subsequent stories from Africa, Sicily, Italy and France earned him a Pulitzer Prize in 1944. Pyle was killed by enemy machine-gun fire on Ie Shima, off Okinawa.

SAN FRANCISCO 1906 EARTHQUAKE: ANNIVERSARY. Apr 18. Business section of San Francisco, some 10,000 acres, destroyed by earthquake, Apr 18, 1906. First quake at 5:13 AM, followed by fire. Nearly 4,000 lives lost.

SPACE MILESTONE: *TITAN 34-D* **ROCKET FAILURE.** Apr 18. Launched from Vandenburg Air Force Base, CA, on Apr 18, 1986, the $65 million *Titan* exploded when it was only a few hundred feet into flight, destroying the $500 million KH-11 reconnaissance satellite payload. Poisonous fumes were released by the explosion, causing concern for the safety of persons in nearby communities.

SURRENDER AT DURHAM STATION: ANNIVERSARY. Apr 18. On Apr 18, 1865, Union General William Tecumseh Sherman and Confederate General Joseph Johnston signed a broad political peace agreement at Durham Station, NC. The agreement promised a general amnesty for all Southerners and pledged Federal recognition of all Southern state governments after their officials took an oath of allegiance to the United States. Sherman is roundly criticized for his role in drawing up the agreement, although he based it on an earlier conversation with Lincoln and Grant. The agreement was rejected by President Andrew Johnson and Sherman and Johnston were forced

April 1995

S	M	T	W	T	F	S
						1
2	3	4	5	6	7	8
9	10	11	12	13	14	15
16	17	18	19	20	21	22
23	24	25	26	27	28	29
30						

Chase's 1995 Calendar of Events — Apr 18-20

to reach a new agreement with terms virtually the same as those given Robert E. Lee.

"THIRD WORLD" DAY: 40th ANNIVERSARY. Apr 18. Anniversary of the first use of the phrase "third world," in the opening speech, by Indonesia's President Sukarno, at the Bandung Conference, Apr 18, 1955. Representatives of nearly 30 African and Asian countries (2,000 attendees) heard Sukarno praise the American war of independence, "the first successful anticolonial war in history." More than half the world's population, he said, was represented at this "first intercontinental conference of the so-called colored peoples, in the history of mankind." The phrase and the idea of a "third world" rapidly gained currency, generally signifying the aggregate of nonaligned peoples and nations—generally the nonwhite and underdeveloped portion of the world.

ZIMBABWE: INDEPENDENCE DAY. Apr 18. National holiday.

BIRTHDAYS TODAY

Nate Archibald, 47, basketball Hall of Famer, born at the Bronx, NY, Apr 18, 1948.
Ed Garvey, 55, lawyer, union official, labor negotiator, born at Burlington, WI, Apr 18, 1940.
Barbara Hale, 73, actress (Della Street on "Perry Mason"), born at Dekalb, IL, Apr 18, 1922.
John James, 39, actor ("Search for Tomorrow," "Dynasty"), born at Minneapolis, MN, Apr 18, 1956.
Dorothy Lyman, 48, actress (Opal in "All My Children"; "Mama's Family"), director, born at Minneapolis, MN, Apr 18, 1947.
Wilber Marshall, 33, football player, born at Titusville, FL, Apr 18, 1962.
Hayley Mills, 49, actress (*Pollyana, The Parent Trap, The Moon Spinners*), born at London, England, Apr, 18, 1946.
Clive Revill, 65, actor (*Bunny Lake Is Missing*, "Wizards and Warriors"), born at Wellington, NZ, Apr 18, 1930.
Eric Roberts, 39, actor, born at Biloxi, MS, Apr 18, 1956.
James Woods, 48, actor (*Holocaust, The Onion Field*), born at Vernal, UT, Apr 18, 1947.

APRIL 19 — WEDNESDAY
109th Day — Remaining, 256

EXPLOSION ON THE USS IOWA. Apr 19. In one of the worst naval disasters since the war in Vietnam, a freak explosion rocked the battleship, USS *Iowa*, killing 47 sailors, on Apr 19, 1989. The explosion occurred in the No. 2 gun turret as the *Iowa* was participating in gunnery exercises about 300 miles northeast of Puerto Rico.

GARFIELD, LUCRETIA RUDOLPH: BIRTH ANNIVERSARY. Apr 19. Wife of James Abram Garfield, 20th president of the US, born at Hiram, OH, Apr 19, 1832. Died Mar 14, 1918.

JOHN PARKER DAY: 220th ANNIVERSARY. Apr 19. Remembering John Parker's order, at Lexington Green, Apr 19, 1775: "Stand your ground. Don't fire unless fired upon; but if they mean to have a war, let it begin here." John Parker, revolutionary soldier, captain of minutemen, born at Lexington, MA, July 13, 1729. Died Sept 17, 1775.

◆ **MODEL UNITED NATIONS OF THE FAR WEST.** Apr 19-23. Hyatt Embarcadero, San Francisco, CA. In conjunction with the 50th anniversary of the UN, San Francisco State University will host the 1995 Model UN conference in which college students from around the country serve as delegates in simulated UN sessions. Sponsor: San Francisco State University. Contact: Dr. John Aviel, (415) 338-2055. For info: UN50 Committee, 312 Sutter St, Ste 610, San Francisco, CA 94108. Phone: (415) 989-1995.

NETHERLANDS-UNITED STATES: DIPLOMATIC ANNIVERSARY. Apr 19. Anniversary of establishment of America's oldest continuously peaceful diplomatic relations. On Apr 19, 1782, the States General of the Netherlands United Provinces admitted John Adams (later to become second president of the US) as minister plenipotentiary of the young American republic. This was the second diplomatic recognition of the US as an independent nation. Adams succeeded in bringing about, on Oct 8, 1782, the signing at The Hague of the first Treaty of Amity and Commerce between the two countries.

SHERMAN, ROGER: BIRTH ANNIVERSARY. Apr 19. American statesman, member of the Continental Congress (1774-1781 and 1783-1784), signer of the Declaration of Independence and of the Constitution, was born at Newton, MA, Apr 19, 1721 (Old Style). He also calculated astronomical and calendar information for an almanac. Sherman died at New Haven, CT, July 23, 1793.

SIERRA LEONE: NATIONAL HOLIDAY. Apr 19. Sierra Leone became a republic on Apr 19, 1971.

SPACE MILESTONE: *SALYUT 7* (USSR). Apr 19. New space station launched from Tyuratam, USSR, to replace the aging *Salyut 6*. Apr 19, 1982.

BIRTHDAYS TODAY

Don Adams, 68, actor (Emmy for "Get Smart"), born at New York, NY, Apr 19, 1927.
Kenneth (Kenneth Everette Battelle), 68, hairdresser, born at Syracuse, NY, Apr 19, 1927.
Wilfried Martens, 59, Prime Minister of Belgium, born at Sleidinge, Belgium, Apr 19, 1936.
Dudley Moore, 60, actor (*Arthur, Bedazzled*), composer, born at London, England, Apr 19, 1935.
Hugh O'Brian (Hugh J. Krampe), 65, actor ("The Life and Legend of Wyatt Earp"), born at Rochester, NY, Apr 19, 1930.
Alan Price, 53, singer, songwriter, born at Fairfield, England, Apr 19, 1942.
Frank John Viola, Jr., 35, baseball player, born at Hempstead, NY, Apr 19, 1960.

APRIL 20 — THURSDAY
110th Day — Remaining, 255

BIRMINGHAM FESTIVAL OF ARTS' SALUTE TO THE NETHERLANDS. Apr 20-30. Birmingham, AL. A salute to the arts and culture of the Netherlands. This festival, now in its 45th year, celebrates the arts and culture of a different country each year. The Netherlands will be the focus country for 1995. Est attendance: 200,000. For info: Sara Crowder, Exec Dir, Festival of Arts, 2027 First Ave N, Ste 910, Birmingham, AL 35203. Phone: (205) 252-7652.

FRENCH, DANIEL CHESTER: BIRTH ANNIVERSARY. Apr 20. American sculptor born Apr 20, 1850. Died Oct 7, 1931.

◆ **HITLER, ADOLPH: BIRTH ANNIVERSARY.** Apr 20. German dictator, frustrated artist, son of Alois (Schicklgruber) Hitler, obsessed with superiority of the "Aryan race" and the evil of Marxism (which he saw as a Jewish plot). Born at Braunau am Inn, Austria, on Apr 20, 1889. Turning to politics, despite a five-year prison sentence (writing *Mein Kampf* during the nine months he served), his rise was predictable and, on Aug 19, 1934, a German plebiscite vested sole executive power in Führer Adolph Hitler. Facing certain defeat by the Allied Forces, on Apr 30, 1945, he shot himself while his mistress, Eva Braun, took poison in a Berlin bunker where they had been hiding for more than three months.

Apr 20-21 ☆ Chase's 1995 Calendar of Events ☆

LUDLOW MINE INCIDENT: ANNIVERSARY. Apr 20. Miners struggling for recognition of their United Mine Workers Union were attacked in Ludlow, CO, by National Guard troops on Apr 20, 1914. The Guardsmen were paid by the mining company. A tent colony was destroyed, five men and one boy were killed by machine-gun fire, and eleven children and two women were burned to death.

NATIONAL WHISTLERS CONVENTION. Apr 20-23. Louisburg, NC. Assemblage of professional and amateur whistlers and whistle collectors. Est attendance: 2,500. Sponsor: Franklin County Arts Council, Inc, Allen DeHart, Project Dir, PO Box 758, Louisburg, NC 27549. Phone: (919) 496-2521.

PUYALLUP SPRING FAIR. Apr 20-23. Puyallup Fairgrounds, Puyallup, WA. Fair to celebrate spring, including exhibits, animals, flowers, rides, demonstrations, gardening, recycling, lots of entertainment, food and much more. Est attendance: 100,000. For info: Puyallup Spring Fair, PO Box 430, Puyallup, WA 98371. Phone: (206) 841-5045.

◆ **RUSSIANS FIRE ON BERLIN: 50th ANNIVERSARY.** Apr 20. At 11 AM on Adolph Hilter's 56th birthday, the Red Army opened fire on Berlin, where the German leader was holed up in a bunker 50 feet below the Reich Chancellery.

SMITH, HOLLAND: BIRTH ANNIVERSARY. Apr 20. Considered the father of amphibious warfare, Holland "Howling Mad" Smith was born Apr 20, 1882, at Hatchechubie, AL. Smith developed techniques for amphibious assaults that involved coordination of land, sea and air forces. During WWII he led troops in assaults in the Marshall and Mariana Islands and also directed forces at Guam, Iwo Jima and Okinawa. Smith died Jan 12, 1967, at San Diego, CA.

STEVENS, JOHN PAUL: 75th BIRTHDAY. Apr 20. Associate justice of the Supreme Court of the US, nominated by President Ford on Dec 1, 1975. (Sworn in Dec 19, 1975.) Justice Stevens was born at Chicago, IL, on Apr 20, 1920.

TAURUS, THE BULL. Apr 20-May 20. In the astronomical/astrological zodiac that divides the sun's apparent orbit into 12 segments, the period Apr 20-May 20 is identified, traditionally, as the sun sign of Taurus, the Bull. The ruling planet is Venus.

TEXAS STATE CHAMPIONSHIP FIDDLERS FROLICS. Apr 20-23. Halletsville, TX. Old-time fiddling championship of Texas. Proceeds to various charities. Phone: (512) 798-2311 or (512) 798-5934. Est attendance: 8,000. Sponsor: Knights of Columbus, Box 46, Hallettsville, TX 77964.

WILLLIAM INGE FESTIVAL. Apr 20-22. Independence Community College, Independence, KS. Seminars, presentations, lectures and banquet with nationally known playwrights honoring the Pulitzer Prize- and Academy Award-winning author of *Picnic* and *Splendor in the Grass*. Annually, the third Thursday and Friday of April. Est attendance: 1,000. For info: Ken Fienen, CVB Dir, Independence Chmbr Comm, PO Box 386, Independence, KS 67301. Phone: (316) 331-1890.

April 1995

S	M	T	W	T	F	S
						1
2	3	4	5	6	7	8
9	10	11	12	13	14	15
16	17	18	19	20	21	22
23	24	25	26	27	28	29
30						

BIRTHDAYS TODAY

Nina Foch, 71, actress (*Scaramouche*), born at Leyden, Holland, Apr, 20, 1924.
Jessica Lange, 46, actress (Oscar for *Tootsie*; *Frances, Country, Sweet Dreams*), born Cloquet, MN, Apr 20, 1949.
Don Mattingley, 33, baseball player, born at Evansville, IN, Apr 20, 1962.
Ryan O'Neal, 54, actor ("Peyton Place," *Love Story, Paper Moon*), born Los Angeles, CA, Apr 20, 1941.
John Paul Stevens, 75, Associate Justice of the US Supreme Court, born at Chicago, IL, Apr 20, 1920.
Luther Vandross, 44, musician, songwriter ("Never Too Much"), born at New York, NY, Apr 20, 1951.

APRIL 21 — FRIDAY
111th Day — Remaining, 254

ALFERD G. PACKER DAY. Apr 21. University Memorial Center, Boulder, CO. Annual observance since 1968. Joyous celebration in honor of Colorado's and the US's only convicted cannibal. 27th annual observance. Sponsor: Program Council and University Memorial Ctr Food Service. Est attendance: 2,000. For info: Food Service Dir, Campus Box 202, University of Colorado, Boulder, CO 80309. Phone: (303) 492-8833.

BIG 8 MEN'S AND WOMEN'S TENNIS CHAMPIONSHIPS. Apr 21-23. Will Rogers Tennis Center, Oklahoma City, OK. Men and women team championships will be determined with the winning teams qualifying for the NCAA Regional. Est attendance: 3,000. For info: Stanley Draper, Jr, Exec Sec'y, Oklahoma City All Sports Assn, 100 W Main, Ste 287, Oklahoma City, OK 73102. Phone: (405) 236-5000.

BRAZIL: TIRADENTES DAY. Apr 21. National holiday commemorating execution of national hero, dentist Jose de Silva Xavier, nicknamed Tiradentes (tooth-puller), a conspirator in revolt against the Portuguese in 1789.

BRONTE, CHARLOTTE: BIRTH ANNIVERSARY. Apr 21. English novelist born Apr 21, 1816, at Hartshead, Yorkshire, England. "Conventionality," she wrote in the preface to *Jane Eyre*, "is not morality. Self-righteousness is not religion. To attack the first is not to assail the last." She died on Mar 31, 1855.

CONNECTICUT STORYTELLING FESTIVAL. Apr 21-23. Connecticut College, New London, CT. Performances by Connecticut storytellers and nationally known guest artists. Est attendance: 300. For info: Connecticut Storytelling Ctr, Connecticut College, Barbara Reed, Dir, Educ Dept, New London, CT 06320. Phone: (203) 439-2764.

CULINARY FESTIVAL '95. Apr 21-23. Scottsdale Center for the Arts, Scottsdale, AZ. To benefit the arts in Arizona, and to expose the public to the latest in the food and beverage business. Est attendance: 60,000. For info: Scottsdale Culinary Festival, Robyn Lee, Box 9431, Scottsdale, AZ 85252.

EARTH DAY CELEBRATION. Apr 21. Chimney Rock Park, Chimney Rock, NC. Environmental programs, speakers demonstrations and free seedlings for guests at this 4th annual celebration. Est attendance: 600. Sponsor: Chimney Rock Park, Mary Jaeger-Gale, Mktg Mgr, PO Box 39, Chimney Rock, NC 28720. Phone: (800) 277-9611.

FESTIVAL OF RIDVAN. Apr 21-May 2. Annual Baha'i festival commemorating the 12 days (Apr 21-May 2, 1863) when Baha'u'llah, the prophet-founder of the Baha'i Faith, resided in a garden called Ridvan (Paradise) in Baghdad, at which time He publicly proclaimed His mission as God's messenger for this age. The first, ninth (Apr 29 in 1995) and twelfth days are celebrated as holy days and are three of the nine days of the year when Baha'is suspend work. For info: Natl Spiritual Assembly of the US, 1320 Nineteenth St, NW, Ste 701, Washington, DC 20036. Phone: (202) 833-8990.

☆ Chase's 1995 Calendar of Events ☆   Apr 21

FIESTA SAN ANTONIO. Apr 21–30. San Antonio, TX. Ten days of culture, heritage, beauty and remembrance. Parades, carnivals, sports, fireworks, music, ethnic feasts, art exhibits, dances—more than 150 events. This colorful Fiesta originated in 1891 with the Battle of Flowers parade honoring the memory of Texas heroes who fought against Genl Santa Ana for Texan independence at the Alamo and San Jacinto. Est attendance: 3,200,000. For info: Fiesta San Antonio Commission, Inc, Mktg Coord, 122 Heiman, San Antonio, TX 78205. Phone: (210) 227-5191.

GLOBAL RESTORATION FAIR. Apr 21–June 5. Presidio Park, San Francisco, CA. This fair will highlight projects designed to restore and renew the environment. For info: Bill Buck, 300 Broadway, Ste 28, San Francisco, CA 94133. Phone: (415) 981-3247.

INDONESIA: KARTINI DAY. Apr 21. Republic of Indonesia. Honors Raden Adjeng Kartini, pioneer in the emancipation of the women of Indonesia.

ITALY: BIRTHDAY OF ROME. Apr 21. National celebration of the founding of Rome. Traditionally in 753 BC.

KALEIDOSCOPE INTERNATIONAL FAIR. Apr 21–23. Five Flags Civic Center, Dubuque, IA. Ethnic food, global art and craft bazaar, continuous ethnic entertainment. The 1995 theme is "Bend in the River: a New Vision." Annually, the fourth weekend in April. Sponsor: Kaleidoscope, An Inter cultural Program. Fax: (319) 588-9751. Est attendance: 10,000. For info: Kaleidoscope, Ruth Nash, Promo Dir, 422 Loras Blvd, Dubuque, IA 52001. Phone: (319) 588-8035.

KENTUCKY DERBY FESTIVAL. Apr 21–May 7. Louisville, KY. Civic celebration. About 70 events, two-thirds of which are free to the public. Est attendance: 1,500,000. For info: Kentucky Derby Festival, Inc, 137 W Muhammad Ali Blvd, Louisville, KY 40202. Phone: (502) 584-6383.

KINDERGARTEN DAY. Apr 21. A day to recognize the importance of play, games and "creative self-activity" in child education, and to note the history of the kindergarten. Observed on anniversary of the birth of Friedrich Froebel (Apr 21, 1782) who established the first kindergarten in 1837. German immigrants brought Froebel's ideas to the US in the 1840s. The first kindergarten in a public school in the US was started in 1873, at St. Louis, MO.

MOON PHASE: LAST QUARTER. Apr 21. Moon enters Last Quarter phase at 11:18 PM, EDT.

MUIR, JOHN: BIRTH ANNIVERSARY. Apr 21. American naturalist, explorer, conservationist and author for whom the 550-acre Muir Woods National Monument (near San Francisco, CA) is named. Muir, born at Dunbar, Scotland, on Apr 21, 1838, emigrated to the US in 1849, where he urged establishment of national parks and profoundly influenced US forest conservation. Died at Los Angeles, CA, on Dec 24, 1914.

MUSTER. Apr 21. Texas A&M University, College Station, TX, and various sites around the world. Muster brings together Aggies of all ages to remember their days at Texas A&M and to honor those who have died since the last muster. It is a way for Aggies to renew their loyalty, unity, friendship for one another and devotion to the university. The largest Muster is held on the Texas A&M campus where more than 8,000 students and former students gather. As part of the ceremonies, a roll call of the absent is read and a comrade answers "here" for the deceased person. It includes a candle-lighting ceremony and three rifle volleys fired by the Ross Volunteers, an honor society within the University's Corps of Cadets. The first Muster was held in 1883. Apr 21 was later selected as a way to also remember San Jacinto Day, the day on which Texas won its independence from Mexico in 1836. Annually, Apr 21. For info on the Off-campus Muster, write to David Wilkinson, Dir, Club Programs, Assn of Former Students, Texas A&M University, PO Box 7368, College Station, TX 77844-7368. For info: Campus Muster Chair, Patricia Wilder, Student Govt Office, Texas A&M University, College Station, TX 77843-1264.

NATIONAL ATLANTIC CITY ARCHERY CLASSIC. Apr 21–23. Convention Center, Atlantic City, NJ. Est attendance: 6,000. For info: Greater Atlantic City Conv and Visitors Bureau, 2314 Pacific Ave, Atlantic City, NJ 08401. Phone: (609) 348-7100.

NATIONAL SCIENCE AND TECHNOLOGY WEEK. Apr 21–29. National Science and Technology Week is sponsored by the Office of Legislative and Public Affairs of the National Science Foundation to promote awareness of science and technology to the general public and especially to children. NSTW learning materials are available free of charge to formal and informal educators by calling or writing. Annually, the last week in April. Sponsor: Natl Science Fdtn, Attn: NSTW'95, 4201 Wilson Blvd, Alexandria, VA 20165.

RED BARON SHOT DOWN: ANNIVERSARY. Apr 21. German flying ace Baron Manfred von Richtofen was shot down and killed during the battle of the Somme on Apr 21, 1918. The "Red Baron," so named for the color of his Fokker triplane, was credited with 80 kills in less than two years. Royal Flying Corp pilots recovered his body and the Allies buried him with full military honors. Asked about his fighting philosophy he was quoted as saying, "I am a hunter. My brother Lothar is a butcher. When I have shot down an Englishman, my hunting passion is satisfied for a quarter of an hour."

SAN JACINTO DAY. Apr 21. Texas. Commemorates Battle of San Jacinto, Apr 21, 1836, in which Texas won independence from Mexico. A 570-foot monument, dedicated on the 101st anniversary of the battle, marks the site on the banks of the San Jacinto River, about 20 miles from present city of Houston, TX, where General Sam Houston's Texans decisively defeated the Mexican forces led by Santa Anna in the final battle between Texas and Mexico.

SHOWCASE '95 TRADE AND HOME SHOW. Apr 21–23. Zion, IL. 100 exhibitors of home improvement products and services, door prizes and entertainment. Est attendance: 6,000. Sponsor: Zion Park District, Richard Walker, 2400 Dowie Memorial Dr, Zion, IL 60099. Phone: (708) 746-5500.

SPACE MILESTONE: *COPERNICUS, OAO 4 (US)*. Apr 21. Orbiting Astronomical Observer, named in honor of Polish astronomer. Launched Apr 21, 1972.

SPRINGFEST. Apr 21–23. Heber Springs, AR. 8th annual festival features arts and crafts, children's activities, food concessions and more. Est attendance: 5,000. For info: Chamber of Commerce, 1001 W Main, Heber Springs, AR 72543. Phone: (501) 362-2444.

◆ **STATE OF THE WORLD FORUM.** Apr 21–23. Masonic, Auditorium, San Francisco, CA. With Mikhail Gorbachev as host, world leaders involved with the resolution of the Cold War will convene to explore new foundations of global security and discuss with youth leaders critical issues for the 21st Century. Sponsor: Gorbachev Foundation. Contact: John Balbach, (415) 771-4567. For info: UN50 Committee, 312 Sutter St, Ste 610, San Francisco, CA 94108. Phone: (415) 989-1995.

VERMONT MAPLE FESTIVAL. Apr 21. St. Albans, VT. Parade, maple supper, pancake breakfasts, fiddlers, talent shows, sugar-on-snow, maple harvest exhibits, 8.5-mile run from village to village, craft show and antique show. Est attendance: 25,000. For info: Vermont Maple Festival, PO Box 255, St. Albans, VT 05478. Phone: (802) 524-5800.

Apr 21-22 ☆ Chase's 1995 Calendar of Events ☆

BIRTHDAYS TODAY

David Boren, 54, US Senator (D, Oklahoma), born at Washington, DC, Apr 21, 1941.
Tony Danza, 44, actor ("Taxi"), born at Brooklyn, NY, Apr 21, 1951.
Queen Elizabeth II, 69, Queen of the United Kingdom, born at London, England, Apr 21, 1926.
Charles Grodin, 60, actor (*Midnight Run, Beethoven*), director, born at Pittsburgh, PA, Apr 21, 1935.
Patti LuPone, 46, actress (*Evita*; "Life Goes On"), born at Northport, NY, Apr 21, 1949.
Elaine May, 63, actress, writer (comedy with Mike Nichols), director (*A New Leaf*), born at Philadephia, PA, Apr 21, 1932.
Iggy Pop, 48, singer, born at Ann Arbor, MI, Apr 21, 1947.
Anthony Quinn, 79, actor (Oscars for *Viva Zapata, Lust for Life*), born at Chihuahua, Mexico, Apr 21, 1916.

APRIL 22 — SATURDAY
112th Day — Remaining, 253

ARBOR DAY. Apr 22. Nebraska. First observance of Arbor Day was in this state on Apr 10, 1872. Now observed on many different dates in many states and countries.

◆ **BABE'S PITCHING DEBUT: ANNIVERSARY.** April 22, 1914. Babe Ruth made his professional pitching debut, playing for the Baltimore Orioles in his own hometown. Allowing just six hits and contributing two singles himself, Ruth shut out the Buffalo Bisons, 6-0.

BLOCK HOUSE STEEPLECHASE RACES. Apr 22. Foothills Equestrian Nature Center, Tryon, NC. 49th annual running of the Block House Steeplechase. Est attendance: 15,000. For info: Tryon Riding & Hunt Club, Mitzi Lindsey, PO Box 1095, Tryon, NC 28782. Phone: (800) 438-3681.

BRAZIL: DISCOVERY OF BRAZIL DAY. Apr 22. Commemorates discovery by Pedro Alvarez Cabral, 1500.

BUFFALO BEANO KITE FLY AND FRISBEE FLING. Apr 22-23. Burl Huffman Athletic Complex, Lubbock, TX. K-9 local Ashley Whippet competition, kite games for kids, single and dual line games for adults, PRO and AMA frisbee events and 2 national sport kite conference competitions highlight a weekend of fun in the sun and wind. Admission: $2 per car. Annually, the last weekend in April. Est attendance: 12,500. Sponsor: Buffalo Beano Co and KFMX Radio, 801 University Ave, Lubbock, TX 79401-2419. Phone: (800) 788-2326.

COINS STAMPED "IN GOD WE TRUST": ANNIVERSARY. Apr 22. By Act of Congress, the phrase "In God We Trust" began to be stamped on all US coins on Apr 22, 1864.

◆ **CONSTANCE STUART LARRABEE: WORLD WAR II PHOTO JOURNAL.** Apr 22-May 28. North Carolina Museum of History, Raleigh, NC. From July 1944 to March 1945, this photographer followed the Allied invasion of Europe through Egypt, France and Italy. This exhibition consists of her documentary photographs complemented by entries from her private journal and correspondence. Call or write the Smithsonian for other dates and venues. For info: Smithsonian Institution Traveling Exhibition Service, 1100 Jefferson Dr SW, Ste 3146, Washington, DC 20560. Phone: (202) 357-2700.

CRAFT FAIR USA: INDOOR SHOW. Apr 22-23. Wisconsin State Fair Park, Milwaukee, WI. Sale of handcrafted items—jewelry, pottery, weaving, leather, wood, glass and sculpture. Est attendance: 13,000. For info: Dennis R. Hill, Dir, 3233 S Villa Circle, Milwaukee, WI 53227. Phone: (414) 321-2100.

DAFFODIL FESTIVAL PARADE. Apr 22. Tacoma, Puyallup, Sumner and Orting, WA. 60th annual parade, featuring a myriad of floats decorated with freshly cut flowers, primarily daffodils. For info: Daffodil Fest, Inc, Lee Fink, Office Mgr, PO Box 1824, Tacoma, WA 98401. Phone: (206) 627-6176.

◆ **EARTH DAY: 25th ANNIVERSARY.** Apr 22. First observed Apr 22, 1970, with message "Give Earth a Chance" and attention to reclaiming the purity of the air, water and living environment. Special 10th anniversary observance on Apr 22, 1980, with assistance from US Environmental Protection Agency. Annually, Apr 22. Note: "Earth" days have been observed by some groups and on various dates. The vernal equinox has been chosen by some for this observance.

EARTH DAY. Apr 22. Burritt Museum and Park, Huntsville, AL. An all-day festival of workshops, games, hikes and exhibits. Discuss global environmental issues, learn new life-style alternatives and be immersed in nature itself. This festival stresses how each individual can impact the environment. Adults and children are encouraged to explore the wonders and workings of this fascinating planet we call "home." Admission charged. Annually, the fourth Saturday in April. Est attendance: 700. For info: Jeff Hughes, Curator of Natural History, Burritt Museum and Park, 3101 Burritt Dr, Huntsville, AL 35801. Phone: (205) 536-2882.

EARTH DAY COMMUNITY FESTIVAL. Apr 22-23. St. Louis, MO. An educational event with hundreds of exhibtitors related to environmental issues and their solutions. For info: Jerry Klamon, EarthWays, 3617 Grandel Sq, St Louis, MO 63108. Phone: (314) 531-1995.

EARTH DAY FAMILY FESTIVAL. Apr 22. Plaza Park and Civic Auditorium, San Jose, CA. More than 150 exhibitors participate in this environmentally oriented fest. Est attendance: 17,500. For info: World Environmental Network, PO Box 53589, San Jose, CA 95153. Phone: (408) 281-7100.

EARTH DAY TAMPA BAY. Apr 22. Tampa, FL. A festival designed to bring awareness to local and regional environmental programs and initiatives. 50-75 exhibits serve as main focus of the event. Entertainment, refreshments and voter registration. Trees are planted and many activities are planned for children. For info: Sally Thompson, EDTB, PO 2800, Tampa, FL 33601-2800. Phone: (813) 274-5890.

FAMILY FOLKLORE. Apr 22-May 21. Skokie Heritage Museum, Skokie, IL. See Jan 4 for full description. Call or write the Smithsonian for other dates and venues. For info: Smithsonian Institution Traveling Exhibition Service, 1100 Jefferson Dr SW, Ste 3146, Washington, DC 20560.

GEORGIA RENAISSANCE SPRING FESTIVAL. Apr 22-June 11. (Saturdays, Sundays and Memorial Day.) Atlanta, GA. Theme park features "the entertainmnet, romance and art of the Renaissance." More than 100 performances daily on 10 stages, thousands of unique craft items. Phone; (404) 964-8575 or (404) 455-0553. Est attendance: 180,000. For info: Georgia Renaissance Festival, Sarah Waters, PO Box 986, Fairburn, GA 30213. Phone: (404) 964-8575.

◆ **GERMANS TRY TO DEAL: 50th ANNIVERSARY.** Apr 22, 1945. Hoping to divide the Allied Big Three, Heinrich Himmler met with British Count Folke Bernadotte to offer Germany's surrender to the Western Allies only. President Truman's swift reply demanded unconditional surrender on all fronts and to all Allied countries.

April 1995	S	M	T	W	T	F	S
							1
	2	3	4	5	6	7	8
	9	10	11	12	13	14	15
	16	17	18	19	20	21	22
	23	24	25	26	27	28	29
	30						

Chase's 1995 Calendar of Events — Apr 22–23

GIRL SCOUT LEADER'S DAY. Apr 22. To recognize the people who make Girl Scouting possible. An opportunity for girls involved in Girl Scouting to honor their troop leaders. Sponsor: Girl Scouts of the USA, Media Services, 420 Fifth Ave, New York, NY 10018.

GOOSE POND COLONY DOGWOOD FESTIVAL. Apr 22–23. Goose Pond Colony, Scottsboro, AL. 2 days of fun for everyone including a golf tournament, food, entertainment, boat show and RV show, arts and crafts, skydivers, hot air balloon rides, Dogwood Festival Ball, and much more. Est attendance: 5,000. For info: Rick Roden, Tourism Dir, PO Box 973, Scottsboro, AL 35768. Phone: (800) 259-5508.

HISTORIC GARDEN WEEK IN VIRGINIA. Apr 22–29. This annual statewide event, celebrating its 62nd anniversary, is billed as "America's Largest Open House." The 8-day program showcases more than 250 of Virginia's finest homes, gardens, plantations and landmarks on more than 30 separate tours. A brochure is available with dates and tour locations. A 200-page guidebook will be available in early March 1995. Please mail a contribution of $2.00 to cover postage and handling to Historic Garden Week, 12 E Franklin St, Richmond, VA 23219. Est attendance: 40,000. For info: Garden Club of Virginia, 12 E Franklin St, Richmond, VA 23219. Phone: (804) 644-7776.

JUST PRAY NO: WORLDWIDE WEEKEND PRAYER. Apr 22–23. Churches throughout the world. Prayer breakfasts, lunches, street rallies and marches to gain media attention. Bible studies and sermons concerning alcoholism and drug abuse and revival meetings aimed at those bound by addiction. Sponsor: "Just Pray No," Ltd, 124 Garfield Pl, E Rockaway, NY 11518. Phone: (516) 599-7399.

KENDUSKEAG STREAM CANOE RACE. Apr 22. Bangor, ME. 16.5 mile white water open canoe race. Est attendance: 1,500. For info: Bangor Parks and Recreation Dept, 647 Main St, Bangor, ME 04401. Phone: (207) 947-1018.

LENIN, NIKOLAI: 125th BIRTH ANNIVERSARY. Apr 22. Russian socialist and revolutionary leader (real name: Vladimir Ilyitch Ulyanov), ideological follower of Karl Marx. Born at Simbirst, on the Volga, Russia, Apr 22, 1870 (New Style) (Apr 10, 1870, Old Style). Leader of the Great October Socialist Revolution of 1917. Lenin died at Gorky, near Moscow, on Jan 21, 1924. His embalmed body, in a glass coffin at the Lenin Mausoleum, has been viewed by millions of visitors to Moscow's Red Square.

LIVINGSTON RAILROAD SWAP MEET. Apr 22. Northern Pacific Passenger Depot, Livingston, MT. Railroad hobby items and collectibles for sale at this 5th annual swap. Est attendance: 750. For info: Cynthia Moses Nesmith, Livingston Depot Center, Box 1319, Livingston, MT 59047. Phone: (406) 222-2300.

MOSSY CREEK BARNYARD FESTIVAL. Apr 22–23. (Also Oct 21–22.) Warner Robins, GA. Arts and crafts chosen from best in the nation; heritage crafts, country and folk music, tales in relaxed atmosphere. Semiannually, the third weekend of April (usually) and of October. Est attendance: 20,000. For info: Mossy Creek Barnyard Festival, Inc, 106 Anne Dr, Warner Robins, GA 31093. Phone: (912) 922-8265.

NATIONAL HEADACHE FOUNDATION FUNDRAISER. Apr 22. New York, NY. Annual black tie silent auction and dinner serves as the major fund raiser for the National Headache Foundation. Proceeds from the evening are used for research, education and service. Est attendance: 150. For info: Natl Headache Fdtn, Suzanne E. Simons, Dir Admin and Dvmt, 5252 N Western Ave, Chicago, IL 60625. Phone: (312) 907-6232.

OKLAHOMA DAY. Apr 22. Oklahoma.

OKLAHOMA LAND RUSH: ANNIVERSARY. Apr 22. At twelve o'clock noon on Apr 22, 1889, a gun shot signaled the start of the Oklahoma land rush as thousands of settlers rushed into the territory to claim land. Under pressure from cattlemen, the federal government opened 1,900,000 acres of central Oklahoma that had been bought from the Creek and Seminole tribes.

PENNSYLVANIA MAPLE FESTIVAL. Apr 22–23. (Also Apr 28–30.) Meyersdale, PA. To celebrate the miracle of the maple. Est attendance: 50,000. Sponsor: Pennsylvania Maple Festival, Box 222, Meyersdale, PA 15552. Phone: (814) 634-0213.

POLK COUNTY RAMP TRAMP FESTIVAL. Apr 22. Polk County 4-H Camp, Benton, TN. A tribute to the ramp, a wild onionlike plant that grows only in the Appalachian Mountains. Bluegrass music and feast of the ramps. Est attendance: 800. For info: Don Ledford, Box 189, Benton, TN 37307. Phone: (615) 338-4504.

RITES OF SPRING. Apr 22. Patapsco State Park, MD. Some 300–500 volunteers will spend Earth Day reclaiming more than 200 acres of eroded farm soil. Turning this area back to woodland will protect the North Branch of the Patapsco River which flows into Chesapeake Bay. For info: Wally Orlinsky, Tree-Mendous Maryland, Tawes State Office Bldg, Forestry Program, E-1, Annapolis, MD 21401. Phone: (410) 974-3776.

SPRING FLING. Apr 22–23. Wichita Falls, TX. A celebration of the arts. Artists from all over the US display their art. Food booths, demonstrations and entertainment. Est attendance: 21,000. For info: Wichita Falls Conv and Visitors Bureau, PO Box 1860, Wichita Falls, TX 76307. Phone: (817) 723-2741.

TASTE OF HEALTH. Apr 22–23. Miami, FL. Celebrating Earth Day, this educational event is designed to encourage dietary common sense and increase awareness of sound environmental practices. Est attendance: 60,000. For info: A Taste of Health, 2911 Grand Ave, Miami, FL 33133. Phone: (305) 892-2252.

TOUR de CURE. Apr 22–June 4. (Tentative) Thousands of cyclists participate in the American Diabetes Association's annual biking event to raise money to help find a cure for diabetes and to provide information and resources to improve the lives of all people affected by diabetes. Tours are held in communities across America, combining fun and fitness with the chance to help people with diabetes. Contact your local affiliate. Sponsor: American Diabetes Assn, Natl HQ, 1660 Duke St, Alexandria, VA 22314. Phone: (703) 549-1500.

WARBIRDS IN ACTION AIR SHOW. Apr 22–23. Minter Field, Shafter, CA. 14th annual airshow features flybys of up to 100 WWII trainers, fighters, bombers and reconaissance aircraft. Also includes aerobatics, air racers, pilots' barbecue and rides. Annually, the fourth weekend in April. Est attendance: 20,000. For info: Neil Keyzers, Curator, Minter Field Air Museum, 401 Vultee St, Shafter, CA 93263. Phone: (805) 393-0291.

BIRTHDAYS TODAY

Eddie Albert (Edward Albert Heimberger), 87, actor ("Green Acres," *The Teahouse of the August Moon*), born at Rock Island, IL, Apr 22, 1908.

Glen Campbell, 60, singer ("Gentle on My Mind," "By the Time I Get to Phoenix," "Country Boy," "Rhinestone Cowboy"), born at Billstown, AR, Apr 22, 1935.

Peter Frampton, 45, singer ("Show Me the Way," "Do You Feel Like We Do"), songwriter, born at Beckenham, England, Apr 22, 1950.

Chris Makepeace, 31, actor (*My Bodyguard*), born at Montreal, Quebec, Canada, Apr 22, 1964.

Yehudi Menuhin, 79, violinist, born at New York, NY, Apr 22, 1916.

Jack Nicholson, 59, actor (*The Last Detail, One Flew Over the Cuckoo's Nest*), producer, born at Neptune, NJ, Apr 22, 1936.

APRIL 23 — SUNDAY
113th Day — Remaining, 252

ALLONS MANGER '95. Apr 23. Belle Rose, LA. To illustrate the unique natural resources of Louisiana bayous: Cajun cooking, craftsmanship, music and culture. Annually, the first Sunday after Easter. Est attendance: 10,000. Sponsor: St. Jules–St. Martin Church, Father M. Jeffery Bayhi, P P Box 38, Belle Rose, LA 70341. Phone: (504) 473-8569.

Apr 23 ☆ Chase's 1995 Calendar of Events ☆

BERMUDA: PEPPERCORN CEREMONY: ANNIVERSARY. Apr 23. St. George. Commemorates the payment of one peppercorn in 1816 to the governor of Bermuda for rental of Old State House by the Masonic Lodge.

BIG BROTHERS/BIG SISTERS APPRECIATION WEEK. Apr 23–29. To honor the men and women who serve on a ONE-TO-ONE® basis as Big Brother and Big Sister volunteers who provide much needed guidance and support to at-risk children, primarily from one parent families, in need of an adult mentor and role model. Sponsor: Big Brothers/Big Sisters of America, Viola W. Bostic, Asst Natl Exec Dir, 230 N 13th St, Philadelphia, PA 19107. Phone: (215) 567-7000.

BUCHANAN, JAMES: BIRTH ANNIVERSARY. Apr 23. The 15th president of the US, born at Cove Gap, PA, on Apr 23, 1791, was the only president who never married. He served one term in office, Mar 4, 1857–Mar 3, 1861, and died at Lancaster, PA, on June 1, 1868.

CANADA–UNITED STATES GOODWILL WEEK. Apr 23–29. To bring about a better understanding of the American and Canadian ways of life and to recognize the anniversary of the signing of the Rush-Bagot treaty Apr 28, 1817, which limited US and British armaments on the Great Lakes. Annually, the full week including Apr 28. Sponsor: Kiwanis Intl, Program Development Dept, 3636 Woodview Trace, Indianapolis, IN 46268.

CERVANTES SAAVEDRA, MIGUEL DE: DEATH ANNIVERSARY. Apr 23. Spanish poet, playwright and novelist died in his 69th year, at Madrid, Spain, on Apr 23, 1616 (also the day of William Shakespeare's death). The exact date of Cervante's birth at Alcala de Henares is unknown, but he was baptized Oct 9, 1547, and some believe he was born on Michaelmas, Sept 29. As soldier and tax collector, Cervantes traveled widely. He spent more than five years in prisons in Spain, Italy and North Africa. His greatest creation was Don Quixote, the immortal Knight of La Mancha whose profession was chivalry. Riding his nag, Rozinante, and accompanied by Squire Sancho Panza, Don Quixote tilts at windmills of the mind in the world's best-known novel. Nearly a thousand editions of *Don Quixote* (a bestseller since its first appearance in 1605) have been published, and it has been translated into more languages than any other book except the Bible.

DAYS OF REMEMBRANCE. Apr 23–30. Days of Remembrance are to commemorate the victims of the Holocaust. Est attendance: 800. For info: United States Holocaust Memorial Museum, 100 Raoul Wallenberg Place SW, Washington, DC 20024. Phone: (202) 488-0400.

EARTH DAY REGIONAL FESTIVAL. Apr 23. Baton Rouge, LA. Festival includes Earth for the Children/Children for the Earth activities and the All Species Parade, "From the River to the Gulf." Est attendance: 80,000. For info: Susan Hamilton, Baton Rouge Earth Day, Inc, PO Box 1471, Baton Rouge, LA 70821. Phone: (504) 389-3113.

EARTH FAIR '95. Apr 23. San Diego, CA. A free public event featuring more than 250 exhibitors representing non-profit, for-profit and government organizations. Events foster public education and awareness of contemporary environmental issues. Included are five stages of live entertainment, a special edition "Earth Times" newspaper, speakers, natural food and a popular Kids Area with numerous activities. Fair will be kicked-off by the Walk 'n Roll Walk-a-thon and Earth Parade. Est attendance: 250,000. For info: San Diego Earth Day, PO Box 9827, San Diego, CA 92169. Phone: (619) 272-7370.

EARTHFEST. Apr 23. Cleveland, OH. Ohio's largest environmental education event. Est attendance: 75,000. For info: Scott Sanders, Cleveland Earth Day Coalition, 3606 Bridge Ave, Cleveland, OH 44113. Phone: (216) 281-6468.

ENGLAND: HARROGATE SPRING FLOWER FESTIVAL. Apr 23–26. Valley Gardens, Harrogate, North Yorkshire. Spectacular exhibits and displays at Britain's largest spring show. For info: Alan Ravenscroft, Show Dir, North of England Horticultural Soc, 4A South Park Road, Harrogate, North Yorkshire, England HG1 5QU.

GOSPEL MUSIC '95. Apr 23–27. Nashville, TN. Seminars, workshops, spectacular concerts and the 26th annual Dove Awards. Est attendance: 2,000. For info: Gospel Music Assn, Bruce Koblish, Exec Dir, 7 Music Circle N, Nashville, TN 37203. Phone: (615) 242-0303.

GRANGE WEEK. Apr 23–29. State and local recognition for Grange's contribution to rural/urban America. Celebrated at National Headquarters in Washington, DC, and in all states with local, county, and state Granges. Begun in 1867, the National Grange is the oldest organized agricultural movement in the US. For info: Judy T. Massabny, Dir of Info, The Natl Grange, 1616 H St NW, Washington, DC 20006. Phone: (202) 628-3507.

HELENA RAILROAD FAIR. Apr 23. Civic Center, Helena, MT. Largest railroad hobby event in Montana features a mix of scale and tin-plate trains. Also railroad memorabilia and collectibles, real life train watching at the MRL Helena depot. Est attendance: 5,000. For info: Bob Solomon, 161 Fairway Dr, Helena, MT 59601. Phone: (406) 442-6118.

◆ **HITLER TAKES COMMAND OF BERLIN: 50th ANNIVERSARY.** Apr 23, 1945. Enlisting the police force, members of the Hilter youth and old men and women to defend the city against the approaching Red Army, Adolf Hitler took personal command of Berlin's defense.

INDUSTRY DAY. Apr 23. 2nd and Court St, Beatrice, NE. Features industries of Gage County and their products. Antique gas engine demonstrations, exhibits of stationary engine equipment. Est attendance: 300. For info: Kent Wilson, Admin, Gage County Historical Soc, PO Box 793, Beatrice, NE 68310. Phone: (402) 228-1679.

INTERGENERATIONAL WEEK. Apr 23–29. Homewood's Corporate Complex, Williamsport, MD. This week was set aside to recognize the unique friendships between young children and older adults and to educate the public on the positive results of intergenerational programs. This week includes an intergenerational seminar, Open House, picnic and parade, and Arts and Crafts Show. Annually, the last week in April. For info: Nancy Higgs, Dir, Down Home Adult Day Care, Denise Feeser, Dir, Choo Choo Child Care, PO Box 250, Williamsport, MD 21795. Phone: (301) 582-4894.

MICHAEL FORBES TROLLEY RUN. Apr 23. Kansas City, MO. 4-mile run and walk to benefit the Children's Center for the Visually Impaired. Kansas city's largest run course ends on the Country Club Plaza with a gigantic After-Bash block party with food and live music. Annually, the fourth Sunday in April. Est attendance: 10,000. For info: Judy Miller, Mktg Chair, c/o Trolley Run, 400 W 57th St, Kansas City, MO 64113. Phone: (816) 333-3166.

	S	M	T	W	T	F	S
April							1
	2	3	4	5	6	7	8
	9	10	11	12	13	14	15
1995	16	17	18	19	20	21	22
	23	24	25	26	27	28	29
	30						

☆ Chase's 1995 Calendar of Events ☆ Apr 23

NATIONAL GIVE-A-SAMPLE WEEK. Apr 23–29. A reminder to consumer-oriented companies that one of the very best sales devices is the offering of free product samples or coupons for products or services, and that the coming summer months are ideal for such programs. Annually, the last full week in April. For info: Robert Jackson, Pres, Mktg Services Consultant, 322 First Ave N, Ste 201, Minneapolis, MN 55401. Phone: (612) 375-0141.

NATIONAL LINGERIE WEEK. Apr 23–29. Celebrating the glamour, allure and importance of lingerie through a series of special events and promotions. For info: MiMi Field, Intimate Apparel Council, c/o The Bromley Group, 150 Fifth Ave, New York, NY 10011. Phone: (212) 807-0878.

NATIONAL MICROBREWERS CONFERENCE AND TRADE SHOW. Apr 23–26. Austin, TX. To educate brewers and potential brewers on the technical and business aspects of small scale professional microbrewing and pubbrewing. Est attendance: 900. Sponsor: The Institute for Brewing Studies, Box 1679, Boulder, CO 80306. Phone: (303) 447-0816.

NATIONAL VOLUNTEER WEEK. Apr 23–29. Natl Volunteer Week honors those who reach out to others through community service and calls attention to the need for more community services for individuals, groups and families, and emphasizes that real social progress depends on the selflessness of caring people. Fax: (202) 223-9256. For info: Points of Light Fdtn, 1737 H Street NW, Washington, DC 20006. Phone: (202) 223-9186.

NEWSPAPER ASSOCIATION OF AMERICA ANNUAL CONVENTION. Apr 23–26. New Orleans, LA. Reviews through a three-day program of speakers, panels and discussion sessions those matters that bear upon present and future operations of newspaper publishing. Est attendance: 1,400. For info: Newspaper Assn of America, Nancy J. Jones, Public Affairs Dir, The Newspaper Center, 11600 Sunrise Valley Dr, Reston, VA 22091. Phone: (703) 648-1117.

ORTHODOX EASTER SUNDAY. Apr 23. Observed by Eastern Orthodox Churches on this date.

PARADE FOR THE PLANET™. Apr 23. New York City, NY. This parade, which is expected to attract 1.5 million people along its route, will celebrate several anniversaries—the 25th of Earth Day, the 50th of the UN, 100th of the Zoological Society and NY Public Library and the 125th of the Natural History Museum and Metropolitan Museum of Art. EDNY will also host a walk-a-thon, Clean/Green Up and EarthFair. For info: Pamela Lippe, Earth Day New York, 10 E 39th St, Ste 601, New York, NY 10016. Phone: (212) 686-4905.

PROFESSIONAL SECRETARIES WEEK. Apr 23–29. Acknowledgment of the contributions of all secretaries to the vital roles of business, industry, education, government and the professions. Annually, the last full week (from Sunday–Saturday) in April. Professional Secretaries Day is observed on Wednesday of this week (Apr 26 in 1995). Sponsor: Professional Secretaries Intl, 10502 NW Ambassador Dr, PO Box 20404, Kansas City, MO 64195-0404. Phone: (816) 891-6600.

READING IS FUN WEEK. Apr 23–29. To highlight the importance of reading. Sponsor: Reading Is Fundamental, Inc, 600 Maryland Ave SW, Rm 600, Washington, DC 20560. Phone: (202) 287-3371.

ST. GEORGE FEAST DAY. Apr 23. Martyr and patron saint of England, who died Apr 23, AD 303. Hero of the George and the Dragon legend. The story says that his faith helped him to slay a vicious dragon that demanded daily sacrifice after the king's daughter became the intended victim.

ST. GEORGE'S DAY ANNUAL CELEBRATION. Apr 23. New York, NY. Celebration of the birth of St. George, the patron saint of England, at an authentic British pub featuring coloring contests, prizes and appearances of St. George and the dragon. Annually, Apr 23. Est attendance: 200. For info: Deven Black, Genl Mgr, North Star Pub, 93 South St, New York, NY 10038. Phone: (212) 509-6757.

SHAKESPEARE, WILLIAM: BIRTH AND DEATH ANNIVERSARY. Apr 23. England's most famous and most revered poet and playwright. He was born at Stratford-on-Avon, England, Apr 23, 1564, baptized there three days later and died there on his birthday, Apr 23, 1616. Author of at least 36 plays and 154 sonnets, Shakespeare created the most influential and lasting body of work in the English language, an extraordinary exploration of human nature. His epitaph: "Good frend for Jesus sake forbeare, To digg the dust enclosed heare. Blese be ye man that spares thes stones, And curst be he that moves my bones."

SKY AWARENESS WEEK. Apr 23–29. A celebration of the sky and an opportunity to appreciate its natural beauty, to understand sky and weather processes and to work together to protect the sky as a natural resource (it's the only one we have). Events are held at schools, nature centers, etc, all across the US. Annually, during Natl Science and Technology Week. Sponsor: THINK WEATHER, Inc., Barbara G. Levine, Pres, 1522 Baylor Ave, Rockville, MD 20850. Phone: (301) 762-7669.

SPACE MILESTONE: SOYUZ 10 (USSR). Apr 23. Cosmonauts V.A. Shatalov, A.S. Yeliseyev and N.N. Rukavishnikov docked Apr 24 with Salyut 1 orbital space station. Return Earth landing at Kazakhstan, USSR, Apr 24. Launched Apr 23, 1971.

SPAIN: BOOK DAY AND LOVER'S DAY. Apr 23. Barcelona. Saint George's Day and the anniversary of the death of Spanish writer Miguel de Cervantes have been observed with special ceremonies in the Palacio de la Disputacion and throughout the city since 1714. Book stands are set up in the plazas and on street corners. This is Spain's equivalent of Valentine's Day. Women give books to men; men give roses to women.

TASTE OF LOUISIANA FOOD FESTIVAL. Apr 23. An array of booths set up to provide a "Taste of Louisiana" good cooking. Sponsor: NE LA Restaurant Assn. For info: Judy M Lowentritt, Tourism Sales Mgr, Monroe-W Monroe CVB, PO Box 6054, Monroe, LA 71211. Phone: (318) 387-5746.

TURKEY: NATIONAL SOVEREIGNTY AND CHILDREN'S DAY. Apr 23. Commemorates Grand National Assembly's inauguration on Apr 23, 1923.

WEEK OF THE YOUNG CHILD. Apr 23–29. To focus on the importance of quality early childhood education. Sponsor: Natl Assn for the Educ of Young Children, Information Services, 1509 16th St, NW, Washington, DC 20036. Phone: (800) 424-2460.

WOODS, GRANVILLE T.: BIRTH ANNIVERSARY. Apr 23. Granville T. Woods was born in Columbus, OH, on Apr 23, 1856. He invented the induction telegraph known as the Synchronous Multiplex Railway Telegraph which allowed communication between dispatchers and trains while the trains were in motion. The communication avenues opened by his invention decreased the number of train accidents. In addition, Woods is credited with a number of other electrical inventions and was compared favorably to Thomas Edison.

BIRTHDAYS TODAY

Valerie Bertinelli, 35, actress ("One Day at a Time," Silent Witness), born at Wilmington, DE, Apr 23, 1960.

Shirley Temple Black, 67, former ambassador, '30s child star with dimples and curls (Heidi, Little Miss Marker, Curly Top, The Littlest Rebel), born Santa Monica, CA, Apr 23, 1928.

Sandra Dee (Alexandra Zuck), 53, actress (Gidget), born Bayonne, NJ, Apr 23, 1942.

Joyce Dewitt, 46, actress ("Three's Company"), born at Wheeling, WV, Apr 23, 1949.

Phil Esposito, 53, former hockey player, born at Sault Ste. Marie, Ontario, Canada, Apr 23, 1942.
Lee Majors, 55, actor ("Six Million Dollar Man," *Gary Francis Powers*), born at Wyandotte, MI, Apr 23, 1940.
Bernadette Devlin McAliskey, 48, political activist, born at Cooks-town, Northern Ireland, Apr 23, 1947.

APRIL 24 — MONDAY
114th Day — Remaining, 251

ARMENIAN MARTYRS DAY. Apr 24. Commemorates the massacre of Armenians under the Ottoman Turks in 1915. Also called Armenian Liberation Day. In 1985, an international controversy concerning this event was revived when it was reported that Adolph Hitler, in a speech at Obersalzberg on Aug 22, 1939, said, "Who today remembers the Armenian extermination?" in an apparent justification of genocide.

BASCOM, GEORGE N.: BIRTH ANNIVERSARY. Apr 24, 1836. After an 1861 raid on an Arizona ranch, West Point Graduate Lt George N. Bascom was assigned to search out Apache chief Cochise, believed to be responsible. He arrested Cochise at Apache Pass, but the chief escaped and declared war, launching a reign of terror known as the Apache Wars. Bascom was born at Owingsville, KY, and died the year following his Apache adventure when he became a casualty of the Civil War battle at Fort Craig, Valverde, NM, on Feb 21, 1862. For info: Fort Bowie National Historic Site, PO Box 276, Bowie, AZ 85605.

CARTWRIGHT, EDMUND: BIRTH ANNIVERSARY. Apr 24. English cleric and inventor (developed the power loom and other weaving inventions) was born at Nottinghamshire, England, Apr 24, 1743. He died at Hastings, Sussex, England, Oct 30, 1823.

CONFEDERATE MEMORIAL DAY IN ALABAMA AND MISSISSIPPI. Apr 24. Annually, last Monday in April. Observed on other dates in some states: Apr 26 in Florida and Georgia, May 10 in North Carolina and South Carolina, last Monday in May in Virginia.

EGYPT: SHAM EL-NESIM. Apr 24. (tentative date.) Sporting Holiday. This feast has been celebrated by all Egyptians since Pharoanic time; people go out and spend the day in parks and along Nile's banks. For info: Egyptian Tourist Authority, 645 N Michigan Ave, Ste 829, Chicago, IL 60611. Phone: (312) 280-4666.

FAST DAY IN NEW HAMPSHIRE. Apr 24. Dates from the days of "public humiliation, fasting and prayer" proclaimed by royal governors during the 17th-century settlement of New England. Annually, the fourth Monday in April.

FORDYCE ON THE COTTON BELT FESTIVAL. Apr 24-29. Fordyce, AR. 15th annual railroad event includes arts and crafts, quilt show and sale, train rides, musical entertainment, rodeo, parade, railroad displays, beauty pageants, gospel singing and more. Est attendance: 10,000. For info: Chamber of Commerce, PO Box 588, Fordyce, AR 71742. Phone: (501) 352-3520.

FORT MOORE ESTABLISHED: ANNIVERSARY. Apr 24, 1847. At the conclusion of the Mexican War, the Morman Battlion of the Army of the West established Fort Moore overseeing the pueblo of Los Angeles. The fort was named in honor of their captain who had perished in the Battle of San Pascual.

	S	M	T	W	T	F	S
April 1995							1
	2	3	4	5	6	7	8
	9	10	11	12	13	14	15
	16	17	18	19	20	21	22
	23	24	25	26	27	28	29
	30						

LIBRARY OF CONGRESS: ANNIVERSARY. Apr 24. Congress approved, on Apr 24, 1800, an act providing "for the purchase of such books as may be necessary for the use of Congress... and for fitting up a suitable apartment for containing them." Thus began one of the world's greatest libraries.

NATIONAL PC/TYPING CONTEST. Apr 24-28. Held in participating offices of Western Temporary Services and chosen locations to honor the secretarial profession and to determine who is the fastest typist in the nation. National first-place prize winner is awarded a word processor and a dream trip for two. Annually, during Professional Secretaries Week. For more info phone office nearest you. Est attendance: 1,000. For info: Western Temp Services, PR Dir, 220 N Wiget Lane, Walnut Creek, CA 94598. Phone: (510) 930-5340.

NEWFOUNDLAND: ST. GEORGE'S DAY. Apr 24. Holiday observed in Newfoundland on Monday nearest Feast Day (Apr 23) of St. George.

SPACE MILESTONE: *CHINA 1* (PEOPLE'S REPUBLIC OF CHINA). Apr 24. China becomes fifth nation to orbit satellite with its own rocket. Broadcast Chinese song "Tang Fang Hung" ("The East Is Red") and telemetric signals. Apr 24, 1970.

THOMAS, ROBERT BAILEY: BIRTH ANNIVERSARY. Apr 24. Founder and editor of *The Farmer's Almanac* (first issue for 1793) was born at Grafton, MA, Apr 24, 1766. Thomas died May 19, 1846, while working on the 1847 edition.

TROLLOPE, ANTHONY: BIRTH ANNIVERSARY. Apr 24. English novelist, born at London, England, Apr 24, 1815. Died there Dec 6, 1882. "Of the needs a book has," he wrote in his autobiography, "the chief need is that it be readable."

WARREN, ROBERT PENN: 90th BIRTH ANNIVERSARY. Apr 24. American poet, novelist, essayist and critic. America's first official poet laureate, 1986-88, Robert Penn Warren was born at Guthrie, KY, on Apr 24, 1905. Warren was awarded the Pulitzer Prize for his novel *All the King's Men*, as well as for his poetry in 1958 and 1979. He died of cancer Sept 15, 1989, at Stratton, VT.

BIRTHDAYS TODAY

Eric Balfour, 18, actor/musician, born at Los Angeles, CA, Apr 24, 1977.
Eric Bogosian, 42, actor, playwright, performance artist, born at Boston, MA, Apr 24, 1953.
Vince Ferragamo, 41, former football player, born at Torrance, CA, Apr 24, 1954.
Stanley J. Kauffmann, 79, critic, born at New York, NY, Apr 24, 1916.
Shirley MacLaine, 61, author, actress (Emmy for "Gypsy in My Soul"; *Irma La Douce*), born at Richmond, VA, Apr 24, 1934.
Barbra Streisand, 53, singer, actress (*Funny Girl*), director (*Prince of Tides*), born New York, NY, Apr 24, 1942.
Rudy Tomjanovich, 47, basketball coach and former player, born at Hamtramck, MI, Apr 24, 1948.

☆ Chase's 1995 Calendar of Events ☆   Apr 25-26

APRIL 25 — TUESDAY
115th Day — Remaining, 250

ABORTION FIRST LEGALIZED: ANNIVERSARY. Apr 25. The first law legalizing abortion in the US was signed on Apr 25, 1967, by Colorado Governor John Arthur Love. The law allowed therapeutic abortions in cases in which a three-doctor panel unanimously agreed.

ANZAC DAY. Apr 25. Australia, New Zealand and Western Samoa. Memorial day and veterans' observance, especially to mark World War I Anzac landing at Gallipoli, Turkey, on Apr 25, 1915 (ANZAC: Australia and New Zealand Army Corps).

◆ **EAST MEETS WEST: 50th ANNIVERSARY.** Apr 25, 1945. US Army Lt Albert Kotzebue encountered a single Soviet soldier near the German village of Lechwitz, 75 miles south of Berlin. Patrols of Gen Lenoard Gerow's V Corps saluted the advance guard of Marshall Ivan Konev's Soviet 58 Division. Soldiers of both nations embraced and exchanged toasts. The Allied armies of East and West had finally met.

EGYPT: SINAI DAY. Apr 25. National holiday celebrating the liberation of Sinai after the peace treaty between Egypt and Israel. For info: Egyptian Tourist Authority, 645 N Michigan Ave, Ste 829, Chicago, IL 60611. Phone: (312) 280-4666.

ENGLAND: BRITISH INTERNATIONAL ANTIQUES FAIR. Apr 25-30. National Exhibition Centre, Birmingham, West Midlands. Major exhibition now in its 12th year. A wide range of quality antiques. For info: Linda Colban, Exhibitions Division, National Exhibition Centre Ltd, Birmingham, England B40 1NT.

FESTIVAL INTERNATIONAL DE LOUISIANE. Apr 25-30. Lafayette, LA. Celebrates the cultural connections between French Louisiana and the French-speaking world. Visual and performing artists of Louisiana and several francophone countries provide open air concerts, exhibits, theater, film and international cuisine. Est attendance: 150,000. For info: Festival Intl de Louisiane, PO Box 4008, Lafayette, LA 70502. Phone: (318) 232-8086.

FREDERICKSBURG DAY. Apr 25. Fredericksburg, VA. To fund one historic garden restoration in Virginia per year. Spring tour of historic homes and gardens. Est attendance: 1,600. For info: Visitor Center, 706 Caroline St, Fredericksburg, VA 22401. Phone: (800) 678-4748.

◆ **LIBERATION OF MILAN: 50TH ANNIVERSARY.** Apr 25, 1945. One day after the Italian Committee for National Liberation called for a general uprising, the Italian partisans liberated Milan.

MACAU: ANNIVERSARY OF THE PORTUGUESE REVOLUTION. Apr 25. Anniversary of the 1974 revolution.

MARCONI, GUGLIELMO: BIRTH ANNIVERSARY. Apr 25. Inventor of wireless telegraphy (1895) born at Bologna, Italy, on this date, 1874. Died at Rome, July 20, 1937.

PORTUGAL'S DAY. Apr 25. Portugal. Public holiday.

SPACE MILESTONE: HUBBLE SPACE TELESCOPE DEPLOYED (US): 5th ANNIVERSARY. Apr 25. Deployed on Apr 25, 1990, by *Discovery*, the telescope is the largest on-orbit observatory to date and is capable of imaging objects up to 14 billion light years away. The resolution of images is expected to be seven to ten times greater than images from Earth-based telescopes, since the Hubble Space Telescope is not hampered by Earth's atmospheric distortion. Launched Apr 12, 1990, from Kennedy Space Center, FL.

BIRTHDAYS TODAY

William J. Brennan, Jr, 89, former Associate Justice of the US Supreme Court, born at Newark, NJ, Apr 25, 1906.

David Corzine, 39, former basketball player, born at Arlington Heights, IL, Apr 25, 1956.

◆ **Ella Fitzgerald**, 77, singer ("Mack The Knife," "Bill Bailey, Won't You Please Come Home"), born Newport News, VA, Apr 25, 1918.

Meadowlark Lemon, 63, former basketball player, born at Lexington, SC, Apr 25, 1932.

Anthony Lukas, 62, author, journalist (*Common Ground*), born at New York, NY, Apr 25, 1933.

Paul Mazursky, 65, director (*Harry & Tonto*, *An Unmarried Woman*), born at Brooklyn, NY, Apr 25, 1930.

Al Pacino, 55, actor (Oscar *Scent of a Woman*; *Dog Day Afternoon*, *Godfather*), born East Harlem, NY, Apr 25, 1940.

Talia Shire, 49, actress (Connie in the Godfather movies and Adrian in the Rocky movies), born at Jamaica, NY, Apr 25, 1946.

APRIL 26 — WEDNESDAY
116th Day — Remaining, 249

AUDUBON, JOHN JAMES: BIRTH ANNIVERSARY. Apr 26. American artist and naturalist, born at Haiti, Apr 26, 1785. Died Jan 27, 1851 at New York, NY.

CHERNOBYL NUCLEAR REACTOR DISASTER: ANNIVERSARY. Apr 26, 1986. At 1:23 AM, local time, an explosion occurred at the Chernobyl atomic power station at Pripyat in the Ukraine. The resulting fire burned for days sending radioactive material into the atmosphere. More than 100,000 persons were evacuated from a 300-square-mile area around the plant. Three months after the explosion 31 persons (including six firefighters) were reported to have died, and thousands of others had been exposed to dangerous levels of radiation. Estimates projected an additional 1,000 cancer cases in the 12 European community nations as a result of this disaster. The plant was encased in a concrete tomb in an effort to prevent the still-hot reactor from overheating again and to minimize further release of radiation.

CONFEDERATE MEMORIAL DAY IN FLORIDA AND GEORGIA. Apr 26. See also: Confederate Memorial Day entries for Apr 24, May 10 and May 29.

GUERNICA MASSACRE: ANNIVERSARY. Apr 26. Late in the afternoon of Monday, Apr 26, 1937, the ancient Basque town of Guernica, in northern Spain, was attacked without warning by German-made airplanes. Three hours of intensive bombing left the town in flames, and citizens who fled to the fields and ditches around Guernica were machine-gunned from the air. This atrocity inspired Pablo Picasso's mural *Guernica*. Responsibility for the bombing was never officially established, but the suffering and anger of the victims and their survivors are still evident at anniversary demonstrations.

HESS, RUDOLF: BIRTH ANNIVERSARY. Apr 26. One of the most bizarre figures of World War II Germany, Walter Richard Rudolf Hess was born at Alexandria, Egypt, on Apr 26, 1894. He was a close friend, confidant and personal secretary to Adolph Hitler who had dictated much of *Mein Kampf* to Hess while both were prisoners at Landsberg Prison. Third in command in Nazi Germany, Hess surprised the world on May 10, 1941, by flying alone to Scotland and parachuting from his plane on what he called a "mission of humanity"—offering peace to Britain if she would join Germany in attacking the Soviet Union. He was immediately taken prisoner of war. At the Nuremberg Trials (1946), after questions about his sanity, he was convicted and sentenced to life imprisonment at Spandau Allied War Crimes Prison in Berlin, Germany. Outliving all other prisoners there, he was the only inmate from 1955 until he succeeded (in his fourth attempt) in committing suicide (as had Hitler, Josef Goebbels, Heinrich Himmler and Hermann Goring before him). He died at age 93, on Aug 17, 1987.

Apr 26-27 ☆ *Chase's 1995 Calendar of Events* ☆

HUG AN AUSTRALIAN DAY. Apr 26. To show our great appreciation for all the love and support the Aussies have given us over the years. Phone: (212) 388-8673 or (717) 274-8451. For info: Wellness Permission League, Tom or Ruth Roy, 2105 Water St, Lebanon, PA 17046.

KGBX TYPEWRITER TOSS. Apr 26. KGBX AM/FM Radio, Springfield, MO. Participating secretaries toss a typewriter from a lift-truck nearly 50-feet in the air. The typewriter landing closest to the bulls-eye wins an array of prizes. Definitely a "smashing" success! Annually, Secretaries Day. 6th annual toss. Est attendance: 100. For info: Mark Phillips, KGBX Radio, 840 S Glenstone, Springfield, MO 65802. Phone: (417) 869-1059.

LOOS, ANITA: BIRTH ANNIVERSARY. Apr 26. American author and playwright, born at Sisson, CA, on Apr 26, 1893. She is best remembered for her book *Gentlemen Prefer Blondes*, published in 1925. Loos, a brunette, died at New York, NY, on Aug 18, 1981.

MONTGOMERY WARD SEIZED: ANNIVERSARY. Apr 26. Montgomery Ward Chairman Sewell Avery was physically removed from his office on Apr 26, 1944, when federal troops seized Ward's Chicago offices after the company refused to obey President Franklin D. Roosevelt's order to recognize a CIO union. Government control ended May 9, shortly before the National Labor Relations Board announced the United Mail Order Warehouse and Retail Employees Union had won an election to represent the company's workers.

OLMSTED, FREDERICK LAW: BIRTH ANNIVERSARY. Apr 26. American landscape architect who participated in design of Yosemite National Park, New York City's Central Park and parks for Boston, Hartford and Louisville. Born at Hartford, CT, Apr 26, 1822; died at Waverly, MA, Aug 28, 1903.

PROFESSIONAL SECRETARIES DAY. Apr 26. Annually, the Wednesday of Professional Secretaries Week. Sponsor: Professional Secretaries Intl, PO Box 20404, 10502 NW Ambassador Dr, Kansas City, MO 64195-0404. Phone: (816) 891-6600.

RACKING HORSE SPRING CELEBRATION. Apr 26-29. Decatur, AL. Racking horses from throughout the country are shown each night, culminating with the selection of award winners. For info: Decatur Conv and Visitors Bureau, 719 6th Ave SE, PO Box 2349, Decatur, AL 35602. Phone: (205) 350-2028.

READ ME DAY. Apr 26. National and local celebrities and other volunteers read in classrooms wearing readable clothing with school appropriate messages. For info: Lee Fairbend, Exec Dir, Book'Em!, 2012 21st Ave S, Nashville, TN 37212. Phone: (615) 834-7323.

RICHTER SCALE DAY. Apr 26. A day to recognize the importance of Charles Francis Richter's research and his work in development of the earthquake magnitude scale that is known as the Richter scale. Richter, an American author, physicist and seismologist, was born on Apr 26, 1900, near Hamilton, OH. An Earthquake Awareness Week was observed in recognition of his work. Richter died at Pasadena, CA, on Sept 30, 1985.

April 1995

S	M	T	W	T	F	S
						1
2	3	4	5	6	7	8
9	10	11	12	13	14	15
16	17	18	19	20	21	22
23	24	25	26	27	28	29
30						

RUNNING OF THE RODENTS. Apr 26. Louisville, KY. To celebrate the pre-finals "Rat Race" that occurs each year, to develop community spirit and to unofficially kick off the Kentucky Derby festivities. Est attendance: 250. Sponsor: Spalding University, Emily Whalin, 851 S Fourth St, Louisville, KY 40203. Phone: (502) 585-7140.

SAN FRANCISCO REVISITED. Apr 26-June 26. M.H. de Young Museum, San Francisco, CA. A two-month exhibition of major artifacts related to the founding of the United Nations. For info: UN50 Committee, 312 Sutter St, Ste 610, San Francisco, CA 94108. Phone: (415) 989-1995.

SECRETARY'S DAY TEA. Apr 26. Rockwood Museum, Wilmington, DE. Barbara Darlin presents a one-woman fashion show, "The Gibson Girl Friday," examining the 19th century world of suffragettes, secretaries and ladies of leisure. For info: Rockwood Museum, 610 Shipley Rd, Wilmington, DE 19809. Phone: (302) 761-4340.

TANZANIA: UNION DAY. Apr 26. Celebrates union between mainland Tanzania (formerly Tanganyika) and the islands of Zanzibar and Pemba, in 1964.

UNITED STATES HOLOCAUST MUSEUM: TWO YEAR ANNIVERSARY. Apr 26. Washington, DC. The United States Holocaust Museum opened to the public on Apr 26, 1993. During its first year of operation over two million visitors toured the permanent exhibition.

BIRTHDAYS TODAY

Carol Burnett, 59, actress (*Once Upon a Mattress*, "Garry Moore Show," "Carol Burnett Show," *The Four Seasons*), born San Antonio, TX, Apr 26, 1936.

Duane Eddy, 57, musician, born at Corning, NY, Apr 26, 1938.

Bambi Linn, 69, dancer, born at Brooklyn, NY, Apr 26, 1926.

Bobby Rydell, 53, singer ("Wild One," "Volare"), born at Philadelphia, PA, Apr 26, 1942.

Gary Wright, 52, musician, born at Englewood, NJ, Apr 26, 1943.

APRIL 27 — THURSDAY
117th Day — Remaining, 248

AFGHANISTAN: SAUR REVOLUTION ANNIVERSARY. Apr 27. National holiday.

AMERICAN QUILTER'S SOCIETY QUILT SHOW. Apr 27-30. Paducah, KY. More than 400 quilts are exhibited with $75,000 awarded in prizes. Seminars, workshops. Est attendance: 30,000. For info: American Quilter's Soc, PO Box 3290, Paducah, KY 42002. Phone: (502) 898-7903.

◈ **BABE RUTH DAY: ANNIVERSARY.** Apr 27, 1947. Babe Ruth Day was celebrated in every ballpark in organized baseball in the US as well as Japan. Mortally ill with throat cancer, Ruth appeared at Yankee Stadium to thank his former club for the honor.

BIG TEN MEN'S TENNIS TEAM CHAMPIONSHIP. Apr 27-30. Indiana University, Bloomington, IN. For info: Big Ten Conference, 1500 W Higgins Rd, Park Ridge, IL 60068-6300. Phone: (708) 696-1010.

DOVE AWARDS PRESENTATION. Apr 27. Telecast live on The Family Channel, the Gospel Music Association presents the 26th annual Dove Awards for gospel music in a 2-hour show. Est attendance: 4,000. For info: Gospel Music Assn (GMA), 7 Music Circle N, Nashville, TN 37203. Phone: (615) 242-0303.

ENGLAND: WESTMINSTER ANTIQUE FAIR. Apr 27-30. Horticultural Hall, Vincent Sq, London. Quality antique fair with everything authenticated and for sale. All furniture pre 1870, most other items pre 1900. For info: Caroline Penman, Penman Antiques Fair, PO Box 114, Haywards Health, Sussex, England RH16 2YU.

☆ Chase's 1995 Calendar of Events ☆   Apr 27

FESTIVAL OF NATIONS 1995. Apr 27–30. St. Paul Civic Center, St. Paul, MN. Celebration by more than 95 ethnic groups presenting food specialties, folk dances and folk arts to the public. Est attendance: 95,000. For info: Intl Institute of Minnesota, 1694 Como Ave, St. Paul, MN 55108. Phone: (612) 647-0191.

GIBBON, EDWARD: BIRTH ANNIVERSARY. Apr 27. English historian and author. His *History of the Decline and Fall of the Roman Empire* remains a model of literature and history. From his description of the Roman emperor Gordianus II: "Twenty-two acknowledged concubines, and a library of sixty-two thousand volumes, attested the variety of his inclinations; and from the productions which he left behind him, it appears that the former as well as the latter were designed for use rather than for ostentation." Born on Apr 27, 1737, at Putney, Surrey, England, Gibbon died at London, on Jan 6, 1794. The remains of his magnificent library, left by Gibbon at Lausanne, Switzerland, were sold at auction in London on Dec 20, 1934. The 274 lots sold fetched £1,577.10s.

GIRLS INCORPORATED NATIONAL CONFERENCE. Apr 27–30. Sheraton New York, New York, NY. Girls Incorporated was formerly Girls Clubs of America. Est attendance: 400. For info: Girls Incorporated, Amy Sutnick Plotch, 30 E 33rd St, New York, NY 10016. Phone: (212) 689-3700.

GODWIN, MARY WOLLSTONECRAFT: BIRTH ANNIVERSARY. Apr 27. English writer whose best-known book was *Vindication of the Rights of Women*, published in 1792. Born at London, England, Apr 27, 1759, and died there Sept 10, 1797. Her daughter, Mary, was the wife of poet Percy Bysshe Shelley but is best remembered as the author of *Frankenstein*, published in 1818.

GRANT, ULYSSES SIMPSON: BIRTH ANNIVERSARY. Apr 27. Eighteenth president of the US (Mar 4, 1869–Mar 3, 1877), born Hiram Ulysses Grant at Point Pleasant, OH, Apr 27, 1822. He graduated from the US Military Academy in 1843. President Lincoln promoted Grant to lieutenant general in command of all the Union armies on Mar 9, 1864. On Apr 9, 1865, Grant received General Robert E. Lee's surrender, at Appomattox Court House, VA, which he announced to the secretary of war as follows: "General Lee surrendered the Army of Northern Virginia this afternoon on terms proposed by myself. The accompanying additional correspondence will show the conditions fully." Nicknamed "Unconditional Surrender Grant," he died at Mount McGregor, NY, July 23, 1885, just four days after completing his memoirs. He was buried in Riverside Park, New York, NY, where Grant's Tomb was dedicated in 1897.

INTERNATIONAL HOME FURNISHINGS MARKET. Apr 27–May 5. (Also Oct 19–27.) High Point and Thomasville, NC. The largest wholesale home furnishings market in the world. (Not open to the general public.) Est attendance: 69,000. Sponsor: Intl Home Furnishings Mktg Assn, PO Box 5687, High Point, NC 27262. Phone: (910) 889-0203.

MAGELLAN, FERDINAND: DEATH ANNIVERSARY. Apr 27. Portuguese explorer Ferdinand Magellan was probably born near Oporto, Portugal, about 1480, but neither the place nor the date is certain. Usually thought of as the first man to circumnavigate the earth, he died before completing the voyage; thus his co-leader, Basque navigator Juan Sebastian de Elcano, became the world's circumnavigator. The westward, 'round-the-world expedition began on Sept 20, 1519, with five ships and about 250 men. Magellan was killed by natives of the Philippine island of Mactan, Apr 27, 1521.

MATANZAS MULE DAY. Apr 27. In one of the first naval actions of the Spanish-American War, US naval forces, on Apr 27, 1898, bombarded the Cuban village of Matanzas. It was widely reported that the only casualty of the bombardment was one mule. "The Matanzas Mule" became instantly famous and remains a footnote in the history of the Spanish-American War.

MORSE, SAMUEL FINLEY BREESE: BIRTH ANNIVERSARY. Apr 27. American artist and inventor, after whom the Morse Code is named, was born at Charlestown, MA, Apr 27, 1791, and died at New York, NY, Apr 2, 1872. Morse conceived the idea of an electromagnetic telegraph while on shipboard, returning from art instruction in Europe in 1832, and he proceeded to develop his idea. With financial assistance approved by the Congress, the first telegraph line in the US was constructed, between Washington, DC, and Baltimore, MD. The first message was tapped out by Morse from the Supreme Court Chamber at the US Capitol building on May 24, 1844, was: "What hath God wrought?"

NEW BEGINNING FESTIVAL. Apr 27–29. Coffeyville, KS. A multi-state arts and crafts festival; cheese festival, carnival, entertainment. Est attendance: 10,000. For info: Chamber of Commerce, Box 457, Coffeyville, KS 67337. Phone: (316) 251-2550.

SIERRA LEONE: INDEPENDENCE DAY. Apr 27. National Day.

SPRING WILDFLOWER PILGRIMAGE. Apr 27–29. Gatlinburg, TN. Nature studies, guided photo trips, motor and hiking tours to explore the spring flora of the Great Smoky Mountains National Park. For info: Gatlinburg Chamber of Commerce, PO Box 527, Gatlinburg, TN 37738. Phone: (800) 568-4748.

***SULTANA* STEAMSHIP EXPLOSION: ANNIVERSARY.** Apr 27. Early in the morning of Apr 27, 1865, America's worst steamship disaster occurred. The *Sultana*, heavily overloaded with an estimated 2,300 passengers, exploded in the Mississippi River, just north of Memphis, en route to Cairo, IL. Most of the passengers were Union soldiers who had been prisoners of war and were eagerly returning to their homes. Although there was never an accurate accounting of the dead, estimates range from the Naval History Division's 1,450 to nearly 2,000. Cause of the explosion was not determined, but the little-known event is unparalleled in US history.

TAKE OUR DAUGHTERS TO WORK DAY. Apr 27. A national public education campaign sponsored by the Ms Foundation for Women. It is designed to focus attention on the neglected needs of girls. TODTWD has succeeded in mobilizing parents, educators, employers and other caring adults to take action to redress the inequalities in girls' lives and focus national attention on the concerns, hopes and dreams of girls. Annually, the fourth Thursday in April. For info: Gail Maynor, Dir, TODTWD, Ms Foundation for Women, 120 Wall St, 33rd Fl, New York, NY 10005. Phone: (800) 436-1800.

TOGO: INDEPENDENCE DAY. Apr 27. National holiday.

WASHINGTON STATE APPLE BLOSSOM FESTIVAL. Apr 27–May 7. Wenatchee, WA. To celebrate the spring blossoms of one of our most important crops, apples. Parades, arts and crafts, gem and mineral show, theatrical productions, exhibit in Riverfront Park, jet ski race and carnival. More than 40 events. Toll-free phone: (800) 57-APPLE. Est attendance: 100,000. Sponsor: Washington State Apple Blossom Festival Assn, Box 850, Wenatchee, WA 98807. Phone: (509) 662-3616.

YOM HASHOAH (HOLOCAUST DAY) 50th Anniversary. Apr 27. Hebrew calendar date: Nisan 27, 5755. A day established by Israel's Knesset as a memorial to the Jewish dead of World War II. Anniversary in Jewish calendar of Nisan 27, 5705 (corresponding to Apr 10, 1945, in the Gregorian calendar), the day on which Allied troops liberated the first Nazi concentration camp, one of the most notorious, Buchenwald, north of Weimar, Germany, established in 1937, where about 56,000 prisoners, many of them Jewish, perished.

BIRTHDAYS TODAY

Anouk Aimee, 61, actress (*A Man and a Woman*, *The Golden Salamander*), born at Paris, France, Apr 27, 1934.

Sheena Easton (Sheena Shirley Orr), 36, singer ("Morning Train"), born at Bellshill, Scotland, Apr 27, 1959.

Coretta Scott King, 68, lecturer, writer, widow of Martin Luther King, Jr, born at Marion, AL, Apr 27, 1927.

Jack Klugman, 73, actor ("Odd Couple," "Quincy, M.E."), born at Philadelphia, PA, Apr 27, 1922.

APRIL 28 — FRIDAY
118th Day — Remaining, 247

ARBOR DAY FESTIVAL. Apr 28–30. Arbor Day Farm, Nebraska City, NE. To celebrate Arbor Day, the tree planter's holiday, in the hometown of J.S. Morton, the founder of Arbor Day. Events include more than 50 artists' and craftsmen's booths, children's environmental festival, a petting barn, old-time fiddlers contest, golf tournament, roving entertainers and tree planting demonstrations. Est attendance: 20,000. For info: The Natl Arbor Day Fdtn, Mary Wolf, 211 N 12th St, #501, Lincoln, NE 68508. Phone: (402) 474-5655.

A(UGUSTA) BAKER'S DOZEN—A CELEBRATION OF STORIES. Apr 28–29. Columbia, SC. Annual event honors Augusta Baker and recognizes her distinguished career as librarian, storyteller, teacher and author. It features outstanding authors, illustrators and storytellers. Sponsors: Richland County Public Library and the College of Library and Information Science, Univ of South Carolina. Est attendance: 4,000. For info: Judy McClendon, PR Librarian, Richland County Public Library, 1431 Assembly St, Columbia, SC 29201. Phone: (803) 929-3440.

BARRYMORE, LIONEL: BIRTH ANNIVERSARY. Apr 28. Famed American actor of celebrated acting family, Lionel Barrymore was born Lionel Blythe, at Philadelphia, PA, Apr 28, 1878. He died at Van Nuys, CA, Nov 15, 1954. US Postal Service stamp issued in 1982 honored Ethel, John and Lionel Barrymore.

BIRDWATCHING WEEKEND. Apr 28–30. Madison, MN. More than 150 birdwatchers check out the birds at Salt Lake, a small salt water lake just west and south of Madison where more than 150 species of birds have been spotted in past years. Sponsor: Minnesota Ornithologists' Union. Housing sponsored by Madison Ambassadors and Saturday evening meal sponsored by Sons of Norway in Madison. Continental breakfast on Saturday sponsored by American Legion Club of Marietta. Annually, the last weekend in April. Est attendance: 150. Sponsor: Donna Ventrella, Chamber of Commerce, 404 6th Ave, Madison, MN 56256. Phone: (612) 598-7373.

BURLINGTON CRUISE NIGHTS. Apr 28. (Also May 26, June 30, July 28, Aug 25 and Sept 29). Fairway Center, Burlington, IA. Relive the past through your classic convertible, Model "T", Highboy, etc. Music and dance of the '50s and '60s. Annually, the last Friday of month, April–September. Sponsored by KCPS Radio, Burlington, IA, 1150 AM. Est attendance: 3,000. For info: Chip Giannettino–KCPS Radio, 408 N Main St, Burlington, IA 52601. Phone: (319) 754-6698.

CANADA: GUELPH SPRING FESTIVAL. Apr 28–May 14. Guelph, Ontario. Music, theatre, dance, opera and jazz with celebrated performers. Est attendance: 12,500. For info: Guelph Spring Festival, Box 1718, Guelph, Ont, Canada N1H 6Z9. Phone: (519) 821-3210.

CANADA: NATIONAL DAY OF MOURNING. Apr 28. A national day of mourning for workers killed or injured on the job in Canada. The Canadian Labour Congress first officially recognized the day in 1986. Pointing to the nearly one million workplace injuries each year in Canada, the CLC has called for stricter health and safety regulations and for annual recognition of this day throughout Canada. Federal legislation (Bill D-223) first recognized the day in 1991.

CHIMNEY ROCK HILLCLIMB 50th ANNIVERSARY. Apr 28–30. Chimney Rock Park, Chimney Rock, NC. Most challenging sports car hillclimb (on asphalt) in the US. The 1.8-mile course includes 19 curves—13 hairpin—through scenic tourist attraction. Grand celebration of 50th Anniversary with former winners and more! Sponsors: Sports Car Club of America (SCCA), Yokohama. Annually: the last full weekend in April. Est attendance: 5,000. For info: Chimney Rock Park, Mary Jaeger-Gale, Dir, PO Box 39, Chimney Rock, NC 28720. Phone: (800) 277-9611.

CIVIL WAR REENACTMENT. Apr 28–30. Rand Park, Keokuk, IA. Battle reenactment, military ball, theatre production, historic encampment, parade. Est attendance: 15,000. For info: Keokuk Area Conv and Tourism Bureau, M. Pohorsky, Dir, 401 Main, Keokuk, IA 52632. Phone: (800) 383-1219.

DAFFODIL FESTIVAL. Apr 28–30. Nantucket Island, MA. Daffodil show, shop window displays, antique car parade, tailgate picnic, flower show (Sunday and Monday). For info: Nantucket Island Chamber of Commerce, 48 Main St, Nantucket, MA 02554-3595. Phone: (508) 228-1700.

DOGWOOD ARTS AND CRAFTS FESTIVAL. Apr 28–30. Huntington Civic Center, Huntington, WV. 25th annual. Craft and food booths plus live entertainment for all ages. Est attendance: 15,000. For info: Shelly Ridgeway, Huntington Civic Center, PO Box 2767, Huntington, WV 25727. Phone: (304) 696-5990.

ELKS NATIONAL HOOP SHOOT. Apr 28–30. Indianapolis, IN. Boy and girl finalists, ages 8 through 13, compete for national hoop shoot title. Est attendance: 700. For info: Emile J. Brady, PO Box 153, Danville, PA 17821. Phone: (717) 275-5355.

GIBBS, MIFFLIN WISTER: BIRTH ANNIVERSARY. Apr 28. Mifflin Wister Gibbs was born at Philadelphia, PA, Apr 28, 1828. In 1873 he became the first black man to be elected a judge in the US, winning an election for City Judge in Little Rock, AR.

GREAT POETRY READING DAY. Apr 28. Read some great poetry/The world better to see. Sponsor: Robert A. Stevens, Clio Area High School, One Mustang Dr, Clio, MI 48420.

HOMER, LOUISE: BIRTH ANNIVERSARY. Apr 28. American operatic singer, born at Sewickley, PA, on Apr 28, 1871. Admired for the beauty and purity of her voice, she sang for major opera companies of Europe and America, including 19 seasons at the Metropolitan Opera House (1900–1919). She died at Winter Park, FL, on May 6, 1947.

JUBILEE—A GRAND CELEBRATION. Apr 28–29. Chattanooga Christian School, Chattanooga, TN. A please touch, see and do festival for the entire family. Auction, crafts, quilts, music, food. Lots of hands-on fun. Est attendance: 30,000. For info: Sharman Sherfey, Chattanooga Christian School, 3354 Broad St, Chattanooga, TN 37409. Phone: (615) 265-6411.

KISS-YOUR-MATE DAY. Apr 28. Show your mate how much you care. Share the pleasure of a kiss when he or she least expects it. Annually, Apr 28. Sponsor: Alan W. Brue, 399 SW 8th St, #1, Boca Raton, FL 33432-5732.

MONROE, JAMES: BIRTH ANNIVERSARY. Apr 28. The 5th president of the US was born at Westmoreland County, VA, Apr 28, 1758, and served two terms in that office (Mar 4, 1817–Mar 3, 1825). Monrovia, the capital city of Liberia, is named after him, as is the Monroe Doctrine, which he enun-

	S	M	T	W	T	F	S
April 1995							1
	2	3	4	5	6	7	8
	9	10	11	12	13	14	15
	16	17	18	19	20	21	22
	23	24	25	26	27	28	29
	30						

ciated at Washington, DC on Dec 2, 1823. Last of three presidents to die on US Independence Day, Monroe died at New York, NY, July 4, 1831.

◈ **MUSSOLINI EXECUTED: 50th ANNIVERSARY.** Apr 28, 1945. Italian partisans shot Benito Mussolini near the lakeside village of Dongo. Leaders of the Fascist Party, several of his friends and his mistress Clara Petacci also were executed. The 23-year-long Fascist rule of Italy was ended.

NATIONAL ARBOR DAY. Apr 28. The Committee for National Arbor Day has as its goal the observance of Arbor Day in all states on the same day. The last Friday in April was selected mainly to secure the support of all media areas—newspapers, magazines, radio and television—on this unified Arbor Day date, which would provide all our citizenry with the opportunity to better learn of the importance of trees to our way of life. This unified Arbor Day date was also found to be a good planting date for many of the states throughout the country. National Arbor Day has been observed in 1970, 1972, 1988, 1990, 1991 and 1993 by Presidential Proclamation. However, Congress did not legislate continuing authority, so the Committee is working to increase support for a permanent observance. More than half the states now observe Arbor Day on the proposed April Friday. Sponsors include: International Society of Arboriculture; Society of Municipal Arborists; American Assn of Nurserymen; National Arborist Assn, Inc; National Recreation and Park Assn; Arborists Assn of New Jersey. For info: Committee for Natl Arbor Day, 640 Eagle Rock Ave, West Orange, NJ 07052. Phone: (201) 731-0594.

NATIONAL DREAM HOTLINE. Apr 28–Apr 30. The National Dream Hotline is sponsored by the School of Metaphysics as an educational service to people throughout the world. Faculty and staff of the College and School of Metaphysics Centers throughout the midwest will offer the benefits of 30 years of research into the significance and meaning of dreams by manning the hotline phones from 6 PM CDT Friday until midnight Sunday. Annually, the last weekend in April. For info: Dr. Barbara Condron, Natl Advisor, School of Metaphysics, HCR 1, Box 15, Windyville, MO 65783. Phone: (417) 345-8411.

NEW ORLEANS JAZZ AND HERITAGE FESTIVAL. Apr 28–May 7. New Orleans, LA. Ten-day festival with thousands of musicians playing. Evening concerts, outdoor daytime activities, Louisiana specialty foods and handmade crafts. Est attendance: 360,000. Sponsor: New Orleans Jazz and Heritage Festival, Chairperson, PO Box 53407, New Orleans, LA 70153-3407.

◈ **PEACE, SECURITY AND THE PACIFIC RIM.** Apr 28. University of San Francisco, San Francisco, CA. This symposium will examine the United Nations' role in security, peacekeeping and human rights issues in the Asia Pacific. Sponsor: Center for the Pacific Rim. Contact: Barbara Bundy (415) 666-6357. For info: UN50 Committee, 312 Sutter St, Ste 610, San Francisco, CA 94108. Phone: (415) 989-1995.

POSITIVE POWER OF HUMOR AND CREATIVITY CONFERENCE. Apr 28–30. Saratoga Springs, NY. Participants will enjoy themselves while learning practical ideas they can apply both personally and on-the-job. Recent conferences have featured special appearances by Jay Leno, Victor Borge, Steve Allen, Sid Caesar The Capital Steps and The Smothers Brothers. More than 8,500 attendees from all 50 states and abroad; Japan, Russia, Saudi Arabia, Sweden, England, Australia, Brazil, South Africa, etc. 10th annual conference. Est attendance: 1,200. Sponsor: Dr. Joel Goodman, Dir, The Humor Project, 110 Spring St, Saratoga Springs, NY 12866. Phone: (518) 587-8770.

RATTLESNAKE DERBY. Apr 28–30. Mangum, OK. Hunters stalk these wily reptiles and attempt to bring in the most snakes and the longest snake. Snake skins and meat will be sold, and entertainment will include live music, a carnival and flea market. A herpetologist will be on hand to educate festival-goers. Est attendance: 40,000. For info: Chamber of Commerce, 222 W Jefferson, Mangum, OK 73554. Phone: (405) 782-2444.

SANTA FE TRAIL DAY. Apr 28. Las Animas, CO. The oldest Student Council-sponsored event in the country. Includes parade and square dance competition. Sponsor: Las Animas High School, Student Council, 300 Grove Ave, Las Animas, CO 81054.

SCOTLAND: MAYFEST 1995. Apr 28–May 20. Various venues, Glasgow, Strathclyde. Annual festival including theater, dance, rock and pop music, street entertainment and classical concerts. Est attendance: 50,000. For info: Billy Kelly, Genl Mgr, 18 Albion St, Glasgow, Scotland G1 1LH.

SOUTH CAROLINA FESTIVAL OF ROSES. Apr 28–30. Edisto Memorial Gardens, Orangeburg, SC. To celebrate the beauty of the roses and the gardens. Est attendance: 25,000. For info: Orangeburg County Chamber of Commerce, Carol P. Whisenhunt, PO Box 328, Orangeburg, SC 29116-0328. Phone: (803) 534-6821.

STAMP EXPO '95: SOUTH. Apr 28–30. Grand Hotel, Anaheim, CA. Fax: (818) 988-4337. Est attendance: 4,000. Sponsor: Intl Stamp Collectors Soc, PO Box 854, Van Nuys, CA 91408. Phone: (818) 997-6496.

SUGARLOAF CRAFT FESTIVAL. Apr 28–30. Maryland State Fairgrounds, Timonium, MD. More than 335 professional artists and craftspeople, demonstrations, live music, children's entertainment and delicious food. Est attendance: 23,000. Sponsor: Sugarloaf Mt Works, Inc, 200 Orchard Ridge Dr, #215, Gaithersburg, MD 20878. Phone: (301) 990-1400.

WORKERS MEMORIAL DAY. Apr 28. Held in memory of workers injured or killed on their jobs. Also, to draw attention to and educate people about their rights involving safety in the workplace. Annually, April 28. For info: Sharolyn Rosier, AFL-CIO, 815 16th St NW, Washington, DC 20000. Phone: (202) 637-5010.

BIRTHDAYS TODAY

Ann-Margret (Ann-Margaret Olsson), 54, actress (Oscar nominations for *Carnal Knowledge* and *Tommy*), born at Stockholm, Sweden, Apr 28, 1941.

James A. Baker, III, 65, former US Secretary of State in Bush administration, born at Houston, TX, Apr 28, 1930.

Barry Louis Larkin, 31, baseball player, born at Cincinnati, OH, Apr 28, 1964.

Harper Lee (Nelle Harper), 69, author (*To Kill a Mockingbird*), born at Monroeville, AL, Apr 28, 1926.

Jay Leno, 45, TV talk show host ("Tonight Show"), comedian, born at New Rochelle, NY, Apr 28, 1950.

Marcia Strassman, 47, actress (Julie Kotter in "Welcome Back Kotter"), born at New York, NY, Apr, 28, 1948.

APRIL 29 — SATURDAY
119th Day — Remaining, 246

APRIL IN TRENCHTOWN. Apr 29–30. St. Charles, MO. Street fair on 2nd St featuring antiques, artists, food and entertainment. For info: Convention and Visitors Bureau, St. Charles, MO, 230 S Main St, Po Box 745, St. Charles, MO 63302. Phone: (800) 366-2427.

ARIZONA STATE CHAMPIONSHIP CHILI COOKOFF. Apr 29. Bullhead City Rotary Park, Bullhead City, AZ. 23rd annual fundraiser for charity. Sanctioned by the International Chili Society, Inc. Food and entertainment. Sponsors: Budweiser and Laughlin casinos. Annually the last Saturday in April. Fax: (602) 763-3545. Est attendance: 9,000. For info: Fred R. Eck, Pres/Founder, PO Box 401, Bullhead City, AZ 86430. Phone: (602) 763-5885.

CLOVER LEAF PLANTING. Apr 29. Site TBA, MD. More than 500 volunteers plant at least 2,000 trees and 15,000–20,000 bulbs and perennials to beautify a major Maryland highway intersection. For info: Wally Orlinsky, Tree-Mendous Maryland, Tawes State Office Bldg, Forrestry Programs, E-1, Annapolis, MD 21401. Phone: (410) 974-3776.

Apr 29 ☆ Chase's 1995 Calendar of Events ☆

DAN BEARY'S SPRINGTIME CAVALCADE OF TEDDY BEARS. Apr 29–30. O'Neill Center, Western CT State University, Danbury, CT. Teddy Bear Show & Sale featuring more than 100 of America's leading teddy bear designers and makers includes bear repair and restoration clinic, appraisals, antique bears and dealers selling famous-name teddy bears from manufacturers throughout the world. Free door prizes, charity raffle, refreshments and free parking. Annually, the fourth full weekend in April. Est attendance: 2,500. For info: John K. Pringle, Pres, Pringle Productions, Ltd, PO Box 757, Bristol, CT 06011-0757. Phone: (203) 585-9940.

ELLINGTON, "DUKE" (EDWARD KENNEDY ELLINGTON): BIRTH ANNIVERSARY. Apr 29. "Duke" Ellington, one of the most influential individuals in jazz history, was born Apr 29, 1899, at Washington, DC. Ellington's professional career began when he was 17, and by 1923 he was leading a small group of musicians at the Kentucky Club in New York City who became the core of his big band. Ellington is credited with being one of the founders of big band jazz. He used his band as an instrument for composition and orchestration to create big band pieces, film scores, operas, ballets, Broadway shows and religious music. Ellington was responsible for more than 1,000 musical pieces. He drew together instruments from different sections of the orchestra to develop unique and haunting sounds such as that of his famous "Mood Indigo." "Duke" Ellington died May 24, 1974, at New York City.

ELLSWORTH, OLIVER: 250th BIRTH ANNIVERSARY. Apr 29. Third chief justice of the US Supreme Court, born at Windsor, CT, on Apr 29, 1745. Died there, Nov 26, 1807.

FIRST BLOOM FESTIVAL. Apr 29–May 7. American Rose Center, Shreveport, LA. Annual festival celebrating the first Spring Bloom of more than 20,000 roses, camellias, azaleas, daffodils and other flowering plants. Entertainment, mini-rose sale and refreshments. Est attendance: 5,000. Sponsor: American Rose Center, Stuart Parkerson, PO Box 30000, Shreveport, LA 71130-0030. Phone: (318) 938-5402.

FIRST THANKSGIVING FESTIVAL. Apr 29–30. Chamizal National Memorial, El Paso, TX. In April 1598 Juan de Oñate arrived at the Rio Grande with 600 colonists and 8,000 head of livestock. After what had been a four-month journey across harsh desert, Oñate ordered a feast of thanksgiving. This festival, in commemoration of that event (which preceded the Pilgrims' thanksgiving meal by 23 years), includes colorful reenactments with conquistadors, Spanish soldiers, monks, women and children. Vendors, food, arts and crafts. Est attendance: 10,000. For info: El Paso Mission Trail Assn, 1 Civic Center Plaza, El Paso, TX 79901. Phone: (915) 534-0677.

FLEMINGTON SPEEDWAY RACING SEASON. Apr 29–Nov 4. (Saturday night.) Flemington Fairgrounds. Racing by various types of stock cars and other vehicles. 6:00 PM starting time. Est attendance: 5,000. For info: Paul or Rick Kuhl, Flemington Fairgrounds, PO Box 293, Rt 31, Flemington, NJ 08822. Phone: (908) 782-2413.

FOXFIELD RACES. Apr 29. (Also Sept 24.) Charlottesville, VA. Steeplechase horse racing. Est attendance: 20,000. For info: Foxfield Racing Assn, W. Patrick Butterfield, Racing Mgr, PO Box 5187, Charlottesville, VA 22905. Phone: (804) 293-9501.

FRIENDSHIP SEES NO COLOR WEEK. Apr 29–May 5. A week to "see beyond color" and to make at least one friend of another race. Dates set to coincide with the anniversary of the Los Angeles riots of 1992. For info: Brian Harris, PO Box 74, Stanton, CA 90680. Phone: (714) 236-0805.

	S	M	T	W	T	F	S
April							1
1995	2	3	4	5	6	7	8
	9	10	11	12	13	14	15
	16	17	18	19	20	21	22
	23	24	25	26	27	28	29
	30						

GREAT CARDBOARD BOAT REGATTA. Apr 29. SIUC Campus Lake, Carbondale, IL. Teams and individuals design, build, and race person-powered boats made of corrugated cardboard. "Titanic Award" for most spectacular sinking plus trophies for team spirit, most creative use of cardboard, best-dressed team, most spectacular-looking boat. Prizes for top finishers in three boat classes: Class I, propelled by oars or paddles; Class II, propelled by mechanical means such as paddlewheels, propellors; Class III, "Instant Boats" made from "Secret Kits" by spectators-turned-participants. Registration 10 AM; races begin 12 noon. 22nd annual regatta. Additional competition at other locations across the country on various dates. Est attendance: 13,500. For info: School of Art and Design, Southern Illinois University, Carbondale, IL 62901. Phone: (618) 867-2346.

HEARST, WILLIAM RANDOLPH: BIRTH ANNIVERSARY. Apr 29. American newspaper editor and publisher, born at San Francisco, CA, Apr 29, 1863. Died at Beverly Hills, CA, Aug 14, 1951.

HERB FESTIVAL. Apr 29. Mattoon, IL. Fresh herbs, everlasting plants and scented geraniums, herb-related seminars, craft demonstrations and antiques. Annually, the last Saturday in April. Sponsors: Coles County Antiques and Craft Dealer Association. Est attendance: 7,000. For info: Mattoon Chamber of Commerce, 1701 Wabash, Mattoon, IL 61938. Phone: (217) 235-5661.

HIROHITO MICHI-NO-MIYA, EMPEROR: BIRTH ANNIVERSARY. Apr 29. Emperor of Japan. Born Apr 29, 1901, at Tokyo, Japan. Hirohito's death on Jan 27, 1989, ended the reign of the world's longest ruling monarch. He became the 124th in a line of monarchs when he ascended to the Chrysanthemum Throne in 1926. Hirohito presided over perhaps the most dynamic and cataclysmic years in the 2,500 years of recorded Japanese history, including the attempted military conquest of Asia; launching an attack on the US, bringing that country into World War II, leading to Japan's ultimate defeat after the US dropped atomic bombs on Hiroshima and Nagasaki; and the amazing economic restoration following the war that led Japan to a preeminent position of economic strength and influence. Although he opposed initiating hostilities with the US, he signed a declaration of war, allowing Japan's militarist Prime Minister, Hideki Tojo, to begin the fateful campaign. During the war's final days he overruled Tojo and advocated surrender. Hirohito, through his allies, survived an assassination plot, and broadcast a taped message to the Japanese people to stop fighting and "endure the unendurable." This radio message was the first time the emperor's voice had ever been heard outside the imperial household and inner circle of government. After the war, Hirohito was allowed to remain on his throne. He denounced his divinity in 1946, bestowed upon him by Japanese law, and became a "symbol of the state" in Japan's new parliamentary democracy. Hirohito turned his energies to his real passion, marine biology, becoming a recognized world authority in the field.

ISLE OF EIGHT FLAGS SHRIMP FESTIVAL. Apr 29–May 1. Fernandina Beach, FL. Commemorates Fernandina's role as the birthplace of the modern shrimping industry. Multi-event festival includes art show, entertainment, antiques, crafts and food. Est attendance: 150,000. For info: Chamber of Commerce, Box 472, Fernandina Beach, FL 32034. Phone: (904) 261-3248.

KITEFEST. Apr 29–30. River Oaks Park, Kalamazoo, MI. Family-oriented kite-flying event includes children's kite-making workshop, kite competitions and lots of kite flying. Est attendance: 5,000. For info: Kalamazoo County Parks Dept, John D. Cosby, 2900 Lake St, Kalamazoo, MI 49001. Phone: (616) 383-8778.

☆ Chase's 1995 Calendar of Events ☆ Apr 29-30

◆ **LIBERATION OF DACHAU: 50th ANNIVERSARY.** Apr 29, 1945. The Charlie Battery of the 522nd Field Artillery Battalion liberated the concentration camp at Dachau, Germany. The 522nd, part of the legendary 442nd (Go for Broke) regimental combat team, was made up of Nisei—second-generation Japanese Americans. The 442nd was the most decorated regiment in the history of the American military as well as the regiment with the highest casualties.

LOS ANGELES RIOTS: ANNIVERSARY. Apr 29. On Apr 29, 1992, a jury in Simi Valley, CA, failed to convict four Los Angeles police officers accused in the video-taped beating of Rodney King, providing the spark that set off rioting, looting and burning in South Central Los Angeles, CA, and other areas across the country. The anger unleashed during and after the violence was attributed to widespread racism, lack of job opportunities and the resulting hopelessness of inner-city poverty.

MARCH OF DIMES WALKAMERICA. Apr 29-30. The March of Dimes' largest fundraiser takes place in communities nationwide with more than one million volunteer walkers participating, who recruit donation based on the distance they complete. Funds raised support research and educational programs to prevent birth defects and to help lower the rate of premature births and infant mortality. Call your local Chapter to learn how to participate. For info: March of Dimes Birth Defects Fdtn, Natl HQ, 1275 Mamaroneck Ave, White Plains, NY 10605. Phone: (914) 428-4574.

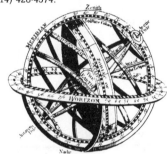

MOON PHASE: NEW MOON. Apr 29. Moon enters New Moon phase at 1:36 PM, EDT.

POMPANO BEACH SEAFOOD FESTIVAL. Apr 29-30. Pompano Beach, FL. Fresh seafood prepared by Broward's finest restaurants plus live music and arts and crafts. Est attendance: 150,000. For info: Pompano Beach Chamber of Commerce, 2200 E Atlantic Blvd, Pompano Beach, FL 33062. Phone: (305) 941-2940.

QUADLINGS OF OZ CONVENTION. Apr 29. Tulsa, OK. To celebrate Dorothy's meeting with Glinda, who told her how the magic shoes would take her home to Kansas. Est attendance: 40. Sponsor: Intl Wizard of Oz Club, Inc, Fred M. Meyer, Secy, 220 N 11th St, Escanaba, MI 49829.

REDBUD TRAIL RENDEZVOUS. Apr 29-30. Rochester, IN. Reenactment of a pre-1840 gathering to trade furs on the Tippecanoe River, tepee village, traditional crafts, pioneer and Indian dances, foods cooked over wood fires. Museum, round barn and Living History Village at north end of grounds. Est attendance: 2,000. For info: Fulton County Historical Soc, 37 E 375 N, Rochester, IN 46975. Phone: (219) 223-4436.

SAND CREEK FOLK LIFE FESTIVAL. Apr 29-30. Athletic Park, Newton, KS. Presents the history, music, arts and crafts of Kansas and the midwest. Craft demonstrations, Civil War reenactment, entertainment, mountainman campsite, arts and crafts show, food booths. Est attendance: 4,000. Sponsor: Harvey County Historical Soc, Mike Smurr, Chair, PO Box 4, Newton, KS 67114. Phone: (316) 283-2221.

SOLAR ECLIPSE. Apr 29. Annular eclipse of the sun. Central eclipse begins at 11:42 AM, EDT, reaches greatest eclipse at 1:23 PM, EDT and ends at 3:23 PM, EDT. The eclipse will be visible over the Pacific Ocean, Central America, South America except extreme southern tip, Atlantic Ocean and West Africa.

SOUTHERN MARYLAND CELTIC FESTIVAL. Apr 29. Jefferson Patterson Park, St. Leonard, MD. Scottish fiddling championship, bagpipe competition, Scottish Heptathlon, Highland Dancing Competition, Celtic marketplace and crafts, parade of clans and nations, Celtic harp workshop, Celtic folk music and dance demonstrations and Celtic foods. Est attendance: 5,000. For info: Mary Beth Dent, Celtic Soc of Southern Maryland, PO Box 209, Prince Frederick, MD 20678. Phone: (410) 257-9003.

SPACE MILESTONE: *CHALLENGER* (US). Apr 29. *Challenger* launched from Kennedy Space Center, FL, on Apr 29, 1985, with crew of seven and animal menagerie including monkeys and rats. Landed after 111 orbits of Earth on May 6, 1985, at Edwards AFB, CA.

TAIWAN: CHENG CHENG KUNG LANDING DAY. Apr 29. Commemorates landing in Taiwan on this day in 1661 of Ming Dynasty loyalist Cheng Cheng Kung (Koxinga), who ousted Dutch colonists who had occupied Taiwan for 37 years. Main ceremonies held at Tainan, in south Taiwan, where Dutch had their headquarters and where Cheng is buried. Cheng's birthday is also joyously celebrated, but according to the lunar calendar—on the 14th day of the seventh moon, Aug 9 in 1995.

◆ **UNITED NATIONS TEACHER'S WORKSHOP.** Apr 29. M.H. de Young Museum, San Francisco, CA. This afternoon's workshop will assist teachers in preparing their students for the archival exhibit on the founding of the UN and for use of the UN curriculum being developed. Sponsor: M.H. de Young Museum, SPICE and UN50 Committee. Contact: Educ Dept, (415) 750-3640. For info: UN50 Committee, 312 Sutter St, Ste 610, San Francisco, CA 94108. Phone: (415) 989-1995.

BIRTHDAYS TODAY

Andre Agassi, 25, tennis player, born at Las Vegas, NV, Apr 29, 1970.
George Allen, 73, former football coach, broadcaster, born at Detroit, MI, Apr 29, 1922.
Daniel Day-Lewis, 38, actor (Oscar for *My Left Foot*; *Unbearable Lightness of Being*), born London, England, Apr 29, 1957.
Robert Gottlieb, 64, editor, born at New York, NY, Apr 29, 1931.
Celeste Holm, 76, actress (*All About Eve*; Oscar for *Gentleman's Agreement*), born at New York, NY, Apr 29, 1919.
Rod McKuen, 62, poet, singer, born at San Francisco, CA, Apr 29, 1933.
Zubin Mehta, 59, conductor, born at Bombay, India, Apr 29, 1936.
Kate Mulgrew, 40, actress (Kate Ryan on "Ryans Hope"; Kate Columbo on "Mrs. Columbo"), born at Dubuque, IA, Apr 29, 1955.
Michelle Pfeiffer, 37, actress (*Dangerous Liasons*, *The Fabulous Baker Boys*), born Orange County, CA, Apr 29, 1958.
Jerry Seinfeld, 41, comedian ("The Jerry Seinfeld Show"), born Brooklyn, NY, Apr 29, 1954.

APRIL 30 — SUNDAY
120th Day — Remaining, 245

ASTRONOMY WEEK. Apr 30-May 6. To take astronomy to the people. Astronomy Week is observed during the calendar week in which Astronomy Day falls. See also: "Astronomy Day" (May 6).

BELTANE. Apr 30. (Also called Bealtaine, May Eve, Walpurgis Night, Cyntefyn, Roodmass and Cethsamhain.) One of the "Greater Sabbats" during the Wiccan year, it celebrates the union or marriage of the Goddess and God. In Scotland, Beltane was one of the quarter days or terms when rents were due and debts settled. On the eve of Beltane, two fires were built close together and cattle driven between them to ward off disease prior to putting them out to pasture for the new season. Annually, on Apr 30.

Apr 30 ★ Chase's 1995 Calendar of Events ★

BIRD RECORDED: ANNIVERSARY. Apr 30, 1941. The first commercially recorded work of Charlie (Bird) Parker, alto saxophonist and originator of the bebop style of modern jazz, was cut this date at Decca Records. During the recording session, picking up from the last two bars of "Swingmatism," Parker took off into a flowing improvisation that included "Hootie Blues," an example of a Parker blues chorus complete with a characteristic riff figure.

BLOSSOMTIME FESTIVAL OF SOUTHWESTERN MICHIGAN. Apr 30-May 7. Benton Harbor, MI. To promote the abundant fruit industry and tourism in Michigan. Spring comes alive as the blossoms announce the forthcoming arrival of the season's most delectable fruits. Est attendance: 250,000. Sponsor: Blossomtime, Inc, Gretchen Schalon, Exec Dir, 151 E Napier Ave, Benton Harbor, MI 49022. Phone: (616) 926-7397.

BRAUN, EVA: 50th DEATH ANNIVERSARY. Apr 30. Adolf Hitler's mistress from 1932 until their assumed suicide Apr 30, 1945, Eva Braun was born Feb 6, 1912, at Munich, Germany. She met Hitler while working as an assistant to his staff photographer and took up residence in his house though few people knew of her existence. On two occasions she attempted suicide because of jealousy stemming from Hitler's affairs with other women. As a sign of gratitude for her loyalty, he married her on Apr 29, 1945, as it became clear that Germany had lost the war. It generally is believed, though no bodies were found, that they took their own lives the next day in a Berlin bunker.

COPPERAS COVE CRIME STOPPERS FUN DOG SHOW. Apr 30. Copperas Cove Civic Center, Copperas Cove, TX. A fun dog show with fun, conformation and obedience classes. Serves as a practice show for AKC-sanctioned dog shows later in the season and major fundraiser for Cove Crime Stoppers. Annually, the last Sunday in April. Est attendance: 325. Sponsor: Copperas Cove Crime Stoppers, Inc, PO Box 1743, Copperas Cove, TX 76522. Phone: (817) 547-1111.

HARRISON, MARY SCOTT LORD DIMMICK: BIRTH ANNIVERSARY. Apr 30. Second wife of Benjamin Harrison, 23rd president of the US, born at Honesdale, PA, Apr 30, 1858. Died Jan 5, 1948.

INTERNATIONAL SCHOOL SPIRIT SEASON. Apr 30-Sept 30. To recognize everyone who has helped to make school spirit better and to provide time to plan improved spirit ideas for the coming school year. Sponsor: Pepsters, Jim Hawkins, Committee for More School Spirit, Box 2652, San Diego, CA 92112. Phone: (619) 280-0999.

ITALY: FEAST OF SAN PELLEGRINO. Apr 30-May 1. Gualdo Tadino (near Perugia). Ceremony with elements of history and folklore, called also "Festa del Maggio" (May Festival), which dates back to the 15th century and is celebrated on the night of April 30-May 1. At the end of several rituals, a huge poplar, formerly barked, is heaved with quick and expert movements in the middle of the square. For info: Italian Govt Travel Office, 500 N Michigan Ave, Chicago, IL 60611. Phone: (312) 644-0990.

JEWISH HERITAGE WEEK. Apr 30-May 7. Fosters intergroup understanding and greater appreciation of each other's culture, history and heritage. Celebrated by public and private schools, communities and organizations throughout the country. Phone: (212) 983-4800 ext 142. Sponsor: Jewish Community Relations Council NY, 711 Third Ave, New York, NY 10017.

LILLY, WILLIAM: BIRTH ANNIVERSARY. Apr 30. English astrologer, author and almanac compiler, born at Diseworth, Leicestershire, England, on Apr 30, 1602. His almanacs were among the most popular in Britain from 1644 until his death, June 9, 1681, at Hersham, Surrey, England.

LOUISIANA: ADMISSION DAY: ANNIVERSARY. Apr 30. Became 18th state on this day in 1812.

MUHAMMAD ALI STRIPPED OF TITLE: ANNIVERSARY. Apr 30. Muhammad Ali was stripped of his world heavyweight boxing championship Apr 30, 1972, when he refused to be inducted into military service. Said Ali, "I have searched my conscience, and I find I cannot be true to my belief in my religion by accepting such a call." He has claimed exemption as a minister of the Black Muslim religion.

NATIONAL HONESTY DAY (WITH PRESENTATION OF HONEST ABE AWARDS). Apr 30. To celebrate honesty and those who are honest and honorable in their dealings with others. Nominations accepted for most honest people and companies. Winners to be awarded "Honest Abe" Awards and given "Abies" on National Honesty Day. Annually, Apr 30. For info: M. Hirsh Goldberg, Author of *The Book of Lies*, 3103 Szold Dr, Baltimore, MD 21208. Phone: (410) 486-4150.

NETHERLANDS: QUEEN'S BIRTHDAY. Apr 30. A public holiday in celebration of the Queen's birthday, and the Dutch National Day.

NORTHERN EUROPE: WALPURGIS NIGHT. Apr 30. Witches' Sabbath. Eve of May Day. Celebrated particularly by university students.

SEIZURE OF IRANIAN EMBASSY IN LONDON: 15th ANNIVERSARY. Apr 30. On Apr 30, 1980, three dissident Arabs from Khuzestan Province of Iran seized the Iranian Embassy in London, England, and took about 20 people hostage. British commandos of the Special Air Service (SAS) stormed the embassy on May 5, 1980, freeing the 19 remaining hostages and killing or capturing the perpetrators of the seizure.

SMITH, MICHAEL J.: 50th BIRTH ANNIVERSARY. Apr 30. Michael J. Smith, 40-year-old pilot of the Space Shuttle *Challenger* on Jan 28, 1986. It was to have been Commander Smith's first space flight. Born at Beaufort, NC, on Apr 30, 1945, Smith perished with all others on board when the Space Shuttle *Challenger* exploded on Jan 28, 1986. See also: "*Challenger* Space Shuttle Explosion Anniversary" (Jan 28).

SWEDEN: FEAST OF VALBORG. Apr 30. An evening celebration in which Sweden "sings in the Spring" by listening to traditional hymns to the Spring, often around community bonfires. Also known as Walpurgis Night, the Feast of Valborg occurs annually on Apr 30.

WASHINGTON, GEORGE: PRESIDENTIAL INAUGURATION ANNIVERSARY. Apr 30. In New York, NY, on Apr 30, 1789, George Washington was inaugurated as the first president of the US under the new Constitution. Robert R. Livingston administered the oath of office to Washington on the balcony of Federal Hall, at the corner of Wall and Broad streets.

WSBA and WARM 103 SPRING CRAFT SHOW. Apr 30. York Fairgrounds, York, PA. More than 250 crafts, from country to contemporary, Victorian and southwestern, handcrafted furniture, wood carvings, dolls, jewelry, pottery, collectibles, quilts, baskets and much more. Admission fee. Est attendance: 6,000. For info: Gina M. Koch, Special Events Dir, PO Box 910, York, PA 17405. Phone: (717) 764-1155.

BIRTHDAYS TODAY

Lawton Chiles, 65, Governor of Florida (D), born at Lakeland, FL, Apr 30, 1930.
Jill Clayburgh, 51, actress (*An Unmarried Woman, Luna*), born at New York, NY, Apr 30, 1944.
Gary Collins, 57, actor, born at Boston, MA, Apr 30, 1938.
Cloris Leachman, 65, actress (Oscar *The Last Picture Show*; Phyllis on "The Mary Tyler Moore Show"), born Des Moines, IA, Apr 30, 1930.
Willie Nelson, 62, singer (platinum *Always On My Mind, Honeysuckle Rose*), born Abbott, TX, Apr 30, 1933.
Al Toon, 32, football player, born at Newport News, VA, Apr 30, 1963.

☆ Chase's 1995 Calendar of Events ☆ May 1

Maye.

MAY 1 — MONDAY
121st Day — Remaining, 244

ADDISON, JOSEPH: BIRTH ANNIVERSARY. May 1. English essayist born at Milston, Wiltshire, England, on May 1, 1672. Died at London, June 17, 1719. "We are," he wrote in *The Spectator*, "always doing something for Posterity, but I would fain see Posterity do something for us."

AMERICAN BOWLING CONGRESS BUD LIGHT MASTERS TOURNAMENT. May 1–6. Reno, NV. Top professional and nonprofessional bowlers from across the country and several foreign nations compete for $235,000 in prize money and one of the sport's most prestigious titles. Each player rolls 10 qualifying games before a cut is made to the top 120 bowlers. Then another 5 qualifying games cut the field to 63, who join the defending champion in a unique, 3-game, double-elimination match play format. Est attendance: 10,000. For info: American Bowling Congress, 5301 S 76th St, Greendale, WI 53129-1127. Phone: (414) 421-6400.

AMERICAN LUNG ASSOCIATION CLEAN AIR CAMPAIGN 1995. May 1–31. Throughout the month of May, the American Lung Association will focus public attention on air pollution, its effect on lung diseases such as asthma, emphysema and lung cancer and what individuals can do to fight for cleaner air. Local associations will designate their Clean Air Week®, holding public outreach activities as well as Clean Air Challenge™ fundraisers. Sponsor: American Lung Assn, Communications Div, 1740 Broadway, New York, NY 10019-4374. Phone: (212) 315-6473.

BELGIUM: PLAY OF ST. EVERMAAR. May 1. Annual performance (for more than 1,000 years) by the village inhabitants of a real "mystery" play, in its original form.

BETTER SLEEP MONTH. May 1–31. Emphasizes the importance of good sleep and encourages Americans to reevaluate their bedtime habits and check bedding for signs of old age. Sponsor: The Better Sleep Council. Sponsor: Ms Andrea Herman, Dir, Better Sleep Council, 333 Commerce St, Alexandria, VA 22314. Phone: (703) 683-8371.

CARPET CARE IMPROVEMENT WEEK. May 1–6. To encourage homeowners, tenants and landlords to improve the appearance, longevity and indoor air quality (for health) through carpet cleaning. Carpet is one of the largest investments inside a home or office. Clean your carpet for health's sake! Sponsor: Carpet and Fabricare Institute, Roger Pierce, Exec Dir, PO Box 149, Vancouver, WA 98666. Phone: (800) 227-7389.

May 1995

S	M	T	W	T	F	S
	1	2	3	4	5	6
7	8	9	10	11	12	13
14	15	16	17	18	19	20
21	22	23	24	25	26	27
28	29	30	31			

CHECKERFEST. May 1–31. Union Station, Indianapolis, IN. Third annual celebration enhances the Indy "500" experience for the community and visitors by offering a multitude of events during the month of May. Events include free concerts, special appearances by IndyCar drivers, children's activities and more. Annually, the month of May. For info: Union Station, PR Dept, 39 W Jackson Pl, Indianapolis, IN 46225. Phone: (317) 267-0700.

CHICAGO DAY AT THE WORLD'S FAIR: ANNIVERSARY. May 1. On May 1, 1893, the gates to Chicago's Columbian Exposition were opened to the public for free. Legend has it that 70,000 of Chicago's 100,000 residents attended that day.

CHILDCARE AWARENESS WEEK. May 1–7. Wilmington, DE. Features childcare seminars and training, a nanny social gathering and a family picnic. Annually, May 1–7. For info: A Choice Nanny, 111 Continental Dr, Ste 108, Newark, DE 19713. Phone: (302) 292-0626.

CLARK, MARK: BIRTH ANNIVERSARY. May 1. US general who served in both World Wars, Mark Clark was born at Madison Barracks, NY, May 1, 1896. In November 1942, he commanded the US forces taking part in the invasion of North Africa, and in January 1943, he became commander of the US Fifth Army, which invaded Italy in September 1943, taking Rome in June of 1944. After the Germans capitulated in Italy, Clark was appointed commander of US occupation forces in Austria. Clark died at Charleston, SC, Apr 17, 1984.

CORRECT POSTURE MONTH. May 1–31. To increase public awareness of the importance of correct posture. Sponsor: American Chiropractic Assn, 1701 Clarendon Blvd, Arlington, VA 22209. Phone: (703) 276-8800.

DATE YOUR MATE MONTH. May 1–31. May is the month to make your romance sizzle. Get back into the dating game by making a date with your mate this month. Annually, in May. Sponsor: Married Mistress & Monogamous Male Assns, Rose Smith, Pres, 4900 Mesa Bonita Ct NW, Albuquerque, NM 87120. Phone: (505) 899-3121.

DENMARK: TIVOLI GARDENS SEASON. May 1–Sept 17. Copenhagen. World-famous for its variety of entertainment, symphony concerts, pantomime and ballet. Beautiful flower arrangements and excellent restaurants. Traditional season: May 1 until the third Sunday in September.

FAMILY SUPPORT MONTH. May 1–31. The purpose of this observance is to support families with children during divorce, separation and custody issues. Promotes respect for mother/child relationships. Sponsor: Children Hurt in Legal Disputes (CHILD), Kathleen Quin, PO Box 241, Wilmette, IL 60091-0241.

FREEDOM SHRINE MONTH. May 1–31. To bring America's heritage of freedom to public attention through presentations or rededications of Freedom Shrine displays of historic American documents by Exchange Clubs. Sponsor: The Natl Exchange Club, 3050 Central Ave, Toledo, OH 43606-1700. Phone: (419) 535-3232.

FUNGAL INFECTION AWARENESS MONTH. May 1–31. To create media attention and heighten public awareness of skin, nail and scalp infections created by fungus, yeasts and mold in order to protect the skin during outdoor, summer activities. Annually, the month of May. Sponsor: Dave Chapman, The Mycology Institute, c/o FCG, 30 Lanidex Plaza W, Parsippany, NJ 07054. Phone: (201) 884-2200.

May 1 ☆ *Chase's 1995 Calendar of Events* ☆

GAZPACHO AFICIONADO TIME. May 1–Oct 31. A time to appreciate one of Spain's finest contributions to international cuisine: gazpacho, a nutritious cold soup made from fresh tomatoes and other vegetables. Observed while tomatoes are ripe.

ITALY: FESTIVAL OF ST. EFISIO. May 1–4. Cagliari. Said to be one of the biggest and most colorful processions in the world. Several thousand pilgrims on foot, in carts and on horseback wearing costumes dating from the 17th century accompany the statue of the saint through the streets.

◆ **JONES, MARY HARRIS (MOTHER JONES): BIRTH ANNIVERSARY.** May 1. Irish-born American labor leader. After the death of her husband and four children (during the Memphis yellow fever epidemic of 1867) and loss of her belongings in Chicago Fire, 1871, she devoted her energies and her life to organizing and advancing the cause of labor. It seemed she was present wherever there were labor troubles. She gave her last speech on her 100th birthday. Born at Cork, Ireland, May 1, 1830. Died Nov 30, 1930.

★ **LAW DAY.** May 1. Presidential Proclamation issued each year for May 1 since 1958 at request. (PL87–20 of Apr 7, 1961.)

LAW DAY USA. May 1. To advance equality and justice under law; to encourage citizen support of law observance and law enforcement; and to foster respect for law and understanding of its essential place in the life of every citizen of the US. Sponsor: American Bar Assn, Marcia L. Kladder, Dir, Special Events, 750 N Lake Shore Dr, Chicago, IL 60611. Phone: (312) 988-6133.

LEI DAY. May 1. Hawaii. On this special day—the Hawaiian version of May Day—leis are made, worn, given, displayed and entered in lei-making contests. One of the most popular Lei Day celebrations takes place in Honolulu at Kapiolani Park in Waikiki. Includes the state's largest lei contest, the crowning of the Lei Day Queen, Hawaiian music, hula and flowers galore.

◆ **LEO SOWERBY CENTENNIAL AND "JubiLEO."** May 1. The centennial of Leo Sowerby's birth will be commemorated throughout the nation during 1995. Orchestras, chamber groups, individual musicians, choral ensembles, church choirs and other organizations will perform many of Sowerby's more than 500 compositions. Sponsor: The Leo Sowerby Foundation, JubiLEO Steering Committee, Keuka College, Keuka, NY 14478.

★ **LOYALTY DAY.** May 1. Presidential Proclamation issued annually for May 1 since 1959 at request. (PL85–529 of July 18, 1958.) Note that an earlier proclamation was issued in 1955.

May 1995

S	M	T	W	T	F	S
	1	2	3	4	5	6
7	8	9	10	11	12	13
14	15	16	17	18	19	20
21	22	23	24	25	26	27
28	29	30	31			

LUCY STONE MARRIED: ANNIVERSARY. May 1, 1855. When nationally known public speaker and feminist Lucy Stone married Henry Blackwell, a marriage contract written by the bride and groom was read at the wedding that disavowed the gross inequity married women suffered under American law, and the word "obey" was omitted from their marriage vows. A year after the ceremony the bride further shocked society by taking back her maiden name, which she kept for the rest of her life.

MAY DAY. May 1. The first day of May has been observed as a holiday since ancient times. Spring festivals, maypoles and maying still are common, but political content of May Day has grown since the 1880s, when it became a workers' day in the US. Now widely observed in socialist countries as a workers' holiday. More recently Loyalty Day and Law Day observances have been encouraged in the US (by presidential and other proclamations) on May 1, contrasting strongly with the workers' demonstrations abroad. In most European countries, when May Day falls on Saturday or Sunday, the Monday following is observed as a holiday, with bank and store closings, parades and other festivities.

MAY DAY. May 1. Lumpkin, GA. Celebration of the planting season with traditional May Day activities and the May Pole Dance. Est attendance: 1,000. For info: Westville Village, Patty Cannington, PR Dir, PO Box 1850, Lumpkin, GA 31815. Phone: (912) 838-6310.

MAY IN MONTCLAIR. May 1–31. Montclair Township, NJ. 16th annual month-long festival celebrating the wealth of cultural, artistic, historic and recreational opportunities in Montclair for residents, merchants and visitors. Est attendance: 4,000. For info: Jean H. Kidd, Coord, May in Montclair, 91 Central Ave, Montclair, NJ 07042. Phone: (201) 744-7660.

MAY IS BETTER HEARING AND SPEECH MONTH. May 1–31. A nationwide public information campaign held each May to inform the 41 million Americans with hearing and speech problems that help is available. Annually, the month of May. For info: Harry Massey, Natl Grange, 1616 H St NW, Washington, DC 20006. Phone: (202) 628-3507.

MAY IS BETTER HEARING MONTH. May 1–31. To educate the public on hearing loss and its effects. Sponsor: Texas Hearing Aid Assn, Jim Wilson, Exec Dir, 222 N Riverside Dr, Fort Worth, TX 76111. Phone: (817) 831-0592.

MENTAL HEALTH MONTH. May 1–31. To heighten public awareness of mental health. Annually, the month of May. Fax: (703) 684-5968. Sponsor: Natl Mental Health Assn, 1021 Prince St, Alexandria, VA 22314. Phone: (703) 684-7722.

MOTHER GOOSE DAY. May 1. To re-appreciate the old nursery rhymes. Motto is "Either alone or in sharing, read childhood nursery favorites and feel the warmth of Mother Goose's embrace." Annually, May 1. Sponsor: Mother Goose Soc, Gloria T. Delamar, Founder, 7303 Sharpless Rd, Melrose Park, PA 19027. Phone: (215) 782-1059.

MULTICULTURAL SUBSTANCE ABUSE AWARENESS WEEK. May 1–5. All nationalities come together to celebrate the non-use of drugs and to help others of ethnic background to get off and stay off of drugs. Annually, the first week in May starting Monday. For info: Multicultural Partnership, Eduardo Hernandez, Dir, 601 E Montecito St, Santa Barbara, CA 93140. Phone: (805) 564-6778.

NATIONAL ALLERGY/ASTHMA AWARENESS MONTH. May 1–31. Kit of materials available for $15 from this non-profit organization. Fax: (415) 332-1832. For info: Pharmacist Planning Services, Inc, Frederick S. Mayer, Pres, c/o Allergy Council of America (ACA), PO Box 1336, Sausalito, CA 94966. Phone: (415) 332-4066.

NATIONAL ARTHRITIS MONTH. May 1–31. Increases awareness of the more than 100 diseases known as arthritis and increases support for the 37 million Americans with arthritis. Sponsor: Arthritis Foundation, Dennis Bowman, 1314 Spring St NW, Atlanta, GA 30309. Phone: (404) 872-7100.

Chase's 1995 Calendar of Events — May 1

NATIONAL BARBECUE MONTH. May 1–31. To encourage people to start enjoying barbecuing early in the season when Daylight-Saving Time lengthens the day. Annually, the month of May. Sponsor: Barbecue Industry Assn. For info: DHM Group, Inc, PO Box 767, Holmdel, NJ 07733.

NATIONAL BIKE MONTH. May 1–31. 39th annual celebration of bicycling for recreation and transportation. Local activities sponsored by bicycling organizations, environmental groups, PTAs, police departments, health organizations and civic groups. Annually, the month of May. For info: Educ Dir for League of American Bicyclists, 190 W Ostend St, Ste 120, Baltimore, MD 21230. Phone: (410) 539-3399.

NATIONAL EGG MONTH. May 1–31. Dedicated to the versatility, convenience, economy and good nutrition of "The incredible edible egg." Annually, the month of May. Sponsor: American Egg Board, Linda Braun, Consumer Services Mgr, 1460 Renaissance Dr, Park Ridge, IL 60068.

NATIONAL GOOD CAR-KEEPING MONTH. May 1–31. To promote increased safety and value through good car maintenance. Sponsor: Good Car-Keeping Institute, Sander Allen, 230 N Michigan Ave, Ste 1020, Chicago, IL 60601.

NATIONAL HAMBURGER MONTH. May 1–31. To pay tribute to one of America's favorite foods. With or without condiments, on or off a bun or bread, hamburgers have grown in popularity since the early 1920s and are now an American meal mainstay. Est attendance: 930. For info: White Castle System, Inc, Marketing Dept, 555 W Goodale St, Columbus, OH 43215-1171. Phone: (614) 228-5781.

NATIONAL HIGH BLOOD PRESSURE EDUCATION MONTH. May 1–31. To promote the control and treatment of high blood pressure. Annually, the month of May. Sponsor: Natl High Blood Pressure Education Program Info Ctr, Natl Heart, Lung and Blood Institute, PO Box 30105, Bethesda, MD 20824-0105. Phone: (301) 251-1222.

NATIONAL PHYSICAL FITNESS AND SPORTS MONTH. May 1–31. Encourages individuals and organizations to promote fitness activities and programs. New Presidential Fitness Partners in May campaign brings together all those groups promoting physical activity during May for mental health, older adults, etc. Events may be educational and involve demonstration and participation. Sponsor: President's Council on Physical Fitness and Sports, 701 Pennsylvania Ave NW, Ste 250, Washington, DC 20004. Phone: (202) 272-3427.

NATIONAL SALAD MONTH. May 1–31. Americans celebrate salads and their role in today's healthy lifestyle. Annually, the month of May. Sponsor: The Assn for Dressings and Sauces, 5775 Peachtree-Dunwoody Rd, Ste 500-G, Atlanta, GA 30342. Phone: (404) 252-3663.

NATIONAL SENIOR TRAVEL MONTH. May 1–31. Combining the spirit of Older Americans Month (May) and National Tourism Week (May 2–8), this month celebrates the many opportunities for seniors to travel. Sponsor: Pam Powell, Promo Dir, New Choices Magazine, 28 W 23rd St, New York, NY 10010.

NATIONAL SIGHT-SAVING MONTH. May 1–31. The spring season brings with it a flurry of yard work and home improvement activities. To reduce the number of eye injuries that occur in and around the home, Prevent Blindness America (formerly National Society to Prevent Blindness), is warning homeowners and "do-it-yourselfers" to watch out for hidden dangers at home. Sponsor: Prevent Blindness America, Marita Gomez, Media Relations, 500 E Remington Rd, Schaumburg, IL 60173. Phone: (800) 331-2020.

NATIONAL STRAWBERRY MONTH. May 1–31. Although strawberries are available year-round, the growing season for California strawberries is at its peak during May, when the volume of the harvest can reach 8 million pounds of strawberries per day. Annually, the month of May. For info: California Strawberry Commission, PO Box 269, Watsonville, CA 95077-0269. Phone: (408) 724-1301.

NATIONAL TUBEROUS SCLEROSIS AWARENESS MONTH. May 1–31. A presidential proclamation in 1974 declared the month of May to be National Tuberous Sclerosis Awareness Month, and it has been observed as such annually. Tuberous sclerosis is a genetic disease that affects 1 in 5,500 Americans every year. Contact the Natl HQ for a listing of current activities. Sponsor: Natl Tuberous Sclerosis Assn (NTSA), 8000 Corporate Dr, Ste 120, Landover, MD 20785. Phone: (800) 225-6872.

★ **OLDER AMERICANS MONTH.** May 1–31. Presidential Proclamation from 1963 through 1973 this was called "Senior Citizens Month." In May 1974 it became Older Americans Month. In 1980 the title included Senior Citizens Day, which was observed on May 8, 1980. Always has been issued since 1963.

PEN-FRIENDS WEEK INTERNATIONAL. May 1–7. To encourage everyone to have one or more pen-friends not only in their own country but in other countries. For complete information on how to become a good pen-friend and information about how to write good letters, send $2 to cover expense of printing, handling and postage. Annually, May 1–7. Sponsor: Intl Soc of Friendship and Good Will, Dr. Stanley J. Drake, Pres, 9538 Summerfield St, San Diego County, CA 91977-2852. Phone: (619) 466-8882.

PERSONAL HISTORY AWARENESS MONTH. May 1–31. To educate individuals and families about the importance of compiling a personal history. Write or call for newsletter or more info. Annually, each May. Sponsor: Margaret L. Ingram, Memories PLUS, PO Box 1339, Albany, OR 97321. Phone: (503) 928-4798.

PHILIPPINES: FEAST OF OUR LADY OF PEACE AND GOOD VOYAGE. May 1–31. Pilgrimage to the shrine of Nuestra Sra de la Paz y Buen Viaje in Antipolo, Rizal.

PHILIPPINES: SANTACRUZAN. May 1–31. Maytime pageant-procession that recalls the quest of Queen Helena and Prince Constantine for the Holy Cross.

PROJECT SAFE BABY MONTH. May 1–31. Dedicated to raising awareness of correct child safety seat use, to increasing the availability of child safety seats and to educating parents, teachers and health professionals about child safety in the car. Sponsor: Midas Intl, Inc. For info: Project Safe Baby Team, Golin/Harris Communications, Inc, 500 N Michigan Ave, Chicago, IL 60611. Phone: (312) 836-7100.

PUBLIC SERVICE RECOGNITION WEEK. May 1–7. Take this opportunity to thank the "Unsung Heroes and Heroines" of the public work force who perform a range of vital services. Public employees are scientists and police officers, teachers and doctors, astronauts and zoologists, engineers and food inspectors, forest rangers and claims representatives, researchers and foreign service agents. Resource materials to promote the celebration available. Annually, the first Monday–Sunday in May. For info: Public Employees Roundtable, PO Box 14270, Washington, DC 20044-4270. Phone: (202) 927-5000.

REACT MONTH. May 1–31. To encourage correct use of CB Radio Emergency Channel 9 in securing help for travel emergencies, etc; and to highlight REACT's role in volunteer public service radio communications. Fax: (316) 263-2118. For info: REACT Intl, Inc, PO Box 998, Wichita, KS 67201. Phone: (316) 263-2100.

REOPENING OF THE 1743 PALATINE HOUSE MUSEUM. May 1–Oct 31. Schoharie, NY. Open Thursday through Monday 1:00–5:00 PM, May through October. The Palatine House is a living museum and is State marked as the oldest existing building in Schoharie County. Est attendance: 900. Sponsor: Schoharie Colonial Heritage Assn, PO Box 554, Schoharie, NY 12157. Phone: (518) 295-7585.

May 1 ☆ Chase's 1995 Calendar of Events ☆

REVISE YOUR WORK SCHEDULE MONTH. May 1-31. To increase awareness, exploration and implementation of nontraditional work schedules such as flextime, telecommuting, job sharing and four-day work weeks. Annually, the month of May. For info: Maggi Payment, Dir, Center for Worktime Options, 1043 University Ave, #192, San Diego, CA 92103. Phone: (619) 232-0404.

ST. TAMENEND'S DAY. May 1. Colonial American joyous May Day celebration, recently revived. Legendary American Indian sage, Chief Tamenend, canonized by fun-loving young colonists asserting independence from old-world patrons. Modern celebrants, tired of overly serious political observances, identify selves by pinning dollar bill to their jackets. For info: Dr. Nicholas Varga, Loyola College, Baltimore, MD 21210.

SAVE THE RHINO DAY. May 1. May Day! May Day! Rhinos are *still* on the verge of extinction! Get involved with local, national and international conservation efforts to stop the senseless slaughter of these gentle pachyderms. Call your local zoo or write Really, Rhinos! For info: Really, Rhinos!, Judyth Lessee, Founder, Box 1285, Tucson, AZ 85702-1285.

SENIOR CITIZENS MONTH. May 1-31. Massachusetts.

SMITH, KATE: BIRTH ANNIVERSARY. May 1. One of America's most popular singers. Kate Smith, who never took a formal music lesson, recorded more songs than any other performer (more than 3,000), made more than 15,000 radio broadcasts and received more than 25 million fan letters. On Nov 11, 1938, she introduced a new song during her regular radio broadcast, written especially for her by Irving Berlin: "God Bless America." It soon became the unofficial national anthem. Born Kathryn Elizabeth Smith at Greenville, VA, on May 1, 1909. She began her radio career on May 1, 1931, with "When the Moon Comes Over the Mountain," a song identified with her throughout her career. She died at Raleigh, NC, June 17, 1986.

SOWERBY, LEO: BIRTH ANNIVERSARY. May 1, 1895. Pulitzer prize-winning composer of more than 550 compositions, born at Grand Rapids, MI, and died July 7, 1968, at Port Clinton, OH.

★ **STEELMARK MONTH.** May 1-31. Presidential Proclamation always May since 1967. Proclamation 3778, Apr 8, 1967, covers all succeeding years. (PL 89-703 of Nov 2, 1966.)

STROKE AWARENESS MONTH. May 1-31. American Heart Assn and other organizations across the country work together to conduct awareness campaigns about stroke, the leading cause of disability in the US. Observance is designed to alert the public about stroke's warning signs, how it might be prevented and resources available to help the stroke survivor. Call local AHA offices at (800) AHA-USA1.

TEILHARD DE CHARDIN, PIERRE: BIRTH ANNIVERSARY. May 1. French Jesuit author, paleontologist and philosopher, born at Sarcenat, France, May 1, 1881. He died at New York, NY, Apr 10, 1955.

UNITED KINGDOM: MAY DAY BANK HOLIDAY. May 1. Bank and public holiday in England, Wales, Scotland and Northern Ireland.

U-2 INCIDENT: 35th ANNIVERSARY. May 1. On the eve of a summit meeting between US President Dwight D. Eisenhower and Soviet Premier Nikita Khrushchev, a U-2 espionage plane flying at about 60,000 feet was shot down over Sverdlovsk, in central USSR, May 1, 1960. The pilot, CIA agent Francis Gary Powers, survived the crash, as did large parts of the aircraft, a suicide kit and sophisticated surveillance equipment. The sensational event, which US officials described as a weather reconnaissance flight gone astray, resulted in cancellation of the summit meeting. Powers was tried, convicted and sentenced to 10 years in prison, in a Moscow court. In 1962 he was returned to the US in exchange for an imprisoned Soviet spy but found an unfriendly American public, which apparently believed he should have used his suicide kit. He died in a helicopter crash in 1977. See also: "Powers, Francis Gary: Birth Anniversary" (Aug 17).

VEGETARIAN RESOURCE GROUP'S ESSAY CONTEST FOR KIDS. May 1. Baltimore, MD. Children ages 18 and under are encouraged to submit 2-3 page essays on topics related to vegetarianism. Essays accepted beginning May 1. Winners will be announced Sept 15, and each will receive a $50 savings bond. Sponsor: The Vegetarian Resource Group, PO Box 1463, Baltimore, MD 21203. Phone: (410) 366-8343.

WILLIAMS, ARCHIE: BIRTH ANNIVERSARY. May 1. Archie Williams, along with Jesse Owens and others, debunked Hitler's theory of the superiority of the Aryan athletes at the 1936 Berlin Olympics. As a black member of the US team Williams won a gold medal by running the 400-meter in 46.5 seconds (.4 second slower than his own record of earlier that year). Williams, who was born May 1, 1915, at Oakland, CA, earned a degree in mechanical engineering from the University of California-Berkeley in 1939, but had to dig ditches for a time because they weren't hiring black engineers. In time Williams became an airplane pilot and for 22 years he trained Tuskegee Institute pilots including the black air corp of WWII. He joined the Army Air Corps in 1942. When asked during a 1981 interview about his treatment in the hands of the Nazis during the 1936 Olympics, he replied, "Well, over there at least we didn't have to ride in the back of the bus." Archie Williams died June 24, 1993, at Fairfax, CA.

WOMEN'S INTERNATIONAL BOWLING CONGRESS ANNUAL MEETING. May 1-3. Tucson, AZ. Est attendance: 3,500. For info: WIBC, 5301 S 76th St, Greendale, WI 53129. Phone: (414) 421-9000.

ZAMBIA: LABOR DAY. May 1. Dedicated to "Freedom and Labor," the motto of Zambia's only political party, UNIP. A day of mobilization for maximum productivity. Annually, the first Monday in May.

BIRTHDAYS TODAY

Steve Cauthen, 35, jockey, born at Covington, KY, May 1, 1960.
Judy Collins, 56, singer ("Both Sides Now," "Chelsea Morning"), born Seattle, WA, May 1, 1939.
Rita Coolidge, 50, singer ("[Your Love Has Lifted Me] Higher and Higher," "We're All Alone"), born Nashville, TN, May 1, 1945.
Glenn Ford, 79, actor (*The Blackboard Jungle, The Fastest Gun Alive*, "The Family Holvak"), born Quebec, Canada, May 1, 1916.
Joseph Heller, 72, writer (*Catch-Twenty-Two, God Knows*), born at Brooklyn, NY, May 1, 1923.
Sonny James (Jimmy Loden), 66, singer ("Young Love"), born at Hackleburg, AL, May 1, 1929.
Bobbie Ann Mason, 55, writer (*In Country, Spence and Lila*), born at Mayfield, KY, May 1, 1940.
Jack Paar, 77, entertainer, born at Canton, OH, May 1, 1918.

May 1995

S	M	T	W	T	F	S
	1	2	3	4	5	6
7	8	9	10	11	12	13
14	15	16	17	18	19	20
21	22	23	24	25	26	27
28	29	30	31			

☆ Chase's 1995 Calendar of Events ☆ May 2–3

MAY 2 — TUESDAY
122nd Day — Remaining, 243

◈ BERLIN SURRENDERS: 50th ANNIVERSARY. May 2, 1945. At 6:45 AM, Soviet Marshall Georgi Zhukov accepted the surrender of Berlin, the German capital. The victory came at a terrible cost for the Red Army, with 304,887 men killed, wounded or missing—10% of its soldiers. About 125,000 Berliners died in the siege, many by suicide.

CARTOON ART APPRECIATION WEEK. May 2–6. To create a greater public awareness of and appreciation for cartoon art and its creators, the Museum schedules a variety of special events that celebrate the contribution of cartoons to the artistic, social and historic fabric of Americana. Also, to promote literacy using cartoon art as an educational tool, classroom activity kits for all ages are available to teachers and parents. Est attendance: 300. For info: Cartoon Art Museum, CAAW Director, 665 Third St, Ste 412, San Francisco, CA 94107. Phone: (415) 546-3922.

CONTRABAND DAYS (PIRATE FESTIVAL). May 2–14. Lake Charles, LA. A major 13-day festival celebrating Jean Lafitte, "The Gentleman Pirate," and the contraband treasures he is supposed to have hidden along the shores of Lake Charles. Events include a mock invasion by pirates, a night boat parade, concerts, fireworks, races and a host of other activities. Fax: (318) 436-1126. Est attendance: 300,000. For info: Contraband Days, Inc, Lana Brunet, PO Box 679, Lake Charles, LA 70602. Phone: (318) 436-5508.

CROSBY, HARRY LILLIS "BING": BIRTH ANNIVERSARY. May 2. American singer, composer and actor, born at Tacoma, WA, May 2, 1904. Died while playing golf near Madrid, Spain, Oct 14, 1977.

DA VINCI, LEONARDO: DEATH ANNIVERSARY. May 2. Italian artist, scientist and inventor. Painter of the famed *The Last Supper*, perhaps the first painting of the High Renaissance, and of the *Mona Lisa*. Inventor of the first parachute. Died May 2, 1519, at age 67.

LEE, PINKY (Pincus Leff): BIRTH ANNIVERSARY. May 2. Pinky Lee was born at St. Paul, MN, May 2, 1907. When young, Lee had dreams of becoming an attorney, but abandoned the idea when classmates laughed at his lisp. His show business debut was in burlesque in the 1930s. He is best remembered for "The Pinky Lee Show" which telecast from Los Angeles in the early 1950s. Pinky Lee died Apr 3, 1993, at Mission Viejo, CA.

NATIONAL ONLINE MEETING AND IOLS '95. May 2–4. New York Hilton, New York, NY. A forum for communication among database producers, online vendors, and users and producers of electronic information services. Papers presented, product reviews and exhibits including online products, CD-ROMs and library systems on display for demonstrations and hands-on experience. Approximate attendance: 5,750. Info submitted by Carol Nixon, Learned Information, Inc, 143 old Marlton Pike, Medford, NJ 08055; phone: (609) 654-6266; fax: (609) 654-4309. Also submitted by: Natl Online Meeting, Prof Martha E. Williams, U of I Coordinated Science Lab, 1308 W Main, Urbana, IL 61801. Phone: (217) 333-1074.

POPE LEO XIII: BIRTH ANNIVERSARY. May 2. Giocchino Vincenzo Pecci, 256th pope of the Roman Catholic Church, born at Carpineto, Italy, May 2, 1810. Elected pope Feb 20, 1878. Died July 20, 1903.

REWARD OFFERED FOR JEFFERSON DAVIS: 130th ANNIVERSARY. May 2. On May 2, 1865, President Andrew Johnson offered a $100,000 reward for the capture of Confederate Jefferson Davis, accusing the Confederate government of complicity in the murder of Abraham Lincoln. Davis and his cabinet had fled southward from Richmond and arrived in Abbeville, SC. After the rejection of the Sherman-Johnston peace treaty, he and the rest of his fugitive cabinet were at odds over future strategy. Davis argued for continuing the struggle from west of the Mississippi while his cabinet officers preferred fleeing to a foreign country.

ROBERT'S RULES DAY. May 2. Anniversary of the birth of Henry M. Robert (General, US Army), author of *Robert's Rules of Order*, a standard parliamentary guide. Born May 2, 1837. Died May 11, 1923.

BIRTHDAYS TODAY

Theodore Bikel, 71, singer, actor, born at Vienna, Austria, May 2, 1924.
Larry Gatlin, 46, singer, songwriter ("Broken Lady," "All the Gold in California"), born at Odessa, TX, May 2, 1949.
Lesley Gore, 49, singer ("I'll Cry If I Want To"), born at Tenafly, NJ, May 2, 1946.
Bianca Jagger, 50, actress, political activist, ex-wife of Mick Jagger, born at Managua, Nicaragua, May 2, 1945.
Benjamin Spock, 92, pediatrician, author, born at New Haven, CT, May 2, 1903.

MAY 3 — WEDNESDAY
123rd Day — Remaining, 242

HONG KONG: TIN HAU FESTIVAL. May 3. This is the festival of the heavenly Queen, the Goddess of fishermen. Fishermen decorate their boats and gather at the temples to worship her and to ask for good catches during the coming year. A celebration at Joss House Bay and Yuen Long and the traditional rites of the temple are also featured. For info: Hong Kong Tourist Assn, 590 Fifth Ave, 5th Fl, New York, NY 10036-4706. Phone: (212) 869-5008.

JAPAN: CONSTITUTION MEMORIAL DAY. May 3. National holiday.

LUMPY RUG DAY. May 3. To encourage the custom of teasing bigots and trigots for shoving unwelcome facts under the rug. The legend of Thiri is that when many cans of worms have been shoved under the rug, the defenders of the status quo obtain a new rug high enough to cover the unwanted facts. Annually, May 3. Sponsor: Puns Corps, Robert L. Birch, Coord, Box 2364, Falls Church, VA 22042-0364. Phone: (703) 533-3668.

MEXICO: DAY OF THE HOLY CROSS. May 3. Celebrated especially by construction workers and miners, a festive day during which anyone who is building must give a party for the workers. A flower-decorated cross is placed on every piece of new construction in the country.

NATIONAL PUBLIC RADIO: FIRST BROADCAST ANNIVERSARY. May 3. National noncommercial radio network, financed by Corporation for Public Broadcasting, began programming on May 3, 1971.

POLAND: CONSTITUTION DAY (SWIETO TRZECIEGO MAJO). May 3. National Day. Celebrates ratification of Poland's first constitution, 1794.

ST. LOUIS STORYTELLING FESTIVAL. May 3–6. St. Louis, MO. A gathering of 40–50 storytellers who celebrate and perpetuate stories and storytelling. Est attendance: 26,000. Sponsor: University of Missouri–St. Louis, 318 Lucas Hall, 8001 Natural Bridge Rd, St. Louis, MO 63121. Phone: (314) 553-5961.

May 3-4 ☆ Chase's 1995 Calendar of Events ☆

SPACE MILESTONE: *DELTA 3914* ROCKET FAILURE. May 3. Launched from Cape Canaveral, FL, on May 3, 1986. The rocket failed, flew out of control and was intentionally destroyed by explosives 90 seconds after launch to avoid the risk of having it land in a populated area.

SUNFEST '95. May 3-7. Flagler Dr, downtown West Palm Beach, FL. Florida's largest music, art and water events festival features some of the best acts in jazz, pop, blues and more. In addition, this family-oriented event includes a juried art show, hand-made crafts, fireworks, water and youth park activities and fabulous foods. 24-hour hotline: (407) 659-5992; Fax: (407) 659-3567. Est attendance: 300,000. For info: SunFest of Palm Beach County, Inc, PO Box 279, West Palm Beach, FL 33402-0279. Phone: (407) 659-5980.

UNITED NATIONS: WORLD PRESS FREEDOM DAY. May 3. A day to recognize that a free, pluralistic and independent press is an essential component of any democratic society and to promote press freedom in the world.

BIRTHDAYS TODAY

James Brown, 62, singer, songwriter ("Papa's Got a Brand New Bag"), born at Augusta, GA, May 3, 1933.
Christopher Cross, 44, musician, songwriter, born at Anston, TX, May 3, 1951.
Doug Henning, 48, magician, born at Fort Gary, Manitoba, Canada, May 3, 1947.
Jeffrey John Hornacek, 32, basketball player, born at Elmhurst, IL, May 3, 1963.
Engelbert Humperdinck (Gerry Dorsey), 59, singer ("Release Me [And Let Me Love Again]," "After The Lovin'"), born Madras, India, May 3, 1936.
David Roderick, 53, corporation executive, born at Pittsburgh, PA, May 3, 1942.
Pete Seeger, 76, folksinger, songwriter ("Blowin' in the Wind"), born at New York, NY, May 3, 1919.
Frankie Valli, 58, singer ("Can't Take My Eyes Off You," "Grease"), born Newark, NJ, May 3, 1937.

MAY 4 — THURSDAY
124th Day — Remaining, 241

CHINA: YOUTH DAY. May 4. Annual public holiday "recalls the demonstration on May 4, 1919, by thousands of patriotic students in Beijing's Tiananmen Square to protest imperialist aggression in China."

CINCO DE MAYO FESTIVAL. May 4-7. World Trade Center and Tom McCall Waterfront Park. Portland, OR. A Mexican celebration with ethnic foods, continuous entertainment, dancing, arts and crafts booths. Part of Portland's Sister City program with Guadalajara, Mexico (See Oct 7). Est attendance: 200,000. For info: Portland-Guadalajara Sister City Assn, PO Box 728, Portland, OR 97207. Phone: (503) 222-2223.

May 1995

S	M	T	W	T	F	S
	1	2	3	4	5	6
7	8	9	10	11	12	13
14	15	16	17	18	19	20
21	22	23	24	25	26	27
28	29	30	31			

CURACAO: MEMORIAL DAY. May 4. Victims of World War II are honored on this day. Military ceremonies at the War Monument. Not an official public holiday.

DISCOVERY OF JAMAICA BY CHRISTOPHER COLUMBUS: ANNIVERSARY. May 4. Christopher Columbus discovered Jamaica on May 4, 1494. The Arawak Indians were its first inhabitants.

ENGLAND: BADMINTON HORSE TRIALS. May 4-7. Badminton, Avon. Famous international horse trials consisting of showjumping, cross country and dressage. For the Mitsubishi Motors Trophy. Est attendance: 200,000. For info: Jane Tuckwell, Badminton Horse Trials, Badminton, Avon, England GL9 1DF.

HEPBURN, AUDREY (EDDA van HEEMSTRA HEPBURN-RUSTEN): BIRTH ANNIVERSARY. May 4. Audrey Hepburn, whose first major movie role in *Roman Holiday* (1953) won her an Academy Award as best actress, was born May 4, 1929, near Brussels, Belgium. Hepburn made 26 movies during her career and received four additional Oscar nominations. During the latter years of her life Hepburn served as spokesperson for the United Nations Children's Fund. She traveled worldwide trying to raise money for the organization. Audrey Hepburn died Jan 20, 1993, at Tolochenaz, Switzerland.

ISLE OF WIGHT: INTERNATIONAL OBOE COMPETITION. May 4-7. Various venues, Isle of Wight. One of the very few international oboe competitions. For info: Norman Thurston, Administrator, 32 Gregory Ave, Pondwell, Ryde PO33 1PZ, Isle of Wight, Great Britain.

KENT STATE STUDENTS' MEMORIAL DAY: 25th ANNIVERSARY. May 4. Services in memory of four students (Allison Krause, 19; Sandra Lee Scheuer, 20; Jeffrey Glenn Miller, 20; and William K. Schroeder, 19) killed by National Guard during demonstrations against the Vietnam War at Kent (Ohio) State University, on May 4, 1970, and of all other students martyred in the cause of human rights.

LOYALTY DAYS AND SEAFAIR FESTIVAL. May 4-7. Newport, OR. Celebration of loyalty to America. Parade, queen and court, sports car races, Navy ships in port and other activities. Est attendance: 5,000. For info: Chamber of Commerce, 555 SW Coast Hwy, Newport, OR 97365. Phone: (503) 265-8801.

MANN, HORACE: BIRTH ANNIVERSARY. May 4. American educator, author, public servant, known as the "father of public education in the US," was born at Franklin, MA, on May 4, 1796. Founder of Westfield (MA) State College and editor of the influential *Common School Journal*. Mann died at Yellow Springs, OH, on Aug 2, 1859.

MARTIN Z. MOLLUSK DAY. May 4. Moorlyn Terrace Beach, Ocean City, NJ. If Martin Z. Mollusk, a hermit crab, sees his shadow at 11 AM, EST, summer comes a week early—if he doesn't, summer begins on time. 1995 will be the 21st annual observance. Est attendance: 300. For info: Mark Soifer, 9th St and Asbury Ave, Ocean City, NJ 08226. Phone: (609) 399-6111.

★ **NATIONAL DAY OF PRAYER.** May 4. Presidential Proclamation always the first Thursday in May since 1981. (PL100-307 of May 5, 1988.) Beginning in 1957, a day in October was designated, except in 1972 and 1975 through 1977.

NATIONAL WEATHER OBSERVER'S DAY. May 4. For those people, amateurs and professionals alike, who love to follow the everyday phenomenon known as weather. Annually, May 4. Sponsor: Alan W. Brue, 399 SW 8th St, #1, Boca Raton, FL 33432-5732.

PAT BOONE CELEBRITY SPECTACULAR. May 4-5. Bethel Bible Village, Chattanooga, TN. The annual Spectacular will host more than 60 celebrities of the worlds of sports and entertainment. It brings together one of the largest groups of entertainers to Chattanooga at one time for concerts, golfing tournament, luncheons and more. Sponsor: Bethel Bible Village. For info: Shannon Craig, Bethel Bible Village, 3001 Hamill Rd, Hixson, TN 37343. Phone: (615) 842-5757.

RELATIONSHIP RENEWAL DAY. May 4. To salute and strengthen committed couples who value change and acceptance in the context of an ongoing relationship. Celebrants will mutually cite the challenges and changes met in the past year and offer each other well-deserved congratulations. Sponsor: Peter M. Rosenzweig, PhD, Cheerleader, Nondisposable Relationships, 713 Golf Mill Professional Bldg, Niles, IL 60648. Phone: (708) 297-5750.

RHODE ISLAND INDEPENDENCE DAY. May 4. Rhode Island abandoned allegiance to Great Britain on this day, 1776.

SHENANDOAH APPLE BLOSSOM FESTIVAL. May 4–7. Winchester, VA. Springtime extravaganza celebrating the blooming of apple trees. Est attendance: 300,000. For info: Shenandoah Apple Blossom Festival, 135 N Cameron St, Winchester, VA 22601. Phone: (703) 662-3863.

SPACE MILESTONE: *ATLANTIS* (US). May 4. First American planetary expedition in 11 years. Space shuttle was launched May 4, 1989. *Atlantis* was on 65th orbit when it landed May 8, 1989. Accomplished major objective of deploying the *Magellan* spacecraft on its way to Venus to map the planet's surface.

TYLER, JULIA GARDINER: BIRTH ANNIVERSARY. May 4. Second wife of John Tyler, 10th president of the US, born at Gardiners Island, NY, May 4, 1820. Died July 10, 1889.

WINSTON SELECT 500 RACE WEEK. May 4–7. Talladega, AL. Talladega Superspeedway hosts top-rated stock car races of the year. Sponsor: RJ Reynolds Tobacco Company's Winston brand. For info: Speedway Press Dept, PO Box 777, Talladega, AL 35160. Phone: (205) 362-2261.

BIRTHDAYS TODAY

Nickolas Ashford, 53, singer, songwriter ("Ain't No Mountain High Enough" with Valerie Simpson), born at Fairfield, SC, May 4, 1942.
Howard Da Silva, 86, actor (*The Lost Weekend*, "For the People"), born Cleveland, OH, May 4, 1909.
Maynard Ferguson, 67, musician, born at Verdun, Quebec, Canada, May 4, 1928.
Jackie Jackson (Sigmund Esco Jackson), 44, singer (Jackson 5), born at Gary, IN, May 4, 1951.
Randy Travis, 36, country western musician ("Forever and Ever, Amen"), born at Marshville, NC, May 4, 1959.
George F. Will, 54, editor, columnist, born at Champaign, IL, May 4, 1941.
Tammy Wynette (Virginia Wynette Pugh), 53, singer ("Stand by Your Man"), born Red Bay, MS, May 4, 1942.

MAY 5 — FRIDAY

125th Day — Remaining, 240

ASSOCIATION OF COLLEGE UNIONS: INTERNATIONAL BOWLING CHAMPIONSHIPS. May 5. Reno, NV. Top regional bowlers vie for doubles, singles and all-events titles, with the all-events champion earning a berth in the TEAM USA National Finals. Est attendance: 1,000. For info: American Bowling Congress, 5301 S 76th St, Greendale, WI 53129-1127. Phone: (414) 421-6400.

BASEBALL'S FIRST PERFECT GAME: ANNIVERSARY. May 5. On May 5, 1904, Denton T. "Cy" Young pitched baseball's first perfect game, not allowing a single opposing player to reach first base. Young's outstanding performance led the Boston Americans in a 3–0 victory over Philadelphia in the American League. The Cy Young Award for pitching was named in his honor.

BESSIE SMITH TRADITIONAL JAZZ FESTIVAL. May 5–7. Chattanooga Choo Choo, Chattanooga, TN. A traditional jazz festival with nationally acclaimed jazz bands presented by the Choo Choo City Jazz Society in the newly renovated Chattanooga Choo Choo Station House. For info: Mike Griffin, Bessie Smith Jazz Festival, 1425 Heritage Landing Dr, Chattanooga, TN 37405. Phone: (615) 266-0944.

BICYCLE TREK CLASSIC®. May 5–7. Spearfish, SD. The Bicycle Trek is a 2½-day adventure hosted by the American Lung Assn of South Dakota. Pedal through the Black Hills. Est attendance: 250. For info: Kathleen Wiebers, American Lung Assn of South Dakota, 208 E 13th, Sioux Falls, SD 57102. Phone: (605) 336-7222.

BIG TEN WOMEN'S GOLF CHAMPIONSHIP. May 5–7. University of Michigan, Ann Arbor, MI. For info: Big Ten Conference, 1500 W Higgins Rd, Park Ridge, IL 60068-6300. Phone: (708) 696-1010.

BLY, NELLY: BIRTH ANNIVERSARY. May 5. Nelly Bly, American journalist and women's rights advocate, whose real name was Elizabeth Cochrane Seaman, was born at Armstrong County, PA, May 5, 1867. Her career is said to have started as the result of her heated response to a newspaper article entitled "What Girls Are Good For." Called "the best reporter in America," she courageously wrote on the then-dangerous subjects of divorce, insanity, mashers, factory conditions, poverty and capital punishment. Died of pneumonia, at New York, NY, Jan 27, 1922. A marker was finally placed on her grave 56 years later, June 22, 1978.

BONZA BOTTLER DAY™. May 5. To celebrate when the number of the day is the same as the number of the month. Bonza Bottler Day™ is an excuse to have a party at least once a month. T-shirts available. For info: Elaine Fremont, 203 Waddell Rd, Taylors, SC 29687. Phone: (803) 244-2023.

BREAUX BRIDGE CRAWFISH FESTIVAL. May 5–7. Parc Hardy, Breaux Bridge, LA. During the first full weekend in May, the small town of Breaux Bridge bursts into a frenzy of excitement and celebration. This festival, which attracts over 200,000 people annually, includes live Cajun music and street dancing, parades, contests, arts and crafts and thousands of pounds of crawfish cooked every way imaginable. For info: BBCFA, PO Box 25, Breaux Bridge, LA 70517. Phone: (800) 346-1958.

CANADA: VANCOUVER INTERNATIONAL WILDLIFE ART SHOW. May 5–7. Vancouver Trade and Convention Center, Exhibit Hall A, Vancouver, British Columbia. Largest exhibition and sale of wildlife art in Western Canada features thousands of pieces of fine art originals, prints and sculpture. Free educational seminars and live animal and bird demonstrations. Est attendance: 7,000. For info: Robert Farrelly, Wes Productions, PO Box 11225, Tacoma, WA 98411. Phone: (206) 596-6728.

CERTIFIED PROFESSIONAL SECRETARY EXAMINATION. May 5–6. The examinations are given at various locations throughout the US. For info: Professional Secretaries Intl, PO Box 20404, 10502 NW Ambassador Dr, Kansas City, MO 64195-0404. Phone: (816) 891-6600.

CRASH OF 1893: ANNIVERSARY. May 5. Wall Street stock prices took a sudden drop May 5, 1893. By the end of the year 600 banks had closed. The Philadelphia and Reading, the Erie, the Northern Pacific, the Union Pacific and also the Atchison, Topeka and Santa Fe railroads had gone into receivership. 15,000 other businesses went into bankruptcy. Other than the "Great Depression" of the 1930s, this was the worst economic crisis in US history. 15–20 percent of the work force was unemployed.

☆ Chase's 1995 Calendar of Events ☆

May 5

FEST-I-FUN. May 5–7. Fort Mill, SC. Celebration of spring in downtown Fort Mill. Annually, the first full weekend in May. Est attendance: 15,000. For info: Michael W. Chase, Chamber of Commerce, Box 1357, Fort Mill, SC 29716. Phone: (803) 547-5900.

INTERNATIONAL TUBA DAY. May 5. To recognize tubists in musical organizations around the world who have to go through the hassle of handling a tuba. Annually, the first Friday in May. Est attendance: 300. For info: Sy Brandon, Music Dept, Millersville Univ, Millersville, PA 17551. Phone: (717) 872-3439.

ISRAEL: YOM HA'ATZMA'UT (INDEPENDENCE DAY). May 5. Hebrew calendar date: Iyar 5, 5755. Celebrates proclamation of independence from British mandatory rule by Palestinian Jews and establishment of the state of Israel and the provisional government on May 14, 1948 (Hebrew calendar date: Iyar 5, 5708).

JAPAN: CHILDREN'S DAY. May 5. National holiday. Observed on the fifth day of the fifth month each year.

JOHNSON, AMY: 65th FLIGHT ANNIVERSARY. May 5. On May 5, 1930, Yorkshire-born Amy Johnson began the first successful solo flight by a woman from England to Australia. Leaving Croydon Airport in a de Havilland D.H. 60 Gypsy Moth named *Jason*, she flew 9,960 miles to Port Darwin, Australia, arriving May 28. The song "Amy, Wonderful Amy" celebrated the fame of this "wonder girl of the air," who became a legend in her own lifetime. Amy Johnson, working as an air ferry pilot during World War II, was lost over the Thames Estuary in 1941.

JOSEY'S WORLD CHAMPION JUNIOR BARREL RACE. May 5–7. Josey's Ranch, Marshall, TX. Youth barrel-racing competition. Annually, the first weekend in May. Est attendance: 2,000. For info: Pam Whisenant, Dir of Conv and Visitor Development, Marshall Chamber of Commerce, PO Box 520, Marshall, TX 75671. Phone: (903) 935-7868.

KOREA: CHILDREN'S DAY. May 5. A time for families to take their children on excursions. The various parks and children's centers throughout the country are packed with excited and colorfully dressed children. A national holiday since 1975.

LASERSHOW '95. May 5–Aug 2. (Weekends through October.) Stone Mountain Park, Stone Mountain, GA. Brilliant laser beams bounce and dance across the mountain's face creating dramatic stories, comical characters and graphic images choreographed to popular music. Those attending will enjoy sitting on the lawn. Bring a blanket. Est attendance: 42,500. For info: Georgia's Stone Mountain Park, PR Dept, PO Box 778, Stone Mountain, GA 30086. Phone: (404) 498-5702.

LOBSTER RACE AND OYSTER PARADE. May 5. Aiken, SC. The world's only thoroughbred lobster races and oyster party. The lobsters are raced in a unique sea-salt-water-filled track called "Lobster Downs." Beach music, sand, gourmet seafood available. Oyster Parade at the "Mardi Claw" (our version of Mardi Gras) to highlight local landlocked maritime costumes will be held in Aiken's historical alley section. 11th annual Running of the Lobsters in Aiken. Est attendance: 10,000. For info: Greater Aiken Chamber of Commerce, Chuck Martin, PO Box 892, Aiken, SC 29802. Phone: (803) 641-1111.

LOUISIANA PASSION PLAY. May 5–Sept 30. Ruston, LA. Held every Friday and Saturday nights in May and September and every Thursday, Friday and Saturday nights in June, July and August. An outdoor theatre presentation/interpretation of some of the major events from the life of Christ—His birth, His teachings and His death and resurrection. For info: Louisiana Passion Play, 3010 S Vienna, Ruston, LA 71270. Phone: (318) 255-6277.

MAY FELLOWSHIP DAY. May 5. Theme: "All Have a Place at the Table." Luncheon worship services, based on the biblical account of Martha and Mary, provide an opportunity to explore and examine women's various roles. An ecumenical event that responds to local community needs. Annually, the first Friday in May. Est attendance: 64,000. Sponsor: Church Women United, 475 Riverside Dr, Rm 812, New York, NY 10115. Phone: (212) 870-2347.

MEXICO: CINCO DE MAYO. May 5. Mexican national holiday recognizing anniversary of Battle of Puebla, May 5, 1862, in which Mexican troops under General Ignacio Zaragoza, outnumbered three to one, defeated invading French forces of Napoleon III. Anniversary is observed by Mexicans everywhere with parades, festivals, dances and speeches.

NETHERLANDS: LIBERATION DAY. May 5. Marks liberation of the Netherlands from Nazi Germany, May 5, 1945.

POWER, TYRONE: BIRTH ANNIVERSARY. May 5. American actor, best known for his motion-picture action-adventure roles, Tyrone Power was born May 5, 1914, at Cincinnati, OH. Power died Nov 15, 1958, at Madrid, Spain.

PROJECT ACES DAY. May 5. A celebration of fitness and unity worldwide when All Children Exercise Simultaneously. "The World's Largest Exercise Class" takes place as schools across the nation and around the world hold fitness classes, assemblies and other fitness education events. Conducted in cooperation with the President's Council on Physical Fitness and Sports during National Physical Fitness and Sports Month. Sponsor: Yoohoo Chocolate. For info: Youth Fitness Coalition, Inc, PO Box 6452, Jersey City, NJ 07306-0452.

ST. JO HERITAGE TOUR. May 5–7. St. Joseph, MO. Tour 13 St. Joseph museums for one low price. Sponsor: Stan Harris, AKMA, 2818 Frederick, St. Joseph, MO 64506. Phone: (816) 233-7003.

SPACE MILESTONE: *FREEDOM 7* (US). May 5. First US astronaut, second man in space, Alan Shepard, Jr, projected 115 miles into space in suborbital flight reaching a speed of more than 5,000 miles per hour, May 5, 1961.

TENNESSEE CRAFTS FAIR. May 5–7. Centennial Park, Nashville, TN. Festival of fine crafts featuring 160 selected Tennessee craftspeople. 24th annual fair. Annually, the first full weekend in May. Est attendance: 50,000. Sponsor: Tennessee Assn of Craft Artists, PO Box 120066, Nashville, TN 37212. Phone: (615) 665-0502.

THAILAND: CORONATION DAY: ANNIVERSARY. May 5. Thailand.

TOAD SUCK DAZE. May 5–7. Downtown, Conway, AR. 14th annual event features toad jumping contests, concerts, parade, street dancing, tug-o-war, arts and crafts and more. Annually, the first full weekend in May. Est attendance: 100,000. For info: Shannon Jeffery, c/o Conway Chamber of Commerce, PO Box 1492, Conway, AR 72033. Phone: (501) 327-7788.

VALLEY OF FLOWERS FESTIVAL. May 5–7. Florissant, MO. Knights of Columbus Park carnival includes game booths, food, beer garden, entertainment, pony rides, and book fair. Civic Center includes craft fair, flea market plant sale, wine and cheese garden, barbecue, snack foods, pretty baby photo contest, high school band festival, antique and classic car show and Discoverland (for children). Annually, first full weekend in May. Sponsor: Valley of Flowers Committee, Dottie Bertolino, Coord, 475 N Hwy 67, Florissant, MO 63031. Phone: (314) 837-0033.

May 1995

S	M	T	W	T	F	S
	1	2	3	4	5	6
7	8	9	10	11	12	13
14	15	16	17	18	19	20
21	22	23	24	25	26	27
28	29	30	31			

WILLA CATHER SPRING CONFERENCE. May 5–6. Red Cloud, NE. A 1½-day event devoted to noted American author Willa Cather. Includes church service, tour of Cather Country, lunch, panel discussion, banquet, entertainment and keynote speaker. Paper session and evening entertainment on Friday; main conference on Saturday. Est attendance: 250. Sponsor: Willa Cather Pioneer Memorial & Educational Fdtn, 326 N Webster, Red Cloud, NE 68970. Phone: (402) 746-2653.

WORLD CHAMPIONSHIP CRIBBAGE TOURNAMENT. May 5–6. Plumas County Fairgrounds, Quincy, CA. Founded in 1972, the tournament now draws entrants from all over the country during the two-day event. Annually, the first weekend in May. Est attendance: 300. For info: Mike Taborski, Tournament Chair, PO Box B, Quincy, CA 95971. Phone: (916) 283-0800.

BIRTHDAYS TODAY

Pat Carroll, 68, actress (Emmy for "Caesar's Hour"; "The Ted Knight Show"), born at Shreveport, LA, May 5, 1927.

Alice Faye (Ann Leppert), 80, actress (*In Old Chicago, Lillian Russell*), born at New York, NY, May 5, 1915.

Michael Palin, 52, actor, comedian ("Monty Python's Flying Circus," *Life of Brian*), born Sheffield, Yorkshire, England, May 5, 1943.

Roger Rees, 51, actor, born at Aberystwyth, Wales, May 5, 1944.

Tina Yothers, 22, actress (Jennifer on "Family Ties"), born at Whittier, CA, May 5, 1973.

MAY 6 — SATURDAY
126th Day — Remaining, 239

AMERICAN BOWLING CONGRESS SENIORS TOURNAMENT. May 6–7. Reno, NV. State and provincial senior champions compete for titles in four divisions on the 40 specially installed lanes at the American Bowling Congress Tournament site. For info: ABC, Tom New, 5301 S 76th St, Greendale, WI 53129. Phone: (414) 421-6400.

ANN ARBOR SPRING ART FAIR. May 6–7. Ann Arbor, MI. To give the public the opportunity to view and invest in the finest arts and crafts. 17th annual. Est attendance: 15,000. For info: Audree Levy, 10629 Park Preston, Dallas, TX 75230. Phone: (214) 369-4345.

ASTRONOMY DAY. May 6. To take astronomy to the people. International Astronomy Day is observed on a Saturday near the first quarter moon between mid-April and mid-May. Co-sponsored by 15 astronomical organizations. (Note: Address below is effective 1-1-95; previous to that, 54 Jefferson Ave SE, ZIP 49503.) For info: Astronomy Day Headquarters, Gary E. Tomlinson, Coord, c/o Chaffee Planetarium, 272 Pearl NW, Grand Rapids, MI 49504. Phone: (616) 456-3987.

◈ **BABE RUTH'S FIRST MAJOR LEAGUE HOME RUN: 80th ANNIVERSARY.** May 6. On May 6, 1915, George Herman "Babe" Ruth, of the Boston Red Sox, hit his first major league home run in a game against the New York Yankees in New York.

BASKETBALL HALL OF FAME AWARDS DINNER. May 6. Springfield Marriott, Springfield, MA. National awards dinner hosted by the Basketball Hall of Fame. Annually, the Saturday before Hall of Fame enshrinement. Est attendance: 500. For info: Robin Jonathan Deutsch, Dir, Mktg/PR, Basketball Hall of Fame, 1150 W Columbus Ave, Springfield, MA 01101-0179. Phone: (413) 781-6500.

CINCO DE MAYO CELEBRATION. May 6. Leadville, CO. Annual celebration of Mexican heritage and culture. A colorful gathering with entertainment, parade, many special events and Mexican dinner and dance. Toll-free phone: (800) 933-3901. For info: James Chavez, PO Box 861, Leadville, CO 80461. Phone: (719) 486-3900.

DANDELION MAY FEST. May 6. Der Marktplatz, Dover, OH. 46th old-fashioned, family-oriented festival which includes the finals in a nationwide dandelion recipe contest judged by celebrities, a dandelion-picking contest, storytelling and presentations about dandelions and food booths featuring dishes made from dandelions, including dandelion coffee ice cream, dandelion pizza, and dandelion bread. Also, dandelion wine and dandelion jelly tasting, both country and German music, 5K fun run, volleyball tournament. Fax: (216) 343-8290. Est attendance: 8,000. For info: Anita Davis, Coord, Der Marktplatz-Breitenbach Wine Cellars, 5934 Old Route 39 NW, Dover, OH 44622. Phone: (216) 343-3603.

DEPOT LANE SINGERS. May 6 and Dec 2. Schoharie, NY. 75–100 singers celebrate the seasons in song. Est attendance: 500. Sponsor: Schoharie Colonial Heritage Assn, PO Box 554, Schoharie, NY 12157. Phone: (518) 295-7505.

EAST COAST CRAFT TRADE SHOW. May 6–7. (Workshops begin May 5.) Valley Forge Convention Center, King of Prussia, PA. A trade event, including seminars, workshops and exhibits. Exhibitors will be manufacturers and wholesalers of craft industry supplies/products; buyers will be wholesalers and craft/hobby store buyers. The show is jointly sponsored by the Mid-Atlantic Craft & Hobby Assn, Southeast Craft & Hobby Industries Assn, Northeast Craft & Hobby Assn and Hobby Industries Association. HIA is organizing the event. Est attendance: 1,600. For info: HIA, 319 E 54th St, PO Box 348, Elmwood Park, NJ 07407. Phone: (201) 794-1133.

EASTPARK CZECH FESTIVAL. May 6. Eastpark Plaza, Lincoln, NE. Czech music, singing and dancing; baked goods for sale. Est attendance: 300. For info: Denise Spale, Pres, 2345 Wildwood Pl, Lincoln, NE 68512. Phone: (402) 421-6233.

ENGLAND: HELSTON FURRY DANCE. May 6. The world-famous Helston Furry Dance is held each year on May 8 (except when the 8th is a Sunday or Monday, in which case it is held on the previous Saturday). Dancing around the streets of Helston, Cornwall, begins early in the morning and continues throughout the day. The "Furry" dance leaves Guildhall at the stroke of noon and winds its way in and out of many of the larger buildings. Final dance at 5:00 PM.

GILLIKINS OF OZ CONVENTION. May 6. To celebrate the birth of L. Frank Baum, Royal Historian of Oz. Est attendance: 20. Sponsor: Intl Wizard of Oz Club Inc, Fred M. Meyer, Secy, 220 N 11th St, Escanaba, MI 49829.

HALFWAY POINT OF SPRING. May 6. At 7:24 AM, EDT, on May 6, 1995, 46 days, 9 hours and 10 minutes of spring will have elapsed, and the equivalent will remain before June 21, 4:34 PM, EDT, which is the summer solstice and the beginning of summer.

May 6 ☆ *Chase's 1995 Calendar of Events* ☆

HINDENBURG DISASTER: ANNIVERSARY. May 6. At 7:20 PM, on May 6, 1937, the dirigible *Hindenburg* exploded as it approached the mooring mast at Lakehurst, NJ, after a trans-Atlantic voyage. Of its 97 passengers and crew, 36 died in the accident, which ended the dream of mass transportation via dirigible.

JEAN LAKE FESTIVAL. May 6–7. Pike Pioneer Museum, Troy, AL. An arts and crafts show with over 80 vendors displaying oil and watercolor paintings along with woodworking and textiles, food and entertainment. Annually, the first full weekend in May. Est attendance: 3,500. Sponsor: Jean Lake Festival, PO Box 972, Troy, AL 36081.

KENTUCKY DERBY. May 6. Churchill Downs, Louisville, KY. The running of "America's premier" thoroughbred horse race, inaugurated in 1875. First jewel in the "Triple Crown," traditionally followed by the Preakness (the second Saturday after Derby) and the Belmont Stakes (the fifth Saturday after Derby). Annually, the first Saturday in May. Est attendance: 130,000. For info: Churchill Downs, 700 Central Ave, Louisville, KY 40208. Phone: (502) 636-4400.

LANSING DAZE. May 6. Lansing High School and Lansing City Hall, Lansing, KS. Features pancake breakfast, ice cream social, car show, craft show, quilt show, yard sale, square dancing and entertainment. Annually, the first weekend in May. Est attendance: 5,000. For info: Leavenworth Conv and Visitors Bureau, 518 Shawnee, Leavenworth, KS 66048. Phone: (800) 844-4114.

LEE-JACKSON LACROSSE CLASSIC. May 6. Virginia Military Institute, Lexington, VA. Community-wide event when the Washington and Lee University lacrosse team meets the Virginia Military Institute team. 2 PM. For info: Lexington Visitors Bureau, 106 E Washington St, Lexington, VA 24450. Phone: (703) 463-3777.

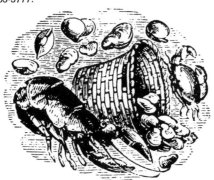

LOW COUNTRY SHRIMP FESTIVAL. May 6. McClellanville, SC. Seafood, arts, crafts, civic display, entertainment and blessing of the fleet. Annually, the first Saturday in May. Est attendance: 2,000. For info: The Archibald Rutledge Academy, PO Box 520, McClellanville, SC 29458. Phone: (803) 887-3323.

MALTA: CARNIVAL. May 6–7. Valletta. Festival dates from 1535 when Knights of St. John introduced Carnival in Malta. Dancing, bands, decorated trucks and grotesque masks. Annually, the first weekend after May 1.

MID-ATLANTIC STUNT KITE CHAMPIONSHIPS. May 6–7. On the beach, Ocean City, MD. World-class kite flyers vie for top honors in Olympic-style kiting events. Sat 9–9, Sun 9–5. Annually, the first weekend in May. Est attendance: 50,000. For info: Roger Chewning, Pres, Sky Festival Productions, 1459 Makefield Rd, Yardley, PA 19067.

May 1995

S	M	T	W	T	F	S
	1	2	3	4	5	6
7	8	9	10	11	12	13
14	15	16	17	18	19	20
21	22	23	24	25	26	27
28	29	30	31			

MIDWEST DOLL SHOW AND SALE. May 6–7. Zion, IL. Doll-related items displayed by dealers and collectors from throughout the Midwest. Est attendance: 1,200. Sponsor: Zion Park District, Richard Walker, Special Events Coord, 2400 Dowie Memorial, Zion, IL 60099. Phone: (708) 746-5500.

MILLFEST. May 6. Old Mill Museum Complex, Lindsborg, KS. Annual running of the Mill (we do not grind wheat) with guided tours of the Mill (visitors must be over 12 years old to tour the Mill), plus entertainment and arts and crafts demonstrations on the grounds. Chamber of Commerce puts on a noon Bar-B-Que. Charge for Bar-B-Que and Museum/Mill tours, entertainment is free. Annually, the first Saturday in May. Est attendance: 400. Sponsor: McPherson County Old Mill Museum, PO Box 94, 120 Mill St, Lindsborg, KS 67456. Phone: (913) 227-3595.

NATIONAL HOMEBREW DAY. May 6. A national celebration of more than 1.5 million amateur homebrewers. National Homebrew Day will be celebrated in conjunction with the 17th annual National Homebrew Competition. Sponsor: American Homebrewers Assn, Box 1679, Boulder, CO 80306. Phone: (303) 447-0816.

NATIONAL NURSES WEEK. May 6–12. A week to honor the outstanding efforts of nurses everywhere to strengthen the health of the nation. Annually, beginning on May 6, National Nurses Day, and ending on May 12, Florence Nightingale's birthday. For info: Amer Nurses Assn, 600 Maryland Ave SW, Ste 100W, Washington, DC 20024.

PEARY, ROBERT E.: BIRTH ANNIVERSARY. May 6. Arctic explorer Robert E. Peary was born at Cresson, PA, on May 6, 1856. After graduating college, he served as a cartographic draftsman in the US Coast and Geodetic Survey for two years prior to joining the US Navy's Corp of Civil Engineers in 1881. The discoverer of the North Pole began his life's work as an explorer in tropical climates as he served as subchief of the Inter-Ocean Canal Survey in Nicaragua. After reading a report on the inland ice of Greenland, Peary became attracted to the Arctic. He organized and led eight Arctic expeditions and is credited with the verification of Greenland's island formation, proving that the polar ice cap extended beyond 82 degrees north latitude, and the discovery of the Melville meteorite on Melville Bay, in addition to his famous discovery of the North Pole on Apr 6, 1909. Peary died on Feb 20, 1920, at Washington, DC.

PENN, JOHN: BIRTH ANNIVERSARY. May 6. Signer of the Declaration of Independence, born at Caroline County, VA, May 6, 1740. Died Sept 14, 1788.

PHILIPPINES: ARAW NG KAGITINGAN. May 6. Legal holiday in the Philippines.

PROGRESSIVE MUSIC SPECTACULAR (PMS). May 6. Historic City Market, Roanoke, VA. An event dedicated to Alternative Music in the Mid-Atlantic Region. Annually, the first Saturday in May. Est attendance: 12,500. Sponsor: E. Laban Johnson, Special Events Coord, City of Roanoke Virginia, 210 Reserve Ave, SW, Roanoke, VA 24016. Phone: (703) 981-2889.

ROAN MOUNTAIN WILDFLOWER TOURS AND BIRDWALKS. May 6–7. Roan Mountain State Park, Roan Mountain, TN. Wildflower and nature identification and inspiration. Annually, the first full weekend in May. Est attendance: 1,200. For info: Jennifer Wilson, Ranger Naturalist, Roan Mtn State Park, Rte 1, Box 236, Roan Mountain, TN 37687. Phone: (615) 772-3303.

SACRED HARP SINGING. May 6. Burritt Museum and Park, Huntsville, AL. Singers from across Alabama and neighboring states will participate in sacred harp singing, a type of religious folk music that began in the 1700s. For info: Pat Robertson, Dir, 3101 Burritt Dr, Huntsville, AL 35801. Phone: (205) 536-2882.

SONGS OF MY PEOPLE. May 6–Jun 18. Fort Wayne Museum of Art, Fort Wayne, IN. See Mar 4 for full description. Call or write the Smithsonian for other dates and venues. For info: Smithsonian Institution Traveling Exhibition Service, 1100 Jef-

☆ Chase's 1995 Calendar of Events ☆ May 6–7

ferson Dr SW, Ste 3146, Washington, DC 20560. Phone: (202) 357-2700.

SPRINGFEST. May 6. Means and Clark Park, Weatherford, OK. Craft fair with concession booths and entertainment all day. Annually, first Saturday in May. Est attendance: 1,000. For info: William Hancock, Exec Dir/Cathy Burbank, Admin Asst, Weatherford Chamber of Commerce, PO Box 729, Weatherford, OK 73096. Phone: (800) 725-7744.

TAGORE, RABINDRANATH: BIRTH ANNIVERSARY. May 6. Hindu poet, mystic and musical composer was born at Calcutta, India, May 6, 1861. Received Nobel Prize (literature) in 1913. Died at Calcutta, Aug 7, 1941. His birthday is observed in Bangladesh on the 25th day of the Bengali month of Baishakha (second week of May), when the poet laureate is honored with songs, dances and discussions of his works.

TOWSONTOWN SPRING FESTIVAL. May 6–7. Towson, MD. Five stages with continuous entertainment. 400 food, craft and display vendors on the street. Art and photography exhibit, antique auto display, trading card show and flea market. Annually, the first Saturday and Sunday in May. Est attendance: 250,000. For info: Towsontown Spring Fest, PO Box 10115, Towson, MD 21285. Phone: (410) 825-1144.

◈ **VALENTINO, RUDOLPH: 100th BIRTH ANNIVERSARY.** May 6, 1895. Rodolpho Alfonzo Rafaello Pietro Filiberto Guglielmi Di Valentina D'Antonguolla, whose professional name was Rudolph Valentino, was born at Castelleneta, Italy. Popular cinema actor. Press reports claim that "at least one weeping veiled woman in black has brought flowers to his tomb" (at Hollywood Memorial Park) on the anniversary every year since his death at New York, NY, Aug 23, 1926.

VIRGINIA GOLD CUP. May 6. Great Meadow, The Plains, VA. Steeplechasing, which began in Ireland in 1762 when two horsemen held a cross-country match race to a far away church steeple, is today one of the fastest growing spectator sports in the US. Great Meadow is the largest steeplechase course in the country with a spectacular hillside amphitheater. The Virginia Gold Cup race, sponsored by BMW, is run over a challenging 4-mile post and rail course of 23 four-ft fences. Advance tickets only—phone: (703) 347-2612. Conducted by the Virginia Gold Cup Assn for benefit of free year-round use of Great Meadow by non-profit community activities. 45 minutes from Washington D.C. Annually, the first Saturday in May. Est attendance: 50,000. For info: Virginia Gold Cup Assn, Box 840, Warrenton, VA 22186. Phone: (703) 347-2612.

WINE & ROSES FEST '95 WITH GRAPE STOMP COMPETITION. May 6. Messina Hof Wine Cellars, Bryan, TX. Join the Grape Stomp Competition and get purple feet! Relax on a vineyard hayride, take a winery tour and taste award-winning Messina Hof Wines. See the 12th annual Texas Artist Competition Exhibit. Arts and Crafts and live entertainment. Est attendance: 2,000. For info: Amy Ping, Mktg Dir, 4545 Old Reliance Rd, Bryan, TX 77808. Phone: (409) 778-9463.

BIRTHDAYS TODAY

Willie Mays, 64, baseball Hall of Famer, born at Fairfield, AL, May 6, 1931.

Bob Seger, 50, musician, singer ("Night Moves," "Travelin' Man"), born Ann Arbor, MI, May 6, 1945.

Richard C. Shelby, 61, US Senator (D, Alabama), born at Birmingham, AL, May 6, 1934.

MAY 7 — SUNDAY

127th Day — Remaining, 238

ANTEC '95. May 7–11. The Hynes Convention Center, Boston, MA. Held in conjunction with the Society of Plastics Engineers' 53nd Annual Technical Conference (ANTEC) to promote national awareness of the significant benefits this diverse family of materials brings to all facets of society. Est attendance: 4,900. Sponsor: James P. Toner, Society of Plastics Engineers, 14 Fairfield Dr, Brookfield, CT 06804-0403. Phone: (203) 775-0471.

BATTLE OF DIEN BIEN PHU ENDS: ANNIVERSARY. After 55 days the Battle of Dien Bien Phu ended on May 7, 1954, when Vietnamese insurgents overran the French forces. This defeat brought French colonial rule in Indochina to an end.

BE KIND TO ANIMALS WEEK. May 7–13. To promote kindness and humane care toward animals. Annually, the first full week of May. Sponsor: American Humane Assn, 63 Inverness Dr E, Englewood, CO 80112. Phone: (303) 792-9900.

BEAUFORT SCALE DAY. May 7. A day to honor the British naval officer, Sir Francis Beaufort, who devised (1805) a scale of wind force from 0 (calm) to 12 (hurricane) that was based on observation, not requiring any special instruments. The scale was adopted for international use in 1874 and has since been enlarged and refined. Beaufort was born at Flower Hill, Meath, Ireland, on May 7, 1774, and died at Brighton, England, on Dec 17, 1857.

BRAHMS, JOHANNES: BIRTH ANNIVERSARY. May 7. Regarded as one of the greatest composers of 19th-century music, Johannes Brahms was born at Hamburg, Germany, May 7, 1833. Brahms wrote in most areas of music and is especially noted for his vocal *Requiem*, *Symphony No 1 in C minor*, and *Symphony No 4 in E minor*. Brahms had the misfortune of falling in love with Robert Schumann's wife after Schumann lost his mind and was institutionalized. Faithful to her husband both in life and after his death, Clara Schumann and Brahms reached an understanding and remained fast friends for the rest of their lives. Brahms never married. Johannes Brahms died at Vienna, Apr 3, 1897.

BROWNING, ROBERT: BIRTH ANNIVERSARY. May 7. English poet, born May 7, 1812. One of the most famous poets of England. Known for his dramatic monologues. Died at Venice, Italy, Dec 12, 1889.

BUCKS COUNTY KITE DAY. May 7. Core Creek Park, Langhorne, PA. Bring your lawn chairs and blanket, pack a lunch and spend the day. Visit the information table next to the stage for information on the various kite activities and clubs. Homemade kites, award-winning kites, stunters, team fliers, trains and much more. Tips and techniques from some of the best. Est attendance: 15,000. Sponsor: Bucks Cnty Dept of Pks and Rec, 901 E Bridgetown Pike, Langhorne, PA 19047. Phone: (215) 757-0571.

CINCO DE MAYO FIESTA. May 7. Traders Village, Grand Prairie, TX. Salute to Mexican-American heritage with folklorico dancers, music, food. Est attendance: 49,000. For info: Traders Village, 2602 Mayfield Rd, Grand Prairie, TX 75052. Phone: (214) 647-2331.

CONSERVE WATER/DETECT-A-LEAK WEEK. May 7–13. To help everyone learn why it is important to conserve our water and how to help accomplish this goal. Sponsor: American Leak Detection, c/o S&S Public Relations, Inc, 400 Skokie Blvd, Ste 200, Northbrook, IL 60062. Phone: (708) 291-1616.

COOPER, GARY: BIRTH ANNIVERSARY. May 7. Frank James Cooper was born on May 7, 1901, at Helena, MT. He changed his name to Gary at the start of his movie career. His first major role was in *The Winning of Barbara Worth* in 1926. Many films followed, including *Wings*, *The Virginian*, *Desire*, *The Plainsman*, *Beau Geste*, *Meet John Doe* and *Sergeant York* (for which he won his first Academy Award), *High Noon* (winning his second Oscar for best actor), *The Court Martial of Billy Mitchell* and *The Wreck of the Mary Deare* among others. He died May 13, 1961, at Hollywood, CA.

May 7 ☆ Chase's 1995 Calendar of Events ☆

CORN PLANTING CEREMONY. May 7. Museum of Indian Culture, Allentown, PA. Lenape (Delaware Indian) ceremony, arts, crafts, demonstrations, foods and family fun. Est attendance: 3,000. Sponsor: Museum of Indian Culture, Lenni Lenape Historical Soc, RD# 2, Fish Hatchery Rd, Allentown, PA 18103-9801. Phone: (610) 797-2121.

FIRST PRESIDENTIAL INAUGURAL BALL: ANNIVERSARY. May 7. Commemorating the inauguration of George Washington, the first Presidential Inaugural Ball was held in New York, NY, May 7, 1789.

FLEXIBLE WORK ARRANGEMENTS WEEK. May 7–13. To promote experimentation with rearranged work schedules and working at home or at a neighborhood work center. Annually, the week beginning with the first Sunday in May. For info: Maggi Payment, Dir, Center for Worktime Options, 1043 University Ave, # 192, San Diego, CA 92103. Phone: (619) 232-0404.

FRANKENMUTH SKYFEST. May 7. Frankenmuth, MI. To encourage participation in a healthy outdoor sport that adapts to all age groups. 1995 will be the 14th annual Skyfest. Annually, the first Sunday in May. Est attendance: 4,500. For info: Audrey Fischer, Kite Kraft/School Haus Square, Frankenmuth, MI 48734. Phone: (517) 652-2961.

◆ **GERMANY'S FIRST SURRENDER: 50th ANNIVERSARY.** May 7, 1945. Russian, American, British and French ranking officers crowded into a second-floor recreation room of a small redbrick schoolhouse (which served as Eisenhower's headquarters) in Reims, Germany. Representing Germany, Field Marshall Alfred Jodl signed an unconditional surrender of all German fighting forces. After a signing that took almost 40 minutes, Jodl was ushered into Eisenhower's presence. The American general asked the German if he fully understood what he had signed and informed Jodl that he would be held personally responsible for any deviation from the terms of the surrender, including the requirement that German commanders sign a formal surrender to the USSR at a time and place determined by that government.

GOODWILL INDUSTRIES WEEK. May 7–13. To call national attention to Goodwill Industries as a leader in Job Training and employment services for people with disabilities and for others with special needs. Annually, the first full week in May. Fax: (301) 530-1516. Sponsor: Goodwill Industries International, Communications Dept, 9200 Wisconsin Ave, Bethesda, MD 20814. Phone: (301) 530-6500.

GOVERNOR'S BAY BRIDGE RUN. May 7. Sandy Point State Park, Annapolis, MD. "Maryland's Most Spectacular Run" 10K foot race across Chesapeake Bay Bridge, dramatic views of the bay and historic Annapolis; designer premium for all finishers; limited to 3,000 entries. Annually, the first Sunday in May. Since 1985. Est attendance: 3,000. For info: Annapolis Striders, Inc, PO Box 187, Annapolis, MD 21404-0187. Phone: (410) 268-1165.

May 1995

S	M	T	W	T	F	S
	1	2	3	4	5	6
7	8	9	10	11	12	13
14	15	16	17	18	19	20
21	22	23	24	25	26	27
28	29	30	31			

HOMESPUN HISTORY DAY. May 7. General Crook House Museum, Fort Omaha Campus MCC College, Omaha, NE. Open House of 1880s Commander General George Crook and wife Mary. Restored Victorian mansion; costumed greeters; living history activities; library/archives photo and paper memorabilia displays; genealogy and research materials; exhibits of homesteader communication along the Oregon Trail; cookies being baked from old-time recipes for sampling enjoyment; 1st Regiment Nebraska Volunteers Civil War Encampment Group serve as greeters. Annually, the first Sunday in May. Sponsor: Historical Soc of Douglas County, PO Box 11398, CAE, Omaha, NE 68111-0398. Phone: (402) 455-9990.

HONG KONG: BIRTHDAY OF LORD BUDDHA. May 7. Religious observances are held in Buddhist temples, and Buddha's statue is bathed. Po Lin Monastery and the other monasteries on Lantau Island are visited by many worshipers during the festival. Annually, the eighth day of fourth lunar month.

***LUSITANIA* SINKING: 80th ANNIVERSARY.** May 7. British passenger liner *Lusitania*, on return trip from New York to Liverpool, carrying nearly 2,000 passengers, was torpedoed by German submarine off the coast of Ireland, sinking within minutes on May 7, 1915; 1,198 lives lost. US President Wilson sent note of protest to Berlin on May 13, but Germany, which had issued warning in advance, pointed to *Lusitania*'s cargo of ammunition for Britain. US maintained "neutrality," for the time being.

MOON PHASE: FIRST QUARTER. May 7. Moon enters First Quarter phase at 5:44 PM, EDT.

NATIONAL FAMILY WEEK. May 7–13. Traditionally the first Sunday and the first full week in May are observed as National Family Week in many Christian churches.

NATIONAL HOSPITAL WEEK. May 7–13. To focus public attention on the many contributions hospitals make to their communities. Sponsor: American Hospital Assn, PO Box 1100, Sacramento, CA 95812-1100. Phone: (916) 552-7504.

NATIONAL PET WEEK. May 7–13. To promote public awareness of veterinary medical service for animal health and care. Annually, the first full week in May. Sponsor: Auxiliary to the American Veterinary Medical Assn, Chris Kanalas, Exec Secy, 1931 N Meacham Rd, Schaumburg, IL 60173. Phone: (708) 397-6651.

NATIONAL POSTCARD WEEK. May 7–13. To advertise use of picture postcards for correspondence and collecting. Annually, the first full week of May. Sponsor: Postcard Historical Soc, John H. McClintock, Box 1765, Manassas, VA 22110. Phone: (703) 368-2757.

NATIONAL SELF-HELP BOOK WEEK. May 7–13. Promote your own self-improvement in the areas of mental and physical health by reading and following advice in an appropriate self-help book. Further, to acknowledge contributions made to self-improvement by authors and publishers of these books. Sponsor: Deaconess Press, Service of Fairview Medical Center. For info: Robert Jackson, Pres, Mktg Services Consultant, 322 First Ave N, Ste 201, Minneapolis, MN 55401. Phone: (612) 375-0141.

NATIONAL TOURISM WEEK. May 7–13. To promote the awareness of the economic, social and cultural importance of tourism. For info: Natl Travel and Tourism Awareness Council, Two Lafayette Ctr, 1133 21st St NW, Ste 800, Washington, DC 20036.

O. HENRY PUN-OFF (WORLD CHAMPIONSHIP). May 7. The O. Henry Museum, Austin, TX. Pundits and punographers match wits for a wordy cause in two separate pun-filled competitions (Punniest of Show and High Lis & Low Puns). Sponsors: Friends of the O. Henry Museum and Punsters United Nearly Yearly (PUNY). Est attendance: 2,000. For info: O. Henry Museum, 409 E Fifth St, Austin, TX 78701. Phone: (512) 472-1903.

PALM HARBOR DAY ARTS, CRAFTS AND MUSIC FESTIVAL. May 7. "Pop" Stansell Park, Palm Harbor, FL. To celebrate the 117th birthday of Palm Harbor. Juried arts and crafts show. Est attendance: 3,000. For info: Palm Harbor Chamber of Commerce, 31954 US 19N, Palm Harbor, FL 34684. Phone: (813) 784-4287.

PASTE-UP DAY. May 7. Honors all paste-up artists who work for newspaper, magazine and book publishers. For info: Foto News, Julie K. Sjuggerud, PO Box 606, Merrill, WI 54452.

PLOWING MATCH. May 7. Woodstock, VT. This annual rite of spring features a horse- and ox-drawn plowing competition and demonstrations of various plowing techniques. Est attendance: 9,500. Sponsor: Billings Farm Museum, PO Box 489, Woodstock, VT 05091. Phone: (802) 457-2355.

PTA TEACHER APPRECIATION WEEK. May 7–13. PTAs across the country conduct activities to strengthen respect and support for teachers and the teaching profession. Sponsor: Natl PTA, 330 N Wabash Ave, Ste 2100, Chicago, IL 60611. Phone: (312) 670-6782.

SMALL BUSINESS WEEK. May 7–13. To honor the 20 million small businesses in the US. Annually, the first full week in May. For info: Small Business Administration, Info Services, 409 3rd St SW, 7th Fl, Washington, DC 20416. Phone: (800) 232-0082.

SOUTHERN APPALACHIAN DULCIMER FESTIVAL. May 7. Tannehill Historical State Park, McCalla, AL. Festival highlighting old-time dulcimer music. Est attendance: 2,000. For info: Vicki Gentry, Tannehill Historical State Park, 12632 Confederate Parkway, McCalla, AL 35111. Phone: (205) 477-5711.

UNIVERSAL FAMILY WEEK. May 7–13. To stress the importance of the fundamental role of good families in strengthening humankind. For complete info, send $1 to cover expense of printing, handling and postage. Annually, the second full week in May. For info: Intl Soc of Friendship and Goodwill, Dr. Stanley Drake, Pres, 9538 Summerfield St, Spring Valley, San Diego Cty, CA 91977-2852. Phone: (619) 466-8882.

WINTERTHUR POINT-TO-POINT. May 7. Winterthur Museum, Garden, and Library, Winterthur, DE. A day of amateur steeplechase racing. Includes a tailgate picnic competition and an antique carriage parade. Annually, the Sunday of the first full weekend in May. Est attendance: 15,000. For info: Winterthur Point-to-Point, Winterthur, DE 19735. Phone: (302) 888-4600.

BIRTHDAYS TODAY

Theresa Brewer, 64, singer ("[Open Up Your Heart and] Let the Sun Shine In"), born Toledo, OH, May 7, 1931.
Pete V. Domenici, 63, US Senator (R, New Mexico), born at Albuquerque, NM, May 7, 1932.
Amy Heckerling, 41, filmmaker, born at New York, NY, May 7, 1954.
Edwin H. Land, 86, inventor, born at Bridgeport, CT, May 7, 1909.
Darren McGavin, 73, actor (*The Man with the Golden Arm*, "Mickey Spillane's Mike Hammer"), born at Spokane, WA, May 7, 1922.
Johnny Unitas, 62, former football player, born at Pittsburgh, PA, May 7, 1933.

MAY 8 — MONDAY
128th Day — Remaining, 237

BASKETBALL HALL OF FAME ENSHRINEMENT CEREMONIES. May 8. Springfield, MA. New electees to the Basketball Hall of Fame will be enshrined. Est attendance: 1,500. For info: Basketball Hall of Fame, PO Box 179, Springfield, MA 01101-0179. Phone: (413) 781-6500.

DUNANT, JEAN HENRI: BIRTH ANNIVERSARY. May 8. Swiss author and philanthropist, founder of the Red Cross Society, was born at Geneva, Switzerland, May 8, 1828. Nobel prize winner in 1901. Died at Heiden, Switzerland, Oct 30, 1910.

◆ **GERMANY'S SECOND SURRENDER: 50th ANNIVERSARY.** May 8, 1945. Stalin refused to recognize the document of unconditional surrender signed at Reims, so a second signing was held at Berlin. The event was turned into an elaborate formal ceremony by the Soviets who after all had lost some ten million lives during the war. As in the Reims document, the end of hostilities was set for 12:01 AM local time on May 9.

LAVOISIER, ANTOINE LAURENT: EXECUTION ANNIVERSARY. May 8. French chemist and the "father of modern chemistry." Especially noted for having first explained the real nature of combustion and for showing that matter is not destroyed in chemical reactions. Born at Paris, France on Aug 26, 1743, Lavoisier was guillotined on May 8, 1794, at the Place de la Revolution for his former position as a tax collector. The Revolutionary Tribunal is reported to have responded to a plea to spare his life with the statement: "We need no more scientists in France."

NATIONAL HERB WEEK. May 8–14. A time to focus on the use and history of herbs—past and present. Annually, the Monday–Sunday in May ending on Mother's Day. Sponsor: Intl Herb Growers & Marketers Assn (IHGMA), 1202 Allanson Rd, Mundelein, IL 60060. Phone: (708) 949-4372.

NO SOCKS DAY. May 8. If we give up wearing socks for one day, it will mean a little less laundry, thereby contributing to the betterment of the environment. Besides, we will all feel a bit freer, at least for one day. Annually, May 8. Phone: (212) 388-8673 or (717) 274-8451. Sponsor: Wellness Permission League, Thomas and Ruth Roy, 2105 Water St, Lebanon, PA 17046.

SEATTLE INTERNATIONAL CHILDREN'S FESTIVAL. May 8–14. Seattle Center Urban Park, Seattle, WA. Performing arts companies from around the world will perform for young audiences. All disciplines encompassed, including music, dance, drama, comedy, clowning, puppetry, mime, acrobatics, juggling. Est attendance: 65,000. For info: Marilyn Raichle, Exec Dir, Seattle Intl Children's Festival, 305 Harrison, Seattle, WA 98109. Phone: (206) 684-7338.

SPALDING ENSHRINEMENT CELEBRITY GOLF TOURNAMENT. May 8. Crestview Country Club, Agawam, MA. Features Basketball Hall of Famers, celebrities and invited guests in four-man teams. For info: Basketball Hall of Fame, Box 179, 1150 W Columbus Ave, Springfield, MA 01101-0179.

TRUMAN, HARRY S.: BIRTH ANNIVERSARY. May 8. The 33rd president of the US, succeeded to that office upon the death of Franklin D. Roosevelt, Apr 12, 1945, and served until Jan 20, 1953. Born at Lamar, MO, May 8, 1884, Truman was the last of the nine US presidents who did not attend college. Affectionately nicknamed "Give 'em Hell Harry" by admirers. The only employment of atomic bombs on populated areas during war occurred during his presidency (Aug 6 and Aug 9, 1945, at Hiroshima and Nagasaki, Japan). Truman died at Kansas City, MO, on Dec 26, 1972.

◆ **V-E DAY: 50th ANNIVERSARY.** May 8, 1945. Victory in Europe Day commemorates unconditional surrender of Germany to Allied Forces. The surrender document was signed by German representatives at General Dwight D. Eisenhower's headquarters at Reims to become effective, and hostilities to end, at one minute past midnight on May 9, 1945. President Harry S. Truman on May 8 declared May 9, 1945, to be "V-E Day." A separate German surrender to the USSR was signed at Karlshorst, near Berlin, on May 8. See also: "Russia: Victory Day" (May 9).

May 8-10 ☆ Chase's 1995 Calendar of Events ☆

WORLD RED CROSS DAY. May 8. A day commemorating the birth of Jean Henri Dunant, the Swiss founder of the international Red Cross movement in 1863, and recognizing the humanitarian work of the Red Cross around the world. For info on activities in your area, contact your local Red Cross chapter. Sponsor: Public Inquiry/Historical Resources, American Red Cross Natl Headquarters, Washington, DC 20006-5399. Phone: (202) 737-8300.

BIRTHDAYS TODAY

David Attenborough, 69, author, naturalist (*Life on Earth, Trials of Life*), born at London, England, May 8, 1926.
Peter Benchley, 55, author, journalist (*Rummies, Jaws*), born at New York, NY, May 8, 1940.
Dennis DeConcini, 58, former US Senator (D, Arizona), born at Tucson, AZ, May 8, 1937.
Melissa Gilbert, 31, actress (Laura on "Little House on the Prairie"; Helen Keller in the *The Miracle Worker* remake), born Los Angeles, CA, May 8, 1964.
Ronnie (Ronald Mandel) Lott, 36, football player, born Albuquerque, NM, May 8, 1959.
Thomas Pynchon, 58, writer (*V., Slow Learner: Early Stories*), born at Glen Cove, NY, May 8, 1937.
Don Rickles, 69, comedian, actor (*Blazing Saddles*, "The Dean Martin Show"), born New York, NY, May 8, 1926.
Toni Tennille, 52, singer (along with husband Daryl Dragon made up singing duo Captain and Tennille; "Love Will Keep Us Together"), born at Montgomery, AL, May 8, 1943.
James R. Thompson, 59, former Governor of Illinois (R), born at Chicago, IL, May 8, 1936.

MAY 9 — TUESDAY
129th Day — Remaining, 236

BOYD, BELLE: BIRTH ANNIVERSARY. May 9. Notorious Confederate spy who later became an actress and lecturer was born at Martinsburg, VA, on May 9, 1843. Author of the book *Belle Boyd in Camp and Prison*, she died June 11, 1900.

BROWN, JOHN: BIRTH ANNIVERSARY. May 9. Abolitionist leader born at Torrington, CT, May 9, 1800. Hanged Dec 2, 1859. Leader of attack on Harpers Ferry, Oct 16, 1859, which was intended to give impetus to movement for escape and freedom for slaves. His aim was frustrated and in fact resulted in increased polarization and sectional animosity. Legendary martyr of the abolitionist movement.

EGYPT: GRAND BAIRAM HOLIDAY. May 9-13. Muslims celebrate the sacrifice by the Prophet Abraham of killing a ram in place of his son, Ismail. Preceded by pilgrimage to Mecca. Dates vary according to the Muslim Lunar Calendar. For info: Egyptian Tourist Authority, 645 N Michigan Ave, Ste 829, Chicago, IL 60611. Phone: (312) 280-4666.

EUROPEAN COMMUNITIES: ANNIVERSARY OBSERVANCE. May 9. European Union, commemorates the May 9, 1950, announcement by French statesman Robert Schuman of the "Schuman Plan" for establishing a single authority for production of coal, iron and steel in France and Germany. For info: European Community Info Service, Delegation Commission European Community, 2100 M St NW, Ste 707, Washington, DC 20037.

NATIONAL BIKE TO WORK DAY. May 9, 1995. At the state or local level, Bike to Work events are conducted by small and large businesses, city governments, bicycle clubs and environmental groups. Annually, the second Tuesday in May. Sponsor: League of American Bicyclists, Susan Jones, Education Dir, 190 W Ostend St, Ste 120, Baltimore, MD 21230-3755. Phone: (410) 539-3399.

NATIONAL TEACHER DAY. May 9. To pay tribute to American educators, sponsored by the National Education Association, Teacher Day falls during the National PTA's Teacher Appreciation Week. Local communities and organizations are encouraged to use this opportunity to honor those who influence and inspire the next generation through their work. Annually, first Tuesday of the first full week in May. Sponsor: Natl Education Assn (NEA), 1201 16th St NW, Washington, DC 20036. Phone: (202) 822-7200.

◆ **RUSSIA: VICTORY DAY: 50th ANNIVERSARY.** May 9. National holiday observed annually to commemorate the Allied Forces defeat of Nazi Germany in World War II and to honor the 20 million Soviet people who died in that war. Among the traditional rituals of this celebration, veterans renew wartime friendships, reminisce and swap stories, sing some of the frontline songs and visit national cemeteries. Hostilities ceased and the German surrender became effective at one minute after midnight on May 9, 1945. See also: "V-E Day" (May 8).

SCHOOL FAMILY DAY. May 9. Some states observe Teacher Day as School Family Day to salute and thank the entire education family—administrators, teachers, support personnel, volunteers and parents. Community groups are urged to plan local observances. Sponsor: Natl Education Assn (NEA), 1201 16th St NW, Washington, DC 20036. Phone: (202) 822-7200.

UNITED WE DANCE. May 9-14. War Memorial Opera House, San Francisco, CA. The San Francisco Ballet presents an international festival celebration of global similarities and differences expressed through the art of dance. Sponsor: San Francisco Ballet. Contact: (415) 861-1177. For info: UN50 Committee, 312 Sutter St, Ste 610, San Francisco, CA 94108. Phone: (415) 989-1995.

BIRTHDAYS TODAY

Candice Bergen, 49, actress (*Starting Over*, "Murphy Brown"), daughter of ventriloquist Edgar Bergen, born Beverly Hills, CA, May 9, 1946.
Albert Finney, 59, actor (*Tom Jones, Shoot the Moon, Annie, The Dresser*), Salford, England, May 9, 1936.
Tony Gwynn, 35, baseball player, born at Los Angeles, CA, May 9, 1960.
Glenda Jackson, 58, actress (Oscars for *Women in Love* and *Touch of Class*), born at Cheshire, England, May 9, 1937.
Billy Joel, 46, singer, composer ("It's Still Rock and Roll to Me," "Just the Way You Are"), born Hicksville, NY, May 9, 1949.
Mike Wallace, 77, TV journalist ("60 Minutes"), born at Brookline, MA, May 9, 1918.

MAY 10 — WEDNESDAY
130th Day — Remaining, 235

BONNIE BLUE NATIONAL HORSE SHOW. May 10-13. Virginia Horse Center, Lexington, VA. Major all-breed event, "A"-rated show of the American Horse Show Association. For info: Lexington Visitors Bureau, 106 E Washington St, Lexington, VA 24450. Phone: (703) 463-3777.

May 1995

S	M	T	W	T	F	S
	1	2	3	4	5	6
7	8	9	10	11	12	13
14	15	16	17	18	19	20
21	22	23	24	25	26	27
28	29	30	31			

☆ Chase's 1995 Calendar of Events ☆ May 10-11

CONFEDERATE MEMORIAL DAY IN NORTH AND SOUTH CAROLINA. May 10. See also Apr 24, Apr 26 and May 29 for Confederate Memorial Day observances in other southern states.

DAY OF THE TEACHER (EL DIA DEL MAESTRO). May 10. The Association of Mexican American Educators and the California Teachers Assn annually sponsor this tribute to all teachers for their lasting influence on children's lives. Designated by the California legislature. Annually, the second Wednesday in May. For info: California Teachers Assn, 1705 Murchison Dr, Burlingame, CA 94010. Phone: (415) 696-1400.

ENGLAND: EUROPEAN JUDO CHAMPIONSHIPS. May 10-14. Birmingham, West Midlands. Forty countries will be competing. For info: National Indoor Arena, Box Office, King Edwards Rd, Birmingham, England B1 2AA.

ENGLAND: ROYAL WINDSOR HORSE SHOW. May 10-14. Home Park, Windsor, Berkshire. Major annual show-jumping event with royal pageantry and color. For bookings: Royal Windsor Horse Show Box Office, Daniel's Department Store, 120 Peascod St, Windsor, Berkshire SL4 1DP, England. Est attendance: 35,000. For info: Penelope Henderson, Sec'y, Royal Windsor Horse Show, The Royal Mews, Windsor Castle, Windsor, Berkshire, England SL4 1NG.

GOLDEN SPIKE DRIVING: ANNIVERSARY. May 10. Anniversary of the meeting of Union Pacific and Central Pacific railways, at Promontory Point, UT, May 10, 1869. On that day a golden spike valued at about $400 was driven by Leland Stanford, president of the Central Pacific, to celebrate the linkage. (It is said that he missed the first stroke!) The golden spike was promptly removed for preservation. Long called the final link in the ocean-to-ocean railroad, this event cannot be accurately described as completing the transcontinental railroad, but it did complete continuous rail tracks between Omaha and Sacramento. See also: "Transcontinental US Railway Completion: Anniversary" (Aug 15).

GOLDEN SPIKE REENACTMENT. May 10. Promontory, UT. Music and speakers, celebrating the Golden Spike ceremony in 1869 that completed the transcontinental railroad. Est attendance: 14,000. Sponsor: Golden Spike Natl Historic Site, PO Box 897, Brigham City, UT 84302-0923. Phone: (801) 471-2209.

HOLLAND TULIP TIME FESTIVAL. May 10-20. Holland, MI. To promote the tulip and to preserve the Dutch cultural heritage in the city of Holland. Phone: (616) 396-4221 or (800) 822-2770. Est attendance: 500,000. Sponsor: Holland Tulip Time Festival, Inc, 171 Lincoln Ave, Holland, MI 49423.

JEFFERSON DAVIS CAPTURED: 130th ANNIVERSARY. May 10. On May 10, 1865, Confederate Jefferson Davis, his wife and cabinet officials were captured in Irwinville, GA, by the 4th Michigan Calvary. The prisoners were taken to Nashville, TN, and later sent to Richmond, VA.

MANDELA INAUGURATION: ANNIVERSARY. May 10, 1994. In a dramatic and historic exchange of power, former political prisoner Nelson Mandela was inaugurated as President of South Africa. Long the focal point of apartheid foes' attempts to end the enforced policy of discrimination in South Africa, Mandela handily won the first free election in South Africa despite many attempts by various political factions to either stop the electoral process or alter the outcome.

MUSLIM FESTIVAL: ID AL-HAJJ. May 10-11. Muslim calendar date: Dhu al-Hijja 10, 1415. Festival ordinarily continues for several days. Different methods for calculating the Gregorian date are used by different Muslim sects or groups. *Chase's* dates are based on astronomical calculations of the visibility of the new moon crescent at Mecca; EST date may vary.

NATIONAL RECEPTIONIST DAY. May 10. Celebrated with a champagne toast and a bouquet or corsage of fresh flowers, the red carpet is optional. Day of recognition for our nation's front line personnel in business, because you only get one chance to make a good first impression. Receptionists may go by other names such as host/hostess, maitre d', front desk clerk, operator, customer service representative, information desk personnel or any personnel responsible for creating or maintaining a favorable image for the company by greeting clients and guests. There are some 892,000 receptionists in the US. Annually, the second Wednesday in May. For info: Jennifer Alexander, Dir, Natl Receptionist Soc, 51 Oakwood Trail, Sparta, NJ 07871. Phone: (201) 729-1903.

NATIONAL THIRD SHIFT WORKERS DAY. May 10. To show appreciation for and to honor those often-forgotten workers who toil through the night to keep countless companies and businesses running smoothly. Annually, the second Wednesday in May. For info: Jeff Corbett, Dir of Mktg, PO Box 1823, Statesville, NC 28687. Phone: (704) 873-0281.

NATIONAL TOURIST APPRECIATION DAY. May 10. Emphasizes the fact that tourism is the second-largest employer in the US. Annually, a Wednesday early in May. Sponsor: Natl Travel and Tourism Awareness Council, Two Lafayette Center, 1133 21st St NW, Ste 800, Washington, DC 20036.

ROSS, GEORGE: BIRTH ANNIVERSARY. May 10. Lawyer and signer of the Declaration of Independence, born at New Castle, DE, May 10, 1730. Died at Philadelphia, PA, July 14, 1779.

VANCOUVER, GEORGE: DEATH ANNIVERSARY. May 10. English navigator, explorer and author for whom Vancouver Island and the cities of Vancouver (British Columbia and Washington) are named was born in 1758 (exact date unknown) and joined the navy at the age of 13. He surveyed the coasts of Australia, New Zealand and western North America and sailed with Captain James Cook to the Arctic in 1780. Vancouver died at Petersham, Surrey, England, on May 10, 1798, just as he was correcting the final pages of his *Journal*, which was published in three volumes at London later that year.

BIRTHDAYS TODAY

Bono, 35, singer (U2), born at Dublin, Ireland, May 10, 1960.
T. Berry Brazleton, 77, pediatrician, author, born at Waco, TX, May 10, 1918.
Judith Jamison, 51, dancer and choreographer, born at Philadelphia, PA, May 10, 1944.
Dave Mason, 49, singer, musician, songwriter ("We Just Disagree," "So High"), born at Worcester, England, May 10, 1946.
Gary Owens, 59, actor ("Rowan & Martin's Laugh-In," "The Gong Show"), born at Mitchell, SD, May 10, 1936.
Ara Parseghian, 72, former football coach, sportscaster, born at Akron, OH, May 10, 1923.
Rony F. Seikaly, 30, basketball player, born at Athens, Greece, May 10, 1965.

MAY 11 — THURSDAY
131st Day — Remaining, 234

AMERICAN BIBLE SOCIETY ANNUAL MEETING. May 11. New York, NY. A gathering of national religious and civic leaders to raise awareness and interest in the work of the American Bible Society. Annually, the second Thursday in May. For info: Mr William P. Cedfeldt, Dir of PR, American Bible Soc, 1865 Broadway, New York, NY 10023. Phone: (212) 408-1200.

AMERICAN INDIAN CELEBRATION. May 11-13. Belleville, IL. Blessed Kateri Tekakwitha's feast will be celebrated with the local American Indian community. Native American music, ceremonies and dance performances. Est attendance: 2,000. For info: Shrine of Our Lady of the Snows, 9500 W Illinois, Hwy 15, Belleville, IL 62223. Phone: (618) 397-6700.

201

May 11-12 ☆ Chase's 1995 Calendar of Events ☆

BATTLE OF HAMBURGER HILL BEGINS: ANNIVERSARY. May 11, 1969. One of the most infamous battles that signified the growing frustration with America's involvement in the Viet Nam war began. Attempting to sieze Dong Ap Bia mountain, American troops repeatedly scaled the hill over a 10-day period, often engaging in bloody hand-to-hand combat with the North Vietnamese. After finally securing the objective, American military decision makers chose to abandon it and the North Vietnamese retook it shortly thereafter. The heavy casualties in the struggle to take the hill inspired the name "Hamburger Hill."

BERLIN, IRVING: BIRTH ANNIVERSARY. May 11. Songwriter. Born Israel Isidore Baline at Tyumen, Russia, on May 11, 1888. Irving Berlin moved to New York, NY, with his family when he was four years old. After the death of his father, he began singing in saloons and on street corners in order to help his family and worked as a singing waiter as a teenager. Berlin became one of America's most prolific songwriters, authoring such songs as "Alexander's Ragtime Band," "White Christmas," "God Bless America," "There's No Business Like Show Business," "Doin' What Comes Naturally," "Puttin' On the Ritz," "Blue Skies" and "Oh! How I Hate to Get Up in the Morning" among others. He could neither read nor write musical notation. Berlin died at the age of 101 on Sept 22, 1989, at New York, NY.

DALI, SALVADOR: BIRTH ANNIVERSARY. May 11. Born at Figueras, Spain, on May 11, 1904. A leading painter in the Surrealist movement, Salvador Dali was equally well known for his sometimes baffling antics and attempts to shock his audiences. The largest collection of his works resides in the Salvador Dali Museum in St. Petersburg, FL. Dali died on Jan 23, 1989, at his hometown of Figueras.

FAIRBANKS, CHARLES WARREN: BIRTH ANNIVERSARY. May 11. Twenty-sixth vice president of the US (1905-1909) born at Unionville Center, OH, May 11, 1852. Died at Indianapolis, IN, June 4, 1918.

GRAHAM, MARTHA: BIRTH ANNIVERSARY. May 11. Martha Graham was born on May 11, 1894, at Allegheny, PA, and became one of the giants of the modern dance movement in the US. She began her dance career at the comparatively late age of 22 and joined the Greenwich Village Follies in 1923. Her new ideas began to surface in the late '20s and '30s, and by the mid-1930s she was incorporating the rituals of the southwestern American Indians in her work. She is credited with bringing a new psychological depth to modern dance by exploring primal emotions and ancient rituals in her work. She performed until the age of 75, and she premiered her 180th ballet, *The Maple Leaf Rag*, in the fall of 1990. Martha Graham died Apr 1, 1991 at the age of 96 at New York, NY.

HART, JOHN: DEATH ANNIVERSARY. May 11. Signer of the Declaration of Independence, farmer and legislator, born about 1711 (exact date unknown), at Stonington, CT. Died on May 11, 1779.

JAMAICA: BOB MARLEY DAY. May 11. Anniversary of the death, at Miami, FL, on May 11, 1981, of Bob Marley. Marley, with his musical group, The Wailers, was one of the most popular and influential performers of reggae music, an "off-beat-accented Jamaican" music closely associated with the political/religious Rastafarian movement (admirers of the late Ethiopian emperor Haile Selassie, who was formerly called Ras Tafari). Marley, who died of cancer, was born Feb 6, 1945, at Rhoden Hall in northern Jamaica. The anniversary of his death has been observed as Bob Marley Day in Jamaica.

JAPAN: CORMORANT FISHING FESTIVAL. May 11-Oct 15. Cormorant fishing on the Nagara River, Gifu. "This ancient method of catching Ayu, a troutlike fish, with trained cormorants, takes place nightly under the light of blazing torches."

MINNESOTA: ADMISSION DAY. May 11. Became 32nd state on this day in 1858.

PELLA TULIP TIME FESTIVAL. May 11-13. Pella, IA. To pay homage to the founders of this predominantly Dutch community. Activities include coronation of queen, parades, flowers, Dutch singing and dancing. Est attendance: 100,000. Sponsor: Pella Historical Soc, 507 Franklin, Pella, IA 50219. Phone: (515) 628-2409.

POKE SALAD FESTIVAL. May 11-13. Blanchard, LA. A parade, beauty pageant, carnival, craft show, fiddling contest, cloggers, country and western music and dinner with poke salad. Est attendance: 5,000. For info: Carolyn Thomas, Sec, PO Box 207, Blanchard, LA 71009. Phone: (318) 929-2839.

BIRTHDAYS TODAY

Stanley Elkin, 65, university professor, writer (*Mac Guffin, Criers & Kibitzers & Criers*), born at Brooklyn, NY, May 11, 1930.
Louis Farrakhan, 62, leader Nation of Islam, born at New York, NY, May 11, 1933.
Robert Jarvik, 49, physician, inventor of artificial heart that went into Barney Clark, born at Midland, MI, May 11, 1946.
James M. Jeffords, 61, US Senator (R, Vermont) up for reelection in Nov '94, born at Rutland, VT, May 11, 1934.
Doug McClure, 60, actor (*Beau Geste*, "The Virginian"), born Glendale, CA, May 11,1935.
Denver Pyle, 75, actor ("The Life and Legend of Wyatt Earp"; Uncle Jesse on "The Dukes of Hazzard"), born Bethune, CO, May 11, 1920.
Mort Sahl, 68, actor (*Don't Make Waves, Doctor You've Got to be Kidding*), born at Montreal, Quebec, Canada, May 11, 1927.

MAY 12 — FRIDAY
132nd Day — Remaining, 233

ALBANY TULIP FESTIVAL. May 12-14. Albany, NY. To celebrate Albany's Dutch heritage. Reenactment of old world tradition of scrubbing the streets, tulip flower show, Kinderkermis (children's fair in Dutch) and Pinksterfest (celebrating spring). Annually, Mother's Day weekend. For info: Convention and Visitors Bureau, 52 S Pearl St, Albany, NY 12207. Phone: (800) 258-3582.

BIG TEN MEN'S GOLF CHAMPIONSHIP. May 12-14. Penn State University, University Park, PA. For info: Big Ten Conference, 1500 W Higgins Rd, Park Ridge, IL 60068-6300. Phone: (708) 696-1010.

BOYNE CITY NATIONAL MUSHROOM HUNTING CHAMPIONSHIP. May 12-14. Boyne City, MI. Est attendance: 15,000. For info: Chamber of Commerce, 28 S Lake St, Boyne City, MI 49712. Phone: (616) 582-6222.

DENMARK: COMMON PRAYER DAY. May 12. Public holiday. The fourth Friday after Easter, known as "Store Bededag," is a day for prayer and festivity.

FASHION SHOW. May 12-13. Mount Mary College, Milwaukee, WI. Student designer fashion show. Est attendance: 1,500. For info: Mary Cain, Mount Mary College, 2900 N Menomonee River Parkway, Milwaukee, WI 53222. Phone: (414) 258-4810.

May 1995

S	M	T	W	T	F	S
	1	2	3	4	5	6
7	8	9	10	11	12	13
14	15	16	17	18	19	20
21	22	23	24	25	26	27
28	29	30	31			

☆ Chase's 1995 Calendar of Events ☆ May 12

INTERNATIONAL BAR-B-Q FESTIVAL. May 12-13. Owensboro, KY. The barbecue capital of the world dishes out a delicious diversion complete with team cooking competitions, pie-eating contest, keg-throwing contest, horseshoe contest, arts and crafts, sack races, country western and bluegrass music. Est attendance: 75,000. For info: BBQ Festival, PO Box 434, Owensboro, KY 42302. Phone: (502) 926-6938.

KENTUCKY SCOTTISH WEEKEND. May 12-14. General Butler State Resort Park, Carrollton, KY. Scottish bagpipe and pipe band competition, highland dancing, athletic demonstrations, bonniest knees contest, weight and hammer throwing and Ceilidh (talent show). Wee Scots Day on Friday, May 12 for school children ages 6-12 (1st-6th grades). Est attendance: 11,000. For info: Genl Butler State Resort Pk, Chet Mitchell, Carrolltown, KY 41008. Phone: (502) 732-4384.

LEAR, EDWARD: BIRTH ANNIVERSARY. May 12. English artist and author, best remembered for his light verse and limericks. Lear was born at Highgate, England, on May 12, 1812, and died at San Remo, Italy, Jan 29, 1888. See also: "Limerick Day" (May 12).

LILAC FESTIVAL. May 12-21. Highland Park, Rochester, NY. Developed by renowned park designer Frederick Law Olmstead, Highland Park is the site of the Lilac Festival, the largest celebration of its kind in North America. In addition to the spectacle of more than 500 varieties of lilacs in bloom, the festival provides free entertainment, a concert series with the Rochester Philharmonic Orchestra, a parade, a 10K race, an antique and collectibles show, a senior citizens day, a juried art show and fireworks. Est attendance: 500,000. For info: Lilac Festival, 171 Resevoir Ave, Rochester, NY 14620. Phone: (716) 256-4960.

LIMERICK DAY. May 12. Observed on the birthday of one of its champions, Edward Lear. The limerick, which dates from the early 18th century, has been described as the "only fixed verse form indigenous to the English language." It gained its greatest popularity following the publication of Edward Lear's *Book of Nonsense* (and its sequels). Write a limerick today! Example: There was a young poet named Lear/Who said, it is just as I fear/Five lines are enough/For this kind of stuff/Make a limerick each day of the year. See also: "Lear, Edward: Birth Anniversary" (May 12).

◆ **McAFEE, MILDRED: BIRTHDAY.** May 12, 1900. Mildred McAfee, born at Parkville, MO, was president of Wellesley College when President Franklin D. Roosevelt appointed her the first director of the US Navy WAVES (Women Accepted for Voluntary Emergency Service) in 1942. During the next three years McAfee's organization trained 78,000 enlisted women and 8,000 officers, and she received the Distinguished Service Medal in 1945.

MISS NEW HAMPSHIRE FINALS. May 12. Practical Arts Auditorium, Manchester, NH. Est attendance: 1,240. For info: Don Anderson, PO Box 9555, Manchester, NH 03108. Phone: (603) 668-4321.

◆ **NIGHTINGALE, FLORENCE: 175th BIRTH ANNIVERSARY.** May 12. English nurse and public health activist who, through her unselfish devotion to nursing, contributed perhaps more than any other single person to the development of modern nursing procedures and dignity of nursing as a profession. Founder of the Nightingale training school for nurses. Author of *Notes on Nursing*. Born at Florence, Italy, May 12, 1820. Died on Aug 13, 1910.

PANOPLY '95. May 12-14. Big Spring International Park, Huntsville, AL. Comprehensive arts festival celebrating the performing arts through opera, symphonic and choral concerts, chamber music, jazz, theater, ballet and modern dance. Visual arts are presented in a juried art show. Fax: (205) 533-3811. Est attendance: 180,000. For info: The Arts Council, 700 Monroe St, Huntsville, AL 35801. Phone: (205) 533-6565.

PONCA CITY RANCH RODEO. May 12-13. Ponca City, OK. Real cowboys compete in events based on real world cowboy tasks. Est attendance: 8,000. For info: Ponca City Rodeo Fdtn, Scott Klososky, Today's Computers, 205 W Hartford, Ponca City, OK 74602. Phone: (405) 765-2340.

PORTUGAL: PILGRIMAGE TO FATIMA. May 12-13. Commemorates first appearance of the Virgin of the Rosary to little shepherd children May 13, 1917. Pilgrims come to Cova da Iria, religious center, candlelit procession, Mass of the sick, for annual observance.

SPITTING PROHIBITION: ANNIVERSARY. May 12. The New York City Department of Health passed a health ordinance on May 12, 1896, prohibiting spitting on sidewalks or in other public places.

SPRING GHOST TALES. May 12. Long Run Park, Louisville, KY. Performance of tales of the supernatural. Est attendance: 3,000. Sponsor: Intl Order of EARS, Inc, Joy Pennington, Pres, 12019 Donohue Ave, Louisville, KY 40243. Phone: (502) 245-0643.

VIRGIN ISLANDS: NURSE OF THE YEAR AWARD. May 12. Annually, on anniversary of birth of Florence Nightingale (May 12, 1820), the Virgin Islands Nurses' Assn selects two nurses for special recognition. Awards are named in honor of deceased outstanding Virgin Island nurses.

WOODMEN RANGER'S DAY. May 12. To celebrate the founding of the Woodmen Rangers® in 1903. For info: John S. Manna, Woodmen of the World, Fraternal Activities Mgr, 1700 Farnam, Omaha, NE 68102. Phone: (402) 271-7258.

BIRTHDAYS TODAY

Burt Bacharach, 66, composer ("Walk On By," "Close to You," "Raindrops Keep Fallin' on My Head"; many film scores), born at Kansas City, MO, May 12, 1929.

Yogi Berra, 70, baseball manager, coach, Hall of Famer, born at St. Louis, MO, May 12, 1925.

Bruce Boxleitner, 44, actor (*How the West Was Won*, "Scarecrow and Mrs. King"), born Elgin, IL, May 12, 1951.

George Carlin, 58, comedian ("That Girl," "The George Carlin Show"), born at New York, NY, May 12, 1937.

Emilio Estevez, 33, actor (*Breakfast Club, Repo Man*), born at New York, NY, May 12, 1962.

Kim Fields, 26, actress (Tootie Ramsey on "The Facts of Life"), born at Los Angeles, CA, May 12, 1969.

Katharine Hepburn, 88, actress (Oscars [only woman to win 4] for *Morning Glory, Guess Who's Coming to Dinner?, The Lion in Winter, On Golden Pond*), born Hartford, CT, May 12, 1907 (per her statement).

Samuel Daniel Nujoma, 66, President of Namibia, born at Ongandjeia, Namibia, May 12, 1929.

Howard K. Smith, 81, journalist, born at Ferriday, LA, May 12, 1914.

Tom Snyder, 59, broadcast journalist, TV personality, born at Milwaukee, WI, May 12, 1936.

Frank Stella, 59, artist (Hard-Edge Abstraction, *Empress of India*), born at Malden, MA, May 12, 1936.

Louis Rodman Whitaker, 38, baseball player, born at Brooklyn, NY, May 12, 1957.

Steve Winwood, 47, musician, singer, born at Birmingham, England, May 12, 1948.

May 13 ☆ *Chase's 1995 Calendar of Events* ☆

MAY 13 — SATURDAY
133rd Day — Remaining, 232

ATTEMPTED ASSASSINATION OF POPE JOHN PAUL II: ANNIVERSARY. May 13. On May 13, 1981, Pope John Paul II was shot twice at close range while riding in an open automobile at St. Peter's Square in Rome, Italy. Two other persons also were wounded. An escaped terrorist, Mehmet Ali Agca (already under sentence of death for the murder of a Turkish journalist), was arrested immediately and was convicted July 22, 1981, of attempted murder of the pope. After convalescence Pope John Paul II was pronounced recovered by his doctors on Aug 14, 1981.

AUTOMOTION '95. May 13-14. Wisconsin Dells, WI. 800 cars on display, including street machines, classics and antiques. Event includes swap meet, antique flea market and "Car Cruise." Est attendance: 15,000. For info: Wisconsin Dells Visitor and Conv Bureau, Box 390, Wisconsin Dells, WI 53965. Phone: (800) 223-3557.

BLACK HILLS BALLOON RALLY. May 13-14. Sturgis, SD. Between 15 and 20 hot-air balloons rise with majestic colors from the Ft Meade Parade Grounds on Saturday and Sunday mornings. Est attendance: 2,500. For info: David McPherson, Black Hills Balloon Rally, Box 126, Sturgis, SD 57785. Phone: (605) 347-5666.

BRAILLE INSTITUTE-OPTIMIST TRACK AND FIELD OLYMPICS. May 13. Braille Institute Youth Center, Los Angeles, CA. To offer athletic competition to blind and visually impaired youths. Annually, the second Saturday in May. Cosponsored by the Optimist Clubs of Southern California. Est attendance: 800. Sponsor: Braille Institute, Communications Dept, 741 N Vermont Ave, Los Angeles, CA 90029. Phone: (213) 663-1111.

CATFISH FESTIVAL. May 13. Eudora, AR. Racing catfish, catfish industry expo, catfish dinners, catfish cookoff, the Jack Pierce Memorial 5K Run and Walk, quilt show, arts, crafts, food, bands. 9th annual festival. Annually, Mother's Day weekend. Est attendance: 5,000. For info: Chamber of Commerce, PO Box 325, Eudora, AR 71640. Phone: (501) 355-8443.

CELEBRATE '95! May 13. Lawrence University, Appleton, WI. Celebrate spring with music, entertainment, food and crafts. Five stages and more than 200 booths. Est attendance: 40,000. For info: Lawrence Univ, Linda Fuerst, PO Box 599, Appleton, WI 54912. Phone: (414) 832-6600.

CHESAPEAKE ANTIQUE FIRE APPARATUS MUSTER. May 13. Frederick, MD. Muster and flea market. Est attendance: 5,000. For info: CAFAA, C.R. Fox, 21 Park Ave, Westminster, MD 21157. Phone: (410) 848-9112.

May 1995

S	M	T	W	T	F	S
	1	2	3	4	5	6
7	8	9	10	11	12	13
14	15	16	17	18	19	20
21	22	23	24	25	26	27
28	29	30	31			

COUNTRY FAIR AND BURRO BARBECUE. May 13. Bullhead Community Park, Bullhead City, AZ. Water fights, entertainment, booths, deep-pit beef barbecue dinner and activities for all ages. Est attendance: 5,000. For info: Leigh Verley, Pres, Rotary Club, PO Box 21026, Bullhead City, AZ 86439. Phone: (602) 758-7676.

CRANBERRY MOUNTAIN SPRING NATURE TOUR. May 13. Monongahela National Forest Visitor Center, Richwood, WV. Guided tours in the Cranberry Glades area with recognized naturalists as leaders. Est attendance: 500. For info: Richwood Chamber of Commerce, Maxine Corbett, 50 Oakford Ave, Richwood, WV 26261. Phone: (304) 846-6790.

ELECTRA GOAT BBQ COOK-OFF. May 13. Electra Rodeo Arena, Electra, TX. Goat Cook-Off, Cow Patty Bingo, live band, horseshoe tournament, tug-o-war, eating contest, children's games, and crafts. Dance 9-1 after cook-off. Annually, the Saturday before Mother's Day. Est attendance: 1,500. Sponsor: Electra Chamber of Commerce, 112 W Cleveland, Electra, TX 76360. Phone: (817) 495-3577.

EXIT GLACIER RUN, 5K AND 10K. May 13. Seward, AK. 13th annual run. Races along scenic Resurrection Road in the Kenai Fjords National Park. T-shirts for all starters; ribbons awarded in each division. Est attendance: 350. For info: Seward Chamber of Commerce, PO Box 749, Seward, AK 99664. Phone: (907) 224-8051.

THE GOOD, THE BAD, AND THE CUDDLY. May 13-Sept 24. Shedd Aquarium, Chicago, IL. See Feb 18 for full description. For info: Smithsonian Institution Traveling Exhibition Service, 1100 Jefferson Dr SW, Ste 3146, Washington, DC 20560. Phone: (202) 357-2700.

GRAND PRIX CATFISH RACES. May 13. Greenville, MS. To recognize the noble racing heart of the thoroughbred pond-raised Mississippi catfish. 11th annual races. Est attendance: 10,000. Sponsor: *Delta Democrat-Times*, Ken Cazalas, Ed, Box 1618, Greenville, MS 38701. Phone: (601) 335-1155.

GUM TREE FESTIVAL. May 13-14. Tupelo, MS. A juried art show with arts, crafts and live entertainment. 10K run. Annually, the second weekend in May. Est attendance: 15,000. For info: Tupelo Conv and Visitors Bureau, Box 1485, Tupelo, MS 38802. Phone: (800) 533-0611.

IROQUOIS STEEPLECHASE. May 13. Percy Warner Park, Nashville, TN. Est attendance: 35,000. For info: Iroquois Steeplechase Race Committee, Henry Hooker, 44 Vantage Way, Ste 500, Nashville, TN 37228. Phone: (615) 256-8755.

JAMESTOWN LANDING DAY. May 13. Jamestown Settlement, Williamsburg, VA. Celebration of the anniversary of the establishment of America's first permanent English settlement at Jamestown, VA in 1607. Militia presentations and sailing demonstrations. Est attendance: 1,200. For info: Media Relations, Jamestown-Yorktown Fdtn, PO Box JF, Williamsburg, VA 23187. Phone: (804) 253-4838.

JAMESTOWN WEEKEND CELEBRATION. May 13-14. Jamestown, VA. Anniversary of the founding of Jamestown, first permanent English colony in North America, in May of 1607. Est attendance: 2,000. For info: Colonial Natl Historical Park, Box 210, Yorktown, VA 23690. Phone: (804) 898-3400.

JUBILEE. May 13. Bennettsville, SC. Arts and crafts festival with entertainment, children's events and juried art show. Annually, the second Saturday in May. Est attendance: 6,000. For info: Ken Harmon, 226 Radio Rd, Bennettsville, SC 29512. Phone: (803) 479-7121.

LOUIS, JOE: BIRTH ANNIVERSARY. May 13. World heavyweight boxing champion, 1937-1949, nicknamed the "Brown Bomber," Joseph Louis Barrow was born near Lafayette, AL, May 13, 1914. He died Apr 12, 1981, at Las Vegas, NV. Burial at Arlington National Cemetery. (Louis's burial there, by presidential waiver, was the 39th exception ever to the eligibility rules for burial in Arlington National Cemetery.)

MACAU: PROCESSION OF OUR LADY OF FATIMA. May 13. To commemorate the miracle of 1917, when the Madonna appeared to residents of Fatima, Portugal, an image of Our Lady is carried in pilgrimage from Sao Domingos Church to Penha Church, generally around 6 PM.

MARKET SQUARE FAIR. May 13. Fredericksburg, VA. To create an awareness of Fredericksburg's historic heritage. Arts, crafts, food and entertainment. Est attendance: 5,500. For info: Visitor Center, 706 Caroline St, Fredericksburg, VA 22401. Phone: (800) 678-4748.

MOTHER'S DAY ANNUAL RHODODENDRON SHOW. May 13–14. Crystal Springs Rhododendron Gardens, Portland, OR. Spectacular display of rhododendron and azalea blooms and plant sale. Est attendance: 6,000. For info: Portland, Oregon Visitors Assn, Three World Trade Center, 26 SW Salmon, Portland, OR 97204-3299. Phone: (503) 771-8386.

NETHERLANDS: NATIONAL WINDMILL DAY. May 13. Second Saturday in May each year. About 950 windmills still survive, and some 300 still are used occasionally and have been designated national monuments by the government. As many windmills as possible are in operation on National Windmill Day for the benefit of tourists.

OHIO RIVER FESTIVAL FOR THE ARTS. May 13–14. Evansville, IN. 26th annual festival with 75 artist from 10 states, stage featuring more than 30 live performances, children's arts activities, 12K River Run and mouthwatering foods. Annually, Mother's Day weekend. Est attendance: 25,000. For info: Kim Setzer, Program Coord, Arts Council of SW Indiana, 123 NW 4th St, Ste 312, Evansville, IN 47708. Phone: (812) 422-2111.

PARADE OF WHEELS. May 13. Rehoboth Beach, DE. Parade anything non-motorized that rolls. Followed by Kids Convention, Teddy Bear Contest, games and activities. Est attendance: 2,500. For info: Carol Everhart, Festival Dir, Rehoboth Beach-Dewey Beach Chambr of Com, PO Box 216, Rehoboth Beach, DE 19971. Phone: (800) 441-1329.

PHILADELPHIA POLICE BOMBING: 10th ANNIVERSARY. May 13. On May 13, 1985, during the siege of a "radical group" in Philadelphia, PA, police in a helicopter reportedly dropped a bomb containing the powerful military plastic explosive C-4 on the building in which the group was housed. The bomb and the resulting fire left 11 persons dead (including 4 children) and destroyed 61 homes.

PRATER'S MILL COUNTRY FAIR. May 13–14. (Also Oct 7–8.) Dalton, GA. Arts and crafts show in the atmosphere of an old-fashioned country fair. Est attendance: 25,000. For info: Prater's Mill Fdtn, Inc, Judy Alderman/Jane Harrell, 848 Shugart Rd, Dalton, GA 30720-2429. Phone: (706) 275-6455.

ROBIDOUX RENDEZVOUS. May 13–14. Robidoux Row Museum, St. Joseph, MO. A celebration of life as it was in the 1800's. Historic reenactment and pioneer crafts are featured. Est attendance: 5,000. Sponsor: Robidoux Poulin, 3rd and Poulin, St. Joseph, MO 64501. Phone: (816) 232-5861.

ST. LAWRENCE SEAWAY ACT: ANNIVERSARY. May 13. President Dwight D. Eisenhower on May 13, 1954, signed legislation authorizing US-Canadian construction of a waterway that would make it possible for ocean-going ships to reach the Great Lakes.

SULLIVAN, ARTHUR: BIRTH ANNIVERSARY. May 13. English composer best known for light operas (with Sir William Gilbert), born May 13, 1842. Died Nov 22, 1900.

TRILLIUM FESTIVAL. May 13–14. Hoffmaster State Park, Muskegon, MI. In celebration of Michigan wildflowers. Annually, Mother's Day weekend. Est attendance: 42,000. For info: Hoffmaster State Park, 6585 Lake Harbor Rd, Muskegon, MI 49441. Phone: (616) 798-3573.

TRY THIS ON—A HISTORY OF CLOTHING, GENDER AND POWER. May 13–June 25. Rotch-Jones-Duff House & Garden Museum, New Bedford, MA. A social history about clothing and appearance in the US over the past 200 years. See Jan 7 for full description. For info: Smithsonian Institution Traveling Exhibition Service, 1100 Jefferson Dr SW, Ste 3146, Washington, DC 20560. Phone: (202) 357-2700.

US-MEXICO WAR DECLARATION: ANNIVERSARY. May 13. Although the war had been in progress since March, when General Zachary Taylor crossed the Rio Grande and established Fort Brown on Mexican territory, the declaration of war by the US did not take place until May 13, 1846.

WELLS, MARY: BIRTH ANNIVERSARY. May 13. Motown's first big star, Mary Wells, was born May 13, 1943 at Detroit, MI. Wells was known for such hits as "The One Who Really Loves You," "You Beat Me to the Punch," "Two Lovers" and her signature song, "My Guy." She was one of a group of black artists of the 60s, along with Smokey Robinson, the Supremes, the Temptations and the Four Tops, who helped end musical segregation by getting played on white radio stations. Mary Wells died July 26, 1992, at Los Angeles.

BIRTHDAYS TODAY

Franklyn Ajaye, 46, actor ("Keep on Truckin," *Car Wash*), born Brooklyn, NY, May 13, 1949.

Beatrice Arthur (Bernice Frankel), 69, actress (*Mame, Fiddler on the Roof*, "Maude," "Golden Girls"), born New York, NY, May 13, 1926.

Clive Barnes, 68, critic, born at London, England, May 13, 1927.

Peter Gabriel, 45, singer (lead singer of Genesis), born at London, England, May 13, 1950.

Dennis Keith (Worm) Rodman, 34, basketball player, born Trenton, NJ, May 13, 1961.

Herbert Ross, 68, actor, choreographer, director on stage and screen, born at New York, NY, May 13, 1927.

Stevie Wonder (Steveland Morris Hardaway), 44, singer, musician (16 Grammy Awards; "I Just Called to Say I Love You"), born Saginaw, MI, May 13, 1951.

MAY 14 — SUNDAY
134th Day — Remaining, 231

ALCOHOL AND OTHER DRUG-RELATED BIRTH DEFECTS WEEK. May 14–20. To increase public awareness of the risks associated with alcohol and other drug use during pregnancy. For info: Natl Council on Alcoholism and Drug Dependence, Inc, Rebecca Fenson, 12 W 21st St, New York, NY 10010. Phone: (212) 206-6770.

CAPE MAY MUSIC FESTIVAL. May 14–June 25. Cape May, NJ. The world's finest chamber ensembles and special guest artists perform in the elegant Victorian setting of Cape May. Est attendance: 5,000. For info: Mid-Atlantic Center for the Arts, PO Box 340, Cape May, NJ 08204. Phone: (609) 884-5404.

FAHRENHEIT, GABRIEL DANIEL: BIRTH ANNIVERSARY. May 14. German physicist whose name is attached to one of the major temperature measurement scales. He introduced the use of mercury in thermometers and greatly improved their accuracy. Born at Danzig, Germany, on May 14, 1686, he died at Amsterdam, Holland, on Sept 16, 1736.

GAINSBOROUGH, THOMAS: BIRTH ANNIVERSARY. May 14. English landscape and portrait painter. Among his most remembered works: *The Blue Boy, The Watering Place* and *The Market Cart*. Born at Sudbury, Suffolk, England, on May 14, 1727, he died at London, on Aug 2, 1788.

May 14 ☆ Chase's 1995 Calendar of Events ☆

GIRLS INCORPORATED WEEK. May 14–20. To focus national and local attention on the goals of Girls Incorporated as an organization for the rights and needs of girls. Annually, beginning the second Sunday in May. Sponsor: Girls Inc, Amy Sutnick-Plotch, 30 E 33rd St, 7th Fl, New York, NY 10016. Phone: (212) 689-3700.

JAMESTOWN, VIRGINIA: FOUNDING ANNIVERSARY. May 14. The first permanent English settlement in what is now the US took place at Jamestown, VA (named for England's King James I), on May 14, 1607. Captains John Smith and Christopher Newport were among the leaders of the group of royally chartered Virginia Company settlers who had traveled from Plymouth, England, in three small ships: *Susan Constant, Godspeed* and *Discovery.*

KIWANIS PRAYER WEEK. May 14–20. Encourages Kiwanis Clubs to promote religious activities throughout their communities and to recognize individuals for their contributions to spiritual welfare. Annually, the second full week in May. Sponsor: Kiwanis Intl, Program Development Dept, 3636 Woodview Trace, Indianapolis, IN 46268.

MALAWI: KAMUZU DAY. May 14. Public holiday in Malawi.

MOON PHASE: FULL MOON. May 14. Moon enters Full Moon phase at 4:48 PM, EDT.

MOTHER OCEAN DAY. May 14. To celebrate the wonder, the vastness and beauty of the ocean. Casting of roses into the sea from the beach and from the water. Annually, on Mother's Day. Sponsor: Natl Week of the Ocean, Inc, Cynthia Hancock, PO Box 179, Ft Lauderdale, FL 33302. Phone: (305) 462-5573.

★ **MOTHER'S DAY.** May 14. Presidential Proclamation always issued for the second Sunday in May. (Pub Res No. 2 of May 8, 1914.)

MOTHER'S DAY. May 14. Observed first in 1907 at the request of Anna Jarvis of Philadelphia, PA, who asked her church to hold service in memory of all mothers on anniversary of her mother's death. Annually, the second Sunday in May.

NATIONAL EDUCATIONAL BOSSES WEEK. May 14–20. A special week to honor bosses in the field of education such as principals and school superintendents. Annually, the third week in May. For info: National Assn of Educational Office Personnel, PO Box 12619, Wichita, KS 67277.

NATIONAL HISTORIC PRESERVATION WEEK. May 14–20. To draw public attention to historic preservation including neighborhoods, districts, landmark buildings, open space and maritime heritage. Annually, the second full week in May. Sponsor: ATTN: Lori Kendis, Natl Trust for Historic Preservation, 1785 Massachusetts Ave NW, Washington, DC 20036. Phone: (202) 673-4141.

May 1995

S	M	T	W	T	F	S
	1	2	3	4	5	6
7	8	9	10	11	12	13
14	15	16	17	18	19	20
21	22	23	24	25	26	27
28	29	30	31			

NATIONAL NURSING HOME WEEK. May 14–20. A community outreach program designed to familiarize the public with long-term care facilities and the services they provide. Activities are conducted locally by individual long-term care facilities. Annually, Mother's Day through the following Saturday. Sponsor: American Health Care Assn, 1201 L St NW, Washington, DC 20005. Phone: (202) 842-4444.

NATIONAL POLICE WEEK. May 14–20. See also "Peace Officer Memorial Day" (May 15). For info: American Police Hall of Fame and Museum, 3801 Biscayne Blvd, Miami, FL 33137. Phone: (305) 573-0202.

NATIONAL SENIOR SMILE WEEK. May 14–20. To heighten public awareness of the importance of dental care and the availability of dental services to older adults. For info: American Dental Assn, Natl Senior Smile Week, 211 E Chicago Ave, Ste 2038, Chicago, IL 60611.

NATIONAL TOWN MEETING ON MAIN STREET. May 14–17. Excelsior Hotel, Little Rock, AR. The only nationwide conference dedicated to preservation-based downtown and neighborhood business district revitalization combines the more than 10 years of experience of the National Trust for Historic Preservation's National Main Street Center, the latest research, potential partners and a network of more than 800 members, downtown experts and local practitioners to discuss issues. Also a large trade show. Est attendance: 700. For info: Linda Donavan Harper, Program Mgr, NMSC Conferences NTHP, 1785 Massachusetts Ave NW, Washington, DC 20036. Phone: (202) 673-4221.

★ **NATIONAL TRANSPORTATION WEEK.** May 14–20. Presidential Proclamation issued for week including third Friday in May since 1960. (PL 86–475 of May 20, 1960, first requested; PL87–449 of May 14, 1962, requested an annual proclamation.)

NORWAY: MIDNIGHT SUN AT NORTH CAPE. May 14–July 30. North Cape. First day of the season with around-the-clock sunshine. At North Cape, the sun never dips below the horizon from May 14 to July 30, but the night is bright long before and after these dates.

OWEN, ROBERT: BIRTH ANNIVERSARY. May 14. English progressive owner of spinning works, philanthropist, Utopian socialist, founder of New Harmony, IN, born at Newtown, Wales, May 14, 1771. Died there Nov 17, 1858.

PARAGUAY: INDEPENDENCE DAY. May 14–15. Two-day celebration begins, commemorating independence from Spain, attained on May 14, 1811.

PHILIPPINES: CARABAO FESTIVAL. May 14–15. Pulilan, Bulacan; Nueva Ecija; Angono, Rizal. Parade of farmers to honor their patron saint, San Isidro, with hundreds of "dressed up" carabaos participating.

★ **POLICE WEEK.** May 14–20. Presidential Proclamation 3537 of May 4, 1963, covers all succeeding years. (PL87–726 of Oct 1, 1962.) Always the week including May 15 since 1962.

SHEEP SHEARING FESTIVAL. May 14. North Andover, MA. Festival features sheep shearing, sheepdog demonstrations, crafts fair, spinning bee, live music, rides, games, museum tours, food booths and sheep-to-shawl demonstrations. Est attendance: 10,000. Sponsor: Museum of American Textile History, 800 Massachusetts Ave, North Andover, MA 01845. Phone: (508) 686-0191.

SPACE MILESTONE: *SOYUZ 40* (USSR). May 14. Two cosmonauts (L. Popov and, from Rumania, D. Prunariu) docked at *Salyut 6* space station on May 15. Returned to Earth on May 22. Launched May 14, 1981.

"THE STARS AND STRIPES FOREVER" DAY. May 14. Anniversary of the first public performance of John Philip Sousa's march, "The Stars and Stripes Forever," at Philadelphia, PA, on May 14, 1897. The occasion was the unveiling of a statue of George Washington, and President William McKinley was present. A bill was introduced in the Congress in 1985 to make "The Stars and Stripes Forever" the official national march.

Chase's 1995 Calendar of Events — May 14-15

UNDERGROUND AMERICA DAY. May 14. Underground America Day is one man's (Malcolm Wells) attempt to get others to think of designing and building structures underground. Each year, Mr. Wells awards prizes and publishes illustrations and humorous suggestions for celebrating Underground America Day. Annually, May 14. Sponsor: Malcolm Wells, 673 Satucket Rd, Brewster, MA 02631. Phone: (508) 896-6850.

VEAL BAN ACTION DAY. May 14. (Mother's Day) To publicize the inhumane and unsanitary conditions under which "milk-fed veal" calves are raised. Local actions include picketing of veal restaurants, leafletting and information tables. Sponsor: Farm Animal Reform Movement, Box 30654, Bethesda, MD 20824. Phone: (301) 530-1737.

WAAC: ANNIVERSARY. May 14. During WWII women became eligible to enlist for noncombat duties in the Women's Auxiliary Army Corps (WAAC) by an act of Congress on May 14, 1942. Women also served as Women Appointed for Voluntary Emergency Service (WAVES), Women's Auxiliary Ferrying Squadron (WAFS), and Coast Guard or Semper Paratus Always Ready Service (SPARS), the Women's Reserve of the Marine Corp.

WOMEN'S INTERNATIONAL BOWLING CONGRESS QUEEN'S TOURNAMENT. May 14-18. Tucson, AZ. For info: WIBC, 5301 S 76th St, Greendale, WI 53129. Phone: (414) 421-9000.

★ **WORLD TRADE WEEK.** May 14-20. Presidential Proclamation has been issued each year since 1948 for the third week of May with three exceptions: 1949, 1955 and 1966.

BIRTHDAYS TODAY

Walter Berry, 31, former basketball player, born at Harlem, NY, May 14, 1964.
Jack Bruce, 52, musician, born at Glasgow, Scotland, May 14, 1943.
David Byrne, 43, singer (with Talking Heads), composer (songs, film scores and *The Catherine Wheel* ballet), born at Dumbarton, Scotland, May 14, 1952.
Byron L. Dorgan, 53, US Senator (D, North Dakota), born at Dickinson, ND, May 14, 1942.
Jim Folsom, Jr., 46, Governor of Alabama (D), born at Montgomery, AL, May 14, 1949.
George Lucas, 51, filmmaker (*Star Wars* trilogy) director, (*American Graffiti*), born Modesto, CA, May 14, 1944.
Jose Dennis Martinez, 40, baseball player, born at Granada, Nicaragua, May 14, 1955.
Patrice Munsel, 70, singer, born at Spokane, WA, May 14, 1925.
Ralph Neas, 49, Executive Director of the Leadership Conference on Civil Rights, born at Brookline, MA, May 14, 1946.
Richard John Neuhas, 59, Lutheran pastor, born at Pembroke, Ontario, Canada, May 14, 1936.
Mike Quick, 36, football player, born at Hamlet, NC, May 14, 1959.

MAY 15 — MONDAY
135th Day — Remaining, 230

BAUM, LYMAN FRANK: BIRTH ANNIVERSARY. May 15. American newspaperman who wrote the Wizard of Oz stories was born at Chittenango, NY, on May 15, 1856. Although *The Wonderful Wizard of Oz* is the most famous, Baum also wrote many other books for children, including more than a dozen about Oz. He died at Hollywood, CA, May 6, 1919.

COTTON, JOSEPH: BIRTH ANNIVERSARY. May 15, 1905. Stage and screen star Joseph Cotton was born at Petersburg, VA. Among Cotton's movie credits were *Citizen Kane*, *The Magnificent Ambersons* and *The Third Man*. Among his most noted performances on Broadway were *The Philadelphia Story* and *Once More With Feeling*. Joseph Cotton died Feb 6, 1994, at Los Angeles.

EASTERN PACIFIC HURRICANE SEASON. May 15-Nov 30. Eastern Pacific defined as: Coast to 140 degrees west longitude. Info from: US Dept of Commerce, Natl Oceanic and Atmospheric Admin, Rockville, MD 20852.

FRISCH, MAX: BIRTH ANNIVERSARY. May 15. Max Frisch was one of Europe's leading post-World War II literary figures. His work includes the novels *Homo Faber*, *I'm Not Stiller*, *Juerg Reinhardt* and plays *The Firebugs*, *Andorra* and many others. In addition to his writing, he was a controversial critic of his native Switzerland, its government and its people. Frisch was born on May 15, 1911, and died on Apr 4, 1991.

GEORGE WALLACE SHOT: ANNIVERSARY. May 15. George Wallace was shot by Arthur Bremer May 15, 1972, while Wallace was in Laurel, MD, campaigning for the US presidency. For the remainder of his life Wallace was paralyzed from the waist down. On Aug 4, 1972, Bremer was sentenced to 67 years in prison for the shooting.

INTERNATIONAL CHILDREN'S ART EXHIBIT. May 15-July 15. San Francisco, CA. In commemoration of the 50th anniversary of the UN, "The Global Family" and "Children's Issues" are the themes for this year's international children's art exhibit representing works received from an art and writing exchange involving 16,000 children from around the globe. Sponsor: Paintbrush Diplomacy. Contact Germain Juneau, (415) 255-7478. For info: UN50 Committee, 312 Sutter St, Ste 610, San Francisco, CA 94108. Phone: (415) 989-1995.

JAPAN: AOI MATSURI (HOLLYHOCK FESTIVAL). May 15. Kyoto. The festival features a pageant reproducing imperial processions of ancient times that paid homage to the shrine of Shimogamo and Kamigamo.

MEXICO: SAN ISIDRO DAY. May 15. Day of San Isidro Labrador celebrated widely in farming regions to honor St. Isidore, the Plowman. Livestock gaily decorated with flowers. Celebrations usually begin about May 13 and continue for about a week.

NATIONAL SALVATION ARMY WEEK. May 15-21. To commemorate the many compassionate services of The Salvation Army to the needy, involving more than one million volunteers annually. For info: The Salvation Army, Natl HQ, 615 Slaters Lane, Alexandria, VA 22313. Phone: (703) 684-5521.

NYLON STOCKINGS: ANNIVERSARY. May 15. On May 15, 1940, nylon hose went on sale at stores throughout the country. Competing producers bought their nylon yarn from E.J. du Pont de Nemours. W.H. Carothers of Du Pont developed nylon, called "Polymer 66," in 1935. It was the first totally man-made fiber and over time substituted for other materials and came to have widespread application.

OCONALUFTEE INDIAN VILLAGE. May 15-Oct 25. Cherokee Indian Reservation, Cherokee, NC. To portray the Cherokee lifestyle of the 1750 period. Also featuring "Unto These Hills" (mid-June-late-August), a drama portraying history of eastern band of Cherokees. Est attendance: 150,000. For info: Cherokee Historical Assn, Margie Douthit, PR, PO Box 398, Cherokee, NC 28719. Phone: (704) 497-2111.

★ **PEACE OFFICER MEMORIAL DAY.** May 15. Presidential Proclamation 3537, of May 4, 1963, covers all succeeding years. (PL87-726 of Oct 1, 1962.) Always May 15 of each year since 1963; however, first issued in 1962 for May 14.

PEACE OFFICER MEMORIAL DAY. May 15. An event honored by some 21,000 police departments nationwide. Memorial ceremonies at 10 AM in American Police Hall of Fame and Museum, Miami, FL, and at Congressional Park, Washington, DC. See also: "National Police Week" (May 14). Sponsor: Natl Assn of Chiefs of Police. Est attendance: 1,000. For info: American Police Hall of Fame and Museum, 3801 Biscayne Blvd, Miami, FL 33137. Phone: (305) 573-0202.

May 15-16 ☆ *Chase's 1995 Calendar of Events* ☆

PERIGEAN SPRING TIDES. May 15-16. Spring tides, the highest possible tides, occur when New Moon or Full Moon falls within 24 hours of the moment the Moon is nearest Earth (perigee) in its monthly orbit on May 15, at 11:00 AM, EDT.

SCHNITZLER, ARTHUR: BIRTH ANNIVERSARY. May 15. Austrian playwright, novelist and medical doctor, Arthur Schnitzler was born at Vienna, Austria, on May 15, 1862. Noted for his psychoanalytical examination of Viennese society. Schnitzler died Oct 21, 1931.

SPACE MILESTONE: *FAITH 7* (US). May 15. Major Gordon Leroy Cooper orbited the Earth 22 times on May 15, 1963.

UNITED NATIONS: INTERNATIONAL DAY OF FAMILIES. May 15. The general assembly (res 47/237) on Sept 20, 1993, voted this as an annual observance beginning in 1994.

WILSON, ELLEN LOUISE AXSON: BIRTH ANNIVERSARY. May 15. First wife of Woodrow Wilson, 28th president of the US, born at Savannah, GA, May 15, 1860. Died Aug 6, 1914.

BIRTHDAYS TODAY

Anna Marie Alberghetti, 59, actress (*Cinderfella*), born at Pesaro, Italy, May 15, 1936.
Eddy Arnold, 77, singer ("Make the World Go Away"), born at Henderson, TN, May 15, 1918.
Richard Avedon, 72, photographer, born at New York, NY, May 15, 1923.
George Brett, 42, former baseball player, born at Moundsville, WV, May 15, 1953.
Joey Matthew Browner, 35, football player, born at Warren, OH, May 15, 1960.
David Cronenberg, 52, filmmaker, born at Toronto, Ont, Canada, May 15, 1943.
Jasper Johns, 65, artist (Neo-Dada Encaustic and collage composition *Flag*), born at Augusta, GA, May 15, 1930.
Trini Lopez, 58, actor, singer (*Marriage on the Rocks*, *The Dirty Dozen*), born at Dallas, TX, May 15, 1937.
Paul Zindel, 59, writer (*Effect of Gamma Rays on Man-in-the-Moon Marigolds*), born at New York, NY, May 15, 1936.

MAY 16 — TUESDAY

136th Day — Remaining, 229

BIOGRAPHERS DAY. May 16. Anniversary of the meeting, in London, England, on May 16, 1763, of James Boswell and Samuel Johnson, beginning history's most famous biographer-biographee relationship. Boswell's *Journal of a Tour to the Hebrides* (1785) and his *Life of Samuel Johnson* (1791) are regarded as models of biographical writing. Thus, this day is recommended as one on which to start reading or writing a biography.

ENGLAND: INTERNATIONAL FESTIVAL OF FLOWERS WITH MUSIC. May 16-20. Westminster Cathedral, London. Festival celebrating the cathedal's 100th anniversary and showing the creative talents of many of the world's finest flower artists. For info: Margaret Ferguson, Dir, Cropley Grove, Ousden, Newmarket, Suffolk, England CB8 8TL.

FONDA, HENRY: 90th BIRTH ANNIVERSARY. May 16. American stage and screen actor, Motion Picture Academy award winner, born Henry Jaynes Fonda, May 16, 1905, at Grand Island, NE. Began his acting career at the Omaha (NE) Playhouse. Fonda died at Los Angeles, CA, Aug 12, 1982.

	S	M	T	W	T	F	S
May		1	2	3	4	5	6
	7	8	9	10	11	12	13
1995	14	15	16	17	18	19	20
	21	22	23	24	25	26	27
	28	29	30	31			

GWINNETT, BUTTON: DEATH ANNIVERSARY. May 16. Signer of the Declaration of Independence, born at Down Hatherley, Gloucestershire, England, about 1735 (exact date unknown). Died following a duel, May 16, 1777.

JOHNSON, MARV: DEATH ANNIVERSARY. May 16. Soul music pioneer. Born at Detroit, MI, and died on May 16, 1993, at Columbia, SC.

MARTIN, BILLY: BIRTH ANNIVERSARY. May 16. Baseball player and manager. Born at Berkeley, CA, on May 16, 1928. Billy Martin's baseball career included managerial stints with five major league teams: the New York Yankees, Minnesota Twins, Detroit Tigers, Texas Rangers and the Oakland Athletics. After a successful playing career, he compiled a record of 1,258 victories to 1,018 losses in his 16 seasons as a manager. His combative and fiery style both on and off the field kept him in the headlines, and he will long be remembered for his on-again/off-again relationship with Yankees' owner George Steinbrenner, for whom he managed the Yankees five different times. Billy Martin died in an automobile accident on Dec 25, 1989.

MORTON, LEVI PARSONS: BIRTH ANNIVERSARY. May 16. Twenty-second vice president of the US (1889-1893) born at Shoreham, VT, May 16, 1824. Died at Rhinebeck, NY, May 16, 1920.

◈ **PEABODY, ELIZABETH PALMER: BIRTH ANNIVERSARY.** May 16, 1804. Elizabeth P. Peabody, born at Billerica, MA, became an innovative educator, author and publisher. She opened her first school in Lancaster, MA, when only 16, her second two years later in Boston, where she became active in the intellectual community. In 1839 Peabody opened a bookstore that quickly became the intellectuals' hangout. With her own printing press Peabody became the first woman publisher in Boston and possibly the US. She published three of her brother-in-law Nathaniel Hawthorne's earliest books. For two years she published and wrote for *The Dial*, the literary magazine and voice of the Transcendental movement. Peabody's enduring accomplishment was the establishment of the first kindergarten in the US, in 1860 in Boston. She organized other kindergartens and promoted the concept vigorously, even creating a magazine, *Kindergarten Messenger*, in 1873. Elizabeth Peabody died Jan 3, 1894, at Jamaica Plain, MA.

SEWARD, WILLIAM HENRY: BIRTH ANNIVERSARY. May 16. American statesman, secretary of state under Lincoln and Andrew Johnson. Seward negotiated the purchase of Alaska from Russia for $7,200,000. At the time some felt the price was too high and referred to the purchase as "Seward's Folly." Born on May 16, 1801, at Florida, NY, he died at Auburn, NY, on Oct 10, 1872.

WEAR PURPLE FOR PEACE DAY: THE FIRST INTERGLACTICAL [SIC] HOLIDAY. May 16. The Moderns theorize that people of space will not communicate with people of Earth, because of our violent nature. This day is an attempt to convey the idea to the people of space that we, the people of Earth, are making an attempt to attain more peace on our planet. The Moderns believe this is essential to making contact with the people of space. Sponsor: The Moderns, Jon E Mod, 715 49th St., Kenosha, WI 53140. Phone: (414) 653-0230.

BIRTHDAYS TODAY

Pierce Brosnan, 43, actor ("The Manions of America," "Remington Steele"), born at County Meath, Ireland, May 16, 1952.
Dan Coats, 52, US Senator (R, Indiana), born at Jackson, MI, May 16, 1943.

Two ways to simplify your life...

CHASE'S 1996

People who rely on *Chase's Annual Events* need to be in the know every day of the year—and like to plan ahead. Now it's easier than ever to make sure you receive the newest edition of *Chase's* as soon as it's available.

☆ Simply return one of the Pre-Publication Order Forms below. Your copy of *Chase's 1996* will be reserved immediately and shipped just as soon as it's off press in October 1995.

☆ To receive *Chase's* **automatically every year**, take advantage of the Standing Order Authorization. There's no chance of missing out on *Chase's*—each year's new edition will be shipped and billed automatically. (Of course, you may change or cancel your Standing Order at any time.)

Both convenient order options carry our unconditional guarantee—you may return *Chase's* for any reason within 10 days of receipt for a full refund. Why not order today and be sure of starting 1996 with *Chase's Annual Events* at your fingertips!

Mail to: Contemporary Books, Dept. C,
Two Prudential Plaza, Suite 1200, Chicago, IL 60601-6790

YES! Send me _____ copies of
Chase's Annual Events 1996 at $49.95 each $ _____
Quantity Discounts:
  Deduct 10% per copy for 3–9 copies
  Deduct 20% per copy for 10 or more copies $ _____
Sales Tax:
  Add applicable tax in AL, CA, FL, IL, NC, NY,
  OH, PA, TX $ _____
Shipping and Handling:
  Add $4.00 for first copy, $2.50 for each
  additional copy $ _____

☐ Check ☐ Money Order
(payable to Contemporary Books)   TOTAL $ _____
☐ VISA ☐ MasterCard

Acct. _____ Exp. ____/____

X _____
Signature if charging to bank card

Name (please print) _____

SHIP TO:

Name _____

Address _____

Address _____

City _____ State _____ Zip _____

STANDING ORDER/Authorization
To ensure that I receive each year's new edition of *Chase's Annual Events*, please accept this Standing Order Authorization to ship me _____ copies of each annual edition, beginning with the 1996 edition, and to bill me at the address shown above.

Signature _____
                                      Date
Name _____
         (please print)
Phone (____) _____

F95

☆ Chase's 1995 Calendar of Events ☆ May 16-17

Tracey Gold, 26, actress ("Shirley," "Goodnight Beantown," "Growing Pains"), born at New York, NY, May 16, 1969.
James B. Hunt, Jr., 58, Governor of North Carolina (D), born at Greensboro, NC, May 16, 1937.
Janet Jackson, 29, singer ("What Have You Done for Me Lately"), actress ("Fame"), born at Gary, IN, May 16, 1966.
Lainie Kazan, 55, singer, born at New York, NY, May 16, 1940.
Olga Korbut, 40, former Olympian gymnast, born at Grodno, USSR, May 16, 1955.
Jack (John Scott) Morris, 40, baseball player, born at St. Paul, MN, May 16, 1955.
Gabriela Sabatini, 25, tennis player, born at Buenos Aires, Argentina, May 16, 1970.
John Thomas Salley, 31, basketball player, born at Brooklyn, NY, May 16, 1964.
Studs Terkel (Louis Terkel), 83, author, journalist (*Hard Times, Working*), born New York, NY, May 16, 1912.
Lowell P. Weicker, Jr., 64, Governor of Connecticut (CP), born at Paris, France, May 16, 1931.
Debra Winger, 40, actress (*Terms of Endearment, Shadowlands*), born Columbus, OH, May 16, 1955.

MAY 17 — WEDNESDAY
137th Day — Remaining, 228

BROWN v BOARD OF EDUCATION: ANNIVERSARY. May 17. The US Supreme Court ruled unanimously May 17, 1954, that segregation of public schools "solely on the basis of race" denied black children "equal educational opportunity" even though "physical facilities and other 'tangible' factors may have been equal. Separate educational facilities are inherently unequal." The case was argued before the Court by Thurgood Marshall, who would go on to become the first black appointed to the Supreme Court.

CANADA: CANADIAN TULIP FESTIVAL. May 17-22. Ottawa, Ontario. The world's largest tulip festival with 3 million tulips in bloom. See a parade of decorated boats, outdoor concerts and fireworks. This "once-in-a-lifetime" mega-event celebrates the 50th anniversary of the end of WWII and the start of the Festival: "Canada and Holland—The Friendship That Flowered." Est attendance: 600,000. For info: Canadian Tulip Festival, 360 Albert St, Ste 1720, Ottawa, Ont, Canada K1R 7X7. Phone: (613) 567-4447.

DENVER INTERNATIONAL CHILDREN'S FESTIVAL. May 17-21. The Plex, Downtown Denver, CO. A five-day showcase of performing arts geared for school children grades K-12, teachers and families. Young audiences will have an opportunity to enjoy theater, music, dance, puppetry and more. Sponsors: NEWS 4, The Denver Post, J. Walter Thompson, Rothgerber Appel, Powers & Johnson. Produced by Downtown Denver Partnership, Inc. Est attendance: 50,000. For info: Denver Intl Children's Fest, c/o Downtown Denver Partnership, 511 16th St, Ste 200, Denver, CO 80202. Phone: (303) 534-6161.

DUBUQUEFEST/VERY SPECIAL ARTS '95. May 17-21. Dubuque, IA. A celebration of folk and fine arts. Dance, opera, music, mime, arts, crafts, poetry, drama and historic architecture. Annually, the third weekend in May, Wednesday-Sunday. Est attendance: 50,000. For info: DubuqueFest, 335 W 2nd, Dubuque, IA 52001. Phone: (319) 583-1217.

FIRST KENTUCKY DERBY: ANNIVERSARY. May 17, 1875. The first running of the Kentucky Derby took place at Churchill Downs, Louisville, KY. Jockey Oliver Lewis rode the horse Aristides to a winning time of 2:37¼.

NAIA SOFTBALL CHAMPIONSHIP. May 17-20. Site to be determined. 15th annual competition. Est attendance: 2,000. For info: Natl Assn of Intercollegiate Athletics, 6120 S Yale Ave, Ste 1450, Tulsa, OK 74136. Phone: (918) 494-8828.

NATIONAL EMPLOYEE HEALTH AND FITNESS DAY. May 17. To focus on the importance of fitness and healthy lifestyles at the worksite. Est attendance: 6,000. Sponsor: Natl Assn of Governor's Councils on Physical Fitness/Sports, 201 S Capitol Ave, Ste 560, Indianapolis, IN 46225-1072. Phone: (317) 237-5630.

NEW YORK STOCK EXCHANGE ESTABLISHED: ANNIVERSARY. May 17. On May 17, 1792, some two dozen merchants and brokers agreed to establish what is now known as the New York Stock Exchange. In fair weather they operated under a buttonwood tree on Wall Street, in New York, NY. In bad weather they moved to the shelter of a coffeehouse to conduct their business.

NORWAY: CONSTITUTION DAY OR INDEPENDENCE DAY. May 17. National holiday. Constitution signed and Norway separated from Denmark on this day, 1814. Parades and children's festivities.

NORWEGIAN 17TH OF MAY FESTIVAL. May 17. Ballard, Seattle, WA. Celebration of Norwegian Constitution Day, May 17, 1814, with luncheon, speeches, parade, dance and entertainment. Est attendance: 10,000. For info: Norwegian 17th of May Festival, PO Box 70433, Seattle, WA 98107. Phone: (206) 784-7894.

PHILIPPINES: FERTILITY RITES. May 17-19. Obando, Bulacan. A triple religious fete in honor of San Pascual, Santa Clara and the Virgin of Salambao marked by dancing of childless couples.

PREAKNESS FROG HOP. May 17. Broadway Market Square, Baltimore, MD. Event determines Baltimore's entry to the international frog jumping contest. Frogs must be four inches in length from the nose to where the tail should be (not counting the legs). Trophies awarded to the 1st-, 2nd-, and 3rd-place winners, plus the 1st-place frog only (not the owner) is flown to California to represent the City of Baltimore in the Calaveras County Frog Jumping Jubilee. Est attendance: 400. For info: Office of Adventures in Fun, Clarence "Du" Burns Arena, 1301 S Ellwood Ave, Baltimore, MD 21224. Phone: (410) 396-9177.

SANTA CRUZ BEACH BOARDWALK GIANT DIPPER: ANNIVERSARY. May 17. The Giant Dipper roller coaster opened at Santa Cruz Beach Boardwalk in Santa Cruz, CA, on May 17, 1924, and quickly became the park's most popular ride. The Dipper was built by Arthur Looff, the son of master carousel horse-carver Charles I.D. Looff. In June of 1987, the Giant Dipper and the Looff carousel were designated National Historic Landmarks by the US National Park Service.

UFF DA DAY. May 17. Reeder, ND. An annual celebration of this Norwegian holiday. Events include a parade, program, Norwegian dinner, Norwegian lefse and rosettes and a dance. Est attendance: 400. For info: V.R. Honeyman, Reeder, ND 58649. Phone: (701) 853-9396.

UNITED NATIONS: WORLD TELECOMMUNICATION DAY. May 17. A day to draw attention to the necessity and importance of further development of telecommunications in the global community. Info from: United Nations, Dept of Public Info, New York, NY 10017.

May 17-18 ⋆ Chase's 1995 Calendar of Events ⋆

US NATIONAL SENIOR SPORTS CLASSIC V: THE SENIOR OLYMPICS. May 17-24. San Antonio, TX. Senior citizens compete nationwide at the local level in eighteen events (track and field, swimming, golf, tennis, archery, table tennis, triathlon, road races, race walking and others). Those placing first through third in each event are eligible for the national classic. More than 7,000 people competed in this biannual event in 1993. Competition is in five year groupings: 55-60, 60-65, etc, with a category for 100+. Yes, some people have competed on the local level in the 100+ category. Est attendance: 20,000. Sponsor: US Natl Senior Sports Organization, 14323 S Outer 40 Rd, Chesterfield, MO 63014. Phone: (314) 878-4900.

USS STARK: ATTACK ANNIVERSARY. May 17. The US Navy's guided missile frigate, *Stark*, sailing off the Iranian coast, in the Persian Gulf, was struck and set afire by two Exocet sea-skimming missiles fired from an Iraqi warplane at 2:10 PM, EDT, May 17, 1987. Also struck was a Cypriot flag tanker. At least 28 American naval personnel were killed in the attack on the *Stark*. Only hours before the attack a Soviet oil tanker had struck a mine in the Persian Gulf.

WEST VIRGINIA STRAWBERRY FESTIVAL. May 17-21. Buckhannon, WV. A family event with activities including 3 parades, carnival, strawberry auctions, band competitions, Party Gras, Concert, and special entertainment for all. Est attendance: 100,000. For info: Scott Jenkins, Pres, Box 117, Buckhannon, WV 26201. Phone: (304) 472-9036.

BIRTHDAYS TODAY

Dennis Hopper, 59, actor (*Easy Rider, Rebel Without a Cause*), born Dodge City, KS, May 17, 1936.
Christian Lacroix, 45, French couturier, born at Arles, France, May 17, 1950.
Sugar Ray Leonard (Ray Charles Leonard), 39, former boxer, born at Washington, DC, May 17, 1956.
Daniel Ricardo Manning, 29, basketball player, born at Hattiesburg, MS, May 17, 1966.
E. Benjamin Nelson, 54, Governor of Nebraska (D), born at McCook, NE, May 17, 1941.
Hazel O'Leary, 58, US Secretary of Energy in Clinton administration, born at Newport News, VA, May 17, 1937.
Maureen O'Sullivan, 84, actress (Jane in the Tarzan films; *Hannah and Her Sisters*), born Voyle, Ireland, May 17, 1911.

MAY 18 — THURSDAY
138th Day — Remaining, 227

BIRTHDAY OF MOTHER'S WHISTLER. May 18. To celebrate the delightful sound of whistling, especially from the beaks of birds and from the nose of Mother's Whistler (preeminent authority on whistling and on the music of rare, exotic and even extinct birds). Sponsor: Mother's Whistler, Warfield and Twin Silo Lanes, Huntingdon Valley, PA 19006. Phone: (215) 947-7007.

CALAVERAS COUNTY FAIR AND JUMPING FROG JUBILEE. May 18-21. Calaveras Fairgrounds, Angels Camp, CA. County fair and reenactment of Mark Twain's "Celebrated Jumping Frog of Calaveras County." This "Superbowl" of the sport of frog jumping attracts more than 3,000 frogs annually from around the world. Est attendance: 45,000. Sponsor: State of California, 39th District Agricultural Assn, Box 489, S Highway 49, Angels Camp, CA 95222. Phone: (209) 736-2561.

	S	M	T	W	T	F	S
May 1995		1	2	3	4	5	6
	7	8	9	10	11	12	13
	14	15	16	17	18	19	20
	21	22	23	24	25	26	27
	28	29	30	31			

CAPRA, FRANK: BIRTH ANNIVERSARY. May 18. The Academy Award-winning director whose movies were suffused with affectionate portrayals of the common man and the strengths and foibles of American democracy. Capra was born in 1897 at Palermo, Sicily. He bluffed his way into silent movies in 1922 and, despite total ignorance of movie making, directed and produced a profitable one-reeler. He was the first to win three directorial Oscars—for *It Happened One Night* (1934), *Mr. Deeds Goes to Town* (1936) and *You Can't Take It with You* (1938). The motion picture academy voted the first and third of these as best picture. Capra said his favorite film was *It's a Wonderful Life* (1946). He died Sept 3, 1991.

ENGLAND: FESTIVAL OF ARTS AND CULTURE—PAIGNTON 700. May 18-21. Various venues, Paignton, Devon. A week of special events to mark the 700th anniversary of Paignton. For info: Simon Bradley, Mktg/Promotion Dir, Riviera Tourist Board, The Tourist Centre, Vaughan Parade, Torquay, England TQ2 5JG.

FONTEYN, MARGOT: BIRTH ANNIVERSARY. May 18. Born Margaret Hookman in Reigate, Surrey, England, May 18, 1919, Dame Margot Fonteyn thrilled ballet audiences for 45 years. She emerged from the Sadler's Wells company during the 1930s and 1940s as a solo artist and followed those successes by partnering with Soviet exile Rudolph Nureyev in the 1960s. She died Feb 21, 1991, in Panama City, Panama.

GATLINBURG SCOTTISH FESTIVAL AND HIGHLAND GAMES. May 18-20. Gatlinburg, TN. Celebration of Scotland, its people and traditions. Music, food and fun. For info: Gatlinburg Chamber of Commerce, PO Box 527, Gatlinburg, TN 37738. Phone: (800) 568-4748.

GRAND PRAIRIE WESTERN DAYS. May 18-21. Grand Prairie, TX. Tribute to western heritage featuring Professional Rodeo Cowboys Association rodeo performances, arts and crafts show, food fair, western art show, Old West gun fights, parade, barbecue, C & W dance and free lunches. Est attendance: 60,000. For info: Traders Village, 2602 Mayfield Rd, Grand Prairie, TX 75052. Phone: (214) 647-2331.

HAITI: FLAG AND UNIVERSITY DAY. May 18. Public holiday.

HARTSCAPADES. May 18-21. Hartsville, SC. Dedicated to family fun—a full weekend of day and evening events. Est attendance: 10,000. For info: Greater Hartsville Chamber of Commerce, Robin Allen, Admin Asst, PO Box 578, Hartsville, SC 29551. Phone: (803) 332-6401.

INTERNATIONAL MUSEUM DAY. May 18. To pay tribute to museums of the world. "Museums are an important means of cultural exchange, enrichment of cultures and development of mutual understanding, cooperation and peace among people." Annually, May 18. Sponsor: Intl Council of Museums, Paris, France. For info: AAM/ICOM, 1225 Eye St NW, Ste 200, Washington, DC 20005. Phone: (202) 289-1818.

LAG B'OMER. May 18. Hebrew calendar date: Iyar 18, 5755. Literally, the 33rd day of the omer (harvest time), the 33rd day after the beginning of Passover. Traditionally a joyous day for weddings, picnics and outdoor activities.

Chase's 1995 Calendar of Events — May 18–19

LITTLE NORWAY FESTIVAL. May 18–21. Petersburg, AK. Celebration of Norwegian Independence Day. The majestic natural beauty that gives Petersburg its nickname ("Little Norway"), the rich Norwegian traditions and spirit and the combination of visitors from near and far with friendly residents form the backdrop to the gala celebration. Est attendance: 3,000. For info: Petersburg Chamber of Commerce, Box 649, Petersburg, AK 99833. Phone: (907) 772-3646.

MARQUETTE, JACQUES: DEATH ANNIVERSARY. May 18. Father Jacques Marquette (Père Marquette), Jesuit missionary-explorer of the Great Lakes region. Died May 18, 1675, near Ludington or Frankfort, MI.

MEN'S BIG 8 BASEBALL CHAMPIONSHIP. May 18–21. All Sports Stadium, Oklahoma City, OK. "Most successful baseball conference championship in the country." Est attendance: 45,000. For info: OK City All Sports Assn, 100 W Main, Ste 287, Oklahoma City, OK 73102. Phone: (405) 236-5000.

MOUNT ST. HELENS ERUPTION: 15th ANNIVERSARY. May 18. A major eruption of Mount St. Helens volcano, in southwestern Washington, on May 18, 1980, blew steam and ash more than 11 miles into the sky. First major eruption of Mount St. Helens since 1857, though on Mar 26, 1980, there had been a warning eruption of smaller magnitude.

POPE JOHN PAUL II: 75th BIRTHDAY. May 18. Karol Wojtyla, 264th pope of the Roman Catholic Church, born at Wadowice, Poland, May 18, 1920. Elected pope on Oct 16, 1978. He was the first non-Italian to be elected pope in 456 years (since the election of Pope Adrian VI, in 1522) and the first Polish pope.

SPACE MILESTONE: APOLLO 10 (US). May 18. Colonel Thomas Stafford and Commander Eugene Cernan brought lunar module (LM) "Snoopy" within 9 miles of moon's surface, May 22. Apollo 10 circled moon 31 times and returned to Earth May 26. Launched May 18, 1969.

VISIT YOUR RELATIVES DAY. May 18. A day to renew family ties and joys by visiting often-thought-of-seldom-seen relatives. Annually, May 18. Sponsor: A.C. Moeller, Box 71, Clio, MI 48420.

BIRTHDAYS TODAY

Perry Como (Pierino Como), 83, singer ("Catch a Falling Star"), actor, born at Canonsburg, PA, May 18, 1912.
Reggie Jackson, 49, baseball Hall of Famer, born at Wyncote, PA, May 18, 1946.
Yannick Noah, 35, tennis player, born at Sedan, France, May 18, 1960.
Pope John Paul II (Karol Wojtyla), 75, Roman Catholic leader, born at Wadowice, Poland, May 18, 1920.
Pernell Roberts, 67, actor (Ride Lonesome, "Bonaza," "Trapper John, M.D."), born Waycross, GA, May 18, 1928.
Warren Bruce Rudman, 65, lawyer, former US Senator (R, New Hampshire), born at Boston, MA, May 18, 1930.
James Stephens, 44, actor ("The Paper Chase"), born Mt Kisco, NY, May 18, 1951.

MAY 19 — FRIDAY
139th Day — Remaining, 226

AMBRIDGE NATIONALITY DAYS. May 19–21. Ambridge, PA. A fest that encompasses the ethnic attributes of the area. Features include crafts, entertainment and ethnic foods. Est attendance: 150,000. For info: Ambridge Area Chamber of Commerce, 719 Merchant St, 2nd Fl, Ambridge, PA 15003.

BIG TEN MEN'S/WOMEN'S OUTDOOR TRACK AND FIELD CHAMPIONSHIPS. May 19–20. Purdue University, West Lafayette, IN. For info: Big Ten Conference, 1500 W Higgins Rd, Park Ridge, IL 60068-6300. Phone: (708) 696-1010.

CHEHAW NATIONAL INDIAN FESTIVAL. May 19–21. Albany GA. Largest cultural festival in the southeast features more than 14 nations and 30 tribes celebrating Mother Earth and paying honor to the Creek Nation. Real Indian food, demonstrations of basketry, pottery making, flint knapping, bow making, flute music, hide tanning and leatherwork. For info: Grace Bannister, Chehaw Natl Indian Festival, PO Box 3492, Albany, GA 31707. Phone: (912) 436-1625.

DARK DAY IN NEW ENGLAND: ANNIVERSARY. May 19. At midday on May 19, 1780, near-total darkness unaccountably descended on much of New England. Candles were lit, fowls went to roost and many fearful persons believed that doomsday had arrived. At New Haven, CT, Colonel Abraham Davenport opposed adjournment of the town council in these words: "I am against adjournment. The day of judgment is either approaching or it is not. If it is not, there is no cause for an adjournment. If it is, I choose to be found doing my duty. I wish therefore that candles may be brought." No scientifically verifiable cause for this widespread phenomenon was ever discovered.

DEPOT DAYS. May 19–21. Leonard Park, Gainesville, TX. Continuous live entertainment, arts and crafts, bike rallies, historic home tours, miniature horse showing, carnival, carriage rides, antique car show, fun run, concessions, Frank Buck Zoo, street dance Friday night and much more. Est attendance: 8,500. For info: Judy Day, Dir, Community Revitalization, Inc, 200 S Rusk, Gainesville, TX 76240. Phone: (817) 665-8871.

DERMOTT'S ANNUAL CRAWFISH FESTIVAL. May 19–20. Dermott, AR. To publicize and popularize crawfish as a delicacy, to promote the area and to raise funds for industrial expansion. Family fun, arts, crafts, exotic foods, carnival, live music and street dances. Annually, the third weekend in May. Est attendance: 25,000. Sponsor: Dermott Area Chamber of Commerce, Herman Foster, Box 147, Dermott, AR 71638. Phone: (501) 538-5656.

ENGLAND: BATH INTERNATIONAL FESTIVAL. May 19–June 4. Various venues, Bath, Avon. International arts festival which will include jazz and opera performances, concerts, talks and exhibitions. Est attendance: 50,000. For info: Bath Festival Office, Linley House, 1 Pierrepoint Place, Bath, Avon, England BA1 1JY.

FISHING HAS NO BOUNDARIES—HAYWARD EVENT. May 19–21. Chippewa Camp Grounds, Hayward, WI. A three-day fishing experience for disabled persons. Any disability, age, sex, race, etc, eligible. Fishing with experienced guides on one of the best fishing waters in Wisconsin, attended by 250 participants and 750 volunteers. Est attendance: 750. For info: Fishing Has No Boundaries, PO Box 375, Hayward, WI 54843. Phone: (715) 634-3185.

HANSBERRY, LORRAINE: 65th BIRTH ANNIVERSARY. May 19. American playwright Lorraine Hansberry was born May 19, 1930, at Chicago, IL. For her now classic play A Raisin in the Sun (1959), she became the youngest American and first black to win the Best Play Award from the New York Critics' Circle. The play, based on an incident in Hansberry's life and titled after the Langston Hughes poem, deals with issues such as racism, cultural pride and self-respect and was the first stage production written by a black woman to appear on Broadway (1959). To Be Young, Gifted, and Black, a book of excerpts from her journals, letters, speeches and plays, was published posthumously in 1969. Tragically, Lorraine Hansberry died of cancer on Jan 12, 1965, at the age of 34.

HARRY'S HAY DAYS. May 19–21. Truman Corners Shopping Center, Grandview, MO. For only a buck the public can go back in time to the era when Harry S. Truman was farming his family homestead. You will have an opportunity to purchase artisans' handcrafted items, participate in Billy Goat Bingo contests, as well as enjoy old-fashioned barbecue and other taste treats. Annually, the third weekend in May. Est attendance: 18,000. For info: Mary Everitt, Fest Coord, 12500 S 71 Hwy, Grandview, MO 64030. Phone: (816) 761-6505.

May 19 ☆ Chase's 1995 Calendar of Events ☆

HO CHI MINH: BIRTH ANNIVERSARY. May 19. Vietnamese leader and first president of the Democratic Republic of Vietnam, born in central Vietnamese village of Kim Lien (Nghe An Province), probably on May 19, 1890. His original name was Nguyen That Thanh. Died at Hanoi, Vietnam, Sept 3, 1969.

HOLY TRINITY'S BRITISH FESTIVAL. May 19. Seville Square, Pensacola, FL. To promote authentic British food, arts and crafts. Est attendance: 4,000. Sponsor: Holy Trinity Episcopal Church, PO Box 3068, Pensacola, FL 32516. Phone: (904) 456-5474.

INTERNATIONAL PICKLE WEEK. May 19–29. To give national recognition to the world's most humorous vegetable. Sponsor: Pickle Packers Intl, Inc. For info: DHM Group, Inc, P.O. Box 767, Holmdel, NJ 07733.

MAIFEST '95. May 19–21. MainStrasse Village, Covington, KY. MainStrasse celebrates the German tradition of welcoming the first spring wines and the beginning of the festival season. Artist and craftsman exhibits, international food and drink, live music and entertainment. Est attendance: 275,000. For info: MainStrasse Village, Cindy Scheidt, 616 MainStrasse, Covington, KY 41011. Phone: (606) 491-0458.

MALCOLM X: 70th BIRTH ANNIVERSARY. May 19. Black nationalist and civil rights activist Malcolm X was born Malcolm Little, May 19, 1925, at Omaha, NE. While serving a prison term he resolved to transform his life. On his release in 1952 he changed his name to Malcolm X and worked for the Nation of Islam until he was suspended by Black Muslim leader Elijah Muhammed on Dec 4, 1963. Malcolm X later (Mar 12, 1964) delivered a "Declaration of Independence" statement to a press conference in New York, NY, and announced that he was forming the Organization of American Unity. He was assassinated as he spoke to a meeting at the Audubon Ballroom in New York, NY, on Feb 21, 1965.

MAY RAY DAY. May 19. To celebrate the beginning of the warm outside days the sun gives us. Also, a day for people named Ray. Annually, May 19. Sponsor: Richard Ankli, The Fifth Wheel Tavern, 639 Fifth St, Ann Arbor, MI 48103.

MILES CITY JAYCEE BUCKING HORSE SALE. May 19–21. Miles City, MT. Sale of bucking horses and bulls, plus a parade, pari-mutuel horse racing, rodeo, dances, barbecue, wild horse race, bareback riding, saddle bronc riding, bull riding and more. Est attendance: 3,200. For info: Miles City Chamber of Commerce, 901 Main St, Miles City, MT 59301. Phone: (406) 232-2890.

MINT JULEP SCALE MEET. May 19–21. Rough River Dam State Resort Park, Falls of Rough, KY. A weekend for model airplane enthusiasts. Est attendance: 400. For info: Rough River Dam State Resort Pk, Tom DeHaven, Falls of Rough, KY 40119. Phone: (502) 257-2311.

MORMON TRAIL RIDE. May 19–21. Las Flores Ranch, Summit Valley, CA. Open trail ride retraces the trails of the Mormon pioneers who crossed the deserts and mountains of California and settled the valley of San Bernardino in 1851. Annually, the third weekend of May. For info: Lou Dudney, Trail Boss, Mormon Trail Ride Assn, Apple Valley Sheriff's Posse, PO Box 338, Victorville, CA 92392. Phone: (619) 247-6911.

★ **NATIONAL DEFENSE TRANSPORTATION DAY.** May 19. Presidential Proclamation customarily issued as "National Defense Transportation Day and National Transportation Week." Issued each year for the third Friday in May since 1957. (PL85-32 of May 16, 1957.)

NATIONAL MEMO DAY. May 19. This is an event cherished by all of us who have to deal with daily memos from the boss. Today is the day for all office people to put up their own memos about having too many memos, memos concerning the latest office gossip or a memo just to give yourself the satisfaction of having posted a memo. For info: Dom Testa, KMJI-Radio, 5359 S Roslyn St, Ste 210, Englewood, CO 80111. Phone: (303) 741-5654.

ORANGE COUNTY DELAWARE RIVER FESTIVAL. May 19–21. Port Jervis, NY. Celebrates the beauty of the Delaware River at the "Tri-States" where New York, New Jersey and Pennsylvania all meet. Street fair, outdoor expo, crafters, art show, shad fishing contest, 5K race and walk, street dancing, music, entertainment, carnival, fireworks, car show, pony rides, petting zoo and much more. For info: Port Jervis Tourism Committee, PO Box 636, Port Jervis, NY 12771. Phone: (914) 856-6600.

PEALE, SARAH MIRIAM: BIRTH ANNIVERSARY. May 19. American portrait painter, member of famous early American family of painters, born at Philadelphia, PA, on May 19, 1800. Died there on Feb 4, 1885.

PICKLE FEST. May 19–20. Atkins, AR (home of the fried dill pickle). Festivities include World's Champion Pickle Juice Drinking Contest, World's Champion Pickle Eating Contest, a pickle tasting booth, tours of the pickle plant, the Pickle Pageant and pickle contest, arts and crafts and live entertainment. Annually, the third weekend in May. For info: People for a Better Atkins, Brian Miller, Chmn, 611 N Church, Atkins, AR 72823. Phone: (501) 641-2785.

RHODODENDRON FESTIVAL. May 19–21. Florence, OR. Abundance of springtime rhododendron blooms celebrated with a Grand Floral parade, arts and crafts show and vendor's fair. Annually, the third weekend in May. Est attendance: 35,000. For info: Chamber of Commerce, Box 26000, Florence, OR 97439. Phone: (503) 997-3128.

RIVER ROAST 1995. May 19–20. Ross's Landing, Chattanooga, TN. The Kidney Foundation's annual barbecue cookoff competition. Two days filled with lots of entertainment, a large volleyball tournament and skulling contest. Est attendance: 20,000. For info: Susan Lockemann, River Roast 1995, 620 Cherokee Blvd, Ste LL1, Chattanooga, TN 37405. Phone: (615) 265-4397.

SCOBEE, FRANCIS R.: BIRTH ANNIVERSARY. May 19. Commander of the ill-fated space shuttle *Challenger*, 46-year-old pilot Francis R. Scobee had been in the astronaut program since 1978 and had been pilot of the *Challenger* in 1984. Born at Cle Elum, WA, on May 19, 1939, Scobee perished with all others on board when the *Challenger* exploded on Jan 28, 1986. See also: "*Challenger* Space Shuttle Explosion Anniversary" (Jan 28).

SENIOR FESTIVAL AND EXPO. May 19–20. Rockford, IL. Only show in Midwest with products and services specifically for seniors. Entertainment and health education. Sponsor: Winnebago County Council on Aging. Est attendance: 10,000. For info: Rockford Area Conv and Visitors Bureau, Memorial Hall, 211 N Main St, Rockford, IL 61101. Phone: (800) 521-0849.

SIMPLON TUNNEL OPENING: ANNIVERSARY. May 19. Tunnel officially opened on this day in 1906. Construction started in 1898. From Brig, Switzerland, to Iselle, Italy.

May 1995

S	M	T	W	T	F	S
	1	2	3	4	5	6
7	8	9	10	11	12	13
14	15	16	17	18	19	20
21	22	23	24	25	26	27
28	29	30	31			

☆ Chase's 1995 Calendar of Events ☆ May 19–20

SPACE MILESTONE: *MARS 2* AND *MARS 3* (USSR). May 19 and 28. Entered Martian orbits on Nov 27 and Dec 2, respectively. *Mars 3* sent down a TV-equipped capsule that soft-landed and transmitted pictures for 20 seconds. Launch dates: May 19 and 28, 1971.

SPRING MUSIC FESTIVAL. May 19–20. Georgia Mountain Fairgrounds, Hiawassee, GA. A weekend of country music with amateur talent. Open invitation to any single performer or band wishing to participate. Est attendance: 10,000. For info: Dale Thurman, Mgr, PO Box 444, Hiawassee, GA 30546. Phone: (706) 896-4191.

STUBB'S EDDY RIVER RENDEZVOUS. May 19–21. Lindsay Park, Davenport, IA. To reenact the years of 1830 to 1860, Buckskinners and traders will camp and demonstrate hawk and knife throws and many contests. Women's and children's games of that period. Est attendance: 4,000. For info: Village of East Davenport Assn, 2215 E 12th St, Davenport, IA 52803. Phone: (319) 322-1860.

SUGARLOAF'S SPRING SOMERSET CRAFTS FESTIVAL. May 19–21. Garden State Exhibit Center, Somerset, NJ. Nearly 300 professional artists and craftspeople, demonstrations and children's entertainment. Annually, the third Friday–Sunday in May. For info: Deann Verdier, Dir, 200 Orchard Ridge Dr, #215, Gaithersburg, MD 20878. Phone: (301) 990-1400.

TURKEY: YOUTH AND SPORTS DAY. May 19. Public holiday commemorating beginning of national movement for independence in 1919, led by Mustafa Kemal Ataturk.

WILLAMETTE VALLEY FOLK FESTIVAL. May 19–21. Eugene, OR. 25th annual folk festival with multiple stages of musical performances, workshops and folk dancing. Sponsor: EMU Cultural Forum. Free. Est attendance: 4,000. For info: Debby Martin, Program Asst, EMU Cultural Forum, University of Oregon, Eugene, OR 97403. Phone: (503) 346-4373.

BIRTHDAYS TODAY

Rick Cerone, 41, former baseball player, born at Newark, NJ, May 19, 1954.
Nora Ephron, 54, writer (*Heartburn* [made into a movie]), born New York, NY, May 19, 1941.
David Hartman, 58, actor (Emmy for "Good Morning America"; *Hello Dolly*), born at Pawtucket, RI, May 19, 1937.
Grace Jones, 43, model, singer ("Sorry," "I Need a Man"), actress, born at Spanishtown, Jamaica, May 19, 1952.
Bill (William) Laimbeer, Jr., 38, former basketball player, born at Boston, MA, May 19, 1957.
James Lehrer, 61, journalist (co-anchor "MacNeil/Lehrer NewsHour"), born at Wichita, KS, May 19, 1934.
Peter Townshend, 50, musician (Who; "I Can See for Miles," "Let My Love Open the Door"), born at London, England, May 19, 1945.

MAY 20 — SATURDAY
140th Day — Remaining, 225

AMELIA EARHART ATLANTIC CROSSING: ANNIVERSARY. May 20. Leaving Harbor Grace, Newfoundland, at 7 PM on May 19, 1932, Amelia Earhart landed near Londonderry, Ireland, on May 20. The 2,026-mile flight took 13 hours and 30 minutes. She was the first woman to fly solo across the Atlantic. Earhart, along with her navigator Fred Noonan, disappeared on July 2, 1937, between Lae, New Guinea, and Howland Island while trying to fly her Lockheed twin-engine plane around the equator to gather scientific data.

AN AMERICAN HERITAGE FESTIVAL. May 20–21. Yorktown Victory Center, Yorktown, VA. Family-centered multicultural festival portrays American lifestyles, crafts and trades of the 17th, 18th and 19th centuries. Est attendance: 1,500. For info: Media Relations, Jamestown-Yorktown Fdtn, PO Box JF, Williamsburg, VA 23187. Phone: (804) 253-4838.

★ **ARMED FORCES DAY.** May 20. Presidential Proclamation 5983, of May 17, 1989, covers the third Saturday in May in all succeeding years. Originally proclaimed as "Army Day" for Apr 6, beginning in 1936 (S.Con.Res. 30 of Apr 2, 1936). S.Con.Res. 5 of Mar 16, 1937, requested annual Apr 6 issuance, which was done through 1949. Always the third Saturday in May since 1950. Traditionally issued once by each Administration.

ATWOOD EARLY ROD RUN. May 20–21. Atwood, KS. Classic car show with several states represented. Includes downtown festival, remote-control car races, barbecue and '50s dance. Est attendance: 1,500. For info: Atwood Ambassadors, 411 Page, Atwood, KS 67730. Phone: (913) 626-9630.

BALZAC, HONORE DE: BIRTH ANNIVERSARY. May 20. French novelist born at Tours, France, May 20, 1799. "It is easier," Balzac wrote in 1829, "to be a lover than a husband for the simple reason that it is more difficult to be witty every day than to say pretty things from time to time." Died at Paris, on Aug 18, 1850.

BEVERLY HILLS: AFFAIRE IN THE GARDENS. May 20–21. (Also Oct 21–22.) Beverly Gardens Park at Rodeo Drive, Beverly Hills, CA. To foster an appreciation of arts in the community. Juried show; only handcrafted items considered. Garden setting. Bi-annually, the third weekends in May and October. Est attendance: 50,000. Sponsor: Beverly Hills Recreation and Parks Dept, Brad Meyerowitz, Sr Rec Supr, 8400 Gregory Way, Beverly Hills, CA 90211. Phone: (310) 550-4628.

CALIFORNIA STRAWBERRY FESTIVAL. May 20–21. Strawberry Meadows of College Park, Oxnard, CA. This weekend event features a variety of gourmet strawberry foods, live musical entertainment, a 200-booth juried arts and crafts show, a 10K race and 2K family fun run, Strawberryland for children and the Strawberry Shortcake Eating Competition. Annually, the third weekend in May. Est attendance: 70,000. For info: Bill Garlock, Festival Mgr, 1621 Pacific Ave, #127, Oxnard, CA 93033. Phone: (805) 385-7578.

CAMEROON: NATIONAL HOLIDAY. May 20. Republic of Cameroon. Commemorates declaration of the republic on May 20, 1972.

CANADA: UPPER CANADA VILLAGE. May 20–Oct 9. Morrisburg, Ontario. Visitors enter the world of the 1860s. Costumed interpreters bring history to life in this fully operational rural community as they perform period activities and entertain as townsfolk or tradesmen applying their skills in restored buildings ranging from homes to trade shops. Many special events and programs offered throughout the season. Est attendance: 200,000. For info: Upper Canada Village, Parks of the St. Lawrence, RR 1, Morrisburg, Ont, Canada K0C 1X0. Phone: (613) 543-3704.

CELEBRATION OF QUILTS IX. May 20–21. Chardon, OH. Quilting guilds and shops set up in Century Village homes to demonstrate a variety of quilting techniques. Antique, contemporary and Amish quilts and quilting supplies. Lectures; museum quilts on display. Est attendance: 3,200. Sponsor: Geauga County Historical Soc, Joan Campbell, PO Box 462, Chardon, OH 44024. Phone: (216) 286-3763.

CHEROKEE ROSE FESTIVAL AND TOURS. May 20–21. Gilmer, TX. Est attendance: 10,000. For info: Chamber of Commerce, Box 854, Gilmer, TX 75644. Phone: (903) 843-2413.

May 20 ☆ Chase's 1995 Calendar of Events ☆

COUNCIL OF NICAEA I: 1,670th ANNIVERSARY. May 20–Aug 25. First ecumenical council of Christian Church, called by Constantine I, first Christian emperor of Roman Empire. Nearly 300 bishops are said to have attended this first of 21 ecumenical councils (latest, Vatican II, began Sept 11, 1962), which was held at Nicaea, Bithynia, in Asia Minor in the year 325. Dates and attendance are approximate. The council condemned Arianism (which denied divinity of Christ), formulated the Nicene Creed and fixed the date of Easter.

DESCENDANTS OF SIMON GAY FAMILY REUNION. May 20–21. Jones Bldg, Pavo Hwy, Moultrie, GA. Events all day both days. Approximate attendance: 1,000. For info: Rondo Gay, Simon Gay Family Reunion, Rt 1 Box 70-A, Moultrie, GA 31768. Phone: (912) 941-5359.

DON MacLEOD COOKOUT (WITH ELK ANTLER AUCTION). May 20. Town Square, Jackson, WY. Hot barbecue beef, hot dogs and baked goods in conjunction with the annual Elk Antler Auction. Boy Scouts harvest the dropped antlers and the proceeds from the auction benefit the BSA and the Natl Elk Refuge. Annually, the third Saturday in May. Est attendance: 800. Sponsor: Teton County Historical Center, Box 1005, Jackson, WY 83001. Phone: (307) 733-9605.

DULCIMER DAYS. May 20–21. Coshocton, OH. Dulcimer competition, exhibits, open stage, jam sessions, Saturday evening concert and more. Est attendance: 1,400. For info: Roscoe Village, Sally Gerycz, 381 Hill St, Coshocton, OH 43812. Phone: (800) 877-1830.

EIGHTEENTH-CENTURY SPRING MARKET FAIR. May 20–21. (Also July 15–16, Oct 21–22.) McLean, VA. Learn 18th-century crafts, music and games. Period wares for sale. Est attendance: 2,500. For info: Claude Moore Colonial Farm at Turkey Run, 6310 Georgetown Pike, McLean, VA 22101. Phone: (703) 442-7557.

ELIZA DOOLITTLE DAY. May 20. To honor Miss Doolittle (heroine of Bernard Shaw's *Pygmalion*) for demonstrating the importance and the advantage of speaking one's native language properly. Est attendance: 15. Sponsor: Doolittle Day Committee, 2460 Devonshire Rd, Ann Arbor, MI 48104.

FRED E. MILLER: PHOTOGRAPHER OF THE CROWS. May 20–Sept 3. Carnegie Museum of Natural History, Pittsburgh, PA. See Jan 14 for full description. Call or write the Smithsonian for other dates and venues. For info: Smithsonian Institution Traveling Exhibition Service, 1100 Jefferson Dr SW, Ste 3146, Washington, DC 20560. Phone: (202) 357-2700.

GETTYSBURG OUTDOOR ANTIQUE SHOW. May 20. (Also Sept 23.) Gettysburg, PA. Features 180 dealers in antiques with exhibits and displays. Annually, the third Saturday in May and the fourth Saturday in September. Est attendance: 25,000. For info: Gettysburg Travel Council, 35 Carlisle St, Gettysburg, PA 17325. Phone: (717) 334-6274.

GIGLI, BENIAMINO: BIRTH ANNIVERSARY. May 20. Celebrated Italian tenor born at Recanati, Italy, on May 20, 1890. Died at Rome, on Nov 30, 1957.

HARLEM: PHOTOGRAPHS BY AARON SISKIND. May 20–July 16. Montclair Art Museum. Montclair, NJ. For full description see Mar 4. Call or write the Smithsonian for other dates and venues. For info: Smithsonian Institution Traveling Exhibition Service, 1100 Jefferson Dr SW, Ste 3146, Washington, DC 20560. Phone: (202) 357-2700.

INTERNATIONAL SPRING FESTIVAL. May 20–21. Micke Grove Park and Zoo, Lodi, CA. Honors ethnic diversity of our country and features stage shows, native dances, costumes, crafts and food. Est attendance: 5,000. For info: San Joaquin County Parks and Recreation, Intl Spring Festival Committee, 11793 N Micke Grove Rd, Lodi, CA 95240. Phone: (209) 953-8800.

IRIS FESTIVAL. May 20–21. 5th & Grand, Ponca City, OK. Arts, crafts, food, entertainment. Est attendance: 12,000. For info: Ponca City Main Street Authority, PO Box 2532, Ponca City, OK 74602. Phone: (405) 763-8082.

LAFAYETTE DAY. May 20. Massachusetts.

LEWIS AND CLARK RENDEZVOUS. May 20–21. St. Charles, MO. Authentic reenactment of Lewis and Clark's encampment in 1804 prior to embarking on the exploration of the Louisiana Purchase. Activities include parades with fife and drum corps, a frontier dinner and ball, court martial reenactment, black powder shoot, church service and 19th-century crafts. Annually, the third weekend in May. Est attendance: 12,000. For info: St. Charles Conv and Visitors Bureau, 230 S Main, St. Charles, MO 63301. Phone: (800) 366-2427.

LINDBERGH FLIGHT: ANNIVERSARY. May 20–21. Anniversary of the first solo trans-Atlantic flight. Captain Charles Augustus Lindbergh, 25-year-old aviator, departed from rainy, muddy Roosevelt Field, Long Island, NY, alone at 7:52 AM, May 20, 1927, in a Ryan monoplane named *Spirit of St. Louis*. He landed at Le Bourget airfield, Paris, at 10:24 PM Paris time (5:24 PM, NY time), on May 21, winning a $25,000 prize offered by Raymond Orteig for the first nonstop flight between New York City and Paris, France (3,600 miles). The "flying fool" as he had been dubbed by some doubters became "Lucky Lindy," an instant world hero. See also: "Lindbergh, Charles Augustus: Birth Anniversary" (Feb 4).

MADISON, DOLLY (DOROTHEA) DANDRIDGE PAYNE TODD: BIRTH ANNIVERSARY. May 20. Wife of James Madison, 4th president of the US, born at Guilford County, NC, May 20, 1768. Died July 12, 1849.

MAIFEST 1995. May 20–21. Hermann, MO. Winery tours, wine tasting, house tours, stage show, volksplatz (craft area), parade each day, museum tours, oompah bands, carnival and beer gardens. Annually, the third weekend of May. Est attendance: 20,000. For info: Visitor Info Center, 306 Market Street, Hermann, MO 65041. Phone: (314) 486-2744.

MECKLENBURG DAY. May 20. North Carolina. Commemorates claimed signing of a declaration of independence from England by citizens of Mecklenburg County on this day, 1775.

MENTONE RHODODENDRON FESTIVAL. May 20–21. Mentone, AL. The festival includes boat rides on the river to view rhododendron and mountain laurel. Crafts, beauty pageant, gala parade and good food. Annually, the third weekend in May. For info: Mentone Rhododendron Festival, PO Box 50, Mentone, AL 35984. Phone: (205) 634-4541.

MIFFLIN-JUNIATA ARTS FESTIVAL. May 20–21. Lewistown Rec Park, Lewistown, PA. 75 juried arts and craft vendors; professional and amateur performances; children's crafts and special exhibits. Admission free. Est attendance: 12,000. For info: Jenny Landis, c/o Mifflin-Juniata Arts Council, PO Box 1126, Lewistown, PA 17044. Phone: (717) 248-8711.

May 1995

S	M	T	W	T	F	S
	1	2	3	4	5	6
7	8	9	10	11	12	13
14	15	16	17	18	19	20
21	22	23	24	25	26	27
28	29	30	31			

☆ Chase's 1995 Calendar of Events ☆ May 20

MOTOR VOTER BILL SIGNED: ANNIVERSARY. May 20, 1993. The latest effort to remove barriers to voter registration resulted in the passage of the Motor Voter Bill, which was signed into law by President William Clinton on this date. This bill requires the states to allow voter registration by mail or when a citizen applies for or renews a driver's license.

NATIONAL PIKE FESTIVAL. May 20–21. Washington County, PA. Festival is historical in nature, celebrating the construction in 1818 of US Route 40, starting at West Alexander, Washington County, PA, and extending 87 miles eastward. Annually, the third weekend in May. Est attendance: 100,000. For info: Washington Co Tourism, 144A McClelland Rd, Canonsburg, PA 15317. Phone: (800) 531-4114.

NATIONAL SAFE BOATING WEEK. May 20. Brings boating safety to the public's attention, decreases the number of boating fatalities and makes the waterways safer for all boaters. Sponsor: US Coast Guard and Natl Safe Boating Council. For info: Natl Safe Boating Council, Inc, Jo Calkin, Commandant (G-NAB-3), US Coast Guard, 2100 Second St SW, Washington, DC 20593. Phone: (800) 368-5647.

◆ **NEW YORK PUBLIC LIBRARY CENTENNIAL CELEBRATION.** May 20. (Tentative.) Central Research Library, 5th Ave and 42nd St, New York, NY. A full day of festivities for the whole family including musical performances, celebrity readings, ribbon-cutting and evening light/laser show. Fax: (212) 768-7439. For info: New York Public Library, Nancy Donner, PR Mgr, 5th Ave and 42nd St, New York, NY 10018-2788. Phone: (212) 221-7676.

"PEC THING." May 20–21. Pecatonica, IL. Also Sept 16–17. Semi-annual antique show with more than 400 exhibitors. Est attendance: 15,000. For info: Winnebago Country Fair Assn, PO Box 38, Pecatonica, IL 61063. Phone: (815) 239-1188.

POLE PEDAL PADDLE. May 20. Bend, OR. Five-stage ironman-type race for singles, teams and couples; includes downhill skiing, nordic skiing, biking, canoeing and running. Annually, the third Saturday in May. Sponsors: US Bank, Pepsi, Bud Light, Teva, Cellular One, Z21TV. Est attendance: 12,000. For info: Mt Bachelor Ski Education Foundation, PO Box 388, Bend, OR 97709. Phone: (503) 388-0002.

PREAKNESS STAKES. May 20. Pimlico Race Course, Baltimore, MD. Running of the Preakness Stakes, middle jewel in the Triple Crown, was inaugurated in 1873. Annually, the third Saturday in May—two Saturdays after the Kentucky Derby—and followed, three Saturdays later, by the Belmont Stakes. Est attendance: 100,000. For info: Maryland Jockey Club, Pimlico Race Course, Baltimore, MD 21215. Phone: (410) 542-9400.

RANCH RODEO. May 20. Fairgrounds, Liberal, KS. Contestants are real working cowboys (they work for local ranches or feedlots). Events include a Wild Cow Milking, Calf Dressing and Wake-Up Call. Annually, the third Saturday in May. Est attendance: 1,500. For info: Debra Huddleston, Box 420, Liberal, KS 67905-0420. Phone: (316) 624-3712.

RHUBARB FESTIVAL. May 20. Intercourse, PA. A lighthearted celebration honoring rhubarb, which grows in abundance in the Pennsylvania Dutch country. Food, games and contests all featuring rhubarb. Est attendance: 10,000. For info: Kitchen Kettle Village, Mel Hauser, Box 380, Intercourse, PA 17534. Phone: (800) 732-3538.

ROUSSEAU, HENRI JULIEN FELIX: BIRTH ANNIVERSARY. May 20. Henri Rousseau, nicknamed Le Douanier because of his onetime post as customs toll-keeper. Celebrated French painter born at Laval, Mayenne, France, May 20, 1844. Clarinet player in a regimental band, painter of deceptively "primitive" pictures of exotic foliage, flowers and fruit of the jungle, with stilted human and animal figures. Died at Hospital Necker, Paris, Sept 4, 1910.

SEND A KID TO KAMP RADIOTHON. May 20. Church of God State Campground, Lexington, KY. Annually, the third Saturday in May since 1988. Fax: (606) 245-1806. Est attendance: 15,000. For info: Dennis Smith, Genl Mgr, 770 AM, WCGW, Lexington Green, Ste 600, 3191 Nicholasville, Lexington, KY 40503. Phone: (606) 885-9119.

SPACE MILESTONE: *PIONEER VENUS I* (US). May 20. Became first Venus orbiter on Dec 4. Launched May 20, 1978.

SPRING BREAKING-UP AT GIANTS RIDGE. May 20–21. Biwabik, MN. Bicycling race, rollerblade. Est attendance: 250. For info: John Filander, PO Box 190, Biwabik, MN 55708. Phone: (800) 688-7669.

STAGECOACH DAYS. May 20–21. Marshall, TX. Arts and craft booths, Wild West shootouts, stagecoach rides, historic home tours, children's events, parade, Miss Loose Caboose Contest, live entertainment. Annually, the third weekend in May. Est attendance: 30,000. For info: Pam Whisenant, Dir of Conv & Visitor Development, PO Box 520, Marshall, TX 75671. Phone: (903) 935-7868.

TRINIDAD: PAN RAMAJAY. May 20–June 3. National Flour Mills, Port of Spain. Steel band competition. Using the steel pan, one of the world's newest musical instruments, bands play music ranging from calypso to classical in this annual competition. The event emphasizes the virtuosity of the players. Preliminaries: May 20–21; Semi-Finals: May 27; Finals: June 3. For info: Trinidad and Tobago TDA, 25 W 43rd St, Ste 1508, New York, NY 10036. Phone: (800) 232-0082.

WEBSTER COUNTY WOODCHOPPING FESTIVAL WITH WOODCHOPPING AND TURKEY CALLING CHAMPIONSHIP CONTESTS. May 20–28. Webster Springs, WV. South Eastern United States World Championship Woodchopping Contest and State Championship Turkey Calling Contest. Est attendance: 10,000. Sponsor: Woodchopping Festival Committee, PO Box 227, Webster Springs, WV 26288. Phone: (304) 847-7666.

WEIGHTS AND MEASURES DAY. May 20. Anniversary of international treaty, signed May 20, 1875, providing for the establishment of an International Bureau of Weights and Measures. The bureau was founded on international territory at Sevres, France.

WHO'S IN CHARGE? WORKERS AND MANAGERS IN THE UNITED STATES. May 20–July 2. Ohio Historical Society, Youngstown, OH. 150 years of conflict and change in the relationships between workers and managers. See Jan 14 for full description. For info: Smithsonian Institution Traveling Exhibition Service, 1100 Jefferson Dr SW, Ste 3146, Washington, DC 20560. Phone: (202) 357-2700.

WILD WEST DAYS. May 20–21. Rockford, IL. Stage coach rides, bank robbery re-enactment, cowboy music, roping demonstrations, gunfights, line dancing and great grub. Est attendance: 5,000. For info: Midway Village, 6799 Guilford Rd, Rockford, IL 61107. Phone: (815) 397-9112.

WILDFLOWER FESTIVAL OF THE ARTS. May 20–21. Dahlonega Town Square, Dahlonega, GA. Springtime in the North Georgia Mountains in a historic 1800s mining town. Visual and performing artists, programs and exhibits on wildflowers of the area, children's art area where children can create their own art. 9th annual fair. Est attendance: 25,000. For info: Dahlonega Welcome Center, 101 S Park St, Dahlonega, GA 30533. Phone: (706) 864-3711.

WRIGHT PLUS. May 20. Oak Park, IL. The Frank Lloyd Wright Home and Studio Foundation's annual housewalk features guided tours of the interiors of 10 buildings designed by Frank Lloyd Wright and his architectural contemporaries, including Unity Temple & Wright's own home and studio. Reservations are required. Tickets limited. (Tickets available Mar 1.) Annually, third Saturday in May. Est attendance: 3,000. For info: Frank Lloyd Wright Home and Studio, 951 Chicago Ave, Oak Park, IL 60302. Phone: (708) 848-1500.

May 20-21 ☆ Chase's 1995 Calendar of Events ☆

BIRTHDAYS TODAY

Cher (Cherilyn Sarkisian), 49, singer ("Half Breed," "Dark Lady"), actress (*Moonstruck*), born El Centro, CA, May 20, 1946.

Joe Cocker, 51, singer ("You Are So Beautiful"), born at Sheffield, England, May 20, 1944.

John R. McKernan, Jr, 47, Governor of Maine (R), born at Bangor, ME, May 20, 1948.

Bronson Pinchot, 36, actor ("Perfect Strangers," *Beverly Hills Cop*), born New York, NY, May 20, 1959.

Ronald Prescott Reagan, 37, dancer, talk show host, son of the former President, born Los Angeles, CA, May 20, 1958.

James "Jimmy" Stewart, 87, actor (*Rear Window, Philadelphia Story, It's a Wonderful Life*), born at Indiana, PA, May 20, 1908.

Constance Towers, 62, actress, born at Whitefish, MT, May 20, 1933.

MAY 21 — SUNDAY
141st Day — Remaining, 224

BURR, RAYMOND WILLIAM STACY: BIRTH ANNIVERSARY. May 21. Stage, film and TV actor best known for the role of Perry Mason in the series of the same name. He was born at New Westminster, British Columbia on May 21, 1917, and died near Healdsburg, CA, on Sept 12, 1993.

CURTISS, GLENN HAMMOND: BIRTH ANNIVERSARY. May 21. American inventor and aviator, born at Hammondsport, NY, on May 21, 1878. The aviation pioneer died at Buffalo, NY, on July 23, 1930.

D.C. BOOTH DAY (WITH FISH CULTURE HALL OF FAME). May 21. D.C. Booth Historic Fish Hatchery, Spearfish, SD. Antique auto show, musical entertainment, refreshments, free tours of the historic Booth home, free fly-tying demonstration, installation of new members into the Fish Culture Hall of Fame. Est attendance: 1,200. For info: Arden Trandahl, D.C. Booth Historic Fish Hatchery, 423 Hatchery Circle, Spearfish, SD 57783. Phone: (605) 642-7730.

DURER, ALBRECHT: BIRTH ANNIVERSARY. May 21. German painter and engraver, one of the greatest artists of the Renaissance, was born at Nuremberg, Germany, May 21, 1471, and died there Apr 6, 1528.

EXAMINER BAY TO BREAKERS RACE. May 21. San Francisco, CA. Largest foot race in the world attracts 100,000 runners each year, from world-class athletes to fun runners; post-race festival, live concert, food and beverage. Annually, the third Sunday in May. Phone: (415) 512-5000, ext 2222. Est attendance: 200,000. For info: *Examiner* Bay to Breakers, PO Box 42900, San Francisco, CA 94142.

FRY, ELIZABETH GURNEY: BIRTH ANNIVERSARY. May 21. English reformer who dedicated her life to improving the condition of the poor and especially of women in prison, born at Earlham, Norfolk, England, May 21, 1780. Died at Ramsgate, on Oct 12, 1845.

May 1995	S	M	T	W	T	F	S
		1	2	3	4	5	6
	7	8	9	10	11	12	13
	14	15	16	17	18	19	20
	21	22	23	24	25	26	27
	28	29	30	31			

GEMINI, THE TWINS. May 21–June 20. In the astronomical/astrological zodiac, which divides the sun's apparent orbit into 12 segments, the period May 21–June 20 is identified, traditionally, as the sun sign period of Gemini, the Twins. The ruling planet is Mercury.

HUMMEL, SISTER MARIA INNOCENTIA: BIRTH ANNIVERSARY. Born at Massing, Bavaria, on May 21, 1909, Sister Maria Innocentia Hummel cultivated an early interest in drawing and attended Munich's Academy of Fine Arts. Although she was eager to learn, she was reluctant to become involved in student life. She entered Siessen Convent run by the Sisters of the Third Order of St. Francis, a teaching order, began teaching art to kindergarten children and was ordained in 1933. Her drawings attained nationwide attention at a symposium of kindergarten teachers, and in 1934 Franz Goebel obtained an exclusive license to translate her drawings into three-dimensional figurines. The first M.I. Hummel figurines were displayed at the Leipzig Trade Fair in 1935; they made their first appearance in the American market in May 1935. She died on Nov 6, 1946, at Siessen, Germany. Many M.I. Hummel Clubs across the country commemorate her birth date with special events and fund-raisers for local charities.

MOON PHASE: LAST QUARTER. May 21. Moon enters Last Quarter phase at 7:36 AM, EDT.

NATIONAL SURGICAL TECHNOLOGIST WEEK. May 21–27. A week sponsored by the Association of Surgical Technologists to recognize the important role of surgical technologists in the provision of quality surgical patient care. Annually, beginning with the third Sunday in May. For info: Michelle Armstrong, Communications Mgr, Assn of Surgical Technology, 7108-C S Alton Way, Englewood, CO 80112. Phone: (303) 694-9130.

NATIONAL WAITRESSES DAY/(ALSO, P.C.) WAITRONS DAY. May 21. A day for restaurant managers and patrons to recognize and to express their appreciation for the many fine and dedicated waitresses and waiters. For info: Gaylord F. Ward, 1505 E Bristol Rd, Burton, MI 48529-2214.

NO LAUGHING MATTER: POLITICAL CARTOONISTS ON THE ENVIRONMENT. May 21–July 23. Sonoma County Museum, Santa Rosa, CA. See Feb 1 for full description. Call or write the Smithsonian for other dates and venues. For info: Smithsonian Institution Traveling Exhibition Service, 1100 Jefferson Dr SW, Ste 3146, Washington, DC 20560. Phone: (202) 357-2700.

"PIETA" ATTACKED: ANNIVERSARY. May 21. May 21, 1972, Lazlo Toth, a Hungarian native, attacked Michelangelo's centuries-old sculpture "The Pieta" while screaming, "I am Jesus Christ!" The Madonna's left arm was severed, her veil and nose were smashed and her left eye was disfigured.

POPE, ALEXANDER: BIRTH ANNIVERSARY. May 21. English poet born at London, England, May 21, 1688. "A man," Pope wrote in 1727, "should never be ashamed to own he has been in the wrong, which is but saying, in other words, that he is wiser today than he was yesterday." Died at Twickenham, May 30, 1744.

RAJIV GANDHI ASSASSINATED: ANNIVERSARY. May 21. Former Indian Prime Minister Rajiv Gandhi was assassinated on May 21, 1991. Gandhi was in the midst of a reelection campaign. He was killed when a bomb, hidden in a bouquet of flowers given by admirers, exploded as he approached a dais to begin a campaign rally. He had served as prime minister between 1984 and 1989 after succeeding his mother, Indira Gandhi, who was assassinated in 1984.

RED CROSS: FOUNDING ANNIVERSARY. May 21. Commemorates the founding of the Red Cross in 1881 by Clara Barton, its first president. The organization is a voluntary, not-for-profit organization governed and directed by volunteers and provides disaster relief at home and abroad. 1.1 million volunteers are involved in community services such as collecting and distributing donated blood and blood products, teaching health

✯ Chase's 1995 Calendar of Events ✯ May 21-22

and safety classes and acting as a medium for emergency communication between Americans and their armed forces.

RESEARCH/STUDY TEAM ON NONVIOLENT LARGE SYSTEMS CHANGE. May 21-23. George Williams College, Williams Bay, WI. Discussions and recommendations on better ways of solving problems without violence. We bring people from Eastern Europe and The Middle East and South Africa and other areas in conflict to the conferences. Presentations are invited. Registration is free. Est attendance: 60. For info: Dr. Donald W. Cole, RODC, The Org Dev Institute, 11234 Walnut Ridge Rd, Chesterland, OH 44026. Phone: (216) 729-7419.

ROGATION SUNDAY. May 21. The fifth Sunday after Easter is the beginning of Rogationtide (Rogation Sunday and the following three days before Ascension Day). Rogation Day rituals date from the fifth century.

RURAL LIFE SUNDAY. May 21. Also known as Soil Stewardship Sunday. With an increase in ecological and environmental concerns, Rural Life Sunday emphasizes the concept that Earth belongs to God, who has granted humanity the use of it, along with the responsibility of caring for it wisely. At the suggestion of the International Association of Agricultural Missions, Rural Life Sunday was first observed in 1929, under plans adopted by the Home Missions Council of North America and the Federal Council of Churches. The day is observed annually by churches of many Christian denominations and includes pulpit exchanges by rural and urban pastors. Under the auspices of the National Association of Soil and Water Conservation Districts, the week beginning with Rural Life Sunday is now widely observed as Soil Stewardship Week, with the Sunday itself alternatively termed Soil Stewardship Sunday. As environmental issues loom larger, observances now feature calls for individual action on the problems of air, water and noise pollution, in addition to problems associated with the conservation of open spaces. Traditionally, Rural Life Sunday is the Sunday preceding Ascension Day (Rogation Sunday).

ST. LOUIS WALK OF FAME INDUCTION CEREMONY. May 21. St. Louis, MO. Outdoor free-admission ceremony with ragtime band. Famous St. Louisans are honored. Nonprofit organization provides a showcase for the cultural heritage of St. Louis. In addition to stars in the sidewalks are plaques describing the achievements and contributions each creative St. Louisan made to our country's culture. (Chuck Berry, Miles Davis, T.S. Eliot, Betty Grable, Vincent Price, Tennessee Williams, Scott Joplin, Charles Lindbergh, Buddy Ebsen, etc.) Annually, the third Sunday in May. Est attendance: 1,200. For info: Joe Edwards, Chair, St. Louis Walk of Fame, 6504 Delmar, St. Louis, MO 63130. Phone: (314) 727-7827.

SAKHAROV, ANDREI DMITRIYEVICH: BIRTH ANNIVERSARY. May 21. Soviet physicist, human rights activist and environmentalist Andrei Sakharov was born at Moscow, Russia, on May 21, 1921. A collaborator in producing the first Soviet atomic bomb, and later the hydrogen bomb, Sakharov later denounced shortcomings of his country's government and was exiled to Gorky, Russia, 1980-1986. He was a formulator of the reform and restructuring concept known as *perestroika* and of *glasnost* (freedom). He was named to the Soviet Congress of Peoples Deputies eight months before his death at Moscow, Russia, on Dec 14, 1989. As a physicist, he was the developer of destructive weapons; as a humanitarian, he was courageous as a dissident from militarism and super-governments and an advocate of human rights and conservation.

SALVATION ARMY ADVISORY ORGANIZATIONS SUNDAY. May 21. A day to recognize and honor those community people whose efforts as board members or volunteers are crucial to The Salvation Army's work. Annually, the last day of National Salvation Army Week. See May 15. For info: Salvation Army, Natl HQ, 615 Slaters Lane, Alexandria, VA 22313. Phone: (703) 684-5500.

SHELBURNE MUSEUM PRESENTS LILAC SUNDAY. May 21. Shelburne Museum, Shelburne, VT. Explore the Museum's acclaimed lilac gardens and grounds. Enjoy 19th-century-style activities including a Victorian picnic, croquet, carriage rides, music and dancing. Wear period costume or a summer bonnet and be admitted at half-price. The American Bus Association has named Lilac Sunday one of the "1993 Top 100 Events in North America." Adults: $15. Students: $9. Children 6-14: $6. Children under 6: Free. Group discounts available. AAA discount. Near Burlington, VT. Est attendance: 3,000. For info: Shelburne Museum, Shelburne, VT 05482. Phone: (802) 985-3346.

BIRTHDAYS TODAY

Peggy Cass, 71, actress (Elinore Hathaway on "The Hathaways"; panelist on "To Tell the Truth"), born Boston, MA, May 21, 1924.

Robert Creeley, 69, author, poet (*Have a Heart, Windows*), born at Arlington, MA, May 21, 1926.

Janet Dailey, 51, writer and novelist (romance novels—*Tangled Vines*), born Storm Lake, IA, May 21, 1944.

Heinz Holliger, 56, oboist, composer, conductor, born at Langenthal, Switzerland, May 21, 1939.

William "Spike" O'Dell, 42, Chicago radio personality, born at Moline, IL, May 21, 1953.

Harold Robbins, 79, author (*The Piranhas, Stiletto*), born at New York, NY, May 21, 1916.

Leo Sayer, 47, singer ("You Make Me Feel Like Dancing"), songwriter, born at Shoreham, England, May 21, 1948.

Mr. T (Lawrence Tero or Tureaud), 43, actor (*Rocky III*, "The A-Team"), born Chicago, IL, May 21, 1952.

MAY 22 — MONDAY
142nd Day — Remaining, 223

CANADA: VICTORIA DAY. May 22. Commemorates the birth of Queen Victoria on May 24, 1819. Observed annually on the first Monday preceding May 25.

◈ **CASSATT, MARY: BIRTH ANNIVERSARY.** May 22. Leading American artist of the Impressionist school, Mary Cassatt was born May 22, 1844 (some sources give 1845) at Allegheny City, PA (now part of Pittsburgh). She settled in Paris in 1874 where she was influenced by Degas and the Impressionists. She was later instrumental in their works becoming well known in the United States. The majority of her paintings and pastels were based on the theme of mother and child. After 1900 her eyesight began to fail, and by 1914 she was no longer able to paint. Cassatt died at Chateau de Beaufresne near Paris, France, June 14, 1926.

CONAN DOYLE, ARTHUR: BIRTH ANNIVERSARY. May 22. English physician Sir Arthur Conan Doyle is best remembered as a detective story writer, especially for the creation of Sherlock Holmes and Dr. Watson. Conan Doyle was born at Edinburgh, Scotland, on May 22, 1859. He was deeply interested in and lectured on the subject of spiritualism. Conan Doyle died at Crowborough, Sussex, England, on July 7, 1930.

May 22-23 ☆ Chase's 1995 Calendar of Events ☆

FRENCH WEST INDIES: SLAVERY ABOLITION DAY. May 22. Martinique. Public holiday celebrated with picnics and parties. Phone: (212) 757-1125, ext 228. For info: Martinique Development & Promotion Bureau, Ms Muriel Wilford, 610 5th Ave, New York, NY 10020.

NAIA MEN'S NATIONAL TENNIS CHAMPIONSHIPS. May 22-27. Shadow Mountain Tennis Club, Tulsa, OK. 44th annual competition. Fax: (918) 494-8841. Est attendance: 1,500. For info: Natl Assn of Intercollegiate Athletics, 6120 S Yale Ave, Ste 1450, Tulsa, OK 74136. Phone: (918) 494-8828.

NAIA WOMEN'S TENNIS CHAMPIONSHIPS. May 22-27. Tulsa Southern Tennis Club, Tulsa, OK. 15th annual tournament. Individuals compete for All-America honors in 9 divisions, while teams compete for the national championship. Fax: (918) 494-8841. Est attendance: 1,500. For info: Natl Assn of Intercollegiate Athletics, 6120 S Yale Ave, Ste 1450, Tulsa, OK 74136. Phone: (918) 494-8828.

★ **NATIONAL MARITIME DAY.** May 22. Presidential Proclamation always issued for May 22 since 1933. (Pub Res No. 7 of May 20, 1933.)

NATIONAL MARITIME DAY. May 22. Anniversary of departure for first steamship crossing of Atlantic from Savannah, GA, to Liverpool, England, by steamship *Savannah*, 1819.

NEW YORK PUBLIC LIBRARY CENTENNIAL GALA. May 22. New York, NY. A spectacular black-tie benefit is planned with 100s of celebrities and friends of the library gathering for an evening of dancing, entertainment and tribute. Fax: (212) 768-7439. For info: New York Public Library, Nancy Donner, PR Mgr, 5th Ave and 42nd St, New York, NY 10018-2788. Phone: (212) 221-7676.

RA, SUN: BIRTH ANNIVERSARY. May 22, 1914. Born Herman (Sonny) Blount, Sun Ra was a pioneering and innovative jazz musician whose avant garde performances mixed elements of theater with his surreal compositional and performance style. Born at Birmingham, AL, and died there on May 30, 1993.

SPRINGSTEEN'S FIRST ALBUM. May 22. On this day in 1966, Bruce Springsteen recorded his first album with his band, the Castilles. The album was never released.

SWITZERLAND: PACING THE BOUNDS. May 22. Liestal. Citizens set off at 8 AM and march along boundaries to the beating of drums and firing of pistols and muskets. Occasion for fetes. Annually, the Monday before Ascension Day.

"THERE WENT JOHNNY!" NIGHT: ANNIVERSARY. May 22. After almost 30 years as host of the *Tonight* show, Johnny Carson hosted his last show on May 22, 1992. Carson became host of the late-night talk show, which began as a local New York program hosted by Steve Allen, on Oct 1, 1962. Over the years Carson occasionally made headlines with such extravaganzas as the marriage of Tiny Tim and Miss Vicki. Johnny received Emmys for his work four years in a row, 1976-79. Ed McMahon, his sidekick of 30 years, and Doc Severinsen, longtime bandleader, left the show with Carson. Jay Leno, the show's exclusive guest host, became the new regular host.

	S	M	T	W	T	F	S
May		1	2	3	4	5	6
1995	7	8	9	10	11	12	13
	14	15	16	17	18	19	20
	21	22	23	24	25	26	27
	28	29	30	31			

YEMEN: NATIONAL DAY. May 22. Public holiday.

BIRTHDAYS TODAY

Charles Aznavour, 71, singer, actor, born at Paris, France, May 22, 1924.
Richard Benjamin, 57, actor (*Goodbye Columbus, Diary of a Mad Housewife*, "He and She"), born New York, NY, May 22, 1938.
Judith Crist, 73, critic, born at New York, NY, May 22, 1922.
Tommy John, 52, baseball Hall of Famer, born at Terre Haute, IN, May 22, 1943.
Michael Sarrazin, 55, actor (*The Flim Flam Man, The Reincarnation of Peter Proud*), born at Quebec City, Que, Canada, May 22, 1940.
Susan Strasberg, 57, actress ("The Marriage," "Toma," *Picnic*), born New York, NY, May 22, 1938.

MAY 23 — TUESDAY
143rd Day — Remaining, 222

DECLARATION OF THE BAB. May 23. Baha'i commemoration of May 23, 1844, when the Bab, the prophet-herald of the Baha'i Faith, announced in Shiraz, Persia, that he was the herald of a new messenger of God. One of the nine days of the year when Baha'is suspend work. For info: Natl Spiritual Assembly of the US, 1320 Nineteenth St, NW, Ste 701, Washington, DC 20036. Phone: (202) 833-8990.

ENGLAND: CHELSEA FLOWER SHOW. May 23-26. (Two of these days will be for members only.) Royal Hospital, Chelsea, London SW3. Britain's major flower show with specially designed gardens and spectacular flower displays. Est attendance: 170,000. For info: Shows Dept, Royal Horticultural Soc, Vincent Sq, London, England SW1P 2PE.

FAIRBANKS, DOUGLAS ELTON: BIRTH ANNIVERSARY. May 23. Douglas Fairbanks was born at Denver, CO, May 23, 1883. He made his professional debut as an actor in Richmond, VA, on Sept 10, 1900, in *The Duke's Jester*. His theatrical career turned to Hollywood, and he became a movie idol appearing in such films as *The Americano, He Comes Up Smiling, The Mollycoddle, The Mark of Zorro, The Three Musketeers, Robin Hood, The Thief of Bagdad, The Black Pirate* and *The Gaucho*. He married "America's Sweetheart," Mary Pickford, in 1918, and in 1919 he joined with D.W. Griffith and Charlie Chaplin to form the production company United Artists. He died at Santa Monica, CA, on Dec 12, 1939.

FULLER, MARGARET: BIRTH ANNIVERSARY. May 23. Journalist and author of the first book on feminism by an American writer. Born Sarah Margaret Fuller, May 23, 1810, at Cambridgeport, MA, she began reading Virgil at age six and Shakespeare at age eight. Her conversational powers won her the admiration of the most intellectual students at Harvard University, and although she left Cambridgeport at age 23, she still communicated with the Boston intellectual community and befriended Ralph Waldo Emerson. In 1839 she returned to Boston and held "conversation" classes, designed to emancipate women from their traditional intellectual subservience to men. She shared editorial duties with Emerson on the Transcendentalist quarterly *The Dial* after a move to New York in 1844, and she was hired by Horace Greeley as literary critic for the *New York Tribune*. Her book, *Women in the Nineteenth Century*, the first feminist statement by an American writer, brought her international acclaim. In 1846 she began duty as a foreign correspondent for the *Tribune*, became caught up in the Italian revolutionary movement, and secretly married a young Roman nobleman, the Marquis Giovanni Angelo Ossoli. She directed a hospital during the siege of Rome by the French. After the fall of Rome, she and her husband found refuge with the English-speaking colony in Florence, Italy and were forced to leave by Tuscan authorities. Fuller, her husband and her child died on July 19, 1850, when their ship wrecked off Fire Island near New York, NY.

Chase's 1995 Calendar of Events — May 23-24

◆ **MANSFIELD, ARABELLA: BIRTH ANNIVERSARY.** May 23, 1846. Arabella Mansfield, born Belle Aurelia Babb near Burlington, IA, was the first woman admitted to the legal profession in the US. In 1869 while teaching at Iowa Wesleyan College, Mansfield was certified as an attorney and admitted to the Iowa bar. According to the examiners, "she gave the very best rebuke possible to the imputation that ladies cannot qualify for the practice of law." Mansfield never did practice law, however, continuing her career as an educator. She joined the faculty of DePauw University, in Greencastle, IN, where she became dean of the schools of art and music. One of the first woman college professors and administrators in the US, Mansfield was also instrumental in the founding of the Iowa Woman Suffrage Society in 1870. She died Aug 2, 1911, at Aurora, IL.

NAIA MEN'S GOLF CHAMPIONSHIP. May 23-26. Tulsa, OK. 44th annual. Fax: (918) 494-8841. Est attendance: 1,000. For info: Natl Assn of Intercollegiate Athletics, 6120 S Yale Ave, Ste 1450, Tulsa, OK 74136. Phone: (918) 494-8828.

NATIONAL GEOGRAPHY BEE: NATIONAL FINALS. May 23-24. National Geographic Society Headquarters, Washington, DC. The first place winner from each state-level competition on Apr 7 advances to the national level. Alex Trebek of "Jeopardy!" fame moderates the finals which are televised on PBS stations. They compete for scholarships and prizes totaling more than $50,000. Est attendance: 400. For info: Natl Geography Bee, Natl Geographic Soc, 1145 17th St NW, Washington, DC 20036. Phone: (202) 828-6635.

◆ **NEW YORK PUBLIC LIBRARY: 100th ANNIVERSARY.** May 23, 1895. New York's then-governor Samuel J. Tilden was the driving force that resulted in the combining of the private Astor and Lenox libraries with a $2 million endowment and 15,000 volumes from the Tilden Trust to become the New York Public Library, incorporated this day in 1895.

ORGANIZATION DEVELOPMENT INSTITUTE ANNUAL INFORMATION EXCHANGE. May 23-26. George Williams College, Williams Bay, WI. Organization development professionals exchange ideas on changes and new developments in the field at this 25th annual conference. FAX: (216) 7209-9319. Est attendance: 120. For info: Dr. Donald W. Cole, RODC, The Org Dev Institute, 11234 Walnut Ridge Rd, Chesterland, OH 44026. Phone: (216) 729-7419.

SOUTH CAROLINA CONSTITUTION RATIFICATION: ANNIVERSARY. May 23. By a vote of 149 to 73, South Carolina became the eighth state to ratify the Constitution, on May 23, 1788.

SWEDEN: LINNAEUS DAY. May 23. Stenbrohult. Commemorates birth, May 23, 1707, of Carolus Linnaeus (Carl von Linne), Swedish naturalist. Died at Uppsala, Sweden, Jan 10, 1778.

WILLIAM CARNEY RECEIVES CONGRESSIONAL MEDAL OF HONOR: ANNIVERSARY. May 23. Sergeant William H. Carney, of the 54th Massachusetts Colored Infantry, was the first black to win the Congressional Medal of Honor. He was cited for his efforts, although wounded twice, during the Battle of Fort Wagner, SC, on June 18, 1863. He was issued the medal on May 23, 1900.

BIRTHDAYS TODAY

Rosemary Clooney, 67, singer ("Mangos"), born at Maysville, KY, May 23, 1928.
Joan Collins, 62, actress (Alexis Carrington Colby on "Dynasty"), born at London, England, May 23, 1933.
Marvelous Marvin Hagler, 43, boxer, born at Newark, NJ, May 23, 1952.
Robert Moog, 61, inventor, born at Flushing, NY, May 23, 1934.
Artie Shaw, 85, musician, born at New York, NY, May 23, 1910.

MAY 24 — WEDNESDAY
144th Day — Remaining, 221

ANTI-SALOON LEAGUE FOUNDED: ANNIVERSARY. May 24. Anti-Saloon League was founded by Howard H. Russell at Oberlin, OH, on May 24, 1893. Efforts in that state were so successful the Anti-Saloon League of America was organized in 1895. The League's permanent home became Otterbein College in Westerville, OH, in 1909.

BASEBALL FIRST PLAYED UNDER THE LIGHTS: 60th ANNIVERSARY. May 24. The Cincinnati Reds defeated the Philadelphia Phillies by a score of 2-1, as more than 20,000 fans enjoyed the first night baseball game in the major leagues on May 24, 1935. The game was played at Crosley Field, Cincinnati, OH.

BELIZE: COMMONWEALTH DAY. May 24. Public holiday.

BROOKLYN BRIDGE: OPENING ANNIVERSARY. May 24. Nearly 14 years in construction, the $16 million Brooklyn Bridge over the East River opened May 24, 1883. Designed by John A. Roebling, the steel suspension bridge has a span of 1,595 feet.

BULGARIA: ENLIGHTENMENT AND CULTURE DAY. May 24. National holiday festively celebrated by schoolchildren, students, people of science and art. Manifestations and concerts to express love for education and culture.

CANADA: GOLDEN SHEAF AWARDS/YORKTON SHORT FILM AND VIDEO FESTIVAL. May 24-28. Yorkton, Saskatchewan. The prestigious "Golden Sheaf" awards are presented at North America's longest continuous running short film festival. Free public screenings, film and video marketplace for distributors and buyers, conferences and workshops. Est attendance: 1,700. For info: Tourism Saskatchewan, Saskatchewan Trade and Conv Center, 1919 Saskatchewan Dr, Regina, Sask, Canada S4P 3V7. Phone: (800) 667-7191.

COPERNICUS, NICOLAUS: DEATH ANNIVERSARY. May 24. The Polish astronomer who is considered the founder of modern astrology, Nicolaus Copernicus revolutionized scientific thought in the 16th century with his theory that the sun was the center of our solar system with the planets revolving around it, rather than Earth, as was the accepted view of officialdom of his day. Born at Torun, Poland, Feb 19, 1473, he died in East Prussia, May 24, 1543.

ERITREA: INDEPENDENCE DAY. May 24. National Day.

JULIA A. MOORE POETRY FESTIVAL. May 24. Flint Public Library, Flint, MI. The public is invited to submit bad poems to this poetry festival and contest dedicated to the memory of America's worst poet. Winners' poems will be read at the library May 24. See also: "Moore, Julia A.: Birth Anniversary" (Dec 1). Est attendance: 130. For info: Julia A. Moore Preservation Society, 1130 Lafayette St, Flint, MI 48503. Phone: (810) 767-6312.

LEUTZE, EMANUEL: BIRTH ANNIVERSARY. May 24. Obscure itinerant painter, born in Germany, May 24, 1816, came to the US when he was 9 years old, began painting by age 15. Painted some of most famous American works, such as *Washington Crossing the Delaware*, *Washington Rallying the Troops at Monmouth* and *Columbus Before the Queen*. Died July 18, 1868.

May 24–25 ☆ Chase's 1995 Calendar of Events ☆

MORSE OPENS FIRST US TELEGRAPH LINE: ANNIVERSARY. May 24. On May 24, 1844, the first US telegraph line was formally opened between Baltimore, MD, and Washington, DC. Samuel F.B. Morse sent the first officially telegraphed words "What hath God wrought?" from the Capitol building to Baltimore. Earlier messages had been sent along the historic line during testing, and one, sent on May 1, contained the news that Henry Clay had been nominated as president by the Whig party, meeting in Baltimore. This message reached Washington one hour prior to a train carrying the same news.

NCAA WOMEN'S GOLF CHAMPIONSHIPS. May 24–27. University of North Carolina, Wilmington, NC. For info: NCAA, 6201 College Blvd, Overland Park, KS 66211-2422. Phone: (913) 339-1906.

NEWHOUSE, SAMUEL I.: 100th BIRTH ANNIVERSARY. May 24. Mysterious multimillionaire businessman who built family publishing and communications empire. Born to immigrant parents in a New York City tenement on May 24, 1895, Newhouse became "America's most profitable publisher." He accumulated 31 newspapers, 7 magazines, 6 television stations, 5 radio stations and 20 cable television systems. His success with the "bottom line" in publishing and communications was without parallel. He died at New York, NY, Aug 29, 1979.

NORWAY: BERGEN INTERNATIONAL FESTIVAL. May 24–June 4. Bergen. World-class music festival features artists from Norway and abroad. One of the largest annual musical events in Scandinavia. For info: Norwegian Tourist Board, 655 Third Ave, New York, NY 10017. Phone: (212) 949-2333.

PALMER, LILLI: BIRTH ANNIVERSARY. May 24. Stage, screen and television actress Lilli Palmer was born Lillie Marie Peiser, at Poznan, Poland, on May 24, 1914. She also painted and was the author of several novels and an autobiography titled *Change Lobsters—And Dance*. She died at Los Angeles, CA, on Jan 27, 1986.

PEALE, JAMES: DEATH ANNIVERSARY. May 24. American portrait and miniature painter (painted portraits of George and Martha Washington and General Sir Thomas Shirley) was born at Chestertown, MD, in 1749 (exact date unknown) and died May 24, 1831.

SPACE MILESTONE: AURORA 7 MERCURY SPACE CAPSULE (US). May 24. Scott Carpenter becomes 2nd American to orbit Earth. Three orbits. Launched May 24, 1962.

BIRTHDAYS TODAY

Gary Burghoff, 61, actor (Emmy "M*A*S*H"), born Bristol, CT, May 24, 1934.
Jane Margaret Burke Byrne, 61, former mayor of Chicago, born at Chicago, IL, May 24, 1934.
Roger Caras, 67, nature writer, born at Methuen, MA, May 24, 1928.
Joe Dumars III, 32, basketball player, born Shreveport, LA, May 24, 1963.

Bob Dylan (Robert Zimmerman), 54, composer, singer, born at Duluth, MN, May 24, 1941.
Frank Oz, 51, puppeteer, born at Hereford, England, May 24, 1944.
Priscilla Beaulieu Presley, 49, actress (Jenna on "Dallas"; Jane in *Naked Gun*), born Brooklyn, NY, May 24, 1946.
Coleman Alexander Young, 49, former mayor of Detroit, MI, born at Tuscaloosa, AL, May 24, 1946.

MAY 25 — THURSDAY
145th Day — Remaining, 220

AFRICAN FREEDOM DAY: ANNIVERSARY. May 25. Public holiday in Chad, Zambia and some other African states. Members of the Organization for African Unity (formed May 25, 1963) commemorate their independence from colonial rule. Sports contests, political rallies and tribal dances.

ARGENTINA: INDEPENDENCE DAY. May 25. Anniversary of establishment of independent republic, following revolt of the provinces against Spanish rule, May 25, 1810.

ASCENSION DAY. May 25. Commemorates Christ's ascension into heaven. Observed since AD 68. Ascension Day is the 40th day after the Resurrection, counting Easter as the 1st day.

◈ **BABE'S 714th BIG ONE: ANNIVERSARY.** May 25, 1935. George Herman Ruth could barely run and could no longer hit like he used to, but occasionally the Babe could still put on a show with his bat. On May 25, Ruth, playing for the Boston Braves, hit three home runs before a crowd of only 10,000 at Pittsburgh's Forbes Field. His last home run of the day—his 714th in regular season play—proved to be Babe's last major league home run as well as his last big-league hit.

BELGIUM: PROCESSION OF THE HOLY BLOOD. May 25. Religious historical procession. Recalls heroic, adventurous crusaders, including Count Thierry of Alsace, who carried back relic of the Holy Blood. Always on Ascension Day.

CANADA: ANNAPOLIS VALLEY APPLE BLOSSOM FESTIVAL. May 25–29. Windsor to Digby, Nova Scotia. Annual festival with barbecues, sports events, art show, Princess Tea, coronation ceremonies, dances, concerts, fireworks, craft fair, children's parade and Grand Street Parade. Est attendance: 125,000. For info: Festival Office, 37 Cornwallis St, Kentville, NS, Canada B4N 2E2. Phone: (902) 678-8322.

CARVER, RAYMOND: BIRTH ANNIVERSARY. May 25. American poet and short story writer who chronicled the lives of America's working poor. Born on May 25, 1938, at Clatskanie, OR, died Aug 2, 1988, at his home at Port Angeles, WA, soon after finishing a book of poetry entitled *A New Path to the Waterfall*.

CONSTITUTIONAL CONVENTION: ANNIVERSARY. May 25. At Philadelphia, PA, on May 25, 1787, the delegates from seven states, forming a quorum, opened the Constitutional Convention, which had been proposed by the Annapolis Convention on Sept 11–14, 1786. Among those who were in attendance: George Washington, Benjamin Franklin, James Madison, Alexander Hamilton and Elbridge Gerry. See also: "Annapolis Convention Anniversary" (Sept 11).

DAVIS, MILES: BIRTH ANNIVERSARY. May 25. Jazz trumpeter Miles Davis was born at Alton, IL, on May 25, 1926. He was influenced by the be-bop music style of Charlie Parker and Dizzy Gillespie and ended up leaving the Julliard School of Music to join Parker's quintet in 1945 at the age of 19. He experimented with different styles throughout his career, exploring new voicings in jazz with arranger Gil Evans and musicians John Coltrane and Red Garland, delving into "modal" music with Tony Williams and Wayne Shorter and moving into a fusion sound in the '60s. His career was beset with bouts of drug addiction, but in the 1970s his return to the music scene found him creating a sound that melded his be-bop origins, modal chord progressions and driving rock rhythms. He died on Sept 28, 1991.

May 1995	S	M	T	W	T	F	S
		1	2	3	4	5	6
	7	8	9	10	11	12	13
	14	15	16	17	18	19	20
	21	22	23	24	25	26	27
	28	29	30	31			

☆ Chase's 1995 Calendar of Events ☆ May 25

ELIZABETH BOWER DAY. May 25. Oklahoma. To honor the Nursing Home Queen, who promoted nursing homes statewide. (She passed away in Dec 1994.) Annually, May 25. Est attendance: 40. Sponsor: McMahon Tomlinson Nursing Center, Cheryl Adams, Social Services, 3126 Arlington, Lawton, OK 73505. Phone: (405) 357-3240.

EMERSON, RALPH WALDO: BIRTH ANNIVERSARY. May 25. American author and philosopher born at Boston, MA, May 25, 1803. Died there Apr 27, 1882. It was Emerson who wrote (in his essay "Self-Reliance," 1841), "A foolish consistency is the hobgoblin of little minds, adored by little statesmen and philosophers and divines. With consistency a great soul has simply nothing to do."

GRUBSTAKE DAYS. May 25–28. Yucca Valley, CA. Includes parade, carnival, PCRA rodeo, dances, tug-of-war, horseshoe tournament, food and community booths, arts and craft booths and breakfasts offered by local service organizations. Annually, Memorial Day weekend. Est attendance: 30,000. For info: Yucca Valley Chamber of Commerce, 56300 29 Palms Hwy, Ste D, Yucca Valley, CA 92284. Phone: (619) 365-6323.

JORDAN: INDEPENDENCE DAY. May 25. National holiday. Commemorates treaty of May 25, 1946, proclaiming autonomy and establishing monarchy.

KODIAK CRAB FESTIVAL. May 25–29. Kodiak, AK. A celebration of spring and the Emerald Isle. Featured are parades, carnival booths and midway, running events, a golf tournament, bicycle and survival suit races, a blessing of the fleet ceremony and concerts. Annually, Memorial Day weekend. Est attendance: 15,000. Sponsor: Kodiak Chamber of Commerce, Box 1485, Kodiak, AK 99615. Phone: (907) 486-5557.

MEMORY DAYS. May 25–29. Grayson, KY. Parade, art show and horse show. Est attendance: 6,000. For info: Chamber of Commerce, Robert L. Caummisar, 301 W Main St, Grayson, KY 41143. Phone: (606) 474-9522.

MURRAY, PHILIP: BIRTH ANNIVERSARY. May 25. American labor leader and founder of the Congress of Industrial Organizations, also active in and a leader of the United Mine Workers, was born near Blantyre, Scotland, on May 25, 1886. Murray died at San Francisco, CA, on Nov 9, 1952.

NAIA MEN'S AND WOMEN'S OUTDOOR TRACK AND FIELD CHAMPIONSHIPS. May 25–27. Azusa Pacific University, Azusa, CA. 44th annual men's and 15th annual women's competition. Fax: (918) 494-8841. Est attendance: 2,500. For info: Natl Assn of Intercollegiate Athletics, 6120 S Yale Ave, Ste 1450, Tulsa, OK 74136. Phone: (918) 494-8828.

NATIONAL MISSING CHILDREN'S DAY. May 25. To promote awareness of the problem of missing children; to offer a forum for change; and to offer safety information for children in school and community. National toll-free phone numbers: 1-(800)-I-AM-LOST, 1-(800)-A-WAY-OUT. Annually, May 25. Sponsor: Child Find of America, Inc, PO Box 277, New Paltz, NY 12561-0277. Phone: (914) 255-1848.

NATIONAL TAP DANCE DAY. May 25. To celebrate this unique American art form that represents a fusion of African and European cultures and to transmit tap to succeeding generations through documentation and archival and performance support. Held on the anniversary of the birth of Bill "Bojangles" Robinson to honor his outstanding contribution to the art of tap dancing on stage and in films through the unification of diverse stylistic and racial elements.

NCAA WOMEN'S COLLEGE SOFTBALL WORLD SERIES. May 25–29. Amateur Softball Association Hall of Fame Stadium, Oklahoma City, OK. Eight of the best collegiate teams compete for the national championships. Double Elimination. Est attendance: 25,000. For info: OK City All Sports Assn, 100 W Main, Ste 287, Oklahoma City, OK 73102. Phone: (405) 236-5000.

ROANOKE FESTIVAL IN THE PARK '95. May 25–June 4. Roanoke, VA. A celebration of life and the arts. Est attendance: 375,000. For info: Roanoke Festival in the Park, PO Box 8276, Roanoke, VA 24014. Phone: (703) 342-2640.

ROBINSON, BILL "BOJANGLES": BIRTH ANNIVERSARY. May 25. "The King of Tap Dancers," Bill "Bojangles" Robinson was born on May 25, 1878, at Richmond, VA. After a career as a vaudeville and cabaret performer, he debuted on Broadway in *Blackbirds* in 1927, and in 1932 he was the star of *Harlems Heaven*, the first all-black talking film. His film credits included *The Littlest Rebel*, *In Old Kentucky*, *Rebecca of Sunnybrook Farm*, *Stormy Weather*, *One Mile From Heaven* and teaching Shirley Temple the stair dance in *The Little Colonel*. Bill Robinson died penniless on Nov 25, 1949, at New York, NY.

SIKORSKY, IGOR: BIRTH ANNIVERSARY. May 25. Aeronautical engineer best remembered for his development of the first successful helicopter in 1939. Also pioneered in multi-engine airplanes and large flying boats that made transoceanic air transportation possible. Born May 25, 1889, at Kiev, Russia, and died Oct 26, 1972, at Easton, CT.

SPACE MILESTONE: *SKYLAB 2* (US). May 25. Joseph P. Kerwin, Paul J. Weitz and Charles (Pete) Conrad, Jr, spent 28 days in space experimentation. Pacific splashdown occurred June 22. Launched May 25, 1973.

THREE RIVERS FESTIVAL AND RIVER REGATTA. May 25–29. Fairmont, WV. An entertaining and educational festival celebrates river activities. Est attendance: 30,000. Sponsor: Three Rivers Festival and Regatta, PO Box 1604, Fairmont, WV 26554. Phone: (304) 363-2625.

TITO (JOSIP BROZ): BIRTH ANNIVERSARY. May 25. Josip Broz, Yugoslavian soldier and political leader, born near Zagreb, Yugoslavia, May 25, 1892. Died May 4, 1980, and was interred in garden of his home at Belgrade. Funeral was attended by leaders of most of the world's major nations.

TUNNEY, JAMES JOSEPH (GENE): BIRTH ANNIVERSARY. May 25. Heavyweight boxing champion, business executive. The famous "long count" occurred in seventh round of Jack Dempsey-Gene Tunney world championship fight, Sept 22, 1927, at Soldier Field, Chicago, IL. Tunney was born at New York, NY, May 25, 1898. Died at Greenwich, CT, Nov 7, 1978.

BIRTHDAYS TODAY

Claude Akins, 77, actor (*Inherit the Wind*, "Lobo"), born Nelson, GA, May 25, 1918.

Dixie Carter, 56, actress (Julia Sugarbaker on "Designing Women"), born at McLemoresville, TN, May 25, 1939.

Jessi Colter (Miriam Johnson), 48, singer, songwriter, born at Phoenix, AZ, May 25, 1947.

Tom T. Hall, 59, singer, songwriter ("The Homecoming"), born at Olive Hill, KY, May 25, 1936.

K.C. Jones, 63, former basketball coach and player, born at Taylor, TX, May 25, 1932.

Connie Sellecca, 40, actress ("Hotel"), born at the Bronx, NY, May 25, 1955.

Beverly Sills, 66, singer, born at Brooklyn, NY, May 25, 1929.

Leslie Uggams, 52, singer ("Sing Along with Mitch"), actress, born at New York, NY, May 25, 1943.

Karen Valentine, 48, actress ("Room 222"), born Sebastopol, CA, May 25, 1947.

MAY 26 — FRIDAY
146th Day — Remaining, 219

ALMA HIGHLAND FESTIVAL AND GAMES. May 26–28. Alma, MI. To host, promote and preserve the piping, drumming, dancing, culture and tradition of the Scottish heritage. 27th annual festival. Annually, Memorial Day weekend. Est attendance: 70,000. For info: Chamber of Commerce, 110 W Superior St, PO Box 516, Alma, MI 48801. Phone: (517) 463-8979.

ANTIQUE AIRPLANE FLY-IN. May 26–28. Amelia Earhart Airport, Atchison, KS. More than 100 antique airplanes from the Kansas City area fly in for awards and a banquet. Held in conjunction with the Riverbend Art Fair in downtown Atchison. 29th annual fly-in. Annually, Memorial Day weekend. Est attendance: 1,000. For info: Steve Lawlor, Rte 4, Box 180, St. Joseph, MO 64507. Phone: (816) 238-2161.

BROOKINGS-HARBOR AZALEA FESTIVAL. May 26–29. Brookings, OR. Parade, street fair, art shows, seafood, 10K run, Bonsai exhibit, regional quilt show, crafts fair, Coast Guard demonstration, old-time fiddlers, live music. 56th annual festival. Annually, Memorial Day weekend. Est attendance: 10,000. For info: Brookings-Harbor Chamber of Commerce, PO Box 940, Brookings, OR 97415. Phone: (800) 535-9469.

CANAAN VALLEY HOLIDAY WEEKEND. May 26–28. Appalachian Base Camp, Terra Alta, WV. Camping weekend with outdoor field trips to learn about and identify birds. Led by experts; appropriate for all levels of experience. Est attendance: 30. For info: Oglebay Institute, Nature/Environmental Ed Dept, Oglebay Park, Wheeling, WV 26003. Phone: (304) 242-6855.

ENGLAND: ENGLISH RIVIERA DANCE FESTIVAL. May 26–June 10. Torquay, Devon. Demonstrations by world champions and participatory events including Modern, Ballroom, Disco and Latin American dance styles. Est attendance: 2,000. For info: Philip Wylie, 73 Hoylake Crescent, Ickenham, Middlesex, England UB10 8JQ.

FEAST OF ST. AUGUSTINE OF CANTERBURY. May 26. Pope Gregory sent Augustine to convert the pagan English. Augustine became the first archbishop of Canterbury. Augustine died May 26, 604 AD.

FLORIDA FOLK FESTIVAL. May 26–27. Stephen Foster State Folk Culture Center, White Springs, FL. To celebrate Florida's folk heritage with music, song, dance and stories. Est attendance: 25,000. Sponsor: Florida Dept of State, Bureau of Florida Folklife, Box 265, White Springs, FL 32096. Phone: (904) 397-2192.

GEORGIA: INDEPENDENCE DAY. May 26. National Day.

GUTHRIE JAZZ BANJO FESTIVAL. May 26–28. Downtown Guthrie, OK. Jazz banjo bands and soloists from all over the US perform music from the gay '90s, as well as the roaring '20s and contemporary selections at various venues around historic Guthrie. Also antique autos on parade. Annually, on Memorial Day weekend. Sponsor: Mr. Brady Hunt/Guthrie CVB. Est attendance: 10,000. For info: Guthrie Conv & Visitors Bureau, Kay Hunt, Dir, PO Box 995, Guthrie, OK 73044. Phone: (405) 282-1947.

JOLSON, AL: BIRTH ANNIVERSARY. May 26. Actor, singer (born Asa Yoelson, St. Petersburg, Russia, May 26, 1886). Died at San Francisco, CA, Oct 23, 1950.

MEMORIAL WEEKEND SALMON DERBY. May 26–29. Petersburg, AK. Est attendance: 400. For info: Petersburg Chamber of Commerce, PO Box 649, Petersburg, AK 99833. Phone: (907) 772-3646.

MONTAGU, LADY MARY WORTLEY: BAPTISM ANNIVERSARY. May 26. English author baptized May 26, 1689. Died Aug 21, 1762.

MOUNTAINFILM '95. May 26–29. Telluride, CO. 17th annual film festival devoted to mountain sports, adventure and the environment. Est attendance: 550. For info: Telluride Central Reservations, PO Box 653, Telluride, CO 81435. Phone: (800) 525-3455.

MUSICAL INSTRUMENT PREVIEW AND AUCTION. May 26–29. Seattle Center, Seattle, WA. 15th annual auction of musical instruments, paraphernalia and memorabilia at the Northwest Folklife Festival. Preview of the items to be auctioned will be on the 3 preceding days, May 26–28. Est attendance: 10,000. For info: Sandy Bradley, Auctioneer, 313 18th Ave, Seattle, WA 98122. Phone: (206) 548-9622.

NAIA BASEBALL WORLD SERIES. May 26–June 1. Site to be determined. 39th annual competition. Fax: (918) 494-8841. Est attendance: 17,500. For info: Natl Assn of Intercollegiate Athletics, 6120 S Yale Ave, Ste 1450, Tulsa, OK 74136. Phone: (918) 494-8828.

NATIONAL DEATH BUSTERS DAY. May 26. On this day every state is encouraged to set a goal of having no traffic fatalities. Everybody is requested to tie a white ribbon to their vehicle radio antenna to show support for National Death Busters Day. Annually, the Friday of Memorial Day weekend. Sponsor: Joseph C. Kearley, 2129 Benbrook Dr, Carrollton, TX 75007. Phone: (214) 245-6192.

◆ **NORMAN ROCKWELL'S HOME-COMING SOLDIER COVER FOR THE SATURDAY EVENING POST: 50th ANNIVERSARY.** May 26. On May 7, 1945, German officials signed a document of unconditional surrender ending the war in Europe. Less than three weeks later, May 26, Norman Rockwell chose as the subject of his Post cover the subject closest to the hearts of the American people—a picture of an American soldier returning home. This illustration, "Home-Coming Soldier," was selected by the US Treasury as the official poster for the eighth war bond drive.

NORTHWEST FOLKLIFE FESTIVAL. May 26–29. Seattle Center, Seattle, WA. Traditional arts event celebrating the many local Northwest cultures and cultures of the world. Includes live music, children's programs, concerts, workshops and handmade crafts. More than 6,000 performers. Annually, Friday-Monday on Memorial Day weekend. Est attendance: 200,000. For info: Northwest Folklife Fest, 305 Harrison St, Seattle, WA 98109-4695. Phone: (206) 684-7300.

OLD WEST DAYS. May 26–29. Jackson, WY. Annual celebration of the Old West in Jackson Hole. Activities include a parade, special rodeo, mule show, street dances, amateur western swing contest, stagecoach rides, a cowboy poetry gathering, plus an authentic Mountain Man Rendezvous. Annually, Memorial Day weekend. For info: Jackson Hole Chamber of Commerce, Box E, Jackson, WY 83001. Phone: (307) 733-3316.

POSTCARD SALE. May 26–27. Howard Johnson Hotel, Hagerstown, MD. 36 tables filled with antique and modern postcards. Est attendance: 200. Sponsor: Postcard History Soc, John H. McClintock, Dir, PO Box 1765, Manassas, VA 22110. Phone: (703) 368-2757.

PUSHKIN, ALEXANDER: BIRTH ANNIVERSARY. May 26. Russian author born May 26, 1799. Died Jan 29, 1837.

☆ Chase's 1995 Calendar of Events ☆ May 26–27

◈ **RIDE, SALLY KRISTEN: BIRTHDAY.** May 26. Dr. Sally Ride, one of first seven women in the US astronaut corps and the first American woman in space, was born at Encino, CA, May 26, 1951. She was married in 1982 to another astronaut, Steve Hawley. Her flight aboard the space shuttle, *Challenger*, was launched from Cape Canaveral, FL, on June 18 and landed at Edwards Air Force Base, CA, on June 24, 1983. The six-day flight was termed "nearly a perfect mission."

RIVERFEST. May 26–28. Riverfront Park, Little Rock, AR. 18th annual outdoor festival of the visual and performing arts with non-stop entertainment, food vendors, visual artists, kid stuff. Est attendance: 150,000. For info: Jane Rogers, PO Box 3232, Little Rock, AR 72203. Phone: (501) 376-4781.

SACRAMENTO JAZZ JUBILEE. May 26–29. Sacramento, CA. More than 100 bands from the US, Canada and foreign countries perform American traditional jazz music in approximately 40 venues around Sacramento at one of the world's largest traditional jazz festivals. Est attendance: 100,000. Sponsor: Sacramento Traditional Jazz Soc, Roger Krum, Exec Dir, 2787 Del Monte Blvd, Sacramento, CA 95691. Phone: (916) 372-5277.

SOUTH SHORE SINGLES FOUNDING ANNIVERSARY. May 26. First Parish Church, Norwell, MA. Nonprofit singles group started in a living room in 1979 and now has more than 600 members. Provides support, social and educational programs to singles, new and seasoned. Celebrated with dinner/dancing at South Shore Country Club, Hingham, MA. Annually, the fourth Friday in May. Sponsor: South Shore Singles. Est attendance: 250. For info: Terry Kleine, Newsletter Ed, 15 Franklin Rd, Hanover, MA 02339. Phone: (617) 878-3289.

SPACE MILESTONE: *SOYUZ 36* (USSR). May 26. Cosmonauts Bertalan Farkas, of Hungary, and Valery Kubasov docked at *Salyut 6* on May 27 for week-long visit with Ryumin and Popov before returning to Earth. Launched May 26, 1980.

SPRING FOLK DANCE CAMP. May 26–29. Camp Russel, Wheeling, WV. This nationally recognized Folk Dance Camp features national and international dances by dance leaders in their field. Est attendance: 100. Sponsor: Oglebay Institute Visual and Creative Arts Dept, Stifel Fine Arts Ctr, 1330 National Rd, Wheeling, WV 26003. Phone: (304) 242-7700.

TRI-STATE ANTI-DRUG AIR SHOW. May 26–27. Tri-State Airport, Kenova, WV. Aerial acts, tactical demonstrations, skydiving, hot-air balloons, military aircraft and much more. Est attendance: 30,000. For info: Director Dan Hieronimus, PO Box 364, Ironton, OH 45638. Phone: (614) 377-9989.

WALES: HAY-ON-WYE FESTIVAL OF LITERATURE. May 26–June 4. Hay-on-Wye, Powys. Largest annual festival of literature takes place in the beautiful market town of Hay-on-Wye in the Black Mountains of the Welsh Marches. Est attendance: 25,000. For info: Festival of Literature, Hay-on-Wye, Powys HR3 5BX, Wales HR3 5BX.

WAYNE, JOHN: BIRTH ANNIVERSARY. May 26. American motion picture actor, born Marion Michael Morrison, at Winterset, IA, on May 26, 1907. Died at Los Angeles, CA, June 11, 1979. "Talk low, talk slow and don't say too much" was his advice on acting.

WORLD CHAMPIONSHIP OLD-TIME PIANO PLAYING CONTEST. May 26–29. Holiday Inn, Decatur, IL. Competition and festival of ragtime, honky-tonk and old-time music. Annually, Memorial Day weekend. Sponsor: Old-Time Music Preservation Assn (OMPA) Inc. Est attendance: 1,200. For info: Judy Leschewski, PO Box 4714, Decatur, IL 62525. Phone: (217) 428-2403.

BIRTHDAYS TODAY

James Arness, 72, actor ("Gunsmoke," *How the West was Won*"), born at Minneapolis, MN, May 26, 1923.

Darrell Evans, 48, former baseball player, born at Pasadena, CA, May 26, 1947.

Genie Francis, 33, actress (Laura Vining on "General Hospital"), born Englewood, NJ, May 26, 1962.

Peggy Lee (Norma Delores Egstrom), 75, singer ("Fever," "Is That All There Is"), actress (*The Jazz Singer*, *Pete Kelly's Blues*), born at Jamestown, ND, May 26, 1920.

Brent Musburger, 56, sportscaster, born at Portland, OR, May 26, 1939.

Stevie Nicks, 47, singer (Fleetwood Mac, "Don't Stop") songwriter ("Edge of Seventeen"), born at Phoenix, CA, May 26, 1948.

◈ *Sally K. Ride*, 44, astronaut, first American woman in space, born at Los Angeles, CA, May 26, 1951.

Philip Michael Thomas, 46, actor ("Miami Vice," *Hair*), born Los Angeles, CA, May 26, 1949.

Wesley Walker, 40, football player, born at San Bernadino, CA, May 26, 1955.

Hank Williams, Jr, 46, singer ("All for the Love of Sunshine," "I Fought the Law"), born at Shreveport, LA, May 26, 1949.

MAY 27 — SATURDAY
147th Day — Remaining, 218

ALABAMA JUBILEE. May 27–29. Point Mallard, Decatur, AL. Hot air balloon races, arts, crafts, antique cars, water and air shows. Annually, Memorial Day weekend. Est attendance: 100,000. For info: Decatur Conv and Visitors Bureau, Box 2349, 719 6th Ave SE, Decatur, AL 35602. Phone: (205) 350-2028.

◈ **BLOOMER, AMELIA JENKS: BIRTH ANNIVERSARY.** May 27. American social reformer and women's rights advocate, born at Homer, NY, May 27, 1818. Her name is remembered especially because of her work for more sensible dress for women and her recommendation of a costume that had been introduced about 1849 by Elizabeth Smith Miller but came to be known as the "Bloomer Costume," or "Bloomers." Amelia Bloomer died at Council Bluffs, IA, Dec 30, 1894.

◈ **CARNAVAL '95.** May 27–29. Mission District, San Francisco, CA. In conjunction with commemoration of the 50th anniversary of the United Nations, "World Peace" is the theme of the 16th annual celebration of world music and dance. This year's carnaval will feature the first Carnaval International Parade. Sponsor: M.E.C.A. Contact: Roberto Hernandez (415) 826-1401. For info: UN50 Committee, 312 Sutter Ste, Ste 610, San Francisco, CA 94108. Phone: (415) 989-1995.

◈ **CARSON, RACHEL (LOUISE): BIRTH ANNIVERSARY.** May 27. American scientist and author, born at Springdale, PA, May 27, 1907. Author of *Silent Spring* (1962), a book that provoked widespread controversy over the use of pesticides. Died Apr 14, 1964.

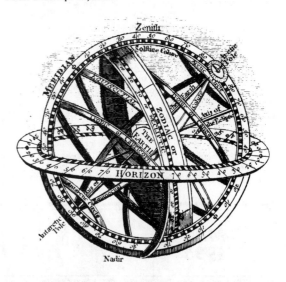

May 27 ☆ Chase's 1995 Calendar of Events ☆

CHESTERTOWN TEA PARTY FESTIVAL. May 27. Chestertown, MD. Reenactment of 1774 tea dumping, crafts, music, food and games. Est attendance: 10,000. Sponsor: Chestertown Tea Party Festival, Inc, Box 526, Chestertown, MD 21620. Phone: (410) 778-0416.

CHICAGOLAND GEMS AND MINERALS ASSOCIATION SHOW. May 27–29. Du Page County Fairgrounds, Wheaton, IL. Exhibits, programs and working demonstrations in jewelry-making hobbies, including fossil preparation and rock identifications, rock swap and children's corner. Annually, Memorial Day weekend. Est attendance: 5,000. For info: VP Gems and Minerals Assn, 1220 W Henderson, Chicago, IL 60657. Phone: (312) 549-3612.

CIVIL WAR WEEKEND. May 27–29. Yorktown Battlefield, Yorktown, VA. 100 reenactors present a recreated Civil War encampment, Confederate field hospital and military drills performed on the battlefield in Colonial National Historical Park. Est attendance: 3,000. For info: Colonial Natl Historical Park, PO Box 210, Yorktown, VA 23690. Phone: (804) 898-3400.

DAKOTA COWBOY POETRY GATHERING. May 27–28. Medora, ND. Local and national cowboy poets reunion. Includes western art show and crafts display. Sponsor: Theodore Roosevelt Medora Fdtn, PO Box 198, Medora, ND 58645. Phone: (701) 224-2525 or (800) 435-5663. Est attendance: 4,000. For info: North Dakota Tourism, Liberty Memorial Bldg, State Capitol Grounds, Bismarck, ND 58505.

◆ **DUNCAN, ISADORA: BIRTH ANNIVERSARY.** May 27. American-born interpretive dancer who revolutionized the entire concept of dance. Bare-footed, freedom-loving, liberated woman and rebel against tradition, experienced worldwide professional success and profound personal tragedy (her two children drowned, her marriage failed and she met a bizarre death when the long scarf she was wearing caught in a wheel of the open car in which she was riding, strangling her). Born at San Francisco, CA, May 27, 1878. Died at Nice, France, Sept 14, 1927.

FARMINGTON INVITATIONAL BALLOON FESTIVAL. May 27–28. Farmington, NM. Hot air balloons launch off the banks of Farmington Lake. Famous Splash and Dash and Hare and Hound races included in the two-day event. Est attendance: 5,000. For info: Farmington Conv and Visitors Bureau, 203 W Main, Ste 401, Farmington, NM 87401. Phone: (800) 448-1240.

FESTIVAL OF FLAGS. May 27–28. Killeen Special Events Ctr and Rodeo Grounds. Killeen, TX. The Festival of Flags is a celebration of Killeen's multi-cultural ethnic heritages. The Festival includes an arts and crafts show, international food booths, PRCA rodeo, street dance, golf tournament, car show, motorcycle display, parade, military static display, carnival, petting zoo, and much, much more. Annually, Memorial Day weekend. Est attendance: 80,000. For info: Wendy Kouba, Greater Killeen Chamber of Commerce, PO Box 548, Killeen, TX 76540. Phone: (800) 869-8265.

FIRST RUNNING OF PREAKNESS: ANNIVERSARY. May 27, 1873. The first running of the Preakness Stakes in Pimlico, MD, was won by Survivor with a time of 2:43. The winning jockey was G. Barbee.

FRENCH AND INDIAN WAR RENDEZVOUS. May 27–28. Fort Frederick State Park, Big Pool, MD. Demonstrations of frontier, tactical battle and cannon-firing skills. Period wares sold by sutlers. Annually, the last full weekend in May. Est attendance: 4,000. For info: Fort Frederick State Park, 11100 Fort Frederick Rd, Big Pool, MD 21711. Phone: (301) 842-2155.

May 1995

S	M	T	W	T	F	S
	1	2	3	4	5	6
7	8	9	10	11	12	13
14	15	16	17	18	19	20
21	22	23	24	25	26	27
28	29	30	31			

FUNFEST. May 27–29. Thompson Park, Amarillo, TX. Community festival with entertainment, food booths, games, free children's area. Special events—volleyball, funtug, golf tournament and talent show. Annually, Memorial Day weekend. Sponsor: Junior League of Amarillo and City of Amarillo Parks and Recreation Dept. Est attendance: 45,000. For info: Junior League of Amarillo, 1700 Polk, Amarillo, TX 79102. Phone: (806) 374-0802.

GASPEE DAYS. May 27–June 11. Paw Tuxet Village, Warwick and Cranston, RI. Month of events commemorating the 1772 burning of the HMS *Gaspee* by Rhode Island colonists. This first blow for freedom led the colonies to form the Committees of Correspondence that led eventually to the Continental Congress and the Declaration of Independence. Annually, the Saturday of Memorial Day weekend to the second Sunday in June. Est attendance: 250,000. For info: Councilman Scott Avedisian, Pres, Gaspee Days, PO Box 1772, Warwick, RI 02888. Phone: (401) 781-1772.

GREAT MISSISSIPPI RIVER ARTS AND CRAFTS FESTIVAL. May 27–28. Mark Twain Historic District, Hannibal, MO. 9th annual arts and crafts show featuring the works of nearly 55 artisans from all over the midwest. Concession stands with an array of All-American home cooking, a variety of entertainment groups and creative learning fun in the children's area. Est attendance: 15,000. For info: Hannibal Visitors Bureau, PO Box 624, Hannibal, MO 63401. Phone: (314) 221-2477.

GREAT MONTEREY SQUID FESTIVAL. May 27–28. Monterey, CA. For those with a taste for history as well as seafood, The Great Monterey Squid Festival is both a tribute to the commercial fishing industry of the Monterey Bay and succulent celebration of the star of the sea, the incredible, edible squid! Annually, the Saturday and Sunday of Memorial Day weekend. Est attendance: 30,000. For info: Bostrom Monterey, Robert Massaro, Dir, 2600 Garden Rd #208, Monterey, CA 93940. Phone: (408) 649-6547.

HALFWAY PARK DAYS. May 27–28. Martin L. (Marty) Snook Memorial Park, Hagerstown, MD. Family fun in the park with arts and crafts, free entertainment, food and children's rides. Annually, Memorial Day weekend. Est attendance: 8,000. For info: Stanley D. Johnson, Lions Club of Halfway, 10923 Holly Terrace, Hagerstown, MD 21740-7804. Phone: (301) 739-3219.

HAMMETT, (SAMUEL) DASHIELL: BIRTH ANNIVERSARY. May 27. The man who brought realism to the genre of mystery writing, Dashiell Hammett was born May 27, 1894, at St. Marys County, MD. Hammett's first two novels, *Red Harvest* (1929) and *The Dain Curse* (1929), were based on his eight years spent as a Pinkerton detective. Hammett is recognized as the founder of the "hard-boiled" school of detective fiction. Three of his novels have been made into films: *The Maltese Falcon* (1930), considered by many to be his finest work; *The Thin Man* (1932), which provided the basis for a series of five movies starring William Powell and Myrna Loy; and *The Glass Key* (1931). Hammett, who refused to name members of an alleged subversive organization during House Un-American Activities Committee hearings, was jailed for six months in 1951, and later was charged with tax delinquency by the Bureau of Internal Revenue. Hammett never resumed writing and died Jan 10, 1961, at New York City.

HEAD-OF-THE-MON-RIVER HORSESHOE TOURNAMENT. May 27-29. Fairmont, WV. Open to horseshoe pitchers with a 1995 State/National Horseshoe Pitchers Assn membership card. Est attendance: 300. For info: Tri-County Horseshoe Club Dir, Davis "Catfish" Woodward, 1133 Sunset Dr, Fairmont, WV 26554. Phone: (304) 366-3819.

HUMPHREY, HUBERT HORATIO: BIRTH ANNIVERSARY. May 27. Born at Wallace, SD, May 27, 1911. Thirty-eighth vice president of the US. Died at Waverly, MN, Jan 13, 1978.

IROQUOIS ARTS SHOWCASE I. May 27-28. Iroquois Indian Museum, Howes Cave, NY. Demonstrations of Iroquois arts and crafts including beadwork, cornhusk dolls, pottery and more. Many items for sale by the artists. Children's activities. Iroquois social dancing, storytelling and nature walks. Annually, Memorial Day weekend. Est attendance: 2,500. Sponsor: Iroquois Indian Museum, PO Box 7, Caverns Rd, Howes Cave, NY 12092. Phone: (518) 296-8949.

ITALIAN FESTIVAL. May 27-28. McAlester, OK. Coal-rich Pittsburg County lured Italian miners to Indian Territory in the 1880s, and now descendants celebrate their heritage at the fairgrounds with folk music, dances, native costumes, arts and crafts and ethnic culinary delights. Est attendance: 8,000. For info: Italian Festival Fdtn, McAlester, OK 74502. Phone: (918) 426-2055.

LITTLE 500. May 27. Anderson, IN. This 46th annual "Granddaddy of all sprint car races" concludes a week-long festival of events in Anderson and serves as an appetizer for the Indianapolis 500 held the next day. Drivers compete for more than $100,000 in prize money in this 500-lap sprint car race. Annually, the day before the Indy 500. Est attendance: 9,000. For info: Dan McFeely, PR, Amer Speed Assn, 202 S Main St, Pendleton, IN 46064. Phone: (317) 778-8088.

LOBSTERFEST. May 27-29. Mystic Seaport, Mystic, CT. A New England lobster bake on the banks of the Mystic River over the Memorial Day weekend. Est attendance: 10,000. For info: Mystic Seaport, 75 Greenmanville Ave, PO Box 6000, Mystic, CT 06355. Phone: (203) 572-5315.

LONGWOOD GARDENS FESTIVAL OF FOUNTAINS. May 27-Sept 30. Kennett Square, PA. A magical nighttime mixture of rainbow-hued fountains, alfresco garden concerts, fireworks, and leisurely evenings in the conservatory. Est attendance: 250,000. For info: Colvin Randall, PR Mgr, PO Box 501, Kennett Square, PA 19348-0501. Phone: (610) 388-6741.

MIAMI/FT. LAUDERDALE HOME SHOW. May 27-June 4. Miami Beach Convention Center, Miami Beach, FL. The nation's largest show of its kind features thousands of home improvement products, services, furnishings and accessories from more than 25 countries. Est attendance: 150,000. For info: Norman H. Cooper, 6915 Red Rd, #228, Coral Gables, FL 33143. Phone: (305) 666-5944.

NATIONAL RIVERS HALL OF FAME INDUCTIONS/AWARDS. May 27. National Rivers Hall of Fame, Dubuque, IA. The ceremonies will honor men and women important in the history and lore of America's rivers. The event will feature the organization's annual meeting, lectures, lunch aboard a paddle-wheeled dredge boat and awards ceremony. Est attendance: 100. For info: Jerry Enzler, Exec Dir, Mississippi River Museum, PO Box 266, Dubuque, IA 52004-0266. Phone: (319) 557-9545.

PAINT-A-CAN CONTEST. May 27. On the grounds of the Mississippi Coast Coliseum and Convention Center, Biloxi, MS. Amateur and professional artists paint 50-gallon cans with attractive decals and other depictions. The cans are then used as trash receptacles along the white sand Mississippi beach of the Gulf of Mexico. Competition in pro and amateur categories. Free souvenir t-shirts to participants and more than $1,000 in cash and other prizes. Annually, the Saturday of the Memorial Day weekend, conducted as part of the annual Mississippi Coast Fair & Expo. Est attendance: 2,500. Sponsor: John McFarland or Gena Collins, c/o Marketing Dept, The Sun Herald, PO Box 4567, Biloxi, MS 39535. Phone: (601) 896-2434.

PRICE, VINCENT: BIRTH ANNIVERSARY. May 27, 1911. Actor, best known for his portrayal of sinister villains in horror films and as host for the TV series "Mystery." Born at St. Louis, MO, and died at Los Angeles, CA, on Oct 25, 1993.

RIVERSPREE FESTIVAL. May 27-28. Elizabeth City, NC. A celebration of life on the river. Sporting events, food, crafts, entertainment. Annually, Memorial Day weekend. Est attendance: 75,000. For info: Chamber of Commerce, Box 426, Elizabeth City, NC 27907. Phone: (919) 335-4365.

RMS QUEEN MARY: MAIDEN VOYAGE ANNIVERSARY. May 27. Anniversary of the maiden voyage from Southampton, England, to New York Harbor, NY, on May 27, 1936.

SENIOR PRO RODEO. May 27-28. Payson Rodeo Grounds, Payson, AZ. Sanctioned by the National Old Timer's Rodeo Association. Contestants have to be 40 years old or older to compete. Some of the best veterans of professional rodeo. Proceeds of the Rodeo benefit the Gary Hardt Athletic Scholarship fund and underprivileged and needy children of Gila County. Est attendance: 3,500. For info: Payson Chamber of Commerce, PO Box 1380, Payson, AZ 85547. Phone: (800) 672-9766.

TASTE OF CINCINNATI. May 27-29. Cincinnati, OH. Greater Cincinnati is famous for its fine food, from elegant five-star dining to five-way chili. This popular eating extravaganza presents a taste of the most delicious culinary delights available. Est attendance: 400,000. For info: Downtown Council, 300 Carew Tower, 441 Vine, Cincinnati, OH 45202. Phone: (513) 579-3191.

TRASH FEST®. May 27-29. Toledo Bend Reservoir, Anacoco, LA. Scuba divers gather to retrieve more than two tons of debris from bottom of lake. Queen coronation, bands, spearfishing tournament, crawfish boil and fireworks display. Est attendance: 300. Sponsor: Toledo Divers Assn, Dick Wilgus, 261 Tibbitt Rd, Anacoco, LA 71403. Phone: (318) 286-5565.

UTICA OLD-FASHIONED ICE CREAM FESTIVAL. May 27-29. Utica, OH. Saluting "America's favorite dessert," ice cream, with a weekend of fun and entertainment—parade, queen contest, arts and crafts, antique gas engines, sheep herding with border collies and plenty of delicious ice cream. Est attendance: 26,000. For info: Utica Sertoma, Howard Stone, Box 303, Utica, OH 43080. Phone: (614) 892-3462.

BIRTHDAYS TODAY

John Barth, 65, author (*Last Voyage of Somebody the Sailor, Letters*), born at Cambridge, MD, May 27, 1930.

Pat Cash, 30, tennis player, born at Melbourne, Australia, May 27, 1965.

Christopher J. Dodd, 51, US Senator (D, Connecticut), born at Willimantic, CT, May 27, 1944.

Lou Gossett, Jr, 59, actor (Emmy for "Roots"; "Sadat," *Enemy Mine*; Oscar nomination *An Officer and a Gentleman*), Brooklyn, NY, May 27, 1936.

Henry Kissinger, 72, former secretary of state, author, born at Fuerth, Germany, May 27, 1923.

May 27-28 ☆ Chase's 1995 Calendar of Events ☆

Christopher Lee, 73, actor (*Dracula, The Mummy*), born at London, England, May 27, 1922.

Lee Meriwether, 60, actress (Cat Woman in TV *Batman*; former Miss America ('55), born Los Angeles, CA, May 27, 1935.

Yasuhiro Nakasone, 78, former Japanese prime minister, born at Takasaki, Gumma Province, Japan, May 27, 1917.

William Sessions, 65, former director of the FBI, born at Fort Smith, AK, May 27, 1930

Bruce Weitz, 52, actor ("Hill Street Blues," "Death of a Centerfold: The Dorothy Stratton Story"), born at Norwalk, CT, May 27, 1943.

Herman Wouk, 80, writer (*Marjorie Morningstar, The Winds of War*), born at New York, NY, May 27, 1915.

MAY 28 — SUNDAY
148th Day — Remaining, 217

AGASSIZ, JEAN LOUIS RODOLPHE: BIRTH ANNIVERSARY. May 28. Swiss geologist, teacher and author, born May 28, 1807. Died at Cambridge, MA, Dec 14, 1873. "The eye of the trilobite," Agassiz wrote in 1870, "tells us that the sun shone on the old beach where he lived; for there is nothing in nature without a purpose, and when so complicated an organ was made to receive the light, there must have been light to enter it."

AZERBAIJAN: NATIONAL DAY. May 28. Public holiday.

CANADA: ROSSLAND PUBLIC MARKET. May 28. (Also June 25, July 30, Aug 27, Sept 24.) Esling Park, Rossland, British Columbia. Gifts, crafts, homemade goods and clothing, produce and preserves, and children's entertainment at this monthly public market. Est attendance: 300. For info: Rossland Chamber of Commerce, Box 1385, Rossland, BC, Canada V0G 1Y0. Phone: (604) 362-5390.

DIONNE QUINTUPLETS: BIRTHDAY. May 28. Five daughters (Marie, Cecile, Yvonne, Emilie and Annette) were born to Oliva and Elzire Dionne, near Callander, Ont, Canada, on May 28, 1934. They were the first quints known to have lived for more than a few hours after birth.

FOXTAIL DRIVE-IN CAR SHOW. May 28-29. Kruse Auction Complex, Auburn, IN. Features muscle cars, street rods and trucks of all kinds. Model and toy show. Fun for the whole family. Annually, on Memorial Day weekend. Est attendance: 1,500. Sponsor: Auburn-Cord-Duesenberg Festival, Inc, Sharon Vick, Publicist, PO Box 271, Auburn, IN 46706. Phone: (219) 925-3600.

GUILLOTIN, JOSEPH IGNACE: BIRTH ANNIVERSARY. May 28. French physician and member of the Constituent Assembly who urged the use of a machine that was sometimes called the "Maiden" for the execution of death sentences—in a less painful, more certain way of dispatching those sentenced to death. The guillotine was first used on Apr 25, 1792, for the execution of a highwayman, Nicolas Jacques Pelletier. Other machines for decapitation had been in use from time to time in other countries since the Middle Ages. Guillotin was born at Saintes, France, May 28, 1738. Died at Paris, Mar 26, 1814.

INDIANAPOLIS 500-MILE RACE. May 28. Indianapolis, IN. Recognized as the world's largest single-day sporting event. First race was in 1911. Annually, the Sunday of Memorial Day weekend. For info: Indianapolis Motor Speedway Corp, 4790 W 16th St, Speedway, IN 46224. Phone: (317) 248-6750.

May 1995

S	M	T	W	T	F	S
	1	2	3	4	5	6
7	8	9	10	11	12	13
14	15	16	17	18	19	20
21	22	23	24	25	26	27
28	29	30	31			

ITALY: PALIO DEI BALESTRIERI. May 28. Gubbio. The last Sunday in May is set aside for a medieval crossbow contest between Gubbio and Saensepolcro; medieval costumes, arms.

MAD-CITY MARATHON. May 28. Madison, WI. 26.2-mile footrace through the streets of Madison. Other events include a 10K run, 5K walk, Health & Fitness Expo, carbo-load dinner and Finish Line Festival. Annually, the Sunday of Memorial Day weekend. Est attendance: 4,000. For info: Madison Festivals, Inc, Kerry Nolen, Race Dir, 615 E Washington Ave, Madison, WI 53703. Phone: (800) 373-6376.

MELON CITY CRITERIUM. May 28. Muscatine, IA. 17th annual day-long criterium at Weed Park has eight races for the public and nine race categories for licensed USCF members. Prize money of approximately $6,000. Est attendance: 3,000. For info: Greg Harper, c/o Harper's Cycling & Fitness, 1106 Grandview Ave, Muscatine, IA 52761. Phone: (319) 263-4043.

MINI GRAND PRIX. May 28-31. Downtown, Fond du Lac, WI. 48 business-sponsored miniature Indy-style 3hp cars maneuver a series of 5 curves, hairpin turns and straight-aways on the streets of downtown Fond du Lac. An entire weekend of family-oriented entertainment, live bands, special attractions and food concessions for young and old race fans. Fond du Lac innkeepers offer special rates for the weekend. For info: Mary Burnett, Dir Tourism Sales, Fond du Lac Conv Bureau, 19 W Scott St, Fond du Lac, WI 54935. Phone: (414) 923-3010.

PITT, WILLIAM: BIRTH ANNIVERSARY. May 28. British prime minister from 1783 to 1801 and from 1804 to 1806. Influenced by Adam Smith's economic theories, he reduced England's large national debt caused by the American Revolution. Born on May 28, 1759, at Hayes, Kent, England. Died Jan 23, 1806, at Putney. He was the son of William Pitt, first earl of Chatham, for whom the city of Pittsburgh was named.

ST. BERNARD OF MONTJOUX FEAST DAY. May 28. Patron saint of mountain climbers, founder of Alpine hospices of the Great and Little St. Bernard, died at age 85, probably on May 28, 1081.

SLUGS RETURN FROM CAPISTRANO. May 28. It's a little known secret that slimy slugs spend their winters in lovely Capistrano and return to our patios and gardens on this date. Bare feet not a good idea now through first frost. Phone: (212) 388-8673 or (717) 274-8451. Sponsor: Wellness Permission League, Tom or Ruth Roy, 2105 Water St, Lebanon, PA 17046.

THORPE, JAMES FRANCIS: BIRTH ANNIVERSARY. May 28. Jim Thorpe, distinguished American athlete, winner of pentathlon and decathlon events at the 1912 Olympic Games, professional baseball and football player, American Indian, born near Prague, OK, May 28, 1888. Died at Lomita, CA, Mar 28, 1953.

BIRTHDAYS TODAY

Caroll Baker, 64, actress (*Baby Doll, Harlow*), born at Johnstown, PA, May 28, 1931.

Barry Commoner, 78, biologist, politician, born at Brooklyn, NY, May 28, 1917.

Kirk Gibson, 38, baseball player, born at Pontiac, MI, May 28, 1957.

Armon Louis Gilliam, 31, basketball player, born Pittsburgh, PA, May 28, 1964.

Rudolph Giuliani, 51, Mayor of New York City, born at Brooklyn, NY, May 28, 1944.

☆ Chase's 1995 Calendar of Events ☆ May 28-29

Gladys Knight, 51, singer (and the Pips; "Neither One of Us [Wants to Be the First to Say Goodbye]," "If I Were Your Woman"), born at Atlanta, GA, May 28, 1944.

Sondra Locke, 48, actress (*The Heart Is a Lonely Hunter, Bronco Billy*), director (*Ratboy*), born Shelbyville, TN, May 28, 1947.

MAY 29 — MONDAY
149th Day — Remaining, 216

AMNESTY ISSUED FOR SOUTHERN REBELS: 130th ANNIVERSARY. May 29. On May 29, 1865, President Andrew Johnson issued a proclamation giving a general amnesty to all who participated in the rebellion against the United States. High ranking members of the Confederate government and military and those who owned more than $20,000 worth of property were excepted and had to apply individually to the President for a pardon. Once an oath of allegience was taken all former property rights, exept those in slaves, were returned to the former owners.

ASCENSION OF BAHA'U'LLAH: ANNIVERSARY. May 29. Baha'i observance of the anniversary of the death in exile of Baha'u'llah (prophet-founder of the Baha'i Faith), May 29, 1892. One of the nine days of the year when Baha'is suspend work. For info: Natl Spiritual Assembly of the US, 1320 Nineteenth St, NW, Ste 701, Washington, DC 20036. Phone: (202) 833-8990.

BOLDER BOULDER 10K. May 29. Boulder, CO. A 10K race of walkers, joggers and world-class runners through the streets of Boulder. Annually, on Memorial Day. Est attendance: 35,000. For info: Bolder Boulder, PO Box 9125, Boulder, CO 80301-9125. Phone: (303) 444-7223.

CHARLES II: RESTORATION AND BIRTH ANNIVERSARY. May 29. Restoration of Charles II to English throne, May 29, 1660. Also his birthday (May 29, 1630). English monarchy restored after Commonwealth period under Cromwell.

CHESTERTON, GILBERT KEITH: BIRTH ANNIVERSARY. May 29. English author. May 29, 1874–June 14, 1936.

CONFEDERATE MEMORIAL DAY IN VIRGINIA. May 29. Annually, the last Monday in May.

ENGLAND: OAK-APPLE DAY. May 29. Anniversary of Charles II's entry into Whitehall, May 29, 1660. Oak-Apple Day actually commemorates the adventures and concealment of Charles in Boscobel's famous oak tree, in 1651. Wearing of an oak twig or leaf or of an oak-apple on this day formerly was popular, along with Maypole ceremonies. Oak-Apple Day is sometimes called Royal Oak Day or Shick-Shack Day.

HENRY, PATRICK: BIRTH ANNIVERSARY. May 29. American revolutionary leader and orator, born at Studley, VA, May 29, 1736. Died near Brookneal, VA, June 6, 1799. Especially remembered for his speech (Mar 23,1775) for arming the Virginia militia, at St. Johns Church, Richmond, VA, when he declared: "I know not what course others may take, but as for me, give me liberty or give me death."

KENNEDY, JOHN FITZGERALD: BIRTH ANNIVERSARY. May 29. Thirty-fifth president of the US, born at Brookline, MA, May 29, 1917. Assassinated while riding in an open automobile, in Dallas, TX, Nov 22, 1963. (Accused assassin Lee Harvey Oswald was killed at the Dallas police station by a gunman, Jack Rubenstein [Ruby], two days later.) Kennedy was the youngest man ever elected to the presidency, the first Roman Catholic, and the first president to have served in the US Navy. He was the fourth US president to be killed by an assassin, and the second to be buried at Arlington National Cemetery (first was William Howard Taft).

KWIN KITE FLY IN. May 29. Oak Grove Regional Park, Stockton, CA. Hundreds of colorful kites are flown. Competition for highest flying, largest and smallest kites. Live entertainment and crafts. Est attendance: 8,000. For info: San Joaquin County Dept of Parks and Rec, 11793 N Micke Grove Rd, Lodi, CA 95240. Phone: (209) 953-8800.

MEMORIAL DAY. May 29. Legal public holiday. (PL90-363 sets Memorial Day on last Monday in May. Applicable to federal employees and District of Columbia.) Also known as Decoration Day. Most countries designate a day each year for decorating graves with flowers and for other memorial tributes to the dead. Especially an occasion for honoring those who have died in battle. (Observance dates from Civil War years in US: first documented observance at Waterloo, NY, May 5, 1865.) See also: "Confederate Memorial Day" (Apr 24, Apr 26, May 10 and May 29).

MEMORIAL DAY PARADE AND SERVICES. May 29. Gettysburg National Cemetery, Gettysburg, PA. 2,000 schoolchildren scatter flowers over the unknown graves. Memorial services follow parade. For info: Gettysburg Travel Council, 35 Carlisle St, Gettysburg, PA 17325. Phone: (717) 334-6274.

★ **MEMORIAL DAY, PRAYER FOR PEACE.** May 29. Presidential Proclamation issued each year since 1948. PL81-512 of May 11, 1950, asks President to proclaim annually this day as a day of prayer for permanent peace. PL90-363 of June 28, 1968, requires that beginning in 1971 it will be observed the last Monday in May. Often entitled "Prayer for Peace Memorial Day," and traditionally requests the flying of the flag at half-staff "for the customary forenoon period."

MOON PHASE: NEW MOON. May 29. Moon enters New Moon phase at 5:27 AM, EDT.

MOSCOW COMMUNIQUE: ANNIVERSARY. May 29. On May 29, 1972, President Richard Nixon and Soviet Party leader Leonid Brezhnev released a joint communique after Nixon's week-long visit to Moscow. During the visit the two men acknowledged their major differences on the Vietnam War, signed a treaty on antiballistic missile systems as well as an interim agreement on limitation of strategic missiles and an agreement for a joint space flight in 1975. This was the first visit ever to Moscow by a US president (May 22-30, 1972).

NATIONAL DESIGN DRAFTING WEEK. May 29–June 2. To bring attention to the services provided by those in the profession of design and drafting. Sponsor: American Design Drafting Assn, PO Box 799, Rockville, MD 20848-0799. Phone: (301) 460-6875.

NATIONAL FROZEN YOGURT WEEK. May 29–June 4. To inform the public of the benefits and colorful history of frozen yogurt, one of America's favorite desserts. Annually, the first week in June. Sponsor: TCBY, Tracy Pritts, Acct Exec, S&S Public Relations, 400 Skokie Blvd, Ste 200, Northbrook, IL 60062.

RHODE ISLAND: RATIFICATION DAY. May 29. Thirteenth state to ratify Constitution, on this day in 1790.

SOCCER TRAGEDY: 10th ANNIVERSARY. May 29. A riot at Heysel stadium in Brussels, Belgium, on May 29, 1985, killed 39 people. Fans attending the European Cup Final, between Liverpool and Juventus of Turin, clashed before the match started. Some 400 persons were injured in the riot. The incident was televised and viewed by millions throughout Europe. More than two years later, Sept 2, 1987, the British government announced that 26 British soccer fans (identified from television tapes) would be extradited to Belgium for trial. Hooliganism at soccer matches became the target of increased security measures for most of England's more than 90 professional teams following the 1985 tragedy.

May 29-30 ☆ *Chase's 1995 Calendar of Events* ☆

SPENGLER, OSWALD: BIRTH ANNIVERSARY. May 29. German historian, author of *The Decline of the West*, born at Blankenburg-am-Harz, Germany, on May 29, 1880. Died at Munich, Germany, May 8, 1936.

STOCK EXCHANGE HOLIDAY (MEMORIAL DAY). May 29. The holiday schedules for the various exchanges are subject to change if relevant rules, regulations or exchange policies are revised. If you have questions, phone: American Stock Exchange (212) 306-1212; Chicago Board of Options Exchange (312) 786-7760; Chicago Board of Trade (312) 435-3500; New York Stock Exchange (215) 656-2065; Pacific Stock Exchange (415) 393-4000; Philadelphia Stock Exchange (215) 496-5000.

TOUR OF SOMERVILLE. May 29. Somerville, NJ. The oldest continuously run major bicycle race in America. 1995 marks the 52nd running. Attracts more than 600 top amateur cyclists for seven events. Annually, on Memorial Day. Est attendance: 40,000. For info: Dan Puntillo, Admin, PO Box 125, Somerville, NJ 08876. Phone: (908) 725-0461.

TRADITIONAL INDIAN DANCES. May 29-Sept 4. Red Rock State Park, Gallup, NM. Educational, entertaining, memorable nightly performances by the world-famous Indian Dancers. Sample traditional foods of the area. Est attendance: 5,000. For info: Inter-Tribal Indian Ceremonial Assn, PO Box 1, Church Rock, NM 87311. Phone: (800) 233-4528.

UNITED KINGDOM: SPRING BANK HOLIDAY. May 29. Bank and public holiday in England, Wales, Scotland and Northern Ireland.

VIRGIN ISLANDS: MEMORIAL DAY. May 29. Parades and ceremonies take place throughout the islands. Yacht races featured on St. Croix. Observed annually on last Monday in May.

"VIRGINIA PLAN" PROPOSED: ANNIVERSARY. May 29. Just five days after the Constitutional Convention met at Philadelphia, PA, in 1787, the "Virginia Plan" was proposed. It called for establishment of a new governmental organization consisting of a legislature with two houses, an executive (chosen by the legislature) and a judicial branch.

WISCONSIN: ADMISSION DAY. May 29. Became 30th state on this day in 1848.

BIRTHDAYS TODAY

Eric Davis, 33, baseball player, born at Los Angeles, CA, May 29, 1962.

Paul Ehrlich, 63, population biologist, born at Philadelphia, PA, May 29, 1932.

Anthony Geary, 47, actor (Luke on "General Hospital"), born at Coalville, UT, May 29, 1948.

Bob Hope (Leslie Townes), 92, comedian ("Road" movies with Bing Crosby; entertains US troops abroad; former longtime emcee for Oscars), born Eltham, England, May 29, 1903.

Felix Rohatyn, 67, investment banker (developed strategy to keep New York City solvent), born at Vienna, Austria, May 29, 1928.

Al Unser, 56, auto racer, born at Albuquerque, NM, May 29, 1939.

Fay (Francis) Thomas Vincent, Jr, 57, former Commissioner of Major League Baseball, born at Waterbury, CT, May 29, 1938.

MAY 30 — TUESDAY
150th Day — Remaining, 215

CROATIA: NATIONAL DAY. May 30. Public holiday.

FIRST AMERICAN DAILY NEWSPAPER PUBLISHED: ANNIVERSARY. May 30. *The Pennsylvania Evening Post* became the first daily newspaper published in the US on May 30, 1783. The paper was published in Philadelphia, PA, by Benjamin Towne.

GERMANY: MARBURG UNIVERSITY FOUNDING: ANNIVERSARY. May 30. University of Marburg was founded May 30, 1527.

HALL OF FAME FOR GREAT AMERICANS: OPENING ANNIVERSARY. May 30. The Hall of Fame for Great Americans at New York University, New York, NY, was dedicated and opened to the public on May 30, 1901.

LOOMIS DAY. May 30. To honor Mahlon Loomis, a Washington, DC, dentist who received a US patent on wireless telegraphy in 1872 (before Marconi was born). Titled "An Improvement in Telegraphing," the patent described how to do without the wires; this patent was backed up by experiment on the Massanutten mountains of Virginia. Annually, May 30. Sponsor: Puns Corps, Robert L. Birch, Box 2364, Falls Church, VA 22042-0364. Phone: (703) 533-3668.

MEMORIAL DAY (TRADITIONAL). May 30. This day honors the tradition of making memorial tributes to the dead, especially remembering those who have died in battle. Observed as a legal public holiday on the last Monday in May.

MUSLIM NEW YEAR. May 30-31. Year 1416 of the Islamic era, or the Era of the Hegira, begins at sunset on this day. Muslim calendar date: Muharram 1, 1416. Different methods for calculating the Gregorian date are used by different Muslim sects or groups. *Chase's* dates are based on astronomical calculations of the visibility of the new moon crescent at Mecca; EST date may vary.

PAPAL "NO" ON ORDAINING WOMEN: ANNIVERSARY. May 30, 1994. Pope John Paul II issued an Apostolic Letter declaring that the "church has no authority whatsoever to confer priestly ordination on women." The issue of a woman's role in the church is very divisive, especially in America, and the Pope's attempt to end debate instead raised the issue to a more passionate level in some quarters.

ST. JOAN OF ARC: FEAST DAY. May 30. French heroine and martyr, known as the Maid of Orleans, led French against English invading army. Captured, found guilty of heresy and burned at the stake in 1431 (at age 19). Innocence declared in 1456. Canonized in 1920.

SPACE MILESTONE: *ARIANE-2* (ESA). May 30. European Space Agency unmanned rocket carrying Intelsat V communications satellite, launched from Kourou space center in French Guiana on May 30, 1986. It was destroyed during launch because of a malfunction.

SPACE MILESTONE: *MARINER 9* (US). May 30. Unmanned spacecraft entered Martian orbit Nov 13, studied temperature and gravitational fields and sent back photographs. First spacecraft to orbit another planet. Launched May 30, 1971.

SPACE MILESTONE: *NASA ATS-6* (US). May 30. Communications satellite with expected six-year life. Stationary orbit. Launched May 30, 1974.

BIRTHDAYS TODAY

Bob Evans, 77, restaurant executive, born at Sugar Ridge, OH, May 30, 1918.

Michael J. Pollard, 56, actor (*Bonnie and Clyde*, "Leo & Liz in Beverly Hills"), born Passaic, NJ, May 30, 1939.

Gale Sayers, 55, former football player, born at Wichita, KS, May 30, 1940.

Clint Walker, 68, actor (*The Dirty Dozen*, "Cheyenne"), born at Hartford, IL, May 30, 1927.

May 1995

S	M	T	W	T	F	S
	1	2	3	4	5	6
7	8	9	10	11	12	13
14	15	16	17	18	19	20
21	22	23	24	25	26	27
28	29	30	31			

MAY 31 — WEDNESDAY
151st Day — Remaining, 214

AMECHE, DON: BIRTH ANNIVERSARY. May 31, 1908. Film, stage and TV actor. Born Dominic Felix Amici at Kenosha, WI, and died Dec 6, 1993, at Scottsdale, AZ.

CANADA: THE NATIONAL. May 31-June 4. Spruce Meadows, Calgary, Alberta. The National Tournament features the Canadian Show Jumping Championship, including a world cup qualifier and Canadian Team Selection Trial and the Royal Bank World Cup. Also the Festival of Music in Spruce Meadows Plaza. Live entertainment and activities daily. Est attendance: 85,000. For info: Spruce Meadows, RR #9, Calgary, AB, Canada T2J 5G5. Phone: (403) 254-3200.

JOHNSTOWN FLOOD: ANNIVERSARY. May 31. Heavy rains in May, 1889, caused the Connemaugh River Dam to burst on May 31 of that year. At nearby Johnstown, PA, the resulting flood killed more than 2,300 persons and destroyed the homes of thousands more. Nearly 800 unidentified drowning victims were buried in a common grave at Johnstown's Grandview Cemetery. So devastating was the flood and so widespread the sorrow for its victims that "Johnstown Flood" entered the language as a phrase to describe a disastrous event. The valley city of Johnstown, in the Allegheny Mountains, has been damaged repeatedly by floods, but that of 1889 took the greatest number of lives. Floods in 1936 (25 deaths) and 1977 (85 deaths) were next most destructive.

NATIONAL SENIOR HEALTH AND FITNESS DAY. May 31. Local sites nationwide. First national event to promote the value of fitness and exercise for older adults. During this day—as part of Older Americans Month activities—seniors across the country will be involved in local exercise and fitness activities. Call the toll-free number for further info and schedule of local activities. To be held annually, the last Wednesday in May. 1995 is the second year for this event; 10,000 seniors are expected to be involved. For info: Mature Market Resource Center, Mary Gay Kay, Program Coord, 621 E Park Ave, Libertyville, IL 60048. Phone: (800) 828-8225.

NATIONAL SPELLING BEE FINALS. May 31-June 1. Washington, DC. Newspapers across the country sponsor 235-250 youngsters in the finals in Washington, DC. Annually, Wed and Thurs of Memorial Day week. Est attendance: 1,000. Sponsor: Scripps-Howard, Reta Rose, Dir Natl Spelling Bee, PO Box 5380, Cincinnati, OH 45201. Phone: (513) 977-3028.

NCAA DIVISION I MEN'S AND WOMEN'S OUTDOOR TRACK CHAMPIONSHIPS. May 31-June 3. University of Tennessee, Knoxville, TN. Est attendance: 40,000. For info: NCAA, 6201 College Blvd, Overland Park, KS 66211-2422. Phone: (913) 339-1906.

NCAA DIVISION I MEN'S GOLF CHAMPIONSHIP. May 31-June 3. Ohio State Scarlet Course, Columbus, OH. Est attendance: 3,000. For info: NCAA, 6201 College Blvd, Overland Park, KS 66211-2422. Phone: (913) 339-1906.

POPE PIUS XI: BIRTH ANNIVERSARY. May 31. Ambrogio Damiano Achille Ratti, 259th pope of the Roman Catholic Church, born at Desio, Italy, May 31, 1857. Elected pope Feb 6, 1922. Died Feb 10, 1939.

SOUTH AFRICA: COMRADES MARATHON. May 31. Durban-Pietermaritzburg, Natal. A 90K marathon run from Pietermaritz to Durban. Entry is open to all runners. For direct info: Comrades Committee, Box 100621, Scottsville, 3209 South Africa. Est attendance: 13,000. For info: South African Tourism Board, 500 Fifth Ave, New York, NY 10110. Phone: (212) 730-2929.

SOUTH AFRICA: REPUBLIC DAY. May 31. National holiday. On May 31, 1910, the Union of South Africa was established. On May 31, 1961, it became the Republic of South Africa.

US INTL FILM AND VIDEO FESTIVAL AWARDS PRESENTATIONS. May 31-June 1. Chicago, IL. World's largest awards competition honoring sponsored, business, television and industrial productions. Founded in 1968. FAX: (708) 834-5565. Est attendance: 300. For info: US Intl Film & Video Festival, J.W. Anderson, Chair, Patricia Meyer, Exec Dir, 841 N Addison Ave, Elmhurst, IL 60126-1291. Phone: (708) 834-7773.

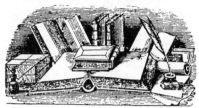

WHITMAN, WALT: BIRTH ANNIVERSARY. May 31. American poet, born May 31, 1819. Following a short-lived and largely unsuccessful career in journalism, Whitman in 1855 published the collection of poetry for which he is now famous, *Leaves of Grass*. Widely criticized in the US and only marginally received abroad, its rhythmically rambling discourse and mystical imagery proved incomprehensible to most early readers. Although Whitman insisted that *Leaves* was only a celebration of himself as an average man, it was in fact a celebration of "the self," as defined by the inseparableness of body and soul. His exploration of the communion of body and soul and his presentation of love as longing and death as satisfaction of longing resulted in what was to his 19th-century audience decidedly indecent physiological imagery. When he died on Mar 26, 1892, his obituary was published in newspapers worldwide, finally marking his place in the history of great American literature. "No really great song," Whitman wrote, "can ever attain full purport till long after the death of its singer—till it has accrued and incorporated the many passions, many joys and sorrows, it has aroused."

WORLD NO-TOBACCO DAY. May 31. An observance of the World Health Organization. For info: World No-Tobacco Day, American Assn for World Health, 1129 20th St NW, Ste 400, Washington, DC 20036. Phone: (202) 466-5883.

BIRTHDAYS TODAY

Tom Berenger, 45, actor ("One Life to Live," *If Tomorrow Comes*), born Chicago, IL, May 31, 1950.

Clint Eastwood, 65, actor (spaghetti westerns, *Dirty Harry*), director (Oscar *Unforgiven*), former mayor of Carmel, CA, born San Francisco, CA, May 31, 1930.

Sharon Gless, 52, actress (Emmy for Det. Cagney on "Cagney and Lacey"), born Los Angeles, CA, May 31, 1943.

Gregory Harrison, 45, actor ("Logan's Run," "Trapper John, M.D."), born at Avalon, Catalina Island, CA, May 31, 1950.

Joe Namath, 52, Pro Football Hall of Famer, actor, born at Beaver Falls, PA, May 31, 1943.

Johnny Paycheck (Don Lytle), 54, singer ("Mr Lovemaker," "Take This Job and Shove It"), songwriter, born Greenfield, OH, May 31, 1941.

Brooke Shields, 30, actress (*Pretty Baby*, *The Blue Lagoon*), born at New York, NY, May 31, 1965.

Terry Waite, 56, Church of England special envoy, former hostage in Lebanon ('87-'91), born May 31, 1939.

Peter Yarrow, 57, composer, singer (Peter, Paul and Mary), born at New York, NY, May 31, 1938.

☆ Chase's 1995 Calendar of Events ☆

June.

JUNE 1 — THURSDAY
152nd Day — Remaining, 213

ADOPT-A-SHELTER-CAT MONTH. June 1-30. To promote the adoption of homeless kittens and cats from local shelters, the ASPCA sponsors this important observance. For further information, send a SASE. Phone: (212) 876-7700, ext 4655. Sponsor: American Soc for the Prevention of Cruelty to Animals (ASPCA), 424 E 92nd St, New York, NY 10128.

ALL DRESSED UP. June 1-Aug 27. Clinton, MD. Special exhibit on 19th-century costuming. Sponsor: Surratt House and Tavern, 9110 Brandywine Rd, Clinton, MD 20735. Phone: (301) 868-1121.

AMERICAN RIVERS MONTH. June 1-30. To focus attention on the outstanding scenic, recreational and aesthetic benefits offered by our natural rivers and streams. Sponsor: American Rivers Inc, Mary-Ellen Kirkbride, 801 Pennsylvania Ave SE, Ste 400, Washington, DC 20003. Phone: (202) 547-6900.

ATLANTIC, CARIBBEAN AND GULF HURRICANE SEASON. June 1-Nov 30. Info from: US Dept of Commerce, Natl Oceanic and Atmospheric Admin, Rockville, MD 20852.

CANADA: MOSAIC FESTIVAL OF CULTURES. June 1-3. Regina, Saskatchewan. City-wide festival with 25 pavilions featuring the crafts of skilled artisans, energetic folk dances, lively music and song, traditional costumes and food from around the world. Est attendance: 30,000. For info: Tourism Saskatchewan, Saskatchewan Trade and Conv Center, 1919 Saskatchewan Dr, Regina, Sask, Canada S4P 3V7. Phone: (800) 667-7191.

CANADA: SHELBURNE COUNTY LOBSTER FESTIVAL. June 1-4. Shelburne County, Nova Scotia. Four days of activities in celebration of the lobster-fishing industry. Local community groups and businesses throughout the county host lobster suppers, sporting events, craft shows, yacht and boat races and much more. "Shelburne County—The Lobster Capital of Canada." Annually, the first weekend in June. Est attendance: 10,000. For info: Marilyn Johnston, Lobster Fest Secy, PO Box 280, Shelburne, NS, Canada B0T 1W0. Phone: (902) 875-3544.

CANCER IN THE SUN MONTH. June 1-30. To promote education and awareness of the dangers of skin cancer from too much exposure to the sun. Kit of materials available for $15 from this non-profit organization. Fax: (415) 332-1832. For info: Pharmacy Council on Dermatology (PCD), Frederick Mayer, Pres, PO Box 1336, Sausalito, CA 94966. Phone: (415) 332-4066.

June 1995

S	M	T	W	T	F	S
				1	2	3
4	5	6	7	8	9	10
11	12	13	14	15	16	17
18	19	20	21	22	23	24
25	26	27	28	29	30	

CAREER NURSE ASSISTANTS DAY. June 1. Recognizes those nurse assistants who provide care to all ill, elderly and longterm patients for 5 or more years. Annually, the first Thursday in June. Sponsor: Genevieve Gipson, RN M Ed, Dir, Career Nurse Assistant Programs, 3577 Easton Rd, Barberton, OH 44203. Phone: (216) 825-9342.

CENTRAL PACIFIC HURRICANE SEASON. June 1-Oct 31. Central Pacific is defined as 140 West Longitude to the International Date Line (180 West Longitude). Info from: US Department of Commerce, Natl Oceanic and Atmospheric Admin, Rockville, MD 20852.

ENGLAND: ROCHESTER DICKENS FESTIVAL. June 1-4. Various venues, Rochester, Kent. World famous festival with displays, competitions and street entertainment all connected with Charles Dickens. For info: Tourist Info Centre, Eastgate Cottage, High St, Rochester, Kent, England ME1 1EW.

EVERETT SALTY SEA DAYS. June 1-4. Everett, WA. Family events to celebrate Everett's waterfront, carnival, food, novelty and commercial booths, live entertainment, fireworks, 3 on 3 Basketball Tournament, classic car show, outboard boat races, Hole-In-One Shoot Out, various nautical events. Sponsor: City of Everett, Pepsi, Rainier Beer, Budweiser Beer, Key Bank, Ramada Inn. Est attendance: 200,000. For info: Salty Sea Days Assn, Marion Pope, Managing Dir, PO Box 7050, Everett, WA 98201. Phone: (206) 339-1113.

FIREWORKS SAFETY MONTH. June 1-30. Activities during this month are designed to warn and educate parents and children about the dangers of playing with fireworks. Prevent Blindness America (formerly known as the National Society to Prevent Blindness), will offer suggestions for safer ways to celebrate the Fourth of July. Materials that can easily be posted or distributed to the community will be provided. Sponsor: Prevent Blindness America, Marita Gomez, Media Relations, 500 E Remington Rd, Schaumburg, IL 60173. Phone: (800) 331-2020.

GAY AND LESBIAN BOOK MONTH. June 1-30. Celebrating the publication of gay and lesbian literary works. For info: Michele Karlsberg, 585 N Railroad Ave, 6G, Staten Island, NY 10304. Phone: (718) 351-9599.

HOWARD, LESLIE: DEATH ANNIVERSARY. June 1. During the return trip from a British-government sponsored tour of Spain, on June 1, 1943, a plane transporting 50-year-old actor Leslie Howard was shot down by German raiders. Rumors that he was serving on a spy mission for his government circulated at the time. In her biography about her father (*A Quite Remarkable Father*) his daughter expressed doubt that her father was the sort to get involved in espionage. Howard's most remembered film role is that of Ashley Wilkes in *Gone with the Wind*.

INTERNATIONAL VOLUNTEERS WEEK. June 1-7. To honor men and women throughout the world who serve as volunteers, rendering valuable service without compensation to the communities in which they live, and to honor nonprofit organizations dedicated to making the world a better place in which to live. For complete info, send $1 to cover expense of printing, handling and postage. Annually, the first seven days of June. Sponsor: Intl Soc of Friendship and Good Will, Dr. Stanley Drake, Pres, 9538 Summerfield St, Spring Valley, San Diego County, CA 91977-2852. Phone: (619) 466-8882.

JUNE DAIRY MONTH. June 1-30. Since 1937, the dairy industry has set aside June as a time to salute American dairy producers who provide quality dairy products all year long. Sponsor: American Dairy Assn, Dir PR, 10255 W Higgins Rd, Ste 900, Rosemont, IL 60018.

JUNE IS TURKEY LOVERS' MONTH. June 1-30. Month-long campaign to promote awareness and increase turkey consumption at a non-holiday time. Annually, the month of June. Sponsor: Natl Turkey Federation, 11319 Sunset Hills Rd, Reston, VA 22090. Phone: (703) 435-7209.

☆ Chase's 1995 Calendar of Events ☆   June 1

KENTUCKY: ADMISSION DAY: ANNIVERSARY. June 1. Became 15th state on this day in 1792.

KENYA: MADARAKA DAY. June 1. Madaraka Day (Self-Rule Day) is observed as a national public holiday.

LITTLE, CLEAVON: BIRTH ANNIVERSARY. June 1. Best known for his role as the black sheriff who cleaned up a town of bumbling redneck toughs in the movie *Blazing Saddles*, Cleavon Little was born June 1, 1939, in Chickasha, OK. Little was the winner of a Tony award for the 1970 musical *Purlie* and an Emmy in 1989 for a guest appearance on the television series "Dear John." Little died Oct 22, 1992, near Sherman Oaks, CA.

MISSISSIPPI BROILER FESTIVAL. June 1-3. Gaddis Recreational Park, Forest, MS. A variety of entertainment, arts, crafts. Also, tennis tournaments, youth league baseball tournament, carnival, coronation of Miss Broiler Festival Queen, Mississippi Track Grand Prix Race, concert and fireworks. Est attendance: 8,000. For info: Memorie Dickson, Forest Area Chamber of Commerce, PO Box 266, Forest, MS 39074. Phone: (601) 469-4332.

MISSOURI RIVER EXPO. June 1-4. Missouri Valley Fairgrounds, Bismarck, ND. An annual 5-day exciting expo with various grandstand shows, demonstrations, displays, and carnival. For info: Bismarck Conv and Visitors Bureau, Box 2274, Bismarck, ND 58502. Phone: (701) 222-4308.

MONROE, MARILYN: BIRTH ANNIVERSARY. June 1. American actress (Norma Jean Mortenson) born at Los Angeles, CA, June 1, 1926. Died Aug 5, 1962.

NATIONAL ACCORDION AWARENESS MONTH. June 1-30. To increase public awareness of this multicultural instrument and its influence and popularity in today's music. E-Mail: Tommy@crl.com. Sponsor: Those Darn Accordions!, 2269 Chestnut St, Ste 183, San Francisco, CA 94123. Phone: (415) 346-5862.

NATIONAL DREAMWORK MONTH. June 1-30. This celebration encourages us to pay attention to our dreams, keep a dream journal, tune into the symbolic language of dreams and recognize the empowerment of the subtle messages sent by the subconscious to the conscious mind. For info: Jean Benedict Raffa, 17 S Osceola Ave, Orlando, FL 32801. Phone: (407) 426-7077.

NATIONAL FRESH FRUIT AND VEGETABLE MONTH. June 1-30. To illustrate to Americans the abundance, variety, good taste, good value and importance to good health of fresh fruits and vegetables. For info: Marketing Manager, United Fresh Fruit and Vegetable Assn, 727 N Washington St, Alexandria, VA 22314. Phone: (703) 836-3410.

NATIONAL FROZEN YOGURT MONTH. June 1-30. To inform the public of the benefits and colorful history of frozen yogurt, one of America's new favorite desserts. Annually, the month of June. Sponsor: TCBY, Tracy Pritts, Acct Exec, S&S PR, 400 Skokie Blvd, Ste 200, Northbrook, IL 60062.

NATIONAL ICED TEA MONTH. June 1-30. To celebrate one of the most widely consumed beverages in the world and one of nature's most perfect beverages, and to encourage Americans to refresh themselves with this all-natural, low-calorie, refreshing thirst-quencher. Sponsor: The Tea Council of the USA, 230 Park Ave, New York, NY 10169. Phone: (212) 986-6998.

NATIONAL PEST CONTROL MONTH. June 1-30. To recognize the fine work of the professional pest control operators who do so much to give us clean, safe homes, workplaces, hospitals, restaurants and recreational areas. For info: Natl Pest Control Assn, 8100 Oak St, Dunn Loring, VA 22027. Phone: (703) 573-8330.

NATIONAL ROSE MONTH. June 1-30. To recognize American grown roses, our national floral emblem. America's favorite flower is grown in all 50 states, and more than 1.2 billion fresh cut roses are sold at retail each year. Sponsor: Roses Inc, Box 99, Haslett, MI 48840. Phone: (517) 339-9544.

NATIONAL SCLERODERMA AWARENESS MONTH. June 1-30. To educate and make the public and patients aware of the disease scleroderma. Phone: (800) 722-HOPE. For info: United Scleroderma Fdtn, Inc, 21 Brennan St, Ste 21, Box 399, Watsonville, CA 95077-0399.

NATIONAL WOMEN'S MUSIC FESTIVAL. June 1-4. Indiana University, Bloomington, IN. 21st annual 4-day event includes 9 concerts, a crafts area, fine arts exhibit, theatre presentations, dances and more than 250 workshops. Annually, the weekend after the last Monday in May. Est attendance: 3,500. For info: NWMF, PO Box 1427, Indianapolis, IN 46202. Phone: (317) 927-9355.

ORTHODOX ASCENSION DAY. June 1. Observed by Eastern Orthodox Churches on this date.

OWN YOUR SHARE OF AMERICA MONTH. June 1-30. A month-long annual educational campaign to encourage individual investment in stocks and to focus attention on individual investors' concerns. Annually, the month of June. Sponsor: Natl Assn of Investors Corp (NAIC), Thomas E. O'Hara, Chair, 711 Thirteen Mile Rd, Madison Heights, MI 48071. Phone: (609) 778-0380.

PEOPLE'S REPUBLIC OF CHINA: INTERNATIONAL CHILDREN'S DAY. June 1. Shanghai.

PORTLAND ROSE FESTIVAL. June 1-25. (Also July 14-16.) Portland, OR. Celebration includes more than 70 events featuring grand floral parade, hot air balloons, band festivals, auto and ski races, carnival, air show and Navy ship visits. Est attendance: 2,000,000. Sponsor: Portland Rose Festival Assn, Paula Fasano, 220 NW Second Ave, Portland, OR 97209. Phone: (503) 227-2681.

SAMOA: INDEPENDENCE DAY. June 1. National Day, public holiday.

SCOTT JOPLIN RAGTIME FESTIVAL. June 1-3. Historic downtown Sedalia, MO. Features both free and admission events indoors and out, including concerts, symposiums, tea dances, ragtime ball, sheet music swap, antiques and collectibles, piano rolls and player pianos, historic walking tours. All events package $90.00. Annually, beginning the first Wednesday in June. Est attendance: 2,500. For info: Scott Joplin Fdtn, 116 E Main, Sedalia, MO 65301. Phone: (816) 826-2271.

June 1-2 ☆ Chase's 1995 Calendar of Events ☆

SOAP OPERA FAN FAIR. June 1-5. Mackinaw City, MI. 50-75 of the country's top Soap Opera Stars will be on hand. Fans who've always dreamed about meeting their favorite Soap Opera stars, getting autographs, posing for photos and home videos will have their dreams fulfilled. There will also be luncheons and dinners with the stars, ferry boat races and moonlight cruises, parade, contests and prizes, great publicity and a 25,000 square foot tent for booths, dances and special events being finalized. Annually, beginning Thursday after Memorial Day weekend. Est attendance: 50,000. For info: Mackinaw Area Tourist Bureau, 708 S Huron Ave, Mackinaw City, MI 49701. Phone: (800) 817-7627.

SPACE MILESTONE: SOYUZ 9 (USSR). June 1. Cosmonauts Nikolayev and Sevastyanov set space endurance record of 17 days, 16 hours and 59 minutes. Launched June 1, 1970.

SURIMI SEAFOOD MONTH. June 1-30. To promote the convenience, taste and versatility of imitation crabmeat, lobster and shrimp. For info: Surimi Seafood Education Center, 1525 Wilson Blvd, Ste 500, Arlington, VA 22209.

TENNESSEE ADMISSION DAY. June 1. Became 16th state on this day in 1796.

WOLF POINT'S HOTTEST CHILI WEEKEND IN JUNE. June 1-4. Sherman Park, Stampede Ground, Wolf Point, MT. International Chili Society sanctioned chili cookoff; car show. Annually, first weekend in June. Est attendance: 4,000. Sponsor: Wolf Point Chamber of Commerce, PO Box 237, Wolf Point, MT 59201. Phone: (406) 653-2012.

YOUNG, BRIGHAM: BIRTH ANNIVERSARY. June 1. Mormon church leader born at Whittingham, VT, June 1, 1801. He died at Salt Lake City, UT, on Aug 29, 1877, and was survived by 17 wives and 47 children. Utah observes, as a state holiday, the anniversary of his entrance into the Salt Lake Valley, July 24, 1847.

ZAM! ZOO AND AQUARIUM MONTH. June 1-30. A national celebration to focus public attention on the role of zoos and aquariums in wildlife education and conservation. Held at 164 AAZPA member institutions in the US and Canada. Sponsor: American Assn of Zoological Parks and Aquariums. For info, contact your local zoo or aquarium.

BIRTHDAYS TODAY

James Hadley Billington, 66, Librarian of Congress, born at Bryn Mawr, PA, June 1, 1929.

Pat Boone, 61, singer, actor ("Bernadine," *State Fair*), author, born Jacksonville, FL, June 1, 1934.
Pat Corley, 65, actor ("Bay City Blues"; Phil on "Murphy Brown"), born Dallas, TX, June 1, 1930.
Morgan Freeman, 58, stage and film actor (*Driving Miss Daisy*), born Memphis, TN, June 1, 1937.
Andy Griffith, 69, actor (*No Time for Sergeants*, "The Andy Griffith Show"), born at Mt Airy, NC, June 1, 1926.
Lisa Hartman, 39, actress (Tabitha Stephens in "Tabitha"; "Knots Landing"), born Houston, TX, June 1, 1956.
Ron Wood, 48, musician (2nd drummer with Rolling Stones), born at London, England, June 1, 1947.

JUNE 2 — FRIDAY
153rd Day — Remaining, 212

BAHAMAS: LABOR DAY. June 2. Public holiday. First Friday in June celebrated with parades, displays and picnics.

BEEF EMPIRE DAYS. June 2-11. Finney County Fairgrounds, Garden City, KS. 27th annual celebration of the beef industry. Live and carcass show, PRCA rodeo (2 nights), parade, beef tasting for the public, cowboy poetry, professional Western art sale, feedlot roping and riding, stock dogs demonstration, walk/run 2mi, 10K, golf tournament, softball tournament, tennis tournament, children's events. Annual event in June. Phone: (316) 275-6807 or (800) 879-9803. Est attendance: 100,000. For info: Beef Empire Days, Sara Robinson, 1511 E Fulton Ter, Garden City, KS 67846.

BELGIUM: PROCESSION OF THE GOLDEN CHARIOT. June 2. Mons. Horse-drawn coach carrying a reliquary of St. Waudru circles the town of Mons. Procession commemorates delivery of Mons from the plague in 1349. In the town square, in afternoon, St. George fights the dragon. Lumecon symbolizes triumph of good over evil, and when victory is won the spectators sing traditional songs.

BILLY BOWLEGS PIRATE FESTIVAL. June 2-10. Fort Walton Beach, FL. Captain Billy Bowlegs and his krewe, portrayed by local businessmen, storm the city in the pirate ship *Blackhawk* for a week of fun and frolicking. Two treasure hunts for valuable prizes will take place during the festival. Fireworks on Friday night. Est attendance: 10,000. Sponsor: Chamber of Commerce, PO Box 640, Ft Walton Beach, FL 32549. Phone: (904) 244-8191.

BULGARIA: HRISTO BOTEV DAY. June 2. Poet and national hero Histro Botev fell fighting Turks, 1876.

CALIFORNIA SENIOR GAMES-SACRAMENTO. June 2-4. Athletic competition for men and women age 50 and older. Compete in five-year age divisions. Sports vary between locations. Must qualify at local meet to compete in state finals. Est attendance: 1,000. For info: Alan Boyd, Coord, 6005 Folsom Blvd, Sacramento, CA 95819. Phone: (916) 277-6077.

CHINA: DRAGON BOAT FESTIVAL. June 2. An important Chinese observance, the Dragon Boat Festival commemorates a hero of ancient China, poet Qu Yuan, who drowned himself in protest against injustice and corruption. It is said that rice dumplings were cast into the water to lure fish away from the body of the martyr, and this is remembered by the eating of zhong zi, glutinous rice dumplings filled with meat and wrapped in bamboo leaves. Dragon boat races are held on rivers. The Dragon Boat Festival is observed in many countries by their Chinese population. Also called Fifth Month Festival or Summer Festival. Annually, the fifth day of the fifth lunar month.

CHURCH POINT BUGGY FESTIVAL. June 2-4. Church Point, LA. Pageants, carnival, arts and crafts, Cajun food, Cajun French music and authentic parade including buggies, horse drawn carriages and antique cars. Annually, the first weekend in June. Est attendance: 8,000. For info: "Teasie" Cary, Mgr, Church Point Buggy Festival, 837 E Ebey, Church Point, LA 70525. Phone: (318) 684-2739.

June 1995	S	M	T	W	T	F	S
					1	2	3
	4	5	6	7	8	9	10
	11	12	13	14	15	16	17
	18	19	20	21	22	23	24
	25	26	27	28	29	30	

☆ Chase's 1995 Calendar of Events ☆ June 2

COTTONWOOD PRAIRIE FESTIVAL. June 2-4. Brickyard Park, Hastings, NE. A celebration of fine arts/crafts, antiques, food and entertainment. Nebraska's largest outdoor sculpture invitational. Annually, first weekend of June. Est attendance: 25,000. For info: Wendy Keele, Coord, PO Box 941, Hastings, NE 68902-0941. Phone: (800) 967-2189.

CRANESVILLE SWAMP WEEKEND. June 2-4. Appalachian Base Camp, Terra Alta, WV. Camping weekend to identify and learn about reptiles and amphibians: specimens collected and studied (later they are set free) and field trips led by experts. Appropriate for all levels of expertise. Est attendance: 30. Sponsor: Oglebay Institute Nature/Environmental Education, Tom Shepard, Oglebay Park, Wheeling, WV 26003. Phone: (304) 242-6855.

CURWOOD FESTIVAL. June 2-4. Owosso, MI. Homecoming celebration commemorating James Oliver Curwood, Owosso-born author and conservationist (June 12, 1878-Aug 13, 1927). The Curwood castle was built for a studio. Open to the public. 50 events including parades, races and music. Annually, the first full weekend in June. Est attendance: 100,000. For info: Curwood Festival, Box 461, Owosso, MI 48867. Phone: (517) 723-2161.

DONUT DAY. June 2-3. Chicago, IL. Founded in 1938 by the Salvation Army for fundraising during the Great Depression, Donut Day is now an annual tradition. Recalling the donuts served to doughboys by the Salvation Army during World War I, symbolic paper "donuts" are given to contributors. Annually, the first Friday and Saturday in June. For info: The Salvation Army, Metro Div HQ, Robert A. Bonesteel, Dir of Communications, 5040 N Pulaski, Chicago, IL 60630. Phone: (312) 725-1100.

FIFTIES REVIVAL. June 2-4. Marshall, MN. Fifties music, games, auto display, poker run, dance, car show and concessions. Est attendance: 2,500. For info: Curt Anton, Shades of the Past Car Club, RR 4, Box 434, Marshall, MN 56258. Phone: (507) 532-6773.

FISHING HAS NO BOUNDARIES EAGLE RIVER. June 2-4. Dock Park, Eagle River, WI. A three-day fishing experience for disabled persons. Any disability, age, sex, race, etc, eligible. Attended by 60 participants and 150 volunteers. For info: Fishing Has No Boundaries, PO Box 2200, Eagle River, WI 54521. Phone: (800) 261-3474.

GOVERNOR'S CUP MARATHON. June 2-3. Helena, MT. Montana's premier running event. Includes a marathon, marathon relay 20K, 10K and 5K. Corporate entries. Approximately 7,000 runners. Annually, the first weekend in June. Sponsor: Blue Cross/Blue Shield. For info: Hal Rawson, Blue Cross & Blue Shield, Box 451, Helena, MT 59624. Phone: (406) 444-8983.

HARBORFEST. June 2-4. Norfolk, VA. Live entertainment, sailboat races, water demonstrations, tall sailing ships from around the world, aerobatic stunt-flying teams, military displays and air precision-flying teams, Chesapeake Bay Seafood, special entertainment for children and seniors, and a spectacular fireworks display. Annually, the first full weekend in June. Est attendance: 400,000. For info: Harborfest Office, 123 Granby St, Norfolk, VA 23510. Phone: (804) 627-5329.

HARVARD MILK DAY FESTIVAL. June 2-4. Harvard, IL. This salute to the dairy farmer includes a parade, evening entertainment, arts and crafts fair, sports and classic car show, farm tours, pedal tractor pull, toy show, farm safety programs, tractor pull, carnival, fireworks, laser shows, prince and princess contest, Milk Run (2-mile and 10K races), cattle show, talent show and Wee Farm. Est attendance: 70,000. For info: Harvard Milk Day Office, PO Box 310, Harvard, IL 60033. Phone: (815) 943-4614.

ITALY: REPUBLIC DAY. June 2. National holiday. Commemorates referendum on June 2, 1946, in which republic status was selected instead of return to monarchy.

KOREA: TANO DAY. June 2. Fifth day of fifth lunar month. Summer food offered at the household shrine of the ancestors. Also known as Swing Day, since girls dressed in their prettiest clothes often compete in swinging matches. The Tano Festival usually lasts from the third through eighth day of the fifth lunar month; May 31–June 5.

LAKE CHAMPLAIN BALLOON FESTIVAL. June 2-4. Champlain Valley Exposition, Essex Jct, VT. Fifty hot air balloons, including spectacular special shapes, will participate in five scheduled launches. The festival also features a juried arts and crafts show, an "evening Balloon Glow" on Saturday at dusk, fireworks and daily skydiving exhibitions. Est attendance: 70,000. For info: Lake Champlain Balloon and Craft Festival, PO Box 83, Underhills Ctr, VT 05490. Phone: (802) 899-2993.

MAINE LAW: ANNIVERSARY. June 2. America's first state-wide statute prohibiting the sale of alcoholic beverages was enacted in the state of Maine, June 2, 1851. The following Independence Day the mayor of Bangor showed his support of the new law by smashing 10 kegs of confiscated booze.

MAMOU CAJUN MUSIC FESTIVAL. June 2-3. Mamou, LA. Festival celebration with various authentic games, a Cajun Queen contest and live Cajun music on Friday night from 6:00 PM until 12:00 midnight. On Saturday, live Cajun music is provided all day from 9:00 AM until 10:00 PM, while various authentic Cajun games are held. Annually, the first Friday and Saturday in June. Est attendance: 5,000. For info: Eric M. Fontenot, RR1, Box 84E, Mamou, LA 70554. Phone: (318) 468-0188.

MARQUIS DE SADE: BIRTH ANNIVERSARY. June 2. Donatien Alphonse François, Comte de Sade, was born at Paris, France, June 2, 1740. French military man, governor-general and author, who spent much of his life in prison because of his acts of cruelty and violence, outrageous behavior and debauchery. The word *sadism* was created from his name to describe cruelty and gratification in inflicting pain. He died near Paris, at the Charenton lunatic asylum, on Dec 2, 1814.

MIAMI/BAHAMAS GOOMBAY FESTIVAL. June 2-4. Miami, FL. A celebration of the black culture and heritage of Bahamian settlers in Miami's Coconut Grove area in the 1800s. Largest black heritage special event in the US. More than 400 vendor booths. Pre-festival events include golf tournament, beauty pageant, sailing regatta. Annually, the first weekend in June. Est attendance: 550,000. Sponsor: Miami/Bahamas Goombay Festival, Susan Neuman, 555 NE 15th St, #25-K, Miami, FL 33132. Phone: (305) 372-9966.

NONGAME WILDLIFE WEEKEND. June 2-4. Blackwater Falls State Park, Davis, WV. Presentations by professionals in the wildlife field regarding nongame wildlife and endangered species. The highlight of the weekend occurs on Saturday with morning workshops and afternoon field trips such as birdwatching, visiting a snake den, interpretive plant walks, caving, forest ecology, streamlife surveys and how to attract wildlife to the backyard. Annually, the first weekend in June. Est attendance: 225. For info: Kathleen C. Leo, Project Leader, WV Nongame Wildlife Program, WV Dept of Natural Resources, PO Box 67, Ward Rd, Elkins, WV 26241. Phone: (304) 637-0245.

OAKLAND SWEDISH FESTIVAL. June 2-4. Downtown, Oakland, NE. Award winning ethnic festival complete with entertainment, dancers, exhibits, demonstrations, music, Smorgasbord, kid's games and activities, tours of Oakland, Troll Stroll, dignitaries, concerts, ethnic food and more! Classic Car Show Sunday; craft show all weekend. First weekend in June in odd-numbered years. Sponsor: City of Oakland and Oakland Chamber of Commerce. Est attendance: 15,000. For info: John Thomas, F&M Bank, 212 N Oakland Ave, Oakland, NE 68045. Phone: (402) 685-5621.

June 2-3 ★ Chase's 1995 Calendar of Events ★

OLEPUT FESTIVAL. June 2-3. Downtown Tupelo, MS. A Mardi Gras festival with music, entertainment, food, parade. Annually, the first weekend in June. Est attendance: 25,000. For info: Tupelo Conv and Visitors Bureau, PO Box 1485, Tupelo, MS 38802-1485. Phone: (800) 533-0611.

ORIGINAL OZARK FESTIVAL. June 2-3. Westside Park, Ozark, AR. Festival celebrates the oldest US city incorporated with the name Ozark. Features old-time artists and craftsmen demonstrating various "lost arts" such as shingle-making, rail-splitting, quilting, soap-making and more. Food, fun music contests, events. Annually, the first Friday night and Saturday of June. Sponsor: Ozark Area Chamber of Commerce, Diana Wilson, Dir, PO Box 283, Ozark, AR 72949. Phone: (501) 667-2525.

ROGUE RIVER JET BOAT MARATHON. June 2-4. Gold Beach, OR. Watch jet boats ply the twisted, rushing whitewater rapids of the mighty Rogue at speeds faster than your eyes can focus, starting from Jot's Resort at the mouth of the Rogue in Gold Beach, about 30 miles upriver to Agness, and back again. Other events include hydro-plane racing and sprint-boat manuevers. Est attendance: 1,000. For info: Gold Beach Chamber of Commerce, 1225 S Ellensburg #3, Gold Beach, OR 97444. Phone: (800) 525-2334.

◆ **RUTH RETIRES: ANNIVERSARY.** June 2, 1935. Three days after he benched himself from his last game (May 30), George Herman "Babe" Ruth announced his retirement from major league baseball.

ST. PIUS X: BIRTH ANNIVERSARY. June 2. Giuseppe Melchiorre Sarto, 257th pope of the Roman Catholic Church, born June 2, 1835, at Riese, Italy. Elected pope Aug 4, 1903. Died Aug 20, 1914. Canonized May 29, 1954.

SCOTLAND: ROYAL SCOTTISH AUTOMOBILE CLUB INTERNATIONAL SCOTTISH RALLY. June 2-4. Throughout Scotland with base at Perth. Scotland's only international rally which attracts many of the world's leading drivers. Est attendance: 200,000. For info: Jonathan Lord, Royal Scottish Automobile Club, 11 Blythswood Sq, Glasgow, Scotland G2 4AG.

SEASPACE '95. June 2-4. Hyatt Regency Houston Hotel, Houston, TX. Scuba diving symposium featuring seminars, photo course, photo exhibit, film festival, hall of exhibits and receptions. Sponsor: Houston Underwater Club. For more info send self-addressed stamped envelope. Est attendance: 10,000. For info: Seaspace, PO Box 3753, Houston, TX 77253-3753. Phone: (713) 973-9300.

SITKA SUMMER MUSIC FESTIVAL. June 2-23. (Tuesdays and Fridays, plus Saturday June 17.) Sitka, AK. Sitka hosts a highly acclaimed chamber music festival that attracts performers and spectators from all over the world. Est attendance: 4,000. For info: Sitka Summer Music Festival, PO Box 201988, Anchorage, AK 99520. Phone: (907) 277-4852.

SOUND AND LIGHT SPECTACULAR: *THE IMMORTAL SHOWBOAT*. June 2-Aug 31. Wilmington, NC. An outdoor drama that tells the story of the World War II battleship through spoken word, music, sound effects and lighting. Est attendance: 13,000. For info: USS *North Carolina* Battleship Memorial, Box 480, Wilmington, NC 28402. Phone: (910) 251-5797.

SUGAR VALLEY RALLY AND WEEKEND. June 2-4. Scottsbluff/Gering, NE, and thru North Platte Valley. Sanctioned computerized rally for pre-1943 antique vehicles. Participants come from throughout the US and foreign countries to compete for $5,000+ in prizes. "Second largest event of this type in US (first is Great American Race)." Also, at Pioneer Park, Scottsbluff, Sugar Valley Arts & Crafts Festival with more than 100 craftsmen/artisans selling and demonstrating their work. Live entertainment and more. Annually, the first weekend in June. Est attendance: 5,500. For info: Pam Hill, Rally Secy, Sugar Valley Rally, 1517 Broadway, Scottsbluff, NE 69361. Phone: (308) 632-2133.

SUN FUN FESTIVAL. June 2-7. Myrtle Beach, SC. Air Show, beauty contests, sporting events, live television shows and sandcastle building contest, parade, greet visitors along the Grand Strand. Annually, beginning the weekend after Memorial Day, lasting six days. Est attendance: 300,000. For info: Marilyn Chewning, Myrtle Beach Area Visitors Bureau, PO Box 2115, Myrtle Beach, SC 29578. Phone: (803) 626-7444.

THREE RIVERS ARTS FESTIVAL. June 2-18. Pittsburgh, PA. Celebration of visual and performing arts. Juried visual arts exhibitions, musical performances, artists market, children's activities and food. Est attendance: 600,000. Sponsor: Three Rivers Arts Festival, 207 Sweetbriar St, Pittsburgh, PA 15211. Phone: (412) 481-7040.

YELL "FUDGE" AT THE COBRAS IN NORTH AMERICA DAY. June 2. Anywhere north of the Panama Canal. In order to keep poisonous cobra snakes out of North America, all citizens are asked to go outdoors at noon local time and yell "Fudge." Fudge makes cobras gag, and the mere mention of it makes them skeedaddle. Annually, June 2. Phone: (212) 388-8673 or (717) 274-8651. For info: Wellness Permission League, c/o Thomas Roy, 2105 Water St, Lebanon, PA 17046.

BIRTHDAYS TODAY

Diana Canova, 42, actress (Corrine on "Soap"; "I'm a Big Girl Now"), born at West Palm Beach, FL, June 2, 1953.
Charles Haid, 52, actor (Renko on "Hill Street Blues"; "Delvecchio"), producer, born San Francisco, CA, June 2, 1943.
Marvin Hamlisch, 51, composer (Oscars for scores *The Sting*, *The Way We Were*; Tony for score *A Chorus Line*), born at New York, NY, June 2, 1944.
Stacy Keach, Jr, 54, actor (*Conduct Unbecoming*, "Mickey Spillane's Mike Hammer"), born at Savannah, GA, June 2, 1941.
Sally Kellerman, 59, actress (*M*A*S*H*, *Back to School*), born Long Beach, CA, June 2, 1936.
Jerry Mathers, 47, actor ("Leave It to Beaver"), born Sioux City, IA, June 2, 1948.
Charlie Watts, 54, musician (drummer with Rolling Stones; "[I Can't Get No] Satisfaction"), born Islington, England, June 2, 1941.

JUNE 3 — SATURDAY
154th Day — Remaining, 211

AMERICAN BOOKSELLERS ASSOCIATION TRADE EXHIBIT AND CONVENTION. June 3-6. Chicago, IL. Publishers display fall titles for attending booksellers or all interested in reaching the retail bookseller. Book-related items also on display. For info: American Booksellers Assn, 828 S Broadway, Tarrytown, NY 10591. Phone: (914) 591-2665.

	S	M	T	W	T	F	S
June 1995					1	2	3
	4	5	6	7	8	9	10
	11	12	13	14	15	16	17
	18	19	20	21	22	23	24
	25	26	27	28	29	30	

☆ Chase's 1995 Calendar of Events ☆ June 3

ART FAIRE. June 3-4. Peddler's Village, Lahaska, PA. 2-day event featuring local artists with works in oil, watercolor, pen and ink, charcoal and sculpture. Artists compete for more than $2,000 in prizes. Plus musical entertainment, mimes, face painting. Children's educational activities. Est attendance: 5,000. For info: Jason Assoc, Peddler's Village, Box 218, Lahaska, PA 18931. Phone: (215) 794-4000.

BUFFALO DAYS CELEBRATION (WITH BUFFALO CHIP THROWING). June 3-4. Luverne, MN. Parade, Arts in the Park, free barbecued buffalo burgers (while they last) and unique buffalo chip throwing contest. Annually, the first weekend in June. Est attendance: 8,000. For info: Luverne Area Chamber of Commerce, 102 E Main, Luverne, MN 56156. Phone: (507) 283-4061.

CAPITOL HILL PEOPLE'S FAIR. June 3-4. Civic Center Park, Denver, CO. More than 500 arts and crafts and other exhibit booths; live entertainment featuring local talent on 6 stages. Est attendance: 250,000. For info: Capitol Hill United Neighborhoods, 1490 Lafayette, Denver, CO 80218. Phone: (303) 837-1839.

CHILDREN'S MIRACLE NETWORK. June 3-4. Anaheim, CA. The largest television fund raiser in history. More than $120 million was raised in 1994. Benefits more than 160 hospitals for children. Live from Disneyland. For info: Children's Miracle Network, 4525 S 2300 E, Ste 202, Salt Lake City, UT 84117. Phone: (801) 278-8900.

CHIMBORAZO DAY. June 3. To bring the shape of the earth into focus by publicizing the fact that Mount Chimborazo, in Ecuador near the equator, pokes further out into space than any other mountain on earth, including Mt Everest. (The distance from sea level at the equator to the center of the earth is 13 miles greater than the radius to sea level at the north pole. This means that New Orleans is about 6 miles further from the center of the earth than is Lake Itaska at the headwaters of the Mississippi. So the Mississippi flows uphill.) Annually, June 3. Sponsor: Puns Corps, Robert L. Birch, Box 2364, Falls Church, VA 22042-0364. Phone: (703) 533-3668.

COUNTRYSIDE VILLAGE ART FAIR. June 3-4. Omaha, NE. Exhibits by 140 artists from 15 states. Juried fine arts show. 25th annual show. Est attendance: 10,000. For info: Countryside Merchants Assn, Judy Drawbaugh, 3268 S 130th Circle, Omaha, NE 68144.

COWS ON THE CONCOURSE. June 3. Madison, WI. Pet a cow on the capitol square to celebrate June Dairy Month. All major breeds of cows will be on display. Ample samplings of all types of dairy products made in Wisconsin. Antique tractors, dairy-related crafts demonstrations and displays. Annually, the first Saturday in June. Est attendance: 20,000. For info: Norman Maier, 6101 Queens Way, Madison, WI 53716. Phone: (608) 221-8698.

DAIRY DAY. June 3. Old City Park, Dallas, TX. Enjoy free samples of ice cream, yogurt, cheese and milk during this tasty tribute to the Texas Dairy Industry. Kids enjoy petting the calves and goats; city folk can try their hand at milking a cow; anyone can participate in the mooing contest; 10-3. Annually, the first Saturday in June. Est attendance: 4,000. For info: Old City Park, Dallas Co Heritage Soc, 1717 Gano, Dallas, TX 75215. Phone: (214) 421-5141.

DARE DAY. June 3. Manteo, NC. Arts, crafts, military band, clogging, food festival and gospel groups. Annually, the first Saturday in June. Est attendance: 9,000. Sponsor: Town of Manteo and County of Dare, Box 1000, Manteo, NC 27954. Phone: (919) 473-1101.

DAVIS, JEFFERSON: BIRTH ANNIVERSARY. June 3. American statesman, US senator, only president of the Confederate States of America, imprisoned May 10, 1865-May 13, 1867, but never brought to trial, deprived of rights of citizenship after the Civil War. Davis was born at Todd County, KY, June 3, 1808, and died at New Orleans, LA, Dec 6, 1889. His citizenship was restored, posthumously, on Oct 17, 1978, when President Carter signed an Amnesty Bill. Carter stated: "Our nation needs to clear away the guilts and enmities and recriminations of the past, to finally set at rest the divisions that threatened to destroy our nation and to discredit the great principles on which it was founded." This bill, he said, "officially completes the long process of reconciliation that has reunited our people following the tragic conflict between the states." Davis's birth anniversary is observed in some states on the first Monday in June.

DENMARK: EEL FESTIVAL. June 3-4. Jyllinge (near Roskilde). Festival celebrated since 1968. Every restaurant and pub in town serves delicious fried eel. Other entertainments include theater, sports, tattoo bands, sailing competitions, flea markets and fireworks. Annually, the first weekend in June.

DEWHURST, COLLEEN: BIRTH ANNIVERSARY. June 3. Colleen Dewhurst was born at Quebec, Canada, on June 3, 1924. Her 40-year career as an actress spanned stage, screen and television. After making her Broadway debut in Eugene O'Neill's *Desire Under the Elms* in 1952, she became the actress most associated with O'Neill's works in the later part of this century, also performing in *Long Day's Journey into Night*, *Mourning Becomes Electra*, *Ah, Wilderness* and *Moon for the Misbegotten*, for which she won her second Tony Award. At the time of her death on Aug 22, 1991, she was president of Actor's Equity Assn, the union for professional actors and stage managers.

DO DAH PARADE. June 3. Kalamazoo, MI. Off-beat entries such as a grill team (complete with spatulas) and a lawn mower drill team. Annually, the first Saturday in June. Est attendance: 50,000. For info: WKMI/WKFR Radio, Attn: Lisa Theisen, 4154 Jennings Dr, Kalamazoo, MI 49005. Phone: (616) 344-0111.

DUKE OF WINDSOR: MARRIAGE ANNIVERSARY. June 3. The Duke of Windsor who, as King Edward VIII, had abdicated the British throne on Dec 11, 1936, was married to Mrs Wallis Warfield Simpson of Baltimore, MD. The marriage took place in Monts, France, on June 3, 1937.

ENGLAND: INTERNATIONAL TT MOTORCYCLE RACES. June 3-9. Isle of Man. World famous motorcycle road races held over a 38-mile course. Practice sessions will be held from May 29-June 2. Est attendance: 40,000. For info: Road Race Dept, Auto Cycle Union, Wood St, Rugby, Warwickshire, England CV21 2YX.

FIRST WOMAN RABBI IN US: ANNIVERSARY. June 3. Sally Jan Priesand was ordained a rabbi on June 3, 1972, making her the first woman rabbi in the US. On Aug 1, 1972, she became assistant rabbi at the Stephen Wise Free Synagogue, New York City.

FORT SISSETON HISTORICAL FESTIVAL. June 3-4. Fort Sisseton State Park, Lake City, SD. Fort Sisseton comes alive the first weekend in June every year. See life as it was in 1864 when the fort was established. Cavalry drills, military costume ball, Indian dancing, Dakota Dan's medicine show, Dutch oven cookoff, muzzleloader shoot, rendezvous, draft horse pulls, fiddling contests, square dancing, melodramas and frontier crafts displays are popular features of the festival. Est attendance: 46,000. For info: Fort Sisseton State Park, RR2 Box 94, Lake City, SD 57247-9704. Phone: (605) 448-5701.

FRESH FISH FEST. June 3-4. Breckenridge, CO. Free fishing, fish dish contest, fish face, fish tale, picnic, fun for under a fin. Est attendance: 8,000. For info: John Hendryson, Dir Sales and Spec Events, Box 7868, Breckenridge, CO 80424. Phone: (303) 453-2913.

June 3 ☆ *Chase's 1995 Calendar of Events* ☆

FRONTIER DAYS. June 3-4. Culbertson, MT. Two days of parades, rodeos, a barn dance, talent show and genuine Western hospitality. Annually, the first weekend in June. Est attendance: 2,000. For info: Ila Mae Forbregd, 1 Broadway Ave, Culbertson, MT 59218.

GOD'S COUNTRY MARATHON XXI. June 3. Coundersport, PA. A 26.2-mile marathon from Galeton High School to Coudersport Area Recreation Park, crossing over Denton Hill Mountain, the eastern Continental Divide of the US. Finish medals to all who complete the course in addition to awards. Fax: (814) 274-8926. For info: Potter County Recreation, Inc, Race Dir, RD 3, Box 272A, Coudersport, PA 16915. Phone: (814) 274-9109.

GREAT CHUNKY RIVER RAFT RACE. June 3. Chunky, MS. Raft builders compete in four raft divisions in race held along the Chunky River. Cash prizes for most creative rafts. Live entertainment noon-4:00. Raft divisions: flat bottom boat, canoe, inflatable raft (store bought), homemade raft. Est attendance: 3,000. For info: Lisa Cowart, Special Project/Market Coord, PO Box 4177, Meridian, MS 39304. Phone: (601) 482-6161.

GREAT WISCONSIN DELLS BALLOON RALLY. June 3-4. Wisconsin Dells, WI. More than 90 vividly colored balloons compete in this annual event. Est attendance: 100,000. For info: Wisconsin Dells Visitor and Conv Bureau, Box 390, Wisconsin Dells, WI 53965. Phone: (800) 223-3557.

GRUNDLOVSFEST. June 3-4. Mill St and City Park, Dannebrog, NE. Danish celebration honoring Denmark's first free Constitution Day, June 5, 1849. Danish dancers, Food Court, arts and crafts, parade, reunions, bicycle tours, children's activities, concerts, Danish Pastries, ballgames, community church and sing-a-long. Annually, the first full weekend in June. Sponsor: Dannebrog Area Booster Club. Est attendance: 1,000. For info: Shirley Johnson, Pub Chair, 522 E Roger Welsch Ave, Dannebrog, NE 68831. Phone: (308) 226-2237.

HOBART, GARRET AUGUSTUS: BIRTH ANNIVERSARY. June 3. Twenty-fourth vice president of the US (1897-1899) born at Long Branch, NJ, June 3, 1844. Died at Patterson, NJ, Nov 21, 1899.

HONE, WILLIAM: BIRTH ANNIVERSARY. June 3. English author and bookseller born at Bath, England, June 3, 1780. Died at Tottenham, England, Nov 6, 1842. Compiler of *The Every-Day Book; or Everlasting Calendar of Popular Amusements* (1826). It was William Hone who said: "A good lather is half the shave."

JACK JOUETT'S RIDE: ANNIVERSARY OBSERVANCE. June 3. Jack Jouett made a heroic 45-mile ride on horseback during the night of June 3-4, 1781, to warn Virginia's Governor Thomas Jefferson and the legislature that the British were coming. Jouett rode from a tavern in Louisa County to Charlottesville, VA, in about 6½ hours, arriving at Jefferson's home at about dawn on June 4. Lieutenant Colonel Tarleton's British forces raided Charlottesville, but Jouett's warning gave the Americans time to escape. Jouett was born at Albemarle County, VA, on Dec 7, 1754, and died at Bath, KY, in 1822 (exact date unknown). Reportedly, Jack Jouett Day, commemorating the historic and heroic ride, is observed annually on the first Saturday in June in the Charlottesville area.

KAHN'S KIDS FEST. June 3. Bicentennial Commons, Sawyer Point, Cincinnati, OH. More than 16 venues line ¾-mile stretch along the banks of the Ohio River with over 100 activities. Annually, the first Saturday in June. Est attendance: 120,000. For info: Leslie Keller, Dir of Events, Cincinnati Recreation Commission, 805 Central Ave, 2 Centennial Plaza, Cincinnati, OH 45202. Phone: (513) 352-1608.

KIA ART FAIR. June 3. Bronson Park, Kalamazoo, MI. Juried art fair, 150+ artists from several states. Food, entertainment, children's art fair, silent auction. Annually, the first Saturday in June. Est attendance: 65,000. Sponsor: Kalamazoo Institute of Arts, Exec Dir, 314 S Park St, Kalamazoo, MI 49007. Phone: (616) 349-7775.

MARBLE MEET AT AMANA. June 3-4. Holiday Inn, Amana, IA. Seminar, banquet and exhibits. Collectors buy, sell and trade marbles. Est attendance: 600. Sponsor: Marbles Collectors Unltd, Box 206, Northboro, MA 01532. Phone: (319) 642-3891.

McHENRY HIGHLAND FESTIVAL. June 3. Garrett County Fairgrounds, Route 219, McHenry (Deep Creek Lake), MD. Traditional Scottish event with bagpipe bands, sheepdog exhibitions, open Scottish athletic and solo piping competitions, shops, foods and more. Est attendance: 2,000. Sponsor: Deep Creek Lake-Garrett County, Promotion Council, Courthouse, 200 S Third St, Oakland, MD 21550. Phone: (301) 334-1948.

MIGHTY CASEY HAS STRUCK OUT: ANNIVERSARY. June 3. The famous comic baseball ballad "Casey at the Bat" was printed in the *San Francisco Examiner* Sunday, June 3, 1888. Appearing anonymously, it was written by Ernest L. Thayer. Recitation of "Casey at the Bat" became part of the repertoire of actor William DeWolf Hopper. The recitation took 5 minutes and 40 seconds. Hopper claimed to have recited it more than 10,000 times, the first being at Wallack's Theater in New York, NY, in 1888. The Library of Congress reissued "Casey at the Bat" in text and recording in 1986. See also: "Thayer, Ernest Lawrence: Birth Anniversary" (Aug 14).

MISSION SAN CARLOS BORROMEO DE CARMELO FOUNDING: 225th ANNIVERSARY. June 3. California mission to the Indians founded June 3, 1770.

PEACHTREE JUNIOR. June 3. Atlanta, GA. A 3K non-competitive run for children ages 7-12. Entries limited to 2,500. Send self-addressed stamped envelope. Est attendance: 2,500. For info: Atlanta Track Club, 3097 E Shadowlawn Ave, Atlanta, GA 30305. Phone: (404) 231-9065.

PET PARADE. June 3. LaGrange, IL. Children and their pets parade through the streets of LaGrange in costume, accompanied by clowns, floats, celebrities and marching bands. Trophies awarded to the most original entries in 10 costume categories. Est attendance: 80,000. For info: West Suburban Chamber of Commerce, PO Box 187, LaGrange, IL 60525. Phone: (708) 352-0494.

REOPENING OF THE RAILROAD CAR MUSEUM. June 3. Depot Lane Complex, Schoharie, NY. Open each weekend, 1-5, June through October. Est attendance: 500. Sponsor: Schoharie Colonial Heritage Assn, PO Box 554, Schoharie, NY 12157. Phone: (518) 295-7505.

REPEAT DAY. June 3. Repeat Day is a day during which we each try to learn a new vocabulary word by repeating it in different sentences as many times as possible throughout the day. Created by Gransby, CT, 5th grader Danny Rubalcaba in response to a *Hartford Courant* challenge. For info: Kenton Robinson, Features Writer, *The Hartford Courant*, 285 Broad St, Hartford, CT 06115. Phone: (203) 241-3947.

☆ Chase's 1995 Calendar of Events ☆ June 3-4

SHAVUOT BEGINS AT SUNDOWN. June 3. Jewish Pentecost. See "Shavuot" (June 4).

SMALL CRAFT WEEKEND. June 3-4. Mystic Seaport, Mystic, CT. The 26th annual weekend when small craft enthusiasts gather at the Museum with their boats. Traditional small boats of every type sail from docks of Mystic Seaport on the Mystic River. Annually, first weekend in June. Est attendance: 4,000. For info: Mystic Seaport, Box 6000, Mystic, CT 06355. Phone: (203) 572-0711.

SOUTH JERSEY CANOE AND KAYAK CLASSIC. June 3. Ocean County Park, Rt 88, Lakewood, NJ. Canoe and Kayak vendors from around the country set up on a beach to show the public the thrill of watersports. You may test paddle the boats of your choice and attend a clinic about canoeing or kayaking throughout the day. $1.00 per car. Annually, the first Saturday in June. Est attendance: 3,000. Sponsor: Wells Mills County Park, Lillian Hoey, Recreation Leader, 905 Wells Mills Rd, Waretown, NJ 08758. Phone: (609) 971-3085.

SPACE MILESTONE: GEMINI 4 (US). June 3. Major McDivitt and Major White made 66 orbits. White took spacewalk and maneuvered 20 minutes outside capsule. Launched June 3, 1965.

THE TONGASS—ALASKA'S MAGNIFICENT RAIN FOREST. June 3-Sept 24. Alaska State Museum, Juneau, AK. This exhibit explores the intricate harmony of the largest non-equatorial rainforest on earth. See Feb 4 for full description. For info: Smithsonian Institution Traveling Exhibition Service, 1100 Jefferson Dr SW, Ste 3146, Washington, DC 20560. Phone: (202) 357-2700.

TOPPENISH MURAL SOCIETY'S "MURAL-IN-A-DAY." June 3. Pioneer Park, Toppenish, WA. 12 professional artists paint a complete, historically authentic mural, 14' x 48', in 8 hours. Starting at 9 AM they work until finished, usually until 4 PM. Accompanied by an arts and crafts show and an ethnic food fair. Annually, the first Saturday in June. Est attendance: 8,000. For info: Toppenish Mural Society, PO Box 1172, Toppenish, WA 98948. Phone: (509) 865-6516.

TURTLE RACES. June 3. Eastern Illinois Fairgrounds, Danville, IL. More than 100 turtles compete in 31st annual races throughout the day. Concessions available. Food & fun. Proceeds go to help the area handicapped. Annually, the first Saturday in June. For info: Ralph Sargent, Turtle Club, PO Box 1332, Danville, IL 61834. Phone: (217) 446-5327.

WORLD'S LARGEST GARAGE SALE. June 3. South Bend, IN. Annually, the first Saturday in June. Est attendance: 12,000. For info: The City of South Bend, Office of Community Affairs, 501 W South St, South Bend, IN 46601. Phone: (219) 235-9951.

ZOO BABIES. June 3-July 2. Cincinnati Zoo and Botanical Garden, Cincinnati, OH. The adorable new offspring are made easy to find with 8-foot storks placed at baby animal areas. Special weekend family entertainment. Est attendance: 180,000. For info: Cincinnati Zoo and Botanical Garden, Events/Promo Dept, 3400 Vine St, Cincinnati, OH 45220. Phone: (513) 281-4701.

BIRTHDAYS TODAY

Chuck Barris, 66, TV producer ("Dating Game," "Newlywed Game," "Gong Show"), born Philadelphia, PA, June 3, 1929.
Tony Curtis (Bernard Schwartz), 70, actor ("Vegas$," Some Like It Hot), born New York, NY, June 3, 1925.
Maurice Evans, 94, actor (Macbeth, Planet of the Apes, "Bewitched"), born Dorchester, England, June 3, 1901.
Allen Ginsberg, 69, poet, born at Newark, NJ, June 3, 1926.
Charles Hart, 34, lyricist, composer, born at London, England, June 3, 1961.
Curtis Mayfield, 53, singer, songwriter ("Freddie's Dead"; score Superfly), born at Chicago, IL, June 3, 1942.
Scott Valentine, 37, actor ("Family Ties"), born at Saratoga Springs, NY, June 3, 1958.
Deniece Williams, 44, singer ("Free," "It's Gonna Take a Miracle"), born Gary, IN, June 3, 1951.

JUNE 4 — SUNDAY
155th Day — Remaining, 210

AMERICAN DESIGN DRAFTING ASSOCIATION CONVENTION. June 4-6. Atlanta, GA. Est attendance: 150. For info: American Design Drafting Assn, R. Howard, Exec Dir, PO Box 799, Rockville, MD 20848-0799. Phone: (301) 460-6875.

ANNUAL OPOLIS REUNION PICNIC. June 4. Opolis City Park, Opolis, KS. To welcome present and former Opolis residents and friends of Opolis. Annually, the first Sunday in June. Sponsor: Opolis Reunion Committee. Est attendance: 150. For info: Mrs Helen Seal, Opolis, KS 66760. Phone: (417) 238-8816.

FINLAND: FLAG DAY. June 4. Finland's armed forces honor the birth anniversary of Carl Gustaf Mannerheim, born June 4, 1867.

FIRST FREE FLIGHT BY A WOMAN: ANNIVERSARY. June 4. Marie Thible, of Lyon, France, accompanied by a pilot (Monsieur Fleurant), on June 4, 1784, became the first woman in history to fly in a free balloon. According to her pilot, Madame Thible gave voice to her high spirits by "singing like a bird" as she drifted across Lyon in a balloon named Le Gustave (for King Gustave III, of Sweden, who was watching the ascent). The balloon reached a height of 8,500 feet in a flight that lasted about 45 minutes. The event occurred one day short of a year after the first flight in history by a man. See also: "First Balloon Flight: Anniversary" (June 5).

GEORGE III: BIRTH ANNIVERSARY. June 4. The English king against whom the American Revolution was directed. Born June 4, 1738; died Jan 29, 1820.

INTERNATIONAL PBX TELECOMMUNICATIONS WEEK. June 4-10. To honor the switchboard operators of private businesses, to acknowledge their dedicated professionalism and the vital role they play in the modern business world. Annually, the first full week in June. Sponsor: Intl PBX Telecommunicators Clubs, Mary Bohl, 2505 Paul St, Eau Claire, WI 54701. Phone: (715) 839-6075.

ITALY: GIOCO DEL PONTE. June 4. Pisa. The first Sunday in June is set aside for the Battle of the Bridge, a medieval parade and contest for possession of the bridge.

JAPAN: DAY OF THE RICE GOD. June 4. Chiyoda. Annual rice-transplanting festival observed on first Sunday in June. Centuries-old rural folk ritual revived in 1930s and celebrated with colorful costumes, parades, music, dancing and prayers to the Shinto rice god, Wbai-sama.

NATIONAL FROZEN YOGURT DAY. June 4. To inform the public of the benefits and colorful history of frozen yogurt, one of America's favorite desserts. Annually, the first Sunday in June. Sponsor: TCBY, Tracy Pritts, Acc Exec, S&S PR, 400 Skokie Blvd, Ste 200, Northbrook, IL 60062.

★ **NATIONAL SAFE BOATING WEEK.** June 4-10. Presidential Proclamation issued since 1981 for the first week in June (PL96-376 of Oct 3, 1980). From 1958 through 1977, issued for a week including July 4 (PL85-445 of June 4, 1958). Not issued from 1978 through 1980.

PENTECOST. June 4. The Christian feast of Pentecost commemorates descent of the Holy Spirit unto the Apostles, 50 days after Easter. Observed on the seventh Sunday after Easter. Recognized since the third century. See also: "Whitsunday" (below).

June 4-5 ☆ Chase's 1995 Calendar of Events ☆

RED CLOUD INDIAN ART SHOW. June 4–Aug 13. Pine Ridge, SD. To encourage Native American artists and give them a chance for exposure. Cash awards will be given. Est attendance: 12,500. Sponsor: Red Cloud Indian School, Brother C.M. Simon, Pine Ridge, SD 57770.

SHAVUOT or FEAST OF WEEKS. June 4. Observed on the following day also. Jewish Pentecost holy day. Hebrew date, Sivan 6, 5755. Celebrates giving of Torah (The Law) to Moses on Mt Sinai.

STATE PARKS OPEN HOUSE AND FREE FISHING DAY. June 4. Wisconsin. Free admission to state parks and forests. No fishing license is required to fish in inland lakes and Lake Michigan accessed through Wisconsin. (Free to in-state as well as out-of-state residents.) Annually, the first Sunday in June. Est attendance: 10,000. For info: Bureau of Parks and Recreation, Dept of Natural Resources, PO Box 7921, Madison, WI 53707. Phone: (608) 266-2621.

TEACHER "THANK YOU" WEEK. June 4-10. Write or call teachers and professors who influenced your life. Tell them about it. Send, or better yet, take a big red apple. Annually, the first full week in June. Sponsor: Lake Superior State Univ, Sault Ste. Marie, MI 49783. Phone: (906) 635-2315.

TEACHER'S DAY IN MASSACHUSETTS. June 4. Annually, the first Sunday in June.

TIANANMEN SQUARE MASSACRE: ANNIVERSARY. June 4, 1989. After almost a month and a half of student demonstrations for democracy, The People's Army, on orders from the government, opened fire on unarmed protestors to clear them from the square.

TONGA: NATIONAL DAY. June 4. National holiday.

WHITSUNDAY. June 4. Whitsunday, the seventh Sunday after Easter, is a popular time for baptism. "White Sunday" is named for the white garments formerly worn by the candidates for baptism and occurs at the Christian feast of Pentecost. See also: "Pentecost" (above).

BIRTHDAYS TODAY

Gene Barry (Eugene Klass), 74, actor (*War of the Worlds*, "Bat Masterson"), born New York, NY, June 4, 1921.
Eldra DeBarge, 34, singer, musician (lead singer of the family group DeBarge), born at Grand Rapids, MI, June 4, 1961.
Bruce Dern, 59, actor (*Coming Home*, *The Burbs*), born at Chicago, IL, June 4, 1936.
Bettina Gregory, 49, journalist, born at New York, NY, June 4, 1946.
Andrea Jaeger, 30, tennis player, born at Chicago, IL, June 4, 1965.

	S	M	T	W	T	F	S
June					1	2	3
	4	5	6	7	8	9	10
1995	11	12	13	14	15	16	17
	18	19	20	21	22	23	24
	25	26	27	28	29	30	

Xavier Maurice McDaniel, 32, basketball player, born at Columbia, SC, June 4, 1963.
Robert Merrill, 76, singer, born at New York, NY, June 4, 1919.
Howard M. Metzenbaum, 78, former US Senator (D, Ohio), born at Cleveland, OH, June 4, 1917.
Michelle Gilliam Phillips, 50, singer (with The Mamas and the Papas; "California Dreamin'"; ex-wife John Phillips), born Long Beach, CA, June 4, 1945.
Dennis Weaver, 71, actor ("Gunsmoke," "McCloud"), born Joplin, MO, June 4, 1924.

JUNE 5 — MONDAY
156th Day — Remaining, 209

BRAIN TUMOR AWARENESS WEEK. June 5-11. Series of educational activities and mailings designed to create an awareness and understanding of brain tumors which strike more than 40,000 people annually in the US. Sponsor: American Brain Tumor Assn, Naomi Berkowitz, Exec Dir, 2720 River Rd, Des Plaines, IL 60018. Phone: (708) 827-9910.

COVERED BRIDGE CELEBRATION. June 5-11. Elizabethton, TN. To celebrate the birthday of the city's covered bridge. Est attendance: 25,000. Sponsor: Elizabethton/Carter County Chamber of Commerce, Box 190, Elizabethton, TN 37644. Phone: (615) 543-2122.

DENMARK: CONSTITUTION DAY. June 5. National holiday. Offices and stores close at noon.

ENGLAND: BRITISH AMATEUR CHAMPIONSHIPS. June 5-10. Royal Liverpool Golf Club, Hoylake. This long-established championship is one of the major events on the golfing calendar. Sponsor: Royal and Ancient Golf Club, St Andrews, Fife, Scotland KY16 9JD.

ENGLAND: DICING FOR BIBLES. June 5. An old Whitmonday ceremony at All Saints Church, St. Ives, Huntingdonshire. A bequest (in 1675) with the intent of providing Bibles for poor children of the parish required winning them at a dice game played in the church. In recent years the dicing has been moved from the altar to a "more suitable" place. Six Bibles are given on Whitmonday each year.

FIRST BALLOON FLIGHT: ANNIVERSARY. June 5. The first public demonstration of a hot-air balloon flight took place at Annonay, France, on June 5, 1783, where the co-inventor brothers, Joseph and Jacques Montgolfier, succeeded in launching their 33-foot-diameter *globe aerostatique*. It rose an estimated 1,500 feet and traveled, windborne, about 7,500 feet before landing after the 10-minute flight—the first sustained flight of any object achieved by man.

INTERNATIONAL COUNTRY MUSIC FAN FAIR. June 5-11. State Fairgrounds and Opryland, Nashville, TN. A weeklong musical celebration to give fans an inside look at Music City USA. Stage shows and other musical events are featured. Sponsors: Grand Ole Opry and Country Music Assn. Est attendance: 24,000. For info: Fan Fair, 2804 Opryland Dr, Nashville, TN 37214. Phone: (615) 889-7503.

KENNEDY, ROBERT F.: ASSASSINATION ANNIVERSARY. June 5. Senator Kennedy was shot while campaigning for the Democratic presidential nomination in Los Angeles, CA, June 5, 1968; he died the following day. Sirhan Sirhan was convicted of his murder.

NATIONAL BATHROOM READING WEEK. June 5-11. To promote America's favorite reading room as the foremost seat of learning. Sponsor: Red-Letter Press, Jack Kreismer, Pres, Box 393, Saddle River, NJ 07458. Phone: (201) 652-4402.

NATIONAL FRAGRANCE WEEK. June 5-9. 13 cities across the country. Opening Day Ceremonies to kick off the Week. Mayoral Proclamation presented in each city. Special fragrance events at participating retail and other noted locations. Annually, the first full week in June. For info: The Fragrance Fdtn, 145 E 32nd St, New York, NY 10016-6002.

Chase's 1995 Calendar of Events — June 5–6

SMITH, ADAM: BIRTH ANNIVERSARY. June 5. Scottish economist and philosopher, author of *An Enquiry into the Nature and Causes of the Wealth of Nations* (published in 1776), born June 5, 1723, at Kirkaldy, Fifeshire, Scotland. Died at Edinburgh, Scotland, July 17, 1790. "Consumption," he wrote, "is the sole end and purpose of production; and the interest of the producer ought to be attended to only so far as it may be necessary for promoting that of the consumer."

SPACE MILESTONE: *SOYUZ T-2* (USSR). June 5. Launched on June 5, 1980, cosmonauts Yuri Malyshev and Vladimir Aksenov docked at *Salyut 6* on June 6, returned to Earth June 9.

UNITED NATIONS: WORLD ENVIRONMENT DAY. June 5. Observed annually on June 5, the anniversary of the opening of the UN Conference on the Human Environment held in Stockholm in 1972, which led to establishment of UN Environment Programme, based in Nairobi. The General Assembly has urged marking the day with activities reaffirming concern for the preservation and enhancement of the environment. Info from: United Nations, Dept of Public Info, New York, NY 10017.

WHITMONDAY. June 5. The day after Whitsunday is observed as a public holiday in many countries.

WORLD CAMPAIGN FOR THE BIOSPHERE. June 5. In observance of World Environment Day. To awaken people to their responsibilities towards The Biosphere, our planetary life-support system, of which we humans constitute an integral part and yet threaten increasingly by our ever-expanding population and profligate activities. Meetings subsequently of the World Council For The Biosphere and its sponsoring organization. Sponsor: The Foundation for Environmental Conservation, Dr. Nicholas Polunin, Pres, 7 Chemin Taverney, 1218 Grand-Saconnex, Geneva, Switzerland.

BIRTHDAYS TODAY

Bosin Blackbear, 74, artist, born at Anadarko, OK, June 5, 1921.
Jacques Demy, 64, filmmaker (*The Umbrellas of Cherbourg*), born Pont-Chateau, France, June 5, 1931.
Robert Lansing, 66, actor (*Under the Yum Yum Tree*, "The Man Who Never Was"), born San Diego, CA, June 5, 1929.
Bill Moyers, 61, journalist ("Bill Moyers' Journal"), born Hugo, OK, June 5, 1934.

JUNE 6 — TUESDAY
157th Day — Remaining, 208

BONZA BOTTLER DAY™. June 6. To celebrate when the number of the day is the same as the number of the month. Bonza Bottler Day™ is an excuse to have a party at least once a month. T-shirts available. For info: Elaine Fremont, 203 Waddell Rd, Taylors, SC 29687. Phone: (803) 244-2023.

CANADA: WINNIPEG INTERNATIONAL CHILDREN'S FESTIVAL. June 6–11. (Tentative.) The Forks, Winnipeg, Manitoba. Festival features song, dance, theatre, mime, puppetry and music by local, national and international artists. For info: Winnipeg Intl Children's Fest, Cathy Wawrykow, 300 — 112 Market Ave, Winnipeg, Canada R3B 0P4. Phone: (204) 958-4735.

D-DAY (OPERATION OVERLORD): ANNIVERSARY. June 6. Allied Expeditionary Force landed in Normandy, France, on this day in 1944, opening a second major European front in the battle against the Nazis in World War II.

FISHING HAS NO BOUNDARIES—THERMOPOLIS. June 6–11. Boysen Reservoir, Thermopolis, WY. A three-day fishing experience for disabled persons. Any disability, age, sex, race, etc, eligible. Fishing with experienced guides, attended by 80 participants and 200 volunteers. For info: Fishing Has No Boundaries, HSC C of C, 250 Earapahoe St, Box 768, Thermopolis, WY 82443. Phone: (307) 864-3192.

GERMANY: WALDCHESTAG. June 6. Frankfurt. Since the 19th century Frankfurters have spent the Tuesday after Whitsunday in their forest. See also: "Whitsunday" (June 4).

HALE, NATHAN: 240th BIRTH ANNIVERSARY. June 6. American patriot Nathan Hale was born at Coventry, CT, on June 6, 1755. During the battles for New York during the American Revolution, he volunteered to seek military intelligence behind enemy lines and was captured on the night of Sept 21, 1776. In an audience before General William Howe, Hale admitted he was an American officer and was ordered hanged the following morning. Although some question them, his dying words, "I only regret that I have but one life to lose for my country," have become a famous symbol of American patriotism. He was hanged on Sept 22, 1776, at Manhattan, NY.

KHACHATURIAN, ARAM (ILICH): BIRTH ANNIVERSARY. June 6. Armenian musician and composer, noted for compositions based on folk music and legend, born at Tbilisi, Georgia, USSR, June 6, 1903. Died May 1, 1978.

KOREA: MEMORIAL DAY. June 6. Nation pays tribute to the war dead, and memorial services are held at the National Cemetery in Seoul. Legally recognized Korean holiday.

MOON PHASE: FIRST QUARTER. June 6. Moon enters First Quarter phase at 6:26 AM, EDT.

NATIONAL PATRIOTS MONTH. June 6–July 4. To encourage all citizens, during this month, to wear red, white and blue each day, to fly the US flag each day, to buy US-made products when possible, to decorate their homes, both indoors and outdoors, as patriotically as possible and to otherwise show their love for and loyalty to our country. Annually, June 6 (Nathan Hale's birthday)–July 4. Sponsor: Patriots' Club of America, Jack L. Coulter, 316 SE 2nd St, Fort Meade, FL 33841-3629.

PET APPRECIATION WEEK (PAW)™. June 6–12. A time to thoughtfully remember our pet companions who share so much with us. Do something special for your own pet(s) or pets in general this week. Sponsor: Pawsters United, Adrienne Sioux Koopersmith, 1437 W Rosemont, #1W, Chicago, IL 60660. Phone: (312) 743-5341.

PROPOSITION 13: ANNIVERSARY. June 6. California voters (65 percent of them) supported a primary election ballot initiative, June 6, 1978, to cut property taxes 57 percent. Regarded as possible omen of things to come across country, a taxpayer's revolt against high taxes and government spending.

SPACE MILESTONE: *SOYUZ 11* (USSR). June 6. Cosmonauts G.T. Dobrovolsky, V.N. Volkov, V.I. Patsayev died during return landing after 24-day space flight, June 30, 1971. *Soyuz 11* had docked at *Salyut* orbital space station June 7–29, where scientific experiments were conducted. First humans to die in space. Launched on June 6, 1971.

◆ **SUSAN B. ANTHONY FINED FOR VOTING: ANNIVERSARY.** June 6. Seeking to test for women the citizenship and voting rights extended to black males under the 14th and 15th amendments, Susan B. Anthony led a group of women who registered and voted in a Rochester, NY, election in 1872. She was arrested, tried and sentenced to pay a fine. She refused to do so and was allowed to go free by a judge who feared she would appeal to a higher court.

SWEDEN: FLAG DAY. June 6. Commemorates the day upon which Gustavus I (Gustavus Vasa) ascended the throne of Sweden in 1523.

UPPERVILLE COLT AND HORSE SHOW. June 6–11. A week long "A-rated" horse show involving hundreds of horse and rider combinations from 8–10-year-old children in the pony divisions to leading Olympic and World Cup riders and horses in the Hunter, Jumper and Grand Prix divisions. Sunday's highlight is the prestigious $50,000 Budweiser/Upperville Jumper Classic sponsored by Budweiser. Annually, beginning the first Tuesday in June. Phone during event: (703) 592-3858. Est attendance: 6,000. For info: L Beth Kearns or Diane Jones, Upperville Colt and Horse Show, PO Box 1288, Warrenton, VA 22186. Phone: (703) 347-5863.

WOODMEN OF THE WORLD FOUNDERS DAY. June 6. To celebrate the founding of Woodmen of the World Life Insurance Society in 1890. For info: Woodmen of the World, Scott J. Darling, Communications Dept, 1700 Farnam, Omaha, NE 68102. Phone: (402) 271-7211.

BIRTHDAYS TODAY

Gary U.S. Bonds (Gary Anderson), 56, singer ("Quarter to Three"), songwriter, born at Jacksonville, FL, June 6, 1939.

Bjorn Borg, 39, tennis player, born at Sodertalje, Sweden, June 6, 1956.

George Deukmejian, 67, Governor of California (R), born at Menands, NY, June 6, 1928.

Marian Wright Edelman, 56, president of Children's Defense Fund, lawyer, civil rights activist, born at Bennettsville, SC, June 6, 1939.

Ruben Mayes, 32, football player, born North Battleford, Saskatchewan, Canada, June 6, 1963.

Billie Whitelaw, 63, actress ("The Dressmaker," *Masterpiece Theater*), born at Coventry, England, June 6, 1932.

JUNE 7 — WEDNESDAY
158th Day — Remaining, 207

BOONE DAY. June 7. Each year on June 7, the Kentucky Historical Society celebrates the anniversary of the day in 1767 when Daniel Boone, America's most famous frontiersman, reportedly first sighted the land that would become Kentucky. The June 7 date is taken from the book, *The Discovery, Settlement and Present State of Kentucky,* by John Filson, published in 1784, with an appendix titled "The Adventures of Col. Daniel Boone." The information in the appendix supposedly originated with Boone, although Filson is the actual author. The work is not considered completely reliable by historians.

BRUMMELL, GEORGE BRYAN "BEAU": BIRTH ANNIVERSARY. June 7. Born at London, England, on June 7, 1778, Beau Brummel was, early in his life, a popular English men's fashion leader, the "arbiter elegantarium" of taste in dress. His extravagance and lack of tact (it was he who reportedly said—indicating the Prince of Wales, later George IV—"Who's your fat friend?") led him from wealth and popularity to poverty and disrepute. Once imprisoned for debt, he became careless of dress and personal appearance. He died in a charitable asylum at Caen, France, on Mar 30, 1840.

COCHISE: DEATH ANNIVERSARY. June 7, 1874. Born around 1810 in the Chiricahua Mountains of Arizona, Cochise became a fierce and courageous leader of the Apache. After his arrest in 1861, he escaped and launched the Apache War which lasted for 25 years. He died 13 years later near his stronghold in southeastern Arizona. For info: Edwin R. Sweeney, Cochise Historian, 323 White Pine, St. Charles, MO 63304. Phone: (314) 441-9157.

	S	M	T	W	T	F	S
June					1	2	3
	4	5	6	7	8	9	10
1995	11	12	13	14	15	16	17
	18	19	20	21	22	23	24
	25	26	27	28	29	30	

CYPRUS: KATAKLYSMOS. June 7. Festivities in all the seaside towns. Celebration of the "flood." Unique and colorful Cypriot celebration associated with the Pentecost, as well as with sea games.

ENGLAND: HORSERACING: THE DERBY. June 7–10. Epsom Racecourse, Epsom, Surrey. "The most famous and prestigious horse race in the world was devised at a noble dinner party in 1779 and named after one of the diners—Lord Derby." Note: The Derby is followed by the Coronation Cup on June 9 and the Oaks Stakes on June 10, all at Epsom Racecourse. Est attendance: 100,000. For info: United Racecourses Ltd, Racecourse Paddock, Epsom Downs, Surrey, England KT18 5LQ.

GAUGUIN, (EUGENE HENRI) PAUL: BIRTH ANNIVERSARY. June 7. French painter born at Paris, France, on June 7, 1848. Formerly a stockbroker, he became a painter in his middle age, and three years later Gauguin renounced his life in Paris to move to Tahiti. He is remembered best for his broad, flat tones and bold colors. Gauguin died May 9, 1903, at Atoana on the island of Hiva Oa in the Marquesas.

INDY SENIOR OLYMPICS. June 7–10. Indianapolis, IN. A multi-sports event for people ages 55 and older. Qualifiers advance to State Competition. Est attendance: 800. For info: Indy Parks, 1502 W 16th St, Indianapolis, IN 46202. Phone: (317) 327-7201.

NEWSPAPER IN EDUCATION AND LITERACY CONFERENCE. June 7–9. Seattle, WA. An event to bring together newspaper executives, educational services literacy program managers and leaders in education to exchange ideas and information. Est attendance: 350. For info: Betty Sullivan, Dir Educ/Services, Newspaper Assn of America Fdtn, 11600 Sunrise Valley Dr, Reston, VA 22091. Phone: (703) 648-1051.

RITTENHOUSE SQUARE ART ANNUAL. June 7–11. Philadelphia, PA. Open-air gallery with more than 20,000 works of fine art. Artists from Tri-State area only (PA, NJ, DE). Est attendance: 20,000. For info: Rittenhouse Sq Fine Arts Annual, PO Box 3752, Philadelphia, PA 19125. Phone: (215) 634-5060.

TEXAS MUSICAL DRAMA. June 7–Aug 19. Palo Duro Canyon State Park, Canyon, TX. Paul Green outdoor drama set beneath 600-foot cliffs. History of area told by cast of 80, accented by song, dance and latest sound and light. 30th season. Nightly except Sundays at 8:30 PM. Sponsor: Texas Panhandle Heritage Fdtn, Inc. Est attendance: 100,000. For info: Patty Bryant, TEXAS, Box 268, Canyon, TX 79015. Phone: (806) 655-2181.

VIVIEN KELLEMS MEMORIAL DAY. June 7. Celebrates birth, June 7, 1896, of industrialist Vivien Kellems. She protested what she considered unfair taxation when she refused to withhold taxes from her employees' wages unless the government named her an agent of the Internal Revenue Department, paid her a salary and reimbursed her expenses in collecting taxes.

BIRTHDAYS TODAY

◆ *Gwendolyn Brooks*, 78, poet, author (Pulitzer for *Annie Allen*), born Topeka, KS, June 7, 1917.

Tom Jones (Thomas Woodward), 55, singer ("It's Not Unusual," "She's A Lady"), born at Pontypridd, Wales, June 7, 1940.

☆ Chase's 1995 Calendar of Events ☆   June 7–8

Bill Kreutzmann, 49, singer, musician (Grateful Dead drummer), born at Palo Alto, CA, June 7, 1946.
Prince (Prince Rogers Nelson), 37, musician, singer, born at Minneapolis, MN, June 7, 1958.

JUNE 8 — THURSDAY
159th Day — Remaining, 206

AMERICAN HEROINE REWARDED: ANNIVERSARY. June 8. On Mar 16, 1697, in an attack on Haverhill, MA, Indians captured Hannah Duston and killed her one-week-old baby, in addition to killing or capturing 39 others. After being taken to an Indian camp, she escaped on Apr 29 after killing 10 Indians with a tomahawk and scalping them as proof of her deed. On June 8, 1697, her husband was awarded, on her behalf, the sum of twenty-five pounds for her heroic efforts, the first public award to a woman in America.

ATTACK ON THE USS *LIBERTY*: ANNIVERSARY. June 8. At 2:00 PM local time on June 8, 1967, the unescorted US intelligence ship USS *Liberty*, sailing in international waters off the Egyptian coast, was attacked without warning by Israeli jet planes and three Israeli torpedo boats. She was strafed and hit repeatedly by rockets, cannon, napalm and finally a torpedo. Casualties: out of a crew of 294 Americans, there were 34 dead and 171 wounded. Israel apologized, claiming mistaken identity, but surviving crew members charged deliberate attack by Israel and cover-up by US authorities.

BILL OF RIGHTS PROPOSAL: ANNIVERSARY. June 8. Bill of Rights, which led to the first 10 amendments to the US Constitution, was first proposed by James Madison on June 8, 1789.

BOLIVIAN EARTHQUAKE: ANNIVERSARY. June 8, 1994. A "deep focus" earthquake registering 8.2 on the Richter scale erupted 400 miles beneath the earth's surface in a remote area of Bolivia. The quake is believed to be the largest ever recorded in the Bolivian area and the giant tremblor was felt as far north as Minneapolis, MN.

BRADLEY COUNTY PINK TOMATO FESTIVAL. June 8–10. Warren, AR. Celebrates and promotes the tomato of Bradley County. Tomato eating contest, dunking booth, barbecue chicken cookoff, arts and crafts. Est attendance: 20,000. For info: Chamber of Commerce, Harry L. McCaskill, Exec Dir, 104 N Myrtle, Warren, AR 71671. Phone: (501) 226-5225.

BUSH, BARBARA PIERCE: BIRTHDAY. June 8. Wife of George Herbert Walker Bush, 41st president of the US, was born at Rye, NY, June 8, 1925. They were married on Jan 6, 1945. Active in the country's campaign for increased literacy.

COWBOY STATE SUMMER GAMES. June 8–11. Casper, WY. The Summer Games feature more than 25 different events including everything from archery to wrestling, basketball to volleyball. Participation is open to Wyoming's amateur athletes of all ages. The Summer Games also feature a spectacular Opening Ceremony that will take place Friday evening, June 9. Est attendance: 5,200. For info: Cowboy State Games, Eileen Ford, PO Box 3485, Casper, WY 82602. Phone: (307) 577-1125.

FESTIVAL OF THE BLUEGRASS. June 8–11. Lexington, KY. More than 25 bands perform bluegrass music in an outdoor setting. Est attendance: 10,000. For info: Jean Cornett, Box 644, Georgetown, KY 40324.

GLENN MILLER BIRTHPLACE SOCIETY FESTIVAL. June 8–10. Clarinda, IA. To commemorate Glenn Miller's contribution to big band music through exhibits, music, films, performances by winners of Glenn Miller scholarships and a big band dance. Annually, the second weekend in June. Phone: (712) 542-2461 and 542-3881. Est attendance: 3,000. Sponsor: Glenn Miller Birthplace Soc, Wilda Martin, PO Box 61, Clarinda, IA 51632.

McGOWAN, WILLIAM: DEATH ANNIVERSARY. June 8. William McGowan, the chair of MCI Communications Corporation, changed the shape of the telecommunications industry by bringing lower-cost long-distance service into the marketplace. MCI's 1974 antitrust suit against AT&T, which was joined by the Justice Department, led to the breakup of AT&T 10 years later. McGowan died on June 8, 1992, at Washington, DC.

McKINLEY, IDA SAXTON: BIRTH ANNIVERSARY. June 8. Wife of William McKinley, 25th president of the US, born at Canton, OH, June 8, 1847. Died May 26, 1907.

MUSLIM HOLIDAY: ASHURA. June 8–9. Commemorates death of Muhammad's grandson and the Battle of Karbala. Processions, drama and special costumes. Muslim calendar date: Muharram 10, 1416. Different methods for calculating the Gregorian date are used by different Muslim sects or groups. *Chase's* dates are based on astronomical calculations of the visibility of the new moon crescent at Mecca; Gregorian EST date may vary.

NAPA VALLEY WINE AUCTION. June 8–11. Meadowood Resort, Napa Valley, CA. The largest charity wine auction in the US, the Napa Valley Wine Auction includes four days of vintner hosted hospitality events, the vitners' dinner created by celebrity chefs, the auction itself and winery open houses. Annually, the second weekend in June. Est attendance: 1,600. Sponsor: Napa Valley Vintners Assn, PO Box 141, St. Helena, CA 94574. Phone: (707) 963-5246.

OK MOZART INTERNATIONAL FESTIVAL. June 8–18. Bartlesville, OK. Festival features world-class artists performing with Solisti New York Orchestra. Est attendance: 40,000. For info: OK Mozart Intl Festival, Box 2344, Bartlesville, OK 74005. Phone: (918) 336-9900.

SPACE MILESTONE: *VENERA 9* AND *10* (USSR): 20th ANNIVERSARY. June 8 and 14. Launched on June 8 and June 14, 1975. Venus exploration vehicles landed on Venus Oct 22 and 25. Sent first pictures ever transmitted from another planet, atmospheric analysis, temperature (905°F) and other data.

WRIGHT, FRANK LLOYD: BIRTH ANNIVERSARY. June 8. American architect born at Richland Center, WI, on June 8, 1867. In his autobiography Wright wrote: "No house should ever be *on* any hill or on anything. It should be *of* the hill, belonging to it, so hill and house could live together each the happier for the other." Wright died at Phoenix, AZ, Apr 9, 1959.

WYTHE, GEORGE: DEATH ANNIVERSARY. June 8. Signer of the Declaration of Independence. Born at Elizabeth County, VA, about 1726 (exact year unknown). Died at Richmond, VA, June 8, 1806.

BIRTHDAYS TODAY

Barbara Pierce Bush, 70, former First Lady, born at Rye, NY, June 8, 1925.
James Darren, 59, singer ("Goodbye Cruel World"), actor (*Gidget*), born at Philadelphia, PA, June 8, 1936.
Griffin Dunne, 40, actor (*Straight Talk*), producer, born at New York, NY, June 8, 1955.
Don Grady, 51, actor ("My Three Sons," "Mickey Mouse Club"), born at San Diego, CA, June 8, 1944.
Sara Paretsky, 48, writer (*Killing Orders*, *Burn Marks*), born at Ames, IA, June 8, 1947.
Joan Rivers, 58, comedienne, talk show host, TV shopping host, born at New York, NY, June 8, 1937.
Boz Scaggs, 51, singer, musician, songwriter (*Silk Degrees*, *Middle Man*), born at Dallas, TX, June 8, 1944.

June 8-9 ☆ *Chase's 1995 Calendar of Events* ☆

Nancy Sinatra, 55, singer ("These Boots Are Made for Walking," "Something Stupid"), born Jersey City, NJ, June 8, 1940.
Jerry Stiller, 66, actor (Shakespearian actor; "Tattinger's"), born Brooklyn, NY, June 8, 1929.
Byron Raymond White, 78, former Associate Justice of the US Supreme Court, born at Fort Collins, CO, June 8, 1917.

JUNE 9 — FRIDAY
160th Day — Remaining, 205

AMERICAN BRAIN TUMOR ASSOCIATION SYMPOSIUM. June 9-11. Ramada Hotel O'Hare, Rosemont, IL. Prominent researchers from leading medical centers nationwide will present cutting edge information on brain tumors and their treatment. Workshops, exhibits, discussion groups. Sponsor: American Brain Tumor Assn, 2720 River Rd, Des Plaines, IL 60018. Phone: (708) 827-9910.

BADGER STATE SUMMER GAMES. June 9-11. (Finals June 23-25.) Regional competition in 8 Wisconsin communities, finals in Madison. 11th annual Olympic-style competition for Wisconsin residents of all ages and abilities, featuring 25 sports and opening ceremonies. Major sponsors: AT&T, Ameritech, GTE, Wisconsin Milk Marketing Board and WPS Insurance Corp. Member of the National Congress of State Games. Est attendance: 60,000. For info: Otto Breitenbach, Exec Dir, or Rick For, PR Dir, Badger State Gamess, PO Box 1377, Madison, WI 53701-1377. Phone: (608) 251-3333.

BRAZOS NIGHTS. June 9-July 15. (Fridays and Saturdays only.) Indian Spring Park, Waco, TX. A family concert series with music of all types on two stages, and historic laser light show at dark. Annually, Friday and Saturday nights for six weeks during summer. Est attendance: 60,000. For info: Sally Gavlik, Rec Super, City of Waco Dept of Leisure Services, PO Box 2570, Waco, TX 76702. Phone: (817) 750-5980.

CENTER OF THE NATION ALL-CAR RALLY. June 9-11. Belle Fourche, SD. This 3-day event includes Friday evening "Cruise Night" when '50s & '60s cars parade through at least two communities; Saturday's big "Show 'n' Shine," held all day in Hermann Park, plus food booths, vendors, live '50s & '60s music and in the evening a '50s Dance at the Community Hall; and on Sunday Poker Run, Demolition Derby and Drag Races. Annually, the second weekend in June. Est attendance: 3,000. For info: Kathy M Wainman, Exec Dir, Belle Fourche Chamber of Commerce, 415 5th Ave, Belle Fourche, SD 57717. Phone: (605) 892-2676.

CHISHOLM TRAIL ROUND-UP FESTIVAL. June 9-11. Ft. Worth, TX. Events include two-day trail ride, Quanah Parker Comanche Indian Pow-Wow and Honor Dance, Tarrant County Fair, Street Fair, chili and barbecue cook-offs, chuckwagon cooks' races, cowboy poetry, live armadillo and pig races, gunfighters, parade, indoor PRCA rodeo, children's area with Western theme carnival rides and 5 live music stages with Nashville music greats. Est attendance: 150,000. For info: Chisholm Trail Round-Up, PO Box 4815, Fort Worth, TX 76164-0815. Phone: (817) 625-7005.

DELMARVA CHICKEN FESTIVAL. June 9-10. Federalsburg, MD. A family event focusing on chicken, the leading agricultural enterprise on the Delmarva Peninsula. Food, entertainment and consumer info are featured. Est attendance: 25,000. For info: Connie Parvis, Delmarva Poultry Industry, Inc, Rd 6, Box 47, Georgetown, DE 19947-9622. Phone: (302) 856-9037.

DONALD DUCK: BIRTHDAY. June 9. Donald Duck was "born" June 9, 1934.

FINLAND: KUOPIO DANCE AND MUSIC FESTIVAL. June 9-16. Kuopio. Some 70 events centering on classical ballet, dance theatre and jazz dance, plus other types depending on the theme for the festival. World-renowned dance and music companies; seminars, lectures, exhibits and films; selection of dance courses with internationally known dance teachers. Est attendance: 40,000. For info: Finnish Tourist Board, 655 Third Ave, New York, NY 10017. Phone: (212) 949-2333.

HONG KONG: LEASE ANNIVERSARY. June 9. Hong Kong, consisting of about 400 square miles (islands and mainland) with more than 5 million persons, has been administered as a British Crown Colony since a 99-year lease was signed on June 9, 1898. In 1997, Hong Kong's sovereignty will revert to the People's Republic of China.

INTERNATIONAL OLD-TIME FIDDLERS CONTEST. June 9-10. International Peace Garden, Dunseith, ND. The International Old Time Fiddlers Contest draws young and old fiddlers alike from both Canada and the US. Contestants compete for 2 days performing a required list of fiddle music in pursuit of cash prizes and trophies. Est attendance: 3,000. For info: Joseph T. Alme, 1725 11th St SW, Minot, ND 58701. Phone: (701) 838-8472.

INTERNATIONAL STRANGE MUSIC WEEKEND. June 9-10. Carter Caves States Resort Park, Olive Hill, KY. Two evening concerts in an outdoor amphitheater feature "strange musicians" who play music on anything from disassembled washing machines to vacuum cleaners and stethoscopes. Daytime activities include a strange musical instrument demonstration and a "Strange Music Hall of Fame" exhibit. Est attendance: 500. For info: John Tierney, Carter Caves State Resort Park, Rte 5, Box 1120, Olive Hill, KY 41164. Phone: (606) 286-4411.

KNIM RADIO'S BIG FISH 15. June 9-10. Nodaway Lake, Maryville, MO. The world's largest free fishing contest attracts hundreds of fishermen from a wide area. Participants in this 15th annual, 18-hour contest compete for thousands of dollars in prizes, including a grand prize boat, motor and trailer. Annually, the second weekend in June. Est attendance: 2,000. For info: Jerry Lutz, Program Dir, KNIM Radio, 1618 S Main, Maryville, MO 64468. Phone: (816) 582-2151.

KUTNER, LUIS: BIRTH ANNIVERSARY. June 9. Human rights attorney Luis Kutner was born June 9, 1908, at Chicago, IL. Kutner, who was responsible for the release of many unjustly confined prisoners, came to be known as "The Springman." He helped free Hungarian Cardinal Josef Mindszenty, poet Ezra Pound and former Congo President Moise Tshombe. He was the author of the living will and he founded the World Habeas Corpus. Kutner was nominated nine times for the Nobel Peace Prize. Luis Kutner died Mar 1, 1993, at Chicago.

LOLOMA, CHARLES: DEATH ANNIVERSARY. June 9. Charles Loloma was a major influence on modern Indian art and was famous for changing the look of American Indian jewelry. A painter, sculptor, potter, he was best known for his jewelry which broke tradition with previous Indian styles using bold mosaics and materials such as coral, fossilized ivory, pearls, diamonds, charolite and sugilite. One of his trademarks was to line the inside of a bracelet or ring with stones more

	S	M	T	W	T	F	S
June					1	2	3
1995	4	5	6	7	8	9	10
	11	12	13	14	15	16	17
	18	19	20	21	22	23	24
	25	26	27	28	29	30	

valuable than those on the outside. Charles Loloma was born in 1921 at Hotevilla on the Hopi Indian Reservation and died on June 9, 1991, at Scottsdale, AZ.

LOUISIANA PEACH FESTIVAL. June 9–18. Ruston, LA. The 45th Annual Peach Festival will feature hot air balloon races, rodeo, parade, concerts, cooking contests and more. Est attendance: 50,000. For info: Ruston/Lincoln Chamber of Commerce, Sue Edmunds, PO Box 150, Ruston, LA 71273-0150. Phone: (318) 255-2031.

MARKET SQUARE DAYS CELEBRATION. June 9–10. Portsmouth, NH. A barbecue and benefit auction, historic tours, arts and crafts, a 10K road race, open air concerts and fireworks are all part of the festivities. One of the biggest celebrations held each June on New Hampshire's picturesque seacoast. Free admission. Charge for barbecue and road race. Est attendance: 100,000. For info: Pro Portsmouth, PO Box 1008, Portsmouth, NH 03802. Phone: (603) 431-5388.

MEDORA MUSICAL. June 9–Sept 3. Medora, ND. Folks whoop it up at this Broadway-class variety show held nightly in a natural amphitheatre in the Bad Lands. Spectacular sunsets on painted buttes and ravines compete for attention with Western song and dance, all paying tribute to our Roughriding Conservation President, Theodore Roosevelt. Est attendance: 80,000. For info: North Dakota Tourism, Pat Hertz, Liberty Memorial Bldg, State Capitol Grounds, Bismarck, ND 58505. Phone: (800) 435-5663.

MINNESOTA INVENTORS CONGRESS. June 9–11. Redwood Falls, MN. To promote creativity and development of ideas into marketable products, to educate inventors and to bring the inventor to experts in the appropriate field, including a test market of up to 10,000 viewers. Highlights include more than 170 inventions from around the world on display, 100 student inventors, the Minnesota Inventors Hall of Fame and the Market Place with inventions for sale. Est attendance: 10,000. For info: Minnesota Inventors Congress, Box 71, Redwood Falls, MN 56283. Phone: (507) 637-2344.

NATIONAL ASPARAGUS FESTIVAL. June 9–11. Hart, MI. Celebrate the harvest of Asparagus. Arts and crafts fair, Kid's Parade, Parade Royale, asparagus luncheon, asparagus dinner, square dance, beer tent with live music and more. Phone days (616) 873-2129 or evenings (616) 861-6228. For info: Sheryl Sallgren, Pres, Natl Asparagus Festival, PO Box 117, Shelby, MI 49455.

NIELSEN, CARL: BIRTH ANNIVERSARY. June 9. Danish composer born June 9, 1865. Died Oct 3, 1931. Funen Quartet plays birthday concert in the composer's childhood home at Lyndelse, near Odense, Denmark.

OLDSMOBILE BALLOON CLASSIC. June 9–11. Vermilion County Airport, Danville, IL. 125 hot-air balloons in 5 races, continuous entertainment, familyland, Balloon Glo, road vendors and activities all day long. Annually, the second weekend in June. Sponsor: Oldsmobile. Est attendance: 100,000. For info: Danville Area Conv & Visitors Bureau, Jeanie Cooke, Exec Dir, Towne Centre, PO Box 992, Danville, IL 61832. Phone: (217) 442-2096.

PAYNE, JOHN HOWARD: BIRTH ANNIVERSARY. June 9. American author, actor, diplomat, born at New York, NY, June 9, 1791. Died at Tunis, Apr 9, 1852. Author of opera libretto (*Clari, or, The Maid of Milan*) that contained the song "Home, Sweet Home."

PORTER, COLE: BIRTH ANNIVERSARY. June 9, 1891. Cole Porter published his first song "The Bobolink Waltz" at the age of ten. His career as a composer and lyricist for Broadway was launched in 1928 when five of his songs were used in the musical play *Let's Do It*. His prolific contributions to the Broadway stage include *Fifty Million Frenchmen, Wake Up and Dream, The Gay Divorcée, Anything Goes, Leave It to Me, Du Barry Was a Lady, Something for the Boys, Kiss Me Kate, Can Can* and *Silk Stockings*. Porter was born at Peru, IN, and died at Santa Monica, CA, Oct 15, 1964.

POTOMAC RIVER FESTIVAL. June 9–11. Colonial Beach, VA. To promote the attractions of Colonial Beach and historic Westmoreland County. Est attendance: 18,000. Sponsor: Chamber of Commerce, 2 Boundary St, PO Box 475, Colonial Beach, VA 22443. Phone: (804) 224-7531.

RED EARTH NATIVE AMERICAN CULTURAL FESTIVAL. June 9–11. Oklahoma City, OK. Thousands of Native Americans representing more than 100 tribes from across North America celebrate their proud heritage at one of the largest intertribal gatherings in the world. Dance competitions, art exhibits, film and video showings, lectures, storytelling and parade. Est attendance: 150,000. For info: Red Earth, Inc, 2100 NE 52, Oklahoma City, OK 73111. Phone: (405) 427-5228.

RIVERFEST. June 9–10. Cape Girardeau, MO. Celebration of Mississippi River heritage. Includes professional entertainment, regional food, working crafters, children's activities, fireworks and contests. All activities take place on the streets of historic downtown on the riverfront. Annually, second weekend in June. Est attendance: 50,000. For info: Cape Girardeau Riverfest Assn, 1005 N Kingshighway, #204, Cape Girardeau, MO 63701. Phone: (314) 335-1388.

SEA MUSIC FESTIVAL. June 9–11. Mystic, CT. World-famous folksingers and musicians performing from Mystic Seaport's tall ships or in formal concert. Est attendance: 7,500. For info: Mystic Seaport, 75 Greenmanville Ave, Box 6000, Mystic, CT 06355. Phone: (203) 572-5315.

SENIOR CITIZENS DAY. June 9. Oklahoma.

SHOW ME STATE BARBECUE COOK-OFF. June 9–10. Civic Center Park, St. Joseph, MO. A variety of barbecue delicacies are prepared by some of Missouri's grill masters. For info: Sharon Ritchey, Krug Park Castle, St. Joseph, MO 64505. Phone: (816) 271-5500.

SOONER STATE SUMMER GAMES. June 9–11. Oklahoma City, OK. (Also, June 16–18.) Oklahoma amateur sports festival. More than 15,000 athletes of all ages from every county in the state. More than 40 sports competitions. Est attendance: 35,000. For info: Sooner State Games, 100 W Main, Ste 287, Oklahoma City, OK 73102. Phone: (405) 235-4222.

STEPHENSON, GEORGE: BIRTH ANNIVERSARY. June 9. English inventor, developer of the steam locomotive, born near Newcastle, England, June 9, 1781. Died near Chesterfield, England, Aug 12, 1848.

STRAWBERRY 100 BIKE TOUR AND FESTIVAL. June 9–11. Crawfordsville, IN. This bike tour is a 50-mile per day, 2-day event, which covers the Crawfordsville community and Montgomery County, IN. Includes 11 historical sites along scenic Sugar Creek Valley. Festival at Lane Place includes 2½ days of arts and crafts, food, music and children's activities. All-day entertainment; car show (Sunday); softball and tennis tournaments; antique tractor exhibits; 10K run. All city museums open. Est attendance: 30,000. For info: Montgomery County VCB, 412 E Main St, Crawfordsville, IN 47933. Phone: (800) 866-3973.

TECUMSEH!: THE EPIC OUTDOOR DRAMA. June 9–Sept 2. Chillicothe, OH. Witness the spectacular reenactment of the life and death of the great Shawnee leader Tecumseh. Held in the large, tiered amphitheater nestled in the hardwood forest of Sugarloaf Mountain. Take a backstage tour, visit the Prehistoric Museum, dine in the open-air Tecumseh Restaurant Terrace. Phone: (614) 775-0700. Est attendance: 85,000. For info: Tecumseh, PO Box 73, Chillicothe, OH 45601-0073.

June 9–10 ☆ Chase's 1995 Calendar of Events ☆

BIRTHDAYS TODAY

George Axelrod, 73, writer (screenplays *Bus Stop, Breakfast at Tiffany's*), born at New York, NY, June 9, 1922.
Michael J. Fox, 34, actor ("Family Ties"; *Back to the Future* films), born Edmonton, Alta, Canada, June 9, 1961.
Marvin Kalb, 65, educator, journalist, born at New York, NY, June 9, 1930.
Jackie Mason (Yacov Moshe Maza), 61, comedian ("Chicken Soup," *The World According to Me*), born Sheyboygan, WI, June 9, 1934.
Robert S. McNamara, 79, banker, former cabinet member, born at San Francisco, CA, June 9, 1916.
David Gene Parker, 44, former baseball player, born at Jackson, MS, June 9, 1951.
Les Paul, 79, musician, singer (with the late Mary Ford; "Hummingbird"), born Waukesha, WI, June 9, 1916.
Wayman Tisdale, 31, basketball player, born at Tulsa, OK, June 9, 1964.

JUNE 10 — SATURDAY
161st Day — Remaining, 204

ALCOHOLICS ANONYMOUS: 60th FOUNDING ANNIVERSARY. June 10. On June 10, 1935, in Akron, OH, Dr. Robert Smith completed his first day of permanent sobriety. "Doctor Bob" and William G. Wilson are considered to have founded Alcoholics Anonymous on that day.

ANTIQUE AUTO SWAP MEET. June 10–11. Grand Prairie, TX. More than 500 vendors from across the US gather to sell, trade or buy automobiles, parts and accessories. Est attendance: 65,000. For info: Traders Village, 2602 Mayfield Rd, Grand Prairie, TX 75052. Phone: (214) 647-2331.

BELMONT STAKES. June 10. Belmont Park, NY. Final race of the "Triple Crown" was inaugurated in 1867. Traditionally run on the fifth Saturday after Kentucky Derby (third Saturday after Preakness). For info: New York Racing Assn, Press Office, PO Box 90, Jamaica, NY 11417. Phone: (718) 641-4700.

BETTY PICNIC. June 10. All Sports Park, Grants Pass, OR. To celebrate the Bettys of this world for their vivacity, impulsiveness and similarities. Annually, the second Saturday in June. Send stamped envelope for info since this is a nonprofit organization. Est attendance: 55. Sponsor: Betty Wilder and Betty Patterson, c/o Betty Picnic, 1012 NE Madrone, Grants Pass, OR 97526. Phone: (503) 476-4104.

CONSERVATION FESTIVAL. June 10–11. Mill Mountain Zoological Park, Roanoke, VA. A festival dedicated to making everyone aware of the fragile state of the environment. Nationally known conservationists conduct seminars and programs that both entertain and inform. Annually, the second weekend in June. Sponsors: The Roanoke Special Events Committee and the Mill Mountain Zoo. Est attendance: 7,500. For info: E. Laban Johnson, Special Events Coord, City of Roanoke, 210 Reserve Ave SW, Roanoke, VA 24016. Phone: (703) 981-2889.

June 1995

S	M	T	W	T	F	S
				1	2	3
4	5	6	7	8	9	10
11	12	13	14	15	16	17
18	19	20	21	22	23	24
25	26	27	28	29	30	

CZECHOSLOVAKIA: RAPE OF LIDICE: ANNIVERSARY. June 10. On June 10, 1942, Nazi German troops executed, by shooting, all male inhabitants of the Czechoslovakian village of Lidice (total population about 500 persons), burned every house and deported the women and children to Germany for "re-education." One of the most remembered atrocities of World War II. June 10, Lidice Memorial Day, is observed in New Jersey.

DONNA REED FESTIVAL FOR THE PERFORMING ARTS. June 10–17. Denison, IA. Performing Arts festival in Donna Reed's hometown. Focus of the festival is educational workshops in various areas of the performing arts taught by Hollywood and New York professionals. Also included are celebrity golf tournament, street fair and theatrical performances for the general public. Est attendance: 8,000. For info: Donna Reed Foundation, Patricia Fleshner, Exec Dir, PO Box 122, Denison, IA 51442. Phone: (800) 336-4692.

ENGLAND: TROOPING THE COLOUR—THE QUEEN'S OFFICIAL BIRTHDAY PARADE. June 10. (Tentative—may be changed to June 17.) Horse Guards Parade, Whitehall, London. Colorful ceremony with music and pageantry during which Her Majesty The Queen takes the salute. Starts at 11 AM. When requesting info, send stamped, self-addressed envelope. Est attendance: 10,000. For info: The Ticket Office, HQ Household Division, 1 Chelsea Barracks, London, England SW1H 8RF.

FRANKENMUTH BAVARIAN FESTIVAL. June 10–17. Frankenmuth, MI. To celebrate the German heritage of Frankenmuth with good food, great entertainment events for all ages, parades, craft show, children's activities, fun special events, "Michigan's Largest Chicken Dance," "Wurst Day at the Festival" and much more. Est attendance: 350,000. Sponsor: Frankenmuth Civic Events Council, Carrie Schlobohm, Exec Dir, 637 S Main St, Frankenmuth, MI 48734. Phone: (517) 652-8155.

GARLAND, JUDY: BIRTH ANNIVERSARY. June 10. American actress and singer, born Frances Gumm, at Grand Rapids, MI, June 10, 1922. Died at London, England, June 22, 1969.

GUS MACKER 3-ON-3 CHARITY BASKETBALL TOURNAMENT. June 10–11. Chillicothe, OH. To benefit Junior Achievement of Ross County, the one-and-only original in-the-streets, in-your-face, call-your-own-fouls, 3-on-3 outdoor backyard, the-way-it-was-meant-to-be-played, Gus Macker Basketball Tournament. Annually, the second weekend in June. Est attendance: 25,000. For info: Junior Achievement Conv and Visitor Bureau, PO Box 150, Chillicothe, OH 45601.

HERB DAY. June 10. New Hampshire Farm Museum, Rte 125, Milton, NH. Herb-related activities: sales of plants, herbal wreaths, soaps, oils, craft items made with herbs. Demonstrations of a herb garden and uses of herbs. Samples of herbal foods and drinks for tasting. Est attendance: 450. For info: Susie McKinley, Admin Asst, PO Box 644, Milton, NH 03851. Phone: (603) 652-7840.

HERITAGE DAY CELEBRATION. June 10. Wisconsin Dells, WI. A return to the past featuring arts and crafts fair, plus old-fashioned exhibits and displays and antique flea market. Annually, weekend following Great Wisconsin Dells Balloon Rally. Est attendance: 2,500. For info: Wisconsin Dells Visitor & Conv Bureau, 701 Superior St, PO Box 390, Wisconsin Dells, WI 53965. Phone: (800) 223-3557.

HERITAGE DAY CELEBRATION. June 10. Delaware and Riverfront Area, Leavenworth, KS. Celebration of Leavenworth's 140-year history—the oldest city in the state! Ethnic foods, square and street dancing, melodrama, barbershoppers, children's events and more. Est attendance: 5,000. For info: Leavenworth Conv and Visitors Bureau, 518 Shawnee, Leavenworth, KS 66048. Phone: (800) 844-4114.

HONG KONG: INTERNATIONAL DRAGON BOAT FESTIVAL. June 10–11. A day recognizing the death of Qu Yuan, 4th-century BC poet and former minister of state who threw himself into the river in protest of the corruption of the court. The local dragon boat races are held on the 5th day of the 5th lunar month—June 2, 1995, on the Gregorian calendar. The international races will be held June 10-11. For info: Hong Kong Tourist Assn, 590 Fifth Ave, 5th Fl, New York, NY 10036-4706. Phone: (212) 869-5008.

JOHN HULL OPENS FIRST MINT IN AMERICA: ANNIVERSARY. June 10. In defiance of English colonial law, John Hull, a silversmith, established the first mint in America, on June 10, 1652. The first coin issued was the Pine Tree Shilling, designed by Hull.

JORDAN: GREAT ARAB REVOLT AND ARMY DAY: ANNIVERSARY. June 10. Commemorates the beginning of the Great Arab Revolt on June 10, 1916. National holiday.

JOYFUL SUMMER CELEBRATION. June 10. To recognize rare June days (à la James R. Lowell) for those unable to wait for the summer solstice. Plan picnics, softball games, canoeing or croquet. Cake and ice cream also desirable. Sponsor: R.H. Plummers, 2222 Fuller Rd, #1115A, Ann Arbor, MI 48105.

LEAGUE OF WOMEN VOTERS COUNCIL, 1995. June 10-13. Washington, DC. 200 members of the League will meet to review national and legislative concerns and activities. Special note—Day on the Hill, when League members lobby their representatives. For info: Public Affairs Dept, League of Women Voters, 1730 M St NW, Washington, DC 20036. Phone: (202) 429-1965.

MAJOR LEAGUE/MINOR LEAGUE: PHOTOGRAPHS OF AMERICA'S BASEBALL STADIUMS BY JIM DOW. June 10-July 23. Everhart Museum, Scranton, PA. See Feb 2 for full description. Call or write the Smithsonian for other dates and venues. For info: Smithsonian Institution Traveling Exhibition Service, 1100 Jefferson Dr SW, Ste 3146, Washington, DC 20560. Phone: (202) 357-2700.

McDANIEL, HATTIE: BIRTH ANNIVERSARY. June 10. Hattie McDaniel was the first African-American to win an Academy Award. She won it in 1940 for her role in the 1939 film *Gone With the Wind*. Her career spanned radio and vaudeville in addition to her screen roles in *Judge Priest, The Little Colonel, Showboat* and *Saratoga*, among others. She was born on June 10, 1889, at Wichita, KS, and died Oct 26, 1952.

MELROSE PLANTATION ARTS AND CRAFTS FESTIVAL. June 10-11. Natchitoches, LA. Features quality handcrafted items from juried exhibitors. Food, including the famous meatpies, and soft drinks will be available. Annually, the second full weekend in June. Est attendance: 20,000. For info: Natchitoches Parish Tourist Commission, PO Box 411, Natchitoches, LA 71458. Phone: (318) 352-8072.

MIDDLE OF NOWHERE TRAIL RIDE. June 10-11. Ainsworth, NE. Trail ride along the Niobrara River. Limit 100 horses; wagons and buggies are invited. Place to keep horses available. 7th annual ride. Est attendance: 120. For info: Mrs Mary Jo Curtis, Organizer, 340 W 3rd, Ainsworth, NE 69210. Phone: (402) 387-2488.

NATIONAL YO-YO DAY. June 10. Arcade, NY. Celebrations of yo-yo playing marked by yo-yo contests, demonstrations, promotions and give-aways across the country. Est attendance: 500. Sponsor: Hummingbird Toy Co, PO Box 276, 32 Water St, Arcade, NY 14009. Phone: (716) 492-5120.

NEW JERSEY FRESH SEAFOOD FESTIVAL. June 10-11. Historic Gardner's Basin Nautical Park, Atlantic City, NJ. Educational festival with maritime exhibits and creative activities for children. Arts and crafts, food booths and two days of live entertainment. Est attendance: 20,000. For info: Greater Atlantic City Conv and Visitors Bureau, 2314 Pacific Ave, Atlantic City, NJ 08401. Phone: (609) 348-7100.

OLD TOWN ART FAIR/CHICKEN TERIYAKI LUNCHES. June 10-11. Chicago, IL. Chicken Teriyaki lunches at $6 per plate. Annually, during Old Town Art Fair. Sponsor: Midwest Buddhist Temple. Est attendance: 1,500. For info: Midwest Buddhist Temple, 435 W Menomonee St, Chicago, IL 60614. Phone: (312) 943-7801.

OPERATION YOUTH. June 10-17. Xavier University, Cincinnati, OH. To provide a hands-on exercise in democracy for teenagers. Students elect, from their ranks, city officials who join their real counterparts in a Cincinnati council meeting. Speakers from education, industry and government. Fax: 513-745-4383. Est attendance: 100. For info: William Smith, CPA, Prof of Accounting, Dir of Operation Youth, Xavier University, 3800 Victory Pkwy, Cincinnati, OH 45207-5161. Phone: (513) 745-3504.

PIER 39 STREET PERFORMERS FESTIVAL. June 10. Pier 39, San Francisco, CA. The Bay Area's best street performers converge in June for the 15th annual Pier 39 Street Performers Festival. A San Francisco tradition, these multi-talented tricksters perform their zany acts at the Pier's Entrance Plaza and Crystal Geyser Center Stage. Free and open to the public; performers pass the hat following each performance to raise money for a local charity. Est attendance: 10,000. For info: Alicia Vargas, PR Dir, Pier 39, PO Box 193730, San Francisco, CA 94119. Phone: (415) 705-5500.

PORTUGAL: DAY OF PORTUGAL. June 10. National holiday. Anniversary of the death of Portugal's national poet, Luis Vas de Camoes (Camoens), born in 1524 (exact date unknown) at either Lisbon or possibly Coimbra. Died at Lisbon, Portugal, June 10, 1580.

POSTCARD SHOW. June 10-11. Convention Center, Ft. Washington, VA. Est attendance: 600. For info: Postcard History Soc, John H. McClintock, PO Box 1765, Manassas, VA 22110. Phone: (703) 368-2757.

SODBUSTER FEST. June 10-11. Museum of Wildlife, Science and Industry, Webster, SD, West Highway 12. Annual pageant, model trains, handcrafts and arts, parade, contests, demonstrations of household duties, ethnic foods, melodrama, tractors, meal served on the premises. Largest collection of antique tractors in the state. Annually, the second weekend of June. Sponsor: Museum of Wildlife, Science & Industry. Est attendance: 3,000. For info: MWSI/NESD, RR 2, Box 141A, Webster, SD 57274. Phone: (605) 345-4751.

SPECIAL LIBRARIES ASSOCIATION ANNUAL CONFERENCE. June 10-15. Montreal, Quebec, Canada. Networking, educational activities, annual board meeting for SLA. Includes exhibitors and vendors of library materials and equipment. Est attendance: 5,000. For info: Alisa N. Cooper, Dir Mktg and Sales, SLA, 1700 18th St NW, Washington, DC 20009. Phone: (202) 234-4700.

TAKE A KID FISHING WEEKEND. June 10-11. St. Paul, MN. Resident adults may fish without a license on these days when fishing with a child under age 16. For info: Richard Hassinger, DNR, 500 Lafayette Rd, Box 12, St. Paul, MN 55146. Phone: (612) 296-3325.

TERRITORIAL DAYS. June 10-11. Prescott, AZ. A celebration of arts, crafts and music and historic home tour. Toll-free phone: (800) 266-7534. Est attendance: 15,000. Sponsor: Chamber of Commerce, PO Box 1147, Prescott, AZ 86302. Phone: (602) 445-2000.

WORLD'S GREATEST YARD SALE. June 10-11. Fairgrounds, York, PA. Fundraising for York Unit, American Cancer Society. Est attendance: 25,000. For info: American Cancer Soc, 226 E Market St, York, PA 17403. Phone: (717) 848-1841.

June 10-11 ☆ *Chase's 1995 Calendar of Events* ☆

BIRTHDAYS TODAY

F. Lee Bailey, 62, lawyer, born at Waltham, MA, June 10, 1933.
Jeff Greenfield, 52, author, journalist, born at New York, NY, June 10, 1943.
Nat Hentoff, 70, music critic, journalist, born at Boston, MA, June 10, 1925.
J. Bennett Johnston, 63, US Senator (D, Louisiana), born at Shreveport, LA, June 10, 1932.
Doug McKeon, 29, actor (*On Golden Pond*), born at Pomptain Plains, NJ, June 10, 1966.
Rose Mofford, 73, former Governor of Arizona (D), born at Globe, AZ, June 10, 1922.
Prince Philip, 74, Duke of Edinburgh, husband of Queen Elizabeth II, born at Corfu, Greece, June 10, 1921.
Maurice Sendak, 67, author, illustrator (*Chicken Soup with Rice, Where the Wild Things Are*), born at Brooklyn, NY, June 10, 1928.
Samuel Knox Skinner, 57, former White House Chief of Staff (Bush administration), born Chicago, IL, June 10, 1938.

JUNE 11 — SUNDAY
162nd Day — Remaining, 203

CHILDREN'S DAY IN MASSACHUSETTS. June 11. Annually, the second Sunday in June.

CHILDREN'S SUNDAY. June 11. Traditionally the second Sunday in June is observed as Children's Sunday in many Christian churches.

CONSTABLE, JOHN: BIRTH ANNIVERSARY. June 11. English landscape painter. Born at East Bergholt, Suffolk, England, on June 11, 1776. He died at London, on Mar 31, 1837.

ENGLAND: THE BRISTOL TO BOURNEMOUTH VINTAGE VEHICLE RUN. June 11. Starts at Ashton Court Estate, Bristol, Avon; finishes at Undercliff Drive, Bournemouth, Dorset. More than 350 pre-1940 cars, motorcycles and light commercial vehicles will take part in this scenic 97-mile run. Est attendance: 7,000. For info: Alan Davidson, 63 Abbots Way, Yeovil, Somerset, England BA21 3HX.

FESTIVAL OF THE RED ROSE. June 11. Manheim, PA. The annual rental payment is one rose to an heir of Baron Henry Williams Stiegel, for land deeded to the Zion Lutheran Church in 1772. Annually, the second Sunday in June. Sponsor: Zion Lutheran Church. Est attendance: 200. For info: John Kendig, 65 S Main St, Manheim, PA 17545. Phone: (717) 665-2308.

GERMANFEST. June 11-17. Fort Wayne, IN. A celebration of our German heritage with folk music, folk dancing, beer tents, German food, beer and wine tastings, Mannerchor (men's choir) singing, Gottesdienste (masses), genealogy workshops, classical organ music, German films, lectures and demonstrations. Est attendance: 40,000. Sponsor: German Heritage Soc, PO Box 12651, Fort Wayne, IN 46864. Phone: (800) 767-7752.

GRANT WOOD ART FESTIVAL. June 11. Stone City-Anamosa, IA. Juried art exhibits, demonstrations, stage entertainment, tours of historic Stone City, quilt show and more! Est attendance: 8,000. Sponsor: Grant Wood Art Festival, Inc, Marguerite Stoll, Media Coord, 124 E Main St, Anamosa, IA 52205. Phone: (319) 462-4267.

June 1995

S	M	T	W	T	F	S
				1	2	3
4	5	6	7	8	9	10
11	12	13	14	15	16	17
18	19	20	21	22	23	24
25	26	27	28	29	30	

INTERNATIONAL MUSIC CAMP. June 11-Aug 1. International Peace Garden, Dunseith, ND. In its 40th season, more than 3,000 students from 25 countries enjoy the many different programs of IMC. Music camps include concert band, piping and drumming, jazz, choir, vocal jazz, orchestra, piano, guitar, electronic music. Also visual arts, dance, drama and creative writing sessions. Est attendance: 3,000. For info: Intl Music Camp, Joseph T. Alme, Dir, 1725 11th St SW, Minot, ND 58701. Phone: (701) 838-8472.

JEFF COOK SUPER BASS CLASSIC. June 11. Goose Pond Colony, Scottsboro, AL. Thousands of dollars worth of cash and prizes. Proceeds go to charity. Admission is free to weigh-in. Sponsor: Jeff Cook of the country music group Alabama. Est attendance: 5,000. For info: Rick Roden, Tourism Dir, PO Box 973, Scottsboro, AL 35768. Phone: (800) 259-5508.

JONSON, BEN: BIRTH ANNIVERSARY. June 11. English playwright and poet, born June 11, 1572. "Talking and eloquence," he wrote, "are not the same: to speak and to speak well, are two things." Jonson died at London, England, on Aug 6, 1637. The epitaph, written on his tombstone in Westminster Abbey: "O rare Ben Jonson."

KING KAMEHAMEHA I DAY. June 11. Designated state holiday in Hawaii honors memory of Hawaiian monarch (1737-1819). Governor appoints state commission to plan annual celebration.

KORN KLUB MUD DRAGS. June 11. (Also Sept. 10.) River Rd, Plattsmouth, NE. Modified cars race in pits of mud in this annual fundraiser. Est attendance: 500. For info: Patricia Baburek, Coord, Kass Kounty King Korn Klub, PO Box 40, Plattsmouth, NE 68048. Phone: (402) 296-4155.

MILK JUG REGATTA. June 11. Minocqua, WI. Boat race featuring boats made from plastic milk jugs. Est attendance: 6,000. For info: Greater Minocqua Chamber of Commerce, PO Box 1006, Minocqua, WI 54548. Phone: (800) 446-6784.

MOUNT PINATUBO ERUPTS IN PHILIPPINES: ANNIVERSARY. June 11. On June 11, 1991, long dormant volcano, Mount Pinatubo in the Philippines erupted with a violent explosion spewing ash and gases into the air that could be seen for more than 60 miles. The surrounding areas were covered with ash and mud created by rainstorms. US Military bases Clark and Subic Bay were also damaged. This came in the midst of negotiations between the US and the Philippines over a future lease for the bases, prompting speculation that the US would not continue to use the bases in the future. On July 6, 1992, Ellsworth Dutton of the National Oceanic and Atmospheric Administration's Climate Monitoring and Diagnostics Laboratory announced that a layer of sulfuric acid droplets released into the Earth's atmosphere by the eruption had cooled the planet's average temperature by about 1 degree Fahrenheit. The greatest difference was noted in the Northern Hemisphere with a drop of 1.5 degrees. Although the temperature drop is temporary and should end within five years, the climate trend will make determining the effect of greenhouse warming on the Earth more difficult through its duration.

★ **NATIONAL FLAG WEEK.** June 11-17. Presidential Proclamation issued each year since 1966 for the week including June 14. (PL89-443 of June 9, 1966.) In addition, the President often calls upon the American people to participate in public ceremonies in which the Pledge of Allegiance is recited.

☆ Chase's 1995 Calendar of Events ☆ June 11-12

ORTHODOX PENTECOST. June 11. Observed by Eastern Orthodox churches on this date.

RACE UNITY DAY. June 11. Baha'i-sponsored observance promoting racial harmony and understanding and the essential unity of mankind. Annually, the second Sunday in June. Established in 1957 by the Baha'is of the US. For info: Natl Spiritual Assembly of the US, 1320 Nineteenth St, NW, Ste 701, Washington, DC 20036. Phone: (202) 833-8990.

◆ **RANKIN, JEANNETTE: BIRTH ANNIVERSARY.** June 11. First woman elected to the US Congress, a reformer, feminist and pacifist, was born at Missoula, MT, June 11, 1880. She was the only member of Congress to vote against a declaration of war against Japan in December 1941. Died May 18, 1973.

ROSE SHOW. June 11. Planting Fields Arboretum, Oyster Bay, Long Island, NY. More than 1,000 roses, entered by amateur growers in competition for awards and an educational display staffed by experts to answer questions. Est attendance: 2,000. Sponsor: Long Island Rose Soc, Inc, Robert Ardini, 30 Barstow Rd, Great Neck, NY 11021. Phone: (516) 773-7936.

"SPACE ODDITY" SONG RELEASE: ANNIVERSARY. June 11, 1969. This single recorded by David Bowie was released to coincide with the *Apollo 11*'s trip to the moon, during which Neil Armstrong and Edwin Aldrin, Jr, landed and walked on the surface of the moon.

STRAUSS, RICHARD GEORG: BIRTH ANNIVERSARY. June 11. German composer, musician and conductor whose best remembered works are *Till Eulenspiegel* (1895), *Also Sprach Zarathustra* (1896) and *Don Quixote* (1898). Born at Munich, Germany, on June 11, 1864, he died at Garmisch-Partenkirchen, Germany, after a heart attack on Sept 8, 1949.

TRINITY SUNDAY. June 11. Christian Holy Day on the Sunday after Pentecost commemorates the Holy Trinity, the three divine persons—Father, Son and Holy Spirit—in one God. See also: "Pentecost" (June 4).

WALES: CARDIFF SINGER OF THE WORLD COMPETITION. June 11-17. St. David's Hall, The Hayes, Cardiff, South Glamorgan. One of the most famous singing competitions in the world. Biannual. For info: Box Office, St. David's Hall, The Hayes, Cardiff, Wales CF1 2SH,.

BIRTHDAYS TODAY

Adrienne Barbeau, 50, actress ("Maude"), born at Sacramento, CA, June 11, 1945.
Henry Cisneros, 48, US Secretary of Housing and Urban Development in Clinton administration, born at San Antonio, TX, June 11, 1947.
Chad Everett (Raymond Cramton), 59, actor ("The Dakotas," "Medical Center"), born South Bend, IN, June 11, 1936.
Gary Fencik, 41, Pro Football Hall of Famer, born Chicago, IL, June 11, 1954.
Joe Montana, 39, football player, born at Monongahela, PA, June 11, 1956.
Jackie Stewart, 56, auto racer, born at Dunbartonshire, Scotland, June 11, 1939.
William Styron, 70, author (*The Confessions of Nat Turner, Sophie's Choice*), born at Newport News, VA, June 11, 1925.
Gene Wilder, 56, actor (*The Producers, Willie Wonka, Blazing Saddles*), born at Milwaukee, WI, June 11, 1939.

JUNE 12 — MONDAY
163rd Day — Remaining, 202

BUSH, GEORGE HERBERT WALKER: BIRTHDAY. June 12. Forty-third vice president of the US under Reagan, and 41st president, elected in 1988. Born at Milton, MA, June 12, 1924. Married to Barbara Pierce, Jan 6, 1945.

FIRST MAN-POWERED FLIGHT ACROSS ENGLISH CHANNEL: ANNIVERSARY. June 12. Bryan Allen, 26-year-old Californian, pedaled the 70-pound *Gossamer Albatross* 22 miles across the English Channel, from Folkestone, England, to Cape Gris-Nez, France, in 2 hours, 49 minutes on June 12, 1979, winning (with the craft's designer, Paul MacCready of Pasadena, CA) the £100,000 prize that had been offered Nov 30, 1977, by British industrialist Henry Kremer, for the first man-powered flight across the English Channel.

***LOVING v VIRGINIA*: ANNIVERSARY.** June 12. On June 12, 1967, the US Supreme Court decision in *Loving v Virginia* swept away all 16 remaining state laws prohibiting interracial marriages.

NATIONAL BASEBALL HALL OF FAME: ANNIVERSARY. June 12. The National Baseball Hall of Fame and Museum, Inc, was dedicated at Cooperstown, NY, June 12, 1939. Nearly 200 individuals have been honored for their contributions to the game of baseball by induction into the Baseball Hall of Fame. The first players chosen for membership (1936) were Ty Cobb, Honus Wagner, Babe Ruth, Christy Mathewson and Walter Johnson. Relics and memorabilia from the history of baseball are housed at this "shrine" of America's national sport.

★ **NATIONAL LITTLE LEAGUE BASEBALL WEEK.** June 12-18. Presidential Proclamation 3296, of June 4, 1959 covers all succeeding years. Always the week beginning with the second Monday in June. (H.Con.Res. 17 of June 1, 1959.) Proclamations 3296 of June 4, 1959, and 3407 of Apr 18, 1961.

PERIGEAN SPRING TIDES. June 12-13. Spring tides, the highest possible tides, occur when New Moon or Full Moon falls within 24 hours of the moment the Moon is nearest Earth (perigee) in its monthly orbit, on June 12 at 9:00 PM, EDT.

PHILIPPINES: INDEPENDENCE DAY. June 12. National holiday. Declared independence from Spain on this day, 1898.

PORTUGAL: ST. ANTHONY'S EVE AND FESTIVAL. June 12. Lisbon. The capital city of Portugal, the birthplace of St. Anthony, honors its favorite saint (the patron saint of young lovers) with an impressive show of "marchas" (walking groups of singers and musicians) from almost all typical quarters of Lisbon, parading along Avenida da Liberdale on the evening of June 12, while in various districts of the city the festival goes on with music, dances, bonfires, etc. For info: Portuguese Natl Tourist Office, 590 Fifth Ave, New York, NY 10036. Phone: (212) 354-4403.

SPACE MILESTONE: *VENERA 4* (USSR). June 12. Launched on June 12, 1967, this instrumental capsule landed on Venus by parachute on Oct 18 and reported a temperature of 536°F.

BIRTHDAYS TODAY

Marv Albert, 52, TV station director, sportscaster, born at New York, NY, June 12, 1943.
George Herbert Walker Bush, 71, 41st US President, born at Milton, MA, June 12, 1924.
Chick Corea, 54, musician, born at Chelsea, MS, June 12, 1941.
Vic Damone (Vito Farinola), 67, singer ("On the Street Where You Live"), born New York, NY, June 12, 1928.
James Houston, 74, author (*Ghost Fox*), illustrator, designer, filmmaker, born Toronto, Ont, Canada, June 12, 1921.
Jim Nabors, 63, actor ("Andy Griffith Show," "Gomer Pyle, U.S.M.C."), born Sylacauga, AL, June 12, 1932.
David Rockefeller, 80, banker, born at New York, NY, June 12, 1915.
Rory Sparrow, 37, former basketball player, born at Suffolk, VA, June 12, 1958.

JUNE 13 — TUESDAY
164th Day — Remaining, 201

BOLLES, DON: DEATH ANNIVERSARY. June 13. Don Bolles, investigative reporter for *The Arizona Republic*, died June 13, 1976, as a result of injuries received when a bomb exploded in his automobile, June 2, 1976, while he was engaged in journalistic investigation of alleged Mafia story. Bolles was awarded, posthumously, the University of Arizona's John Peter Zenger Award, Dec 9, 1976.

BURLINGTON STEAMBOAT DAYS/AMERICAN MUSIC FESTIVAL. June 13-18. Mississippi Riverfront at Port of Burlington, IA. Week-long event offers community and visitors a chance to enjoy top-name entertainment and a carnival setting. Est attendance: 95,000. For info: Steamboat Days, Sandy Schafer, PO Box 271, Burlington, IA 52601. Phone: (319) 754-4334.

ENGLAND: ROYAL ASCOT. June 13-16. Ascot Racecourse, Ascot, Berkshire. World-famous horseracing attended by members of the British Royal Family. Est attendance: 70,000. For info: The Secretary, Grand Stand Office, Ascot Racecourse, Ascot, Berkshire, England SL5 7JN.

EVERS, MEDGAR: ASSASSINATION ANNIVERSARY. June 13. Civil rights leader Medgar Wiley Evers was active in seeking integration of schools and voter registration. He was assassinated by Byron de la Beckwith on June 13, 1963. The public outrage following his death was one of the factors that led President John F. Kennedy to propose a comprehensive civil rights law.

HOME OWNERS LOAN ACT: ANNIVERSARY. June 13. The Federal Savings and Loan Association was authorized on June 13, 1933, with the passage of the Home Owners Loan Act. The purpose of the legislation was to provide a convenient place for investment and to lend money on first mortgages. The first association was the First Federal Savings and Loan Association of Miami, FL, which was chartered on Aug 8, 1933.

MISSION SAN LUIS REY DE FRANCIA FOUNDING: ANNIVERSARY. June 13. California mission to the Indians founded June 13, 1798. Abandoned by 1846; restoration begun in 1892.

MOON PHASE: FULL MOON. June 13. Moon enters Full Moon phase at 12:03 AM, EDT.

NATIONAL CLAY WEEK. June 13-17. City Park, Uhrichsville, OH. The fest commemorates the heritage of the valley as the "Clay Center of the World." Midway and parade. For info: Barbara Roberts, 918 N 3rd St, Dennison, OH 44621. Phone: (614) 922-3028.

NEBRASKALAND DAYS AND BUFFALO BILL RODEO. June 13-19. North Platte, NE. To relive the Old West. Parades, contests, shoot-outs, art shows, frontier revue, top country/western stars. Est attendance: 100,000. For info: Nebraskaland Days, Box 706, North Platte, NE 69103. Phone: (308) 532-7939.

ST. ANTHONY OF PADUA: FEAST DAY. June 13. Born at Lisbon, Portugal, Aug 15, 1195, St. Anthony is patron of the illiterate and the poor. Died at Padua, June 13, 1231. Public holiday, Lisbon.

SCOTT, WINFIELD: BIRTH ANNIVERSARY. June 13. American army general, negotiator of peace treaties with Indians and twice nominated for president (1848 and 1852). Leader of brilliant military campaign in Mexico in 1847. Scott was born at Petersburg, VA, on June 13, 1786, and died at West Point, NY, on May 29, 1866.

SUMMER STORYTELLER-IN-RESIDENCE. June 13-Aug 19. Wren's Nest, Atlanta, GA. Wonderful storyteller, puppeteer and poet Akbar Imhotep will be featured at the Wren's Nest during the summer months. He will share with his audience the wonderful African-American folk tales about Br'er Rabbit and Br'er Fox and their antics. This program attracts visitors from around the world. ($1 additional fee over regular admission charge.) Est attendance: 5,000. For info: Karen Kelly, Wren's Nest, 1050 Ralph David Abernathy Blvd SW, Atlanta, GA 30310. Phone: (404) 753-7735.

VIEIRA da SILVA, MARIA-HELENA: BIRTH ANNIVERSARY. June 13. Portuguese-born painter, Maria Helena Vieira da Silva was born on June 13, 1908, at Lisbon. Although a French citizen, she was considered by many as Portugal's greatest contemporary artist. Her work spanned the Parisian School and Abstract Expressionism. She died Mar 6, 1992, at Paris, France.

◆ **YANKEE STADIUM SILVER JUBILEE ANNIVERSARY.** June 13, 1948. The New York Yankees celebrated the 25th anniversary of the House That Ruth Built. Ruth's famous uniform number 3 was retired that day as Ruth made his final appearance at Yankee Stadium. He died just two months later.

YEATS, WILLIAM BUTLER: 130th BIRTH ANNIVERSARY. June 13. Nobel prize-winning Irish poet and dramatist, born at Dublin, Ireland, on June 13, 1865. He once wrote: "If a poet interprets a poem of his own he limits its suggestibility." Yeats died in France on Jan 28, 1939. After World War II his body was returned, as he had wished, for reburial in a churchyard at Drumcliff, Ireland.

BIRTHDAYS TODAY

Bettina Bunge, 32, tennis player, born at Adliswick, Switzerland, June 13, 1963.
Christo (Christo Javacheff), 60, conceptual artist (*Running Fence, Valley Curtin*), born Babrovo, Bulgaria, June 13, 1935.
Sarunas Marciulionis, 31, basketball player, born at Kaunas, Lithuania, June 13, 1964.
Malcolm McDowell, 52, actor (*A Clockwork Orange, O Lucky Man*), born at Leeds, England, June 13, 1943.
Ally (Alexandra Elizabeth) Sheedy, 33, actress (*St. Elmo's Fire, Breakfast Club, War Games*), born New York, NY, June 13, 1962.
Richard Thomas, 44, actor ("The Waltons," "Roots: The Next Generation"), born at New York, NY, June 13, 1951.
Jim Guy Tucker, Jr., 52, Governor of Arkansas (D), born at Oklahoma City, OK, June 13, 1943.

JUNE 14 — WEDNESDAY
165th Day — Remaining, 200

ALZHEIMER, ALOIS: BIRTH ANNIVERSARY. June 14. The German psychiatrist and pathologist Alois Alzheimer was born at Markbreit am Mainz, Germany, on June 14, 1864. In 1907 an article by Alzheimer appeared in *Allgemeine Zeitschrift fur Psychiatrie*, first describing the disease that was named for him. It was thought of as a kind of presenile dementia, usually beginning at age 40-60, but whether Alzheimer's disease is an entity separate from senile dementia remains unanswered. Alzheimer died at Breslau, Germany, on Dec 19, 1915.

June 1995

S	M	T	W	T	F	S
				1	2	3
4	5	6	7	8	9	10
11	12	13	14	15	16	17
18	19	20	21	22	23	24
25	26	27	28	29	30	

ANTIQUE AUTO SHOW AND SWAP MEET. June 14-18. Petit Jean Mountain, Morrilton, AR. 37th annual show and meet with more than 100 antique and classic cars competing for awards, from turn-of-the-century to 1970 models. More than 1,000 vendor spaces filled with antique cars, parts and related items. Also, arts and crafts. Est attendance: 90,000. For info: Museum of Automobiles, Rt 3, Box 306, Morrilton, AR 72110. Phone: (501) 727-5427.

◈ **BARTLETT, JOHN: 175th BIRTH ANNIVERSARY.** June 14. American editor and compiler (Bartlett's *Familiar Quotations* [1855] and other works) was born at Plymouth, MA, on June 14, 1820. Though he had little formal education, he created one of the most used reference works of the English language. No quotation of his own is among the more than 22,000 listed today, but in the preface to the first edition he wrote that the object of this work "originally made without any view of publication" was to show "the obligation our language owes to various authors for numerous phrases and familiar quotations which have become 'household words.'" Bartlett died at age 85 at Cambridge, MA, on Dec 3, 1905.

BOURKE-WHITE, MARGARET: BIRTH ANNNIVERSARY. June 14, 1906. Margaret Bourke was born at New York City. One of the original photojournalists, she developed her personal style while photographing the Krupp Iron Works in Germany and the Soviet Union during the first Five-Year Plan. Bourke-White was one of the four original staff photographers for *Life* magazine in 1936. The first woman attached to the US armed forces during World War II, she covered the Italian campaign, the siege of Moscow and the American soldiers' crossing of the Rhine into Germany, and she shocked the world with her photographs of survivors and corpses in the concentration camps. Bourke-White photographed Mahatma Gandhi and covered the migration of millions of people after the Indian subcontinent was divided into Hindu India and Muslim Pakistan. She served as a war correspondent during the Korean War. Among her several books, the most famous was probably her collaboration with her second husband, novelist Erskine Caldwell, a study of rural poverty in the American South, called *You Have Seen Their Faces*. Margaret Bourke-White died Aug 27, 1971, at Stamford, CT.

CANADA: PROVINCIAL EX. June 14-18. Keystone Center, Brandon, Manitoba. Voted Canadian Regional Fair of the Year. Held on the Conklin Midway. Events include the Bud Pro Rodeo, Mid-Canada Truck Show, 7 free stages, summer saloon, kids world and petting zoo, dairy show, trade show, and family fair. Est attendance: 92,000. For info: Provincial Exhibition of Manitoba, #3 - 1175 18th St, Brandon, Man, Canada R7A 7C5. Phone: (204) 726-3590.

FAMILY HISTORY DAY. June 14. Every summer family reunions are so busy with games and activities that most of us forget the true purpose: to share the folklore, legends and myths that bind us together. Each participant should share at least one good recollection (fact or fiction). Don't forget the hot dogs and lemonade. Phone: (212) 388-8673 or (717) 274-8451. Sponsor: Wellness Permission League, Tom or Ruth Roy, 2105 Water St, Lebanon, PA 17046.

FINLAND: JYVASKYLA ARTS FESTIVAL. June 14-21. Jyvaskyla. Annual festival begun in 1955 for the purpose of stimulating discussion of contemporary philosophical and social concerns and building a bridge between different national and ethnic traditions. About 50 events including seminars, music, visual arts and film. Est attendance: 10,000. For info: Finnish Tourist Board, 655 Third Ave, New York, NY 10017. Phone: (212) 949-2333.

FIRST NONSTOP TRANSATLANTIC FLIGHT: ANNIVERSARY. June 14-15. Captain John Alcock and Lieutenant Arthur W. Brown flew a Vickers Vimy bomber 1,900 miles nonstop from St. Johns, Newfoundland, to Clifden, County Galway, Ireland, June 14-15, 1919. In spite of their crash landing in an Irish peat bog, their flight inspired public interest in aviation and led to many other flights. See also: "Lindbergh Flight Anniversary" (May 20).

FIRST US BREACH OF PROMISE SUIT: ANNIVERSARY. June 14. The first breach of promise suit in the US was filed on June 14, 1623, in the Virginia Council of State, in Charles City, VA. Reverend Greville Pooley brought suit against Cicely Jordan, who had jilted him in favor of another man.

★ **FLAG DAY.** June 14. Presidential Proclamation issued each year for June 14. Proclamation 1335, of May 30, 1916, covers all succeeding years. Has been issued annually since 1941. (PL81-203 of Aug 3, 1949.) Customarily issued as "Flag Day and National Flag Week," as in 1986; the president usually mentions "a time to honor America," Flag Day to Independence Day (89 Stat. 211). See also: "National Flag Day USA: Pause for the Pledge" (this date).

FLAG DAY: ANNIVERSARY OF THE STARS AND STRIPES. June 14. On this day in 1777, John Adams introduced the following resolution before the Continental Congress, meeting at Philadelphia, PA: "Resolved, That the flag of the thirteen United States shall be thirteen stripes, alternate red and white; that the union be thirteen stars, white on a blue field, representing a new constellation." Legal holiday in Pennsylvania.

FLAG DAY CEREMONIES. June 14. Betsy Ross House, Philadelphia, PA. 11:00 AM program followed by band concert in Atwater Kent Gardens of Betsy Ross House. Est attendance: 1,000. For info: Betsy Ross House, 239 Arch St, Philadelphia, PA 19106. Phone: (215) 627-5343.

JAPAN: RICE PLANTING FESTIVAL. June 14. Osaka. Ceremonial transplanting of rice seedlings in paddyfield at Sumiyashi Shrine, Osaka.

KIAMICHI OWA-CHITO FESTIVAL OF THE FOREST. June 14-17. Broken Bow, OK. Lovely Beavers Bend State Park provides the setting for a celebration of American Indian culture and of the forest industry. Compete in Ax Throwing, Cross Buck Sawing, Pole Climbing, Pole Filling and Logging to become the "Bull of the Woods." Other festival activities include kids games, food booths, princess contest, arts and crafts show, photography contest, battle of the bands, golf tournament, canoe races, archery contest, turkey calling contest, 5K road race, talent contest and more. Est attendance: 50,000. For info: Chamber of Commerce, PO Box 249, Broken Bow, OK 74728. Phone: (405) 584-3393.

NATIONAL FLAG DAY USA: PAUSE FOR THE PLEDGE. June 14. Held simultaneously across the country at 7:00 PM, EDT. Public law 99-54 recognizes the Pause for the Pledge as part of National Flag Day ceremonies. The concept of the Pause for the Pledge of Allegiance was conceived as a way for all citizens to share a patriotic moment. National ceremony at Fort McHenry National Monument and Historic Shrine.

SPACE MILESTONE: *MARINER 5* (US). June 14. Interplanetary probe of Venus established 72½-87½ percent carbon dioxide content of atmosphere on Oct 18 flyby of planet. Launched June 14, 1967.

SPACE MILESTONE: *VOSTOK 5* (USSR). June 14. Lieutenant Colonel Valery Bykovsky orbits Earth 81 times, 2,046,000 miles. Landed June 19. Launched June 14, 1963.

June 14-15 ☆ *Chase's 1995 Calendar of Events* ☆

UNIVAC COMPUTER: BIRTHDAY. June 14. Univac 1, the world's first commercial computer, designed for the US Bureau of the Census, was unveiled, demonstrated and dedicated at Philadelphia, PA, June 14, 1951. Though this milestone of the computer age was the first commercial electronic computer, it had been preceded by ENIAC (Electronic Numeric Integrator and Computer) completed under the supervision of J. Presper Eckert, Jr, and John W. Mauchly, at the Moore School of Electrical Engineering, University of Pennsylvania, in 1946.

US ARMY ESTABLISHED BY CONGRESS: ANNIVERSARY. June 14. Anniversary of Congressional Resolution of June 14, 1775, establishing the army as the first US military service.

WARREN G. HARDING BECOMES FIRST PRESIDENT TO BROADCAST ON RADIO. June 14. On June 14, 1922, Warren G. Harding became the first president to broadcast a message over the radio. The event was the dedication of the Francis Scott Key Memorial in Baltimore, MD. The first official government message was broadcast on Dec 6, 1923.

BIRTHDAYS TODAY

Boy George (B. George Alan O'Dowd), 34, singer (lead singer Culture Club; "Do You Really Want to Hurt Me?"), born London, England, June 14, 1961.

Marla Gibbs, 49, actress (Mary on "227"; Florence on "The Jeffersons"), born Chicago, IL, June 14, 1946.

Steffi Graf, 26, tennis player, born at Bruhl, West Germany, June 14, 1969.

Eric Heiden, 37, Olympic speed skater, born at Madison, WI, June 14, 1958.

Burl Ives (Burl Icle Ivanhoe Ives), 86, singer ("A Little Bitty Tear"), actor (Oscar *The Big Country*; *Cat on a Hot Tin Roof*), born Hunt, IL, June 14, 1909.

Dorothy McGuire, 77, actress (*A Tree Grows in Brooklyn*, *Gentlemen's Agreement*, "Rich Man, Poor Man"), born Omaha, NE, June 14, 1918.

Thomas F. (Mack) McLarty, 49, Counsel to Pres Clinton, born at Prescott, AR, June 14, 1946.

Samuel Bruce Perkins, 34, basketball player, born at Brooklyn, NY, June 14, 1961.

JUNE 15 — THURSDAY
166th Day — Remaining, 199

ARKANSAS: ADMISSION DAY. June 15. Became 25th state on this day in 1836.

CANADA: SAM STEELE DAYS. June 15-18. Cranbrook, British Columbia. Parade, Lord Strathcona Musical Ride, Sweetheart Pageant, banquet and ball, loggers' sports, elephant hunt, railway museum tours and much more. Est attendance: 19,000. For info: Laura Kennedy, Sam Steele Society, PO Box 115, Cranbrook, BC, Canada V1C 4H6. Phone: (604) 426-4161.

CORPUS CHRISTI. June 15. Roman Catholic festival celebrated in honor of the Eucharist. A solemnity observed on the Thursday following Trinity Sunday since 1246. In the US Corpus Christi is celebrated on the Sunday following Trinity Sunday. See also: "Corpus Christi (US Observance)" (June 18).

FIRST FATAL AVIATION ACCIDENT: ANNIVERSARY. June 15. Two French aeronauts, Jean Francois Pilatre de Rozier and P.A. de Romain, attempting to cross the English Channel from France to England in a balloon, were killed June 15, 1785, when their balloon caught fire and crashed to the ground. Pilatre de Rozier, the first man to fly, thus became a fatality in the first fatal accident in aviation history.

FORT UNION TRADING POST RENDEZVOUS. June 15-18. Williston, ND. Re-creation of the fur trade era. Tomahawk throws, buffalo chip tosses, pit sawing, frying pan throws, blacksmith and craft demonstrations. Sponsor: National Park Service, Fort Union Trading Post NHS, Buford Rte, RR3, PO Box 71, Williston, ND 58801. Est attendance: 6,000. For info: North Dakota Tourism, Liberty Memorial Bldg, State Capitol Grounds, Bismarck, ND 58505. Phone: (800) 435-5663.

GRIEG, EDVARD: BIRTH ANNIVERSARY. June 15, 1843. Pianist, composer, conductor, and teacher, the first Scandinavian to compose nationalistic music. Born at Bergen, Norway, and died there Sept 4, 1907.

HUG HOLIDAY. June 15-22. To honor, recognize and express our appreciation for one another through the simple form of a hug. Hug Pledge Day June 15 and Hug Holiday Week June 15-22. 12th annual observance. Phone: (714) 832-HUGS. Sponsor: Jo Lindberg, Hugs for Health Foundation, PO Box 1704, Tustin, CA 92681.

JACKSON, RACHEL DONELSON ROBARDS: BIRTH ANNIVERSARY. June 15. Wife of Andrew Jackson, 7th president of the US, born at Halifax County, NC, June 15, 1767. Died Dec 22, 1828.

MAGNA CARTA DAY: 780th ANNIVERSARY. June 15. Anniversary of King John's sealing, in 1215, of the Magna Carta "in the meadow called Ronimed between Windsor and Staines on the fifteenth day of June in the seventeenth year of our reign." This document is regarded as the first charter of English liberties and one of the most important documents in the history of political and human freedom as understood today. Four original copies of the 1215 charter survive.

MARATHON READING OF *ULYSSES*. June 15-16. Irish Times Pub, Washington, DC. Annual marathon reading of James Joyce's *Ulysses* starts around 11 AM on June 15 and ends in the early morning of June 16—the same as the setting of the novel. For info: Irish Times Pub, 14 F St NW, Washington, DC 20001. Phone: (202) 543-5433.

MISS LOUISIANA PAGEANT. June 15-17. Monroe Civic Center, Monroe, LA. Preliminaries followed by the main event of crowning the new Miss Louisiana. For info: Judy M. Lowentritt, Tourism Sales Mgr, Monroe-W Monroe CVB, PO Box 6054, Monroe, LA 71211. Phone: (318) 329-2225.

OWN YOUR SHARE OF AMERICA. June 15. A day to promote and encourage individual investment in stocks of American companies and focus attention on individual investor's concerns and the individual's value to the securities markets. Annually, June 15. For info: Natl Assn of Investors Corp, Thomas E. O'Hara, Chmn, PO Box 220, Royal Oak, MI 48068. Phone: (810) 543-0612.

	S	M	T	W	T	F	S
June 1995					1	2	3
	4	5	6	7	8	9	10
	11	12	13	14	15	16	17
	18	19	20	21	22	23	24
	25	26	27	28	29	30	

✯ Chase's 1995 Calendar of Events ✯ June 15–16

PAPILLION DAYS. June 15–18. City Park, Papillion, NE. Carnival, arts and crafts, food, games, fireworks, kiddie parade, road rally, baby races, grand parade, adult dance, barbecue, pancake breakfast and free entertainment under a "Big Top Tent." Band concert and Ice Cream Social conclude the weekend festivities. '95 French theme for entire festival. Annually, Father's Day Weekend. Est attendance: 35,000. Sponsor: Loren Johnson, Exec Dir, Papillion Chamber of Commerce, 122 E Third St, Papillion, NE 68046. Phone: (402) 339-3050.

QUARTERLY ESTIMATED FEDERAL INCOME TAX PAYERS' DUE DATE. June 15. For those individuals whose fiscal year is the calendar year and who make quarterly estimated federal income tax payments, today is one of the due dates. (Jan 15, Apr 15, June 15 and Sept 15, 1995.)

SPACE MILESTONE: *SOYUZ 29* (USSR). June 15. Launched on June 15, 1978, cosmonauts Vladimir Kovalyonok and Alexander Ivanchenkov docked with *Salyut 6* space station on June 17. Returned to Earth in *Soyuz 31*, Nov 2, after 4½ months' stay and visits by two pairs of cosmonauts while there. Return landing in Kazakhstan, USSR.

SUMMERFEST ART FAIRE. June 15–17. Tabernacle Square, Logan, UT. A celebration of the arts. Events include live musical theater, a fine arts exhibit, benefit evening and a three-day art faire, featuring various arts demonstrations, exhibiting artists and artisans, jazz festival and children's activities. Annually, the third weekend in June. Est attendance: 20,000. For info: Cache Valley Health Care Foundation, c/o Logan Regional Hospital, 1400 N 500 East, Logan, UT 84321. Phone: (801) 750-5430.

SUPERMAN CELEBRATION. June 15–18. Metropolis, IL. Weekend full of Super activities. Antique and classic car show, bicycle ride, fun run, Superman drama, flea market, entertainment, museum and collector stamp. Est attendance: 52,000. For info: Massac County Chamber of Commerce, PO Box 188, Metropolis, IL 62960. Phone: (618) 524-2714.

SUWANNEE RIVER GOSPEL JUBILEE. June 15–17. Spirit of the Suwannee, Live Oak, FL. Great gospel music by America's top gospel artists in an outdoor concert setting. For info: Jean Cornett, Rt 1, Box 98, Live Oak, FL 32060. Phone: (904) 364-1683.

TWELFTH AMENDMENT TO US CONSTITUTION RATIFIED. June 15. The 12th Amendment to the Constitution was ratified on June 15, 1804. It changed the method of electing the president and vice-president after a tie in the electoral college during the election of 1800. Rather than each elector voting for two candidates with the candidate receiving the most votes elected president and the second place candidate elected vice-president, each elector was now required to designate his choice for president and vice-president, respectively.

***UNTO THESE HILLS* CHEROKEE INDIAN DRAMA.** June 15–Aug 26. Cherokee Indian Reservation, Cherokee, NC. Nightly (except Sundays), to portray the history of the Eastern Band of Cherokee Indians. Est attendance: 90,000. Sponsor: Cherokee Historical Assn, Margie Douthit, PR, PO Box 398, Cherokee, NC 28719. Phone: (704) 497-2111.

US OPEN (GOLF) CHAMPIONSHIP. June 15–18. Shinnecock, Hills Country Club, Southampton, NY. Sponsor: US Golf Assn, Championship Dept, Golf House, Far Hills, NJ 07931. Phone: (908) 234-2300.

BIRTHDAYS TODAY

Jim Belushi, 41, actor ("NBC's Saturday Night Live," *Men at Work*), born Chicago, IL, June 15, 1954.
Wade Boggs, 37, baseball player, born at Omaha, NE, June 15, 1958.
Brett Morgan Butler, 38, baseball player, born at Los Angeles, CA, June 15, 1957.
Mario M. Cuomo, 63, Governor of New York (D), up for reelection Nov '94, born at Queens, NY, June 15, 1932.
Helen Hunt, 32, actress ("Mad About You," *Peggy Sue Got Married*), born Los Angeles, CA June 15, 1963.

Waylon Jennings, 58, singer ("Amanda," "Luckenbach, Texas"), born Littlefield, TX, June 15, 1937.
Saul Steinberg, 81, artist, cartoonist, born at Rimnicu-Sarat, Romania, June 15, 1914.

JUNE 16 — FRIDAY
167th Day — Remaining, 198

BAYOU BOOGALOO AND CAJUN FOOD FESTIVAL. June 16–18. Town Point Park, Norfolk, VA. Town Point Park transforms itself into the big bayou complete with hot cajun and zydeco music, spicy foods and even hotter dancing and entertainment. Est attendance: 25,000. For info: Festevents, Promotions Dir, 120 W Main St, Norfolk, VA 23510.

BLOOMSDAY. June 16. Anniversary of events in Dublin (June 16, 1904) recorded in James Joyce's *Ulysses*, whose central character is Leopold Bloom.

BUTCH CASSIDY OUTLAW TRAIL RIDE. June 16–18. Diamond Mountain, Vernal, UT. Three days of historic trails, great food, programs around the camp fire. Bring your own camper, tent, sleeping bag, horse and gear. Two short rides and one full-day ride through beautiful scenery. What every horse lover wants to do on his horse. Deadline for 1996 Centennial Ride is May 1, 1996. Sponsors: Dinosaur Travel Board, Uintah Arts Council, USU, IGA, McDonalds. For info call (800) 477-5559 or locally (801) 789-6932. Est attendance: 130. For info: Dinaland Travel, 25 E Main, Vernal, UT 84078.

CANADA NORTH: THE YELLOWKNIFE MIDNIGHT GOLF CLASSIC. June 16–17. Yellowknife, Northwest Territories. Thieving ravens stealing well-placed chip shots off the sand "greens" is one of the legendary hazards awaiting golfers at this all-night event. Join local golf nuts and visiting celebrities in this rollicking social event that starts at 1:00 PM and ends around 6:00 AM Saturday. Annually, the Friday closest to June 21. Est attendance: 250. For info: Yellowknife Golf Club, Secy, Box 388, Yellowknife, NWT, Canada X1A 2N3. Phone: (403) 920-5647.

CANADA: RAVEN MAD DAZE. June 16. Yellowknife, Northwest Territories. Entertainment in the streets and late night sales to celebrate the summer solstice. Fax: (403) 920-4640. Est attendance: 10,000. For info: Chamber of Commerce, #6 4807 49th St, Yellowknife, NWT, Canada X1A 3T5. Phone: (403) 920-4944.

CITY STAGES MUSIC FESTIVAL. June 16–18. Linn Park, Birmingham, AL. Music and heritage festival featuring local and nationally known talent on 11 outdoor stages, folklife festival, children's festival, jazz camp, songwriter's workshop and classical music oasis. Annually, Father's Day weekend. Est attendance: 235,000. For info: City Stages, PO Box 2266, Birmingham, AL 35201. Phone: (205) 251-1272.

CZECH DAYS. June 16–17. Tabor, SD. Czechs dressed in their festive costumes gather with people from all parts of the world in this gala celebration. Fine Czech foods, dancing, music and entertainment. Est attendance: 15,000. For info: Tabor Area Chamber of Commerce, Inc, Box 21, Tabor, SD 57063. Phone: (605) 463-2476.

DER SCHNITZER. June 16–18. Amana, IA. A woodcraft show with displays, products, demonstrations, equipment, supplies, entertainment, food and fellowship. Annually, the third weekend in June. Sponsors: Personalized Wood Products, Inc, and Millstream Brewing Co. Est attendance: 3,000. For info: R.C. Eichacker, Coord, Box 193, Amana, IA 52203. Phone: (319) 622-3100.

June 16 ☆ Chase's 1995 Calendar of Events ☆

DOLLARS AGAINST DIABETES (DAD'S) DAY. June 16-18. DAD's Day is a national fundraising event conducted in more than 300 cities to help the Diabetes Research Institute find a permanent cure for the disease. Annually, on Father's Day weekend. Sponsor: Building and Construction Trades Dept of the AFL-CIO. Est attendance: 30,000. For info: Bob Bonitati, Sr, VP, The Kamber Group, 1920 L St NW, Ste 700, Washington, DC 20036. Phone: (800) 804-3237.

FUN FLIGHT. June 16-18. (Tentative; call for firm info.) Model Hobby Park, Alamagordo, NM. Friday—a workshop where kids (no age limit) build model rockets (which they get to keep); on Saturday they launch them in the 4th annual Fun Flight. Phone in New Mexico: (505) 437-2840. Est attendance: 250. For info: Bob Turner, Space Center, PO Box 533, Alamagordo, NM 88311-0533. Phone: (800) 545-4021.

GOLDEN RAINTREE FESTIVAL. June 16-18. New Harmony, IN. Midway, craft demonstrations, parades, chicken dinner, craft show and flea market. Times vary. Annually, the third Friday-Sunday in June. Est attendance: 3,000. For info: Julie Rutherford, Special Projects Coord, PO Box 579, New Harmony, IN 47631. Phone: (812) 682-3730.

GREAT AMERICAN BRASS BAND FESTIVAL. June 16-18. Centre College Campus, Danville, KY. Brass bands and ensembles from throughout the country in concert Saturday and Sunday. Free to the public. Est attendance: 35,000. Sponsor: George Foreman, Chair, Great American Brass Band Festival, PO Box 429, Danville, KY 40423. Phone: (606) 236-4692.

GRIFFIN, JOHN HOWARD: 75th BIRTH ANNIVERSARY. June 16. American author and photographer deeply concerned about racial problems in US. To better understand blacks in the American South, Griffin blackened his skin by the use of chemicals and ultraviolet light, keeping a journal as he traveled through the South, resulting in his best-known book, *Black Like Me*. Born at Dallas, TX, June 16, 1920. Died at Fort Worth, TX, Sept 9, 1980.

HARBORFEST. June 16-18. Nantucket Island, MA. Island-wide activities. Blessing of the fleet, boat parade, children's pirate parade, best chowder contest, windsurfing, kayak events, boat tour, marine exhibitions, demonstrations and tours. Sponsor: Nantucket Island Chamber of Commerce, 48 Main St, Nantucket, MA 02554. Phone: (508) 228-1700.

INTERNATIONAL WIZARD OF OZ CLUB CONVENTION. June 16-18. Holiday Inn, Rosemont, IL. Est attendance: 150. Sponsor: Wizard of Oz Club, Inc, Fred M. Meyer, Secy, 220 N 11th St, Escanaba, MI 49829.

LADIES' DAY INITIATED IN BASEBALL: ANNIVERSARY. June 16. The New York Giants hosted the first Ladies' Day baseball game on June 16, 1883. Both escorted and unescorted ladies were admitted to the game free.

LAKEFRONT FESTIVAL OF ARTS. June 16-18. Milwaukee Art Museum, Milwaukee, WI. Juried art fair with 185 artists from around the country. Artist demonstrations, children's activities, food and more. Fundraiser for Milwaukee Art Museum. Museum is open admission-free during the festival. Est attendance: 50,000. For info: Friends of Art, Beth Hoffman, 750 N Lincoln Memorial Dr, Milwaukee, WI 53202. Phone: (414) 224-3850.

	S	M	T	W	T	F	S
June					1	2	3
1995	4	5	6	7	8	9	10
	11	12	13	14	15	16	17
	18	19	20	21	22	23	24
	25	26	27	28	29	30	

LAST DUSKY SEASIDE SPARROW: DEATH ANNIVERSARY. June 16. The last survivor of dusky seaside sparrows, whose habitat was a 10-mile stretch of marshland on Florida's east coast (near Titusville), died at age 12, on June 16, 1987. The last specimen, named "Orange Band," lived its last days in a cage at Walt Disney World. For possible experimental cloning its heart and lungs were frozen and preserved.

LOUDON CAMEL CLASSIC. June 16-18. NH International Speedway, Loudon, NH. 72nd anniversary running of the oldest AMA grand National Championship Road Race in America. Gate admission, but advance tickets available at a savings. Est attendance: 40,000. For info: New Hampshire Intl Speedway, PO Box 7888, Loudon, NH 03301. Phone: (603) 783-4931.

OLD FORT RIVER FESTIVAL. June 16-18. Fort Smith, AR. Multi-faceted festival including juried arts and crafts, large all-free children's area, 2 stages with continuous entertainment and sporting events including men's 3-on-3 basketball tourney, 2-mile walk/run, 30-mile bike ride and bike critierium. Annually, Father's Day weekend. Sponsor: Old Fort River Festival, Inc. Est attendance: 50,000. For info: Kathy Liggett, Exec Dir, Old Fort River Fest, PO Box 3025, Fort Smith, AR 72913. Phone: (501) 783-6363.

QUECHEE BALLOON FESTIVAL AND CRAFTS FAIR. June 16-18. Village Green, Quechee, VT. 20 hot air balloons, juried crafts fair (over 60 craftspeople), balloon rides, old-fashioned BBQ, live music and entertainment, 5-library book sale, volleyball tournament, children's tent, food court and much more. 16th annual fair. Annually, the third weekend in June. Est attendance: 13,500. Sponsor: Quechee Chamber of Commerce, PO Box 106, Quechee, VT 05059. Phone: (800) 295-5451.

RENO RODEO. June 16-25. Reno, NV. Top professional cowboys ride in one of the west's richest rodeos. Ticket Number: (800) 842-7633. Est attendance: 120,000. For info: Reno Rodeo Assn, Box 12335, Reno, NV 89510. Phone: (702) 329-3877.

RIVERBEND FESTIVAL 1995. June 16-24. Ross' Landing, Chattanooga, TN. Chattanooga's urban celebration presenting the arts, fantastic sporting events, children's activities and musical events featuring nationally acclaimed artists. Nine days of attractions celebrating Ross's Landing Park. Daily activities including the unique Bessie Smith Strut. On Monday, June 19, a New Orleans-style block party will be held with entertainment on several stages down M.L. King Blvd. One of the highlights of the festival, it celebrates the cultural background in Chattanooga. Lots of food concessions from Cajun and ribs to Mexican and more. Est attendance: 500,000. For info: Friends of the Festival, Attn: Carla Sanderlin, PO Box 886, Chattanooga, TN 37401. Phone: (615) 756-2211.

ROUND BARN FESTIVAL. June 16-18. Rochester, IN. Crafts, foods, kiddy rides and contests, free entertainment all day, 10-10. Tours of round barns. Fulton County Museum and Round Barn Museum open 9-5. Indian crafts demonstrated. Parade 8 PM Friday. Regatta, Balloon Assention, Golf Tournament, Karaoke, and much more. Annually, the third weekend in June. Est attendance: 6,000. For info: Round Barn Festival Inc, 617 Main St, Rochester, IN 46975. Phone: (219) 223-6773.

SPACE MILESTONE: FIRST WOMAN IN SPACE: VOSTOK 6 (USSR): ANNIVERSARY. June 16. Valentina Tereshkova, 26, former cotton mill worker, born on collective farm near Yaroslavl, USSR, became the first woman in space when her 10,300-lb spacecraft, *Vostok 6*, took off from the Tyuratam launch site on June 16, 1963. She manually controlled *Vostok 6* during the 70.8-hour flight through 48 orbits of Earth, and landed by parachute (separate from her cabin) on June 19, 1963. In November 1963 she married cosmonaut Andrian Nikolayev, who had piloted *Vostok 3* through 64 earth orbits, Aug 11-15, 1962. Their child Yelena (1964) was the first born to space-traveler parents.

SQUEEZE PLAY FIRST USED: ANNIVERSARY. June 16. In a game against Princeton, two members of the Yale baseball team executed the first squeeze play on June 16, 1894. The squeeze play is executed with a runner on third, and less than two outs. The batter bunts the ball slowly to the infield, allowing the runner on third to score safely. Clark Griffith, manager of the New York Highlanders in the American League, introduced the play into the major leagues in 1904.

TIMBERFEST. June 16–18. Woodruff, WI. A celebration of the old logging days in Northern Wisconsin. Lumberjack shows, arts & crafts, petting zoo. Est attendance: 13,000. For info: Al Hanley, Minocqua-ArborVitae-Woodruff Area C of C, PO Box 1006, Minocqua, WI 54548. Phone: (800) 446-6784.

UNITED CEREBRAL PALSY'S CASUAL DAY. June 16. In return for a small contribution to the United Cerebral Palsy fund, employers across the country permit their employees to come to work in casual clothes. Funds raised help UCP affiliates provide programs and services to people with cerebral palsy and other disabilities. Annually, the Friday before the first day of summer. Est attendance: 500,000. For info: United Cerebral Palsy Assn (UCPA), Special Events Dept, 1522 K St NW, Ste 1112, Washington, DC 20005. Phone: (800) 872-5827.

WONAGO WORLD CHAMPIONSHIP RODEO. June 16–18. Milwaukee, WI. More than 150 of North America's best cowboys and cowgirls compete in six professionally sanctioned contests ranging from bronc riding to bull riding for top prize and world championship points. Theme of the production changes annually. Additional attractions include special animal trainers, trick ropers, trick riders and professional clowns. Held in conjunction with West Allis Western Days and North America's Largest Horse Drawn Parade on June 16. Est attendance: 10,000. For info: W. Bruce Lehrke, Pres, Longhorn Rodeo, Natl HQ, PO Box 70159, Nashville, TN 37207. Phone: (615) 876-1016.

BIRTHDAYS TODAY

Billy "Crash" Craddock, 56, singer ("Don't Destroy Me," "Ruby, Baby"), born at Greensboro, NC, June 16, 1939.
Roberto Duran, 44, former boxer, born at Chorillo, Panama, June 16, 1951.
Neil Goldschmidt, 55, former Governor of Oregon (D), born at Eugene, OR, June 16, 1940.
◈ *Katharine Graham*, 78, newspaper executive (*The Washington Post*), born at New York, NY, June 16, 1917.
Joyce Carol Oates, 57, writer (*Triumph of the Spider Monkey, The Time Traveler*), born at Lockport, NY, June 16, 1938.
Irving Penn, 78, photographer, born at Plainfield, PA, June 16, 1917.
Wayne ("Tree") Rollins, 40, basketball player, born at Winter Haven, FL, June 16, 1955.
Erich Segal, 58, author (*Acts of Faith; Man, Woman and Child*), born at Brooklyn, NY, June 16, 1937.
Joan Van Ark, 52, actress (Val in "Knots Landing"), born at New York, NY, June 16, 1943.

JUNE 17 — SATURDAY
168th Day — Remaining, 197

AFRICA'S LEGACY IN MEXICO: PHOTOGRAPHS BY TONY GLEATON. June 17–July 30. Witle Museum of Science and History, San Antonio, TX. See Feb 11 for full description. Call or write the Smithsonian for other dates and venues. For info: Smithsonian Institution Traveling Exhibition Service, 1100 Jefferson Dr SW, Ste 3146, Washington, DC 20560. Phone: (202) 357-2700.

◈ **BORMANN, MARTIN: BIRTH ANNIVERSARY.** June 17. Known as Hitler's "Evil Genius," Martin Bormann was born at Halberstadt, Saxony, Germany, June 17, 1900. Bormann was named secretary to the Führer in April 1943, and in that capacity became the guard at Hitler's door, preventing all from entering unless he wished them to. It has been suggested that some of Hitler's worst mistakes might actually have been policies put forth by Bormann. Martin Bormann was last seen before he escaped from Hitler's bunker May 1, 1945. He was reported killed on the street outside. Bormann was declared dead by the German government in 1973.

BUNKER HILL DAY. June 17. Suffolk County, MA. Legal holiday in the county in commemoration of the Battle of Bunker Hill that took place on this day in 1775.

CANADA: BAZAART. June 17. MacKenzie Art Gallery, Regina, Saskatchewan. Juried outdoor art show and sale with 150 booths. Stage entertainment, pottery, stained glass and oil and watercolor paintings. Crafts and food. Est attendance: 20,000. For info: Tourism Saskatchewan, Saskatchewan Trade and Conv Center, 1919 Saskatchewan Dr, Regina, Sask, Canada S4P 3V7. Phone: (800) 667-7191.

CANADA: CELEBRATION '95/MULTICULTURAL FESTIVAL. June 17–19. Halifax, Nova Scotia. 3 days of cultural events to include 9 tents housing exhibits, food booths, children's tent, performances, beer tent with live bands, fashion show and parade. 11th annual. Annually, the third weekend in June. Est attendance: 40,000. For info: Barbara Campbell, Exec Dir, Multicultural Assn of Nova Scotia, 1809 Barrington St, Ste 901, Halifax, NS, Canada B3J 3K8. Phone: (902) 423-6534.

CHAUTAUQUA FESTIVAL IN THE PARK. June 17–25. Elizabeth Brown Memorial Park, Wytheville, VA. Parade, music, education, performing arts, art shows, photography, quilt and needlework, creative writing contests, antique show and sale, flower show, hot air balloons, children's activities, cooking workshops, 5K run and much more. Free. Annually, in June. Est attendance: 100,000. For info: June Kerss, Wythe Arts Council, PO Box 911, Wytheville, VA 24382. Phone: (703) 228-4833.

ENGLAND: BROADSTAIRS DICKENS FESTIVAL. June 17–24. Broadstairs, Kent. A week of festivities commemorating the association of the famous novelist Charles Dickens to the town of Broadstairs. For info: Mrs Lee Ault, Honorary Festival Organiser, Rooftops, 58 High St, Broadstairs, Kent, England CT10 1JT.

FAMILY FROLIC. June 17–18. Roscoe Village, Coshocton, OH. A festival of fun and games dedicated to the child in all of us. Games and activities for the entire family. Est attendance: 5,000. For info: Roscoe Village Fdtn, 381 Hill St, Coshocton, OH 43812. Phone: (800) 877-1830.

FARM DAYS WEEKEND. June 17–18. Wheeling, WV. A celebration of farming and old traditions from yesteryear—music, dance, arts, crafts and food. Sponsor: Oglebay's Good Children's Zoo. Est attendance: 4,500. For info: Good Children's Zoo, John Hargleroad, Dir, Oglebay Park, Wheeling, WV 26003. Phone: (304) 243-4030.

FISHING HAS NO BOUNDARIES—PIERRE. June 17–18. Lake Oahe, Pierre, SD. A two-day fishing experience for disabled persons. Any disability, age, sex, race, etc, eligible. Fishing with experienced guides attended by 150 participants and 400 volunteers. For info: Fishing Has No Boundaries, Pierre Area C of C, PO Box 548, Pierre, SD 57501. Phone: (800) 962-2034.

June 17 ☆ Chase's 1995 Calendar of Events ☆

GERMANY: DAY OF UNITY. June 17. Public holiday throughout Germany.

HOG DAY. June 17. Hillsborough, NC. Barbecue cooking contest, arts and crafts. A variety of entertainment. Est attendance: 16,500. Sponsor: Hillsborough Area Chamber of Commerce, 150 E King St, Hillsborough, NC 27278. Phone: (919) 732-8156.

HOOPER, WILLIAM: BIRTH ANNIVERSARY. June 17. Signer of the Declaration of Independence, born at Boston, MA, June 17, 1742. Died Oct 14, 1790.

HOT AIR BALLOON SHOW AND AERONAUTICAL EXTRAVAGANZA. June 17–18. Falls City, NE. 3 balloon lift-offs at airport (Brenner Field) plus ethnic entertainment and food in town all day Saturday, and craft show; model airplanes, parachutists, fly-in breakfast, stunt pilot, food concessions at Brenner Field Sunday. Annually, mid-June. Est attendance: 2,000. Sponsor: Chamber of Commerce, PO Box 146, Falls City, NE 68355. Phone: (402) 245-4228.

ICELAND: INDEPENDENCE DAY. June 17. Anniversary of founding of republic is major festival, especially in Reykjavik. Parades, competitions, street dancing.

INTERNATIONAL FREEDOM FESTIVAL. June 17–July 4. Detroit, MI, and Windsor, Ontario, Canada. A transborder festival featuring more than 100 events on both sides of the Detroit River. The Hudson's Freedom Festival Fireworks is claimed to be the largest pyrotechnics display in North America. Est attendance: 4,500,000. For info: Iana Guida, Dir, PR, The Parade Co, 9600 Mt Elliott, Detroit, MI 48211. Phone: (313) 923-8259.

LAKESTRIDE HALF-MARATHON. June 17. Ludington, MI. Half-marathon race that begins in downtown Ludington and takes runners along a scenic course through the wooded trails and sand dunes of Lake Michigan at Ludington State Park. Annually, the day before Father's Day. Est attendance: 10,000. For info: Ludington Area Conv and Visitors Bureau, Sue Brillhart, Exec Dir, 5827 W US 10, Ludington, MI 49431. Phone: (800) 542-4600.

MIDNIGHT SUN FESTIVAL. June 17–18. Nome, AK. A celebration of the summer solstice, which is when Nome experiences the midnight sun with more than 22 hours of direct sunlight. The official "Midnight Sun Button" is designed by the school children of Nome. The festival usually includes a street dance, Eskimo dances and barbecue. Est attendance: 500. For info: Nome Chamber of Commerce, PO Box 240, Nome, AK 99762. Phone: (907) 443-5535.

NATIONAL JOUSTING HALL OF FAME JOUSTING TOURNAMENT. June 17. Natural Chimneys Regional Park, Mt Solon, VA. Ring jousting for novice, amateur, semi-professional and professional jousters. Medieval jousting and Robin Hood skit. Annually, the third Saturday in June. Est attendance: 500. For info: Upper Valley Regional Park Authority, Box 478, Grottoes, VA 24441. Phone: (703) 350-2510.

NATIONAL JUGGLING DAY. June 17. Juggling clubs affiliated with the International Jugglers Association in cities all over North America hold local festivals to demonstrate, teach and celebrate their art. Sponsor: Intl Jugglers' Assn, Paul Kyprie, Mktg Dir, 650 Hidden Valley #315, Ann Arbor, MI 48104. Phone: (313) 994-0368.

NEW OXFORD FLEA MARKET AND ART AND CRAFT SHOW. June 17. New Oxford, PA. Arts, crafts, antiques and flea market. Annually, the third Saturday in June. Est attendance: 20,000. For info: Gettysburg Travel Council, 35 Carlisle St, Gettysburg, PA 17325. Phone: (717) 334-6274.

June 1995

S	M	T	W	T	F	S
				1	2	3
4	5	6	7	8	9	10
11	12	13	14	15	16	17
18	19	20	21	22	23	24
25	26	27	28	29	30	

OPERA FESTIVAL OF NEW JERSEY. June 17–July 16. Kirby Arts Center, The Lawrenceville School, Lawrenceville, NJ. A professional organization founded in 1984 to present opera productions, in English, of high artistic caliber to audiences of all ages and backgrounds. 1995 season will feature the operas *Don Pasquale, La Traviata* and the world premiere of Peter Westergaard's *The Rarest Progress*. Est attendance: 12,500. For info: Opera Festival of New Jersey, 55 Princeton-Hightstown Rd, Princeton Jnct, NJ 08550. Phone: (609) 936-1505.

POCOMOKE RIVER CANOE CHALLENGE. June 17. Pocomoke River, Snow Hill, MD. 12½-mile race for canoes and kayaks. Starts at 8 AM. Annually, the third Saturday in June. For info: Paula Sparrow, Pocomoke City, PO Box 29, Pocomoke City, MD 21851. Phone: (410) 957-1333.

PRINTERS ROW BOOK FAIR. June 17–18. Chicago, IL. More than 150 booksellers and publishers from all over the US and Canada fill the street with new, used, rare and antique books for sale; demonstrations of paper marbling, paper making and book binding; food, music and readings by authors and poets. Est attendance: 70,000. For info: Near South Planning Bd, 1727 S Indiana, #104, Chicago, IL 60616. Phone: (312) 987-1980.

SEAFOOD IN SEASIDE. June 17–18. Seaside Heights, NJ. 4th annual seafood festival on the beautiful bayfront, 10–6. Seafood vendors, foods for all, arts and crafts, entertainment. Est attendance: 30,000. For info: Seaside Business Assn, PO Box 98, Seaside Heights, NJ 08751. Phone: (908) 793-1510.

SOUTH AFRICA REPEALS LAST APARTHEID LAW: ANNIVERSARY. June 17. On June 17, 1991, the Parliament of South Africa repealed the Population Registration Act, removing the law that was the foundation of apartheid. The law, first enacted in 1950, required the classification by race of all South Africans at birth. It established four compulsory racial categories: white, mixed race, Asian and black. Although this marked the removal of the last of the apartheid laws, blacks in South Africa still could not vote.

STRAVINSKY, IGOR FEDOROVICH: BIRTH ANNIVERSARY. June 17. Russian composer and author, born at Oranienbaum (near Leningrad), June 17 (Gregorian), 1882. Died at New York, NY, on Apr 6, 1971.

VIRGINIA INDIAN HERITAGE FESTIVAL. June 17. Jamestown Settlement, Williamsburg, VA. Native Americans from Virginia and other states assemble to celebrate their culture with dance, storytelling and demonstrations and sales of traditional crafts and foods. Est attendance: 3,500. For info: Media Relations, Jamestown-Yorktown Fdtn, PO Box JF, Williamsburg, VA 23187. Phone: (804) 253-4838.

WATERGATE DAY. June 17. Anniversary of arrests, at Democratic Party Headquarters (in Watergate complex, Washington, DC) on June 17, 1972, which led to revelations of political espionage, threats of imminent impeachment of the president and on Aug 9, 1974, the resignation of President Richard M. Nixon.

BIRTHDAYS TODAY

Elroy "Crazylegs" Hirsch, 72, pro football Hall of Famer, born at Wausau, WI, June 17, 1923.

Guy Hunt, 62, Governor of Alabama (R), born at Holly Pond, AL, June 17, 1933.

Mark Linn-Baker, 42, actor ("Perfect Strangers," *My Favorite Year*), born St. Louis, MO, June 17, 1953.

Barry Manilow, 49, singer, songwriter ("Mandy," "I Write the Songs"), born at Brooklyn, NY, June 17, 1946.

Dean Martin (Dino Crocetti), 78, singer, actor, born at Steubenville, OH, June 17, 1917.

Joe Piscopo, 44, comedian (former "Saturday Night Live" regular), born at Passaic, NJ, June 17, 1951.

JUNE 18 — SUNDAY
169th Day — Remaining, 196

AJL '95. June 18–21. Marriott Downtown, Chicago, IL. The Association of Jewish Libraries was created in 1965 to support the production, collection, organization and dissemination of Jewish resources and library/media services. The annual convention, with three exciting and informative days of sessions on all aspects of Jewish library concerns, helps keep the Judaica librarians informed and skilled for the performance of their work while providing a forum for sharing and networking. For info: Shoshanah Seidman, Co-chair, AJL '95, Assn of Jewish Libraries, 9056 Tamaroa Terr, Skokie, IL 60076.

AMATEUR RADIO WEEK. June 18–25. To bring amateur radio to the attention of the public. Annually the week culminating in the fourth weekend of June. Sponsor: American Radio Relay League, PR Coor, 225 Main St, Newington, CT 06111. Phone: (203) 666-1541.

ASRT ANNUAL CONFERENCE. June 18–22. Conference for radiologic sciences professionals offering technical exhibits, continuing education courses, business meetings, house of delegates, scientific displays and many other special events. Est attendance: 1,200. For info: American Soc of Radiologic Technologists, 15000 Central Ave SE, Albuquerque, NM 87123-3917. Phone: (505) 298-4500.

CAHN, SAMMY: BIRTH ANNIVERSARY. June 18. Tin Pan Alley legend Sammy Cahn was born Samuel Cohen, June 18, 1913, at New York City. Cahn's initial ambition was to play the fiddle in vaudeville, but the best he did was some semi-professional work as a teenager. His first song writing hit was "Rhythm Is Our Business" which he wrote in collaboration with Saul Chaplin. He was nominated for 26 Academy Awards and won four times for "Three Coins in the Fountain" (1954), "All the Way" (1957), "High Hopes" (1959) and "Call Me Irresponsible" (1963). In the late 1940s he began working with composer Jimmy Van Heusen, and the two in essence were the personal songwriting team for Frank Sinatra for a time. Cahn wrote the greatest number of Sinatra hits, including "Love and Marriage," "The Second Time Around," "High Hopes" and "The Tender Trap." His autobiography, *I Should Care*, was published in 1974. Sammy Cahn died Jan 15, 1993, at Los Angeles.

CARPENTER ANT AWARENESS WEEK. June 18–24. Wood-destroying organisms cause Americans to spend $3.5 billion annually. This week will focus attention on the identification, biology and habits of carpenter ants, and provide consumers with information on the elimination of these costly pests. Annually, the last full week of June. For info: Jerry Batzner, Pres, Batzner Pest Management, Inc, 16700 W Victor Rd, New Berlin, WI 53151. Phone: (414) 797-4160.

CORPUS CHRISTI (US OBSERVANCE). June 18. A movable Roman Catholic celebration commemorating the institution of the Holy Eucharist. The solemnity has been observed on the Thursday following Trinity Sunday since 1246, except in the US, where it is observed on the Sunday following Trinity Sunday.

EGYPT: EVACUATION DAY. June 18. Public holiday celebrating the anniversary of the withdrawal of the British Army from the Suez Canal area of Egypt in 1954. For info: Egyptian Tourist Authority, 645 N Michigan Ave, Ste 829, Chicago, IL 60611. Phone: (312) 280-4666.

EIGHTEENTH-CENTURY WHEAT HARVEST. June 18. The Claude Moore Colonial Farm at Turkey Run, McLean, VA. Help the colonial farm family cut and bind wheat, the farmer's second most important cash crop. Light refreshment and 18th-century games. Annually, the third Sunday in June. Est attendance: 300. For info: Gretchen Brodtman, The Claude Moore Colonial Farm, Turkey Run, 6310 Georgetown Pike, McLean, VA 22101. Phone: (703) 442-7557.

★ **FATHER'S DAY.** June 18. Presidential Proclamation issued for third Sunday in June in 1966 and annually since 1971. (PL 92-278 of Apr 24, 1972.)

FATHER'S DAY. June 18. Recognition of the third Sunday in June as Father's Day occurred first at the request of Mrs John B. Dodd of Spokane, WA, on June 19, 1910. It was proclaimed for that date by the mayor of Spokane and recognized by the governor of Washington. The idea was publicly supported by President Calvin Coolidge in 1924, but not presidentially proclaimed until 1966. It was assured of annual recognition by Public Law 92-278 of Apr 1972.

FINLAND: JOENSUU FESTIVAL. June 18–24. Joensuu. International event with concerts of symphonic and popular music, and solo concerts, dance and theater performances, art exhibitions, films. Est attendance: 30,000. For info: Finnish Tourist Board, 655 Third Ave, New York, NY 10017. Phone: (212) 949-2333.

FIRST AMERICAN WOMAN IN SPACE: ANNIVERSARY. June 18. Dr. Sally Ride, 32-year-old physicist and pilot, functioned as a "mission specialist" and became the first American woman in space on June 18, 1983, when she began a six-day mission aboard the space shuttle *Challenger*. The "near perfect" mission was launched from Cape Canaveral, FL, and landed, June 24, 1983, at Edwards Air Force Base, CA. See also: "Ride, Sally Kristen: Birthday" (May 26) and "Woman in Space Anniversary" (June 16).

FOLGER, HENRY CLAY, JR: BIRTH ANNIVERSARY. June 18. American businessman and industrialist who developed one of the finest collections of Shakespeareana in the world and bequeathed it (The Folger Shakespeare Library, Washington, DC) to the American people. Born at New York, NY, on June 18, 1857. Died June 11, 1930.

FROG JUMPING CONTEST. June 18. Old Forge, NY. Annual event on Father's Day. Est attendance: 400. For info: Adirondack Assn, Bob Hall, PO Box 68, Old Forge, NY 13420. Phone: (315) 369-6983.

HARD ROCK CAFE 5K. June 18. Hard Rock Cafe, 63 W Ontario, Chicago, IL. This 5th annual 5K (3.1 miles) is a family fun run with proceeds benefitting a to-be-designated charity. The race begins and ends at Hard Rock, and all runners are invited to join post-race festivities. Annually, on Father's Day. Sponsor: Hard Rock Cafe, Chicago, IL. Est attendance: 3,800. For info: Hard Rock Cafe 5K Dir, PO Box 10597, Chicago, IL 60610-0597. Phone: (312) 527-2200.

KYSER, KAY: BIRTH ANNIVERSARY. June 18. American bandleader whose band, "Kay Kyser's Kollege of Musical Knowledge," enjoyed immense popularity in the swing era. He was born James King Kern Kyser at Rocky Mount, NC, on June 18, 1906. Kyser, a shrewd showman and performer, said he never learned to read music or play an instrument. Among his hit recordings were "Three Little Fishes" and "Praise the Lord and Pass the Ammunition," a World War II favorite. Kyser died at Chapel Hill, NC, July 23, 1985.

NOME RIVER RAFT RACE. June 18. Nome, AK. Homemade rafts paddle their way down the 1-2 mile course on the Nome River. The victorious team claims 1st place recognition and the ownership of the fur-lined Honey-Bucket which is handed down from year to year. This event draws the entire town out for a fun afternoon at Nome's largest summer event. Annually, Sunday closest to Summer Solstice. Approximate attendance: 20 rafts. For info: Bering Sea Lions Club, Box 326, Nome, AK 99762. Phone: (907) 443-5172.

June 18-19 ☆ Chase's 1995 Calendar of Events ☆

ORTHODOX FESTIVAL OF ALL SAINTS. June 18. Observed by Eastern Orthodox churches on the Sunday following Orthodox Pentecost (June 11 in 1995). Marks the end of the 18-week Triodion cycle.

ROARING CAMP RAILROAD RIDE. June 18. Roaring Camp, Felton, CA. Treat dad to a chuckwagon barbecue, model train displays and chance to win engineer-in-cab drawing. Sponsor: Jeanette Guire, Roaring Camp Depot, Felton, CA 95018. Phone: (408) 335-4400.

SEYCHELLES: CONSTITUTION DAY. June 18. National holiday.

SPACE MILESTONE: *STS-7* (US). June 18. Shuttle *Challenger*, launched from Kennedy Space Center, FL, June 18, 1983, with crew of five, including Sally K. Ride, first American woman in space, Robert Crippen, Norman Thagard, John Fabian and Frederick Houck. Landed at Edwards Air Force Base, CA, on June 24 after near-perfect 6-day mission.

WAR OF 1812: DECLARATION ANNIVERSARY. June 18. After much debate in Congress between "hawks," such as Henry Clay and John Calhoun, and "doves," such as John Randolph, Congress issued a declaration of war on Great Britain on June 18, 1812. The action was prompted primarily by Britain's violation of America's rights on the high seas and British incitement of Indian warfare on the frontier. Less than a cause for the war, but fueling the fire, was a desire by some to acquire Florida and Canada. The hostilities ended with the signing of the Treaty of Ghent on Dec 24, 1814, at Ghent, Belgium.

WAUSAU AIR SHOW AND FLY-IN. June 18. Wausau Municipal Airport, Wausau, WI. Airplanes on display, pancake breakfast and air show fly-in. Nationally famous aerobatic performers. Est attendance: 10,000. Sponsor: Wausau Airshow Committee, 715 Woods Pl, Wausau, WI 54403. Phone: (715) 848-6000.

BIRTHDAYS TODAY

Roger Joseph Ebert, 53, film critic ("Siskel and Ebert"), born at Urbana, IL, June 18, 1942.
Carol Kane, 43, actress (*Hester Street, The Princess Bride*, "Taxi"), born Cleveland, OH, June 18, 1952.

	S	M	T	W	T	F	S
June					1	2	3
	4	5	6	7	8	9	10
1995	11	12	13	14	15	16	17
	18	19	20	21	22	23	24
	25	26	27	28	29	30	

Donald Keene, 73, literary critic, translator, educator, born at New York, NY, June 18, 1922.
Paul McCartney, 53, singer, songwriter (The Beatles), born Liverpool, England, June 18, 1942.
Sylvia Porter, 82, journalist, author (*Sylvia Porter's A Home of Your Own*), born at Patchogue, NY, June 18, 1913.
John D. Rockefeller, IV, 58, US Senator (D, West Virginia), born at New York, NY, June 18, 1937.
Isabella Rossellini, 43, model, actress, born at Rome, Italy, June 18, 1952.
Tom Wicker, 69, journalist, author (*One of US: Richard Nixon & the American Dream*), born at Hamlet, NC, June 18, 1926.

JUNE 19 — MONDAY
170th Day — Remaining, 195

BASCOM, EARL W.: BIRTH ANNIVERSARY. June 19, 1906. Rodeo showman and pioneer, Earl W. Bascom was born at Vernal, UT. During his career he developed the first side-delivery rodeo chute (1916), the first hornless bronc saddle (1922) and the first one-handed bare-back rigging (1924). He produced the first rodeo in Mississippi and also produced the first rodeo performed at night under electric lights (1935).

EMANCIPATION DAY IN TEXAS. June 19. In honor of the emancipation of the slaves in Texas on June 19, 1865.

FIRST RUNNING OF THE BELMONT STAKES: ANNIVERSARY. June 19. The first running of the Belmont Stakes took place on June 19, 1867 at Jerome Park, NY. The team of jockey J. Gilpatrick and his horse Ruthless finished in a time of 3:05. The Belmont Stakes continued at Jerome Park until 1889, then moved to Morris Park, NY between 1890 and 1905, and settled at Belmont Park, NY in 1906 where it has continued to the present day. The Belmont stakes is the oldest of horse racing's Triple Crown.

FORTAS, ABE: BIRTH ANNIVERSARY. June 19. Abe Fortas was born June 19, 1910, at Memphis, TN. He was appointed to the Supreme Court by President Lyndon Johnson in 1965. Prior to his appointment he was known as a civil libertarian, having argued cases for government employees and other individuals accused by Senator Joe McCarthy of having communist affiliations. He argued the 1963 landmark Supreme Court case of *Gideon v Wainwright*, which established the right of indigent defendants to free legal aid in criminal prosecutions. In 1968, he was nominated by Johnson to succeed Chief Justice Earl Warren, but his nomination was withdrawn after much conservative opposition in the Senate. In 1969 Fortas became the first Supreme Court Justice to be forced to resign after revelations about questionable financial dealings were made public. Fortas died Apr 5, 1982, at Washington, DC.

GARFIELD: BIRTHDAY. June 19. America's favorite lasagna-loving cat is 17. *Garfield*, a modern classic comic strip created by Jim Davis first appeared on June 19, 1978, and has brought laughter to millions. For info: Paws, Inc, Kim Campbell, 5440 E Co Rd 450 N, Albany, IN 47320.

GEHRIG, LOU: BIRTH ANNIVERSARY. June 19. Baseball great Henry Louis Gehrig (lifetime batting average of .341), who played in seven World Series, was born at New York, NY, on June 19, 1903, and died there June 2, 1941, from the degenerative muscle disease alateral sclerosis, which has become known as Lou Gehrig's Disease.

◆ **GLOBAL YOUTH PROJECT.** June 19–30. San Francisco, CA. In commemoration of the 50th anniversary of the United Nations, Swedish filmmaker Staffan Hildebrand profiles the lives of young artists from around the world. The film is interspersed with music and dance performance and is intended to appeal to youth and their families to foster greater appreciation across cultures and across generations. Sponsors: UN50 Secretariat-NY and Government of Sweden. Contact: Matthew Inge, (415) 434-5310. For info: UN50 Committee, 312 Sutter St, Ste 610, San Francisco, CA 94108. Phone: (415) 989-1995.

☆ Chase's 1995 Calendar of Events ☆ June 19-20

HUBBARD, ELBERT: BIRTH ANNIVERSARY. June 19. Born at Bloomington, IL, on June 19, 1856, Elbert Green Hubbard, aspiring American author and craftsman, founded the Roycroft Press at East Aurora, NY. Best known of his writings were *A Message to Garcia* and a series of essays entitled *Little Journeys*. Hubbard lost his life with the sinking of the *Lusitania*, May 7, 1915.

JUNETEENTH: 130 YEARS. June 19. Celebrated in Texas and other parts of the Deep South in memory of this day in 1865 when Union General Granger proclaimed the slaves of Texas free.

◆ **A MOMENT IN HISTORY.** June 19-30. War Memorial Veterans Building, San Francisco, CA. An exhibit of archival materials on the 1945 United Nations Conference on International Organization. For info: UN50 Committee, 312 Sutter St, Ste 610, San Francisco, CA 94108. Phone: (415) 989-1995.

MOON PHASE: LAST QUARTER. June 19. Moon enters Last Quarter phase at 6:01 PM, EDT.

ROANOKE VALLEY HORSE SHOW. June 19-25. Salem Civic Center, Salem, VA. Multi-breed horse show. Est attendance: 30,000. For info: Salem Civic Ctr, John Saunders, Box 886, Salem, VA 24153. Phone: (703) 375-3004.

ROSENBERG EXECUTION: ANNIVERSARY. June 19. Anniversary of the electrocution of the only married couple ever executed together in the US. Julius (35) and Ethel (37) Rosenberg were executed, for espionage, at Sing Sing Prison, Ossining, NY, June 19, 1953. Time for the execution was advanced several hours to avoid conflict with the Jewish Sabbath.

SPACE MILESTONE: *ARIANE* (ESA). June 19. Launched from Kourou, French Guiana, June 19, 1981. *Ariane* carried two satellites into orbit (*Meteostat 2*, ESA weather satellite, and *Apple*, a geostationary communications satellite for India, to be stationed over Sumatra).

WORLD SAUNTERING DAY. June 19. A day to revive the lost art of Victorian sauntering and discourage jogging, lollygagging, sashaying, fast walking and trotting. [Originated by the late W.T. Rabe of Saute Ste. Marie, MI.]

BIRTHDAYS TODAY

Paula Abdul, 33, singer, dancer, choreographer ("Forever Your Girl"), born at Los Angeles, CA, June 19, 1962.

Alan Cranston, 81, former US Senator (D, California), born at Palo Alto, CA, June 19, 1914.

Charles Gwathmey, 57, architect, born at Charlotte, NC, June 19, 1938.

Howell Thomas Heflin, 74, US Senator (D, Alabama), born at Poulan, GA, June 19, 1921.

Pauline Kael, 76, critic, born at Sonoma County, CA, June 19, 1919.

Nancy Marchand, 67, actress (Mrs. Pynchon on "Lou Grant"; *Brain Donors*), born at Buffalo, NY, June 19, 1928.

Mildred Natwick, 87, actress (*Barefoot in the Park*, *Daisy Miller*), born at Baltimore, MD, June 19, 1908.

Marisa Pavan, 63, actress (*The Diary of Anne Frank*), born at Cagliari, Sardinia, June 19, 1932.

Phylicia Rashad, 47, actress (Claire Huxtable on "The Cosby Show"), born at Houston, TX, June 19, 1948.

Gena Rowlands, 59, actress (Adrienne Van Leyden in "Peyton Place"; *A Woman Under the Influence*), born at Cambria, WI, June 19, 1936.

Salman Rushdie, 48, author (*The Jaguar Smile*, *Midnight's Children*), born at Bombay, India, June 19, 1947.

Kathleen Turner, 41, actress (*Body Heat*, *Peggy Sue Got Married*, *Romancing the Stone*), born Springfield, MO, June 19, 1954.

Ann Wilson, 44, singer, musician (lead singer Heart), born at San Diego, CA, June 19, 1951.

JUNE 20 — TUESDAY
171st Day — Remaining, 194

ENGLAND: CITY OF LONDON FESTIVAL. June 20-July 6. London. Annual arts festival held in some of the city's most historically interesting buildings, including St. Paul's Cathedral and the Tower of London. For info: City Arts Trust, Bishopgate Hall, 230 Bishopgate, London, England EC2M 4QM.

FIRST BALLOON HONEYMOON: ANNIVERSARY. June 20. Roger Burnham and Eleanor Waring took the first balloon honeymoon on June 20, 1909, ascending at 12:40 PM in the balloon "Pittsfield." They began their trip at Woods Hole, Cape Cod, MA, and landed at 4:30 PM in an orchard at Holbrook, MA.

FIRST BANK CHARTERED BY CONGRESS. June 20. The National Bank of Philadelphia at Phildelphia, PA, was chartered on June 20, 1863.

◆ **FIRST DOCTOR OF SCIENCE DEGREE EARNED BY A WOMAN: 100th ANNIVERSARY.** June 20. On June 20, 1895, Caroline Willard Baldwin became the first woman to earn a doctor of science degree at Cornell University, Ithaca, NY.

GREAT SEAL OF THE UNITED STATES: ANNIVERSARY. June 20. Charles Thomson, first official recordkeeper of the US, submitted his report to Congress, recommending a design for the Great Seal, on June 20, 1782. The Congress adopted his report on the same day. See also: "Great Seal of the US: Anniversary" (Jan 28, July 4 and Sept 16).

◆ **REFLECTIONS ON THE FUTURE OF THE UN.** June 20-21. Herbst Theater, San Francisco, CA. This meeting of former US Ambassadors will explore the UN's changing role as a vehicle for US foreign policy. Sponsors: UN50 Committee and Southern Center for Intl Studies. Contact: Stephanie Rapp, (415) 989-1995. For info: UN50 Committee, 312 Sutter St, Ste 610, San Francisco, CA 94108. Phone: (415) 989-1995.

SPANISH-AMERICAN WAR SURRENDER OF GUAM TO US: ANNIVERSARY. June 20. Having not known that a war was in progress and having no ammunition on the island, the Spanish commander of Guam surrendered to Capt Glass of the USS *Charleston* on June 20, 1898.

VIRGIN ISLANDS: ORGANIC ACT DAY: ANNIVERSARY. June 20. Commemorates the enactment by the US Congress, on July 22, 1954, of the Revised Organic Act, under which the government of the Virgin Islands is organized. Observed annually on the third Monday in June.

WEST VIRGINIA: ADMISSION DAY. June 20. Became 35th state on this day in 1863.

WEST VIRGINIA DAY CELEBRATION. June 20. Ninth St Plaza, Huntington, WV. Live entertainment, free birthday cake, wagon rides, jugglers, dunking booth, food, crafts and much more. Est attendance: 1,500. For info: Rick Abel, Greater Huntington Parks & Recreation, PO Box 9361, Huntington, WV 25704. Phone: (304) 696-5954.

WOMAN RUNS THE HOUSE: ANNIVERSARY. June 20. Miss Alice Robertson of Oklahoma became the first woman to preside in the US House of Representatives on June 20, 1921. Miss Robertson presided for half an hour.

BIRTHDAYS TODAY

Danny Aiello, Jr, 62, actor ("Lady Blue," *Do the Right Thing*), born New York, NY, June 20, 1933.

Chet Atkins, 71, musician (guitarist who has played all styles; country music legend), born at Luttrell, TN, June 20, 1924.

Olympia Dukakis, 64, actress, theatrical director (Oscar *Moonstruck*; *Steel Magnolias*), born Lowell, MA, June 20, 1931.

John Goodman, 43, actor ("Roseanne," *The Flintstones*), born Afton, MO, June 20, 1952.

Cyndi Lauper, 42, singer ("Girls Just Want to Have Fun"), born at Brooklyn, NY, June 20, 1953.

Anne Murray, 50, singer (gold album *Country*; Glen Campbell's "Goodtime Hour"), born Springhill, NS, Canada, June 20, 1945.
Lionel Richie, 46, singer ("Truly"), songwriter, born at Tuskegee, AL, June 20, 1949.
Andre Watts, 49, musician, born at Nuremburg, Germany, June 20, 1946.
Brian Wilson, 53, singer, songwriter (Beach Boys), born at Hawthorne, CA, June 20, 1942.

JUNE 21 — WEDNESDAY
172nd Day — Remaining, 193

◆ **BATTLE OF OKINAWA ENDED: 50th ANNIVERSARY.** June 21, 1945. With American grenades exploding in the background, inside the Japanese command cave at Mabuni the battle for Okinawa was ended when Maj Gen Isamu Cho and Lt Gen Mitsuru Ushijima killed themselves in the ceremonial rite of hara-kiri. In the long battle which had begun Apr 1, the American death toll reached enormous proportions by Pacific battle standards—7,613 died on land and 4,907 in the air or from kamikaze attacks. A total of 36 US warships were sunk. More than 70,000 Japanese and 80,000 civilian Okinawans died in the course of the battle.

CANADA: SHORTEST NIGHT. June 21. Burwash Landing, Yukon. Dance, midnight horseshoe tournament, live music, BBQ. For info: Yukon Tourism, Box 2703, Whitehorse, Yukon, Canada Y1A 2C6. Phone: (403) 841-4441.

CANADA: SUMMER SOLSTICE DANCE. June 21. Whitehorse, Yukon. 9th annual celebration of the longest day of the year with all-night music and dancing. Est attendance: 600. For info: Tourism Yukon, Box 2703, Whitehorse, Yukon, Canada Y1A 2C6. Phone: (403) 668-4921.

CANADA: WESTERN CANADA FARM PROGRESS SHOW. June 21-24. Regina Exhibition Park, Regina, Saskatchewan. Newest innovations in agriculture and technology for farmers, manufacturers and distributors. Commercial displays, lectures and demonstrations, seminars, fashion shows. Est attendance: 40,000. For info: Tourism Saskatchewan, Saskatchewan Trade and Conv Center, 1919 Saskatchewan Dr, Regina, Sask, Canada S4P 3V7. Phone: (800) 667-7191.

CANCER, THE CRAB. June 21-July 22. In the astronomical/astrological zodiac, which divides the sun's apparent orbit into 12 segments, the period June 21-July 22 is identified, traditionally, as the sun sign of Cancer, the Crab. The ruling planet is the moon.

CHICKEN CLUCKING CONTEST. June 21. Broadway Market Square, Baltimore, MD. Amateur chicken cluckers compete for trophies and "poultry related" prizes. Country music band performs throughout the event. Est attendance: 250. Sponsor: Dept of Recreation and Parks, Office of Adventures in Fun, Clarence "Du" Burns Arena, 1301 S Ellwood Ave, Baltimore, MD 21224. Phone: (410) 396-9177.

June 1995

S	M	T	W	T	F	S
				1	2	3
4	5	6	7	8	9	10
11	12	13	14	15	16	17
18	19	20	21	22	23	24
25	26	27	28	29	30	

CIVIL RIGHTS WORKERS DISAPPEAR: ANNIVERSARY. June 21. James Chaney, Andrew Goodman and Michael Schwerner left Meridian, MS, at 9 AM on June 21, 1964, to investigate a church burning at Philadelphia, MS. They were expected back by 4 PM. They failed to return and an anxious search was begun. Their murdered bodies were found on Aug 4. The three young men were workers on the Mississippi Summer Project to increase black voter registration organized by SNCC (Student Nonviolent Coordinating Committee).

◆ **HEARTLAND TRAIN/KANSAS CITY UN50.** June 21-26. San Francisco, CA (and other locations). This traveling museum will trace President Truman's whistle stop tour of the US. Rail cars will focus on different themes including the environment, human rights and peacekeeping, and the train will carry a UN photographic exhibit and UN post office. Sponsor: Center for Global Community. Contact: Jody Edgerton, (816) 531-3976. For info: UN50 Committee, 312 Sutter St, Ste 610, San Francisco, CA 94108. Phone: (415) 989-1995.

THE JEFFERSON AWARDS. June 21. US Supreme Court, Washington, DC. The Jefferson Awards were founded in 1972 by Jacqueline Kennedy Onassis and Senator Robert Taft, Jr, to honor the highest ideals and achievements in the field of public service in the United States. Each year in the US Supreme Court nine winners are selected in the following categories: Greatest Public Service by an Elected or Appointed Official, Greatest Public Service Performed by a Private Citizen, Greatest Public Service Benefiting the Disadvantaged, Greatest Public Service by an Individual Thirty-Five Years or Under and Outstanding Public Service Benefiting Local Communities. One winner receives the award in each of the first four categories, and five unsung heroes are selected in the last category. Sponsor: American Institute for Public Service, Samuel S. Beard, Pres, 1025 Connecticut Ave NW, Ste 307, Washington, DC 20036.

MIDNIGHT SUN BASEBALL GAME. June 21. Fairbanks, AK. To celebrate the summer solstice. Game is played without artificial lights at 10:35 PM. Est attendance: 4,000. For info: Alaska Goldpanners, Box 71154, Fairbanks, AK 99707. Phone: (907) 451-0095.

MIDSUMMER. June 21. One of the "Lesser Sabbats" during the Wiccan year, celebrating the peak of the Sun-God in his annual cycle. Annually, on the summer solstice.

NATIONAL SQUARE DANCE CONVENTION. June 21-24. Birmingham-Jefferson Civic Center, Birmingham, AL. Dancers from all over the world gather together for dancing enjoyment, educational panels, sewing clinics, fashion show and fellowship at the 44th annual world's largest square dance event. Annually, in June. Approximate attendance: 25,000. For info: David and Sara Meadows, Registration and Housing Chairmen, PO Box 610409, Birmingham, AL 35261-0409. Phone: (205) 833-6732.

OUTDOOR SUMMER THEATER. June 21-Aug 20. Farmington, NM. Featuring *Anasazi, The Ancient Ones*, plus George M. Cohan's *The Tavern*. Productions will be on alternating nights, Tues-Sun. They are performed in a natural sandstone amphitheater at the Lion's Wilderness park, with an optional southwest-style dinner served prior to each performance. Est attendance: 9,000. For info: Farmington Conv and Visitors Bureau, 203 W Main, Ste 401, Farmington, NM 87401. Phone: (800) 448-1240.

SARTRE, JEAN PAUL: 90th BIRTH ANNIVERSARY. June 21. French philosopher, "father of existentialism," born at Paris, France, June 21, 1905. In 1964, Sartre rejected the Nobel Prize for Literature when it was awarded to him. He died at Paris, Apr 15, 1980. In *Being and Nothingness*, he wrote: "Man can will nothing unless he has first understood that he must count on no one but himself; that he is alone, abandoned on earth in the midst of his infinite responsibilities, without help, with no other aim than the one he sets for himself, with no other destiny than the one he forges for himself on this earth."

Chase's 1995 Calendar of Events — June 21-22

SCANDINAVIAN HJEMKOMST FESTIVAL. June 21-25. Civic Center and Trollwood Park, Fargo, ND, and Hjemkomst Center, Moorhead, MN. (Plus various mini-fests in both cities.) Celebration of Scandinavian heritage with traditional Norwegian, Icelandic, Finnish, Swedish and Danish crafts and foods. Authentic folk costume Style Show. Grandparent-Grandchild Fun Tour with passports, prizes and pictures as elves. Annual folk art juried exhibit featuring weaving at Concordia College. Music and dance performed and taught by guest artists from Scandinavia and the US. 1995 feature at Civic Center: Finland: Yesterday and Today Expo with cooking lessons, sauna, skits, roots and history, demonstrations and displays. Reindeer to pet. Annually, the last full weekend in June. Est attendance: 30,000. For info: Bev Paulson, Pres, 3107 S Rivershore Dr, Moorhead, MN 56560-4963. Phone: (800) 235-7654.

SONOMA-MARIN FAIR. June 21-25. Petaluma Fairgrounds, Petaluma, CA. Annual country fair with livestock exhibitions, flowers, carnival and entertainment. Est attendance: 60,000. For info: Sonoma-Marin Fair, PO Box 182, Petaluma, CA 94953. Phone: (707) 763-0931.

SUMMER. June 21-Sept 23. In the Northern Hemisphere summer begins today with the summer solstice, at 4:34 PM, EDT. Note that in the Southern Hemisphere today is the beginning of winter. Anywhere between the Equator and Arctic Circle, the sun rises and sets farthest north on the horizon for the year, and length of daylight is maximum (12 hours, 8 minutes at Equator, increasing to 24 hours at Arctic Circle).

TANNER, HENRY OSSAWA: BIRTH ANNIVERSARY. June 21. Henry Ossawa Tanner was one of the first black artists to be exhibited in galleries in the US. His works were exhibited in New York City, Washington, DC, Chicago, New Orleans and Louisville, among other cities. He was born on June 21, 1859, at Pittsburgh, PA. He died May 25, 1937, at Paris.

TOMPKINS, DANIEL D.: BIRTH ANNIVERSARY. June 21. Sixth vice president of the US (1817-1825), born at Fox Meadows, NY, June 21, 1774. Died at Staten Island, NY, June 11, 1825.

US WOMEN'S AMATEUR PUBLIC LINKS (GOLF) CHAMPIONSHIP. June 21-25. Hominy Hill Golf Course, Colts Neck, NJ. Sponsor: US Golf Assn, Championship Dept, Golf House, Far Hills, NJ 07931. Phone: (908) 234-2300.

WASHINGTON, MARTHA DANDRIDGE CUSTIS: BIRTH ANNIVERSARY. June 21. Wife of George Washington, first president of the US, born at New Kent County, VA, June 21, 1731. Died May 22, 1802.

WHITEWATER WEDNESDAY. June 21. Kernville, CA. One-hour raft trips on the wild and scenic Kern River. Lunch included. Est attendance: 700. For info: Kernville Chamber of Commerce, Rose Ann Viton, PO Box 397, Kernville, CA 93238. Phone: (619) 376-2629.

BIRTHDAYS TODAY

Meredith Baxter, 48, actress ("Bridget Loves Bernie," "Family," "Family Ties"), born at Los Angeles, CA, June 21, 1947.
Benazir Bhutto, 42, Pakistani political leader, born at Karachi, Pakistan, June 21, 1953.
Tom Chambers, 36, former basketball player, born Ogden, UT, June 21, 1959.
Derrick Coleman, 28, basketball player, born at Mobile, AL, June 21, 1967.
Joe Flaherty, 55, writer, actor ("Second City TV," "SCTV Network 90"), born Pittsburgh, PA, June 21, 1940.
Michael Gross, 48, actor ("Family Ties"), born at Chicago, IL, June 21, 1947.
Mariette Hartley, 54, actress ("Peyton's Place," Polaroid commercials/James Garner), born New York, NY, June 21, 1941.
Bernie Kopell, 62, actor ("Get Smart," "The Love Boat"), born New York, NY, June 21, 1933.
Nils Lofgren, 44, musician, singer, songwriter, born at Chicago, IL, June 21, 1951.
Stephen Merrill, 49, Governor of New Hampshire (R), born at Norwich, CT, June 21, 1946.
Jane Russell, 74, actress (*The Outlaw, Gentlemen Prefer Blondes*; Howard Hughes designed bra for her), born Bemidji, MN, June 21, 1921.
Maureen Stapleton, 70, actress (Oscar for *Reds; Little Foxes*), born at Troy, NY, June 21, 1925.
Rick Sutcliffe, 39, baseball player, born at Independence, MO, June 21, 1956.
Prince William (William Philip Arthur Louis), 13, of Wales, son of Prince Charles and Princess Diana, born at London, England, June 21, 1982.

JUNE 22 — THURSDAY
173rd Day — Remaining, 192

ANTIQUE AUTO SHOW AND COLLECTOR CAR FESTIVAL. June 22-24. St. Ignace, MI. Parade, cruise night and a nostalgia concert. Entries from 40 states and Canada. Fax: (906) 643-9784. Est attendance: 80,000. For info: Edward K. Reavie, 268 Hillcrest Blvd, St. Ignace, MI 49781. Phone: (906) 643-8087.

ASPEN MUSIC FESTIVAL. June 22-Aug 30. Aspen, CO. Nine weeks of concerts performed by highly acclaimed artists. Est attendance: 100,000. Sponsor: Music Associates of Aspen, Inc, Box AA, Aspen, CO 81612. Phone: (303) 925-3254.

BOISE RIVER FESTIVAL. June 22-25. Boise, ID. International award-winning citywide celebration of the river, parks, greenbelt and people of Boise. Features more than 300 events including entertainment, children's activities, sporting competitions, food and specialty events. Fax: (208) 383-7397. Est attendance: 600,000. Sponsor: Steven Wood Schmader, CFE, Exec Dir, Boise River Festival, 205 N 10th, Ste 210, Boise, ID 83702. Phone: (208) 383-7397.

CHESAPEAKE-LEOPARD AFFAIR: ANNIVERSARY. June 22. One of the events leading to the War of 1812 occurred June 22, 1807, about 40 miles east of Chesapeake Bay. The US frigate *Chesapeake* was fired upon and boarded by the crew of the British man-of-war *Leopard*. The *Chesapeake*'s commander, James Barron, was court-martialed and convicted of not being prepared for action. Later Barron killed one of the judges (Stephen Decatur) in a duel fought at Bladensburg, MD, Mar 22, 1820.

CLARKSON CZECH FESTIVAL. June 22-25. Main St, Clarkson, NE. Czech food, entertainment, music, polkas, cooking, demonstrations, carnival, arts and crafts. Annually, the last Thursday through Sunday in June. Sponsor: Clarkson Commercial Club. Est attendance: 10,000. For info: Robert Brabec, 515 Elm St, Clarkson, NE 68629. Phone: (402) 892-3331.

DAHLONEGA BLUEGRASS FESTIVAL. June 22-24. Blackburn Park, Dahlonega, GA. Top names in bluegrass music; barbecue and craft booths. Selected as a "Top 20 Event of the Southeast" by the Southeast Tourism Society. 21st annual festival. For info: Dahlonega Bluegrass Festival, c/o Norman Adams, PO Box 98, 112 N Park Ave, Dahlonega, GA 30533. Phone: (706) 864-7203.

HOMESTEAD DAYS. June 22-25. 2nd & Court St, Beatrice, NE. This community-wide celebration recognizes the importance of the Homestead Act of 1862 to the settlement of Nebraska. Entertainment and special museum exhibits. Est attendance: 10,000. For info: Kent Wilson, Admin, Gage County Historical Society, PO Box 793, Beatrice, NE 68310. Phone: (402) 228-1679.

June 22-23 ☆ Chase's 1995 Calendar of Events ☆

JAZZ IN JUNE. June 22-25. Norman, OK. Four outdoor, free jazz concerts featuring national headliners and regional, up-and-coming groups. Annually, the weekend beginning with the fourth Thursday in June. Est attendance: 23,000. Sponsor: Jazz in June, c/o KGOU Radio, The University of Oklahoma, Norman, OK 73019.

JOE LOUIS-BRADDOCK AND SCHMELING FIGHT ANNIVERSARIES. June 22. At Chicago's Comiskey Park, on June 22, 1937, Joe Louis won the World Heavyweight Championship title by knocking out James J. Braddock (eighth round). Louis retained the title until his retirement in 1949. Exactly one year after the Braddock fight, on June 22, 1938, Louis met Germany's Max Schmeling, at New York City's Yankee Stadium. Louis knocked out Schmeling in the first round.

LINDBERGH, ANNE MORROW: BIRTHDAY. June 22. American author and aviator, born June 22, 1907. In *Gift from the Sea*, she wrote: "By and large, mothers and housewives are the only workers who do not have regular time off. They are the great vacationless class."

LITTLE BIG HORN DAYS. June 22-25. Hardin, MT. To celebrate the history of the Old West. This annual celebration commemorates the anniversary of Custer's Last Stand. Activities include carnival, Custer's Last Stand reenactment, parade, rodeo, shows and dances. Est attendance: 10,000. For info: Hardin Area Chamber of Commerce, 219 N Center Ave, Hardin, MT 59034. Phone: (406) 665-1672.

PAPP, JOSEPH: BIRTH ANNIVERSARY. June 22. Born Joseph Papirofsky at Brooklyn, NY, on June 22, 1921, Joe Papp became one of the leading figures in American theatre. At the helm of the New York Public Theatre, Papp produced a wide range of work from the classical to that of the newest American dramatists, including *Hair, Two Gentlemen of Verona, The Pirates of Penzance, The Mystery of Edwin Drood, Sticks and Bones, That Championship Season, Cuba and His Teddy Bear* and *A Chorus Line*, among others. He began his rise in 1954 with the Shakespearean Theatre Workshop (which became the New York Shakespeare Festival), operating from a church basement to take touring productions around the city on a flatbed truck. When the truck broke down in Central Park, Papp turned his touring company into Shakespeare-in-the-Park. His career encompassed producing and directing more than 400 productions, many of which garnered prestigious awards, including 3 Pulitzer Prizes, 6 New York Critics Circle Awards and 28 Tonys. Papp died on Oct 31, 1991.

◆ **REDISCOVERING JUSTICE.** June 22-25. Univ of San Francisco, CA. As part of the commemoration of the UN's 50th anniversary this conference will enable students to explore the role and responsibility of world religions on issues of peace, the environment and religious and ethnic freedom. Sponsors: Univ of San Francisco and Grace Cathedral. Contact: Akab Ziajka, (415) 666-6848. For info: UN50 Committee, 312 Sutter St, Ste 610, San Francisco, CA 94108. Phone: (415) 989-1995.

SHAKESPEARE ON THE GREEN. June 22-25. (Also June 29-July 2; July 6-9.) Elmwood Park, Univ of Nebraska, Omaha, NE. Non-profit professional presentations of the works of William Shakespeare in a beautiful outdoor setting for the families of the Great Plains region. One of a handful of "free" festivals across the country. Includes preshow seminars and workshops. Picnic area and concessions, Elizabethan entertainment featuring music, dancing, singing, juggling and acrobatics. Help to support the Festival is welcome. Annually, the last weekend in June and the first two weekends in July. Est attendance: 25,000. Sponsor: Michael Markey, Managing Dir, Nebraska Shakespeare Festival, c/o Dept of Fine Arts, Creighton Univ, Omaha, NE 68178. Phone: (402) 280-2391.

◆ **US DEPARTMENT OF JUSTICE: 125th ANNIVERSARY.** June 22. Established by act of June 22, 1870, the Department of Justice is headed by the attorney general. Prior to 1870, the attorney general (whose office had been created Sept 24, 1789) had been a member of the president's cabinet but had not been the head of a department.

V-MAIL DELIVERY: ANNIVERSARY. June 22. The first V-Mail (V for victory) was dispatched from New York on June 22, 1942. This system was devised during WW II to conserve cargo space for war materials and supplies. Special paper was used for writing the letters. At post offices, the letters were opened, censored and photographed in reduced proportions. The film was then transported overseas. A complete roll of film contained 1,600 letters.

WATERMELON THUMP (WITH WORLD CHAMPION SEED-SPITTING CONTEST). June 22-24. Luling, TX. Features World Champion Seed-Spitting Contest, street dance each night, giant parade on Saturday, free live entertainment in the Beer Garden, car rally, champion melon auction, arts and crafts exhibit and sales, food, games, rides, beer garden. Annually, the last Thursday, Friday, and Saturday in June. Est attendance: 35,000. For info: Susan H. Ward, Luling Watermelon Thump Assn, Box 710, Luling, TX 78648. Phone: (210) 875-3214.

BIRTHDAYS TODAY

Bill Blass, 73, fashion designer, born at Fort Wayne, IN, June 22, 1922.
Ed Bradley, 54, broadcast journalist ("60 Minutes"), born at Philadelphia, PA, June 22, 1941.
Klaus Maria Brandauer, 51, actor, born at Altausse, Austria, June 22, 1944.
Clyde Drexler, 33, basketball player, born at Houston, TX, June 22, 1962.
Diane Feinstein, 62, US Senator (D, California) up for reelection in Nov '94, born at San Francisco, CA, June 22, 1933.
Kris Kristofferson, 59, singer, actor (*Alice Doesn't Live Here Anymore, A Star Is Born*), born at Brownsville, TX, June 22, 1936.
Todd Rundgren, 47, singer (*Something/Anything*), producer, born at Upper Darby, PA, June 22, 1948.
Meryl Streep, 46, actress (Oscars for *Kramer vs. Kramer* and *Sophie's Choice*), born at Summit, NJ, June 22, 1949.
Lindsay Wagner, 46, actress (Jaime Sommers on "The Bionic Woman"; *The Paper Chase*), born at Los Angeles, CA, June 22, 1949.

JUNE 23 — FRIDAY
174th Day — Remaining, 191

CANADA: CHANGING THE GUARD. June 23-Aug 27. Ottawa, Ontario. Colorful precision drill performed by Canada's two oldest regiments, daily 10 to 10:30 AM on Parliament Hill, weather permitting. Fax: (613) 991-5744. For info: Ceremonial Guard, CFB Ottawa South, Bldg 16, Ottawa, Ont, Canada K1A 0K5. Phone: (613) 991-2117.

CANADA: MIDNIGHT MADNESS. June 23. Inuvik, Northwest Territories. Celebrates the Midnight Sun with a variety of community events including the Midnight Sun Fun Run, music and dancing in Jim Koe Park and late night sidewalk sales. Annually, the Friday closest to summer solstice. Fax: (403) 979-2071. Est attendance: 1,500. For info: Peggy Curtis, Coord, Box 1160, Inuvik, NWT, Canada X0E 0T0. Phone: (403) 979-2607.

June 1995

S	M	T	W	T	F	S
				1	2	3
4	5	6	7	8	9	10
11	12	13	14	15	16	17
18	19	20	21	22	23	24
25	26	27	28	29	30	

Chase's 1995 Calendar of Events — June 23

CANADA: RIVERFEST. June 23–July 2. Brockville, Ontario. A variety of water-related events on the St. Lawrence River. Est attendance: 250,000. For info: Riverfest, PO Box 742, Brockville, Ont, Canada K6V 5V8. Phone: (613) 342-8975.

CANADA: SASKATCHEWAN JAZZ FESTIVAL. June 23–July 2. Saskatoon, Saskatchewan. 9th annual festival emphasizes mainstream jazz and a wide variety of other styles from Dixieland and blues to contemporary fusion and gospel. More than 370 musicians and 125 performances (most free), plus workshops, seminars and videos. Live jazz and blues in the city's nightclubs and lounges too. Est attendance: 30,000. For info: Tourism Saskatchewan, Saskatchewan Trade and Conv Center, 1919 Saskatchewan Dr, Regina, Sask, Canada S4P 3V7. Phone: (800) 667-7191.

CANADA: YUKON INTERNATIONAL STORYTELLING FESTIVAL. June 23–25. Whitehorse, Yukon. Storytellers from all over Canada and abroad. Fax: (403) 667-7556. Est attendance: 5,000. For info: Yukon Intl Storytelling Fest, PO Box 5029, Whitehorse, Yukon, Canada Y1A 4S2. Phone: (403) 633-7550.

CLEARWATER CHAMBER OF COMMERCE RODEO. June 23–25. West Hwy 275, Clearwater, NE. 8 PM each day. Annually, the last full weekend of June. Est attendance: 5,000. Sponsor: Clearwater Chamber of Commerce, Box 201, Clearwater, NE 68726.

COBB, IRVIN S.: BIRTH ANNIVERSARY. June 23. American writer and humorist born June 23, 1876. Died Mar 10, 1944.

COLORADO SHAKESPEARE FESTIVAL. June 23–Aug 13. Boulder, CO. One of three top Shakespeare festivals in the country. This is the 38th annual season. Est attendance: 50,000. For info: Patti McFerran, Colorado Shakespeare Festival, Campus Box 261, Boulder, CO 80309-0261. Phone: (303) 492-2783.

CUSTER'S LAST STAND REENACTMENT. June 23–25. Hardin, MT. Reenactment of the Battle of the Little Bighorn based on an historical outline by Joe Medicine Crow, anthropologist and historian of the Crow Indian Tribe. Reenacted by more than 200 Indian and cavalry riders in an outdoor arena near the site of the original battle. Est attendance: 10,000. For info: Charlene Steinmetz, 219 N Center Ave, Hardin, MT 59034. Phone: (406) 665-1672.

DALESBURG MIDSUMMER FESTIVAL. June 23. Dalesburg Lutheran Church, rural Vermillion, SD. Celebration of Scandinavian heritage. A dance to raise the Midsummer Pole, a smorgasbord, Scandinavian arts and crafts area, band concert and more. Est attendance: 900. For info: Ronald Johnson, Midsummer Committee, RR1 Box 243, Vermillion, SD 57069-9533. Phone: (605) 253-2575.

DENMARK: MIDSUMMER EVE. June 23. Celebrated all over the country with bonfires and merrymaking.

DENMARK: VIKING FESTIVAL. June 23–July 2. Frederiksund (about 25 miles northwest of Copenhagen). Famous outdoor plays based on Danish legends. Annually, the next to last Friday in June through the first Sunday in July.

ENGLAND: SHREWSBURY INTERNATIONAL MUSIC FESTIVAL. June 23–July 2. Shrewsbury, Shropshire. Noncompetitive festival that brings together music and dance groups from all over the world. Est attendance: 2,700. For info: Concertworld (UK) Ltd, 150 Waterloo Rd, London, England SE1 8SB.

FOSSE, ROBERT LOUIS (BOB): BIRTH ANNIVERSARY. June 23. Bob Fosse was born June 23, 1927, at Chicago, IL. The son of a vaudeville singer, he began his show business career at the age of 13. He was the only director in history to win an Oscar, an Emmy and a Tony for his work. As a choreographer and director he was known for his unique dance style that focused on explosive angularity of the human body in its movement. His body of work included the plays *Pippin, Sweet Charity, Pajama Game, Chicago, Dancin', Redhead* and *Damn Yankees*. His films included *Cabaret, Lenny, Star 80,* and the autobiographical *All That Jazz*. Fosse died Sept 23, 1987, at Washington, DC.

GMC TRUCK–MT WASHINGTON HILLCLIMB—THE CLIMB TO THE CLOUDS. June 23–25. Gorham, NH. The "Climb to the Clouds" Hillclimb is America's oldest Hillclimb, originating in 1904. An SCCA-Sanctioned Sportscar Hillclimb, which benefits the DARE programs in Northern New Hampshire Center. Annually, the last weekend in June. Est attendance: 5,000. For info: Climb to the Clouds Registrar, PO Box 278, Gorham, NH 03581. Phone: (603) 466-3988.

HAMPTON JAZZ FESTIVAL. June 23–25. Hampton Coliseum, Hampton, VA. 28th annual music extravaganza offering the best of jazz, blues and contemporary music, the "Grandfather of Jazz Festivals." Admission charged. Est attendance: 29,000. For info: Hampton Jazz Festival, PO Box 126, Hampton, VA 23669. Phone: (804) 838-4203.

HELEN KELLER FESTIVAL. June 23–25. Tuscumbia, AL. Commemorates the remarkable life of Helen Keller with stage shows for all ages, arts and crafts fair, free musical entertainment, races, historic tours of Helen Keller Birthplace and other beautiful homes and much more! *Miracle Worker* play performed evenings during festival and for 5 weekends following. Annually, the last weekend in June. Est attendance: 100,000. For info: Virginia Ware Gilluly, Tourism Coord, 104 S Pine St, Florence, AL 35630. Phone: (205) 764-4661.

INDIANA FIDDLERS' GATHERING. June 23–25. Tippecanoe Battlefield, Battle Ground, IN. Under the huge trees that surround the Tippecanoe Battlefield, clusters of musicians will gather. Enjoy listening to these impromptu concerts, or bring your instruments and play along. Stage performances are scheduled throughout the weekend, and draw musicians and audiences from all around the Midwest. There's music, food and fun for all ages. Annually the last weekend in June. Est attendance: 8,000. Sponsor: Indiana Fiddlers' Gathering Inc, PO Box 49, Battle Ground, IN 47920. Phone: (317) 742-1419.

KIWANIS INTERNATIONAL CONVENTION. June 23–27. New Orleans, LA. Est attendance: 14,000. For info: Kiwanis Intl, Conv Dept, 3636 Woodview Trace, Indianapolis, IN 46268.

LAST FORMAL SURRENDER OF CONFEDERATE TROOPS: 130th ANNIVERSARY. June 23. On June 23, 1865, the last formal surrender of Confederate troops took place in the Oklahoma Territory. Cherokee leader and Confederate Brigadier General Waite surrendered his command of a battalion formed by Indians.

LUXEMBOURG: NATIONAL HOLIDAY. June 23. Official birthday of His Royal Highness Grand Duke Jean. Also, Luxembourg's independence is celebrated on this day.

MICHIGAN SUGAR FESTIVAL. June 23–25. Sebewaing, MI. In recognition of the importance of sugar beet growing and processing to Michigan and the US, this 31st annual festival offers: 3 days of parades, competition for the Michigan Sugar Queen (the "sweetest girl in the world"), 3-on-3 basketball, road race and walkathon, historical events recognition, live entertainment, day and evening carnival, music, food demonstration and sales and contest. Est attendance: 50,000. For info: Melvin Kuhl, Chair, 1600 Ridge Rd, Sebewaing, MI 48759. Phone: (517) 883-2150.

June 23-24 ☆ *Chase's 1995 Calendar of Events* ☆

MIDSUMMER DAY. June 23. Celebrates the beginning of summer with maypoles, music, dancing and bonfires. Observed mainly in northern Europe, especially Scandinavian countries. Day of observance is sometimes St. John's Day (June 24), with celebration on St. John's Eve (June 23) as well. Time approximates the summer solstice. See also: "Summer" (June 21).

NEWFOUNDLAND: DISCOVERY DAY. June 23.

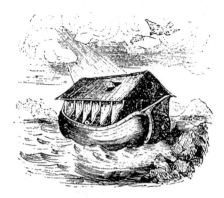

◆ **NOAH'S FLOOD.** June 23-24. Congregation Church, San Francisco, CA. In commemoration of the 50th anniversary of the UN and as a gift to the City of San Francisco from Canada, 120 children will perform Benjamin Britten's short opera which celebrates the diversity of the ethnic and cultural background of the children. This event will be presented by the San Francisco Opera Assn. Sponsor: UN50, Canada. Contact: Nicholas Goldschmidt, (416) 397-5727. For info: UN50 Committee, 312 Sutter St, Ste 610, San Francisco, CA 94108. Phone: (415) 989-1995.

NORWAY: MIDSUMMER NIGHT. June 23. Celebrated with bonfires, fireworks and open-air dancing throughout the country. For info: Norwegian Tourist Board, 655 Third Ave, New York, NY 10017. Phone: (212) 949-2333.

OREGON BACH FESTIVAL. June 23-July 8. Hult Center for the Performing Arts and the University of Oregon, Eugene, OR. Artistic Director Helmuth Rilling commemorates the 50th anniversary of the end of World War II with the theme "War, Peace and Reconciliation," performing the choral-orchestral music of Bach, Britten, Penderecki and Haydn. Forty concerts, events and programs spanning a range of musical genres. Est attendance: 30,000. For info: Oregon Bach Fest, 961 E 18th St, Eugene, OR 97403. Phone: (503) 346-5666.

PORTUGAL: ST. JOHN'S FESTIVAL. June 23-24. Porto. Although other great festivals in honor of Saint John are held in many other towns, it is in Porto that the festivities are most colorful. On every corner of the city there are "cascatas" (arrangements of religious motifs), bonfires and groups of merrymakers singing and dancing all night. Most events held at the great amusement fair of Fontainhas. For info: Portuguese Natl Tourist Office, 590 Fifth Ave, New York, NY 10036. Phone: (212) 354-4403.

SALMON TOURNAMENT WEEK. June 23-29. Lake Michigan, Manistee County, MI. Powder Puff, ProAm and "10 Grand" tournaments held during the week. Fish Lake Michigan for cash prizes. Est attendance: 2,000. For info: Capt. Fred MacDonald, Pres, Manistee Cnty Sport Fishing Assn, 180 Harrison St, Manistee, MI 49660. Phone: (616) 723-7975.

SAND PLUM FESTIVAL. June 23-25. Downtown Guthrie, OK. Guthrie celebrates the ripening of the Sand Plum, used in Victorian cooking since the turn of the century. Juried fine arts show, children's activities, live entertainment, storytelling and more. Annually, the last weekend in June. Est attendance: 10,000. For info: Guthrie Conv & Visitors Bureau, Kay Hunt, Dir, PO Box 995, Guthrie, OK 73044. Phone: (800) 299-1889.

SNOWMASS HOT AIR BALLOON FESTIVAL. June 23-25. Snowmass Village, CO. Forty-five beautifully colored hot air balloons fill the sky over Snowmass Village while the pilots compete in a race and other contests. Est attendance: 6,000. For info: Snowmass Village Resort Assn, PO Box 5566, Snowmass Village, CO 81615. Phone: (303) 923-2000.

STERNWHEELER DAYS. June 23-25. Port Marina Park, Cascade Locks, OR. Return of the 599-passenger sternwheeler "Columbia Gorge" to home port for the summer. Salmon Bake, Mountain Men Encampment, Bingo, Children's rides, parade, food, crafts, live music. Annually, the fourth weekend in June. Est attendance: 10,000. Sponsor: Columbia Gorge Lions, PO Box 522, Cascade Locks, OR 97014. Phone: (503) 374-8313.

SWEDEN: MIDSUMMER. June 23-24. Celebrated throughout Sweden. Maypole dancing, games and folk music.

TELLURIDE WINE FESTIVAL. June 23-25. Telluride, CO. This 22nd annual festival presents the best of international wines and local delicacies. Not just for connoisseurs. Annually, the last weekend in June. Est attendance: 350. For info: Telluride Central Reservations, PO Box 653, Telluride, CO 81435. Phone: (800) 525-3455.

◆ **VISIONS OF PEACE.** June 23-24. Herbst Theater, San Francisco, CA. A forum of Nobel Peace Prize winners convenes to express their hopes and visions for the world. Laureates will offer their views of the UN's role in the next 50 years. Contact: Stephanie Rapp, (415) 989-195. For info: UN50 Committee, 312 Sutter St, Ste 610, San Francisco, CA 94108. Phone: (415) 989-1995.

WATER SKI DAYS. June 23-25. Lake City, MN. Gala water ski shows, grand parade, Venetian sailboat parade, stage shows, arts and crafts show, classic car show and carnival. Est attendance: 20,000. For info: Lake City Area Chamber of Commerce, 212 S Washington St, Box 150, Lake City, MN 55041. Phone: (800) 369-4123.

BIRTHDAYS TODAY

June Carter Cash, 66, singer (Grammy with husband Johnny "Jackson"), born at Maces Spring, VA, June 23, 1929.
James Levine, 52, conductor, born at Cincinnati, OH, June 23, 1943.
Wilma Rudolph, 55, track athlete, born at Saint Bethlehem, TN, June 23, 1940.
Ted Schackelford, 49, actor ("Knots Landing," "Dallas"), born at Oklahoma City, OK, June 23, 1946.
Clarence Thomas, 47, Supreme Court Justice, born at Pinpoint, GA, June 23, 1948.

JUNE 24 — SATURDAY
175th Day — Remaining, 190

AEBLESKIVER DAYS. June 24. Tyler, MN. A celebration of the city's Danish heritage with Danish food, crafts and folk dancing. Est attendance: 5,000. Sponsor: Tyler Area Chamber of Commerce, Box Q, Tyler, MN 56178. Phone: (507) 247-3905.

ARTS AND CRAFTS SHOW. June 24-25. (Also Sept 2-3.) East Tawas, MI. Features crafters from around the state and midwest. Est attendance: 5,000. For info: Tawas Area Chamber of Commerce, Box 608, Tawas City, MI 48764. Phone: (800) 558-2927.

BEARTOOTH RUN. June 24. Red Lodge, MT. 8.2-mile or 4.4-mile foot race up scenic Beartooth Pass. Race will start at 7,000 feet and finish at 9,000 feet. Annually, last Saturday in June. Est

June 1995

S	M	T	W	T	F	S
				1	2	3
4	5	6	7	8	9	10
11	12	13	14	15	16	17
18	19	20	21	22	23	24
25	26	27	28	29	30	

attendance: 350. For info: Joan Cline, Box 988, Red Lodge, MT 59068. Phone: (406) 446-1718.

BEECHER, HENRY WARD: BIRTH ANNIVERSARY. June 24. Famous American clergyman and orator was born at Litchfield, CT, June 24, 1813. Died Mar 8, 1887. His dying words were "Now comes the mystery."

BERLIN AIRLIFT: ANNIVERSARY. June 24. In the early days of the Cold War the Soviet Union challenged the West's right of access to Berlin. The Soviets created a blockade June 24, 1948, and an airlift to supply some 2,250,000 people resulted. The airlift lasted a total of 321 days and brought into Berlin 1,592,787 tons of supplies. Joseph Stalin finally backed down and the blockade ended May 12, 1949.

BULLWHACKER DAYS. June 24-25. Mahaffie Farmstead, Olathe, KS. A celebration of Olathe's Santa Fe Trail heritage with 1800s period demonstrations, music, crafts, children's games and covered wagon rides. The Mahaffie Farmstead on the Santa Fe Trail near the Oregon Trail served warm meals for travelers as a stagecoach stop from 1863 to 1869. "Bullwhackers" were the men who drove teams of oxen. Annually, the last weekend in June. Est attendance: 7,000. For info: Michelle Caron, Mahaffie Farmstead, PO Box 768, Olathe, KS 66061. Phone: (913) 782-6972.

CANADA: FRIENDSHIP FESTIVAL. June 24. (Also June 28-30 and July 1-4.) Fort Erie, Ontario, and Buffalo, NY. Stage entertainment, highland games, horse show, air show, sports tournaments, arts and crafts, custom cars, fireworks, midway, food court and children's festival. Phone: (905) 871-6454 or from US (800) 268-0180. Est attendance: 450,000. For info: Friendship Festival, 516 Garrison Rd, Unit 2, Fort Erie, Ont, Canada L2A 1N2.

CARAMOOR INTERNATIONAL MUSIC FESTIVAL. June 24-Aug 20. 50th Anniversary Celebratory Season. Yo-Yo Ma performs with Orchestra of St. Lukes, André Previn conducting. Season features greatest international stars including Andre Watts, Kathleen Battle, Yitzhak Perlman, Jessie Norman and Misha Dichter. Also Jazz series, children's programs and tours of the house, museum and grounds. Est attendance: 40,000. For info: Libby Alson, Caramoor, PO Box 816, Katonah, NY 10536. Phone: (914) 232-5035.

CELEBRATION OF THE SENSES. June 24. Treat yourself to a stimulation of the five senses—taste, touch, scent, sight and sound—and you may experience the elevation known to many mystics as the elusive sixth sense. Phone: (212) 388-8673 or (717) 274-8451. Sponsor: Wellness Permission League, Tom or Ruth Roy, 2105 Water St, Lebanon, PA 17046.

CHURCH AND SYNAGOGUE LIBRARY ASSOCIATION CONFERENCE. June 24-27. Houghton College, Houghton, NY. Est attendance: 275. Sponsor: Judith Janzen, Exec Sec'y, Church and Synagogue Library Assn, Box 19357, Portland, OR 97280-0357. Phone: (503) 244-6919.

CIARDI, JOHN: BIRTH ANNIVERSARY. June 24. American poet, critic, translator, teacher, etymologist and author of children's books, born at Boston, MA, on June 24, 1916. John Anthony Ciardi's criticism and other writings were often described as honest and sometimes as harsh. Ciardi died at Edison, NJ, on Mar 30, 1986.

CORVETTE AND HIGH PERFORMANCE SUMMER MEET. June 24-25. Puyallup, WA. Buy, sell and show cars and parts—new, used and reproductions. Est attendance: 3,250. For info: Larry or Karen Johnson, Show Organizers, PO Box 7753, Olympia, WA 98507. Phone: (206) 786-8844.

◆ **DEMPSEY, JACK: 100th BIRTH ANNIVERSARY.** June 24. William Harrison Dempsey, known as "The Manassa Mauler," was world heavyweight boxing champion from 1919 to 1926. Following his boxing career Dempsey became a successful New York restaurant operator. Born on June 24, 1895, at Manassa, CO, Dempsey died on May 31, 1983, at New York, NY.

EASTERN MUSIC FESTIVAL. June 24-Aug 5. Guilford College, Greensboro, NC. 34th annual summer festival of classical concerts and recitals performed by resident professionals and a corps of talented young students from the United States and abroad. Est attendance: 63,000. For info: Juanita Lawson-Haith, Dir of Mktg, Eastern Music Festival, PO Box 22026, Greensboro, NC 27420. Phone: (910) 333-7450.

ENGLAND: BOURNEMOUTH MUSICMAKERS FESTIVAL. June 24-July 8. Various venues, including Winter Gardens Theatre, Exeter Road, Bournemouth, Dorset. Amateur bands, choirs and orchestras from around the world gather for this annual festival. For info: Paul B. Buck, Bournemouth Tourism, Westover Rd, Bournemouth, England BH1 2BU.

ENGLAND: NATIONAL MUSIC DAY. June 24-25. Various venues throughout England. A memorable celebration of music with a wide variety of entertainment. For info: National Music Day, Avon House, 360 Oxford St, London, England W1N 9HA.

FISHING HAS NO BOUNDARIES—BEMIDJI. June 24-25. Paul Bunyon Park, Bemidji, MN. A two-day fishing experience for disabled persons. Any disability, age, sex, race, etc, eligible. Fishing with experienced guides attended by 50 participants and 130 volunteers. For info: Fishing Has No Boundaries, Bemidji C of C, 300 Bemidji Ave, PO Box 850, Bemidji, MN. Phone: (800) 458-2223.

FLYING SAUCER SIGHTING: ANNIVERSARY. June 24. First reported sighting of "flying saucers" over Mt Rainier, WA, on June 24, 1947, by Kenneth Arnold of Boise, ID.

FRONTIER ARMY DAYS. June 24-25. Fort Abraham Lincoln State Park, Mandan, ND. Frontier Army re-creation groups re-enact frontier military life with demonstrations and living history by infantry, cavalry and "old Scouts" groups, along with others who populated the fort during its heyday. Est attendance: 3,000. For info: Chuck Erickson, Fort Abraham Lincoln State Park, Rt 2, Box 139, Mandan, ND 58554. Phone: (701) 663-9571.

GALESBURG RAILROAD DAYS. June 24-25. Galesburg, IL. 18th annual festival celebrating the city's railroad heritage that dates back to 1854. Carnival, street fair, railroad exhibits and displays, railyard tours, 5K run, concerts and basketball. Includes more than 40 events. Est attendance: 55,000. For info: Galesburg Area CVB, PO Box 631, Galesburg, IL 61402-0631. Phone: (309) 343-1194.

HIGH WIND CLASSIC. June 24-25. Hood River Event site, Hood River, OR. Hood River's oldest slalom sailboarding event invites amateurs to race in this benefit for the American Heart Association. Est attendance: 1,000. Sponsor: Hood River Windsurfing, 101 Oak Ave, Hood River, OR 97031. Phone: (503) 386-5787.

HOOSIER STATE GAMES: REGIONALS. June 24-25. Various locations, IN. Statewide amateur multi-sports festival. Athletes of all ages and skill levels compete in 19 sports. Finals held July 14-16 at Indianapolis. For info: Indiana Sports Corp, 201 S Capitol Ave, Ste 1200, Indianapolis, IN 46225. Phone: (317) 237-5000.

ITALY: GIOCO DEL CALCIO. June 24-28. Florence. Revival of a 16th-century football match in medieval costumes. Fireworks also on June 24.

KANSAS STATE BARBEQUE CHAMPIONSHIP. June 24. Sar-Ko-Par Trails Park, Lenexa, KS. 5,000 pounds of barbeque smoked over an open flame by over 160 teams. Contestants use a variety of grills and smokers to cook beef, poultry, pork and those meats that fall in the miscellaneous category like alligator, buffalo and rattlesnake. Children's games and activities and great food and drink round out this summer feast. Sponsors: Kingsford Charcoal and K.C. Masterpiece Barbeque. Annually, the fourth Saturday in June. Est attendance: 40,000. Sponsor: Lenexa Conv and Visitors Bureau, PO Box 15626, Lenexa, KS 66215. Phone: (800) 950-7867.

KAUFMAN, IRVING: BIRTH ANNIVERSARY. June 24. Irving R. Kaufman gained national attention in 1951 as the judge who sentenced Julius and Ethel Rosenberg to die in the electric chair after their conviction on charges of espionage. Although he is mostly remembered for this celebrated case, Kaufman was a respected jurist who wrote landmark decisions in cases involving the 1st Amendment, antitrust laws and civil rights. In 1971, his dissenting opinion in *US v New York Times* over the publication of the Pentagon Papers argued against prior restraint in that case. The Supreme Court agreed with Kaufman and reversed the decision. In 1961 he ordered the desegregation of all-black Lincoln School in New Rochelle, NY, in *Taylor v Board of Education*. His opinion stated that "compliance with the Supreme Court's edict was not to be less forthright in the North than in the South." Kaufman was born at New York, NY on June 24, 1910. He died there on Feb 3, 1992.

MACAU: MACAU DAY. June 24. Celebrates defeat of the Dutch invasion of 1622 and pays homage to patron saint of Macau, St. John the Baptist.

MARGARET BRENT DEMANDS A POLITICAL VOICE: ANNIVERSARY. June 24. On June 24, 1647, Margaret Brent made her claim as America's first feminist by demanding a voice and vote for herself in the Maryland colonial assembly. Brent came to America in 1638 and was the first woman to own property in Maryland. At the time of her demands she was serving as secretary to Governor Leonard Calvert. She was ejected from the meetings, but when Calvert died she became his executor and became acting governor, presiding over the General Assembly.

MODEL AIRPLANE SHOW: "MEETING OF THE GIANTS." June 24-25. Danville, VA. Model airplane fly-in featuring airplanes with wing span of 60" or more. IMAA and AMA-sanctioned event. Sponsor: Danville Aeromodelers. Danville Parks & Recreation Dept. Est attendance: 3,000. For info: Kim Davis, Danville Parks and Recreation, PO Box 3300, Danville, VA 24543. Phone: (804) 799-5216.

MONTEREY BAY BLUES FESTIVAL. June 24-25. Monterey Fairgrounds, Monterey, CA. One of the most popular festivals in California, the Blues Festival offers quality performers, food and arts and crafts. Past performers have included Etta James, Dr. John, John Mayall, Albert King, Ruth Brown, Betty Wright, Clarence Carter, Charles Brown, Robert Cray, B.B. King, Neville Brothers and many others. This Festival is dedicated to preserving Blues as an American artform. Annually, last Saturday and Sunday in June. Est attendance: 28,000. For info: Bostrom Management, 2600 Garden Rd, Ste 208, Monterey, CA 93940. Phone: (408) 394-2652.

MUSEUM COMES TO LIFE DAY. June 24. Court House Park, Ainsworth, NE. Cowboy poets, an original pioneer play, demonstrations and vendors. Pioneer dress suggested at this 3rd annual festival. Annually, the last Saturday in June. Est attendance: 400. For info: Sellors-Barton Museum Committee, Mary Jo Curtis, Chair, 340 W Third St, Ainsworth, NE 69210. Phone: (402) 387-2488.

NATIONAL FORGIVENESS DAY. June 24. Anyone holding resentments against another person is asked to communicate with that person to talk over the problem. For info: Center of AWESOME Love, 1014 McKinley St, Fremont, OH 43420. Phone: (419) 355-0810.

ONIZUKA, ELLISON S.: BIRTH ANNIVERSARY. June 24. Lieutenant Colonel Ellison S. Onizuka, 39-year-old aerospace engineer, was mission specialist aboard the Space Shuttle *Challenger* when it exploded on Jan 28, 1986 (killing all aboard). Onizuka was born on June 24, 1946, at Kealakekua, Kona, HI. See also: "*Challenger* Space Shuttle Explosion Anniversary" (Jan 28).

PERU: COUNTRYMAN'S DAY. June 24. Half-day public holiday.

POLISH FEST. June 24-26. Summerfest Gounds, Milwaukee, WI. Polish Fest is the first in a series of summer ethnic festivals on Milwaukee's lakefront. Featured are top national Polish and polka bands, historical and cultural exhibits, pageants, dancing, food, fireworks and other events with special contemporary stage featuring Blood, Sweat and Tears and the Turtles. Est attendance: 65,000. Sponsor: Polish Festivals, Inc, 7128 W Rawson Ave, Franklin, WI 53132. Phone: (414) 529-2140.

ROUTE 66 DAYS. June 24. Centennial Park, Ash Fork, AZ. Entertainment, art and craft booths, parade, swap meet, contests, Ash Fork Lottery, street dance on "Old 66", car show. Annually, last Saturday in June. Est attendance: 1,000. For info: Joan Zumbo, Pres, Ash Fork Chamber of Commerce, PO Box 494, Ash Fork, AZ 86320. Phone: (602) 637-2442.

ST. JOHN THE BAPTIST DAY. June 24. Celebrates birth of saint.

SEWANEE CONCERTS AND FESTIVAL. June 24-July 30. Sewanee, TN. Intensive training for serious students in orchestra and chamber music with private instruction. Also composition, theory and conducting. Fax: (615) 598-1145. Est attendance: 800. For info: Sewanee Summer Music Ctr, Martha McCrory, Dir, Univ of the South, 735 Univ Ave, Sewanee, TN 37375-1000. Phone: (800) 598-1225.

SNAKE HUNT. June 24-25. Cross Fork, PA. To raise funds for the fire company. Est attendance: 4,500. For info: Kettle Creek Hose Co #1, Barry Gipe, Chmn, Box 264, Cross Fork, PA 17729. Phone: (717) 923-0848.

SPACE MILESTONE: *SOYUZ T-6* (USSR). June 24. Three-man crew (V. Dzhanibekov, A. Ivanchenkov and Jean-Loup Chretien) docked at *Salyut 7*, visiting two other cosmonauts in residence there before returning to Earth on July 2. Launched June 24, 1982.

STERNWHEEL DAYS. June 24-25. Augusta, KY. (18th annual.) Sternwheelers grace the shore of the Ohio River. Antique car show, Kentucky arts and crafts market. On stage and strolling entertainment all weekend. Fun for the whole family. Est attendance: 8,000. For info: City of Augusta, Sternwheel Days Committee, Judy Bonar, PO Box 85, Augusta, KY 41002. Phone: (606) 756-2183.

☆ Chase's 1995 Calendar of Events ☆ June 24–25

TECHNICAL EXPOSITION AND CONFERENCE. June 24–28. Atlanta, GA. Five days of the largest annual trade show of newspaper systems and equipment with a conference program designed to help improve newspaper production operations. Future technology is featured. Est attendance: 13,500. For info: Nancy J. Jones, Public Affairs Dir, The Newspaper Center, 11600 Sunrise Valley Dr, Reston, VA 22091. Phone: (703) 648-1117.

TREK FEST. June 24. Riverside, IA. Riverside welcomes all for a small town celebration with a Star Trek theme. Parade at 10 AM. Est attendance: 7,500. Sponsor: Riverside Area Community Club, Box 55, Riverside, IA 52327. Phone: (319) 648-5475.

◆ **THE UN ROCKS THE WORLD.** June 24–25. Golden Gate Park, San Francisco, CA. A popular musical celebration featuring world music and art. Sponsors: UN50 Committee and UN50, NY. Contact: Stephanie Rapp. For info: UN50 Committee, 312 Sutter St, Ste 610, San Francisco, CA 94108. Phone: (415) 989-1995.

WORLD CHAMPIONSHIP ROTARY TILLER RACE AND PURPLEHULL PEA FESTIVAL. June 24. Emerson, AR. World Championship Rotary Tiller Race, International Pea Shelling Championship, domino tournament, concessions, arts, crafts, entertainment, children's games, 5K run, Queen's pageant (various ages), antique farm equipment, 3-on-3 basketball. 6th annual festival. Est attendance: 10,000. For info: Mayor Joe Mullins, PO Box 1, Emerson, AR 71740. Phone: (501) 547-2476.

YOUTH SING PRAISE PERFORMANCE. June 24. Belleville, IL. Talented high school students from all over the country gather at the Shrine to perform a musical in the Outdoor Amphitheatre. Saturday, 7 PM. Est attendance: 2,000. For info: Shrine of Our Lady of the Snows, 9500 W Illinois, Hwy 15, Belleville, IL 62223. Phone: (618) 397-6700.

BIRTHDAYS TODAY

Nancy Allen, 45, actress (*Carrie, Blow Out, Robo Cop*), born at New York, NY, June 24, 1950.
Claude Chabrol, 65, filmmaker (*La femme Infidèle, The Cousins*), born at Sardent, France, June 24, 1930.
Mick Fleetwood, 53, musician (drummer with Fleetwood Mac; "Dreams," "Don't Stop"), born at Cornwall, England, June 24, 1942.
Phyllis George, 46, sports announcer, former Miss America, born at Denton, TX, June 24, 1949.
Michele Lee, 53, actress (Karen Fairgate MacKenzie in "Knots Landing"), born at Los Angeles, CA, June 24, 1942.
Robert Reich, 49, US Secretary of Labor in Clinton administration, born at Scranton, PA, June 24, 1946.

JUNE 25 — SUNDAY
176th Day — Remaining, 189

ARNOLD, HENRY H. "HAP": BIRTH ANNIVERSARY. June 25. US general and commander of the Army Air Force in all theaters throughout WWII, Arnold was born June 25, 1886, in Gladwyne, PA. Although no funds were made available, as early as 1938 Arnold was persuading the US aviation industry to step up manufacturing of airplanes. Production grew from 6,000 to 262,000 per year from 1940-44. He supervised pilot training and by 1944 Air Force personnel strength had grown to 2 million from a pre-war high of 21,000. He served on the US Joint Chiefs of Staff Committee and the Combined Chiefs of Staff Committee for the Allies. A popular individual with the brass and his subordinates, his nickname of "Hap" was appropriate. Made a full general in 1944, he became the US Army Air Force's first Five-Star General when the Air Force was made a separate military branch equal to the Army and Navy. Arnold died Jan 15, 1950, at Sonoma, CA.

ASH LAWN–HIGHLAND SUMMER OPERA FESTIVAL. June 25–Aug 14. Charlottesville, VA. Opera and musical theater performed in the estate's boxwood gardens and sung in English. Annually, Mid-June–Mid-August. The Summer Festival is owned by the College of William and Mary. Est attendance: 10,000. For info: Judith H. Walker, Genl Mgr, Rt 2, Box 37, Charlottesville, VA 22902. Phone: (804) 293-4500.

BATTLE OF LITTLE BIG HORN: ANNIVERSARY. June 25. Lieutenant Colonel George Armstrong Custer, leading military force of more than 200 men, attacked an encampment of Sioux Indians led by Chiefs Sitting Bull and Crazy Horse, near Little Bighorn River, MT. Custer and all men in his immediate command were killed. Brief battle (about two hours) of Little Bighorn occurred on Sunday, June 25, 1876. One horse, named Comanche, is said to have been the only survivor among Custer's forces.

CIVIL WAR IN YUGOSLAVIA: ANNIVERSARY. June 25. In an Eastern Europe freed from the iron rule of Communism and the USSR, separatist and nationalist tensions suppressed for decades rose to a violent boiling point. On June 25, 1991, the Yugoslavian republics of Croatia and Slovenia declared their independence, sparking a fractious and bitter war that spread throughout what was formerly Yugoslavia. Ethnic rivalries between Serbians and Croatians began the military conflicts that spread to Slovenia, and in 1992 fighting began in Bosnia-Herzegovina between Serbians and ethnic Muslims. Although the new republics were recognized by the UN and sanctions passed to stop the fighting, it still raged on despite the efforts of UN peacekeeping forces in the various regions. At presstime (1994), the UN was planning to pull its forces out and end the weapons embargo.

GAY AND LESBIAN PRIDE PARADE AND MUSIC FEST/RALLY. June 25. Chicago, IL. Chicago's 26th annual parade begins at 2 PM; the rally begins at 3:30 PM. Est attendance: 150,000. For info: Gay and Lesbian Pride Parade, PO Box 14131, Chicago, IL 60614. Phone: (312) 348-8243.

GILLARS, MILDRED "AXIS SALLY" E.: DEATH ANNIVERSARY. June 25. Mildred E. Gillars received the nickname "Axis Sally" during World War II, when she broadcast Nazi propaganda to US troops in Europe. An American citizen, she was arrested after the war and tried and convicted of treason. She was sentenced to 10 to 30 years in prison and fined $10,000. She was released after 12 years and later taught music in a convent school in Columbus, OH. She died on June 25, 1988.

GREEN MOUNTAIN NATIONALS CAR SHOW. June 25. Willow Park, Bennington, VT. Open Sunday Car Show: judged event, more than 27 classes, car corral, vendors, displays, winners' parade at 3. Annually, fourth weekend in June. Est attendance: 1,600. Sponsor: BASIC Car Club, PO Box 1102, Bennington, VT 05201. Phone: (802) 447-3311.

INTERNATIONAL SIT-ON-THE-FRONT-PEW SUNDAY. June 25. An event that can be fun for the whole family, designed to fill churches from the front pew back, so that clergy will feel encouraged that their people really do care and do want to see and hear them. Annually, the fourth Sunday in June. Sponsor: KJIL—Great Plains Christian Radio, Don Hughes, Genl Mgr, or Rebecca Ottun, Dir of Communications, PO Box 991, Meade, KS 67864. Phone: (316) 873-2991.

LOG CABIN DAY. June 25. Michigan. Commemorates log cabins with tours, open houses and special festivities throughout the state. Est attendance: 11,000. Sponsor: Log Cabin Soc of Michigan, 3503 Edwards Rd, Sodus, MI 49126. Phone: (616) 944-5719.

MASTERS OF THE GRILL. June 25. Mangia Trattoria, Kenosha, WI. Owner/Chef Tony Mantuano will gather five of the Midwest's best chefs to cook everything from appetizers to dessert on the grill. Attendees can sit on the back patio and chat with the chefs, gathering valuable culinary tips. Reservations: (414) 652-4285. Cost: $55 per person for the meal and wine, plus tax and gratuity not included. Sponsor: Mangia Trattoria. For info: Cindy Kurman, Pres, Kurman Communications, Inc., 213 N Institute Pl, Ste 406, Chicago, IL 60610. Phone: (312) 944-6444.

June 25-26 ☆ *Chase's 1995 Calendar of Events* ☆

MOZAMBIQUE: NATIONAL DAY. June 25. National holiday.

O'NEILL, ROSE CECIL: BIRTH ANNIVERSARY. June 25. Rose O'Neill was born on June 25, 1874, at Wilkes-Barre, PA. Her career included work as an illustrator, author and doll designer, the latter gaining her commercial success with the Kewpie Doll. In 1910, *The Ladies Home Journal* devoted a full page to her Kewpie Doll designs, which turned into a marketing phenomenon for three decades. Died at Springfield, MO, on Apr 6, 1944.

ORWELL, GEORGE: BIRTH ANNIVERSARY. June 25. English satirist, author of *Animal Farm, 1984*, and other works was born at Motihari, Bengal, on June 25, 1903. George Orwell was the pseudonym of Eric Arthur Blair. Died at London, England, Jan 21, 1950.

SINGING ON THE MOUNTAIN. June 25. Grandfather Mountain, Linville, NC. Modern and traditional gospel music featuring top groups and nationally known speakers. 71st annual sing. Annually, the fourth Sunday in June. Free admission. Est attendance: 12,000. For info: Grandfather Mountain, Harris Prevost, Box 995, Linville, NC 28646. Phone: (704) 733-4337.

SLOVENIA: NATIONAL DAY. June 25. Public holiday.

SUPREME COURT BANS OFFICIAL PRAYER: ANNIVERSARY. June 25. On June 25, 1962, the US Supreme Court ruled that a prayer read aloud in public schools violated the 1st Amendment's separation of church and state. On June 1, 1985, the court again struck down a law pertaining to the 1st Amendment when it disallowed an Alabama law that permitted a daily one-minute period of silent meditation or prayer in public schools. (Vote 6 to 3)

◆ **UNITED NATIONS ASSOCIATION OF THE USA ANNUAL CONVENTION.** June 25. Fairmont Hotel, San Francisco, CA. As part of the commemoration of the 50th anniversary of the UN, UNA-USA hold their annual national convention. Sponsor: UNA-USA. Contact: Jim Olson, (212) 697-3232. For info: UN50 Committee, 312 Sutter St, Ste 610, San Francisco, CA 94108. Phone: (415) 989-1995.

◆ **UNITED NATIONS 50th ANNIVERSARY CELEBRATION-SAN FRANCISCO.** June 25. San Francisco, CA. Locations and events throughout the day: Grace Cathedral/interfaith service; M.H. de Young Museum/reception for delegates; Davies Symphony Hall/special command performance (Royal Philharmonic Orchestra as part of a world tour celebrating the 50th anniversary); War Memorial Veterans Building/official ceremony (an international celebration of the signing of the UN Charter); Moscone Ballroom/Black Tie Gala (closing ceremonial dinner marking the end of the official period of celebration and commemoration). For info: UN50 Committee, 312 Sutter St, Ste 610, San Francisco, CA 94108. Phone: (415) 989-1995.

June 1995

S	M	T	W	T	F	S
				1	2	3
4	5	6	7	8	9	10
11	12	13	14	15	16	17
18	19	20	21	22	23	24
25	26	27	28	29	30	

US FORCES INVADE KOREA: ANNIVERSARY. June 25. Invasion began June 25, 1950, and US ground forces entered conflict June 30. Armistice signed at Panmunjom, South Korea, on July 27, 1953.

VIRGINIA RATIFICATION DAY. June 25. Tenth state to ratify Constitution, on this day in 1788.

BIRTHDAYS TODAY

Dorothy Gilman, 72, author (*Incident at Madamya*; Mrs. Polifax novels), born New Brunswick, NJ, June 25, 1923.

June Lockhart, 70, actress (mom in 2nd "Lassie" series and in "Lost in Space"), born New York, NY, June 25, 1925.

Sidney Lumet, 71, director (*12 Angry Men, Serpico, Deathtrap, Dog Day Afternoon, Network*), born Philadelphia, PA, June 25, 1924.

George Michael, 32, singer (Wham!; "Wake Me Up Before You Go-Go"), born at Radlett, England, June 25, 1963.

Willis Reed, 53, coach, former basketball player, born at Hico, LA, June 25, 1942.

Carly Simon, 50, singer ("You're So Vain," "Nobody Does It Better"), songwriter, born at New York, NY, June 25, 1945.

Jimmie Walker, 47, actor, comedian ("Good Times," "B.A.D. Cats"), born at New York, NY, June 25, 1948.

JUNE 26 — MONDAY
177th Day — Remaining, 188

◆ **BUCK, PEARL SYDENSTRICKER: BIRTH ANNIVERSARY.** June 26. American author, noted authority on China. Nobel Prize winner for *The Good Earth*. Born at Hillsboro, WV, June 26, 1892. Died Mar 6, 1973.

CN TOWER: OPENING ANNIVERSARY. June 26. Birthday of the world's tallest free-standing, self-supporting structure, the CN Tower, 1,815 feet, 5 inches high, in Toronto, Ontario, Canada, which opened on June 26, 1976. For info: CN Tower, 301 Front St W, Toronto, Ont, Canada M5V 2T6. Phone: (416) 360-8500.

DOUBLEDAY, ABNER: BIRTH ANNIVERSARY. June 26. Abner Doubleday served in the US Army during the Mexican War and against the Seminole Indians in Florida prior to his service in the American Civil War. He was stationed at Charleston, SC, where he manned the first of Fort Sumter's guns to fire back at the South after its initial bombardments. His service found him at the battle of Second Bull Run, Antietam, Fredricksburg, and as a major general commanding a division at Gettysburg. As a school boy in Cooperstown, NY, he constantly organized ball games. He was credited with inventing the game of baseball in the year of 1839, and this fact was confirmed by a committee set up to investigate the origins of baseball by sporting goods manufacturer Albert Spaulding. Abner Doubleday was born at Ballston Spa, NY, June 26, 1819, and died at Mendham, NJ, on Jan 26, 1893.

ENGLAND: LAWN TENNIS CHAMPIONSHIPS AT WIMBLEDON. June 26-July 9. Wimbledon, London. World famous men's and women's singles and doubles championships for the most coveted titles in tennis. Advanced booking required—application forms for ticket lottery due by Dec 1994. For info: All England Lawn Tennis and Croquet Club, Church Rd, Wimbledon, London, England SW19 5AE.

FINLAND: SATA-HAME ACCORDION FESTIVAL. June 26-July 2. Ikaalinen. Est attendance: 48,000. For info: Finnish Tourist Board, 655 Third Ave, New York, NY 10017. Phone: (212) 949-2333.

MADAGASCAR: INDEPENDENCE DAY. June 26. National holiday.

MIDDLETON, ARTHUR: BIRTH ANNIVERSARY. June 26. American Revolutionary leader and signer of the Declaration of Independence, born near Charleston, SC, June 26, 1742. Died at Goose Creek, SC, Jan 1, 1787.

Chase's 1995 Calendar of Events — June 26–27

MONTGOLFIER, JOSEPH MICHEL: DEATH ANNIVERSARY. June 26. French merchant and inventor, born at Vidalonlez-Annonay, France, in 1740 (exact date unknown) who, with his brother Jacques Etienne in Nov 1782 conducted experiments with paper and fabric bags filled with smoke and hot air, which led to invention of the hot air balloon and man's first flight. Died at Balaruc-les-Bains, France, June 26, 1810. See also: "Montgolfier, Jacques Etienne: Birth Anniversary" (Jan 7), "First Balloon Flight: Anniversary" (June 5) and "Aviation History Month" (Nov 1).

PIONEER WEEK. June 26–30. Historic Jefferson College, Washington, MS. A hands-on program for children to learn through talks, demonstrations, and activities how children lived 150–200 years ago (fee $25). Annually, usually in June. Est attendance: 20. For info: Historic Jefferson College, Anne L. Gray, Historian, PO Box 700, Washington, MS 39190. Phone: (601) 442-2901.

PRE-OPERA LECTURE SERIES. June 26–July 23. Ash Lawn–Highland Summer Festival, Charlottesville, VA. Scholars present lectures during the Opera season on the season's productions. Annually, June–August. Est attendance: 10,000. For info: Judith H. Walker, Genl Mgr, Rt 6, Box 37, Charlottesville, VA 22902. Phone: (804) 293-4500.

ST. LAWRENCE SEAWAY: DEDICATION ANNIVERSARY. June 26. President Dwight D. Eisenhower and Queen Elizabeth II jointly dedicated the St. Lawrence Seaway in formal ceremonies held at St. Lambert, Quebec, Canada, on June 26, 1959. A project undertaken jointly by Canada and the US, the waterway (which provides access between the Atlantic Ocean and the Great Lakes) had been opened to traffic on Apr 25, 1959.

SIGNING OF THE FEDERAL CREDIT UNION ACT: ANNIVERSARY. June 26, 1934. Commemorates signing by President Franklin Delano Roosevelt of the Federal Credit Union Act, thus enabling the formation of credit unions anywhere in the US.

◆ **UNITED NATIONS CHARTER SIGNED: 50th ANNIVERSARY.** June 26, 1945. The Charter of the United Nations, drafted in San Francisco during the United Nations Conference on International Organization, was signed by representatives of the 50 attending states. By Oct 24, the required number of nations had ratified the charter and the UN officially came into existence. Oct 24 has since been celebrated as United Nations Day.

UNITED NATIONS: INTERNATIONAL DAY AGAINST DRUG ABUSE AND ILLICIT TRAFFICKING. June 26. Following a recommendation of the 1987 International Conference on Drug Abuse and Illicit Trafficking, the United Nations General Assembly on June 26, 1987 (Res 42/112), expressed its determination to strengthen action and cooperation for an international society free of drug abuse and proclaimed June 26 as an annual observance to raise public awareness. Info from: Dept of Public Info, New York, NY 10017.

VIRGINIA STATE HORSE SHOW ALL-BREED EVENT. June 26–July 2. Fairgrounds on Strawberry Hill, Richmond, VA. Competition and hunters, jumpers, Arabians, Half-Arabians, Morgans, Saddlebreds, Walking Horses, Driving. The most inclusive all-breed event in the state with the most divisions of competition. "A" rated by the American Horse Shows Association and recognized by various other breed associations. Highlights include "Championship Night" with its "Parade of Breeds." Fax: (804) 228-3252. Est attendance: 10,000. For info: Sue Mullins, Equine Dir, PO Box 26805, Richmond, VA 23261. Phone: (804) 228-3238.

◆ **ZAHARIAS, MILDRED "BABE" DIDRIKSON: BIRTH ANNIVERSARY.** June 26, 1914. Born Mildred Ella Didrikson at Port Arthur, TX, the great athlete was nicknamed Babe because of her baseball-playing; she was named to the women's All-America basketball team when she was 16; at the 1932 Olympic Games, she won two gold medals—and also set world records—in the javelin throw and the 80-meter high hurdles; only a technicality prevented her from obtaining the gold in the high jump. Didrikson married professional wrestler George Zaharias in 1938, six years after she began playing golf casually. In 1946 Babe won the US Women's Amateur tournament, and in 1947 she won 17 straight golf championships and became the first American winner of the British Ladies' Amateur Tournament. Turning professional in 1948, she won the US Women's Open in 1950 and 1954, the same year she won the All-American Open. Babe also excelled in softball, baseball, swimming, figure skating, billiards—even football. In a 1950 Associated Press poll she was named the woman athlete of the first half of the 20th century. A victim of cancer, she died at age 42 on Sept 27, 1956, at Galveston, TX.

BIRTHDAYS TODAY

Claudio Abbado, 62, conductor, born at Milan, Italy, June 26, 1933.

Jerome Kersey, 33, basketball player, born at Clarksville, VA, June 26, 1962.

Greg LeMond, 34, cyclist, born at Los Angeles, CA, June 26, 1961.

Eleanor Parker, 73, actress (Baroness in *The Sound of Music*), born Cedarsville, OH, June 26, 1922.

Charles Robb, 56, US Senator (D, Virginia) up for reelection in Nov '94, born at Phoenix, AZ, June 26, 1939.

Charlotte Zolotow, 80, author of children's books (*The Moon Was the Best, Peter and the Pigeons*), born Norfolk, VA, June 26, 1915.

JUNE 27 — TUESDAY
178th Day — Remaining, 187

DJIBOUTI: NATIONAL HOLIDAY. June 27.

GRAND TETON MUSIC FESTIVAL. June 27–Aug 19. Walk Festival Hall, Teton Village, WY. Full symphony orchestra concerts on Friday and Saturday evenings, Chamber Music Tuesday through Thursday, performed by professional musicians from America's finest symphony orchestras with internationally acclaimed guest soloist and conductor. Est attendance: 700. For info: Grand Teton Music Festival, Box 490, Teton Village, WY 83025. Phone: (800) 959-4863.

HEARN, LAFCADIO: BIRTH ANNIVERSARY. June 27. Author, born on the Greek island of Santa Maura, June 27, 1850. Hearn, who had been a newspaper reporter in Cincinnati, OH, and in New Orleans, LA, went to Japan in 1890 as a magazine writer. Deeply attracted to the country and to the Japanese people, he stayed there as a writer and teacher until his death at Okubo, Japan, on Sept 26, 1904. Though his writings are little remembered in America, he remains a popular figure in Japan, where his books are still used, especially in language classes. His home at Matsue is a tourist shrine.

HILL, MILDRED J.: BIRTH ANNIVERSARY. June 27. Mildred J. Hill, American musician, composer and teacher, was born at Louisville, KY, on June 27, 1859. She composed the melody for "Good Morning to All," a song intended to be sung by teachers and students in school classrooms. It was first published in *Song Stories for the Sunday School* in 1893. The lyrics were written by Mildred Hill's younger sister, Patty Smith Hill, and were amended in 1924 to include a stanza beginning "Happy Birthday to You." Mildred Hill died at Chicago, IL, at midnight, June 4/5, 1916, without knowing that her melody would become the world's most popular. See also: "Happy Birthday to 'Happy Birthday to You,'" (June 27) and "Hill, Patty Smith: Birth Anniversary" (Mar 27).

◆ **KELLER, HELEN: BIRTH ANNIVERSARY.** June 27. American writer, worldwide advocate of help for the blind and prevention of blindness. Born at Tuscumbia, AL, June 27, 1880. Blind and deaf from 19 months. Died at Easton, CT, June 1, 1968. Yearly observances promoted by organization she helped found in 1915. Special aim is conquest of blindness worldwide by year 2000. For info: Helen Keller Intl, PR Officer, 90 Washington St, 15th Fl, New York, NY 10006. Phone: (212) 943-0890.

LEFT-HANDED GOLFERS 59th NATIONAL AMATEUR CHAMPIONSHIP. June 27–30. Palm Beach Garden, FL. 72 holes of competition to determine open, senior, super senior and women's division left-handed champions. Est attendance: 300. For info: Dave Mccall, Exec Secy/Treas, Natl Assn of Left-Handed Golfers, 1307 N Orchard, Espanola, NM 87532. Phone: (800) 844-6254.

MOON PHASE: NEW MOON. June 27. Moon enters New Moon phase at 8:50 PM, EDT.

NATIONAL COLUMNIST'S DAY. June 27. Newspaper columnists, who bring you joy all year long, deserve to be celebrated by their readers at least once a year. Now you can send your favorite columnists, local or nationally syndicated, your own wishes for a Happy Columnist's Day and make him or her feel wonderful. Annually, the fourth Tuesday in June. For info: Jim Six, Columnist, The Gloucester County Times, 309 S Broad St, Woodbury, NJ 08096. Phone: (609) 845-3300.

PARNELL, CHARLES STEWART: BIRTH ANNIVERSARY. June 27. Irish nationalist leader and home rule advocate born at Avondale, County Wicklow, Ireland, on June 27, 1846. Politically ruined as a result of an affair with Katherine O'Shea, the estranged wife of a member of Parliament. She was divorced by her husband (who named Parnell correspondent), and on June 25, 1891, she married Parnell. Less than a month later Parnell was defeated in a by-election. He made his last public speech on Sept 27, 1891, and died in the arms of his wife, at Brighton on Oct 6, 1891. Reportedly he was given "a magnificent funeral" by the city of Dublin, where he was buried. The anniversary of Parnell's death is observed by some as Ivy Day when a sprig of ivy is worn on the lapel to remember him. See also: "Ivy Day" (Oct 6).

SPACE MILESTONE: SOYUZ 30 (USSR). June 27. Launched on June 27, 1978, cosmonauts Pyotr Klimuk and Miroslav Hernaszewski (from Poland) linked with *Salyut 6* on June 28, greeting crew of *Soyuz 29*. Returned to Earth July 5.

SPACE MILESTONE: SOYUZ T-9 (USSR). June 27. Launched from Tyuratam June 27, 1983, with two cosmonauts (V. Lyakhov and A. Aleksandrov). The 40-ton *Soyuz T-9* docked at *Salyut 7* the next day and returned to Earth nearly five months later, Nov 23.

SPACE MILESTONE: STS-4 (US). June 27. Shuttle *Columbia* launched from Kennedy Space Center, Florida, on June 27, 1982, with astronauts K. Mattingly and Henry Hartsfield along with 22,000 lbs of cargo, landed at Edwards Air Force Base, California, on July 4.

BIRTHDAYS TODAY

Bruce Babbitt, 57, US Secretary of Interior in Clinton administration, born at Los Angeles, CA, June 27, 1938.
Julia Duffy, 44, actress ("Newhart," "Designing Women"), born at St. Paul, MN, June 27, 1951.
Craig Anthony Hodges, 35, basketball player, born Park Forest, IL, June 27, 1960.
Brereton C. Jones, 56, Governor of Kentucky (D), born at Point Pleasant, WV, June 27, 1939.
Norma Kamali, 50, fashion designer, born at New York, NY, June 27, 1945.
Captain Kangaroo (Bob Keeshan), 68, TV personality, born at Lynbrook, NY, June 27, 1927.
Anna Moffo, 61, opera singer, born at Wayne, PA, June 27, 1934.
H. Ross Perot, 65, philanthropist, businessman, '92 presidential candidate (United We Stand), born at Texarkana, TX, June 27, 1930.
Chuck Connors Person, 31, basketball player, born at Brantley, AL, June 27, 1964.
Rico Petrocelli, 52, former baseball player, born Brooklyn, NY, June 27, 1943.

JUNE 28 — WEDNESDAY
179th Day — Remaining, 186

CHILDREN'S DAY. June 28. University Cultural Center, Detroit, MI. Activities include storytelling, face painting, theatre, film, music, dance, visual art, make-and-take crafts, youth entertainment, food and Cinema Canada presentations. Est attendance: 10,000. For info: Sue Mosey, University Cultural Center Assn, 4735 Cass, Detroit, MI 48202. Phone: (313) 577-5088.

CLARA MAASS DAY. June 28. New Jersey. Commemorates the birth in 1876 of Clara Louise Maass, the heroic nurse who gave her life in the yellow fever experiments of 1901. Sponsor: Clara Maass Foundation, 1 Franklin Ave, Belleville, NJ 07109. Phone: (201) 450-2277.

CYPRUS: ST. PAUL'S FEAST. June 28–29. Kato Paphos, Cyprus. Religious festivities at Kato Paphos at which the archbishop officiates. Procession of the Icon of St. Paul through the streets.

ENGLAND: HENLEY ROYAL REGATTA. June 28–July 2. Henley-on-Thames, Oxfordshire. International rowing event that is one of the big social events of the year. Est attendance: 250,000. For info: The Secretary, Henley Royal Regatta, Regatta Headquarters, Henley-on-Thames, Oxfordshire, England RG9 2LY.

June 1995	S	M	T	W	T	F	S
					1	2	3
	4	5	6	7	8	9	10
	11	12	13	14	15	16	17
	18	19	20	21	22	23	24
	25	26	27	28	29	30	

Chase's 1995 Calendar of Events — June 28–29

FRANCE: FESTIVAL OF THE TARASQUE. June 28. Tarascon (Bouches du Rhone). Reenactment of legendary monstrous beast that lurked near banks of the Rhone, capsizing boats and devouring flocks and men. St. Martha is credited with having tamed the monster.

HUTCHFEST '95. June 28–July 4. Hutchinson, KS. Family festival with free entertainment, games, activities, arts and crafts, prairie heritage and "name entertainment." Est attendance: 60,000. Sponsor: HutchFest '95 Board of Directors, Downtown Dvmt Dept, 20 E 1st, Hutchinson, KS 67501. Phone: (316) 694-2677.

◆ **MAYER, MARIA GOEPPERT: BIRTH ANNIVERSARY.** June 28, 1906. German-American physicist, Maria Goeppert Mayer was born at Kattowitz, Germany. A participant in the Manhattan Project, she worked on the separation of uranium isotopes for the atomic bomb. Mayer became the first American woman to win the Nobel Prize when she shared the 1963 prize for physics with J. Hans Daniel Jensen and Eugene P. Wigner for their explanation of the atomic nucleus, known as the nuclear shell theory. Maria Mayer died Feb 20, 1972, at San Diego, CA.

MONDAY HOLIDAY LAW: ANNIVERSARY. June 28. On June 28, 1968, President Lyndon B. Johnson approved Public Law 90-363, which amended section 6103(a) of title 5, United States Code, establishing Monday observance of Washington's Birthday, Memorial Day, Labor Day, Columbus Day and Veterans Day. The new holiday law took effect on Jan 1, 1971. Veterans Day observance subsequently reverted to its former observance date, Nov 11. See individual holidays for further detail.

MONTANA TRADITIONAL JAZZ FESTIVAL. June 28–July 2. Helena, MT. Top-name Dixieland jazz bands from all over the nation take over Helena's historic Last Chance Gulch. Features concerts, a jazz mass, sidewalk art sale and food fair. Est attendance: 6,500. For info: Don West, PO Box 856, Helena, MT 59624. Phone: (406) 449-7969.

ROUSSEAU, JEAN JACQUES: BIRTH ANNIVERSARY. June 28. Philosopher, born at Geneva, Switzerland, June 28, 1712. Died July 2, 1778. "Man is born free," he wrote in *The Social Contract*, "and everywhere he is in chains."

RUBENS, PETER PAUL: BIRTH ANNIVERSARY. June 28. Flemish painter and diplomat born June 28, 1577, at Siegen, Westphalia. Died of gout at Antwerp, Belgium, May 30, 1640.

WINDJAMMER DAYS. June 28–29. Boothbay Harbor, ME. Windjammer Days, the premier maritime event along the coast of Maine. Parades, concerts, street dance, fireworks, visiting military vessels, windjammers sailing into harbor under full sail and much more. Fun for the whole family. Est attendance: 20,000. Sponsor: Boothbay Harbor Chamber of Commerce, PO Box 356, Boothbay, ME 04538. Phone: (207) 633-2353.

WORLD WAR I: BEGINNING AND ENDING ANNIVERSARY. June 28. Archduke Francis Ferdinand and his wife assassinated at Sarajevo, Bosnia, on June 28, 1914, touching off the conflict that became World War I. Also, the anniversary of signing of the Treaty of Versailles, on June 28, 1919, formally ending the war.

BIRTHDAYS TODAY

Eric Ambler, 86, author (*The Dark Frontier, Journey into Fear*), born at London, England, June 28, 1909.

Kathy Bates, 47, actress (Oscar for *Misery; Fried Green Tomatoes*), born at Memphis, TN, June 28, 1948.

Don Baylor, 46, former baseball player, born at Austin, TX, June 28, 1949.

Danielle Brisebois, 26, actress (Stephanie on "All in the Family"; Mary-Frances on "Knots Landing"), born Brooklyn, NY, June 28, 1969.

Mel Brooks (Melvyn Kaminsky), 67, actor, director (*The Producers, Blazing Saddles, Young Frankenstein, Silent Movie, Life Stinks!*), born at New York, NY, June 28, 1928.

John Elway, 35, football player, born at Port Angeles, WA, June 28, 1960.

Mark Eugene Grace, 31, baseball player, born at Winston-Salem, NC, June 28, 1964.

Carl Levin, 61, US Senator (D, Michigan), born at Detroit, MI, June 28, 1934.

Leon Panetta, 57, White House Chief of Staff, former congressman from CA and former director OMB, born Monterery, CA, June 28, 1938.

JUNE 29 — THURSDAY
180th Day — Remaining, 185

CANADA: NANISIVIK MIDNIGHT SUN MARATHON AND ROAD RACES. June 29–July 3. Nanisivik, Northwest Territories. Runners challenge the treeless pass that links the Inuit Village of Artic Bay to the mining community of Nanisivik north of the Artic Circle on the northern shore of Canada's Baffin Island. A variety of events includes 10K, 32K and 42K races. At heights from 2m to 535m and in temperatures ranging from −5 degrees C to +10 degrees C, runners take on what is considered the toughest marathon in the world. The Ultra, which to date has been conquered by only a handful of men and women, doubles the already grueling 42K course. Entrants stay with miners' families or are placed in Government houses in Nanisivik. Participation is limited to 100 runners. For info: Linda Brunner, Midnight Sun Marathon, Strathcoma Mineral Service, Ltd, 12th fl, 20 Toronto St, Toronto, Ont, Canada, M5C 2B8. Phone: (416) 869-0772.

DUBACH CHUCK WAGON RACES. June 29–July 2. Dubach, LA. Wagon racing, horse race, barn dance and crawfish creole dinner; family camping and horse trail ride. Races begin at 5 on Friday and 2 on Saturday and Sunday. Annually, Fourth of July weekend. For info: Ruel Jean Roberson, Rt 1, Box 161 A, Dubach, LA 71235. Phone: (318) 255-6277.

GOETHALS, GEORGE WASHINGTON: BIRTH ANNIVERSARY. June 29. American engineer and army officer, chief engineer of the Panama Canal, and first civil governor of the Canal Zone, born at Brooklyn, NY, June 29, 1858. Died at New York, NY, Jan 21, 1928.

GRENADA: FISHERMAN'S BIRTHDAY. June 29. Gouyave. Celebration includes the blessing of the boats and nets and boat races, ending with a mini-carnival with street dancing. Est attendance: 5,000. For info: Grenada Tourist Info Office, 820 Second Ave, Ste 900-D, New York, NY 10017. Phone: (800) 927-9554.

◆ **LATHROP, JULIA C.: BIRTH ANNIVERSARY.** June 29, 1858. A pioneer in the battle to establish child-labor laws, Julia C. Lathrop was the first woman member of the Illinois State Board of Charities and in 1900 was instrumental in establishing the first juvenile court in the US. In 1912, President Taft named Lathrop chief of the newly created Children's Bureau, then part of the US Dept of Commerce and Labor. In 1925 she became a member of the Child Welfare Committee of the League of Nations. Julia Lathrop died Apr 15, 1932, at Rockford, IL.

MAYO, WILLIAM JAMES: BIRTH ANNIVERSARY. June 29. American surgeon, one of the Mayo brothers, establishers of the Mayo Foundation, born at LeSeuer, MN, June 29, 1861. Died July 28, 1939.

June 29–30 ☆ Chase's 1995 Calendar of Events ☆

PETER AND PAUL DAY. June 29. Feast day for St. Peter and St. Paul. Commemorates dual martyrdom of Christian apostles Peter (by crucifixion) and Paul (by beheading) during persecution by Roman Emperor Nero. Observed since 3rd century.

SOUTH AFRICA: STANDARD BANK NATIONAL ARTS FESTIVAL. June 29–July 8. Grahamstown, South Africa. Important cultural event showcasing South Africa's top creative talent. Included is a feast of drama, dance, music and visual art. For info: Standard Bank Natl Arts Fest, PO Box 304, Grahamstown, 6140, South Africa. Phone: (212) 730-2929.

TATANKA FESTIVAL. June 29–July 2. Jamestown, ND. A celebration honoring the buffalo, the native animal that once roamed this region in vast herds. The world's largest buffalo, built at Jamestown, is a tribute to these animals. A live buffalo herd, a frontier village, and a buffalo museum are also located on the grounds. Est attendance: 7,000. For info: Jeanette, Chamber of Commerce, PO Box 1530, Jamestown, ND 58402-1530. Phone: (701) 252-4830.

TURKEY: ST. PETER'S DAY. June 29. Antakya. Peter first preached Christianity at this place. Ceremonies at St. Peter's Grotto, early Christian cave near Antakya.

US SENIOR OPEN (GOLF) CHAMPIONSHIP. June 29–July 2. Congressional Country Club, Bethesda, MD. Sponsor: US Golf Assn, Championship Dept, Golf House, Far Hills, NJ 07931. Phone: (908) 234-2300.

"WORLD'S OLDEST RODEO." June 29–July 4. Prescott, AZ. PRCA-approved rodeo featuring saddle and bareback bronc riding, bull riding, calf roping, steer wrestling, team roping and wild horse race. Est attendance: 30,000. For info: Prescott Frontier Days, Inc, Box 2037, Prescott, AZ 86302. Phone: (800) 358-1888.

BIRTHDAYS TODAY

Gary Busey, 51, actor, musician ("The Texas Wheelers," *The Buddy Holly Story*), born Goose Creek, TX, June 29, 1944.

Elizabeth Hanford Dole, 59, President, American Red Cross, born at Salisbury, NC, June 29, 1936.

Fred Grandy, 47, congressman (R, Iowa), former actor (Gopher on "Love Boat"), born Sioux City, IA, June 29, 1948.

Pedro Guerrero, 39, former baseball player, born at San Pedro de Macoris, Dominican Republic, June 29, 1956.

Harmon Killebrew, 59, former baseball player, born at Payette, ID, June 29, 1936.

Slim Pickens (Louis Bert Lindley), 76, actor ("The Outlaws," "Hee Haw," *Dr. Strangelove, Blazing Saddles*), born at Kingsberg, CA, June 29, 1919.

Kwame Toure (Stokeley Carmichael), 54, civil rights leader, born at Port of Spain, British West Indies, June 29, 1941.

Ruth Warrick, 80, actress (*Citizen Kane*; Phoebe Tyler on "All My Children"), born at St. Louis, MO, June 29, 1915.

JUNE 30 — FRIDAY
181st Day — Remaining, 184

CIVIL WAR HERITAGE DAYS. June 30–July 9. Gettysburg, PA. Living history encampments with both Union and Confederate army campsites, reenactment of the Civil War Battle, lectures, band concerts, firefighter's festival and encampment church services. Est attendance: 37,500. For info: Gettysburg Travel Council, 35 Carlisle, Gettysburg, PA 17325. Phone: (717) 334-6274.

June 1995

S	M	T	W	T	F	S
				1	2	3
4	5	6	7	8	9	10
11	12	13	14	15	16	17
18	19	20	21	22	23	24
25	26	27	28	29	30	

DIXON PETUNIA FESTIVAL. June 30–July 4. Dixon, IL. Carnival, bingo, arts and crafts show, children's events, car show, Taste of Sauk, beer garden, tennis tournament, bike race, ducky derby, parade and fireworks. Est attendance: 55,000. Sponsor: Petunia Festival Corp, 74 Galena Ave, Dixon, IL 61021. Phone: (815) 284-3361.

ENGLAND: EXETER FESTIVAL. June 30–July 16. Various venues, Exeter, Devon. Major annual music and arts festival featuring the best of British arts and culture and celebrating the centenary of the National Trust. For info: Lesley Maynard, Fest Organizer, Exeter City Council, Festival Office, Dix's Field, Exeter, England EX1 1JN.

ETHNIC FESTIVAL. June 30–July 2. Downtown South Bend, IN. An urban festival including food, crafts, display and diversity of ethnic entertainment. Annually, Fourth of July weekend. For info: Rebecca Jasinski, Events Coord, Office of Community Affairs, 501 W South St, South Bend, IN 46601. Phone: (219) 235-9951.

FIDDLERS JAMBOREE AND CRAFTS FESTIVAL. June 30–July 1. Smithville, TN. Home of Joe L. Evins Appalachian Regional Crafts Center and Edgar Evins State Park. Clogging, buck dancing, dobro guitar, dulcimer, banjo, gospel singing and old-time fiddle band. More than 250 craft booths. Est attendance: 62,500. For info: Chamber of Commerce, Box 64, Smithville, TN 37166. Phone: (615) 597-4163.

GETTYSBURG CIVIL WAR RELIC AND COLLECTOR'S SHOW. June 30–July 2. Gettysburg, PA. Accoutrements, weapons, uniforms and personal effects from American military history, 1865 and earlier. Leading collectors and dealers of Civil War material. Est attendance: 2,000. For info: Gettysburg Travel Council, 35 Carlisle St, Gettysburg, PA 17325. Phone: (717) 334-6274.

GUATEMALA: ARMED FORCES DAY. June 30. Guatemala observes public holiday.

"THE GUIDING LIGHT": ANNIVERSARY. June 30. Previously on radio, "The Guiding Light" joined the ranks of television soap operas on June 30, 1952. It holds the title of longest-lasting daytime show and longest-lasting series. The Bauer family members were played by Charita Bauer as Bertha (Bert) Bauer, Theo Goetz as Papa Bauer and Lyle Sudro as Bill Bauer. The Grants were Susan Douglas as Kathy Grant, James Lipton as Dr. Richard Grant and Alice Yourman as Laura Grant. Jone Allison played Meta Banning and Herb Nelson played the part of Joe Roberts.

LEAP SECOND ADJUSTMENT TIME. June 30. June 30 is one of the times that has been favored for the addition or subtraction of a second from our clock time (to coordinate atomic and astronomical time). The determination to adjust is made by the Bureau International de l'Heure, in Paris, France. See also: "Note about Leap Seconds" (see Contents).

☆ Chase's 1995 Calendar of Events ☆   June 30

MANISTEE NATIONAL FOREST FESTIVAL. June 30–July 4. Manistee, MI. "The largest Fourth of July Celebration in the state." Grand Parade, a Venetian boat parade, fireworks over Lake Michigan and children's parade. Tours of historic buildings, Lake Bluff Audubon Center, the Big Manistee River by canoe, North Country Trail and other local attractions. Juried arts and crafts show, band concerts, entertainment, dancing, flea market, races, sports tournament and many other special events. Annually, on the 4th of July and a weekend. Est attendance: 40,000. For info: Manistee Area Chamber of Commerce, 11 Cypress St, Manistee, MI 49660. Phone: (616) 723-2575.

MISSISSIPPI DEEP SEA FISHING RODEO. June 30–July 4. Gulfport, MS. 48th annual celebration, featuring fishing events, entertainment, rides, games and family fun. Sponsor: MS Deep Sea Rodeo, Inc. Est attendance: 150,000. For info: Ken Ernst, 2266 Sunkist Country Club Rd, Biloxi, MS 39532. Phone: (601) 863-2713.

MONROE, ELIZABETH KORTRIGHT: BIRTH ANNIVERSARY. June 30. Wife of James Monroe, fifth president of the US, born at New York, NY, June 30, 1768. Died Sept 23, 1830.

MOON PHASE: LAST QUARTER. June 30. Moon enters Last Quarter phase at 3:31 PM, EDT.

NATIONAL TOM SAWYER DAYS (WITH FENCE PAINTING CONTEST). June 30–July 4. Hannibal, MO. Frog jumping, mud volleyball, Tom and Becky Contest parade, arts & crafts show and fireworks launched from the banks of the Mississippi River. Highlight is the National Fence Painting Contest. Sponsor: Hannibal Jaycees. Est attendance: 100,000. For info: Hannibal Visitors Bureau, Box 624, Hannibal, MO 63401. Phone: (314) 248-1119.

NORTHERN IRELAND: GAME AND COUNTRY FAIR. June 30–July 2. Shane's Castle, Antrim, County Antrim. Ireland's premier field sports event covering all aspects of outdoor sports and country life. Est attendance: 35,000. For info: Mr John Beach, Secy, Estate Office, Shane's Castle, Randalstown Rd, Antrim, Northern Ireland BT41 4NE.

RIVERFEST. June 30–July 4. Riverside Park, Lacrosse, WI. Now in it's 12th year, Riverfest has become the city's premier summer event! Top name entertainers, river activities and continuous entertainment highlight this fest. Free musical entertainment all day, every day! A food fair, beverage tent, Venetian Parade, clowns, children's activities and many athletic events are held daily and much more. Est attendance: 40,000. For info: Riverfest, Inc, PO Box 1745, Lacrosse, WI 54602. Phone: (608) 782-6000.

SALEM FAIR AND EXPOSITION. June 30–July 9. Salem Civic Center Complex, Salem, VA. Fastest growing and second largest state fair in Virginia. Wholesome family fun. Featuring agricultural exhibits, crafts, food, games and carnival rides. Annually, beginning on Friday preceding July 4. Est attendance: 350,000. For info: Salem Civic Center, John Saunders, Box 886, Salem, VA 24153. Phone: (703) 375-3004.

SIBERIAN EXPLOSION ANNIVERSARY. June 30. Early on the morning of June 30, 1908, a spectacular explosion occurred over central Siberia. The seismic shock, fire storm, ensuing "black rain," and the illumination that was reportedly visible for hundreds of miles led to speculation about whether a meteorite or an extraterrestrial visitor was the most probable cause. Said to have been the most powerful explosion in history.

TWENTY-SIXTH AMENDMENT RATIFIED. June 30. The 26th Amendment to the Constitution granted the right to vote in all federal, state and local elections to all persons 18 years or older. On the date of ratification, June 30, 1971, the United States gained an additional 11 million voters.

WALES: WELSH INTERNATIONAL FESTIVAL OF STORYTELLING. June 30–July 2. St. Donats, Llantwit Major, South Glamorgan. Traditional stories and music in the grounds of a medieval castle by the sea. For info: David Ambrose, St. Donats Art Centre, St. Donats, Llantwit Major, South Glamorgan, Wales CF6 9WF.

WHEELER, WILLIAM ALMON: BIRTH ANNIVERSARY. June 30. Nineteenth vice president of the US (1877–1881), born at Malone, NY, June 30, 1819. Died there June 4, 1887.

ZAIRE: INDEPENDENCE DAY. June 30. National holiday.

BIRTHDAYS TODAY

Nancy Dussault, 59, actress ("Too Close for Comfort," "The Ted Knight Show"), born at Pensacola, FL, June 30, 1936.

Lena Horne, 78, singer, actress (*Stormy Weather, Jamaica, Death of a Gunfighter*), born at Brooklyn, NY, June 30, 1917.

Mitch (Mitchell James) Richmond, 30, basketball player, born at Fort Lauderdale, FL, June 30, 1965.

Mike Tyson, 29, former boxer, born at Brooklyn, NY, June 30, 1966.

David Wayne, 79, actor (*Adam's Rib, Wait Till the Sun Shines Nellie*, "The Good Life"), born Traverse City, MI, June 30, 1916.

Iulye.

JULY 1 — SATURDAY
182nd Day — Remaining, 183

ANTI-BOREDOM MONTH. July 1-31. Tenth annual sponsorship of a "self-awareness" event to encourage people to examine whether they, co-workers, family or friends are experiencing "an extended period of boredom" in their lives. The Boring Institute identifies this as "a warning sign" of problems that include depression, self-destructive behavior, and even suicide. Advice is offered on how to avoid and overcome boredom. Sponsor: The Boring Institute, Alan Caruba, Founder, Box 40, Maplewood, NJ 07040. Phone: (201) 763-6392.

BATTLE OF GETTYSBURG: ANNIVERSARY. July 1. After the Southern success at Chancellorsville, VA, Confederate General Robert E. Lee led his forces on an invasion of the North, initially targeting Harrisburg, PA. As Union forces moved to counter the invasion, the battle lines were eventually formed at Gettysburg, PA, in one of the Civil War's most crucial battles, beginning on July 1, 1863. On the climactic third day of the battle (July 3), Lee ordered an attack on the center of the Union line, later to be known as Pickett's Charge. The 15,000 rebels were repulsed, ending the Battle of Gettysburg. After the defeat, Lee's forces retreated back to Virginia, listing more than one-third of its men as casualties in the failed invasion. Union General George Meade initially failed to pursue the retreating rebels, allowing Lee's army to escape across the rain-swollen Potomac River.

BIG ISLAND 6th ANNUAL GYM BODYBUILDING CLASSIC. July 1. Royal Waikoloan Hotel, Island of Hawaii. Bodybuilding and fitness contest. Produced by Big Island Gym. Qualifier for National Fitness Competition. Features local and international bodybuilding champions and guest posers. Admission charge. Est attendance: 1,000. For info: Sue Lovell, Big Island Gym, 74-5603 Alapa St, Kailua-Kona, HI 96740. Phone: (808) 329-9432.

BIGGEST ALL NIGHT GOSPEL SINGING IN THE WORLD. July 1. Memorial Field, Bonifay, FL. Sponsor: Bonifay Kiwanis Club. Est attendance: 10,000. For info: Rickey Callahan, PO Box 425, Bonifay, FL 32425. Phone: (904) 547-3613.

BLERIOT, LOUIS: BIRTH ANNIVERSARY. July 1. Louis Bleriot, aviation pioneer and first man to fly an airplane across the English Channel (July 25, 1909), was born at Cambrai, France, July 1, 1872. He died at Paris, Aug 2, 1936.

July 1995

S	M	T	W	T	F	S
						1
2	3	4	5	6	7	8
9	10	11	12	13	14	15
16	17	18	19	20	21	22
23	24	25	26	27	28	29
30	31					

BUILD A SCARECROW DAYS. July 1-2. New Hampshire Farm Museum, Rte 125, Milton, NH. Visitors can make their own scarecrow and place it either in the museum garden or their own. Supplies are furnished by the museum. Annually, 4th of July weekend. Sponsor: New Hampshire Farm Museum. Est attendance: 250. For info: Susie McKinley, Admin Asst, PO Box 644, Milton, NH 03851. Phone: (603) 652-7840.

BUREAU OF INTERNAL REVENUE ESTABLISHED. July 1. The Bureau of Internal Revenue was established on July 1, 1862, by act of Congress.

BURUNDI: INDEPENDENCE DAY. July 1. National holiday. Anniversary of establishment of independence, July 1, 1962.

CANADA: CANADA DAY. July 1. National holiday. Canada's national day, formerly known as Dominion Day. Observed on following day when July 1 is a Sunday. Commemorates the confederation of Upper and Lower Canada and some of the Maritime Provinces into the Dominion of Canada on July 1, 1867.

CANADA: CANADA DAY CELEBRATION. July 1. Jubilee Park & Downtown Bridgetown, Nova Scotia. Celebration includes sidewalk sales, pancake breakfast, opening ceremonies, church service, firemen's competition, live entertainment and fireworks. Annually, July 1. Est attendance: 3,000. For info: Bob Powell, PO Box 609, Bridgetown, NS, Canada B0S 1C0. Phone: (902) 665-2938.

CANADA: CANADA DAY CELEBRATION. July 1. Watson Lake, Yukon. Outdoor concert, triathlon, softball and horseshoe tournaments, water slide, races and games, birthday cake and lots more. Est attendance: 900. For info: Tracy MacKay, Watson Lake Canada Day Celebration, Watson Lake, Yukon, Canada Y1A 2C6. Phone: (403) 536-2246.

CANADA: CANADA DAY WEEKEND PARTY IN THE PARK. July 1-3. Bancroft, Ontario. Fireworks on Canada Day, July 1. Annually, weekend closest to July 1. Est attendance: 3,000. For info: Bancroft and Dist Chmbr of Commerce, PO Box 539, Bancroft, Ont, Canada K0L 1C0. Phone: (613) 332-1513.

CANADA: NOVA SCOTIA INTERNATIONAL TATTOO. July 1-7. Halifax, Nova Scotia. The Tattoo combines international military and civilian performers in bands, singing, dancing, marching, gymnastics and comedy. Annually, July 1-7. Est attendance: 60,000. For info: The Nova Scotia Intl Tattoo, Box 3233 South, Halifax, NS, Canada B3J 3H5. Phone: (902) 420-1114.

CANADA: RAYMOND STAMPEDE. July 1. Ray Knight Arena, Raymond, Alberta. Annual rodeo held at the home of Canada's first and oldest rodeo, started in 1902 by Ray Knight, founder of the town of Raymond. Est attendance: 3,500. For info: Al Heggie, Raymond Stampede Committee Chair, PO Box 335, Raymond, Alta, Canada. Phone: (403) 752-3322.

CANADA: YUKON GOLD PANNING CHAMPIONSHIPS. July 1. Dawson City, Yukon. Contestants compete for the honor of Territorial Champion Gold Panner or various other categories. Dawson visitors can join in and compete for the Cheechako Award. Est attendance: 2,000. For info: Klondike Visitors Assn, Box 389, Dawson City, YT, Canada Y0B 1G0. Phone: (403) 993-5575.

CHILDREN'S COLONIAL DAYS FAIR. July 1-2. Yorktown Victory Center, Yorktown, VA. Hoop rolling, sack races, pie-eating contests and crafts to make and take home are among a variety of activities planned for young people. Est attendance: 1,500. For info: Media Relations, Jamestown-Yorktown Fdtn, PO Box JF, Williamsburg, VA 23187. Phone: (804) 253-4838.

CLEMSON, THOMAS GREEN: BIRTH ANNIVERSARY. July 1. The man for whom Clemson University was named was born at Philadelphia, PA, on July 1, 1807. The mining engineer and agriculturist married John C. Calhoun's daughter, Anna. Clemson bequeathed the old Calhoun plantation to South Carolina, and Clemson Agricultural College (now Clemson Univer-

Chase's 1995 Calendar of Events — July 1

sity) was founded there in 1889. Clemson died at Clemson, SC, Apr 6, 1888.

COURT TV DEBUT: ANNIVERSARY. July 1. The continuing evolution of home television brought on by the advent of cable television added another twist on July 1, 1991, with the advent of Court TV. Trials are broadcast in their entirety, with occasional commentary from the channel's anchor desk and switching between several trials in progress. Trials with immense popular interest, such as the William Kennedy Smith rape trial, the sentencing hearing of Marlon Brando's son and the Jeffrey Dahmer trial are broadcast along with more low-profile cases.

CREATIVE ICE CREAM FLAVOR DAY AND CONTEST. July 1-20. Chicago, IL. To promote individuality and personal independence through the medium of homemade American ice cream and alternative mix-ins and toppings—from Chocolate Sweet Potato to Peachberry Rhubarb. Contest runs from July 1-20. Est attendance: 500. For info: Cool Temptations Inc, 2808 N Halsted, Chicago, IL 60657. Phone: (312) 348-5865.

DAYS OF '47 CELEBRATION. July 1-31. Salt Lake City, UT. State holiday commemorating the arrival of the Mormon Pioneers into the Salt Lake Valley on July 24, 1847. Annually, third largest parade in US on July 24th. Days of '47 Rodeo held July 17-24. Nearly every community throughout the state holds similar celebrations, rodeos and parades. Sponsor: Days of '47 Committee. For info: David Porter, Pub Dir, Utah Travel Council, Council Hall/Capitol Hill, Salt Lake City, UT 84114. Phone: (801) 538-1030.

DENMARK: AALBORG AND REBILD FESTIVAL (AMERICAN INDEPENDENCE DAY CELEBRATION). July 1-4. Aalborg. This celebration of the American Independence Day, at the Rebild National Park, Aalborg, Denmark, is described as "the largest single gathering for this occasion in the world." Guest speakers and Danish and American entertainment. Est attendance: 10,000. Sponsor: 4 July Committee, Aalborg Tourist og Kongres Bureau A/S, Osteragade 8, PO Box 1862, DK-9100 Aalborg, Denmark.

DEPOT FESTIVAL OF THE ARTS. July 1-2. Livingston, MT. Juried outdoor art show held each year during rodeo week. Open to all arts and crafts, jury and booth fees charged. Annually, July 4th weekend. Est attendance: 3,000. For info: Cynthia Moses NeSmith, Dir, Depot Center, PO Box 1319, Livingston, MT 59047. Phone: (406) 222-2300.

DIXON, WILLIE: BIRTH ANNIVERSARY. July 1. Blues legend Willie Dixon was born in Vicksburg, MI, on July 1, 1915. He moved to Chicago, IL, in 1936 and began his career as a musician with the Big Three Trio. With the advent of instrument amplification Dixon migrated away from his acoustic upright bass into producing and song writing with Chess Studios. His style of writing and producing established him as one of the primary architects of the classic Chicago sound in the 1950s. His songs were performed by Elvis Presley, the Everly Brothers, the Rolling Stones, Led Zeppelin, the Doors, Cream, the Yardbirds, Aerosmith, Jimi Hendrix and the Allman Brothers, among many others. Willie Dixon died Jan 29, 1992, at Burbank, CA.

DORSEY, THOMAS A.: BIRTH ANNIVERSARY. July 1. Thomas A. Dorsey, who was widely known as the father of gospel music, was born July 1, 1899, at Villa Rica, GA. Originally a blues composer and pianist, Dorsey eventually combined blues and religious music to develop the gospel music for which he became well known. It was Dorsey's composition "Take My Hand, Precious Lord" that Rev. Dr. Martin Luther King, Jr, had asked to have performed just moments before his assasination. Dorsey, who composed more than 1,000 gospel songs and hundreds of blues songs in his lifetime, died Jan 23, 1993, at Chicago, IL.

ENGLAND: CHELTENHAM INTERNATIONAL FESTIVAL OF MUSIC. July 1-16. Cheltenham, Gloucestershire. The best of contemporary British music including symphony and chamber music, opera, dance, and jazz. Est attendance: 32,000. For info: Mr Kim Sargeant, Festival Organiser, Town Hall, Imperial Square, Cheltenham, Gloucestershire, England GL50 1QA.

FARM SANCTUARY HOEDOWN. July 1-2. Farm Sanctuary, Watkins Glen, NY. America's only shelter for farm animals hosts a fun and educational weekend of farm tours, hayrides, lectures, exhibits, music, food and more, all located on a beautiful 175-acre working farm with hundreds of cows, pigs, turkeys and other animals. Est attendance: 250. For info: Farm Sanctuary, PO Box 150, Watkins Glen, NY 14891. Phone: (607) 583-2225.

FIRST ADHESIVE US POSTAGE STAMPS ISSUED. July 1. On July 1, 1847, the first adhesive US postage stamps were issued by the US Postal Service.

FREEDOM DAYS. July 1-4. Farmington, NM. Celebration with a variety of special events, including spectacular fireworks, free outdoor concert, food fair, auction, street dance, parade, triathlon. Est attendance: 65,000. For info: Farmington Conv and Visitors Bureau, 203 W Main, Ste 401, Farmington, NM 87401. Phone: (800) 448-1240.

GHANA: REPUBLIC DAY. July 1. National holiday.

GREAT CARDBOARD BOAT REGATTA. July 1. Rock Island, IL. Teams and individuals design, build, and race person-powered boats made of corrugated cardboard. "Titantic Award" for most spectacular sinking plus trophies for team spirit, most creative use of cardboard and most spectacular-looking boat. Prizes for top finishers in three boat classes: propelled by oars or paddles; propelled by mechanical means such as paddlewheels or propellers; "Instant Boats" made from "Secret Kits" by spectators-turned-participants. Registration 7:30; races begin 9 AM. Est attendance: 2,000. For info: Jerry Tutskey, Rock Island Parks and Recreation, 1320 24th St, Rock Island, IL 61201. Phone: (309) 788-7275.

GREAT HOUSEBOAT/CRUISER PARADE. July 1. Rough River State Resort Park, Falls of Rough, KY. Decorated boats, cruisers and houseboats, fireworks and parade. Est attendance: 18,000. For info: Rough River Dam State Resort Pk, Tom DeHaven, Rec Supr, Falls of Rough, KY 40119. Phone: (502) 257-2311.

GUATEMALA: BANKER'S DAY. July 1. National holiday.

INTERNATIONAL CHERRY PIT-SPITTING CONTEST. July 1. Tree-Mendus Fruit Farm, Eau Claire, MI. A nutritious sport—is there a better way to dispose of the pits once you have eaten the cherry? Entrants eat a cherry and then spit the pit as far as possible on a blacktop surface. The pit that goes the farthest including the roll is the champ. Annually, the first Saturday in July. Est attendance: 1,000. Sponsor: Tree-Mendus Fruit Farm, Herb Teichman, owner, East Eureka Rd, Eau Claire, MI 49111. Phone: (616) 782-7101.

JONESBOROUGH DAYS. July 1-4. Jonesborough, TN. Celebrate Tennessee's oldest town with a parade, arts and crafts exhibits, live music and entertainment, fireworks and more. Est attendance: 75,000. For info: Jonesborough Civic Trust, Debbie Haney, Steering Committee Chair, PO Box 451, Jonesborough, TN 37659. Phone: (615) 753-5281.

July 1 ☆ Chase's 1995 Calendar of Events ☆

KEY CLUB INTERNATIONAL CONVENTION. July 1-5. Anaheim, CA. Est attendance: 2,800. For info: Key Club Intl, 3636 Woodview Trace, Indianapolis, IN 46268-3196. Phone: (317) 875-8755.

KSNT GO 4th. July 1-2. Topeka, KS. A fund-raiser for non-profit groups and a fireworks display synchronized to music. Est attendance: 150,000. For info: KSNT-TV, PO Box 2700, Topeka, KS 66601. Phone: (913) 582-4000.

LINCOLN SIGNS INCOME TAX: ANNIVERSARY. July 1. President Abraham Lincoln signed into law a bill levying a 3% income tax on annual incomes of $600–$10,000, and 5% on incomes over $10,000. This tax law of July 1, 1862, actually went into effect, unlike an earlier law passed in August, 1861, making it the first income tax levied by the United States.

LONE TREE DAYS. July 1-4. Central City, NE. A celebration of the Lone Tree Monument. The original Lone Tree was a giant, solitary cottonwood which was visible for 20 miles in every direction. The tree was a resting stop along the old California Trail. The Monument is a Nebraska Historic Landmark. The Chamber of Commerce in cooperation with many of its area clubs and organizations have built a 4th of July celebration around the history of the area. Includes flea market, sidewalk sale, Chuckwagon Bar-B-Q, Fireman's Dance, Beer Garden, fireworks display, annual "duck regatta," classic and antique car fest, baking contest, golf tournament, water fights and much more. Est attendance: 1,500. For info: Sherry Johnson, Exec Dir, Central City Area Chamber of Commerce, PO Box 418, Central City, NE 68826. Phone: (308) 946-3897.

MANDAN RODEO DAYS. July 1-4. Mandan, ND. 4 days of wild west action featuring the country's top cowboys. Other activities include an old-fashioned parade, street dances, nightly concerts, ethnic foods, and a huge Art-in-the-Park Festival. Annually, July 4 weekend. Sponsor: Mandan. For info: Sara Coleman, Bismarck/Mandan Conv & Visitors Bureau, Box 2274, Bismarck, ND 58502. Phone: (800) 767-3555.

MARYLAND SYMPHONY AT ANTIETAM. July 1. Antietam Battlefield, Sharpsburg, MD. Maryland Symphony Orchestra features patriotic music and light classics in its salute to independence. Followed by fireworks at dusk. Annually, the Saturday closest to July 4. Est attendance: 35,000. For info: Maryland Symphony Orchestra, 12 Rochester Pl, Hagerstown, MD 21740. Phone: (301) 797-4000.

NATIONAL BAKED BEAN MONTH. July 1-31. To pay tribute to one of America's favorite and most healthful and nutritious foods, baked beans, made with dry or canned beans. Sponsor: Michigan Bean Commission, Dale Kuenzli, Exec Dir, 1031 S US27, St. Johns, MI 48879.

NATIONAL HOT DOG MONTH. July 1-31. Celebrates one of America's favorite hand-held foods with fun facts and new topping ideas. More than 16 billion hot dogs per year are sold in the US. Spnsor: Deli/Prepared Meats Committee. For info: Natl Live Stock and Meat Bd, Susan Lamb Parenti, Asst Dir Test Kitchens/Edit, 444 N Michigan Ave, Chicago, IL 60611. Phone: (312) 670-9253.

July 1995

S	M	T	W	T	F	S
						1
2	3	4	5	6	7	8
9	10	11	12	13	14	15
16	17	18	19	20	21	22
23	24	25	26	27	28	29
30	31					

NATIONAL ICE CREAM MONTH. July 1-31. To promote America's favorite dessert, ice cream. Sponsor: Intl Ice Cream Assn, Attn: Lynn Bagorazzi, 1250 H St NW, Ste 900, Washington, DC 20005. Phone: (202) 296-4250.

NATIONAL JULY BELONGS TO BLUEBERRIES MONTH. July 1-31. To make the public aware that this is the peak month for fresh blueberries. Sponsor: North American Blueberry Council, Box 1036, Folsom, CA 95763.

NATIONAL PURPOSEFUL PARENTING MONTH. July 1-31. This celebration encourages parents to incorporate "purpose" in their parenting. The effort is designed to elevate the level of parental effectiveness by building awareness and providing interested participants with tips for positive, conscientious parenting. For further info, send a self-addressed stamped envelope to PWOP. Sponsor: Parenting Without Pressure (PWOP), Teresa Langston, 1330 Boyer St, Longwood, FL 32750. Phone: (407) 767-2524.

NATIONAL RECREATION AND PARKS MONTH. July 1-31. To showcase and invite community participation in quality leisure activities for all segments of the population. Sponsor: National Recreation and Park Assn, 2775 S Quincy St, Ste 300, Arlington, VA 22206. Phone: (703) 820-4940.

NATIONAL TENNIS MONTH. July 1-31. To promote tennis at the grassroots level. Players can participate in tournaments, clinics and special events at more than 2,000 tennis facilities nationwide. Presented by *Tennis Magazine* and World Team-Tennis. For info: Natl Tennis Month, Attn: Marilyn Wilkes, 5520 Park Ave, Box 395, Trumbull, CT 06611. Phone: (203) 373-7123.

OLD-FASHIONED 4TH OF JULY. July 1-4. Shakopee, MN. How 4th of July was celebrated 100 years ago. Est attendance: 950. For info: Murphy's Landing, 2817 Hwy 101, Shakopee, MN 55379. Phone: (612) 445-6900.

ORO CITY. July 1-2. (Also July 8-9.) Leadville, CO. Rebirth of a miners' camp with entertainment and yarn spinners. Gathering with entertainment and demonstrations. For info: Chamber of Commerce, PO Box 861, Leadville, CO 80461. Phone: (800) 933-3901.

PEPSI 400 NASCAR WINSTON CUP STOCK CAR RACE. July 1. Daytona International Speedway, Daytona Beach, FL. Sponsor: Pepsi. For info: Daytona Intl Speedway, Glenn Barber, PR, Box 2801, Daytona Beach, FL 32120-2801. Phone: (904) 254-6782.

PICNIC ON THE POINT. July 1-2. Central riverfront, Cincinnati, OH. "Old-fashioned" family picnic, national talent and spectacular fireworks. Est attendance: 150,000. For info: Leslie Keller, 2 Centennial Plaza, 805 Central Ave, Cincinnati, OH 45202. Phone: (513) 352-1608.

PLAY TACOMA CELEBRATION. July 1-31. Tacoma, WA. July is packed full of fun, including swimming, sailing, dancing, camping and musical entertainment. Annually, the month of July. Est attendance: 500,000. For info: Metropolitan Park Dist, Rec Dept, 4702 S 19th St, Tacoma, WA 98405. Phone: (206) 305-1036.

PUNXSUTAWNEY GROUNDHOG FESTIVAL. July 1-8. Punxsutawney, PA. Provides residents and visitors a festive week of summer celebration. Music, contests, crafts, food, theater and entertainers. Est attendance: 8,000. Sponsor: Groundhog Festival Committee, Mrs Judy Freed, Secy, RD 2, Box 115, Punxsutawney, PA 15767. Phone: (814) 938-7687.

RHODES, CECIL JOHN: BIRTH ANNIVERSARY. July 1. English-born, South African millionaire politician. Said to have controlled at one time 90% of the world's diamond production. His will founded the Rhodes Scholarships at Oxford University (for superior scholastic achievers, aged 18-25). Rhodesia was named for him. Born July 1, 1853, at Bishop's Stortford, Hertfordshire, Rhodes died Mar 26, 1902, at Cape Town, South Africa.

ROCKPORT ART FESTIVAL. July 1-2. Rockport Ski Basin, Rockport, TX. The work of top artists from across the country line the waterfront for two days. Jazz and plenty of good food.

☆ Chase's 1995 Calendar of Events ☆ July 1-2

For info: Rockport Fulton Area Chamber of Commerce, 404 Broadway, Rockport, TX 78382. Phone: (512) 729-5519.

RUSSIAN RUBLE BECOMES CONVERTIBLE: ANNIVERSARY. July 1. On July 1, 1992, the Russian Ruble became convertible with other currencies worldwide. The step towards convertibility had long been held to be an important first move towards bringing the Russian economy into the mainstream of the world's economy. The official rate on July 1 was approximately 125 rubles to the dollar and authorities hoped that it would eventually stabilize at approximately 80 rubles to the dollar. Previously it was illegal to convert rubles although they traded briskly on the black market.

SAND, GEORGE: BIRTH ANNIVERSARY. July 1. French novelist, author of more than 100 volumes, whose real name was Amandine Aurore Lucile (Dupin) Dudevant, was born at Paris, France, on July 1, 1804. Died at Nohant, France, on June 8, 1876. She is better remembered for having been a liberated woman during a romantic epoch than for her literary works.

SAUERKRAUT SALAD AND SANDWICH SEASON. July 1-Aug 31. To launch the outdoor eating season with light and tasty sauerkraut salads and sandwiches. Sponsor: Natl Kraut Packers Assn. For info: DHM Group, Inc, PO Box 767, Holmdel, NJ 07733.

SHINDIG-ON-THE-GREEN. July 1. (Also July 8, 15, 22, Aug 12, 19, 26, Sept 2.) Downtown Asheville, NC. A free open-air festival held on the City-County Plaza. Clogging and bluegrass bands, dulcimer players and balladeers, banjo players and all kinds of traditional mountain entertainment. Shindig begins each evening at 7:30. 26th annual. For info: Folk Heritage, Asheville Chamber of Commerce, PO Box 1010, Asheville, NC 28802-1010. Phone: (800) 257-1300.

SOUTH AFRICA: ROTHMANS JULY HANDICAP. July 1. Durban. Premier horse racing event in South Africa with stakes of over one million rand. Annually, the first Saturday in July. Est attendance: 60,000. For info: Durban Turf Club, PO Box 924, Durban, 4000, South Africa,. Phone: (212) 730-2929.

SPACE MILESTONE: KOSMOS 1383 (USSR). July 1. First search and rescue satellite—equipped to hear distress calls from aircraft and ships—launched in cooperative project with the US and France, July 1, 1982.

SPIDERS! July 1-Sept 24. Cranbrook Institute of Science, Bloomfield Hill, MI. An exhibit to "spider-wise" visitors on how spiders go about their daily lives. See Mar 11 for full description. For info: Smithsonian Institution Traveling Exhibition Service, 1100 Jefferson Dr SW, Ste 3146, Washington, DC 20560. Phone: (202) 357-2700.

SUNDOWN SALUTE. July 1-4. Junction City, KS, with fireworks over Milford Lake. Free Independence Day Celebration includes Coors 10K Freedom Run, parade, Veterans Ceremony, family activities, music, crafts and much more. Sundown Salute Inc is a non-profit organization made up of citizens in the interest of freedom. The Celebration is run completely from corporate, private and individual donations. Est attendance: 35,000.

For info: Geary County Conv & Visitors Bureau, Mayumi Ameku, Exec Dir, 425 N Washington, PO Box 1846, Junction City, KS 66441-1846. Phone: (800) 528-2489.

TAHITI AND HER ISLANDS AWARENESS MONTH. July 1-31. To celebrate the exotic paradise of the 115 islands that comprise French Polynesia. Throughout this month, Tahitians celebrate their heritage during the colorful Heiva I Tahiti annual festival, which includes traditional singing and dancing, as well as outrigger canoe and fruit-carrying races. Known for their turquoise lagoons and white sand beaches, Tahiti and her islands have been the inspiration of many authors, including Herman Melville, Jack London and James A. Michener. For info: Tahiti Tourism Board, 300 N Continental, Ste 180, El Segundo, CA 90245. Phone: (310) 471-6170.

US FOOD RATIONING: ANNIVERSARY. July 1. Allowing each individual 8 ounces a week, it was announced July 1, 1918, that the US would initiate sugar rationing for the duration of World War I. Herbert Hoover, US Food Administrator, asked additionally for voluntary wheatless Mondays and Wednesdays, meatless Tuesdays, and porkless Thursdays and Fridays as well as asking for an increase in the use of dark "Victory bread." All rationing regulations were suspended in late December after the war had ended.

US's FIRST ZOO: ANNIVERSARY. July 1. The Philadelphia Zoological Society, the first US zoo, opened July 1, 1874. On that day, 3,000 visitors traveled by foot, horse and carriage and steamboat to visit the exhibits. Price of admission was 25 cents for adults and 10 cents for children. There were 1,000 animals in the zoo when it opened.

BIRTHDAYS TODAY

Dan Akroyd, 43, actor ("Saturday Night Live," *Ghostbusters, The Blues Brothers, Dragnet, Driving Miss Daisy, My Girl*), born Ottawa, Ontario, Canada, July 1, 1952.
Karen Black, 53, actress (*Five Easy Pieces, Nashville*), born at Park Ridge, IL, July 1, 1942.
Leslie Caron, 64, actress (*Gigi, An American in Paris*), dancer, born at Paris, France, July 1, 1931.
Olivia DeHavilland, 79, actress (Oscar for *To Have and Have Not; Gone With the Wind*), born Tokyo, Japan, July 1, 1916.
Princess Diana, 34, Princess of Wales, born at Sandringham, England, July 1, 1961.
Jamie Farr (Jameel Farah), 59, actor (Klinger on "M*A*S*H"; *The Blackboard Jungle*), born Toledo, OH, July 1, 1936.
Estee Lauder, 87, cosmetics executive, born at New York, NY, July 1, 1908.
Carl Lewis, 34, track athlete, born at Birmingham, AL, July 1, 1961.
Jean Marsh, 61, writer, actress (wrote and played Rose in *Upstairs, Downstairs; 9 to 5, Master of the Game*), born at Stoke Newington, England, July 1, 1934.
Sydney Pollack, 61, filmmaker (*Tootsie, The Way We Were, The Fabulous Baker Boys*), born at South Bend, IN, July 1, 1934.
Twyla Tharp, 54, dancer, choreographer, born at Portland, IN, July 1, 1941.

JULY 2 — SUNDAY
183rd Day — Remaining, 182

ALABAMA IMPACT: EXHIBITION. July 2-Sept 3. Huntsville Museum of Art, Huntsville, AL. Organized by the Fine Arts Museum of the South in Mobile, this exhibition features work in various media by contemporary Alabama artists of significant regional and national renown. Est attendance: 2,500. For info: Marylyn Coffey, Publicist, Huntsville Museum of Art, 700 Monroe St SW, Huntsville, AL 35801. Phone: (205) 535-4350.

BE NICE TO NEW JERSEY WEEK. July 2-8. A time to recognize the assets of the state most maligned by American comedians. Annually, the first full week of July. For info: Lone Star Publications of Humor, Attn: Lauren Barnett, Box 29000, Ste 103, San Antonio, TX 78229.

July 2 ☆ Chase's 1995 Calendar of Events ☆

BLACK HILLS ROUNDUP. July 2-4. Roundup Grounds, Belle Fourche, SD. This traditional Independence Day celebration includes an exciting PRCA rodeo each day, western South Dakota's largest fireworks display at dusk on the shore of Orman Dam, a big Fourth of July parade downtown at 10 AM and a thrilling carnival and midway. Est attendance: 5,000. For info: Kathy Wainman, Dir, Chamber of Commerce, 415 Fifth Ave, Belle Fourche, SD 57717. Phone: (605) 892-2676.

CACTUS PETE'S CARL HAYDEN DAZE CELEBRATION. July 2-3. Jackpot, NV. The late Carl Hayden, Cactus Pete's former publicist, is honored by a writer's contest, an off-road race and home-built airplane race. Annually, the Fourth of July weekend. Sponsor: Cactus Pete's Resort Casino. Est attendance: 200. For info: Marcus Prates, PR, PO Box 508, Jackpot, NV 89825. Phone: (702) 755-2321.

CANADA: CULTURES CANADA. July 2-Sept 5. National Capital Region, Ottawa-Hull. A series of concerts expressing the ethnic and regional diversity of Canadian culture. Traditional performances and new forms of art derived from traditional roots. Annually, summer-long, on 5 stages along Confederation Blvd. In-state phone: (800) 465-1867. Est attendance: 80,000. For info: Natl Capital Commission, 7th Fl, 161 Laurier Ave W, Ottawa, Ont, Canada K1P 6J6. Phone: (613) 239-5000.

CANADA: WOOD MOUNTAIN WAGON TREK. July 2-7. Willow Bunch, Saskatchewan. 10th annual wagon trek from Willow Bunch to Wood Mountain. Must supply own horse. For info: Tourism Saskatchewan, Saskatchewan Trade and Conv Center, 1919 Saskatchewan Dr, Regina, Sask, Canada S4P 3V7. Phone: (800) 667-7191.

CIVIL RIGHTS ACT OF 1964 PASSED: ANNIVERSARY. July 2. On July 2, 1964, President Lyndon Johnson signed the Voting Rights Act of 1964 into law, prohibiting discrimination on the basis of race in public accommodations, in publicly owned or operated facilities, in employment and union membership and in the registration of voters. The bill included Title VI, which allowed for the cutoff of federal funding in areas where discrimination persisted.

CRANMER, THOMAS: BIRTH ANNIVERSARY. July 2. English clergyman, reformer and martyr, born at Aslacton, Nottinghamshire, England, July 2, 1489. One of the principal authors of *The English Book of Common Prayer*. Archbishop of Canterbury. Tried for treason and burned at the stake, Oxford, England, Mar 21, 1556.

DECLARATION OF INDEPENDENCE RESOLUTION: ANNIVERSARY. July 2. Anniversary of adoption by the Continental Congress, Philadelphia, PA, July 2, 1776, of a resolution introduced on June 7, 1776, by Richard Henry Lee of Virginia: "Resolved, That these United Colonies are, and of right ought to be, free and independent States, that they are absolved from all allegiance to the British Crown, and that all political connection between them and the State of Great Britain is, and ought to be, totally dissolved. That it is expedient forthwith to take the most effectual measures for forming foreign Alliances. That a plan of confederation be prepared and transmitted to the respective Colonies for their consideration and approbation." This resolution prepared the way for adoption, on July 4, 1776, of the Declaration of Independence. See also: "Declaration of Independence: Anniversary" (July 4).

FOURTH OF JULY WEEKEND/TORCHLIGHT PROCESSION. July 2-4. Winston-Salem, NC. Military encampment, demonstrations, re-enactments of events in early Salem. A special weekend of activities. Torchlight Procession: Re-enactment of America's first Fourth of July celebration by legislative proclamation. 200 costumed participants, narration and music. Sponsor: Old Salem Inc. Est attendance: 5,000. For info: Linda Georgitis, PR, Box F, Salem Station, Winston-Salem, NC 27108. Phone: (910) 721-7345.

HALFWAY POINT OF 1995. July 2. At noon on July 2, 1995, 182½ days of the year will have elapsed and 182½ will remain before Jan 1, 1996.

HOT PROSPECTS. July 2-Aug 27. (Sunday afternoons at 3 PM.) Concert Grove at Prospect Park, Brooklyn, NY. Prospect Park's summer performing series, presenting family-oriented programs of dance, music, clowns, puppets, magic and all kinds of summer fun. Est attendance: 5,250. For info: Public Info Office, Prospect Park, 95 Prospect Park W, Brooklyn, NY 11215. Phone: (718) 965-8999.

INDEPENDENCE SUNDAY IN IOWA. July 2. Sunday preceding July 4, by proclamation of the governor.

MARSHALL, THURGOOD: BIRTH ANNIVERSARY. July 2. Thurgood Marshall, the first African-American on the US Supreme Court, was born July 2, 1908, at Baltimore, MD. For more than 20 years, he served as director-counsel of the NAACP Legal Defense and Educational Fund. He experienced his greatest legal victory on May 17, 1954, when the Supreme Court decision on *Brown v Board of Education* declared an end to the "separate but equal" system of racial segregation in public schools in 21 states. Marshall argued 32 cases before the Supreme Court, winning 29 of them, before becoming a member of the high court himself. Nominated by President Lyndon Johnson, he began his 24-year career on the high court Oct 2, 1967, becoming a voice of dissent in an increasingly conservative court. Marshall announced his retirement June 27, 1991, and he died Jan 24, 1993, at Washington, DC.

MUSIC FOR LIFE WEEK. July 2-8. To be observed to promote functional music in individuals, communities, cities, states, encouraging everyone to use music for consolation and comfort at times of loss; to implement humor; to accompany and ease work; and for inspiration and nurture. Est attendance: 75. Sponsor: Ann Fabe Isaacs, Pres, Composer, Music For Life, 8080 Springvalley Dr, Cincinnati, OH 45236-1395. Phone: (513) 631-1777.

NATIONAL CANNED LUNCHEON MEAT WEEK. July 2-8. To create public awareness of the enduring popularity, great taste and versatlity of canned luncheon meat. The observance kicks off in the canned meat capital of the world, Austin, MN, with festival that includes food, games and family activities. Annually, the first week in July. Sponsor: Hormel Foods. For info: Canned Luncheon Meat Council, PO Box 800, Austin, MN 55912. Phone: (507) 437-5355.

NATIONAL CHERRY FESTIVAL. July 2-9. Traverse City, MI. Family civic celebration featuring sports of all sorts, parades and band competitions. Family fun, food and free entertainment. Est attendance: 500,000. For info: Natl Cherry Festival, Thomas Kern, Exec Dir, Box 141, Traverse City, MI 49685.

OLD-FASHIONED FROG JUMP JAMBOREE. July 2. The Inn on Lake Waramaug, New Preston, CT. Kids under 16 years of age are welcome to enter their favorite frogs in an old-fashioned frog-jumping contest. Prizes; no entry fee. 95 frogs entered. Est attendance: 500. For info: Nancy Conant, The Inn on Lake Waramaug, 107 N Shore Rd, New Preston, CT 06777. Phone: (203) 868-0563.

	S	M	T	W	T	F	S
July 1995	2	3	4	5	6	7	1 8
	9	10	11	12	13	14	15
	16	17	18	19	20	21	22
	23	24	25	26	27	28	29
	30	31					

PRAIRIE PIONEER DAYS. July 2-4. Arapahoe, NE. Parade, games for kids, food booths, free swimming, watermelon bust, quilt show, horseshoe pitching, black powder shoot, model airplane event, baseball games, fireworks and more. For info: Arapahoe Chamber of Commerce, Ariel McNamara, Secy, PO Box 624, Arapahoe, NE 68922. Phone: (308) 962-5203.

RED LODGE HOME OF CHAMPIONS RODEO AND PARADE. July 2-4. Red Lodge, MT. This rodeo is part of the Professional Rodeo Cowboys Assn circuit and brings nearly all of the national champions to Red Lodge each year. At noon each day the Home of Champions Parade struts its way through downtown Red Lodge with colorful floats, antique cars, horses, dancing girls, wagons and more. Annually, July 2-4. Sponsor: Red Lodge Rodeo Assn. Est attendance: 6,000. For info: Red Lodge Area Chamber of Commerce, Box 988, Red Lodge, MT 59068. Phone: (406) 446-1718.

ROARING CAMP JUMPING FROG CONTEST. July 2. Felton, CA. Contestants may bring a frog or rent one; prizes for longest jump and fastest frog. For info: Jeanette Guire, Roaring Camp and Big Trees Railroad, Box G-1, Felton, CA 95018. Phone: (408) 335-4484.

ST. LOUIS RACE RIOTS: ANNIVERSARY. July 2. On July 2, 1917, between 20 and 75 blacks were killed in a race riot in St. Louis, MO; hundreds more were injured. To protest this violence against blacks, W.E.B. DuBois and James Weldon Johnson, of the NAACP, led a silent march down Fifth Avenue in New York City.

SOCIETY FOR THE PRESERVATION & ENCOURAGEMENT OF BARBERSHOP QUARTET SINGING IN AMERICA, INC (SPEBSQSA) INTERNATIONAL CONVENTION. July 2-8. Miami, FL. More than 10,000 members attend a week of shows, meetings and chorus and quartet contests. Est attendance: 10,000. For info: Brian Lynch, Communications Specialist, 6315 3rd Ave, Kenosha, WI 53143-5199. Phone: (414) 653-8440.

BIRTHDAYS TODAY

Jose Canseco, 31, baseball player, born at Havana, Cuba, July 2, 1964.
Polly Holliday, 58, actress (Flo Castleberry in "Alice" and in "Flo"), born at Jasper, AL, July 2, 1937.
Rene Lacoste, 90, former tennis player, born at Paris, France, July 2, 1905.
Cheryl Ladd, 43, actress (Kris Munroe on "Charlie's Angels"; "Grace Kelly"), born at Huron, SD, July 2, 1952.
Jimmy McNichol, 34, actor ("The Fitzpatricks," "California Fever"), born at Los Angeles, CA, July 2, 1961.
John H. Sununu, 56, former Governor of New Hampshire, born at Havana, Cuba, July 2, 1939.
Dave Thomas, 63, founder of Wendy's (stars in TV ads), Horatio Alger Award Winner, adoption advocate, born Camden, NJ, July 2, 1932.

JULY 3 — MONDAY
184th Day — Remaining, 181

AIR CONDITIONING APPRECIATION DAYS. July 3-Aug 15. Northern Hemisphere. During Dog Days, the hottest time of the year in the Northern Hemisphere, to acknowledge the contribution of air conditioning to a better way of life. Annually, July 3-Aug 15. For info: Air-Conditioning and Refrig Institute, 4301 N Fairfax Dr, Ste 425, Arlington, VA 22203. Phone: (703) 524-8800.

CANADA: KIMBERLEY INTERNATIONAL OLD-TIME ACCORDION CHAMPIONSHIPS. July 3-8. Kimberley, British Columbia. Jam sessions, play-downs, finals and trophy presentation. Social evening dance and entertainment; pancake breakfast. Annually, the second weekend in June. Est attendance: 5,000. For info: Jean Irvine, P O Box 473, Kimberley, BC, Canada V1A 3B9.

CARIBBEAN OR CARICOM DAY. July 3. The anniversary of the treaty establishing the Caribbean Community (also called the Treaty of Chaguaramas), signed by the prime ministers of Barbados, Guyana, Jamaica and Trinidad and Tobago on July 4, 1973. Observed as a public holiday by the participating nations. Annually, the first Monday in July.

COMPLIMENT-YOUR-MIRROR DAY. July 3. Participation consists of complimenting your mirror on having such a wonderful owner and keeping track of whether other mirrors you meet during the day smile at you. Annually, July 3. Sponsor: Puns Corps, c/o Bob Birch, Grand Punscorpion, Box 2364, Falls Church, VA 22042-0364. Phone: (703) 533-3668.

DOG DAYS. July 3-Aug 15. Hottest days of the year in Northern Hemisphere. Usually about 40 days, but variously reckoned at 30-54 days. Popularly believed to be an evil time "when the sea boiled, wine turned sour, dogs grew mad, and all creatures became languid, causing to man burning fevers, hysterics and phrensies" (from Brady's *Clavis Calendarium*, 1813). Originally the days when Sirius, the Dog Star, rose just before or at about the same time as sunrise (no longer true owing to precession of the equinoxes). Ancients sacrificed a brown dog at beginning of Dog Days to appease the rage of Sirius, believing that star was the cause of the hot, sultry weather.

EARTH AT APHELION. July 3. At approximately 10:00 PM, EDT, planet Earth will reach aphelion, that point in its orbit when it is farthest from the sun (about 94,510,000 miles). The Earth's mean distance from the sun (mean radius of its orbit) is reached early in the months of April and October. Note that Earth is farthest from the sun during Northern Hemisphere summer. See also: "Earth at Perihelion" (Jan 4).

ENNIS RODEO AND PARADE. July 3-4. Ennis, MT. Billed as the fastest two-day rodeo in Montana, this is a non-stop weekend of excitement. Parade with clowns, bucking broncos and everything imaginable. Annually, July 3-4. Est attendance: 4,000. For info: Pat Hamilton, Ennis Rodeo Club, PO Box 236, Ennis, MT 59729. Phone: (406) 682-4700.

FIRST BANK OPENS IN US: ANNIVERSARY. July 3. The Bank for Savings in New York, NY, the first bank in the US, opened on July 3, 1819. On the first day of business $2,807.00 was deposited.

FORT NECESSITY MEMORIAL. July 3. Fort Necessity National Battlefield, Farmington, PA. In the battle on July 3, 1754, soldiers and Native Americans gave their lives for a cause. The park takes this opportunity to present a special program to remember those who fought here 241 years ago. Annually, July 3. Est attendance: 100. Sponsor: Fort Necessity NB, Rd 2, Box 528, Farmington, PA 15437. Phone: (412) 329-5512.

GEORGE WASHINGTON TAKES COMMAND OF THE CONTINENTAL ARMY. July 3. On July 3, 1775, George Washington took command of the Continental Army in Cambridge, MA.

IDAHO: ADMISSION DAY. July 3. Became 43rd state on this day in 1890.

MAN WATCHERS' COMPLIMENT WEEK. July 3-8. Since men aren't used to receiving compliments, make it a point to find something nice to say to a man. P.S. Women need compliments too. Sponsor: Man Watchers, Inc, Suzy Mallery, Pres, 12308 Darlington Ave, Los Angeles, CA 90049. Phone: (310) 826-9101.

NUDE RECREATION WEEK. July 3-9. To promote acceptance of the body and understanding of the nude recreation movement as a natural solution to many problems of modern living. Sponsor: The Naturist Soc, Box 132, Oshkosh, WI 54902. Phone: (414) 231-9950.

July 3-4 ☆ Chase's 1995 Calendar of Events ☆

OLD-FASHIONED PICNIC IN THE PARK. July 3-4. Lincoln Park, Pittsburg, KS. Old-fashioned picnic with families gathering for contests, games, food, carnival, live music, night golf with lighted golf balls and a giant fireworks display set to music. Annually, July 3-4. Est attendance: 15,000. Sponsor: Pittsburg Parks & Recreation Dept, PO Box 688, Pittsburg, KS 66762. Phone: (316) 231-8310.

SPACE MILESTONE: SOYUZ 14 (USSR). July 3. Launched July 3, 1974, cosmonauts P. Popovich and Y. Artyukhin linked with *Salyut 3* space station during 15-day mission.

SPIRIT OF AMERICA (WITH FREEDOM AND AUDIE MURPHY PATRIOTISM AWARDS). July 3-4. Pt Mallard, Decatur, AL. Event was started in 1966 to lift patriotic spirits during the tragedy of the Vietnam War. Presentation of Freedom Award and Audie Murphy Patriotism Award. Est attendance: 50,000. For info: Denise Martin, Decatur Conv and Visitors Bureau, 719 6th Ave SE, PO Box 2349, Decatur, AL 35602. Phone: (205) 350-2028.

STAY OUT OF THE SUN DAY. July 3. For health's sake, give your skin a break today. Phone: (212) 388-8673 or (717) 274-8451. Sponsor: Tom and Ruth Roy, Wellness Permission League, 2105 Water St, Lebanon, PA 17046.

VIRGIN ISLANDS: DANISH WEST INDIES EMANCIPATION DAY. July 3. Commemorates freeing of slaves in the Danish West Indies in 1848. Ceremony in Frederiksted, St. Croix, where actual proclamation was first read by Governor-General Peter Von Scholten.

ZAMBIA: HEROES DAY. July 3. First Monday in July is Zambian national holiday—memorial day for Zambians who died in the struggle for independence. Political rallies stress solidarity.

BIRTHDAYS TODAY

Lamar Alexander, 55, former US Secretary of Education in Bush administration, born July 3, 1940.

Betty Buckley, 48, actress (Sandra Sue Abbott "Abby" Bradford in "Eight Is Enough"), born Big Springs, TX, July 3, 1947.

Tom Cruise, 33, actor (*Taps, Rain Man, Color of Money, Born on the Fourth of July*), born Syracuse, NY, July 3, 1962.

Pete Fountain, 65, jazz musician, born at New Orleans, LA, July 3, 1930.

Stavros Spyros Niarchos, 86, shipping executive, born at Athens, Greece, July 3, 1909.

Tom Stoppard, 58, playwright (*Travesties, On the Razzle, The Real Thing*), born Zlin, Czechoslovakia, July 3, 1937.

JULY 4 — TUESDAY
185th Day — Remaining, 180

AIR SHOW AND FIREWORKS DISPLAY. July 4. Aransas County Airport and Rockport Beach Park, Rockport, TX. A display of military, antique and aerobatic planes on display at the Aransas County Airport and a dazzling show of aerobatic flying over Beach Park. In the evening a beautiful fireworks show will be reflected off the serene waters at Beach Park. Est attendance: 40,000. For info: Rockport-Fulton Area Chamber of Commerce, 404 Broadway, Rockport, TX 78382. Phone: (512) 729-6445.

◆ **AMERICA THE BEAUTIFUL PUBLISHED; 100th ANNIVERSARY.** July 4. The poem "America the Beautiful" by Katherine Lee Bates, a Wellesley College professor, was first published in the *Congregationalist*, a church publication, on July 4, 1895.

July 1995

S	M	T	W	T	F	S
						1
2	3	4	5	6	7	8
9	10	11	12	13	14	15
16	17	18	19	20	21	22
23	24	25	26	27	28	29
30	31					

AMERICAN MENSA ANNUAL MEETING. July 4-9. Adam's Mark, St. Louis, MO. Gathering of members of the international high IQ society. Est attendance: 1,500. For info: Mensa, Sheila Skolnik, Exec Dir, 2626 E 14th St, Brooklyn, NY 11235. Phone: (718) 934-3700.

AMERICAN REDNECK DAY. July 4. To celebrate the work-hard, play-hard, independent spirit of the rural working class. Est attendance: 350. For info: American Redneck Trading Post, Ed Mason, Natl Dir, 317 Fogwell Rd, Centreville, MD 21617. Phone: (410) 758-0777.

ANVIL MOUNTAIN RUN. July 4. Nome, AK. At 8 AM the 17K run up 1,134-ft Anvil Mountain and return to the city of Nome starts the day's activities. Annually, July 4. Fax/Msg: (907) 443-5777. For info: Rasmussen's Music Mart, PO Box 2, Nome, AK 99762-0002. Phone: (907) 443-2798.

ARMSTRONG, LOUIS: NOT BORN THIS DAY. July 4. Although he often said he was born July 4, 1900, according to personnel at the Louis Armstrong Archives of Queens College in Flushing, NY, documents in their collection indicate that Armstrong was actually born Aug 4, 1901.

BOOM BOX PARADE. July 4. Main St, Willimantic, CT. Connecticut's unique people's parade. Anyone can march, enter a float or watch; only requirement—bring a radio. No "real" bands allowed. (Marching music broadcast WILI-AM Radio and played by "boom boxes" along the parade route.) 11 AM. Est attendance: 10,000. Sponsor: WILI-AM, 720 Main St, Willimantic, CT 06226. Phone: (203) 456-1111.

BRISTOL CIVIC, MILITARY AND FIREMEN'S PARADE. July 4. Bristol, RI. The nation's oldest 4th of July parade. Features floats, bands, veteran and patriotic organizations, and military units. Patriotic exercises, a tradition dating to 1785, are held prior to the parade. Annually, July 4. Est attendance: 175,000. For info: Bristol County Chamber of Commerce, 654 Metacom Ave, Warren, RI 02885. Phone: (401) 245-0750.

COOLIDGE, CALVIN: BIRTH ANNIVERSARY. July 4. The 30th president of the US was born John Calvin Coolidge at Plymouth, VT, July 4, 1872. He succeeded to the presidency Aug 3, 1923, following the death of Warren G. Harding. Coolidge was elected president once, in 1924, but did "not choose to run for president in 1928." Nicknamed Silent Cal, he is reported to have said, "If you don't say anything, you won't be called on to repeat it." Coolidge died at Northampton, MA, on Jan 5, 1933.

DECLARATION OF INDEPENDENCE: APPROVAL AND SIGNING ANNIVERSARY. July 4. On July 4, 1776, the Declaration of Independence was approved by the Continental Congress: "Signed by Order and in Behalf of the Congress, John Hancock, President, Attest, Charles Thomson, Secretary." The official signing occurred on Aug 2, 1776. The manuscript journals of the Congress for that date state: "The declaration of independence being engrossed and compared at the table was signed by the members."

FAMILY DAY CELEBRATION. July 4. Dahlonega, GA. Voted as "Top 20 Event" by the Southeast Tourism Society. Craft booths of all kinds, continuous music all day, food booths, kids fun booths with prizes, clogging and buckdancing. Fireworks after dark at the high school football field. Est attendance: 18,000. For info: Dahlonega Welcome Center, 101 S Park St, Dahlonega, GA 30533. Phone: (706) 864-3711.

FIREWORKS CELEBRATION. July 4. Demopolis, AL. Each year the Demopolis Area Chamber of Commerce presents a spectacular array of fireworks in celebration of Independence Day. The fireworks celebration is held at the Demopolis City Landing and begins at 9 PM. Est attendance: 10,000. For info: Jane Gross, Box 667, Demopolis, AL 36732. Phone: (205) 289-0270.

FOSTER, STEPHEN: BIRTH ANNIVERSARY. July 4. Stephen Collins Foster, one of America's most famous and best-loved song writers, was born at Lawrenceville, PA, on July 4, 1826. Among his nearly 200 songs: "Oh! Susanna," "Camptown

Races," "Old Folks at Home" ("Swanee River"), "Jeanie with the Light Brown Hair," "Old Black Joe," "Beautiful Dreamer." Foster died in poverty at Bellevue Hospital at New York, NY, on Jan 13, 1864. The anniversary of his death has been observed as Stephen Foster Memorial Day by Presidential Proclamation since 1952.

FREDERICKSBURG HERITAGE FESTIVAL. July 4. Fredericksburg, VA. To create an awareness of Fredericksburg's historic heritage. Parade, raft race, hayrides, country music, fireworks and Civil War battle reenactment. Est attendance: 15,000. For info: Visitor Center, 706 Caroline St, Fredericksburg, VA 22401. Phone: (800) 678-4748.

FREEDOM FEST. July 4. Lake of the Woods, Mahomet, IL. Our annual celebration of patriotism. Family activities capped off by a spectacular fireworks display. Admission charged. Est attendance: 7,000. For info: Champaign Co Forest Preserve District, PO Box 1040, Mahomet, IL 61853. Phone: (217) 586-3360.

FREEDOM WEEK. July 4–10. To disseminate throughout the world information about freedom and liberty. For complete info and many famous quotations about freedom and liberty, send $1 to cover expense of printing, handling and postage. Annually, July 4–10. Sponsor: Intl Soc of Friendship and Good Will, Dr. Stanley Drake, Pres, 9538 Summerfield St, Spring Valley, San Diego County, CA 91977-2852. Phone: (619) 466-8882.

GATLINBURG'S JULY 4TH MIDNIGHT PARADE. July 4. Gatlinburg, TN. The first July 4th parade in the nation, held each year beginning at 12:01 AM, features thousands of lights, music, patriotic celebrations and many unusual floats. For info: Gatlinburg Chamber of Commerce, PO Box 527, Gatlinburg, TN 37738. Phone: (615) 436-7333.

GREAT CARDBOARD BOAT REGATTA. July 4. Rotary Riverview Park, Sheboygan, WI. The most spectacular and hilarious races of somewhat seaworthy craft ever launched in Wisconsin. Person-powered cardboard boats compete in various classes for prizes. Awards for the most spirited team, the most beautiful boats, boats following a theme and the most spectacular sinking (Titanic Award). Prizes for top finishers in three boat classes: propelled by oars or paddles; propelled by mechanical means such as paddlewheels, propellers; "Instant Boats" made from "Secret Kits" by spectators-turned-participants. Sponsor: School of Art and Design, Southern Illinois University, Carbondale, IL. Est attendance: 16,000. For info: John Michael Kohler Arts Center, 608 New York Ave, PO Box 489, Sheboygan, WI 53082-0489. Phone: (414) 458-6144.

HOOD RIVER OLD-FASHIONED FOURTH OF JULY. July 4. Port Marina Park, Hood River, OR. Innertube races, parade, barbecue, fun run, entertainment and fireworks over the Columbia River. Sponsor: Hood River Chamber of Commerce, Port Marina Park, Hood River, OR 97031. Phone: (800) 366-3530.

ICE CREAM SOCIAL. July 4. Indianapolis, IN. Independence Day celebration. Music, lawn games and living history. Est attendance: 2,000. For info: President Benjamin Harrison Home, PR Dept, 1230 N Delaware St, Indianapolis, IN 46202. Phone: (317) 631-1898.

INDEPENDENCE DAY (FOURTH OF JULY). July 4. The US commemorates adoption of the Declaration of Independence by the Continental Congress on July 4, 1776. The nation's birthday. Legal holiday in all states and territories.

INDEPENDENCE DAY CELEBRATION. July 4. Washington, DC. The nation's capital celebrates the 4th of July with a parade past many historic monuments on the Mall. National Symphony Orchestra concert in the evening followed by spectacular fireworks display over the Washington Monument. For info: Washington, DC Conv and Visitors Assn, 1212 New York Ave NW, Washington, DC 20005.

INDEPENDENCE DAY CHALLENGE RUN. July 4. Boyne City, MI. Eight-mile course includes 1¼-mile uphill grade and several other hills on paved and gravel roads overlooking Lake Charlevoix. Est attendance: 25,000. For info: Chamber of Commerce, 28 S Lake St, Boyne City, MI 49712. Phone: (616) 582-6832.

LEWISTOWN 4TH PARADE AND CELEBRATION. July 4. Lewistown, MT. Parade, breakfast, bingo, barbecue, baseball and games. The Shrine Circus (two performances), fireworks. Est attendance: 1,000. For info: Lewistown Chamber of Commerce, PO Box 818, Lewistown, MT 59457. Phone: (406) 538-5436.

MOUNT MARATHON RACE. July 4. Seward, AK. Grueling footrace begins in downtown Seward, then ascends and descends 3,022-ft Mt Marathon. Race began as a wager between two sourdoughs. 68th running. Est attendance: 10,000. For info: Seward Chamber of Commerce, PO Box 749, Seward, AK 99664. Phone: (907) 224-8051.

MOUNT RUSHMORE 4TH OF JULY CELEBRATION. July 4. Mt Rushmore National Memorial, SD. Features the Hill City Fife and Drum Corps; a military band and the dramatic sculpture-lighting program conclude the celebration. Est attendance: 2,000. For info: Mount Rushmore Natl Memorial, PO Box 268, Keystone, SD 57751. Phone: (605) 574-2523.

MYSTIC SEAPORT: INDEPENDENCE DAY CELEBRATION. July 4. Mystic, CT. Visitors can participate in a recreation of an 1870's Fourth of July with costumed roleplayers. There are patriotic ceremonies and a parade of the "Antiques and Horribles." Kids old-fasioned spelling bee. Est attendance: 3,500. For info: Mystic Seaport, 75 Greenmanville Ave, Box 6000, Mystic, CT 06355. Phone: (203) 572-5315.

NATIONAL COUNTRY MUSIC DAY. July 4. A day for those who love to listen to America's favorite kind of music—country music. Annually, July 4. For info: Alan W. Brue, 399 SW 8th St #1, Boca Raton, FL 33432-5732.

OATMAN SIDEWALK EGG FRYING CONTEST. July 4. Oatman, AZ. Contest held at high noon on the downtown streets of old Rt 66. Also, Old West gun fights, wild burros roaming the streets, food, entertainment and more. "Note: Only solar heat may be used to fry the eggs." 30-minute time limit. Annually, July 4. Est attendance: 2,500. For info: Spirit Mountain Productions, PO Box 2755, Bullhead City, AZ 86442. Phone: (602) 763-5885.

OLD VERMONT FOURTH. July 4. Billings Farm & Museum, Woodstock, VT. A traditional 4th of July with patriotic speeches, "1890" flags, a spelling bee for adults, ice cream making, sack races, and more. The costumed Ed Larkin dancers will perform after lunch. Est attendance: 650. For info: Billings Farm and Museum, PO Box 489, Woodstock, VT 05091. Phone: (804) 457-2355.

OLD-FASHIONED FOURTH. July 4. Old City Park, Dallas, TX. The "All-Join-In-Parade," sack races, games and special entertainment guests are a festive way to honor the birth of our nation. Annually, July 4. Est attendance: 2,500. For info: Old City Park, Dallas Co Heritage Soc, 1717 Gano, Dallas, TX 75215. Phone: (214) 421-5141.

July 4 ☆ Chase's 1995 Calendar of Events ☆

PEACHTREE ROAD RACE. July 4. Atlanta, GA. 10K run. 50,000-runner limit; advance registration only. Send self-addressed stamped envelope by Mar 1, 1995. 40,000 entrants on first-come basis and 10,000 selected by lottery from other entries postmarked in March 1995. Est attendance: 50,000. For info: Atlanta Track Club, 3097 E Shadowlawn Ave, Atlanta, GA 30305. Phone: (404) 231-9065.

PHILIPPINES: FIL-AMERICAN FRIENDSHIP DAY. July 4. National. Formerly the National Independence Day, now celebrated as Fil-American Friendship Day.

ROAN MOUNTAIN STATE PARK'S FIREWORKS JAMBOREE. July 4. Roan Mountain, TN. Bluegrass music, clogging and fireworks. Est attendance: 15,000. For info: Jennifer Wilson, Ranger Naturalist, Roan Mtn State Park, Rte 1, Box 236, Roan Mountain, TN 37687. Phone: (615) 772-3303.

SANDBLAST. July 4. Fort Lauderdale, FL. A sand-sculpture competition on Fort Lauderdale beach for amateurs and professionals, featuring live music, fireworks and food. Annually, July 4. Sponsor: *Sun Sentinel*. For info: Greater Fort Lauderdale Conv/Visitors Bureau, 200 E Las Olas Blvd, Ste 1500, Fort Lauderdale, FL 33301. Phone: (305) 765-4466.

SHELBURNE MUSEUM'S OLD-TIME FARM DAY AND GRAND OLD FOURTH. July 4. Celebrate the Fourth of July by paying tribute to Vermont's rural farm traditions. Farm animals, horse-drawn wagon rides, music and dance, crafts, games and storytelling featured. Cow-milking, butter-churning, ice-cream-making and pie-baking contests all day. Annually, July 4. Est attendance: 3,000. For info: Shelburne Museum, US Rte 7, Shelburne, VT 05482. Phone: (802) 985-3346.

SPIRIT OF FREEDOM FESTIVAL. July 4. Florence, AL. Celebrate America's birthday on the banks of the beautiful Tennessee River at McFarland Park. Enjoy sunning, swimming, bicycling, golf and entertainment. Playground, picnic tables and campground available. Annually, on July 4. Est attendance: 60,000. For info: Jimmy Ray Stanfield, Florence Parks & Recreation Dept, PO Box 2040, Florence, AL 35630. Phone: (205) 760-6416.

STOCK EXCHANGE HOLIDAY (INDEPENDENCE DAY). July 4. The holiday schedules for the various exchanges are subject to change if relevant rules, regulations or exchange policies are revised. If you have questions, phone: American Stock Exchange (212) 306-1212; Chicago Board of Options Exchange (312) 786-7760; Chicago Board of Trade (312) 435-3500; New York Stock Exchange (212) 656-2065; Pacific Stock Exchange (415) 393-4000; Philadelphia Stock Exchange (215) 496-5000.

TEN THOUSAND CRESTONIANS. July 4. Downtown and McKinley Park, Creston, IA. Parade, fireworks, flea market, talent show, historical village, carnival and food. All to celebrate US founding. Est attendance: 10,000. For info: Creston Chamber of Commerce, PO Box 471, Creston, IA 50801. Phone: (515) 782-7021.

WALES: LLANGOLLEN INTERNATIONAL MUSICAL EISTEDDFOD. July 4-9. Eisteddfod Field, Llangollen, Clwyd. For one week the little Welsh town of Llangollen becomes the most cosmopolitan place in Britain. Thousands of singers and folk dancers from more than 30 countries take part in this annual international music festival. Friendly rivalry among amateur groups performing amid the Welsh rivers and mountains. Est attendance: 120,000. For info: Llangollen Intl Musical Eisteddfod, Llangollen, Clwyd, Wales, LL20 8NG.

July 1995

S	M	T	W	T	F	S
						1
2	3	4	5	6	7	8
9	10	11	12	13	14	15
16	17	18	19	20	21	22
23	24	25	26	27	28	29
30	31					

WOOD RADIO/SMITHS INDUSTRIES FIREWORKS GALA. July 4. Ah-Nab-Awen Park (Bicentennial Park), downtown Grand Rapids, MI. The 13th annual fireworks gala held for the Grand Rapids community. The biggest and brightest fireworks display in Grand Rapids. Family and musical entertainment provided for spectators. Sponsors: Smiths Industries, WOOD Radio. Est attendance: 350,000. For info: Juli Agacinski, Promo Dir, WOOD Radio, 180 N Division, Grand Rapids, MI 49503. Phone: (616) 459-1919.

WORLD'S GREATEST LIZARD RACE. July 4. Chaparral Park, Lovington, NM. Children cheer as their lizards race down a 16-ft ramp; winners are awarded trophies. Entertainment and other games are also featured. Annually, July 4. Est attendance: 9,000. For info: Lovington Chamber of Commerce, PO Box 1347, Lovington, NM 88260. Phone: (505) 396-5311.

WROK/WZOK ANYTHING THAT FLOATS ROCK RIVER RAFT RACE. July 4. Rock River, Rockford, IL. More than 100 rafts compete for prizes in speed and creativity. Est attendance: 60,000. Sponsor: WROK/WZOK Radios, Jan Thorpe, Box 6186, Rockford, IL 61125. Phone: (815) 399-2233.

WSB-TV SALUTE 2 AMERICA PARADE. July 4. Atlanta, GA. One of the largest 4th of July parades in the US. 34th annual parade steps off at 1 PM; "Let Freedom Sing." Est attendance: 300,000. Sponsor: WSB-TV Salute 2 America Parade, 1601 W Peachtree St NE, Atlanta, GA 30309. Phone: (404) 897-7385.

ZAMBIA: UNITY DAY. July 4. First Tuesday in July. Memorial day for Zambians who died in the struggle for independence. Political rallies stressing solidarity throughout country.

ZOOBALEE. July 4. Lee Richardson Zoo, Garden City, KS. 2K and 10K runs, pie-eating contest, celebrity dunk tank, ethnic foods, bands, turtle races, animal mimic contest (sound of the animal), big wheel races, animal trick demonstrations and much, much more. Est attendance: 4,000. For info: Friends of Lee Richardson Zoo, Dianne Duncan, Development Dir, Box 1638, Garden City, KS 67846. Phone: (316) 276-6243.

BIRTHDAYS TODAY

Al Davis, 66, football executive, born at Brockton, MA, July 4, 1929.
Harvey Grant, 30, basketball player, born at Augusta, GA, July 4, 1965.
Horace Junior Grant, 30, basketball player, born at Augusta, GA, July 4, 1965.
Leona Helmsley, 75, former hotel executive who went to jail, born at Brooklyn, NY, July 4, 1920.
Ann Landers (Esther Pauline Friedman), 77, advice columnist, born at Sioux City, IA, July 4, 1918.
Gina Lollobrigida, 67, actress, photographer (*Belles de Nuit; Bread, Love and Dreams*), born at Auviaco, Italy, July 4, 1928.
Jawann Oldhem, 38, basketball player, born Seattle, WA, July 4, 1957.
Geraldo Rivera, 52, journalist, talk show host ("Geraldo"), author (*Exposing Myself*), born at New York, NY, July 4, 1943.
Eva Marie Saint, 71, actress (Oscar for *On the Waterfront*; *North by Northwest, Exodus, The Sandpiper, The Stalking Moon*), born Newark, NJ, July 4, 1924.
Pam Shriver, 33, tennis player, born at Baltimore, MD, July 4, 1962.
Neil Simon, 68, playwright (*The Odd Couple, Barefoot in the Park*), born at New York, NY, July 4, 1927.
George Steinbrenner, 65, principal owner of the NY Yankees, former general manager, born at Rocky River, OH, July 4, 1930.

Abigail Van Buren (Pauline Esther Friedman), 77, advice columnist, born at Sioux City, IA, July 4, 1918.

JULY 5 — WEDNESDAY
186th Day — Remaining, 179

AMERICAN ASSOCIATION OF HANDWRITING ANALYSTS/AMERICAN HANDWRITING ANALYSIS FOUNDATION ANNUAL CONVENTION. July 5-9. Vancouver, BC, Canada. Speakers, books, workshops and social gatherings of graphologists and handwriting enthusiasts from around the world at this joint convention of two of the largest American handwriting analysis organizations. Est attendance: 200. For info: Amer Assn Handwriting Analysts, PO Box 6201, San Jose, CA 95150. Phone: (708) 325-2266.

CANADA: THE NORTH AMERICAN. July 5-9. Spruce Meadows, Calgary, Alberta. Show jumping tournament featuring the Spruce Meadows North American Show Jumping Championships. Est attendance: 80,000. For info: Spruce Meadows, RR #9, Calgary, AB, Canada T2J 5G5. Phone: (403) 254-3200.

CAPE VERDE: NATIONAL DAY. July 5.

CHESAPEAKE TURTLE DERBY. July 5. Baltimore, MD. Ten races with winner of each race competing in special grand sweepstakes race to determine the grand champ. Trophies and prizes awarded to the winners. Est attendance: 400. For info: Office of Adventures in Fun, Clarence "Du" Burns Arena, 1301 S Ellwood Ave, Baltimore, MD 21224. Phone: (410) 396-9177.

ISLE OF MAN: TYNWALD DAY. July 5. Tynwald Hill at St. John's. Traditionally, on July 5, Old Midsummer Day, the island's parliament of Tynwald assembles at the meeting place of the Vikings, to promulgate new laws.

MOON PHASE: FIRST QUARTER. July 5. Moon enters First Quarter phase at 4:02 PM, EDT.

NATIONAL LABOR RELATIONS ACT OF 1935 (THE WAGNER ACT): ANNIVERSARY. July 5. This bill, signed into law July 5, 1935, guaranteed workers the right to organize and bargain collectively with their employers. It also prohibited the formation of company unions. An enforcement agency, the National Labor Relations Board, was created by the Act.

SCOTLAND: BELL'S SCOTTISH OPEN. July 5-8. Gleneagles Hotel Golf Club, Auchterarder, Tayside. Scotland's principal professional golf tournament. For info: PR, United Distillers UK plc, Cherrybank, Perth, Scotland PH2 ONG.

VENEZUELA: INDEPENDENCE DAY. July 5. National holiday. Commemorates Proclamation of Independence from Spain on this date in 1811.

WORKAHOLICS DAY. July 5. A day to recognize this often misunderstood group of people. Let's salute these hard-working people who are too busy to take the time out to give themselves a much deserved pat on the back for a job well done. Annually, July 5 For info: Workman Publishing Co., Attn: Publicity Dept, 708 Broadway, New York, NY 10003. Phone: (800) 722-7202.

ZETKIN, CLARA: BIRTH ANNIVERSARY. July 5. Women's rights advocate, born at Wiederau, Germany, July 5, 1857. Zetkin has been credited with being the initiator of International Women's Day, which has been observed on Mar 8 at least since 1910. She died at Arkhangelskoe, Russia, June 20, 1933. See also: "International (Working) Women's Day" (Mar 8).

BIRTHDAYS TODAY

Eliot Feld, 53, dancer, born at Brooklyn, NY, July 5, 1942.
Richard "Goose" Gossage, 44, baseball player, born at Colorado Springs, CO, July 5, 1951.
Katherine Helmond, 61, actress (*The House of Blue Leaves*, "Soap," "Who's the Boss"), born Galveston, TX, July 5, 1934.
Huey Lewis (Hugh Anthony Cregg III), 44, singer (lead of Huey Lewis and the News; "Do You Believe in Love"), born at San Francisco, CA, July 5, 1951.

James Lofton, 39, football player, born at Fort Ord, CA, July 5, 1956.
Robbie Robertson, 51, musician (guitarist with The Band or Hawks), born at Toronto, Ontario, Canada, July 5, 1944.
Janos Starker, 71, musician, born at Budapest, Hungary, July 5, 1924.

JULY 6 — THURSDAY
187th Day — Remaining, 178

BABE RUTH COMMEMORATIVE STAMP: ANNIVERSARY. July 6, 1983. The US Postal Service issued the Babe Ruth Commemorative postage stamp, which was a graphic rendering of a photograph of Ruth in his classic follow-through of a typical home run swing—bat held high and body twisted.

BELGIUM: OMMEGANG PAGEANT. July 6. Splendid historic festival of medieval pageantry at the illuminated Grand-Palace in Brussels. The annual event (first Thursday in July) re-creates an entertainment given in honor of Charles V and his court.

CANADA: QUEBEC INTERNATIONAL SUMMER FESTIVAL. July 6-16. Quebec City, Quebec. Artists from North America, Africa and Europe perform in the streets and parks of Old Quebec. Est attendance: 2,500,000. For info: Quebec Intl Summer Festival, 160 Saint-Paul St, CP 24, Succ B, Quebec, PQ, Canada G1K 7A1. Phone: (418) 692-4540.

CANADA: SHAKESPEARE ON THE SASKATCHEWAN FESTIVAL. July 6-Aug 20. Saskatoon, Saskatchewan. Innovative and distinctly "Canadian" Shakespearean productions at this award-winning festival. Tentative 1995 plans include Shakespeare's *Henry IV*, *Comedy of Errors* and an opera. Est attendance: 10,000. For info: Tourism Saskatchewan, Saskatchewan Trade and Conv Center, 1919 Saskatchewan Dr, Regina, Canada, S4P 3V7. Phone: (800) 667-7191.

COMOROS: INDEPENDENCE DAY: 20th ANNIVERSARY. July 6. Federal and Islamic Republic of Comoros commemorates Declaration of Independence on July 6, 1975.

ETTELBRUCK, LUXEMBOURG: REMEMBRANCE DAY: 50th ANNIVERSARY. July 6. In honor of US General George Patton, Jr, liberator of the Grand-Duchy in 1945, who is buried at the American Military Cemetery in Hamm, Germany, among 5,100 soldiers of his famous Third Army.

FINLAND: TIME OF MUSIC. July 6-12. Viitasaari. Festival in a town of 300 lakes focuses on contemporary music with top international and Finnish artists performing more than 100 works from some of the most interesting composers of our time, many being performed for the first time ever. Also, a sizeable program for children. Est attendance: 8,000. For info: Finnish Tourist Board, 655 Third Ave, New York, NY 10017. Phone: (212) 949-2333.

FIRST AIRSHIP CROSSING OF ATLANTIC: ANNIVERSARY. July 6. The first airship crossing of the Atlantic was completed as a British dirigible landed at New York's Roosevelt Field on July 6, 1919.

FIRST BLACK US STATE'S ATTORNEY. July 6. On July 6, 1961, Cecil Francis Poole became the first black US attorney when he was sworn in as US attorney for the Northern District of California. He served until his retirement on Feb 3, 1970.

FIRST SUCCESSFUL ANTIRABIES INOCULATION: ANNIVERSARY. July 6. On July 6, 1885, Louis Pasteur gave the first successful antirabies inoculation to a boy who had been bitten by an infected dog.

MAJOR LEAGUE BASEBALL HOLDS FIRST ALL-STAR GAME: ANNIVERSARY. July 6. Prior to the summer of 1933, All-Star contests consisted of pre- and postseason exhibitions that often found teams made up of a few stars playing beside journeymen and even minor leaguers. On July 6, 1933, the first midsummer All-Star Game was held at Comiskey Park, Chicago, IL. Babe Ruth led the American League, with a home run, as they defeated the National League 4-2.

★ Chase's 1995 Calendar of Events ★

July 6–7

MALAWI: INDEPENDENCE DAY. July 6. National holiday. Commemorates attainment of independence from Britain on July 6, 1964. Malawi became a republic on July 6, 1966.

MARION COUNTY FAIR. July 6–9. State Fairgrounds, Salem, OR. Carnival, entertainment, commercial exhibits, open class exhibits, demonstrations, contests and livestock. Est attendance: 40,000. For info: Mary Boedigheimer, Coord, PO Box 703, Salem, OR 97308. Phone: (503) 585-9998.

MONTANA GOVERNOR'S CUP WALLEYE TOURNAMENT. July 6–8. Fort Peck, MT. Two-person team event, limited to 200 teams. 10% of the teams entering receive cash awards of up to $10,000. $260.00 entry fee. Kids' fishing event also. Annually, the 2nd weekend in July. Est attendance: 1,000. For info: Glasgow Area Chamber of Commerce & Agriculture, Box 325, Glasgow, MT 59230. Phone: (406) 228-2222.

◆ **OPERATION OVERCAST: 50th ANNIVERSARY.** July 6, 1945. As the end of the war approached, the US Army had begun to move the German scientific equipment and scientists from the German territory designated for Russian occupation. On this date, the American Joint Chiefs of Staff authorized Operation Overcast, under which 350 German and Austrian scientists were transported to the US in a manner of months.

POTTER, (HELEN) BEATRIX: BIRTH ANNIVERSARY. July 6. Creator of the Peter Rabbit stories for children, born at London, England, July 6, 1866. Died at Sawrey, Lancashire, Dec 22, 1943.

REPUBLICAN PARTY FORMED: ANNIVERSARY. July 6. The Republican Party formally originated at a convention in Ripon, WI, on July 6, 1854.

SPACE MILESTONE: SOYUZ 21 (USSR). July 6. Two cosmonauts, Colonel B. Volynov and Lieutenant Colonel V. Zholobov, traveled to Salyut 5 space station (launched June 22, 1976) to study Earth's surface and conduct zoological-botanical experiments. Forty-eight-day stay on space station. Return landing on Aug 24. Launch date was July 6, 1976.

TURKEY RAMA. July 6–8. McMinnville, OR. This annual event began as a result of many years' development of turkey business in the area. In 1962 McMinnville was crowned Turkey Capital of the World. Activities include traditional turkey BBQ, biggest turkey contest, street sales, fun runs, athletic tournaments, carnival, art and car shows, entertainment. Est attendance: 27,000. For info: Greater McMinnville Chamber of Commerce, 417 N Adams, McMinnville, OR 97128. Phone: (503) 472-6196.

BIRTHDAYS TODAY

Allyce Beasley, 41, actress (Agnes on "Moonlighting"), born Brooklyn, NY, July 6, 1954.
Ned Beatty, 58, actor (*Deliverance*, "Szysznyk," *Hear My Song*), born in Lexington, KY, July 6, 1937.
Grant Goodeve, 43, actor ("Eight is Enough," "Dynasty"), born at New Haven, CT, July 6, 1952.
Merv Griffin, 70, TV host, business executive, born at San Mateo, CA, July 6, 1925.
Wojciech Jaruzelski, 72, Polish army officer and political leader, born at Kurow, Poland, July 6, 1923.
Janet Leigh, 68, actress (*That Forsyte Woman, Psycho*), born at Merced, CA, July 6, 1927.
Nancy Davis Reagan, 74, former First Lady, born at New York, NY, July 6, 1921.
Della Reese (Deloreese Patricia Early), 63, singer ("Don't You Know," "And That Reminds Me"), born at Detroit, MI, July 6, 1932.
Sylvester Stallone, 49, actor (*Rocky* films, *Cliffhanger*), director (*Staying Alive*), born at New York, NY, July 6, 1946.
Burt Ward, 50, actor (Robin in "Batman"), born at Los Angeles, CA, July 6, 1945.

JULY 7 — FRIDAY
188th Day — Remaining, 177

BENNINGTON MUSEUM ANTIQUE SHOW AND SALE. July 7–9. Monument Elementary School, Bennington, VT. 15th annual benefit for the Bennington Museum featuring 35 fine antique dealers. Est attendance: 3,500. For info: Bennington Museum, W Main Street, Bennington, VT 05201. Phone: (802) 447-1571.

BONZA BOTTLER DAY™. July 7. To celebrate when the number of the day is the same as the number of the month. Bonza Bottler Day is an excuse to have a party at least once a month. T-shirts. For info: Elaine Fremont, 203 Waddell Rd, Taylors, SC 29687. Phone: (803) 244-2023.

CANADA: CALGARY EXHIBITION AND STAMPEDE. July 7–16. Calgary, Alberta. Billed as the "Greatest Outdoor Show on Earth!" The world's top professional cowboys compete for supremacy during the Calgary Stampede Half Million Dollar Rodeo. Each evening, nine heart-stopping chuckwagon races explode in an all-out dash to the finish. Outdoor stage spectaculars, International Stock Show, midway, casino and a city-wide celebration are what you can expect, jam-packed into 10 of the most varied and exciting days you'll ever experience. Est attendance: 1,200,000. For info: Calgary Stampede Assn, PO Box 1060 Stn M, Calgary, AB, Canada T2P 2K8. Phone: (800) 661-1260.

CANADA: DRUMMONDVILLE WORLD FOLKLORE FESTIVAL. July 7–16. Drummondville, Quebec. Groups from more than 20 countries participate in the singing and dancing of folkloric songs and themes. Fax: (819) 474-6585. Est attendance: 1,000,000. For info: Drummondville World Folklore Festival, 405 Saint-Jean St, Drummondville, PQ, Canada J2B 5L7. Phone: (819) 472-1184.

CANADA: THE GREAT RENDEZVOUS. July 7–16. Old Fort Williams, Thunder Bay, Ontario. Join the celebration of the arrival of the partners, clerks and voyageurs to the inland headquarters of the Northwest Company with historical re-enactors camping around the palisade, voyageur games and something for everyone to enjoy. Annually, the first Friday of July through the third Sunday of July. Est attendance: 14,000. For info: Laura J. Craig, Reservations, Sales and Customer Service, Vickers Heights Post Office, Thunder Bay, Ont, Canada P0T 2Z0. Phone: (807) 577-8461.

ELKHART GRAND PRIX AND MOTOR SPORTS WEEKEND. July 7–9. Elkhart, IN. The largest karting event race in the world follows a 4,900-ft course through downtown streets. More than 700 competitors from all over the world. Also features top name race car drivers, antique and hot rod car show and swap meet, racing car displays, carnival and celebrity entertainment at the 7th annual holding of this event. Annually, the second weekend in July. Est attendance: 60,000. Sponsor: Curt Paluzzi, Natl Kart News, 51535 Bittersweet Rd, Granger, IN 46530. Phone: (219) 277-0033.

July 1995

S	M	T	W	T	F	S
						1
2	3	4	5	6	7	8
9	10	11	12	13	14	15
16	17	18	19	20	21	22
23	24	25	26	27	28	29
30	31					

☆ Chase's 1995 Calendar of Events ☆ July 7-8

ENGLAND: BRITISH GRAND PRIX. July 7-9. Silverstone Circuit, Towcester, Northamptonshire. Britain's round of the Formula One World Championship is the highlight of the motor racing calendar. For info: Silverstone Circuit Ltd, Towcester, Northamptonshire, England NN12 8TN.

HISTORIC FARM DAYS. July 7-9. Middle Fork River Forest Preserve, Penfield, IL. This weekend-long event features antique farm machinery displays, agricultural demonstrations, square dancing, a tractor parade, children's pedal tractor pull, food and entertainment. Admission charged. Est attendance: 3,000. For info: Champaign County Forest Preserve District, PO Box 1040, Mahomet, IL 61853. Phone: (217) 586-3360.

JAPAN: TANABATA (STAR FESTIVAL). July 7. As an offering to the stars, children set up bamboo branches to which colorful strips of paper bearing poems are tied.

KLOWN KARNIVAL. July 7-8. Plainview, NE. Bar-B-Que, clowns everywhere, dance, home town carnival, parade, horseshoe tournament, ice cream social, sand volleyball, Klown Band. Annually, the second weekend in July. Est attendance: 1,800. Sponsor: Plainview Chamber of Commerce, PO Box 813, Plainview, NE 68769. Phone: (402) 582-4433.

LINCOLN ASSASSINATION CONSPIRATORS: 130th HANGING ANNIVERSARY. July 7. Four persons convicted of complicity with John Wilkes Booth in the assassination of President Abraham Lincoln (Apr 14, 1865) were hanged at Washington, DC, on July 7, 1865. The four: Mary E. Surratt, Lewis Payne, David E. Harold and George A. Atzerodt.

SANTA FE CHAMBER MUSIC FESTIVAL. July 7-Aug 21. St. Francis Auditorium, Museum of Fine Arts, Santa Fe, NM. Highly acclaimed chamber music festival draws on international talent, featuring works from the baroque, romantic and classical periods, including world premiere performances. Annually, beginning during the first full week in July through the third week in August. Est attendance: 14,000. For info: Santa Fe Chamber Music Fest, Gretchen Grogan, Dir of PR, PO Box 853, Santa Fe, NM 87504. Phone: (505) 983-2075.

SLOW PITCH SOFTBALL TOURNAMENT. July 7-9. Elm Park, Williamsport, PA. 22nd annual charitable tournament with 64 teams. Sponsor: Miller Lite. Est attendance: 8,500. For info: Williamsport Beverage Co, Inc, Don Phillips, 532 Sylvan Dr, South Williamsport, PA 17701. Phone: (717) 322-3331.

SOLOMON ISLANDS: NATIONAL HOLIDAY. July 7.

STATE GAMES OF OREGON. July 7-9. Portland, OR. Oregon's Olympic style amateur sports festival. Annually, the first weekend after July 4th. Est attendance: 16,000. For info: Ron Allen, Exec Dir, 4840 SW Western Ave, Beaverton, OR 97005. Phone: (503) 520-1319.

SURRATT, MARY: EXECUTION ANNIVERSARY. July 7. Mary E. Surratt, who had been convicted by a military commission of conspiracy in the assassination of President Abraham Lincoln, was hanged on July 7, 1865, becoming the first woman executed for a crime in the US. Her conviction was and is a subject of controversy, as the only crime she appeared to have committed was to own the boarding house where John Wilkes Booth planned the assassination.

TALKEETNA MOOSE DROPPING FESTIVAL (WITH MOOSE DROPPING TOSS GAME). July 7-9. Talkeetna, AK. Parade, booths, entertainment, 5K fun run, Mountain Mother contest and the famous moose dropping toss game. Annually, the second weekend in July. Est attendance: 2,000. For info: Talkeetna Historical Soc, PO Box 76, Talkeetna, AK 99676. Phone: (907) 733-2487.

TANZANIA: SABA SABA DAY. July 7. Marks the day in 1954 when Tanzania mainland's ruling party, TANU, was formed.

WILD HORSE STAMPEDE. July 7-9. Wolf Point, MT. The "Granddaddy" of all Montana rodeos features a wild horse race, three rodeos, three parades and Native American culture. Est attendance: 9,500. For info: Wolf Point Chamber of Commerce, Box 237, Wolf Point, MT 59201. Phone: (406) 653-2012.

WINKIES OF OZ CONVENTION. July 7-9. Escanaba, MI. Celebrates Dorothy's melting of the Wicked Witch of the West. Est attendance: 150. Sponsor: The Intl Wizard of Oz Club, Inc, Fred M. Meyer, Secy, 220 N 11th St, Escanaba, MI 49829.

BIRTHDAYS TODAY

Pierre Cardin, 73, fashion designer, born at Venice, Italy, July 7, 1922.

Alan J. Dixon, 68, former US Senator (D, Illinois), born at Belleville, IL, July 7, 1927.

Shelley Duvall, 46, actress (Olive Oyl in *Popeye*; *Nashville*, *Roxanne*), born Houston, TX, July 7, 1949.

Jessica Hahn, 36, secretary, born at Massapeque, NY, July 7, 1959.

William M. Kunstler, 76, lawyer, born at New York, NY, July 7, 1919.

Gian Carlo Menotti, 84, composer, born at Cadigliano, Italy, July 7, 1911.

Wally Phillips, 70, Chicago radio personality, born at Portsmouth, OH, July 7, 1925.

Ralph Lee Sampson, 35, former basketball player, born at Harrisonburg, VA, July 7, 1960.

Doc Severinsen, 68, composer, conductor, musician (band leader on "The Tonight Show"), born Arlington, OR, July 7, 1927.

Ringo Starr (Richard Starkey), 55, singer, musician (drummer with The Beatles), born at Liverpool, England, July 7, 1940.

Matt Suhey, 37, former football player, born at Bellefonte, PA, July 7, 1958.

JULY 8 — SATURDAY
189th Day — Remaining, 176

ASPINWALL CROSSES US ON HORSEBACK: ANNIVERSARY. July 8. On July 8, 1911, Nan Jane Aspinwall rode into New York City carrying a letter to Mayor William Jay Gaynor from San Francisco Mayor Patrick Henry McCarthy, becoming the first woman to cross the US on horseback. She began her trip in San Francisco on Sept 1, 1910, and covered 4,500 miles in 301 days.

BIG LAKE REGATTA WATER FESTIVAL. July 8-9. (Also July 15-16.) Big Lake, AK. Family fun such as a decorated boat parade, water and jet skiing and a big top tent with games. The following weekend is a triathlon of biking, swimming and running. Annually, the second and third weekends in July. Est attendance: 300. For info: Big Lake Chamber of Commerce, Box 520067, Big Lake, AK 99652. Phone: (907) 892-7144.

BON ODORI "FESTIVAL OF THE LANTERNS." July 8. Chicago, IL. 500 participants, most clad in colorful kimonos, dance in celebration to music of different prefectures of Japan. The beat of the huge *taiko* (drum) helps keep tempo. Dances are performed outdoors, and public participation is encouraged. 8 PM. Annually, the second Saturday in July. Est attendance: 750. For info: Midwest Buddhist Temple, 435 W Menomonee St, Chicago, IL 60614. Phone: (312) 943-7801.

CANADA: HARRISON FESTIVAL OF THE ARTS. July 8-15. Harrison Hot Springs, British Columbia. African roots theme with music (African, Caribbean, Latin, blues, gospel, jazz) on three stages. Theater and visual art exhibits; crafts market. Various venues throughout the village. Variety of activities for the entire family. Est attendance: 25,000. For info: Harrison Fest, Box 399, Harrison Hot Springs, BC, Canada V0M 1K0. Phone: (604) 796-3664.

July 8 ☆ *Chase's 1995 Calendar of Events* ☆

CANADA: SASKATCHEWAN AIR SHOW '95. July 8–9. The largest annual air show on the prairies includes an impressive array of military and civilian aeronautics performances and ground displays. Est attendance: 100,000. For info: Captain Andrew Mckenzie, Deputy Marketing Coordinator, 15th Wing, PO Box 5000, Moose Jaw, Sask, Canada S6H 7Z8. Phone: (800) 461-7469.

CANADA: SASKATOON EXHIBITION. July 8–15. Exhibition Centre, Saskatoon, Saskatchewan. Free stage entertainment, concessions, horse racing, midway, parade, 4-H Jr activities, casino, exhibits, beef, dairy, sheep and swine shows, Saskatchewan Youth Talent Search. Est attendance: 200,000. For info: Tourism Saskatchewan, Saskatchewan Trade and Conv Center, 1919 Saskatchewan Dr, Regina, Sask, Canada S4P 3V7. Phone: (800) 667-7191.

DECLARATION OF INDEPENDENCE: FIRST PUBLIC READING ANNIVERSARY. July 8. Colonel John Nixon read the Declaration of Independence to the assembled residents at Philadelphia's Independence Square, July 8, 1776.

EASTER IN JULY LILY FESTIVAL. July 8–9. Smith River, CA. Gala celebration of the lily bulb industry in the Easter lily capitol of the world. Annually, the second weekend in July. Est attendance: 9,000. For info: Rowdy Creek Fish Hatchery, PO Box 328, Smith River, CA 95567. Phone: (707) 487-3443.

ECKSTINE, BILLY: BIRTH ANNIVERSARY. July 8, 1914. Band leader and bass-baritone singer Billy Eckstine was born William Clarence Eckstein at Pittsburgh, PA. After performing with the Earl Hines band for almost 20 years, Eckstine formed his own band in 1944. At one time or another the band's ranks included Charlie Parker, Dizzy Gillespie, Miles Davis, Fats Navarro, Dexter Gordon, Gene Ammons, Art Blakey and vocalist Sarah Vaughan—some of the greatest be-bop musicians of all time. Among Eckstine's hits were "Fools Rush In," "Everything I Have Is Yours," "I Apologize," "My Foolish Heart," "Blue Moon" and "Body and Soul." Billy Eckstine died Mar 8, 1993, at Pittsburgh.

ENGLAND: STRATFORD-UPON-AVON FESTIVAL. July 8–22. Various venues, Stratford-upon-Avon, Warwickshire. Events will include music, theatre, art and dance. For info: J. Hargreaves, Stratford-upon-Avon Festival, Ryon Hill House, Warwick Rd, Stratford-upon-Avon, Warwickshire, England CV37 0NZ.

FINLAND: SAVONLINNA OPERA FESTIVAL. July 8–Aug 5. Savonlinna. One of Europe's most important and best known music festivals. Est attendance: 60,000. For info: Savonlinna Opera Festival, Box Office, Olavinkatu 35, SF-57130 Savonlinna, Finland,.

FISHING HAS NO BOUNDARIES—SANDUSKY. July 8–9. Lake Erie, Sandusky, OH. A two-day fishing experience for disabled persons. Any disability, age, sex, race, etc, eligible. Fishing with experienced guides attended by 75 participants and 200 volunteers. For info: Fishing Has No Boundaries, 3807 Deer Path Dr, Sandusky, OH 44870. Phone: (419) 626-6211.

July 1995

S	M	T	W	T	F	S
						1
2	3	4	5	6	7	8
9	10	11	12	13	14	15
16	17	18	19	20	21	22
23	24	25	26	27	28	29
30	31					

HANDCAR RACES AND STEAM FESTIVAL. July 8–9. Felton, CA. Teams race railroad handcars along 300 meters of track. Festival displays rare, old steam- and gas-powered farm and railroad equipment. For info: Jeanette Guire, Roaring Camp & Big Trees Railroad, Box G-1, Felton, CA 95018. Phone: (408) 335-4484.

HEALTHY EXCHANGES ANNUAL POTLUCK. July 8. Fairgrounds, DeWitt, IA. A low-fat, low-sugar potluck for everyone in the US who uses Healthy Exchanges recipes. Old-time fun, games and music for the entire family. Est attendance: 750. For info: JoAnna M. Lund, Ed/Founder, Healthy Exchanges Food Newsletter, PO Box 124, DeWitt, IA 52742. Phone: (319) 659-8234.

IROQUOIS ARTS SHOWCASE II. July 8–9. Iroquois Indian Museum, Howes Cave, NY. Demonstrations of Iroquois arts and crafts including beadwork, cornhusk dolls, pottery and more. Many items for sale by the artists. Children's activities. Iroquois social dancing, storytelling and nature walks. Annual. Est attendance: 2,500. Sponsor: Iroquois Indian Museum, PO Box 7, Caverns Rd, Howes Cave, NY 12092. Phone: (518) 296-8949.

THE KIDS BRIDGE. July 8–Oct 1. Pittsburgh Children's Museum, Pittsburgh, PA. Using the 46-foot-long bridge constructed inside The Children's Museum in Boston as a model, this exhibition, like the original, teaches multiculturalism through activity. The bridge symbolically spans cultural and ethnic differences and marks the entrance to this interactive exhibit where children play hopscotch from different countries, listen to cross-cultural lullabies, and learn common phrases in 5 different languages. Call or write the Smithsonian for other dates and venues. For info: Smithsonian Institution Traveling Exhibition Service, 1100 Jefferson Dr SW, Ste 3146, Washington, DC 20560. Phone: (202) 357-2700.

NEWPORT MUSIC FESTIVAL. July 8–23. Newport, RI. Unique chamber music programs, American debuts and international artists with special guest artists from the Festival's 27-year history. Three concerts daily held in Newport's fabled mansions. Morning and afternoon concerts $25, evening concerts $30. Box office phone after June 1: (401) 849-0700. Sponsor: Yamaha Corp of America. Est attendance: 22,000. For info: The Newport Music Festival, Mark P. Malkovich III, PO Box 3300, Newport, RI 02840-0993. Phone: (401) 846-1133.

OLD-FASHIONED ICE CREAM FESTIVAL. July 8–9. Rockwood Museum, Wilmington, DE. Old-fashioned ice cream social. Hot air balloons, craft show, antique show, baby parade, fashion show, old-time music, high-wheeled bicycles, games and homemade ice cream. Annually, the weekend after the 4th of July weekend. Est attendance: 14,000. For info: Rockwood Museum, 610 Shipley Rd, Wilmington, DE 19809. Phone: (302) 761-4340.

OLIVE BRANCH PETITION: ANNIVERSARY. July 8. Representatives of New Hampshire, Massachusetts Bay, Rhode Island, Providence, Connecticut, New York, New Jersey, Pennsylvania, Delaware, Maryland, Virginia, North Carolina and South Carolina signed, on July 8, 1775, a petition from the Congress to the King (George III), a final attempt by moderates in the Second Continental Congress to avoid a complete break with England.

SETTLERS DAYS. July 8–9. Bedford, PA. Native American drumming and singing will accompany exhibition and competition dancing. American Indian arts and crafts and native foods. Est attendance: 5,000. For info: Old Bedford Village, PO Box 1976, Bedford, PA 15522. Phone: (814) 623-1156.

SODBUSTER DAYS. July 8–9. Fort Ransom State Park, Valley City, ND. Remember the way things were done in rural North Dakota during the early 1920s with shelling corn by hand, rope weaving, horse-drawn plowing and haying. Est attendance: 6,000. For info: Ft Ransom State Park, RR1, Box 20A, Ft Ransom, ND 58033. Phone: (701) 973-4331.

Chase's 1995 Calendar of Events — July 8–9

SONGS OF MY PEOPLE. July 8–Aug 20. Fairpark Culture Series, Shreveport, LA. See Mar 4, for full description. Call or write the Smithsonian for other dates and venues. For info: Smithsonian Institution Traveling Exhibition Service, 1100 Jefferson Dr SW, Ste 3146, Washington, DC 20560. Phone: (202) 357-2700.

SUNG, KIM IL: DEATH ANNIVERSARY. July 8, 1994. President Kim Il Sung, the only leader in the history of North Korea, died just a few weeks before an historic summit with the president of South Korea was to take place. A Stalinist-styled dictator, Sung had created a god-like personality cult surrounding himself and his son and presumed heir apparent Kim Jong Il. His death came at a crucial time in world politics. North Korea and the US had recently cooled rhetoric regarding North Korea's nuclear program and had begun further talks just hours prior to the announcement of Sung's death. The North–South Summit and the US–North Korean talks were postponed.

THREE RIVERS FESTIVAL. July 8–16. Ft Wayne, IN. A city-wide extravaganza of more than 250 events including parades, food, hot air balloons, concerts, competitions and fireworks. Annually, beginning the Saturday after the first Thursday after the 4th of July. Est attendance: 1,500,000. For info: Three Rivers Festival, 2301 Fairfield Ave, Ste 107, Fort Wayne, IN 46807. Phone: (219) 745-3378.

TIVOLI—A BITE OF SCANDINAVIA. July 8–9. Nordic Heritage Museum, Seattle, WA. A two-day outdoor Scandinavian festival featuring crafts, foods, entertainment and children's activities. Swedish pancake breakfast in morning, barbecued salmon in afternoon and food booths throughout both days. Est attendance: 3,500. For info: Marianne Forssblad, Dir, Nordic Heritage Museum, 3014 W 67th, Seattle, WA 98117. Phone: (206) 789-5707.

TUPPER LAKE WOODSMEN'S DAYS (WITH CHAINSAW SCULPTURING CONTEST). July 8–9. Tupper Lake, NY. Competitions for woodsmen and lumberjacks. Includes the 5th annual Northeast Regional Chainsaw Sculpturing Contest. Annually, the second weekend in July. Est attendance: 33,000. For info: Tupper Lake Woodsmen's Assn, 19 Front St, PO Box 759, Tupper Lake, NY 12986. Phone: (518) 359-9444.

UGLY TRUCK CONTEST. July 8. Pelican Rapids, MN. Farm trucks and beat-up pick-ups, the worse the better, are put on display and judged. The only requirement is that the vehicle must travel under its own power. Owner of the ugliest truck wins a cash prize. Est attendance: 100. For info: Ugly Truck Contest, c/o Len Zierke, PO Box 435, Pelican Rapids, MN 56572. Phone: (218) 863-6693.

WAYNE CHICKEN SHOW. July 8. Wayne, NE. To allow humankind to pay tribute to chickenkind (without laying the proverbial egg). Fun run and parade. Annually, the second Saturday in July. Est attendance: 10,000. Sponsor: The Wayne Chicken Show Committee, Jane O'Leary, Box 262, Wayne, NE 68787. Phone: (402) 375-3729.

WESTERN HERITAGE DAYS. July 8–9. Grant-Kohrs Ranch NHS (National Historic Site), Deer Lodge, MT. Demonstrations of blacksmithing, chuckwagon cooking, and draft horse use. Calf branding and special programs for children. Free admission. Annually, the second full weekend in July. Est attendance: 1,200. For info: Superintendent, Grant-Kohrs Ranch NHS, PO Box 790, Deer Lodge, MT 59722. Phone: (406) 846-2070.

WORLD FOLKFEST. July 8–15. Springville, UT. Large international folk dance event. No dance performances on Sunday, July 9. Est attendance: 26,000. Sponsor: Springville World Folkfest, 50 S Main, PO Box 306, Springville, UT 84663. Phone: (801) 489-2726.

BIRTHDAYS TODAY

Roone Arledge, 64, TV executive, born at Forest Hills, NY, July 8, 1931.
Raffi Cavoukian, 47, children's singer and songwriter, born at Cairo, Egypt, July 8, 1948.
Kim Darby, 47, actress ("Rich Man, Poor Man," *True Grit*), born Los Angeles, CA, July 8, 1948.
Phil Gramm, 53, US Senator (R, Texas), born at Fort Benning, GA, July 8, 1942.
Cynthia Gregory, 49, ballerina, born at Los Angeles, CA, July 8, 1946.
Anjelica Huston, 44, actress (Oscar for *Prizzi's Honor*; *The Grifters, Addams Family*), born Los Angeles, CA, July 8, 1951.
Walter Kerr, 82, drama critic, born at Evanston, IL, July 8, 1913.
Steve Lawrence (Sidney Liebowitz), 60, singer ("Party Doll," "Go Away Little Girl"), born at New York, NY, July 8, 1935.
George Wilcken Romney, 88, former business executive, Governor of Michigan and presidential candidate, born at Chihuahua, Mexico, July 8, 1907.
Alyce Faye Wattleton, 52, former executive director of Planned Parenthood Federation of America, born at St. Louis, MO, July 8, 1943.

JULY 9 — SUNDAY
190th Day — Remaining, 175

BARN DAY. July 9. Two miles south of Filley, NE. Threshing demonstration, antique farm equipment, entertainment. The barn is on the National Register of Historic Sites and is the largest limestone barn in NE. Annually, the second Sunday in July. Est attendance: 700. For info: Kent Wilson, Admin, Gage County Historical Soc, PO Box 793, Beatrice, NE 68310. Phone: (402) 228-1679.

CHILDREN'S FESTIVAL. July 9–11. Jacksonville, OR. A complete cultural event for children between the ages of 2 and 12. Features performing arts, crafts, animal petting farm, music, storytelling, hands-on science and magic. All performers are volunteer and no sales for profit occur during the 5 sessions. Est attendance: 10,000. For info: The Storytelling Guild, Jackson County Library System, 413 W Main St, Medford, OR 97501. Phone: (503) 776-7288.

FOURTEENTH AMENDMENT TO US CONSTITUTION RATIFIED: ANNIVERSARY. July 9. The 14th Amendment, ratified on July 9, 1868, defined United States citizenship and provided that no State shall have the right to abridge the rights of any citizen without due process and equal protection under the law. Coming three years after the Civil War, the 14th Amendment also included provisions for barring individuals who assisted in any rebellion or insurrection against the United States from holding public office, and releasing federal and state governments from any financial liability incurred in the assistance of rebellion or insurrection against the United States.

FUTURE HOMEMAKERS OF AMERICA NATIONAL LEADERSHIP MEETING. July 9–13. Washington, DC. This meeting is a unique oopportunity to gain a national perspective on FHA/HERO activities and issues, elect officers, receive specialized leadership training and enhance chapter activities. FHA/HERO will be celebrating its 50th Anniversary. Sponsor: Cathy Burdette, Future Homemakers of America, Inc, 1910 Association Dr, Reston, VA 22091. Phone: (703) 476-4900.

HOWE, ELIAS: BIRTH ANNIVERSARY. July 9. American inventor of the sewing machine. Born on July 9, 1819, at Spencer, MA. He died on Oct 3, 1867, at Brooklyn, NY.

July 9–10 ☆ *Chase's 1995 Calendar of Events* ☆

MARTYRDOM OF THE BAB. July 9. Baha'i observance of the anniversary of the execution by a firing squad, July 9, 1850, in Tabriz, Persia, of the 30-year-old Mirza Ali Muhammed, the Bab (prophet-herald of the Baha'i Faith). One of the nine days of the year when Baha'is suspend work. For info: Natl Spiritual Assembly of the US, 1320 Nineteenth St, NW, Ste 701, Washington, DC 20036. Phone: (202) 833-8990.

NATIONAL THERAPEUTIC RECREATION WEEK. July 9–15. To increase awareness of therapeutic recreation programs and services, and to expand leisure opportunities for individuals with disabilities in their local communities. Annually, the second week in July. Sponsor: Natl Therapeutic Recreation Society, 2775 S Quincy St, Ste 300, Arlington, VA 22206. Phone: (703) 578-5548.

OLD CRAFTS DAY. July 9. Galloway House and Village, Fond du Lac, WI. To demonstrate crafts of late 1800s—as many as 30 different craftsmen. Annually, the second Sunday in July. Est attendance: 800. Sponsor: Fond du Lac County Historical Soc, 332 14th St, Fond du Lac, WI 54935. Phone: (414) 922-6390.

RADCLIFFE, ANN WARD: BIRTH ANNIVERSARY. July 9. English novelist famous for her gothic novels (fiction works especially popular in the late 18th and early 19th centuries). Among her works were *The Romance of the Forest*, *The Mysteries of Udolpho* and *The Italian*. She was born on July 9, 1764, at London, England, and died there on Feb 7, 1823.

RESPIGHI, OTTORINO: BIRTH ANNIVERSARY. July 9. Italian composer born at Bologna, Italy, on July 9, 1879. He died at Rome, on Apr 18, 1936.

SOVIET ROCKET THREAT: ANNIVERSARY. July 9. On this day in 1960, Soviet Premier Nikita Krushchev threatened military action against the US if Washington were to attempt an overthrow of the Castro regime in Cuba.

SPECIAL RECREATION DAY. July 9. A day to focus attention on the recreation rights, needs, aspirations and abilities of disabled people. For info: John A. Nesbitt, Pres, Special Recreation Inc, 362 Koser Ave, Iowa City, IA 52246-3038. Phone: (319) 337-7578.

SPECIAL RECREATION WEEK. July 9–15. To focus attention on the recreation rights, needs, aspirations and abilities of people who are disabled. Sponsor: Special Recreation, Inc, John A. Nesbitt, Pres, 362 Koser Ave, Iowa City, IA 52246-3038. Phone: (319) 337-7578.

BIRTHDAYS TODAY

Brian Dennehy, 57, actor ("Star of the Family"), born at Bridgeport, CT, July 9, 1938.
Margaret Gillis, 42, dancer, choreographer, born at Montreal, Quebec, Canada, July 9, 1953.
James Hampton, 59, actor ("F Troop," "Love, American Style"), born at Oklahoma City, OK, July 9, 1936.
Tom Hanks, 39, actor (*Forrest Gump*, *Sleepless in Seattle*; Oscar for *Philadelphia*), born Concord, CA, July 9, 1956.
David Hockney, 58, artist, born at Bradford, England, July 9, 1937.
Mathilde Krim, 69, geneticist, philanthropist, born at Como, Italy, July 9, 1926.
Leonard Pennario, 71, musician, composer, born at Buffalo, NY, July 9, 1924.
O.J. Simpson, 48, Pro Football Hall of Famer, born at San Francisco, CA, July 9, 1947.
Jimmy Smits, 37, actor (*Glitz*, "L.A. Law"), born New York, NY, July 9, 1958.

	S	M	T	W	T	F	S
July							1
	2	3	4	5	6	7	8
1995	9	10	11	12	13	14	15
	16	17	18	19	20	21	22
	23	24	25	26	27	28	29
	30	31					

JULY 10 — MONDAY
191st Day — Remaining, 174

ASHE, ARTHUR: BIRTH ANNIVERSARY. July 10. Arthur Ashe, who was renowned for his list of firsts as a black tennis player, was born July 10, 1943, at Richmond, VA. Ashe was chosen for the US Davis Cup team in 1963 and was named to captain the team in 1980. He won the US men's singles championship and US Open in 1968. In 1975 he won the men's singles title at Wimbledon. Ashe won a total of 33 career titles. In 1985 he was inducted into the International Tennis Hall of Fame. He retired from tennis in 1979 at the age of 36 after he had quadruple bypass surgery resulting from a massive heart attack he suffered earlier that year. A social activist both before and after his retirement, Ashe worked to eliminate racism and stereotyping. He was arrested numerous times for his activism, including once in South Africa while protesting that country's policy of apartheid. In 1988 he helped create inner-city tennis programs for youths. That same year his three-volume *A Hard Road to Glory: A History of the African-American Athlete* was published. Aware that *USA Today* intended to publish an article revealing that he was infected by the AIDS virus, he announced at a press conference on Apr 8, 1992, that he probably contracted HIV through a transfusion of tainted blood during a second round of bypass surgery in 1983. In September of 1992 he began a $5 million fund-raising effort on behalf of the Arthur Ashe Foundation for the Defeat of AIDS. During the last year of his life Ashe campaigned for public awareness regarding the AIDS epidemic. Arthur Ashe died at New Yory City on Feb 6, 1993, from pneumonia.

BAHAMAS: INDEPENDENCE DAY: ANNIVERSARY. July 10. Public holiday. At 12:01 AM, on July 10, 1973, the Bahamas gained their independence after 250 years as a British Crown Colony.

◆ **BETHUNE, MARY McLEOD: BIRTH ANNIVERSARY.** July 10, 1875. Mary Jane McLeod Bethune was born in Mayesville, SC, the first in her family to be born free. Bethune became a teacher and in 1904 founded her own school in Florida, the Daytona Normal and Industrial School for Negro Girls. In 1931, the school merged with a local men's college, Cookman Institute, and was renamed Bethune-Cookman College. An adviser on minority affairs under President Franklin D. Roosevelt, she directed the Division of Negro Affairs of the National Youth Administration. Bethune died May 18, 1955 at Daytona Beach, FL.

BORIS YELTSIN INAUGURATED AS RUSSIAN PRESIDENT: ANNIVERSARY. July 10. On July 10, 1991, Boris Yeltsin took the oath of office as the first popularly elected president in the 1,000-year history of the Republic of Russia. Yeltsin defeated the Communist Party candidate resoundingly, establishing himself as a powerful political counterpoint to Mikhail Gorbachev, the president of the Soviet Union, of which Russia was the largest republic. Yeltsin had been dismissed from the Politburo in 1987 and resigned from the Communist Party in 1989. His popularity forced Gorbachev to make concessions to the republics in the new union treaty forming the Confederation of Independent States.

CHILE: DIA DEL BIBLIOTECARIO. July 10. "Day of the Librarian" commemorates the government decree of July 10, 1969, creating the Colegio de Bibliotecarios de Chile (Chilean Association of Librarians).

Chase's 1995 Calendar of Events — July 10-11

CLERIHEW DAY. July 10. A day recognized in remembrance of Edmund Clerihew Bentley, journalist and author of the celebrated detective thriller *Trent's Last Case* (1912), but perhaps best known for his invention of a popular humorous verse form, the clerihew, consisting of two rhymed couplets of unequal length:
> Edmund's middle name was Clerihew
> A name possessed by very few,
> But verses by Mr. Bentley
> Succeeded eminently.

Bentley was born at London on July 10, 1875, and died there on Mar 30, 1956.

DALLAS, GEORGE MIFFLIN: BIRTH ANNIVERSARY. July 10. Eleventh Vice President of the US (1845-49), born at Philadelphia, PA, July 10, 1792. Died there Dec 31, 1864.

GWYNNE, FREDERICK HUBBARD: BIRTH ANNIVERSARY. July 10, 1926. Stage, screen and TV actor, best known for the TV roles Herman Munster in "The Munsters" and Officer Muldoon in "Car 54, Where Are You?" Gwynne was born at New York, NY, and died at Taneytown, MD, on July 2, 1993.

HONG KONG: BIRTHDAY OF LU PAN. July 10. The birthday of Lu Pan, the Master Builder, is a holiday for everybody connected with the building trades. Ceremonies sponsored by the Builders' Guilds are held at Lu Pan Temple in Kennedy Town. Festive dinner parties. Annually, the 13th day of 6th lunar month.

PROUST, MARCEL: BIRTH ANNIVERSARY. July 10. Anniversary of birth of famed French author (July 10, 1871-Nov 19, 1922). "Happiness," he wrote in *The Past Recaptured*, "is beneficial for the body but it is grief that develops the powers of the mind."

◆ ***RAINBOW WARRIOR* SINKING: 10th ANNIVERSARY.** July 10. The 160-ft ship, the *Rainbow Warrior*, operated by Greenpeace, an environmentalist organization, was sunk and a photographer aboard was killed while the ship was at Marsden Wharf, Auckland, New Zealand, on July 10, 1985. Reportedly it was caused by a bomb attached to the underside of the ship by saboteurs. The ship had been scheduled for use in a protest against nuclear tests in the South Pacific Ocean by the French government.

SPACE MILESTONE: *TELSTAR* (US). July 10. First privately owned satellite (American Telephone and Telegraph Company) and first satellite to relay live TV pictures across the Atlantic. Launched on July 10, 1962.

SWITZERLAND: SEMPACH BATTLE COMMEMORATION. July 10. On the morning of the first Monday after July 4, the Lucerne government, military and student delegations and historical groups make their way in solemn procession to the battlefield of 1386. Commemorative address, battle report and solemn service in the chapel. Also an evening procession.

WHISTLER, JAMES ABBOTT McNEILL: BIRTH ANNIVERSARY. July 10. American painter (especially known for painting of his mother), born at Lowell, MA, July 10, 1834. Died at London, England, July 17, 1903. When a woman declared that a landscape reminded her of Whistler's paintings, he reportedly said, "Yes, madam, Nature is creeping up."

WILDFLOWER FESTIVAL. July 10-16. Crested Butte, CO. The wildflower capitol of Colorado celebrates the beauty and abundance of the flowers surrounding Crested Butte. Workshops and lectures. Est attendance: 450. For info: Dana Spencer, Crested Butte Chamber of Commerce, PO Box 216, Crested Butte, CO 81224. Phone: (800) 545-4505.

WYOMING: ADMISSION DAY. July 10. Became 44th state on this day in 1890.

BIRTHDAYS TODAY

Saul Bellow, 80, author (*Herzog*, *The Bellarosa Connection*), born at Lachine, Quebec, Canada, July 10, 1915.
David Brinkley, 75, ABC TV journalist, born at Wilmington, NC, July 10, 1920.
Roger Timothy Craig, 35, football player, born at Davenport, IA, July 10, 1960.
Andre Dawson, 41, baseball player, born at Miami, FL, July 10, 1954.
David Norman Dinkins, 68, former and first black Mayor of New York City (D), born at Trenton, NJ, July 10, 1927.
Arlo Guthrie, 48, singer (Woody Guthrie's son; "The City of New Orleans"), born Brooklyn, NY, July 10, 1947.
Jerry Herman, 62, composer, lyricist, born at New York, NY, July 10, 1933.
Jean Kerr, 72, author (*Finishing Touches*, *Please Don't Eat the Daisies*), born at Scranton, PA, July 10, 1923.
Virginia Wade, 50, former tennis player, born at Bournemouth, England, July 10, 1945.

JULY 11 — TUESDAY
192nd Day — Remaining, 173

ADAMS, JOHN QUINCY: BIRTH ANNIVERSARY. July 11. Sixth president of the US, and the son of the second president, John Quincy Adams was born at Braintree, MA, July 11, 1767. After his single term as president, he served 17 years as a member of Congress, from Plymouth, MA. He died Feb 23, 1848, at the House of Representatives (in the same room in which he had taken the presidential Oath of Office on Mar 4, 1825). John Quincy Adams was the only president whose father had also been president of the US.

ANTHONY WAYNE DAY. July 11. To recognize the contributions of American Revolutionary War general Anthony Wayne (Jan 1, 1745-Dec 15, 1796). Observed in Michigan since governor's proclamation of 1971. Est attendance: 30,000. Sponsor: Wayne State University, Pat Borninski, MR Info Officer, Detroit, MI 48202. Phone: (313) 577-2150.

◆ **BABE'S DEBUT IN MAJORS: ANNIVERSARY.** July 11, 1914. Babe Ruth made his debut in major league baseball when he took the mound in Fenway Park for the Boston Red Sox against the Cleveland Indians. Ruth was relieved for the last two innings but was the winning pitcher in a 4-3 game.

BOWDLER'S DAY. July 11. A day to remember the prudish medical doctor, Thomas Bowdler, born near Bath, England, on July 11, 1754. He gave up the practice of medicine and undertook the cleansing of the works of Shakespeare by removing all the words and expressions he considered to be indecent or impious. His *Family Shakespeare*, in 10 volumes, omitted all those words "which cannot with propriety be read aloud in a family." He also "purified" Edward Gibbon's *History of the Decline and Fall of the Roman Empire* and selections from the Old Testament. So offensive was his censorship that his name became synonymous with self-righteous expurgation, and the word *bowdlerize* has become part of the English language. Bowdler died at Rhyddings, in South Wales, Feb 24, 1825.

DAY OF THE FIVE BILLION: ANNIVERSARY. July 11. An eight-pound baby boy, Matej Gaspar, born at 1:35 AM, EST, July 11, 1987, at Zagreb, Yugoslavia, was proclaimed the 5 billionth inhabitant of Earth. The United Nations Fund for Population Activities, hoping to draw attention to population growth, proclaimed July 11 as "Day of the Five Billion," noting that 150 babies are born each minute. The US Census Bureau has estimated that the world's population will reach 6.2 billion by the year 2000.

MONGOLIAN PEOPLE'S REPUBLIC: NATIONAL HOLIDAY. July 11. Commemorates establishment of Mongolian Communist government, on July 11, 1921.

July 11-12 ★ *Chase's 1995 Calendar of Events* ★

◆ NAPALM USED: 50th ANNIVERSARY. July 11, 1945. The US dropped several thousand pounds of the recently developed weapon napalm on Japanese forces still holed up on Luzon in the Philippines. Napalm, which was later used heavily as a defoliant in Vietnam, was a thickener consisting of a mixture of aluminum soaps used to jell gasoline.

OLIVIER, LAURENCE: DEATH ANNIVERSARY. July 11. Actor, director and theater manager. Born at Dorking, England, on May 22, 1907. Thought by many to be the most influential actor of this century, Laurence Olivier's theatrical and film career in many ways shaped the art forms within which he participated. Honored with nine Academy Award nominations, three Oscars and five Emmy awards, his repertoire included most of the prime Shakespearean roles, Archie Rice in John Osborne's *The Entertainer* and roles in such films as *Rebecca, Pride and Prejudice, The Boys from Brazil, Sleuth, Marathon Man* and *Wuthering Heights*. Olivier was an innovative and tireless theater manager with London's Old Vic company and the National Theatre of Great Britain. The National Theatre's largest auditorium and Britain's equivalent of Broadway's Tony awards carry his name. He was knighted in 1947 and made a peer of the throne in 1970. Olivier died at Ashurst, England, on July 11, 1989.

PERIGEAN SPRING TIDES. July 11-12. Spring tides, the highest possible tides, occur when New Moon or Full Moon takes place within 24 hours of the moment the Moon is nearest Earth (perigee) in its monthly orbit, on July 11, at 6:00 AM, EST.

SMITH, JAMES: DEATH ANNIVERSARY. July 11. Signer of the Declaration of Independence, born in Ireland about 1719 (exact date unknown). Died July 11, 1806.

SPACE MILESTONE: *SKYLAB* (US): FALLS TO EARTH. July 11. Eighty-two-ton spacecraft that was launched May 14, 1973, reentered Earth's atmosphere on July 11, 1979. Expectation was that 20-25 tons probably would survive to hit Earth, including one piece of about 5,000 pounds. Intense international public interest in where it would fall. Chance that some person would be hit by a piece of *Skylab* was calculated at one in 152. Targets were drawn, *Skylab* parties held, and there was broad media coverage of the largest man-made object to fall to Earth from orbit. *Skylab* broke up and fell to Earth in a shower of pieces over the Indian Ocean and Australia, with no known or reported casualties.

TWO THOUSAND DAYS BEFORE "2000"™. July 11. Today is the 1,635th day before the year 2000. Now is the time to do all those things you've wanted to do, but put off for way too long. Although the countdown began on July 11, 1994, it is not too late to participate as this timely theory is "worked" by just taking a moment each day to do something you ordinarily wouldn't. That way, when the clock strikes midnight on Dec 31, 1999, you can say you've accomplished 2000 things before 2000 and 1,635 new, enriching feats and gestures of goodwill and brotherhood . . . if you start today! For info: Adrienne Sioux Koopersmith, 1437 W Rosemont, 1W, Chicago, IL 60660. Phone: (312) 743-5341.

UNITED NATIONS: WORLD POPULATION DAY. July 11. In June 1989, the Governing Council of the United Nations Development Programme recommended that July 11 be observed by the international community as World Population Day. An outgrowth of the Day of Five Billion (July 11, 1987), the Day seeks to focus public attention on the urgency and importance of population issues, particularly in the context of overall development plans and programs, and the need to create solutions to these problems. Info from: United Nations, Dept of Public Info, New York, NY 10017.

BIRTHDAYS TODAY

Harold Bloom, 65, literary critic, born at New York, NY, July 11, 1930.
Nicolai Gedda, 70, opera singer, born at Stockholm, Sweden, July 11, 1925.
Debbie Harry, 50, singer (lead singer of Blondie; "The Tide is High"), born at Miami, FL, July 11, 1945.
Tab Hunter (Arthur Gelien), 64, actor (*Damn Yankees, Judge Roy Bean*, "The Tab Hunter Show"), born New York, NY, July 11, 1931.
Theodore Maiman, 68, physicist, developed first working laser, born at Los Angeles, CA, July 11, 1927.
Bonnie Pointer, 44, singer (Pointer Sisters; "Steam Heat"), born at East Oakland, CA, July 11, 1951.
Leon Spinks, 42, former boxer, born at St. Louis, MO, July 11, 1953.

JULY 12 — WEDNESDAY
193rd Day — Remaining, 172

CANADA: BIG VALLEY JAMBOREE. July 12-16. Craven, Saskatchewan. The largest country music festival in the world, featuring top-name entertainers, with more than 60 acts. CCMA semi-finalists, beer gardens, live entertainment, tradeshows, unserviced camping and parking. Est attendance: 50,000. For info: Glenda Mohr, Mktg and Advertising, Big Valley Dvmts, Inc, Box 200, Regina, Saskatchewan, Canada S4P 2Z6. Phone: (306) 721-6060.

CENTRAL PENNSYLVANIA FESTIVAL OF THE ARTS. July 12-16. University Park, State College, PA. A celebration of the arts—visual, performing, film—with 325 artists on hand. Est attendance: 150,000. For info: Central Pennsylvania Festival of the Arts, Box 1023, State College, PA 16804. Phone: (814) 237-3682.

CHOCTAW INDIAN FAIR. July 12-15. Philadelphia, MS. The fair presents cultural programs demonstrating native arts, crafts, social dancing, stickball, archery, blowgun and rabbit stick competition. Also native foods are prepared and served on the reservation. Annually, the second Wednesday through Saturday after July 4. Est attendance: 60,000. For info: Connie Sampsell, Dir, Philadelphia-Neshoba County Ch of Comm, PO Box 51, Philadelphia, MS 39350. Phone: (601) 656-1742.

HOG CALLING CONTEST. July 12. Broadway Market Square, Baltimore, MD. Amateur hog callers compete for prizes. Country music entertainment. Live hogs on display. Est attendance: 250. For info: Office of Adventures in Fun, Clarence "Du" Burns Arena, 1301 S Ellwood Ave, Baltimore, MD 21224. Phone: (410) 396-9177.

INTERNATIONAL TINNITUS SEMINAR. July 12-15. Portland, OR. A quadrennial event for Tinnitus researchers and hearing professionals to share the most recent treatment and management of the condition. Tinnitus is a ringing in the head or ears caused most commonly by noise exposure. Est attendance: 400. For info: American Tinnitus Assn, Patricia Daggett, PO Box 5, Portland, OR 97207.

July 1995

S	M	T	W	T	F	S
						1
2	3	4	5	6	7	8
9	10	11	12	13	14	15
16	17	18	19	20	21	22
23	24	25	26	27	28	29
30	31					

Chase's 1995 Calendar of Events — July 12-13

IRELAND: ORANGEMEN'S DAY. July 12. Annual observance commemorates Battle of Boyne, July 1 (Old Style), 1690, in which the forces of King William III of England, Prince of Orange, defeated those of James II, at Boyne River in Ireland. Ordinarily observed on July 12. If July 12 is a Saturday or a Sunday the holiday observance is on the following Monday.

KIRIBATI: INDEPENDENCE DAY. July 12. Republic of Kiribati attained independence July 12, 1979.

MARINERFEST. July 12-16. Tawas Bay, MI. Mayor's fish boil, carnival, auto and boat show, breakfast and lunch "in the park," sidewalk sale. Phone: (800) 55-TAWAS. Est attendance: 10,000. For info: Chamber of Commerce, Box 608, Tawas City, MI 48764.

MOON PHASE: FULL MOON. July 12. Moon enters Full Moon phase at 6:49 AM, EDT.

NORTHERN IRELAND: ORANGEMAN'S DAY. July 12. National holiday.

OSLER, SIR WILLIAM: BIRTH ANNIVERSARY. July 12. Anniversary of birth at Tecumseh, Ontario, Canada, on July 12, 1849, of William Osler, physician, teacher and author of *Principles and Practice of Medicine*. Osler died at Oxford, England, Dec 29, 1919.

SAO TOME AND PRINCIPE: NATIONAL DAY. July 12. National holiday observed.

SPACE MILESTONE: *PHOBOS 2* **(USSR).** July 12. Sent back the first close-up photos of Phobos, one of two small moons of Mars. Launched from Soviet space probe in central Asia on July 12, 1988.

THOREAU, HENRY DAVID: BIRTH ANNIVERSARY. July 12. American author and philosopher, born at Concord, MA, July 12, 1817. Died there May 6, 1862. In *Walden* he wrote "I frequently tramped eight or ten miles through the deepest snow to keep an appointment with a beechtree, or a yellow birch, or an old acquaintance among the pines."

VIDEO GAMES DAY. July 12. A day for kids who love video games to celebrate the fun they have playing them and to thank their parents for all the cartridges and quarters they have provided to indulge this enthusiasm. Sponsor: David Earle, Pres, Kid Video Warriors.

WEDGEWOOD, JOSIAH: BIRTH ANNIVERSARY. July 12. Famed English pottery designer and manufacturer, born at Burslem, Staffordshire, England, July 12, 1730. Died at Etruria, Staffordshire, England, Jan 3, 1795.

BIRTHDAYS TODAY

Milton Berle (Milton Berlinger), 87, comedian, actor (Uncle Miltie—Mr. Television; "The Milton Berle Show"), born New York, NY, July 12, 1908.

Van Cliburn (Harvey Lavan, Jr), 61, pianist, born at Shreveport, LA, July 12, 1934.

Bill Cosby, 57, comedian, actor (Emmys "I Spy," "The Cosby Show"), born Philadelphia, PA, July 12, 1938.

Mark O. Hatfield, 73, US Senator (R, Oregon), born at Dallas, OR, July 12, 1922.

Christine McVie, 52, singer, musician (member Fleetwood Mac; "Got a Hold On Me"), born Birmingham, England, July 12, 1943.

Richard Simmons, 47, TV personality, weight loss guru, author, born New Orleans, LA, July 12, 1948.

Roger Smith, 70, former chair of General Motors, born at Columbus, OH, July 12, 1925.

Kristi Yamaguchi, 24, figure skater (Olympic medalist), born at Hayward, CA, July 12, 1971.

JULY 13 — THURSDAY
194th Day — Remaining, 171

BASTILLE DAYS FESTIVAL (LA KERMESSE DE LA BASTILLE). July 13-16. Milwaukee, WI. Celebrated "on the streets" of East Town in the heart of downtown Milwaukee. "Storm the Bastille" 5K run, sidewalk cafes, bicycle races, French marketplace, continuous entertainment on 4 stages, waiters'/waitresses' race, New France encampment, dancing in the streets. Admission is free. Est attendance: 250,000. For info: East Town Assn, 770 N Jefferson, Milwaukee, WI 53202. Phone: (414) 271-1416.

FORREST, NATHAN BEDFORD: BIRTH ANNIVERSARY. July 13. Confederate cavalry commander whose birthday is observed as a holiday in Tennessee, Forrest was also one of the founders of the short-lived original Ku Klux Klan. Forrest was born July 13, 1821, at Bedford County, TN, and died Oct 29, 1877, at Memphis, TN.

FRANCE: NIGHT WATCH OR LA RETRAITE AUX FLAMBEAUX. July 13. France. Celebrates eve of the Bastille's fall.

JAPAN: BON FESTIVAL (FEAST OF LANTERNS). July 13-15. Religious rites throughout Japan in memory of the dead who, according to Buddhist belief, revisit Earth during this period. Lanterns are lighted for the souls. Spectacular bonfires in the shape of the character *dai* are burned on hillsides on last day of the Bon or O-Bon Festival, bidding farewell to the spirits of the dead.

"LIVE AID" CONCERTS: 10th ANNIVERSARY. July 13. Concerts at Philadelphia, PA, and London, England (Kennedy and Wembley Stadiums) on July 13, 1985, were seen by 162,000 attendees and an estimated 1.5 billion television viewers. Organized to raise funds for African famine relief; the musicians performed without fee, and nearly $100 million was pledged toward aid to the hungry.

LOGGER DAYS. July 13-16. Libby, MT. Celebration of logging in the community. Est attendance: 4,000. For info: Libby Chamber of Commerce, PO Box 704, Libby, MT 59923. Phone: (406) 293-4167.

MICAJAH. July 13-15. (Also July 20-22 and 27-29.) Autryville, NC. To promote patriotism, local history and heritage. *Micajah* is an outdoor drama depicting the life of Micajah Autry, local hero of the Battle of the Alamo. Est attendance: 400. Sponsor: Micajah Autry Soc, Donna N. Cashwell, Pres, Box 52, Autryville, NC 28318. Phone: (919) 525-4621.

NEW ORLEANS WINE AND FOOD EXPERIENCE. July 13-16. New Orleans, LA. Vintner dinners, wine tasting in art and antique shops, wine seminars, cooking demonstations and grand tastings featuring more than 150 wineries and 80 local restaurants. Est attendance: 1,000. For info: Amy Hymel, 1008 N Peters St, New Orleans, LA 70116. Phone: (504) 522-5730.

NORTHWEST ORDINANCE: ANNIVERSARY. July 13. On July 13, 1787, the Northwest Ordinance, providing for government of the territory north of the Ohio River, became law. The ordinance guaranteed freedom of worship and the right to trial by jury, and it prohibited slavery.

OREGON TRAIL DAYS. July 13-16. Gering, NE. Oldest continuing celebration in state of Nebraska commemorating Oregon Trail. Parades, barbecues, street dances, Intl Food Fair, concert, Nebraska State CASI Chili Cookoff, musical plays and a western art show highlight the 74th annual celebration. Annually, the second full weekend in July. Est attendance: 25,000. For info: Nadine Sieb, Event Cood, PO Box 222, Gering, NE 69341.

Chase's 1995 Calendar of Events

SINCLAIR LEWIS DAYS. July 13-16. Sauk Centre, MN. Parade, park activities, park dance, Miss Sauk Centre pageant, Fun Golf Tournament, flea market, craft sale, softball tournament, kiddie caravan and volleyball tournament in Sinclair Lewis's hometown. Est attendance: 1,200. For info: Chamber of Commerce, Box 222, Sauk Centre, MN 56378. Phone: (612) 352-5201.

US WOMEN'S OPEN (GOLF) CHAMPIONSHIP. July 13-16. Broadmoor Golf Club, Colorado Springs, CO. Sponsor: US Golf Assn, Championship Dept, Golf House, Far Hills, NJ 07931. Phone: (908) 234-2300.

◈ **WOOLLEY, MARY E.: BIRTH ANNIVERSARY.** July 13, 1863. Mary Woolley was born at South Norwalk, CT. The first woman to attend classes and graduate from Brown University (1894), Woolley pursued a career in education, serving on the faculty of Wellesley College. In 1901 she was named president of Mount Holyoke College, where her impact was enormous as she not only upgraded the school's faculty and curriculum but raised the funds to add 16 major buildings to the campus. In 1932 she became the first female representative of the US at a major diplomatic conference, serving as a delegate to the Conference on Reduction and Limitation of Armaments. Woolley died Sept 5, 1947, at Westport, CT.

BIRTHDAYS TODAY

Harrison Ford, 53, actor (*American Graffiti*; *Star Wars* and *Indiana Jones* movies), born at Chicago, IL, July 13, 1942.
Robert Forster, 54, actor (*Reflection in a Golden Eye*, "Banyon"), born at Rochester, NY, July 13, 1941.
Jack Kemp, 60, former US Secretary of Housing and Urban Development in Bush administration, born at Los Angeles, CA, July 13, 1935.
Cheech Marin, 49, writer (*My Name Is Cheech the School Bus Driver*), actor (Cheech & Chong movies), born Los Angeles, CA, July 13, 1946.
Roger McGuinn (James Joseph McGuinn), 53, musician, born at Chicago, IL, July 13, 1942.
Erno Rubik, 51, inventor of the Rubik's Cube, born at a hospital air raid shelter, Budapest, Hungary, July 13, 1944.
Wole Soyinka, 61, author (*The Lion and the Jewel*, *The Strong Breed*), born Abeokuta, Nigeria, July 13, 1934.
David Storey, 62, author, playwright (*The Performance of Small Firms*), born at Wakefield, England, July 13, 1933.
Spud (Anthony Jerome) Webb, 32, basketball player, born at Dallas, TX, July 13, 1963.

JULY 14 — FRIDAY
195th Day — Remaining, 170

AEROSPACE AMERICA '95. July 14-16. Will Rogers World Airport, Oklahoma City, OK. Considered one of the top five air shows in the world. Voted number one event in Oklahoma for 1990. More than 100 military aircraft, 85 warbirds, top aerobatic performers in the world, trade show. Est attendance: 100,000. For info: Oklahoma City All Sports Assn, 100 W Main, Ste 287, Oklahoma City, OK 73102. Phone: (405) 236-5000.

BALLOON RALLY AND ARTS AND CRAFTS SHOW. July 14-16. Angel Fire, NM. This beautiful event lifts everyone's spirits as colorful balloons fly over our majestic valley. Arts and crafts to please everyone. Est attendance: 500. For info: Angel Fire Chamber of Commerce, PO Box 547, Angel Fire, NM 87710. Phone: (800) 446-8117.

BASCOM, FLORENCE: BIRTH ANNIVERSARY. July 14, 1862. After receiving her third bachelor's degree from the University of Wisconsin in 1884 and a master's degree in 1887, Florence Bascom entered Johns Hopkins University and became the first woman to receive a doctorate in 1893. She taught at Ohio State and became a professor at Bryn Mawr. She also was the first woman appointed a geologist with the US Geological Survey, was Associate Editor of "American Geologist" (1890-1905) and became the first woman elected a Fellow of the Geological Survey of America. She died on June 18, 1945.

BASTILLE DAY CELEBRATION. July 14. Boston, MA. A street celebration with typical Parisian joie de vivre; dining and dancing under the stars. A special celebration as The French Library and Cultural Center is celebrating its 50th Anniversary in 1995! Est attendance: 2,500. For info: The French Library, 53 Marlborough St, Boston, MA 02116-2099. Phone: (617) 266-4351.

BASTILLE DAY MOONLIGHT GOLF TOURNAMENT. July 14. Greenleaf Point Golf Course, Fort McNair, Washington, DC. After a cook-out, teams of 4 tee-off wearing glow-in-the-dark necklaces and hitting glow-in-the-dark balls. Participants are given membership into the Natl Capital Moonlight Golf Assn. Annually, July 14. For info: Ft McNair Sports Center, Bldg 17, 4th & P St SW, Washington, DC 20319-5050.

BEREA CRAFT FESTIVAL. July 14-16. Berea, KY. Craftspeople from 20 states gather to exhibit, demonstrate and sell their work. Est attendance: 12,000. For info: Berea Craft Enterprises, Sandy Chowning, Secy, Box 128, Berea, KY 40403. Phone: (606) 986-2258.

BIG SKY STATE GAMES. July 14-16. Billings, MT. An Olympic-styled festival for Montana citizens. This statewide multi-sport program is designed to inspire people of all ages and skill levels to develop their physical and competitive abilities to the height of their potential through participation in fitness activities. Annually, the third weekend in July. Est attendance: 12,000. For info: Big Sky State Games, PO Box 2318, Billings, MT 59103.

BLISSFESTIVAL. July 14-16. Cross Village, MI. Traditional, accoustic and folk music, 2 stages, workshops, camping, kids' area and activities. Music styles from jazz to bluegrass. Est attendance: 4,000. For info: Blissfest, Box 441, Harbor Springs, MI 49740. Phone: (616) 348-2815.

CANADA: CANADIAN TURTLE DERBY. July 14-16. Boissevain, Manitoba. A fun-filled family weekend, adjacent to the famous International Peace Garden. Live turtle races. Curling Bonspiel, ball and volleyball tourneys, vintage auto and farm machinery show, mini-triathlon, fireworks and children's entertainment. Ample camping available. 24th annual derby. Est attendance: 5,000. For info: Canadian Turtle Derby, Ivan Strain, PO Box 122, Boissevain, Man, Canada R0K 0E0. Phone: (204) 534-6000.

CANADA: HARVEST MOON. July 14-30. Station Arts Centre/Seager Wheeler Place. Rosthern, Saskatchewan. Presentation of an original script of good humour and old-fashioned sentiment depictions the past, present and future of agriculture in Saskatchewan through the eyes of Seager Wheeler, internationally acclaimed for his innovative agriculture techniques. Est attendance: 10,000. For info: Tourism Saskatchewan, Saskatchewan Trade and Conv Center, 1919 Saskatchewan Dr, Regina, Sask, Canada S4P 3V7. Phone: (800) 667-7191.

	S	M	T	W	T	F	S
July							1
	2	3	4	5	6	7	8
1995	9	10	11	12	13	14	15
	16	17	18	19	20	21	22
	23	24	25	26	27	28	29
	30	31					

CANADA: JULYFEST. July 14–16. Kimberley, British Columbia. Platzl entertainment, sports events (soccer, bocce ball), refreshment garden at arena, events for all ages at Rotary Park; parade. Annually, the third weekend in July. Est attendance: 3,000. For info: Jill Hamacher, Mgr, Kimberley Bavarian Soc, 350 Ross St, Kimberley, BC, Canada V1A 2Z9. Phone: (604) 427-3666.

CANADA: SASKATCHEWAN HANDCRAFT FESTIVAL. July 14–16. Battleford Arena and Alex Dillabough Centre, Battleford, Saskatchewan. Juried craft show, children's pavilion, craft market and demonstrations and an ethnic food pavilion. Est attendance: 7,000. For info: Tourism Saskatchewan, Saskatchewan Trade and Conv Center, 1919 Saskatchewan Dr, Regina, Sask, Canada S4P 3V7. Phone: (800) 667-7191.

CANADA: SENIORS EXPO: "LIFE IS WHAT YOU MAKE IT." July 14–17. World Trade and Convention Center, Halifax, Nova Scotia. More than 170 exhibitors display their products and services for seniors and their families. Consumer show features professional and amateur entertainment, hobbies, crafts, art and educational seminars. Annually, the third weekend in July. Est attendance: 27,000. Sponsor: Senior Citizens Secretariat, 4th Fl, Dennis Bldg, 1740 Granville St, PO Box 2065, Halifax, NS, Canada B3J 2Z1. Phone: (902) 424-7957.

CATFISH DAYS. July 14–16. East Grand Forks, MN. The catfish-catching tournament is held on the Red and Red Lake Rivers. 150 teams from all over the US participate. Other weekend events held on the riverbank include a bike race, horseshoe tournament, volleyball and frog-kissing contest. Evening entertainment includes live music by various bands. Annually, the third Saturday and Sunday in July. Est attendance: 3,000. For info: EGF Chamber of Commerce, 218 4th St NW, East Grand Forks, MN 56721. Phone: (218) 773-7481.

CHILDREN'S PARTY AT GREEN ANIMALS. July 14. Green Animals Topiary Gardens, Portsmouth, RI. Annual party for children and adults at Green Animals, a delightful topiary garden. Party includes pony rides, merry-go-round, games, clowns, refreshments, hot dogs, hamburgers and more. Annually, July 14. Est attendance: 2,000. For info: The Preservation Soc of Newport County, 424 Bellevue Ave, Newport, RI 02840. Phone: (401) 847-1000.

FORD, GERALD RUDOLPH: BIRTHDAY. July 14. Thirty-eighth president of the US. Born Leslie King, at Omaha, NE, July 14, 1913. Ford became 41st vice president of the US Dec 6, 1973, by appointment following the resignation of Spiro T. Agnew from that office on Oct 10, 1973. Ford became president on Aug 9, 1974, following the resignation from that office on that day of Richard M. Nixon. He was the first nonelected vice president and president of the US.

FRANCE: BASTILLE DAY or FETE NATIONAL. July 14. Public holiday commemorating the fall of the Bastille at the beginning of the French Revolution, on July 14, 1789. Also celebrated or observed in many other countries.

GUTHRIE, WOODROW WILSON "WOODY": BIRTH ANNIVERSARY. July 14. American folksinger, songwriter ("This Land Is Your Land," "Union Maid," "Hard Traveling"), July 14, 1912–Oct 3, 1967.

HOOSIER STATE GAMES: FINALS. July 14–16. Indianapolis, IN. Finals competition in Indiana's amateur multi-sports festival. All ages and skill levels in 19 sports. For info: Indiana Sports Corp, 201 S Capitol Ave, Ste 1200, Indianapolis, IN 46225. Phone: (317) 237-5000.

HORSE AND CARRIAGE WEEKEND. July 14–15. Mystic Seaport, Mystic, CT. 9th annual gathering of horse-drawn antique carriages. Costumed drivers offer rides along Mystic Seaport waterfront and down its village streets for a small fee. The variety of carriages range from workday to deluxe. At noon on both days see a rally on the village green as the horses are put through their paces. Est attendance: 5,500. For info: Mystic Seaport, 75 Greenmanville Ave, PO Box 6000, Mystic, CT 06355. Phone: (203) 572-5315.

ITALY DECLARES WAR ON JAPAN: 50th ANNIVERSARY. July 14, 1945. Italy, no longer in the control of the Fascists, dramatically ended its Axis partnership by declaring war on Japan.

MIDNIGHT SUN FESTIVAL. July 14–21. Yellowknife, NWT. More than 40 events over a seven-day period with visual and performing artists from across the Northwest Territory gather in Yellowknife to demonstrate their work. For info: Vicki Tompkins, Fest Coord, Box 995, Yellowknife, NWT, Canada X1A 2N7. Phone: (403) 920-7118.

MIDSUMMER NIGHTS' FAIR. July 14–15. Lions Park, Norman, OK. Juried and invited artists and craftspeople display their work for show and sale in booths under the stars. A pottery auction, dalmatian mascot contest, festival foods, free entertainment and activities for the whole family are set for this 18th annual event, 6PM–midnight. Est attendance: 6,000. Sponsor: Firehouse Art Center, 444 S Flood, Norman, OK 73069. Phone: (405) 329-4523.

MINNEAPOLIS AQUATENNIAL. July 14–23. Minneapolis, MN. Major civic water festival to celebrate Minnesota's 10,000 lakes and quality of life. Some 50 events including sports competitions, musical entertainment, pageants and parades. Traditional family fun and community spirit in a completely volunteer-run and sponsor-driven festival. Annually, the third week in July. Est attendance: 1,000,000. For info: Paula Beadle, Exec Dir, Riverplace, 43 Main St SE, Ste 145, Minneapolis, MN 55414. Phone: (612) 331-8371.

NATCHITOCHES/NORTHWESTERN STATE UNIVERSITY FOLK FESTIVAL. July 14–16. Northwestern State University, Natchitoches, LA. The festival is a "purist" folk festival in that folk artists who are reviving a traditional Louisiana folk art or still working a Louisiana tradition are invited. Music, food, crafts and storytellers. Annually, the third weekend in July. Est attendance: 24,000. For info: Don Hatley, Louisiana Folklife Center, Box 3663, NSU, Natchitoches, LA 71497. Phone: (318) 357-4332.

NETHERLANDS: NORTH SEA JAZZ FESTIVAL. July 14–16. Nederlands Congresgebouw (Dutch Congress Bldg), The Hague. Live jazz festival with top name entertainment. Annually, the second weekend in July. Est attendance: 70,000. For info: North Sea Jazz Fest, PO Box 87919, 2508 DH The Hague, Netherlands.

POTOMAC SOUTH BRANCH WEEKEND. July 14–16. Appalachian Base Camp, Terra Alta, WV. This weekend is for the amateur or professional canoeist who wants to learn the flora & fauna of West Virginia while canoeing. Est attendance: 30. Sponsor: Oglebay Institute Nature/Environmental Educ Dept, Tom Shepherd, Oglebay Institute, Wheeling, WV 26003. Phone: (304) 242-6855.

SEAFAIR. July 14–Aug 6. Seattle, WA. Seafair hosts more than 55 events, including the Torchlight Parade, Unlimited Hydroplane Race and Airshow on Lake Washington, Kent Cornucopia Days, Milk Carton Derby, Chinatown festival, Kiddies Parade, Beach Party, relay team swim marathon, Soul Festival, Bon Odori, several other parades and much, much more. Est attendance: 1,000,000. Sponsor: Seafair, 2001 6th Ave, Ste 2800, Seattle, WA 98121-2574. Phone: (206) 728-0123.

SHERWOOD ROBIN HOOD FESTIVAL. July 14–15. Sherwood, OR. Parade, archery contest, hot-air balloon rally, dance, music, beer garden, game booths, food and arts and crafts. Annually, the third weekend in July. Est attendance: 2,500. For info: Robin Hood Festival Assn, PO Box 496, Sherwood, OR 97140. Phone: (503) 625-7537.

July 14-15 ☆ Chase's 1995 Calendar of Events ☆

STAMP EXPO. July 14-16. Elks Lodge, Pasadena, CA. Fax: (818) 988-4337. Est Attendance: 4,000. For info: Intl Stamp Collectors Soc, PO Box 854, Van Nuys, CA 91408. Phone: (818) 997-6496.

STEAMBOAT'S RAINBOW WEEKEND. July 14-16. Steamboat Springs, CO. 15th annual Hot Air Balloon Rodeo; 21th annual Art-in-the-Park; Strings in the Mountains Chamber Music Festival; PRCA Rodeo; kids' tent and entertainment. Est attendance: 15,000. For info: Steamboat Springs Chamber Resort Assn, PO Box 774408, Steamboat Sprgs, CO 80477. Phone: (303) 879-0880.

SUMMER SUNFEST. July 14-16. MainStrasse Village, Covington, KY. Annual family fun festival will feature games and special activities, international food and drink, arts and crafts, an International Bazaar, continuous live entertainment that highlights the '50s through the '90s. Hours: Friday 6-11 PM; Saturday noon-11:30 PM; Sunday noon-9 PM. Est attendance: 100,000. For info: MainStrasse Village, Cindy Scheidt, 616 Main St, Covington, KY 41011. Phone: (606) 491-0458.

VIRGINIA LAKE FESTIVAL. July 14-16. Clarksville, VA. Fun-filled weekend with Pig Pickin, opening ceremonies on Friday; arts and crafts show, flea market, balloons, 5-mile run, fish fry and fireworks on Saturday; beach music festival, 2nd day of arts and crafts on Sunday. Annually, the third weekend in July (Friday-Sunday). Est attendance: 50,000. For info: Virginia Lake Country Chamber of Commerce, Box 1017, 325 Virginia Ave, Clarksville, VA 23927. Phone: (804) 374-2436.

WALES: INTERNATIONAL POTTERS FESTIVAL. July 14-16. Arts Centre, Penglais Hill, Aberystwyth, Dyfed. Potters from all over the world demonstrate and exhibit their crafts. For info: Alan Hewson, Arts Centre, Penglais Hill, Aberystwyth, Dyfed, Wales SY23 3DE.

◆ **WOMEN'S RIGHTS CONVENTION CALLED: ANNIVERSARY.** July 14, 1848. An announcement placed by Elizabeth Cady Stanton and Lucretia Mott Appeared in the *Seneca Falls Courier*. The notice concerned a women's rights convention to be held at the Wesleyon Chapel, Seneca Falls, NY, on July 19-20.

WRONG DAYS. July 14-15. Wright, MN. 30th annual festival with teen and adult dances, coronation, amateur contest, smorgasbord, flea and crafts market, softball and horseshoe tournaments, dunk tank, parade, kids' games and more. Est attendance: 2,000. For info: Bonnie Adams, Wrong Days in Wright, Rt 1, Wright, MN 55798. Phone: (218) 357-2911.

YARMOUTH CLAM FESTIVAL. July 14-16. Yarmouth, ME. Family-oriented festival. Annually, the third weekend in July. Est attendance: 150,000. Sponsor: Yarmouth Chamber of Commerce, Charles W. Nulle, Exec Dir, 16 US Route 1, Yarmouth, ME 04096. Phone: (207) 846-3984.

BIRTHDAYS TODAY

Polly Bergen, 65, actress ("To Tell the Truth," *The Winds of War*), singer, born Knoxville, TN, July 14, 1930.
Ingmar Bergman, 77, filmmaker (*The Seventh Seal, Wild Strawberries, Cries and Whispers*), born at Uppsala, Sweden, July 14, 1918.
John Chancellor, 68, broadcast journalist, born at Chicago, IL, July 14, 1927.
Douglas Edwards, 78, broadcast journalist, born at Ada, OK, July 14, 1917.
Gerald R. Ford (Leslie King), 82, 38th US president, born at Omaha, NE, July 14, 1913.

	S	M	T	W	T	F	S
July 1995							1
	2	3	4	5	6	7	8
	9	10	11	12	13	14	15
	16	17	18	19	20	21	22
	23	24	25	26	27	28	29
	30	31					

Missy Gold, 25, actress (Katie Gatling on "Benson"), born at Great Falls, MT, July 14, 1970.
Frances Lear, 72, editor and publisher, born at Hudson, NY, July 14, 1923.
Robert Stephens, 64, actor (*The Private Life of Sherlock Holmes*), born at Bristol, England, July 14, 1931.
Steve Stone, 48, sportscaster, former baseball player, born at Cleveland, OH, July 14, 1947.

JULY 15 — SATURDAY
196th Day — Remaining, 169

ABRAHAM LINCOLN BIRTHPLACE NATIONAL HISTORIC SITE: FOUNDER'S DAY WEEKEND. July 15-16. Hodgenville, KY. Special pioneer demonstrations, drama and musical programs commemorating the anniversary of the park and the founding of the National Park Service. Annually, weekend closest to July 17. Est attendance: 400. For info: Gary V. Talley, Natl Park Service, 2995 Lincoln Farm Rd, Hodgenville, KY 42748. Phone: (502) 358-3137.

ANTI-CRUELTY SOCIETY DOGWASH. July 15. Chicago, IL. A fun annual event for all dog owners in the city to have their dogs washed by Society volunteers. Cost depends on size of dog. Phone: (312) 644-8338, ext 311. Est attendance: 150. For info: The Anti-Cruelty Soc, 157 W Grand Ave, Chicago, IL 60610.

CANADA: ROTHESAY CRAFT FESTIVAL. July 15-16. Rothesay Common, Rothesay, New Brunswick. Major outdoor craft sale and entertainment by New Brunswick Performers. Est attendance: 7,000. For info: New Brunswick Crafts Council Inc, PO Box 1231, Fredericton, NB, Canada E3B 5C8. Phone: (506) 450-8989.

CANADA: ST. SWITHUN'S SOCIETY ANNUAL CELEBRATION. July 15. Toronto, Ontario. To encourage the celebration of Saint Swithun's Day and to promote feelings of goodwill. Organizes spontaneous celebrations throughout the year and "creates annual honorary members." Affiliated with the Friends of Winchester Cathedral. Publishes "The Water Spout" newsletter, free upon request. Est attendance: 100. Sponsor: St. Swithun's Soc, Norman A. McMullen, Pres, 427 Lynett Crescent, Richmond Hill, Ont, Canada L4C 2V6. Phone: (905) 883-0984.

CANADA: WINNIPEG FRINGE FESTIVAL. July 15-23. Winnipeg, Manitoba. A 9-day, noon/to/midnight, non-stop theatrical smorgasbord with more than 100 companies from around the world. Est attendance: 90,000. For info: The Winnipeg Fringe Festival, 174 Market Ave, Winnipeg, Manitoba, Canada R3B 0P8. Phone: (204) 956-1340.

COLTON COUNTRY DAY. July 15. Main St, Colton, NY. Annual flea market, live entertainment, Museum exhibits. Special programs to be announced. Annually, the third weekend in

July. Sponsor: Colton Historical Soc. Est attendance: 1,200. For info: Dennis Eickhoff, Town Historian, PO Box 109, Colton, NY 13625. Phone: (315) 262-2800.

COMMERCIAL AIR FLIGHT BETWEEN THE US AND USSR BEGINS: ANNIVERSARY. July 15. On July 15, 1968, a Soviet Aeroflot jet landed at Kennedy Airport in New York, NY, marking the start of direct commerical air flight between the US and the then USSR.

EIGHTEENTH-CENTURY SUMMER MARKET FAIR. July 15-16. McLean, VA. 18th-century games, music and crafts. Militia will drill. Est attendance: 2,500. Sponsor: Claude Moore Colonial Farm at Turkey Run, 6310 Georgetown Pike, McLean, VA 22101. Phone: (703) 442-7557.

FINLAND: INTERNATIONAL JAZZ FESTIVAL. July 15-23. Pori. 30th international festival presenting the jazz music of today with top name performances. 12 different venues; some go on all night. Est attendance: 90,000. For info: Finnish Tourist Board, 655 Third Ave, New York, NY 10017. Phone: (212) 949-2333.

FINLAND: KAUSTINEN FOLK MUSIC FESTIVAL. July 15-23. Kaustinen. Scandinavia's largest annual international festival of folk music and dance. Thousands of Finnish and hundreds of foreign artists perform. Every visitor has opportunity to join in the music, dance and song. First organized in 1968. Est attendance: 100,000. For info: Finnish Tourist Board, 655 Third Ave, New York, NY 10017. Phone: (212) 949-2333.

FRIENDSVILLE FIDDLE AND BANJO CONTEST. July 15. Friendsville, MD. Fiddlers from Garrett County and surrounding region participate in event. Old-time bluegrass entertainment. Est attendance: 1,000. For info: Deep Creek Lake-Garrett County, Promotion Council, Court House, 200 S Third St, Oakland, MD 21550. Phone: (301) 334-1948.

HARVEST WEEKENDS. July 15-Aug 6. (Saturdays and Sundays only.) Messina Hof Vineyards, Bryan, TX. Join the Messina Hof Family Pickers Club in bucolic surroundings as you and your friends pick grapes and follow the harvest through production. Conclude your morning with a toe-tingling grape stomp and a Vintner's lunch. Est attendance: 1,000. Sponsor: Messina Hof Vineyards, 4545 Old Reliance Rd, Bryan, TX 77808. Phone: (409) 778-9463.

HIGHLIGHTS FOUNDATION WRITERS WORKSHOP AT CHAUTAUQUA. July 15-22. Chautauqua, NY. A seven-day immersion in study, peer exchange and individual professional guidance for those interested in writing for children. Est attendance: 150. For info: Jan Keen, Conf Dir, Highlights Fdtn, Inc, 814 Court St, Honesdale, PA 18431. Phone: (717) 253-1192.

KANSAS RIVER VALLEY ART FAIR. July 15-16. Gage Park, Topeka, KS. 16th annual juried fine arts festival. More than 100 artists from several state area. Children's activities. Est attendance: 7,000. For info: Anita Wolgast, Historic Ward-Meade Park, 124 NW Fillmore, Topeka, KS 66606. Phone: (913) 295-3888.

KOHLER OUTDOOR ARTS FESTIVAL. July 15-16. Sheboygan, WI. A multiarts extravaganza featuring the works of 130 artists, demonstrations, live entertainment, children's workshops and exhibition tours. Est attendance: 16,000. For info: John Michael Kohler Arts Center, 608 New York Ave, PO Box 489, Sheboygan, WI 53082-0489. Phone: (414) 458-6144.

LOGGERS/SAWDUST FESTIVAL: STATE CHAMPIONSHIP. July 15-16. Rodeo Grounds, Payson, AZ. Logging skills such as cross-cutting, bucking, log rolling, ax throwing, tree-topping and greased-pole climbs plus Blues Fesitval and Arts and Craft show. Est attendance: 3,500. For info: Chamber of Commerce, PO Box 1380, Payson, AZ 85547. Phone: (800) 672-9766.

MICAJAH AUTRY DAY. July 15. Autryville, NC. Patriotic festival with parade, arts and crafts, food, personalities, music and dance. Est attendance: 2,500. For info: Micajah Autry Soc, Box 52, Autryville, NC 28318. Phone: (919) 525-4621.

MOORE, CLEMENT CLARKE: BIRTH ANNIVERSARY. July 15. American author and teacher, best remembered for his popular verses, "A Visit from Saint Nicholas" (" 'Twas the Night Before Christmas"), which was first published anonymously and without Moore's knowledge in a newspaper on Dec 23, 1823. Moore was born at New York, NY, July 15, 1779, and died at Newport, RI, on July 10, 1863.

OREGON COAST MUSIC FESTIVAL. July 15-29. Coos Bay, OR. Musical celebration features regional, national and international artists performing symphonic and chamber music. Est attendance: 10,000. For info: Oregon Coast Music Festival, Box 663, Coos Bay, OR 97420. Phone: (503) 267-0938.

PRESCOTT/BLUEGRASS FESTIVAL. July 15-16. Watson Lake Park, Prescott, AZ. Music, food, drink and fun featuring a wide variety of bands in the hills of Prescott. Two full days of music and entertainment, 12-6 PM daily. Phone: (602) 445-2000 or toll free: (800) 266-7534. Est attendance: 6,000. For info: Prescott Chamber of Commerce, PO Box 1147, Prescott, AZ 86302-1147.

RESPECT CANADA DAY. July 15. To show our friends to the north that we really do know they are not just some strange northern province of America. Sponsor: Wellness Permission League. Phone: (212) 388-8673 or (717) 274-8451. For info: Thomas and Ruth Roy, Wellness Permission League, 2105 Water St, Lebanon, PA 17046.

ST. FRANCES XAVIER CABRINI: BIRTH ANNIVERSARY. July 15. First American saint, last of 13 children, founder of schools, orphanages, convents and hospitals, born at Lombardy, Italy, July 15, 1850. Died of malaria at Chicago, IL, Dec 22, 1917. Canonized July 7, 1946.

ST. SWITHIN'S DAY. July 15. Swithun (Swithin), Bishop of Winchester (AD 852-862), died July 2, 862. Little is known of his life, but his relics were transferred into Winchester Cathedral on July 15, 971, a day on which there was a heavy rainfall. According to old English belief, it will rain for 40 days thereafter when it falls on this day. "St. Swithin's Day, if thou dost rain, for 40 days it will remain; St. Swithin's Day, if thou be fair, for 40 days, 'twill rain nea mair."

SCOTLAND: CUTTY SARK TALL SHIPS RACE. July 15-18. Leith, Lothian. Fleet visit and pre-race festivities in Leith. Est attendance: 750,000. For info: The Sail Training Assn, 5 Mumby Rd, Gosport, Hampshire, England PO12 1AA.

SPACE MILESTONE: *APOLLO-SOYUZ* TEST PROJECT (US, USSR): 20th ANNIVERSARY. July 15. After three years of planning, negotiation and preparation, the first US-USSR joint space project reached fruition with the link-up in space of *Apollo 18* (crew: T. Stafford, V. Brand, D. Slayton; landed in Pacific Ocean July 24, during 136th orbit) and *Soyuz 19* (crew: A.A. Leonov, V.N. Kubasov; landed July 21, after 96 orbits). *Apollo 18* and *Soyuz 19* were linked 47 hours (July 17-19) while joint experiments and transfer of personnel and materials back and forth between craft took place. Launch date was July 15, 1975.

SUMMERFAIR. July 15-16. Rocky Mountain College Campus, Billings, MT. Major regional craft fair with arts and crafts, food, music, dance, drama and children's activities on the campus green. Est attendance: 20,000. Sponsor: Yellowstone Art Center, Trina Cullinan, 401 N 27th, Billings, MT 59101. Phone: (406) 256-6804.

July 15-16 ☆ *Chase's 1995 Calendar of Events* ☆

TEDDY BEAR'S PICNIC. July 15-16. Lahaska, PA. Bring your teddy bear for the festivities: competitions, craftspeople and vendors, puppet shows, parades and jugglers, live entertainment. Est attendance: 3,000. For info: Peddler's Village, PO Box 218, Lahaska, PA 18931. Phone: (215) 794-4000.

TOBAGO: TOBAGO HERITAGE FESTIVAL. July 15-Aug 1. A two-week festival celebrating indigenous cultural art forms features different villages presenting a variety of activities from the "ole time" wedding and Belmana Riots to the pulsating rhythms of the music and dance. For info: Trinidad and Tobago Tourist Bd, 25 W 43rd St, Ste 1508, New York, NY 10036. Phone: (800) 232-0082.

TRY THIS ON—A HISTORY OF CLOTHING, GENDER AND POWER. July 15-Aug 27. Tallahassee Museum, Tallahassee, FL. A social history about clothing and appearance in the US over the past 200 years. See Jan 7 for full description. For info: Smithsonian Institution Traveling Exhibition Service, 1100 Jefferson Dr SW, Ste 3146, Washington, DC 20560. Phone: (202) 357-2700.

WOODCRAFT FESTIVAL. July 15-16. Grand Rapids, MN. Northern Minnesota woodcrafters demonstrate traditional woodcrafting techniques, discussions with artisans and hands-on activities. Est attendance: 2,000. For info: R.M. Drake, Forest History Center, 2609 Cnty Rd 76, Grand Rapids, MN 55744. Phone: (218) 327-4482.

BIRTHDAYS TODAY

Willie Aames, 35, actor ("Eight Is Enough," "Charles in Charge"), born at Newport Beach, CA, July 15, 1960.
Kim Alexis, 35, model, born at Lockport, NY, July 15, 1960.
Julian Bream, 62, musician (classical guitar, lute), born London, England, July 15, 1933.
Alex(ander) Karras, 60, former football player, actor (dad in "Webster"), born at Gary, IN, July 15, 1935.
Ken Kercheval, 60, actor ("Dallas," "Search for Tomorrow") born at Wolcottville, IN, July 15, 1935.
Linda Ronstadt, 49, singer ("Heart Like a Wheel," "Simple Dreams"), songwriter, born at Tucson, AZ, July 15, 1946.
Jan-Michael Vincent, 51, actor (*The Winds of War*, "Airwolf"), born Denver, CO, July 15, 1944.
George V. Voinovich, 59, Governor of Ohio (R), born at Cleveland, OH, July 15, 1936.

JULY 16 — SUNDAY
197th Day — Remaining, 168

◆ **ATOMIC BOMB TESTED: 50th ANNIVERSARY.** July 16, 1945. In the New Mexican Desert at Alamongordo Air Base, 125 miles southeast of Albuquerque, the experimental atomic bomb was set off at 5:30 AM. Dubbed "Fat Boy" by its creator, the plutonium bomb vaporized the steel scaffolding holding it as the immense fireball rose 8,000 feet in a fraction of a second—ultimately creating a mushroom cloud to a height of 41,000 feet. At ground zero the bomb emitted heat three times the temperature of the interior of the sun. All plant and animal life for a mile around had ceased to exist. When informed by President Truman at Potsdam of the successful experiment, Winston Churchill responded, "It's the Second Coming in wrath!"

BOLIVIA: LA PAZ DAY. July 16. Founding of city, now capital of Bolivia, on this day, 1548.

BRAILLE RALLYE. July 16. Braille Institute Youth Center, Los Angeles, CA. Blind or visually impaired students from the Braille Institute use braille or large print instructions to navigate sighted drivers through more than 80 miles of LA streets. Celebrity drivers from television, radio and film help make this a special event. Annually, in July. For info: Braille Institute Communications Dept, 741 N Vermont Ave, Los Angeles, CA 90029. Phone: (213) 663-1111.

★ **CAPTIVE NATIONS WEEK.** July 16-22. Presidential Proclamation issued each year since 1959 for the third week of July. (PL86-90 of July 17, 1959.)

CODMAN HOUSE ANTIQUE VEHICLE MEET. July 16. Codman Estate, Lincoln, MA. Display of 200 antique cars, trucks and fire engines from 1969 and earlier. Entertainment, tours of the historic Codman House, refreshments. 10 AM-3 PM. Admission charged. Annually, the second Sunday after Independence Day/Independence Day weekend. Est attendance: 1,800. For info: Hetty Startup, Site Admin, Soc for Prsvn of New Eng Antiquities, 141 Cambridge St, Boston, MA 02114. Phone: (617) 227-3956.

COMET CRASHES INTO JUPITER: ANNIVERSARY. July 16, 1994. The first fragment of the comet Shoemaker-Levy crashed into the planet Jupiter beginning a series of spectacular collisions, each unleashing more energy than the combined effect of an explosion of all our world's nuclear arsenal. Video imagery from earth-bound telescopes as well as the Hubble telescope provided vivid records of the explosions and their aftereffects. In 1993 the comet had shattered into a series of about a dozen large chunks that resembled "pearls on a string" after its orbit brought it within the gravitational effects of our solar system's largest planet.

DISTRICT OF COLUMBIA: ESTABLISHING LEGISLATION ANNIVERSARY. July 16. On July 16, 1790, George Washington signed legislation that selected the District of Columbia as the permanent capital of the US. Boundaries of the district were established in 1792. Plans called for the government to remain housed at Philadelphia, PA, until 1800, when the new national capital would be ready for occupancy.

EARTHQUAKE JOLTS PHILIPPINES: 5th ANNIVERSARY. July 16. An earthquake measuring 7.7 on the Richter scale struck the Philippines on July 16, 1990, killing an estimated 1,621 persons and leaving approximately 1,000 missing. The quake struck in an area north of Manila, and heavy damage was reported in Cabanatuan, Baguio, and on Luzon island. The quake was the worst in the Philippines in 14 years.

FAST OF TAMMUZ. July 16. Jewish holiday. Hebrew calendar date: Tammuz 18, 5755 (because Tammuz 17 falls on the Sabbath in 1995). Shiva Asar B'Tammuz begins at first light of day and commemorates the first century Roman siege that breached the walls of Jerusalem. Begins a three-week time of mourning.

ITALY: FEAST OF THE REDEEMER. July 16. Venice. Procession of gondolas and other craft commemorating the end of the epidemic of 1575. Annually, the third Sunday in July.

LUXEMBOURG: BEER FESTIVAL. July 16. At Diekirch an annual beer festival is held on the third Sunday in July.

MAINE POTATO BLOSSOM FESTIVAL. July 16-23. Fort Fairfield, ME. Promotes and observes the importance of Maine's prime agricultural product. Est attendance: 15,000. For info: Fort Fairfield Chamber of Commerce, 121 Main St, Fort Fairfield, ME 04742. Phone: (207) 472-3802.

	S	M	T	W	T	F	S
July							1
	2	3	4	5	6	7	8
	9	10	11	12	13	14	15
1995	16	17	18	19	20	21	22
	23	24	25	26	27	28	29
	30	31					

★ Chase's 1995 Calendar of Events ★ July 16-17

MISSION SAN DIEGO DE ALCALA FOUNDING: ANNIVERSARY. July 16. First of 21 California missions to the Indians, founded July 16, 1769.

MORMON BATTALION RECRUITMENT DAY: ANNIVERSARY. July 16, 1846. At Council Bluffs, IA, a collection of 500 Mormon men, following the advice of their leader Brigham Young, joined the Army of the West to do battle in the war with Mexico. Known as the US Mormon Battalion, these men, along with 50 women and children, marched 2,000 miles to California, in the longest march in modern military history. They established the first wagon road into Southern California from Santa Fe, NM. They arrived in San Diego, CA, Jan 27, 1847, and their arrival is commemorated each year in San Diego's Old Town with a military parade.

NATIONAL ICE CREAM DAY. July 16. To promote America's favorite dessert, ice cream, on "Sundae Sunday." Annually, the third Sunday in July. For info: Lynn Bagorazzi, 1250 H St NW, Ste 900, Washington, DC 20005. Phone: (202) 296-4250.

NATIONAL SPORTING ASSOCIATION WORLD SPORTS EXPO '95. July 16-18. McCormick Place, Chicago, IL. The international kick-off of the exposition where sporting goods retailers will see 1996 products. FAX: (708) 439-0111. Est attendance: 85,000. Sponsor: Natl Sporting Goods Assn, Tom Drake, Dir of Mktg and PR, 1699 Wall St, Mt Prospect, IL 60056-5780. Phone: (708) 439-4000.

ST. CYRIL'S PARISH FESTIVAL. July 16. Kiwanis Park, Sheboygan, WI. Slovenian foods, entertainment, authentically costumed dancers, games, family activities. 10:30-7. Polka Mass at 10:30, $2,000 bingo at 1:00. Est attendance: 8,000. Sponsor: Cyril & Methodius Parish, 822 New Jersey Ave, Sheboygan, WI 53081. Phone: (414) 457-9330.

SPACE MILESTONE: *APOLLO 11* (US): MAN SENT TO THE MOON. July 16. Launched on July 16, 1969, and resulted in man's first moon landing. Lunar module *Eagle* landed on moon, 4:17 PM EDT, July 20. Commander Neil Armstrong descended from *Eagle* to moon's surface, followed shortly by Colonel Edwin Aldrin, Jr. After rejoining spaceship *Columbia*, piloted by Lieutenant Colonel Michael Collins, astronauts returned to Earth, July 24, bringing firsthand reports of lunar surface, photographs and rock samples. Man's first landing on an extraterrestrial body.

SPACE WEEK. July 16-22. The calendar week containing July 20 has been observed in a number of communities and states as Space Week, commemorating the July 20, 1969, landing on the moon by two US astronauts, Neil Alden Armstrong and Edwin Eugene Aldrin, Jr. See also: "Space Milestone: Moon Day" (July 20).

SPACEWEEK. July 16-24. Spaceweek International seeks to expand public support for the nation's space program by creating awareness, involvement and interest among young people in America's space program. By focusing on the nine-day period coinciding with the epic flight of Apollo 11 that first put a man on the moon, the Assn has succeeded in reaching nearly one-half million Americans annually, operating or consulting on more than 500 programs across the country. Some 20 others countries are official Spaceweek Intl partners. Annually, July 16-24 For info: Spaceweek Intl Assn, 1110 NASA Rd 1, Houston, TX 77058.

STANWYCK, BARBARA: BIRTH ANNIVERSARY. July 16. Actress Barbara Stanwyck was born Ruby Stevens in the Flatbush section of Brooklyn, NY, on July 16, 1907. She began her career at the age of 15 as a chorus girl, and at the age of 18 she won a leading role in the Broadway melodrama *Noose*, appearing for the first time as Barbara Stanwyck. She appeared in 82 films including *Stella Dallas, Double Indemnity, Sorry, Wrong Number, The Lady Eve* and the television series "The Big Valley." As her popularity grew, she demanded increasingly higher salaries in the 1930s and early '40s. In 1944, the government listed her as the nation's highest paid woman, earning $400,000 per year. Stanwyck died at Santa Monica, CA, on Jan 21, 1990.

SUMMER SOCIAL. July 16. Woodstock, VT. Antique bicycling, games, traditional music, amusements and hands-on activities for the whole family. Bring a picnic; horse-drawn wagons will take you to the riverbank for lunch. Est attendance: 700. For info: Billings Farm and Museum, PO Box 489, Woodstock, VT 05091. Phone: (802) 457-2355.

◆ **WELLS, IDA B.: BIRTH ANNIVERSARY.** July 16, 1862. African-American journalist and anti-lynching crusader Ida B. Wells was born the daughter of slaves at Holly Springs, MS, during the Civil War and grew up as Jim Crow and lynching were becoming prevalent. Wells argued that lynchings occurred not to defend white women but because of whites' fear of economic competition from blacks. She traveled extensively, founding antilynching societies and black women's clubs. Wells's *Red Record* (1895) was one of the first accounts of lynchings in the South. She died Mar 25, 1931, at Chicago, IL.

WSBA and WARM 103 SUMMER CRAFT SHOW. July 16. York Fairgrounds, York, PA. More than 250 crafts, from country to contemporary, Victorian and southwestern, handcrafted furniture, wood carvings, dolls, jewelry, pottery, collectibles, quilts, baskets and much more. Admission fee. Est attendance: 4,000. For info: Gina M. Koch, Special Events Dir, PO Box 910, York, PA 17405. Phone: (717) 764-1155.

BIRTHDAYS TODAY

Richard H. Bryan, 58, US Senator (D, Nevada) up for reelection in Nov '94, born at Washington, DC, July 16, 1937.

Stewart Copeland, 43, musician (drummer with Police), songwriter, born at McLean, VA, July 16, 1952.

Margaret Court, 53, former tennis player, born at Albury, Australia, July 16, 1942.

Barnard Hughes, 80, actor (*Sisters*, "Doc"), born at Bedford Hills, NY, July 16, 1915.

Bess Myerson, 71, former Miss America ('45), former government official, born at New York, NY, July 16, 1924.

Ginger Rogers (Virginia Katherine McNath), 84, dancer/actress (many films with Fred Astaire; Oscar for *Kitty Foyle*), born Independence, MO, July 16, 1911.

Barry Sanders, 27, football player, born at Wichita, KS, July 16, 1968.

Richard L. Thornburgh, 63, former US Attorney General, born at Pittsburgh, PA, July 16, 1932.

Pinchas Zukerman, 47, violinist, born at Tel Aviv, Israel, July 16, 1948.

JULY 17 — MONDAY
198th Day — Remaining, 167

ABBOTT, BERENICE: BIRTH ANNIVERSARY. July 17. Berenice Abbott was born at Springfield, OH, on July 17, 1898, and went on to become a pioneer of American photography. She is best remembered for her black and white photographs of New York City in the 1930s, many of which appeared in the book *Changing New York*. After publishing this collection she began photographing scientific experiments that illustrated the laws and processes of physics. She died at Monson, ME, on Dec 11, 1991.

July 17-18 ☆ Chase's 1995 Calendar of Events ☆

CANADA: CALGARY FOLK MUSIC FESTIVAL. July 17-23. Calgary, Alberta. Celebration of local, national and international folk music. Est attendance: 21,000. For info: Folk Festival Soc of Calgary, PO Box 2897 Stn. M, Calgary, AB, Canada T2P 3C3. Phone: (403) 233-0904.

GARDNER, ERLE STANLEY: BIRTH ANNIVERSARY. July 17. American author of detective fiction, born at Malden, MA, July 17, 1889. Best remembered for his Perry Mason detective story series about lawyer-detective, Gardner also wrote novels under the pen name A.A. Fair. Gardner died at Temecula, CA, Mar 11, 1970.

GERRY, ELBRIDGE: BIRTH ANNIVERSARY. July 17. Fifth vice president of the US (1813-1814) born at Marblehead, MA, July 17, 1744. Died at Washington, DC, Nov 23, 1814. His name became part of the language (gerrymander) after he signed a redistricting bill, while governor of Massachusetts, in 1812.

HEMINGWAY DAYS FESTIVAL (WITH HEMINGWAY LOOK-ALIKE CONTEST). July 17-23. Key West, FL. Tribute to Ernest Hemingway and his heirs features short story, trivia, storytelling, arm-wrestling and fishing contests, and a writer's workshop and conference, a 5K sunset run, a sailing regatta and a concert at Hemingway's Key West home. Est attendance: 7,500. For info: Hemingway Days Festival, Michael Whalton, Dir, Box 4045, Key West, FL 33041. Phone: (305) 294-4440.

IRAQ: NATIONAL DAY. July 17. National holiday.

KANSAS CITY HOTEL DISASTER: ANNIVERSARY. July 17. Anniversary of the collapse of aerial walkways at the Hyatt Regency Hotel in Kansas City, MO, on July 17, 1981. About 1,500 people were attending the popular weekly tea dance when, at about 7:00 PM, two concrete and steel skywalks that were suspended from the ceiling of the hotel's atrium broke loose and fell on guests in the crowded lobby, killing 114 people. In 1986, a state board revoked the licenses of two engineers convicted of gross negligence for their part in designing the hotel.

KOREA: CONSTITUTION DAY: ANNIVERSARY. July 17. Legal national holiday. Commemorates the proclamation of the constitution of the republic of Korea on July 17, 1948. Ceremonies at Seoul's capitol plaza and all major cities.

◆ **POTSDAM CONFERENCE: 50th ANNIVERSARY.** July 17-Aug 2, 1945. The Allied Big Three (the US, Soviet Union and Great Britain) met in a palace at Potsdam, to discuss Germany's future. On Aug 1, they issued a 6,000-word communique laying out how Germany would be disarmed, the Nazi party abolished and the country divided into four sectors (with France designated the fourth country to occupy a zone). Reparations were set with the Soviet Union as the country that suffered the greatest under Germany's hand, receiving the greatest share. Trials were set for Nazi leaders charged with authorizing or committing war crimes. Territorial claims, where directly addressed, were particularly favorable to the Soviet Union.

PUERTO RICO: MUNOZ-RIVERA DAY. July 17. Public holiday on the anniversary of the birth of Luis Munoz-Rivera. The Puerto Rican patriot, poet and journalist was born at Barranquitas, Puerto Rico, on July 17, 1859. He died at Santurce, a suburb of San Juan, Puerto Rico, Nov 15, 1916.

SAINT ANN'S SOLEMN NOVENA. July 17-26. Scranton, PA. Religious festival in honor of St. Ann, the mother of the Virgin Mary, the grandmother of Jesus. Some 10,000 people come to St. Ann's shrine each day of the Solemn Novena. Mass is televised each day on national cable network (The Faith and Values Channel). Phone (800)-THE-MASS or (717) 941-0100. Fax (717) 941-0185. For info: Rev Peter Grace, Dir Media, Box 111, Scranton, PA 18504.

SPACE MILESTONE: *SOYUZ T-12* (USSR). July 17. Launched on July 17, 1984, cosmonaut Svetlana Savitskaya became the first woman to walk in space (July 25) and the first woman to make more than one space voyage. With cosmonauts V. Dzhanibekov and I. Volk. Docked at *Salyut 7* on July 18 and returned to Earth July 29.

US AMATEUR PUBLIC LINKS (GOLF) CHAMPIONSHIP. July 17-22. Stow Acres Country Club, Stow, MA. Sponsor: US Golf Assn, Championship Dept, Golf House, Far Hills, NJ 07931. Phone: (908) 234-2300.

"WRONG WAY" CORRIGAN DAY. July 17. On July 17, 1938, Douglas Groce Corrigan, an unemployed airplane mechanic, left Brooklyn, NY's Floyd Bennett field, ostensibly headed for Los Angeles, CA, in a 1929 Curtiss Robin monoplane. He landed 28 hours, 13 minutes later at Dublin, Ireland's Baldonnell airport, after 3,150-mile nonstop flight without radio or special navigation equipment and in violation of American and Irish flight regulations. Born at Galveston, TX, on Jan 22, 1907, Corrigan received hero's welcome home, was nicknamed "Wrong Way" Corrigan because he claimed he accidentally followed the wrong end of his compass needle. The 31-year-old bachelor commented, on return, "I can't get over the number of girls who seem to think because I flew the Atlantic I would make a perfect husband."

BIRTHDAYS TODAY

Diahann Carroll (Carol Diahann Johnson), 60, actress, singer ("Julia," "Dynasty," *Porgy and Bess*, *Hurry Sundown*), born at New York, NY, July 17, 1935.

Phyllis Diller (Phyllis Driver), 78, actress ("The Beautiful Phyllis Diller Show," *Boy Did I Get the Wrong Number*), born Lima, OH, July 17, 1917.

David Hasselhoff, 43, actor ("The Young and the Restless," "Knight Rider"), born at Baltimore, MD, July 17, 1952.

Phoebe Snow, 43, singer ("Poetry Man"), born at New York, NY, July 17, 1952.

Donald Sutherland, 60, actor (*M*A*S*H*, *Klute*) born at St. John, New Brunswick, Canada, July 17, 1935.

Robert Thomas Thigpen, 32, baseball player, born Tallahassee, FL, July 17, 1963.

JULY 18 — TUESDAY

199th Day — Remaining, 166

CHICAGO GOLF CLUB: ANNIVERSARY. July 18. The first 18-hole golf course in America, laid out by Charles Blair MacDonald, was incorporated on July 18, 1893, in Wheaton, IL. MacDonald was the architect of many of the early US courses which he attempted to model on the best in Scotland and England. It was his belief that at each tee a golfer should face a hazard at the average distance of his shot.

FULLER, BOBBY: DEATH ANNIVERSARY. July 18. On this day in 1966, Bobby Fuller, leader of the rock group Bobby Fuller Four, was found dead at his car in Los Angeles, CA. No definite cause of death was ever proven. The Bobby Fuller Four is best remembered for their 1966 hit song "I Fought the Law," which was written by Sonny Curtis, a member of Buddy Holly's Crickets.

July 1995

S	M	T	W	T	F	S
						1
2	3	4	5	6	7	8
9	10	11	12	13	14	15
16	17	18	19	20	21	22
23	24	25	26	27	28	29
30	31					

☆ Chase's 1995 Calendar of Events ☆ July 18–19

MANDELA, NELSON: BIRTHDAY. July 18. South African President Nelson Rolihlahla Mandela was born the son of a Tembu tribal chieftain on July 18, 1918, at Mbhashe, near Umtata, in the Transkei territory of South Africa. Giving up his hereditary rights, Mandela chose to become a lawyer and earned his degree at the University of South Africa. He joined the African National Congress (ANC) in 1944, eventually becoming deputy national president in 1952. His activities in the struggle against apartheid resulted in his conviction for sabotage in 1964. During his 28 years in jail, Mandela remained a symbol of hope to South Africa's nonwhite majority, the demand for his release a rallying cry for civil rights activists. That release finally came on Feb 11, 1990, as millions watched via satellite television. In 1994 Mandela was elected President of South Africa in the first all-race election there. See also: "Mandela, Nelson: Prison Release Anniversary" (Feb 11).

NATIONAL BABY FOOD FESTIVAL. July 18–23. Fremont, MI. One of Western Michigan's largest family festivals. Featured throughout the week will be arts and crafts, sporting events, spectacular midway, nationally renowned entertainment. A summertime festival highlighting the Great American lifestyle. Sponsor: Fremont Chamber of Commerce. Est attendance: 165,000. For info: Fremont Chamber of Commerce, 33 W Main, Fremont, MI 49412. Phone: (616) 924-2270.

ODETTS, CLIFFORD: BIRTH ANNIVERSARY. July 18. Clifford Odetts began his writing career as a poet before turning to acting. He helped found the Group Theatre in 1931. In 1935 he returned to writing with works for the Group Theatre such as *Waiting for Lefty, Awake and Sing, The Golden Boy*. His proletarian views helped make him a popular playwright during the Depression years. Other plays include *The Big Knife, The Country Girl*, and *The Flowering Peach*. Clifford Odetts was born July 18, 1906, at Philadelphia, PA, and died at Los Angeles, CA, on Aug 15, 1963.

PRESIDENTIAL SUCCESSION ACT: ANNIVERSARY. July 18. On July 18, 1947, President Harry S. Truman signed an Executive Order determining the line of succession should the president be temporarily incapacitated or die in office. The speaker of the house and president pro tem of the senate are next in succession after the vice president. This line of succession became the 25th Amendment to the Constitution, which was ratified on Feb 10, 1967.

PROSPECT PARK FISHING CONTEST. July 18–22. Prospect Park, Brooklyn, NY. A contest for young anglers, 15 and under, with prizes for the largest or most fish. At the Rustic Shelter near the Wollman Rink parking lot. Groups must register ahead of time. Est attendance: 700. For info: Public Info Office, Prospect Park, 95 Prospect Park W, Brooklyn, NY 11215. Phone: (718) 965-8999.

RUTLEDGE, JOHN: DEATH ANNIVERSARY. July 18. American statesman, associate justice on the Supreme Court, born at Charleston, SC, in Sept 1739. Nominated second Chief Justice of the Supreme Court, to succeed John Jay, and served as Acting Chief Justice until his confirmation was denied because of his opposition to the Jay Treaty. He died at Charleston, SC, on July 18, 1800.

SPACE MILESTONE: *ROHINI 1* (INDIA). July 18. First successful launch from India, orbited 77-lb satellite. July 18, 1980.

WHITE, GILBERT: 275th BIRTH ANNIVERSARY. July 18. Born at Selborne, Hampshire, England, on July 18, 1720. Gilbert White has been called the "father of British naturalists." His book, *The Natural History of Selborne*, published in 1788, enjoyed immediate success and is said to have never been out of print. White died near his birthplace, June 26, 1793. His home survives as a museum.

BIRTHDAYS TODAY

Dick Button, 66, sportscaster, former figure skater, born at Englewood, NJ, July 18, 1929.

Hume Cronyn, 84, actor (*The Seventh Cross, Sunrise at Campobello*), director, born London, Ont, Canada, July 18, 1911.

Dion Di Mucci, 56, singer (Dion and the Belmonts), born at the Bronx, NY, July 18, 1939.

John Glenn, 74, US Senator (D, Ohio), first American astronaut to orbit Earth, born at Cambridge, OH, July 18, 1921.

Elizabeth McGovern, 34, actress (*Ordinary People, Racing with the Moon*), born at Evanston, IL, July 18, 1961.

Calvin Peete, 52, golfer, born at Detroit, MI, July 18, 1943.

Martha Reeves, 54, singer (lead of Martha & The Vandellas; "Power of Love"), born at Detroit, MI, July 18, 1941.

Ricky Skaggs, 41, musician (bluegrass guitar), singer ("I Don't Care"), born at Cordell, KY, July 18, 1954.

Red Skelton, 82, comedian, actor ("The Red Skelton Show"), born at Vincennes, IN, July 18, 1913.

Hunter S. Thompson, 56, journalist, editor, born at Louisville, KY, July 18, 1939.

Yevgeny Alelesandrovich Yevtushenko, 62, poet, born at Zima, USSR, July 18, 1933.

JULY 19 — WEDNESDAY
200th Day — Remaining, 165

ANN ARBOR SUMMER ART FAIR AND STREET FAIR. July 19–22. Ann Arbor, MI. Juried art fair with more than 560 of the nation's finest artists and contemporary craftspeople. Free family art activity area, specialty shops and restaurants. Also, the 36th annual Street Fair of visual and performing arts. For Street Fair info: Ann Arbor Street Art/Street Fair, Inc, Susan L. Froelich, Coord, Box 1352, Ann Arbor, MI 48106; phone: (313) 994-5260. For Summer Art Fair, call The Guild. Est attendance: 500,000. For info: Shary Brown, Art Fairs Dir, The Guild, 118 N Fourth Ave, Ann Arbor, MI 48104-1402. Phone: (313) 662-3382.

ATTACK ON FORT WAGNER: ANNIVERSARY. July 19. In a second attempt to capture Fort Wagner on June 19, 1863, outside Charleston, SC, Federal troops were repulsed after losing 1,515 men as opposed to Southern losses of only 174. The attack was led by the 54th Massachusetts Colored Infantry, commanded by Colonel Robert Gould Shaw, who was killed in the action. This was the first use of black troops in the war. The film *Glory* was based on the Massachusetts 54th and this was the attack featured in the film. Fort Wagner was never taken by the Union.

CRAB AND CLAM BAKE. July 19. Somers Cove Marina, Crisfield, MD. One of Maryland's finest seafood festivals, featuring steamed crabs and clams, fried fish, raw clams, fried clams and side dishes. This event raises funds to support the Somerset County museum. Annually, the third Wednesday in July. Est attendance: 5,000. For info: Crisfield Area Chamber of Commerce, PO Box 292, Crisfield, MD 21817. Phone: (800) 782-3913.

DEGAS, EDGAR: BIRTH ANNIVERSARY. July 19. French Impressionist painter, especially noted for his paintings of dancers in motion, was born at Paris, France, July 19, 1834. Degas died Sept 26, 1917.

DOLL SHOW. July 19. Broadway Market Square, Baltimore, MD. Open to girls and boys 14 years of age and under. 12 different categories of dolls. Trophies and prizes for winners. Overall Doll Show Queen & King are selected from the winners of the different doll categories. The Doll Show King & Queen are presented additional prizes and crowned at the conclusion of the show. Est attendance: 300. For info: Office of Adventures and Fun, Clarence "Du" Burns Arena, 1301 S Ellwood St, Baltimore, MD 21224. Phone: (410) 396-9177.

MAYO, CHARLES HORACE: BIRTH ANNIVERSARY. July 19. American surgeon, one of the Mayo brothers, founders of the Mayo Clinic and Mayo Foundation, born at Rochester, MN, July 19, 1865. Died at Chicago, IL, May 26, 1939.

July 19–20 ☆ Chase's 1995 Calendar of Events ☆

MOON PHASE: LAST QUARTER. July 19. Moon enters Last Quarter phase at 7:10 AM, EDT.

NICARAGUA: NATIONAL LIBERATION DAY. July 19. Following the National Day of Joy (July 17—anniversary of date in 1979 when dictator Anastasio Somoza Debayle fled Nicaragua) is annual July 19 observance of National Liberation Day, anniversary of day the National Liberation Army claimed victory over the Somoza dictatorship.

ST. VINCENT DE PAUL: OLD FEAST DAY. July 19. A day remembering the founder of the Vincentian Congregation and the Sisters of Charity, born in France in 1581 (exact date unknown). He died Sept 27, 1660. His Feast Day was formerly observed on July 19 but is now observed on the anniversary of his death, Sept 27.

UNITED AIRLINES FLIGHT 232 CRASHES IN IOWA CORNFIELD: ANNIVERSARY. July 19. On July 19, 1989, Chicago-bound, United Airlines flight 232 crashed into a cornfield in Sioux City, IA. The DC-10 aircraft was carrying 293 people when it slammed into the runway and flipped over, bursting into a fireball. Miraculously, 181 of the passengers and crew survived the crash. Emergency personnel at the airport and in Sioux City had been alerted to the situation approximately one-half hour before the crash when the pilot reported an uncontained engine failure and a complete hydraulic system failure. The rear engine failed when a turbine disintegrated sending shards of metal through the tail cone section, which dislodged from the aircraft and was later found in Alta, IA, approximately 70 miles east of the crash site.

♦ **WOMEN'S CONVENTION HELD AT SENECA FALLS: ANNIVERSARY.** July 19. A convention concerning the rights of women, called by Lucretia Mott and Elizabeth Cady Stanton, was held at Seneca Falls, NY, on July 19–20, 1848. The issues discussed included voting, property rights and divorce. The convention drafted a "Declaration of Sentiments" that paraphrased the Declaration of Independence, addressing man instead of King George, and called for women's "immediate admission to all the rights and privileges which belong to them as citizens of the United States." This convention was the beginning of an organized women's rights movement in the US. The most controversial issue was Stanton's demand for women's right to vote.

July 1995

S	M	T	W	T	F	S
						1
2	3	4	5	6	7	8
9	10	11	12	13	14	15
16	17	18	19	20	21	22
23	24	25	26	27	28	29
30	31					

♦ **YALOW, ROSALYN: BIRTHDAY.** July 19, 1921. Medical physicist Rosalyn Yalow was born at New York City. Along with Andrew V. Schally and Roger Guillemin, in 1977 Yalow was awarded the Nobel Prize for Physiology or Medicine. Through her research on medical applications of radioactive isotopes, Yalow developed RIA, a sensitive and simple technique used to measure minute concentrations of hormones and other substances in blood or other body fluids. First applied to the study of insulin concentration in the blood of diabetics, RIA was soon used in hundreds of other applications.

BIRTHDAYS TODAY

Philip Agee, 60, former CIA agent, author (*Inside the Company: CIA Diary*), born at Tacoma Park, FL, July 19, 1935.
Natalya Bessmertnova, 54, Bolshoi prima ballerina, born at Moscow, USSR, July 19, 1941.
Vikki Carr (Florencia Bisenta deCasilla), 54, singer ("It Must Be Him," "With Pen In Hand"), born at El Paso, TX, July 19, 1941.
Roosevelt "Rosey" Grier, 63, former football player, actor, born at Linden, NJ, July 19, 1932.
George Stanley McGovern, 73, former US senator and '72 Democratic presidential nominee, born at Avon, SD, July 19, 1922.
Ilie Nastase, 49, former tennis player, born at Bucharest, Romania, July 19, 1946.

JULY 20 — THURSDAY
201st Day — Remaining, 164

AMERICAN ROWING NATIONAL CHAMPIONSHIPS. July 20–23. Lake Shawnee, Topeka, KS. 124th annual championships. Sponsor: US Rowing Assn. Fax: (913) 233-9952 on 9th ring. Est attendance: 44,000. For info: Don Craig, 4336 SE 25th St Terrace, Topeka, KS 66605. Phone: (913) 233-9951.

BASEBALL DECLARED NON-ESSENTIAL OCCUPATION: ANNIVERSARY. July 20. Secretary of War Newton D. Baker ruled on July 20, 1918, that baseball was a non-essential occupation. He stated that all players of draft age should seek "employment to aid successful prosecution of the war or shoulder guns and fight." On July 26, Baker allowed baseball to continue until Sept 1. Nearly 250 ballplayers entered the armed services.

BRENNAN, WILLIAM: SUPREME COURT RESIGNATION: 5th ANNIVERSARY. July 20. US Supreme Court Associate Justice William J. Brennan resigned on July 20, 1990, creating not only a vacancy on the high court but setting the stage for a legal watershed. Brennan was appointed to the Supreme Court in 1956 by President Dwight Eisenhower, taking his seat in a recess appointment and being confirmed the following year. His liberal leanings and judicial activism raised the ire of many conservatives, even leading Eisenhower to call his appointment his "worst mistake." Brennan believed that the courts should go beyond strict constructionism and original intent doctrines in interpreting the law. In a 1985 speech, he answered his critics by saying, "We current justices read the Constitution in the only way we can: as 20th century Americans.... The ultimate question must be, what do the words of the text mean in our time?" He was responsible for many landmark decisions in the last half of the 20th century including the decision requiring the Little Rock, AR, schools to desegregate that included the controversial assertion that it is not only the Constitution that is the "supreme law of the land" and must be obeyed, but the Supreme Court's interpretation of it as well. His legacy also includes major decisions upholding affirmative action, a losing battle to declare the death penalty unconstitutional, decisions broadening free speech and free press guarantees, expansion of the due process guarantees under the 14th Amendment, and protection of flagburning as a form of expression. Brennan's resignation provided President George Bush with an opportunity to swing the balance of the Supreme Court to a more conservative, constructionist majority for the first time since the days of Franklin D. Roosevelt's presidency.

CANADA: JUST FOR LAUGHS: THE MONTREAL INTERNATIONAL COMEDY FESTIVAL. July 20-30. Montreal, Quebec. 13th annual international festival devoted to humor. More than 250 comics from Canada, England, US, Australia, New Zealand, France, Belgium, Scotland and Italy. Past years have included Tim Allen, Jerry Seinfeld, John Candy, Bob Newhart, Mary Tyler Moore, Jay Leno, Graham Chapman, Steve Allen, Lily Tomlin, Jerry Lewis, Ben Elton and Sinbad. Est attendance: 635,000. For info: Just for Laughs Comedy Festival, 2101 Boul Saint-Laurent, Montreal, Quebec H2X 2T5. Phone: (514) 845-3155.

CANADA: KLONDIKE DAYS. July 20-29. Edmonton, Alberta. Edmonton's early days as a frontier community and gateway to the Yukon in the rush for gold sets the stage for the city's annual Klondike Days festival. Parades, pancake breakfasts, gambling casinos, entertainment, gold panning and the World Championship Sourdough Raft Race. Est attendance: 765,000. For info: Edmonton Northlands, R.J. Gray, Major Events Mgr, Box 1480, Edmonton, Alta, Canada T5J 2N5. Phone: (403) 471-7210.

CANADA: SHELBURNE FOUNDERS' DAYS. July 20-23. Shelburne, Nova Scotia, Canada. To commemorate the settlement of Shelburne by the United Empire Loyalists in 1783. Community groups and volunteers organize canoe and yacht races, canoe jousting, greased pole games, fireworks, canteens, craft demonstrations, musical entertainment, pony and wagon rides and a Loyalist Garden Party. Annually, the weekend closest to July 21. Est attendance: 7,000. For info: Jerry Locke, Parks/Recreation Dept Dir, PO Box 699, Shelburne, NS, Canada B0T 1W0. Phone: (902) 875-3873.

COLOMBIA: INDEPENDENCE DAY. July 20. National holiday. Gained independence from Spain, 1819.

COUNTRY JAM USA. July 20-23. Eau Claire, WI. The country stars will shine. Today's hottest stars playing the music that is the heart and soul of America. Food and beverage tents, campgrounds. Phone: (800) 7800-JAM. Est attendance: 120,000. For info: Summer Festivals, Inc, Paul Novitzke, 1711 S Hastings Way, Eau Claire, WI 54701.

CRAFT FAIR OF THE SOUTHERN HIGHLANDS. July 20-23. (Also Oct 19-22.) Asheville Civic Center, Asheville, NC. In 1948 this fair started the crafts revival so evident today. More than a craft fair, this event includes demonstrations, children's workshops and traditional mountain music—all in addition to the 175 craft exhibits by the finest craftspeople in both traditional and contemporary crafts. Est attendance: 35,000. Sponsor: Becky Orr, Admin, Southern Highland Handicraft Guild, PO Box 9545, Asheville, NC 28815. Phone: (704) 298-7928.

DESTINY IN DAYTON. July 20-23. Bryan College, Dayton, TN. Five performances of an original drama written from the actual transcript of the Scopes trial—"the world's most famous court trial." Also, an arts and crafts fair, an antique car show and much more. Est attendance: 13,000. For info: Bryan College, Attn: Tom Davis, PO Box 7000, Dayton, TN 37321-7000. Phone: (615) 775-2041.

ENGLAND: SOUTHERN CATHEDRALS FESTIVAL. July 20-23. Chichester Cathedral, Chichester, Sussex. Festival of church music and concerts performed by the Cathedral Choirs of Chichester, Salisbury and Winchester. Est attendance: 1,000. For info: Alan Thurlow, 2 St. Richard's Walk, Chichester, Sussex, England PO19 1QA.

FOLKMOOT USA: THE NORTH CAROLINA INTERNATIONAL FOLK FESTIVAL. July 20-30. Waynesville, NC. A festival of international folk dance featuring groups from 10 countries. Est attendance: 75,000. For info: Folkmoot USA, PO Box 523, Waynesville, NC 28786. Phone: (704) 452-2997.

GENEVA ACCORDS. July 20. An agreement covering cessation of hostilities in Vietnam, signed at Geneva, Switzerland, July 20, 1954, on behalf of the commanders-in-chief of French forces in Vietnam and the People's Army of Vietnam. A further declaration of the Geneva Conference was released July 21, 1954. Partition, foreign troop withdrawal and elections for a unified government, within two years, were among provisions.

HILLARY, SIR EDMUND PERCIVAL: BIRTHDAY. July 20. Explorer, mountaineer born at Auckland, New Zealand, on July 20, 1919. With Tenzing Norgay, a Sherpa guide, became first to ascend summit of highest mountain in the world, Mt Everest (29,028 ft), at 11:30 AM on May 29, 1953. "We climbed because nobody climbed it before," he said.

HOT DOG NIGHT. July 20. Luverne, MN. More than 12,000 hot dogs are served free of charge; free orange drink is also provided. Various demonstrations. Est attendance: 5,000. Sponsor: Luverne Area Chamber of Commerce, 102 E Main, Luverne, MN 56156. Phone: (507) 283-4061.

JOHNSON COUNTY PEACH FESTIVAL. July 20-23. Clarksville, AR. Oldest outdoor festival in Arkansas. Princess Elberta, Queen Elberta and Miss Arkansas Valley pageants; horseshoe pitching, frog jump, terrapin derby, peach eating contest, peach cobbler contest, peach jam & jellie contest, square dancing, 4-mile race, diaper derby, greased pig chase, parade, water balloon toss, fiddling jamboree, street dance and bass tournament. 54th annual festival. Est attendance: 20,000. For info: Johnson County Chamber of Commerce, PO Box 396, Clarksville, AR 72830. Phone: (501) 754-2340.

LAS VEGAS NIGHT. July 20. Minocqua, WI. Shoppers play Las Vegas-type games to determine discounts on purchases. Est attendance: 3,000. For info: Greater Minocqua Chamber of Commerce, PO Box 1006, Minocqua, WI 54548. Phone: (800) 446-6784.

MIDSUMMER NIGHTS' STROLL. July 20-21. New Harmony, IN. Small groups will be led to historic homes serving period food and refreshments. Annually, the third Thursday and Friday in July. Est attendance: 300. For info: Midsummer Nights' Stroll, Special Projects Coord, PO Box 579, New Harmony, IN 47631. Phone: (812) 682-4488.

RIOT ACT: ANNIVERSARY. July 20. To "read the riot act" now usually means telling children to be quiet or less boisterous, but in 18th-century England reading the riot act was a more serious matter. On July 20, 1715, the Riot Act took effect. By law, in England, if 12 or more persons were unlawfully assembled to the disturbance of the public peace an authority was required "with a loud voice" to command silence and read the riot act proclamation: "Our sovereign lord the king chargeth and commandeth all persons, being assembled, immediately to disperse themselves, and peaceably to depart to their habitations, or to their lawful business, upon the pains contained in the act made in the first year of King George, for preventing tumults and riotous assemblies. God save the king." Any persons who failed to obey within one hour were to be seized, apprehended and carried before a justice of the peace.

SCOTLAND: GOLF OPEN CHAMPIONSHIP. July 20-23. Old Course, St. Andrews, Fife. Major event on the golfing calendar. Est attendance: 180,000. For info: The Secretary, Royal and Ancient Golf Club, St. Andrews, Fife, Scotland, KY16 9JD. Phone: (033) 472-112.

July 20–21 ☆ Chase's 1995 Calendar of Events ☆

SPACE MILESTONE: MOON DAY. July 20. Anniversary of man's first landing on moon. Two US astronauts (Neil Alden Armstrong and Edwin Eugene Aldrin, Jr) landed lunar module *Eagle* at 4:17 PM, EDT, July 20, 1969, and remained on lunar surface 21 hours, 36 minutes and 16 seconds. The landing was made from the *Apollo XI*'s orbiting command and service module, code named *Columbia*, whose pilot Michael Collins, remained aboard. Armstrong was first to set foot on moon. Armstrong and Aldrin were outside spacecraft, walking on Moon's surface, approximately 2¼ hours.

BIRTHDAYS TODAY

Kim Carnes, 49, singer ("Bette Davis Eyes"), songwriter (co-wrote score *Flashdance*), born at Hollywood, CA, July 20, 1946.

Judy Chicago (Judy Cohen), 56, artist, feminist, born at Chicago, IL, July 20, 1939.

Larry E. Craig, 50, US Senator (R, Idaho), born at Council, ID, July 20, 1945.

Chuck Daly, 62, former basketball coach, born at St. Mary's, PA, July 20, 1933.

Nelson Doubleday, 62, baseball executive, publisher, born at Long Island, NY, July 20, 1933.

Sir Edmund Hillary, 76, explorer (first to climb Mt Everest), born at Auckland, New Zealand, July 20, 1919.

Sally Ann Howes, 61, actress (*Dead of Night, The History of Mr. Polly*), singer, born London, England, July 20, 1934.

Mike Ilitch, 66, owner of the Detroit Red Wings and Little Caesar's pizza franchises, born at Detroit, MI, July 20, 1929.

Barbara Ann Mikulski, 59, US Senator (D, Maryland), born at Baltimore, MD, July 20, 1936.

Diana Rigg, 57, actress (Mrs. Emma Peel on "The Avengers"; Tony for *Medea*; *King Lear, Witness for the Prosecution*), born Doncaster, Yorkshire, England, July 20, 1938.

Carlos Santana, 48, musician, born at Autlan, Mexico, July 20, 1947.

JULY 21 — FRIDAY
202nd Day — Remaining, 163

ARCADIA DAZE. July 21–23. Arcadia MI. The scenic village of Arcadia is the setting for this mid-summer event. Activities include a parade on Sunday at 1:30 PM on Lake Street. An art fair, steak fry, fishing contest, games for the children, pancake breakfast, 5K and 10K running race on Saturday and a street dance. An old-fashioned good time for the whole family. Annually, the fourth weekend in July. Sponsor: Arcadia Lion's Club. Est attendance: 2,500. For info: Wesley Hall, 3269 Lake St, Arcadia, MI 49613. Phone: (616) 889-5555.

ARTS IN THE PARK. July 21–23. Kalispell, MT. 26th Annual juried art show and fair, food and entertainment. Est attendance: 30,000. For info: Hockaday Center for the Arts, PO Box 83, Kalispell, MT 59903. Phone: (406) 755-5268.

BELGIUM: NATIONAL HOLIDAY. July 21. Marks accession of first Belgian king, Leopold I, at independence from Netherlands on this day, 1831.

BLUE RIVER FESTIVAL. July 21–23. Crete, NE. Weekend of activities including arts and craft booths, sidewalk sales, children's parade and games, music, car show, food, airport open house, fireworks, high school alumni banquet and class reunions. Annually, the weekend of the fourth Saturday in July. Est attendance: 5,500. Sponsor: Marilyn McElravy, Mgr, Crete Chamber of Commerce, PO Box 264, Crete, NE 68333. Phone: (402) 826-2136.

	S	M	T	W	T	F	S
July 1995							1
	2	3	4	5	6	7	8
	9	10	11	12	13	14	15
	16	17	18	19	20	21	22
	23	24	25	26	27	28	29
	30	31					

CANADA: DAWSON CITY MUSIC FESTIVAL. July 21–23. Dawson City, Yukon. Entertainers and artists from Canada and the US, featuring workshops, concerts and dances at this 17th annual festival. Tickets should be purchased in advance. Fax: (403) 993-5510. Est attendance: 750. Sponsor: Dawson City Music Festival Assn, Box 456, Dawson City, Yukon, Canada Y0B 1G0. Phone: (403) 993-5584.

CANADA: FOLK ON THE ROCKS. July 21–23. Yellowknife, Northwest Territories. Weekend extravaganza of music and cultural arts. Fax: (403) 873-3654. Est attendance: 1,000. For info: Folk on the Rocks, Box 326, Yellowknife, NT, Canada X1A 2N3. Phone: (403) 920-7806.

CANADA: GREAT NORTHERN ARTS FESTIVAL. July 21–30. Inuvik, Northwest Territories. Exhibition and sale by more than 75 Northern artists. Workshops, demonstrations and seminars on printmaking, drawing, painting, tapestry and carving. Nightly entertainment and special presentations—drumming and dancing, storytelling, throat singing, fashion show, jam sessions—leading up to the Great Northern MusicFest (July 28–30). Est attendance: 6,000. For info: Charlene Alexander, Great Northern Arts Festival, Box 2921, Inuvik, NWT, Canada X0E 0T0. Phone: (403) 979-3536.

CENTRAL NEBRASKA ETHNIC FESTIVAL. July 21–23. Downtown Grand Island, NE. The streets come alive with color, music and foods from all ethnic cultures in the Central Nebraska area. Singing, dancing, eating, laughter and enjoyment for the whole family. Annually, the fourth weekend in July. Phone: (308) 385-5444 ext. 239. Est attendance: 20,000. For info: Dianne Kelley, Downtown Dvmt Dir, PO Box 1306, Grand Island, NE 68802.

CHEYENNE FRONTIER DAYS. July 21–30. Frontier Park, Cheyenne, WY. Held annually since 1897 the world's largest outdoor rodeo is the "Daddy of 'em All" with nine rodeos, nine night shows, three free pancake breakfasts, four parades, chuckwagon racing, carnival midway and exhibitors. Annually, the last full week in July. Est attendance: 210,000. For info: Cheyenne Frontier Days, PO Box 2477, Cheyenne, WY 82003. Phone: (800) 227-6336.

CLEVELAND, FRANCES FOLSOM: BIRTH ANNIVERSARY. July 21. Wife of Grover Cleveland, 22nd president of the US, born at Buffalo, NY, July 21, 1864. Died Oct 29, 1947.

COHO FAMILY FESTIVAL. July 21–23. Sheboygan, WI. Fishing competition features a main tournament, kids' tournament and super tournament. Tons of fish are caught and registered. Daily, category and bag limit prizes. Exhibits, entertainment, basketball shoot-out for all ages, sand volleyball tournament and sand sculpturing contest, fish display and food at Derby Headquarters. Sponsor: Sheboygan Rotary, PO Box 1172, Sheboygan, WI 53082-1172. Est attendance: 27,500. For info: Conv and Visitors Bureau, 712 Riverfront Dr, Ste 101, Sheboygan, WI 53081. Phone: (414) 457-9495.

DOWNTOWN DENVER INTERNATIONAL BUSKERFEST. July 21–23. Denver, CO. A four-day European-style street festival featuring performing artists from around the world. Colorful, lively outdoor performances by jugglers, acrobats, fire-eaters, musicians, mimes, tightrope walkers, puppeteers, street comics, theatre troupes and more on the 16th

Chase's 1995 Calendar of Events — July 21

Street Mall in Downtown Denver. 11–dusk. Est attendance: 100,000. Sponsor: Downtown Denver Partnership, Inc, 511 16th St, Ste 200, Denver, CO 80202-4250. Phone: (303) 534-6161.

FAIRBANKS SUMMER ARTS FESTIVAL. July 21–Aug 6. University of Alaska, Fairbanks, AK. Two weeks of workshops and concerts involving music, dance, theater, opera theatre, ice skating theater and the visual arts. More than 60 guest artists and instructors from across the nation are featured. Fax: (907) 479-4329. Est attendance: 12,500. For info: Jo Scott, Fairbanks Summer Arts Fest, Box 80845, Fairbanks, AK 99708. Phone: (907) 479-6778.

FIRST ROBOT HOMICIDE: ANNIVERSARY. July 21. The first reported killing of a human by a robot occurred at Jackson, MI. On July 21, 1984, a robot turned and caught a 34-year-old worker between it and a safety bar, crushing him. He died of the injuries on July 26. According to the National Institute for Occupational Safety and Health, it was "the first documented case of a robot-related fatality in the US."

FRENCH WEST INDIES: SCHOELCHER DAY. July 21. Guadeloupe, St. Martin and St. Barthelemy. Public holiday honoring Victor Schoelcher, the French parliamentarian who led the campaign to abolish slavery in 1848. For info: French West Indies Tourist Board, 610 5th Ave, New York, NY 10020.

GUAM: LIBERATION DAY. July 21. Guam ceded to US by Spain, 1898. US forces returned to Guam on this day, 1944.

HEMINGWAY, ERNEST: BIRTH ANNIVERSARY. July 21. American short story writer and novelist born at Oak Park, IL, July 21, 1899. Made his name with such works as *The Sun Also Rises* (1926), *A Farewell to Arms* (1929), *For Whom the Bell Tolls* (1940) and *The Old Man and the Sea* (1952). He was awarded the Nobel Prize in 1954 and wrote little thereafter; he shot himself on July 2, 1961, at Ketchum, ID, having been seriously ill for some time.

KENTUCKY STATE CHAMPIONSHIP OLD-TIME FIDDLER'S CONTEST. July 21–22. Rough River Dam State Resort Park, Falls of Rough, KY. Old-time and bluegrass music competition, Governor's cup trophy is awarded to the champion fiddler. Annually, the third Friday and Saturday in July. Est attendance: 3,000. For info: Brent L. Miller, PO Box 4042, Leitchfield, KY 42755. Phone: (502) 259-3578.

McLUHAN, MARSHALL: BIRTH ANNIVERSARY. July 21. (Herbert) Marshall McLuhan, university professor and author, called "the Canadian sage of the electronic age," was born at Edmonton, Alberta, Canada, July 21, 1911. *Understanding Media* and *The Medium Is the Massage* (not to be confused with his widely quoted aphorism: "The medium is the message"), among other books, were widely acclaimed for their fresh view of communication. McLuhan is reported as saying: "Most people are alive in an earlier time, but you must be alive in our own time." He died at Toronto, Ontario, Dec 31, 1980.

MOUNTAIN MAN RENDEZVOUS. July 21–30. Red Lodge, MT. Mountain men, cavalry, Indians, traders, scouts, buffalo hunters, whiskey runners and horse traders. Annual. Est attendance: 6,000. For info: Joan Cline, Red Lodge Chamber of Commerce, PO Box 998, Red Lodge, MT 59068. Phone: (406) 446-1718.

NATIONAL WOMEN'S HALL OF FAME: ANNIVERSARY. July 21. Seneca Falls, NY. Founded to honor American women whose contributions "have been of the greatest value in the development of their country." Located in community known as "birthplace of women's rights," where first Women's Suffrage Movement convention was held in 1848. Dedicated with 23 inductees, on July 21, 1979. Earlier National Women's Hall of Fame, honoring "Twenty outstanding women of the Twentieth Century," was dedicated at New York World's Fair, on May 27, 1965.

NORTH AMERICAN REGATTA. July 21–23. Lake Sakakawea, ND. Held on massive Lake Sakakawea, the Regatta features Hobie sailing and sailboarding, and offers seminars and free sailboarding lessons for spectators. Est attendance: 10,000. For info: North Dakota Parks and Recreation, 604 East Blvd, Bismarck, ND 58505. Phone: (800) 435-5663.

NORTH DAKOTA STATE FAIR. July 21–29. Minot, ND. For 9 days the State Fair features the best in big-name entertainment, farm and home exhibits, displays, the Midway and NDRA rodeo. Est attendance: 244,000. For info: North Dakota State Fair, Box 1796, Minot, ND 58702. Phone: (701) 857-7620.

PRO FOOTBALL HALL OF FAME FESTIVAL. July 21–Aug 5. Canton, OH. To honor the 1995 Class of Enshrinees inducted into the Hall of Fame. Est attendance: 500,000. Sponsor: Pro Football Hall of Fame Fest, Janice C. Meyer, CFE, Festival Exec Dir, 229 Wells Ave NW, Canton, OH 44703. Phone: (216) 456-7253.

SALEM ART FAIR & FESTIVAL. July 21–23. Bush's Pasture Park, Salem, OR. A celebration of the arts with 200 arts/crafts booths, artist demonstrations, continuous performing arts, food, children's parade and art activities, historic Bush House Museum, special exhibition, 5K run, street painting and folk arts. Annually, the third full weekend in July. Est attendance: 100,000. For info: Salem Art Assn, 600 Mission St SE, Salem, OR 97302. Phone: (503) 581-2228.

SCARBOROUGH-INDIANAPOLIS PEACE GAMES. July 21–24. Indianapolis, IN. Annual multi-sport event competition between sister cities in the US and Ontario, Canada. Est attendance: 3,000. For info: Indianapolis-Scarborough Peace Games, 1502 W 16th St, Indianapolis, IN 46202. Phone: (317) 327-7201.

SHOW ME STATE GAMES. July 21–23. Columbia, MO. (Also July 28–30.) An Olympic-style festival for Missouri citizens. This statewide multi-sport program is designed to inspire Missourians of every age and skill level to develop their physical and competitive abilities to the height of their potential through participation in fitness activities. Annually, the last two full weekends in July. Est attendance: 40,000. For info: Gary Filbert, Exec Dir, Show Me State Games, 404 Jesse Hall, Columbia, MO 65211. Phone: (314) 882-2101.

SONG OF HIAWATHA PAGEANT. July 21–23. (Also July 28–30, Aug 4–6. Pipestone, MN. 47th annual presentation of pageant based on Longfellow's poem, held in a natural outdoor amphitheater. Elaborate lighting, lovely costumes and cast of 200, help make an unforgettable event. Annually, last two weekends in July and first weekend in August. Sponsor: The Hiawatha Club. For info: Mick Myers, Exec Dir, Pipestone Chmbr Comm, PO Box 8 CAE, Pipestone, MN 56164. Phone: (507) 825-4126.

TWIN-O-RAMA. July 21–23. Cassville, WI. Celebration honoring twin and other multiple births. Annually. Evening phone: (608)725-5348. Est attendance: 7,500. For info: Twin-O-Rama, Inc, PO 545, Cassville, WI 53806. Phone: (608) 725-5037.

BIRTHDAYS TODAY

Les Aspin, 57, former US Secretary of Defense in Clinton administration, born at Milwaukee, WI, July 21, 1938.

Norman Jewison, 69, producer, director (*Moonstruck, Fiddler on the Roof*), born at Toronto, Ontario, Canada, July 21, 1926.

Don Knotts, 71, actor, comedian ("The Andy Griffith Show," *The Ghost and Mr. Chicken*), born Morgantown, WV, July 21, 1924.

Jon Lovitz, 38, comedian ("Saturday Night Live," *A League of Their Own*), born at Tarzana, CA, July 21, 1957.

Janet Reno, 57, US Attorney General in Clinton administration, born at Miami, FL, July 21, 1938.

Isaac Stern, 75, violinist, born at Kreminiecz, USSR, July 21, 1920.

Cat Stevens (Stephen Demetri Georgiou), 47, singer, songwriter, chosen Muslim name is Yusuf Islam, born at London, England, July 21, 1948.

July 21-22 ☆ Chase's 1995 Calendar of Events ☆

Paul D. Wellstone, 51, US Senator (D, Minnesota), born at Washington, DC, July 21, 1944.
Robin Williams, 43, actor (*The World According to Garp, The Fisher King*), comedian, born Chicago, IL, July 21, 1952.

JULY 22 — SATURDAY
203rd Day — Remaining, 162

ANTIQUE AND CLASSIC BOAT RENDEZVOUS. July 22. Mystic Seaport, Mystic, CT. Pre-1952 power and sailing yachts on view for Mystic Seaport visitors. Mystic River parade. Est attendance: 4,000. For info: Mystic Seaport, 75 Greenmanville Ave, Box 6000, Mystic, CT 06355. Phone: (203) 527-5315.

CANADA: GREAT CARIBOO RIDE. July 22-30. Cariboo Country (Cache Creek to 100 Mile House), British Columbia. Annual horse trek exploring pioneer trails and discovering new ones. Riding during day, camping at night. Est attendance: 60. For info: Great Cariboo Ride Soc, PO Box 1025, 100 Mile House, BC, Canada V0K 2E0. Phone: (604) 395-2753.

CENTRAL MAINE EGG FEST. July 22. Pittsfield, ME. Features "World's Largest Egg" competition, an early morning breakfast, parade, games, crafts, fireworks. Annually, the fourth Saturday in July. Est attendance: 25,000. For info: Central Maine Egg Festival Committee, Box 82, Pittsfield, ME 04967. Phone: (207) 487-5416.

CHILDREN'S DAY. July 22. New Hampshire Farm Museum, Rte 125, Milton, NH. Hands-on activities for the children, Teddy Bear Picnic, small animals for petting, rides, games, face painting, wood-working table. Annually, the fourth Saturday in July. Sponsor: New Hampshire Farm Museum. Est attendance: 1,500. For info: Susie McKinley, Admin Asst, PO Box 644, Milton, NH 03851. Phone: (603) 652-7840.

CRAFT FAIR USA: INDOOR SHOW. July 22-23. Wisconsin State Fair Park, Milwaukee, WI. Sale of handcrafted items: jewelry, pottery, weaving, leather, wood, glass, yulecraft and sculpture. Est attendance: 13,000. For info: Dennis R. Hill, Dir, 3233 S Villa Circle, Milwaukee, WI 53227. Phone: (414) 321-2100.

DILLINGER, JOHN: DEATH ANNIVERSARY. July 22. Bank robber, murderer, prison escapee and the first person to receive the FBI's appellation "Public Enemy No. 1" (July 1934). After nine years in prison (1924-1933), Dillinger traveled, in 1933-1934, through Indiana, Illinois, Ohio, Wisconsin, Minnesota and Iowa, leaving a path of violent crimes. Reportedly betrayed by the "Lady in Red," he was killed by FBI agents as he left Chicago's Biograph movie theater (where he had watched *Manhattan Melodrama*, starring Clark Gable), on July 22, 1934. He was born at Indianapolis, IN, on June 28, 1902.

FREMONT CAR SHOW. July 22. Gerber Grounds, Fremont, MI. A wide variety of vehicles ranging from vintage models to car-crushing monster trucks. Annually, the six weekends prior to Labor Day. Sponsor: Fremont Auto Parts Center/Auto Value. Est attendance: 25,000. For info: Casper and Terri Braafhart, Exec Dirs, 3810 W 72nd St, Newaygo, MI 49337. Phone: (616) 924-4899.

GREAT CARDBOARD BOAT RACE. July 22. Sandy Beach, Heber Springs, AR. Cardboard boats which defy natural laws float and race on Greers Ferry Lake; open to all comers; winner competes in the Annual America's International Cardboard Cup Challenge; other activities scheduled all day. 8th annual race. Est attendance: 6,500. For info: Chamber of Commerce, 1001 W Main, Heber Springs, AR 72543. Phone: (501) 362-2444.

	S	M	T	W	T	F	S
July 1995							1
	2	3	4	5	6	7	8
	9	10	11	12	13	14	15
	16	17	18	19	20	21	22
	23	24	25	26	27	28	29
	30	31					

INTERNATIONAL CHILDBIRTH EDUCATION AWARENESS DAY. July 22. ICEA Convention, Phoenix, AZ. A day designed to promote the benefits of education concerning the childbearing year. Expectant couples can find information on birthing processes, coping strategies, support techniques, breastfeeding, infant care and more. Such information has been directly linked with healthier outcomes for mother and baby and increased satisfaction with the birth experience. Est attendance: 1,000. Sponsor: Sam Cook, RNC, ICCE, Dir PR, The Intl Childbirth Education Assn, 3603 Nortree St, San Jose, CA 95148. Phone: (408) 238-6162.

OLD BEDFORD VILLAGE BLUEGRASS FESTIVAL. July 22-23. Bedford, PA. A host of the finest bluegrass bands will entertain you with your favorite knee-slapping, foot-stomping music. Delicious foods including a chicken Bar-B-Q and pig roast. For info: Old Bedford Village, Box 1976, Bedford, PA 15522. Phone: (814) 623-1156.

PIED PIPER OF HAMELIN ANNIVERSARY. July 22. According to legend, the German town of Hamelin, plagued with rats, bargained with a piper who promised to, and did, pipe the rats out of town and into the Weser River. Refused payment for his work, the piper then piped the children out of town and into a hole in a hill, never to be seen again. All on July 22, 1376, according to 16th-century accounts. More recent historians suggest that the event occurred in 1284 when young men of Hamelin left the city on colonizing adventures.

RAT-CATCHERS DAY. July 22. A day to recognize the rat-catchers who labor to exterminate members of the genus *Rattus*, disease-carrying rodents that infest most of the "civilized" world. Observed on anniversary of the extraordinary feat of the Pied Piper of Hamelin on July 22, 1376 (according to 16th-century chronicler Richard Rowland Verstegen).

SPACE MILESTONE: *SOYUZ TM-3 (USSR)*. July 22. Two Soviet cosmonauts, Aleksandr Viktorenko and Aleksandr Aleksandrov, along with the first Syrian space traveler, Mohammed Faris, were launched on a projected 10-day mission, on July 22, 1987. Launched from the Baikonur base in Central Asia, the spacecraft orbited Earth for two days before linking with Soviet space station *Mir*.

SPOONER'S DAY (WILLIAM SPOONER BIRTH ANNIVERSARY). July 22. A day named for the Reverend William Archibald Spooner (born at London, England, July 22, 1844, warden of New College, Oxford, 1903-1924, died at Oxford, England, Aug 29, 1930), whose frequent slips of the tongue led to coinage of the term spoonerism to describe them. A day to remember and emulate the scholarly man whose accidental transpositions gave us blushing crow (for crushing blow), tons of soil (for sons of toil), queer old dean (for dear old queen), swell foop (for fell swoop), and half-warmed fish (for half-formed wish).

VERMONT FORESTRY EXPO. July 22-23. Fairgrounds, Rutland, VT. This two-day event features amateur and professional lumberjack competitions plus an equipment expo and parade, tug-o-war, children's events and games. Educational displays and videos provide the visitor with information about Vermont's timber industry. Fax: (802)533-9222. Est attendance: 2,500. For info: Vermont Foresty Expo, 24 Sparhawk Rd, Greensboro Bend, VT 05824. Phone: (802) 533-9212.

VIRGINIA SCOTTISH GAMES. July 22-23. Alexandria, VA. Bagpipes, Highland dance, drumming, fiddling, celtic harp competition, animal trials and Scottish athletic games competition and British antique car show. Scottish food and gifts are sold. Annually, the fourth weekend in July. Phone for events line: (703) 838-5005. Est attendance: 30,000. For info: Virginia Scottish Games, c/o Alexandria Conv & Visitors Bureau, 221 King St, Alexandria, VA 22314. Phone: (703) 838-4200.

Chase's 1995 Calendar of Events — July 22-24

WHO'S IN CHARGE? WORKERS AND MANAGERS IN THE UNITED STATES. July 22–Sept 3. Elmhurst Historical Museum, Elmhurst, IL. 150 years of conflict and change in the relationships between workers and managers. See Jan 14 for full description. For info: Smithsonian Institution Traveling Exhibition Service, 1100 Jefferson Dr SW, Ste 3146, Washington, DC 20560. Phone: (202) 357-2700.

BIRTHDAYS TODAY

Orson Bean (Dallas Frederick Burroughs), 67, actor ("To Tell the Truth," "Mary Hartman, Mary Hartman"), born at Burlington, VT, July 22, 1928.
Albert Brooks (Albert Einstein), 48, comedian, actor (*Network News, Lost in America*), born at Los Angeles, CA, July 22, 1947.
Willem Dafoe, 40, actor, born at Appleton, WI, July 22, 1955.
Oscar De La Renta, 63, fashion designer, born at Santo Domingo, Dominican Republic, July 22, 1932.
Robert J. Dole, 72, US Senator (R, Kansas), born at Russell, KS, July 22, 1923.
Jim Edgar, 49, Governor of Illinois (R), up for reelection Nov '94, born at Vinita, OK, July 22, 1946.
Danny Glover, 48, actor ("Chiefs," *Lethal Weapon, The Color Purple*), born San Francisco, CA, July 22, 1947.
Don Henley, 48, musician (drummer; The Eagles), songwriter (co-wrote "The Boys of Summer"), born Linden, TX, July 22, 1947.
Kay Bailey Hutchison, 52, US Senator (R, Texas) up for reelection in Nov '94, born at Galveston, TX, July 22, 1943.
Rose Kennedy, 105, matriarch (mother of JFK), born at Boston, MA, July 22, 1890.
William V. Roth, Jr, 74, US Senator (R, Delaware), up for reelection in Nov '94, born at Great Falls, MT, July 22, 1921.
Bobby Sherman, 50, singer, actor, born at Santa Monica, CA, July 22, 1945.
Margaret Whiting, 71, singer ("The Money Tree," "The Wheel of Hurt"), born at Detroit, MI, July 22, 1924.

JULY 23 — SUNDAY
204th Day — Remaining, 161

DRYSDALE, DON: BIRTH ANNIVERSARY. July 23, 1936. Elected to the Baseball Hall of Fame in 1984, Don Drysdale was a pitcher for the Brooklyn and Los Angeles Dodgers from 1956 to 1969, compiling a won-lost record of 209-166 with a career ERA of 2.95. Following his playing career he became a successful and popular broadcast announcer for the Chicago White Sox and then for the Los Angeles Dodgers. He was born at Van Nuys, CA, and died at Montreal, Canada, on July 3, 1993.

EGYPT, ARAB REPUBLIC OF: NATIONAL DAY. July 23. Anniversary of the Revolution of July 23, 1952, which was launched by army officers and changed Egypt from a monarchy to a republic.

FIRST US SWIMMING SCHOOL: OPENING ANNIVERSARY. July 23. The first swimming school in the US opened on July 23, 1827, in Boston, MA. Some of its pupils included John Quincy Adams and James Audubon.

JAPAN: SOMA NO UMAOI (WILD HORSE CHASING). July 23–25. Hibarigahara, Haramachi, Fukushima Prefecture, Japan. 1,000 horsemen clad in ancient armor compete for possession of three shrine flags shot aloft on Hibarigahara Plain, and men in white costumes attempt to catch wild horses corralled by the horsemen.

LEO, THE LION. July 23–Aug 22. In the astronomical/astrological zodiac, which divides the sun's apparent orbit into 12 segments, the period July 23–Aug 22 is identified, traditionally, as the sun sign of Leo, the Lion. The ruling planet is the sun.

NEWPORT MARITIME TEDDY BEAR FESTIVAL I. July 23. Islander Doubletree Hotel, Newport, RI. Teddy Bear Show and Sale featuring dozens of America's leading teddy bear designers and makers includes bear repair and restoration clinic, appraisals, antique bears and dealers selling famous-name teddy bears from manufacturers throughout the world. Free door prizes, charity raffle, refresments and free parking. Est attendance: 2,500. For info: John K. Pringle, Pringle Productions, Ltd., PO Box 757, Bristol, CT 06011-0757. Phone: (203) 585-9940.

PERSEID METEOR SHOWERS. July 23–Aug 20. Among the best-known and most spectacular meteor showers are the Perseids, peaking about Aug 10–12. As many as 50–100 may be seen in a single night. Wish upon a "falling star"!

PROFESSIONAL SECRETARIES INTERNATIONAL: ANNUAL CONVENTION. July 23–26. Seattle, WA. Est attendance: 1,700. For info: Professional Secretaries Intl, Dale Shuter, CPS, Conv Mgr, PO Box 20404, 10502 NW Ambassador Dr, Kansas City, MO 64195-0404. Phone: (816) 891-6600.

REGISTER'S ANNUAL GREAT BICYCLE RIDE ACROSS IOWA. July 23–29. A week-long bicycle ride across Iowa with 7,500 riders from around the country. Annually, the last full week of July. After Nov 1 and before Mar 1 send a business-size self-addressed-stamped envelope to address below. Sponsor: *Des Moines Register*. Est attendance: 7,500. For info: RAGBRAI, Box 622, Des Moines, IA 50303-0622. Phone: (515) 284-8282.

ST. APOLLINARIS: FEAST DAY. July 23. First bishop of Ravenna, and a martyr, of unknown date. Observed July 23.

SPACE MILESTONE: SOYUZ 37 (USSR). July 23. Launched on July 23, 1980, cosmonauts Viktor Gorbatko and, the first non-Caucasian in space, Lieutenant Colonel Pham Tuan (Vietnam), docked at *Salyut 6* on July 24. Returned to Earth July 31.

SWITZERLAND: DORNACH BATTLE COMMEMORATION. July 23. The victory at Dornach in 1499 is remembered on the battlefield and in the city of Solothurn on the Sunday nearest to July 22. Dornach observes commemorative festival every five years.

BIRTHDAYS TODAY

Gloria DeHaven, 70, actress (*Two Girls and a Sailor*, "Nakia"), born at Los Angeles, CA, July 23, 1925.
Nicholas Gage, 56, journalist, film producer, writer (*Eleni*), born at Lia, Greece, July 23, 1939.
Arata Isozaki, 64, architect, born at Oita, Japan, July 23, 1931.
Anthony M. Kennedy, 59, Supreme Court Justice, born at Sacramento, CA, July 23, 1936.
Belinda Montgomery, 45, actress ("Man From Atlantis"), born at Winnipeg, Manitoba, Canada, July 23, 1950.
Gary Dwayne Payton, 27, basketball player, born at Oakland, CA, July 23, 1968.

JULY 24 — MONDAY
205th Day — Remaining, 160

CENTRAL MONTANA FAIR. July 24–29. Lewistown, MT. Draft and open-to-all-class horseshow, three sessions of rodeo, carnival, 4-H and open exhibits, night shows, demolition derby. Annually, last full week in July. Est attendance: 25,000. Sponsor: Central Montana Fair, PO Box 1098, Lewistown, MT 59457. Phone: (406) 538-8841.

DETROIT'S BIRTHDAY. July 24. Anniversary of the landing, on July 24, 1701, at the site of Detroit, by Antoine de la Mothe Cadillac, in the service of Louis XIV of France. Fort Pontchartrain du Detroit was first settlement on site.

July 24-25 ☆ Chase's 1995 Calendar of Events ☆

DUMAS, ALEXANDRE: BIRTH ANNIVERSARY. July 24. French playwright and novelist, born at Villers-Cotterets, France, July 24, 1802. He is said to have managed the production of more than 1,200 volumes, including *The Count of Monte Cristo* and *The Three Musketeers*. Father of Alexandre Dumas (Dumas fils), also a novelist and playwright (1824–1895). Dumas died near Dieppe, France, Dec 5, 1870.

◆ **EARHART, AMELIA: BIRTH ANNIVERSARY.** July 24. American aviatrix lost on flight from New Guinea to Howland Island, in the Pacific Ocean, July 3, 1937. First woman to cross the Atlantic solo and fly solo across the Pacific from Hawaii to California. Born at Atchison, KS, July 24, 1898.

MOZART FESTIVAL. July 24–Aug 6. San Luis Obispo, CA. Concerts celebrating the spirit of Wolfgang Amadeus Mozart. Est attendance: 5,000. Sponsor: Mozart Festival Assn, Pat Martin, Exec Dir, PO Box 311, San Luis Obispo, CA 93406. Phone: (805) 781-3008.

NATIONAL ASSOCIATION OF SCIENTIFIC MATERIALS MANAGERS ANNUAL CONFERENCE AND TRADE SHOW. July 24–28. Palmer House, Chicago, IL. Members of NAOSMM (Nay-o-sum) hold their 22nd annual conference and trade show. Scientific materials managers are involved in the handling of scientific materials—from purchasing agents to stockroom managers in situations ranging from colleges and universities to commercial firms and plants to hospitals, testing labs or research labs. Conference will offer seminars on a wide variety of professional subjects. Est attendance: 250. For info: NAOSMM Info, Dept of Chemistry, Oregon St University, Gilbert Hall 154, Corvallis, OR 97331-4003. Phone: (503) 737-6708.

NATIONAL HIGH SCHOOL FINALS RODEO. July 24–30. Cam-Plex, Gillette, Wyoming. More than 1,400 teenaged contestants from North America will compete in rodeo events patterned after official Pro-Rodeo Association guidelines. Annually, the last Monday–Sunday in July. Toll-free phone: (800) 46-NHSRA. Est attendance: 95,000. For info: Natl High School Rodeo Assn, 11178 N Huron, Ste 7, Denver, CO 80234. Phone: (303) 452-0820.

PIONEER DAY. July 24. Utah. Commemorates first settlement on this day, 1847, by Brigham Young.

◆ **POTSDAM DECLARATION: 50th ANNIVERSARY.** July 24, 1945. As the Potsdam Conference came to a close in Germany, Churchill, Truman and China's representatives fashioned a communique to Japan offering them an opportunity to end the war. It demanded that Japan completely disarm, allowed them sovereignty to the four main islands and to minor islands to be determined by the Allies, and insisted that all Japanese citizens be given immediate and complete freedom of speech, religion and thought. The Japanese would be allowed to continue enough industry to maintain their economy. The communique concluded with a demand for unconditional surrender. Unaware these demands were backed up by an atomic bomb, on July 26 Japanese Prime Minister Admiral Kantaro Suzuki rejected the Potsdam Declaration.

VIRGIN ISLANDS: HURRICANE SUPPLICATION DAY. July 24. Legal holiday. Population attends churches to pray for protection from hurricanes. Annually, the fourth Monday in July.

WALES: ROYAL WELSH SHOW. July 24–27. Royal Welsh Showground, Llanelwedd, Builth Wells, Powys. National agricultural show incorporating exhibitions of livestock, farm machinery, horticulture and forestry displays. For info: The Secretary, Royal Welsh Agricultural Soc, Llanelwedd, Builth Wells, Powys, Wales LD2 3SY.

WORLD HORSESHOE TOURNAMENT. July 24–Aug 5. Georgia National Fairgrounds, Perry, GA. Horseshoe pitchers from around the globe converge on Perry for two weeks of competition. Cash prizes and trophies will be awarded at the 77th annual tournament. Est attendance: 5,000. For info: Donnie Roberts, Box 7927, Columbus, OH 43207. Phone: (614) 444-8510.

BIRTHDAYS TODAY

Bella Abzug, 75, lawyer, feminist, former congresswoman (hats), born at New York, NY, July 24, 1920.
Kevin Butler, 33, football player, born at Savannah, GA, July 24, 1962.
Ruth Buzzi, 59, comedienne, actress ("Rowan & Martin's Laugh-In"), born at Westerly, RI, July 24, 1936.
Carroll A. Campbell, Jr, 55, Governor of South Carolina (R), born at Greenville, SC, July 24, 1940.
Joe Barry Carroll, 37, basketball player, born Pine Bluff, AR, July 24, 1958.
Lynda Carter, 44, actress ("Wonder Woman," "Partners in Crime"; former Miss World-USA), singer, born Phoenix, AZ, July 24, 1951.
Robert Hays, 48, actor (*Airplane!*), born at Bethesda, MD, July 24, 1947.
Julie Krone, 32, jockey, born at Benton Harbor, MI, July 24, 1963.
Karl Malone, 32, basketball player, born at Mount Sinai, LA, July 24, 1963.
Pat Oliphant, 60, cartoonist, born at Adelaide, Australia, July 24, 1935.
Marc Racicot, 47, Governor of Montana (R), born at Thompson Falls, MT, July 24, 1948.
Peter Serkin, 48, musician, born at New York, NY, July 24, 1947.
Billy Taylor, 74, musician, born at Greenville, NC, July 24, 1921.

JULY 25 — TUESDAY
206th Day — Remaining, 159

CAVE MAN NEVER DAYS. July 25–Aug 9. A 16-day period when you get to act like a cave man, walk and talk like a cave man, cook and look, move and groove, shimmy and shake, read and plead like a cave man because "after all, there is a little Neanderthal in all of us." [Taken from the sponsor's cartoon "Cave Man Never" © 1990.] For info: Adrienne Sioux Koopersmith, 1437 W Rosemont, 1W, Chicago, IL 60660. Phone: (312) 743-5341.

FIRST AIRPLANE CROSSING OF ENGLISH CHANNEL: ANNIVERSARY. July 25. On Sunday, July 25, 1909, Louis Bleriot, after asking from the cockpit, "Where is England?" took off from Les Baraques (near Calais), France, and landed on English soil at Northfall Meadow, near Dover, where he was greeted first by English police and customs officers. This, the world's first international overseas airplane flight was accomplished in a 28-hp monoplane with wingspan of 23 ft. Bleriot was born at Cambrai, France, July 1, 1872. See also: "Bleriot, Louis: Birth Anniversary" (July 1).

	S	M	T	W	T	F	S
July							1
1995	2	3	4	5	6	7	8
	9	10	11	12	13	14	15
	16	17	18	19	20	21	22
	23	24	25	26	27	28	29
	30	31					

Chase's 1995 Calendar of Events — July 25–26

GARFIELD COUNTY FAIR. July 25–29. Burwell, NE. County fair, carnival and flea market. Est attendance: 10,000. For info: Garfield County Frontier Fair Assn, Peggy Haskell, Box 747, Burwell, NE 68823. Phone: (308) 346-5210.

GERMANY: BAYREUTHER FESTSPIELE. July 25–Aug 28. (Tentative.) Bayreuth. Music festival. For info: Bayreuther Festspiele, Richard Wagner, GmbH, Presseburo, Postfach 100262, D-95402 Bayreuth, Germany.

HARRISON, ANNA SYMMES: BIRTH ANNIVERSARY. July 25. Wife of William Henry Harrison, ninth president of the US, born at Morristown, NJ, July 25, 1775. Died Feb 25, 1864.

PUERTO RICO: CONSTITUTION DAY. July 25. Also called Commonwealth Day or Occupation Day. Commemorates proclamation of constitution on July 25, 1952.

PUERTO RICO: LOIZA ALDEA FIESTA. July 25–28. Best known of Puerto Rico's patron saint festivities. Villagers of Loiza Aldea, 20 miles east of San Juan, don devil masks and colorful costumes for a variety of traditional activities.

ROCKBRIDGE REGIONAL FAIR. July 25–29. Lexington, VA. Held at the Virginia Horse Center. 4-H Show, rodeo, carnival, exhibits (commercial and crafts), activities of all kinds. For info: Lexington Visitors Bureau, 106 E Washington St, Lexington, VA 24450. Phone: (703) 463-3777.

TEST-TUBE BABY: 17th BIRTHDAY. July 25. Anniversary of the birth of Louise Brown, on this date in 1978, at Oldham, England. First documented birth of a baby conceived outside the body of a woman. Parents: Gilbert John and Lesley Brown, of Bristol, England. Physicians: Patrick Christopher Steptoe and Robert Geoffrey Edwards.

US JUNIOR AMATEUR (GOLF) CHAMPIONSHIP. July 25–29. Fargo Country Club, Fargo, ND. Sponsor: US Golf Assn, Championship Dept, Golf House, Far Hills, NJ 07931. Phone: (908) 234-2300.

WORLD FOOTBAG CHAMPIONSHIPS. July 25–30. Golden, CO. Six-day sports event spotlights competition of footskills—the Super Bowl of footbag! Now in its 15th year, it attracts the world's top footbag competitors from the US and 8 other countries. Prize money exceeds $15,000. Sponsor: Wham-O Inc., I Dig Footbag, Sipa Sipa Footbags, Holiday Inn and The World Footbag Assn. Est attendance: 3,000. For info: Bruce Guettich, Dir, World Footbag Assn, 1317 Washington Ave, Ste 7, Golden, CO 80401. Phone: (800) 878-8797.

BIRTHDAYS TODAY

Louise Joy Brown, 17, first test-tube baby, born at Oldham, England, July 25, 1978.

Stanley Dancer, 68, harness racer, born at New Egypt, NY, July 25, 1927.

Midge Decter, 68, journalist, born at St. Paul, MN, July 25, 1927.

Douglas Dean Drabek, 33, baseball player, born at Victoria, TX, July 25, 1962.

Estelle Getty, 71, actress (Sophia Petrillo in "Golden Girls"), born New York, NY, July 25, 1924.

Walter Payton, 41, Pro Football Hall of Famer, born at Columbia, MS, July 25, 1954.

Nate Thurmond, 54, basketball Hall of Famer, born at Akron, OH, July 25, 1941.

JULY 26 — WEDNESDAY
207th Day — Remaining, 158

ATOMIC BOMB DELIVERED: 50th ANNIVERSARY. July 26, 1945. The US cruiser *Indianapolis* arrived at Tinian Island in the Marianas with a deadly cargo. Aboard were the makings of the atomic bomb. On the island waited scientists prepared to complete the assembly.

CANADA: CRANBROOK AIR FAIR. July 26. Cranbrook Airport, Cranbrook, British Columbia. International aerobatic acts, aircraft displays, parachute jumpers, helicopter rides, water bombing, search and rescue, classic cars, food concessions. Est attendance: 10,000. For info: Karin Penner, Mgr, Cranbrook Chamber of Commerce, PO Box 84, Cranbrook, BC, Canada V1C 4H6. Phone: (604) 426-5914.

CANADA: TRIAL OF LOUIS RIEL. July 26–Aug 31. Shumiatcher Theatre at the MacKenzie Art Gallery, Regina, Saskatchewan. In its 29th season this play is a reenactment of the dramatic events surrounding Louis Riel, the Metis leader in the North West Rebellion of 1885. One of the longest running stage shows in Canada. Est attendance: 3,000. For info: Tourism Saskatchewan, Saskatchewan Trade and Conv Center, 1919 Saskatchewan Dr, Regina, Sask, Canada S4P 3V7. Phone: (800) 667-7191.

CATLIN, GEORGE: BIRTH ANNIVERSARY. July 26, 1796. American artist famous for his paintings of Native American life. In 1832 he toured North and South American tribes, recording their lives in his work. He died on Dec 23, 1872 at Jersey City, NJ.

CHINCOTEAGUE PONY PENNING. July 26–27. Chincoteague Island, VA. To round up the 150 wild ponies living on Assateague Island and swim them across the inlet to Chincoteague, where about 50–60 of them are sold. Est attendance: 50,000. For info: Chamber of Commerce, Jacklyn Russell, Box 258, Chincoteague, VA 23336. Phone: (804) 336-6161.

CLINTON, GEORGE: BIRTH ANNIVERSARY. July 26. Fourth vice president of the US (1805–1812), born at Little Britain, NY, July 26, 1739. Died at Washington, DC, Apr 20, 1812.

CUBA: NATIONAL HOLIDAY: ANNIVERSARY OF REVOLUTION. July 26. Anniversary of 1953 beginning of Fidel Castro's revolutionary "26th of July Movement."

CURAÇAO: CURAÇAO DAY. July 26. "Although not officially recognized by the government as a holiday, various social entities commemorate the fact that on this day Alonso de Ojeda, a companion of Christopher Columbus, discovered the Island of Curaçao in 1499, sailing into Santa Ana Bay, the entrance of the harbor of Willemstad." Festivities on this day.

HOOD RIVER COUNTY FAIR. July 26–30. Hood River, OR. From 4-H activities to the excitement of the carnival, this annual old-fashioned country fair is bustling with things to do! For info: Peggy Packer, Hood River Chamber of Commerce. Phone: (503) 354-2865.

HUXLEY, ALDOUS: BIRTH ANNIVERSARY. July 26. English author, satirist, mystic and philosopher, Aldous Leonard Huxley was born at Godalming, Surrey, England, on July 26, 1894. Best known of his works are *Brave New World*, *Crome Yellow* and *Point Counter Point*. Huxley died at Los Angeles, CA, Nov 22, 1963.

JOHN HUNT MORGAN CAPTURED: ANNIVERSARY. July 26. After harrassing Union forces in Tennessee and Ohio throughout the Civil War, Confederate raider John Hunt Morgan was captured at New Lisbon, OH on July 26, 1863. Morgan was imprisoned in the Ohio Penitentiary from which he later escaped.

LIBERIA: INDEPENDENCE DAY. July 26. National holiday. Became republic on this day, 1847, under aegis of the US societies for repatriating Negroes in Africa.

July 26–27 ☆ Chase's 1995 Calendar of Events ☆

MALDIVES: NATIONAL DAY. July 26. National holiday is observed in Maldives.

NEW YORK RATIFICATION DAY. July 26. Eleventh state to ratify Constitution, on this day in 1788.

PIGEON FARMERS' FESTIVAL. July 26–30. Pigeon, MI. To recognize and honor farmers for their unique contribution to rural life in Michigan's "thumb area." Features parades, barbecues, stage shows, carshow, crafts and sidewalk sales. Sponsor: Pigeon Chamber of Commerce. Est attendance: 3,500. For info: Sally Rummel, c/o The News Weekly, Pigeon, MI 48755. Phone: (517) 453-3100.

PUBLICATION OF FIRST ESPERANTO BOOK: ANNIVERSARY. July 26. To commemorate the anniversary of the publication of Dr. Zamenhof's first textbook about the international language, Esperanto, July 26, 1887. Sponsor: Intl Soc of Friendship and Good Will (ISFGW), Dr Stanley J. Drake, 9538 Summerfield St, Spring Valley, San Diego County, CA 91977-2852. Phone: (619) 466-8882.

SHAW, GEORGE BERNARD: BIRTH ANNIVERSARY. July 26. Irish playwright, essayist, vegetarian, socialist, antivivisectionist and, he said, "... one of the hundred best playwrights in the world." Born at Dublin, Ireland, on July 26, 1856. Died at Ayot St. Lawrence, England, Nov 2, 1950.

SPACE MILESTONE: APOLLO 15 (US). July 26. Launched on July 26, 1971, astronauts David R. Scott and James B. Irwin landed on Moon (lunar module *Falcon*) while Alfred M. Worden piloted command module *Endeavor*. Rover 1, a fourwheel vehicle, was used for further exploration. Departed Moon Aug 2, after nearly three days. Pacific landing Aug 7.

US ARMY FIRST DESEGREGATION: ANNIVERSARY. In 1944, during World War II, the US Army ordered desegregation of its training camp facilities. Later the same year black platoons were assigned to white companies in a tentative step toward integration of the battlefield. However, it was not until after the War—July 26, 1948—that President Harry Truman signed an order officially integrating the armed forces.

BIRTHDAYS TODAY

Blake Edwards, 73, producer, writer, director (*Victor/Victoria, The Pink Panther*), born at Tulsa, OK, July 26, 1922.

Susan George, 45, actress (*Straw Dogs*), born London, England, July 26, 1950.

Mick (Michael Philip) Jagger, 52, singer, songwriter, musician (lead singer of The Rolling Stones; "Satisfaction," "Ruby Tuesday," "Honky Tonk Woman"), born Dartford, England, July 26, 1943.

July 1995

S	M	T	W	T	F	S
						1
2	3	4	5	6	7	8
9	10	11	12	13	14	15
16	17	18	19	20	21	22
23	24	25	26	27	28	29
30	31					

JULY 27 — THURSDAY
208th Day — Remaining, 157

ANNIE OAKLEY DAYS. July 27–30. Greenville, OH. Festival to keep alive the memory of Annie Oakley. Visit the newly dedicated Annie Oakley Memorial Park. Est attendance: 50,000. Sponsor: Annie Oakley Days Committee, Inc, Rodney Oda, Pres, P O Box 436, Greenville, OH 45331. Phone: (513) 548-3492.

ATLANTIC TELEGRAPH CABLE LAID: ANNIVERSARY. July 27. Cable laying successfully completed on this day, 1866.

BIX BEIDERBECKE MEMORIAL JAZZ FESTIVAL. July 27–30. Davenport, IA. Annually, the last full weekend of July. For info: Bix Beiderbecke Memorial Soc, PO Box 3688, Davenport, IA 52808. Phone: (319) 324-7170.

DUMAS, ALEXANDRE (DUMAS FILS): BIRTH ANNIVERSARY. July 27. French novelist and playwright, as was his father. Author of *La Dame aux Camélias*. Dumas fils was born at Paris, France, July 27, 1824, and died at Marly-le-Roi, France, on Nov 27, 1895.

DUROCHER, LEO: 90th BIRTH ANNIVERSARY. July 27. Leo Durocher was born on July 27, 1905, at West Springfield, MA. He began his major league baseball career with the New York Yankees in 1925. He also played for the St. Louis Cardinals' "Gashouse Gang" and the Brooklyn Dodgers, where he first served as player-manager in 1939. It was during that season that he used the phrase "Nice guys finish last," which would become his trademark. As a manager, he guided the New York Giants into two world series, defeating Cleveland in 1954. Following a 5-year period away from baseball, he resurfaced as a coach with the now Los Angeles Dodgers in 1961. In 1966 he signed with the Chicago Cubs as manager. He was at the helm in the ill-fated 1969 season when the Cubs were overtaken by the Miracle Mets during the last month of the season. After leaving the Cubs, he spent one season with the Houston Astros, then retired from baseball in 1973. He died on Oct 7, 1991.

EAA INTERNATIONAL FLY-IN CONVENTION AND SPORT AVIATION EXHIBITION. July 27–Aug 2. Wittman Regional Airport, Oshkosh, WI. "World's largest and most significant annual aviation event." Est attendance: 850,000. Sponsor: Experimental Aircraft Assn and EAA Fdtn, John Burton, PR Dir, PO Box 3086, Oshkosh, WI 54903-3086. Phone: (414) 426-4800.

FAIRFEST '95. July 27–31. Adams County Fairgrounds, Hastings, NE. Annual county fair. Est attendance: 60,000. For info: James Gleason, Genl Mgr, PO Box 342, Hastings, NE 68902. Phone: (402) 462-3247.

GREAT TEXAS MOSQUITO FESTIVAL. July 27–29. Clute, TX. More than 100 booths: arts and crafts, food, entertainment, novelty-games, and carnival. Meet "Willie Man Chew," a 25-ft inflatable mosquito dressed in cowboy boots and hat. Annually, the last Thursday-Saturday in July. Est attendance: 40,000. Sponsor: City of Clute Pks and Rec Dept, PO Box 997, Clute, TX 77531. Phone: (409) 265-8392.

KOREAN WAR ARMISTICE: ANNIVERSARY. July 27. Armistice agreement ending war that had lasted 3 years and 32 days was signed at Panmunjom, Korea, July 27, 1953 (July 26, US time), by US and North Korean delegates. Both sides claimed victory at conclusion of 2 years, 17 days of truce negotiations.

LAST CHANCE STAMPEDE AND FAIR. July 27–30. Fairgrounds, Helena, MT. Fast-paced Montana rodeo action along with exhibits, carnival rides, 4-H animal judging and Demo-Derby. Annually, the last full weekend in July. Est attendance: 12,000. For info: Loren Davis, 1429 Helena Ave, Helena, MT 59601. Phone: (406) 442-1098.

MOON PHASE: NEW MOON. July 27. Moon enters New Moon phase at 11:13 AM, EDT.

NEBRASKA'S BIG RODEO. July 27-29. Fairgrounds, Burwell, NE. 74th professional rodeo. Contestants compete in 4 exciting performances in historic, outdoor rodeo arena. Added thrills—chuckwagon races, wild horse races, Dinnerbell Derby, bull fighting-Burwell style, quilt and art shows, parade, flea market, llama and longhorn cattle show, c/w music and dancing and Miss Burwell Rodeo Pageant. Est attendance: 10,000. Sponsor: Garfield County Frontier Fair Assn, Peggy Haskell, Secy, Box 747, Burwell, NE 68823. Phone: (308) 346-5210.

PEOPLE'S REPUBLIC OF CHINA: FESTIVAL OF HUNGRY GHOSTS. July 27-Aug 25. Important Chinese festival, also known as Ghosts Month. According to Chinese legend, during the 7th lunar month, the souls of the dead are released from purgatory to roam the Earth. Joss sticks are burnt in homes; prayers, food and "ghost money" are offered to appease the ghosts. Market stallholders combine to hold celebrations to ensure that their businesses will prosper in coming year. Wayang (Chinese street opera) and puppet shows are performed, and fruit and Chinese delicacies are offered to the spirits of the dead. Chung Yuan (All Souls' Day) is observed on the 15th day of the 7th lunar month: Aug 10 in 1995.

SWANTON SUMMER FESTIVAL. July 27-30. Village Green, Swanton, VT. Amusement rides, food concessions, free entertainment, giant parade on Sunday, arts and crafts. Annually, last weekend of July. Est attendance: 10,000. Sponsor: Swanton Chamber of Commerce, c/o Marie Speer, 20 S River St, Swanton, VT 05488. Phone: (802) 868-7200.

TAKE YOUR HOUSEPLANTS FOR A WALK DAY. July 27. Walking your plants around the neighborhood enables them to know their environment, thereby providing them with a sense of knowing, bringing on wellness. Phone: (212) 388-8673 or (717) 274-8451. For info: Wellness Permission League, Tom or Ruth Roy, 2105 Water St, Lebanon, PA 17046.

US DEPARTMENT OF STATE BIRTHDAY. July 27. The first presidential cabinet department, called the Department of Foreign Affairs, was established by the Congress on July 27, 1789. Later the name was changed to Department of State.

BIRTHDAYS TODAY

Peggy Fleming, 47, figure skater (Olympic medalist), born at San Jose, CA, July 27, 1948.
Bobbie Gentry (Roberta Streeter), 53, singer, songwriter ("Ode to Billie Joe"), born at Chicasaw County, MS, July 27, 1942.
Norman Lear, 73, TV scriptwriter, producer ("All in the Family"), born at New Haven, CT, July 27, 1922.
Maureen McGovern, 46, singer ("The Morning After"), actress, born at Youngstown, OH, July 27, 1949.
Betty Thomas, 47, director, actress ("Hill Street Blues"), born at St. Louis, MO, July 27, 1948.

JULY 28 — FRIDAY
209th Day — Remaining, 156

ANTIQUE POWER EXHIBITION. July 28-30. Burton, OH. More than 100 old-time engines toot whistles and puff smoke as they demonstrate yesterday's feats such as wood sawing and grain threshing. Daily parades, slow races, tractor pulls. Est attendance: 2,600. For info: Geauga County Historical Soc, Marlene F. Collins, PO Box 153, Burton, OH 44021. Phone: (216) 834-4012.

BAGELFEST '95 (WITH BAGEL BUGGY DERBY). July 28-29. Mattoon, IL. The World's Biggest Bagel Breakfast with free bagels and toppings, entertainment, Bagel Buggy Derby, Bagel Beauties, arts and crafts, bingo and a street dance. Annually, the last Friday and Saturday in July. Est attendance: 40,000. For info: Mattoon Chamber of Commerce, 1701 Wabash, Mattoon, IL 61938. Phone: (217) 235-5661.

BELE CHERE. July 28-30. Asheville, NC. A community celebration featuring a variety of food, music, children's activities and events. Annually, the last full weekend in July. Est attendance: 300,000. For info: Bele Chere Festival, PO Box 7148, Asheville, NC 28802. Phone: (704) 253-1009.

BLACKBEARD PIRATE JAMBOREE. July 28-29. Town Point Park, Norfolk, VA. Gather all ye scoundrels, scalawags, and rogues. Don your pirate garb for 2 days of concerts, pirate festivities and adventurous children's activities. This waterfront event features a parade of sail, costume contest, coconut bowling, limbo dancing and more. Est attendance: 20,000. For info: Norfolk Festevents, Ltd, Promotions Dir, 120 W Main St, Norfolk, VA 23510. Phone: (804) 627-7809.

CANADA: CANADA'S NATIONAL UKRAINIAN FESTIVAL. July 28-30. Dauphin, Manitoba. Celebrates freedom and the flavor of the old traditions and culture of the Ukraine. There's song, dance, costume and Ukrainian cuisine. Fax: (204) 638-5851. For info: Canada's National Ukrainian Festival, Pat Maksymchuk, 119 Main St, Dauphin, Man, Canada R7N 1K4. Phone: (204) 638-5645.

CANADA: GREAT NORTHERN MUSIC FESTIVAL. July 28-30. Jim Koe Park, Inuvik, Northwest Territory. A three-day music festival featuring more than 50 musicians from across Northern Canada, as well as many guest performers from the South. Join us for a weekend of fiddling, drumming, jigging, country, rock, folk, and blues in the land of the midnight sun. Est attendance: 500. For info: Bob Mumford, Dir, Box 2921, Inuvik, NWT, Canada X0E 0T0. Phone: (403) 979-3536.

CANADA: LEACOCK HERITAGE FESTIVAL. July 28-Aug 7. Downtown waterfront parks, Orillia, Ontario. A celebration of culture, humor and history wrapped in good old-fashioned family fun. The festival highlights Stephen Leacock and the Orillia he knew at the turn of the century. 11 days of music and theater in the park, humorous song at the Aquatheatre, children's festival, waterfront park picnic, Leacock Medal of Humour readings on lawns of the Leacock Home, old-fashioned street dance, costume contest, literary contests and much more. Fax: (705) 325-7666. Est attendance: 50,000. For info: Doug Little, Leacock Heritage Festival, PO Box 2305, Orillia, Ont, Canada L3V 6S3. Phone: (705) 325-3261.

CANADA: NOVA SCOTIA BLUEGRASS AND OLDTIME MUSIC FESTIVAL. July 28-30. Ardoise, Nova Scotia. 24th annual family event featuring acoustic music by groups from the US and Canada's Atlantic area. Annually, the last full weekend in July. Est attendance: 3,000. For info: Jerry Murphy, Dir, Downeast Bluegrass Oldtime Music Soc, PO Box 546, Elmsdale, NS, Canada B0N 1M0. Phone: (902) 883-7189.

CEDAR GROVE HOLLAND FEST. July 28-29. Cedar Grove, WI. Residents celebrate with street scrubbing, klompen dancing, films, wooden shoe races for children, parade, art fair and traditional Dutch foods. The Holland Fest Players perform each evening. Also Holland Festival Ron/Walk. Annually, the last Friday and Saturday in July. Est attendance: 6,000. For info: Cedar Grove Holland Fest, PO Box HF, Cedar Grove, WI 53013. Phone: (414) 457-9495.

CELEBRATION OF OUR LADY OF THE SNOWS. July 28-Aug 5. Belleville, IL. This annual prayer event invites the public to come and celebrate the patroness of the Shrine at the Outdoor Amphitheatre. A candlelight procession will close the services each evening. Est attendance: 2,500. For info: Shrine of Our Lady of the Snows, 9500 W Illinois, Hwy 15, Belleville, IL 62223. Phone: (618) 397-6700.

July 28 ☆ *Chase's 1995 Calendar of Events* ☆

DODGE CITY DAYS. July 28–Aug 6. Dodge City, KS. Western heritage celebration with concerts, arts and crafts, parades, PRCA rodeo, street dances, cookouts, staged train robbery, art show, antique car show and county fair. Est attendance: 75,000. For info: Dodge City Conv & Visitors Bureau, PO Box 1474, Fourth & Spruce, Dodge City, KS 67801. Phone: (316) 225-8186.

GERMAN FEST. July 28–30. Milwaukee, WI. Featuring bands from Milwaukee, the US and Germany. Cultural exhibition with a genealogy section, movies and travelogue of Germany, a bookstore and Trachtenschau, a "style show" of ethnic costumes. Food including Spanferkel (roasted chicken), rollbraten and tortes from our Konditorei. Est attendance: 100,000. For info: German Fest Milwaukee, Inc, Attn: Linda Schmitz, 8229 W Capitol Dr, Milwaukee, WI 53222. Phone: (414) 464-9444.

GILROY GARLIC FESTIVAL. July 28–30. Gilroy, CA. Midsummer harvest celebration in the "Garlic Capital of the World." Great garlic recipe contest/cookoff. Ethnic foods, continuous entertainment on four stages, arts, crafts and garlic queen pageant. Est attendance: 130,000. Sponsor: Gilroy Garlic Festival Assn, Inc, Box 2311, Gilroy, CA 95021. Phone: (408) 842-1625.

KALAMAZOO COUNTY FLOWERFEST. July 28–30. Bronson Park, Library Lane, Kalamazoo and Portage, MI. Celebration of color and beauty, highlighting the Kalamazoo County bedding plant industry, the largest producer of bedding plants in the nation. The festival offers three-dimensional sculptures, floral mounds, flower show, landscape seminars, AAS garden winners and entertainment. Annually, the fourth week in July. Est attendance: 60,000. Sponsor: Kalamazoo County Flowerfest, Inc, PO Box 986, Portage, MI 49081-0986. Phone: (616) 381-3597.

LOGGING MUSEUM FESTIVAL DAYS. July 28–29. Rangeley, ME. Bean-hole beans, Logger's Hall of Fame, Miss Woodchip Contest, parade, logging competition. Sponsor: Logging Museum. Est attendance: 1,500. For info: Steve Richard, Box 154, Rangeley, ME 04970. Phone: (207) 864-5595.

MID AMERICA ALL-INDIAN CENTER POWWOW. July 28–30. Sedgwick County Park, Wichita, KS. Each year different tribes of Indians from all over the US come to Wichita to hold ceremonies celebrating their heritage. Authentic arts, crafts and food. Indians perform traditional dances and hold Indian "Give Aways," a gift-giving tradition to honor friends and relatives. Annually, the last full weekend in July. Est attendance: 30,000. For info: Mid America All-Indian Center, 650 N Seneca, Wichita, KS 67203. Phone: (316) 262-5221.

NATIONAL BALLOON CLASSIC. July 28–Aug 6. Indianola, IA. A spectator-oriented balloon extravaganza involving fun events utilizing up to 100 balloons. It is held on the Classics specially designed balloon field with a natural amphitheater for perfect viewing. Balloons fly morning and evening, weather permitting. The Classic stage features live local & regional entertainment. Est attendance: 65,000. For info: Kristie Wildung, PO Box 346, Indianola, IA 50125. Phone: (515) 961-8415.

NESHOBA COUNTY FAIR. July 28–Aug 4. Philadelphia, MS. Billed as "Mississippi's Giant Houseparty," it is one of the nation's last Campground Fairs. At this 106th annual fair, harness racing, state and national political speaking, crafts, music and amusement are the order each day. Est attendance: 175,000. For info: Connie Sampsell, Exec Dir, Philadelphia-Neshoba Co Chamber of Comm, PO Box 51, Philadelphia, MS 39350. Phone: (601) 656-1742.

ONASSIS, JACQUELINE LEE BOUVIER KENNEDY: BIRTH ANNIVERSARY. July 28, 1929. Editor, widow of John Fitzgerald Kennedy (35th president of the US), born at Southampton, NY. Later married (Oct 20, 1968) Greek shipping magnate Aristotle Socrates Onassis, who died Mar 15, 1975. The widely admired and respected former First Lady died May 19, 1994, at New York City.

OZARK EMPIRE FAIR. July 28–Aug 6. Springfield, MO. Regional fair with carnival, exhibits, livestock and entertainment. Est attendance: 230,000. For info: Ozark Empire Fair, PO Box 630, Springfield, MO 65801. Phone: (417) 833-2660.

PERU: NATIONAL INDEPENDENCE DAY. July 28. At defeat of Spain by Simon Bolivar, Peru became independent, 1824.

SINGING TELEGRAM: ANNIVERSARY. July 28. Anniversary of the first singing telegram, said to have been delivered to singer Rudy Vallee on his 32nd birthday, July 28, 1933. Early singing telegrams often were delivered in person by uniformed messengers on bicycle. Later they were usually sung over the telephone.

SPACE MILESTONE: *RANGER 7* (US). July 28. Televised back to Earth 4,308 close-up photographs of moon. Launched July 28, 1964.

SPACE MILESTONE: *SKYLAB 3* (US). July 28. Launched on July 28, 1973, Alan L. Bean, Owen K. Garriott, Jack R. Lousma started record 59-day mission to test man's space flight endurance. Pacific splashdown Sept 25.

TERRY FOX DAY: BIRTH ANNIVERSARY. July 28. Birthday of Terrence Stanley Fox, Canadian youth who captured the hearts and admiration of millions during his brief life. Stricken with cancer, requiring amputation of the athlete's right leg at age 18, Fox determined to devote his life to a fight against the disease. His "Marathon of Hope," a planned 5,200-mile run westward across Canada, started Apr 12, 1980, at St. John's, Newfoundland, and continued 3,328 miles to Thunder Bay, Ontario, Sept 1, 1980, when he was forced by spread of the disease to stop. During the run (on an artificial leg) he raised $24 million for cancer research and inspired millions with his courage. Terry Fox was born at Winnipeg, Manitoba, July 28, 1958, and died at New Westminster (near Vancouver), British Columbia, Canada, June 28, 1981.

UFO DAYS. July 28–30. Elmwood, WI. Starts Friday afternoon at 5:00; ends Sunday evening at 8:00. Annually, the last full weekend in July. Est attendance: 3,000. For info: UFO Days, N5427 County Rd S, Elmwood, WI 54740-8022. Phone: (715) 639-5732.

VETERANS BONUS ARMY EVICTION: ANNIVERSARY. July 28. Some 15,000 unemployed veterans of World War I marched on Washington, DC, in the summer of 1932, demanding payment of a war bonus. After two months' encampment in Washington's Anacostia Flats, eviction of the bonus marchers by the US Army was ordered by President Herbert Hoover. Under the leadership of General Douglas MacArthur, Major Dwight D. Eisenhower and Major George S. Patton, Jr (among others), cavalry, tanks and infantry attacked. Fixed bayonets, tear gas and the burning of the veterans' tents hastened the end of the confrontation. One death was reported.

July 1995

S	M	T	W	T	F	S
						1
2	3	4	5	6	7	8
9	10	11	12	13	14	15
16	17	18	19	20	21	22
23	24	25	26	27	28	29
30	31					

Chase's 1995 Calendar of Events — July 28–29

BIRTHDAYS TODAY

Bill Bradley, 52, US Senator (D, New Jersey), born at Crystal City, MO, July 28, 1943.
Darryl Hickman, 64, actor ("The Many Loves of Dobie Gillis," "The Americans"), born at Los Angeles, CA, July 28, 1931.
Linda Kelsey, 49, actress (Billie on "Lou Grant"), born Minneapolis, MN, July 28, 1946.
Jacques Piccard, 73, inventor, explorer, born at Brussels, Belgium, July 28, 1922.
Sally Anne Struthers, 47, actress (Gloria on "All in the Family"), born Portland, OR, July 28, 1948.
Rick Wright, 50, singer, musician (keyboard with Pink Floyd), born London, England, July 28, 1945.

JULY 29 — SATURDAY
210th Day — Remaining, 155

CLASSIC CAR SHOW IN THE PARK. July 29–30. Ta-Ha-Zouka Park, Norfolk, NE. More than 350 cars on display. Saturday night is cruise night, listen to "rockin' oldies" show, '50s and '60s dance; on Sunday, look at cars in the park competing for prizes. Also model car show, live band, Nebraska radio control truck pullers, special attraction cars and door prizes. Est attendance: 10,000. For info: Dave Eilerps, 116 Morningside Dr, Norfolk, NE 68701.

CUSTER STATE PARK VOLKSMARCH. July 29–30. Custer, SD. Promotes health and recreation with a 6.2-mile hike on marked trail. Est attendance: 330. Sponsor: Custer State Park, c/o Sally Svenson and Jon Corey, HC 83, Box 70, Custer, SD 57730. Phone: (605) 255-4515.

ENGLAND: COWE'S WEEK. July 29–Aug 5. Cowes, Isle of Wight. Yachting festival covering all classes of yacht racing. Est attendance: 14,000. For info: Cowes Week Organizers, Cowes Combined Clubs, 18 Bath Rd, Cowes, Isle of Wight, England PO31 7QN.

FESTIVAL AT CARHENGE. July 29. Alliance, NE. Carhenge is a replica of Stonehenge in dimension and orientation, but built with old cars. Event features a pageantry-filled ceremony with the adding of more "stones." The basic circle consists of 26 cars planted trunk down with 7 others as lintels. Events during the day include a road rally, car arts show and a sculpture made of car parts. Est attendance: 750. For info: Friends of Carhenge, PO Box 464, Alliance, NE 69301. Phone: (308) 762-4954.

FIRST ASSEMBLY DAY. July 29–30. Jamestown, Colonial Natl Historical Park, Yorktown, VA. Living history interpreters commemorate the beginning of representative government in the new world. Est attendance: 2,000. Sponsor: Colonial Natl Historical Park, PO Box 210, Yorktown, VA 23690. Phone: (804) 898-3400.

HANOVER DUTCH FESTIVAL. July 29. Hanover, PA. Handmade crafts, ethnic foods, music, entertainment, children's carnival, petting zoo, antique car show. 12th annual festival. Est attendance: 7,000. For info: Hanover Area Chamber of Commerce, 146 Broadway, Hanover, PA 17331. Phone: (717) 637-6130.

◆ **INDIANAPOLIS SUNK: 50th ANNIVERSARY.** July 29, 1945. After delivering the atomic bomb to Tinian Island, the American cruiser *Indianapolis* was headed for Okinawa to train for the invasion of Japan when it was torpedoed by a Japanese submarine. Of 1,196 crew members, over 350 were immediately killed in the explosion or went down with the ship. There were no rescue ships nearby and those fortunate enough to survive endured the next 84 hours in ocean waters. By the time they were spotted by air on Aug 2, only 318 sailors remained alive, the others either having drowned or been eaten by sharks. This is the US Navy's worst ever loss at sea.

LOUIS ARMSTONG: CULTURAL LEGACY. July 29–Oct 8. Gibbs Museum of Art, Charleston, SC. See Jan 28 for full description. For info: Smithsonian Institution Traveling Exhibition Service, 1100 Jefferson Dr SW, Ste 3146, Washington, DC 20560. Phone: (202) 357-2700.

MILITARY FIELD DAYS. July 29. Big Pool, MD. Reactivated military units from French and Indian War, Revolutionary War and Civil War set up encampments, perform tactical demonstrations and do living history programs. Period sutlers offer their wares for sale. Entrance fee. Annually, the last full weekend in July. For info: Ralph Young, Ft Frederick State Park, 11100 Fort Frederick Rd, Big Pool, MD 21711. Phone: (301) 842-2155.

MONTANA STATE FAIR. July 29–Aug 5. Great Falls, MT. Horse racing, petting zoo, carnival, discount days, nightly entertainment and plenty of food. Fax: (406) 452-8955. Est attendance: 202,000. For info: Bill Ogg, State Fair, Box 1888, Great Falls, MT 59403. Phone: (406) 727-8900.

◆ **MUSSOLINI, BENITO: BIRTH ANNIVERSARY.** July 29. The anniversary of birth of Italian Fascist leader, at Dovia, Italy, July 29, 1883. Self-styled "Il Duce" (the leader), Mussolini governed Italy, first as prime minister and later as absolute dictator, 1922–1943. Reportedly, under his regime "the trains ran on time." It was Mussolini who said: "War alone . . . puts the stamp of nobility upon the peoples who have the courage to face it." But military defeat of Italy in World War II was Mussolini's downfall. Repudiated and arrested by the Italian government, he was temporarily rescued by German paratroops in 1943. Later, as they attempted to flee in disguise to Switzerland, he and his mistress, Clara Petacci, were shot and killed by Italian partisans near Lake Como, Italy, Apr 28, 1945.

NORWAY: OLSOK EVE. July 29. Commemorates Norway's Viking king St. Olav, who fell in battle at Stiklestad near Trondheim, Norway, July 29, 1030. Bonfires, historical pageants.

QUAYLE, MARILYN TUCKER: BIRTHDAY. July 29. Wife of the 44th Vice-President of the US, J. Danforth Quayle, born at Indianapolis, IN, July 29, 1949.

QUILT SHOW. July 29. Woodstock, VT. A month-long juried showing of quilts made by Windsor County quilters, displayed with selected 19th-century Vermont quilts. Daily quilting demonstrations and activities. Est attendance: 9,700. For info: Billings Farm and Museum, PO Box 489, Woodstock, VT 05091. Phone: (802) 457-2355.

RAIN DAY IN WAYNESBURG, PENNSYLVANIA. July 29. Legend has it that rain will fall in Waynesburg, PA, on July 29 as it has most years for the last century, according to local records in this community, which was laid out in 1796 and incorporated in 1816.

RISING WATER FALLING WATER POWWOW. July 29–30. The Show Place Exhibition Center, Richmond, VA. Traditional Powwow, traders, crafts, food. Sponsor: Rising Water Dancers and Falling Water Drum. Est attendance: 3,500. For info: Nokomis Lemons, Coord, Rt #2, Box 107B, Bruington, VA 23023.

ST. SPYRIDON GREEK ORTHODOX FESTIVAL. July 29–30. Kiwanis Park, Sheboygan, WI. Greek music and dancing. Festival foods include Greek pastries, shish kabobs, gyros. Est attendance: 10,000. Sponsor: St. Spyridon Greek Orthodox Church, c/o 2903 S 11th St, Sheboygan, WI 53081. Phone: (414) 452-3096.

TARKINGTON, BOOTH: BIRTH ANNIVERSARY. July 29. American novelist, born at Indianapolis, IN, July 29, 1869. Died there May 19, 1946.

VIRGINIA HIGHLANDS FESTIVAL. July 29–Aug 13. Abingdon, VA. Featuring one of the largest antique markets in the USA. Entertainment, arts, crafts, workshops, historic tours, youth events, wine tasting, bike race, marathon, hot air balloons, lectures and much more! Named one of the top 100 events in North America by the ABA. For info: Abingdon Convention and Visitors Bureau; phone: (800) 435-3440.

WHITE RIVER WATER CARNIVAL. July 29–Aug 5. Batesville, AR. Public receptions, Miss White River pageant, bass classic tournament, 4-mile run, 4-ball golf tournament, arts, crafts, parade, tour of historic homes, children's activities, entertainment in the park, ski show. Est attendance: 9,000. For info: Chamber of Commerce, 409 Vine, Batesville, AR 72501. Phone: (501) 793-2378.

WOOD CARVERS FESTIVAL. July 29. Blackduck, MN. Exhibits and sales by woodcarvers from region, US and Canada. Food, music and drawings. Est attendance: 3,000. For info: Jim Schram, Blackduck Woodcarvers, HCR 3, Box 221, Blackduck, MN 56630. Phone: (218) 835-4669.

ZAMBIA: MUTOMBOKO CEREMONY. July 29. Ancient annual ceremony to honor Senior Chief, the traditional leader of the Luunda peoples of the Luapula Province in north-central Zambia. Joyous communal party and cultural get-together.

BIRTHDAYS TODAY

Melvin Belli, 88, lawyer (*Everybody's Guide to the Law*), born at Sonora, CA, July 29, 1907.
Peter Jennings, 57, journalist (anchorman "ABC Evening News"), born at Toronto, Ontario, Canada, July 29, 1938.
Nancy Landon Kassebaum, 63, US Senator (R, Kansas), born at Topeka, KS, July 29, 1932.
Marilyn Tucker Quayle, 46, former Second Lady, born at Indianapolis, IN, July 29, 1949.
Patty Scialfa, 39, singer, born at Deal, NJ, July 29, 1956.
Michael Spinks, 39, boxer, born St. Louis, MO, July 29, 1956.
Paul Taylor, 65, dancer, choreographer, born at Allegheny, NY, July 29, 1930.
David Warner, 54, actor ("Holocaust," *Tron*), born at Manchester, England, July 29, 1941.

JULY 30 — SUNDAY

211th Day — Remaining, 154

BRONTE, EMILY: BIRTH ANNIVERSARY. July 30. English novelist, one of the Brontë sisters, best known for *Wuthering Heights*. Born July 30, 1818. Died Dec 19, 1848.

	S	M	T	W	T	F	S
July 1995							1
	2	3	4	5	6	7	8
	9	10	11	12	13	14	15
	16	17	18	19	20	21	22
	23	24	25	26	27	28	29
	30	31					

COMEDY CELEBRATION DAY. July 30. Golden Gate Park, San Francisco, CA. A comedy extravaganza from local and nationally known comics to say thank you to the city that's nurtured so many comedic artists. Annually, the last Sunday in July. Sponsors: *The San Francisco Chronicle*, Robin Williams, local comedy clubs. Est attendance: 50,000. For info: Jose Simon, Exec Producer, 468 Dellbrook Ave, San Francisco, CA 94131. Phone: (415) 566-3042.

DICKENS UNIVERSE. July 30–Aug 5. Kresge College, University of California, Santa Cruz, CA. A lively gathering of Dickens enthusiasts and scholars. Lectures, workshops, films, exhibits, performances, dances, parties. Theme for 1995: *Great Expectations, Jane Eyre* and Subjectivity. Est attendance: 250. Sponsor: John O. Jordan, Dir, The Dickens Project, 354 Kresge College, University of California, Santa Cruz, CA 95064. Phone: (408) 459-2103.

DIETARY MANAGERS ASSOCIATION ANNUAL MEETING AND EXPO. July 30–Aug 3. Park Plaza Hotel, Boston, MA. 35th annual. Est attendance: 800. For info: Susan Wrona, Mktg and Sales, Dietary Mgrs Assn, One Pierce Pl, Ste 1220W, Itasca, IL 60143-1277. Phone: (708) 775-9200.

FRENCH WEST INDIES: TOUR DES YOLES RONDES. July 30–Aug 6. Martinique. An exciting and beautiful spectator sport, the weeklong race of yawls (traditional sailboats used by Martinique fishermen) depart Francois on the Atlantic at 10 AM, then sail north to cover the island in seven stopovers. For info: Martinique Development & Promotion Bureau, Ms Muriel Wiltord, 610 5th Ave, New York, NY 10020. Phone: (212) 757-1125.

HILL, ANITA FAYE: BIRTHDAY. July 30. Law professor, Anita Hill, born on an Oklahoma farm, July 30, 1956, shocked the nation and forever changed public attitudes toward sexual harassment of women in October 1991. Her nationally televised testimony accused now-Justice Clarence Thomas of sexual harassment before the all-male Senate Judiciary Committee during that committee's confirmation hearings concerning President Bush's nomination of Thomas as a justice of the Supreme Court. From a rural upbringing in relative obscurity, youngest of 13 children, Yale graduate, she abruptly became one of the nation's best-known and most-admired women as a result of her courageous televised testimony. The Hill-Thomas hearings galvanized American women to strive for feminine representation in the US Senate.

HOFFA, JAMES: DISAPPEARANCE ANNIVERSARY. July 30. Former Teamsters Union leader, 62-year-old James Riddle Hoffa was last seen July 30, 1975, outside a restaurant in Bloomfield Township, near Detroit, MI. His 13-year federal prison sentence had been commuted by former President Richard M. Nixon in 1971. On Dec 8, 1982, seven years and 131 days after his disappearance, an Oakland County judge declared Hoffa officially dead as of July 30, 1982.

JUNIOR NATURE CAMP. July 30–Aug 12. Camp Giscowheco, Wheeling, WV. This camp is designed for the young camper, 11–15 years of age, and focuses on nature and the environment. Est attendance: 120. Sponsor: Oglebay Institute, Nature/Environmental Educ Dept, Jeff Donahue, Oglebay Institute, Wheeling, WV 26003. Phone: (304) 242-6855.

RANCH RODEO. July 30. Burwell, NE. A new concept in rodeo where ranch hands compete against the clock while performing tasks that are part of their everyday life. Est attendance: 500. For info: Conv and Visitors Bureau, PO Box 747, Burwell, NE 68823. Phone: (308) 346-5210.

STEILACOOM SALMON BAKE. July 30. Sunnyside Beach, Steilacoom, WA. Grilled salmon prepared on the shores of Puget Sound complemented by corn on the cob, clam nectar, fruit pies, salads and beverages. Special activities include canoe and kayak races and music. Annually, the last Sunday in July. Est attendance: 1,000. Sponsor: Steilacoom Historical Museum Assn, PO Box 88016, Steilacoom, WA 98388. Phone: (206) 584-4133.

★ Chase's 1995 Calendar of Events ★ July 30–31

SUMMERFEST. July 30. Hillandale Park, Harrisonburg, VA. To provide a showcase of talent in western Virginia to highlight the uniqueness of the history, nature and arts of the Shenandoah Valley. Arts and crafts, drama, dance and music. Annually, the last Sunday in July (10–5). Est attendance: 6,000. For info: Dept of Parks and Recreation, 305 S Dogwood Dr, Harrisonburg, VA 22801. Phone: (703) 433-9168.

VANUATU: INDEPENDENCE DAY: 15th ANNIVERSARY. July 30. Vanuatu became an independent republic on July 30, 1980, and observes its national holiday on this day.

VEBLEN, THORSTEIN: BIRTH ANNIVERSARY. July 30. American economist, born at Valders, WI, July 30, 1857. Died at Menlo Park, CA, Aug 3, 1929. "Conspicuous consumption," he wrote in *The Theory of the Leisure Class*, "of valuable goods is a means of reputability to the gentleman of leisure."

BIRTHDAYS TODAY

Paul Anka, 54, singer, songwriter ("Diana"; "My My Way" for Frank Sinatra), born Ottawa, Ontario, July 30, 1941.

Peter Bogdanovich, 56, producer, director (*The Last Picture Show, Paper Moon*), born Kingston, NY, July 30, 1939.

Delta Burke, 39, actress (played Suzanne Sugarbaker on "Designing Women"; former Miss Florida), born at Orlando, FL, July 30, 1956.

Kate Bush, 37, singer ("The Man with the Child in His Eyes"), songwriter, born at Lewisham, England, July 30, 1958.

Edd Byrnes, 62, actor ("77 Sunset Strip," *Darby's Rangers*), born at New York, NY, July 30, 1933.

Bill (James William) Cartwright, 38, former basketball player, born at Lodi, CA, July 30, 1957.

Anita Faye Hill, 39, law professor (the Clarence Thomas hearings), born at an Oklahoma farm, July 30, 1956.

Christopher Paul Mullin, 32, basketball player, born at New York, NY, July 30, 1963.

Ken Olin, 41, actor ("Hill Street Blues," "Falcon Crest"), born at Chicago, IL, July 30, 1954.

David Sanborn, 50, saxophonist, composer, born at Tampa, FL, July 30, 1945.

Arnold Schwarzenegger, 48, bodybuilder, actor (*Terminator, Twins*), born Graz, Austria, July 30, 1947.

Eleanor Marie Cutri Smeal, 56, feminist (former president of NOW), born at Ashtabula, OH, July 30, 1939.

JULY 31 — MONDAY
212th Day — Remaining, 153

CANADA: CHOCOLATE FESTIVAL 1995. July 31–Aug 5. St. Stephen, New Brunswick. Events include chocolate meals, chocolate teas, chocolate-eating contests, "Choctail hour" and the ever-popular Ganong Chocolate Factory tours. (Tours must be booked in advance.) Est attendance: 5,000. For info: Chocolate Fest '95, Doug Dougherty, PO Box 5002, St. Stephen, NB, Canada E3L 2X5. Phone: (506) 465-4193.

FEAST OF ST. IGNATIUS OF LOYOLA. July 31. 1491–1556. Founder of the Society of Jesus and Jesuits. Canonized in 1622.

FIRST US GOVERNMENT BUILDING. July 31. The cornerstone of the Mint, the first US government building, was laid on this day in 1792.

KENNEDY-CHENG UNITED NATIONS MEMBERSHIP SUMMIT: ANNIVERSARY. July 31. President John F. Kennedy agreed during talks held with General Chen Cheng July 31–Aug 1, 1961, to support Nationalist China in its bid for UN membership and oppose the admission of Communist China to the United Nations.

KENNEDY INTERNATIONAL AIRPORT DEDICATION: ANNIVERSARY. July 31. New York's International Airport at Idlewild Field was dedicated by President Harry S. Truman on July 31, 1948. It was later renamed John F. Kennedy International Airport.

MOBY DICK MARATHON. July 31. Mystic Seaport, Mystic, CT. Marathon reading of the classic *Moby Dick* in celebration of Herman Melville's birthday. Reading takes place on deck and in foc's'le of the last wooden whaler, *Charles W. Morgan*. Annually, beginning on July 31. Est attendance: 6,000. For info: Mystic Seaport, 75 Greenmanville Ave, PO Box 6000, Mystic, CT 06355-0990. Phone: (203) 572-5315.

REEVES, JIM: DEATH ANNIVERSARY. July 31. Country music star Jim Reeves died at age 41, July 31, 1964, when the single-engine plane in which he was traveling crashed in a dense fog. Reeves's biggest hit was "He'll Have to Go" (1959), and he was inducted into the Country Music Hall of Fame in 1967.

US GIRLS' JUNIOR (GOLF) CHAMPIONSHIP. July 31–Aug 5. Longmeadow Country Club, Longmeadow, MA. Sponsor: US Golf Assn, Championship Dept, Golf House, Far Hills, NJ 07931. Phone: (908) 234-2300.

US PATENT OFFICE OPENS: ANNIVERSARY. July 31. On July 31, 1790, the first US Patent Office opened its doors, and the first US patent was issued to Samuel Hopkins of Vermont for a new method of making pearlash and potash. The patent was signed by George Washington and Thomas Jefferson.

BIRTHDAYS TODAY

Geraldine Chaplin, 51, actress (*Nashville, Roseland, Chaplin*), born at Santa Monica, CA, July 31, 1944.

Milton Friedman, 83, economist, journalist, born at Brooklyn, NY, July 31, 1912.

Evonne Goolagong, 44, former tennis player, born at Barellan, Australia, July 31, 1951.

Curt Gowdy, 76, sports commentator, born at Green River, WY, July 31, 1919.

Irv Kupcinet, 83, former TV talk-show host, columnist, born at Chicago, IL, July 31, 1912.

Don Murray, 66, actor (*Bus Stop*, "Knots Landing"), born at Hollywood, CA, July 31, 1929.

William F. Weld, 50, Governor of Massachusetts (R), born at Smithtown, NY, July 31, 1945.

☆ Chase's 1995 Calendar of Events ☆

Aug 1

August.

AUGUST 1 — TUESDAY
213th Day — Remaining, 152

BENIN: NATIONAL DAY. Aug 1. Public holiday.

BURK, MARTHA (CALAMITY JANE): DEATH ANNIVERSARY. Aug 1. Known as a frontierswoman and companion to Wild Bill Hickock, Calamity Jane Burk was born Martha Jane Canary in Princeton, MO, probably in the year 1852. As a young girl living in Montana, she became an excellent markswoman. She went to the Black Hills of South Dakota as a scout for a geological expedition in 1875. Several opposing traditions account for her nickname, one springing from her kindness to the less fortunate, while another attributes it to the harsh warnings she would give men who would offend her. She died on Aug 1, 1903, and was buried at Deadwood, SD, next to Wild Bill Hickock.

CANADA: BUFFALO DAYS. Aug 1-6. Regina Exhibition Park, Regina, Saskatchewan. Midway rides, grandstand shows, livestock exhibits, casino, fashion show, homecraft displays, parade, horse racing. Est attendance: 200,000. For info: Tourism Saskatchewan, Saskatchewan Trade and Conv Center, 1919 Saskatchewan Dr, Regina, Sask, Canada S4P 3V7. Phone: (800) 667-7191.

DIARY OF ANNE FRANK: THE LAST ENTRY: ANNIVERSARY. Aug 1. To escape deportation to concentration camps, the Jewish family of Otto Frank hid for two years in the back-room office and warehouse of his food products business in Amsterdam. Gentile friends smuggled in food and other supplies during their confinement. Thirteen-year-old Anne Frank, who kept a journal during the time of their hiding, penned her last entry in the diary Aug 1, 1944: "[I] keep on trying to find a way of becoming what I would like to be, and what I could be, if . . . there weren't any other people living in the world." Three days later (Aug 4, 1944) Grüne Polizei raided the "Secret Annex" where the Frank family was hidden. Anne and her sister were sent to Bergen-Belsen concentration camp where Anne died at age 15, two months before the liberation of Holland. Young Anne's diary, later found in the family's hiding place, has been translated into 30 languages and has become a symbol of the indomitable strength of the human spirit.

EMANCIPATION OF 500: ANNIVERSARY. Aug 1. Virginia planter Robert Carter III confounded his family and friends by filing a deed of emancipation for his 500 slaves Aug 1, 1791. One of the wealthiest men in the state, Carter owned 60,000 acres over 18 plantations. The deed included the following words, "I have for some time past been convinced that to retain them in Slavery is contrary to the true principles of Religion and Justice and therefore it is my duty to manumit them." The document established a schedule by which 15 slaves would be freed each Jan 1, over a 21-year period plus slave children would be freed at age 18 for females and 21 for males. It is believed this was the largest act of emancipation in US history and predated the Emancipation Proclamation by 70 years.

FINLAND: LAHTI ORGAN FESTIVAL. Aug 1-7. Lahti. Annual festival emphasizing organ music. Also, choirs, orchestras, instrumental groups and solo artists. Plus master classes, seminars and panel discussions. Est attendance: 10,000. For info: Finnish Tourist Board, 655 Third Ave, New York, NY 10017. Phone: (212) 949-2333.

FOOT HEALTH MONTH. Aug 1-31. To educate people on the importance of taking care of their feet. Eighty percent of adults suffer from foot problems at some point in their lives. Efforts throughout the month are dedicated to making life more comfortable for people on their feet. Tips on how to prevent problems and treatment alternatives are available. Sponsor: Dr. Scholl's. For info: Andy Horrow, P/N Ste 1214, 303 E Wacker Dr, Chicago, IL 60601. Phone: (312) 856-8826.

INTERNATIONAL CLOWN WEEK. Aug 1-7. To call public attention to the charitable activities of clowns and the wholesome entertainment they provide. Annually, August 1-7. Phone: (218) 834-8415 (W) or (218) 834-6406 (H). For info: Clowns of America Intl, Inc, Thomas S. Oswald, Intl Chair, PO Box 306, Two Harbors, MN 55616.

JAMAICA: ABOLITION OF SLAVERY. Aug 1. National day. Spanish settlers introduced the slave trade into Jamaica in 1509 and sugar cane in 1640. Slavery continued until Aug 1, 1838, when it was abolished.

LUGHNASADH. Aug 1. (Also called August Eve, Lammas Eve, Lady Day Eve and Feast of Bread.) One of the "Greater Sabbats" during the Wiccan year, Lughnasadh marks the first harvest. Annually, Aug 1.

MELVILLE, HERMAN: BIRTH ANNIVERSARY. Aug 1. American author, best known for his novel *Moby Dick*, born at New York, NY, Aug 1, 1819. Died Sept 28, 1891.

◆ **MITCHELL, MARIA: BIRTH ANNIVERSARY.** Aug 1, 1818. An interest in her father's hobby and an ability for mathematics resulted in Maria Mitchell becoming the first female professional astronomer. In 1847, while assisting her father in a survey of the sky for the US Coast Guard, Mitchell discovered a new comet and determined its orbit. She received many honors because of this, including being elected to the American Academy of Arts and Sciences—its first woman. Mitchell joined the staff at Vassar Female College in 1865—the first US female professor of astronomy—and in 1873 was a cofounder of the Association for the Advancement of Women. Born at Nantucket, MA, Mitchell died June 28, 1889, at Lynn, MA.

NATIONAL CATFISH MONTH. Aug 1-31. To increase awareness of genuine US farm-raised catfish. For info: Loyal Order of Catfish Lovers, c/o Golin/Harris, 666 Third Ave, 2nd Fl, New York, NY 10017.

NATIONAL NIGHT OUT. Aug 1. Designed to heighten crime prevention awareness and to promote police-community partnerships. Annually, the first Tuesday in August. Phone: (610) 649-7055 or (800) 648-3688. For info: Matt A. Peskin, Dir, Natl Assn of Town Watch, PO Box 303, Wynnewood, PA 19096.

NATIONAL WATER QUALITY MONTH. Aug 1-31. To increase the awareness of water as a precious resource and the importance of quality water in our everyday lives. Sponsor: Culligan Intl, One Culligan Pkwy, Northbrook, IL 60062. Phone: (708) 205-6000.

OAK RIDGE ATOMIC PLANT BEGUN: ANNIVERSARY. Aug 1. Ground was broken in Oak Ridge, TN, on Aug 1, 1943, for the first plant built to manufacture the uranium-235 needed to build an atomic bomb. The plant was largely completed by

August 1995

S	M	T	W	T	F	S
		1	2	3	4	5
6	7	8	9	10	11	12
13	14	15	16	17	18	19
20	21	22	23	24	25	26
27	28	29	30	31		

☆ Chase's 1995 Calendar of Events ☆   Aug 1-2

July of 1944 at a final cost of $280 million. By August 1945 the total cost for development of the A-bomb ran to $1 billion.

ROMANCE AWARENESS MONTH. Aug 1-31. To educate couples about how romance can improve their relationships. To encourage couples to display romance throughout the year rather than just Valentine's Day and anniversaries. Phone: (800) 368-7978 or (714) 459-7620. Sponsor: Celebrate Romance, Eileen Buchheim, 5199 E Pacific Coast Hwy, Ste 303A, Long Beach, CA 90804.

ROUNDS RESOUNDING DAY. Aug 1. To sing rounds, catches and canons in folk contrapuntal tradition. Motto: "As rounds re-sound and resound all the world's joined in a circle of harmony." Annually, Aug 1. Sponsor: Rounds Resounding Soc, Gloria T. Delamar, Founder, 7303 Sharpless Rd, Melrose Park, PA 19027. Phone: (215) 782-1059.

SWITZERLAND: NATIONAL DAY. Aug 1. Anniversary of the founding of the Swiss Confederation. Commemorates a pact made in 1291. Parades, patriotic gatherings, bonfires and fireworks. Young citizens' coming-of-age ceremonies. Observed since 600th anniversary of Swiss Confederation was celebrated in 1891.

TRINIDAD AND TOBAGO: EMANCIPATION DAY. Aug 1. Public holiday.

TRINIDAD: BOAT RACING. Aug 1. Starting from the National Stadium Sea Front at 8 AM, an international regatta racing in celebration of the 25th anniversary of "The Great Race." For info: Trinidad and Tobago TDA, 25 W 43rd St, Ste 1508, New York, NY 10036. Phone: (800) 232-0082.

US CUSTOMS: ANNIVERSARY. Aug 1. "The first US customs officers began to collect the revenue and enforce the Tariff Act of July 4, 1789, on Aug 1, 1789. Since then, the customhouse and the customs officer have stood as symbols of national pride and sovereignty at ports of entry along the land and sea borders of our country." (From Presidential Proclamation 4306.)

BIRTHDAYS TODAY

Tempestt Bledsoe, 22, actress (Vanessa on "The Cosby Show"), born Chicago, IL, Aug 1, 1973.
Ronald H. Brown, 54, US Secretary of Commerce in Clinton administration, born at Washington, DC, Aug 1, 1941.
Robert Cray, 42, singer, guitarist, songwriter, born at Columbus, GA, Aug 1, 1953.
Alfonse M. D'Amato, 58, US Senator (R, New York), born at Brooklyn, NY, Aug 1, 1937.
Dom DeLuise, 62, comedian, actor, born at Brooklyn, NY, Aug 1, 1933.
Jerry Garcia, 53, singer, musician (leader the Grateful Dead), born at San Francisco, CA, Aug 1, 1942.
Giancarlo Giannini, 53, actor (*Swept Away . . . By an Unusual Destiny in the Blue Sea of August*), born Spezia, Italy, Aug 1, 1942.
Arthur Hill, 73, actor (*Harper*, *The Andromeda Strain*, "Owen Marshall"), born Melfort, Sask, Canada, Aug 1, 1922.
Yves Saint Laurent, 59, fashion designer, born at Oran, Algeria, Aug 1, 1936.
Tom Wilson, 64, cartoonist, born at Grant Town, WV, Aug 1, 1931.

AUGUST 2 — WEDNESDAY
214th Day — Remaining, 151

ABBOTT'S MAGIC GET-TOGETHER (WITH TALENT CONTESTS). Aug 2-5. Colon, MI. Magic convention. Est attendance: 1,200. Sponsor: Abbott's Magic Co, Greg Bordner, 124 St. Joseph St, Colon, MI 49040. Phone: (616) 432-3235.

ALBERT EINSTEIN'S ATOMIC BOMB LETTER: ANNIVERSARY. Aug 2. Albert Einstein, world-famous scientist, a refugee from Nazi Germany, wrote a letter to US President Franklin D. Roosevelt, Aug 2, 1939, first mentioning a possible "new phenomenon . . . chain reactions . . . vast amounts of power" and "the construction of bombs." "A single bomb of this type," he wrote, "carried by boat and exploded in a port, might very well destroy the whole port together with some of the surrounding territory." An historic letter that marked the beginning of atomic weaponry. Six years and four days later, Aug 6, 1945, the Japanese port of Hiroshima was destroyed by the first atomic bombing of a populated place.

BALDWIN, JAMES ARTHUR: BIRTH ANNIVERSARY. Aug 2. Black American author noted for descriptions of black life in the US. Born Aug 2, 1924, at New York, NY. His best known work, *Go Tell It on the Mountain*, was published in 1953. Died at Saint Paul-de-Vence, France, Nov 30, 1987.

CANADA: MAHONE BAY WOODEN BOAT FESTIVAL. Aug 2-6. Mahone Bay, Nova Scotia. Five-day festival celebrating Mahone Bay's wooden boat building heritage. Events include workshops, demonstrations, water activities, boat races, children's activities, boat yard tours, musical entertainment, street dancing, a parade, fireworks, food and a nautical trade fair. Free admission. Annually, the Wednesday through Sunday prior to the first Monday in August. Phone: (800) 565-0000, operator 503. Est attendance: 25,000. For info: Mahone Bay Wooden Boat Festival Soc, Festival Coord, PO Box 609, Mahone Bay, NS, Canada B0J 2E0.

DECLARATION OF INDEPENDENCE: OFFICIAL SIGNING ANNIVERSARY. Aug 2. Contrary to widespread misconceptions, the 56 signers did not sign as a group and did not do so on July 4, 1776. John Hancock and Charles Thompson signed only draft copies that day, the official day the Declaration was adopted by Congress. The signing of the official declaration occurred on Aug 2, 1776, when 50 men probably took part. Later that year, five more apparently signed separately and one added his name in a subsequent year. (From "Signers of the Declaration...," US Dept of the Interior, 1975.) See also: "Declaration of Independence" (July 4).

EDSEL OWNERS CLUB ANNUAL CONVENTION. Aug 2-6. (Tentative.) Oakland, CA. Nov 19, 1995, marks the 36th anniversary of Ford Motor Company's announcement that it was discontinuing the manufacture of the Edsel. Overall production of the automobile was only 110,810 units in three years. Today the Edsel is a collectors' item. Convention will include a judging, banquet and general membership meeting. Est attendance: 300. For info: Edsel Owners Club, 9211 Portland Ave S, Bloomington, MN 55420-3839. Phone: (612) 884-3091.

GEORGIA MOUNTAIN FAIR. Aug 2-13. Georgia Mountain Fairgrounds, Hiawassee, GA. Authentic mountain demonstrations including corn millin', board splittin', soap and hominy makin', and others. Pioneer village with re-created one-room school, log cabin, barn and corncrib. Nashville talent, clogging, midway and much more. Est attendance: 125,000. For info: Dale Thurman, Mgr, Georgia Mountain Fair, PO Box 444, Hiawassee, GA 30546. Phone: (706) 896-4191.

IRAQ INVADES KUWAIT: 5th ANNIVERSARY. Aug 2. President Saddam Hussein of Iraq ordered the invasion of Kuwait on Aug 2, 1990. Hussein claimed that Kuwait presented a serious threat to Iraq's economic existence by overproducing oil and driving prices down on the world market. After conquering the capital, Kuwait City, Hussein installed a military government in Kuwait, prior to annexing it to Iraq on the claim that Kuwait was historically part of Iraq. The US and most other nations immediately condemned the aggression and the United

☆ Chase's 1995 Calendar of Events ☆

Nations passed measures calling for broad economic sanctions against Iraq. As Iraqi forces began to mass along the border with Saudi Arabia, the US and other nations sent troops to Saudi Arabia to protect that country from invasion with an operation named Desert Shield. The multinational force included troops from other Arab countries such as Egypt, Syria and Morocco in addition to forces from Western governments with large economic interests in the region. Approximately 21,000 foreign nationals from several countries were detained by Iraq and were transported to various strategic locations to deter possible retaliatory attacks. The US military action was the largest mobilization of forces since the Vietnam War. The following January Desert Shield became Operation Desert Storm as the Allied forces went to war against Iraq.

KIDS' DAY. Aug 2. Broadway Market Square, Baltimore, MD. All kids 12 years of age and under enjoy an hour of free fun, games, refreshments and entertainment. Est attendance: 300. For info: Office of Adventures and Fun, Clarence "Du" Burns Arena, 1301 S Ellwood Ave, Baltimore, MD 21224. Phone: (410) 396-9177.

L'ENFANT, PIERRE CHARLES: BIRTH ANNIVERSARY. Aug 2. The architect, engineer and Revolutionary War officer who designed the plan for the city of Washington, DC, Pierre Charles L'Enfant was born at Paris, France, on Aug 2, 1754. He died at Prince Georges County, MD, on June 14, 1825.

LOY, MYRNA (WILLIAMS): 90th BIRTH ANNIVERSARY. Aug 2, 1905. Actress known for her film roles in the "Thin Man" series and for speaking out against the House Committee on UnAmerican Activities. Born near Helena, MT, and died at New York City on Dec 14, 1993.

MISS CRUSTACEAN USA BEAUTY PAGEANT AND OCEAN CITY CREEP. Aug 2. State Beach, Ocean City, NJ. Participants are hermit tree crabs. To determine most beautiful tree crab and fastest tree crab on earth. Begins at 1:00 PM, EST. Phone: (609) 399-6111, ext 222 or (609) 399-0272 (direct line). Est attendance: 500. For info: City of Ocean City, Mark Soifer, PR Dir, City Hall, Ocean City, NJ 08226.

ST. ELIAS DAY (ILLINDEN): ANNIVERSARY OF MACEDONIAN UPRISING. Aug 2. Most sacred, honored and celebrated day of the Macedonian people. Anniversary of the uprising, on Aug 2, 1903, of Macedonians against Turkey. Turkish reprisals against the insurgents were ruthless, including the destruction of 105 villages and the execution of more than 1,700 noncombatants.

SCOTLAND: ABERDEEN INTERNATIONAL YOUTH FESTIVAL. Aug 2-12. Aberdeen, Gampian. Talented young people from all areas of the performing arts come from around the world to participate in this festival. For info: Nicola Willis, 3 Nutborn House, Clifton Rd, London, England SW19 4QT.

August 1995

S	M	T	W	T	F	S
		1	2	3	4	5
6	7	8	9	10	11	12
13	14	15	16	17	18	19
20	21	22	23	24	25	26
27	28	29	30	31		

BIRTHDAYS TODAY

Ron de Lugo, 65, US Delegate (D, Virgin Islands), born at Englewood, NJ, Aug 2, 1930.
Linda Fratianne, 35, figure skater, born at Los Angeles, CA, Aug 2, 1960.
Carroll O'Connor, 71, actor (*Marlowe*, "All in the Family," "In the Heat of the Night"), born New York, NY, Aug 2, 1924.
Peter O'Toole, 62, actor (*Lawrence of Arabia, Becket*), born at Connemara, Ireland, Aug 2, 1933.
Beatrice Straight, 77, actress (Oscar for *Network*), born at Old Westbury, NY, Aug 2, 1918.

AUGUST 3 — THURSDAY
215th Day — Remaining, 150

CANADA: HALIFAX INTERNATIONAL BUSKERFEST. Aug 3-13. Halifax, Nova Scotia. Street performers and artists from around the world, vaudeville nights, an international street food fair, and a 9-day special children's program. Annually, beginning the first Thursday in August. Est attendance: 700,000. For info: Dale Thompson, Exec Producer, G.M. Buskers' Fest, 1652 Barrington St, Halifax, NS, Canada B3J 2A2. Phone: (902) 425-4329.

CANADA: ROCKHOUND GEMBOREE. Aug 3-6. Bancroft, Ontario. Daily expeditions to prime mineral locations, dealers, demonstrations and displays, swapping. Est attendance: 12,000. For info: Bancroft and District Chamber of Commerce, PO Box 539, Bancroft, Ont, Canada K0L 1C0. Phone: (613) 332-1513.

GUINEA-BISSAU: COLONIZATION MARTYR'S DAY. Aug 3. National holiday is observed.

MAINE FESTIVAL OF THE ARTS. Aug 3-6. Brunswick, ME. More than 600 artists and craftspeople on the site. Folk Arts Village highlighting music and crafts traditional to Maine. Festival stage featuring the best contemporary music and dance. Storytelling, vaudeville and many special children's events. Downeast gourmet foods! Daily ticket costs will vary. Annually, the first weekend in August. Est attendance: 18,000. For info: Maine Arts, Inc, 582 Congress St, Portland, ME 04101. Phone: (207) 772-9012.

MAINE LOBSTER FESTIVAL. Aug 3-6. Rockland, ME. Celebration and promotion of the lobster industry featuring lobster dinners, arts, crafts, exhibits, live entertainment, parade, contests and a road race. Est attendance: 50,000. For info: Rockland Thomaston Area Chamber of Commerce, PO Box 508, Rockland, ME 04841. Phone: (207) 596-0376.

MOON PHASE: FIRST QUARTER. Aug 3. Moon enters First Quarter phase at 11:16 PM, EDT.

MOUNTAIN DANCE AND FOLK FESTIVAL. Aug 3-5. Asheville, NC. More than 400 performers compete in this three-day fest, which celebrates the cultural heritage of the Southern Appalachian Mountains. Includes dance teams, mountain fiddlers, banjo pickers and dulcimer sweepers. 68th annual festival. Oldest of its kind, founded in 1927 by Bascom Lamar Lunsford. Annually, the first Thurs-Sat in August. Est attendance: 2,000. For info: Zack Allen, Folk Heritage Dir, Asheville Area Chamber of Commerce, PO Box 1010, Asheville, NC 28802. Phone: (800) 257-5583.

PYLE, ERNEST TAYLOR: BIRTH ANNIVERSARY. Aug 3. Ernie Pyle was born Aug 3, 1900, at Dana, IN, and began his career in journalism in 1923. After serving as managing editor of the Washington *Daily News*, he returned to his first journalistic love of working as a roving reporter in 1935. His column was syndicated by nearly 200 newspapers and often focused on figures behind the news. His reports of the bombing of London in 1940 and subsequent reports from Africa, Sicily, Italy and France earned him a Pulitzer Prize in 1944. He was killed by machine gun fire at the Pacific island of Ie Shima Apr 18, 1945.

☆ *Chase's 1995 Calendar of Events* ☆   Aug 3-4

ROCKINGHAM OLD HOME DAYS. Aug 3-6. Bellows Falls and Rockingham, VT. Village Square closed Friday and Saturday. Sidewalk sales Thursday thru Saturday. Road races. Fireworks on Saturday evening. Sunday Pilgrimage to the Old Rockingham Meeting House with ceremonies and speeches. Annually, the first weekend in August. Est attendance: 2,000. For info: Great Falls Chamber of Commerce, PO Box 554, Bellows Falls, VT 05101. Phone: (802) 463-4280.

SANTA CRUZ BEACH BOARDWALK LOOFF CAROUSEL: ANNIVERSARY. Aug 3. Danish woodcarver Charles I.D. Looff delivered the classic carousel on Aug 3, 1911. A furniture-maker by trade, Looff began carving carousel animals as a hobby after immigrating to America. His first carousel was installed at Coney Island in New York in 1875. The Boardwalk carousel features jeweled horses and a 342-pipe Ruth band organ built in 1894. The carousel and the park's Giant Dipper roller coaster were designated National Historic Landmarks by the US National Park Service in June of 1987.

SCOPES, JOHN T.: BIRTH ANNIVERSARY. Aug 3. Central figure in a cause célèbre (the "Scopes Trial" or the "Monkey Trial"), John Thomas Scopes was born Aug 3, 1900, at Paducah, KY. An obscure 24-year-old schoolteacher at the Dayton, TN, high school in 1925, he became the focus of world attention. Scopes never uttered a word at his trial, which was a contest between two of America's best-known lawyers (William Jennings Bryan and Clarence Darrow). The trial, July 10-21, 1925, resulted in Scopes's conviction. He was fined $100 "for teaching evolution" in Tennessee. The verdict was upset on a technicality, and the statute he was accused of breaching was repealed in 1967. Scopes died at Shreveport, LA, Oct 21, 1970.

STEINBECK FESTIVAL XVI. Aug 3-6. Salinas Community Center, Salinas, CA. Features lectures, presentations, films, tours of Steinbeck Country, special trips to Steinbeck locales, dinner events and Educator's Seminar. Annually, the first weekend in August. Est attendance: 700. For info: Amanda Piper, Festival Coord, Steinbeck Foundation, 371 Main St, Salinas, CA 93901. Phone: (408) 753-6411.

TEXAS FOLKLIFE FESTIVAL. Aug 3-6. San Antonio, TX. To provide historic understanding of the crafts, art, food, music, history and heritage of the more than 30 different cultural and ethnic groups who helped to settle and develop the state of Texas. Est attendance: 70,000. Sponsor: Institute of Texan Cultures, 801 S Bowie St, San Antonio, TX 78205-3296. Phone: (210) 558-2300.

TUNISIA: BOURGUIBA HARIB BIRTHDAY. Aug 3. Public holiday commemorating birth, Aug 3, 1902, of Tunisia's first president.

WISCONSIN STATE FAIR. Aug 3-13. State Fair Park, Milwaukee, WI. Wisconsin celebrates its rural heritage at the state's largest and oldest annual event: livestock judging, concessions, giant midway and top-name entertainment. Est attendance: 930,000. For info: Wisconsin State Fair Park, Dept C, PO Box 14990, West Allis, WI 53214-0990. Phone: (414) 266-7000.

BIRTHDAYS TODAY

Tony Bennett (Anthony Dominick Benedetto), 69, singer ("I Left My Heart in San Francisco," "In the Middle of an Island"), born New York, NY, Aug 3, 1926.
Delores Del Rio (Delores Ansunsolo), 90, actress (*Flying Down to Rio*), born at Durango, Mexico, Aug 3, 1905.
Maggie Kuhn, 90, founder of the Gray Panthers, born at Buffalo, NY, Aug 3, 1905.
Martin Sheen (Ramon Estevez), 55, actor (*Badlands*, *Platoon*), born at Dayton, OH, Aug 3, 1940.
Leon Uris, 71, author (*Mitla Pass*, *Exodus*), born at Baltimore, MD, Aug 3, 1924.

AUGUST 4 — FRIDAY
216th Day — Remaining, 149

AMERICOVER '95. Aug 4-6. The American First Day Cover Society will hold their 40th annual convention and exhibition, Americover '95, at the Irvine Marriott at John Wayne Airport in Irvine, CA. The AFDCS is the only society of philatelists (stamp collectors) dedicated exclusively to First Day Covers and First Day Cover Collecting. Sponsor: American First Day Cover Society, PO Box 1335, Maplewood, NJ 07040-0456.

ARMSTRONG, LOUIS: BIRTH ANNIVERSARY. Aug 4. Jazz musician born this day at New Orleans, LA, 1901. Died at New York, NY, July 6, 1971. Asked to define jazz, Armstrong reportedly replied, "Man, if you gotta ask, you'll never know."

BLUEBERRY FESTIVAL. Aug 4-5. Library Lawn and Village Green, Montrose, PA. 16th annual fundraiser for the Susquehanna County Library and Historical Society. Two days filled with food, fun and festivity. Raffles, book sale, children's games, silent auction, handstitched quilt, commemorative items, and more. Annually, the first Friday and Saturday in August. Est attendance: 5,000. For info: Susquehanna Co Historical Soc & Free Library Assn, 2 Monument Sq, Montrose, PA 18801. Phone: (717) 278-1881.

BOOM DAYS CELEBRATION. Aug 4-6. Leadville, CO. Celebration of the mining of the early years in Leadville. Held in connection with Leadville's International Championship Burro Race. For info: Chamber of Commerce, Box 861, Leadville, CO 80461. Phone: (800) 933-3901.

BRATWURST DAYS. Aug 4-5. Sheboygan, WI. Sheboygan celebrates bratwurst at Kiwanis Park with music, entertainment, Bratozica, circus, parade, flea market and, of course, bratwurst-eating contests. Since 1953. Annually, the first Friday and Saturday in August. Est attendance: 40,000. Sponsor: Sheboygan Jaycees, PO Box 561, Sheboygan, WI 53082. Phone: (414) 457-9495.

BURKINA FASO: NATIONAL DAY. Aug 4. Anniversary of the day (Aug 4, 1984) on which the Republic of Upper Volta changed its name to Burkina Faso (or Bourkina Fasso). The Republic of Upper Volta formerly commemorated attainment of autonomy, Dec 11, 1958, as its National Day.

CANADA: AGRIFAIR. Aug 4-7. Abbotsford, British Columbia. It's great fun for the young, and young-at-heart with attractions like: draft horses, dairy, beef, poultry and smaller animals; hands-on milking display, maternity pen, antique farm display bursting with antique toys, a children's midway, plus everything from a tractor pull to an exhilarating rodeo to nightly fireworks. Est attendance: 40,000. Sponsor: Central Fraser Valley Fairs, PO Box 2334, Clearbrook, BC, Canada V2T 4X2. Phone: (604) 852-6674.

CANADA: ICELANDIC FESTIVAL OF MANITOBA (ISLENDINGADAGURINN). Aug 4-6. Gimli, Manitoba. Celebration of Gimli's Icelandic heritage features parade, poetry, music and traditional Icelandic cuisine. For info: Icelandic Festival of Manitoba, Art Kilgour, Box 1871, Gimli, Man, Canada R0C 1B0. Phone: (204) 642-7417.

CANADA: PIONEER DAYS '95. Aug 4-7. Steinbach, Manitoba. Mennonite pioneer demonstrations and related activities in a heritage village setting. Est attendance: 10,000. For info: Harv Klassen, Exec Dir, Box 1136, Steinbach, Manitoba, Canada R0A 2A0. Phone: (204) 326-9661.

CARNATION CITY FESTIVAL. Aug 4-13. Alliance, OH. To honor the scarlet carnation, Ohio's state flower. Alliance is home of the carnation. 10 days of activities. Fax: (216) 823-4434. Est attendance: 100,000. For info: Carnation Fest Bd of Dir, 210 E Main St, Alliance, OH 44601. Phone: (216) 823-6260.

Aug 4   ☆ *Chase's 1995 Calendar of Events* ☆

CIVIL RIGHTS WORKERS FOUND SLAIN: ANNIVERSARY. Aug 4. After disappearing on June 21, three civil rights workers were found murdered and buried in an earthen dam outside Philadelphia, MS, on Aug 4, 1964. Prior to their disappearance, James Chaney, Andrew Goodman and Michael Schwerner were detained for six hours by Neshoba County police on charges of speeding. When their car was found, burned, on June 23, President Johnson ordered an FBI search for the three men.

COAST GUARD DAY. Aug 4. Celebrates anniversary of founding of the US Coast Guard, Aug 4, 1790.

CUNNINGHAM, GLENN: BIRTH ANNIVERSARY. Aug 4. Glenn Clarence Cunningham, the "Kansas Ironman," American track athlete and 1934-37 world record holder for the mile, member of the US Olympic teams in 1932 and 1936, was born at Atlanta, KS, on Aug 4, 1909. On June 16, 1934, at Princeton, NJ, Cunningham set a world record for the mile (4:06.7 min). Cunningham died at Menifee, AR, on Mar 10, 1988.

DECATUR CELEBRATION. Aug 4-6. Decatur, IL. A free family street festival with 22 city blocks of fun with 13 entertainment stages and much more. Est attendance: 300,000. For info: Decatur Celebration, 142 E Prairie, Ste 201, Decatur, IL 62523. Phone: (217) 423-4222.

DOLLY SODS WEEKEND. Aug 4-7. Appalachian Base Camp, Terra Alta, WV. Focuses on the study of rocks and fossils by day and the heavens by night. Est attendance: 30. Sponsor: Oglebay Institute Nature/Environmental Educ Dept, Tom Shepherd, Oglebay Institute, Wheeling, WV 26003. Phone: (304) 242-6855.

FESTIVAL IN THE PINES. Aug 4-6. Coconino County Fairgrounds, Flagstaff, AZ. Features 250 national artists and craftspeople, ethnic and traditional foods, continuous entertainment on three stages and a children's activity area. Annually, the first weekend in August. Est attendance: 75,000. For info: Mill Avenue Merchants Assn, 520 S Mill Ave, Ste 201, Tempe, AZ 85281. Phone: (602) 967-4877.

FRENCH WEST INDIES: TOUR DE LA GUADELOUPE. Aug 4-13. Guadeloupe. 10-day international bicycle race covers the island. For info: French West Indies Tourist Bd, 610 5th Ave, New York, NY 10020. Phone: (900) 990-0040.

FROG DAYS. Aug 4-6. Melvina, WI. True Americana served up family style with Men's 30 and over 18 team softball tournament, a Sunday chicken barbecue and frog racing for all ages. Est attendance: 5,000. For info: Frog Days, Rt 1, Box 90, Cashton, WI 54619.

GREAT ARKANSAS PIG-OUT. Aug 4-6. Morrilton, AR. Focus is on food of all kinds: barbecue, chicken, homemade ice cream, catfish, funnel cakes, hamburgers, spaghetti dinners. Activities include pig chase, hog calling, children's fairy tale park, rides, arts, crafts, entertainment, games, street dance, gospel singing. 6th annual pig-out. Est attendance: 20,000. For info: Chamber of Commerce, 118 N Moose, Morrilton, AR 72110. Phone: (501) 354-2393.

August 1995

S	M	T	W	T	F	S
		1	2	3	4	5
6	7	8	9	10	11	12
13	14	15	16	17	18	19
20	21	22	23	24	25	26
27	28	29	30	31		

MANDELA, NELSON: POLICE CAPTURE ANNIVERSARY. Aug 4. Nelson Rolihlahla Mandela, charismatic black South African leader, was born in 1918, the son of the Tembu tribal chief, at Umtata, Transkei territory of South Africa. A lawyer and political activist, Mandela, who in 1952 established the first black law partnership in South Africa, had been in conflict with the white government there much of his life. Acquitted of a treason charge after a trial that lasted from 1956 to 1961, he was apprehended again by security police on Aug 4, 1962. The subsequent trial, widely viewed as an indictment of white domination, resulted in Mandela's being sentenced to five years in prison. In 1963 he was taken from the Pretoria prison to face a new trial—for sabotage, high treason and conspiracy to overthrow the government—and in June 1964 he was sentenced to life in prison. See also: "Mandela, Nelson: Prison Release Anniversary" (Feb 11).

MICHIGAN FESTIVAL. Aug 4-13. East Lansing, MI. Performing arts festival. Est attendance: 375,000. For info: The Michigan Festival, Inc, 1331 E Grand River, Ste 113, East Lansing, MI 48823. Phone: (517) 351-6620.

MUNCHKINS OF OZ CONVENTION. Aug 4-6. Wilmington, DE. Celebrates Dorothy's arrival over the rainbow in Oz. Est attendance: 150. Sponsor: Intl Wizard of Oz Club, Inc, Fred M. Meyer, Secy, 220 N 11th St, Escanaba, MI 49829.

NORWAY: OSLO CHAMBER MUSIC FESTIVAL. Aug 4-12. Oslo. Chamber Music Festival. Special emphasis on Norwegian Culture with Edvard Grieg as a central figure. Annually. For info: Norwegian Tourist Bd, 655 Third Ave, New York, NY 10017. Phone: (219) 949-2333.

SCALIGER, JOSEPH JUSTUS: BIRTH ANNIVERSARY. Aug 4. French scholar who has been called the founder of scientific chronology. Born at Agen, France, Aug 4, 1540, the 10th son of classical scholar Julius Caesar Scaliger. In 1582 he suggested a new system for measuring time and numbering years. His "Julian Period" (named for his father) consisted of 7,980 consecutive years (beginning Jan 1, 4713 BC), the relationship of which was relatively easily understood. His Julian Period is still in use. He died at Leiden, Netherlands, Jan 21, 1609.

SCHUMAN, WILLIAM HOWARD: BIRTH ANNIVERSARY. Aug 4. American composer who won the first Pulitzer Prize for composition and founded the Julliard School of Music, was born Aug 4, 1910, in New York. His compositions include *American Festival Overture, Symphony No. 3, New England Triptych*, the baseball opera *The Mighty Casey* and *On Freedom's Ground* which was written for the centennial of the Statue of Liberty in 1986. He was instrumental in the conception of the Lincoln Center for the Performing Arts and served as its first president during the 1960s. In 1985 he was awarded a special Pulitzer Prize for his contributions and he also received a National Medal of Arts in 1985 and a Kennedy Center Honor in 1989. Mr. Schuman died at New York City, NY, on Dec 12, 1992.

SCOTLAND: EDINBURGH MILITARY TATTOO: THE MAIN EVENT. Aug 4-26. Edinburgh Castle, Edinburgh, Lothian. Display of military color and pageantry held at night on the floodlit esplanade of Edinburgh Castle. A unique blend of music, ceremony, entertainment and theatre, the Tattoo is probably Edinburgh's most popular annual event. Phone: (031) 225-1188, Fax: (031) 225-8627. Est attendance: 200,000. For info: The Edinburgh Military Tattoo, The Tattoo Office, 22 Market St, Edinburgh, Scotland EH1 1QB.

SQUARE FAIR '95. Aug 4-6. Lima Town Square, Lima, OH. Community outdoor arts festival. Est attendance: 100,000. For info: Council for the Arts of Greater Lima, Sherry Krouse, Asst Dir, Box 1124, Lima, OH 45802. Phone: (419) 222-1096.

SUSSEX COUNTY FARM AND HORSE SHOW. Aug 4-13. Augusta, NJ. "New Jersey's Best Fair." The state's largest livestock and horse show also includes educational exhibits, amusements, commercial exhibits and entertainment. Located off Rt 206 Plains Rd. Gate opens 5 PM Fri and closes 5 PM Sun.

Chase's 1995 Calendar of Events — Aug 4–5

Est attendance: 200,000. For info: Warren Welsh, Exec Secy, Box M, Branchville, NJ 07826. Phone: (201) 948-0540.

TANANA VALLEY FAIR. Aug 4–12. Tanana Valley Fairgrounds, Fairbanks, AK. The oldest fair in Alaska, including agricultural products such as 40+-pound cabbages, crafts and entertainment, rodeo, midway, contests and magic and musical performances. Est attendance: 130,000. For info: Tanana Valley Fair Assn, 1800 College Rd, Fairbanks, AK 99709. Phone: (907) 452-3750.

TAWAS BAY WATERFRONT FINE ART SHOW. Aug 4–6. Tawas City Park, Tawas City, MI. Phone: (800) 55-TAWAS. Est attendance: 9,000. For info: Chamber of Commerce, Box 608, Tawas City, MI 48764-0608.

TELLURIDE JAZZ CELEBRATION. Aug 4–6. Telluride, CO. Some of Jazz's hottest rising stars and most accomplished musicians team up in the intimate nightclubs of Telluride. Annually, the first weekend in August. Sponsor: American Express. Est attendance: 1,500. For info: Telluride Chamber Resort Assn, 666 W Colorado Ave, PO Box 653, Telluride, CO 81435. Phone: (800) 525-3455.

TWINS DAY FESTIVAL. Aug 4–6. Glenn Chamberlin Park, Twinsburg, OH. A festival celebrating the heritage of Twinsburg (the only city named after twins—early settlers Aaron and Moses Wilcox), twins and others of multiple birth. Parade, twins contests, talent show, entertainment, food, fun, games, exhibitions, Golf Tournament, twin research, arts and craft show and fireworks. Admission: $1 for non-twins (twins free), shuttle bus available. Est attendance: 92,500. For info: The Twins Days Committee, PO Box 29, Twinsburg, OH 44087. Phone: (216) 425-3652.

VIRGIN ISLANDS: NICOLE ROBIN DAY. Aug 4. Commemorates the safe return of the *Nicole Robin* and her crew on Aug 4, 1973, after a harrowing 23-day experience. Nicole Robin Day is celebrated by appropriate ceremonies and festivities throughout the Virgin Islands.

VIRGINIA STATE HORSE SHOW—QUARTER HORSE DIVISION. Aug 4–6. Fairgrounds on Strawberry Hill, Richmond, VA. Quarter horse competition plus youth and amateur clinic. Est attendance: 1,000. For info: Sue Mullins, Equine Dir, PO Box 26805, Richmond, VA 23261. Phone: (804) 228-3238.

WALLENBERG, RAOUL: BIRTH ANNIVERSARY. Aug 4. Swedish architect Raoul Gustaf Wallenberg was born at Stockholm, Sweden, on Aug 4, 1912. He was the second person in history (Winston Churchill was the first) to be voted honorary American citizenship (US House of Representatives 396-2, Sept 22, 1981). Little is known of his architectural achievements but he is credited with saving 100,000 Jews from almost certain death at the hands of the Nazis during World War II. Wallenberg was arrested by Soviet troops at Budapest, Hungary, on Jan 17, 1945, and according to the official Soviet press agency Tass, died in prison in 1947 (exact date undisclosed).

WEST VIRGINIA GLASS FESTIVAL. Aug 4–6. Oglebay, Wheeling, WV. Features dozens of glass artisans and glass-related crafts: candles, flowers, WV gourmet foods. Lots of entertainment and wonderful festival foods. Annually, the first weekend of August. Est attendance: 6,300. For info: John Hargleroad, Oglebay, Wheeling, WV 26003. Phone: (304) 243-4028.

WEST VIRGINIA SQUARE AND ROUND DANCE CONVENTION. Aug 4–6. Buckhannon, WV. Promotes the fun and fellowship of western-style square dancing, rounds, contra and clogging. Annually, the first full weekend in August. Sponsor: WV Square and Round Dance Federation. Est attendance: 1,000. For info: Frank and Sheila Landis, 175 Fairmor Dr, Westover, WV 26505.

WHITE OAK RENDEZVOUS. August 4–6. Deer River, MN. A reenactment of fur trade history placed in a setting of 80 acres of pasture and wilderness. 200 tipis and white canvas lodges placed around a NW Co Fur Post from 1798. Blacksmith shop, root cellar, clerk's quarters and store, birchbark canoe-building shed, Ojibwe village. Music, dance, puppetry and 1-mile long nature trail with plants labeled to explain their use in 1798 for medicine, shelter or food. 18th-century craft demonstrations and sales. Also, specialty foods from the fur trade period. Est attendance: 12,000. For info: Perry Vining, Rendezvous Bourgeois, Box 306, Deer River, MN 56636. Phone: (218) 246-9393.

WORLD FREEFALL CONVENTION. Aug 4–13. Qunicy, IL. Skydivers converge in Quincy and fill the skies with their brilliant colored parachutes. Spectators can enjoy the sights and take part in helicopter rides, bi-plane rides and hot air balloon rides. Those wishing to skydive may purchase a tandem skydive or take lessons and skydive on their own. Est attendance: 20,000. For info: Rob Ebbing, World Freefall Convention, 1515 Kentucky, Quincy, IL 62301. Phone: (217) 222-5867.

BIRTHDAYS TODAY

Wesley Addy, 82, actor (*Whatever Happened to Baby Jane?*, *Seconds*), born Omaha, NE, Aug 4, 1913.

Roger Clemens, 33, baseball player, born at Dayton, OH, Aug 4, 1962.

Elizabeth, the Queen Mother, 95 (Elizabeth Angela Marguerite), 95, born at Hertfordshire, England, Aug 4, 1900.

Maurice Richard, 74, hockey player, born at Montreal, Quebec, Canada, Aug 4, 1921.

Kristofer Tabori, 43, actor ("Seventh Avenue," "Chicago Story"), born at Los Angeles, CA, Aug, 4, 1952.

Helen Thomas, 75, journalist (long-time White House correspondent), born at Winchester, KY, Aug 4, 1920.

AUGUST 5 — SATURDAY
217th Day — Remaining, 148

AIKEN, CONRAD: BIRTH ANNIVERSARY. Aug 5. American poet, short-story writer, critic and Pulitzer Prize winner (poetry, 1916). He was born at Savannah, GA, on Aug 5, 1899, and died there Aug 17, 1973.

AMHERST'S TEDDY BEAR RALLY. Aug 5. Amherst, MA. Teddy bear contest, official bear appraiser, coloring event, Teddy Bear Hospital, UMass Oompah Band, Winnie-the-Pooh readings, and more than 165 teddy bear dealers and exhibits. Entertainment and refreshments; free. 13th annual rally. Est attendance: 20,000. For info: Amherst Rotary Club, PO Box 542, Amherst, MA 01004. Phone: (413) 256-8983.

BEARD, MARY R.: BIRTH ANNIVERSARY. Aug 5. American historian Mary Ritter Beard was born at Indianapolis, IN, on Aug 5, 1876. Many of her books were written in collaboration with her husband, Charles A. Beard. She died at Phoenix, AZ, on Aug 14, 1958.

CALIFORNIA DRY BEAN FESTIVAL. Aug 5–6. Tracy, CA. To promote the California dry bean industry, to raise funds for nonprofit organizations and to educate the public. Est attendance: 50,000. For info: Tracy Chamber of Commerce, Tom Hawkins, Coord, 223 E 10th St, Tracy, CA 95376. Phone: (209) 835-2131.

CANADA: RED DEER INTERNATIONAL AIR SHOW. Aug 5–6. Red Deer, Alberta. All eyes will be on the skies as the best in the airshow world perform at one of Canada's most spectacular aerial events. Est attendance: 50,000. For info: Red Deer Air Show Assn, #208, 4911 51st St, Canada T4N 6V4. Phone: (403) 886-5050.

CANADA: YORKTON THRESHERMEN'S SHOW AND SENIORS' FESTIVAL. Aug 5–6. Western Development Museum Grounds, Yorkton, Saskatchewan. Old-fashioned entertainment, arts and crafts; demonstrations of steam gas engines, threshing, stoking, bag-tying, clay-oven baking and antique cars. Est attendance: 8,000. For info: Tourism Saskatchewan, Saskatchewan Trade and Conv Center, 1919 Saskatchewan Dr, Regina, Canada S4P 3V7. Phone: (800) 667-7191.

Aug 5 ☆ Chase's 1995 Calendar of Events ☆

CHEROKEE DAYS OF RECOGNITION (WITH BLOWGUN CONTEST). Aug 5-6. Red Clay State Park, Cleveland, TN. Traditional and contemporary Cherokee dances, storytelling, Native American food, Cherokee arts and crafts and the annual blowgun contest are featured. Annually, the first weekend in August. Sponsor: Tennessee Dept of Environment and Conservation. Est attendance: 15,000. For info: Lois I. Osborne, Park Mgr, Red Clay State Hist Park, 1140 Red Clay Park Rd SW, Cleveland, TN 37311. Phone: (615) 478-0339.

CHICAGO PEACE AND MUSIC FESTIVAL. Aug 5-6. Cricket Hill, Chicago, IL. An awareness campaign focusing on social, humanitarian and environmental issues. Poetry readings, ecological exhibits. Jazz, blues, rock, pop, reggae and alternative bands will be featured. Est attendance: 15,000. For info: Chicago Peace and Music Festival, PO Box 477409, Chicago, IL 60647-7409. Phone: (312) 252-9150.

CIRCLE K INTERNATIONAL CONVENTION. Aug 5-9. Phoenix, AZ. Est attendance: 1,000. For info: Circle K Intl, 3636 Woodview Trace, Indianapolis, IN 46268-3196. Phone: (317) 875-8755.

CROP DAY. Aug 5. Greenwood, MS. Features cotton row tours—remembering when cotton was king. Civil War brigade encampment, cannon shoot, bed races, canoe races, hovercraft rally. Also art, crafts sales, entertainment and barbecue. Annually, the first Saturday in August. Est attendance: 15,000. For info: Janice Moor, Dir, Leflore County Chamber of Commerce, PO Box 848, Greenwood, MS 38930. Phone: (601) 453-4152.

DICK EVANS MEMORIAL CYCLE RACE. Aug 5-6. Oahu, HI. Three-stage cycling race—toad race, sprints and criterium. Open to the public. For info: Event Marketing, Inc., 1001 Bishop St, #477, Pauahi Tower, Honolulu, HI 96813. Phone: (808) 521-4322.

ELIOT, JOHN: BIRTH ANNIVERSARY. Aug 5. John Eliot, American "Apostle to the Indians," translator of the Bible into an Indian tongue (the first Bible to be printed in America), was born at Hertfordshire, England, on Aug 5, 1604. He died at Roxbury, MA, on May 21, 1690.

FESTIVAL OF NATIONS. Aug 5-13. Red Lodge, MT. A nine-day extravaganza. All events are free, following the philosophy of Festival founders that the Festival should be a fun, educational experience including information on cooking, crafts, customs, dances and languages. Annually, the first Saturday of August through the following weekend. Est attendance: 15,000. For info: Joan Cline, Exec Secretary, Red Lodge Chamber of Commerce, PO Box 988, Red Lodge, MT 59068. Phone: (406) 446-1718.

FIRST ENGLISH COLONY IN NORTH AMERICA: FOUNDING ANNIVERSARY. Aug 5. Sir Humphrey Gilbert, English navigator and explorer, aboard his sailing ship, the *Squirrel*, sighted the Newfoundland coast and took possession of the area around St. Johns harbor in the name of the Queen, Aug 5, 1583, thus establishing the first English colony in North America. Gilbert was lost at sea, in a storm off the Azores, on his return trip to England.

August 1995

S	M	T	W	T	F	S
		1	2	3	4	5
6	7	8	9	10	11	12
13	14	15	16	17	18	19
20	21	22	23	24	25	26
27	28	29	30	31		

FORGOTTEN PAST STEAM AND GAS SHOW. Aug 5-6. Benton, KY. Antique steam and gas engines, fully preserved and working. Crafts, flea market and swap meet. Annually, the first Saturday and Sunday in August. For info: Bill Luebker, US Hwy 68 East, Benton, KY 42025.

GRENADA: CARRIACOU REGATTA. Aug 5-8. Annual weekend of yacht and boat races, swimming, cultural shows including Big Drum dancing. Est attendance: 2,000. For info: Grenada Tourist Info Office, 820 Second Ave, Ste 900-D, New York, NY 10017. Phone: (800) 927-9554.

HOME OF THE HAMBURGER CELEBRATION. Aug 5. Seymour, WI. Giant parade, Bun Run, live music, hamburger cook-off, Hamburger Olympics. Est attendance: 10,000. For info: Home of the Hamburger Inc, PO Box 173, Seymour, WI 54165. Phone: (414) 833-6800.

INTERNATIONAL FESTIVAL WEEK. Aug 5-13. Calais, ME, and St. Stephen, New Brunswick, Canada. Festival of international cooperation between St. Stephen and Calais with the theme "Hands across the Border." Celebrating the friendship of two countries joining as one. Est attendance: 14,000. For info: Intl Fest Committee, PO Box 773, Calais, ME 04619. Phone: (207) 454-3216.

JONATHAN HAGER FRONTIER CRAFT DAYS. Aug 5-6. Jonathan Hager House, City Park, Key Street, Hagerstown, MD. Featuring dozens of demonstrating craftsmen, great bluegrass music and great food. Annually, the first weekend in August. Est attendance: 10,000. For info: John Nelson, Jonathan Hager House and Museum, 19 Key St, Hagerstown, MD 21740. Phone: (301) 739-8393.

LEAGUE OF NEW HAMPSHIRE CRAFTSMEN'S FAIR. Aug 5-13. Mt Sunapee State Park, Sunapee, NH. "America's oldest crafts fair"—62nd annual fair. More than 150 crafts booths, Living with Crafts exhibit, daily performing arts, children's activities and more. Admission: $7 adults, $4 seniors. Phone: (800) 639-1610 or (603) 224-3375. Est attendance: 53,000. For info: League of NH Craftsmen, 205 N Main St, Concord, NH 03301.

LYNCH, THOMAS: BIRTH ANNIVERSARY. Aug 5. Signer, Declaration of Independence, born Prince George's Parish, SC, Aug 5, 1749. Died 1779 (lost at sea, exact date of death unknown).

MUD VOLLEYBALL TOURNAMENT. Aug 5. Rockford, IL. More than 300 teams from the Midwest participate in the world's largest one-day, one-site mud volleyball tournament. Benefit for Epilepsy Association. Est attendance: 5,000. For info: Epilepsy Assn, 321 W State St, Ste 208, Rockford, IL 61101. Phone: (815) 964-2689.

NATIONAL CERTIFIED REGISTERED NURSE ANESTHETIST WEEK. Aug 5-10. To provide recognition for the nation's certified registered nurse anesthetists (CRNAs) who administer more than 65% of the anesthesia in the US each year. Sponsor: Amer Assn of Nurse Anesthetists, Betty Colitti-Stuffers, PR Dir, 222 S Prospect, Park Ridge, IL 60068.

NATIONAL MUSTARD DAY. Aug 5. Mustard lovers across the nation pay tribute to the condiment of kings—and the king of condiments—by lathering their favorite spread on hot dogs, pretzels, and even ice cream (an acquired taste)! The Mount Horeb Mustard Museum ("If you collect us, they will come") contains the world's largest collection of prepared mustards and mustard memorabilia. Annually, Aug 5. Est attendance: 400. Sponsor: The Mount Horeb Mustard Museum, Barry M. Levenson, Curator, 109 E Main St, Mount Horeb, WI 53572. Phone: (608) 437-3986.

OKLAHOMA ALL NIGHT SINGING. Aug 5. Veterans' Memorial Park, Konawa, Oklahoma. Famous quartets from across the nation entertain thousands in this 32nd annual Gospel Music Homecoming! An Oklahoma summer tradition, campers begin arriving on Thursday and Friday, others bring blankets and picnic baskets. The park is usually filled by noon on Saturday for the evening presentation that concludes Gospel Music

Week. Food available, parking and admission free. Est attendance: 20,000. Sponsor: Oklahoma All-Night Gospel Singing Enterprises, Inc, Konawa, OK 74849. Phone: (405) 925-3434.

OWENSBORO SHOW. Aug 5-6. Owensboro, KY. To promote the art and culture of Prehistoric Native America and to display outstanding collections of relics; buying, selling and trading. Est attendance: 3,000. For info: Kathy Gerber, Box 7, Tell City, IN 47586. Phone: (812) 547-4881.

PENNSYLVANIA RENAISSANCE FAIRE. Aug 5–Oct 8. Cornwall, PA. Recreation of 16th-century Elizabethan village. Lords and ladies, mongers, merchants, jousting, human chess match, medieval foods and crafts. Faire takes place Saturday–Monday from Aug 5 until Labor Day; Saturdays and Sundays only thereafter. Est attendance: 150,000. For info: Mount Hope Estate and Winery, PO Box 685, Cornwall, PA 17016. Phone: (717) 665-7021.

PONY GIRLS SOFTBALL NATIONAL CHAMPIONSHIPS. Aug 5-7. (Tentative.) Site TBA. Slow and fast pitch National Championship for girls ages 9 and 18. Annually, beginning the first weekend in August. Fast pitch phone: Marilyn Fisher, (815) 726-1836. Slow pitch phone: Lynn Parnell, (817) 292-9391. Est attendance: 10,700. Sponsor: Pony Baseball, PO Box 225, Washington, PA 15301.

STREETSCENE '95. Aug 5. Covington, VA. Car Show, open to all types of vehicles! 10 AM-4 PM. Entertainment throughout the day. Annually, the first Saturday in August. Est attendance: 12,500. For info: Kars Unlimited, Inc, PO Box 851, Covington, VA 24426. Phone: (703) 962-3642.

SURFSIDE'S INTERNATIONAL FIESTA. Aug 5. Surfside, FL. To salute our international visitors. Pool party, ethnic foods, outdoor dance music of all types by live band. Games and prizes for children and adults. Arts and crafts displays. Est attendance: 2,000. Sponsor: Surfside Tourist Bd, 9301 Collins Ave, Surfside, FL 33154. Phone: (305) 864-0722.

WALES: ROYAL NATIONAL EISTEDDFOD. Aug 5-12. Colwyn Bay, North Wales. The Royal National Eisteddfod of Wales (Eisteddfod Genedlaethol Frenhinol Cymru) is a cultural event with competitive festivals of music, drama, literature, arts and crafts. All events conducted in Welsh with simultaneous translation into English available. Est attendance: 160,000. For info: Royal Natl Eisteddfod, 40 Parc Ty Glas, Llanishen, Cardiff, South Glamorgan, Wales, CF4 5WU.

WYATT EARP HOMECOMING WESTERN DAYS. Aug 5. Monmouth, IL. Luncheons with "Wyatt Earp." OK Corral Reenactment and Life and Wyatt Earp Shows. Birthplace tours, Pioneer Cemetery tours, stagecoach and pony rides, music, exhibitors. Phone for Chamber of Commerce: (309) 734-3181. Est attendance: 250. For info: Wyatt Earp Birthplace Historic House Museum, c/o 1020 E Detroit Ave, Monmouth, IL 61462. Phone: (309) 734-6419.

BIRTHDAYS TODAY

Loni Anderson, 49, actress ("WKRP in Cincinnati," *The Jayne Mansfield Story*), born St. Paul, MN, Aug 5, 1946.
Neil Alden Armstrong, 65, former astronaut, first man to walk on moon, born at Wapakoneta, OH, Aug 5, 1930.
Patrick Ewing, 33, basketball player, born at Kingston, Jamaica, Aug 5, 1962.
Samantha Sang, 42, singer ("Emotion"), born at Melbourne, Australia, Aug 5, 1953.
Erika Slezak, 49, actress ("One Life to Live"), born at Los Angeles, CA, Aug 5, 1946.

Sammi Smith, 52, singer ("Help Me Make It Through the Night"), born Orange, CA, Aug 5, 1943.

AUGUST 6 — SUNDAY
218th Day — Remaining, 147

AMERICAN FAMILY DAY. Aug 6. Observed on the first Sunday in August. The observance date is designated by statute in Arizona and Michigan.

BALL, LUCILLE: BIRTH ANNIVERSARY. Aug 6. Film and television pioneer and comedian. Born at Butte, MT, on Aug 6, 1911. In addition to her many other film and television credits, Lucille Ball always will be remembered for her role in the 1950s CBS sitcom *I Love Lucy*. As Lucy Riccardo, the wife of band leader Ricky Riccardo (her real-life husband Desi Arnaz), her comedic style became a trademark of early television comedy. Ball appeared in three additional sitcoms. She became the first woman to head a major motion picture and TV studio when she purchased Arnaz's share of Desilu Productions. Ball and Arnaz were divorced in 1960; she later married Gary Morton. She died on Sept 26, 1989.

BOLIVIA: INDEPENDENCE DAY. Aug 6. National holiday. Gained freedom from Spain, 1825.

BURRO RACE. Aug 6. Leadville, CO. For info: Chamber of Commerce, Box 861, Leadville, CO 80461. Phone: (719) 486-3900.

CANADA: DIGBY SCALLOP DAYS FESTIVAL. Aug 6-13. Digby, Nova Scotia. A celebration of our sea-faring heritage, honoring the fishermen of the world famous Digby Scallop Fleet. Savour the marvelous mollusks of the deep. Cheer the scallop shuckers in competition. Enjoy local talent. Take in our grand street parade and children's events. Annually, the first week in August. Est attendance: 12,000. For info: Digby Scallop Days Assn, PO Box 983, Digby, NS, Canada B0V 1A0.

CANADA: FOLKLORAMA—CANADA'S CULTURAL CELEBRATION. Aug 6-19. Winnipeg, Manitoba. Over 40 pavilions representing various cultures offer food, entertainment and cultural displays. Begun in 1970, Folklorama has become the largest multicultural festival of its kind in the world. US Toll-free phone: (800) 665-0234. Est attendance: 500,000. For info: Claudette Leclerc, Exec Dir, Folklorama, 375 York Ave, Winnipeg, Man, Canada R3C 3J3. Phone: (204) 982-6221.

CANOE JOUST. Aug 6. Sturgis Park, Snow Hill, MD. Three-person canoe jousting teams attempt to knock opponents from their canoe platforms in this wet and wild event. Noon-5. Annually, the first Sunday of August. For info: Kathy Fisher, PO Box 207, Snow Hill, MD 21863.

CELEBRATION OF PEACE DAY. Aug 6. Library Lawn, Concord, MA. To bring people of all shades of opinion together in unity to reflect on peace and justice, and to celebrate life. Annually, the first Sunday in August. Est attendance: 150. Sponsor: Natl Peace Day Celebrations, Inc, Marie M. Strain, Founder and Pres, 93 Pilgrim Rd, Concord, MA 01742. Phone: (508) 369-3751.

CHERRY RIVER FESTIVAL. Aug 6-13. Richwood, WV. A land-locked Navy of Admirals parade as a climax to this week-long world-famous event which features pageants, contests, athletic events, parades and dances. Est attendance: 4,000. For info: Maxine Corbett, Board of Directors, Cherry River Festival, Box 482, Richwood, WV 26261. Phone: (304) 846-6790.

ELECTROCUTION FIRST USED TO CARRY OUT DEATH PENALTY: ANNIVERSARY. Aug 6. On Aug 6, 1890, at Auburn Prison, Auburn, NY, William Kemmler of Buffalo, NY, became the first man to be executed by electrocution. He had been convicted of the hatchet murder of his common-law wife, Matilde Ziegler, on Mar 28, 1889. This first attempt at using electrocution to carry out the death penalty was a botched affair. As reported by George Westinghouse, Jr, "It has been a brutal affair. They could have done better with an axe."

Aug 6 ☆ Chase's 1995 Calendar of Events ☆

◆ **FIRST ATOMIC BOMB DROPPED ON INHABITED AREA: 50th ANNIVERSARY.** Aug 6. Hiroshima, Japan. At 8:15 AM, local time, on Aug 6, 1945, an American B-29 bomber, the *Enola Gay*, dropped an atomic bomb named "Little Boy" over the center of the city of Hiroshima, Japan. The bomb exploded about 1,800 feet above the ground, killing more than 105,000 civilians and destroying the city. It is estimated that another 100,000 persons were injured and died subsequently as a direct result of the bomb and the radiation it produced. This was the first time in history that such a devastating weapon had been used by any nation.

FRIENDSHIP DAY. Aug 6. A day to heed the advice of 18th-century English philosopher Samuel Johnson, "A man should keep his friendships in constant repair." A time to acknowledge old friends and celebrate new ones. For info: Linda Gorin, The Best To You, 7920 Silverton Ave #F, San Diego, CA 92126. Phone: (619) 578-2740.

GREAT CARDBOARD BOAT REGATTA. Aug 6. Leon, IA. Person-powered water craft are designed, built of corrugated cardboard and raced over 200-yd course. Trophies awarded for 3 places in 4 divisions, plus 6 special awards. Est attendance: 4,500. For info: Dennis Moon, RR 1, Box 18-B, Leon, IA 50144. Phone: (515) 446-7859.

◆ **HIROSHIMA DAY: 50 YEARS.** Aug 6. Memorial observances in many places for victims of first atomic bombing of populated place, which occurred at Hiroshima, Japan, Aug 6, 1945.

INTERNATIONAL AEROBATIC CHAMPIONSHIPS. Aug 6-11. Start your day at the airport on Sunday with pancake breakfast and plane rides. Opening ceremonies, parade of nations and Fond du Lac Cup Sunday at 2. Week-long competition follows featuring free forums and exhibits at FDL County Airport. Phone: (800) 937-9123, ext 95. For info: Fond du Lac Conv Bureau, 19 W Scott St, Fond du Lac, WI 54935.

ITALY: JOUST OF THE QUINTANA. Aug 6. Ascoli/Piceno. The first Sunday in August is set aside for the Torneo della Quintana, an historical pageant with 15th-century costumes.

◆ **JAPAN: PEACE FESTIVAL: 50 YEARS.** Aug 6. Hiroshima, Japan. The festival held annually at Peace Memorial Park is observed in memory of the victims of the Aug 6, 1945, atomic bomb explosion there.

JUDGE CRATER DAY. Aug 6. Anniversary of mysterious disappearance at age 41, on Aug 6, 1930, of Joseph Force Crater, justice of the New York State Supreme Court. Never seen or heard from after disappearance on this date. Declared legally dead in 1939.

NATIONAL PSYCHIATRIC TECHNICIAN WEEK. Aug 6-12. To honor psychiatric technicians who work in hospital ward settings, caring for patients in psychiatric hospitals. Sponsor: Ellen Adcock, Dir Special Projects, Lakeshore Mental Health Institute, 5908 Lyons View Dr, Knoxville, TN 37919. Phone: (615) 584-1561.

NEARING, SCOTT: BIRTH ANNIVERSARY. Aug 6. American sociologist, antiwar crusader, back-to-the-land advocate and author, with his wife Helen, of *Living the Good Life* (1954). He was born at Morris Run, PA, Aug 6, 1883. He died a century later at his farm at Harborside, ME, on Aug 24, 1983.

OLD-FASHIONED ICE CREAM SOCIAL. Aug 6. Galloway House and Village, Fond du Lac, WI. Generous portions of ice cream served at old-fashioned prices. Turn-of-the-century village, 30-room Victorian mansion. Annually, the first Sunday in August. Est attendance: 1,000. Sponsor: Fond du Lac County Historical Soc, 332 14th St, Fond du Lac, WI 54935. Phone: (414) 922-6390.

PYROTECHNICS GUILD INTERNATIONAL FIREWORKS CONVENTION. Aug 6-11. (Tentative.) Stevens Point, WI. The world's finest fireworks are displayed and members compete for the title of Grand Master during this annual international pyrotechnic convention. The Pyrotechnics Guild promotes the design, construction and safe display of the highest quality pyrotechnics. Annually, the first week after the first Friday in August. Est attendance: 2,000. Sponsor: Pyrotechnics Guild International Inc, Mark Wray, 11144 Clare Ave, Northridge, CA 91326-2333. Phone: (818) 363-6277.

SCOTTISH BAGPIPE AND HIGHLAND DANCE FESTIVAL. Aug 6. Lake Waramaug, New Preston, CT. The festival will feature pipe and drum corps, folkdancers, authentic costumes and a Sunday brunch. Annually, the first Sunday in August. Est attendance: 1,000. For info: The Inn on Lake Waramaug, Nancy Conant, 107 N Shore Rd, New Preston, CT 06777. Phone: (203) 868-0563.

SPACE MILESTONE: *VOSTOK 2* (USSR). Aug 6. Launched on Aug 6, 1961, Gherman Titov orbited Earth 17 times over period of 25 hours, 18 minutes. Titov broadcast messages in passage over countries, controlled spaceship manually for two hours.

TENNYSON, ALFRED LORD: BIRTH ANNIVERSARY. Aug 6. English poet born at Somersby, Lincolnshire, England, Aug 6, 1809. Appointed English poet laureate in succession to William Wordsworth. Died at Aldworth, England, Oct 6, 1892.

TISHA B'AV or FAST OF AB. Aug 6. Hebrew calendar date: Ab 10, 5755 (because Ab 9 falls on the Sabbath in 1995). Commemorates and mourns the destruction of the first and second Temples in Jerusalem (586 BC and AD 70).

VOLKSFEST. Aug 6. New Glarus, WI. Celebration of Swiss Independence Day. Annually, the first Sunday in August. Est attendance: 2,000. For info: Volksfest, Box 713, New Glarus, WI 53574. Phone: (608) 527-2095.

◆ **VOTING RIGHTS ACT OF 1965 SIGNED: 30th ANNIVERSARY.** Aug 6, 1965. Signed into law by Pres Lyndon Johnson, the Voting Rights Act of 1965 was designed to thwart attempts to discriminate against minorities at the polls. The act suspended literacy and other disqualifying tests, authorized appointment of federal voting examiners and provided for judicial relief on the federal level to bar discriminatory poll taxes. Congress has voted to extend the Act in 1975, 1984 and 1991.

W.C. HANDY FESTIVAL. Aug 6-12. Florence, AL. A weeklong street-strutting, toe-tapping and hand-clapping celebration of the musical heritage of Florence native W.C. Handy—"the father of the blues"—culminating in a spectacular Saturday evening concert. Est attendance: 110,000. For info: Music Preservation Soc, PO Box 1827, Florence, AL 35631. Phone: (205) 766-7642.

BIRTHDAYS TODAY

Peter Bonerz, 57, actor ("Bob Newhart Show," "9 to 5"), director, born at Portsmouth, NH, Aug 6, 1938.

Soleil Moon Frye, 19, actress ("Punky Brewster"), born at Glendora, CA, Aug 6, 1976.

Catherine Hicks, 44, actress (Marilyn Monroe in TV movie *Marilyn*; *Peggy Sue Got Married*), born Scottsdale, AZ, Aug 6, 1951.

Freddie Laker, 73, former airline executive, born at Canterbury, England, Aug 6, 1922.

Robert Mitchum, 78, actor (*The Night of the Hunter*, *The Story of G.I. Joe*), born Bridgeport, CT, Aug 6, 1917.

David Maurice Robinson, 30, basketball player, born at Key West, FL, Aug 6, 1965.

August 1995

S	M	T	W	T	F	S
		1	2	3	4	5
6	7	8	9	10	11	12
13	14	15	16	17	18	19
20	21	22	23	24	25	26
27	28	29	30	31		

☆ Chase's 1995 Calendar of Events ☆ Aug 7

AUGUST 7 — MONDAY
219th Day — Remaining, 146

ANTIGUA AND BARBUDA: AUGUST MONDAY. Aug 7–8. The first Monday in August and the day following form the August Monday Public Holiday in Antigua and Barbuda.

AUSTRALIA: PICNIC DAY. Aug 7. The first Monday in August is a bank holiday in New South Wales and Picnic Day in Northern Territory, Australia.

BAHAMAS: EMANCIPATION DAY. Aug 7. Public holiday in Bahamas. Annually, the first Monday in August.

BUNCHE, RALPH JOHNSON: BIRTH ANNIVERSARY. Aug 7. American statesman, UN official, Nobel Peace prize recipient, born at Detroit, MI, Aug 7, 1904. Died Dec 9, 1971.

BUSH ORDERS COMMENCEMENT OF DESERT SHIELD: 5th ANNIVERSARY. Aug 7. On Aug 7, 1990, five days after the Iraqi invasion of Kuwait, President George Bush ordered the military buildup that would become known as Desert Shield, to prevent further Iraqi invasion.

CANADA: CIVIC HOLIDAY. Aug 7. The first Monday in August is observed as a holiday in seven of Canada's 10 provinces. Civic Holiday in Manitoba, New Brunswick, Northwest Territories, Ontario and Saskatchewan; British Columbia Day in British Columbia; and Heritage Day in Alberta.

COLORADO DAY. Aug 7. Colorado. Annually, the first Monday in August. Commemorates Admission Day, Aug 1, 1876, when Colorado became the 38th state.

DON'T WAIT—CELEBRATE! WEEK. Aug 7–13. To encourage frequent festivities acknowledging small but significant accomplishments such as team wins, good grades, completed projects, new neighbors, braces off, balanced checkbooks. Gathering for these mini-galas will enhance and nurture relationships while raising the self-esteem of the honorees. Annually, the second week in August. Sponsor: Celebration Creations, 7817 67th Ave N, Minneapolis, MN 55428. Phone: (612) 879-4592.

FIRST PICTURE OF EARTH FROM SPACE: ANNIVERSARY. Aug 7. US satellite *Explorer VI* transmitted the first picture of Earth from space Aug 7, 1959, and for the first time man had a likeness of his planet based on more than Earth measurements, projections and conjectures.

GRENADA: EMANCIPATION DAY. Aug 7. Grenada observes public holiday annually on the first Monday in August.

GULF OF TONKIN RESOLUTION: ANNIVERSARY. Aug 7. Congress, on Aug 7, 1964, approved the "Gulf of Tonkin Resolution," pertaining to the war in Vietnam, which gave President Lyndon Johnson authority "to take all necessary measures to repel any armed attack against the forces of the United States and to prevent further aggression."

HALFWAY POINT OF SUMMER. Aug 7. At 11:23 AM, EDT, on Aug 7, 1995, 46 days, 19 hour and 49 minutes will have elapsed and the equivalent will remain before 8:13 AM EDT, Sept 22, 1995, the autumnal equinox and the beginning of autumn.

ICELAND: SHOP AND OFFICE WORKERS' HOLIDAY. Aug 7. In Iceland an annual holiday for shop and office workers is observed on the first Monday in August.

ISING, RUDOLF C.: BIRTH ANNIVERSARY. Aug 7. Rudolf C. Ising, co-creator with Hugh Harmon of "Looney Tunes" and "Merrie Melodies," was born Aug 7, 1903, at Kansas City, MO. Ising and Harmon's initial production, *Bosko the Talk-Ink Kid* (1929), was the first talkie cartoon synchronizing dialogue on the soundtrack with the action on screen. Ising received an Academy Award in 1948 for *Milky Way*, a cartoon about three kittens. Ising, who emphasized the musical element of cartoons, was the primary developer of "Merrie Melodies." His greatest strengths were in writing and producing rather than illustration. During WWII he headed the animation division for the Army Air Forces movie unit developing training films. Rudolf Ising died July 18, 1992, at Newport Beach, CA.

JAMAICA: INDEPENDENCE DAY. Aug 7. National holiday observing achievement of Jamaican independence on Aug 6, 1962. Annually, the first Monday in August.

KID'S SWAP SHOP. Aug 7. Broadway Market Square, Baltimore, MD. For children to trade any used item (games, puzzles, toys, books) they no longer wish to keep. Youngsters 14 years of age and under permitted to participate. Est attendance: 250. For info: Office of Adventures and Fun, Clarence "Du" Burns Arena, 1301 S Ellwood Ave, Baltimore, MD 21224. Phone: (410) 396-9177.

NATIONAL SMILE WEEK. Aug 7–13. "Share a smile and it will come back to you, bringing happiness to you and the giver." Annually, the first Monday in August through the following Sunday. Fax: (210) 434-6473. Sponsor: Heloise, Newspaper Columnist, Box 795000, San Antonio, TX 78279.

PGA CHAMPIONSHIP. Aug 7–13. Riviera Country Club, Pacific Palisades, CA. The 77th championship competition conducted by the Professional Golfers' Association of America. Fax: (918) 491-6300. For ticket info: (800) PGA-IN95. Est attendance: 180,000. For info: PGA Championship, 1250 Capri Dr, Pacific Palisades, CA 90272. Phone: (310) 573-7780.

PSYCHIC WEEK. Aug 7–11. To utilize the power of the psyche to bring peace, find lost individuals and concentrate "psychic power" on beneficial causes. Annually, the first week in August (Mon–Fri). [Created by the late Richard R. Falk.]

PURPLE HEART: ANNIVERSARY. Aug 7. At Newburgh, NY, on Aug 7, 1782, General George Washington ordered the creation of a Badge of Military Merit. The badge consisted of a purple cloth heart with silver braided edge. Only three are known to have been awarded during the Revolutionary War. The award was reinstituted on the bicentennial of Washington's birth, Feb 22, 1932, and recognizes those wounded in action.

SCOTLAND: SUMMER BANK HOLIDAY. Aug 7. Bank and public holiday in Scotland.

STURGIS RALLY AND RACES. Aug 7–13. Sturgis, SD. The grandaddy of all motorcycle Rally and Races. For 55 years the small community of Sturgis has welcomed motorcycle enthusiasts from around the world to a week of varied cycle racing, tours of the beautiful Black Hills, old and new cycles at the National Motorcycle Museum, trade shows and thousands of bikes on display. Annually, beginning the Monday after the first full weekend in August. Est attendance: 150,000. For info: Sturgis Rally & Races, Inc, PO Box 189, Sturgis, SD 57785. Phone: (605) 347-6570.

US WAR DEPARTMENT: ESTABLISHMENT ANNIVERSARY. Aug 7. The second presidential cabinet department. The War Department was established by Congress on Aug 7, 1789.

US WOMEN'S AMATEUR (GOLF) CHAMPIONSHIP. Aug 7–12. The Country Club, Brookline, MA. Sponsor: US Golf Assn, Championship Dept, Golf House, Far Hills, NJ 07931. Phone: (908) 234-2300.

ZAMBIA: YOUTH DAY. Aug 7. First Monday in August. National holiday. Youth activities are order of the day. Focal point is Lusaka's Independence Stadium.

BIRTHDAYS TODAY

Lana Cantrell, 52, singer, actress, born at Sydney, Australia, Aug 7, 1943.

Edwin W. Edwards, 68, Governor of Louisiana (D), born at Marksville, LA, Aug 7, 1927.

Aug 7–9 ☆ *Chase's 1995 Calendar of Events* ☆

Stan Freberg, 69, satirist, born at Pasadena, CA, Aug 7, 1926.
Garrison Keillor, 53, producer (host "The Prairie Home Companion"), writer (*Lake Wobegon Days*), born Anoka, MN, Aug 7, 1942.
Alberto Salazar, 38, marathon runner, born at Havana, Cuba, Aug 7, 1957.
B.J. Thomas (Billy Joe Thomas), 53, singer ("Hooked on a Feeling," "Raindrops Keep Falling on My Head"), born Houston, TX, Aug 7, 1942.

AUGUST 8 — TUESDAY
220th Day — Remaining, 145

BAHAMAS: FOX HILL DAY. Aug 8. Nassau. Annually, the second Tuesday in August.

BONZA BOTTLER DAY™. Aug 8. (Tenth anniversary today.) To celebrate when the number of the day is the same as the number of the month. Bonza Bottler Day is an excuse to have a party at least once a month. T-shirts available. For info: Elaine Fremont, 203 Waddell Rd, Taylors, SC 29687. Phone: (803) 244-2023.

CANADA: THE FLIGHT BEGINS. Aug 8–13. Abbotsford, British Columbia. During the week preceding the Airshow, local businesses take flight with decorations depicting aeronautical theme. The entire downtown area comes with airshow frivolity including Flight Begins Breakfast, Flight Begins Bus Tour, First Night Spectacular featuring Tug-of-War, old-fashioned picnic with Wacky Waiters Race, Abbotsford Concert Band and mascots and a Hot Air Balloon Festival and Paper Jet Competition. Est attendance: 14,000. For info: Abbotsford-Matsqui Chamber of Commerce, 2462 McCallum Rd, Abbotsford, BC, Canada V2S 3P9, BC.

HENSON, MATTHEW A.: BIRTH ANNIVERSARY. Aug 8. American black explorer, born Aug 8, 1866, at Charles County, MD. Orphaned as a youth, Henson, at the age of 12, served as a cabin boy on the sailing ship *Katie Hines*. He met Robert E. Peary while working in a Washington, DC, store in 1888, and was hired to be Peary's valet. He accompanied Peary on his seven subsequent Arctic expeditions. During the successful 1908–1909 expedition to the North Pole, Henson and two of the four Eskimo guides reached their destination on Apr 6, 1909. Peary arrived minutes later and verified the location. Henson's account of the expedition, *A Negro Explorer at the North Pole* was published in 1912. In addition to the Congressional medal awarded all members of the North Pole expedition, Henson also received the Gold Medal of the Geographical Society of Chicago and, at 81, was made an honorary member of the Explorers Club in New York, NY.

INTERTRIBAL INDIAN CEREMONIAL. Aug 8–13. Red Rock State Park, Gallup, NM. A major Indian festival with more than 50 tribes from the US and Mexico. Parades, Indian dances, rodeos, arts and crafts and foods. Est attendance: 32,500. For info: Intertribal Indian Ceremonial Assn, Box 1, Church Rock, NM 87311. Phone: (800) 233-4528.

August 1995

S	M	T	W	T	F	S
		1	2	3	4	5
6	7	8	9	10	11	12
13	14	15	16	17	18	19
20	21	22	23	24	25	26
27	28	29	30	31		

MONDALE, JOAN ADAMS: 65th BIRTHDAY. Aug 8. Wife of 42nd vice president of the US, Walter F. Mondale, born at Eugene, OR, on Aug 8, 1930. Married Dec 27, 1955.

◆ **MORRIS, ESTHER HOBART McQUIGG: BIRTH ANNIVERSARY.** Aug 8, 1814. Esther Hobart McQuigg Morris was born at Tioga County, NY, but eventually moved to the Wyoming Territory, where she worked in the women's rights movement and had a key role in getting a women's suffrage bill passed. Morris became justice of the peace of South Pass City, WY, in 1870, one of the first times a woman held public office in the US. She represented Wyoming at the national suffrage convention in 1895. She died Apr 2, 1902, at Cheyenne, WY.

ODIE: 17th BIRTHDAY. Aug 8. Commemorates the birthday of Odie, Garfield's sidekick, who first appeared in the "Garfield" comic strip on Aug 8, 1978. For info: Paws, Inc., Kim Campbell, 5440 E. Co. Rd., 450 N, Albany, IN 47320.

SHILTS, RANDY: BIRTH ANNIVERSARY. Aug 8, 1951. Journalist known for his reporting on the AIDS epidemic. One of the first openly homosexual journalists to work for a mainstream newspaper and the author of *And the Band Played On: Politics, People and the AIDS Epidemic*. Born at Davenport, IA, and died at Guerneville, CA, on Feb 17, 1994.

SNEAK SOME ZUCCHINI ONTO YOUR NEIGHBORS' PORCH NIGHT. Aug 8. Due to overzealous planting of zucchini, citizens are asked to drop off baskets of the squash on neighbors' doorsteps. Annually, Aug 8. Phone: (212) 388-8673 or (717) 274-8451. For info: Wellness Permission League, Tom or Ruth Roy, 2105 Water St, Lebanon, PA 17046.

SPACE MILESTONE: *PIONEER VENUS* MULTIPROBE (US). Aug 8. Second craft in *Pioneer Venus* program. Split into five and probed Venus atmosphere Dec 9. Launched on Aug 8, 1978.

BIRTHDAYS TODAY

Beatrice, 7, Princess of York, born at London, England, Aug 8, 1988.
Keith Carradine, 45, actor (*Nashville, Will Rogers Follies*), singer, born at San Mateo, CA, Aug 8, 1950.
Benny Carter, 88, jazz musician, composer, born at New York, NY, Aug 8, 1907.
Dino DeLaurentiis, 76, producer, born at Torre Annunziata, Italy, Aug 8, 1919.
The Edge, 34, musician, born at Dublin, Ireland, Aug 8, 1961.
Dustin Hoffman, 58, actor (Oscars *Rain Man, Kramer vs Kramer; The Graduate, Midnight Cowboy, Hero*), born Los Angeles, CA, Aug 8, 1937.
Joan Adams Mondale, 65, author, art patron, born at Eugene, OR, Aug 8, 1930.
Edward T. Schafer, 49, Governor of North Dakota (R), born at Bismark, ND, Aug 8, 1946.
Connie Stevens, 57, actress (Cricket Blake on "Hawaiian Eye"), born at Brooklyn, NY, Aug 8, 1938.
Mel Tillis, 63, singer, songwriter, born at Pahokee, FL, Aug 8, 1932.
Esther Williams, 72, actress, swimmer (*Take Me Out to the Ball Game, Dangerous When Wet*), born Los Angeles, CA, Aug 8, 1923.

AUGUST 9 — WEDNESDAY
221st Day — Remaining, 144

INDIANA STATE FAIR. Aug 9–20. Indiana State Fairgrounds Event Center, Indianapolis, IN. Top-rated livestock exhibition, world-class harness racing, top country music, giant Midway and Pioneer Village. Est attendance: 690,000. For info: Jeff Fites, Publicity Coord, Indiana State Fair, 1202 E 38th St, Indianapolis, IN 46205-2869. Phone: (317) 927-7500.

◆ **JAPAN: MOMENT OF SILENCE: 50 YEARS.** Aug 9. Nagasaki. Memorial observance held at Peace Memorial Park for victims of second atomic bombing (Nagasaki, Aug 9, 1945).

☆ Chase's 1995 Calendar of Events ☆ Aug 9–10

JAPANESE REJECT SURRENDER: 50th ANNIVERSARY. Aug 9, 1945. At the same time the atomic bomb was being dropped on the city of Nagasaki, the Japanese Supreme War Direction Council was meeting in Tokyo to discuss the Potsdam Declaration, which called for unconditional surrender. The six generals were evenly divided. Foreign Minister Shigenori Togo and Prime Minister Admiral Suzuki were for surrender. Minister of War General Anami was adamant in his opposition to surrender. An impasse resulted even though news of the devastation at Nagasaki reached the gathering.

KAHOKA FESTIVAL OF BLUEGRASS MUSIC. Aug 9–12. Kahoka, MO. Top Bluegrass bands, clogging workshops and shows, Fiddlers Frolic, jam sessions, Miss Kahoka Pageant, clogging contest and old fashioned square dance. 23rd annual festival. Est attendance: 4,000. For info: Delbert Spray, Dir, RR 1, Kahoka, MO 63445. Phone: (314) 853-4344.

MUSLIM FESTIVAL: BIRTH OF PROPHET MUHAMMAD. Aug 9–10. Mulid al-Nabi (Birth of the Prophet Muhammad) is observed on Muslim calendar date, Rabi al-Awal 12, 1416. Different methods for calculating the Gregorian date are used by different Muslim sects or groups. Chase's dates are based on astronomical calculations of the visibility of the new moon crescent at Mecca. EST date may vary.

MUSTANG LEAGUE WORLD SERIES. Aug 9–15. International youth baseball World Series for players league age 9 and 10. Est attendance: 3,000. Sponsor: Pony Baseball, Inc, PO Box 225, Washington, PA 15301. Phone: (412) 225-1060.

OLD FIDDLERS' CONVENTION. Aug 9–12. Galax, VA. Dance, folksongs, oldtime and bluegrass music competition. Est attendance: 35,000. For info: Edward F. Carico, Box 655, Galax, VA 24333. Phone: (703) 236-8541.

PALOMINO LEAGUE WORLD SERIES. Aug 9–15. (Tentative.) Greensboro, NC. International young adult baseball World Series for players of league ages 17 and 18. Annually, beginning in the second week of August. Est attendance: 6,650. Sponsor: Pony Baseball, Inc, Box 225, Washington, PA 15301. Phone: (412) 225-1060.

PRESIDENTIAL RESIGNATION: ANNIVERSARY. Aug 9. Effective at noon, Aug 9, 1974, the resignation from the presidency of the US by Richard Milhous Nixon had been announced in a speech to the American people on Thursday evening, Aug 8. Nixon, under threat of impeachment as a result of the Watergate scandal, became the first person to resign the presidency.

SECOND ATOMIC BOMB DROPPED ON INHABITED AREA: 50th ANNIVERSARY. Aug 9. Nagasaki, Japan. On Aug 9, 1945, three days after the atomic bombing of Hiroshima, an American B-29 bomber named *Bock's Car* left its base on Tinian Island carrying a plutonium bomb nicknamed "Fat Man." Its target was the Japanese city of Kokura, but because of clouds and poor visibility the bomber headed for a secondary target, Nagasaki, where at 11:02 AM, local time, it dropped the bomb, killing an estimated 70,000 persons and destroying about half the city. In 1985, it was said that the bombardier who pushed the button releasing the bomb would like to apologize to the survivors (called *hibakusha*), but the mayor of Nagasaki declined, saying he could not find it in his heart to meet him. Memorial services are held annually at Nagasaki and also at Kokura, where those who were spared because of the bad weather also grieve for those at Nagasaki who suffered in their stead.

SINGAPORE: INDEPENDENCE DAY. Aug 9. Most festivals in Singapore are Chinese, Indian or Malay, but celebration of national day is shared by all to commemorate achievement of independence in 1965. Music, parades, dancing.

SWEDEN: CRAYFISH PREMIERE. Aug 9. Crayfish may be sold and served in restaurants the following day after the season opens. Annually, the second Wednesday in August.

TAIWAN: CHENG CHENG KUNG BIRTH ANNIVERSARY. Aug 9. Joyous celebration of birth of Cheng Cheng Kung (Koxinga), the Ming Dynasty loyalist who ousted the Dutch colonists from Taiwan in 1661. Dutch landing is commemorated annually on Apr 29, but Cheng's birthday is honored on 14th day of seventh moon according to the Chinese lunar calendar.

WALTON, IZAAC: BIRTH ANNIVERSARY. Aug 9. English author of classic treatise on fishing, *The Compleat Angler*, published in 1653, was born at Stafford, England, Aug 9, 1593. Died at Winchester, England, Dec 15, 1683. "Angling," Walton wrote, "may be said to be so like the mathematics, that it can never be fully learnt."

WEBSTER-ASHBURTON TREATY SIGNED. Aug 9. The treaty delimiting the eastern section of the Canadian-American border was negotiated in 1842 by the US Secretary of State, Daniel Webster, and Alexander Baring, president of the British Board of Trade. The treaty established the boundaries between the St. Croix and Connecticut rivers, between Lake Superior and the Lake of the Woods, and between Lakes Huron and Superior. The treaty was signed in Washington, DC, on Aug 9, 1842.

BIRTHDAYS TODAY

Sam Elliott, 51, actor ("Mission Impossible," *Gettysburg*), born at Sacramento, CA, Aug 9, 1944.
J. James Exon, 74, US Senator (D, Nebraska), born at Geedes, SD, Aug 9, 1921.
Melanie Griffith, 38, actress (*Working Girl, Night Moves, Smile*), born at New York, NY, Aug 9, 1957.
Whitney Houston, 32, singer ("And I Will Always Love You"), born at Newark, NJ, Aug 9, 1963.
Brett Hull, 31, hockey player, born at Belleview, Ontario, Canada, Aug 9, 1964.
Rod Laver, 57, former tennis player, born at Queensland, Australia, Aug 9, 1938.
Ken Norton, 50, former boxer, born at Jacksonville, FL, Aug 9, 1945.
David Steinberg, 53, comedian ("The Music Scene," "The David Steinberg Show"), born at Winnipeg, Manitoba, Canada, Aug 9, 1942.

AUGUST 10 — THURSDAY
222nd Day — Remaining, 143

BABY PARADE. Aug 10. Ocean City, NJ. Oldest baby parade on East Coast. Begins at 11:15 AM, EST. Annually, the second Thursday in August. Est attendance: 14,000. For info: PR Dept, City Hall, Ocean City, NJ 08226. Phone: (609) 399-0272.

CAMPBELL, ANGUS: BIRTH ANNIVERSARY. Aug 10. Professor of psychology and sociology, author and director of the Institute for Social Research at the University of Michigan, called a "man with a scientist's mind and a humanist's heart," born Aug 10, 1910, at Leiters, IN. He was one of the principal researchers in studies of social and racial problems and attitudes. Campbell died Dec 15, 1980, at Ann Arbor, MI.

CANADA: CARDSTON STAMPEDE. Aug 10–12. Cardston, Alberta. Part of Cardston's annual Heritage Days celebrations. Rodeo's first hornless bronc saddle is said to have debuted here in 1922. For info: Cardston Heritage Days, PO Box 1322, Cardston, Alta, Canada T0K 0K0.

CANADA: EDMONTON FOLK MUSIC FESTIVAL. Aug 10–13. Gallagher Park, Edmonton, Alberta. Folk music and fun for the entire family highlighting blues, country, Celtic, traditional folk and bluegrass music, arts and crafts displays and a food fair. Est attendance: 50,000. For info: Folk Music Festival, PO Box 4130, Edmonton, AB, Canada T6E 4T2. Phone: (403) 429-1899.

Aug 10–11 ☆ Chase's 1995 Calendar of Events ☆

COLT LEAGUE WORLD SERIES. Aug 10–16. (Tentative.) Lafayette, IN. International young adult baseball World Series for players of league ages 15 and 16. Annually, beginning in the second week of August. Est attendance: 2,336. Sponsor: Pony Baseball, PO Box 225, Washington, PA 15301. Phone: (412) 225-1060.

ECUADOR: INDEPENDENCE DAY. Aug 10. National holiday. Celebrates attainment of independence Aug 10, 1809.

FIDDLEFEST. Aug 10–12. Galax, VA. Street Festival with arts and crafts, music, food and games. Coincides with 60th annual Old Fiddlers Convention. Annually, the second weekend in August. Sponsor: Galax Downtown Assoc, PO Box 544, Galax, VA 24333. Phone: (703) 236-2184.

HOOVER, HERBERT CLARK: BIRTH ANNIVERSARY. Aug 10. The 31st president of the US was born at West Branch, IA, Aug 10, 1874. Hoover was the first president born west of the Mississippi River and the first to have a telephone on his desk (installed Mar 27, 1929). "Older men declare war. But it is youth that must fight and die," he said in Chicago, IL, at the Republican National Convention, June 27, 1944. Hoover died at New York, NY, Oct 20, 1964. The Sunday nearest Aug 10 is observed in Iowa as Herbert Hoover Day.

IOWA STATE FAIR. Aug 10–20. Iowa State Fairgrounds, Des Moines, IA. One of America's oldest and largest state fairs with one of the world's largest livestock shows. Ten-acre carnival, superstar grandstand stage shows, track events, spectacular free entertainment. 160-acre campgrounds. Est attendance: 894,000. For info: Iowa State Fair, Statehouse, 400 E 14th St, Des Moines, IA 50319-0198. Phone: (515) 262-3111.

◆ **JAPAN'S UNCONDITIONAL SURRENDER: 50th ANNIVERSARY.** Aug 10, 1945. Less than 24 hours after a meeting to discuss the acceptance of the Potsdam Declaration ended in stalemate, another gathering to discuss the surrender terms took place in Emperor Hirohito's bomb shelter. As on the previous day [see Aug 9], the participants were stalemated. War Genl Anami continued to express the belief that "we may be able to reverse the situation in our favor, pulling victory out of defeat." Hirohito came down on the side of Suzuki and Togo. He believed continuation of the war would only result in further loss of Japanese lives. A message was transmitted from Tokyo to Japanese ambassadors in Switzerland and Sweden to accept the terms issued at Potsdam on July 26, 1945, except that the prerogative of the Japanese emperor's sovereign rule must be maintained. The Allies devised a plan under which the emperor and the Japanese government would administer under the rule of the Supreme Commander of the Allied Powers, and the Japanese surrendered.

MISSOURI: ADMISSION DAY. Aug 10. Became 24th state on this day in 1821.

MOON PHASE: FULL MOON. Aug 10. Moon enters Full Moon phase at 2:15 PM, EDT.

NATIONAL HOBO CONVENTION. Aug 10–12. Britt, IA. The 95th annual convention and accompanying events honors the hobo—a traveling laborer who pursues jobs via the rail lines. Visit the Hobo Jungle and the Hobo Museum. Join the Hoboes for storytelling, parade, the traditional Mulligan Stew and the Coronation of the Hobo King and Queen, as well as entertainment, carnival, gospel concert, 6-block craft and flea market, 5K walk, 5 & 10K runs, fine arts show, classic and antique car show and more. Annually, the weekend of the second Saturday in August. Est attendance: 30,000. For info: Britt Chamber of Commerce, PO Box 63, Britt, IA 50423. Phone: (515) 843-3867.

August 1995

S	M	T	W	T	F	S
		1	2	3	4	5
6	7	8	9	10	11	12
13	14	15	16	17	18	19
20	21	22	23	24	25	26
27	28	29	30	31		

RIBFEST. Aug 10–12. Kalamazoo, MI. Features the smell of sizzling ribs as rib-burners from throughout the United States tantalize the taste buds of West Michigan. Festival will feature live entertainment, family-oriented events, food booths and the "Peoples Choice Award" for best ribs. Annually, the second weekend in August. Est attendance: 22,000. Sponsor: Comm Advocates for Persons with Developmental Disabilities, Laurie DeHaven, 806 S Westnedge St, Kalamazoo, MI 49008-1162. Phone: (616) 342-9801.

SPACE MILESTONE: *DISCOVERER 13* (US). Aug 10. Ejected space capsule, first object recovered after orbiting. Launched Aug 10, 1960.

BIRTHDAYS TODAY

Ian Anderson, 48, musician, singer (lead of Jethro Tull; "Bungle in the Jungle"), born at Blackpool, England, Aug 10, 1947.

Rosanna Arquette, 36, actress (*Desperately Seeking Susan, New York Stories*), born at New York, NY, Aug 10, 1959.

Jimmy Dean (Seth Ward), 67, singer ("Big Bad John," "P.T. 109"), born Plainview, TX, Aug 10, 1928.

Eddie Fisher, 67, singer ("Heart," "Cindy, Oh Cindy"), born Philadelphia, PA, Aug 10, 1928.

Rhonda Fleming (Marilyn Lewis), 72, actress ("Stage Door," *The Best of Broadway*), born Los Angeles, CA, Aug 10, 1923.

Betsy Johnson, 53, fashion designer, born at Wethersfield, CT, Aug 10, 1942.

Benjamin Ward, 69, police administrative official, born at Brooklyn, NY, Aug 10, 1926.

AUGUST 11 — FRIDAY
223rd Day — Remaining, 142

ATCHISON, DAVID R.: BIRTH ANNIVERSARY. Aug 11. Missouri legislator who was president of the US for one day. Born at Frogtown, KY, on Aug 11, 1807, Atchison's strong pro-slavery opinions made his name prominent in legislative debates. He served as president pro tempore of the Senate a number of times, and he became president of the US for one day—Sunday, Mar 4, 1849—pending the swearing in of President-elect Zachary Taylor on Monday, Mar 5, 1849. The city of Atchison, KS, and the county of Atchison, MO, are named for him. He died at Gower, MO, on Jan 26, 1886.

BEATLES' NATIONAL APPLE WEEK: 30th ANNIVERSARY. Aug 11–18. The Beatles proclaimed this week to be "National Apple Week" in 1965, when they launched their Apple recording label. Presentation boxes of "Our First Four" releases were sent to the Queen, the Queen Mother and Princess Margaret. In return, the Queen Mother sent a thank-you note saying that she was "greatly touched by this kind thought from the Beatles" and "much enjoyed listening to these recordings."

BOND, CARRIE JACOBS: BIRTH ANNIVERSARY. Aug 11. American composer of well-known songs, including "I Love You Truly" and "A Perfect Day," and of scores for motion pictures, Carrie Jacobs Bond was born at Janesville, WI, on Aug 11, 1862. She died at Hollywood, CA, at age 84, on Dec 28, 1946.

CANADA: ABBOTSFORD INTERNATIONAL AIRSHOW. Aug 11–13. Abbotsford Airport, Abbotsford, British Columbia. International and national aerobatic flying teams including the Canadian Snowbirds, USAF Thunderbirds, the Stealth bomber and more. Annually, the second weekend in August. Sponsor: Abbotsford Intl Airshow Society. Phone: (604) 852-8511 or (604) 328-JETS. Est attendance: 250,000. For info: Ministry of Tourism and Culture, 1117 Wharf St, Victoria, BC, Canada V8W 2Z2. Phone: (604) 852-8511.

☆ Chase's 1995 Calendar of Events ☆ Aug 11

CANADA: CANADIAN OPEN FIDDLE CONTEST. Aug 11–12. Shelburne, Ontario. 45th annual competition; annually on the weekend after Canada Day Civic Holiday. Est attendance: 8,000. For info: Shelburne Rotary Club, PO Box 27, Shelburne, Ont, Canada L0N 1S0. Phone: (519) 925-3013.

CANADA: DOCKSIDE FESTIVAL. Aug 11–13. Gravenhurst, Ontario, Canada. Relaxing weekend of arts, crafts and entertainment overlooking beautiful Lake Muskoka. Features more than 130 juried Canadian artisans showing and selling their work in Sagamo Park. Ongoing live daily entertainment, Saturday evening Dance in the Park. Annually, the second weekend in August. Est attendance: 12,000. For info: Muskoka Winter Carnival Assn, PO Box 129, Gravenhurst, Ont, Canada P1P 1T5. Phone: (705) 687-8432.

CAROLINA CRAFTSMEN'S SUMMER CLASSIC. Aug 11–13. Convention Center, Myrtle Beach, SC. Arts and crafts. Est attendance: 20,000. For info: The Carolina Craftsmen, 1240 Oakland Ave, Greensboro, NC 27403. Phone: (910) 274-5550.

CHAD: INDEPENDENCE DAY. Aug 11. National holiday.

ELVIS WEEK. Aug 11–16. Memphis, TN. To remember, honor and celebrate the life and career of Elvis Aaron Presley. Three dozen events at Graceland and throughout the city of Memphis. Est attendance: 35,000. Sponsor: Graceland, Div of Elvis Presley Enterprises, Inc, Box 16508, Memphis, TN 38186-0508. Phone: (800) 238-2000.

FINLAND: TURKU MUSIC FESTIVAL. Aug 11–20. Turku. One of Finland's oldest music festivals in its oldest city. Music ranges from the medieval era to present day, performed by world famous artists and groups in halls with fine acoustics, churches and even museums. Est attendance: 15,000. For info: Finnish Tourist Board, 655 Third Ave, New York, NY 10017. Phone: (212) 949-2333.

FIRST FOREIGN-BORN OFFICER APPOINTED CHAIR OF JOINT CHIEFS: ANNIVERSARY. Aug 11, 1993. Pres Bill Clinton appointed Army General John Shalikashvili to succeed Colin Powell as Chairman of the Joint Chiefs of Staff. Shalikashvili was born in Poland, but his family fled to Germany in 1944 to escape advancing Soviet troops. After moving to the USA, his family lived in Peoria, IL. "General Shali" has a distinguished military record and is a Vietnam War veteran.

◆ **FREDERICK DOUGLASS SPEAKS: ANNIVERSARY.** Aug 11, 1841. Having escaped from slavery only three years earlier, Frederick Douglass was legally a fugitive when he first spoke before an audience. At an antislavery convention on Nantucket Island Douglass spoke simply but eloquently about his life as a slave. His words were so moving that he was asked to become a full-time lecturer for the Massachusetts Anti-Slavery Society. Douglass became a brilliant orator, writer and abolitionist who championed the rights of blacks as well as the rights of all humankind.

GOLD COAST ART FAIR. Aug 11–13. Chicago, IL. Features the work of 600 artists of painting and sculpture in a large outdoor exhibit. Est attendance: 830,000. Sponsor: Gold Coast Art Fair, c/o Near North News, 222 W Ontario, Ste 502, Chicago, IL 60610-3695.

GRENADA: CARNIVAL. Aug 11–15. Grenada's annual Carnival celebration with pageant, steelband, calypso and street "jump up" party. Aug 14 is Carnival Monday, a holiday in Grenada. Est attendance: 5,000. For info: Grenada Tourist Info Office, 820 Second Ave, Ste 900-D, New York, NY 10017. Phone: (800) 927-9554.

HALEY, ALEX PALMER: BIRTH ANNIVERSARY. Aug 11. Born at Ithaca, NY, Aug 11, 1921, Alex Palmer Haley was raised by his grandmother in Henning, TN. In 1939 he entered the US Coast Guard and served as a cook, but eventually he became a writer and college professor. His interview with Malcolm X for *Playboy* led to his first book, *The Autobiography of Malcolm X*, which sold 6 million copies and was translated into 8 languages. *Roots*, his Pulitzer Prize–winning novel published in 1976, sold millions, was translated into 37 languages and was made into an 8-part TV miniseries in 1977. The story also generated an enormous interest in family ancestry. Haley died in 1992.

LITTLE LEAGUE BASEBALL: BIG LEAGUE WORLD SERIES. Aug 11–19. Floyd Hull Stadium, Ft Lauderdale, FL. Eleven teams, foreign and regional, play for world championship. Est attendance: 25,000. For info: Floyd Hull, Tournament Dir, 1000 SE 9th Ave, Ft Lauderdale, FL 33316. Phone: (305) 566-4395.

NO MAN'S LAND CELEBRATION. Aug 11–13. Breckenridge, CO. Legend has it that Breckenridge was left off the map in several historic treaties. One weekend a year Breckenridge is not a part of America. Celebrated with a parade, dances, barbecue and run. Est attendance: 25,000. For info: Breckenridge Resort Chamber of Commerce, John Hendryson, PO Box 1909, Breckenridge, CO 80424. Phone: (303) 453-2913.

◆ **NORMAN ROCKWELL'S THE SWIMMER COVER FOR THE *SATURDAY EVENING POST*: 50th ANNIVERSARY.** Aug 11. This *Post* cover of a man happily submerged up to his neck on a hot day was a fooler. Published on Aug 11, 1945, it was actually painted in March when it's much too cold to take to the water in Vermont. Model George Zimmer posed for Rockwell in a cozy indoor studio.

PRESIDENTIAL JOKE DAY: ANNIVERSARY. Aug 11. A day to recall presidential jokes. Anniversary of President Ronald Reagan's voice-test joke of Aug 11, 1984. In preparation for a radio broadcast, during a thought-to-be-off-the-record voice level test, instead of counting "one, two, three ...," the president said: "My fellow Americans, I am pleased to tell you I just signed legislation which outlaws Russia forever. The bombing begins in five minutes." The statement was picked up by live television cameras and microphones and was seen and heard by millions worldwide. The incident provoked national and international reactions, including a news network proposal of new ground rules concerning the use of "off-the-record" remarks.

ST. CLARE OF ASSISI: FEAST DAY. Aug 11. Chiara Favorone di Offreduccio, a religious leader inspired by St. Francis of Assisi, was the first woman to write her own religious rule. Born at Assisi, Italy in 1193, she died there Aug 11, 1253. A "Privilege of Poverty" freed her order from any constraint to accept material security, making the "Poor Clares" totally dependent on God.

SHANTY DAYS: CELEBRATION OF THE LAKE. Aug 11–13. Legion Park, Algoma, WI. Three day festival to celebrate lakeshore heritage. Entertainment, ethnic food, coaster cars, arts and crafts, beach volleyball, 5K Walk/Run, fishing contest, kid's area, minnow racing, book sale, community parade, venetian boat parade, photo contest and fireworks finale. Annually, the second full weekend in Aug. Est attendance: 20,000. Sponsor: Carol Wiese, Exec Dir, Algoma Area Chmbr of Com, 1226 Lake St, Algoma, WI 54201. Phone: (414) 487-2041.

SOUTHEASTERN SPORTS-A-RAMA. Aug 11–13. The 9th Annual Hunting Show and Sale featuring more than 100 booths of hunting and fishing supplies, equipment and services for sale, pre-season prices, amateur Deer Head exhibit, along with the annual Indoor Archery Tournament and daily seminars. Chattanooga-Hamilton County Convention and Trade Center, Chattanooga, TN. Sponsor: Esau Inc. Est attendance: 8,000. For info: Danyl Hobby, Southeastern Sports A-Rama, PO Box 50096, Knoxville, TN 37950. Phone: (615) 588-1233.

SPACE MILESTONE: *VOSTOK 3* (USSR). Aug 11. Launched Aug 11, 1962, Andrian Nikolayev orbits Earth 64 times over a period of 94 hours, 25 minutes, covering distance of 1,242,500 miles. Achieved radio communication with *Vostok 4* and telecast from spacecraft.

Aug 11-12 ☆ Chase's 1995 Calendar of Events ☆

STAMP EXPO '95: CALIFORNIA. Aug 11-13. Grand Hotel, Anaheim, CA. Fax: (818) 988-4337. Est attendance: 4,000. Sponsor: Intl Stamp Collectors Society, PO Box 854, Van Nuys, CA 91408. Phone: (818) 997-6496.

TETONKAHA RENDEZVOUS (WITH STATE WILD GAME COOKOFF). Aug 11-13. Bruce, SD. Experience the fur-trading atmosphere of the 1840s. Taste foods from the State Wild Game cookoff. Est attendance: 300. For info: Lee Kratochvil, RR #2, Box 42, Bruce, SD 57220. Phone: (605) 627-5441.

WATTS RIOT: 30th ANNIVERSARY. Aug 11. On Aug 11, 1965, a minor clash between the California Highway Patrol and two young blacks set off six days of riots in the Watts area of Los Angeles. Thirty-four deaths were reported and more than 3,000 people were arrested. Damage to property was listed at $40 million. The less-immediate cause of the disturbance and the others that followed was racial tension between whites and blacks in American society.

WYANDOTTE WATERFEST. Aug 11-13. Bishop Park, Wyandotte, MI. Outdoor boat show, Great Lakes jet ski competition, fireworks, parade of boats and professional band. Annually, the second full weekend in August. For info: Leslie Lupo, Dir, 3131 Biddle Ave, Wyandotte, MI 48192. Phone: (313) 246-4505.

BIRTHDAYS TODAY

Joanna Coles, 51, children's author, born at Newark, NJ, Aug 11, 1944.
Mike Douglas (Michael Delaney Dowd, Jr), 70, TV host, singer, born at Chicago, IL, Aug 11, 1925.
Jerry Falwell, 62, clergyman (head of Moral Majority PAC), born at Lynchburg, VA, Aug 11, 1933.
Hulk Hogan (Terry Gene Bollea), 42, wrestler, actor (*Suburban Commando, Adventure in Paradise*), born at Augusta, GA, Aug 11, 1953.
Joe Jackson, 40, musician, songwriter, born at Burton-On-Trent, England, Aug 11, 1955.
Anna Massey, 58, actress (*Bunny Lake Is Missing, A Doll's House*), born Thankeham, England, Aug 11, 1937.
Carl Rowan, 70, journalist, author (*Breaking Barriers*), born at Ravenscraft, TN, Aug 11, 1925.

AUGUST 12 — SATURDAY
224th Day — Remaining, 141

ALL-AMERICAN SOAP BOX DERBY. Aug 12. (Tentative.) Derby Downs, Akron, OH. A week-long festival culminating in world championship race by regional champs from USA, Canada, Germany, Ireland, Australia and Philippines. 58th annual derby. Est attendance: 15,000. Sponsor: Intl Soap Box Derby, Inc, Jeff Iula, Genl Mgr, PO Box 7233, Derby Downs, Akron, OH 44306. Phone: (216) 733-8723.

ANTIQUE AUTO SHOW AND FLEA MARKET. Aug 12-13. Boyne City, MI. Est attendance: 10,000. For info: Chamber of Commerce, 28 S Lake St, Boyne City, MI 49712. Phone: (616) 582-6222.

BLUEBERRY FESTIVAL. Aug 12. State Office Building, Methodist Church and Main Street Theatre, Ketchikan, AK. A street fair featuring arts and crafts, food, games, and contests for all ages. Annually, the second Saturday in August. Est attendance: 2,500. For info: Ketchikan Area Arts and Humanities Council, 338 Main St, Ketchikan, AK 99901. Phone: (907) 225-2211.

August 1995

S	M	T	W	T	F	S
		1	2	3	4	5
6	7	8	9	10	11	12
13	14	15	16	17	18	19
20	21	22	23	24	25	26
27	28	29	30	31		

BRONCO LEAGUE WORLD SERIES. Aug 12-17. (Tentative.) Monterey, CA. International youth baseball World Series for players of league ages 11 and 12. Annually, beginning in the second week of August. Est attendance: 4,350. Sponsor: Pony Baseball, PO Box 225, Washington, PA 15301. Phone: (412) 225-1060.

BUD BILLIKEN PARADE. Aug 12. Chicago, IL. A parade especially for children begun in 1929 by Robert S. Abbott. The second largest parade in the United States, it features bands, floats, drill teams and celebrities. Annually, the second Saturday in August. Sponsor: Chicago Defender Charities, Michael Brown, PR Dir, 2400 S Michigan, Chicago, IL 60616. Phone: (312) 225-2400.

CANADA: DISCOVERY DAYS CELEBRATION. Aug 12-15. Watson Lake, Yukon Territory. Celebration of the discovery of gold in the Yukon includes pool party, wild game barbecue, whipped cream slide, tug o' war, Peterbilt Truck Pull, square dancing, mini-triathlon, parade, bubble gum-blowing contest, fancy hatmaking, children's events and more. Est attendance: 900. For info: Tracy MacKay, Recreation Programmer, Town of Watson Lake Recreation Dept, Box 590, Watson Lake, YT, Canada Y0A 1C0. Phone: (403) 536-2246.

CANADA: SAINT-JEAN-sur-RICHELIEU HOT AIR BALLOON FESTIVAL. Aug 12-20. Saint-Jean-sur-Richelieu, Quebec. Activities surrounding the demonstrations of fiesta-style hot-air balloons. Est attendance: 100,000. For info: Daniel Béland, Aéroport Municipal, Bureau #1, Saint-Jean-sur-Richelieu, Que, Canada J3B 7B5.

CANTINFLAS: BIRTH ANNIVERSARY. Aug 12. Mexico's most famous comic actor, Cantinflas, was born Aug 12, 1911, at Mexico City. His real name was Mario Moreno Reyes. Particularly popular with the poor of Mexico because he most often portrayed the underdog, Cantinflas got his start in Mexico City *carpas*, the equivalent of vaudeville. He became internationally known for his role in *Around the World in 80 Days*. The name Cantinflas was invented by the comic to prevent his parents from learning he was in show business which was considered a shameful endeavor. Cantinflas died on Apr 20, 1993, at Mexico City.

CHESTER COUNTY OLD FIDDLERS' PICNIC. Aug 12. Hibernia County Park, Coatesville, PA. Old-time musicians gather to make their traditional music. Annually, the second Saturday in August. Est attendance: 7,000. For info: Chester County Parks and Recreation Dept, Chester County Govt Services Center, 601 Westtown Rd, Ste 160, West Chester, PA 19382-4534. Phone: (610) 344-6415.

CIVIL WAR DAYS. Aug 12-13. Rockford, IL. Troops from Union and Confederate armies meet at Midway Village performing military drills, skirmishes and battle reenactment, camp-life demonstrations and period fashion show. Est attendance: 16,000. For info: Midway Village and Museum Center, 6799 Guilford Rd, Rockford, IL 61107. Phone: (815) 397-9112.

COLUMBUS DAYS. Aug 12-20. Columbus, NE. City-wide. Band concert, prayer breakfast, coronation ball, baby show, arts and crafts, archery, horseshoes, lip sync, bed race, talent show, community picnic, quilt show, teen dance, book sale, big wheel races, volleyball, tennis, pie-eating contest, etc. Parade 6:00 Sunday. Est attendance: 12,500. Sponsor: Dale Collinsworth, Exec VP, Columbus Area Chamber of Commerce, PO Box 515, Columbus, NE 68602-0515. Phone: (402) 564-2769.

COUNTRY SIDEWALK SALE. Aug 12-13. Peddler's Village, Lahaska, PA. A great weekend of bargains. All shops will be offering an array of incredible bargains displayed on country-decorated booths, carts and tables. Est attendance: 5,000. For info: Peddlers Village, PO Box 218, Lahaska, PA 18931. Phone: (215) 794-4000.

FAMILY DAYS CELEBRATION. Aug 12-13. Farmington, PA. Special hands-on family activities. Dress in military uniforms, make musket cartridges, sign your name with a feather pen, learn about the Native Americans or enlist with Washington's troops—these are just some of the activities planned. Sponsor: Natl Park Service, Ft Necessity Natl Battlefield, RD 2, Box 528, Farmington, PA 15437. Phone: (412) 329-5512.

FRENCH WEST INDIES: FETE DES CUISINIERES. Aug 12. Guadeloupe. Women in Creole costumes carry baskets of decorated stuffed lobsters and other island specialties. For info: French West Indies Tourist Bd, 610 5th Ave, New York, NY 10020. Phone: (900) 990-0040.

INDIAN DAY. Aug 12. Massachusetts.

INTER-STATE FAIR AND RODEO. Aug 12-20. Coffeyville, KS. "Largest outdoor fair and rodeo event in southeast Kansas and northeast Oklahoma." Est attendance: 60,000. Sponsor: Montgomery County Fair Assn, Box 457, Coffeyville, KS 67337. Phone: (316) 251-2550.

JOUR DE FETE. Aug 12-13. St. Genevieve, MO. Area's largest craft fair with more than 700 exhibitors, entertainment and food. Est attendance: 40,000. For info: Tourist Info Office, 66 S Main, St. Genevieve, MO 63670. Phone: (314) 883-7097.

KING PHILIP ASSASSINATION: ANNIVERSARY. Aug 12. Native American, Philip, son of Massasoit, chief of the Wampanog tribe, was killed Aug 12, 1676, near Mt Hope, RI, by a renegade Indian of his own tribe, bringing to an end the first and bloodiest war between American Indians and white settlers of New England, a war that had raged for nearly two years and was known as King Philip's War.

KRXL CAR CRUISE. Aug 12. Kirksville, MO. Motorcycles, cars, antiques, hot rods, classics, originals and street rods. Est attendance: 3,000. Sponsor: Steve Lloyd, Station Mgr, KRXL, 1308 N Baltimore St, Kirksville, MO 63501. Phone: (816) 665-9841.

LEITERSBURG PEACH FESTIVAL. Aug 12-13. Leitersburg Ruritan Community Park, Leitersburg, MD. This peach-oriented event features crafts, horse and carriage rides, bluegrass and country music and a quilt raffle. Annually, the second weekend in August. Est attendance: 10,000. For info: Leitersburg Ruritan Club, 21378 Leiters Mill Rd, Hagerstown, MD 21742. Phone: (301) 797-6387.

MATHEWSON, CHRISTY: BIRTH ANNIVERSARY. Aug 12. Famed American baseball player, Christopher (Christy) Mathewson, one of the first players named to Baseball's Hall of Fame, was born at Factoryville, PA, Aug 12, 1880. Died at Saranac Lake, NY, Oct 7, 1925. He pitched three complete games during the 1905 World Series without allowing opponents to score a run. In 17 years he won 372 games while losing 188 and striking out 2,499 players.

MICHIGAN RENAISSANCE FESTIVAL. Aug 12-Sept 24. (Weekends and Labor Day only.) Holly, MI. Continuous entertainment on 7 stages featuring comedy, drama, bawdy and classical music and folk dance. More than 130 craft shops; games, food and much more. For info: Michigan Renaissance Fest, PR Dir, 700 E Maple, Birmingham, MI 48009. Phone: (800) 966-8215.

MIDDLE CHILDREN'S DAY. Aug 12. To salute the middle-born children whose youthful activities were limited due to their always being "too young or too old." Today, they are JUST RIGHT! Sponsor: Mid-Kid Co, 402 Oak Ave, Sebring, FL 33870.

MINNESOTA RENAISSANCE FESTIVAL. Aug 12-Sept 25. (Weekends and Labor Day only.) Shakopee, MN. A celebration of 16th-century Renaissance Europe with entertainment, food, arts and crafts, games and armored jousting. Est attendance: 310,000. For info: Minnesota Renaissance Festival, 3525 W 145th St, Shakopee, MN 55379. Phone: (800) 966-8215.

MODEL RAILROAD SHOW. Aug 12-13. Kingswood Regional High School, Wolfeboro, NH. Operating layouts, railroad model contest, ladies railroad motif handcraft contest, dealers and test track. Annually, the second weekend in August. Est attendance: 1,000. For info: Railroad Information, RR 2, Box 500, Wolfeboro, NH 03894. Phone: (603) 569-4876.

MONTANAFAIR. Aug 12-19. MetraPark, Billings, MT. Montana's biggest event featuring exhibits, livestock events, carnival, rodeo, entertainment, indoor motorcross and parimutuel horse racing. Est attendance: 240,000. For info: MetraPark, PO Box 2514, Billings, MT 59103. Phone: (406) 256-2400.

NATIONAL DOLLHOUSE AND MINATURES TRADE SHOW. Aug 12-14. Hyatt At Union Station, St. Louis, MO. Manufacturers, importers, wholesalers and hand crafters of dollhouses and miniatures exhibit their wares for buyers from retail and wholesale companies only. For info: Patty Parrish, Assn/Trade Show Mgr, PO Box 2188, Zanesville, OH 49702-2188. Phone: (614) 452-4541.

NO LAUGHING MATTER: POLITICAL CARTOONISTS ON THE ENVIRONMENT. Aug 12-Nov 26. Grace Hudson Museum, Ukiah, CA. See Feb 1 for full description. Call or write the Smithsonian for other dates and venues. For info: Smithsonian Institution Traveling Exhibition Service, 1100 Jefferson Dr SW, Ste 3146, Washington, DC 20560. Phone: (202) 357-2700.

O-BON FESTIVAL. Aug 12. Morikami Museum, Delray Beach, FL. Traditional Japanese summer festival welcomes the ancestral spirits back to Earth with folk dancing, folk music, games and amusements, culminating with the floating of paper lanterns on Morikami pond at dusk and a fireworks display. Est attendance: 5,000. Sponsor: The Morikami Museum, Larry Rosensweig, Dir, 4000 Morikami Park Rd, Delray Beach, FL 33446. Phone: (407) 495-0233.

OLD BEDFORD VILLAGE GOSPEL FESTIVAL. Aug 12-13. Bedford, PA. Come and hear some of the finest gospel groups. For info: Old Bedford Village, PO Box 1976, Bedford, PA 15522. Phone: (814) 623-1156.

OLD-TIME FARM DAY. Aug 12. New Hampshire Farm Museum, Milton, NH. Museum presents 17th annual "Festival of Farming as Practiced from the 1780s to the 1930s" with more than 60 demonstrations and exhibits, wagon rides, country music, chicken barbecue and tours of the historic farm buildings. Est attendance: 2,000. For info: NH Farm Museum, PO Box 644, Milton, NH 03851. Phone: (603) 652-7840.

POLISH AMERICAN WEEKEND. Aug 12-13. Penn's Landing, Christopher Columbus Blvd and Chestnut St in historic Philadelphia, PA. Two-day festival of Polish food, music and dance. For info: Polish American Congress, Eastern Pennsylvania District, 308 Walnut St, Philadelphia, PA 19106. Phone: (215) 739-3408.

PONY LEAGUE WORLD SERIES. Aug 12-19. (Tentative.) Washington, PA. International youth baseball World Series for teams of players ages 13 and 14. Annually, beginning in the second week of August. Est attendance: 16,700. Sponsor: Pony Baseball, PO Box 225, Washington, PA 15301. Phone: (412) 225-1060.

RAILROADER'S FESTIVAL (WITH SPIKE-DRIVING CONTEST). Aug 12. Promontory, UT. Visitors relive the rush to complete the transcontinental railroad. Professionals vie for the world's record in a spike-driving contest. Annually, the second Saturday in August. Sponsor: Golden Spike National Historic Site, PO Box 897, Brigham City, UT 84302. Phone: (801) 471-2209. For info: Dave Porter, Publicity Dir, Utah Travel Council, Council Hall, Capitol Hill, Salt Lake City, UT 84114. Phone: (801) 538-1030.

Aug 12-13 ☆ *Chase's 1995 Calendar of Events* ☆

ROCK CITY FAIRY TALE FESTIVAL. Aug 12-13. (Also Aug 19-20, 26-27.) Rock City Gardens, Lookout Mountain, TN. Memories are made and dreams come true each summer as Rock City holds its annual Fairy Tale Festival and storybook characters come to life. Cinderella, Little Red Riding Hood, Sleeping Beauty and others will greet visitors as they enter the gardens. Est attendance: 40,000. For info: Todd Smith, See Rock City Inc, PO Box 246, Lookout Mountain, TN 37350. Phone: (706) 820-2531.

ST. JOHNS MINT FESTIVAL. Aug 12-13. St. Johns, MI. To honor area mint farmers. Annually, the second full weekend in August. Est attendance: 60,000. For info: St. Johns Area Chamber of Commerce, Linda Curtis, Box 61, St. Johns, MI 48879. Phone: (517) 224-7248.

SCOTLAND: EDINBURGH BOOK FESTIVAL. Aug 12-28. Charlotte Square Gardens, Edinburgh, Lothian. A biennial event which is Britain's largest book display and sale giving the public a chance to meet authors, hear readings and buy both new and old titles. For info: Ms Shona Munro, Edinburgh Book Festival, Scottish Book Centre, 137 Dundee St, Edinburgh, Scotland EH111BG.

SCOTLAND: EDINBURGH INTERNATIONAL FILM FESTIVAL. Aug 12-27. Edinburgh, Lothian. Festival includes documentaries, special events and lectures. For info: Edinburgh International Film Festival, Filmhouse, 88 Lothian Rd, Edinburgh, Scotland EH3 9BZ.

SEWARD SILVER SALMON DERBY. Aug 12-20. Seward, AK. Alaska's largest salmon derby. Fishermen vie for cash prizes, including tagged fish, daily top fish awards and sweepstakes drawing. 40th annual. Est attendance: 5,000. For info: Seward Chamber of Commerce, PO Box 749, Seward, AK 99664. Phone: (907) 224-8051.

SPACE MILESTONE: *ECHO I* (US). Aug 12. First successful communications balloon. Launched Aug 12, 1960.

SPACE MILESTONE: *ENTERPRISE* (US). Aug 12. Reusable orbiting vehicle (space shuttle) makes first successful flight on its own within Earth's atmosphere. Launched from Boeing 747 on Aug 12, 1977.

STATE WILD GAME COOKOFF. Aug 12. Oakwood Lakes State Park, SD. Open to any cook with a pot to cook in. Categories include porridge and roasted meat. Contestants are also judged on their best "true fabrication" of how they caught the critter for the pot. Est attendance: 100. For info: Brookings Renegade Muzzleloaders, Rte 2, Box 29, White, SD 57276. Phone: (605) 693-4589.

THAILAND: BIRTHDAY OF THE QUEEN. Aug 12. The entire kingdom of Thailand celebrates the birthday of Queen Sirikit.

VICTORIAN GARDEN WALK. Aug 12-13. General Crook House Museum, Fort Omaha Campus, MCC College, Omaha, NE. Authentic Victorian Garden featuring china doll, bugbane, spiderwort, goat's beard, prairie grasses; brochure and costumed guides; exhibit: Homesteaders Garden; guided historic Fort Omaha walking or trolley tours. Lemonade. Annually, the second full weekend in August. Sponsor: Historical Society of Douglas County, PO Box 11398, CAE, Omaha, NE 68111-0398. Phone: (402) 455-9990.

August 1995

S	M	T	W	T	F	S
		1	2	3	4	5
6	7	8	9	10	11	12
13	14	15	16	17	18	19
20	21	22	23	24	25	26
27	28	29	30	31		

WATERMELON FESTIVAL. Aug 12. Rush Springs, OK. Beginning with watermelon judging and ending with the crowning of the festival queen, the highlight of the celebration is the serving of 50,000 pounds of free watermelon at Jeff Davis Park. Est attendance: 20,000. For info: Chamber of Commerce, Box 298, Rush Springs, OK 73082. Phone: (405) 476-3277.

WORLD WHIMMY DIDDLE COMPETITION. Aug 12. Folk Art Center, Asheville, NC. Gee Haw Whimmy Diddle Competitions with awards, demonstrations, storytelling and live music. Est attendance: 900. Sponsor: Southern Highland Handicraft Guild, Milepost 382, Blue Ridge Parkway, Box 9545, Asheville, NC 28815. Phone: (704) 298-7928.

BIRTHDAYS TODAY

Dale Bumpers, 70, US Senator (D, Arkansas), born at Charleston, AR, Aug 12, 1925.

John Derek (Dereck Harris), 69, actor (*All the King's Men, Prince of Players*), born at Hollywood, CA, Aug 12, 1926.

William Goldman, 64, writer (*The Princess Bride, Marathon Man*), born at Chicago, IL, Aug 12, 1931.

George Hamilton, 56, actor (*Love at First Bite, Act One*, "The Survivors"), born Memphis, TN, Aug 12, 1939.

Michael Kidd (Milton Greenwald), 76, choreographer, born at Brooklyn, NY, Aug 12, 1919.

"Buck" (Alvis Edgar) Owens, 66, singer ("Hee Haw" show, "Act Naturally"), songwriter, born at Sherman, TX, Aug 12, 1929.

John Poindexter, 59, former national security advisor, born at Washington, IN, Aug 12, 1936.

Mstislav Leopoldovich Rostropovich, 68, musician, born at Baku, USSR, Aug 12, 1927.

Fife Symington, 50, Governor of Arizona (R), born at New York City, Aug 12, 1945.

Porter Wagoner, 65, singer ("The Carroll County Accident"), born at West Plains, MO, Aug 12, 1930.

Jane Wyatt, 82, actress (mother in "Father Knows Best"; Spock's mom), born New York, NY, Aug 12, 1913.

AUGUST 13 — SUNDAY
225th Day — Remaining, 140

AMERICAN SOCIETY OF ASSOCIATION EXECUTIVES ANNUAL MEETING AND EXPOSITION. Aug 13-16. Washington, DC. Major meeting for ASAE members, nonmembers and suppliers including education sessions, board meetings and awards presentations. Est attendance: 5,000. For info: Lorri Lee, American Soc of Assn Executives, 1575 Eye St NW, Washington, DC 20005-1168. Phone: (202) 626-2798.

BERLIN WALL ERECTED: ANNIVERSARY. Aug 13. Early in the morning of Sunday, Aug 13, 1961, the East German government closed the border between east and west sectors of Berlin with barbed wire fence to discourage further population movement to the west. Telephone and postal services were interrupted, and, later in the week, a concrete wall was built to strengthen the barrier between official crossing points. The dismantling of the Wall was begun on Nov 9, 1989. See also: "Berlin Wall: Dismantling Anniversary" (Nov 9).

CANADA: OPASQUIAK INDIAN DAYS. Aug 13-19. The Pas, Manitoba. Celebrates the Native culture and its traditional skills and sports with fiddling and jigging, an Indian Princess pageant, square dance competitions and a 97K/60-mile canoe race. For info: Opasquiak Indian Days, Caroline Constant, Box 297, The Pas, Man, Canada R9A 1K4. Phone: (204) 623-5483.

CATALINA WATER SKI RACE. Aug 13. Long Beach, CA. International water ski racing teams race 62 miles across open seas, round-trip from Long Beach to Catalina Island and back. Est attendance: 9,000. For info: Long Beach Boat and Ski Club, PO Box 2370, Long Beach, CA 90801. Phone: (714) 894-3498.

CAXTON, WILLIAM: BIRTH ANNIVERSARY. Aug 13. First English printer, born Aug 13, 1422. Died, at London, England, 1491. Caxton produced first book printed in English

Chase's 1995 Calendar of Events — Aug 13–14

(while he was still at Bruges), the *Recuyell of the Histories of Troy*, in 1476, and in the autumn of 1476 set up print shop in Westminster, becoming the first printer in England.

CENTRAL AFRICAN REPUBLIC: INDEPENDENCE DAY: 35 YEARS. Aug 13. Commemorates Proclamation of Independence of the Central African Republic, on Aug 13, 1960.

COUNTRY JAMBOREE. Aug 13. Trollwood Park, Fargo, ND. A full day of country music featuring the region's finest country music performers. Sponsor: K100-FM. Est attendance: 9,000. For info: Carolyn Altmann, Dir of Trollwood Park, PO Box 1796, Fargo, ND 58107. Phone: (701) 241-8160.

ELVIS PRESLEY REMEMBERED. Aug 13. St. Louis, MO. Singer Elvis Presley, his entire family and Col Tom Parker are impersonated in a live, display window show entitled "From Tupelo to Graceland" to mark anniversary week of his death. Later a two-hour live music show featuring Steve Davis and the Memphis Mafia will be held in the Elvis Room at Blueberry Hill. Est attendance: 800. For info: Blueberry Hill, 6504 Delmar, St. Louis, MO 63130. Phone: (314) 727-0880.

FAMILY DAY. Aug 13. To focus attention on family solidarity and its potential as the best teacher of basic beliefs and values. Annually, the second Sunday in August. Sponsor: Kiwanis Intl, Program Dvmt Dept, 3636 Woodview Trace, Indianapolis, IN 46268.

FIVE STATE FREE FAIR. Aug 13–21. Fairgrounds, Liberal, KS. Because Kansas, Oklahoma, New Mexico, Texas and Colorado are so close together, residents of each state enjoy one big fair. KRCA Finals Rodeo, Aug 19–21. Country-western entertainment, a carnival, kid's day and stock car races also featured. Annually, the third full week of August. Est attendance: 30,000. For info: Debra Huddleston, Box 420, Liberal, KS 67905. Phone: (316) 624-3712.

HERBERT HOOVER DAY. Aug 13. Iowa. Annually, the Sunday nearest Aug 10.

HITCHCOCK, ALFRED (JOSEPH): BIRTH ANNIVERSARY. Aug 13. English film director, master of suspense, was born at London, Aug 13, 1899. He produced a string of classics including *Rebecca, Suspicion, Notorious, Rear Window, To Catch a Thief, The Birds, Psycho* and *Frenzy*, in addition to his TV series "Alfred Hitchcock Presents." He died on Apr 29, 1980, at Beverly Hills, CA.

INTERNATIONAL LEFTHANDERS DAY. Aug 13. To recognize the needs and frustrations of lefthanders and celebrate the good life of lefthandedness. For info: Lefthanders Intl, Box 8249, Topeka, KS 66608. Phone: (913) 234-2177.

ITALY: PALIO DEL GOLFO. Aug 13. La Spezia. A rowing contest over a 2,000-meter course is held on the second Sunday in August.

KRUPP, von BOHLEN und HALBACH, ALFRIED: BIRTH ANNIVERSARY. Aug 13. As sole owner of the massive Krupp industries, Alfried Krupp took over the factories of German-occupied countries and used them for the Nazi war machine. Sometimes he had complete facilities dismantled and then reassembled inside Germany. He used prisoners of war, civilians from occupied countries and inmates of concentration camps as forced labor in his factories. Found guilty as a war criminal by the Military Court at Nuremberg in 1948, he regained his property after serving three years of a twelve-year sentence. He was named Alfried von Bohlen und Halback at birth but the family was authorized by Emperor Wilhelm II to add the mother's maiden name of Krupp to their own. Alfried was born Aug 13, 1907, at Essen, Germany, and died July 30, 1967, at Essen.

NATIONAL RECREATIONAL SCUBA DIVING WEEK. Aug 13–19. During this week, scuba retailers and instructors all over the US and the Caribbean will be staging events of interest to both divers and non-divers. For certified divers there are underwater treasure events to participate in. For non-divers there are many "Discover Scuba" experiences where you can hop in a swimming pool with a certified instructor for a fun and relaxing first scuba experience. For info: PADI, 1251 E Dyer Road, #100, Santa Ana, CA 92705-5605. Phone: (714) 540-7234.

OAKLEY, ANNIE: BIRTH ANNIVERSARY. Aug 13. Annie Oakley was born at Darke County, OH, Aug 13, 1860. She developed an eye as a markswoman early as a child. She became so proficient that she was able to pay off the mortgage on her family farm by selling the game she killed. A few years after defeating vaudeville marksman Frank Butler in a shooting match, she married him and they toured as a team until joining Buffalo Bill's Wild West Show in 1885. She was one of the star attractions for 17 years. She died Nov 2, 1926, at Greenville, OH.

RATHKAMP MATCHCOVER SOCIETY CONVENTION. Aug 13–20. Airport Hilton, Newark, NJ. To promote matchcover collecting as an inexpensive hobby for all age groups. Est attendance: 400. Sponsor: Tri-State Cardinal Matchcover Club, John C. Williams, 1359 Surrey Rd, Dept AE, Vandalia, OH 45377-1646. Phone: (513) 890-8684.

ROASTING EARS OF CORN FOOD FEST. Aug 13. The Museum of Indian Culture, Allentown, PA. Traditional American Indian event with activities for the whole family. Est attendance: 3,000. For info: Lenni Lenape Historical Soc, RD 2, Fish Hatchery Rd, Allentown, PA 18103-9801. Phone: (610) 797-2121.

SCOTLAND: EDINBURGH INTERNATIONAL FESTIVAL/EDINBURGH FESTIVAL FRINGE. Aug 13–Sept 2. Edinburgh, Lothian. The Festival is one of the world's largest arts festivals attracting many international stars. For info: Edinburgh Festival Society, 21 Market St, Edinburgh, Scotland EH1 1BW. The Festival Fringe consists of the more than 10,000 performances of more than 1,000 shows in 150 venues, including theatre, dance, music and children's shows, that take place at the same time—on the fringes of—the main festival. Est attendance: 620,000. For info: Edinburgh Festival Fringe Society, 180 High St, Edinburgh, Scotland EH1 1QS.

SCOTTS BLUFF COUNTY FAIR. Aug 13–21. County Fairgrounds, Mitchell, NE. County fair including stock displays and shows, special events such as large name country western concert, largest amateur rodeo in Nebraska, chain-saw artists, rubber check races, demolition derby. Beef sale on Monday, Aug 21. Est attendance: 35,000. For info: Leon Meyer, Fair Mgr, Box 157, Mitchell, NE 69357. Phone: (308) 623-1828.

TUNISIA: WOMEN'S DAY. Aug 13. General holiday. Celebration of independence of women.

BIRTHDAYS TODAY

Fidel Castro, 68, President of Cuba, born at Mayari, Oriente Province, Cuba, Aug 13, 1927.

Quinn Cummings, 28, actress (*The Goodbye Girl*, Annie on "Family"), born Los Angeles, CA, Aug 13, 1967.

M. Joycelyn Elders, M.D., Surgeon General of the US, born Schaal, AR, Aug 13, 1933.

Dan Fogelberg, 44, composer, singer ("Same Old Lang Syne," "Leader of the Band"), born at Peoria, IL, Aug 13, 1951.

Pat Harrington, Jr, 66, actor, comedian ("The Jack Paar Show," "One Day at a Time"), born at New York, NY, Aug 13, 1929.

Don Ho, 65, singer ("Tiny Bubbles"), born at Kakaako, HI, Aug 13, 1930.

Ben Hogan, 83, former golfer, born at Dublin, TX, Aug 13, 1912.

Rex Humbard, 76, evangelist, born at Little Rock, AR, Aug 13, 1919.

AUGUST 14 — MONDAY
226th Day — Remaining, 139

ATLANTIC CHARTER SIGNING: ANNIVERSARY. Aug 14. The eight-point agreement was signed on Aug 14, 1941, by US President Franklin D. Roosevelt and British Prime Minister Winston S. Churchill. The charter grew out of a three-day conference aboard ship in the Atlantic Ocean, off the Newfoundland coast, and stated policies and hopes for the future agreed to by the two nations.

Aug 14–15 ☆ *Chase's 1995 Calendar of Events* ☆

CANADA: YUKON DISCOVERY DAY. Aug 14. In the Klondike region of the Yukon, at Bonanza Creek (formerly known as Rabbit Creek), George Washington Carmack discovered gold on Aug 16 or 17, 1896. During the following year more than 30,000 people joined the gold rush to the area. Anniversary is celebrated as a holiday (Discovery Day) in the Yukon, on nearest Monday.

COLOGNE CATHEDRAL: COMPLETION ANNIVERSARY. Aug 14. The largest Gothic church in northern Europe, the Cologne Cathedral in Cologne, Germany, was completed Aug 14, 1880, just 632 years after rebuilding began on Aug 14, 1248. In fact, there had been a church on its site since 873, but a fire in 1248 made rebuilding necessary. The cathedral was again damaged, by bombing, during World War II.

◆ **IMPERIAL PALACE ATTACKED: 50th ANNIVERSARY.** Aug 14, 1945. The Japanese news broadcast an overseas radio bulletin that the emperor would soon accept the Potsdam Proclamation. Hirohito in fact had already recorded a statement to that effect to be released to the Japanese public. That evening, more than a thousand Japanese soldiers attacked the Imperial Palace with hopes of seizing the recording and preventing the release of the message. Troops faithful to the emperor managed to drive them away, and the message was released the following day.

LIBERTY TREE DAY. Aug 14. Massachusetts.

NATIONAL AVIATION WEEK. Aug 14–20. A celebration of flight designed to increase public awareness, knowledge and appreciation of aviation. Annually, the week of Aug 19, Orville Wright's birthday. Fax: (318) 268-5464. Est attendance: 200. For info: Lafayette Natural History Museum and Planetarium, David Hostetter, Curator of Planetarium, 637 Girard Park Dr, Lafayette, LA 70503. Phone: (318) 268-5544.

SOCIAL SECURITY ACT: ANNIVERSARY. Aug 14. The Congress approved, on Aug 14, 1935, the Social Security Act, which contained provisions for the establishment of a Social Security Board to administer federal old-age and survivors insurance in the US. By signing the bill into law, President Franklin D. Roosevelt was fulfilling a 1932 campaign promise.

◆ **SOVIET ADVANCE IN MANCHURIA: 50th ANNIVERSARY.** Aug 14, 1945. Despite Japan's acceptance of the surrender terms of the Potsdam Declaration, Soviet troops continued advancing into Manchuria, while additional Soviet troops invaded Sakhalin and the Kurile islands.

THAYER, ERNEST LAWRENCE: BIRTH ANNIVERSARY. Aug 14. The man who wrote the famous comic baseball ballad "Casey at the Bat" was born at Lawrence, MA, on Aug 14, 1863. The *magna cum laude* graduate of Harvard wrote a series of comic ballads for the *San Francisco Examiner*, of which "Casey at the Bat" was the last. It was published on Sunday, June 3, 1888, and Thayer received $5 in payment for it. Ernest L. Thayer, who regarded the ballad's fame as a nuisance, and whose other writings are widely forgotten, died at Santa Barbara, CA, on Aug 21, 1940.

August 1995

S	M	T	W	T	F	S
		1	2	3	4	5
6	7	8	9	10	11	12
13	14	15	16	17	18	19
20	21	22	23	24	25	26
27	28	29	30	31		

365-INNING SOFTBALL GAME: ANNIVERSARY. Aug 14–15. The Gager's Diner softball team played the Bend'n Elbow Tavern in a 365-inning softball game in 1976. Starting at 10 AM Aug 14, the game was called because of rain and fog at 4 PM, Aug 15. The 70 players, including 20 women, raised $4,000 for construction of a new softball field and for the Monticello, NY, Community General Hospital. The Gagers beat the Elbows 491–467. To date, this remains the longest softball game on record.

◆ **VICTORY DAY or V-J DAY: 50 YEARS.** Aug 14. Anniversary of President Truman's announcement, on Aug 14, 1945, that Japan had surrendered to the Allies, setting off celebrations across the nation. Official ratification of surrender occurred aboard the USS *Missouri* in Tokyo Bay, on Sept 2 (Far Eastern time).

WEIRD CONTEST WEEK. Aug 14–18. 11 AM in front of Music Pier, Boardwalk and Moorlyn Terr, Ocean City, NJ. One contest daily. Contests include artistic pie eating, chewing something meaningful out of a TastyKake Pie, Salt Water Taffy Sculpting, French Fry Sculpting, Wet T-Shirt Throwing, Animal and Celebrity Impersonation. Annually, the third week in August. Sponsors: The City of Ocean City, TastyKake Baking Company, Shriver's Saltwater Taffy. Phone: (609) 399-0272 or (609) 399-6111, Ext 222. Est attendance: 4,500. For info: City of Ocean City, Mark Soifer, PR Dir, City Hall, Ocean City, NJ 08226.

BIRTHDAYS TODAY

Neal Anderson, 31, football player, born at Graceville, FL, Aug 14, 1964.
Russell Baker, 70, journalist, author, born at Loudoun County, VA, Aug 14, 1925.
Julia Child, 83, food authority, author (*The French Chef*), born at Pasadena, CA, Aug 14, 1912.
David Crosby, 54, singer (of Crosby, Stills & Nash), songwriter, born at Los Angeles, CA, Aug 14, 1941.
Alice Ghostley, 69, actress (Bernice on "Designing Women"; Esmeralda on "Bewitched"), born at Eve, MO, Aug 14, 1926.
Buddy Greco, 69, singer ("Mr. Lonely"), composer, musician, born Philadelphia, PA, Aug 14, 1926.
Earvin "Magic" Johnson, 36, former basketball player, born at Lansing, MI, Aug 14, 1959.
Arthur Betz Laffer, 55, economist, born at Youngstown, OH, Aug 14, 1940.
Gary Larson, 45, cartoonist, born at Tacoma, WA, Aug 14, 1950.
Steve Martin, 50, comedian, actor ("Saturday Night Live," *All of Me, Roxanne, LA Story, Parenthood, Father of the Bride*), born Waco, TX, Aug 14, 1945.
Susan Saint James, 49, actress ("MacMillan and Wife," "Kate and Allie"), born Long Beach, CA, Aug 14, 1946.
Robyn Smith, 51, jockey, born at San Francisco, CA, Aug 14, 1944.
Danielle Steel, 48, author (*Vanished, Wanderlust*), born at New York, NY, Aug 14, 1947.

AUGUST 15 — TUESDAY
227th Day — Remaining, 138

ASSUMPTION OF THE VIRGIN MARY. Aug 15. Greek and Roman Catholic churches celebrate Mary's ascent to Heaven.

BARRYMORE, ETHEL: BIRTH ANNIVERSARY. Aug 15. Celebrated award-winning actress of stage, screen and television, born Ethel Blythe, at Philadelphia, PA, Aug 15, 1879. Died at Beverly Hills, CA, June 18, 1959. US Postal Service stamp was issued in 1982 featuring Ethel, John and Lionel Barrymore.

BONAPARTE, NAPOLEON: BIRTH ANNIVERSARY. Aug 15. Anniversary of birth of French emperor, Napoleon Bonaparte, on island of Corsica, Aug 15, 1769. He died in exile, at 5:49 PM, May 5, 1821, on island of St. Helena. Public holiday in Corsica, France.

330

☆ Chase's 1995 Calendar of Events ☆   Aug 15-16

CANADA: WESTERN CANADA GAMES. Aug 15-20. Abbotsford, B.C. The sixth annual Western Canada Summer Games. Competition in 23 sports at venues throughout Abbotsford-Matsqui. Competitors will be young athletes from Western Canada. Along with five full days of keen sport competition, this "Celebration of Excellence" will feature an opening and closing ceremony, plus a variety of entertainment and activities at the Games community festival. Est attendance: 8,000. For info: Abbotsford-Matsqui Chamber of Commerce, 2462 McCallum Rd, Abbottsford, BC, Canada V2S 3P9. Phone: (604) 855-1995.

CONGO: NATIONAL HOLIDAY. Aug 15. Congolese national day.

DORMITION OF THEOTOKOS. Aug 15. Orthodox Church observance of Assumption of the Virgin Mary depends on use of Old or New Calendar. According to New Calendar (Gregorian), the Dormition Fast is observed Aug 1-14, followed by Dormition of Theotokos on Aug 15.

FERBER, EDNA: BIRTH ANNIVERSARY. Aug 15. Edna Ferber was born at Kalamazoo, MI, on Aug 15, 1887. She wrote her first novel, *Dawn O'Hara* in 1911 and became a prolific writer, producing many popular magazine stories. Her novel *So Big* brought her a huge commercial success in 1924 as well as a Pulitzer Prize. Her other novels include *Show Boat, Cimarron, Saratoga Trunk, Giant* and *Ice Palace*, all of which were made into successful films. Ferber collaborated with George Kaufman in writing for the stage on *The Royal Family, Dinner At Eight, Stage Door, The Land Is Bright* and *Bravo*. Miss Ferber died at New York, NY, on Apr 16, 1968.

FINLAND: TAMPERE INTERNATIONAL THEATRE FESTIVAL. Aug 15-20. Tampere. 27th annual festival includes approximately 50 productions from abroad and Finland and 160 performances, plus about 150 other events. Seminars, open stage, exhibitions, etc. Est attendance: 80,000. For info: Tampere Intl Theatre Festival Office, Leena Vihola, Exec Dir, Tullikamarinaukio 2, FIN-33100, Tampere, Finland.

HARDING, FLORENCE KLING DEWOLFE: BIRTH ANNIVERSARY. Aug 15. Wife of Warren Gamaliel Harding, 29th president of the US, born at Marion, OH, Aug 15, 1860. Died Nov 21, 1924.

HIROHITO'S RADIO ADDRESS: 50th ANNIVERSARY. Aug 15, 1945. At noon Japanese radio broadcast the Japanese national anthem, followed by a prerecorded statement by Emperor Hirohito announcing Japan's decision to surrender, citing the devastating power of the new atomic bomb. This was the first time most Japanese citizens had heard the voice of their emperor.

INDIA: INDEPENDENCE DAY. Aug 15. Anniversary of Indian independence (1947).

KOREA: LIBERATION DAY: 50 YEARS. Aug 15. National holiday commemorates acceptance by Japan of Allied terms of surrender in 1945, thereby freeing Korea from 36 years of Japanese domination. Also marks formal proclamation of the Republic of Korea in 1948. Military parades and ceremonies throughout country.

LIECHTENSTEIN: NATIONAL DAY. Aug 15. Public holiday.

NATIONAL RELAXATION DAY. Aug 15. Everyone should have a special day during the year for total relaxation. Today is a time to think about and encourage new ideas for enjoyment of a relaxation day. Annually, Aug 15. Sponsor: Sean Moeller, 12079 Belann Ct, Clio, MI 48420.

PANAMA: PANAMA CITY FOUNDATION DAY. Aug 15. Traditional annual cultural observance recognizes foundation of Panama City.

PENN STATE'S AG PROGRESS DAYS. Aug 15-17. The Larson Agricultural Research Center, Rock Springs, PA. To provide the public with the latest information on agricultural industries and developments Penn State has made in the field of agriculture. More than 300 commercial exhibitors. Est attendance: 45,000. For info: Robert Oberheim, Penn State Univ, 431 Agricultural Admin Bldg, University Park, PA 16802. Phone: (814) 865-2081.

PORTUGAL: OUR LADY OF THE "MONTE" FESTIVAL. Aug 15. Funchal, Madeira Island. Held on one of Madeira's many hills, the "Monte" being one of the closest to Funchal, this festival includes the procession in honor of Our Lady and a fun fair in the evening with burning charcoal fires preparing the "espetadas" (meat on spits) in the open. For info: Portuguese Tourist Office, 590 Fifth Ave, New York, NY 10036. Phone: (212) 354-4403.

SCOTT, SIR WALTER: BIRTH ANNIVERSARY. Aug 15. Anniversary of birth in Edinburgh, Scotland, of famed poet and novelist (Aug 15, 1771–Sept 21, 1832). "But no one shall find me rowing against the stream," he wrote in the introduction to *The Fortunes of Nigel*, "I care not who knows it—I write for the general amusement."

TRANSCONTINENTAL US RAILWAY COMPLETION: 125th ANNIVERSARY. Aug 15. The Golden Spike ceremony at Promontory Point, UT, May 10, 1869, was long regarded as the final link in a transcontinental railroad track reaching from an Atlantic port to a Pacific port. In fact, that link occurred unceremoniously on another date in another state. Diaries of engineers working at the site establish "the completion of a transcontinental track at a point 928 feet east of today's milepost 602, or 3,812 feet east of the present Union Pacific depot building at Strasburg (formerly Comanche)," CO. The final link was made at 2:53 PM, on Aug 15, 1870 (see: Robert A. LeMassena's "... The True Transcontinental Hook-up" in *Kansas Quarterly*, Vol III, No 2). Annual celebration at Strasburg, CO, on a weekend in August. See also: "Golden Spike Day" (May 10).

WEDDING OF THE SEA. Aug 15. Convention Center, Atlantic City, NJ. Begins with mass and interdenominational service; participants proceed to the beach and into the ocean for the blessing of all sea vessels including the lifeguard boats. Annually, Aug 15. For info: Greater Atlantic City Conv and Visitors Bureau, 2314 Pacific Ave, Atlantic City, NJ 08401. Phone: (609) 348-7100.

WOODSTOCK: ANNIVERSARY. Aug 15. The Woodstock Music and Art Fair opened on this day in 1969 on an alfalfa field on or near Yasgur's Farm at Bethel, NY. The three-day concert featured 24 bands and drew a crowd of more than 400,000 people.

BIRTHDAYS TODAY

Mike Connors (Krekor Ohanian), 70, actor ("Mannix"), born at Fresno, CA, Aug 15, 1925.
Linda Ellerbee, 51, journalist, born at Bryan, TX, Aug 15, 1944.
Vernon Eulion Jordan, Jr, 60, civil rights leader, born at Atlanta, GA, Aug 15, 1935.
Phyllis Stewart Schlafly, 71, antifeminist, author, born at St. Louis, MO, Aug 15, 1924.
Gene (Eugene) Upshaw, Jr, 50, union executive and former football player, born at Robstown, TX, Aug 15, 1945.
Kathryn Whitmire, 49, first woman mayor of Houston, TX, born at Houston, TX, Aug 15, 1946.

AUGUST 16 — WEDNESDAY
228th Day — Remaining, 137

ARTISTS IN THE PARK. Aug 16. Cate Park, Wolfeboro, NH. Juried exhibit and sale including 42 artists and craftspeople, demonstrations throughout the day, family entertainment shows at 12 noon, 1:30 and 3:00 PM. Held rain or shine 10 AM to 5 PM. Sponsors: Governor Wentworth Arts Council. For info: Deborah Hopkins, Co-Chair, PO Box 1322, Wolfeboro, NH 03894. Phone: (603) 569-4994.

Aug 16 ☆ *Chase's 1995 Calendar of Events* ☆

BABE RUTH DIES: ANNIVERSARY. Aug 16, 1948. Baseball fans of all ages and all walks of life mourned when the great Bambino died of cancer at New York City at the age of 53. The left-handed pitcher and "Sultan of Swat" hit 714 home runs in 22 major league seasons of play and had played in 10 World Series. His body lay in state at the main entrance of Yankee Stadium where people waited in line for hours to march past the coffin. On Aug 19, countless numbers of people surrounded St. Patrick's Cathedral for the funeral mass and lined the streets along the route to the cemetery as America bade farewell to one of baseball's greatest legends.

BATTLE OF CAMDEN: ANNIVERSARY. Aug 16. Revolutionary War battle fought near Camden, SC, on Aug 16, 1780. American troops led by General Horatio Gates suffered disastrous losses. Nearly 1,000 Americans killed and another 1,000 captured by the British. British losses about 325. One of America's worst defeats in the war.

BEGIN, MENACHEM: BIRTH ANNIVERSARY. Aug 16. Menachem Begin was born at Brest-Litovsk, Poland, on Aug 16, 1913. A militant Zionist and anti-Communist, he fled to Russia in 1939 ahead of the advancing Nazis, where he was soon arrested and sent to Siberia. Freed in 1941, Begin found his way to Palestine and quickly became a leader in the Jewish underground fighting for Israel's independence; by 1943 he headed the national military organization. While he was successful on the radical periphery of power, Begin's election as Prime Minister of Israel in 1977 quickly introduced him to the necessary compromises of governing a nation, and over time he became less outspoken and dramatic. In 1979 at Camp David, MD, he signed a peace treaty between Israel and Egypt with President Jimmy Carter and President Anwar el Sadat of Egypt. Once the peace with Egypt finally was cemented, Begin became embroiled in the continuing international controversy over the future of the Israeli-occupied West Bank and Gaza Strip. Begin spent the last 8 years of his life as a virtual recluse, never fully recovering from the death of his wife, Aliza. He died Mar 9, 1992.

BENNINGTON BATTLE DAY: ANNIVERSARY. Aug 16. Anniversary of battle fought Aug 16, 1777, is legal holiday in Vermont.

CANADA: ROYAL RED ARABIAN HORSE SHOW. Aug 16–20. Regina Exhibition Park, Regina, Saskatchewan. Royal Red Canadian International Arabian Horse Show showcases some of the best Arabian and Half-Arabian horses in North America. Est attendance: 15,000. For info: Tourism Saskatchewan, Saskatchewan Trade and Conv Center, 1919 Saskatchewan Dr, Regina, Canada, S4P 3V7. Phone: (800) 667-7191.

DOMINICAN REPUBLIC: RESTORATION OF THE REPUBLIC. Aug 16. The anniversary of the Restoration of the Republic is celebrated as an official public holiday.

ELVIS PRESLEY SALUTE. Aug 16. Broadway Market Square, Baltimore, MD. Salutes rock and roll singer Elvis Presley. Elvis impersonator, live musical entertainment, Elvis memorabilia on display and available for purchase and "Let Me Be Your Teddy Bear" and Elvis Look-Alike contests featured. Est attendance: 400. For info: Office of Adventures and Fun, Clarence "Du" Burns Arena, 1301 S Ellwood Ave, Baltimore, MD 21224. Phone: (410) 396-9177.

ENGLAND: MANCHESTER MASSACRE OR BATTLE OF PETERLOO: ANNIVERSARY. Aug 16. Anniversary of demonstration by more than 50,000 persons protesting unemployment, starvation wages, overcrowding, high costs and British government policies. The mass meeting, Aug 16, 1819, was held in St. Peter's Fields, Manchester, England. Police and cavalry charged the unarmed crowd with sabres. Casualty estimates for the 10-minute battle varied widely, but several deaths and up to 500 injuries were claimed.

JOE MILLER'S JOKE DAY. Aug 16. A day to tell a joke in honor of the English comic actor Joseph (or Josias) Miller, who was born in 1684 (exact date unknown). Miller acted at the Drury Lane Theatre in London, England, starting in 1709, and was a popular favorite for many years. It is probable that his family had established "Miller's Droll Booth" at Bartholomew's Fair. Joe Miller is remembered mainly because of a book with which he had no direct connection: *Joe Miller's Jests*, compiled by John Mottley, and first published in 1739. It consisted of only 70 pages, containing 247 jokes. Republished, revised and expanded hundreds of times it reached more than 1,500 jokes in the ensuing two centuries. Joe Miller died at London on Aug 16, 1738, never having heard of the celebrated book of *Joe Miller's Jests*. From *Joe Miller's Jests*, London, 1739: "A melting Sermon being preached in a country Church, all fell a weeping but one Man, who being asked, why he did not weep with the rest? O! said he. I belong to another Parish."

MacFADDEN, BERNARR: BIRTH ANNIVERSARY. Aug 16. Physical culture enthusiast and publisher, born at Mill Springs, MO, Aug 16, 1868. He was once publisher of *Physical Culture Magazine*, *True Story Magazine*, *True Romances*, *True Detective Mysteries Magazine* and many others. MacFadden made parachute jumps on his 81st, 83rd and 84th birthdays. He died at Jersey City, NJ, of jaundice, following a three-day fast, on Oct 12, 1955.

101 WILD WEST RODEO. Aug 16–19. Rodeo Arena, Ponca City, OK. PRCA Rodeo. Est attendance: 18,000. For info: Ponca City Rodeo Fdtn, Scott Klososky, Today's Computers, 205 W Hartford, Ponca City, OK 74602. Phone: (405) 765-2340.

PRECANEX 1995: STAMP EXHIBIT/CONVENTION. Aug 16–20. Recreation Center, Wildwood, NJ. Postage stamp exhibit held in conjunction with annual convention of the national association. 46th annual National Stamp Show by the Sea. Est attendance: 5,000. Sponsor: Natl Assn of Precancel Collectors, Inc, Glen W. Dye, Secy, 5121 Park Blvd, Wildwood, NJ 08260-0121.

PRESLEY, ELVIS: DEATH ANNIVERSARY. Aug 16. One of America's most popular singers, Elvis Presley was pronounced dead at the Memphis Baptist Hospital at 3:30 PM, on Aug 16, 1977. He was 42 years old. The 10th anniversary (1987) of his death was an occasion for pilgrimages by admirers to Graceland, his home and gravesite at Memphis, TN, when an estimated 20,000 persons participated in a candlelight tribute to Presley. See also: "Presley, Elvis: Birth Anniversary" (Jan 8).

SWITZERLAND: INTERNATIONAL FESTIVAL OF MUSIC. Aug 16–Sept 9. Lucerne. A classical music festival considered to have the most spectacular scenery of any music festival in Europe. Features 50 events including symphony concerts, chamber music, concerts presenting young artists and choral concerts. Annually, in August and September. For info: Intl Festival of Music, Rosmarie Hohler-Welti, Mgr Press/PR, Postfach CH-6002, Lucerne, Switzerland.

TONTITOWN GRAPE FESTIVAL. Aug 16–20. (Tentative.) Tontitown, AR. More than 6,000 lbs of homemade pasta and sauce are served at the Italian spaghetti dinners; more than 100 crafts exhibitors; carnival. 97th annual festival. Est attendance: 25,000. For info: Tontitown Grape Festival, PO Box 39, Tontitown, AR 72770. Phone: (501) 361-2612.

BIRTHDAYS TODAY

Robert Culp, 65, actor (Kelly Robinson on "I Spy"; *Bob and Carol and Ted and Alice*), born Berkeley, CA, Aug 16, 1930.

Frank Gifford, 65, sportscaster, born Santa Monica, CA, Aug 16, 1930.

August 1995

S	M	T	W	T	F	S
		1	2	3	4	5
6	7	8	9	10	11	12
13	14	15	16	17	18	19
20	21	22	23	24	25	26
27	28	29	30	31		

Eydie Gorme (Edith Gormezano), 63, singer ("Blame It on the Bossa Nova"), born at New York, NY, Aug 16, 1932.
Timothy Hutton, 35, actor (*Taps, Made in Heaven*), born Malibu, CA, Aug 16, 1960.
Madonna (Madonna Louise Veronica Ciccone), 37, singer ("Material Girl"), actress, born at Bay City, MI, Aug 16, 1958.
Carol Moseley-Braun, 48, US Senator (D, Illinois), born at Chicago, IL, Aug 16, 1947.
Julie Newmar, 60, actress (Cat Woman on "Batman"; *Li'l Abner*), born at Hollywood, CA, Aug 16, 1935.
Christian Okoye, 34, football player, born at Enugu, Nigeria, Aug 16, 1961.
Fess Parker, 68, actor ("Daniel Boone," *Davy Crockett*), born Ft Worth, TX, Aug 16, 1927.
Lesley Ann Warren, 49, actress (*Victor/Victoria, Life Stinks!*), born New York, NY, Aug 16, 1946.

AUGUST 17 — THURSDAY
229th Day — Remaining, 136

BALLOON CROSSING OF ATLANTIC OCEAN: ANNIVERSARY. Aug 17. Three Americans—Max Anderson, 44; Ben Abruzzo, 48; and Larry Newman, 31—all of Albuquerque, NM, became first to complete transatlantic trip in a balloon. Starting from Presque Isle, ME, Aug 11, they traveled some 3,200 miles in 137 hours, 18 minutes, landing at Miserey, France (about 60 miles west of Paris), in their craft, named the *Double Eagle II*, on Aug 17, 1978.

COBBLESTONE FESTIVAL. Aug 17-20. Falls City, NE. Includes games, races, contests, carnival rides, flea market, food concessions, car show, demolition derby, tractor pull, parade, craft demos and more. Annually, mid-August. Est attendance: 2,500. Sponsor: Chamber of Commerce, PO Box 146, Falls City, NE 68355. Phone: (402) 245-4228.

CROCKETT, DAVID "DAVY": BIRTH ANNIVERSARY. Aug 17. American frontiersman, adventurer and soldier, born Aug 17, 1786, in Hawkins County, TN. Died during final heroic defense of the Alamo, Mar 6, 1836. In his *Autobiography* (1834), Crockett wrote, "I leave this rule for others when I'm dead, Be always sure you're right—then go ahead."

ELWOOD GLASS FESTIVAL. Aug 17-19. Elwood, IN. Glass factory tours, parade, craft market, flea market, quilt show. Est attendance: 25,000. For info: Chamber of Commerce, 108 S Anderson St, Elwood, IN 46036. Phone: (317) 552-0180.

FORT SUMTER SHELLED BY NORTHERN FORCES: ANNIVERSARY. Aug 17. In what would become a long siege, Union forces began shelling Fort Sumter in Charleston, SC, on Aug 17, 1863. The site of the first shots fired during the Civil War, Sumter endured the siege for a year and a half before being returned to Union hands.

450-MILE OUTDOOR FESTIVAL. Aug 17-20. The World's Largest Outdoor Sale Festival. 127-corridor outdoor Sale Festival stretches from Covington, KY, covering 450 miles and ends in Gadsden, AL. The road comes alive and draws 35,000 cars carrying 80,000 people in search of antique cars, quilts, primitive furniture, arts and crafts and musical entertainment. Sponsor: Fentress County. For info: Toni Salyer, PO Box 1128, Jamestown, TN 38556. Phone: (615) 879-7713.

GABON: NATIONAL DAY. Aug 17. National holiday.

GOLDWYN, SAMUEL: BIRTH ANNIVERSARY. Aug 17. Motion picture producer and industry pioneer, born Samuel Goldfish, at Warsaw, Poland, Aug 17, 1882. Goldwyn died at Los Angeles, CA, Jan 31, 1974. Attributed to Goldwyn is the observation: "Anybody who goes to see a psychiatrist ought to have his head examined."

HOPE WATERMELON FESTIVAL. Aug 17-20. Hope, AR. An inexpensive event the entire family can participate in while promoting the city of Hope and having fun. Hope is the birthplace of 42nd President of USA Bill Clinton. Annually the third weekend in August. Est attendance: 85,000. Sponsor: Hope-Hempstead County Chamber of Commerce, 108 W 3rd, Hope, AR 71801. Phone: (501) 777-3640.

INDONESIA: INDEPENDENCE DAY: 50 YEARS. Aug 17. National holiday. Republic proclaimed on this day, 1945, upon withdrawal of Japanese.

KENTUCKY STATE FAIR (WITH WORLD CHAMPIONSHIP HORSE SHOW). Aug 17-27. Kentucky Fair and Expo Center, Louisville, KY. Midway, concerts by nationally known artists and the World's Championship Horse Show. For info: Harold Workman, KY Fair and Expo Ctr, Box 37130, Louisville, KY 40233. Phone: (502) 367-5000.

LITTLESTOWN GOOD OLD DAYS CELEBRATION. Aug 17-19. Gettysburg, PA. Antique show, arts and crafts, flea market, music and entertainment. Est attendance: 20,000. For info: Gettysburg Travel Council, 35 Carlisle St, Gettysburg, PA 17325. Phone: (717) 334-6274.

MISSOURI RIVER FESTIVAL OF THE ARTS. Aug 17-26. Thespian Hall, Boonville, MO. Performing arts festival with major symphony, opera, jazz, children's program, Broadway music and folk music. Est attendance: 5,000. For info: Judy Shields, Admin, Friends of Historic Boonville, PO Box 1776, Boonville, MO 65233. Phone: (816) 882-7977.

MISSOURI STATE FAIR. Aug 17-27. Sedalia, MO. Livestock shows, commercial and competitive exhibits, horse show, car races, tractor pulls, carnival and headline musical entertainment. Economical family entertainment. Est attendance: 400,000. For info: Missouri State Fair, Dianne Larkin, Publicity Dir, Box 111, Sedalia, MO 65302. Phone: (816) 530-5600.

MOON PHASE: LAST QUARTER. Aug 17. Moon enters Last Quarter phase at 11:04 PM, EDT.

POWERS, FRANCIS GARY: BIRTH ANNIVERSARY. Aug 17. One of America's most famous aviators, Francis Gary Powers was born at Jenkins, KY, Aug 17, 1929. The CIA agent, pilot of a U-2 overflight across the Soviet Union, was shot down May 1, 1960, near Sverdlovsk, USSR. He was tried, convicted and sentenced to 10 years imprisonment, at Moscow, USSR, in Aug 1960. Returned to the US in 1962, in exchange for an imprisoned Soviet spy (Colonel Rudolf Abel), he found an unwelcoming homeland. Powers died in a helicopter crash near Los Angeles, CA, on Aug 2, 1977.

SOLDIERS' REUNION CELEBRATION. Aug 17. Newton, NC. Parade climaxes the 106th annual soldiers' celebration—"oldest patriotic event of its kind in the US, honoring all veterans." Annually, the third Thursday in August. Concerts, arts, crafts and games. Est attendance: 20,000. For info: Soldiers' Reunion Committee, Box 267, Newton, NC 28658. Phone: (704) 464-2383.

SUN PRAIRIE'S SWEET CORN FESTIVAL. Aug 17-20. Sun Prairie, WI. Family-oriented fun. Carnival, midget auto races, parade, beer, brats, food, exhibits and all the hot buttered sweet corn you can eat. Est attendance: 100,000. Sponsor: Chamber of Commerce, 109 E Main, Sun Prairie, WI 53590. Phone: (608) 837-4547.

SWEDEN: SOUR HERRING PREMIERE. Aug 17. By ordinance, the year's supply of sour herring may begin to be sold on the third Thursday in August.

Aug 17-18 ☆ *Chase's 1995 Calendar of Events* ☆

BIRTHDAYS TODAY

Belinda Carlisle, 37, singer ("Mad About You"), born at Hollywood, CA, Aug 17, 1958.

Harrison V. Chase, 82, associate professor emeritus, Florida St Univ, co-founder and co-editor of *Chase's Annual Events* for its first 13 years, born Big Rapids, MI, Aug 17, 1913.

Robert DeNiro, 52, actor (Oscars *Raging Bull, The Godfather II; Taxi Driver*), born New York, NY, Aug 17, 1943.

Maureen O'Hara, 75, actress (*The Quiet Man, Miracle on 34th Street, The Hunchback of Notre Dame, Only the Lonely*), born Dublin, Ireland, Aug 17, 1920.

Sean Penn, 35, actor (*Fast Times at Ridgemont High, Casualties of War*), born at Santa Monica, CA, Aug 17, 1960.

Nelson Piquet, 43, auto racer, born at Brasilia, Brazil, Aug 17, 1952.

Larry Rivers, 72, artist, born at New York, NY, Aug 17, 1923.

Guillermo Vilas, 43, tennis player, born at Mar del Plata, Argentina, Aug 17, 1952.

AUGUST 18 — FRIDAY
230th Day — Remaining, 135

ALABAMA PRO NATIONAL TRUCK AND TRACTOR PULL. Aug 18–19. Lexington, AL. Five different classes compete. Pro National 7500 Superstock, Pro National 6200 2-Wheel Drive Trucks, Pro National 7200 Modified Tractors (multi-engine), State Level 6200 4-Wheel Drive Truck, Local Level 4x4 Street Class and Local Level 15, 200 Farm Class. Annually, the third Friday and Saturday in August. Est attendance: 12,000. For info: Shoals Chamber of Commerce, Tourism Coord, 104 S Pine St, Florence, AL 35630. Phone: (205) 764-4661.

AMERICAN NEUTRALITY APPEAL: ANNIVERSARY. Aug 18. President Woodrow Wilson, on Aug 18, 1914, followed his Aug 4th Proclamation of Neutrality with an appeal to the American people to remain impartial in thought and deed with respect to the war that was raging in Europe (World War I).

BAD POETRY DAY. Aug 18. After all the "good" poetry you were forced to study in school, here's a chance for a payback. Invite some friends over, compose some really rotten verse and send it to your old high school English teacher. Phone: (212) 388-8673 or (717) 274-8451. Sponsor: Wellness Permission League, Tom or Ruth Roy, 2105 Water St, Lebanon, PA 17046.

BLACK HILLS STEAM AND GAS THRESHING BEE. Aug 18–20. Sturgis, SD. Held annually one mile east of the Sturgis Airport. The pioneer heritage of our country comes alive with old steam and gas tractors, threshing machine demonstrations, flea market, parade, antique automobiles and dozens of other demonstrations related to yesteryear. Est attendance: 3,200. For info: Chamber of Commerce, Box 504, Sturgis, SD 57785. Phone: (605) 347-2556.

CANADA: DISCOVERY DAYS. Aug 18–21. Dawson City, Yukon. Gold was discovered in the Klondike Aug 16 (some believe 17), 1896, and each year Dawson commemorates the event with a festival including parade, raft and canoe races, Demo Derby, mud bog, ball tournament, dances and more. 1995 is also the 100th anniversary of the Royal Canadian Mounted Police in the Yukon. Est attendance: 3,000. For info: Discovery Days, PO Box 308, Dawson City, Yukon, Canada Y0B 1G0. Phone: (403) 993-5434.

CANADA: FRINGE THEATER EVENT. Aug 18–27. Edmonton, Alberta. A ten-day extravaganza of new plays, old plays, dance, music, mime and street entertainment. More than 800 performances of 150 productions in 18 theaters, in the parks and on the streets. Performers from around the world. Regarded as the largest and most exciting festival of alternative theater in North America. Est attendance: 420,000. Sponsor: Fringe Festival, c/o Judy Lawrence, Chinook Theatre, 10329-83 Ave, Edmonton, Alta, Canada T6E 2C6. Phone: (403) 448-9000.

CANADA: OWEN SOUND SUMMERFOLK FESTIVAL. Aug 18–20. Kelso Beach Park, Owen Sound, Ontario. A celebration of traditional and contemporary folk music, crafts, workshops, concerts. Canadian and international artists. Annually, the third weekend in August. Fax: (519) 371-2973. Est attendance: 10,000. For info: Georgian Bay Folk Soc, Box 521, Owen Sound, Ont, Canada N4K 5R1. Phone: (519) 371-2995.

CANADA: SOURDOUGH RENDEZVOUS GOLD RUSH BATHTUB RACE. Aug 18–21. Beginning in Whitehorse, Yukon. 4th annual 462-mile bathtub race from Whitehorse to Dawson. Fax: (403) 668-6755. Est attendance: 400. For info: Yukon Sourdough Rendezvous Society, Box 5108, Whitehorse, YT, Canada Y1A 4S3. Phone: (403) 667-2148.

COSHOCTON CANAL FESTIVAL. Aug 18–20. Coshocton, OH. Celebrates the arrival of first canal boat in Roscoe. Crafts, parades and old-time entertainment. Annually, the third weekend in August. Est attendance: 8,000. For info: Roscoe Village Fdtn, 381 Hill St, Coshocton, OH 43812. Phone: (800) 877-1830.

DANISH FESTIVAL. Aug 18–19. Greenville, MI. To honor the Danish heritage of the area. Fax: (616) 754-4710. Est attendance: 100,000. Sponsor: Danish Festival Inc, Kathy Jo VanderLaan, 116 E Washington, Greenville, MI 48838. Phone: (616) 754-6369.

DARE, VIRGINIA: BIRTH ANNIVERSARY. Aug 18. Virginia Dare, the first child of English parents to be born in the New World, was born to Ellinor and Ananias Dare, at Roanoke Island, NC, Aug 18, 1587. When a ship arrived to replenish their supplies in 1591, the settlers (including Virginia Dare) had vanished, without leaving a trace of the settlement.

FESTIVAL OF THE LITTLE HILLS. Aug 18–20. Frontier Park and South Main, St. Charles, MO. The largest festival of the year, activities include demonstrations by craftspeople and artisans. Annually, the third weekend in August. Est attendance: 300,000. For info: St. Charles Conv and Visitors Bureau, 230 S Main, St. Charles, MO 63301. Phone: (800) 366-2427.

GINZA HOLIDAY: JAPANESE CULTURAL FESTIVAL. Aug 18–20. Midwest Buddhist Temple, Chicago, IL. See the Waza, 300 years of Edo craft tradition come alive as master craftsmen from Tokyo demonstrate their arts. Japanese folk and classical dancing, martial arts, taiko (drums), flower arrangements and cultural displays. Chicken Teriyaki, sushi, udon, shaved ice, corn-on-the-cob and refreshments. Annually, the third weekend in August. Est attendance: 5,000. For info: Midwest Buddhist Temple, 435 W Menomonee St, Chicago, IL 60614. Phone: (312) 943-7801.

GREAT PIKES PEAK COWBOY POETRY GATHERING. Aug 18–20. Pikes Peak Center, Colorado Springs, CO. Fundraising event for the Library District featuring poetry, a Western Trade Show, children's programs, The Great Chow Down Barbecue, street dance, pancake toss contest, and a concert with special guest. Phone: (719) 531-6333 ext 1150. Est attendance: 24,000. For info: Judy Evans, Coord, Pikes Peak Library Dist, 5550 N Union Blvd, Colorado Springs, CO 80918.

HAWAII ADMISSION DAY HOLIDAY. Aug 18. The third Friday in August is observed as a state holiday each year, recognizing the anniversary of Hawaii's statehood. Hawaii became the 50th state on Aug 21, 1959.

HOLZFEST. Aug 18–20. Amana, IA. Woodcrafters of all types display and sell their products. Fellowship, entertainment, food and demonstrations. Annually, the third weekend in August. Est attendance: 11,000. Sponsor: Personalized Wood Products, R.C. Eichacker, Box 193, Amana, IA 52203. Phone: (319) 622-3100.

August 1995

S	M	T	W	T	F	S
		1	2	3	4	5
6	7	8	9	10	11	12
13	14	15	16	17	18	19
20	21	22	23	24	25	26
27	28	29	30	31		

Chase's 1995 Calendar of Events — Aug 18-19

LEWIS, MERIWETHER: BIRTH ANNIVERSARY. Aug 18. American explorer (of Lewis and Clark expedition), born Aug 18, 1774. Died Oct 11, 1809.

MILWAUKEE IRISH FEST. Aug 18-20. Milwaukee, WI. World's largest and most comprehensive Irish music and cultural event, featuring dozens of major Irish and Irish American entertainers. Activities include sports, drama, contests, displays, food, market place, dance and children's activities. Weeklong summer school series of lectures and demonstrations in Irish music, dance and culture precedes the festival. Annually, the third weekend in August. Est attendance: 95,000. For info: Milwaukee Irish Fest, 515 N Glenview Ave, Milwaukee, WI 53213. Phone: (414) 476-3378.

MONTANA COWBOY POET GATHERING. Aug 18-19. Lewistown, MT. Gather cowboy poets to Lewistown to share lyric and poetry. Est attendance: 900. Sponsor: Cowboy Poetry Committee, Lewistown Area Chamber of Commerce, Attn: Kathy Thompson, PO Box 818, Lewistown, MT 59457. Phone: (406) 538-5436.

◈ NINETEENTH AMENDMENT TO US CONSTITUTION RATIFIED: 75th ANNIVERSARY. Aug 18. The 19th Amendment, ratified on August 18, 1920, extended the right to vote to women.

TEXAS RANCH ROUNDUP. Aug 18-19. Wichita Falls, TX. Cowboys from prestigious ranches in Texas compete in events that make up their daily work. Cattle roping and penning and other rodeo-type events plus ranch cooking contests, ranch talent contests. Est attendance: 23,000. For info: Wichita Falls Conv and Visitors Bureau, Paula Moers, Services Coord, PO Box 1860, Wichita Falls, TX 76307. Phone: (817) 723-2741.

TRAILS WEST!. Aug 18-20. Civic Center Park, St. Joseph, MO. Arts festival celebrating St. Joseph's unique cultural heritage. Sponsor: Allied Arts Council, 118 S 8th, St. Joseph, MO 64501. Phone: (816) 233-0231.

WORLD'S OLDEST CONTINUOUS PRCA RODEO. Aug 18-20. Payson, AZ. The Payson Rodeo has been held continuously since 1884. Named #1 small outdoor rodeo in America! Est attendance: 20,000. For info: Payson Chamber of Commerce, Box 1380, Payson, AZ 85547. Phone: (800) 672-9766.

YANKTON RIVERBOAT DAYS AND SUMMER ARTS FESTIVAL. Aug 18-20. Yankton, SD. Parades, water ski show and competition, tug-o-war, mud volleyball, antique tractor pull, continuous entertainment, juried arts fair, food vendors, rodeo and square dancing. Est attendance: 75,000. For info: Yankton Chamber of Commerce, PO Box 588, Yankton, SD 57078-0588. Phone: (605) 665-3636.

BIRTHDAYS TODAY

Rosalynn (Eleanor) Smith Carter, 68, former First Lady, wife of President Jimmy Carter, born at Plains, GA, Aug 18, 1927.
Walter J. Hickel, 76, Governor of Alaska (I), born at Claflin, KS, Aug 18, 1919.
Fat (Lafayette) Lever, 35, basketball player, born at Pine Bluff, AR, Aug 18, 1960.
Luc Montagnier, 63, virologist, born at Chabris, France, Aug 18, 1932.
Martin Mull, 52, actor, comedian ("Mary Hartman, Mary Hartman," "Roseanne"), born Chicago, IL, Aug 18, 1943.
Roman Polanski, 62, filmmaker (*Rosemary's Baby, MacBeth, Chinatown*), born at Paris, France, Aug 18, 1933.
Robert Redford, 58, actor (*Butch Cassidy and the Sundance Kid, The Sting*), director (*A River Runs Through It*), born Santa Monica, CA, Aug 18, 1937.

Patrick Swayze, 41, actor, dancer ("North and South," *Dirty Dancing*), born at Houston, TX, Aug 18, 1954.
Malcolm Jamal Warner, 25, actor ("The Cosby Show"), born at Jersey City, NJ, Aug 18, 1970.
Caspar Willard Weinberger, 78, former Secretary of Defense, born at San Francisco, CA, Aug 18, 1917.
Shelley Winters (Shelly Schrift), 73, actress (Oscars *A Patch of Blue, The Diary of Anne Frank; The Poseidon Adventure*), born St. Louis, MO, Aug 18, 1922.

AUGUST 19 — SATURDAY
231st Day — Remaining, 134

AFGHANISTAN: INDEPENDENCE DAY. Aug 19. National day.

AFRICAN-AMERICAN CULTURAL FESTIVAL. Aug 19. Miller Plaza, Chattanooga, TN. Local dancers, music choral, gospel, jazz, poets, writers. Also African marketplace with clothing, food, jewelry, crafts, drama and other special attractions. Sponsor: Chattanooga African-American Museum. Est attendance: 20,000. For info: Chatt. African American Museum, Vilma Fields, 730 M. L. King Blvd, Chattanooga, TN 37403. Phone: (615) 267-1076.

AFRICA'S LEGACY IN MEXICO: PHOTOGRAPHS BY TONY GLEATON. Aug 19-Oct 1. Harrison Museum of African American Culture, Roanoke, VA. See Feb 11 for full description. Call or write the Smithsonian for other dates and venues. For info: Smithsonian Institution Travelling Exhibition Service, 1100 Jefferson Dr SW, Ste 3146, Washington, DC 20560. Phone: (202) 357-2700.

ANTIQUE MARINE ENGINE EXPOSITION. Aug 19-20. Mystic Seaport, Mystic, CT. Collectors from across the US and Canada gather for the 3rd annual exposition of pre-World War II marine engines and engine models. Unique engines power watercraft in a boat parade on the Mystic River. For info: Mystic Seaport, 75 Greenmanville Ave, PO Box 6000, Mystic, CT 06355-0990. Phone: (203) 572-5315.

BATTLE OF BLUE LICKS CELEBRATION. Aug 19-20. Blue Licks Battlefield State Park, Mount Olivet, KY. Commemorates the anniversary of Revolutionary War Battle of Blue Licks (which involved Daniel Boone). Living history demonstrations, arts, crafts, games, competitions and battle reenactment. Annually, the third weekend in August. Est attendance: 3,000. For info: Blue Licks Battlefield State Pk, Bill Stevens, Pk Mgr, Mt. Olivet, KY 41064. Phone: (606) 289-5507.

BIKE VAN BUREN IX. Aug 19-20. Van Buren County, PA. A laid-back bicycle tour of the villages, landmarks, and landscape of this rural Iowa county. The "red carpet of hospitality" is rolled out for the bikers as they pass through. Est attendance: 400. Sponsor: Villages of Van Buren, Inc, Mary E. Muir, Exec Dir, PO Box 9, Keosauqua, IA 52565. Phone: (800) 868-7822.

BONANZAVILLE USA PIONEER DAYS. Aug 19-20. Bonanzaville USA, West Fargo, ND. The 43 buildings in the Victorian-era pioneer village come to life. Ice cream and lemonade are sold at the drug store, log cabin residents turn out lefse and jellies for sample and sale, a country market sells garden produce, threshing machines crank up, church services are performed in both German and Norwegian, and the highlights of the weekend are the old-time vehicle parades. Est attendance: 12,000. For info: Margo Lang, Bonanzaville USA, 1351 W Main, West Fargo, ND 58078. Phone: (701) 282-2822.

CANADA: OJIBWAY KEESHIGUNUN. Aug 19-20. Old Fort William, Thunder Bay, Ontario. A celebration of Old Fort William's native culture. Taste historic foods, enjoy unique demonstrations and join in crafts and games. Experience the atmosphere of a powwow and the complexity of native living. Annually, the second to last weekend in August. Est attendance: 6,000. For info: Laura Craig, Vickers Heights PO, Thunder Bay, Ont, Canada P0T 2Z0. Phone: (807) 577-8461.

Aug 19 ☆ Chase's 1995 Calendar of Events ☆

CANDLELIGHT VIGILS FOR HOMELESS ANIMALS' DAY. Aug 19. Humane societies, SPCAs, animal welfare and animal rights organizations nationwide take part in this 4th annual vigil to increase public awareness of the millions of young, healthy dogs and cats (including many purebreds) killed in shelters each year for lack of a home. The motto of the day is "Spay/Neuter . . . It Stops the Killing!" If your local shelter is not participating, contact the event's national coordinator for the location of the vigil closest to you or for detailed information on holding a vigil of your own. A special "Vigil Packet" is available for $5.00. For info: Intl Soc for Animal Rights, 421 S State St, Clarks Summit, PA 18411. Phone: (800) 543-4727.

COLORADO STATE FAIR. Aug 19–Sept 4. State Fairgrounds, Pueblo, CO. One of the nation's oldest western fairs, it is also Colorado's largest single event, drawing more than a million visitors. Family fun, top-name entertainment, lots of food and festivities. For info: Colorado State Fair, Jerry Robbe, Pres/Genl Mgr, Pueblo, CO 81004. Phone: (719) 561-8484.

COUP ATTEMPT IN THE SOVIET UNION: ANNIVERSARY. Aug 19. On Aug 19, 1991, a Soviet coalition of hard-line communists staged a coup d'état, removing Soviet President Mikhail Gorbachev from power. The hard-liners, who included Vice President Gennady Yanayev, the Soviet Defense Minister, the head of the KGB and the Soviet Interior Minister, claimed that Gorbachev was removed because of his ill health. Yanayev was installed as president and a six-month state of emergency was declared. Gorbachev had been on vacation in the Crimea and was due to return to Moscow the next day to sign the historic union treaty that would have taken much of the power away from the Central USSR government and given it to the republics signing the treaty. On the third day, Aug 21, in the face of the massive public resistance, the conspirators gave up, flew to the Crimea hoping to negotiate with Gorbachev and were taken into custody. A shaken and chastened Gorbachev returned to Moscow in an extremely weakened position. The coup began a chain reaction of events involving the independence of the republics, the dissolution of the Soviet Union and the loss of power by Gorbachev.

DRAGON BOAT RACES. Aug 19–20. Keokuk, IA. 25-member teams row 43-foot long Chinese Dragon Boats on a course set on the Mississippi River below the dam in Keokuk. Est attendance: 12,000. For info: Keokuk Area Conv and Tourism Bureau, Jeff White, 401 Main, Keokuk, IA 52632. Phone: (800) 383-1219.

ENGLAND: THREE CHOIRS FESTIVAL. Aug 19–26. Gloucester Cathedral, Gloucester, Gloucestershire. Europe's oldest music festival, with performances by the three cathedral choirs of Hereford, Worcester and Gloucester. For info: Mr Anthony Boden, Secretary, Three Choirs Festival, Community House, College Green, Gloucester, England GL1 2LZ.

FORBES, MALCOLM: BIRTH ANNIVERSARY. Aug 19. Publisher, born Aug 19, 1919, at New York, NY. Malcolm Forbes was an unabashed proponent of capitalism and his beliefs led to his colorful and successful climb to the top of the magazine publishing industry. Known as much for his lavish lifestyle as his publishing acumen, Forbes was also an avid motorcyclist and hot air balloonist. He died on Feb 24, 1990, at Far Hills, NJ.

GERMAN PLEBISCITE: ANNIVERSARY. Aug 19. In a plebiscite, Aug 19, 1934, 89.9 percent of German voters approved giving Chancellor Adolf Hitler the additional office of president, placing the Fuhrer in uncontestable and supreme command of that country's destiny.

August 1995

S	M	T	W	T	F	S
		1	2	3	4	5
6	7	8	9	10	11	12
13	14	15	16	17	18	19
20	21	22	23	24	25	26
27	28	29	30	31		

HI-AYH MIDNIGHT MADNESS FUN BICYCLE RIDE. Aug 19. San Diego, CA. 22-mile ride along the San Diego Bay. Begins at the stroke of midnight. Est attendance: 2,000. For info: San Diego Council, American Youth Hostels, 335 W Beech St, San Diego, CA 92101. Phone: (619) 338-9981.

HOMETOWN DAYS. Aug 19–20. Strasburg, CO. To remind people of the first continuous chain of rails from an Atlantic to a Pacific port. The rails were joined Aug 15, 1870, at Comanche, which was later renamed Strasburg. Annually, the third weekend in August. Est attendance: 1,000. For info: Sandy Miller, Rt 1, Box 85, Strasburg, CO 80136. Phone: (303) 622-4890.

HOOD RIVER APPLE JAM. Aug 19. Hood River, OR. The Gorge's premier performing arts festival with music, food and entertainment for the entire family. Est attendance: 2,500. For info: Showman Inc, PO Box 13243, Portland, OR 97213. Phone: (503) 387-7529.

LEADVILLE 100-MILE ULTRA RUN. Aug 19–20. Leadville, CO. 100-mile course through the Rocky Mountains with aid stations. For info: Chamber of Commerce, Box 861, Leadville, CO 80461. Phone: (800) 933-3901.

MAINE HIGHLAND GAMES. Aug 19. Thomas Point Beach, Brunswick, ME. Presented by the Saint Andrew's Society of Maine. Bagpipe Bands, Highland and Scottish dancing, Scottish arts and crafts fair, folksingers, Scottish fiddling, children's games; adult athletics including Tossing of the Caber, Wheat Sheaf Toss and Putting of the Stone; border collie herding demonstrations, Highland cattle and individual piping contests. American and Scottish foods galore. The only Scottish event of its kind held in Maine! Scots and non-Scots will enjoy the color, pageantry and friendly atmosphere. 17th annual games. 9–5. Adults $8, children under 12 $4. For info: Thomas Point Beach, Meadow Rd, Box 5419, Brunswick, ME 04011. Phone: (207) 725-6009.

MID-SUMMER OCEAN SWIM AND BEACH BARBECUE. Aug 19. 12th & Ocean Ave, Seaside Park, NJ. Swim 1½ miles in the ocean (usually a northward direction parallel to the coastline). Barbecue follows the swim. Bring beach blankets or chairs. Annually, third Saturday in August. Est attendance: 300. Sponsor: Maureen Rohrs, Ocean County YMCA, PO Box 130, Toms River, NJ 08754. Phone: (908) 341-9622.

NASH, OGDEN: BIRTH ANNIVERSARY. Aug 19. American writer, best remembered for his humorous verse. Born at Rye, NY, Aug 19, 1902, and died May 19, 1971. "Undeniably brash/Was young Ogden Nash/Whose notable verse/Was admirably terse/And written with panache."

★ **NATIONAL AVIATION DAY.** Aug 19. Presidential Proclamation 2343, of July 25, 1939, covers all succeeding years. Always Aug 19 of each year since 1939. Observed annually on anniversary of birth of Orville Wright who piloted "first self-powered flight in history," Dec 17, 1903. First proclaimed by President Franklin D. Roosevelt.

NATIONAL INVITATIONAL FUN FLY AND MODEL AIRCRAFT SHOW. Aug 19. Chimney Rock Park, Chimney Rock, NC. Academy of Aeronautics (AMA) members from across the country are invited to fly, compete and display radio-controlled aircraft in this 3rd annual show. Est attendance: 2,000. For info: Mary Jaeger-Gale, Mktg Mgr, PO Box 39, Chimney Rock, NC 28720. Phone: (800) 277-9611.

CHASE'S 1996

Two ways to simplify your life . . .

People who rely on *Chase's Annual Events* need to be in the know every day of the year—and like to plan ahead. Now it's easier than ever to make sure you receive the newest edition of *Chase's* as soon as it's available.

☆ Simply return one of the Pre-Publication Order Forms below. Your copy of *Chase's 1996* will be reserved immediately and shipped just as soon as it's off press in October 1995.

☆ To receive *Chase's* **automatically every year**, take advantage of the Standing Order Authorization. There's no chance of missing out on *Chase's*—each year's new edition will be shipped and billed automatically. (Of course, you may change or cancel your Standing Order at any time.)

Both convenient order options carry our unconditional guarantee—you may return *Chase's* for any reason within 10 days of receipt for a full refund. Why not order today and be sure of starting 1996 with *Chase's Annual Events* at your fingertips!

Mail to: Contemporary Books, Dept. C,
Two Prudential Plaza, Suite 1200, Chicago, IL 60601-6790

YES! Send me _____ copies of
Chase's Annual Events 1996 at $49.95 each $ _______

Quantity Discounts:
Deduct 10% per copy for 3–9 copies
Deduct 20% per copy for 10 or more copies $ _______

Sales Tax:
Add applicable tax in AL, CA, FL, IL, NC, NY, OH, PA, TX $ _______

Shipping and Handling:
Add $4.00 for first copy, $2.50 for each additional copy $ _______

☐ Check ☐ Money Order
(payable to Contemporary Books) TOTAL $ _______
☐ VISA ☐ MasterCard

Acct. ______________________ Exp. ___/___

X ______________________
Signature if charging to bank card

____
Name (please print)

SHIP TO:

____
Name

____
Address

____
Address

____
City State Zip

STANDING ORDER/Authorization

To ensure that I receive each year's new edition of ***Chase's Annual Events***, please accept this Standing Order Authorization to ship me _______ copies of each annual edition, beginning with the 1996 edition, and to bill me at the address shown above.

Signature ______________________
Date

Name ______________________
(please print)

Phone (   ) ______________________

M95

Mail to: Contemporary Books, Dept. C,
Two Prudential Plaza, Suite 1200, Chicago, IL 60601-6790

YES! Send me _____ copies of
Chase's Annual Events 1996 at $49.95 each $ _______

Quantity Discounts:
Deduct 10% per copy for 3–9 copies
Deduct 20% per copy for 10 or more copies $ _______

Sales Tax:
Add applicable tax in AL, CA, FL, IL, NC, NY, OH, PA, TX $ _______

Shipping and Handling:
Add $4.00 for first copy, $2.50 for each additional copy $ _______

☐ Check ☐ Money Order
(payable to Contemporary Books) TOTAL $ _______
☐ VISA ☐ MasterCard

Acct. ______________________ Exp. ___/___

X ______________________
Signature if charging to bank card

____
Name (please print)

SHIP TO:

____
Name

____
Address

____
Address

____
City State Zip

STANDING ORDER/Authorization

To ensure that I receive each year's new edition of ***Chase's Annual Events***, please accept this Standing Order Authorization to ship me _______ copies of each annual edition, beginning with the 1996 edition, and to bill me at the address shown above.

Signature ______________________
Date

Name ______________________
(please print)

Phone (   ) ______________________

M95

Chase's 1995 Calendar of Events — Aug 19-20

NATURAL CHIMNEYS JOUSTING TOURNAMENT. Aug 19. Natural Chimneys Regional Park, Mt Solon, VA. A modern version of a medieval contest, believed to be the oldest continuously held sporting event in America. Annually, the third Saturday in August. Est attendance: 550. For info: Upper Valley Regional Park Authority, Box 478, Grottoes, VA 24441. Phone: (703) 350-2510.

NORTHEASTERN WISCONSIN ANTIQUE POWER ASSOCIATION THRESHEREE. Aug 19-20. Geisel Farm, Sturgeon Bay, WI. Displays of operating antique farm equipment. Horse and antique tractor pulls. Barn dance. Crafts tables and games for all kids. Annuallly, the third full weekend in August. Est attendance: 3,000. For info: Josephine Bochek, 4376 Rudy Rd, Sturgeon Bay, WI 54235.

PETALUMA RIVER FESTIVAL. Aug 19. Petaluma, CA. Celebrate Petaluma's river heyday with crafts and food booths; steamboat and sternwheeler rides. Sea chanteys aboard 1891 scow schooner. Est attendance: 25,000. Sponsor: Petaluma River Festival Assn, PO Box 2013, Petaluma, CA 94953. Phone: (707) 762-5331.

RODDENBERRY, GENE: BIRTH ANNIVERSARY. Aug 19. The creator of the popular TV series "Star Trek," Gene Roddenberry was born at El Paso, TX, on Aug 19, 1921. Turning from his first career as an airline pilot to writing, he created one of the most successful TV science fiction series ever. The original series, which ended its run in 1969, lives on in re-runs, and the "Star Trek: The Next Generation" and "Star Trek: Deep Space Nine" series have continuing popularity. Six films also have spawned from the original concept. Roddenberry died on Oct 24, 1991, at Santa Monica, CA.

SPACE MILESTONE: *SOYUZ T-7* (USSR). Aug 19. Launched from Tyuratam, USSR, on Aug 19, 1982, with second woman in space (test pilot Svetlana Savitskaya) and two other cosmonauts. Docked at *Salyut 7* and visited the cosmonauts in residence there for the three previous months before returning to Earth on Aug 27 in the *Soyuz T-5* vehicle that had been docked there. The *Soyuz T-7* returned to Earth Dec 10 with the remaining two cosmonauts.

SPACE MILESTONE: *SPUTNIK 5* (USSR): 35th ANNIVERSARY. Aug 19. Space menagerie satellite launched on Aug 19, 1960. The dog passengers, Belka and Strelka, became first living organisms recovered from orbit, Aug 19, 1960.

WALES: LLANDRINDOD WELLS VICTORIAN FESTIVAL. Aug 19-27. Llandrindod Wells, Powys. The whole town turns the clocks back to the Victorian age. For info: Festival Office, Old Town Hall, Temple St, Llandrindod Wells, Powys, Wales LD1 5DL.

WRIGHT, ORVILLE: BIRTH ANNIVERSARY. Aug 19. Aviation pioneer born at Dayton, OH, on Aug 19, 1871, and died there Jan 30, 1948.

BIRTHDAYS TODAY

Morten Anderson, 35, football player, born at Struer, Denmark, Aug 19, 1960.

William Jefferson Clinton, 49, 42nd US president, born at Hope, AR, Aug 19, 1946.

Ronald Maurice Darling, Jr, 35, baseball player, born at Honolulu, HI, Aug 19, 1960.

David F. Durenberger, 61, former US Senator (R, Minnesota), born at St. Cloud, MN, Aug 19, 1934.

Ring Lardner, Jr, 80, writer (*You Know Me, Haircut & Other Stories*), born at Chicago, IL, Aug 19, 1915.

Cindy Nelson, 40, skier, born at Lutsen, MN, Aug 19, 1955.

Jill St. John (Jill Oppenheim), 55, actress (*Diamonds Are Forever*), born at Los Angeles, CA, Aug 19, 1940.

Willie Shoemaker, 64, former jockey, born at Fabens, TX, Aug 19, 1931.

John Stamos, 32, actor ("General Hospital," "Full House"), born at Los Angeles, CA, Aug 19, 1963.

AUGUST 20 — SUNDAY
232nd Day — Remaining, 133

EXETER ROAD RALLY. Aug 20. Exeter City Park, Exeter, NE. Participants decipher clues which take them throughout the countryside and into neighboring communities. Evening meal served. Registration 1 PM. 11th annual rally. Annually, the third Sunday in August. Est attendance: 150. For info: Norene Fitzgerald, Box 368, Exeter, NE 68351. Phone: (402) 759-4910.

GUEST, EDGAR ALBERT: BIRTH ANNIVERSARY. Aug 20. Newspaperman and author of folksy, homespun verse that enjoyed great popularity and was syndicated in more than 100 newspapers. Born at Birmingham, England, Aug 20, 1881, died at Detroit, MI, Aug 5, 1959. "Eddie Guest Day" usually proclaimed on birth anniversary in Detroit.

HARRISON, BENJAMIN: BIRTH ANNIVERSARY. Aug 20. The 23rd president of the US, born at North Bend, OH, Aug 20, 1833, was the grandson of William Henry Harrison, ninth president of the US. His term of office, Mar 4, 1889–Mar 3, 1893, was preceded and followed by the presidential terms of Grover Cleveland (who thus became the 22nd and 24th president of the US). Harrison died at Indianapolis, IN, Mar 13, 1901.

HUNGARY: NATIONAL DAY. Aug 20. National holiday.

LITTLE LEAGUE BASEBALL WORLD SERIES. Aug 20-26. Williamsport, PA. Eight teams from the US and foreign countries compete for the World Championship. Est attendance: 100,000. For info: Little League Baseball HQ, Box 3485, Williamsport, PA 17701. Phone: (717) 326-1921.

LOVECRAFT, H. P.: BIRTH ANNIVERSARY. Aug 20. H(oward) P(hillips) Lovecraft, American author of horror tales of the supernatural, a pioneering science fiction writer and a notable epistolarian, was born at Providence, RI, on Aug 20, 1890. He died there Mar 15, 1937.

LUXEMBOURG: OUR LADY OF GIRSTERKLAUS PROCESSION. Aug 20. Rosport. Tradition since 1328. Annually, the Sunday after Aug 15.

LUXEMBOURG: SCHUEBERMESS SHEPHERD'S FAIR. Aug 20-Sept 2. Fair dates from 1340. (Two weeks beginning on the next to last Sunday of August.)

MOSQUITO AWARENESS (WITH WORLD CHAMPIONSHIP MOSQUITO-CALLING CONTEST). Aug 20. Crowley's Ridge State Park, Walcott, AR. World Championship Mosquito-calling contest, 2nd annual Great Mosquito Cook-Off for recipes using ½ cup of mosquitos, Mr. and Ms. Mosquito Leggs competition, Great Mosquito Race, mosquito in hay stack contest, mosquito-swatting contest, soda-pop-sucking contest, demonstrations of mosquito-trapping techniques and a few serious programs on insects. 9th annual weekend. Est attendance: 3,000. For info: Crowley's Ridge State Park, PO Box 97, Walcott, AR 72474-0097. Phone: (501) 573-6751.

NATIONAL HOMELESS ANIMALS DAY. Aug 20. A day to call attention to the fact that from 12–17 million healthy dogs and cats are killed each year at animal shelters because of overpopulation—a problem that has a solution: spay/neuter! Also a day to memorialize the animals killed due to owner irresponsibility and to sympathize with the caring shelter personnel who must take the lives of the animals. The day will be marked by Candlelight Vigils sponsored by animal protection organizations throughout the US. Sponsor: Intl Society for Animal Rights, Inc, 421 S State St, Clarks Summit, PA 18411. Phone: (717) 586-2200.

NATIONAL RELIGIOUS SOFTWARE WEEK. Aug 20-26. To celebrate the increasing role computer software, such as BibleSource and MacBible, is playing in biblical studies. Annually, the last full week of August. For info: Jonathan Petersen, Dir, Media Relations, Zondervan Publishing House, 5300 Patterson Ave SE, Grand Rapids, MI 49530. Phone: (616) 698-3417.

NEWPORT MARITIME TEDDY BEAR FESTIVAL. Aug 20. Islander Doubletree Hotel, Newport, RI. Show and sale featuring "dozens of America's leading Teddy Bear designers and makers." Also, dealers selling name-brand bears; bear repair and restoration clinic; appraisals; antique bears; door prizes, raffle and more. For info: John K. Pringle, Pres, Pringle Productions, Ltd, PO Box 757, Bristol, CT 06011-0757. Phone: (203) 585-9940.

O'HIGGINS, BERNARDO: BIRTH ANNIVERSARY. Aug 20. First ruler of Chile after its declaration of independence. Called the "Liberator of Chile." Born at Chillan, Chile, Aug 20, 1778. Died at Lima, Peru, Oct 24, 1842.

OLD ADOBE FIESTA. Aug 20. Petaluma, CA. Craftspeople will be on hand to show visitors how to make candles, weave baskets, bake bread, churn butter. Entertainment will be a live Hispanic band, square dancers and contests for costumes, whiskers and kids. Local non-profit organizations will provide refreshment booths. Take a self-guided tour of General Vallejo's Casa Grande, now a historical museum. 10–5. Send self-addressed stamped envelope. Est attendance: 3,000. For info: Old Adobe Assn, PO Box 631, Petaluma, CA 94953. Phone: (707) 762-4871.

PLUTONIUM FIRST WEIGHED: ANNIVERSARY. Aug 20. University of Chicago scientist Glenn Seaborg and his colleagues first weighed plutonium, the first man-made element, on Aug 20, 1942.

PRESIDENT BENJAMIN HARRISON'S 162nd BIRTHDAY CELEBRATION. Aug 20. Indianapolis, IN, Harrison's hometown. Also free tours of the Victorian mansion. Est attendance: 500. For info: President Benjamin Harrison Home, PR Dept/Debbie Browning, 1230 N Delaware St, Indianapolis, IN 46202. Phone: (317) 631-1898.

SAARINEN, (GOTTLIEB) ELIEL: BIRTH ANNIVERSARY. Aug 20. Famed architect. Born at Helsinki, Finland, Aug 20, 1873. Died at Bloomfield Hills, MI, July 1, 1950.

SPACE MILESTONE: *VIKING 1* AND *2* (US). Aug 20–Sept 9. Sister ships launched toward Mars from Cape Canaveral, FL, on Aug 20 and Sept 9, 1975. *Viking 1*'s lander touched down on Mars July 20, 1976, and *Viking 2*'s lander on Sept 3, 1976. Sent back to Earth high-quality photographs, analysis of atmosphere, weather information and results of sophisticated experiments intended to determine whether life may be present on Mars.

SPACE MILESTONE: *VOYAGER 2* (US). Aug 20. Launched Aug 20, 1977, unmanned spacecraft starts trip to Jupiter (1979), Saturn (1981), Uranus (1986) and Neptune (1989).

TOBACCO HARVEST. Aug 20. McLean, VA. Help cut and hang tobacco to dry. Enjoy 18th-century games and light refreshments afterwards. Est attendance: 300. Sponsor: Claude Moore Colonial Farm at Turkey Run, 6310 Georgetown Pike, McLean, VA 22101. Phone: (703) 442-7557.

TURKEY: VICTORY DAY. Aug 20. Nationwide. Military parades, performing of the Mehtar band, the world's oldest military band, fireworks.

XEROX 914 DONATED TO SMITHSONIAN: ANNIVERSARY. Aug 20. On Aug 20, 1985, the original Xerox 914 copying machine (which had been introduced to the public 25 years earlier—in March 1960) was formally presented to the Smithsonian Institution's National Museum of American History, in Washington, DC. Invented by Chester Carlson, a patent lawyer, the quick and easy copying of documents by machine revolutionized the world's offices.

BIRTHDAYS TODAY

Connie Chung (Constance Yu-Hwa), 49, journalist (co-anchor "CBS Evening News"), born at Washington, DC, Aug 20, 1946.
Carla Fracci, 59, dancer, born at Milan, Italy, Aug 20, 1936.
Isaac Hayes, 53, musician, singer, songwriter, composer (scores for *Shaft*, *Tough Guys*, *Truck Turner*), born at Covington, TN, Aug 20, 1942.
Donald King, 64, boxing promoter, born at Cleveland, OH, Aug 20, 1931.
Mark Edward Langston, 35, baseball player, born at San Diego, CA, Aug 20, 1960.
George John Mitchell, 62, former US Senator (D, Maine), born at Waterville, ME, Aug 20, 1933.
Graig Nettles, 51, former baseball player, born at San Diego, CA, Aug 20, 1944.
Ron Paul, 60, former congressman from Texas, Libertarian Party candidate for president of US, born at Pittsburgh, PA, Aug 20, 1935.
Robert Plant, 47, singer, born at Bromwich, England, Aug 20, 1948.

AUGUST 21 — MONDAY
233rd Day — Remaining, 132

AMERICAN BAR ASSOCIATION FOUNDING: ANNIVERSARY. Aug 21. Organized at Sarasota, NY, Aug 21, 1878.

AQUINO, BENIGNO: ASSASSINATION ANNIVERSARY. Aug 21. Filipino opposition leader Benigno S. Aquino, Jr, was shot and killed at the Manila airport on his return to the Philippines on Aug 21, 1983. The killing precipitated greater anti-Marcos feeling and figured significantly in the Feb 7, 1986, election that brought about the collapse of the government administration of Ferdinand E. Marcos and the inauguration of Corazon C. Aquino, widow of the slain man, as president.

HAWAII STATEHOOD: ANNIVERSARY. Aug 21, 1959. Pres Dwight Eisenhower signed a proclamation admitting Hawaii to the Union. The statehood bill had passed the previous March with a stipulation that statehood should be approved by a vote of Hawaiian residents. The referendum passed by a huge margin in June and Eisenhower proclaimed Hawaii the 50th state on Aug 21.

OZMA'S BIRTHDAY. Aug 21. Celebration of the birth of the Queen of Oz. Sponsor: Intl Wizard of Oz Club, Inc, Fred M. Meyer, Secy, 220 N 11th St, Escanaba, MI 49829.

SPACE MILESTONE: *GEMINI 5* (US). Aug 21. Launched on Aug 21, 1965, Lieutenant Colonel Cooper and Lieutenant Commander Conrad orbit Earth 128 times for new international record of 8 days. First launch of a satellite from a manned spacecraft.

WEST MICHIGAN FAIR. Aug 21–27. West Michigan Fairgrounds, Ludington, MI. Traditional county fair, complete with carnival rides and games, livestock judging, 4-H exhibits, live entertainment and food. Annually, the third week in August. Est attendance: 20,000. For info: Ludington Area Conv and Visitors Bureau, Sue Brillhart, Exec Dir, 5827 W US 10, Ludington, MI 49431. Phone: (800) 542-4600.

WYOMING STATE FAIR. Aug 21–26. Douglas, WY. To recognize the products, achievements and cultural heritage of the people of Wyoming by bringing together rural and urban residents for an inexpensive, entertaining and educational experience. Est attendance: 100,000. For info: Wyoming State Fair, Drawer 10, Douglas, WY 82633. Phone: (307) 358-2398.

BIRTHDAYS TODAY

Wilton Norman Chamberlain, 59, basketball Hall of Famer, born at Philadelphia, PA, Aug 21, 1936.
Jackie DeShannon, 51, singer, songwriter ("Put a Little Love in Your Heart"), born at Hazel, KY, Aug 21, 1944.

✯ Chase's 1995 Calendar of Events ✯ Aug 22-23

Princess Margaret, 65, Countess of Snowdon, sister of Queen Elizabeth II, born at Glamis, Scotland, Aug 21, 1930.
Jim McMahon, 36, football player, born at Jersey City, NJ, Aug 21, 1959.
Kenny Rogers, 57, singer ("Lucille," "Lady"), born at Houston, TX, Aug 21, 1938.
Clarence Williams, III, 56, actor ("The Mod Squad"), born at New York, NY, Aug 21, 1939.

AUGUST 22 — TUESDAY
234th Day — Remaining, 131

BE AN ANGEL DAY. Aug 22. A day to do "one small act of service for someone. Be a blessing in someone's life." Annually, Aug 22. Sponsor: Angelic Alliance, Jane M Howard, PO Box 95, Upperco, MD 21155. Phone: (410) 833-6912.

DEBUSSY, CLAUDE: BIRTH ANNIVERSARY. Aug 22. (Achille) Claude Debussy, French musician and composer, especially remembered for his impressionistic "tone poems," was born at St. Germain-en-Laye, France, on Aug 22, 1862. He died at Paris, France, on Mar 25, 1918.

INTERNATIONAL YACHT RACE: ANNIVERSARY. Aug 22. A silver trophy (then known as the "Hundred Guinea Cup," and offered by the Royal Yacht Squadron) was won in a race around the Isle of Wight, Aug 22, 1851, by the US yacht *America*. The trophy, later turned over to the New York Yacht Club, became known as the America's Cup.

JUDSON, EMILY CHUBBOCK: BIRTH ANNIVERSARY. Aug 22. American poet and author Emily C. Judson, born at Eaton, NY, on Aug 22, 1817. She wrote under the pseudonym Fanny Forester. Judson died at Hamilton, NY, on June 1, 1854.

KGB FOUNDER STATUE DISMANTLED: ANNIVERSARY. Aug 22. In the wake of the popular revolt that smashed the right-wing Soviet coup, a crowd of 10,000 Muscovites watched on Aug 22, 1991, as cranes dismantled a 14-ton statue of Felix Dzerzhinsky, a Polish intellectual tapped by Vladimir Lenin to organize the fledgling Soviet Union's secret police. After trucks had hauled away the massive likeness of Dzerzhinsky, Moscow residents adorned the statue's pedestal and the nearby KGB headquarters with graffiti.

LANGLEY, SAMUEL PIERPONT: BIRTH ANNIVERSARY. Aug 22. American astronomer, physicist and aviation pioneer for whom Langley Air Force Base, VA, is named. Born at Roxbury, MA, Aug 22, 1834, Langley died at Aiken, SC, on Feb 27, 1906.

MICHIGAN STATE FAIR. Aug 22-Sept 4. State Fairgrounds, Detroit, MI. Est attendance: 400,000. For info: State of Michigan, Dept of Commerce, 1120 W State Fair Ave, Detroit, MI 48203. Phone: (313) 368-1000.

US AMATEUR (GOLF) CHAMPIONSHIP. Aug 22-27. Newport Country Club, Newport, RI, Wanumetonomy Golf & Country Club, Middletown, RI. Sponsor: US Golf Assn, Championship Dept, Golf House, Far Hills, NJ 07931. Phone: (908) 234-2300.

◈ **VIETNAM CONFLICT BEGINS: 50th ANNIVERSARY.** Aug 22, 1945. Less than a week after the Japanese surrender, a team of Free French parachuted into southern Indo-China in response to a successful coup by a Communist guerrilla named Ho Chi Minh in a land called Vietnam.

WILLARD, ARCHIBALD M.: BIRTH ANNIVERSARY. Aug 22. American artist, best known for his painting, *The Spirit of '76*, was born at Bedford, OH, on Aug 22 or 26, 1836. Willard died at Cleveland, OH, Oct 11, 1918.

BIRTHDAYS TODAY

Ray Bradbury, 75, author (*The Toynbee Convector, Farenheit 451*), born Waukegan, IL, Aug 22, 1920.
Gerald Paul Carr, 63, former astronaut, born at Denver, CO, Aug 22, 1932.
Henri Cartier-Bresson, 87, photographer, born at Chanteloup, France, Aug 22, 1908.
Valerie Harper, 54, actress (Rhoda on "The Mary Tyler Moore Show" and "Rhoda"), born Suffern, NY, Aug 22, 1941.
John Lee Hooker, 78, singer ("Boom, Boom"), born at Clarksdale, MS, Aug 22, 1917.
Paul Leo Molitor, 39, baseball player, born at St. Paul, MN, Aug 22, 1956.
Bill Parcell, 54, football coach, born at Englewood, NY, Aug 22, 1941.
Norman H. Schwarzkopf, 61, retired army general, born at Trenton, NJ, Aug 22, 1934.
Cindy Williams, 47, actress (*American Graffiti*, "Laverne & Shirley"), born Van Nuys, CA, Aug 22, 1948.
Carl Yastrzemski, 56, baseball Hall of Famer, born at Southampton, NY, Aug 22, 1939.

AUGUST 23 — WEDNESDAY
235th Day — Remaining, 130

EAST COAST SURFING CHAMPIONSHIPS AND SPORTS FESTIVAL. Aug 23-27. Oceanfront to 12th St, Virginia Beach, VA. 33rd annual championship. Pro and amateur surfing and volleyball, 5K run, bike races, skimboarding, catamaran races, golf tournament, windsurfing, team tug-of-war, bands, food. Est attendance: 100,000. For info: Virginia Beach Jaycees, PO Box 62041, Virginia Beach, VA 23466. Phone: (804) 499-8822.

FIRST MAN-POWERED FLIGHT ANNIVERSARY. Aug 23. At Schafter, CA, Aug 23, 1977, Bryan Allen pedaled the 70-lb *Gossamer Condor* for a mile at a "minimal altitude of two pylons," in a flight certified by the Royal Aeronautical Society of Britain, winning a £50,000 prize offered by British industrialist Henry Kremer. See also: "First Man-Powered Flight Across English Channel: Anniversary" (June 12).

ITALY: STRESA MUSICAL WEEKS. Aug 23-Sept 18. Stresa. International festival includes concerts by symphonic orchestras, chamber music, recitals and a series by young winners of international musical contests. 34th annual festival. For info: Assoc Settimane Musicali di Stresa, Via R Bonghi 4, 28049 Stresa (Lago Maggiore), Italy.

MASTERS, EDGAR LEE: BIRTH ANNIVERSARY. Aug 23. American poet, author of the *Spoon River Anthology*, was born at Garnett, KS, Aug 23, 1869. He died at Melrose Park, PA, Mar 5, 1950.

NEVADA STATE FAIR. Aug 23-27. Reno Livestock Events Center, Reno, NV. State entertainment and carnival, with home arts, agriculture and commercial exhibits. Est attendance: 85,000. For info: Nevada State Fair, Gary Lubra, Exec Dir, 1350-A N Wells Ave, Reno, NV 89512. Phone: (702) 688-5767.

PERRY, OLIVER HAZARD: BIRTH ANNIVERSARY. Aug 23. American naval hero. Born Aug 23, 1785. Died Aug 23, 1819. Best remembered is his announcement of victory at the Battle of Lake Erie on Sept 10, 1813: "We have met the enemy, and they are ours."

SACCO-VANZETTI MEMORIAL DAY. Aug 23. Nicola Sacco and Bartolomeo Vanzetti were electrocuted at the Charlestown, MA, prison on Aug 23, 1927. Convicted of a shoe factory payroll robbery during which a guard had been killed, Sacco and Vanzetti maintained their innocence to the end. Six years of appeals marked this American cause célèbre during which substantial evidence was presented to show that both men were elsewhere at the time of the crime. Even a confession from another person failed to save them from public anger. However, on the 50th anniversary of their execution, Massachusetts governor Michael S. Dukakis proclaimed Aug 23, 1977, a memorial day, noting that the 1921 trial had been "permeated by prejudice."

SPACE MILESTONE: INTELSAT-4 F-7 (US). Aug 23. International Communications Satellite Consortium—*Intelsat*—to relay communications from North and South America to Europe and Africa. Launched Aug 23, 1973.

VIRGO, THE VIRGIN. Aug 23–Sept 22. In the astronomical/astrological/zodiac, which divides the sun's apparent orbit into 12 segments, the period Aug 23–Sept 22 is identified, traditionally, as the sun sign of Virgo, the Virgin. The ruling planet is Mercury.

BIRTHDAYS TODAY

Michael James Boddicker, 38, baseball player, born at Cedar Rapids, IA, Aug 23, 1957.
Barbara Eden (Barbara Huffman), 61, actress ("I Dream of Genie," *The Wonderful World of the Brothers Grimm*), born at Tucson, AZ, Aug 23, 1934.
Gene Kelly, 83, actor, dancer (*American in Paris, Singin' in the Rain*), born at Pittsburgh, PA, Aug 23, 1912.
Shelley Long, 46, actress (Diane Chambers on "Cheers"; *Irreconcilable Differences*), born at Fort Wayne, IN, Aug 23, 1949.
Patricia McBride, 53, dancer, born at Teaneck, NJ, Aug 23, 1942.
Vera Miles, 65, actress (*The Wrong Man, Psycho*), born Boise City, OK, Aug 23, 1930.
Mark Russell (Mark Ruslander), 63, political comedian ("Real People"), born at Buffalo, NY, Aug 23, 1932.
Richard Sanders, 55, actor ("WKRP in Cincinnati," "Berrengers"), born at Harrisburg, PA, Aug 23, 1940.
Rick Springfield, 46, singer, actor, born at Sydney, Australia, Aug 23, 1949.

AUGUST 24 — THURSDAY
236th Day — Remaining, 129

ACTON FAIR. Aug 24–27. Acton, ME. A country fair featuring horse and ox pulls, firemen's muster, 4-H projects, flowers, arts and crafts, stage shows and handicraft. Est attendance: 20,000. For info: Lista C. Staples, PO Box 75, Shapleigh, ME 04076. Phone: (207) 636-2026.

CRIM FESTIVAL OF RACES. Aug 24–26. Flint, MI. Festival includes international 10-mile road race, 5K and 8K runs, 5K and 8K walks, 1-mile run and Teddy Bear Trot (¼-mile run) for children ages 4–12. Pledges go to Special Olympics. Sports and Fitness Expo on Aug 24–26; Pasta Party on Aug 25 at University Pavillion Rink; Food Festival on Aug 26; Cycling Classic, a leg of the Tour de Michigan also on Aug 26. Est attendance: 25,000. For info: Crim Festival of Races, Box 981, Flint, MI 48501. Phone: (810) 235-3398.

HOTTER 'N HELL HUNDRED WEEKEND. Aug 24–27. Wichita Falls, TX. To encourage bicycling and to benefit the city of Wichita Falls, its parks and greenspaces. Spaghetti dinner, consumer show, and speakers on Friday in conjunction with packet pick-up. Also festival and consumer show all day Saturday. The largest sanctioned century ride in the USA—more than 10,000 riders. On Saturday, Aug 26, rides of 25, 50, 100 miles and 10K or 100K (6.2 and 62 miles) and a 100-mile USCF Road Race. USCF criterium races on Aug 27. Est attendance: 15,000. For info: Wichita Falls Bicycling Club, Inc, PO Box 2096, Wichita Falls, TX 76307. Phone: (817) 692-2925.

ITALY: VESUVIUS DAY. Aug 24. Anniversary of the eruption of Vesuvius, an active volcano in southern Italy, on Aug 24, AD 79, which destroyed the cities of Pompeii, Stabiae and Herculaneum.

JARVIS, GREGORY B.: BIRTH ANNIVERSARY. Aug 24. Gregory B. Jarvis, a civilian engineer with Hughes Aircraft Co, was born in Detroit, MI, on Aug 24, 1944. Jarvis was the 41-year-old payload specialist who perished with other crew members and Christa McAuliffe in the Space Shuttle *Challenger* explosion on Jan 28, 1986. See also: "*Challenger* Space Shuttle Explosion: Anniversary" (Jan 28).

MINNESOTA STATE FAIR. Aug 24–Sept 4. St. Paul, MN. Major entertainers, agriculture displays, arts, crafts, food, carnival rides, animal judging and performances. Est attendance: 1,600,000. For info: Minnesota State Fair, 1265 Snelling Ave N, St. Paul, MN 55108. Phone: (612) 642-2200.

NEW YORK STATE FAIR. Aug 24–Sept 4. Empire Expo Center, Syracuse, NY. Agriculture and livestock competitions, top-name entertainment, the International Horse Show, business and industrial exhibits, the midway and ethnic presentations. Est attendance: 900,000. For info: NY State Fair, Joseph LaGuardia, Dir of Mktg, NYS Fair, Empire Expo Ctr, Syracuse, NY 13209. Phone: (315) 487-7711.

OREGON STATE FAIR. Aug 24–Sept 4. Salem, OR. Exhibits, products and displays illustrate Oregon's role as one of the nation's major agricultural and recreational states. Floral gardens, carnival, big-name entertainment, horse show and food. Est attendance: 730,000. For info: Oregon State Fair, 2330 17th St NE, Salem, OR 97310. Phone: (503) 378-3247.

ST. BARTHELEMY: PATRON SAINT DAY. Aug 24. The festival of St. Barthelemy is celebrated for several days, beginning on Aug 24. The "look and feeling of a French country fair."

ST. BARTHOLOMEW'S DAY MASSACRE: ANNIVERSARY. Aug 24. Anniversary of the massacre in Paris and throughout France of thousands of Huguenots, as ordered by King Charles IX (and approved by his mother, the regent, Catherine de Médici). The massacre began when the church bells tolled at dawn on St. Bartholomew's Day, Aug 24, 1572, and continued for several days. Pope Gregory XIII ordered a medal struck to commemorate the event, but Protestant countries abhorred the killings, estimated at 2,000 to 70,000.

SOUTHERN CYCLONE: ANNIVERSARY. Aug 24. A hurricane hit Savannah, GA, and Charleston, SC, on Aug 24, 1893, killing between 1,000 and 2,000 people.

SPACE MILESTONE: VOYAGER 2 (US). Aug 24. Launched in 1977, *Voyager* had its first close encounter with Neptune on Aug 24, 1989.

TELLURIDE MUSHROOM FESTIVAL. Aug 24–27. Telluride, CO. To educate people about the types of wild mushrooms—edible, poisonous, psychoactive—and their cultivation. Est attendance: 150. For info: Fungophile, Inc, Box 480503, Denver, CO 80248-0503. Phone: (303) 296-9359.

☆ Chase's 1995 Calendar of Events ☆   Aug 24-25

TENNESSEE WALKING HORSE NATIONAL CELEBRATION. Aug 24–Sept 2. Celebration Grounds, Shelbyville, TN. This 57th annual show determines which of two thousand or more entries will be crowned world champion. Ten days and nights of high-stepping excitement. Est attendance: 250,000. For info: Tennessee Walking Horse Natl Celebration, PO Box 1010, Shelbyville, TN 37160. Phone: (615) 684-5915.

UKRAINE: INDEPENDENCE DAY. Aug 24. National day.

WASHINGTON, DC: INVASION ANNIVERSARY. Aug 24–25. British forces briefly invaded and raided Washington, DC, burning the capitol, the president's house and most other public buildings on Aug 24–25, 1814. President James Madison and other high US government officials fled to safety until British troops (not knowing the strength of their position) departed the city two days later.

BIRTHDAYS TODAY

Jim Capaldi, 51, musician, born at Evesham, England, Aug 24, 1944.
Gerry Cooney, 39, former boxer, born at Brooklyn, NY, Aug 24, 1956.
Steve Guttenberg, 37, actor ("Billy," "No Soap, Radio"), born at Brooklyn, NY, Aug 24, 1958.
Marlee Matlin, 30, actress (Oscar *Children of a Lesser God*), born Morton Grove, IL, Aug 24, 1965.
Reggie (Reginald Wayne) Miller, 30, basketball player, born at Riverside, CA, Aug 24, 1965.
Cal Ripken, Jr, 35, baseball player, born at Havre de Grace, MD, Aug 24, 1960.
Louis Teicher, 71, pianist, composer, born at Wilkes-Barre, PA, Aug 24, 1924.
Mason Williams, 57, composer, born at Abilene, TX, Aug 24, 1938.

AUGUST 25 — FRIDAY
237th Day — Remaining, 128

ALASKA STATE FAIR. Aug 25–Sept 4. Palmer, AK. Cows and critters, music and dancing, rides, excitement and family fun at the state's largest summer extravaganza. More than 500 events including demonstrations, high-caliber entertainment, rodeos, horse shows, homemaking and agricultural exhibits. Est attendance: 235,000. For info: Alaska State Fair, Inc, 2075 Glenn Hwy, Palmer, AK 99645. Phone: (907) 745-4827.

BE KIND TO HUMANKIND WEEK. Aug 25–31. Make the world a nicer place one person at a time! Conscientious effort is the key with caring the very essence of life itself. Share your care and make life worth while! Sacrifice Our Wants for Others' Needs Sunday. Motorist Consideration Monday. Touch-a-heart Tuesday. Willing to Lend a Hand Wednesday. Thoughtful Thursday. Forgive Your Foe Friday. Speak Kind Words Saturday. Sponsor: Lorraine Jara, PO Box 586, Island Heights, NJ 08732-0586.

CANADA: MORDEN CORN AND APPLE FESTIVAL. Aug 25–27. Morden, Manitoba. It's fun and it's free!! Fax: (204) 822-6494. Est attendance: 35,000. For info: Cheryl Howdle, Mgr, Chamber of Commerce, 102-195 Stephen St, Morden, Manitoba, Canada R6M 1V3. Phone: (204) 822-5630.

CHARLESTON STERNWHEEL REGATTA FESTIVAL. Aug 25–Sept 4. Charleston, WV. A 10-day festival celebrating Charleston's river heritage. Est attendance: 400,000. Sponsor: Charleston Festival Commission, Inc, Susie Salisbury, #3 Capitol St, Charleston, WV 25301. Phone: (304) 348-6419.

CHARLOTTE OBSERVER MOONRIDE AND MOONLIGHT CRITERIUM. Aug 25. (Tentative.) Charlotte, NC. The Criterium at 8 PM is a US Cycling Fed-sanctioned professional bicycle race, followed at midnight by the Moonride, a family leisure ride. Est attendance: 3,700. For info: Observer Moonride, Promotion Dept, Box 30294, Charlotte, NC 28230. Phone: (704) 358-5800.

ENGLAND: ARUNDEL FESTIVAL. Aug 25–Sept 3. Arundel, West Sussex. Festival of the arts with some open air performances of Shakespeare, opera and dance in the beautiful setting of Arundel Castle. Est attendance: 500. For info: Julie Young, Arundel Festival, The Mary Gate, Arundel, West Sussex, England BN18 9AT.

ENGLAND: BEATLES FESTIVAL. Aug 25–29. Liverpool, Merseyside. Music and entertainment connected with the Beatles. For info: Mr Bill Heckle, Cavern City Tours Ltd, The Cavern Club, 10 Mathew St, Liverpool, Merseyside, England L2 6RE.

GREAT AMERICAN DUCK RACE. Aug 25–27. Deming, NM. "World's richest duck race" for a $7,500 purse. Also Duck Queen and Darling Duckling contests, Best Dressed Duck, Tortilla Toss, parade and other festivities. Est attendance: 22,000. For info: The Great American Duck Race, Box 8, Deming, NM 88030. Phone: (505) 544-2213.

HARTE, BRET: BIRTH ANNIVERSARY. Aug 25. Francis Bret(t) Harte, journalist, poet, printer, teacher and novelist, especially remembered for his early stories of California ("The Luck of Roaring Camp," "The Outcasts of Poker Flat" and "How Santa Claus Came to Simpson's Bar") was born at Albany, NY, Aug 25, 1836. He died at London, England, May 5, 1902.

KELLY, WALT: BIRTH ANNIVERSARY. Aug 25. American cartoonist and creator of the comic strip "Pogo" was born at Philadelphia, PA, on Aug 25, 1913. It was Kelly's character Pogo who paraphrased Oliver Hazard Perry to say "We has met the enemy, and it is us." Kelly died at Hollywood, CA, Oct 18, 1973. See also: "Perry, Oliver Hazard: Birth Anniversary" (Aug 23).

KISS-AND-MAKE-UP-DAY. Aug 25. A day to make amends and for relationships that need mending! For info: Jacqueline V. Milgate, Jay Inc, 170 Linden Oaks Dr, Rochester, NY 14625.

MEXICAN FIESTA. Aug 25–27. Henry Maier Festival Park, Milwaukee, WI. Hottest festival in Milwaukee brings Mexico to the lakefront. Three full days of fun, food and fiesta! Jalapeño-eating contest, national and international entertainment. Est attendance: 60,000. For info: Mexican Fiesta, 1030 W Mitchell St, Milwaukee, WI 53204. Phone: (414) 383-7066.

NATIONAL PARK SERVICE ANNIVERSARY OBSERVANCE. Aug 25. Colonial Natl Historical Park, Jamestown and Yorktown, VA. Patriotic band concert at Yorktown in evening. Est attendance: 5,000. Sponsor: Colonial National Historical Park, PO Box 210, Yorktown, VA 23690. Phone: (804) 898-3400.

PINKERTON, ALLAN: BIRTH ANNIVERSARY. Aug 25. Scottish-born American detective, founder of detective agency in Chicago, IL, in 1850, first chief of US Army's secret service, remembered now because of his strike-breaking employments and his lack of sympathy for working people. Pinkerton was born at Glasgow, Scotland, Aug 25, 1819, and died at Chicago, IL, July 1, 1884.

SMITH, SAMANTHA: 10th DEATH ANNIVERSARY. Aug 25. American schoolgirl whose interest in world peace drew praise and affection from people around the world. In 1982, the ten-year-old wrote a letter to Soviet leader Yuri Andropov asking him "Why do you want to conquer the whole world, or at least our country?" The letter was widely publicized in the USSR and Andropov replied personally to her. Samantha Smith was invited to visit and tour the Soviet Union. On Aug 25, 1985, the airplane on which she was riding crashed in Maine killing all aboard, including Samantha and her father. In 1986, minor planet No 3147, an asteroid between Mars and Jupiter, was named Samantha Smith in her memory.

Aug 25-26 ☆ Chase's 1995 Calendar of Events ☆

STIFTUNGSFEST. Aug 25-27. Young America, MN. Continuous German musical entertainment under the big tent in Willkommen Park. Roast beef dinners, famous seasoned hamburgers, brats and kraut, beer garden, carnival, softball tournament, arts and crafts, flower show, kiddie parade and more. Est attendance: 10,000. For info: J.C. Ingebrand, Stiftungsfest 1995, PO Box 133, Young America, MN 55397. Phone: (612) 467-3365.

TOWN POINT JAZZ FESTIVAL. Aug 25-27. Town Point Park, Norfolk, VA. One of the hottest free jazz festivals on the East Coast, this three-day spectacular takes place on the banks of the Elizabeth River and features top jazz artists of today. Est attendance: 75,000. For info: Norfolk Festevents, Ltd, Promotions Dir, 120 W Main St, Norfolk, VA 23510. Phone: (804) 627-7809.

URUGUAY: INDEPENDENCE DAY. Aug 25. National holiday. Declared independence from Brazil on this day, 1825.

BIRTHDAYS TODAY

Cecil D. Andrus, 64, Governor of Idaho (D), born at Hood River, OR, Aug 25, 1931.
Anne Archer, 48, actress ("Falcon Crest," *A Couple of White Chicks Sitting Around Talking*), born Los Angeles, CA, Aug 25, 1947.
Cornelius O'Landa Bennett, 29, football player, born Birmingham, AL, Aug 25, 1966.
Sean Connery, 65, actor (James Bond movies; *The Man Who Would Be King*), born Edinburgh, Scotland, Aug 25, 1930.
Elvis Costello (Declan McManus), 41, musician, songwriter ("Oliver's Army"), born London, England, Aug 25, 1954.
Don DeFore, 78, actor ("The Adventures of Ozzie & Harriet," "Hazel"), born at Cedar Rapids, IA, Aug 25, 1917.
Mel Ferrer, 78, actor (*Scaramouche, The Sun Also Rises*), born Elberon, NJ, Aug 25, 1917.
Monty Hall, 72, former TV host ("Let's Make a Deal"), born at Winnipeg, Manitoba, Canada, Aug 25, 1923.

AUGUST 26 — SATURDAY
238th Day — Remaining, 127

CANADA: WELLING PIONEER DAY. Aug 26. Welling, Alberta. The town of Welling is noted in rodeo history as the home of the first side-delivery rodeo chute and arena, constructed in 1916. (Side-delivery chutes are now standard at all rodeos world-wide.) Pioneer Day—a day of fun and activities—is celebrated with their annual cowboy boot throwing contest. Annually, the last Saturday of August. Sponsor: Richard Wilde, Pres, Welling Recreation Society, Box 57, Welling, Alta, Canada, T0K 2N0. Phone: (403) 752-4576.

CHAMPLAIN VALLEY FAIR. Aug 26-Sept 4. Essex Junction, VT. Vermont's largest fair. Agricultural exhibits and competitions, variety of entertainment, arts and crafts, midway rides, commercial exhibits, great food and much more. Est attendance: 300,000. For info: George Rousseau, Dir of Sales and Mktg, PO Box 209, Essex Junction, VT 05453. Phone: (802) 878-5545.

CHILDREN'S DAY. Aug 26. Woodstock, VT. Traditional farm activities from corn shelling to sawing firewood—19th-century games, traditional spelling bee, ice cream and butter making, wagon rides. Sponsor: Billings Farm Museum, PO Box 489, Woodstock, VT 05091. Phone: (802) 457-2355.

CHRISTMAS BAZAAR. Aug 26-27. Trollwood Park, Fargo, ND. A two-day arts and crafts festival featuring more than 300 booths. All items handmade. Concessions available. Strolling and staged entertainment is also scheduled. Annually, the last full weekend in August. Est attendance: 35,000. For info: Carolyn Altmann, Dir of Trollwood Park, PO Box 1796, Fargo, ND 58107. Phone: (701) 241-8160.

CORN FESTIVAL. Aug 26. Shippensburg, PA. Entertainment, children's activities, demonstrations, food. More than 275 crafts stands. Est attendance: 40,000. For info: Corn Festival, Box F, Shippensburg, PA 17257. Phone: (717) 532-5509.

DANKFEST. Aug 26-27. Harmony Museum, Harmony, PA. Old crafts, historic district tours, German food and entertainment. Est attendance: 3,000. For info: Historic Harmony, Sharon Anno, Admin Asst, PO Box 524, Main and Mercer Sts, Harmony, PA 16037. Phone: (412) 452-7341.

DeFOREST, LEE: BIRTH ANNIVERSARY. Aug 26. American inventor of the electron tube, radio knife for surgery, the photoelectric cell, and a pioneer in the creation of talking pictures and television. Born at Council Bluffs, IA, on Aug 26, 1873, DeForest was holder of hundreds of patents but perhaps best remembered by the moniker he gave himself in the title of his autobiography, *Father of Radio*, published in 1950. So unbelievable was the idea of wireless radio broadcasting that DeForest was accused of fraud and arrested for selling stock to underwrite the invention that later was to become an essential part of daily life in America and throughout the world. DeForest died at Hollywood, CA, on June 30, 1961.

DENMARK: HO SHEEP MARKET. Aug 26. The village of Ho, near Esbjerg, holds its annual sheep market on the last Saturday in August, when some 50,000 people visit the fair.

FERRARO, GERALDINE ANNE: 60th BIRTHDAY. Aug 26. Geraldine A. Ferraro, of Queens, NY, the first woman to be nominated as candidate of a major political party for the office of US vice president, was born Aug 26, 1935, at Newburgh, NY. Married to New York businessman John A. Zaccaro in 1960, she continued to use her maiden name. The former schoolteacher and attorney was first elected to the Congress in 1978 and was nominated for the vice presidency at the Democratic National Convention, San Francisco, CA, July 1984.

FIRST BASEBALL GAMES TELEVISED: ANNIVERSARY. Aug 26. WXBS television, in New York City, broadcast the first major league baseball games on Aug 26, 1939—a doubleheader between the Cincinnati Reds and the Brooklyn Dodgers at Ebbets Field. Announcer Red Barber interviewed Leo Durocher, manager of the Dodgers, and William McKechnie, manager of the Reds, between games.

ICE CREAM SOCIAL. Aug 26. Shane Hill Nursing Home, Rockford, OH. Employees and families, and residents and their families gather for games, music and entertainment at family night ice cream social. Annually, the fourth Saturday of August. Est attendance: 500. For info: Dorothy Trisel, Shane Hill Nursing Home, 10731 State Rte 118, Rockford, OH 45882-0159.

ISHERWOOD, CHRISTOPHER: BIRTH ANNIVERSARY. Aug 26. Author of short stories, plays and novels, Christopher William Isherwood was born at High Lane, Cheshire, England, on Aug 26, 1904. The play and motion picture *I Am a Camera* and the musical *Cabaret* were based on a short story, "Sally Bowles," in his collection from the 1930s entitled *Goodbye to Berlin*, which contained the line "I am a camera with its shutter

	S	M	T	W	T	F	S
August			1	2	3	4	5
1995	6	7	8	9	10	11	12
	13	14	15	16	17	18	19
	20	21	22	23	24	25	26
	27	28	29	30	31		

open, quite passive, recording, not thinking." Isherwood, who urged the acceptance of homosexuality as "entirely natural," died at Santa Monica, CA, on Jan 4, 1986.

KALEVA DAYS. Aug 26–27. Kaleva, MI. Grand parade, children's activities, flea market, baseball tournament, arts and crafts, horseshoe pitching and beer tent. Est attendance: 3,000. For info: Kaleva Days Committee, Margaret Peterson, PO Box 367, Kaleva, MI 49645. Phone: (616) 362-3941.

KRAKATOA: ERUPTION ANNIVERSARY. Aug 26. Anniversary of the biggest explosion in historic times. The eruption of the Indonesian volcanic island, Krakatoa (Krakatau), on Aug 26, 1883, was heard 3,000 miles away, created tidal waves 120 feet high (killing 36,000 persons), hurled five cubic miles of earth fragments into the air (some to a height of 50 miles) and affected the oceans and the atmosphere for years.

MAKE YOUR OWN LUCK DAY. Aug 26. A day to take affirmative actions in your life to direct events and take control of your destiny for a happier, more productive and successful life by making your own luck and creating opportunities. Remember, the welcome mat is always out at the door of opportunity. For info: J. Richard Falls, PO Box 165090, Irving, TX 75016-5090. Phone: (214) 252-9026.

MARYLAND RENAISSANCE FESTIVAL. Aug 26–Oct 15. (Weekends only). Annapolis, MD. A 16th-century English festival with Henry VIII, swordswallowers, magicians, authentic jousting, juggling, music, theatre, games, food and crafts. Est attendance: 215,000. For info: Jules Smith, Maryland Renaissance Festival, PO Box 315, Crownsville, MD 21032. Phone: (410) 266-7304.

MARYLAND STATE FAIR. Aug 26–Sept 4. Timonium, MD. Home arts, agricultural and livestock presentations, midway rides, live entertainment and thoroughbred horse racing. Est attendance: 600,000. For info: Max Mosner, State Fairgrounds, PO Box 188, Timonium, MD 21094. Phone: (410) 252-0200.

MOON PHASE: NEW MOON. Aug 26. Moon enters New Moon phase at 12:31 AM, EDT.

PARADISE SPRING TREATIES COMMEMORATION. Aug 26–27. Paradise Spring Historic Park, Wabash, IN. Held on the original treaty grounds on the bank of the Wabash River, this living history festival commemorates the signing of land treaties with the Miami and Potawatomi Indians in October 1826. Pre-1860 re-enactors in authentic period clothing demonstrate traditional pioneer and Indian crafts. Festival also features merchants, blanket traders, period food purveyors and militia. Est attendance: 3,000. For info: Paradise Spring, Inc, PO Box 353, Wabash, IN 46992. Phone: (219) 563-1168.

ROANOKE BEACH PARTY. Aug 26. Victory Stadium, Roanoke, VA. Annual beach party featuring sand volleyball shells, live music. 1995 will have special activities for children, including a toddlers race/crawl. Est attendance: 20,000. Sponsor: Roanoke Special Events Committee, 210 Reserve Ave SW, Roanoke, VA 24016. Phone: (703) 981-2889.

ROCKBRIDGE COMMUNITY FESTIVAL. Aug 26. Lexington, VA. Fun for all ages—crafts, exhibits, live music, food and games. For info: Lexington Visitors Bureau, 106 E Washington St, Lexington, VA 24450. Phone: (703) 463-3777.

SOUTH DAKOTA STATE FAIR. Aug 26–Sept 3. Huron, SD. Est attendance: 260,000. For info: South Dakota State Fair, PO Box 1275, Huron, SD 57350-1275. Phone: (605) 352-1431.

SPACE MILESTONE: SOYUZ 15 (USSR). Aug 26. Launched on Aug 26, 1974, cosmonauts G. Sarafanov and L. Demin returned to Earth Aug 28, making an emergency night landing.

SPACE MILESTONE: SOYUZ 31 (USSR). Aug 26. Launched on Aug 26, 1978, Valery Bykovsky and Sigmund Jaehn docked at Salyut 6 on Aug 27, stayed for a week, then returned to Earth in Soyuz 29 vehicle, leaving their Soyuz 31 docked at space station. Earth landing on Sept 3.

SUSAN B. ANTHONY DAY. Aug 26. Massachusetts.

WICC GREATEST BLUEFISH TOURNAMENT ON EARTH. Aug 26–27. Long Island Sound, NY, CT. One of the nation's largest fishing tournaments of its kind. More than $50,000 in prizes for the biggest fish at this 13th annual tourney. Annually, the last weekend in August. Est attendance: 7,000. For info: Megan E. O'Connell, Tourn Dir, 2 Lafayette Sq, BPT, CT 06604. Phone: (203) 366-2583.

◈ **WOMEN'S EQUALITY DAY.** Aug 26. Anniversary of certification as part of US Constitution, in 1920, of the 19th Amendment, prohibiting discrimination on the basis of sex with regard to voting. Congresswoman Bella Abzug's bill to designate Aug 26 of each year as "Women's Equality Day" in August 1974 became Public Law 93-382.

★ **WOMEN'S EQUALITY DAY.** Aug 26. Presidential Proclamation issued in 1973 and 1974 at request and since 1975 without request.

WOMEN'S SUFFRAGE 75th ANNIVERSARY CELEBRATION. Aug 26. The League of Women Voters and women's organizations around the country will celebrate the 75th anniversary of the 19th amendment. For info: Public Affairs Dept, League of Women Voters, 1730 M St NW, Washington, DC 20036. Phone: (202) 429-1965.

BIRTHDAYS TODAY

Benjamin Crowninshield Bradlee, 74, journalist, editor, born at Boston, MA, Aug 26, 1921.
Geraldine Anne Ferraro, 60, first woman vice-presidential candidate, born at Newburgh, NY, Aug 26, 1935.
Ronny Graham, 76, actor, composer ("Chico and the Man"), born at Philadelphia, PA, Aug 26, 1919.
Irving R. Levine, 73, broadcast journalist, born at Pawtucket, RI, Aug 26, 1922.
Branford Marsalis, 35, musician, born at New Orleans, LA, Aug 26, 1960.

AUGUST 27 — SUNDAY
239th Day — Remaining, 126

ANNAPOLIS RUN. Aug 27. Navy-Marine Corps Memorial Stadium, Annapolis, MD. Maryland's Premier Ten-Mile Foot Race, through historic Annapolis and along Naval Academy seawalls; designer premium for all finishers; entries limited to 3,500. Annually, last Sunday of August since 1976. Est attendance: 3,500. Sponsor: Annapolis Striders, Inc, PO Box 187, Annapolis, MD 21404-0187. Phone: (410) 268-0187.

BARNEGAT BAY CRAB RACE AND SEAFOOD FESTIVAL. Aug 27. Seaside Heights, NJ. Crab race, seafood festival and flea market. Est attendance: 4,000. Sponsor: Lucy Greene, Exec VP, Barbara Morgan, Admin Asst, Toms River-Ocean Co Chmbr of Com, 1200 Hooper Ave, Toms River, NJ 08753. Phone: (908) 349-0220.

Aug 27-28 ☆ *Chase's 1995 Calendar of Events* ☆

BELGIUM: WEDDING OF THE GIANTS. Aug 27. Ath. Traditional cultural observance. Annually, the fourth Sunday in August.

BOULEVARD EVENTS/TOYOTA TIMES!. Aug 27–Sept 4. Charleston, WV. Nine days of events with fun for everyone. Gymnastics, parades, clowns, magicians, musicians, story time, eating contest and various demonstrations. For info: Debra Sampson, 227-2 Locust Ave, South Charleston, WV 25303. Phone: (304) 744-8666.

DAWES, CHARLES GATES: BIRTH ANNIVERSARY. Aug 27. Thirtieth vice president of the US (1925-1929), born at Marietta, OH, Aug 27, 1865. Died at Evanston, IL, Apr 23, 1951.

"THE DUCHESS" WHO WASN'T DAY. Aug 27. At least once on Aug 27 (her birthdate in 1850), repeat from the novel *Molly Bawn*, the following quotation, which has passed into the English language, is in books of quotations and is used almost daily by writers, sports reporters, comic creators, etc: "Beauty is in the eye of the beholder." Margaret Wolfe Hungerford often wrote under the pseudonym "The Duchess," which was the title of her most popular novel—hence the name of this event. Her novels were read in Ireland, England, Canada, USA, Australia and Continental Europe. A popular romance novelist with about 40 books published, Hungerford was born at Rosscarbery, County Cork, Ireland, on Aug 27, 1850; she died at Bandon, County Cork in 1897. For info: Peggy Shirley, 1261 Quail Run, Columbia, SC 29206. Phone: (803) 782-0560.

ENGLAND: NOTTING HILL CARNIVAL. Aug 27-28. London. Annual multi-cultural celebration...streets of Notting Hill and Ladbroke Grove. 3-mile route for spectacular costume bands, steelbands, calypsonians and soca-on-the-move. 50 static sound systems, 3 live stages featuring top national and international musicians. Hundreds of stalls selling food from all over the world, and arts and crafts. Free event. Est attendance: 1,500,000. For info: Notting Hill Carnival Ltd, 332 Ladbroke Grove, London, England W10 5AH.

FIRST COMMERCIAL OIL WELL: ANNIVERSARY. Aug 27. On Aug 27, 1859, W. A. "Uncle Billy" Smith discovered oil in a shaft being sunk in western Pennsylvania. Drilling had reached 69 feet, 6 inches when Smith saw a dark film floating on the water below the derrick floor. Soon 20 barrels of crude were being pumped each day.

FIRST PLAY PRESENTED IN NORTH AMERICAN COLONIES. Aug 27. Acomac, VA, was the site of the first play presented in the North American colonies on Aug 27, 1655. The play was *Ye Bare and Ye Cubb*, by Phillip Alexander Bruce. Three local residents were arrested and fined for acting in the play. At the time, most colonies had laws prohibiting public performances; Virginia, however, had no such ordinance.

HAMLIN, HANNIBAL: BIRTH ANNIVERSARY. Aug 27. Fifteenth vice president of the US (1861-1865) born at Paris, ME, Aug 27, 1809. Died at Bangor, ME, July 4, 1891.

JOHNSON, LYNDON BAINES: BIRTH ANNIVERSARY. Aug 27. The 36th president of the US succeeded to the presidency following the assassination of John F. Kennedy. Johnson's term of office: Nov 22, 1963–Jan 20, 1969. In 1964, he said: "The challenge of the next half-century is whether we have the wisdom to use [our] wealth to enrich and elevate our national life—and to advance the quality of American civilization." Johnson was born Aug 27, 1908, near Stonewall, TX, and died at San Antonio, TX, Jan 22, 1973.

MOUNTBATTEN, LOUIS: ASSASSINATION ANNIVERSARY. Aug 27. Lord Mountbatten (Louis Francis Albert Victor Nicholas Mountbatten), celebrated British war hero, cousin of Queen Elizabeth II, last viceroy of India, son of Prince Louis of Battenberg (Battenberg family changed name to Mountbatten and renounced German titles during World War I), was killed by bomb, along with his 14-year-old grandson and two others, while on his yacht in Donegal Bay off the coast of Ireland, on Aug 27, 1979. Provisional Irish Republican Army claimed responsibility for the explosion and for the killing of 18 British soldiers later the same day, deepening the crisis and conflict between Protestants and Catholics and between England and Ireland. Lord Mountbatten was born at Windsor, England, June 25, 1900.

PONY EXPRESS FESTIVAL. Aug 27. Hollenberg Pony Express Station, Hanover, KS. Re-enactment of Pony Express ride with mochila exchange, pioneer living history demonstrations, 1860s historic dress group, circuit-rider church service, buffalo-burger noon meal on the grounds and a re-enactment of the arrival of an Oregon Trail wagon train. Annually, the last Sunday in August. Sponsor: Friends of Hollenberg Station. Est attendance: 4,000. For info: Duane Durst, Curator, Hollenberg Pony Express Station, RR1, Box 183, Hanover, KS 66945. Phone: (913) 337-2635.

REPUBLIC OF MOLDOVA: NATIONAL DAY. Aug 27. Public holiday.

BIRTHDAYS TODAY

Mangosuthu Gatsha Buthelezi, 67, Chief Minister of Kwazulu, South Africa, born at Mahlabatini, Natal, South Africa, Aug 27, 1928.

Darryl Dragon, 53, musician, songwriter, born at Studio City, CA, Aug 27, 1942.

J. Robert Kerrey, 52, US Senator (D, Nebraska) up for reelection in Nov '94, born at Lincoln, NE, Aug 27, 1943.

Mother Teresa (Agnes Gonxha Bojaxhiu), 85, nun, missionary, born at Skopje, Yugoslavia, Aug 27, 1910.

Martha Raye (Margaret Theresa Yvonne Ree), 79, comedienne, singer, actress (*Rhythm on the Range*, "Martha Raye Show"), tireless USO touring, born Butte, MT, Aug 27, 1916.

Tommy Sands, 58, singer ("Teen-Age Crush," "Goin' Steady"), born at Chicago, IL, Aug 27, 1937.

Tuesday Weld (Susan Kerr), 52, actress ("The Many Loves of Dobie Gillis," *Looking for Mr. Goodbar*), born New York, NY, Aug 27, 1943.

AUGUST 28 — MONDAY
240th Day — Remaining, 125

BOYER, CHARLES: BIRTH ANNIVERSARY. Aug 28. American film star born at Figeac, France, Aug 28, 1889. Died at Scottsdale, AZ, Aug 26, 1978.

COMMERCIAL RADIO BROADCASTING: ANNIVERSARY. Aug 28. In 1922 broadcasters realized radio could earn profits from the sale of advertising time. WEAF in New York ran a commercial "spot" on Aug 28, which was sponsored by the Queensboro Realty Corporation of Jackson Heights in New York City to promote Hawthorne Court, a group of apartment buildings in Queens. The commercial rate was $100 for ten minutes.

	S	M	T	W	T	F	S
August 1995			1	2	3	4	5
	6	7	8	9	10	11	12
	13	14	15	16	17	18	19
	20	21	22	23	24	25	26
	27	28	29	30	31		

DREAM DAY. Aug 28. One of the three Kingdom Respect Days Color Commemorated by wearing Black and White Colors to celebrate humanity in the Spirit of WAO (way-o, WeAreOne), on the anniversaries of three culturally historic events: the Aug 28, 1955, kidnapping/lynching of 14-year-old Chicagoan Emmett Till that prompted the first March on Washington after 600,000 viewed his remains and ignited the Civil Rights Renaissance of the '60s; the Aug 28, 1963, March on Washington led by Rev. Dr. Martin Luther King, Jr, that prompted the passage of the Civil Rights Bill of 1964; and the Aug 28, 1818, death of Chicago's "under acclaimed" Black Founding Father, Jean Baptiste Pointe DuSable. Dream Day includes a Quest/parade to reinforce the Vision of love, liberty, peace and joy by playing the amplified "I Have a Dream" speech in its entirety during the 4-block Quest to show visible support for those who have overcome the evil era of racism and are motivating others to change. It ends Freedom Colors Dream Week. See also: "Humanitarian Day" (Jan 15) and "Victims of Violence Holy Day" (Apr 4). Phone: (312) RESPECT (737-7328). Sponsor: Global Committee Commemorating Kingdom Respect Days, PO Box 21050, Chicago, IL 60621.

ENGLAND: WAYS WITH WORDS LITERATURE FESTIVAL. Aug 28–Sept 4. Dartington Hall, Dartington, Devon. Around one hundred writers give lectures, seminars, interviews, discussions and readings. Book stalls, workshops and plays are also included in this festival. For info: K. Dunbar, Dir, Ways With Words Literature Festival, Droridge Farm, Dartington, Totnes, Devon, England TQ9 6JQ.

EVERLY BROTHERS HOMECOMING. Aug 28–Sept 2. Central City, KY. The homecoming celebration features top artists. Est attendance: 20,000. For info: Everly Brothers Foundation, PO Box 309, Central City, KY 42330. Phone: (502) 754-9603.

FEAST OF ST. AUGUSTINE. Aug 28. Bishop of Hippo, author of *Confessions* and *The City of God*, died Aug 28, 430 AD.

GOETHE, JOHANN WOLFGANG: BIRTH ANNIVERSARY. Aug 28. German author-philosopher born Aug 28, 1749. Died Mar 22, 1832.

HAYES, LUCY WARE WEBB: BIRTH ANNIVERSARY. Aug 28. Wife of Rutherford Birchard Hayes, 19th president of the US, born at Chilicothe, OH, Aug 28, 1831. Died June 25, 1889. She was nicknamed "Lemonade Lucy" because she and the president, both abstainers, served no alcoholic beverages at White House receptions.

HONG KONG: LIBERATION DAY. Aug 28. Public holiday. Annually, the last Monday in August.

NORTHERN IRELAND: OUL' LAMMAS FAIR. Aug 28–29. Ballycastle, County Antrim. The oldest of Ireland's traditional fairs, probably dating back much further than its charter of 1606. For info: Tourist Info Center, Sheskburn House, 7 Mary St, Ballycastle, County Atrim BT54 6QH, Northern Ireland.

SETON, ELIZABETH ANN BAYLEY: BIRTH ANNIVERSARY. Aug 28. First American-born saint was born on this day in 1774. See also: "Seton, Elizabeth Ann Bayley: Feast Day" (Jan 4).

UNITED KINGDOM: SUMMER BANK HOLIDAY. Aug 28. Bank and public holiday in England, Wales and Northern Ireland. (Scotland not included.)

BIRTHDAYS TODAY

William S. Cohen, 55, US Senator (R, Maine), born at Bangor, ME, Aug 28, 1940.
Ben Gazzara, 65, actor (*Anatomy of a Murder*, "Run for Your Life"), born New York, NY, Aug 28, 1930.
Ron Guidry, 45, former baseball player, born at Lafayette, LA, Aug 28, 1950.
Scott Hamilton, 37, figure skater (Olympic medalist), born at Haverford, PA, Aug 28, 1958.
Donald O'Connor, 70, actor, dancer ("Donald O'Connor Texaco Show," *Singin' in the Rain*), born Houston, TX, Aug 28, 1925.

Emma Samms (Emma Samuelson), 35, actress ("General Hospital"; Fallon Carrington Colby on "Dynasty"), born London, England, Aug 28, 1960.
David Soul, 49, actor (Hutch on "Starsky and Hutch"; *Salem's Lot*), born Chicago, IL, Aug 28, 1946.

AUGUST 29 — TUESDAY
241st Day — Remaining, 124

"ACCORDING TO HOYLE" DAY (EDMOND HOYLE DEATH ANNIVERSARY). Aug 29. A day to remember Edmond Hoyle, and a day for fun and games *according to the rules*. For many years Hoyle lived in London, England, and gave instructions in the playing of games. His name became synonymous with the idea of correct play according to the rules, and the phrase "according to Hoyle" became a part of the English language. Hoyle died at London on Aug 29, 1769.

BEHEADING OF ST. JOHN THE BAPTIST. Aug 29. Commemorates the martyrdom of Saint John the Baptist, beheaded upon order from King Herod, about 29 AD.

BERGMAN, INGRID: 80th BIRTH ANNIVERSARY. Aug 29. One of cinema's greatest actresses. Bergman was born at Stockholm, Sweden, Aug 29, 1915, and died at London, England, on her 67th birthday, Aug 29, 1982. Three times a winner of Motion Picture Academy Awards. Controversy over her personal life made her and her films unpopular to American audiences during an interval of several years between periods of awards and adulation.

FLEMINGTON AGRICULTURAL FAIR. Aug 29–Sept 4. Flemington, NJ. Est attendance: 225,000. For info: Karlene Pisarcik, Flemington Fair Grounds, PO Box 293, Rt 31, Flemington, NJ 08822. Phone: (908) 782-2413.

HOLMES, OLIVER WENDELL: BIRTH ANNIVERSARY. Aug 29. Physician and author, born at Cambridge, MA, Aug 29, 1809. Died at Boston, MA, Oct 7, 1894. "A moment's insight," he wrote, "is sometimes worth a life's experience."

LARBAUD, VALERY NICOLAS: BIRTH ANNIVERSARY. Aug 29. Novelist, essayist and translator of English literature. Born at Vichy, France, Aug 29, 1881; died there Feb 2, 1957.

MORE HERBS, LESS SALT DAY. Aug 29. It's healthier, zestier and lustier! Phone: (212) 388-8673 or (717) 274-8451. Sponsor: Wellness Permission League, Tom or Ruth Roy, 2105 Water St, Lebanon, PA 17046.

PARKER, CHARLIE: 75th BIRTH ANNIVERSARY. Aug 29. Jazz saxophonist Charlie Parker was born Aug 29, 1920, at Kansas City, KS. He earned the nickname "Yardbird" (later "Bird") from his habit of sitting in the backyard of speakeasies fingering his saxophone. His career as a jazz saxophonist took him from jam sessions in Kansas City to New York where he met Dizzy Gillespie, Chubby Jackson and others who were creating a style of music that would become known as bop or bebop. Although his musical genius was unquestioned, Parker's addiction to heroin (begun at age 15) constantly haunted his life. He died at Rochester, NY on Mar 12, 1955, at the age of 34.

SOVIET COMMUNIST PARTY SUSPENDED: ANNIVERSARY. Aug 29. The Soviet parliament suspended all activities of the Communist Party on Aug 29, 1991, seizing its property and bringing to an end the institution that ruled the Soviet Union for nearly 75 years. The action followed an unsuccessful coup Aug 19-21 that sought to overthrow the government of Soviet President Mikhail Gorbachev but instead prompted a sweeping wave of democratic change. See also: "Coup Attempt in the Soviet Union: Anniversary" (Aug 19).

BIRTHDAYS TODAY

Sir Richard Attenborough, 72, filmmaker (*In Which We Serve, The Great Escape*), born Cambridge, England, Aug 29, 1923.
Carl Banks, 33, football player, born at Flint, MI, Aug 29, 1962.
Jim Florio, 58, former Governor of New Jersey (D), born at Brooklyn, NY, Aug 29, 1937.
William Friedkin, 56, filmmaker (Oscar for *The French Connection, The Exorcist*), born at Chicago, IL, Aug 29, 1939.
Elliott Gould (Elliott Goldstein), 57, actor (*M*A*S*H, The Long Good-Bye*), born at Brooklyn, NY, Aug 29, 1938.
Michael Jackson, 37, singer, song writer ("We Are the World" with Lionel Richie; *Bad, Thriller, Beat It*, "The Girl Is Mine," "Billie Jean," "Don't Stop 'Til You Get Enough"), born Gary, IN, Aug 29, 1958.
John Sidney McCain, III, 59, US Senator (R, Arizona), born in Panama Canal Zone, Aug 29, 1936.
Mark Morris, 39, choreographer, dancer, born at Seattle, WA, Aug 29, 1956.
William Edward Perdue, 30, basketball player, born at Melbourne, FL, Aug 29, 1965.
David H. Pryor, 61, US Senator (D, Arkansas), born at Camden, AR, Aug 29, 1934.
Barry Sullivan, 83, actor (*The Bad and the Beautiful*, "The Road West"), born New York, NY, Aug 29, 1912.

AUGUST 30 — WEDNESDAY
242nd Day — Remaining, 123

ARTHUR, ELLEN LEWIS HERNDON: BIRTH ANNIVERSARY. Aug 30. Wife of Chester Alan Arthur, 21st president of the US, born at Fredericksburg, VA, Aug 30, 1837. Died Jan 12, 1880.

CANADA: GRAND FORKS INTERNATIONAL BASEBALL TOURNAMENT. Aug 30-Sept 4. James Donalson Park, Grand Forks, British Columbia. 12-team invitational tournament draws from the four corners of North America and from the Pacific Rim. Annually, Labor Day weekend, beginning Wednesday. Est attendance: 27,500. For info: Larry Seminoff, Box 1214, Grand Forks, BC, Canada V0H 1H0. Phone: (604) 442-2110.

	S	M	T	W	T	F	S
August			1	2	3	4	5
	6	7	8	9	10	11	12
1995	13	14	15	16	17	18	19
	20	21	22	23	24	25	26
	27	28	29	30	31		

CENTRAL STATES THRESHERMEN'S REUNION. Aug 30-Sept 4. Pontiac, IL. Large steam show includes steam engines, antique tractors and gas engines and other farm equipment. Many special displays including blacksmiths, old-time general store, arts and crafts, flea market, farmer's market and quilt show. Daily parade of power. Est attendance: 14,000. For info: Carl Sellmyer, RR2, Box 161-A, Pontiac, IL 61764. Phone: (815) 844-3474.

FIRST WHITE HOUSE PRESIDENTIAL BABY: BIRTH ANNIVERSARY. Aug 30. Frances Folsom Cleveland (Mrs. Grover Cleveland) was the first presidential wife to have a baby at the White House when she gave birth to a baby girl (Ester) on Aug 30, 1893. The first child ever born in the White House was a granddaughter to Thomas Jefferson in 1806.

HUEY P. LONG DAY. Aug 30. Louisiana.

LENIN ASSASSINATION ATTEMPT: ANNIVERSARY. Aug 30. As Vladimir Ilyich Lenin emerged from a Moscow meeting hall after addressing a labor rally on Aug 30, 1918, three shots were fired at him by Fanny (Fanya) Dora Kaplan, a member of the Socialist-Revolutionary Party. Two of the bullets hit Lenin—in the neck and the collarbone. Though seriously wounded, he resumed his political activities after a brief convalescence. The bullet in his neck never was removed; the one in his collarbone was removed several years later in an attempt to alleviate the headaches from which he'd begun to suffer. Fanny Kaplan was executed Sept 3, 1918, for her attempt on Lenin's life.

MacMURRAY, FRED: BIRTH ANNIVERSARY. Aug 30. Fred MacMurray was born at Kankakee, IL on Aug 30, 1908. His film and television career included a wide variety of roles, ranging from comedy (*The Absent Minded Professor, Son of Flubber, The Shaggy Dog, The Happiest Millionaire*) to serious drama (*The Caine Mutiny, The Miracle of the Bells, Fair Wind to Java, Double Indemnity*). From 1960 to 1972 he portrayed Steven Douglas, the father on "My Three Sons," which was second only to "Ozzie and Harriet" as network TV's longest running family sit-com. He died on Nov 5, 1991.

PERU: SAINT ROSE OF LIMA'S DAY. Aug 30. Saint Rose of Lima was the first saint of the western hemisphere. She lived at the time of the colonization by Spain in the 16th century. Patron saint of the Americas and the Philippines. Public holiday in Peru.

SOUTH MOUNTAIN FAIR. Aug 30-Sept 3. Gettysburg, PA. Display of agricultural products, arts, crafts and industrial and agricultural exhibits. Est attendance: 12,500. For info: Gettysburg Travel Council, 35 Carlisle St, Gettysburg, PA 17325. Phone: (717) 334-6274.

SPACE MILESTONE: *DISCOVERY*. Aug 30. Space shuttle *Discovery* made its maiden flight with six-member crew. Launched from Kennedy Space Center, FL, on Aug 30, 1984. Deployed three satellites and used robot arm before landing, Sept 5, 1984, at Edwards Air Force Base, CA.

SPACE MILESTONE: *STS-8* (US). Aug 30. Shuttle *Challenger* with five astronauts (Richard Truly, Daniel Brandenstein, Guion Bluford, Jr, Dale Garner and William Thornton) was launched from Kennedy Space Center, FL, on Aug 30, 1983. Landed at Edwards AFB, CA, on Sept 5.

SWEET CORN FESTIVAL. Aug 30-Sept 2. Millersport, OH. To raise funds for community projects and civic improvements. Food, games and crafts. Est attendance: 225,000. Sponsor: Millersport Lions Club, Joan Elliott, 13495 Oak Rd NE, Thornville, OH 43076. Phone: (614) 246-6217.

WILKINS, ROY: BIRTH ANNIVERSARY. Aug 30. Roy Wilkins, grandson of a Mississippi slave, civil rights leader, active in the National Association for the Advancement of Colored People (NAACP), retiring as its executive director in 1977. Born at St. Louis, MO, Aug 30, 1901. Died at New York, NY, Sept 8, 1981.

★ Chase's 1995 Calendar of Events ★ Aug 30–31

BIRTHDAYS TODAY

Elizabeth Ashley (Elizabeth Ann Cole), 54, actress (*Agnes of God, Cat on a Hot Tin Roof*, "Evening Shade"), born Ocala, FL, Aug 30, 1941.
Timothy Bottoms, 44, actor (*Last Picture Show, The Paper Chase*), born at Santa Barbara, CA, Aug 30, 1951.
Jean-Claude Killy, 52, skier, born at Saint Cloud, France, Aug 30, 1943.
Robert L. Parish, 42, basketball player, born at Shreveport, LA, Aug 30, 1953.
Kitty Wells (Muriel Deason), 76, singer ("Jealousy"), born at Nashville, TN, Aug 30, 1919.
Ted Williams, 77, baseball Hall of Famer, born at San Diego, CA, Aug 30, 1918.

AUGUST 31 — THURSDAY
243rd Day — Remaining, 122

AUBURN-CORD-DUESENBERG FESTIVAL. Aug 31–Sept 5. Auburn, IN. Classic car parade, antiques show, decorators showcase and classic car auction. Est attendance: 300,000. For info: Auburn-Cord-Duesenberg Festival, Box 271, Auburn, IN 46706. Phone: (219) 925-3600.

BLUE HILL FAIR. Aug 31–Sept 4. Blue Hill, ME. A "down-to-earth" country fair. Annually, Labor Day weekend. Est attendance: 50,000. For info: Dwight Webber, Pres, East Blue Hill, ME 04629. Phone: (207) 374-9976.

CHARLESTON EARTHQUAKE: ANNIVERSARY. Aug 31. Charleston, SC. The first major earthquake in the recorded history of the eastern US occurred Aug 31, 1886. It is believed that about 100 persons perished in the quake centered near Charleston but felt up to 800 miles away. The first shock was at 9:51 PM, EST. Though a number of smaller eastern US quakes had been described and recorded since 1638, this was the most terrible and affected persons living in an area of some 2 million square miles.

COUNT YOUR LOSSES. Aug 31–Nov 26. Clinton, MD. Special exhibit on historic sites lost to growth in the surrounding county. Est attendance: 1,800. Sponsor: Surratt House and Tavern, 9110 Brandywine Rd, Clinton, MD 20735. Phone: (301) 868-1121.

KYRGYZSTAN: INDEPENDENCE DAY. Aug 31. National holiday.

LOUISIANA SHRIMP AND PETROLEUM FESTIVAL AND FAIR. Aug 31–Sept 4. Morgan City, LA. To recognize and celebrate importance of the shrimp and oil industry to the area. Street fair and dance, arts, crafts, parade, Cajun culinary classic, antique show, shrimp cook-off, music in the park, unique children's village (a magical adventureland), gospel tent, coronation pageant and ball, blessing of the fleet. Est attendance: 150,000. Sponsor: Louisiana Shrimp and Petroleum Festival and Fair Assn, Box 103, Morgan City, LA 70381. Phone: (504) 385-0703.

MALAYSIA: NATIONAL DAY. Aug 31. National holiday.

NATIONAL HOT ROD ASSOCIATION US NATIONALS. Aug 31–Sept 4. Indianapolis, IN. Est attendance: 152,000. For info: Indianapolis Raceway Park, Lex Dudas, Gen Mgr, PO Box 34300, Indianapolis, IN 46234. Phone: (317) 291-4090.

NATIONAL SWEETCORN FESTIVAL. Aug 31–Sept 4. McFerron Park, Hoopeston, IL. 51st annual festival includes 29 tons of free corn on the cob, nationally sanctioned beauty pageant, carnival, flea market, horse show, demolition derby, bands, and talent shows. Est attendance: 50,000. For info: Jeanie Cook, Ex Dir, Danville Area Conv/Visitors Bureau, PO Box 992, Danville, IL 61834. Phone: (217) 442-2096.

POLAND: SOLIDARITY FOUNDING 15th ANNIVERSARY. Aug 31. The Polish trade union, Solidarity, was formed at the Baltic Sea port of Gdansk, Poland, on Aug 31, 1980. The union was outlawed by the government and many of its leaders arrested. Led by Lech Walesa, Solidarity persisted in its opposition to the communist-controlled government and, on Aug 19, 1989, Polish president Wojcieck Jaruzelski astonished the world by nominating for the post of Prime Minister Tadeusz Mazowiecki, "a longtime Catholic activist," a deputy in the Polish Assembly, 1961–1972, and editor-in-chief of *Tygodnik Solidarnosc*, Solidarity's weekly newspaper, bringing to an end 42 years of Communist Party domination.

SAROYAN, WILLIAM: BIRTH ANNIVERSARY. Aug 31. American writer of Armenian descent, author of *The Human Comedy* and of Pulitzer Prize-winning play *The Time of Your Life*, was born at Fresno, CA, Aug 31, 1908, and died there on May 18, 1981. In April 1981, he gave reporters a final statement for publication after his death: "Everybody has got to die, but I have always believed an exception would be made in my case. Now what?"

SHAWN, WILLIAM: BIRTH ANNIVERSARY. Aug 31. William Shawn, editor of *The New Yorker* for 35 years, was born at Chicago, IL, Aug 31, 1907. A quiet, considerate and courteous man, Shawn was virtual dictator of editorial policy for the magazine, which in turn had an impact on the literary and reportorial styles of writers throughout the country. Non-fiction pieces in *The New Yorker* contributed to public opinion on important issues during Shawn's tenure. William Shawn died Dec 8, 1992, at New York, NY.

TRINIDAD AND TOBAGO: INDEPENDENCE DAY. Aug 31. National holiday. Became Commonwealth nation Aug 31, 1962.

BIRTHDAYS TODAY

James Coburn, 67, actor ("Klondike," *The Devil and Miss Jones*; Oscar *The More the Merrier*), born Laurel, NE, Aug 31, 1928.
Richard Gere, 46, actor (*An Officer and a Gentleman, Pretty Woman*), born at Philadelphia, PA, Aug 31, 1949.
Debbie Gibson, 25, singer ("Only in My Dreams," "Foolish Beat"), born at Brooklyn, NY, Aug 31, 1970.
Buddy Hackett (Leonard Hacker), 71, comedian, actor (*The Love Bug, The Music Man*; cartoon voices), born New York, NY, Aug 31, 1924.
Van Morrison, 50, singer, songwriter ("Brown Eyed Girl," "Domino"), born at Belfast, Northern Ireland, Aug 31, 1945.
Edwin Moses, 40, track athlete, born at Dayton, OH, Aug 31, 1955.
Itzhak Perlman, 50, violinist, born at Tel Aviv, Israel, Aug 31, 1945.
Frank Robinson, 60, baseball executive, former manager, Hall of Famer, born at Beaufort, TX, Aug 31, 1935.
Daniel Schorr, 79, journalist, born at New York, NY, Aug 31, 1916.
Glenn Tilbrook, 38, singer, musician, born at London, England, Aug 31, 1957.

Sept 1 ☆ Chase's 1995 Calendar of Events ☆

September.

SEPTEMBER 1 — FRIDAY
244th Day — Remaining, 121

BABY SAFETY MONTH. Sept 1-30. The Juvenile Products Manufacturers Assn, Inc (JPMA), a national trade organization of juvenile product manufacturers devoted to helping parents keep baby safe, is disseminating information to parents, grandparents and other child caregivers about baby safety. The information from JPMA pertains to safe selection of juvenile products through the Association's Safety Certification Program and tips on correct use of products such as cribs, car seats, infant carriers and decorative accessories. For a free copy of JPMA's brochure "Safe and Sound for Baby," send a stamped, self-addressed envelope. For info: Juvenile Products Manufacturers Assn, Inc, Two Greentree Centre, Ste 225, PO Box 955, Marlton, NJ 08053.

BE KIND TO EDITORS AND WRITERS MONTH. Sept 1-30. A time for editors and writers to show uncommon courtesy toward each other. For info: Lone Star Publications of Humor, Attn: Lauren Barnett, Box 29000, Ste 103, San Antonio, TX 78229.

BOARD AND CARE RECOGNITION MONTH: SHARE THE CARE. Sept 1-30. New Jersey. To recognize owners and operators of board and care facilities around the state (in the 21 counties). They care for elderly, psychiatric or other vulnerable residents, many of whom have no relatives. Annually since 1988, by Proclamation of the Governor. For info: Nancy O. Heydt, Chair, Board and Care Convenors Assn, c/o MCDSS, PO Box 3000, Freehold, NJ 07728.

BRAZIL: INDEPENDENCE WEEK. Sept 1-7. The independence of Brazil is commemorated with civic and cultural ceremonies promoted by federal, state and municipal authorities. On Sept 7 a grand military parade takes place and the National Defense League organizes the Running Race in Honor of the Symbolic Torch of the Brazilian Nation.

BRITT DRAFT HORSE SHOW. Sept 1-3. Hancock County Fairgrounds, Britt, IA. Largest draft horse hitch show in North America, featuring 16 draft horse hitches from the US and Canada representing the very best of the Belgian, Percheron and Clydesdale performance horses. Annually, Labor Day weekend. Est attendance: 10,000. Sponsor: Britt Draft Horse Assn, Randel or Melodie Hiscocks, PO Box 312, Britt, IA 50423. Phone: (515) 843-4181.

September 1995

S	M	T	W	T	F	S
					1	2
3	4	5	6	7	8	9
10	11	12	13	14	15	16
17	18	19	20	21	22	23
24	25	26	27	28	29	30

BUMBERSHOOT, THE SEATTLE ARTS FESTIVAL. Sept 1-4. Seattle Center, Seattle, WA. Celebrates the arts in every genre, includes kids' activities. Annually, Labor Day weekend. Est attendance: 250,000. For info: One Reel, PO Box 9750, Seattle, WA 98109. Phone: (206) 682-4386.

CARTIER, JACQUES: DEATH ANNIVERSARY. Sept 1. French navigator and explorer who sailed from St. Malo, France, Apr 20, 1534, in search of a northwest passage to the Orient. Instead, he discovered the St. Lawrence River, explored Canada's coastal regions and took possession of the country for France. Cartier was born at St. Malo, about 1491 (exact date unknown) and died there on Sept 1, 1557.

CHICAGO JAZZ FESTIVAL. Sept 1-3. Chicago, IL. World's largest free outdoor jazz festival. Traditional jazz, avante-garde quintets, Latin ensembles and hard-bop sextets—the whole gamut. Annually, Labor Day weekend. Est attendance: 300,000. Sponsor: Mayor's Office of Special Events, Jennifer Johnson, 121 N LaSalle St, Chicago, IL 60602. Phone: (312) 744-3315.

CHICKEN BOY'S BIRTHDAY. Sept 1. Chicken Boy is a 22-foot-tall statue of a boy with a chicken's head, holding a bucket of chicken. Formerly the signage for the restaurant for which he is named, he was rescued from destruction when the restaurant went out of business by Future Studio of Los Angeles, a graphic design studio. Chicken Boy has since become a pop culture icon (some call him the Statue of Liberty of Los Angeles). Sponsor: Amy Inouye, Owner of Future Studio, PO Box 292000, Los Angeles, CA 90029. Phone: (213) 660-0620.

CHILD INJURY PREVENTION WEEK. Sept 1-7. Tot Lot Children's Center, New Hyde Park, NY. We will host a child safety fair to provide information on injury prevention, safety issues, complimentary life saving information and to serve as a resource for activity and media information for all interested groups on a national level. Fax: (516) 482-8122. Est attendance: 1,200. Sponsor: Safety By Design, Ltd, PO Box 4312, Great Neck, NY 11023. Phone: (516) 488-5395.

CHILDREN'S EYE HEALTH AND SAFETY MONTH. Sept 1-30. Prevent Blindness America (formerly known as National Society to Prevent Blindness) directs its educational efforts to common causes of eye injuries and common eye problems among children. Materials that can easily be posted or distributed to the community will be provided. Sponsor: Prevent Blindness America, Marita Gomez, Media Relations, 500 E Remington Rd, Schaumburg, IL 60173. Phone: (800) 331-2020.

CHILE: NATIONAL MONTH. Sept 1-30. A month of special significance in Chile: arrival of spring, Independence of Chile anniversary (proclaimed Sept 18, 1810), anniversary of the armed forces rising of Sept 11, 1973, to overthrow the government and celebration of the 1980 Constitution and Army Day, Sept 19.

EMMA M. NUTT DAY. Sept 1. A day to honor the first woman telephone operator, Emma M. Nutt, who reportedly began that professional career at Boston, MA, on Sept 1, 1878, and continued working as a telephone operator for some 33 years.

ENGLAND: BLACKPOOL ILLUMINATIONS. Sept 1-Nov 5. The Promenade, Blackpool, Lancashire. "A five-mile spectacle of lighting." Est attendance: 8,000,000. For info: Tourism and Services Dept, 1 Clifton St, Blackpool, Lancashire, England FY1 1LY.

FALL FOLK DANCE CAMP. Sept 1-4. Camp Russel, Wheeling WV. Nationally-recognized folk dance instructors lead participants in ethnic, international and national dances. Est attendance: 100. Sponsor: Oglebay Institute Visual and Creative Arts Dept, Oglebay Institute, Stifel Fine Arts Ctr, 1330 National Rd, Wheeling, WV 26003. Phone: (304) 242-7700.

☆ Chase's 1995 Calendar of Events ☆   Sept 1

FARISH STREET FESTIVAL. Sept 1-2. Jackson, MS. The Farish Street Festival features outdoor performances of jazz, blues, gospel and reggae music. The "Crystal Palace" dance/concert held where Duke Ellington, Count Basie, Ella Fitzgerald and Sara Vaughan once performed. Est attendance: 10,000. For info: Earline Strickland, Mississippi Cultural Arts Coalition, PO Box 22668, Jackson, MS 39225. Phone: (601) 355-2787.

FESTIVAL OF MOUNTAIN AND PLAIN: A TASTE OF COLORADO. Sept 1-4. Civic Center Park, Denver, CO. City's largest celebration featuring culinary delights from rural restaurants. Plus, arts and crafts, children's area, exhibits. Free concerts by top-name entertainers on 6 stages. Est attendance: 350,000. For info: Taste of Colorado, c/o Downtown Denver Partnership, 511 16th St, Ste 200, Denver, CO 80202. Phone: (303) 534-6161.

FIELD OF DREAMS EXTRAVAGANZA. Sept 1-3. Field of Dreams, Dyersville, IA. Included are a fantasy camp, celebrity baseball game and parade. Tickets go on sale in spring. Annually, Labor Day weekend. Est attendance: 18,000. For info: Karen Kramer, Tourism Dir, Dyersville Area Chamber of Commerce, PO Box 187, Dyersville, IA 52040. Phone: (319) 875-2311.

HOISINGTON CELEBRATION. Sept 1-4. BiCentennial Park, Hoisington, KS. 99th annual event includes dances, demolition derby, horseshoe tournament, parade, talent contest, car show, carnival, kiddie events, royalty and baby contests, float contest. Annually, Labor Day weekend. Est attendance: 35,000. For info: Hoisington Labor Day Committee, 123 N Main, Hoisington, KS 67544. Phone: (316) 653-4311.

INTERNATIONAL GAY SQUARE DANCE MONTH. Sept 1-30. Emphasis on square dancing as a healthy, fun recreational activity. For info: Intl Assn of Gay Square Dance Clubs (IAGSDC), PO Box 15428, Crystal City, VA 22215-0428. Phone: (800) 835-6462.

ITALIAN STREET FAIR. Sept 1-4. Maryland Farms Office Complex, Brentwood, TN. 41st annual affair benefits the Nashville Symphony Orchestra. Annually, Labor Day weekend. Est attendance: 100,000. For info: Nashville Symphony Guild, 208 23rd Ave N, Nashville, TN 37203. Phone: (615) 329-3033.

JAPAN: KANTO EARTHQUAKE MEMORIAL DAY. Sept 1. Day to remember the 57,000 people who died during Japan's greatest earthquake which took place on Sept 1, 1923.

KOREAN AIR LINES FLIGHT 007 DISASTER: ANNIVERSARY. Sept 1. Korean Air Lines Flight 007, en route from New York, NY, to Seoul, Korea, reportedly strayed more than 100 miles off course, flying over secret Soviet military installations on the Kamchatka Peninsula and Sakhalin Island. Two and one half hours after it was said to have entered Soviet airspace, at 3:26 AM, Korean time (Aug 31 at 1:26 PM, EST), a Soviet interceptor plane destroyed the Boeing 747 with 269 persons on board (240 passengers and 29 crew members) which then crashed in the Sea of Japan. There were no survivors. President Reagan, in Proclamation 5093, appointed Sunday, Sept 11, 1983, as a National Day of Mourning, and recommended "homage to the memory of those who died."

LIBRARY CARD SIGN-UP MONTH. Sept 1-30. National effort to sign up every child for a library card. Annually, the month of September. Sponsor: American Library Assn, 50 E Huron St, Chicago, IL 60611. Phone: (312) 280-5041.

LIBYAN ARAB JAMAHIRIYA: REVOLUTION DAY. Sept 1. Commemorates the revolution of Sept 1, 1969. National holiday.

LITTLE BALKANS DAYS/PAACA LITTLE BALKANS FOLKLIFE FESTIVAL. Sept 1-3. Pittsburg, KS. This event attracts over 100 arts and crafts booths, is spread all over Pittsburg from the mall on the south to the Historical Museum on the north and in the parks in between. Contests, tournaments, food, live entertainment, street dances, displays, hot-air balloon races, and much more than space will allow. Annually, Labor Day weekend. Est attendance: 40,000. Sponsor: Pittsburg Area Festival Assn, Diane James, Pres, PO Box 1115, Pittsburg, KS 66762. Phone: (316) 231-1000.

MARFA LIGHTS FESTIVAL. Sept 1-3. Marfa, TX. Family-oriented festival with big-name music stars, arts and crafts and more. Annually, Friday-Sunday of Labor Day weekend. Est attendance: 2,500. Sponsor: Marfa Chamber of Commerce, PO Box 635, Marfa, TX 79843. Phone: (915) 729-4942.

MISSION SAN LUIS OBISPO DE TOLOSA: FOUNDING ANNIVERSARY. Sept 1. California mission to the Indians founded Sept 1, 1772.

MONTREUX DETROIT JAZZ FESTIVAL. Sept 1-4. Detroit, MI. Celebration for jazz lovers and festival fans. Features Detroit groups as well as international stars and reflects Detroit's roots in the development of American jazz. Est attendance: 350,000. For info: Detroit Jazz Festival, Music Hall Ctr, 350 Madison, Detroit, MI 48226. Phone: (313) 963-7622.

NATIONAL BED CHECK MONTH. Sept 1-30. To remind the public to check their mattresses and foundations for signs of wear and tear at least once a year and replace beds every 8 to 10 years for optimum sleep comfort and support. Sponsor: The Better Sleep Council. For info: Andrea Herman, 333 Commerce St, Alexandria, VA 22314. Phone: (703) 683-8371.

NATIONAL CHAMPIONSHIP CHUCKWAGON RACES. Sept 1-3. Clinton, AR. Five divisions of chuckwagon races; bronc fanning race; Snowy River Race; live entertainment, barn dance, western show, Gold Rush Race, western art, saddles, tack clothing vendors. 10th annual races. Annually, Labor Day weekend. Est attendance: 20,000. For info: Dan Eoff, Rt 6, Box 187-1, Clinton, AR 72031. Phone: (501) 745-8407.

NATIONAL CHICKEN MONTH. Sept 1-30. Focuses consumer attention on chicken as the most nutritious, convenient, economical, and versatile food available. Fax: (202) 293-4005. Sponsor: Natl Broiler Council, Bill Roenigk, 1155 15th St NW, Ste 614, Washington, DC 20005. Phone: (202) 296-2622.

NATIONAL CHOLESTEROL EDUCATION AND AWARENESS MONTH. Sept 1-30. To increase the public's consciousness of the dangers of elevated levels of cholesterol and develop public awareness of the beneficial role of proper nutrition, exercise, stopping smoking and prescribed medical treatment in the control of cholesterol. Kit of materials available for $15. FAX: (415) 332-1832. For info: Pharmacists Planning Service, Inc, Frederick S. Mayer, Pres, PO Box 1336, Sausalito, CA 94966. Phone: (415) 332-4066.

NATIONAL COURTESY MONTH. Sept 1-30. Are we a nation of rude dawgs? Whatever happened to mutual respect? Our moms say, "It's nice to be nice." Don't believe it? Try it! Sponsor: Tom Danaher, PO Box 1778, Las Vegas, NV 89125.

NATIONAL HARD CRAB DERBY AND FAIR. Sept 1-3. Crisfield, MD. Crab picking and crab cooking contest, parade, beauty pageant and crab races. Est attendance: 20,000. For info: Crisfield Chamber of Commerce, Box 292, Crisfield, MD 21817. Phone: (800) 782-3913.

NATIONAL HONEY MONTH. Sept 1-30. To honor the United States's 211,600 beekeepers, and 4 million colonies of honey bees, which produce more than 230 million pounds of honey each year. For info: Natl Honey Bd, Mary Humann, 421 21st Ave, #203, Longmont, CO 80501. Phone: (303) 776-2337.

Sept 1 ☆ *Chase's 1995 Calendar of Events* ☆

NATIONAL LITERACY MONTH. Sept 1-30. Events are held in numerous cities nationwide during this month, which is designated in order to raise awareness that 40 million Americans are functionally illiterate. Sponsor: Literacy Volunteers of America. For info: Schenkein/Sherman, 1125 17th St, Ste 1400, Denver, CO 80202. Phone: (800) 626-4601.

NATIONAL MIND MAPPING MONTH. Sept 1-30. To promote creativity and innovation in business through the use of mind mapping, a visual form of outlining. For info: Joyce Wycoff, Pres, Mind Play, 1103 Habor Hills Dr, Santa Barbara, CA 93109. Phone: (805) 962-9933.

NATIONAL ORGANIC HARVEST MONTH. Sept 1-30. National education and promotion campaign about organic food and agriculture. Local and regional events organized by retailers, manufacturers, distributors and consumer groups include food fairs, tastings, farm tours, cooking demonstrations and meet-the-farmer days. Annually, the month of September. For info: Anne Day, PR Director, PO Box 1078, Greenfield, MA 01302. Phone: (703) 768-1861.

NATIONAL PAPAYA MONTH. Sept 1-30. To celebrate the peak of the Hawaiian papaya season and encourage consumers to enjoy Hawaiian papaya for its taste and nutritious value. Est attendance: 1,000. Sponsor: The Papaya Admin Committee, 230 Kekuanaoa St, Hilo, HI 96720. Phone: (808) 969-1160.

NATIONAL PIANO MONTH. Sept 1-30. Recognizes America's most popular instrument and its more than 20 million players; also encourages piano study by people of all ages. Sponsor: Natl Piano Fdtn, 4020 S McEwen, Ste 105, Dallas, TX 75244-5019. Phone: (214) 233-9107.

NATIONAL RICE MONTH. Sept 1-30. To help focus attention on the importance of rice to the American diet and to salute the entire US rice industry. For info: USA Rice Council, PO Box 740121 Dept NRM, Houston, TX 77274. Phone: (713) 270-6699.

NATIONAL SCHOOL SUCCESS MONTH. Sept 1-30. Today's young people have many distractions from school and are sometimes overwhelmed when it comes to academics. Parents are often unskilled at effectively redirecting the attention of their children, especially their teenagers. This observance is to recognize parents who want to support and encourage their children to succeed in school and to explore ways to do that. Annually, the month of September. For info: Teresa Langston, Parenting Without Pressure, 1330 Boyer St, Longwood, FL 32750. Phone: (407) 767-2524.

OATMEAL FESTIVAL. Sept 1-2. Bertram/Oatmeal, TX. To celebrate the town of Oatmeal being put back on the state map years after being left off. Fun, food and foolishness. Annually, the Friday and Saturday of Labor Day weekend. Est attendance: 5,000. For info: Oatmeal Fest, PO Box 70, Bertram/Oatmeal, TX 78605. Phone: (512) 355-2197.

OKLAHOMA STATE PRISON RODEO. Sept 1-2. McAlester, OK. This PRCA rodeo held entirely behind the walls of the Oklahoma State Penitentiary features prison inmates who try their luck at such unusual contests as wild cow milking, bull and horse racing, and "Money the Hard Way," as well as traditional rodeo contests. Est attendance: 15,000. For info: Chamber of Commerce, PO Box 759, McAlester, OK 74502. Phone: (918) 423-2550.

ON THE WATERFRONT. Sept 1-3. Rockford, IL. Largest food and music festival in the region. Eight music stages, more than 60 specialty foods, dozens of special events. Est attendance: 400,000. For info: Mark Miller, Exec Dir, On the Waterfront, Inc, PO Box 437, Rockford, IL 61105-0437. Phone: (815) 968-5600.

PLEASURE YOUR MATE MONTH. Sept 1-30. To promote love and show appreciation to your mate. Look for new ways to create happiness together. Use this event to establish a life-long habit of sharing pleasure. Annually, the month of September. Sponsor: Donald Etkes, MFCC, 112 Harvard Ave, Ste 148, Claremont, CA 91711. Phone: (909) 624-0451.

PROJECT AWARE MONTH. Sept 1-30. Throughout the month, scuba divers, instructors and dive stores across the US promote environmental and marine conservation in conjunction with AWARE (Aquatic World Awareness and Education). Events include underwater and beach clean-ups, and seminars stressing the importance of protecting the Earth's aquatic resources. For info: Scott D. Jones, PADI, Professional Assn of Diving Instructors, 1251 E Dyer Rd, #100, Santa Ana, CA 92705-5605. Phone: (714) 540-7234.

PTA MEMBERSHIP ENROLLMENT MONTH IN TEXAS. Sept 1-30. Texas. Texas PTA membership recruitment theme, September 1995: "Put Your Heart in PTA—For All Children." Texas PTA is the largest child advocacy organization in Texas with more than 800,000 members. National PTA is the largest child advocacy organization in the nation with more than 7 million members. For info: Texas PTA, 408 W 11th St, Austin, TX 78701. Phone: (512) 476-6769.

REUTHER, WALTER PHILIP: BIRTH ANNIVERSARY. Sept 1. American labor leader who began work in a steel factory at age 16 and later became president of the United Automobile Workers (UAW) and the Congress of Industrial Organizations (CIO). Born at Wheeling, WV, on Sept 1, 1907, Reuther spent several years traveling in his twenties and worked for two years in a Russian automobile factory. Often at the center of controversy, he was the target of an assassin in 1948. Reuther and his wife died in an airplane crash on May 9, 1970, near the UAW Family Education Center at Black Lake, MI. The Family Education Center, a project which he had cherished, was later named for Walter and May Reuther.

SANTA-CALI-GON DAYS. Sept 1-4. Independence, MO. To celebrate the period of time in the mid-1800s when wagon trains left Independence, MO, to travel west along the Santa Fe, California and Oregon trails. Est attendance: 175,000. Sponsor: Chamber of Commerce, Box 1077, Independence, MO 64051. Phone: (816) 252-4745.

SELF-IMPROVEMENT MONTH. Sept 1-30. To disseminate information about the importance of life-long learning and self-improvement. For complete information and lists of books, material and cassettes, send $1 to cover expense of printing, handling and postage. Annually, the month of September. Sponsor: Intl Soc of Friendship and Good Will, Dr. Stanley Drake, Pres, 9538 Summerfield St, Spring Valley, San Diego County, CA 91977-2852. Phone: (619) 466-8882.

SELF-UNIVERSITY WEEK. Sept 1-7. Reminds adults (in or out of school) that each of us has a responsibility to help shape the future by pursuing life-long education. Annually, the first seven days of September. For info: Autodidactic Press, PO Box 872749, Wasilla, AK 99687. Phone: (907) 376-2932.

SEPTEMBER SKIRMISH. Sept 1-3. Point Mallard, Decatur, AL. Staged in honor of local Confederate Generals John Hunt

September 1995

S	M	T	W	T	F	S
					1	2
3	4	5	6	7	8	9
10	11	12	13	14	15	16
17	18	19	20	21	22	23
24	25	26	27	28	29	30

Morgan and Joe Wheeler. More than 200 authentically clad Yankee and Rebel soldiers meet in daily battles. Visitors may join a young recruit and follow his life as a soldier during the Candlelight Camp Tour on Friday night or join the Civil War roundtable discussion. Annually, Labor Day weekend. For info: Decatur Conv/Visitors Bureau, 719 6th Ave SE, PO Box 2349, Decatur, AL 35602. Phone: (205) 350-2028.

SIGOURNEY, LYDIA: BIRTH ANNIVERSARY. Sept 1. Prolific American author, Lydia Howard Huntley Sigourney was born Sept 1, 1791, at Norwich, CT. Her writings, mainly moral and religious works, included such titles as *How to Be Happy, Letters to Young Ladies* and *Pleasant Memories of Pleasant Lands.* More than 65 books came from her pen before her death on June 10, 1865, at Hartford, CT.

SLOVAKIA: NATIONAL DAY. Sept 1. Anniversary of the adoption of the Constitution of the Slovak Republic.

SOUTHERN GOSPEL MUSIC MONTH. Sept 1-30. To promote the growth, enjoyment and awareness nationwide of Southern Gospel music, a cherished American art form, by promoting radio airplay, concert attendance and retail awareness. Fax: (502) 587-6153. Sponsor: Southern Gospel Music Guild, One Riverfront Plaza, Ste 1710, Louisville, KY 40202. Phone: (502) 587-9653.

STOP THE VIOLENCE AND SAVE OUR KIDS MONTH. Sept 1-30. To encourage a nationwide effort to tackle the problems plaguing our youth such as drugs, gangs, crime, violence and vandalism, and to challenge others to take part in turning around young lives. Annually, in September. Sponsor: Paul Bernstein, Tune Inc, 748 Miller Dr SE, PO Box 5000, Leesburg, VA 22075.

THOMAS POINT BEACH BLUEGRASS FESTIVAL. Sept 1-3. Brunswick, ME. Featuring world class bluegrass musicians on the southern coast of Maine. Swimming, playground, picnic area, snack bar, free camping, Sunday morning worship service on the beach. Fun for the whole family! 18th annual festival. For tickets and info call (207) 725-6009. Enjoy an end-of-summer family outing and terrific bluegrass music entertainment. Est attendance: 4,000. For info: Thomas Point Beach, Meadow Rd, Box 5419, Brunswick, ME 04011. Phone: (207) 725-6009.

TOPEKA RAILROAD DAYS FESTIVAL. Sept 1-4. Topeka, KS. A celebration of a century of Topeka's railroad heritage with exhibits, music, food and a 60-mile excursion train trip complete with dinner and entertainment. Annually, Labor Day weekend. Est attendance: 75,000. For info: Topeka Railroad Days, Inc, 605 S Kansas Ave, Topeka, KS 66603. Phone: (913) 232-5533.

TOTAH FESTIVAL. Sept 1-3. Farmington, NM. Native American juried fine arts and crafts show and marketplace—highlighted by an Indian rug auction. Est attendance: 10,000. For info: Farmington Conv and Visitors Bureau, 203 W Main, Ste 401, Farmington, NM 87401. Phone: (800) 448-1240.

TWITTY, CONWAY: BIRTH ANNIVERSARY. Sept 1, 1933. Country and Western music star who began his career as a rock and roll performer in the style of Elvis Presley. He was born at Friars Point, MS, and died June 5, 1993, at Springfield, MO.

UZBEKISTAN: INDEPENDENCE DAY. Sept 1. National holiday.

VERMONT STATE FAIR. Sept 1-10. Fairgrounds, Rutland, VT. Est attendance: 100,000. For info: Vermont State Fair, 175 S Main St, Rutland, VT 05701. Phone: (802) 775-5200.

VINTAGE AUTO RACE AND CONCOURS D'ELEGANCE. Sept 1-4. Steamboat Springs, CO. Street race and display of classic vintage automobiles. Annually, Labor Day weekend. Sponsors: Continental, American Express and Valvoline. For info: Janet Nichols, Special Events Coord, Steamboat Springs Chamber Resort Assn, PO Box 774408, Steamboat Springs, CO 80477. Phone: (303) 879-0882.

WEST VIRGINIA ITALIAN HERITAGE FESTIVAL. Sept 1-3. Clarksburg, WV. Est attendance: 150,000. For info: Italian Heritage Festival Office, Box 1632, Clarksburg, WV 26301. Phone: (304) 622-7314.

WISCONSIN COW CHIP THROW. Sept 1-2. Prairie du Sac, WI. Cow Chip Throw, 5K & 10K runs, arts & crafts fair, live music, parade, Tour du Chips, Cow Pie Eating Contest. Est attendance: 30,000. For info: Wisconsin Cow Chip Throw, PO Box 3, Prairie du Sac, WI 53578. Phone: (608) 643-4317.

WOMEN OF ACHIEVEMENT MONTH. Sept 1-30. Westchester, Rockland and Putnam counties, NY. A month-long celebration led by *Women's News* and local women's organizations, to focus on health and wealth for business and professional women, and on the annual Women of Achievement awards, which will be presented on Sept 28. For info: Women's News, 33 Halstead Ave, PO Box 829, Harrison, NY 10528. Phone: (914) 835-5400.

WOODSTOCK FAIR. Sept 1-4. Woodstock, CT. Annually, Labor Day weekend. Est attendance: 188,000. For info: Woodstock Fair, PO Box 1, South Woodstock, CT 06267. Phone: (203) 928-3246.

BIRTHDAYS TODAY

Yvonne DeCarlo (Peggy Yvonne Middleton), 71, actress (Lily Munster on "The Munsters"; *Salome*), born at Vancouver, British Columbia, Canada, Sept 1, 1924.

Barry Gibb, 49, singer (with Bee Gees; "Stayin' Alive"), songwriter, born at Manchester, England, Sept 1, 1946.

Vinnie (Vincent) Johnson, 39, basketball player, born at Brooklyn, NY, Sept 1, 1956.

Seiji Ozawa, 60, conductor, born at Hoten, Japan, Sept 1, 1935.

Ann W. Richards, 62, Governor of Texas (D), up for reelection in Nov '94, born at Waco, TX, Sept 1, 1933.

Lily Tomlin, 56, actress ("Laugh-In," *The Search for Signs of Intelligent Life in the Universe*), born Detroit, MI, Sept 1, 1939.

SEPTEMBER 2 — SATURDAY
245th Day — Remaining, 120

ARTQUAKE 1995. Sept 2-4. Portland, OR. The festival has an outstanding array of visual arts, music, theatre and dance, outdoor craft market booths, literary arts, film and hands-on activity. Est attendance: 200,000. For info: Arts Celebrations, Inc, PO Box 9100, Portland, OR 97207. Phone: (503) 227-2787.

BISON-TEN-YELL DAY. Sept 2. Annually, Sept 2. Sponsor: Puns Corps, Bob Birch, Grand Punscorpion, Box 2364, Falls Church, VA 22042-0364. Phone: (703) 533-3668.

BOX CAR DAYS. Sept 2-4. Tracy, MN. To celebrate the city's heritage as a railroad community. Annually, Labor Day weekend. Est attendance: 10,000. For info: Chamber of Commerce, Con Rettmer, Exec Dir, 372 Morgan St, Tracy, MN 56175. Phone: (507) 629-4021.

CALENDAR ADJUSTMENT DAY: ANNIVERSARY. Sept 2. Pursuant to the British Calendar Act of 1751, Britain (and the American colonies) made the "Gregorian Correction" on Sept 2, 1752. The Act proclaimed that the day following Wednesday, Sept 2, should become Thursday, Sept 14, 1752. There was rioting in the streets by those who felt cheated and who demanded the eleven days back. The Act also provided that New Year's Day (and the change of year number) should fall on Jan 1 (instead of Mar 25) in 1752 and every year thereafter. See also: "Gregorian Calendar Adjustment: Anniversary" (Feb 24, Oct 4).

Sept 2 ☆ Chase's 1995 Calendar of Events ☆

CANADA: MACTAQUAC CRAFT FESTIVAL. Sept 2-3. Mactaquac Provincial Park, Fredericton, New Brunswick. Major outdoor craft sale and entertainment by New Brunswick performers. Est attendance: 13,000. For info: NB Crafts Council, Box 1231, Fredericton, NB, Canada E3B 5C8. Phone: (506) 450-8989.

CAPITAL DAY. Sept 2. The Saturday before Labor Day is designated to honor the American savers and investors who directly or indirectly provide the funds for our capitalistic economy to grow and to progress. Sponsor: Albert L. Maguire, 1094 Maple St, Arroyo Grande, CA 93420. Phone: (805) 489-2645.

CAPITAL DISTRICT SCOTTISH GAMES. Sept 2. Altamont, NY. Celtic festival features the Northeastern US Pipe Band and Open Highland Dancing Championships, Highland athletics, Scottish dogs, country dancers, Celtic goods, Scottish/American food/beverages. Est attendance: 10,000. For info: Donald Martin, Capital District Scottish Games, 7 Lori Lane, Latham, NY 12110. Phone: (518) 785-5951.

CAROLINA CRAFTSMEN'S LABOR DAY CLASSIC. Sept 2-4. Benton Convention Center, Winston-Salem, NC. Arts and crafts. Est attendance: 20,000. For info: The Carolina Craftsmen, 1240 Oakland Ave, Greensboro, NC 27403. Phone: (910) 274-5550.

CHARLESTON DISTANCE RUN. Sept 2. Charleston, WV. To provide an amateur 15-mile race of professional quality for residents and visitors to Charleston. A 5K (3.1 miles) race also will be held. Est attendance: 1,500. Sponsor: Charleston Festival Commission, Inc, Danny Wells, Race Dir, #3 Capitol St, Charleston, WV 25301. Phone: (304) 744-8666.

CLEVELAND NATIONAL AIR SHOW. Sept 2-4. Burke Lakefront Airport, Cleveland, OH. Country's oldest air show, featuring extensive military and foreign aircraft participation. Est attendance: 200,000. Sponsor: Cleveland Natl Air Show, Charles K. Newcomb, Burke Lakefront Airport, Cleveland, OH 44114. Phone: (216) 781-0747.

CLOTHESLINE FAIR. Sept 2-4. Prairie Grove Battlefield State Park, Prairie Grove, AR. 44th annual fair with more than 200 arts and crafts exhibitors; parade; folk, bluegrass, gospel and country music, square dancing exhibitions and competitions; living history programs and tours, 5K run. Annually, Labor Day Weekend. Est attendance: 45,000. For info: Prairie Grove Battlefield State Park, PO Box 306, Prairie Grove, AR 72753. Phone: (501) 846-2990.

COMMONWHEEL LABOR DAY WEEKEND ARTS AND CRAFTS FESTIVAL. Sept 2-4. Memorial Park, Manitou Springs, CO. 21st annual juried arts and crafts festival, featuring 100 fine artists and craftsmen and a variety of foods with continuous live entertainment ranging from Celtic harp music to jazz to mime. There is a special children's art area. There is no admission charged. Est attendance: 20,000. Sponsor: Commonwheel Artists Fair, PO Box 42, Manitou Springs, CO 80829. Phone: (719) 685-1008.

DAYS OF MARATHON: ANNIVERSARY. Sept 2-9. Anniversary of the events in 490 BC from which the marathon race is derived. Phidippides, "an Athenian and by profession and practice a trained runner," according to Herodotus, was dispatched from Marathon to Sparta (26 miles), Sept 2 (Metageitnion 28), to seek help in repelling the invading Persian army. Help being unavailable by religious law until after the next full moon, Phidippides ran the 26 miles back to Marathon on Sept 4. Under the leadership of Miltiades, and without Spartan aid, the Athenians defeated the Persians at the Battle of Marathon on Sept 9.

September 1995

S	M	T	W	T	F	S
					1	2
3	4	5	6	7	8	9
10	11	12	13	14	15	16
17	18	19	20	21	22	23
24	25	26	27	28	29	30

According to legend Phidippides carried the news of the battle to Athens and died as he spoke the words "Rejoice, we are victorious." The marathon race was revived at the 1896 Olympic Games in Athens to commemorate Phidippides' heroism. Course distance, since 1924, is 26 miles, 385 yards. Oldest in the US is the Boston Marathon, an annual event since 1897. See also: "Battle of Marathon: Anniversary" (Sept 9).

DENMARK: AARHUS FESTIVAL WEEK. Sept 2-11. Observed from the first Saturday in September and for nine days after, since 1965, with theater, ballet, opera, sports, exhibitions and special programs for children.

ENGLAND: GREAT FIRE OF LONDON: ANNIVERSARY. Sept 2-5. The fire generally credited with bringing about our system of fire insurance started Sept 2, 1666, in the wooden house of a baker named Farryner, in London's Pudding Lane, near the Tower. During the ensuing three days more than 13,000 houses were destroyed, though it is believed that only six lives were lost in the fire. London had experienced three disastrous fires previously, in 798, 982 and 1212.

FAIR AT NEW BOSTON. Sept 2-3. George Rogers Clark Park, Springfield, OH. An authentic recreation of an 18th-century trades fair with period crafts, military encampments, food and entertainment. Annually, the Saturday and Sunday of Labor Day weekend. Est attendance: 10,000. For info: Fairmaster, PO Box 1251, Springfield, OH 45501. Phone: (513) 864-2526.

FORTEN, JAMES: BIRTH ANNIVERSARY. Sept 2. James Forten was born of free black parents on Sept 2, 1766, at Philadelphia, PA. As a powder boy on an American Revolutionary warship, he escaped being sold as a slave when his ship was captured due to the intervention of the son of the British commander. During the year in England he became involved with abolitionists. On his return to Philadelphia, he became an apprentice to a sailmaker and eventually purchased the company for which he worked. He was active in the abolition movement, and in 1816, his support was sought by the American Colonization Society for the plan to settle American blacks in Liberia. He rejected their ideas and their plans to make him the ruler of the colony. From the large profits of his successful sailmaking company, he contributed heavily to the abolitionist movement and was a supporter of William Lloyd Garrison's anti-slavery journal, *The Liberator*. James Forten died at Philadelphia on Mar 4, 1842.

GIANTS RIDGE MOUNTAIN BIKE FESTIVAL. Sept 2-3. Biwabik, MN. Trail rides, uphill-downhill, cross-country. Est attendance: 400. For info: John Filander, PO Box 190, Biwabik, MN 55708. Phone: (800) 688-7669.

IROQUOIS INDIAN FESTIVAL. Sept 2-3. Iroquois Indian Museum, Caverns Rd, Howes Cave, NY. Performances of social dancing by the "Jim Sky Dancers" in traditional Iroquois clothing. Demonstrations of crafts such as moccasin making, basket weaving, pottery making, silversmithing, and stone carving. Archaeology and art exhibits, Iroquois foods, films and videos, children's activities and lectures. Est attendance: 8,000. Sponsor: Iroquois Indian Museum, PO Box 7, Howes Cave, NY 12092. Phone: (518) 296-8949.

JUBILEE DAYS FESTIVAL. Sept 2-4. Zion, IL. 47th annual community-wide festival featuring an arts and crafts festival, Queen's Pageant, Illinois' largest Labor Day Parade and fireworks. Annually, Labor Day weekend. Est attendance: 10,000. Sponsor: Jubilee Days Fest, Inc, Richard Walker, Exec Dir, PO Box 23, Zion, IL 60099. Phone: (708) 746-5500.

★ Chase's 1995 Calendar of Events ★   Sept 2

KANSAS CITY RENAISSANCE FESTIVAL. Sept 2–Oct 15. (Weekends & Labor Day.) Bonner Springs, KS. Re-creation 16th-century harvest faire featuring more than 150 artisans selling handcrafted wares and continuous entertainment on 7 stages. Food, special events and the Children's Realm are highlights. Est attendance: 160,000. For info: The Kansas City Renaissance Festival, 207 Westport Rd, #206, Kansas City, MO 64111. Phone: (800) 966-8215.

LABOR DAY IN THE PINES. Sept 2–3. Angel Fire, NM. Two days of Labor Day fun during this most colorful time of year. Music, arts and crafts and an old-fashioned pancake breakfast await young and old. Est attendance: 300. For info: Angel Fire Chamber of Commerce, PO Box 547, Angel Fire, NM 87710. Phone: (800) 446-8117.

McAULIFFE, CHRISTA: BIRTH ANNIVERSARY. Sept 2. Christa McAuliffe, a 37-year-old Concord, NH, high school teacher, was to have been the first "ordinary citizen" in space. Born Sharon Christa Corrigan at Boston, MA, on Sept 2, 1948, she perished with six crew members in the Space Shuttle *Challenger* explosion on Jan 28, 1986. See also: "Challenger Space Shuttle Explosion Anniversary" (Jan 28).

MOON PHASE: FIRST QUARTER. Sept 2. Moon enters First Quarter phase at 5:03 AM, EDT.

MOUNTAIN EAGLE INDIAN FESTIVAL. Sept 2–4. Hunter, NY. Representatives of 40 Indian nations, authentic Indian handcrafts, Native American food, dance competitions, lectures and storytelling. Est attendance: 15,000. For info: Indian Festival Committee, PO Box 295, Hunter, NY 12442. Phone: (518) 263-3800.

NATIONAL FRISBEE DISC FESTIVAL. Sept 2. Washington Monument grounds, Washington, DC. "Frisbee for the Family." Exhibitions of throwing and catching skills, frisbee-catching dogs and clinics for the audience. Est attendance: 10,000. Sponsor: Natl Frisbee Disc Festival, 6150 Bryantown Dr, Bryantown, MD 20617-2227. Phone: (301) 645-5043.

OMAHA CACTUS AND SUCCULENT SOCIETY SHOW AND SALE. Sept 2–3. Southroads Mall, Bellevue, NE. Hundreds of plants on display Saturday and Sunday, with a huge selection of unusual succulents for sale both days. Club members' plant entries are evaluated by accredited judges. Annually, Labor Day weekend. Evening phone: (402) 346-7377. Est attendance: 1,000. For info: Marcella Newman, 1306 S 35th St, Omaha, NE 68105.

OREGON TRAILS RODEO. Sept 2–4. Hastings, NE. PRCA-sponsored rodeo. Annually, Labor Day weekend. Est attendance: 3,000. For info: James Gleason, Genl Mgr, PO Box 342, Hastings, NE 68902. Phone: (402) 462-3247.

PIONEER DAYS CELEBRATION. Sept 2–4. Bedford, PA. Rollicking fun fills Old Bedford Village. Pioneer games, a pig roast, corn boil, wagon rides, live entertainment, demonstrations and skirmishes by militiamen. Performances by Native Americans. For info: Old Bedford Village, Jamie Nesbit, PO Box 1976, Bedford, PA 15522. Phone: (814) 623-1156.

PIQUA HERITAGE FESTIVAL. Sept 2–4. Piqua, OH. To promote and celebrate our local history and that of Ohio. Annually, Labor Day weekend. Est attendance: 160,000. Sponsor: Piqua Heritage Fest, Inc, Box 1418, Piqua, OH 45356. Phone: (513) 778-0441.

POWERS' CROSSROADS COUNTRY FAIR AND ART FESTIVAL. Sept 2–4. Newnan, GA. 300 top US and Canadian artists and craftsmen. Old plantation skills revived including using a grist mill and sorghum syrup mill. Patriotic entertainment, country music and down home country cooking. Location: 12 miles west of Newnan on GA State Hwy 34 W. Est attendance: 60,000. Sponsor: Coweta Festivals, Inc, Box 899, Newnan, GA 30264. Phone: (404) 253-2011.

PRAIRIE ARTS FESTIVAL. Sept 2. West Point, MS. 5K run, 600 arts and crafts exhibitors, juried fine arts competition, kiddsville, train rides, four stages with entertainment and evening concert in the park. 9–5. Annually, the Saturday before Labor Day. Est attendance: 30,000. Sponsor: Prairie Arts Festival, Louise Campbell, Coord, PO Box 177, West Point, MS 39773. Phone: (601) 494-5121.

RIVERFEST. Sept 2–3. Cincinnati, OH. Three-city end-of-summer celebration, national talent on 6 stages. The event is capped off with one of the country's best fireworks displays. Est attendance: 600,000. For info: Leslie Keller, 2 Centennial Plaza, 805 Central Ave, Cincinnati, OH 45202. Phone: (513) 352-1608.

SCOTLAND: BRAEMAR ROYAL HIGHLAND GATHERING. Sept 2. Princess Royal and Duke of Fife Memorial Park, Braemar, Grampian. Kilted clansmen from all over the world gather. Traditional activities including tossing cabers, dancing and playing bagpipes. Est attendance: 17,000. For info: Mr. W.A. Meston, Secretary, Coilacriech, Ballater, Aberdeenshire, Scotland.

SEWARD SILVER SALMON 10K RUN. Sept 2. Seward, AK. Exhilirating road race begins on Nash Road, ends near center of town. T-shirts for all starters; trophies and ribbons in each division. 14th annual. Est attendance: 100. For info: Seward Chamber of Commerce, PO Box 749, Seward, AK 99664. Phone: (907) 224-8051.

STREET ROD/MUSCLE CAR SHOW. Sept 2–3. Veteran's Memorial Park, Boyne City, MI. Showing of muscle cars and street rodders with music, prizes and more. Est attendance: 12,000. For info: Tim Van Alstine, 9 East St, Boyne City, MI 49712.

TASTE OF MADISON. Sept 2–3. Capitol Concourse, Madison, WI. Food and entertainment fest, including booths from more than 60 restaurants, five stages, waiters' race and Kiddie Korner. Est attendance: 100,000. For info: Greater Madison Conv and Visitors Bureau, Kerry Nolen, Events Coord, 615 E Washington Ave, Madison, WI 53703. Phone: (800) 373-6376.

TUNA TOURNAMENT. Sept 2–4. Galilee, Narragansett, RI. More than 100 boats and up to 400 anglers registered. 37th annual tournament. Phone: (800) 556-2484 or (401) 737-8845. For info: Rhode Island Tourism, 7 Jackson Walkway, Providence, RI 02903.

TWIN CITIES MODEL TRAIN SHOW. Sept 2–3. Monroe Civic Center, Monroe, LA. A show to present model trains. For info: Monroe-West Monroe CVB, PO Box 6054, Monroe, LA 71211. Phone: (318) 387-2372.

US TREASURY DEPARTMENT: ANNIVERSARY. Sept 2. The third presidential cabinet department, the Treasury Department, was established by Congress on Sept 2, 1789.

◆ **VICTORY DAY OR V-J DAY: 50 YEARS.** Sept 2. Official ratification of Japanese surrender to the Allies occurred aboard the USS *Missouri* in Tokyo Bay on Sept 2 (Far Eastern time) in 1945, thus prompting President Truman's declaration of this day as Victory-over-Japan Day. Japan's initial, informal agreement of surrender was announced by Truman and celebrated in the US on Aug 14.

VIETNAM: INDEPENDENCE DAY: 50 YEARS. Sept 2. On Sept 2, 1945, Ho Chi Minh formally proclaimed the independence of Vietnam and the establishment of the Democratic Republic of Vietnam. National holiday.

WESTFEST. Sept 2–3. West, TX. West celebrates its Czech heritage with folk dances, Czech pastries, sausage, polka music, arts & crafts, children's area, 5K run, parade. Est attendance: 40,000. For info: Westfest, Box 65, West, TX 76691. Phone: (817) 826-5058.

Sept 2-3 ☆ *Chase's 1995 Calendar of Events* ☆

WILHELM TELL FESTIVAL. Sept 2-4. New Glarus, WI. Presentation of Schiller's Wilhelm Tell drama in German and English. Alpine festival with Swiss entertainment. Annually, Labor Day weekend. Est attendance: 2,000. For info: Wilhelm Tell Guild, Box 456, New Glarus, WI 53574. Phone: (608) 527-2095.

WORLD CHAMPIONSHIP BARBECUE GOAT COOK-OFF AND ARTS AND CRAFTS FAIR. Sept 2. Richards Park, Brady, TX. 22nd annual cookoff to promote Brady and the sheep and goat industry. Arts and crafts fair featuring local and statewide artists. Est attendance: 17,000. Sponsor: Chamber of Commerce, 101 E First St, Brady, TX 76825. Phone: (915) 597-3491.

WORLD SHEEP AND CRAFTS FESTIVAL. Sept 2-4. Bethel, MO. To promote the sheep and wool industries. Annually, Labor Day weekend. Est attendance: 11,000. For info: World Sheep and Crafts Fest, PO Box 107, Bethel, MO 63434. Phone: (314) 633-2652.

BIRTHDAYS TODAY

Cleveland Amory, 78, author (*The Cat and the Curmudgeon, The Cat Who Came for Christmas*), born at Nahant, MS, Sept 2, 1917.
Terry Bradshaw, 47, Pro Football Hall of Famer, sports color broadcaster, born at Shreveport, LA, Sept 2, 1948.
Jimmy Connors, 43, tennis player, born at East St. Louis, IL, Sept 2, 1952.
Eric Dickerson, 35, football player, born at Sealy, TX, Sept 2, 1960.
Mark Harmon, 44, actor ("St. Elsewhere," "Eleanor and Franklin: The White House Years"), born at Burbank, CA, Sept 2, 1951.
Linda Purl, 40, actress (Ashley Pfister on "Happy Days"; Charlene Matlock on "Matlock"), born Greenwich, CT, Sept 2, 1955.
Alan K. Simpson, 64, US Senator (R, Wyoming), born at Denver, CO, Sept 2, 1931.
Peter Ueberroth, 58, businessman, born at Evansville, IL, Sept 2, 1937.

SEPTEMBER 3 — SUNDAY
246th Day — Remaining, 119

AMATI, NICOLO: BIRTH ANNIVERSARY. Sept 3. Celebrated Italian violin maker. Sept 3, 1596–Aug 12, 1684.

BEGINNING OF THE PENNY PRESS: ANNIVERSARY. Sept 3. Benjamin H. Day launched the *New York Sun* on Sept 3, 1833. This was the first truly successful penny newspaper to originate in the US. The *Sun* was sold on sidewalks by newspaper boys. By 1836 the paper was the largest seller in the country with a circulation of 30,000. It was possibly Day's concentration on human-interest stories and sensationalism that made his publication a success while abortive efforts at penny papers in Philadelphia, New York and Boston had failed.

CANADA: GREAT KLONDIKE OUTHOUSE RACE. Sept 3. Dawson City, Yukon. Crazy race of outhouses on wheels over a 1.5-mile course through the streets of downtown Dawson City. Awards presentation at Diamond Tooth Gertie's gambling hall following the race. Est attendance: 3,000. For info: Klondike Visitors Assn, Box 389, Dawson City, Yukon, Canada Y0B 1G0. Phone: (403) 993-5575.

CHEETAH RUN. Sept 3. Throughout the Cincinnati Zoo in Ohio. The Cheetah Run is a 2.5-mile course throughout the beautiful Zoo grounds. The race is followed by a Fun Run for children around our Swan Lake. An awards ceremony follows the race. Sponsor: Frito Lay, Inc. Est attendance: 1,500. For info: Dan Behnke, Events and Promotions Specialist, The Cincinnati Zoo, 3400 Vine St, Cincinnati, OH 45220. Phone: (513) 281-4701.

CRANDALL, PRUDENCE: BIRTH ANNIVERSARY. Sept 3. American schoolteacher who sparked controversy in the 1830s with her efforts to educate black girls. Born on Sept 3, 1803, to a Quaker family at Hopkinton, RI. When her private academy for girls was boycotted because she admitted a black girl, she started a school for "young ladies and misses of colour." Died on Jan 28, 1890.

DELICATO CHARITY GRAPE STOMP. Sept 3. Delicato Vineyards, Manteca, CA. Wine, food and entertainment. Annually, the Sunday of Labor Day weekend. Fax: (209) 825-6207. Est attendance: 7,000. For info: Delicato Vineyards, Dorothy Indelicato, 12001 S Hwy 99, Manteca, CA 95336-9209. Phone: (209) 825-6213.

◆ ESCAPE TO FREEDOM: ANNIVERSARY. Sept 3, 1838. Dressed as a sailor and carrying identification papers borrowed from a retired merchant seaman, Frederick Douglass boarded a train in Baltimore, MD, a slave state, and rode to Wilmington, DE, where he caught a steamboat to the free city of Philadelphia. He then transferred to a train headed for New York City where he entered the protection of the underground railway network. Douglass later became a great orator and one of the leaders of the antislavery struggle.

ITALY: HISTORICAL REGATTA. Sept 3. Venice. Traditional competition among two-oar racing gondolas, preceded by a procession of Venetian ceremonial boats of the epoch of the Venetian Republic. Annually, the first Sunday in September.

ITALY: JOUST OF THE SARACEN. Sept 3. Arezzo. The first Sunday in September is set aside for the Giostra del Saracino, a tilting contest of the 13th century, with knights in armor.

JEWISH RENAISSANCE FAIR. Sept 3. Morristown, NJ. Simultaneous stages of theater, music and comedy all day long, the artists' quarter, cafes and restaurants, arts and crafts expo, carnival games, golf course and storybook hay ride. Star-studded show on main stage late in the day. Annually, the Sunday of Labor Day weekend. Est attendance: 10,000. Sponsor: Rabbi Boruch Klar, Dir of Community Outreach, Rabbinical College of America, 226 Sussex Ave, PO Box 1996, Morristown, NJ 07960. Phone: (201) 267-9404.

NATIONAL FINANCIAL SERVICES WEEK. Sept 3-9. Take a moment to say "thanks" to all those financial folks who keep us financially healthy. Sponsor: Tuality Healthcare Hospital, Joyce Curran, Human Resource Mgr, Box 309, Hillsboro, OR 97123.

NATIONAL RELIGIOUS REFERENCE BOOKS WEEK. Sept 3-9. To commemorate the role reference books, such as concordances, commentaries, atlases, and dictionaries, play in biblical studies. Annually, the first week of September. For info: Jonathan Petersen, Dir Media Relations, Zondervan Publishing House, 5300 Patterson Ave SE, Grand Rapids, MI 49530. Phone: (616) 698-3417.

September 1995

S	M	T	W	T	F	S
					1	2
3	4	5	6	7	8	9
10	11	12	13	14	15	16
17	18	19	20	21	22	23
24	25	26	27	28	29	30

POLAND: INTERNATIONAL ORATORIO-CANTATA FESTIVAL "WRATISLAVIA CANTANS." Sept 3-16. Wroclaw. Vocal-instrumental and vocal music: oratorio, symphonic and choral concerts; chamber music, vocal recitals, opera, "Festival School" (lectures, meetings, discussions, exhibitions), ballet, "Music and Fine Arts," concerts/exhibition openings, tourney of counter-tenors and open-air concerts. Course for oratorio music interpretation, scholarly session, exhibitions. For info: Intl Oratorio-Cantata Festival "Wratislavia Cantans," Arts and Culture Center, Rynek-Ratusz 24, PL-50101 Wroclaw, Poland.

QATAR: INDEPENDENCE DAY. Sept 3. National holiday.

"RUNNING OF THE SHEEP" SHEEP DRIVE. Sept 3. Reedpoint, MT. Hundreds of Montana bred woolies charge down the six blocks of Main Street. Also a parade, Egg Toss and contests, such as the ugliest sheep and the prettiest ewe. Annually, Sunday of Labor Day weekend. Est attendance: 8,500. For info: Marian Cain, Reedpoint Community Club, Reedpoint, MT 59069. Phone: (406) 326-2193.

SAN MARINO: NATIONAL DAY. Sept 3. Public holiday.

STARS ACROSS AMERICA JERRY LEWIS MDA LABOR DAY WEEKEND. Sept 3-4. The annual TV broadcast to raise money for neuromuscular disease research, medical and support services and public health education. For info: Muscular Dystrophy Assn, 3300 E Sunrise Dr, Tucson, AZ 85718-3208. Phone: (602) 529-2000.

TREATY OF PARIS: SIGNING ANNIVERSARY. Sept 3. Treaty between Britain and the United States, ending the Revolutionary War, signed at Paris, France, Sept 3, 1783. American signatories: John Adams, Benjamin Franklin and John Jay.

WORLD WAR II: DECLARATION ANNIVERSARY. Sept 3. British ultimatum to Germany, demanding halt to invasion of Poland (which had started at dawn on Sept 1), expired at 11 AM, GMT, Sept 3, 1939. At 11:15 AM, in a radio broadcast, Prime Minister Neville Chamberlain announced the declaration of war against Germany. France, Canada, Australia, New Zealand and South Africa quickly issued separate declarations of war. Winston Churchill was named First Lord of the Admiralty.

BIRTHDAYS TODAY

Eileen Brennan, 58, actress (*The Last Picture Show, Private Benjamin*), born Los Angeles, CA, Sept 3, 1937.

Kitty Carlisle, 80, actress (*A Night at the Opera*; "To Tell the Truth"), singer, born at New Orleans, LA, Sept 3, 1915.

Anne Jackson, 69, actress (*Lovers and Other Strangers*; many stage roles), born at Allegheny, PA, Sept 3, 1926.

Valerie Perrine, 52, actress (*Lenny, W.C. Fields and Me*), born Galveston, TX, Sept 3, 1943.

Charlie Sheen (Carlos Irwin Estevez), 30, actor (*Platoon, Hot Shots*), born New York, NY, Sept 3, 1965.

Mort Walker (Mortimer Walker Addison), 72, cartoonist, born at El Dorado, KS, Sept 3, 1923.

SEPTEMBER 4 — MONDAY

247th Day — Remaining, 118

◆ **ANTHRACITE COAL MINERS DAY.** Sept 4. Girard Park, Shenandoah, PA. Parade in honor of all anthracite coal miners and the unveiling of the Anthracite Miners Memorial Statue. Est attendance: 10,000. Sponsor: Shenandoah Area Chamber of Commerce, PO Box 606, Shenandoah, PA 17976. Phone: (800) 755-1942.

BANGOR LABOR DAY ROAD RACE. Sept 4. Bass Park, Bangor, ME. Five-mile road race. Annually, on Labor Day. Est attendance: 200. Sponsor: Craig R. Orff, Parks and Rec, 647 Main St, Bangor, ME 04401. Phone: (207) 947-1018.

BRUCKNER, ANTON: BIRTH ANNIVERSARY. Sept 4. Austrian composer born at Ansfelden, Austria, Sept 4, 1824. Died at Vienna, Austria, Oct 11, 1896.

BUHL DAY. Sept 4. Buhl Farm, Sharon, PA. To honor the laboring man. Observed annually on Labor Day. Est attendance: 25,000. For info: Nancy Emmett, Box 709, Sharon, PA 16146. Phone: (412) 981-5522.

BURNHAM, DANIEL: BIRTH ANNIVERSARY. Sept 4. American architect and city planner, Daniel Hudson Burnham, born Sept 4, 1846, at Henderson, NY. Burnham was an advocate of the tall, fireproof buildings, probably the first to be called "sky-scrapers." In 1909 he proposed a long-range city plan for Chicago, IL, that was a key factor in the "forever open, clear and free" policy which resulted in Chicago having the most beautiful lakefront of any major urban city in the US.

CHATEAUBRIAND, FRANCOIS RENE DE: BIRTH ANNIVERSARY. Sept 4. French poet, novelist, historian, explorer and statesman, witness to the French Revolution. Born at St. Malo, France, Sept 4, 1768. Died at Paris, France, July 4, 1848.

COLUMBIA RIVER CROSS CHANNEL SWIM. Sept 4. Hood River, OR. The annual swim across the mighty Columbia River draws 350 contestants each year to swim the approximately one mile distance for fun. Est attendance: 400. For info: Hood River County Chamber of Commerce, Columbia River Cross Channel Swim, Port Marina Park, Hood River, OR 97031. Phone: (800) 366-3530.

CURACAO: ANIMALS' DAY. Sept 4. In Curaçao the Association for the Protection of Animals organizes an animal show for this day and the best kept animals are awarded prizes.

FULL EMPLOYMENT WEEK. Sept 4-10. An extension of Labor Day to include the full week beginning with that holiday. Meetings, rallies, etc, were held during 1977 Full Employment Week to demonstrate that "Americans want work, and there is plenty of work to do."

GREAT BATHTUB RACE. Sept 4. Nome, AK. Lets people know that participants bathe at least once a year. Further, when politicians participate they can let their constituents know that they clean up their act yearly. Takes place at noon on Labor Day. Bathtubs mounted on wheels are raced down Front Street. Each team has 5 members, one in the tub with bubbles apparent in the bath water. Must be full of water at beginning and have at least 10 gallons at the finish line. Other 4 team members must wear large brim hats and suspenders. Person in tub carries one bar of soap, towel and bath mat throughout entire race. No entry fee. Winning team claims trophy, a statue of Miss Piggy and Kermit taking a bath, which is handed down from year to year. Annually, on Labor Day. For info: Rasmussen's Music Mart, PO Box 2, Nome, AK 99762-0002. Phone: (907) 443-2798.

HORTON BAY LABOR DAY BRIDGE WALK. Sept 4. Horton Bay, MI. The walk is a spoof of the well-known walk across the Mackinac Bridge. The walkers traverse a culvert 16 feet and 3.75 inches long by candlelight. Speeches will follow including propositions for a twenty dollar per hour minimum wage and appearances by celebrity impersonators. Annually, on Labor Day. For info: Boyne Country Conv and Visitors Bureau, Peter Fitzsimons, PO Box 694, Petoskey, MI 49770.

HUCKLEBERRY FINN RAFT RACE. Sept 4. The Inn on Lake Waramaug, New Preston, CT. Dozens of homemade rafts worth no more than $25.00 will compete in the 12th annual Inn on Lake Waramaug Huck Finn Raft Race. Huck Finn era costumes are encouraged. Entrants must paddle or sail (no motors allowed) one-half mile on the lake and then back to the finish line. Prizes awarded; barbecue and other refreshments available lakeside. Est attendance: 350. For info: Nancy Conant, The Inn on Lake Waramaug, 107 N Shore Rd, New Preston, CT 06777. Phone: (203) 868-0563.

Sept 4–5 ★ Chase's 1995 Calendar of Events ★

LABOR DAY. Sept 4. Legal public holiday. (Public Law 90-363 sets Labor Day on the first Monday in September.) Observed on this day in all states and in Canada. First observance believed to have been a parade at 10 AM, on Tuesday, Sept 5, 1882, in New York, NY, probably organized by Peter J. McGuire, a carpenters and joiners union secretary. In 1883, a union resolution declared "the first Monday in September of each year a Labor Day." By 1893 more than half of the states were observing Labor Day on one or another day, and a bill to establish Labor Day as a federal holiday was introduced in Congress. On June 28, 1894, President Grover Cleveland signed into law an act making the first Monday in September a legal holiday for federal employees and the District of Columbia.

LOS ANGELES, CALIFORNIA: BIRTHDAY. Sept 4. Los Angeles founded, by decree, on Sept 4, 1781, and called "El Pueblo de Nuestra Senora La Reina de Los Angeles de Porciuncula."

MACKINAC BRIDGE WALK. Sept 4. St. Ignace, MI. By tradition Labor Day is the only day of the year pedestrians are permitted to walk across the five-mile-long span, one of the world's longest suspension bridges, connecting Michigan's two peninsulas. Walk is from St. Ignace to Mackinaw City. Est attendance: 55,000. Sponsor: Mackinac Bridge Authority, PO Box 217, St. Ignace, MI 49781. Phone: (906) 643-7600.

NATIONAL MIND MAPPING IN SCHOOLS WEEK. Sept 4–10. To promote better learning and creativity through mind mapping, a visual form of outlining. For info: Pat Barnes, Buzan Ctr, 415 Federal Hwy, Lake Park, FL 33403. Phone: (407) 881-0188.

NEWSPAPER CARRIER DAY. Sept 4. Anniversary of the hiring of the first "newsboy" in the US, 10-year-old Barney Flaherty, who is said to have answered the following classified advertisement which appeared in *The New York Sun*, Sept 4, 1833: "To the Unemployed—a number of steady men can find employment by vending this paper. A liberal discount is allowed to those who buy to sell again."

POLK, SARAH CHILDRESS: BIRTH ANNIVERSARY. Sept 4. Wife of James Knox Polk, 11th president of the US. Born at Murfreesboro, TN, Sept 4, 1803, and died, Aug 14, 1891.

STOCK EXCHANGE HOLIDAY (LABOR DAY). Sept 4. The holiday schedules for the various exchanges are subject to change if relevant rules, regulations or exchange policies are revised. If you have questions, phone: American Stock Exchange (212) 306-1212; Chicago Board of Options Exchange (312) 786-7760; Chicago Board of Trade (312) 435-3500; New York Stock Exchange (212) 656-2065; Pacific Stock Exchange (415) 393-4000; Philadelphia Stock Exchange (215) 496-5000.

WAIKIKI ROUGHWATER SWIM. Sept 4. Waikiki Beach, Honolulu, HI. The 26th annual swim is 2.4 miles from Sans Souci Beach to Duke Kahanamoku Beach. "The World's Largest Open Water Swimming Event." Preregistration is required. Annually, Labor Day. Est attendance: 1,200. For info: Jim Anderson, Pres, 3176 E Manoa Rd, Honolulu, HI 96822.

YORK, DICK: BIRTH ANNIVERSARY. Sept 4. Actor Dick York was born on Sept 4, 1928, at Fort Wayne, IN, and began his acting career as a child performing on radio in Chicago. At the age of 15 he starred on the network show, "That Brewster Boy." His Broadway credits include *Tea and Sympathy* and *Bus Stop*. He played the schoolteacher who prompted the Scopes Monkey Trial in *Inherit the Wind* among other films, but it was as Darrin Stephens, the husband of the nose-twitching witch Samantha in the comedy "Bewitched," that he became most famous. York performed the role from the show's inception in 1964 until 1969 when he had to leave due to complications from a back injury sustained while filming the movie *They Came to Cordura* in 1958. He spent his remaining years supporting and soliciting money for charity groups despite his physical handicap. York died on Feb 20, 1992, at Grand Rapids, MI.

BIRTHDAYS TODAY

Carlos Romero Barcelo, 63, US Resident Commissioner (D, Commonwealth of Puerto Rico), born at San Juan, PR, Sept 4, 1932.

Craig Claiborne, 75, editor, author, born at Sunflower, MS, Sept 4, 1920.

Dawn Fraser, 58, swimmer, born at Balmain, Australia, Sept 4, 1937.

Mitzi Gaynor (Franchesca Mitzi Marlene de Charney von Gerber), 64, singer, dancer, actress (*South Pacific*), born Chicago, IL, Sept 4, 1931.

Paul Harvey, 77, broadcaster/commentator ("The Rest of the Story"), born at Tulsa, OK, Sept 4, 1918.

Alexander Liberman, 83, editor, painter, photographer, sculptor, born at Kiev, USSR, Sept 4, 1912.

Donald Petersen, 69, business executive, born at Pipestone, MN, Sept 4, 1926.

Jennifer Salt, 51, actress (Eunice Tate on "Soap"), born at Los Angeles, CA, Sept 4, 1944.

Tom Watson, 46, golfer, born at Kansas City, MO, Sept 4, 1949.

SEPTEMBER 5 — TUESDAY
248th Day — Remaining, 117

◆ **BABE RUTH'S FIRST PRO HOMER: ANNIVERSARY.** Sept 5, 1914. Babe Ruth hit his first home run as a professional while playing for Providence in the International League, a sort of minor league affiliate of the Boston Red Sox. He pitched a one-hit shutout against Toronto.

BE LATE FOR SOMETHING DAY. Sept 5. To create a release from the stresses and strains resulting from a consistent need to be on time. Sponsor: Procrastinators' Club of America, Inc, Les Waas, Pres, Box 712, Bryn Athyn, PA 19009. Phone: (215) 947-9020.

FIRST CONTINENTAL CONGRESS ASSEMBLY: ANNIVERSARY. Sept 5. The first assembly of this forerunner of the US Congress took place at Philadelphia, PA, on Sept 5, 1774. Peyton Randolph, delegate from Virginia, was elected president.

FRALEY FAMILY MUSIC WEEKEND. Sept 5–10. Carter Caves State Resort Park, Olive Hill, KY. A weekend devoted to preserving the mountain music of eastern Kentucky. Musical concerts and instrument workshops. Est attendance: 500. For info: Carter Caves State Resort Pk, John Tierney, Naturalist, Olive Hill, KY 41164. Phone: (606) 286-4411.

GERALD FORD: ASSASSINATION ATTEMPTS: 20th ANNIVERSARY. Sept 5. Lynette A. "Squeaky" Fromme, a follower of convicted murderer Charles Manson, attempted to shoot President Gerald Ford on Sept 5, 1975. On September 22 of the same year another attempt on Ford's life took place when Sara Jane Moore shot at him.

September 1995

S	M	T	W	T	F	S
					1	2
3	4	5	6	7	8	9
10	11	12	13	14	15	16
17	18	19	20	21	22	23
24	25	26	27	28	29	30

ISRAELI OLYMPIAD MASSACRE: ANNIVERSARY. Sept 5. Eleven members of the Israeli Olympic Team were killed in an attack on the Olympic Village in Munich and attempted kidnapping of team members Sept 5-6, 1972. Four of seven guerrillas, members of the Black September faction of the Palestinian Liberation Army, were also killed. In retaliation Israeli jets bombed Palestinian positions in Lebanon and Syria Sept 8, 1972.

MICHIGAN'S GREAT FIRE OF 1881: ANNIVERSARY. Sept 5. According to Michigan Historical Commission "Small fires were burning in the forests of the 'Thumb area of Michigan,' tinder-dry after a long, hot summer, when a gale swept in from the southwest on Sept 5, 1881. Fanned into an inferno, the fire raged for three days. A million acres were devastated in Sanilac and Huron counties alone. At least 125 persons died, and thousands more were left destitute. The new American Red Cross won support for its prompt aid to the fire victims. This was the first disaster relief furnished by this great organization."

NIELSEN, ARTHUR CHARLES: BIRTH ANNIVERSARY. Sept 5. Marketing research engineer, founder of A.C. Nielsen Co, in 1923, known for radio and TV audience surveys, was born Sept 5, 1897, and died at Chicago, IL, June 1, 1980.

SPACE MILESTONE: *VOYAGER 1* (US). Sept 5. Launched on Sept 5, 1977, twin of *Voyager 2* which was launched Aug 20.

SWITZERLAND: ST. GOTTHARD AUTOMOBILE TUNNEL: OPENING ANNIVERSARY. Sept 5. The longest underground motorway in the world, the St. Gotthard Auto Tunnel in Switzerland, was opened to traffic on Sept 5, 1980. More than 10 miles long, requiring $417,000,000 and 10 years for construction, it became the most direct route from Switzerland to the southern regions of the continent. The St. Gotthard Pass, the main passage since the Middle Ages, was closed much of every year by massive snow drifts.

WESTERN SAMOA: NATIONAL HOLIDAY. Sept 5.

ZANUCK, DARRYL F.: BIRTH ANNIVERSARY. Sept 5. Born at Wahoo, NE, on Sept 5, 1902, Darryl F. Zanuck became a celebrated—and controversial—movie producer. He was also a co-founder of Twentieth Century Studios that later merged with Fox. His film credits include *The Jazz Singer* (the first sound picture), *Forever Amber*, *The Snake Pit* and *The Grapes of Wrath*. He died on Dec 21, 1979, at Palm Springs, CA.

BIRTHDAYS TODAY

John Claggett Danforth, 59, former US Senator (R, Missouri) born at St. Louis, MO, Sept 5, 1936.
William Devane, 56, actor (*From Here to Eternity*, "Knots Landing"), born at Albany, NY, Sept 5, 1939.
Willie Gault, 35, football player, born at Griffin, GA, Sept 5, 1960.
Cathy Lee Guisewite, 45, cartoonist ("Cathy"), born at Dayton, OH, Sept 5, 1950.
Arthur Koestler, 90, author (*The Thirteenth Tribe*, *Scum of the Earth*), born at Budapest, Hungary, Sept 5, 1905.
Carol Lawrence (Carol Maria Laraia), 60, singer, actress, born at Melrose Park, IL, Sept 5, 1935.
Bob Newhart, 66, comedian ("The Bob Newhart Show," "Newhart"), born at Chicago, IL, Sept 5, 1929.
Raquel Welch, 53, actress (*Three Musketeers*, *Woman of the Year*), model, born Chicago, IL, Sept 5, 1942.

☆   ☆   ☆

SEPTEMBER 6 — WEDNESDAY
249th Day — Remaining, 116

ADDAMS, JANE: BIRTH ANNIVERSARY. Sept 6. American worker for peace, social welfare, rights of women, founder of Hull House (Chicago), co-winner of Nobel Prize, 1931. Born at Cedarville, IL, Sept 6, 1860. Died May 21, 1935, at Chicago, IL.

BALTIC STATES' INDEPENDENCE RECOGNIZED: ANNIVERSARY. Sept 6. The Soviet government recognized the independence of the Baltic states—Latvia, Estonia and Lithuania—on Sept 6, 1991. The action came 51 years after the Baltic states were annexed by the Soviet Union. All three Baltic states had earlier declared their independence, and many nations had already recognized them diplomatically, including the US on Sept 2, 1991.

BEECHER, CATHARINE ESTHER: BIRTH ANNIVERSARY. Sept 6, 1800. Catharine Esther Beecher was born at East Hampton, NY. In addition to teaching herself mathematics, philosophy and Latin, Beecher had been formally educated in art and music. An early advocate for equal education for women, she founded the Hartford Female Seminary, which was widely recognized for its advanced curriculum. She was also instrumental in the founding of women's colleges in Iowa, Illinois and Wisconsin. Beecher died May 12, 1878, at Elmira, NY.

CANADA: THE MASTERS. Sept 6-10. Spruce Meadows, Calgary, Alberta. World-class show jumping competition, along with Equi-Fair, Breeds for the World and Festival of Nations. Feature events are the Bank of Montreal Nations' Cup and the $680,000 du Maurier Ltd. International, the world's richest Grand Prix. Est attendance: 135,000. For info: Spruce Meadows, RR #9, Calgary, AB, Canada T2J 5G5. Phone: (403) 254-3200.

DALTON, JOHN: BIRTH ANNIVERSARY. Sept 6. English chemist, physicist, teacher and developer of atomic theory, was born at Eaglesfield, near Cockermouth, England, on Sept 6, 1766. Dalton died at Manchester, England, on July 27, 1844.

DEFEAT OF JESSE JAMES DAYS. Sept 6-11. Northfield, MN. Bank raid reenactment, 5 and 15K runs, arts, crafts, bike race, parade and professional rodeo. Est attendance: 150,000. For info: Northfield Chamber of Commerce, PO Box 198, Northfield, MN 55057. Phone: (507) 645-5604.

MICHIGAN WINE AND HARVEST FESTIVAL. Sept 6-10. Kalamazoo and Paw Paw, MI. Celebration of the grape harvest. Grape stomping, live entertainment, ethnic food fair, wine tasting, vineyard and winery tours, art fairs, midway rides and steamboat rides. Annually, the weekend after Labor Day. Est attendance: 220,000. Sponsor: Michigan Wine and Harvest Festival, Inc, 128 N Kalamazoo Mall, Kalamazoo, MI 49007. Phone: (616) 381-4003.

NATIONAL DO IT! DAY (AKA FIGHT PROCRASTINATION DAY). Sept 6. A day for the organizationally challenged to just "Do It!" and get delayed chores done. Annually, the first Wednesday after Labor Day. For info: Ethel M. Cook, Organizational Consultant, 4 Hilda Rd, Bedford, MA 01730. Phone: (617) 275-2326.

NATIONAL STEARMAN FLY-IN. Sept 6-10. Galesburg, IL. The largest gathering of Stearman airplanes—the biplane trainers that gave wings to more military pilots than any other series of aircraft in the world. Est attendance: 7,500. For info: Galesburg Area CVB, PO Box 631, Galesburg, IL 61402-0631. Phone: (309) 343-1194.

Sept 6–7 ☆ *Chase's 1995 Calendar of Events* ☆

PRIME BEEF FESTIVAL, WYATT EARP DAYS. Sept 6–9. Downtown Monmouth, Monmouth College, Monmouth Park, Wyatt Earp Birthplace, Monmouth, IL. Princess Pageant, Parade (6th); Carnival from noon-midnight, Beef and Pork Show, judging (8th); evening shows, Wyatt Earp Birthplace Open House, authentic O.K. Corral reenactment and Old West shows (9th); music, food, exhibitors. Annually, the weekend beginning with the Wednesday after Labor Day. Sponsor: Monmouth Area Chamber of Commerce and the Wyatt Earp Birthplace Association, phone: (309) 734-6419. Est attendance: 5,000. For info: Monmouth Chamber of Commerce, PO Box 857, 620 S Main St, Monmouth, IL 61462. Phone: (309) 734-3181.

ROSE, BILLY: BIRTH ANNIVERSARY. Sept 6. Billy Rose (William S. Rosenberg), American theatrical producer, author, songwriter and husband of Fanny Brice, was born at New York, NY, on Sept 6, 1899. His songs include: "That Old Gang of Mine," "Me and My Shadow," "Without a Song," "It's Only a Paper Moon" and hundreds of others. Rose died at Montego Bay, Jamaica, Feb 10, 1966.

ST. PETERSBURG NAME RESTORED: ANNIVERSARY. Sept 6. Russian legislators on Sept 6, 1991, voted to restore the name St. Petersburg to the nation's second largest city. The city had been known as Leningrad for 67 years in honor of the Soviet Union's founder, Vladimir I. Lenin. The city, founded in 1703 by Peter the Great, has had three names in the 20th century with Russian leaders changing its German-sounding name to Petrograd at the beginning of World War I in 1914 and Soviet Communist leaders changing its name to Leningrad in 1924 following their leader's death.

SWAZILAND: INDEPENDENCE DAY: ANNIVERSARY. Sept 6. Commemorates attainment of national independence on Sept 6, 1968. National holiday.

BIRTHDAYS TODAY

Jane Curtin, 48, actress (original on "Saturday Night Live"; "Kate and Allie"), comedienne, born Cambridge, MS, Sept 6, 1947.

Swoosie Kurtz, 51, actress (*The World According to Garp*; Tony for *The House of Blue Leaves*), born Omaha, NE, Sept 6, 1944.

Jo Anne Worley, 58, comedienne, actress ("Rowan & Martin's Laugh-In"), born Lowell, IA, Sept 6, 1937.

SEPTEMBER 7 — THURSDAY
250th Day — Remaining, 115

BRAZIL: INDEPENDENCE DAY. Sept 7. Declared independence from Portugal on this day, 1822. National holiday.

	S	M	T	W	T	F	S
September						1	2
1995	3	4	5	6	7	8	9
	10	11	12	13	14	15	16
	17	18	19	20	21	22	23
	24	25	26	27	28	29	30

CANADA: TORONTO INTERNATIONAL FILM FESTIVAL. Sept 7–16. Toronto, Ontario. International film festival at various mid-town theaters. Est attendance: 25,000. For info: Festival of Festivals, 70 Carlton St, Toronto, Ont, Canada M5B 1L7. Phone: (416) 968-3456.

CORBETT-SULLIVAN PRIZE FIGHT: ANNIVERSARY. Sept 7. John L. Sullivan was knocked out by James J. Corbett in the 21st round of a prize fight at New Orleans, LA, Sept 7, 1892. It was the first major fight under the Marquess of Queensberry Rules.

GRANDMA MOSES DAY. Sept 7. Anna Mary (Robertson) Moses, modern primitive American painter born at Greenwich, NY, Sept 7, 1860. Started painting at the age of 78. Her 100th birthday was proclaimed Grandma Moses Day in New York state. Died at Hoosick Falls, NY, Dec 13, 1961.

GREAT PEANUT TOUR. Sept 7–10. Skippers, VA. Assorted bicycle rides from 13 to 125 miles. Special peanut tour ride to examine peanuts growing, method of harvesting and a sampling of more than 40 peanut goodies. Annually, the weekend following Labor Day. Est attendance: 1,500. For info: Emporia Bicycle Club, Robert C. Wrenn, PO Box 668, Emporia, VA 23847. Phone: (804) 348-4215.

HOLLY, BUDDY: BIRTH ANNIVERSARY. Sept 7. American popular music performer, composer and band leader. Called one of the most innovative and influential musicians of his time, he was a pioneer of rock 'n' roll. His hits included "That'll Be the Day" and "Peggy Sue." Born Charles Harden Holley, at Lubbock, TX, on Sept 7, 1936, Buddy Holly died at age 22, in an airplane crash near Mason City, IA, on Feb 3, 1959.

HUMMER/BIRD CELEBRATION. Sept 7–10. Rockport and Fulton, TX. To celebrate the spectacular fall migration of the ruby-throated hummingbird from its summer nesting grounds in the north along the eastern Gulf Coast on the way to its winter grounds in Mexico and Central America, and its 500-mile journey across the Gulf. There will be programs, workshops, booths, concessions and bus and boat tours. Est attendance: 4,000. For info: Hummer/Bird Celebration, PO Box 1055, Rockport, TX 78382. Phone: (800) 242-0071.

KASS KOUNTY KING KORN KARNIVAL AND MUD DRAGS. Sept 7–10. Plattsmouth, NE. Krowning of a King and Queen of Kornland, 3 large parades. Free street entertainment—fire department Wall of Water, flower show, Korn Palace, museum exhibits, Hauf Brau Garten, Ugly Pickup Contest, Cow Chip Bingo, fun run, scarecrow contest, flea market, go cart, two-wheel bicycle race, Mud Drag races on Sunday and much more. Annually, the second weekend in September. Est attendance: 30,000. For info: Patricia Baburek, Coord, 141 S 3rd St, PO Box 40, Plattsmouth, NE 68048. Phone: (402) 296-4155.

MARION POPCORN FESTIVAL. Sept 7–9. Marion, OH. Performances by nationally known entertainers every evening; parade, athletic competition, arts and crafts. 11 AM to midnight daily. Annually, the Thursday, Friday and Saturday after Labor Day. Est attendance: 350,000. For info: Marion Popcorn Festival, PO Box 1101, Marion, OH 43301-1101. Phone: (614) 387-3378.

NEITHER SNOW NOR RAIN DAY. Sept 7. Anniversary of the opening to the public, on Labor Day, 1914, of the New York Post Office Building at Eighth Avenue between 31st and 33rd Streets. On the front of this building was an inscription supplied by William M. Kendall of the architectural firm that planned the building. The inscription, a free translation from Herodotus: "Neither snow nor rain nor heat nor gloom of night stays these couriers from the swift completion of their appointed rounds." Long believed to be the motto of the US Post Office and Postal Service, they have, in fact, no motto ... but the legend remains. Info from: New York Post Office Public Info Office and US Postal Service.

☆ Chase's 1995 Calendar of Events ☆ Sept 7-8

QUEEN ELIZABETH I: BIRTH ANNIVERSARY. Sept 7. Queen of England, daughter of Henry VIII and Anne Boleyn, after whom the "Elizabethan Era" was named, was born in Greenwich Palace on Sept 7, 1533. She ascended the throne in 1558 at the age of 25. During her reign, the British defeated the Spanish Armada in July 1588, the Anglican Church was essentially established and England became a world power. She died at Richmond, England, Mar 24, 1603.

SOUTHERN UTAH FOLKLIFE FESTIVAL. Sept 7-9. Springdale, UT. Displays of arts, crafts and cultural history of the Southern Utah Paiute Indians and early pioneers. Storytelling, water witching, herbal remedies, stilt walking. Est attendance: 5,000. Sponsor: Linda Sappington, 3015 Provost Rd, St. George, UT 84770. Phone: (801) 673-6290.

WILLIAMSBURG OLD-FASHIONED TRADING DAYS. Sept 7-9. Courthouse Square, Williamsburg, KY. Arts and crafts, gospel singing, 8K run, antique car show and more. Annually, the first Thursday, Friday and Saturday after Labor Day. Est attendance: 20,000. For info: Theresa Estes, Coord, 522 Main St, Williamsburg, KY 40769.

BIRTHDAYS TODAY

Corbin Bernsen, 41, actor ("L.A. Law," "Ryan's Hope"), born at North Hollywood, CA, Sept 7, 1954.
Taylor Caldwell, 95, novelist (*Dear and Glorious Physician*), born Manchester, England, Sept 7, 1900.
Michael DeBakey, 87, distinguished heart surgeon, born at Lake Charles, LA, Sept 7, 1908.
Michael Feinstein, 39, singer, pianist, born at Columbus, OH, Sept 7, 1956.
Arthur Ferrante, 74, pianist (Ferrante and Teicher; "Exodus," "Tonight," "Midnight Cowboy"), composer, born at New York, NY, Sept 7, 1921.
Chrissie Hynde, 44, singer, songwriter (lead singer of The Pretenders), born Akron, OH, Sept 7, 1951.
Daniel Ken Inouye, 71, US Senator (D, Hawaii), born at Honolulu, HI, Sept 7, 1924.
Julie Kavner, 44, actress (*Radio Days*, "Rhoda"; Marge Simpson's voice on "The Simpsons"), born Los Angeles, CA, Sept 7, 1951.
Elia Kazan (Elia Kazanjoglou), 86, filmmaker (*On the Waterfront, East of Eden, Viva Zapata!*), born at Constantinople, Turkey, Sept 7, 1909.
John Philip Law, 58, actor (*The Russians Are Coming, Barbarella*), born Hollywood, CA, Sept 7, 1937.

SEPTEMBER 8 — FRIDAY
251st Day — Remaining, 114

AUTUMN MAGIC (WITH TOM MIX FESTIVAL). Sept 8-10. Guthrie, OK. Features the International Tom Mix festival, antique/collectible market, Western Legends memorabilia market, classic car show, western re-enactors, live entertainment. Est attendance: 12,000. For info: Kay Hunt, Dir, Guthrie Conv & Visitors Bureau, PO Box 995, Guthrie, OK 73044. Phone: (800) 299-1889.

BALD IS BEAUTIFUL CONVENTION. Sept 8-10. Morehead City, NC. To cultivate a sense of pride for all bald-headed men (folks) everywhere and eliminate the vanity associated with the loss of one's hair. As seen on national television; Bald is Beautiful Contests; family event. Annually, the second weekend in September. Fax: (919) 726-6061. Est attendance: 200. Sponsor: Bald Headed Men of America, John T. Capps III, Founder, 102 Bald Dr, Morehead City, NC 28557. Phone: (919) 726-1855.

CARRY NATION FESTIVAL: 22nd ANNUAL. Sept 8-10. Downtown Holly, MI. 22nd annual festival re-creates the historical visit of Carry Nation, the Kansas City saloon smasher. Includes pageant, parade, entertainment tent, carnival, street dance and craft show. Annually, the weekend after Labor Day. Est attendance: 7,000. For info: Holly Chamber of Commerce, PO Box 214, Holly, MI 48442. Phone: (810) 634-1900.

CATHEDRAL STATE PARK WEEKEND. Sept 8-10. Appalachian Base Camp, Terra Alta, WV. A camping weekend led by experts in the study, identification, collection and tasting of mushrooms. Est attendance: 30. For info: Oglebay Institute Nature/Environmental Educ Dept, Oglebay Park, Wheeling, WV 26003. Phone: (304) 242-6855.

CHILI AND BLUEGRASS FESTIVAL. Sept 8-10. Tulsa, OK. Free concerts from the best in bluegrass and country music and tantalizing chili and BBQ cooking during the International Chili Society Mid-America Regional Chili & BBQ Cookoff. Open car show, children's Kiddie Korral. Annually, Friday, Saturday and Sunday following Labor Day. Est attendance: 65,000. For info: Chili and Bluegrass Festival, 201 W 5th, Ste 450, Tulsa, OK 74103. Phone: (918) 583-2617.

CLINE, PATSY: BIRTH ANNIVERSARY. Sept 8. Country and western singer, born Virginia Patterson Hensley on Sept 8, 1932, at Winchester, VA. Patsy Cline got her big break in 1957 when she won an Arthur Godfrey Talent Scout show, singing "Walking After Midnight." Her career took off and she became a featured singer at the Grand Ole Opry, attaining the rank of top female country singer. She died in a plane crash on Mar 5, 1963, at Camden, TN, along with singers Hawkshaw Hawkins and Cowboy Copas.

CORN PALACE FESTIVAL. Sept 8-17. Mitchell, SD. Celebration of the harvest and the annual redecoration of the Corn Palace (with ears of corn). Midway and carnival rides, games, agricultural displays, outdoor exhibits and 10 days of entertainment on the stage of the Corn Palace. Est attendance: 30,000. For info: Corn Palace Committee, Box 250, Mitchell, SD 57301. Phone: (605) 996-7738.

GERMAN VILLAGE OKTOBERFEST. Sept 8-10. Columbus, OH. Willkommen! The German Village Oktoberfest has something for everyone—oompah-pah bands, great food, polka dancing, the best in traditional and contemporary music, strolling entertainers, arts and crafts and The Kinderplatz (children's place) with free kiddie rides and family entertainment. Annually, the weekend after Labor Day. Est attendance: 75,000. Sponsor: Aaron Leventhal, Dir, German Village Oktoberfest, 624 S Third St, Columbus, OH 43206. Phone: (614) 224-4300.

GREEK FESTIVAL. Sept 8-10. Hellenic Cultural Center, Salt Lake City, UT. An annual showcase of Greek culture and heritage. Encompasses church tours, tour of Hellenic Museum, dancing by Greek youth groups and ala carte food and pastries with a boutique and gourmet shop featured. Annually, the first weekend after Labor Day. Est attendance: 30,000. Sponsor: Greek Orthodox Community of Salt Lake City, William (Bill) Kandas, Festival Chair, 279 S 300 West, Salt Lake City, UT 84101.

HARVEST MOON. Sept 8. So called because the full moon nearest the autumnal equinox extends the hours of light into the evening and helps the harvester with his long day's work. Moon enters Full Moon phase at 11:37 PM, EDT.

HUFF 'N PUFF HOT AIR BALLOON RALLY. Sept 8-10. Topeka, KS. 20th annual hot air balloon rally. Annually, the second weekend in September. Est attendance: 10,000. For info: Great Plains Balloon Club, Box 1093, Topeka, KS 66601. Phone: (913) 582-5869.

INDIAN HERITAGE FESTIVAL. Sept 8-9. Martinsville, VA. Native American heritage, crafts, food and more. Annually, usually the second Saturday in September. No competitions. Gathering atmosphere. Est attendance: 5,000. Sponsor: Virginia Museum of Natural History, Martinsville, VA 24112. Phone: (703) 666-8600.

INDIAN SUMMER FESTIVAL. Sept 8-10. Maier Festival Park, Milwaukee, WI. Festival dedicated to promoting the unique culture of the American Indian. Cultural events and exhibits, competition powwow, arts and crafts, traditional and contemporary entertainment and traditional food. Est attendance: 75,000. For info: Indian Summer Festivals, Inc, 7441 W Greenfield Ave, Ste 109, Milwaukee, WI 53214. Phone: (414) 774-7119.

Sept 8 ☆ *Chase's 1995 Calendar of Events* ☆

JIMMY DOOLITTLE QUARTER MIDGET NATIONAL TROPHY RACE. Sept 8–10. Rough River Dam State Resort Park, Falls of Rough, KY. Remote-controlled model airplane maneuvering reminiscent of old barnstorming days. Est attendance: 300. For info: Rough River Dam State Resort, Tom DeHaven, Rec Supt, Falls of Rough, KY 40119. Phone: (502) 257-2311.

KANSAS STATE FAIR. Sept 8–17. Hutchinson, KS. Commercial and competitive exhibits, entertainment, carnival, car racing and other special attractions. Annually, beginning the first Friday after Labor Day. Est attendance: 335,500. For info: Kansas State Fair, Robert A. Gottschalk, Gen Mgr, 2000 N Poplar, Hutchinson, KS 67502. Phone: (316) 669-3600.

LONG, HUEY P.: ASSASSINATION ANNIVERSARY. Sept 8. Louisiana senator Huey P. Long was shot at Baton Rouge, LA, Sept 8, 1935. The assassin, Dr. Carl A. Weiss, was killed by Long's bodyguards.

LOS ANGELES COUNTY FAIR. Sept 8–Oct 1. Pomona, CA. World's Largest County Fair features free stage acts and attractions, livestock, thoroughbred racing, carnival, monorail, flower and garden show, home arts, nightly fireworks, horse shows, commercial exhibits, photography and much more. Est attendance: 1,400,000. For info: Sid Robinson, Communications Mgr, Box 2250, Pomona, CA 91769. Phone: (909) 623-3111.

MALTA: SIEGE BROKEN: ANNIVERSARY. Sept 8. "Two Sieges and Regatta Day" festivities now commemorate victory over the Turks, Sept 8, 1565, when the siege that began in May 1565 was broken by the Maltese and the Knights of St. John after a loss of nearly 10,000 lives. Also commemorated is survival of the 1943 siege by the Axis Powers. Parades, fireworks, boat races, etc, especially at the capital, Valleta, and the Grand Harbour.

MARIGOLD FESTIVAL. Sept 8–10. Pekin, IL. Parade, flower show, arts, crafts, paddle boat rides, golf, cardboard boat races, Civil War reenactment, carnival and other family-oriented activities. Fax: (309) 346-2104. Est attendance: 100,000. For info: Chamber of Commerce, PO Box 636, Pekin, IL 61555-0636. Phone: (309) 346-2106.

MISS AMERICA FIRST CROWNED: ANNIVERSARY. Sept 8. Margaret Gorman of Washington, DC, was crowned the first Miss America at the end of a two-day pageant in Atlantic City, NJ, on September 8, 1921.

MISSION SAN GABRIEL ARCHANGEL: FOUNDING ANNIVERSARY. Sept 8. California mission to the Indians founded Sept 8, 1771.

MONTANA TRAPPERS ASSOCIATION STATE CONVENTION. Sept 8–10. Lewistown, MT. Gathering of trappers—show, trade, yearly awards, banquet. Open to all. Annually, the second weekend in September. Est attendance: 400. Sponsor: Montana Trappers Assn, Fran Buell, PO Box 133, Gildford, MT 59525. Phone: (406) 376-3178.

MOON PHASE: FULL MOON. Sept 8. Moon enters Full Moon phase at 11:37 PM, EDT.

MUSKIES INC INTERNATIONAL MUSKIE TOURNAMENT. Sept 8–10. Northcentral, MN. 28th annual fundraiser for nonprofit sportsman's organization. Proceeds go toward muskie stocking, rearing and research projects along with Dept of Natural Resources fisheries improvements. This tournament strongly encourages "catch and release." Annually, the Friday, Saturday and Sunday after Labor Day. The $40 entry fee entitles the contestant to participate in the Sunday banquet and compete for thousands of dollars in prizes. Grand Prize in the "Release Division" is a boat, motor and trailer with a retail value of approximately $18,000. We welcome corporate sponsorship inquiries. Est attendance: 600. For info: Dave Griffin, Twin Cities Chapter of Muskies, Inc, 4434 Dorchester Rd, Mound, MN 55364. Phone: (612) 472-6039.

NATIONAL CHAMPIONSHIP INDIAN POWWOW. Sept 8–10. Traders Village, Grand Prairie, TX. Hundreds of Indians gather for colorful traditional dance contests, Indian arts and crafts shows, homemade tepee competition and Indian food. Est attendance: 75,000. For info: Dallas-Fort Worth Inter-Tribal Assn, Traders Village, 2602 Mayfield Rd, Grand Prairie, TX 75052. Phone: (214) 647-2331.

NORDICFEST. Sept 8–10. Libby, MT. Scandinavian festival with food booths, craft shows, quilt show, art show, parade, folk dance workshops, big name performances and a Fjord horse show. Est attendance: 30,000. For info: Floyd Van Weelden, 314 Norman Ave, Libby, MT 59923. Phone: (406) 293-6838.

OHIO RIVER STERNWHEEL FESTIVAL. Sept 8–10. Ohio River Levee, Marietta, OH. A three-day riverfront extravaganza. More than two dozen sternwheelers line the Ohio River shore in Marietta, OH. Continuous musical entertainment for all ages, food concessions, clowns, queen coronation, sternwheel races, fireworks. Annually, the weekend following Labor Day. Est attendance: 75,000. For info: Ohio River Sternwheel Fest Committee, 316 Third St, Marietta, OH 45750. Phone: (614) 373-5178.

OKTOBERFEST '95. Sept 8–10. MainStrasse Village, Covington, KY. The tradition of the celebration of the German "storybook wedding reception" kicks off with a beer-tapping ceremony. Features include German and American food, live Bavarian music and dancing, arts and crafts, contests, children's rides and much more. Est attendance: 275,000. For info: MainStrasse Village, Cindy Scheidt, 616 Main St, Covington, KY 41011. Phone: (606) 491-0458.

OUTRIGGER HOTEL'S ANNUAL HAWAIIAN OCEANFEST. Sept 8–17. Oahu, HI. Includes: on Sept 8–9 the Hawaiian International Ocean Challenge, an international competition between six-person teams of the world's best lifeguards in kayak, surf rescue, swim, paddleboard, outrigger canoe races; on Sept 8 the Hawaiian Roughwater Swim at Makapuu Beach; on Sept 10 the Outrigger's Waikiki Kings Race, a four-part run, kayak, swim, paddleboard on Waikiki Beach; on Sept 10 the Diamond Head Wahine Windsurfing Classic, 24 professional female windsurfers in slalom and wave jumping competition at Diamond Head Beach (PBA members only); on Sept 16 the Diamond Head Biathlon, a 1.1-mile swim along Waikiki Beach followed by a 5-mile run up and around Diamond Head Crater finishing at Kapiolani Park; and on Sept 17 the Waikiki Beach Volleyball Shootout featuring men's and women's teams competing for the championship. Some events are open to the public. Est attendance: 10,000. For info: Event Marketing, 1001 Bishop St, Ste 477, Honolulu, HI 96813. Phone: (808) 521-4322.

September 1995

S	M	T	W	T	F	S
					1	2
3	4	5	6	7	8	9
10	11	12	13	14	15	16
17	18	19	20	21	22	23
24	25	26	27	28	29	30

Chase's 1995 Calendar of Events — Sept 8

PARDON DAY: ANNIVERSARY. Sept 8. Anniversary of the "full, free, and absolute pardon unto Richard Nixon, for all offenses against the United States which he, Richard Nixon, has committed or may have committed or taken part in during the period from January 20, 1969, through August 9, 1974." (Presidential Proclamation 4311, Sept 8, 1974, by Gerald R. Ford.)

PEPPER, CLAUDE DENSON: 95th BIRTH ANNIVERSARY. Sept 8. US Representative and former Senator, born near Dudleyville, AL, on Sept 8, 1900. Claude Pepper's career in politics spanned 53 years and 10 presidents, and he became the champion for America's senior citizens. His political career began in the Florida Legislature in 1928, and he was elected to the US Senate in 1936, where he was a principal architect of many of the nation's "safety net" social programs including Social Security, the minimum wage and medical assistance for the elderly and for handicapped children. After a 14-year career in the Senate, he returned to Congress in the House of Representatives where he served 14 terms. A staunch advocate for senior citizens, he served as chairman of the House Select Committee on Aging, drafted legislation banning forced retirement and fought against cutting Social Security benefits to elderly retired workers, 50 years after campaigning for the establishment of retirement programs. Pepper died at Washington, DC, on May 30, 1989.

POLISH FEST. Sept 8-10. Wisconsin Dells, WI. Traditional Polish celebration with festive and colorful dance, dress, food and live music. Polish art exposition and sale. Est attendance: 6,000. For info: Wisconsin Dells Visitor and Conv Bureau, 701 Superior St, Box 390, Wisconsin Dells, WI 53965-0390. Phone: (800) 223-3557.

POSTCARD SHOW. Sept 8-9. Budget Motor Lodge, Mt Laurel, NJ. For info: Postcard History Society, John McClintock, PO Box 1765, Manassas, VA 22110. Phone: (703) 368-2757.

RIVERFEST '95 IN MANCHESTER. Sept 8-10. Arms Park, Manchester, NH. Festival on the bank of the Merrimack River includes entertainment for the whole family, well-known national music groups, arts and crafts, special events, a canoe race, rowing regatta and other water activities, plus a fireworks extravaganza. Admission charged. Est attendance: 100,000. For info: Riverfest, Inc, PO Box 21, Manchester, NH 03105. Phone: (603) 623-2623.

SELLERS, PETER (RICHARD HENRY): 70th BIRTH ANNIVERSARY. Sept 8. Award-winning British comedian and film star, especially remembered for his role as the bumbling character Inspector Clouseau. Born Sept 8, 1925. Died at London, England, July 24, 1980.

"STAR TREK": FIRST BROADCAST ANNIVERSARY. Sept 8. The first of 79 episodes of the TV series "Star Trek" was aired on the NBC network on Sept 8, 1966. Although the science fiction show set in the future only lasted a few seasons, it has remained enormously popular through syndication reruns. It has been given new life through 6 motion pictures, a cartoon TV series and the very popular TV series "Star Trek: The Next Generation" and "Star Trek: Deep Space Nine." It has consistently ranked among the biggest titles in the motion picture, television, home video and licensing divisions of Paramount Pictures.

STEAMBOAT DAYS FESTIVAL. Sept 8-10. Jeffersonville, IN. Nonprofit community festival. Est attendance: 100,000. For info: PO Box Steamboat, Jeffersonville, IN 47131-0764. Phone: (812) 284-2628.

SUGARLOAF CRAFT FESTIVAL. Sept 8-10. Prince William County Fairgrounds, Manassas, VA. More than 335 professional artists and craftspeople, demonstrations, live music and children's entertainment. Est attendance: 30,000. Sponsor: Sugarloaf Mt Works, Inc, 200 Orchard Ridge Dr #215, Gaithersburg, MD 20878. Phone: (301) 990-1400.

TENNESSEE STATE FAIR. Sept 8-17. Fairgrounds, Nashville, TN. Est attendance: 325,000. For info: Edd R. Townsend, Mgr, Box 40208, Nashville, TN 37204. Phone: (615) 862-8980.

TOSAFEST 1995. Sept 8-10. Wauwatosa, WI. A 3-day food, music and sports festival held in Wauwatosa's historic village includes soccer, tennis and softball tournaments; 5K run, the Great Tosa Bake-off and continuous entertainment. Proceeds benefit community organizations. Est attendance: 30,000. For info: Tosafest, Maureen Badding, 1837 N 68th St, Wauwatosa, WI 53213. Phone: (414) 476-3238.

TRAVERSE DES SIOUX COMMEMORATIVE ENCAMPMENT. Sept 8-10. Traverse des Sioux Park, St. Peter, MN. A Rendezvous/Native American event marking the Treaty of Traverse des Sioux of 1851, which opened 24 million acres in four states to settlement and ended a great Native American culture. Traders, buckskinners, memorabilia, Indian demonstrations, 19th-century crafts, food, music and fun. Annually, the first weekend after Labor Day. Est attendance: 7,000. Sponsor: John Hans, Dir, Nicollet County Historical Society, 1851 North Minnesota Ave, St. Peter, MN 56082. Phone: (507) 931-2160.

UNITED NATIONS: INTERNATIONAL LITERACY DAY. Sept 8. An international day observed by the organizations of the United Nations system. Info from: United Nations, Dept of Public Info, New York, NY 10017.

USA INTERNATIONAL DRAGON BOAT FESTIVAL. Sept 8-10. Dubuque, IA. Teams of 25 enthusiastic paddlers race ornately carved and painted dragon boats on the Mississippi River. Pageantry, competition, international fellowship. Held in conjunction with River Fest. Annually, the second weekend in September. Sponsor: Dubuque Chapter of the American Dragon Boat Assn. Est attendance: 20,000. For info: Earl Brimeyer, 2595 Rhomberg Ave, Dubuque, IA 52001. Phone: (319) 583-6345.

UTAH STATE FAIR. Sept 8-17. Salt Lake City, UT. Est attendance: 400,000. For info: Utah State Fair Park, 155 N 1000 W, Salt Lake City, UT 84116. Phone: (801) 538-8440.

WESTERN WASHINGTON FAIR. Sept 8-24. Puyallup, WA. Entertainment, animals, rides, displays and food. Est attendance: 1,300,000. For info: Western Washington Fair, PO Box 430, Puyallup, WA 98371. Phone: (206) 841-5045.

WHEELCHAIR TENNIS TOURNAMENT. Sept 8-10. Oka Hester Park, Greensboro, NC. Oka Hester Park hosts the North Carolina Wheelchair Sports Association Tennis Tournament. Est attendance: 25. For info: Wheelchair Tennis Tournament, 915 Barker Rd, Bear Creek, NC 27407. Phone: (919) 837-5001.

WYANDOTTE HERITAGE DAYS. Sept 8-10. Bishop Park Area, Wyandotte, MI. Outdoor craft show, living history encampments, Teddy Bear picnic, costume show, Historic Home and Church Tours and colonial dinners. For info: Mark Partin, Museum Dir, 3131 Biddle Ave, Wyandotte, MI 48192. Phone: (313) 246-4520.

YELLOW DAISY FESTIVAL. Sept 8-10. Georgia's Stone Mountain Park, Stone Mountain, GA. 26th annual. Arts and crafts festival with more than 450 exhibitors. Continuous entertainment and foods. Annually, the weekend after Labor Day. Est attendance: 225,000. For info: K. Thweatt, PR Office, Georgia's Stone Mountain Park, PO Box 778, Stone Mountain, GA 30086. Phone: (404) 498-5702.

BIRTHDAYS TODAY

Sid Caesar, 73, comedian, actor ("Your Show of Shows"), born at Yonkers, NY, Sept 8, 1922.

Maurice Cheeks, 39, former basketball player, born at Chicago, IL, Sept 8, 1956.

Wendell Hampton Ford, 71, US Senator (D, Kentucky), born at Davies County, KY, Sept 8, 1924.

Lyndon H. Larouche, Jr, 73, political activist, born at Rochester, NH, Sept 8, 1922.
Marilyn (Williamson) Mims, 41, opera singer, born at Collins, MS, Sept 8, 1954.
Sam Nunn, 57, US Senator (D, Georgia), born at Perry, GA, Sept 8, 1938.
Heather Thomas, 38, actress (Jody Banks on "The Fall Guy"), born at Greenwich, CT, Sept 8, 1957.
Rogie Vachon, 50, hockey player, born at Palmarolle, Quebec, Canada, Sept 8, 1945.

SEPTEMBER 9 — SATURDAY
252nd Day — Remaining, 113

BATTLE OF MARATHON: ANNIVERSARY. Sept 9. On the day of the ninth month's full moon (Boedromion 6) in the year 490 BC, the numerically superior invading army of Persia was met and defeated on the Plain of Marathon by the Athenian army, led by Miltiades. More than 6,000 men died in the day's battle which drove the Persians to the sea, and the mound of earth covering the dead is still visible at the site. See also: "Days of Marathon: Anniversary"(Sept 2) for the legendary running of Phidippides and the origin of the marathon race. Info from: Greswell's *Origines Kalendariae Hellenicae*.

BONZA BOTTLER DAY™. Sept 9. To celebrate when the number of the day is the same as the number of the month. Bonza Bottler Day is an excuse to have a party at least once a month. T-shirts available. For info: Elaine Fremont, 203 Waddell Rd, Taylors, SC 29687. Phone: (803) 244-2023.

BOONESBOROUGH DAYS. Sept 9-10. Shafer Memorial Park, Boonsboro, MD. Crafts, antiques, living history, demonstrations, and food. Est attendance: 10,000. For info: Deborah Miller, 21109 Reno Monument Rd, Boonsboro, MD 21713. Phone: (301) 432-2792.

BULGARIA: SOCIALIST REVOLUTION ANNIVERSARY. Sept 9. National holiday in Bulgaria.

CALIFORNIA: ADMISSION DAY. Sept 9. Became 31st state on this day in 1850.

CANADA: BINDER TWINE FESTIVAL. Sept 9. Kleinburg, Ontario. Outdoor arts and crafts show, pioneer skills demonstrations, old-fashioned food and entertainment. Est attendance: 20,000. For info: Binder Twine Festival Inc, PO Box 6, Kleinburg, Ont, Canada L0J 1C0.

CANADA: TRAIL OF '98 INTERNATIONAL ROAD RELAY. Sept 9-10. Dawson City, Yukon. Mixed, men, ladies and masters categories with teams consisting of six to ten members, race the 176 km (110 mi) from tidewater in Skagway to Whitehorse. For info: Yukon Tourism, Box 2703, Whitehorse, Yukon, Canada Y1A 2C6. Phone: (403) 668-1236.

CEDARHURST CRAFT FAIR. Sept 9-10. Mitchell Museum, Mount Vernon, IL. National juried invitational show featuring more than 150 artists. Est attendance: 20,000. For info: Cedarhurst Craft Fair, Glenna Brandt, Mitchell Museum, PO Box 923, Mount Vernon, IL 62864. Phone: (618) 242-1236.

September 1995

S	M	T	W	T	F	S
					1	2
3	4	5	6	7	8	9
10	11	12	13	14	15	16
17	18	19	20	21	22	23
24	25	26	27	28	29	30

CHADDS FORD DAYS. Sept 9-10. Chadds Ford, PA. Open-air colonial fair with 18th-century craft demonstrations, Brandywine Valley art, live old-time music, country rides and games, Colonial crafts for sale, good food. Annually, the second weekend in September. Est attendance: 11,500. For info: Susan Hauser, Chadds Ford Historical Society, Box 27, Chadds Ford, PA 19317. Phone: (610) 388-7376.

CHINESE MOON FESTIVAL. Sept 9-10. Chinatown, Los Angeles, CA. The Moon Festival celebrates the end of the summer harvest—a fun family holiday. Stage shows, food, arts and crafts, vendor booths, carnival, ceremonies, mooncakes, telescopes and more! Annually, on the full moon of the eighth month of the lunar calendar since at least 100 BC on the closest week-end. Est attendance: 25,000. For info: Chinese Chamber of Commerce of Los Angeles, 977 N Broadway, Ste E, Los Angeles, CA 90012. Phone: (213) 617-0396.

CIVIL WAR REENACTMENT. Sept 9-10. Bedford, PA. Visiting militia portray civil war lifestyles through an authentic encampment, parades, drills, demonstrations and flag-raising ceremonies. For info: Old Bedford Village, PO Box 1976, Bedford, PA 15522. Phone: (814) 623-1156.

★ **FEDERAL LANDS CLEANUP DAY.** Sept 9. Presidential Proclamation 5521, of Sept 5, 1986, covers all succeeding years. The first Saturday after Labor Day. (PL99-402 of Aug 27, 1986.)

FESTIVAL '95: FALL FESTIVAL OF ARTS AND CRAFTS. Sept 9-10. Dalton, GA. Juried works of more than 250 artists and craftspersons. Indoor and outdoor exhibits. More than $14,000 in Cash Awards, Corporate Purchase Awards available to artists. Entertainment, regional and ethnic foods, and a children's art market. 32nd annual. Est attendance: 7,000. For info: Creative Arts Guild, Box 1485, Dalton, GA 30722-1485. Phone: (706) 278-0168.

FESTIVAL OF THE ARTS. Sept 9. Means & Clark Park, Weatherford, OK. More than 100 booths involving artists from all over the country as well as crafters from surrounding states. Concession booths and live entertainment all day long. 22nd annual. Annually, the second Saturday in September. Est attendance: 2,000. For info: William Hancock, Exec Dir or Helen Blair, Admin Asst, Weatherford Chamber of Commerce, PO Box 729, Weatherford, OK 73096. Phone: (800) 725-7744.

FLAX SCUTCHING FESTIVAL. Sept 9-10. Stahlstown, PA. Demonstrations of the art of making linen from the flax plant. Second oldest continuous complete flax demonstration festival in the world. Annually, the second weekend in September. Est attendance: 15,000. For info: Flax Scutching Festival, Frank Newell, RD #1, Box 216, Stahlstown, PA 15687. Phone: (412) 593-2119.

GREAT KINETIC SCULPTURE RACE. Sept 9. Omaha, NE. Starting Saturday at 12 noon, Heartland of America Park (8th & Douglas St). Handmade artistic, all-terrain, people-powered vehicles race across a treacherous route of mud, water, pavement, gravel, sand and tall dandelions. Cheating is encouraged and the last entry across the finish line can win the race. Named a "Best Bet" by the Intl Events Group. Annually, the second Saturday in September. Est attendance: 10,000. For info: Evan Mills, Self-Appointed Person-in-Charge, 1308 Park Ave, Ste 9, Omaha, NE 68103. Phone: (402) 341-5150.

HERITAGE DAY FESTIVAL. Sept 9. (Rain date Sept 10.) Bay Blvd at Philadelphia Ave, Lavallette, NJ. Clowns, antique cars, WWII vehicles, bands, games, food, children's games, pony rides, train rides and finale with Ocean County String Band performing. 10-dusk. Annually, the second Saturday in September. Est attendance: 20,000. For info: Joy Grosko, Dir, Heritage Committee Inc, 13 Camden Ave, Lavallette, NJ 08735. Phone: (908) 793-3652.

INDIAN DAY. Sept 9. Oklahoma. Observed on the first Saturday after the Full Moon in September.

JOHN WILKES BOOTH ESCAPE ROUTE TOUR. Sept 9. Clinton, MD. (Also Sept 23.) A 12-hour bus tour over the route used by Lincoln's assassin. Est attendance: 45. Sponsor: Surratt House and Tavern, PO Box 427, Clinton, MD 20735. Phone: (301) 868-1121.

KID'RIFIC. Sept 9–10. Hartford, CT. A two-day children's festival of learning and fun, held under colorful tents on the streets of downtown Hartford. Hands-on art and science activities, entertainment, storytelling, face painting, a petting zoo with a 13-foot giraffe, motorized train ride, horse-drawn hayride and food are featured. Annually, the first weekend after Labor Day. Est attendance: 50,000. For info: Steven A. Lazaroff, Asst Dir for Events Programming, Hartford Downtown Council, 250 Constitution Plaza, Hartford, CT 06103. Phone: (203) 728-3089.

KING TURKEY DAYS (WITH TURKEY RACE). Sept 9–16. Worthington, MN. Festival to thank farm people for their patronage. On Sept 16 there are a parade of live turkeys and a turkey race down the 150-yard Main St race course. Est attendance: 20,000. Sponsor: King Turkey Days, Inc, Box 608, Worthington, MN 56187. Phone: (800) 279-2919.

KOREA: CHUSOK. Sept 9. Gala celebration by Koreans everywhere. Autumn harvest thanksgiving moon festival. Observed on 15th day of eighth lunar month each year. Koreans pay homage to ancestors and express gratitude to guarding spirits for another year of rich crops. A time to visit tombs, leave food, and prepare for coming winter season. Traditional food is "moon cake," made on eve of Chusok, with rice, chestnuts and jujube fruits. Games, dancing and gift exchanges. Observed since Silla Dynasty (beginning of First Millennium).

LENEXA SPINACH FESTIVAL. Sept 9. Sar-Ko-Par Trails Park, Lenexa, KS. Celebrating the city's reputation as Spinach Capital of the World, this tongue-in-cheek festival boasts a variety of spinach cuisine, a spinach recipe contest, arts and crafts booths, a briefcase throw and a water balloon launching. Est attendance: 7,000. For info: Kristy Iseman, Tourism Sales Mgr, Lenexa Conv & Visitors Bureau, 11900 W 87th St Pkwy, Ste 115, Lenexa, KS 66215. Phone: (800) 950-7867.

LITTLE FALLS SIDEWALK ARTS AND CRAFTS FAIR. Sept 9–10. Little Falls, MN. Artists, craftspeople and hobbyists displaying and selling their items. Est attendance: 40,000. For info: Chamber of Commerce, 200 NW First St, Little Falls, MN 56345. Phone: (612) 632-5155.

LOON LAKE HARVEST FESTIVAL. Sept 9–10. Aurora, MN. To promote old and new farm equipment, farmers market, crafts, food, classic car show, ugly farmer contest and draft horse demonstration. Annually, the second weekend in September. Est attendance: 2,000. For info: Bruce Petron, Pres, Loon Lake Harvest Festival, 3816 Hwy 100, Aurora, MN 55705. Phone: (218) 638-2725.

LUXEMBOURG: ANNIVERSARY LIBERATION CEREMONY. Sept 9. Petange. Commemoration of liberation of Grand-Duchy by the Allied forces in 1944. Ceremony at monument of the American soldier.

MARYLAND RECREATIONAL VEHICLE SHOW. Sept 9–17. Timonium State Fairgrounds, Timonium, MD. Fourth annual outdoor show of motorhomes, trailers, 5-wheel trailers and pickup campers, campground booths, accessories and related items. Annually, beginning first Saturday after Labor Day. Est attendance: 8,000. Sponsor: Richard T. Albright, Pres, Maryland Rec Vehicle Dealers Assn, Inc, 8332 Pulaski Hwy, Baltimore, MD 21237. Phone: (410) 687-7200.

MID-AMERICAN SPORT KITE CLASSIC. Sept 9–10. River Oaks Park, Kalamazoo, MI. A family-oriented sport kite competition, with events for Junior, Novice, Intermediate, Experienced and Master classes. Competitions for Individuals, Teams and Pairs in Precision and Ballet. Annually, the weekend after Labor Day. Sponsors: Kazoo Stringfellows and Battle Creek Cloud Cutters Kite Clubs. Est attendance: 1,500. For info: John Cosby, Coord, PO Box 2241, Kalamazoo, MI 49003. Phone: (616) 345-5432.

MONTANA STATE CHOKECHERRY FESTIVAL. Sept 9. Lewistown, MT. Queen contest, tasting and judging of jam, jelly, wine; pancake breakfast, race and run-walk, parade, farmers market, Mill Ditch duck race, art festival and street dance. Annually, first weekend after Labor Day. Est attendance: 1,500. For info: Lewistown Chamber of Commerce, PO Box 818, Lewistown, MT 59457. Phone: (406) 538-5436.

MOON FESTIVAL OR MID-AUTUMN FESTIVAL. Sept 9. This festival, observed on the 15th day of the eighth moon of the lunar calendar year, is called by different names in different places, but is widely recognized throughout the Far East, including People's Republic of China, Taiwan, Korea, Singapore and Hong Kong. An important harvest festival at the time the moon is brightest, it is also a time for homage to ancestors. Special harvest foods are eaten, especially "moon cakes."

MOUNTAINEER FOLK FESTIVAL. Sept 9–10. Fall Creek Falls State Park, Pikeville, TN. Handmade crafts, traditional mountain music, and pioneer skills demonstrations are just part of the fun. Country cooking, bluegrass music and different exhibits round out the jam-packed schedule. Annually, the weekend after Labor Day. Co-sponsored by Van Buren County Historical Society. Est attendance: 5,000. Sponsor: Betty Dunn Nature Center, Fall Creek Falls State Park, Rte 3, Pikeville, TN 37367. Phone: (615) 881-5708.

MYSTIC SEAPORT PHOTO WEEKEND. Sept 9–10. Mystic Seaport, Mystic, CT. An entire weekend designed for photographers: 19th-century costumed models, scenic ships and grounds. Est attendance: 4,000. For info: Mystic Seaport, 75 Greenmanville Ave, Box 6000, Mystic, CT 06355-5315. Phone: (203) 572-5315.

NANTICOKE INDIAN POWWOW. Sept 9–10. Millsboro, DE. Annual gathering of Native Americans during which a program of Indian dances and music is produced. Indian food, arts and crafts are sold during this event. Time: Dance sessions noon–5 on Saturday; Sunday worship service from 11–12 and dance 2–4:30. Annually, the second weekend in September. Est attendance: 40,000. Sponsor: Nanticoke Indian Assn, Rte 24, Millsboro, DE 19966. Phone: (302) 945-3400.

NATIVE AMERICAN HERITAGE FESTIVAL. Sept 9–10. Patrick Henry Community College, Martinsville, VA. 11th annual traditional powwow, crafts and restricted dancing. Annually, the second weekend in September. For info: Virginia Museum of Natural History, Bonnie Wilson, 1001 Douglas Ave, Martinsville, VA 24112. Phone: (703) 666-8600.

NORDICFEST PARADE. Sept 9. Libby, MT. Parade showcases a large contingent of Norwegian fjord horses plus a variety of floats and entertainments, many of which celebrate the ethnic origins of the area's earliest settlers. Annually, the second Saturday in September. For info: Shirley Wasco, Parade Chmn, Nordicfest, Inc, 351 S Central Rd, Libby, MT 59923. Phone: (406) 293-8815.

NORTH KOREA: NATIONAL DAY. Sept 9. National holiday in the Democratic People's Republic of [North] Korea.

OCEAN COUNTY FOLK ARTS FESTIVAL. Sept 9. Ocean County Park, Rte 88, Lakewood, NJ. This event highlights music and crafts from Ocean County. Folk musicians, cloggers, square dancers and fine area artists; music workshops for adults and children; canoeing and delicious food. Est attendance: 6,500. For info: Ocean County Park & Recreation, 1198 Bandon Rd, Toms River, NJ 08753. Phone: (908) 506-9090.

PEOPLE'S REPUBLIC OF CHINA: MID-AUTUMN FESTIVAL. Sept 9. To worship the moon god, and according to folk legend this day is also the birthday of the earth god T'u-ti Kung. The festival indicates the year's hard work in the fields will soon end with the harvest. People express gratitude to heaven as represented by the moon and earth as symbolized by the earth god for all good things from the preceding year. 15th day of eighth month of Chinese lunar calendar.

Sept 9 ☆ Chase's 1995 Calendar of Events ☆

POTATO BOWL. Sept 9-16. Grand Forks, ND. Potato Bowl USA salutes the potato featuring queen contest, concert, football game, a parade and other events. For info: Grand Forks Conv and Visitors Bureau, 202 N 3rd St, Ste 200, Grand Forks, ND 58203. Phone: (800) 866-4566.

PUBLIC LANDS DAY. Sept 9. To involve citizen volunteers in cleaning and maintaining public lands. Annually, the Saturday after Labor Day. Est attendance: 1,000,000. Sponsor: Keep America Beautiful, Inc, 9 W Broad St, Stamford, CT 06902.

QUADRANGLE FESTIVAL. Sept 9-10. (Tentative date.) Downtown Texarkana, TX and AR. 14th annual festival features 5K run around Texarkana Historical Museum that sits on the TX-AR state line. Also, antiques, crafts, collectibles; antique autos, pet shows and turtle races; country/western, traditional and contemporary music; folk artists and craftspeople; street dancing and food vendors. Est attendance: 50,000. For info: Texarkana Museum System, PO Box 2343, Texarkana, TX 75504. Phone: (903) 793-4831.

ROAN MOUNTAIN NATURALISTS RALLY. Sept 9-10. Roan Mountain State Park, Elizabethton, TN. Provides both professional and amateur naturalists an annual opportunity to gather together to enjoy the mountains. Always held weekend after Labor Day. For info: Elizabethton/Carter County Chamber of Commerce, PO Box 190, Elizabethton, TN 37643. Phone: (615) 543-2122.

ROCVALE BALLOON RALLY. Sept 9-10. Rockford, IL. Officially sanctioned competition and festival with 3 balloon launches, entertainment, arts and crafts exhibitors and food. Sponsor: Rocvale Children's Home. Est attendance: 26,000. For info: Rockford Area Conv and Visitors Bureau, Memorial Hall, 211 N Main St, Rockford, IL 61101. Phone: (800) 521-0849.

SANDERS, COLONEL HARLAND DAVID: 150th BIRTH ANNIVERSARY. Sept 9. Founder of Kentucky Fried Chicken, born near Henryville, IN, Sept 9, 1890. Died Dec 16, 1980.

SANTA ROSALIA FESTIVAL. Sept 9-10. Monterey Fairgrounds, Monterey, CA. A 58-year tradition dedicated to Santa Rosalia, patron saint of the Sicilian fishermen in Monterey, the Festival includes parade, outdoor mass, procession on Fisherman's Wharf, blessing of the fleet, entertainment, arts and crafts and Italian food. Benefits the Italian Heritage Society of the Monterey Peninsula. Annually, the second Saturday and Sunday in September. Est attendance: 7,500. For info: Bostrom Management, 2600 Garden Rd, Ste 208, Monterey, CA 93940. Phone: (408) 649-6544.

SONGS OF MY PEOPLE. Sept 9-Oct 22. Winston-Salem State University, Winston-Salem, NC. See Mar 4 for full description. Call or write the Smithsonian for other dates and venues. For info: Smithsonian Institution Traveling Exhibition Service, 1100 Jefferson Dr SW, Ste 3146, Washington, DC 20560. Phone: (202) 357-2700.

TAJIKISTAN: INDEPENDENCE DAY. Sept 9. National holiday.

UNITED TRIBES POWWOW. Sept 9-10. United Tribes College, Bismarck, ND. The United Tribes Powwow draws many contestants for Native American singing and dancing competition. Colorful costumes, Indian foods and artifacts add to the festivities. Phone: (701) 224-2525 or (800) 435-5663. Est attendance: 50,000. For info: North Dakota Tourism, Pat Hertz, Liberty Memorial Bldg, State Capitol Grounds, Bismarck, ND 58505.

VACAVILLE ONION FESTIVAL. Sept 9-10. Vacaville, CA. Celebrity entertainment, cooking demonstrations, cuisine center, kiddie korner, onion patch, arts and crafts and more to promote the pungent bulb. Est attendance: 14,000. For info: Colleen Duke, PO Box 5263, Vacaville, CA 95696. Phone: (707) 448-6424.

WALES: CARDIFF FESTIVAL OF MUSIC. Sept 9-Oct 7. Cardiff, Wales. The festival takes the theme of "Stories and Legends" and includes performances by the BBC National Orchestra of Wales. For info: Marketing Office, St. David's Hall, The Hayes, Cardiff, Wales, UK CF1 2SH.

WALES: WALKER CUP GOLF CHAMPIONSHIP. Sept 9-10. Royal Porthcawl Golf Course, Porthcawl. For info: US Golf Assn, Golf House, Far Hills, NJ 07913. Phone: (908) 234-2300.

WILLIAM, THE CONQUEROR: DEATH ANNIVERSARY. Sept 9. Anniversary of the death of William I, The Conqueror, King of England and Duke of Normandy, whose image is portrayed in the Bayeux Tapestry. Victorious over Harold at the Battle of Hastings (the Norman Conquest) in 1066, William was crowned King of England at Westminster Abbey on Christmas Day of that year. Later, at age 60, while waging war in France, William met his death. Having burnt the town of Nantes, he was riding victoriously through the ruins when his horse stepped on a hot coal and reared up abruptly, causing internal injury to King William, who died of the injury at Rouen, on Sept 9, 1087.

WILLOW TREE FESTIVAL. Sept 9-10. Gordon, NE. This festival is a cornucopia of arts and crafts, 3 quality performing stages including a children's stage, children's activities, fun and food. Annually, the second weekend in September. Est attendance: 4,000. For info: Willow Tree Festival, PO Box 303, Gordon, NE 69343. Phone: (308) 282-1185.

WYATT EARP PRIME BEEF DAY. Sept 9. Monmouth, IL. Birthplace tours, costumed hosts and Earp cousins. Held during the Warren County Prime Beef Festival which includes a parade, beef contest and auction, midway, pig, chicken and beef scrambles, fly-in breakfast, evening shows and children's activities. Phone for Chamber of Commerce: (309) 734-3181. Est attendance: 4,000. For info: Wyatt Earp Birthplace Historic House Museum, c/o 1020 E Detroit Ave, Monmouth, IL 61462. Phone: (309) 734-6419.

BIRTHDAYS TODAY

B.J. (Benjamin Roy) Armstrong, 28, basketball player, born at Detroit, MI, Sept 9, 1967.

Michael Keaton (Michael Douglas), 44, actor ("Report to Murphy," *Batman*), born at Pittsburgh, PA, Sept 9, 1951.

Daniel Lewis Majerlee, 30, basketball player, born Traverse City, MI, Sept 9, 1965.

Kristy McNichol, 33, actress (Letitia "Buddy" Lawrence on "Family"; "Empty Nest"), born Los Angeles, CA, Sept 9, 1962.

Billy Preston, 49, musician, singwriter, singer ("Will It Go Round in Circles," "Nothing from Nothing"), born Houston, TX, Sept 9, 1946.

Cliff Robertson, 70, actor ("Falcon Crest," *Brainstorm, Charly, PT-109*), born at La Jolla, CA, Sept 9, 1925.

September 1995

S	M	T	W	T	F	S
					1	2
3	4	5	6	7	8	9
10	11	12	13	14	15	16
17	18	19	20	21	22	23
24	25	26	27	28	29	30

Chase's 1995 Calendar of Events — Sept 9–10

James "Jimmy the Greek" Snyder (Demetrius George Synodinos), 72, ex-oddsmaker, born at Steubenville, OH, Sept 9, 1923.

Joe Theisman, 46, Pro Football Hall of Famer, born at New Brunswick, NJ, Sept 9, 1949.

John Thompson, 54, college basketball coach, born at Washington, DC, Sept 9, 1941.

SEPTEMBER 10 — SUNDAY
253rd Day — Remaining, 112

AFRICAN-AMERICAN HERITAGE FESTIVAL. Sept 10. Pocomoke River State Park, Snow Hill, MD. Celebrates the role played by African-Americans in shaping history and culture in Worcester County. 10–6. Annually, the second Sunday in September. Est attendance: 2,000. For info: Gabriel and Diana Purnell, 10214 Old Ocean City Blvd, Berlin, MD 21811. Phone: (410) 632-2600.

BELGIUM: HISTORICAL PROCESSION. Sept 10. Tournai. Traditional cultural observance. Annually, the Sunday closest to Sept 8.

BELIZE: ST. GEORGE'S CAYE DAY. Sept 10. Public holiday celebrated in honor of the battle between the European Baymen Settlers and the Spaniards for the territory of Belize.

BRAXTON, CARTER: BIRTH ANNIVERSARY. Sept 10. American revolutionary statesman and signer of the Declaration of Independence. Born on Sept 10, 1736, at Newington, VA. Died on Oct 10, 1797, at Richmond, VA.

BUFFALO BILL DAYS. Sept 10–17. Leavenworth, KS. Arts and crafts, outhouse races, 450-foot-long banana split, parade, aircraft display and rides and children's carnival. Grab your western garb and join the fun. Annually, beginning the second Sunday in September. Est attendance: 20,000. For info: Leavenworth-Lansing Conv/Visitors Bureau, Connie Boyd, Dir, 518 Shawnee, Box 44, Leavenworth, KS 66048. Phone: (800) 844-4114.

CANADA: INTERNATIONAL DART TOURNAMENT. Sept 10–11. Dawson City, Yukon. Held at Gertie's. Singles and doubles. For info: Yukon Tourism, Box 2703, Whitehorse, Yukon, Canada Y1A 2C6. Phone: (403) 993-5575.

CANDA: INTERNATIONAL TEXAS HOLDEM POKER TOURNAMENT. Sept 10. Dawson City, Yukon. Most Northern Holdem Tournament hosted by White Ram bed and breakfast. Annually, the Sunday after Labor Day. For info: Yukon Tourism, Box 2703, Whitehorse, Yukon, Canada Y1A 2C6. Phone: (403) 993-5772.

CODMAN HOUSE ARTISANS' FAIR OF CRAFTS. Sept 10. Codman Estate, Lincoln, MA. Outdoor craft fair with 100 artisans selling their wares. Crafts include spinning and weaving, silver and pewter, baskets, pottery, toys and more. Demonstrations, entertainment, children's activities, tours of historic Codman House, refreshments. 10 AM–5 PM. Admission charged. Annually, the Sunday after Labor Day. Est attendance: 4,000. For info: Hetty Startup, Site Admin Codman House, Society Preservation New Eng Antiquities, 141 Cambridge St, Boston, MA 02114. Phone: (617) 227-3956.

ENCHANTED CIRCLE CENTURY BIKE TOUR. Sept 10. Red River, NM. 100-mile scenic ride and one of the longest and most difficult bicycle tours in the Southwest. Est attendance: 1,000. For info: Red River Chamber of Commerce, PO Box 870, Red River, NM 87558. Phone: (800) 348-6444.

ENGLAND: BATTLE OF BRITAIN WEEK. Sept 10–16. Annually, the week containing Battle of Britain Day (Sept 15).

FALL HAT WEEK. Sept 10–17. A week of celebration during which the straw hat is put aside in favor of the felt hat by both men and women. Local businesses are encouraged to plan hat-related activities. Annually, in September. For info: Casey Push, Exec Dir, Millinery Info Bureau, 302 W 12 St, PH-C, New York, NY 10014. Phone: (212) 627-8333.

FESTIVAL-IN-THE-PARK. Sept 10. (Rain date Sept 24.) Memorial Park, Nutley, NJ. 22nd annual festival, a craft and collectibles show to benefit the Nutley Historical Society and Historic Restoration Trust. Annually, the first Sunday after Labor Day. Est attendance: 15,000. For info: Festival-in-the-Park, Douglas J. Eisenfelder, 51 Enclosure, Nutley, NJ 07110. Phone: (201) 667-3013.

FUNK, ISAAC KAUFFMAN: BIRTH ANNIVERSARY. Sept 10. American publisher born at Clifton, OH, Sept 10, 1839. Partner in Funk and Wagnalls. Died at Montclair, NJ, Apr 4, 1912.

GUNSTON HALL CAR SHOW. Sept 10. Gunston Hall Plantation, Lorton, VA. 29th annual show. Trophies for the oldest car, best in class, most popular and best in show. A trophy for Best of Marque will be awarded to a car from each club entering at least five representative cars. Est attendance: 4,000. For info: Special Events, Gunston Hall, 10709 Gunston Rd, Lorton, VA 22079. Phone: (703) 550-9220.

HOMESTEADER HARVEST FESTIVAL. Sept 10. Beaver Creek Nature Area, Brandon, SD. This is a re-creation of the harvest and celebration that went with it back in the pioneer days. Annually, the second Sunday in September. Est attendance: 1,200. For info: Palisades State Park, 25495 485th Ave, Garretson, SD 57030-6117. Phone: (605) 594-3824.

INTERFAITH FELLOWSHIP IN NEW YORK CITY: ANNIVERSARY CELEBRATION. Sept 10. 165 W. 57th St, New York, NY. In the belief that love is the core of all Spiritual Paths, Interfaith Fellowship was founded to bring people of all different Faiths together in Fellowship and Celebration every Sunday morning in NYC. Founded Sept 10, 1989, it was legally incorporated as a church in the state of New York in August 1990. Celebration held annually, the first Sunday after Labor Day. Sponsor: Diane Berke. Est attendance: 250. For info: Rev Jon Mundy, 459 Carol Dr, Monroe, NY 10950. Phone: (914) 783-0383.

ITALY: GIOSTRA DELLA QUINTANA. Sept 10. Foligno. A revival of a 17th-century joust of the Quintana, featuring 600 knights in full costume. Annually, the second Sunday in September.

JCBC CENTURY RIDE. Sept 10. Junction City, Milford Lake, Fort Riley, KS. 100-mile bicycle ride (in loops) that features various course routes and optional stopping points. Sponsor: Bicycle Club. Est attendance: 200. For info: Casey Thomas, 2206 Prospect Circle, Junction City, KS 66441. Phone: (913) 762-3310.

LISCO OLD-TIMERS DAY. Sept 10. Lisco, NE. Local talent puts on program and skits about an honored old-timer or couple. Parade, barbecue (free will offering), horseshoe tournament, horse pull, team roping and more. Annually, Sunday after Labor Day. Sponsor: Lisco Old-timers Day Committee. Est attendance: 2,000. For info: Bill Vogler, c/o Lisco State Bank, Lisco, NE 69148. Phone: (308) 772-3226.

MARRIAGE CELEBRATION. Sept 10. Belleville, IL. An evening honoring married couples celebrating 25, 40, 50 and 60 years of marriage in 1995. Sunday, 7 PM. 23rd annual. Est attendance: 700. For info: Shrine of Our Lady of the Snows, 9500 W Illinois, Hwy 15, Belleville, IL 62223. Phone: (618) 397-6700.

MT WASHINGTON AUTO ROAD BICYCLE HILLCLIMB. Sept 10. Gorham, NH. A USCF-Sanctioned Bicycle Hillclimb, which benefits the Tin Mountain Conservation Center, Jackson, NH. Race registration required. Annually, the second Sunday in September. Est attendance: 800. For info: Bicycle Hillclimb Registrar, PO Box 1170, Conway, NH 03818. Phone: (603) 447-6991.

★ **NATIONAL GRANDPARENTS DAY.** Sept 10. Presidential Proclamation 4679, of Sept 6, 1979, covers all succeeding years. First Sunday in September following Labor Day (PL96–62 of Sept 6, 1979). First issued in 1978 (Proc 4580 of Aug 3, 1978), requested by Public Law 325 of July 28, 1978.

Sept 10-11 ☆ *Chase's 1995 Calendar of Events* ☆

NATIONAL GRANDPARENTS DAY. Sept 10. To honor grandparents, to give grandparents an opportunity to show love for their children's children and to help children become aware of the strength, information and guidance older people can offer. Annually, the first Sunday after Labor Day. For info: Marian McQuade, Founder, 140 Main St, Oak Hill, WV 25901. Phone: (304) 469-6884.

NATIONAL HOUSEKEEPERS WEEK. Sept 10-16. A week to recognize housekeeping staff; let them know that they are appreciated and that the work they do is worthwhile, vital, commendable and acknowledged by all. Annually, the second full week in September. Sponsor: NEHA, 1001 Eastwind Dr, Ste 301, Westerville, OH 43081-3361. Phone: (614) 895-7166.

NATIONAL PET MEMORIAL DAY. Sept 10. To memorialize a person's deceased pet and to recollect the impact that pet had on the owner's family. Annually, the second Sunday in September. Est attendance: 70. Sponsor: Intl Assn of Pet Cemeteries, 5055 US11, Ellenburg Depot, NY 12935. Phone: (518) 594-3000.

NORTHEAST MISSOURI TRIATHLON CHAMPIONSHIP. Sept 10. Thousand Hills State Park, Kirksville, MO. Swim ¾ mile, bike 18 miles, run 5 miles. USA Triathlon Federation certified. Qualifier for International Course Nationals. Annually, the Sunday after Labor Day. Est attendance: 800. For info: KRXL Radio, Box 130, Kirksville, MO 63501. Phone: (816) 626-2213.

SWAP IDEAS DAY. Sept 10. To encourage people to explore ways in which their ideas can be put to work for the benefit of humanity, and to encourage development of incentives that will encourage use of creative imagination. Annually, Sept 10. Sponsor: Puns Corps, c/o Robert L. Birch, Publicity Chmn, Box 2364, Falls Church, VA 22042-0364. Phone: (703) 533-3668.

WERFEL, FRANZ: BIRTH ANNIVERSARY. Sept 10. Austrian author, born at Prague, Czechoslovakia, on Sept 10, 1890. Died at Hollywood, CA, Aug 26, 1945.

BIRTHDAYS TODAY

Cap (Casper N.) Boso, 32, football player, born Kansas City, MO, Sept 10, 1963.

Jose Feliciano, 50, singer, musician ("Light My Fire," "Hi-Heel Sneakers"), born at Larez, Puerto Rico, Sept 10, 1945.

Amy Irving, 42, actress (*Carrie, Honeysuckle Rose*; singing voice of Jessica Rabbit), born at Palo Alto, CA, Sept 10, 1953.

Charles Bishop Kuralt, 61, journalist (retired from "Sunday Morning"), born at Wilmington, NC, Sept 10, 1934.

Arnold Daniel Palmer, 66, golfer, born at Youngstown, PA, Sept 10, 1929.

Yma Sumac, 67, singer, born at Ichocan, Peru, Sept 10, 1928.

Robert Wise, 81, filmmaker (*The Curse of the Cat People, The Sound of Music*), born Winchester, IN, Sept 10, 1914.

September 1995

S	M	T	W	T	F	S
					1	2
3	4	5	6	7	8	9
10	11	12	13	14	15	16
17	18	19	20	21	22	23
24	25	26	27	28	29	30

SEPTEMBER 11 — MONDAY
254th Day — Remaining, 111

ANNAPOLIS CONVENTION: ANNIVERSARY. Sept 11-14. Twelve delegates from New York, New Jersey, Delaware, Pennsylvania and Virginia met at Annapolis, MD, Sept 11-14, 1786, to discuss commercial matters of mutual interest. The delegates voted, on Sept 14, to adopt a resolution prepared by Alexander Hamilton asking all states to send representatives to a convention at Philadelphia, PA, in May 1787 "to render the constitution of the Federal Government adequate to the exigencies of the Union."

FEARLESS FORECASTS OF TV'S FALL FLOPS. Sept 11. An annual forecast of which new television shows will not last through the network's fall season. Uncannily accurate. Sponsor: The Boring Institute, Alan Caruba, Founder, Box 40, Maplewood, NJ 07040. Phone: (201) 763-6392.

GNAWED, CLAWED, BATTERED, BUSTED, BROKEN BIRD FEEDER FESTIVAL. Sept 11-24. Wild Bird Maketplace stores. People who maintain wild bird feeders bring in their Gnawed, Clawed, Battered, Busted, Broken Bird Feeders to display damage inflicted by squirrels, storms or other phenomena. Sponsor: John F. Gardner, Pres, Wild Bird Marketplace, 710 W Main St, PO Box 1184, New Holland, PA 17557. Phone: (717) 354-2841.

LAWRENCE, DAVID HERBERT: BIRTH ANNIVERSARY. Sept 11. English novelist, author of *Lady Chatterley's Lover*. Born at Eastwood, Nottinghamshire, England, Sept 11, 1885. Died on Mar 2, 1930.

LIND, JENNY: US PREMIERE: ANNIVERSARY. Sept 11. Jenny Lind, the "Swedish Nightingale," gave her first American performance in the Castle Garden Theatre, New York, NY, on this day in 1850.

NATIONAL BOSS/EMPLOYEE EXCHANGE DAY. Sept 11. To help bosses and employees appreciate each other by sharing each other's point of view for a day. Annually, the first Monday after Labor Day. Sponsor: A.C. Moeller, Box 71, Clio, MI 48420.

NATIONAL MIND MAPPING FOR PROJECT MANAGEMENT WEEK. Sept 11-16. An ideal time to introduce this brainstorming process to employees at any business or organization. Mind mapping is a creative, idea-generating technique which utilizes free writing or note taking. Make use of a lunch period or plan a brainstorming break to plan projects using this fun technique. For info: Tim Richardson, Pres, Total Development Resources, Inc, 363-6 Atlantic Blvd, Ste 201, Atlantic Beach, FL 32233-5251. Phone: (800) 226-4473.

NO NEWS IS GOOD NEWS DAY. Sept 11. Don't read, listen to or watch the news today and you'll feel better tonight. Phone: (212) 388-8673 or (717) 274-8451. Sponsor: Tom and Ruth Roy, Wellness Permission League, 2105 Water St, Lebanon, PA 17046.

O. HENRY: BIRTH ANNIVERSARY. Sept 11. William Sydney Porter, American author, who wrote under the pen name O. Henry. Best known for his short stories, including "Gift of the Magi." Born at Greensboro, NC, Sept 11, 1862. Died at New York, NY, June 5, 1910.

PAKISTAN: FOUNDER'S DEATH ANNIVERSARY. Sept 11. Pakistan recognizes observance of the death anniversary in 1948 of Quaid-i-Azam Mohammed Ali Jinnah (founder of Pakistan) with public holiday.

SCARECROW CONTEST AND DISPLAY. Sept 11-Oct 28. Peddler's Village, Lahaska, PA. Contestants compete for $5500 in cash prizes. Categories include: "A Scarecrow Whirligig"—a scarecrow that makes noise and moves with the wind; "An Extraordinary Contemporary Scarecrow"—an imaginative piece created to give a good scare in the garden; "A Traditional Scarecrow"—an outstanding example of the American Scarecrow; and "The Amateur Scarecrow"—a traditional, contemporary or whirligig example of the art. Public votes for their favorites. Free admission. Est attendance: 75,000. For info: Peddler's Village, PO Box 218, Lahaska, PA 18931. Phone: (215) 794-4000.

TYLER'S CABINET RESIGNS: ANNIVERSARY. Sept 11. In protest of President John Tyler's veto of the Banking Bill all of his cabinet except Secretary of State Daniel Webster resigned on this day in 1841.

BIRTHDAYS TODAY

Daniel K. Akaka, 71, US Senator (D, Hawaii) up for reelection in Nov '94, born at Honolulu, HI, Sept 11, 1924.
Brian DePalma, 55, filmmaker (*The Untouchables, Bonfire of the Vanities, Carrie*), born Newark, NJ, Sept 11, 1940.
Lola Falana, 52, actress ("The New Bill Cosby Show," "Ben Vereen—Comin' At Ya"), born Camden, NJ, Sept 11, 1943.
William Xavier Kienzle, 67, author (*Body Count, The Rosary Murders*), former priest, born at Detroit, MI, Sept 11, 1928.
Thomas Wade Landry, 71, Pro Football Hall of Famer, born at Mission, TX, Sept 11, 1924.
Bob Packwood, 63, US Senator (R, Oregon), born at Portland, OR, Sept 11, 1932.
Alfred Slote, 69, author (*Finding Buck McHenry, The Trading Game*), born at Brooklyn, NY, Sept 11, 1926.

SEPTEMBER 12 — TUESDAY
255th Day — Remaining, 110

CAPE VERDE: INDEPENDENCE DAY. Sept 12. National holiday.

CHARLES LEROUX'S LAST JUMP: ANNIVERSARY. Sept 12. American aeronaut of French extraction, born in New York, NY, about 1857, claimed by *Soviet Life* to have been a nephew of President Abraham Lincoln, achieved world fame as a parachutist. After his first public performance (Philadelphia, PA, 1887) he toured European cities where his parachute jumps attracted wide attention. Credited with 238 successful jumps. On Sept 12, 1889, he jumped from a balloon over Tallinn, Estonia, and perished in the Bay of Reval. A monument to his memory was erected at Tallinn five years after his death.

ENGLAND: CHELSEA ANTIQUES FAIR. Sept 12–23. Chelsea Old Town Hall, King's Rd, London. Antiques fair with a wide range of pre-1830 furniture and other items with a pre-1851 dateline for sale. Est attendance: 12,000. For info: Penman Antiques Fair, PO Box 114, Haywards Heath, West Sussex, England RH16 2YU.

GUINEA-BISSAU: NATIONAL HOLIDAY. Sept 12. Amilcar Cabral's birthday, Sept 12, is observed as national holiday.

MARYLAND: DEFENDERS DAY. Sept 12. Maryland. Annual reenactment of bombardment of Ft McHenry in 1814 which inspired Francis Scott Key to write the "Star Spangled Banner."

MENCKEN, HENRY LOUIS: BIRTH ANNIVERSARY. Sept 12. American newspaperman, lexicographer, and critic, "the Sage of Baltimore" was born at Baltimore, MD, Sept 12, 1880, and died there Jan 29, 1956. "If, after I depart this vale," he wrote in 1921 (Epitaph, *Smart Set*), "you ever remember me and have thought to please my ghost, forgive some sinner and wink your eye at some homely girl."

OWENS, JESSE: BIRTH ANNIVERSARY. Sept 12. James Cleveland (Jesse) Owens, American athlete, winner of four gold medals at the 1936 Olympic Games in Berlin, Germany, was born at Oakville, AL, Sept 12, 1913. Died at Tucson, AZ, Mar 31, 1980. Owens set 11 world records in track and field. During one track meet, at Ann Arbor, MI, Owens, representing Ohio State University, broke five world records and tied a sixth in the space of 45 minutes, on May 23, 1935.

SPACE MILESTONE: *LUNA 2* (USSR). Sept 12. First spacecraft to land on moon. Launched on Sept 12, 1959.

SPACE MILESTONE: *LUNA 16* (USSR). Sept 12. First unmanned spacecraft to land on moon (Sea of Fertility), collect samples and return to Earth. Launched on Sept 12, 1970.

WARNER, CHARLES DUDLEY: BIRTH ANNIVERSARY. Sept 12. American newspaperman, born at Plainfield, MA, Sept 12, 1829. Author of many works, perhaps best remembered for a single sentence (in an editorial, *Hartford Courant*, Aug 24, 1897): "Everybody talks about the weather, but nobody does anything about it." The quotation is often mistakenly attributed to his friend, Mark Twain. Died at Hartford, CT, Oct 20, 1900.

BIRTHDAYS TODAY

Deron Leigh Cherry, 36, football player, born at Riverside, NJ, Sept 12, 1959.
Irene Dailey, 75, actress (the 4th Liz Matthews in "Another World"), born at New York, NY, Sept 12, 1920.
Linda Gray, 54, actress (Sue Ellen on "Dallas"), born Santa Monica, Ca, Sept 12, 1941.
Timothy Duane Hardaway, 29, basketball player, born Chicago, IL, Sept 12, 1966.
George Jones, 64, singer ("White Lightning," "Race Is On," "He Stopped Loving Her Today"; CMA and Grammy Awards, the man Tammy didn't stand by), born at Saratoga, TX, Sept 12, 1931.
Michael Stephen (Mickey) Lolich, 55, former baseball player, born at Portland, OR, Sept 12, 1940.
Maria Muldaur, 52, singer ("Midnight at the Oasis," "I'm a Woman"), born New York, NY, Sept 12, 1943.
Peter Scolari, 41, actor ("Bosom Buddies," "Newhart"), born at New Rochelle, NY, Sept 12, 1954.
Barry White, 51, singer ("Can't Get Enough of Your Love, Babe," "You're the First, the Last, My Everything"), born at Galveston, TX, Sept 12, 1944.

SEPTEMBER 13 — WEDNESDAY
256th Day — Remaining, 109

ANDERSON, SHERWOOD: BIRTH ANNIVERSARY. Sept 13. American author and newspaper publisher, born at Camden, OH, Sept 13, 1876. His best remembered book is *Winesburg, Ohio*. Anderson died at Colon, Panama, Mar 8, 1941.

BARRY, JOHN: DEATH ANNIVERSARY. Sept 13. Commemorates day in 1803 on which Commodore John Barry died. First American commodore who fought in Revolutionary War.

MORTON PUMPKIN FESTIVAL. Sept 13–16. Morton, IL. Carnival, parade, entertainment and fantastic food to celebrate the pumpkin in the "Pumpkin Capital of the World." Est attendance: 50,000. For info: Chamber of Commerce, 415 W Jefferson St, Morton, IL 61550. Phone: (309) 263-2491.

PENDLETON ROUND-UP. Sept 13–16. Pendleton, OR. America's classic rodeo. An 85-year-old western tradition, with participating Indian tribes from the Pacific Northwest. Plus historical Happy Canyon, an outdoor pageant each evening. Est attendance: 54,000. For info: Pendleton Round-up Assn, Box 609, Pendleton, OR 97801. Phone: (800) 457-6336.

PERSHING, JOHN J.: BIRTH ANNIVERSARY. Sept 13. US Army general who commanded the American Expeditionary Force (AEF) during World War I, Pershing was born Sept 13, 1860, at Laclede, MO. The AEF, as part of the inter-Allied offensive, successfully assaulted the Saint-Mihiel salient in Sept 1918 and later that month quickly regrouped for the Meuse-Argonne operation that led to the Armistice of Nov 11, 1918. Pershing died July 15, 1948, at Washington, DC.

Sept 13-14 ☆ Chase's 1995 Calendar of Events ☆

REED, WALTER: BIRTH ANNIVERSARY. Sept 13. American army physician (especially known for his Yellow Fever research). Born at Gloucester County, VA, on Sept 13, 1851. Served as an army surgeon for more than 20 years and as a professor at the Army Medical College. He died at Washington, DC, Nov 22, 1902. The US Army's general hospital in Washington, DC, is named in his honor.

"STAR-SPANGLED BANNER" INSPIRED: ANNIVERSARY. Sept 13-14. On the night of Sept 13, 1814, Francis Scott Key was aboard a ship that was delayed in Baltimore harbor by the British attack there on Fort Henry. Key had no choice but to anxiously watch the battle. That experience and seeing the American flag still flying over the fort the next morning inspired him to pen the verses that, coupled with the tune of a popular drinking song, became our national anthem.

STEAMBOAT MOTORCYCLE WEEKEND. Sept 13-17. Steamboat Springs, CO. Vintage and Modern street races. Preweekend includes motorcross, trials race and dirt track. Display of classic motorcycles. 14th annual. Est attendance: 10,000. For info: Steamboat Springs Chamber Resort Assn, Janet Nichols, Special Event Coord, PO Box 774408, Steamboat Sprgs, CO 80477. Phone: (303) 879-0882.

US CAPITAL ESTABLISHED AT NEW YORK CITY: ANNIVERSARY. Sept 13. On Sept 13, 1788, the Congress picked New York, NY, as the location of the new US government.

US SENIOR WOMEN'S AMATEUR (GOLF) CHAMPIONSHIP. Sept 13-15. Somerset Country Club, St. Paul, MN. Sponsor: US Golf Assn, Championship Dept, Golf House, Far Hills, NJ 07931. Phone: (908) 234-2300.

BIRTHDAYS TODAY

Jacqueline Bisset, 51, actress (*Rich & Famous, The Deep*), born at Weybridge, England, Sept 13, 1944.

Ernest L. Boyer, 67, educator, born at Dayton, OH, Sept 13, 1928.

Nell Carter, 47, actress (Tony for *Ain't Misbehavin'*; "Gimme a Break") born Birmingham, AL, Sept 13, 1948.

Peter Cetera, 51, singer (lead singer of Chicago for a while; solo hit "Glory of Love"; "The Next Time I Fall" with Amy Grant), songwriter, born Chicago, IL, Sept 13, 1944.

Claudette Colbert (Claudette Chauchoin), 90, actress (Oscar *It Happened One Night, I Met Him in Paris*), born Paris, France, Sept 13, 1905.

Robert Indiana, 67, artist (pop artist, *As I Opened Fire*), born New Castle, IA, Sept 13, 1928.

Judith Martin, 57, author, journalist ("Miss Manners"), born at Washington, DC, Sept 13, 1938.

Fred Silverman, 58, TV producer, born at New York, NY, Sept 13, 1937.

Larry Speakes, 56, former government official, born at Cleveland, MS, Sept 13, 1939.

Mel Torme, 70, singer ("Comin' Home Baby"), composer ("The Christmas Song"), born at Chicago, IL, Sept 13, 1925.

SEPTEMBER 14 — THURSDAY
257th Day — Remaining, 108

CANADA: ARTS FESTIVAL XII. Sept 14-17. Annapolis Royal, Nova Scotia. Performance, workshops, art exhibits, a costume arts ball and readings by famous Canadian writers. Annually, the third weekend in September. Est attendance: 2,500. For info: Susan Tileston, Exec Dir, ARCAC, Annapolis Royal Community Arts Council, Box 534, Annapolis Royal, NS, Canada B0S 1A0. Phone: (902) 532-7069.

DANTE ALIGHIERI: DEATH ANNIVERSARY. Sept 14. Italian poet, author of the *Divine Comedy*, died Sept 14, 1321 at Ravenna, Italy. He was born in May 1265 (exact date unknown) at Florence, Italy.

FLORIDA STATE BOAT AND SPORTS SHOW. Sept 14-17. Florida State Fairgrounds, Tampa, FL. Sponsor: Natl Marine Manufacturers Assn. For info: Cathy Johnston, 14502 N Dale Mabry, Ste 332, Tampa, FL 33618. Phone: (813) 264-0490.

GEMAYEL, BASHIR: ASSASSINATION ANNIVERSARY. Sept 14. Although first government reports described his rescue and quoted him as saying "Thank God I survived this one," later news dispatches indicated that Bashir Gemayel, 34-year-old president-elect of Lebanon, died along with at least eight other persons when a bomb exploded at his Phalange party headquarters in Beirut, Lebanon, on Sept 14, 1982. On Sept 21, the Lebanese parliament elected his 39-year-old brother, Amin Gemayel, to succeed him.

INTERNATIONAL CROSS-CULTURE DAY. Sept 14. People all over the world are asked to observe this day in remembrance of cultural differences that give their nations their important distinctive attributes and heritages. Also, people should renew efforts to know and understand differences of other cultures as a means of promoting international goodwill. Sponsor: Window on the World, Robert Jackson, 322 First Ave N, #202, Minneapolis, MN 55401. Phone: (612) 375-0141.

LODI GRAPE FESTIVAL AND HARVEST FAIR. Sept 14-17. 413 E Lockeford St, Lodi, CA. County fair featuring competitive exhibits in art, photography, floriculture, viticulture, agriculture, home arts and crafts; commercial exhibits, concessions, carnival, live entertainment, grape stomps and a parade. Est attendance: 80,000. For info: Mark A. Armstrong, Genl Mgr, Box 848, Lodi, CA 95241. Phone: (209) 369-2771.

McKINLEY, WILLIAM: DEATH ANNIVERSARY. Sept 14. President William McKinley was shot at Buffalo, NY, Sept 6, 1901. He died eight days later on Sept 14, 1901. Assassin Leon Czolgosz was executed Oct 29, 1901.

◆ **MOTLEY, CONSTANCE BAKER: BIRTH.** Sept 14, 1921. New York's first black woman state senator and federal judge, and the first woman elected borough president of Manhattan, Constance Baker Motley became interested in law and civil rights when she was barred from a public beach at age 15. She went on to become one of the top civil rights lawyers of the '50's and '60's. She presented arguments before the US Supreme Court for seven cases and won them all. Motley was born at New Haven, CT.

NATIONAL ANTHEM DAY. Sept 14. Maryland.

NATIONAL GUITAR FLAT-PICKING CHAMPIONSHIPS AND WALNUT VALLEY FESTIVAL. Sept 14-17. Cowley County Fairgrounds, Winfield, KS. The Walnut River is the site of this 24th annual family event featuring 4 stages with 8 contests, at least 14 workshops and many first-class concerts. The Walnut Valley Arts and Crafts Festival features handmade instruments and a large variety of art and craft items, both ornamental and functional. All-weather facilities. Est attendance: 12,000. Sponsor: Walnut Valley Association, Bob Redford, PO Box 245, Winfield, KS 67156. Phone: (316) 221-3250.

September 1995

S	M	T	W	T	F	S
					1	2
3	4	5	6	7	8	9
10	11	12	13	14	15	16
17	18	19	20	21	22	23
24	25	26	27	28	29	30

OGALLALA'S INDIAN SUMMER RENDEZVOUS. Sept 14–17. Downtown Ogallala, NE. Festival includes dances, a parade, local and featured entertainment, craft show, kids' games, automobile show, refreshment garden and food festival. The annual Rendezvous breakfast and the start of the Cowboy Capital Trail Ride to the River City Round-up in Omaha are on Saturday. Annually, third weekend in September. Est attendance: 5,000. For info: Ogallala/Keith County Chamber of Commerce, PO Box 628, Ogallala, NE 69153. Phone: (800) 658-4390.

PENN'S LANDING IN-WATER BOAT SHOW. Sept 14–17. Penn's Landing Boat Basin, Philadelphia, PA. Sponsor: Natl Marine Manufacturers Assn. Est attendance: 20,000. For info: Jim Ranieri, 514 Harriet Lane, Havertown, PA 19083. Phone: (215) 449-9910.

PRINCESS GRACE: DEATH ANNIVERSARY. Sept 14. Princess Grace of Monaco (neé Grace Kelly) was killed in an automobile accident Sept 14, 1982, when her car plunged off a mountain road in Monte Carlo. Her daughter Stephanie, who was with her at the time, was treated for shock and bruises. Grace Kelly met Prince Rainier III while she was making her third Hitchcock film, *To Catch a Thief*. They were married in 1956.

ROCHESTER FAIR. Sept 14–24. Rochester, NH. Agricultural Fair. Est attendance: 170,000. Sponsor: Rochester Fair Assn, 72 Lafayette St, Rochester, NH 03867. Phone: (603) 332-6585.

◈ **SANGER, MARGARET (HIGGINS): BIRTH ANNIVERSARY.** Sept 14. Feminist, nurse and founder of the birth control movement in the US. Born at Corning, NY, Sept 14, 1879. (Note: birth year not entirely certain because, apparently, Sanger often used a later date when obliged to divulge her birthday. Best evidence now points to Sept 14, 1879, rather than the frequently used 1883 date.) She died at Tucson, AZ, Sept 6, 1966.

SETON, ELIZABETH ANN: 20th CANONIZATION ANNIVERSARY. Sept 14. Elizabeth Ann Seton became the first native-born American to be canonized on Sept 14, 1975. She was declared a saint in 1974 by Pope Paul VI.

SOLO TRANSATLANTIC BALLOON CROSSING: ANNIVERSARY. Sept 14–18. Joe W. Kittinger, 56-year-old balloonist left Caribou, ME, in a ten-story-tall helium-filled balloon named *Rosie O'Grady's Balloon of Peace* on Sept 14, 1984, crossed the Atlantic Ocean and reached the French coast, above the town of Capbreton, in bad weather on Sept 17 at 4:29 PM, EDT. He crash landed amid wind and rain near Savone, Italy, at 8:08 AM, EDT, Sept 18. Kittinger suffered a broken ankle when he was thrown from the balloon's gondola during the landing. His nearly 84-hour flight, covering about 3,535 miles, was the first solo balloon crossing of the Atlantic Ocean and a record distance for a solo balloon flight.

WEST VIRGINIA OIL AND GAS FESTIVAL. Sept 14–17. Sistersville, WV. To honor the oil and gas industry—past, present and future. 27th annual. Est attendance: 20,000. For info: West Virginia Oil and Gas Festival, Stewart Bradfield, Box 25, Sistersville, WV 26175.

BIRTHDAYS TODAY

Allan Bloom, 65, author (*The Closing of the American Mind, Love and Friendship*), born at Indianapolis, IN, Sept 14, 1930.

Zoe Caldwell, 62, actress (*Medea, The Prime of Miss Jean Brodie*), born Melbourne, Australia, Sept 14, 1933.

Mary Crosby, 36, actress (Kristin Shepard on "Dallas" [she shot J.R.]), born at Los Angeles, CA, Sept 14, 1959.

Joey Heatherton, 51, actress (*Cry Baby, Bluebeard*, "Dean Martin and the Golddiggers"), born Rockville Centre, NY, Sept 14, 1944.

Kate Millett (Katherine Murray Millett), 61, feminist, writer, sculptor, born at St. Paul, MN, Sept 14, 1934.

Clayton Moore, 81, actor ("The Lone Ranger"), born at Chicago, IL, Sept 14, 1914.

SEPTEMBER 15 — FRIDAY
258th Day — Remaining, 107

ACUFF, ROY: BIRTH ANNIVERSARY. Sept 15. Grand Ole Opry "King of Country Music" Roy Acuff was born Sept 15, 1903, at Maynardville, TN. Singer and fiddler Acuff (who was co-founder of Acuff-Rose Publishing Company, the leading publisher of country music) was a regular host on weekly Grand Ole Opry broadcasts. He frequently appeared at the Opry with his group, the Smoky Mountain Boys. In December of 1991 Acuff became the first living member elected to the Country Music Hall of Fame. Some of his more famous songs were "The Wabash Cannonball" (his theme song), "Pins and Needles (In My Heart)," and "Night Train to Memphis." Roy Acuff died Nov 23, 1992, at Nashville, TN.

ALOHA WEEK FESTIVALS. Sept 15–24. Oahu, Hawaii. (Also Sept 22–Oct 1, Hawaii; Sept 29–Oct 8, Molokai; Oct 6–15, Maui; Oct 13–22, Lanai; Oct 20–29, Kauai.) Celebration of Hawaiian culture, traditions and customs. Fax: (808) 941-4753. Est attendance: 1,500,000. For info: Aloha Festivals, Exec Dir, 1649 Kalakaua Ave, Honolulu, HI 96826. Phone: (808) 944-8857.

ANTIQUE AND CLASSIC CAR SHOW. Sept 15–17. Willow Park, Bennington, VT. Brass cars, Woodies, costume judging and events of skill and dexterity in handling these wonderful machines of yesteryear are all part of this car show. A flea market with auto-related parts and memorabilia entices collectors seeking that elusive fender or gas running lamp. A display and demonstration of antique tractor and farm machinery are also featured. Est attendance: 10,000. Sponsor: Michael Williams, Bennington Area Chamber of Commerce, Veterans Memorial Dr, Bennington, VT 05201. Phone: (802) 447-3311.

THE BIG E. Sept 15–Oct 1. West Springfield, MA. New England's Great State Fair and one of the nation's largest fairs. Each September, The Big E features all free entertainment including top-name talent, a big-top circus and horse show. Other offerings include children's attractions, a daily parade, historic village, the Avenue of States, Better Living Center and much more. Annually, beginning the second Friday after Labor Day. Est attendance: 1,000,000. For info: Catherine Pappas, Communications Mgr, Eastern States Exposition, 1305 Memorial Ave, West Springfield, MA 01089. Phone: (413) 737-2443.

BISMARCK FOLKFEST. Sept 15–17. State Capitol Grounds, Bismarck, ND. An annual celebration of the Bavarian kind for the harvest season. Folkfest features parades, dances, exhibits, and a Folkfest Queen. Phone: (701) 224-2525 or (800) 435-5663. Est attendance: 45,000. For info: North Dakota Tourism, Pat Hertz, Liberty Memorial Bldg, State Capitol Grounds, Bismarck, ND 58505.

BUDWEISER BLUEGRASS FESTIVAL & BBQ RIB BURN-OFF. Sept 15–17. Town Point Park, Norfolk, VA. This three-day hoe down features backporch workshops, clogging, fierce barbecue competition featuring national cookers from the US and Canada and nationally known bluegrass recording artists. For info: Promo Dir, Festevents, 120 W Main St, Norfolk, VA 23510. Phone: (804) 627-7809.

CHARLESTON MARITIME FESTIVAL. Sept 15–17. Charleston, SC. A celebration of maritime heritage, featuring marine paintings, carvings, sculpture with many exhibits and activities throughout historic Charleston. Est attendance: 25,000. For info: Southeastern Management Co, 211 Meeting St, Charleston, SC 29401. Phone: (800) 221-5273.

Sept 15 ☆ *Chase's 1995 Calendar of Events* ☆

CHRISTIE, AGATHA: 105th BIRTH ANNIVERSARY. Sept 15. English author of nearly a hundred books (mysteries, drama, poetry and nonfiction), born at Torquay, England, Sept 15, 1890. Died at Wallingford, England, Jan 12, 1976. "Every murderer," she wrote, in *The Mysterious Affair at Styles*, "is probably somebody's old friend."

CIVIL WAR MUSTER AND MERCANTILE EXPOSITION. Sept 15-17. Lindsay Park, Davenport, IA. To reenact the Civil War battle and perpetuate Victorian customs by demonstrations. Est attendance: 10,000. For info: Village of East Davenport, 2215 E 12th St, Davenport, IA 52803. Phone: (319) 322-1860.

COOPER, JAMES FENIMORE: BIRTH ANNIVERSARY. Sept 15. American novelist, historian and social critic, born at Burlington, NJ, on Sept 15, 1789, James Fenimore Cooper was one of the earliest American writers eager to develop a native American literary tradition. His most popular works are the five novels comprising *The Leatherstocking Tales*, featuring the exploits of one of the truly unique American fictional characters, Natty Bumppo. These novels, *The Deerslayer, The Last of the Mohicans, The Pathfinder, The Pioneers* and *The Prairie*, chronicle Natty Bumppo's continuing flight away from the rapid settlement of America. Other works, including *The Monikins* and *Satanstoe*, reveal him as an astute critic of American life both socially and politically. He died Sept 14, 1851, at Cooperstown, NY, the town founded by his father.

CORN ISLAND STORYTELLING FESTIVAL: 20th ANNIVERSARY. Sept 15-16. Louisville, KY. Homecoming for more than 100 storytellers who have appeared over the last 19 years. Festival includes an "olio," mixture of tales, "Fest of Storytelling" and "ghost tales" told at Long Run Park. Est attendance: 16,000. For info: Joy Pennington, Intl Order of EARS, Inc, 12019 Donohue Ave, Louisville, KY 40243. Phone: (502) 245-0643.

COSTA RICA: INDEPENDENCE DAY. Sept 15. National holiday. Gained independence from Spain on this day, 1821.

DETROIT FESTIVAL OF ARTS. Sept 15-17. Detroit, MI. Detroit Cultural Center and Wayne State University become a 15-block live entertainment forum with more than 300 stage and street performers, 100 visual artists and 200,000 guests. Est attendance: 250,000. For info: Pat Borninski, MR Officer, Wayne State Univ, 3222 FAB, Detroit, MI 48202. Phone: (313) 577-2150.

DISCOVERY LAUNCHES SATELLITE OF DISCOVERY: ANNIVERSARY. Sept 15. A satellite that was launched from the space shuttle *Discovery* on Sept 15, 1991, discovered large windstorms in Earth's upper atmosphere. The satellite's 10 instruments became functional on Nov 7 and detected the continent-sized windstorms, which measured up to 200 mile-per-hour velocities in areas that are 600 to 6,000 miles wide in the mesosphere. The largest storm was discovered in the Southern hemisphere and reached from western Australia eastward to points halfway across the Atlantic Ocean.

September 1995

S	M	T	W	T	F	S
					1	2
3	4	5	6	7	8	9
10	11	12	13	14	15	16
17	18	19	20	21	22	23
24	25	26	27	28	29	30

EISENFEST. Sept 15-17. Amana, IA. A metal craft event with displays, products, demonstrations, equipment, supplies, contests and entertainment, food and fellowship. Annually, third weekend in September. For info: R.C. Eichacker, Coord, Box 193, Amana, IA 52203. Phone: (319) 622-3100.

EL SALVADOR: INDEPENDENCE DAY. Sept 15. National holiday. Gained independence from Spain on this day, 1821.

ENGLAND: BATTLE OF BRITAIN DAY. Sept 15. Commemorates end of biggest daylight bombing raid of Britain by German Luftwaffe, on Sunday, Sept 15, 1940. Said to have been the turning point against Hitler's siege of Britain in World War II.

FALL ARTS FESTIVAL. Sept 15-Oct 1. Jackson, WY. Featuring Arts for the Parks Awards Ceremony, special shows by more than 35 major galleries, program presentations, working demonstrations, special exhibits, concerts, dance and theater. Annually, the third Friday in September through the first Sunday in October. Est attendance: 15,000. For info: Jackson Hole Chamber of Commerce, Box E, Jackson, WY 83001. Phone: (307) 733-3316.

FIRST NATIONAL CONVENTION FOR BLACKS: ANNIVERSARY. Sept 15. The first national convention for blacks was held at Bethel Church, Philadelphia, PA, on Sept 15, 1830. The convention was called to find ways to better the condition of black people and was attended by delegates from seven states. Bishop Richard Allen was elected as the first convention president.

GOAT DAYS. Sept 15-16. USA Stadium, Millington, TN. Goat contests with large cash prizes and trophies, children's area camping, fishing, goat races, dancing, music, food, rodeo, crafts and more. No admission or parking fees. Annually, the third weekend in September. Est attendance: 5,000. For info: Goat Days Intl, 4880 Navy Rd, Millington, TN 38053. Phone: (901) 872-4559.

GREENPEACE FOUNDED: ANNIVERSARY. Sept 15. The environmental organization Greenpeace, committed to a green and peaceful world, was founded Sept 15, 1971, by 12 members of the Don't Make a Wave committee of Vancouver, BC, Canada, when the boat *Phyllis Cormack* sailed to Amchitka, AK, to protest US nuclear testing. Greenpeace's basic principle is "that determined individuals can alter the actions and purposes of even the overwhelmingly powerful by 'bearing witness'—drawing attention to an environmental abuse through their mere unwavering presence, whatever the risk."

GUATEMALA: INDEPENDENCE DAY. Sept 15. National holiday. Gained independence from Spain on this day in 1821.

HARVEST FESTIVAL. Sept 15. Intercourse, PA. A fall fest featuring special decorations, food, games and strolling entertainers. Benefits the local fire department. Est attendance: 7,500. For info: Kitchen Kettle Village, Mel Hauser, Box 380, Intercourse, PA 17534. Phone: (800) 732-3538.

HONDURAS: INDEPENDENCE DAY. Sept 15. National holiday. Gained independence from Spain on this day, 1821.

JAPAN: OLD PEOPLE'S DAY OR RESPECT FOR THE AGED DAY. Sept 15. National holiday.

LEBANON BOLOGNA FEST. Sept 15-17. Lebanon, PA. Free entertainment and rides with paid admission. Food, arts and crafts, antique auto show & more. Est attendance: 20,000. For info: Lebanon Valley Tourist and Visitors Bureau, Box 329, Lebanon, PA 17042. Phone: (717) 272-8555.

MARITIME FESTIVAL. Sept 15-17. Charleston, SC. Festival features art related to Charleston's maritime history—fine art, paintings, sculpture, wood carvings and more. For info: Southeastern Management Co, PO Box 20159, Charleston, SC 29413-0159. Phone: (800) 723-1748.

MELON DAYS. Sept 15-16. Green River, UT. Arts and crafts, parade, activities in city park and free watermelon. Sponsor: Melon Days, Office of the Mayor, Box 620, Green River, UT 84525. Phone: (801) 564-3448.

☆ Chase's 1995 Calendar of Events ☆   Sept 15

MONTEREY JAZZ FESTIVAL. Sept 15–17. Monterey, CA. Celebrating its 38th year, the country's oldest jazz festival features the sounds of some of the world's finest jazz musicians. Est attendance: 30,000. For info: Monterey Jazz Festival, PO Box Jazz, Monterey, CA 93942. Phone: (408) 373-3366.

★ **NATIONAL HISPANIC HERITAGE MONTH.** Sept 15–Oct 15. Presidential Proclamation. Beginning in 1989, always issued for Sept 15–Oct 15 of each year (PL 100-402 of Aug 17, 1988). Previously issued each year for the week including Sept 15 and 16 since 1968 at request (PL90-498 of Sept 17, 1968).

NEW HAMPSHIRE HIGHLAND GAMES. Sept 15–17. Loon Mountain, Lincoln, NH. From the tossing of the caber to the lilting melodies of the clarsach plus massed pipe bands on parade, there's something for everyone at New Hampshire's 20th annual Highland Games: a 3-day Scottish festival crammed with music, dance, crafts, athletic events, Scottish food and more. For those of Scottish heritage, there's also a chance to look up one's clan connection, as more than 60 Scottish clans and societies have tents with displays. Admission charged. Est attendance: 25,000. For info: New Hampshire Highland Games, PO Box 495, Dublin, NH 03444-0495. Phone: (603) 563-8801.

NICARAGUA: INDEPENDENCE DAY. Sept 15. National holiday. Gained independence from Spain on this day, 1821.

NORTHERN APPALACHIAN STORYTELLING FESTIVAL. Sept 15–16. Straughn Hall, Mansfield University, Mansfield, PA. Showcases the talent of the nation's top storytellers who share through their performances a sense of roots and cultural diversity. There are performances Friday and Saturday evening, plus a ghost story session late Friday night. In addition, there are storytelling master classes Saturday afternoon and a workshop on Thursday and Friday. 15th annual. Annually, the third weekend in September. Est attendance: 1,700. For info: Dr. A. Vernon Lapps, Dir, N Appalachian Storytelling Fest, PO Box 117, Mansfield, PA 16933. Phone: (717) 662-4782.

OFFICE OLYMPICS. Sept 15. Downtown Shreveport, LA. A 1-day event that spotlights the office employee! 100 teams of 5 (men and women) office workers compete in such zany events as The Water Break Relay, The Office Chair Roll-off, Toss the Boss, Memo Mania, The Human Post-it-Note, Musical Office Chairs. Sponsor: KITT Radio/Ferris Office Furnishings, Inc. Est attendance: 2,000. For info: Melinda R. Coyer, Office Olympics Founder, KITT Radio, PO Box 20007, Shreveport, LA 71120. Phone: (318) 425-8692.

OKTOBERFEST. Sept 15–17. (Also Sept 22–24.) Denver, CO. Colorado's largest German festival saluting the state's German heritage. Schuhplattler, bratwurst, authentic German entertainment, pretzels and beer are featured in this re-creation of the famous Munich festival. Est attendance: 200,000. For info: Larimer Square, 1429 Larimer St, Denver, CO 80202. Phone: (303) 534-2367.

ON THE WATERFRONT SWAP MEET AND CAR CORRAL. Sept 15–17. Downtown St. Ignace, MI. Car show, sports car events, toys, Corvette and Chevy judging. Plus Mustangs and GTOs on display. Fax: (906) 643-9784. Est attendance: 4,000. For info: Ed Reavie, 268 Hillcrest Blvd, St. Ignace, MI 49781. Phone: (906) 643-8087.

QUARTERLY ESTIMATED FEDERAL INCOME TAX PAYER'S DUE DATE. Sept 15. For those individuals whose fiscal year is the calendar year and who make quarterly estimated federal income tax payments, today is one of the due dates. (Jan 15, Apr 15, June 15 and Sept 15, 1995.)

ROCKFORD STORYTELLING FESTIVAL. Sept 15–16. Midway Village & Museum Center, Rockford, IL. A two-day family affordable event featuring nationally known storytellers from around the US. The festival is set in Midway Village, a recreated 19th-century town, and offers tales tall and true for people young and old. Annually, the second weekend after Labor Day. 7th annual. Est attendance: 2,500. For info: Rockford Area Conv & Visitors Bureau, 211 N Main St, Rockford, IL 61101. Phone: (800) 521-0849.

SPACE MILESTONE: *ARIANE-3* (ESA). Sept 15. European Space Agency rocket carrying two (Australian and European) communications satellites into earth orbit marked re-entry of western nations into commercial space projects. Launched from Kourou, French Guiana, with Arianespace, a private company operating the rocket for the 13-nation European Space Agency on Sept 15, 1987.

SPACE MILESTONE: *SOYUZ 22* (USSR). Sept 15. Launched on Sept 15, 1976, Cosmonauts V. Bykovsky and V. Asenov entered Earth orbit to study "geological and geographical characteristics on Earth's surface in the interests of the national economy." Returned to Earth on Sept 23.

STAMP EXPO. Sept 15–17. Grand Hotel, Anaheim, CA. Fax: (818) 988-4337. Est attendance: 4,000. Sponsor: Intl Stamp Collectors Society, PO Box 854, Van Nuys, CA 91408. Phone: (818) 997-6496.

STATE FAIR OF OKLAHOMA. Sept 15–Oct 2. Oklahoma City, OK. International exhibitors, horse and livestock shows, PRCA rodeo, Ringling Bros and Barnum & Bailey Circus, Walt Disney's World on Ice, FFA, 4-H, grandstand concerts, car racing, tractor and truckpulls, automobile show, flower and garden exhibits. Est attendance: 1,800,000. For info: State Fair of Oklahoma, Box 74943, Oklahoma City, OK 73147. Phone: (405) 948-6700.

TAFT, WILLIAM HOWARD: BIRTH ANNIVERSARY. Sept 15. The 27th president of the US was born at Cincinnati, OH, Sept 15, 1857. His term of office: Mar 4, 1909–Mar 3, 1913. Following his presidency he became a law professor at Yale University until his appointment as Chief Justice of the US Supreme Court in 1921. Believed to have been the heaviest US president (weighed 225 pounds at age 20). Died at Washington, DC, Mar 8, 1930, and was buried at Arlington National Cemetery.

WO-ZHA-WA FALL FESTIVAL. Sept 15–17. Wisconsin Dells, WI. Celebrates the beginning of the fall season. Arts, crafts, 100-unit parade and carnival. Est attendance: 100,000. For info: Visitor and Convention Bureau, Box 390, Wisconsin Dells, WI 53965. Phone: (800) 223-3557.

"WORLD'S LARGEST" WEATHER VANE ANNIVERSARY. Sept 15. At the edge of White Lake in Montague, MI, on Sept 15, 1984, the "world's largest" weather vane was dedicated. Forty-eight feet high with a 26-foot wind arrow, the weather vane weighing 3,500 lbs is adorned with a 14-foot replica of a 19th-century Great Lakes schooner.

BIRTHDAYS TODAY

Ernest Byner, 33, football player, born at Midgeville, GA, Sept 15, 1962.
Jackie Cooper, 73, actor (*Our Gang* shorts, Skippy; "The People's Choice"), producer, born Los Angeles, CA, Sept 15, 1922.
Norm Crosby, 68, comedian (hosted "The Comedy Shop"), born at Boston, MA, Sept 15, 1927.
Edward J. Derwinski, 69, former US Secretary of Veterans Affairs in Bush administration, born at Chicago, IL, Sept 15, 1926.
Sherman Douglas, 29, basketball player, born Washington, DC, Sept 15, 1966.
Prince Harry (Henry Charles Albert David), 11, of Wales, born at London, England, Sept 15, 1984.
Tommy Lee Jones, 49, actor (Oscar *The Fugitive; Coalminer's Daughter*), born at San Saba, TX, Sept 15, 1946.
Dan Marino, 34, football player, born at Pittsburgh, PA, Sept 15, 1961.

Sept 15-16 ★ Chase's 1995 Calendar of Events ★

Joe Morris, 35, football player, born at Fort Bragg, NC, Sept 15, 1960.
Merlin Olsen, 55, former football player, actor, born at Logan, UT, Sept 15, 1940.
Gaylord Perry, 57, baseball Hall of Famer, born at Williamston, NC, Sept 15, 1938.
Bobby Short (Robert Waltrip), 69, singer-pianist (as child called "The Midget King of Swing"; cafe song stylist of prodigious repertoire), born Danville, IL, Sept 15, 1926.
Oliver Stone, 49, motion picture director (*Platoon, JFK, Wall Street*), screenwriter, born New York, NY, Sept 15, 1946.

SEPTEMBER 16 — SATURDAY
259th Day — Remaining, 106

APPLEJACK FESTIVAL. Sept 16–17. Nebraska City, NE. To promote local orchards and their abundant apple harvest. Est attendance: 30,000. Sponsor: Nebraska City Chamber of Commerce, 806 First Ave, Nebraska City, NE 68410. Phone: (402) 873-3000.

ARTS AND CRAFTS FESTIVAL. Sept 16–17. Veterans Memorial Dr, Bennington, VT. Annual Arts and Crafts Festival proceeds benefit local art programs. Annually, second weekend after Labor Day. Sponsor: Mountain Valley Artists Assn. Est attendance: 5,000. For info: Muriel Bohne, Apple Hill Rd, Bennington, VT 05201. Phone: (802) 442-9624.

BARNESVILLE BUGGY DAYS. Sept 16–17. Barnesville-Lamar County, GA. A weeklong festival commemorating the days of spoke wheels and surrey tops. Included in festivities are entertainment, arts and crafts, food, footrace and parade. Est attendance: 50,000. For info: Chamber of Commerce, PO Box 506, Barnesville, GA 30204-0506. Phone: (404) 358-2732.

BELLE MEADE PLANTATION FALL FEST. Sept 16–17. Nashville, TN. Est attendance: 20,000. For info: Cindy Graham, Belle Meade Plantation, 5025 Harding Rd, Nashville, TN 37205. Phone: (615) 356-0501.

BIG WHOPPER LIAR'S CONTEST. Sept 16. Murphy Auditorium, New Harmony, IN. Thirty "story tellers" compete to see who can tell the BIGGEST Whopper. Annually, the third Saturday in September. Est attendance: 425. For info: Tim Rutherford, PO Box 644, New Harmony, IN 47631. Phone: (812) 682-3730.

BUDWEISER OKTOBERFEST-ZINZINNATI. Sept 16–17. Cincinnati, OH. Events featuring oom-pah-pah bands, dancing, German foods, pastries and more. Est attendance: 500,000. For info: Downtown Council, 300 Carew Tower, 441 Vine, Cincinnati, OH 45202. Phone: (513) 579-3191.

September 1995

S	M	T	W	T	F	S
					1	2
3	4	5	6	7	8	9
10	11	12	13	14	15	16
17	18	19	20	21	22	23
24	25	26	27	28	29	30

CANADA: SHEARWATER INTERNATIONAL AIR SHOW. Sept 16–17. Canadian Forces Base Shearwater, Nova Scotia. Atlantic Canada's major air show featuring air display of world-class military and civilian aerobatic performers, static aircraft display and ground exhibits. Est attendance: 95,000. For info: Executive Dir, Shearwater Intl Air Show, PO Box 218, Shearwater, NS, Canada B0J 3A0. Phone: (902) 465-2725.

CANDLELIGHT WALKING TOUR OF HISTORIC CHESTERTOWN. Sept 16. Chestertown, MD. To view the interiors of historic buildings and houses in a colonial town. Annually, the third Saturday in September. Est attendance: 1,500. For info: Historical Society of Kent County, Nancy Nunn, Curator, PO Box 665, Chestertown, MD 21620. Phone: (410) 778-3499.

CELEBRATION OF AMERICA'S BOUNTY. Sept 16–17. McLean, VA. Agricultural fair and food festival. Exhibits, farm animals, children's games, food and rides. Fair held rain or shine in the Pavilions of Turkey Run. Sponsors: Claude Moore Colonial Farm and the Agricultural Research Institute. Est attendance: 3,500. For info: Claude Moore Colonial Farm at Turkey Run, 6310 Georgetown Pike, McLean, VA 22101. Phone: (703) 442-7557.

CHAUTAUQUA OF THE ARTS. Sept 16–17. Columbus, IN. Fine artists and craftsmen gather to demonstrate and sell their works. Lodging info: (800) 468-6564. Est attendance: 40,000. For info: Dixie McDonough, PO Box 2624, Columbus, IN 47202-2624. Phone: (812) 265-5080.

CHEROKEE STRIP CELEBRATION. Sept 16. Perry, OK. To commemorate the opening of the Cherokee Strip to settlement on Sept 16, 1893. Annually, the Saturday or weekend nearest Sept 16. Est attendance: 10,000. Sponsor: Chamber of Commerce, Cheryle Leach, Box 426, Perry, OK 73077. Phone: (405) 336-4684.

CHEROKEE STRIP DAY: ANNIVERSARY. Sept 16. Optional holiday, Oklahoma. Greatest "run" for Oklahoma land in 1893.

COVERED BRIDGE FESTIVAL. Sept 16–17. Washington County, PA. Arts and crafts, entertainment at 8 different covered bridges. Annually, the third weekend in September. Est attendance: 140,000. Sponsor: Washington Co Tourism, 144A McClelland Rd, Canonsburg, PA 15317. Phone: (800) 531-4114.

CURTIS FALL FESTIVAL. Sept 16–17. Curtis City Park, Curtis, NE. A fun-filled weekend, including a parade, rides, booths, band competition, barbeque, flea market, rodeo, quilt show, demolition derby, street dance and many other exciting and varied events. Annually, the third weekend in September. Est attendance: 1,500. Sponsor: Medicine Creek Chamber, c/o Diane Wetzel, Box 45, Curtis, NE 69025. Phone: (308) 367-4500.

ENGLAND: SEVEN FLORENTINE HEADS EXHIBITION. Sept 16–Oct 15. The Gallery, Windsor Castle, Windsor. 15th-century metalpoint drawings from the Royal Library. Est attendance: 50,000. For info: Exhibitions Dept, The Royal Library, Windsor Castle, Windsor, Berkshire, England SL4 1NJ.

FALL SORGHUM AND HARVEST FESTIVAL. Sept 16. Burritt Museum and Park, Huntsville, AL. Living History presentation of sorghum sweetener process using mule-powered press and boiling furnace. Also, trades and skills of 19th-century rural Alabama. Annually, the third Saturday in September. For info: Charles Pautler, Curator of History, Burritt Museum and Park, 3101 Burritt Dr, Huntsville, AL 35810. Phone: (205) 536-2882.

FESTIVAL OF THE SEA. Sept 16–17. Pt Pleasant Beach, NJ. Sample a variety of local cuisine from seafood to pizza. Arts and crafts exhibits, health exhibits, pony rides, games for kids, entertainment and much more. Also, a five-mile run preceeding the festival. The next day we have a zany parade and an Ocean Inner Tube Race. Annually, the third weekend in September. Est attendance: 60,000. Sponsor: Greater Point Pleasant Area Chamber of Commerce, 517 A Arnold Ave, Pt Pleasant Bch, NJ 08742. Phone: (908) 899-2424.

Chase's 1995 Calendar of Events — Sept 16

GENERAL MOTORS: FOUNDING ANNIVERSARY. Sept 16. The giant automobile manufacturing company was founded Sept 16, 1908, by William Crapo "Billy" Durant, a Flint, MI, entrepreneur.

GREAT SEAL OF THE UNITED STATES: FIRST USE ANNIVERSARY. Sept 16. On Sept 16, 1782, the Great Seal of the United States was, for the first time, impressed upon an official document. That document authorized George Washington to negotiate a prisoner of war agreement with the British. See also: "Great Seal of the United States Anniversary" (Jan 28, June 20 and July 4).

HARVEST SHOW. Sept 16–17. Horticulture Center, Fairmount Park, Philadelphia, PA. More than 700 gardeners enter more than 350 horticultural and artistic categories. Educational exhibits and samples of freshley harvested crops and homemade preserved products judged. A series of special events, including live music and a children's activity tent. Est attendance: 5,000. For info: The Pennsylvania Horticultural Soc, 325 Walnut St, Philadelphia, PA 19106-2777. Phone: (215) 625-8250.

HODAG MUSKIE CHALLENGE. Sept 16–17. Rhinelander, WI. $15,500 catch and release Muskie Tournament. Annually, the third weekend in September. Est attendance: 200. For info: Rhinelander Area Chamber of Commerce, PO Box 795, Rhinelander, WI 54501. Phone: (800) 236-4386.

JERSEY SHORE SEA KAYAK AND BAY CANOEING SHOW. Sept 16. Bayville, NJ. Canoe and sea kayak vendors from around the country set up on a beach and share the joys of watersport with the public. Test paddling is allowed all day and there are clinics to teach about canoeing and kayaking. At this show you may sign up for a trip for the following day. There is no fee. Annually, the third Saturday in September. Est attendance: 1,000. For info: Lillian Hoey, Rec Leader, Wells Mills County Park, 905 Wells Mills Rd, Waretown, NJ 08758. Phone: (609) 971-3085.

JOHNNY APPLESEED FESTIVAL. Sept 16–17. Fort Wayne, IN. A return to the pioneer spirit of the early 1800s, at the gravesite of John Chapman, known as the "Johnny Appleseed" who planted hundreds of apple orchards along the early Indiana frontier. Crafts, music, food and storytelling. Annually, the third weekend in September. Est attendance: 300,000. For info: Parks and Recreation Dept, 705 E State Blvd, Fort Wayne, IN 46805. Phone: (219) 483-0057.

KUNSTFEST. Sept 16–17. New Harmony, IN. German craft demonstrations and exhibits, homemade German food, historic tours of homes. Annually, the third weekend in September. Est attendance: 2,000. For info: Julie Rutherford, Tour/Promo Coord, PO Box 579, New Harmony, IN 47631. Phone: (812) 682-3730.

LOG SHOW BY THE SEA. Sept 16. Port of Brookings-Harbor, Brookings. Open competition, trophies, cash prizes, food, drinks, music, free admission; big party on Oregon's Banana Belt. Annually, the third Saturday in September. Est attendance: 4,000. For info: Brookings-Harbor Chamber of Commerce, PO Box 940, Brookings, OR 97415. Phone: (800) 535-9469.

◆ **MAYFLOWER DAY: 375th ANNIVERSARY.** Sept 16. Anniversary of the departure of the *Mayflower* from Plymouth, England, Sept 16, 1620, with 102 passengers and a small crew. Vicious storms were encountered en route which caused serious doubt about the wisdom of continuing, but she reached Provincetown, MA, on Nov 21, and discharged the Pilgrims at Plymouth, MA, on Dec 26, 1620.

MEXICO: INDEPENDENCE DAY. Sept 16. National Day.

MIDDLEMARK, MARVIN: BIRTH ANNIVERSARY. Sept 16. Marvin Middlemark was born on Sept 16, 1919, at Long Island, NY. His passion for inventing and tinkering led to many inventions, most of which enjoyed little commercial success. It was as the inventor of a device to improve TV reception, known as "rabbit ears," that he became successful. He died on Sept 14, 1989 at Old Westbury, NY.

MISS AMERICA PAGEANT FINALE. Sept 16. Convention Center, Atlantic City, NJ. Crowning of Miss America 1996. The pageant is tentatively scheduled to take place Sept 12–16 with the crowning on the 16th, but this is subject to change. NOTE: Date is "tentative but probable," per sponsor. Est attendance: 60,000. For info: Greater Atlantic City Conv and Visitors Bureau, 2314 Pacific Ave, Atlantic City, NJ 08401. Phone: (609) 348-7100.

MOON PHASE: LAST QUARTER. Sept 16. Moon enters Last Quarter phase at 5:09 PM, EDT.

OLD BEDFORD VILLAGE SENIOR DAYS. Sept 16–17. Bedford, PA. Old Bedford Village will feature two special days full of activities geared toward our older visitors. Escorted wagon rides, specialty foods, "pure western" revue by Ted & Ruth Reinhart. For info: Old Bedford Village, Jamie Nesbit, PO Box 1976, Bedford, PA 15522. Phone: (814) 623-1156.

PALESTINIAN MASSACRE: ANNIVERSARY. Sept 16. Christian militiamen (the Phalangists) on Sept 16, 1982, entered Sabra and Shatila, two Palestinian refugee camps in West Beirut. They began open shooting and by Sept 18 hundreds of Palestinians, including elderly men, women and children, were dead. Phalangists had demanded blood of Palestinians since the assassination of their president, Bashir Gemayel on Sept 14. Survivors of the massacre said they had not seen Israeli forces inside the camp; however, they claimed Israelis sealed off boundaries to the camps and allowed Christian militiamen to enter.

PANIZZI, ANTHONY: BIRTH ANNIVERSARY. Sept 16. Sir Anthony Panizzi, the only librarian ever hanged in effigy, was born Antonio Genesio Maria Panizzi at Brescello, Italy, on Sept 16, 1797. As a young man he joined a forbidden Italian patriotic society which advocated the overthrow of the oppressive Austrians who then controlled most of northern Italy. Tried in absentia by an Austrian court in 1820, he was sentenced to death, hanged in effigy, and all his property was confiscated. He fled to England in 1823, learned the language and by 1831 was employed in the British Museum. In spite of his continuing outspoken and unpopular enthusiasm for Italian politics, the naturalized Englishman was, in 1856, named principal librarian of the British Museum. Later described as the "prince of librarians," Panizzi died at London, England, on Apr 8, 1879.

PAPUA NEW GUINEA: INDEPENDENCE DAY. Sept 16. National holiday.

PARKMAN, FRANCIS: BIRTH ANNIVERSARY. Sept 16. American historian, author of *The Oregon Trail*, was born at Boston, MA, on Sept 16, 1823, and died there on Nov 8, 1893.

ROAN MOUNTAIN STATE PARK FALL FESTIVAL. Sept 16–17. Roan Mountain, TN. Crafts, exhibits, music and clogging. Est attendance: 20,000. For info: Jennifer Wilson, Ranger Naturalist, Roan Mt State Park, Rte 1, Box 236, Roan Mountain, TN 37687. Phone: (615) 772-3303.

RUSSIAN RIVER JAZZ FEST. Sept 16–17. Johnson's Beach, Guerneville, CA. A variety of top jazz performers in the majestic setting of the Russian River and redwoods. 19th annual festival. Est attendance: 12,000. For info: Russian River Jazz Festival, PO Box 1913, Guerneville, CA 95446. Phone: (707) 869-3940.

SCARECROW FESTIVAL. Sept 16–17. Peddler's Village, Lahaska, PA. Weekend festival includes scarecrow making, pumpkin painting workshops, jack-o'-lantern and gourd art contest, musical entertainment and scarecrow competition display. Free admission. Charge for workshops. Est attendance: 7,000. Sponsor: Peddler's Village, PO Box 218, Lahaska, PA 18931. Phone: (215) 794-4000.

Sept 16–17 ★ Chase's 1995 Calendar of Events ★

SOUTHWEST IOWA PROFESSIONAL HOT AIR BALLOON RACES. Sept 16–17. Creston, IA. Hare and hound races held at sunrise and sunset; art and book fairs, parade and marching band contest and much more. Annually, the third weekend in September. Est attendance: 12,000. For info: Chamber of Commerce, Box 471, Creston, IA 50801. Phone: (515) 782-7021.

STAY AWAY FROM SEATTLE DAY. Sept 16. Observed worldwide, except in Seattle, WA, to give America's "Best Place to Live" city a break from the influx of people moving to the area. On this day every year, the rest of us will try to keep the appeal of Seattle from haunting us with its siren call. Phone: (212) 388-8673 or (717) 274-8451. Sponsor: Thomas and Ruth Roy, Wellness Permission League, 2105 Water St, Lebanon, PA 17046.

SUN HERALD SAND SCULPTURE CONTEST. Sept 16. Biloxi, MS. To conduct the "nation's largest" sand sculpture contest and promote the art of sand sculpting. More than $4,000 in prizes. 20th annual contest. Annually, the third Saturday in September. Est attendance: 3,500. For info: *The Sun Herald*, Mktg Dept, Box 4567, Biloxi, MS 39535. Phone: (601) 896-2434.

TORQUEMADA, TOMAS DE: DEATH ANNIVERSARY. Sept 16. One of history's most malevolent persons, feared and hated by millions. As Inquisitor-General of Spain, he ordered burning at the stake for more than 10,000 persons, and burning in effigy for another 7,000 (according to 18th-century estimates). Torquemada persuaded Ferdinand and Isabella to rid Spain of the Jews. More than a million families were driven from the country and Spain suffered a commercial decline from which it never recovered. Torquemada was born at Valladolide, Spain, in 1420 (exact date unknown) and died at Avila, Spain, Sept 16, 1498.

TRAIL OF COURAGE LIVING-HISTORY FESTIVAL. Sept 16–17. Rochester, IN. Portrayal of life in frontier Indiana when it was Indian territory. Historic skits, music, dancing, wigwam and tepee villages, historic encampments for Revolutionary War, French and Indian War, Voyageurs, Western Fur Trade, and Plains Indians; recreated 1832 Chippeway Village, also Woodland Indian village, pioneer foods and crafts, muzzleloading and tomahawk contests. Museum, round barn and Living History Village on grounds. Est attendance: 16,000. Sponsor: Fulton County Historical Society, 37 E 375N, Rochester, IN 46975. Phone: (219) 223-4436.

TRY THIS ON—A HISTORY OF CLOTHING, GENDER AND POWER. Sept 16–Oct 29. Sloan Museum, Flint, MI. A social history about clothing and appearance in the US over the past 200 years. See Jan 7 for full description. For info: Smithsonian Institution Traveling Exhibition Service, 1100 Jefferson Dr SW, Ste 3146, Washington, DC 20560. Phone: (202) 357-2700.

US MID-AMATEUR (GOLF) CHAMPIONSHIP. Sept 16–21. Caves Valley Golf Course, Owings Mill, MD; Woodholme Country Club, Pikesville, MD. Sponsor: US Golf Assn, Championship Dept, Golf House, Far Hills, NJ 07931. Phone: (908) 234-2300.

◆ **WALES: NORTH WALES MUSIC FESTIVAL.** Sept 16–23. Various venues, North Wales. Events include recitals, music dramas, orchestras and a Purcell tercentenary programme. For info: Geraint Lewis, Artistic Dir, Newton Court, Monmouth, Gwent, Wales.

WINGS 'N' WATER FESTIVAL. Sept 16–17. Wetlands Institute, Stone Harbor, NJ. A wildfowl arts festival, spread throughout the resort towns of Avalon and Stone Harbor, featuring wildlife arts, carvings, decoys, Duck Stamps, crafts, music, seafood and children's activities. Est attendance: 8,000. Sponsor: Wetlands Institute, Cindy O'Connor, Exec Dir, 1075 Stone Harbor Blvd, Stone Harbor, NJ 08247-1424. Phone: (609) 368-1211.

BIRTHDAYS TODAY

Lauren Bacall (Betty Joan Perske), 71, actress (*Applause, Woman of the Year, Key Largo, How to Marry a Millionaire*), born at New York, NY, Sept 16, 1924.

Ed Begley, Jr., 46, actor ("St. Elsewhere"), born at Los Angeles, CA, Sept 16, 1949.

Charlie Byrd, 70, musician ("Meditation"; "Desafinado" with Stan Getz), born at Chuckatuck, VA, Sept 16, 1925.

David Copperfield (Kotkin), 39, illusionist, born at Metuchen, NJ, Sept 16, 1956.

Peter Falk, 68, actor (*The Great Race*, "Columbo"), born at New York, NY, Sept 16, 1927.

Allen Funt, 81, TV producer, host ("Candid Camera"), born at Brooklyn, NY, Sept 16, 1914.

Orel Leonard Hershiser IV, 37, baseball player, born at Buffalo, NY, Sept 16, 1958.

B.B. King, 70, singer ("Rock Me Baby," "The Thrill is Gone"), born at Itta Bena, MS, Sept 16, 1925.

John Knowles, 69, writer (*Backcasts: Memories & Recollections of Seventy Years as a Sportsman*), born Farmont, WV, Sept 16, 1926.

Richard Marx, 32, singer, born at Chicago, IL, Sept 16, 1963.

Mark McEwen, 41, weatherman, music editor, born at San Antonio, TX, Sept 16, 1954.

Janis Paige, 72, singer, actress (*The Pajama Game* on stage; *Silk Stockings*), born at Tacoma, WA, Sept 16, 1923.

Timothy Raines, 36, baseball player, born at Sanford, FL, Sept 16, 1959.

Robin Yount, 40, former baseball player, born at Danville, IL, Sept 16, 1955.

SEPTEMBER 17 — SUNDAY
260th Day — Remaining, 105

ARTS/QUINCY RIVERFEST. Sept 17. Quincy, IL. Celebration of the arts and the Mississippi River in the river front parks. Fine arts and crafts, Dixieland Jazz stage, Country-Western stage, along with children's area featuring hands-on activities and performances. Tours of boats including a working tugboat. Annually, the third Sunday in September. Est attendance: 20,000. For info: Vicki Ebbing, Coord, Quincy Soc of Fine Arts, 428 Maine, Quincy, IL 62301. Phone: (217) 222-3432.

BEL AIR FESTIVAL FOR THE ARTS. Sept 17. Bel Air, MD. Arts, crafts, photography exhibits and entertainment. Application deadline May 1. Est attendance: 40,000. For info: Bel Air Rec Committee, Donna Clauer, 1909 Wheel Rd, Bel Air, MD 21015. Phone: (410) 836-2395.

CHILDREN'S FESTIVAL. Sept 17. Tinicum Park, Erwinna, PA. Performing and visual activities providing opportunity for expression and creativity. Est attendance: 5,000. Sponsor: Bucks County Dept of Parks and Recreation, 901 E Bridgetown Pike, Langhorne, Bucks County, PA 19047. Phone: (215) 757-0571.

September 1995

S	M	T	W	T	F	S
					1	2
3	4	5	6	7	8	9
10	11	12	13	14	15	16
17	18	19	20	21	22	23
24	25	26	27	28	29	30

CHIMNEY ROCK HILLFALL. Sept 17. Chimney Rock Park, Chimney Rock, NC. Gravity-powered tub race with Walter Mitty-type vehicles that incorporate a No. 2 washtub in seat or chassis. Run on 1½-mile winding course down side of mountain. Clowns, llamas, rope climbers and traditional mountain music. Annually, the third Sunday in September. Est attendance: 2,500. For info: Chimney Rock Park, Mary Jaeger-Gale, PO Box 39, Chimney Rock, NC 28720. Phone: (800) 277-9611.

★ **CITIZENSHIP DAY.** Sept 17. Presidential Proclamation always issued for September 17 at request (PL82-261 of Feb 29, 1952). Customarily issued as "Citizenship Day and Constitution Week."

CONSTITUTION OF THE US: ANNIVERSARY. Sept 17. On Sept 17, 1787, delegations from twelve states at the Constitutional Convention in Philadelphia, PA, voted unanimously to approve the proposed document. Thirty-nine of the 42 delegates present signed it and the Convention adjourned, after drafting a letter of transmittal to the Congress. The proposed constitution stipulated that it would take effect when ratified by nine states.

★ **CONSTITUTION WEEK.** Sept 17-23. Presidential Proclamation always issued for the period of September 17-23 each year since 1955 (PL 84-915 of Aug 2, 1956).

DULCIMER FESTIVAL. Sept 17. Burritt Museum and Park, Huntsville, AL. Dulcimer makers and players from several states will provide a blend of informal jam sessions, performances, demonstrations and displays. Annually, the third Sunday in September. For info: Pat Robertson, Dir, 3101 Burritt Dr, Huntsville, AL 35801. Phone: (205) 536-2882.

FOSTER, ANDREW (RUBE): BIRTH ANNIVERSARY. Sept 17. Rube Foster's efforts in baseball earned him the title of "The Father of Negro Baseball." He was a manager and star pitcher, pitching 51 victories in one year. After playing for the Chicago Lelands in 1907 and leading that team to a record of 110 wins and 10 losses, in 1908 he formed the Chicago American Giants who won 129 games and lost only 6 in their first season. In 1919, he called a meeting of black baseball owners and organized the first Negro League. He served as its president until his death in 1930. Foster was born Sept 17, 1879 at Galveston, TX.

GRAND DETOUR ART FAIR. Sept 17. John Deere Historic Site, Grand Detour, IL. An invitational fine arts and craft show with $1600 in cash awarded. Held around the perimeter of the John Deere Historic Site grounds from 10-4 with 60 to 70 artists exhibiting. Children's art exhibit and food. Annually, the third Sunday in September. Sponsor: Arts Alliance of Ogle County. Est attendance: 6,000. For info: The Grand Detour Art Fair, c/o Linda Freeh, PO Box 65, Oregon, IL 61061.

HENDRICKS, THOMAS ANDREWS: BIRTH ANNIVERSARY. Sept 17. Twenty-first vice president of the US (1885) born Muskingum County, OH, Sept 17, 1819. Died at Indianapolis, IN, Nov 25, 1885.

"HEY RUBE GET A TUBE." Sept 17. Ocean Ave and Atlantic Ocean, Pt Pleasant Beach, NJ. A parade featuring zany floats, bands, clowns and more, followed by an Ocean Inner Tube Race. Contestants dash into the ocean and paddle backwards from one beach to another. Annually, the third Sunday in September. Est attendance: 20,000. Sponsor: Lions Club, PO Box 444, Pt Pleasant, NJ 08742. Phone: (908) 899-3306.

INTERNATIONAL PRIORITIES WEEK. Sept 17-23. A week to focus on managing personal and business priorities in order to increase productivity by identifying and improving personal and organizational development while maintaining a balance in one's life. Annually, two weeks after Labor Day. For info: Tee Houston-Aldridge, Priority Management Systems, 500 108th Ave NE, Ste 1740, Bellevue, WA 98004. Phone: (800) 221-9031.

NATIONAL ADULT DAY CARE CENTER WEEK. Sept 17-23. A time to acknowledge the centers, the staff and the contributions that these programs make to society in meeting the needs of older adults who live alone and of families that seek to care for their relatives. Festivals, open houses, health fairs, recognition events and other community events will be celebrated throughout the week. Annually, the third full week in September. Sponsor: Natl Institute on Adult Daycare, Betty Ransom, 409 Third St SW, Washington, DC 20024. Phone: (202) 479-1200.

NATIONAL CHIROPRACTIC WEEK. Sept 17-23. A week to promote the discovery of chiropractic, Sept 15, 1895. Chiropractic focuses on keeping the human nerve system open allowing ultimate personal potential. This is chiropractics 100-year anniversary. For info: Hilpisch Chiropractic Clinic, 1401 Helmo Ave N, Oakdale, MN 55128. Phone: (612) 731-8211.

NATIONAL CONSTITUTION DAY. Sept 17. To celebrate and commemorate the signing of the US Constitution on September 17, 1787. The Center gives away pocket Constitutions free of charge. Fax: (215) 923-1749. For info: Natl Constitution Center, The Bourse, 111 S Independence Mall East, Ste 560, Philadelphia, PA 19106. Phone: (215) 923-0004.

NATIONAL FARM ANIMALS AWARENESS WEEK. Sept 17-23. A week to promote awareness of farm animals and their natural behaviors. Each day of the week is dedicated to learning about a specific group of farm animals and to appreciating their many interesting and unique qualities. Annually, the third full week in September. Sponsor: The Humane Society of the US, Farm Animal Section, 2100 L St NW, Washington, DC 20037. Phone: (202) 452-1100.

★ **NATIONAL FARM SAFETY WEEK.** Sept 17-23. Presidential Proclamation issued since 1982 for the third week in September. Previously, from 1944, for one of the last two weeks in July.

NATIONAL LAUNDRY WORKERS WEEK. Sept 17-23. To promote public awareness of the contribution and importance of laundry service employees. Annually, the third week in September. For info: Nemaha Co Good Samaritan Ctr, Rte 1, Box 4, Auburn, NE 68305.

NATIONAL REHABILITATION WEEK. Sept 17-23. An opportunity to celebrate the determination of more than 43 million people with disabilities in America. It is also a time to salute the dedicated health care professionals who provide rehabilitation care, and an occasion to call attention to the unmet needs of our nation's disabled citizens. Est attendance: 300. For info: Allied Services, Corporate Programs Dept, PO Box 1103, Scranton, PA 18501-1103. Phone: (717) 348-1498.

NATIONAL SINGLES WEEK. Sept 17-23. To celebrate single life; to recognize singles and their associations; to present dignified options for finding a partner or living fully without one. Sponsor: Singles Press Assn, Box 6243, Scottsdale, AZ 85261-6243. Phone: (602) 788-6001.

SELFRIDGE, THOMAS E.: DEATH ANNIVERSARY. Sept 17. Lieutenant Thomas E. Selfridge, 26-year-old passenger in 740-lb biplane piloted by Orville Wright, was killed when, after four minutes in the air, the plane fell from a height of 75 feet. Nearly 2,000 spectators witnessed the crash at Fort Myer, VA, on Sept 17, 1908. The plane was being tested for possible military use by the Army Signal Corps. Orville Wright was seriously injured in the crash. Selfridge Air Force Base, MI, was named after the young lieutenant, a West Point graduate, who was the first fatality of powered airplane travel.

SPACE MILESTONE: *PEGASUS 1* (US). Sept 17. 23,000-pound research satellite, launched Feb 16, 1965, broke up over Africa and fell to Earth. Major pieces believed to have fallen into Atlantic Ocean off the coast of Angola on Sept 17, 1978.

STRATTON ARTS FESTIVAL. Sept 17–Oct 15. Base Lodge, Stratton Ski Resort, Stratton, VT. This festival, in its 31st year, is a month-long celebration of the arts. The festival's showcase exhibition and sale offers two floors of refined exhibition space, weekend demonstrations and an education program. The festival sells the highest quality work, eclectic in discipline and style, of both established and emerging Vermont artists and craftspeople. Sponsor: Stratton Arts Festival. Est attendance: 9,100. For info: Roderic Cain, PO Box 576, Stratton Mt, VT 05155. Phone: (802) 297-3265.

TOLKIEN WEEK. Sept 17–23. To promote appreciation and enjoyment of the works of J.R.R. Tolkien. Sponsor: American Tolkien Society, Attn: Phil Helms, Box 373, Highland, MI 48357-0373. Phone: (813) 585-0985.

WILLIAMS, HANK, SR: BIRTH ANNIVERSARY. Sept 17. Hiram King Williams, country and western singer, born at Georgia, AL, on Sept 17, 1923. He achieved his first major hit with "Lovesick Blues," which brought him a contract with the Grand Ole Opry. His string of hits included "Cold, Cold Heart," "Honky Tonk Blues," "Jambalaya," "Your Cheatin' Heart," "Take These Chains From My Heart" and "I'll Never Get Out Of This Life Alive," which was released prior to his death on Jan 1, 1953, at Oak Hill, VA.

WOOL DAY: SHEEP TO SHAWL AND BORDER COLLIES. Sept 17. Billings Farm and Museum, Woodstock, VT. This day-long event focuses on the many aspects of wool production, including a "sheep to shawl" demonstration and sheep herding, as well as many hands-on activities for all ages to enjoy. These activities include carding wool, drop spindle spinning, weaving, knitting and rug hooking. Border Collie demonstrations will show how these natural shepherds will efficiently round up sheep, drive them from one area to another, and corral them. Est attendance: 1,200. Sponsor: Billings Farm & Museum, PO Box 489, Woodstock, VT 05091. Phone: (802) 457-2355.

BIRTHDAYS TODAY

Anne Bancroft (Anna Maria Italiano), 64, actress (Tony and Oscar for *The Miracle Worker*; *The Graduate, To Be or Not to Be, The Turning Point*), born at New York, NY, Sept 17, 1931.
Paul Benedict, 57, actor ("The Jeffersons"; many stage roles), born at Silver City, NM, Sept 17, 1938.
Warren E. Burger, 88, former Chief Justice of the US (retired), born at St. Paul, MN Sept 17, 1907.
Anthony Carter, 35, football player, born Riviera Beach, FL, Sept 17, 1960.
Charles Ernest Grassley, 62, US Senator (R, Iowa), born at New Hartford, IA, Sept 17, 1933.
Chaim Herzog, 77, President of Israel, born at Belfast, Northern Ireland, Sept 17, 1918.
Phil Jackson, 50, basketball coach, former player, born at Deer Lodge, MT, Sept 17, 1945.
Ken Kesey, 60, author (*One Flew Over the Cuckoo's Nest*), born LaHunta, CO, Sept 17, 1935.
Dorothy Loudon, 62, actress, singer ("The Garry Moore Show"), born at Boston, MA, Sept 17, 1933.
Jeff MacNelly, 48, cartoonist ("Shoe"), born at New York, NY, Sept 17, 1947.
Roddy McDowall, 67, actor (*How Green Was My Valley, My Friend Flicka, Planet of the Apes*; Emmy for "Not Without Honor" [*Our American Heritage*]), born London, England, Sept 17, 1928.
Cassandra Peterson, 44, actress (Elvira), born at Manhattan, KS, Sept 17, 1951.

John Ritter, 47, actor (Emmy "Three's Company"; *Problem Child*), born Burbank, CA, Sept 17, 1948.
David H. Souter, 56, associate justice of the US Supreme Court, born at Melrose, MA, Sept 17, 1939.

SEPTEMBER 18 — MONDAY
261st Day — Remaining, 104

AMERICAN TINNITUS ASSOCIATION ADVISORS MEETING. Sept 18. New Orleans, LA. For info: American Tinnitus Assn, Patricia Daggett, Box 5, Portland, OR 97207.

CHILE: INDEPENDENCE DAY: ANNIVERSARY. Sept 18. National holiday. Gained independence from Spain, 1818.

DIEFENBAKER, JOHN: BIRTH ANNIVERSARY. Sept 18. Canadian lawyer, statesman and Conservative prime minister (1957–1963). Born at Normandy Township, Ontario, Canada, on Sept 18, 1895. Died at Ottawa, Ontario, Aug 16, 1979. Diefenbaker was a member of the Canadian Parliament from 1940 until his death.

GEIGER, RAY: BIRTH ANNIVERSARY. Sept 18. Born in Irvington, NJ, on Sept 18, 1910, Ray Geiger became editor of the *Farmer's Almanac* in 1934, shortly after graduating from Notre Dame. He remained editor until completing the 1994 edition in the fall of 1993 (when his son Peter became editor), the longest running almanac editor in American history. He died Apr 1, 1994, at Auburn, ME. *The Farmer's Almanac* offices are in Lewiston, ME.

IRON HORSE OUTRACED BY HORSE: ANNIVERSARY. Sept 18. In a widely celebrated race, the first locomotive built in America, the Tom Thumb, lost to a horse, on Sept 18, 1830. Mechanical difficulties plagued the steam engine over the nine-mile course between Riley's Tavern and Baltimore, MD, and a boiler leak prevented the locomotive from finishing the race. In the early days of trains, engines were nicknamed "Iron Horses."

JOHNSON, SAMUEL: BIRTH ANNIVERSARY. Sept 18. English lexicographer and literary lion, creator of the first great dictionary of the English language (1755) and author of poems and essays. Less well known is his novel *Rasselas: Prince of Abyssinia* (1759) written to pay for his mother's funeral. It begins with what has been called "the most beautiful sentence ever written": "Ye who listen with credulity to the whispers of fancy, and pursue with eagerness the phantoms of hope; who expect that age will perform the promises of youth, and that the deficiencies of the present day will be supplied by the morrow; attend to the history of Rasselas, Prince of Abyssinia." Johnson was born at Lichfield, in Staffordshire, England, Sept 18, 1709, and died at London, England, on Dec 13, 1784.

NATIONAL MIND MAPPING AND BRAINSTORMING WEEK. Sept 18–24. To promote creativity and innovation in business and organizations through brainstorming and mind mapping, a visual form of outlining. For info: Joyce Wycoff, Pres, Mind Play, 1103 Harbor Hills Dr, Santa Barbara, CA 93109. Phone: (805) 962-9933.

NATIONAL QUARTET CONVENTION. Sept 18–23. Freedom Hall, Kentucky Fair and Exposition Center, Louisville, KY. Six-day event with focus on the nightly Southern Gospel concerts. Daytime activities include special seminars and record company showcases, celebrity golf tournament, daily Bible study and chapel service and three-hour cruise on the Belle of Louisville. Annually, beginning the third Monday in September. Est attendance: 15,000. For info: Clarke Beasley, Exec Dir, 1 Riverfront Plaza, Ste 1710, Louisville, KY 40202. Phone: (502) 587-9071.

September 1995

S	M	T	W	T	F	S
					1	2
3	4	5	6	7	8	9
10	11	12	13	14	15	16
17	18	19	20	21	22	23
24	25	26	27	28	29	30

☆ Chase's 1995 Calendar of Events ☆ Sept 18-19

NATIONAL YOUTH OF THE YEAR. Sept 18-20. Washington, DC. Each year a Boys and Girls Club member is selected by a panel of judges from among five regional finalists to be the National Youth of the Year and spokesperson for Boys and Girls Clubs of America. Selections are made on the basis of leadership qualities and service exhibited to home and family, spiritual values, service to community and Club, excellence in school and obstacles overcome. Winners are presented at a Congressional breakfast and to the president at the White House. For info: Boys and Girls Clubs of America, 1230 W Peachtree St NW, Atlanta, GA 30309.

READ, GEORGE: BIRTH ANNIVERSARY. Sept 18. Lawyer and signer of the Declaration of Independence, born at Cecil County, MD, Sept 18, 1733. Died on Sept 21, 1798.

SPACE MILESTONE: SOYUZ 38 (USSR). Sept 18. Launched on Sept 18, 1980, Cosmonauts Arnaldo Tamayo Mendes (Cuba) and Yuri Romanenko docked at Salyut 6 for week-long mission, returning to Earth Sept 26.

STORY, JOSEPH: BIRTH ANNIVERSARY. Sept 18. Associate justice of the US Supreme Court (1811-1845) was born at Marblehead, MA, on Sept 18, 1779. "It is astonishing" he wrote a few months before his death, "how easily men satisfy themselves that the Constitution is exactly what they wish it to be." Story died on Sept 10, 1845, having served 33 years on the Supreme Court bench.

US AIR FORCE ESTABLISHED: BIRTHDAY. Sept 18. The US Air Force became a separate military service on Sept 18, 1947. Responsible for providing an Air Force that is capable, in conjunction with the other armed forces, of preserving the peace and security of the United States, the department is separately organized under the Secretary of the Air Force and operates under the authority, direction and control of the Secretary of Defense.

US TAKES OUT ITS FIRST LOAN: ANNIVERSARY. Sept 18. The first loan taken out by the US was negotiated and secured by Alexander Hamilton on Feb 17, 1790. After beginning negotiations with the Bank of New York and the Bank of North America, on Sept 18, 1789, Hamilton obtained the sum of $191,608.81 from the two banks in what became known as the Temporary Loan of 1789. The loan was obtained without authority of law and was used to pay the salaries of the president, senators, representatives and officers of the first Congress. Repayment was completed on June 8, 1790.

US WOMEN'S MID-AMATEUR (GOLF) CHAMPIONSHIP. Sept 18-23. Essex Country Club, Manchester, MA. Sponsor: US Golf Assn, Championship Dept, Golf House, Far Hills, NJ 07931. Phone: (908) 234-2300.

WHITE WOMAN MADE AMERICAN INDIAN CHIEF: ANNIVERSARY. Sept 18. On Sept 18, 1891, Harriet Maxwell Converse was made a chief of the Six Nations Tribe at the Tonawanda Reservation, NY. She was given the name Ga-is-wa-noh, which means "The Watcher." She had been adopted as a member of the Seneca tribe in 1884 in appreciation of her efforts on behalf of the tribe.

BIRTHDAYS TODAY

Frankie Avalon, 55, singer ("Why," "Venus," "Dede Dinah"), actor (teen flicks with Annette Funicello), born at Philadelphia, PA, Sept 18, 1940.

Robert F. Bennett, 62, US Senator (R, Utah), born at Salt Lake City, UT, Sept 18, 1933.

Robert Blake (Michael Gubitosi), 57, actor ("Baretta," *In Cold Blood, Little Rascals*), born at Nutley, NJ, Sept 18, 1938.

Rossano Brazzi, 79, singer, actor (*South Pacific*, "The Survivors"), born Bologna, Italy, Sept 18, 1916.

Ryne Sandberg, 36, former baseball player, born at Spokane, WA, Sept 18, 1959.

Jack Warden, 75, actor ("N.Y.P.D.," "The Bad News Bears," "Crazy Like a Fox"; Emmy for *Brian's Song*), born at Newark, NJ, Sept 18, 1920.

SEPTEMBER 19 — TUESDAY
262nd Day — Remaining, 103

BROUGHAM, HENRY PETER: BIRTH ANNIVERSARY. Sept 19. Scotch jurist and orator born at Edinburgh, Scotland, Sept 19, 1778. Died at Cannes, France, May 7, 1868. Brougham carriage named after him. "Education," he said, "makes a people easy to lead, but difficult to drive; easy to govern, but impossible to enslave."

CARROLL, CHARLES: BIRTH ANNIVERSARY. Sept 19. American Revolutionary leader and signer of the Declaration of Independence, born at Annapolis, MD, Sept 19, 1737. Died (the last surviving signer of the Declaration) Nov 14, 1832.

GOLDING, SIR WILLIAM: BIRTH ANNIVERSARY. Sept 19, 1911. Born at Columb Minor in Cornwall, England, this celebrated author was recognized for his contributions to literature with a Nobel Prize in 1983. His first and most popular novel was *Lord of the Flies*. He died June 19, 1993, near Truro, Cornwall.

GRANT COUNTY HOME PRODUCTS DINNER. Sept 19. Grant County Civic Center, Ulysses, KS. 33rd annual public dinner grown, prepared and served by Grant County residents. Approximately 1,800 people attend including the Governor of Kansas and many state officials. Annually, the third Tuesday of September. For info: Dennis Zimmerman, Grant County Chamber of Commerce, 115 W Grant, Ulysses, KS 67880. Phone: (316) 356-4700.

JACKSON COUNTY APPLE FESTIVAL. Sept 19-23. Jackson County, OH. Mountains of apples and barrels of cider. Homemade apple butter, apple pies and candy apples. Est attendance: 210,000. For info: Jackson County Apple Festival, Inc, PO Box 8, Jackson, OH 45640-0008. Phone: (614) 286-1339.

LUYTS, JAN: BIRTH ANNIVERSARY. Sept 19. Dutch scholar, physicist, mathematician and astronomer, Jan Luyts was born at Hoorn in western Netherlands on Sept 19, 1655. Little remembered except for his books: *Astronomica Institutio ...* (1689), and *Introductio ad Geographiam ...* (1690).

MEXICO CITY EARTHQUAKE: 10th ANNIVERSARY. Sept 19-20. Nearly 10,000 persons perished in the earthquakes (8.1 and 7.5 respectively, on the Richter Scale) that devastated Mexico City Sept 19 and 20, 1985. Damage to buildings was estimated at more than $1 billion, and 100,000 homes were destroyed or severely damaged.

NETHERLANDS: PRINSJESDAG. Sept 19. Official opening of parliament at The Hague. On the third Tuesday in September, the queen of the Netherlands, by tradition, rides in a golden coach to the hall of knights for the annual opening of parliament.

POWELL, LEWIS F., JR: BIRTHDAY. Sept 19. Retired associate justice of the Supreme Court of the United States, nominated by President Nixon on Oct 21, 1971. (Took office on Jan 7, 1972.) Justice Powell was born at Suffolk, VA, on Sept 19, 1907. In 1987, he announced his retirement from the Court.

ST. CHRISTOPHER (ST. KITTS) AND NEVIS: INDEPENDENCE DAY. Sept 19. National holiday.

ST. JANUARIUS (GENNARO): FEAST DAY. Sept 19. Fourth-century bishop of Benevento, martyred near Naples, Italy, whose relics in the Naples Cathedral are particularly famous because on his feast days the blood in glass vial is said to liquefy in response to prayers of the faithful. In Sept 1979, the Associated Press reported that some 5,000 persons gathered at the cathedral at dawn, and that "the blood liquefied after 63 minutes of prayers." This phenomenon is said to occur also on the first Saturday in May.

Sept 19-21 ☆ *Chase's 1995 Calendar of Events* ☆

TITAN II MISSILE EXPLOSION: ANNIVERSARY. Sept 19. The third major accident involving America's most powerful single weapon occurred near Damascus, AR, Sept 19, 1980. The explosion, at 3 AM, came nearly eleven hours after the fire had started in the missile silo. The multi-megaton nuclear warhead (a hydrogen bomb) reportedly was briefly airborne, but came to rest a few hundred feet away. One dead, 21 injured in accident. Previous major Titan Missile accidents: Aug 9, 1965, near Searcy, AR (53 dead); and Aug 24, 1978, near Rock, KS (2 dead, 29 injured).

UNITED NATIONS: INTERNATIONAL DAY OF PEACE/ OPENING DAY OF GENERAL ASSEMBLY. Sept 19. The United Nations General Assembly, on Nov 30, 1981, declared "that the third Tuesday of September, the opening day of the regular sessions of the General Assembly, shall be officially proclaimed and observed as International Day of Peace and shall be devoted to commemorating and strengthening the ideals of peace both within and among all nations and peoples." An International Year of Peace was proclaimed for 1986. A Peace Month, and a University for Peace also have been proposed and are under consideration. Info from: United Nations, Dept of Public Info, New York, NY 10017.

BIRTHDAYS TODAY

Jim Abbott, 28, baseball player, born at Flint, MI, Sept 19, 1967.
Clifton Daniel, 83, journalist, born at Zebulon, NC, Sept 19, 1912.
Jeremy Irons, 47, actor (*Moonlighting, Dead Ringers*), born at Cowes, Isle of Wight, England, Sept 19, 1948.
Joan Lunden, 44, broadcast journalist (co-host of "Good Morning America"), born at Sacramento, CA, Sept 19, 1951.
Randolph Mantooth, 50, actor ("Emergency"), born at Sacramento, CA, Sept 19, 1945.
Lewis Franklin Powell, Jr, 88, Associate Justice of the US Supreme Court (retired), born at Suffolk, VA, Sept 19, 1907.
Mike Royko, 63, journalist, author (*Boss, Slats Grobnick*), born at Chicago, IL, Sept 19, 1932.
Twiggy (Leslie Hornby), 46, actress (*The Boy Friend, The Blues Brothers, My One and Only*), born London, England, Sept 19, 1949.
Adam West, 67, actor ("Batman," "The Last Precinct") born at Walla Walla, WA, Sept 19, 1928.
Paul Williams, 55, singer, composer ("Love Boat" theme song), born Omaha, NE, Sept 19, 1940.

SEPTEMBER 20 — WEDNESDAY
263rd Day — Remaining, 102

BILLIE JEAN KING WINS THE "BATTLE OF THE SEXES": ANNIVERSARY. Sept 20. Billie Jean King defeated Bobby Riggs in the nationally televised "Battle of the Sexes" tennis match in three straight sets on Sept 20, 1973.

EQUAL RIGHTS PARTY FOUNDING: ANNIVERSARY. Sept 20. San Francisco, CA. On Sept 20, 1884, the Equal Rights Party was formed. Their candidate for president, nominated in convention, was Mrs Belva Lockwood. The vice-presidential candidate was Marietta Stow.

FINANCIAL PANIC OF 1873: ANNIVERSARY. Sept 20. For the first time in its history the New York Stock Exchange was forced to close, Sept 20, 1873, because of a banking crisis. Although the worst of the panic and crisis was over within a week, the psychological effect on businessmen, investors and the nation at large was more lasting.

September 1995

S	M	T	W	T	F	S
					1	2
3	4	5	6	7	8	9
10	11	12	13	14	15	16
17	18	19	20	21	22	23
24	25	26	27	28	29	30

GOLDEN ASPEN MOTORCYCLE RALLY. Sept 20-24. Ruidoso, NM. Trade show, bike shows, riding tours and skill events. Est attendance: 9,000. Sponsor: Golden Aspen Rally Assn, PO Box 1458, Ruidoso, NM 88345. Phone: (800) 452-8045.

MORTON, FERDINAND "JELLY ROLL": 110th BIRTH ANNIVERSARY. Sept 20. American jazz pianist, composer, singer and orchestra leader, was born at New Orleans, LA, Sept 20, 1885. Ferdinand Joseph (Jelly Roll) Morton, subject of a biography entitled *Mr. Jelly Roll* (by Alan Lomax), died July 10, 1941, at Los Angeles, CA.

NATIONAL RESEARCH COUNCIL: FIRST MEETING ANNIVERSARY. Sept 20. Anniversary of first meeting of National Research Council, at New York, NY, Sept 20, 1916. Formed at request of President Woodrow Wilson for the purpose of "... encouraging the investigation of natural phenomena ..." for American business and national security. Sept 20 also marks the anniversary of the first meeting of the American Association for the Advancement of Science at Philadelphia, PA, in 1848.

RIVER CITY ROUNDUP. Sept 20-24. Omaha, NE. Take a moment to reflect on the Midwest's proud past. A celebration of Omaha's agricultural and western heritage; world championship rodeos, hot air balloon races, barbeque and chili contests, trail rides, the world's largest 4-H Livestock Expo and a downtown parade. Est attendance: 400,000. For info: Danelle Myer, Mktg Dir, River City Roundup, PO Box 6253, Omaha, NE 68106. Phone: (402) 554-9610.

SINCLAIR, UPTON (BEALL): BIRTH ANNIVERSARY. Sept 20. American novelist and politician born at Baltimore, MD, Sept 20, 1878. Died at Bound Brook, NJ, Nov 25, 1968. He worked for political and social reforms, and his best-known novel, *The Jungle*, prompted one of the nation's first pure food laws.

BIRTHDAYS TODAY

Arnold "Red" Auerbach, 78, basketball Hall of Famer, former coach, born at Brooklyn, NY, Sept 20, 1917.
Donald A. Hall, 67, author (*Lucy's Christmas, Oxcart Man*), born New Haven, CT, Sept 20, 1928.
Guy LaFleur, 44, hockey player, born at Thurso, Quebec, Canada, Sept 20, 1951.
Sophia Loren (Sofia Scicoloni), 61, actress (Oscar for *Two Women*; *The Black Orchid, Marriage Italian Style*, "Brief Encounter"), born Rome, Italy, Sept 20, 1934.
Anne Meara, 71, actress ("Fame"), comedienne (Stiller and Meara), born at New York, NY, Sept 20, 1924.

SEPTEMBER 21 — THURSDAY
264th Day — Remaining, 101

ARMENIA: NATIONAL DAY. Sept 21. Public holiday.
BELIZE: INDEPENDENCE DAY. Sept 21. National holiday.
BIOSPHERE DAY. Sept 21. A day to remind all humanity annually of the fragility of our only life-support, with the consequent need to safeguard it our foremost human imperative. The Bio-

Chase's 1995 Calendar of Events — Sept 21-22

sphere is that layer of our planet's periphery (solid, liquid and gaseous) in which any form of life exists naturally. Participants are resolved to spread fundamental information worldwide concerning The Biosphere and how to preserve it and foster it. The motto and guiding principle is "Every Day a Biosphere Day"—for it is The Biosphere that is gravely threatened by human overpopulation and profligacy, not the inert and more solid Planet Earth. A major Biosphere fund and substantial Biosphere Prizes are planned to support Biosphere Day and spread its vital message, and also small Biosphere Clubs such as it is hoped will ultimately form a world-wide network. Sponsor: The Foundation for Environmental Conservation, Dr. Nicholas Polunin, Pres, 7 Chemin Taverney, 1218 Grand-Saconnex, Geneva, Switzerland.

HONG KONG: BIRTHDAY OF CONFUCIUS. Sept 21. Religious observances are held by the Confucian Society at Confucius Temple in Causeway Bay. Observed on 27th day of 8th lunar month.

HURRICANE HUGO HITS AMERICAN COAST: ANNIVERSARY. Sept 21. After ravaging the Virgin Islands, Hurricane Hugo hit the American coast at Charleston, SC, on Sept 21, 1989. In its wake, Hugo left destruction totaling at least eight billion dollars.

LAUREL HERBSTFEST. Sept 21-24. Laurel, MT. A four-day celebration of the town's German heritage. Activities include a pageant, arts and crafts fair, dances, parade, fun run, two full days of musical entertainment, a variety of food booths, a beer garden with lots of polka music and dancing. Est attendance: 8,000. For info: Jean Carroll Thompson, PO Box 1192, Laurel, MT 59044. Phone: (406) 628-8306.

LAVITSEF. Sept 21-24. City-wide, Norfolk, NE. Concerts, pet show, health fair, flower show, pancake feed, 10K run, cycling series, craft fair, softball tournament, ice cream social, baby contests. Parade at 10 AM Saturday. Est attendance: 5,000. For info: Holly Swanson, PO Box 386, Norfolk, NE 68702-0386. Phone: (402) 371-2932.

MALTA: INDEPENDENCE DAY. Sept 21. National Day.

NORWALK INTERNATIONAL IN-WATER BOAT SHOW. Sept 21-24. Norwalk Cove Marina, Norwalk, CT. Boats and marine products display and sales. For info: Natl Marine Manufacturers Assn, 600 Third Ave, New York, NY 10016. Phone: (212) 922-1212.

STATE FAIR OF VIRGINIA ON STRAWBERRY HILL. Sept 21-Oct 1. Richmond, VA. The pride of Virginia's industry of agriculture can be seen in more than 3,000 exhibitions, competitions and shows. Virginia's greatest annual educational and entertainment event. Est attendance: 600,000. For info: Kieth T. Hessey, Genl Mgr, 600 E Laburnum Ave, Richmond, VA 23222. Phone: (804) 228-3200.

TAYLOR, MARGARET SMITH: BIRTH ANNIVERSARY. Sept 21. Wife of Zachary Taylor, 12th president of the US, born at Calvert County, MD, Sept 21, 1788. Died on Aug 18, 1852.

WATTICISM DAY: ANNIVERSARY. Sept 21. Anniversary of speech by then US Interior Secretary James Watt to trade association executives at the US Chamber of Commerce, Sept 21, 1983. Referring to his advisory committee, Watt said: "We have every kind of mixture you can have. I have a black, I have a woman, two Jews and a cripple. And we have talent." He later apologized for an "unfortunate choice of words."

WELLS, HERBERT GEORGE: BIRTH ANNIVERSARY. Sept 21. English novelist and historian, born at Bromley, in Kent, England, on Sept 21, 1866. Among his books: *The Time Machine, The Invisible Man, The War of the Worlds* and *The Outline of History*. Wells died at London, Aug 13, 1946. "Human history," he wrote, "becomes more and more a race between education and catastrophe."

WORLD GRATITUDE DAY. Sept 21. To unite all people in a positive emotion of gratitude, creating a world community. Gratitude gatherings, any assemblage aware of the spirit of gratitude, are encouraged. Annual international children's poster exhibit. Sponsor: World Gratitude Day, Edna Lemle, Pres, 132 W 31st St, New York, NY 10001.

BIRTHDAYS TODAY

Leonard Cohen, 61, singer, songwriter, born at Montreal, Quebec, Canada, Sept 21, 1934.
Cecil Grant Fielder, 32, baseball player, born at Los Angeles, CA, Sept 21, 1963.
Henry Gibson, 60, comedian ("Rowan and Martin's Laugh-In"), actor (*Nashville*), born Germantown, PA, Sept 21, 1935.
Artis Gilmore, 46, former basketball player, born at Chipley, FL, Sept 21, 1949.
Larry Hagman, 64, actor ("I Dream of Jeannie," "Dallas"), born at Fort Worth, TX, Sept 21, 1931.
Stephen King, 48, author (*Christine, Pet Sematary, The Duel, Misery*), born at Portland, ME, Sept 21, 1947.
Bill Kurtis, 55, journalist (Chicago CBS anchorman), born at Pensacola, FL, Sept 21, 1940.
Sidney Moncrief, 38, former basketball player, born at Little Rock, AR, Sept 21, 1957.
Rob Morrow, 33, actor ("Northern Exposure"), born at New Rochelle, NY, Sept 21, 1962.
Bill Murray, 45, comedian ("Saturday Night Live"), actor (*Ghostbusters, Groundhog Day, What About Bob?, Mad Dog and Glory*), born Evanston, IL, Sept 21, 1950.

SEPTEMBER 22 — FRIDAY
265th Day — Remaining, 100

AMERICAN BUSINESS WOMEN'S DAY. Sept 22. A day on which all Americans can recognize the important contributions more than 55 million American working women have made and are continuing to make to this nation. Annually, Sept 22. Sponsor: American Business Women's Assn, PR Dept, 9100 Ward Pkwy, Kansas City, MO 64114. Phone: (816) 361-6621.

CHILE CHALLENGE OFF-ROAD BIKE RACE. Sept 22-24. Angel Fire, NM. Annual bike race held on the ski mountain in this beautiful alpine setting. All skill levels. Est attendance: 200. For info: Angel Fire Resort, PO Drawer B, Angel Fire, NM 87710. Phone: (800) 633-7463.

COMMON GROUND COUNTRY FAIR. Sept 22-24. Windsor Fairgrounds, Windsor, ME. Old-time country fair celebrating rural life with the revival of forgotten skills and demonstrations of technology appropriate for the future. Features Maine-produced food, crafts, entertainment, farming demonstrations and talks, a very special Children's Area with daily and ongoing participatory activities. Annually, the third weekend after Labor Day. Sponsor: Maine Organic Farmers and Gardeners Assn. Est attendance: 60,000. For info: Susan Pierce, Special Events Dir, PO Box 2176, Augusta, ME 04338. Phone: (207) 623-5115.

DEAR DIARY DAY. Sept 22. Put it on paper. You'll feel better. No need to be a professional writer. Phone: (212) 388-8673 or (717) 274-8451. Sponsor: Wellness Permission League, Tom or Ruth Roy, 2105 Water St, Lebanon, PA 17046.

DURHAM FAIR. Sept 22–24. Durham, CT. 75th annual fair features agricultural exhibits, demonstrations and competitions; name CW entertainers; antiques and craft show; rides, amusements and food. Est attendance: 275,000. For info: Durham Agricultural Fair Assn, Inc, Box 225, Durham, CT 06422.

FABULOUS 1890s WEEKEND. Sept 22–24. Mansfield, PA. Night football in America began in 1892 with a game between Mansfield University and Wyoming Seminary. Annually, Mansfield celebrates a "Fabulous 1890s Weekend" to celebrate the event. Motorless parade, period exhibits, crafts and other events, including the re-creation of the first night football game. Sponsors: Mansfield University of Pennsylvania and Greater Area Mansfield Chamber of Commerce. Est attendance: 10,000. For info: Dennis Miller, Dir PR, Mansfield University, Mansfield, PA 16933. Phone: (717) 662-4293.

FAIRMOUNT MUSEUM DAYS FESTIVAL/REMEMBERING JAMES DEAN. Sept 22–24. Fairmount, IN. Fairmount, the town where James Dean grew up, honors Dean and other celebrated former citizens such as Jim Davis, creator of Garfield, journalist Phil Jones and Robert Sheets of the National Hurricane Center. The Fairmount Museum boasts "the Authentic James Dean Exhibit" of memorabilia and personal items of Dean's and it sponsors the festival that also includes a parade, James Dean Look-Alike Contest, custom car show featuring the James Dean Run for pre-1970 autos, Garfield Cat Photo Contest, 10K run, carnival, games, live entertainment and more. Est attendance: 33,000. For info: Fairmount Historical Museum, Inc, 203 E Washington St, PO Box 92, Fairmount, IN 46928. Phone: (317) 948-4555.

FARADAY, MICHAEL: BIRTH ANNIVERSARY. Sept 22. English scientist and early experimenter with electricity, born Sept 22, 1791. Died Aug 25, 1867.

FESTIVAL OF ADVENTURES. Sept 22–24. Aitkin, MN. This 4th annual festival celebrates the area's fur trading history. Rendezvous at Depot Museum with trappers, traders, ethnic dancers, bluegrass music and food. Phone: (800) 526-8342 or (218) 927-2316. Est attendance: 5,000. For info: Aitkin Area Chamber of Commerce, PO Box 127, Aitkin, MN 56431. Phone: (218) 927-2316.

FIRST ALL-WOMAN JURY EMPANELED IN COLONIES. Sept 22. The General Provincial Court in Patuxent, MD, empaneled the first all-woman jury in the colonies on Sept 22, 1656. The jury heard the case of Judith Catchpole, accused of murdering her child. The defendant claimed she had never even been pregnant, and after all the evidence was heard, the jury acquitted her.

FLATLANDERS FALL CLASSIC; FLATLANDERS FALL FESTIVAL; PIONEER FOLK FESTIVAL; EARLY IRON CAR CLUB ROD RUN. Sept 22–24. Goodland, KS. Classic on Fri and Sat features mini-sprint stock car races; Fall Festival on Sat on Main St is a street festival featuring crafts, food, entertainment; Sat and Sun Pioneer Folk Festival at High Plains Museum highlights crafts and chores of the 19th century; also Sat and Sun, the Early Iron Car Club Rod Run on Highways 24 and 27. For info on Early Iron Rod Run, write to PO Box 628, Goodland, KS 67735; phone: (913) 899-3483. Est attendance: 5,000. For other info: Goodland Chamber of Commerce, 104 W 11th, Goodland, KS 66603. Phone: (913) 899-7130.

GREAT AMERICAN DULCIMER CONVENTION. Sept 22–23. Pine Mountain State Resort Park, Pineville, KY. Talented performers on lap and hammered dulcimers will appear in concert along with singers and other musicians. Dulcimer lessons, crafts and craft demonstrations. Est attendance: 1,000. For info: Pine Mountain State Resort Pk, Dean M. Henson, Naturalist, 1050 State Park Rd, Pineville, KY 40977. Phone: (800) 325-1712.

HOBBIT DAY. Sept 22. To commemorate the birthdays of Frodo and Bilbo Baggins and their creator J.R.R. Tolkien. Sponsor: American Tolkien Society, Box 373, Highland, MI 48357-0373. Phone: (813) 585-0985.

ICE CREAM CONE: BIRTHDAY. Sept 22. Italo Marchiony emigrated from Italy in the late 1800s and soon thereafter went into the portable restaurant business in New York, NY, with a pushcart dispensing lemon ice. Success soon led to a small fleet of pushcarts, and the inventive Marchiony was inspired to develop a cone, first made of paper, later of pastry, to hold the tasty delicacy. On Sept 22, 1903, his application for a patent for his new mold was filed, and US Patent No. 746971 was issued to him on Dec 15, 1903.

◈ **LONG COUNT DAY.** Sept 22. Anniversary of world championship boxing match between Jack Dempsey and Gene Tunney, at Soldier Field, Chicago, IL, on Sept 22, 1927. It was the largest fight purse ($990,446) in the history of boxing to that time. Nearly half the population of the US is believed to have listened to the radio broadcast of this fight. In the seventh round of the 10-round fight, Tunney was knocked down. Following the rules, Referee Dave Barry interrupted the count when Dempsey failed to go to the farthest corner. The count was resumed and Tunney got to his feet at the count of nine. Stopwatch records of those present claimed the total elapsed time from the beginning of the count until Tunney got to his feet at 12–15 seconds. Tunney, generally awarded seven of the ten rounds, won the fight and claimed the world championship. Dempsey's appeal was denied and he never fought again. Tunney retired the following year after one more (successful) fight.

MALI: PROCLAMATION OF THE REPUBLIC: 35th ANNIVERSARY. Sept 22. National holiday commemorating independence from France on Sept 22, 1960. Mali, in West Africa, was known as the French Sudan while a colony of France.

MID-SOUTH FAIR. Sept 22–Oct 1. Fairgrounds, Memphis, TN. Regional fair featuring concerts, free entertainment, midway, rodeo, livestock, exhibits and special events. Est attendance: 500,000. For info: Mid-South Fair, Verna Jones, 340 Early Maxwell Blvd, Memphis, TN 38104. Phone: (901) 274-8800.

NATIONAL CENTENARIANS DAY. Sept 22, 1995. A day to recognize and honor elderly individuals who have lived a century or longer. A day not only to recognize these individuals, but to listen to them discuss the memories—filled with historical information—that they have of their rich lives. Take time today to listen to a Centenarian. Special celebration held at Williamsport Retirement Village. Annually. For info: Meg Cliber, Mktg Dir, Williamsport Retirement Village, A Division of Brooks Grove Fdtn, 154 N Artizan St, Williamsport, MD 21795. Phone: (301) 223-7971.

NATIONAL LAUNDRY WORKERS DAY. Sept 22. To promote public awareness of the contribution and importance of laundry service employees. Annually, the Friday of the third full week in September. For info: Nemaha Co Good Samaritan Ctr, Rte 1, Box 4, Auburn, NE 68305.

PACIFIC RIM WILDLIFE ART SHOW. Sept 22–24. Tacoma Dome Exhibition Hall, Tacoma, WA. Wildlife fine art by some of the most renowned artists from North America. The show offers booth sales, competition and auction pieces and informative seminars for the entire family. Annually, the last weekend in September. Est attendance: 9,000. Sponsor: WES Productions,

September 1995

S	M	T	W	T	F	S
					1	2
3	4	5	6	7	8	9
10	11	12	13	14	15	16
17	18	19	20	21	22	23
24	25	26	27	28	29	30

Bob Farrelly, Pres, PO Box 11225, Tacoma, WA 98411. Phone: (206) 596-6728.

RACKING WORLD CELEBRATION. Sept 22-30. Decatur, AL. Week-long event featuring racking horses from across the nation. The highlight of the 75-class event is the crowning of the World Grand Racking Horse Champion on the last night. Annually, the last full week in September. Est attendance: 68,000. For info: Decatur Conv and Visitors Bureau, 719 6th Ave SE, PO Box 2349, Decatur, AL 35602. Phone: (205) 350-2028.

ST. FRANCOIS RIVER RENDEZVOUS. Sept 22-24. Farmington, MO. Black powder shoot, blanket traders, 1840s food booths, Native American Indian Powwow and competitive dancing. Both modern and primitive campgrounds provided. Est attendance: 5,000. For info: Farmington Chamber of Commerce, N Washington St, Farmington, MO 63640. Phone: (314) 756-3615.

SNOW HILL HERITAGE WEEKEND. Sept 22-24. Snow Hill, MD. Events include concerts, art show, antique show, historic home tours, river cruises, games, food and more. Annually, the third weekend after Labor Day. Est attendance: 3,000. For info: Snow Hill Area Chamber of Commerce, PO Box 176, Snow Hill, MD 21863. Phone: (800) 852-0335.

SOCIETY FOR THE ERADICATION OF TELEVISION CONVENTION. Sept 22. San Diego, CA. To discuss the problems of our television-obsessed society and plan activities to combat them. Est attendance: 32. Sponsor: Society for the Eradication of Television, Steve Wagner, Dir, Box 10491, Oakland, CA 94610-0491. Phone: (510) 763-8712.

STANHOPE, PHILIP DORMER: BIRTH ANNIVERSARY. Sept 22. Philip Dormer Stanhope, the 4th Earl of Chesterfield, was born at London, England, on Sept 22, 1694. His brilliant career in politics gave many opportunities for demonstration of his extraordinary skill as an orator. On Feb 20, 1751, he brought a bill into the House of Lords that caused the "New Style" Gregorian calendar to replace the "Old Style" Julian calendar in 1752. His influential political career was eclipsed by the fame of the letters he wrote to his natural son, Philip. Beginning when his son was five years old, they continued until the son's death at age 36, giving shrewd counsel on manners, morals and the ways of the world. Published less than a year after his death at London, Mar 24, 1773, the *Letters* became immensely popular, were translated and republished in many editions and in many countries. The Chesterfield, a kind of sofa, is said to be named for him.

TACY RICHARDSON'S RIDE: ANNIVERSARY. Sept 22. Remembers early morning ride, Sept 22, 1777, of 23-year-old Tacy Richardson (Jan 1, 1754-June 18, 1807) who rode her favorite horse, "Fearnaught," several perilous miles from the family farm (near the meeting of the Perkiomen and the Schuylkill, Montgomery County, PA) to the James Vaux mansion to warn General George Washington of the approach of British troops led by General William Howe. In reality, the British crossing of the Schuylkill at Gordon's Ford was a feint to deceive Washington who indeed hastily withdrew to Pottstown. General Howe spent that night in the same quarters Washington had occupied only a few hours earlier. Poems and family tradition memorialize the bravery of Tacy's ride.

US POSTMASTER GENERAL ESTABLISHED: ANNIVERSARY. Sept 22. Congress established office of postmaster general—Sept 22, 1789, following the departments of state, war and treasury.

WILD WEST FILM FEST. Sept 22-24. Tuolumne County, CA. A Friday night gala honors Western movie stars, followed by two days of Western film showings, celebrity/film making panel discussions, autograph signing, live entertainment, film actor guests, arts and crafts and other family-oriented activities held in an historic 1850 gold rush community. A professional rodeo, BBQ and barn dance on Saturday evening. Est attendance: 10,000. For info: Tuolumne County Visitors Bureau, PO Box 4020, Sonora, CA 95370. Phone: (800) 446-1333.

BIRTHDAYS TODAY

Wally (Walter Wayne) Backman, 36, baseball player, born at Hillsboro, OR, Sept 22, 1959.
Scott Baio, 34, actor ("Happy Days," "Stoned," "Charles in Charge"), born Brooklyn, NY, Sept 22, 1961.
Shari Belafonte-Harper, 41, model, actress, born at New York, NY, Sept 22, 1954.
Debbie Boone, 39, singer ("You Light Up My Life," "Baby, I'm Yours"), born Hackensack, NJ, Sept 22, 1956.
Joan Jett, 35, singer ("I Love Rock 'n' Roll," "Crimson and Clover"), musician, born at Philadelphia, PA, Sept 22, 1960.
Tom Lasorda, 68, baseball manager (LA Dodgers), former player, born Norristown, PA, Sept 22, 1927.
Catherine Oxenberg, 34, actress (Amanda in "Dynasty"), born New York, NY, Sept 22, 1961.
Eugene Roche, 67, actor ("Soap," "Webster"), born at Boston, MA, Sept 22, 1928.
Mike Sullivan, 56, Governor of Wyoming (D), born at Omaha, NE, Sept 22, 1939.

SEPTEMBER 23 — SATURDAY
266th Day — Remaining, 99

AL'S RUN AND WALK FOR KIDS. Sept 23. Milwaukee, WI. Choose from an 8K run, 4-mile walk or 2.5-mile walk along the lake and through the streets of Milwaukee. Named after Al McGuire, network basketball announcer. Net proceeds benefit Children's Hospital of Wisconsin. Finish line party on Summerfest grounds with free entertainment and fitness expo. Est attendance: 22,000. For info: Children's Hospital Foundation, Al's Run and Walk for Kids, PO Box 1997 MS#3060, Milwaukee, WI 53201. Phone: (414) 266-6320.

AUTUMN. Sept 23-Dec 22. In the Northern Hemisphere, autumn begins today with the autumnal equinox, at 8:13 AM, EDT. Note that in the Southern Hemisphere today is the beginning of spring. Everywhere on Earth (except near the poles) the sun rises due east and sets due west, and daylight length is nearly identical—about 12 hours, 8 minutes.

BANNED BOOKS WEEK—CELEBRATING THE FREEDOM TO READ. Sept 23-29. Brings to the attention of the general public the importance of the freedom to read and the harm censorship causes to our society. Sponsors: (1) American Library Assn, (2) American Booksellers Assn, (3) American Society of Journalists and Authors, (4) Assn of American Publishers, (5) Natl Assn of College Stores, (6) American Assn of University Presses. For info: American Library Assn, Judith F. Krug, Office for Intellectual Freedom, 50 E Huron St, Chicago, IL 60611. Phone: (312) 280-4223.

BASEBALL'S GREATEST DISPUTE: ANNIVERSARY. Sept 23. On Sept 23, 1908, in the decisive game between the Chicago Cubs and the New York Giants, the National League pennant race erupted in controversy during the bottom of the ninth with the score tied 1-1, at the Polo Grounds, New York, NY. New York was at bat with two men on. The batter hit safely to center field, scoring the winning run. Chicago claimed that the runner on first, Fred Merkle, seeing the winning run score, headed toward the dugout without advancing to second base, thus invalidating the play. The Chicago second baseman, Johnny Evers, attempted to get the ball and tag Merkle out, but was prevented by the fans streaming onto the field. Days later Harry C. Pulliam, head of the National Commission of Organized Baseball, decided to call the game a tie. The teams were forced to play a post-season playoff game, which the Cubs won 4-2. Fans invented the terms "boner" and "bonehead" in reference to the play and it has gone down in baseball history as "Merkle's Boner."

Sept 23 ☆ *Chase's 1995 Calendar of Events* ☆

CHECKERS DAY. Sept 23. A day to recognize the important role of dogs in American politics. Anniversary of the nationally televised "Checkers Speech" by then vice-presidential candidate Richard M. Nixon, on Sept 23, 1952. Nixon was found "clean as a hound's tooth" in connection with a private fund for political expenses, and he declared he would never give back the cocker spaniel dog, Checkers, which had been a gift to his daughters. Other dogs prominent in American politics: Abraham Lincoln's dog, Fido; Franklin D. Roosevelt's much-traveled terrier, Fala; Harry S. Truman's dogs, Mike and Feller; Dwight D. Eisenhower's dog, Heidi; Lyndon Johnson's beagles, Him and Her; Ronald Reagan's dogs, Lucky and Rex; and George Bush's dog, Millie.

CHICKAHOMINY FALL FESTIVAL POWWOW. Sept 23-24. Tribal grounds near Roxbury, VA. Native American dancing, ceremonies, awards, crafts, food, and camping. 44th annual festival. Usually the weekend of the fourth Saturday in September. Est attendance: 3,000. Sponsor: Chickahominy Indian Tribe, Powhatan Red Cloud-Owen, 4803 Charles City Rd, Charles City, VA 23030. Phone: (804) 966-7043.

CLOWNFEST '95. Sept 23-24. Boardwalk, Seaside Heights, NJ. Clowns from all over the country converge on Seaside Heights. Big Top Circus Shows, Clown Strolling, and Huge Clown Parade on the Boardwalk, Sunday, Sept 24th at 2 PM. There is also professional & amateur Clown competition. 500 clowns and 600 general public clowns-for-a-day. Fax: (908) 747-3841. Sponsor: WOBM Radio. Est attendance: 175,000. For info: Natl Clown Arts Council, Fred Collins, 49 Dodd St, Ste One, Bloomfield, NJ 07003. Phone: (201) 429-7592.

COPPER MAGNOLIA FESTIVAL. Sept 23-24. Washington, MS. Demonstration and sale of handmade crafts, family entertainment. Annually, the last full weekend in September. Est attendance: 3,000. For info: Historic Jefferson College, Anne L. Gray, Historian, Box 700, Washington, MS 39190. Phone: (601) 442-2901.

CRAFT FAIR USA: INDOOR SHOW. Sept 23-24. Wisconsin State Fair Park, Milwaukee, WI. Sale of handcrafted items: jewelry, pottery, weaving, leather, wood, glass, yulecraft and sculpture. Est attendance: 15,000. For info: Dennis R. Hill, Dir, Craft Fair USA, 3233 S Villa Circle, Milwaukee, WI 53227. Phone: (414) 321-2100.

DYERSVILLE FESTIVAL OF THE ARTS. Sept 23-24. Beckman High School, Dyersville, IA. The art and craft festival in the "Farm Toy Capital of the World," near the film site of the movie *Field of Dreams*. Features the Dyersville Quilt Show, with more than 100 quilted items on display, including award winners from the Iowa State Fair. Midwestern crafts booths, art and crafts demonstrations, art on display and for sale, food and entertainment. Annually, the last full weekend in September. Est attendance: 5,000. For info: Dyersville Area Chamber of Commerce, PO Box 187, Dyersville, IA 52040. Phone: (319) 875-2311.

FALL CLASSIC HORSESHOE TOURNAMENT. Sept 23-24. Fairmont, WV. This event closes out our summer outdoor activities for the calendar year. Est attendance: 100. For info: Tri-County Horseshoe Club, Davis "Catfish" Woodward, 1133 Sunset Dr, Fairmont, WV 26554. Phone: (304) 366-3819.

FALL FOLIAGE FESTIVAL AND BREW FEST. Sept 23. Steamboat Springs, CO. German Oktoberfest with tasting of "home brews" from the surrounding four state area. Also includes a Kindergarten, polka bands, arts and crafts fair, food fair and more. Est attendance: 5,000. Sponsor: Steamboat Springs Chamber Resort Assn, Janet Nichols, Special Event Coord, PO Box 774408, Steamboat Springs, CO 80477. Phone: (303) 879-0882.

FERDINAND VI OF SPAIN: BIRTH ANNIVERSARY. Sept 23. King of Spain from 1746 to 1759, his greatest claim to fame was keeping Spain at peace. Born at Madrid, Spain, on Sept 23, 1713, he died at Villaviciosa de Odon, Spain, on Aug 10, 1759.

FORT ATKINSON RENDEZVOUS. Sept 23-24. Fort Atkinson, Fort Atkinson, IA. Buckskinner, military, artisans and food vendors depicting life on the frontier from 1840-1849. Annually, last full weekend in September. Est attendance: 10,000. For info: Scot Michelson, Ranger, Volga River Recreation Area, Fayette, IA 52142.

GENEVA AREA GRAPE JAMBOREE. Sept 23-24. Geneva, OH. Grape harvest and products. 32nd annual Jamboree. Annually, the last full weekend in September. Est attendance: 250,000. For info: Geneva Grape Jamboree, Box 92, Geneva, OH 44041. Phone: (216) 466-5262.

GETTYSBURG OUTDOOR ANTIQUE SHOW. Sept 23. Gettysburg, PA. More than 175 dealers displaying their wares on the sidewalk. Annually, the fourth Saturday in September. Est attendance: 25,000. For info: Gettysburg Travel Council, 35 Carlisle St, Gettysburg, PA 17325. Phone: (717) 334-6274.

GOLDEN LEAF FESTIVAL. Sept 23-24. Smith Haven Park, Mullins, SC. Family-oriented event includes the Great Parade, the Original Golden Leaf Husband Holler, Golden Leaf car show, Tot Trot, crafts, music and entertainment. Annually, the fourth weekend in September. Est attendance: 20,000. For info: Golden Leaf Festival Dir, PO Box 691, Mullins, SC 29574. Phone: (803) 464-6651.

GOVERNOR'S INVITATIONAL FIRELOCK MATCH. Sept 23-24. Ft Frederick State Park, Big Pool, MD. Reactivated 18th-century military units from several states compete in both individual and team competition. Annually, the last full weekend in September. Est attendance: 2,000. For info: Fort Frederick State Park, 11100 Fort Frederick Rd, Big Pool, MD 21711. Phone: (301) 842-2155.

KIWANIS KIDS' DAY. Sept 23. To honor and assist youth—our greatest resource. Annually, the fourth Saturday in September. Sponsor: Kiwanis Intl, Program Dvmt Dept, 3636 Woodview Trace, Indianapolis, IN 46268.

LIBRA, THE BALANCE. Sept 23-Oct 22. In the astronomical/astrological zodiac that divides the sun's apparent orbit into twelve segments, the period Sept 23-Oct 22 is identified traditionally as the sun sign of Libra, the Balance. The ruling planet is Venus.

LIPPMANN, WALTER: BIRTH ANNIVERSARY. Sept 23. American journalist, political philosopher and author. Born at New York, NY, on Sept 23, 1889, he died there on Dec 14, 1974. As a syndicated newspaper columnist he was the foremost and perhaps the most influential commentator in the nation. "Without criticism," he said in an address to the International Press Institute in 1965, "and reliable and intelligent reporting, the government cannot govern."

MABON. Sept 23. (Also called Alban Elfed.) One of the "Lesser Sabbats" during the Wiccan year, Mabon marks the second harvest as Nature prepares for the coming of winter. Annually, on the autumnal equinox.

MARION COUNTY COUNTRY HAM DAYS. Sept 23-24. Lebanon, KY. Country ham breakfast, served in the streets of Lebanon. Pokey pig run, pigasus parade and other specialties.

September 1995

S	M	T	W	T	F	S
					1	2
3	4	5	6	7	8	9
10	11	12	13	14	15	16
17	18	19	20	21	22	23
24	25	26	27	28	29	30

Est attendance: 50,000. For info: Lebanon-Marion County Chamber of Commerce, Joan M. Osbourn, 107A W Main St, Lebanon, KY 40033. Phone: (502) 692-2661.

McGUFFEY, WILLIAM HOLMES: BIRTH ANNIVERSARY. Sept 23. American educator and author of the famous *McGuffey Readers*, born at Washington County, PA, Sept 23, 1800. Died at Charlottesville, VA, May 4, 1873.

★ **NATIONAL HUNTING AND FISHING DAY.** Sept 23. Presidential Proclamation 4682, Sept 11, 1979, covers all succeeding years. The fourth Saturday of September of each year.

NORTHEAST MONTANA THRESHING BEE AND ANTIQUE SHOW. Sept 23-24. Culbertson, MT. See how grandma and grandpa worked and lived. At the threshing grounds one mile southeast of town. Annually, the fourth full weekend in Sept. Est attendance: 2,000. For info: Rodney Iverson or Clifford Johnson, Culbertson, MT 59218. Phone: (406) 787-5265.

OLD-TIME BARNEGAT BAY DECOY AND GUNNING SHOW. Sept 23-24. Pinelands High & Middle Schools and Tip Seaman County Park, Tuckerton, NJ. Gathering to celebrate the local waterfowling heritage. Emphasizes traditional skills such as decoy carving, working decoy rigs, sneakbox building, gunning, retrieving and goose calling contests. More than 400 vendors. Est attendance: 40,000. Sponsor: Michael T Mangum, Chief Naturalist, Wells Mills County Park, Box 905, Wells Mills Rd, Waretown, NJ 08758. Phone: (609) 971-3085.

OLD-TIME FIDDLERS' CONTEST. Sept 23-24. Payson Rodeo Grounds, Payson, AZ. Toe tappin' musical playoffs for the state championships. The winners here will go to the national finals. Annually, the last weekend in September. Est attendance: 3,000. For info: Town of Payson, 303 N Beeline Hwy, Payson, AZ 85541. Phone: (602) 474-5242.

PALATINE MUSEUM WOOL DAY. Sept 23. Palatine House Museum, Spring St, Schoharie, NY. Palatine House was built in 1743. This event is fun for the whole family. Est attendance: 200. For info: Schoharie Colonial Heritage Assn, Box 554, Schoharie, NY 12157. Phone: (518) 295-7505.

PANCAKE DAY. Sept 23. Centerville, IA. Free pancakes, entertainment and two-mile long parade. Est attendance: 30,000. Sponsor: Chamber of Commerce, 128 N 12th, Centerville, IA 52544. Phone: (515) 437-4102.

PAULUS, FRIEDRICH: BIRTH ANNIVERSARY. Sept 23. The German commander of the Sixth Army who led the advance on Stalingrad in 1942, Friedrich von Paulus was born Sept 23, 1890, at Breitenau, Germany. Paulus' troops succeeded in taking most of Stalingrad in November 1942, but eventually became trapped within the city they had captured. Paulus surrendered to the Russians Jan 31, 1943, the same day that Hitler promoted him to field marshal. He appeared as a key witness for the Soviet prosecution at the Nuremberg trials. Paulus died on Feb 1, 1957, at Dresden, West Germany.

PLANET NEPTUNE: DISCOVERY ANNIVERSARY. Sept 23. First observed on Sept 23, 1846. Neptune is 2,796,700,000 miles from the sun (about 30 times as far from the sun as Earth). Eighth planet from the sun, Neptune takes 164.8 years to revolve around the sun. Diameter is about 31,000 miles compared to Earth at 7,927 miles.

PROPOSAL DAY®. Sept 23. Proposal Day honors unmarried adults everywhere who are seeking marriage. Both men and women are encouraged to propose marriage to their true love on Proposal Day. List released today of the ten current most eligible singles in the world, according to American Singles. Annually, on the first day of autumn (Autumnal Equinox). For info: John Michael O'Loughlin, 1333 W Campbell #125, Richardson, TX 75080. Phone: (214) 721-9975.

RAINBOW OF ARTS CRAFT FESTIVAL. Sept 23-24. Rockwood Park, Chesterfield, VA. 18th annual fest includes crafts, children's Imagination Station and live entertainment. Est attendance: 12,000. For info: Laura Dalton, Chesterfield County Parks and Rec, PO Box 40, Chesterfield, VA 23832. Phone: (804) 748-1623.

RHYTHM ON THE RIVER. Sept 23-25. Riverfront Park, Little Rock, AR. A three day fun filled music festival with spectacular entertainment, international dance, food and costumes. Est attendance: 60,000. Sponsors: Arkansas Power & Light, First Commercial Bank and Budweiser. For info: Beth Miller, Exec Dir, Arkansas River Festival, Inc, PO Box 3232, Little Rock, AR 72203. Phone: (501) 370-3284.

SAUDI ARABIA: KINGDOM UNIFICATION. Sept 23. National holiday. Commemorates Sept 23, 1932.

TEXAS HERITAGE DAY. Sept 23. Old City Park, Dallas, TX. Learn about life on the prairie and become a pioneer for the day as you try your hand at the skills that built our country. Demonstrations of blacksmithing and cooking over an open hearth. Celebrate Texas culture as fiddlers and banjo players perform the music, song and dance of the 19th century. Est attendance: 1,500. For info: Old City Park, Dallas Co Heritage Soc, 1717 Gano St, Dallas, TX 75215. Phone: (214) 421-5141.

◆ **WOODHULL, VICTORIA CHAFLIN: BIRTH ANNIVERSARY.** Sept 23. American feminist, reformer and first female candidate for the presidency of the US. Born at Homer, OH, on Sept 23, 1838; died at Norton Park, Bremmons, Worcestershire, England, on June 10, 1927.

BIRTHDAYS TODAY

Ray Charles (Robinson), 65, singer ("Georgia on My Mind," "I Can't Stop Loving You," "What'd I Say," "America the Beautiful," "Drown in My Own Tears"; Kennedy Center Award), composer, musician, born Albany, GA, Sept 23, 1930.

Julio Iglesias, 52, singer ("To All the Girls I've Loved Before" with Willie Nelson), songwriter, born at Madrid, Spain, Sept 23, 1943.

Tony Joseph Mandarich, 29, football player, born Oakville, Ontario, Canada, Sept 23, 1966.

Les McCann, 60, musician, singer ("Compared to What"), born at Lexington, KY, Sept 23, 1935.

Larry Mize, 37, golfer, born at Augusta, GA, Sept 23, 1958.

Mary Kay Place, 48, writer, actress (Loretta Haggers on "Mary Hartman, Mary Hartman"), born at Tulsa, OK, Sept 23, 1947.

Mickey Rooney (Joe Yule, Jr), 75, actor (the Andy Hardy movies; *The Human Comedy, The Black Stallion*), born at Brooklyn, NY, Sept 23, 1920.

Bruce Springsteen, 46, singer (lead of E Street Band; the Boss), songwriter ("Born in the USA," "Reason to Believe," "Born to Run," "Dancing in the Dark," "Hungry Heart"), born Freehold, NJ, Sept 23, 1949.

SEPTEMBER 24 — SUNDAY
267th Day — Remaining, 98

ARRIVAL OF BULLWINKLE: FIRST BROADCAST ANNIVERSARY. Sept 24. On this day in 1961 "The Bullwinkle Show" featuring Bullwinkle the Moose and Rocky the Flying Squirrel first aired.

◆ **BABE RUTH'S FAREWELL TO THE YANKEES: ANNIVERSARY.** Sept 24. Babe Ruth played his last game with the New York Yankees at Yankee Stadium on this day in 1934.

FANEUIL HALL OPENED TO THE PUBLIC: ANNIVERSARY. Sept 24. On Sept 24, 1742, Faneuil Hall in Boston, MA, opened to the public. Designed by painter John Smibiert, it was enlarged in 1805 according to plans by Charles Bulfinch.

Sept 24-25 ☆ *Chase's 1995 Calendar of Events* ☆

FESTIFALL. Sept 24. Friendship Hill National Historic Site, Point Marion, PA. To celebrate the 19th-century arts, crafts and music of the Allegheny Plateau. Historic foods are available for purchase. Tours of the Albert Gallatin house also provided. Annually, the last Sunday in September. Est attendance: 2,000. For info: Friendship Hill Assn and Natl Park Service, RD 1 Box 149-A, Point Marion, PA 15474. Phone: (412) 725-9190.

FITZGERALD, F. SCOTT: BIRTH ANNIVERSARY. Sept 24. American short story writer and novelist; author of *This Side of Paradise, The Great Gatsby* and *Tender Is the Night*. Was born Francis Scott Key Fitzgerald, at St. Paul, MN, on Sept 24, 1896. He died at Hollywood, CA, Dec 21, 1940.

GEISEL, THEODOR "DR. SEUSS": DEATH ANNIVERSARY. Sept 24. Theodor Seuss Geisel, the creator of *The Cat in the Hat, If I Ran the Zoo, The Grinch Who Stole Christmas, Green Eggs and Ham, Fox in Sox, Horton Hears a Who* and dozens of other wonderful books for children, died at the age of 87 on Sept 24, 1991, at La Jolla, CA. See also: "Geisel, Theodor 'Dr. Seuss': Birth Anniversary" (Mar 2).

★ **GOLD STAR MOTHER'S DAY.** Sept 24. Presidential Proclamation always for last Sunday of each September since 1936. Proclamation 2424 of Sept 14, 1940, covers all succeeding years.

GUINEA-BISSAU: INDEPENDENCE DAY. Sept 24. National holiday is observed.

HENSON, JIM: BIRTH ANNIVERSARY. Sept 24. Puppeteer, born Sept 24, 1936, at Greensville, MS. Jim Henson created a unique brand of puppetry known as the Muppets. Kermit the Frog, Big Bird, Bert and Ernie, Gonzo, Animal, Miss Piggy and Oscar the Grouch are a few of the puppets that captured the hearts of children and adults alike in television and film in productions including "Sesame Street," "The Jimmy Dean Show," "The Muppet Show," *The Muppet Movie, The Muppets Take Manhattan, The Great Muppet Caper* and *The Dark Crystal*. Henson began his career in 1954 as producer of the TV show "Sam and Friends" in Washington, DC. He gave birth to the Muppets in 1956. His creativity was rewarded with eighteen Emmy Awards, seven Grammy awards, four Peabody awards and five ACE awards from the National Cable Television Association. Henson died unexpectedly at the age of 53 on May 16, 1990, at New York, NY.

MARSHALL, JOHN: BIRTH ANNIVERSARY. Sept 24. Fourth Chief Justice of Supreme Court, born at Germantown, VA, on Sept 24, 1755. Served in House of Representatives and as secretary of state under John Adams. Appointed by President Adams to the position of chief justice in January 1801 and became known as "The Great Chief Justice." Marshall's court was largely responsible for defining the role of the Supreme Court and basic organizing principles of government in the early years after adoption of the Constitution, in such cases as *Marbury v. Madison, McCulloch v. Maryland, Cohens v. Virginia* and *Gibbons v. Ogden*. He died at Philadelphia, PA, on July 6, 1835.

MILLIGAN MINI-POLKA DAY. Sept 24. Milligan Auditorium, Milligan, NE. Four bands play polka music from 11–11. Roast pork, dumpling, kraut dinner served all day. Beer, kolaches and dancing. Est attendance: 500. For info: Scott T. Oliva, Corp Board Member, PO Box 102, Milligan, NE 68406. Phone: (402) 629-4446.

MOON PHASE: NEW MOON. Sept 24. Moon enters New Moon phase at 12:55 PM, EDT.

NATIONAL BLUEBIRD OF HAPPINESS DAY. Sept 24. Beauregard Jones, Las Vegas's Bluebird of Happiness, and Las Vegas resident, encourages all Bluebirds to sing out today about the joys of life! Sponsor: Tom Danaher, PO Box 1778, Las Vegas, NV 89125.

NATIONAL DOG WEEK. Sept 24–30. To promote the relationship of dogs to mankind and emphasize the need for the proper care and treatment of dogs. Annually, the last full week in September. Sponsor: Dogs on Stamps Study Unit (DOSSU), c/o Morris Raskin, Secy, 3208 Hana Rd, Edison, NJ 08817. Phone: (908) 248-1865.

NATIONAL GOOD NEIGHBOR DAY. Sept 24. To build a nation and world that cares by increasing appreciation and understanding of our fellow man, beginning next door. Annually, the fourth Sunday in September. Sponsor: Good Neighbor Day Fdtn, Rebecca E. Mattson, Box 379, Lakeside, MT 59922. Phone: (406) 844-3000.

RELIGIOUS FREEDOM WEEK. Sept 24–Oct 1. As proclaimed by Congressional Resolution and Presidential Proclamation the week of Sept 25th each year is celebrated as Religious Freedom Week. This date commemorates the anniversary of the Bill of Rights and the right to believe and practice the religion of one's own choice as laid out in the First Amendment. Annually, to include anniversary date. For info: Rev Susan Taylor, Religious Freedom Week Committee, 400 C St NE, Washington, DC 20002. Phone: (202) 543-6404.

ROSH HASHANAH BEGINS AT SUNDOWN. Sept 24. Jewish New Year. See "Rosh Hashanah" (Sept 25).

SCHWENKFELDER THANKSGIVING. Sept 24. On this day in 1734 members of the Schwenkfelder Society gave thanks for their deliverance from Old World persecution as they prepared to take up new lives in the Pennsylvania-Dutch counties of Pennsylvania. Still celebrated.

TRINIDAD: REPUBLIC DAY. Sept 24. Port of Spain. Public holiday with ceremonial and other observances islandwide. For info: Trinidad and Tobago TDA, 25 W 43rd St, Ste 1508, New York, NY 10036. Phone: (800) 232-0082.

BIRTHDAYS TODAY

Arne H. Carlson, 61, Governor of Minnesota (R), born at New York City, Sept 24, 1934.
Sheila MacRae, 72, singer, actress, born at London, England, Sept 24, 1923.
Linda McCartney, 53, singer ("Admiral Halsey" with Paul McCartney), photographer, born New York, NY, Sept 24, 1942.
Jim McKay, 74, sportscaster, born at Philadelphia, PA, Sept 24, 1921.
Anthony Newley, 64, actor, singer (*Oliver Twist, No Time to Die*), born at London, England, Sept 24, 1931.
Rafael Corrales Palmeiro, 31, baseball player, born at Havana, Cuba, Sept 24, 1964.

SEPTEMBER 25 — MONDAY

268th Day — Remaining, 97

FIRST AMENDMENTS TO THE US CONSTITUTION SUBMITTED TO STATES: ANNIVERSARY. Sept 25. Twelve amendments to the Constitution (in response to requests of immediate changes by five of the state ratifying conventions) were submitted to the states for consideration on Sept 25, 1789.

September 1995	S	M	T	W	T	F	S
						1	2
	3	4	5	6	7	8	9
	10	11	12	13	14	15	16
	17	18	19	20	21	22	23
	24	25	26	27	28	29	30

Chase's 1995 Calendar of Events — Sept 25-26

FIRST AMERICAN NEWSPAPER: PUBLICATION ANNIVERSARY. Sept 25. The first (and only) edition of *Publick Occurrences Both Foreign and Domestick* was published on Sept 25, 1690, by Benjamin Harris, at the London-Coffee-House, Boston, MA. Authorities considered this first newspaper published in the US offensive and ordered immediate suppression.

MAJOR LEAGUE BASEBALL'S FIRST DOUBLE HEADER. Sept 25. On Sept 25, 1882, the first major league baseball double header was played between the Providence and Worcester teams.

NATIONAL FOOD SERVICE EMPLOYEES WEEK. Sept 25-29. To promote public awareness of the contributions and importance of food service employees to life in America. Annually, the fourth week in September. Sponsor: Women and Infants Hospital of Rhode Island, Dietary Dept, 101 Dudley St, Providence, RI 02908.

NATIONAL MIND MAPPING FOR PROBLEM SOLVING WEEK. Sept 25-30. An ideal time to introduce this brainstorming process to employees at any business or organization. Mind mapping is a creative, idea-generating technique which utilizes free writing or note taking. Make use of a lunch period or plan a brainstorming break to solve problems using this fun technique. For info: Tim Richardson, Pres, Total Development Resources, Inc, 363-6 Atlantic Blvd, Ste 201, Atlantic Beach, FL 32233-5251. Phone: (800) 226-4473.

NATIONAL ONE-HIT WONDER DAY. Sept 25. Honors the one-hit wonders of rock-n-roll. Anyone who ever had a hit single deserves eternal remembrance. For info: *One Shot* Magazine, Steven Rosen, Editor and Publisher, 11667 Elkhead Range Rd, Littleton, CO 80127. Phone: (303) 979-2318.

PACIFIC OCEAN DISCOVERED: ANNIVERSARY. Sept 25. On Sept 25, 1513, Vasco Nuñez de Balboa, a Spanish conquistador, stood high atop a peak in the Darien, becoming the first European to look upon the Pacific Ocean, claiming it as the South Sea in the name of the King of Spain.

RAMEAU, JEAN PHILLIPPE: BIRTH ANNIVERSARY. Sept 25. Birthday of French composer Jean Phillippe Rameau. Baptised at Dijon, France, Sept 25, 1683. Called by some the greatest French composer and musical theorist of the 18th century, Rameau died at Paris, France, Sept 12, 1764.

ROSH HASHANAH OR JEWISH NEW YEAR. Sept 25. Jewish holy day; observed on following day also. Hebrew calendar date: Tishri 1, 5756. Rosh Hashanah is beginning of ten days of repentance and spiritual renewal. (Began at sundown of previous day.)

SHOSTAKOVICH, DMITRI: BIRTH ANNIVERSARY. Sept 25. Russian composer born at St. Petersburg (now Leningrad), USSR, Sept 25, 1906. Died at Moscow, USSR, Aug 9, 1975.

SMITH, WALTER WESLEY "RED": 90th BIRTH ANNIVERSARY. Sept 25. Pulitzer Prize-winning sports columnist and newspaperman for 54 years, Walter Wesley (Red) Smith was born at Green Bay, WI, on Sept 25, 1905. Called the "nation's most respected sportswriter," Smith's columns appeared in some 500 newspapers. He died at Stamford, CT, Jan 15, 1982.

WALES: SWANSEA FESTIVAL OF MUSIC AND THE ARTS. Sept 25-Nov 5. Swansea, West Glamorgan. Major professional arts festival in Wales, including concerts, jazz opera, dance and art exhibitions. For info: Swansea Festival Office, The Guildhall, Swansea, West Glamorgan, Wales SA1 4PA.

BIRTHDAYS TODAY

Michael Douglas, 51, actor ("The Streets of San Francisco," *Fatal Attraction*), director, born New York, NY, Sept 25, 1944.
Mark Hamill, 44, actor ("General Hospital"; Luke in *Star Wars*), born Oakland, CA, Sept 25, 1951.
Heather Locklear, 34, actress ("T.J. Hooker," "Dynasty," "Melrose Place"), born at Los Angeles, CA, Sept 25, 1961.
Scottie Pippen, 30, basketball player, born at Hamburg, AR, Sept 25, 1965.
Juliet Prowse, 59, dancer, actress, born at Bombay, India, Sept 25, 1936.
Christopher Reeve, 43, actor (*Superman*), born at New York, NY, Sept 25, 1952.
Robert Walden, 52, actor ("Lou Grant," *All the King's Men*), born New York, NY, Sept 25, 1943.
Barbara Walters, 64, journalist, interviewer, former "Today" host, born at Boston, MA, Sept 25, 1931.

SEPTEMBER 26 — TUESDAY
269th Day — Remaining, 96

APPLESEED, JOHNNY: BIRTH ANNIVERSARY. Sept 26. John Chapman, better known as Johnny Appleseed, believed to have been born at Leominster, MA, on Sept 26, 1774. Died at Allen County, IN, Mar 11, 1847. Planter of orchards and friend of wild animals, he was regarded as a great medicine man by the Indians.

BARTOK, BELA: 50th DEATH ANNIVERSARY. Sept 26, 1945. Hungarian composer, born Mar 25, 1881, at Nagyszentmiklos (now in Romania), died at New York City.

BEATLES LAST ALBUM RELEASED: ANNIVERSARY. Sept 26. The Beatles' 13th album, *Abbey Road*, was released in the United Kingdom on Sept 26, 1969. The album zoomed to number one on the record charts and stayed there for 11 weeks. It was the last album The Beatles made together.

COMAL COUNTY FAIR. Sept 26-Oct 1. Comal County Fairgrounds, New Braunfels, TX. Local competition of livestock, arts and crafts, antiques, horseshoe pitching, agricultural products, poultry, handwork, baked goods and plants. 102nd annual fair. Est attendance: 148,000. For info: Jan Jochec, 457 Landa St, New Braunfels, TX 78130. Phone: (210) 625-1505.

ELIOT, T.S.: BIRTH ANNIVERSARY. Sept 26. Thomas Stearns Eliot, Nobel Prize winner, poet, playwright and critic, was born at St. Louis, MO, on Sept 26, 1888. "There never was a time," he believed, "when those that read at all, read so many more books by living authors than books by dead authors; there never was a time so completely parochial, so shut off from the past." Eliot died at London, England, on Jan 4, 1965.

ENGLAND: HORSE OF THE YEAR SHOW. Sept 26-Oct 1. Wembley Arena, Wembley, London. Some of the world's top showjumpers compete on tough courses in this annual indoor event. For info: Show Secretary, British Showjumping Assn, British Equestrian Centre, Stoneleigh, Coventry, England CV8 2LR.

GERSHWIN, GEORGE: BIRTH ANNIVERSARY. Sept 26. American composer remembered for his many enduring songs and melodies, including: "The Man I Love," "Strike Up the Band," "Funny Face," "I Got Rhythm" and the opera *Porgy and Bess*. Many of his works were in collaboration with his brother, Ira. Born at Brooklyn, NY, on Sept 26, 1898, he died of a brain tumor, at Beverly Hills, CA, July 11, 1937. See also: "Gershwin, Ira: Birth Anniversary" (Dec 6).

POPE PAUL VI: BIRTH ANNIVERSARY. Sept 26. Giovanni Battista Montini, 262nd pope of the Roman Catholic Church, born at Concesio, Italy, on Sept 26, 1897. Elected pope June 21, 1963. Died at Castel Gandolfo, near Rome, Italy, Aug 6, 1978.

Sept 26–28 ☆ *Chase's 1995 Calendar of Events* ☆

SHAMU'S BIRTHDAY. Sept 26. Shamu was born at Sea World in Orlando, FL, on Sept 26, 1985, and is the first killer whale born in captivity to survive. Shamu is now living at Sea World's Texas park.

US SENIOR AMATEUR (GOLF) CHAMPIONSHIP. Sept 26–Oct 1. Prairie Dunes Country Club, Hutchinson, KS. Sponsor: US Golf Assn, Championship Dept, Golf House, Far Hills, NJ 07931. Phone: (908) 234-2300.

WCVL/WIMC SENIOR CITIZENS HEALTH FAIR & EXPO. Sept 26. Montgomery County Fairgrounds, Crawfordsville, IN. An expo and trade show highlighting products and services of interest to seniors. Nearly 50 display booths, free health screenings for glaucoma, blood pressure, cholesterol, etc. Food service, travel logs. Annually, the last Tuesday in September. Est attendance: 2,000. For info: Mr. Dick Munro, Gen Mgr, WCVL/WIMC Radio, PO Box 603, Crawfordsville, IN 47933. Phone: (317) 362-8200.

YEMEN ARAB REPUBLIC: NATIONAL HOLIDAY. Sept 26. Commemorates proclamation of the republic on Sept 26, 1962.

BIRTHDAYS TODAY

Lynn Anderson, 48, singer ("Rose Garden"), born at Grand Forks, ND, Sept 26, 1947.
Melissa Sue Anderson, 33, actress (Mary Ingalls Kendall on "Little House on the Prairie"), born Berkeley, CA, Sept 26, 1962.
Bryan Ferry, 50, singer (lead with Roxy Music; "Heart on My Sleeve"), songwriter, born Durham, England, Sept 26, 1945.
Julie London, 69, singer ("Cry Me a River"), actress (Dixie on "Emergency"), born Santa Rosa, CA, Sept 26, 1926.
Olivia Newton-John, 47, actress (*Grease*), singer ("Physical," "Heart Attack"), born Cambridge, England, Sept 26, 1948.
Christine T. Whitman, 49, Governor of New Jersey (R), raised at Hunterdon County, NJ, born Sept 26, 1946.

SEPTEMBER 27 — WEDNESDAY
270th Day — Remaining, 95

ANCESTOR APPRECIATION DAY. Sept 27. A day to learn about and appreciate one's forebears. Sponsor: A.A.D. Assn, Box 3, Montague, MI 49437.

CONRAD, WILLIAM: 75th BIRTH ANNIVERSARY. Sept 27, 1920. Actor, best known for his roles in the TV series "Cannon" and "Jake and the Fat Man." Born at Louisville, KY and died at North Hollywood, CA, Feb 11, 1994.

CRUIKSHANK, GEORGE: BIRTH ANNIVERSARY. Sept 27. English illustrator, especially known for caricatures and for illustration of Charles Dickens's books. Born Sept 27, 1792, and died Feb 1, 1878.

FAST OF GEDALYA. Sept 27. Jewish holiday. Hebrew calendar date: Tishri 3, 5756. Tzom Gedalya begins at first light of day and commemorates the sixth century BC assassination of Gedalya Ben Achikam.

FOOD SERVICE EMPLOYEES DAY. Sept 27. To promote public awareness of the contributions and importance of food service employees to life in America. Annually, the Wednesday of Food Service Employees Week. Sponsor: Women and Infants Hospital of Rhode Island, Dietary Dept, 101 Dudley St, Providence, RI 02908.

NAST, THOMAS: BIRTH ANNIVERSARY. Sept 27. American political cartoonist born Sept 27, 1840. Died Dec 7, 1902.

ST. VINCENT DE PAUL: FEAST DAY. Sept 27. French priest, patron of charitable organizations, and co-founder of the Sisters of Charity. Canonized 1737 (lived 1581?–1660).

SPACE MILESTONE: *SOYUZ 12* (USSR). Sept 27. Two Soviet cosmonauts (V.G. Lazarev and O.G. Makarov) made two-day flight. Launched Sept 27, 1973.

WARREN COMMISSION REPORT: ANNIVERSARY. Sept 27. On this day in 1964, the Warren Commission issued a report stating that Lee Harvey Oswald acted alone in the assassination of President John F. Kennedy on Nov 23, 1963. Congress reopened the investigation and in 1979 the House Select Committee on Assassinations issued a report stating a conspiracy was most likely involved. See also: "Committee on Assassinations Report: Anniversary" (Mar 29).

BIRTHDAYS TODAY

Wilford Brimley, 61, actor (*Cocoon*, "Our House"), born Salt Lake City, UT, Sept 27, 1934.
Shaun Cassidy, 36, singer ("Da Doo Ron Ron," "That's Rock 'n' Roll"), actor ("The Hardy Boys," "General Hospital"), born at Los Angeles, CA, Sept 27, 1959.
Stephen Douglas Kerr, 30, basketball player, born at Beirut, Lebanon, Sept 27, 1965.
Meat Loaf (Marvin Lee Aday), 48, singer, musician (Eddie in *The Rocky Horror Picture Show*), born Dallas, TX, Sept 27, 1947.
Greg Morris, 61, actor ("Mission Impossible," "Vega$"), born at Cleveland, OH, Sept 27, 1934.
Arthur Heller Penn, 73, filmmaker (*Bonnie and Clyde, Alice's Restaurant, The Miracle Worker*), born at Philadelphia, PA, Sept 27, 1922.
Mike Schmidt, 46, former baseball player, born at Dayton, OH, Sept 27, 1949.
Sada Thompson, 66, actress (*Twigs*; Kate Lawrence on "Family"), born Des Moines, IA, Sept 27, 1929.

SEPTEMBER 28 — THURSDAY
271st Day — Remaining, 94

BADOGLIO, PIETRO: BIRTH ANNIVERSARY. Sept 28. A principal force behind the downfall of Benito Mussolini in 1943, Field Marshall Pietro Badoglio was born Sept 28, 1871, in Grazzano Monferrato, Italy. Badoglio was a veteran of the Italian colonial wars and made a reputation during World War I fighting the Austrians on Isonzo. Appointed Chief of Staff when Italy entered WWII, he resigned this post after the failure of the 1940 invasion of Greece. Having long opposed Mussolini, he became the first prime minister of an anti-fascist government when the dictator was deposed and signed the 1943 Italian armistice with the Allies. Badoglio died Oct 31, 1956, in Grazzano Monferrato.

CABRILLO DAY: ANNIVERSARY OF DISCOVERY OF CALIFORNIA. Sept 28. California. Commemorates discovery of California on Sept 28, 1542, by Portuguese navigator Juan Rodriguez Cabrillo who reached San Diego Bay on that date. Cabrillo died at San Miguel Island, CA, Jan 3, 1543. His birth date is unknown. The Cabrillo National Monument marks his landfall and Cabrillo Day is still observed in California (in some areas on the Saturday nearest Sept 28).

September 1995

S	M	T	W	T	F	S
					1	2
3	4	5	6	7	8	9
10	11	12	13	14	15	16
17	18	19	20	21	22	23
24	25	26	27	28	29	30

CAPP, AL: BIRTH ANNIVERSARY. Sept 28. American satirical cartoonist Al Capp (born Alfred Gerald Caplin), creator of "Li'l Abner" and originator of Sadie Hawkins Day, was born at New Haven, CT, Sept 28, 1909. He died at Cambridge, MA, Nov 5, 1979.

CLAN DONALD USA ANNUAL GENERAL MEETING. Sept 28–Oct 1. William Penn Hotel, Pittsburgh, PA. The annual gathering of members of Clan Donald, USA. Also, the annual meeting of the Clan Donald Foundation. Est attendance: 425. For info: Beth Gay, Natl PR Dir, Box 1110, Moultrie, GA 31768. Phone: (912) 985-6540.

ENGLAND: BOUCHERCON 26—THE WORLD MYSTERY CONVENTION. Sept 28–Oct 1. Nottingham, Nottinghamshire. Largest mystery event in the world attracting up to two thousand enthusiasts. For info: Sarah Boiling, Broadway Media Centre, 14 Broad St, Nottingham, England NG1 3AL.

ENGLAND: SOHO JAZZ FESTIVAL. Sept 28–Oct 7. Various venues, Soho, London. Various events in Soho's clubs, pubs, theatres and churches. For info: Mr Robert Guterman, 29 Romilly St, Soho, London, England W1V 6HP.

FIRST NIGHT FOOTBALL GAME: ANNIVERSARY. Sept 28. Mansfield, PA. Marks the first night football game in America played Sept 28, 1892, between Mansfield State Normal School (now Mansfield University) and Wyoming Seminary. Est attendance: 10,000. For info: Steve McCloskey, Sports Info Dir, Mansfield University, Mansfield, PA 16933. Phone: (717) 662-4845.

PRESTON COUNTY BUCKWHEAT FESTIVAL. Sept 28–Oct 1. Kingwood, WV. Celebrating the fall harvest, with coronations, parades, exhibits, music and Buckwheat Cake Dinners. 53rd annual festival. Annually, beginning the last Thursday in September. Est attendance: 100,000. Sponsor: Kingwood Volunteer Fire Dept, Lucille H. Crogan, Festival Secy, PO Box 74, Kingwood, WV 26537. Phone: (304) 329-0021.

STATE FAIR OF TEXAS. Sept 28–Oct 22. Fair Park, Dallas, TX. Exposition features a Broadway musical, college football games, fireworks shows, concerts, livestock events and traditional fair events and entertainment including exhibits, creative arts and parades. Est attendance: 3,200,000. For info: State Fair of Texas, PO Box 150009, Dallas, TX 75315. Phone: (214) 421-8716.

TAIWAN: CONFUCIUS'S BIRTHDAY AND TEACHERS' DAY. Sept 28. National holiday, designated as Teachers' Day. Confucius is the Latinized name of Kung-futzu, born in Shantung province on the 27th day of the tenth moon (lunar calendar) in the 22nd year of Kuke Hsiang of Lu (551 BC). He died at age 72, having spent some 40 years as a teacher. Teachers' Day is observed annually on Sept 28.

VAN BUREN COUNTY LITERACY DAY. Sept 28. Storytellers visit Elementary Centers. Children from three local school districts will make and distribute bookmarks throughout the county. Annually, Sept 28. Est attendance: 200. For info: Tomorrow's Literate Children (TLC), A Van Buren Co Reading Project, Rte 2, Box 95A, Keosauqua, IA 52565.

WIGGIN, KATE DOUGLAS: BIRTH ANNIVERSARY. Sept 28. Kate Wiggin was born Kate Douglas Smith Sept 28, 1856, at Philadelphia, PA. She helped organize the first free kindergarten on the west coast in 1878 in San Francisco, and in 1880 she and her sister established the California Kindergarten Training School. After moving back to the east coast she devoted herself to writing, producing a number of children's books including *The Birds' Christmas Carol, Timothy's Quest, Polly Oliver's Problem* and *Rebecca of Sunnybrook Farm*. She died at Harrow, England, on Aug 24, 1923.

WILLARD, FRANCES ELIZABETH CAROLINE: BIRTH ANNIVERSARY. Sept 28. American educator and reformer, president of the Women's Christian Temperance Union, 1879–1898, and women's suffrage leader, born at Churchville, NY, Sept 28, 1839. Died at New York, NY, Feb 18, 1898.

BIRTHDAYS TODAY

Brigitte Bardot (Camille Javal), 61, actress (*And God Created Woman, Viva Maria*; animal rights activist), born Paris, France, Sept 28, 1934.
Jerry Clower, 69, comedian ("Nashville on the Road"), born at Liberty, MS, Sept 28, 1926.
Johnny Earl Dawkins, Jr, 32, basketball player, born Washington, DC, Sept 28, 1963.
Ben E. King, 57, singer, musician (wrote lyrics to "There Goes My Baby"), born at Henderson, NC, Sept 28, 1938.
Madeleine M. Kunin, 62, former Governor of Vermont (D), born at Zurich, Switzerland, Sept 28, 1933.
Steve Largent, 41, former football player, born at Tulsa, OK, Sept 28, 1954.
Marcello Mastroianni, 71, actor (*White Nights; Yesterday, Today and Tomorrow*), born Fontana Liri, Italy, Sept 28, 1924.
William Windom, 72, actor (Emmy for "My World and Welcome to It"; "Murder She Wrote"), born New York, NY, Sept 28, 1923.

SEPTEMBER 29 — FRIDAY
272nd Day — Remaining, 93

AMERICAN INDIAN CEREMONIAL DANCING. Sept 29–30. Taos, NM. Sundown dance is performed at dusk each Sept 29, followed the next day (San Geronimo's Day) with foot races, high pole climb, clowning, feasting and dancing.

CANADA: OKANAGAN WINE FESTIVAL. Sept 29–Oct 8. Penticton, Kelowna, Vernon, Oliver. A 10-day annual celebration surrounding the annual wine grape harvest: winemasters dinners, wine tastings, seminars, picnics, sports and family fun events throughout scenic Okanagan Valley. Est attendance: 10,000. For info: Okanagan Wine Fest Soc, 185 Lakeshore Dr, Penticton, BC, Canada V2A 1B7. Phone: (604) 490-8866.

ENGLAND: SCOTLAND YARD: FIRST APPEARANCE ANNIVERSARY. Sept 29. The first public appearance of Greater London's Metropolitan Police occurred on Sept 29, 1829, amid jeering and abuse from disapproving political opponents. Public sentiment turned to confidence and respect in the ensuing years. The Metropolitan Police had been established by act of Parliament in June 1829, at the request of Home Secretary Sir Robert Peel, after whom the London police officers became more affectionately known as "bobbies." Scotland Yard, the site of their first headquarters near Charing Cross, soon became the official name of the force.

MICHAELMAS. Sept 29. The feast of St. Michael and All Angels in the Greek and Roman Catholic Churches.

NELSON, HORATIO: BIRTH ANNIVERSARY. Sept 29. English naval hero of the Battle of Trafalgar born Sept 29, 1758. Died Oct 21, 1805.

OHIO SWISS FESTIVAL. Sept 29–30. Sugarcreek, OH. Swiss music, yodeling, games, costumes, parades. Continuous entertainment and tons of Swiss cheese. Annually, the fourth Friday and Saturday following Labor Day. For info: Ohio Swiss Festival, Box 158, Sugarcreek, OH 44681. Phone: (216) 852-4113.

Sept 29-30 ☆ *Chase's 1995 Calendar of Events* ☆

OKTOBERFEST. Sept 29–Oct 1. Amana Colonies, Amana, IA. Traditional German festival with ethnic food and drink, craft displays and entertainment. Est attendance: 20,000. For info: Amana Colonies Conv & Visitors Bureau, PO Box 303, Amana, IA 52203. Phone: (800) 245-5465.

OKTOBERFEST. Sept 29. LaCrosse, WI. German fall festival featuring family events, live entertainment, lots of food, Torchlight Parade on Thursday night and the Big Maple Parade. Fun for the whole family. For info: LaCrosse Area Conv & Visitors Bureau, 410 E Veterans Memorial Dr, LaCrosse, WI 54601. Phone: (608) 784-3378.

SPACE MILESTONE: *DISCOVERY* (US). Sept 29. Space Shuttle *Discovery*, after numerous reschedulings, launched from Kennedy Space Center, FL, on Sept 29, 1988, with a five-member crew on board, and landed on Oct 3, 1988, at Edwards AFB, CA. It marked the first American manned flight since the Challenger tragedy in 1986. See also: "Challenger, Space Shuttle Explosion Anniversary" (Jan 28).

SPACE MILESTONE: *SALYUT 6* (USSR). Sept 29. Soviet space station launched on Sept 29, 1977. Burned up when it re-entered Earth's atmosphere after nearly five years, July 29, 1982.

SUGARLOAF'S FALL SOMERSET CRAFTS FESTIVAL. Sept 29–Oct 1. Garden State Exhibit Center, Somerset, NJ. Nearly 300 professional artists and craftspeople, demonstrations and children's entertainment. Annually, the last Friday–Sunday in September. For info: Deann Verdier, Dir, Sugarloaf Mountain Works, Inc, 200 Orchard Ridge Dr, #215, Gaithersburg, MD 20878. Phone: (301) 990-1400.

TYLENOL DEATHS: ANNIVERSARY. Sept 29. On Sept 29, 1982, the first of seven deaths, including that of a 10-year-old child, occurred as a result of the individuals unknowingly taking Tylenol capsules that had been deliberately contaminated with cyanide. After a California man was poisoned taking Tylenol laced with strychnine, Johnson and Johnson, the manufacturer of the product, recalled all capsules of the pain-reliever, some 264,000 bottles. Many lawsuits resulted. The killer has never been identified.

VIRGINIA-CAROLINA CRAFTSMEN'S FALL CLASSIC. Sept 29–Oct 1. Roanoke Civic Center, Roanoke, VA. Arts and crafts. Est attendance: 20,000. For info: The Virginia-Carolina Craftsmen, 1240 Oakland Ave, Greensboro, NC 27403. Phone: (910) 274-5550.

WINDY CITY CLASSIC CONVENTION. Sept 29–31. Hyatt Woodfield, Schaumburg, IL. Second annual convention held by Chicago's Windy City Jitterbug Club to promote and exchange steps and styling among dancers throughout the country. Hotline: (800) 562-7919. For info: CWCJC, PO Box 713, Franklin Park, IL 60131. Phone: (312) 725-7230.

BIRTHDAYS TODAY

Michelangelo Antonioni, 83, filmmaker (*Blowup, Zabriskie Point*), born Ferrara, Italy, Sept 29, 1912.
Gene Autry, 88, actor, singer ("Gene Autry Show," "Rudolf, the Red-Nosed Reindeer"), born Tioga, TX, Sept 29, 1907
Anita Ekberg, 64, actress (*La Dolce Vita*), born at Malmo, Sweden, Sept 29, 1931.
Bryant Gumbel, 47, TV host ("Today"), sportscaster, born at New Orleans, LA, Sept 29, 1948.
Hersey R. Hawkins Jr, 30, basketball player, born at Chicago, IL, Sept 29, 1965.

Madeline Kahn, 53, actress (Tony for *The Sisters Rosensweig*; *Blazing Saddles*), born Boston, MA, Sept 29, 1942.
Jerry Lee Lewis, 60, singer, musician ("Whole Lot of Shakin' Goin On," "Great Balls of Fire"), born Ferriday, LA, Sept 29, 1935.
Larry Linville, 56, actor (Frank on "M*A*S*H"; "Grandpa Goes to Washington"), born Ojai, CA, Sept 29, 1939.
John MacBeth Paxson, 35, former basketball player, born at Dayton, OH, Sept 29, 1960.

SEPTEMBER 30 — SATURDAY
273rd Day — Remaining, 92

◆ **BABE SETS HOME RUN RECORD: ANNIVERSARY.** Sept 30, 1927. George Herman "Babe" Ruth hit his 60th home run of the season off Tom Zachary, of the Washington Senators. Ruth's record for the most homers in a single season stood for 24 years—until Roger Maris hit 61 in 1961.

◆ **BABE'S LAST GAME AS YANKEE: ANNIVERSARY.** Sept 30, 1934. On Sept 30, 1934, Babe Ruth played his last game for New York Yankees. While in St. Louis to watch the fifth game of the World Series (between the St. Louis Cardinals and Detroit Tigers), and angry that he was not to be named Yankees manager, Ruth told Joe Williams, sports editor of the Scripp-Howard newspapers, that after 15 seasons he would no longer be playing for the Yankees.

BEEF-A-RAMA/ARTS AND CRAFTS SHOW. Sept 30. Minocqua, WI. Merchants roast over 1,200 pounds of beef and sell sandwiches in the park. Snacks, bargains, arts, crafts, live music. Est attendance: 8,500. For info: Greater Minocqua-Arbor Vitae-Woodcruff Area Chmbr of Comm, PO Box 1006, Minocqua, WI 54548. Phone: (800) 446-6784.

BOTSWANA: INDEPENDENCE DAY ANNIVERSARY. Sept 30. National holiday. The former Bechuanaland Protectorate (British Colony) became the independent Republic of Botswana on Sept 30, 1966.

BUFFALO ROUND-UP BICYCLE RALLY. Sept 30–Oct 1. Custer State Park, Custer, SD. Bicycle races, hikebike. For info: Custer Co Chamber of Commerce, 447 Crook St, Custer, SD 57730. Phone: (605) 673-2244.

BURLINGTON OLD-FASHIONED APPLE HARVEST FESTIVAL. Sept 30–Oct 1. Burlington, WV. To celebrate the apple harvest in the beautiful Potomac Highlands of West Virginia. Est attendance: 35,000. Sponsor: Auxiliary of Burlington United Methodist Family Services, PO Box 96, Burlington, WV 26710-0096. Phone: (304) 289-3511.

CAPOTE, TRUMAN: BIRTH ANNIVERSARY. Sept 30. American novelist and literary celebrity, was born Truman Streckfus Persons at New Orleans, LA, on Sept 30, 1924. He later took the name of his stepfather and became Truman Capote. Among his best remembered books: *Other Voices, Other Rooms; Breakfast at Tiffany's* and *In Cold Blood*. He was working on a new novel, *Answered Prayers*, at the time of his death at Los Angeles, CA, on Aug 25, 1984.

EVERYBODY'S DAY FESTIVAL. Sept 30. Thomasville, NC. A true hometown street festival for "everybody." Crafts, food vendors and live entertainment. Est attendance: 35,000. For info: Thomasville Area Chamber of Commerce, Box 727, Thomasville, NC 27360.

September 1995

S	M	T	W	T	F	S
					1	2
3	4	5	6	7	8	9
10	11	12	13	14	15	16
17	18	19	20	21	22	23
24	25	26	27	28	29	30

Chase's 1995 Calendar of Events — Sept 30

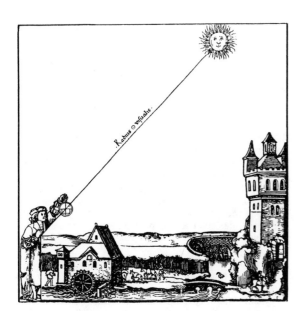

FEAST OF ST. JEROME. Sept 30. Patron saint of scholars and librarians.

FIRST ANNUAL FAIR IN AMERICA: ANNIVERSARY. Sept 30. According to the Laws and Ordinances of New Netherlands (now New York and New Jersey), on Sept 30, 1641, authorities declared that "henceforth there shall be held annually at Fort Amsterdam" a Cattle Fair (Oct 15) and a Hog Fair (Nov 1), and that "whosoever hath any things to sell or buy can regulate himself accordingly."

FIRST CRIMINAL EXECUTION IN AMERICAN COLONIES: ANNIVERSARY. Sept 30. John Billington, one of the first pilgrims to land in America, was hanged for murder on Sept 30, 1630, becoming the first criminal to be executed in the American Colonies.

HAITI MILITARY DEPOSES ARISTIDE: ANNIVERSARY. Sept 30, 1991. A military coup, led by Brigadier Gen Raoul Cedras toppled Haiti's first freely elected president, Jean-Bertrand Aristide. The deposed president was later deported and efforts by various countries and the United Nations, most notably the US, to return him to power included economic sanctions and the threat of military action.

LONG BEACH ISLAND CHOWDER COOK-OFF. Sept 30–Oct 1. Bayfront Park, Beach Haven, NJ. Weekend-long festival featuring unlimited tasting of up to 30 different red/white clam chowders prepared by area restaurants. Entertainment; other food/beverages available. Annually, the first weekend in October. Est attendance: 9,000. For info: Jeanne Di Paola, Exec Dir, Long Beach Island Chamber of Commerce, 265 W 9th St, Ship Bottom, NJ 08008. Phone: (800) 292-6372.

MEREDITH ENROLLS AT OLE MISS: ANNIVERSARY. Sept 30. Rioting broke out on Sept 30, 1962, when James Meredith became the first black to enroll in the all-white University of Mississippi. President Kennedy sent US troops to the area to force compliance with the law. Three people died in the fighting and 50 were injured. On June 6, 1966, Meredith was shot while participating in a civil rights march in Mississippi. On June 25 Meredith, barely recovered, rejoined the marchers near Jackson, MS.

MOUNTAIN STATE FOREST FESTIVAL. Sept 30–Oct 8. Elkins, WV. Promotes the natural resources of the area with emphasis on forests. Est attendance: 100,000. For info: Mountain State Forest Festival, Box 369, Elkins, WV 26241. Phone: (304) 636-1824.

NECKER, JACQUES: BIRTH ANNIVERSARY. Sept 30. French banker and statesman, born at Geneva, Switzerland, Sept 30, 1732. His dismissal from his post as head of France's Department of Finance was the immediate cause of the storming of the Bastille, July 14, 1789. Necker died near Geneva, Apr 9, 1804.

PIONEER POWER DAY THRESHING BEE. Sept 30–Oct 1. Lewistown, MT. An old-fashioned Threshing Bee. Old steam engines at work, threshing machines, all types of steam power machines and other old restored engines. Barbecue served. Est attendance: 500. For info: Central MT Flywheelers Assn, Bryan Sallee, 1015 W Washington, Lewistown, MT 59457.

SEPTEMBER FEST. Sept 30–Oct 1. DeSoto Caverns Park, Childersburg, AL. Regional artists and craftspeople's work. Music, clogging, good food and Civil War artillery campsites and maneuvers. Est attendance: 10,000. For info: DeSoto Caverns Park, Rebecca Grevas, DeSoto Caverns Pkwy, Childersburg, AL 35044. Phone: (205) 378-7252.

TRI-STATE BAND FESTIVAL. Sept 30. Luverne, MN. 45th annual festival with more than 2,500 high school students from Minnesota, South Dakota and Iowa; trophies awarded in four classes. Annually, the last Saturday in September. Est attendance: 10,000. For info: Norma De Jongh, Luverne Area Chamber of Commerce, 102 E Main, Luverne, MN 56156. Phone: (507) 283-4061.

VOLKSMARCH. Sept 30–Oct 1. Angel Fire, NM. This worldwide popular "walking" sport is enjoyed by people of all ages. Participate for medals and have your "passport" stamped while enjoying our incredible scenic beauty. Est attendance: 50. For info: Angel Fire Chamber of Commerce, PO Box 547, Angel Fire, NM 87710. Phone: (800) 446-8117.

WALKTOBERFEST. Sept 30–Oct 1. Each October hundreds of thousands of Americans participate in the American Diabetes Association's annual walk-a-thon to raise money to help find a cure for diabetes and to provide information and resources to improve the lives of all people affected by the disease. Walks are held in communities across America, combining fun and fitness with the chance to help people with diabetes. Sponsor: American Diabetes Assn, Natl HQ, 1660 Duke St, Alexandria, VA 22314. Phone: (703) 549-1500.

BIRTHDAYS TODAY

Deborah Allen, 42, singer ("Baby I Lied"), songwriter, born at Memphis, TN, Sept 30, 1953.

Angie Dickinson (Angeline Brown), 64, actress (Emmy for "Police Woman"; *Rio Bravo, Dressed to Kill*), born at Kulm, ND, Sept 30, 1931.

Deborah Kerr, 74, actress (*From Here to Eternity, The King and I*), born at Helensburgh, Scotland, Sept 30, 1921.

Lester Garfield Maddox, 80, former Govenor of Georgia, born at Atlanta, GA, Sept 30, 1915.

Johnny Mathis, 60, singer ("It's Not For Me to Say," "Chances Are"), born San Francisco, CA, Sept 30, 1935.

Marilyn McCoo, 52, singer (with The Fifth Dimension; "Up Up and Away," "Wedding Bell Blues," "One Less Bell to Answer"), actress, born at Jersey City, NJ, Sept 30, 1943.

James Ralph Sasser, 59, US Senator (D, Tennessee) up for reelection in Nov '94, born at Memphis, TN, Sept 30, 1936.

Oct 1 ☆ *Chase's 1995 Calendar of Events* ☆

October.

OCTOBER 1 — SUNDAY
274th Day — Remaining, 91

ADOPT-A-SHELTER-DOG MONTH. Oct 1-31. To promote the adoption of puppies and dogs from local shelters, the ASPCA sponsors this important observance. For more information, send SASE. Phone: (212) 876-7700, ext 4655. Sponsor: American Soc for the Prevention of Cruelty to Animals, (ASPCA), 424 E 92nd St, New York, NY 10128.

AMERICAN SAMOA: SWARM OF PALOLO OF THE PACIFIC. Oct 1-Nov 30. Pago Pago, American Samoa. October-November is the swarm of palolo (caviar) of the Pacific. This coral worm floats from the reef and locals turn out en masse for harvest. For info: Office of Tourism, Pago Pago, Amer Samoa 96799. Phone: (684) 633-1091.

APPLE FESTIVAL. Oct 1. Historic Ward-Meade Park, Topeka, KS. Celebration of harvest and heritage of Kansas located in 5½-acre historical park. Ethnic foods, live entertainment, special shows, turn-of-the century town square activities, more than 100 craft booths. Annually, the first Sunday in October. Est attendance: 10,000. For info: Anita Wolgast, Historic Ward-Meade Park, 124 NW Fillmore, Topeka, KS 66606. Phone: (913) 295-3888.

AUTO BATTERY SAFETY MONTH. Oct 1-31. Knowing how to properly jump-start a dead vehicle battery should be a learning requirement for students in drivers' education classes. Prevent Blindness America (formerly known as National Society to Prevent Blindness) will offer safety tips and instructions to new and beginning drivers. Sponsor: Prevent Blindness America, Marita Gomez, Media Relations, 500 E Remington Rd, Schaumburg, IL 60173. Phone: (800) 331-2020.

◆ **BABE CALLS HIS SHOT? ANNIVERSARY.** Oct 1, 1932. On Oct 1, 1932, in the fifth inning of Game 3 of the World Series, with a count of two balls and two strikes and with hostile Cubs fans shouting epithets at him, Babe Ruth pointed to the center-field bleachers in Chicago's Wrigley Field and followed up by hitting a soaring home run high above the very spot to which he had just gestured. With that homer Ruth squashed the Chicago Cubs' hopes of winning the game, and the Yankees went on to sweep the Series with four straight victories. For over half a century the question has remained: Did Ruth actually call his shot that day? Even eyewitnesses disagree. Joe Williams of *The New York Times* wrote, "In no mistaken motions, the Babe notified the crowd that the nature of his retaliation would be a wallop right out of the confines of the park."

But Cubs pitcher Charlie Root said, "Ruth did *not* point at the fence before he swung. If he'd made a gesture like that, I'd have put one in his ear and knocked him on his ass. Ruth's daughter has said that he denied it. But the Babe himself also claimed he had. Fact or folklore? Either way, legend!

◆ **BABE'S LAST PITCHING: ANNIVERSARY.** Oct 1, 1933. Babe Ruth had pitched only once in 12 years when he took the mound for the final game of the season. Ruth hurled all nine innings, hit a home run in the fifth and beat his original team, the Boston Red Sox, 6-5. It was a fitting finish to his 20th season in the majors: because this turned out to be his last pitching appearance, his 1-0 record in 1933 meant that Ruth never had a losing season as a pitcher.

BAYOU HURRICANE: ANNIVERSARY. Oct 1-2. Approximately 2,000 persons died Oct 1-2, 1893, when the Louisiana Bayou country was submerged in a storm that raged from the Gulf of Mexico. The unexpected arrival of this storm caught thousands of residents off guard. The coast was swept by a 10-12-ft tidal wave.

BIG "C"ATTUS DAY. Oct 1. Cattus Island County Park, Toms River, NJ. Environmental organizations fair with natural history programs throughout the day. Est attendance: 500. Sponsor: Shaun O'Rourke, Cooper Environmental Center, 1170 Cattus Island Blvd, Toms River, NJ 08753. Phone: (908) 270-6960.

BLESSING OF THE FISHING FLEET. Oct 1. Church of Saints Peter and Paul and Fisherman's Wharf, San Francisco, CA. Annually, the first Sunday in October.

BRAZIL: FESTIVAL OF PENHA. Oct 1-31. Rio de Janeiro. Pilgrimages, especially on Saturdays during October, to the Church of Our Lady of Penha, which is built on top of a rock, requiring a climb of 365 steps (representing the days of the year), or ride in car on inclined plane (for children, invalids and aged), for those troubled and sick who seek hope or cure.

BUFFALO WALLOW CHILI COOKOFF. Oct 1. Custer State Park, Custer, SD. Musical entertainment, hayrides, a scavenger hunt, and lots of chili cooking and tasting. Annually, the day before the CSP Buffalo Roundup (See Oct 2). Est attendance: 3,000. For info: Custer Co Chamber of Commerce, 447 Crook St, Custer, SD 57730. Phone: (605) 673-2244.

CAMPAIGN FOR HEALTHIER BABIES MONTH. Oct 1-31. A month-long concentrated effort to focus attention on the March of Dimes Birth Defects Foundation and its community health programs, public awareness messages, advocacy and fundraising efforts. This month is designated to showcase infant mortality prevention efforts. Sponsor: March of Dimes Birth Defects Fdtn, 1275 Mamaroneck Ave, White Plains, NY 10605. Phone: (914) 428-7100.

CARTER, JIMMY: BIRTHDAY. Oct 1. James Earl Carter, Jr (he had his name legally changed to Jimmy), 39th president of the US, was born at Plains, GA, Oct 1, 1924. His term of office: Jan 20, 1977–Jan 20, 1981. In a 1976 speech, Carter said: "The first step in providing economic equality for women is to ensure a stable economy in which every person who wants to work can work." Most remembered for the Camp David Accord, the peace treaty between Israel and Egypt.

CHRISTMAS SEAL® CAMPAIGN 1995. Oct 1-Dec 31. An American tradition dating back to 1907 when the first Christmas Seals® were made available in the US, the annual campaign is a major support of American Lung Association programs dedicated to fighting lung diseases such as asthma, emphysema, tuberculosis and lung cancer, as well as their causes. Sponsor: American Lung Assn, Communications Div, 1740 Broadway, New York, NY 10019-4374. Phone: (212) 315-6468.

COMPUTER LEARNING MONTH. Oct 1-31. A month-long focus of events and activities for learning new uses of computers and software, sharing ideas and helping others gain the benefits of computers and software. Numerous national contests are held to recognize students, educators and parents for their innovative ideas; computers and software are awarded to winning entries. Annually, the month of October. For info: Sally

October 1995

S	M	T	W	T	F	S
1	2	3	4	5	6	7
8	9	10	11	12	13	14
15	16	17	18	19	20	21
22	23	24	25	26	27	28
29	30	31				

390

Bowman Alden, Exec Dir, Computer Learning Fdtn, Dept CHS, PO Box 60005, Palo Alto, CA 94306. Phone: (415) 327-3347.

CONSUMER INFORMATION MONTH. Oct 1-31. To promote awareness of reliable information sources that will help consumers acquire quality products and services and/or resolve related problems. Related resources available. Sponsor: Consumers Index to Product Evaluations and Info Sources, Pierian Press, PO Box 1808, Ann Arbor, MI 48106.

CO-OP AWARENESS MONTH. Oct 1-31. Every year an estimated $5.1 billion in co-op advertising funds is left unspent by US retailers. Co-op Awareness Month was created to remind retailers of their co-op funds in time for them to research and spend these remaining funds by the end of the calendar year. Sponsor: Sales Development Services, Kelly M. Smith, Dir Client Services, 1335 Dublin Rd, Ste 200A, Columbus, OH 43235. Phone: (614) 481-3530.

CUTS AND CURLS FOR CHARITY MONTH. Oct 1-31. *Modern Salon Magazine* invities participation from salons nationwide to promote breast cancer awareness by donating a portion of sales to the American Cancer Society. Annually, the month of October to coincide with Breast Cancer Awareness Month. Est attendance: 5,000. For info: Kathleen Quinn, Dir, PR, TFA Communications, 9450 W Bryn Mawr, Ste 300, Rosemont, IL 60018. Phone: (708) 671-2900.

CYPRUS: INDEPENDENCE DAY. Oct 1. National holiday.

ENERGY AWARENESS MONTH. Oct 1-31. An annual observance to promote a greater understanding and awareness of energy sources, how they can be used wisely, and the importance of energy to the United States. Organizations are encouraged to sponsor or participate in local activities to promote energy awareness. Sponsor: Attn: Bonnie Winsett, Dept of Energy, PA-5, 1000 Independence Ave SW, Washington, DC 20585. Phone: (202) 586-6827.

ENERGY MANAGEMENT IS A FAMILY AFFAIR—IMPROVE YOUR HOME. Oct 1-Mar 31, 1996. Replace energy-consuming units with new efficient home conveniences and remodel to prevent heating and cooling loss. For info: Home Improvement Time, J.A. Stewart, Sr, Program Admin, 7425 Steubenville Pike, Oakdale, PA 15071.

FAMILY HISTORY MONTH. Oct 1-31. To celebrate and publicize family history: the challenge of the research; the fascination of gathering family stories, customs and traditions; the fun of involving family members; its importance as an academic discipline; the value of passing it on to our children. Annually, the month of October. Sponsor: Monmouth County Genealogy Club, Monmouth County Historical Assn, 70 Court St, Freehold, NJ 07728.

FIREPUP'S BIRTHDAY. Oct 1. Firepup spends his time teaching fire safety awareness to children in a fun-filled and non-threatening manner. Sponsor: Natl Fire Safety Council, Inc, PO Box 378, Michigan Center, MI 49254-0378. Phone: (517) 764-2811.

FRYEBURG FAIR. Oct 1-8. Rte 5, Fryeburg, ME. Agricultural exposition, draft horse competitions, oxen and horse pulling, midway, country shows each evening, harness racing, Woodsmen's Day (always Monday), tractor pulling, baking contests, Forestry Resource Center, Firemans Muster, sheep dog trials. Annually, the first week in October that includes the first Wednesday in October. Est attendance: 300,000. For info: June Hammond, PO Box 78, Fryeburg, ME 04037. Phone: (207) 935-3268.

GERMANY: ERNTEDANKFEST. Oct 1. A harvest thanksgiving festival, or potato harvest festival, Erntedankfest or Erntedanktag, is generally observed on the first Sunday in October.

GET ORGANIZED WEEK. Oct 1-7. This is an opportunity for those who want to lower stress and create more time for themselves by making their life a little simpler, a little easier, a little more manageable. It's a time to get organized! Annually, the first week in October. Sponsor: National Assn of Professional Organizers, Deborah J Barnett, Exec Dir, 1604 N Country Club, Tucson, AZ 85716. Phone: (602) 322-9753.

GRANDPARENTS' DAY IN MASSACHUSETTS. Oct 1. Annually, the first Sunday in October.

HARRISON, CAROLINE LAVINIA SCOTT: BIRTH ANNIVERSARY. Oct 1. First wife of Benjamin Harrison, 23rd president of the US, born at Oxford, OH, Oct 1, 1832. Died Oct 25, 1892.

HISPANIC HERITAGE FESTIVAL. Oct 1-31. Dade County, Miami, FL. Commemorates the discovery of America by Christopher Columbus and the Hispanic contribution to the economic and cultural development of Florida through a series of special, cutural and educational events. Est attendance: 500,000. For info: Eloy Vazquez, Hispanic Heritage Council, Inc, 4011 W Flagler St, #505, Miami, FL 33134. Phone: (305) 541-5023.

HOROWITZ, VLADIMIR: BIRTH ANNIVERSARY. Oct 1. Virtuoso pianist, born at Berdichev, Russia, on Oct 1, 1904 (Sept 18, old style). Vladimir Horowitz is widely hailed as one of the world's greatest pianists, renowned for his masterful technique. His first public debut was in Kiev in 1920, and at the age of 20 he played a series of 23 recitals in Leningrad, performing a total of more than 200 works with no duplications. He made his US debut in 1928 with the New York Philharmonic. He settled in the US in 1940 and became a citizen in 1944. His career swung full circle on Apr 20, 1986, when he performed his first concert in his native Russia after a self-imposed absence of 60 years. He died on Nov 5, 1989 at New York, NY.

HUNGER AWARENESS MONTH. Oct 1-31. Hunger Awareness Month, sponsored by the Food Industry Crusade Against Hunger, is designed to educate the nation on the rapid rise of unnecessary hunger both in the US and abroad. FICAH and other food industry organizations will present to the public causes and statistics relating to hunger, possible solutions to the problem and avenues for the public to be part of the solution. FICAH believes hunger is 100% curable. Sponsor: Food Industry Crusade Against Hunger, 800 Connecticut Ave NW, Washington, DC 20006. Phone: (202) 429-4555.

INTERNATIONAL ASSOCIATION OF CULINARY PROFESSIONALS' COOKBOOK MONTH. Oct 1-31. A month to salute excellence in cookbook writing. 30 top cookbooks receive the prestigious Julia Child Cookbook Awards. Annually, the month of October. Fax: (502) 589-3602. For info: Daniel Maye, Exec Dir, Intl Assn of Culinary Professionals, 304 W Liberty, Ste 201, Louisville, KY 40202. Phone: (502) 581-9786.

INTERNATIONAL BOOK FAIR MONTH. Oct 1-31. A fun, school-based event that provides students and their families access to affordable books and encourages leisure reading while raising funds for the school. Sponsored annually by School Book Fairs/Trumpet Book Fairs. For info: Sharon Rice, Mktg Mgr, School Book Fairs, 801 94th Ave N, St Petersburg, FL 33702. Phone: (813) 578-7600.

JAPAN: NEWSPAPER WEEK. Oct 1-7. During this week newspapers make an extensive effort to acquaint the public with their functions and attempts to carry out the role of a newspaper in a free society. Annually, the first week in October.

Oct 1 ☆ *Chase's 1995 Calendar of Events* ☆

KENTUCKY APPLE FESTIVAL. Oct 1-7. Paintsville, KY. Apple blossom beauty pageants, apple auction, country music show, arts and crafts, flea market, antique car show, Corvette show, Apple Bowl, Terrapin Trot, amusement rides, and an AKC-sanctioned dog show. Est attendance: 75,000. For info: Kentucky Apple Festival, Inc, Ray Tosti, Chmn, PO Box 879, Paintsville, KY 41240. Phone: (606) 789-4355.

LAWRENCE, JAMES: BIRTH ANNIVERSARY. Oct 1. Brilliant American naval officer, whose last battle was a defeat, but whose dying words became a most honored naval motto. Lawrence, born at Burlington, NJ, on Oct 1, 1781, was commander of the USS *Hornet*, and later captain of the *Chesapeake* when she engaged in a naval duel with HMS *Shannon*, off Boston, on June 1, 1813. The *Chesapeake* was captured and towed to Halifax as a British prize. Lawrence was mortally wounded by a musket ball during the engagement and uttered his famous last words, "Don't give up the ship," as he was being carried off the ship's deck.

LONGWOOD GARDENS AUTUMN'S COLORS. Oct 1-27. Kennett Square, PA. Outdoor display features a harvest rainbow of brilliant foliage on 1,050 acres. Est attendance: 50,000. For info: Colvin Randall, PR Mgr, Longwood Gardens, PO Box 501, Kennett Sq, PA 19348-0501. Phone: (610) 388-6741.

LUPUS AWARENESS MONTH. Oct 1-31. To promote public awareness of lupus symptoms to aid in early diagnosis and treatment of this disease. Sponsor: Lupus Foundation of America, John M. Huber, Exec Dir, 4 Research Pl, Ste 180, Rockville, MD 20850-3226. Phone: (301) 670-9292.

MENTAL ILLNESS AWARENESS WEEK. Oct 1-7. To increase public awareness of the causes, symptoms and treatments for mental illnesses. Annually, the first full week in October. Sponsor: American Psychiatric Assn, Div of Public Affairs, 1400 K St NW, Washington, DC 20005. Phone: (202) 682-6220.

★ **MINORITY ENTERPRISE DEVELOPMENT WEEK.** Oct 1-7. Presidential Proclamation issued since 1983 without request for the first full week in October except in 1991 when issued for Sept 22-28 and in 1992 for Sept 27-Oct 3, to coincide with the National Conference. Issued for Oct 3-10 in 1993.

MOON PHASE: FIRST QUARTER. Oct 1. Moon enters First Quarter phase at 10:36 AM, EDT.

NATIONAL AIDS AWARENESS MONTH. Oct 1-31. To educate consumers, patients, students and professionals on the prevention of AIDS and sexually transmitted diseases. Kit of materials available for $15. Fax: (415) 332-1832. Sponsor: Pharmacists Planning Service, Inc, Frederick Mayer, PO Box 1336, Sausalito, CA 94966. Phone: (415) 332-4066.

NATIONAL APPLE JACK MONTH. Oct 1-31. A month during which Apple Jack will be honored as America's oldest native distilled spirit, established in the early 1700s. Sponsor: Laird's Apple Jack. For info: Ted Worner, 7064 Charleston Point Dr, Lake Worth, FL 33467. Phone: (407) 439-4937.

NATIONAL CAR CARE MONTH. Oct 1-31. To educate motorists about the importance of maintaining their cars in an effort to improve air quality, highway safety and fuel conservation. Sponsor: Car Care Council, One Grande Lake Dr, Port Clinton, OH 43452. Phone: (419) 734-5343.

NATIONAL CLOCK MONTH. Oct 1-31. To promote the importance and increased use of clocks by US consumers. Annually, the month of October. Sponsor: Clock Mfrs and Mktg Assn, 710 E Ogden Ave, Ste 113, Naperville, IL 60563. Phone: (708) 369-2406.

October 1995	S	M	T	W	T	F	S
	1	2	3	4	5	6	7
	8	9	10	11	12	13	14
	15	16	17	18	19	20	21
	22	23	24	25	26	27	28
	29	30	31				

NATIONAL COMMUNICATE WITH YOUR KID MONTH. Oct 1-31. The generation gap—parents and teens are often on different wavelengths and complain that they cannot talk to each other. The purpose of this observance is to open the doors to better communication between parents and teens and to build positive teen/parent relationships. Annually, the month of October. For info: Teresa Langston, Parenting Without Pressure, 1330 Boyer St, Longwood, FL 32750. Phone: (407) 767-2524.

NATIONAL COSMETOLOGY MONTH. Oct 1-31. Encourages better personal appearance in order to increase self-awareness, self-worth and self-development. Sponsor: Natl Cosmetology Assn, PR Coord, 3510 Olive St, St. Louis, MO 63103. Phone: (314) 534-7980.

NATIONAL DENTAL HYGIENE MONTH. Oct 1-31. To increase public awareness of the importance of preventive oral health care and the dental hygienist's role as the preventive professional. Annually, during the month of October. Sponsor: American Dental Hygienists' Assn, Public Relations, 444 N Michigan Ave, Ste 3400, Chicago, IL 60611. Phone: (312) 440-8900.

NATIONAL DEPRESSION EDUCATION AND AWARENESS MONTH. Oct 1-31. A non-profit campaign to educate patients, the elderly and professionals about depression disorders. Kit of materials available for $15. Annually, the month of October. Fax: (415) 332-1832. For info: Pharmacists Planning Service, Inc, Frederick Mayer, PO Box 1336, Sausalito, CA 94966. Phone: (800) 421-4211.

NATIONAL DESSERT MONTH. Oct 1-31. Held each year to focus on the pleasures of sweet and fanciful desserts. Sponsor: Martini & Rossi Asti, c/o The Alden Group, 52 Vanderbilt Ave, New York, NY 10017. Phone: (212) 867-6400.

★ **NATIONAL DISABILITY EMPLOYMENT AWARENESS MONTH.** Oct 1-31. Presidential Proclamation issued for the month of October (PL100-630, Title III, Sec 301a of Nov 7, 1988). Previously issued as "National Employ the Handicapped Week" for a week beginning during the first week in October since 1945.

NATIONAL DISABILITY EMPLOYMENT AWARENESS MONTH. Oct 1-31. To encourage the full integration of people with disabilities into the workforce. Phone for deaf: (202) 376-6205; fax: (202) 376-6219. Sponsor: US Pres Committee on Employment of People with Disabilities, 1331 F St NW, 3rd Fl, Washington, DC 20004. Phone: (202) 376-6200.

NATIONAL FAMILY SEXUALITY EDUCATION MONTH. Oct 1-31. A national coalition effort to support parents as the first and primary sexuality educators of their children by providing information for parents and young people. Sponsor: Planned Parenthood Federation of America, Trish Torruella, Dir of Educ, 810 Seventh Ave, New York, NY 10019. Phone: (212) 261-4633.

NATIONAL HEALTH CARE FOOD SERVICE WEEK. Oct 1-7. A week in recognition of the contributions of health care food service workers in providing food and nutrition services to patients who are hospitalized or institutionalized. Annually, the first full week in October. Sponsor: American Society for Hospital Food Service Administrators of the American Hospital Assn. For info: American Hospital Assn, ASHFA, Avis Criss, 840 N Lake Shore Dr, Chicago, IL 60611. Phone: (312) 280-6000.

NATIONAL INFERTILITY AWARENESS WEEK. Oct 1-7. Increases public awareness of infertility and informs people about their choices. Sponsor: RESOLVE, Inc, 1310 Broadway, Somerville, MA 02144-1731. Phone: (617) 623-1156.

NATIONAL KITCHEN AND BATH MONTH. Oct 1-31. To facilitate the exchange of information about the two most frequently remodeled rooms in the house. Activities include kitchen and bath tours, remodeling seminars and kitchen/bath dealer open houses. Sponsor: Natl Kitchen and Bath Assn, 687 Willow Grove St, Hackettstown, NJ 07840. Phone: (800) 367-6522.

NATIONAL LIVER AWARENESS MONTH. Oct 1-31. To increase understanding of the importance of liver functions, to promote healthful practices and to encourage research into the causes and cures of liver disease. Annually, the month of October. Sponsor: American Liver Fdtn, Ari Maravel, Dir PR, 1425 Pompton Ave, Cedar Grove, NJ 07009. Phone: (800) 223-0179.

NATIONAL PASTA MONTH. Oct 1-31. To promote the nutritional value of pasta while educating the public about healthy, easy ways to prepare it. Annually, the month of October. Sponsor: Natl Pasta Assn, Christina Salcedo, Fleishman-Hillard, 1330 Ave of the Americas, New York, NY 10019. Phone: (212) 265-9150.

NATIONAL PIZZA MONTH. Oct 1-31. To promote the good fun and nutrition of pizza—"America's Number One Fun Food." *Pizza Festival*, an international pizza recipe contest, is held during October, and participants are encouraged to submit recipes in categories ranging from the exotic to the traditional. Winners will be announced at the Pizza Expo in March 1996. Sponsor: *Pizza Today* Magazine and the Natl Assn of Pizza Operators. For info: Natl Assn of Pizza Operators, Gerry Durnell, Exec Dir, PO Box 1347, New Albany, IN 47151. Phone: (812) 949-0909.

NATIONAL POPCORN POPPIN' MONTH. Oct 1-31. To celebrate the wholesome, economical, natural food value of popcorn, America's native snack. Sponsor: The Popcorn Institute, 401 N Michigan Ave, Chicago, IL 60611-4267. Phone: (312) 644-6610.

NATIONAL PORK MONTH. Oct 1-31. The National Pork Producers Council, in cooperation with state pork producers associations, marks October as National Pork Month. While promotions are conducted throughout the year, special emphasis is placed on Pork, The Other White Meat®, during October. For info: Charles Harness, VP Communications, Natl Pork Producers Council, PO Box 10383, Des Moines, IA 50306. Phone: (515) 223-2600.

NATIONAL SARCASTICS AWARENESS MONTH. Oct 1-31. To help people everywhere understand the positive and negative aspects of sarcasm. For info: Sarcastics Anonymous, Virginia Tooper, Dir of Barbs, Box 10944, Pleasanton, CA 94588. Phone: (510) 786-4567.

NATIONAL SEAFOOD MONTH. Oct 1-31. To promote the taste, variety and nutrition of fish and shellfish. Sponsor: Natl Fisheries Institute, Emily Holt, 1525 Wilson Blvd, Ste 500, Arlington, VA 22209.

NATIONAL SUDDEN INFANT DEATH SYNDROME AWARENESS MONTH. Oct 1-31. Month-long focus on Sudden Infant Death Syndrome (also called crib death, the nation's major cause of death for infants beyond one week of age) promotes the SAVE OUR BABIES CAMPAIGN to increase public awareness and funds available for medical research and family services. Sponsor: SIDS Alliance, Phipps Y. Cohe, Public Affairs Dir, 10500 Little Patuxent Pkwy, Ste 420, Columbia, MD 21044. Phone: (800) 221-7437.

NATIONAL YOUTH AGAINST TOBACCO MONTH. Oct 1-31. Santa Barbara, CA. Summit, Oct 28. A national network of youth summits will culminate October activities encouraging teen smokers to kick the habit and acknowledging teens who have never used tobacco products. Annually, the month of October. Summit is the last Saturday in October. Est attendance: 500. For info: Janet Benner, PhD, Regional Coord, Tobacco Education and Prevention, 1806 Cliff Dr #B, Santa Barbara, CA 93109. Phone: (805) 899-3300.

NIGERIA: INDEPENDENCE DAY: 35 YEARS. Oct 1. National holiday. Became independent on this day in 1960, and a republic on this day in 1963.

PEOPLE'S REPUBLIC OF CHINA: NATIONAL DAY. Oct 1. Commemorates the founding of the People's Republic of China, Oct 1, 1949.

POLISH AMERICAN HERITAGE MONTH. Oct 1-31. A national celebration of Polish history, culture and pride, in cooperation with the Polish American Congress and Polonia Across America. For info: Polish American Cultural Center, Natl HQ, Michael Blichasz, Chair, 308 Walnut St, Philadelphia, PA 19106. Phone: (215) 922-1700.

POLISH AMERICAN HERITAGE MONTH CELEBRATION. Oct 1-31. Philadelphia and Eastern Pennsylvania. A month-long national celebration commemorating the contributions of Poles and Polish Americans to the US, world history and culture in such areas as science, art, music, politics, the military and religion. Exhibitions, lectures, music and more. For info: Eastern Pennsylvania District Office, Polish American Congress, Inc, 308 Walnut St, Philadelphia, PA 19106. Phone: (215) 922-1700.

PULASKI DAY PARADE. Oct 1. Philadelphia, PA. Parade honoring the Polish patriot known as the "Father of the American Cavalry." Begins at 20th St and the Benjamin Franklin Pkwy to 17th St, then east to Independence Mall. For info: Polish American Congress, Eastern Pennsylvania District, 308 Walnut St, Philadelphia, PA 19106. Phone: (215) 739-3408.

QUILTING IN THE TETONS. Oct 1-7. Jackson, WY. Quilting exhibits and workshops following the Fall Arts Festival. For info: Teton County Extension Office-Quilting in the Tetons, Box 1708, Jackson, WY 83001. Phone: (307) 733-3087.

RALLY 'ROUND BUCKS COUNTY. Oct 1. Langhorne, PA. A fun rally with question-answer format that takes participants through beautiful nature scenery. Est attendance: 350. For info: Bucks County Dept of Parks and Recreation, 901 E Bridgetown Pike, Langhorne, PA 19047.

SA'FEST 95 (SAFEty FESTival): CARTOONISTS AGAINST CRIME INTERNATIONAL CARTOON ART COMPETITION. Oct 1-31. Chicago, IL. Held in conjunction with National Crime Prevention Month and Domestic Violence Awareness Month, competition is designed to draw attention to mankind's vulnerability to random and violent crimes both in the US and abroad. The emphasis is on "How to Play It Safe® in the 90s." Cartoons propose unique safety measures, precautions and tips for avoiding crime, showing cartoons can be educational as well as entertaining. Exhibition will be held in downtown Chicago. For info: Adrienne Sioux Koopersmith, Cartoonists Against Crime™, 1437 W Rosemont, 1W, Chicago, IL 60660. Phone: (312) 743-5341.

STOCKTON, RICHARD: BIRTH ANNIVERSARY. Oct 1. Lawyer and signer of the Declaration of Independence, born at Princeton, NJ, Oct 1, 1730. Died there, Feb 8, 1781.

UNITED NATIONS: INTERNATIONAL DAY FOR THE ELDERLY. Oct 1. On Dec 14, 1990, the General Assembly designated Oct 1 as the International Day for the Elderly. It appealed for contributions to the Trust Fund for Aging (which supports projects in developing countries in implementation of the Vienna International Plan of Action on Aging adopted at the 1982 World Assembly on Aging) and endorsed an action program on aging for 1992 and beyond as outlined by the Secretary-General. (Resolution 45/106.) Info from: United Nations, Dept of Public Info, New York, NY 10017.

UNITED STATES 1996 FEDERAL FISCAL YEAR BEGINS. Oct 1, 1995-Sept 30, 1996.

☆ Chase's 1995 Calendar of Events ☆

Oct 1-2

UNIVERSAL CHILDREN'S WEEK. Oct 1-7. To disseminate throughout the world info on the needs of children and to distribute copies of the Declaration of the Rights of the Child. For complete info, send $1 to cover expense of printing, handling and postage. Annually, the first seven days of October. Sponsor: Intl Soc of Friendship and Good Will, Dr. Stanley Drake, Pres, 9538 Summerfield St, Spring Valley, San Diego County, CA 91977-2852. Phone: (619) 466-8882.

UNIVERSITY OF CHICAGO FIRST DAY OF CLASSES: ANNIVERSARY. Oct 1. Chicago, IL. On Oct 1, 1892, the University of Chicago opened with an enrollment of 594 and a faculty of 103, including eight former college presidents.

VEGETARIAN AWARENESS MONTH. Oct 1-31. To advance awareness of the many surprising environmental, economic, ethical, health, humanitarian and other benefits of a meatless lifestyle. To promote personal and planetary healing with respect for all life. Sponsor: Vegetarian Awareness Network/VEGANET, PO Box 321, Knoxville, TN 37901. Phone: (800) 872-8343.

WORLD VEGETARIAN DAY. Oct 1. Celebration of vegetarianism's benefits to humans, animals and our planet. Sponsor: North American Vegetarian Society, Box 72, Dolgeville, NY 13329. Phone: (518) 568-7970.

ZOOFEST '95. Oct 1-29. (Saturdays and Sundays only.) Asheboro, NC. A celebration of animals, autumn and the arts. Est attendance: 20,000. Sponsor: North Carolina Zoological Park, 4401 Zoo Pkwy, Asheboro, NC 27203. Phone: (800) 488-0444.

BIRTHDAYS TODAY

Julie Andrews (Julia Wells), 60, actress, singer (Emmy "Julie Andrews Hour"; Oscar *Mary Poppins*; stage *My Fair Lady, Camelot*), born Walton-on-Thames, England, Oct 1, 1935.
Tom Bosley, 68, actor ("Happy Days," "Murder She Wrote"), born at Chicago, IL, Oct 1, 1927.
Jimmy Carter (James Earl Carter, Jr), 71, 39th US president, born at Plains, GA, Oct 1, 1924.
Richard Harris, 65, singer ("MacArthur Park"), actor (*Camelot, Hawaii, Juggernaut*), born Limerick, Ireland, Oct 1, 1930.
Walter Matthau, 75, actor (*The Odd Couple, Kotch, Grumpy Old Men*), born New York, NY, Oct 1, 1920.
Randy Quaid, 45, actor (*The Last Picture Show, Dead Solid Perfect*), born Houston, TX, Oct 1, 1950.
William Hubbs Rehnquist, 71, Chief Justice of the US Supreme Court, born at Milwaukee, WI, Oct 1, 1924.
Stella Stevens, 59, actress (Jane Hancock on "Ben Casey"; Lute-Mac Sanders in "Flamingo Road"), born Hot Coffee, MS, Oct 1, 1936.
Grete Waitz, 42, track athlete, born at Oslo, Norway, Oct 1, 1953.

OCTOBER 2 — MONDAY
275th Day — Remaining, 90

BOOK IT!/NATIONAL YOUNG READER'S DAY. Oct 2. The Pizza Hut BOOK IT! Program and the Library of Congress celebrate the annual start of the largest reading incentive program in America; more than 19 million children in grades K-6 participate via public/private school classrooms. Schools, libraries, families and communities nationwide use this day to celebrate the joys of reading in a variety of creative and educational ways. This event is now a combination of two previous celebratory days and will be marked of the first school day of October, beginning this year. Phone: (800)-4-BOOK-IT. For info: Creamer Dickson Basford, Camela Stuby, 1633 Broadway, 27th Fl, New York, NY 10019.

BUFFALO ROUND-UP. Oct 2-14. Custer, SD. To round up, brand and separate 1,500 buffalo before auction in November. Est attendance: 1,600. For info: Craig Pugsley, Custer State Park, HC 83, Box 70, Custer, SD 57730. Phone: (605) 255-4515.

CHARLIE BROWN AND SNOOPY: BIRTHDAY. Oct 2. On Oct 2, 1995, Charles M. Schulz's PEANUTS comic strip celebrates its 45th anniversary. Syndicated by United Feature Syndicate, the PEANUTS comic strip now appears in 2,400 newspapers and is translated into 26 languages in 68 countries. For info: United Feature Syndicate, Nancy Nicolelis, PR Dir, 200 Park Ave, New York, NY 10166.

★ **CHILD HEALTH DAY.** Oct 2. Presidential Proclamation always for first Monday of October. Proclamation has been issued since 1928. In 1959 Congress changed celebration day from May 1 to the present observance (Pub Res No. 46 of May 18, 1928, and PL86-352 of Sept 22, 1959).

FALL FOLIAGE FESTIVAL. Oct 2-8. Walden, Cabot, Plainfield, Peacham, Barnet, Groton and Marshfield, VT. Seven towns welcome visitors during Vermont's famous fall foliage season. Send self-addressed stamped envelope. Est attendance: 900. For info: Fall Festival Committee, Box 38, West Danville, VT 05873.

FRENCH LIBRARY AND CULTURAL CENTER 50th ANNIVERSARY CONFERENCE: THE GLOBALIZATION OF VENTURE CAPITAL. Oct 2-3. The Four Seasons Hotel, Boston, MA. Corporate convention will feature workshops on specific fields and regions. Part of 50th anniversary festivities in deference to the many contributions made to the center by Gen Georges F. Doriot, one of the founders of the field of venture capital, a former Harvard School of Business faculty member and former president of the French Library. Conference chaired by Richard J. Testa of Testa, Hurwitz & Thibeault. Attendance limited. Fax: (617) 266-1780. For info: Phyllis Dohanian, Exec Dir, The French Library and Cultural Center, 53 Marlborough St, Boston, MA 02116-2099. Phone: (617) 266-4351.

GANDHI, MOHANDAS KARAMCHAND (MAHATMA): BIRTH ANNIVERSARY. Oct 2. Indian political and spiritual leader who achieved world honor and fame for his advocacy of nonviolent resistance as a weapon against tyranny was born Oct 2, 1869, at Porbandar, India. He was assassinated in the garden of his home at New Delhi, on Jan 30, 1948. On the anniversary of Gandhi's birth (Gandhi Jayanti) thousands gather at the park on the Jumna River in Delhi where Gandhi's body was cremated. Hymns are sung, verses from the Gita, the Koran and the Bible are recited, and cotton thread is spun on small spinning wheels (one of Gandhi's favorite activities). Other observances held at his birthplace and throughout India.

GUINEA: INDEPENDENCE DAY. Oct 2. National Day.

GUNN, MOSES: BIRTH ANNIVERSARY. Oct 2, 1929. The 1981 winner of the NAACP Image Award for his performance as Booker T. Washington in the film *Ragtime* was born in St. Louis, MO, and grew up to be one of America's finest actors. His appearances on stage ranged from the title role in *Othello* to Jean Genet's *The Blacks*. He received an Emmy nomination for his role in *Roots* and was awarded several Obies for off-Broadway performances. In film he appeared in *Shaft* and *The Great White Hope*. He died Dec 17, 1993, at Guilford, CT.

	S	M	T	W	T	F	S
October 1995	1	2	3	4	5	6	7
	8	9	10	11	12	13	14
	15	16	17	18	19	20	21
	22	23	24	25	26	27	28
	29	30	31				

Chase's 1995 Calendar of Events — Oct 2-3

HULL, CORDELL: BIRTH ANNIVERSARY. Oct 2. American statesman who served in both houses of the Congress and as secretary of state, was born Oct 2, 1871, at Pickett County, TN. Noted for his contributions to the "Good Neighbor" policies of the US with regard to countries of the Americas, and to the establishment of the United Nations, Hull died at Bethesda, MD, July 23, 1955.

LONDON BRIDGE DAYS. Oct 2-8. Lake Havasu City, AZ. Celebration of the dedication of London Bridge to Lake Havasu City, AZ. For info: Visitors and Conv Bureau, 1930 Mesquite Ave, Ste 3, Lake Havasu City, AZ 86403. Phone: (800) 242-8278.

MARSHALL, THURGOOD: SWORN IN TO SUPREME COURT: ANNIVERSARY. Oct 2. On Oct 2, 1967, Thurgood Marshall was sworn in as the first Black associate justice to the US Supreme Court. On June 27, 1991, he announced his resignation, effective upon the confirmation of his successor.

McFARLAND, GEORGE (SPANKY): BIRTH ANNIVERSARY. Oct 2, 1928. Chubby child star of the "Our Gang" comedy film shorts. Born at Dallas, TX, and died at Grapevine, TX, June 30, 1993.

NAME YOUR CAR DAY. Oct 2. A day to honor your car by giving it a pet name. In honor of John Pertzborn's 1955 Buick "Elvis," a car that shakes, rattles and rolls. For info: John H. Pertzborn, Feature Reporter, KSDK-TV, 1000 Market St, St. Louis, MO 63101. Phone: (314) 444-5119.

NATIONAL CUSTOMER SERVICE WEEK. Oct 2-6. A week-long opportunity for businesses to focus on the importance of the customer and the customer service profession. Sponsor: Intl Customer Service Assn, Elizabeth Gleason, 401 N Michigan Ave, Chicago, IL 60611-4267. Phone: (312) 321-6800.

NATIONAL SPINNING AND WEAVING WEEK. Oct 2-8. To celebrate the timeless craft of weaving and spinning and to honor craftsmen and women past and present who perpetuate a legacy of fine handmade textiles. Fax: (612) 646-0806. For info: Handweavers Guild of America, Inc, 2402 University Ave, Ste 702, St. Paul, MN 55114. Phone: (612) 646-0802.

PHILEAS FOGG'S WAGER DAY. Oct 2. Anniversary, from Jules Verne's *Around the World in Eighty Days*, of the famous wager of Oct 2, 1872, upon which the book is based: "I will bet twenty thousand pounds against any one who wishes, that I will make the tour of the world in eighty days or less." Then, consulting a pocket almanac, Phileas Fogg said: "As today is Wednesday, the second of October, I shall be due in London, in this very room of the Reform Club, on Saturday, the twenty-first of December, at a quarter before nine PM; or else the twenty thousand pounds ... will belong to you." See also: "Phileas Fogg Wins a Wager Day" (Dec 21).

PUMPKIN CORNERS PUMPKIN DAY. Oct 2. Stamford, NE. Pick your own pumpkins. Free hay rides and 50-cent hot dogs and cider. Est attendance: 300. For info: Raylene Stephens, Rte 1, Box 7, Stamford, NE 68977.

SPACE MILESTONE: RECORD TIME IN SPACE. Oct 2. On Oct 2, 1984, three Soviet cosmonauts returned to Earth after setting a record of 237 days in space—at *Salyut 7* space station, since Feb 9.

◆ **STREETER, RUTH CHENEY: 100th BIRTH ANNIVERSARY.** Oct 2, 1895. Born at Brookline, MA, Ruth Cheney Streeter was the first director of the US Marine Corps Women's Reserve. She was active in unemployment relief, public health, welfare and old-age assistance in New Jersey during the 1930s. A student of aeronautics, she learned to fly while serving as an adjutant of a flight group in the Civil Air Patrol during the early years of World War II and became the only female member of the Committee on Aviation of the New Jersey Defense Council. Ruth Cheney Streeter died Sept 30, 1990, at Morristown, NJ.

SUPREME COURT 1995-1996 TERM BEGINS. Oct 2. Traditionally, the Supreme Court's annual term begins on the first Monday in October and continues with seven two-week sessions of oral arguments. Between the sessions are six recesses during which the opinions are written by the Justices. Ordinarily, all cases are decided by the following June or July.

"THE TWILIGHT ZONE": THE FIRST EPISODE. Oct 2. On Oct 2, 1959, "The Twilight Zone" went on the air with these now-familiar surreal words: "There is a fifth dimension, beyond that which is known to man. It is a dimension as vast as space and as timeless as infinity. It is the middle ground between light and shadow, between science and superstition, and it lies between the pit of man's fear and the summit of his knowledge. This is the dimension of imagination. It is an area which we call The Twilight Zone." The program ran five seasons for 154 installments, with a one-year hiatus between the third and fourth seasons. It now is considered to have had some of the best drama ever to appear on television. It is noted for its scripts with twists of fate and fine lines between reality and illusion.

UNITED NATIONS: UNIVERSAL CHILDREN'S DAY. Oct 2. First Monday in October is designated by the United Nations General Assembly as Universal Children's Day. First observance was in 1953. A time to honor children with special ceremonies and festivals and to make children's needs known to governments. Observed on different days and in different ways in more than 120 nations.

UNITED NATIONS: WORLD HABITAT DAY. Oct 2. The United Nations General Assembly, by a resolution of Dec 17, 1985, has designated the first Monday of October each year as World Habitat Day. The first observance of this day, Oct 5, 1986, marked the 10th anniversary of the first international conference on the subject. (Habitat: United Nations Conference on Human Settlements, Vancouver, Canada, 1976.) Info from: United Nations, Dept of Public Info, New York, NY 10017.

WORLD FARM ANIMALS DAY. Oct 2. (Gandhi's birthday.) To memorialize the needless suffering and death of billions of farm animals each year through mock funerals, memorial services, vigils and similar activities. Fax: (301) 530-5747. Sponsor: Farm Animal Reform Movement, Box 30654, Bethesda, MD 20824. Phone: (301) 530-1737.

BIRTHDAYS TODAY

Clay S. Felker, 67, publisher, born at St. Louis, MO, Oct 2, 1928.
Don McLean, 50, singer ("Crying"), songwriter ("American Pie," "Vincent"), born at New Rochelle, NY, Oct 2, 1945.
Rex Reed, 56, movie critic, born at Fort Worth, TX, Oct 2, 1939.
Sting (Gordon Sumner), 44, musician, singer (lead with The Police; "Roxanne"), songwriter ("Every Breath You Take"), actor (*Dune*, *Bring on the Night*), born at London, England, Oct 2, 1951.

OCTOBER 3 — TUESDAY
276th Day — Remaining, 89

"ANDY GRIFFITH SHOW": FIRST BROADCAST ANNIVERSARY. Oct 3. Marks the airing of the first episode on Oct 3, 1960. The 12,000+ members of "The Andy Griffith Show" Rerun Watchers Club and others celebrate this day with festivities every year.

BANCROFT, GEORGE: BIRTH ANNIVERSARY. Oct 3. American historian, born at Worcester, MA, Oct 3, 1800. Died at Washington, DC, Jan 27, 1891.

GERMANY: REUNIFICATION: 5th ANNIVERSARY. Oct 3. East and West Germany reunited on Oct 3, 1990, just four days short of East Germany's 41st founding anniversary (Oct 7, 1949). Originally scheduled to coincide with the all-German parliamentary elections on Dec 2, 1990, reunification was rescheduled for the earlier date by East Germany's parliament in response to the deterioration of the economy.

HONDURAS: FRANCISCO MORAZAN HOLIDAY. Oct 3. Public holiday in honor of Francisco Morazan, national hero, who was born in 1799.

KOREA: NATIONAL FOUNDATION DAY. Oct 3. National holiday also called Tangun Day as it commemorates day when legendary founder of the Korean nation, Tangun, established his kingdom of Chosun in 2333 BC.

KURTZMAN, HARVEY: BIRTH ANNIVERSARY. Oct 3. Cartoonist and founder of *Mad* magazine, Harvey Kurtzman was born on Oct 3, 1902, at Brooklyn, NY. At the age of 14 he had his first cartoon published and he began his career in comic books in 1943. His career led him to EC Comics and with the support of William Gaines, he created *Mad* magazine, which first appeared in 1952. He died Feb 21, 1993, at Mount Vernon, NY.

MANSON, PATRICK: BIRTH ANNIVERSARY. Oct 3. British parasitologist and surgeon sometimes called the "father of tropical medicine." Sir Patrick's research into insects as carriers of parasites was instrumental in later understanding of mosquitoes as transmitters of malaria. Born in Scotland on Oct 3, 1844, Manson died Apr 9, 1922.

MRS W.H. FELTON BECOMES FIRST WOMAN US SENATOR. Oct 3. Mrs. W.H. Felton, 87, of Cartersville, GA, was appointed on Oct 3, 1922, by Governor Thomas Hardwick of Georgia to the Senate seat vacated by the death of Thomas E. Watson.

NATIONAL CONVENTION OF THE AMERICAN BUSINESS WOMEN'S ASSOCIATION. Oct 3-8. Portland, OR. Businesswomen gather to learn, network and elect ABWA's national board of directors for the coming year. Seminars, speakers and the announcement of the Top Ten Business Women of ABWA and the American Business Woman of ABWA are featured. Est attendance: 5,000. Sponsor: American Business Women's Assn, PR Dept, 9100 Ward Parkway, Kansas City, MO 64114. Phone: (816) 361-6621.

NETHERLANDS: RELIEF OF LEIDEN DAY. Oct 3. Celebration of the liberation of Leiden in 1574.

NOBEL CONFERENCE XXXI. Oct 3-4. Gustavus Adolphus College, St. Peter, MN. Annual two-day scientific symposium (31st year), and the only one sanctioned by the Nobel Foundation, Stockholm. Annually, the first Tuesday and Wednesday in October. Est attendance: 4,000. For info: Dean Wahlund, Dir of Public Affairs, Gustavus Adolphus College, St. Peter, MN 56082. Phone: (507) 933-7550.

NORWAY: PAGEANTRY IN OSLO. Oct 3. The Storting (Norway's Parliament) convenes on first weekday in October, when it decides date for the ceremonial opening of the Storting—usually the following weekday—and the parliamentary session is then opened by King Olav V in the presence of Corps Diplomatique, preceded and followed by a military procession between the Royal Palace and the Storting.

ROBINSON NAMED BASEBALL'S FIRST BLACK MAJOR LEAGUE MANAGER: ANNIVERSARY. Oct 3. The only major league player selected most valuable player in both the American and National Leagues, Frank Robinson was hired by the Cleveland Indians as baseball's first black major league manager on Oct 3, 1974. During his playing career Robinson represented the American League in four World Series playing for the Baltimore Orioles and led the Cincinnati Reds to a National League Pennant. Robinson hit 586 home runs in 21 years of play.

YOM KIPPUR BEGINS AT SUNDOWN. Oct 3. Jewish Day of Atonement. See "Yom Kippur" (Oct 4).

BIRTHDAYS TODAY

Jeff Bingaman, 52, US Senator (D, New Mexico) up for reelection in Nov '94, born at El Paso, TX, Oct 3, 1943.

Lindsey Buckingham, 48, musician, singer, songwriter (with Fleetwood Mac; "Go Your Own Way"), born at Palo Alto, CA, Oct 3, 1947.

Chubby Checker (Ernest Evans), 54, musician, singer ("The Twist"), born at South Philadelphia, PA, Oct 3, 1941.

Dennis Lee Eckersley, 41, baseball player, born at Oakland, CA, Oct 3, 1954.

Madlyn Rhue, 61, actress (*It's a Mad Mad Mad Mad World*, "Executive Suite"), born Washington, DC, Oct 3, 1934.

Gore Vidal, 70, writer (*Burr, Myra Breckinridge*), born West Point, NY, Oct 3, 1925.

David Winfield, 44, baseball player, born at St. Paul, MN, Oct 3, 1951.

OCTOBER 4 — WEDNESDAY
277th Day — Remaining, 88

GREGORIAN CALENDAR ADJUSTMENT ANNIVERSARY. Oct 4. Pope Gregory XIII, on Feb 24, 1582, issued a bulletin which decreed that the day following Tuesday, Oct 4, 1582, should be Friday, Oct 15, 1582, thus correcting the previously used Julian Calendar, then 10 days out of date. This reform was effective in most Catholic countries, though the Julian Calendar continued in use in Britain and the American colonies until 1752. See also: "Gregorian Calendar Adjustment Anniversary" (Feb 24, Sept 2).

HAYES, RUTHERFORD BIRCHARD: BIRTH ANNIVERSARY. Oct 4. Rutherford Birchard Hayes, 19th president of the US (Mar 4, 1877–Mar 3, 1881), was born at Delaware, OH, Oct 4, 1822. In his inaugural address, Hayes said: "He serves his party best who serves the country best." He died at Fremont, OH, Jan 17, 1893.

JOHNSON, ELIZA McCARDLE: BIRTH ANNIVERSARY. Oct 4. Wife of Andrew Johnson, 17th president of the US, born at Leesburg, TN, Oct 4, 1810. Died, Jan 15, 1876.

LESOTHO: NATIONAL DAY. Oct 4. National holiday.

MISSISSIPPI STATE FAIR. Oct 4-15. Jackson, MS. Features nightly professional entertainment, livestock show, midway carnival, domestic art exhibits. Est attendance: 601,000. For info: Mississippi Fair Commission, PO Box 892, Jackson, MS 39205. Phone: (601) 961-4000.

REMINGTON, FREDERIC S.: BIRTH ANNIVERSARY. Oct 4, 1861. Born in Canton, NY, Remington studied art at Yale Art School. He travelled extensively throughout North America. He is best known for the work he did in the western US, capturing life on the plains—Indians, cowboys, horses and more. He owned a ranch near Peabody, KS. He died Dec 26, 1909, age 48, at Ridgefield, CT, following an appendectomy. For info: Lowell McAllister, Dir, Frederic Remington Art Museum, 303 Washington St, Ogdensburg, NY 13669. Phone: (315) 393-2425.

RUNYAN, DAMON: BIRTH ANNIVERSARY. Oct 4. American newspaperman and author, born at Manhattan, KS, Oct 4, 1884. Died at New York, NY, Dec 10, 1946. "... always try to rub up against money," he wrote, "for if you rub up against money long enough, some of it may rub off on you."

ST. FRANCIS OF ASSISI: FEAST DAY. Oct 4. Giovanni Francesco Bernardone, religious leader, founder of the Friars Minor (Franciscan Order), born at Assisi, Umbria, Italy, in 1181. Died there on Oct 3, 1226. One of the "most attractive and best-loved saints of all time."

SPACE MILESTONE: *LUNA 3* (USSR). Oct 4. First satellite to photograph moon's distant side. Launched Oct 4, 1959.

October 1995

S	M	T	W	T	F	S
1	2	3	4	5	6	7
8	9	10	11	12	13	14
15	16	17	18	19	20	21
22	23	24	25	26	27	28
29	30	31				

Chase's 1995 Calendar of Events — Oct 4–5

SPACE MILESTONE: *SPUTNIK* ANNIVERSARY. Oct 4. Anniversary of launching of first successful man-made earth satellite. *Sputnik I* ("fellow traveller of earth") weighing 184 lbs was fired into orbit from the USSR's Tyuratam launch site on Oct 4, 1957. Transmitted radio signal for 21 days, decayed Jan 4, 1958. Beginning of Space Age and man's exploration beyond Earth.

STRATEMEYER, EDWARD L.: BIRTH ANNIVERSARY. Oct 4. American author of children's books, Stratemeyer was born at Elizabeth, NJ, on Oct 4, 1862. He created numerous series of popular children's books including "The Bobbsey Twins," "The Hardy Boys," "Nancy Drew" and "Tom Swift." He and his Stratemeyer Syndicate, using 60 or more pen names, produced more than 800 books. More than four million copies were in print in 1987. Stratemeyer died at Newark, NJ, on May 10, 1930.

TEN-FOUR DAY. Oct 4. The fourth day of the 10th month is a day of recognition for radio operators, whose code words, "Ten-Four," signal an affirmative reply.

WORLD DAIRY EXPO. Oct 4–8. Madison, WI. World's largest dairy trade show features more than 600 exhibits and 1,700 head of cattle shown in six shows at the Dane County Exposition Center. About 200 animals offered for sale. Educational forums. Open to the public. Est attendance: 61,000. For info: World Dairy Expo, 122 E Olin Ave, Suite 270, Madison, WI 53713. Phone: (608) 251-3976.

YELTSIN SHELLS WHITE HOUSE: ANNIVERSARY. Oct 4, 1993. In a violent spectacle, the Russian military loyal to Pres Boris Yeltsin shelled the Russian White House, the symbol of Russian independence from communist rule. The parliament building had been seized by Yeltsin opponents led by Ruslan Khasbulatov and Alexander Rutskoi, who in the August 1992 coup had stood with Yeltsin on the White House balcony as he proclaimed that Russia should not submit to the coup plotters' attempts to turn back history. These former-allies-turned-foes reacted to Yeltsin's Sept 21 decision to dissolve the Russian parliament by seizing the parliament building. Yeltsin and his supporters succeeded in isolating them inside, cutting electricity and the flow of information to the building. But Rutskoi urged his followers to seize the Kremlin and other institutions, precipitating a day of bloody fighting in the streets of Moscow before the attackers retreated back into the White House.

YOM KIPPUR or DAY OF ATONEMENT. Oct 4. Holiest Jewish observance. A day for fasting, repentance and seeking forgiveness. Hebrew calendar date: Tishri 10, 5756.

BIRTHDAYS TODAY

Mike Adamle, 46, broadcast journalist, born at Euclid, OH, Oct 4, 1949.
Clifton Davis, 50, singer, actor ("That's My Mama," "Amen"), composer, born at Chicago, IL, Oct 4, 1945.
Charlton Heston, 73, actor (*The Ten Commandments, Ben-Hur, Planet of the Apes*), born Evanston, IL, Oct 4, 1922.
Patti LaBelle (Patricia Louise Holte), 51, singer ("Since I Don't Have You,"), born at Philadelphia, PA, Oct 4, 1944.
Jan Murray (Murry Janofsky), 78, comedian (emcee of "Dollar a Second," "Treasure Hunt"), born New York, NY, Oct 4, 1917.
Buddy Roemer, 52, former Governor of Louisiana (D), born at Shreveport, LA, Oct 4, 1943.
Susan Sarandon (Susan Tomaling), 49, actress (*Atlantic City, Thelma and Louise, Lorenzo's Oil*), born New York, NY, Oct 4, 1946.

Alvin Toffler, 67, author (*Future Shock, Power Shift*), born at New York, NY, Oct 4, 1928.

OCTOBER 5 — THURSDAY
278th Day — Remaining, 87

ARTHUR, CHESTER ALAN: 165th BIRTH ANNIVERSARY. Oct 5. The 21st president of the US, Chester Alan Arthur, was born at Fairfield, VT, Oct 5, 1830, and succeeded to the presidency following the death of James A. Garfield. Term of office: Sept 20, 1881–Mar 3, 1885. Arthur was not successful in obtaining the Republican Party's nomination for the following term. He died at New York, NY, Nov 18, 1886.

CHIEF JOSEPH SURRENDER: ANNIVERSARY. Oct 5. After a 1,700-mile retreat, Chief Joseph and the Nez Perce Indians surrendered to US Cavalry troops at Bear's Paw near Chinook, MT, on Oct 5, 1877. Chief Joseph made his famous speech of surrender, "From where the sun now stands, I will fight no more forever."

CIVIL WAR "SUBMARINE" ATTACK: ANNIVERSARY. Oct 5. In an attempt to disrupt the Union blockade of Charleston Harbor, the Confederate semi-submersible *David* rammed the Federal ironclad *New Ironsides* with a spar torpedo on Oct 5, 1863. This was the first successful Southern attack using a submersible craft. Although both sides experimented with submarine warfare during the Civil War the results were far from encouraging as the submarines caused more fatalities to their own crews than to the opposing side.

COWBOY HALL OF FAME CEREMONY AND BANQUET. Oct 5. Willcox Community Center, Willcox, AZ. Est attendance: 150. For info: Willcox Chamber of Commerce, Harold Skinner, Chairman; 1500 N Circle I Rd, Willcox, AZ 85643. Phone: (602) 384-2272.

ENGLAND: NOTTINGHAM GOOSE FAIR. Oct 5–7. Forest Recreation Ground, Nottingham. Held annually since 1284 (except during the Great Plague in 1665 and the two World Wars), the fair formerly lasted three weeks and boasted as many as 20,000 geese on display. Now lasting three days, the Nottingham Goose Fair always begins on the first Thursday in October. A traditional fair with modern amusements.

ENRICO FERMI ATOMIC POWER PLANT: ACCIDENT ANNIVERSARY. Oct 5. A radiation alarm and Class I alert at 3:09 PM, EST, on Oct 5, 1966, signaled a problem at the Enrico Fermi Atomic Power Plant, Lagoona Beach, near Monroe, MI. The accident was contained, but nearly a decade was required to complete the decommissioning and disassembly of the plant.

GODDARD, ROBERT HUTCHINGS: BIRTH ANNIVERSARY. Oct 5. The "father of Space Age." Born at Worcester, MA, on Oct 5, 1882. Largely ignored or ridiculed during his lifetime because of his dreams of rocket travel, including travel to other planets. Launched a liquid-fuel-powered rocket on Mar 16, 1926, at Auburn, MA. Died Aug 10, 1945. See also: "Goddard Day" (Mar 16) and "Anniversary Day" (Oct 19).

HASENFUS, EUGENE: CAPTURE ANNIVERSARY. Oct 5. A cargo plane carrying arms intended for the Nicaraguan rebels (contras) was shot down over Nicaragua on Oct 5, 1986. All crew members except Eugene Hasenfus, ex-US Marine, of Marinette, WI, perished. Hasenfus parachuted from the doomed plane and was captured the following day. The US administration at first denied "any link" to the mission. Hasenfus was put on trial for violating Nicaraguan security laws. He was convicted and sentenced to 30 years in prison on Nov 15, 1986. On request of president Daniel Ortega, the Nicaraguan National Assembly pardoned Hasenfus on Dec 17, 1986. He was released that day and flown back to the US. This episode commenced the unravelling and exposure of covert operations and possible violations of US law which in turn led to Congressional "Iran Contra" hearings and investigation by a presidentially-appointed special prosecutor.

Oct 5-6 ☆ *Chase's 1995 Calendar of Events* ☆

ORPHAN TRAIN HERITAGE SOCIETY OF AMERICA, INC, ANNUAL REUNION. Oct 5-7. (Tentative.) Clarion Inn, Fayetteville, AR. Between 1854 and 1929 more than 150,000 homeless children were transported out of New York City, Boston, Chicago and Philadelphia aboard trains accompanied by "agents." The agents arranged for mid-western families to take the children. Some children were taken by the adults signing a contract, others were indentured. Annual Reunion brings Orphan Train Riders and their families from all over America to talk about their personal experiences. Fax: (501) 756-0769. Est attendance: 125. For info: Mary Ellen Johnson, PO Box 496, Johnson, AR 72744. Phone: (501) 756-2780.

RILEY FESTIVAL. Oct 5-8. Greenfield, IN. To celebrate the birth of Hoosier poet James Whitcomb Riley. Est attendance: 75,000. For info: Riley Fest Assn, Nancy Alldredge, PO Box 554, Greenfield, IN 46140. Phone: (317) 462-4188.

SPACE MILESTONE: *CHALLENGER*. Oct 5. Space shuttle *Challenger* makes sixth mission with crew of seven, including two women. Launched from and landed at Kennedy Space Center, FL (Oct 5-13, 1984). Kathryn D. Sullivan became the first American woman to walk in space.

STONE, THOMAS: DEATH ANNIVERSARY. Oct 5. Signer of the Declaration of Independence, born 1743 (exact date unknown) in Charles County, MD. Died at Alexandria, VA, Oct 5, 1787.

TECUMSEH: DEATH ANNIVERSARY. Oct 5. Shawnee Indian Chief and orator, born at Old Piqua near Springfield, OH, in March 1768. Tecumseh is regarded as one of the greatest of American Indians. He came to prominence between the years 1799 and 1804 as a powerful orator, defending his people against whites. He denounced as invalid all treaties by which Indians ceded their lands and condemned the chieftains who had entered into such agreements. His contention was that the land was owned in common for hunting and not by individual tribes. With his brother Tenskwatawa, the Prophet, he established a town on the Tippecanoe River near Lafayette, IN, and then embarked on a mission to organize an Indian confederation to stop white encroachment. Although he advocated peaceful methods and negotiation, he did not rule out war as a last resort as he visited tribes throughout the country. While he was away, William Henry Harrison defeated the Prophet at the Battle of Tippecanoe on Nov 7, 1811, and burned the town. Tecumseh organized a large force of Indian warriors and assisted the British in the War of 1812. Tecumseh was defeated and killed at the Battle of the Thames on Oct 5, 1813.

BIRTHDAYS TODAY

Bill Dana, 71, actor, comedian ("The Steve Allen Show," "The Bill Dana Show"), born at Quincy, MA, Oct 5, 1924.

Bob Geldof, 44, singer, songwriter (lead singer and songwriter of Boomtown Rats), born at Dublin, Ireland, Oct 5, 1951.

Vaclav Havel, 59, dramatist, former President of the Republic of Czechoslovakia, born at Prague, Czechoslovakia, Oct 5, 1936.

Glynis Johns, 72, actress (*Mary Poppins, The Ref, A Little Night Music*), born Pretoria, South Africa, Oct 5, 1923.

Steve Miller, 52, musician, singer (leader Steve Miller Band; "The Joker," "Abracadabra"), born at Dallas, TX, Oct 5, 1943.

Walter Dale Miller, 70, Governor of South Dakota (R), born at New Underwood, SD, Oct 5, 1925.

Donald Pleasence, 76, actor (*The Caretaker, The Madwoman of Chaillot, The Great Escape, Halloween*), born Worksop, England, Oct 5, 1919.

OCTOBER 6 — FRIDAY
279th Day — Remaining, 86

AMERICAN LIBRARY ASSOCIATION: FOUNDING ANNIVERSARY. Oct 6. Founded at Philadelphia, PA, Oct 6, 1876.

ARKANSAS STATE FAIR AND LIVESTOCK SHOW. Oct 6-15. Barton Coliseum and State Fairground, Little Rock, AR. Est attendance: 400,000. For info: Arkansas State Fair, PO Box 166660, Little Rock, AR 72216. Phone: (501) 372-8341.

BLACK HILLS JEEP JAMBOREE. Oct 6-8. Sturgis and Deadwood, SD. Annual jamboree for 4-wheel drive enthusiasts, who try their luck on some of the roughest terrain in the Black Hills. For info: Sturgis Chamber of Commerce, PO Box 504, Sturgis, SD 57785. Phone: (605) 347-2556.

CANADA: INUVIK LIONS CLUB DELTA DAZE. Oct 6-9. Inuvik, NWT. Community fall celebration includes kids' carnival, steak barbecue, raffle of 20-oz gold bar, dances, beer gardens, helicopter rides, prince and princess crowning and casino. Est attendance: 550. Sponsor: Inuvik Lions Club, Vicki Boudreau, Box 1215, Inuvik, NWT, Canada X0E 0T0. Phone: (403) 979-2612.

CANADA: OKTOBERFEST. Oct 6-14. Kitchener and Waterloo, Ontario. "The second largest Oktoberfest in the world." More than 70 events including one of the premier parades in Canada on Canadian Thanksgiving morning. Est attendance: 700,000. For info: Kitchener-Waterloo Oktoberfest Inc, PO Box 1053, Kitchener, Ont, Canada N2G 4G1. Phone: (519) 570-4267.

COLUMBUS DAY FESTIVAL AND HOT AIR BALLOON REGATTA. Oct 6-8. Downtown square and Industrial Park, Columbus, KS. Hot Air Balloon Regatta; Balloon Glow on Friday evening; races for prizes among the balloons on Saturday and Sunday. Also, car show, arts and crafts fair, entertainment, children's festival, chili cook off and more. Est attendance: 9,000. Sponsor: Columbus Chamber of Commerce, Jean Pritchett, 320 E Maple, Columbus, KS 66725. Phone: (316) 429-1492.

COME AND TAKE IT DAY. Oct 6-Oct 8. Gonzales, TX. This celebration commemorating the first shot fired for Texas independence in 1835 is named for the defiant battle cry of the colonists when the Mexican military demanded the return of a cannon. Est attendance: 30,000. For info: Chamber of Commerce, Box 134, Gonzales, TX 78629. Phone: (210) 672-6532.

EGYPT: ARMED FORCES DAY. Oct 6. The Egyptian Army celebrates crossing into Sinai in 1973. For info: Egyptian Tourist Authority, 645 N Michigan Ave, Ste 829, Chicago, IL 60611. Phone: (312) 280-4666.

EL-SADAT, ANWAR, ASSASSINATION: ANNIVERSARY. Oct 6. Egyptian president and Nobel Peace Prize recipient, Anwar el-Sadat, was killed by assassins at Cairo on Oct 6, 1981, while he was reviewing a military parade commemorating the 1973 Egyptian-Israeli War. At least eight other persons were reported killed in the attack on Sadat. Anwar el-Sadat was born Dec 25, 1918, at Mit Abu Al-Kom, a village near the Nile River delta.

	S	M	T	W	T	F	S
October							
1995	1	2	3	4	5	6	7
	8	9	10	11	12	13	14
	15	16	17	18	19	20	21
	22	23	24	25	26	27	28
	29	30	31				

☆ Chase's 1995 Calendar of Events ☆ Oct 6

ENGLAND: *THE DAILY TELEGRAPH* CHELTENHAM FESTIVAL OF LITERATURE. Oct 6-15. Cheltenham, Gloucestershire. An annual festival of literature and related events including a drama festival, book and poetry readings, talks and discussions by writers and literary people. Est attendance: 32,000. For info: Ms. Nicki Bennett, Festival Organizer, Town Hall, Imperial Square, Cheltenham, Gloucestershire, England GL50 1QA.

FALL CELEBRATION (WITH OLE TIME FIDDLERS CONVENTION). Oct 6-15. Georgia Mountain Fairgrounds, Hiawassee, GA. The Fall Music Festival and the Ole Time Fiddlers Convention have been combined into a 10-day event called the Fall Celebration. The first weekend will be the Fall Music Festival with country music, arts and crafts and a carnival. Open invitation to any single performers or bands who wish to participate. Amateur talent. Arts, crafts and carnival continue during the week with the Ole Time Fiddlers Convention the last weekend. Est attendance: 75,000. For info: Dale Thurman, Mgr, PO Box 444, Hiawassee, GA 30546. Phone: (706) 896-4191.

FALL POWWOW AND FESTIVAL. Oct 6-8. James E. Ward Agricultural Center, Lebanon, TN. Traditional crafts, music and dance. For info: Tennessee Tourist Development, PO Box 23170, Nashville, TN 37202. Phone: (615) 741-2158.

FIREANT FESTIVAL. Oct 6-8. Marshall, TX. Arts and crafts, chili cook-off, Tour de FireAnt bike ride, 5K run, fireant calling contest, fireant round-up, rubber chicken chunking, gurning contest, parade and street dance. Sponsored by American Cyanamid. Est attendance: 50,000. For info: Pam Whisenant, Marshall Chamber of Commerce, PO Box 520, Marshall, TX 75671. Phone: (903) 935-7868.

GREAT AMERICAN BEER FESTIVAL. Oct 6-7. Denver, CO. To celebrate the depth, variety and distinctiveness of beers brewed by American breweries. More than 900 beers from most of America's breweries. 14th annual festival. Est attendance: 17,000. Sponsor: The Assn of Brewers, Box 1679, Boulder, CO 80306. Phone: (303) 447-0816.

INTERNATIONAL ASSOCIATION OF PET CEMETERIES FALL MEETING. Oct 6-8. Boston, MA. Est attendance: 70. Sponsor: Intl Assn of Pet Cemeteries, 5055 US11, Ellenburg Depot, NY 12935. Phone: (518) 594-3000.

IRELAND: IVY DAY. Oct 6. The anniversary of the death of Irish nationalist leader and Home Rule advocate Charles Stewart Parnell (q.v.) is observed, especially in Ireland, as Ivy Day. A sprig of ivy is worn on the lapel to remember Parnell. James Joyce's short story, "Ivy Day in the Committee Room," published in the collection entitled *Dubliners*, addresses this event.

LIND, JENNY: 175th BIRTH ANNIVERSARY. Oct 6. Swedish opera singer born at Stockholm, Sweden, Oct 6, 1820. She died at Malvern, England, Nov 2, 1887.

LITERALLY HAUNTED HOUSE. Oct 6-7. (Also Oct 13-14, 20-21, 27-28 & 31.) New Albany, IN. Friends of Culbertson Mansion fundraiser. Est attendance: 6,000. Sponsor: Culbertson Mansion, 914 E Main, New Albany, IN 47150. Phone: (812) 944-9600.

NATIONAL GERMAN-AMERICAN DAY. Oct 6. Celebration of German heritage, and contributions German-Americans have made to the building of the nation. A Presidential Proclamation has been issued each year since 1987. Annually, Oct 6. Sponsor: Society for German-American Studies, 3418 Boudinot Ave, Cincinnati, OH 45211.

NATIONAL PICKLED PEPPER WEEK. Oct 6-16. To recognize the zest added to the American menu by the bell, banana, cherry, chili, peperoncini and other members of the hot, mild and sweet pickled pepper family. Sponsor: Pickle Packers Intl, Inc. For info: DHM Group, Inc, PO Box 767, Holmdel, NJ 07733.

NATIONAL STORYTELLING FESTIVAL. Oct 6-8. Jonesborough, TN. The 3-day celebration showcases storytellers, stories and traditions from across America and around the world. Est attendance: 9,000. Sponsor: Natl Storytelling Assn (formerly NAPPS), PO Box 309, Jonesborough, TN 37659. Phone: (800) 525-4514.

OZARK TRAIL FESTIVAL. Oct 6-8. Heber Springs, AR. Arts and crafts, pioneer parade, 5K race, horseshoe pitch, children's activities, food concessions and much more. 30th annual festival. Est attendance: 10,000. For info: Chamber of Commerce, 1001 W Main, Heber Springs, AR 72543. Phone: (501) 362-2444.

PAUL BUNYAN SHOW. Oct 6-8. Hocking College Campus, Nelsonville, OH. Live demonstrations of forestry equipment, lumberjack contests, professional timber harvester competitions, forest industry trade show, chainsaw sculptors, Robbins Crossing interpretive history program and activities, and steam show exhibits. Est attendance: 50,000. For info: Hocking Technical College, Ohio Forestry Assn, Judy Sinnott, Public Info Dir, Ext 2102, Nelsonville, OH 45764. Phone: (614) 753-3591.

PHYSICIAN ASSISTANT (PA) DAY. Oct 6. To acknowledge the unique contribution of Physician Assistants to the delivery of health care on the anniversary of the graduation of the first class of PAs from Duke University. For info: American Academy of Physician Assistants, 950 N Washington St, Alexandria, VA 22314-1534. Phone: (703) 836-2272.

RED WING SHEEP DOG TRIAL. Oct 6-8. Red Wing, MN. Farmers' and ranchers' dogs compete. Est attendance: 400. For info: Red Wing Sheep Dog Trial Assn, Charles O'Reilly, Course Dir, Rte 4, Box 33, Red Wing, MN 55066. Phone: (800) 852-4422.

REX ALLEN DAYS. Oct 6-8. Willcox, AZ. Annual celebration honors hometown boy, Rex Allen, who gained fame as a singer, cowboy movie star and narrator for Walt Disney Productions. Activities include golf tournaments, parade, country fair, PRCA rodeo, General Willcox International Turtle Race, arts and crafts, country western concert, carnival, softball tournament, cowboy dances and miniature art show. Est attendance: 20,000. For info: Willcox Chamber of Commerce, 1500 N Circle I Rd, Willcox, AZ 85643. Phone: (602) 384-2272.

SOUTH ALABAMA STATE FAIR. Oct 6-14. Garrett Coliseum/Fairgrounds, Montgomery, AL. A midway filled with exciting rides and games, arts and crafts, exhibits, a circus, petting zoo, food and entertainment. Est attendance: 219,000. For info: Hazel Ashmore, PO Box 3304, Montgomery, AL 36109-0304. Phone: (205) 272-6831.

SWAPPIN' MEETIN'. Oct 6-7. Southeast Community College, Cumberland, KY. A celebration of the rich heritage of the mountain people. Handmade goods such as quilts and woodwork are displayed; special events include photo contest, hayride and folk singing. Phone (606) 589-2145 ext 2102 or 2013. Est attendance: 3,000. For info: Michael Corriston, 700 College Rd, Cumberland, KY 40823.

SZYMANOWSKI, KAROL: BIRTH ANNIVERSARY. Oct 6. Birthday of one of Poland's outstanding composers, whose art has played a role in Polish 20th-century music. Born at Timoshovka, Ukraine, on Oct 6, 1882. Died on Mar 28, 1937.

TENNYSON, ALFRED LORD: DEATH ANNIVERSARY. Oct 6. English poet, born at Somersby, Lincolnshire, England, on Aug 6, 1809, the fourth son of a rector. At Trinity College, Cambridge, he became part of a group of young men including Arthur Hallam, whose early death prompted Tennyson to write his elegiac poem "In Memoriam." His most celebrated works include the poems "The Lady of Shalott" and "The Lotus-eaters" and the verse novelettes *Maud, Enoch Arden, Locksley Hall Sixty Years After* and *The Idylls of the King*. Tennyson was named Poet Laureate of England in 1850 to succeed William Wordsworth, and he became a peer in 1884. Died at Aldworth, Sussex, Oct 6, 1892.

Oct 6-7 ☆ *Chase's 1995 Calendar of Events* ☆

VICTORIAN WEEK. Oct 6-15. Cape May, NJ. A 10-day celebration of Cape May's Victorian heritage, with self-guided tours of Victorian homes, antiques and crafts shows, Victorian theatre, boisterous sing-alongs, entertaining and educational lectures. Annually, the 10 days beginning the Friday before Columbus Day. Est attendance: 10,000. For info: Mid-Atlantic Center for the Arts, PO Box 340, Cape May, NJ 08204. Phone: (609) 884-5404.

WEST VIRGINIA PUMPKIN FESTIVAL. Oct 6-8. Little League Complex, Fairgrounds Rd, Milton, WV. Crafts and demonstrations, pumpkin contests, exhibits, pageants, entertainment and outstanding concessions. Est attendance: 25,000. For info: West Virginia Pumpkin Festival, PO Box 358, Milton, WV 25541. Phone: (304) 743-9222.

WESTINGHOUSE, GEORGE: BIRTH ANNIVERSARY. Oct 6. American inventor born on Oct 6, 1846. Died on Mar 12, 1914.

BIRTHDAYS TODAY

Shana Alexander, 70, journalist (formerly of "60 Minutes" Point-Counterpoint), author, born at Boston, MA, Oct 6, 1925.
Britt Ekland, 53, actress (*The Night They Raided Minsky's*), born at Stockholm, Sweden, Oct 6, 1942.
Thor Heyerdahl, 81, anthropologist, explorer, author (*Kon Tiki*), born at Larvik, Norway, Oct 6, 1914.
Ruben Angel Sierra (Garcia), 30, baseball player, born at Rio Piedras, Puerto Rico, Oct 6, 1965.
Fred Travalena, 53, actor ("Keep On Truckin'," "ABC Comedy Hour"), born New York, NY, Oct 6, 1942.
Stephanie Zimbalist, 39, actress (Laura Holt in "Remington Steele"), born at Encino, CA, Oct 6, 1956.

OCTOBER 7 — SATURDAY
280th Day — Remaining, 85

ALZHEIMER'S ASSOCIATION MEMORY WALK. Oct 7-8. Memory Walk is a fun, family-oriented event developed to raise funds and generate awareness for Alzheimer's Association Chapters across the country. More than 200 communities participate. Est attendance: 80,000. For info: Alzheimer's Assn, Mindy Leonard, 919 N Michigan Ave, Ste 1000, Chicago, IL 60611. Phone: (312) 335-5738.

AMERICAN HEART WALK. Oct 7-8. This annual event is an opportunity to promote exercise as well as to educate people about heart disease and stroke. Participants in more than 900 US cities and towns ask family and friends for a flat donation to the American Heart Association for walking the course. 200,000 participants nationwide. Annually, the first weekend in October. Call local AHA offices at (800) AHA-USA1.

ANNUAL ARTS AND CRAFTS FAIR. Oct 7-8. Base Lodge, Jay Peak Ski Resort, Jay, VT. This annual event features more than 60 statewide artisans. The show is non-juried providing many local crafters the opportunity to display their wares. Foliage tramway rides, face painting and music are all part of this fall foliage event. Est attendance: 6,000. For info: Jay Peak Resort, Rte 242, Jay, VT 05859. Phone: (802) 988-2611.

APPLE BUTTER FESTIVAL. Oct 7-8. Burton, OH. Enjoy the beautiful Geauga autumn with the aroma of fresh apple butter simmering in real copper kettles over outdoor fires, while visiting the antique and craft shows. Est attendance: 8,300. For info: Geauga County Historical Soc, Marlene F. Collins, PO Box 153, Burton, OH 44021. Phone: (216) 834-4012.

October 1995

S	M	T	W	T	F	S
1	2	3	4	5	6	7
8	9	10	11	12	13	14
15	16	17	18	19	20	21
22	23	24	25	26	27	28
29	30	31				

APPLE HARVEST FESTIVAL. Oct 7-8. (Also Oct 14-15.) South Mountain Fairgrounds, Gettysburg, PA. Celebration includes tours of orchards, apple-butter boiling, and antique cider press. Est attendance: 100,000. For info: Gettysburg Travel Council, 35 Carlisle St, Gettysburg, PA 17325. Phone: (717) 334-6274.

ART IN THE PARK. Oct 7. Demopolis, AL. The first Saturday in October, Confederate Square in historic downtown Demopolis comes alive for the annual arts and crafts show. Craftsmen from several states display their hand-made wares for the day-long show and sale. Sponsor: Demopolis Chamber of Commerce. Est attendance: 15,000. For info: Jane Gross, Box 667, Demopolis, AL 36732. Phone: (205) 289-0270.

BIG ISLAND RENDEZVOUS. Oct 7-8. Bancroft Bay City Park, Albert Lea, MN. Fur-trade era festival with bluegrass and country-old-time music, black powder shoot, tipi village, traders' row, workshops, traditional and ethnic foods. Named one of the top 25 festivals in Minnesota by Office of Tourism. Est attendance: 13,000. Sponsor: Big Island Rendezvous and Festival, Inc, Box 686, Albert Lea, MN 56007. Phone: (800) 658-2526.

CANADA: ATLANTIC WINTER FAIR. Oct 7-9. (Also Oct 13-15.) Exhibition Park, Halifax, Nova Scotia. Showcase of livestock, agricultural displays, barrel racing and pole bending contests. Est attendance: 85,000. For info: David Coombes, Special Events Coord, Atlantic Winter Fair, Box 3, Prospect Ct, RR 2, Armdale, NS, Canada B3L 4J2. Phone: (902) 876-8222.

CHOCOLATE HARVEST. Oct 7. Mattoon, IL. All chocolate-related festival including antiques, crafts and gifts, demonstrations, antique and custom car show, boulevard scarecrow contest, outdoor market exhibitors, chocolate goods and concessions. Annually, the first Saturday in October. Sponsors: Coles County Antique and Craft Dealer Association. Est attendance: 7,000. For info: Mattoon Chamber of Commerce, 1701 Wabash, Mattoon, IL 61938. Phone: (217) 235-5661.

CHOWDERFEST. Oct 7-9. Mystic Seaport, Mystic, CT. Columbus Day Weekend. A riverfront festival of New England chowders. Est attendance: 9,000. For info: Mystic Seaport, 75 Greenmanville Ave, PO Box 6000, Mystic, CT 06355-0990. Phone: (203) 572-5315.

CIRCLE CITY CLASSIC. Oct 7. Hoosier Dome, Indianapolis, IN. "Bowl"-style football game between two predominantly black universities is preceded by several days of related activities, including concerts, Princess Pageant, College Fair and parade. Est attendance: 65,000. For info: Indiana Sports Corp, 201 S Capitol Ave, Ste 1200, Indianapolis, IN 46225. Phone: (317) 237-5000.

CITY STAGE. Oct 7-8. Greensboro, NC. A free two-day street festival featuring the performing and visual arts, with food vendors, arts and crafts exhibitors and demonstrators, local, regional and national performers on six stages, kiddie rides, beer gardens, concessions, souvenirs and two warm-up events—block party and the 10K run and 5K walk for the arts. Annually, the first full weekend in October. Est attendance: 200,000. For info: United Arts Council of Greensboro, PO Box 877, Greensboro, NC 27402. Phone: (910) 333-7440.

COUNTRY FAIR ON THE SQUARE AND "ALL CAR SHOW." Oct 7-8. Downtown Square, Gainesville, TX. More than 100 vintage cars, classic cars, street rods and pickups, all makes and models on display. Entertainment, quilt show, food, arts display, barbecue cook-off, children's activities and more. Annually, the first weekend in October. Est attendance: 10,000. Sponsor: Community Revitalization, Gainesville Antique Car Club, Judy Day, 200 S Rusk, Gainesville, TX 76240. Phone: (817) 665-8871.

DALTON DEFENDERS DAY. Oct 7-8. Coffeyville, KS. Event to honor citizens killed during Dalton Gang's robbery of two banks on Oct 5, 1892. Est attendance: 10,000. For info: Conv and Visitors Bureau, PO Box 457, Coffeyville, KS 67337. Phone: (800) 626-3357.

FALL HARVEST FESTIVAL. Oct 7-8. Veteran's Memorial Park, Boyne City, MI. Historic encampment, arts and crafts show, large farmers' market. Est attendance: 17,000. For info: Deborah Thompson, Fall Harvest Festival, 28 S Lake, Boyne City, MI 49712.

FEAST OF THE HUNTERS' MOON. Oct 7-8. Fort Ouiatenon Historic Park, Lafayette, IN. Re-creation of French and Indian life at mid-1700s fur trading outpost. 5,200 participants. Est attendance: 65,000. For info: Tippecanoe County Historical Assn, 909 South St, Lafayette, IN 47901. Phone: (317) 742-8411.

FELL'S POINT FUN FESTIVAL. Oct 7-8. Fells's Point National Historic District, Baltimore, MD. Popular outdoor street festival held in Baltimore's original seaport. 300 arts and crafts vendors, flea market, five stages, Family & Children's Area, two beer gardens, International Food Row and more. Annually, the first full weekend in October. Sponsor: The Preservation Society. Est attendance: 325,000. For info: Bea Haskins, Festival Coord, 812 S Ann St, Baltimore, MD 21231. Phone: (410) 675-6756.

GEORGIA RENAISSANCE FALL FESTIVAL. Oct 7-29. (Saturdays and Sundays only.) Atlanta, GA. Theme park presents "the entertainment, romance and art of the Renaissance." More than 100 performances daily on 10 stages, thousands of unique craft items. Holiday shopping, Haunted Castle and Renaissance Trick or Treating. Phone: (404) 964-8575 or (404) 455-0553. Est attendance: 120,000. For info: Georgia Renaissance Festival, Sarah Waters, PO Box 986, Fairburn, GA 30213.

GOLDSTAR CHILI FEST. Oct 7-8. Cincinnati, OH. The Midwest's fastest growing family festival featuring sizzlin' national country acts in addition to 25 vendors serving up spicy fare & chili cook-offs. Est attendance: 80,000. For info: Downtown Council, 300 Carew Tower, 441 Vine, Cincinnati, OH 45202. Phone: (513) 579-3191.

GOPHER HILL FESTIVAL. Oct 7. Old School Grounds, Ridgeland, SC. Celebrate Gopher Hill (area known as Ridgeland), named for gopher tortoise that lives here and is on South Carolina's endangered species list. Est attendance: 12,000. For info: Jasper County Chamber of Commerce, Box 1267, Ridgeland, SC 29936. Phone: (803) 726-8126.

HARVEST DAYS. Oct 7-8. New Hampshire Farm Museum, Rte 125, Milton, NH. Gathering harvest from museum garden, pumpkin decorating, pie contest, harvest figure making, corn husking demonstration and corn husk doll making. Sponsor: New Hampshire Farm Museum. Est attendance: 500. For info: Susie McKinley, Admin Asst, PO Box 644, Milton, NH 03851. Phone: (603) 652-7840.

HARVEST OF HARMONY. Oct 7. Downtown Grand Island, NE. Harvest of Harmony, observing its 54nd annual event, is the largest high school band parade competition (120 bands last year) in the state of Nebraska. 52 float entries and 52 queen candidates also in the parade. Stadium competition for bands continues at the high school all day. Annually, the first Saturday in October. Sponsor: Grand Island Area Chamber of Commerce. Est attendance: 15,000. For info: Jan Schmidt, Coord, 309 W 2nd St, PO Box 1486, Grand Island, NE 68802. Phone: (308) 382-9210.

HIMMLER, HEINRICH: BIRTH ANNIVERSARY. Oct 7, 1900. Considered the most notorious leader of the Third Reich, second only to Hitler, Heinrich Himmler was born at Munich, Germany. His ruthless pursuit of power gained him appointment as the head of the SS, thereby in charge of criminal and political police, the concentration camp system and a private army, the Waffen SS. Himmler also held the positions Minister of the Interior, Commander of the Replacement Army and Army Group Commander of the Eastern Front by the end of the war. Himmler ingested poison, and died by his own hand on May 23, 1945, at Lüneburg, Germany.

INTERNATIONAL SPACE HALL OF FAME INDUCTION. Oct 7. Space Center, Alamogordo, NM. International recognition and induction of space pioneers and explorers into the International Space Hall of Fame. Annually, the first Saturday in October. Sponsor: Intl Space Hall of Fame Fdtn. Est attendance: 500. For info: Public Affairs, Space Center, PO Box 533, Alamogordo, NM 88311. Phone: (505) 437-2840.

JOHNNY APPLESEED DAYS. Oct 7-8. Lake City, MN. Apple pie, apple pancake breakfast, scarecrow-building contest, arts and crafts fair, apple rolling, giant maze, games. Est attendance: 5,000. For info: Lake City Area Chamber of Commerce, 212 S Washington St, Lake City, MN 55041. Phone: (800) 369-4123.

KERNVILLE STAMPEDE RODEO. Oct 7-8. (Tentative.) Kernville, CA. This is a yearly PRCA rodeo. Est attendance: 1,750. Sponsor: Recreation Complex Committee, Kernville Chamber of Commerce, PO Box 1072, Weldon, CA 93283. Phone: (619) 378-3157.

KNOX COUNTY SCENIC DRIVE. Oct 7-8. (Also Oct 14-15.) Knox County, IL. A self-conducted 100-mile driving tour through rural Spoon River Valley resplendent with fall colors. Different attractions at every stop feature food, crafts, art, antiques, fresh produce, old skills demonstrations, flea markets and games. Est attendance: 75,000. For info: Galesburg Area CVB, PO Box 631, Galesburg, IL 61402-0631. Phone: (309) 343-1194.

LAPIC WINEFEST. Oct 7. Lapic Winery, New Brighton, PA. Enjoy wine stomping, music, hay rides, ethnic food and crafts and tours of the winery at this annual festival. Annually, the first Saturday in October. For info: Beaver County Tourist Promotion Agency, April Koehler, Exec Dir, 215B Ninth St, Monaca, PA 15061. Phone: (412) 846-2031.

MAKOTI THRESHING SHOW. Oct 7-8. Makoti, ND. To acquire, rebuild and maintain antique farm machinery and motor vehicles. Threshing and other demonstrations. Annually, the first weekend in October. Est attendance: 8,000. Sponsor: Makoti Threshers, Inc, Loren Quandt, PO Box 124, Makoti, ND 58756. Phone: (701) 726-5649.

MID-CONTINENT RAILWAY STEAM TRAIN AUTUMN COLOR TOURS. Oct 7-8. (Also Oct 14-15.) Mid-Continent Railway Museum, North Freedom, WI. Authentic steam train tours amid the brilliant hues of autumn. Annually, the first and second weekends in October. Est attendance: 3,500. For info: Wisconsin Dells Visitor & Conv Bureau, PO Box 390, 701 Superior St, Wisconsin Dells, WI 53965. Phone: (800) 223-3557.

MORRO BAY HARBOR FESTIVAL. Oct 7-8. Morro Bay, CA. Celebrates a working waterfront at play! Officially launches "October Is National Seafood Month" on the west coast. Showcases seafood, fishing industry and diversity of marine life and coastal lifestyles. Features Seafood Faire and winetasting. Annually, the first full weekend in October. Phone in California: (800) 231-0592. Est attendance: 40,000. Sponsor: Morro Bay Harbor Festival, Inc, Festival Coord, PO Box 1869, Morro Bay, CA 93443. Phone: (805) 772-1155.

NATCHEZ FALL PILGRIMAGE. Oct 7-27. Natchez, MS. Tours of 24 antebellum homes furnished with fine antiques and surrounded by lovely gardens. Carriage and bus sightseeing tours, evening entertainment. Est attendance: 18,000. Sponsor: Natchez Pilgrimage Tours, PO Box 347, Corner of Canal and State Streets, Natchez, MS 39121. Phone: (800) 647-6742.

Oct 7 ☆ Chase's 1995 Calendar of Events ☆

NORTHEAST MARBLE MEET. Oct 7–8. Radisson Inn, Marlborough, MA. Auction, lunch and exhibits; collectors buy, sell and trade marbles. Annually, Columbus Day weekend. Est attendance: 500. For info: Bert Cohen, 169 Marlborough St, Boston, MA 02116. Phone: (617) 247-4754.

OCTOBERFEST. Oct 7–8. Ponca City, OK. At the Marland Estate Grounds, 901 Monument, 10 AM–9 PM Saturday, 11 AM–5 PM Sunday. Phone: (405) 765-8901 after 5 PM & weekends. Est attendance: 15,000. For info: Marland Estate Commission, Bettie J. Marsh, 1700 N Woodland Rd, Ponca City, OK 74604.

OKTOBERFEST. Oct 7–8. (Also Oct 14–15, 21–22, 28–29.) (Always four weekends.) Hermann, MO. Winery Tours, winetasting, music and entertainment at all four wineries. Also, craft demonstrations, beer gardens, museum tours, German food, bands and dancing. Sponsored by Vitners Assn. Est attendance: 70,000. For info: Hermann Visitor Center, Dolores Smith, Coord, 306 Market St, Hermann, MO 65041. Phone: (314) 486-2744.

OKTOBERFEST. Oct 7–8. St. Charles, MO. A city-wide celebration of French and German heritage. Activities include a parade, German bands, foods, and costumes. Annually, the first full weekend in Oct. For info: Convention and Visitors Bureau, St. Charles, MO, 230 S Main St, PO Box 745, St. Charles, MO 63302. Phone: (800) 366-2427.

OLD SEDGWICK COUNTY FAIR. Oct 7–8. Old Cowtown Museum, Wichita, KS. Re-creation of an 1870s fair with horse-drawn carriages, craft demonstrations and sales, music and reenactments. Est attendance: 10,000. For info: Elizabeth D. Kennedy, Assoc Dir, Old Cowtown Museum, 1871 Sim Park Dr, Wichita, KS 67203. Phone: (316) 264-0671.

OYSTER FESTIVAL. Oct 7. Chincoteague Island, VA. Oysters fixed every way possible—all you can eat. Annually, Saturday of Columbus Day weekend. Tickets can be obtained in advance by contacting Chincoteague Chamber of Commerce. Est attendance: 2,500. Sponsor: Chincoteague Chamber of Commerce, Box 258, Chincoteague Island, VA 23336. Phone: (804) 336-6161.

RILEY, JAMES WHITCOMB: BIRTH ANNIVERSARY. Oct 7. American "Hoosier" poet, born at Greenfield, IN, on Oct 7 probably in 1853, but possibly several years earlier. Riley died at Indianapolis on July 22, 1916.

ROCKPORT SEAFAIR. Oct 7–8. Ski Basin area, Rockport, TX. Features fresh-from-the-bay seafood, Gumbo cook-off, ongoing live musical entertainment, sailing regatta, "anything-that-floats-but-a-boat race," crab races, crab beauty contest, arts and crafts booths and land parade on Saturday. Annually, Columbus Day weekend. Est attendance: 40,000. For info: Rockport Seafair, PO Box 2256, Rockport, TX 78381-2256. Phone: (512) 729-3312.

ST. SIMONS BY THE SEA: AN ARTS FEST. Oct 7–8. St. Simons Island, GA. Sponsored by the Coastal Center for the Arts (formerly Island Art Center), now in its 47th year. Free admission for visitors; "juried to a quality level." For info: Mittie B. Hendrix, Exec Dir, Coastal Center for the Arts, 2012 Demere Rd, St. Simons Is, GA 31522. Phone: (912) 634-0404.

October 1995

S	M	T	W	T	F	S
1	2	3	4	5	6	7
8	9	10	11	12	13	14
15	16	17	18	19	20	21
22	23	24	25	26	27	28
29	30	31				

SIPPIN' CIDER DAYS. Oct 7–8. Leavenworth Landing, Leavenworth, KS. Musicals, juggling, dancing, crafts, food, chili cook-off and children's activities including Oz characters and child-sized replica of the Emerald City—celebrate harvest time at this family event. Est attendance: 8,000. For info: Leavenworth Conv & Visitors Bureau, 518 Shawnee, Leavenworth, KS 66048. Phone: (800) 844-4114.

VERMONT APPLE FESTIVAL. Oct 7–8. Riverside Middle School, Springfield, VT. Craft fair, apples, apple cider press and other entertainment. Annually, Columbus Day weekend. Est attendance: 5,500. For info: Springfield Chamber of Commerce, 14 Clinton St, Springfield, VT 05156. Phone: (802) 885-2779.

VIRGINIA CHILDREN'S FESTIVAL. Oct 7. Town Point Park, Norfolk, VA. An all-day family program hosted by nationally famous children's entertainers, costumed characters and five stages of entertainment. Also, magic, giant puppets, creative dance and many other activities for a day of fantasy and fun. Est attendance: 45,000. For info: Norfolk Festevents, Ltd, Promotions Dir, 120 W Main St, Norfolk, VA 23510. Phone: (804) 627-7809.

WALLACE, HENRY AGARD: BIRTH ANNIVERSARY. Oct 7. Thirty-third vice president of the US (1941–1945) born at Adair County, IA, Oct 7, 1888. Died at Danbury, CT, Nov 18, 1965.

WISE, THOMAS JAMES: BIRTH ANNIVERSARY. Oct 7. English bibliophile and literary forger, born at Gravesend, Oct 7, 1859. One of England's most distinguished bibliographic experts, he was revealed, in 1934, to have forged dozens of "first editions" and "unique" publications over a period of more than 20 years. Many of them had been sold at high prices to collectors and libraries. The forgeries in some cases purported to predate the real first editions. Wise, whose health was broken when the exposure came, died at Hampstead, England, May 13, 1937.

WOLLERSHEIM WINERY GRAPE STOMP FESTIVAL. Oct 7–8. Wollersheim Winery, Prairie du Sac, WI. Old-world tradition featuring wine tasting, grape-spitting contests, vineyard and wine cellar tours, and the "La Feet Classique Grape Stomp." Annually, the first weekend in October. Est attendance: 2,500. For info: Wisconsin Dells Visitor and Conv Bureau, 701 Superior St, Box 390, Wisconsin Dells, WI 53965. Phone: (800) 223-3557.

WROK/WZOK ANNUAL CHILI SHOOT-OUT. Oct 7. Rockford, IL. Area cooks see who can make the best chili in the county. Chili-pepper-eating and cow-chip-throwing contests. Sponsor: WROK/WZOK-FM. Est attendance: 8,000. For info: WROK/WZOK, Jan Thorpe, 3901 Brendenwood Lane, Rockford, IL 61107. Phone: (815) 399-2233.

BIRTHDAYS TODAY

June Allyson (Ella Geisman), 78, actress (*Little Women, The Shrike*, "The June Allyson Show"), born Lucerne, NY, Oct 7, 1917.

Imamu Amiri Baraka (Leroi Jones), 61, poet, dramatist, born at Newark, NJ, Oct 7, 1934.

Charles Dufoit, 59, Swiss conductor, born at Lausanne, Switzerland, Oct 7, 1936.

Thomas Keneally, 60, novelist, born at New South Wales, Australia, Oct 7, 1935.

R.D. Laing (Ronald David Laing), 68, psychiatrist, author, born at Glasgow, Scotland, Oct 7, 1927.

Yo-Yo Ma, 40, musician, born at Paris, France, Oct 7, 1955.

Al Martino (Alfred Cini), 68, actor, singer (*Hello Dolly, Phantom of the Opera*), born Philadephia, PA, Oct 7, 1927.

John Cougar Mellencamp, 44, singer (*American Fool, Uh-Huh* [first album under his full name]), born Seymour, IN, Oct 7, 1951.

Oliver Laurence North, 52, US Marine Corps Lieutenant Colonel, center of Iran-Contra controversy, candidate for senate (VA), born at San Antonio, TX, Oct 7, 1943.

☆ Chase's 1995 Calendar of Events ☆   Oct 8

OCTOBER 8 — SUNDAY
281st Day — Remaining, 84

ALVIN C. YORK DAY. Oct 8. On this day in 1918, Sergeant Alvin C. York (in the Argonne Forest, France, and separated from his patrol) killed 20 enemy soldiers and captured a hill, 132 enemy soldiers and 35 machine guns. He was awarded the US Medal of Honor and French Croix de Guerre. Interestingly, York had petitioned for exemption from the draft as a conscientious objector but was turned down by his local draft board.

AMERICAN BEER WEEK. Oct 8-14. To celebrate and recognize the significance of quality beer and its contribution to life in America. Sponsor: Assn of Brewers, Box 1679, Boulder, CO 80306. Phone: (303) 447-0816.

AMERICAN INDIAN TIME OF THANKSGIVING. Oct 8. Allentown, PA. American Indian ceremony with activities for the whole family. Annually, the second Sunday in October. Est attendance: 3,000. For info: Museum of Indian Culture, Lenni Lenape Historical Soc, RD# 2, Fish Hatchery Rd, Allentown, PA 18103-9801. Phone: (610) 797-2121.

AMERICAN SAMOA: WHITE SUNDAY. Oct 8. Second Sunday in October is a "children's day" on the island. Children demonstrate skits, prayers, songs and special presentations for parents, friends and relatives. A feast is prepared by the parents and served to the children.

ASRT RADIATION THERAPY CONFERENCE IN CONJUNCTION WITH THE AMERICAN SOCIETY FOR THERAPEUTIC RADIOLOGY AND ONCOLOGY. Oct 8-12. Premier radiation therapy conference for radiation therapists, dosimetrists, educators, managers and students. Included in the program are a vast array of continuing education and refresher courses, business meetings, career opportunities, technical exhibits, poster displays and an essay competition. Additionally, an honary presentation is scheduled to recognize a radiation therapist who has made a significant contribution to the profession of radiation. Est attendance: 1,600. Sponsor: American Soc of Radiologic Technologists, 15000 Central Ave SE, Albuquerque, NM 87123-3917. Phone: (505) 298-4500.

BRAZIL: CIRIO DE NAZARE. Oct 8-21. Greatest festival of northern Brazil, the Feast of Cirio starts on second Sunday of October in city of Belem (St. Mary of Bethlehem), capital of the state of Para. Festival lasts two weeks.

BURGOO FESTIVAL. Oct 8. Downtown, North Utica, IL. Only the "Burgoomeister" knows the secret recipe for this pioneer stew, burgoo, served outdoors at this annual Utica festival. Other events include arts and crafts, antiques, food, a reenactment of Civil War living conditions by uniformed volunteers and the firing of a Civil War cannon. Annually, the second Sunday in October. Est attendance: 20,000. For info: LaSalle Co Historical Museum, PO Box 278, Utica, IL 61373. Phone: (815) 667-4861.

CHICAGO FIRE: ANNIVERSARY. Oct 8. Great fire of Chicago began, according to legend, when Mrs O'Leary's cow kicked over the lantern in her barn on DeKoven St, on this day in 1871. The fire leveled 3½ sq miles, destroying 17,450 buildings and leaving 98,500 people homeless and about 250 people dead. Financially, the loss was $200 million. On the same day a fire destroyed the entire town of Peshtigo, WI, killing an estimated 600 people.

★ **FIRE PREVENTION WEEK.** Oct 8-14. (Approximate.) Presidential Proclamation issued annually for the first or second week in October since 1925. For many years prior to 1925, National Fire Prevention Day was observed in October.

FIRE PREVENTION WEEK. Oct 8-14. To increase awareness of the dangers of fire and to educate the public on how to stay safe from fire. Sponsor: Natl Fire Protection Assn, Public Affairs and Educ Dept, One Batterymarch Park, Quincy, MA 02269. Phone: (617) 770-3000.

HARVEST CELEBRATION. Oct 8. Woodstock, VT. Traditional celebration of the harvest featuring a husking bee, barn dance and the arrival of the giant pumpkins. Also, farm harvest activities including food preservation, traditional toy making and cider pressing. Est attendance: 1,300. Sponsor: Billings Farm and Museum, PO Box 489, Woodstock, VT 05091. Phone: (802) 457-2355.

HOME-BASED BUSINESS WEEK. Oct 8-14. Conference and other special events planned to celebrate, recognize and promote the home-based entrepreneur. Annually, the week including the second Tuesday of October. For info: American Assn of Home-Based Businesses, Beverley Williams, 7601 Dew Wood Dr, Derwood, MD 20855. Phone: (301) 963-9153.

HUNTER'S MOON. Oct 8. The full moon following Harvest Moon. So called because the moon's light in evening extends day's length for hunters. Moon enters Full Moon phase at 11:52 AM, EDT.

MOON PHASE: FULL MOON. Oct 8. Moon enters Full Moon phase at 11:52 AM, EDT.

NATIONAL METRIC WEEK. Oct 8-14. To maintain an awareness of the importance of the metric system as the primary system of measurement for the United States. Sponsor: Natl Council of Teachers of Mathematics, Communications Mgr, 1906 Association Dr, Reston, VA 22091. Phone: (703) 620-9840.

★ **NATIONAL SCHOOL LUNCH WEEK.** Oct 8-14. Presidential Proclamation issued for the week beginning with the second Sunday in October since 1962 (PL87-780 of Oct 9, 1962). Note: Not issued in 1981.

OZZIE AND HARRIET SHOW RADIO DEBUT: ANNIVERSARY. Oct 8. On their ninth wedding anniversary, Oct 8, 1944, Ozzie and Harriet Nelson made their CBS Radio debut in "The Adventures of Ozzie and Harriet." Although their sons David and Ricky were referred to frequently on air and eventually played by others, it was not until Feb 20, 1949, that David (age 12) and Rick (age 8) first appeared playing themselves on the show. "The Adventures of Ozzie and Harriet" hit television airwaves Oct 3, 1952, on ABC. The show was cancelled at the end of the 1965-66 season after 435 episodes, 409 of which were in black and white and 26 in color.

PERU: DAY OF THE NAVY. Oct 8. Public holiday in Peru.

PESHTIGO FOREST FIRE ANNIVERSARY. Oct 8. One of the most disastrous forest fires in history began at Peshtigo, WI, on this day in 1871. It burned across six counties, killing more than 1,100 persons.

RICKENBACKER, EDWARD V.: BIRTH ANNIVERSARY. Oct 8. American aviator, auto racer, war hero. Born Oct 8, 1890. Died July 23, 1973.

SCHUTZ, HEINRICH: BIRTH ANNIVERSARY. Oct 8. German musician and composer sometimes called the father of German music. Born at Kostritz, Saxony, on Oct 8, 1585. Schutz died at Dresden on Nov 6, 1672. His works enjoyed renewed attention on the occasions of the bicentennials (1885) and tricentennials (1985) of two of his most devoted followers: George Frederick Handel and Johann Sebastian Bach.

SUKKOT BEGINS AT SUNDOWN. Oct 8. Jewish Feast of Tabernacles. See "Sukkot" (Oct 9).

WESTERN SAMOA: WHITE SUNDAY. Oct 8. The second Sunday in October. For the children of Samoa, this is the biggest day of the year. On this day, traditional roles are reversed, as children lead church services, are served special foods and receive gifts of new church clothes and other special items. All the children dress in white. The following Monday is an official holiday.

Oct 8–9 ☆ *Chase's 1995 Calendar of Events* ☆

BIRTHDAYS TODAY

Rona Barrett, 59, gossip columnist, born at New York, NY, Oct 8, 1936.
David Carradine, 55, actor (*Boxcar Bertha*; "Kung-Fu" series), born Hollywood, CA, Oct 8, 1940.
Chevy Chase (Cornelius Crane), 52, comedian, actor ("Saturday Night Live," *Caddyshack, Memoirs of an Invisible Man*), born New York, NY, Oct 8, 1943.
Clodagh (Clodagh Aubry), 58, designer, born at Galway, Ireland, Oct 8, 1937.
Billy Conn, 78, former light-heavyweight champion of the world, referee, born at Pittsburgh, PA, Oct 8, 1917.
Jesse Jackson, 54, civil rights leader ("I am somebody"; "Keep hope alive"; founder of Operation PUSH [formerly Operation Bread Basket]), DC "Shadow" Senator, born Greenville, NC, Oct 8, 1941.
Sarah Purcell, 47, TV personality ("Real People"), born at Richmond, IN, Oct 8, 1948.
Sigourney Weaver, 46, actress (*Ghostbusters, Gorillas in the Mist, Aliens*), born New York, NY, Oct 8, 1949.

OCTOBER 9 — MONDAY
282nd Day — Remaining, 83

BANNEKER, BENJAMIN: DEATH ANNIVERSARY. Oct 9. American astronomer, mathematician, clockmaker, surveyor and almanac author. Called "first black man of science." Born near Baltimore, MD, in 1736. Died at Baltimore County, MD, Oct 9, 1806. Took part in original survey of city of Washington. Benjamin Banneker's *Almanac* was published in 1792–1797. A fire that started during his funeral destroyed his home, library, notebooks, almanac calculations, clocks and virtually all belongings and documents related to his life.

CANADA: THANKSGIVING DAY. Oct 9. Observed on second Monday in October each year.

★ **COLUMBUS DAY.** Oct 9. Presidential Proclamation always the second Monday in October. Observed on Oct 12 from 1934–1970 (Pub Res No 21 of Apr 30, 1934). PL90-363 of June 28, 1968 required that beginning in 1971 it would be observed on the second Monday in October.

COLUMBUS DAY OBSERVANCE. Oct 9. Public Law 90-363 sets observance of Columbus Day on the second Monday in October. Applicable to federal employees and to the District of Columbia, but observed also in most states on this day. Commemorates the landfall of Columbus in the New World, Oct 12, 1492. See also: "Columbus Day (Traditional)" (Oct 12).

DISCOVERERS' DAY IN HAWAII. Oct 9. Honors all discoverers, including Pacific and Polynesian navigators. Second Monday in October.

FESTIVAL IN THE SCHOOLS. Oct 9–13. Nashville, Tennessee and Davidson Counties, TN public and private schools. An artist-in-residence program brings area authors, storytellers, poets, songwriters, illustrators and cartoonists into classrooms to talk to children about their work. Est attendance: 6,000. For info: Marilyn Friedlander, Coord, Tennessee Humanities Council, 1003 18th Ave S, Nashville, TN 37212. Phone: (615) 320-7001.

ICELAND: LEIF ERIKSON DAY: 995 YEARS. Oct 9. Celebrates discovery of North America in the year 1000 by Norse explorer.

KOREA: ALPHABET DAY (HANGUL). Oct 9. Celebrates anniversary of promulgation of Hangul (24-letter phonetic alphabet) by King Sejong of the Yi Dynasty, in 1446.

October 1995

S	M	T	W	T	F	S
1	2	3	4	5	6	7
8	9	10	11	12	13	14
15	16	17	18	19	20	21
22	23	24	25	26	27	28
29	30	31				

★ **LEIF ERIKSON DAY.** Oct 9. (Tentative.) Pesidential Proclamation always issued for Oct 9 since 1964 (PL88-566 of Sept 2, 1964) at request. May be changed in 1995 to avoid falling on Columbus Day.

LENNON, JOHN: 55th BIRTH ANNIVERSARY. Oct 9. John Winston Lennon, English composer, musician, member of "The Beatles," a sensationally popular group of musical performers who captivated audiences first in England and Germany, and later throughout the world. Born at Liverpool, England, Oct 9, 1940. Assassinated at New York City, NY, Dec 8, 1980.

MISSION DELORES: FOUNDING ANNIVERSARY. Oct 9. The oldest building in San Francisco, CA. Formerly known as Mission San Francisco de Asis, the mission survived the great earthquake and fire of 1906. Founded Oct 9, 1776.

NATIONAL NEWSPAPER WEEK. Oct 9–15. To highlight newspapers' role in our daily lives. Sponsor: Newspaper Assn Managers Inc, Mary Ann Gentile, PO Box 28875, Atlanta, GA 30358. Phone: (404) 256-0444.

NATIONAL PET PEEVE WEEK. Oct 9–13. A chance for people to make others aware of all the little things in life they find so annoying, in the hope of changing some of them. Annually, the second full week of October. Fax: (616) 235-6853. Sponsor: Ad★America Advertising, Kelly D. Fleming, Pres, Two Fountain Pl, Ste 260, Grand Rapids, MI 49503. Phone: (616) 235-6926.

NATIONAL SCHOOL LUNCH WEEK. Oct 9–13. To celebrate good nutrition and wholesome, low-cost school lunches. Annually, the second full week in October. Sponsor: American School Food Service Assn, 1600 Duke St, 7th Fl, Alexandria, VA 22314. Phone: (703) 739-3900.

NATIVE AMERICANS' DAY (SOUTH DAKOTA). Oct 9. Observed in the state of South Dakota as a legal holiday, dedicated to the remembrance of the great Native American leaders who contributed so much to the history of South Dakota. Annually, the second Monday in October. For info: Indian Affairs Office, Public Safety Building, Pierre, SD 57501. Phone: (605) 773-3415.

SPACE MILESTONE: *SOYUZ 25* (USSR). Oct 9. Cosmonauts Vladimir Kovalyonok and Valery Ryumin launched on Oct 9, 1977. Intended link with *Salyut 6* unsuccessful. Craft returned to Earth Oct 11.

SUKKOT, SUCCOTH, or FEAST OF TABERNACLES, FIRST DAY. Oct 9. Hebrew calendar date: Tishri 15, 5756, begins nine-day festival in commemoration of Jewish people's 40 years of wandering in the desert, and thanksgiving for the fall harvest. This high holiday season closes with Shemini Atzeret (see entry on Oct 16) and Simchat Torah (see entry on Oct 17).

UGANDA: INDEPENDENCE DAY. Oct 9. National holiday commemorating achievement of autonomy on Oct 9, 1962.

UNITED NATIONS: WORLD POST DAY. Oct 9. An annual special observance of Postal Administrations of the Universal Postal Union (UPU). Info from: United Nations, Dept of Public Info, New York, NY 10017.

VIRGIN ISLANDS-PUERTO RICO FRIENDSHIP DAY. Oct 9. Columbus Day (second Monday in October) also celebrates historical friendship between peoples of Virgin Islands and Puerto Rico.

Chase's 1995 Calendar of Events — Oct 9–11

BIRTHDAYS TODAY

Jackson Browne, 45, singer, songwriter ("Running on Empty," "Lawyers in Love"), born Heidelberg, Germany, Oct 9, 1950.
Martin Gottfried, 62, drama critic, born at New York, NY, Oct 9, 1933.
Trent Lott, 54, US Senator (R, Mississippi) up for reelection in '94, born at Duck Hill, MS, Oct 9, 1941.
Russell Myers, 57, cartoonist, born at Pittsburg, KS, Oct 9, 1938.
Joe Pepitone, 55, former baseball player, born at Brooklyn, NY, Oct 9, 1940.
Donald Sinden, 72, actor (*The Day of the Jackal*), born at Plymouth, England, Oct 9, 1923.
Mike Singletary, 37, former football player ("Samurai"), born at Houston, TX, Oct 9, 1958.

OCTOBER 10 — TUESDAY
283rd Day — Remaining, 82

AGNEW RESIGNATION: ANNIVERSARY. Oct 10. On Oct 10, 1973, Spiro Theodore Agnew became the second person to resign the office of Vice President of the United States. Agnew entered a plea of no contest to a charge of income tax evasion (on contract kickbacks received while he was governor of Maryland and after he became vice president). He was sentenced to pay a $10,000 fine and serve three years probation. On Apr 27, 1981, following a class-action suit, a Maryland circuit court ruled that Agnew must pay the state $248,735 ($147,500 in kickbacks plus $101,235 in interest) for violating the public trust. Agnew was elected vice president twice, serving under President Richard M. Nixon.

BONZA BOTTLER DAY™. Oct 10. To celebrate when the number of the day is the same as the number of the month. Bonza Bottler Day™ is an excuse to have a party at least once a month. T-shirts available. For info: Elaine Fremont, 203 Waddell Rd, Taylors, SC 29687. Phone: (803) 244-2023.

DOUBLE TENTH DAY. Oct 10. Tenth day of 10th month, Double Tenth Day, is observed by many Chinese as anniversary of the outbreak of the revolution against the imperial Manchu dynasty, Oct 10, 1911. Sun Yat-sen and Huan Hsing were among the revolutionary leaders.

FIJI: INDEPENDENCE DAY. Oct 10. National holiday.

HAYES, HELEN: BIRTH ANNIVERSARY. Oct 10, 1900. Actress Helen Hayes, often called the First Lady of the American theater, was born at Washington, DC. Haye's greatest stage triumph was her role as the long-lived British monarch Queen Victoria in the play *Victoria Regina*. Her first great success was in *Coquette* (1927). She won an Academy Award for best actress for her first major film role in *The Sin of Madelon Claudet* (1931) and also won best supporting actress for her role in *Airport* (1971). Helen Hayes died Mar 17, 1993, at Nyack, NY.

JAPAN: HEALTH-SPORTS DAY. Oct 10. National holiday to encourage physical activity for building sound body and mind. Created in 1966 to commemorate the day of the opening of the 18th Olympic Games in Tokyo on Oct 10, 1964.

KRUGER, PAUL: BIRTH ANNIVERSARY. Oct 10. Stephanus Johannes Paulus Kruger, former president of the South African Republic and a leader of the Boers, was born Oct 10, 1825. His birthday is commemorated in South Africa. Kruger died in Switzerland on July 14, 1904.

NORSK HOSTFEST. Oct 10–14. State Fairgrounds, Minot, ND. The Northern Plains' biggest ethnic festival draws thousands of people to Minot for Scandinavian and American entertainment, Scandinavian delicacies, arts and crafts exhibits, and dignitaries representing Sweden, Norway, Denmark, Iceland and Finland with big name entertainment nightly. Toll-free phone: (800) 264-2626. Est attendance: 65,000. For info: Bruce Nelson, Minot Conv & Visitors Bureau, 1015 S Broadway, PO Box 2066, Minot, ND 58702. Phone: (701) 857-8206.

OKLAHOMA HISTORICAL DAY. Oct 10. Oklahoma.

VERDI, GIUSEPPI: BIRTH ANNIVERSARY. Oct 10. Italian composer. Born at Le Roncole, Italy, on Oct 10, 1813. His 26 operas include *Rigoletto*, *Il Trovatore*, *La Traviata* and *Aida*, and are among the most popular of all operatic music today. Died at Milan, Italy, on Jan 27, 1901.

BIRTHDAYS TODAY

Wilhelmina Holladay, 73, philanthropist, museum founder and president, interior designer, born at Elmira, NY, Oct 10, 1922.
Martina Navratilova, 39, former tennis player, born at Prague, Czechoslovakia, Oct 10, 1956.
Harold Pinter, 65, dramatist, director (*Butley*), born at London, England, Oct 10, 1930.
David Lee Roth, 40, singer, musician (lead with Van Halen; "Jump," *Eat 'Em and Smile*), born Bloomington, IN, Oct 10, 1955.
Tanya Tucker, 37, singer ("Delta Dawn," "Lizzie and the Rainman"), born at Seminole, TX, Oct 10, 1958.
Ben Vereen, 49, actor, singer, dancer (Tony for *Pippin*; *Roots*, *All That Jazz*, "Webster"), born at Miami, FL, Oct 10, 1946.

OCTOBER 11 — WEDNESDAY
284th Day — Remaining, 81

★ **GENERAL PULASKI MEMORIAL DAY.** Oct 11. Presidential Proclamation always issued for Oct 11 since 1929. Requested by Congressional Resolution each year from 1929–1946. (Since 1947 has been issued by custom.) Note: Proclamation 4869, of Oct 5, 1981, covers all succeeding years but proclamations have been issued in each successive year.

HOPPE, WILLIE: BIRTH ANNIVERSARY. Oct 11. Celebrated champion American billiards player, William F. Hoppe, won tournaments from age 18 until age 64. Born Oct 11, 1887, at Cronwall on the Hudson. Hoppe died at Miami, FL, on Feb 1, 1959.

MUSEUM OF APPALACHIA TENNESSEE FALL HOMECOMING. Oct 11–15. Museum of Appalachia, Norris, TN. Mountain, folk music and craft festival featuring several hundred old-time mountain craftsmen and musicians. Est attendance: 50,000. Sponsor: Museum of Appalachia, John Rice Irwin, PO Box 0318, Norris, TN 37828. Phone: (615) 494-7680.

NATIONAL COMING OUT DAY. Oct 11. International day of visibility for the lesbian and gay community since 1988. Local community groups sponsor activities and events which in the past have included "coming out" dances, rallies and demonstrations, educational films, fairs and workshops, literature drops, fund raisers, and religious blessings of lesbian and gay couples and families. Annually, Oct 11. For info: Human Rights Campaign Fund, PO Box 34640, Washington, DC 20043-4640. Phone: (800) 866-6263.

PARSON WEEMS: BIRTH ANNIVERSARY. Oct 11. Mason Locke Weems, 19th child of David Weems, and father of 10, was born in Anne Arundel County, MD, on Oct 11, 1759. An Episcopal clergyman and travelling bookseller, Weems is remembered for the fictitious stories he presented as historical fact. Best known of his "fables" is the story describing George Washington cutting down his father's cherry tree with a hatchet. Weems's fictionalized histories, however, delighted many readers who accepted them as true. They became immensely popular and were best sellers for many years. Weems died May 23, 1825, at Beaufort, SC.

PATENT ISSUED FOR FIRST ADDING MACHINE: ANNIVERSARY. Oct 11. On Oct 11, 1887, a patent was granted to Dorr Eugene Felt for the Comptmeter, which was the first adding machine known to be absolutely accurate at all times.

RELICS AND RODS RUN TO THE SUN. Oct 11–15. Ramada London Bridge Resort, Lake Havasu City, AZ. Pre-1960 car show. For info: Visitors and Conv Bureau, 1930 Mesquite Ave, Ste 3, Lake Havasu City, AZ 86403. Phone: (800) 242-8278.

ROBINSON, ROSCOE, JR: BIRTH ANNIVERSARY. Oct 11, 1928. The first black American to achieve the Army rank of four-star general. Born at St. Louis, MO, and died at Washington, DC, on July 22, 1993.

◆ **ROOSEVELT, ANNA ELEANOR: BIRTH ANNIVERSARY.** Oct 11. Wife of Franklin Delano Roosevelt, 32nd president of the US, was born at New York, NY, Oct 11, 1884. She led an active and independent life and was the first wife of a president to give her own news conference in the White House (1933). Widely known throughout the world, she was affectionately called "the first lady of the world." She served as US delegate to the United Nations General Assembly for a number of years before her death at New York, NY, Nov 7, 1962. A prolific writer, she wrote in *This Is My Story*, "No one can make you feel inferior without your consent."

STONE, HARLAND FISKE: BIRTH ANNIVERSARY. Oct 11. Former associate justice and later chief justice of the US Supreme Court who wrote more than 600 opinions and dissents for that court, was born at Chesterfield, NH, on Oct 11, 1872. Stone served on the Supreme Court from 1925 until his death, at Washington, DC, on Apr 22, 1946.

UNITED NATIONS: INTERNATIONAL DAY FOR NATURAL DISASTER REDUCTION. Oct 11. The General Assembly made this designation for the second Wednesday of October each year as part of its efforts to foster international cooperation in reducing the loss of life, property damage and social and economic disruption caused by natural disasters. Info from: United Nations, Dept of Public Info, New York, NY 10017.

BIRTHDAYS TODAY

Joseph Alsop, 85, journalist, author, born at Avon, CT, Oct 11, 1910.
Art Blakey, 76, musician, born at Pittsburgh, PA, Oct 11, 1919.
Robert Gale, 50, physician, co-founder of the International Bone Marrow Registry, born at Brooklyn Heights, NY, Oct 11, 1945.
Daryl Hall, 47, singer, musician (Hall and Oates), born at Pottstown, PA, Oct 11, 1948.
Patty Murray, 45, US Senator (D, Washington), born at Seattle, WA, Oct 11, 1950.
William Perry, 68, US Secretary of Defense in Clinton administration, born at Vandergrift, PA, Oct 11, 1927.
Jerome Robbins, 77, choreographer, born at New York, NY, Oct 11, 1918.
Roy Scheider, 60, actor (*Jaws*, "Seaquest"), born Orange, NJ, Oct 11, 1935.

OCTOBER 12 — THURSDAY
285th Day — Remaining, 80

AUTUMN GLORY FESTIVAL. Oct 12-15. Oakland, MD. Foliage celebration with state banjo and fiddle contests, parade, arts and crafts and antique show. Est attendance: 50,000. Sponsor: Deep Creek Lake-Garrett County Promotion Council, Festival HQ, Court House, 200 S Third St, Oakland, MD 21550. Phone: (301) 334-1948.

COLUMBUS DAY (TRADITIONAL). Oct 12. Public holiday in most countries in the Americas and in most Spanish-speaking countries. Observed under different names and on different dates (most often, as in US, on the second Monday in October). Anniversary of Christopher Columbus's arrival, Oct 12, 1492, after a dangerous voyage across "shoreless Seas," at the Bahamian Island of Guanahani, which he renamed El Salvador and claimed in the name of the Spanish crown. In his *Journal*, he wrote: "As I saw that they (the natives) were friendly to us, and perceived that they could be much more easily converted to our holy faith by gentle means than by force, I presented them with some red caps, and strings of beads to wear upon the neck, and many other trifles of small value, wherewith they were much delighted, and becamed wonderfully attached to us." See also: "Columbus Day Observance" (Oct 11).

EQUATORIAL GUINEA: INDEPENDENCE DAY. Oct 12. National holiday.

FAIR OF 1850. Oct 12-Nov 12. Westville Village, Lumpkin, GA. Special demonstrations of harvest-time activities, cane grinding and syrup making. The last remaining antebellum animal-powered cotton gin will be in operation. Est attendance: 10,000. Sponsor: Westville Village, Patty Cannington, PO Box 1850, Lumpkin, GA 31815. Phone: (912) 838-6310.

INTERNATIONAL MOMENT OF FRUSTRATION SCREAM DAY. Oct 12. To share any or all of our frustrations, all citizens of the world will go outdoors at twelve hundred hours Greenwich time and scream for thirty seconds. We will all feel better, or Earth will go off its orbit. Annually, Oct 12. Phone: (212) 388-8673 or (717) 274-8451. For info: Thomas and Ruth Roy, Wellness Permission League, 2105 Water St, Lebanon, PA 17046.

McNAIR, RONALD E.: 45th BIRTH ANNIVERSARY. Oct 12. Ronald E. McNair, a 35-year-old physicist, was the second black American astronaut in space (Feb 1984). He was born at Lake City, SC, on Oct 12, 1950. As mission specialist for the crew, he perished in the space shuttle *Challenger* explosion on Jan 28, 1986. See also: "Challenger Space Shuttle Explosion Anniversary" (Jan 28).

MEXICO: DIA DE LA RAZA. Oct 12. Columbus Day is observed as the "Day of the Race," a fiesta time to commemorate the discovery of America as well as the common interests and cultural heritage of the Spanish and Indian peoples and the Hispanic nations.

NATIONAL DESSERT DAY. Oct 12. Held each year on the second Thursday in October to focus on the pleasures of sweet and fanciful desserts. Est attendance: 250. Sponsor: Martini & Rossi Asti, c/o The Alden Group, 52 Vanderbilt Ave, New York, NY 10017. Phone: (212) 867-6400.

PORTUGAL: LAST PILGRIMAGE TO FATIMA. Oct 12-13. Crowds of pilgrims from Portugal and all over the world travel to Fatima to celebrate the last apparition of the Virgin to the little shepherds on Oct 12, 1917. For info: Portuguese Natl Tourist Office, 590 Fifth Ave, New York, NY 10036. Phone: (212) 354-4403.

RACIAL BRAWL ON *KITTY HAWK*: ANNIVERSARY. Oct 12-13. On Oct 12-13, 1972, 46 black and caucasian sailors were injured in a racial fight that broke out on the *Kitty Hawk* while the aircraft carrier was stationed off North Vietnam. According to the *New York Times*, Oct 14, 1972, "The Navy has had the reputation in the black community of being a service in which blacks are limited to stewards' jobs." The previous year, 1971, the Navy had begun a five-year program to increase the numbers of black recruits to match their 12% level in the general population.

SPAIN: NATIONAL HOLIDAY. Oct 12.

TRUMBULL, JONATHAN: BIRTH ANNIVERSARY. Oct 12. American patriot, counselor and friend of George Washington, governor of Connecticut Colony, born Oct 12, 1710. Died, Aug 17, 1785.

Chase's 1995 Calendar of Events — Oct 12-13

WORLD SCRABBLE CHAMPIONSHIP. Oct 12-15. England. Players compete for the world championship in the popular game invented by unemployed architect Alfred Butts in 1931. Est attendance: 64. Sponsor: Natl Scrabble Assn, Kathy Hummel, c/o Williams & Co, Box 700, Front St Garden, Greenport, NY 11944. Phone: (516) 477-0033.

BIRTHDAYS TODAY

Susan Anton, 45, singer ("Killin' Time" with Fred Knoblock), actress (*Goldengirl*), born Yucaipa, CA, Oct 12, 1950.
Kirk Cameron, 25, actor ("Growing Pains"), born at Panorama City, CA, Oct 12, 1970.
John Engler, 47, Governor of Michigan (R), born at Mt. Pleasant, MI, Oct 12, 1948.
Dick Gregory, 63, comedian, author, activist, born at St. Louis, MO, Oct 12, 1932.
Tony Kubek, 59, sportscaster, born at Milwaukee, WI, Oct 12, 1936.
Jean Nidetch, 72, founder of Weight Watchers, born at Brooklyn, NY, Oct 12, 1923.
Luciano Pavarotti, 60, opera singer, actor, born at Modena, Italy, Oct 12, 1935.
Adam Rich, 27, actor (Nicholas Bradford on "Eight Is Enough"), born at Brooklyn, NY, Oct 12, 1968.

OCTOBER 13 — FRIDAY
286th Day — Remaining, 79

AMERICA'S SEXY WIVES CONTEST. Oct 13-14. Holiday Inn Pyramid, Albuquerque, NM. Contest to pick the sexiest wife in America. Wives will be judged on their commitment to their marriages; the reason they feel they represent sexiness; and their charm, poise and creative expression. Annually, the second weekend in October. For info: Rose Smith, Pres, Married Mistress & Monogamous Male Assn, 4900 Mesa Bonita Ct NW, Albuquerque, NM 87120. Phone: (505) 899-3121.

BOB EVANS FARM FESTIVAL. Oct 13-15. Rio Grande, OH. More than 150 craftspersons, clothed in authentic costumes and using tools and techniques developed more than 100 years ago, demonstrate and sell their crafts. Continuous live entertainment, contests and homemade food. Est attendance: 60,000. Sponsor: Bob Evans Farms, Inc, PR Dept, 3776 S High St, Columbus, OH 43207. Phone: (614) 245-5305.

BROWN, JESSE LEROY: BIRTH ANNIVERSARY. Oct 13. Jesse Leroy Brown was the first black American naval aviator and also the first black naval officer to lose his life in combat when he was shot down over Korea on Dec 4, 1950. On Mar 18, 1972, USS *Jesse L. Brown* was launched as the first ship to be named in honor of a black naval officer. Brown was born on Oct 13, 1926, at Hattiesburg, MS.

BUTTERFIELD OVERLAND STAGE DAYS. Oct 13-15. Benson, AZ. Commemorates when Benson was a stop for the Overland Stage and the Southern Pacific Railroad came through to make Benson a "Hub City" for three major rail lines. Parade, stage coach rides, auto show, entertainment, game and food booths, exhibitions, dinner/dance, chili cookoff, beard growing contests, essay contests, ropings, 10K run. 10th annual. Annually, the second weekend in October. Est attendance: 5,000. For info: Kay Daggett, Program Admin, Benson-San Pedro Valley C of C, PO Box 2255, Benson, AZ 85602. Phone: (602) 586-2842.

CRANBERRY HARVEST WEEKEND. Oct 13-15. Nantucket Island, MA. Guided tours of bogs, cookery contest, craft exhibitions at this 5th annual festival. Sponsor: Nantucket Island Chamber of Commerce, 48 Main St, Nantucket, MA 02554-3595.

DANVILLE HARVEST JUBILEE AND WORLD TOBACCO AUCTIONEERING CHAMPIONSHIP. Oct 13-15. Auctioneers Park, Danville, VA. A celebration of the harvest season with arts, crafts, entertainment, contests and more. Auctioneering championship held to recognize the art of tobacco auctioneering. Modern-day tobacco auction system used around the US and other countries originated in Danville, VA, in the 1850s. Auctioneers can sell a pile of tobacco at a speed of 400 to 500 words a minute. Est attendance: 40,000. For info: Danville Harvest Jubilee, Tish Lindsey, Box 3300, Danville, VA 24543. Phone: (804) 799-5200.

FRIDAY THE THIRTEENTH. Oct 13. Variously believed to be a lucky or unlucky day. Every year has at least one Friday the 13th, but never more than three. Two Fridays in 1995 fall on the 13th day, one in January and one in October. Fear of the number 13 is known as triskaidekaphobia.

GUMBO FESTIVAL. Oct 13-15. Bridge City, LA. To promote Cajun-French culture and provide the opportunity for people from everywhere to enjoy continuous Cajun entertainment on an outdoor stage and cuisine. Gumbo cooking contests and 5K Bridge Run. Annually, the second weekend in October. Est attendance: 150,000. Sponsor: Holy Guardian Angels Church, Rev Msgr J. Anthony Luminais, Pastor, Box 9069, Bridge City, LA 70094. Phone: (504) 436-4712.

HALLOWEEN FRIGHT NIGHT. Oct 13-15. (Also Oct 20-22, 26-31.) At Mt Hope's Scaregrounds fearsome creatures roam the village streets, and a suspenseful stage show with visual and lighting effects gives you the scare of your life. Est attendance: 25,000. For info: Jennifer L Wenrich, Mgr PR, PO Box 685, Cornwall, PA 17016. Phone: (717) 665-7021.

HAUNTED TRAIL AND KIDDIE TRAIL. Oct 13-15. (Also Oct 20-22, 27-29.) Ritter Park Amphitheater, Huntington, WV. A 15-minute haunted walk on a scary trail. Also featured, a spooky trail for small kids. Est attendance: 8,000. For info: Rick Abel, Greater Huntington Park and Rec District, PO Box 9361, Huntington, WV 25704. Phone: (304) 696-5954.

HOOD RIVER VALLEY HARVEST FEST. Oct 13-15. Hood River, OR. The valley welcomes visitors for two days of continuous entertainment, crafts, fresh locally-grown produce and fruits. 13th annual festival. Est attendance: 30,000. For info: Hood River Chamber of Commerce, Port Marina Park, Hood River, OR 97031. Phone: (800) 366-3530.

KING BISCUIT BLUES FESTIVAL. Oct 13-14. Downtown Historic District, Helena, AR. Two days of the best in Delta Blues. Many of the Blues "Greats" got their start in Helena on the 1940s "King Biscuit Time" radio show and return to perform. "King Biscuit Time" first aired on station KFFA in Helena in 1941. Other activities include gospel music stage, barbecue cook-off, art gallery and games. 10th annual festival. Annually, the second weekend in October. Est attendance: 50,000. For info: Main Street Helena, PO Box 247, Helena, AR 72342. Phone: (501) 338-9144.

NATIONAL SCHOOL CELEBRATION. Oct 13. The National School Celebration will provide a high-profile celebration uniting our nation's youth during regular school hours for a patriotic observance. Every school in the nation is invited to participate. This event perpetuates the original spirit of the 1892 National School Celebration declared by Pres Benjamin Harrison, for which the first pledge of allegiance was written. Annually, the second Friday in October. For info: Celebration USA, 17853 Santiago Blvd, Ste 107, Villa Park, CA 92667. Phone: (714) 283-1892.

Oct 13　★ *Chase's 1995 Calendar of Events* ★

NATIONAL WILD TURKEY CALLING CONTEST AND TURKEY TROT FESTIVAL. Oct 13-14. Town Square, Yellville, AR. 50th annual wild turkey calling contest, turkey trivia, turkey shoot, Miss Turkey Trot pageant, turkey dinner, arts, crafts, Mr. & Ms. Drumsticks, musical entertainment, Turkey Trot 5K run, parade. Annually, the second weekend in October. Est attendance: 14,000. For info: Chamber of Commerce, PO Box 369, Yellville, AR 72687. Phone: (501) 449-4676.

NEMONT/AGRIBITION. Oct 13-14. Wolf Point, MT. Annual bull sale and agribusiness show: two days of exhibits and entertainment ending with a major bull sale. Est attendance: 1,750. For info: Wolf Point Chamber of Commerce, Box 237, Wolf Point, MT 59201. Phone: (406) 653-2012.

NORMAN ROCKWELL'S HOME-COMING MARINE COVER FOR THE *SATURDAY EVENING POST*: 50th ANNIVERSARY. Oct 13. Norman Rockwell used a marine by the name of Peters as the central character for his Oct 13, 1945 *Post* cover depicting a beribboned young marine in a mechanic's garage, surrounded by young and old neighborhood friends. The ribbons Peters wore on his chest for this illustration were "earned the hard way" according to Rockwell. The October cover appeared shortly after the Japanese signed the act of surrender aboard the battleship *Missouri* off Tokyo, Sept 2, 1945.

NORTH CAROLINA STATE FAIR. Oct 13-22. State Fairgrounds, Raleigh, NC. Agricultural Fair with livestock, arts and crafts, home arts, entertainment and carnival. Est attendance: 675,000. For info: Sam Rand, Mgr, North Carolina State Fair, 1025 Blue Ridge Blvd, Raleigh, NC 27607. Phone: (919) 821-7400.

OKTOBERFEST AND THANKS CANADA! CELEBRATION. Oct 13-15. Fairgrounds, Helena, MT. German-style celebration "held in a very large tent" with German food, beer and high quality free entertainment. A variety of side events as well. Annually, the second weekend in October. Est attendance: 30,000. For info: Don Gilbreath, 920 E Lyndale, Helena, MT 59601. Phone: (406) 442-6449.

PARKE COUNTY COVERED BRIDGE FESTIVAL. Oct 13-22. Rockville, IN. 32 covered bridges, turn-of-the-century theme. Guided bus tours on bridge routes to many rural villages. Hundreds of booths of arts, crafts, demonstrations, old-fashioned homemade foods and a Farmers' Market. Billie Creek Village. Est attendance: 1,000,000. Sponsor: Parke County, Inc, Anne Lynk, PO Box 165, Rockville, IN 47872-0165. Phone: (317) 569-5226.

October 1995

S	M	T	W	T	F	S
1	2	3	4	5	6	7
8	9	10	11	12	13	14
15	16	17	18	19	20	21
22	23	24	25	26	27	28
29	30	31				

PITCHER, MOLLY: BIRTH ANNIVERSARY. Oct 13. "Molly Pitcher," heroine of the American Revolution, was a water carrier at the Battle of Monmouth (Sunday, June 28, 1778) where she distinguished herself by loading and firing a cannon after her husband, John Hays, was wounded. Affectionately known as "Sergeant Molly" after General Washington issued her a warrant as a noncommissioned officer. Her real name was Mary Hays McCauley (née Ludwig). Born near Trenton, NJ, Oct 13, 1754. Died at Carlisle, PA, on Jan 22, 1832.

ST. EDWARD, THE CONFESSOR: FEAST DAY. Oct 13. King of England, 1042-1066, Edward was the son of King Ethelred the Unready. Born in 1003, he died on Jan 5, 1066. On Oct 13, 1163, his remains were transported in a ceremony which was of national interest. Since then Oct 13 has been observed as his principal feast day.

SCOTLAND: BORDERS FESTIVAL. Oct 13-29. Various venues throughout the Scottish Borders. A widely varied program of drama, music, visual and community arts created from the rich heritage and traditions in the Scottish Borders. For info: Ian Yates, Administrator, Leisure and Rec Dept, Ettrick and Lauderdale District Council, Paton St, Galashiels, Scotland TD1 3DL.

SOUTHERN FESTIVAL OF BOOKS: A CELEBRATION OF THE WRITTEN WORD. Oct 13-15. Legislative Plaza, Nashville, TN. To promote reading, writing, the literary arts and a broader understanding of the language and culture of the South, this 7th annual festival will feature readings, talks and panel discussions by up to 150 authors; exhibit booths of publishing companies and bookstores; autographing sessions; a comprehensive children's program; and the Cafe Stage, a performance corner for authors, storytellers and musicians. Est attendance: 25,000. For info: Marilyn Friendlander, Coord, Tennessee Humanities Council, 1003 18th Ave S, Nashville, TN 37212. Phone: (615) 320-7001.

SPACE MILESTONE: US NATIONAL COMMISSION ON SPACE. Oct 13. President Reagan signed executive order creating a National Commission on Space to prepare 20-year agenda for civilian space program. Oct 13, 1984.

STAMP EXPO '95: ANAHEIM. Oct 13-15. Grand Hotel, Anaheim, CA. Fax: (818) 988-4337. Est attendance: 4,000. Sponsor: Intl Stamp Collectors Society, PO Box 854, Van Nuys, CA 91408. Phone: (818) 997-6496.

STATE CRAFT FESTIVAL. Oct 13-15. Tyler State Park, Richboro, PA. State-wide juried craft festival. Est attendance: 25,000. For info: Pennsylvania Designer Craftsmen, Box 238, State College, PA 16804. Phone: (215) 579-5997.

SUGARLOAF CRAFT FESTIVAL. Oct 13-15. Maryland State Fairgrounds, Timonium, MD. More than 400 professional artists and craftspeople, demonstrations, children's entertainment and live music. Est attendance: 35,000. Sponsor: Sugarloaf Mt Works, Inc, 200 Orchard Ridge Dr #215, Gaithersburg, MD 20878. Phone: (301) 990-1400.

TENNESSEE RIVER FIDDLERS CONVENTION. Oct 13-14. McFarland Park, Florence, Al. The lazy Tennsse River will be the backdrop to the sounds of banjos, dobros, guitars and fiddles during the Bluegrass celebration. Musicians, both young and old, compete for over $3500 in prize money in 15 categories. Annually, the second full weekend in October. Est attendance: 10,000. Sponsor: City of Florence, Park and Recreation Dept, 2500 Chisholm Rd, PO Box 2040, Florence, AL 35630. Phone: (205) 760-6416.

US NAVY: AUTHORIZATION ANNIVERSARY. Oct 13. Commemorates legislation passed by Second Continental Congress Oct 13, 1775, authorizing the acquisition of ships and establishment of a navy.

VIRCHOW, RUDOLF: BIRTH ANNIVERSARY. Oct 13. German political leader, scientist, teacher and author. Called "the founder of cellular pathology." Born at Schivelbein, Prussia, on Oct 13, 1821. Died at Berlin, Sept 5, 1902.

VIRGINIA CHRISTMAS SHOW. Oct 13–15. Prince William County Fairgrounds, Manassas, VA. 250 artisans, crafters, specialty Christmas food shops, legendary "Sgt Santa" displaying and selling. Craft demonstrations, entertainment and more. 5th annual show. Est attendance: 25,000. For info: Virginia Show Productions, PO Box 305, Chase City, VA 23924. Phone: (804) 372-3996.

BIRTHDAYS TODAY

Sammy Hagar, 46, singer, musician ("Your Love Is Driving Me Crazy"), born at Monterrey, CA, Oct 13, 1949.
Marie Osmond, 36, actress, singer (Donny and Marie, "Ripley's Believe It or Not"), born Ogden, UT, Oct 13, 1959.
Jerry Lee Rice, 33, football player, born at Starkville, MS, Oct 13, 1962.
Glenn Anton (Doc) Rivers, 34, basketball player, born at Maywood, IL, Oct 13, 1961.
Nipsey Russell, 71, comedian, actor, born at Atlanta, GA, Oct 13, 1924.
Paul Simon, 54, singer/songwriter (with Simon and Garfunkel; "The Sounds of Silence," "Mrs. Robinson," "Bridge over Troubled Water"; solo: "50 Ways to Leave Your Lover," "Gracelend"), born Newark, NJ, Oct 13, 1941.
Margaret Hilda Roberts Thatcher, 70, former Prime Minister of England, born at Grantham, England, Oct 13, 1925.
Reggie Theus, 38, former basketball player, born at Inglewood, CA, Oct 13, 1957.

OCTOBER 14 — SATURDAY
287th Day — Remaining, 78

A IS FOR APPLES WEEKEND. Oct 14–15. Chadds Ford, PA. 18th-century apple cookery will be demonstrated at the Barns-Brinton House the weekend of the Apple Wine Festival at the Chaddsford winery. Visitors stroll the footpath between the winery and the historic house that was once William Barns' Tavern. Tour the tavern and have samples of apple-based foods. Annually, the middle weekend in October. Est attendance: 500. For info: Chadds Ford Historical Society, Box 27, Chadds Ford, PA 19317. Phone: (610) 988-7376.

ALASKA DAY CELEBRATION. Oct 14–18. Sitka, AK. Celebration of the transfer ceremony in which the Russian flag was lowered and the Stars and Stripes raised, formally transferring the ownership of Alaska to the United States on Oct 18, 1867. Annually, Oct 14–18. For info: Sitka Conv and Visitors Bureau, Box 1226, Sitka, AK 99835. Phone: (907) 747-5940.

ALMA COLLEGE: ANNIVERSARY. Oct 14. Formal establishment of Alma College, at Alma, MI, was achieved by a Presbyterian Synod resolution on Oct 14, 1886.

BE BALD AND BE FREE DAY. Oct 14. For those who are bald and who either do wear or do not wear a wig or toupee, this is the day to go "shiny" and be proud. Annually, Oct 14. Phone: (212) 388-8673 or (717) 274-8451. For info: Thomas or Ruth Roy, Wellness Permission League, 2105 Water St, Lebanon, PA 17046.

BELIZE: COLUMBUS DAY. Oct 14. Public Holiday.

BENTONSPORT ARTS FESTIVAL. Oct 14–15. Bentonsport National Historic District, IA. Juried art show held as part of a countywide scenic drive. Variety of media represented. Held on river bank in historic and picturesque restored 1840s village. Food and lodging available locally. Guided tours via horse-drawn wagon. Demonstrations. Annually, the second full weekend in October. Est Attendance: 12,000. For info: Greef General Store, Bentonsport, PO Box 9, Keosauqua, IA 52565. Phone: (319) 592-3579.

CATOCTIN COLORFEST ARTS AND CRAFTS SHOW. Oct 14–15. Thurmont, MD. Annually, the second weekend in October. Est attendance: 100,000. For info: Catoctin ColorFest, Inc, Beverly Zienda, Pres, Box 33, Thurmont, MD 21788. Phone: (301) 271-4432.

DOZENAL SOCIETY OF AMERICA: ANNUAL MEETING. Oct 14. Garden City, Long Island, NY. Est attendance: 24. For info: Prof. Gene Zirkel, Nassau Community College, Garden City, Long Island, NY 11530. Phone: (516) 669-0273.

EISENHOWER, DWIGHT DAVID: 105th BIRTH ANNIVERSARY. Oct 14. The 34th president of the US, Dwight David Eisenhower, was born at Denison, TX, Oct 14, 1890. Serving two terms as president, Jan 20, 1953–Jan 20, 1961, Eisenhower was the first president to be baptized after taking office (Sunday, Feb 1, 1953). Nicknamed "Ike," he held the rank of five-star general of the army (resigned in 1952, and restored by act of Congress in 1961). He served supreme commander of the Allied forces in western Europe during WWII. In his Farewell Address (Jan 17, 1961), speaking about the "conjunction of an immense military establishment and a large arms industry," he warned: "In the councils of government, we must guard against the acquisition of unwarranted influence, whether sought or unsought, by the military-industrial complex. The potential of the disastrous rise of misplaced power exists and will persist." An American hero, Eisenhower died at Washington, DC, Mar 28, 1969.

FOREST CRAFTS/SCENIC DRIVE FESTIVAL. Oct 14–15. Van Buren County, IA. Scenic landscapes, historic architecture, wood carvers, buckskinners camp, flea market, quilt show, arts festival, farm toy show, country music show, food fair, crafts and more. Annually, the second full weekend in October. Est attendance: 12,000. Sponsor: Villages of Van Buren, Inc, Mary E. Muir, Exec Dir, PO Box 9, Keosauqua, IA 52565. Phone: (800) 868-7822.

GAUTIER MULLET FESTIVAL. Oct 14. Jackson County Community College Campus, Gautier, MS. Arts and crafts booths, local talent and entertainment, mullet cookoff, food booths, exhibitions, barrel painting, poster contest, free hayride and other events. Annually, the second Saturday in October. Est attendance: 18,000. Sponsor: City of Gautier, 3330 Hwy 90, Gautier, MS 39553. Phone: (601) 497-1878.

INTERPLANETARY CONFEDERATION DAYS. Oct 14–15. A celebration to promote worldwide recognition of our brother planets in the Milky Way Galaxy—32 worlds waiting to join ours in an Interplanetary Confederation. FAX: (619) 444-9637. Est attendance: 350. Sponsor: Unarius Academy of Science, 145 S Magnolia Ave, El Cajon, CA 92020-4522. Phone: (619) 444-7062.

JAPAN: MEGA KENKA MATSURI OR ROUGHOUSE FESTIVAL. Oct 14–15. Himeji. Palanquin bearers jostle one another to demonstrate their skill and balance in handling their burdens.

KING AWARDED NOBEL PEACE PRIZE: ANNIVERSARY. Oct 14. Martin Luther King, Jr became the youngest recipient of the Nobel Peace Prize when awarded the honor on Oct 14, 1964. Dr. King donated the entire $54,000 prize money to furthering the causes of the civil rights movement.

MAJOR LEAGUE/MINOR LEAGUE: PHOTOGRAPHS OF AMERICA'S BASEBALL STADIUMS BY JIM DOW. Oct 14–Nov 26. Henderson Cnty Library Auditorium, Hendersonville, NC. See Feb 2 for full description. Call or write the Smithsonian for other dates and venues. For info: Smithsonian Institution Traveling Exhibition Service, 1100 Jefferson Dr SW, Ste 3146, Washington, DC 20560.

MISSISSINEWA 1812. Oct 14–15. Marion, IN. Largest War of 1812 living history event in United States includes reenactment of battle. Military, trappers and woodland Indians living as they did 183 years ago. Est attendance: 30,000. For info: Mississinewa Battlefield Society, 402 S Washington, Ste 509, PO Box 1324, Marion, IN 46952. Phone: (317) 662-0096.

MOUNTAIN GLORY FESTIVAL. Oct 14. Marion, NC. A celebration of mountain heritage in western North Carolina. Arts, crafts, children's arena and continuous entertainment. Annually, the second Saturday in October. Est attendance: 20,000. For info: Chamber of Commerce, John R. Birdsong, Exec Dir, 17 N Garden St, Marion, NC 28752. Phone: (704) 652-4240.

Oct 14–15 ☆ Chase's 1995 Calendar of Events ☆

NIGHT IN OLD LULING. Oct 14. Luling, TX. Antique car parade, open class car show, costumes of yesteryear, cake auction, food and game booths, free live entertainment on stage throughout the evening. Proceeds benefit civic projects, including the creation of an oil museum. Annually, the second Saturday in October. Est attendance: 1,500. Sponsor: Night in Old Luling Assn, PO Box 710, Luling, TX 78648. Phone: (210) 875-3214.

OKTOBERFEST. Oct 14–15. Traders Village, Grand Prairie, TX. Bavarian biergarten setting featuring polka bands, German, Polish, Norwegian and Czech folk dancing and plenty of German food and beer. Est attendance: 60,000. For info: Traders Village, 2602 Mayfield Rd, Grand Prairie, TX 75052. Phone: (214) 647-2331.

PEACE CORPS: 35th BIRTHDAY. Oct 14. At the improbable hour of 2:00 AM, on Oct 14, 1960, then presidential candidate John F. Kennedy spoke impromptu to several thousand university students from the steps of the University of Michigan Union building. He asked: "How many of you who are going to be doctors are willing to spend your days in Ghana? How many of you (technicians and engineers) are willing to work in the Foreign Service?" The response was quick and favorable, and 19 days later, in San Francisco, Kennedy formally proposed the Peace Corps, which was to draw more than 100,000 volunteers into service during the next three decades.

PENN, WILLIAM: BIRTH ANNIVERSARY. Oct 14. Founder of Pennsylvania, born at London, England, on Oct 14, 1644. Penn died on July 30, 1718. "Men are generally more careful," Penn wrote, "of the breed of their horses and dogs than of their children." Presidential Proclamation 5284 of Nov 28, 1984, conferred honorary citizenship of the United States upon William Penn and his second wife, Hannah Callowhill Penn. They were the third and fourth persons to receive honorary US citizenship (following Winston Churchill and Raoul Wallenberg).

ROARING CAMP HARVEST FAIRE. Oct 14–15. (Also Oct 21–22.) Felton, CA. Demonstrations of 1880s skills and crafts. For info: Jeanette Guire, Roaring Camp and Big Trees Railroad, Box G-1, Felton, CA 95018. Phone: (408) 335-4484.

ST. CHARLES SCARECROW FESTIVAL. Oct 14–15. St. Charles, IL. Est attendance: 40,000. For info: St. Charles Conv and Visitors Bureau, Jean Becker, Exec Dir, PO Box 11, St. Charles, IL 60174. Phone: (800) 777-4373.

SEDONA ARTS FESTIVAL. Oct 14–15. Sedona, AZ. Annual celebration of visual, performing and culinary arts. A special gathering of creative artisans presenting a unique collection of fine arts, crafts, foods, music and entertainment in a festive showcase surrounded by the scenic grandeur of towering red rocks. Featured events include Patrons Purchase Program. Admission $5.00; children 12 and under free with paid adult. Proceeds will benefit the cultural arts of Sedona. Est attendance: 7,000. For info: Sedona Arts Festival, PO Box 2729, Sedona, AZ 86339. Phone: (602) 282-8949.

SPACE MILESTONE: *SOYUZ 23* (USSR). Oct 14. Launched Oct 14, 1976. Cosmonauts V. Zudov and V. Rozhdestvensky unable to dock at *Salyut 5* space station as planned. Return landing on Earth, Oct 16.

TAMARACK TIME! Oct 14. Bigfork, MT. Old-fashioned village celebration of harvest and autumn. Local chefs prepare their specialties for a taste treat. Est attendance: 1,000. For info: Elna Darrow, Box 400, Bigfork, MT 59911. Phone: (406) 837-4848.

THE TONGASS—ALASKA'S MAGNIFICENT RAIN FOREST. Oct 14–Nov 26. Anchorage Museum of History and Art, Anchorage, AK. This exhibit explores the intricate harmony of the largest non-equatorial rain forest on earth. See Feb 4 for full description. For info: Smithsonian Institution Traveling Exhibition Service, 1100 Jefferson Dr SW, Ste 3146, Washington, DC 20560. Phone: (202) 357-2700.

TOUR OF SOUTHERN GHOSTS. Oct 14–31. Stone Mountain, GA. A Tour of Southern Ghosts set on the grounds of the Antebellum Plantation at Stone Mountain Park. A celebration of Southern ghostlore presented in the grand tradition of good, old-fashioned storytelling. Tour fee. Est attendance: 100,000. For info: Georgia's Stone Mountain Park, PR Dept, PO Box 778, Stone Mountain, GA 30086. Phone: (404) 498-5702.

TRINIDAD: PAN IS BEAUTIFUL VIII. Oct 14–Nov 4. Jean Pierre Complex, Port of Spain. Bi-annual steel band music festival, a major cultural activity of Trinidad and Tobago. Famous name steel bands and soloists in competition. For info: Trinidad and Tobago TDA, 25 W 43rd St, Ste 1508, New York, NY 10036. Phone: (800) 232-0082.

WORLD WRISTWRESTLING CHAMPIONSHIPS. Oct 14. Petaluma, CA. Nationally recognized event with more than 500 entrants vying for the title of World Wristwrestling Champion. Est attendance: 1,000. For info: Bill Soberanes, c/o *Argus Courier*, 830 Petaluma Blvd N, Petaluma, CA 94952. Phone: (707) 778-1430.

YORKTOWN DAY WEEKEND. Oct 14–15. Yorktown Victory Center, Yorktown, VA. An encampment of Revolutionary War reenactors features military demonstrations, music, games and contests to salute the anniversary of Washington's victory at Yorktown in October 1781. Est attendance: 1,500. For info: Jamestown-Yorktown Fdtn, PO Box JF, Williamsburg, VA 23187. Phone: (804) 253-4838.

BIRTHDAYS TODAY

John Dean, 57, lawyer (White House counsel during Watergate), born at Akron, OH, Oct 14, 1938.
Greg Evigan, 42, actor ("B.J. and the Bear," "Masquerade"), born South Amboy, NJ, Oct 14, 1953.
Gary Graffman, 67, musician, born at New York, NY, Oct 14, 1928.
Charles Everett Koop, 79, former US Surgeon General, born at Brooklyn, NY, Oct 14, 1916.
Ralph Lauren, 56, designer, born at the Bronx, NY, Oct 14, 1939.
Roger Moore, 67, actor (James Bond movies; "The Saint"), born London, England, Oct 14, 1928.

OCTOBER 15 — SUNDAY
288th Day — Remaining, 77

CHINA: CANTON AUTUMN TRADE FAIR. Oct 15–Nov 15. The Guangzhou (Canton) Autumn Trade Fair is held during the same dates each year.

CREDIT UNION WEEK. Oct 15–21. Worldwide observance to recognize the contribution of credit unions to the development and practice of democracy. For info: Credit Union Natl Assn, Steve Bosack, Box 431, Madison, WI 53701.

CROW RESERVATION OPENED FOR SETTLEMENT: ANNIVERSARY. Oct 15. By presidential proclamation 1.8 million acres of Crow Indian reservation were opened to settlers on Oct 15, 1892. The government had induced the Crow to give up a portion of their land in the mountainous western area in the state of Montana. The Crow received 50 cents per acre for their land.

October 1995

S	M	T	W	T	F	S
1	2	3	4	5	6	7
8	9	10	11	12	13	14
15	16	17	18	19	20	21
22	23	24	25	26	27	28
29	30	31				

GETTING THE WORLD TO BEAT A PATH TO YOUR DOOR WEEK. Oct 15–21. To focus attention on improving "public relationships" in order to create success for companies, products and individuals. Free self-evaluation available. Annually, the third week in October. May call Fax-On-Demand: (805) 96FAX-IT. For info: Barbara Gaughen, Pres, Gaughen Public Relations, 226 E Canon Perdido, Ste B, Santa Barbara, CA 93101. Phone: (805) 965-8482.

IACOCCA, LEE A.: BIRTHDAY. Oct 15. Lido Anthony Iacocca, son of Italian immigrant parents, was born at Allentown, PA, on Oct 15, 1924, and, in the 1980s, he was called "America's first corporate folk hero." His best-selling autobiography, *Iacocca*, sold more than 5 million copies. The mechanical engineer-turned-automobile executive (former president of Ford Motor Company, and later chairman of Chrysler Corporation) served as chairman of the (centennial rehabilitation) Statue of Liberty-Ellis Island foundation.

MANN, MARTY: BIRTH ANNIVERSARY. Oct 15. American social activist and author was born at Chicago, IL, on Oct 15, 1904. She was founder, in 1944, of the National Committee for Education on Alcoholism and author of *A New Primer on Alcoholism*. She died at Bridgeport, CT, on July 22, 1980.

MAPLE LEAF FESTIVAL. Oct 15–22. Carthage, MO. Brilliant fall foliage gives this Victorian city the perfect backdrop. A 150-plus unit parade, four-state marching band competition, statewide car show, historic homes tour, arts and craft show plus entertainment and much more. Annually, the third week in October. Est attendance: 70,000. For info: Chamber of Commerce, 107 E 3rd St, Carthage, MO 64836. Phone: (417) 358-2373.

MATA HARI: EXECUTION ANNIVERSARY. Oct 15. Possibly history's most famous spy, Mata Hari refused a blindfold and threw a kiss to the firing squad at her execution, Oct 15, 1917. Early estimates claimed her espionage for Germany in World War I was responsible for 50,000 Allied deaths. More recent studies suggest that she was a double agent (secret code name: "H-21") and that her "intelligence" was of little value to either side. Born Margaret Gertrude Zelle, at Leewarden, Netherlands, her life as a strip-teaser, courtesan, blackmailer and spy—and her liaisons with high officials in the British, French and German governments—fascinated millions of readers. See also: "Mata Hari: Birth Anniversary" (Aug 7).

★ **NATIONAL FOREST PRODUCTS WEEK.** Oct 15–21. Presidential Proclamation always issued for the week beginning with the third Sunday in October since 1960 (PL86-753 of Sept 13, 1960).

NATIONAL GROUCH DAY. Oct 15. Honor a grouch; all grouches deserve a day to be recognized. Annually, Oct 15. Sponsor: Alan R. Miller, Carter Middle School, 300 Upland Dr, Clio, MI 48420. Phone: (313) 686-0503.

NATIONAL SHUT-IN VISITATION DAY. Oct 15. A day set aside for the purpose of caring for and visiting the sick, incapacitated and elderly. Annually, the third Sunday in October. Sponsor: Holy Rosary Church. For info: Natl Society for Shut-Ins, Rev. Msgr. Felix A. Losito, PO Box 1392, Reading, PA 19603. Phone: (610) 374-2930.

NIETZSCHE, FRIEDRICH WILHELM: BIRTH ANNIVERSARY. Oct 15. Influential German philosopher born at Rocken on Oct 15, 1844. Especially remembered among his philosophical beliefs are contempt for the weak and expected ultimate triumph of a superman. Nietzsche died at Weimar on Aug 25, 1900, a decade after becoming insane.

STEILACOOM APPLE SQUEEZE. Oct 15. Steilacoom, WA. People are invited to experience the "good old days" by bringing their own apples and making cider. Twenty presses are set up for this activity. Also includes a demonstration press, an apple identifier and booths featuring apple products for sale. Est attendance: 1,500. For info: Steilacoom Historical Museum Assn, PO Box 88016, Steilacoom, WA 98388-0016. Phone: (206) 584-4133.

★ **WHITE CANE SAFETY DAY.** Oct 15. Presidential Proclamation always issued for Oct 15 since 1964 (PL88-628 of Oct 6, 1964).

WILSON, EDITH BOLLING GALT: BIRTH ANNIVERSARY. Oct 15. Second wife of Woodrow Wilson, 28th president of the US, born at Wytheville, VA, Oct 15, 1872. Died Dec 28, 1961.

WODEHOUSE, PELHAM GRENVILLE: BIRTH ANNIVERSARY. Oct 15. English author, humorist, creator of Jeeves, born at Guildford, Surrey, England, Oct 15, 1881. Died at Southampton, Long Island, NY, Feb 14, 1975.

WSBA/WARM 103 HOLIDAY CRAFT SHOW. Oct 15. York Fairgrounds, York, PA. More than 250 crafts, from country to contemporary, Victorian and southwestern, handcrafted furniture, wood carvings, dolls, jewelry, pottery, collectibles, quilts, baskets and much more. Admission fee. Est attendance: 7,000. For info: Gina M. Koch, Special Events Dir, PO Box 910, York, PA 17405. Phone: (717) 764-1155.

BIRTHDAYS TODAY

Trace (Raymond Lester) Armstrong, 30, football player, born Bethesda, MD, Oct 15, 1965.
John Kenneth Galbraith, 87, economist, diplomat, author, born at Iona Station, Ontario, Canada, Oct 15, 1908.
Lee Iacocca, 71, former automobile executive (Ford and Chrysler), born at Allentown, PA, Oct 15, 1924.
Tito Jackson (Toriano Adaryll Jackson), 42, singer, musician (Jackson 5), born at Gary, IN, Oct 15, 1953.
Linda Lavin, 56, actress (Tony for *Broadway Bound*; "Alice"), born Portland, ME, Oct 15, 1939.
Penny Marshall, 53, director (*Big*), actress ("Laverne and Shirley"), born the Bronx, NY, Oct 15, 1942.
Ned Ray McWherter, 65, Governor of Tennessee (D), born at Palmersville, TN, Oct 15, 1930.
Jim Palmer, 50, sportscaster, born at New York, NY, Oct 15, 1945.
Mario Puzo, 75, author (*The Godfather, Fourth K.*), born New York, NY, Oct 15, 1920.
Sarah (Ferguson), 36, Duchess of York (estranged wife of Prince Andrew), born at London, England, Oct 15, 1959.
Arthur Meier Schlesinger, Jr, 78, historian, author, born at Columbus, OH, Oct 15, 1917.
Roscoe Tanner, 44, tennis player, born at Lookout Mountain, TN, Oct 15, 1951.

OCTOBER 16 — MONDAY
289th Day — Remaining, 76

BEN-GURION, DAVID: BIRTH ANNIVERSARY. Oct 16. First prime minister of the state of Israel. Born at Plonsk, Poland, on Oct 16, 1886. Died at Tel Aviv, Israel, on Dec 1, 1973.

BIRTH CONTROL CLINIC OPENED: ANNIVERSARY. Oct 16. Margaret Sanger, Fania Mindell and Ethel Burne opened the first birth control clinic in the US on Oct 16, 1916, at 46 Amboy St, Brooklyn, NY. Sanger believed that the poor should be able to control the size of their families.

DICTIONARY DAY. Oct 16. The birthday of Noah Webster, American teacher and lexicographer, is occasion to encourage every person to acquire at least one dictionary—and to use it regularly.

Oct 16-17 ☆ *Chase's 1995 Calendar of Events* ☆

DOUGLAS, WILLIAM ORVILLE: BIRTH ANNIVERSARY. Oct 16. American jurist, world traveler, conservationist, outdoorsman and author. Born at Maine, MN, Oct 16, 1898. Served as justice of the US Supreme Court longer than any other (36 years). Died at Washington, DC, Jan 19, 1980.

JAMAICA: NATIONAL HEROES DAY. Oct 16. National holiday established in 1969. Always observed on third Monday of October.

JOHN BROWN'S RAID: ANNIVERSARY. Oct 16. On Oct 16, 1859, fanatical abolitionist John Brown, with a band of about 20 men, seized the US Arsenal at Harpers Ferry, WV. Brown was captured and the insurrection put down by Oct 19. Brown was hanged at Charles Town, WV, Dec 2, 1859.

MAINTENANCE PERSONNEL DAY. Oct 16. To promote the public awareness of the contribution and importance of maintenance employees in hospitals, nursing homes and other health care facilities. For info: Dick Armstrong, Maintenance Super, Heartland Health Care Center, 512 Draper Dr, Temple, TX 76501.

MARIE ANTOINETTE: EXECUTION ANNIVERSARY. Oct 16. Queen Marie Antoinette, whose extravagance and "let them eat cake" attitude toward the starving French underclass made her a symbol and target of the French Revolution, was beheaded on Oct 16, 1793.

MOON PHASE: LAST QUARTER. Oct 16. Moon enters Last Quarter phase at 12:26 PM, EDT.

NATIONAL BOSS DAY. Oct 16. For all employees to honor their bosses. Annually, Oct 16. For info: Mrs Patricia Bays Haroski, Originator, 2871-F-Walnut View Ct, Winston-Salem, NC 27103.

NATIONAL BUSINESS WOMEN'S WEEK. Oct 16-20. Declared annually by the President of the United States, National Business Women's Week establishes a week of special recognition of the role of the working woman in American society, the economy and in the family. It is commemorated nationwide by special activities. Annually, starting the third Monday in October. Sponsor: Business and Professional Women's Clubs. For info: Liana Sayer, Issues Research Mgr, BPW/USA, 2012 Massachusetts Ave NW, Washington, DC 20036. Phone: (202) 293-1100.

NORTHERN INTERNATIONAL LIVESTOCK EXPOSITION. Oct 16-21. MetraPark, Billings, MT. Rodeo, trade show exhibits, cattle and horse sales. Fax: (406) 256-2494. Est attendance: 65,600. For info: Kevin Dawe, Genl Mgr, NILE Office, PO Box 1981, Billings, MT 59103. Phone: (406) 256-2495.

O'NEILL, EUGENE GLADSTONE: BIRTH ANNIVERSARY. Oct 16. American playwright, recipient of Pulitzer and Nobel Prize. Born at New York, NY, Oct 16, 1888. Died at Boston, MA, Nov 27, 1953.

PEACE WITH JUSTICE WEEK. Oct 16-24. A national week to witness for peace based on justice. The National Council of Churches which sponsors the Week, seeks to network people of faith to work on justice issues. Linking World Food Day, Peace Sabbath, Children's Sabbaths and Unitd Nations/Disarmament Day, the week is the nation's largest annual multi-issue interfaith event. Est attendance: 50,000. For info: Peace with Justice Week, NCC Office for World Community, 475 Riverside Dr, Room 670, New York, NY 10115-0122. Phone: (212) 870-2424.

SHEMINI ATZERET. Oct 16. Hebrew calendar date: Tishri 22, 5756. The eighth day of Solemn Assembly, part of the Sukkot Festival (see entry on Sept 9), with memorial services and cycle of Biblical readings in the synagogue.

UNITED NATIONS: WORLD FOOD DAY. Oct 16. Annual observance to heighten public awareness of the world food problem and to strengthen solidarity in the struggle against hunger, malnutrition and poverty. Date of observance is anniversary of founding of Food and Agriculture Organization (FAO), Oct 16, 1945, at Quebec, Canada. Info from: United Nations, Dept of Public Info, New York, NY 10017.

VIRGIN ISLANDS: HURRICANE THANKSGIVING DAY. Oct 16. Third Monday of October is legal holiday celebrating end of hurricane season.

WEBSTER, NOAH: BIRTH ANNIVERSARY. Oct 16. American teacher and journalist whose name became synonymous with the word "dictionary" after his compilations of the earliest American dictionaries of the English language. Born at West Hartford, CT, on Oct 16, 1758. Died at New Haven, CT, on May 28, 1843.

WILDE, OSCAR: BIRTH ANNIVERSARY. Oct 16. Irish poet and playwright, Oscar (Fingal O'Flahertie Wills) Wilde, was born at Dublin, Ireland, on Oct 16, 1854. At the height of his career he was imprisoned for two years on a morals offense, during which time he wrote "A Ballad of Reading Gaol." Best known of his plays is *The Importance of Being Earnest*. "There is only one thing in the world worse than being talked about," he wrote in his *Picture of Dorian Gray*, "and that is not being talked about." Wilde died, self-exiled, at Paris, France, on Nov 30, 1900. His dying words are said to have been: "This wallpaper is killing me; one of us has got to go."

YALE UNIVERSITY FOUNDED: ANNIVERSARY. Oct 16. On October 16, 1701, the Collegiate School was founded in Killingworth, CT. The school was started by Congregationalists dissatisfied with the growing liberalism at Harvard. In 1745, the school was moved to New Haven, CT, where it became Yale College, named after Elihu Yale, son of one of the founders of New Haven, and a governor of India for the East India Company. Yale became a university in 1887. The first degrees were awarded in 1716.

BIRTHDAYS TODAY

Melissa Belote, 39, swimmer, born at Washington, DC, Oct 16, 1956.

Manute Bol, 33, basketball player, born at Gogrial, Sudan, Oct 16, 1962.

Gunter Grass, 68, author (*The Tin Drum, Dog Years*), born at Danzig, Germany, Oct 16, 1927.

Angela Lansbury, 70, actress (*National Velvet, Sweeny Todd*, "Murder She Wrote"), born London, England, Oct 16, 1925.

Suzanne Somers, 49, actress ("Three's Company," *American Graffiti*), born San Bruno, CA, Oct 16, 1946.

OCTOBER 17 — TUESDAY
290th Day — Remaining, 75

BLACK POETRY DAY. Oct 17. To recognize the contribution of black poets to American life and culture, and to honor Jupiter Hammon, first black in America to publish his own verse. Jupiter Hammon of Huntington, Long Island, NY, was born Oct 17, 1711. Est attendance: 200. Sponsor: Black Poetry Day Committee, Multicultural Affairs, Affirmative Action Office, State University College, Plattsburgh, NY 12901.

October 1995

S	M	T	W	T	F	S
1	2	3	4	5	6	7
8	9	10	11	12	13	14
15	16	17	18	19	20	21
22	23	24	25	26	27	28
29	30	31				

HAMMON, JUPITER: BIRTH ANNIVERSARY. Oct 17. America's first published black poet, whose birth anniversary, Oct 17, is celebrated annually as Black Poetry Day, was born in 1711, probably at Long Island, NY. Though born into slavery he was taught to read, and as a trusted servant, was allowed to use his master's library. "With the publication on Christmas Day, 1760, of the 88-line broadside poem 'An Evening Thought,' Jupiter Hammon, then 49, became the first black in America to publish poetry." Hammon died in 1790. The exact date and place of his death and burial are as yet unknown.

JOHNSON, RICHARD MENTOR: BIRTH ANNIVERSARY. Oct 17. Ninth vice president of the US (1837-1841). Born at Floyd's Station, KY, Oct 17, 1780. Died at Frankfort, KY, Nov 19, 1850.

POPE JOHN PAUL I: BIRTH ANNIVERSARY. Oct 17. Albino Luciani, 263rd pope of the Roman Catholic Church. Born at Forno di Canale, Italy, Oct 17, 1912. Elected pope Aug 26, 1978. Died at Rome, 34 days after his election, Sept 28, 1978. Shortest papacy since Pope Leo XI (Apr 1-27, 1605).

SAN FRANCISCO EARTHQUAKE OF 1989: ANNIVERSARY. Oct 17. The San Francisco Bay area was rocked by an earthquake registering 7.1 on the Richter scale on Oct 17, 1989, at 5:04 PM, PDT, just as the nation's baseball fans settled in to watch the 1989 World Series. The quake caused damage estimated at 10 billion dollars and killed 67 people, many of whom were caught in the collapse of the double decked Interstate 80, in Oakland, CA. A large World Series audience was tuned in to the pre-game coverage when the quake hit and knocked the broadcast off the air.

SIMCHAT TORAH. Oct 17. Hebrew calendar date: Tishri 23, 5756. Rejoicing in the Torah concludes the nine-day Sukkot Festival (see entry on Sept 9). Public reading of the Pentateuch is completed and begun again, symbolizing the need for ever-continuing study.

UNITED NATIONS: INTERNATIONAL DAY FOR THE ERADICATION OF POVERTY. Oct 17. The General Assembly proclaimed this observance (Res 47/196) to promote public awareness of the need to eradicate poverty and destitution in all countries, particularly the developing nations.

BIRTHDAYS TODAY

Beverly Garland, 69, actress ("My Three Sons," "Scarecrow and Mrs. King"), born at Santa Cruz, CA, Oct 17, 1926.
Margot Kidder, 47, actress (Lois Lane in *Superman* movies, born Yellowknife, NWT, Canada, Oct 17, 1948.
Evel Knievel (Robert Craig), 57, motorcycle stunt performer, born at Butte, MT, Oct 17, 1938.
Michael McKean, 48, actor ("Laverne & Shirley," *This Is Spinal Tap*), born New York, NY, Oct 17, 1947.
Steve Douglas McMichael, 38, football player, born Houston, TX, Oct 17, 1957.
Arthur Miller, 80, dramatist (*Death of a Salesman, Rhinoceros*), born at New York, NY, Oct 17, 1915.
George Wendt, 47, actor (Norm Peterson on "Cheers"), born at Chicago, IL, Oct 17, 1948.

OCTOBER 18 — WEDNESDAY
291st Day — Remaining, 74

ALASKA DAY: ANNIVERSARY. Oct 18. Alaska. Anniversary of transfer of Alaska from Russia to the US. Became official on Sitka's Castle Hill, Oct 18, 1867.

ANDREE, SALOMON AUGUSTE: BIRTH ANNIVERSARY. Oct 18. Swedish explorer and balloonist was born at Grenna, Sweden, on Oct 18, 1854. His North Pole expedition of 1897 attracted world attention but ended tragically. With two companions, Nils Strindberg and Knut Frankel, Andree left Spitzbergen on July 11, 1897, in a balloon, hoping to place the Swedish flag at the North Pole. The last message from Andree, borne by carrier pigeons, was dated at noon, July 13, 1897. The frozen bodies of the explorers were found 33 years later by another polar expedition in the summer of 1930. Taken back to Stockholm they were cremated on Oct 10, 1930. Diaries, maps and exposed photographic negatives also were found. The photos were developed successfully, providing a pictorial record of the ill-fated expedition.

BAZAARFEST. Oct 18-21. West Acres Shopping Center, Fargo, ND. Open to non-profit organizations and churches to sell their handcrafted items. Est attendance: 68,000. For info: Dee Lander, Promo/Mktg, West Acres Office, West Acres Shopping Center, Fargo, ND 58103. Phone: (701) 282-2222.

BERGSON, HENRI: BIRTH ANNIVERSARY. Oct 18. French philosopher, Nobel Prize winner and author of *Creative Evolution*, born Oct 18, 1859. Died Jan 4, 1941.

BROOKS, JAMES DAVID: BIRTH ANNIVERSARY. Oct 18. James Brooks was born at St. Louis, MO, on Oct 18, 1906. During the depression Brooks worked as a muralist in the Federal Art Project of the Works Progress Administration. His best known work of that period was "Flight," a 235-foot mural on the rotunda of the Marine Air Terminal at La Guardia National Airport in New York. It was painted over during the 1950s but restored in 1980. After leaving the WPA he worked as an art correspondent with the US Army from 1942 to 1945. After World War II, Brooks returned to New York, his interest shifting to abstraction. He came to be considered one of the greatest abstract expressionists. His paintings were exhibited in the historic "Ninth Street Exhibition," as a part of the Museum of Modern Art's exhibits "Twelve Americans" and "New American Painting," among others. He died on Mar 8, 1992, at Brookhaven, NY.

CANADA: PERSONS DAY. Oct 18. A day to commemorate the anniversary of the 1929 ruling that declared women to be persons in Canada. Prior to this ruling English common law prevailed ("Women are persons in matters of pains and penalties, but are not persons in matters of rights and privileges"). The celebrated cause, popularly known as the "Persons Case," was brought by five women of Alberta, Canada; leader of the courageous "Famous Five" was Emily Murphy (1868-1933). This ruling by the Judicial Committee of England's Privy Council, Oct 18, 1929, overturned a 1928 decision of the Supreme Court of Canada. Fifty years after the Persons Case decision, in 1979, the Governor General's Awards in Commemoration of the Persons Case were established to recognize deserving persons who have made outstanding contributions to the quality of life of women in Canada.

CIRCLEVILLE PUMPKIN SHOW. Oct 18-21. Circleville, OH. More than 100,000 pounds of pumpkins, squash and gourds. Est attendance: 300,000. Sponsor: Pumpkin Show Inc, Hugh Dresbach, Secy, 216 N Court St, Circleville, OH 43113. Phone: (614) 474-7000.

◆ **GERMANY: WAR CRIMES TRIAL: 50th ANNIVERSARY.** Oct 18, 1945. The first session of the German war crimes trials started at Berlin with indictments against 24 former Nazi leaders. Later sessions were held at Nuremberg, starting Nov 20, 1945. One defendant committed suicide during the trial and another was excused because of his physical and mental condition. The trial lasted more than 10 months and delivery of the judgment was completed on Oct 1, 1946. Twelve were sentenced to death by hanging, 3 to life imprisonment, 4 to lesser prison terms and 3 were acquitted. It was hoped that the war crimes trials would serve as a deterrent to future world leaders who might consider crimes against humanity.

Oct 18–19 ☆ *Chase's 1995 Calendar of Events* ☆

LIEBLING, A.J.: BIRTH ANNIVERSARY. Oct 18. American journalist and author who said "Freedom of the press belongs to those who own one." Abbott Joseph Liebling was born at New York, NY, on Oct 18, 1904. Died there Dec 28, 1963.

MERCOURI, MELINA (MARIA AMALIA): BIRTH ANNIVERSARY. Oct 18, 1922[?]. Greek film actress who later successfully moved into a political career. She appeared in more than 70 films and plays, most famously in *Never on Sunday* in 1960. She was elected to the Greek parliament in 1977 and named Minister of Culture in 1981. She was born in either 1922 or 1925 at Athens, Greece, and died Mar 6, 1994, at New York City.

MISSOURI DAY. Oct 18. The third Wednesday of October each year. Observed by teachers and pupils of schools with appropriate exercises throughout state of Missouri.

ST. LUKE: FEAST DAY. Oct 18. Patron saint of doctors and artists, himself a physician and painter, author of the third Gospel and Acts of the Apostles. Died about AD 68. Legend says that he painted portraits of Mary and Jesus.

TRUDEAU, PIERRE ELLIOTT: BIRTHDAY. Oct 18. Canadian politician, born at Montreal, Que, Canada, Oct 18, 1919. Prime Minister of Canada, Apr 20, 1968–June 4, 1979, and Mar 3, 1980–June 30, 1984.

WATER POLLUTION CONTROL ACT: ANNIVERSARY. Oct 18. Overriding President Nixon's veto, on Oct 18, 1972, Congress passed a $25 billion Water Pollution Control Act. Industry was to be required to halt discharges by 1985 and industrywide standards were to be established.

BIRTHDAYS TODAY

Chuck Berry (Charles Edward Anderson), 69, singer, songwriter ("Johnny B. Goode," "Roll Over Beethoven," "Sweet Little Sixteen," "Back in the USA," "Maybellene," "Brown-Eyed Handsome Man"), musician (famous guitar riff), born St. Louis, MO, Oct 18, 1926.
Pam Dawber, 44, actress (Mindy on "Mork & Mindy"; Samantha on "My Sister Sam"), born Farmington, MI, Oct 18, 1951.
Mike Ditka, 56, Pro football Hall of Famer, former coach, broadcaster, born Carnegie, PA, Oct 18, 1939.
Tommy Hearns, 37, boxer, born at Detroit, MI, Oct 18, 1958.
Jesse Helms, 74, US Senator (R, North Carolina), born at Monroe, NC, Oct 18, 1921.
Wynton Marsalis, 34, jazz musician, born at New Orleans, LA, Oct 18, 1961.
Erin Moran, 34, actress (Joanie Cunningham on "Happy Days" and "Joanie Loves Chachi"), born at Burbank, CA, Oct 18, 1961.
Laura Nyro, 48, singer ("I Am the Blues," "Save the Country"), composer, born Bronx, NY, Oct 18, 1947.
George C. Scott, 68, actor (Oscar for *Patton* [which he refused]; "East Side/West Side") born at Wise, VA, Oct 18, 1927.
Ntozake Shange (Paulette L. Williams), 47, dramatist, poet, born at Trenton, NJ, Oct 18, 1948.
Wendy Wasserstein, 45, writer (*The Heidi Chronicles, Bachelor Girls*), born at Brooklyn, NY, Oct 18, 1950.

October 1995

S	M	T	W	T	F	S
1	2	3	4	5	6	7
8	9	10	11	12	13	14
15	16	17	18	19	20	21
22	23	24	25	26	27	28
29	30	31				

OCTOBER 19 — THURSDAY
292nd Day — Remaining, 73

ANNIVERSARY DAY. Oct 19. So named and observed by Robert Hutchings Goddard, rocket pioneer, as anniversary of day (Oct 19, 1899) when he first began to speculate about a space ship that could travel to Mars.

BROWNE, THOMAS: BIRTH ANNIVERSARY. Oct 19. Physician, scholar and author, Thomas Browne was born at London, England, on Oct 19, 1605. He wrote: "I could never divide myself from any man upon the difference of opinion, or be angry with his judgment for not agreeing with me in that from which perhaps within a few days I should dissent myself." At age 55, he wrote: "The long habit of living indisposeth us for dying." His most famous work, *Religio Medici*, was published in 1642. Browne died at Norwich, England, on his 77th birthday, Oct 19, 1682.

EAST TEXAS YAMBOREE. Oct 19–21. Gilmer, TX. Est attendance: 100,000. For info: Upshur County Chamber of Commerce, Box 854, Gilmer, TX 75644. Phone: (903) 843-2413.

EVALUATE YOUR LIFE DAY. Oct 19. To encourage everyone to check and see if they're really headed where they want to be. Sponsor: Wellness Permission League. Phone: (212) 388-8673 or (717) 274-8451. For info: Tom and Ruth Roy, Wellness Permission League, 2105 Water St, Lebanon, PA 17046.

FALL GOSPEL JUBILEE. Oct 19–21. Spirit of The Suwannee, Live Oak, FL. For info: Jean Cornett, Rte 1, Box 98, Live Oak, FL 32060. Phone: (904) 364-1683.

JEFFERSON, MARTHA WAYLES SKELTON: BIRTH ANNIVERSARY. Oct 19. Wife of Thomas Jefferson, 3rd president of the US. Born at Charles City County, VA, on Oct 19, 1748. Died on Sept 6, 1782.

LITERACY VOLUNTEERS OF AMERICA CONFERENCE. Oct 19–21. Hyatt Regency, Buffalo, NY. An annual gathering of literacy providers, volunteer tutors and adult learners. Featuring: workshops, keynote and guest speakers, awards banquet, training programs and networking with experts in the adult literacy field. Annually, in the fall. Est attendance: 1,000. Sponsor: Literacy Volunteers of America, Margaret Price, Dir Field Svc, 5795 Widewaters Pkwy, Syracuse, NY 13214-1846. Phone: (315) 445-8000.

NORTH AMERICAN CONFERENCE ON THE FAMILY. Oct 19–22. Winnipeg, Man. Family Service America and Family Service Canada sponsor this joint conference with in-depth presentations and workshops on "Championing Change," "Developing Tools for Tomorrow," "Effective Responses to Family Needs" and "Visioning for a Brighter Future." Est attendance: 1,200. Sponsor: North Amer Conf on the Family, Conf Coord, Family Service Amer, Inc, 11700 W Lake Park Dr, Milwaukee, WI 53224. Phone: (414) 359-1040.

OKTOBERFEST. Oct 19–22. Tulsa, OK. This colorful ethnic festival reflects an authentic German flavor in food, music and entertainment with biergartens located in River West Festival Park. Attractions include bands, dancing, sing-alongs, contests, visual arts and crafts area, children's entertainment. Est attendance: 100,000. For info: River Parks, 707 S Houston, Ste 202, Tulsa, OK 74127. Phone: (918) 596-2001.

◆ **PECK, ANNIE S.: BIRTH ANNIVERSARY.** Oct 19, 1850. World-renowned mountain climber Annie S. Peck was born in 1850. She won an international following in 1895 when she climbed the Matterhorn in the Swiss Alps. Peck climbed the Peruvian peak Huascaran (21,812 feet), giving her the record for the highest peak climbed in the Western Hemisphere by an American man or woman, and at age 61 she climbed Mt Coropuna in Peru (21,250 feet) and placed a "Votes for Women" banner at its pinnacle. Annie Peck died July 18, 1935, at New York City.

PIONEER DAY. Oct 19. Washington, MS. Through talks and demonstrations, youngsters learn how children lived 150–200 years ago, Annually, the third Thursday in October. Est attendance: 1,500. For info: Historic Jefferson College, Anne L. Gray, Historian, P O Box 700, Washington, MS 39190. Phone: (601) 442-2901.

SCOTTISH TATTOO, FESTIVAL AND HIGHLAND GAMES. Oct 19. (Also 21–22.) Stone Mountain, GA. A regal celebration of the Scottish tradition including the Tattoo on Oct 19 and the Festival and Highland Games on Oct 21–22. Kilted clans compete in fun-filled Highland athletic events and drumming competitions. Pipe bands, folk dancing, parades and pageantry. Admission charged. 21st annual festival. Est attendance: 150,000. For info: Georgia's Stone Mountain Park, PR Dept, PO Box 778, Stone Mountain, GA 30086. Phone: (404) 498-5702.

WAR EAGLE FAIR. Oct 19–22. War Eagle Mills Farm, War Eagle, AR. 42nd annual fair with more than 350 booths of baskets, quilts, woodwork, pottery, furniture, wood carving, toys, leather, books, candles, weaving, jams and jellies and food under circus style tents on the historic farm grounds. "One of the most highly respected shows in the country." Est attendance: 135,000. For info: Shirley Sutton, War Eagle Fair, Rt 1, Box 157, Hindsville, AR 72738. Phone: (501) 789-5398.

WORLD'S LARGEST HALLOWEEN PARTY. Oct 19–22. (Also Oct 26–29.) Louisville, KY. Est attendance: 50,000. Sponsor: Joyce Compton, Coord Spec Events, PO Box 37250, Louisville, KY 40233. Phone: (502) 459-2181.

YORKTOWN DAY. Oct 19. More than 7,000 English and Hessian troops, led by British General Lord Cornwallis, surrendered to General George Washington at Yorktown, VA, on Oct 19, 1781. The event effectively ended the war between Britain and her American colonies. There were no more major battles, but the provisional treaty of peace was not signed until Nov 30, 1782, and the final Treaty of Paris, on Sept 3, 1783.

YORKTOWN DAY: "AMERICA'S REAL INDEPENDENCE DAY." Oct 19. Yorktown, VA. Representatives of US, France and other nations involved in American Revolution gather to celebrate anniversary of the victory which assured American independence. Parade and commemorative ceremonies. Annually, Oct 19. Est attendance: 2,000. For info: The Yorktown Day Assn, Box 210, Yorktown, VA 23690. Phone: (804) 898-3400.

BIRTHDAYS TODAY

Jack Anderson, 73, journalist, columnist, author, (*Japan Conspiracy, Stormin' Norman*), born at Long Beach, CA, Oct 19, 1922.

Bern Bennett, 74, former CBS staff announcer, born at Rochester, NY, Oct 19, 1921.

Bradley Lee Daugherty, 30, basketball player, born at Black Mountain, NC, Oct 19, 1965.

Patricia Ireland, 50, feminist, social activist, president of National Organization for Women, born at Oak Park, IL, Oct 19, 1945.

John LeCarre (David John Moore Cornwell), 64, author (*The Russia House, A Small Town in Germany*), born at Poole, England, Oct 19, 1931.

John Lithgow, 50, actor (*Twilight Zone—The Movie, I'm Dancing As Fast As I Can*), born at Rochester, NY, Oct 19, 1945.

Peter Max, 58, artist, designer, born at Berlin, Germany, Oct 19, 1937.

LaWanda Page, 75, actress (Aunt Esther on "Sanford and Son"), born Cleveland, OH, Oct 19, 1920.

OCTOBER 20 — FRIDAY
293rd Day — Remaining, 72

AMERICAN THEATRICAL PRODUCTIONS SUSPENDED: ANNIVERSARY. Oct 20. American theatrical productions were brought to an abrupt halt on Oct 20, 1774, along with other entertainments when the Continental Congress passed an order proclaiming that the colonies "discountenance and discourage all horse racing and all kinds of gaming, cock fighting, exhibitions of shows, plays, and other expensive diversions and entertainments."

ARKANSAS CRAFT GUILD FALL SHOW AND SALE (ON WAR EAGLE WEEKEND). Oct 20–22. Four Runners Inn Convention Center, Eureka Springs, AR. More than 70 exhibitors of original handcrafts, including woodworks, pottery, stained glass, metalworks, musical instruments, jewelry, baskets, quilts and more at 30th annual weekend craft show. Free admission and parking. Est attendance: 5,000. For info: Arkansas Craft Guild, PO Box 800, Mountain View, AR 72560. Phone: (501) 269-3897.

BIRTH OF THE BAB. Oct 20. Baha'i observance of anniversary of the birth, Oct 20, 1819, in Shiraz, Persia, of Siyyid Ali Muhammad, who later took the title of "the Bab"; the Bab was the prophet-herald of the Baha'i Faith. One of the nine days of the year when Baha'is suspend work. For info: Natl Spiritual Assembly of the US, 1320 Nineteenth St, NW, Ste 701, Washington, DC 20036. Phone: (202) 833-8990.

DAYTONA FALL MOTORCYCLE RACES. Oct 20–22. Daytona International Speedway, Daytona Beach, FL. For info: Daytona Intl Speedway, Glenn Barber, PR, Box 2801, Daytona Beach, FL 32120-2801. Phone: (904) 254-6782.

DEWEY, JOHN: BIRTH ANNIVERSARY. Oct 20. American psychologist, philosopher and educational reformer was born at Burlington, VT, on Oct 20, 1859. His philosophical views of education have been termed pragmatism, instrumentalism and experimentalism. Died at New York, NY, on June 1, 1952.

FALL FESTIVAL OF LEAVES. Oct 20–22. Bainbridge, Ross County, OH. Celebration of beauty of the season and region. Folk arts, crafts, music, antique car show, log sawing contest, flea markets and parades. To obtain a map of self-guided scenic tours send SASE to sponsor. Sponsor: Fall Festival of Leaves, Box 571, Bainbridge, Ross County, OH 45612.

FANTASY FEST '95. Oct 20–29. Key West, FL. Week-long adult costume festival with street parties, masked balls and night-time grand parade. Est attendance: 50,000. Sponsor: Fantasy Fest '95, Ann Dickinson, Box 230, Key West, FL 33041. Phone: (305) 296-1817.

FESTIVAL OF THE HORSE. Oct 20–29. Oklahoma City, OK. Celebrating Oklahoma's horse industry with a showcase of breeds, thoroughbred racing at Remington Park, Prairie Circuit Finals Rodeo at Lazy E Arena and more than 20 equine sports and top entertainment events. Est attendance: 120,000. For info: Oklahoma Events, Inc, Ste 1212, Equity Tower, 1601 NW Expressway, Oklahoma City, OK 73118. Phone: (405) 842-4141.

GUATEMALA: REVOLUTION DAY. Oct 20. Public holiday in Guatemala.

HALLOWEEN HARVEST OF HORRORS. Oct 20. Tom Sawyer State Park, Louisville, KY. Storytelling performance of ghost tales. Est attendance: 2,000. For info: Intl Order of EARS, Inc, Lee Pennington, 12019 Donohue Ave, Louisville, KY 40243. Phone: (502) 245-0643.

HELLDORADO DAYS. Oct 20–22. Tombstone, AZ. To relive the historic events of Tombstone. Reenactments of early day gunfights, 1880 fashion show, melodrama, carnival and parade. Est attendance: 7,000. Sponsor: Helldorado Inc and Tombstone Lions Club, Box 297, Tombstone, AZ 85638. Phone: (602) 457-3966.

KENYA: KENYATTA DAY. Oct 20. Kenya observes this day as a public holiday.

Oct 20-21 ☆ *Chase's 1995 Calendar of Events* ☆

KLAN MEMBERS CONVICTED OF CONSPIRACY: ANNIVERSARY. Oct 20. A federal jury convicted seven members of the Ku Klux Klan of conspiracy in the 1964 murders of three civil rights workers, James Chaney, Andrew Goodman and Michael Schwerner, in Mississippi. Eight others were acquitted and no verdict was reached on three others.

MANN, JAMES ROBERT: BIRTH ANNIVERSARY. Oct 20. American lawyer and legislator, born near Bloomington, IL, Oct 20, 1856. Republican member of Congress from Illinois, from 1896 until his death, Nov 30, 1922. Mann was the author and sponsor of the "White Slave Traffic Act," also known as the "Mann Act," passed by Congress on June 25, 1910. The act prohibited, under heavy penalties, the interstate transportation of women for immoral purposes.

MOSCOW SOCCER TRAGEDY: ANNIVERSARY. Oct 20. The world's worst soccer disaster occurred at Moscow on Oct 20, 1982, when 340 sports fans were killed during a game between Soviet and Dutch players. Details of the event, blaming police for the tragedy in which spectators were crushed to death in an open staircase, were not published until nearly seven years later (July 1989) in *Sovietsky Sport*. In 1985, three soccer game disasters, in England and Belgium, took 93 lives and injured nearly 800 persons. In April 1989, 95 persons perished in a crush at a soccer match in Sheffield, England.

NEEWOLLAH. Oct 20-28. Independence, KS. Halloween spelled backwards means a broadway musical presentation (local talent); Queen Neelah talent and coronation ceremonies, carnival, wide range of concessions, street acts, kiddie parade, grand parade, nationally known country western stars and concerts. Annually, the last full week in October. Est attendance: 80,000. Sponsor: Neewollah, Inc, Ken Fienen, CVB Dir, PO Box 386, Independence, KS 67301. Phone: (316) 331-1890.

SATURDAY NIGHT MASSACRE: ANNIVERSARY. Oct 20. Anniversary of dramatic turning point in the Watergate affair. The swiftly moving events of Oct 20, 1973: White House announcement (8:24 PM, EDT) that President Richard M. Nixon had discharged Archibald Cox (Special Watergate Prosecutor) and William B. Ruckelshaus (Deputy Attorney General). The Attorney General, Elliot L. Richardson, resigned. Immediate and widespread demands for impeachment of the president ensued and were not stilled until President Nixon resigned, on Aug 9, 1974.

STATE FAIR OF LOUISIANA. Oct 20-29. Fairgrounds, Shreveport, LA. Educational, agricultural, commercial exhibits, entertainment. Fax: (318) 631-4909. Est attendance: 250,000. For info: Sam Giordano, Louisiana State Fairgrounds, PO Box 38327, Shreveport, LA 71133. Phone: (318) 635-1361.

WREN, CHRISTOPHER: BIRTH ANNIVERSARY. Oct 20. Sir Christopher Wren, English architect, astronomer and mathematician, born on Oct 20, 1632. Died on Feb 25, 1723. His epitaph, written by his son, is inscribed over the interior of the north door in St. Paul's Cathedral, London: "Si monumentum requiris, circumspice." (If you would see his monument, look about you.)

BIRTHDAYS TODAY

Joyce Brothers, 67, psychologist, author, born at New York, NY, Oct 20, 1928.
Art Buchwald, 70, columnist, author (*While Reagan Slept*), born at Mount Vernon, NY, Oct 20, 1925.
William Christopher, 63, actor ("M*A*S*H," *With Six You Get Eggroll*), born Evanston, IL, Oct 20, 1932.

Arlene Francis (Arlene Kazanjian), 87, actress, panelist ("What's My Line?"), born Boston, MA, Oct 20, 1908.
Keith Hernandez, 42, former baseball player, born at San Francisco, CA, Oct 20, 1953.
Mickey Mantle, 64, baseball Hall of Famer, born at Spavinaw, OK, Oct 20, 1931.
Tom Petty, 43, musician, singer (with the Heartbreakers; "Refugee"; later with Traveling Wilburys), born Gainesville, FL, Oct 20, 1952.

OCTOBER 21 — SATURDAY
294th Day — Remaining, 71

AMERICAN HEART ASSOCIATION'S HEARTFEST. Oct 21-28. American Heart Association volunteers, in cooperation with supermarkets, restaurants, schools, and corporate and hospital cafeterias across the country, work together to bring heart-healthful nutrition information to consumers. Call local AHA Offices at (800) AHA-USA1.

APPLE BUTTER STIRRIN'. Oct 21-22. Coshocton, OH. Kettles of apple butter simmer over open fires; arts, crafts, contests, in Roscoe Village, a restored canal town. Est attendance: 11,000. For info: Roscoe Village Fdtn, 381 Hill Street, Coshocton, OH 43812. Phone: (800) 877-1830.

AUTUMN HISTORIC FOLKLIFE FESTIVAL. Oct 21-22. Downtown Historic District, Hannibal, MO. The Hannibal Arts Council sponsors its 19th annual festival celebrating Hannibal's cultural heritage and tradition. More than 50 artisans demonstrate the crafts of the mid-1800s, 20 vendors prepare food and drink over wood fires and street performers play traditional tunes. Also Rendezvous, Living History and children's areas with many activities. Est attendance: 25,000. For info: Hannibal Arts Council, PO Box 1202, Hannibal, MO 63401. Phone: (314) 221-6545.

BATTLE OF TRAFALGAR: ANNIVERSARY. Oct 21. "This famous naval action, on Oct 21, 1805, between the British Royal Navy and the combined French and Spanish fleets, removed forever the threat of Napoleon's invasion of England. The British victory, off Trafalgar, set the seal of eternal fame on Viscount Horatio Nelson who died in the moment of victory." Inscription under copy of painting in Main Post Office, near Charing Cross, London.

BOO AT THE ZOO. Oct 21-31. Good Children's Zoo, Oglebay, Wheeling, WV. Safe trick-or-treating, Halloween cartoons and train ride. Est attendance: 17,000. For info: Good Children's Zoo Office, John Hargleroad, Operations Manager, Oglebay, Wheeling, WV 26003. Phone: (304) 243-4028.

BROWN'S CROSSING FALL CRAFTSMEN FAIR. Oct 21-22. Milledgeville, GA. Arts and crafts sale. 26th annual sale. Est attendance: 15,000. For info: Carole Sirmans, 400 Browns Crossing Rd NW, Milledgeville, GA 31061. Phone: (912) 452-9327.

CARLETON, WILL: 150th BIRTH ANNIVERSARY. Oct 21. Anniversary of the birth of poet Will Carleton, Oct 21, 1845, observed (by 1919 statute) in Michigan schools where poems of Carleton must be read on this day. Best known of his poems: "Over the Hill to the Poorhouse." Carleton died in 1912.

COLERIDGE, SAMUEL TAYLOR: BIRTH ANNIVERSARY. Oct 21. English poet and essayist born Oct 21, 1772. Died July 25, 1834. In *Table Talk*, he wrote: "I wish our clever young poets would remember my homely definitions of prose and poetry; that is, prose = words in their best order; poetry = the *best* words in the best order."

✯ Chase's 1995 Calendar of Events ✯ Oct 21

A COUNTRY AFFAIR. Oct 21–22. Ben Franklin School, Menomonee Falls, WI. Community League's 13th annual fair features antiques, folk art and country collectibles. More than 90 exhibitors from 5 states; a Country Luncheon of homemade food and desserts; a "Country Pantry" and silent auction. Admission charged. Est attendance: 6,500. For info: Jeanne Verbsky, Publicity, Community League, Inc, N82 W18380 Independence Lane, Menomonee Falls, WI 53051. Phone: (414) 251-9062.

EIGHTEENTH-CENTURY AUTUMN MARKET FAIR. Oct 21–22. McLean, VA. Crafts, games, music and dancing. Period food and wares. Est attendance: 2,500. Sponsor: Claude Moore Colonial Farm at Turkey Run, 6310 Georgetown Pike, McLean, VA 22101. Phone: (703) 442-7557.

FALL COLOR CRUISE AND FOLK FESTIVAL. Oct 21–22. (Also Oct 28–29.) Shellmound Recreation Area and Ross's Landing, Chattanooga, TN. One of the largest boating events in the South, a major arts and crafts show and a large folk festival. The fall foliage can be enjoyed by car, bus or the Southern Belle Riverboat. Alhambra Shriners and other corporate sponsors. Est attendance: 130,000. For info: Alabama Temple, Attn: Rayford McLaurin, 1000 Alhambra Dr, Chattanooga, TN 37421. Phone: (615) 238-4626.

FALL FIESTA OF THE ARTS. Oct 21–22. Swan Lake Gardens, Sumter, SC. Features visual and performing arts, concerts and choral groups. Annually, the third weekend in October. Est attendance: 6,000. For info: Sumter Cultural Commission, 10 Mood Ave, Sumter, SC 29150. Phone: (803) 436-2258.

FILLMORE, CAROLINE CARMICHAEL McINTOSH: BIRTH ANNIVERSARY. Oct 21. Second wife of Millard Fillmore, 13th president of the US, born at Morristown, NJ, Oct 21, 1813. Died Aug 11, 1881.

GILLESPIE, JOHN BIRKS "DIZZY." Oct 21. Dizzy Gillespie, trumpet player, composer, band leader and one of the founding fathers of modern jazz, was born Oct 21, 1917, at Cheraw, SC. In the early 1940s Gillespie and alto saxophonist Charlie (Yardbird) Parker created be-bop. In the late '40s he created a second music revolution by incorporating Afro-Cuban music into jazz. In 1953 someone fell on Gillespie's trumpet and bent it. Finding he could hear the sound better, he kept it that way, his puffed cheeks and bent trumpet becoming his trademarks. Gillespie won a Grammy in 1975 for *Oscar Peterson and Dizzy Gillespie* and again in 1991 for *Live at the Royal Festival Hall*. His autobiography, *To Be or Not to Bop*, was published in 1979. Dizzy Gillespie died Jan 6, 1993, at Englewood, NJ.

GREAT PUMPKIN FESTIVAL. Oct 21–22. Bedford, PA. See record-breaking pumpkins. Bring in your own pumpkin for weigh-in (cash prizes). Carve or paint a jack-o-lantern, live entertainment, costume parade, make a scarecrow, and pumpkin-pie eating contest. Sponsor: Old Bedford Village, Box 1976, Bedford, PA 15522. Phone: (814) 623-1156.

GREAT TEDDY BEAR JAMBOREE SHOW AND SALE. Oct 21–22. Bristol, CT. To bring together teddy bear lovers, collectors and others to view thousands of bears, old and new. Hungry bear snack bar, teddy bear hospital, children's storytelling and coloring corner. Annually, the third weekend in October. Fax: (203) 584-5063. Est attendance: 10,000. Sponsor: Friends of Bristol Senior Ctr, Cheryl Yetke, Office Mgr, 240 Stafford Ave, Bristol, CT 06010. Phone: (203) 583-2562.

INCANDESCENT LAMP DEMONSTRATED: ANNIVERSARY. Oct 21. Thomas A. Edison on Oct 21, 1879, demonstrated the first incandescent lamp that could be used economically for domestic purposes. This prototype, developed at his Menlo Park, NJ, laboratory, could burn for 13½ hours.

INDIAN HERITAGE FESTIVAL. Oct 21. Huntsville, AL. The Indian Heritage Festival is a day-long event to explore the history, culture, traditions, and skills of Alabama's Indian people. Come explore the history of these first Alabamians and the different cultures from the habitation of caves through the Mississippian period cultures of the Mound builders and the tribal civilizations at the time of DeSoto's arrival, to the Indians of today. Admission charged. Annually, the third Saturday in October. Est attendance: 3,000. For info: Jeff Hughes, Curator of Natural History, Burritt Museum and Park, 3101 Burritt Dr, Huntsville, AL 35801. Phone: (205) 536-2882.

INTERNATIONAL GOLD CUP. Oct 21. Great Meadow, The Plains, VA. A day of steeplechasing in the heart of Virginia's hunt country. Gates open at 10 AM for special events and activities. Corporate and chalet entertainment packages available. Conducted by the Virginia Gold Cup Assn for benefit of free year-round use of Great Meadow for non-profit community activities. Est attendance: 50,000. For info: Virginia Gold Cup Assn, PO Box 840, Warrenton, VA 22186. Phone: (703) 347-2612.

MAKE A DIFFERENCE. Oct 21. The national day of community service is sponsored by USA Weekend, (a Sunday supplement delivered in over 400 newspapers). Volunteer projects completed on the day are judged by well known celebrities and 10 winning projects receive $2,000 each to donate to their charity of choice. Winners are honored in April during National Volunteer Week at a special Make A Difference Day awards luncheon and at the White House. The Points of Light Foundation is a partner. Approximately 500,000 people nationwide participate. For info: Pam Brown, Make A Difference Day Editor USA Weekend, 1000 Wilson Blvd, Arlington, VA 22229-0012. Phone: (703) 276-6445.

MARSHALL ISLANDS: COMPACT DAY. Oct 21. National holiday.

ST. MARY'S COUNTY OYSTER FESTIVAL. Oct 21–22. Fairgrounds, Leonardtown, MD. Oysters served every style, national oyster shucking contest and national oyster cookoff. Est attendance: 25,000. For info: Oyster Fest Office, David L. Taylor, Admin, Box 766, California, MD 20619-0766. Phone: (301) 863-5015.

SOMALIA: NATIONAL DAY. Oct 21. National holiday observed in Somalia.

SPIDERS! Oct 21–Jan 14. Royal Ontario Museum, Toronto, Ont. An exhibit to "spider-wise" visitors to how spiders go about their daily lives. See Mar 11 for full description. For info: Smithsonian Institution Traveling Exhibition Service, 1100 Jefferson Dr SW, Ste 3146, Washington, DC 20560. Phone: (202) 357-2700.

SWEETEST DAY. Oct 21. Observed annually on the third Saturday in October. For info: Detroit Sweetest Day Committee, John M. Sanders, Chair, 901 Wilshire Dr, Ste 360, Troy, MI 48084. Phone: (810) 362-3223.

TAIWAN: OVERSEAS CHINESE DAY. Oct 21. Thousands of overseas Chinese come to Taiwan for this and other occasions that make October a particularly memorable month.

TOWN POINT VIRGINIA WINE FESTIVAL. Oct 21. Town Point Park, Norfolk, VA. Sampling of the finest wines from Virginia's best wineries. Also, a variety of Virginia food and wares from area and regional gourmet specialty shops and restaurants will be showcased. 7th annual. Est attendance: 10,000. For info: Norfolk Festevents, Ltd, Promotions Dir, 120 W Main St, Norfolk, VA 23510. Phone: (804) 627-7809.

WYATT EARP O.K. CORRAL ANNIVERSARY. Oct 21. Monmouth, IL. Lunch at local restaurants with Wyatt Earp, birthplace and pioneer cemetery tours, videos of corral reenactments. Phone for Chamber of Commerce: (309) 734-3181. Est attendance: 100. For info: Wyatt Earp Birthplace Historic House Museum, c/o 1020 E Detroit Ave, Monmouth, IL 61462. Phone: (309) 734-6419.

Oct 21-22 ☆ Chase's 1995 Calendar of Events ☆

BIRTHDAYS TODAY

Sir Malcolm Arnold, 74, composer, born at Northampton, England, Oct 21, 1921.
George Bell, 36, former baseball player, born at San Pedro de Macoris, Dominican Republic, Oct 21, 1959.
Elvin Bishop, 53, musician, born at Tulsa, OK, Oct 21, 1942.
Carrie Fisher, 39, actress (Princess Leia in *Star Wars*; *Shampoo*), novelist (*Postcards From the Edge* [made into film]), born Beverly Hills, CA, Oct 21, 1956.
Frances Fitzgerald, 55, journalist, author, born at New York, NY, Oct 21, 1940.
Edward Charles "Whitey" Ford, 67, baseball Hall of Famer, born at New York, NY, Oct 21, 1928.
Ursula LeGuin, 66, author (*The Wind's Twelve Quarters, A Wizard of Earthsea*), born at Berkeley, CA, Oct 21, 1929.
Sir Georg Solti, 83, conductor (formerly of Chicago Symphony Orchestra), born at Budapest, Hungary, Oct 21, 1912.

OCTOBER 22 — SUNDAY
295th Day — Remaining, 70

BEADLE, GEORGE: BIRTH ANNIVERSARY. Oct 22. George Beadle was born on a farm near Wahoo, NE, on Oct 22, 1903. He began his professional career as as assistant professor in genetics at Harvard, moving up finally to become president of the University of Chicago. Dr. Beadle won many international prizes, including the Nobel Prize for Medicine in 1958 for his work in genetic research, as well as the National Award of the American Cancer Society in 1959 and the Kimber Genetica Award of the National Academy of Science in 1960. Beadle demonstrated how the genes control the basic chemistry of the living cell. Because of his work, he has been termed "the man who did most to put modern genetics on its chemical basis." Beadle died June 9, 1989, at Pomona, CA.

CUBAN MISSILE CRISIS ANNIVERSARY. Oct 22. President John F. Kennedy, in a nationwide television address on Oct 22, 1962, demanded the removal from Cuba of Soviet missiles, launching equipment and bombers, and imposed a naval "quarantine" to prevent further weaponry from reaching Cuba. On Oct 28, the USSR announced it would remove the weapons in question.

FOREST GROVE ARTS AND CRAFTS SHOW. Oct 22. Forest Grove Community Club, Forest Grove, MT. 15th annual craft show and sale, indoors, tables available, 11 AM to 4 PM. Annually, the first Sunday of the Hunting Season. Est attendance: 650. Sponsor: Forest Grove Community Club, May Charbonneau, Box 6, Forest Grove, MT 59441. Phone: (406) 538-3510.

GROLIER, JEAN: 430th DEATH ANNIVERSARY. Oct 22. The celebrated French bibliophile, Jean Grolier de Servieres, whose exact birthday at Lyon, France, in 1479 is unknown, died at Paris, Oct 22, 1565. A government official, Grolier's consuming interest was books and he assembled one of the world's finest collections—a library of more than 3,000 elegantly bound volumes. The Grolier Club of New York City is named for him.

October 1995

S	M	T	W	T	F	S
1	2	3	4	5	6	7
8	9	10	11	12	13	14
15	16	17	18	19	20	21
22	23	24	25	26	27	28
29	30	31				

HOLY SEE: NATIONAL HOLIDAY. Oct 22. The state of Vatican City and the Holy See observe Oct 22 as a national holiday.

LAKESIDE HISTORICAL SOCIETY HOMETOWN REUNION. Oct 22. Lakeside, CA. 17th hometown reunion. Est attendance: 450. For info: Lakeside Historical Society, Shirley Anderson, Box 1886, Lakeside, CA 92040. Phone: (619) 561-1886.

LISZT, FRANZ: BIRTH ANNIVERSARY. Oct 22. Hungarian composer. Oct 22, 1811–July 31, 1886.

METROPOLITAN OPERA HOUSE: OPENING ANNIVERSARY. Oct 22. Grand opening of the original New York Metropolitan Opera House, on Oct 22, 1883, was celebrated with a performance of Gounod's *Faust*.

MOTHER-IN-LAW DAY. Oct 22. Traditionally, the fourth Sunday in October is occasion to honor mothers-in-law for their contribution to the success of families and for their good humor in enduring bad jokes.

NATIONAL CONSUMERS WEEK. Oct 22-28. Consumer education, including the protection of consumers against fraud and misrepresentation of products and services. This year's theme: know your consumer rights. Annually, the last full week in October. For info: White House Office of Consumer Affairs, 1620 L St NW, Ste 700, Washington, DC 20036. Phone: (202) 634-4329.

NATIONAL SAVE YOUR BACK WEEK. Oct 22-28. To educate the population on proper back care. Sponsor: Daniel S. Romm, MD, Medical Arts Bldg, 890 Poplar Church Rd, Ste 305, Camp Hill, PA 17011. Phone: (717) 975-9994.

NEW INTERNATIONAL VERSION OF THE BIBLE WEEK. Oct 22-28. To mark the anniversary of date (Oct 27, 1978) of the publication of the New International Version of the Bible, the most popular contemporary-English Bible translation today, with more than 90 million copies distributed worldwide. For info: Judy Waggoner, Mgr, Print Media, Zondervan Publishing House, 5300 Patterson Ave SE, Grand Rapids, MI 49530. Phone: (616) 698-3209.

RANDOLPH, PEYTON: DEATH ANNIVERSARY. Oct 22. First president of the Continental Congress died, Oct 22, 1775. Born about 1721 (exact date unknown), probably at Williamsburg, VA.

WORLD'S END DAY: ANNIVERSARY. Oct 22. Anniversary of the day, Oct 22, 1844, set as the day on which the world would end, by followers of William Miller, religious leader and creator of a movement known as Millerism. Stories about followers disposing of all earthly possessions and climbing to high places on that date are believed to be apocryphal. (Miller was born at Pittsfield, MA, Feb 15, 1782. Died at Low Hampton, NY, Dec 20, 1849.)

BIRTHDAYS TODAY

Brian Anthony Boitano, 32, Olympic Men's Figure Skating Champion ('88), born at Mountain View, CA, Oct 22, 1963.
John Hubbard Chafee, 73, US Senator (R, Rhode Island) up for reelection in '94, born Providence, RI, Oct 22, 1922.
Catherine Deneuve (Dorleac), 52, actress (*Repulsion, Indochine*), born Paris, France, Oct 22, 1943.
Annette Funicello, 53, actress, singer ("Mickey Mouse Club"; Beach Party movies), born at Utica, NY, Oct 22, 1942.
Jeff Goldblum, 43, actor ("Tenspeed and Brown Shoe," *The Fly*), born Pittsburgh, PA, Oct 22, 1952.
Derek Jacobi, 57, actor ("I Claudius," *The Day of the Jackal*), born London, England, Oct 22, 1938.
Timothy Francis Leary, 75, psychologist, chemist/philosopher born at Springfield, MA, Oct 22, 1920.
Christopher Lloyd, 57, actor ("Taxi," *Back to the Future, Star Trek III, Who Framed Roger Rabbit, The Dream Team*), born Stamford, CT, Oct 22, 1938.
Robert Rauschenberg, 70, artist (construction, *Monogram*, born Port Arthur, TX, Oct 22, 1925.

☆ Chase's 1995 Calendar of Events ☆ Oct 23-24

OCTOBER 23 — MONDAY
296th Day — Remaining, 69

APPERT, NICOLAS: BIRTH ANNIVERSARY. Oct 23. Also known as "Canning Day," this is the anniversary of the birth of French chef, chemist, confectioner, inventor and author, Nicolas Appert, at Chalons-Sur-Marne, on Oct 23, 1752. Appert, who also invented the bouillon tablet, is best remembered for devising a system of heating foods and sealing them in air-tight containers. Known as the "father of canning," Appert won a prize of 12,000 francs from the French government in 1809, and the title "Benefactor of Humanity" in 1812, for his work and inventions which revolutionized our previously seasonal diet. Appert died at Massy, June 3, 1841, but the methods he developed for preserving food continue in ever greater use.

BEIRUT TERRORIST ATTACK: ANNIVERSARY. Oct 23. A suicidal terrorist attack on American forces in Beirut, Lebanon, killed 240 US personnel on Oct 23, 1983, when a truck loaded with TNT was driven into and exploded at US Headquarters there. At the same time a similar attack on French forces killed scores more.

EDERLE, GERTRUDE: BIRTH ANNIVERSARY. Oct 23. American swimming champion, born at New York, NY, Oct 23, 1906. Gertrude Caroline Ederle was the first woman to swim the English Channel (from Cape Gris-nez, France to Dover, England). At age 19 she broke the previous world record by swimming the 35-mile distance in 14 hours, 31 minutes, on Aug 6, 1926. During her swimming career she broke many other records and was a gold medal winner at the 1924 summer Olympic Games.

HUNGARY DECLARES ITS INDEPENDENCE: ANNIVERSARY. Oct 23. On Oct 23, 1989, Hungary declared itself an independent republic, 33 years after Russian troops crushed a popular revolt against Soviet rule. The announcement followed a week-long purge by Parliament of the Stalinist elements from Hungary's 1949 constitution, which defined the country as a socialist people's republic. Acting head of state Matyas Szuros made the declaration in front of tens of thousands of Hungarians in Parliament Square, speaking from the same balcony from which Imre Nagy addressed rebels 33 years earlier. Nagy was hanged for treason after Soviet intervention. Free elections held in March 1990 placed the Alliance of Free Democrats in control of a coalition government, removing the Communist Party to the ranks of the opposition for the first time in four decades.

NATIONAL MOLE DAY. Oct 23. Celebrated on Oct 23 each year from 6:02 AM–6:02 PM in observance of the "mole." The "mole" is a way of counting the Avogadro number, 6.02×10 to the 23rd power of anything (just like a "dozen" is a way of counting 12 of anything). Mole Day owes its existence to an early 19th-century Italian physics professor named Amadeo Avogadro. He discovered that the number of molecules in a mole is the same for all substances. Because of this, chemists are able to precisely measure quantities of chemicals in the laboratory. Mole Day is celebrated to help all persons, especially chemistry students, become enthused about chemistry, which is the central science. Individual teachers develop their own ways of observing Mole Day. 1995 theme: "Let's celebrate Mole-di Gras." Fax: (608) 326-2333. E-mail: PRAIRIMO @LLWISC.LL.PBS.ORG. For info: National Mole Day Fdtn, Maurice Oehler, Exec Dir, 1220 S 5th St, Prairie du Chien, WI 53821.

ST. JOHN OF CAPISTRANO: DEATH ANNIVERSARY. Oct 23. Giovanni da Capistrano, Franciscan lawyer, educator and preacher, was born at Capistrano, Italy, in 1386. He died, of plague, on Oct 23, 1456. Feast Day is Mar 28.

SCORPIO, THE SCORPION. Oct 23–Nov 22. In the astronomical/astrological zodiac that divides the sun's apparent orbit into 12 segments, the period Oct 23–Nov 22 is identified, traditionally, as the sun-sign of Scorpio, the Scorpion. The ruling planet is Pluto or Mars.

SOLAR ECLIPSE. Oct 23. Total eclipse of the sun. Central eclipse begins at 10:53 PM (EDT), Oct 23, reaches greatest eclipse at 12:23 AM (EDT), Oct 24, ends at 2:12 AM (EDT). The eclipse will be visible over extreme northeast Africa, central and eastern Asia except northeastern part, northern Indian Ocean, northern half of Australia and western Pacific Ocean.

STEVENSON, ADLAI EWING: BIRTH ANNIVERSARY. Oct 23. Twenty-third vice president of the US (1893–1897) born at Christian County, KY, Oct 23, 1835. Died at Chicago, IL, June 14, 1914. He was grandfather of Adlai E. Stevenson, Democratic candidate for president in 1952 and 1956. See also: "Stevenson, Adlai Ewing: Birth Anniversary" (Feb 5).

SWALLOWS DEPART FROM SAN JUAN CAPISTRANO. Oct 23. Traditional date for swallows to depart, for the winter, from old mission of San Juan Capistrano, CA. See also: "Swallows Return to San Juan Capistrano" (Mar 19).

THAILAND: CHULALONGKORN DAY. Oct 23. Annual commemoration of the death of King Chulalongkorn the Great who died Oct 23, 1910, after a 42-year reign. Special ceremonies with floral tributes and incense at the foot of his equestrian statue in front of Bangkok's National Assembly Hall. It was King Chulalongkorn who abolished slavery in Thailand.

TV TALK SHOW HOST DAY. Oct 23. To celebrate the many TV talk show hosts whose personalities and intellects enable them to bring out the best in their guests. Sponsor: Glenn Rothenberger, Blue Collar Show, 541 Clinton St, Ste 2B, Brooklyn, NY 11231. Phone: (718) 802-1689.

BIRTHDAYS TODAY

Jim Bunning, 64, former baseball player, born at Southgate, KY, Oct 23, 1931.

Johnny Carson, 70, former TV talk show host ("The Tonight Show"), born at Corning, IA, Oct 23, 1925.

Michael Crichton, 53, writer (*Jurassic Park, Rising Sun*), born at Chicago, IL, Oct 23, 1942.

Diana Dors (Diana Fluck), 64, actress (*Oliver Twist, Yield to the Night*), born at Swindon, England, Oct 23, 1931.

Doug Flutie, 33, football player, born Manchester, MD, Oct 23, 1962.

Pele (Edson Arantes do Nascimento), 55, soccer player, born at Tres Coracoes, Brazil, Oct 23, 1940.

Juan "Chi-Chi" Rodriguez, 61, golfer, born at Rio Piedras, Puerto Rico, Oct 23, 1934.

Mike Tomczak, 33, football player, born Calumet City, IL, Oct 23, 1962.

Alfred Matthew "Weird Al" Yankovic, 36, singer, satirist, born at Lynwood, CA, Oct 23, 1959.

OCTOBER 24 — TUESDAY
297th Day — Remaining, 68

◆ **LOCKWOOD, BELVA A. BENNETT: BIRTH ANNIVERSARY.** Oct 24. Belva Lockwood, an educator, lawyer and advocate for women's rights, was born in 1830 at Royalton, NY. In 1879 she was admitted to practice before the US Supreme Court—the first woman to do so. While practicing law in Washington, DC, she secured equal property rights for women. By adding amendments to statehood bills, Lockwood helped to provide voting rights for women in Oklahoma, New Mexico and Arizona. In 1884 she was the first woman formally nominated for the US presidency. Belva Lockwood died May 19, 1917, at Washington, DC.

Oct 24-25 ☆ Chase's 1995 Calendar of Events ☆

MOON PHASE: NEW MOON. Oct 24. Moon enters New Moon phase at 12:36 AM, EDT.

SHERMAN, JAMES SCHOOLCRAFT: BIRTH ANNIVERSARY. Oct 24. Twenty-seventh vice president of the US (1909-1912) born at Utica, NY, Oct 24, 1855. Died there Oct 30, 1912.

STOCK MARKET PANIC: ANNIVERSARY. Oct 24. After several weeks of downward trend in stock prices, investors began panic selling on Black Thursday, Oct 24, 1929. More than 13 million shares dumped. Desperate attempts to support the market brought brief rally. See also: "Stock Market Collapse Anniversary" (Oct 29).

★ **UNITED NATIONS DAY.** Oct 24. Presidential Proclamation. Always issued for Oct 24 since 1948. (By unanimous request of the UN General Assembly.)

UNITED NATIONS: DISARMAMENT WEEK. Oct 24-30. Annual observance begins on anniversary of founding of United Nations. In December 1978, the United Nations General Assembly invited all states to implement measures exposing the danger of the arms race and the need for its cessation. As requested by the Assembly, the secretary-general in 1979 drew up the elements of a model program for the week that might assist states in developing their own observances. Info from: United Nations, Dept of Public Info, New York, NY 10017.

◆ **UNITED NATIONS DAY: 50th ANNIVERSARY OF FOUNDING.** Oct 24. Official United Nations holiday commemorates founding of the United Nations and effective date of the United Nations Charter, Oct 24, 1945. Info from: United Nations, Dept of Public Info, New York, NY 10017.

UNITED NATIONS: WORLD DEVELOPMENT INFORMATION DAY. Oct 24. Anniversary of adoption by United Nations General Assembly, in 1970, of the International Development Strategy for the Second United Nations Development Decade. Object is to "draw the attention of the world public opinion each year to development problems and the necessity of strengthening international cooperation to solve them." Info from: United Nations, Dept of Public Info, New York, NY 10017.

ZAMBIA: INDEPENDENCE DAY. Oct 24. Zambia. National holiday commemorates the Instruments of Independence signed on this day in 1964. Celebrations in all cities, but main parades of military, labor and youth organizations are at capital, Lusaka.

BIRTHDAYS TODAY

F. Murray Abraham, 55, actor (Oscar *Amadeus*), born El Paso, TX, Oct 24, 1940.
Kevin Kline, 48, actor (*A Fish Called Wanda*, *Silverado*), born St. Louis, MO, Oct 24, 1947.
Denise Levertov, 72, poet, born at Ilford, England, Oct 24, 1923.
David Nelson, 59, actor ("The Adventures of Ozzie and Harriet"), born New York, NY, Oct 24, 1936.

October 1995

S	M	T	W	T	F	S
1	2	3	4	5	6	7
8	9	10	11	12	13	14
15	16	17	18	19	20	21
22	23	24	25	26	27	28
29	30	31				

Y.A. Tittle (Yelberton Abraham Tittle), 69, Pro Football Hall of Famer, born at Marshall, TX, Oct 24, 1926.
Bill Wyman, 54, musician, born at London, England, Oct 24, 1941.

OCTOBER 25 — WEDNESDAY
298th Day — Remaining, 67

CARTOONISTS AGAINST CRIME DAY™. Oct 25. A day in honor of all those cartoonists, graphic designers and illustrators, who join together to promote the prevention of crime through the art and medium of cartooning. Motto: CARTOONISTS AGAINST CRIME: WE DRAW CARTOONS—NOT GUNS! CAC's Slogan is to TOON OUT CRIME by employing comic art as an educational tool for implanting seeds of safety prevention into the minds of viewers/readers. Sponsor: Cartoonists Against Crime, Adrienne Sioux Koopersmith, 1437 W Rosemont, #1W, Chicago, IL 60660. Phone: (312) 743-5341.

CENTER CITY CHURCH CONFERENCE. Oct 25-26. First Presbyterian Church, Aurora, IL. Interdenominational conference of churches from America's center cities (pop. 50,000-500,000) with church consultant Lyle Schaller. Workshops to discuss urban issues. Cost $150. Sponsor: First Presbyterian Church and Lyle Schaller. Est attendance: 450. For info: Rev Noel Allen, First Presbyterian Church, 325 E Downer Pl, Aurora, IL 60505. Phone: (708) 844-0050.

CHAUCER, GEOFFREY: 595th DEATH ANNIVERSARY. Oct 25. English poet and the best-known English writer of the Middle Ages, was born at London, England, probably about 1340. His greatest work, *Canterbury Tales*, consists of some 17,000 poetic lines. Unfinished at his death, it tells the stories of 23 pilgrims. Among his lesser known prose writings was a treatise on the Astrolabe entitled *Brede and Milke for Children* (1387), written for "little Lewis, my son." Chaucer died at London on Oct 25, 1400, and is buried at Westminster Abbey.

FIRST FEMALE FBI AGENTS: ANNIVERSARY. Oct 25. The first women to become FBI agents completed training at Quantico, VA, Oct 25, 1972. The new agents, Susan Lynn Roley and Joanne E. Pierce graduated from the 14-week course with a group of 45 men.

GRENADA INVADED BY US: ANNIVERSARY. Oct 25. On Oct 25, 1983, some 2,000 US Marines and Army Rangers invaded the Caribbean island of Grenada, taking control after a political coup the previous week had made the island a "Soviet-Cuban colony," according to President Reagan.

LOUISIANA YAMBILEE. Oct 25-29. Opelousas, LA. Sweet potato, corn, rice and soybean shows, cooked foods contest, yam auction, "yum-yum contest," costume contest, carnival, diaper derby, grand parade and the crowning of King Will Yam and Queen Marigold. Est attendance: 80,000. For info: Louisiana Yambilee, Inc, Sheryl Badeaux, Box 444, Opelousas, LA 70570. Phone: (318) 948-8848.

MACAULAY, THOMAS BABINGTON: BIRTH ANNIVERSARY. Oct 25. English essayist and historian, born Oct 25, 1800. Died, Dec 28, 1859. "Nothing," he wrote, "is so useless as a general maxim."

NATIONAL MAGIC WEEK. Oct 25-31. To promote brotherly love through magic performances at hospitals and nursing homes. More than 8,000 members nationwide participate. Sponsor: Society of American Magicians, Inc, Anthony D. Murphy, Esq, 11 Angel Rd, North Reading, MA 01864. Phone: (617) 523-6434.

PEACE, FRIENDSHIP AND GOOD WILL WEEK. Oct 25-31. To encourage and foster international understanding, good human relations, friendship, good will and peace throughout the world. For complete info, send $1 to cover expense of printing, handling and postage. Annually, the last seven days in October. Sponsor: Intl Soc of Friendship and Good Will, Dr. Stanley Drake, Pres, 9538 Summerfield St, Spring Valley, San Diego County, CA 91977-2852. Phone: (619) 466-8882.

PICASSO, PABLO RUIZ: BIRTH ANNIVERSARY. Oct 25. Called by many the greatest artist of the 20th century, Pablo Picasso excelled as a painter, sculptor and engraver. He is said to have commented once: "I am only a public entertainer who has understood his time." Born at Malaga, Spain, Oct 25, 1881, Picasso died Apr 8, 1973.

SOUREST DAY. Oct 25. To emphasize the balance of things in nature. A day for sour (Sauer) people. Sponsor: Richard Ankli, The Fifth Wheel Tavern, 639 Fifth St, Ann Arbor, MI 48103.

SOUTHERN CALIFORNIA FIRESTORMS: ANNIVERSARY. Oct 25, 1993. The Southern California fire season began viciously when fires swept from the celebrity-studded beachfront homes of Malibu to the Mexican border. Blown out of the desert by the fierce Santa Anna winds, the fires destroyed suburban enclaves south of Los Angeles in Laguna Beach and northeast of LA in Altadena. As winds died down, firefighters appeared to gain control as the flames reached the Santa Monica Mountains, but the winds roared again spreading the fire into Malibu—often jumping the Pacific Coast Highway to destroy the beachfront homes of the wealthy celebrities who lived there. Damage from the fires was estimated at over $1 billion.

◆ **TAIWAN: RETROCESSION DAY: 50 YEARS.** Oct 25. Commemorates restoration of Taiwan to Chinese rule, on Oct 25, 1945, after half a century of Japanese occupation.

BIRTHDAYS TODAY

Anthony Franciosa (Anthony Papaleo), 67, actor ("The Name of the Game," "Wheels"), born at New York, NY, Oct 25, 1928.

Hanna Gray, 65, educator (former president of University of Chicago), born at Heidelberg, Germany, Oct 25, 1930.

Brian Kerwin, 46, actor ("Lobo," "The Blue and the Gray"), born at Chicago, IL, Oct 25, 1949.

Bob Knight, 55, basketball coach, born at Massillon, OH, Oct 25, 1940.

Helen Reddy, 53, singer ("I Am Woman"), songwriter, born Melbourne, Australia, Oct 25, 1942.

Marion Ross, 59, actress (Marion Cunningham on "Happy Days"), born Albert Lea, MN, Oct 25, 1936.

Anne Tyler, 54, author (*The Accidental Tourist, Searching for Caleb*), born Minneapolis, MN, Oct 25, 1941.

OCTOBER 26 — THURSDAY
299th Day — Remaining, 66

THE ARC (A NATIONAL ORGANIZATION ON MENTAL RETARDATION): ANNUAL CONVENTION. Oct 26-28. Hyatt Regency, Indianapolis, IN. The Arc was formerly named Association for Retarded Citizens of the United States. Est attendance: 1,500. For info: The Arc Natl HQ, Mike Meyers, 500 E Border St, Ste 300, Arlington, TX 76010. Phone: (817) 261-6003.

AUSTRIA: NATIONAL DAY. Oct 26. National holiday observed.

EDGAR ALLAN POE FESTIVAL. Oct 26-31. (Also Nov 2-5, Nov 9-12.) Mount Hope Estate, Cornwall, PA. An evening of suspense featuring the spine-chilling short stories of Edgar Allan Poe. Professionals from the Pennsylvania Renaissance Faire Actors Conservatory perform, wine served. Est attendance: 8,000. For info: Mount Hope Estate and Winery, PO Box 685, Cornwall, PA 17016. Phone: (717) 665-7021.

ERIE CANAL: ANNIVERSARY. Oct 26. The Erie Canal, first US man-made major waterway, was opened on Oct 26, 1825, providing a water route from Lake Erie to the Hudson River. Started on July 4, 1817, the canal cost $7,602,000. Cannons fired and celebrations held all along the route for the opening.

FORT LAUDERDALE INTERNATIONAL BOAT SHOW. Oct 26-30. Greater Ft Lauderdale/Broward Co Conv Center. Everything from small boats to mega-yachts to boating equipment. Visitors attend from all over the world. For info: Greater Ft Lauderdale Conv/Visitors Bureau, 200 E Las Olas Blvd, Ste 1500, Fort Lauderdale, FL 33301. Phone: (305) 765-5900.

THE GREAT PUMPKIN CARVE. Oct 26. Chadds Ford, PA. Local artists carve huge pumpkins on the grounds of the Chadds Ford Historical Society, 5-9 PM. Attractions include judging and food. Annually, the last Thursday before Halloween. Est attendance: 1,500. For info: Chadds Ford Historical Soc, Box 27, Chadds Ford, PA 19317. Phone: (610) 388-7376.

HANSOM, JOSEPH: BIRTH ANNIVERSARY. Oct 26. English architect and inventor Joseph Aloysius Hansom registered his "Patent Safety Cab" in 1834. The two-wheeled, one-horse, enclosed cab, with driver seated above and behind the passengers, quickly became a familiar and favorite vehicle for public transportation. Hansom was born at York, England, on Oct 26, 1803, and died at London, on June 29, 1882.

HORSELESS CARRIAGE DAY. Oct 26. From James Boswell's *Life of Samuel Johnson*: Oct 26, 1769 "... we dined together at the Mitre tavern. ... We went home to his house to tea. ... There was a pretty large circle this evening. Dr. Johnson was in very good humour, lively, and ready to talk upon all subjects. Mr. Ferguson, the self-taught philosopher, told him of a new-invented machine which went without horses: a man who sat in it turned a handle, which worked a spring that drove it forward. 'Then, Sir (said Johnson), what is gained is, the man has his choice whether he will move himself alone, or himself and the machine too.' "

LIMOUSINE SCAVENGER HUNT. Oct 26. Tacoma, WA. Costumed adults form teams and test their endurance, creativity and willpower in a crazy scavenger hunt, travelling around town in limousines in pursuit of items. Annually, the Thursday before Halloween. Est attendance: 200. Sponsor: Metropolitan Park District, Rec Dept, 4702 S 19th St, Tacoma, WA 98405. Phone: (206) 305-1036.

MULE DAY. Oct 26. Anniversary of the first importation of Spanish jacks to the US, a gift from King Charles III of Spain. Mules are said to have been bred first in this country by George Washington from a pair delivered in Boston on Oct 26, 1785.

ROCKEFELLER, ABBY GREENE ALDRICH: BIRTH ANNIVERSARY. Oct 26, 1874. A philanthropist and art patron, Abby Rockefeller was one of the three founders of the New York Museum of Modern Art in 1929. Born at Providence, RI, she died Apr 5, 1948 at New York City.

SCARLATTI, DOMENICO: BIRTH ANNIVERSARY. Oct 26. Italian composer, born Oct 26, 1685. Died, July 23, 1757.

SOUTHWEST LOUISIANA STATE FAIR AND EXPOSITION. Oct 26-29. Civic Center, Lake Charles, LA. Merchant exhibits and pageants, fireworks and special events. Est attendance: 43,000. For info: Southwest Louisiana State Fair, PO Box 809, Lake Charles, LA 70602. Phone: (318) 436-7575.

SPACE MILESTONE: *SOYUZ 3* (USSR). Oct 26. Launched on Oct 26, 1968. Colonel Georgi Beregovoy orbited Earth 64 times, rendezvousing with unmanned *Soyuz 2*, launched Oct 25, 1968. Both vehicles returned to Earth under ground control.

VIRGINIA FESTIVAL OF AMERICAN FILM. Oct 26-29. University of Virginia, Charlottesville, VA. This 8th annual 4-day program celebrates and explores the unique character of American film past and present, bringing together leading filmmakers, performers, writers, scholars and the public in a setting that fosters serious discussion and academic exchange. Phone: (800) UVA-FEST. Est attendance: 30,000. For info: VA Festival of American Film, 104 Midmont Lane, University of Virginia, Charlottesville, VA 22903.

Oct 26–27 ☆ Chase's 1995 Calendar of Events ☆

BIRTHDAYS TODAY

Hillary Rodham Clinton, 49, First Lady, attorney, born at Park Ridge, IL, Oct 26, 1946.
Chuck Forman, 45, former football player, born at Frederick, MO, Oct 26, 1950.
Bob Hoskins, 53, actor (*Mona Lisa, Who Framed Roger Rabbit*), born Bury St. Edmonds, Suffolk, England, Oct 26, 1942.
Francois Mitterand, 79, President of France, born at Jarnac, France, Oct 26, 1916.
Ivan Reitman, 49, filmmaker (*Ghostbusters* movies), born Komarno, Czechoslovakia, Oct 26, 1946.
Pat Sajak, 49, TV personality ("Wheel of Fortune"), born Chicago, IL, Oct 26, 1946.
Jaclyn Smith, 48, actress (Kelly Garrett on "Charlie's Angels"; former Breck Girl) born Houston, TX, Oct 26, 1947.

OCTOBER 27 — FRIDAY
300th Day — Remaining, 65

ALBEMARLE CRAFTSMAN'S FAIR. Oct 27–29. Knobbs Creek Recreation Center, NC. 37th Annual Albemarle Craftsman's Fair, one of the oldest demonstrating craft shows in the nation and one of the most unique and prestigious shows of its kind. Features woodworkers, carvers, potters, quilters, weavers, sterling silver jewelry makers and other talented artisans. Annually, the last Friday–Sunday of October. Est attendance: 6,000. For info: President, Albermarle Craftman's Fair, PO Box 1301, Elizabeth City, NC 27906-1301. Phone: (919) 338-3954.

CANADA: FALL FAIR. Oct 27–29. Keystone Centre, Brandon, Manitoba. Featuring AG EX—Manitoba's largest livestock show and sale and the Manitoba Rodeo Championship finals. Fax: (204) 725-0202. Est attendance: 35,000. For info: Provincial Exhibition of Manitoba, #3-1175 18th St, Brandon, MB, Canada R7A 7C5. Phone: (204) 726-3590.

COOK, JAMES: BIRTH ANNIVERSARY. Oct 27. English sea captain and explorer born Oct 27, 1728. Died, Feb 14, 1779.

FEDERALIST PAPERS: ANNIVERSARY. Oct 27. The first of the 85 "Federalist" papers appeared in print in a New York City newspaper, Oct 27, 1787. These essays, written by Alexander Hamilton, James Madison and John Jay argued in favor of adoption of the new Constitution and the new form of federal government. The last of the essays was completed Apr 4, 1788.

GHOST TALES AROUND THE CAMPFIRE. Oct 27. Jefferson College, Washington, MS. Storytellers weave their spells of mystery, surprise and suspense as they tell tales around a bonfire. Est attendance: 300. For info: Historic Jefferson College, Anne L. Gray, Historian, PO Box 700, Washington, MS 39190. Phone: (601) 442-2901.

GRENADINES: NATIONAL DAY. Oct 27. National holiday observed in Grenadines.

HALLOWEEN NIGHT WALKS. Oct 27–28. A.B. Brooks Nature Center, Oglebay, Wheeling, WV. A unique Halloween adventure: a walk through the woods at night with stops at stations to learn about nature's night creatures. The evening concludes with songs, stories and nature nuggets around a campfire. Est attendance: 800. Sponsor: Oglebay Institute Nature/Environmental Educ Dept, Oglebay, Wheeling, WV 26003. Phone: (304) 242-6855.

NAVY DAY. Oct 27. Observed on this date since 1922.

NBA DAY IN BASKETBALL CITY, USA. Oct 27. (Tentative.) Springfield Civic Center, Springfield, MA. The Boston Celtics will meet another NBA team. Est attendance: 9,000. For info: Basketball Hall of Fame, Box 179, 1150 W Columbus Ave, Springfield, MA 01101. Phone: (413) 781-6500.

NEW YORK CITY SUBWAY: ANNIVERSARY. Oct 27. Running from City Hall to West 145th St, the New York City subway began operation on Oct 27, 1904. It was the first underground and underwater rail system in the world.

PAGANINI, NICOLO: BIRTH ANNIVERSARY. Oct 27. Hailed as the greatest violin virtuoso of all time, Paganini was born at Genoa, Italy, Oct 27, 1782. Unusually long arms contributed to his legendary Mephistophelian appearance—and probably to his unique skills as a performer. His immensely popular concerts brought him great wealth, but his compulsive gambling repeatedly humbled the genius. Paganini died at Nice, France, May 27, 1840.

ROOSEVELT, THEODORE: BIRTH ANNIVERSARY. Oct 27. Theodore Roosevelt, 26th president of the US, succeeded to the presidency on the death of William McKinley. His term of office: Sept 14, 1901–Mar 3, 1909. Roosevelt was the first president to ride in an automobile (1902), to submerge in a submarine (1905) and to fly in an airplane (1910). Although his best-remembered quote was perhaps, "Speak softly and carry a big stick," he also said: "The first requisite of a good citizen in this Republic of ours is that he shall be able and willing to pull his weight." Born at New York, NY, Oct 27, 1858, Roosevelt died at Oyster Bay, NY, Jan 6, 1919. His last words: "Put out the light."

SAINT VINCENT AND THE GRENADINES: INDEPENDENCE DAY. Oct 27. National Day.

SERVETUS, MICHAEL: EXECUTION ANNIVERSARY. Oct 27. Spanish theologian and physician. Servetus was condemned to death on Oct 26, 1553, for blasphemy and burned at the stake at Geneva, Switzerland, on Oct 27, 1553. Servetus was born at Tudela, Navarre, Spain, in 1511 (exact date unknown).

TAMBO, OLIVER REGINALD: BIRTH ANNIVERSARY. Oct 27, 1917. Former president of South Africa's anti-apartheid African National Congress, Tambo was born at Transkei, South Africa, and died Apr 24, 1993, at Johannesburg.

THOMAS, DYLAN MARLAIS: BIRTH ANNIVERSARY. Oct 27. Welsh poet and playwright, born at Swansea, Walsea, Oct 27, 1914. Died at New York, NY, Nov 9, 1953.

TURKMENISTAN: INDEPENDENCE DAY. Oct 27. National holiday.

UGLY PICKUP PARADE AND CONTEST. Oct 27. Chadron, NE. Honors beat-up old pickups, US manufacturing prowess and ingenuity, and selects the ugliest pickup in all the land. Ugly pickup queen contest held prior to the parade. Est attendance: 2,500. For info: *Chadron Record*, PO Box 1141, Chadron, NE 69337. Phone: (308) 432-5511.

October 1995	S	M	T	W	T	F	S
	1	2	3	4	5	6	7
	8	9	10	11	12	13	14
	15	16	17	18	19	20	21
	22	23	24	25	26	27	28
	29	30	31				

✯ Chase's 1995 Calendar of Events ✯ Oct 27-28

BIRTHDAYS TODAY

Warren Christopher, 70, US Secretary of State in Clinton administration, born at Scranton, ND, Oct 27, 1925.
John Cleese, 56, actor (*Life of Brian*, "Fawlty Towers"), writer ("Monty Python's Flying Circus," *A Fish Called Wanda*), born Weston-Super-Mare, England, Oct 27, 1939.
Floyd Cramer, 62, musician, singer/songwriter ("Last Date," "On the Rebound"), born at Shreveport, LA, Oct 27, 1933.
Frederick De Cordova, 85, producer (third for "Tonight Show"), director, born at New York, NY, Oct 27, 1910.
Ruby Dee, 71, actress ("Ossie and Rubie," *Zora Is My Name, Do the Right Thing*), born Cleveland, OH, Oct 27, 1924.
Nanette Fabray, 75, actress (Emmy "Caesar's Hour"; "One Day at a Time"; *Our Gang* comedies), born San Diego, CA, Oct 27, 1920.
Simon LeBon, 37, singer, born at Bushey, England, Oct 27, 1958.
Roy Lichtenstein, 72, artist, born at New York, NY, Oct 27, 1923.
Carrie Snodgress, 49, actress (*Diary of a Mad Housewife*), born Chicago, IL, Oct 27, 1946.

OCTOBER 28 — SATURDAY
301st Day — Remaining, 64

ALABAMA RENAISSANCE FAIRE. Oct 28–29. Florence, AL. Celebration in grand 16th-century style with music, arts and crafts, costumes, theatre and dance. Listen to minstrels, dulcimers, and autoharps, watch as knights in shining armor transform Wilson Park into "Fountain-on-the-Green," the scene of a 16th-century faire. Annually, the fourth weekend in October. Est attendance: 25,000. For info: Virginia Ware Gilluly, Tourism Coord, Shoals Chamber of Commerce, 104 S Pine St, Florence, AL 35630. Phone: (205) 764-4661.

ANN ARBOR WINTER ART FAIR. Oct 28–29. Ann Arbor, MI. Fine art and selected craft show. Some of the best artists and craftsmen in the country. 22th annual fair. Est attendance: 15,000. For info: Audree Levy, 10629 Park Preston, Dallas, TX 75230. Phone: (214) 369-4345.

BOO AT THE ZOO. Oct 28–29. Asheboro, NC. Halloween Carnival and Haunted Tram. Est attendance: 20,000. For info: North Carolina Zoological Park, Events and Projects, 4401 Zoo Pkwy, Asheboro, NC 27203. Phone: (800) 488-0444.

BREEDERS' CUP CHAMPIONSHIP. Oct 28. Belmont Park, Elmont, NY. Join the excitement of the 12th annual Breeders' Cup, the Super Bowl of horse racing. Est attendance: 60,000. For info: Breeders Cup Ltd, PO Box 4230, Lexington, KY 40544. Phone: (606) 223-5444.

CHAMPIONSHIP CAT SHOW. Oct 28–29. Indiana State Fairgrounds, Indianapolis, IN. Exhibition and judging of longhair and shorthair purebred cats and kittens and mixed breed household pets. Est attendance: 2,000. Sponsor: Indy Cat Club, Cat Fanciers Assn, Maribeth Echard, 8507 N Illinois, Indianapolis, IN 46260. Phone: (317) 251-4486.

CRAFT FAIR USA: INDOOR SHOW. Oct 28–29. Wisconsin State Fair Park, Milwaukee, WI. Sale of handcrafted items including jewelry, pottery, weaving, leather, wood, glass and sculpture. Est attendance: 19,000. For info: Dennis R. Hill, Dir, 3233 S Villa Circle, Milwaukee, WI 53227. Phone: (414) 321-2100.

CZECH REPUBLIC: FOUNDATION OF THE REPUBLIC. Oct 28. National Day, anniversary of the bloodless revolution in Prague on Oct 28, 1918 (after which the Czechs and Slovaks united to form Czechoslovakia, a union they dissolved bloodlessly in 1993).

DICKINSON, ANNA ELIZABETH: BIRTH ANNIVERSARY. Oct 28. Influential American orator and author of the Civil War era was born at Philadelphia, PA, on Oct 28, 1842. As an advocate of abstinence, abolition and woman suffrage, she earned the nickname "American Joan of Arc." She died on Oct 22, 1932.

ERASMUS, DESIDERIUS: BIRTH ANNIVERSARY. Oct 28. Dutch author and scholar, Desiderius Erasmus, was born at Rotterdam, probably on Oct 28, 1467. Best known of his writings is *Encomium Moriae* (In Praise of Folly). Erasmus died at Basel, Switzerland, on July 12, 1536.

ESCOFFIER, GEORGES AUGUSTE: BIRTH ANNIVERSARY. Oct 28. Celebrated French chef and author, inventor of the peche Melba (honoring the operatic singer, Dame Nellie Melba), was born on Oct 28, 1846. Known as the "king of chefs and the chef of kings," Escoffier was awarded the Legion d'Honneur in recognition of his contribution to the international reputation of French cuisine. His service at the Savoy and Carlton hotels in London, England, brought him world fame. He died at Monte Carlo, Monaco, on Feb 12, 1935.

FINGER LAKES CRAFTSMEN CHRISTMAS IN OCTOBER ARTS AND CRAFTS SHOW. Oct 28–29. Monroe County Fairgrounds Dome Arena, Rochester, NY. All media, all categories including photography and prints. 16th annual show. Est attendance: 12,000. For info: Finger Lakes Craftsmen Shows, Ronald L. Johnson, 1 Freshour Dr, Shortsville, NY 14548. Phone: (716) 289-9439.

FIRST WOMAN US AMBASSADOR APPOINTED: ANNIVERSARY. Oct 28. Helen Eugenie Moore Anderson became the first woman to hold the post of US ambassador when she was sworn in by President Harry S. Truman on Oct 28, 1949. She served as Ambassador to Denmark.

GREAT PUMPKIN WEEKEND IN CHADDS FORD. Oct 28–29. The John Chads House, Chadds Ford, PA. The Chadds Ford Historical Society invites visitors to come find out more about pumpkin or "pompion," a staple of the early American diet. Guides in colonial garb will offer tours of the circa 1725 house and springhouse, as well as samples of pumpkin bread. Pumpkins carved by local artists will be on display all weekend across the road. Annually, the last weekend in October. Est attendance: 500. For info: Patricia Casey, Admin, Chadds Ford Historical Soc, Box 27, Chadds Ford, PA 19317. Phone: (610) 388-7376.

GREECE: "OHI DAY." Oct 28. National holiday commemorating Greek resistance and refusal to open her borders when Mussolini's Italian troops attacked Greece on Oct 28, 1940. "Ohi" means no! Celebrated with military parades especially in Athens and Thessaloniki.

GUAVAWEEN CELEBRATION. Oct 28. Ybor City, Tampa, FL. A Halloween celebration based on Tampa's nickname, "The Big Guava." Each year the zany celebration features the mythical Mama Guava, who has sworn to take the "bore" out of Ybor City, and a parade of her devoted followers. Events include the "Mama Guava Stumble" (a satirical parade), street parties, music, live entertainment, dancing and food. Est attendance: 120,000. For info: Cookie Ellis, Ybor City Chamber of Commerce, 1800 E 9th Ave, Tampa, FL 33605. Phone: (813) 248-3712.

HALLOWEEN CELEBRATION. Oct 28. Shakopee, MN. Celebration of Halloween at Murphy's Landing. By reservation only. Est attendance: 2,000. For info: Murphy's Landing, 2817 Highway 101, Shakopee, MN 55379. Phone: (612) 445-6900.

Oct 28 ★ Chase's 1995 Calendar of Events ★

HALLOWEEN FESTIVAL AND HAUNTED WALK. Oct 28. Prospect Park, Brooklyn, NY. Carnival games, musicians, storytellers and a haunted walk through the woods with many scary surprises. Est attendance: 15,000. For info: Public Info Office, Prospect Park, 95 Prospect Park W, Brooklyn, NY 11215. Phone: (718) 965-8999.

HALLOWEEN HAPPENING. Oct 28–29. Children can trick-or-treat at the Zoo from 14 different treat stations. Other activities include hay rides, pumpkin giveaways, a pumpkin carving contest and the city's largest costume contest. Est attendance: 23,000. For info: Cincinnati Zoo and Botanical Garden, Events and Promo Dept, 3400 Vine St, Cincinnati, OH 45220. Phone: (513) 281-4701.

HANSON, HOWARD: BIRTH ANNIVERSARY. Oct 28. Born Oct 28, 1896, at Wahoo, NE, where he began his study of music, Howard Hanson in 1921 became the first American to win the Priz de Rome. In 1924 he became head of the Eastman School of Music at the University of Rochester, NY, where he served for 40 years. Best known for the music he composed, Hanson was awarded the Pulitzer Prize as outstanding contemporary composer in 1944 for his composition *Symphony No. 4*, the George Foster Peabody award in 1946, the Laurel Leaf of the American Composers Alliance in 1957 and the Huntington Hartford Foundation Award in 1959. He died at Rochester on Feb 26, 1981.

HARVARD UNIVERSITY: FOUNDING ANNIVERSARY. Oct 28. Harvard University founded at Cambridge, MA, when, on Oct 28, 1636, the Massachusetts General Court voted to provide 400 pounds for a "schoale or colledge."

HUNGARY: ANNIVERSARY OF 1956 REVOLUTION. Oct 28. National holiday.

THE KIDS BRIDGE. Oct 28–Jan 21. Children's Museum of Houston, Houston, TX. See July 8 for full description. Call or write the Smithsonian for other dates and venues. For info: Smithsonian Institution Traveling Exhibition Service, 1100 Jefferson Dr SW, Ste 3146, Washington, DC 20560. Phone: (202) 357-2700.

LONGWOOD GARDENS CHRYSANTHEMUM FESTIVAL. Oct 28–Nov 19. Kennett Square, PA. 15,000 chrysanthemums and amazing topiaries are featured in indoor displays accompanied by performances and activities throughout the month of November. Est attendance: 50,000. For info: Colvin Randall, PR Mgr, Longwood Gardens, PO Box 501, Kennett Square, PA 19348-0501. Phone: (610) 388-6741.

LOUIS ARMSTRONG: CULTURAL LEGACY. Oct 28–Jan 7. New Orleans Museum of Art, New Orleans, LA. See Jan 28 for full description. For info: Smithsonian Institution Traveling Exhibition, 1100 Jefferson Dr SW, Ste 3146, Washington, DC 20560. Phone: (202) 357-2700.

MORE THAN MEETS THE EYE. Oct 28–Jan 7. Fernbank Museum of Natural History, Atlanta, GA. An original approach to the topics of sight, visual impairment and blindness. See Jan 7 for full description. For info: Smithsonian Institution Traveling Exhibition Service, 1100 Jefferson Dr SW, Ste 3146, Washington, DC 20560. Phone: (202) 357-2700.

PIONEER AND INDIAN FESTIVAL. Oct 28. Mississippi Crafts Center, Natchez Trace Parkway, Ridgeland, MS. Arts and crafts of the pioneer era (basket weaving, pottery, spinning), Indian stickball, dances, falconry, music and food. Annually, the fourth Saturday in October. Sponsor: Craftsmen's Guild of Mississippi, Inc. Est attendance: 5,000. For info: Mississippi Crafts Center, Martha Garrott, PO Box 69, Ridgeland, MS 39158. Phone: (601) 856-7546.

ST. JUDE'S DAY. Oct 28. St. Jude, the saint of hopeless causes, was martyred along with St. Simon in Persia and their feast is celebrated jointly. St. Jude was supposedly the brother of Jesus, and like his brother a carpenter by trade. He is most popular with those who attempt the impossible and with students, who often ask for his help on exams.

SALK, JONAS: BIRTHDAY. Oct 28. Dr. Jonas Salk, developer of the Salk Polio Vaccine, born at New York, NY, Oct 28, 1914. Salk announced his development of a successful vaccine in 1953, the year after a polio epidemic claimed some 3,300 lives in the US.

SEA WITCH HALLOWEEN FESTIVAL. Oct 28–29. Rehoboth Beach-Dewey Beach, DE. Sea Witch Hunt, bicycle hayrides on Rehoboth Ave, broom tossing contest on beach, scarecrow making, mounted horse shows on the beach, pumpkin glow on boardwalk, haunted house, costume parade, square-dancing and live entertainment. Annually, the last weekend in October. Est attendance: 20,000. For info: Carol Everhart, Festival Dir, PO Box 216, Rehoboth Beach, DE 19971. Phone: (800) 441-1329.

SORGHUM DAY FESTIVAL. Oct 28. Wewoka, OK. The autumn air fills with the sweet-smelling aroma of sorghum during this old-time festival and fall tradition. Visitors can witness sorghum-making, a quilt show, antique car show, pioneer demonstrations and more than 100 craft booths. Annually, the fourth Saturday in October. Est attendance: 20,000. For info: Wewoka Chamber of Commerce, PO Box 719, Wewoka, OK 74884. Phone: (405) 257-5485.

SPACE MILESTONE: INTERNATIONAL SPACE RESCUE AGREEMENT. Oct 28. US and USSR officials agreed upon space rescue cooperation, Oct 28, 1970.

STATUE OF LIBERTY: DEDICATION ANNIVERSARY. Oct 28. Frederic Auguste Bartholdi's famous sculpture, the statue of *Liberty Enlightening the World*, on Bedloe's Island in New York Harbor, was dedicated on Oct 28, 1886. Groundbreaking for the structure was in April 1883. A sonnet by Emma Lazarus, inside the pedestal of the statue, contains the words: "Give me your tired, your poor, your huddled masses yearning to breathe free, the wretched refuse of your teeming shore. Send these, the homeless, tempest-tost to me, I lift my lamp beside the golden door!"

WILD FOODS DAY. Oct 28. Fall Creek Falls State Park, Pikeville, TN. Field trips demonstrating edible wild plants; workshop on preparing wild meats; and a fish canning workshop. Est attendance: 75. Sponsor: Nature Center, Fall Creek Falls State Park, Pikeville, TN 37367. Phone: (615) 881-5708.

BIRTHDAYS TODAY

Jane Alexander (Jane Quigley), 56, actress (*The Great White Hope, Kramer vs. Kramer*), chair of Natl Endowment for Arts, born Boston, MA, Oct 28, 1939.

Charlie Daniels, 59, musician, singer, songwriter ("Devil Went Down to Georgia"), born Wilmington, NC, Oct 28, 1936.

Dennis Franz, 51, actor ("Hill Street Blues," "NYPD Blue"), born at Maywood, IL, Oct 28, 1944.

Jami Gertz, 30, actress (Muffy on "Square Pegs"), born Chicago, IL Oct 28, 1965.

Telma Hopkins, 47, singer, actress, born at Louisville, KY, Oct 28, 1948.

Bruce Jenner, 46, track athlete, sportscaster, born at Mount Kisco, NY, Oct 28, 1949.

Bowie Kuhn, 69, former baseball commissioner, born at Tacoma Park, MD, Oct 28, 1926.

Julia Roberts, 28, actress (*Steel Magnolias, Pretty Woman, Mystic Pizza*), born Smyrna, GA, Oct 28, 1967.

Jonas Edward Salk, 81, scientist, immunologist (the Salk polio vaccine), born at New York, NY, Oct 28, 1914.

October 1995

S	M	T	W	T	F	S
1	2	3	4	5	6	7
8	9	10	11	12	13	14
15	16	17	18	19	20	21
22	23	24	25	26	27	28
29	30	31				

☆ Chase's 1995 Calendar of Events ☆ Oct 29-30

OCTOBER 29 — SUNDAY
302nd Day — Remaining, 63

ALOHA CLASSIC PRO WINDSURFING WORLD CUP FINAL. Oct 29–Nov 11. Ho'okipa Beach, Maui, HI. 12th annual windsurfing competition. 34 countries represented. $200,000 prize money. Est attendance: 10,000. For info: Ehman Productions, PO Box 479, Paia, Maui, HI 96779. Phone: (808) 575-9151.

BOSWELL, JAMES: BIRTH ANNIVERSARY. Oct 29. Scottish biographer, born at Edinburgh, Scotland, Oct 29, 1740. Died, London, England, May 19, 1795. "I think," he wrote in his monumental biography, the *Life of Samuel Johnson*, "no innocent species of wit or pleasantry should be suppressed: and that a good pun may be admitted among the smaller excellencies of lively conversation."

CHICAGO MARATHON. Oct 29. Daley Plaza, Chicago, IL. Both the 18th annual marathon (26.2 miles) and the 5th annual 5K (3.1 miles) have reputations for attracting an international field of top athletes, and the marathon is ranked as one of the fastest in the world. Annually, the last Sunday in October. Est attendance: 12,000. For info: Chicago Marathon, Mike Nishi, Genl Mgr, PO Box 10597, Chicago, IL 60610-0597. Phone: (312) 527-2200.

EMMETT, DANIEL DECATUR: BIRTH ANNIVERSARY. Oct 29. Creator of words and music for the song "Dixie," which became a fighting song for Confederate troops and unofficial "national anthem" of the South. Emmett was born on Oct 29, 1815, at Mount Vernon, OH, where he died on June 28, 1904.

◆ **GOEBBELS, PAUL JOSEF: BIRTH ANNIVERSARY.** Oct 29. German Nazi leader who became Hitler's minister of propaganda. Originally rejected from the military because of a limp caused by infantile paralysis. Born on Oct 29, 1897. Killed himself and his family on May 1, 1945.

REFORMATION SUNDAY. Oct 29. Many Protestant churches commemorate Reformation Day (Oct 31—anniversary of the day on which Martin Luther nailed his 95 theses to the door of Wittenberg's Palace church, protesting the sale of papal indulgences, in 1517), on the Sunday preceding Oct 31, each year.

STANDARD TIME RESUMES. Oct 29–Apr 6, 1996. Standard Time resumes at 2:00 AM, on the last Sunday in October in each time zone, as provided by the Uniform Time Act of 1966 (as amended in 1986 by Public Law 99-359). Many use the popular rule: "Spring forward, Fall back" to remember which way to turn their clocks. See also: "Daylight-Saving Time" (Apr 2).

STOCK MARKET CRASH: ANNIVERSARY. Oct 29. New York, NY. Prices on the Stock Exchange plummeted and virtually collapsed four days after President Herbert Hoover had declared "The fundamental business of the country ... is on a sound and prosperous basis." More than 16 million shares were dumped and billions of dollars were lost on Oct 29, 1929. The boom was over and the nation faced nearly a decade of depression. Some analysts had warned the buying spree, with prices 15 to 150 times above earnings, had to stop at some point. Frightened investors ordered their brokers to sell at whatever price. The resulting Great Depression, which lasted till about 1939, involved North America, Europe and other industrialized countries. In 1932 one out of four US workers was unemployed.

TURKEY: NATIONAL HOLIDAY. Oct 29. Anniversary of the founding of republic in 1923.

BIRTHDAYS TODAY

Jesse Lee Barfield, 36, former baseball player, born at Joliet, IL, Oct 29, 1959.
Michael D'Andrea Carter, 35, football player, born at Dallas, TX, Oct 29, 1960.
Richard Dreyfus, 48, actor (*American Graffiti, Jaws*; Oscar *The Goodbye Gye Girl*), born Brooklyn, NY, Oct 29, 1947.
Randy Jackson (Steven Randall Jackson), 34, singer, born at Gary, IN, Oct 29, 1961.
Kate Jackson, 47, actress ("The Rookies," "Charlie's Angels," "Scarecrow and Mrs. King"), born at Birmingham, AL, Oct 29, 1948.
Dirk Kempthorne, 44, US Senator (R, Idaho), born at San Diego, CA, Oct 29, 1951.
Connie Mack, 55, US Senator (R, Florida) up for reelection in '94, born at Philadelphia, PA, Oct 29, 1940.
Bill Mauldin, 74, political cartoonist, born at Mountain Park, NM, Oct 29, 1921.
Melba Moore, 50, singer ("You Stepped into My Life"), actress ("Melba"), born at New York, NY, Oct 29, 1945.
John Haley "Zoot" Sims, 70, musician, born at Inglewood, CA, Oct 29, 1925.
J.T. Smith, 40, football player, born at Leonard, TX, Oct 29, 1955.

OCTOBER 30 — MONDAY
303rd Day — Remaining, 62

ADAMS, JOHN: BIRTH ANNIVERSARY. Oct 30. Second president of the US (term of office: Mar 4, 1797–Mar 3, 1801), had been George Washington's vice president, and was the father of John Quincy Adams (6th president of the US). Born at Braintree, MA, Oct 30, 1735, he once wrote in a letter to Thomas Jefferson: "You and I ought not to die before we have explained ourselves to each other." John Adams and Thomas Jefferson died on the same day, July 4, 1826, the 50th anniversary of adoption of the Declaration of Independence. Adams's last words: "Thomas Jefferson still survives." Jefferson's last words: "Is it the fourth?"

ATLAS, CHARLES: BIRTH ANNIVERSARY. Oct 30. Charles Atlas (ex-97-lb weakling), whose original name was Angelo Siciliano, was born Oct 30, 1893, at Acri, Calbria, Italy. A bodybuilder and physical culturist, he created a popular mail-order bodybuilding course. The legendary sand-kicking episode used later in advertising for his course occurred at Coney Island when a lifeguard kicked sand in Atlas's face and stole his girlfriend. Three generations of comic books carried his advertisements. He died on Dec 24, 1972, at Long Beach, NY.

CASSIUS CLAY NAMED HEAVYWEIGHT CHAMP: ANNIVERSARY. Oct 30. Twenty-two-year-old Cassius Clay, who later changed his name to Muhammed Ali, became world heavyweight boxing champion by defeating Sonny Liston, Oct 30, 1964. Ali was well-known for both his fighting ability and his personal style. His most famous saying was, "I am the greatest!" Convicted of violating the Selective Service Act in 1967, he was stripped of his title; the Supreme Court reversed the decision, though, in 1971. Ali is the only fighter to win the heavyweight fighting title three separate times, and he defended that title nine times.

425

Oct 30–31 ☆ *Chase's 1995 Calendar of Events* ☆

DEVIL'S NIGHT. Oct 30. Formerly a "Mischief Night" on the evening before Halloween and an occasion for harmless pranks, chiefly observed by children. However, in some areas of the US, the destruction of property and endangering of lives has led to the imposition of dusk-to-dawn curfews during the last two or three days of October. Not to be confused with "Trick or Treat," or "Beggar's Night," usually observed on Halloween. See also: "Hallowe'en" (Oct 31).

HALSEY, WILLIAM "BULL" FREDERICK: BIRTH ANNIVERSARY. Oct 30. American admiral and fleet commander who played a leading role in the defeat of the Japanese in the Pacific naval battles of WWII, William Halsey was born at Elizabeth, NJ, Oct 30, 1882. In April 1942, aircraft carriers under his command ferried Jimmy Doolittle's B-25s to within several hundred miles of Japan's coast. From that location the aircraft were launched from the decks of the carriers for a raid on Tokyo. In October 1942, as commander of all the South Pacific area, Halsey led naval forces in the defeat of Japan at Guadalcanal and in November 1943, he directed the capture of Bougainville. He supported the landings in the Philippines in June 1944. In the great naval battle of Leyte (Oct 23–25, 1944) he assisted in an overwhelming defeat of the Japanese. On Sept 2, 1945, Japan's final instrument of surrender was signed in Tokyo Bay aboard Halsey's flagship, the *USS Missouri*. Halsey died at Fishers Island, NY, on Aug 16, 1959.

MOON PHASE: FIRST QUARTER. Oct 30. Moon enters First Quarter phase at 4:17 PM EST.

ONLINE/CD-ROM '95 CONFERENCE. Oct 30–Nov 1. Palmer House, Chicago, IL. Est attendance: 3,000. For info: Online Inc, 462 Danbury Rd, Wilton, CT 06897-2126. Phone: (203) 761-1466.

POUND, EZRA LOOMIS: 110th BIRTH ANNIVERSARY. Oct 30. American poet born at Hailey, ID, Oct 30, 1885. As result of his pro-Fascist radio broadcasts from Italy, Pound was indicted for treason July 26, 1943, and arrested near Genoa, by US Army, May 5, 1945. Confined to St. Elizabeth's Hospital, Washington, DC, from 1946–1958 as being mentally unable to stand trial, he was never tried for treason. Died in Italy, Nov 1, 1972.

SHERIDAN, RICHARD BRINSLEY: BIRTH ANNIVERSARY. Oct 30. Dramatist, born at Dublin, Ireland, Oct 30, 1751. Died, London, England, July 7, 1816. Sheridan is said to have extended the following invitation to a young lady: "Won't you come into the garden? I would like my roses to see you."

SISLEY, ALFRED: BIRTH ANNIVERSARY. Oct 30. French impressionist painter, born at Paris, France, Oct 30, 1839. One of the most influential artists of his time, he died near Fontainbleau, Jan 29, 1899.

"WAR OF THE WORLDS": BROADCAST ANNIVERSARY. Oct 30. On Oct 30, 1938, as part of a series of radio dramas based on famous novels, Orson Welles (born Kenosha, WI, May 6, 1915) with the Mercury Players produced H.G. Wells's *War of the Worlds*. Near panic resulted when listeners believed the simulated news bulletins, which described a Martian invasion of New Jersey, to be real.

BIRTHDAYS TODAY

Nestor Almendros, 65, cinematographer, film director, born at Barcelona, Spain, Oct 30, 1930.
Fred Friendly, 80, broadcast journalist, born at New York, NY, Oct 30, 1915.
Dick Gautier, 58, actor (Conrad Birdie in *Bye Bye Birdie*; "Here We Go Again"), born Los Angeles, CA, Oct 30, 1937.
Harry Hamlin, 44, actor ("L.A. Law," "Studs Lonigan"), born at Pasadena, CA, Oct 30, 1951.
Louis Malle, 63, filmmaker (*Pretty Baby, Atlantic City, Au Revoir Les Enfants*), born at Thumeries, France, Oct 30, 1932.
Diego Maradona, 35, Argentinian soccer player, born at Lanus, Argentina, Oct 30, 1960.
Gordon Parks, 83, photographer, author, born at Fort Scott, KS, Oct 30, 1912.
Grace Slick, 56, singer (with Jefferson Airplane; "White Rabbit"), born Evanston, IL, Oct 30, 1939.
Henry Winkler, 50, actor ("Happy Days," "Henry Winkler Meets William Shakespeare," "An American Christmas Carol"), co-producer ("MacGyver"), born at New York, NY, Oct 30, 1945.

OCTOBER 31 — TUESDAY
304th Day — Remaining, 61

CANDY, JOHN FRANKLIN: BIRTH ANNIVERSARY. Oct 31, 1950. Comedic actor who got his start in Second City improvisation at Toronto and graduated to film stardom. Born at Toronto, Ontario, Canada, and died Mar 4, 1994, while on location for a film at Chupederos, Mexico.

CHIANG KAI-SHEK: BIRTH ANNIVERSARY. Oct 31. Chinese soldier and statesman, was born at Chekiang on Oct 31, 1887. He died at Taipei, Taiwan, on Apr 5, 1975.

HALLOWE'EN OR ALL HALLOW'S EVE. Oct 31. An ancient celebration combining Druid autumn festival and Christian customs. Hallowe'en (All Hallow's Eve) is the beginning of Hallowtide, a season that embraces the Feast of All Saints (Nov 1) and the Feast of All Souls (Nov 2). The observance, dating from the sixth or seventh centuries, has long been associated with thoughts of the dead, spirits, witches, ghosts and devils. In fact, the ancient Celtic Feast of Samhain, the festival that marked the beginning of winter and of the New Year, was observed on Nov 1. See also: "Trick or Treat or Beggar's Night" (Oct 31).

HALLOWEEN. Oct 31. Arapahoe, NE. Merchants and business personnel dress in Halloween costume, free soup and sandwiches at noon in Ella Missina Community Building, children's parade and prizes, haunted house, chili cook-off and auction. For info: Ariel McNamara, Sec'y, Chamber of Commerce, PO Box 624, Arapahoe, NE 68922. Phone: (308) 962-5203.

HALLOWEEN FROLIC. Oct 31. Hiawatha, KS. 81st annual celebration for townspeople and school children at Halloween to promote Hiawatha and surrounding area. Est attendance: 12,000. Sponsor: Hiawatha Chamber of Commerce, 413 Oregon, Hiawatha, KS 66434. Phone: (913) 742-7136.

HALLOWEEN PARADE. Oct 31. (Rain date: Nov 1.) Toms River, NJ. Largest Halloween parade in state with 3,000 participants, 75 prizes and 60 thousand spectators. Parade covers a one-mile route. For info: Brian Kubiel, 26 Robbins St, Toms River, NJ 08753. Phone: (908) 349-1621.

HOUDINI, HARRY: DEATH ANNIVERSARY. Oct 31. Harry Houdini (whose real name was Ehrich Weisz), magician, illusionist and escape artist, died of peritonitis following a blow to the abdomen, at 10:30 PM, 1926, at Grace Hospital, Detroit, MI. Last words reported to have been "Robert Ingersoll," name of famed agnostic. Houdini's death anniversary, on Halloween, is occasion for meetings of magicians. See also: "Houdini, Harry: Birth Anniversary" (Mar 24).

October 1995

S	M	T	W	T	F	S
1	2	3	4	5	6	7
8	9	10	11	12	13	14
15	16	17	18	19	20	21
22	23	24	25	26	27	28
29	30	31				

Chase's 1995 Calendar of Events — Oct 31

◆ **KEATS, JOHN: 200th BIRTH ANNIVERSARY.** Oct 31. One of England's greatest poets, born at London, England, on Oct 31, 1795, and died (of consumption) at the age of 25, at Rome, Italy, on Feb 23, 1821. Keats wrote, to Fanny Brawne (in 1820): "If I should die ... I have left no immortal work behind me—nothing to make my friends proud of my memory—but I have loved the principle of beauty in all things, and if I had had time I would have made myself remembered."

LANDON, MICHAEL: BIRTH ANNIVERSARY. Oct 31. Actor and TV producer Michael Landon was born Eugene Orowitz on Oct 31, 1936, at Forest Hills, NY. Known for his portrayals of Little Joe Cartwright on "Bonanza" and Charles Ingalls on "Little House on the Prairie," Landon's career also included writing, directing and producing for TV. After being diagnosed with cancer in April 1991, he went public with this information and received support from thousands of fans and admirers prior to his death on July 1, 1991.

LOW, JULIETTE GORDON: BIRTH ANNIVERSARY. Oct 31. Founded Girl Scouts of the USA Mar 12, 1912, in Savannah, GA. Oct 31, 1860–Jan 17, 1927.

MILL AVENUE MASQUERADE ADVENTURE. Oct 31. Tempe, AZ. A hauntingly fun time for the entire family, the Masquerade features a costume parade through the streets of Old Town Tempe. Live entertainment and a costume contest. Annually, Oct 31. Est attendance: 5,000. For info: Mill Avenue Merchants Assn, 520 S. Mill Ave, Ste 201, Tempe, AZ 85281. Phone: (602) 967-4877.

MOUNT RUSHMORE COMPLETION: ANNIVERSARY. Oct 31. The Mount Rushmore National Memorial was completed on Oct 31, 1941, after 14 years of work. First suggested by Jonah Robinson of the South Dakota State Historical Society, the memorial was dedicated in 1925 and work began in 1927. The memorial contains sculptures of the heads of Presidents George Washington, Thomas Jefferson, Abraham Lincoln and Theodore Roosevelt. The 60-foot tall sculptures represent, respectively, the nation's founding, political philosophy, preservation and expansion and conservation.

MULTIMEDIA SCHOOLS '95 CONFERENCE. Oct 31–Nov 1. Palmer House, Chicago, IL. Est attendance: 500. For info: Online, Inc, 462 Danbury Rd, Wilton, CT 06897-2126. Phone: (203) 761-1466.

NATIONAL MAGIC DAY. Oct 31. Traditionally observed on anniversary of death of Harry Houdini in 1926.

★ **NATIONAL UNICEF DAY.** Oct 31. Presidential Proclamation 3817, of Oct 27, 1967, covers all succeeding years. Always Oct 31. Free educational materials about children in the developing world accompany fundraising material: there are a wide variety of materials for every age level. Info from: Group Programs Dept, US Committee for UNICEF, 331 38th St, New York, NY 10016.

NEVADA: ADMISSION DAY. Oct 31. Became 36th state on this day in 1864.

REFORMATION DAY. Oct 31. Anniversary of the day on which Martin Luther nailed his 95 theses to the door of Wittenberg's Palace church, denouncing the selling of papal indulgences, on Oct 31, 1517—the beginning of the Reformation in Germany. Observed by many Protestant churches on Reformation Sunday, the Sunday before Oct 31, each year (Oct 29 in 1995).

SAMHAIN. (Also called November Eve, Hallowmas, Hallowe'en, All Hallow's Eve, Feast of Souls, Feast of the Dead, Feast of Apples and Calan Gaeaf.) One of the "Greater Sabbats" during the Wiccan year, Samhain marks the death of the Sun-God, who then awaits his rebirth from the Mother Goddess at Yule (Dec 22 in 1995). In the Celtic tradition, the feast of Samhain was also celebrated as New Year's Eve, as their new year began on Nov 1. Annually, Oct 31.

SLEIDANUS, JOHANNES: DEATH ANNIVERSARY. Oct 31. German historian, born at Schleiden in 1506, died at Strasbourg, Oct 31, 1556. His *Famous Chronicle of Oure Time*, called *Sleidanes Comentaires*, was first translated into English in 1560. The translator spoke thus to the book: "Go forth my painful Boke, Thou art no longer mine. Eche man may on thee loke, The Shame or praise is thine."

SWAN, JOSEPH WILSON: BIRTH ANNIVERSARY. Oct 31. English scientist and inventor born at Sunderland, Durham, England, Oct 31, 1828. Pioneer in photographic chemistry, incandescent electric lamp and man-made fibres. Died at Warlingham, Surrey, England, May 27, 1914.

TAIWAN: CHIANG KAI-SHEK DAY. Oct 31. National holiday to honor memory of Generalissimo Chiang Kai-Shek, born Oct 31, 1887, the first constitutional president of the Republic of China.

TRICK OR TREAT OR BEGGAR'S NIGHT. Oct 31. A popular custom on Hallowe'en, in which children wearing costumes visit neighbors' homes, calling out "Trick or Treat," and "begging" for candies or gifts to place in their beggars' bags. In recent years there has been increased participation by adults, often parading in elaborate or outrageous costumes and also requesting candy.

WATERS, ETHEL: BIRTH ANNIVERSARY. Oct 31. Married at thirteen, Ethel Waters began her singing career at the urging of friends. At age 17 she was singing at Baltimore's Lincoln Center, billing herself as Sweet Mama Stringbean. Her career took her to New York, where she divided her work between the stage, nightclubs and films. She made her Broadway debut in 1927 in the revue *Africana*, and her other stage credits included *Blackbirds, Rhapsody in Black, Thousands Cheer, At Home Abroad* and *Mamba's Daughters*. Her memorable stage roles in *Cabin in the Sky* and *A Member of the Wedding* (for which she won the Drama Critics' Award) were recreated for film. Born Oct 31, 1900, at Chester, PA, she died on Sept 9, 1977, at Chatsworth, GA.

YOUTH HONOR DAY. Oct 31. Iowa and Massachusetts.

BIRTHDAYS TODAY

Michael Collins, 64, astronaut, born in Rome, Italy, Oct 31, 1931.
Dale Evans, 83, actress, evangelist ("The Roy Rogers Show"), born Uvalde, TX, Oct 31, 1912.
Deidre Hall, 47, actress ("Our House," "Days of Our Lives"), born Lake Worth, FL, Oct 31, 1948.
Frederick Stanley McGriff, 32, baseball player, born at Tampa, FL, Oct 31, 1963.
Larry Mullen, 34, musician (drummer with U2; Grammy for album *The Joshua Tree*), born at Dublin, Ireland, Oct 31, 1961.
Jane Pauley, 45, TV personality, born at Indianapolis, IN, Oct 31, 1950.
Dan Rather, 64, journalist (co-anchor "CBS Evening News"), born at Wharton, TX, Oct 31, 1931.
Roy Romer, 67, Governor of Colorado (D), born at Garden City, KS, Oct 31, 1928.
David Ogden Stiers, 53, actor (Charles on "M*A*S*H"; *North and South*), born Peoria, IL, Oct 31, 1942.

Nouember.

NOVEMBER 1 — WEDNESDAY
305th Day — Remaining, 60

ALGERIA: REVOLUTION ANNIVERSARY. Nov 1. National holiday.

ALL-HALLOMAS or ALL HALLOWS or ALL SAINTS DAY. Nov 1. Roman Catholic Holy Day of Obligation. Commemorates the blessed, especially those who have no special feast days. Observed on Nov 1 since Pope Gregory IV set the date of recognition in 835.

ANTIGUA and BARBUDA: NATIONAL HOLIDAY. Nov 1.

AVIATION HISTORY MONTH. Nov 1-30. Anniversary of aeronautical experiments in November 1782 (exact dates unknown), by Joseph Michel Montgolfier and Jacques Etienne Montgolfier, brothers living at Annonay, France. Inspired by Joseph Priestly's book *Experiments Relating to the Different Kinds of Air*, the brothers experimented with filling paper and fabric bags with smoke and hot air, leading to the invention of the hot air balloon, man's first flight and the entire science of aviation and flight.

BALD EAGLE MIGRATION. Nov 1-Dec 15. (Dates approximate.) Below Canyon Ferry Dam, 15 miles east of Helena, MT. Approximately 400 Bald Eagles come to feed on spawning salmon during this 6-week period every year. For info: Helena Area Chamber of Commerce, 201 E Lyndale, Helena, MT 59601. Phone: (406) 442-4121.

CHILD SAFETY AND PROTECTION MONTH. Nov 1-30. To increase awareness of adult responsibility in ensuring the safety and protection of children. Sponsor: Natl PTA, Attn: Program Outreach Div, 330 N Wabash Ave, Ste 2100, Chicago, IL 60611-3690. Phone: (312) 670-6782.

DIABETIC EYE DISEASE MONTH. Nov 1-30. Can people with diabetes prevent the onset of diabetic eye disease? During this observance, Prevent Blindness America (formerly known as National Society to Prevent Blindness) tells how good/poor control of diabetes can affect diabetic retinopathy, cataracts and glaucoma. Sponsor: Prevent Blindness America, Marita Gomez, Media Relations, 500 E Remington Rd, Schaumburg, IL 60173. Phone: (800) 331-2020.

ENGLAND: LONDON INTERNATIONAL ENVIRONMENTAL FILM FESTIVAL. Nov 1-5. Royal Geographical Society, Kensington Gore, London. A showcase for a unique selection of some of the most effective and imaginative environmental films ever made. For info: Victoria Cliff Hodges, Dir, Green Screen, 45 Shelton St, Covent Garden, London, England WC2H 9JH.

November 1995

S	M	T	W	T	F	S
			1	2	3	4
5	6	7	8	9	10	11
12	13	14	15	16	17	18
19	20	21	22	23	24	25
26	27	28	29	30		

GUATEMALA: KITE FESTIVAL OF SANTIAGO SACATEPEQUEZ. Nov 1. Long ago when evil spirits disturbed the good spirits in the local cemetery, a magician told the townspeople a secret way to get rid of the evil spirits—by flying kites (because the evil spirits were frightened by the noise of wind against paper). Since then the kite festival has been held at the cemetery each year on Nov 1 or Nov 2, and it is said that "to this day no one knows of bad spirits roaming the streets or the cemetery of Santiago Sacatepequez," a village about 20 miles from Guatemala City. Nowadays the youth of the village work for many weeks to make the elaborate and giant kites to fly on All Saints Day (Nov 1) or All Souls Day (Nov 2).

HONG KONG: CHUNG YEUNG FESTIVAL. Nov 1. This festival relates to the old story of the Han Dynasty, when a soothsayer advised a man to take his family to a high place on the ninth day of the ninth moon for 24 hours in order to avoid disaster. The man obeyed and found, on returning home, that all living things had died a sudden death in his absence. Part of the celebration is climbing to high places.

INTERNATIONAL CREATIVE CHILD AND ADULT MONTH. Nov 1-30. To encourage people to realize ordinary, everyday creativity as well as the high-order creativity present in gifted-talented populations. Creativity can be improved and developed, reducing stress, increasing wellbeing and benefiting individuals personally and all peoples collectively. Sponsor: Natl Assn for Creative Children and Adults, Ann F. Isaacs, 8080 Springvalley Dr, Cincinnati, OH 45236-1395. Phone: (513) 631-1777.

INTERNATIONAL DRUM MONTH. Nov 1-30. To celebrate the world-wide popularity of all types of drums. Annually, the month of November. Sponsor: Percussive Arts Society, Jerome Hershman, IDM Committee, 38 W 21st St, 5th Fl, New York, NY 10010-6906. Phone: (212) 924-9175.

MEDICAL SCHOOL FOR WOMEN OPENED IN BOSTON: ANNIVERSARY. Nov 1. Founded by Samuel Gregory, a pioneer in medical education for women, the Boston Female Medical School opened on Nov 1, 1848, as the first medical school exclusively for women. The original enrollment was twelve students. In 1874 the school merged with the Boston University School of Medicine and formed one of the first co-ed medical schools in the world.

MEXICO: DAY OF THE DEAD. Nov 1-2. Observance begins during last days of Oct when "Dead Men's Bread" is sold in bakeries—round loaves, decorated with sugar skulls. Departed souls are remembered not with mourning but with a spirit of friendliness and good humor. Cemeteries are visited and graves decorated.

MISSION SAN JUAN CAPISTRANO: FOUNDING ANNIVERSARY. Nov 1. California mission founded Nov 1, 1776, collapsed during 1812 earthquake. The swallows of Capistrano nest in ruins of the old mission church, departing each year on Oct 23, and returning the following year on or near St. Joseph's Day (Mar 19).

NASHVILLE'S COUNTRY HOLIDAYS. Nov 1-Jan 1. Nashville, TN. A Nashville winter tradition, this citywide festival is comprised of more than 60 events. Free brochure. For info: Nashville Conv & Visitors Bureau, 161 Fourth Ave N, Nashville, TN 37219. Phone: (615) 259-4700.

NATIONAL ALZHEIMER'S DISEASE MONTH. Nov 1-30. To increase awareness of the disease and what the Alzheimer's Association is doing to help persons with Alzheimer's Disease, their families and their caregivers. Sponsor: Alzheimer's Assn, Niles Frantz, 919 N Michigan, Ste 1000, Chicago, IL 60611. Phone: (800) 272-3900.

NATIONAL AUTHORS' DAY. Nov 1. This observance was adopted by the General Federation of Women's Clubs in 1929 and in 1949 was given a place on the list of special days, weeks and months prepared by the US Dept of Commerce. The Resolution states: "by celebrating an Authors' Day as a nation, we would not only show patriotism, loyalty, and appreciation of the men and women who have made American literature possible, but would also encourage and inspire others to give of themselves in making a better America. . . ." It was also resolved "that we commemorate an Authors' Day to be observed on November First each year; . . . [and] that we salute these heroes with waving flags, the emblem of the National existence they have striven to perfect." Originated and first resolution submitted by Mrs Nellie Verne Burt McPherson, Pres, Bement (Illinois) Woman's Club on Oct 30, 1928. Note: There is NO printed material available. Submitted by: Mrs Sue Cole, Granddaughter of Originator.

NATIONAL DIABETES MONTH. Nov 1-30. The American Diabetes Association sponsors events and activities across the country to raise awareness of the information and services available for the 13 million people with diabetes. For further information contact your local affiliate. Sponsor: American Diabetes Assn, Natl HQ, 1660 Duke St, Alexandria, VA 22314.

NATIONAL EPILEPSY AWARENESS MONTH. Nov 1-30. To increase understanding of epilepsy and public awareness of the fact that most people with the disorder can lead normal healthy lives. Annually, the month of November. Sponsor: Epilepsy Foundation of America, Ann Scherer, Dir Pub Info and Educ, 4351 Garden City Dr, Landover, MD 20785. Phone: (800) 332-1000.

NATIONAL FIG WEEK. Nov 1-7. To celebrate the completion of the California fig harvest and encourage consumers to use California figs as part of their diet for the taste, high fiber and nutritional value. Sponsor: California Fig Advisory Board, PO Box 709, Fresno, CA 93712. Phone: (213) 938-3300.

NATIONAL HOSPICE MONTH. Nov 1-30. A celebration of hospice care for terminally ill people and their families. Recognition of the professionals and volunteers who deliver this unique type of care. For info: David R. Schneider, Mktg Mgr, Natl Hospice Organization, 1901 N Moore St, Ste 901, Arlington, VA 22209. Phone: (703) 243-5900.

PEANUT BUTTER LOVER'S MONTH. Nov 1-30. Celebration of America's favorite food and #1 sandwich. Sponsor: Peanut Advisory Board, 1950 N Park Pl, Ste 525, Atlanta, GA 30339.

PLAN YOUR EPITAPH DAY. Nov 1. (Also, Apr 6.) Dedicated to the proposition that a forgettable gravestone is a fate worse than death, and that everyone can be in the same league with Shakespeare and W.C. Fields. Semiannually, coincides with Ching Ming Festival and the Day of the Dead. For info: Lance Hardie, Dead or Alive, PO Box 4595, Arcata, CA 95521. Phone: (707) 822-6924.

PROJECT RED RIBBON. Nov 1-Jan 1. Nationwide. Encourages Americans to "Tie One On For Safety" by attaching a ribbon to a visible location on their vehicle. Motorists demonstrate their commitment to drive safe and sober and remind other motorists to do the same. Sponsor: Sally O'Connor, Dir of Programs, MADD, Natl Office, 511 E John Carpenter Frwy #700, Irving, TX 75062. Phone: (214) 744-6233.

REAL JEWELRY MONTH. Nov 1-30. The national association for consumer education and information about fine jewelry, Jewelers of America, annually hosts jewelry symposiums, fund-raising events, consumer press events and national ads during the month of November. Sponsor: Jewelers of America, Inc., Eileen Farrell, Dir of Comm, 1185 Ave of the Americas, 30th Fl, New York, NY 10036. Phone: (212) 768-8777.

THANKSGIVING CANNED GOODS DRIVE. Nov 1-22. (Weekdays only.) Donations of canned goods and non-perishable food items collected for needy families. For info: Office of Adventures in Fun, Clarence "Du" Burns Arena, 1301 S Ellwood Ave, Baltimore, MD 21224. Phone: (410) 396-9177.

TREASURES BY THE SEA. Nov 1-Feb 28. Myrtle Beach, SC. Longest continuous display of thematic lights in the US, featuring special events and affordable accommodations, shopping. Est attendance: 100,000. For info: Marilyn Chewning, Myrtle Beach Area Chamber of Commerce, PO Box 2115, Myrtle Beach, SC 29578. Phone: (803) 626-7444.

VIRGIN ISLANDS: LIBERTY DAY. Nov 1. Officially "D. Hamilton Jackson Memorial Day," commemorating establishment of the first press in Virgin Islands, in 1915.

WINTER FESTIVAL OF LIGHTS. Nov 1-Jan 8. (Also weekends thru Feb 4.) Oglebay, Wheeling, WV. More than 500,000 lights sparkle over breathtaking landscapes covering 300 acres of this nationally famous park. Enjoy the free drive through the lighted trees, outlined buildings and gigantic, animated theme displays, plus special activities and fine lodging packages. Light show runs in conjunction with Wheeling's "City of Lights." Est attendance: 1,000,000. For info: Visitor Services, Oglebay, Wheeling, WV 26003. Phone: (800) 624-6988.

WORLD COMMUNICATION WEEK. Nov 1-7. To stress the importance of communication among the more than five billion human beings in the world who speak more than 3,000 languages and to promote communication by means of the international language Esperanto. For complete info, send $1 to cover expense of printing, handling and postage. Annually, the first seven days of November. Sponsor: Intl Society of Friendship and Goodwill, Dr. Stanley Drake, Pres, 9538 Summerfield St, Spring Valley, San Diego County, CA 91977-2852. Phone: (619) 466-8882.

BIRTHDAYS TODAY

Larry Claxton Flynt, 53, publisher, born at Magoffin County, KY, Nov 1, 1942.

James Jackson Kilpatrick, 75, journalist (conservative side of "60 Minutes" Point-Counterpoint), born at Oklahoma City, OK, Nov 1, 1920.

Betsy Palmer (Patricia Bromek), 69, actress (panelist on "I've Got a Secret"; "Knots Landing," "Today"), born East Chicago, IA, Nov 1, 1926.

Gary Player, 60, golfer, born at Johannesburg, South Africa, Nov 1, 1935.

Fernando Valenzuela, 35, baseball player, born at Fundicion, Mexico, Nov 1, 1960.

NOVEMBER 2 — THURSDAY
306th Day — Remaining, 59

ALL-SOULS DAY. Nov 2. Commemorates the faithful departed. Catholic observance.

BOONE, DANIEL: BIRTH ANNIVERSARY. Nov 2. American frontiersman, explorer and militia officer, born in Berks County, near Reading, PA, on Nov 2, 1734 (New Style). In February 1778, he was captured at Blue Licks, KY, by Shawnee Indians, under Chief Blackfish, whose adopted son he became when inducted into the tribe as "Big Turtle." Boone escaped after five months, and in 1781 was captured briefly by the British. He experienced a series of personal and financial disasters during his life, but continued a rugged existence, hunting until his 80s. Boone died at St. Charles County, MO, Sept 26, 1820. The bodies of Daniel Boone and his wife, Rebecca, were moved to Frankfort, KY, in 1845.

Nov 2-3 ☆ Chase's 1995 Calendar of Events ☆

FALL POSTAGE STAMP MEGA EVENT. Nov 2-5. Madison Square Garden, New York, NY. 150 dealers offering stamps and postal history materials for sale. Display of private collections, seminars, lectures, free appraisal service. Event highlights US, UN and foreign postal administrations. Exhibits on printing stamps. Stamp clubs and societies attend. Free admission. Est attendance: 12,000. For info: Joseph B. Savarese, Exec VP, Amer Stamp Dealers Assn, 3 School St, Glen Cove, NY 11542. Phone: (516) 759-7000.

HARDING, WARREN GAMALIEL: BIRTH ANNIVERSARY. Nov 2. The 29th president of the US and the first to have a radio, was born at Corsica, OH, Nov 2, 1865. His term of office: Mar 4, 1921–Aug 2, 1923 (died in office). His undistinguished administration was tainted by the "Teapot Dome Scandal," and his sudden death while on a western speaking tour (San Francisco, CA, Aug 2, 1923) prompted many dark rumors. In 1927, the book *The President's Daughter* by Nan Britton, dedicated "to all unwedded mothers, and to their innocent children whose fathers are usually not known to the world," added doubts about morality in high political places.

NEW YORK SUBWAY ACCIDENT: ANNIVERSARY. Nov 2. The Brighton Beach Express, exceeding its speed limit five times over (going 30 mph) while approaching the station near Malbone Street tunnel in Brooklyn, jumped the tracks Nov 2, 1918, killing 97 people and injuring 100. The supervisor-engineer, taking the place of a striking motorman of the Brotherhood of Locomotive Engineers, was tried and acquitted of charges of negligence.

NORTH DAKOTA: ADMISSION DAY. Nov 2. Became 39th state on this day in 1889.

◆ **POLK, JAMES KNOX: 200th BIRTH ANNIVERSARY.** Nov 2. The 11th president of the US was born at Mecklenburg County, NC, Nov 2, 1795. His term of office: Mar 4, 1845–Mar 3, 1849. A compromise candidate at the 1844 Democratic Party convention, Polk was awarded the nomination on the ninth ballot. He declined to be a candidate for a second term, and declared himself to be "exceedingly relieved" at the completion of his presidency. He died shortly thereafter at Nashville, TN, June 15, 1849.

SAMPSON COUNTY EXPO. Nov 2. Clinton, NC. To promote products and services in Sampson County and provide entertainment, good food and information. Est attendance: 7,000. Sponsor: Chamber of Commerce, Box 467, Clinton, NC 28328.

◆ **SENECA FALLS CONVENTION SURVIVOR VOTES: 75th ANNIVERSARY.** Nov 2. The only woman who attended the historic Seneca Falls Women's Rights Convention in 1848 who lived long enough to exercise her right to vote under the 19th Amendment, Charlotte Woodward voted in Philadelphia in the general election Nov 2, 1920.

SOUTH DAKOTA: ADMISSION DAY. Nov 2. Became 40th state on this day in 1889.

SPRUCE GOOSE FLIGHT: ANNIVERSARY. Nov 2. The mammoth flying boat, H-2, *Hercules*, then the world's largest airplane, was designed, built and flown (once) by Howard Hughes. Its first and only flight—about one mile and at an altitude of 70 feet—over Long Beach Harbor, CA, was made on Nov 2, 1947. The $25 million, 200-ton plywood craft was nicknamed the "Spruce Goose." It is now displayed near the *Queen Mary* at Long Beach, CA.

TOUR O' HAWAII. Nov 2-5. Oahu, HI. Concluding World Cup amateur series which attracts the top federation teams from around the world. Stages are 14K time trials, 90K road sprint, 160K road race, 11K hill climb, 250-yard sprints and criterium consisting of 35 laps around a 1.2K course. For info: Event Marketing, Inc, 1001 Bishop St, #477, Pauahi Tower, Honolulu, HI 96813. Phone: (808) 521-4322.

VIRGINIA CHRISTMAS SHOW. Nov 2-5. Showplace Exhibition Center, Richmond, VA. Featuring 450 artisans and crafters, Christmas food shops and Christmas Holiday Theatre, entertainment, legendary "Sgt Santa." 9th annual show. Est attendance: 40,000. For info: Virginia Show Productions, Patricia Wagstaff, PO Box 305, Chase City, VA 23924. Phone: (804) 372-3996.

BIRTHDAYS TODAY

Shere Hite, 53, author (*The Hite Report, Women and Love*), born at St. Joseph, MO, Nov 2, 1942.
Burt Lancaster, 82, actor (*From Here to Eternity*; Oscar for *Elmer Gantry*), born New York, NY, Nov 2, 1913.
Ann Rutherford, 75, actress (*Andy Hardy* pictures; "Leave It to the Girls"), born Toronto, Ont, Canada, Nov 2, 1920.
William D. Schaefer, 74, Governor of Maryland (D), born at Baltimore, MD, Nov 2, 1921.
David Stockton, 54, golfer, born at San Bernardino, CA, Nov 2, 1941.
Alfre Woodard, 42, actress (*Cross Creek, Miss Firecracker, Grand Canyon*), born at Tulsa, OK, Nov 2, 1953.

NOVEMBER 3 — FRIDAY
307th Day — Remaining, 58

AUSTIN, STEPHEN FULLER: BIRTH ANNIVERSARY. Nov 3. A principal founder of Texas, for whom its capital city was named, Stephen Fuller Austin was born at Wythe County, VA, on Nov 3, 1793. He first visited Texas in 1821 and established a settlement there the following year, continuing a colonization project started by his father, Moses Austin. Thrown in prison when he advocated formation of a separate state (Texas still belonged to Mexico), he was freed in 1835, lost a campaign for the presidency (of the Republic of Texas) to Sam Houston (q.v.) in 1836, and died (while serving as Texas secretary of state), on Dec 27, 1836.

BRYANT, WILLIAM CULLEN: BIRTH ANNIVERSARY. Nov 3. American poet born at Cummington, MA, Nov 3, 1794. Died at New York, NY, June 12, 1878.

CANADA: BANFF FESTIVAL OF MOUNTAIN FILMS. Nov 3-5. Banff, Alberta. Mountain and adventure films from around the world will be entered for this award competition at the Banff Centre. A weekend of film, seminars and exhibits. Est attendance: 6,000. For info: Banff Festival of Mountain Film, Banff Centre, PO Box 1020, Banff, AB, Canada T0L 0C0. Phone: (403) 762-6125.

DOMINICA: NATIONAL DAY. Nov 3. National holiday.

FEDERATED STATES OF MICRONESIA: INDEPENDENCE DAY. Nov 3. National holiday.

November 1995

S	M	T	W	T	F	S
			1	2	3	4
5	6	7	8	9	10	11
12	13	14	15	16	17	18
19	20	21	22	23	24	25
26	27	28	29	30		

Chase's 1995 Calendar of Events — Nov 3-4

GREAT NY STATE SNOW EXPO. Nov 3-5. Empire State Plaza, Albany, NY. One of the longest continuous-running snow expos in the country, the 34th annual show features ski deck shows, seminars, videos, bargain bazaar, auction and fashion shows. Annually, the first weekend in November. Est attendance: 20,000. Sponsor: Heather Schlachter, Show Promoter, Ed Lewi Assn, 8 Wade Rd, Latham, NY 12110. Phone: (518) 783-1333.

JAPAN: CULTURE DAY. Nov 3. National holiday.

LONGHORN WORLD CHAMPIONSHIP RODEO. Nov 3-5. Joel Coliseum, Winston-Salem, NC. More than 200 of North America's best cowboys and cowgirls compete in 6 professionally sanctioned contests ranging from bronc riding to bull riding for top prize money and world championship points. The theme of the production changes annually. Additional attractions include special animal trainers, trick ropers, trick riders and professional clowns. 26th annual. Est attendance: 18,000. For info: Longhorn World Chmpshp Rodeo, Inc, PO Box 70159, Nashville, TN 37207. Phone: (615) 876-1016.

NATIONAL FARM TOY SHOW. Nov 3-5. Beckman HS, Dyersville, IA. Features tours of the Ertl, Spec Cast and Scale Models toy companies, a sanctioned Pedal Pull for kids, collector social, auction and the Mercy Hospital Bazaar held at St. Francis Xavier Basilica. Annually, the first weekend in November. Est attendance: 15,000. Sponsor: Karen Kramer, Tourism Dir, Dyersville Area Chamber of Commerce, PO Box 187, Dyersville, IA 52040. Phone: (319) 875-2311.

PANAMA: NATIONAL HOLIDAY. Nov 3, 1903. Independence Day. Panama declared itself independent of Colombia.

POLKA FEST. Nov 3-5. Wisconsin Dells, WI. Popular Midwest polka bands playing song and dance favorites at Holiday Inn. Ethnic specialties. Est attendance: 1,500. For info: Wisconsin Dells Visitor and Conv Bureau, 701 Superior St, Box 390, Wisconsin Dells, WI 53965. Phone: (800) 223-3557.

SANDWICH DAY. Nov 3. To celebrate the birthday of John Montague, Fourth Earl of Sandwich. Montague created the world's first fast food, the sandwich. Sponsor: Ziploc Sandwich Bags. Est attendance: 100. For info: DowBrands, Mary Anne Surber, PO Box 68511, Indianapolis, IN 46268.

SANDWICH DAY: BIRTH ANNIVERSARY OF JOHN MONTAGUE. Nov 3. A day to recognize the inventor of the sandwich, John Montague, Fourth Earl of Sandwich, who was born on Nov 3, 1718. England's first lord of the admiralty, secretary of state for the northern department, postmaster general, the man after whom Capt Cook named the Sandwich Islands in 1778. A rake and a gambler, he is said to have invented the sandwich as a time-saving nourishment while engaged in a 24-hour-long gambling session in 1762. He died at London, England, on Apr 30, 1792.

SPACE MILESTONE: SPUTNIK 2 (USSR): ANNIVERSARY. Nov 3. Dog, Laika, first animal projected by man into space, Nov 3, 1957. Radiation measurements. 1,121 lbs. Nicknamed "Muttnik" by the American press.

VIRGINIA-CAROLINA CRAFTSMEN'S CHRISTMAS CLASSIC. Nov 3-5. State Fairgrounds, Richmond, VA. Arts and crafts. Est attendance: 35,000. For info: The Carolina Craftsmen, 1240 Oakland Ave, Greensboro, NC 27403. Phone: (910) 274-5550.

WALSH INVITATIONAL RIFLE TOURNAMENT. Nov 3-5. (Also Nov 10-12 and 17-19.) Xavier University, Cincinnati, OH. To promote marksmanship and sportsmanship in the competitive spirit of collegiate athletics. International smallbore rifle and air rifle match open to all competitors. Recognized as "the largest indoor international rifle match in the nation." Sponsor: Xavier University Athletic Dept. Est attendance: 300. For info: Alan Joseph, Dept of Athletics, O'Conner Sports Ctr, Xavier University, 3800 Victory Pkwy, Cincinnati, OH 45207.

WESTERN SAMOA: ARBOR DAY. Nov 3. The first Friday in November is observed as Arbor Day in Western Samoa.

WHITE, EDWARD DOUGLASS: 150th BIRTH ANNIVERSARY. Nov 3. Ninth Chief Justice of the Supreme Court. Born at La Fourche Parish, LA, on Nov 3, 1845. During the Civil War, he served in the Confederate Army after which he returned to New Orleans to practice law. Elected to the US Senate in 1891, he was appointed to the Supreme Court by Grover Cleveland in 1894. He became Chief Justice under President William Taft in 1910 and served until 1921. He died at Washington, DC, on May 19, 1921.

WILL ROGERS DAYS. Nov 3-5. Claremore, OK. A birthday celebration to commemorate Will Rogers's birth. Birthday party at 1879 Living History Ranch, arts and crafts and country fair. Est attendance: 50,000. For info: Will Rogers Memorial, Joseph Carter, Box 157, Claremore, OK 74018. Phone: (800) 828-9643.

WORLD COMMUNITY DAY. Nov 3. Theme: "Preserving the Fruits of God's Labor." An ecumenical event that affirms a national commitment to peace and justice within the US. "The service lifts up children as one of God's greatest fruits, which we are to nourish for the benefit of each child and our community." Annually, the first Friday of November. Est attendance: 52,000. Sponsor: Church Women United, 475 Riverside Dr, Rm 812, New York, NY 10115. Phone: (212) 870-2347.

WURSTFEST. Nov 3-12. Landa Park, New Braunfels, TX. To honor and celebrate German heritage. Music, dancing, food, arts and crafts, historical exhibits, sporting events, and special demonstrations. Est attendance: 100,000. For info: Wurstfest Assn, PO Box 310309, New Braunfels, TX 78131. Phone: (210) 625-9167.

BIRTHDAYS TODAY

Adam Ant (Stewart Goddard), 41, singer ("Goody Two Shoes"), born at London, England, Nov 3, 1954.
Roseanne Arnold, 42, comedienne, actress ("Roseanne," *She Devil*), born Salt Lake City, UT, Nov 3, about 1953.
Ken Berry, 62, actor ("F Troop," "Mayberry RFD," "Mama's Family"), singer, dancer, born Moline, IL, Nov 3, 1933.
Charles Bronson (Charles Buchinsky), 73, actor (*The Dirty Dozen, The Valachi Papers*), born Ehrenfeld, PA, Nov 3, 1922.
Michael S. Dukakis, 62, former Governor of Massachusetts (D), born at Brookline, MA, Nov 3, 1933.
Steve Landesberg, 50, actor ("Barney Miller," "Friends and Lovers"), born Bronx, NY, Nov 3, 1945.
James Reston, 86, journalist, born at Clydebank, Scotland, Nov 3, 1909.
Phil Simms, 39, football player, born at Lebanon, KY, Nov 3, 1956.
Louis Wade Sullivan, 62, former US Secretary of Health and Human Services in Bush administration, born at Blakely, GA, Nov 3, 1933.
Monica Vitti (Monica Luisa Ceciarelli), 62, actress (*The Red Desert*), born Rome, Italy, Nov 3, 1933.

NOVEMBER 4 — SATURDAY
308th Day — Remaining, 57

CANADA: NORTHLANDS FARMFAIR. Nov 4-12. Northlands Agricom, Edmonton, Alberta. International Purebred Livestock Show and sale. Est attendance: 40,000. For info: Edmonton Northlands, Leroy Emerson, Agricultural Mgr, Box 1480, Edmonton, Alta, Canada T5J 2N5. Phone: (403) 471-7210.

CHRISTMAS GIFT AND HOBBY SHOW. Nov 4-12. Indiana State Fairgrounds, Indianapolis, IN. 46th annual event. Selling show of arts, crafts, collectibles and other gift items. Est attendance: 82,000. For info: Patrick Buchen, Pres, HSI Show Productions, Box 20760, Indianapolis, IN 46220. Phone: (317) 255-4151.

Nov 4-5   ☆ *Chase's 1995 Calendar of Events* ☆

DAN BEARY'S FALL CAVALCADE OF TEDDY BEARS. Nov 4-5. O'Neill Center, Western Connecticut State Univ, Danbury, CT. More than 100 of America's leading Teddy Bear designers and makers, Bear Repair and Restoration Clinic, appraisals, antique bears, dealers selling famous-name Teddy Bears, free door prizes and more. Annually, the first weekend in November. Est attendance: 2,500. For info: John K. Pringle, Pres, Pringle Productions, Ltd, PO Box 757, Bristol, CT 06011-0757. Phone: (203) 585-9940.

GREAT LAKES INVITATIONAL BARBERSHOP SHOW. Nov 4. Grand Rapids, MI. Presented by Grand Rapids Chapter of Society for the Preservation and Encouragement of Barber Shop Quartet Singing in America. Est attendance: 2,000. For info: J. Schneider, 1311 Northlawn NE, Grand Rapids, MI 49505. Phone: (616) 361-6820.

MISCHIEF NIGHT. Nov 4. Observed in England, Australia and New Zealand. Nov 4, the eve of Guy Fawkes Day is occasion for bonfires and firecrackers to commemorate failure of the plot to blow up the Houses of Parliament on Nov 5, 1605. See also: "England: Guy Fawkes Day" (Nov 5).

NATIONAL SERVICE DAY. Nov 4. Each year all chapters of Alpha Phi Omega conduct service projects related to a specific area of concern. The area for emphasis is determined at the national convention in December for the following year. 1994 projects concentrated on "Chemical Dependency and Eating Disorders Education and Awareness." Annually, the first Saturday in November. Sponsor: Patrick W. Burke, Natl Exec Dir, Alpha Phi Omega, 14901 E 42nd St, Independence, MO 64055. Phone: (816) 373-8667.

PANAMA: FLAG DAY. Nov 4. Public holiday.

PHILLPOTTS, EDEN: BIRTH ANNIVERSARY. Nov 4. English novelist, poet and playwright, born in India on Nov 4, 1862. A friend of Arnold Bennett, Phillpotts wrote more than a hundred novels. He died near Exeter, England, on Dec 29, 1960.

RE-IMAGING CONFERENCE CONTROVERSY: ANNIVERSARY. Nov 4-7, 1993. In Minneapolis, MN, an interfaith conference was held to begin a "re-imaging" of many basic tenets of Christian theology as related to women. Controversy arose following the conference when it was learned that parts of the conference featured participation in rituals that deviated from church doctrine and included references to God as "Sofia," a biblical reference to the feminine aspects of God. About 200 Presbyterian church women participated in the event which drew the ire of many in the Presbyterian Church when it was learned that church funds had been used to support the event.

ROGERS, WILL: BIRTH ANNIVERSARY. Nov 4. William Penn Adair Rogers, American writer, actor, humorist and grass-roots philosopher, born at Oologah, Indian Territory (now Oklahoma), Nov 4, 1879. With aviator Wiley Post, he was killed in an airplane crash near Point Barrow, AK, Aug 15, 1935. "My forefathers," he said, "didn't come over on the *Mayflower*, but they met the boat."

SADIE HAWKINS DAY. Nov 4. Widely observed in US, usually on the first Saturday in November. Tradition established in "Li'l Abner" comic strip in 1930s by cartoonist Al Capp. Popularly an occasion when women and girls are encouraged to take the initiative in inviting the man or boy of their choice for a date. A similar tradition is associated with Feb 29 in Leap Years.

SEIZURE OF US EMBASSY IN TEHERAN: ANNIVERSARY. Nov 4. On Nov 4, 1979, about 500 Iranian "students" seized the US Embassy in Teheran, taking some 90 hostages, of whom about 60 were Americans. They vowed to hold the hostages until the former Shah, Mohammed Reza Pahlavi (in the US for medical treatments) was returned to Iran for trial. The Shah died July 27, 1980, in an Egyptian military hospital near Cairo. The remaining 52 American hostages were released and left Teheran on Jan 20, 1981, after 444 days of captivity. The release occurred on America's Presidential Inauguration Day, and during the hour in which the American Presidency was transferred from Jimmy Carter to Ronald Reagan.

SOUTHEASTERN NEW MEXICO ARTS AND CRAFTS FESTIVAL. Nov 4-5. Lea County Fairgrounds, Lovington, NM. Displays from more than 100 local and regional artists. No commercially manufactured items allowed. 18th annual festival. Annually, the first weekend in November. Est attendance: 10,000. For info: Lovington Chamber of Commerce, PO Box 1347, Lovington, NM 88260. Phone: (505) 396-5311.

STEEPLECHASE AT CALLAWAY GARDENS. Nov 4. Pine Mountain, GA. A 7-race steeplechase "meet" where riders match their horses for speed and split-second timing over bush jumps. Box seating and infield tailgating spaces available. Est attendance: 12,000. For info: The Steeplechase at Callaway Gardens, PO Box 2311, Columbus, GA 31902. Phone: (706) 324-6252.

SWEDEN: ALL SAINT'S DAY. Nov 4. Honors the memory of deceased friends and relatives. Annually, the Saturday following Oct 30.

WILL ROGERS DAY. Nov 4. Oklahoma.

BIRTHDAYS TODAY

Martin Balsam, 76, actor ("Archie Bunker's Place," *Twelve Angry Men*), born New York, NY, Nov 4, 1919.

Art Carney, 77, actor (Oscar *Harry and Tonto*; 6 Emmys for "The Honeymooners"), born Mt Vernon, NY, Nov 4, 1918.

Walter Leland Cronkite, Jr, 79, journalist (former anchorman "CBS Evening News"), born at St. Joseph, MO, Nov 4, 1916.

Alfred Heineken, 72, brewery executive, born at Amsterdam, Netherlands, Nov 4, 1923.

Ralph Macchio, 33, actor ("Eight Is Enough," *Karate Kid*), born at Huntington, NY, Nov 4, 1962.

Andrea McArdle, 32, singer, actress (Broadway's *Annie*), born Philadelphia, PA, Nov 4, 1963.

Cameron Mitchell, 77, actor ("The High Chaparral," "Swiss Family Robinson"), born at Dallastown, PA, Nov 4, 1918.

Markie Post, 45, actress ("Night Court," "Hearts Afire"), born Palo Alto, CA, Nov 4, 1950.

Loretta Swit, 58, actress ("M*A*S*H"), born at Passaic, NJ, Nov 4, 1937.

NOVEMBER 5 — SUNDAY
309th Day — Remaining, 56

DEBS, EUGENE VICTOR: BIRTH ANNIVERSARY. Nov 5. American politician, first president of the American Railway Union, founder of the Social Democratic Party of America, and Socialist Party candidate for president of the US in 1904, 1908, 1912 and 1920, sentenced to 10-year prison term in 1918 (for sedition) and pardoned by President Harding in 1921. Debs was born at Terre Haute, IN, Nov 5, 1855, and died at Elmhurst, IL, Oct 20, 1926.

November 1995

S	M	T	W	T	F	S
			1	2	3	4
5	6	7	8	9	10	11
12	13	14	15	16	17	18
19	20	21	22	23	24	25
26	27	28	29	30		

Chase's 1995 Calendar of Events — Nov 5

DURANT, WILL: 110th BIRTH ANNIVERSARY. Nov 5. American author and popularizer of history and philosophy. Among his books: *The Story of Philosophy* and *The Story of Civilization* (a 10-volume series of which the last four were co-authored by his wife, Ariel). Born at North Adams, MA, on Nov 5, 1885, and died Nov 7, 1981 at Los Angeles, CA.

ENGLAND: GUY FAWKES DAY: 390 YEARS. Nov 5. United Kingdom. Anniversary of the "Gunpowder Plot." Conspirators planned to blow up the Houses of Parliament and King James I, on Nov 5, 1605. Twenty barrels of gunpowder, which they had secreted in a cellar under Parliament, were discovered on the night of Nov 4, the very eve of the intended explosion, and the conspirators were arrested. They were tried and convicted, and on Jan 31, 1606, eight survivors (including Guy Fawkes) were beheaded and their heads displayed on pikes at London Bridge. Though there were at least 11 conspirators, Guy Fawkes is most remembered. In 1606, the Parliament, which was to have been annihilated, enacted a law establishing Nov 5 as a day of public thanksgiving. It is still observed, and on the night of Nov 5, "the whole country lights up with bonfires and celebration." "Guys" are burned in effigy; and the old verses repeated: "Remember, remember the fifth of November,/Gunpowder treason and plot;/I see no reason why Gunpowder Treason/Should ever be forgot."

ENGLAND: RAC LONDON TO BRIGHTON VETERAN CAR RUN. Nov 5. London. A 57-mile run for maximum of 400 veteran cars, along the A23 road from Serpentine Row, Hyde Park, London, to Madiera Drive, Brighton, England. Celebrates emancipation—the abolition of English law requiring that a man walk in front of motor vehicles carrying a red flag. Annually, the first Sunday in November. For info: RAC Motor Sports Assn Ltd, Motor Sports House, Riverside Park, Colnbrook, Slough, England, SL3 OHG.

HUG-A-BEAR SUNDAY. Nov 5. Hug-a-Bears are soft, cuddly bears hand-sewn by the Telephone Pioneers of America to comfort frightened or hurt children or adults faced with difficult situations. Day will include distribution of thousands of bears to law enforcement and emergency agencies who deal with people in traumatic situations. Phone: (303) 571-9270 or (303) 595-3983. Sponsor: Telephone Pioneers of America, Sue Saunders, 930 15th St, Rm 1249, Denver, CO 80202.

KEY CLUB INTERNATIONAL WEEK. Nov 5-11. To recognize service projects of more than 160,000 school members. Annually, the first full week of November. Est attendance: 2,800. Sponsor: Key Club Intl, 3636 Woodview Trace, Indianapolis, IN 46268-3196. Phone: (317) 875-8755.

LOEWY, RAYMOND: BIRTH ANNIVERSARY. Nov 5. Raymond Fernand Loewy, the "father of streamlining," an inventor, engineer and industrial designer whose ideas changed the look of 20th-century life, was born at Paris, France, Nov 5, 1893. His designs are evident in almost every area of modern life—the US Postal Service logo, the president's airplane, *Air Force One*, in streamlined automobiles, trains, refrigerators and pens. Loewy was, he once said, "amazed at the chasm between the excellent quality of much of American production and its gross appearance, clumsiness and noise." "Between two products equal in price, function and quality," he said, "the better looking will outsell the other." Loewy died at Monte Carlo on July 14, 1986.

NATIONAL CHEMISTRY WEEK. Nov 5-11. To celebrate the contributions of chemistry to modern life, and to help the public understand that chemistry affects every part of our lives. Activities include an array of outreach programs such as open houses, contests, workshops, exhibits and classroom visits. 7 million participants nationwide. Sponsor: American Chemical Society, Denise L. Creech, 1155 16th St NW, Washington, DC 20036. Phone: (202) 872-6078.

NATIONAL NOTARY PUBLIC WEEK. Nov 5-11. To recognize the fundamental contributions made by notaries to the law and the people of the US. Annually, the full week containing Nov 7. Sponsor: American Society of Notaries, Eugene E. Hines, Exec Dir, 918 16th St NW, Washington, DC 20006. Phone: (202) 955-6162.

NATIONAL OSTEOPATHIC MEDICINE WEEK. Nov 5-11. To salute osteopathic physicians, osteopathic medical students and the healthcare professionals who work in osteopathic hospitals. Sponsors: American Osteopathic Assn, Auxiliary to the American Osteopathic Assn, American Osteopathic Hospital Assn and Association of Osteopathic State Executive Directors. For info: American Osteopathic Assn, Sharon K. Mellor, Dir Communications, 142 E Ontario, Chicago, IL 60611. Phone: (800) 621-1773.

NATIONAL RADIOLOGIC TECHNOLOGY WEEK. Nov 5-11. Increases public awareness of the health professionals who utilize medical radiation and diagnostic imaging techniques to aid in the diagnosis and treatment of disease. Fax: (505) 298-5063. Sponsor: American Soc of Radiologic Technologists, 15000 Central Ave SE, Albuquerque, NM 87123-3917. Phone: (505) 298-4500.

NEW YORK WEEKLY JOURNAL: FIRST ISSUE ANNIVERSARY. Nov 5, 1733. John Peter Zenger, colonial American printer and journalist, published the first issue of the *New York Weekly Journal* newspaper. He was arrested and imprisoned on Nov 17, 1734, for libel. The trial remains an important landmark in the history of the struggle for freedom of the press. See also: "Zenger, John P.: Arrest Anniversary" (Nov 17).

◆ **TARBELL, IDA M.: BIRTH ANNIVERSARY.** Nov 5. American writer, editor and historian, born at Erie County, PA, Nov 5, 1857. Died at Bethel, CT, Jan 6, 1944.

UNITED NATIONS: INTERNATIONAL WEEK OF SCIENCE AND PEACE. Nov 5-11. In 1988, the General Assembly proclaimed this week to generate and increase public interest in this topic and to encourage activities conducive to the study and dissemination of information on the links between progress in science and technology and the maintenance of peace and security. To be celebrated each year during the week in which Nov 11 falls. Info from: United Nations, Dept of Public Info, New York, NY 10017.

VIRGINIA THANKSGIVING FESTIVAL. Nov 5. Berkeley Plantation, Charles City County, VA. Promotes awareness of the historical significance of Thanksgiving Day by re-creation of the original Thanksgiving celebration, which took place near Richmond, VA, at Berkeley Plantation in 1619. Est attendance: 4,000. For info: Virginia Thanksgiving Festival, Shirley J. Baber, Exec Dir, PO Box 5132, Richmond, VA 23220. Phone: (804) 272-3226.

BIRTHDAYS TODAY

Bryan Adams, 36, singer ("Heaven," "Summer of '69"), songwriter ("Everything I Do"), born Vancouver, British Columbia, Canada, Nov 5, 1959.

Arthur Garfunkel, 54, singer (formerly of Simon and Garfunkel; solo "All I Know,"), actor (*Carnal Knowledge, Catch-22, Bad Timing*), born Forest Hills, NY, Nov 5, 1941.

Lloyd Anthony Moseby, 36, former baseball player, born at Portland, AR, Nov 5, 1959.

Tatum O'Neal, 32, actress (Oscar for *Paper Moon* [youngest recipient]; *Bad News Bears*), daughter of Ryan O'Neal, born Los Angeles, CA, Nov 5, 1963.

Nov 5-7

★ Chase's 1995 Calendar of Events ★

Roy Rogers (Leonard Slye), 83, singer, actor (King of the Cowboys; "The Roy Rogers Show,"), born at Cincinnati, OH, Nov 5, 1912.
Sam Shepard (Samuel Shepard Rogers), 52, author, actor (*A Lie of the Mind, Buried Child*), born at Ft Sheridan, IL, Nov 5, 1943.
Elke Sommer (Elke Schletze), 54, actress (*A Shot in the Dark, The Prize*), born Berlin, Germany, Nov 5, 1941.
Ike Turner, 64, singer (Ike and Tina Turner Revue), born at Clarksdale, MS, Nov 5, 1931.

NOVEMBER 6 — MONDAY
310th Day — Remaining, 55

AUSTRALIA: RECREATION DAY. Nov 6. The first Monday in November is observed as Recreation Day in Northern Tasmania, Australia.

BLACK SOLIDARITY DAY. Nov 6. Yale University, New Haven, CT. Students participate in a speak-out, share a common meal, relate important events regarding their heritage and enjoy music and other expressions of talent. Annually, the first Monday in November. Est attendance: 250. For info: Kimberly Goff-Crews, Esq, Asst Dean and Dir, Afro-American Cultural Ctr, Yale University, PO Box 3439, New Haven, CT 06520. Phone: (203) 432-4131.

NAISMITH, JAMES: BIRTH ANNIVERSARY. Nov 6. Inventor of the game of basketball was born at Almonte, Ontario, Canada, Nov 6, 1861. Died at Lawrence, KS, Nov 28, 1939. Basketball became an Olympic sport in 1936.

NORTHERN IRELAND: BELFAST FESTIVAL AT QUEEN'S. Nov 6-25. Queen's University, Belfast, County Antrim. International festival of the arts which includes cinema, drama, opera and all types of music from folk to classical. Est attendance: 80,000. For info: Mr. Robert Agnew, Fest House, 25 College Gardens, Belfast, Northern Ireland, BT9 6BS.

PADEREWSKI, IGNACE JAN: 135th BIRTH ANNIVERSARY. Nov 6. Polish composer, pianist, patriot born at Kurylowka, Podolia, Poland, Nov 6, 1860. He died at New York, NY, June 29, 1941. When Poland fell into the hands of the Soviets after WWII his family decided he would remain buried in Arlington National Cemetary. In May 1963, President John F. Kennedy dedicated a plaque to Paderewski's memory and declared that the pianist would rest in Arlington until Poland was free. Paderewski's remains were returned to his native country on June 29, 1992, the 51st anniversary of his death, after Poland held its first parliamentary election following its independence from the Soviet Union.

SAXOPHONE DAY (ADOLPHE SAX BIRTH ANNIVERSARY). Nov 6. A day to recognize the birth anniversary of Adolphe Sax, Belgian musician and inventor of the saxophone and the saxotromba. Born at Dinant, Belgium on Nov 6, 1814, Antoine Joseph Sax, later known as Adolphe, was the eldest of 11 children of a musical instrument builder. Sax contributed an entire family of brass wind instruments for band and orchestra use. He was accorded fame and great wealth, but business misfortunes led to bankruptcy. Sax died in poverty at Paris on Feb 7, 1894.

SOUSA, JOHN PHILIP: BIRTH ANNIVERSARY. Nov 6. American composer and band conductor, remembered for many stirring marches such as "Stars and Stripes Forever," "Semper Fidelis," "El Capitan," born at Washington, DC, Nov 6, 1854. Died at Reading, PA, Mar 6, 1932. See also: "The Stars and Stripes Forever" (May 14).

	S	M	T	W	T	F	S
November				1	2	3	4
1995	5	6	7	8	9	10	11
	12	13	14	15	16	17	18
	19	20	21	22	23	24	25
	26	27	28	29	30		

SWEDEN: GUSTAVUS ADOLPHUS DAY. Nov 6. Honors Sweden's King and military leader killed in 1632.

TCHAIKOVSKY, PETER ILICH: DEATH ANNIVERSARY. Nov 6. Ranked among the outstanding composers of all time, Peter Ilich Tchaikovsky was born at Vatkinsk, Russia, May 7, 1840. His obvious musical talent was not encouraged and he embarked upon a career in jurisprudence, not studying music seriously until 1861. Among his famous works are the three-act ballet *Sleeping Beauty*, two-act ballet *Nutcracker*, and the symphony *Pathetique*. In an effort to hide his homosexuality, he married a former fellow student from the Moscow Conservatory of Music. The marriage fell apart almost immediately. Tchaikovsky tried to commit suicide at the time, and later his former wife attempted to blackmail him. Mystery surrounds his death. At that time it was believed he'd caught cholera from contaminated water, but 20th-century scholars believe he probably committed suicide to avoid his homosexuality being revealed. He died at St. Petersburg, Nov 6, 1893.

YOSEMITE VINTNERS' HOLIDAYS. Nov 6–Dec 8. (Sundays through Thursdays.) The Ahwahnee Hotel, Yosemite National Park, CA. Seminars on winemaking and tasting amid Yosemite's natural beauty, featuring 4 vitners per session from California's most prestigious wineries. For info: Yosemite Park and Curry Co, Yosemite Reservations, Vintners' Holidays, 5410 E Home Ave, Fresno, CA 93727. Phone: (209) 372-8839.

BIRTHDAYS TODAY

Ray Conniff, 79, bandleader, born at Attleboro, MA, Nov 6, 1916.
Sally Field, 49, actress (Oscars *Norma Rae, Places in the Heart*; Emmy *Sybil*), born Pasadena, CA, Nov 6, 1946.
Glen Frey, 47, musician, songwriter, singer (with Eagles "Hotel California"; solo "The Heat Is On," "You Belong to the City"), born Detroit, MI, Nov 6, 1948.
Lance Kerwin, 35, actor ("James at 15," "The Family Holvak"), born at Newport Beach, CA, Nov 6, 1960.
Mike Nichols (Michael Igor Peschkowsky), 64, comedian, actor, theater producer/director, filmmaker (Oscar *The Graduate*), born Berlin, Germany, Nov 6, 1931.
Maria Owings Shriver, 40, broadcast journalist ("Today"; Mrs. Arnold Schwarzenegger), born at Chicago, IL, Nov 6, 1955.

NOVEMBER 7 — TUESDAY
311th Day — Remaining, 54

CAMUS, ALBERT: BIRTH ANNIVERSARY. Nov 7. French writer and philosopher, winner of the Nobel Prize for Literature in 1957, was born at Mondavi, Algeria, Nov 7, 1913. "The struggle to reach the top is itself enough to fulfill the heart of man. One must believe that Sisyphus is happy," he wrote, in *Le Mythe de Sisyphe*. Camus was killed in an automobile accident in France, Jan 4, 1960.

CANADIAN PACIFIC RAILWAY: TRANSCONTINENTAL COMPLETION ANNIVERSARY. Nov 7. At 9:30 AM, on Nov 7, 1885, the last spike was driven at Craigellachie, British Columbia, completing the Canadian Pacific Railway's 2,980-mile transcontinental railroad track between Montreal, Quebec, in the east and Port Moody, British Columbia, in the west.

CURIE, MARIE SKLODOWSKA: BIRTH ANNIVERSARY. Nov 7. Polish chemist and physicist, born at Warsaw, Poland, Nov 7, 1867. Died July 4, 1934.

Chase's 1995 Calendar of Events — Nov 7-8

FIRST BLACK GOVERNOR ELECTED: ANNIVERSARY. Nov 7. On Nov 7, 1989, L. Douglas Wilder was elected governor of Virginia, becoming the first elected black governor in US history. Wilder had previously served as lieutenant governor of Virginia.

GENERAL ELECTION DAY. Nov 7. Always the first Tuesday after the first Monday in November. Many state and local government elections on this day, as well as presidential and congressional elections in the appropriate years. All US Congressional seats and one-third of US senatorial seats are up for election in even-numbered years. Presidential elections are held in even-numbered years that can be divided equally by 4.

GREAT OCTOBER SOCIALIST REVOLUTION: ANNIVERSARY. Nov 7. This holiday in the old Soviet Union was observed for two days with parades, military displays and appearances by Soviet leaders. According to the old Russian calendar, the revolution took place on Oct 25, 1917. Soviet calendar reform causes observance to fall on Nov 7 (Gregorian). The Bolshevik Revolution began in Petrograd, Russia, on the evening of Nov 6 (Gregorian), 1917. A new government headed by Nikolai Lenin took office the following day under the name Council of People's Commissars. Leon Trotsky was commissar for foreign affairs and Josef Stalin became commissar of national minorities. No longer celebrated as such since the dissolution of the USSR in 1991 after 74 years of Communist rule.

HALFWAY POINT OF AUTUMN. Nov 7. At 5:15 AM, EDT, on Nov 7, 1995, 44 days, 22 hours and 2 min of autumn will have elapsed and the equivalent will remain before Dec 22, 3:17 AM, EST, which is the winter solstice and the beginning of winter.

MOON PHASE: FULL MOON. Nov 7. Moon enters Full Moon phase at 2:20 AM, EST.

NATIONAL NOTARY PUBLIC DAY. Nov 7. To recognize the fundamental contributions made by notaries to the law and the people of the US. Annually, Nov 7. Sponsor: American Society of Notaries, Eugene E. Hines, Exec Dir, 918 16th St NW, Washington, DC 20006. Phone: (202) 955-6162.

NATIONAL SPLIT PEA SOUP WEEK. Nov 7-13. To promote the use of split peas in split pea soup. Annually, the second week in November. Sponsor: USA Dry Pea and Lentil Industry, Victoria Scalise, 5071 Highway 8, Moscow, ID 83843. Phone: (208) 882-3023.

REPUBLICAN SYMBOL: ANNIVERSARY. Nov 7. Thomas Nast used an elephant to represent the Republican Party in a satirical cartoon in *Harper's Weekly* Nov 7, 1874, and still today the elephant is a well-recognized symbol for the Republican Party in political cartoons.

ROOSEVELT ELECTED TO FOURTH TERM: ANNIVERSARY. Nov 7. Defeating Thomas Dewey on Nov 7, 1944, Franklin D. Roosevelt became the first, and only, person elected to four terms as President of the US. Roosevelt was inaugurated the following Jan 20 but died in office Apr 12, 1945, serving only 53 days of the fourth term.

BIRTHDAYS TODAY

Billy (William Franklin) Graham, 77, evangelist, born at Charlotte, NC, Nov 7, 1918.
Al Hirt, 73, jazz musician, born New Orleans, LA, Nov 7, 1922.
Joni Mitchell (Roberta Joan Anderson), 52, singer ("Help Me"; album *Blue*), songwriter ("Both Sides Now," "Woodstock"), born McLeod, Alberta, Canada, Nov 7, 1943.
Dana Plato, 31, actress (Kimberly Drummond on "Diff'rent Strokes"), born at Maywood, CA, Nov 7, 1964.
Joan Sutherland, 69, opera singer, born Sydney, Australia, Nov 7, 1926.
Mary Travers, 58, composer, singer (Peter, Paul and Mary, "Blowin' in the Wind"), born Louisville, KY, Nov 7, 1937.

NOVEMBER 8 — WEDNESDAY
312th Day — Remaining, 53

ABET AND AID PUNSTERS DAY. Nov 8. Laugh instead of groan at incredibly dreadful puns. All-time greatest triple pun: "Though he's not very humble, there's no police like Holmes." Register worst puns with Punsters Unlimited. (Originated by Earl Harris, retired, and the late William Rabe.)

BARNARD, CHRISTIAAN NEETHLING: BIRTHDAY. Nov 8. South African surgeon and medical pioneer who performed the first known human heart transplant was born at Beaufort West, South Africa, on Nov 8, 1922. The first human heart transplant took place at a Cape Town hospital in 1967. The organ donor had been killed in an automobile accident and the recipient lived less than a month after the surgery. In the following year more than a hundred such transplants were performed, but the number declined in succeeding years. In his book *Good Life, Good Death: A Doctor's Case for Euthanasia and Suicide*, Dr. Barnard wrote: "I have learned from my life in medicine that death is not always an enemy ... it achieves what medicine cannot achieve—it stops suffering."

CANADA: CANADIAN FINALS RODEO. Nov 8-12. Northlands Coliseum, Edmonton, Alberta. Professional rodeo championships of Canada. Est attendance: 79,000. For info: Edmonton Northlands, D Craig, Event Mgr, Box 1480, Edmonton, Alta, Canada T5J 2N5. Phone: (403) 471-7335.

HALLEY, EDMUND: BIRTH ANNIVERSARY. Nov 8. English astronomer and mathematician born at London, England, Nov 8, 1656. Astronomer Royal, 1721-1742. Died at Greenwich, England, Jan 14, 1742. He observed the great comet of 1682 (now named for him), first conceived its periodicity and wrote his *Synopsis of Comet Astronomy*: "... I may venture to foretell that this Comet will return again in the year 1758." It did, and Edmund Halley's memory is kept alive by the once-every-generation appearance of Halley's Comet. There have been 28 recorded appearances of this comet since 240 BC. Average time between appearances is 76 years. Often regarded as a harbinger of disaster, Halley's Comet is next expected to be visible in 2061.

MERCHANT SAILING SHIP PRESERVATION DAY. Nov 8. On this day in 1941 the whaler *Charles W. Morgan* arrived in Mystic, CT, to be restored. This was the beginning of the modern era of preservation of merchant sailing ships.

MITCHELL, MARGARET: BIRTH ANNIVERSARY. Nov 8. American novelist, author of the bestselling novel in US history, *Gone with the Wind*, published in 1936, for which she received the Pulitzer Prize for fiction in 1937. Born at Atlanta, GA, Nov 8, 1900, she died there, after being struck by an automobile, on Aug 16, 1949. *Gone with the Wind* was her first and only novel.

MONTANA: ADMISSION DAY. Nov 8. Became 41st state on this day in 1889.

NATIONAL ASSOCIATION FOR GIFTED CHILDREN CONVENTION. Nov 8-12. Tampa, FL. Educational training for administrators, parents and teachers. Est attendance: 2,300. Sponsor: Natl Assn for Gifted Children, 1155 15th St NW #1002, Washington, DC 20005. Phone: (202) 785-4268.

◆ **X-RAY DISCOVERY DAY: 100th ANNIVERSARY.** Nov 8. Physicist Wilhelm Conrad Roentgen (q.v.) discovered X-rays on Nov 8, 1895, beginning a new era in physics and medicine. Although X-rays had been observed previously, it was Roentgen, a professor at the University of Wurzburg (Germany), who successfully repeated X-ray experimentation and who is credited with the discovery.

435

BIRTHDAYS TODAY

Christiaan Barnard, 73, heart surgeon (performed first heart transplant), born at Beaufort West, South Africa, Nov 8, 1922.
Mary Hart, 44, TV host, born at Madison, SD, Nov 8, 1951.
June Havoc, 79, actress (*Brewster's Millions*, "Willy"), born Seattle, WA, Nov 8, 1916.
Christie Hefner, 43, business executive (head of Playboy), daughter of Hugh Hefner, born at Chicago, IL, Nov 8, 1952.
Ricki Lee Jones, 41, singer, musician ("Chuck E.'s In Love"), born Chicago, IL, Nov 8, 1954.
Patti Page (Clara Ann Fowler), 68, singer ("Let Me Go Lover," "Allegheny Moon"), born at Clarence, OK, Nov 8, 1927.
Bonnie Raitt, 46, singer ("Sweet Forgiveness"), actress, daughter of John Raitt, born Los Angeles, CA, Nov 8, 1949.
Morley Safer, 64, newscaster, born at Toronto, Ontario, Canada, Nov 8, 1931.

NOVEMBER 9 — THURSDAY
313th Day — Remaining, 52

AGNEW, SPIRO THEODORE: BIRTHDAY. Nov 9. Thirty-ninth vice president of the US, born at Baltimore, MD, Nov 9, 1918. Twice elected vice president (1968 and 1972), Agnew, on Oct 10, 1973, became the second person to resign that office. Agnew entered a plea of no contest to a charge of income tax evasion (on contract kickbacks received while he was governor of Maryland and after he became vice president), and was sentenced to pay a $10,000 fine and serve three years probation. On Apr 27, 1981, following a class-action suit, a Maryland circuit court ruled that Agnew must pay the state $248,735 ($147,500 in kickbacks plus $101,235 in interest) for violating the public trust. Agnew did not testify at the trial. See also: "Vice Presidential Resignation: Anniversary" (Dec 28) and "Calhoun, John Caldwell: Birth Anniversary" (Mar 18).

ATHABASCAN FIDDLING FESTIVAL. Nov 9–12. Fairbanks, AK. The festival is celebrated to preserve the tradition of Athabascan fiddling and dancing. Activities including demonstrations of dancing, guitar-playing and fiddling, as well as performances by village representatives. Est attendance: 4,500. For info: Institute of Alaska Native Arts, Box 70769, Fairbanks, AK 99707. Phone: (907) 456-7491.

BAHAMAS: REMEMBRANCE DAY. Nov 9. The governor general, prime minister, members of the diplomatic corps and civic leaders lay wreaths in memory of the war dead at service in the Garden of Remembrance, downtown Nassau.

BERLIN WALL OPENED: ANNIVERSARY. Nov 9, 1989. After 28 years as a symbol of the Cold War, the Berlin Wall was opened.

November 1995

S	M	T	W	T	F	S
			1	2	3	4
5	6	7	8	9	10	11
12	13	14	15	16	17	18
19	20	21	22	23	24	25
26	27	28	29	30		

CAMBODIA: INDEPENDENCE DAY. Nov 9. National Day.

EAST COAST BLACKOUT: 30th ANNIVERSARY. Nov 9. Massive electric power failure starting in western New York state at 5:16 PM, Nov 9, 1965, cut electric power to much of northeastern US and Ontario and Quebec in Canada. More than 30 million persons in an area of 80,000 square miles were affected. The experience provoked studies of the vulnerability of 20th-century societal technology.

FRENCH LIBRARY AND CULTURAL CENTER 50th ANNIVERSARY GALA. Nov 9. Boston, MA. Celebration of the founding of The French Library and Cultural Center and its many contributions to the Greater Boston and French-speaking communities. It is the largest such center in New England. Attendance limited. Fax: (617) 266-1780. For info: Phyllis Dohanian, Exec Dir, The French Library and Cultural Center, 53 Marlborough St, Boston, MA 02116-2099. Phone: (617) 266-4351.

GRIMES, LEONARD ANDREW: BIRTH ANNIVERSARY. Nov 9. Reverend Leonard Grimes was born at Leesburg, VA, to parents who were free on Nov 9, 1815. As a free black man living in Washington, DC, he despised slavery and became active in assisting fugitive slaves to escape. He was caught and imprisoned in Richmond, VA. After his imprisonment he founded and became the first minister of the Twelfth Street Baptist Church in Boston, MA, where he served until his death on Mar 14, 1874.

INTERNATIONAL GIFT FESTIVAL. Nov 9–11. Fairfield, PA. Traditional handcrafts from more than 30 countries. For times call or write Gettysburg Travel Council. Est attendance: 1,750. For info: Gettysburg Travel Council, 35 Carlisle St, Gettysburg, PA 17325. Phone: (717) 334-6274.

KRISTALLNACHT (CRYSTAL NIGHT): ANNIVERSARY. Nov 9–10. During the evening of Nov 9 and into the morning of Nov 10, 1938, mobs in Germany destroyed thousands of shops and homes carrying out a pogrom against Jews. Synagogues were burned down or demolished. There were bonfires in every Jewish neighborhood, fueled by Jewish prayer books, Torah scrolls and volumes of philosophy, history and poetry. More than 30,000 Jews were arrested and 91 killed. The night got its name from the smashing of glass store windows.

LOVEJOY, ELIJAH P.: BIRTH ANNIVERSARY. Nov 9. American newspaper publisher and abolitionist born Nov 9, 1802. Died Nov 7, 1837.

MINIATURE GOLF NATIONAL CHAMPIONSHIP. Nov 9–12. Mountasia Regency, Jacksonville, FL. Miniature golf players from North America compete in four age groups—grade school, high school and college, adults and seniors (60 and up) for the best in each division. National championship is televised on ESPN. Sponsor: City of Jacksonville, FL. Est attendance: 1,000. Sponsor: Skip Laun, Exec Dir, Miniature Golf Assn of Amer, PO Box 32353, Jacksonville, FL 32237. Phone: (904) 781-4653.

NATIONAL CHILD SAFETY COUNCIL: FOUNDING ANNIVERSARY. Nov 9. National Child Safety Council (NCSC) was founded Nov 9, 1955, in Jackson, MI. NCSC is the oldest and largest nonprofit organization in the United States dedicated solely to the personal safety and well-being of young children. Sponsor: NCSC, Box 1368, Jackson, MI 49204-1368. Phone: (517) 764-6070.

NATIONAL FFA CONVENTION. Nov 9–11. Kansas City, MO. Est attendance: 32,000. For info: Natl FFA Ctr, Communications Resources Team, 5632 Mt Vernon Memorial Hwy, Box 15160, Alexandria, VA 22309-0160. Phone: (703) 360-3600.

NORSEFEST. Nov 9–11. Madison, MN. Lutefisk supper and lutefisk eating contest (current record: 8 pounds—we're looking for competitors from out of the area). Norwegian arts and craft show and out house race and more. Est attendance: 2,500. For info: Donna Ventrella, Chamber of Commerce, 404 Sixth Ave, Madison, MN 56256. Phone: (612) 598-7373.

Chase's 1995 Calendar of Events — Nov 9-10

RETURN DAY. Nov 9. Georgetown, DE. The day when officially tabulated election returns are read from the balcony of Georgetown's red brick, Greek Revival courthouse to the throngs of voters assembled below. Always the second day after a general election. An official "half-holiday" in Sussex County. Reportedly Return Day has become so popular "that it is for all intents and purposes a state holiday as well."

SOUTHERN CHRISTMAS SHOW. Nov 9-19. Charlotte, NC. Show filled with exquisite art and crafts from nationally known artisans. Ideas for decorating trees, mantels and doors, festive foods for holiday celebrations from fresh hot strudel to plum pudding. Gifts for everyone on your holiday list. Plus cooking demonstrations, Christmas Tree Lane, Santa Claus and Olde Towne. Fax: (704) 849-0248. For info: Southern Shows, Inc, PO Box 36859, Charlotte, NC 28236. Phone: (800) 849-0248.

VIETNAM VETERANS MEMORIAL STATUE UNVEILING: ANNIVERSARY. Nov 9. The Vietnam Veterans Memorial was completed by the addition of a statue, "Three Servicemen" (sculpted by Frederick Hart), which was unveiled on Nov 9, 1984. The statue faces the black granite wall on which are inscribed the names of more than 58,000 Americans who were killed or missing in action in the Vietnam War.

WHITE, STANFORD: BIRTH ANNIVERSARY. Nov 9. American architect who designed the old Madison Square Garden, the Washington Arch, and the Players, Century and Metropolitan Clubs in New York City, Stanford White was born there on Nov 9, 1853. White was shot to death on the roof of the Madison Square Garden by Harry Thaw on June 25, 1906.

BIRTHDAYS TODAY

Spiro T. Agnew, 77, former vice president who had to resign over taxes on kickbacks, born Baltimore, MD, Nov 9, 1918.
Lou Ferrigno, 44, actor (*Pumping Iron*, "The Incredible Hulk"; twice Mr. Universe), born Brooklyn, NY, Nov 9, 1951.
Bob Gibson, 60, baseball Hall of Famer, born at Omaha, NE, Nov 9, 1935.
Bob Graham, 59, US Senator (D, Florida), born at Coral Gables, FL, Nov 9, 1936.
Carl Sagan, 61, astronomer, biologist, author (*Broca's Brain, Cosmos*), born New York, NY, Nov 9, 1934.
Tom Weiskopf, 53, golfer, born at Massillon, OH, Nov 9, 1942.

NOVEMBER 10 — FRIDAY
314th Day — Remaining, 51

BURTON, RICHARD: 70th BIRTH ANNIVERSARY. Nov 10. Welsh-born stage and film actor. Richard Burton was never knighted and never an Oscar-winner, but he was generally regarded as the possessor of one of the great acting talents of his time. Born Richard Jenkins at Pontrhydyfen, South Wales, on Nov 10, 1925, the son of a coal miner later took the name of his guardian, schoolmaster Philip Burton. An intense and tempestuous personal life and career suggested a self-destructive bent. Burton died at Geneva, Switzerland, of a cerebral hemorrhage, at age 58, on Aug 5, 1984.

EDMUND FITZGERALD MEMORIAL BEACON LIGHTING. Nov 10. 4 PM-6:00 PM, Two Harbors, MN. Slide presentation on *Edmund Fitzgerald* and other shipwrecks on Lake Superior; beacon lighting; lighthouse tour held at Split Rock Lighthouse. Est attendance: 325. For info: Lee Radzak, Split Rock Lighthouse, 2010 Hwy 61 E, Two Harbors, MN 55616. Phone: (218) 226-4372.

EDMUND FITZGERALD SINKING: 20th ANNIVERSARY. Nov 10. The ore carrier *Edmund Fitzgerald* broke in two during a heavy storm in Lake Superior (near Whitefish Point) Nov 10, 1975. There were no survivors of this, the worst Great Lakes ship disaster of the decade, that took the lives of 29 crew members.

FOUR CORNER STATES BLUEGRASS FESTIVAL IN WICKENBURG. Nov 10-12. Wickenburg, AZ. Old-time fiddle, banjo, mandolin, flat-pick guitar championships. Clogging and gospel music. Special entertainment by nationally known bands as well as 14 competitive events. Annually, the second weekend in November. Est attendance: 5,000. Sponsor: Chamber of Commerce, Drawer CC, Wickenburg, AZ 85358. Phone: (602) 684-5479.

GOLDSMITH, OLIVER: BIRTH ANNIVERSARY. Nov 10. Irish writer, author of the play *She Stoops to Conquer*. Born Nov 10, 1728, and died Apr 4, 1774. "A book may be amusing with numerous errors," he wrote (Advertisement to *The Vicar of Wakefield*), "or it may be very dull without a single absurdity."

HOGARTH, WILLIAM: BIRTH ANNIVERSARY. Nov 10. English painter and engraver, famed for his satiric series of engravings (*A Harlot's Progress, A Rake's Progress, Four Stages of Cruelty*, etc), born at London, England, Nov 10, 1697, and died there on Oct 26, 1764.

HOLIDAY AT THE BENJAMIN HARRISON HOME. Nov 10-Dec 31. Indianapolis, IN. (Closed Thanksgiving, Christmas Eve and Christmas Day.) Daily guided tours of the 23rd President's home decorated in seasonal, Victorian style. Est attendance: 4,500. For info: President Benjamin Harrison Home, PR Dept, 1230 N Delaware St, Indianapolis, IN 46202-2598. Phone: (317) 631-1898.

HOLIDAY HIGHLIGHTS. Nov 10-11. Ella Missing Community Center, Arapahoe, NE. Display of craft items hand-made and hand-crafted by the hobbyist. (No commercial crafts accepted.) Annually, the second weekend (Friday and Saturday) in November. To insure space, get entries in early. Sponsor: Arapahoe Chamber of Commerce. Est attendance: 600. For info: Ariel McNamara, Secy, Arapahoe Chamber of Commerce, PO Box 624, Arapahoe, NE 68922. Phone: (308) 962-5203.

KIRTLAND, JARED: BIRTH ANNIVERSARY. Nov 10. American physician and naturalist, Dr. Jared Potter Kirtland (for whom Kirtland's Warbler is named) was born at Wallingford, CT, on Nov 10, 1793. The first of the now rare Kirtland's Warblers to be identified and studied was found on his farm near Cleveland, OH, in 1851. Dr. Kirtland died at Rockport, near Cleveland, on Dec 10, 1877.

LONGHORN WORLD CHAMPIONSHIP RODEO. Nov 10-12. Memorial Auditorium, Greenville, SC. More than 200 of North America's best cowboys and cowgirls compete in 6 professionally sanctioned contests ranging from bronc riding to bull riding for top prize money and world championship points. The theme of the production changes annually. Additional attractions include special animal trainers, trick ropers, trick riders and professional clowns. 26th annual. Est attendance: 13,000. For info: Longhorn World Chmpshp Rodeo, Inc, PO Box 70159, Nashville, TN 37207. Phone: (615) 876-1016.

LUTHER, MARTIN: BIRTH ANNIVERSARY. Nov 10. Augustinian monk who was a founder and leader of the Reformation and of Protestantism was born at Eisleben, Saxony, Nov 10, 1483. Luther tacked his 95 Theses "On the Power of Indulgences" on the door of Wittenburg's castle church, on Oct 31, 1517, the eve of All Saints' Day. Luther asserted that the Bible was the sole authority of the church, called for reformation of abuses by the Roman Catholic Church, and denied the supremacy of the Pope. Tried for heresy by the Roman Church, threatened with excommunication and finally banned by a papal bull (Jan 2, 1521), he responded by burning the bull. In 1525 he married Katherine von Bora, one of nine nuns who had left the convent due to his teaching. Luther died near his birthplace, at Eisleben, Feb 18, 1546.

Nov 10-11 ☆ *Chase's 1995 Calendar of Events* ☆

MARINE CORPS BIRTHDAY. Nov 10. Commemorates the Marine Corps' establishment in 1775. Originally part of the navy, it became a separate unit on July 11, 1789.

MARRIAGE ENRICHMENT WEEKEND. Nov 10-12. Belleville, IL. Join other couples as they gather to discuss, reflect, pray, relax and share their marriage journey. The weekend includes lodging at the Pilgrim's Inn, meals, program materials and refreshments. Registration is limited. Est attendance: 80. For info: Shrine of Our Lady of the Snows, 9500 W Illinois, Hwy 15, Belleville, IL 62223. Phone: (618) 397-6700.

PLANNED PARENTHOOD BOOK FAIR. Nov 10-12. Montgomery County Fairgrounds Coliseum, Dayton, OH. More than 250,000 used books, magazines and records on display. Excellent collectibles and antiques. Annually, the second weekend in November. Est attendance: 9,000. For info: Friends of Planned Parenthood, Dev Dir, 224 N Wilkinson St, Dayton, OH 45402. Phone: (513) 226-0780.

SOUTHWEST RHINOCEROS EXHIBITION. Nov 10-12. Tucson, AZ. Rhinoceros aficionadoes gather to buy, sell and exchange rhino collectibles. Sponsor: Really, Rhinos!, Judyth Lessee, Founder, Box 1285, Tucson, AZ 85702. Phone: (602) 327-9048.

SPACE MILESTONE: *LUNA 17* (USSR). Nov 10. Launched on Nov 10, 1970, this unmanned spacecraft landed and released *Lunakhod 1* (8-wheel, radio-controlled vehicle) on Moon's Sea of Rains, Nov 17, explored lunar surface, sending data back to Earth.

STAMP EXPO. Nov 10-12. Elks Lodge, Pasadena, CA. Fax: (818) 988-4337. Est attendance: 4,000. Sponsor: Intl Stamp Collectors Society, Box 854, Van Nuys, CA 91408. Phone: (818) 997-6496.

TEXAS PTA CONVENTION. Nov 10-12. Austin Conv Center, Austin, TX. This 86th annual business meeting of the Texas PTA. Election of officers, adoption of legislative positions and resolutions are part of the general business. More than 70 workshops are presented on issues such as legislation, year-round schools, AIDS, parent involvement and site-based decision making. Est attendance: 3,000. Sponsor: Texas PTA, Texas Congress of Parents and Teachers, 408 W 11th St, Austin, TX 78701. Phone: (512) 476-6769.

VETERANS DAY OBSERVED. Nov 10. For federal employees, when a holiday falls on a Saturday, it is observed on the preceding Friday.

WATERFOWL FESTIVAL (WITH GOOSE-CALLING CONTEST). Nov 10-12. Easton, MD. Three-day festival of wildlife art. Five hundred of the country's most prestigious wildlife exhibitors present the finest in wildlife art, carvings, sculpture, photography, books, gifts, decoys and antique guns. Retriever Demonstrations, World Championship Goose and Regional Duck Calling Contests, a Decoy Auction and seminars are also featured. Annually, the second full weekend in November. Proceeds contributed to conservation. Est attendance: 20,000. Sponsor: Waterfowl Festival, Box 929, Easton, MD 21601. Phone: (410) 822-4567.

BIRTHDAYS TODAY

Jack Anthony Clark, 40, former baseball player, born at New Brighton, PA, Nov 10, 1955.
Donna Fargo (Yvonne Vaughan), 46, singer ("Funny Face"), songwriter, born Mt Airy, NC, Nov 10, 1949.
Russell Charles Means, 55, Native American rights activist, born at Pine Ridge, SD, Nov 10, 1940.

November 1995

S	M	T	W	T	F	S
			1	2	3	4
5	6	7	8	9	10	11
12	13	14	15	16	17	18
19	20	21	22	23	24	25
26	27	28	29	30		

MacKenzie Phillips, 36, actress (Julie on "One Day at a Time"), daughter of John and Michelle Phillips of the Mamas and Papas, born Alexandria, VA, Nov 10, 1959.
Ann Reinking, 46, dancer, actress (*Pippin*, born Seattle, WA, Nov 10, 1949.
Tim Rice, 51, lyricist (co-wrote musicals *Evita* and *Jesus Christ Superstar* with Andrew Lloyd Webber), born Amersham, England, Nov 10, 1944.

NOVEMBER 11 — SATURDAY
315th Day — Remaining, 50

ADAMS, ABIGAIL SMITH: BIRTH ANNIVERSARY. Nov 11. Wife of John Adams, 2nd president of the US, born at Weymouth, MA, Nov 11, 1744. Died Oct 28, 1818.

ALDRICH, THOMAS BAILEY: BIRTH ANNIVERSARY. Nov 11. American author and editor. Best known for his book *The Story of a Bad Boy* (1870), an autobiographical work. Aldrich was born at Portsmouth, NH, on Nov 11, 1836, and died at Boston, MA, on Mar 19, 1907.

ANGOLA: NATIONAL DAY. Nov 11. National holiday observed.

ARMISTICE DAY. Nov 11. Anniversary of Armistice between Allied and Central Powers, signed at 5 AM, Nov 11, 1918, in Marshal Foch's railway car in the Forest of Compiegne, France. Hostilities ceased at 11 AM. Also recognized in some places as Remembrance Day, Veterans Day, Victory Day and World War I Memorial Day. Many places observe silent memorial at the 11th hour of the 11th day of the 11th month each year. See also: "Veterans Day" (Nov 11).

BONZA BOTTLER DAY™. Nov 11. To celebrate when the number of the day is the same as the number of the month. Bonza Bottler Day™ is an excuse to have a party at least once a month. T-shirts available. For info: Elaine Fremont, 203 Waddell Rd, Taylors, SC 29687. Phone: (803) 244-2023.

CANADA: REMEMBRANCE DAY. Nov 11. Public holiday.

CHRISTMAS AT BILTMORE. Nov 11-Dec 31. Biltmore Estate, Asheville, NC. Christmas is celebrated in much the same fashion as it was in 1895 when George Vanderbilt formally opened Biltmore House. Special features include holiday music and candlelight evenings available on select dates. Closed Nov 23, Dec 25 and Jan 1. Est attendance: 150,000. For info: The Biltmore Co, 1 N Pack Square, Asheville, NC 28801. Phone: (800) 543-2961.

CIRCLE K INTERNATIONAL SERVICE DAY. Nov 11. This day is set aside for all Circle K clubs worldwide to perform a campus and community service project. Circle K is a college student service organization sponsored by Kiwanis. Sponsor: Circle K Intl, 3636 Woodview Trace, Indianapolis, IN 46268-3196. Phone: (317) 875-8755.

CIVIL WAR LIVING HISTORY WEEKEND. Nov 11-12. Burritt Museum and Park, Huntsville, AL. The 19th Alabama Infantry and the 42nd Indiana Infantry will sponsor this Civil War Living History encampment. The original 19th Alabama formed in Huntsville in 1861 commanded by General Joe Wheeler. Their first engagement was the Battle of Shiloh. The 42nd Indiana was also active in this area and at the Battle of Chickamauga they allegedly captured the 19th Alabama's flag. Camp life, various drills and a mock skirmish will be re-enacted each day. The highlight of the event will be the candlelight tour on Saturday night. Annually, the second weekend in November. For info: Pat Robertson, Dir, 3101 Burritt Dr, Huntsville, AL 35810. Phone: (205) 536-2882.

DOSTOYEVSKY, FYODOR MIKHAILOVICH: BIRTH ANNIVERSARY. Nov 11. Russian novelist, author of *The Brothers Karamazov*, *Crime and Punishment*, and *The Idiot*, was born at Moscow, Nov 11, 1821 (New Style) and died at St. Petersburg, Feb 9, 1881 (New Style). A political revolutionary, he was tried, convicted and sentenced to death, but instead of execution he served a sentence in a Siberian prison and later served in the army there.

ENGLAND: LORD MAYOR'S PROCESSION AND SHOW. Nov 11. The City, London. Each year a colorful parade takes place from the Guildhall to the Royal Courts of Justice to mark the inauguration of the new Lord Mayor. For info: Mr Reid, Pageantmaster, 5 The Green, London, England N14 7EG.

FRENCH WEST INDIES: CONCORDIA DAY. Nov 11. St. Martin. Public holiday. Parades and joint ceremony by French and Dutch officials at the obelisk Border Monument commemorating the long-standing peaceful coexistence of both countries. For info: Ms Michel Coutosiev, Mktg Challenges Int'l, 10 E 21st St, New York, NY 10010. Phone: (212) 529-9069.

"GOD BLESS AMERICA" FIRST PERFORMED: ANNIVERSARY. Nov 11. Irving Berlin wrote this song especially for Kate Smith. She first sang it during her regular radio broadcast on Nov 11, 1938. It quickly became a great patriotic favorite of the nation and one of Smith's most requested songs.

KALAMAZOO HOLIDAY PARADE. Nov 11. Kalamazoo, MI. With more than 3,000 participants, the Kalamazoo Holiday Parade is the largest in west Michigan and will be remembered for its giant helium balloons, colorful floats, spirited marching bands and delightful fantasy characters, clowns, musicians and multi-talented performers. Annually the second Saturday in November. Est attendance: 100,000. For info: Stanley Lauderdale, Downtown Kalamazoo Inc, 141 E Michigan Ave, Ste 301, Kalamazoo, MI 49007. Phone: (616) 344-0795.

MARTINMAS. Nov 11. The Feast Day of St. Martin of Tours, who lived about 316–397 AD. A bishop, who became one of the most popular saints of the Middle Ages. The period of warm weather often occurring about the time of his feast day is sometimes called St. Martin's Summer (especially in England).

MONCTON SALE OF FINE CRAFT. Nov 11–12. Moncton Arena, Moncton, NB. All exhibitors must be juried members of MBCC. Est attendance: 8,000. For info: New Brunswick Crafts Council Inc, PO Box 1231, Fredericton, NB E3B 5C8. Phone: (506) 450-8989.

MUSEUM OPEN HOUSE. Nov 11–12. Clinton, MD. 6th annual open house features free tours, special gift shop sales, refreshments. Est attendance: 500. Sponsor: Surratt House and Tavern, 9110 Brandywine Rd, Clinton, MD 20735. Phone: (301) 868-1121.

◆ **PATTON, GEORGE S., JR: 110th BIRTH ANNIVERSARY.** Nov 11. American military officer, graduate of West Point (1909), George Smith Patton, Jr, was born at San Gabriel, CA, on Nov 11, 1885. An undistinguished student, Patton was ambitious and flamboyant. He lived for combat and was bored by what he described as the "horrors of peace." His career included action in the punitive expedition into Mexico (1916), in Europe in World War I, and bold and brilliant soldiering in North Africa and Europe in World War II. He received world attention and official censure in 1943 for slapping a hospitalized shell-shocked soldier. As a full general, owing to his critical public statements, he was relieved of his command in 1945. One of the best-known generals of World War II, Patton is remembered in his star-laden helmet and wearing ivory-handled pistols. He died at Heidelberg, Germany, on Dec 21, 1945, of injuries received in an automobile accident.

SPACE MILESTONE: GEMINI 12 (US). Nov 11. Last Project Gemini manned Earth orbit. Nov 11, 1966.

SPACE MILESTONE: STS-5 (US). Nov 11. Shuttle Columbia launched from Kennedy Space Center, FL, Nov 11, 1982, with four astronauts: Vance Brand, Robert Overmyer, William Lenoir and Joseph Allen. "First operational mission" delivered two satellites into orbit for commercial customers. Columbia landed at Edwards Air Force Base, CA, Nov 16, 1982.

SWEDEN: ST. MARTIN'S DAY. Nov 11. Originally in memory of St. Martin of Tours; also associated with Martin Luther, who is celebrated the day before. Marks the end of the autumn's work and the beginning of winter activities.

SWITZERLAND: MARTINMAS GOOSE (MARTINIGIANS). Nov 11. Sursee, Canton Lucerne. At 3 PM, on Martinmas (the day on which interest is due), the "Gansabhauet" is staged in front of Town Hall. Blindfolded participants try to bring down, with a single sword stroke, a dead goose suspended on a wire.

TEDDY BEAR CONVENTION. Nov 11. First Christian Church, Family Life Center, Jenks, OK. Teddy Bear Show and Sale, supplies for bear making, teddy bear benefit drawing. Annually, the second Saturday before Thanksgiving. Sponsor: Classic Traditions. Est attendance: 2,000. For info: Monica Murray, PO Box 728, Jenks, OK 74037. Phone: (918) 299-5416.

★ **VETERANS DAY.** Nov 11. Presidential Proclamation. Formerly called "Armistice Day" and proclaimed each year since 1926 for Nov 11. PL83-380 of June 1, 1954, changed the name to "Veterans Day." PL90-363 of June 28, 1968, required that beginning in 1971 it would be observed the fourth Monday in October. PL 94-97 of Sept 18, 1975, required that effective Jan 1, 1978, the observance would revert to Nov 11.

VETERANS DAY. Nov 11. Veterans Day was observed on Nov 11 from 1919 through 1970. Public Law 90-363, the "Monday Holiday Law," provided that, beginning in 1971, Veterans Day would be observed on "the fourth Monday in October." This movable observance date, which separated Veterans Day from the Nov 11 anniversary of World War I Armistice, proved unpopular. State after state moved its observance back to the traditional Nov 11 date, and finally Public Law 94-97 of Sept 18, 1975, required that, effective Jan 1, 1978, the observance of Veterans Day revert to Nov 11. See also: "Armistice Day" (Nov 11).

VETERANS DAY CELEBRATION. Nov 11. Mamou, LA. Memorial services, a parade, live Cajun music and a dance that evening. Annually, on Veterans Day. Sponsor: American Legion Post 123. Est attendance: 1,000. For info: Gerald J. Fontenot, Chamber of Commerce, PO Box 34, Mamou, LA 70554. Phone: (318) 468-3272.

WASHINGTON: ADMISSION DAY. Nov 11. Became 42nd state on this day in 1889.

WINEFEST '95. Nov 11. Messina Hof Wine Cellars, Bryan, TX. Visit tasting booths of Brazos Valley restaurants and Texas wineries. Sample the foods and wines Texas is famous for. See beautiful culinary art and chef's demonstrations. Be a part of the premier of a new Messina Hof Vintage at our Premier Dinner. Est attendance: 2,500. For info: Amy Ping, Mktg Dir, Messina Hof Wine Cellars, 4545 Old Reliance Rd, Bryan, TX 77808. Phone: (409) 778-9463.

WORLD OF CHRISTMAS. Nov 11–Jan 8. Herbert Hoover Library/Museum, West Branch, IA. This year's exhibit focuses on some of the many "worlds" of Christmas. Est attendance: 10,000. For info: Herbert Hoover Library/Museum, PO Box 488, West Branch, IA 52358. Phone: (319) 643-5301.

Nov 11-12 ☆ Chase's 1995 Calendar of Events ☆

BIRTHDAYS TODAY

Bibi Andersson (Birgitta Anderson), 60, actress (*Story of a Woman*), born Stockholm, Sweden, Nov 11, 1935.
Barbara Boxer, 55, US Senator (D, California), born at Brooklyn, NY, Nov 11, 1940.
Philip McKeon, 31, actor (Tommy in "Alice"), older brother of Nancy McKeon, born Westbury, NY, Nov 11, 1964.
Demi Moore, 33, actress ("General Hospital," *Ghost, Indecent Proposal*), born Rosewell, NM, Nov 11, 1962.
Daniel Ortega Saavedra, 50, former president of Nicaragua, born at La Libertad, Chontales, Nicaragua, Nov 11, 1945.
Kurt Vonnegut, Jr, 73, novelist (*Slaughterhouse Five, Cat's Cradle*), born Indianapolis, IN, Nov 11, 1922.
Jonathan Winters, 70, comedian, actor ("The Jonathan Winters Show," "Mork & Mindy"), born at Dayton, OH, Nov 11, 1925.
Frank Urban "Fuzzy" Zoeller, 44, golfer, born at New Albany, IN, Nov 11, 1951.

NOVEMBER 12 — SUNDAY
316th Day — Remaining, 49

★ **AMERICAN EDUCATION WEEK.** Nov 12-18. Presidential Proclamation 5403, of Oct 30, 1985, covers all succeeding years. Always the first full week preceding the fourth Thursday in November. Issued from 1921-1925 and in 1936, sometimes for a week in December and sometimes as National Education Week. After an absence of a number of years, this proclamation was issued each year from 1955-1982 (issued in 1955 as a prelude to the White House Conference on Education). Previously, Proclamation 4967, of Sept 13, 1982, covered all succeeding years as the second week in November.

AMERICAN EDUCATION WEEK. Nov 12-18. Focuses attention on the importance of public education and all that it stands for. Annually, the week preceding the weekend in which Thanksgiving occurs. For info: Natl Education Assn (NEA), 1201 16th St NW, Washington, DC 20036. Phone: (202) 822-7200.

BIRTH OF BAHA'U'LLAH. Nov 12. Baha'i observance of anniversary of the birth of Baha'u'llah (born Mirza Husayn Ali) on Nov 12, 1817, in Nur, Persia. Baha'u'llah was prophet-founder of the Baha'i Faith. One of the nine days of the year when Baha'is suspend work. For info: Natl Spiritual Assembly of the US, 1320 Nineteenth St, NW, Ste 701, Washington, DC 20036. Phone: (202) 833-8990.

BLACKMUN, HARRY A.: BIRTHDAY. Nov 12. Associate justice of the Supreme Court of the United States, nominated by President Nixon on Apr 14, 1970. (Qualified, June 9, 1970.) Justice Blackmun was born at Nashville, IL, on Nov 12, 1908.

ENGLAND: REMEMBRANCE DAY SERVICE AND PARADE. Nov 12. Cenotaph, Whitehall, London. Wreath-laying ceremony to commemorate the dead of both world wars by Her Majesty The Queen, members of the Royal Family, government and service organizations. For info: Public Info Office, HQ London District Military, Chelsea Barracks, London, England SW1H 8RF.

FIRST PROFESSIONAL FOOTBALL PLAYER: ANNIVERSARY. Nov 12. On Nov 12, 1892, William "Pudge" Heffelfinger became the first professional football player when he was paid $25 for expenses and a cash bonus of $500. It was the cash bonus that made him professional. Scoring the winning touchdown for the Allegheny Athletic Association, his team beat the Pittsburgh Athletic Club 4-0. Info from: Joe Harrigan, Historian, Football Hall of Fame, 2121 George Halas Dr, Canton, OH 44708.

KELLY, GRACE PATRICIA: BIRTH ANNIVERSARY. Nov 12. American award-winning actress who became Princess Grace of Monaco when she married that country's ruler, Prince Rainier III, in 1956. Born at Philadelphia, PA, Nov 12, 1929. She died of injuries sustained in an automobile accident, Sept 12, 1982.

NATIONAL CULINARY WEEK. Nov 12-18. To recognize and celebrate culinarians and culinary excellence at all levels, honoring the profession, the diversity of cuisine in America and fundamental importance of good food in people's lives. Annually, the Sunday-Saturday before Thanksgiving. Fax: (904) 825-4758. Sponsor: The American Culinary Fed, Brent Frei, Dir of Communications, 10 San Bartola Dr, St. Augustine, FL 32085. Phone: (904) 824-4468.

NATIONAL GEOGRAPHY AWARENESS WEEK. Nov 12-18. Focus public awareness on the importance of the knowledge of geography. To be on the mailing list for a free Geography Awareness Week packet of teaching materials, send your name and mailing address to Geography Education Program, Natl Geographic Society, PO Box 37138, Washington, DC 20013-7138. More than 225,000 packets mailed out last year. For info: Natl Geographic Society, Geography Education Program, 1145 17th St NW, Washington, DC 20036.

NEW YORK CITY MARATHON. Nov 12. New York, NY. 25,000 runners from all over the world gather to compete in the largest spectator event with more than 2 million spectators watching from the sidelines. Sponsor: NY Road Runners Club, 9 E 89th St, New York, NY 10128. Phone: (212) 860-4455.

OPERATING ROOM NURSE WEEK. Nov 12-18. To inform health care consumers that the nurse in the operating room cares for patients before, during and after surgery. Annually, the week including Nov 14. Sponsor: Assn of Operating Room Nurses, Member Services, 2170 S Parker Rd, Ste 300, Denver, CO 80231-5711. Phone: (303) 755-6300.

POLISHFEST '95. Nov 12. Soldiers' and Sailors' Memorial Hall, Pittsburgh, PA. A family-oriented celebration of the cultural heritage of Poland. Arts, crafts, music, song, dance, food and fun! Annually, the second Sunday in November. Est attendance: 4,500. Sponsor: Pittsburgh Polishfest Comm, 808 Phineas St, Pittsburgh, PA 15212. Phone: (412) 231-1493.

RODIN, AUGUSTE: BIRTH ANNIVERSARY. Nov 12. French sculptor born at Paris, France, Nov 12, 1840. Died Nov 17, 1917.

SPACE MILESTONE: STS-2 (US). Nov 12. Shuttle *Columbia*, launched from Kennedy Space Center, FL, on Nov 12, 1981, with Joe Engle and Richard Truly on board, became first spacecraft launched from Earth for a second orbiting mission. Landed at Edwards Air Force Base, CA, Nov 14, 1981.

◈ **STANTON, ELIZABETH CADY: BIRTH ANNIVERSARY.** Nov 12. American woman suffragist and reformer, Elizabeth Cady Stanton was born at Johnstown, NY, on Nov 12, 1815. "We hold these truths to be self-evident," she said at the first Women's Rights Convention, in 1848, "that all men and women are created equal." She died at New York, NY, on Oct 26, 1902.

SUN YAT-SEN: BIRTH ANNIVERSARY (TRADITIONAL). Nov 12. Although his actual birth date in 1866 is not

November 1995

S	M	T	W	T	F	S
			1	2	3	4
5	6	7	8	9	10	11
12	13	14	15	16	17	18
19	20	21	22	23	24	25
26	27	28	29	30		

known, Dr. Sun Yat-Sen's traditional birthday commemoration is held on Nov 12. Heroic leader of China's 1911 revolution, he died at Peking on Mar 12, 1925. The death anniversary is also widely observed. See also: "Sun Yat-Sen: Death Anniversary" (Mar 12).

TUNISIA: TREE FESTIVAL. Nov 12. National agricultural festival. Annually, the second Sunday in November.

TYLER, LETITIA CHRISTIAN: BIRTH ANNIVERSARY. Nov 12. First wife of John Tyler, 10th president of the US, born at New Kent County, VA, Nov 12, 1790. Died Sept 10, 1842.

WASSAIL TEA. Nov 12. Greensboro, NC. Old-fashioned Wassail teas and Christmas cookies welcome the holiday season at the many shops along State Street Station. Strolling musicians, Santa with candy and official lighting of the State Street Christmas tree. No admission charge. 1–5 PM. Phone: (910) 230-0623 or (910) 275-3969. Est attendance: 3,000. For info: Betsy Seale, State Street Station, PO Box 13563, Greensboro, NC 27415.

BIRTHDAYS TODAY

Harry A. Blackmun, 87, Associate Justice of the US Supreme Court (retired), born at Nashville, IL, Nov 12, 1908.
Nadia Comaneci, 34, gymnast, Olympic medalist, born at Onesti, Romania, Nov 12, 1961.
Kim Hunter (Janet Cole), 73, actress (Nola on "The Edge of Night"; *The Planet of the Apes*), born Detroit, MI, Nov 12, 1922.
Stefanie Powers, 53, actress (Jennifer Hart on "Hart to Hart"), born at Hollywood, CA, Nov 12, 1942.
Neil Young, 50, singer (with Buffalo Springfield, "For What It's Worth," and Crosby, Stills, Nash & Young, album *Deja Vu*), songwriter, born at Toronto, Ontario, Canada, Nov 12, 1945.

NOVEMBER 13 — MONDAY
317th Day — Remaining, 48

BOOTH, EDWIN (THOMAS): BIRTH ANNIVERSARY. Nov 13. Famed American actor and founder of the Players Club, born near Bel Air, MD, Nov 13, 1833. Died at New York, NY, June 7, 1893.

BRANDEIS, LOUIS DEMBITZ: BIRTH ANNIVERSARY. Nov 13. American jurist, associate justice of US Supreme Court (1916–1939), born at Louisville, KY, Nov 13, 1856. Died at Washington, DC, Oct 5, 1941.

FINLAND: INTERNATIONAL CHILDREN'S FILM FESTIVAL. Nov 13–19. Oulu. Est attendance: 10,000. For info: Finnish Tourist Board, 655 Third Ave, New York, NY 10017. Phone: (212) 949-2333.

HOLLAND TUNNEL ANNIVERSARY. Nov 13. On Nov 13, 1927, the Holland Tunnel, running under the Hudson River between New York, NY, and Jersey City, NJ, was opened to traffic. The tunnel was built and operated by the New York–New Jersey Bridge and Tunnel Commission. Comprised of two tubes, each large enough for two lanes of traffic, the Holland was the first underwater tunnel built in the US.

MAXWELL, JAMES CLERK: BIRTH ANNIVERSARY. Nov 13. British physicist noted for his work in the field of electricity and magnetism. Born at Edinburgh, Scotland, on Nov 13, 1831, he died of cancer on Nov 5, 1879, at Cambridge, England.

NATIONAL CHILDREN'S BOOK WEEK. Nov 13–19. To encourage the enjoyment of reading for children. Sponsor: The Children's Book Council, Inc, 568 Broadway, Ste 404, New York, NY 10012. Phone: (212) 966-1990.

STEVENSON, ROBERT LOUIS: 145th BIRTH ANNIVERSARY. Nov 13. Scottish author, born at Edinburgh, Scotland, Nov 13, 1850. Died at Samoa, Dec 3, 1894.

STOKES BECOMES FIRST BLACK MAYOR IN US: ANNIVERSARY. Nov 13. Carl Burton Stokes became the first black in the US elected mayor when he won the Cleveland, OH, mayoral election on Nov 13, 1967.

TRELAWNEY, EDWARD JOHN: BIRTH ANNIVERSARY. Nov 13. English traveler and author, friend of Shelley and Byron, born at London, Nov 13, 1792. He died at Sompting, Sussex, Aug 13, 1881, and was buried at Rome, next to Shelley.

BIRTHDAYS TODAY

Garry Marshall, 61, producer, director (*Beaches, Pretty Woman*), born New York, NY, Nov 13, 1934.
Richard Mulligan, 63, actor ("Empty Nest," *S.O.B*), born New York, NY, Nov 13, 1932.
Dack Rambo, 54, actor ("All My Children," "The Guns of Will Sonnett"), born at Delano, CA, Nov 13, 1941.
Madeline Sherwood, 73, actress (Oscar *The Year of Living Dangerously*), born Montreal, Quebec, Canada, Nov 13, 1922.
Vinny (Vincent Frank) Testaverde, 32, football player, born Brooklyn, NY, Nov 13, 1963.
Charlie Tickner, 42, figure skater, born at Oakland, CA, Nov 13, 1953.

NOVEMBER 14 — TUESDAY
318th Day — Remaining, 47

AROUND THE WORLD IN 72 DAYS: ANNIVERSARY. Nov 14. Newspaper reporter Nellie Bly (pen name used by Elizabeth Cochrane Seaman) set off on Nov 14, 1889, to attempt to break Jules Verne's imaginary hero Phileas Foggs' record of voyaging around the world in 80 days. She did beat Foggs' record, taking 72 days, 6 hours, 11 minutes and 14 seconds to make the trip.

BLOOD TRANSFUSION: ANNIVERSARY. Nov 14. Samuel Pepys, diarist and Fellow of the Royal Society, wrote in his diary for Nov 14, 1666: "Dr. Croone told me ... there was a pretty experiment of the blood of one dog let out, till he died, into the body of another on one side, while all his own run out on the other side. The first died upon the place, and the other very well and likely to do well. This did give occasion to many pretty wishes, as of the blood of a Quaker to be let into an Archbishop, and such like; but, as Dr. Croone says, may, if it takes, be of mighty use to man's health, for the amending of bad blood by borrowing from a better body." Two days later, Nov 16, Pepys noted: "This noon I met with Mr. Hooke, and he tells me the dog which was filled with another dog's blood, at the College the other day, is very well, and like to be so as ever, and doubts not its being found of great use to men...."

COOR'S LIGHT TRIPLE CROWN OF SURFING. Nov 14–Dec 9. The 13th annual Triple Crown is comprised of three professional big wave surf meets on Oahu's North shore that mark the conclusion of the year-long Association of Surfing Professionals (ASP) world tour. Following the Triple Crown the ASP Men's and Women's Champions are crowned. Phone: (808) 638-7266 or Ocean Promotion (808) 325-7400. Est attendance: 10,000. For info: Randy Rarick, 59-063-A Hoalua St, Haleiwa, HI 96712.

DOW-JONES TOPS 1,000: ANNIVERSARY. Nov 14. On Nov 14, 1972, the Dow-Jones Index of 30 major industrial stocks closed at 1,003.16—topping the 1,000 mark for the first time.

EISENHOWER, MAMIE DOUD: BIRTH ANNIVERSARY. Nov 14. Wife of Dwight David Eisenhower, 34th president of the US, born at Boone, IA, Nov 14, 1896. Died Nov 1, 1979, at Washington, DC.

FULTON, ROBERT: BIRTH ANNIVERSARY. Nov 14. Inventor of the steamboat, born Nov 14, 1765. Died Feb 24, 1815.

Nov 14–15 ☆ *Chase's 1995 Calendar of Events* ☆

GUINEA-BISSAU: RE-ADJUSTMENT MOVEMENT'S DAY. Nov 14. National holiday.

INDIA: CHILDREN'S DAY. Nov 14. Holiday observed throughout India.

JOHN GILPIN'S RIDE: PUBLICATION ANNIVERSARY. Nov 14. William Cowper's popular and memorable ballad: "The diverting History of John Gilpin, Showing How He Went Farther Than He Intended, and Came Safe Home Again," was first published, anonymously, in *The Public Advertiser*, London, Nov 14, 1782.

JORDAN: KING HUSSEIN'S 60th BIRTHDAY. Nov 14. H.M. King Hussein's birthday is nationally honored each year on the anniversary of his birth, Nov 14, 1935.

MONET, CLAUDE: 155th BIRTH ANNIVERSARY. Nov 14. French painter born Nov 14, 1840. Died, Dec 5, 1926.

NATIONAL COMMUNITY EDUCATION DAY. Nov 14. To recognize and promote strong relationships between public schools and the communities they serve and to help schools develop new relationships with parents, community members, local organizations and agencies. Annually, the Tuesday of American Education Week. Sponsor: Natl Community Education Assn, 3929 Old Lee Hwy, Ste 91-A, Fairfax, VA 22030-2401. Phone: (703) 359-8973.

NEHRU, JAWAHARLAL: BIRTH ANNIVERSARY. Nov 14. Indian leader and first prime minister after independence. Born Nov 14, 1889. Died May 27, 1964.

SPACE MILESTONE: *APOLLO 12* (US). Nov 14. Launched on Nov 14, 1969, this was the second manned lunar landing—in Ocean of Storms. Astronauts Conrad, Bean and Gordon. First pinpoint landing. Astronauts visited *Surveyor 3* and took samples. Earth splashdown Nov 24.

BIRTHDAYS TODAY

Boutros Boutros-Ghali, 73, Secretary-General of the United Nations, born at Cairo, Egypt, Nov 14, 1922.

Charles, 47, Prince of Wales, heir to British throne, born at London, England, Nov 14, 1948.

Brian Keith, 74, actor (*Nevada Smith*, "Family Affair," "Hardcastle and McCormick"), born Bayone, NJ, Nov 14, 1921.

Jack Sikma, 40, former basketball player, born at Kankakee, IL, Nov 14, 1955.

McLean Stevenson, 66, actor ("M*A*S*H," "Hello, Larry"), born at Bloomington, IL, Nov 14, 1929.

Don Stewart, 60, actor, singer (Michael Bauer on "Guiding Light"), born at Staten Island, NY, Nov 14, 1935.

NOVEMBER 15 — WEDNESDAY
319th Day — Remaining, 46

AMERICAN ENTERPRISE DAY. Nov 15. To celebrate the achievements of America's free market economy and to focus on the challenges facing American business and industry in a changing world. Sponsor: Future Business Leaders of America, Phi Beta Lambda, Inc, 1912 Association Dr, Reston, VA 22091. Phone: (703) 860-3334.

AMERICAN FEDERATION OF LABOR: FOUNDING ANNIVERSARY. Nov 15. Anniversary of the founding, at Pittsburgh, PA, on Nov 15, 1881, of the Federation of Organized Trades and Labor Unions of the United States and Canada which, reorganized in 1886, became the American Federation of Labor.

November 1995

S	M	T	W	T	F	S
			1	2	3	4
5	6	7	8	9	10	11
12	13	14	15	16	17	18
19	20	21	22	23	24	25
26	27	28	29	30		

BELGIUM: DYNASTY DAY. Nov 15. National holiday in honor of Belgian monarchy.

BRAZIL: REPUBLIC DAY. Nov 15. Commemorates the Proclamation of the Republic on Nov 15, 1889.

FIRST BLACK PROFESSIONAL HOCKEY PLAYER: ANNIVERSARY. Nov 15. When Arthur Dorrington signed a contract to play hockey with the Atlantic City Seagulls of the Eastern Amateur League on Nov 15, 1950, he became the first black man to play organized hockey in the US. He played for the Seagulls during the 1950 and 1951 seasons.

GEORGE SPELVIN DAY. Nov 15. Believed to be the anniversary of George Spelvin's theatrical birth—in Charles A. Gardiner's play *Karl the Peddler* on Nov 15, 1886, in a production at New York, NY. Spelvin, a fictitious creation, is said to have appeared in more than 10,000 Broadway performances. The name (or equivalent Georgina, Georgetta, etc) is used in play programs to conceal the fact that an actor is performing in more than one role. See also: "England: Walter Plinge Day" (Dec 2) for British equivalent.

GINGERBREAD HOUSE COMPETITION. Nov 15–Jan 6. Peddler's Village, Lahaska, PA. Design, bake and decorate a Christmas Gingerbread House and compete for cash prizes. Six different categories including Amateurs and Children. Gingerbread Seminar on Oct 18. Winning entrees displayed in Gazebo. Open daily. Free admission. Est attendance: 75,000. For info: Peddler's Village, PO Box 218, Lahaska, PA 18931. Phone: (215) 794-4000.

JAPAN: SHICHI-GO-SAN. Nov 15. Annual children's festival. The *Shichi-Go-San* (Seven-Five-Three) rite, observed on Nov 15, is "the most picturesque event in the autumn season." On this day, parents take their three-year-old children of either sex, five-year-old boys and seven-year-old girls to the parish shrines dressed in their best clothes. There the guardian spirits are thanked for the healthy growth of the children and prayers are offered for their further development.

MOON PHASE: LAST QUARTER. Nov 15. Moon enters Last Quarter phase at 6:40 AM, EST.

NATIONAL CLEAN OUT YOUR REFRIGERATOR DAY. Nov 15. With the holidays just around the corner, now is the time to clean out those leftovers and make way for turkey, stuffing and pumpkin pie. An organized, clean refrigerator is the perfect way to start the holiday season. Sponsor: Whirlpool Corp, Carol Sizer, Mgr, Media/Community Relations, Admin Cntr, 2000 N State Rt 63, Benton Harbor, MI 49022. Phone: (616) 923-3231.

NATIONAL EDUCATIONAL SUPPORT PERSONNEL DAY. Nov 15. A mandate of the delegates to the 1987 National Education Association Representative Assembly called for a special day during American Education Week to honor the contributions of school support employees. Local associations and school districts salute support staff on this 9th annual observance, the Wednesday of American Education Week. For info: Natl Education Assn (NEA), 1201 16th St NW, Washington, DC 20036. Phone: (202) 822-7200.

O'KEEFFE, GEORGIA: BIRTH ANNIVERSARY. Nov 15. Described as one of the greatest American artists of the 20th century, Georgia O'Keeffe was born at Sun Prairie, WI, Nov 15, 1887. In 1924 she married the famous photographer, Alfred Stieglitz. His more than 500 photographs of her have been called "the greatest love poem in the history of photography." She died at Santa Fe, NM, on Mar 6, 1986, at age 98.

ROMMEL, ERWIN: BIRTH ANNIVERSARY. Nov 15. German field marshal and commander of the German Afrika Korps in WWII, Erwin Rommel was born Nov 15, 1891, at Heidenheim, in Wurttemberg, Germany. Rommel commanded the 7th Panzer Division in the Battle of France. Considered an excellent commander, Rommel's early success in Africa made him a legend as the "Desert Fox," but in early 1943 he was outmaneuvered by Field Marshal Bernard Montgomery and Germany surrendered Tunis in May of that year. Implicated in July 1944 in an attempted assassination of Hitler, he was given the choice of suicide or a trial and chose the former. Rommel died by his own hand on Oct 14, 1944, near Ulm, Germany.

SMOKY MOUNTAIN LIGHTS AND WINTERFEST. Nov 15–Feb 28. Gatlinburg, TN. A magical celebration of more than two million twinkling lights in animated motion. The entire city of Gatlinburg is lit up with exciting displays featuring up to 60-ft animated displays, for the entire family to enjoy. More than 100 contemporary and traditional events such as yule log burnings, living Christmas tree, festival of trees and Christmas Faire with a taste of Dickens. Drive-thru and walk-thru tours available. Annually, mid-November through the end of February. Est attendance: 1,000,000. Sponsor: Deana Davenport Ivey, Dir of Communications, Gatlinburg Chamber of Commerce, PO Box 527, Gatlinburg, TN 37738. Phone: (800) 568-4748.

SPACE MILESTONE: *BURAN* (USSR). Nov 15. The Soviet Union's first reusable space plane, *Buran*, completed a smooth, unmanned mission at approximately 1:25 AM, EST, on Nov 15, 1988, after orbiting the Earth twice in 3 hours, 25 minutes. Launched at Baikonur, Soviet central Asia, the importance of this mission was in its computer-controlled lift-off and return.

BIRTHDAYS TODAY

Ed Asner, 66, actor ("The Mary Tyler Moore Show," "Lou Grant," "Roots"), born Kansas City, MO, Nov 15, 1929.
Daniel Barenboim, 53, musician, conductor (with Chicago Symphony), born Buenos Aires, Argentina, Nov 15, 1942.
Joanna Barnes, 61, actress (Katie in "The Trials of O'Brien"), born Boston, MA, Nov 15, 1934.
Petula Clark, 63, singer ("Downtown," "I Know a Place," "This Is My Song"), actress (*Finian's Rainbow, Goodbye Mr Chips*), born Ewell, Surrey, England, Nov 15, 1932.
John Coleman, 60, TV meteorologist, originator of The Weather Channel, born at Champaign, IL, Nov 15, 1935.
Joseph Wapner, 76, television personality ("People's Court"), retired judge, born Los Angeles, CA, Nov 15, 1919.
Sam Waterston, 55, actor (*Friendly Fire*, "I'll Fly Away," *The Great Gatsby*), born Cambridge, MA, Nov 15, 1940.

NOVEMBER 16 — THURSDAY
320th Day — Remaining, 45

AMERICAN SPEECH-LANGUAGE-HEARING ASSOCIATION CONVENTION. Nov 16–20. Cincinnati, OH. Scientific sessions held on language, speech disorders, hearing science and hearing disorders and matters of professional interest to speech-language pathologists and audiologists. Est attendance: 10,000. For info: American Speech-Language-Hearing Assn, Cheryl Russell, Conv Dir, 10801 Rockville Pike, Rockville, MD 20852-3279. Phone: (301) 897-5700.

CANADA: SASKATOON FALL FAIR AND MEXABITION. Nov 16–19. Prairieland Exhibition Center, Saskatoon, Saskatchewan. Major livestock show features 7 breeds beef cattle, commercial cattle show and sale, Cinderella Classic, prospect calf show and sale, Saskatchewan Jr Classic, 4-H grooming competition. Agricultural trade show features new farm equipment and advances in agricultural practices. Est attendance: 20,000. For info: Tourism Saskatchewan, Saskatchewan Trade and Conv Center, 1919 Saskatchewan Dr, Regina, Sask, Canada S4P 3V7. Phone: (800) 667-7191.

EXPANDED NEW ENGLAND KINDERGARTEN CONFERENCE. Nov 16–18. Lombardo's-Lantana's-Holiday Inn Complex, Randolph, MA. Annual meeting dedicated to providing services for professionals working in early childhood programs. Consideration of current issues, research reviews, curriculum suggestions, parent/teacher collaboratives. Sponsor: Lesley College, Cambridge, MA. Est attendance: 2,500. For info: Mary Mindess, Conf Coord or Julie Wack, Conf Mgr, Expanded New England Kindergarten Conf, 29 Mellen St, Cambridge, MA 02138. Phone: (617) 349-8922.

GREAT AMERICAN SMOKEOUT. Nov 16. A day observed annually to encourage smokers to kick the habit for at least 24 hours. Annually, the third Thursday in November. Sponsor: American Cancer Society, 1180 Avenue of the Americas, New York, NY 10036. Phone: (212) 382-2169.

HANDY, WILLIAM CHRISTOPHER: BIRTH ANNIVERSARY. Nov 16. American composer, bandleader, "Father of the Blues," W.C. Handy was born at Florence, AL, Nov 16, 1873. He died at New York, NY, Mar 28, 1958.

◆ **HINDEMITH, PAUL: 100th BIRTH ANNIVERSARY.** Nov 16. Prolific composer and teacher, born at Hanau, Germany, Nov 16, 1895. Became a resident and citizen of the US during World War II. Died at Frankfurt, Germany, Dec 28, 1963.

NEW STAR RISING BLUEGRASS FESTIVAL. Nov 16–19. Spirit of The Suwannee, Live Oak, FL. A new concept in music festivals, this one features up and coming bands and new talent. For info: Jean Cornett, Rt 1, Box 98, Live Oak, FL 32060. Phone: (904) 364-1683.

OKLAHOMA: ADMISSION DAY: ANNIVERSARY. Nov 16. Became 46th state on this day in 1907.

RIEL, LOUIS: HANGING ANNIVERSARY. Nov 16. Born at St. Boniface, Manitoba, Canada, on Oct 23, 1844, Louis Riel, leader of the Metis (French/Indian mixed ancestry) was elected to Canada's House of Commons in 1873 and 1874, but never seated. Confined to asylums for madness (feigned or falsely charged, some said), Riel became a US citizen in 1883. In 1885 he returned to western Canada to lead the North West Rebellion. Defeated, he surrendered, was tried for treason, convicted and hanged, at Regina, NWT, Canada, on Nov 16, 1885. Seen as a true patriot and protector of French culture in Canada, Riel's life and death became a legend and a symbol of the problems between French and English Canadians.

ST. EUSTATIUS, WEST INDIES: STATIA AND AMERICA DAY. Nov 16. St. Eustatius, Leeward Islands. To commemorate the first salute to an American flag by a foreign government, from Fort Oranje in 1776. Festivities include sports events and dancing. During the American Revolution St. Eustatius was an important trading center and a supply base for the colonies.

SPACE MILESTONE: *SKYLAB 4* (US). Nov 16. Thirtieth manned US space flight was launched on Nov 16, 1973. Three astronauts, G.P. Carr, W.R. Page and E.G. Gibson. Space walks. Returned to Earth after 84 days, on Feb 8, 1974.

SPACE MILESTONE: *VENERA 3* (USSR). Nov 16. Launched Nov 16, 1965, it crashed into Venus, Mar 1, 1966. First manmade object on another planet.

UNIVERSITY OF CHICAGO'S FIRST FOOTBALL VICTORY: ANNIVERSARY. Nov 16. The University of Chicago, which played to a 0–0 tie with Northwestern in its first-ever football game, Oct 22, 1892, won its first game for Coach Amos Alonzo Stagg 10-4 against Illinois in Chicago on Nov 16, 1892. Formerly a founding member of the Big Ten, Chicago now competes in the University Athletic Association against the likes of New York University, Emory and Washington University in St. Louis.

Nov 16–17 ☆ *Chase's 1995 Calendar of Events* ☆

BIRTHDAYS TODAY

Lisa Bonet, 28, actress ("The Cosby Show," "A Different World," *Angel Heart*), born San Francisco, CA, Nov 16, 1967.
Elizabeth Drew, 60, journalist, born at Cincinnati, OH, Nov 16, 1935.
Bob Gibson, 64, singer (folk music), musician, born at New York, NY, Nov 16, 1931.
Dwight Gooden, 31, baseball player, born at Tampa, FL, Nov 16, 1964.
Harvey Martin, 45, former football player, born Dallas, TX, Nov 16, 1950.
Burgess Meredith, 86, actor (*Rocky*, "Gloria"; Emmy for *Tail Gunner Joe*), born Cleveland, OH, Nov 16, 1909.
Martine Van Hammel, 50, ballerina, born at Brussels, Belgium, Nov 16, 1945.
Jo Jo White, 49, former basketball player, born at St. Louis, MO, Nov 16, 1946.

NOVEMBER 17 — FRIDAY
321st Day — Remaining, 44

BUCK FEVER DAYS AND CHILI FEED. Nov 17–18. Monocqua, WI. Area businesses compete for "Chili Champ." Free chili welcoming hunters. For info: Al Handley, Minocqua Area Chamber of Commerce, PO Box 1006, Minocqua, WI 54548. Phone: (800) 446-6784.

ENGLAND: HENRY PURCELL TERCENTENARY TUDELEY. Nov 17–25. All Saints' Church, Tudeley, Kent. A series of concerts and events in celebration of the tercentenary of the death of Henry Purcell, who died Nov 25, 1695. For info: Stephen Coles, Artistic Dir, Tudeley Festival, Postern Park Oast, Tonbridge, Kent, England TN11 0QT.

HOLIDAY FOLK FAIR. Nov 17–19. MECCA, Milwaukee, WI. Multi-ethnic festival, featuring costumes, dancing, entertainment, exhibits, workshops, gifts and food from every corner of the globe. Annually, the weekend prior to Thanksgiving Day. Est attendance: 65,000. For info: Intl Institute of Wisconsin, Holiday Folk Fair, 1110 N Old World Third St, Milwaukee, WI 53203. Phone: (414) 225-6220.

HOMEMADE BREAD DAY. Nov 17. A day for the family to remember and enjoy the making, baking and eating of nutritious homemade bread. Sponsor: Homemade Bread Day Committee, Box 3, Montague, MI 49437.

LONGHORN WORLD CHAMPIONSHIP RODEO. Nov 17–19. Municipal Auditorium, Nashville, TN. More than 250 of North America's best cowboys and cowgirls compete in 6 professionally sanctioned contests ranging from bronc riding to bull riding for top prize money and world championship points. The theme of the production changes annually. Additional attractions include special animal trainers, trick ropers, trick riders and professional clowns. 30th annual. Est attendance: 18,000. For info: Longhorn World Chmpshp Rodeo, Inc, PO Box 70159, Nashville, TN 37207. Phone: (615) 876-1016.

MOBIUS, AUGUST: BIRTH ANNIVERSARY. Nov 17. German astronomer, mathematician, teacher and author, August Ferdinand Mobius, was born at Schulpforte, Germany, on Nov 17, 1790. Mobius was a pioneer in the field of topology, and first described the Mobius net and the Mobius strip. He died at Leipzig, on Sept 26, 1868.

MONTGOMERY, BERNARD LAW: BIRTH ANNIVERSARY. Nov 17. Bernard Law Montgomery, who commanded the British Eighth Army to victory at El Alamein in north Africa in 1943, was born Nov 17, 1887, at St. Mark's Vicarage, Kennington Oval, London, England. He also led the Eighth Army in the Sicilian and Italian campaigns and commanded all ground forces in the 1944 Normandy landing. Montgomery died Mar 24, 1976 at Alton, Hampshire, England.

NATIONAL BIBLE WEEK INAUGURAL LUNCHEON. Nov 17. (Friday before NBW) in New York, NY. Armed Service Chaplain honored with Witherspoon Chaplains Award. Information upon request. Est attendance: 500. Sponsor: Laymen's Natl Bible Assn, Inc, 1865 Broadway, 12th Fl, New York, NY 10023. Phone: (212) 408-1390.

★ **NATIONAL FARM-CITY WEEK.** Nov 17–23. Presidential Proclamation issued for a week in November since 1956, customarily for the week ending with Thanksgiving Day. Requested by congressional resolutions from 1956–1958; since 1959 issued annually without request.

PEALE, TITIAN RAMSEY: BIRTH ANNIVERSARY. Nov 17. American artist, naturalist, son of Charles Willson Peale, born at Philadelphia, PA, on Nov 17, 1799. Died there on Mar 13, 1885.

QUEEN ELIZABETH I: ASCENSION ANNIVERSARY. Nov 17. Anniversary of ascension of Elizabeth I to English throne, Nov 17, 1558; celebrated as a holiday in England for more than a century after her death in 1603.

SILVER BELLS IN THE CITY. Nov 17. Lansing, MI. Michigan's capital city sparkles with hospitality when more than 2,000 candles line the streets of downtown Lansing's business district for this celebration of lights, music and holiday cheer including the lighting of the State of Michigan Holiday Tree. Annually, the Friday before Thanksgiving. Coordinated by Arts Council Center of Greater Lansing, Inc. Est attendance: 20,000. For info: Arts Council Center of Greater Lansing, Inc., Center for the Arts, 425 S Grand Ave, Lansing, MI 48933. Phone: (517) 372-4636.

STAMP EXPO: FALL. Nov 17–19. Wilshire Ebell Convention Complex, Los Angeles, CA. Fax: (818) 988-4337. Est attendance: 4,000. Sponsor: Intl Stamp Collectors Society, PO Box 854, Van Nuys, CA 91408. Phone: (818) 997-6496.

SUEZ CANAL ANNIVERSARY. Nov 17. Formal opening of the Suez Canal, on Nov 17, 1869.

SUGARLOAF CRAFT FESTIVAL. Nov 17–19. Montgomery County Fairgrounds, Gaithersburg, MD. More than 480 professional artists and craftspeople, demonstrations, children's entertainment. Est attendance: 50,000. Sponsor: Sugarloaf Mt Works, Inc, 200 Orchard Ridge Dr, #215, Gaithersburg, MD 20878. Phone: (301) 990-1400.

TELLABRATION! AN EVENING OF STORYTELLING FOR GROWN-UPS. Nov 17–18. Many sites throughout Connecticut. Simultaneous storytelling concerts for adults. Annually, the weekend before Thanksgiving. Est attendance: 1,500. For info: Barbara Reed, Dir, Connecticut Storytelling Center, Dept of Education, Connecticut College, New London, CT 06320. Phone: (203) 439-2764.

ZENGER, JOHN PETER: ARREST ANNIVERSARY. Nov 17. Colonial printer and journalist who established the *New York Weekly Journal* (first issue, Nov 5, 1733). Zenger was arrested on Nov 17, 1734, for libels against the colonial governor, but continued to edit his newspaper from jail. Trial was held

November 1995

S	M	T	W	T	F	S
			1	2	3	4
5	6	7	8	9	10	11
12	13	14	15	16	17	18
19	20	21	22	23	24	25
26	27	28	29	30		

during August 1735. Zenger's acquittal was an important early step toward freedom of the press in America. Zenger was born in Germany in 1697, came to the US in 1710 and died July 28, 1746, at New York City.

BIRTHDAYS TODAY

Terry E. Branstad, 49, Governor of Iowa (R), born at Leland, IA, Nov 17, 1946.
Peter Cook, 58, comedian, actor (*Bedazzled*), author, born Devonshire, England, Nov 17, 1937.
Howard Dean, 47, Governor of Vermont (D), born at East Hampton, NY, Nov 17, 1948.
Danny Devito, 51, actor ("Taxi," *Twins, Batman 2*), director (*Throw Mama from the Train* and acted in), born Neptune, NJ, Nov 17, 1944.
Shelby Foote, 79, writer, historian (*The Civil War*), born Greenville, MS, Nov 17, 1916.
Lauren Hutton, 51, model, actress (*American Gigolo*), born Charleston, SC, Nov 17, 1944.
Gordon Lightfoot, 57, singer ("Sundown"), songwriter ("Early Morning Rain"), born Orillia, Ont, Canada, Nov 17, 1938.
Bob Mathias, 65, former track athlete, born at Tulare, CA, Nov 17, 1930.
Martin Scorsese, 53, director (*Mean Streets, Color of Money, Raging Bull, Goodfellas*), born Flushing, NY, Nov 17, 1942.
Tom Seaver, 51, baseball Hall of Famer, born at Fresno, CA, Nov 17, 1944.

NOVEMBER 18 — SATURDAY
322nd Day — Remaining, 43

ASSOCIATED PUBLISHERS, INC, FOUNDED: ANNIVERSARY CELEBRATION. Nov 18. Washington, DC. The Associated Publishers, Inc, was organized in Washington, DC, Nov 18, 1920, for the purpose of publishing books on black history and culture. The founders were Carter Godwin Woodson, John W. Davis and Louis R. Mehlinger. A celebration is held each year on the Saturday of the week that Nov 18 falls in, Nov 18 in 1995. The celebration includes workshops, exhibits and a luncheon. Est attendance: 350. Sponsor: Associated Publishers, Inc, W. Leanna Miles, 1407 14th St, NW, Washington, DC 20005-3704. Phone: (202) 265-1441.

CAROLINA MODEL RAILROADERS' OPEN HOUSE. Nov 18. (Also, Nov 25.) Greensboro, NC. Annual open house on two Saturdays from 10-4 at the old Southern Railway passenger station, corner of Washington and Church Streets, downtown Greensboro. This display features HO and N Scale trains running on the clubs' large layouts. Admission: $2.00 for adults with a maximum of $5.00 per family. Est attendance: 1,200. For info: Ben Stemkowski, PR, c/o Carolina Model Railroaders, PO Box 313, Jamestown, NC 27282. Phone: (910) 656-7968.

CHRISTMAS WALK. Nov 18-Dec 30. Pella Historical Village, Pella, IA. To celebrate an old-fashioned Christmas. Closed Sundays and Christmas Day. Est attendance: 1,000. For info: Pella Historical Society, 507 Franklin, Pella, IA 50219. Phone: (515) 628-2409.

COOSA RIVER CHRISTMAS. Nov 18. Rome, GA. Parade of pontoon boats decorated with Christmas scenes in lights. The night is topped off with a spectacular fireworks display. Annually, the Saturday before Thanksgiving. Est attendance: 45,000. For info: Coosa River Christmas, Johnette Chambers, PO Box 5823, Rome, GA 30162-5823. Phone: (706) 295-5576.

CUSTER STATE PARK BUFFALO AUCTION. Nov 18. Custer, SD. A live sale at 10:00 AM, MST, of 300-400 surplus buffalo (calves, yearlings, mature cows and two-year-old bulls). Est attendance: 600. For info: Ron Walker, Custer State Park, HC 83, Box 70, Custer, SD 57730. Phone: (605) 255-4515.

DAGUERRE, LOUIS JACQUES MANDE: BIRTH ANNIVERSARY. Nov 18. French tax collector, theatre scene-painter, physicist and inventor, was born at Cormeilles-en-

Parisis, France, on Nov 18, 1789. He is remembered for his invention of the daguerreotype photographic process—one of the earliest to permit a photographic image to be chemically fixed to provide a permanent picture. The process was presented to the French Academy of Science on Jan 7, 1839. Daguerre died near Paris, France, on July 10, 1851.

FESTIVAL OF TREES. Nov 18-26. (Closed Thanksgiving Day.) River Center, Davenport, IA. Outdoors, the lights of the season are spectacular and can be enjoyed from horse-drawn carriages. Indoors, hundreds of spectacular holiday creations for the home are on display at The River Center. Additional events include a holiday concert by the Quad City Symphony, Festival luncheons and a Holiday Hoedown. On Nov 18, The Festival of Trees Holiday Parade, the largest inflatable character balloon parade in the country. Est attendance: 57,000. For info: Sue Gerdes, Quad City Arts, 1715 Second Ave, Rock Island, IL 61201. Phone: (309) 793-1213.

GAINESVILLE DOWNTOWN FESTIVAL AND ART SHOW. Nov 18-19. (Tentative.) 14th annual festival, downtown, Gainesville, FL. 200 artists and craftsmen display work for purchase. Two full days of music from Plaza stage, children's activity area, food vendors, hands on art opportunities Est attendance: 80,000. Sponsor: City of Gainesville-Cultural Affairs, Downtown Redevelopment Agency, PO Box 490-Station 30, Gainesville, FL 32602. Phone: (904) 334-2197.

GALLI-CURCI, AMELITA: BIRTH ANNIVERSARY. Nov 18. Italian born operatic soprano, made US debut Nov 18, 1916, in Chicago, IL. Born at Milan, Italy, Nov 18, 1889. She died at La Jolla, CA, Nov 26, 1963.

GILBERT, SIR WILLIAM SCHWENCK: BIRTH ANNIVERSARY. Nov 18. English author of librettos for the famed Gilbert and Sullivan comic operas, born at London, England, Nov 18, 1836. Died May 29, 1911, as a result of a heart attack experienced while saving a woman from drowning.

GINGERBREAD VILLAGE. Nov 18-Dec 22. Mormon Pioneer Visitors Center, 34th & State St, Omaha, NE. View more than 100 gingerbread houses of every shape and size delightfully decorated for the holidays. Gingerbread houses given to charities before Christmas. Est attendance: 14,500. For info: Kenneth R. Barker, Gingerbread Village, 3215 State St, Omaha, NE 68112. Phone: (402) 453-9372.

HAITI: ARMY DAY. Nov 18. Commemorates the Battle of Vertieres, Nov 18, 1803, at which Haitians defeated the French.

HOLIDAY RHAPSODY IN LIGHTS. Nov 18-Jan 2. Northeast Community College, Norfolk, NE. Holiday light display features over 300,000 lights. 12 life size animated displays reminiscent of department store windows of the '40s and '50s. Scenes include: Candy Cane Express, Elves workshop, Reindeer house, Miracle on 34th Street, Santa's mailroom and others. Controlled by computer relays—there are 120 soldiers marching and 160 gingerbread men and women skating on the lake under their own London Bridge. Display activated between 5:30 PM and midnight each evening. For info: Holly Swanson, PO Box 386, Norfolk, NE 68702-0386. Phone: (402) 371-2932.

Nov 18 ☆ *Chase's 1995 Calendar of Events* ☆

HONDA STARLIGHT CELEBRATION. Nov 18–Jan 2. Bicentennial Commons Sawyer Point, Cincinnati, OH. This 6th annual free holiday lighting event features 10 large displays and more than 300,000 lights. The Kick-off is Nov 18 at 3 PM and includes live stage entertainment, carolers, roller blading, clowns, storytellers, magicians and an appearance by Santa Claus. Sponsors: Honda, Cincinnati Recreation Commission, Star 64 Television. Annually, the Saturday before Thanksgiving. Est attendance: 28,000. For info: Ginny Buzzell, STAR 64, 5177 Fishwick Dr, Cincinnati, OH 45216. Phone: (513) 641-4400.

JEWISH BOOK MONTH. Nov 18–Dec 17. To promote interest in Jewish books. Phone: (212) 532-4949 ext 297. Sponsor: Jewish Book Council, 15 E 26th St, New York, NY 10010.

LA POSADA de KINGSVILLE: A CELEBRATION OF LIGHTS. Nov 18. (Also Dec 2.) Kingsville, TX. Celebration of lights recaptures the joy and spirit of Christmas. Businesses and neighborhoods twinkle with lights from the third weekend in November through Dec 31. Events include a fiesta market in historic district, 5K jingle bell run, street dance, lighted procession of Mary and Joseph's search for an inn, special activities for children, nighttime parade with lighted floats and holiday music and more. Call for dates of specific events. Annually, the third Saturday in November and the first Saturday in December. Est attendance: 10,000. For info: Kingsville Visitor Center, 101 N 3rd, Box 1562, Kingsville, TX 78364-1562. Phone: (800) 333-5032.

LATVIA: INDEPENDENCE DAY. Nov 18. National holiday.

LOMBROSO, CESARE: BIRTH ANNIVERSARY. Nov 18. Italian founder of criminology, born at Verona, Italy, Nov 18, 1836. A professor of psychiatry, Lombroso believed that criminality could be identified with certain physical types of people. He died at Turin on Oct 19, 1909.

MERCER, JOHN HERNDON (JOHNNY): BIRTH ANNIVERSARY. Nov 18. American songwriter, singer, radio performer and actor, born Nov 18, 1909, at Savannah, GA. Johnny Mercer wrote lyrics (and often the music) for some of the great American popular music from the 1930s through the 1960s, including "Moon River," "Autumn Leaves," "One for My Baby," "Charade," "Satin Doll," "On the Achison, Topeka, and the Santa Fe," "You Must Have Been a Beautiful Baby," "Come Rain or Come Shine," "Hooray for Hollywood," "Jeepers Creepers" and countless more. Mercer died on June 25, 1976, and was buried at Savannah.

MICKEY MOUSE'S BIRTHDAY. Nov 18. The comical activities of squeaky-voiced Mickey Mouse first appeared Nov 18, 1928, on the screen of the Colony Theatre in New York City. The film, Walt Disney's "Steamboat Willie," was the first animated cartoon talking picture.

NORDIC YULE FEST. Nov 18–19. Nordic Heritage Museum, Seattle, WA. Annual Christmas fair featuring crafts, holiday foods, music and dance and children's activities. Lucia procession on Sunday afternoon. Est attendance: 4,500. For info: Nordic Heritage Museum, 3014 NW 67th St, Seattle, WA 98117. Phone: (206) 789-5707.

OMAN: NATIONAL HOLIDAY. Nov 18. Sultanate of Oman celebrates its national day.

PINCHBECK, CHRISTOPHER: DEATH ANNIVERSARY. Nov 18. English inventor, jeweler and clockmaker. Inventor of the copper and zinc alloy which looked like gold but became synonymous with cheapness. Noted manufacturer of automated musical clocks and instruments. Born at Clerkenwell, London, England, about 1670 (exact date unknown). Died at London, England, Nov 18, 1732.

RANCH HAND BREAKFAST. Nov 18. King Ranch, Kingsville, TX. A chuck wagon breakfast cooked and served outdoors at the world-famous King Ranch. See Longhorn cattle and real cowboys on horse back. Annually, the third Saturday in November. Est attendance: 5,000. For info: Kingsville Visitor Center, 101 N 3rd, Box 1562, Kingsville, TX 78364-1562. Phone: (800) 333-5032.

SHRINE OYSTER BOWL GAME. Nov 18. Foreman Field, Norfolk, VA. VMI vs Georgia Southern. Benefits the Shriners Hospitals for Crippled and Burned Children. Est attendance: 20,000. For info: Oyster Bowl Office, Box 11063, Norfolk, VA 23517.

SOUTHERN WILDLIFE FESTIVAL. Nov 18–19. Calhoun College, Decatur, AL. Event features original paintings, photography and competition in color and black and white. Vendors specialize in wildlife items and wildlife carvings. Annually, the third weekend of November. For info: Decatur Conv and Visitors Bureau, 719 6th Ave SE, PO Box 2349, Decatur, AL 35602. Phone: (205) 350-2028.

THAILAND: ELEPHANT ROUND-UP AT SURIN. Nov 18. Elephant demonstrations in morning, elephant races and tug-of-war between 100 men and one elephant. Observed since 1961 on third Saturday in November. Special trains from Bangkok on previous day.

TRY THIS ON—A HISTORY OF CLOTHING, GENDER AND POWER. Nov 18–Dec 31. Clarksville-Montgomery County History Museum, Clarksville, TN. A social history about clothing and appearance in the US over the past 200 years. See Jan 7 for full description. For info: Smithsonian Institution Traveling Exhibition Service, 1100 Jefferson Dr SW, Ste 3146, Washington, DC 20560. Phone: (202) 357-2700.

UNITED STATES UNIFORM TIME ZONE PLAN: ANNIVERSARY. Nov 18. Charles Ferdinand Dowd, a Connecticut school teacher and one of the early advocates of uniform time, proposed a time zone plan of the US (four zones of 15 degrees), which he and others persuaded the railroads to adopt and place in operation, on Nov 18, 1883. Info from Natl Bureau of Standards Monograph 155. See also: "US Standard Time Act: Anniversary" (Mar 19).

BIRTHDAYS TODAY

Margaret Eleanor Atwood, 56, author (*Cat's Eye, Dancing Girls & Other Stories*), born at Ottawa, Ontario, Canada, Nov 18, 1939.

Linda Evans, 53, actress (Crystal Carrington on "Dynasty"; "Bachelor Father"), born Hartford, CT, Nov 18, 1942.

Warren Moon, 39, football player, born Los Angeles, CA, Nov 18, 1956.

Alan Shepard, 72, former astronaut, first American in space, born at East Derry, NH, Nov 18, 1923.

Ted Stevens, 72, US Senator (R, Alaska), born at Indianapolis, IN, Nov 18, 1923.

Susan Sullivan, 51, actress ("It's a Living," "Falcon Crest"), born at New York, NY, Nov 18, 1944.

Brenda Vaccaro, 56, actress (*Cactus Flower, How Now Dow Jones, The Goodbye People*), born Brooklyn, NY, Nov 18, 1939.

November 1995

S	M	T	W	T	F	S
			1	2	3	4
5	6	7	8	9	10	11
12	13	14	15	16	17	18
19	20	21	22	23	24	25
26	27	28	29	30		

NOVEMBER 19 — SUNDAY
323rd Day — Remaining, 42

ALASCATTALO DAY. Nov 19. Anchorage, AK. To honor humor in general and Alaskan humor in particular. Event is named after "alascattalo," said to be the genetic cross between a moose and a walrus. Est attendance: 30. Sponsor: Parsnackle Press, Steven C. Levi, Ed, 8512 E Fourth Ave, Anchorage, AK 99504. Phone: (907) 337-2021.

BELIZE: GARIFUNA DAY. Nov 19. Public holiday celebrating the first arrival of Black Caribs from St. Vincent and Rotan to Southern Belize.

CAMPANELLA, ROY: BIRTH ANNIVERSARY. Nov 19. Roy Campanella, one of the first black major leaguers and a star of one of baseball's greatest teams, the Brooklyn Dodgers' "Boys of Summer," was born at Philadelphia, PA, Nov 19, 1921. Campy, as he was often called, was named the National League MVP three times in his 10 years of play, in 1951, 1953 and 1955. Campanella had his highest batting average in 1951 (.325) and in 1953 he established three single-season records for a catcher—most putouts (807), most home runs (41) and most runs batted in (142)—as well as having a batting average of .312. His career was cut short on Jan 28, 1958, when an automobile accident left him paralyzed. Campy gained even more fame after his accident as an inspiration and spokesman for the handicapped. He was named to the Baseball Hall of Fame in 1969. Roy Campanella died June 26, 1993, at Woodland Hills, CA.

CLARK, GEORGE ROGERS: BIRTH ANNIVERSARY. Nov 19. American soldier and frontiersman, born in Albemarle County, VA, on Nov 19, 1752. Died at Louisville, KY, on Feb 13, 1818.

EIGHTEENTH-CENTURY THRESHING DAY. Nov 19. McLean, VA. Help the farm family thresh wheat, make yeast cakes and celebrate the end of the season with period games and light refreshment. Est attendance: 500. Sponsor: Claude Moore Colonial Farm at Turkey Run, Gretchen Brodtman, 6310 Georgetown Pike, McLean, VA 22101. Phone: (703) 442-7557.

FIRST PRESIDENTIAL LIBRARY: ANNIVERSARY. Nov 19. On Nov 19, 1939, President Franklin D. Roosevelt laid the cornerstone for his presidential library at Hyde Park, NY. He donated the land, but public donations provided funds for the building which was dedicated on June 30, 1941.

GARFIELD, JAMES ABRAM: BIRTH ANNIVERSARY. Nov 19. The 20th president of the US (and the first left-handed president) was born at Orange, OH, Nov 19, 1831. Term of office: Mar 4–Sept 19, 1881. While walking into the Washington, DC, railway station on the morning of July 2, 1881, Garfield was shot by disappointed office seeker Charles J. Guiteau. He survived, in very weak condition, until Sept 19, 1881, when he succumbed to blood poisoning at Elberon, NJ (where he had been taken for recuperation). Guiteau was tried, convicted and, on June 30, 1882, hanged at the jail in Washington.

GERMANY: VOLKSTRAUERTAG. Nov 19. Memorial Day and national day of mourning in all German states. Observed on the Sunday before Totensonntag. See also: "Germany: Totensonntag" (Nov 26).

GETTYSBURG ADDRESS MEMORIAL CEREMONY. Nov 19. Gettysburg, PA. 132nd anniversary of Lincoln's Gettysburg Address is celebrated with brief memorial services at the Soldiers' National Monument in Gettysburg National Cemetery at 2 PM. Est attendance: 3,000. For info: Gettysburg Travel Council, 35 Carlisle St, Gettysburg, PA 17325. Phone: (717) 334-6274.

HAVE A BAD DAY DAY. Nov 19. For those who are filled with revulsion at being told endlessly to "have a nice day," this day is a brief respite. Store and business owners are to ask workers to tell customers to "have a bad day." Annually, Nov 19. Phone: (212) 388-8673 or (717) 274-8451. For info: Wellness Permission League, Thomas or Ruth Roy, 2105 Water St, Lebanon, PA 17046.

LINCOLN'S GETTYSBURG ADDRESS: ANNIVERSARY. Nov 19. On this day in 1863, 17 acres of the battlefield at Gettysburg, PA, were dedicated as a national cemetery. Noted orator Edward Everett spoke for two hours. The address that Lincoln delivered in less than two minutes was later recognized as one of the most eloquent of the English language. Five manuscript copies in Lincoln's hand survive, including the rough draft begun in ink at the Executive Mansion in Washington and concluded in pencil at Gettysburg on the morning of the dedication (kept at the Library of Congress).

LIVE TURKEY "OLIMPIKS." Nov 19. The INN on Lake Waramaug, New Preston, CT. Live turkeys in athletic competition of turkey skills: high jump, sprinting, fastest eater and fastest turkey through a zig-zag slalom course. Also contests for heaviest turkey and best-dressed turkey. Refreshments available. Est attendance: 1,500. For info: Nancy Conant, The INN on Lake Waramaug, 107 N Shore Rd, New Preston, CT 06777. Phone: (800) 525-3466.

MONACO: NATIONAL HOLIDAY. Nov 19.

MOTHER GOOSE PARADE. Nov 19. El Cajon, CA. "A gift to the children." Floats depict Mother Goose rhymes and fairy tales. Bands, equestrians and clowns. Traditionally, the Sunday before Thanksgiving. Est attendance: 250,000. For info: Mother Goose Parade Assn, Box 1155, El Cajon, CA 92022-1155. Phone: (619) 444-8712.

NATIONAL ADOPTION WEEK. Nov 19–25. To commemorate the success of three kinds of adoption—infant, special needs and intercountry—through a variety of special events. Annually, the week of Thanksgiving. Sponsor: Natl Council for Adoption, Paul Denhalter, Exec VP, 1930 17th St NW, Washington, DC 20009-6207. Phone: (202) 328-1200.

NATIONAL BIBLE SUNDAY. Nov 19. An ecumenical observance by churches across the country in celebration of the Bible. Annually, the Sunday before Thanksgiving. For info: American Bible Society, Mary Cook or Roy Bickley, Natl Program Development, 1865 Broadway, New York, NY 10023. Phone: (212) 408-1200.

NATIONAL BIBLE WEEK. Nov 19–26. An interfaith campaign to promote reading and study of the Bible. Resource packets available. Annually, from the Sunday preceding Thanksgiving to the following Sunday. Fax: (212) 408-1448. Sponsor: Laymen's Natl Bible Assn, Inc, 1865 Broadway, 12th Fl, New York, NY 10023. Phone: (212) 408-1390.

PUERTO RICO: DISCOVERY DAY: ANNIVERSARY. Nov 19. Public holiday. Columbus discovered Puerto Rico in 1493 on his second voyage to the New World.

◆ **SUFFRAGISTS' VOTING ATTEMPT: ANNIVERSARY.** Nov 19. Testing the wording of the 14th Amendment that says "no State shall make or enforce any law which shall abridge the privileges or immunities of citizens of the United States," 172 New Jersey suffragists, including 4 black women, attempted to vote in the presidential election on Nov 19, 1868. Denied, they cast their votes instead into a women's ballot box overseen by 84-year-old Quaker Margaret Pryer.

BIRTHDAYS TODAY

Dick Cavett, 59, entertainer (TV interview show), former comedy writer ("The Tonight Show"), born at Gibbon, NE, Nov 19, 1936.

Jodie Foster, 33, actress (Oscars *The Accused*, *Silence of the Lambs*; *Taxi Driver*; Joey in "Courtship of Eddie's Father"), director (*Little Man Tate*), born Los Angeles, CA, Nov 19, 1962.

Thomas R. Harkin, 56, US Senator (D, Iowa), born at Cumming, IA, Nov 19, 1939.
Jeane Kirkpatrick, 69, political scientist, former US United Nations ambassador, born at Duncan, OK, Nov 19, 1926.
Calvin Klein, 53, fashion designer (made bluejeans expensive), born at New York, NY, Nov 19, 1942.
Ahmad Rashad (Bobby Moore), 46, sportscaster, born at Portland, OR, Nov 19, 1949.
Tommy G. Thompson, 54, Governor of Wisconsin (R), born at Elroy, WI, Nov 19, 1941.
Ted Turner, 57, cable TV executive (CNN, TNN), born Cincinnati, OH, Nov 19, 1938.
Garrick Utley, 56, journalist, born at Chicago, IL, Nov 19, 1939.
John Francis Welch, Jr, 60, chairman and chief executive officer of General Electric Co, born at Salem, MA, Nov 19, 1935.

NOVEMBER 20 — MONDAY

324th Day — Remaining, 41

BELLO, ANDRES: BIRTH ANNIVERSARY. Nov 20. Venezuelan diplomat, author and humanist, was born at Caracas, Venezuela, Nov 20, 1781. Bello died at Santiago, Chile, Oct 15, 1865.

BILL OF RIGHTS: ANNIVERSARY OF FIRST STATE RATIFICATION. Nov 20. New Jersey on Nov 20, 1789, became the first state to ratify 10 of the 12 amendments to the US Constitution proposed by Congress on Sept 25. These 10 amendments came to be known as the Bill of Rights. See also: "First Proposed Amendments to the US Constitution: Anniversary" (Sept 25).

CHATTERTON, THOMAS: BIRTH ANNIVERSARY. Nov 20. English poet Thomas Chatterton was born at Bristol, England, on Nov 20, 1752, and killed himself at age 17 by taking arsenic in his London garret on the night of Aug 24, 1770. A gifted but lonely child, before he reached his teens Chatterton had created a fantasy poet-priest, Thomas Rowley, who lived in the 16th century. With his own pen, Chatterton created enough verses "by" Rowley to fill more than 600 printed pages. Chatterton's fantasy-forgery poems attracted little attention during his short life, but in spite of their youthful deficiencies they were later admired by Wordsworth, Coleridge, Shelley, Keats, Byron and many others. In addition, he became the subject of at least one play, an opera and a novel.

GOULD, CHESTER: BIRTH ANNIVERSARY. Nov 20. Creator of the comic strip "Dick Tracy," was born at Pawnee, OK, on Nov 20, 1900. Gould's "Dick Tracy," which first appeared Oct 4, 1931 in *The Detroit Daily Mirror*, later was syndicated in nearly 1,000 newspapers worldwide and it was said that Dick Tracy's name was more widely known than that of the president of the US. Among the villains over whom Dick Tracy triumphed were: Pruneface, Flattop, Flyface, Mole, and 88 Keys. Notable among the innovations appearing in the comic strip before they were used in real life were the two-way wrist radio and television and the use of closed-circuit TV lineups. Gould drew and wrote the comic strip himself from 1931 until his retirement in 1977. He died at Woodstock, IL, May 11, 1985.

HUBBLE, EDWIN POWELL: BIRTH ANNIVERSARY. Nov 20. American astronomer Edwin Hubble, was born at Marshfield, MO, on Nov 20, 1889. His discovery and development of the concept of an expanding universe had been described as the "most spectacular astronomical discovery" of the twentieth century. As a tribute, the Hubble Space Telescope, deployed Apr 25, 1990 from US Space Shuttle *Discovery*, was named for him. The Hubble Space Telescope, with a 240-centimeter mirror, was to allow man to see farther into space than he had ever seen from telescopes on Earth. Hubble died at San Marino, CA, on Sept 28, 1953.

INTERNATIONAL FESTIVAL OF LIGHTS. Nov 20–Dec 31. Throughout Greater Battle Creek, MI. Holiday lighting and entertainment festival featuring more than one million lights, a 100-unit parade and a wide variety of performing arts. Starts the Monday before Thanksgiving and ends New Year's Eve. Est Attendance: 800,000. For info: Jeffrey T. Nowicki, Special Projects Dir, Cereal City Dev Corp, 2 W Michigan Ave, Ste 208, Battle Creek, MI 49017. Phone: (616) 968-1515.

KENNEDY, ROBERT FRANCIS: 70th BIRTH ANNIVERSARY. Nov 20. US Senator and brother of John F. Kennedy (35th president), was born at Brookline, MA, Nov 20, 1925. An assassin shot him in Los Angeles, CA, June 5, 1968, while he was campaigning for the presidential nomination. He died on June 6, 1968. Sirhan Sirhan was convicted of his murder.

LAGERLOF, SELMA: BIRTH ANNIVERSARY. Nov 20. Swedish author, member of the Swedish Academy and the first woman to receive the Nobel Prize for literature (1909) was born in Sweden's Varmland Province on Nov 20, 1858. She died there Mar 16, 1940.

LAURIER, SIR WILFRED: BIRTH ANNIVERSARY. Nov 20. Canadian statesman (premier, 1896–1911) born Nov 20, 1841. Died Feb 17, 1919.

MARRIAGE OF ELIZABETH AND PHILIP: ANNIVERSARY. Nov 20. The Princess Elizabeth Alexandra Mary was wed to Philip Mountbatten on Nov 20, 1947. Elizabeth was the first child of the Duke and Duchess of York who became King George VI and Queen Elizabeth in 1936 upon the abdication of Edward VIII. Philip, the former Prince Philip of Greece, had become a British subject nine months earlier and the title Duke of Edinburgh was bestowed on him. The bride later became Elizabeth II, Queen of the United Kingdom of Great Britain and Northern Ireland and Head of the Commonwealth, upon the death of her father on Feb 6, 1952, her coronation taking place at Westminster Abbey on June 2, 1953.

MEXICO: 85th REVOLUTION ANNIVERSARY. Nov 20. Anniversary of the social revolution launched by Francisco I. Madero in 1910. National holiday.

TIERNEY, GENE: 75th BIRTH ANNIVERSARY. Nov 20. Known best for the title role in the film *Laura*, actress Gene Tierney was born on Nov 20, 1920, at Brooklyn, NY. Her other films include *Leave Her to Heaven, Belle Starr, Heaven Can Wait, A Bell for Adano, Dragonwyck, Advice and Consent* and her last film *The Pleasure Seekers*. She died on Nov 6, 1991.

UNITED NATIONS: AFRICA INDUSTRIALIZATION DAY. Nov 20. The General Assembly proclaimed this day for the purpose of mobilizing the commitment of the international community to the industrialization of the continent. (Res 44/237, Dec 22, 1989.) Info from: United Nations, Dept of Public Info, New York, NY 10017.

WOLCOTT, OLIVER: BIRTH ANNIVERSARY. Nov 20, 1726. Signer of the Declaration of Independence, Governor of Connecticut, born at Windsor, CT; died on Dec 1, 1797.

BIRTHDAYS TODAY

Joseph Robinette Biden, Jr, 53, US Senator (D, Delaware), born at Scranton, PA, Nov 20, 1942.
Robert C. Byrd, 78, US Senator (D, West Virginia) up for reelection in '94, born at North Wilkesboro, NC, Nov 20, 1917.

☆ Chase's 1995 Calendar of Events ☆ Nov 20-22

Steve Dahl, 41, Chicago radio personality, born La Canada, CA, Nov 20, 1954.

Richard Dawson, 63, actor, TV game show host ("Laugh-In"; Emmy "Family Feud"), born Gosport, England, Nov 20, 1932.

Bo Derek (Cathleen Collins), 39, actress (*10, Bolero, Tarzan, A Change of Season*), born Long Beach, CA, Nov 20, 1956.

Mark Gastineau, 39, former football player, born at Ardmore, OK, Nov 20, 1956.

Nadine Gordimer, 72, writer, (*July's People, Lifetimes Under Apartheid*), born at Springs, the Transvaal, South Africa, Nov 20, 1923.

Veronica Hamel, 52, actress (Joyce in "Hill Street Blues"), born Philadelphia, PA, Nov 20, 1943.

Ruth Laredo, 58, concert pianist, born at Detroit, MI, Nov 20, 1937.

Dick Smothers, 57, comedian (the one with bass fiddle), singer (social commentary in variety show led to being fired by CBS), born New York, NY, Nov 20, 1938.

Beryl Sprinkel, 72, economist, born at Richmond, MO, Nov 20, 1923.

David Walters, 44, Governor of Oklahoma (D), born at Canute, OK, November 20, 1951.

Judy Woodruff, 49, journalist, author, born at Tulsa, OK, Nov 20, 1946.

NOVEMBER 21 — TUESDAY
325th Day — Remaining, 40

ALTAMONT FAIR FESTIVAL OF LIGHTS. Nov 21–Jan 1. Altamont Fairgrounds, Altamont, NY. Drive-through holiday event featuring larger-than-life illuminated sculptures and fantasy caricatures. For info: Conv and Visitors Bureau, 52 S Pearl St, Albany, NY 12207. Phone: (800) 258-3582.

BEAUMONT, WILLIAM: BIRTH ANNIVERSARY. Nov 21. US Army surgeon whose contribution to classic medical literature and world fame resulted from another man's shotgun wound. When Canadian fur trapper Alexis St. Martin received an apparently mortal wound on June 6, 1822—a nearly point blank blast to the abdomen—Dr. Beaumont began observing his stomach and digestive processes through an opening in his abdominal wall. His findings were published in 1833 in *Experiments and Observations on the Gastric Juice and the Physiology of Digestion*. St. Martin returned to Canada in 1834 and resisted Beaumont's efforts to have him return for further study. He outlived his doctor by 20 years and was buried at a depth of eight feet to discourage any attempt at posthumous examination. Beaumont, born at Lebanon, CT, on Nov 21, 1785, died Apr 25, 1853, at St. Louis, MO.

FRENCHMAN ROWS ACROSS PACIFIC: ANNIVERSARY. Nov 21. Gerard d'Aboville completed a four-month solo journey across the Pacific Ocean on Nov 21, 1991. D'Aboville began rowing across the Pacific on July 11 when he left Choshi, Japan. His journey ended at Ilwaco, WA.

GREEN, HETTY: BIRTH ANNIVERSARY. Nov 21. Henrietta Howland Robinson Green, better known as Hetty Green, reported to have been the richest woman in America, was born at New Bedford, MA, Nov 21, 1835. She was an able financier who managed her own wealth, which was estimated to have been in excess of 100 million dollars. Died at New York, NY, July 3, 1916.

MAN'S FIRST FREE FLIGHT (BALLOON): ANNIVERSARY. Nov 21. Jean Francois Pilatre de Rozier and the Marquis Francois Laurent d'Arlandes became the first men to fly when they ascended in a Montgolfier hot air balloon at Paris, France, on Nov 21, 1783, less than six months after the first public balloon flight demonstration (June 5, 1783), and only a year after the first experiments with small paper and fabric balloons by the Montgolfier brothers, Joseph and Jacques, at Annonay, France, in November 1782. The first manned free flight lasted about 25 minutes and carried the passengers nearly six miles at a height of about 300 feet, over the city of Paris. Benjamin Franklin was one of the spectators at the flight.

NORTH CAROLINA: RATIFICATION DAY. Nov 21. Twelfth state to ratify Constitution, on this day in 1789.

POPE BENEDICT XV: BIRTH ANNIVERSARY. Nov 21. Giacomo dela Chiesa, 258th pope of the Roman Catholic Church, born at Pegli, Italy, Nov 21, 1854, elected pope Sept 3, 1914. Died Jan 22, 1922.

VOLTAIRE, JEAN FRANCOIS MARIE: BIRTH ANNIVERSARY. Nov 21. French author and philosopher to whom is attributed (perhaps erroneously) the statement: "I disapprove of what you say, but I will defend to the death your right to say it." Born Nov 21, 1694. Died May 30, 1778.

WORLD HELLO DAY. Nov 21. "Everyone who participates greets 10 people. People in 171 countries have participated in this annual activity for advancing peace through personal communication. Heads of state of 114 countries have expressed approval of the event." 23rd annual observance. Sponsor: The McCormack Brothers, Box 993, Omaha, NE 68101.

WORLD'S CHAMPIONSHIP DUCK CALLING CONTEST AND WINGS OVER THE PRAIRIE FESTIVAL. Nov 21–25. Stuttgart, AR. Annually, Thanksgiving weekend. Duck calling contests, duck gumbo cookoff, carnival, 10K race, arts and crafts, beauty pageant, concessions, sporting collectibles, Sportsman's Dinner and Dance, commercial exhibitors, fun shoot. Fax: (501) 673-1604. Est attendance: 60,000. For info: Stuttgart Chamber of Commerce, 507 S Main, Stuttgart, AR 72160. Phone: (501) 673-1602.

BIRTHDAYS TODAY

James (Anderson) DePreist, 59, conductor Oregon Symphony, born at Philadelphia, PA, Nov 21, 1936.

George Kenneth Griffey, Jr, 26, baseball player, born at Donora, PA, Nov 21, 1969.

Goldie Hawn, 50, actress ("Laugh-In"; Oscar *Cactus Flower*; *Private Benjamin*), born Washington, DC, Nov 21, 1945.

David Hemmings, 54, actor (*Blow Up, The Charge of the Light Brigade*), born Guildford, England, Nov 21, 1941.

Lorna Luft, 43, actress ("Trapper John, M.D."), daughter of Judy Garland, born Los Angeles, CA, Nov 21, 1952.

Natalia Makarova, 55, ballerina, born at Leningrad, USSR, Nov 21, 1940.

Stan Musial, 75, baseball Hall of Famer, born at Donora, PA, Nov 21, 1920.

Marlo Thomas, 57, actress ("That Girl"; author, *Free to be . . . You and Me*), born at Detroit, MI, Nov 21, 1938.

NOVEMBER 22 — WEDNESDAY
326th Day — Remaining, 39

BRITTEN, (EDWARD) BENJAMIN: BIRTH ANNIVERSARY. Nov 22. English composer born, Lowestoft, Suffolk, England, Nov 22, 1913. Lord Britten, Baron Britten of Aldeburgh, died at Aldeburgh, Dec 4, 1976.

CARMICHAEL, HOAGIE: BIRTH ANNIVERSARY. Nov 22. Hoagland Howard Carmichael, attorney who gave up the practice of law to become an actor and songwriter, was born at Bloomington, IN, Nov 22, 1899. Among his many popular songs: "Stardust," "Lazybones," "Two Sleepy People" and "Skylark." Carmichael died at Rancho Mirage, CA, Dec 27, 1981.

***CHINA CLIPPER* ANNIVERSARY.** Nov 22. A Pan American Martin 130 "flying boat" called the *China Clipper* began regular trans-Pacific mail service on Nov 22, 1935. The plane, powered by four Pratt and Whitney Twin Wasp engines and piloted by Captain Edwin Musick, took off from San Francisco. It reached Manila, Philippines, 59 hours and 48 minutes later. About 20,000 persons watched the historic take-off. Commercial passenger service was established the following year (Oct 21, 1936).

Nov 22 ☆ *Chase's 1995 Calendar of Events* ☆

De GAULLE, CHARLES ANDRE MARIE: BIRTH ANNIVERSARY. Nov 22. President of France from December 1958 until his resignation in April 1969, Charles de Gaulle was born at Lille, France, Nov 22, 1890. A military leader, he wrote *The Army of the Future* (1934) in which he predicted just the type of armored warfare that was used against his country by Nazi Germany in WWII. After France's defeat at the hands of the Germans, he declared the existence of "Free France" and made himself head of that organization. When the French Vichy government began to collaborate openly with the Germans, the French citizenry looked to De Gaulle for leadership. His greatest moment of triumph was when he entered liberated Paris on Aug 26, 1944. De Gaulle died at Colombey-les-Deux-Eglises, France, Nov 19, 1970.

ELIOT, GEORGE: BIRTH ANNIVERSARY. Nov 22. English novelist George Eliot, whose real name was Mary Ann Evans, was born at Chilvers Coton, Warwickshire, England, on Nov 22, 1819. She died at Chelsea, Dec 22, 1880.

GARNER, JOHN NANCE: BIRTH ANNIVERSARY. Nov 22. Thirty-second vice president of US (1933–1941) born at Red River County, TX, Nov 22, 1868. Died at Uvalde, TX, Nov 7, 1967.

GERMANY: BUSS UND BETTAG. Nov 22. Buss und Bettag (Repentance Day) is observed on the Wednesday before the last Sunday of the church year. A legal public holiday in all German states except Bavaria (where it is observed only in communities with predominantly Protestant population).

GREAT ALASKA SHOOTOUT. Nov 22–25. Sullivan Area, Anchorage, AK. Top NCAA basketball action as eight men's division I teams from around the country compete. Est attendance: 35,000. For info: Univ of Alaska, Anchorage, Athletic Dept, 3211 Providence Dr, Anchorage, AK 99508. Phone: (907) 786-1230.

KENNEDY, JOHN F.: ASSASSINATION ANNIVERSARY. Nov 22. President John F. Kennedy was slain by a sniper while riding in an open automobile at Dallas, TX, Nov 22, 1963. Accused Lee Harvey Oswald was killed in police custody awaiting trial as assassin.

LEBANON: INDEPENDENCE DAY. Nov 22. National Day.

MOON PHASE: NEW MOON. Nov 22. Moon enters New Moon phase at 10:43 AM, EST.

NATIONAL STOP THE VIOLENCE DAY. Nov 22. Radio and television stations across the nation are encouraged to promote "Peace on the Streets" and help put an end to gang (and other) violence through Stop the Violence Day. Participating stations unite to call for a one-day cease fire, the idea being, "If we can stop the violence for one day, we can stop the violence everyday, one day at a time." Stations also encourage listeners/viewers to wear and display white ribbons that day and drive with their headlights on as show of peace. Many stations hold peace rallies with local community leaders and also conduct a moment of silence on the air in honor of the year's victims of violence. Begun in 1990. Annually, on the anniversary of President John F. Kennedy's assassination. For info: Cliff Berkowitz, Pres, Promotional Rescue, Paradigm Radio, 309 O ST, Eureka, CA 95501. Phone: (707) 443-9842.

POST, WILEY: BIRTH ANNIVERSARY. Nov 22. Barnstorming aviator, stunt parachutist and adventurer, Wiley Post was born at Grand Plain, TX, on Nov 22, 1898. Post, who taught himself to fly, and his plane, the *Winnie Mae*, were the center of world attention in the 1930s. He was co-author (with his navigator, Harold Gatty) of *Around the World in Eight Days*. In 1935, Post and friend Will Rogers started on flight to the Orient. Plane crashed near Point Barrow, AK, on Aug 15, 1935; both were killed.

ST. CECILIA: FEAST DAY. Nov 22. Roman virgin, Christian martyr and patron of music and musicians lived during third century. Survived sentences of burning and beheading. Subject of poetry and musical compositions and her feast day is still an occasion for musical events.

SCOTLAND: SCOTTISH OPEN BADMINTON CHAMPIONSHIP. Nov 22–26. Kelvin Hall International Sports Arena, Argyle Street, Glasgow. An International Badminton Federation 2 Star Grand Prix International Tournament. Prize fund $35,000. Est attendance: 7,000. For info: Scottish Badminton Union, Cockburn Centre, 40 Bogmoor Pl, Glasgow, Scotland G51 4TQ.

TEXAS AGGIE BONFIRE. Nov 22. Polo field, Texas A&M University, College Station, TX. Bonfire is held each year before the Texas A&M University vs. University of Texas at Austin football game. Construction of the bonfire, which traditionally stands some 55 feet tall and has a circumference of more than 135 feet, begins in October. The bonfire tradition at Texas A&M dates back to 1909. Est attendance: 40,000. For info: Dept of Student Actvities, Texas A&M University, College Station, TX 77843-1236.

WONDERLAND OF LIGHTS. Nov 22–Dec 30. Marshall, TX. More than 6 million tiny white lights cover the city. Features the living Christmas tree, JC's lighted Christmas parade, candlelight home tours. Live entertainment in square on Fridays and Saturdays. Annually, from Thanksgiving to New Year's Day. Est attendance: 750,000. For info: Pam Whisenant, Dir of Conv & Visitors Dvmt, Greater Marshall Chamber of Commerce, PO Box 520, Marshall, TX 75671. Phone: (903) 935-7868.

BIRTHDAYS TODAY

Boris Becker, 28, tennis player, born at Liemen, Germany, Nov 22, 1967.

Tom Conti, 54, actor, born at Paisley, Scotland, Nov 22, 1941.

Jamie Lee Curtis, 37, actress (Hannah on "Anything But Love"; *Love Letters, A Fish Named Wanda*), born Los Angeles, CA, Nov 22, 1958.

Rodney Dangerfield (Jacob Cohen), 74, actor, comedian (*Easy Money, Back to School, Caddyshack*, "The Dean Martin Show"), born Babylon, NY, Nov 22, 1921.

Harry Edwards, 53, sports sociologist, born at St. Louis, MO, Nov 22, 1942.

Terry Gilliam, 55, actor, writer ("Monty Python's Flying Circus," *The Life of Brian*), director (*Brazil*), born Minneapolis, MN, Nov 22, 1940.

Lew Hays, 81, one of the founders of PONY League baseball for 13- and 14-year-olds, born at Butler, PA, Nov 22, 1914.

Mariel Hemingway, 34, actress (*Manhattan, Personal Best, Superman IV*; caused controversy by kissing Roseanne on TV), born Ketchum, ID, Nov 22, 1961.

Billy Jean King, 52, tennis player (beat Bobby Riggs in the battle of the sexes), born at Long Beach, CA, Nov 22, 1943.

Claiborne Pell, 77, US Senator (D, Rhode Island), born at New York, NY, Nov 22, 1918.

Robert Vaughn, 63, actor ("The Man From U.N.C.L.E.," *The Magnificent Seven*), born New York, NY, Nov 22, 1932.

November 1995

S	M	T	W	T	F	S
			1	2	3	4
5	6	7	8	9	10	11
12	13	14	15	16	17	18
19	20	21	22	23	24	25
26	27	28	29	30		

NOVEMBER 23 — THURSDAY
327th Day — Remaining, 38

AMERICAN BICYCLE ASSOCIATION GRAND NATIONALS. Nov 23-28. Myriad Convention Center, Oklahoma City, OK. This World Championship BMX (bike motocross) competition draws more than 10,000 contestants and out-of-state spectators, from at least 15 countries, making it the largest out-of-state visitor attraction in the nation to be held during the Thanksgiving holidays. For info: OK City All Sports Assn, 100 W Main, Ste 287, Oklahoma City, OK 73102. Phone: (405) 236-5000.

ASHFORD, EMMETT LITTLETON: BIRTH ANNIVERSARY. Nov 23. Emmett Littleton Ashford, born Nov 23, 1914, at Los Angeles, CA, was the first black to officiate at a major league baseball game. Ashford began his pro career calling games in the minors in 1951 and went to the majors in 1966. He was noted for his flamboyant style when calling strikes and outs as well as for his dapper dress which included cufflinks with his uniform. He died Mar 1, 1980, at Marina del Rey, CA.

ATLANTA MARATHON AND HALF MARATHON. Nov 23. Atlanta, GA. 26.2-mile and 13.1-mile races. Advance registration only; entry forms available in July. Send SASE. Est attendance: 8,000. For info: Atlanta Track Club, 3097 E Shadowlawn Ave, Atlanta, GA 30305. Phone: (404) 231-9065.

BILLY THE KID: BIRTH ANNIVERSARY. Nov 23. Legendary outlaw of western US. Probably named Henry McCarty at birth, at New York, NY, Nov 23, 1859, he was better known as William H. Bonney. Ruthless killer, a failure at everything legal, at age 21, he escaped from jail while under sentence of hanging. Recaptured at Stinking Springs, NM, and returned to jail, he again escaped, only to be shot through heart by pursuing Lincoln County Sheriff Pat Garrett, at Fort Sumner, NM, during night of July 14, 1881. His last words, answered by two shots, reportedly were "Who is there?"

CHANNEL 6 THANKSGIVING DAY PARADE. Nov 23. Philadelphia, PA. Est attendance: 300,000. For info: Valerie Lagauskas, Parade Dir, WPVI-TV, 4100 City Line Ave, Philadelphia, PA 19131. Phone: (215) 581-4529.

CHRISTMAS AT THE CAPITOL. Nov 23-Dec 27. State Capitol Rotunda, Pierre, SD. More than 50 SD evergreens are decorated with handcrafted ornaments, each with its own theme. Est attendance: 50,000. For info: Dottie J Howe, 114 Lee Hill Rd, Pierre, SD 57501. Phone: (605) 224-2048.

"DR. WHO" PREMIERE: ANNIVERSARY. Nov 23, 1963. First episode of "Dr. Who" premiered on British TV with William Hartnell as the first doctor. Traveling through time and space in the TARDIS (an acronym for Time and Relative Dimensions in Space), the doctor and his companions found themselves in mortal combat with creatures such as the Daleks. "Dr. Who" didn't air in the U.S. until almost 10 years later.

FIRST PLAY-BY-PLAY FOOTBALL GAME BROADCAST: ANNIVERSARY. Nov 23. The first play-by-play football game broadcast in the United States took place on this day in 1919. Texas A&M–7, University of Texas–0.

FOODS AND FEASTS IN 17th-CENTURY VIRGINIA. Nov 23-25. Jamestown Settlement, Williamsburg, VA. Daily activities devoted to the foods of the Jamestown colonists and their Powhatan Indian neighbors. Est attendance: 4,500. For info: Media Relations, Jamestown-Yorktown Fdtn, Box JF, Williamsburg, VA 23187. Phone: (804) 253-4838.

JAPAN: LABOR THANKSGIVING DAY. Nov 23. National holiday.

LONGWOOD GARDENS CHRISTMAS DISPLAY. Nov 23-Jan 1. Kennett Sq, PA. Indoor conservatory display of 2,300 poinsettias, decorated trees, daily concerts. Outdoors, 200,000 lights and holiday fountain displays. Est attendance: 150,000. For info: Colvin Randall, PR Mgr, Longwood Gardens, Kennett Square, PA 19348-0501. Phone: (610) 988-6741.

MACY'S THANKSGIVING DAY PARADE. Nov 23. New York, NY. Starts at 9 AM EST in Central Park West. A part of everyone's Thanksgiving, the parade grows bigger and better each year. Featuring floats, giant balloons, marching bands and famous stars, the parade is televised for the whole country. 69th annual parade. For info: New York Conv/Vis Bureau, Macy's Thanksgiving Day Parade, 2 Columbus Circle, New York, NY 10019. Phone: (212) 397-8222.

MARX, HARPO: BIRTH ANNIVERSARY. Nov 23. Harpo (Adolph Arthur) Marx was born at New York, NY, on Nov 23, 1893. He was the second born of the famed Marx brothers who became a popular comedy team of stage, screen and radio for 30 years. Harpo wore a blond curly wig and pretended to be a mute who communicated by honking a horn. He was an expert player of the harp, though the comedy he brought to his playing tended to distract the viewer from noticing the quality of his musicianship. He died on Sept 28, 1964, at Hollywood, CA. Other family members who participated in the comedy team were Groucho (Julius), Chico (Leonard) and, briefly, Zeppo (Herbert) and Gummo (Milton).

MICHIGAN THANKSGIVING PARADE. Nov 23. Woodward Ave, Detroit, MI. The 68th annual parade kicks off the holiday season with nearly 100 units marching. Annually, on Thanksgiving morning. Est attendance: 1,200,000. For info: Iano Guida, Dir PR, The Parade Co, 9600 Mt Elliott, Detroit, MI 48211. Phone: (313) 923-7400.

OLD TYME FARMERS' DAYS. Nov 23-25. Spirit of the Suwannee, Live Oak, FL. Thanksgiving dinner, cane grinding, Old Tyme Farm contests, mule contests, covered wagon and buggy rides. Special bluegrass concert. For info: Jean Cornett, Rt 1 Box 98, Live Oak, FL 32060. Phone: (904) 364-1683.

PIERCE, FRANKLIN: BIRTH ANNIVERSARY. Nov 23. The 14th president of the US, was born at Hillsboro, NH, Nov 23, 1804. Term of office: Mar 4, 1853-Mar 3, 1857. Not nominated until the 49th ballot at the Democratic party convention in 1852, he was refused his party's nomination, in 1856, for a second term. Pierce died at Concord, NH, Oct 8, 1869.

POARCH CREEK INDIAN THANKSGIVING DAY POWWOW AND RUN/WALK. Nov 23. Atmore, AL. A day-long powwow filled with dance competitions and dance exhibitions by Indian dance teams from throughout the southeast. Princess contests, more than 100 craft booths, games and plenty of food. The Poarch Creek Indians extend the hand of friendship to share the rich American Indian culture with the general public. Also, a 5K Run/Walk over the scenic reservation is held; includes 11 age categories with each entrant receiving a specially designed T-shirt. Unique awards given to top 3 male and female winners and to the top 3 winners in each category. Annually, on Thanksgiving Day. Est attendance: 20,000. Sponsor: Poarch Creek Indians, Susan Wicker, Community Relations Coord, HCR 69A, Box 85B, Atmore, AL 36502. Phone: (205) 368-9136.

RUTLEDGE, EDWARD: BIRTH ANNIVERSARY. Nov 23. Signer of the Declaration of Independence, governor of South Carolina, born at Charleston, SC, Nov 23, 1749. Died there Jan 23, 1800.

SAGITTARIUS, THE ARCHER. Nov 23-Dec 21. In the astronomical/astrological zodiac that divides the sun's apparent orbit into 12 segments, the period Nov 22-Dec 21 is identified, traditionally, as the sun-sign of Sagittarius, the Archer. The ruling planet is Jupiter.

STOCK EXCHANGES HOLIDAY (THANKSGIVING DAY). Nov 23. The holiday schedules for the various exchanges are subject to change if relevant rules, regulations or exchange policies are revised. If you have questions, phone: American Stock Exchange (212) 306-1212; Chicago Board of Options Exchange (312) 786-7760; Chicago Board of Trade (312) 435-3500; New York Stock Exchange (212) 656-2065; Pacific Stock Exchange (415) 393-4000; Philadelphia Stock Exchange (215) 496-5000.

★ **THANKSGIVING DAY.** Nov 23. Presidential Proclamation. Always issued for the fourth Thursday in November. See also: "Anniversary of First US Holiday by Presidential Proclamation" (Nov 26).

THANKSGIVING DAY. Nov 23. Legal public holiday. (Public Law 90-363 sets Thanksgiving Day on the fourth Thursday in November). Observed on this day in all states.

THANKSGIVING HUNT WEEKEND (WITH BLESSING OF THE HOUNDS). Nov 23-25. Cismont, VA. "Blessing of the Hounds," Thanksgiving feast, site seeing (historic sites), ballooning, tours of vineyards, afternoon Foot Hunt and Hunt Tea are only part of the traditional Virginia Thanksgiving celebration. For info: The Boar's Head Inn and Sports Club, PO Box 5307, Charlottesville, VA 22905. Phone: (804) 476-1988.

TURKEY TROT. Nov 23. Parkersburg City Park Pavilion, Parkersburg, WV. Three-mile fun run/walk on Thanksgiving morning. Drawings held for frozen turkeys and each participant receives a long-sleeved T-shirt. Est attendance: 650. For info: St. Joseph's Hospital, Sports Medicine Dept, PO Box 327, Parkersburg, WV 26102. Phone: (304) 424-4393.

BIRTHDAYS TODAY

Jerry Bock, 67, composer, born at New Haven, CT, Nov 23, 1928.
Ellen Drew, 80, actress (*Hollywood Boulevard*), born Kansas City, MO, Nov 23, 1915.
Krzysztof Penderecki, 62, composer, born at Debica, Poland, Nov 23, 1933.
Andrew Toney, 38, former basketball player, born at Birmingham, AL, Nov 23, 1957.
Maurice Zolotow, 82, author (*Billy Wilder in Hollywood*), born New York, NY, Nov 23, 1913.

NOVEMBER 24 — FRIDAY

328th Day — Remaining, 37

BARKLEY, ALBEN WILLIAM: BIRTH ANNIVERSARY. Nov 24. Thirty-fifth vice president of the US (1949-1953), born at Graves County, KY, Nov 24, 1877. Died at Lexington, VA, Apr 30, 1956.

BATTLE OF CHATTANOOGA: ANNIVERSARY. Nov 24. After reinforcing the besieged Union army at Chattanooga, TN, General Ulysess S. Grant launched the Battle of Chattanooga on Nov 24, 1863. Falsely secure in the knowledge that his troops were in an impregnable position on Lookout Mountain, Confederate General Braxton Bragg and his army were overrun by the Union forces, Bragg himself barely escaping capture. The battle is famous for the Union Army's spectacular advance up a heavily fortified slope into the teeth of the enemy guns.

BELSNICKEL CRAFT SHOW. Nov 24-25. Boyertown, PA. Sale of quality crafts with folk art emphasis. Annually, the first Friday and Saturday after Thanksgiving. Est attendance: 4,700. For info: Boyertown Area Historical Society, Holly K. Green, 341 W Philadelphia Ave, Boyertown, PA 19512. Phone: (610) 367-9843.

BLACK FRIDAY. Nov 24. The traditional beginning of the Christmas shopping season on the Friday following Thanksgiving.

CANADA: WINTERGREEN '95. Nov 24-26. Regina, Saskatchewan. Craft market features the finest in Saskatchewan crafts. Est attendance: 7,000. For info: Tourism Saskatchewan, Saskatchewan Trade and Conv Center, 1919 Saskatchewan Dr, Regina, Sask, Canada S4P 3V7. Phone: (800) 667-7191.

CARNEGIE, DALE: BIRTH ANNIVERSARY. Nov 24. American inspirational lecturer and author, Dale Carnegie was born at Maryville, MO, Nov 24, 1888. His best known book, *How to Win Friends and Influence People*, published in 1936, sold nearly five million copies and was translated into 29 languages. Carnegie died at New York, NY, Nov 1, 1955.

CAROLINA CRAFTSMEN'S CHRISTMAS CLASSIC. Nov 24-26. State Fairgrounds, Columbia, SC. Arts and crafts. Est attendance: 35,000. For info: The Carolina Craftsmen, 1240 Oakland Ave, Greensboro, NC 27403. Phone: (910) 274-5550.

CAROLINA CRAFTSMEN'S CHRISTMAS CLASSIC. Nov 24-26. Coliseum Exhibition Building, Greensboro, NC. Arts and crafts. Est attendance: 35,000. For info: The Carolina Craftsmen, 1240 Oakland Ave, Greensboro, NC 27403. Phone: (910) 974-5550.

A CHARLES DICKENS VICTORIAN CHRISTMAS. Nov 24-26. (Also Dec 1-3, 8-10, 15-17, 22-24 and 26-27.) Mount Hope Estate and Winery, Cornwall, PA. An open house in colorfully decorated Mount Hope Mansion, wine sampling in the billiards room, actors portraying such Dickens favorites as Tiny Tim, Oliver Twist and Ebeneezer Scrooge. Est attendance: 7,000. For info: Mount Hope Estate and Winery, PO Box 685, Cornwall, PA 17016. Phone: (717) 665-7021.

"D.B. COOPER" HIJACKING: ANNIVERSARY. Nov 24-25. A middle-aged man whose plane ticket was made out to "D.B. Cooper" parachuted from a Northwest Airlines 727 jetliner on Nov 25, 1971, carrying $200,000 which he had collected from the airline as ransom for the plane and passengers as a result of threats he made during his Nov 24th flight from Portland, OR, to Seattle, WA. He jumped from the plane over an area of wilderness south of Seattle, and was never apprehended. Several thousand dollars of the marked ransom money turned up in February 1980, along the Columbia River, near Vancouver, WA.

DAYTONA BEACH FALL '95 SPEEDWAY SPECTACULAR. Nov 24-26. Daytona International Speedway, Daytona Beach, FL. Featuring the annual Turkey Rod Run car show of all makes and models collector vehicles. Show includes display of classics, sportscars, muscle cars, race cars, custom and special interest vehicles on the speedway infield all three days with a large swap meet of auto parts and accessories and car sales corral. Also crafts sale. Annually, Thanksgiving weekend. Est attendance: 50,000. For info: Daytona Beach Racing and Recreational Facilities District, Rick D'Louhy, Exec Dir, PO Box 1958, Daytona Beach, FL 32115-1958. Phone: (904) 255-7355.

DICKENS FESTIVAL. Nov 24-Dec 9. Utah State Fairpark, Salt Lake City, UT. Christmas entertainment, food, booths and Father Christmas in an old English setting with period costumes. Annually, the Friday after Thanksgiving. Est attendance:

November 1995

S	M	T	W	T	F	S
			1	2	3	4
5	6	7	8	9	10	11
12	13	14	15	16	17	18
19	20	21	22	23	24	25
26	27	28	29	30		

☆ Chase's 1995 Calendar of Events ☆   Nov 24

75,000. For info: Utah State Fairpark, 155 North 1000 West, Salt Lake City, UT 84116. Phone: (801) 538-8440.

DICKENS OLDE FASHIONED CHRISTMAS FESTIVAL. Nov 24-26. Holly, MI. (Also, Dec 2-3, 9-10, 16-17, 23-24.) Circa 1850 comes to life in downtown Holly. Bah humbug with Scrooge, encourage Tiny Tim, sing with the carolers, banter with the street vendors. Enjoy delicacies like roasted chestnuts, open-flame baked potatoes and plum pudding. Children's fantasy parades daily and entertainment on the hour. Annually, Thanksgiving weekend and each Saturday and Sunday until Christmas. Est attendance: 12,000. For info: Holly Area Chamber of Commerce, PO Box 214, Holly, MI 48442. Phone: (810) 634-1900.

FESTIVAL OF CAROLS. Nov 24. Kansas Capitol Grounds, Topeka, KS. The ringing in of the Christmas season. Musical program, arrival of Santa Claus, lighting of Christmas decorations, hayrides, ethnic dances and music. Annually, the Friday after Thanksgiving beginning at 4 PM. Est attendance: 4,250. For info: Marsha J. Sheahan, VP PR, Greater Topeka Chamber of Commerce, 120 SE 6th Ave, Ste 110, Topeka, KS 66603-3515. Phone: (913) 234-2644.

FESTIVAL OF LIGHTS. Nov 24-Jan 3. Cincinnati, OH. Exquisite lighting, entertainment, Santa Claus and carolers. Also, skating outside on Fountain Square under the Holiday tree. For info: Downtown Council, 300 Carew Tower, 441 Vine, Cincinnati, OH 45202. Phone: (513) 579-3191.

FINGER LAKES CRAFTSMEN THANKSGIVING WEEKEND CHRISTMAS ARTS AND CRAFTS SHOW. Nov 24-26. Monroe County Fairgrounds Dome Arena, Rochester, NY. All media, all categories including photography and prints. 26th annual show. Est attendance: 16,000. For info: Finger Lakes Craftsmen Show, Ronald L. Johnson, 1 Freshour Dr, Shortsville, NY 14548. Phone: (716) 289-9439.

FISH HOUSE PARADE. Nov 24. Aitkin, MN. 5th annual fancy parade of different fish houses used for ice fishing during the winter. Annually, the Friday after Thanksgiving. Est attendance: 2,000. For info: Carroll Kukowski, Exec Dir, Aitkin Area Chamber of Commerce, PO Box 127, Aitkin, MN 56431. Phone: (800) 526-8342.

FOLKWAYS OF CHRISTMAS. Nov 24-Dec 23. Also Dec 27-29. Shakopee, MN. Ethnic holiday traditions, presented by costumed interpreters, at historic Murphy's Landing. Open weekdays by reservation, 10-5 pm. Annually, beginning the Friday after Thanksgiving. Est attendance: 7,350. For info: Murphy's Landing, 2187 E Hwy 101, Shakopee, MN 55379. Phone: (612) 220-3988.

GROSSE POINTE SANTA CLAUS PARADE. Nov 24. Grosse Pointe, MI. 19th annual parade includes more than 100 units with 1,500 participants. Annually, the day after Thanksgiving. Est attendance: 10,000. For info: Jane Coté, 20913 Hawthorne, Harper Woods, MI 48225.

HOLIDAY CRAFT AND GIFT SHOW. Nov 24-26. Wisconsin State Fair Park, Milwaukee, WI. Combined show with 200 commercial gift exhibitors and 400 craftsmen. Est attendance: 48,000. For info: Dennis R. Hill, Dir, 3233 S Villa Circle, Milwaukee, WI 53227. Phone: (414) 321-2100.

JOPLIN, SCOTT: BIRTH ANNIVERSARY. Nov 24. American musician and composer famed for his piano rags, born at Texarkana, TX, Nov 24, 1868. Died at New York, NY, Apr 1, 1917.

JULE FEST. Nov 24-26. Elk Horn, IA. Danish Christmas festival. Est attendance: 3,000. For info: Danish Windmill, Lisa Riggs, PO Box 245, Elk Horn, IA 51531. Phone: (712) 764-7472.

MANSTEIN, ERICH von: BIRTH ANNIVERSARY. Nov 24. Considered by many to be the greatest strategist of World War II, Erich von Manstein was born on Nov 24, 1887, at Berlin, Germany. His plan for the invasion of France in 1940 was a complete success. He was dismissed by Hitler in March 1944. Manstein died at Irschenhausen, Germany, on June 10, 1973.

MOUNTAIN MAN RENDEZVOUS. Nov 24-26. Felton, CA. A living history demonstration of fur trappers and traders of the American West during the 1830s and 1840s. Native American dances, axe throwing and authentic teepees in a mountain meadow. For info: Jeanette Guire, Roaring Camp and Big Trees Railroad, Box G-1, Felton, CA 95018. Phone: (408) 335-4484.

NATIONAL GAME AND PUZZLE WEEK. Nov 24-30. To increase appreciation of games and puzzles while conserving the tradition of investing time with family and friends. Annually, the last week in November. Sponsor: Tim "Game Boy" Walsh, Patch Products, PO Box 268, Beloit, WI 53512-0268. Phone: (608) 362-6896.

NATIONAL "RED KETTLE" KICK OFF FOR THE SALVATION ARMY'S CHRISTMAS SEASON. Nov 24. Thousands of volunteers will man these collection kettles to fund holiday relief efforts for more than seven million needy Americans. Sponsor: The Salvation Army, Natl HQ, 615 Slaters Lane, Alexandria, VA 22313. Phone: (703) 684-5500.

NEW ENGLAND FIELD DAYS. Nov 24-26. Mystic Seaport, Mystic, CT. Families enjoy special Thanksgiving weekend activities including wagon rides, food, entertainment and outdoor games on the green. Est attendance: 3,000. For info: Mystic Seaport Museum, PO Box 6000, Mystic, CT 06355. Phone: (203) 572-5315.

OLD GERMAN CHRISTMAS. Nov 24-25. Sauter House, Papillion, NE. Old Town community celebration of Christmas with table Christmas displays, luncheon, carols, German food, the lighting of the tree, live Nativity pageant, brass ensemble and drawing for dollars. Santa arrives on horse-drawn sleigh at 4:30 PM. Annually, the Friday after Thanksgiving. Est attendance: 600. For info: Papillion Chamber of Commerce, 122 E Third St, Papillion, NE 68046. Phone: (402) 339-3050.

SANTA SPECTACULAR. Nov 24-Dec 24. In front of Union Station, Indianapolis, IN. The annual celebration tracking Santa's journey and arrival to Indianapolis is presented in an originally choreographed show with a cast of dozens of entertainers and costumed characters. Grand finale features the lighting of the exterior of the building and a firework display. Annually, from the Friday after Thanksgiving until Christmas. Est attendance: 2,000. For info: Union Station, PR Dept, 39 W Jackson, Indianapolis, IN 46225. Phone: (317) 267-0700.

SANTALAND USA. Nov 24-Dec 23. Downtown, Madison, MN. Santaland USA is a group of child-sized buildings and figures, including a church, schoolhouse, barn, bakery, toy shop, Santa and Mrs Claus' house, Santa's Workshop, Santa's sleigh and reindeer, a mouse house, camels and a manger scene. All are built, furnished and decorated by volunteers and along with other scenes, such as a railroad train and pond, take up most of the mezzanine area of a large retail store. Santaland is loved by children and adults alike. Each year a new building is added. Santaland is sponsored by the Madison Chamber; free admission. Est attendance: 2,000. For info: Donna C. Ventrella, 404 6th Ave, Madison, MN 56256. Phone: (612) 598-7373.

SINKIE DAY. Nov 24. SINKIES (people who occasionally dine over the kitchen sink) are encouraged to participate in and celebrate this time-honored, casual-yet-tasteful cuisine culture. This is a particularly appropriate day to become acquainted with the SINKIE style of dining. Christmas shopping and Thanksgiving leftovers provide the perfect reasons to enjoy a quick meal. Sponsor: Intl Assn of People Who Dine Over the Kitchen Sink, Norm Hankoff, Founder, 1579 Farmers Lane, #252, Santa Rosa, CA 95405.

SPINOZA, BARUCH: BIRTH ANNIVERSARY. Nov 24. Dutch philosopher, born at Amsterdam, Nov 24, 1632. Died Feb 21, 1677. "Peace is not an absence of war," wrote Spinoza, in 1670, "it is a virtue, a state of mind, a disposition for benevolence, confidence, justice."

Nov 24-25 ☆ *Chase's 1995 Calendar of Events* ☆

STAMP EXPO '95: AMERICA. Nov 24-26. Grand Hotel, Anaheim, CA. Fax: (818) 988-4337. Est attendance: 4,000. Sponsor: Intl Stamp Collectors Society, PO Box 854, Van Nuys, CA 91408. Phone: (818) 997-6496.

STARTER HALL OF FAME TIP-OFF CLASSIC. Nov 24. (Date tentative.) Springfield Civic Center, Springfield, MA. Official opening game of intercollegiate basketball season. Est attendance: 9,000. For info: Robin Deutsch, Dir Mktg, Basketball Hall of Fame, Box 179, 1150 W Columbus Ave, Springfield, MA 01101. Phone: (413) 781-6500.

STERNE, LAURENCE: BIRTH ANNIVERSARY. Nov 24. Novelist, author, born Nov 24, 1713. Died at London, England, Mar 18, 1768. In his Dedication to *Tristram Shandy*, Sterne wrote: "I live in a constant endeavour to fence against the infirmities of ill health, and other evils of life, by mirth; being firmly persuaded that every time a man smiles,—but much more so, when he laughs, that it adds something to this Fragment of Life."

TAYLOR, ZACHARY: BIRTH ANNIVERSARY. Nov 24. The soldier who became 12th president of the US was born at Orange County, VA, Nov 24, 1784. Term of office: Mar 4, 1849-July 9, 1850. Nominated at the Whig party convention in 1848, but the story goes, he did not accept the letter notifying him of his nomination because it had postage due. He cast his first vote in 1846, when he was 62 years old. Becoming ill on July 4, 1850, he died at the White House, July 9. His last words: "I am sorry that I am about to leave my friends."

US MILITARY LEAVES PHILIPPINES: ANNIVERSARY. Nov 24. The Philippines became a US colony at the turn of the century when it was taken over from Spain after the Spanish American War. Though President Franklin D. Roosevelt signed a bill Mar 24, 1934, granting the Philippines independence to be effective July 4, 1946, before that date Manila and Washington signed a treaty allowing the US to lease military bases on the island. In 1991 the Philippine Senate voted to reject a renewal of that lease, and Nov 24, 1992, after almost 100 years of military presence on the island, the last contingent of US marines left Subic Base.

VICTORIAN CHRISTMAS WALK. Nov 24-Dec 24. Larimer Square, Denver, CO. Christmas walk in an old-fashioned tradition, turning back the hands of time to the Victorian era. Carolers, Dickens, Father Christmas, carriage rides and more. Est attendance: 100,000. For info: Larimer Square, 1429 Larimer St, Denver, CO 80202. Phone: (303) 534-2367.

WAY OF LIGHTS. Nov 24-Jan 7. Belleville, IL. Remember the birth of Christ while journeying along a path illuminated with 150,000 white lights to the traditional crib. Indoor activities include children's show, choirs and an International Christmas Tree display, all expressing the message of Christ's peace. Est attendance: 350,000. For info: Shrine of Our Lady of the Snows, 9500 W Illinois, Hwy 15, Belleville, IL 62223. Phone: (618) 397-6700.

YOU'RE WELCOMEGIVING DAY. Nov 24. To create a four-day weekend. Sponsor: Richard Ankli, The Fifth Wheel Tavern, 639 Fifth St, Ann Arbor, MI 48103-4840.

ZAIRE: NEW REGIME ANNIVERSARY. Nov 24. National holiday.

BIRTHDAYS TODAY

William Frank Buckley, 70, editor (*The National Review*), author (*God and Man at Yale*), born at New York, NY, Nov 24, 1925.

Marlin Fitzwater, 53, former principal press secretary in the Bush administration, White House spokesman, born at Salinas, KS, Nov 24, 1942.

Garson Kanin, 83, writer (*A Gift of Time, Tracy and Hepburn*), director (*Adam's Rib, Pat and Mike*), husband of Ruth Gordon, born Rochester, NY, Nov 24, 1912.

Alfredo Kraus, 68, operatic tenor, born at Las Palmas, Canary Islands, Nov 24, 1927.

Stanley Livingston, 45, actor ("My Three Sons"), born at Los Angeles, CA, Nov 24, 1950.

☆ ☆ ☆
NOVEMBER 25 — SATURDAY
329th Day — Remaining, 36

AUTOMOBILE SPEED REDUCTION: ANNIVERSARY. Nov 25. Anniversary of the presidential order, Nov 25, 1973, requiring a cutback from the 70 mile-per-hour speed limit. The 55 mile-per-hour National Maximum Speed Limit (NMSL) was established by the Congress in January 1974 (PL 93-643). The National Highway Traffic Administration reported that "analysis of available data shows that the 55 mph NMSL forestalled 48,310 fatalities through 1980. There were also reductions in crash-related injuries and property damage." Motor fuel savings were estimated at 2.4 billion gallons per year. Notwithstanding, in 1987, the Congress permitted states to increase speed limits on rural interstate highways to 65 miles per hour.

BRACH'S HOLIDAY CHRISTMAS PARADE. Nov 25. Michigan Ave, Chicago, IL. Syndicated across the country with floats by more than 30 corporations. Sponsor: Brach's Candy. Est attendance: 600,000. For info: Production Contractors, Inc, 1711 W Fullerton Ave, Chicago, IL 60614. Phone: (312) 935-8747.

CANADA: CANADIAN WESTERN AGRIBITION. Nov 25-Dec 1. Regina, Saskatchewan. Canada's premiere agricultural event is the world's largest indoor livestock show and marketplace. More than 4,000 cattle, sheep, swine and horses are brought to Agribition. The event also features a pedigreed seed show and agricultural trade and technology show. Livestock producers from around the world come to Agribition to compare breeding stocks and programs, perhaps importing new blood lines based on what they discover. Winning at a livestock show of Agribition's status brings recognition as a producer of quality breeding stock. The general public can view the impressive livestock brought to the show. Also part of Agribition is a variety of entertainment, including the province's biggest indoor professional rodeo. Est attendance: 160,000. For info: Tourism Saskatchewan, 1919 Saskatchewan Dr, Regina, Sask, Canada S4P 3V7. Phone: (800) 667-7191.

CANADA: SAINT JOHN SALE OF FINE CRAFT. Nov 25-26. Saint John, NB. All exhibitors must be juried members of the NBCC. Est attendance: 6,000. For info: New Brunswick Crafts Council Inc, PO Box 1231, Fredericton, NB, Canada E3B 5C8. Phone: (506) 450-8989.

CARNEGIE, ANDREW: BIRTH ANNIVERSARY. Nov 25. American financier, philanthropist and benefactor of more than 2,500 libraries, was born at Dunfermline, Scotland, Nov 25, 1835. Carnegie Hall, Carnegie Foundation, and the Carnegie Endowment for International Peace are among his gifts. Carnegie wrote, in 1889, "Surplus wealth is a sacred trust which its possessor is bound to administer in his lifetime for the good of the community ... The man who dies ... rich dies disgraced." Carnegie died at his summer estate, "Shadowbrook," MA, on Aug 11, 1919.

CHRISTMAS FAIR AND GALLERY. Nov 25-Dec 28. Firehouse Art Center, Norman, OK. Shoppers find gifts for friends, children and family, from stocking stuffers to "executive" gifts! Original works and limited production pieces by juried and invited artists and craftspeople. Est attendance: 3,500. For info: Firehouse Art Ctr, 444 S Flood, Norman, OK 73069. Phone: (405) 329-4523.

November 1995

S	M	T	W	T	F	S
			1	2	3	4
5	6	7	8	9	10	11
12	13	14	15	16	17	18
19	20	21	22	23	24	25
26	27	28	29	30		

CHRISTMAS IN OLD DODGE CITY. Nov 25–Dec 24. Citywide, Dodge City, KS. Home tours, bazaar, musical productions, 19th-century Christmas decorations in Old Dodge City. Annually, after Thanksgiving through Christmas. Est attendance: 10,000. For info: Dodge City Conv & Visitors Bureau, PO Box 1474, Dodge City, KS 67801. Phone: (316) 225-8186.

CHRISTMAS IN ROSELAND. Nov 25–Dec 31. American Rose Center, Shreveport, LA. Holiday magic transforms the American Rose Center with lights, holiday scenes, music and light sculptures. Est attendance: 40,000. For info: American Rose Center, 8877 Jefferson-Paige Rd, Shreveport, LA 71119. Phone: (318) 938-5402.

CHRISTMAS INNS OF CAPE MAY. Nov 25–26. (Also Dec 2-3, 9-10, 16-17, 23-30) Cape May, NJ. A unique guided tour of 5 restored Victorian inns (all listed on the National Register) in the downtown Historic District. Visitors spend 15–20 minutes at each inn, entertained by the authentically costumed hosts who portray original owners of homes. Each specially decorated inn presents a different aspect of a Victorian Christmas. Est attendance: 5,000. For info: John Dunwoody, 719 Columbia Ave, Cape May, NJ 08204. Phone: (609) 884-8075.

ELECTRIC LIGHT PARADE. Nov 25. Downtown Lovington, NM. Christmas shines in Lovington with more than 60 entries including floats, motorhomes, cars and motorcycles decorated with Christmas lights. Annually, the first Saturday following Thanksgiving at 8 PM. Est attendance: 7,000. For info: Lovington Chamber of Dvmt & Commerce, Inc, PO Box 1347, Lovington, NM 88260. Phone: (505) 396-5311.

GERMANY: FRANKFURT CHRISTMAS MARKET. Nov 25–Dec 23. "Weinachtsmarkt auf dem Romerberg," the Christmas market in Frankfurt, is one of Germany's best. Bells are rung simultaneously from nine downtown churches. Glockenspiels sounded by hand and trumpets blown from the old St. Nicolas Church.

HMONG NEW YEAR CELEBRATION. Nov 25. Sheboygan, WI. Sheboygan's Hmong community invites the public to share in its culture. Hmong dance performances, games for children, speakers, traditional music. Est attendance: 2,000. For info: Hmong Mutual Assistance Assn of Sheboygan, Inc, 1504 N 13th St, Sheboygan, WI 53081. Phone: (414) 458-0808.

"THE LIGHT OF THE WORLD" CHRISTMAS PAGEANT. Nov 25. (Also Dec 3 and 10). Courthouse square, Minden, NE. Pageant presented on three sides of the courthouse square with around 115 local citizens performing with beautiful costumes. At the climax some 10,000 Christmas lights are turned on the courthouse. 7 PM; free admission. Annually, the first Saturday after Thanksgiving and the first two Sundays in December. Est attendance: 10,000. Sponsor: Marjorie Madsen, Mgr, Minden Chamber of Commerce, PO Box 375, Minden, NE 68959. Phone: (308) 832-1811.

NATION, CARRY AMELIA MOORE: BIRTH ANNIVERSARY. Nov 25. American temperance leader, famed as hatchet-wielding smasher of saloons, born at Garrard County, KY, Nov 25, 1846. Died at Leavenworth, KS, June 9, 1911.

POPE JOHN XXIII: BIRTH ANNIVERSARY. Nov 25. Angelo Roncalli, 261st pope of the Roman Catholic Church, born at Sotte il Monte, Italy, on Nov 25, 1881. Elected pope, Oct 28, 1958. Died June 3, 1963.

PURCELL, HENRY: 300th DEATH ANNIVERSARY. Nov 25, 1695. English composer of the early Baroque period was born at London circa 1659. Purcell's work includes more than 100 songs, the opera *Dido and Aeneas* and *The Fairy Queen*, incidental music for a version of Shakespeare's *A Midsummer Night's Dream*. Purcell died at London.

ST. CATHERINE'S DAY. Nov 25. Patron saint of maidens, mechanics and philosophers, as well as of all who work with wheels.

SHOPPING REMINDER DAY. Nov 25. A reminder to shoppers that there are only 24 more shopping days (excluding Sundays and Christmas Eve) after today until Christmas, and that one month from today a new countdown will begin for Christmas 1996.

SURINAME: INDEPENDENCE DAY. Nov 25. Holiday.

SWINE TIME. Nov 25. Climax, GA. This homecoming for past residents promotes swine and raises funds for the Climax Community Club. A day of special events for people of all ages, including parade, clogging, music of all kind and a variety of food. Est attendance: 25,000. For info: Climax Community Club, Clifford Wells, PO Box 131, Climax, GA 31734. Phone: (912) 246-0910.

BIRTHDAYS TODAY

Kathryn Crosby, 62, actress born at Houston, TX, Nov 25, 1933.

Russell Earl "Bucky" Dent, 44, former baseball player, manager, born at Savannah, GA, Nov 25, 1951.

Joe Jackson Gibbs, 55, professional football coach, born at Mocksville, NC, Nov 25, 1940.

Bernie Kosar, 32, professional football player, born at Boardman, OH, Nov 25, 1963.

John Larroquette, 48, actor (Emmy for "Night Court," "The John Larroquette Show"), born at New Orleans, LA, Nov 25, 1947.

NOVEMBER 26 — SUNDAY
330th Day — Remaining, 35

ANNIVERSARY OF FIRST US HOLIDAY BY PRESIDENTIAL PROCLAMATION. Nov 26. President George Washington proclaimed Nov 26, 1789, to be Thanksgiving Day. Both Houses of Congress, by their joint committee, had requested him to recommend a day of public thanksgiving and prayer, to be observed by acknowledging with grateful hearts the many and signal favors of Almighty God, especially by affording them an opportunity to peaceably establish a form of government for their safety and happiness. Proclamation issued Oct 3, 1789.

CASABLANCA PREMIERE: ANNIVERSARY. Nov 26, 1942. Due to the landing of the Allies in Casablanca and other points in North Africa on Nov 8, the premiere and release of the film were moved up from previously planned June 1943. The film was premiered in New York City on Thanksgiving Day with the general nationwide release following on Jan 23, 1943, during the Roosevelt-Churchill conferences in Casablanca.

CHRISTMAS AT UNION STATION. Nov 26–Dec 24. Omaha, NE. Celebration of the Christmas holiday around a giant 50-ft Christmas tree in the splendor of Omaha's old Union Station. Est attendance: 20,000. For info: Western Heritage Museum, 801 S 10th St, Omaha, NE 68108. Phone: (402) 444-5072.

CUSTER BATTLEFIELD BECOMES LITTLE BIGHORN BATTLEFIELD: ANNIVERSARY. Nov 26. The US Congress approved a bill on Nov 26, 1991, renaming Custer Battlefield National Monument as Little Bighorn Battlefield National Monument. The bill also authorized the construction of a memorial to the Native Americans who fought and died at the battle known as Custer's Last Stand. Introduced by Representative Ben Nighthorse Campbell, the only Native American in Congress, the bill was signed into law by President George Bush.

GERMANY: TOTENSONNTAG. Nov 26. In Germany, Totensonntag is the Protestant population's day for remembrance of the dead. It is celebrated on the last Sunday of the church year (the Sunday before Advent).

GRIMKE, SARAH MOORE: BIRTH ANNIVERSARY. Nov 26. American antislavery and women's rights advocate along with her sister Angelina. Born on Nov 26, 1792 at Charleston, SC, and died on Dec 23, 1873 at Hyde Park, MA.

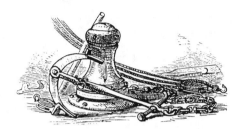

JAPAN AGREES TO END USE OF DRIFT NETS: ANNIVERSARY. Nov 26. On Nov 26, 1991, Japan agreed to comply with a 1989 United Nations moratorium on the use of huge fishing nets in the Northern Pacific Ocean. The large nets extend up to 40 miles and have been criticized as "walls of death," causing wide-spread destruction of marine life, including whales, turtles, birds and many varieties of fish. Japan agreed to end half of its driftnet fishing by the June 30, 1992, deadline, and the remainder by the end of 1992.

JOHN F. KENNEDY DAY IN MASSACHUSETTS. Nov 26. Annually, the last Sunday in November.

JOHN HARVARD DAY. Nov 26. English clergyman and scholar, founder of Harvard College. Born Nov 26, 1607 and died Sept 24, 1638.

MERRY PRAIRIE CHRISTMAS. Nov 26–Dec 31. Fargo-Moorhead, ND. Events include a holiday parade, Christmas concerts on the historic Fargo Theatre's Mighty Wurlitzer Pipe Organ, Santa Village, the Red River Dance and Performing Company holiday show and the Bonanzaville, USA Pioneer Village staging Christmas as it was 100 years ago. For info: Fargo-Moorhead Conv and Visitors Bureau, 2001 44th St SW, Fargo, ND 58103. Phone: (800) 235-7654.

NATIONAL HOME CARE WEEK. Nov 26–Dec 2. To educate the public and commemorate providers. Annually, the week after Thanksgiving week. For info: Natl Assn for Home Care, 519 C St NE, Stanton Park, Washington, DC 20002. Phone: (202) 547-7424.

PASADENA DOO DAH PARADE. Nov 26. Pasadena, CA. No theme, no judging, no prizes, no order of march, no motorized vehicles and no music. Annually, the Sunday following Thanksgiving Day.

◆ **TRUTH, SOJOURNER: DEATH ANNIVERSARY.** Nov 26. A former slave who had been sold four different times, Sojourner Truth became an evangelist who argued for abolition and women's rights. After a troubled early life, she began her evangelical career in 1843, traveling through New England until she discovered the utopian colony called the Northampton Association of Education and Industry. It was there she was exposed to, and became an advocate for, the cause of abolition, working with Frederick Douglass, Wendall Phillips, William Lloyd Garrison and others. In 1850, she befriended Lucretia Mott, Elizabeth Cady Stanton and other feminist leaders and actively began supporting calls for women's rights. In 1870, she attempted to petition Congress to create a "Negro State" on public lands in the west, but withdrew it because she felt her petition had too few names. Born in Ulster County, NY, about 1790, she died on Nov 26, 1883, at Battle Creek, MI.

WALKER, MARY EDWARDS: BIRTH ANNIVERSARY. Nov 26. American physician and women's rights leader, born at Oswego, NY, Nov 26, 1832. First female surgeon in US Army (Civil War). Spent four months in Confederate prison. First and only woman ever to receive Medal of Honor (Nov 11, 1865). Two years before her death, on June 3, 1916, a government review board asked that her award be revoked. She continued to wear it, in spite of official revocation, until her death, Feb 21, 1919. On June 11, 1977, the secretary of the army posthumously restored the Medal of Honor to Dr. Walker. Special commemorative US postage stamp was issued in 1982, marking 150th anniversary of her birth.

BIRTHDAYS TODAY

Robert Goulet (Stanley Applebaum), 62, singer ("My Love, Forgive Me"), actor (Sir Lancelot in *Camelot* on stage; *I'd Rather Be Rich*, "The Blue Light"), born at Lawrence, MA, Nov 26, 1933.
Johnny Hector, 35, football player, born at Lafayette, LA, Nov 26, 1960.
Richard Caruthers Little, 57, impressionist, born at Ottawa, Ontario, Canada, Nov 26, 1938.
Charles Monroe Schulz, 73, cartoonist ("Peanuts"), born at Minneapolis, MN, Nov 26, 1922.
Tina Turner, 57, singer (with Ike: "It's Gonna Work Out Fine," "A Fool in Love"; solo: "What's Love Got to Do with It," "Better Be Good to Me"), born Nutbush, TN, Nov 26, 1938.

NOVEMBER 27 — MONDAY
331st Day — Remaining, 34

BEARD, CHARLES A.: BIRTH ANNIVERSARY. Nov 27. American historian Charles Austin Beard who wrote many books in collaboration with his wife, Mary R. Beard, was born near Knightstown, IN, on Nov 27, 1874. He died at New Haven, CT, Sept 1, 1948.

BENEDICT, JULIUS: BIRTH ANNIVERSARY. Nov 27. The German-born musician and composer Sir Julius Benedict was born at Stuttgart, Germany, on Nov 27, 1804. Knighted in 1871, Benedict died at London, England, June 5, 1885.

DUBCEK, ALEXANDER: BIRTH ANNIVERSARY. Nov 27, 1921. The man who attempted to give his country "socialism with a human face," Alexander Dubcek was born at Uhrocev, a village in western Slovakia. As first secretary of the Czechoslovak Communist Party during the "Prague Spring" of 1968, he moved to achieve the "widest possible democratization" and to loosen the dominant influence of the USSR. As a result Czechoslovakia was invaded by armed forces of the Warsaw Pact on Aug 21, 1968. Dubcek died Nov 7, 1992, at Prague.

ENGLAND: BONNIE PRINCE CHARLIE 250th ANNIVERSARY. Nov 27–Dec 10. City Centre, Derby, Derbyshire. Various commemorative events including guided tours, displays, street entertainment, reenactments and musical events. Admission free. For info: Bob Thorne, Dir, Business Development Assoc, Derwent Business Centre, Clarke St, Derby, England DE1 2BU.

KEMBLE, FANNY: BIRTH ANNIVERSARY. Nov 27. Frances Anne Kemble, English actress, born at London, England, Nov 27, 1809. Died Jan 15, 1893.

LIVINGSTON, ROBERT R.: BIRTH ANNIVERSARY. Nov 27. Member of the Continental Congress, farmer, diplomat and jurist, was born at New York City, Nov 27, 1746. It was Livingston who administered the oath of office to President George Washington in 1789. He died at Clermont, NY, Feb 26, 1813.

☆ Chase's 1995 Calendar of Events ☆ Nov 27-29

REVEREND FRANCIS GASTRELL'S EJECTMENT: ANNIVERSARY. Nov 27. On Nov 27, 1759, the Stratford-upon-Avon town corporation gave orders to bring an "action of Ejectment" against the Rev Francis Gastrell, Vicar of Frodsham, who lived in William Shakespeare's home. Gastrell, it is said, had cut down the 150-year-old mulberry tree that had been planted by William Shakespeare. Gastrell maliciously felled the tree because he was annoyed by the many Shakespeare enthusiasts who came to look at it. He sold the tree for firewood, but it was recovered by a jeweler-woodcarver, Thomas Sharp, who fashioned hundreds of relics from it. Gastrell was ejected from Stratford "amid the ragings and cursings of its people, a citizen well lost"—for one of "the meanest petty infamies in our annals." The Rev Gastrell's wife, Jane, is said to have been an accomplice in this celebrated arborcide.

SPACE MILESTONE: SOYUZ T-3 (USSR). Nov 27. Launched on Nov 27, 1980, three cosmonauts, O. Makarov, L. Kizim and G. Strekalov, docked at Salyut 6 space station on Nov 28. Returned to Earth, Dec 10, 1980.

SWITZERLAND: ONION MARKET (ZIBELEMARIT). Nov 27. Berne. Best known and most popular of Switzerland's many autumn markets. Great heaps of onions in front of Federal Palace. Fourth Monday in November commemorates granting of market right to people after great fire of Berne in 1405.

WEIZMANN, CHAIM: BIRTH ANNIVERSARY. Nov 27. Israeli statesman born near Pinsk, Byelorussia, on Nov 27, 1874. He played an important role in bringing about the British government's Balfour Declaration, calling for the establishment of a national home for Jews in Palestine. He died at Tel Aviv, Israel, on Nov 9, 1952.

BIRTHDAYS TODAY

Buffalo Bob (Bob Smith), 78, TV personality ("The Howdy Doody Show"), born Buffalo, NY, Nov 27, 1917.
Robin Givens, 31, actress ("Head of the Class," A Rage in Harlem), born New York, NY, Nov 27, 1964.
Boris Grebenshikov, 42, Soviet rock musician, born at Leningrad, USSR, Nov 27, 1953.
David Merrick, 83, producer (Hello, Dolly!; Beckett), born St. Louis, MO, Nov 27, 1912.
Eddie Rabbitt, 51, singer ("I Love a Rainy Night"), songwriter ("Kentucky Rain"), born at Brooklyn, NY, Nov 27, 1944.
Gail Henion Sheehy, 58, author, journalist (The Silent Passage: Menopause; Pathfinders), born at Mamaroneck, NY, Nov 27, 1937.
Mona Washbourne, 92, actress (Mrs. Pearce, My Fair Lady), born Birmingham, England, Nov 27, 1903.

NOVEMBER 28 — TUESDAY
332nd Day — Remaining, 33

ALBANIA: NATIONAL DAY. Nov 28. National holiday observed.

BLAKE, WILLIAM: BIRTH ANNIVERSARY. Nov 28. English poet and artist, born at London, England, Nov 28, 1757. Died there Aug 12, 1827.

BUNYAN, JOHN: BIRTH ANNIVERSARY. Nov 28. English cleric and author of A Pilgrim's Progress, born Nov 28, 1628 at Elstow, Bedfordshire. Died at London, Aug 31, 1688.

DESERT STORM: UN DEADLINE RESOLUTION: 5th ANNIVERSARY. Nov 28. On Nov 28, 1990, the United Nations passed the twelfth in a series of resolutions concerning the Iraqi invasion of Kuwait. Resolution 678 authorized states "to use all necessary means" against Iraq unless it withdrew its forces from Kuwait by Jan 15, 1991. Iraq did not comply, and the Allied Forces began their attack within hours of the expiration of the deadline.

LULLY, JEAN BAPTISTE: BIRTH ANNIVERSARY. Nov 28. Versatile musician and composer, born at Florence, Italy, on Nov 28, 1632, who chose France for his homeland. Noted for his quick temper, it is said that he struck his own foot with a baton while in a rage. The resulting wound led to blood poisoning, from which he died, at Paris, Mar 22, 1687.

MAURITANIA: INDEPENDENCE DAY. Nov 28. National holiday. Attained sovereignty Nov 28, 1960.

PANAMA: INDEPENDENCE FROM SPAIN. Nov 28. Public holiday in Panama.

SPACE MILESTONE: MARINER 4 (US). Nov 28. Launched Nov 28, 1964 toward Mars. On July 14, 1965, after 228 days, approached within 6,118 miles of Mars. Took photographs.

SPACE MILESTONE: STS-9 (US). Nov 28. Shuttle Columbia launched from Kennedy Space Center, FL, Nov 28, 1983 with five astronauts (John Young, Brewster Shaw, Jr, Owen Garriot, Robert Parker, Byron Lichtenberg) and German physicist Ulf Merbold. Landed at Edwards Air Force Base, CA, on Dec 8.

BIRTHDAYS TODAY

Alexander Godunov, 46, ballet dancer, born at Sakhalin, USSR, Nov 28, 1949.
Berry Gordy, Jr, 66, record and motion picture executive (founder of Motown), born at Detroit, MI, Nov 28, 1929.
Ed Harris, 45, actor (The Right Stuff), born at Englewood, NJ, Nov 28, 1950.
Gary Hart (Gary Hartpence), 59, former senator, presidential candidate, born at Ottawa, KS, Nov 28, 1936.
Hope Lange, 62, actress ("The Ghost and Mrs. Muir," Bus Stop), born Reading Ridge, CT, Nov 28, 1933.
Randy Newman, 52, singer, songwriter ("Love Story," "Short People"), composer (scores for Ragtime and The Natural), born Los Angeles, CA, Nov 28, 1943.
Paul Shaffer, 46, band leader (David Letterman's music man), comedian, born at Thunder Bay, Ontario, Canada, Nov 28, 1949.
Roy Tarpley, 31, basketball player, born at New York, NY, Nov 28, 1964.

NOVEMBER 29 — WEDNESDAY
333rd Day — Remaining, 32

ALCOTT, LOUISA MAY: BIRTH ANNIVERSARY. Nov 29. American author, born at Philadelphia, PA, Nov 29, 1832. Died at Boston, MA, Mar 6, 1888. Her most famous novel was Little Women, the classic story of Meg, Jo, Beth and Amy.

◆ **BERKELEY, BUSBY: 100th BIRTH ANNIVERSARY.** Nov 29. William Berkeley Enos was born Nov 29, 1895, at Los Angeles, CA. After serving in World War I as an entertainment officer, he changed his name to Busby Berkeley and began a career in show business as an actor. He turned to directing in 1921 and his lavish Broadway and Hollywood creations included Forty-Second Street, Gold Diggers of 1933, Footlight Parade, Hollywood Hotel, Stage Struck, Gold Diggers in Paris, Babes in Arms, Strike Up the Band, Girl Crazy and Take Me Out to the Ball Game, among many others. He retired in 1962 and returned to Broadway in 1970 to supervise a revival of No, No, Nanette. He died Mar 14, 1976, at Palm Springs, CA.

Nov 29-30 ☆ Chase's 1995 Calendar of Events ☆

CZECHOSLOVAKIA ENDS COMMUNIST RULE: ANNIVERSARY. Nov 29. On Nov 29, 1989, Czechoslovakia ended 41 years of one-party communist rule when the Czechoslovak parliament voted unanimously to repeal the constitutional clauses giving the Communist Party a guaranteed leading role in the country and promoting Marxism-Leninism as the state ideology. The vote came at the end of a 12-day revolution sparked by the beating of protestors on Nov 17. Although the Communist party remained in power, the tide of reform led to its ouster by the Civic Forum, headed by playwright Vaclav Havel. The Civic Forum demanded free elections with equal rights for all parties, a mixed economy and support for foreign investment. In the first free elections in Czechoslovakia since World War II, Vaclav Havel was elected president. He resigned in 1993 when it became clear that Czechoslovakia would split into two separate nations.

MOON PHASE: FIRST QUARTER. Nov 29. Moon enters First Quarter phase at 1:28 AM, EST.

PHILLIPS, WENDELL: BIRTH ANNIVERSARY. Nov 29. American women's suffrage, anti-slavery, prison reform leader, born Nov 29, 1811. Died Feb 2, 1884.

◆ **ROSS, NELLIE TAYLOE: BIRTH ANNIVERSARY.** Nov 29, 1876. Nellie Tayloe Ross became the first female governor in the US when she was chosen to serve out the last month and two days of her husband's term as governor of Wyoming after he died in office. She was elected in her own right in the Nov 4, 1924, election but lost the 1927 race. Ross was appointed vice chairman of the Democratic National Committee in 1926 and named director of the US Mint by President Franklin D. Roosevelt in 1933. She served in that capacity for 20 years. Ross died Dec 20, 1977, at Washington, DC.

THOMSON, CHARLES: BIRTH ANNIVERSARY. Nov 29. America's first official record keeper. Chosen secretary of the First Continental Congress on Sept 5, 1774, Thomson recorded proceedings for 15 years and delivered his journals together with tens of thousands of records to the federal government in 1789. Born in Ireland, Nov 29, 1729. Died on Aug 16, 1824. It was Thomson who notified George Washington of his election as president.

UNITED NATIONS: INTERNATIONAL DAY OF SOLIDARITY WITH THE PALESTINIAN PEOPLE. Nov 29. Annual observance proclaimed by UN General Assembly in 1977. At request of Assembly, observance is organized by secretary-general in consultation with Committee on the Exercise of the Inalienable Rights of the Palestinian People. Recommendations include a plan for return of the Palestinians to their homes and the establishment of an "independent Palestinian entity." Info from: United Nations, Dept of Public Info, New York, NY 10017.

WAITE, MORRISON R.: BIRTH ANNIVERSARY. Nov 29. Seventh Chief Justice of the Supreme Court. Born Nov 29, 1816, at Lyme, CT. Appointed Chief Justice by President Ulysses S. Grant on Jan 19, 1874. The Waite Court is remembered for its controversial rulings that did much to rehabilitate the idea of states' rights after the Civil War and early Reconstruction years. Waite died at Washington, DC, on Mar 23, 1888.

BIRTHDAYS TODAY

Jacques Rene Chirac, 63, French political leader, born at Paris, France, Nov 29, 1932.

Madeleine L'Engle, 77, writer (*A Wrinkle in Time, Summer of the Great-Grandmother*), born New York, NY, Nov 29, 1918.

Diane Ladd (Rose Diane Ladner), 63, actress (*Alice Doesn't Live Here Anymore, Ramblin' Rose, The Cemetery Club*), born Meridian, MS, Nov 29, 1932.

Chuck Mangione, 55, musician, composer (Grammy for "Bellavia"), born Rochester, NY, Nov 29, 1940.

John Mayall, 62, musician, band leader (The Bluesbreakers), born Manchester, England, Nov 29, 1933.

Cathy Moriarty, 35, actress (*Raging Bull, The Mambo Kings*), born Bronx, NY, Nov 29, 1960.

Vin Scully, 68, sportscaster, born at the Bronx, NY, Nov 29, 1927.

Garry Shandling, 46, comedian ("The Larry Sanders Show"), born at Chicago, IL, Nov 29, 1949.

Paul Simon, 67, US Senator (D, Illinois), born at Eugene, OR, Nov 29, 1928.

NOVEMBER 30 — THURSDAY
334th Day — Remaining, 31

ARTICLES OF PEACE BETWEEN GREAT BRITAIN AND THE UNITED STATES: ANNIVERSARY. Nov 30. These provisional articles of peace, which were to end America's War of Independence, were signed at Paris, France, Nov 30, 1782. The refined and definitive treaty of peace between Great Britain and the US was signed at Paris, on Sept 3, 1783. In it "His Britannic Majesty acknowledges the said United States ... to be free, sovereign and independent states; that he treats them as such; and for himself, his heirs and successors, relinquishes all claims to the government, propriety and territorial rights of the same, and every part thereof...."

BARBADOS: INDEPENDENCE DAY. Nov 30. National holiday.

CHRISTMAS ON THE RIVER. Nov 30–Dec 2. Demopolis, AL. A day of fun with arts and crafts, children's parade, AL BBQ Cook-off, and a river boat parade. Est attendance: 40,000. For info: Jane Gross, Box 667, Demopolis, AL 36732. Phone: (205) 289-0270.

CHURCHILL, WINSTON: BIRTH ANNIVERSARY. Nov 30. Winston Leonard Spencer Churchill, British statesman and the first man to be made an honorary citizen of the US (by an act of Congress, Apr 9, 1963), born (prematurely) Nov 30, 1874, at Blenheim Palace, Oxfordshire, England. Died Jan 24, 1965, at London, England. Dedicated to Britain and total victory over Germany, Churchill as minister of defense and prime minister was considered a strong leader during WWII.

CLEMENS, SAMUEL LANGHORNE (MARK TWAIN): 160th BIRTH ANNIVERSARY. Nov 30. Celebrated American author, whose books include: *The Adventures of Tom Sawyer, The Adventures of Huckleberry Finn*, and *The Prince and the Pauper*. Born at Florida, MO, Nov 30, 1835. Twain is quoted as saying "I came in with Halley's Comet in 1835. It is coming again next year, and I expect to go out with it." He did. Twain died at Redding, CT, on Apr 21, 1910 (just one day after Halley's Comet perihelion).

COMPUTER SECURITY DAY. Nov 30. The use of computers increases daily. This annual event reminds people to protect their computers, programs and data at home and at work. For info: Special Interest Group for Security Audit Control (SIG-SAC), Assn for Computing Machinery, PO Box 39110, Washington, DC 20016.

November 1995

S	M	T	W	T	F	S
			1	2	3	4
5	6	7	8	9	10	11
12	13	14	15	16	17	18
19	20	21	22	23	24	25
26	27	28	29	30		

☆ Chase's 1995 Calendar of Events ☆ Nov 30

ENGLAND: WESTMINSTER ANTIQUES FAIR. Nov 30–Dec 3. Horticultural Hall, Vincent Sq, London. Quality antiques fair with everything authenticated and for sale. Furniture pre-1870, most other items pre-1900. Annually, April and December. Est Attendance: 5,000. For info: Penman Antiques Fairs, PO Box 114, Haywards Heath, Sussex, England RH16 2YU.

FESTIVAL OF TREES. Nov 30–Dec 3. Agriculture Hall, Kansas Expocentre, Topeka, KS. Elaborately decorated Christmas trees create centerpiece of celebration. Trees range from Dickens-style traditional trees to avant garde creations representing new generation. Also included are a boutique, concessions and entertainment. Proceeds benefit Sheltered Living, Inc, a program that assists disabled adults in acquiring independent living skills. Admission free. Annually, the first weekend in December. For info: Merilee Larson, Sheltered Living, 2044 SW Fillmore St, Topeka, KS 66604. Phone: (913) 233-2566.

FRENCH WEST INDIES: INTERNATIONAL JAZZ FESTIVAL. Nov 30–Dec 10. Martinique. World renowned performers, concerts and study classes. For info: Martinique Devmt and Promo Bureau, Ms Muriel Wiltord, 610 5th Ave, New York, NY 10020. Phone: (212) 757-1125.

MORAVIAN CANDLE TEA. Nov 30–Dec 2. (Also Dec 7–9.) Single Brothers' House, Old Salem, Winston-Salem, NC. To share traditions of the Moravian Church and prepare the visitors for the coming of the Christmas season. Est attendance: 13,700. Sponsor: Women's Fellowship, Home Moravian Church, Tamra Thomas, 660 Kingsbury Circle, Winston-Salem, NC 27106-5711. Phone: (910) 722-6119.

PHILIPPINES: BONIFACIO DAY. Nov 30. Also known as National Heroes' Day. Commemorates birth of Andres Bonifacio, leader of the 1896 revolt against Spain. Bonifacio was born Nov 30, 1863.

ST. ANDREW'S DAY. Nov 30. Feast day of the apostle and martyr, Andrew, who died about 60 AD. Patron saint of Scotland.

ST. OLAF CHRISTMAS FESTIVAL. Nov 30–Dec 3. St. Olaf College, Northfield, MN. This celebration of the Christmas season brings together 500 student musicians (a 90-piece symphony orchestra and 450 singers) to perform sacred and folk songs from around the world. Annually, the first full weekend in December. Est attendance: 16,000. For info: Dan Jorgensen, Dir PR, St. Olaf College, 1520 St. Olaf Ave, Northfield, MN 55057-1098. Phone: (507) 646-3002.

SIDNEY, PHILIP: BIRTH ANNIVERSARY. Nov 30. English poet, statesman and soldier was born at Penshurst, Kent, England, on Nov 30, 1554. Best known of his poems is *Arcadia* (1580). Mortally wounded as he led an English detachment aiding the Dutch near Zutphen, on Sept 22, 1586, Sidney gave his water bottle to another dying soldier with the words "Thy necessity is yet greater than mine." He died at Arnheim on Oct 17, 1586, and all of England mourned the death of this man who had given new meaning to English patriotism.

STATUE OF RAMSES II UNEARTHED: ANNIVERSARY. Nov 30. Egyptian construction workers in the ancient provincial town of Akhimim, 300 miles south of Cairo, unearthed a statue of Ramses II on Nov 30, 1991. Akhimim was an important provincial district that included the city of Ipu, a mecca for worshippers of the fertility god Min. The statue was uncovered during an excavation to prepare a foundation for a post office. An additional statue was uncovered 33 feet away, but the identity of its subject was still unknown.

STAY HOME BECAUSE YOU'RE WELL DAY. Nov 30. So we can call in "well," instead of faking illness, and stay home from work. Phone: (212) 388-8673 or (717) 274-8451. For info: Wellness Permission League, Thomas or Ruth Roy, 2105 Water St, Lebanon, PA 17046.

◈ SWIFT, JONATHAN: BIRTH ANNIVERSARY. Nov 30. Clergyman and satirist born at Dublin, Ireland, Nov 30, 1667. Died there Oct 19, 1745. "I never saw, heard, nor read," Swift wrote in *Thoughts on Religion* "that the clergy were beloved in any nation where Christianity was the religion of the country. Nothing can render them popular but some degree of persecution."

UKRAINIAN FAMINE FILM BROADCAST. Nov 30. In the rapidly changing former Soviet Union, the film *Famine 33* produced by Oles Yanchuk, was broadcast on republic-wide TV on Nov 30, 1991. The film chronicled the forced collectivization of the agriculture industry in 1933 and the resulting famine which led to the death of more than 7 million Ukrainians. The famine was not officially recognized until 1990, when the Central Committee of the Ukrainian Communist Party first acknowledged that the millions of deaths were caused by the seizure of crops. The airing of the film heralded a significant departure from prior Soviet handling of history.

VICTORIAN CHRISTMAS SLEIGHBELL PARADE. Nov 30–Dec 3. Manistee, MI. Re-creation of Manistee history. No motorized vehicles, no amplification. Horse-drawn entries, walking entries, singers, animals and St. Nick in historic garb. Est attendance: 5,000. For info: Manistee Area Chamber of Commerce, 11 Cypress St, Manistee, MI 49660. Phone: (616) 723-2575.

BIRTHDAYS TODAY

Maryon Pittman Allen, 70, former US Senator, born at Meridian, MS, Nov 30, 1925.

◈ *Shirley Chisholm*, 71, author, former congresswoman, born at Brooklyn, NY, Nov 30, 1924.

Dick Clark, 66, entertainer (TV's "American Bandstand"), producer, born at Mount Vernon, NY, Nov 30, 1929.

Mike Espy, 42, US Secretary of Agriculture in Clinton administration, born at Yazoo, MS, Nov 30, 1953.

Robert Guillaume, 68, actor ("Soap," "Benson") born at St. Louis, MO, Nov 30,1927.

Billy Idol, 40, singer ("Mony Mony," "Eyes Without a Face"), songwriter, born at Surrey, England, Nov 30, 1955.

Bo Jackson (Vincent Edward Jackson), 33, baseball player, former football player, born at Bessemer, AL, Nov 30, 1962.

G. Gordon Liddy, 65, Watergate participant, born at New York, NY, Nov 30, 1930.

David Mamet, 48, dramatist, director (*House of Games, Things Change*), born Chicago, IL, Nov 30, 1947.

Virginia Mayo, 73, actress (*The Best Years of Our Lives*), born at St. Louis, MO, Nov 30, 1922.

Mandy Patinkin, 43, actor (*Evita, Sunday in the Park with George*), born Chicago, IL, Nov 30, 1952.

Paul Stookey, 58, singer, songwriter (Paul of Peter, Paul and Mary), born Baltimore, MD, Nov 30, 1937.

Efrem Zimbalist, Jr, 72, actor ("The F.B.I.," *Airport*), born New York, NY, Nov 30, 1923.

Dec 1 ☆ *Chase's 1995 Calendar of Events* ☆

December.

DECEMBER 1 — FRIDAY
335th Day — Remaining, 30

ARKANSAS CRAFT GUILD CHRISTMAS SHOWCASE AND SALE. Dec 1-3. Statehouse Convention Center, Excelsior Hotel, Little Rock, AR. 17th annual. 100 exhibitors of original handcrafted gifts including woodworks, pottery, stained glass, metalworks, musical instruments, jewelry, baskets and quilts. Admission charged ($2/adult). Est attendance: 5,000. For info: Arkansas Craft Guild, PO Box 800, Mountain View, AR 72560. Phone: (501) 269-3897.

BECKY THATCHER DAY. Dec 1. Northeastern Missouri. A day in honor of all girls and women who have written or inspired great literature. Traditionally observed on Dec 1, the birthday of Laura Hawkins (b. 1836), girl friend of young Sam Clemens, and model for the character of Becky Thatcher in Mark Twain's writings. Sponsor: Becky Thatcher Area Girl Scout Council, 512 Church St, Hannibal, MO 63401. Phone: (314) 221-0339.

BINGO'S BIRTHDAY MONTH. Dec 1-31. To celebrate the innovation and manufacture of the game of Bingo in 1929 by Edwin S. Lowe, which today has grown to a five-billion-dollar-a-year charitable fundraiser. Sponsor: Bingo Bugle, Inc, Roger Snowden, Pres, Box 527, Vashon, WA 98070. Phone: (800) 327-6437.

BONANZAVILLE'S CHRISTMAS ON THE PRAIRIE. Dec 1-2. Bonanzaville USA, West Fargo, ND. An old-fashioned Christmas celebration including caroling, special entertainment, hay rides, live animal displays, Santa and Mrs Claus and more. Est attendance: 5,000. For info: Margo Lang, Bonanzaville USA, W Main Ave, West Fargo, ND 58078. Phone: (701) 282-2822.

CANADA: SUNDOG HANDCRAFT FAIRE. Dec 1-3. Saskatchewan Place, Saskatoon, Saskatchewan. More than 150 artisans, artists and specialty food producers display their work at this, the largest event of its kind in western Canada. Entertainment, craft demonstrations, free children's activity center and a multicultural food court. Est attendance: 15,000. For info: Tourism Saskatchewan, Saskatchewan Trade and Conv Center, 1919 Saskatchewan Dr, Regina, Sask, Canada S4P 3V7. Phone: (800) 667-7191.

CANADA: YUKON ORDER OF PIONEERS: ANNIVERSARY. Dec 1. The Yukon Order of Pioneers held its founding meeting on Dec 1, 1894, at Fortymile, Yukon, Canada. It began as a vigilante police force to deter claim jumping, and later inaugurated Discovery Day (Aug 17), a statutory Yukon holiday commemorating the discovery of gold on Bonanza Creek.

December 1995

S	M	T	W	T	F	S
					1	2
3	4	5	6	7	8	9
10	11	12	13	14	15	16
17	18	19	20	21	22	23
24	25	26	27	28	29	30
31						

CENTRAL AFRICAN REPUBLIC: NATIONAL DAY. Dec 1. National holiday.

CHAMPAGNE DINNER. Dec 1. Rosecliff, Newport, RI. Dinner and tastings of imported champagnes held in the beautiful ballroom of Rosecliff, a mansion modelled after the Grand Trianon at Versailles. Reservations required. Annually, the first Friday in December. Est attendance: 200. For info: The Preservation Society of Newport County, 424 Bellevue Ave, Newport, RI 02840. Phone: (401) 847-1000.

CHRISTMAS GREENS SHOW. Dec 1-3. Jackman-Long Bldg, State Fairgrounds, Salem, OR. 60,000 square feet of Christmas ideas. Half the building houses decorated trees, large Christmas displays, 100s of floral arrangements, wreaths and wall hangings. The other half houses 106 handcrafters and their wares. A wide variety of crafts, also fresh greens and wreaths. Food available in restaurant area. Proceeds benefit community projects. Annually, the second Friday, Saturday and Sunday after Thanksgiving. Est attendance: 15,000. Sponsor: Linda Nelson, Willamette Christmas Assn, 1856 Dearborn Ave, NE, Salem, OR 97303. Phone: (503) 393-4439.

CHRISTMAS ON THE PRAIRIE. Dec 1-3. Saunders County Museum, Wahoo, NE. Old-fashioned Christmas featuring entertainment by local groups, lots of period costumes, special postal cancellation, children's activities common to the 1800s and demonstrations in the historical village decorated in the 1800s style. Annually, the first weekend after Thanksgiving weekend (1 full week after Thanksgiving). Sponsor: Christmas on the Prairie Steering Committee. Est attendance: 300. For info: Deb Playfair, Saunders County Museum, 240 N Walnut, Wahoo, NE 68066-1858. Phone: (402) 443-3090.

CHRISTMAS PAST AT AUDUBON ACRES. Dec 1-2. Audubon Acres, 900 Sanctuary Rd, Chattanooga, TN. Enjoy the sights, tastes, smells and excitement of Christmas long ago; see demonstrations and reenactments of early skills; enjoy a stroll to the swinging bridge to observe signs of nature in the winter. Sponsor: Chattanooga Audubon Society. Est attendance: 20,000. For info: Stacy Tilley, Audubon Acres, 900 N Sanctuary Rd, Chattanooga, TN 37421. Phone: (615) 892-1499.

CHRISTMAS SHOWCASE AND SALE. Dec 1-3. Statehouse Convention Center, Excelsior Hotel, Little Rock, AR. High-quality juried holiday craft show. Annually, the first weekend in December. Est attendance: 5,000. For info: Arkansas Craft Guild, PO Box 800, Mountain View, AR 72560. Phone: (501) 269-3897.

CHRISTMAS STROLL. Dec 1. Lewistown, MT. Christmas parade, tree decorating contest, Santa comes to Lewistown, food booths, horse and wagon (sled) rides and Christmas button for a chance on prizes. Annually, the first Friday after Thanksgiving weekend. Est attendance: 1,500. For info: Lewistown Chmbr of Commerce, PO Box 818, Lewistown, MT 59457. Phone: (406) 538-5436.

CHRISTMAS STROLL. Dec 1-3. Nantucket Island, MA. Christmas trees, costumed carolers, theatrical performances. Annually, the first weekend in December. For info: Nantucket Island Chamber of Commerce, 48 Main St, Nantucket, MA 02554-3595. Phone: (508) 228-1700.

COLORECTAL CANCER EDUCATION AND AWARENESS MONTH. Dec 1-31. To educate consumers, patients, and professionals regarding the need for early diagnosis and treatment of colorectal cancer. Sponsor: Pharmacists Planning Service, Inc, Frederick S. Mayer, Pres, PO Box 1336, Sausalito, CA 94966. Phone: (415) 332-4066.

COUNTRY CHRISTMAS. Dec 1-31. (Except Dec 25.) Coshocton, OH. Nineteenth-century holiday atmosphere, 51-room country inn, hot-mulled cider in visitor center, occasional cooking demonstrations in doctor's house, candlelighting ceremony (Dec 2, 9, 16), special weekend activities including Christmas wagon rides first three weekends in December. Est attendance: 8,000. For info: Roscoe Village Fdtn, 381 Hill St, Coshocton, OH 43812. Phone: (800) 877-1830.

★ Chase's 1995 Calendar of Events ★   Dec 1

COWBOY CHRISTMAS. Dec 1-3. Wickenburg, AZ. Gathering of cowboy poets and singers who carry on the workings of ranch life and traditional cowboy life in verse and song. Annually, the first weekend in December. Est attendance: 1,500. For info: Wickenburg Chamber of Commerce, PO Drawer CC, Wickenburg, AZ 86358. Phone: (602) 684-5479.

CREOLE CHRISTMAS IN THE FRENCH QUARTER. Dec 1-31. New Orleans, LA. Nostalgic caroling in Jackson Square, candlelight tours of 19th-century houses in holiday dress, holiday concerts, ice carving competition, cooking demonstrations, Celtic Christmas program, gingerbread house exhibits, tours to celebration in the Oaks, levee bonfire excursions and other family activities. For info: Sandra Dartus, Exec Dir, 1008 N Peters St, New Orleans, LA 70116. Phone: (800) 673-5725.

ENGLAND: WINMAU WORLD DARTS CHAMPIONSHIPS. Dec 1-2. London. Major event in the darts calendar with top players from all over the world competing. For info: Mr Olly Craft, Dir, British Darts Organisation, 2 Pages Lane, London, England N10 1PS.

FLOSSIE BEADLE WEEK. Dec 1-7. Lakeside, CA. Honoring the first historian of Lakeside. Est attendance: 100. Sponsor: Lakeside Historical Society, PO Box 1886, Lakeside, CA 92040. Phone: (619) 561-1886.

GETTYSBURG YULETIDE FESTIVAL. Dec 1-3. (Also Dec 9-10.) Gettysburg, PA. Tours of decorated historic homes, live nativity scene, caroling and handbell choirs, Christmas parade, Breakfast with Santa, community concerts, holiday dessert tasting and candlelight walking tour. Annually, the first and second weekends in December. For info: Gettysburg Travel Council, 35 Carlisle St, Gettysburg, PA 17325. Phone: (717) 334-6274.

GINGERBREAD AND LACE: A CHRISTMAS CELEBRATION. Dec 1. (Also Dec 8.) Ash Lawn-Highland, Charlottesville, VA. Customs of 1870s: caroling, ornament-making, tree-trimming and refreshments. Est attendance: 100. For info: Ash Lawn-Highland, Rte 6, Box 37, Charlottesville, VA 22902. Phone: (804) 293-9539.

ICELAND: UNIVERSITY STUDENTS' CELEBRATION: ANNIVERSARY. Dec 1. Marks the day in 1918 when Iceland became an independent state from Denmark (but still remained under the king of Denmark).

LEGENDS OF CHRISTMAS. Dec 1-Jan 1. Rock City Gardens, Lookout Mountain, GA (near Chattanooga, TN). See the Legends of Christmas depicted throughout the gardens. Discover the stories behind Christmas wreaths, the first Christmas tree, Christmas stockings or why we put candles in our windows. Selected Top 20 event by Southeast Tourism Society. Annually, December thru New Year's Day. Est attendance: 10,500. For info: Shelda Spencer Rees, Mktg, 1400 Patten Rd, Lookout Mountain, GA 30750. Phone: (706) 820-2531.

MACAU: RESTORATION OF INDEPENDENCE DAY. Dec 1. Macau is Chinese territory under Portuguese administration. Portugal is due to return Macau to China in December 1999. First colonized by Portugal in 1557.

MITTEN TREE KICK-OFF CEREMONY. Dec 1. Abel Wolman Municipal Building, Baltimore, MD. To collect mittens and gloves for needy families. Kick-off ceremony includes holiday musical entertainment and a visit from Santa Claus and his elves. Public is invited to hang their donations on ten-foot live fir tree. Donations will be collected throughout the month of December. Sponsor: The Dept of Rec and Parks, Office of Adventures in Fun, Clarence "Du" Burns Arena, 1301 S Ellwood Ave, Baltimore, MD 21224. Phone: (410) 396-9177.

MOORE, JULIA A. DAVIS: BIRTH ANNIVERSARY. Dec 1. Julia Moore, known as the "Sweet Singer of Michigan," was born in a log cabin in Kent County, MI, on Dec 1, 1847. A writer of homely verse and ballads, Moore enjoyed remarkable popularity and gave many public readings before realizing that her public appearances were occasions for laughter and ridicule. One newspaper account of her recitation of a tragic ballad declared: "A wild ovation startled the poet. Pandemonium broke loose in the galleries. The crowd shouted and stamped its feet." Her poems were said to be "so bad, her subjects so morbid and her naivete so genuine" that they were actually gems of humorous genius. At her final public appearance she told her audience: "You people paid 50 cents to see a fool, but I got $50 to look at a house full of fools." Moore died in 1920.

NATIONAL STRESS-FREE FAMILY HOLIDAYS MONTH. Dec 1-31. The holidays are so fraught with busy schedules that families often miss out on quality time together because outside demands have left them virtually drained. This observance is a reminder for parents to strive for more stress-free holidays for their families. Annually, the month of December. For info: Teresa Langston, Parenting Without Pressure, 1330 Boyer St, Longwood, FL 32750. Phone: (407) 767-2524.

NEWS 4 PARADE OF LIGHTS. Dec 1-2. Denver, CO. This evening holiday parade features dazzling lighted floats, giant helium-filled holiday balloons, magical costumed characters, the area's best marching bands, high-stepping equestrian units and much more. Friday at 7:30, Saturday at 7:00. Also, the Olive Garden Reindeer Dash, a 3K family fun run, will be held Saturday at 6:40 just before the parade. Est attendance: 350,000. For info: Downtown Denver Partnership, Inc, 511 16th St, Ste 200, Denver, CO 80202. Phone: (303) 534-6161.

OLD-FASHIONED CHRISTMAS CELEBRATION. Dec 1-3. (Also Dec 8-10.) Bedford, PA. The spirit of an old-fashioned Christmas recaptured—pathways lit with candle luminaries, live nativity scene in a log barn, original log houses decorated with old-fashioned ornaments and choral groups singing in the log church. Evenings 5-9. For info: Old Bedford Village, Jamie Nesbit, PO Box 1976, Bedford, PA 15522. Phone: (814) 623-1156.

OLD TOWN TEMPE FALL FESTIVAL OF THE ARTS. Dec 1-3. Tempe, AZ. Featuring more than 500 artists and craftspeople, traditional and ethnic foods, continuous entertainment on three stages and a children's activity area. Annually, the first weekend in December. Est attendance: 225,000. For info: Mill Avenue Merchants Assn, 520 S Mill Ave, Ste 201, Tempe, AZ 85281. Phone: (602) 967-4877.

PORTUGAL: INDEPENDENCE DAY. Dec 1. Public holiday.

ROMANIA: NATIONAL DAY. Dec 1. National holiday.

◆ **ROSA PARKS DAY: 40th ANNIVERSARY OF ARREST.** Dec 1. Anniversary of the arrest of Rosa Parks, Dec 1, 1955, in Montgomery, AL, for refusing to give up her seat and move to the back of a municipal bus. Her arrest triggered a year-long boycott of the city bus system and led to legal actions which ended racial segregation on municipal buses throughout the southern US. The event has been called the birth of the modern civil rights movement. Rosa McCauley Parks was born at Tuskegee, AL, on Feb 4, 1913.

SAFE TOYS AND GIFTS MONTH. Dec 1-31. What are the top 10 dangerous toys to children's eyesight? Prevent Blindness America (formerly known as National Society to Prevent Blindness) issues a list of toys hazardous to eyesight. Tips on how to choose age-appropriate toys will be distributed. Sponsor: Prevent Blindness America, Marita Gomez, Media Relations, 500 E Remington Rd, Schaumburg, IL 60173. Phone: (800) 331-2020.

Dec 1-2 ☆ Chase's 1995 Calendar of Events ☆

SPACE MILESTONE: *ATLANTIS* (US). Dec 1. Deployed a new-generation spy satellite able to peer down on the Soviet Union in darkness and through clouds. All-military crew: Navy Cmdr Robert Gibson, Air Force Col Guy Gardner, Air Force Col Richard Mullane, Air Force Lt Col Jerry Ross and Navy Cmdr William Shepherd. Launched Dec 1, 1988. Landed Dec 6, 1988.

UNITED NATIONS: WORLD AIDS DAY. Dec 1. In 1988, the World Health Organization of the United Nations declared Dec 1 as World AIDS Day, an international day of awareness and education about AIDS. The WHO is the leader in global direction and coordination of AIDS prevention, control, research and education. Also see World AIDS Day entry below for information address in US.

UNIVERSAL HUMAN RIGHTS MONTH. Dec 1-31. To disseminate throughout the world information about human rights and distribute copies of the Universal Declaration of Human Rights in English and other languages. Please send $1 to cover expense of printing, handling and postage. Annually, the month of December. Sponsor: Intl Society of Friendship & Good Will, Dr Stanley Drake, Pres, 9538 Summerfield St, Spring Valley, San Diego County, CA 91977-2852. Phone: (619) 466-8882.

VICTORIAN CHRISTMAS WALK. Dec 1-2. Village of East Davenport, Davenport, IA. 50 village shops come alive with costumed people recalling Christmas traditions of the Victorian era. Est attendance: 15,000. For info: Village of East Davenport Assoc, 2215 E 12th St, Davenport, IA 52803.

WORLD AIDS DAY. Dec 1. To focus world attention on the fight against AIDS. For info: World AIDS Day, Amer Assn for World Health, 1129 20th St NW, Ste 400, Washington, DC 20036. Phone: (202) 466-5883.

YULETIDE TRADITIONS. Dec 1-31. Special events at Ash Lawn-Highland, Monticello and Historic Michie Tavern. For info: Charlottesville/Albemarle Visitors Bureau, Box 161 Dept YT, Charlottesville, VA 22902. Phone: (804) 977-1783.

ZOOLIGHTS. Dec 1-31. (Closed Dec 24-25.) Point Defiance Zoo & Aquarium, Point Defiance Park, Tacoma, WA. A spectacular display of holiday lights to be seen from a walk-through tour of the zoo. More than 450,000 lights form animal silhouettes. Est attendance: 136,000. For info: Point Defiance Zoo & Aquarium, 5400 N Pearl St, Tacoma, WA 98407-3218. Phone: (206) 591-5337.

BIRTHDAYS TODAY

Woody Allen (Allen Stewart Konigsburg), 60, actor, writer, producer (Oscar for *Annie Hall*; *Sleeper, Manhattan Murder Mystery*), born Brooklyn, NY, Dec 1, 1935.
Carol Alt, 35, model, born at Queens, NY, Dec 1, 1960.
Bette Midler, 50, singer ("You Are the Wind Beneath My Wings"), actress (*Beaches, For the Boys, The Rose, Big Business, Outrageous Fortune*), Honolulu, HI, Dec 1, 1945.
Richard Pryor, 55, actor, comedian (*Blue Collar, Stir Crazy*, "The Richard Pryor Show"), born Peoria, IL, Dec 1, 1940.
Lou Rawls, 60, blues singer ("A Natural Man," "You'll Never Find Another Love Like Mine," "You've Made Me So Very Happy"), actor, born Chicago, IL, Dec 1, 1935.
Lee Trevino, 56, golfer, born at Dallas, TX, Dec 1, 1939.

December 1995

S	M	T	W	T	F	S
					1	2
3	4	5	6	7	8	9
10	11	12	13	14	15	16
17	18	19	20	21	22	23
24	25	26	27	28	29	30
31						

DECEMBER 2 — SATURDAY
336th Day — Remaining, 29

ANTIETAM NATIONAL BATTLEFIELD MEMORIAL ILLUMINATION. Dec 2. Antietam National Battlefield, Sharpsburg, MD. 7th annual memorial illumination to the 23,110 casualties at the Battle of Antietam. Nearly 500 volunteers light 23,110 luminaries (candles in bags) that are placed along a 4.5-mile route through the Battlefield. Annually, the first Saturday in December; rain/snow date the following Saturday (Dec 9). Free admission; donations accepted. Sponsor: Little Heiskell Charter Chapter of American Business Women's Assn (ABWA). Fax:(301) 733-7385. Est attendance: 14,000. For info: Georgene Charles, Chair, ABWA, PO Box 188, 11850 National Pike Rd, Clear Spring, MD 21722. Phone: (301) 842-2722.

APPALACHIAN POTTERS MARKET. Dec 2. McDowell High School commons area, Marion, NC. One-day display and sale of clay work only. Annually, the first Saturday in December. Sponsor: McDowell Arts and Crafts Assn. Est attendance: 1,500. Sponsor: Nancy Greenlee, PO Box 1387, Marion, NC 28752. Phone: (704) 652-8610.

ARMY-NAVY FOOTBALL GAME. Dec 2. Veterans Stadium, Philadelphia, PA. To possibly determine winner of Commander-in-Chief's trophy. Est attendance: 67,000. For info: Office of the Dir of Intercollegiate Athletics, Building 639, USMA, West Point, NY 10996. Phone: (914) 938-3512.

ARTIFICIAL HEART TRANSPLANT: ANNIVERSARY. Dec 2. On Dec 2, 1982, Barney C. Clark, 61, became the first recipient of a permanent artificial heart. The operation was performed at the University of Utah Medical Center in Salt Lake City. Near death at the time of the operation, Clark survived almost 112 days after the implantation. He died on Mar 23, 1983.

BACHELOR SOCIETY WILDERNESS WOMAN CONTEST AND BACHELOR AUCTION. Dec 2-3. Talkeetna, AK. A bachelor get-together including "Wilderness Woman" contest with women competing in a variety of Alaskan "bush" activity events, followed by Bachelors Ball at which bachelors are auctioned off for drinks with highest-bidding women. Est attendance: 400. For info: Talkeetna Bachelor Society, PO Box 258, Talkeetna, AK 99676. Phone: (907) 733-2423.

BROWN, JOHN: EXECUTION ANNIVERSARY. Dec 2. Abolitionist leader who is remembered for his raid on the US Arsenal at Harper's Ferry was hanged for treason at Charleston, WV, Dec 2, 1859.

CALLAS, MARIA: BIRTH ANNIVERSARY. Dec 2. American opera singer born at New York City, Dec 2, 1923. Died at Paris, Sept 16, 1977.

CELEBRATIONS AROUND THE WORLD '95: EXHIBITION. Dec 2-Jan 7. Huntsville Museum of Art, Huntsville, AL. An educational exhibition of art, artifacts, toys, costumes and decorations having to do with festivals around the world. Each holiday season the museum concentrates on a different country or region. Fun and colorful for both children and adults. Annual. Est attendance: 3,000. For info: Marylyn Coffey, Publicist or Debra Taylor, Educ Dir, Huntsville Museum of Art, Von Braun Civic Center, 700 Monroe St SW, Huntsville, AL 35801. Phone: (205) 535-4350.

CHIEFTAINS RUN. Dec 2. Berry College Campus, Rome, GA. This run is AAU sanctioned and certified. 5K, 10K, 15K and open to males and females. Held on the beautiful 28,000-acre Berry College campus off Highway 27 North. Annually, the first Saturday in December. For info: Berry College, Mt. Berry, GA 30149. Est attendance: 700. Sponsor: Junior Service League of Rome, PO Box 1003, Rome, GA 30161. Phone: (706) 291-6960.

CHRISTMAS CRAFT SHOW. Dec 2-3. H.O. Weeks Recreation Center, Aiken, SC. This show consists of more than 200 Christmas exhibits and other items from the southeast's top craftsmen. The displays range from woodwork to fine porcelain sculpture. 25th annual show. Est attendance: 13,000. For info:

Anne Hagelston, PO Box 1177, Aiken, SC 29802. Phone: (803) 642-7630.

CHRISTMAS FESTIVAL. Dec 2. Natchitoches, LA. Known as Festival of Lights, featuring junior and main parades, fireworks, food, entertainment and Christmas lighting. Annually, the first Saturday in December. Toll-free phone: (800) 259-1714. Est attendance: 150,000. For info: Natchitoches Parish Tourist Commission, Box 411, Natchitoches, LA 71458. Phone: (318) 352-4411.

CHRISTMAS IN BEAR LAKE. Dec 2. Downtown Bear Lake, MI. Horse-drawn hayrides, Santa in his workshop, crafts and goodies, Christmas caroling, a community sing-along and a live Nativity scene. Sponsor: Bear Lake Promoters. Est attendance: 600. For info: Marge Six, 6511 S Shore Rd, Bear Lake, MI 49614. Phone: (616) 864-3038.

CHRISTMAS ON THE RIVER. Dec 2. Ross's Landing, Chattanooga, TN. The Tennessee River comes alive with the splendid colors of decorated boats as they cruise past Ross's Landing. Area groups perform Christmas carols. Evening concludes with a spectacular fireworks show. Hours: 10 AM–9 PM. Annually, the first Saturday in December. Sponsor: Downtown Partnership. Est attendance: 10,000. For info: Carla Watson, Event Coord, Christmas on the River, Downtown Partnership, 701 Broad St, Ste LL1, Chattanooga, TN 37402. Phone: (615) 765-0771.

CUT YOUR OWN CHRISTMAS TREE. Dec 2–3. (Also Dec 9–10, 16–24.) Charlottesville, VA. Est attendance: 100. For info: Ash Lawn–Highland, Rte 6, Box 37, Charlottesville, VA 22902. Phone: (804) 293-9539.

DICKENS ON THE STRAND. Dec 2–3. Galveston, TX. To focus on the 19th-century architecture of Galveston's Strand and ties to Charles Dickens's 19th-century London. Annually, the first Saturday–Sunday in December. Est attendance: 60,000. For info: Galveston Historical Foundation, 2016 Strand, Galveston, TX 77550. Phone: (409) 765-7834.

ENGLAND: WALTER PLINGE DAY. Dec 2. A day to recognize Walter Plinge, said to have been a London pub landlord in 1900. His generosity to actors led to the use of his name as an actor, in play programs, usually to conceal the fact that an actor was playing more than one role. See also: "George Spelvin Day" (Nov 15) for US equivalent.

FIREMEN'S CARNIVAL. Dec 2. Nome, AK. Young and old alike enjoy games of chance, concession stands, bingo and cakewalks. Annually, the first Saturday in December. Sponsor: Nome Volunteer Fire Department. Est attendance: 1,000. For info: Robert K. Lewis, Nome Volunteer Fire Dept, Box 82, Nome, AK 99762. Phone: (907) 443-2439.

FIRST SELF-SUSTAINING NUCLEAR CHAIN REACTION: ANNIVERSARY. Dec 2. On Dec 2, 1942, physicist Enrico Fermi led a team of scientists at the University of Chicago in producing the first controlled, self-sustaining nuclear chain reaction. Their first, simple nuclear reactor was built under the stands of the University's football stadium.

HERITAGE CHRISTMAS. Dec 2. (Also Dec 9.) Old Mill Museum Complex, Lindsborg, KS. A celebration of Christmas past with the music, drama and costumes of a traditional pioneer Christmas on the Kansas prairie. Admission by donation. Annually, the first two Saturday evenings in December. Est attendance: 1,200. Sponsor: McPherson County Old Mill Museum, PO Box 94, Lindsborg, KS 67456. Phone: (913) 227-3595.

KRISTKINDL MARKT. Dec 2–3. Stone Hill Winery Pavilion, Hermann, MO. A 600-year-old German tradition with juried crafts, gifts, German soups, sausages and many interesting booths. Christmas carolers and entertainment in a wonderland setting. Annually, the first weekend in December. For info: Dolores Smith, Coord, HTG Visitor Center, 306 Market St, Hermann, MO 65041. Phone: (314) 486-2744.

LAO PEOPLE'S DEMOCRATIC REPUBLIC: NATIONAL DAY. Dec 2. National holiday.

LIGHTED BOAT PARADE ON THE TENN-TOM WATERWAY. Dec 2. Columbus, MS. Boat owners from throughout the area decorate their boats in holiday themes and parade at Lock and Dam area in Columbus. Grand Finale fireworks show lights up the sky and is reflected in the water. Est attendance: 20,000. For info: Columbus Conv and Visitors Bureau, PO Box 789, Columbus, MS 39703. Phone: (800) 327-2686.

MADD'S NATIONAL CANDLELIGHT VIGIL OF REMEMBRANCE AND HOPE. Dec 2. Whitehouse, Washington, D.C. Families from all across the nation whose loved ones have been killed or injured by a drunk driver come together for a poignant remembrance of music, candlelight and spoken words. The event gives victim families a special way to remember their loved one at holiday time and to seek solutions for a safer America. Est attendance: 500. Sponsor: Janice Lord, Dir of Victim Service, MADD Natl Office, 511 E John Carpenter Freeway, Ste 700, Irving, TX 75062. Phone: (214) 744-6233.

MADRIGAL DINNER AND CONCERT. Dec 2–3. Mount Mary College, Milwaukee, WI. Saturday dinner concert and silent auction. Sunday dessert concert and silent auction. Est attendance: 300. For info: Mary Cain, Mount Mary College, 2900 N Menomonee River Pkwy, Milwaukee, WI 53222. Phone: (414) 258-4810.

MONROE DOCTRINE: ANNIVERSARY. Dec 2. President James Monroe, in his annual message to the Congress, Dec 2, 1823, enunciated the doctrine which bears his name and which was long hailed as a statement of US policy. "... In the wars of the European powers in matters relating to themselves we have never taken any part ... we should consider any attempt on their part to extend their system to any portion of this hemisphere as dangerous to our peace and safety...."

NORWEGIAN CHRISTMAS. Dec 2–3. Brooklyn Park, MN. Old-fashioned farm Christmas with turn-of-the-century decorations, carolers, making of traditional gifts, lefse and other Norwegian delicacies, visit by St. Nicholas and sleigh rides. Est attendance: 3,500. For info: Kay Grotenhuis, Brooklyn Park Historical Farm, 4345 101st Ave N, Brooklyn Park, MN 55443. Phone: (612) 493-4604.

OLD-FASHIONED DANISH CHRISTMAS. Dec 2. Dannebrog, NE. Day filled with Christmas Tree Fantasy, living nativity scene, crafters/working artists expo, Danish buffet luncheon, Danish pastry, Danish tree ornament cut-outs demonstration and more. Annually, the first Saturday in December. For info: Shirley Johnson, Treasurer, Dannebrog Area Booster Club, 522 E Roger Welsch Ave, Dannebrog, NE 68831. Phone: (308) 226-2237.

PALM HARBOR ART, CRAFT AND MUSIC FESTIVAL. Dec 2–3. Palm Harbor, FL. Exhibitors from all over the US. Juried arts and craft show. Est attendance: 40,000. For info: Palm Harbor Chamber of Commerce, 33451 US 19N, Ste 300, Palm Harbor, FL 34684-2699. Phone: (813) 784-4287.

★ **PAN AMERICAN HEALTH DAY.** Dec 2. Presidential Proclamation 2447, of Nov 23, 1940, covers all succeeding years. Always Dec 2. The 1940 Pan American Conference of National Directors of Health adopted a resolution recommending that a "Health Day" be held annually in the countries of the Pan American Union.

PERIGEAN SPRING TIDES. Dec 2–3. Spring tides, the highest possible tides, occur when New Moon or Full Moon takes place within 24 hours of the moment the Moon is nearest Earth (perigee) in its monthly orbit, on Dec 2 at 7:00 AM EST.

SANTA BY STAGE COACH PARADE. Dec 2. El Centro, CA. Annually, the first Saturday in December. Est attendance: 25,000. Sponsor: El Centro Chamber of Commerce, Box 3006, El Centro, CA 92244. Phone: (619) 352-3681.

SCOTTISH CHRISTMAS WALK. Dec 2. Alexandria, VA. Alexandria's annual salute to its 18th-century Scottish founders. Parade of kilted pipers, highland dancers. Festivities include concerts, church services, sales, crafts, performances, and children's events. Annually, the first Saturday in December. Est attendance: 30,000. For info: The Campagna Center, 418 S Washington St, Alexandria, VA 22314. Phone: (703) 549-0111.

SEURAT, GEORGES PIERRE: BIRTH ANNIVERSARY. Dec 2. French painter born at Paris, France, Dec 2, 1859. Died there Mar 29, 1891.

SNOWFLAKE FESTIVAL. Dec 2–10. Klamath Falls, OR. Celebration of winter season in the community. Est attendance: 25,000. For info: Klamath Falls City Parks & Rec, 226 S 5th St, Klamath Falls, OR 97601. Phone: (503) 883-5371.

SPACE MILESTONE: *SOYUZ 16* (USSR). Dec 2. Six-day mission began Dec 2, 1974, with cosmonauts A.V. Flipchenko and N.N. Rukavishnikov. Rehearsal for US-USSR link-up in July 1975.

TWELVE VILLAGES OF CHRISTMAS. Dec 2–3. (Also Dec 9–10, 16–17 and 22–23.) Each of 12 cities in Washington County, KS, organizes events including special lighting of entire county, craft festivals, drawings, special musical programs, home tour, retail open houses, youth Christmas card contest; some towns have a Christmas tree in every yard; large Christmas stocking give-aways; and much more. Annually, the three weekends before Christmas in December. Est attendance: 6,000. Sponsor: Washington County Economic Development, Mrs Billie Jo Smart, Court House, Washington, KS 66968. Phone: (913) 325-2116.

UNITED ARAB EMIRATES: NATIONAL DAY. Dec 2. Anniversary of the day when a federation of seven sheikdoms declared independence (Dec 2, 1971) and became known as the United Arab Emirates.

VICTORIAN CHRISTMAS HOME TOUR. Dec 2. Leadville, CO. Catch a glimpse of Leadville's grand and glorious past with a tour of six historic Victorian homes decked for the season. An elegant brunch and period costumes also featured. For info: Chamber of Commerce, Box 861, Leadville, CO 80461. Phone: (800) 933-3901.

WICHITA WINTERFEST. Dec 2. East Douglas from Topeka to the Arkansas River, Wichita, KS. 4th annual festival for all ages to kick off the winter in holiday style. Carnival rides, games and horse-drawn carriage rides. 5k race for running, rolling or walking. Holiday food court, street carolers. Fireworks to Christmas music over the river, accompainied by the official holiday lighting ceremony, followed by the Snow Ball Dance. Annually, first Saturday of December. Est attendance: 35,000. For info: Downtown United, 350 W Douglas, Wichita, KS 67202. Phone: (316) 268-1130.

A YORKTOWN CHRISTMAS. Dec 2–16. Yorktown Victory Center, Yorktown, VA. In the museum's re-created 18th-century farmsite and Continental Army encampment, costumed interpreters prepare traditional holiday fare and tell stories of Christmas during the Revolutionary War era. Est attendance: 2,500. Sponsor: Jamestown-Yorktown Foundation, Media Rel, PO Box JF, Williamsburg, VA 23187. Phone: (804) 253-4838.

BIRTHDAYS TODAY

Randy Gardner, 37, figure skater, born at Marina del Rey, CA, Dec 2, 1958.
Adolph Green, 80, actor, composer, born at New York, NY, Dec 2, 1915.
Alexander Meigs Haig, Jr, 71, general, presidential candidate, former secretary of state in Reagan administration ("I am in control."), born at Philadelphia, PA, Dec 2, 1924.
Julie Harris, 70, actress ("Knots Landing," *I Am a Camera, Member of the Wedding*), born Grosse Pointe, MI, Dec 2, 1925.
Edwin Meese, III, 64, former US attorney general, born at Oakland, CA, Dec 2, 1931.
Garry Meier, 46, Chicago radio personality, born Chicago, IL, Dec 2, 1949.
Harry Reid, 56, US Senator (D, Nevada), born at Searchlight, NV, Dec 2, 1939.
Charlie Ventura, 79, musician, born at Philadelphia, PA, Dec 2, 1916.
William Wegman, 52, artist/photographer, born at Holyoke, MA, Dec 2, 1943.

DECEMBER 3 — SUNDAY
337th Day — Remaining, 28

ADVENT, FIRST SUNDAY. Dec 3. Advent includes the four Sundays before Christmas, Dec 3, Dec 10, Dec 17 and Dec 24 in 1995.

AMERICAN SOCIETY OF ASSOCIATION EXECUTIVES MANAGEMENT CONFERENCE. Dec 3–6. Chicago, IL. Annual conference with emphasis on peer-to-peer educational programs; includes non-hospitality suppliers. Est attendance: 2,000. For info: Lorri Lee, American Society of Assn Executives, 1575 Eye St NW, Washington, DC 20005-1168. Phone: (202) 626-2798.

BHOPAL POISON GAS DISASTER: ANNIVERSARY. Dec 3. At Bhopal, India, on the night of Dec 3, 1984, a leak of deadly gas (methyl isocyanate) at a Union Carbide Corporation plant killed more than 2,000 persons and injured more than 200,000 in the world's worst industrial accident.

CHRISTMAS CANDLELIGHT TOUR. Dec 3. Fredericksburg, VA. To instill the spirit of Christmases past by opening historic homes to the public. Eighteenth-century music played throughout the tour. Est attendance: 6,000. For info: Visitor Center, 706 Caroline St, Fredericksburg, VA 22401. Phone: (800) 678-4748.

CHRISTMAS TO REMEMBER. Dec 3. Laurel, MT. To officially open the Christmas season in Laurel, this day-long celebration includes the arrival of Santa, a community bazaar, trolley rides, children's craft activities, musical entertainment, a children's matinee and fireworks. Annually, the first Sunday of December. Sponsor: Christmas to Remember, Inc. Est attendance: 4,000. For info: Jean Carroll Thompson, PO Box 463, Laurel, MT 59044. Phone: (406) 628-4508.

CONRAD, JOSEPH: BIRTH ANNIVERSARY. Dec 3. Polish novelist, born Dec 3, 1857; died Aug 3, 1924. Author of *Lord Jim* and *Heart of Darkness*, among others.

December 1995	S	M	T	W	T	F	S
						1	2
	3	4	5	6	7	8	9
	10	11	12	13	14	15	16
	17	18	19	20	21	22	23
	24	25	26	27	28	29	30
	31						

Chase's 1995 Calendar of Events — Dec 3–4

FIRST HEART TRANSPLANT: ANNIVERSARY. Dec 3. Dr. Christiaan Barnard, a South African surgeon, performed the world's first successful heart transplantation on Dec 3, 1967 at Cape Town, South Africa. See also: "Barnard, Christiaan Neethling: Birthday" (Nov 8).

ILLINOIS ADMISSION DAY: ANNIVERSARY. Dec 3. Became 21st state on this day in 1818.

A JAMESTOWN CHRISTMAS. Dec 3–31. Jamestown Settlement, Williamsburg, VA. Recalls traditional English Christmas celebrations with decorations, holiday food preparation and the Lord of Misrule. A video program compares 17th-century holiday traditions to Christmas at Jamestown. Est attendance: 12,000. For info: Media Relations, Jamestown-Yorktown Fdtn, Box JF, Williamsburg, VA 23187. Phone: (804) 253-4838.

MONTOYA, CARLOS: BIRTH ANNIVERSARY. Dec 3. Guitarist and composer renowned for popularizing flamenco guitar music. His solo performances of the Spanish folk form lifted flameco from its traditional accompaniment role. Montoya never learned to read music and relied on the traditional improvisational nature of flamenco rooted in the Andalusian Gypsy form of music that stressed rhythms and harmonic patterns. He was born Dec 3, 1903, at Madrid, Spain, and died on March 3, 1993, at Wainscott, NY.

NATIONAL ALL BREED DOG SHOW. Dec 3. Convention Center, Atlantic City, NJ. One of the largest all breed dog shows in the country with more than 2,200 entrants from Canada and across the US. Est attendance: 5,000. For info: Greater Atlantic City Conv and Visitors Bureau, 2314 Pacific Ave, Atlantic City, NJ 08401. Phone: (609) 348-7100.

NETHERLANDS: MIDWINTER HORN BLOWING. Dec 3–Jan 6, 1996. Twente, and several other areas in the Netherlands. Midwinter horn blowing, folklore custom of announcing the birth of Christ, begins with Advent and continues until Epiphany (Jan 6) of the following year.

POLISH-AMERICAN CHRISTMAS GALA. Dec 3. Cicero, IL. Traditional and new Polish-American carols, narrated in English; spirited Polish and American folk songs; Polish folk and court dances; Polish holiday customs; colorful folk costumes. Annually, the first Sunday in December. Est attendance: 1,500. For info: The Lira Singers, 3750 W Peterson Ave, Chicago, IL 60659. Phone: (800) 547-5472.

STUART, GILBERT CHARLES: BIRTH ANNIVERSARY. Dec 3. American portrait painter whose most famous painting is that of George Washington. He also painted portraits of Madison, Monroe, Jefferson and other important Americans. Stuart was born near Narragansett, RI, on Dec 3, 1755, and died July 9, 1828, at Boston, MA.

UNITED NATIONS: INTERNATIONAL DAY OF DISABLED PERSONS. Dec 3. With Resolution 47/3 on Oct 14, 1992, at the end of the Decade of Disabled Persons, the General Assembly proclaimed Dec 3 to be an annual observance to promote the continuation of integrating the disabled into general society.

WSBA/WARM 103 CHRISTMAS CRAFT SHOW. Dec 3. York Fairgrounds, York, PA. More than 250 crafts, from country to contemporary, Victorian and southwestern, handcrafted furniture, wood carvings, dolls, jewelry, pottery, collectibles, quilts, baskets and much more. Admission fee. Est attendance: 6,000. For info: Gina M. Koch, Special Events Dir, PO Box 910, York, PA 17405. Phone: (717) 764-1155.

BIRTHDAYS TODAY

Jean Luc Godard, 65, filmmaker (*Breathless*, *Weekend*), born at Paris, France, Dec 3, 1930.

Ferlin Husky, 68, singer ("Gone," "On the Wings of a Dove"), born Flat River, MO, Dec 3, 1927.

Rick Mears, 44, auto racer, born at Wichita, KS, Dec 3, 1951.

Jaye P. Morgan, 63, singer ("That's All I Want From You," "The Longest Walk"), born at New York, NY, Dec 3, 1932.

Sven Vilhem Nykvist, 73, Swedish cinematographer, born at Moheda, Sweden, Dec 3, 1922.

Ozzy Osbourne, 47, singer, songwriter (originally lead singer for heavy metal Black Sabbath), born at Birmingham, England, Dec 3, 1948.

Andy Williams, 65, singer (platinum album *Love Story*, 13 gold albums), born Wall Lake, IA, Dec 3, 1930.

DECEMBER 4 — MONDAY
338th Day — Remaining, 27

BUTLER, SAMUEL: BIRTH ANNIVERSARY. Dec 4. English author, born Dec 4, 1835. Died London, June 18, 1902.

CARLYLE, THOMAS: 200th BIRTH ANNIVERSARY. Dec 4. Scottish essayist and historian, born Dec 4, 1795. Died at London, Feb 4, 1881. "A well-written Life is almost as rare as a well-spent one," Carlyle wrote in his *Critical and Miscellaneous Essays*.

CENTRAL AFRICAN REPUBLIC: NATIONAL DAY OBSERVED. Dec 4. Commemorates Proclamation of the Republic on Dec 1, 1958. Usually observed on the first Monday in December.

CHASE'S ANNUAL EVENTS 38th BIRTHDAY. Dec 4. First copies of first edition of *Chase's Calendar of Annual Events* (for the year 1958) were delivered by the printer at Flint, MI, Dec 4, 1957. Two thousand copies, consisting of 32 pages and listing 364 events, were printed. The 1995 edition consists of 592 pages listing more than 10,000 events.

DAY OF THE ARTISANS. Dec 4. Honors the nation's workers.

GIANT CHRISTMAS TREE AT ROCKEFELLER CENTER. Dec 4. (APPROXIMATE.) New York, NY. Tree-lighting of the 75-foot Norway spruce signals the opening of the holiday season in New York City. Date is approximate—always a weekday during the first week of December.

LAST AMERICAN HOSTAGE RELEASED IN LEBANON: ANNIVERSARY. Dec 4. A sad chapter of US history came to a close on Dec 4, 1991, when Terry Anderson, an Associated Press correspondent, became the final American hostage held in Lebanon to be freed. Anderson had been held since Mar 16, 1985, one of 15 Americans who were held hostage for from 2 months to as long as 6 years and 8 months. Three of the hostages, William Buckley, Peter Kilburn and Lt. Col. William Higgins, were killed during their captivity. The other hostages, released previously one or two at a time, were Jeremy Levin, Benjamin Weir, the Rev. Lawrence Martin Jenco, David Jacobsen, Thomas Sutherland, Frank Herbert Reed, Joseph Cicippio, Edward Austin Tracy, Alan Steen, Jesse Turner and Robert Polhill.

NATIONAL GRANGE FOUNDING. Dec 4. The 128th anniversary of the National Grange, which set in motion the first organized agricultural movement in the US.

RUSSELL, LILLIAN: BIRTH ANNIVERSARY. Dec 4. American singer and actress who in 1881 gained fame in the comic opera *The Great Mogul*. Born Helen Louise Leonard on Dec 4, 1861 at Clinton, IA. Married four times, and died on June 6, 1922 at Pittsburgh, PA.

ST. BARBARA'S DAY. Dec 4. On this day a young girl places a twig from a cherry tree in a glass of water. If it blooms by Christmas Eve, she is certain to marry the following year.

☆ Chase's 1995 Calendar of Events ☆

Dec 4-5

TANNEHILL VILLAGE CHRISTMAS. Dec 4-8. (Also Dec 11-15.) Tannehill Historical State Park, McCalla, AL. 1800s village with cabins decorated in old-fashioned Christmas style and staffed by volunteers wearing period costumes. Est attendance: 7,000. Sponsor: Tannehill Historical State Park, Iron and Steel Museum, Vicki Gentry, Dir, 12632 Confederate Pkwy, McCalla, AL 35111. Phone: (205) 477-5711.

BIRTHDAYS TODAY

Max Baer, Jr, 58, actor, producer ("The Beverly Hillbillies"; produced *Ode to Billy Joe*), born at Oakland, CA, Dec 4, 1937.
Jeff Bridges, 46, actor (*The Fisher King, The Last Picture Show*), born at Los Angeles, CA, Dec 4, 1949.
Helen M. Chase, 71, homemaker and former chronicler of contemporary civilization as co-editor of *Chase's Annual Events*, born at Whitehall, MI, Dec 4, 1924.
Deanna Durbin, 74, actress (*It Started with Eve, Can't Help Singing*), born at Winnipeg, Manitoba, Canada, Dec 4, 1921.
Chris Hillman, 53, musician (with The Byrds; "Mr Tambourine Man," "Turn, Turn, Turn"), born at Los Angeles, CA, Dec 4, 1942.
Stewart Rawlings Mott, 58, philanthropist, born at Flint, MI, Dec 4, 1937.

DECEMBER 5 — TUESDAY
339th Day — Remaining, 26

AFL-CIO: FOUNDING ANNIVERSARY. Dec 5. The American Federation of Labor and the Congress for Industrial Organizations joined together on this day in 1955, following 20 years of rivalry, to become the nation's leading advocate for trade unions.

HAITI: DISCOVERY DAY: ANNIVERSARY. Dec 5. Commemorates the discovery of Haiti by Christopher Columbus in 1492. Public holiday.

◆ MONTGOMERY BUS BOYCOTT BEGINS: 40th ANNIVERSARY. Dec 5, 1955. On Dec 1, 1955, Rosa Parks was arrested in Montgomery, AL, for refusing to give up her seat on a bus to a white man. In support of Parks, and to protest the arrest, the black community of Montgomery organized a boycott of the bus system. The boycott lasted from Dec 5, 1955, to Dec 20, 1956, when a US Supreme Court ruling was implemented in Montgomery, integrating the public transportation system.

MOZART, WOLFGANG AMADEUS: DEATH ANNIVERSARY. Dec 5. Born on Jan 27, 1756, at Salzburg, Austria, Johann Chrysostom Wolfgang Amadeus Mozart is regarded as one of the greatest musical geniuses of all time. As he gained proficiency at the harpsichord and the violin, he began composing at the age of 5. He later concentrated his efforts on the pianoforte and under the direction of his father, Mozart and his sister, Maria Anna, began touring Europe in 1762. Throughout the tours, Mozart continued composing and in 1775 he began service under the Archbishop in Salzburg. While in Salzburg he composed *Idomeneo*, the first of his mature operas. Mozart broke with the Archbishop in 1781 and settled in Vienna, risking his existence without official patronage. After years of waiting, he was engaged in the service of the emperor, Joseph II, as a chamber composer, but at far less wages than his predecessor. Although he enjoyed artistic and popular success, Mozart's finances were never secure. An ardent Freemason, he borrowed extensively from his Masonic brothers, despite the successes of his operas *The Marriage of Figaro, Don Giovanni, Cosi fan tutti* and *The Magic Flute* and his extensive work for the piano. Mozart became seriously ill in the fall of 1791, and his rapid decline led to speculation that he had been poisoned, although this later was proved false. Mozart died a pauper on Dec 5, 1791, at Vienna, Austria.

◆ PROHIBITION REPEAL: ANNIVERSARY. Dec 5. Congress proposed repeal of Amendment XVIII ("... the manufacture, sale, or transportation of intoxicating liquors, within, the importation thereof into, or the exportation thereof from the United States and all territory subject to the jurisdiction thereof, for beverage purposes is hereby prohibited ...") on Feb 20, 1933. By Dec 5, 1933, the repeal amendment had been ratified by the required 36 states and went into effect immediately as Amendment XXI to the US Constitution.

THAILAND: KING'S BIRTHDAY AND NATIONAL DAY. Dec 5. Celebrated throughout the kingdom with colorful pageantry. Stores and houses decorated with spectacular illuminations at night. Public holiday.

TWENTY-FIRST AMENDMENT TO THE US CONSTITUTION RATIFIED: ANNIVERSARY. Dec 5. Prohibition ended with the repeal of the Eighteenth Amendment, as the Twenty-First Amendment was ratified on Dec 5, 1933.

UNITED NATIONS: INTERNATIONAL VOLUNTEER DAY FOR ECONOMIC AND SOCIAL DEVELOPMENT. Dec 5. In a resolution of Dec 17, 1985, the United Nations General Assembly recognized the desirability of encouraging the work of all volunteers. It invited governments to observe, annually on Dec 5, the "International Volunteer Day for Economic and Social Development, urging them to take measures to heighten awareness of the important contribution of volunteer service." A day commemorating the establishment in December 1970 of the UN Volunteers program, and inviting world recognition of volunteerism in the international development movement. Info from United Nations, Dept of Public Info, New York, NY 10017.

VAN BUREN, MARTIN: BIRTH ANNIVERSARY. Dec 5. The eighth president of the US (term of office: Mar 4, 1837–Mar 3, 1841) was the first to have been born a citizen of the US. A widower for nearly two decades before he entered the White House, his daughter-in-law, Angelica, served as White House hostess during an administration troubled by bank and business failures, depression and unemployment. Van Buren was born at Kinderhook, NY, Dec 5, 1782, and died there July 24, 1862, leaving an estate of about $250,000.

◆ WHEATLEY, PHILLIS: DEATH ANNIVERSARY. Dec 5. Born in Africa in about 1753, Phillis Wheatley was brought to the US in 1761 and purchased as a slave by a Boston tailor named John Wheatley. She was allotted unusual privileges for a slave, including being allowed to learn to read and write. She wrote her first poetry at age 14, and her first work was published in 1770. Wheatley's fame as a poet spread throughout Europe as well as the US after her *Poems on Various Subjects, Religious and Moral* was published in England in 1773. She was invited to visit George Washington's army headquarters after he read a poem she'd written about him in 1776. Phillis Wheatley died Dec 5, 1784, at Boston.

BIRTHDAYS TODAY

Morgan Brittany (Suzanne Cupito), 45, actress (Katherine Wentworth on "Dallas"; Vivien Leigh in "Moviola"), born at Hollywood, CA, Dec 5, 1950.

December 1995

S	M	T	W	T	F	S
					1	2
3	4	5	6	7	8	9
10	11	12	13	14	15	16
17	18	19	20	21	22	23
24	25	26	27	28	29	30
31						

Joan Didion, 61, author, journalist (*After Henry, Run River*), born at Sacramento, CA, Dec 5, 1934.
Carrie Hamilton, 32, actress (Reggie in "Fame"; *Tokyo Pop*), daughter of Carol Burnett, born New York, NY, Dec 5, 1963.
Jim Messina, 48, singer (with Loggins and Messina; "Your Mama Won't Dance"), songwriter, born Maywood, CA, Dec 5, 1947.
Chad Mitchell, 59, singer (leader Chad Mitchell trio; "Lizzie Borden"), born at Portland, OR, Dec 5, 1936.
Art Monk, 38, football player, born at White Plains, NY, Dec 5, 1957.
Strom Thurmond, 93, US Senator (R, South Carolina), born at Edgefield, SC, Dec 5, 1902.
Calvin Trillin, 60, author (*American Stories, Remembering Denny*), born at Kansas City, MO, Dec 5, 1935.

DECEMBER 6 — WEDNESDAY
340th Day — Remaining, 25

ALTAMONT CONCERT ANNIVERSARY. Dec 6. A free concert featuring performances by the Rolling Stones; Jefferson Airplane; Santana; Crosby, Stills, Nash and Young; and the Flying Burrito Brothers turned into tragedy on this day in 1969. The "thank-you" concert for 300,000 fans was marred by overcrowding, drug overdoses and the fatal stabbing of a spectator by a member of the Hell's Angels motorcycle gang, who had been hired as security guards for the event. The concert was held at the Altamont Speedway, Livermore, CA.

ECUADOR: DAY OF QUITO. Dec 6. Commemorates founding of city of Quito by Spaniards in 1534.

FINLAND: INDEPENDENCE DAY. Dec 6. National holiday. Declaration of independence from Russia on this day, 1917.

GERALD FORD SWEARING-IN AS VICE-PRESIDENT: ANNIVERSARY. Dec 6. On this day in 1973, Gerald Ford was sworn in as vice president under Richard Nixon, following the resignation of Spiro Agnew after pleading no contest to a charge of income tax evasion. See also "Agnew, Spiro Theodore: Birthday" (Nov 9) and "Ford, Gerald Rudolph: Birthday" (July 14).

GERSHWIN, IRA: BIRTH ANNIVERSARY. Dec 6. Pulitzer Prize-winning American lyricist and author who collaborated with his brother, George, and with many other composers. Among his Broadway successes: *Lady Be Good, Funny Face, Strike Up the Band*, and such songs as: "The Man I Love," "Someone to Watch Over Me," "I Got Rhythm" and hundreds of others. Born at New York City, Dec 6, 1896, died at Beverly Hills, CA, Aug 17, 1983.

HALIFAX, NOVA SCOTIA, DESTROYED: ANNIVERSARY. Dec 6. On Dec 6, 1917, 1,654 people were killed in Halifax when the Norwegian ship *Imo* plowed into the French munitions ship *Mont Blanc*. *Imo* was loaded with supplies for war-torn Europe, and *Mont Blanc* was loaded with 4,000 tons of TNT, 2,300 tons of picric acid, 61 tons of other explosives and a deck of highly flammable benzene, which ignited and touched off an explosion. In addition to those killed, 1,028 were injured. A tidal wave caused by the explosion washed much of the city's remains out to sea.

KEEP AMERICA BEAUTIFUL, INC, ANNUAL MEETING (WITH NATIONAL AWARDS). Dec 6-9. Washington, DC. National convention for organization and its affiliates. Training programs, solid waste education workshops; National Awards presented on Dec 8. Sponsor: Keep America Beautiful, Inc, Denise Harkin, Exec Asst, 9 W Broad St, Stamford, CT 06902. Phone: (203) 323-8987.

KILMER, JOYCE (ALFRED): BIRTH ANNIVERSARY. Dec 6. American poet most famous for the poem "Trees," which was published in 1913, was born Dec 6, 1886. Kilmer was killed in action near Ourcy, France, in World War I, on July 30, 1918. Camp Kilmer was named for him.

LEVINE, CHARLES A.: DEATH ANNIVERSARY. Dec 6. Charles A. Levine, whose efforts to beat Charles Lindbergh across the Atlantic by plane were stymied by a lawsuit, nevertheless became the first air passenger to cross the Atlantic Ocean. Levine's 225-horsepower plane, *The Columbia*, was grounded when one of his co-pilots filed a suit hours after Lindbergh took off from Roosevelt Field. Not to be overshadowed by Lindbergh's success, Levine announced that his flight, leaving June 4, 1927, would fly beyond Paris and to Berlin, with himself as a passenger. Piloted by Clarence Chamberlin, the plane exhausted its fuel and landed in Eisleben, Germany, on June 6, 100 miles short of his goal. The flight set a new record of 3,911 miles in 43 hours of non-stop flight, besting Lindbergh by approximately 300 miles. Levine was born at North Adams, MA, in 1897 and died at Washington, DC, on Dec 6, 1991.

MOON PHASE: FULL MOON. Dec 6. Moon enters Full Moon phase at 8:27 PM, EST.

ST. NICHOLAS' DAY. Dec 6. One of the most venerated saints of both eastern and western Christian churches, of whose life little is known, except that he was Bishop Myra in the fourth century, and that from early times he has been one of the most often pictured saints, especially noted for his charity. Santa Claus and the presentation of gifts is said to derive from St. Nicholas.

THIRTEENTH AMENDMENT TO THE US CONSTITUTION RATIFIED: 130th ANNIVERSARY. Dec 6. On Dec 6, 1865, the Thirteenth Amendment to the Constitution was ratified abolishing slavery in the United States. The Thirteenth, Fourteenth and Fifteenth amendments are considered the Civil War Amendments.

BIRTHDAYS TODAY

Steve Bedrosian, 38, baseball player, born at Methuen, MA, Dec 6, 1957.
Dave Brubeck, 75, musician, born at Concord, CA, Dec 6, 1920.
Alfred Eisenstaedt, 97, photographer, born at Dirschau, Germany, Dec 6, 1898.
Thomas Hulce, 42, actor (*Amadeus, The Inner Circle*), born at Plymouth, MI, Dec 6, 1953.
Don Nickles, 47, US Senator (R, Oklahoma), born at Ponca City, OK, Dec 6, 1948.
Walter Perkins, 54, football coach, born at Olive, MS, Dec 6, 1941.

DECEMBER 7 — THURSDAY
341st Day — Remaining, 24

AMERICAN SPORTSCASTERS HALL OF FAME DINNER. Dec 7. New York, NY. Awards to the greats in sportscasting—Sportscaster of the Year, Graham McNamee Awardee, Hall of Fame inductees, Sports Personality of the Year, and Sports Legend Award. Est attendance: 600. For info: Louis O. Schwartz, Pres, ASA, 5 Beekman St, New York, NY 10038. Phone: (212) 227-8080.

ANGEL FIRE SKI AREA OPENING. Dec 7. Angel Fire, NM. Bring a can of food and receive a free lift ticket. Food goes to needy homes for Christmas. Est attendance: 1,000. For info: Angel Fire Chamber of Commerce, PO Box 547, Angel Fire, NM 87710. Phone: (800) 633-7463.

BELGIUM: LOVER'S FAIR. Dec 7. Arlon, Belgium. Traditional cultural observance. First Thursday in December.

Dec 7-8 ☆ *Chase's 1995 Calendar of Events* ☆

CANDLELIGHT. Dec 7-10. Old City Park, Dallas, TX. Turn-of-the-century decorations, horse-drawn surrey rides and traditional entertainment. Also featuring the Children's Wonderland with old fashioned crafts and games, bell choirs, orchestras, choirs and dancers. Annually, the second weekend in December. Est attendance: 14,000. For info: Old City Park, Dallas Co Heritage Soc, 1717 Gano St, Dallas, TX 75215. Phone: (214) 421-5141.

CATHER, WILLA: BIRTH ANNIVERSARY. Dec 7. American author born at Winchester, VA, Dec 7, 1873. Died at New York, NY, Apr 24, 1947.

CHRISTMAS AT PIONEER VILLAGE. Dec 7-8. Worthington, MN. Many different activities are held including: Christmas carolers, sleigh rides, Santa and Mrs. Claus, their elves, refreshments and the beautiful decorations. Free admission. For info: Worthington Area Chamber of Commerce, Darlene Macklin, Exec VP, 1121 Third Ave, PO Box 608, Worthington, MN 56187-0608. Phone: (507) 372-2919.

CHRISTMAS TOUR OF HOMES. Dec 7-8. Pella, IA. Four homes decorated for the holidays by Pella Garden Club members. Open to the public. Christmas tea at Pella Historical Village. Est attendance: 3,000. For info: Pella Historical Society, 507 Franklin, Pella, IA 50219. Phone: (515) 628-2409.

COTE D'IVOIRE: NATIONAL DAY. Dec 7.

DELAWARE RATIFIES CONSTITUTION: ANNIVERSARY. Dec 7. Delaware became the first state to ratify the proposed Constitution on Dec 7, 1787. It did so by unanimous vote.

NATIONAL FIRE SAFETY COUNCIL: FOUNDING ANNIVERSARY. Dec 7. Founded Dec 7, 1979, to promote fire prevention awareness. Sponsor: Natl Fire Safety Council, Inc, PO Box 378, Michigan Center, MI 49254-0378. Phone: (517) 764-2811.

PEARL HARBOR DAY: ANNIVERSARY. Dec 7. At 7:55 AM (local time) on Dec 7, 1941, "a date that will live in infamy," nearly 200 Japanese aircraft attacked Pearl Harbor, Hawaii, long considered the US "Gibraltar of the Pacific." The raid, which lasted little more than one hour, left nearly 3,000 dead. Nearly the entire US Pacific Fleet was at anchor there and few ships escaped damage. Several were sunk or disabled, while 200 US aircraft on the ground were destroyed. The attack on Pearl Harbor brought about immediate US entry into World War II, a Declaration of War being requested by President Franklin D. Roosevelt, and approved by the Congress on Dec 8, 1941.

SPACE MILESTONE: *APOLLO 17* (US). Dec 7. Launched on Dec 7, 1972, three-man crew: Eugene A. Cernan, Harrison H. Schmidt, Ronald E. Evans, explored moon, Dec 11-14. Lunar landing module named *Challenger*. Pacific splashdown, Dec 19.

TUSSAUD, MARIE GROSHOLTZ: BIRTH ANNIVERSARY. Dec 7. Creator of Madame Tussaud's waxwork museum born at Strasbourg, France, Dec 7, 1761. Many of the wax figures she created are still on view at Madame Tussaud's in London. She died at London, Apr 15, 1850.

WASSAIL CELEBRATION. Dec 7-10. Woodstock, VT. Activities include a horse rider and carriage parade, Santa comes to town, Wassail Dance, concert by Revere Handbell Choir, caroling and burning the Yule Log on the Village Green. Est attendance: 3,000. Sponsor: Woodstock Area Chamber of Commerce, PO Box 486, Woodstock, VT 05091. Phone: (802) 457-3555.

December 1995

S	M	T	W	T	F	S
					1	2
3	4	5	6	7	8	9
10	11	12	13	14	15	16
17	18	19	20	21	22	23
24	25	26	27	28	29	30
31						

BIRTHDAYS TODAY

Gregg Allman, 48, singer, musician ("Midnight Rider"; with Allman Brothers Band "Ramblin Man"), born Nashville, TN, Dec 7, 1947.

Johnny Bench, 48, baseball Hall of Famer, born at Oklahoma City, OK, Dec 7, 1947.

Larry Bird, 39, former basketball player, born at West Baden, IN, Dec 7, 1956.

Ellen Burstyn (Edna Rae Gilhooley), 63, actress (Catherine Martell in "Twin Peaks"; *The Hustler, The Exorcist, The Cemetery Club*), born at Detroit, MI, Dec 7, 1932.

Thad Cochran, 58, US Senator (R, Mississippi), born at Pontotoc, MS, Dec 7, 1937.

C. Thomas Howell, 29, actor ("Two Marriages," *Soul Man, Tank*), born at Los Angeles, CA, Dec 7, 1966.

Jacob Kainen, 86, artist, curator, born at Waterbury, CT, Dec 7, 1909.

Victor Kermit Kiam II, 69, business executive, born at New Orleans, LA, Dec 7, 1926.

Tom Waits, 46, singer (album *Foreign Affairs*, songwriter ("I Never Talk to Strangers"), born at Pomona, CA, Dec 7, 1949.

Eli Wallach, 80, actor (*The Tiger Makes Out*; Emmy "The Poppy Is Also a Flower"), born New York, NY, Dec 7, 1915.

DECEMBER 8 — FRIDAY
342nd Day — Remaining, 23

BERGALIS, KIMBERLY: DEATH ANNIVERSARY. Dec 8. Kimberly Bergalis, the first patient believed to have contracted the AIDS virus from a health care professional, died on Dec 8, 1991, at Ft. Pierce, FL. Her case sparked controversy over how the disease is transmitted, and incited calls to ban infected health care professionals from the workplace.

CAROLS BY CANDLELIGHT. Dec 8-10. Gunston Hall Plantation, Lorton, VA. The Candlelit Hall will be decorated for Christmas in the 18th-century manner, and period music will be played. Caroling and refreshments in the reception center. Est attendance: 600. For info: Special Events, Gunston Hall, 10709 Gunston Rd, Lorton, VA 22079. Phone: (703) 550-9220.

DAVIS, SAMMY, JR: 70th BIRTH ANNIVERSARY. Dec 8. Born Dec 8, 1925, at New York, NY, Sammy Davis, Jr, was the son of vaudevillians and first appeared on the stage at the age of four. Often described as one of the greatest entertainers in the world, his first film appearance was in *Rufus Jones for President* in 1931. He then joined the Will Mastin Trio. The popular song and dance team featured his father and his adopted uncle, but as Davis matured, his singing, dancing and impersonations soon became the center of the act. After military service in World War II, the trio became popular on the fashionable night club circuit. Davis began performing on his own in the 1950s headlining club engagements, appearing on television variety shows and making numerous popular records. His Broadway debut came in 1956 in the hit musical *Mr. Wonderful,* and in the late '50s and early '60s he starred in a number of films, including a series with Frank Sinatra and the Rat Pack. Davis continued to perform in clubs, concerts and on TV until he was stricken with throat cancer, from a lifetime of chain-smoking, in 1987. He died at Los Angeles, CA, on May 16, 1990.

☆ Chase's 1995 Calendar of Events ☆ Dec 8-9

DURANT, WILLIAM CRAPO: BIRTH ANNIVERSARY. Dec 8. "Billy" Durant, a leading producer of carriages in Flint, MI, promoter of the Buick car, co-founder of Chevrolet, was founder, in 1908, of General Motors. He lost, regained, and again lost control of GM, after which he founded Durant Motors, went bankrupt in the Depression, and operated a Flint bowling alley in his last working years. Durant was born at Boston, MA, on Dec 8, 1861, and died at New York, NY, Mar 18, 1947.

FEAST OF THE IMMACULATE CONCEPTION. Dec 8. Roman Catholic Holy Day of Obligation.

FIRST STEP TOWARD A NUCLEAR-FREE WORLD: ANNIVERSARY. Dec 8. On Dec 8, 1987, the former Soviet Union and the United States signed a treaty in Washington eliminating medium-range and shorter-range missiles. This was the first treaty completely doing away with two entire classes of nuclear arms. These 500- to 5,500-kilometer range missiles were to be scrapped under strict supervision within three years of the signing.

GUAM: LADY OF CAMARIN DAY. Dec 8. Declared a legal holiday by Guam legislature on Mar 2, 1971.

HOBAN, JAMES: DEATH ANNIVERSARY. Dec 8. Irish-born architect who designed the US President's Executive Mansion, later known as The White House. He was born in 1762 (exact date unknown) and died Dec 8, 1831. The cornerstone for the White House, Washington's oldest public building, was laid in 1792.

MORRISON, JIM: BIRTH ANNIVERSARY. Dec 8. Singer, songwriter, known as "The Lizard King," lead singer of The Doors, Jim Morrison is considered to be one of the fathers of contemporary rock. Born Dec 8, 1943, at Melbourne, FL, and died at Paris, France, July 3, 1971.

RIVERA, DIEGO: BIRTH ANNIVERSARY. Dec 8. Mexican painter whose murals became center of political controversy and demands for banishment, born at Guanajuato, Mexico, Dec 8, 1886. Died in his studio at San Angel, near Mexico City, Nov 25, 1957.

SOVIET UNION DISSOLVED: ANNIVERSARY. Dec 8. On Dec 8, 1991, the Union of Socialist Soviet Republics ceased to exist, as the republics of Russia, Byelorussia and Ukraine signed an agreement in Minsk, Byelorussia, creating the Commonwealth of Independent States. The remaining republics, with the exception of Georgia, joined in the new Commonwealth as it began the slow and arduous process of creating a new confederation while removing the yoke of Communism and dealing with strong separatist and nationalist movements within the various republics.

SUGARLOAF CRAFT FESTIVAL. Dec 8-10. Montgomery County Fairgrounds, Gaithersburg, MD. More than 300 artists and craftspeople, demonstrations, prize drawings and delicious food. Est attendance: 18,000. Sponsor: Sugarloaf Mt Works, Inc, 200 Orchard Ridge Dr, #215, Gaithersburg, MD 20878. Phone: (301) 990-1400.

THURBER, JAMES: BIRTH ANNIVERSARY. Dec 8. James Grover Thurber, American humorist and artist, born at Columbus, OH, Dec 8, 1894. Died at New York, NY, Nov 2, 1961.

BIRTHDAYS TODAY

Kim Basinger, 42, actress (*The Natural, The Getaway, My Stepmother Is an Alien*), born Athens, GA, Dec 8, 1953.
Red Berenson, 54, former hockey player, born at Regina, Saskatchewan, Canada, Dec 8, 1941.
James Galway, 56, musician, born at Belfast, Northern Ireland, Dec 8, 1939.
James MacArthur, 58, actor ("Hawaii Five-O,"), born at Los Angeles, CA, Dec 8, 1937.
Sinead O'Connor, 29, singer, songwriter, born at Dublin, Ireland, Dec 8, 1966.
Maximilian Schell, 65, actor (*Judgment at Nuremberg, The Odessa File*), producer, born at Vienna, Austria, Dec 8, 1930.

Flip Wilson, 62, actor, comedian (*Uptown Saturday Night*, "The Flip Wilson Show"), born Jersey City, NJ, Dec 8, 1933.

DECEMBER 9 — SATURDAY
343rd Day — Remaining, 22

AMERICA'S FIRST FORMAL CREMATION: ANNIVERSARY. Dec 9. The first formal cremation of a human body in America took place near Charleston, SC, on Dec 9, 1792. Henry Laurens, colonial statesman and signer of the Treaty of Paris, ending the Revolutionary War, in his Will, provided: "I do solemnly enjoin it on my son, as an indispensable duty, that as soon as he conveniently can, after my decease, he cause my body to be wrapped in twelve yards of tow cloth and burned until it be entirely consumed, and then, collecting my bones, deposit them wherever he may think proper." Laurens died Dec 8, 1792, at his plantation, and was cremated there.

ARMY AND NAVY UNION DAY IN MASSACHUSETTS. Dec 9. Second Saturday in December.

BIRDSEYE, CLARENCE: BIRTH ANNIVERSARY. Dec 9. American industrialist who developed a way of deep-freezing foods. He was marketing frozen fish by 1925 and was one of the founders of General Foods Corp. Born on Dec 9, 1886 at Brooklyn, NY. Died at New York City, Oct 7, 1956.

CHRISTMAS CANDLELIGHT TOUR. Dec 9-10. Old Town Alexandria, VA. Travel by trolley between three Alexandria museums and an eighteenth-century tavern conducting tours by candlelight with food, music, and entertainment of the period. Est attendance: 1,500. Sponsor: Alexandria Conv and Visitors Bureau, 221 King St, Alexandria, VA 22314. Phone: (703) 838-4200.

CHRISTMAS CRAFT FAIR USA: INDOOR SHOW. Dec 9-10. Wisconsin State Fair Park, Milwaukee, WI. Sale of handcrafted items—jewelry, pottery, weaving, leather, wood, glass and sculpture. Est attendance: 11,887. For info: Dennis R. Hill, Dir, 3233 S Villa Circle, Milwaukee, WI 53227. Phone: (414) 321-2100.

CHRISTMAS ON THE COOSA. Dec 9. Wetumpka, AL. Riverboat parade with colorfully decorated boats in Christmas theme cruising up the river. Est attendance: 25,000. For info: Susan Williams, PO Box 480, Wetumpka, AL 36092. Phone: (205) 567-5147.

FOXX, REDD: BIRTH ANNIVERSARY. Dec 9. Born John Elroy Sanford on Dec 9, 1922, in St. Louis, MO, Redd Foxx plied his comedic trade on vaudeville stages, in nightclubs, television, films and record albums. His talents reached a national audience with the TV sitcom "Sanford and Son." He died after collapsing during a rehearsal for a new TV sitcom, "The Royal Family," on Oct 11, 1992.

GENOCIDE CONVENTION: ANNIVERSARY. Dec 9. The United Nations General Assembly unanimously approved the Convention on Prevention and Punishment of the Crime of Genocide on Dec 9, 1948. It took effect Jan 12, 1951, when ratification by 20 nations had been completed. President Truman sent it to the US Senate for approval on June 16, 1949; it was supported by Presidents Kennedy, Johnson, Nixon, Ford, Carter and Reagan. Thirty-seven years after its submission, and after approval by more than 90 nations, the Senate approved it, Feb 19, 1986, by a vote of 83 to 11.

HARRIS, JOEL CHANDLER: BIRTH ANNIVERSARY. Dec 9. American author, creator of the "Uncle Remus" stories, born Dec 9, 1848. Died July 3, 1908.

Dec 9–10 ☆ *Chase's 1995 Calendar of Events* ☆

HISTORICAL SOCIETY'S CHRISTMAS CANDLE-LIGHT TOUR. Dec 9–10. Crawfordsville, IN. A walking tour of fine Victorian homes in Crawfordsville. Horse sleigh rides, hot cider, cookies and musical entertainment: dulcimer players, choral groups and brass quartets. Sponsor: Montgomery County Historical Society. Est attendance: 1,500. For info: Montgomery County VCB, 412 E Main St, Crawfordsville, IN 47933. Phone: (800) 866-3973.

◆ **HOPPER, GRACE: BIRTH ANNIVERSARY.** Dec 9. Grace Hopper was born Dec 9, 1906, at New York City. When she retired from the US Navy at the age of 79, she was the oldest naval officer ever on active duty. She attained the rank of Rear Admiral and was a leader in the computer revolution, having developed the computer language COBOL. Grace Hopper died Jan 1, 1992 at Arlington, WV.

MILTON, JOHN: BIRTH ANNIVERSARY. Dec 9. English poet and defender of freedom of the press born in Bread Street, Cheapside, London, on Dec 9, 1608. Died from gout, Nov 8, 1674. "No man who knows aught," he wrote, "can be so stupid to deny that all men naturally were born free."

MOUNT WASHINGTON TAVERN CANDLELIGHT TOURS. Dec 9–10. Fort Necessity National Battlefield, Farmington, PA. Annually, the second weekend in December. Est attendance: 500. Sponsor: Ft Necessity Natl Battlefield, Farmington, PA 15437. Phone: (412) 329-5512.

OLDE TIME CHRISTMAS AT MOUNT WASHINGTON TAVERN. Dec 9–10. Mt Washington Tavern, Fort Necessity National Battlefield, Farmington, PA. Candlelit and decorated for the holidays, the Tavern imparts the sights, smells and tastes of a festive 1800s Christmas. 2:30–7:30. Park admission: $2 for ages 17 and up; Tavern admission free for this event. For info: Natl Park Service, Fort Necessity Natl Battlefield, RD 2, Box 528, Farmington, PA 15437. Phone: (412) 329-5512.

O'NEILL, THOMAS PHILIP, II (TIP): BIRTH ANNIVERSARY. Dec 9, 1912. Democratic congressman from Massachusetts 1953–87, Speaker of the House of Representatives 1977–87, Tip O'Neill was born at Cambridge, MA, and died Jan 5, 1994, at Boston.

OZCANABANS OF OZ CONVENTION. Dec 9. Escanaba, MI. Celebrates Christmas in Oz. Est attendance: 20. Sponsor: The Wizard of Oz Club, Inc, Fred M. Meyer, Secy, 220 N 11th St, Escanaba, MI 49829.

SANDYS, EDWIN: BIRTH ANNIVERSARY. Dec 9. Sir Edwin Sandys, English statesman and one of the founders of the Virginia Colony (treasurer, the Virginia Company, 1619–1620), born at Worcestershire, Dec 9, 1561. Died in October 1629 (exact date unknown).

34th STREET EXPRESS. Dec 9. Chartered Amtrak train leaves from Boston, MA. Christmas shopping special to New York City also takes in the Radio City Christmas Spectacular. Annually, the second Saturday in December. Est attendance: 500. For info: Mystic Valley Railway Society, Inc, PO Box 486, Hyde Park, MA 02136-0486. Phone: (617) 361-4445.

UNITED REPUBLIC OF TANZANIA: INDEPENDENCE AND REPUBLIC DAY. Dec 9. On Dec 9, 1961 Tanganyika became independent of Britain and on Dec 9, 1962 it became a republic within the commonwealth. The republics of Tanganyika and Zanzibar joined to become one state (Apr 27, 1964) renamed (Oct 29, 1964) the United Republic of Tanzania.

VICTORIAN CHRISTMAS BY CANDLELIGHT. Dec 9–11. Clinton, MD. Candlelight tours with Christmas greenery, antique toys, cards, ornaments, music and Father Christmas. Est attendance: 800. Sponsor: Surratt House and Tavern, 9110 Brandywine Rd, Clinton, MD 20735. Phone: (301) 868-1121.

WINTERFEST BOAT PARADE. Dec 9. (Tentative.) More than 100 decorated yachts sail up Fort Lauderdale's intracoastal waterway starting at Port Everglades. Est attendance: 75,000. For info: Greater Ft. Lauderdale Conv/Visitors Bureau, 200 E Las Olas Blvd, Ste 1500, Fort Lauderdale, FL 33301. Phone: (305) 767-0686.

BIRTHDAYS TODAY

Joan Armatrading, 45, singer (album *Me, Myself, I*), songwriter, born at Saint Kitts, West Indies, Dec 9, 1950.

Otis Birdsong, 40, former basketball player, born at Winter Haven, FL, Dec 9, 1955.

Beau Bridges, 54, actor ("James Brady Story," *Fabulous Baker Boys*), born at Los Angeles, CA, Dec 9, 1941.

Dick Butkus, 53, actor, former football player ("Rich Man, Poor Man," "Half Nelson"), born at Chicago, IL, Dec 9, 1942.

Thomas Andrew Daschle, 48, US Senator (D, South Dakota), born at Aberdeen, SD, Dec 9, 1947.

Kirk Douglas (Issur Danielovich Demsky), 79, actor, author (*Lonely Are the Brave, Lust for Life*), born Amsterdam, NY, Dec 9, 1916.

Douglas Fairbanks, Jr, 86, actor, buisnessman, (*The Prisoner of Zenda, State Secret*), born at New York, NY, Dec 9, 1909.

Tom Kite (Thomas O. Kite, Jr), 46, golfer, born at Austin, TX, Dec 9, 1949.

John Malkovich, 42, actor (*The Killing Fields, The Sheltering Sky*), filmmaker, born Christopher, IL, Dec 9, 1953.

Donny Osmond, 38, actor, singer ("Donny and Marie," *Joseph and the Amazing Technicolor Dreamcoat*), born Ogden, UT, Dec 9, 1957.

Dick Van Patten, 67, actor ("Eight Is Enough," "Mama"), born Richmond Hill, NY, Dec 9, 1928.

DECEMBER 10 — SUNDAY
344th Day — Remaining, 21

ANIMALS' MESSIAH. Dec 10. Chicago, IL. A do-it-yourself Messiah sung to the animals in the society's shelter. It begins at noon and is free though a donation would be appreciated. Members of local youth orchestras perform and soloists also are included. A bake sale follows. Est attendance: 100. For info: The Anti-Cruelty Society, 157 W Grand Ave, Chicago, IL 60610. Phone: (312) 644-8338.

DEWEY, MELVIL: BIRTH ANNIVERSARY. Dec 10. American librarian and inventor of the Dewey decimal book classification system was born at Adams Center, NY, on Dec 10, 1851. Born Melville Louis Kossuth Dewey, he was an advocate of spelling reform, urged use of the metric system and was interested in many other education reforms. Dewey died at Highlands County, FL, on Dec 26, 1931.

DICKINSON, EMILY: 165th BIRTH ANNIVERSARY. Dec 10. One of America's greatest poets, Emily Dickinson was born at Amherst, MA, on Dec 10, 1830. She was reclusive, mysterious and frail in health. Neither she nor her poems were known to more than a few persons. Seven of her poems were published during her life, but after her death her sister, Lavinia, discovered almost 2,000 more poems written on the backs of envelopes and other scraps of paper—locked in her bureau. They were published gradually, over 50 years, beginning in 1890. She died at 6 PM, on May 15, 1886, at Amherst, MA. The little-known Emily Dickinson who was born, lived and died at Amherst, now is recognized as one of the most original poets of the English-speaking world.

	S	M	T	W	T	F	S
December 1995						1	2
	3	4	5	6	7	8	9
	10	11	12	13	14	15	16
	17	18	19	20	21	22	23
	24	25	26	27	28	29	30
	31						

Chase's 1995 Calendar of Events — Dec 10-11

FIRST US HEAVYWEIGHT CHAMP DEFEATED IN ENGLAND: ANNIVERSARY. Dec 10. Tom Molineaux, the first unofficial heavyweight champion of the US, was a freed slave from Virginia. He was beaten in the 40th round by Tom Cribb, the English champion, in a boxing match at Copthall Common in London on Dec 10, 1810.

GALLAUDET, THOMAS HOPKINS: BIRTH ANNIVERSARY. Dec 10. American teacher who established, in 1817, the first school for the deaf in the US. Born at Philadelphia, PA, on Dec 10, 1787. Gallaudet died at Hartford, CT, Sept 9, 1851.

HANDBELL CHOIR CONCERT AND TREE LIGHTING. Dec 10. The Inn on Lake Waramaug, New Preston, CT. A handbell choir will perform two identical half-hour concerts of Christmas songs. Spectators are welcome to sing along to some of the songs. Est attendance: 350. For info: Nancy Conant, The Inn on Lake Waramaug, 107 North Shore Rd, New Preston, CT 06777. Phone: (203) 868-0563.

★ **HUMAN RIGHTS DAY.** Dec 10. Presidential Proclamation 2866, of Dec 6, 1949, covers all succeeding years. Customarily issued as "Bill of Rights Day, Human Rights Day and Week."

★ **HUMAN RIGHTS WEEK.** Dec 10-16. Presidential Proclamation issued since 1958 for the week of Dec 10-16, except in 1986. See also: "Human Rights Day" (Dec 10) and "Bill of Rights Day" (Dec 15).

MISSISSIPPI: ADMISSION DAY: ANNIVERSARY. Dec 10. Became 20th state on this day in 1817.

NOBEL, ALFRED BERNHARD: DEATH ANNIVERSARY. Dec 10. Swedish chemist and engineer who invented dynamite was born at Stockholm, Sweden, Oct 21, 1833, and died at San Remo, Italy, Dec 10, 1896. His will established the Nobel Prize.

NOBEL PRIZE AWARDS CEREMONIES. Dec 10. Oslo, Norway and Stockholm, Sweden. Alfred Nobel, Swedish chemist and inventor of dynamite who died in 1896, provided in his will that income from his $9 million estate should be used for annual prizes—to be awarded to people who are judged to have made the most valuable contributions to the good of humanity. The Nobel Peace Prize is awarded by a committee of the Norwegian parliament and the presentation is made at the Oslo City Hall. Five other prizes, for physics, chemistry, medicine, literature and economics, are presented in a ceremony at Stockholm, Sweden. Both ceremonies traditionally are held on the anniversary of the death of Alfred Nobel. The current value of each prize is about $200,000. See also "Nobel, Alfred Bernhard: Death Anniversary" (Dec 10).

QUINCY PRESERVES CHRISTMAS CANDLELIGHT TOUR. Dec 10. Qunicy, IL. 5-8 PM. Tour one of Quincy's historic neighborhoods decked out in its finest holiday splendor. Luminaries line paths from house to house. Inside each home ticket holders are awed by the architectural splendors set off by beautiful holiday decorations. Annually, the second Sunday in December. Est attendance: 7,000. For info: Vicki Ebbing, Quincy Preserves, 428 Maine, Ste 270, Quincy, IL 62301. Phone: (217) 222-3432.

RALPH BUNCHE AWARDED NOBEL PEACE PRIZE: ANNIVERSARY. Dec 10. On Dec 10, 1950, Dr. Ralph Johnson Bunche became the first black man awarded the Nobel Peace Prize. Bunche was awarded the prize for his efforts in mediation between Israel and neighboring Arab states in 1949.

RED CLOUD: DEATH ANNIVERSARY. Dec 10. Sioux Indian chief Red Cloud was born in 1822 (exact date unknown), near North Platte, NE. A courageous leader and defender of Indian rights, Red Cloud was the son of Lone Man and Walks as She Thinks. His unrelenting determination caused US abandonment of the Bozeman trail and of three forts that interfered with Indian hunting grounds. Red Cloud died at Pine Ridge, SD, on Dec 10, 1909. The US Postal Service issued a postage stamp honoring him in its Great Americans Series in 1987.

SPACE MILESTONE: SOYUZ 26 (USSR). Dec 10. Launched on Dec 10, 1977, Cosmonauts Yuri Romanenko and Georgi

Grechko linked *Soyuz 26* with *Salyut 6* space station on Dec 11. Returned to Earth in *Soyuz 27*, Mar 16, 1978, after record-setting 96 days in space.

THAILAND: CONSTITUTION DAY. Dec 10. A public holiday throughout Thailand.

UNITED NATIONS: HUMAN RIGHTS DAY. Dec 10. Official United Nations observance day. Date is the anniversary of adoption of the "Universal Declaration of Human Rights" in 1948. The Declaration sets forth basic rights and fundamental freedoms to which all men and women everywhere in the world are entitled. Info from: United Nations, Dept of Public Info, New York, NY 10017.

UNITED NATIONS: INTERNATIONAL DECADE OF THE WORLD'S INDIGENOUS PEOPLE: YEAR TWO. Dec 10, 1995-Dec 9, 1996. The goal of this decade is to strengthen international cooperation to resolve problems faced by indigenous people in such areas as human rights, the environment, development, education and health.

UNITED NATIONS: THIRD DECADE TO COMBAT RACISM AND RACIAL DISCRIMINATION: YEAR THREE. Dec 10, 1995-Dec 9, 1996. See Jan 1 for details.

WREN'S NEST CHRISTMAS OPEN HOUSE. Dec 10. Wren's Nest, Atlanta, GA. This event will feature food prepared from genuine Victorian recipes, carols sung by local choirs and storytelling. The beautiful Queen Anne Victorian home will be decorated in authentic period style. ($1 contribution requested.) Est attendance: 500. For info: Karen Kelly, Wren's Nest, 1050 Ralph David Abernathy Blvd, SW, Atlanta, GA 30310. Phone: (404) 753-7735.

BIRTHDAYS TODAY

Mark Aguirre, 36, former basketball player, born at Chicago, IL, Dec 10, 1959.

Susan Dey, 43, actress, model ("The Partridge Family," "L.A. Law," "Love and War"), born at Pekin, IL, Dec, 10, 1952.

Harold Gould, 72, actor ("Rhoda," "Under One Roof"), born at Schenectady, NY, Dec 10, 1923.

Dorothy Lamour, 81, actress, singer (*The Hurricane, Road to Singapore*), born at New Orleans, LA, Dec 10, 1914.

Gloria Loring, 49, singer, actress (Liz Chandler DiMera in "Days of Our Lives"), born at New York, NY, Dec 10, 1946.

Dennis Morgan, 85, actor ("21 Beacon Street"), born Prentice, WI, Dec 10, 1910.

DECEMBER 11 — MONDAY

345th Day — Remaining, 20

ALASKA'S FIRST BROADCAST TELEVISION STATION: ANNIVERSARY. Dec 11. At 6:00 PM, on Dec 11, 1953, KTVA, Channel 11, at Anchorage, Alaska, signed on the air, becoming Alaska's first broadcast station.

BUELL, MARJORIE H.: BIRTH ANNIVERSARY. Dec 11, 1904. Cartoonist, creator of comic strip character Little Lulu, Marjorie Buell was considered a pioneer for creating a female character that outsmarted the neighborhood boys. She was born at Philadelphia, PA, and died May 30, 1993, at Elyria, OH.

Dec 11-12 ☆ Chase's 1995 Calendar of Events ☆

CANNON, ANNIE JUMP: BIRTH ANNIVERSARY. Dec 11. American astronomer and discoverer of five stars, was born at Dover, DE on Dec 11, 1863. Author and winner of the National Academy of Science Draper Medal, she died at Cambridge, MA, Apr 13, 1941.

EDWARD VIII: ABDICATION ANNIVERSARY. Dec 11. Christened Edward Albert Christian George Andrew Patrick David, King Edward VIII was born at Richmond Park, England, on June 12, 1894, and became Prince of Wales in July 1911. He ascended to the English throne upon the death of his father, George V, on Jan 20, 1936, but coronation never took place. He abdicated on Dec 11, 1936, in order to marry "the woman I love," twice-divorced American Wallis Warfield Simpson. They were married in France, June 3, 1937. Edward was created Duke of Windsor by his brother-successor, George VI. The Duke died at Paris, May 28, 1972, but was buried in England, near Windsor Castle.

HILLHAVEN HO HO HOTLINE. Dec 11-24. Seattle, WA. Skilled nursing facility residents play Santa and Mrs Claus for children to call. About 2,500 calls received. Sponsor: The Hillhaven Corporation, Tacoma, WA. Est attendance: 2,500. For info: Donna Albers, Natl Mktg Dir, Caller Service 2264, Tacoma, WA 98401-2264. Phone: (206) 756-4731.

INDIANA: ADMISSION DAY. Dec 11. Became 19th state on this day in 1816.

"MAGNUM, P.I." SERIES PREMIERE: ANNIVERSARY. Dec 11, 1980. The acclaimed television series premiered on CBS starring Tom Selleck in the title role of Thomas Sullivan Magnum, Vietnam vet turned private eye in Hawaii. Other cast regulars were Jonathan Hillerman as Jonathan Quayle Higgins III, Roger E. Mosley as T.C. (Theodore Calvin), Larry Manetti as Orville "Rick" Wright and Orson Welles supplied a voice from 1980 to 1985 for the unseen Robin Masters. May be the only TV series to kill off the star due to expected cancellation and explain that away as a dream the following season. The theme music was composed by Mike Post and Pete Carpenter. Final episode aired May 1, 1988.

MOST BORING CELEBRITIES OF THE YEAR. Dec 11. Eleventh annual list of celebrities chosen because of "massive media over-exposure" during the year. For info: The Boring Institute, Alan Caruba, Founder, Box 40, Maplewood, NJ 07040. Phone: (201) 763-6392.

UNITED NATIONS: UNICEF ANNIVERSARY. Dec 11. Anniversary of the establishment by the United Nations General Assembly, on Dec 11, 1946, of the United Nations International Children's Emergency Fund (UNICEF). Info from: United Nations, Dept of Public Info, New York, NY 10017.

BIRTHDAYS TODAY

Max Baucus, 54, US Senator (D, Montana), born at Helena, MT, Dec 11, 1941.
Teri Garr, 46, actress (*Young Frankenstein, Tootsie, The Black Stallion*), born at Lakewood, OH, Dec, 11, 1949.
David Gates, 55, singer, songwriter, born at Tulsa, OK, Dec 11, 1940.
Tom Hayden, 55, journalist, activist, politician (ex-husband of Jane Fonda), born at Royal Oak, MI, Dec 11, 1940.
Jermaine Jackson, 41, singer, musician (Jackson 5; "Daddy's Home," "Let's Get Serious"), born at Gary, IN, Dec 11, 1954.
John F. Kerry, 52, US Senator (D, Massachusetts), born at Denver, CO, Dec 11, 1943.
Brenda Lee (Brenda Mae Tarpley), 51, singer ("I'm Sorry," "All Alone Am I"), born at Atlanta, GA, Dec 11, 1944.

December 1995

	S	M	T	W	T	F	S
						1	2
	3	4	5	6	7	8	9
	10	11	12	13	14	15	16
	17	18	19	20	21	22	23
	24	25	26	27	28	29	30
	31						

Donna Mills, 52, actress (Abby on "Knots Landing"), born Chicago, IL, Dec 11, 1943.
Rita Moreno, 64, singer, actress (Oscar *West Side Story*; Tony *The Ritz*), born at Hunacao, Puerto Rico, Dec 11, 1931.
Carlo Ponti, 82, producer, born at Milan, Italy, Dec 11, 1913.
Susan Seidelman, 43, filmmaker (*Desperately Seeking Susan*), born at Philadelphia, PA, Dec 11, 1952.
Aleksandr Isayevich Solzhenitsyn, 77, author (*One Day in the Life of Ivan Denisovich*), born at Kislovodsk, USSR, Dec 11, 1918.
Curtis Williams, 33, musician, singer (with The Penguins; "Earth Angel"), born at Buffalo, NY, Dec 11, 1962.

DECEMBER 12 — TUESDAY
346th Day — Remaining, 19

BONZA BOTTLER DAY™. Dec 12. To celebrate when the number of the day is the same as the number of the month. Bonza Bottler Day™ is an excuse to have a party at least once a month. T-shirts available. For info: Elaine Fremont, 203 Waddell Rd, Taylors, SC 29687. Phone: (803) 244-2023.

CHILDREN'S VICTORIAN CHRISTMAS CELEBRATION. Dec 12-27. Washington, MS. Display of 14 or more Christmas trees decorated by fifth grade students. Trees and decorations remain on display throughout the holiday season. Closed Christmas weekend. Est attendance: 2,000. For info: Historic Jefferson College, Anne L. Gray, Historian, Box 700, Washington, MS 39190. Phone: (601) 442-2901.

DAY OF OUR LADY OF GUADALUPE. Dec 12. The legend of Guadalupe tells how in December 1533, an Indian, Juan Diego, saw the Virgin Mother on a hill near Mexico City, who instructed him to go to the bishop and have him build a shrine to her on the site of the vision. Initially rebuffed, the Virgin Mother appeared to Juan Diego three days later. She instructed him to pick roses growing on a stony and barren hillside nearby and take them to the bishop as proof. Although flowers do not normally bloom in December, Juan Diego found the roses and took them to the bishop. As he opened his mantle to drop the roses on the floor, an image of the Virgin Mary appeared among them. The bishop built the sanctuary as instructed. Our Lady of Guadalupe became the patroness of Mexico City, by 1746 was the patron saint of all New Spain and by 1910 of all Latin America.

FIRST BLACK SERVES IN US HOUSE OF REPRESENTATIVES: 125th ANNIVERSARY. Dec 12. Joseph Hayne Rainey of Georgetown, SC, was sworn in on Dec 12, 1870, as the first black to serve in the US House of Representatives. Rainey filled the seat of Benjamin Franklin Whittemore, that had been declared vacant by the House. He served until Mar 3, 1879.

FLAUBERT, GUSTAVE: BIRTH ANNIVERSARY. Dec 12. French author whose works include one of the greatest French novels, *Madame Bovary*, was born at Rouen on Dec 12, 1821. Flaubert died at Croisset, France, on May 8, 1880.

GARRISON, WILLIAM LLOYD: BIRTH ANNIVERSARY. Dec 12. American anti-slavery leader, poet and journalist, was born at Newburyport, MA, on Dec 12, 1805. Garrison died at New York, NY, on May 24, 1879.

JAY, JOHN: 250th BIRTH ANNIVERSARY. Dec 12. American statesman, diplomat and first chief justice of the US Supreme Court (1789-1795), co-author (with Alexander Hamilton and James Madison) of the influential *Federalist* papers, was born at New York, NY, on Dec 12, 1745. Jay died at Bedford, NY, on May 17, 1829.

Chase's 1995 Calendar of Events — Dec 12–14

KENYA: JAMHURI DAY. Dec 12. Jamhuri Day (Independence Day) is Kenya's official National Day, commemorating proclamation of the republic on Dec 12, 1963.

MEXICO: GUADALUPE DAY. Dec 12. One of Mexico's major celebrations. Honors the "Dark Virgin of Guadalupe," the republic's patron saint. Parties and pilgrimages, with special ceremonies at the Shrine of Our Lady of Guadalupe, in Mexico City.

PENNSYLVANIA RATIFIES CONSTITUTION ANNIVERSARY. Dec 12. Pennsylvania became the second state to ratify the US Constitution, by a vote of 46 to 23, on Dec 12, 1787.

POINSETTIA DAY (JOEL ROBERTS POINSETT BIRTH ANNIVERSARY). Dec 12. A day to enjoy poinsettias and to honor Dr. Joel Roberts Poinsett, the American diplomat who introduced the Central American plant which is named for him into the US. Poinsett was born at Charleston SC, Mar 2, 1799. He also served as a member of Congress and as secretary of war. He died near Statesburg, SC, Dec 12, 1851. The poinsettia has become a favorite Christmas season plant.

RUSSIAN ELECTION SURPRISE: ANNIVERSARY. Dec 12, 1993. In a shocking rebuke to Boris Yeltsin's attempts to stabilize the country, Russian voters gave Vladimir Zhironovsky and his misnamed Liberal Democratic Party 28% of the vote, granting them a large block of power in the new Russian Parliament and the world's imagination. Zhironovsky's platform, which has been called racist and fascist, tapped into a groundswell of nationalist fervor with calls to return Russia to the glory days of its empire and emphasizing harsh treatment for minorities and those who would stand in his way.

BIRTHDAYS TODAY

Tracy Austin, 33, tennis player, born at Rolling Hills, CA, Dec 12, 1962.
Bob Barker, 72, TV personality, game show host (Emmy "The Price Is Right"), born at Darrington, WA, Dec 12, 1923.
Shelia E (Sheila Escoveda), 36, singer, musician ("The Glamorous Life"), born at San Francisco, CA, Dec 12, 1959.
Connie Francis (Constance Franconero), 57, singer ("My Happiness," "Where the Boys Are"), born Newark, NJ, Dec 12, 1938.
Edward Irwin Koch, 71, former mayor of New York City, born at New York, NY, Dec 12, 1924.
Cathy Rigby, 43, former gymnast, born at Long Beach, CA, Dec 12, 1952.
Frank Sinatra (Francis Albert), 80, singer ("All the Way," "Strangers in the Night,"), actor (*From Here to Eternity*), born Hoboken, NJ, Dec 12, 1915.
Dionne Warwick, 54, singer ("I Say a Little Prayer for You," "What Do You Get When You Fall in Love," "This Girl's in Love With You"), born East Orange, NJ, Dec 12, 1941.
Grover Washington, 52, musician, born at Buffalo, NY, Dec 12, 1943.

DECEMBER 13 — WEDNESDAY
347th Day — Remaining, 18

BROOKS, PHILLIPS: BIRTH ANNIVERSARY. Dec 13. American clergyman and composer born at Boston, MA, Dec 13, 1835. Perhaps best remembered for his lyrics for the Christmas carol "O Little Town of Bethlehem." Brooks died at Boston, Jan 23, 1893.

ENTOMOLOGICAL SOCIETY OF AMERICA ANNUAL MEETING. Dec 13–17. Dallas, TX. Est attendance: 3,000. For info: ESA, 9301 Annapolis Rd, Lanham, MD 20706-3115. Phone: (301) 731-4535.

HEINE, HEINRICH: BIRTH ANNIVERSARY. Dec 13. German author, born at Dusseldorf, Dec 13, 1797. Died at Paris, Feb 17, 1856.

LINCOLN, MARY TODD: BIRTH ANNIVERSARY. Dec 13. Wife of Abraham Lincoln, 16th president of the US, born at Lexington, KY, Dec 13, 1818. Died July 16, 1882.

MALTA: REPUBLIC DAY. Dec 13. National holiday. Malta became a republic on Dec 13, 1974.

NOEL NIGHT. Dec 13. University Cultural Center, Detroit, MI. Holiday celebration with caroling, carriage rides, yuletide treats, face-painting, vocal and instrumental performances, puppet shows, pictures with Santa, festive suppers and children's activities. Est attendance: 5,000. For info: University Cultural Center Assn, Sue Mosey, 4735 Cass, Detroit, MI 48202. Phone: (313) 577-5088.

NORTH AND SOUTH KOREA END WAR: ANNIVERSARY. Dec 13. On Dec 13, 1991, North and South Korea signed a treaty of reconciliation and nonaggression, formally ending the Korean War—38 years after fighting ceased in 1953. This agreement was not hailed as a peace treaty, and the armistice that was signed on July 27, 1953, between the United Nations and North Korea, was to remain in effect until it could be transformed into a formal peace.

SWEDEN: SANTA LUCIA DAY. Dec 13. Nationwide celebration of festival of light, honoring St. Lucia. Many hotels have their own Lucia, a young girl attired in a long flowing white gown, who serves guests coffee and lussekatter (saffron buns) in the early morning.

BIRTHDAYS TODAY

Tim Conway, 62, actor, comedian ("McHale's Navy," "The Carol Burnett Show"), born at Willoughby, OH, Dec 13, 1933.
John Davidson, 54, singer, actor, born at Pittsburgh, PA, Dec 13, 1941.
Richard Lamar Dent, 35, football player, born Atlanta, GA, Dec 13, 1960.
Ted Nugent, 46, singer (with Amboy Dukes: "Journey to the Center of the Mind"; solo: "Cat Scratch Fever"), born Detroit, MI, Dec 13, 1949.
Christopher Plummer, 66, actor (Emmy *The Moneychangers*; *The Sound of Music*), born Toronto, Ont, Canada, Dec 13, 1929.
Dick Van Dyke, 70, actor, comedian (*Mary Poppins*, "The Dick Van Dyke Show"), born West Plains, MO, Dec 13, 1925.

DECEMBER 14 — THURSDAY
348th Day — Remaining, 17

ALABAMA: ADMISSION DAY: ANNIVERSARY. Dec 14. Became 22nd state on this day in 1819.

DOOLITTLE, JAMES HAROLD: BIRTH ANNIVERSARY. Dec 14, 1896. American aviator and World War II hero, General James Doolittle was born at Alameda, CA. A Lieutenant General in the US Army Air Force, he was the first person to fly across North America in less than a day. On Apr 18, 1942, Doolittle led a squadron of 16 B-25 bombers, launched from aircraft carriers, on the first US aerial raid on Japan of WWII. Bombs were dropped on the Japanese cities of Tokyo, Yokohama, Osaka, Kobe and Nagoya. He was awarded the Congressional Medal of Honor for this accomplishment. Doolittle also headed the Eighth Air Force during the Normandy invasion. He died Sept 27, 1993, at Pebble Beach, CA.

Dec 14-15 ☆ *Chase's 1995 Calendar of Events* ☆

EGYPT: MARITIME DISASTER: ANNIVERSARY. Dec 14. The ferry *Salem Express* sank off the port city of Safaga, Egypt, on Dec 14, 1991, claiming the lives of 462 passengers and crew members. 180 people survived the disaster, the worst in modern Egypt's maritime history.

INTERNATIONAL FINALS RODEO. Dec 14-17. Myriad Convention Center, Oklahoma City, OK. The top 15 money-winning IPRA cowboys and cowgirls compete for world championships in seven contests. Prize money more than $250,000. Trade show, dances, bucking stock sale. Annually, the third weekend in December. Sponsor: OK City All Sports Assn, 100 W Main, Ste 287, Oklahoma City, OK 73102. Est attendance: 40,000. For info: Stanley Draper, Jr, Exec Secy, Oklahoma City All Sports Assn, 100 W Main, Ste 287, Oklahoma City, OK 73102. Phone: (405) 236-5000.

NOSTRADAMUS: BIRTH ANNIVERSARY. Dec 14. French physician, best remembered for his astrological predictions (written in rhymed quatrains), was born Michel de Notredame, at St. Remy, Provence, France, Dec 14, 1503. Many believed that his book of prophecies actually foretold the future. Nostradamus died at Salon, France, July 2, 1566.

SMITH, MARGARET CHASE: BIRTHDAY. Dec 14. First woman to be elected to both houses of the US Congress, born at Skowhegan, ME, Dec 14, 1897.

SOUTH POLE: DISCOVERY ANNIVERSARY. Dec 14. The elusive object of many expeditions dating from the 7th century, the South Pole was located and visited Dec 14, 1911, by Roald Amundsen with four companions and 52 sled dogs. All five men and 12 of the dogs returned to base camp safely. Next to visit the South Pole, on Jan 17, 1912, was a party of five led by Captain Robert F. Scott, all of whom perished during the return trip. A search party found their frozen bodies 11 months later. See also: "Amundsen, Roald: Birth Anniversary" (July 16).

WASHINGTON, GEORGE: DEATH ANNIVERSARY. Dec 14. The first President of the United States of America died at Mount Vernon, VA, on Dec 14, 1799.

BIRTHDAYS TODAY

Leonardo Boff, 57, Brazilian Catholic theologian, born at Concordia, Brazil, Dec 14, 1938.
Bill Buckner, 46, former baseball player, born at Vallejo, CA, Dec 14, 1949.
Patty Duke, 49, actress (Oscar *The Miracle Worker*; Emmy *My Sweet Charlie*), born at New York, NY, Dec 14, 1946.
Don Hewitt, 73, TV news producer, born at New York, NY, Dec 14, 1922.
Charlie Rich, 63, singer, musician ("Behind Closed Doors," "The Most Beautiful Girl"), born at Forrest City, AR, Dec 14, 1932.
Stan Smith, 49, tennis player, born at Pasadena, CA, Dec 14, 1946.

December 1995

S	M	T	W	T	F	S
					1	2
3	4	5	6	7	8	9
10	11	12	13	14	15	16
17	18	19	20	21	22	23
24	25	26	27	28	29	30
31						

DECEMBER 15 — FRIDAY
349th Day — Remaining, 16

BILL OF RIGHTS: ANNIVERSARY. Dec 15. The first ten amendments to the US Constitution, known as the Bill of Rights, became effective following ratification by Virginia on Dec 15, 1791. The anniversary of ratification and of effect is observed as Bill of Rights Day.

★ **BILL OF RIGHTS DAY.** Dec 15. Presidential Proclamation. Has been proclaimed each year since 1962, but was omitted in 1967 and 1968. (Issued in 1941 and 1946 at Congressional request and in 1947 without request.) Since 1968 has been included in Human Rights Day and Week Proclamation.

CURACAO: KINGDOM DAY AND ANTILLEAN FLAG DAY. Dec 15. This day commemorates the Dec 15, 1954, Charter of Kingdom, signed in the Knight's Hall in The Hague, granting the Netherlands Antilles complete autonomy. Solemn meeting of the "Staten," the Parliament of the Netherlands Antilles. The Antillean Flag was hoisted for the first time on this day in 1959.

EIFFEL, ALEXANDRE GUSTAVE: BIRTH ANNIVERSARY. Dec 15. Eiffel, the French engineer who designed the thousand-foot-high, million-dollar, open-lattice wrought iron Eiffel Tower, and who participated in designing the Statue of Liberty, was born at Dijon, France, on Dec 15, 1832. The Eiffel Tower, weighing more than 7,000 tons, was built for the Paris International Exposition of 1889. Eiffel died at Paris on Dec 23, 1923.

HALCYON DAYS. Dec 15-29. Traditionally, the seven days before and the seven days after the winter solstice. To the ancients a time when fabled bird (called the halcyon) calmed the wind and waves—a time of calm and tranquility.

HANDEL'S MESSIAH. Dec 15-17. Centennial Hall, Rock Island, IL. Traditional holiday favorite performed by the 350 voices of the Handel Oratorio Society and the Augustana College Choir with the Augustana Symphony Orchestra. For info: Augustana College Ticket Office, Rock Island, IL 61201. Phone: (309) 794-7306.

INTERNATIONAL LANGUAGE WEEK. Dec 15-21. To disseminate information about mankind's quest for an international language to solve the communication problem of humans, and to supply information about the international language Esperanto. Esperanto was created in 1887 by Dr. L.L. Zamenhof as a solution to the world's language problem. For complete info, send $1 to cover expense of printing, handling and postage. Annually, Dec 15-21. Sponsor: Intl Society of Friendship and Good Will, Dr. Stanley Drake, Pres, 9538 Summerfield St, Spring Valley, San Diego Count, CA 91977-2852. Phone: (619) 466-8882.

MILITARY DICTATORSHIP ENDED IN CHILE: ANNIVERSARY. Dec 15. In an election on Dec 15, 1989, Patricio Aylwin defeated General Augusto Pinochet's former finance minister, Hernan Buchi, bringing the military dictatorship of Pinochet to an end. Fourteen months previously, Pinochet suffered defeat in a national plebiscite on eight more years of his rule. This defeat prompted democratic elections and crippled the Pinochet regime. Pinochet came to power when he toppled Marxist president Salvador Allende in a 1973 coup. Patricio Aylwin avoided a two-candidate runoff by achieving 55.2 percent of the vote. He was inaugurated on March 11, 1990.

MOON PHASE: LAST QUARTER. Dec 15. Moon enters Last Quarter phase at 12:31 AM, EST.

PUERTO RICO: NAVIDADES. Dec 15-Jan 6, 1996. Traditional Christmas season begins mid-December and ends on Three Kings Day. Elaborate nativity scenes, carolers, special Christmas foods. Trees from Canada and US. Gifts on Christmas Day and on Three Kings Day.

SITTING BULL: DEATH ANNIVERSARY. Dec 15. (c. 1834-1890) Famous Sioux Indian leader, medicine man and warrior of the Hunkpapa Teton band. Known also by his native

474

name, Tatanka-yatanka, Sitting Bull was born on the Grand River, SD. He first accompanied his father on the warpath at the age of 14 against the Crow and thereafter rapidly gained influence within his tribe. He became known in 1886 when he led a raid on Fort Buford. His steadfast refusal to go to a reservation led General Phillip Sheridan to initiate a campaign against him which led to the massacre of Lieutenant Colonel George Custer's men at the Little Bighorn, after which he fled to Canada, remaining there until 1881. Although many in his tribe surrendered on their return, Sitting Bull remained hostile until his death in a skirmish with the US soldiers along the Grand River on Dec 15, 1890.

SPACE MILESTONE: VEGA 1 (USSR). Dec 15. Craft launched Dec 15, 1984, to rendezvous with Halley's Comet in March 1986. *Vega 2*, launched Dec 21, 1984, was part of same mission which, in cooperation with the US, carried US-built "comet-dust" detection equipment.

UNDERDOG DAY. Dec 15. To salute, before the year's end, all of the underdogs and unsung heroes—the Number Two people who contribute so much to the Number One people we read about. (Sherlock Holmes's Dr. Watson and Robinson Crusoe's Friday are examples.) Annually, the third Friday in December. Sponsor: P. Moeller, Chief Underdog, Box 71, Clio, MI 48420.

ZAMENHOF, DR. L.L.: BIRTH ANNIVERSARY. Dec 15. Lazarus Ludovic Zamenhof was born Dec 15, 1859, in Bialystok, near the borders of Lithuania, Poland and Byelorussia, where many different languages were spoken. Zamenhof realized the need for a common tongue while a child and later developed the International Language, Esperanto, which means "he who hopes." Sponsor: Intl Society of Friendship and Good Will (ISFGW), Dr Stanley Drake, Pres, 9538 Summerfield St, Spring Valley, San Diego County, CA 91977-2852. Phone: (619) 466-8882.

BIRTHDAYS TODAY

Nick Buoniconti, 55, former football player, born at Springfield, MA, Dec 15, 1940.
Dave Clark, 53, musician (leader Dave Clark Five; "I Like It Like That"), born London, England, Dec 15, 1942.
Friedrich Hundertwasser, 67, artist, born at Vienna, Austria, Dec 15, 1928.
Don Johnson (Donald Wayne), 46, actor ("Miami Vice," *A Boy and His Dog*), born Flatt Creek, MO, Dec 15, 1949.
Edna O'Brien, 64, author (*Country Girls Trilogy, Time and Tide*), born at Tuamgraney, Ireland, Dec 15, 1931.
Daryl Turner, 34, football player, born at Wadley, GA, Dec 15, 1961.

DECEMBER 16 — SATURDAY
350th Day — Remaining, 15

AUSTEN, JANE: BIRTH ANNIVERSARY. Dec 16. English novelist born Dec 16, 1775. Died July 18, 1817.

BAHRAIN: INDEPENDENCE DAY. Dec 16. National holiday.

BANGLADESH: VICTORY DAY. Dec 16. National holiday in the People's Republic of Bangladesh.

◆ **BEETHOVEN, LUDWIG VAN: 225th BIRTH ANNIVERSARY.** Dec 16. Regarded by many as the greatest orchestral composer of all time, Ludwig van Beethoven was born at Bonn, Germany, Dec 16, 1770. Impairment of his hearing began before he was 30, but even total deafness did not halt his composing and conducting. His last appearance on the concert stage was to conduct the premiere of his Ninth Symphony, at Vienna, on May 7, 1824. He was unable to hear either the orchestra or the applause. Often in love, he never married. Of a stormy temperament, he is said to have died during a violent thunderstorm on Mar 26, 1827, at Vienna. ". . . I have spent no less than the whole morning and the whole of yesterday afternoon," he wrote to Carl Holz in 1825, "over the correction of the two pieces, and am quite hoarse with swearing and stamping."

BOSTON TEA PARTY: ANNIVERSARY. Dec 16. Anniversary of Dec 16, 1773, Boston patriots' boarding of British vessel at anchor in Boston Harbor. Contents of nearly 350 chests of tea were dumped into the harbor.

EAT WHAT YOU WANT DAY. Dec 16. Here's a day you may actually enjoy yourself. Ignore all those on-again/off-again warnings. Sponsor: Wellness Permission League. Phone: (212) 388-8673 or (717) 274-8451. For info: Tom and Ruth Roy, Wellness Permission League, 2105 Water St, Lebanon, PA 17046.

KAZAKHSTAN: DAY OF THE REPUBLIC. Dec 16. National Day.

MAN WILL NEVER FLY SOCIETY MEETING. Dec 16. Kitty Hawk, NC. Group meets to prove that man will never fly. Society motto: "Birds fly, men drink." Annually on Dec 16, the night before the anniversary of the Wright Brothers first flight. Est attendance: 450. For info: Al Jones, Box 1903, Kill Devil Hill, NC 27948. Phone: (919) 441-2424.

◆ **MEAD, MARGARET: BIRTH ANNIVERSARY.** Dec 16. American anthropologist and author, especially known for her studies of primitive peoples of southwest Pacific Ocean area, and for her forthright manner in speaking and writing. Born at Philadelphia, PA, Dec 16, 1901. Died at New York, NY, Nov 15, 1978.

MEXICO: POSADAS. Dec 16-24. A nine-day annual celebration throughout Mexico. Processions of "pilgrims" knock at doors asking for posada (shelter), commemorating the search by Joseph and Mary for a shelter in which the infant Jesus might be born. Invited inside, fun and merrymaking ensue with blindfolded guests trying to break a "piñata" (papier mache decorated earthenware utensil filled with gifts and goodies) suspended from the ceiling. Once broken the gifts are distributed and celebration continues.

NCAA DIVISION I-AA FOOTBALL CHAMPIONSHIP. Dec 16. Site to be determined. National college football championship game in 1-AA division. Several social events scheduled for visitors, coaches and community. Est attendance: 29,835. Sponsor: NCAA, 6201 College Blvd, Overland Park, KS 66211-2422. Phone: (913) 339-1906.

***NEW WORLD SYMPHONY* PREMIERE: ANNIVERSARY.** Dec 16. Anton Dvorak's *New World Symphony* premiered Dec 16, 1893, at the newly erected Carnegie Hall with the New York Philharmonic playing. The composer attended and enjoyed enthusiastic applause from the audience. The symphony contains snatches from black spirituals and American folk music. Dvorak, a Bohemian, had been in the US only a year when he composed it as a greeting to his friends in Europe.

PHILIPPINES: PHILIPPINE CHRISTMAS OBSERVANCE. Dec 16-Jan 6. Philippine Islands. Said to be world's longest Christmas celebration.

PHILIPPINES: SIMBANG GABI. Dec 16-25. Nationwide. A nine-day novena of pre-dawn masses, also called "Misa de Gallo." One of the traditional Filipino celebrations of the holiday season.

POLISH CHRISTMAS OPEN HOUSE. Dec 16. Polish American Cultural Center Museum, Philadelphia, PA. Sw. Mikolaj (Polish St. Nicholas) will greet everyone with gifts for the children. Polish Christmas Tree and entertainment. Free Admission. For info: Polish American Cultural Center Museum, Michael Blichasz, Chair, 308 Walnut St, Philadelphia, PA 19106. Phone: (215) 922-1700.

SALEM CHRISTMAS. Dec 16. Winston-Salem, NC. Re-creation of Christmas season as Moravian records indicate that early 19th-century Salem might have observed it. Choral and church band music, craft and trades demonstrations and special decorations. Est attendance: 5,000. For info: Old Salem Inc, Linda Georgitis, Box F, Winston-Salem, NC 27108. Phone: (910) 721-7300.

SANTAYANA, GEORGE: BIRTH ANNIVERSARY. Dec 16. Philosopher and author born at Madrid, Spain, on Dec 16, 1863. At the age of nine he emigrated to the US where he attended and later taught at Harvard University. In 1912 he returned to Europe and traveled extensively. It was Santayana who said "Those who cannot remember the past are condemned to repeat it." He died at Rome, on Sept 26, 1952.

THE TONGASS—ALASKA'S MAGNIFICENT RAIN FOREST. Dec 16–Jan 28. Natural History Museum of Los Angeles, Los Angeles, CA. This exhibit explores the intricate harmony of the largest non-equatorial rain forest on earth. See Feb 4 for full description. For info: Smithsonian Institution Traveling Exhibition Service, 1100 Jefferson Dr SW, Ste 3146, Wasington, DC 20560. Phone: (202) 357-2700.

UNITED NATIONS REVOKES RESOLUTION ON ZIONISM. Dec 16. On Dec 16, 1991, the United Nations voted 111 to 25 to revoke Resolution 3379, which equated Zionism with racism. Resolution 3379 was approved Nov 10, 1975, with 72 countries voting in favor, 35 against and 32 abstentions. The largest block of changed votes came from the former Soviet Union and Eastern Europe.

BIRTHDAYS TODAY

Bruce N. Ames, 67, biochemist, cancer researcher, born at New York, NY, Dec 16, 1928.
Steven Bochco, 52, TV writer, producer, creator ("Hill Street Blues," "NYPD Blue"), born New York, NY, Dec 16, 1943.
Arthur Charles Clarke, 78, author, scientist (*2001: A Space Odyssey, Islands in the Sky*), born Minehead, England, Dec 16, 1917.
Mike Flanagan, 44, former baseball player, born at Manchester, NH, Dec 16, 1951.
William "The Refrigerator" Perry, 33, football player, born at Aiken, SC, Dec 16, 1962.
Clifton Ralph Robinson, 29, basketball player, born at Buffalo, NY, Dec 16, 1966.
Lesley Stahl, 54, journalist ("60 Minutes," former White House correspondent), born at Lynn, MA, Dec 16, 1941.
Liv Johanne Ullmann, 56, actress (*The Immigrants, Scenes From a Marriage*), born at Tokyo, Japan, Dec 16, 1939.

DECEMBER 17 — SUNDAY
351st Day — Remaining, 14

AZTEC CALENDAR STONE DISCOVERY: ANNIVERSARY. Dec 17. On Dec 17, 1790, one of the wonders of the western hemisphere—the Aztec Calendar, or Solar, Stone—was found beneath the ground by workmen repairing Mexico City's Central Plaza. The centuries-old, intricately carved stone, 11 ft, 8 in in diameter and weighing nearly 25 tons, proved to be a highly developed calendar monument to the sun. Believed to have been carved in the year 1479, this extraordinary time-counting basalt tablet stood originally in the Great Temple of the Aztecs. Buried, along with other Aztec idols, soon after the Spanish conquest in 1521, it remained hidden until 1790. Its 52-year cycle had regulated many Aztec ceremonies, including grisly human sacrifices to save the world from destruction by the gods.

BHUTAN: NATIONAL DAY. Dec 17. National holiday observed.

BOLIVAR, SIMON: DEATH ANNIVERSARY. Dec 17. Commemorated in Venezuela and other Latin American countries. Bolivar, called "The Liberator," was born July 24, 1783, at Caracas, Venezuela, and died on Dec 17, 1830, at Santa Marta, Colombia.

EIGHTEENTH-CENTURY CHRISTMAS WASSAIL. Dec 17. McLean, VA. Greet the winter solstice and Christmas season with a toast to the apple trees. Caroling and warm refreshments. Bring pots or other noisemakers to frighten off evil spirits threatening next year's apple crop. Est attendance: 500. Sponsor: Claude Moore Colonial Farm at Turkey Run, 6310 Georgetown Pike, McLean, VA 22101. Phone: (703) 442-7557.

FIRST FLIGHT TRADITIONAL ANNIVERSARY CELEBRATION. Dec 17. Kill Devil Hills, NC. Each year since 1928, on the anniversary of the Wright Brothers' first successful heavier-than-air flight at Kitty Hawk, NC, Dec 17, 1903, a celebration has been held at the Wright Brothers National Memorial, with wreaths, flyover and other observances—regardless of weather.

FLOYD, WILLIAM: BIRTH ANNIVERSARY. Dec 17. Signer of the Declaration of Independence, member of Congress, born at Brookhaven, Long Island, Dec 17, 1734. Died at Westernville, NY, Aug 4, 1821.

HAWAIIAN CHRISTMAS LOOONG DISTANCE INVITATIONAL ROUGH-H_2O SWIM. Dec 17. Waikiki Beach, Honolulu, HI. 7K (4.33 mi) swim across Waikiki Bay and return. Pre-registration is required. Hawaii's longest open ocean race. Annually, the week following Honolulu Marathon. Est attendance: 90. For info: Jim Anderson, 3176 E Manoa Rd, Honolulu, HI 96822.

INTERNATIONAL SHAREWARE DAY. Dec 17. A day to take the time to reward the efforts of thousands of computer programmers who trust that if we try their programs and like them, we will pay for them. Unfortunately, very few payments are received, thus stifling the programmers' efforts. This observance is meant to prompt each of us to inventory our PCs and Macs, see if we are using any shareware, and then take the time in the holiday spirit to write payment checks to the authors. Hopefully this will keep shareware coming. Annually, the third Sunday in December. Sponsor: David Lawrence, Host, Online Tonight, PO Box 10, Glenwood, MD 21738. Phone: (301) 854-5459.

LIBBY, WILLARD FRANK: BIRTH ANNIVERSARY. Dec 17. American educator, chemist, atomic scientist, and Nobel Prize winner was born at Grand Valley, CO, Dec 17, 1908. He was the inventor of the Carbon-14 "atomic clock" method for dating ancient and prehistoric plant and animal remains and minerals. Died at Los Angeles, CA, Sept 8, 1980.

★ **PAN AMERICAN AVIATION DAY.** Dec 17. Presidential Proclamation 2446, of Nov 18, 1940, covers all succeeding years (Pub Res No. 105 of Oct 10, 1940).

REENACTMENT OF THE BOSTON TEA PARTY. Dec 17. Congress Street Bridge, Boston, MA. Reenactment of "Boston's most notorious protest, the single most important event leading to the American Revolution." Annually, the Sunday closest to Dec 16. Starts at Old South Meeting House 5:30 p.m. Est attendance: 1,000. For info: Boston Tea Party Ship, Reneé, Congress St Bridge, Boston, MA 02210. Phone: (617) 338-1773.

◈ **SAMPSON, DEBORAH: BIRTH ANNIVERSARY.** Dec 17, 1760. Born at Plympton, MA, Deborah Sampson spent her childhood as an indentured servant. In 1782, wishing to participate in the Revolutionary War, Sampson disguised herself as a man and enlisted in the Continental Army's 4th Massachusetts Regiment under the name Robert Shurtleff. She received both musket and sword wounds, but it was an attack of fever that

December 1995

S	M	T	W	T	F	S
					1	2
3	4	5	6	7	8	9
10	11	12	13	14	15	16
17	18	19	20	21	22	23
24	25	26	27	28	29	30
31						

Chase's 1995 Calendar of Events — Dec 17-19

unmasked her identity and led to her dismissal from the army in 1783. In 1802 Sampson became perhaps the first woman to lecture professionally in the US when she began giving public speeches on her experiences. Full military pension was provided for her heirs by act of Congress in 1838. Deborah Sampson died Apr 29, 1827, at Sharon, MA.

SATURNALIA. Dec 17-23. Anniversary of ancient Roman festival honoring Saturn, the planter god. Approximates the winter solstice. Most festive period of the year in Rome, and immensely popular. Gifts, lights and the closing of businesses mark the period. Some say that the date for the observance of the nativity of Jesus was selected by the early Christian church leaders to fall on Dec 25 partly to counteract the popular but disapproved of pre-Christian Roman festival of Saturnalia.

TELL SOMEONE THEY'RE DOING A GOOD JOB WEEK. Dec 17-23. Every day this week tell someone "you're doing a good job." Sponsor: Joe Hoppel, Radio Station WCMS, 900 Commonwealth Place, Virginia Beach, VA 23464. Phone: (804) 424-1050.

WHITTIER, JOHN GREENLEAF: BIRTH ANNIVERSARY. Dec 17. American author born at Haverhill, MA, Dec 17, 1807. Died Sept 7, 1892.

★ **WRIGHT BROTHERS DAY.** Dec 17. Presidential Proclamation always issued for Dec 17 since 1963 (PL88-209 of Dec 17, 1963). Issued twice earlier at Congressional request in 1959 and 1961.

WRIGHT BROTHERS FIRST POWERED FLIGHT: ANNIVERSARY. Dec 17, 1903. Orville and Wilbur Wright, brothers, bicycle shop operators, inventors and aviation pioneers, after three years of experimentation with kites and gliders, achieved the first documented successful powered and controlled flights of an airplane. The flights, near Kitty Hawk, NC, piloted first by Orville then by Wilbur Wright, were sustained for less than one minute but represented man's first powered airplane flight, and the beginning of a new form of transportation. Orville Wright was born at Dayton, OH, Aug 19, 1871, and died there Jan 30, 1948. Wilbur Wright was born at Millville, IN, Apr 16, 1867, and died at Dayton, OH, May 30, 1912.

BIRTHDAYS TODAY

Bob Guccione, 65, publisher, born at Brooklyn, NY, Dec 17, 1930.
Albert King, 36, former basketball player, born at Brooklyn, NY, Dec 17, 1959.
Eugene Levy, 49, comedian, writer ("Second City TV," "SCTV Network 90"), born Hamilton, Ont, Canada, Dec 17, 1946.
Sy Oliver, 85, music arranger, composer, born at Battle Creek, MI, Dec 17, 1910.
Gene Rayburn, 78, former TV game show host ("Match Game"), comic ("Steve Allen Show"), born at Christopher, IL, Dec 17, 1917.
William Safire, 66, author, journalist (*Coming to Terms, Words of Wisdom*), born at New York, NY, Dec 17, 1929.
Tommy Steele, 59, actor (*The Happiest Millionaire, Half a Sixpence*), born at London, England, Dec 17, 1936.

DECEMBER 18 — MONDAY
352nd Day — Remaining, 13

BELGIUM: NUTS FAIR. Dec 18. Bastogne. Traditional cultural observance. Third Monday in December.

CHANUKAH. Dec 18-25. Feast of Lights, or Feast of Dedication. Festival lasting eight days, beginning on Kislev 25. Commemorates victory of Maccabees over Syrians (165 BC) and rededication of Temple of Jerusalem. Begins on Hebrew calendar date Kislev 25, 5756.

COBB, TYRUS RAYMOND "TY": BIRTH ANNIVERSARY. Dec 18. Famed American baseball player born at Narrows, GA, Dec 18, 1886. Died at Atlanta, GA, July 17, 1961. Lifetime batting average of .367 compiled over 24 years during which he played in more than 3,000 games.

GRIMALDI, JOSEPH: BIRTH ANNIVERSARY. Dec 18. Known as the "greatest clown in history" and the "king of pantomime," Joseph Grimaldi began his stage career at age two. He was an accomplished singer, dancer and acrobat. Born at London on Dec 18, 1778, he is best remembered as the original "Joey the Clown," and for the innovative humor he brought to the clown's role in theater. Illness forced his early retirement in 1823, and he died at London on May 31, 1837.

MEXICO: FEAST OF OUR LADY OF SOLITUDE. Dec 18. Oaxaca. Pilgrims venerate the patron of the lonely.

NEW JERSEY RATIFIES CONSTITUTION ANNIVERSARY. Dec 18. New Jersey, on Dec 18, 1787, became the third state to ratify the Constitution (following Delaware and Pennsylvania). It did so unanimously.

NIGER: REPUBLIC DAY. Dec 18. National holiday.

SPACE MILESTONE: *SOYUZ 13* (USSR). Dec 18. Two Soviet cosmonauts (P.I. Klimuk and V.V. Lebedev) began, Dec 18, 1973, 8-day orbit of Earth. Returned to Earth on Dec 26, 1973.

STRADIVARI, ANTONIO: DEATH ANNIVERSARY. Dec 18. Celebrated Italian violin maker was born probably in the year 1644, and died at Cremona, Dec 18, 1737.

THIRTEENTH AMENDMENT TO THE US CONSTITUTION: ANNIVERSARY. Dec 18. On Dec 18, 1865, ratification of the 13th Amendment was proclaimed, abolishing slavery in the nation. "Neither slavery nor involuntary servitude, save as a punishment for crime whereof the party shall have been duly convicted, shall exist within the United States, or any place subject to their jurisdiction." See also: "Emancipation Proclamation Anniversary" (Jan 1) for Abraham Lincoln's proclamation freeing slaves in the rebelling states.

WEBER, CARL MARIA VON: BIRTH ANNIVERSARY. Dec 18. Composer, "founder of German romantic school," was born at Eutin, Germany, on Dec 18, 1786. Member of a musical family, he is remembered mainly for his operas, especially the immensely popular *Der Freischutz* (1821). He died at London, England, on June 5, 1826.

BIRTHDAYS TODAY

(William) Ramsey Clark, 68, former attorney general, born at Dallas, TX, Dec 18, 1927.
Ossie Davis, 78, actor (*Raisin in the Sun, Grumpy Old Men*, "Evening Shade"), born Cogdell, GA, Dec 18, 1917.
Janie Frickie, 45, singer (Country Music Assn Female Vocalist of Year '82, '83; "It Ain't Easy"), born Whitney, IN, Dec 18, 1950.
Leonard Maltin, 45, movie critic, author (*Maltin's Guide*) born at New York, NY, Dec 18, 1950.
Charles Oakley, 32, basketball player, born at Cleveland, OH, Dec 18, 1963.
Keith Richards, 52, musician, singer (with The Rolling Stones; "Happy"), born at Dartford, England, Dec 18, 1943.
Steven Spielberg, 48, filmmaker (*E.T., Indiana Jones, Close Encounters of the Third Kind, The Color Purple*, Oscar for *Schindler's List*), born Cincinnati, OH, Dec 18, 1947.

DECEMBER 19 — TUESDAY
353rd Day — Remaining, 12

CHRISTMAS GREETINGS FROM SPACE: ANNIVERSARY. Dec 19. At 3:15 PM, EST, Dec 19, 1958, the US Earth satellite *Atlas* transmitted the first radio voice broadcast from

Dec 19-20 ☆ Chase's 1995 Calendar of Events ☆

space, a 58-word recorded Christmas greeting from President Dwight D. Eisenhower: "to all mankind America's wish for peace on earth and good will toward men everywhere." Satellite had been launched from Cape Canaveral Dec 18.

FISKE, MINNIE MADDERN: BIRTH ANNIVERSARY. Dec 19. American theater actress with a long, distinguished career. First stage appearance at the age of three as "Little Minnie Maddern." Born on Dec 19, 1865 at New Orleans, LA, and died on Feb 15, 1932.

LIVERMORE, MARY ASHTON: BIRTH ANNIVERSARY. Dec 19. American reformer and women's suffrage leader, born Dec 19, 1821. Died May 23, 1905.

MUSLIM FESTIVAL: LAILAT AL MIRAJ. Dec 19-20. Commemorates ascent of the Prophet Muhammad into Heaven. Muslim calendar date: Rajab 27, 1416. Different methods for calculating the Gregorian date are used by different Muslim sects or groups. *Chase's* dates are based on astronomical calculations of the visibility of the new moon crescent at Mecca; EST date may vary.

PARRY, WILLIAM: BIRTH ANNIVERSARY. Dec 19. British explorer, Sir William Edward Parry, was born at Bath, England, on Dec 19, 1790. Remembered for his Arctic expeditions and for his search for a Northwest Passage, Parry died at Ems, Germany, July 8, 1855.

SPACE MILESTONE: *COSMOS 1614*. Dec 19. Unmanned space shuttle made one orbit before splashdown in Black Sea. Dec 19, 1984.

SPACE MILESTONE: *INTELSAT 4 F-3* (US). Dec 19. Communications satellite launched by NASA on contract with COMSAT. Mission involved intercontinental relay phone and TV communications. Dec 19, 1971.

UNITED STATES INVASION OF PANAMA: ANNIVERSARY. Dec 19. On Dec 19, 1989, the US launched operation "Just Cause," invading Panama in an attempt to seize Manuel Noriega and bring him to justice for narcotics trafficking. Seven months after Noriega had ruled unfavorable election results null and void, the US toppled the Noriega government and oversaw the installation of Guillermo Endara as president, restoring a democratic form of government. Although the initial military action was declared a success, Noriega eluded capture and eventually sought sanctuary in the Papal Nunciature in Panama City. He surrendered to US troops on Jan 4, 1990, and was flown to the US to stand trial.

BIRTHDAYS TODAY

Jennifer Beals, 32, actress (*Flashdance, The Bride* [Frankenstein with Sting], *Into the Soup*), born Chicago, IL, Dec 19, 1963.

Richard E. Leakey, 51, anthropologist, born at Nairobi, Kenya, Dec 19, 1944.

Kevin McHale, 38, former basketball player, born at Hibbing, MN, Dec 19, 1957.

Alyssa Milano, 23, actress (Samantha Micelli on "Who's the Boss"), born at Brooklyn, NY, Dec 19, 1972.

Tim Reid, 51, actor ("Frank's Place," "WKRP in Cincinnati"), born Norfolk, VA, Dec 19, 1944.

Cicely Tyson, 56, actress (Emmy for "The Autobiography of Miss Jane Pittman"; *Sounder*), born at New York, NY, Dec 19, 1939.

Robert Urich, 50, actor ("Marcus Welby," "Spencer: For Hire"), born Toronto, Ont, Canada, Dec 19, 1945.

December 1995

S	M	T	W	T	F	S
					1	2
3	4	5	6	7	8	9
10	11	12	13	14	15	16
17	18	19	20	21	22	23
24	25	26	27	28	29	30
31						

DECEMBER 20 — WEDNESDAY
354th Day — Remaining, 11

AMERICAN POET LAUREATE ESTABLISHMENT: 10th ANNIVERSARY. Dec 20. A bill empowering the Librarian of Congress to name, annually, a Poet Laureate/Consultant in Poetry was signed into law by President Ronald Reagan on Dec 20, 1985. In return for a $10,000 stipend as Poet Laureate and a salary (about $35,000) as the Consultant in Poetry, the person named will present at least one major work of poetry and will appear at selected national ceremonies. The first Poet Laureate of the US was Robert Penn Warren, appointed to that position by the Librarian of Congress on Feb 26, 1986. See also: "Warren, Robert Penn: Birthday" (Apr 24).

FIRESTONE, HARVEY S.: BIRTH ANNIVERSARY. Dec 20. American industrialist, businessman and founder of the Firestone Tire and Rubber Co. Harvey Samuel Firestone was born at Columbiana County, OH, Dec 20, 1868. A close friend of Henry Ford, Thomas Edison and John Burroughs, Firestone was also author of two books about rubber. He died at Miami Beach, FL, Feb 7, 1938.

FIRST JOINT SIBERIAN-AMERICAN MUSICAL THEATRE PRODUCTION: 5th ANNIVERSARY. Dec 20. On Dec 20, 1990, the first joint Siberian-American production of an American Musical premiered in the Siberian city of Omsk, Russia. The musical was *Charlie Bar* by Americans Jane Boyd and Gregg Opelka. It was staged by an American production team using a Russian cast performing the play in Russian. The production was a joint effort between the Musical Comedy Theatre of Omsk and New Tuners Theatre of Chicago.

FIRST US SCIENTIST RECEIVES NOBEL PRIZE: ANNIVERSARY. Dec 20. University of Chicago professor Albert Michelson, eminent physicist known for his research on the speed of light and optics, on Dec 20, 1907, became the first US scientist to receive the Nobel Prize.

LANGER, SUSANNE K.: 100th BIRTH ANNIVERSARY. Dec 20. Susanne Langer, a leading American philosopher, author of *Philosophy in a New Key: A Study in the Symbolism of Reason, Rite, and Art*, was born at New York, NY, on Dec 20, 1895. Her studies of esthetics and art exerted a profound influence on thinking in the fields of psychology, philosophy and the social sciences. She died at Old Lyme, CT, on July 17, 1985.

LOUISIANA PURCHASE DAY. Dec 20. One of the greatest real estate deals in history was completed on Dec 20, 1803, when more than a million square miles of the Louisiana Territory were turned over to the US by France, for a price of about $20 per square mile.

◆ **MONTGOMERY BUS BOYCOTT ENDS: ANNIVERSARY.** Dec 20, 1956. The US Supreme Court ruling of Nov 13, 1956, calling for integration of the Montgomery, AL, public bus system was implemented. Since Dec 5, 1955, the black community of Montgomery had refused to ride on the segregated buses. The boycott was in reaction to the Dec 1, 1955, arrest of Rosa Parks for refusing to relinquish her seat on a Montgomery bus to a white man.

MUDD DAY. Dec 20. A day to remember Dr. Samuel A. Mudd (born near Bryantown, MD, Dec 20, 1833), sentenced to life

478

☆ Chase's 1995 Calendar of Events ☆ Dec 20-21

imprisonment for giving medical aid to disguised John Wilkes Booth, fleeing assassin of Abraham Lincoln. Imprisoned four years before being pardoned by President Andrew Johnson. Died on Jan 10, 1883.

◆ **SACAGAWEA: DEATH ANNIVERSARY.** Dec 20. As a young Shoshone Indian woman, Sacagawea in 1805 (with her 2-month-old boy strapped to her back) traveled with the Lewis and Clark Expedition, serving as an interpreter. It is said that the expedition could not have succeeded without her aid. She was born about 1787, and died at Fort Manuel on the Missouri River, Dec 20, 1812. Few women have been so often honored. There are statues, fountains and memorials of her, and her name has been given to a mountain peak. Few facts about her life are firmly established and some legends have her living to nearly a hundred years of age.

SOUTH CAROLINA: SECESSION ANNIVERSARY. Dec 20. South Carolina's legislature voted, Dec 20, 1860, to secede from the United States of America, the first state to do so.

VIRGINIA COMPANY EXPEDITION TO AMERICA: ANNIVERSARY. Dec 20. Three small ships, the *Susan Constant*, the *Godspeed* and the *Discovery*, commanded by Captain Christopher Newport, departed London, England, on Dec 20, 1606, bound for America, where the royally chartered Virginia Company's approximately 120 persons established the first permanent English settlement in what is now the United States at Jamestown, VA, on May 14, 1607.

BIRTHDAYS TODAY

Jenny Agutter, 43, actress (Emmy "The Snow Goose"), born London, England, Dec 20, 1952.
Uri Geller, 49, psychic, clairvoyant, born at Tel Aviv, Israel, Dec 20, 1946.
George Roy Hill, 73, filmmaker (*Butch Cassidy and the Sundance Kid*), born at Minneapolis, MN, Dec 20, 1922.
John Hillerman, 63, actor ("The Betty White Show," "Magnum, P.I."), born Denison, TX, Dec 20, 1932.
Sidney Hook, 93, philosopher, educator, born at New York, NY, Dec 20, 1902.
David Levine, 69, artist, caricaturist, born at Brooklyn, NY, Dec 20, 1926.
Mahathir bin Mohamed, 70, prime minister of Malaysia, born at Alor Star, Malaysia, Dec 20, 1925.

DECEMBER 21 — THURSDAY
355th Day — Remaining, 10

BOLL, HEINRICH: BIRTH ANNIVERSARY. Dec 21. German novelist, winner of the 1972 Nobel Prize for Literature, author of some twenty books including *Billiards at Half-Past Nine*, *The Clown* and *Group Portrait With Lady*, was born Dec 21, 1917, at Cologne, Germany. He died near Bonn, Germany, on July 16, 1985.

DISRAELI, BENJAMIN: BIRTH ANNIVERSARY. Dec 21. British novelist and statesman, born Dec 21, 1804. Died Apr 19, 1881. "No government," he wrote, "can be long secure without a formidable opposition."

FOLDES, ANDOR: BIRTH ANNIVERSARY. Dec 21. Pianist Andor Foldes was born at Budapest, Hungary, on Dec 21, 1913. A child prodigy born to a musical family, he played a Mozart piano concerto with the Budapest Philharmonic when he was 8. He moved to the US in 1939 and is renowned for introducing the work of composer Bela Bartok to the US. He died on Feb 9, 1992, at Zurich, Switzerland.

FOREFATHERS' DAY: 375 YEARS. Dec 21. Observed mainly in New England in commemoration of landing at Plymouth Rock on this day in 1620.

HUMBUG DAY. Dec 21. Allows everyone preparing for Christmas to vent their frustrations. 12 "humbugs" allowed. Phone: (212) 388-8673 or (717) 274-8451. Sponsor: Wellness Permission League, Tom or Ruth Roy, 2105 Water St, Lebanon, PA 17046.

MOON PHASE: NEW MOON. Dec 21. Moon enters New Moon phase at 9:22 PM, EST.

NATIONAL FLASHLIGHT DAY. Dec 21. The longest night of the year, the night when people could use a flashlight the most. (Winter Solstice occurs during the night at 3:18 in the morning of Dec 22.) For info: James Morgan, 589 Elma St, Akron, OH 44310.

PAN AMERICAN FLIGHT 103 EXPLOSION: ANNIVERSARY. Dec 21. Pan Am World Airways Flight 103 exploded in mid-air and crashed into the heart of Lockerbie, Scotland, on Dec 21, 1988, the result of a terrorist bombing. The 258 passengers and crew members and a number of persons on the ground were killed in the disaster. The tragedy raised questions about security and the notification of passengers in the event of threatened flights. In the resultant investigation it was revealed that government agencies and the airline had known that the flight was possibly the target of a terrorist attack.

PERIGEAN SPRING TIDES. Dec 21-22. Spring Tides, the highest possible tides, occur when New Moon or Full Moon falls within 24 hours of the moment the Moon is nearest Earth (perigee) in its monthly orbit on Dec 22, at 5:00 AM, EST.

PHILEAS FOGG WINS A WAGER DAY. Dec 21. Anniversary, from Jules Verne's *Around the World in Eighty Days*, of the winning of Phileas Fogg's wager, on Dec 21, 1872, when Fogg walked into the saloon of the Reform Club in London, announcing "Here I am gentlemen!" exactly 79 days, 23 hours, 59 minutes and 59 seconds after starting his trip "around the world in 80 days," to win his £20,000 wager. See also: "Phileas Fogg's Wager Day" (Oct 2).

◆ **PILGRIM LANDING: 375th ANNIVERSARY.** Dec 21. According to Governor William Bradford's *History of Plymouth Plantation*, "On Munday," (Dec 21, 1620, New Style) the Pilgrims, aboard the *Mayflower*, reached Plymouth, MA, "sounded ye harbor, and founde it fitt for shipping; and marched into ye land, & founde diverse cornfields, and ye best they could find, and ye season & their presente necessitie made them glad to accepte of it.... And after wards tooke better view of ye place, and resolved wher to pitch their dwelling; and them and their goods." Plymouth Rock, the legendary place of landing since it first was "identified" nearly 150 years after the landing (in 1769), has been a historic shrine since. The landing anniversary is observed in much of New England as Forefathers' Day. See also: "Forefathers' Day" (Dec 21).

SPACE MILESTONE: *APOLLO 8* (US). Dec 21. First moon voyage launched on Dec 21, 1968, and manned by Colonel Frank Borman, Captain James A. Lovell, Jr, Major William A. Anders, orbited moon on Dec 24, returned to Earth on Dec 27. First men to see the side of the moon away from Earth.

STALIN, JOSEPH VISSARIONOVICH: BIRTH ANNIVERSARY. Dec 21. Russian dictator whose family name was Dzhugashvili, was born at Gori, Georgia, Dec 21, 1879. One of the most powerful and most feared men of the 20th century, Stalin died (of a stroke) at the Kremlin, in Moscow, Mar 5, 1953.

SZOLD, HENRIETTA: BIRTH ANNIVERSARY. Dec 21. Teacher, writer, scholar, social worker, organizer and pioneer Zionist, Henrietta Szold is best remembered as founder and first president of Hadassah, the Women's Zionist Organization of America, Inc. Born at Baltimore, MD, on Dec 21, 1860, she was influenced by her father Rabbi Benjamin Szold, an active and vocal abolitionist. She established the first "Night School" in Baltimore and one of the first of its kind in the US to focus on teaching English and job skills to immigrants. Her trip to Palestine in 1910 sparked the genesis of Hadassah. While there, Szold was alarmed by the lack of social, medical and educational services and returned with the idea that a national women's Zionist organization must be formed to carry out practical projects. As the "Mother of Social Service in Palestine," Szold viewed volunteerism as one of the greatest human endeavors. See also: "Hadassah: Anniversary" (Feb 24). For info: Wendy Hirschhorn, Dir Pub Affairs, Hadassah, 50 W 58th St, New York, NY 10019. Phone: (212) 303-8153.

YALDA. Dec 21. This Iranian observance originates out of the concept of Light and Good versus Darkness and Evil. As the longest night of the year, the winter solstice was considered extremely unlucky, and efforts were made to ward off the imminent Evil. Families kept vigil through the night, fires burning brightly to help the sun (and Goodness) in the battle against darkness (and Evil). Also an occasion of thanksgiving for the previous harvest and prayers for the success of the next one. For info: Yassaman Djalali, Librarian, West Valley Branch Library, 1243 San Tomas Aquino Rd, San Jose, CA 95117. Phone: (408) 244-4766.

ZAPPA, FRANK: BIRTH ANNIVERSARY. Dec 21, 1949. Rock musician and composer, Zappa was noted for his satire and as a leading advocate against censorship of contemporary music. Born at Baltimore, MD, he died Dec 4, 1993, at Los Angeles, CA.

BIRTHDAYS TODAY

Phil Donahue, 60, TV talk show host ("Donahue," "Donahue, Posner"), born at Cleveland, OH, Dec 21, 1935.
Chris Evert, 41, tennis player, born at Fort Lauderdale, FL, Dec 21, 1954.
Jane Fonda, 58, actress (Oscars for *Klute* and *Coming Home*; *Barbarella, The China Syndrome, Julia, On Golden Pond*), wife of Ted Turner, born New York, NY, Dec 21, 1937.
Florence Griffith Joyner, 36, track athlete, born at Los Angeles, CA, Dec 21, 1959.
Barbara Roberts, 59, Governor of Oregon (D), born at Corvallis, Oregon, Dec 21, 1936.
Andrew James Van Slyke, 35, baseball player, born at Utica, NY, Dec 21, 1960.
Kurt Waldheim, 77, former President of Austria, former UN secretary-general, born at Woerdern, Austria, Dec 21, 1918.
Paul Winchell, 73, ventriloquist, actor, born at New York, NY, Dec 21, 1922.

DECEMBER 22 — FRIDAY
356th Day — Remaining, 9

BRACEBRIDGE DINNER. Dec 22. (Also Dec 24–25.) The Ahwahnee Hotel, Yosemite National Park, CA. A pageant held each Christmastime. (Seats are allocated by lottery.) Sponsor: Yosemite Park and Curry Co, 5410 E Home Ave, Fresno, CA 93727. Phone: (209) 252-4848.

CALIFORNIA KIWIFRUIT DAY. Dec 22. National campaign to educate Americans about the nutritional benefits of kiwifruit (they provide twice the vitamin C of oranges); ways to enjoy kiwifruit; kiwifruit's colorful history. Annually, the first day of winter. For info: California Kiwifruit Commission, 1540 River Park Dr, Ste 110, Sacramento, CA 95815.

CAPRICORN, THE GOAT. Dec 22–Jan 19. In the astronomical and astrological zodiac that divides the sun's apparent orbit into 12 segments, the period Dec 22–Jan 19 is identified, traditionally, as the sun-sign of Capricorn, the Goat. The ruling planet is Saturn.

December 1995

S	M	T	W	T	F	S
					1	2
3	4	5	6	7	8	9
10	11	12	13	14	15	16
17	18	19	20	21	22	23
24	25	26	27	28	29	30
31						

ELLERY, WILLIAM: BIRTH ANNIVERSARY. Dec 22. Signer of the Declaration of Independence, born, Dec 22, 1727. Died, Feb 15, 1820.

FIRST GORILLA BORN IN CAPTIVITY: ANNIVERSARY. Dec 22. On this day in 1956, "Colo" was born at the Columbus, OH, zoo, weighing in at 3¼ pounds.

JOHNSON, CLAUDIA ALTA "LADY BIRD" TAYLOR: BIRTHDAY. Dec 22. Widow of Lyndon Baines Johnson, 36th president of the US, born at Karnack, TX, Dec 22, 1912.

OGLETHORPE, JAMES EDWARD: BIRTH ANNIVERSARY. Dec 22. English general, author and colonizer of Georgia. Founder of the city of Savannah. Oglethorpe was born at London on Dec 22, 1696. He died June 30, 1785.

PUCCINI, GIACOMO: BIRTH ANNIVERSARY. Dec 22. Italian composer, born Dec 22, 1858. Died Nov 29, 1924.

ROBINSON, EDWIN ARLINGTON: BIRTH ANNIVERSARY. Dec 22. Three-time Pulitzer Prize winner best known for his short dramatic poems, including "Richard Cory" and "Miniver Cheevy." Born Dec 22, 1869, at Head Tide, ME, and died at Los Angeles, CA, on Apr 6, 1935.

WINTER. Dec 22–Mar 20. In the Northern Hemisphere winter begins today with the winter solstice, at 3:17 AM, EST. Note that in the Southern Hemisphere today is the beginning of summer. Between Equator and Arctic Circle the sunrise and sunset points on the horizon are farthest south for the year, and daylight length is minimum (ranging from 12 hours, 8 minutes at the equator to zero at the Arctic Circle).

YULE. Dec 22. (Also called Alban Arthan.) One of the "Lesser Sabbats" during the Wiccan year, Yule marks the death of the Sun-God and his rebirth from the Earth Goddess. Annually, on the winter solstice.

BIRTHDAYS TODAY

Barbara Billingsley, 73, actress (June Cleaver on "Leave It to Beaver;" *Airplane!*), born Los Angeles, CA, Dec 22, 1922.
Steve Carlton, 51, baseball Hall of Famer, born at Miami, FL, Dec 22, 1944.
Steve Garvey, 47, former baseball player, born at Tampa, FL, Dec 22, 1948.
Claudia Alta "Lady Bird" Johnson, 83, former first lady, born at Karnack, TX, Dec 22, 1912.
Diane K. Sawyer, 49, journalist (staff asst to Richard Nixon at time of his resignation; "60 Minutes,") born at Glasgow, KY, Dec 22, 1946.
Jan Stephenson, 44, golfer, born at Sydney, Australia, Dec 22, 1951.

DECEMBER 23 — SATURDAY
357th Day — Remaining, 8

FEDERAL RESERVE SYSTEM: ANNIVERSARY. Dec 23. Established pursuant to authority contained in the Federal Reserve Act of Dec 23, 1913, the system serves as the nation's central bank, has responsibility for execution of monetary policy. It is called on to contribute to the strength and vitality of the US economy, in part by influencing the lending and investing activities of commercial banks and the cost and availability of money and credit.

FIRST NONSTOP FLIGHT AROUND THE WORLD WITHOUT REFUELING. Dec 23. In 1987, Dick Rutan and Jeana Yeager set a new world record of 216 hours of continuous flight, breaking their own record of 111 hours set on July 15, 1986. The aircraft *Voyager* departed from Edwards Air Force Base in California on Dec 14, 1987 and landed on Dec 23, 1987. The journey covered 24,986 miles at an official speed of 115 miles per hour.

JAPAN: BIRTHDAY OF THE EMPEROR. Dec 23. National Day.

☆ Chase's 1995 Calendar of Events ☆ Dec 23-24

MERRIE OLDE ENGLAND CHRISTMAS FESTIVAL. Dec 23-26. Boar's Head Inn, Charlottesville, VA. The highlight of the festival, the "Feast Before Forks," is a multi-course Medieval style banquet enjoyed without the use of forks, including cuisine and entertainment reminiscent of the 16th century. Other activities include puppet shows, jugglers, wine tastings, folk dances and storytelling. Est attendance: 200. For info: The Boar's Head Inn & Sports Club, PO Box 5307, Charlottesville, VA 22905. Phone: (800) 476-1988.

METRIC CONVERSION ACT: 20th ANNIVERSARY. Dec 23. The Congress of the United States, on Dec 23, 1975, passed Public Law 94-168, known as the Metric Conversion Act of 1975. This act declares that the SI (International System of Units) will be this country's basic system of measurement and establishes the United States Metric Board which is responsible for the planning, coordination and implementation of the nation's voluntary conversion to SI. (Congress had authorized the metric system as a legal system of measurement in the US by an act passed on July 28, 1866. In 1875, the US became one of the original signers of the Treaty of the Metre, which established an international metric system.)

MEXICO: FEAST OF THE RADISHES. Dec 23. Oaxaca. Figurines of people and animals cleverly carved out of radishes are sold during festivities.

MONROE, HARRIET: BIRTH ANNIVERSARY. Dec 23. American poet, editor and founder of *Poetry* magazine. Born Dec 23, 1860, at Chicago, IL. Died on Sept 26, 1936.

SMILES, SAMUEL: BIRTH ANNIVERSARY. Dec 23. Scottish writer, born Dec 23, 1812. Died at London, Apr 17, 1904. "A place for everything," he wrote in *Thrift*, "and everything in its place."

SMITH, JOSEPH: BIRTH ANNIVERSARY. Dec 23. Mormon prophet born at Sharon, VT, Dec 23, 1805. Killed by a mob while in Carthage, IL, jail, June 27, 1844.

TOJO HIDEKI: EXECUTION ANNIVERSARY. Dec 23. Tojo Hideki, prime minister of Japan from Oct 16, 1941, until his resignation on July 19, 1944. After Japan's surrender in Aug 1945, Tojo was arrested as a war criminal, tried by a military tribunal and sentenced to death on Nov 12, 1948. Born at Tokyo on Dec 30, 1884, Tojo was hanged (with six other Japanese wartime military leaders) at Sugamo Prison in Tokyo on Dec 23, 1948, the sentence being carried out by the US Eighth Army.

BIRTHDAYS TODAY

Akihito, 62, Emperor of Japan, born at Tokyo, Japan, Dec 23, 1933.
Robert Bly, 69, author (*Iron John: A Book About Men; What Have I Ever Lost By Dying?*), born Madison, MN, Dec 23, 1926.
Jose Greco, 77, dancer, born at Abruzzi, Italy, Dec 23, 1918.
James Gregory, 84, actor ("Barney Miller," *The Manchurian Candidate*), born at New York, NY, Dec 23, 1911.
Jim (James Joseph) Harbaugh, 32, football player, born at Toledo, OH, Dec 23, 1963.
Susan Lucci, 46, actress (Erica Kane on "All My Children"; *Mafia Princess*), born at Westchester, NY, Dec 23, 1949.
Gerald O'Loughlin, 74, actor ("The Rookies," "Our House"), born at New York, NY, Dec 23, 1921.
Ruth Roman, 72, actress (Minnie Littlejohn in *The Long Hot Summer*; Sylvia Lean in "Knots Landing"), born Boston, MA, Dec 23, 1923.

DECEMBER 24 — SUNDAY
358th Day — Remaining, 7

ARNOLD, MATTHEW: BIRTH ANNIVERSARY. Dec 24. English poet and essayist, born Dec 24, 1822. Died Apr 15, 1888. "One has often wondered," he wrote in *Culture and Anarchy*, "whether upon the whole earth there is anything so unintelligent, so unapt to perceive how the world is really going, as an ordinary young Englishman of our upper class."

AUSTRIA: "SILENT NIGHT, HOLY NIGHT" CELEBRATIONS. Dec 24. Oberndorf, Hallein and Wagrain, Salzburg, Austria. Commemorating the creation of the Christmas carol here in 1818.

CARSON, CHRISTOPHER "KIT": BIRTH ANNIVERSARY. Dec 24. American frontiersman, soldier, trapper, guide and Indian agent best known as Kit Carson. Born on Dec 24, 1809, in Madison County, KY, he died at Fort Lyon, CO, on May 23, 1868.

CHRISTMAS AT THE ZOO. Dec 24-26. North Carolina Zoological Park, Asheboro, NC. Christmas tree for the birds; free admission. Est attendance: 5,000. Sponsor: North Carolina Zoological Park, Tracy Hall, Events and Projects Coord, 4401 Zoo Pkwy, Asheboro, NC 27203. Phone: (800) 488-0444.

CHRISTMAS BELLS RING AGAIN IN ST. BASIL'S: 5th ANNIVERSARY. Dec 24. For the first time since the death of Lenin, the bells of St. Basil's Cathedral, on Red Square in Moscow, rang to celebrate Christmas on Dec 24, 1990.

CHRISTMAS EVE. Dec 24. Family gift-giving occasion in many Christian countries.

FIRST SURFACE-TO-SURFACE GUIDED MISSILE: ANNIVERSARY. Dec 24. On Dec 24, 1942, German rocket engineer Wernher von Braun launched the first surface-to-surface guided missile. Buzz bombs, a form of guided missile, were used by Germany against Great Britain starting Sept 8, 1944. On Feb 24, 1949, the first rocket to reach outer space (an altitude of 25 miles) was fired. The two-stage rocket, a Wac Corporal set in the nose of a German V-2, was launched from the White Sands Proving Grounds, NM, by a team of scientists headed by Dr. Wernher von Braun.

HUGHES, HOWARD ROBARD: BIRTH ANNIVERSARY. Dec 24. Wealthy American recluse born Dec 24, 1905. Died in airplane en route from Acapulco, Mexico, to Houston, TX, on Apr 5, 1976.

JOULE, JAMES PRESCOTT: BIRTH ANNIVERSARY. Dec 24. English physicist and inventor after whom Joule's Law was named, was born at Salford, Lancashire, England, on Dec 24, 1818. Joule died at Cheshire on Oct 11, 1889.

NATIONAL ROOF-OVER-YOUR-HEAD DAY. Dec 24. To draw attention to the plight of the homeless across America and to generate support for local and national efforts on their behalf. Varies annually. No further info available. Sponsor: There but for the Grace Society.

RUSH, BENJAMIN: 250th BIRTH ANNIVERSARY. Dec 24. Physician, patriot and humanitarian of the American Revolution, born on a plantation at Byberry, PA, Dec 24, 1745. A signer of the Declaration of Independence, his writings on mental illness earned him the title "Father of Psychiatry." His tract, *Inquiry*, attacked the common wisdom of the time that alcohol was a positive good. He was the first American to call alcoholism a chronic disease. Benjamin Rush died at Philadelphia, PA, Apr 19, 1813.

BIRTHDAYS TODAY

Jill Bennett, 64, actress (*Lust for Life, The Nanny*), born Penang, Federated Malay States, Dec 24, 1931.
Anthony Fauci, 55, health administrator, physician, born at Brooklyn, NY, Dec 24, 1940.
A.P. Lutali, 76, Governor of American Samoa (D), born at Au-nu'u, American Samoa, Dec 24, 1919.
Nicholas Meyer, 50, author, director (*The Canary Trainer: From the Memoirs of John H. Watson*), born New York, NY, Dec 24, 1945.

DECEMBER 25 — MONDAY
359th Day — Remaining, 6

BARTON, CLARA: BIRTH ANNIVERSARY. Dec 25. Clarissa Harlowe Barton, American nurse and philanthropist, founder of the American Red Cross, was born at Oxford, MA, Dec 25, 1821. In 1881, she became first president of the American Red Cross (founded May 21, 1881). She died at Glen Echo, MD, Apr 12, 1912.

BOGART, HUMPHREY: BIRTH ANNIVERSARY. Dec 25. American stage and screen actor, Humphrey DeForest Bogart was born at New York, NY, on Dec 25, 1899. Among his best remembered films were: *The African Queen*, *The Maltese Falcon*, *Casablanca* and *To Have and Have Not*. Bogart died Jan 14, 1957.

BOOTH, EVANGELINE CORY: BIRTH ANNIVERSARY. Dec 25. Salvation Army general, active in England, Canada and the US. Author and composer of songs, Booth was born at London, England on Dec 25, 1865. She died at Hartsdale, NY, on July 17, 1950.

CEAUSESCU, NICOLAE: DEATH ANNIVERSARY. Dec 25. On Christmas evening, 1989, a broadcast of a Christmas symphony on state-run television was interrupted with the report that Romanian president Nicolae Ceausescu and his wife had been executed, bringing to an end the last hard-line regime in the Soviet bloc. In a brutally fast uprising, the Romanian people ousted the Ceausescu regime, and the National Salvation Front, an ad hoc pro-democracy coalition, took charge of the country. Ceausescu's downfall began when he ordered members of his black-shirted state police, the Securitate, to use force to quell a disturbance in the town of Timisorara. The brutal crackdown led to estimates of as many as 4,500 killed. Ceausescu's rule was marked by corruption, deprivation and terror. Although he broke from Moscow on many issues, his ruling methods adhered to Stalinist techniques. Sparked by the revolt and massacre in Timisorara, the Romanian revolution was the sixth and only violent overthrow of a Communist government in the last six months of 1989.

CHRISTMAS. Dec 25. Christian festival commemorating the birth of Jesus of Nazareth. Most popular of Christian observances, Christmas as a Feast of the Nativity dates from the 4th century. Although Jesus's birth date is not known, the Western church selected Dec 25 for the feast, possibly to counteract the non-Christian festivals of that approximate date. Many customs from non-Christian festivals (Roman Saturnalia, Mithraic sun's birthday, Teutonic yule, Druidic and other winter solstice rites) have been adopted as part of the Christmas celebration (lights, mistletoe, holly and ivy, holiday tree, wassailing and gift-giving, for example). Some Orthodox Churches celebrate Christmas on Jan 7 based on the "old calendar" (Julian). Theophany (recognition of the divinity of Jesus) is observed on this date and also on Jan 6, especially by the Eastern Orthodox Church.

CHRISTMAS FIRESIDE CHAT WARNING: ANNIVERSARY. Dec 25. In his Dec 25, 1943, Christmas message to the American people, Franklin D. Roosevelt warned, "The war is now reaching the stage when we shall have to look forward to large casualty lists—dead, wounded and missing. War entails just that. There is no easy road to victory. And the end is not yet in sight."

JINNAH, MOHAMMED ALI: BIRTH ANNIVERSARY. Dec 25. The founder of the Islamic Republic of Pakistan, Mohammed Ali Jinnah, was born at Karachi, Dec 25, 1876. When Pakistan became an independent political entity (Aug 15, 1947), Jinnah became its first governor general. He died at Karachi, Sept 11, 1948.

KELLY TIRE BLUE-GRAY ALL STAR FOOTBALL CLASSIC. Dec 25. Cramton Bowl, Montgomery, AL. College seniors from northern schools compete against their southern counterparts. Sponsors: Montgomery Lion's Club and Kelly Tire. Est attendance: 22,000. For info: Charles W. Jones, Exec Dir, Box 94, Montgomery, AL 36101. Phone: (205) 265-1266.

NEWTON, ISAAC: BIRTH ANNIVERSARY. Dec 25. Sir Isaac Newton, mathematician, scientist and author, was born near Grantham, Lincolnshire, England, on Dec 25, 1642. Best known of his works is the *Philosophiae Naturalis Principia Mathematica* (1687) in which he enunciated the famed three laws of motion. Newton died at London, England, on Mar 20, 1727, and was buried at Westminster Abbey.

STOCK EXCHANGE HOLIDAY (CHRISTMAS DAY). Dec 25. The holiday schedules for the various exchanges are subject to change if relevant rules, regulations or exchange policies are revised. If you have questions, phone: American Stock Exchange (212) 306-1212; Chicago Board of Options Exchange (312) 786-7760; Chicago Board of Trade (312) 435-3500; New Stock Exchange (212) 656-2065; Pacific Stock Exchange (415) 393-4000; Philadelphia Stock Exchange (215) 496-5000.

UNITED KINGDOM: CHRISTMAS HOLIDAY. Dec 25. Bank and public holiday in England, Wales, Scotland and Northern Ireland.

WASHINGTON CROSSING THE DELAWARE REENACTMENT. Dec 25. Washington Crossing, PA. Reenactment pageant of George Washington's crossing of the Delaware, Dec 25, 1776, which led to the victory at Trenton, turning point of the American Revolution. Est attendance: 12,000. Sponsor: Washington Crossing Fdtn, Emma Adams, PO Box 17, Washington Crossing, PA 18977. Phone: (215) 493-4076.

WEST, REBECCA: BIRTH ANNIVERSARY. Dec 25. English author, literary critic, prize-winning journalist and noted feminist, Dame Rebecca West was born Cicely Isabel Fairfield at London, on Dec 25, 1892. She died there Mar 15, 1983.

BIRTHDAYS TODAY

Jimmy Buffett, 49, singer ("Margaritaville"), songwriter, born at Pascagoula, MS, Dec 25, 1946.
Cab Calloway, 88, band leader, singer ("Minnie the Mermaid"), born at Rochester, NY, Dec 25, 1907.
Larry Csonka, 49, Pro Football Hall of Famer, born at Akron, OH, Dec 25, 1946.
Rickey Henderson, 37, baseball player, born at Chicago, IL, Dec 25, 1958.
Annie Lennox, 41, singer (singer for Eurythmics; "Sweet Dreams Are Made of This"), songwriter, born at Aberdeen, Scotland, Dec 25, 1954.
Barbara Ann Mandrell, 47, singer ("I Was Country When Country Wasn't Cool"), born Houston, TX, Dec 25, 1948.
Little Richard (Penniman), 60, singer ("Tutti Frutti," "Long Tall Sally"), songwriter, born Macon, GA, Dec 25, 1935.
Gary Sandy, 50, actor ("All That Glitters;" Andy Travis in "WKRP in Cincinnati"), born Dayton, OH, Dec 25, 1945.
Sissy (Mary Elizabeth) Spacek, 46, actress (Oscar for *Coal Miner's Daughter*; *Missing*, *The River*), born at Quitman, TX, Dec 25, 1949.

☆ Chase's 1995 Calendar of Events ☆   Dec 26-27

DECEMBER 26 — TUESDAY
360th Day — Remaining, 5

AMERICAN YOUTH HOSTELS: CHRISTMAS TRIP. Dec 26-31. San Diego, CA. A six-day bicycling adventure through the mountains, deserts and coastal areas of southern California. Begins and ends in San Diego. Annually, Dec 26-31. Est attendance: 100. Sponsor: San Diego Council, American Youth Hostels, Inc, 335 W Beech St, San Diego, CA 92101. Phone: (619) 338-9981.

BAHAMAS: JUNKANOO. Dec 26. Kaleidoscope of sound and spectacle combining a bit of Mardi Gras, mummer's parade and ancient African tribal rituals. Revelers in colorful costumes parade through the streets to sounds of cowbells, goat skin drums and many other homemade instruments. Always on Boxing Day.

BOXING DAY. Dec 26. Ordinarily observed on the first day after Christmas. Now a legal holiday in Canada, the United Kingdom and many other countries. Formerly (according to Robert Chambers) a day when Christmas gift boxes were "regularly expected by a postman, the lamplighter, the dustman, and generally by all those functionaries who render services to the public at large, without receiving payment therefore from any individual." When Boxing Day falls on a Saturday or Sunday, the Monday or Tuesday immediately following may be proclaimed or observed as a bank or public holiday.

BOXING DAY ANNUAL CELEBRATION. Dec 26. North Star Pub, New York, NY. Annual observation of Boxing Day at an authentic British pub featuring free Bangers and Mash (sausages and mashed potatoes), yard-of-ale contests, Christmas crackers, games and prizes. Collection of food and clothing for homeless people. Est attendance: 500. For info: Deven Black, Genl Mgr, North Star Pub, 93 South St, New York, NY 10038. Phone: (212) 509-6757.

HILLHAVEN THANK YOU SANTA HOTLINE. Dec 26-27. Seattle, WA. Skilled nursing facility residents play Santa and Mrs Claus for children to call to thank Santa for the gifts they received. About 2,500 calls received. For info: Donna Albers, Natl Mktg Dir, Caller Service 2264, Tacoma, WA 98401-2264. Phone: (206) 756-4731.

IRELAND: DAY OF THE WREN. Dec 26. Dingle Peninsula. Masked revelers and musicians go from door to door asking for money. Traditional day and night of public merrymaking.

KWANZAA. Dec 26-Jan 1, 1996. American black family observance (since 1966) in recognition of traditional African harvest festivals. Stresses unity of the black family, with community-wide harvest feast (karamu) on seventh day. Kwanzaa means "first fruit" in Swahili. An optional observance to avoid commercialization of Christmas traditions.

LOVERA, JUAN: BIRTH ANNIVERSARY. Dec 26. Venezuelan "Artist of Independence," whose best-known canvases commemorate the independence dates of Apr 19, 1810, and July 5, 1811, and who is known as the founder of historical painting in Venezuela, was born Dec 26, 1778. Died in 1841 (exact date unknown).

LUXEMBOURG: BLESSING OF THE WINE. Dec 26. Greiveldange. Winemakers parade to the church, where a barrel of wine is blessed.

MAO TSE-TUNG: BIRTH ANNIVERSARY. Dec 26. Chinese librarian, teacher, communist-revolutionist and "founding father" of the People's Republic of China, born in Hunan Province, China, on Dec 26, 1893. Died at Peking, Sept 9, 1976.

MILLER, HENRY (VALENTINE): BIRTH ANNIVERSARY. Dec 26. Controversial American novelist born at New York, NY, on Dec 26, 1891. Died at Pacific Palisades, CA, on June 7, 1980.

NATIONAL WHINER'S DAY™. Dec 26. A day dedicated to whiners, especially those who return Christmas gifts and need lots of attention on this day. The most famous whiner(s) of the year will be announced. Nominations accepted through Dec 20. For more information, please send S.A.S.E. Sponsor: Rev. Kevin C. Zaborney, Box 123, Clio, MI 48420. Phone: (313) 982-8436.

NELSON, THOMAS: BIRTH ANNIVERSARY. Dec 26. Merchant and signer of the Declaration of Independence, born at Yorktown, VA, Dec 26, 1738. Died at Hanover County, VA, Jan 4, 1789.

ST. STEPHEN'S DAY. Dec 26. One of the seven deacons, named by the apostles to distribute alms, died during 1st century. Feast Day is Dec 26 and is observed as a public holiday in Austria.

SECOND DAY OF CHRISTMAS. Dec 26. Observed as holiday in many countries.

UNITED KINGDOM: BOXING DAY BANK HOLIDAY. Dec 26. Bank and public holiday in England, Wales, Scotland and Northern Ireland.

WKA DAYTONA CLASSIC DIRT KARTING RACES. Dec 26-29. Daytona Beach Municipal Stadium, Daytona Beach, FL. For info: Daytona Intl Speedway, Glenn Barber, PR, Box 2801, Daytona Beach, FL 32120-2801. Phone: (904) 254-6782.

BIRTHDAYS TODAY

Steve Allen, 74, entertainer, TV pioneer (the first "Tonight" show), composer, author, born at New York, NY, Dec 26, 1921.
Evan Bayh, 40, Governor of Indiana (D), born at Terre Haute, IN, Dec 26, 1955.
Susan Butcher, 41, sled-dog racer, born at Cambridge, MA, Dec 26, 1954.
Carlton Fisk, 48, former baseball player, born at Bellows Falls, VT, Dec 26, 1947.
Alan King (Irwin Kniberg), 68, comedian, author ("Seventh Avenue," *Help! I'm a Prisoner in a Chinese Bakery*), born New York, NY, Dec 26, 1927.
Lynn Martin, 56, former US Secretary of Labor in Bush administration, born at Evanston, IL, Dec 26, 1939.
Ozzy Smith, 41, baseball player, born at Mobile, AL, Dec 26, 1954.
Phil Spector, 55, music producer, born at New York, NY, Dec 26, 1940.
Richard Widmark, 81, actor (*Kiss of Death, Madigan*), born Sunrise, MN, Dec 26, 1914.

DECEMBER 27 — WEDNESDAY
361st Day — Remaining, 4

CAYLEY, GEORGE: BIRTH ANNIVERSARY. Dec 27. Aviation pioneer Sir George Cayley, English scientist and inventor, was a theoretician who designed airplanes, helicopters and gliders. He is credited as the father of aerodynamics, and he was the pilot of the world's first manned glider flight. Born at Scarborough, Yorkshire, England, Dec 27, 1773. Died at Brompton Hall, Yorkshire, Dec 15, 1857.

DIETRICH, MARLENE: BIRTH ANNIVERSARY. Dec 27. Marlene Dietrich was born at Berlin, Germany on Dec 27, 1901. After her hopes of becoming a concert violinist ended with a wrist injury, she enrolled in Max Reinhardt's drama school and went on to become a screen legend. Her ascent to stardom began in 1930 when Josef Von Sterberg cast her as Lola-Loala in *The Blue Angel*, the first film with sound produced in Germany. A year later, she and Sterberg moved to Hollywood and began a string of six US films together with *Morocco*, the only film for which she received an Academy Award nomination. Some of her other film roles include *Destry Rides Again, Kismet, 80 Days Around the World, Touch of Evil, Judgement at Nuremberg* and *Witness for the Prosecution*. During the 1950s she began a career as a cabaret singer in a stage revue that toured the globe including a hostile homecoming in Berlin in 1960. Her later years were spent away from the spotlight. She died on May 6, 1992.

Dec 27–28 ☆ *Chase's 1995 Calendar of Events* ☆

KEPLER, JOHANNES: BIRTH ANNIVERSARY. Dec 27. One of the world's greatest astronomers, called "the father of modern astronomy," German mathematician Johannes Kepler was born at Wurttemberg on Dec 27, 1571. Died at Regensburg, Nov 15, 1630.

PASTEUR, LOUIS: BIRTH ANNIVERSARY. Dec 27. French chemist-bacteriologist born at Dole, Jura, France, Dec 27, 1822. Died at Villeneuve l'Etang, France, Sept 28, 1895. Discoverer of prophylactic inoculation against rabies. Pasteurization process named for him.

RADIO CITY MUSIC HALL: ANNIVERSARY. Dec 27. Radio City Music Hall, in New York City, opened Dec 27, 1932.

ST. JOHN, APOSTLE-EVANGELIST: FEAST DAY. Dec 27. Son of Zebedee, Galilean fisherman, and Salome. Died about AD 100. Roman Rite Feast Day is Dec 27. (Observed on May 8 by Byzantine Rite.)

SOUNDS OF THE SEASON: A HOLIDAY CONCERT. Dec 27–29. Ash Lawn–Highland, Charlottesville, VA. Madrigal singers, yule log and cider. Annually, Dec 27–29. Est attendance: 150. For info: Ash Lawn–Highland, James Monroe Pkwy, Charlottesville, VA 22902. Phone: (804) 293-9539.

WORLD ENDURO CHAMPIONSHIP KART RACES. Dec 27–30. Daytona International Speedway, Daytona Beach, FL. For info: Daytona Intl Speedway, Glenn Barber, PR, Box 2801, Daytona Beach, FL 32120-2801. Phone: (904) 253-6782.

BIRTHDAYS TODAY

Tovah Feldshuh, 43, actress (*Holocaust*), born at New York, NY, Dec 27, 1952.
Bernard Lanvin, 60, fashion designer, born at Neuilly, France, Dec 27, 1935.
William Howell Masters, 80, physician, born at Cleveland, OH, Dec 27, 1915.
Anna Russell, 84, comedienne, born at London, England, Dec 27, 1911.
Lee Salk, 69, psychologist, author, born at New York, NY, Dec 27, 1926.

DECEMBER 28 — THURSDAY

362nd Day — Remaining, 3

AUSTRALIA: PROCLAMATION DAY. Dec 28. Observed in South Australia.

HOLY INNOCENTS DAY (CHILDERMAS). Dec 28. Commemoration of the massacre of children in Bethlehem, ordered by King Herod who wanted to destroy, among them, the infant Savior. Early and medieval accounts claimed as many as 144,000 victims, but more recent writers, noting that Bethlehem was a very small town, have revised the estimates of the number of children killed to between 6 and 20.

IOWA: ADMISSION DAY. Dec 28. Became 29th state on this day in 1846.

MESSINA EARTHQUAKE: ANNIVERSARY. Dec 28. Messina, Sicily. The ancient town of Messina was struck by an earthquake on Dec 28, 1908. Nearly 80,000 persons died in the disaster and half of the town's buildings were destroyed.

MOLSON, JOHN: BIRTH ANNIVERSARY. Dec 28. John Molson, an orphan, left his home at Lincolnshire, England, to settle in Montreal, in 1782. He soon acquired a brewery and became patriarch of the Molson brewery family. Born at Lincolnshire, Dec 28, 1763. Died at Montreal, Que, Canada, Jan 11, 1836.

December 1995

S	M	T	W	T	F	S
					1	2
3	4	5	6	7	8	9
10	11	12	13	14	15	16
17	18	19	20	21	22	23
24	25	26	27	28	29	30
31						

MOON PHASE: FIRST QUARTER. Dec 28. Moon enters First Quarter phase at 2:06 PM, EST.

NEPAL: NATIONAL HOLIDAY. Dec 28. The birthday of His Majesty the King is a public holiday.

POOR RICHARD'S ALMANACK: ANNIVERSARY. Dec 28. The *Pennsylvania Gazette*, on Dec 28, 1732, carried the first known advertisement for the first issue of *Poor Richard's Almanack*, by Richard Saunders (Benjamin Franklin) for the year 1733. The advertisement promised "many pleasant and witty verses, jests and sayings... new fashions, games for kisses ... men and melons... breakfast in bed, &c." America's most famous almanac, *Poor Richard V's* was published through the year 1758, and has been imitated many times since.

"In 1732 I first publish'd my Almanack, under the name of *Richard Saunders*; it was continu'd by me about twenty-five years, commonly call'd *Poor Richard's Almanack*. I endeavor'd to make it both entertaining and useful, and it accordingly came to be in such demand, that I reap'd considerable profit from it, vending annually near ten thousand. And observing that it was generally read, scarce any neighborhood in the province being without it, I consider'd it as a proper vehicle for conveying instruction among the common people, who bought scarcely any other books; I therefore filled all the little spaces that occurr'd between the remarkable days in the calendar with proverbial sentences, chiefly such as inculcated industry and frugality, as the means of procuring wealth, and thereby securing virtue; it being more difficult for a man in want, to act always honestly, as, to use here one of those proverbs, *it is hard for an empty sack to stand upright*."
—*The Autobiography of Benjamin Franklin*

WILSON, WOODROW: BIRTH ANNIVERSARY. Dec 28. The 28th president of the US was born Thomas Woodrow Wilson at Staunton, VA, Dec 28, 1856. Twice elected president (1912 and 1916), it was Wilson who said "The world must be made safe for democracy," as he asked the Congress to declare war on Germany, Apr 2, 1917. His first wife, Ellen, died Aug 6, 1914, and he married Edith Bolling Galt on Dec 18, 1915. He suffered a paralytic stroke Sept 16, 1919, never regaining his health. There were many dark speculations about who (possibly Mrs. Wilson?) was running the government during his illness. His second term of office ended Mar 3, 1921, and he died at Washington, DC, Feb 3, 1924.

BIRTHDAYS TODAY

John Akers, 61, business executive, born at Boston, MA, Dec 28, 1934.
Carlos Carson, 37, football player, born at Lake Worth, FL, Dec 28, 1958.
Hubie Green, 49, golfer, born at Birmingham, AL, Dec 28, 1946.
Lou Jacobi, 82, actor on stage and screen (*Irma La Douce*), born Toronto, Ontario, Canada, Dec 28, 1913.
David Peterson, 52, Canadian political leader, born at Toronto, Ontario, Canada, Dec 28, 1943.
Manuel Puig, 63, Argentinian writer (*Kiss of the Spider Woman*), born at General Villegas, Argentina, Dec 28, 1932.

Maggie Smith, 61, actress (*Prime of Miss Jean Brodie*; Tony for *Lettice & Lovage*), born Ilford, England, Dec 28, 1934.
Denzel Washington, 41, actor ("St. Elsewhere," *Glory, Malcolm X*), born Mt Vernon, NY, Dec 28, 1954.
Edgar Winter, 49, singer, musician (*Edgar Winter's White Trash, They Only Come Out At Night*), born Beaumont, TX, Dec 28, 1946.

DECEMBER 29 — FRIDAY
363rd Day — Remaining, 2

ALL-COLLEGE BASKETBALL TOURNAMENT. Dec 29–30. Myriad Convention Center, Oklahoma City, OK. The oldest college basketball tournament in the country outdating the NCAA, NAIA and the NIT. Host school is Oklahoma University. Rest of field to be announced; 153 universities have competed. Annually, between Christmas and New Year's. Est attendance: 25,000. For info: Stanley Draper, Jr, OK City All Sports Assn, 100 W Main, Ste 287, Oklahoma City, OK 73102. Phone: (405) 236-5000.

CASALS, PABLO: BIRTH ANNIVERSARY. Dec 29. Famed cellist, Pablo Carlos Salvador Defillio de Casals, born at Venrell, Spain, Dec 29, 1876. Died at Rio Pedros, Puerto Rico, Oct 22, 1973.

GLADSTONE, WILLIAM EWART: BIRTH ANNIVERSARY. Dec 29. English statesman and author for whom the Gladstone (luggage) bag was named. Inspiring orator, eccentric individual, intensely loved or hated by all who knew him (cheered from the streets and jeered from the balconies), Gladstone is said to have left more writings (letters, diaries, journals, books) than any other major English politician. However, his preoccupation with the charitable rehabilitation of prostitutes was perhaps easily misunderstood. Born at Liverpool, England, on Dec 29, 1809, he was four times Britain's prime minister. Gladstone died at Hawarden, Wales, on May 19, 1898.

JOHNSON, ANDREW: BIRTH ANNIVERSARY. Dec 29. Seventeenth president of the US, Andrew Johnson, proprietor of a tailor shop in Laurens, SC, before he entered politics, was born Dec 29, 1808, at Raleigh, NC. Upon Abraham Lincoln's assassination Johnson became president. He was the only US president to be impeached, and he was acquitted Mar 26, 1868. After his term of office as president (Apr 15, 1865–Mar 3, 1869) he made several unsuccessful attempts to win public office. Finally he was elected to the US Senate from Tennessee, and served in the Senate from Mar 4, 1875, until his death, at Carter's Station, TN, on July 31, 1875.

MOST DUBIOUS NEWS STORIES OF THE YEAR. Dec 29. An annual review of news stories based on bogus scientific and other claims, reported in 1995 by the nation's media. Sponsor: Alan Caruba, Founder, Natl Anxiety Center, Box 40, Maplewood, NJ 07040. Phone: (201) 763-6392.

NEPAL: BIRTHDAY OF HIS MAJESTY THE KING. Dec 29. National holiday of Nepal. Three-day celebration with huge public rally at Tundikkel, gay pageantry, musical bands and illumination in the towns at night.

ST. THOMAS OF CANTERBURY: FEAST DAY: 825th DEATH ANNIVERSARY. Dec 29. Thomas, Archbishop of Canterbury, was born at London, in 1118, and was murdered at the Canterbury Cathedral, Dec 29, 1170.

◆ **TEXAS: ADMISSION DAY: 150th ANNIVERSARY.** Dec 29. Became 28th state on this day in 1845.

WOUNDED KNEE MASSACRE: 105th ANNIVERSARY. Dec 29. Anniversary of the massacre of more than 200 Native American men, women and children by the US Seventh Cavalry on Dec 29, 1890, at Wounded Knee Creek, SD. Government efforts to suppress a ceremonial religious practice, the Ghost Dance (which called for a messiah who would restore the bison to the plains, make the white men disappear and bring back the old Native American way of life), had resulted in the death of Sitting Bull on Dec 15, 1890, which further inflamed the disgruntled Native Americans and culminated in the slaughter at Wounded Knee on Dec 29.

BIRTHDAYS TODAY

Tom Bradley, 78, former mayor of Los Angeles, CA, born Calvert, TX, Dec 29, 1917.
Ted Danson, 48, actor (Sam on "Cheers"; *Three Men and a Baby*), born at San Diego, CA, Dec 29, 1947.
Marianne Faithfull, 49, singer ("As Tears Go By," "Summer Nights"), actress, born at London, England, Dec 29, 1946.
Alexander A. Farrelly, 72, Governor of the Virgin Islands (D), born at St. Croix, Virgin Islands, Dec 29, 1923.
Mervyn Fernandez, 36, football player, born Merced, CA, Dec 29, 1959.
Ed Flanders, 61, actor (Emmy for "A Moon for the Misbegotten"; "St. Elsewhere"), born Minneapolis, MN, Dec 29, 1934.
William Gaddis, 73, novelist, born at Manhattan, NY, Dec 29, 1922.
Thomas Edwin Jarriel, 61, broadcast journalist, born at LaGrange, GA, Dec 29, 1934.
Mary Tyler Moore, 59, actress (2 Emmys "Dick Van Dyke Show"; 3 Emmys "Mary Tyler Moore Show"; *Ordinary People*), born Brooklyn, NY, Dec 29, 1936.
Jon Voight, 57, actor (*Midnight Cowboy, Deliverance*), born Yonkers, NY, Dec 29, 1938.

DECEMBER 30 — SATURDAY
364th Day — Remaining, 1

ENGLAND: MODEL ENGINEER EXHIBITION. Dec 30–Jan 7. Olympia, Hammersmith Road, Kensington W14 8UX, London. Exhibition of working models, static displays and demonstrations of model making. Est attendance: 45,000. For info: Exhibition Mgr, Argus Specialist Exhibitions, Argus House, Boundary Way, Hemel Hempstead, Herefordshire, England HP2 7ST.

GUGGENHEIM, SIMON: BIRTH ANNIVERSARY. Dec 30. American capitalist and philanthropist, born at Philadelphia, PA, Dec 30, 1867. He established, in memory of his son, the John Simon Guggenheim Memorial Foundation, in 1925. Died on Nov 2, 1941.

HOLIDAY BOWL PARADE AND GAME. Dec 30. Along Harbor Drive, San Diego, CA. The morning of the Holiday Bowl football game (between a Western Athletic Conference and a Big Ten team). This televised parade features floats, marching bands, giant balloons and unique specialty units. Parade at 10 AM, game in afternoon. Annually, Dec 30. Est attendance: 25,000. For info: Stephanie Saatoff, The Holiday Bowl, PO Box 601400, San Diego, CA 92160-1400. Phone: (619) 283-5808.

KIPLING, RUDYARD: BIRTH ANNIVERSARY. Dec 30, 1865. English poet, novelist and short story writer, Nobel prize laureate, Kipling was born at Bombay, India. Kipling became seriously ill while visiting New York City and the news media made a great event of the sickness. There were frequent bulletins from the sickroom, interviews with doctors and stories about the crowds of people who watched outside Kipling's hotel. After his recovery Kipling is said to have remarked that the American people "never forgave him for not dying in New York." He died at London, on Jan 18, 1936.

LEACOCK, STEPHEN: BIRTH ANNIVERSARY. Dec 30. Canadian economist and humorist, born Dec 30, 1869. Died Mar 28, 1944. "Lord Ronald ...," he wrote in *Nonsense Novels*, "flung himself upon his horse and rode madly off in all directions."

MADAGASCAR: NATIONAL HOLIDAY. Dec 30. Anniversary of the Democratic Republic of Madagascar.

MONITOR SINKS: ANNIVERSARY. Dec 30. The Union ironclad ship USS *Monitor* (which achieved fame after the Battle of the *Monitor* and the *Merrimac*) sank on Dec 30, 1862, off Cape Hatteras during a storm. Sixteen of her crew were lost.

Dec 30-31 ☆ *Chase's 1995 Calendar of Events* ☆

PARKS, BERT: BIRTH ANNIVERSARY. Dec 30. Bert Parks was born at Atlanta, GA, on Dec 30, 1914. As an actor whose career spanned radio, film, television and Broadway stages, his name became synonymous with the Miss America pageant which he emceed for 25 years. He was fired from the Miss America post in 1980 when pageant officials wanted to acquire a younger look. The firing generated an outpouring of sympathy and a massive letter-writing campaign spearheaded by Johnny Carson. Parks made a special return appearance for the 1990 pageant, once again singing his signature song "There She Is." He got his big break in show business in 1945 as the emcee for the radio quiz show "Break the Bank" and later as the host of "Stop the Music." When both shows changed mediums to television they did so with Parks at the microphone, launching a television career that included hosting a variety of quiz shows and guest appearances on dramatic series. Bert Parks died on Feb 2, 1992, at La Jolla, CA.

PHILIPPINES: RIZAL DAY. Dec 30. The Philippines. Commemorates martyrdom of Dr. Jose Rizal on this day in 1896.

◆ **TOJO HIDEKI: BIRTH ANNIVERSARY.** Dec 30. Japanese prime minister during World War II. Born on Dec 30, 1884, at Tokyo. Arrested in August 1945 as a war criminal and hanged on Dec 23, 1948. See also: "Tojo Hideki: Execution Anniversary" (Dec 23).

BIRTHDAYS TODAY

Bo Diddley (Ellas McDaniel), 67, singer, songwriter ("Who Do You Love," "Say Man," "I'm a Man"), musician, born at McCombs, MS, Dec 30, 1928.
Ben Johnson, 34, sprinter, born at Falmouth, Jamaica, Dec 30, 1961.
Davy Jones, 49, actor, singer (with The Monkees; "Daydream Believer"), born at Manchester, England, Dec 30, 1946.
Sandy Koufax, 60, baseball Hall of Famer, sportscaster, born at Brooklyn, NY, Dec 30, 1935.
Jack Lord, 65, actor ("Hawaii Five-O," *God's Little Acre*), born New York, NY, Dec 30, 1930.
Michael Nesmith, 53, singer, songwriter and director (member of the Monkees), born at Houston, TX, Dec 30, 1942.
Tracey Ullman, 36, actress, singer ("The Tracey Ullman Show"), born Buckinghamshire, England, Dec, 30, 1959.
Jo Van Fleet, 73, actress (Oscar *East of Eden*; "Cinderella"), born Oakland, CA, Dec 30, 1922.

DECEMBER 31 — SUNDAY

365th Day — Remaining, 0

ANNUAL WORLD PEACE MEDITATION. Dec 31. An opportunity for people around the world to focus their thoughts and energy on peace. The event is observed internationally, beginning at noon Greenwich mean time and lasting one hour. (7AM–8AM EST.) For info: Rhea Giffin, Coord, PO Box 1151, Coeur d'Alene, ID 83816-1151. Phone: (208) 664-1691.

CHECK THE SMOKE ALARMS DAY. Dec 31. Celebrate life and the coming new year by checking your smoke detectors. Fire safety is a New Year's resolution you and your family deserve. An ounce of prevention is worth more than a pound of regret. For info: Judith Bates-Gorman, 1731 Morton St, Layayette, IN 47904. Phone: (317) 742-4480.

COWBELLION HERD NEW YEAR'S EVE ESCAPADE AND REVEL. Dec 31. Mobile, AL. To honor Michael Krafft who gave birth to the country's first "mystic society," the Cowbellion de Rakin Society in 1830. This led to the formation of all the later "mystic societies" and "krewes," which stage the famed Mardi Gras parades and balls in Mobile and New Orleans. Annually, New Year's Eve. Est attendance: 20,000. For info: Resurrected Cowbellion de Rakin Society (Michael Krafft Memorial Assn, Inc), PO Box 16564, Mobile, AL 36616-0564. Phone: (205) 626-2694.

CZECH-SLOVAK DIVORCE: ANNIVERSARY. Dec 31. The 74-year-old state of Czechoslovakia died at the stroke of midnight Dec 31, 1992. In the republic of Slovakia a celebration was held in the streets of Bratislava amid fireworks, bell ringing, singing of the new country's national anthem and the raising of the Slovak flag. In the Czech Republic no official festivities took place. Rather, on Jan 1, 1993, the Czechs celebrated with a solemn oath by their parliament. The nation of Czechoslovakia ended peacefully though polls showed that most Slovaks and Czechs would have preferred that Czechoslovakia survive. Before the split Czech Prime Minister Vaclav Klaus and Slovak Prime Minister Vladimir Meciar reached agreement on dividing everything from army troops, gold reserves and the art on government building walls.

ENGLAND: ALLENDALE BAAL FESTIVAL. Dec 31. Allendale, Northumberland. Villagers parade in costume with burning tar barrels on their heads to celebrate the New Year. For info: Mr B. Thompson, Heathcote, Allendale, Hexham, Northumberland, England NE47.

FIRST NIGHT HARTFORD. Dec 31. Hartford, CT. Day-long celebration beginning at 2:00 PM and ending at midnight. Visual and performing arts, procession, theater, music, activities for children, parties and fireworks to welcome the New Year. Est attendance: 30,000. For info: Kathy Butler, Deputy Dir, Hartford Downtown Council, 250 Constitution Plaza, Hartford, CT 06103. Phone: (203) 728-3089.

FIRST NIGHT 1996. Dec 31. Boston, MA. A New Year's Eve Arts Festival that magically transforms the streets of Boston from an everyday workplace into a once-a-year, one-of-a-kind community of festivity and awe. 250 presentations of the visual and performing arts provide a community experience and energetic environment by 1,000 artists using the streets and buildings of Boston as their theater. Highlights include afternoon Family & Children's Festival, a Grand Procession at 5:30 and fireworks at midnight. Est attendance: 1,200,000. For info: First Night, Inc, 20 Park Plaza, Ste 927, Boston, MA 02116. Phone: (617) 542-1399.

FIRST NIGHT ROANOKE. Dec 31. Downtown, Roanoke, VA. An alcohol-free New Year's Eve Celebration of the Lively Arts. First-rate entertainment and hands-on activities for the entire community. 12 site locations. Held annually on New Year's Eve. Est attendance: 10,000. For info: Wendi Schultz, CFE Exec Dir, Roanoke Festival in the Park, PO Box 8276, Roanoke, VA 24014. Phone: (703) 342-2640.

FIRST NIGHT WAYNESBORO. Dec 31. Waynesboro, VA. An alcohol/drug-free community celebration of performing and visual arts for the whole family with more than 100 performances and/or exhibits at over 25 sites in the Waynesboro Downtown District. Kicked off with a Grand Processional. Special activities for children. Performances followed by a Grand Finale at midnight. Entry to any and all events and exhibits via entry button ($5.00 per person; children under 5 free). Activities run from 2 PM until midnight. Est attendance: 6,000. For info: First Night Waynesboro, PO Box 716, Waynesboro, VA 22980. Phone: (703) 943-2988.

GHANA: REVOLUTION DAY. Dec 31. National holiday.

JAPAN: NAMAHAGE. Dec 31. In evening, groups of "Namahage" men disguised as devils make door-to-door visits, growling "Any good-for-nothing fellow hereabout?" The object of this annual event is to give sluggards an opportunity to change their minds and become diligent. Otherwise, according to legend, they will be punished by devils. Oga Peninsula, Akita Prefecture, Japan.

	S	M	T	W	T	F	S
December						1	2
	3	4	5	6	7	8	9
1995	10	11	12	13	14	15	16
	17	18	19	20	21	22	23
	24	25	26	27	28	29	30
	31						

☆ Chase's 1995 Calendar of Events ☆   Dec 31

KING ORANGE JAMBOREE PARADE. Dec 31. Miami, FL. Annual New Year's Eve parade for the past 60 years. Nationally televised by NBC, parade moves 2.2 miles through downtown Miami's Biscayne Blvd by the bay. Est attendance: 500,000. For info: Orange Bowl Committee, Lisa Franson, 601 Brickell Key Dr, Ste 206, Miami, FL 33131. Phone: (305) 371-4600.

LEAP SECOND ADJUSTMENT TIME. Dec 31. Dec 31 is one of the times which has been favored for the addition or subtraction of a second from clock time (to coordinate atomic and astronomical time). The determination to adjust is made by the Bureau International de l'Heure, in Paris. See also: "Leap Seconds" (see Contents).

MAKE UP YOUR MIND DAY. Dec 31. A day for all those people who have a hard time making up their minds. Make a decision today and follow through with it! Annually, Dec 31. Sponsor: A.C. Moeller and M.A. Dufour, Box 71, Clio, MI 48420.

MARSHALL, GEORGE CATLETT: BIRTH ANNIVERSARY. Dec 31. Chairman of the newly formed Joint Chiefs of Staff Committee throughout the US's involvement in WWII, General George Marshall was born Dec 31, 1880, in Uniontown, PA. He accompanied Roosevelt or represented the US at most Allied war conferences. He served as secretary of state and was designer of the Marshall Plan after the war. George Marshall died Oct 16, 1959, at Washington, DC.

NEW YEAR'S EVE. Dec 31.

NEW YEAR'S EVE CELEBRATION. Dec 31. New York, NY. Fireworks and midnight run in Central Park; lighted "Big Apple" in Times Square; fireworks at Prospect Park in Brooklyn and in South Street Seaport. For info: New York Conv and Visitors Bureau, 2 Columbus Circle, New York, NY 10019.

NEW YEAR'S EVE FIREWORKS. Dec 31. North end of the long meadow, Prospect Park, Brooklyn, NY. Fireworks and live music to ring in the New Year. Annually, New Year's Eve. Est attendance: 10,000. For info: Public Info Office, Prospect Park, 95 Prospect Park W, Brooklyn, NY 11215. Phone: (718) 965-8999.

NEW YEAR'S FEST. Dec 31. Kalamazoo, MI. New Year's Eve non-alcoholic celebration of the arts. Fireworks at midnight, live entertainment throughout the evening. Est attendance: 6,000. For info: New Year's Fest, Inc, 128 N Kalamazoo Mall, Kalamazoo, MI 49007. Phone: (616) 381-4003.

1996 FIRST NIGHT STATE COLLEGE. Dec 31–Jan 1. State College, PA. A non-alcoholic family celebration of the visual and performing arts held in conjunction with the Borough of State College's Centennial in 1996. Est attendance: 10,000. For info: Central Pennsylvania Festival of the Arts, Box 1023, State College, PA 16804. Phone: (814) 237-3682.

NIXON, JOHN: DEATH ANNIVERSARY. Dec 31. Commander of the Philadelphia City Guard, born 1733 (exact date unknown). Died at Philadelphia, PA, Dec 31, 1808. Appointed to conduct the first public reading of the Declaration of Independence, July 8, 1776. Nixon, a revolutionary patriot and businessman, was the son of Richard and Sarah Nixon.

NO RESOLUTION DAY. Dec 31. A celebration of not making New Year's resolutions, because they set people up for frustration and waste energy in ill-chosen directions. Instead of setting goals, tune in to your intuition and capitalize on your energy moment to moment. For info: Marcia Yudkin, Creative Ways, PO Box 1310, Boston, MA 02117. Phone: (617) 266-1613.

NOON YEAR'S EVE PARTY. Dec 31. Albuquerque, NM. A New Year's Eve party for early risers who have difficulty staying awake until midnight. Celebration includes countdown to 12 (noon) and "Happy Noon Year!" greeting. Est attendance: 700. Sponsor: KOB Radio, Larry Ahrens, Morning Show Host, 77 Broadcast Plaza SW, Albuquerque, NM 87103.

PORTUGAL: ST. SYLVESTER'S FESTIVAL. Dec 31. Funchal, Madeira Island. A spectacular show that transforms the whole city into a vast fairground and culminates with a spectacular fireworks display over the beautiful bay. For info: Portuguese Natl Tourist Office, 590 Fifth Ave, New York, NY 10036. Phone: (212) 354-4403.

ST. SYLVESTER'S DAY. Dec 31. Observed in Belgium, Germany, France, Switzerland. Commemorates death of Pope Sylvester I in 1335. Feasting, particularly upon "St. Sylvester's Carp."

SCOTLAND: HOGMANAY. Dec 31. Scottish celebration of New Year's Eve.

TEMPE FIESTA BOWL BLOCK PARTY. Dec 31. Tempe, AZ. A family-oriented celebration of the New Year, the Block Party features continuous entertainment, ethnic and traditional foods, an area for children, a New Year's Eve pep rally and spectacular fireworks displays to usher in the New Year. Annually, New Year's Eve. Est attendance: 125,000. For info: Mill Avenue Merchants Assn, 520 S Mill Ave, Ste 201, Tempe, AZ 85281. Phone: (602) 967-4877.

WATERFRONT NEW YEAR'S EVE CELEBRATION. Dec 31. Town Point Park, Norfolk, VA. Live music and dancing under a big-top tent. Count down the New Year with Norfolk's own 12th annual "little bit of Times Square" silver ball and a dazzling fireworks display at the stroke of midnight. Est attendance: 55,000. For info: Norfolk Festevents, Ltd, Promotions Dept, 120 W Main St, Norfolk, VA 23510. Phone: (804) 627-7809.

WESTERN SAMOA: SAMOAN FIRE DANCE. Dec 31. New Year's Eve is occasion for Samoan bamboo fireworks, singing and traditional performances such as the Samoan Fire Dance.

YOU'RE ALL DONE DAY. Dec 31. Acknowledge all that you have accomplished in the past year and savor the satisfaction of finishing a long task. [No further info available.] Sponsor: Long Haul Committee.

BIRTHDAYS TODAY

John Denver (Henry John Deutschendorf), 52, singer, songwriter ("Sunshine on My Shoulder," "Annie's Song"), actor (*Oh God*), born Rosewell, NM, Dec 31, 1943.

Ben Kingsley (Krishna Bhanji), 52, actor (Oscar *Gandhi; Bugsy*), born Yorkshire, England, Dec 31, 1943.

Tim Matheson, 47, actor (*Animal House*, "The Virginian," "Bonanza"), born Los Angeles, CA, Dec 31, 1948.

Odetta (Odetta Homes Felious Gordon), 65, folksinger, musician, born at Birmingham, AL, Dec 31, 1930.

Donna Summer (LaDonna Andrea Gaines), 47, singer ("Bad Girls"), born at Boston, MA, Dec 31, 1948.

Diane Halfin von Furstenberg, 49, fashion designer, author, born at Brussels, Belgium, Dec 31, 1946.

See ya next year!

☆ Chase's 1995 Calendar of Events ☆

PRESIDENTIAL PROCLAMATIONS ISSUED, JANUARY 1, 1993–JUNE 30, 1994
(See Index and Chronological Text for Expected 1995 Presidential Proclamations.)

No. Title, Observance Dates, (Date of signing).

No. Title, Observance Dates, (Date of signing).

—————————— 1993 ——————————

6521 National Sanctity of Human Life Day: Jan 17, 1993 (Jan 4, 1993)
6522 Braille Literacy Week: Jan 4–10, 1993 (Jan 5, 1993)
6523 National Law Enforcement Training Week: Jan 3–9, 1993 (Jan 5, 1993)
6524 Martin Luther King, Jr, Federal Holiday: Jan 18, 1993 (Jan 11, 1993)
6525 National Day of Fellowship and Hope: Jan 22, 1993 (Jan 20, 1993)
6526 Death of Thurgood Marshall (Jan 24, 1993)
6527 National Women and Girls in Sports Day: Feb 4, 1993 (Feb 3, 1993)
6528 American Heart Month: Feb 1–28, 1993 (Feb 14, 1993)
6529 National Visiting Nurse Associations Week: Feb 14, 1993 (Feb 18, 1993)
6530 American Wine Appreciation Week: Feb 21–27, 1993 (Feb 23, 1993)
6531 National FFA Organization Awareness Week: Feb 21–27, 1993 (Feb 25, 1993)
6532 Save Your Vision Week: Mar 7–13, 1993 (Mar 5, 1993)
6533 Irish-American Heritage Month: Mar 1–31, 1993 (Mar 6, 1993)
6534 To Revoke Proclamation No. 6491 of Oct 14, 1992 (Mar 6, 1993)
6535 American Red Cross Month: Mar 1–31, 1993 (Mar 17, 1993)
6536 National Poison Prevention Week: Mar 21–27, 1993 (Mar 17, 1993)
6537 Women's History Month: Mar 1–31, 1993 (Mar 19, 1993)
6538 National Agriculture Day: Mar 20, 1993 (Mar 20, 1993)
6539 Greek Independence Day: A National Day of Celebration of Greek and American Democracy: Mar 25, 1993 (Mar 25, 1993)
6540 Education and Sharing Day, U.S.A.: Apr 2, 1993 (Apr 2, 1993)
6541 National Former Prisoner of War Recognition Day: Apr 9, 1993 (Apr 9, 1993)
6542 National Preschool Immunization Week: Apr 18–24, 1993 (Apr 9, 1993)
6543 To Extend Special Rules of Origin Applicable to Certain Textile Articles Woven or Knitted in Canada (Apr 9, 1993)
6544 To Modify Duty-Free Treatment Under the Andean Trade Preferences Act, To Modify the Generalized System of Preferences, and for Other Purposes (Apr 13, 1993)
6545 Pan American Day and Pan American Week: Apr 14, 1993, and Apr 11–17, 1993 (Apr 14, 1993)
6546 National Volunteer Week: Apr 18–24, 1993 (Apr 17, 1993)
6547 National Credit Education Week: Apr 18–24, 1993 (Apr 22, 1993)
6548 Nancy Moore Thurmond National Organ and Tissue Donor Awareness Week: Apr 18–24, 1993 (Apr 23, 1993)
6549 Cancer Control Month: Apr 1–30, 1993 (Apr 23, 1993)
6550 Jewish Heritage Week: Apr 25–May 2, 1993, and Apr 10–17, 1993 (Apr 28, 1993)
6551 National Crime Victims' Rights Week: Apr 25–May 1, 1993 (Apr 28, 1993).

6552 Death of Cesar Chavez (Apr 28, 1993)
6553 National Day of Prayer: May 6, 1993 (Apr 30, 1993)
6554 National Arbor Day: Apr 30, 1993 (Apr 30, 1993)
6555 Law Day, U.S.A.: May 1, 1993 (Apr 30, 1993)
6556 Loyalty Day: May 1, 1993 (May 1, 1993)
6557 Asian/Pacific American Heritage Month: May 1–30, 1993 (May 3, 1993)
6558 National Walking Week: May 3–8, 1993 (May 6, 1993)
6559 Mother's Day: May 9, 1993 (May 7, 1993)
6560 Be Kind to Animals and National Pet Week: May 2–8, 1993 (May 7, 1993)
6561 Small Business Week: May 9–15, 1993 (May 14, 1993)
6562 National Defense Transportation Day and National Transportation Week: May 21, 1993, and May 16–22, 1993 (May 19, 1993)
6563 World Trade Week: May 16–22, 1993 (May 19, 1993)
6564 National Maritime Day: May 22, 1993 (May 21, 1993)
6565 Older Americans Month: May 1–30, 1993 (May 25, 1993)
6566 Prayer for Peace, Memorial Day: May 31, 1993 (May 28, 1993)
6567 Emergency Medical Services Week: May 23–29, 1993, and May 15–21, 1994 (May 28, 1993)
6568 Time for the National Observance of the Fiftieth Anniversary of World War II: May 30–June 7, 1993 (May 31, 1993)
6569 Suspension of Entry as Immigrants and Nonimmigrants of Persons Who Formulate or Implement Policies That Are Impeding the Negotiations Seeking the Return to Constitutional Rule in Haiti (June 3, 1993)
6570 National Safe Boating Week: June 6–12, 1993 (June 4, 1993)
6571 Lyme Disease Awareness Week: June 6–12, 1993, and June 5–11, 1994 (June 4, 1993)
6572 Flag Day and National Flag Week: June 14, 1993, and June 13–19, 1993 (June 14, 1993)
6573 Father's Day: June 20, 1993 (June 17, 1993)
6574 Suspension of Entry as Immigrants and Nonimmigrants of Persons Who Formulate or Implement Policies That Are Impeding the Transition to Democracy in Zaire or Who Benefit from Such Policies (June 21, 1993)
6575 To Modify Duty-Free Treatment Under the Generalized System of Preferences and for Other Purposes (June 25, 1993)
6576 National Youth Sports Program Day, 1993: July 1, 1993 (July 1, 1993)
6577 Agreement on Trade Relations Between the United States of America and Romania (July 2, 1993)
6578 National Literacy Day, 1993 and 1994: July 2, 1993, July 2, 1994 (July 2, 1993)
6579 To Implement An Accelerated Tariff Schedule of Duty Elimination and to Modify Rules of Origin Under the United States-Canada Free-Trade Agreement (July 4, 1994)
6580 Captive Nations Week, 1993: July 11–17, 1993 (July 15, 1993)
6581 National Veterans Golden Age Games Week, 1993: July 17–23, 1993 (July 22, 1993)
6582 40th Anniversary of the Korean Armistice July 27, 1993 (July 27, 1993)
6583 Death of General Matthew B. Ridgway (July 29, 1993)

☆ Chase's 1995 Calendar of Events ☆

PRESIDENTIAL PROCLAMATIONS (cont'd)
ISSUED, JANUARY 1, 1993-JUNE 30, 1994

(See Index and Chronological Text for Expected 1995 Presidential Proclamations.)

No. Title, Observance Dates, (Date of signing).

6584 Helsinki Human Rights Day, 1993: Aug 1, 1993 (Aug 1, 1993)
6585 To designate Peru as a Beneficiary Country for Purposes of the Andean Trade Preference Act (Aug 11, 1993)
6586 Women's Equality Day, 1993: Aug 26, 1993 (Aug 18, 1993)
6587 National POW/MIA Recognition Day, 1993: Sept 10, 1993 (Sept 3, 1993)
6588 National D.A.R.E. Day, 1993 and 1994: Sept 9, 1993 (Sept 9, 1993)
6589 Commodore John Barry Day, 1993: Sept 13, 1993 (Sept 13, 1993)
6590 Gold Star Mother's Day, 1993: Sept 26, 1993 (Sept 13, 1993)
6591 Minority Enterprise Development Week, 1993: Oct 3-9, 1993 (Sept 13, 1993)
6592 National Hispanic Heritage Month, 1993: Sept 15-Oct 15, 1993 (Sept 15, 1993)
6593 Citizenship Day and Constitution Week, 1993: Sept 17, 1993, Sept 17-23, 1993 (Sept 17, 1993)
6594 National Historically Black Colleges and Universities Week, 1993: Sept 19-25, 1993 (Sept 21, 1993)
6595 National Farm Safety and Health Week, 1993: Sept 19-25, 1993 (Sept 21, 1993)
6596 National Rehabilitation Week, 1993 and 1994: Sept 19-25, 1993, Sept 18-24, 1994 (Sept 22, 1993)
6597 Energy Awareness Month, 1993: October (Sept 22, 1993)
6598 Death of General James H. Doolittle (Sept 30, 1993)
6599 To Amend the Generalized System of Preferences (Sept 30, 1993)
6600 National Breast Cancer Awareness Month, 1993: October (Sept 30, 1993)
6601 Fire Prevention Week, 1993: Oct 3-9, 1993 (Sept 30, 1993)
6602 Child Health Day, 1993: Oct 4, 1993 (Oct 4, 1993)
6603 Mental Illness Awareness Week, 1993: Oct 3-9, 1993 (Oct 5, 1993)
6604 German-American Day, 1993: Oct 6, 1993 (Oct 5, 1993)
6605 National Disability Employment Awareness Month, 1993: October (Oct 6, 1993)
6606 Country Music Month, 1993: October (Oct 7, 1993)
6607 Leif Erikson Day, 1993: Oct 9, 1993 (Oct 8, 1993)
6608 Columbus Day, 1993: Oct 11, 1993 (Oct 8, 1993)
6609 National School Lunch Week, 1993: Oct 10-16, 1993 (Oct 8, 1993)
6610 General Pulaski Memorial Day, 1993: Oct 11, 1993 (Oct 9, 1993)
6611 National Down Syndrome Awareness Month, 1993: October (Oct 14, 1993)
6612 White Cane Safety Day, 1993: Oct 15, 1993 (Oct 15, 1993)
6613 World Food Day, 1993 and 1994: Oct 16, 1993; Oct 16, 1994 (Oct 16, 1993)
6614 National Forest Products Week, 1993: Oct 17-23, 1993 (Oct 16, 1993)
6615 National Mammography Day, 1993: Oct 19, 1993 (Oct 18, 1993)
6616 National Biomedical Research Day, 1993: Oct 21, 1993 (Oct 20, 1993)
6617 National Consumers Week, 1993: Oct 24-30, 1993 (Oct 21, 1993)

No. Title, Observance Dates, (Date of signing).

6618 United Nations Day, 1993: Oct 24, 1993 (Oct 23, 1993)
6619 National Domestic Violence Awareness Month, 1993 and 1994: October (Oct 28, 1993)
6620 National Health Information Management Week, 1993: Oct 31-Nov 6, 1993 (Nov 3, 1993)
6621 Veterans Day, 1993: Nov 11, 1993 (Nov 5, 1993)
6622 National Women Veterans Recognition Week, 1993: Nov 7-13, 1993 (Nov 10, 1993)
6623 Geography Awareness Week, 1993 and 1994: Nov 14-20, 1993; Nov 13-19, 1994 (Nov 14, 1993)
6624 National Farm-City Week, 1993: Nov 19-25, 1993 (Nov 16, 1993)
6625 Thanksgiving Day, 1993: Nov 25, 1993 (Nov 17, 1993)
6626 National Children's Day, 1993: Nov 21, 1993 (Nov 18, 1993)
6627 National Military Families Recognition Day, 1993: Nov 22, 1993 (Nov 18, 1993)
6628 National Family Week, 1993 and 1994: Nov 21-27, 1993; Nov 20-26, 1994 (Nov 22, 1993)
6629 National Adoption Week, 1993 and 1994: Nov 21-27, 1993; Nov 20-26, 1994 (Nov 24, 1993)
6630 National Hospice Month, 1993 and 1994: November (Nov 29, 1993)
6631 National Home Care Week, 1993 and 1994: Nov 28-Dec 4, 1993; Nov 27-Dec 3, 1994 (Nov 29, 1993)
6632 World AIDS Day, 1993: Dec 1, 1993 (Nov 30, 1993)
6633 National Drunk and Drugged Driving Prevention Month, 1993: December (Dec 3, 1993)
6634 International Year of the Family, 1994 (Dec 6, 1993)
6635 To Amend the Generalized System of Preferences (Dec 9, 1993)
6636 Suspension of Entry as Immigrants and Nonimmigrants of Persons Who Formulate, Implement, or Benefit from Policies that are Impeding the Transition to Democracy in Nigeria (Dec 10, 1993)
6637 Human Rights Day, Bill of Rights Day, and Human Rights Week, 1993: Dec 10, 1993, Dec 15, 1993, Dec 10-16, 1993 (Dec 10, 1993)
6638 Wright Brothers Day, 1993: Dec 17, 1993 (Dec 10, 1993)
6639 National Firefighters Day, 1993: Dec 15, 1993 (Dec 14, 1993)
6640 Modification of Import Limitations on Certain Dairy Products (Dec 15, 1993)
6641 To Implement the North American Free Trade Agreement and for Other Purposes (Dec 15, 1993)
6642 Fifth Anniversary Day of Remembrance for the Victims of the Bombing of Pan AM Flight 103: Dec 21, 1993 (Dec 17, 1993)
6643 National Law Enforcement Training Week, 1994: Jan 2-8, 1994 (Dec 21, 1993)

---------- 1994 ----------

6644 Death of Thomas P. O'Neill, Jr. (Jan 6, 1994)
6645 Martin Luther King, Jr., Federal Holiday, 1994: Jan 17, 1994 (Jan 14, 1994)
6646 Religious Freedom Day, 1994: Jan 16, 1994 (Jan 14, 1994)
6647 National Good Teen Day, 1994: Jan 16, 1994 (Jan 14, 1994)
6648 American Heart Month, 1994: February (Feb 3, 1994)
6649 National Women and Girls in Sports Day, 1994: Feb 3, 1994 (Feb 3, 1994)

☆ Chase's 1995 Calendar of Events ☆

PRESIDENTIAL PROCLAMATIONS (cont'd) ISSUED, JANUARY 1, 1993–JUNE 30, 1994
(See Index and Chronological Text for Expected 1995 Presidential Proclamations.)

No. Title, Observance Dates, (Date of signing).

6650 To Amend the Generalized System of Preferences and for Other Purposes (Feb 16, 1994)
6651 National Poison Prevention Week, 1994: Mar 20–26, 1994 (Mar 1, 1994)
6652 Save your Vision Week, 1994: Mar 6–12, 1994 (Mar 2, 1994)
6653 American Red Cross Month, 1994: March (Mar 2, 1994)
6654 Women's History Month, 1994: March (Mar 2, 1994)
6655 To Amend the Generalized System of Preferences (Mar 3, 1994)
6656 Irish-American Heritage Month, 1994: March (Mar 8, 1994)
6657 National Agriculture Day, 1994: Mar 20, 1994 (Mar 18, 1994)
6658 Education and Sharing Day, U.S.A., 1994: Mar 23, 1994 (Mar 23, 1994)
6659 Greek Independence Day: A National Day of Celebration of Greek and American Democracy, 1994: Mar 25, 1994 (Mar 25, 1994)
6660 Small Family Farm Week, 1994: Mar 20–26, 1994 (Mar 25, 1994)
6661 National Day of Reconciliation, Apr 4, 1994 (Apr 3, 1994)
6662 Transfer of Functions of the ACTION Agency to the Corporation for National and Community Service (Apr 4, 1994)
6663 National Former Prisoner of War Recognition Day, 1994: Apr 9, 1994 (Apr 6, 1994)
6664 Cancer Control Month, 1994: April (Apr 7, 1994)
6665 Jewish Heritage Week, 1994: Apr 10–17, 1994 (Apr 8, 1994)
6666 Pan American Day and Pan American Week, 1994: Apr 14, 1994, Apr 10–16, 1994 (Apr 8, 1994)
6667 National Public Safety Telecommunicators Week, 1994: Apr 11–17, 1994 (Apr 12, 1994)
6668 National Day of Prayer, 1994: May 5, 1994 (Apr 12, 1994)
6669 251st Anniversary of the Birth of Thomas Jefferson: Apr 13, 1994 (Apr 13, 1994)
6670 National Park Week, 1994: May 23–29, 1994 (Apr 14, 1994)
6671 Death of Those Aboard American Helicopters in Iraq (Apr 14, 1994)
6672 Nancy Moore Thurmond National Organ and Tissue Donor Awareness Week, 1994: Apr 17–23, 1994 (Apr 15, 1994)
6673 National Volunteer Week, 1994: Apr 17–23, 1994 (Apr 15, 1994)
6674 National Youth Service Day, 1994 and 1995: Apr 19, 1994, Apr 18, 1995 (Apr 19, 1994)
6675 National Infant Immunization Week: Apr 24–30, 1994 (Apr 20, 1994)
6676 To Amend the Generalized System of Preferences (Apr 21, 1994)
6677 Announcing the Death of Richard Milhous Nixon (Apr 22, 1994)
6678 National Crime Victims' Rights Week, 1994: Apr 24–30, 1994 (Apr 25, 1994)
6679 Law Day, U.S.A., 1994: May 1, 1994 (Apr 30, 1994)
6680 Loyalty Day, 1994: May 1, 1994 (Apr 30, 1994)
6681 Small Business Week, 1994: May 1–7, 1994 (Apr 30, 1994)

No. Title, Observance Dates, (Date of signing).

6682 Public Service Recognition Week, 1994: May 2–8, 1994 (May 3, 1994)
6683 Mother's Day, 1994: May 8, 1994 (May 5, 1994)
6684 National Walking Week, 1994: May 1–7, 1994 (May 6, 1994)
6685 Suspension of Entry of Aliens Whose Entry is Barred Under United Nations Security Council Resolution 917 or Who Formulate, Implement, or Benefit from Policies That Are Impeding the Negotiations Seeking the Return to Constitutional Rule in Haiti (May 7, 1994)
6686 Asian/Pacific American Heritage Month, 1994: May (May 9, 1994)
6687 Older Americans Month, 1994: May (May 9, 1994)
6688 Labor History Month, 1994: May (May 10, 1994)
6689 National Defense Transportation Day and National Transportation Week, 1994: May 20, 1994, May 15–21, 1994 (May 16, 1994)
6690 World Trade Week, 1994: May 22–28, 1994 (May 18, 1994)
6691 National Trauma Awareness Month, 1994: May (May 18, 1994)
6692 National Maritime Day, 1994: May 22, 1994 (May 19, 1994)
6693 Armed Forces Day, 1994: May 21, 1994 (May 21, 1994)
6694 Pediatric and Adolescent AIDS Awareness Week, 1994: May 29–June 4, 1994 (May 25, 1994)
6695 National Safe Boating Week, 1994: June 5–11, 1994 (May 27, 1994)
6696 Prayer for Peace, Memorial Day, 1994: May 30, 1994 (May 30, 1994)
6697 D-Day National Remembrance Day and Time for the National Observance of the Fiftieth Anniversary of World War II, 1994: June 6, 1994 (May 30, 1994)
6698 National Women in Agriculture Day, 1994: June 9, 1994 (May 31, 1994)
6699 Flag Day and National Flag Week, 1994: June 14, 1994, June 12–18, 1994 (June 10, 1994)
6700 National Men's Health Week, 1994: June 12–19, 1994 (June 10, 1994)
6701 Father's Day, 1994: June 19, 1994 (June 14, 1994)
6702 National Housing Week, 1994: June 20–27, 1994 (June 21, 1994)
6703 50th Anniversary of the GI Bill of Rights: June 22, 1994 (June 21, 1994)

A Word About Presidential Proclamations

A presidential proclamation is generally issued very close to the date of the observance being proclaimed, so we don't know in advance what they will be. In order to give readers a sense of what kinds of observances are proclaimed, we include the most recent full year of proclamations and those issued during the year when we are preparing our next edition, cutting off as we get close to press time. Proclamations for which legislation has been passed authorizing the proclaiming of an observance for ensuing years are listed in the chronology (day-by-day) section and indicated by a star. The text of a proclamation does not include information on what congresspersons or organizations were involved. Your congressperson's office may be able to help you. Read "The Congressional Process . . ." on the following page to learn how an observance becomes proclaimed.

☆ Chase's 1995 Calendar of Events ☆

THE CONGRESSIONAL PROCESS FOR DECLARING SPECIAL OBSERVANCES

From the first proclamation of a public holiday (George Washington's proclamation of Nov 26, 1789, as Thanksgiving Day) to the present, the federal government has been active in seeing that special observances be set aside to commemorate people, events, ideas and activities worthy of national recognition. These have ranged from the proclamation of official holidays to the recognition or commemoration of subjects through special days, weeks, months or years.

The first usual step in having something so commemorated is the introduction of a bill in the House of Representatives or in the Senate calling for a day, week, month or year to be designated for special recognition.

Because of the large number of such resolutions introduced, there are much more restrictive rules applied to these resolutions than are applied to other legislative measures. In the House, such a resolution needs at least 218 co-sponsors (a majority of the members) before it will be considered by the Post Office and Civil Service Committee. Once the resolution has its 218 co-sponsors, it is taken up by the Subcommittee on Census and Population, and usually routinely referred to the whole Committee, which then reports the resolution to the floor of the full House. It is voted upon at that point, but from the time the resolution has its 218 co-sponsors there is little question about its eventual success. After House passage, the resolution also must be voted upon by the Senate. If it passes there, it goes to the President for signing.

On the Senate side, commemorative legislation goes directly to the Senate Judiciary Committee. It will not be considered by the Committee, however, unless it has at least 50 co-sponsors with at least 20 of them being from each party. The filing of commemorative resolutions is restricted to three months: February, June and October. After being passed by the full Committee, the process then is the same as for House resolutions, with passage required by the House and signing by the President.

Under unusual circumstances, this process can be circumvented by the chairpersons of the House and Senate committees. However, in nearly every case, this procedure is followed.

Both committees also have policies on what types of commemorative legislation they will consider. Neither committee, for example, will report for floor consideration commemorative legislation concerning a commercial enterprise, specific product or political organization. Neither committee also will report any proposal, other than official holidays, to be a recurring annual commemoration. In addition, neither committee will consider commemorative legislation regarding living persons or governmental units.

The President also has the authority to declare by presidential proclamation any commemorative event, but this is rarely done and virtually all commemorative celebrations come about by going through the congressional legislative process, after which they are issued as presidential proclamations.

History records a total of 6,703 presidential proclamations, mostly commemorative or hortatory, issued to date—from George Washington's proclamation of Thanksgiving Day in 1789 to Bill Clinton's proclamation concerning the 50th anniversary of the GI Bill of Rights, signed on June 21, 1994.

Based on information provided by The Hon. Dale E. Kildee, Member of Congress, 7th District, Michigan, with revisions to include recent changes.

DECADE OF THE BRAIN PROCLAMATION

Proclamation 6158 of July 17, 1990

Decade of the Brain, 1990-1999

By the President of the United States of America

A Proclamation

The human brain, a 3-pound mass of interwoven nerve cells that controls our activity, is one of the most magnificent—and mysterious—wonders of creation. The seat of human intelligence, interpreter of senses, and controller of movement, this incredible organ continues to intrigue scientist and layman alike.

Over the years, our understanding of the brain—how it works, what goes wrong when it is injured or diseased—has increased dramatically. However, we still have much more to learn. The need for continued study of the brain is compelling: millions of Americans are affected each year by disorders of the brain ranging from neurogenetic diseases to degenerative disorders such as Alzheimer's, as well as stroke, schizophrenia, autism, and impairments of speech, language, and hearing.

Today, these individuals and their families are justifiably hopeful, for a new era of discovery is dawning in brain research. Powerful microscopes, major strides in the study of genetics, and advanced brain imaging devices are giving physicians and scientists ever greater insight into the brain. Neuroscientists are mapping the brain's biochemical circuitry, which may help produce more effective drugs for alleviating the suffering of those who have Alzheimer's or Parkinson's disease. By studying how the brain's cells and chemicals develop, interact, and communicate with the rest of the body, investigators are also developing improved treatments for people incapacitated by spinal cord injuries, depressive disorders, and epileptic seizures. Breakthroughs in molecular genetics show great promise of yielding methods to treat and prevent Huntington's disease, the muscular dystrophies, and other life-threatening disorders.

Research may also prove valuable in our war on drugs, as studies provide greater insight into how people become addicted to drugs and how drugs affect the brain. These studies may also help produce effective treatments for chemical dependency and help us to understand and prevent the harm done to the preborn children of pregnant women who abuse drugs and alcohol. Because there is a connection between the body's nervous and immune systems, studies of the brain may also help enhance our understanding of Acquired Immune Deficiency Syndrome.

Many studies regarding the human brain have been planned and conducted by scientists at the National Institutes of Health, the National Institute of Mental Health, and other Federal research agencies. Augmenting Federal efforts are programs supported by private foundations and industry. The cooperation between these agencies and the multidisciplinary efforts of thousands of scientists and health care professionals provide powerful evidence of our Nation's determination to conquer brain disease.

To enhance public awareness of the benefits to be derived from brain research, the Congress by House Joint Resolution 174 has designated the decade beginning January 1, 1990, as the "Decade of the Brain" and has authorized and requested the President to issue a proclamation in observance of this occasion.

NOW, THEREFORE, I, GEORGE BUSH, President of the United States of America, do hereby proclaim the decade beginning January 1, 1990, as the Decade of the Brain. I call upon all public officials and the people of the United States to observe that decade with appropriate programs, ceremonies, and activities.

IN WITNESS WHEREOF, I have hereunto set my hand this seventeenth day of July, in the year of our Lord nineteen hundred and ninety, and of the Independence of the United States of America the two hundred and fifteenth.

George Bush

From *The Federal Register*, 55 F.R. 29555

☆ *Chase's 1995 Calendar of Events* ☆

THE NATIONAL EDUCATION GOALS

1. By the year 2000, all children in America will start school ready to learn.

2. By the year 2000, the high school graduation rate will increase to at least 90 percent.

3. By the year 2000, American students will leave grades four, eight, and twelve having demonstrated competency in challenging subject matter including English, mathematics, science, foreign languages, civics and government, economics, art, history, and geography; and every school in America will insure that all students learn to use their minds well, so they may be prepared for responsible citizenship, further learning, and productive employment in our nations modern economy.

4. By the year 2000, the nation's teaching force will have access to programs for the continued improvement of their professional skills and the opportunity to acquire the knowledge and skills needed to instruct and prepare all American students for the next century.

5. By the year 2000, U.S. students will be first in the world in science and mathematics achievement.

6. By the year 2000, every adult American will be literate and will possess the knowledge and skills necessary to compete in a global economy and exercise the rights and responsibilities of citizenship.

7. By the year 2000, every school in America will be free of drugs, violence, and the unauthorized presence of firearms and alcohol and will offer a disciplined environment conducive to learning.

8. By the year 2000, every school will promote partnerships that will increase parental involvement and participation in promoting the social, emotional, and academic growth of children.

Established in 1990 by the nation's governors and endorsed by Presidents George Bush and William Clinton, the National Education Goals provide a focal point for reform and renewal in education. The goals set a series of targets that America's elected leaders hold as a vision for the nation to achieve by the year 2000.

☆ *Chase's 1995 Calendar of Events* ☆

NATIONAL DAYS OF THE WORLD FOR 1995

(Compiled from publications of the U.S. Department of State, the United Nations and from information received from the countries listed.)

Most nations set aside one or more days each year as National Public Holidays, often recognizing the anniversary of the attainment of independence, or the birthday of the country's ruler. Below, the National Days are listed alphabetically. It should be noted that in some countries the Gregorian Calendar date of observance varies from year to year. See the Index and the main chronology for further details of observance, and for numerous holidays in addition to the National Days listed here.

Afghanistan	Aug 19
Albania	Nov 28
Algeria	Nov 1
Angola	Nov 11
Antigua and Barbuda	Nov 1
Argentina	May 25
Armenia	Sept 21
Australia	Jan 26
Austria	Oct 26
Azerbaijan	May 28
Bahamas	July 10
Bahrain	Dec 16
Bangladesh	Mar 26
Barbados	Nov 30
Belgium	July 21
Belize	Sept 21
Benin	Aug 1
Bhutan	Dec 17
Bolivia	Aug 6
Botswana	Sept 30
Brazil	Sept 7
Brunei Darussalam	Feb 23
Bulgaria	Mar 3
Burkina Faso	Aug 4
Burundi	July 1
Belarus	July 27
Cambodia	Nov 9
Cameroon	May 20
Canada	July 1
Cape Verde	July 5
Central African Republic	Dec 1
Chad	Aug 11
Chile	Sept 18
China	Oct 1
Colombia	July 20
Comoro Islands	July 6
Congo	Aug 15
Costa Rica	Sept 15
Cote D'Ivoire	Dec 7
Croatia	May 30
Cuba	Jan 1
Cyprus	Oct 1
Czech Republic	Oct 28
Democratic Yemen	Oct 14
Denmark	Apr 16
Djibouti	June 27
Dominica	Nov 3
Dominican Republic	Feb 27
Ecuador	Aug 10
Egypt	July 23
El Salvador	Sept 15
Equatorial Guinea	Oct 12
Eritrea	May 24
Estonia	Feb 24
Ethiopia	Sept 12
Fiji	Oct 10
Finland	Dec 6
France	July 14
Gabon	Aug 17
Gambia	Feb 18
Georgia	May 26
Germany	Oct 3
Ghana	Mar 6
Great Britain	June 11
Greece	Mar 25
Grenada	Feb 7
Grenadines	Oct 27
Guatemala	Sept 15
Guinea	Oct 2
Guinea-Bissau	Sept 24
Guyana	Feb 23
Haiti	Jan 1
Holy See	Oct 22
Honduras	Sept 15
Hungary	Aug 20
Iceland	June 17
India	Jan 26
Indonesia	Aug 17
Iran	Feb 11
Iraq	July 17
Ireland	Mar 17
Israel	Apr 14
Italy	June 2
Jamaica	Aug 1
Japan	Dec 23
Jordan	May 25
Kazakhstan	Dec 16
Kenya	Dec 12
Kiribati	July 12
Korea, Democratic People's Republic of	Sept 9
Korea, Republic of	Aug 15
Kuwait	Feb 25
Kyrgyzstan	Aug 31
Lao People's Democratic Republic	Dec 2
Latvia	Nov 18
Lebanon	Nov 22
Lesotho	Oct 4
Liberia	July 26
Libyan Arab Jamahiriya	Sept 1
Liechtenstein	Aug 15
Lithuania	Feb 16
Luxembourg	June 23
Madagascar	June 26
Malawi	July 6
Malaysia	Aug 31
Maldives	July 26
Mali	Sept 22
Malta	Sept 21
Marshall Islands	Oct 21
Mauritania	Nov 28
Mauritius	Mar 12
Mexico	Sept 16
Micronesia (Federated States of)	Nov 3
Moldova, Republic of	Aug 27
Monaco	Nov 19
Mongolia	July 11
Morocco	Mar 3
Mozambique	June 25
Myanmar (formerly Burma)	Jan 4
Namibia, Republic of	Mar 21
Nauru	Jan 31
Nepal	Dec 28
Netherlands	Apr 30
New Zealand	Feb 6
Nicaragua	Sept 15
Niger	Dec 18
Nigeria	Oct 1
Norway	May 17
Oman	Nov 18
Pakistan	Mar 23
Panama	Nov 3
Papua New Guinea	Sept 16
Paraguay	May 15
Peru	July 28
Philippines	June 12
Poland	May 3
Portugal	June 10
Qatar	Sept 3
Romania	Dec 1
Rwanda	July 1
Saint Christopher (St Kitts) and Nevis	Sept 19
Saint Lucia	Feb 22
Saint Vincent and the Grenadines	Oct 27
Samoa	June 1
San Marino	Sept 3
Sao Tome and Principe	July 12
Saudi Arabia	Sept 23
Senegal	Apr 4
Seychelles	June 18
Sierra Leone	Apr 27
Singapore	Aug 9
Slovakia	Sept 1
Slovenia	June 25
Solomon Islands	July 7
Somalia	Oct 21
South Africa	May 31
Spain	Oct 12
Sri Lanka	Feb 4
Sudan	Jan 1
Suriname	Nov 25
Swaziland	Sept 6
Sweden	June 6
Switzerland	Aug 1
Syria	Apr 17
Tajikistan	Sept 9
Tanzania, United Republic of	Apr 26
Thailand	Dec 5
The Former Yugoslav Republic of Macedonia	Aug 2
Togo	Apr 27
Tonga	July 4
Trinidad and Tobago	Aug 31
Tunisia	Mar 20
Turkey	Oct 29
Turkmenistan	Oct 27
Tuvalu	Oct 2
Uganda	Oct 9
Ukraine	Aug 24
United Arab Emirates	Dec 2
United Republic of Tanzania	Apr 26
United States of America	July 4
Uruguay	Aug 25
Uzbekistan	Sept 1
Vanuatu	July 30
Venezuela	July 5
Vietnam	Sept 2
Western Samoa	June 1
Yemen	May 22
Zaire	June 30
Zambia	Oct 24
Zimbabwe	Apr 18

☆ Chase's 1995 Calendar of Events ☆

CALENDAR INFORMATION FOR THE YEAR 1995

Time shown is Eastern Standard Time. All dates are given in terms of the Gregorian calendar.
(Based in part on information prepared by the Nautical Almanac Office, US Naval Observatory.)

ERAS	YEAR	BEGINS
Byzantine	7504	Sept 14
Jewish (AM)*	5755	Sept 15
Chinese (Year of the Pig)	4693	Jan 31
Roman (AUC)	2748	Jan 14
Nabonassar	2744	Apr 25
Japanese	2655	Jan 1
Grecian (Selucidae)	2307	Sept 14 (or Oct 14)
Indian (Saka)	1917	Mar 22
Diocletian	1712	Sept 12
Islamic (Hegira)*	1416	June 20

*Year begins at sunset.

RELIGIOUS CALENDARS

Epiphany	Jan 6
Shrove Tuesday	Feb 28
Ash Wednesday	Mar 1
Lent	Mar 1-Apr 15
Palm Sunday	Apr 9
Good Friday	Apr 14
Easter Day	Apr 16
Ascension Day	May 25
Whit Sunday (Pentecost)	June 4
Trinity Sunday	June 11
First Sunday in Advent	Dec 3
Christmas Day (Monday)	Dec 25

Eastern Orthodox Church Observances

Christmas	Jan 7
Great Lent begins	Mar 6
Pascha (Easter)	Apr 23
Ascension	June 1
Pentecost	June 11

Jewish Holy Days

Purim	Mar 16
Passover (1st day)	Apr 15
Yom Hashoah	Apr 27
Yom Ha'atsma'ut	May 4
Shavuot	June 4-5
Tisha B'av	Aug 6
Rosh Hashanah (New Year)	Sept 25-26
Yom Kippur	Oct 4
Succoth	Oct 9-17
Chanukah	Dec 18-25

Islamic Holy Days

First Day of Ramadan	Feb 1
Islamic New Year (1416)	May 30

CIVIL CALENDAR—USA—1995

New Year's Day	Jan 1
Martin Luther King's Birthday (obsvd)	Jan 16
Lincoln's Birthday	Feb 12
Washington's Birthday (obsvd)/Presidents' Day	Feb 20
Memorial Day (obsvd)	May 29
Independence Day	July 4
Labor Day	Sept 4
Columbus Day	Oct 9
General Election Day	Nov 7
Veterans Day	Nov 11
Thanksgiving Day	Nov 23

Other Days Widely Observed in US—1995

Groundhog Day (Candlemas)	Feb 2
St Valentine's Day	Feb 14
Mother's Day	May 14
Children's Day	June 11
Flag Day	June 14
Father's Day	June 18
Hallowe'en	Oct 31
National Grandparents Day	Sept 10

CIVIL CALENDAR—CANADA—1995

Commonwealth Day	Mar 13
Victoria Day	May 22
Canada Day	July 1
Labor Day	Sept 4
Thanksgiving Day	Oct 9
Remembrance Day	Nov 11
Boxing Day	Dec 26

CIVIL CALENDAR—MEXICO—1995

New Year's Day	Jan 1
Constitution Day	Feb 5
Benito Juarez Birthday	Mar 21
Labor Day	May 1
Battle of Puebla Day	May 5
Independence Day	Sept 16
Dia de La Raza	Oct 12
Mexican Revolution Day	Nov 20
Guadalupe Day	Dec 12

CIVIL CALENDAR—UNITED KINGDOM—1995

Accession of Queen Elizabeth II	Feb 6
St. David (Wales)	Mar 1
Commonwealth Day	Mar 13
St. Patrick (Ireland)	Mar 17
Birthday of Queen Elizabeth II	Apr 21
St. George (England)	Apr 23
Coronation Day	June 2
Birthday of Prince Philip, Duke of Edinburgh	June 10
The Queen's Official Birthday	June 10
Remembrance Sunday	Nov 12
Birthday of the Prince of Wales	Nov 14
St. Andrew (Scotland)	Nov 30

BANK AND PUBLIC HOLIDAYS—UNITED KINGDOM—1995

(Observed during 1995 in England and Wales, Scotland and Northern Ireland unless otherwise indicated)

New Year	Jan 2
Bank Holiday (Scotland)	Jan 2
St Patrick's Day (Northern Ireland)	Mar 17
Good Friday	Apr 14
Easter Monday (except Scotland)	Apr 17
May Day Bank Holiday	May 1
Spring Bank Holiday	May 29
Orangeman's Day (Battle of the Boyne) (Northern Ireland)	July 12
Bank Holiday (Scotland)	Aug 7
Summer Bank Holiday (except Scotland)	Aug 28
Christmas Day Holiday	Dec 25
Boxing Day Holiday	Dec 26

SEASONS

Spring (Vernal Equinox)	Mar 20, 9:15 PM, EST
Summer (Summer Solstice)	June 21, 4:35 PM, EDT
Autumn (Autumnal Equinox)	Sept 23, 8:14 AM, EDT
Winter (Winter Solstice)	Dec 22, 3:18 AM, EST

DAYLIGHT SAVING TIME SCHEDULE—1995

Sunday, Apr 2, 2:00 AM–Sunday, Oct 29, 2:00 AM—in all time zones.

CHRONOLOGICAL CYCLES

Dominical Letter	A
Epact	29
Golden Number (Lunar Cycle)	XX
Julian Period (year of)	6708
Roman Indiction	3
Solar Cycle	16

ECLIPSES

Partial eclipse of the Moon	Apr 15
Annular eclipse of the Sun	Apr 29
Penumbral eclipse of the Moon	Oct 8
Total eclipse of the Sun	Oct 24

☆ Chase's 1995 Calendar of Events ☆

CALENDAR INFORMATION FOR THE YEAR 1996

Time shown is Eastern Standard Time. All dates are given in terms of the Gregorian calendar.
(Based in part on information prepared by the Nautical Almanac Office, US Naval Observatory.)

ERAS	YEAR	BEGINS
Byzantine	7505	Sept 14
Jewish (AM)*	5757	Sept 14
Chinese (Year of the Rat)	4694	Feb 19
Roman (A.U.C.)	2749	Jan 14
Nabonassar	2745	Apr 24
Japanese	2656	Jan 1
Grecian (Selucidae)	2308	Sept 14 (or Oct 14)
Indian (Saka)	1918	Mar 21
Diocletian	1713	Sept 11
Islamic (Hegira)*	1417	May 17

*Year begins at sunset.

RELIGIOUS CALENDARS

Epiphany	Jan 6
Shrove Tuesday	Feb 20
Ash Wednesday	Feb 21
Lent	Feb 21–Apr 6
Palm Sunday	Mar 31
Good Friday	Apr 5
Easter Day	Apr 7
Ascension Day	May 16
Whit Sunday (Pentecost)	May 26
Trinity Sunday	June 2
First Sunday in Advent	Dec 1
Christmas Day	Dec 25

Eastern Orthodox Church Observances

Christmas	Jan 7
Great Lent begins	Feb 26
Pascha (Easter)	Apr 14
Ascension	May 23
Pentecost	June 2

Jewish Holy Days

Purim	Mar 5
Passover (1st day)	Apr 4
Yom Hashoah	Apr 16
Yom Ha'atsma'ut	Apr 24
Shavuot	May 24
Tisha B'av	July 25
Rosh Hashanah (New Year)	Sept 14–15
Yom Kippur	Sept 23
Succoth	Sept 28–29
Chanukah	Dec 6–13

Islamic Holy Days

First Day of Ramadan	Jan 21
Islamic New Year (1417)	May 18

CIVIL CALENDAR—USA—1996

New Year's Day	Jan 1
Martin Luther King's Birthday (obsvd)	Jan 15
Lincoln's Birthday	Feb 12
Washington's Birthday (obsvd)	Feb 19
Memorial Day (obsvd)	May 27
Independence Day	July 4
Labor Day	Sept 2
Columbus Day	Oct 14
General Election Day	Nov 5
Veterans Day	Nov 11
Thanksgiving Day	Nov 28

Other Days Widely Observed in US—1996

Groundhog Day (Candlemas)	Feb 2
St Valentine's Day	Feb 14
Mother's Day	May 12
Children's Day	June 9
Flag Day	June 14
Father's Day	June 16
Hallowe'en	Oct 31
National Grandparents Day	Sept 8

CIVIL CALENDAR—CANADA—1996

Commonwealth Day	Mar 11
Victoria Day	May 20
Canada Day	July 1
Labor Day	Sept 2
Thanksgiving Day	Oct 14
Remembrance Day	Nov 11
Boxing Day	Dec 26

CIVIL CALENDAR—MEXICO—1996

New Year's Day	Jan 1
Constitution Day	Feb 5
Benito Juarez Birthday	Mar 21
Labor Day	May 1
Battle of Puebla Day	May 5
Independence Day	Sept 16
Dia de La Raza	Oct 12
Mexican Revolution Day	Nov 20
Guadalupe Day	Dec 12

CIVIL CALENDAR—UNITED KINGDOM—1996

Accession of Queen Elizabeth II	Feb 6
St. David (Wales)	Mar 1
Commonwealth Day	Mar 11
St. Patrick (Ireland)	Mar 17
Birthday of Queen Elizabeth II	Apr 21
St. George (England)	Apr 23
Coronation Day	June 2
Birthday of Prince Philip, Duke of Edinburgh	June 10
The Queen's Official Birthday	June 8
Remembrance Sunday	Nov 10
Birthday of the Prince of Wales	Nov 14
St. Andrew (Scotland)	Nov 30

BANK AND PUBLIC HOLIDAYS—UNITED KINGDOM—1996

(Observed during 1996 in England and Wales, Scotland and Northern Ireland unless otherwise indicated)

New Year	Jan 1
Bank Holiday (Scotland)	Jan 2
St Patrick's Day (Northern Ireland)	Mar 18
Good Friday	Apr 5
Easter Monday (except Scotland)	Apr 8
May Day Bank Holiday	May 6
Spring Bank Holiday	May 27
Orangeman's Day (Battle of the Boyne) (Northern Ireland)	July 12
Bank Holiday (Scotland)	Aug 5
Summer Bank Holiday (except Scotland)	Aug 26
Christmas Day Holiday	Dec 25
Boxing Day	Dec 26

SEASONS

Spring (Vernal Equinox)	Mar 20, 3:04 PM, EST
Summer (Summer Solstice)	June 20, 10:24 PM, EDT
Autumn (Autumnal Equinox)	Sept 22, 2:01 PM, EDT
Winter (Winter Solstice)	Dec 21, 9:07 AM, EST

DAYLIGHT SAVING TIME SCHEDULE—1996

Sunday, Apr 3, 2:00 AM–Sunday, Oct 30, 2:00 AM—in all time zones.

CHRONOLOGICAL CYCLES

Dominical Letter	GF
Golden Number (Lunar Cycle)	XXI
Julian Period (year of)	6709
Roman Indiction	4
Solar Cycle	17

ECLIPSES

Total eclipse of the Moon	Apr 3
Partial eclipse of the Sun	Apr 17
Total eclipse of the Moon	Sept 26
Partial eclipse of the Sun	Oct 12

☆ *Chase's 1995 Calendar of Events* ☆

ASTRONOMICAL PHENOMENA FOR THE YEARS 1995–1997

All dates are given in terms of Eastern Standard or Daylight Time and the Gregorian calendar.
(Based in part on information prepared by the Nautical Almanac Office, US Naval Observatory.)

1995

PRINCIPAL PHENOMENA, 1995
EARTH

Perihelion .... Jan 4
Aphelion .... July 3
Equinoxes .... Mar 20, Sept 23
Solstices .... June 21, Dec 22

PHASES OF THE MOON

New Moon	First Quarter	Full Moon	Last Quarter
Jan 1	Jan 8	Jan 16	Jan 23
Jan 30	Feb 7	Feb 15	Feb 22
Mar 1	Mar 9	Mar 16	Mar 23
Mar 30	Apr 8	Apr 15	Apr 21
Apr 29	May 7	May 14	May 21
May 29	June 6	June 13	June 19
June 27	July 5	July 12	July 19
July 27	Aug 3	Aug 10	Aug 17
Aug 26	Sept 2	Sept 8	Sept 16
Sept 24	Oct 1	Oct 8	Oct 16
Oct 24	Oct 30	Nov 7	Nov 15
Nov 22	Nov 29	Dec 6	Dec 15
Dec 21	Dec 28		

ECLIPSES

Partial eclipse of the Moon .... Apr 15
Annular eclipse of the Sun .... Apr 29
Penumbral eclipse of the Moon .... Oct 8
Total eclipse of the Sun .... Oct 23–24

VISIBILITY OF PLANETS
IN MORNING AND EVENING TWILIGHT

	Morning	Evening
VENUS	Jan 1–July 15	Sept 27–Dec 31
MARS	Jan 1–Feb 12	Feb 12–Dec 29
JUPITER	Jan 1–June 1	June 1–Dec 6
SATURN	Mar 24–Sept 14	Jan 1–Feb 17
		Sept 14–Dec 31

CHRONOLOGICAL CYCLES

Dominical Letter .... A
Epact 29
Golden Number (Lunar Cycle) I
Julian Period (year of) 6708
Roman Indiction 3
Solar Cycle 16

All dates are given in terms of the Gregorian calendar in which 1994 January 14 corresponds to 1994 January 1 of the Julian calendar.

1996

PRINCIPAL PHENOMENA, 1996
EARTH

Perihelion .... Jan 4
Aphelion .... July 5
Equinoxes .... Mar 20, Sept 22
Solstices .... June 20, Dec 21

PHASES OF THE MOON

New Moon	First Quarter	Full Moon	Last Quarter
		Jan 5	Jan 13
Jan 20	Jan 27	Feb 4	Feb 12
Feb 18	Feb 26	Mar 5	Mar 12
Mar 19	Mar 26	Apr 3	Apr 10
Apr 17	Apr 25	May 3	May 10
May 17	May 25	June 1	June 8
June 15	June 24	June 30	July 7
July 15	July 23	July 30	Aug 6
Aug 14	Aug 21	Aug 28	Sept 4
Sept 12	Sept 20	Sept 26	Oct 4
Oct 12	Oct 19	Oct 26	Nov 3
Nov 10	Nov 17	Nov 24	Dec 3
Dec 10	Dec 17	Dec 24	

ECLIPSES

Total eclipse of the Moon .... Apr 3
Partial eclipse of the Sun .... Apr 17
Total eclipse of the Moon .... Sept 26

1997

PRINCIPAL PHENOMENA, 1997
EARTH

Perihelion .... Jan 1
Aphelion .... July 4
Equinoxes .... Mar 20, Sept 22
Solstices .... June 21, Dec 21

PHASES OF THE MOON

New Moon	First Quarter	Full Moon	Last Quarter
			Jan 1
Jan 8	Jan 15	Jan 23	Jan 31
Feb 7	Feb 14	Feb 22	Mar 2
Mar 8	Mar 15	Mar 23	Mar 31
Apr 7	Apr 14	Apr 22	Apr 29
May 6	May 14	May 22	May 29
June 5	June 13	June 20	June 27
July 4	July 12	July 19	July 26
Aug 3	Aug 11	Aug 18	Aug 24
Sept 1	Sept 9	Sept 16	Sept 23
Oct 1	Oct 9	Oct 15	Oct 23
Oct 31	Oct 7	Nov 14	Nov 21
Nov 29	Dec 7	Dec 13	Dec 21
Dec 29			

ECLIPSES

Partial eclipse of the Sun .... Sept 1
Partial eclipse of the Moon .... Mar 23–24
Total eclipse of the Sun .... Mar 8
Total eclipse of the Moon .... Sept 16

☆ Chase's 1995 Calendar of Events ☆

LOOKING FORWARD

1996 • Budapest World's Fair, May 11–Oct 4
• The Games of the XXVIth Olympiad will be held at Atlanta, GA
• Intl Year for Eradication of Poverty
• US presidential election, Nov 5, 1996
• Tennessee Statehood Bicentennial
• Utah Statehood Centennial
• UNESCO 50th Anniversary
• UNICEF 50th Anniversary
• UN Conference on Human Settlements (Habitat (II), June 3–14, Turkey
1997 • British lease of Hong Kong expires
• PTA Centennial
1998 • World Cup in France
• WHO 50th Anniversary
• Yukon Territory Centennial
1999 • Bicentennial of death of George Washington, at Mount Vernon, VA, Dec 14, 1799
• Intl Year of the Elderly
2000 • US presidential election, Nov 7, 2000
• Holy Year. Traditionally observed by Roman Catholic Church at 25-year intervals since 1450, in addition to extraordinary Holy Years
2001 • Beginning of 21st Century and the Third Millennium of the Christian Era
• 250th Birth Anniversary of James Madison
2003 • Ohio Statehood Bicentennial
2004 • US presidential election, Nov 2, 2004
2007 • Oklahoma Statehood Centennial
2008 • 100th Birth Anniversary of Theodore Roosevelt
2009 • Bicentennial of birth of Abraham Lincoln
2050 • World population predicted to reach 10 billion, according to 1984 World Bank estimate (1987 world pop: five billion)
2061 • Halley's comet returns

SELECTED SPECIAL YEARS: 1957–1995

Intl Geophysical Year: July 1, 1957–Dec 31, 1958
World Refugee Year: July 1, 1959–June 30, 1960
Intl Cooperation Year: 1965
Intl Book Year: 1972
World Population Year: 1974
Intl Women's Year: 1975
Intl Year of the Child: 1979
Intl Year for Disabled Persons: 1981
World Communications Year: 1983
Intl Youth Year: 1985
Intl Year of Peace: 1986
Intl Year of Shelter for the Homeless: 1987
Year of the Reader: 1987
Year of the Young Reader: 1989
Intl Literacy Year: 1990
US Decade of the Brain: 1990–99
Intl Space Year: 1992
Intl Year for World's Indigenous Peoples: 1993
Intl Year of the Family: 1994
UN Year for Tolerance: 1995

SPACE OBJECTS BOX SCORE

The Space Objects Box Score, furnished by the Project Operations Branch Code 513 of the NASA/Goddard Space Flight Center, reflects the numbers of objects now in orbit by type—payload or debris—and by owning nation or organization. It is based on data computed at Goddard Space Flight Center, NORAD, or provided by satellite owners. The debris includes pieces launched into space along with the functioning payloads, as well as fragments produced by in-space breakups. Decayed objects are those that either impacted or that have landed on other planets and the moon.

CURRENT AND HISTORICAL STATUS (as of August 11, 1994)

COUNTRY	OBJECTS IN ORBIT			DECAYED OBJECTS		
	SO P/L	SO DEB	TOTAL	DO P/L	DO DEB	TOTAL
ARGENTINA	1	0	1	0	0	0
ARAB SAT. COMM. ORG.	0	0	0	0	0	0
ASIASAT CORP.	0	0	0	0	0	0
AUSTRALIA	6	1	7	1	0	1
BRAZIL	4	0	4	0	0	0
CANADA	16	0	16	1	0	1
CZECH	1	0	1	1	0	1
ESA	25	138	163	4	454	458
ESRO	0	0	0	7	3	10
FRANCE/FRG	2	0	2	0	0	0
FRANCE	23	18	41	7	59	66
FRG	14	2	16	5	5	10
IMSO	3	0	3	0	0	0
INDIA	10	3	13	6	8	14
INDONESIA	6	0	6	1	1	2
ISRAEL	0	0	0	2	2	4
ITALY	4	0	4	5	0	5
ITSO	44	0	44	1	0	1
JAPAN	51	51	102	10	74	84
KOREA	2	0	2	0	0	0
LUXEMBOURG	3	2	5	0	0	0
MEXICO	3	0	3	0	0	0
NATO	7	2	9	0	0	0
NETHERLANDS	0	0	0	1	3	4
PAKISTAN	0	0	0	1	0	1
PORTUGAL	1	0	1	0	0	0
PRC	15	81	96	23	81	104
SAUDI ARABIA	3	0	3	0	0	0
SPAIN	3	2	3	0	0	0
SWEDEN	3	0	3	0	0	0
THAILAND	1	0	1	0	0	0
UK	18	1	19	8	4	12
USA	639	2,745	3,384	649	2,875	3,524
USSR	1,295	2,426	3,721	1,616	9,614	11,230
TOTAL	2,203	5,470	7,673	2,349	13,183	15,532

SO P/L: Space Objects Payload, which are those in Earth's orbit.
SO DEB: Space Objects Debris.
DO P/L: Decayed Objects Payload
DO DEB: Decayed Objects Debris.
ESA: European Space Agency.
ESRO: European Space Research Organization.
FRANCE/FRG: France/Fed. Rep. Ger.
FRG: Federal Republic of Germany.
IMSO: International Maritime Satellite Organization.
ITSO: International Telecommunications Satellite Organization.
PRC: People's Republic of China.
UK: United Kingdom.
USSR: Russia

NOTE: The change in the format of this information reflects a change in what information is released to the public.

☆ *Chase's 1995 Calendar of Events* ☆

UNIVERSAL, STANDARD AND DAYLIGHT TIMES

Universal Time (UT) is also known as Greenwich Mean Time (GMT) and is the standard time of the Greenwich meridian (0° of longitude). A time given in UT may be converted to local mean time by the addition of east longitude (or the subtraction of west longitude), where the longitude of the place is expressed in time-measure at the rate of one hour for every 15°. Local clock times may differ from standard times, especially in summer when clocks are often advanced by one hour ("daylight saving" or "summer" time).

The time used in this book is Eastern Standard Time. The following table provides conversion between Universal Time and all Time Zones in the United States. An asterisk denotes that the time is on the preceding day.

Universal Time	Eastern Daylight Time	Eastern Standard Time and Central Daylight Time	Central Standard Time and Mountain Daylight Time	Mountain Standard Time and Pacific Daylight Time	Pacific Standard Time
0^h	* 8 P. M.	* 7 P. M.	* 6 P. M.	* 5 P. M.	* 4 P. M.
1	* 9	* 8	* 7	* 6	* 5
2	*10	* 9	* 8	* 7	* 6
3	*11 P. M.	*10	* 9	* 8	* 7
4	0 Midnight	*11 P. M.	*10	* 9	* 8
5	1 A. M.	0 Midnight	*11 P. M.	*10	* 9
6	2	1 A. M.	0 Midnight	*11 P. M.	*10
7	3	2	1 A. M.	0 Midnight	*11 P. M.
8	4	3	2	1 A. M.	0 Midnight
9	5	4	3	2	1 A. M.
10	6	5	4	3	2
11	7	6	5	4	3
12	8	7	6	5	4
13	9	8	7	6	5
14	10	9	8	7	6
15	11 A. M.	10	9	8	7
16	12 Noon	11 A. M.	10	9	8
17	1 P. M.	12 Noon	11 A. M.	10	9
18	2	1 P. M.	12 Noon	11 A. M.	10
19	3	2	1 P. M.	12 Noon	11 A. M.
20	4	3	2	1 P. M.	12 Noon
21	5	4	3	2	1 P. M.
22	6	5	4	3	2
23	7 P. M.	6 P. M.	5 P. M.	4 P. M.	3 P. M.

The longitudes of the standard meridians for the standard time zones are:
Eastern..... 75° West   Central..... 90° West   Mountain..... 105° West   Pacific..... 120° West

LEAP SECONDS

The information below is developed by the editors from data supplied by the National Bureau of Standards of the US and the Bureau International de L'Heure.

Because of Earth's slightly erratic rotation and the need for greater precision in time measurement it has become necessary to add a "leap second" from time to time to man's clocks to coordinate them with astronomical time. Rotation of the Earth has been slowing since 1900, making an astronomical second longer than an atomic second. Since 1972, by international agreement, adjustments have been made to keep astronomical and atomic clocks within 0.9 second of each other. The determination to add (or subtract) seconds is made by the Bureau International de l'Heure, in Paris. Preferred times for adjustment have been June 30 and December 31, but any time may be designated by the Bureau International de l'Heure. The first such adjustment was made in 1972, and as of June 30, 1994, a total of 19 leap seconds have been added. The additions have been made at 23:59:60 UTC (Coordinated Universal Time) = 6:59:60 EST (Eastern Standard Time). Leap seconds have been inserted into the UTC time scale on the following dates:

June 30, 1972	Dec 31, 1976	June 30, 1982	Dec 31, 1990
Dec 31, 1972	Dec 31, 1977	June 30, 1983	June 30, 1992
Dec 31, 1973	Dec 31, 1978	June 30, 1985	June 30, 1993
Dec 31, 1974	Dec 31, 1979	Dec 31, 1987	June 30, 1994
Dec 31, 1975	June 30, 1981	Dec 31, 1989	

☆ *Chase's 1995 Calendar of Events* ☆
WORLD MAP OF TIME ZONES

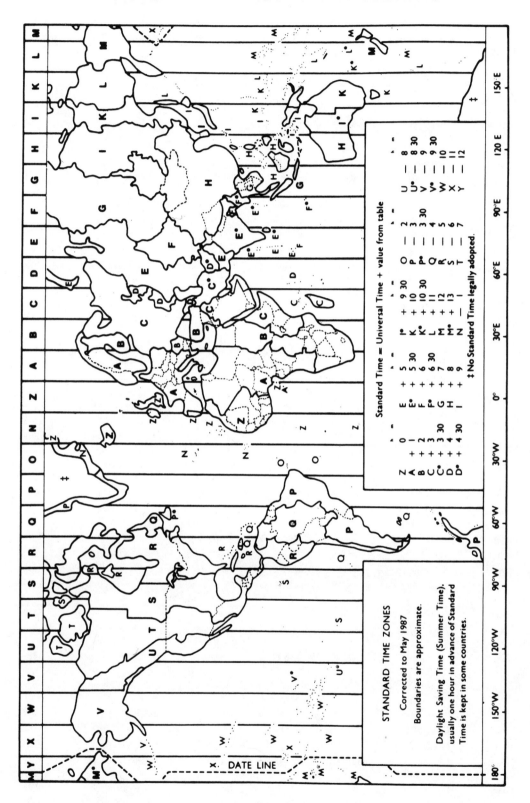

☆ Chase's 1995 Calendar of Events ☆

SOME FACTS ABOUT THE STATES

State	Capital	Popular name	Area (sq. mi.)	State bird	State flower	State tree	Admitted to the Union	Order of admission
Alabama	Montgomery	Cotton or Yellowhammer State; or Heart of Dixie	51,609	Yellowhammer	Camellia	Southern pine (Longleaf pine)	1819	22
Alaska	Juneau	Last Frontier	591,004	Willow ptarmigan	Forget-me-not	Sitka spruce	1959	49
Arizona	Phoenix	Grand Canyon State	114,000	Cactus wren	Saguaro (Giant cactus)	Paloverde	1912	48
Arkansas	Little Rock	Land of Opportunity	53,187	Mockingbird	Apple blossom	Pine	1836	25
California	Sacramento	Golden State	158,706	California valley quail	Golden poppy	California redwood	1850	31
Colorado	Denver	Centennial State	104,091	Lark bunting	Rocky Mountain columbine	Blue spruce	1876	38
Connecticut	Hartford	Constitution State	5,018	Robin	Mountain laurel	White oak	1788	5
Delaware	Dover	First State	2,044	Blue hen chicken	Peach blossom	American holly	1787	1
Florida	Tallahassee	Sunshine State	58,664	Mockingbird	Orange blossom	Cabbage (Sabal) palm	1845	27
Georgia	Atlanta	Empire State of the South	58,910	Brown thrasher	Cherokee rose	Live oak	1788	4
Hawaii	Honolulu	Aloha State	6,471	Nene (Hawaiian goose)	Hibiscus	Kukui	1959	50
Idaho	Boise	Gem State	83,564	Mountain bluebird	Syringa (Mock orange)	Western white pine	1890	43
Illinois	Springfield	Land of Lincoln	56,345	Cardinal	Native violet	White oak	1818	21
Indiana	Indianapolis	Hoosier State	36,185	Cardinal	Peony	Tulip tree or yellow poplar	1816	19
Iowa	Des Moines	Hawkeye State	56,275	Eastern goldfinch	Wild rose	Oak	1846	29
Kansas	Topeka	Sunflower State	82,277	Western meadowlark	Sunflower	Cottonwood	1861	34
Kentucky	Frankfort	Bluegrass State	40,409	Kentucky cardinal	Goldenrod	Kentucky coffeetree	1792	15
Louisiana	Baton Rouge	Pelican State	47,752	Pelican	Magnolia	Bald cypress	1812	18
Maine	Augusta	Pine Tree State	33,265	Chickadee	White pine cone and tassel	White pine	1820	23
Maryland	Annapolis	Old Line State	10,577	Baltimore oriole	Black-eyed Susan	White Oak	1788	7
Massachusetts	Boston	Bay State	8,284	Chickadee	Mayflower	American elm	1788	6
Michigan	Lansing	Wolverine State	58,527	Robin	Apple blossom	White pine	1837	26
Minnesota	St. Paul	Gopher State	84,402	Common loon	Pink and white lady's-slipper	Norway, or red, pine	1858	32
Mississippi	Jackson	Magnolia State	47,689	Mockingbird	Magnolia	Magnolia	1817	20
Missouri	Jefferson City	Show Me State	69,697	Bluebird	Hawthorn	Flowering dogwood	1821	24
Montana	Helena	Treasure State	147,046	Western meadowlark	Bitterroot	Ponderosa pine	1889	41
Nebraska	Lincoln	Cornhusker State	77,355	Western meadowlark	Goldenrod	Cottonwood	1867	37
Nevada	Carson City	Silver State	110,540	Mountain bluebird	Sagebrush	Single-leaf piñon	1864	36
New Hampshire	Concord	Granite State	9,304	Purple finch	Purple lilac	White birch	1788	9
New Jersey	Trenton	Garden State	7,787	Eastern goldfinch	Purple violet	Red oak	1787	3
New Mexico	Santa Fe	Land of Enchantment	121,593	Roadrunner	Yucca flower	Piñon, or nut pine	1912	47
New York	Albany	Empire State	49,108	Bluebird	Rose	Sugar maple	1788	11
North Carolina	Raleigh	Tar Heel State or Old North State	52,669	Cardinal	Dogwood	Pine	1789	12
North Dakota	Bismarck	Flickertail State	70,702	Western meadowlark	Wild prairie rose	American elm	1889	39

☆ Chase's 1995 Calendar of Events ☆

State	Capital	Popular name	Area (sq. mi.)	State bird	State flower	State tree	Admitted to the Union	Order of admission
Ohio	Columbus	Buckeye State	41,330	Cardinal	Scarlet carnation	Buckeye	1803	17
Oklahoma	Oklahoma City	Sooner State	69,956	Scissor-tailed flycatcher	Mistletoe	Redbud	1907	46
Oregon	Salem	Beaver State	97,073	Western meadowlark	Oregon grape	Douglas fir	1859	33
Pennsylvania	Harrisburg	Keystone State	45,308	Ruffed grouse	Mountain laurel	Hemlock	1787	2
Rhode Island	Providence	Ocean State	1,212	Rhode Island Red	Violet	Red maple	1790	13
South Carolina	Columbia	Palmetto State	31,113	Carolina wren	Carolina jessamine	Palmetto	1788	8
South Dakota	Pierre	Sunshine State	77,116	Ring-necked pheasant	American pasqueflower	Black Hills spruce	1889	40
Tennessee	Nashville	Volunteer State	42,114	Mockingbird	Iris	Tulip poplar	1796	16
Texas	Austin	Lone Star State	266,807	Mockingbird	Bluebonnet	Pecan	1845	28
Utah	Salt Lake City	Beehive State	84,899	Sea Gull	Sego lily	Blue spruce	1896	45
Vermont	Montpelier	Green Mountain State	9,614	Hermit thrush	Red clover	Sugar maple	1791	14
Virginia	Richmond	Old Dominion	40,767	Cardinal	Dogwood	Dogwood	1788	10
Washington	Olympia	Evergreen State	68,139	Willow goldfinch	Coast rhododendron	Western hemlock	1889	42
West Virginia	Charleston	Mountain State	24,231	Cardinal	Rhododendron	Sugar maple	1863	35
Wisconsin	Madison	Badger State	56,153	Robin	Wood violet	Sugar maple	1848	30
Wyoming	Cheyenne	Equality State	97,809	Meadowlark	Indian paintbrush	Cottonwood	1890	44

STATE & TERRITORY ABBREVIATIONS: UNITED STATES

Alabama....................AL	Kansas....................KS	Ohio....................OH
Alaska....................AK	Kentucky....................KY	Oklahoma....................OK
Arizona....................AZ	Louisiana....................LA	Oregon....................OR
Arkansas....................AR	Maine....................ME	Pennsylvania....................PA
American Samoa....................AS	Maryland....................MD	Puerto Rico....................PR
California....................CA	Massachusetts....................MA	Rhode Island....................RI
Canal Zone....................CZ	Michigan....................MI	South Carolina....................SC
Colorado....................CO	Minnesota....................MN	South Dakota....................SD
Connecticut....................CT	Mississippi....................MS	Tennessee....................TN
Delaware....................DE	Missouri....................MO	Trust Territories....................TT
District of Columbia....................DC	Montana....................MT	Texas....................TX
Florida....................FL	Nebraska....................NE	Utah....................UT
Georgia....................GA	Nevada....................NV	Vermont....................VT
Guam....................GU	New Hampshire....................NH	Virginia....................VA
Hawaii....................HI	New Jersey....................NJ	Virgin Islands....................VI
Idaho....................ID	New Mexico....................NM	Washington....................WA
Illinois....................IL	New York....................NY	West Virginia....................WV
Indiana....................IN	North Carolina....................NC	Wisconsin....................WI
Iowa....................IA	North Dakota....................ND	Wyoming....................WY

PROVINCE & TERRITORY ABBREVIATIONS: CANADA

Alberta....................Alta	Newfoundland....................Nfld	Quebec....................Que
British Columbia....................BC	Nova Scotia....................NS	Saskatchewan....................Sask
Manitoba....................Man	Ontario....................Ont	Yukon Territory....................Yukon
New Brunswick....................NB	Prince Edward Island....................PEI	Northwest Territories....................NWT

☆ Chase's 1995 Calendar of Events ☆

SOME FACTS ABOUT CANADA

Province/Territory	Capital	*Population	Flower	Land/Fresh Water (sq. mi.)	Total Area
Alberta	Edmonton	2,186,100	Wild rose	400,423/10,437	410,860
British Columbia	Victoria	2,724,900	Pacific dogwood	578,230/11,227	589,458
Manitoba	Winnipeg	1,032,700	Prairie crocus	340,834/63,129	403,964
New Brunswick	Fredericton	711,900	Purple violet	44,797/835	45,633
Newfoundland	St. John's	588,200	Pitcher plant	230,219/21,147	251,367
Northwest Territories	Yellowknife	43,700	Mountain avens	2,017,306/82,829	2,100,136
Nova Scotia	Halifax	859,600	Mayflower	32,835/1,647	34,482
Ontario	Toronto	8,650,300	White trillium	553,788/110,229	664,012
Prince Edward Island	Charlottetown	124,600	Lady's-slipper	3,515/0	3,515
Quebec	Quebec City	6,353,000	White garden lily	843,109/114,269	957,379
Saskatchewan	Regina	983,900	Red lily	354,365/51,347	405,712
Yukon Territory	Whitehorse	22,200	Fireweed	297,050/2,784	299,835

*1984

SOME FACTS ABOUT THE PRESIDENTS

	Name	Birthdate/Place	Party	Tenure	Died	First Lady	Vice President
1.	George Washington	2/22/1732, Westmoreland Cnty, VA	Federalist	1789–1797	12/14/1799	Martha Dandridge Custis	John Adams
2.	John Adams	10/30/1735, Braintree (Quincy), MA	Federalist	1797–1801	7/4/1826	Abigail Smith	Thomas Jefferson
3.	Thomas Jefferson	4/13/1743, Shadwell, VA	Democratic-Republican	1801–1809	7/4/1826	Martha Wayles Skelton	Aaron Burr, 1801–05 George Clinton, 1805–09
4.	James Madison	3/16/1751, Port Conway, VA	Democratic-Republican	1809–1817	6/28/1836	Dolley Payne Todd	George Clinton 1809–12 Elbridge Gerry 1813–14(?)
5.	James Monroe	4/28/1758, Westmoreland Cnty, VA	Democratic-Republican	1817–1825	7/4/1831	Elizabeth Kortright	Daniel D. Tompkins
6.	John Q. Adams	7/11/1767, Braintree (Quincy), MA	Democratic-Republican	1825–1829	2/23/1848	Louisa Catherine Johnson	John C. Calhoun
7.	Andrew Jackson	3/15/1767, Waxhaw Settlement, SC	Democratic-Republican	1829–1837	6/8/1845	Mrs. Rachel Donelson Robards	John C. Calhoun 1829–32 Martin Van Buren 1833–37
8.	Martin Van Buren	12/5/1782, Kinderhook, NY	Democratic-Republican	1837–1841	7/24/1862	Hannah Hoes	Richard M. Johnson
9.	William H. Harrison	2/9/1773, Charles City Cnty, VA	Whig	1841	4/4/1841†	Anna Symmes	John Tyler
10.	John Tyler	3/29/1790, Charles City Cnty, VA	Whig	1841–1845	1/18/1862	Letitia Christian	
11.	James K. Polk	11/2/1795, near Pineville, NC	Democrat	1845–1849	6/15/1849	Sarah Childress	George M. Dallas
12.	Zachary Taylor	11/24/1784, Barboursville, VA	Whig	1849–1850	7/9/1850†	Margaret Mackall Smith	Millard Fillmore
13.	Millard Fillmore	1/7/1800, Locke, NY	Whig	1850–1853	3/8/1874	Abigail Powers Mrs. Caroline Carmichael McIntosh	
14.	Franklin Pierce	11/23/1804, Hillsboro, NH	Democrat	1853–1857	10/8/1869	Jane Means Appleton	William R. D. King
15.	James Buchanan	4/23/1791, near Mercersburg, PA	Democrat	1857–1861	6/1/1868		John C. Breckinridge
16.	Abraham Lincoln	2/12/1809, near Hodgenville, KY	Republican	1861–1865	4/15/1865*	Mary Todd	Hannibal Hamlin, 1861–65 Andrew Johnson, 1865

☆ *Chase's 1995 Calendar of Events* ☆

SOME FACTS ABOUT THE PRESIDENTS (continued)

	Name	Birthdate/Place	Party	Tenure	Died	First Lady	Vice President
17.	Andrew Johnson	12/29/1808, Raleigh, NC	Democrat	1865–1869	7/31/1875	Eliza McCardle	
18.	Ulysses S. Grant	4/27/1822, Point Pleasant, OH	Republican	1869–1877	7/23/1885	Julia Boggs Dent	Schuyler Colfax, 1869–73 Henry Wilson, 1873–75
19.	Rutherford B. Hayes	10/4/1822, Delaware, OH	Republican	1877–1881	1/17/1893	Lucy Ware Webb	William A. Wheeler
20.	James A. Garfield	11/19/1831, Orange, OH	Republican	1881	9/19/1881*	Lucretia Rudolph	Chester Alan Arthur
21.	Chester A. Arthur	10/5/1829, Fairfield, VT	Republican	1881–1885	11/18/1886	Ellen Lewis Herndon	
22.	Grover Cleveland	3/18/1837, Caldwell, NJ	Democrat	1885–1889	6/24/1908	Frances Folsom	Thomas A. Hendricks, 1885
23.	Benjamin Harrison	8/20/1833, North Bend, OH	Republican	1889–1893	3/13/1901	Caroline Lavinia Scott Mrs. Mary Dimmick	Levi P. Morton
24.	Grover Cleveland	3/18/1837, Caldwell, NJ	Democrat	1893–1897	6/24/1908	Frances Folsom	Adlai Stevenson, 1893–97
25.	William McKinley	1/29/1843, Niles, OH	Republican	1897–1901	9/14/1901*	Ida Saxton	Garret A. Hobart, 1897–99 Theodore Roosevelt, 1901
26.	Theodore Roosevelt	10/27/1858, New York City	Republican	1901–1909	1/6/1919	Alice Hathaway Lee Edith Kermit Carow	Charles W. Fairbanks
27.	William H. Taft	9/15/1857, Cincinnati, OH	Republican	1909–1913	3/8/1930	Helen Herron	James S. Sherman
28.	Woodrow Wilson	12/29/1856, Staunton, VA	Democrat	1913–1921	2/3/1924	Ellen Louise Axson Edith Bolling Galt	Thomas R. Marshall
29.	Warren G. Harding	11/2/1865, near Corsica, OH	Republican	1921–1923	8/2/1923†	Florence Kling DeWolfe	Calvin Coolidge
30.	Calvin Coolidge	7/4/1872, Plymouth Notch, VT	Republican	1923–1929	1/5/1933	Grace Anna Goodhue	Charles G. Dawes
31.	Herbert C. Hoover	8/10/1874, West Branch, IA	Republican	1929–1933	10/20/1964	Lou Henry	Charles Curtis
32.	Franklin D. Roosevelt	1/30/1882, Hyde Park, NY	Democrat	1933–1945	4/12/1945†	Eleanor Roosevelt	John N. Garner, 1933–41 Henry A. Wallace, 1941–45 Harry S. Truman, 1945
33.	Harry S. Truman	5/8/1884, Lamar, MO	Democrat	1945–1953	12/26/1972	Elizabeth Virginia (Bess) Wallace	Alben W. Barkley
34.	Dwight D. Eisenhower	10/14/1890, Denison, TX	Republican	1953–1961	3/28/1969	Mamie Geneva Doud	Richard M. Nixon
35.	John F. Kennedy	5/29/1917, Brookline, MA	Democrat	1961–1963	11/22/1963*	Jacqueline Lee Bouvier	Lyndon B. Johnson
36.	Lyndon B. Johnson	8/27/1908, near Stonewall, TX	Democrat	1963–1969	1/22/1973	Caudia Alta (Lady Bird) Taylor	Hubert H. Humphrey
37.	Richard M. Nixon	1/9/1913, Yorba Linda, CA	Republican	1969–1974**	4/22/1994	Thelma Catherine (Pat) Ryan	Spiro T. Agnew, 1969–73 Gerald R. Ford, 1973–74
38.	Gerald R. Ford	7/14/1913, Omaha, NE	Republican	1974–1977		Elizabeth (Betty) Bloomer	Nelson A. Rockefeller
39.	James E. Carter, Jr	10/1/1924, Plaines, GA	Democrat	1977–1981		Rosalynn Smith	Walter F. Mondale
40.	Ronald W. Reagan	2/6/1911, Tampico, IL	Republican	1981–1989		Nancy Davis	George H. W. Bush
41.	George H. W. Bush	6/12/1929, Milton, MA	Republican	1989–1993		Barbara Pierce	J. Danforth Quayle
42.	William J. Clinton	8/19/1946, Hope, AR	Democrat	1993–		Hillary Rodham	Albert Gore, Jr.

* assainated while in office
** resigned Aug 9, 1974
† died while in office—nonviolently

Sources: *World Book*, 1991 Edition; *Encyclopedia Americana*, 1990 Edition; *Collier's Encyclopedia*, 1988 Edition

☆ *Chase's 1995 Calendar of Events* ☆
U.S. SUPREME COURT JUSTICES

Chief Justices:	Term	Appointed By
John Jay	1789–1795	Washington
John Rutledge	1795	Washington
Oliver Ellsworth	1796–1800	Washington
John Marshall	1801–1835	J. Adams
Roger B. Taney	1836–1864	Jackson
Salmon P. Chase	1864–1873	Lincoln
Morrison R. Waite	1874–1888	Grant
Melville W. Fuller	1888–1910	Cleveland
Edward D. White	1910–1921	Taft
William H. Taft	1921–1930	Harding
Charles E. Huges	1930–1941	Hoover
Harlan F. Stone	1941–1946	F. D. Roosevelt
Frederick M. Vinson	1946–1953	Truman
Earl Warren	1953–1969	Eisenhower
Warren E. Burger	1969–1986	Nixon
William H. Rehnquist	1986–	Reagan

Associate Justices:	Term	Appointed By
James Wilson	1789–1798	Washington
John Rutledge	1789–1791	Washington
William Cushing	1790–1810	Washington
John Blair	1790–1796	Washington
James Iredell	1790–1799	Washington
Thomas Johnson	1792–1793	Washington
William Paterson	1793–1806	Washington
Samuel Chase	1796–1811	Washington
Bushrod Washington	1799–1829	J. Adams
Alfred Moore	1800–1804	J. Adams
William Johnson	1804–1834	Jefferson
H. Brockholst Livingston	1807–1823	Jefferson
Thomas Todd	1807–1826	Jefferson
Gabriel Duvall	1811–1835	Madison
Joseph Story	1812–1845	Madison
Smith Thompson	1823–1843	Monroe
Robert Trimble	1826–1828	J. Q. Adams
John McLean	1830–1861	Jackson
Henry Baldwin	1830–1844	Jackson
James M. Wayne	1835–1867	Jackson
Philip P. Barbour	1836–1841	Jackson
John Catron	1837–1865	Van Buren
John McKinley	1838–1852	Van Buren
Peter V. Daniel	1842–1860	Van Buren
Samuel Nelson	1845–1872	Tyler
Levi Woodbury	1845–1851	Polk
Robert C. Grier	1846–1870	Polk
Benjamin R. Curtis	1851–1857	Fillmore
John A. Campbell	1853–1861	Pierce
Nathan Clifford	1858–1881	Buchanan
Noah H. Swayne	1862–1881	Lincoln
Samuel F. Miller	1862–1890	Lincoln
David Davis	1862–1877	Lincoln
Stephen J. Field	1863–1897	Lincoln
William Strong	1870–1880	Grant
Joseph P. Bradley	1870–1892	Grant
Ward Hunt	1873–1882	Grant
John M. Harlan	1877–1911	Hayes
William B. Woods	1881–1887	Hayes
Stanley Matthews	1881–1889	Garfield
Horace Gray	1882–1902	Arthur
Samuel Blatchford	1882–1893	Arthur
Lucius Q. C. Lamar	1888–1893	Cleveland
David J. Brewer	1890–1910	Harrison
Henry B. Brown	1891–1906	Harrison
George Shiras, Jr.	1892–1903	Harrison
Howell E. Jackson	1893–1895	Harrison
Edward D. White	1894–1910	Cleveland
Rufas W. Peckham	1896–1909	Cleveland
Joseph McKenna	1898–1925	McKinley
Oliver W. Holmes, Jr.	1902–1932	T. Roosevelt
William R. Day	1903–1922	T. Roosevelt
William H. Moody	1906–1910	T. Roosevelt
Horace H. Lurton	1910–1914	Taft
Charles E. Hughes	1910–1916	Taft
Willis Van Devanter	1911–1937	Taft
Joseph R. Lamar	1911–1916	Taft
Mahlon Pitney	1912–1922	Taft
James C. McReynolds	1914–1941	Wilson
Louis D. Brandeis	1916–1939	Wilson
John H. Clarke	1916–1922	Wilson
George Sutherland	1922–1938	Harding
Pierce Butler	1923–1939	Harding
Edward T. Sanford	1923–1930	Harding
Harlan F. Stone	1925–1941	Coolidge
Owen J. Roberts	1930–1945	Hoover
Benjamin N. Cardozo	1932–1938	Hoover
Hugo L. Black	1937–1971	F. D. Roosevelt
Stanley F. Reed	1938–1957	F. D. Roosevelt
Felix Frankfurter	1939–1962	F. D. Roosevelt
William O. Douglas	1939–1975	F. D. Roosevelt
Frank Murphy	1940–1949	F. D. Roosevelt
James F. Byrnes	1941–1942	F. D. Roosevelt
Robert H. Jackson	1941–1954	F. D. Roosevelt
Wiley B. Rutledge	1943–1949	F. D. Roosevelt
Harold H. Burton	1945–1958	Truman
Tom C. Clark	1949–1967	Truman
Sherman Minton	1949–1956	Truman
John M. Harlan	1955–1971	Eisenhower
William J. Brennan, Jr.	1956–1990	Eisenhower
Charles E. Whittaker	1957–1962	Eisenhower
Potter Stewart	1958–1981	Eisenhower
Byron R. White	1962–1993	Kennedy
Arthur J. Goldberg	1962–1965	Kennedy
Abe Fortas	1965–1969	Johnson
Thurgood Marshall	1967–1991	Johnson
Harry A. Blackmun	1970–1994	Nixon
Lewis F. Powell, Jr.	1972–1987	Nixon
William H. Rehnquist	1972	Nixon
John P. Stevens	1975–	Ford
Sandra Day O'Connor	1981–	Reagan
Antonin Scalia	1986–	Reagan
Anthony M. Kennedy	1988–	Reagan
David H. Souter	1990–	Bush
Clarence Thomas	1991–	Bush
Ruth Bader Ginsburg	1993–	Clinton
Stephen G. Breyer	1994–	Clinton

☆ Chase's 1995 Calendar of Events ☆

STATE GOVERNORS/US SENATORS

GOVERNORS
Name (Party, State)

Jim Folsom, Jr (D, AL)*
Walter J. Hickel (I, AK)*
Fife Symington (R, AZ)*
Jim Guy Tucker, Jr (D, AR)*
Pete Wilson (R, CA)*
Roy Romer (D, CO)*
Lowell P. Weicker, Jr (CP, CT)*
Tom Carper (D, DE)
Lawton Chiles (D, FL)*
Zell Miller (D, GA)*
John Waihee III (D, HI)*
Cecil D. Andrus, (D, ID)*
Jim Edgar (R, IL)*
Evan Bayh (D, IN)
Terry R. Branstad (R, IA)*
Joan Finney (D, KS)*
Brereton C. Jones (D, KY)
Edwin W. Edwards (D, LA)
John R. McKernan, Jr (R, ME)*
William Donald Schaefer (D, MD)*
William F. Weld (R, MA)*
John Engler (R, MI)*
Arne H. Carlson (R, MN)*
Kirk Fordice (R, MS)
Mel Carnahan (D, MO)
Marc Racicot (R, MT)
E. Benjamin Nelson (D, NE)*
Bob Miller (D, NV)*
Stephen Merrill (R, NH)*
Christine T. Whitman (R, NJ)
Bruce King (D, NM)*
Mario M. Cuomo (D, NY)*
James B. Hunt, Jr (D, NC)
Edward T. Schafer (R, ND)
George V. Voinovich (R, OH)*
David Walters (D, OK)*
Barbara Roberts (D, OR)*
Robert P. Casey (D, PA)*
Bruce Sundlan (D, RI)*
Carroll A. Campbell, Jr (R, SC)*
Walter Dale Miller (R, SD)*
Ned Ray McWherter (D, TN)*
Ann W. Richards (D, TX)*
Mike Leavitt (R, UT)
Howard Dean (D, VT)*
George F. Allen (R, VA)
Mike Lowry (D, WA)*
Gaston Caperton (D, WV)
Tommy G. Thompson (R, WI)*
Michael Sullivan (D, WY)*

SENATORS
Name (Party, State)

Howell Heflin (D, AL)
Richard C. Shelby (D, AL)
Ted Stevens (R, AK)
Frank H. Murkowski (R, AK)
Dennis DeConcini (D, AZ)††
John McCain (R, AZ)
Dale L. Bumpers (D, AR)
David H. Pryor (D, AR)
Diane Feinstein (D, CA)†
Barbara Boxer (D, CA)
Hank Brown (R, CO)
Ben Nighthorse Campbell (D, CO)
Christopher J. Dodd (D, CT)
Joseph I. Lieberman (D, CT)†
William V. Roth, Jr (R, DE)†
Joseph R. Biden (D, DE)
Bob Graham (D, FL)
Connie Mack (R, FL)†
Sam Nunn (D, GA)
Paul D. Coverdell (R, GA)
Daniel K. Inouye (D, HI)
Daniel K. Akaka (D, HI)†
Larry E. Craig (R, ID)
Dirk Kempthorne (R, ID)
Paul Simon (D, IL)
Carol Mosely-Braun (D, IL)
Richard G. Lugar (R, IN)†
Dan Coats (R, IN)
Charles E. Grassley (R, IA)
Tom Harkin (D, IA)
Robert J. Dole (R, KS)
Nancy L. Kassebaum (R, KS)
Wendell H. Ford (D, KY)
Mitch McConnell (R, KY)
J. Bennett Johnston (D, LA)
John B. Breaux (D, LA)
William S. Cohen (R, ME)
George J. Mitchell (D, ME)††
Paul S. Sarbanes (D, MD)†
Barbara A. Mikulski (D, MD)
Edward M. Kennedy (D, MA)†
John F. Kerry (D, MA)
Donald W. Riegle, Jr (D, MI)††
Carl Levin (D, MI)
Dave Durenberger (IR, MN)††
Paul D. Wellstone (DFL, MN)
Thad Cochran (R, MS)
Trent Lott (R, MS)†
John C. Danforth (R, MO)††
Christopher S. Bond (R, MO)

SENATORS
Name (Party, State)

Max S. Baucus (D, MT)
Conrad Burns (R, MT)†
J. James Exon (D, NE)
Bob Kerrey (D, NE)†
Harry M. Reid (D, NV)
Richard H. Bryan (D, NV)†
Robert C. Smith (R, NH)
Judd Gregg (R, NH)
Bill Bradley (D, NJ)
Frank R. Lautenberg (D, NJ)†
Pete V. Domenici (R, NM)
Jeff Bingaman (D, NM)†
Daniel Patrick Moynihan (D, NY)†
Alfonse M. D'Amato (R, NY)
Jesse A. Helms (R, NC)
Lauch Faircloth (R, NC)
Kent Conrad (D, ND)†
Byron L. Dorgan (D, ND)
John Glenn (D, OH)
Howard M. Metzenbaum (D, OH)††
David L. Boren (D, OK)
Don Nickles (R, OK)
Mark O. Hatfield (R, OR)
Bob Packwood (R, OR)
Arlen Specter (R, PA)
Harris Wofford (D, PA)†
Claiborne Pell (D, RI)
John H. Chafee (R, RI)†
Strom Thurmond (R, SC)
Ernest F. Hollings (D, SC)
Larry Pressler (R, SD)
Thomas A. Daschle (D, SD)
Jim Sasser (D, TN)†
Harlan Mathews (D, TN)††
Phil Gramm (R, TX)
Kay Bailey Hutchinson (R, TX)†
Orrin G. Hatch (R, UT)†
Robert F. Bennett (R, UT)
Patrick J. Leahy (D, VT)
James M. Jeffords (R, VT)†
John W. Warner (R, VA)
Charles S. Robb (D, VA)†
Slade Gorton (R, WA)†
Patty Murray (D, WA)
Robert C. Byrd (D, WV)†
John D. Rockefeller (D, WV)
Herb Kohl (D, WI)†
Russell D. Feingold (D, WI)
Malcolm Wallop (R, WY)††
Alan K. Simpson (R, WY)

*Denotes governorship up for reelection at press time.
†Denotes senate seat up for reelection at press time.
††Denotes senators retiring.

★ *Chase's 1995 Calendar of Events* ★
THE NAMING OF HURRICANES

(Compiled from information issued by the US Department of Commerce, National Oceanic and Atmospheric Administration.)

Why are hurricanes named? Experience shows that the use of short, distinctive given names greatly reduces confusion when two or more tropic storms occur at the same time. The use of easily remembered names in written and spoken communications is quicker and less subject to error than the older, more cumbersome latitude-longitude identification methods, advantages which are especially important in exchanging detailed storm information between hundreds of widely scattered stations, airports, coastal bases and ships at sea.

The practice of naming hurricanes began hundreds of years ago, but only relatively recently did they begin to be named solely for women. During World War II forecasters and meteorologists began using female names for storms in weather map discussions, and in 1953 the US weather services adopted the practice, creating a new international phonetic alphabet of women's names from A-W to name hurricanes. The practice came to an end in 1978 when men's names were introduced into the storm lists also.

Because hurricanes affect other nations and are tracked by their weather services, the lists have an international flavor. Names for are selected from library sources and agreed upon during international meetings of the World Meteorological Organization by nations involved, and can be retired and replaced with new names in the event of particularly severe storms.

The National Hurricane Center near Miami, FL keeps a constant watch on oceanic storm-breeding areas for tropical disturbances which may herald the formation of a hurricane. If a disturbance intensifies into a tropical storm—with rotary circulation and wind speeds above 39 miles per hour—the Center will give the storm a name from one of the six lists below. The lists are rotated year by year so that the 1995 set, for example, will be used again to name storms in 2001.

ATLANTIC HURRICANE NAMES

1995	1996	1997	1998	1999	2000
ALLISON	ARTHUR	ANA	ALEX	ARLENE	ALBERTO
BARRY	BERTHA	BILL	BONNIE	BRET	BERYL
CHANTAL	CESAR	CLAUDETTE	CHARLEY	CINDY	CHRIS
DEAN	DOLLY	DANNY	DANIELLE	DENNIS	DEBBY
ERIN	EDOUARD	ERIKA	EARL	EMILY	ERNESTO
FELIX	FRAN	FABIAN	FRANCES	FLOYD	FLORENCE
GABRIELLE	GUSTAV	GRACE	GEORGES	GERT	GORDON
HUMBERTO	HORTENSE	HENRI	HERMINE	HARVEY	HELENE
IRIS	ISIDORE	ISABEL	IVAN	IRENE	ISAAC
JERRY	JOSEPHINE	JUAN	JEANNE	JOSE	JOYCE
KAREN	KYLE	KATE	KARL	KATRINA	KEITH
LUIS	LILI	LARRY	LISA	LENNY	LESLIE
MARILYN	MARCO	MINDY	MITCH	MARIA	MICHAEL
NOEL	NANA	NICHOLAS	NICOLE	NATE	NADINE
OPAL	OMAR	ODETTE	OTTO	OPHELIA	OSCAR
PABLO	PALOMA	PETER	PAULA	PHILIPPE	PATTY
ROXANNE	RENE	ROSE	RICHARD	RITA	RAFAEL
SEBASTIEN	SALLY	SAM	SHARY	STAN	SANDY
TANYA	TEDDY	TERESA	TOMAS	TAMMY	TONY
VAN	VICKY	VICTOR	VIRGINIE	VINCE	VALERIE
WENDY	WILFRED	WANDA	WALTER	WILMA	WILLIAM

EASTERN PACIFIC HURRICANE NAMES

1995	1996	1997	1998	1999	2000
ADOLPH	ALMA	ANDRES	AGATHA	ADRIAN	ALETTA
BARBARA	BORIS	BLANCA	BLAS	BEATRIZ	BUD
COSME	CHRISTINA	CARLOS	CELIA	CALVIN	CARLOTTA
DALILA	DOUGLAS	DOLORES	DARBY	DORA	DANIEL
ERICK	ELIDA	ENRIQUE	ESTELLE	EUGENE	EMILIA
FLOSSIE	FAUSTO	FELICIA	FRANK	FERNANDA	FABIO
GIL	GENEVIEVE	GUILLERMO	GEORGETTE	GREG	GILMA
HENRIETTE	HERNAN	HILDA	HOWARD	HILARY	HECTOR
ISMAEL	ISELLE	IGNACIO	ISIS	IRWIN	ILEANA
JULIETTE	JULIO	JIMENA	JAVIER	JOVA	JOHN
KIKO	KENNA	KEVIN	KAY	KENNETH	KRISTY
LORENA	LOWELL	LINDA	LESTER	LIDIA	LANE
MANUEL	MARIE	MARTY	MADELINE	MAX	MIRIAM
NARDA	NORBERT	NORA	NEWTON	NORMA	NORMAN
OCTAVE	ODILE	OLAF	ORLENE	OTIS	OLIVIA
PRISCILLA	POLO	PAULINE	PAINE	PILAR	PAUL
RAYMOND	RACHEL	RICK	ROSLYN	RAMON	ROSA
SONIA	SIMON	SANDRA	SEYMOUR	SELMA	SERGIO
TICO	TRUDY	TERRY	TINA	TODD	TARA
VELMA	VANCE	VIVIAN	VIRGIL	VERONICA	VICENTE
WALLIS	WINNIE	WALDO	WINIFRED	WILEY	WILLA
XINA	XAVIER	XINA	XAVIER	XINA	XAVIER
YORK	YOLANDA	YORK	YOLANDA	YORK	YOLANDA
ZELDA	ZEKE	ZELDA	ZEKE	ZELDA	ZEKE

If over 24 tropical cyclones occur in a year, then the Greek alphabet will be used following Zeke or Zelda.

☆ *Chase's 1995 Calendar of Events* ☆
THE NAMING OF HURRICANES (cont'd)

The four lists of names for Central Pacific and Western Pacific hurricanes (tropical cyclones) are not rotated on a yearly basis. Meteorologists follow each list until all those names have been used, then go on to the next list. They do go back to List 1 after List 4, and as with Atlantic and Eastern Pacific lists, the name of a particularly severe storm is retired and replaced. Iniki on List 2—the name of the hurricane that devastated Hawaii—has been replaced with Iolona.

CENTRAL PACIFIC TROPICAL CYCLONE NAMES

LIST 1		LIST 2	
AKONI	ah-KOH-nee	AKA	AH-kah
EMA	EH-mah	EKEKA	eh-KEH-kak
HANA	HAH-nah	HALI	HAH-lee
IO	EE-oo	IOLONA	ee-OH-lah-nah
KELI	KEH-lee	KEONI	keh-OH-nee
LALA	LAH-lah	*LI	LEE
MOKE	MOH-keh	MELE	MEH-leh
NELE	NEH-leh	NONA	NOH-nah
OKA	OH-kah	OLIWA	oh-LEE-vah
PEKE	PEH-keh	PAKA	PAH-kah
ULEKI	oo-LEH-kee	UPANA	oo-PAH-nah
WILA	VEE-lah	WENE	WEH-neh

LIST 3		LIST 4	
ALIKA	ah-LEE-kah	ANA	AH-nah
ELE	EH-leh	ELA	EH-lah
HUKO	HOO-koh	HALOLA	hah-LOH-lah
IOKE	ee-OH-keh	IUNE	ee-OO-neh
KIKA	KEE-kah	KIMO	KEE-mo
LANA	LAH-na	LOKE	LOH-keh
MAKA	MAH-kah	MALIA	mah-LEE-ah
NEKI	NEH-kee	NIALA	nee-AH-lah
OLEKA	oh-LEH-kah	OKO	OH-koh
PENI	PEH-nee	PALI	PAH-lee
ULIA	oo-LEE-ah	ULIKA	oo-LEE-kah
WALI	WAH-lee	WALAKA	wah-LAH-kah

All letters in the Hawaiian language are pronounced, including double or triple vowels.
*As of Aug 11, 1994, Central Pacific Tropical Cyclone names had been used through Li.

WESTERN PACIFIC TROPICAL CYCLONE NAMES

LIST 1	LIST 2	LIST 3	LIST 4
ANGELA	ABE	AMY	AXEL
BRIAN	BECKY	BRENDAN	BOBBIE
COLLEEN	CECIL	CAITLIN	CHUCK
DAN	DOT	*DOUG	DEANNA
ELSIE	ED	ELLIE	ELI
FORREST	FLO	FRED	FAYE
GAY	GENE	GLADYS	GARY
HUNT	HATTIE	HARRY	HELEN
IRMA	IRA	IVY	IRVING
JACK	JEANA	JOEL	JANIS
KORYN	KYLE	KINNA	KENT
LEWIS	LOLA	LUKE	LOIS
MARIAN	MANNY	MELISSA	MARK
NATHAN	NELL	NAT	NINA
OFELIA	OWEN	ORCHID	OSCAR
PERCY	PAGE	PAT	POLLY
ROBYN	RUSS	RUTH	RYAN
STEVE	SHARON	SEITH	SIBYL
TASHA	TIM	TERESA	TED
VERNON	VANESSA	VERNE	VAL
WINONA	WALT	WILDA	WARD
YANCY	YUNYA	YURI	YVETTE
ZOLA	ZEKE	ZELDA	ZACK

*As of Aug 11, 1994, Western Pacific Tropical Cyclone names had been used through Doug.

☆ Chase's 1995 Calendar of Events ☆
NATIONAL FILM REGISTRY 1989-1993

The National Film Preservation Act of 1988 (Public Law 100-446) called for the development of a comprehensive national film preservation program. Originally authorized to expire on Sept 27, 1992, it was extended by the 1992 Act (Public Law 102-307) to June 26, 1996. The goal of the National Film Preservation Act is to develop a comprehensive national film preservation program by first studying the current efforts to preserve film nationwide. This is being done in conjunction with the other film archives, educators and film industry representatives. It is the intention of the *National Film Board and Librarian* to assist in the coordination of film preservation after this study.

The Board consists of 18 members including individuals from the film industry who are selecting films for the Registry at the rate of 25 a year. One copy of each film selected is collected "in archival quality" by the Library of Congress for the National Film Registry Collection. Films have been selected on the basis of historical, cultural and aesthetic significance. A film must be at least 10 years old to be considered.

FILMS SELECTED FOR THE NATIONAL FILM REGISTRY, 1989-1993
(Titles listed alphabetically)

1. *Adam's Rib* (1949)
2. *All About Eve* (1950)
3. *All Quiet on the Western Front* (1930)
4. *An American in Paris* (1951)
5. *Annie Hall* (1977)
6. *Badlands* (1973)
7. *The Bank Dick* (1940)
8. *The Battle of San Pietro* (1945)
9. *The Best Years of Our Lives* (1946)
10. *Big Business* (1929)
11. *The Big Parade* (1925)
12. *The Birth of a Nation* (1915)
13. *The Black Pirate* (1926)
14. *Blade Runner* (1982)
15. *The Blood of Jesus* (1941)
16. *Bonnie and Clyde* (1967)
17. *Bringing Up Baby* (1938)
18. *Carmen Jones* (1954)
19. *Casablanca* (1942)
20. *Castro Street* (1966)
21. *Cat People* (1942)
22. *The Cheat* (1915)
23. *Chinatown* (1974)
24. *Chulas Fronteras* (1976)
25. *Citizen Kane* (1941)
26. *City Lights* (1931)
27. *The Crowd* (1928)
28. *David Holzman's Diary* (1968)
29. *Detour* (1946)
30. *Dodsworth* (1936)
31. *Dog Star Man* (1964)
32. *Double Indemnity* (1944)
33. *Dr. Strangelove, (or How I Learned to Stop Worrying and Love the Bomb)* (1964)
34. *Duck Soup* (1933)
35. *Eaux d'Artifice* (1953)
36. *Fantasia* (1940)
37. *Footlight Parade* (1933)
38. *Frankenstein* (1931)
39. *The Freshman* (1925)
40. *The General* (1927)
41. *Gertie the Dinosaur* (1914)
42. *Gigi* (1958)
43. *The Godfather* (1972)
44. *The Godfather, Part II* (1974)
45. *The Gold Rush* (1925)
46. *Gone with the Wind* (1939)
47. *The Grapes of Wrath* (1940)
48. *The Great Train Robbery* (1903)
49. *Greed* (1924)
50. *Harlan County, U.S.A.* (1976)
51. *High Noon* (1952)
52. *High School* (1968)
53. *His Girl Friday* (1940)
54. *How Green Was My Valley* (1941)
55. *I Am a Fugitive from a Chain Gang* (1932)
56. *Intolerance* (1916)
57. *It Happened One Night* (1934)
58. *It's a Wonderful Life* (1946)
59. *The Italian* (1915)
60. *Killer of Sheep* (1977)
61. *King Kong* (1933)
62. *Lassie Come Home* (1943)
63. *Lawrence of Arabia* (1962)
64. *The Learning Tree* (1969)
65. *Letter from an Unknown Woman* (1948)
66. *Love Me Tonight* (1932)
67. *Magical Maestro* (1952)
68. *The Magnificent Ambersons* (1942)
69. *The Maltese Falcon* (1941)
70. *March of Time: Inside Nazi Germany—1938* (1938)
71. *Meshes of the Afternoon* (1943)
72. *Modern Times* (1936)
73. *Morocco* (1930)
74. *Mr. Smith Goes to Washington* (1939)
75. *My Darling Clementine* (1946)
76. *Nanook of the North* (1922)
77. *Nashville* (1975)
78. *The Night of the Hunter* (1955)
79. *A Night at the Opera* (1935)
80. *Ninotchka* (1939)
81. *Nothing But a Man* (1964)
82. *On the Waterfront* (1954)
83. *One Flew Over the Cuckoo's Nest* (1975)
84. *Out of the Past* (1947)
85. *Paths of Glory* (1957)
86. *A Place in the Sun* (1951)
87. *Point of Order* (1964)
88. *The Poor Little Rich Girl* (1917)
89. *Primary* (1960)
90. *The Prisoner of Zenda* (1937)
91. *Psycho* (1960)
92. *Raging Bull* (1980)
93. *Rebel Without a Cause* (1955)
94. *Red River* (1948)
95. *Ride the High Country* (1962)
96. *The River* (1937)
97. *Salesman* (1969)
98. *Salt of the Earth* (1954)
99. *The Searchers* (1956)
100. *Shadow of a Doubt* (1943)
101. *Shadows* (1959)
102. *Shane* (1953)
103. *Sherlock, Jr.* (1924)
104. *Singin' in the Rain* (1952)
105. *Snow White and the Seven Dwarfs* (1937)
106. *Some Like It Hot* (1959)
107. *Star Wars* (1977)
108. *Sullivan's Travels* (1941)
109. *Sunrise* (1927)
110. *Sunset Boulevard* (1950)
111. *Sweet Smell of Success* (1957)
112. *Tevya* (1939)
113. *Top Hat* (1935)
114. *Touch of Evil* (1958)
115. *The Treasure of the Sierra Madre* (1948)
116. *Trouble in Paradise* (1932)
117. *2001: A Space Odyssey* (1968)
118. *Vertigo* (1958)
119. *What's Opera, Doc?* (1957)
120. *Where Are My Children?* (1916)
121. *The Wind* (1928)
122. *Within Our Gates* (1920)
123. *The Wizard of Oz* (1939)
124. *A Woman Under the Influence* (1974)
125. *Yankee Doodle Dandy* (1942)

☆ Chase's 1995 Calendar of Events ☆

THE ACADEMY OF TELEVISION ARTS AND SCIENCES (ATAS) TELEVISION HALL OF FAME

The Academy of Television Arts & Sciences Television Academy Hall of Fame was established in 1984 to recognize the lifelong accomplishments of television's greatest contributors, including, in addition to those who appear before the camera, those who contribute to the industry as writers, executives and producers.

Bronze sculptures and bas reliefs of some of the Hall of Fame's inductees have been erected in the Hall of Fame Plaza. Located in the forecourt of the Television Academy's North Hollywood international headquarters, the plaza's centerpiece is a 27-foot Emmy statue, a replica of the internationally famous Emmy statuette, symbol of television excellence.

1984
Lucille Ball
Milton Berle
Paddy Chayefsky
Norman Lear
Edward R. Murrow
Edward S. Paley
General David Sarnoff

1985
Carol Burnett
Sid Caesar
Walter Cronkite
Joyce C. Hall
Rod Serling
Ed Sullivan
Sylvester "Pat" Weaver

1986
Steve Allen
Fred Coe
Walt Disney
Jackie Gleason
Mary Tyler Moore
Frank Stanton
Burr Tillstrom

1987
Johnny Carson
Jaques-Yves Cousteau
Leonard Goldenson
Jim Henson
Bob Hope
Ernie Kovacs

1988
Jack Benny
George Burns
Gracie Allen
Chet Huntley
David Brinkley
Red Skelton
David Susskind
David Wolper

1989
Roone Arledge
Fred Astaire
Perry Como
Joan Ganz Cooney
Don Hewitt
Carroll O'Connor
Barbara Walters

1990
Desi Arnaz
Leonard Bernstein
James Garner
"I Love Lucy"
Danny Thomas
Mike Wallace

1991
Bill Cosby
Andy Griffith
Ted Koppel
Sheldon Leonard
Dinah Shore
Ted Turner

1992
Dick Clark
John Chancellor
Phil Donahue
Mark Goodson
Bob Newhart
Agnes Nixon
Jack Webb

1993
Alan Alda
Howard Cosell
Barry Diller
Fred Friendly
William Hanna/Joseph Barbera
Oprah Winfrey

☆ Chase's 1995 Calendar of Events ☆

MAJOR AWARDS

1993 ACADEMY AWARDS (OSCARS)

Picture: *Schindler's List*
Actor: Tom Hanks, *Philadelphia*
Actress: Holly Hunter, *The Piano*
Director: Steven Spielberg, *Schindler's List*
Supporting Actor: Tommy Lee Jones, *The Fugitive*
Supporting Actress: Anna Paquin, *The Piano*
Original Screenplay: Jane Campion, *The Piano*
Adapted Screenplay: Steven Zaillian, *Schindler's List*
Cinematography: Janusz Kaminski, *Schindler's List*
Art Direction: Allan Starski and Ewa Braun, *Schindler's List*
Film Editing: Michael Kahn, *Schindler's List*
Foreign Language Film: *Belle Epoque*, Spain
Animated Short Film: Nicholas Park, *The Wrong Trousers*
Documentary Feature: Susan Raymond and Alan Raymond, *I Am a Promise*
Documentary Short Subject: Margaret Lazarus and Renner Wunderlich, *Defending Our Lives*
Costume Design: Gabriella Pescucci, *The Age of Innocence*
Sound Design: Gary Summers, Gary Rydstrom, Shawn Murphy and Ron Judkins, *Jurassic Park*
Visual Effects: Dennis Muren, Stan Winston, Phil Tippett and Michael Lantieri, *Jurassic Park*
Makeup: Greg Cannom, Ve Neill and Yolanda Toussieng, *Mrs. Doubtfire*
Sound Effects Editing: Gary Rydstrom and Richard Hymns, *Jurassic Park*
Live Action Short Film: Pepe Danquart, *Black Rider*
Original Score: John Williams, *Schindler's List*
Original Song: Bruce Springsteen, "Streets of Philadelphia," *Philadelphia*

1993-94 SEASON TONY AWARDS

Play: *Angels in America: Perestroika*
Musical: *Passion*
Revival, play: *An Inspector Calls*
Revival, musical: *Carousel*
Actor, play: Stephen Spinella, *Angels in America: Perestroika*
Actress, play: Diana Rigg, *Medea*
Actor, musical: Boyd Gaines, *She Loves Me*
Actress, musical: Donna Murphy, *Passion*
Book, musical: *Passion*, James Lapine
Score, musical: *Passion*, Stephen Sondheim
Director, play: Stephen Daldry, *An Inspector Calls*
Director, musical: Nicholas Hytner, *Carousel*
Featured Actor, play: Jeffrey Wright, *Angels in America: Perestroika*
Featured Actress, play: Jane Adams, *An Inspector Calls*
Featured Actor, musical: Jarrod Emick, *Damn Yankees*
Featured Actress, musical: Audra Ann McDonald, *Carousel*
Scenic Design: Bob Crowley, *Carousel*
Costume Design: Ann Hould-Ward, *Beauty and the Beast*
Lighting Design: Rick Fisher, *An Inspector Calls*
Choreography: Sir Kenneth MacMillan, *Carousel*
Special Award to a Regional Theater: McCarter Theatre of Princeton, NJ
Special Award for Lifetime Achievement: Jessica Tandy and Hume Cronyn

1993 AMERICAN MUSIC AWARDS
(Awarded Feb 7, 1994)
21ST Annual Awards

POP/ROCK
Male Vocalist: Eric Clapton
Female Vocalist: Whitney Houston
Band, Duo or Group: Aerosmith
Single Record: "I Will Always Love You," Whitney Houston
Album: *The Bodyguard*, Whitney Houston
New Artist: Stone Temple Pilots

SOUL/RHYTHM & BLUES
Male Vocalist: Luther Vandross
Female Vocalist: Whitney Houston
Band, Duo or Group: En Vogue
Single Record: "I Will Always Love You," Whitney Houston
Album: *The Bodyguard*, Whitney Houston
New Artist: Toni Braxton

COUNTRY
Male Vocalist: Garth Brooks
Female Vocalist: Reba McEntire
Band, Duo or Group: Alabama
Single: "Chattahoochee," Alan Jackson
Album: *A Lot About Livin' (And a Little 'Bout Love)*, Alan Jackson
New Artist: John Michael Montgomery

HEAVY METAL/HARD ROCK
Artist: Aerosmith
New Artist: Stone Temple Pilots

RAP/HIP HOP
Artist: Dr. Dre
New Artist: Dr. Dre

ADULT CONTEMPORARY
Artist: Kenny G
New Artist: Toni Braxton
Album: *The Bodyguard*, Whitney Houston

☆ Chase's 1995 Calendar of Events ☆

1993 GRAMMY AWARDS

Record of the Year: "I Will Always Love You," Whitney Houston, David Foster, producer
Album of the Year: *The Bodyguard*, Whitney Houston; David Foster, Narada Michael Walden, La Reid, Babyface, Whitney Houston, David Cole, Robert Clivilles & Bebe Winans, producers
Song of the Year: "A Whole New World," Alan Menken, Tim Rice, songwriters
Best New Artist: Toni Braxton
Best Pop Vocal Performance, Female: "I Will Always Love You," Whitney Houston
Best Pop Vocal Performance, Male: "If I Ever Lose My Faith in You," Sting
Best Pop Performance by a Duo or Group with Vocal: "A Whole New World" (Aladdin's Theme), Peabo Bryson and Regina Belle
Best Traditional Pop Performance: *Steppin' Out*, Tony Bennett
Best Pop Instrumental Performance: "Barcelona Mona," Bruce Hornsby and Branford Marsalis
Best Rock Vocal Performance, Female: Not given (combined with Best Rock Vocal Performance Solo)
Best Rock Vocal Performance, Male: Not given (combined with Best Rock Vocal Performance Solo)
Best Rock Vocal Performance, Solo: "I'd Do Anything for Love (But I Won't Do That)," Meat Loaf
Best Rock Performance by a Duo or Group with Vocal: "Livin' on the Edge," Aerosmith
Best Hard Rock Performance with Vocal: "Plush" (Track from *Core*), Stone Temple Pilots
Best Metal Performance with Vocal: "I Don't Want to Change the World" (Track from *Live & Loud*), Ozzy Osbourne
Best Rock Instrumental Performance: "Sofa" (Track from *Zappa's Universe*—conducted and arranged by Joel Thorne), Zappa's Universe Rock Group featuring Steve Vai
Best Rock Song: "Runaway Train," David Pirner, songwriter
Best Alternative Music Album: *Zooropa*, U2
Best R&B Vocal Performance, Female: *Another Sad Love Song*, Toni Braxton
Best R&B Vocal Performance, Male: *A Song for You*, Ray Charles
Best R&B Performance by a Duo or Group with Vocal: *No Ordinary Love*, Sade
Best R&B Instrumental Performance: Not given
Best Rhythm & Blues Song: "That's the Way Love Goes," Janet Jackson, James Harris, III and Terry Lewis, songwriters
Best Rap Solo Performance: "Let Me Ride," Dr. Dre
Best Rap Performance by a Duo or Group: "Rebirth of Slick (Cool Like Dat)," Digable Planets
Best New Age Album: *Spanish Angel*, Paul Winter Consort
Best Contemporary Jazz Performance, Instrumental: *The Road to You*, Pat Metheny Group
Best Jazz Vocal Performance: *Take a Look*, Natalie Cole
Best Jazz Instrumental Solo: "Miles Ahead" (Track from *So Near, So Far: Musings for Miles*), Joe Henderson
Best Jazz Instrumental Performance, Individual or Group: *So Near, So Far (Musings for Miles)*, Joe Henderson
Best Large Jazz Ensemble Performance: *Miles and Quincy Live at Montreux*, Miles Davis and Quincy Jones
Best Country Vocal Performance, Female: "Passionate Kisses," Mary Chapin-Carpenter
Best Country Vocal Performance, Male: "Ain't That Lonely Yet," Dwight Yoakam
Best Country Performance by a Duo or Group with Vocal: "Hard Workin' Man," Brooks and Dunn
Best Country Vocal Collaboration: "Does He Love You," Reba McEntire and Linda Davis
Best Country Instrumental Performance: "Red Wing," Asleep at the Wheel featuring Eldon Bhamblin, Johnny Gumble, Chet Atkins, Vince Gill, Marty Stuart and Reuben "Lucky Oceans" Gosfield
Best Bluegrass Album: *Waitin' for the Hard Times to Go*, The Nashville Bluegrass Band
Best Country Song: "Passionate Kisses," Lucinda Williams, songwriters
Best Rock Gospel Album: *Free at Last*, DC Talk

Best Pop/Contemporary Gospel Album: *The Live Adventure*, Steven Curtis Chapman
Best Southern Gospel or Bluegrass Gospel Album: *Good News*, Kathy Mattea
Best Traditional Soul Gospel Album: *Stand Still*, Shirley Caesar
Best Contemporary Soul Gospel Album: *All Out*, The Winans
Best Gospel Album by a Choir or Chorus: *Live . . . We Come Rejoicing* (Brooklyn Tabernacle Choir), Carol Cymbala, choir director
Best Latin Pop Album: *Aries*, Luis Miguel
Best Tropical Latin Album: *Mi Tierra*, Gloria Estefan
Best Mexican/American Album: *Live*, Selena
Best Traditional Blues Album: *Blues Summit*, B. B. King
Best Contemporary Blues Album: *Feels Like Rain*, Buddy Guy
Best Traditional Folk Album: *The Celtic Harp*, The Chieftains
Best Contemporary Folk Album: *Other Voices/Other Rooms*, Nanci Griffith
Best Reggae Album: *Bad Boys*, Inner Circle
Best World Music Album: *A Meeting by the River*, Ry Cooder and V.M. Shatt
Best Polka Album: *Accordionally Yours*, Walter Ostanek and His Band
Best Musical Album for Children: *Aladdin* (Original Motion Picture Soundtrack), (various); Alan Menken and Tim Rice, producers
Best Spoken Word Album for Children: *Audrey Hepburn's Enchanted Tales*, Audrey Hepburn, Deborah Raffin and Michael Viner, producers
Best Spoken Word or Non-Musical Album: *On the Pulse of Morning*, Maya Angelou
Best Spoken Comedy Album: *Jammin' in New York*, George Carlin
Best Musical Show Album: *The Who's Tommy—Original Cast Recording*, George Martin, album producer and Pete Townshend, composer and lyricist
Best Instrumental Compositon: "Forever In Love," Kenny G., composer
Best Instrumental Composition Written for a Motion Picture or Television: *Aladdin*, Alan Menken, composer
Best Song Written Specifically for a Motion Picture or for Television: "A Whole New World" (Aladdin's Theme) (From *Aladdin*), Alan Menken and Tim Rice, songwriters
Best Music Video—Short Form: "Steam," Peter Gabriel; Stephen R. Johnson, video director; Prudence Fenton, video producer

Best Music Video—Long Form: "Ten Summoner's Tales," Sting; Doug Nichol, video director; Julie Fong, video producer
Best Arrangement on an Instrumental: "Mood Indigo," Dave Grubin, arranger
Best Instrumental Arrangement Accompanying Vocal(s): "When I Fall in Love" (Track from *Sleepless in Seattle* Soundtrack), Jeremy Lubbock and David Foster, arrangers
Best Recording Package: "The Complete Billie Holiday on Verve 1945-1956," (Billie Holiday) David Lau, art director
Best Album Notes: "The Complete Billie Holiday on Verve 1945-1956," (Billie Holiday) Buck Clayton, Phil Schaap and Joel E. Siegel, album notes writers
Best Historical Album: "The Complete Billie Holiday on Verve 1945-1956," (Billie Holiday) Michael Lang and Phil Schaap, producers
Best Engineered Album (Non-Classical): *Ten Summoner's Tales*, Sting; Hugh Padgham, engineer
Producer of the Year (Non-Classical): David Foster
Best Classical Album: *Bartok: The Wooden Prince and Cantata Profana*, Pierre Boulez, conductor; Chicago Symphony Orchestra and Chorus; John Ale John Tomlinson, baritone; Karl-August Naegler, producer

Continued

☆ Chase's 1995 Calendar of Events ☆

MAJOR AWARDS (cont'd)

Best Orchestral Performance: Bartok: *The Wooden Prince*, Pierre Boulez, conductor; Chicago Symphony Orchestra
Best Opera Recording: Handel: *Semele*; John Nelson, conductor; English Chamber Orchestra and Ambrosian Opera Chorus; Prin. Solos: Battle, Horne, Ramey, Aler, McNair, Chance, Mackie, Doss; Dr. Steven Paul, producer
Best Performance of a Choral Work: Bartok: *Cantata Profana*; Pierre Boulez, conductor; Chicago Symphony Orchestra and Chorus; Margaret Hillis, choral director
Best Classical Performance—Instrumental Soloist(s) (with Orchestra): Berg: Violin Concerto/Rihm: Time Chant; Anne-Sophie Mutter, violin; James Levine, conductor; Chicago Symphony Orchestra
Best Classical Performance—Instrumental Soloist (Without Orchestra): Barber: The Complete Solo Piano Music; John Browning, piano
Best Chamber Music Performance: Ives: String Quartet Nos. 1 & 2/Barber: String Quartet Op. 11; Emerson String Quartet
Best Classical Vocal Performance: *The Art of Arleen Auger* (Larsen, Purcell, Schumann, Mozart); Arleen Auger, soprano
Best Contemporary Composition: Carter: Violin Concerto; Elliott Carter, composer
Best Engineered Recording, Classical: Bartok: *The Wooden Prince and Cantata Profana*; Pierre Boulez, conductor; Chicago Symphony Orchestra, chorus and soloists; Rainer Maillard, engineer
Classical Producer of the Year: Judith Sherman

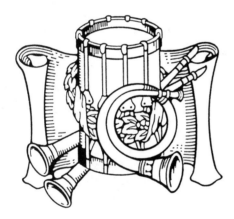

1993-94 AMERICAN BLACK ACHIEVEMENT AWARDS

Lifetime Achievement Award: Ray Charles
Career Achievement Award: Denzel Washington
Thurgood Marshall Black Education Fund Educational Achievement Award: Dr. William DeLauder
Martin Luther King, Jr., Award for Public Service: Dr. Benjamin Chavis
Jackie Robinson Award for Athletics: Michael Jordan
Business and Professional Award: Dr. David Satcher
Dramatic Arts Award: Laurence Fishburne and Angela Bassett
Fine Arts Award: Maya Angelou
Music Award: Janet Jackson
Religion Award: The Rev. Dr. Cecil Murray
Trailblazer Award: Secretary Jesse Brown and Secretary Michael Espy

25TH ANNUAL DOVE AWARDS GOSPEL MUSIC ASSOCIATION

Song of the Year: "In Christ Alone," Shawn Craig and Don Koch
Songwriter of the Year: Steven Curtis Chapman
Male Vocalist of the Year: Michael English
Female Vocalist of the Year: Twila Paris
Group of the Year: 4HIM
Artist of the Year: Michael English
New Artist of the Year: Point of Grace
Producer of the Year: Wayne Kirkpatrick
Rap Recorded Song of the Year: "Socially Acceptable," DC Talk
Metal Recorded Song of the Year: "Psychedelic Super Jesus," Bride
Rock Recorded Song of the Year: "Jesus is Just Alright," DC Talk
Contemporary Recorded Song of the Year: "Go There With You," Steven Curtis Chapman
Inspirational Recorded Song of the Year: "Holding Out Hope To You," Michael English
Instrumental Album of the Year: *Psalms, Hymns and Spiritual Songs*, Kurt Kaiser
Praise and Worship Album of the Year: *Songs from the Loft*, Susan Ashton, Gary Chapman, Ashley Cleveland, Amy Delaine, Amy Grant, Kim Hill, Wes King, Michael James, Donna McElroy and Michael W. Smith
Musical Album of the Year: *God With Us*, Don Moen
Choral Collection Album of the Year: *Al Denson Presents the Youth Chorus Book, Vol. III*, Al Denson
Children's Music Album of the Year: *Come to the Cradle*, Michael Card
Recorded Music Packaging of the Year: *The Wonder Years 1983-1993*, Michael W. Smith; creative artists, D. Rhodes, Buddy Jackson, Beth Middleworth and Mark Tucker
Short Form Video of the Year: "Hand on My Shoulder," Sandi Patti; producers and directors, Jack Clark, Stephen Yake
Long Form Music Video of the Year: "The Live Adventure," Steven Curtis Chapman; producers and directors, Bret Wolcott, Douglas C. Forbes, Michael Salomon
Southern Gospel Recorded Song of the Year: "Satisfied," The Gaither Vocal Band
Country Recorded Song of the Year: "There But for the Grace of God Go I," Paul Overstreet
Contemporary Black Gospel Recorded Song: "Sold Out," Helen Baylor
Traditional Black Gospel Recorded Song of the Year: "Why We Sing," Kirk Franklin
Metal Album of the Year: *Tamplin*, Ken Tamplin
Rock Album of the Year: *Wake-Up Call*, Petra
Contemporary Album of the Year: *Hope*, Michael English
Inspirational Album of the Year: *The Season of Love*, 4HIM
Southern Gospel Album of the Year: *Southern Classics*, The Gaither Vocal Band
Country Album of the Year: *Walk On*, Bruce Carroll
Contemporary Black Gospel Album of the Year: *Start All Over*, Helen Baylor
Traditional Black Gospel Album of the Year: *Kirk Franklin and the Family*, Kirk Franklin

☆ Chase's 1995 Calendar of Events ☆

1993 ACE AWARDS

Dramatic Series: "Avonlea," The Disney Channel
Comedy Series: "The Larry Sanders Show," HBO
Movie or Miniseries: "The Positively True Adventures of the Alleged Texas Cheerleader-Murdering Mom," HBO
Comedy Special: "HBO Comedy Hour: John Leguizamo's 'Spic-o-Rama,' " HBO
Actor in a Comedy Series: Rip Torn, "The Larry Sanders Show," HBO
Actress in a Comedy Series: Wendie Malick, "Dream On," HBO
Writing a Comedy Series: Maya Forbes, Victor Levin, Drake Sather, Garry Shandling and Paul Simms, "The Larry Sanders Show: Larry's Agent," HBO
Directing a Comedy Series: Todd Holland, "The Larry Sanders Show: The Guest Host," HBO
Actor in a Dramatic Series: Gary Oldman, "Fallen Angels: Dead End for Delia," Showtime
Actress in a Dramatic Series: Mariangela Pino, "The Showtime 30-Minute Movie: Evening Class," Showtime
Writing a Dramatic Series: Chris Geroimo, "The Showtime 30-Minute Movie: The Witness," Showtime
Directing a Dramatic Series: Alfonso Cuaron, "Fallen Angels: Murder Obliquely," Showtime
Actor in a Movie or Miniseries: Brian Dennehy, "Foreign Affairs, " TNT
Actress in a Movie or Miniseries: Holly Hunter, "The Positively True Adventures of the Alleged Texas Cheerleader-Murdering Mom," HBO
Supporting Actor in a Movie or Miniseries: Maximillian Schell, "Stalin," HBO
Supporting Actress in a Movie or Miniseries: Juanita Jennings, "Laurel Avenue," HBO
Writing a Movie or Miniseries: Larry Gelbart, "Barbarians at the Gate," HBO
Directing a Movie or Miniseries: David Wheatley, "Hostages," HBO
Performance in a Comedy Special: John Leguizamo, "HBO Comedy Hour: John Leguizamo's 'Spic-o-Rama,' " HBO
Directing a Comedy Special: Peter Askin, "HBO Comedy Hour: John Leguizamo's 'Spic-o-Rama,' " HBO
Informational or Documentary Host: Peter Jepson-Young, "The Broadcast Tapes of Dr. Peter: America Undercover," HBO
Writing an Entertainment Special: John Leguizamo, "HBO Comedy Hour: John Leguizamo's 'Spic-o-Rama,' " HBO
Cultural/Entertainment Documentary or Informational Special: "Mo' Funny: Black Comedy in America," HBO
Talk Show Special or Series: "Larry King Live," CNN
Educational or Instructional Special or Series: "How the West Was Lost," Discovery
Magazine Host: Richard Hart, "Next Step," Discovery
Sports Events Coverage Special: "ESPN Sunday Night NFL," ESPN
Animated Programming Special or Series: "Rugrats," Nickelodeon

PEN/FAULKNER AWARD FOR FICTION (1994)

(An award given by an organization of writers to honor their peers)

Philip Roth, *Operation Shylock*

1993 GOLDEN GLOBE AWARDS

MOVIES
Drama: *Schindler's List*
Musical or Comedy: *Mrs. Doubtfire*
Director: Steven Spielberg, *Schindler's List*
Actor, Drama: Tom Hanks, *Philadelphia*
Actress, Drama: Holly Hunter, *The Piano*
Actor, Musical or Comedy: Robin Williams, *Mrs. Doubtfire*
Actress, Musical or Comedy: Angela Bassett, *What's Love Got to Do With It?*
Supporting Actor: Tommy Lee Jones, *The Fugitive*
Supporting Actress: Winona Ryder, *The Age of Innocence*
Screenplay: Steven Zaillian, *Schindler's List*
Foreign Film: *Farewell My Concubine*
Original Score: Kitaro, *Heaven and Earth*
Original Song: Bruce Springsteen, music and lyrics, "Streets of Philadelphia"
Cecil B. De Mille Award: Robert Redford
Special Achievement Award: Ensemble cast of *Short Cuts*: Andie MacDowell, Bruce Davison, Julianne Moore, Matthew Modine, Anne Archer, Fred Ward, Jennifer Jason Leigh, Chris Penn, Lili Taylor, Robert Downey, Jr., Madeleine Stowe, Tim Robbins, Lily Tomlin, Tom Waits, Frances McDormand, Peter Gallagher, Annie Ross, Lori Singer, Jack Lemmon, Lyle Lovett, Buck Henry, Huey Lewis

TELEVISION
Series, Musical or Comedy: "Seinfeld"
Actor, Series, Musical or Comedy: Jerry Seinfeld, "Seinfeld"
Actress, Series, Musical or Comedy: Helen Hunt, "Mad About You"
Actor, Supporting Role: Beau Bridges, "The Positively True Adventures of the Alleged Texas Cheerleader-Murdering Mom"
Actress, Supporting Role: Julia Louis-Dreyfus, "Seinfeld"
Series, Drama: "NYPD Blue"
Actor, Series, Drama: David Caruso, "NYPD Blue"
Actress, Series, Drama: Kathy Baker, "Picket Fences"
Miniseries or TV Movie: "Barbarians at the Gate"
Actor, Miniseries or TV Movie: James Garner, "Barbarians at the Gate"
Actress, Miniseries or TV Movie: Bette Midler, "Gypsy"

1993 GEORGE POLK AWARDS

Awarded for special achievement in journalism

Radio Commentary: Daniel Schorr, National Public Radio
Magazine Reporting: "An Anthropologist on Mars," Oliver Sacks, *The New Yorker*
Television Reporting: Bosnia Coverage, Christiane Amanpour, CNN
Medical Reporting: "The HMO Maze," Larry Keller and Fred Schulte, *Sun-Sentinel* (Ft. Lauderdale, FL)
Local Reporting: "Illegal Smuggling from China," Ying Chan, *Daily News*
National Reporting: "America's Atomic Victims," Eileen Welsome, *The Albuquerque Tribune*
Foreign Reporting: Somalia Coverage, Keith Richburg, *The Washington Post*
Career Award: Richard Dudman, *St. Louis Post-Dispatch*
Political Reporting: "Cash Transactions," *The State Journal-Register* (Springfield, IL)
Business Reporting: "Coalfield Contracts: Mining at What Price?" Paul Nyden, *The Charleston Gazette* (W. Va)
Financial Reporting: Prudential-Bache Securities Investigation, Scot J. Paltrow, *Los Angeles Times*
Regional Reporting: Mississippi Flood, Isabel Wilkerson, *The New York Times*
Book: *Lenin's Tomb*, David Remnick, Random House

☆ Chase's 1995 Calendar of Events ☆

MAJOR AWARDS (cont'd)

1994 AMERICAN LIBRARY ASSOCIATION AWARDS FOR CHILDREN'S BOOKS
(Announced Feb 4, 1994)

NEWBERY MEDAL
For most distinguished contribution to American literature for children published in 1993:
Lois Lowry, author, *The Giver*
Honors Books
Jane Leslie Conly, author, *Crazy Lady*
Laurence Yep, author, *Dragon's Gate*
Russell Freedman, author, *Eleanor Roosevelt: A Life of Discovery*

CALDECOTT MEDAL
For most distinguished American picture book for children published in 1993:
Allen Say, illustrator and author, *Grandfather's Journey*
Honors Books
Ted Lewin, illlustrator, and Elisa Bartone, author, *Peppe the Lamplighter*
Denise Fleming, illustrator and author, *In the Small, Small Pond*
Kevin Henkes, illustrator and author, *Owen*
Gerald McDermott, illustrator and author, *Raven: A Trickster Tale from the Pacific Northwest*
Chris Raschko, illustrator and author, *Yo! Yes?*

CORETTA SCOTT KING AWARD
For outstanding books by African-American authors and illustrators:
Angela Johnson, author, *Toning the Sweep*
Tom Feelings, illustrator, *Soul Looks Back in Wonder*
Honors Books—Authors
Joyce Carol Thomas, author, *Brown Honey in Broomwheat Tea*
Walter Dean Myers, author, *Malcolm X: By Any Means Necessary*
Honors Books—Illustrators
Floyd Cooper, illustrator, *Brown Honey in Broomwheat Tea*
James Ransome, illustrator, *Uncle Jed's Barbershop*

MARGARET A. EDWARDS AWARD
For lifetime achievement in writing books for young adults:
Walter Dean Myers, recipient

1994 NATIONAL ASSOCIATION OF BROADCASTERS CRYSTAL AWARDS FOR OUTSTANDING COMMUNITY SERVICE

KBHP-FM, Bemidji, MN; KCBS-AM, San Francisco, CA; KLBJ-AM, Austin, TX; KOJM-AM, Havre, MT; KPSN-FM, Phoenix, AZ; KRMG-AM, Tulsa, OK; KSJN-FM, St. Paul, MN; WCCO-AM, Minneapolis, MN; WWTC-AM, Minneapolis, MN; WXYV-FM, Baltimore, MD.

1994 NATIONAL ENDOWMENT FOR THE HUMANITIES AWARD

Jefferson Lecturer in Humanities: Gwendolyn Books, Pulitzer Prize-winning poet. (Highest honor bestowed by the federal government for distinguished intellectual achievement in the humanities)
Charles Frankel Prize: Ricardo E. Alegría, John Hope Franklin, Hanna Holborn Gray, Andrew Heiskell and Laurel T. Ulrich. (To honor the outstanding achievements of Americans who have helped expand the public's understanding of history, literature, philosophy or other subjects in the humanities)

1994 PULITZER PRIZES

THE ARTS
Drama: *Three Tall Women*, Edward Albee
Fiction: *The Shipping News*, E. Annie Proulx
History: No Award
Biography: *W.E.B. DuBois: Biography of a Race 1868–1919*, David Levering Lewis
Poetry: *Neon Vernacular: New and Selected Poems*, Yusef Komunyakaa
General Nonfiction: *Lenin's Tomb: The Last Days of the Soviet Empire*, David Remnick

JOURNALISM
Public Service: *Akron Beacon Journal*
Spot News Reporting: *The New York Times* staff
Investigative Reporting: *The Providence Journal-Bulletin* staff
Explanatory Journalism: Ronald Kotulak, *Chicago Tribune*
Beat Reporting: Eric Freedman and Jim Mitzelfeld, *The Detroit News*
National Reporting: Eileen Welsome, *The Albuquerque Tribune*
International Reporting: *The Dallas Morning News* team
Feature Writing: Isabel Wilkerson, *The New York Times*
Commentary: William Raspberry, *The Washington Post*
Criticism: Lloyd Schwartz, *The Boston Phoenix*
Editorial Writing: R. Bruce Dold, *Chicago Tribune*
Editorial Cartooning: Michael P. Ramirez, *The Commercial Appeal*, Memphis, TN.
Spot News Photography: Paul Watson, *The Toronto Star*
Feature Photography: Kevin Carter, free-lance photographer, for *The New York Times*

MUSIC
Music: *Of Reminiscences and Reflections*, Gunther Schuller

1993–1994 DAYTIME EMMY AWARDS

Outstanding Directing in a Game/Audience Participation Show: Bob Levy, "American Gladiators" (Syndicated)
Outstanding Directing in a Talk Show: Peter Kimball, Joey Ford and Duke Struck, "The Oprah Winfrey Show" (Syndicated)
Outstanding Directing in a Service Show: Russell Morah, "This Old House (PBS)
Outstanding Achievement in Directing—Special Class: Bob McKinnon, "Good Morning America" (ABC)
Outstanding Achievement in Live and Tape Sound Mixing and Sound Effects for a Drama Series: Clyde Kaplan, Jeff Smith, Steve Wacker, Neal Weinstein, Bob Maryon and Jerry Martz, "The Bold and the Beautiful" (CBS)
Outstanding Achievement in Makeup: Ron Wild and Karen Stephens, "Adventures in Wonderland" (Disney Channel)
Outstanding Achievement in Makeup for a Drama Series: Eva Polywka, Gloria Grant and Diane Ford, "As the World Turns" (CBS)

☆ Chase's 1995 Calendar of Events ☆

Outstanding Achievement in Graphics and Title Design: Iraj Paran, Tom Wogatzke and James Hickey, "The Addams Family" (ABC); Billy Pittard, Suzanne Kiley and Dale Everett, "The Ricki Lake Show" (Syndicated)

Outstanding Achievement in Art Direction/Set Decoration/Scenic Design: Victor DiNapoli, Mike Pantuso, Pete Ortiz, Bob Phillips, Nat Mongioi and Michael J. Kelley, "Sesame Street" (PBS)

Outstanding Achievement in Art Direction/Set Decoration/Scenic Design for a Drama Series: Sy Tomashoff, Jack Forrestel, Richard Harvey and Elsa Zamparelli, "The Bold and the Beautiful" (CBS)

Outstanding Writing in a Children's Series: Daryl Busby and Tom J. Astle, "Adventures in Wonderland" (Disney Channel); Norman Stiles, Lou Berger, Molly Boylan, Sara Compton, Judy Freudberg, Tony Geiss, Ian Ellis James, Emily P. Kingsley, David Korr, Sonia Manzano, Joey Mazzarino, Nancy Sans, Luis Santeiro, Josh Selig, Jon Stone, Cathi R. Turow, Belinda Ward and John Weidman, "Sesame Street" (PBS)

Outstanding Writing in an Animated Program: Ray Bradbury, "Halloween Tree" (Syndicated)

Outstanding Achievement in Live and Tape Sound Mixing and Sound Effects: Bob Manahan, Tim Harsh, Alan Porzio and Andrew Sommers, "Beakman's World" (CBS)

Outstanding Achievement in Technical Direction/Electronic Camera/Video Control: Ralph Mensch, Frank Biondo, Dave Driscoll and Dick Sens, "Sesame Street" (PBS)

Outstanding Achievement in Technical Direction/Electronic Camera/Video Control for a Drama Series: Chuck Guzzi, Donna Stock, Gordon T. Sweeney, David G. Navarette, Ted L. Morales, Joel Binger, Jim Velarde, Scha Jani and Roberto Bosio, "The Bold and the Beautiful" (CBS)

Outstanding Achievement in Costume Design: Fred Buchholz, Ed Christie, Bill Kellard, Rollin Krewson, Laurent Linn, Peter MacKennan, Connie Peterson, Terry Roberson, Stephen Rotondaro, Mark Ruffin, Carol Yannuzzi and Mark Zeszotek, "Sesame Street" (PBS)

Outstanding Achievement in Costume Design for a Drama Series: Charles Schoonmaker and Margarita Delgado, "Another World" (NBC)

Outstanding Achievement in Multiple Camera Editing: Robert J. Emerick, Evamarie Keller and John Tierney, "Sesame Street" (PBS)

Outstanding Achievement in Multiple Camera Editing for a Drama Series: Jim Jewell, "The Bold and the Beautiful" (CBS)

Outstanding Achievement in Single Camera Editing: David Greenwald, "Girlfriend" (ABC Afterschool Special) (ABC)

Outstanding Achievement in Music Direction and Composition: Richard Stone and Steve Bernstein, "Animaniacs" (Fox)

Outstanding Achievement in Music Direction and Composition for a Drama Series: Barbara Miller-Gidaly, A. J. Gundell, John Henry, Wes Boatman, Michael Licari, Dominic Messinger, Larry Hold, Richard Hazard and Barry Devorzon, "Guiding Light" (CBS)

Outstanding Original Song: Richard Stone and Tom Ruegger, "Animaniacs (Main Title Theme)" (Fox)

Outstanding Directing in a Children's Series: Ed Wiseman and Mark Mannucci, "Reading Rainbow" (PBS)

Outstanding Directing in a Children's Special: Hank Saroyan, "William Saroyan's the Parsley Garden" (ABC Weekend Special) (ABC)

Outstanding Achievement in Hairstyling: Andre Walker, "The Oprah Winfrey Show" (Syndicated)

Outstanding Achievement in Hairstyling for a Drama Series: Annette Bianco, Stanley Steve Hall and Joyce Sica, "Another World" (NBC)

Outstanding Writing in a Children's Special: Amy Dunkleberger, "Other Mothers" (CBS Schoolbreak Special) (CBS)

Outstanding Achievement in Writing—Special Class: Terrence McDonnell, Steven Dorfman, Kathy Easterling, Debbie Griffin, Frederik Pohl and Steve D. Tamerius, "Jeopardy!" (Syndicated)

Outstanding Achievement in Lighting Direction: Bill Berner, "Sesame Street" (PBS)

Outstanding Achievement in Lighting Direction for a Drama Series: Lauri Moorman, "The Bold and the Beautiful" (CBS); William Roberts and Ray Thompson, "The Young and the Restless" (CBS)

Outstanding Achievement in Single Camera Photography: Randy Drummond, "Reading Rainbow" (PBS)

Outstanding Achievement in Film Sound Editing: Michael A. Gollom, Timothy J. Borquez, Michael Geisler, Tom Jeager, Gregory LaPlante, Kenneth Young, William B. Griggs, Timothy Mertens and Patrick Foley, "Rocko's Modern Life" (Nickelodeon)

Outstanding Achievement in Film Sound Mixing: Ross Davis, John Anderson and Don Summer, "Other Mothers" (CBS Schoolbreak Special)

Outstanding Performer in a Children's Series: Shari Lewis as: Shari, "Lamb Chop's Play-Along" (PBS)

Outstanding Talk Show Host: Oprah Winfrey, "The Oprah Winfrey Show" (Syndicated)

Outstanding Service Show Host: T. Berry Brazelton, "What Every Baby Knows" (Lifetime)

Outstanding Supporting Actor in a Drama Series: Justin Deas as: Buzz Cooper, "Guiding Light" (CBS)

Outstanding Supporting Actress in a Drama Series: Susan Haskell as: Marty Saybrooke, "One Life to Live" (ABC)

Outstanding Children's Special: Frank Doelger and Howard Meltzer, "Dead Drunk: The Kevin Tunell Story" (HBO)

Outstanding Game/Audience Participation Show: Merv Griffin and George Vosburgh, "Jeopardy!" (Syndicated)

Outstanding Younger Actor in a Drama Series: Roger Howarth as: Todd Manning, "One Life to Live" (ABC)

Outstanding Younger Actress in a Drama Series: Melissa Hayden as: Bridget Reardon, "Guiding Light" (CBS)

Outstanding Animated Children's Program: Vanessa Coffey, Gabor Csupo, Arlene Klasky, Paul Germain, Charles Swenson, Mary Harrington, Geraldine Clarke, Howard E. Baker, Norton Virgien and Jim Duffy, "Rugrats" (Nickelodeon)

Outstanding Children's Series: Michael Loman, Lisa Simon, Arlene Sherman and Yvonne Hill-Ogunkoya, "Sesame Street" (PBS)

Outstanding Talk Show: Debra DiMaio, Oprah Winfrey, Dianne Hudson, Mary Kay Clinton, Alice McGee, Ellen Rakieten, David Boul, Rudy Guido, Legrande Green and Dana Newton, "The Oprah Winfrey Show" (Syndicated)

Outstanding Service Show: Russell Morash, Nina Sing Fialkow and Bruce Irving, "This Old House" (PBS)

Outstanding Special Class Program: Dick Schneider, Willard Scott and Katie Couric, "Macy's 67th Annual Thanksgiving Day Parade" (CBS)

Outstanding Drama Series Directing Team: Bruce Barry, Jo Anne Sedwick, Irene Pace, Brian Mertes, John O'Connell, Matthew Lagle, Scott Riggs and Lisa Connor, "Guiding Light" (CBS)

Outstanding Game Show Host: Bob Barker, "The Price Is Right" (CBS)

Outstanding Drama Series Writing Team: Michael Malone, Josh Griffith, Jean Passanante, Susan Bedsow-Horgan, Chris Whitesell, Becky Cole, David Colson, Lloyd Gold and David Smilow, "One Life to Live" (ABC:)

Outstanding Lead Actor in a Drama Series: Michael Zaslow as: Roger Thorpe, "Guiding Light" (CBS)

Outstanding Lead Actress in a Drama Series: Hillary B. Smith as: Nora Gannon, "One Life to Live" (ABC)

Outstanding Drama Series: Felicia Minei Behr, Terry Cacavio, Thomas DeVilliers and Nancy Horwich, "All My Children" (ABC)

… Chase's 1995 Calendar of Events …

ALPHABETICAL INDEX

Events are generally listed under key words; many broad categories have been created, including African-American, Agriculture, Air Shows, Animals, Antiques, Art—Fine and Performing, Arts and Crafts, Auto Shows and Races, Aviation, Balloon (Hot-air), Bicycle, Birds, Boats, Books, Children, Christmas, Cowboys/Old West, Dance, Education, Employment/Occupations/Professions, Environment, Ethnic, Family, Film, Fishing, Flowers, Food and Beverages, Handicapped, Health and Welfare, Home Shows and Tours, Horses, Human Relations, Journalism, Kites, Library/Librarians, Literature, Music, Native American, Parades, Photography, Radio, Railroad/Trains, Renaissance Fairs, Rodeos, Safety, Science, Senior Citizens, Space, Space Milestones, Storytelling, Television, Theater, Women, names of sports, types of music, types of animals, etc. Events that can be attended are also listed under the states or countries where they are to be held. Index indicates only initial date for each event. See chronology for inclusive dates of events lasting more than one day.

A

Aames, Willie: Birth......................July 15
Aardvark Week, Natl......................Mar 5
Aaron, Hank: Birth.......................Feb 5
Aaron, Hank: Home Run Record: Anniv.....Apr 8
Abbado, Claudio: Birth..................June 26
Abbott, Berenice: Birth Anniv...........July 17
Abbott, Jim: Birth......................Sept 19
Abdul, Paula: Birth.....................June 19
Abdul-Jabbar, Kareem: Birth.............Apr 16
Abet and Aid Punsters Day...............Nov 8
Abolition Soc Founded, First American:
  Anniv.................................Apr 14
Abortion First Legalized: Anniv.........Apr 25
Abortion: Roe v Wade Supreme Court Decision:
  Anniv.................................Jan 22
Abraham Baldwin Agri College Homecoming
  (Tifton, GA).........................Mar 30
Abraham, F. Murray: Birth...............Oct 24
Abzug, Bella: Birth.....................July 24
ACBL Bridge Tournament (Yuma, AZ).......Feb 23
Accession of Queen Elizabeth II: Anniv..Feb 6
According to Hoyle Day..................Aug 29
Accordion Awareness Month, Natl.........June 1
Accordion Chmpshp, Kimberley Intl Old-Time
  (Kimberley, BC)......................July 3
Accordion Fest, Sata-Hame (Ikaalinen,
  Finland).............................June 26
Ace, Goodman: Birth Anniv...............Jan 15
Achelis, Elisabeth: Birth Anniv.........Jan 24
Ackland, Joss: Birth....................Feb 29
ACTION Agency Transfer of Functions
  6662.................................Page 488
Activity Professionals Conv, Natl Assn of
  (Nashville, TN)......................Mar 29
Activity Professionals Day, Natl........Jan 27
Activity Professionals Week, Natl......Jan 22 Acuff,
  Roy: Birth Anniv.....................Sept 15
Adamle, Mike: Birth.....................Oct 4
Adams, Abigail: Birth Anniv.............Nov 11
Adams, Bryan: Birth.....................Nov 5
Adams, Don: Birth.......................Apr 19
Adams, Edie: Birth......................Apr 16
Adams, John Quincy: Birth Anniv.........July 11
Adams, John Quincy: Returns to Congress.Mar 4
Adams, John: Birth Anniv................Oct 30
Adams, Louisa Catherine Johnson: Birth
  Anniv................................Feb 12
Adams, Mason: Birth.....................Feb 26
Adams, Richard C.: Death Anniv..........Mar 9
Addams, Jane: Birth Anniv...............Sept 6
Adding Machine, Patent Issued for First:
  Anniv................................Oct 11
Addison, Joseph: Birth Anniv............May 1
Addy, Wesley: Birth.....................Aug 4
Ade, George: Birth Anniv................Feb 9
Admission Day Holiday (Hawaii)..........Aug 18
Adobe Fiesta, Old (Petaluma, CA)........Aug 20
Adolfo: Birth...........................Feb 15
Adopt-a-Shelter-Cat Month...............June 1
Adopt-a-Shelter-Dog Month...............Oct 1
Adoption: Natl Birthparents Week........Apr 2
Adoption Week, Natl.....................Nov 19
Adoption Week, Natl 6629...............Page 488
Adult Day Care Center Week, Natl........Sept 17
Advent, First Sunday of.................Dec 3
Advertising: Co-op Awareness Month......Oct 1
Advertising: Mobius Awards (Chicago, IL).Feb 2
Aebleskiver Days (Tyler, MN)............June 24

Afghanistan
  Independence Day.....................Aug 19
  Saur Revolution Anniv................Apr 27
AFL-CIO Founding Anniv..................Dec 5
AFL-CIO Merger: Anniv...................Feb 9
Africa: Industrial Development Decade for Africa,
  Second...............................Jan 1
Africa: Industrialization Day (UN)......Nov 20
Africa: Transport/Communications Decade,
  Second (UN)..........................Jan 1
African-Americans
  African-American Cultural Festival
    (Chattanooga, TN)..................Aug 19
  African-American Heritage Fest (Snow Hill,
    MD)................................Sept 10
  African-American History and Heritage Tour
    (Dallas, TX).......................Feb 4
  African Cultural Ball (Chattanooga, TN)..Feb 3
  Afro-American Cultural Ctr Conf (New Haven,
    CT)................................Mar 31
  Afro-American History Month..........Feb 1
  Ashford, Emmet: Birth Anniv..........Nov 23
  Associated Publishers, Inc: Founding
    Anniv..............................Nov 18
  Attack on Fort Wagner: Anniv.........July 19
  Black American Arts Festival (Greensboro,
    NC)................................Jan 14
  Black History Month Kickoff (Washington,
    DC)................................Jan 27
  Black History Parade (Monroe, LA)....Feb 25
  Black Poetry Day.....................Oct 17
  Black Solidarity Day (New Haven, CT).Nov 6
  Blacks Ruled Eligible to Vote: Anniv.Apr 3
  Bolin, Jane M.: Birth................Apr 11
  Brown, Jesse Leroy: Birth Anniv......Oct 13
  Bud Billiken Parade (Chicago, IL)....Aug 12
  Civil Rights Act of 1964: Anniv......July 2
  Civil Rights Act of 1965: Anniv......Aug 6
  Civil Rights Act of 1968: Anniv......Apr 11
  Civil Rights Workers Found Slain: Anniv..Aug 4
  Desegregation, US Army First: Anniv..July 26
  Emancipation of 500: Anniv...........Aug 1
  Escape to Freedom (F. Douglass): Anniv.Sept 3
  First American Abolition Soc Founded:
    Anniv..............................Apr 14
  First Black Governor Elected: Anniv..Nov 7
  First Black Pro Hockey Player: Anniv.Nov 15
  First Black Serves in US House Reps:
    Anniv..............................Dec 12
  First Black US Cabinet Member: Anniv.Jan 18
  First Black US State's Attorney: Anniv.July 6
  First National Convention for Blacks:
    Anniv..............................Sept 15
  Forten, James: Birth Anniv...........Sept 2
  Foster, Andrew: Birth Anniv..........Sept 17
  Frederick Douglass Speaks: Anniv.....Aug 11
  Gibbs, Mifflin Wister: Birth Anniv...Apr 28
  Grimes, Leonard Andrew: Birth Anniv..Nov 9
  Harlem: Photographs by Aaron Siskind (Los
    Angeles, CA).......................Mar 4
  Historically Black Coll/Univ Week, Natl
    6594...............................Page 488
  King Awarded Nobel: Anniv............Oct 14
  Kwanzaa Fest.........................Dec 26
  Loving v Virginia: Anniv.............June 12
  Malcolm X: Assassination Anniv.......Feb 21
  McDaniel, Hattie: Birth Anniv........June 10
  Medgar Evers Assassinated: Anniv.....June 13
  Meredith (James) Enrolls at Ole Miss:
    Anniv..............................Sept 30
  Miami/Bahamas Goombay Fest (Miami,
    FL)................................June 2
  Minority Enterprise Development Week (Pres
    Proc)..............................Oct 1
  Minority Enterprise Development Week
    6591...............................Page 488
  Montgomery Boycott Arrests: Anniv....Feb 22
  Montgomery Bus Boycott Begins: Anniv.Dec 5
  Montgomery Bus Boycott Ends: Anniv...Dec 20
  Motley, Constance Baker: Birth.......Sept 14
  Racial Brawl on Kitty Hawk: Anniv....Oct 12
  Ralph Bunche Awarded Nobel Peace Prize:
    Anniv..............................Dec 10
  Robinson Named First Black Manager:
    Anniv..............................Oct 3
  Rosa Parks Day.......................Dec 1
  St. Louis Race Riots: Anniv..........July 2
  Scott Joplin Ragtime Fest (Sedalia, MO).June 1
  Songs of My People (Fort Wayne, IN)..May 6
  Songs of My People (Paterson, NJ,)...Mar 4
  Songs of My People (Shreveport, LA)..July 8
  Songs of My People (Winston-Salem, NC).Sept 9
  South Africa Repeals Last Apartheid Law:
    Anniv..............................June 17
  Stokes Becomes First Black Mayor in US:
    Anniv..............................Nov 13
  SW Black Student Leadership Conf (College
    Station, TX).......................Jan 19
  Tanner, Henry Ossawa: Birth Anniv....June 21
  Truth, Sojourner: Death Anniv........Nov 26
  Wheatley, Phillis: Death Anniv.......Dec 5
  Wm Carney Receives Cong Medal:
    Anniv..............................May 23
  Zora Neale Hurston Fest (Eatonville, FL).Jan 26
Africa's Legacy in Mexico/Photos (Ann Arbor,
  MI)..................................Apr 15
Africa's Legacy in Mexico: Photos (Roanoke,
  VA)..................................Aug 19
Africa's Legacy in Mexico: Photos (San Antonio,
  TX)..................................June 17
Africa's Legacy in Mexico: Photos (Toledo,
  OH)..................................Feb 11
Agassi, Andre: Birth....................Apr 29
Agassiz, Jean: Birth Anniv..............May 28
Agee, Philip: Birth.....................July 19
Aggiecon (College Station, TX)..........Mar 23
Agnew Resignation: Anniv................Oct 10
Agnew, Spiro: Birth Anniv...............Nov 9
Agriculture
  Abraham Baldwin Ag College Homecoming
    (Tifton, GA).......................Mar 30
  Acton Fair (Acton, ME)...............Aug 24
  Agriculture Day, Natl................Mar 20
  Agriculture Day, Natl 6538, 6657....Page 488
  Agriculture Week, Natl...............Mar 20
  Agrifair (Clearbrook, BC, Canada)....Aug 4
  Alabama State Fair, South (Montgomery,
    AL)................................Oct 6
  Alaska State Fair (Palmer, AK).......Aug 25
  Arkansas State Fair (Little Rock, AR).Oct 6
  Atlantic Winter Fair (Halifax, NS, Canada).Oct 7
  Barn Day (Filley, NE)................July 9
  Beef Empire Days (Garden City, KS)...June 2
  Big E (West Springfield, MA).........Sept 15
  Black Hills Steam and Gas Threshing Bee
    (Sturgis, SD)......................Aug 18

516

☆ Chase's 1995 Calendar of Events ☆ Index

Agriculture (cont'd)—Alaska

Bob Evans Farm Fest (Rio Grande, OH) . . **Oct 13**
Build a Scarecrow Days (Milton, NH) . . . . . **July 1**
Canadian Western Agribition (Regina, Sask, Canada) . . . . . **Nov 25**
Central States Threshermen's Reunion (Pontiac, IL) . . . . . **Aug 30**
Champlain Valley Fair (Essex Junction, VT) . . . . . **Aug 26**
Colorado State Fair (Pueblo, CO) . . . . **Aug 19**
Comal County Fair (New Braunfels, TX) . **Sept 26**
Common Ground Country Fair (Windsor, ME) . . . . . **Sept 22**
Corn Palace Fest (Mitchell, SD) . . . . **Sept 8**
Crop Day (Greenwood, MS) . . . . **Aug 5**
Dairy Expo, World (Madison, WI) . . . . **Oct 4**
Delmarva Chicken Fest (Federalsburg, MD) . . . . . **June 9**
Durham Fair (Durham, CT) . . . . **Sept 22**
Eighteenth-Century Threshing Day (McLean, VA) . . . . . **Nov 19**
Fair of 1850 (Westville Village, GA) . . . . **Oct 12**
Fairfest (Hastings, NE) . . . . **July 27**
Fall Fair (Brandon, Man, Canada) . . . . **Oct 27**
Farm Days Weekend (Wheeling) . . . . **June 17**
Farm Safety and Health Week, Natl 6595 . . . . . **Page 488**
Farm Sanctuary Hoedown (Watkins Glen, NY) . . . . . **July 1**
Farm-City Week, Natl (Pres Proc) . . . . **Nov 17**
Farm-City Week, Natl 6624 . . . . **Page 488**
Farmfair (Edmonton, Alta, Canada) . . . . **Nov 4**
Flemington Agricultural Fair (Flemington, NJ) . . . . . **Aug 29**
Florida Citrus Festival (Winter Haven, FL) **Jan 26**
Florida State Fair (Tampa, FL) . . . . **Feb 3**
Fryeburg Fair (Fryeburg, ME) . . . . **Oct 1**
Game and Country Fair (Antrim) . . . . **June 30**
Garfield County Fair (Burwell, NE) . . . . . . **July 25**
Grange Week . . . . **Apr 23**
Grant County Home Products Dinner (Ulysses, KS) . . . . . **Sept 19**
Harvard Milk Day Festival (Harvard, IL) . . **June 2**
Harvest Days (Milton, NH) . . . . **Oct 7**
Historic Farm Days (Penfield, IL) . . . . **July 7**
Hood River County Fair (Hood River, OR) . . . . . **July 26**
Husker Feed Grains and Soybean Conference (Lincoln, NE) . . . . . **Jan 11**
Indiana State Fair (Indianapolis, IN) . . . . **Aug 9**
Iowa State Fair (Des Moines, IA) . . . . **Aug 10**
Kansas State Fair (Hutchinson, KS) . . . . . . **Sept 8**
Kentucky State Fair (Louisville, KY) . . . . . . **Aug 17**
Livestock Show, Rio Grande Valley (Mercedes, TX) . . . . . **Mar 15**
Lodi Grape Festival (Lodi, CA) . . . . **Sept 14**
Loon Lake Harvest Festival (Aurora, MN) . **Sept 9**
Los Angeles County Fair (Pomona, CA) . . **Sept 8**
Louisiana, State Fair of (Shreveport, LA) . **Oct 20**
Makoti Threshing Show (Makoti, ND) . . . . . **Oct 7**
Marion County Fair (Salem, OR) . . . . **July 6**
Maryland State Fair (Timonium) . . . . **Aug 26**
Michigan State Fair (Detroit, MI) . . . . **Aug 22**
Mid-South Fair (Memphis, TN) . . . . **Sept 22**
Minnesota State Fair (St. Paul, MN) . . . . . . **Aug 24**
Mississippi State Fair (Jackson, MS) . . . . . . **Oct 4**
Missouri State Fair (Sedalia, MO) . . . . **Aug 17**
Montana State Fair (Great Falls, MT) . . . . **July 29**
MontanaFair (Billings, MT) . . . . **Aug 12**
Nemont/Agribition (Wolf Point, MT) . . . . . **Oct 13**
Nevada State Fair (Reno, NV) . . . . **Aug 23**
New York State Fair (Syracuse, NY) . . . . . **Aug 24**
North Carolina State Fair (Raleigh, NC) . . **Oct 13**
North Dakota State Fair (Minot, ND) . . . . . **July 21**
Northeast MT Threshing Bee/Antique Show (Culbertson, MT) . . . . . **Sept 23**
Northeast WI Antiq Power Thresheree (Sturgeon Bay, WI) . . . . . **Aug 19**
Northern Illinois Farm Show (Rockford, IL) **Jan 3**
Northern Intl Livestock Expo (Billings, MT) . . . . . **Oct 16**
Of Field and Forest (East Meredith, NY) . . **June 4**
Oklahoma, State Fair of (Oklahoma City, OK) . . . . . **Sept 15**
Old Time Farm Day (Milton, NH) . . . . **Aug 12**
Old Tyme Farmers' Days (Live Oak, FL) . . **Nov 23**
Oregon State Fair (Salem, OR) . . . . **Aug 24**
Organic Harvest Month, Natl . . . . **Sept 1**
Ozark Empire Fair (Springfield, MO) . . . . . **July 28**
Penn State's AG Progress Days (Rock Springs, PA) . . . . . **Aug 15**
Pigeon Farmer's Fest (Pigeon, MI) . . . . . . **July 26**
Pioneer Power Day Threshing Bee (Lewistown, MT) . . . . . **Sept 30**
Plowing Match (Woodstock, VT) . . . . **May 7**
Powers' Crossroads Country Fair/Art Fest (Newnan, GA) . . . . . **Sept 2**
Rice Month, Natl . . . . **Sept 1**
Rochester Fair (Rochester, NH) . . . . **Sept 14**
Rockbridge Regional Fair (Lexington, VA) . . . . . **July 25**
Rotary Tiller Race, W Chmpshp (Emerson, AR) . . . . . **June 24**
Round Barn Fest (Rochester, IN) . . . . **June 16**
Royal Manitoba Winter Fair (Brandon, Man, Canada) . . . . . **Mar 27**
Royal Welsh Show (Powys, Wales) . . . . . . **July 24**
Ruffin, Edmund: Birth Anniv . . . . **Jan 5**
Running of the Sheep Sheep Drive (Reedpoint, MT) . . . . . **Sept 3**
Rural Life Sunday . . . . **May 21**
St. Johns Mint Festival (St. Johns, MI) . . . **Aug 12**
Salem Fair and Exposition (Salem, VA) . . **June 30**
Saskatoon Fall Fair/Mexabition (Saskatoon, Sask) . . . . . **Nov 16**
Scotts Bluff County Fair (Mitchell, NE) . . **Aug 13**
Small Family Farm Week 6660 . . . . **Page 488**
Sodbuster Days (Fort Ransom State Park) **July 8**
Sonoma-Marin Fair (Petaluma, CA) . . . . . **June 21**
Sorghum Day Festival (Wewoka, OK) . . . . . **Oct 28**
South Dakota State Fair (Huron, SD) . . . . **Aug 26**
South Mountain Fair (Gettysburg, PA) . . . **Aug 30**
Southwest Louisiana State Fair/Expo (Lake Charles, LA) . . . . . **Oct 26**
Southwestern Expo Livestock Show/Rodeo (Ft Worth, TX) . . . . . **Jan 20**
Sussex Farm and Horse Show (Augusta, NJ) . . . . . **Aug 4**
Tanana Valley Fair (Fairbanks, AK) . . . . **Aug 4**
Tennessee State Fair (Nashville, TN) . . . . . **Sept 8**
Texas, State Fair of (Dallas, TX) . . . . **Sept 28**
Triumph of Agriculture Expo (Omaha, NE) . . . . . **Mar 14**
Upper Canada Village (Morrisburg, Ont, Canada) . . . . . **May 20**
Utah State Fair (Salt Lake City, UT) . . . . . . **Sept 8**
Vermont State Fair (Rutland, VT) . . . . **Sept 1**
Virginia on Strawberry Hill, State Fair (Richmond, VA) . . . . . **Sept 21**
Webster County Woodchopping Fest (Webster Springs, WV) . . . . . **May 20**
Western Canada Farm Progress Show (Regina, Sask) . . . . . **June 21**
Western Stock Show and Rodeo, Natl (Denver, CO) . . . . . **Jan 10**
Western Washington Fair (Puyallup, WA) . **Sept 8**
Wheat Harvest, 18th Century (McLean, VA) . . . . . **June 18**
Wisconsin Farm Woman of Year Month . . . . **Feb 1**
Wisconsin State Fair (Milwaukee, WI) . . . . . **Aug 3**
Women in Agriculture Day, Natl 6698 . . **Page 488**
Wool Day: Sheep to Shawl/Border Collies (Woodstock, VT) . . . . . **Sept 17**
Wyoming State Fair (Douglas) . . . . **Aug 21**
Aguirre, Mark: Birth . . . . . Dec 10
Agutter, Jenny: Birth . . . . . Dec 20
AIDS Awareness Month, Natl . . . . . Oct 1
AIDS: Bergalis, Kimberly: Death Anniv . . . . . Dec 8
AIDS Day, World (UN) . . . . . Dec 1
AIDS: White, Ryan: Death Anniv . . . . . Apr 8
Aiello, Danny: Birth . . . . . June 20
Aiken, Conrad: Birth Anniv . . . . . Aug 5
Ailey, Alvin: Birth Anniv . . . . . Jan 5
Aimee, Anouk: Birth . . . . . Apr 27
Ainge, Daniel: Birth . . . . . Mar 17
Air Conditioning Appreciation Days . . . . . July 3
Airplanes, Air Shows. See Aviation
Ajaye, Franklyn: Birth . . . . . May 13
Akaka, Daniel K.: Birth . . . . . Sept 11
Akers, John: Birth . . . . . Dec 28
Akihito: Birth . . . . . Dec 23
Akins, Claude: Birth . . . . . May 25
Akroyd, Dan: Birth . . . . . July 1
Alabama
Admission Day . . . . **Dec 14**
Alabama Impact: Exhib . . . . **July 2**
Alabama Jubilee (Decatur) . . . . **May 27**
Alabama Natl Truck/Tractor Pull (Lexington) . . . . . **Aug 18**
Alabama Renaissance Faire (Florence) . . . **Oct 28**
American Bowling Congress Convention (Mobile) . . . . . **Mar 13**
Art in the Park (Demopolis) . . . . **Oct 7**
Birmingham Fest Arts Salute Switzerland (Birmingham) . . . . . **Apr 20**
Birmingham Sport and Boat Show (Birmingham) . . . . . **Jan 31**
Celebrations Around World (Huntsville) . . . **Dec 2**
Christmas on the Coosa (Wetumpka) . . . . . **Dec 9**
Christmas on the River (Demopolis) . . . . . **Nov 30**
City Stages Music Festival (Birmingham) . . . . . **June 16**
Civil War Living History Weekend (Huntsville) . . . . . **Nov 11**
Confederate Memorial Day . . . . **Apr 24**
Cowbellion Herd New Year's Eve Escapade/Revel (Mobile) . . . . . **Dec 31**
Dulcimer Fest (Huntsville) . . . . **Sept 17**
Earth Day (Huntsville) . . . . **Apr 22**
Fall Sorghum/Harvest Fest (Huntsville) . . **Sept 16**
Fireworks Celebration (Demopolis) . . . . **July 4**
First Monday Trade Days (Scottsboro) . . . . **Jan 1**
450-Mile Outdoor Fest . . . . **Aug 17**
Goose Pond Colony Dogwood Fest (Scottsboro) . . . . . **Apr 22**
Helen Keller Fest (Tuscumbia) . . . . **June 23**
Indian Dance and Crafts Fest (Childersburg) . . . . . **Apr 1**
Indian Day (Bridgeport) . . . . **Apr 15**
Indian Heritage Fest (Huntsville) . . . . **Oct 21**
Jean Lake Fest (Troy) . . . . **May 6**
Jeff Cook Super Bass Classic (Scottsboro) . . . . . **June 11**
Joe Cain Procession (Mobile) . . . . **Feb 26**
Kelly Tire Blue-Gray All Star Classic (Montgomery) . . . . . **Dec 25**
Lathe-Turned Objects, Intl: Exhib (Huntsville) . . . . . **Apr 16**
Longhorn World Chmpshp Rodeo (Huntsville) . . . . . **Mar 3**
Mentone Rhododendron Fest (Mentone) **May 20**
Panoply (Huntsville) . . . . **May 12**
Poarch Creek Indian Thanksgiving Day Powwow (Atmore) . . . . . **Nov 23**
Racking Horse Spring Celebration (Decatur) . . . . . **Apr 26**
Racking World Celebration (Decatur) . . . **Sept 22**
Sacred Harp Singing (Huntsville) . . . . **May 6**
Senior Bowl Football Game (Mobile) . . . . . **Jan 21**
September Fest (Childersburg) . . . . **Sept 30**
September Skirmish (Decatur) . . . . **Sept 1**
South Alabama State Fair (Montgomery) . . **Oct 6**
Southern Appalachian Dulcimer Fest (McCalla) . . . . . **May 7**
Southern Wildlife Festival (Decatur) . . . . . **Nov 18**
Spirit of America (Decatur) . . . . **July 3**
Spirit of Freedom Fest (Florence) . . . . **July 4**
Square Dance Conv, Natl (Birmingham) **June 21**
Tannehill Village Christmas (McCalla) . . . . . **Dec 4**
Tennessee River Fiddlers Conv (Florence) . . . . . **Oct 13**
W.C. Handy Fest (Florence) . . . . **Aug 6**
Winston Select 500 Race Week (Talladega) . . . . . **May 4**
YAM Programs for Youth Art Month (Huntsville) . . . . . **Mar 12**
Alamo: Anniv of the Fall . . . . . Mar 6
Alascattalo Day (Anchorage, AK) . . . . . Nov 19
Alaska
Admission Day . . . . **Jan 3**
Alascattalo Day (Anchorage) . . . . **Nov 19**
Alaska Day . . . . **Oct 18**
Alaska Day Celebration (Sitka) . . . . **Oct 14**
Alaska Earthquake: Anniv . . . . **Mar 27**
Alaska's First Broadcast TV Station: Anniv . . . . . **Dec 11**
Anchorage Fur Rendezvous (Anchorage) **Feb 10**
Anvil Mountain Run (Nome) . . . . **July 4**
Athabascan Fiddling Festival (Fairbanks) . **Nov 9**
Bachelor Soc Wild Woman Contest/Auction (Talkeetna) . . . . . **Dec 2**
Bering Sea Ice Golf Classic (Nome) . . . . . **Mar 18**
Big Lake Regatta Water Festival (Big Lake) . . . . . **July 8**
Blueberry Festival (Ketchikan) . . . . **Aug 12**
Exit Glacier Run (Seward) . . . . **May 13**
Fairbanks Summer Arts Fest (Fairbanks) **July 21**
Festival of Native Arts (Fairbanks) . . . . **Feb 23**
Festival of the North (Ketchikan) . . . . **Feb 3**
Firemen's Carnival (Nome) . . . . **Dec 2**
Great Alaska Shootout (Anchorage) . . . . . **Nov 22**
Great Bathtub Race (Nome) . . . . **Sept 4**
Hope Sled Dog Friendship Run (Nome) . . . **Apr 1**
Kodiak Crab Festival (Kodiak) . . . . **May 25**
Little Norway Fest (Petersburg) . . . . **May 18**
Memorial Weekend Salmon Derby (Petersburg) . . . . . **May 26**
Midnight Sun Baseball Game (Fairbanks) . . . . . **June 21**
Midnight Sun Festival (Nome) . . . . **June 17**
Mount Marathon Race (Seward) . . . . **July 4**
Nenana Tripod Raising Festival (Nenana) **Feb 25**
Nome River Raft Race (Nome) . . . . **June 18**

Index ☆ Chase's 1995 Calendar of Events ☆

Alaska (cont'd)

Seward Polar Bear Jump Fest (Seward)..Jan 20
Seward Silver Salmon 10K Run, Annual (Seward)...................Sept 2
Seward Silver Salmon Derby (Seward)..Aug 12
Seward's Day..................Mar 30
State Fair (Palmer)..............Aug 25
Summer Music Fest (Sitka)........June 2
Talkeetna Moose Dropping Festival (Talkeetna)..................July 7
Tanana Valley Fair (Fairbanks)....Aug 4
Tongass: Alaska's Rain Forest (Anchorage)...................Oct 14
Tongass: Alaska's Rain Forest (Fairbanks)..................Apr 8
Tongass: Alaska's Rain Forest (Juneau)....................June 3
Valentine Heart Throb Biathlon (Nome)..Feb 11

Albania

First Miss Albania Crowned: Anniv....Jan 31
National Day..................Nov 28
Albee, Edward: Birth............Mar 12
Alberghetti, Anna Marie: Birth...May 15
Albert, Eddie: Birth.............Apr 22
Albert, Marv: Birth..............June 12
Alcohol and Other Drug-Related Birth Defects Week..........................May 14
Alcohol Awareness Month, Natl...Apr 1
Alcohol-Free Weekend...........Apr 7
Alcoholics Anonymous: Founding Anniv..June 10
Alcott, Amy: Birth...............Feb 22
Alcott, Louisa May: Birth Anniv..Nov 29
Alda, Alan: Birth................Jan 28
Aldrich, Thomas B.: Birth Anniv..Nov 11
Aldrin, Edwin "Buzz": Birth.......Jan 20
Alexander, Jane: Birth............Oct 28
Alexander, Lamar: Birth...........July 3
Alexander, Shana: Birth...........Oct 6
Alexis, Kim: Birth................July 15
Alferd G. Packer Day (Boulder, CO)..Apr 21
Alger, Horatio, Jr.: Birth Anniv...Jan 13
Algeria: Revolution Anniv.........Nov 1
Ali (Cassius Clay) Named Heavyweight Champ: Anniv.........................Oct 30
Ali, Muhammad: Birth..............Jan 17
Ali, Muhammad: Stripped of Title: Anniv..Apr 30
Aliens Barred under UN Res 6685.......Page 488
All Dressed Up (Clinton, MD)......June 1
All Fools' Day....................Apr 1
All Hallows........................Nov 1
All Hallow's Eve..................Oct 31
All Saints' Day...................Nov 1
All States Picnic (Yuma, AZ).......Jan 18
All the News That's Fit to Print: Anniv..Feb 10
All-Hallomas......................Nov 1
All-Souls' Day....................Nov 2
All-Star Game, First Major League Baseball..July 6
Allen, Betty: Birth................Mar 17
Allen, Debbie: Birth...............Jan 16
Allen, Deborah: Birth..............Sept 30
Allen, Elizabeth: Birth............Jan 25
Allen, Ethan: Birth Anniv..........Jan 10
Allen, George F.: Birth............Mar 8
Allen, George: Birth...............Apr 29
Allen, Marcus: Birth...............Mar 26
Allen, Maryon Pittman: Birth.......Nov 30
Allen, Mel: Birth..................Feb 14
Allen, Nancy: Birth................June 24
Allen, Rex, Days (Willcox, AZ)......Oct 6
Allen, Steve: Birth................Dec 26
Allen, Woody: Birth................Dec 1
Allergy/Asthma Awareness Month, Natl..May 1
Alley, Kirstie: Birth..............Jan 12
Allilueva, Svetlana: Birth.........Feb 28
Allman, Gregg: Birth...............Dec 7
Allons Manger (Belle Rose, LA)....Apr 23
Allyson, June: Birth...............Oct 7
Alma College (MI): Anniv..........Oct 14
Almanac, Poor Richard's: Anniv....Dec 28
Almendros, Nestor: Birth..........Oct 30
Aloha Week Festivals (Hawaii).....Sept 15
Alpert, Herb: Birth................Mar 31
Alpha Phi Omega Natl Service Day..Nov 4
Alphabet Day (Korea)..............Oct 9
Alsop, Joseph: Birth...............Oct 11
Alt, Carol: Birth..................Dec 1
Altamont Concert Anniv............Dec 6
Altman, Robert: Birth..............Feb 20
Alzheimer, Alois: Birth Anniv......June 14
Alzheimer's Assn Memory Walk......Oct 7
Alzheimer's Disease Month, Natl...Nov 1
Amateur Athletic Union Sullivan Award Dinner (Indpls, IN)..................Feb 27
Amateur Radio Week................June 18
Amati, Nicolo: Birth Anniv.........Sept 3
Ambler, Eric: Birth................June 28
AMD Awareness Month...............Feb 1

Ameche, Don: Birth Anniv..........May 31
America Day, Underground..........May 14
America, God Bless, 1st Performed: Anniv..Nov 11
America, Spirit of (Decatur, IL)..July 3
America the Beautiful Published...July 4
American. See key word
American Bombing of Libya: Anniv..Apr 14
American Booksellers Assn Exhib/Conv (Chicago, IL).......................June 3
American Crossword Puzzle Tournament (Stamford, CT)....................Mar 17
American Diabetes Alert...........Mar 28
American Education Week..........Nov 12
American Enterprise Day...........Nov 15
American Federation of Labor: Founding Anniv...........................Nov 15
American Heart Assn's Heartfest..Oct 21
American History Month............Feb 1
American Redneck Day..............July 4
American Rivers Month.............June 1

American Samoa

Flag Day..........................Apr 17
Swarm of Palolo...................Oct 1
White Sunday......................Oct 8
Americana Indian and Western Art Show/Sale (Yuma, AZ)..........................Feb 3
Ames, Bruce N.: Birth.............Dec 16
Amis, Kingsley: Birth.............Apr 16
Amory, Cleveland: Birth...........Sept 2
Ampere, Andre: Birth Anniv........Jan 22
Ancestor Appreciation Day.........Sept 27
Andersen, Hans Christian: Birth Anniv..Apr 2
Anderson, Helen E.M.: 1st Woman US Ambassador: Anniv................Oct 28
Anderson, Ian: Birth..............Aug 10
Anderson, Jack: Birth.............Oct 19
Anderson, Lindsay: Birth..........Apr 17
Anderson, Loni: Birth.............Aug 5
Anderson, Lynn: Birth.............Sept 26
Anderson, Marian: Easter Concert: Anniv..Apr 9
Anderson, Melissa Sue: Birth......Sept 26
Anderson, Morten: Birth...........Aug 19
Anderson, Neal: Birth.............Aug 14
Anderson, Richard: Birth..........Jan 23
Anderson, Sherwood: Birth Anniv..Sept 13
Anderson, Sparky: Birth...........Feb 22
Andersson, Bibi: Birth............Nov 11
Andree, Salomon A.: Birth Anniv..Oct 18
Andress, Ursula: Birth............Mar 19
Andretti, Mario: Birth............Feb 28
Andrew, Prince: Birth.............Feb 19
Andrews, Julie: Birth.............Oct 1
Andrus, Cecil D.: Birth...........Aug 25
Andy Griffith Show: First Broadcast Anniv..Oct 3
Anesthetic First Used in Surgery (Anniv)..Mar 30
Angel Day, Be an..................Aug 22
Angelou, Maya: Birth..............Apr 4
Anglund, Joan Walsh: Birth........Jan 3
Angola: National Day..............Nov 11

Animals

Aardvark Week, Natl..............Mar 5
Adopt-a-Shelter-Cat Month........June 1
Adopt-a-Shelter-Dog Month........Oct 1
All-Breed Dog Show, Natl (Atlantic City, NJ)............................Dec 3
Animal's Messiah (Chicago, IL)...Dec 10
Answer Your Cat's Question Day..Jan 22
Anti-Cruelty Society Dogwash (Chicago, IL)..............................July 15
Barnegat Bay Crab Race/Seafood Fest (Seaside, NJ)......................Aug 27
Be Kind to Animals and Natl Pet Week 6560.............................Page 488
Be Kind to Animals Week..........May 7
Boo at the Zoo (Asheboro, NC)....Oct 28
Boo at the Zoo (Wheeling, WV)....Oct 21
Buffalo Auction, Custer State Park (Custer, SD).............................Nov 18
Buffalo Round-Up (Custer, SD)....Oct 2
Burro Race (Leadville, CO).......Aug 6
Candlelight Vigil for Homeless Animals Day................................Aug 19
Carpenter Ant Awareness Week.....June 18
Cat Festival (Belgium)...........Mar 2
Cat Show, Chmpshp (Indianapolis, IN)..Oct 28
Cat Show, Chmpshp (Indianapolis, IN)..Jan 7
Cat Show, Chesapeake Club CFA All-Breed (Baltimore, MD)...................Mar 11
Catfish Races, Grand Prix (Greenville, MS)..............................May 13
Chicken Clucking Contest (Baltimore, MD)..............................June 21
Christmas at the Zoo (Asheboro, NC)..Dec 24
Cobras in North America Day, Yell "Fudge" at

the..............................June 2
Copperas Crime Stoppers Fun Dog Show (Copperas Cove, TX).................Apr 30
Cow Milked While Flying: Anniv...Feb 18
Cows on the Concourse (Madison, WI)..June 3
Cranesville Weekend (Terra Alta, WV)..June 2
Crufts Dog Show (Birmingham, England)..Mar 16
Dixie Natl Livestock Show/Rodeo (Jackson, MS).............................Jan 30
Dog Week, Natl...................Sept 24
Duck Race, Great American (Deming, NM).............................Aug 25
Elm Farm Ollie Day Celebration (WI)..Feb 18
Entomological Soc America, Annual Mtg (Dallas, TX)........................Dec 13
Fancy Rat and Mouse Annual Show (Riverside, CA)....................Jan 28
Farm Sanctuary Hoedown (Watkins Glen, NY).............................July 1
Festival of the Horse (Oklahoma City, OK).............................Oct 20
Field Trial Chmp, Natl (Bird Dogs) (Gr Junction, TN)..............................Feb 13
Frog Days (Melvina, WI)..........Aug 4
Frog Hop, Preakness (Baltimore, MD)..May 17
Frog Jump Jamboree, Old-Fashioned (New Preston, CT)......................July 2
Frog Jumping Contest (Old Forge, NY)..June 18
Frog Jumping Contest, Roaring Camp (Felton, CA)..............................July 2
Goat Days (Millington, TN).......Sept 15
Good, the Bad and the Cuddly (Chicago)..May 13
Good, the Bad and the Cuddly (Royal Oak, MI).............................Feb 18
Gopher Hill Fest (Ridgeland, SC)..Oct 7
Gorilla Born in Captivity, First: Anniv..Dec 22
Grand American Coon Hunt (Orangeburg, SC).............................Jan 6
Ho Sheep Market (Denmark).......Aug 26
Hog Calling Contest (Baltimore, MD)..July 12
Homeless Animals Day, Natl......Aug 20
Humpback Whale Awareness Month (Honolulu, HI).............................Feb 1
Jumping Frog Jubilee/Calaveras Fair (Angel Camp, CA)........................May 18
Lizard Race, World's Greatest (Lovington, NM).............................July 4
Miss Crustacean/Ocean City Creep (Ocean City, NJ)........................Aug 2
Mosquito Awareness Weekend (Walcott, AR).............................Aug 20
Moth-er Day......................Mar 14
Mule Day.........................Oct 26
Mule Day, Matazas................Apr 27
Nongame Wildlife Weekend (Davis, WV)..June 2
Pet Appreciation Week............June 6
Pet Cemeteries Fall Mtg, Intl Assn..Oct 6
Pet Memorial Day, Natl...........Sept 10
Pet Owners Independence Day.....Apr 18
Pet Parade (LaGrange, IL).......June 3
Pet Sitters Week, Natl Prof......Mar 6
Pet Week, Natl...................May 7
Pig Day, Natl....................Mar 1
Pony Penning, Chincoteague (Chincoteague Island, VA).......................July 26
Prevention of Animal Cruelty Month..Apr 1
Rabbit Week (Wheeling, WV)......Apr 15
Race to the Sky Dog Sled Race (Helena, MT).............................Feb 11
Rattlesnake Derby (Mangum, OK)..Apr 28
Rattlesnake Hunt (Waurika, OK)..Apr 7
Red Wing Sheep Dog Trial (Red Wing, MN)..Oct 6
Responsible Pet Owner Month.....Feb 1
Rhino Exhib, Southwest (Tuscon, AZ)..Nov 10
Running of the Rodents (Louisville, KY)..Apr 26
Running of the Sheep Sheep Drive (Reedpoint, MT).............................Sept 3
Saskatoon Fall Fair/Mexabition (Saskatoon, Sask)..........................Nov 16
Save the Rhino Day...............May 1
Seal, Intl Day of the (Washington, DC)..Mar 5
Shamu's Birthday.................Sept 26
Sheep and Crafts Festival, World (Bethel, MO).............................Sept 2
Sheep Shearing Fest (North Andover, MA).............................May 14
Sheep Shearing/Kids Day (Petaluma, CA)..Apr 8
Snake Hunt (Cross Fork, PA).....June 24
Southern Wildlife Festival (Decatur, AL)..Nov 18
Spiders! (Bloomfield, MI).......July 1
Spiders! (New York, NY).........May 11
Spiders! (Toronto, Canada)......Oct 21
Swarm of Palolo (American Samoa)..Oct 1
Swine Time (Climax, GA).........Nov 25

☆ *Chase's 1995 Calendar of Events* ☆ **Index**

Animals (cont'd)—Arkansas

Tatanka Fest (Jamestown, ND) . . . . **June 29**
Thanksgiving Hunt Weekend (Cismont, VA) . . . . **Nov 23**
Turkey Days, King (Worthington, MN) . . . . **Sept 9**
Turkey "Olimpiks," Live (New Preston, CT) . . . . **Nov 19**
Turtle Derby, Canadian (Boissevain, Man, Canada) . . . . **July 14**
Turtle Derby, Chesapeake (Baltimore, MD) . . . . **July 5**
Turtle Races (Danville, IL) . . . . **June 3**
US's First Zoo: Anniv (Philadelphia, PA) . . **July 1**
Veal Ban Action Day . . . . **May 14**
Wayne Chicken Show (Wayne, NE) . . . . **July 8**
Wild Horse Chasing (Japan) . . . . **July 23**
Wild Turkey Calling, Natl/Festival (Yellville, AR) . . . . **Oct 13**
Wool Day: Sheep to Shawl/Border Collies (Woodstock, VT) . . . . **Sept 17**
World Champshp Hog Calling Contest (Weatherford, OK) . . . . **Feb 25**
World Farm Animals Day . . . . **Oct 2**
ZAM! Zoo and Aquarium Month . . . . **June 1**
Zoo Babies (Cincinnati, OH) . . . . **June 3**
Zoo Fling (Asheboro, NC) . . . . **Apr 1**
Zoobalee (Garden City, KS) . . . . **July 4**
ZooLights (Tacoma, WA) . . . . **Dec 1**
Anka, Paul: Birth . . . . July 30
Ann-Margret: Birth . . . . Apr 28
Annapolis Convention: Anniv . . . . Sept 11
Annenberg, Walter: Birth . . . . Mar 13
Annie Oakley Days (Greenville, OH) . . . . . . July 27
Anniversary Day . . . . Oct 19
Announcement of Ten Best Puns of Year . . . . Jan 1
Annunciation, Feast of . . . . Mar 25
Answer Your Cat's Question Day . . . . Jan 22
Ant, Adam: Birth . . . . Nov 3
Antebellum Jubilee (Stone Mountain, GA) . Mar 31
Antec '95 . . . . May 7
Anthem Day, National . . . . Mar 3
Anthony, Susan B.: Birth Anniv . . . . Feb 15
Anthony, Susan B.: Day . . . . Aug 26
Anthony, Susan B.: Death Anniv . . . . Mar 13
Anthony, Susan B.: Fined for Voting: Anniv . June 6
Anthony Wayne Day . . . . July 11
Anti-Boredom Month, Natl . . . . July 1
Anti-Saloon League Founded: Anniv . . . . . . May 24
Antietam Battlefield, Meml Illumination (Sharpsburg, MD) . . . . Dec 2
Antigua and Barbuda: August Monday . . . . . . Aug 7
Antigua: National Holiday . . . . Nov 1
Antiques
Antique Airplane Fly-In (Atchison, KS) . . **May 26**
Antique Auto Show and Flea Market (Boyne City, MI) . . . . **Aug 12**
Antique Auto Show/Collector Car Fest (St. Ignace, MI) . . . . **June 22**
Antique Auto Swap Meet (Grand Prairie, TX) . . . . **June 10**
Antique Marine Engine Exposition (Mystic, CT) . . . . **Aug 19**
Antique Power Exhib (Burton, OH) . . . . . . **July 28**
Antique Valentine Exhibit (Clinton, MD) . . . **Feb 9**
Antique/Classic Boat Rendezvous (Mystic, CT) . . . . **July 22**
Art and Antique Fair (Breda, Holland) . . . . . **Apr 8**
Atlantic City Auction/Flea Market (Atlantic City, NJ) . . . . **Feb 17**
Atlantique City Antiq/Collectibles (Atlantic City, NJ) . . . . **Mar 18**
Auburn-Cord-Duesenberg Fest (Auburn, IN) . . . . **Aug 31**
Bennington Museum Antiques Show/Sale (Bennington, VT) . . . . **July 7**
Black Hills Threshing Bee (Sturgis, SD) . . **Aug 18**
Blackpowder Trade Fair/Gun Show (Albert Lea, MN) . . . . **Feb 11**
British Intl Antiques Fair (Bristol, England) . . . . **Apr 25**
Chelsea Antiques Fair (London, England) . **Mar 7**
Chelsea Fair (London, England) . . . . **Sept 12**
Chesapeake Antique Fire Muster (Westminster, MD) . . . . **May 13**
Codman House Antique Vehicle Meet (Lincoln, MA) . . . . **July 16**
Colonial Heritage Assn Antique Show/Sale (Schoharie, NY) . . . . **Mar 4**
Forgotten Past Steam and Gas Show (Benton, KY) . . . . **Aug 5**
Gettysburg Outdoor Antique Show (Gettysburg, PA) . . . . **May 20, Sept 23**
Littlestown Good Old Days (Gettysburg, PA) . . . . **Aug 17**
Makoti Threshing Show (Makoti, ND) . . . . . **Oct 7**
Old Bedford Village Senior Days (Bedford, PA) . . . . **Sept 16**
Pec Thing (Pecatonica, IL) . . . . **May 20**
Philadelphia Antiques Show (Philadelphia, PA) . . . . **Apr 8**
Snow Hill Heritage Weekend (Snow Hill, MD) . . . . **Sept 22**
Southwest Antique Show/Sale (Yuma, AZ) . . . . **Jan 13**
Virginia Highlands Fest (Abingdon, VA) . . **July 29**
West London Antiques Fair (London, England) . . . . **Jan 12**
Westminster Antiques Fair (London, England) . . . . **Apr 27, Nov 30**
Anton, Susan: Birth . . . . Oct 12
Antonioni, Michelangelo: Birth . . . . Sept 29
Ants: Carpenter Ant Awareness Week . . . . June 18
Anxiety: Chicken Little Awards . . . . Apr 3
Anxiety Month, Natl . . . . Apr 1
Apache Wars Begin: Anniv . . . . Feb 4
Apartheid Law, South Africa Repeals Last: Anniv . . . . June 17
Aphelion, Earth at . . . . July 3
Apollo I: Spacecraft Fire Anniv . . . . Jan 27
Appalachia Museum Tennessee Fall Homecoming (Norris, TN) . . . . Oct 11
Appalachian: Kentucky Hills Weekend (Corbin, KY) . . . . Mar 3
Appert, Nicolas: Birth Anniv . . . . Oct 23
Apple Blossom Fest (Annapolis Valley, NS, Canada) . . . . May 25
Apple Blossom Fest, Shenandoah (Winchester, VA) . . . . May 4
Apple Blossom Fest, Washington State (Wenatchee, WA) . . . . Apr 27
Apple Butter Fest (Burton, OH) . . . . Oct 7
Apple Butter Stirrin' (Coshocton, OH) . . . . . Oct 21
Apple Fest (Topeka, KS) . . . . Oct 1
Apple Fest, Jackson County (Jackson, OH) . . . . Sept 19
Apple Fest, Kentucky (Paintsville, KY) . . . . . . Oct 1
Apple Fest, Vermont (Springfield, VT) . . . . Oct 7
Apple Harvest Fest (Gettysburg, PA) . . . . Oct 7
Apple Harvest Fest, Burlington Old-Fash (Burlington, WV) . . . . Sept 30
Apple Jack Month, Natl . . . . Oct 1
Apple Squeeze, Steilacoom (Steilacoom, WA) . . . . Oct 15
Apple Week, Beatles' Natl: Anniv . . . . Aug 11
Applejack Fest (Nebraska City, NE) . . . . Sept 16
Apples Weekend, A Is for (Chadds Ford, PA) . . . . Oct 14
Appleseed Days, Johnny (Lake City, MN) . . . Oct 7
Appleseed: Johnny Appleseed Day . . . . Mar 11
Appleseed, Johnny: Birth Anniv . . . . Sept 26
April Fools' Day . . . . Apr 1
Aquarius Begins . . . . Jan 20
Aquatennial (Minneapolis, MN) . . . . July 14
Aquatic: Project Aware Month . . . . Sept 1
Aquino, Benigno: Assassination Anniv . . . . Aug 21
Aquino, Corazon: Birth . . . . Jan 25
Arab Oil Embargo Lifted: Anniv . . . . Mar 13
Arbor Day (Arizona) . . . . Feb 3
Arbor Day (Arizona) . . . . Apr 7
Arbor Day: Anniv (Nebraska) . . . . Apr 22
Arbor Day Fest (Nebraska City, NE) . . . . Apr 28
Arbor Day, Natl (Proposed) . . . . Apr 28
Arbor Day, Natl 6554 . . . . Page 488
Arcadia Daze (Arcadia, MI) . . . . July 21
Arcaro, Eddie: Birth . . . . Feb 19
Archer, Anne: Birth . . . . Aug 25
Archery Classic, Natl Atlantic City (Atlantic City, NJ) . . . . Apr 21
Archery: Southeastern Sports-A-Rama (Chattanooga, TN) . . . . Aug 11
Archibald, Nate: Birth . . . . Apr 18
Argentina
Independence Day . . . . **May 25**
National Holiday . . . . **Apr 2**
Aries Begins . . . . Mar 21
Arizona
ACBL Bridge Tournament (Yuma) . . . . **Feb 23**
Admission Day . . . . **Feb 14**
All States Picnic (Yuma) . . . . **Jan 18**
Americana Indian and Western Art Show/Sale (Yuma) . . . . **Feb 3**
Apache Wars Begin: Anniv . . . . **Feb 4**
Arbor Day . . . . **Feb 3, Apr 7**
Arizona Renaissance Fest (Apache Junction) . . . . **Feb 11**
Arizona State Chmpshp Chili Cookoff (Bullhead City) . . . . **Apr 29**
Butterfield Overland Stage Days (Benson) . . . . **Oct 13**
Circle K Intl Convention (Phoenix) . . . . **Aug 5**
Country Fair and Burro Barbecue (Bullhead City) . . . . **May 13**
Country/Bluegrass Fest (Prescott) . . . . . . **July 15**
Cowboy Christmas (Wickenburg) . . . . **Dec 1**
Cowboy Hall of Fame Ceremony (Willcox) . **Oct 5**
Culinary Fest (Scottsdale) . . . . **Apr 21**
Desert Foothills Music Fest (Carefree) . . . **Feb 15**
Doll Show (Yuma) . . . . **Feb 4**
Fest in the Pines (Flagstaff) . . . . **Aug 4**
Four Corner States Bluegrass Fest (Wickenburg) . . . . **Nov 10**
Gold Rush Days (Wickenburg) . . . . **Feb 10**
Helldorado Days (Tombstone) . . . . **Oct 20**
Jazz Fest (Lake Havasu) . . . . **Jan 13**
La Fiesta De Los Vaqueros (Tucson) . . . . . **Feb 22**
Loggers/Sawdust Fest: State Chmpshp (Payson) . . . . **July 15**
London Bridge Days (Lake Havasu City) . . **Oct 2**
Lost Dutchman Days (Apache Junction) . **Feb 24**
Mill Avenue Masquerade Adventure (Tempe) . . . . **Oct 31**
NBA: All-Star Weekend (Phoenix) . . . . **Feb 10**
Oatman Sidewalk Egg Frying Contest (Oatman) . . . . **July 4**
Old Time Fiddlers' Contest (Payson) . . . . **Sept 23**
Old Town Tempe Fall Fest of Arts (Tempe) **Dec 1**
Relics and Rods Run to the Sun (Lake Havasu City) . . . . **Oct 11**
Rex Allen Days (Willcox) . . . . **Oct 6**
Route 66 Days (Ash Fork) . . . . **June 24**
Sedona Arts Fest (West Sedona) . . . . **Oct 14**
Senior Pro Rodeo (Payson) . . . . **May 27**
Sonora Showcase (Yuma) . . . . **Jan 28**
Southwest Antique Show/Sale (Yuma) . . . **Jan 13**
Southwest Rhino Exhib (Tucson) . . . . **Nov 10**
Southwest Senior Chmpshp (Yuma) . . . . . **Jan 17**
SPEBSQSA Mid-Winter Convention (Tucson) . . . . **Jan 22**
Tempe Fiesta Bowl Block Party (Tempe) . **Dec 31**
Territorial Days (Prescott) . . . . **June 10**
WIBC Queen's Tourn (Tucson) . . . . **May 14**
Women's Intl Bowling Congress An Mtg (Tucson) . . . . **May 1**
Women's Intl Bowling Congress Chmpshp (Tucson) . . . . **Mar 30**
World's Oldest Continuous PRCA Rodeo (Payson) . . . . **Aug 18**
World's Oldest Rodeo (Prescott) . . . . **June 29**
Yuma Jaycee's Silver Spur Parade (Yuma) . **Feb 4**
Yuma Jaycee's Silver Spur Rodeo (Yuma) **Feb 10**
Yuma Square and Round Dance Fest (Yuma) . . . . **Mar 3**
Arkansas
Admission Day . . . . **June 15**
Antique Auto Show/Swap Meet (Morrilton) . . . . **June 14**
Arkansas Craft Guild Christmas Show (Little Rock) . . . . **Dec 1**
Arkansas Craft Show/Sale (Eureka Springs) . . . . **Oct 20**
Bradley County Pink Tomato Fest (Warren) . . . . **June 8**
Catfish Fest (Eudora) . . . . **May 13**
Christmas Showcase/Sale (Little Rock) . . . **Dec 1**
Chuckwagon Races, Natl Chmpshp (Clinton) . . . . **Sept 1**
Clothesline Fair (Prairie Grove) . . . . **Sept 2**
Dermott's Annual Crawfish Festival (Dermott) . . . . **May 19**
Duck Calling Contest/Wings Over Prairie Fest (Stuttgart) . . . . **Nov 21**
Eagles Et Cetera (Bismarck) . . . . **Jan 27**
Fordyce on the Cotton Belt Fest (Fordyce) . . . . **Apr 24**
Great Arkansas Pig-Out (Morrilton) . . . . . . **Aug 4**
Great Cardboard Boat Race(Heber Springs) . . . . **July 22**
Hope Watermelon Fest (Hope) . . . . **Aug 17**
Johnson City Peach Festival (Clarksville) **July 20**
King Biscuit Blues Fest (Helena) . . . . **Oct 13**
Mosquito Awareness Weekend (Walcott) . **Aug 20**
Old Fort River Fest (Fort Smith) . . . . **June 16**
Original Ozark Fest (Ozark) . . . . **June 2**
Orphan Train Heritage Society Reunion (Fayetteville) . . . . **Oct 5**
Ozark Trail Festival (Heber Springs) . . . . . . **Oct 6**
Ozark UFO Conf (Eureka Springs) . . . . **Apr 7**
Picklefest (Atkins) . . . . **May 19**
Quadrangle Fest (Texarkana) . . . . **Sept 9**
Rhythm on the River (Little Rock) . . . . . . **Sept 23**
Riverfest (Little Rock) . . . . **May 26**
Rotary Tiller Race, World/Purple Hull Pea Fest (Emerson) . . . . **June 24**

Index ☆ Chase's 1995 Calendar of Events ☆

Arkansas (cont'd)——Arts & Crafts

Spinach Fest (Alma) . . . . Apr 14
Springfest (Heber Springs) . . . . Apr 21
State Fair and Livestock Show (Little Rock) . . . . Oct 6
Toad Suck Daze (Conway) . . . . May 5
Tontitown Grape Fest (Tontitown) . . . . Aug 16
Town Meeting on Main St, Natl (Little Rock) . . . . May 14
War Eagle Fair (War Eagle) . . . . Oct 19
White River Water Carnival (Batesville) . . July 29
Wild Turkey Calling, Natl/Festival (Yellville) . . . . Oct 13

Arkin, Alan: Birth . . . . Mar 26
Arledge, Roone: Birth . . . . July 8
Arlen, Harold: Birth Anniv . . . . Feb 15
Armatrading, Joan: Birth . . . . Dec 9
Armed Forces Day (Egypt) . . . . Oct 6
Armed Forces Day (Pres Proc) . . . . May 20
Armed Forces Day 6693 . . . . Page 488
Armenia: National Day . . . . Sept 21
Armenian Appreciation Day . . . . Apr 3
Armenian Christmas . . . . Jan 6
Armenian Martyrs Day . . . . Apr 24
Armistice Day . . . . Nov 11
Armstrong, BJ: Birth . . . . Sept 9
Armstrong, Louis: Birth Anniv . . . . Aug 4
Armstrong, Louis: Not Born This Day . . . . July 4
Armstrong, Neil Alden: Birth . . . . Aug 5
Armstrong, Trace: Birth . . . . Oct 15
Armstrong-Jones, Anthony: Birth . . . . Mar 7
Army and Navy Union Day (MA) . . . . Dec 9
Army Birthday . . . . June 14
Army Day (Lesotho, South Africa) . . . . Jan 20
Army First Desegregation, US: Anniv . . . . July 26
Army-Navy Football Game (Philadelphia, PA) . . . . Dec 2
Arnaz, Desi, Jr.: Birth . . . . Jan 19
Arness, James: Birth . . . . May 26
Arnold, Benedict: Birth Anniv . . . . Jan 14
Arnold, Eddy: Birth . . . . May 15
Arnold, Henry "Hap": Birth Anniv . . . . June 25
Arnold, Matthew: Birth Anniv . . . . Dec 24
Arnold, Roseanne: Birth . . . . Nov 3
Arnold, Sir Malcolm: Birth . . . . Oct 21
Arquette, Rosanna: Birth . . . . Aug 10
Art: Fine and Performing
Affaire in Gardens, Beverly Hills (Beverly Hills, CA) . . . . May 20
Alabama Impact: Exhib (Huntsville, AL) . . . July 2
Alabama Jubilee (Decatur, AL) . . . . May 27
Americana Indian and Western Art Show/Sale (Yuma, AZ) . . . . Feb 3
Ann Arbor Film Festival (Ann Arbor, MI) . Mar 14
Ann Arbor Spring Art Fair (Ann Arbor, MI) May 6
Ann Arbor Summer Art/Street Fair (Ann Arbor, MI) . . . . July 19
Ann Arbor Winter Art Fair (Ann Arbor, MI) Oct 28
Art and Antique Fair (Breda, Holland) . . . . Apr 8
Art Faire (Lahaska, PA) . . . . June 3
Art in the Park Fest (Oakland Park, FL) . . . Apr 8
Artists' Natl Juried Competition (St. Simon's Is, GA) . . . . Mar 25
Artists' Salute Black History Month (Los Angeles, CA) . . . . Feb 4
Artquake (Portland, OR) . . . . Sept 2
Arts Fest XI (Annapolis, NS, Canada) . . . Sept 14
Arts in the Park (Kalispell, MT) . . . . July 21
Arts/Quincy Fest (Quincy, IL) . . . . Sept 17
Arundel Festival (Arundel, England) . . . . Aug 25
Ballet Introduced to the US . . . . Feb 7
Balloon Rally and Arts and Crafts Show (Angel Fire, NM) . . . . July 14
Bazaart (Regina, Sask, Canada) . . . . June 17
Bel Air Fest for the Arts (Bel Air, MD) . . . Sept 17
Belfast Fest at Queen's (Belfast, Northern Ireland) . . . . Nov 6
Bentonsport Arts Festival (Keosauqua, IA) . . . . Oct 14
Birmingham Arts Salute Switzerland (Birmingham, AL) . . . . Apr 20
Black American Arts Festival (Greensboro, NC) . . . . Jan 14
Brooks, James David: Birth Anniv . . . . Oct 18
Bumbershoot: The Seattle Arts Fest (Seattle, WA) . . . . Sept 1
C.M. Russell Auction Orig Western Art (Great Falls, MT) . . . . Mar 15
Cartoon Art Appreciation Week . . . . May 2
Cartoonists Against Crime Day . . . . Oct 25
Celebrations Around World (Huntsville, AL) . . . . Dec 2
Central PA Fest of Arts (State College, PA) . . . . July 12
Chautauqua of the Arts (Columbus, IN) . Sept 16
Chicago Intl Art Expo (Chicago, IL) . . . . May 12
Children's Festival (Erwinna, PA) . . . . Sept 17
City of London Festival (London, England) . . . . June 20
City Stage (Greensboro, NC) . . . . Oct 7
Coconut Grove Arts Fest (Coconut Grove, FL) . . . . Feb 18
Cottonwood Prairie Fest/Sculpture Invit (Hastings, NE) . . . . June 2
Countryside Village Art Fair (Omaha, NE) June 3
Cultures Canada (Ottawa-Hull, Ont, Canada) . . . . July 2
Dare Day (Manteo, NC) . . . . June 3
Detroit Festival of the Arts (Detroit, MI) . Sept 15
Dogwood Arts & Crafts Fest (Huntington, WV) . . . . Apr 28
Dogwood Arts Fest (Knoxville, TN) . . . . Apr 7
Dogwood Arts Fest (Maryville, TN) . . . . Apr 6
Donna Reed Fest for the Performing Arts (Denison, IA) . . . . June 10
Downtown Denver Intl Buskerfest (Denver, CO) . . . . July 21
Dubuquefest/Very Special Arts (Dubuque, IA) . . . . May 17
Edinburgh Intl Fest/Fest Fringe (Edinburgh, Scotland) . . . . Aug 13
Epstein Young Artists Program: Anniv . . . Mar 24
Fairbanks Summer Arts Fest (Fairbanks, AK) . . . . July 21
Fall Arts Fest (Jackson, WY) . . . . Sept 15
Fall Fiesta of the Arts (Sumter, SC) . . . . Oct 21
Festival of the North (Ketchikan, AK) . . . . Feb 3
First Friday (Milwaukee, WI) . . . . Jan 6
First Night (Boston, MA) . . . . Dec 31
Flowers Exhib (Windsor Castle, Windsor, England) . . . . Apr 8
Folkmoot USA (Waynesvile, NC) . . . . July 20
From George to George: Monceaux (Grand Rapids, MI) . . . . Jan 1
Gainesville Downtown Fest and Art Show (Gainesville, FL) . . . . Nov 18
Gallery Night (Milwaukee, WI) . . . . Jan 20
Gasparilla Sidewalk Art Fest (Tampa, FL) . Mar 4
George Spelvin Day . . . . Nov 15
Gold Coast Art Fair (Chicago, IL) . . . . Aug 11
Grant Wood Art Festival (Stone City-Anamosa, IA) . . . . June 11
Great Northern Arts Fest (Inuvik, NWT, Canada) . . . . July 21
Gum Tree Fest (Tupelo, MS) . . . . May 13
Hood River Apple Jam (Hood River, OR) Aug 19
Hot Prospects (Brooklyn, NY) . . . . July 2
Intl Wildlife Art Expo (Vancouver, BC, Canada) . . . . May 5
Iroquois Arts Showcase I and II (Howes Cave, NY) . . . . May 27, July 8
Joensu Festival (Joensu, Finland) . . . . June 18
Just for Laughs Festival (Montreal, Que, Canada) . . . . July 20
Kansas River Valley Art Fair (Topeka, KS) July 15
Kennedy Center Imagination Celeb (Colorado Springs, CO) . . . . Apr 1
KIA Art Fair (Kalamazoo, MI) . . . . June 3
Kinetic Sculpture Race (Omaha, NE) . . . . Sept 9
Kohler Outdoor Arts Festival (Sheboygan, WI) . . . . July 15
Lakefront Festival of Arts (Milwaukee, WI) . . . . June 16
Las Olas Art Fest (Fort Lauderdale, FL) . . . Mar 4
Lathe-Turned Objects, Intl: Exhib (Huntsville, AL) . . . . Apr 16
Little Falls Sidewalk Arts/Crafts (Little Falls, MN) . . . . Sept 9
Macon Cherry Blossom Fest (Macon, GA) . . . . Mar 17
Maine Fest of the Arts (Brunswick, ME) . . . Aug 3
Maritime Fest (Charleston, SC) . . . . Sept 15
Mayfest (Glasgow, Scotland) . . . . Apr 28
Michigan Fest (East Lansing, MI) . . . . Aug 4
Midsummer Night's Fair (Norman, OK) . . July 14
Missouri River Fest of Arts (Boonville, MO) . . . . Aug 17
Mural-in-a-Day (Toppenish, WA) . . . . June 3
Nautical/Wildlife Art Fest/Craft Show (Ocean City, MD) . . . . Jan 14
New Oxford Flea Market/Art/Craft Show (Gettysburg, PA) . . . . June 17
New Prospects (Brooklyn, NY) . . . . Mar 5
Northwest Folklife Fest (Seattle, WA) . . . . May 26
Ohio River Fest for the Arts (Evansville, IN) . . . . May 13
Pacific Rim Wildlife Art Show (Tacoma, WA) . . . . Sept 22
Panoply (Huntsville, AL) . . . . May 12
Pier 39 Street Performers Festival (San Francisco, CA) . . . . June 10
Pieta Attacked: Anniv . . . . May 21
Pin Ear on the Van Gogh (Ft Wayne, IN) . . Mar 30
Quebec Intl Summer Fest (Quebec, Que, Canada) . . . . July 6
Red Cloud Indian Art Show (Pine Ridge, SD) . . . . June 4
Rittenhouse Square Art Annual (Philadelphia, PA) . . . . June 7
Riverfest (Little Rock, AR) . . . . May 26
Roanoke Festival in the Park '95 (Roanoke, VA) . . . . May 25
Rockport Art Festival (Rockport, TX) . . . . July 1
Royal Natl Eisteddfod (Colwyn Bay, North Wales) . . . . Aug 5
Russell, Charles M.: Birth Anniv . . . . Mar 19
Sa'Fest Intl Cartoon Art Competition . . . . Oct 1
St. Simons Arts Fest (St. Simons Island, GA) . . . . Oct 7
Salvador Dali Museum: Anniv . . . . Mar 7
Savonlinna Opera Fest (Savonlinna, Finland) . . . . July 8
Seven Florentine Heads (Windsor Castle, England) . . . . Sept 16
Shaw Fest (Niagara-on-the-Lake, Ont, Canada) . . . . Apr 18
Snow Hill Heritage Weekend (Snow Hill, MD) . . . . Sept 22
Society of Illustrators Annual Exhib (New York, NY) . . . . Feb 11
Sound/Light Spectacular (Wilmington, NC) . . . . June 2
South Mountain Fair (Gettysburg, PA) . . . Aug 30
Southern Gospel Music Month . . . . Sept 1
Spring Fling (Wichita Falls, TX) . . . . Apr 22
Square Fair (Lima, OH) . . . . Aug 4
Standard Bank Natl Arts Fest (Grahamstown, S Africa) . . . . June 29
Stratford-upon-Avon Fest (England) . . . . July 8
Stratton Arts Festival (Stratton Mt, VT) . . Sept 17
Summerfest Art Faire (Logan, UT) . . . . June 15
Tampere Theatre Fest (Tampere, Finland) Aug 15
Tawas Bay Waterfront Fine Art Show (Tawas City, MI) . . . . Aug 4
Texas Musical Drama (Canyon, TX) . . . . June 7
Three Rivers Arts Fest (Pittsburgh, PA) . . June 2
Totah Festival (Farmington, NM) . . . . Sept 1
Wildflower Fest of the Arts (Dahlonega, GA) . . . . May 20
Wildlife Expo (Orlando, FL) . . . . Feb 3
Winnipeg Fringe Fest (Winnipeg, Man, Canada) . . . . July 15
Winslow Homer engravings exhibit (Austin, TX) . . . . Jan 23
World of Drawings and Watercolours (London, England) . . . . Jan 25
YAM Programs for Youth Art Month (Huntsville, AL) . . . . Mar 12
Yellowstone Art Center's Auction (Billings, MT) . . . . Mar 4
Youth Art Month . . . . Mar 1

Arthritis Foundation National Telethon . . . . Apr 2
Arthritis Month, Natl . . . . May 1
Arthur, Beatrice: Birth . . . . May 13
Arthur, Chester A.: Birth Anniv . . . . Oct 5
Arthur, Ellen: Birth Anniv . . . . Aug 30
Articles of Confederation: Ratification Anniv Mar 1
Articles of Peace Anniv . . . . Nov 30
Artisans, Day of the . . . . Dec 4
Artists' Salute Black History Month (Los Angeles, CA) . . . . Feb 4
Arts and Crafts
Appalachian Potters Market (Marion, NC) . Dec 2
Arkansas Craft Show/Sale (Eureka Springs, AR) . . . . Oct 20
Art in the Park (Dempolis, AL) . . . . Oct 7
Artists in the Park (Wolfeboro, NH) . . . . Aug 16
Arts and Crafts Fair, Annual (Jay, VT) . . . . Oct 7
Arts and Crafts Fest (Bennington, VT) . . Sept 16
Arts and Crafts Show (East Tawas, MI) . June 24
Arts and Crafts Show, Balloon Rally and (Angel Fire, NM . . . . July 14
Beef-a-Rama/Arts and Crafts Show (Minocqua, WI) . . . . Sept 30
Belsnickel Craft Show (Boyertown, PA) . . Nov 24
Berea Craft Fest (Berea, KY) . . . . July 14
Brown's Crossing Fall Crafts Fair (Milledgeville, GA) . . . . Oct 21
Capitol Hill People's Fair (Denver, CO) . . . June 3
Carolina Crafts Labor Day Classic (Winston-Salem, NC) . . . . Sept 2
Carolina Craftsmen's Christmas Classic (Columbia, SC) . . . . Nov 24

☆ Chase's 1995 Calendar of Events ☆ Index

Carolina Craftsmen's Christmas Classic (Greensboro, NC)....................Nov 24
Carolina Craftsmen's Spring Classic (Greensboro, NC)..............Mar 3, Apr 7
Carolina Craftsmen's Summer Classic (Myrtle Beach, SC)..........................Aug 11
Catoctin Colorfest Arts and Crafts Show (Thurmont, MD).....................Oct 14
Cedarhurst Craft Fair (Mt Vernon, IL)....Sept 9
Celebration of Quilts (Chardon, OH)....May 20
Cherished Treasures Quilt Display (Omaha, NE)..................................Feb 14
Christmas Craft Fair USA: Indoor Show (Milwaukee, WI).......................Dec 9
Christmas Craft Show (Aiken, SC)........Dec 2
Christmas Craft Show (York, PA).........Dec 3
Christmas Fair and Gallery (Norman, OK)................................Nov 25
Christmas Showcase/Sale (Little Rock, AR)..................................Dec 1
Clothesline Fair (Prairie Grove, AR)......Sept 2
Codman House Artisans' Fair (Lincoln, MA)................................Sept 10
Commonwheel Arts and Crafts Fest (Manitou Springs, CO).........................Sept 2
Copper Magnolia Fest (Washington, MS) Sept 23
Country Affair, A (Menomonee Falls, WI).Oct 21
Craft Fair of the Southern Highlands (Asheville, NC)................................July 20
Craft Fair USA: Indoor Show (Milwaukee, WI)...........Apr 22, July 22, Sept 23, Oct 28
Craft Month, Natl.......................Mar 1
Dankfest (Harmony, PA)................Aug 26
Depot Fest of the Arts (Livingston, MT)...July 1
Dyersville Festival of the Arts (Dyersville, IA).................................Sept 23
East Coast Craft Trade Show (Elmwood Park, NJ)..................................May 6
Easter Craft Show (York, PA)...........Feb 12
Eisenfest (Amana, IA).................Sept 15
Fells Pt Fun Fest (Baltimore, MD).........Oct 7
Fest in the Pines (Flagstaff, AZ)..........Aug 4
Fest of the Arts (Weatherford, OK).......Sept 9
Festival: Fall Fest of Arts and Crafts (Dalton, GA).................................Sept 9
Festival of Native Arts (Fairbanks, AK)...Feb 23
Festival-in-the-Park (Nutley, NJ)........Sept 10
Fiddlers Jamboree and Crafts Fest (Smithville, TN)................................June 30
Finger Lakes Christmas Craft Show (Rochester, NY).................................Oct 28
Finger Lakes Craftsmen Arts/Crafts Show (Rochester, NY).....................Nov 24
Finger Lakes Spring Arts/Crafts Show (Rochester, NY)......................Apr 8
Flax Scutching Fest (Stahlstown, PA)....Sept 9
Forest Grove Arts and Crafts Show (Forest Grove, MT)..........................Oct 22
Gift Festival, Intl (Fairfield, PA)...........Nov 9
Grand Detour Art Fair (Grand Detour, IL)Sept 17
Great Mississippi R Arts/Crafts Fest (Hannibal, MO)................................May 27
Holiday Craft and Gift Show (Milwaukee, WI).................................Nov 24
Holiday Craft Show (York, PA)...........Oct 15
Holiday Highlights (Arapahoe, NE)......Nov 10
Holzfest (Amana, IA)...................Aug 18
Indian Dance and Crafts Fest (Childersburg, AL)..................................Apr 1
Jean Lake Fest (Troy, AL)................May 6
Jonathan Hager Frontier Craft Days (Hagerstown, MD)....................Aug 5
Jour de Fete (St. Genevieve, MO)........Aug 12
Jubilee (Bennettsville, SC)...............May 13
Knitting/Needlecraft/Design Exhib (Esher, England).............................Jan 19
League of NH Craftsmen's Fair (Sunapee, NH).................................Aug 5
Mactaquac Craft Fest (Fredericton, NB, Canada).............................Sept 2
Market Square Fair (Fredericksburg, VA).May 13
Melrose Plantation Arts/Crafts Fest (Natchitoches, LA).................June 10
Mifflin-Juniata Arts Fest (Lewistown, PA) May 20
Moncton Sale of Fine Craft (Fredericton, NB)...............................Nov 11
Mossy Creek Barnyard Fest (Warner Robins, GA)................................Apr 22
Nautical North American Arts/Crafts Show (Ocean City, MD)....................Jan 14
New Beginning Fest (Coffeyville, KS)....Apr 27
NW Knitting/Needlecraft Exhib (Manchester, England).............................Feb 23
Old Crafts Day (Fond du Lac, WI)........July 9
Old Fort River Fest (Fort Smith, AR)....June 16
Old Town Tempe Fall Fest of Arts (Tempe, AZ)..................................Dec 1
Original Ozark Fest (Ozark, AR).........June 2
Palm Harbor Arts/Crafts/Music Fest (Palm Harbor, FL)..........................Dec 2
Palm Harbor Day Arts/Crafts/Music Fest (Palm Harbor, FL)...........................May 7
Pioneer and Indian Festival (Ridgeland, MS)................................Oct 28
Potters Fest, Intl (Wales)..............July 14
Powers' Crossroads Country Fair/Art Fest (Newnan, GA).......................Sept 2
Prairie Arts Festival (Westpoint, MS).....Sept 2
Prater's Mill Country Fair (Dalton, GA)..May 13
Quilt Competition/Display (Lahaska, PA).Jan 30
Quilting in the Tetons (Jackson, WY)......Oct 1
Rainbow of Arts Craft Fest (Chesterfield, VA)................................Sept 23
Robidoux Rendezvous (St. Joseph, MO).May 13
Rothesay Craft Fest (Rothesay, NB, Canada)............................July 15
Saint John Sale of Fine Craft (St John, NB, Canada)............................Nov 25
Salem Art Fair and Fest (Salem, OR)....July 21
Saskatchewan Handcraft Fest (Battleford, Sask, Canada)............................July 14
Scarecrow Contest (Lahaska, PA)......Sept 11
Southeastern New Mexico Arts/Crafts Fest (Lovington, NM).......................Nov 4
Southeastern Wildlife Expo (Charleston, SC)................................Feb 17
Southern Knitting/Needlecraft Exhibit (Bristol, England)............................Mar 17
Spinning and Weaving Week, Natl.........Oct 2
Spring Arts Fest (Gainesville, FL).........Apr 1
Spring Craft Show (York, PA)...........Apr 30
Spring Craft/Gift Show (Millwaukee, WI).Mar 11
Spruce Bog Craft Fair (Whitehorse, Yukon, Canada)............................Feb 20
State Craft Festival (Richboro, PA).......Oct 13
Sugarloaf Craft Fest (Gaithersburg, MD)...................Apr 7, Nov 17, Dec 8
Sugarloaf Craft Fest (Manassas, VA).....Sept 8
Sugarloaf Craft Fest (Timonium, MD)..........................Apr 28, Oct 13
Sugarloaf's Crafts Fest (Somerset, NJ)........................May 19, Sept 29
Summer Craft Show (York, PA).........July 16
SummerFair (Billings, MT)..............July 15
Summerfest (Harrisonburg, VA).........July 30
Sundog Handcraft Faire (Saskatoon, Sask, Canada).............................Dec 1
Swappin' Meetin' (Cumberland, KY).......Oct 6
Tarpon Springs Crafts Fest (Tarpon Springs, FL)..................................Apr 8
Tennessee Crafts Fair (Nashville, TN)....May 5
Territorial Days (Prescott, AZ)..........June 10
Toms River Wildfowl Art & Decoy Show (Toms River, NJ)............................Feb 4
Totah Festival (Farmington, NM).........Sept 1
Tupper Lake Woodsmen's Days/Chainsaw Sculpturing (NY).....................July 8
Virginia Christmas Show (Manassas, VA).Oct 13
Virginia Craftsmen's Christmas Classic (Richmond, VA)......................Nov 3
Virginia Spring Show (Richmond, VA)....Mar 10
Virginia-Carolina Craftmen's Fall Classic (Roanoke, VA).......................Sept 29
War Eagle Fair (War Eagle, AR)..........Oct 19
Waterfowl Festival (Easton, MD)........Nov 10
West Virginia Glass Fest (Wheeling, WV)..Aug 4
Wintergreen (Regina, Sask, Canada)....Nov 24
Wood Carvers Fest (Blackduck, MN)....July 29
World Chmpshp BBQ Goat Cookoff/Crafts Fair (Brady, TX)..........................Sept 2
Wyandotte Heritage Days (Wyandotte, MI)...................................Sept 8
Year of American Craft 6516..........Page 488
Aruba: Flag Day.........................Mar 18
Ascension Day..........................May 25
Ascension of Baha'u'llah.................May 29
Ash Wednesday...........................Mar 1
Ashe, Arthur: Birth Anniv................July 10
Ashford, Emmet: Birth Anniv..............Nov 23
Ashford, Evelyn: Birth...................Apr 15
Ashford, Nickolas: Birth..................May 4
Ashley, Elizabeth: Birth..................Aug 30
Ashura (Muslim Holiday)..................June 8
Asia and Pacific, UN Transport/Communications Decade for.............................Jan 1
Asian/Pacific American Heritage Month 6557............................Page 488
Asimov, Isaac: Death Anniv................Apr 6
Asner, Ed: Birth.........................Nov 15
Aspin, Les: Birth........................July 21
Aspinwall Crosses US on Horseback: Anniv July 8
Assassination Attempt, Lenin: Anniv......Aug 30
Assassination Attempt: Pope John Paul II: Anniv...............................May 13
Assassinations Report, Committee on: Anniv................................Mar 29
Assembly Day, First (Yorktown, VA).......July 29
Assumption of The Virgin Mary...........Aug 15
Astin, John: Birth.......................Mar 30
Astrological Signs
Aquarius..............................Jan 20
Aries..................................Mar 21
Cancer...............................June 21
Capricorn.............................Dec 22
Gemini...............................May 21
Leo..................................July 23
Libra................................Sept 23
Pisces................................Feb 20
Sagittarius...........................Nov 23
Scorpio...............................Oct 23
Taurus................................Apr 20
Virgo................................Aug 23
Astrology Day, International..............Mar 21
Astronomy Day...........................May 6
Astronomy Week.........................Apr 10
AT&T Divestiture: Anniv....................Jan 8
Atchison, David R.: Birth Anniv...........Aug 11
Atkins, Chet: Birth......................June 20
Atlantic Charter 50th Signing: Anniv......Aug 14
Atlantic Telegraph Cable Laid: Anniv......July 27
Atlas, Charles: Birth Anniv...............Oct 30
Atomic Bomb, 1st Dropped Inhabited Area: Anniv..................................Aug 6
Atomic Bomb, 2nd Dropped Inhabited Area: Anniv..................................Aug 9
Atomic Bomb Tested: Anniv...............July 16
Atomic Plant Begun, Oak Ridge: Anniv.....Aug 1
Attenborough, David: Birth................May 8
Attenborough, Sir Richard: Birth..........Aug 29
Attucks, Crispus: Day.....................Mar 5
Atwood Early Rod Run (Atwood, KS).....May 20
Atwood, Margaret Eleanor: Birth.........Nov 18
Auden, Wystan H.: Birth Anniv............Feb 21
Audubon, John J.: Birth Anniv............Apr 26
Auerbach, Arnold "Red": Birth...........Sept 20
Augusta Futurity (Augusta, GA)............Jan 6
Austen, Jane: Birth Anniv................Dec 16
Austin, Stephen F.: Birth Anniv............Nov 3
Austin, Tracy: Birth......................Dec 12
Australia
Anzac Day............................Apr 25
Australia Day.........................Jan 26
Australia Day Observance...............Jan 30
Canberra Day.........................Mar 19
Eight Hour Day (Labor Day)...............Mar 5
Melbourne Music Fest (Melbourne)......Feb 10
Picnic Day.............................Aug 7
Proclamation Day......................Dec 28
Recreation Day.........................Nov 6
Australians: Hug an Australian Day........Apr 26
Austria
Mozart Week (Salzburg)................Jan 20
National Day..........................Oct 26
Silent Night, Holy Night Celebrations....Dec 24
Authors' Day, Natl........................Nov 1
Auto Shows and Races
Antique and Classic Car Show (Bennington, VT)..............................Sept 15
Antique Auto Show and Flea Market (Boyne City, MI)..........................Aug 12
Antique Auto Show/Collector Car Fest (St. Ignace, MI).........................June 22
Antique Auto Show/Swap Meet (Morrilton, AR)................................June 14
Antique Auto Swap Meet (Grand Prairie, TX)................................June 10
ARCA 200 Late Model Stock Race (Daytona Beach, FL)..........................Feb 12
Atlantique City Antique/Classic Car Auction (Atlantic City, NJ)......................Feb 17
Atwood Early Rod Run (Atwood, KS)....May 20
Auburn-Cord-Dusenberg Fest (Auburn, IN).................................Aug 31
Auto Battery Safety Month...............Oct 1
Auto Show, Intl (Baltimore, MD)..........Feb 4
Automotion (Wisconsin Dells, WI).......May 13
Black Hills Jeep Jamboree (Sturgis, SD)....Oct 6
Braille Rallye (Los Angeles, CA).........July 16
British Grand Prix (Towcestere, England).July 7
Burlington Cruise Nights (Burlington, IA) Apr 28
Busch Clash (Daytona Beach, FL).......Feb 12
Car Care Month, Natl.....................Oct 1

Arts & Crafts (cont'd)—Auto Shows & Races

Index ☆ Chase's 1995 Calendar of Events ☆

Auto Shows & Races (cont'd)—Balloons

Car-Keeping Month, Natl .... **May 1**
Center of Nation All-Car Rally (Belle Fourche, SD) .... **June 9**
Checkerfest (Indianapolis, IN) .... **May 1**
Chimney Rock Hillclimb (Chimney Rock, NC) .... **Apr 28**
Chrysler Opens Detroit Factory .... **Mar 31**
Classic Car Show (Norfolk, NE) .... **July 29**
Codman House Antique Vehicle Meet (Lincoln, MA) .... **July 16**
Corvette and High Performance Meets (Puyallup, WA) .... **Feb 11, June 24**
Country Fair on Square/All Car Show (Gainesville, TX) .... **Oct 7**
Daytona Beach Fall Speed Spectacular (Daytona Bch, FL) .... **Nov 24**
Daytona Beach Spring Speed Spectac (Daytona Bch, FL) .... **Apr 1**
Daytona 500 NASCAR-FIA Winston Cup (Daytona Bch, FL) .... **Feb 19**
Death Busters Day, Natl .... **May 26**
Edsel Owners Club Conv (Oakland, CA) .. **Aug 2**
Elkhart Grand Prix/Motor Sports Weekend (Elkhart, IN) .... **July 7**
Esso Vintage Vehicle Run (Bristol, England) .... **June 11**
55 mph Speed Limit: Anniv .... **Jan 2**
Flatlanders Classic/Early Iron Rod Run (Goodland, KS) .... **Sept 22**
Flemington Speedway Racing Season (Flemington, NJ) .... **Apr 29**
Florida 200 (Daytona Beach, FL) .... **Feb 17**
Foxtail Drive-in Car Show (Auburn, IN) .. **May 28**
Fremont Car Show (Fremont, MI) .... **July 22**
GMC Truck Mt Wash Hillclimb (Gorham, NH) .... **June 23**
Goody's 300/NASCAR Busch Grand Natl (Daytona Beach FL) .... **Feb 18**
Grand Center Sport, Fishing & RV Show (Grand Rapids, MI) .... **Mar 16**
Green Mountain Nationals Car Show (Bennington, VT) .... **June 25**
Gunston Hall Car Show (Lorton, VA) .... **Sept 10**
Hot Rod Assn US Natls (Indianapolis, IN) **Aug 31**
Indy 500-Mile Race (Indianapolis, IN) .... **May 28**
International Race of Champions (Daytona Beach, FL) .... **Feb 17**
Kass Kounty Mud Drags (Plattsmouth, NE) .... **Sept 7**
Korn Klub Mud Drags (Plattsmouth, NE) **June 11**
KRXL Car Cruise (Kirksville, MO) .... **Aug 12**
Little 500 (Anderson, IN) .... **May 27**
Loudon Camel Classic (Loudon, NH) ... **June 16**
Mini Grand Prix (Fond du Lac, WI) ...... **May 28**
Name Your Car Day .... **Oct 2**
NC RV and Camping Show (Charlotte, NC) **Feb 3**
NC RV and Camping Show (Greensboro, NC) .... **Mar 24**
NC RV and Camping Show (Raleigh, NC) **Mar 10**
NC RV and Travel Show (Fayetteville, NC) **Jan 13**
On the Waterfront Swap Meet/Car Corral (St. Ignace, MI) .... **Sept 15**
Pepsi 400 Nascar (Daytona, FL) .... **July 1**
Pontiac Camper, Travel & RV Show (Pontiac, MI) .... **Jan 25**
Project Red Ribbon .... **Nov 1**
Project Safe Baby Month .... **May 1**
RAC London/Brighton Veteran Car Run (London, England) .... **Nov 5**
Rally 'Round Bucks County (Langhorne, PA) .... **Oct 1**
Recreational Vehicle Show (Timonium, MD) .... **Feb 17**
Relics and Rods Run to the Sun (Lake Havasu City, AZ) .... **Oct 11**
Rolex 24 at Daytona (Daytona, FL) .... **Feb 4**
Royal Scottish Auto Club Rally (Scotland) .... **June 2**
Speed Weeks (Daytona Beach, FL) .... **Feb 4**
Street Rod/Muscle Car Show (Boyne City, MI) .... **Sept 2**
Streetscene (Covington, VA) .... **Aug 5**
Sugar Valley Rally (Scottsbluff, NE) ...... **June 2**
Summer Solstice at Carhenge (Alliance, NE) .... **July 29**
Twin 125-Mile Qualifying Races (Daytona Beach, FL) .... **Feb 16**
Vintage Auto Race (Steamboat Springs, CO) .... **Sept 1**
Winston Select 500 Race Week (Talladega, AL) .... **May 4**
WKA Daytona Dirt Karting Races (Daytona Bch, FL) .... **Dec 26**
World Enduro Chmpshp Kart Races (Daytona Beach, FL) .... **Dec 27**
Automobile Speed Reduction: Anniv ...... Nov 25
Autry, Gene: Birth .... Sept 29
Autry, Micajah, Day (Autryville, NC) .... July 15
Autry, Micajah: Drama (Autryville, NC) .... July 13
Autumn Begins .... Sept 23
Autumn Glory Fest (Oakland, MD) .... Oct 12
Autumn, Halfway Point .... Nov 7
Autumn Magic/Tom Mix Festival (Guthrie, OK) .... Sept 8
Autumn Market Fair, Eighteenth-Century (McLean, VA) .... Oct 21
Avalon, Frankie: Birth .... Sept 18
Avedon, Richard: Birth .... May 15
Aviation
Abbotsford Intl Airshow (Abbotsford, BC, Canada) .... **Aug 11**
Aerobatic Chmpshps, Intl (Fon du Lac, WI) .... **Aug 6**
Aerospace America (Oklahoma City, OK) **July 14**
Air Show and Fireworks Display (Rockport, TX) .... **July 4**
Amelia Earhart Atlantic Crossing: Anniv . **May 20**
Antique Airplane Fly-In (Atchison, KS) .. **May 26**
Aviation Day, Natl (Pres Proc) .... **Aug 19**
Aviation History Month .... **Nov 1**
Aviation in America: Anniv .... **Jan 9**
Aviation Week, Natl .... **Aug 14**
Berlin Airlift: Anniv .... **June 24**
Campbell Becomes 1st American Air ACE: Anniv .... **Apr 14**
Cayley, George: Birth Anniv .... **Dec 27**
China Clipper: Anniv .... **Nov 22**
Cleveland Natl Air Show (Cleveland, OH) . **Sept 2**
Commercial Air Flight Between the US and USSR Begins .... **July 15**
Cranbrook Air Fair (Cranbrook, BC, Canada) .... **July 26**
EAA Intl Fly-In Conv/Sport Aviation Exhib (Oshkosh, WI) .... **July 27**
Elm Farm Ollie Day Celebration (WI) ..... **Feb 18**
First Airplane Crossing English Channel: Anniv .... **July 25**
First Airship Crossing of Atlantic: Anniv . . **July 6**
First Balloon Flight Across English Channel: Anniv .... **Jan 7**
First Balloon Flight: Anniv .... **June 5**
First Fatal Aviation Accident: Anniv ..... **June 15**
First Flight Anniv Celeb (Kill Devil Hills, NC) .... **Dec 17**
First Free Flight by a Woman: Anniv ..... **June 4**
First Man-Powered Flight Across English Channel: Anniv .... **June 12**
First Man-Powered Flight: Anniv .... **Aug 23**
First Nonstop Flight Around World Without Refueling .... **Dec 23**
First Nonstop Transatlantic Flight: Anniv **June 14**
Flight Begins (Abbotsford, BC) .... **Aug 8**
Flight 232 Crashes in Iowa Cornfield .... **July 19**
Fun Flight (Alamogordo, NM) .... **June 16**
Fun Fly and Model Aircraft Show, Natl (Chimney Rock, NC) .... **Aug 19**
Jimmy Doolittle Quarter Midget Race (Falls of Rough, KY) .... **Sept 8**
Johnson, Amy: Flight Anniv .... **May 5**
Kennedy Intl Airport Dedication .... **July 31**
Levine, Charles A.: Death Anniv .... **Dec 6**
Lindbergh, Charles Augustus: Birth Anniv . **Feb 4**
Lindbergh Flight: Anniv .... **May 20**
Man Will Never Fly Society Meeting (Kitty Hawk, NC) .... **Dec 16**
Man's First Free Flight (Balloon): Anniv . . **Nov 21**
Military Air Transport Crashes in Indiana: Anniv .... **Feb 6**
Mint Julep Scale Meet (Falls of Rough, KY) .... **May 19**
Model Airplane Show (Danville, VA) .... **June 24**
Montgolfier, Jacques: Birth Anniv .... **Jan 7**
Pan Am Flight 103 Explosion: Anniv ..... **Dec 21**
Pan American Aviation Day (Pres Proc) . . **Dec 17**
Post, Wiley: Birth Anniv .... **Nov 22**
Red Deer Intl Air Show (Red Deer, AB, Canada) .... **Aug 5**
Saskatchewan Air Show '95 (Moose Jaw, Sask, Canada) .... **July 8**
Shearwater Intl Air Show (Shearwater, NS, Canada) .... **Sept 16**
Sikorsky, Igor: Birth Anniv .... **May 25**
Solo TransAtlantic Balloon Crossing: Anniv .... **Sept 14**
Spruce Goose Flight Anniversary .... **Nov 2**
Stearman Fly-In, Natl (Galesburg, IL) ..... **Sept 6**
Streeter, Ruth Cheney: Birth Anniv .... **Oct 2**
Tri-State Anti-Drug Air Show (Kenova, WV) .... **May 26**
US Air Force Established: Birth .... **Sept 18**
Warbirds in Action Air Show (Shafter, CA) .... **Apr 22**
Wausau Air Show and Fly-in (Wausau, WI) .... **June 18**
Wright Brothers First Powered Flight: Anniv .... **Dec 17**
Axelrod, George: Birth .... June 9
Aylwin, Patricio: Military Dictatorship Ended .... Dec 15
Azalea Fest (Muskogee, OK) .... Apr 8
Azalea Fest, Brookings-Harbor (Brookings, OR) .... May 26
Azerbaijan: National Day .... May 28
Aznavour, Charles .... May 22
Aztec Calendar Stone Discovery: Anniv . . . Dec 17

B

Bab. See Baha'i
Babbitt, Bruce: Birth .... June 27
Baby Food Fest, Natl (Fremont, MI) .... July 18
Baby Parade (Ocean City, NJ) .... Aug 10
Baby Safety Month .... Sept 1
Bacall, Lauren: Birth .... Sept 16
Bach, Catherine: Birth .... Mar 1
Bach, Johann Sebastian: Birth Anniv ...... Mar 21
Bacharach, Burt: Birth .... May 12
Bachelor Soc Wild Woman Contest/Auction (Talkeetna, AK) .... Dec 2
Back Week, Natl Save Your .... Oct 22
Backman, Wally: Birth .... Sept 22
Backwards Day .... Jan 27
Bacon, Nathaniel: Birth Anniv .... Jan 2
Bad Day Day, Have a .... Nov 19
Bad Poetry Day .... Aug 18
Baden-Powell, Robert: Birth Anniv .... Feb 22
Badger State Winter Games (Wausau, WI) . . . Feb 3
Badminton Day, Natl .... Mar 30
Badminton, Scottish Open Chmpshp (Glasgow, Scotland) .... Nov 22
Badoglio, Pietro: Birth Anniv .... Sept 28
Baer, Max, Jr: Birth .... Dec 4
Baez, Joan: Birth .... Jan 9
Bagelfest (Mattoon, IL) .... July 28
Baha'i
Ascension of Baha'u'llah .... **May 29**
Baha'i New Year's Day: Naw-Ruz .... **Mar 21**
Birth of Baha'u'llah .... **Nov 12**
Birth of the Bab .... **Oct 20**
Declaration of the Bab .... **May 23**
Fest of Ridvan .... **Apr 21**
Martyrdom of the Bab .... **July 9**
Race Unity Day .... **June 11**
Bahamas
Emancipation Day .... **Aug 7**
Fox Hill Day (Nassau) .... **Aug 8**
Independence Day .... **July 10**
Junkanoo .... **Dec 26**
Labor Day .... **June 2**
Miami/Bahamas Goombay Festival ...... **June 2**
Remembrance Day .... **Nov 9**
Bahrain: Independence Day .... Dec 16
Bailey, F. Lee: Birth .... June 10
Baines, Harold: Birth .... Mar 15
Baio, Scott: Birth .... Sept 22
Baked Bean Month, Natl .... July 1
Baker, Caroll: Birth .... May 28
Baker, James A., III: Birth .... Apr 28
Baker, Joe Don: Birth .... Feb 12
Baker, Russell: Birth .... Aug 14
Baker Street Irregulars Dinner/Cocktail Pty (New York, NY) .... Jan 6
Bakers Month, Natl Retail .... Jan 1
Bakery-Deli Assn, Retailers: Conv/Exhib (St. Louis, MO) .... Mar 11
Bakker, Jim: Birth .... Jan 2
Balanchine, George: Birth Anniv .... Jan 9
Balboa: Pacific Ocean Discovered: Anniv . Sept 25
Bald and Be Free Day, Be .... Oct 14
Bald Is Beautiful Convention (Morehead, NC) .... Sept 8
Baldwin, Alec: Birth .... Apr 3
Baldwin, James Arthur: Birth Anniv .... Aug 2
Baldwin, Roger Nash: Birth Anniv .... Jan 21
Balfour, Eric: Birth .... Apr 24
Balkans Days, Little/Folklife Fest (Pittsburg, KS) .... Sept 1
Ball, Lucille: Birth Anniv .... Aug 6
Ballesteros, Seve: Birth .... Apr 9
Ballet Introduced to the US .... Feb 7
Balloon, Hot Air
Aviation in America: Anniv .... **Jan 9**
Balloon Classic, Natl (Indianola, IA) ..... **July 28**
Balloon Crossing of Atlantic: Anniv ...... **Aug 17**
Black Hills Balloon Rally (Sturgis, SD) . . . **May 13**

☆ Chase's 1995 Calendar of Events ☆ Index

Balloons (cont'd)—Beer

Columbus Fest/Hot Air Balloon Regatta (Columbus, KS) . . . Oct 6
Farmington Invitational Balloon Fest (Farmington, NM) . . . May 27
First Balloon Flight Across English Channel: Anniv . . . Jan 7
First Balloon Flight: Anniv . . . June 5
First Balloon Honeymoon: Anniv . . . June 20
Great Wisc Dells Balloon Rally (WI Dells, WI) . . . June 3
Hot Air Balloon Show/Aero Extravaganza (Falls City, NE) . . . June 17
Hot-Air Balloon Fest (St Jean-sur-Richelieu, Que) . . . Aug 12
Huff 'n Puff Balloon Rally (Topeka, KS) . . . Sept 8
Lake Champlain Balloon Fest (Essex Jct, VT) . . . June 2
Man's First Free Flight: Anniv . . . Nov 21
Monterey Cnty Hot Air Affair (Monterey, CA) . . . Mar 11
Oldsmobile Balloon Classic (Danville, IL) June 9
Quechee Balloon Fest/Crafts Fair (Quechee, VT) . . . June 16
Rocvale Balloon Rally (Rockford, IL) . . . Sept 9
Snowmass Hot Air Balloon Fest (Snowmass Village, CO) . . . June 23
Solo TransAtlantic Balloon Crossing: Anniv . . . Sept 14
Southwest Iowa Prof Hot Air Balloon Races (Creston, IA) . . . Sept 16
Steamboat's Rainbow Weekend (Steamboat Springs, CO) . . . July 14
Winterfest Balloon Rally (Angel Fire, NM) . Feb 3
Balsam, Martin: Birth . . . Nov 4
Balzac, Honore De: Birth Anniv . . . May 20
Banana Season, Aspen/Snowmass (Snowmass Village, CO) . . . Apr 1
Bancroft, Anne: Birth . . . Sept 17
Bancroft, George: Birth Anniv . . . Oct 3
Band Fest, Great American Brass (Danville, KY) . . . June 16
Bandaranaike, Sirimavo: Birth . . . Apr 17
Bands. See Music
Bangerter, Norman H.: Birth . . . Jan 4
Bangladesh
Independence Day . . . Mar 26
Martyrs Day . . . Feb 21
Victory Day . . . Dec 16
Banjo: Guthrie Jazz Banjo Fest (Guthrie, OK) . . . May 26
Bank, First Chartered by Congress . . . June 20
Bank Holiday, Spring (United Kingdom) . . . May 29
Bank Holiday, Summer (Scotland) . . . Aug 7
Bank Holiday, Summer (United Kingdom) . . Aug 28
Bank Opens in US, First: Anniv . . . July 3
Banking: Capital Day . . . Sept 2
Banks, Carl: Birth . . . Aug 29
Banks, Ernie: Birth . . . Jan 31
Banned Books Week . . . Sept 23
Banneker, Benjamin: Death Anniv . . . Oct 9
Bannister, Roger: Birth . . . Mar 23
Bar Assn, American Founding: Anniv . . . Aug 21
Bar-B-Q Festival, Intl (Owensboro, KY) . . . May 12
Barbados: Independence Day . . . Nov 30
Barbeau, Adrienne: Birth . . . June 11
Barbecue Month, Natl . . . May 1
Barber, Red: First Baseball Games Televised: Anniv . . . Aug 26
Barber, Walter "Red": Birth Anniv . . . Feb 17
Barbershop Ballad Contest (Forest Grove, OR) . . . Mar 10
Barbershop Quartet Day . . . Apr 11
Barbershop Show, Great Lakes Invit (Grand Rapids, MI) . . . Nov 4
Barbershop Singing Intl Conv (Miami, FL) . . July 2
Barbershop: SPEBSQSA Convention (Tucson, AZ) . . . Jan 22
Barbuda: National Holiday . . . Nov 1
Barbuda. See Antigua
Bardot, Brigitte: Birth . . . Sept 28
Barenboim, Daniel: Birth . . . Nov 15
Barfield, Jesse: Birth . . . Oct 29
Barker, Bob: Birth . . . Dec 12
Barkin, Ellen: Birth . . . Apr 16
Barkley, Alben: Birth Anniv . . . Nov 24
Barkley, Charles: Birth . . . Feb 20
Barn Fest, Round (Rochester, IN) . . . June 16
Barnard, Christiaan: Birth . . . Nov 8
Barnard, Christiaan: First Heart Transplant . . Dec 3
Barnes, Clive: Birth . . . May 13
Barnes, Joanna: Birth . . . Nov 15
Baron Bliss Day (Belize) . . . Mar 9
Barr: Roseanne Arnold: Birth . . . Nov 3
Barrel Race, Jr, Josey's World Champion (Marshall, TX) . . . May 5
Barrett, Rona: Birth . . . Oct 8
Barris, Chuck: Birth . . . June 3
Barry Day, Commodore John 6589 . . . Page 488
Barry, Gene: Birth . . . June 4
Barry, John: Death Anniv . . . Sept 13
Barry, Marion: Birth . . . Mar 6
Barrymore, Drew: Birth . . . Feb 22
Barrymore, Ethel: Birth Anniv . . . Aug 15
Barrymore, John: Birth Anniv . . . Feb 15
Barrymore, Lionel: Birth Anniv . . . Apr 28
Barth, John: Birth . . . May 27
Bartlett, John: Birth Anniv . . . June 14
Bartok, Bela: Birth Anniv . . . Mar 25
Bartok, Bela: Death Anniv . . . Sept 26
Barton, Clara: Birth Anniv . . . Dec 25
Barton, Clara: Red Cross Founding Anniv . May 21
Baryshnikov, Mikhail: Birth . . . Jan 27
Bascom, Earl W.: Birth Anniv . . . June 19
Bascom, Florence: Birth Anniv . . . July 14
Bascom, George N.: Birth Anniv . . . Apr 24
Bascom, "Texas Rose": Birth Anniv . . . Feb 25
Baseball/Softball
Ashford, Emmet: Birth Anniv . . . Nov 23
Babe Ruth's Farewell to the Yankees: Anniv . . . Sept 24
Babe Ruth's First Major League Home Run: Anniv . . . May 6
Baseball Declared Non-Essential: Anniv . July 20
Baseball Game First Played Under the Lights: Anniv . . . May 24
Baseball Stadiums (Hendersonville) . . . Oct 14
Baseball Stadiums (Scranton) . . . June 10
Baseball's First Perfect Game Anniv . . . May 5
Baseball's Greatest Dispute: Anniv . . . Sept 23
Bronco League World Series (Monterey, CA) . . . Aug 12
Cobb, Ty: Birth Anniv . . . Dec 18
Colt League World Series (Lafayette, IN) Aug 10
Doubleday, Abner: Birth Anniv . . . June 26
Field of Dreams Extravaganza (Dyersville, IA) . . . Sept 1
First Baseball Games Televised: Anniv . . . Aug 26
First Baseball Strike Ends: Anniv . . . Apr 13
Grand Forks Intl Tourn (Grand Forks, BC, Canada) . . . Aug 30
Hall of Fame, Natl Baseball: Anniv . . . June 12
Home Run Record: Anniv . . . Apr 8
Ladies' Day Initiated . . . June 16
Little League Baseball Week, Natl . . . June 12
Little League: Big League World Ser (Ft Lauderdale, FL) . . . Aug 11
Little League World Series (Williamsport, PA) . . . Aug 20
Major League Baseball First All Star Game: Anniv . . . July 6
Major League's First Double Header . . . Sept 25
Major/Minor League: Photos/Stadiums (Hollywood, FL) . . . Feb 2
Major/Minor League: Photos/Stadiums (Naples, FL) . . . Apr 8
Men's Big 8 Baseball Chmpshp (Oklahoma City, OK) . . . May 18
Midnight Sun Baseball Game (Fairbanks, AK) . . . June 21
Mustang League World Series (Irving, TX) Aug 9
NAIA Baseball World Series . . . May 26
NAIA Softball Chmpshp . . . May 17
Palomino League World Series (Greensboro, NC) . . . Aug 9
Peanut-Kids-Baseball Day (Coshohocken, PA) . . . Apr 8
Pony Girls Softball Natl Championships . . Aug 5
Pony League World Series (Washington, PA) . . . Aug 12
Robinson Named First Black Manager: Anniv . . . Oct 3
Squeeze Play First Used: Anniv . . . June 16
Taft Opened Baseball Season: Anniv . . . Apr 14
Basinger, Kim: Birth . . . Dec 8
Basketball
All-College Basketball Tourn (Oklahoma City, OK) . . . Dec 29
All-College Tourn (Oklahoma City, OK) . . . Jan 6
Elks Natl Hoop Shoot (Indianapolis, IN) . . Apr 28
First Women's Collegiate Basketball: Anniv . . . Mar 22
Great Alaska Shootout (Anchorage, AK) . Nov 22
Gus Macker Charity Basketball Tourn (Chillicothe, OH) . . . June 10
Hall of Fame Awards Dinner (Springfield, MA) . . . May 6
Hall of Fame Enshrinement Ceremonies (Springfield, MA) . . . May 8
NAIA Men's Div I Basketball Chmpshp (Tulsa, OK) . . . Mar 14
NAIA Wom Div I Basketball Chmpshp (Jackson, TN) . . . Mar 15
NAIA Wom Div II Basketball Chmpshp Tourn (Monmouth, OR) . . . Mar 9
NBA: All-Star Weekend (Phoenix, AZ) . . . Feb 10
NBA Day in Basketball City, USA (Springfield, MA) . . . Oct 27
NCAA Div I Women's Basketball Chmpshp (Richmond, VA) . . . Apr 1
NCAA Div I Women's Chmpshp (various) Mar 23
NJCAA Div II Natl Basketball Finals (Danville, IL) . . . Mar 16
Sez Who? Fourplay March Madness (Chicago, IL) . . . Mar 12
Spalding Enshrinement Celebr Golf Tourn (Agawam, MA) . . . May 8
Starter Hall of Fame Tip-Off Classic (Springfield, MA) . . . Nov 24
Bastille Day (France) . . . July 14
Bastille Day Celebration (Boston, MA) . . . July 14
Bastille Day Moonlight Golf (Washington, DC) . . . July 14
Bastille Days (Milwaukee, WI) . . . July 13
Bataan Death March: Anniv . . . Apr 10
Bateman, Jason: Birth . . . Jan 14
Bateman, Justine: Birth . . . Feb 19
Bates, Alan: Birth . . . Feb 17
Bates, Kathy: Birth . . . June 28
Bath: Natl Kitchen and Bath Month . . . Oct 1
Bathroom Reading Week, Natl . . . June 5
Bathtub Race, Great (Nome, AK) . . . Sept 4
Battle of Blue Licks Celebration (Mount Olivet, KY) . . . Aug 19
Battle of Britain Day (England) . . . Sept 15
Battle of Britain Week (England) . . . Sept 10
Battle of Dien Bieh Phu Ends: Anniv . . . May 7
Battle of Little Big Horn: Anniv . . . June 25
Baucus, Max: Birth . . . Dec 11
Baum, Lyman F.: Birth Anniv . . . May 15
Bavarian: Bismarck Folkfest (Bismarck, ND) . . . Sept 15
Bavarian Fest (Frankenmuth, MI) . . . June 10
Baxter, Meredith: Birth . . . June 21
Bayh, Evan: Birth . . . Dec 26
Baylor, Don: Birth . . . June 28
Bazaarfest (Fargo, ND) . . . Oct 18
Be an Angel Day . . . Aug 22
Be Bald and Be Free Day . . . Oct 14
Be Kind to Animals Week . . . May 7
Be Kind to Humankind Week . . . Aug 25
Be Late for Something Day . . . Sept 5
Be On-Purpose Month, Natl . . . Jan 1
Beach Party, Roanoke (Roanoke, VA) . . . Aug 26
Beadle, Flossie, Week (Lakeside, CA) . . . Dec 1
Beadle, George: Birth Anniv . . . Oct 22
Beals, Jennifer: Birth . . . Dec 19
Bean, Alan: Birth . . . Mar 15
Bean, Andy: Birth . . . Mar 13
Bean Fest, California Dry (Tracy, CA) . . . Aug 5
Bean Month, Natl Baked . . . July 1
Bean, Orson: Birth . . . July 22
Beard, Charles: Birth Anniv . . . Nov 27
Beard Growing Contest, St. Patrick's Day (Shamrock, TX) . . . Mar 17
Beard, Mary R.: Birth Anniv . . . Aug 5
Beasley, Allyce: Birth . . . July 6
Beatles Last Album Released . . . Sept 26
Beatles Last Public Appearance: Anniv . . . Jan 30
Beatles' Natl Apple Week: Anniv . . . Aug 11
Beatrice, Princess of York: Birth . . . Aug 8
Beatty, Ned: Birth . . . July 6
Beatty, Warren: Birth . . . Mar 30
Beaufort Scale Day . . . May 7
Beaumont, William: Birth Anniv . . . Nov 21
Beauty Contests, Pageants
First Miss Albania Crowned: Anniv . . . Jan 31
First Miss America: Anniv . . . Sept 8
Miss America Pageant Finale (Atlantic City, NJ) . . . Sept 16
Miss Crustacean USA (Ocean City, NJ) . . . Aug 2
Miss Louisiana Pageant (Monroe, LA) . . June 15
Miss New Hampshire Finals (Manchester, NH) . . . May 12
Parks, Bert: Birth Anniv . . . Dec 30
Becker, Boris: Birth . . . Nov 22
Bed Check Month, Natl . . . Sept 1
Bedrosian, Steve: Birth . . . Dec 6
Beecher, Catharine Esther: Birth Anniv . . . Sept 6
Beecher, Henry W: Birth Anniv . . . June 24
Beef-a-Rama/Arts and Crafts Show (Minocqua, WI) . . . Sept 30
Beer Fest (Luxembourg) . . . July 16
Beer Fest, Great American (Denver, CO) . . . Oct 6
Beer: Microbrewers Conf and Trade Show (Austin, TX) . . . Apr 23

Index ☆ Chase's 1995 Calendar of Events ☆

Beer (cont'd)—Blondin

Beer: Natl Homebrew Day .... May 6
Beer Week, American .... Oct 8
Beethoven, Ludwig Van: Birth Anniv .... Dec 16
Beggars' Night .... Oct 31
Begin, Menachem: Birth Anniv .... Aug 16
Begley, Ed, Jr: Birth .... Sept 16
Behan, Brendan: Birth Anniv .... Feb 9
Beirut Terrorist Attack: Anniv .... Oct 23
Belafonte, Harry: Birth .... Mar 1
Belafonte-Harper, Shari: Birth .... Sept 22
Bele Chere (Asheville, NC) .... July 28
Belgium
Cat Festival .... Mar 2
Dynasty Day .... Nov 15
Historical Procession (Tournai) .... Sept 10
Lover's Fair .... Dec 7
National Holiday .... July 21
Nuts Fair (Bastogne) .... Dec 18
Ommegang Pageant .... July 6
Play of St. Evermaar .... May 1
Procession of Golden Chariot .... June 2
Procession of the Holy Blood .... May 25
Wedding of the Giants .... Aug 27
Belize
Baron Bliss Day .... Mar 9
Columbus Day .... Oct 14
Commonwealth Day .... May 24
Garifuna Day .... Nov 19
Independence Day .... Sept 21
St. George's Caye Day .... Sept 10
Bell, Alexander Graham: Birth Anniv .... Mar 3
Bell, George: Birth .... Oct 21
Bellamy, Edward: Birth Anniv .... Mar 26
Belli, Melvin: Birth .... July 29
Bello, Andres: Birth Anniv .... Nov 20
Bellow, Saul: Birth .... July 10
Belmondo, Jean-Paul: Birth .... Apr 9
Belmont Stakes (Belmont Park, NY) .... June 10
Belote, Melissa: Birth .... Oct 16
Beltane .... Apr 30
Belushi, Jim: Birth .... June 15
Benatar, Pat: Birth .... Jan 10
Bench, Johnny: Birth .... Dec 7
Benchley, Peter: Birth .... May 8
Benedict, Julius: Birth Anniv .... Nov 27
Benedict, Paul: Birth .... Sept 17
Benet, William Rose: Birth Anniv .... Feb 2
Ben-Gurion, David: Birth Anniv .... Oct 16
Benin, Peoples Republic of: National Day .... Aug 1
Benjamin, Richard: Birth .... May 22
Bennett, Bern: Birth .... Oct 19
Bennett, Cornelius: Birth .... Aug 25
Bennett, Jill: Birth .... Dec 24
Bennett, Robert F.: Birth .... Sept 18
Bennett, Tony: Birth .... Aug 3
Bennington Battle Day .... Aug 16
Benny, Jack: Birth Anniv .... Feb 14
Benson, George: Birth .... Mar 22
Benson, Robbie: Birth .... Jan 21
Bentley, Edmund Clerihew: Clerihew Day .... July 10
Benton, Thomas Hart: Birth Anniv .... Apr 15
Bentsen, Lloyd: Birth .... Feb 11
Berenger, Tom: Birth .... May 31
Berenson, Marisa: Birth .... Feb 15
Berenson, Red: Birth .... Dec 8
Bergalis, Kimberly: Death Anniv .... Dec 8
Bergen, Candice: Birth .... May 9
Bergen, Edgar: Birth Anniv .... Feb 16
Bergen, Polly: Birth .... July 14
Bergman, Ingmar: Birth .... July 14
Bergman, Ingrid: Birth Anniv .... Aug 29
Bergson, Henri: Birth Anniv .... Oct 18
Berkeley, Busby: Birth Anniv .... Nov 29
Berle, Milton: Birth .... July 12
Berlin Airlift: Anniv .... June 24
Berlin Wall Erected: Anniv .... Aug 13
Berlin Wall Opened: Anniv .... Nov 9
Berlin, Irving: Birth Anniv .... May 11
Berman, Shelley: Birth .... Feb 3
Bermuda: Peppercorn Ceremony .... Apr 23
Bernsen, Corbin: Birth .... Sept 7
Bernstein, Carl: Birth .... Feb 14
Bernstein, Elmer: Birth .... Apr 4
Bernstein, Robert: Birth .... Jan 5
Berra, Yogi: Birth .... May 12
Berry, Chuck: Birth .... Oct 18
Berry, Ken: Birth .... Nov 3
Berry, Walter: Birth .... May 14
Bertholdi, Frederic A.: Birth Anniv .... Apr 2
Bertinelli, Valerie: Birth .... Apr 23
Bertolucci, Bernardo: Birth .... Mar 16
Bessmertnova, Natalya: Birth .... July 19
Bethune, Mary McLeod: Birth Anniv .... July 10
Bethune, Norman: Birth Anniv .... Mar 3
Better Hearing and Speech Month, May Is .... May 1
Betty Picnic (Grants Pass, OR) .... June 10
Bhopal Poison Gas Disaster: Anniv .... Dec 3
Bhutan: National Day .... Dec 17
Bhutto, Benazir: Birth .... June 21
Bible Society, American: Annual Meeting (New York, NY) .... May 11
Bible Sunday, Natl .... Nov 19
Bible Week Inaugural Luncheon, Natl .... Nov 17
Bible Week, Natl .... Nov 19
Bible Week, New Intl Version of .... Oct 22
Bicycle
Amer Bicycle Assn Grand Nationals (Oklahoma City, OK) .... Nov 23
American Youth Hostel: Christmas Trip (San Diego, CA) .... Dec 26
Bicycle Trek Classic (Spearfish, SD) .... May 5
Bike Month, Natl .... May 1
Bike to Work Day, Natl .... May 9
Bike Van Buren (Van Buren County, IA) .... Aug 19
Buffalo Round-Up Bicycle Rally (Custer, SD) .... Sept 30
Charlotte Observer Moonride Criterium (Charlotte, NC) .... Aug 25
Chile Challenge Off-Road Bike Race (Angel Fire, NM) .... Sept 22
Dick Evans Memorial Cycle Race (Oahu, HI) .... Aug 5
Enchanted Circle Century Bike Tour (Red River, NM) .... Sept 10
Giants Ridge Mountain Bike Fest (Biwabik, MN) .... Sept 2
HI-AYH Midnight Madness Fun Ride (San Diego, CA) .... Aug 19
Hotter 'n Hell Hundred Weekend (Wichita Falls, TX) .... Aug 24
JCBC Century Ride (Junction City and Fort Riley, KS) .... Sept 10
Melon City Criterium (Muscatine, IA) .... May 28
Mt Washington Bicycle Hillclimb (Gorham) .... Sept 10
Old Fort River Fest (Fort Smith, AR) .... June 16
Perry's "BRR" (Bike Ride to Rippey) (Perry, IA) .... Feb 4
Pole Pedal Paddle (Bend, OR) .... May 20
Pole, Pedal, Paddle (Jackson, WY) .... Apr 1
Possum Pedal 100 Bicycle Ride/Race (Graham, TX) .... Mar 25
Register's Bicycle Ride Across Iowa (Des Moines, IA) .... July 23
Strawberry 100 Bike Tour/Fest (Crawfordsville, IN) .... June 9
Tour de Cure .... Apr 22
Tour de la Guadeloupe (Guadeloupe, French West Indies) .... Aug 4
Tour of Somerville (Somerville, NJ) .... May 29
Biden, Joseph Robinette, Jr: Birth .... Nov 20
Big Brothers/Big Sisters Appreciation Week .... Apr 23
Big Island Rendezvous (Albert Lea, MN) .... Oct 7
Big Ten Men's and Women's Outdoor Track/Field (West Lafayette, IN) .... May 19
Big Ten Men's Gymnastics Chmpshp .... Mar 25
Big Ten Men's Swimming/Diving Chmpshp .... Feb 23
Big Ten Women's Gymnastics Chmpshp .... Mar 25
Big Whopper Liar's Contest (New Harmony, IN) .... Sept 16
Big Wind: Anniv .... Apr 12
Biggest All Night Gospel Singing (Bonifay, FL) .... July 1
Bigotry: Lumpy Rug Day .... May 3
Bike Month, Natl .... May 1
Bike to Work Day, Natl .... May 9
Bikel, Theodore: Birth .... May 2
Bill of Rights Proposal: Anniv .... June 8
Bill of Rights: Anniv .... Dec 15
Bill of Rights: Anniv of First State Ratification .... Nov 20
Bill of Rights Day (Pres Proc) .... Dec 15
Bill of Rights: Religious Freedom Week .... Sept 24
Billings, John S.: Birth Anniv .... Apr 12
Billingsley, Barbara: Birth .... Dec 22
Billington, James Hadley: Birth .... June 1
Billington, John: First Crim Execution in Colonies: Anniv .... Sept 30
Bills Week, Pay Your .... Feb 17
Billy Bowlegs Fest (Fort Walton, FL) .... June 2
Billy The Kid: Birth Anniv .... Nov 23
Biltmore: Christmas at Biltmore (Asheville, NC) .... Nov 11
Bingaman, Jeff: Birth .... Oct 3
Bingo's Birthday Month .... Dec 1
Biographers Day .... May 16
Biomedical Research Day, Natl 6616 .... Page 488
Biosphere Day .... Sept 21
Biosphere, World Campaign for the .... June 5
Bird Day (IA) .... Mar 21
Bird, Larry: Birth .... Dec 7
Birds
Bald Eagle Appreciation Days (Keokuk) .... Jan 21
Bald Eagle Day (Cassville, WI) .... Jan 28
Bald Eagle Migration (Helena, MT) .... Nov 1
Bird Day (IA) .... Mar 21
Birdwatching Weekend (Madison, MN) .... Apr 28
Birthday of Mother's Whistler .... May 18
Buzzards' Day .... Mar 15
Canaan Valley Holiday Weekend (Terra Alta, WV) .... May 26
Curlew Day .... Mar 16
Duck Calling Contest/Wings Over Prairie (Stuttgart, AR) .... Nov 21
Eagle Days (Junction City, KS) .... Jan 21
Eagles Et Cetera (Bismarck, AR) .... Jan 27
Gnawed, Clawed Bird Feeder Fest (New Holland, PA) .... Sept 11
Homes for Birds Week .... Feb 19
Hummer/Bird Celebration (Rockport, Fulton, TX) .... Sept 7
Kirtland, Jared: Birth Anniv .... Nov 10
Last Dusky Seaside Sparrow: Death Anniv .... June 16
Nottingham Goose Fair (Nottingham, England) .... Oct 5
Old-Time Barnegat Decoy/Gun Show (Tuckerton, NJ) .... Sept 23
Pigeons Return to City-County Bldg (Ft Wayne, IN) .... Mar 20
Reelfoot Eagle Tours (Tiptonville, TN) .... Jan 1
Roan Mtn Wildflower Tours/Birdwalks (Roan Mtn, TN) .... May 6
Sandhill Crane Migration (Grand Island, NE) .... Mar 5
Swallows Return to San Juan Capistrano Mar 19
Waterfowl Festival (Easton, MD) .... Nov 10
Wild Bird Feeding Month, Natl .... Feb 1
Wings Over the Platte (Grand Island, NE) Mar 17
Birdseye, Clarence: Birth Anniv .... Dec 9
Birdsong, Otis: Birth .... Dec 9
Birkebeiner, American, (Cable to Hayward, WI) .... Feb 23
Birth Control Clinic Opened, First: Anniv .... Oct 16
Birth Defects Prevention Month, March of Dimes .... Jan 1
Birth Defects Week, Alcohol and Other Drug-Related .... May 14
Birth of a Nation, The: Premiere Anniv .... Feb 8
Birthday: Chicken Boy's .... Sept 1
Birthparents Week, Natl .... Apr 2
Bishop, Elvin: Birth .... Oct 21
Bishop, Joey: Birth .... Feb 3
Bison-Ten-Yell Day .... Sept 2
Bisset, Jacqueline: Birth .... Sept 13
Blab, Uwe: Birth .... Mar 26
Black American Arts Festival (Greensboro, NC) .... Jan 14
Black Friday .... Nov 24
Black Heritage Festival of Louisiana (Lake Charles, LA) .... Mar 3
Black History Month Kickoff (Washington, DC) .... Jan 27
Black Nazarene, Feast of the (Philippines) .... Jan 9
Black Nazarene Fiesta (Philippines) .... Jan 1
Black Poetry Day .... Oct 17
Black Press Day: Anniv of the First Black Newspaper .... Mar 16
Black, Karen: Birth .... July 1
Black, Shirley Temple: Birth .... Apr 23
Blackbear, Bosin: Birth .... June 5
Blackbeard Pirate Jamboree (Norfolk, VA) July 28
Blackman, Rolando: Birth .... Feb 26
Blackmun, Harry A.: Birth .... Nov 12
Blackstone, William: Birth Anniv .... Mar 5
Blair, Bonnie: Birth .... Mar 18
Blair, Linda: Birth .... Jan 22
Blake, Eubie: Birth Anniv .... Feb 7
Blake, Robert: Birth .... Sept 18
Blake, William: Birth Anniv .... Nov 28
Blakey, Art: Birth .... Oct 11
Blame Someone Else Day .... Jan 13
Blass, Bill: Birth .... June 22
Blatty, William: Birth .... Jan 7
Blaz, Ben: Birth .... Feb 14
Bledsoe, Tempestt: Birth .... Aug 1
Bleriot, Louis: Birth Anniv .... July 1
Blessing of Animals at the Cathdral (Mexico) .... Jan 17
Bliss, Lizzie: Birth Anniv .... Apr 11
Blondin, Charles: Birth Anniv .... Feb 28

✫ Chase's 1995 Calendar of Events ✫   Index

Blood Pressure Education Month, High, Natl.................May 1
Blood Transfusion Anniv.................Nov 14
Bloody Sunday (Northern Ireland): Anniv...Jan 30
Bloom, Allan: Birth.................Sept 14
Bloom, Claire: Birth.................Feb 15
Bloom, Harold: Birth.................July 11
Bloomer, Amelia Jenks: Birth Anniv.......May 27
Bloomsday.................June 16
Blossom Fest, Hood River Valley (Hood River, OR).................Apr 15
Blossomtime Fest of SW Michigan (Benton Harbor, MI).................Apr 30
Blue Hill Fair (Blue Hill, ME).................Aug 31
Blue River Fest (Crete, NE).................July 21
Blueberries Month, Natl July Belongs to....July 1
Blueberry Fest (Montrose, PA).................Aug 4
Blueberry Festival (Ketchikan, AK).......Aug 12
Bluebird of Happiness Day, Natl.........Sept 24
Bluegrass: See also Fiddlers
  Blissfest (Cross Village, MI).................July 14
  Chili and Bluegrass Fest (Tulsa, OK).....Sept 8
  Country/Bluegrass Fest (Prescott, AZ)....July 15
  Dahlonega Bluegrass Fest (Dahlonega, GA).................June 22
  Festival of the Bluegrass (Lexington, KY) June 8
  Four Corner States Bluegrass Fest (Wickenburg, AZ).................Nov 10
  Friendsville Fiddle/Banjo Contest (Friendsville, MD).................July 15
  Kahoka Fest of Bluegrass Music (Kahoka, MO).................Aug 9
  Midwinter Bluegrass Fest (Hannibal, MO) Feb 17
  New Star Rising Bluegrass Fest (Live Oak, FL).................Nov 16
  Nova Scotia Oldtime Music Fest (Ardoise, NS).................July 28
  Old Bedford Village Bluegrass Fest (Bedford, PA).................July 22
  Old Fiddlers' Convention (Galax, VA).....Aug 9
  Polk County Ramp Tramp Fest (Benton, TN).................Apr 22
  Thomas Pt Beach Bluegrass Fest (Brunswick, ME).................Sept 1
Blume, Judy: Birth.................Feb 12
Bly, Nellie: Around the World in 72 Days: Anniv.................Nov 14
Bly, Nellie: Birth Anniv.................May 5
Bly, Robert: Birth.................Dec 23
Blyleven, Bert: Birth.................Apr 6
Boats/Ships/Other Vessels
  American Rowing Natl Chmpshps (Topeka, KS).................July 20
  Antique/Classic Boat Rendezvous (Mystic, CT).................July 22
  Anything that Floats River Raft Race (Rockfork, IL).................July 4
  Big Lake Regatta Water Fest (Big Lake, AK).................July 8
  Birmingham Sport and Boat Show (Birmingham, AL).................Jan 31
  Boat Racing.................Aug 1
  Burlington Steamboat Days/Music Fest (Burlington, IA).................June 13
  Canoe Joust (Snow Hill, MD).................Aug 6
  Cardboard Boat Regatta, Great (Sheboygan, WI).................July 4
  Carriacou Regatta (Grenada).................Aug 5
  Civil War Submarine Attack: Anniv.......Oct 5
  Coshocton Canal Fest (Coshocton, OH) Aug 18
  Cowe's Week (Isle of Wight).................July 29
  Cutty Sark Tall Ships Race (Weymouth, Dorset, Scotland).................July 15
  Devizes/Westminster Intl Canoe Race (Devizes, England).................Apr 14
  Dragon Boat Fest (Hong Kong).................June 10
  Dragon Boat Races (Keokuk, IA).........Aug 19
  Easter Regatta (Grenada).................Apr 14
  Egyptian Maritime Disaster: Anniv......Dec 14
  Everett Salty Sea Days (Everett, WA)....June 1
  Fall Color Cruise and Folk Fest (Chattanooga, TN).................Oct 21
  Florida State Boat and Sports Show (Tampa, FL).................Sept 14
  Frenchman Rows Across Pacific: Anniv..Nov 21
  Fresh Seafood Festival (Atlantic City, NJ).................June 10
  Ft Lauderdale Intl Boat Show (Ft. Lauderdale, FL).................Oct 26
  Ft Lauderdale Spring Boat Show (Ft Lauderdale, FL).................Apr 13
  Grand Center Boat Show (Grand Rapids, MI).................Feb 21
  Great Cardboard Boat Race (Heber Springs, AR).................July 22
  Great Cardboard Boat Regatta (Carbondale, IL).................Apr 29
  Great Cardboard Boat Regatta (Leon, IA).Aug 6
  Great Cardboard Boat Regatta(Rock Island, IL).................July 1
  Great Chunky River Raft Race (Chunky, MS).................June 3
  Great Houseboat/Cruiser Parade (Falls of Rough, KY).................July 1
  Great Plains Rowing Chmpshp (Topeka, KS).................Apr 14
  Halifax, Nova Scotia, Destroyed: Anniv...Dec 6
  Hansa Dusi Canoe Marathon (Natal, S Africa).................Jan 19
  Harborfest (Nantucket Island, MA).....June 16
  Harborfest (Norfolk, VA).................June 2
  Head of the River Race (London, England).................Mar 25
  Henley Royal Regatta (Henley-on-Thames, England).................June 28
  Historical Regatta (Italy).................Sept 3
  Huckleberry Finn Raft Race (New Preston, CT).................Sept 4
  Indianapolis Boat/Sport/Travel Show (Indianapolis, IN).................Feb 17
  Ironclads Battle in Charleston Harbor: Anniv.................Jan 31
  Jersey Shore Kayak/Canoeing Show (Bayville, NJ).................Sept 16
  Kenduskeag Stream Canoe Race (Bangor, ME).................Apr 22
  Lighted Boat Parade (Columbus, MS)....Dec 2
  London Intl Boat Show (London, England) Jan 5
  Loyalty Days and Seafair Fest (Newport, OR).................May 4
  Mahone Bay Wooden Boat Fest (Mahone Bay, NS, Canada).................Aug 2
  Marinerfest (Tawas City, MI).................July 12
  Maritime Day, Natl 6564, 6692.......Page 488
  Merchant Sailing Ship Preservation Day..Nov 8
  Miami Intl Boat/Sailboat Show (Miami Beach, FL).................Feb 16
  Milk Jug Regatta (Minocqua, WI).......June 11
  Milwaukee Boat Show at MECCA (Milwaukee, WI).................Feb 15
  Monitor Sinking: Anniv.................Dec 30
  National Safe Boating Week.................May 20
  New Orleans Boat & Sportfishing Show (New Orleans, LA).................Feb 1
  New York Natl Boat Show (New York, NY).Jan 6
  Newport Yacht Club Frostbite Fleet Race (Newport, RI).................Jan 1
  Nome River Raft Race (Nome, AK).....June 18
  North American Regatta (Lake Sakajawea, ND).................July 21
  Norwalk Intl In-Water boat Show (Norwalk, CT).................Sept 21
  Ohio River Sternwheel Fest (Marietta, OH).................Sept 8
  Outrigger's Hawaiian Ocean Fest (Oahu, HI).................Sept 8
  Oxford-Cambridge Race (London, England).................Apr 1
  Palio del Golfo (Italy).................Aug 13
  Penn's Landing In-Water Boat Show (Philadelphia, PA).................Sept 14
  Philadelphia Boat Show (Philadelphia, PA).................Jan 21
  Pocomo River Canoe Challenge (Snow Hill, MD).................June 17
  Pole, Pedal, Paddle (Jackson, WY).......Apr 1
  Pontiac Boat, Sport & Fishing Show (Pontiac, MI).................Feb 15
  Remember the Maine Day.................Feb 15
  Rogue River Jet Boat Marathon (Gold Beach).................June 2
  Safe Boating Week, Natl 6570, 6695...Page 488
  San Antonio Sport, Boat and RV Show (San Antonio, TX).................Jan 24
  San Diego Boat and Sport Fishing Show (San Diego, CA).................Jan 5
  Sea Music Fest (Mystic, CT).................June 9
  Small Craft Weekend (Mystic, CT).......June 3
  Sound/Light Spectacular (Wilmington, NC).................June 2
  South Jersey Canoe/Kayak Classic (Lakewood, NJ).................June 3
  Spanish War/Maine Memorial Day.......Feb 15
  Steamboat Days Fest (Jeffersonville, IN).Sept 8
  Sternwheel Days (Augusta, KY).........June 24
  Sternwheel Regatta Fest, Charleston (Charleston, WV).................Aug 25
  Sternwheeler Days (Cascade Locks, OR).................June 23
  Sugar Creek Canoe Race and Tri Sport Fest (Crawford, IN).................Apr 23
  Tampa Boat Show (Tampa, FL).........Jan 19
  Three Rivers Fest (Ft Wayne, IN).......July 8
  Tioman Intl Regatta (Pahang, Malaysia)..Sept 3
  Toronto Intl Boat Show (Toronto, Ont, Canada).................Jan 6
  Tour des Yoles Rondes (Martinique, French West Indies).................July 30
  USA Intl Dragon Boat Fest (Dubuque, IA) Sept 8
  Virginia Boat Show (Richmond, VA).....Jan 16
  Wedding of the Sea (Atlantic City, NJ)...Aug 15
  White River Water Carnival (Batesville, AR).................July 29
  Whitewater Wednesday (Kernville, CA)..June 21
  Windjammer Days (Boothbay Harbor, ME).................June 28
  Winterfest Boat Parade (Ft. Lauderdale, FL).................Dec 9
  Wyandotte Waterfest (Wyandotte, MI)...Aug 11
  Yacht Race, Intl: Anniv.................Aug 22
Bochco, Steven.................Dec 16
Bock, Jerry: Birth.................Nov 23
Bocuse, Paul: Birth.................Feb 11
Boddicker, Michael: Birth.................Aug 23
Bodybuilding: Atlas, Charles: Birth Anniv..Oct 30
Bodybuilding, Big Is/Fitness Contest (Kailua-Kona, HI).................July 1
Boff, Leonardo: Birth.................Dec 14
Bogarde, Dirk: Birth.................Mar 28
Bogart, Humphrey: Birth Anniv.........Dec 25
Bogdanovich, Peter: Birth.................July 30
Boggs, Wade: Birth.................June 15
Bogosian, Eric: Birth.................Apr 24
Bogues, Muggsy: Birth.................Jan 9
Boitano, Brian: Birth.................Oct 22
Bol, Manute: Birth.................Oct 16
Bolin, Jane M.: Birth.................Apr 11
Bolivar, Simon: Death Anniv.............Dec 17
Bolivia
  Alacitis Fair.................Jan 24
  Independence Day.................Aug 6
  La Paz Day.................July 16
Boll, Heinrich: Birth Anniv.................Dec 21
Bolles, Don: Death Anniv.................June 13
Bollingen Prize Award: Anniv.............Feb 19
Bologna Fest, Lebanon.................Sept 15
Bombeck, Erma: Birth.................Feb 21
Bon Festival (Feast of Lanterns) (Japan)..July 13
Bon Jovi, Jon: Birth.................Mar 2
Bon Odori Festival of the Lanterns (Chicago, IL).................July 8
Bonanzaville USA Pioneer Days (West Fargo, ND).................Aug 19
Bonaparte, Napoleon: Birth Anniv.......Aug 15
Bond, C.J.: Birth Anniv.................Aug 11
Bond, Christopher Samuel: Birth.........Mar 6
Bond, Julian: Birth.................Jan 14
Bonds, Gary U.S.: Birth.................June 6
Bonerz, Peter: Birth.................Aug 6
Bonet, Lisa: Birth.................Nov 16
Bonheur, Rosa: Birth Anniv.............Mar 22
Bonilla, Bobby: Birth.................Feb 23
Bonnie Blue Natl Horse Show (Lexington, VA).................May 10
Bono: Birth.................May 10
Bono, Chastity: Birth.................Mar 4
Bono, Sonny: Birth.................Feb 16
Bonza Bottler Day...Jan 1, Feb 2, Mar 3, Apr 4, May 5, June 6, July 7, Aug 8, Sept 9, Oct 10, Nov 11, Dec 12
Books. See also Library/Librarians
  American Booksellers Assn Exhib/Conv (Chicago, IL).................June 3
  Associated Publishers, Inc: Founding Anniv.................Nov 18
  Authors' Day, Natl.................Nov 1
  Banned Books Week.................Sept 23
  Bible Sunday, Natl.................Nov 19
  Bible Week, Natl.................Nov 19
  Bible Week, New Intl Version of.........Oct 22
  Book Blitz Month, Natl.................Jan 1
  Book Day (Spain).................Apr 23
  Book Fair Month, Intl.................Oct 1
  Book It/Natl Young Reader's Day.........Oct 2
  Celebrity Read a Book Week (Williamsport, MD).................Jan 22
  Children's Book Day, Intl.................Apr 2
  Children's Book Week, Natl.............Nov 13
  Cookbook Festival (Fargo, ND).........Feb 11
  Culinary Professionals' Cookbook Month, Intl Assn.................Oct 1

Index ☆ Chase's 1995 Calendar of Events ☆
Books (cont'd)—Bulgaria

Edinburgh Book Festival (Edinburgh, Scotland)...Aug 12
Farmer's Almanac editor Ray Geiger: Birth Anniv...Sept 18
Gay and Lesbian Book Month...June 1
Jewish Book Month...Nov 18
Library Week, Natl...Apr 9
Moby Dick Marathon (Mystic, CT)...July 31
Planned Parenthood Book Fair (Dayton, OH)...Nov 10
Printers Row Book Fair (Chicago, IL)...June 17
Read Me Day (TN)...Apr 26
Read to Your Child Day...Feb 14
Reading Is Fun Week...Apr 23
Religious Reference Books Week, Natl...Sept 3
Return the Borrowed Books Week...Mar 1
Self-Help Book Week, Natl...May 7
Southern Festival of Books (Nashville, TN)...Oct 13
State St Station Cookbook Fair (Greensboro, NC)...Mar 25
Talking Book Week...Apr 17
Van Buren Co Literacy Day...Sept 28
Boom Box Parade (Willimantic, CT)...July 4
Boom Days Celebration (Leadville, CO)...Aug 4
Boone, Daniel: Battle Blue Licks Celeb (Mount Olivet, KY)...Aug 19
Boone, Daniel: Birth Anniv...Nov 2
Boone, Daniel: Boone Day...June 7
Boone Day...June 7
Boone, Debbie: Birth...Sept 22
Boone, Pat: Birth...June 1
Boone, Pat, Celebrity Spectacular (Chattanooga, TN)...May 4
Boonesborough Days (Boonsboro, MD)...Sept 9
Boorman, John: Birth...Jan 18
Booth, Edwin: Birth Anniv...Nov 13
Booth, Evangeline: Birth Anniv...Dec 25
Booth, John Wilkes, Escape Route Tour (Clinton, MD)...Apr 15, Sept 9
Booth, William: Birth Anniv...Apr 10
Boren, David: Birth...Apr 21
Borg, Bjorn: Birth...June 6
Borge, Victor: Birth...Jan 3
Borglum, Gutzon: Birth Anniv...Mar 25
Borgnine, Ernest: Birth...Jan 24
Boring Celebrities of the Year, Most...Dec 11
Boring Film Awards, Most...Mar 13
Bork, Robert Heron: Birth...Mar 1
Borman, Frank: Birth...Mar 14
Bormann, Martin: Birth Anniv...June 17
Bosley, Tom: Birth...Oct 1
Boso, Cap: Birth...Sept 10
Boss Day, Natl...Oct 16
Boss/Employee Exchange Day, Natl...Sept 11
Boston Marathon (Boston, MA)...Apr 17
Boston Massacre: Anniv...Mar 5
Boston Tea Party: Anniv...Dec 16
Boston Tea Party, Reenactment of (Boston, MA)...Dec 17
Boswell, James: Birth Anniv...Oct 29
Bosworth, Brian: Birth...Mar 9
Botswana: Independence Day...Sept 30
Bottler Day, Bonza...Jan 1, Feb 2, Mar 3, Apr 4, May 5, June 6, July 7, Aug 8, Sept 9, Oct 10, Nov 11, Dec 12
Bottoms, Timothy: Birth...Aug 30
Boulez, Pierre: Birth...Mar 26
Bourguiba, Habib: Birth...Aug 3
Bourke-White, Margaret: Birth Anniv...June 14
Boutros-Ghali, Boutros: Birth...Nov 14
Bowditch, Nathaniel: Birth Anniv...Mar 26
Bowdler's Day...July 11
Bowie, David: Birth...Jan 8
Bowling
ABC Seniors Tourn (Reno, NV)...May 6
ACU Intl Bowling Chmpshp (Reno, NV)...May 5
Amer Bowling Congress Chmpshp Tourn (Reno, NV)...Feb 4
American Bowling Congress Convention (Mobile, AL)...Mar 13
Amer Bowling Congress Masters Tourn (Reno, NV)...May 1
General Tire PBA Men's Tourn (Fairlawn, OH)...Apr 17
WIBC Annual Meeting (Tucson, AZ)...May 1
WIBC Chmpshp Tourn (Tucson, AZ)...Mar 30
WIBC Queen's Tourn (Tucson, AZ)...May 14
Box Car Days (Tracy, MN)...Sept 2
Boxer, Barbara: Birth...Nov 11
Boxing Day (United Kingdom)...Dec 26
Boxing Day Annual Celebration (New York, NY)...Dec 26
Boxing Day Bank Holiday (UK)...Dec 26
Boxing: First US Heavyweight Champ Defeated: Anniv...Dec 10
Boxing: Muhammad Ali Named Champ: Anniv...Oct 30
Boxing: Muhammad Ali Stripped of Title: Anniv...Apr 30
Boxleitner, Bruce: Birth...May 12
Boy George: Birth...June 14
Boy Scouts: Baden-Powell, Robert: Birth Anniv...Feb 22
Boy Scouts of America Anniv Week...Feb 5
Boy Scouts of America Founded: Anniv...Feb 8
Boy Scouts of America Scout Sunday...Feb 6
Boyd, Belle: Birth Anniv...May 9
Boyer, Charles: Birth Anniv...Aug 28
Boyer, Ernest L.: Birth...Sept 13
Boys and Girls Club Week...Apr 16
Bradbury, Ray: Birth...Aug 22
Bradford, William: Birth Anniv...Mar 19
Bradlee, Benjamin Crowninshield: Birth...Aug 26
Bradley, Bill: Birth...July 28
Bradley, Ed: Birth...June 22
Bradley, Omar Nelson: Birth Anniv...Feb 12
Bradley, Tom: Birth...Dec 29
Bradshaw, Terry: Birth...Sept 2
Brady, Nicholas F.: Birth...Apr 11
Brahms, Johannes: Birth Anniv...May 7
Braille Institute Track/Field Olympics (Los Angeles, CA)...May 13
Braille Literacy Week 6522...Page 488
Braille, Louis: Birth Anniv...Jan 4
Braille Rallye (Los Angeles, CA)...July 16
Brain, Decade of the...Jan 1
Brain Tumor Awareness Week...June 5
Brand, Oscar: Birth...Feb 7
Brandauer, Klaus Maria: Birth...June 22
Brandeis, Louis D: Birth Anniv...Nov 13
Brando, Marlon: Birth...Apr 3
Branstad, Terry E.: Birth...Nov 17
Bratwurst Days (Sheboygan, WI)...Aug 4
Braun, Eva: Birth Anniv...Apr 30
Brawl in US House of Representatives, First: Anniv...Jan 30
Braxton, Carter: Birth Anniv...Sept 10
Brazil
Carnival...Feb 25
Cirio de Nazare...Oct 8
Discovery of Brazil Day...Apr 22
Fest of Penha (Rio de Janeiro)...Oct 1
Independence Day...Sept 7
Independence Week...Sept 1
Nosso Senhor Do Bonfim Fest...Jan 20
Republic Day...Nov 15
San Sebastian's Day...Jan 20
Tiradentes Day...Apr 21
Brazleton, T. Berry: Birth...May 10
Brazzi, Rossano: Birth...Sept 18
Breach of Promise Suit, First US: Anniv...June 14
Bread Day, Homemade...Nov 17
Bream, Julian: Birth...July 15
Breast Cancer Awareness Month, Natl 6600...Page 488
Breast Cancer: Cuts and Curls for Charity Month...Oct 1
Breaux, John B.: Birth...Mar 1
Breckinridge, John Cabell: Birth Anniv...Jan 21
Breeders' Cup Chmpshp (Elmont, NY)...Oct 28
Brennan, Eileen: Birth...Sept 3
Brennan, William J., Jr: Birth...Apr 25
Brennan, Wm: Resigns Supreme Court: Anniv...July 20
Brenner, David: Birth...Feb 4
Brent, Margaret: Demands a Political Voice: Anniv...June 24
Brett, George: Birth...May 15
Brewer, Theresa: Birth...May 7
Brickhouse, Jack: Birth...Jan 24
Bridal Expo (Davenport, IA)...Jan 28
Bridal Expo, Cocoanut Grove (Santa Cruz, CA)...Jan 22
Bridge: ACBL Tournament (Yuma, AZ)...Feb 23
Bridge Celebration, Covered (Elizabethton, TN)...June 5
Bridge Fest, Covered (Washington Cnty, PA)...Sept 16
Bridge Fest, Parke County Covered (Rockville, IN)...Oct 13
Bridge over the Neponset: Anniv...Apr 1
Bridge Walk, Mackinac (St. Ignace, MI)...Sept 4
Bridges, Beau: Birth...Dec 9
Bridges, Jeff: Birth...Dec 4
Bridges, Lloyd: Birth...Jan 15
Brimley, Wilford: Birth...Sept 27
Brinkley, Christie: Birth...Feb 2
Brinkley, David: Birth...July 10
Brisebois, Danielle: Birth...June 28
Brisson, Frederick: Birth...Mar 17
British Fest, Holy Trinity (Pensacola, FL)...May 19
British Museum: Anniv...Jan 15
Brittany, Morgan: Birth...Dec 5
Britten, (Edward) Benjamin: Birth Anniv...Nov 22
Broadbent, Ed: Birth...Mar 21
Broderick, Matthew: Birth...Mar 21
Broiler Fest, Mississippi (Forest, MS)...June 1
Brokaw, Tom: Birth...Feb 6
Bronco League World Series (Monterey, CA)...Aug 12
Bronson, Charles: Birth...Nov 3
Bronte, Charlotte: Birth Anniv...Apr 21
Bronte, Emily: Birth Anniv...July 30
Brook, Peter: Birth...Mar 21
Brooklyn Bridge: Opening Anniv...May 24
Brooks, Albert: Birth...July 22
Brooks, Garth: Birth...Feb 7
Brooks, Gwendolyn: Birth...June 7
Brooks, James David: Birth Anniv...Oct 18
Brooks, Mel: Birth...June 28
Brooks, Phillips: Birth Anniv...Dec 13
Broomball Hockey Tourn, Co-Ed (Sheboygan, WI)...Feb 4
Brosnan, Pierce: Birth...May 16
Brotherhood/Sisterhood Week...Feb 19
Brothers, Joyce: Birth...Oct 20
Brougham, Henry P.: Birth Anniv...Sept 19
Broun, Heywood Hale: Birth...Mar 10
Brown, Hank: Birth...Feb 12
Brown, Helen Gurley: Birth...Feb 18
Brown, James: Birth...May 3
Brown, Jesse: Birth...Mar 27
Brown, Jesse Leroy: Birth Anniv...Oct 13
Brown, Jim: Birth...Feb 17
Brown, John: Birth Anniv...May 9
Brown, John: Execution Anniv...Dec 2
Brown, John: Raid Anniv...Oct 16
Brown, Louise: First Test Tube Baby Birth...July 25
Brown, Louise Joy: Birth...July 25
Brown, Ronald H.: Birth...Aug 1
Browne, Jackson: Birth...Oct 9
Browne, Thomas: Birth Anniv...Oct 19
Browner, Joey: Birth...May 15
Browning, Elizabeth Barrett: Birth Anniv...Mar 6
Browning, Robert: Birth Anniv...May 7
Brownmiller, Susan: Birth...Feb 15
Broz, Josip "Tito": Birth Anniv...May 25
Brubeck, Dave: Birth...Dec 6
Bruce, Jack: Birth...May 14
Bruckner, Anton: Birth Anniv...Sept 4
Brummell, George Bryan "Beau": Birth Anniv...June 7
Brunei: National Day...Feb 23
Bryan, Richard H.: Birth...July 16
Bryan, William Jennings: Birth Anniv...Mar 19
Bryant, Anita: Birth...Mar 25
Bryant, William C.: Birth Anniv...Nov 3
Bryson, Peabo: Birth...Apr 13
Buchanan, James: Birth Anniv...Apr 23
Buchwald, Art: Birth...Oct 20
Buck, Pearl S: Birth Anniv...June 26
Buckingham, Lindsey: Birth...Oct 3
Buckley, Betty: Birth...July 3
Buckley, William Frank: Birth...Nov 24
Buckner, Bill: Birth...Dec 14
Buckwheat Fest, Preston County (Kingwood, WV)...Sept 28
Buddha: Birth Anniv...Apr 8
Buddha: Birthday (Hong Kong)...May 7
Budweiser Fest and BBQ Rib Burn-Off (Norfolk, VA)...Sept 15
Buell, Marjorie H.: Birth Anniv...Dec 11
Buffalo Auction, Custer State Park (Custer, SD)...Nov 18
Buffalo Bill (William F. Cody): Birth Anniv...Feb 26
Buffalo Bob: Birth...Nov 27
Buffalo Days Celebration (Luverne, MN)...June 3
Buffalo Round-Up (Custer, SD)...Oct 2
Buffalo Soldiers Day 6461...Page 488
Buffett, Jimmy: Birth...Dec 25
Buggy Days, Barnesville (Barnesville, GA) Sept 16
Buggy Festival, Church Point (Church Point, LA)...June 2
Buhl Day (Sharon, PA)...Sept 4
Building: Underground America Day...May 14
Buildings Safety Week, Natl...Apr 9
Bulgaria
Enlightenment and Culture Day...May 24
Hristo Botev Day...June 2
Liberation Day...Mar 3
St. Lasarus's Day...Apr 1

☆ Chase's 1995 Calendar of Events ☆ Index

Bulgaria (cont'd)—California

Socialist Revolution: Anniv . . . . **Sept 9**
Vinegrower's Day . . . . **Jan 14**
Viticulturists' Day . . . . **Feb 14**
Bullfinch Exchange Festival (Japan) . . . . Jan 7
Bullnanza (Guthrie, OK) . . . . Feb 3
Bumbershoot: The Seattle Arts Fest (Seattle, WA) . . . . Sept 1
Bumpers, Dale: Birth . . . . Aug 12
Bun Day (Iceland) . . . . Feb 27
Bunche, Ralph: Awarded Nobel Peace Prize: Anniv . . . . Dec 10
Bunche, Ralph J.: Birth Anniv . . . . Aug 7
Bundy, McGeorge: Birth . . . . Mar 30
Bunge, Bettina: Birth . . . . June 13
Bunker Hill Day (Suffolk County, MA) . . . . June 17
Bunning, Jim: Birth . . . . Oct 23
Bunsen Burner Day . . . . Mar 31
Bunsen, Robert: Birth and Bunsen Burner Day . . . . Mar 31
Bunyan, John: Birth Anniv . . . . Nov 28
Buonarroti Simoni, Michelangelo: Birth Anniv . . . . Mar 6
Buoniconti, Nick: Birth . . . . Dec 15
Burbank, Luther: Birth Anniv . . . . Mar 7
Burdon, Eric: Birth . . . . Apr 5
Bureau of Internal Revenue Established . . . . . July 1
Bureaucracy: Tom Sawyer's Cat's Birthday . . Jan 3
Burger, Warren E.: Birth . . . . Sept 17
Burgess, Anthony: Birth . . . . Feb 25
Burghoff, Gary: Birth . . . . May 24
Burgoo Fest (Utica, IL) . . . . Oct 8
Burk, Martha (Calamity Jane): Death Anniv . Aug 1
Burke, Delta: Birth . . . . July 30
Burke, Edmund: Birth Anniv . . . . Jan 12
Burkina Faso: National Day . . . . Aug 4
Burma: Independence Day . . . . Jan 4
Burnett, Carol: Birth . . . . Apr 26
Burnham, Daniel: Birth Anniv . . . . Sept 4
Burns, Conrad: Birth . . . . Jan 25
Burns, George: Birth . . . . Jan 20
Burr, Raymond William: Birth Anniv . . . . May 21
Burro Race (Leadville, CO) . . . . Aug 6
Burroughs, John: Birth Anniv . . . . Apr 3
Bursting Day (Iceland) . . . . Feb 28
Burstyn, Ellen: Birth . . . . Dec 7
Burton, LeVar: Birth . . . . Feb 16
Burton, Richard: Birth Anniv . . . . Nov 10
Burundi: Independence Day . . . . July 1
Buscaglia, Leo: Birth . . . . Mar 31
Busch Clash (Daytona Beach, FL) . . . . Feb 12
Busey, Gary: Birth . . . . June 29
Bush, Barbara Pierce: Birth . . . . June 8
Bush, George H.W.: Birth . . . . June 12
Bush, Kate: Birth . . . . July 30
Business: AT&T Divestiture: Anniv . . . . Jan 8
Business: Give-a-Sample Week, Natl . . . . Apr 23
Business: Home Owners Loan Act: Anniv . June 13
Business, Small, Week 6561 . . . . Page 488
Business Week, Home-Based . . . . Oct 8
Business Week, Small . . . . May 7
Business Week, Small 6681 . . . . Page 488
Business Women's Day, American . . . . Sept 22
Business Women's Week, Natl . . . . Oct 16
Butcher, Susan: Birth . . . . Dec 26
Buthelezi, Mangosuthu Gatsha: Birth . . . . . . Aug 27
Butkus, Dick: Birth . . . . Dec 9
Butler, Brett: Birth . . . . June 15
Butler, Kevin: Birth . . . . July 24
Butler, Samuel: Birth Anniv . . . . Dec 4
Butterfield Overland Stage Days (Benson, AZ) . . . . Oct 13
Button, Dick: Birth . . . . July 18
Buttons, Red: Birth . . . . Feb 5
Buzzards' Day . . . . Mar 15
Buzzi, Ruth: Birth . . . . July 24
Byner, Ernest: Birth . . . . Sept 15
Byrd, Charlie: Birth . . . . Sept 16
Byrd, Robert C.: Birth . . . . Nov 20
Byrne, David: Birth . . . . May 14
Byrne, Jane Margaret Burke: Birth . . . . May 24
Byrnes, Edd: Birth . . . . July 30
Byron, George Gordon: Birth Anniv . . . . Jan 22

C

Caan, James: Birth . . . . Mar 26
Cabinet, US: Perkins, Frances (1st Woman Appointed) . . . . Mar 4
Cabrillo Day (CA) . . . . Sept 28
Cactus/Succulent Society Show/Sale, Omaha (Bellevue, NE) . . . . Sept 2
Caesar, Sid: Birth . . . . Sept 8
Cahn, Sammy: Birth Anniv . . . . June 18
Caine, Michael: Birth . . . . Mar 14
Cajun: Bread Bridge Crawfish Fest (Breaux Bridge, LA) . . . . May 5
Cajun: Mamou Mardi Gras Celebration (Mamou, LA) . . . . Feb 28
Calamity Jane (Martha Burk): Death Anniv . . Aug 1
Caldwell, Sarah: Birth . . . . Mar 6
Caldwell, Taylor: Birth . . . . Sept 7
Caldwell, Zoe: Birth . . . . Sept 14
Calendar Adjustment Day: Anniv . . . . Sept 2
Calendar Day, Gregorian . . . . Feb 24
Calendar Stone, Aztec, Discovery Anniv . . . Dec 17
Calhoun, John C: Birth Anniv . . . . Mar 18
California
Admission Day . . . . **Sept 9**
American Council on Education An Mtg (San Francisco) . . . . **Feb 12**
American Youth Hostels: Christmas Trip (San Diego) . . . . **Dec 26**
America's Intl Dixieland Jazz Fest (Sacramento) . . . . **May 26**
Americover (Irvine) . . . . **Aug 4**
Artists' Salute Black History Month (Los Angeles) . . . . **Feb 4**
Beverly Hills Affaire in Gardens (Beverly Hills) . . . . **May 20**
Beyond Category: Genius/Duke Ellington (Los Angeles) . . . . **Jan 7**
Blessing of the Fishing Fleet (San Francisco) . . . . **Oct 1**
Bob Hope Chrysler Golf Classic (Bermuda Dunes) . . . . **Feb 15**
Bracebridge Dinner (Yosemite Natl Pk) . . **Dec 22**
Braille Institute Track/Field Olympics (Los Angeles) . . . . **May 13**
Braille Rallye (Los Angeles) . . . . **July 16**
Bridal Expo, Cocoanut Grove (Santa Cruz) . . . . **Jan 22**
Bronco League World Series (Monterey) **Aug 12**
Cabrillo Day . . . . **Sept 28**
Calaveras Fair/Jumping Frog Jubilee (Angel Camp) . . . . **May 18**
California Gold Discovery: Anniv . . . . **Jan 24**
California Senior Games (Sacramento) . . **June 2**
California Strawberry Fest (Oxnard) . . . . **May 20**
Carnival '95 (San Francisco) . . . . **May 27**
Catalina Water Ski Race (Long Beach) . . . **Aug 13**
Cherry Blossom Fest (Lodi) . . . . **Apr 8**
Cherry Blossom Fest (San Francisco) . . . **Apr 14**
Children's Children's Art Exhibit Intl (San Francisco) . . . . **May 15**
Children's Miracle Network (Anaheim) . . . **June 3**
Chinese Moon Festival (Los Angeles) . . . . **Sept 9**
Chinese New Year Celebration (Los Angeles) . . . . **Jan 23**
Chinese New Year Festival (San Francisco) . . . . **Jan 31**
Chinese New Year Golden Dragon Parade (Los Angeles) . . . . **Feb 18**
Chinese New Year Golden Dragon Parade (San Francisco) . . . . **Feb 11**
Clam Chowder Cookoff and Chowder Chase (Santa Cruz) . . . . **Feb 25**
Comedy Celebration Day (San Francisco) . . . . **July 30**
Community Services Month . . . . **Apr 1**
Cribbage Tourn, World Chmpshp (Quincy) **May 5**
Delicato Charity Grape Stomp (Manteca) **Sept 3**
Dickens Universe (Santa Cruz) . . . . **July 30**
Dry Bean Fest, California (Tracy) . . . . **Aug 5**
Earth Day Family Fest (San Jose) . . . . **Apr 22**
Earth Fair (San Diego) . . . . **Apr 23**
Easter in July Lily Fest (Smith River) . . . . . **July 8**
Edsel Owners Club Conv (Oakland) . . . . . . **Aug 2**
Examiner Bay to Breaker Race (San Francisco) . . . . **May 21**
Fancy Rat and Mouse Show (Riverside) . . **Jan 28**
Flossie Beadle Week (Lakeside) . . . . **Dec 1**
Folk on the Rocks (Yellowknife, NT) . . . . . **July 21**
Gilroy Garlic Fest (Gilroy) . . . . **July 28**
Global Restoration Fair (San Francisco) . . **Apr 21**
Global Youth Project (San Francisco) . . . **June 19**
Great LA Clean UP (Los Angeles) . . . . **Apr 17**
Great Monterey Squid Fest (Monterey) . . . **May 27**
Grubstake Days (Yucca Valley) . . . . **May 25**
Handcar Races and Steam Fest (Felton) . . **July 8**
Harlem: Photographs by Aaron Siskind (Los Angeles) . . . . **Mar 4**
Heartland Train/Kansas City UN50 (San Francisco) . . . . **June 21**
HI-AYH Midnight Madness Fun Bike Ride (San Diego) . . . . **Aug 19**
Hobby Industry Association Trade Show (Anaheim) . . . . **Jan 29**
Holiday Bowl Parade/Game (San Diego) . **Dec 30**
Imperial Valley Produce Charity Ball (Imperial Valley) . . . . **Feb 11**
John Steinbeck's Birthday (Salinas) . . . . . **Feb 27**
Kernville Stampede Rodeo (Kernville) . . . . . **Oct 7**
Key Club Intl Convention (Anaheim) . . . . . . **July 1**
Kwin Kite Fly In (Stockton) . . . . **May 29**
Lakeside Historical Soc Hometown Reunion (Lakeside) . . . . **Oct 22**
Laura Ingalls Wilder Gingerbread Sociable (Pomona) . . . . **Feb 4**
Lodi Grape Festival (Lodi) . . . . **Sept 14**
Los Angeles County Fair (Pomona) . . . . . . **Sept 8**
Los Angeles Marathon/Family Reunion (Los Angeles) . . . . **Mar 5**
Los Angeles Riots: Anniv . . . . **Apr 29**
Model UN of the Far West (San Francisco) . . . . **Apr 19**
Moment in History (San Francisco) . . . . . **June 19**
Monterey Bay Blues Fest (Monterey) . . . **June 24**
Monterey County Hot Air Affair (Monterey, CA) . . . . **Mar 11**
Monterey Jazz Fest (Monterey) . . . . **Sept 15**
Mormon Trail Ride (Summit Valley) . . . . . . **May 19**
Morro Bay Harbor Fest (Morro Bay) . . . . . . **Oct 7**
Mother Goose Parade (El Cajon) . . . . **Nov 19**
Mountain Man Rendezvous (Felton) . . . . . **Nov 24**
Mozart Festival (San Luis Obispo) . . . . **July 24**
Mt Rubidoux Easter Sunrise Service (Riverside) . . . . **Apr 16**
NAIA Outdoor Track/Field Chmpshps (Azusa) . . . . **May 25**
Napa Valley Wine Auction (Napa Valley) . . **June 8**
Nixon Birthday Holiday (Yorba Linda) . . . . . **Jan 9**
No Laughing Matter: Cartoon/Envi (Yuba City) . . . . **Feb 1**
No Laughing Matter: Cartoon/Environ (San Jacinto) . . . . **Apr 8**
No Laughing Matter: Cartoon/Environ (Santa Rosa) . . . . **May 21**
No Laughing Matter: Cartoon/Environ (Ukiah) . . . . **Aug 12**
Noah's Flood (San Francisco) . . . . **June 23**
Northern CA Cherry Blossom Fest (San Francisco) . . . . **Apr 15**
Old Adobe Fiesta (Petaluma) . . . . **Aug 20**
Online Conference (Chicago) . . . . **Oct 30**
Pasadena Doo Dah Parade (Pasadena) . . **Nov 26**
Peace, Security Pacific Rim (San Francisco) . . . . **Apr 28**
Petaluma River Fest (Petaluma) . . . . **Aug 19**
PGA Golf Chmpshp (Pacific Palisades) . . . **Aug 7**
Pier 39 Street Performers Festival (San Francisco) . . . . **June 10**
Pier 39's Tulipmania (San Francisco) . . . . . . **Mar 4**
Proposition 13: Anniv . . . . **June 6**
Quality of Life Expo (Los Angeles) . . . . **Mar 2**
Rediscovering Justice (San Francisco) . **June 22**
Reflections on Future of UN (San Francisco) . . . . **June 20**
Roaring Camp Harvest Faire (Felton) . . . . **Oct 14**
Roaring Camp Jumping Frog Contest (Felton) . . . . **July 2**
Roaring Camp Railroad Ride (Felton) . . . **June 18**
Rose Bowl Game (Pasadena) . . . . **Jan 2**
Russian River Jazz Fest (Guerneville) . . . **Sept 16**
San Diego Boat and Sport Fishing Show (San Diego) . . . . **Jan 5**
Santa by Stage Coach Parade (El Centro) . **Dec 2**
Santa Rosalia Fest (Monterey) . . . . **Sept 9**
Sea Lion "Haul Out": Anniv (San Francisco) . . . . **Jan 19**
Sheep Shearing/Kids Day (Petaluma) . . . . . **Apr 8**
Snowbird Breakfast (El Centro) . . . . **Jan 14**
Snowfest (Tahoe City) . . . . **Mar 3**
Society for Eradication of Television Conv (San Diego) . . . . **Sept 22**
Sonoma-Marin Fair (Petaluma) . . . . **June 21**
Southern CA Firestorms: Anniv . . . . **Oct 25**
Spring Fest, Intl (Lodi) . . . . **May 20**
Stamp Expo (Anaheim) . . . . **Mar 24, Sept 15**
Stamp Expo (Long Beach) . . . . **Jan 6**
Stamp Expo (Los Angeles) . . . . **Apr 7**
Stamp Expo (Palm Springs) . . . . **Feb 17**
Stamp Expo (Pasadena) . **Feb 17, July 14, Nov 10**
Stamp Expo America (Anaheim) . . . . **Nov 24**
Stamp Expo USA (Anaheim) . . . . **Feb 10**
Stamp Expo: Anaheim (Anaheim) . . . . **Oct 13**
Stamp Expo: California (Anaheim) . . . . **Aug 11**
Stamp Expo: Fall (Los Angeles) . . . . **Nov 17**
Stamp Expo: South (Anaheim) . . . . **Apr 28**
Stamp Expo: West (Anaheim) . . . . **Jan 13**
State of the World Forum (San Francisco) **Apr 21**
Steinbeck Fest (Salinas) . . . . **Aug 3**
Summer Stamp Expo (Anaheim) . . . . **July 14**

Index ☆ Chase's 1995 Calendar of Events ☆

California (cont'd)—Canada

Swallows Depart (San Juan Capistrano) . . **Oct 23**
Teddy Bear Tea (Barstow) . . . . **Feb 11**
Ten Best Censored Stories 1994 (Rohnert Park) . . . . **Apr 1**
Tongas: Alaska's Rain Forest (Los Angles) . . . . **Dec 16**
Tournament of Roses Parade (Pasadena) . . **Jan 2**
UN Assn USA Annual Conv (San Francisco) . . . . **June 25**
UN 50th Anniversary Celeb (San Francisco) . . . . **June 25**
UN Rocks the World (San Francisco, CA) . . . . **June 24**
UN Teacher's Workshop (San Francisco) . **Apr 29**
Vacaville Onion Fest (Vacaville) . . . . **Sept 9**
Visions of Peace (San Francisco) . . . . **June 23**
Warbirds in Action Air Show (Shafter) . . . **Apr 22**
Whiskey Flat Days (Kernville) . . . . **Feb 17**
Whitewater Wednesday (Kernville) . . . . . . **June 21**
Wild West Film Fest (Tuolumne County) . **Sept 22**
World Wristwrestling Chmpshp (Petaluma) . . . . **Oct 14**
Yosemite Chefs' Holidays (Yosemite Natl Park) . . . . **Jan 1**
Yosemite Nordic Holiday Race (Yosemite Natl Park) . . . . **Mar 4**
Yosemite Spring Skifest (Yosemite Natl Park) . . . . **Apr 8**
Yosemite Vintners' Holidays (Yosemite Natl Park) . . . . **Nov 6**
Callas, Maria: Birth Anniv . . . . **Dec 2**
Calle Ocho: Open House (Miami, FL) . . . . . . **Mar 12**
Calloway, Cab: Birth . . . . **Dec 25**
Cambodia: Independence Day . . . . **Nov 9**
Camden, Battle of: Anniv . . . . **Aug 16**
Cameron, Kirk: Birth . . . . **Oct 12**
Cameroon
National Holiday . . . . **May 20**
Youth Day . . . . **Feb 11**
Camp Fire Birthday Sabbath . . . . **Mar 18**
Camp Fire Birthday Sunday . . . . **Mar 19**
Camp Fire Birthday Week . . . . **Mar 13**
Camp Fire Founders Day . . . . **Mar 17**
Camp Week, American . . . . **Mar 5**
Campanella, Roy: Birth Anniv . . . . **Nov 19**
Campbell, Angus: Birth Anniv . . . . **Aug 10**
Campbell, Ben Nighthorse: Birth . . . . **Apr 13**
Campbell, Carroll A., Jr: Birth . . . . **July 24**
Campbell, Douglas: Becomes 1st American Air ACE: Anniv . . . . **Apr 14**
Campbell, Earl: Birth . . . . **Mar 29**
Campbell, Glen: Birth . . . . **Apr 22**
Campbell, Malcolm: Birth Anniv . . . . **Mar 11**
Camus, Albert: Birth Anniv . . . . **Nov 7**
Canada
AAHA/Amer Handwriting Analysis Conv (Vancouver, BC) . . . . **July 5**
Abbotsford Intl Airshow (Abbotsford, BC) . . . . **Aug 11**
Agrifair (Clearbrook, BC) . . . . **Aug 4**
Apple Blossom Fest (Annapolis Valley, NS) . . . . **May 25**
Arts Fest XI (Annapolis, NS) . . . . **Sept 14**
Atlantic Winter Fair (Halifax, NS) . . . . **Oct 7**
Banff Fest of Mountain Films (Banff, AB) . . **Nov 3**
Bazaart (Regina, Sask) . . . . **June 17**
Big Valley Jamboree (Craven, Sask) . . . . . **July 12**
Binder Twine Fest (Kleinburg, Ont) . . . . **Sept 9**
Buffalo Days (Regina, Sask) . . . . **Aug 1**
Calgary Exhibition/Stampede (Calgary, AB) . . . . **July 7**
Calgary Folk Fest (Calgary, AB) . . . . **July 17**
Calgary Winter Fest (Calgary, AB) . . . . **Feb 17**
Canada Day . . . . **July 1**
Canada Day Celeb (Bridgetown, NS) . . . . . **July 1**
Canada Day Celeb (Watson Lake, Yukon) . **July 1**
Canada Day Weekend (Bancroft, Ont) . . . . **July 1**
Canada's Coldest Recorded Temperature: Anniv . . . . **Jan 31**
Canada's National Ukrainian Festival (Dauphin) . . . . **July 28**
Canada-US Goodwill Week . . . . **Apr 23**
Canadian Chmpshp Dog Derby (Yellowknife, NT) . . . . **Mar 24**
Canadian Finals Rodeo (Edmonton, Alta) . **Nov 8**
Canadian Open Fiddle Contest (Shelburne, Ont) . . . . **Aug 11**
Canadian Tulip Fest (Ottawa, Ont) . . . . . . **May 17**
Canadian Turtle Derby (Boissevain, Man) **July 14**
Canadian Western Agribition (Regina, Sask) . . . . **Nov 25**
Cardston Stampede (Cardston, Alta) . . . . **Aug 10**
Celebration '94/Multicultural Fest (Halifax, NS) . . . . **June 17**
Changing the Guard (Ottawa, Ont) . . . . . **June 23**
Chocolate Fest (St. Stephen, NB) . . . . **July 31**
Civic Holiday . . . . **Aug 7**
CN Tower: Anniv . . . . **June 26**
CN Tower New Year Celeb (Toronto, Ont) . **Jan 1**
Commonwealth Day . . . . **Mar 13**
Constitution Act: Anniv . . . . **Apr 18**
Cranbrook Air Fair (Cranbrook, BC) . . . . . **July 26**
Cultures Canada (Ottawa-Hull, Ont) . . . . . . **July 2**
Curling Chmpshps, World (Brandon, MB) . **Apr 8**
Dart Festival, Intl (Dawson City, Yukon) . **Sept 10**
Dawson City Music Fest (Dawson City, Yukon) . . . . **July 21**
Digby Scallop Days Fest (Digby, NS) . . . . . **Aug 6**
Discovery Days (Dawson City, Yukon) . . . **Aug 18**
Discovery Days Celeb (Watson Lake, YT) **Aug 12**
Dockside Festival (Gravenhurst, Ont) . . . **Aug 11**
Drummondville Folk Fest (Drummondville, Que) . . . . **July 7**
Edmonton Folk Music Fest (Edmonton, AB) . . . . **Aug 10**
Elmira Maple Syrup Fest (Elmira, Ont) . . . . **Apr 1**
Fall Fair (Brandon, Man) . . . . **Oct 27**
Farmfair (Edmonton, Alta) . . . . **Nov 4**
Festival Du Voyageur (Winnipeg, Man) . . . **Feb 10**
Flight Begins (Abbotsford, BC) . . . . **Aug 8**
Folklorama—Canada's Cultural Celeb (Winnipeg, Man) . . . . **Aug 6**
Freedom Fest, Intl (Windsor, Ont) . . . . . . **June 17**
Friendship Fest (Ft. Erie/Buffalo, Ont) . . **June 24**
Fringe Theater Event (Edmonton, Alta) . . **Aug 18**
Frostbite Music Fest (Whitehorse, Yukon) **Feb 17**
Grand Forks Intl Baseball Tourn (Grand Forks, BC) . . . . **Aug 30**
Great Cariboo Ride (100 Mile House, BC) . . . . **July 22**
Great Klondike Outhouse Race (Dawson City, Yukon) . . . . **Sept 3**
Great Northern Arts Fest (Inuvik, NWT) . **July 21**
Great Northern Music Fest (Inuvik, NWT) **July 28**
Great Rendezvous (Thunder Bay, Ont) . . . **July 7**
Guelph Spring Fest (Guelph, Ont) . . . . **Apr 28**
Halifax Intl Buskerfest (Halifax, NS) . . . . . . **Aug 3**
Halifax, Nova Scotia, Destroyed: Anniv . . . **Dec 6**
Harrison Fest of Arts (Harrison Hot Sprgs, BC) . . . . **July 8**
Harvest Moon (Rosthern, Sask) . . . . **July 14**
Hot-Air Balloon Fest (St Jean-sur-Richelieu, Que) . . . . **Aug 12**
Icelandic Festival of Manitoba (Gimli) . . . . **Aug 4**
Intl Curling Bonspiel (Dawson City, Yukon) . . . . **Feb 15**
Intl Fest (St. Stephen, NB/Calais, ME) . . . . **Aug 5**
Intl Wildlife Art Expo (Vancouver, BC) . . . . **May 5**
Inuvik Delta Daze (Inuvik, NWT) . . . . **Oct 6**
Jasper in January (Jasper, AB) . . . . **Jan 13**
Julyfest (Kimberley, BC) . . . . **July 14**
Just For Laughs Festival (Montreal, Que) **July 20**
Kimberley Intl Accordion Chmpshp (Kimberley, BC) . . . . **July 3**
Klondike Days (Edmonton, Alta) . . . . **July 20**
Klondyke Centennial Society Casino Night (Dawson City) . . . . **Feb 24**
Leacock Heritage Fest (Orillia, Ont) . . . . . **July 28**
Mactaquac Craft Fest (Fredericton, NB) . . **Sept 2**
Mahone Bay Wooden Boat Fest (Mahone Bay, NS) . . . . **Aug 2**
Maple Fest of Novia Scotia (Cumberland Co, NS) . . . . **Mar 18**
Mardi Gras Fest (Prince George, BC) . . . . . **Feb 9**
Masters (Calgary, AB) . . . . **Sept 6**
Midnight Madness (Inuvik, NWT) . . . . **June 23**
Midnight Sun Festival (Yellowknife, NWT) **July 14**
Minden Sled Dog Derby (Minden, Ont) . . . **Jan 21**
Moncton Sale of Fine Craft (Fredericton, NB) . . . . **Nov 11**
Morden Corn/Apple Fest (Morden, Man) . **Aug 25**
MOSAIC Fest of Cultures (Regina, Sask) . **June 1**
Mourning, Natl Day of . . . . **Apr 28**
Muskoka Winter Carnival (Gravenhurst, Ont) . . . . **Feb 23**
Nanisivik Midnight Sun Marathon (Nanisivik, NWT) . . . . **June 29**
Nationals (Calgary, AB) . . . . **May 31**
North American (Calgary, AB) . . . . **July 5**
North American Conf on the Family (Winnipeg, Man) . . . . **Oct 19**
Northern Manitoba Trappers' Fest (The Pas, Man) . . . . **Feb 15**
Nova Scotia Bluegrass/Oldtime Music Fest (Ardoise, NS) . . . . **July 28**
Nova Scotia Intl Tattoo (Halifax, NS) . . . . . . **July 1**
Objiway Keeshigunun (Thunder Bay, Ont) . . . . **Aug 19**
Okanagan Wine Fest (BC) . . . . **Sept 29**
Oktoberfest (Kitchener/Waterloo, Ont) . . . **Oct 6**
Oktoberfest/Thanks Canada (Helena, MT) . . . . **Oct 13**
Ontario Winter Carnival Bon Soo (Sault Ste Marie, Ont) . . . . **Jan 27**
Opasquiak Indian Days (The Pas) . . . . **Aug 13**
Owen Sound Summerfolk Fest (Owen Sound, Ont) . . . . **Aug 18**
Persons Day . . . . **Oct 18**
Pioneer Days (Steinback, Man) . . . . **Aug 4**
Provincial Ex (Brandon, Man) . . . . **June 14**
Quebec Intl Summer Fest (Quebec, Que) . **July 6**
Raven Mad Daze (Yellowknife, NWT) . . . . **June 16**
Raymond Stampede (Raymond, Alta) . . . . . **July 1**
Red Deer Intl Air Show (Red Deer, AB) . . . **Aug 5**
Remembrance Day . . . . **Nov 11**
Riverfest (Brockville, Ont) . . . . **June 23**
Rockhound Gemboree (Bancroft, Ont) . . . **Aug 3**
Rossland Public Market (Rossland, BC) . . **May 28**
Rossland Winter Carnival (Rossland, BC) **Jan 27**
Rothesay Craft Fest (Rothesay) . . . . **July 15**
Royal Manitoba Winter Fair (Brandon, Man) . . . . **Mar 27**
Royal Red Arabian Horse Show (Regina) **Aug 16**
Saint John Sale of Fine Craft (St John, NB) . . . . **Nov 25**
St. Swithun's Celebration (Richmond Hill, Ont) . . . . **July 15**
Sam Steele Days (Cranbrook, BC) . . . . . . **June 15**
Saskatchewan Air Show '95 (Moose Jaw) . **July 8**
Saskatchewan Handcraft Fest (Battleford, Sask) . . . . **July 14**
Saskatchewan Jazz Fest (Saskatoon, Sask) . . . . **June 23**
Saskatoon Exhibition (Saskatoon, Sask) . . **July 8**
Saskatoon Fall Fair/Mexabition (Saskatoon, Sask) . . . . **Nov 16**
Scarborough-Indianapolis Peace Games (Scarborough, Ont) . . . . **July 21**
Seniors Expo: Life Is What You Make It (Halifax, NS) . . . . **July 14**
Shakespeare on the Saskatchewan Fest (Saskatoon) . . . . **July 6**
Shaw Fest (Niagara-on-the-Lake, Ont) . . . **Apr 18**
Shearwater Intl Air Show (Shearwater, NS) . . . . **Sept 16**
Shelburne County Lobster Fest (Shelburne Co, NS) . . . . **June 1**
Shelburne Founders' Days (Shelburne, NS) . . . . **July 20**
Shortest Night (Burwash Landing, Yukon) . . . . **June 21**
Sourdough Rendezvous Bathtub Race (Whitehorse, YT) . . . . **Aug 18**
Spaghetti Bridge Building Contest (Kelowna, BC) . . . . **Mar 10**
Special Libraries Assn Annual Conference (Montreal) . . . . **June 10**
Spiders! (Toronto) . . . . **Oct 21**
Spruce Bog Craft Fair (Whitehorse, Yukon) . . . . **Feb 20**
Summer Solstice Dance (Whitehorse, Yukon) . . . . **June 21**
Sundog Handcraft Faire (Saskatoon, Sask) . . . . **Dec 1**
Sunrise Fest (Inuvik, NWT) . . . . **Jan 5**
Texas Holdem Poker Tourn, Intl (Dawson City) . . . . **Sept 10**
Thanksgiving Day . . . . **Oct 9**
Toronto Intl Boat Show (Toronto, Ont) . . . . **Jan 6**
Toronto Intl Film Fest (Toronto, Ont) . . . . . **Sept 7**
Trail of '98 Intl Road Relay (Whitehorse, Yukon) . . . . **Sept 9**
Trial of Louis Riel (Regina, Sask) . . . . **July 26**
Upper Canada Village (Morrisburg, Ont) . **May 20**
Victoria Day . . . . **May 22**
W Canada Farm Progress Show (Regina, Sask) . . . . **June 21**
Welling Pioneer Day (Welling, AB) . . . . **Aug 26**
Western Canada Games (Abbotsford, BC) . . . . **Aug 15**
Winnipeg Fringe Fest (Winnipeg, Man) . . **July 15**
Winnipeg Intl Children's Fest (Winnipeg, Man) . . . . **June 6**
Winter Carnival (Bancroft, Ont) . . . . **Feb 10**
Wintergreen (Regina, Sask) . . . . **Nov 24**
Winterlude (Ottawa, Ont) . . . . **Feb 3**
Wood Mountain Wagon Trek (Willow Bunch) . . . . **July 2**
Yellowknife Midnight Golf Classic (Yellowknife, NWT) . . . . **June 16**
Yorkton Film/Video Fest (Yorkton, Sask) . **May 24**
Yorkton Threshermen's Show/Sr Fest (Yorkton,

☆ Chase's 1995 Calendar of Events ☆ Index

Canada (cont'd)—Chernobyl

Sask) ... Aug 5
Yukon Discovery Day ... Aug 14
Yukon Gold Panning (Dawson City, Yukon) ... July 1
Yukon Intl Storytelling Fest (Whitehorse, Yukon) ... June 23
Yukon Order of Pioneers: Anniv ... Dec 1
Yukon Quest Sled Dog Race (Whitehorse, Yukon) ... Feb 11
Canada Day, Respect ... July 15
Canada: Surfside Salutes Canada Week (Surfside, FL) ... Mar 5
Canadian American Days Fest (Myrtle Beach, SC) ... Mar 11
Canadian Pacific RR: Transcontinental Completion Anniv ... Nov 7
Canal Fest, Coshocton (Coshocton, OH) ... Aug 18
Cancer (Zodiac) Begins ... June 21
Cancer in the Sun Month ... June 1
Candlelight Vigil for Homeless Animals Day ... Aug 19
Candlelight Walking Tour of Chestertown (Chestertown, MD) ... Sept 16
Candlemas Day (Presentation of the Lord) ... Feb 2
Candy, John: Birth Anniv ... Oct 31
Canned Food Month ... Feb 1
Canned Luncheon Meat Week Natl ... July 2
Canning: Appert, Nicholas: Birth Anniv ... Oct 23
Cannon, Annie Jump: Birth Anniv ... Dec 11
Cannon, Dyan: Birth ... Jan 4
Canoe Race and Tri Sport Fest, Sugar Creek (Crawford, IN) ... Apr 23
Canova, Diana: Birth ... June 2
Canseco, Jose: Birth ... July 2
Cantinflas: Birth Anniv ... Aug 12
Cantrell, Lana: Birth ... Aug 7
Capaldi, Jim: Birth ... Aug 24
Caperton, Gaston: Birth ... Feb 21
Cape Verde
Independence Day ... Sept 12
National Day ... July 5
Capistrano, Slugs Return from ... May 28
Capital Day ... Sept 2
Capote, Truman: Birth Anniv ... Sept 30
Capp, Al: Birth Anniv ... Sept 28
Capra, Frank: Birth Anniv ... May 18
Capricorn Begins ... Dec 22
Captain Kangaroo: Birth ... June 27
Captive Nations Week (Pres Proc) ... July 16
Captive Nations Week 6580 ... Page 488
Car Care Month, Natl ... Oct 1
Car Day, Name Your ... Oct 2
Car-Keeping Month, Natl ... May 1
Cara, Irene: Birth ... Mar 18
Carabao Fest (Philippines) ... May 14
Caras, Roger: Birth ... May 24
Caraway, Hattie Wyatt: Birth Anniv ... Feb 1
Caray, Harry: Birth ... Mar 1
Carbon-14 Dating Inventor (Libby): Birth Anniv ... Dec 17
Cardboard Boat: See Boats ... Apr 29
Cardin, Pierre: Birth ... July 7
Cardinale, Claudia: Birth ... Apr 15
Care Sunday (England) ... Apr 2
Caribbean or Caricom Day ... July 3
Caricom or Caribbean Day ... July 3
Carleton, Will: Birth Anniv ... Oct 21
Carlin, George: Birth ... May 12
Carlisle, Belinda: Birth ... Aug 17
Carlisle, Kitty: Birth ... Sept 3
Carlson, Arne H: Birth ... Sept 24
Carlton, Steve: Birth ... Dec 22
Carlyle, Thomas: Birth Anniv ... Dec 4
Carmichael, Hoagie: Birth Anniv ... Nov 22
Carnahan, Mel: Birth ... Feb 11
Carnation City Fest (Alliance, OH) ... Aug 4
Carnaval Miami (Miami, FL) ... Mar 3
Carnegie, Andrew: Birth Anniv ... Nov 25
Carnegie, Dale: Birth Anniv ... Nov 24
Carnes, Kim: Birth ... July 20
Carney, Art: Birth ... Nov 4
Carney, Wm, Receives Cong'l Medal: Anniv May 23
Carnival ... Feb 27
Carnival (Grenada) ... Aug 11
Carnival (Malta) ... May 6
Carnival Season ... Jan 6
Carnival Week (Milan, Italy) ... Feb 26
Caroline, Princess: Birth ... Jan 23
Caron, Leslie: Birth ... July 1
Carousel: Santa Cruz Beach Looff Carousel: Anniv ... Aug 3
Carpenter Ant Awareness Week ... June 18
Carper, Tom: Birth ... Jan 23
Carpet Care Improvement Week, Natl ... May 1
Carr, Gerald Paul: Birth ... Aug 22
Carr, Vikki: Birth ... July 19
Carradine, David: Birth ... Oct 8
Carradine, Keith: Birth ... Aug 8
Carrey, Jim: Birth ... Jan 17
Carriacou Carnival (Carriacou, Grenada) ... Feb 27
Carroll, Charles: Birth Anniv ... Sept 19
Carroll, Diahann: Birth ... July 17
Carroll, Joe Barry: Birth ... July 24
Carroll, Lewis: Birth Anniv. See Dodgson, Charles ... Jan 27
Carroll, Pat: Birth ... May 5
Carson, Carlos: Birth ... Dec 28
Carson, Christopher "Kit": Birth Anniv ... Dec 24
Carson, Johnny: Birth ... Oct 23
Carson, Johnny: There Went Johnny Night May 22
Carson, Rachel: Birth Anniv ... May 27
Carson, Rachel: Silent Spring Publication: Anniv ... Apr 13
Cart, Explosion of the (Florence, Italy) ... Apr 16
Carter, Anthony: Birth ... Sept 17
Carter, Benny: Birth ... Aug 8
Carter Caves Crawlathon (Olive Hill, KY) ... Jan 27
Carter, Dixie: Birth ... May 25
Carter, Gary: Birth ... Apr 8
Carter, (William) Hodding, III: Birth ... Apr 7
Carter, Jimmy: Birth ... Oct 1
Carter, Joseph: Birth ... Mar 7
Carter, Lynda: Birth ... July 24
Carter, Michael: Birth ... Oct 29
Carter, Nell: Birth ... Sept 13
Carter, Robert III: Emancipation of 500: Anniv ... Aug 1
Carter, Rosalyn (Eleanor) Smith: Birth ... Aug 18
Cartier, Jacques: Death Anniv ... Sept 1
Cartier-Bresson, Henri: Birth ... Aug 22
Cartoon Art Appreciation Week ... May 2
Cartoonists Against Crime Day ... Oct 25
Cartoonists on Environment, No Laughing (Santa Rosa, CA) ... May 21
Cartoonists on Environment, No Laughing (Ukiah, CA) ... Aug 12
Cartoons on Environment, No Laughing (San Jacinto, CA) ... Apr 8
Cartoons: Garfield Birthday ... June 19
Cartwright, Bill: Birth ... July 30
Cartwright, Edmund: Birth Anniv ... Apr 24
Caruso, Enrico: Birth Anniv ... Feb 25
Carver, George Washington: Death Anniv ... Jan 5
Carver, Raymond: Birth Anniv ... May 25
Carvey, Dana: Birth ... Apr 2
Casablanca Premiere: Anniv ... Nov 26
Casady, Jack: Birth ... Apr 13
Casals, Pablo: Birth Anniv ... Dec 29
Casanova, Giacomo G.: Birth Anniv ... Apr 2
Casey, Mighty: Struck Out ... June 3
Casey, Robert P.: Birth ... Jan 9
Cash, Johnny: Birth ... Feb 26
Cash, June Carter: Birth ... June 23
Cash, Pat: Birth ... May 27
Cass, Peggy: Birth ... May 21
Cassatt, Mary: Birth Anniv ... May 22
Cassidy, David: Birth ... Apr 12
Cassidy, Shaun: Birth ... Sept 27
Cassini, Oleg: Birth ... Apr 11
Castro, Fidel: Birth ... Aug 13
Cat Festival (Belgium) ... Mar 2
Cat Month, Adopt-a-Shelter- ... June 1
Cat Show, Chmpshp (Indianapolis, IN) ... Oct 28
Cat Show, Chmpshp (Indianapolis, IN) ... Jan 7
Cat Show, Chesapeake Club CFA All-Breed (Baltimore, MD) ... Mar 11
Cat's Question Day, Answer Your ... Jan 22
Cataract Awareness Month ... Mar 1
Catfish Days (East Grand Forks, MN) ... July 14
Catfish Fest (Eudora, AR) ... May 13
Catfish Month, Natl ... Aug 1
Catfish Races, Grand Prix (Greenville, MS) May 13
Cathedral State Park Weekend (Terra Alta, WV) ... Sept 8
Cathedrals Fest, Southern (Chichester, Sussex, England) ... July 20
Cather, Willa: Birth Anniv ... Dec 7
Cather, Willa, Spring Conference (Red Cloud, NE) ... May 5
Catholic Educat'l Assn Conv/Expo, Natl (Cincinnati, OH) ... Apr 18
Catlin, George: Birth Anniv ... July 26
Catt, Carrie Lane Chapman: Birth Anniv ... Jan 9
Cauthen, Steve: Birth ... May 1
Cavazos, Lauro F.: Birth ... Jan 4
Cave Man Never Days ... July 25
Caves: Carter Caves Crawlathon (Olive Hill, KY) ... Jan 27
Cavett, Dick: Birth ... Nov 19
Cavoukian, Raffi: Birth ... July 8
Caxton's "Mirror of the World" Translation: Anniv ... Mar 8
Caxton, William: Birth Anniv ... Aug 13
Cayley, George: Birth Anniv ... Dec 27
Ceausescu, Nicolae: Death Anniv ... Dec 25
Ceccato, Aldo: Birth ... Feb 18
Celebrate (Appleton, WI) ... May 13
Celebrate, Don't Wait, Week ... Aug 7
Celebration '94/Multicultural Fest (Halifax, NS, Canada) ... June 17
Celebration of Love Week ... Feb 12
Celebration of the Senses ... June 24
Celebrities, Most Boring of the Year ... Dec 11
Celtic Fest, Southern Maryland (St. Leonard, MD) ... Apr 29
Censored Stories 1994, Ten Best (Rohnert Park, CA) ... Apr 1
Centenarians Day, Natl ... Sept 22
Central African Republic
Independence Day ... Aug 13
National Day ... Dec 1
Cerebral Palsy, Star-athon Telethon ... Jan 21
Cerone, Rick: Birth ... May 19
Cervantes Saavedra, Miguel De: Death Anniv ... Apr 23
Cetera, Peter: Birth ... Sept 13
Chabrol, Claude: Birth ... June 24
Chad
African Freedom Day ... May 25
Independence Day ... Aug 11
Chadds Ford Days (Chadds Ford, PA) ... Sept 9
Chafee, John Hubbard: Birth ... Oct 22
Chaing Kai-Shek: Birth Anniv ... Oct 31
Challenger Space Shuttle Explosion: Anniv Jan 28
Chamberlain, Richard: Birth ... Mar 31
Chamberlain, Wilton Norman: Birth ... Aug 21
Chambers, Tom: Birth ... June 21
Chamorro, Violeta: Birth ... Oct 18
Chancellor, John: Birth ... July 14
Chaney ... June 21
Chaney, Goodman, Schwerner: Civil Rights Workers Slain ... Aug 4
Chaney, James: Klan Members Convicted ... Oct 20
Channing, Carol: Birth ... Jan 31
Channing, Stockard: Birth ... Feb 13
Chanukah ... Dec 18
Chaplin, Charles S.: Birth Anniv ... Apr 16
Chaplin, Geraldine: Birth ... July 31
Chapman, John: Johnny Appleseed Day ... Mar 11
Chapman, John: Johnny Appleseed Fest (Fort Wayne, IN) ... Sept 16
Charisse, Cyd: Birth ... Mar 8
Charles I Execution: Anniv ... Jan 30
Charles II: Restoration and Birth Anniv ... May 29
Charles, Prince: Birth ... Nov 14
Charles, Ray: Birth ... Sept 23
Charleston Earthquake: Anniv ... Aug 31
Charlie Brown and Snoopy: Birthday ... Oct 2
Charlie the Tuna: Sorry Charlie Day ... Apr 1
Charlotte Observer Moonride Criterium (Charlotte, NC) ... Aug 25
Charo: Birth ... Jan 15
Charro Days (Brownsville, TX) ... Feb 23
Chase's Annual Events: Birthday ... Dec 4
Chase's Annual Events Deadline Approach ... Apr 1
Chase, Chevy: Birth ... Oct 8
Chase, Harrison V.: Birth ... Aug 17
Chase, Helen M.: Birth ... Dec 4
Chase, Samuel: Birth Anniv ... Apr 17
Chase, Sylvia: Birth ... Feb 23
Chase, William D.: Birth ... Apr 8
Chateaubriand, Francois Rene de: Birth Anniv ... Sept 4
Chatterton, Thomas: Birth Anniv ... Nov 20
Chaucer, Geoffrey: Death Anniv ... Oct 25
Chautauqua Festival in the Park (Wytheville, VA) ... June 17
Chavez, Cesar, Death of 6552 ... Page 488
Chavez, Cesar Estrada: Birth Anniv ... Mar 31
Check the Smoke Alarms Day ... Dec 31
Checker, Chubby: Birth ... Oct 3
Checkers Day ... Sept 23
Cheeks, Maurice: Birth ... Sept 8
Chefs' Holidays, Yosemite (Yosemite Natl Park, CA) ... Jan 1
Chekhov, Anton Pavlovich: Birth Anniv ... Jan 29
Chemistry: Natl Mole Day ... Oct 23
Chemistry Week, Natl ... Nov 5
Cheney, Richard B.: Birth ... Jan 30
Cheng Cheng Kung: Birth Anniv (Taiwan) ... Aug 9
Cher: Birth ... May 20
Chernobyl Nuclear Reactor Disaster: Anniv Apr 26

Index ☆ Chase's 1995 Calendar of Events ☆

Cherokee—Christmas

Cherokee Rose Fest/Tours (Gilmer, TX)...May 20
Cherokee Strip Celebration (Perry, OK)...Sept 16
Cherokee Strip Day (OK)................Sept 16
Cherry Blossom Fest (Lodi, CA)............Apr 8
Cherry Blossom Fest, Northern CA (San Francisco, CA)........................Apr 15
Cherry, Deron: Birth.....................Sept 12
Cherry Fest, Natl (Traverse City, MI).......July 2
Cherry Month, Natl.......................Feb 1
Cherry Pit Spitting Contest, Intl (Eau Claire, MI)....................................July 1
Cherry River Festival (Richwood, WV)......Aug 6
Chesapeake-Leopard Affair: Anniv.......June 22
Chesterton, Gilbert: Birth Anniversary....May 29
Chiang Kai-Shek Day (Taiwan)............Oct 31
Chicago Day at World's Fair: Anniv.........May 1
Chicago Fire: Anniv.......................Oct 8
Chicago Flood, Great: Anniv..............Apr 13
Chicago, IL: Harold Washington Elected First Black Mayor..........................Apr 11
Chicago Jazz Fest (Chicago, IL)...........Sept 1
Chicago, Judy: Birth....................July 20
Chicago, Univ of, First Day Classes: Anniv..Oct 1
Chicago, Univ of, First Football Game.....Nov 16
Chicken Boy's Birthday..................Sept 1
Chicken Clucking Contest (Baltimore, MD).................................June 21
Chicken Fest, Delmarva (Federalsburg, MD).................................June 9
Chicken Little Awards.....................Apr 3
Chicken Month, Natl......................Sept 1
Chicken Show, Wayne (Wayne, NE)........July 8
Chief Joseph Surrender: Anniv............Oct 5
Child, Julia: Birth.......................Aug 14
Childermas.............................Dec 28
Children
A(ugusta) Baker's Dozen (Columbia, SC) **Apr 28**
Adoption Week, Natl 6507............**Page 488**
All American Soap Box Derby (Akron, OH)................................**Aug 12**
American Camp Week..................**Mar 5**
Andersen, Hans Christian: Birth Anniv....**Apr 2**
Baby Parade (Ocean City, NJ).........**Aug 10**
Baby Safety Month....................**Sept 1**
Big Brothers/Big Sisters Appreciation Week................................**Apr 23**
Book Fair Month, Intl....................**Oct 1**
Book It/Natl Young Reader's Day.........**Oct 2**
Boy Scouts of America Anniv Week.......**Feb 5**
Boy Scouts of America Founded: Anniv...**Feb 8**
Boy Scouts of America Scout Sunday....**Feb 6**
Bud Billiken Parade (Chicago, IL).......**Aug 12**
Camp Fire Birthday Sabbath............**Mar 18**
Camp Fire Birthday Sunday.............**Mar 19**
Camp Fire Birthday Week...............**Mar 13**
Camp Fire Founders Day................**Mar 17**
Campaign for Healthier Babies Month....**Oct 1**
Charlie Brown and Snoopy: Birthday......**Oct 2**
Child Health Day (Pres Proc).............**Oct 2**
Child Health Day 6602...............**Page 488**
Child Injury Prevention Week (New Hyde Park, NY)................................**Sept 1**
Child Safety and Protection Month.......**Nov 1**
Child Safety Council, Natl: Founding Anniv................................**Nov 9**
Childbirth Education Awarenn Day, Intl..**July 22**
Childcare Awareness Week (Wilmington, DE)..................................**May 1**
Childhood Cancer Month 6469........**Page 488**
Children and Hospitals Week............**Mar 19**
Children's Art Exhibit, Intl (San Francisco, CA)................................**May 15**
Children's Book Day, Intl................**Apr 2**
Children's Book Week, Natl..............**Nov 13**
Children's Colonial Days Fair (Yorktown, VA)..................................**July 1**
Children's Day (Detroit, MI)............**June 28**
Children's Day (Japan)..................**May 5**
Children's Day (Korea)..................**May 5**
Children's Day (MA)....................**June 11**
Children's Day (Milton, NH).............**July 22**
Children's Day (Woodstock, VT)........**Aug 26**
Children's Day, Intl (People's Republic of China).................................**June 1**
Children's Day, Missing, Natl............**May 25**
Children's Day, Natl 6590, 6626........**Page 488**
Children's Day/National Sovereignty (Turkey)............................**Apr 23**
Children's Day, Universal (UN)...........**Oct 2**
Children's Dental Health Month, Natl......**Feb 1**
Children's Eye Health and Safety Month..**Sept 1**
Children's Festival (Erwinna, PA).........**Sept 17**
Children's Festival (Jacksonville, OR).....**July 9**
Children's Film Festival, Intl (Oula, Finland)...........................**Nov 13**
Children's Miracle Network (Anaheim, CA)..................................**June 3**
Children's Party at Green Animals (Newport, RI).................................**July 14**
Children's Sunday.....................**June 11**
Children's Victorian Christmas Open House (Washington, MS)....................**Dec 12**
Communicate with Your Kid Month, Natl..**Oct 1**
Craft Month, Natl........................**Mar 1**
Creative Child and Adult Month, Intl......**Nov 1**
Dan Beary's Spring Cav Teddy Bears (Danbury, CT)...................................**Apr 29**
Decade of the Brain......................**Jan 1**
Denver Intl Children's Fest (Denver, CO) **May 17**
Doll Show (Baltimore, MD).............**July 19**
Dr. Seuss (Geisel): Birth Anniv...........**Mar 2**
Dr. Seuss (Geisel): Death Anniv........**Sept 24**
Elks Natl Hoop Shoot (Indianapolis, IN)..**Apr 28**
Epstein Young Artists Program: Anniv...**Mar 24**
Exchange Club Child Abuse Prevention Month.................................**Apr 1**
Expanded New England Kindergarten Conf (Randolph, MA)......................**Nov 16**
Family Support Month...................**May 1**
Firepup's Birthday......................**Oct 1**
Gifted Children Conv, Natl Assn (Tampa, FL)..................................**Nov 8**
Girl Scout Leader's Day.................**Apr 22**
Girl Scout Sabbath......................**Mar 4**
Girl Scout Sunday.......................**Mar 5**
Girl Scout Week.........................**Mar 5**
Girl Scouts Founding: Anniv............**Mar 12**
Girls Incorporated Natl Conference (New York, NY)..................................**Apr 27**
Girls Incorporated Week................**May 14**
Good Teen Day, Natl 6520, 6647.......**Page 488**
Great Teddy Bear Jamboree Show and Sale (Bristol, CT)........................**Oct 21**
Happy Birthday, A.C.! (Salem, OR).......**Feb 18**
Highlights Fdtn Writers Workshop (Chautauqua, NY)................................**July 15**
Hillhaven Thank You Santa Hotline (Seattle, WA).................................**Dec 26**
Hug-a-Bear Sunday.....................**Nov 5**
Intergenerational Week (Williamsport, MD).................................**Apr 23**
Junior Nature Camp (Wheeling, WV)....**July 30**
Kahn's Kids Fest (Cincinnati, OH).......**June 3**
Key Club Intl Week......................**Nov 5**
Keystone Club Conference (Atlanta, GA) **Mar 29**
Kid'rific (Hartford, CT)..................**Sept 9**
Kids Bridge Exhibition (Pittsburgh, PA)...**July 8**
Kids Bridge Exhibtion (Houston, TX).....**Oct 28**
Kids' Day (Baltimore, MD)...............**Aug 2**
Kid's Swap Shop (Baltimore, MD)........**Aug 7**
Kiwanis Kids' Day, Natl................**Sept 23**
Knuckles Down Month, Natl..............**Apr 1**
Library Card Sign-up Month.............**Sept 1**
Little League Baseball Week, Natl......**June 12**
Little League Baseball World Series (Williamsport, PA).....................**Aug 20**
Little League: Big League World Ser (Ft Lauderdale, FL)........................**Aug 11**
Majic Bodacious Bunny Hunt (Ft Wayne, IN)..................................**Apr 14**
Middle Children's Day..................**Aug 12**
Month of the Young Child (MI)...........**Apr 1**
Mother Goose Day.......................**May 1**
Music in Our Schools Month.............**Mar 1**
Natl School Breakfast Week..............**Mar 6**
Natl Youth Service Day.................**Apr 18**
Night of a Thousand Stars..............**Apr 12**
Odie's Birthday..........................**Aug 8**
Orphan Train Heritage Soc Reunion (Fayetteville, AR)......................**Oct 5**
Peachtree Junior (Atlanta, GA)..........**June 3**
Peanut-Kids-Baseball Day (Coshohocken, PA)...................................**Apr 8**
Pediatric and Adolescent AIDS Awareness Wk 6694.................................**Page 488**
Pet Week, Natl..........................**May 7**
Pony Girls Softball Natl Championships..**Aug 5**
Preschool Immunization Week 6542...**Page 488**
Project Safe Baby Month.................**May 1**
Read Me Day (TN).......................**Apr 26**
Read to Your Child Day.................**Feb 14**
Really Rosie, a Musical (Columbia, SC)..**Mar 16**
Rio Grande Valley Livestock Show (Mercedes, TX)..................................**Mar 15**
Safe Toys and Gifts Month..............**Dec 1**
St. Louis Variety Club Telethon (St. Louis, MO)..................................**Mar 4**
School Celebration, Natl................**Oct 13**
School Lunch Week, Natl.................**Oct 9**
Seattle Intl Children's Fest (Seattle, WA)..**May 8**
Send a Kid to Kamp Radiothon (Lexington, KY)..................................**May 20**
Sheep Shearing/Kids Day (Petaluma, CA).**Apr 8**
Spelling Bee Finals, Natl...............**May 31**
Stop Violence/Save Kids Month..........**Sept 1**
Sudden Infant Death Syndrome Awareness Month, Natl............................**Oct 1**
Take a Kid Fishing Weekend (St. Paul, MN).................................**June 10**
Talk With Your Teen About Sex Month, Natl **Mar 1**
Teddy Bear Tea (Barstow, CA)...........**Feb 11**
Teddy Bear's Picnic (Lahaska, PA)......**July 15**
Truancy Law: Anniv.....................**Apr 12**
Universal Children's Week...............**Oct 1**
Vegetarian Resource Group's Essay Contest for Kids.....................................**May 1**
Video Games Day.......................**July 12**
Virginia Children's Fest (Norfolk, VA)......**Oct 7**
WEBE 108 Kids Fest (Trumbull, CT).....**Mar 12**
Week of the Young Child................**Apr 23**
Winnipeg Intl Children's Fest (Winnipeg, Man, Canada).............................**June 6**
YAM Programs for Youth Art Month (Huntsville, AL)...................................**Mar 12**
Youth Against Tobacco Month, Natl.......**Oct 1**
Youth Art Month.........................**Mar 1**
Youth Day (Cameroon)..................**Feb 11**
Youth Day (Taiwan).....................**Mar 29**
Youth Day (Zambia)......................**Aug 7**
Youth of the Year, Natl (Washington, DC) **Sept 18**
Youth Sing Praise Performance (Belleville, IL)..................................**June 24**
Chile
Dia Del Bibliotecario...................**July 10**
Independence Day......................**Sept 18**
Military Dictatorship Ended: Anniv......**Dec 15**
National Month.........................**Sept 1**
Chiles, Lawton: Birth....................Apr 30
Chili Cookoff/Quail-Egg Eat, Prairie Dog (Gr Prairie, TX).............................Apr 1
Chimborazo Day.........................June 3
China Clipper: Anniv....................Nov 22
China, People's Republic of
Canton Autumn Trade Fair..............**Oct 15**
Canton Spring Trade Fair...............**Apr 15**
Ching Ming Fest.........................**Apr 6**
Double 10th Day........................**Oct 10**
Dragon Boat Fest.......................**June 2**
Fest of Hungry Ghosts..................**July 27**
International Children's Day.............**June 1**
Lantern Festival........................**Feb 14**
Mid-Autumn Festival....................**Sept 9**
National Day............................**Oct 1**
Shanghi Communique: Anniv............**Feb 27**
Sun Yat-Sen Birth Anniv................**Nov 12**
Tiananmen Sq Massacre: Anniv.........**June 4**
Youth Day...............................**May 4**
Chinese Lunar New Year Fest (Baltimore, MD).....................................Feb 5
Chinese Moon Festival (Los Angeles, CA)..Sept 9
Chinese New Year........................Jan 31
Chinese New Year Celebration (Los Angeles, CA)...................................Jan 23
Ching Ming Fest (China)..................Apr 6
Chirac, Jacques Rene: Birth.............Nov 29
Chiropractic Week, Natl.................Sept 17
Chisholm, Shirley: Birth................Nov 30
Chocolate Fest (Knoxville, TN)...........Jan 14
Chocolate Fest (Norman, OK).............Feb 11
Chocolate Fest (St. Stephen, NB, Canada).July 31
Chocolate Fest, Galesburg (Galesburg, IL).Feb 11
Chocolate Harvest (Mattoon, IL)...........Oct 7
Chocolate Week, American...............Mar 19
Choctaw Indian Fair (Philadelphia, MS)...July 12
Cholesterol Education and Awareness Month, Natl....................................Sept 1
Chowderfest (Mystic, CT).................Oct 7
Christ, Circumcision of...................Jan 1
Christianity Week, Consicer...............Apr 2
Christie, Agatha: Birth Anniv............Sept 15
Christie, Julie: Birth...................Apr 14
Christmas
A Yorktown Christmas (Yorktown, VA)....**Dec 2**
Altamont Fair Fest of Lights (Altamont, NY).................................**Nov 21**
American Youth Hostel: Christmas Trip (San Diego, CA)..........................**Dec 26**
Animal's Messiah (Chicago, IL)..........**Dec 10**
Antietam Battlefield Illumination (Sharpsburg, MD)..................................**Dec 2**
Arkansas Craft Guild Christmas Show (Little Rock, AR)............................**Dec 1**
Armenian Christmas......................**Jan 6**
Black Friday............................**Nov 24**
Bonanzaville's Christmas on Prairie (W Fargo,

☆ Chase's 1995 Calendar of Events ☆ Index

ND)....Dec 1
Brach's Holiday Christmas Parade (Chicago, IL)....Nov 25
Candlelight (Dallas, TX)....Dec 7
Carolina Craftsmen's Christmas Classic (Columbia, SC)....Nov 24
Carolina Craftsmen's Christmas Classic (Greensboro, NC)....Nov 24
Carols by Candlelight (Lorton, VA)....Dec 8
Charles Dickens Victorian Christmas (Cornwall, PA)....Nov 24
Children's Vict'n Christmas Open House (Washington, MS)....Dec 12
Christmas....Dec 25
Christmas at Biltmore (Asheville, NC)....Nov 11
Christmas at Pioneer Village (Worthington, MN)....Dec 7
Christmas at the Capitol (Pierre, SD)....Nov 23
Christmas at the Zoo (Asheboro, NC)....Dec 24
Christmas at Union Station (Omaha, NE) Nov 26
Christmas Bazaar (Fargo, ND)....Aug 26
Christmas Bells Ring Again (CIS): Anniv .Dec 24
Christmas Candelight Tour (Fredricksburg, VA)....Dec 3
Christmas Candlelight Tour (Alexandria, VA)....Dec 9
Christmas Craft Fair USA: Indoor Show (Milwaukee, WI)....Dec 9
Christmas Craft Show (Aiken, SC)....Dec 2
Christmas Craft Show (York, PA)....Dec 3
Christmas Eve....Dec 24
Christmas Fair and Gallery (Norman, OK)....Nov 25
Christmas Festival (Natchitoches, LA)....Dec 2
Christmas Gift and Hobby Show (Indianapolis, IN)....Nov 4
Christmas Greens Show (Salem, OR)....Dec 1
Christmas Greetings from Space: Anniv .Dec 19
Christmas in Bear Lake (Bear Lake, MI)....Dec 2
Christmas in Old Dodge City (Dodge City, KS)....Nov 25
Christmas in Roseland (Shreveport, LA) .Nov 25
Christmas Inns of Cape May (Cape May, NJ)....Nov 25
Christmas, Old German (Papillion, NE)..Nov 24
Christmas on the Coosa (Wetumpka, AL) .Dec 9
Christmas on the Prairie (Wahoo, NE)....Dec 1
Christmas on the River (Chattanooga, TN) Dec 2
Christmas on the River (Demopolis, AL) .Nov 30
Christmas Past at Audubon Acres (Chattanooga, TN)....Dec 1
Christmas Seal Campaign....Oct 1
Christmas Showcase/Sale (Little Rock, AR)....Dec 1
Christmas Stroll (Lewistown, MT)....Dec 1
Christmas Stroll (Nantucket, MA)....Dec 1
Christmas to Remember (Laurel, MT)....Dec 3
Christmas Tour of Homes (Pella, IA)....Dec 7
Christmas Walk (Pella, IA)....Nov 18
Coosa River Christmas (Rome, GA)....Nov 18
Country Christmas (Coshocton, OH)....Dec 1
Cowboy Christmas (Wickenburg, AZ)....Dec 1
Creole Christmas in French Quarter (New Orleans, LA)....Dec 1
Cut Your Own Christmas Tree (Charlottesville, VA)....Dec 2
Dickens Festival (Salt Lake City, UT)....Nov 24
Dickens Olde Fashioned Christmas Fest (Holly, MI)....Nov 24
Eighteenth-Century Christmas Wassail (McLean, VA)....Dec 17
Electric Light Parade (Lovington, NM)....Nov 25
Festival of Carols (Topeka, KS)....Nov 24
Festival of Lights (Cincinnati, OH)....Nov 24
Festival of Trees (Rock Island, IL)....Nov 18
Folkways of Christmas (Shakopee, MN) .Nov 24
Frankfurt Christmas Market (Frankfurt, Germany)....Nov 25
Garden of Lights (Muskogee, OK)....Jan 1
Gettysburg Yuletide Fest (Gettysburg, PA) Dec 1
Giant Christmas Tree/Rockefeller Center (New York, NY)....Dec 4
Gingerbread and Lace Christmas Celeb (Charlottesville, VA)....Dec 1
Gingerbread House Competition (Lahaska, PA)....Nov 15
Gingerbread Village (Omaha, NE)....Nov 18
Grosse Pte Santa Claus Parade (Grosse Pte, MI)....Nov 24
Handbell Choir Concert/Tree Lighting (New Preston, CT)....Dec 10
Handel's Messiah (Rock Island, IL)....Dec 15
Heritage Christmas (Lindsborg, KS)....Dec 2
Hillhaven Ho Ho Hotline (Seattle, WA)....Dec 11
Hillhaven Thank You Santa Hotline (Seattle, WA)....Dec 26
Historical Soc Christmas Candle Tour (Crawfordsv, IN)....Dec 9
Holiday at Benjamin Harrison Home (Indianapolis, IN)....Nov 10
Holiday Rhapsody in Lights (Norfolk, NE) Nov 18
Honda Starlight Celebration (Cincinnati, OH)....Nov 18
Humbug Day....Dec 21
International Festival of Lights (Battle Creek, MI)....Nov 20
Jamestown Christmas (Williamsburg, VA) .Dec 3
Jule Fest (Elk Horn, IA)....Nov 24
Kristkindl Markt (Hermann, MO)....Dec 2
La Posada de Kingsville/Celeb of Lights (Kingsville, TX)....Nov 18
Legends of Christmas (Lookout Mountain, GA)....Dec 1
Light of the World Christmas Pageant (Minden, NE)....Nov 25
Lighted Boat Parade (Columbus, MS)....Dec 2
Longwood Gardens Christmas Display (Kennett Square, PA)....Nov 23
Macy's Thanksgiving Day Parade (New York, NY)....Nov 23
MADD's Project Red Ribbon....Nov 1
Madrigal Dinner and Concert (Milwaukee, WI)....Dec 2
Merrie Olde England Xmas Fest (Charlottesville, VA)....Dec 23
Merry Prairie Christmas (Fargo-Moorhead, ND)....Nov 26
Navidades (Puerto Rico)....Dec 15
News 4 Parade of Lights (Denver, CO)....Dec 1
Noel Night (Detroit, MI)....Dec 13
Nordic Yule Fest (Seattle, WA)....Nov 18
Norwegian Christmas (Brooklyn Park, MN)....Dec 2
Old Calendar Orthodox Christmas....Jan 7
Olde Christmas/Mt Washington Tavern (Farmington, PA)....Dec 9
Old-Fashioned Christmas Celebration (Bedford, PA)....Dec 1
Old-Fashioned Danish Christmas (Dannebrog, NE)....Dec 2
Ozcanabans of Oz Convention (Escanaba, MI)....Dec 9
Philippines: Christmas Observance....Dec 16
Polish Christmas Open House (Philadelphia, PA)....Dec 16
Polish-American Christmas Gala (St. Francis, WI)....Dec 3
Quincy Preserves Christmas Tour (Quincy, IL)....Dec 10
St. Nicholas' Day....Dec 6
St. Olaf Christmas Fest (Northfield, MN) .Nov 30
Salem Christmas (Winston-Salem, NC)..Dec 16
Santa by Stage Coach Parade (El Centro, CA)....Dec 2
Santa Spectacular (Indinapolis, IN)....Nov 24
Santaland (Madison, WI)....Nov 24
Scottish Christmas Walk (Alexandria, VA) .Dec 2
Shopping Reminder Day....Nov 25
Silent Night, Holy Night Celebrations (Austria)....Dec 24
Smoky Mountain Lights/Winterfest (Gatlinburg, TN)....Nov 15
Sounds of Season: Holiday Concert (Charlottesville, VA)....Dec 27
Southern Christmas Show (Greensboro, NC)....Nov 9
Star of Wonder Planetarium Show (Newport News, VA)....Jan 2
Tannehill Village Christmas (McCalla, AL) .Dec 4
Twelve Villages of Christmas (Washington County, KS)....Dec 2
Victorian Christmas by Candlelight (Clinton, MD)....Dec 9
Victorian Christmas Home Tour (Leadville, CO)....Dec 2
Victorian Christmas Sleighbell Parade (Manistee, MI)....Nov 30
Victorian Christmas Walk (Davenport, IA) .Dec 1
Victorian Christmas Walk (Denver, CO)..Nov 24
Virginia Christmas Show (Manassas, VA) .Oct 13
Virginia Christmas Show (Richmond, VA) .Nov 2
Virginia Craftsmen's Christmas Classic (Richmond, VA)....Nov 3
Wassail Celebration (Woodstock, VT)....Dec 7
Wassail Tea (Greensboro, NC)....Nov 12
Way of Lights (Belleville, IL)....Nov 24
Whiner's Day, Natl....Dec 26
Wichita Winterfest (Wichita, KS)....Dec 2
Wonderland of Lights (Marshall, TX)....Nov 22
World of Christmas (West Branch, IA)....Nov 11
Wren's Nest Christmas Open House (Atlanta, GA)....Dec 10
Wyatt Earp Christmas Walk (Monmouth, IL)....Sept 9
Yuletide Traditions (Charlottesville, VA)...Dec 1
ZooLights (Tacoma, WA)....Dec 1
Christo: Birth....June 13
Christopher, Warren: Birth....Oct 27
Christopher, William: Birth....Oct 20
Chronic Fatigue Syndrome Awareness Month, Natl....Mar 1
Chrysler Opens Detroit Factory....Mar 31
Chuckwagon Races, Natl Chmpshp (Clinton, AR)....Sept 1
Chung, Connie: Birth....Aug 20
Chung Yeung Fest (Hong Kong)....Nov 1
Church Conf, Center City (Aurora, IL)....Oct 25
Churchill, Winston: Birth Anniv....Nov 30
Churchill, Winston: Day....Apr 9
Church/Synagogue Library Assn Conf (Houghton, NY)....June 24
Ciardi, John: Birth Anniv....June 24
Cinco de Mayo (Mexico)....May 5
Cinco de Mayo Fest (Portland, OR)....May 4
Cinco de Mayo Fiesta (Grand Prairie, TX)..May 7
Circle City Classic (Indianapolis, IN)....Oct 7
Circle K Intl Convention (Phoenix, AZ)....Aug 5
Circle K Intl Week....Feb 5
Circle K Service Day, Intl....Nov 11
Circumcision of Christ....Jan 1
Circus: Clown Week, Natl....Aug 1
Confederation of Independent States (formerly USSR)
Baltic States' Independence Recognized: Anniv....Sept 6
Boris Yeltsin Inaugurated: Anniv....July 10
Christmas Bells Ring Again: Anniv....Dec 24
Coup Attempt in the Soviet Union: Anniv Aug 19
Easter Celebrated in Red Square: Anniv...Apr 7
First Joint Siberian-American Musical: Anniv....Dec 20
Great October Socialist Revolution: Anniv Nov 7
KGB Founder Statue Dismantled: Anniv .Aug 22
Moscow Communique: Anniv....May 29
St. Petersburg Name Restored: Anniv....Sept 6
Soviet Communist Party Suspended: Anniv....Aug 29
Soviet Georgia Votes Independence: Anniv....Mar 31
Soviet Union Dissolved: Anniv....Dec 8
Cisneros, Henry: Birth....June 11
Citizenship Day (Pres Proc)....Sept 17
Citizenship Day and Constitution Week 6593....Page 488
Citrus Festival, Florida (Winter Haven, FL) .Jan 26
Citrus Fiesta, Texas (Mission, TX)....Jan 30
Civil Rights Act of 1964 Passed: Anniv....July 2
Civil Rights Act of 1965: Anniv....Aug 6
Civil Rights Act of 1968: Anniv....Apr 11
Civil Rights Workers Disappear: Anniv....June 21
Civil Rights Workers Found Slain: Anniv....Aug 4
Civil War, American
Amnesty Issued for Southern Rebels: Anniv....May 29
Antietam Battlefield Meml Illumination (Sharpsburg, MD)....Dec 2
Attack on Fort Sumter: Anniv....Apr 12
Attack on Fort Wagner: Anniv....July 19
Battle of Chattanooga: Anniv....Nov 24
Battle of Gettysburg: Anniv....July 1
Battle of Waynesborough: Anniv....Mar 2
Battle West Point/Prairie (Columbus/West Pt, MS)....Feb 11
Carolina Campaign Begins: Anniv....Feb 1
Civil War Days (Rockford, IL)....Aug 12
Civil War Ending: Anniv....Apr 9
Civil War Heritage Days (Gettysburg, PA)....June 30
Civil War Living History Weekend (Huntsville, AL)....Nov 11
Civil War Muster and Mercantile Expo (Davenport, IA)....Sept 15
Civil War Peace Talks: Anniversary....Feb 3
Civil War Reenactment (Bedford, PA)....Sept 9
Civil War Reenactment (Keokuk, IA)....Apr 28
Civil War Relic/Collector's Show (Gettysburg, PA)....June 30
Civil War Submarine Attack: Anniv....Oct 5
Civil War Weekend (Yorktown, VA)....May 27
Columbia Surrenders to Sherman: Anniversary....Feb 17

Christmas (cont'd)—Civil War

Index ☆ Chase's 1995 Calendar of Events ☆

Civil War (cont'd)—Computer

Confederate States Congress Adjournment........Mar 18
Defeat at Five Forks: Anniversary........Apr 1
Doubleday, Abner: Birth Anniv........June 26
Fall of Richmond: Anniversary........Apr 3
Fort Sumter Returned to Union Control: Anniv........Feb 17
Fort Sumter Shelled by North: Anniv....Aug 17
Fredericksburg Heritage Fest (Fredericksburg, VA)........July 4
Gettysburg Address Memorial Ceremony (Gettysburg, PA)........Nov 19
Homefront: America in the 1940s (West Branch, IA)........Apr 15
Ironclads Battle in Charleston Harbor: Anniv........Jan 31
Jefferson Davis Captured: Anniv........May 10
John Hunt Morgan Captured: Anniv.....July 26
John Wilkes Booth Escape Route Tour (Clinton, MD)........Apr 15, Sept 9
Johnson Impeachment Proceedings: Anniv........Mar 5
Last Formal Surrender of Confederate Troops: Anniv........June 23
Lee Birthday Celebrations (Alexandria, VA)........Jan 22
Lincoln, Abraham: Assassination Conspirators Hanging........July 7
Lincoln Approves 13th Amendment (Freedom Day)........Feb 1
Lincoln Assassination Anniv........Apr 14
Lincoln Signs Income Tax: Anniv........July 1
Lincoln's Gettysburg Address: Anniv....Nov 19
Marigold Fest (Pekin, IL)........Sept 8
Memorial Day Parade and Services (Gettysburg, PA)........May 29
Monitor Sinking: Anniv........Dec 30
Mt Washington Tavern Candlelight Tours (Farmington, PA)........Dec 9
Peace Overtures in Civil War: Anniv......Mar 2
Reward Offered for Jefferson Davis: AnnivMay 2
Ruffin, Edmund: Birth Anniv........Jan 5
September Fest (Childersburg, AZ).....Sept 30
September Skirmish (Decatur, AL).......Sept 1
Surratt, Mary: Execution Anniv........July 7
Surrender at Durham Station: Anniversary........Apr 18
Tubman, Harriet: Death Anniv........Mar 10
Vote to Impeach Pres Andrew Johnson: Anniv........Feb 24
William Carney Receives Cong Medal: Anniv........May 23
Claiborne, Craig: Birth........Sept 4
Claiborne, Liz: Birth........Mar 31
Clam Chowder Cookoff and Chowder Chase (Santa Cruz, CA)........Feb 25
Clapton, Eric: Birth........Mar 30
Clark, Abraham: Birth Anniversary........Feb 15
Clark, Barney: Artificial Heart Transplant: Anniv........Dec 2
Clark, Barney: Birth Anniv........Jan 21
Clark, Barney: Death Anniv........Mar 23
Clark, Dave: Birth........Dec 15
Clark, Dick: Birth........Nov 30
Clark, George R: Birth Anniv........Nov 19
Clark, Jack: Birth........Nov 10
Clark, Mark: Birth Anniv........May 1
Clark, Petula: Birth........Nov 15
Clark, Ramsey: Birth........Dec 18
Clark, Roy: Birth........Apr 15
Clark, Susan: Birth........Mar 8
Clarke, Arthur Charles: Birth........Dec 16
Clay, Cassius, Jr (Muhammad Ali): Birth...Jan 17
Clay, Cassius, Named Heavyweight Champ: Anniv........Oct 30
Clay Week, Natl (Uhrichsville, OH).......June 13
Clayburgh, Jill: Birth........Apr 30
Clayton, Adam: Birth........Mar 13
Clayton, Mark: Birth........Apr 8
Clean Air Campaign, Amer Lung Assn......May 1
Clean Out Your Computer Day........Feb 13
Clean Out Your Refrigerator Day, Natl.....Nov 15
Clean-Off-Your-Desk Day, Natl........Jan 9
Cleese, John: Birth........Oct 27
Clemency, Grant of Executive 6518......Page 488
Clemens, Roger: Birth........Aug 4
Clemens, Samuel (Mark Twain): Birth AnnivNov 30
Clements, Fr. George Harold: Birth........Jan 26
Clemons, Clarence: Birth........Jan 11
Clemson, Thomas: Birth Anniversary.......July 1
Clerihew Day (Edmund Bentley Clerihew Birth Anniv)........July 10
Cleveland, Ester: First White House Presidential Baby........Aug 30
Cleveland, Frances: Birth Anniv........July 21
Cleveland, Grover: Birth Anniversary......Mar 18
Cliburn, Van: Birth........July 12
Cline, Patsy: Birth Anniv........Sept 8
Clinton, George: Birth Anniv........July 26
Clinton, Hillary Rodham: Birth........Oct 26
Clinton, William Jefferson: Birth........Aug 19
Clock Month, Natl........Oct 1
Clock Tower Jazz Fest (Rockford, IL)......Jan 20
Clodagh: Birth........Oct 8
Clooney, Rosemary: Birth........May 23
Close, Glenn: Birth........Mar 19
Clothesline Fair (Prairie Grove, AR)........Sept 2
Clothing, Gender, Power, Try This On: (New York, NY)........Jan 7
Clothing/Gender/Power, Try This On: (Clarksville, TN)........Nov 18
Clothing/Gender/Power, Try This On: (Flint, MI)........Sept 16
Clothing/Gender/Power, Try This On: (Janesville, WI)........Mar 11
Clothing/Gender/Power, Try This On: (New Bedford, MA)........May 13
Clothing/Gender/Power, Try This On: (Tallahassee, FL)........July 15
Clothing: All Dressed Up (Clinton)........June 1
Clower, Jerry: Birth........Sept 28
Clown Week, Intl........Aug 1
Clowns: Klown Karnival (Plainview, NE)....July 7
Clymer, George: Birth Anniv........Mar 16
CN Tower: Anniv........June 26
Co-op Awareness Month........Oct 1
Coast Guard Day........Aug 4
Coats, Dan: Birth........May 16
Cobain, Kurt: Birth Anniv........Feb 20
Cobb, Irvin S.: Birth Anniv........June 23
Cobb, Tyrus "Ty": Birth Anniv........Dec 18
Cobblestone Fest (Falls City, NE)........Aug 17
Cobras in North America Day, Yell "Fudge" at the........June 2
Coburn, James: Birth........Aug 31
Cochise: Death Anniv........June 7
Cochran, Thad: Birth........Dec 7
Cocker, Joe: Birth........May 20
Cody, William F. "Buffalo Bill": Birth Anniv .Feb 26
Coffee Cup Washing, Dangerous Dan's....Mar 20
Cohen, Leonard: Birth........Sept 21
Cohen, William S.: Birth........Aug 28
Coho Family Festival (Sheboygan, WI)....July 21
Coin Week, Natl........Apr 16
Coins Stamped "In God We Trust": Anniv..Apr 22
Colbert, Claudette: Birth........Sept 13
Cole, Nat "King": Birth Anniv........Mar 17
Cole, Natalie: Birth........Feb 6
Coleman, Dabney: Birth........Jan 3
Coleman, Derrick: Birth........June 21
Coleman, Gary: Birth........Feb 8
Coleman, John: Birth........Nov 15
Coleman, Ornette: Birth........Mar 19
Coleridge, Samuel T: Birth Anniv........Oct 21
Coles, Joanna: Birth........Aug 11
Colfax, Schuyler: Birth Anniv........Mar 23
Collins, Gary: Birth........Apr 30
Collins, Joan: Birth........May 23
Collins, Judy: Birth........May 1
Collins, Michael: Birth........Oct 31
Collins, Phil: Birth........Jan 31
Collinsworth, Cris: Birth........Jan 27
Cologne Cathedral Completion: Anniv....Aug 14
Colonial: Amer Heritage Fest (Yorktown, VA)........May 20
Colonial: First Assembly Day (Yorktown, VA)........July 29
Colonialism, Intl Decade for Eradication of (UN)........Jan 1
Colorado
Abortion First Legalized: Anniv........Apr 25
Alferd G. Packer Day (Boulder)........Apr 21
Aspen Music Festival (Aspen)........June 22
Aspen/Snowmass Banana Season (Snowmass Village)........Apr 1
Aspen/Snowmass Winterskol (Snowmass Village)........Jan 11
Bolder Boulder 10K (Boulder)........May 29
Boom Days Celebration (Leadville).......Aug 4
Burro Race (Leadville)........Aug 6
Capitol Hill People's Fair (Denver).......June 3
Cinco De Mayo Celebration (Leadville)...May 6
Colorado Day........Aug 7
Colorado Shakespeare Fest (Boulder)..June 23
Colorado State Snow Sculpting Chmpshp (Breckenridge)........Jan 18
Commonwheel Arts and Crafts Fest (Manitou Springs)........Sept 2
Day in the Warsaw Ghetto (Denver).......Apr 8
Denver Intl Children's Fest (Denver).....May 17
Downtown Denver Intl Buskerfest (Denver)........July 21
Fall Foliage/Brew Fest (Steamboat Springs)........Sept 23
Fest of Mountain & Plain: A Taste (Denver)........Sept 1
Footbag Chmpshps, World (Golden)....July 25
Fresh Fish Fest (Breckenridge)........June 3
Great American Beer Fest (Denver).......Oct 6
Great Pikes Peak Cowboy Poetry Gath (Colo Springs)........Aug 18
Hendo's Fat Tuesday at Breckenridge (Breckenridge)........Feb 28
Hometown Days (Strasburg)........Aug 19
Kennedy Center Imagination Celeb (Colorado Springs)........Apr 1
Leadville 100-Mile Ultra Run (Leadville)..Aug 19
Leadville's Crystal Carnival (Leadville)....Mar 3
Ludlow Mine Incident: Anniv........Apr 20
MountainFilm (Telluride)........May 26
News 4 Parade of Lights (Denver)........Dec 1
No Man's Land Celebration (Breckenridge)........Aug 11
Oktoberfest (Denver)........Sept 15
Oro City (Leadville)........July 1
Santa Fe Trail Day (Las Animas)........Apr 28
Snowmass Hot Air Balloon Fest (Snowmass Village)........June 23
Snowmass Mardi Gras (Snowmass Village)........Feb 28
State Fair (Pueblo)........Aug 19
Steamboat Motorcycle Weekend (Steamboat Springs)........Sept 13
Steamboat's Rainbow Weekend (Steamboat Springs)........July 14
Telluride Jazz Celebration (Telluride).....Aug 4
Telluride Mushroom Fest (Telluride)....Aug 24
Telluride Wine Festival (Telluride).......June 23
Ullr Fest (Breckenridge)........Jan 9
Victorian Christmas Home Tour (Leadville)Dec 2
Victorian Christmas Walk (Denver)......Nov 24
Vintage Auto Race/Concours Elegance (Steamboat Springs)........Sept 1
Western Stock Show and Rodeo, Natl (Denver)........Jan 10
Wildflower Festival (Crested Butte).....July 10
Colorectal Cancer Education/Awareness Month........Dec 1
Colt League World Series (Lafayette, IN)..Aug 10
Colter, Jessi: Birth........May 25
Columbia: Independence Day........July 20
Columbian Exposition: Chicago Day: Anniv.May 1
Columbus, Christopher
Columbus Day (Observed)........Oct 9
Columbus Day (Traditional)........Oct 12
Columbus Day 6608........Page 488
Columbus Day, Natl (Pres Proc)........Oct 9
Columbus Sails for New World: Anniv.....Aug 3
Discovery of Jamaica by: Anniv........May 4
Columbus, Knights of, Founders' Day.....Mar 29
Columnist's Day, Natl........June 27
Comaneci, Nadia: Birth........Nov 12
Come and Take It Day (Gonzales, TX).......Oct 6
Comedy Celebration Day (San Francisco, CA)........July 30
Comics: Funky Winkerbean: Anniv........Mar 27
Commercial Air Flight Between the US and USSR Begins........July 15
Commercial Radio Broadcasting: Anniv...Aug 28
Commodore Perry Day........Apr 10
Common Prayer Day (Denmark)........May 12
Commoner, Barry: Birth........May 28
Commonwealth Day (Canada)........Mar 13
Commonwealth Day (United Kingdom)....Mar 13
Commonwealth Day, Belize........May 24
Communication Week, World........Nov 1
Communist Party Suspended, Soviet: Anniv........Aug 29
Community Day, World........Nov 3
Community Education Day, Natl........Nov 14
Community Services Month (California)....Apr 1
Como, Perry: Birth........May 18
Comoros: Independence Day........July 6
Compliment-Your-Mirror Day........July 3
Computer
Computer Day, Clean Out Your........Feb 13
Computer Learning Month........Oct 1
Computer Security Day........Nov 30
Intl Shareware Day........Dec 17
Multimedia Schools Conf (Chicago, IL)..Oct 31
National Online Meeting (New York, NY)..May 2
Online Conf (Chicago, IL)........Oct 30

☆ Chase's 1995 Calendar of Events ☆ Index

Religious Software Week, Natl . . . . Aug 20
Conan Doyle, Arthur: Birth Anniv . . . . May 22
Condom Week, Natl . . . . Feb 14
Confederate Heroes Day . . . . Jan 19
Confederate Memorial Day (AL, MS) . . . . Apr 24
Confederate Memorial Day (FL, GA) . . . . Apr 26
Confederate Memorial Day (NC, SC) . . . . . . May 10
Confederate Memorial Day (VA) . . . . May 29
Confederation, Articles of: Ratification Anniv . . . . Mar 1
Confessions Day, True . . . . Mar 15
Confucius: Birthday Observance (Hong Kong) . . . . Sept 21
Confucius's Birthday (Taiwan) . . . . Sept 28
Congo: National Holiday . . . . Aug 15
Congress Assembles (US) . . . . Jan 3
Congress: First Meeting Anniv . . . . Mar 4
Congress (House of Reps) First Quorum: Anniv . . . . Apr 1
Congress: Woman Runs the House: Anniv June 20
Conn, Billy: Birth . . . . Oct 8
Connecticut
Afro-American Cultural Ctr Anniv Conf (New Haven) . . . . Mar 31
American Crossword Puzzle Tournament (Stamford) . . . . Mar 17
Antique Marine Engine Exposition (Mystic) . . . . Aug 19
Antique/Classic Boat Rendezvous (Mystic) . . . . July 22
Black Solidarity Day (New Haven) . . . . Nov 6
Boom Box Parade (Willimantic) . . . . July 4
Chowderfest (Mystic) . . . . Oct 7
Connecticut Storytelling Fest (New London) . . . . Apr 21
Constitution Ratification: Anniv . . . . Jan 9
Dan Beary's Cavalcade of Teddy Bears (Danbury) . . . . Apr 29, Nov 4
Durham Fair (Durham) . . . . Sept 22
First Night Hartford (Hartford) . . . . Dec 31
Great Teddy Bear Jamboree Show and Sale (Bristol) . . . . Oct 21
Handbell Choir Concert and Tree Lighting (New Preston) . . . . Dec 10
Horse and Carriage Weekend (Mystic) . . July 14
Huckleberry Finn Raft Race (New Preston) . . . . Sept 4
Kid'rific (Hartford) . . . . Sept 9
Live Turkey "Olimpiks" (New Preston) . . . Nov 19
Lobsterfest (Mystic) . . . . May 27
Maple Sugaring Fest (New Preston) . . . . . Mar 13
Moby Dick Marathon (Mystic) . . . . July 31
Mystic Seaport Museum: Independence Day Celeb (Mystic) . . . . July 4
Mystic Seaport Photo Weekend (Mystic) . Sept 9
New England Field Days (Mystic) . . . . Nov 24
Norwalk Intl In-Water boat Show (Norwalk) . . . . Sept 21
Old-Fashioned Frog Jump Jamboree (New Preston) . . . . July 2
Plainfield Home and Product Show (Plainfield) . . . . Mar 25
Scottish Bagpipe and Highland Dance Fest (New Preston) . . . . Aug 6
Sea Music Fest (Mystic) . . . . June 9
Seaport Seniors Month (Mystic) . . . . Apr 1
Small Craft Weekend (Mystic) . . . . June 3
Tellabration Eve of Storytelling/Grown-ups (New London) . . . . Nov 17
WEBE 108 Kids Fest (Trumbull) . . . . Mar 12
Who's in Charge: Workers/Managers (Fairfield) . . . . Mar 18
WICC Greatest Bluefish Tourn (Long Island Sound, NY) . . . . Aug 26
Winter Fun Fest (New Preston) . . . . Feb 12
Woodstock Fair (Woodstock) . . . . Sept 1
Connery, Sean: Birth . . . . Aug 25
Conniff, Ray: Birth . . . . Nov 6
Connors, Chuck: Birth Anniv . . . . Apr 10
Connors, Jimmy: Birth . . . . Sept 2
Connors, Mike: Birth . . . . Aug 15
Conrad, Joseph: Birth Anniv . . . . Dec 3
Conrad, Kent: Birth . . . . Mar 12
Conrad, Robert: Birth . . . . Mar 1
Conrad, William: Birth Anniv . . . . Sept 27
Conservation Festival (Roanoke, VA) . . . . . June 10
Conserve Water/Detect-a-Leak Week . . . . May 7
Consider Christianity Week . . . . Apr 2
Constable, John: Birth Anniv . . . . June 11
Constitution, United States
1st Amendments Submitted to States . . . Sept 25
11th Amendment (States' Sovereignty) . . . . Feb 7
12th Amendment Ratified . . . . June 15
13th Amendment Ratified . . . . Dec 6
13th Amendment: Anniv . . . . Dec 18
14th Amendment Ratified . . . . July 9
15th Amendment Ratified . . . . Feb 3
16th Amendment Ratified (Income Tax) . . . Feb 3
17th Amendment Ratified . . . . Apr 8
18th Amendment (Prohibition): Anniv . . . . Jan 16
19th Amendment Ratified . . . . Aug 18
20th Amendment (Inaugural, Congress opening dates) . . . . Jan 23
20th Amendment to US Constitution: Adoption . . . . Feb 6
21st Amendment (Prohibition Repeal) . . . . Dec 5
21st Amendment Ratified . . . . Dec 5
22nd Amendment (Two Term Limit): Ratified . . . . Feb 27
23rd Amendment Ratified . . . . Mar 29
24th Amendment (Eliminated Poll Taxes) . Jan 23
25th Amendment (Pres Succession, Disability) . . . . Feb 10
26th Amendment Ratified . . . . June 30
Bill of Rights: Anniv of First State Ratification . . . . Nov 20
Constitution Day, Natl . . . . Sept 17
Constitution of the US: Anniv . . . . Sept 17
Constitution Week (Pres Proc) . . . . Sept 17
Constitutional Convention: Anniv . . . . May 25
Equal Rights Amendment Sent to States for Ratification . . . . Mar 22
Federalist Papers: Anniv . . . . Oct 27
Presidential Succession Act: Anniv . . . . . . July 18
Women's Suffrage Amendment Introduced: Anniv . . . . Jan 10
Construction: Underground America Day . May 14
Consumer Information Month . . . . Oct 1
Consumer Protection Week, Natl . . . . Oct 22
Consumers Week, Natl 6617 . . . . Page 488
Conti, Tom: Birth . . . . Nov 22
Continental Congress Assembly, First: Anniv . . . . Sept 5
Contraband Days (Lake Charles, LA) . . . . May 2
Converse, Harriet: White Woman Made Indian Chief: Anniv . . . . Sept 18
Conway, Tim: Birth . . . . Dec 13
Cook, James: Birth Anniv . . . . Oct 27
Cook, Peter: Birth . . . . Nov 17
Cookbook Festival (Fargo, ND) . . . . Feb 11
Cooking: Natl Culinary Week . . . . Nov 12
Cookout, Don Macleod (Jackson, WY) . . . . May 20
Coolidge, Calvin: Birth Anniv . . . . July 4
Coolidge, Grace Anna Goodhue: Birth AnnivJan 3
Coolidge, Rita: Birth . . . . May 1
Coon Hunt, Grand American (Orangeburg, SC) . . . . Jan 6
Cooney, Gerry: Birth . . . . Aug 24
Cooper, Alice: Birth . . . . Feb 4
Cooper, D.B.: Hijacking Anniv . . . . Nov 24
Cooper, Gary: Birth Anniv . . . . May 7
Cooper, Jackie: Birth . . . . Sept 15
Cooper, James Fenimore: Birth Anniv . . . . Sept 15
Cooper, L. Gordon: Birth . . . . Mar 6
Cooper, Michael: Birth . . . . Apr 15
Copeland, Stewart: Birth . . . . July 16
Copernicus, Nicolaus: Birth Anniv . . . . Feb 19
Copernicus, Nicolaus: Death Anniv . . . . May 24
Copper Magnolia Fest (Washington, MS) . . Sept 23
Copperfield, David: Birth . . . . Sept 16
Coppola, Francis Ford: Birth . . . . Apr 7
Coray, Melissa Burton: Birth Anniv . . . . Mar 2
Corbett-Sullivan Prize Fight: Anniv . . . . Sept 7
Corea, Chick: Birth . . . . June 12
Corley, Pat: Birth . . . . June 1
Corman, Roger: Birth . . . . Apr 5
Corn Fest (Shippensburg, PA) . . . . Aug 26
Corn Fest, Sun Prairie's Sweet (Sun Prairie, WI) . . . . Aug 17
Corn Fest, Sweet (Millersport, OH) . . . . Aug 30
Corn Palace Fest (Mitchell, SD) . . . . Sept 8
Corpus Christi . . . . June 15
Corpus Christi (US): Observance . . . . June 18
Correct Posture Month . . . . May 1
Corrigan, Mairead: Birth . . . . Jan 27
Corrigan, "Wrong Way" Day . . . . July 17
Cort, Bud: Birth . . . . Mar 29
Corvette. See Auto Shows
Corzine, David: Birth . . . . Apr 25
Cosby, Bill: Birth . . . . July 12
Cosell, Howard: Birth . . . . Mar 25
Cosmetology Month, Natl . . . . Oct 1
Cosmonauts' Day (Alamogordo, NM) . . . . Apr 8
Costa Rica: Independence Day . . . . Sept 15
Costas, Bob: Birth . . . . Mar 22
Costello, Elvis: Birth . . . . Aug 25
Costner, Kevin: Birth . . . . Jan 18
Cote D'Ivoire: National Day . . . . Dec 7
Cotton Bowl Classic, Mobil (Dallas, TX) . . . . . Jan 2
Cotton: Crop Day (Greenwood, MS) . . . . Aug 5
Cotton, Joseph: Birth Anniv . . . . May 15
Council of Nicaea I: Anniv . . . . May 20
Count Your Losses (Clinton, MD) . . . . Aug 31
Country Day, Colton (Colton, NY) . . . . July 15
Country Fair and Burro Barbecue (Bullhead City, AZ) . . . . May 13
Country Ham Days, Marion County (Lebanon, KY) . . . . Sept 23
Country Holidays, Nashville's (Nashville, TN) . . . . Nov 1
Country Jamboree (Fargo, ND) . . . . Aug 13
Country Music Day, Natl . . . . July 4
Country Music Fan Fair, Intl (Nashville, TN) June 5
Country Music Month 6606 . . . . Page 488
Country Sidewalk Sale (Lahaska, PA) . . . . . Aug 12
Coup Attempt in the Soviet Union: Anniv . . Aug 19
Court TV Debut: Anniv . . . . July 1
Court, Margaret: Birth . . . . July 16
Courtenay, Tom: Birth . . . . Feb 25
Courtesy Month, Natl . . . . Sept 1
Coverdell, Paul D.: Birth . . . . Jan 20
Covered Bridge Celebration (Elizabethton, TN) . . . . June 5
Covered Bridge Fest (Washington Cnty, PA) . . . . Sept 16
Cow Chip Throw, Wisconsin (Prairie du Sac, WI) . . . . Sept 1
Cow Chip Throwing Chmpshp, World (Beaver, OK) . . . . Apr 17
Cowbellion Herd New Year's Escapade/Revel (Mobile, AL) . . . . Dec 31
Cowboys/Old West
Annie Oakley Days (Greenville, OH) . . . . . July 27
Bascom, Texas Rose: Birth Anniv . . . . Feb 25
Buffalo Bill Days (Leavenworth, KS) . . . . Sept 10
Bullwhacker Days (Olathe, KS) . . . . June 24
Burk, Martha (Calamity Jane) Death Anniv . . . . Aug 1
Butch Cassidy Outlaw Trail Ride (Vernal, UT) . . . . June 16
Butterfield Overland Stage Days (Benson, AZ) . . . . Oct 13
Cherokee Strip Celebration (Perry, OK) . Sept 16
Chisholm Trail Round-Up Fest (Ft Worth, TX) . . . . June 9
Come and Take It Day (Gonzales, TX) . . . . . Oct 6
Cowboy Christmas (Wickenburg, AZ) . . . . Dec 1
Cowboy Hall of Fame Ceremony (Willcox, AZ) . . . . Oct 5
Cowboy Poet Gathering, Montana (Lewistown, MT) . . . . Aug 18
Cowboy Poetry Gathering, Dakota (Medora, ND) . . . . May 27
Cowboy Poetry Gathering, Texas (Alpine, TX) . . . . Mar 3
Cowboy State Games: Summer, Winter (Casper, WY) . . . . Feb 18, June 18
Custer's Last Stand Reenactment (Hardin, MT) . . . . June 23
Defeat of Jesse James Days (Northfield, MN) . . . . Sept 6
Dodge City Days (Dodge City, KS) . . . . . . July 28
Dubach Chuck Wagon Races (Dubach, LA) . . . . June 29
Eighty-Niner Day Celeb (Guthrie, OK) . . . Apr 18
Fort Atkinson Rendezvous (Fort Atkinson, IA) . . . . Sept 23
Frontier Army Days (Mandon, ND) . . . . . June 24
Gold Rush Days (Wickenburg, AZ) . . . . Feb 10
Great Pikes Peak Cowboy Poetry Gath (Colo Sprngs, CO) . . . . Aug 18
Homestead Days (Beatrice, NE) . . . . June 22
Lone Tree Days (Central City, NE) . . . . July 1
Mountain Man Rendezvous (Red Lodge, MT) . . . . July 21
Nebraskaland Days/Buffalo Bill Rodeo (North Platte, NE) . . . . June 13
Oakley, Annie: Birth Anniv . . . . Aug 13
Oatman Sidewalk Egg Frying Contest (Oatman, AZ) . . . . July 4
Old Sedgwick County Fair (Wichita) . . . . . . Oct 7
Old West Days (Jackson, WY) . . . . May 26
Oregon Train Days (Gering, NE) . . . . July 13
Pendleton Round-Up (Pendleton, OR) . . . Sept 13
Pioneer Days Celebration (Bedford, PA) . . Sept 2
Pony Express Festival (Hanover, KS) . . . . Aug 27
Pony Express-Jesse James Days (St. Joseph, MO) . . . . Apr 1
Reenactment of Cowtown's Last Gunfight (Fort Worth, TX) . . . . Feb 8
Rex Allen Days (Willcox,AZ) . . . . Oct 6
River City Roundup (Omaha, NE) . . . . Sept 20

Computer (cont'd)—Cowboys

Index ★ Chase's 1995 Calendar of Events ★

Cowboys (cont'd)—Dawes

Russell, Charles M.: Birth Anniv........Mar 19
St. Francis River Rendezvous (Farmington, MO)..................................Sept 22
Santa-Cali-Gon Days (Independence, MO)....................................Sept 1
Stagecoach Days (Marshall, TX)........May 20
Territorial Days (Prescott, AZ)........June 10
Texas Ranch Roundup (Wichita Falls, TX) Aug 18
Trails West! (St. Joseph, MO)..........Aug 18
Wagons/Wires/Rails along Oregon Trail (Omaha, NE)......................Mar 12
Wild West Film Fest (Tuolumne County, CA)...................................Sept 22
Wyatt Earp Birthday Celebration (Monmouth, IL)...........................Mar 18
Wyatt Earp Christmas Walk (Monmouth, IL)...................................Sept 9
Wyatt Earp Homecoming Western Days (Monmouth, IL)..................Aug 5
Wyatt Earp OK Corral Anniv (Monmouth, IL)...................................Oct 21
Cows on the Concourse (Madison, WI)....June 3
Coxey's Army, March on Washington: Anniv Mar 25
CPA's Goof-Off Day, Natl..............Apr 18
Crab and Clam Bake (Crisfield, MD)...July 19
Crab Derby and Fair, Natl Hard (Crisfield, MD)..................................Sept 1
Crab Festival, Kodiak (Kodiak, AK)....May 25
Crab: Miss Crustacean/Ocean City Creep (Ocean City, NJ)..................Aug 2
Craddock, Billy "Crash": Birth........June 16
Craft Month, Natl....................Mar 1
Crafts. See Arts and Crafts Shows
Craig, Larry E: Birth................July 20
Craig, Roger: Birth..................July 10
Cramer, Floyd: Birth.................Oct 27
Cranberry Harvest Weekend (Nantucket Island, MA)....................Oct 13
Cranberry Mountain Spring Nature Tour (Richwood, WV)..................May 13
Crandall, Prudence: Birth Anniv......Sept 3
Crane Migration, Sandhill (Grand Island, NE)...................................Mar 5
Cranes: Wings Over the Platte (Grand Island, NE).........................Mar 17
Cranesville Weekend (Terra Alta, WV)....June 2
Cranmer, Thomas: Birth Anniv........July 2
Cranston, Alan: Birth................June 19
Crapper, Thomas: Day................Jan 27
Crash of 1893: Anniv.................May 5
Crater, Judge Joseph F: Day..........Aug 6
Crawfish Festival, Dermott's Annual (Dermott, AR)......................May 19
Crawford, Michael: Birth.............Jan 19
Crawford, Samuel "Wahoo Sam": Birth Anniv...............................Apr 18
Cray, Robert: Birth..................Aug 1
Crayfish Premier (Sweden)............Aug 9
Creative Child and Adult Month, Intl...Nov 1
Creative Romance Month..............Feb 1
Credit Education Week, Natl 6547....Page 488
Credit Union Act: Signing Anniv......June 26
Credit Union Law, First US: Anniv....Apr 6
Credit Union Week....................Oct 15
Creeley, Robert: Birth...............May 21
Cremation, America's First: Anniv....Dec 9
Crenshaw, Ben: Birth.................Jan 11
Creole Christmas in French Quarter (New Orleans, LA).....................Dec 1
Cribb, Tom: First US Heavyweight Defeated: Anniv......................Dec 10
Cribbage Tourn, Northwest (Baker City, OR)......................................Mar 10
Cribbage Tourn, World Chmpshp (Quincy, CA).....................................May 5
Crichton, Michael: Birth.............Oct 23
Crime: See also Safety
Crime Prevention Week, Natl..........Feb 5
Crime Stoppers Fun Dog Show, Copperas (Copperas Cove, TX)..........Apr 30
Crime Victims' Rights Week, Natl 6551, 6678..................................Page 488
Crispus Attucks Day..................Mar 5
Crist, Judith: Birth.................May 22
Croatia: National Day................May 30
Crockett, David: Birth Anniv.........Aug 17
Cronenberg, David: Birth.............May 15
Cronkite, Walter Leland, Jr: Birth...Nov 4
Cronyn, Hume: Birth..................July 18
Crook, Homespun History Day (Omaha, NE) May 7
Crop Day (Greenwood, MS).............Aug 5
Crosby, David: Birth.................Aug 14
Crosby, Harry L. "Bing": Birth Anniv....May 2
Crosby, Kathryn: Birth...............Nov 25
Crosby, Mary: Birth..................Sept 14

Crosby, Norm: Birth..................Sept 15
Cross, Christopher: Birth............May 3
Cross Culture Day, Intl..............Sept 14
Cruelty: Prevention of Animal Cruelty Month Apr 1
Cruikshank, George: Birth Anniv......Sept 27
Cruise, Tom: Birth...................July 3
Crystal, Billy: Birth................Mar 14
Csonka, Larry: Birth.................Dec 25
Cuba
 Cuban Missile Crisis Anniv..........Oct 22
 Liberation Day......................Jan 1
 National Day........................July 26
Cuckoo Dancing Week..................Jan 11
Culinary Fest (Scottsdale, AZ).......Apr 21
Culinary Professionals' Cookbook Month, Intl Assn.....................Oct 1
Culinary Week, Natl..................Nov 12
Culp, Robert: Birth..................Aug 16
Cultural Development, World Decade for (UN)........................Jan 1
Cultural Diversity Day...............Feb 16
Cultures, MOSAIC Fest of (Regina, Sask, Canada)...................June 1
Cummings, Quinn: Birth...............Aug 13
Cummings, Terry: Birth...............Mar 15
Cunningham, Glenn: Birth Anniv.......Aug 4
Cunningham, Merce: Birth.............Apr 16
Cunningham, Randall: Birth...........Mar 27
Cuomo, Mario M.: Birth...............June 15
Curacao
 Animals' Day........................Sept 4
 Curacao Day.........................July 26
 Kingdom Day and Antillean Flag Day..Dec 15
 Memorial Day........................May 4
Curie, Marie Sklodowska: Birth Anniv....Nov 7
Curlew Day...........................Mar 16
Curling Chmpshps, World (Brandon, MB, Canada)......................Apr 8
Curtin, Jane: Birth..................Sept 6
Curtis, Charles: Birth Anniv.........Jan 25
Curtis, Jamie Lee: Birth.............Nov 22
Curtis, Tony: Birth..................June 3
Curtiss, Glenn: Birth Anniv..........May 21
Custer Battlefield Becomes Little Bighorn Battlefield...................Nov 26
Custer, GA: Battle of Waynesborough: Anniv Mar 2
Custer, George: Battle of Little Big Horn: Anniv...............................June 25
Custer: Little Big Horn Days (Hardin, MT) June 22
Custer's Last Stand Reenactment (Hardin, MT).............................June 23
Customer Service Week, Natl..........Oct 2
Customer Service Week, Natl 6485....Page 488
Cuts and Curls for Charity Month.....Oct 1
Cutter Races, Shriner's Invitational (Jackson, WY)......................Feb 18
Cyprus
 Green Monday........................Mar 6
 Independence Day....................Oct 1
 Kataklysmos.........................June 7
 Procession of Icon of St. Lazarus...Apr 15
 St. Paul's Feast....................June 28
Czech Days (Tabor, SD)...............June 16
Czech Fest, Clarkson (Clarkson, NE)..June 22
Czech Fest, Eastpark (Lincoln, NE)...May 6
Czechoslovakia
 Czech-Slovak Divorce: Anniv.........Dec 31
 Czechoslovakia Ends Communist Rule: Anniv..............................Nov 29
 Foundation of the Republic..........Oct 28
 Rape of Lidice: Anniv...............June 10
 Teachers' Day.......................Mar 28
Czechoslovakian: Westfest (West, TX)....Sept 2

D

D'Amato, Alfonse M.: Birth...........Aug 1
D-Day: Anniv.........................June 6
D-Day Natl Remembrance Day/Natl Observance WWII 50th 6697.......Page 488
D.A.R.E. Day, Natl 6588.............Page 488
Da Silva, Howard: Birth..............May 4
Da Vinci, Leonardo: Death Anniv......May 2
Daffodil Fest (Nantucket Is, MA).....Apr 28
Daffodil Fest Parade (Tacoma/Puyallup/Sumner/Orting, WA)..........Apr 22
Dafoe, Willem: Birth.................July 22
Daguerre, Louis: Birth Anniv.........Nov 18
Dahl, Steve: Birth...................Nov 20
Dailey, Irene: Birth.................Sept 12
Dailey, Janet: Birth.................May 21
Dairy Expo, World (Madison, WI)......Oct 4
Dairy Month, June....................June 1
Daisy Festival, Yellow (Stone Mountain, GA) Sept 8
Dali, Salvador: Birth Anniv..........May 11

Dali, Salvador, Museum: Anniv........Mar 7
Dallas, George: Birth Anniv..........July 10
Dalton Defenders Day (Coffeyville, KS)...Oct 7
Dalton, John: Birth Anniv............Sept 6
Dalton, Timothy: Birth...............Mar 21
Daltry, Roger: Birth.................Mar 1
Daly, Chuck: Birth...................July 20
Daly, Tyne: Birth....................Feb 21
Damone, Vic: Birth...................June 12
Dana, Bill: Birth....................Oct 5
Dance
 Ailey, Alvin: Birth Anniv...........Jan 5
 Balanchine, George: Birth Anniv.....Jan 9
 Ballet Introduced to US.............Feb 7
 Chicago's Windy City Jitterbug Dance (Franklin Pk)....................Jan 7
 English Riviera Dance Festival (Torquay, England).....................May 26
 Fall Folk Dance Camp (Wheeling, WV)...Sept 1
 Fonteyn, Margot: Birth Anniv........May 18
 Fosse, Robert Louis: Birth Anniv....June 23
 Gay Square Dance Month, Intl........Sept 1
 Kupio Dance & Music Fest (Kuopio, Finland)..........................June 9
 Merrie Monarch Festival (Hilo, HI)..Apr 16
 Milligan Mini-Polka Day (Milligan, NE)...Sept 24
 Polka Fest (Wisconsin Dells, WI)....Nov 3
 Spring Folk Dance Camp (Wheeling, WV) May 26
 Square Dance Conv, Natl (Birmingham, AL)........................June 21
 Tap Dance Day, Natl.................May 25
 West Virginia Square/Round Dance Conv (Buckhannon, WV)............Aug 4
 Windy City Classic Conv (Schaumburg, IL)..................................Sept 29
 World Folkfest (Springville, UT)....July 8
 Yuma Square and Round Dance Fest (Yuma, AZ)..........................Mar 3
Dancer, Stanley: Birth...............July 25
Danforth, John Claggett: Birth.......Sept 5
Dangerfield, Rodney: Birth...........Nov 22
Dangerous Dan's Annual Cooffee Cup Washing.............................Mar 20
Daniel, Clifton: Birth...............Sept 19
Daniels, Charlie: Birth..............Oct 28
Daniels, William: Birth..............Mar 31
Danish: Aebleskiver Days (Tyler, MN)....June 24
Danish Fest (Greenville, MI).........Aug 18
Danish: Grundlovsfest (Dannebrog, NE)...June 3
Dankfest (Harmony, PA)...............Aug 26
Danson, Ted: Birth...................Dec 29
Dante, Alighieri: Death Anniv........Sept 14
Dantley, Adrian: Birth...............Feb 28
Danza, Tony: Birth...................Apr 21
Darby, Kim: Birth....................July 8
Dare, Virginia: Birth Anniv..........Aug 18
Dark Day in New England: Anniv.......May 19
Darling, Ron: Birth..................Aug 19
Darren, James: Birth.................June 8
Darrow, Clarence: Birth Anniv........Apr 18
Darrow, Clarence, Death Commemoration (Chicago, IL)..................Mar 13
Dart Festival, Intl (Dawson City, Yukon)...Sept 10
Dart Tourn, Blueberry Hill Open (St. Louis, MO)....................Feb 24
Dart Tourn, Chris Stratton (Battle Creek, MI).........................Jan 22
Dart Tourn, St. Patrick's Day (Battle Creek, MI).......................Mar 16
Darts: Winmau World Darts Chmpshp (London, England).....................Dec 1
Darwin, Charles Robert: Birth Anniv..Feb 12
Daschle, Thomas Andrew: Birth........Dec 9
Date Your Mate Month.................May 1
Daugherty, Brad: Birth...............Oct 19
Daumier, Honore: Birth Anniv.........Feb 26
Davidson, John: Birth................Dec 13
Davis, Adelle: Birth Anniv...........Feb 25
Davis, Al: Birth.....................July 4
Davis, Angela: Birth.................Jan 26
Davis, Bette: Birth Anniv............Apr 5
Davis, Clifton: Birth................Oct 4
Davis, Eric: Birth...................May 29
Davis, Geena: Birth..................Jan 21
Davis, Jefferson: Birth Anniv........June 3
Davis, Jefferson, Captured: Anniv....May 10
Davis, Jefferson: Reward Offered for....May 2
Davis, Mac: Birth....................Jan 21
Davis, Miles: Birth Anniv............May 25
Davis, Ossie: Birth..................Dec 18
Davis, Sammy, Jr: Birth..............Dec 8
Davis-Bacon Suspension re: Hurricane Andrew 6491..................Page 488
Dawber, Pam: Birth...................Oct 18
Dawes, Charles G.: Birth Anniv.......Aug 27

☆ Chase's 1995 Calendar of Events ☆ Index

Dawkins—Douglass

Dawkins, Darryl: Birth....Jan 11
Dawkins, Johnny: Birth....Sept 28
Dawson, Andre: Birth....July 10
Dawson, Richard: Birth....Nov 20
Day, Doris: Birth....Apr 3
Day-Lewis, Daniel: Birth....Apr 29
Daylight-Saving Time, US....Apr 2
De Cordova, Frederick: Birth....Oct 27
De Forest, Lee: Birth Anniv....Aug 26
De Gaulle, Charles: Birth Anniv....Nov 22
De Hostos, Eugenio Maria: Birth Anniv....Jan 11
de Klerk, Frederik: Birth....Mar 18
De La Renta, Oscar: Birth....July 22
de Lugo, Ron: Birth....Aug 2
De Montaigne, Michel: Birth Anniv....Feb 28
De Sade, Donatien: Birth Anniv....June 2
Dean, Dizzy: Birth Anniv....Jan 16
Dean, Howard: Birth....Nov 17
Dean, James: Birth Anniv....Feb 8
Dean, James: Fairmount Fest/Remembering (Fairmount, IN)....Sept 22
Dean, Jimmy: Birth....Aug 10
Dean, John: Birth....Oct 14
Dear Diary Day....Sept 22
Death Busters Day, Natl....May 26
DeBakey, Michael: Birth....Sept 7
DeBarge, Eldra: Birth....June 4
DeBerg, Steve: Birth....Jan 18
Debs, Eugene V: Birth Anniv....Nov 5
Debussy, Claude: Birth Anniv....Aug 22
Decade of the Brain....Jan 1
DeCarlo, Yvonne: Birth....Sept 1
Decatur, Stephen: Birth Anniv....Jan 5
Decency, Rally for: Anniv....Mar 23
Decisions: Make Up Your Mind Day....Dec 31
Declaration of Independence
  Approval and Initial Signing Anniv....July 4
  First Public Reading: Anniv....July 8
  Official Signing Anniv....Aug 2
  Resolution Anniv....July 2
Declaration of the Bab....May 23
DeConcini, Dennis: Birth....May 8
Decoy Fest, Chincoteague Easter (Chincoteague Is, VA)....Apr 15
Decter, Midge: Birth....July 25
Dee, Ruby: Birth....Oct 27
Dee, Sandra: Birth....Apr 23
Defense Depot Tracy: Anniv....Jan 1
Defense Transportation Day and Week, Natl 6562, 6689....Page 488
Defense Transportation Day, Natl (Pres Proc)....May 19
DeFore, Don: Birth....Aug 25
Degas, Edgar: Birth Anniv....July 19
DeHaven, Gloria: Birth....July 23
DeHavilland, Olivia: Birth....July 1
Del Rio, Delores: Birth....Aug 3
Delano, Jane: Birth Anniv....Mar 26
DeLaurentiis, Dino: Birth....Aug 8
Delaware
  Childcare Awareness Week (Wilmington)....May 1
  Lincoln, Abraham: Birthday Observance (also OR)....Feb 6
  Munchkins of Oz Convention (Wilmington)....Aug 4
  Nanticoke Indian Powwow (Millsboro)....Sept 9
  Old-Fashioned Ice Cream Fest (Wilmington)....July 8
  Parade of Wheels (Rehoboth Beach)....May 13
  Ratification Day....Dec 7
  Return Day (Georgetown)....Nov 9
  Sea Witch Halloween Fest (Rehoboth Beach/Dewey Beach)....Oct 28
  Secretary's Day Tea (Wilmington)....Apr 26
  Winterthur Point-to-Point (Winterthur)....May 7
Delmonico, Lorenzo: Birth Anniv....Mar 13
DeLorean, John: Birth....Jan 6
DeLuise, Dom: Birth....Aug 1
Democracy Stifled in Peru: Anniv....Apr 6
Dempsey, Jack: Birth Anniv....June 24
Dempsey, Jack: Long Count Day....Sept 22
Demy, Jacques: Birth....June 5
Deneuve, Catherine: Birth....Oct 22
DeNiro, Robert: Birth....Aug 17
Denmark
  Aalborg and Rebild Fest (Aalborg and Rebild)....July 1
  Aarhus Fest Week....Sept 2
  Common Prayer Day....May 12
  Constitution Day....June 5
  Eel Fest (Jyllinge)....June 3
  Ho Sheep Market....Aug 26
  Midsummer Eve....June 23
  Queen Margrethe's Birthday....Apr 16
  Street Urchins' Carnival....Feb 27
  Tivoli Gardens Season (Copenhagen)....May 1
  Viking Festival....June 23
Dennehy, Brian: Birth....July 9
Dent, Richard: Birth....Dec 13
Dent, Russell Earl "Bucky": Birth....Nov 25
Dental Health Month, Natl Children's....Feb 1
Dental Hygiene Week, Natl....Oct 1
Dental School, First Woman to Graduate: Anniv....Feb 21
Denver, Bob: Birth....Jan 9
Denver, John: Birth....Dec 31
DePalma, Brian: Birth....Sept 11
Depot Days (Gainesville, TX)....May 19
DePreist, James (Anderson): Birth....Nov 21
Depression Educ and Awareness Month, Natl....Oct 1
Derek, Bo: Birth....Nov 20
Derek, John: Birth....Aug 12
Dern, Bruce: Birth....June 4
Derwinski, Edward J.: Birth....Sept 15
Descartes, Rene: Birth....Mar 31
Desegregation, US Army First: Anniv....July 26
Desert Shield, Bush Orders Commencement: Anniv....Aug 7
Desert Shield: UN Deadline Resolution: Anniv....Nov 28
Desert Storm: Ground War Begins: Anniv....Feb 23
Desert Storm: Kuwait City Liberated/War Ends: Anniv....Feb 27
Desert Storm: War Against Iraq Begins: Anniv....Jan 16
DeShannon, Jackie: Birth....Aug 21
Design Drafting Assn Convention, American (Atlanta, GA)....June 4
Design Drafting Week, Natl....May 29
Desk Day, Natl Clean-Off-Your....Jan 9
Dessert Day, Natl....Oct 12
Dessert Month, Natl....Oct 1
Detect-a-Leak Week, Conserve Water/....May 7
Detroit (MI): Birth....July 24
Deukmejian, George: Birth....June 6
Devane, William: Birth....Sept 5
Development Decade, Fourth (UN)....Jan 1
Development Information Day, World (UN)....Oct 24
Devil's Night....Oct 30
Devito, Danny: Birth....Nov 17
Dewey, John: Birth Anniv....Oct 20
Dewey, Melvil: Birth Anniv....Dec 10
Dewhurst, Colleen: Birth Anniv....June 3
Dewitt, Joyce: Birth....Apr 23
Dey, Susan: Birth....Dec 10
Di Mucci, Dion: Birth....July 18
Dia de la Raza (Mexico)....Oct 12
Diabetes Month, Natl....Nov 1
Diabetes, Dollars Against (DAD's) Day....June 16
Diabetic Eye Disease Month....Nov 1
Diamond, Neil: Birth....Jan 24
Diana, Princess: Birth....July 1
Diary Day, Dear....Sept 22
Dibble, Robert: Birth....Jan 24
Dicing for Bibles....June 5
Dickens, Charles: Birth Anniv....Feb 7
Dickens, Charles: Dickens Universe (Santa Cruz, CA)....July 30
Dickens, Charles, Victorian Christmas (Cornwall, PA)....Nov 24
Dickens Fest, Broadstairs (Broadstairs, England)....June 17
Dickens Fest, Rochester (Rochester, England)....June 1
Dickens Festival (Salt Lake City, UT)....Nov 24
Dickens on the Strand (Galveston, TX)....Dec 2
Dickerson, Eric: Birth....Sept 2
Dickinson, Angie: Birth....Sept 30
Dickinson, Anna: Birth Anniv....Oct 28
Dickinson, Emily: Birth Anniv....Dec 10
Dictionary Day....Oct 16
Diddley, Bo: Birth....Dec 30
Didion, Joan: Birth....Dec 5
Diefenbaker, John: Birth Anniv....Sept 18
Diego, Jose De: Birth Anniv....Apr 16
Diesel, Rudolph: Birth Anniv....Mar 18
Diet Month, January....Jan 1
Diet: Rid World Fad Diets/Gimmicks Day....Jan 17
Dietary Managers' Pride in Food Service Week....Feb 6
Dietrich, Marlene: Birth Anniv....Dec 27
Diller, Phyllis: Birth....July 17
Dillinger, John: Death Anniv....July 22
Dillman, Bradford: Birth....Apr 14
Dillon, Matt: Birth....Feb 18
Ding Ling: Death Anniv....Mar 4
Dinkins, David: Birth....July 10
Dionne Quintuplets: Birth....May 28
Disability Employment Awareness Month, Natl (Pres Proc)....Oct 1
Disarmament Decade, Third (UN)....Jan 1
Disarmament Week (UN)....Oct 24
Disaster Reduction, Natural, Intl Day For (UN)....Oct 11
Disaster Reduction, Natural, Intl Decade for (UN)....Jan 1
Discoverers' Day (Hawaii)....Oct 9
Discovery Days (Dawson City, Yukon, Canada)....Aug 18
Dishonor List, New Year's....Jan 1
Disraeli, Benjamin: Birth Anniv....Dec 21
Distinguished Service Medal: Anniv....Mar 7
Ditka, Mike: Birth....Oct 18
Divac, Vlade: Birth....Feb 2
Dix, Dorothea L.: Birth Anniv....Apr 4
Dixieland Jazz Fest, America's Intl (Sacramento, CA)....May 26
Dixon, Alan J.: Birth....July 7
Dixon, Jeanne: Birth....Jan 5
Dixon, Willie: Birth Anniv....July 1
Djibouti: National Holiday....June 27
Do It! Day, Natl (aka Fight Procrastination Day)....Sept 6
Dobson, Kevin: Birth....Mar 18
Doctor Who Premiere: Anniv....Nov 23
Doctorow, E.L.: Birth....Jan 6
Doctors' Day....Mar 30
Dodd, Christopher J.: Birth....May 27
Dodgson, Charles Lutwidge: Birth Anniv....Jan 27
Dog. See also Sled Dog
Dog Days....July 3
Dog Month, Adopt-a-Shelter-....Oct 1
Dog sled: Race to the Sky (Helena, MT)....Feb 11
Dog Trial, Red Wing Sheep (Red Wing, MN)....Oct 6
Dog Week, Natl....Sept 24
Dogwash, Anti-Cruelty Society (Chicago, IL)....July 15
Dolby, Ray: Birth....Jan 18
Dole, Elizabeth Hanford: Birth....June 29
Dole, Robert J.: Birth....July 22
Dolenz, Mickey: Birth....Mar 8
Doll Fest (Japan)....Mar 3
Doll, Kewpie: O'Neill, Rose Cecil: Birth Anniv....June 25
Doll Show (Baltimore, MD)....July 19
Doll Show (Yuma, AZ)....Feb 4
Doll Show and Sale, Midwest (Zion, IL)....May 6
Dollars Against Diabetes (DAD's) Day....June 16
Dollhouse/Miniatures Show, Natl (St. Louis, MO)....Aug 12
Dollhouses/Miniatures Trade Show (NYC, NY)....Feb 18
Dolly Sods Weekend (Terra Alta, WV)....Aug 4
Domenici, Pete V.: Birth....May 7
Domingo, Placido: Birth....Jan 21
Dominica: National Day....Nov 3
Dominican Republic
  Independence Day....Feb 27
  National Holiday....Jan 26
  Restoration of the Republic....Aug 16
Domino, Fats: Birth....Feb 26
Domino Tournament, Texas St Chmpshp (Hallettsville, TX)....Jan 22
Donahue, Phil: Birth....Dec 21
Donahue, Troy: Birth....Jan 17
Donald Duck: Birthday....June 9
Donaldson, Sam: Birth....Mar 11
Donovan: Birth....Feb 10
Don't Go to Work Unless It's Fun Day....Apr 3
Don't Wait—Celebrate! Week....Aug 7
Donut Day (Chicago, IL)....June 2
Doolittle, Death of General James H. 6598....Page 488
Doolittle, Eliza: Day....May 20
Doolittle, James Harold: Birth Anniv....Dec 14
Dorgan, Byron L.: Birth....May 14
Dormition of Theotokos....Aug 15
Dornach Battle Commemoration (Switzerland)....July 23
Dorrington, Arthur: First Black Pro Hockey Player: Anniv....Nov 15
Dors, Diana: Birth....Oct 23
Dorsett, Tony: Birth....Apr 7
Dorsey, Thomas A.: Birth Anniv....July 1
Dostoyevsky, Fyodor M: Birth Anniv....Nov 11
Double 10th Day (China)....Oct 10
Doubleday, Abner: Birth Anniv....June 26
Doubleday, Nelson: Birth....July 20
Douglas, Kirk: Birth....Dec 9
Douglas, Michael: Birth....Sept 25
Douglas, Mike: Birth....Aug 11
Douglas, Sherman: Birth....Sept 15
Douglas, William O: Birth Anniv....Oct 16
Douglass, Frederick
  Death Anniv....Feb 20

Index ☆ Chase's 1995 Calendar of Events ☆

Douglass (cont'd)—Education

Escape to Freedom: Anniv .... Sept 3
Frederick Douglass Speaks: Anniv ...... Aug 11
Dove Awards Presentation .... Apr 27
Dow-Jones Tops 1,000: Anniv .... Nov 14
Down Syndrome Awareness Month 6611 Page 488
Downs, Hugh: Birth .... Feb 14
Dozenal Soc of America: Annual Mtg (Garden City, NY) .... Oct 14
Drabek, Doug: Birth .... July 25
Dragon Boat Fest (China) .... June 2
Dragon Boat Fest, Intl (Hong Kong) ...... June 10
Dragon Boat Fest, USA Intl (Dubuque, IA) .. Sept 8
Dragon Boat Races (Keokuk, IA) .... Aug 19
Dragon, Darryl: Birth .... Aug 27
Dream Day .... Aug 28
Dream Hotline, Natl .... Apr 28
Dreamwork Month, Natl .... June 1
Drew, Elizabeth: Birth .... Nov 16
Drew, Ellen: Birth .... Nov 23
Drexler, Clyde: Birth .... June 22
Dreyfus, Richard: Birth .... Oct 29
Drug Abuse, United Nations Decade Against .... Jan 1
Drug Abuse/Illicit Trafficking, Intl Day Against (UN) .... June 26
Drug and Alcohol Awareness Week, Natl PTA .... Mar 5
Drum Month, Intl .... Nov 1
Drysdale, Don: Birth Anniv .... July 23
Dubcek, Alexander: Birth Anniv .... Nov 27
Dubois, W.E.B: Birth Anniv .... Feb 23
Duck Calling Contest, World Chmpshp (Stuttgart, AR) .... Nov 21
Duck Race, Great American (Deming, NM) Aug 25
Duckworth, Kevin: Birth .... Apr 1
Duffy, Julia: Birth .... June 27
Duffy, Patrick: Birth .... Mar 17
Dufoit, Charles: Birth .... Oct 7
Dukakis, Michael S.: Birth .... Nov 3
Dukakis, Olympia: Birth .... June 20
Duke, Patty: Birth .... Dec 14
Dulcimer Days (Coshocton, OH) .... May 20
Dulcimer: Dulcibrrr (Falls of Rough, KY) .... Feb 3
Dulcimer Fest (Huntsville, AL) .... Sept 17
Dulcimer Fest, Southern Appalachian (McCalla, AL) .... May 7
Dumars, Joe: Birth .... May 24
Dumas, Alexandre (Fils): Birth Anniv ...... July 27
Dumas, Alexandre: Birth Anniv .... July 24
Dumb Week (Greece) .... Apr 2
Dunant, Jean Henri: Birth Anniv .... May 8
Dunaway, Faye: Birth .... Jan 14
Duncan, David Douglas: Birth .... Jan 23
Duncan, Isadora: Birth Anniv .... May 27
Duncan, Sandy: Birth .... Feb 20
Dunn, T.R.: Birth .... Feb 1
Dunne, Griffin: Birth .... June 8
Duper, Mark: Birth .... Jan 25
duPont, Pierre S., IV: Birth .... Jan 22
Duran, Roberto: Birth .... June 16
Durang, Christopher: Birth .... Jan 2
Durant, Will: Birth Anniv .... Nov 5
Durant, William: Birth Anniv .... Dec 8
Durbin, Deanna: Birth .... Dec 4
Durenberger, David F.: Birth .... Aug 19
Durer, Albrecht: Birth Anniv .... May 21
Durning, Charles: Birth .... Feb 28
Durocher, Leo: Birth Anniv .... July 27
Dusky Seaside Sparrow, Last: Death Anniv .... June 16
Dussault, Nancy: Birth .... June 30
Duston, Hannah: American Heroine Rewarded: Anniv .... June 8
Dutch Fest, Hanover (Hanover, PA) .... July 29
Duvall, Robert: Birth .... Jan 5
Duvall, Shelley: Birth .... July 7
Dvorak, Anton: New World Symphony Premiere: Anniv .... Dec 16
Dykstra, Lenny: Birth .... Feb 10
Dylan, Bob: Birth .... May 24
Dysart, Richard: Birth .... Mar 30
Dzerzhinsky, Felix: KGB Statue Dismantled: Anniv .... Aug 22

E

E, Sheila: Birth .... Dec 12
Eagle Appreciation Days, Bald (Keokuk, IA) Jan 21
Eagle Tours, Reelfoot (Tiptonville, TN) ...... Jan 1
Eagles Et Cetera (Bismarck, AR) .... Jan 27
Earhart, Amelia, Atlantic Crossing: Anniv .. May 20
Earhart, Amelia: Birth Anniv .... July 24
Early, Jubal: Battle of Waynesborough: Anniv .... Mar 2
Earp, Wyatt, Birthday Celebration (Monmouth, IL) .... Mar 18
Earp, Wyatt: OK Corral Anniv (Monmouth, IL) .... Oct 21
Earth; Earth Day. See also Environment
Earth Day (Celebrate Spring) .... Mar 20
Earth Day (Environment) .... Apr 22
Earth, 1st Picture of, From Space: Anniv .... Aug 7
Earthquake Anniv, Alaska .... Mar 27
Earthquake, Bolivian: Anniv .... June 8
Earthquake Jolts Philippines .... July 16
Earthquake Mexico City Earthquake: Anniv .... Sept 19
Earthquake, San Francisco: Anniv .... Apr 18
Earthquake: San Francisco 1989: Anniv .... Oct 17
Earthquake, Southern California: Anniv .... Jan 17
East Coast Blackout Anniv .... Nov 9
Easter
Chincoteague Easter Decoy Fest (Chincoteague Is, VA) .... Apr 15
City of Baltimore's Easter Celebration (Baltimore, MD) .... Apr 12
Consider Christianity Week .... Apr 2
Curtis Easter Pageant (Curtis, NE) .... Apr 9
Easter Beach Run (Daytona Beach, FL) .. Apr 15
Easter Celebrated in Red Square: Anniv ... Apr 7
Easter Egg Hunt (Rockford, OH) .... Apr 15
Easter Even .... Apr 15
Easter Fest (Kautokeino, Karasjok, Norway) .... Apr 13
Easter in July Lily Fest (Smith River, CA) . July 8
Easter Monday .... Apr 17
Easter Monday Bank Holiday (United Kingdom) .... Apr 17
Easter Monday Celebration (Portage, MI) Apr 17
Easter Parade (Atlantic City, NJ) .... Apr 16
Easter Regatta (Grenada) .... Apr 14
Easter Seal Telethon, Natl .... Mar 4
Easter Sunday .... Apr 16
Easter Sundays Through the Year 2000 .. Apr 16
Easter Sunrise Service (Chimney Rock, NC) .... Apr 16
Good Egg Treasure Hunt (Wheeling, WV) Apr 14
Holy City Easter Pageant (Lawton, OK) .. Apr 15
Holy Humor Month .... Apr 1
Holy Week .... Apr 9
Holy Week Festivities (Portugal) .... Apr 9
Majic Bodacious Bunny Hunt (Ft Wayne, IN) .... Apr 14
Marion Easter Pageant (Marion, IN) ...... Apr 14
Marksville Easter Egg Knocking Contest (Marksvil, LA) .... Apr 16
Moravian Easter Sunrise Service (Winston-Salem, NC) .... Apr 16
Mt Rubidoux Sunrise Service (Riverside, CA) .... Apr 16
Multi-Cultural Easter Egg Display (Belleville, IL) .... Apr 9
Orthodox Easter Sunday .... Apr 23
Passion Week .... Apr 2
Passiontide .... Apr 2
White House Easter Egg Roll (Washington, DC) .... Apr 17
White House Easter Egg Roll: Anniv .... Apr 2
Easton, Sheena: Birth .... Apr 27
Eastwood, Clint: Birth .... May 31
Eat What You Want Day .... Dec 16
Eaton, Robert J.: Birth .... Feb 13
Ebert, Roger Joseph: Birth .... June 18
Ebsen, Buddy: Birth .... Apr 2
Eckersley, Dennis: Birth .... Oct 3
Eckstine, Billy: Birth Anniv .... July 8
Eclipses
Annular Solar Eclipse .... Apr 29
Partial Lunar Eclipse .... Apr 15
Solar (Total) .... Oct 23
Economics: Crash of 1893: Anniv .... May 5
Ecuador
Day of Quito .... Dec 6
Independence Day .... Aug 10
Eddy, Duane: Birth .... Apr 26
Edelman, Marian Wright: Birth .... June 6
Eden, Barbara: Birth .... Aug 23
Ederle, Gertrude: Birth Anniv .... Oct 23
Edgar Allan Poe Fest (Cornwall, PA) ...... Oct 26
Edgar, Jim: Birth .... July 22
Edge, The: Birth .... Aug 8
Edison Festival of Light (Ft Myers, FL) ...... Feb 3
Edison, Thomas: Birth Anniv .... Feb 11
Edison, Thomas: Incandescent Lamp Demonstrated: Anniv .... Oct 21
Edison, Thomas: Record of a Sneeze: Anniv . Feb 2
Editors and Writers Month, Be Kind to ..... Sept 1
Edmund Fitzgerald Memorial Beacon Lighting (Two Harbors, MN) .... Nov 10
Edmund Fitzgerald Sinking: Anniv .... Nov 10
Education
All Class Reunion Party (Roanoke, VA) ... May 6
American Council Education An Mtg (San Francisco, CA) .... Feb 12
American Education Week .... Nov 12
American Education Week (Pres Proc) .. Nov 12
American History Month (MA) .... Feb 1
Backwards Day .... Jan 27
Banned Books Week .... Sept 23
Book Fair Month, Intl .... Oct 1
Book It/Natl Young Reader's Day .... Oct 2
Braille Literacy Week 6522 .... Page 488
Bunsen Burner Day .... Mar 31
Catholic Educ Assn Conv/Expo, Natl (Cincinnati, OH) .... Apr 18
Celebrity Read a Book Week (Williamsport, MD) .... Jan 22
Certified Prof Secretary Exam .... May 5
Chemistry Week, Natl .... Nov 5
Child/Adult Learning Disabil Intl Conf (Orlando, FL) .... Mar 1
Children's Book Day, Intl .... Apr 2
Children's Book Week, Natl .... Nov 13
Collegiate/Future Secretaries Conf (Bloomington, MN) .... Mar 23
Community Education Day, Natl .... Nov 14
Computer Learning Month .... Oct 1
Conference on Student Govt Assns (College Station, TX) .... Feb 18
Day of the Teacher (El Dia Del Maestro) . May 10
Decade of the Brain .... Jan 1
Descartes, Rene: Birth .... Mar 31
Dream Hotline, Natl .... Apr 28
Education and Sharing Day, USA 6540, 6658 .... Page 488
Education First Week, Natl 6509 .... Page 488
Educational Support Personnel Day, Natl Nov 15
Elementary School Teacher Day .... Jan 16
Eliza Doolittle Day .... May 20
Expanded New England Kindergarten Conf (Randolph, MA) .... Nov 16
Family Sexuality Education Month, Natl ... Oct 1
Fashion Show (Milwaukee, WI) .... May 12
Festival in the Schools (Nashville area, TN) Oct 9
Free Paper Week, Natl .... Mar 19
Geography Awareness Week 6623 ..... Page 488
Geography Awareness Week, Natl .... Nov 12
Geography Bee Finals, Natl (Washington, DC) .... May 23
Geography Bee, School Level, Natl .... Jan 2
Geography Bee State Test, Natl .... Apr 7
Gifted Children Conv, Natl Assn (Tampa, FL) .... Nov 8
Harvest of Harmony (Grand Island, NE) ... Oct 7
Historically Black Coll/Univ Week, Natl 6594 .... Page 488
Homemakers of America FHA/Hero Week, Natl Future .... Feb 12
Kindergarten Day .... Apr 21
Library Card Sign-up Month .... Sept 1
Listening Awareness Month .... Apr 1
Literacy Day, Intl (UN) .... Sept 8
Literacy Day, Natl 6578 .... Page 488
Literacy Month, Natl .... Sept 1
Literacy Volunteers Amer Conf (Buffalo, NY) .... Oct 19
Lutheran School Week, Natl .... Mar 5
Mathematics Education Month .... Apr 1
Membership Enrollment Month in Texas .. Sept 1
Metric Week, Natl .... Oct 8
Michigan Science Teachers Assn Conf (Lansing, MI) .... Feb 3
Mind Mapping in Schools Week, Natl ..... Sept 4
Model UN of the Far West (San Francisco, CA) .... Apr 19
Mole Day, Natl .... Oct 23
Multimedia Schools Conference (Chicago, IL) .... Oct 31
Museum Day, Intl .... May 18
Music in Our Schools Month .... Mar 1
Muster (College Station, TX) .... Apr 21
Newspaper in Education & Literacy Conf (Seattle, WA) .... June 7
Newspapers in Education Week .... Mar 6
Night of a Thousand Stars .... Apr 12
Nobel Conference (St. Peter, MN) .... Oct 3
Operation Youth (Cincinnati, OH) ...... June 10
Peabody, Elizabeth Palmer: Birth Anniv .. May 16
Pioneer Day (Washington, MS) .... Oct 19
Preschool Immunization Week 6542 ... Page 488
Project ACES Day .... May 5
PTA Drug and Alcohol Awareness Week,

☆ Chase's 1995 Calendar of Events ☆ Index

Natl . . . . Mar 5
PTA Founder's Day, Natl . . . . Feb 17
PTA Teacher Appreciation Week . . . . May 7
Public School, First in America: Anniv . . . Feb 13
Read Me Day (TN) . . . . Apr 26
Read to Your Child Day . . . . Feb 14
Reading Is Fun Week . . . . Apr 23
Repeat Day (vocabulary) . . . . June 3
School Breakfast Week, Natl . . . . Mar 6
School Celebration, Natl . . . . Oct 13
School Counseling Week, Natl . . . . Feb 6
School Family Day . . . . May 9
School Lunch Week, Natl . . . . Oct 9
School Lunch Week, Natl 6609 . . . . Page 488
School Nurse Day, Natl . . . . Jan 25
School Spirit Season, Intl . . . . Apr 30
School Success Month, Natl . . . . Sept 1
Science and Technology Week, Natl . . . . Apr 21
Self-Improvement Month . . . . Sept 1
Self-University Week . . . . Sept 1
Southern Festival of Books (Nashville, TN) . . . . Oct 13
Spaceweek . . . . July 16
Spelling Bee Finals, Natl . . . . May 31
Student Government Day (MA) . . . . Apr 7
Student Volunteer Day . . . . Feb 20
Sullivan, Anne: Birth Anniv . . . . Apr 14
Teacher Appreciation Day, Natl . . . . Apr 7
Teacher Day, Natl . . . . May 9
Teacher "Thank You" Week . . . . June 4
Teacher's Day (MA) . . . . June 4
Teachers' Day (Czechoslovakia) . . . . Mar 28
Texas PTA Conv (Austin, TX) . . . . Nov 10
Thank You, School Librarian Day . . . . Apr 12
Truancy Law: Anniv . . . . Apr 12
University of Chicago First Day of Classes Oct 1
University of Virginia Founder's Day (Charlottesville, VA) . . . . Apr 13
Week of the Young Child . . . . Apr 23
Women's History Month, Natl . . . . Mar 1
World's Largest Concert . . . . Mar 2
Yale University Founded: Anniv . . . . Oct 16
Educational Bosses Week, Natl . . . . May 14
Edward, Prince: Birth . . . . Mar 10
Edward VIII: Abdication Anniv . . . . Dec 11
Edwards, Blake: Birth . . . . July 26
Edwards, Douglas: Birth . . . . July 14
Edwards, Edwin W.: Birth . . . . Aug 7
Edwards, Harry: Birth . . . . Nov 22
Eel Fest (Jyllinge, Denmark) . . . . June 3
Egg Fest, Central Maine (Pittsfield, ME) . . . July 22
Egg Frying Contest Oatman Sidewalk (Oatman, AZ) . . . . July 4
Egg Month, Natl . . . . May 1
Egg Races (Switzerland) . . . . Apr 17
Egg Roll, White House Easter Egg Roll: Anniv . . . . Apr 2
Egg Salad Week . . . . Apr 17
Eggar, Samantha: Birth . . . . Mar 5
Eggsibit (Phillipsburg, NJ) . . . . Apr 1
Egypt, Arab Republic of
Armed Forces Day . . . . Oct 6
Egyptian Maritime Disaster: Anniv . . . . Dec 14
Evacuation Day . . . . June 18
Grand Bairam Holiday . . . . May 9
National Day . . . . July 23
Sham El-Nesim . . . . Apr 24
Sinai Day . . . . Apr 25
Statue of Ramses II Unearthed: Anniv . . . Nov 30
Ehrlich, Paul: Birth . . . . May 29
Eiffel, Alexandre Gustave: Birth Anniv . . . . Dec 15
Eiffel Tower: Anniv (Paris, France) . . . . Mar 31
Eikenberry, Jill: Birth . . . . Jan 21
Einstein, Albert: Atomic Bomb Letter Anniv . Aug 2
Einstein, Albert: Birth Anniv . . . . Mar 14
Eisenhower, David: Birth . . . . Apr 1
Eisenhower, Dwight David: Birth Anniv . . . . Oct 14
Eisenhower, Mamie Doud: Birth Anniv . . . . Nov 14
Eisenstaedt, Alfred: Birth . . . . Dec 6
Eisner, Michael: Birth . . . . Mar 7
Ekberg, Anita: Birth . . . . Sept 29
Ekland, Britt: Birth . . . . Oct 6
El Salvador
Independence Day . . . . Sept 15
War Officially Ended: Anniv . . . . Jan 31
Elders, M. Joycelyn: Birth . . . . Aug 13
Elections, Caucuses, Conventions
Election Day, General (US) . . . . Nov 7
Return Day (Georgetown, DE) . . . . Nov 9
Ross Perot Answers Question: Anniv . . . . Feb 20
Electricity: Incandescent Lamp Demonstrated: Anniv . . . . Oct 21
Electrocution for Death Penalty, First . . . . Aug 6
Elephant Round-Up at Surin (Thailand) . . . . Nov 18
Eliot, George: Birth Anniv . . . . Nov 22
Eliot, John: Birth Anniv . . . . Aug 5
Eliot, T.S.: Birth Anniv . . . . Sept 26
Elizabeth Bower Day (OK) . . . . May 25
Elizabeth I, Queen: Ascension Anniv . . . . Nov 17
Elizabeth II, Accession of Queen: Anniv . . . . Feb 6
Elizabeth II: Marriage of Elizabeth and Philip: Anniv . . . . Nov 20
Elizabeth II, Queen: Birth . . . . Apr 21
Elizabeth, the Queen Mother: Birth . . . . Aug 4
Elkin, Stanley: Birth . . . . May 11
Ellerbee, Linda: Birth . . . . Aug 15
Ellery, William: Birth Anniv . . . . Dec 22
Ellington, Duke: Birth Anniv . . . . Apr 29
Elliott, Sam: Birth . . . . Aug 9
Elliott, Sean: Birth . . . . Feb 2
Ellis Island Opened: Anniv . . . . Jan 1
Ellsberg, Daniel: Birth . . . . Apr 7
Ellsworth, Oliver: Birth Anniversary . . . . Apr 29
Elm Farm Ollie Day Celebration (WI) . . . . Feb 18
Elvis Presley Remembered (St. Louis, MO) Aug 13
Elvis Presley Salute (Baltimore, MD) . . . . Aug 16
Elvis Week (Memphis, TN) . . . . Aug 11
Elway, John: Birth . . . . June 28
Emaishen (Luxembourg) . . . . Apr 17
Emancipation Day (Texas) . . . . June 19
Emancipation of 500: Anniv . . . . Aug 1
Emancipation Proclamation: Anniv . . . . Jan 1
Embassy Seizure, US, in Teheran: Anniv . . . . Nov 4
Embroidery Month, Intl . . . . Feb 1
Emerson, Ralph Waldo: Birth Anniv . . . . May 25
Emmett, Daniel D: Birth Anniv . . . . Oct 29
Employment/Labor/Occupations/Professions
Activity Professionals Conv, Natl Assn (Nashvil, TN) . . . . Mar 29
Activity Professionals Day, Natl . . . . Jan 27
Activity Professionals Week, Natl . . . . Jan 22
AFL-CIO: Founding Anniv . . . . Dec 5
AFL-CIO Merger: Anniv . . . . Feb 9
Agriculture Day, Natl . . . . Mar 20
Agriculture Week, Natl . . . . Mar 20
Air Conditioning Appreciation Days . . . . July 3
American Business Women's Day . . . . Sept 22
American Enterprise Day . . . . Nov 15
American Redneck Day . . . . July 4
American Soc Assn Executives Mgmt/Mtgs Forum (Nashville, TN) . . . . Mar 12
American Soc Assn Executives Mgmt Conf (Chicago, IL) . . . . Dec 3
American Soc Assn Executives Mtg/Expo (Washington, DC) . . . . Aug 13
Bakery-Deli Assn, Retailers: Conv/Exhib (St. Louis, MO) . . . . Mar 11
Be Kind to Editors and Writers Month . . . . Sept 1
Board and Care Recognition Month (New Jersey) . . . . Sept 1
Boss Day, Natl . . . . Oct 16
Boss/Employee Exchange Day, Natl . . . . Sept 11
Buildings Safety Week, Natl . . . . Apr 9
Business Women's Assn, Amer, Natl Conv of (Portland, OR) . . . . Oct 3
Business Women's Week, Natl . . . . Oct 16
Career Nurse Assistants Day . . . . June 1
Certified Prof Secretary Exam . . . . May 5
Certified Registered Nurse Anesthetist Week, Natl . . . . Aug 5
Chiropractic Week, Natl . . . . Sept 17
Clean-Off-Your-Desk Day, Natl . . . . Jan 9
Coal Miners Day, Anthracite (Shenandoah, PA) . . . . Sept 4
Collegiate/Future Secretaries Conf (Bloomington, MN) . . . . Mar 23
Columnist's Day, Natl . . . . June 27
Community Services Month (California) . . Apr 1
Computer Security Day . . . . Nov 30
Co-op Awareness Month . . . . Oct 1
Cosmetology Month, Natl . . . . Oct 1
CPA's Goof-Off Day, Natl . . . . Apr 18
Credit Union Week . . . . Oct 15
Culinary Professionals' Cookbook Month, Intl Assn . . . . Oct 1
Customer Service Week, Natl . . . . Oct 2
Customer Service Week, Natl 6485 . . . . Page 488
Dairy Day (Dallas, TX) . . . . June 3
Dairy Month, June . . . . June 1
Day of the Artisans . . . . Dec 4
Dental Hygiene Week, Natl . . . . Oct 1
Design Drafting Assn Convention, American (Atlanta, GA) . . . . June 4
Design Drafting Week . . . . May 29
Dietary Managers Assn Mtg/Expo (Boston, MA) . . . . July 30
Dietary Managers' Pride in Food Service Week . . . . Feb 6
Disability Employment Awareness Month, Natl 6476 . . . . Page 488
Disability Employment Awareness Month, Natl 6605 . . . . Page 488
Don't Go to Work Unless It's Fun Day . . . . Apr 3
Embroidery Month, Intl . . . . Feb 1
Emergency Medical Services Week 6567 . . . . Page 488
Employee Health and Fitness Day, Natl . . May 17
Employment Week, Full . . . . Sept 4
Engineers Week, Natl . . . . Feb 19
FFA Convention, Natl (Kansas City, MO) . . Nov 9
FFA Organization Awareness Week, Natl 6531 . . . . Page 488
Financial Services Week, Natl . . . . Sept 3
Firefighters Day, Natl 6639 . . . . Page 488
First Commercial Oil Well: Anniv . . . . Aug 27
First Robot Homicide: Anniv . . . . July 21
Five-Dollar-a-Day Minimum Wage: Anniv . . Jan 5
Flexible Work Arrangements Week . . . . May 7
Food Service Employees Day, Natl . . . . Sept 27
Food Service Employees Week, Natl . . . . Sept 25
Forestry Expo, Vermont (Rutland, VT) . . . July 22
Geographers Annual Meeting, Assn Amer (Chicago, IL) . . . . Mar 14
Get Organized Week . . . . Oct 1
Getting World to Beat a Path to Your Door Week . . . . Oct 15
Goodwill Industries Week . . . . May 7
Health Care Food Service Week, Natl . . . . Oct 1
Hire a Veteran Week 6502 . . . . Page 488
Home Care Week, Natl . . . . Nov 26
Home-Based Business Week . . . . Oct 8
Homemakers of America FHA/Hero Week, Natl Future . . . . Feb 12
Housekeepers Week, Natl . . . . Sept 10
Human Resources Month . . . . Jan 1
Humorists Are Artists Month (HAAM) . . . . Mar 1
Inventors Congress, Minnesota (Redwood Falls, MN) . . . . June 9
KGBX Typewriter Toss (Springfield, MO) . Apr 26
Labor Day . . . . Sept 4
Labor History Month 6688 . . . . Page 488
Laundry Day, Natl . . . . Sept 22
Laundry Week, Natl . . . . Sept 17
Law Day USA . . . . May 1
Law Enforcement Training Week 6643 . Page 488
Logging Museum Festival Days (Rangeley, ME) . . . . July 28
Ludlow Mine Incident: Anniv . . . . Apr 20
Maintenance Day . . . . Jan 18
Maintenance Personnel Day . . . . Oct 16
Manufacturing Week, Natl (Chicago, IL) . . Mar 13
Market Ability Month . . . . Jan 1
Medical Lab Week, Natl . . . . Apr 9
Medical Staff Services Awareness Week, Natl 6500 . . . . Page 488
Memo Day, Natl . . . . May 19
Michigan Science Teacher Assn Conf (Lansing, MI) . . . . Feb 3
Mind Mapping and Brainstorming Week, Natl . . . . Sept 18
Mind Mapping for Problem Solving Week, Natl . . . . Sept 25
Mind Mapping for Project Management Week, Natl . . . . Sept 11
Mind Mapping Month, Natl . . . . Sept 1
Minority Enterprise Development Week 6591 . . . . Page 488
Montana Trappers Assn State Convention (Lewistown, MT) . . . . Sept 8
Montgomery Ward Seized: Anniv . . . . Apr 26
National Labor Relations Act: Anniv . . . . July 5
NE Speakers Assn Building Professional Competence Month . . . . Apr 1
New Jersey Restaurant/Hospitality Expo (Somerset, NJ) . . . . Mar 19
New York Stock Exchange: Anniv . . . . May 17
Northern Illinois Farm Show (Rockford, IL) Jan 3
Notary Public Day, Natl . . . . Nov 7
Notary Public Week, Natl . . . . Nov 5
Nurse of the Year Award . . . . May 12
Nurses Week, Natl . . . . May 6
Nursing Conf on Pediatric Primary Care (Nashville, TN) . . . . Mar 14
Nursing Home Week, Natl . . . . May 14
Occupational Therapy Month, Natl . . . . Apr 1
Office Olympics (Shreveport, LA) . . . . Sept 15
Old Tyme Farmers' Days (Live Oak, FL) . . Nov 23
Operating Room Nurse Week . . . . Nov 12
Organization Development Info Exchange (Wms Bay, WI) . . . . May 23
Organize Your Home Office Day, Natl . . . . Mar 28
Osteopathic Medicine Week, Natl . . . . Nov 5

Education (cont'd)—Employment

Index ☆ Chase's 1995 Calendar of Events ☆

Employment (cont'd)—Environment

Paste-Up Day....May 7
Patent Granted for Hat Blocking Machine: Anniv....Apr 3
Paul Bunyan Show (Nelsonville, OH)....Oct 6
PBX Telecommunications Week, Intl....June 4
Peace Officer Memorial Day, Natl....May 15
Pest Control Month, Natl....June 1
Pet Owners Independence Day....Apr 18
Pet Week, Natl....May 7
Physician Assistant Day....Oct 6
Police Week, Natl....May 14
Printing Ink Day, Natl....Jan 17
Printing Week, Intl....Jan 15
Professional Secretaries Intl Conv (Seattle, WA)....July 23
Psychiatric Technicians Week, Natl....Aug 6
Public Service Recognition Week....May 1
Radiologic Technology Week, Natl....Nov 5
Receptionist Day, Natl....May 10
Rehabilitation Week, Natl....Sept 17
Retail Bakers Month, Natl....Jan 1
Revise Your Work Schedule Month....May 1
Schneiderman, Rose: Birth Anniv....Apr 6
School Counseling Week, Natl....Feb 6
Scientific Materials Mgrs Conf, Natl Assn (Chicago, IL)....July 24
Secretaries Day, Professional....Apr 26
Secretaries Leadership Conf, Prof (Philadelphia, PA)....Feb 10
Secretaries Week, Professional....Apr 23
Secretary's Day Tea (Wilmington, DE)....Apr 26
Shop/Office Workers' Holiday (Iceland)....Aug 7
Show and Tell Day at Work....Jan 9
Small Business Week....May 7
Small Business Week 6561....Page 488
Social Work Month, Natl Professional....Mar 1
Sporting Assn World Sports Expo, Natl (Chicago, IL)....July 16
Stay Home Because You're Well Day....Nov 30
Surgical Technologist Week, Natl....May 21
Take Our Daughters to Work Day....Apr 27
Technical Expo and Conf (Atlanta, GA)....June 24
Tell Someone They're Doing a Good Job Week....Dec 17
Thank You, School Librarian Day....Apr 12
Third Shift Workers Day, Natl....May 10
Timberfest (Woodruff, WI)....June 16
Tin Hau (Fishermen's) Festl (Hong Kong)....May 3
Tourism Week, Natl....May 7
Tupper Lake Woodsmen's Days (Tupper Lake, NY)....July 8
TV Talk Show Host Day....Oct 23
Typing Contest, Natl....Apr 24
United Cerebral Palsy's Casual Day....June 16
Visiting Nurse Associations Week, Natl 6529....Page 488
Visitor Appreciation Day, Natl....May 10
Waitresses Day, Natl....May 21
Washington St Activity Professionals Conv (Walla Walla, WA)....Aug 17
Weather Observer's Day, Natl....May 4
Weatherman's Day....Feb 5
Welding Month, Natl....Apr 1
West Virginia Oil and Gas Fest (Sisterville, WV)....Sept 14
Who's in Charge: Workers/Managers (Elmhurst, IL)....July 22
Who's in Charge: Workers/Managers (Fairfield, CT)....Mar 18
Who's in Charge: Workers/Managers (Grand Rapids)....Jan 14
Who's in Charge: Workers/Managers (Youngstown, OH)....May 20
Wisconsin Farm Woman of Year Month....Feb 1
Woodmen Ranger's Day....May 12
Word Processing Transcriptionist Week, Natl....Jan 8
Workaholic's Day....July 5
Workers Memorial Day....Apr 28
Working Women's Day, Intl....Mar 8
Enberg, Dick: Birth....Jan 9
Encourager® Day, Be an....Feb 1
Energy Awareness Month....Oct 1
Energy Awareness Month 6589, 6597....Page 488
Energy Management Is a Family Affair—Improve Your Home....Oct 1
Engineers Week, Natl....Feb 19
Engines: Antique Power Exhib (Burton, OH)....July 28
England
Accession of Queen Elizabeth II: Anniv....Feb 6
Allendale Baal Fest (Allendale, Northumberland)....Dec 31
Arundel Festival (Arundel)....Aug 25
Badminton Horse Trials (Badminton)....May 4
Bath Intl Fest (Bath)....May 19
Battle of Britain Day....Sept 15
Battle of Britain Week....Sept 10
Beatles Fest (Liverpool)....Aug 25
Blackpool Illuminations (Blackpool, Lancashire)....Sept 1
Bonnie Prince Charlie Anniv....Nov 27
Boucheron 26—The World Mystery Convention (Nottingham)....Sept 28
Bournemouth Musicmakers Fest (Bournemouth)....June 24
British Amateur Golf Chmpshp (Hoylake)....June 5
British Grand Prix (Towcester)....July 7
British Intl Antiques Fair (Bristol)....Apr 25
British Museum: Anniv....Jan 15
Broadstairs Dickens Festival (Broadstairs)....June 17
Care Sunday....Apr 2
Charles II: Restoration and Birth Anniv....May 29
Chelsea Antiques Fair (London)....Mar 7, Sept 12
Chelsea Flower Show (London)....May 23
Cheltenham Festival of Literature (Cheltenham)....Oct 6
Cheltenham Gold Cup Mtg (Prestbury)....Mar 14
Cheltenham Intl Fest of Music (Cheltenham)....July 1
Christmas Holiday....Dec 25
City of London Festival (London)....June 20
Cowe's Week (Isle of Wight)....July 29
Crufts Dog Show (Birmingham)....Mar 16
Devizes/Westminster Intl Canoe Race (Devizes)....Apr 14
Dicing for Bibles....June 5
English Riviera Dance Festival (Torquay)....May 26
Epsom Derby (Surrey)....June 7
Esso Vintage Vehicle Run (Bristol)....June 11
European Judo Chmpshps (Birmingham)....May 10
Exeter Festival (Exeter)....June 30
Fest of Arts and Culture—Paignton 700 (Torquay)....May 18
Flowers Exhib (Windsor Castle, Windsor)....Apr 8
Grand Natl Horseracing Meeting (Liverpool)....Apr 6
Great Fire of London: Anniv....Sept 2
Guy Fawkes Day....Nov 5
Henry Purcell Tercentenary Tudeley Fest (Tudeley, Kent)....Nov 17
Hallaton Bottle Kicking (Hallaton)....Apr 17
Harrogate Intl Youth Music Fest (Harrogate)....Apr 14
Harrogate Spring Flower Fest (Harrogate)....Apr 23
Harveys Wine Museum Anniv (Bristol)....Apr 14
Head of the River Race (London)....Mar 25
Helston Furry Dance....May 6
Henley Royal Regatta (Henley-on-Thames)....June 28
Horse of the Year Show (London)....Sept 26
Intl Fest of Flowers with Music (London)....May 16
Intl TT Motorcycle Race (Isle of Man)....June 3
Jorvik Viking Festival (York)....Feb 10
Knitting/Needlecraft/Design Exhibition (Esher)....Jan 19
Lawn Tennis Chmpshps at Wimbledon (London)....June 26
London Intl Boat Show (London)....Jan 5
London Intl Environmental Film Fest (London)....Nov 1
London Parade....Jan 1
Lord Mayor's Procession/ Show (London)....Nov 11
Manchester Massacre/Battle of Peterloo....Aug 16
Marriage of Elizabeth and Philip: Anniv....Nov 20
Model Engineer Exhibition (London)....Dec 30
Mothering Sunday....Mar 26
National Music Day....June 24
Northwest Knitting/Needlecraft Exhib (Manchester)....Feb 23
Notting Hill Carnival (London)....Aug 27
Nottingham Goose Fair (Nottingham)....Oct 5
Oak-Apple Day....May 29
Oxford-Cambridge Race (London)....Apr 1
Plough Monday....Jan 9
Queen Elizabeth I: Birth Anniv....Sept 7
RAC London/Brighton Veteran Car Run (London)....Nov 5
Remembrance Day Service/Parade (London)....Nov 12
Rochester Dickens Fest (Rochester)....June 1
Royal Ascot (Ascot)....June 13
Royal Palette (Windsor)....Jan 1
Royal Shakespeare Theatre Season (Warwickshire)....Mar 17
Royal Windsor Horse Show (Windsor)....May 10
St. George: Feast Day....Apr 23
Scotland Yard: First Appearance Anniv....Sept 29
Seizure of Iranian Embassy: Anniv....Apr 30
Seven Florentine Heads (Windsor Castle)....Sept 16
Shrewsbury Intl Music Fest (Shropshire)....June 23
Soho Jazz Festival (London)....Sept 28
Southern Cathedrals Fest (Chichester)....July 20
Southern Knitting/Needlecraft Exhibition (Bristol)....Mar 17
Stratford-upon-Avon Fest....July 8
Summer Time....Mar 26
Three Choirs Festival (Gloucester, Glouchestershire)....Aug 19
Trooping Colours/Queen's Official Birthday....June 10
Tynwald Day....July 5
UK Year of Literature/Writing (Swansea)....Jan 1
Walter Plinge Day....Dec 2
Ways With Words Literature Festival (Dartington)....Aug 28
West London Antiques Fair (London)....Jan 12
West Sussex Intl Youth Music Fest (W Sussex)....Apr 13
Westminster Antique Fair (London)....Apr 27
Winmau World Darts Championship (London)....Dec 1
World of Drawings and Watercolours (London)....Jan 25
Engler, John: Birth....Oct 12
English, Alex: Birth....Jan 5
English Channel, First Airplane Crossing: Anniv....July 25
English Channel, First Man-Powered Flight Across: Anniv....June 12
English Colony in North America, First: Anniv....Aug 5
Enterprise Day, American....Nov 15
Entomological Soc of America, Annual Mtg (Dallas, TX)....Dec 13
Environment
American Rivers Month....June 1
Arbor Day, Natl (Proposed)....Apr 28
Biosphere Day....Sept 21
Campaign for the Biosphere, World....June 5
Chernobyl Reactor Disaster: Anniv....Apr 26
Chicago Peace and Music Festival (Chicago, IL)....Aug 5
Clean Air Campaign, Amer Lung Assn....May 1
Clover Leaf Planting (Annapolis, MD)....Apr 29
Conservation Awareness Program (Waimanalo, HI)....Feb 1
Conservation Festival (Roanoke, VA)....June 10
Conserve Water/Detect-a-Leak Week....May 7
Earth Day....Mar 20, Apr 22
Earth Day (Huntsville, AL)....Apr 22
Earth Day Celebration (Chimney Rock, NC)....Apr 21
Earth Day Community Fest (St. Louis, MO)....Apr 22
Earth Day Family Fest (San Jose, CA)....Apr 22
Earth Day Regional Fest (Baton Rouge, LA)....Apr 23
Earth Day Tampa Bay (Tampa, FL)....Apr 22
EarthFair (San Diego, CA)....Apr 23
EARTHFest (Cleveland, OH)....Apr 23
Energy Awareness Month....Oct 1
Energy Awareness Month 6597....Page 488
Environment Day, World (UN)....June 5
Environmental Policy Act, Natl....Jan 1
Exxon Valdez Oil Spill: Anniv....Mar 24
Giant Sequoia in Natl Forests 6457....Page 488
Global Restoration Fair (San Francisco, CA)....Apr 21
Great LA Clean UP (Los Angeles, CA)....Apr 17
Greenpeace Founded: Anniv....Sept 15
Humpback Whale Awareness Month (Honolulu, HI)....Feb 1
Japan Agrees to End Use of Drift Nets: Anniv....Nov 26
Keep America Beautiful Annual Meeting (Washington, DC)....Dec 6
Keep America Beautiful Month, Natl....Apr 1
Last Dusky Seaside Sparrow: Death Anniv....June 16
London Intl Environmental Film Fest (London, England)....Nov 1
Mother Ocean Day....May 14
No Laughing: Political Cartoon/Environ (San Jacinto, CA)....Apr 8
No Laughing: Political Cartoon/Environ (Santa Rosa, CA)....May 21
No Laughing: Political Cartoon/Environ (Ukiah, CA)....Aug 12
No Laughing Matter: Political Cartoon/Environ (Yuba City, CA)....Feb 1

☆ Chase's 1995 Calendar of Events ☆ Index

Environment (cont'd)—Farentino

No Socks Day . . . May 8
Nongame Wildlife Weekend (Davis, WV) . . June 2
Parade for the Planet (NY, NY) . . . Apr 23
Project Aware Month . . . Sept 1
Public Lands Day . . . Sept 9
Recreation and Parks Month, Natl . . . July 1
Rites of Spring (Patapsco St Pk, MD) . . . Apr 22
Rural Life Sunday . . . May 21
Save the Rhino Day . . . May 1
Silent Spring Publication: Anniv . . . Apr 13
Sky Awareness Week . . . Apr 23
Spring Prairie Fest (Olathe, KS) . . . Apr 15
Taste of Health (Miami, FL) . . . Apr 22
Tongass: Alaska's Rain Forest (Anchorage, AK) . . . Oct 14
Tongass: Alaska's Rain Forest (Fairbanks, AK) . . . Apr 8
Tongass: Alaska's Rain Forest (Juneau, AK) . . . June 3
Tongass: Alaska's Rain Forest (Los Angeles, CA) . . . Dec 16
Tongass: Alaska's Rain Forest (Ocala, FL) . Feb 4
Trash Fest (Anacoco, LA) . . . May 27
Tree-Mendous Month (Annapolis, MD) . . . Apr 6
Water Pollution Control Act: Anniv . . . Oct 18
Water Quality Month, Natl . . . Aug 1
Water, World Day for (UN) . . . Mar 22
Week of Ocean Fest Sea-Son, Natl (Ft Lauderdale, FL) . . . Apr 14
Week of the Ocean, Natl . . . Apr 16
Ephron, Nora: Birth . . . May 19
Epilepsy Awareness Month, Natl . . . Nov 1
Epiphany . . . Jan 6
Epiphany (Twelfth Day) . . . Jan 6
Equal Rights Party Founding: Anniv . . . Sept 20
Equatorial Guinea: Independence Day . . . Oct 12
Equinox, Autumn . . . Sept 23
Equinox, Spring . . . Mar 20
Eraser Day, Rubber . . . Apr 15
Erasmus, Desiderius: Birth Anniv . . . Oct 28
Erector set: Happy Birthday. A.C.! (Salem, OR) . . . Feb 18
Erie Canal: Anniv . . . Oct 26
Erikson, Leif: Day (Iceland) . . . Oct 9
Erikson, Leif: Day (Pres Proc) . . . Oct 9
Erikson, Leif: Day 6607 . . . Page 488
Eritrea: Independence Day . . . May 24
Erving, Julius "Doctor J": Birth . . . Feb 22
Escoffier, George: Birth Anniv . . . Oct 28
Esiason, "Boomer": Birth . . . Apr 17
Esperanto: Founder Birth Anniv . . . Dec 15
Esperanto: Publication Anniv . . . July 26
Esposito, Phil: Birth . . . Apr 23
Espy, Mike: Birth . . . Nov 30
Estevez, Emilio: Birth . . . May 12
Estonia: Baltic States' Independence Recognized: Anniv . . . Sept 6
Estonia: Independence Day . . . Feb 24
Estrada, Erik: Birth . . . Mar 16
Ethnic Observances
Aebleskiver Days (Tyler, MN) . . . June 24
African-American Heritage Fest (Snow Hill, MD) . . . Sept 10
Armenian Appreciation Day . . . Apr 3
Asian/Pacific American Heritage Month 6557, 6686 . . . Page 488
Central Nebraska Ethnic Fest (Grand Island, NE) . . . July 21
Chinese Moon Festival (Los Angeles, CA) Sept 9
Cinco De Mayo Celebration (Leadville, CO) . . . May 6
Clarkson Czech Fest (Clarkson, NE) . . . June 22
Cross Culture Day, Intl . . . Sept 14
Cultural Diversity Day . . . Feb 16
Czech Days (Tabor, SD) . . . June 16
Danish Fest (Greenville, MI) . . . Aug 18
Eastpark Czech Fest (Lincoln, NE) . . . May 6
Ethnic Festival (South Bend) . . . June 30
First Thanksgiving Festival (El Paso, TX) . Apr 29
German Christmas, Old (Papillion, NE) . . Nov 24
German-American Day, Natl . . . Oct 6
German-American Day 6604 . . . Page 488
Greek Independence Day 6539, 6659 . . Page 488
Grundlovsfest (Dannebrog, NE) . . . June 3
Hispanic Heritage Month . . . Sept 15
Hispanic Heritage Month, Natl 6592 . . . Page 488
Holiday Folk Fair (Milwaukee, WI) . . . Nov 17
Irish-American Heritage Month 6533, 6656 . . . Page 488
Jewish Heritage Week 6558 . . . Page 488
Julyfest (Bavarian) (Kimberley, BC, Canada) . . . July 14
Kaleidoscope Intl Fair (Dubuque, IA) . . . Apr 21
Kids Bridge Exhibition (Houston, TX) . . . Oct 28
Multicultural Communication Month . . . Apr 1
Multicultural Substance Abuse Awareness Week . . . May 1
Noah's Flood (San Francisco, CA) . . . June 23
Oktoberfest (LaCrosse, WI) . . . Sept 29
Old-Fashioned Danish Christmas (Dannebrog, NE) . . . Dec 2
Polish-American Heritage Month . . . Oct 1
Polish-American Heritage Month 6493 . Page 488
Polish-American Heritage Month Celeb . . . Oct 1
Polish-American Weekend (Philadelphia, PA) . . . Aug 12
Prager Memorial Day: German-Amer Remembrance . . . Apr 5
Pulaski Day Parade (Philadelphia, PA) . . . Oct 1
St. Spyridon Greek Orthodox Fest (Sheboygan, WI) . . . July 29
Scandinavian Hjemkomst Fest (Fargo, ND/ Moorhead, MN) . . . June 21
Strength/Diversity: Japanese Ameri Women (Logan, KS) . . . Apr 15
Thai Heritage Month . . . Apr 1
Westfest (Czech) (West, TX) . . . Sept 2
Europe: Summer Daylight-Saving Time . . . Mar 26
European Communities: Schuman Plan Anniv . . . May 9
Evacuation Day (Boston, MA) . . . Mar 17
Evacuation Day (Egypt) . . . June 18
Evaluate Your Life Day . . . Oct 19
Evans, Bob: Birth . . . May 30
Evans, Dale: Birth . . . Oct 31
Evans, Darrell: Birth . . . May 26
Evans, Heloise Cruse: Birth . . . Apr 15
Evans, Linda: Birth . . . Nov 18
Evans, Maurice: Birth . . . June 3
Everett, Chad: Birth . . . June 11
Everett, Edward: Birth Anniv . . . Apr 11
Everett, Jim: Birth . . . Jan 3
Everly, Don: Birth . . . Feb 1
Everly, Phil: Birth . . . Jan 19
Evers, Medgar, Assassinated: Anniv . . . June 13
Evert, Chris: Birth . . . Dec 21
Everybody's Day Fest (Thomasville, NC) . . Sept 30
Evigan, Greg: Birth . . . Oct 14
Ewing, Patrick: Birth . . . Aug 5
Exchange Club Birthday, Natl . . . Mar 27
Exchange Club Child Abuse Prevention Month . . . Apr 1
Exchange Club: Freedom Shrine Month . . . May 1
Execution: First Criminal in American Colonies: Anniv . . . Sept 30
Executives, Amer Soc of Assn: Mgmt Conf (Chicago, IL) . . . Dec 3
Executives, Amer Soc of Assn: Mtg/Expo (Washington, DC) . . . Aug 13
Exhibitions. See Arts; Photography; key words
Exon, J. James: Birth . . . Aug 9
Explosion: Halifax, Nova Scotia, Destroyed: Anniv . . . Dec 6
Explosion of the Cart (Florence, Italy) . . . Apr 16
Exxon Valdez Oil Spill: Anniv . . . Mar 24
Eye Care Month, Natl . . . Jan 1
Eyes: See Health and Welfare

F

Fabares, Shelley: Birth . . . Jan 20
Fabian: Birth . . . Feb 6
Fabray, Nanette: Birth . . . Oct 27
Fahrenheit, Gabriel D.: Birth Anniv . . . May 14
Fair at New Boston (Springfield, OH) . . . Sept 2
Fair, First Annual, in America: Anniv . . . Sept 30
Fair of 1850 (Westville Village, GA) . . . Oct 12
Fair, Woodstock (Woodstock, CT) . . . Sept 1
Fairbanks, Charles W.: Birth Anniv . . . May 11
Fairbanks, Douglas, Jr: Birth . . . Dec 9
Fairbanks, Douglas: Birth Anniv . . . May 23
Fairchild, Morgan: Birth . . . Feb 3
Faircloth, Lauch: Birth . . . Jan 14
Fairs. See Agriculture
Fairy Tale Fest, Rock City (Rock City, TN) . Aug 12
Faithfull, Marianne: Birth . . . Dec 29
Falana, Lola: Birth . . . Sept 11
Falco: Birth . . . Feb 19
Falk, Peter: Birth . . . Sept 16
Falkland Islands War: Anniv . . . Apr 2
Fall Color Cruise and Folk Fest (Chattanooga, TN) . . . Oct 21
Fall Fest, Belle Meade Plantation (Nashville, TN) . . . Sept 16
Fall Fest, Curtis (Curtis, NE) . . . Sept 16
Fall Fest of Leaves (Ross County, OH) . . . Oct 20
Fall Fest, Roan Mountain State Park (Roan Mountain, TN) . . . Sept 16
Fall Fest, Who-Zha-Wa (Wisconsin Dells, WI) . . . Sept 15
Fall Harvest Fest (Boyne City, MI) . . . Oct 7
Fall Hat Week . . . Sept 10
Fall Homecoming, Museum of Appalachia (Norris, TN) . . . Oct 11
Fall of the Alamo: Anniv . . . Mar 6
Fall Powwow and Fest (Lebanon, TN) . . . Oct 6
Falwell, Jerry: Birth . . . Aug 11
Family
ABCs of Genealogy (Moultrie, GA) . . . Mar 18
Adoption Week, Natl . . . Nov 19
Adoption Week, Natl 6629 . . . Page 488
Adult Day Care Center Week . . . Sept 17
American Family Day . . . Aug 6
Ancestor Appreciation Day . . . Sept 27
Be an Encourager® Day . . . Feb 1
Birthparents Week, Natl . . . Apr 2
Children's Day (MA) . . . June 11
Communicate with Your Kid Month, Natl . . Oct 1
Creative Romance Month . . . Feb 1
Date Your Mate Month . . . May 1
Descendants of Simon Gay Family Reunion (Moultrie, GA) . . . May 20
Domestic Violence Awareness Month 6619 . . . Page 488
Energy Management Is a Family Affair—Improve Your Home . . . Oct 1
Exchange Club Child Abuse Prevention Month . . . Apr 1
Family Day . . . Aug 13
Family Day Celebrataion (Dahlonega, GA) July 4
Family Day, Natl 6628 . . . Page 488
Family Days Celebration (Farmington, PA) . . . Aug 12
Family Folklore (Lexington, KY) . . . Jan 14
Family Folklore (Skokie, IL) . . . Apr 22
Family Frolic (Coshocton, OH) . . . June 17
Family History Begins With Me . . . Jan 1
Family History Day . . . June 14
Family History Month . . . Oct 1
Family Sexuality Education Month, Natl . . . Oct 1
Family Support Month . . . May 1
Family Week, Natl . . . May 7
Father's Day . . . June 18
Father's Day 6573, 6701 . . . Page 488
Grandparents Day (MA) . . . Oct 1
Grandparents Day, Natl . . . Sept 10
Hugging Day, Natl . . . Jan 21
Infertility Awareness Week, Natl . . . Oct 1
Intl Day of Families (UN) . . . May 15
Intl Year of the Family 6634 . . . Page 488
Kiss-Your-Mate Day . . . Apr 28
Loving v Virginia: Anniv . . . June 12
MADD's Natl Candlelight Vigil (Washington, D.C.) . . . Dec 2
Marriage Celebration (Belleville, IL) . . . Sept 10
Marriage Enrichment Weekend (Belleville, IL) . . . Nov 10
Middle Children's Day . . . Aug 12
Military Families Recognition Day, Natl 6627 . . . Page 488
Mother-in-Law Day . . . Oct 22
Mother's Day . . . May 14
Mother's Day 6559 . . . Page 488
New England Field Days (Mystic, CT) . . . Nov 24
No Housework Day . . . Apr 7
North American Conf on Family (Winnipeg, Man) . . . Oct 19
Pet Memorial Day, Natl . . . Sept 10
Pleasure Your Mate Month . . . Sept 1
Proposal Day . . . Mar 20, Sept 23
Purposeful Parenting Month, Natl . . . July 1
Read to Your Child Day . . . Feb 14
Relationship Renewal Day . . . May 4
Roaring Camp Railroad Ride (Felton, CA) . . . June 18
Room of One's Own Day . . . Jan 25
School Success Month, Natl . . . Sept 1
Sippin' Cider Days (Leavenworth, KS) . . . Oct 7
Stress-Free Family Holidays Month, Natl . . Dec 1
Take Our Daughters to Work Day . . . Apr 27
Talk With Your Teen About Sex Month, Natl . . . Mar 1
Twin-O-Rama (Cassville, WI) . . . July 21
Twins Day Fest (Twinsburg, OH) . . . Aug 4
Universal Children's Week . . . Oct 1
Universal Family Week . . . May 7
Visit Your Relatives Day . . . May 18
Weddings Month, Natl . . . Feb 1
World Marriage Day . . . Feb 12
Yours, Mine and Ours Month, Natl . . . Jan 1
Fan Fair, Intl Country Music (Nashville, TN) June 5
Faneuil Opened to the Public: Anniv . . . Sept 24
Fantasy Fest (Key West, FL) . . . Oct 20
Faraday, Michael: Birth Anniv . . . Sept 22
Farentino, James: Birth . . . Feb 24

Index ☆ *Chase's 1995 Calendar of Events* ☆

Fargo——Fitzgerald

Fargo, Donna: Birth ... Nov 10
Farm Animals Day, World ... Oct 2
Farm Days Weekend (Wheeling, WV) ... June 17
Farm Fest, Bob Evans (Rio Grande, OH) ... Oct 13
Farm Safety Week, Natl (Pres Proc) ... Sept 17
Farm Show, Northern Illinois (Rockford, IL) . Jan 3
Farm Toy Show, Natl (Dyersville, IA) ... Nov 3
Farm-City Week, Natl (Pres Proc) ... Nov 17
Farm-City Week, Natl 6477 ... Page 488
Farm/Ranch Machinery Show, Triumph Agri Expo (Omaha, NE) ... Mar 14
Farmer, James: Birth ... Jan 12
Farmer, Philip Jose ... Jan 26
Farmers' Festival, Pigeon (Pigeon, MI) ... July 26
Farmers, Future: See FFA ... Nov 9
Farmfair (Edmonton, Alta, Canada) ... Nov 4
Farr, Jamie: Birth ... July 1
Farrakhan, Louis: Birth ... May 11
Farrell, James T.: Birth Anniv ... Feb 27
Farrell, Mike: Birth ... Feb 6
Farrelly, Alexander: Birth ... Dec 29
Farrow, Mia: Birth ... Feb 9
Fasching (Germany and Austria) ... Feb 27
Fasching Sunday (Germany, Austria) ... Feb 26
Fashion Show (Milwaukee, WI) ... May 12
Fast Day (NH) ... Apr 24
Fast of Esther: Ta'anit Esther ... Mar 15
Fast of Gedalya ... Sept 27
Fat Tuesday, Hendo's (Breckenridge, CO) . . Feb 28
Father's Day ... June 18
Father's Day (Pres Proc) ... June 18
Father's Day 6573 ... Page 488
Fatima, Last Pilgrimage to (Portugal) ... Oct 12
Fatima, Pilgrimage to (Portugal) ... May 12
Fauci, Anthony: Birth ... Dec 24
Fauntroy, Walter Edward: Birth ... Feb 6
Fawcett, Farrah: Birth ... Feb 2
Fawkes, Guy: Day (England) ... Nov 5
Faye, Alice: Birth ... May 5
FBI: First Female FBI Agent: Anniv ... Oct 25
Fearless Forecasts of TV Flops ... Sept 11
Feast of San Pellegrino (Gualdo Tadino, Italy) ... Apr 30
Feast of the Incappucciati (Gradoli, Italy) . . Feb 23
Feast of the Lanterns (Bon Fest) (Japan) . . July 13
Feast of the Ramson (Richwood, WV) ... Apr 15
Feast of the Redeemer (Italy) ... July 16
Federal Lands Cleanup Day (Pres Proc) . . . Sept 9
Federal Reserve System: Anniv ... Dec 23
Federalist Papers: Anniv ... Oct 27
Feiffer, Jules: Birth ... Jan 26
Feingold, Russell D.: Birth ... Mar 2
Feinstein, Diane: Birth ... June 22
Feinstein, Michael: Birth ... Sept 7
Feld, Eliot: Birth ... July 5
Feldon, Barbara: Birth ... Mar 12
Feldshuh, Tovah: Birth ... Dec 27
Feliciano, Jose: Birth ... Sept 10
Felker, Clay S.: Birth ... Oct 2
Fell, Norman: Birth ... Mar 24
Fellini, Federick: Birth Anniv ... Jan 20
Fellowship and Hope, Natl Day of 6525 . . Page 488
Fellowship Day, May ... May 5
Felton, Mrs W.H., Becomes First Woman US Senator ... Oct 3
Feltsman, Vladimir: Birth ... Jan 8
Feminine Empowerment Month, Natl ... Mar 1
Fence Painting Contest: Natl Tom Sawyer (Hannibal, MO) ... June 30
Fencik, Gary: Birth ... June 11
Fenwick, Millicent Hammond: Birth Anniv . . Feb 25
Ferber, Edna: Birth Anniv ... Aug 15
Ferdinand VI of Spain: Birth Anniv ... Sept 23
Ferguson, Maynard: Birth ... May 4
Ferlinghetti, Lawrence: Birth ... Mar 24
Fermi Atomic Power Plant: Accident Anniv . . Oct 5
Fernandez, Mervyn: Birth ... Dec 29
Ferragamo, Vince: Birth ... Apr 24
Ferrante, Arthur: Birth ... Sept 7
Ferraro, Geraldine A.: Birth ... Aug 26
Ferraro, Geraldine Anne: Birth ... Aug 26
Ferrer, Jose: Birth Anniv ... Jan 8
Ferrer, Mel: Birth ... Aug 25
Ferrigno, Lou: Birth ... Nov 9
Ferry, Bryan: Birth ... Sept 26
Fertility Rights (Philippines) ... May 17
Festifall (Point Marion, PA) ... Sept 24
FFA Convention, Natl (Kansas City, MO) ... Nov 9
Fiber Focus Month, Natl ... Feb 1
Fiddlers. See also Bluegrass
  Athabascan Fiddling Festival (Fairbanks, AK) ... Nov 9
  Autumn Glory Fest (Oakland, MD) ... Oct 12
  Canadian Open Fiddle Contest (Shelburne, Ont, Canada) ... Aug 11
  Chester Old Fiddlers' Picnic (Coatesville, PA) ... Aug 12
  Fiddlefest (Galax, VA) ... Aug 10
  Fiddlers Jamboree and Crafts Fest (Smithville, TN) ... June 30
  Friendsville Fiddle/Banjo Contest (Friendsville, MD) ... July 15
  Indiana Fiddlers Gathering (Battle Ground, IN) ... June 23
  International Old-Time Fiddlers Contest (Dunseith, ND) ... June 9
  Kentucky St Fiddler's Chmp (Falls of Rough, KY) ... July 21
  Old Fiddlers' Convention (Galax, VA) ... Aug 9
  Old Time Fiddlers' Contest (Payson, AZ) Sept 23
  Tennessee River Fiddlers Conv (Florence, AL) ... Oct 13
  Texas St Chmpshp Fiddlers Frolics (Hallettsville, TX) ... Apr 20
Field of Dreams Extravaganza (Dyersville, IA) ... Sept 1
Field, Sally: Birth ... Nov 6
Field Trial Chmp (bird dog), Natl (Gr Junction, TN) ... Feb 13
Fielder, Cecil: Birth ... Sept 21
Fields, Kim: Birth ... May 12
Fields, W.C.: Birth Anniv ... Apr 9
Fiesta Bowl Block Party, Tempe (Tempe, AZ) ... Dec 31
Fifties Revival (Marshall, MN) ... June 2
Fig Week, Natl ... Nov 1
Fiji: Independence Day ... Oct 10
Fillmore, Abigail P.: Birth Anniv ... Mar 13
Fillmore, Caroline: Birth Anniv ... Oct 21
Fillmore, Millard: Birth Anniv ... Jan 7
Film
  Ann Arbor Film Festival (Ann Arbor, MI) . Mar 14
  Banff Fest of Mountain Films (Banff, AB, Canada) ... Nov 3
  Birth of a Nation, The: Premiere Anniv ... Feb 8
  Casablanca Premiere: Anniv ... Nov 26
  Children's Film Festival, Intl (Oula, Finland) ... Nov 13
  Edinburgh Intl Film Fest (Lothian, Scotland) ... Aug 12
  Fairmount Fest/Remembering J. Dean (Fairmount, IN) ... Sept 22
  London Intl Environmental Film Fest (London, England) ... Nov 1
  Most Boring Film Awards ... Mar 13
  MountainFilm (Telluride, CO) ... May 26
  Record of a Sneeze: Anniv ... Feb 2
  Sundance Film Festival (Park City, UT) . . Jan 19
  Toronto Intl Film Fest (Toronto, Ont, Canada) ... Sept 7
  US Industrial Film/Video Awards (Chicago, IL) ... May 31
  VA Fest of American Film (Charlottesville, VA) ... Oct 26
  Wild West Film Fest (Tuolumne County, CA) ... Sept 22
  Wyatt Earp Birthday Celebration (Monmouth, IL) ... Mar 18
  Yorkton Film/Video Fest (Yorkton, Sask, Canada) ... May 24
Financial Panic of 1873: Anniv ... Sept 20
Financial Services Week, Natl ... Sept 3
Fine Arts. See Arts
Finger Lakes Christmas Arts/Craft Show (Rochester, NY) ... Nov 24
Finger Lakes Christmas in Oct Craft Show (Rochester, NY) ... Oct 28
Finland
  Flag Day ... June 4
  Independence Day ... Dec 6
  Intl Children's Film Festival (Oulu) ... Nov 13
  Intl Jazz Fest (Pori) ... July 15
  Joensu Festival (Joensu) ... June 18
  Jyvaskyla Arts Fest (Jyvaskyla) ... June 14
  Kaustinen Folk Music Fest (Kaustinen) . . July 15
  Kupio Dance & Music Fest (Kuopio) ... June 9
  Lahti Organ Fest (Lahti) ... Aug 1
  Sata-Hame Accordion Fest (Ikaalinen) . . June 26
  Savonlinna Opera Fest (Savonlinna) ... July 8
  Tampere Theatre Fest (Tampere) ... Aug 15
  Time of Music (Viitasaari) ... July 6
  Turku Music Fest (Turku) ... Aug 11
Finney, Albert: Birth ... May 9
Finney, Joan: Birth ... Feb 12
Fire Apparatus, Chesapeake Antique (Westminster, MD) ... May 13
Fire: Happy Land Social Club Fire ... Mar 25
Fire of 1881, Great Michigan: Anniv ... Sept 5
Fire of London, Great: Anniv ... Sept 2
Fire Prevention Week ... Oct 8
Fire Prevention Week (Pres Proc) ... Oct 8
Fire Prevention Week 6601 ... Page 488
Fire Safety Council, Natl: Anniv ... Dec 7
Fire: Sleep Safety Month ... Feb 1
Fireant Fest (Marshall, TX) ... Oct 6
Firefighters Day, Natl 6639 ... Page 488
Firelock Match, Governor's Invitational (Big Pool, MD) ... Sept 23
Firemen's Carnival (Nome, AK) ... Dec 2
Firepup's Birthday ... Oct 1
Fireside Chat, FDR's First: Anniv ... Mar 12
Firestone, Harvey: Birth Anniv ... Dec 20
Firestorms, Southern CA: Anniv ... Oct 25
Fireworks Conv, Pyrotechnics Guild (Stevens Point, WI) ... Aug 6
Fireworks Jamboree, Roan Mtn St Park's (Roan Mountain, TN) ... July 4
Fireworks Safety Month ... June 1
First American to Circumnavigate the Earth ... Apr 10
First American to Orbit Earth: Anniv ... Feb 20
First Monday Trade Days (Scottsboro, AL) . . Jan 1
First Night (Boston, MA) ... Dec 31
Fiscal Year, US Federal ... Oct 1
Fischer, Bobby: Birth ... Mar 9
Fiscus, Kathy: Death Anniv ... Apr 8
Fish House Parade (Aitkin, MN) ... Nov 24
Fisher, Carrie: Birth ... Oct 21
Fisher, Eddie: Birth ... Aug 10
Fishing
  Blessing of the Fishing Fleet (San Francisco, CA) ... Oct 1
  Coho Family Festival (Sheboygan, WI) . . July 21
  D.C. Booth Day (Spearfish, SD) ... May 21
  Fisherman's Birthday (Gouyave, Grenada) ... June 29
  Fishing Has No Boundaries (Bemidji, MN) ... June 24
  Fishing Has No Boundaries (Eagle River, WI) ... June 2
  Fishing Has No Boundaries (Hayward, WI) ... May 19
  Fishing Has No Boundaries (Pierre, SD) June 17
  Fishing Has No Boundaries (Sandusky, OH) ... July 8
  Fishing Has No Boundaries (Thermopolis, WY) ... June 6
  Game Fishing Tourn (Grenada) ... Jan 25
  Grand Center Sport, Fishing & RV Show (Grand Rapids, MI) ... Mar 16
  Grand Prix Catfish Races (Greenville, MS) ... May 13
  Grand Strand Fishing Rodeo (Myrtle Beach, SC) ... Apr 1
  Hodag Muskie Challenge (Rhinelander, WI) ... Sept 16
  Jeff Cook Super Bass Classic (Scottsboro, AL) ... June 11
  Lake Winnebago Sturgeon Season (Fond du Lac, WI) ... Feb 12
  Memorial Weekend Salmon Derby (Petersburg, AK) ... May 26
  Mississippi Deep Sea Fishing Rodeo (Gulfport, MS) ... June 30
  Montana Governor's Cup Walleye Tourn (Ft Peck, MT) ... July 6
  Morrow Bay Harbor Fest (Morro Bay, CA) . Oct 7
  Muskie Tourn, Intl (Northcentral, MN) ... Sept 8
  Perchville USA (Tawas Bay, MI) ... Feb 3
  Pontiac Boat, Sport & Fishing Show (Pontiac, MI) ... Feb 15
  Prospect Park Fishing Contest (Brooklyn, NY) ... July 18
  Salmon Tournament Week (Manistee, MI) ... June 23
  Santa Rosalia Fest (Monterey, CA) ... Sept 9
  Seward Silver Salmon Derby (Seward, AK) ... Aug 12
  State Parks Open House and Free Fishing Day (Wisconsin) ... June 4
  Take a Kid Fishing Weekend (St. Paul, MN) ... June 10
  Tin Hau Festival (Hong Kong) ... May 3
  Tuna Tournament (Narragansett, RI) ... Sept 2
  Walton, Izaac: Birth Anniv ... Aug 9
  WICC Greatest Bluefish Tourn (Long Is Sound, NY, CT) ... Aug 26
  ZAM! Zoo and Aquarium Month ... June 1
Fisk, Carlton: Birth ... Dec 26
Fiske, Minnie M: Birth Anniv ... Dec 19
Fitch, John: Birth Anniv ... Jan 21
Fitness, Physical, and Sports Month, Natl . . May 1
Fitness: Project ACES Day ... May 5
Fitzgerald, Edward: Birth ... Mar 31
Fitzgerald, Ella: Birth ... Apr 25

☆ Chase's 1995 Calendar of Events ☆ Index

Fitzgerald (cont'd)—Folk

Fitzgerald, F. Scott: Birth Anniv .... **Sept 24**
FitzGerald, Frances: Birth .... **Oct 21**
Fitzwater, Marlin: Birth .... **Nov 24**
Five Billion, Day of the: Anniv .... **July 11**
Five State Free Fair (Liberal, KS) .... **Aug 13**
Flack, Roberta: Birth .... **Feb 10**
Flag Act of 1818: Anniv .... **Apr 4**
Flag Day (American Samoa) .... **Apr 17**
Flag Day (Pres Proc) .... **June 14**
Flag Day and Week, Natl, 6572, 6699 .... **Page 488**
Flag Day: Anniv of the Stars and Stripes . . **June 14**
Flag Day Ceremonies (Philadelphia, PA) . . **June 14**
Flag Day USA, Pause for Pledge, Natl .... **June 14**
Flag Week, Natl (Pres Proc) .... **June 11**
Flags, Festival of (Killeen, TX) .... **May 27**
Flaherty, Joe: Birth .... **June 21**
Flaherty, Robert J.: Birth Anniv .... **Feb 16**
Flanagan, Mike: Birth .... **Dec 16**
Flanders, Ed: Birth .... **Dec 29**
Flashlight Day, Natl .... **Dec 21**
Flaubert, Gustave: Birth Anniv .... **Dec 12**
Flax Scutching Fest (Stahlstown, PA) ...... **Sept 9**
Fleetwood, Mick: Birth .... **June 24**
Fleming, Peggy: Birth .... **July 27**
Fleming, Rhonda: Birth .... **Aug 10**
Flexible Work Arrangements Week .... **May 7**
Flight. See Aviation
Flood, Curt: Birth .... **Jan 18**
Flood, Great Chicago: Anniv .... **Apr 13**
Flood, Johnstown: Anniv .... **May 31**
Florida
Admission Day .... **Mar 3**
ARCA 200 Late Model Stock Car Race (Daytona Beach) .... **Feb 12**
Art in the Park Fest (Oakland Park) .... **Apr 8**
Biggest All Night Gospel Singing (Bonifay) .... **July 1**
Billy Bowlegs Fest (Fort Walton) .... **June 2**
Busch Clash (Daytona Beach) .... **Feb 12**
Calle Ocho: Open House (Miami) .... **Mar 12**
Carnaval Miami (Miami) .... **Mar 3**
Carquest Auto Parts Bowl (Miami) .... **Jan 2**
Child/Adult Learning Disabilities Intl Conf (Orlando) .... **Mar 1**
Coconut Grove Arts Fest (Coconut Grove) .... **Feb 18**
Confederate Memorial Day .... **Apr 26**
Daytona Beach Speed Spectacular (Daytona Bch) .... **Apr 1, Nov 24**
Daytona Fall Motorcycle Races (Daytona Beach) .... **Oct 20**
Daytona 500 NASCAR-FIA Winston Cup (Daytona Bch) .... **Feb 19**
Daytona Motorcycle Week (Daytona Beach) .... **Mar 3**
Daytona Supercross by Honda (Daytona Beach) .... **Mar 11**
Daytona 200 Motorcycle Road Race (Daytona Beach) .... **Mar 12**
Doral-Ryder Golf Open (Miami) .... **Feb 25**
Earth Day Tampa Bay (Tampa) .... **Apr 22**
Easter Beach Run (Daytona Beach) .... **Apr 15**
Edison Festival of Light (Ft Myers) .... **Feb 3**
Fall Gospel Jubilee (Live Oak) .... **Oct 19**
Fantasy Fest (Key West) .... **Oct 20**
Florida Citrus Bowl (Orlando) .... **Jan 2**
Florida Citrus Festival (Winter Haven) .... **Jan 26**
Florida Folk Fest (White Springs) .... **May 26**
Florida State Boat and Sports Show (Tampa) .... **Sept 14**
Florida 200 Auto Race (Daytona Beach) . **Feb 17**
Ft Lauderdale Intl Boat Show (Ft. Lauderdale) .... **Oct 26**
Ft Lauderdale Spring Boat Show (Ft Lauderdale) .... **Apr 13**
Fun-in-the-Sun Postcard Sale (Orlando) . **Jan 21**
Gainesville Downtown Fest and Art Show (Gainesville) .... **Nov 18**
Gasparilla Distance Classic (Tampa) ..... **Feb 18**
Gasparilla Pirate Fest/Invasion/Parade (Tampa) .... **Feb 4**
Gasparilla Sidewalk Art Fest (Tampa) .... **Mar 4**
Gifted Children Conv, Natl Assn (Tampa) . **Nov 8**
Gold Wing Getaway (Daytona Beach) ..... **Mar 3**
Goody's 300/NASCAR Busch Grand Natl Race (Daytona Beach) .... **Feb 18**
Grant Seafood Fest (Grant) .... **Feb 18**
GTE Suncoast (Golf) Classic (Tampa) .... **Feb 6**
Guavaween Celebration (Tampa) .... **Oct 28**
Hall of Fame (Football) Bowl Game (Tampa) .... **Jan 2**
Hatsume Fair (Delray Beach) .... **Feb 25**
Hemingway Days Fest (Key West) .... **July 17**
Hispanic Heritage Fest (Miami) .... **Oct 1**
Hoggetowne Medieval Faire (Gainesville) **Feb 17**
Holy Trinity's British Fest (Pensacola) ... **May 19**
Homestead Rodeo (Homestead) .... **Feb 3**
Honda Golf Classic (Weston Hills) .... **Mar 5**
International Race of Champions (Daytona Beach) .... **Feb 17**
Isle of Eight Flags Shrimp Festival (Fernandina) .... **Apr 29**
King Orange Jamboree Parade (Miami) . . **Dec 31**
Knights of St. Yago-Illuminated Night Parade (Tampa) .... **Feb 11**
Las Olas Art Fest (Fort Lauderdale) ...... **Mar 4**
Left-Handed Golfers Am Chmpshp (Palm Beach Garden) .... **June 27**
Little League Baseball: Big League Series (Ft Lauderdale) .... **Aug 11**
Major/Minor League: Photos/Baseball Stadiums (Hollywood) .... **Feb 2**
Major/Minor League: Photos/Baseball Stadiums (Naples) .... **Apr 8**
McGuire's St. Patrick's Day Run (Pensacola) .... **Mar 11**
Miami Intl Boat/Sailboat Show (Miami Beach) .... **Feb 16**
Miami/Bahamas Goombay Fest (Miami) . . **June 2**
Miami/Ft. Lauderdale Home Show (Miami Beach) .... **May 27**
Mini Golf Natl Chmpshp (Jacksonville) .... **Nov 9**
More Than Meets the Eye (Jacksonville) . **Apr 15**
New Star Rising Bluegrass Fest (Live Oak) .... **Nov 16**
O-Bon Fest (Delray Beach) .... **Aug 12**
Ocean Fest Sea-Son, Natl Week of (Ft Lauderdale) .... **Apr 14**
Old Tyme Farmers' Days (Live Oak) ..... **Nov 23**
Orange Bowl Football Game (Miami) ...... **Jan 2**
Palm Harbor Day Arts/Crafts/Music Fest (Palm Harbor) .... **May 7, Dec 2**
Pepsi 400 Nascar (Daytona) .... **July 1**
PGA Seniors' Chmpshp (Palm Beach Gardens) .... **Apr 13**
Polo: Intl Open/Handicap (Wellington) .... **Mar 1**
Polo: USPA Rolex Gold Cup (Wellington) **Mar 17**
Polo: World Cup (Wellington) .... **Apr 1**
Pompano Beach Seafood Fest (Pompano Beach) .... **Apr 29**
Ponce de Leon Discovers Florida .... **Apr 2**
Rolex 24 at Daytona (Daytona Beach) .... **Feb 4**
St. Petersburg Fest of States (St. Petersburg) .... **Mar 17**
Salvador Dali Museum: Anniv .... **Mar 7**
Sandblast (Ft Lauderdale) .... **July 4**
Scratch Ankle (Milton) .... **Mar 23**
Snow Fest (Pensacola Beach) .... **Jan 28**
Society Barbershop Singing Intl Conv (Miami) .... **July 2**
Southeast Florida Scottish Festival & Games (Key Biscayne) .... **Mar 4**
Speed Weeks (Daytona Beach) .... **Feb 4**
Spiffs Intl Folk Fair (St. Petersburg) ...... **Mar 16**
Spring Arts Fest (Gainesville) .... **Apr 1**
Springtime Tallahassee (Tallahassee) .... **Mar 15**
State Fair (Tampa) .... **Feb 3**
Strawberry Fest/Hillsborough County Fair (Plant City) .... **Mar 2**
SunFest (W Palm Beach) .... **May 3**
Super Bowl XXIX (Miami) .... **Jan 29**
Surfside Salutes Canada Week (Surfside) . **Mar 5**
Surfside's Intl Fiesta (Surfside) .... **Aug 5**
Suwannee River Gospel Jubilee (Live Oak) .... **June 15**
Tampa Boat Show (Tampa) .... **Jan 19**
Tarpon Springs Arts and Crafts Fest (Tarpon Springs) .... **Apr 8**
Taste of Health (Miami) .... **Apr 22**
Tongass: Alaska's Rain Forest (Ocala) .... **Feb 4**
Try This On/History of Clothing, Gender Power (Tallahassee) .... **July 15**
Twin 125-Mile Qualifying Races (Daytona Beach) .... **Feb 16**
Washington Birthday Fest (Eustis) .... **Feb 10**
Wildlife Expo (Orlando) .... **Feb 3**
Winter Equestrian Fest (Tampa) .... **Mar 15**
Winterfest Boat Parade (Ft. Lauderdale) . . **Dec 9**
WKA Daytona Classic Dirt Karting Races (Daytona Bch) .... **Dec 26**
World Enduro Chmpshp Kart Races (Daytona Beach) .... **Dec 27**
Zora Neale Hurston Fest (Eatonville) ..... **Jan 26**
Florio, Jim: Birth .... **Aug 29**
Flossie Beadle Week (Lakeside, CA) .... **Dec 1**
Flower Events
Albany Tulip Fest (Albany, NY) .... **May 12**
Azalea Fest (Muskogee, OK) .... **Apr 8**
Blossomtime Fest of SW Michigan (Benton Harbor, MI) .... **Apr 30**
Camellia Fest (Ft. Valley, GA) .... **Feb 11**
Canadian Tulip Fest (Ottawa, Ont) ...... **May 17**
Carnation City Fest (Alliance, OH) .... **Aug 4**
Chelsea Flower Show (London, England) **May 23**
Cherokee Rose Fest/Tours (Gilmer, TX) . **May 20**
Christmas in Roseland (Shreveport, LA) . **Nov 25**
Daffodil Fest (Nantucket Is, MA) .... **Apr 28**
Daffodil Parade (Tacoma/Puyallup/Sumner/Orting) .... **Apr 22**
Dixon Petunia Fest (Dixon, IL) .... **June 30**
Fest of Flowers (Asheville, NC) .... **Apr 7**
First Bloom Fest (Shreveport, LA) .... **Apr 29**
Flower Fest (Hana Matsuri, Japan) .... **Apr 8**
Flowers with Music, Intl Fest (London, England) .... **May 16**
Harrogate Spring Flower Fest (Harrogate, England) .... **Apr 23**
Holland Tulip Time Fest (Holland, MI) . . . **May 10**
Hood River Valley Blossom Fest (Hood River, OR) .... **Apr 15**
Indiana Flower and Patio Show (Indianapolis, IN) .... **Mar 11**
Iris Fest (Ponca City, Ok) .... **May 20**
Kalamazoo County Flowerfest (Kalamazoo/Portage, MI) .... **July 28**
Lilac Festival (Rochester, NY) .... **May 12**
Lily Fest, Easter in July (Smith River, CA) . **July 8**
Longwood Gardens Acres of Spring (Kennett Square, PA) .... **Apr 1**
Longwood Gardens Christmas Display (Kennett Square, PA) .... **Nov 23**
Longwood Gardens Chrysanthemum Fest (Kennett Square, PA) .... **Oct 28**
Longwood Gardens Welcome Spring (Kennett Square, PA) .... **Jan 21**
Marigold Fest (Pekin, IL) .... **Sept 8**
Maryland Home and Flower Show (Baltimore, MD) .... **Mar 8**
Mentone Rhododendron Fest (Mentone, AL) .... **May 20**
Mother's Day Annual Rhododendron Show (Portland, OR) .... **May 13**
New York Flower Show (New York, NY) . . . **Mar 2**
NW Flower and Garden Show (Seattle, WA) .... **Feb 22**
Pella Tulip Time Fest (Pella, IA) .... **May 11**
Philadelphia Flower Show (Philadelphia, PA) .... **Mar 5**
Pier 39's Tulipmania (San Francisco, CA) . . **Mar 4**
Poinsettia Day .... **Dec 12**
Portland Rose Fest (Portland, OR) ...... **June 1**
Rhododendron Fest (Florence, OR) ..... **May 19**
Rose Month, Natl .... **June 1**
Rose Show (Long Island, NY) .... **June 11**
Shelburne Museum Presents Lilac Sunday (Shelburne, VT) .... **May 21**
Shenandoah Apple Blossom Fest (Winchester, VA) .... **May 4**
Skagit Valley Tulip Fest (Mt Vernon, WA) . **Mar 31**
South Carolina Fest of Roses (Orangeburg, SC) .... **Apr 28**
Southern Spring Show (Charlotte, NC) . . **Feb 25**
Spring Wildflower Pilgrimage (Galinburg, TN) .... **Apr 27**
Tournament of Roses Parade (Pasadena, CA) .... **Jan 2**
Trillium Fest (Muskegon, MI) .... **May 13**
Valley of Flowers Festival (Florissant, MO) **May 5**
Victorian Garden Walk (Omaha, NE) ..... **Aug 12**
Wildflower Festival (Crested Butte, CO) . **July 10**
Yellow Daisy Festival (Stone Mountain, GA) .... **Sept 8**
Floyd, William: Birth Anniv .... **Dec 17**
Flutie, Doug: Birth .... **Oct 23**
Flying Saucer Sighting: Anniv .... **June 24**
Flynt, Larry Claxton: Birth .... **Nov 1**
Foch, Nina: Birth .... **Apr 20**
Fogelberg, Dan: Birth .... **Aug 13**
Fogg, Phileas: Wager Day .... **Oct 2**
Fogg, Phileas: Wins a Wager Day .... **Dec 21**
Foldes, Andor: Birth Anniv .... **Dec 21**
Folger, Henry C: Birth Anniv .... **June 18**
Foliage Fest, Fall (Walden, VT) .... **Oct 2**
Folk Arts Fest, Ocean County (Lakewood, NJ) .... **Sept 9**
Folk Dance Camp, Fall (Wheeling, WV) ..... **Sept 1**
Folk Dance Camp, Spring (Wheeling, WV) . **May 26**
Folk Fair, Holiday (Milwaukee, WI) .... **Nov 17**
Folk Fair, Spiffs Intl (St. Petersburg, FL) . . . **Mar 16**
Folk Fest, Drummondville (Drummondville, Que, Canada) .... **July 7**
Folk Fest, Fall Color Cruise and (Chattanooga, TN) .... **Oct 21**
Folk Fest, Florida (White Springs, FL) ..... **May 26**

Index ☆ Chase's 1995 Calendar of Events ☆

Folk (cont'd)——Food & Beverages

Folk Fest, Natchitoches/NW State U (Natchitoches, LA)....July 14
Folk Life Fest, Sand Creek (Newton, KS)...Apr 29
Folk Music Fest, Kaustinen (Kaustinen, Finland)....July 15
Folk music: Shindig-on-the-Green (Asheville)....July 1
Folkfest, Bismarck (Bismarck, ND)....Sept 15
Folkfest, World (Springville, UT)....July 8
Folklife Fest, Autumn Historic (Hannibal, MO)....Oct 21
Folklife Fest, Southern Utah (Springdale, UT)....Sept 7
Folklife Fest, Texas (San Antonio, TX)....Aug 3
Folklorama—Canada's Cultural Celeb (Winnipeg, Manitoba)....Aug 6
Folkmoot USA: The NC Intl Folk Fest (Waynesville, NC)....July 20
Folkways of Christmas (Shakopee, MN)...Nov 24
Folsom, Jim: Birth....May 14
Fonda, Henry: Birth Anniv....May 16
Fonda, Jane: Birth....Dec 21
Fonda, Peter: Birth....Feb 23
Fonteyn, Margot: Birth Anniv....May 18
Food and Beverages
A Is for Apples Weekend (Chadds Ford, PA)....**Oct 14**
Alferd G. Packer Day (Boulder, CO)....**Apr 21**
Allons Manager (Belle Rose, LA)....**Apr 23**
Anti-Saloon League Founded: Anniv....**May 24**
Appert, Nicholas: Birth Anniv....**Oct 23**
Apple Butter Fest (Burton, OH)....**Oct 7**
Apple Butter Stirrin' (Coshocton, OH)...**Oct 21**
Apple Fest (Topeka, KS)....**Oct 1**
Apple Fest, Kentucky (Paintsville, KY)....**Oct 1**
Apple Fest, Vermont (Springfield, VT)....**Oct 7**
Apple Harvest Fest (Gettysburg, PA)....**Oct 7**
Apple Harvest Fest, Old-Fashioned (Burlington, WV)....**Sept 30**
Apple Jack Month, Natl....**Oct 1**
Apple Squeeze, Steilacoom (Steilacoom, WA)....**Oct 15**
Applejack Fest (Nebraska City, NE)....**Sept 16**
Asparagus Fest, Natl (Hart and Shelby, MI)....**June 9**
Baby Food Fest, Natl (Fremont, MI)....**July 18**
Bagelfest (Mattoon, IL)....**July 28**
Baked Bean Month, Natl....**July 1**
Bakery-Deli Assn, Retailers: Conv/Exhib (St. Louis, MO)....**Mar 11**
Bar-B-Q Festival, Intl (Owensboro, KY)..**May 12**
Barbecue Month, Natl....**May 1**
Bayou Boogalou & Cajun Food Fest (Norfolk, VA)....**June 16**
Beef Empire Days (Garden City, KS)....**June 2**
Beer Fest, Great American (Denver, CO)..**Oct 6**
Beer Week, American....**Oct 8**
Blueberries Month, Natl July Belongs to..**July 1**
Blueberry Festival (Ketchikan, AK)....**Aug 12**
Bracebridge Dinner (Yosemite Natl Park, CA)....**Dec 22**
Bratwurst Days (Sheboygan, WI)....**Aug 4**
Breaux Bridge Crawfish Fest (Breaux Bridge, LA)....**May 5**
Broiler Fest, Mississippi (Forest, MS)....**June 1**
Buck Fever Days and Chili Feed (Monocqua, WI)....**Nov 17**
Buffalo Wallow Chili Cookoff (Custer, SD) **Oct 1**
Bun Day (Iceland)....**Feb 27**
Burgoo Fest (Utica, IL)....**Oct 8**
California Dry Bean Fest (Tracy, CA)....**Aug 5**
California Kiwifruit Day....**Dec 22**
California Strawberry Fest (Oxnard, CA) **May 20**
Canned Food Month....**Feb 1**
Canned Luncheon Meat Week....**July 2**
Catfish Month....**Aug 1**
Celebration of America's Bounty (McLean, VA)....**Sept 16**
Celebration of Chocolate (Oregon, IL)....**Feb 1**
Champagne Dinner (Newport, RI)....**Dec 1**
Cherry Fest, Natl (Traverse City, MI)....**July 2**
Cherry Month, Natl....**Feb 1**
Cherry Pit Spitting Contest, Intl (Eau Claire, MI)....**July 1**
Chicken Festival, Delmarva (Federalsburg, MD)....**June 9**
Chicken Month, Natl....**Sept 1**
Chili and Bluegrass Fest (Tulsa, OK)....**Sept 8**
Chili Cookoff, Arizona Chmpshp (Bullhead City, AZ)....**Apr 29**
Chili Fest, Goldstar (Cincinnati, OH)....**Oct 7**
Chocolate Fest (Knoxville, TN)....**Jan 14**
Chocolate Fest (Norman, OK)....**Feb 11**
Chocolate Fest (St.Stephen, NB, Canada)....**July 31**
Chocolate Harvest (Mattoon, IL)....**Oct 7**
Chocolate Week, American....**Mar 19**
Chowderfest (Mystic, CT)....**Oct 7**
Citrus Festival, Florida (Winter Haven, FL)....**Jan 26**
Clam Chowder Cookoff and Chowder Chase (Santa Cruz, CA)....**Feb 25**
Clam Fest, Yarmouth (Yarmouth, ME)...**July 14**
Clean Out Your Refrigerator Day, Natl...**Nov 15**
Cookbook Festival (Fargo, ND)....**Feb 11**
Cooking in Paradise (St. Barthelemy, French West Indies)....**Apr 1**
Corn Fest (Shippensburg, PA)....**Aug 26**
Country Fair and Burro Barbecue (Bullhead City, AZ)....**May 13**
Country Ham Days, Marion County (Lebanon, KY)....**Sept 23**
Crab and Clam Bake (Crisfield, MD)....**July 19**
Crawfish Festival, Dermott's Annual (Dermott, AR)....**May 19**
Crayfish Premiere (Sweden)....**Aug 9**
Creative Ice Cream Flavor Day and Contest (Chicago, IL)....**July 1**
Culinary Fest (Scottsdale, AZ)....**Apr 21**
Culinary Professionals' Cookbook Month, Intl Assn....**Oct 1**
Culinary Week, Natl....**Nov 12**
Dairy Expo, World (Madison, WI)....**Oct 4**
Dairy Month, June....**June 1**
Delicato Charity Grape Stomp (Manteca, CA)....**Sept 3**
Dessert Day, Natl....**Oct 12**
Dessert Month, Natl....**Oct 1**
Dietary Managers Assn Mtg/Expo (Boston, MA)....**July 30**
Dietary Mgrs' Pride in Food Service Week.**Feb 6**
Digby Scallop Days Fest (Digby, NS, C)...**Aug 6**
Eat What You Want Day....**Dec 16**
Eel Fest (Jyllinge, Denmark)....**June 3**
Egg Month, Natl....**May 1**
Egg Salad Week....**Apr 17**
Feast of the Ramson (Richwood, WV)....**Apr 15**
Fest of Mountain & Plain: A Taste (Denver, CO)....**Sept 1**
Fiber Focus Month, Natl....**Feb 1**
Fig Week, Natl....**Nov 1**
Florida Strawberry Fest (Plant City, FL)...**Mar 2**
Food Day, World (UN)....**Oct 16**
Food Day, World 6613....**Page 488**
Food Rationing, US: Anniv....**July 1**
Food Service Employees Day....**Sept 27**
Food Service Employees Week, Natl....**Sept 25**
Foods/Feasts 17th-Century Virginia (Williamsburg, VA)....**Nov 23**
French West Indies: Fete des Cuisinieres **Aug 12**
Fresh Fish Fest (Breckenridge, CO)....**June 3**
Fresh Fruits and Vegetables Month....**June 1**
Fresh Seafood Festival (Atlantic City, NJ)....**June 10**
Frozen Food Month, Natl....**Mar 1**
Frozen Yogurt Day, Natl....**June 4**
Frozen Yogurt Month, Natl....**June 1**
Frozen Yogurt Week, Natl....**May 29**
Fulton Oysterfest (Fulton, TX)....**Mar 3**
Galesburg Chocolate Fest (Galesburg, IL)....**Feb 11**
Gazpacho Aficionado Time....**May 1**
Gilroy Garlic Fest (Gilroy, CA)....**July 28**
Goat BBQ Cook-Off, Electra (Electra, TX)....**May 13**
Grant County Home Products Dinner (Ulysses, KS)....**Sept 19**
Grant Seafood Fest (Grant, FL)....**Feb 18**
Great American Meatout....**Mar 20**
Great American Pies Month....**Feb 1**
Great American Pizza Bake....**Feb 7**
Great Arkansas Pig-Out (Morrilton, AR)..**Aug 4**
Great Monterey Squid Fest (Monterey, CA)....**May 27**
Great Peanut Tour (Skippers, VA)....**Sept 7**
Gumbo Fest (Bridge City, LA)....**Oct 13**
Hamburger Month, Natl....**May 1**
Hard Crab Derby and Fair, Natl (Crisfield, MD)....**Sept 1**
Harvest Weekends (Bryan, TX)....**July 15**
Harveys Wine Museum Anniv (Bristol, England)....**Apr 14**
Health Care Food Service Week, Natl....**Oct 1**
Healthy Exchanges Annual Potluck (DeWitt, IA)....**July 8**
Herb Day (Milton, NH)....**June 10**
Herb Fest (Mattoon, IL)....**Apr 29**
Herb Week, Natl....**May 8**
Highland County Maple Fest (Highland County, VA)....**Mar 11**
Hog Day (Hillsborough, NC)....**June 17**
Home of the Hamburger Celeb (Seymour, WI)....**Aug 5**
Homebrew Day, Natl....**May 6**
Homemade Bread Day....**Nov 17**
Honey Month, Natl....**Sept 1**
Hong Kong Food Fest....**Mar 4**
Hope Watermelon Fest (Hope, AR)....**Aug 17**
Hot and Spicy Food Intl Day....**Jan 21**
Hot Dog Month, Natl....**July 1**
Hot Dog Night (Luverne, MN)....**July 20**
Hot Tea Month, Natl....**Jan 1**
Hunger Awareness Month....**Oct 1**
Ice Cream Cone: Birth....**Sept 22**
Ice Cream Day, Natl....**July 16**
Ice Cream Fest, Old-Fashioned (Utica, OH)....**May 27**
Ice Cream Fest, Old-Fashioned (Wilmington, DE)....**July 8**
Ice Cream Month, Natl....**July 1**
Ice Cream Social (Indianapolis, IN)....**July 4**
Ice Cream Social (Rockford, OH)....**Aug 26**
Iced Tea Month, Natl....**June 1**
Jackson County Apple Fest (Jackson, OH)....**Sept 19**
Jalapeno Fest (Laredo, TX)....**Feb 17**
Johnson City Peach Festival (Clarksville, AR)....**July 20**
June is Turkey Lovers' Month....**June 1**
Kansas St Barbeque Champship (Lenexa, KS)....**June 24**
Kraut and Frankfurter Week, Natl....**Feb 9**
Lebanon Bologna Fest (Lebanon, PA)...**Sept 15**
Leitersburg Peach Festival (Leitersburg, MD)....**Aug 12**
Lobster Fest, Maine (Rockland, ME)....**Aug 3**
Lobsterfest (Mystic, CT)....**May 27**
Long Beach Island Chowder Cook-Off (Beach Haven, NJ)....**Sept 30**
Louisiana Peach Fest (Ruston, LA)....**June 9**
Maple Fair, Parke County (Rockville, IN) .**Feb 25**
Maple Fest of Nova Scotia (Cumberland Co, NS, Canada)....**Mar 18**
Maple Fest, Pennsylvania (Meyersdale, PA)....**Apr 22**
Maple Fest, Vermont (St. Albans)....**Apr 21**
Maple Sugaring Fest (New Preston, CT) .**Mar 13**
Maple Syrup Fest (Beaver, PA)....**Apr 8**
Maple Syrup Festival (Wakarusa, IN)....**Mar 24**
Maple Syrup Saturday (Appleton, WI)....**Mar 25**
Marion Popcorn Fest (Marion, OH)....**Sept 7**
Masters of the Grill (Kenosha, WI)....**June 25**
Melon Days (Green River, UT)....**Sept 15**
Microbrewers Conference/Trade Show, Natl (Austin, TX)....**Apr 23**
Mint Fest, St. Johns (St. Johns, MI)....**Aug 12**
Morden Corn/Apple Fest (Morden, Man, Canada)....**Aug 25**
More Herbs, Less Salt Day....**Aug 29**
Morton Pumpkin Fest (Morton, IL)....**Sept 13**
Mushroom Fest, Telluride (Telluride, CO) **Aug 24**
Mushroom Hunting Chmpshps, Natl (Boyne City, MI)....**May 12**
Mustard Day, Natl....**Aug 5**
Napa Valley Wine Auction (Napa Valley, CA)....**June 8**
New Jersey Restaurant/Hospitality Expo (Somerset, NJ)....**Mar 19**
Noodle Month, Natl....**Mar 1**
Nuts Fair (Bastogne, Belgium)....**Dec 18**
Oatman Sidewalk Egg Frying Contest (Oatman, AZ)....**July 4**
Oatmeal Month....**Jan 1**
Old Town Art Fair Chicken Teriyaki Lunch (Chicago, IL)....**June 10**
Old-Fashioned Ice Cream Social (Fond du Lac, WI)....**Aug 6**
Onion Market (Zibelemarit, Switzerland) .**Nov 27**
Organic Harvest Month, Natl....**Sept 1**
Oyster Fest (Chincoteague Island, VA)....**Oct 7**
Oyster Festival (Biloxi, MS)....**Mar 16**
Pancake Day (Centerville, IA)....**Sept 23**
Pancake Day, Intl (Liberal, KS)....**Feb 28**
Pancake Week, Natl....**Feb 26**
Papaya Month, Natl....**Sept 1**
Pasta Month, Natl....**Oct 1**
Peanut Butter Lover's Month....**Nov 1**
Peanut Month, Natl....**Mar 1**
Pecan Day....**Mar 25**
Pickle Week, Intl....**May 19**
Pickled Pepper Week, Natl....**Oct 6**
Picklefest (Atkins, AR)....**May 19**
Pie Day, Natl....**Jan 23**
Pink Tomato Fest, Bradley County (Warren, AR)....**June 8**

☆ Chase's 1995 Calendar of Events ☆ Index

Food & Beverages (cont'd)—Fuller

Pizza Expo (Las Vegas, NV) . . . . . Mar 14
Pizza Month, Natl . . . . . Oct 1
Poke Salad Fest (Blanchard, LA) . . . . . May 11
Popcorn Introduced to Colonists: Anniv . Feb 22
Popcorn Poppin' Month, Natl . . . . . Oct 1
Pork Month, Natl . . . . . Oct 1
Poteet Strawberry Fest (Poteet, TX) . . . . . Apr 7
Prairie Dog Chili Cookoff (Grand Prairie, TX) . . . . . Apr 1
Prime Beef Festival, Wyatt Earp Days (Monmouth, IL) . . . . . Sept 6
Prune Breakfast Month, Natl . . . . . Jan 1
Pumpkin Show, Circleville (Circleville, OH) . . . . . Oct 18
Ranch Hand Breakfast (Kingsville, TX) . . . Nov 18
Retail Bakers Month, Natl . . . . . Jan 1
Return Shopping Carts to the Supermarket Month . . . . . Feb 1
Rhubarb Festival (Intercourse, PA) . . . . . May 20
Ribfest (Kalamazoo, MI) . . . . . Aug 10
Rice God, Day of the (Chiyoda, Japan) . . June 4
Rice Month, Natl . . . . . Sept 1
Rice Planting Festival (Japan) . . . . . June 14
River Roast (Chattanooga, TN) . . . . . May 19
Roasting Ears of Corn Food Fest (Allentown, PA) . . . . . Aug 13
Rockport Seafair (Rockport, TX) . . . . . Oct 7
St. Mary's County Oyster Fest (Leonardtown, MD) . . . . . Oct 21
Salad Month, Natl . . . . . May 1
Sand Plum Fest (Guthrie, OK) . . . . . June 23
Sandwich Day . . . . . Nov 3
Sauce Month, Natl . . . . . Mar 1
Sauerkraut Salad and Sandwich Season . . July 1
Schmeckfest (Freeman, SD) . . . . . Mar 30
Sea Lion Suds Fest (Gold Beach, OR) . . . Mar 18
Seafood and Wine Fest, Newport (Newport, OR) . . . . . Feb 24
Seafood Fest, Pompano Beach (Pompano Beach, FL) . . . . . Apr 29
Seafood in Seaside (Seaside Hts, NJ) . . June 17
Seafood Month, Natl . . . . . Oct 1
Sedona Arts Fest (West Sedona, AZ) . . . . Oct 14
Show Me State Barbeque Cook-Off (St. Joseph, MO) . . . . . June 9
Shrimp Fest, Low Country (McClellanville, SC) . . . . . May 6
Sinkie Day (Eat over Kitchen sink) . . . . . Nov 24
Snack Food Month, Natl . . . . . Feb 1
Sneak Some Zucchini onto Your Neighbors' Porch Night . . . . . Aug 8
Soup Month, Natl . . . . . Jan 1
Sour Herring Premiere (Sweden) . . . . . Aug 17
Spinach Festival, Lenexa (Lenexa, KS) . . . Sept 9
Split Pea Soup Week, Natl . . . . . Nov 7
State St Station Cookbook Fair (Greensboro, NC) . . . . . Mar 25
Steilacoom Salmon Bake (Steilacoom, WA) . . . . . July 30
Strawberry Fest, West Virginia (Buckhannon, WV) . . . . . May 17
Strawberry Month, Natl . . . . . May 1
Sun Prairie's Sweet Corn Fest (Sun Prairie, WI) . . . . . Aug 17
Surimi Seafood Month . . . . . June 1
Swarm of Palolo (American Samoa) . . . . . Oct 1
Sweet Corn Fest (Millersport, OH) . . . . . Aug 30
Sweetcorn Fest, Natl (Hoopeston, IL) . . . Aug 31
Taste of Cincinnati (Cincinnati, OH) . . . . . May 27
Taste of Madison (Madison, WI) . . . . . Sept 2
Telluride Wine Festival (Telluride, CO) . . June 23
Texas Citus Fiesta (Mission, TX) . . . . . Jan 30
Tomato Month, Natl Florida . . . . . Apr 1
Town Pt Pk Wine Fest (Norfolk, VA) . . . . . Oct 21
Turkey Rama (McMinnville, OR) . . . . . July 6
Vacaville Onion Fest (Vacaville, CA) . . . . . Sept 9
Veal Ban Action Day . . . . . May 14
Vegetarian Awareness Month . . . . . Oct 1
Vegetarian Day, World . . . . . Oct 1
Washington State Apple Blossom Fest (Wenatchee, WA) . . . . . Apr 27
Watermelon Fest (Rush Springs, OK) . . . . Aug 12
Watermelon Thump (Luling, TX) . . . . . June 22
Whale of Wine Fest/Brewing Up Storm (Gold Bch, OR) . . . . . Jan 14
Wild Foods Day (Pikeville, TN) . . . . . Oct 28
Wine & Roses Fest (wine) (Bryan, TX) . . . . May 6
Wine Fest, Okanagan (Canada) . . . . . Sept 29
Winefest (Bryan, TX) . . . . . Nov 11
Wine/Harvest Fest, Michigan (Kalamazoo/Paw Paw, MI) . . . . . Sept 6
Wolf Point's Hottest Chili Weekend (Wolf Point, MT) . . . . . June 1
Wollersheim Winery Grape Stomp Fest (Prairie du Sac, WI) . . . . . Oct 7
World Beef Expo (Madison, WI) . . . . . Apr 6
World Chmpshp BBQ Goat Cookoff (Brady, TX) . . . . . Sept 2
WROK/WZOK Annual Chili Shoot-Out (Rockford, IL) . . . . . Oct 7
Yambilee, Louisiana (Opelousas, LA) . . . . Oct 25
Yosemite Chefs' Holidays (Yosemite Natl Park, CA) . . . . . Jan 1
Yosemite Vintners' Holidays (Yosemite Natl Park, CA) . . . . . Nov 6
Foot Health Month . . . . . Aug 1
Footbag Chmpshps, World (Golden, CO) . . July 25
Football
Army-Navy Football Game (Philadelphia, PA) . . . . . Dec 2
Carquest Auto Parts Bowl (Miami, FL) . . . . Jan 2
Circle City Classic (Indianapolis, IN) . . . . . Oct 7
Cotton Bowl Classic, Mobil (Dallas, TX) . . . Jan 2
Fabulous 1890s Weekend (Mansfield, PA) Sept 22
First Night Football Game: Anniv (Mansfield, PA) . . . . . Sept 28
First Play-by-Play Football Game Broadcast . . . . . Nov 23
First Professional Football Player: Anniv . Nov 12
Florida Citrus Bowl (Orlando, FL) . . . . . Jan 2
Hall of Fame (Football) Bowl Game (Tampa, FL) . . . . . Jan 2
Holiday Bowl Parade/Game (San Diego, CA) . . . . . Dec 30
Hula Bowl Game (Honolulu, HI) . . . . . Jan 22
Kelly Tire Blue-Gray All Star Classic (Montgomery, AL) . . . . . Dec 25
NCAA Division I-AA Football Chmpshp . . Dec 16
NFL Pro Bowl (Honolulu, HI) . . . . . Feb 5
NFL Pro Bowl Beach Challenge (Oahu, HI) Feb 3
Orange Bowl Football Game (Miami, FL) . . Jan 2
Pro Football Hall of Fame Fest (Canton, OH) . . . . . July 21
Senior Bowl Football Game (Mobile, AL) . Jan 21
Shrine Oyster Bowl Game (Norfolk, VA) . Nov 18
Super Bowl XXIX (Miami, FL) . . . . . Jan 29
Tourn of Roses Game: Anniv . . . . . Jan 1
Univ of Chicago 1st Football Victory: Anniv . . . . . Nov 16
USF&G Sugar Bowl (New Orleans, LA) . . . Jan 2
Foote, Shelby: Birth . . . . . Nov 17
Forbes, Malcolm: Birth Anniv . . . . . Aug 19
Ford, Eileen: Birth . . . . . Mar 25
Ford, Elizabeth (Betty): Birth . . . . . Apr 8
Ford, Gerald: Assassination Attempts: Anniv . . . . . Sept 5
Ford, Gerald: Birthday . . . . . July 14
Ford, Gerald: Vice-Presidential Swearing In: Anniv . . . . . Dec 6
Ford, Glenn: Birth . . . . . May 1
Ford, Harrison: Birth . . . . . July 13
Ford, John: Birth Anniv . . . . . Feb 1
Ford, Wendell Hampton: Birth . . . . . Sept 8
Ford, "Whitey": Birth . . . . . Oct 21
Fordice, Kirk: Birth . . . . . Feb 10
Forefather's Day . . . . . Dec 21
Foreman, George: Birth . . . . . Jan 10
Forest Fest, Mountain State (Elkins, WV) . . Sept 30
Forest Products Week, Natl (Pres Proc) . . . . Oct 15
Forest Products Week, Natl 6614 . . . . . Page 488
Forestry Expo, Vermont (Rutland, VT) . . . . . July 22
Forestry: Paul Bunyan (Nelsonville, OH) . . . . Oct 6
Forgiveness Day, Natl . . . . . June 24
Forman, Chuck: Birth . . . . . Oct 26
Forman, Milos: Birth . . . . . Feb 18
Forrest, Nathan Bedford: Birth Anniv . . . . . July 13
Forster, E.M.: Birth Anniv . . . . . Jan 1
Forster, Robert: Birth . . . . . July 13
Forsythe, John: Birth . . . . . Jan 29
Fort Moore Established: Anniv . . . . . Apr 24
Fort Necessity Memorial (Farmington, PA) . . July 3
Fort Sumter Shelled by North: Anniv . . . . . Aug 17
Fortas, Abe: Birth Anniv . . . . . June 19
Forten, James: Birth Anniv . . . . . Sept 2
Foss, Joseph: Birth . . . . . Apr 17
Fosse, Robert Louis: Birth Anniv . . . . . June 23
Foster, Andrew: Birth Anniv . . . . . Sept 17
Foster, Jodie: Birth . . . . . Nov 19
Foster, Stephen: Birth Anniv . . . . . July 4
Foster, Stephen: Memorial Day (Pres Proc) Jan 13
Foundation Day, Natl (Japan) . . . . . Feb 11
Fountain, Pete: Birth . . . . . July 3
Fountains, Longwood Gardens Fest of (Kennett Square, PA) . . . . . May 27
Fourth of July. See Independence Day
Fox, Michael J.: Birth . . . . . June 9
Fox, Terry: Day . . . . . July 28
Foxfield Races (Charlottesville, VA) . . . . . Apr 29
Foxx, Redd: Birth Anniv . . . . . Dec 9
Foyt, A.J.: Birth . . . . . Jan 16
Fracci, Carla: Birth . . . . . Aug 20
Fragrance Week, Natl . . . . . June 5
Fraley Family Music Weekend (Olive Hill, KY) . . . . . Sept 5
Frampton, Peter: Birth . . . . . Apr 22
France
Bastille Day . . . . . July 14
Battle of Dien Bieh Phu Ends: Anniv . . . . . May 7
Eiffel Tower: Anniv (Paris) . . . . . Mar 31
Festival of the Tarasque . . . . . June 28
Nice Carnival . . . . . Feb 17
Night Watch . . . . . July 13
Franciosa, Anthony: Birth . . . . . Oct 25
Francis, Arlene: Birth . . . . . Oct 20
Francis, Connie: Birth . . . . . Dec 12
Francis, Genie: Birth . . . . . May 26
Frank, Anne, Diary: Last Entry: Anniv . . . . . Aug 1
Frankel, Max: Birth . . . . . Apr 3
Franklin, Aretha: Birth . . . . . Mar 25
Franklin, Benjamin: Birth Anniv . . . . . Jan 17
Franklin, Benjamin: Poor Richard's Almanac Anniv . . . . . Dec 28
Franklin, Bonnie: Birth . . . . . Jan 6
Frann, Mary: Birth . . . . . Feb 27
Franz, Dennis: Birth . . . . . Oct 28
Fraser, Dawn: Birth . . . . . Sept 4
Fratianne, Linda: Birth . . . . . Aug 2
Frazier, Joe: Birth . . . . . Jan 12
Freberg, Stan: Birth . . . . . Aug 7
Free Paper Week, Natl . . . . . Mar 19
Freedom Day . . . . . Feb 1
Freedom Day, Natl (Pres Proc) . . . . . Feb 1
Freedom Fest, Intl (Detroit, MI/Windsor, Ont, Canada) . . . . . June 17
Freedom Fest, Spirit of (Florence, AL) . . . . . July 4
Freedom of Information Day . . . . . Mar 16
Freedom Shrine Month . . . . . May 1
Freedom Week . . . . . July 4
Freedom Week, Religious . . . . . Sept 24
Freefall Conv, World (Quincy, IL) . . . . . Aug 4
Freeman, Al, Jr: Birth . . . . . Mar 21
Freeman, Morgan: Birth . . . . . June 1
French, Daniel C.: Birth Anniv . . . . . Apr 20
French and Indian Rendezvous (Big Pool, MD) . . . . . May 27
French Library Anniv Gala (Boston, MA) . . . . Nov 9
French Quarter Fest (New Orleans, LA) . . . . . Apr 7
French West Indies
Concordia Day (St. Martin) . . . . . Nov 11
Cooking in Paradise (St. Barthelemy) . . . . . Apr 1
Fete des Cuisinieres (Guadeloupe) . . . . . Aug 12
International Jazz Fest (Martinique) . . . . . Nov 30
Schoelcher Day . . . . . July 21
Slavery Abolition Day (Martinique) . . . . . May 22
Tour de la Guadeloupe (Guadeloupe) . . . . . Aug 4
Tour des Yoles Rondes (Martinique) . . . . . July 30
Frenchman Rows Across Pacific: Anniv . . . . Nov 21
Frey, Glen: Birth . . . . . Nov 6
Frickie, Janie: Birth . . . . . Dec 18
Friday the Thirteenth . . . . . Jan 13, Oct 13
Friedan, Betty Naomi: Birth . . . . . Feb 4
Friedkin, William: Birth . . . . . Aug 29
Friedman, Milton: Birth . . . . . July 31
Friendly, Fred: Birth . . . . . Oct 30
Friendship Day . . . . . Aug 6
Friendship: Peace, Friendship, Good Will Week . . . . . Oct 25
Friendship: Secret Pal Day . . . . . Jan 14
Friendship Sees No Color Week . . . . . Apr 29
Friendship Week, Intl . . . . . Feb 19
Fringe Theater Event (Edmonton, Alta, Canada) . . . . . Aug 18
Frisbee Disc Fest, Natl (Washington, DC) . . . Sept 2
Frisch, Max: Birth Anniv . . . . . May 15
Frog Days (Melvina, WI) . . . . . Aug 4
Frog Hop, Preakness (Baltimore, MD) . . . . . May 17
Frog Jump Jamboree, Old-Fashioned (New Preston, CT) . . . . . July 2
Frog Jumping Contest (Old Forge, NY) . . . June 18
Frog Jumping Contest, Roaring Camp (Felton, CA) . . . . . July 2
Frog Jumping Jubilee/Calaveras Fair (Angel Camp, CA) . . . . . May 18
Frontier Days (Culbertson, MT) . . . . . June 3
Frost, David: Birth . . . . . Apr 7
Frost, Robert Lee: Birth Anniv . . . . . Mar 26
Frozen Food Month, Natl . . . . . Mar 1
Frozen Yogurt Day, Natl . . . . . June 4
Frozen Yogurt Month, Natl . . . . . June 1
Frozen Yogurt Week, Natl . . . . . May 29
Fruits and Vegetables Month, Fresh . . . . . June 1
Fry, Elizabeth: Birth Anniv . . . . . May 21
Frye, Soleil Moon: Birth . . . . . Aug 6
Fuller, Alfred Carl: Birth Anniv . . . . . Jan 13
Fuller, Bobby: Death Anniv . . . . . July 18

Index ☆ Chase's 1995 Calendar of Events ☆

Fuller (cont'd)——Ghana

Fuller, Margaret: Birth Anniv..............May 23
Fuller, Melville Weston: Birth Anniv........Feb 11
Fulton, Robert: Birth Anniv...............Nov 14
Fun Flight (Alamogordo, NM)............June 16
Funfest (Amarillo, TX)....................May 27
Fungal Infection Awareness Month.........May 1
Funicello, Annette: Birth..................Oct 22
Funk, Isaac K: Birth Anniv...............Sept 10
Funky Winkerbean: Anniv.................Mar 27
Funt, Allen: Birth.......................Sept 16
Fur Rendezvous, Anchorage (Anchorage, AK)....................................Feb 10
Fur Trade: Festival Du Voyageur (Winnipeg, Man, Canada)...............................Feb 10
Fur Trade: Festival of Adventures (Aitkin, MN)..................................Sept 22
Fur Trade: Ft Union Trade Post Rendezvous (Williston, ND).....................June 15
Fur Trade: White Oak Rendezvous (Deer River, MN)....................................Aug 4
Furnishings Market, Intl Home (High Point, NC)....................................Apr 27
Fuster, Jaime B.: Birth....................Sept 4
Future Homemakers Natl Leadership Mtg (Washington, DC)........................July 9
Future Homemakers of America FHA/Hero Week, Natl...................................Feb 12

G

Gable, Clark: Birth Anniv.................Feb 1
Gabon: National Day.....................Aug 17
Gabor, Eva: Birth.........................Feb 11
Gabor, Zsa Zsa: Birth......................Feb 6
Gabriel, Peter: Birth......................May 13
Gaddis, William: Birth....................Dec 29
Gagarin, Yuri A.: Birth Anniv...............Mar 9
Gage, Nicholas: Birth....................July 23
Gail, Max: Birth..........................Apr 5
Gainsborough, Thomas: Birth Anniv......May 14
Galbraith, John Kenneth: Birth............Oct 15
Gale, Robert: Birth.......................Oct 11
Gallant, Roy: Birth.......................Apr 17
Gallaudet, Thomas: Birth Anniv...........Dec 10
Galli-Curci, Amelita: Birth Anniv.........Nov 18
Gallo, Frank: Birth.......................Jan 13
Galway, James: Birth......................Dec 8
Gambia: Independence Day...............Feb 18
Game and Puzzle Week, Natl.............Nov 24
Games (Multisport Competition)
Badger State Summer Games (WI)......June 9
Badger State Winter Games (Wausau, WI) .Feb 3
Big Sky State Games (Billings, MT)......July 14
Cowboy State Winter Sports Fest and Summer Games (Casper, WY)..........Feb 18, June 8
Hoosier St Games Finals (Indianapolis, IN).................................July 14
Hoosier St Games Regionals (IN).......June 24
Maine Highland Games (Brunswick, ME).Aug 19
Show Me State Games (Columbia, MO)..July 21
Sooner State Summer Games (Oklahoma City, OK)...................................June 9
State Games of Oregon (Portland, OR)...July 7
Western Canada Games (Abbotsford, BC, Canada)............................Aug 15
Winter Games of Oregon (various, OR)..Mar 11
Games Day, Video........................July 12
Games: World Scrabble Chmpshp.........Oct 12
Gandhi, Mohandas: Assassination Anniv...Jan 30
Gandhi, Mohandas: Birth Anniv............Oct 2
Gandhi, Rajiv, Assassinated: Anniv........May 21
Ganzel, Teresa: Birth.....................Mar 23
Garage Sale, World's Largest (Nationwide) May 14
Garage Sale, World's Largest (South Bend, IN)..................................June 3
Garagiola, Joe: Birth.....................Feb 12
Garcia, Jerry: Birth.......................Aug 1
Garcia-Marquez, Gabriel: Birth.............Mar 6
Garden: Michigan Home & Garden Show (Pontiac, MI)............................Mar 2
Garden Month, Natl.......................Apr 1
Garden Week, Historic (Virginia)..........Apr 22
Garden Week, Natl.........................Apr 9
Gardener Day, Master.....................Mar 21
Gardenia, Vincent: Birth Anniv.............Jan 7
Gardner, Erle Stanley: Birth Anniv........July 17
Gardner, Randy: Birth.....................Dec 2
Garfield: Birthday.......................June 19
Garfield, James A: Birth Anniv............Nov 19
Garfield, Lucretia R.: Birth Anniv..........Apr 19
Garfunkel, Arthur: Birth...................Nov 5
Garland, Beverly: Birth...................Oct 17
Garland, Judy: Birth Anniv..............June 10
Garlic Fest, Gilroy (Gilroy, CA)...........July 28
Garner, James: Birth......................Apr 7
Garner, John Nance: Birth Anniv..........Nov 22
Garr, Teri: Birth..........................Dec 11
Garrick, David: Birth Anniv...............Feb 19
Garrison, William Lloyd: Birth Anniv.......Dec 12
Garvey, Ed: Birth.........................Apr 18
Garvey, Steve: Birth......................Dec 22
Gas Explosion, Chicago's Freak...........Jan 17
Gasparilla Pirate Fest/Invasion/Parade (Tampa, FL).....................................Feb 4
Gastineau, Mark: Birth...................Nov 20
Gastrell, Francis: Ejectment Anniv.........Nov 27
Gates, David: Birth.......................Dec 11
Gatlin, Larry: Birth........................May 2
Gauguin, Paul: Birth Anniv...............June 7
Gault, Willie: Birth.......................Sept 5
Gautier, Dick: Birth.......................Oct 30
Gay and Lesbian Book Month..............June 1
Gay and Lesbian Pride Parade/Rally (Chicago, IL)..................................June 25
Gay: Coming Out Day, Natl...............Oct 11
Gay, Simon, Descendants Family Reunion (Moultrie, GA)............................May 20
Gay Square Dance Month, Intl.............Sept 1
Gayle, Crystal: Birth......................Jan 9
Gaynor, Mitzi: Birth.......................Sept 4
Gazpacho Aficionado Time................May 1
Gazzara, Ben: Birth.......................Aug 28
Geary, Anthony: Birth....................May 29
Gedalya, Fast of.........................Sept 27
Gedda, Nicolai: Birth.....................July 11
Gehrig, Lou: Birth Anniv.................June 19
Gehry, Frank: Birth.......................Feb 28
Geiger, Ray: Birth Anniv.................Sept 18
Geisel, Theodor "Dr. Seuss": Birth Anniv...Mar 2
Geisel, Theodor "Dr. Seuss": Death Anniv.Sept 24
Gelbart, Larry: Birth......................Feb 25
Geldof, Bob: Birth.........................Oct 5
Geller, Uri: Birth.........................Dec 20
Gem and Mineral Show (Jackson, MS).....Feb 25
Gemayel, Bashir: Assassination Anniv....Sept 14
Gemini Begins............................May 21
Gems and Minerals Assn Show, Chicagoland (Wheaton, IL)..........................May 27
Genealogy, ABCs of (Moultrie, GA)........Mar 18
Genealogy: Family History Month..........Oct 1
General Election Day (US).................Nov 7
General Motors: Founding Anniv.........Sept 16
Geneva Accords..........................July 20
Genocide Convention: Anniv...............Dec 9
Gentry, Bobbie: Birth.....................July 27
Gentry, Dennis: Birth.....................Feb 10
Geographers Annual Meeting, Assn Amer (Chicago, IL)..............................Mar 14
Geography Awareness Week 6623......Page 488
Geography Awareness Week, Natl.........Nov 12
Geography Bee Finals, Natl (Washington, DC)...................................May 23
Geography Bee School Level, Natl..........Jan 2
Geography Bee State Test, Natl............Apr 7
Geography: Tahiti and Her Islands Awareness Month.................................July 1
George III: Birth Anniv....................June 4
George, Phyllis: Birth....................June 24
George Spelvin Day.......................Nov 15
George, Susan: Birth.....................July 26
Georgia
ABCs of Genealogy (Moultrie)..........Mar 18
Abraham Baldwin Agricultural College Homecoming (Tifton)................Mar 30
After the Revolution: Everyday Life America (Athens)............................Jan 28
Antebellem Jubilee (Stone Mountain)....Mar 31
Artists' Natl Juried Competition (St. Simon's Is)...................................Mar 25
Atlanta Marathon and Half Marathon (Atlanta)............................Nov 23
Augusta Futurity (Augusta)...............Jan 6
Barnesville Buggy Days (Barnesville)...Sept 16
Bluegrass Fest, Dahlonega (Dahlonega) June 22
Brown's Crossing Fall Craftsmen Fair (Milledgeville)........................Oct 21
Camellia Fest (Ft. Valley)................Feb 11
Chehaw Natl Indian Fest' (Albany)......May 19
Chieftains Run (Rome).....................Dec 2
Confederate Memorial Day.................Apr 26
Coosa River Christmas (Rome).........Nov 18
Descendants of Simon Gay Family Reunion (Moultrie).............................May 20
Design Drafting Assn Convention, American (Atlanta)..............................June 4
Fair of 1850 (Westville Village)...........Oct 12
Fall Fair (Hiawassee).....................Oct 6
Fall Fest of Arts and Crafts (Dalton).....Sept 9
Family Day Celebration (Dahlonega).....July 4
Georgia Renaissance Fall Fest (Fairburn)..Oct 7
Georgia Renaissance Spring Fest (Atlanta).............................Apr 22
Jefferson Davis Captured: Anniv........May 10
Keystone Club Conference (Atlanta)....Mar 29
Lasershow (Stone Mountain)............May 5
Legends of Christmas (Lookout Mountain)............................Dec 1
Macon Cherry Blossom Fest (Macon)...Mar 17
May Day (Lumpkin).......................May 1
More Than Meets the Eye (Atlanta)......Oct 28
Mossy Creek Barnyard Fest (Warner Robins)............................Apr 22
Mountain Fair (Hiawassee)...............Aug 2
Okefenokee Spring Fling (Waycross)......Apr 8
Peachtree Junior (Atlanta)..............June 3
Peachtree Road Race (Atlanta)..........July 4
Powers' Crossroads Country Fair/Art Fest (Newnan)...........................Sept 2
Prater's Mill Country Fair (Dalton).......May 13
Pro-Am Snipe Excursion and Hunt (Moultrie)............................Apr 1
Ratification Day..........................Jan 2
Scottish Tattoo/Fest/Highland Games (Stone Mountain)...........................Oct 19
Southern Cyclone: Anniv...............Aug 24
Spring Festival (Lumpkin)................Apr 6
Spring Music Fest (Hiawassee).........May 19
St. Simons Arts Fest (St. Simons Island) ..Oct 7
Steeplechase at Callaway Gardens (Pine Mountain)............................Nov 4
Summer Storyteller-in-Residence (Atlanta)...........................June 13
Swine Time (Climax)....................Nov 25
Technical Expo and Conf (Atlanta).....June 24
Tour of Southern Ghosts (Stone Mountain)............................Oct 14
Voting Rights Act Signed: Anniv..........Aug 6
Wildflower Fest of the Arts (Dahlonega) .May 20
World Horseshoe Tournament (Perry)...July 24
Wren's Nest Christmas Open House (Atlanta)............................Dec 10
WSB-TV Salute 2 America Parade (Atlanta)..............................July 4
Yellow Daisy Festival (Stone Mountain)...Sept 8
Georgia (Europe): Independence Day.....May 26
Georgia (Europe): Soviet Georgia Votes Independence..........................Mar 31
Gerard, Gil: Birth.........................Jan 23
Gere, Richard: Birth......................Aug 31
German. See also Octoberfest; Oktoberfest
German Fest (Milwaukee, WI).............July 28
German: Kunstfest (New Harmony, IN)....Sept 16
German: Laurel Herbstfest (Laurel, MT)...Sept 21
German Plebiscite: Anniv.................Aug 19
German: Schmeckfest (Freeman, SD).....Mar 30
German: Stiftungsfest (Young America, MN)..................................Aug 25
German Village Oktoberfest (Columbus, OH)...................................Sept 8
German: Wurstfest (New Braunfels, TX)....Nov 3
German-American Day 6604............Page 488
Germanfest (Fort Wayne, IN).............June 11
Germany
Bayreuther Festspiele (Bayreuth).......July 25
Buss und Bettag.......................Nov 22
Day of Unity............................June 17
Erntedankfest...........................Oct 1
Frankfurt Christmas Market (Frankfurt)..Nov 25
Kristallnacht: Anniv......................Nov 9
Munich Fasching Carnival................Jan 7
Reunification Anniv......................Oct 3
Totensonntag...........................Nov 26
Volkstrauertag..........................Nov 19
Waldchestag (Frankfurt).................June 6
War Crimes Trial Anniv..................Oct 18
Geronimo: Death Anniv...................Feb 17
Gerry, Elbridge: Birth Anniv..............July 17
Gershwin, George: Birth Anniv...........Sept 26
Gershwin, Ira: Birth Anniv.................Dec 6
Gertz, Jami: Birth........................Oct 28
Gerulaitis, Vitas: Birth...................July 26
Get a Different Name Day................Feb 13
Get Organized Week.......................Oct 1
Getty, Estelle: Birth......................July 25
Gettysburg Address, Lincoln's: Anniv......Nov 19
Gettysburg Address Memorial Ceremony (Gettysburg, PA)........................Nov 19
Gettysburg Outdoor Antique Show (Gettysburg, PA)........................May 20
Ghali, Boutros Boutros-: Birth............Nov 14
Ghana
Independence Day......................Mar 6
Republic Day.............................July 1
Revolution Day..........................Dec 31

☆ Chase's 1995 Calendar of Events ☆ Index

Ghost Tales Around Campfire (Washington, MS) .... Oct 27
Ghost Tales, Spring (Louisville, KY) .... May 12
Ghostley, Alice: Birth .... Aug 14
Ghosts, Fest of Hungry (China) .... July 27
GI Bill of Rights, 50th Anniv 6703 .... Page 488
Giamatti, Angelo Bartlett: Birth Anniv .... Apr 4
Giannini, Giancarlo: Birth .... Aug 1
Giant Dipper, Santa Cruz Beach, Anniv .... May 17
Gibb, Barry: Birth .... Sept 1
Gibbon, Edward: Birth Anniv .... Apr 27
Gibbs, Joe Jackson: Birth .... Nov 25
Gibbs, Marla: Birth .... June 14
Gibbs, Mifflin Wister: Birth Anniv .... Apr 28
Gibson, Bob (baseball): Birth .... Nov 9
Gibson, Bob (musician): Birth .... Nov 16
Gibson, Debbie: Birth .... Aug 31
Gibson, Henry: Birth .... Sept 21
Gibson, Kirk: Birth .... May 28
Gibson, Mel: Birth .... Jan 3
Gielgud, Sir John: Birth .... Apr 14
Gifford, Frank: Birth .... Aug 16
Gift Festival, Intl (Fairfield, PA) .... Nov 9
Gigli, Beniamino: Birth Anniv .... May 20
Gilbert, Alfred C.: Happy Birthday, A.C.! (Salem, OR) .... Feb 18
Gilbert, Melissa: Birth .... May 8
Gilbert, Sir William: Birth Anniv .... Nov 18
Gillars, Mildred E.: Death Anniv .... June 25
Gillespie, John B. "Dizzy": Birth Anniv .... Oct 21
Gilley, Mickey: Birth .... Mar 9
Gilliam, Armon: Birth .... May 28
Gilliam, Terry: Birth .... Nov 22
Gillikins of Oz Convention (Escanaba, MI) .... May 6
Gillis, Margaret: Birth .... July 9
Gilman, Dorothy .... June 25
Gilmore, Artis: Birth .... Sept 21
Gilmour, Dave: Birth .... Mar 6
Gingerbread House Competition (Lahaska, PA) .... Nov 15
Gingerbread Village (Omaha, NE) .... Nov 18
Gingerbread/Lace Christmas Celeb (Charlottesvil, VA) .... Dec 1
Ginsberg, Allen: Birth .... June 3
Ginza Holiday: Japanese Cultural Festival (Chicago, IL) .... Aug 18
Girl Scout Leader's Day .... Apr 22
Girl Scout Sabbath .... Mar 4
Girl Scout Sunday .... Mar 5
Girl Scout Week .... Mar 5
Girl Scouts Founding: Anniv .... Mar 12
Girls Incorporated Natl Conference (New York, NY) .... Apr 27
Girls Incorporated Week .... May 14
Giuliani, Rudolph: Birth .... May 28
Give-a-Sample Week, Natl .... Apr 23
Givenchy, Hubert de: Birth .... Feb 21
Givens, Robin: Birth .... Nov 27
Gladstone, William: Birth Anniv .... Dec 29
Glaser, Paul Michael: Birth .... Mar 25
Glass Fest, Elwood (Elwood, IN) .... Aug 17
Glaucoma Awareness Week, Natl .... Jan 22
Gleason, Jackie: Birth Anniv .... Feb 26
Glenn, John: Birth .... July 18
Glenn, Scott: Birth .... Jan 26
Gless, Sharon: Birth .... May 31
Global Fest (Waco, TX) .... Apr 7
Glover, Danny: Birth .... July 22
Goat Cookoff, World Chmpshp BBQ/Crafts Fair (Brady, TX) .... Sept 2
God Bless America First Performed: Anniv .... Nov 11
Godard, Jean Luc: Birth .... Dec 3
Goddard Day .... Mar 16
Goddard, Robert H.: Birth Anniv .... Oct 5
Godunov, Alexander: Birth .... Nov 28
Godwin, Mary Wollstonecraft: Birth Anniv .... Apr 27
Goebbels, Paul Josef: Birth Anniv .... Oct 29
Goethals, George W: Birth Anniv .... June 29
Goethe, Johann W.: Birth Anniv .... Aug 28
Gogol, Nikolai: Birth Anniv .... Mar 31
Gold Coast Art Fair (Chicago, IL) .... Aug 11
Gold Discovery, California: Anniv .... Jan 24
Gold, Missy: Birth .... July 14
Gold Rush Days (Wickenburg, AZ) .... Feb 10
Gold Rush: Whiskey Flat Days (Kernville, CA) .... Feb 17
Gold Star Mother's Day (Pres Proc) .... Sept 24
Gold Star Mother's Day 6590 .... Page 488
Gold, Tracey: Birth .... May 16
Goldblum, Jeff: Birth .... Oct 22
Golden Aspen Motorcycle Rally (Ruidoso, NM) .... Sept 20
Golden Leaf Festival (Mullins, SC) .... Sept 23
Golden Raintree Festival (New Harmony, IN) .... June 16
Golden Spike Driving: Anniv .... May 10
Golden Spike Reenactment (Promontory, UT) .... May 10
Goldman, William: Birth .... Aug 12
Goldschmidt, Neil: Birth .... June 16
Goldsmith, Oliver: Birth Anniv .... Nov 10
Goldwyn, Samuel: Birth Anniv .... Aug 17
Golf
  Bastille Day Moonlight Golf (Washington, DC) .... July 14
  Bell's Scottish Open .... July 5
  Bering Sea Ice Golf Classic (Nome, AK) .... Mar 18
  Big 10 Men's Golf Chmpshp (University Park, PA) .... May 12
  Big 10 Women's Chmpshp (Ann Arbor, MI) .... May 5
  Bob Hope Chrysler Classic (Bermuda Dunes, CA) .... Feb 15
  British Amateur Golf Chmpshp (Hoylake, England) .... June 5
  Chicago Golf Club: Anniv .... July 18
  Doral-Ryder Open (Miami, FL) .... Feb 25
  Golf Open Chmpshp (St. Andrews, Scotland) .... July 20
  GTE Suncoast Classic (Tampa, FL) .... Feb 6
  Honda Golf Classic (Weston Hills, FL) .... Mar 5
  Left-Handed Golfers Am Chmpshp (Palm Beach Garden, FL) .... June 27
  MCI Heritage Classic (Hilton Head Island, SC) .... Apr 10
  Mini Golf Natl Chmpshp (Jacksonville, FL) .... Nov 9
  NAIA Men's Golf Chmpshp .... May 23
  NCAA Div I Men's Golf Chmpshp (Columbus, OH) .... May 31
  NCAA Women's Golf Chmpshp (Wilmington, NC) .... May 24
  PGA Chmpshp (Pacific Palisades, CA) .... Aug 7
  PGA Seniors' Chmpshp (Palm Beach Gardens, FL) .... Apr 13
  Southwest Senior Chmpshp (Yuma, AZ) .... Jan 17
  Spalding Enshrinement Celebr Golf Tourn (Agawam, MA) .... May 8
  US Amateur Chmpshp (Newport and Middletown, RI) .... Aug 22
  US Amateur Public Links Chmpshp (Stow, MA) .... July 17
  US Girls' Junior Chmpshp (Longmeadow, MA) .... July 31
  US Junior Amateur Chmpship (Fargo, ND) .... July 25
  US Mid-Amateur Chmpshp (Owings Mill/ Pikesville, MD) .... Sept 16
  US Open Chmpshp (Southampton, NY) .... June 15
  US Senior Amateur Chmpshp (Hutchinson, KS) .... Sept 26
  US Senior Open (Golf) Championship (Bethesda, MD) .... June 29
  US Senior Women's Amateur Chmpshp (St. Paul, MN) .... Sept 13
  US Women's Am Public Links Chmpshp (Colts Neck, NJ) .... June 21
  US Women's Amateur Chmpshp (Brookline, MA) .... Aug 7
  US Women's Mid-Amateur Chmpshp (Manchester, MA) .... Sept 18
  US Women's Open Chmpshp (Colorado Springs, CO) .... July 13
  Walker Cup Golf Chmpshp (Porthcawl, Wales) .... Sept 9
  Yellowknife Midnight Classic (Yellowknife, NWT, Canada) .... June 16
  Zaharias, Babe: Birth Anniv .... June 26
Gompers, Samuel: Birth Anniv .... Jan 27
Good Friday .... Apr 14
Good Friday Bank Holiday (United Kingdom) .... Apr 14
Good Samaritan Involvement Day .... Mar 13
Good Will: Peace, Friendship, Good Will Week .... Oct 25
Goodall, Jane: Birth .... Mar 4
Gooden, Dwight: Birth .... Nov 16
Goodeve, Grant: Birth .... July 6
Goodman, Andrew: Civil Rights Workers Disappear .... June 21
Goodman, Andrew: Klan Members Convicted .... Oct 20
Goodman, John: Birth .... June 20
Goodson, Mark: Birth Anniv .... Jan 24
Goodwill Industries Week .... May 7
Goodwill Week, Canada-United States .... Apr 23
Goof-Off Day, Natl .... Mar 22
Goolagong, Evonne: Birth .... July 31
Goombay Festival, Miami/Bahamas .... June 2
Gopher Hill Fest (Ridgeland, SC) .... Oct 7
Gorbachev, Mikhail Sergeyvich: Birth .... Mar 2
Gordimer, Nadine: Birth .... Nov 20
Gordy, Berry, Jr: Birth .... Nov 28
Gore, Albert, Jr: Birth .... Mar 31
Gore, Lesley: Birth .... May 2
Goren, Charles: Birth .... Mar 4
Gorilla Born in Captivity, First: Anniv .... Dec 22
Gorme, Eydie: Birth .... Aug 16
Gortner, Marjoe: Birth .... Jan 14
Gorton, Slade: Birth .... Jan 8
Gospel. See Music
Gospel Jubilee, Fall (Live Oak, FL) .... Oct 19
Gospel Music (Nashville, TN) .... Apr 23
Gospel Singing, Biggest All Night (Bonifay, FL) .... July 1
Gossage, Richard "Goose": Birth .... July 5
Gossett, Lou, Jr: Birth .... May 27
Gottfried, Martin: Birth .... Oct 9
Gottlieb, Robert: Birth .... Apr 29
Gould, Chester: Birth Anniv .... Nov 20
Gould, Elliot: Birth .... Aug 29
Gould, Harold: Birth .... Dec 10
Goulet, Robert: Birth .... Nov 26
Gowdy, Curt: Birth .... July 31
Grace, Mark: Birth .... June 28
Grace, Princess: Death Anniv .... Sept 14
Grady, Don: Birth .... June 8
Graf, Steffi: Birth .... June 14
Graffman, Gary: Birth .... Oct 14
Graham, Bill: Birth .... Jan 8
Graham, Billy: Birth .... Nov 7
Graham, Bob: Birth .... Nov 9
Graham, Calvin ("Baby Vet"): Birth Anniv .... Apr 3
Graham, Katharine: Birth .... June 16
Graham, Martha: Birth Anniv .... May 11
Graham, Martha: Death Anniv .... Apr 1
Graham, Ronny: Birth .... Aug 26
Gramm, Phil: Birth .... July 8
Grand Canyon Natl Park: Anniv .... Feb 26
Grandparents Day (MA) .... Oct 1
Grandparents Day, Natl .... Sept 10
Grandy, Fred: Birth .... June 29
Grange Founding, Natl: Anniv .... Dec 4
Grange Week .... Apr 23
Grant, Harvey: Birth .... July 4
Grant, Horace: Birth .... July 4
Grant, Julia Dent: Birth Anniv .... Jan 26
Grant, U.S.: Battle of Chattanooga: Anniv .... Nov 24
Grant, U.S.: Birth Anniv .... Apr 27
Grant, U.S.: Peace Overtures in Civil War: Anniv .... Mar 2
Grape Jamboree, Geneva Area (Geneva, OH) .... Sept 23
Grape Stomp, Delicato Charity (Manteca, CA) .... Sept 3
Grass, Gunter: Birth .... Oct 16
Grassle, Karen: Birth .... Feb 25
Grassley, Charles Ernest: Birth .... Sept 17
Gratitude Day, World .... Sept 21
Graves, Peter: Birth .... Mar 18
Gray, Erin: Birth .... Jan 7
Gray, Hanna: Birth .... Oct 25
Gray, Linda: Birth .... Sept 12
Gray, Robert .... Apr 10
Grayson, Kathryn: Birth .... Feb 9
Great American Smokeout .... Nov 16
Great Americans, Hall of Fame for: Opening Anniv .... May 30
Great (Holy) Week .... Apr 9
Great NY State Snow Expo (Albany, NY) .... Nov 3
Great Outdoors Show, Northeast (Albany, NY) .... Mar 17
Grebenshikov, Boris: Birth .... Nov 27
Greco, Buddy: Birth .... Aug 14
Greco, Jose: Birth .... Dec 23
Greece
  Dumb Week .... Apr 2
  Independence Day .... Mar 25
  Midwife's Day or Women's Day .... Jan 8
  Ohi Day .... Oct 28
Greek Festival (Salt Lake City, UT) .... Sept 8
Greek Independence Day 6539 .... Page 488
Greek: St. Spyridon Greek Orthodox Fest (Sheboygan, WI) .... July 29
Greeley, Andrew: Birth .... Feb 5
Green, Adolph: Birth .... Dec 2
Green, Al: Birth .... Apr 13
Green, Hetty: Birth Anniv .... Nov 21
Green, Hubie: Birth .... Dec 28
Green, Mark: Birth .... Mar 15
Green Monday (Cyprus) .... Mar 6
Greene, Bob: Birth .... Mar 10
Greene, Shecky: Birth .... Apr 8
Greenfield, Jeff: Birth .... June 10
Greenpeace Founded: Anniv .... Sept 15

Ghost—Greenpeace

Index ☆ Chase's 1995 Calendar of Events ☆

Greenpeace (cont'd)—Handwriting

Greenpeace: Rainbow Warrior Sinking: Anniv ... July 10
Greensboro Sit-in Anniv ... Feb 1
Greenspan, Alan: Birth ... Mar 6
Greer, Germaine: Birth ... Jan 29
Gregg, Judd: Birth ... Feb 14
Gregorian Calendar Adjustment: Anniv ... Oct 4
Gregorian Calendar Day ... Feb 24
Gregory, Bettina: Birth ... June 4
Gregory, Cynthia: Birth ... July 8
Gregory, Dick: Birth ... Oct 12
Gregory, James: Birth ... Dec 23
Grenada
Carnival ... Aug 11
Carriacou Carnival (Carriacou) ... Feb 27
Carriacou Regatta ... Aug 5
Easter Regatta ... Apr 14
Emancipation Day ... Aug 7
Fisherman's Birthday (Gouyave) ... June 29
Game Fishing Tourn ... Jan 25
Independence Day ... Feb 7
Invasion by US: Anniv ... Oct 25
Grenadines and St. Vincent: Independence Day ... Oct 27
Gretzky, Wayne: Birth ... Jan 26
Grey, Jennifer: Birth ... Mar 26
Grey, Joel: Birth ... Apr 11
Grey, Zane: Birth Anniv ... Jan 31
Grieg, Edvard: Birth Anniv ... June 15
Grieg, Edvard: Oslo Music Fest (Norway) ... Aug 4
Grier, Roosevelt "Rosey": Birth ... July 19
Griese, Bob: Birth ... Feb 3
Griffey, George: Birth ... Nov 21
Griffin, John H: Birth Anniv ... June 16
Griffin, Merv: Birth ... July 6
Griffith, Andy, Show: First Broadcast Anniv ... Oct 3
Griffith, Andy: Birth ... June 1
Griffith, David (Lewelyn) Wark: Birth Anniv ... Jan 22
Griffith Joyner, Florence: Birth ... Dec 21
Griffith, Melanie: Birth ... Aug 9
Griffiths, Martha: Birth ... Jan 29
Griffiths, Martha: Speech Against Sex Discrimination ... Feb 8
Grimaldi, Joseph: Birth Anniv ... Dec 18
Grimes, Leonard Andrew: Birth Anniv ... Nov 9
Grimke, Sarah: Birth Anniv ... Nov 26
Grimm, Jacob: Birth Anniv ... Jan 4
Grimm, Wilhelm: Birth Anniv ... Feb 24
Grodin, Charles: Birth ... Apr 21
Groening, Matt: Birth ... Feb 15
Grolier, Jean: Death Anniv ... Oct 22
Gross, Mary: Birth ... Mar 25
Gross, Michael: Birth ... June 21
Grotius, Hugo: Birth Anniv ... Apr 10
Grouch Day, Natl ... Oct 15
Groundhog Day ... Feb 2
Groundhog Day (Punxsutawney, PA) ... Feb 2
Groundhog Day (Sun Prairie, WI) ... Feb 2
Groundhog Fest, Punxsutawney (Punxsutawney, PA) ... July 1
Groundhog Run (Kansas City, MO) ... Feb 5
Grubstake Days (Yucca Valley, CA) ... May 25
Guadalupe Hidalgo, Treaty of ... Feb 2
Guadalupe, Day of Our Lady of ... Dec 12
Guam
Discovery Day ... Mar 6
Lady of Camarin Day ... Dec 8
Liberation Day ... July 21
Magellan Day ... Mar 6
Spanish-American War Surrender to US: Anniv ... June 20
Guatemala
Armed Forces Day ... June 30
Banker's Day ... July 1
Independence Day ... Sept 15
Kite Fest of Santiago Sacatepequez ... Nov 1
Revolution Day ... Oct 20
Guavaween Celebration (Tampa, FL) ... Oct 28
Guccione, Bob: Birth ... Dec 17
Guernica Massacre: Anniv ... Apr 26
Guerrero, Pedro: Birth ... June 29
Guest, Christopher: Birth ... Feb 5
Guest, Edgar A.: Birth Anniv ... Aug 20
Guggenheim, Simon: Birth Anniv ... Dec 30
Guiding Light: Anniv ... June 30
Guidry, Ron: Birth ... Aug 28
Guilford Courthouse, Battle, Observance (Greensboro, NC) ... Mar 15
Guillaume, Robert: Birth ... Nov 30
Guillotin, Joseph: Birth Anniv ... May 28
Guinea: Independence Day ... Oct 2
Guinea-Bissau
Colonization Martyr's Day ... Aug 3
Independence Day ... Sept 24
National Heroes Day ... Jan 20
National Holiday ... Sept 12
Re-Adjustment Movement's Day ... Nov 14
Guisewite, Cathy Lee: Birth ... Sept 5
Guitar Month, Intl ... Apr 1
Gulf of Tonkin Resolution: Anniv ... Aug 7
Gumbel, Bryant: Birth ... Sept 29
Gumbo Fest (Bridge City, LA) ... Oct 13
Gun Shows, Sales, Competitions
Blackpowder Trade Fair/Gun Show (Albert Lea, MN) ... Feb 11
Governor's Invitational Firelock Match (Big Pool, MD) ... Sept 23
NCAA Men's/Women's Rifle Chmpshp ... Mar 2
Old-Time Barnegat Decoy/Gun Show (Tuckerton, NJ) ... Sept 23
Walsh Invitational Rifle Tourn (Cincinnati, OH) ... Nov 3
Gunn, Moses: Birth Anniv ... Oct 2
Guth, Alan: Birth ... Feb 27
Guthrie, Arlo: Birth ... July 10
Guthrie, Janet: Birth ... Mar 7
Guthrie, Woodie: Birth Anniv ... July 14
Guttenberg, Steve: Birth ... Aug 24
Guyana: National Day ... Feb 23
Gwathmey, Charles: Birth ... June 19
Gwinnett, Button: Death Anniv ... May 16
Gwynn, Tony: Birth ... May 9
Gwynne, Fred: Birth Anniv ... July 10
Gymnastics Chmpshp, Big Ten Men's ... Mar 25
Gymnastics Chmpshp, Big Ten Women's ... Mar 25
Gypsy Rose Lee (Rose L Hovick): Birth Anniv ... Feb 9

H

Habib, Philip: Birth ... Feb 25
Habitat Day, World (UN) ... Oct 2
Hackett, Buddy: Birth ... Aug 31
Hackman, Gene: Birth ... Jan 30
Hadassah: Anniv ... Feb 24
Haden, Pat: Birth ... Jan 23
Hagar, Sammy: Birth ... Oct 13
Haggard, Merle: Birth ... Apr 6
Hagler, Marvelous Marvin: Birth ... May 23
Hagman, Larry: Birth ... Sept 21
Hahn, Jessica: Birth ... July 7
Haid, Charles: Birth ... June 2
Haig, Alexander Meigs, Jr: Birth ... Dec 2
Hailey, Arthur: Birth ... Apr 5
Hair: Be Bald and Be Free Day ... Oct 14
Haiti
Ancestors' Day ... Jan 2
Army Day ... Nov 18
Discovery Day Anniv ... Dec 5
Flag and University Day ... May 18
Independence Day ... Jan 1
Military Deposes Aristide: Anniv ... Sept 30
Haitian Immigration 6569 ... Page 488
HAL (computer): Birth ... Jan 12
Halberstam, David: Birth ... Apr 10
Halcyon Days ... Dec 15
Hale, Barbara: Birth ... Apr 18
Hale, Nathan: Birth Anniv ... June 6
Haley, Alex Palmer: Birth Anniv ... Aug 11
Haley, Charles: Birth ... Jan 6
Halfway Point of 1995 ... July 2
Halfway Point of Autumn ... Nov 7
Halfway Point of Spring ... May 6
Halfway Point of Summer ... Aug 7
Halfway Point of Winter ... Feb 4
Halifax Buskerfest, Intl (Halifax, NS, Canada) ... Aug 3
Halifax Independence Day (NC) ... Apr 12
Hall, Anthony Michael: Birth ... Apr 14
Hall, Arsenio: Birth ... Feb 12
Hall, Daryl: Birth ... Oct 11
Hall, Deidre: Birth ... Oct 31
Hall, Donald A.: Birth ... Sept 20
Hall, Monty: Birth ... Aug 25
Hall of Fame for Great Americans: Opening Anniv ... May 30
Hall, Tom T.: Birth ... May 25
Halley, Edmund: Birth Anniv ... Nov 8
Halloween
Boo at the Zoo (Wheeling, WV) ... Oct 21
Devil's Night ... Oct 30
Fantasy Fest (Key West, FL) ... Oct 20
Ghost Tales Around Campfire (Washington, MS) ... Oct 27
Great Pumpkin Fest (Bedford, PA) ... Oct 21
Great Pumpkin Weekend (Chadds Ford, PA) ... Oct 28
Guavaween Celebration (Tampa, FL) ... Oct 28
Hallowe'en or All Hallow's Eve ... Oct 31
Halloween (Arapahoe, NE) ... Oct 31
Halloween Celebration (Shakopee, MN) ... Oct 28
Halloween Festival and Haunted Walk (Brooklyn, NY) ... Oct 28
Halloween Fright Night (Cornwall, PA) ... Oct 13
Halloween Frolic (Hiawatha, KS) ... Oct 31
Halloween Happening (Cincinnati, OH) ... Oct 28
Halloween Harvest of Horrors (Louisville, KY) ... Oct 20
Halloween Night Walks (Oglebay, WV) ... Oct 27
Halloween Parade (Toms River, NJ) ... Oct 31
Haunted Trail and Kiddie Trail (Huntington, WV) ... Oct 13
Literally Haunted House (New Albany, IN) ... Oct 6
Magic Day, Natl ... Oct 31
Magic Week, Natl ... Oct 25
Mill Avenue Masquerade Adventure (Tempe, AZ) ... Oct 31
Neewollah (Independence, KS) ... Oct 20
Pumpkin Corners Pumpkin Day (Stamford, NE) ... Oct 2
Samhain ... Oct 31
Sea Witch Halloween Fest (Rehoboth Beach/Dewey Beach, DE) ... Oct 28
Tour of Southern Ghosts (Stone Mountain, GA) ... Oct 14
Trick or Treat or Beggar's Night ... Oct 31
UNICEF Day, Natl (Pres Proc) ... Oct 31
War of the Worlds Broadcast: Anniv ... Oct 30
World's Largest Halloween Pty (Louisville, KY) ... Oct 19
Halsey, William "Bull": Birth Anniv ... Oct 30
Hamburger Celeb, Home of the (Seymour, WI) ... Aug 5
Hamburger Hill, Battle of: Anniversary ... May 11
Hamburger Month, Natl ... May 1
Hamel, Veronica: Birth ... Nov 20
Hamill, Mark: Birth ... Sept 25
Hamilton, Alexander: Birth Anniv ... Jan 11
Hamilton, Carrie: Birth ... Dec 5
Hamilton, George: Birth ... Aug 12
Hamilton, Scott: Birth ... Aug 28
Hamlin, Hannibal: Birth Anniv ... Aug 27
Hamlin, Harry: Birth ... Oct 30
Hamlisch, Marvin: Birth ... June 2
Hammett, Dahsiell: Birth Anniv ... May 27
Hammon, Jupiter: Birth Anniv ... Oct 17
Hampton, James: Birth ... July 9
Hampton, Lionel: Birth ... Apr 12
Hancock, Herbie: Birth ... Apr 12
Hancock, John: Birth Anniv ... Jan 23
Handbell Choir Concert/Tree Lighting (New Preston, CT) ... Dec 10
Handcar Races and Steam Fest (Felton, CA) ... July 8
Handel, George Frederick: Birth Anniv ... Feb 23
Handell Oratorio Soc Spring Performance (Rock Is, IL) ... Apr 8
Handicapped
Braille Institute Track/Field Olympics (Los Angeles, CA) ... May 13
Disability Employment Awareness Month, Natl (Pres Proc) ... Oct 1
Disability Employment Awareness Month, Natl 6605 ... Page 488
Disabled Persons, Asian/Pacific Decade (UN) ... Jan 1
Disabled Persons, Intl Day of ... Dec 3
Fishing Has No Boundaries (Bemidji, MN) ... June 24
Fishing Has No Boundaries (Eagle River, WI) ... June 2
Fishing Has No Boundaries (Hayward, WI) ... May 19
Fishing Has No Boundaries (Pierre, SD) ... June 17
Fishing Has No Boundaries (Sandusky, OH) ... July 8
Fishing Has No Boundaries (Thermopolis, WY) ... June 6
Goodwill Industries Week ... May 7
Keller, Helen: Birth Anniv ... June 27
More Than Meets the Eye (Albuquerque, NM) ... Jan 7
More Than Meets the Eye (Atlanta, GA) ... Oct 28
More Than Meets the Eye (Jacksonville, FL) ... Apr 15
Rehabilitation Week, Natl 6467 ... Page 488
Special Recreation Day ... July 9
Special Recreation Week ... July 9
Talking Book Week ... Apr 17
Therapeutic Recreation Week, Natl ... July 9
Wheelchair Tennis Tourn (Greensboro, NC) ... Sept 8
White Cane Safety Day 6481 ... Page 488
Handwriting Analysis Fdtn Conv, Amer (Vancouver, BC) ... July 5
Handwriting Analysis Week, Natl ... Jan 23

☆ Chase's 1995 Calendar of Events ☆ Index

Handwriting (cont'd)—Health & Welfare

Handwriting Day, Natl ... Jan 23
Handy, William C: Birth Anniv ... Nov 16
Hangover Awareness Day, Natl ... Feb 7
Hangul (Korea) ... Oct 9
Hanks, Tom: Birth ... July 9
Hansberry, Lorraine: Birth Anniv ... May 19
Hansom, Joseph: Birth Anniv ... Oct 26
Hanson, Howard: Birth Anniv ... Oct 28
Happy Day, I Want You to Be ... Mar 3
Happy Land Social Club Fire ... Mar 25
Harbaugh, Jim: Birth ... Dec 23
Harborfest (Nantucket Island, MA) ... June 16
Harborfest (Norfolk, VA) ... June 2
Hardaway, Tim: Birth ... Sept 12
Harding, Florence: Birth Anniv ... Aug 15
Harding, Warren G.: Birth Anniv ... Nov 2
Harding, Warren G.: First Radio Broadcast ... June 14
Hari, Mata: Execution Anniv ... Oct 15
Harkin, Thomas R.: Birth ... Nov 19
Harmon, Mark: Birth ... Sept 2
Harp Singing, Sacred (Huntsville, AL) ... May 6
Harper, Valerie: Birth ... Aug 22
Harrington, Pat, Jr: Birth ... Aug 13
Harris, Ed: Birth ... Nov 28
Harris, Emmylou: Birth ... Apr 2
Harris, Franco: Birth ... Mar 7
Harris, Joel Chandler: Birth Anniv ... Dec 9
Harris, Julie: Birth ... Dec 2
Harris, Lou: Birth ... Jan 6
Harris, Richard: Birth ... Oct 1
Harrison, Anna: Birth Anniv ... July 25
Harrison, Benjamin: Birth Anniv ... Aug 20
Harrison, Benjamin: Birth Celeb (Indianapolis, IN) ... Aug 20
Harrison, Benjamin, Holiday at Home (Indianapolis, IN) ... Nov 10
Harrison, Caroline L. S.: Birth Anniv ... Oct 1
Harrison, George: Birth ... Feb 25
Harrison, Gregory: Birth ... May 31
Harrison, Mary: Birth Anniv ... Apr 30
Harrison, Rex: Birth Anniv ... Mar 5
Harrison, William Henry: Birth Anniv ... Feb 9
Harry, Debbie: Birth ... July 11
Harry, Prince: Birth ... Sept 15
Harry's Hay Days (Grandview, MO) ... May 19
Hart, Charles: Birth ... June 3
Hart, Gary: Birth ... Nov 28
Hart, John: Death Anniv ... May 11
Hart, Mary: Birth ... Nov 8
Harte, Bret: Birth Anniv ... Aug 25
Hartley, Mariette: Birth ... June 21
Hartman, David: Birth ... May 19
Hartman, Lisa: Birth ... June 1
Hartscapades (Hartsville, SC) ... May 18
Haru-No-Yabuiri (Japan) ... Jan 16
Harvard, John: Day ... Nov 26
Harvard University Founding Anniv ... Oct 28
Harvest Faire, Roaring Camp (Felton, CA) ... Oct 14
Harvest Fest (Intercourse, PA) ... Sept 15
Harvest Fest, Hood River Valley (Hood River, OR) ... Oct 13
Harvest Moon ... Sept 8
Harvest Moon (Rosthern, Sask, Canada) ... July 14
Harvest Show (Philadelphia, PA) ... Sept 16
Harvey, Paul: Birth ... Sept 4
Harwell, Ernie: Birth ... Jan 25
Hasenfus, Eugene: Capture Anniv ... Oct 5
Hasselhoff, David: Birth ... July 17
Hat Blocking Machine, Patent Granted: Anniv ... Apr 3
Hat Week, Fall ... Sept 10
Hat Week, Straw ... Apr 2
Hatch, Orrin Grant: Birth ... Mar 22
Hatcher, Mickey: Birth ... Mar 15
Hate Week ... Apr 4
Hatfield, Mark O.: Birth ... July 12
Hatsume Fair (Delray Beach, FL) ... Feb 25
Hauer, Rutger: Birth ... Jan 23
Haunted House, Literally (New Albany, IN) ... Oct 6
Have a Bad Day Day ... Nov 19
Havel, Vaclav: Birth ... Oct 5
Havlicek, John: Birth ... Apr 8
Havoc, June: Birth ... Nov 8
Hawaii
  Admission Day Holiday ... Aug 18
  Aloha Classic Pro Windsurfing World Chmpshp (Maui) ... Oct 29
  Aloha Week Festivals (various) ... Sept 15
  Big Is Bodybuilding/Fitness Contest (Kailua-Kona) ... July 1
  Conservation Awareness Program (Waimanalo) ... Feb 1
  Dick Evans Memorial Cycle Race (Oahu) ... Aug 5
  Discoverers' Day ... Oct 9
  Hawaii Statehood: Anniv ... Aug 21
  Hawaiian Christmas Rough Water Swim (Honolulu) ... Dec 17
  Hula Bowl Game (Honolulu) ... Jan 22
  Humpback Whale Awareness Month (Honolulu) ... Feb 1
  King Kamehameha I Day ... June 11
  Kuhio Day ... Mar 26
  Lei Day ... May 1
  Merrie Monarch Festival (Hilo) ... Apr 16
  NFL Pro Bowl (Honolulu) ... Feb 5
  NFL Pro Bowl Beach Challenge (Oahu) ... Feb 3
  Outrigger's Hawaiian Ocean Fest (Oahu) ... Sept 8
  Queen Liliuokalani Deposed: Anniv ... Jan 17
  Tour o' Hawaii (Oahu) ... Nov 2
  Triple Crown Surfing: Hawaiian Pro Men's (Oahu) ... Nov 14
  Waikiki Roughwater Swim (Honolulu) ... Sept 4
Hawkins, Hersey: Birth ... Sept 29
Hawn, Goldie: Birth ... Nov 21
Hayden, Tom: Birth ... Dec 11
Haydn, Franz Joseph: Birth Anniv ... Mar 31
Hayes, Helen: Birth Anniv ... Oct 10
Hayes, Ira Hamilton: Birth Anniv ... Jan 12
Hayes, Isaac: Birth ... Aug 20
Hayes, Lucy: Birth Anniv ... Aug 28
Hayes, Rutherford B.: Birth Anniv ... Oct 4
Hays, Lew: Birth ... Nov 22
Hays, Robert: Birth ... July 24
Headache Fdtn Fundraiser, Natl (New York, NY) ... Apr 22
Health and Welfare
  Activity Professionals Conv, Natl Assn (Nashvil, TN) ... Mar 29
  Activity Professionals Day, Natl ... Jan 27
  Adoption Week, Natl ... Nov 19
  Adult Day Care Center Week, Natl ... Sept 17
  AIDS Awareness Month, Natl ... Oct 1
  AIDS Day, World ... Dec 1
  AIDS Day, World (UN) ... Dec 1
  AIDS Day, World 6632 ... Page 488
  Alcohol and Other Drug-Related Birth Defects Week ... May 14
  Alcohol Awareness Month, Natl ... Apr 1
  Alcohol-Free Weekend ... Apr 7
  Alcoholics Anonymous: Founding Anniv June 10
  Allergy/Asthma Awareness Month, Natl ... May 1
  Alzheimer's Assn Memory Walk ... Oct 7
  Alzheimer's Disease, Natl ... Nov 1
  AMD Awareness Month ... Feb 1
  Anesthetic First Used in Surgery (Anniv) Mar 30
  Anti-Boredom Month, Natl ... July 1
  Anxiety Month, Natl ... Apr 1
  ARC (Assn Retarded Citizens) Conv (Indianapolis, IN) ... Oct 26
  Arthritis Foundation National Telethon ... Apr 2
  Arthritis Month, Natl ... May 1
  Artificial Heart Transplant: Anniv ... Dec 2
  Awareness Day, Natl ... Apr 16
  Baby Safety Month ... Sept 1
  Bald is Beautiful Convention (Morehead, NC) ... Sept 8
  Be Kind to Humankind Week ... Aug 25
  Bed Check Month, Natl ... Sept 1
  Better Hearing and Speech Month, May Is May 1
  Better Sleep Month ... May 1
  Big Brothers/Big Sisters Appreciation Week ... Apr 23
  Biomedical Research Day, Natl 6616 ... Page 488
  Birth Control Clinic Opened, First: Anniv ... Oct 16
  Birthparents Week, Natl ... Apr 2
  Board and Care Recognition Month (New Jersey) ... Sept 1
  Boost Your Self-Esteem Month, Natl ... Feb 1
  Braille Rallye (Los Angeles, CA) ... July 16
  Brain Tumor Assn Symposium, Amer (Rosement, IL) ... June 9
  Brain Tumor Awareness Week ... June 5
  Breast Cancer Awareness Month, Natl 6600 ... Page 488
  Buildings Safety Week, Natl ... Apr 9
  Campaign for Healthier Babies Month ... Oct 1
  Cancer Control Month 6549, 6664 ... Page 488
  Cancer in the Sun Month ... June 1
  Career Nurse Assistants Day ... June 1
  Cataract Awareness Month ... Mar 1
  Child Health Day (Pres Proc) ... Oct 2
  Child Health Day 6602 ... Page 488
  Child Injury Prevention Week (New Hyde Park, NY) ... Sept 1
  Child Safety and Protection Month ... Nov 1
  Child/Adult Learning Disabil Intl Conf (Orlando, FL) ... Mar 1
  Childbirth Education Awareness Day, Intl ... July 22
  Childhood Cancer Month 6469 ... Page 488
  Children and Hospitals Week ... Mar 19
  Children's Dental Health Month, Natl ... Feb 1
  Children's Eye Health and Safety Month ... Sept 1
  Children's Miracle Network (Anaheim, CA) ... June 3
  Chiropractic Week, Natl ... Sept 17
  Cholesterol Education/Awareness Month, Natl ... Sept 1
  Christmas Seal Campaign ... Oct 1
  Chronic Fatigue Syndrome Awareness Month, Natl ... Mar 1
  Circle K Service Day, Intl ... Nov 11
  Clean Air Campaign, Amer Lung Assn ... May 1
  Colorectal Cancer Education/Awareness Month ... Dec 1
  Community Services Month (California) ... Apr 1
  Compliment-Your-Mirror Day ... July 3
  Condom Week, Natl ... Feb 14
  Consumer Information Month ... Oct 1
  Consumer Protection Week, Natl ... Oct 22
  Correct Posture Month ... May 1
  Cosmetology Month, Natl ... Oct 1
  Creative Child and Adult Month, Intl ... Nov 1
  Crime Prevention Week, Natl ... Feb 5
  Cuts and Curls for Charity Month ... Oct 1
  Death Busters Day, Natl ... May 26
  Dental Hygiene Week, Natl ... Oct 1
  Depression Educ and Awareness Month, Natl ... Oct 1
  Diabetes Alert, American ... Mar 28
  Diabetes Month, Natl ... Nov 1
  Diabetic Eye Disease Month ... Nov 1
  Dietary Mgrs' Pride in Food Service Week ... Feb 6
  Disability Employment Awareness Month, Natl ... Oct 1
  Doctors' Day ... Mar 30
  Dollars Against Diabetes (DAD's) Day ... June 16
  Domestic Violence Awareness Month 6619 ... Page 488
  Down Syndrome Awareness Month 6611 ... Page 488
  Drunk/Drugged Driving Prevention Month, Natl 6633 ... Page 488
  Easter Seal Telethon, Natl ... Mar 4
  Elizabeth Bower Day (OK) ... May 25
  Emergency Medical Services Week 6567 ... Page 488
  Employee Health and Fitness Day, Natl ... May 17
  Epilepsy Awareness Month, Natl ... Nov 1
  Evaluate Your Life Day ... Oct 19
  Eye Care Month, Natl ... Jan 1
  Family Sexuality Education Month, Natl ... Oct 1
  Fiber Focus Month, Natl ... Feb 1
  Foot Health Month ... Aug 1
  Fungal Infection Awareness Month ... May 1
  Get Organized Week ... Oct 1
  Goof-Off Day, Natl ... Mar 22
  Great American Pizza Bake ... Feb 7
  Great American Smokeout ... Nov 16
  Handwriting Analysis Week, Natl ... Jan 23
  Hangover Awareness Day, Natl ... Feb 7
  Have a Heart Day, Natl ... Feb 14
  Headache Fdtn Fundraiser, Natl (New York, NY) ... Apr 22
  Health Care Food Service Week, Natl ... Oct 1
  Health Education Week in New York State Feb 19
  Health Information Management Week, Natl 6620 ... Page 488
  Health Radiation Therapy Conf, ASRT ... Oct 8
  Healthy Exchanges Annual Potluck (DeWitt, IA) ... July 8
  Healthy Weight, Healthy Look Day ... Jan 19
  Healthy Weight Week ... Jan 15
  Heart Assn's Heartfest, American ... Oct 21
  Heart Month, American ... Feb 1
  Heart Month, American (Pres Proc) ... Feb 1
  Heart Month, American 6649 ... Page 488
  Heart Transplant, First: Anniv ... Dec 3
  Heart Walk, American ... Oct 7
  Help Someone See Week ... Mar 4
  Help Someone See Week Observance (Rockford, OH) ... Mar 1
  High Blood Pressure Education Month, Natl ... May 1
  Home Care Week, Natl ... Nov 26
  Home Care Week, Natl 6631 ... Page 488
  Hospice Month, Natl ... Nov 1
  Hospice Month, Natl 6630 ... Page 488
  Hospital Week, Natl ... May 7
  Hug-a-Bear Sunday ... Nov 5
  Hugging Day, Natl ... Jan 21
  Human Relations Month (New Hanover County, NC) ... Feb 1
  Hunger Awareness Month ... Oct 1

Index ☆ Chase's 1995 Calendar of Events ☆

Health & Welfare (cont'd)—Hirsch

Indy Sr Olympics (Indianapolis, IN)......June 7
Infant Immunization Week, Natl 5575..Page 488
Infertility Awareness Week, Natl..........Oct 1
January Diet Month......................Jan 1
Just Pray No: Worldwide Day of Prayer..Apr 22
Kazoo Day, Natl........................Jan 28
Kiss-and-Make-Up-Day.................Aug 25
Lefthanders Day, Intl...................Aug 13
Lister, Joseph: Birth Anniv..............Apr 5
Literacy Day, Intl (UN)..................Sept 8
Liver Awareness Month, Natl.............Oct 1
Lupus Awareness Month.................Oct 1
Lyme Disease Awareness Week 6571..Page 488
MADD's Natl Candlelight Vigil (Washington, D.C.)................................Dec 2
MADD's Project Red Ribbon.............Nov 1
Maintenance Personnel Day............Oct 16
Make A Difference Day.................Oct 21
Make Your Own Luck Day...............Aug 26
Mammography Day, Natl 6615........Page 488
March of Dimes Birth Defects Prevention Month................................Jan 1
Marfan Syndrome Awareness Month......Feb 1
May Is Better Hearing Month............May 1
Medical Lab Week, Natl..................Apr 9
Men's Health Week, Natl 6700.........Page 488
Mental Health Month....................May 1
Mental Illness Awareness Week...........Oct 1
Mental Illness Awareness Week 6603..Page 488
Mental Retardation Awareness Month.....Mar 1
Missing Children's Day, Natl............May 25
Moment of Frustration Scream Day, Intl..Oct 12
Month of the Young Child (MI)............Apr 1
More Herbs, Less Salt Day..............Aug 29
More Than Meets the Eye (Albuquerque, NM)..................................Jan 7
More Than Meets the Eye (Atlanta, GA)..Oct 28
More Than Meets the Eye (Jacksonville, FL)..................................Apr 15
Multicultural Substance Abuse Awareness Week................................May 1
Music for Life Week.....................July 2
Natl Glaucoma Awareness Week........Jan 22
Natl School Breakfast Week.............Mar 6
No News Is Good News Day............Sept 11
Nude Recreation Week..................July 3
Nurse Anesthetist Week, Natl Certified Registered...........................Aug 5
Nurses Week, Natl......................May 6
Nursing Conf on Pediatric Primary Care (Nashville, TN).......................Mar 14
Nursing Home Week, Natl..............May 14
Nutrition Month, Natl...................Mar 1
Occupational Therapy Month, Natl.......Apr 1
Operating Room Nurse Week...........Nov 12
Osteopathic Medicine Week, Natl.........Nov 5
Pan American Health Day (Pres Proc)....Dec 2
Pediatric and Adolescent AIDS Awareness Wk 6694...............................Page 488
Pest Control Month, Natl................June 1
Pet Peeve Week, Natl....................Oct 9
Pharmacists Day........................Jan 12
Physical Fitness and Sports Month, Natl..May 1
Physician Assistant Day.................Oct 6
Poison Prevention Awareness Month.....Mar 1
Poison Prevention Week, Natl...........Mar 19
Population Day, World (UN).............July 11
Preschool Immunization Week 6542...Page 488
Procrastination Week, Natl..............Mar 6
Project ACES Day.......................May 5
Project Red Ribbon......................Nov 1
Psychiatric Technicians Week, Natl.......Aug 6
PTA Drug and Alcohol Awareness Week, Natl..................................Mar 5
Purposeful Parenting Month, Natl........July 1
Quality of Life Expo (Los Angeles, CA)...Mar 2
Radiologic Technologists Annual Conf ASRT..............................June 18
Radiologic Technology Week, Natl........Nov 5
Rat-Catchers Day......................July 22
REACT Month, Intl.......................May 1
Red Cross: Founding Anniv..............May 21
Red Cross Month........................Mar 1
Red Cross Month, American (Pres Proc)..Mar 1
Red Kettle Kick Off Campaign for Salvation Army, Natl..........................Nov 24
Rehabilitation Week, Natl...............Sept 17
Rehabilitation Week, Natl 6596........Page 488
Relationship Renewal Day..............May 4
Relaxation Day, Natl...................Aug 15
Remembrance Day.....................Apr 15
Rid World Fad Diets/Gimmicks Day......Jan 17
Rosacea Awareness Month.............Mar 1
Safe Boating Week, Natl................May 20
St. Louis Variety Club Telethon (St. Louis, MO)..................................Mar 4
Salk Vaccine: Anniv.....................Apr 12
Salvation Army Week, Natl.............May 15
Sanctity of Human Life Day, Natl 6521.Page 488
Save Your Back Week, Natl.............Oct 22
Save Your Vision Week..................Mar 5
Save Your Vision Week (Pres Proc).......Mar 5
Save Your Vision Week 6532, 6652.....Page 488
School Counseling Week, Natl...........Feb 6
School Lunch Week, Natl.................Oct 9
School Lunch Week, Natl 6609........Page 488
School Nurse Day, Natl.................Jan 25
Scleroderma Awareness Week, Natl.....June 1
Self-Help Book Week, Natl..............May 7
Self-Improvement Month................Sept 1
Senior Health and Fitness Day, Natl.....May 31
Senior Smile Week, Natl...............May 14
Sexually Transmitted Diseases (STDs) Awareness Mnth, Natl..................Apr 1
Shamrocks for Dystrophy...............Mar 17
Sight-Saving Month, Natl................May 1
Sight-Saving Sabbath...................Jan 22
Singles Week, Natl.....................Sept 17
Sleep Safety Month......................Feb 1
Social Security Act: Anniv.............Aug 14
Social Work Month, Natl Professional.....Mar 1
South Shore Singles Founding Anniv (Norwell, MA).........................May 26
Special Recreation Day.................July 9
Special Recreation Week...............July 9
Speech-Lang-Hearing Conv, American (Cincinnati, OH).......................Nov 16
Spitting Prohibition: Anniv.............May 12
Star-athon Telethon for Cerebral Palsy...Jan 21
Stars Across Amer J. Lewis MDA Labor Day Wknd................................Sept 3
Stay Out of the Sun Day................July 3
Stress Awareness Month................Apr 1
Stress-Free Family Holidays Month, Natl..Dec 1
Stroke Awareness Month................May 1
Successful Antirabies Inoculation, First: Anniv.................................July 6
Sudden Infant Death Syndrome Awareness Month, Natl..........................Oct 1
Surgical Technologist Week, Natl.......May 21
Taste of Health (Miami, FL).............Apr 22
Test Tube Baby: Birthday..............July 25
Thanksgiving Canned Goods Drive (Baltimore, MD).................................Nov 1
Therapeutic Recreation Week, Natl.......July 9
Thyroid Disease Awareness Month.......Jan 1
Tinnitus Assn Advisors Mtg, American (New Orleans, CA).......................Sept 18
Tinnitus Seminar, Intl (Portland, OR)....July 12
Top Junk-Food Stories..................Jan 15
Tour de Cure...........................Apr 22
Trauma Awareness Month, Natl 6691..Page 488
Tuberous Sclerosis Awareness Month, Natl.................................May 1
United Nations Decade Against Drug Abuse................................Jan 1
Vegetarian Awareness Month............Oct 1
Volunteers of America Week, Natl........Mar 5
Volunteers Week, Intl...................June 1
WalkAmerica, March of Dimes...........Apr 29
Walktoberfest (Amer Diabetes Assn)...Sept 30
WCVL/WIMC Sr Citizen Health Fair/Expo (Crawfordsville).....................Sept 26
Week of the Young Child...............Apr 23
White Cane Safety Day (Pres Proc)......Oct 15
White, Ryan: (AIDS) Death Anniv.........Apr 8
Women's Heart Health Day..............Feb 1
Workers Memorial Day..................Apr 28
World Food Day (UN)...................Oct 16
World Health Day.......................Apr 7
World No-Tobacco Day.................May 31
World Red Cross Day....................May 8
World's Largest Garage Sale (Nationwide).........................May 14
Hearing. See Health
Hearing Month, May Is Better.............May 1
Hearn, Lafcadio: Birth Anniv............June 27
Hearns, Tommy: Birth...................Oct 18
Hearst, William R.: Birth Anniv...........Apr 29
Heart. See Health
Heart Day, Natl Have a..................Feb 14
Heart Month, American (Pres Proc)........Feb 1
Heatherton, Joey: Birth................Sept 14
Heckerling, Amy: Birth...................May 7
Hector, Johnny: Birth...................Nov 26
Hedin, Sven: Birth Anniv................Feb 19
Heffelfinger, William: First Pro Football Player...............................Nov 12
Heflin, Howell Thomas: Birth............June 19
Hefner, Christie: Birth....................Nov 8
Hefner, Hugh: Birth.......................Apr 9
Heiden, Eric: Birth......................June 14
Heine, Heinrich: Birth Anniv.............Dec 13
Heineken, Alfred: Birth...................Nov 4
Helen Keller Fest (Tuscumbia, AL)........June 23
Helldorado Days (Tombstone, AZ).........Oct 20
Heller, Joseph: Birth.....................May 1
Hello Day, World.........................Nov 21
Hell's Angels: Altamont Concert Anniv......Dec 6
Helmond, Katherine: Birth.................July 5
Helms, Edgar J.: Birth Anniv.............Jan 19
Helms, Jesse: Birth......................Oct 18
Helmsley, Leona: Birth...................July 4
Help Someone See Week..................Mar 4
Hemingway Days Fest (Key West, FL).....July 17
Hemingway, Ernest: Birth Anniv..........July 21
Hemingway, Margaux: Birth...............Feb 19
Hemingway, Mariel: Birth................Nov 22
Hemlock Day.............................Mar 5
Hemmings, David: Birth..................Nov 21
Hemsley, Sherman: Birth..................Feb 1
Henderson, Florence: Birth...............Feb 14
Henderson, Rickey: Birth................Dec 25
Henderson, Skitch: Birth.................Jan 27
Hendricks, Thomas A: Birth Anniv........Sept 17
Henley, Don: Birth......................July 22
Henner, Marilu: Birth.....................Apr 6
Henning, Doug: Birth.....................May 3
Henried, Paul: Death Anniv...............Mar 29
Henry, Patrick: Birth Anniv...............May 29
Henson, Jim: Birth Anniv.................Sept 24
Henson, Matthew A.: Birth Anniv...........Aug 8
Hentoff, Nat: Birth......................June 10
Hepburn, Audrey: Birth Anniv.............May 4
Hepburn, Katharine: Birth................May 12
Herb Fest (Mattoon, IL)..................Apr 29
Herb Week, Natl..........................May 8
Herbstfest, Laurel (Laurel, MT)...........Sept 21
Heritage Day Celebration (Leavenworth, KS)..................................June 10
Heritage Day Celebration (Wisconsin Dells, WI)..................................June 10
Heritage Day Fest (Lavallette, NJ)..........Sept 9
Heritage Fair/Spring Fest (Ashland, VA).....Apr 8
Heritage Fest (New Harmony, IN)..........Apr 17
Heritage Fest, Leacock (Orillia, Ont, Canada)...............................July 28
Heritage Fest, Piqua (Piqua, OH)..........Sept 2
Herman, Jerry: Birth.....................July 10
Hermit, Robert the: Death Anniv............Apr 1
Hernandez, Keith: Birth..................Oct 20
Hersh, Seymour: Birth.....................Apr 8
Hershey, Barbara: Birth...................Feb 5
Hershiser, Orel: Birth...................Sept 16
Herzog, Chaim: Birth....................Sept 17
Hess, Rudolf: Birth Anniv.................Apr 26
Hesseman, Howard: Birth.................Feb 27
Heston, Charlton: Birth....................Oct 4
Hewes, Joseph: Birth Anniv...............Jan 23
Hewitt, Don: Birth.......................Dec 14
Heyerdahl, Thor: Birth....................Oct 6
Hiawatha Pageant, Song of (Pipestone, MN)..................................July 21
Hickel, Walter J.: Birth...................Aug 18
Hickman, Darryl: Birth...................July 28
Hicks, Catherine: Birth....................Aug 6
High Blood Pressure Education Month, Natl May 1
High Wind Classic (Hood River, OR)......June 24
Highland. See Scottish
Hiking: Custer State Park Volksmarch (Custer, SD)..................................July 29
Hilgenberg, Jay: Birth...................Mar 21
Hill, Anita Faye: Birth...................July 30
Hill, Arthur: Birth........................Aug 1
Hill, George Roy: Birth..................Dec 20
Hill, Mildred J.: Birth Anniv.............June 27
Hill, Patty Smith: Birth Anniv.............Mar 27
Hillary, Sir Edmund: Birth................July 20
Hillerman, John: Birth...................Dec 20
Hillhaven Ho Ho Hotline (Seattle, WA).....Dec 11
Hillman, Chris: Birth......................Dec 4
Hills, Carla Anderson: Birth...............Jan 3
Himmler, Heinrich: Birth Anniv.............Oct 7
Hinamatsuri (Japan).......................Mar 3
Hindemith, Paul: Birth Anniv.............Nov 16
Hindenburg Disaster: Anniv................May 6
Hines, Gregory: Birth....................Feb 14
Hirohito Michi-no-Miya, Emperor: Birth Anniv..................................Apr 29
Hiroshima Day............................Aug 6
Hirsch, Elroy "Crazylegs": Birth..........June 17
Hirsch, Judd: Birth.......................Mar 15

☆ Chase's 1995 Calendar of Events ☆ Index

Hirt—Hospital

Hirt, Al: Birth .... Nov 7
Hispanic: Cinco de Mayo Fest (Portland, OR) .... May 4
Hispanic Heritage Fest (Miami, FL) .... Oct 1
Hispanic Heritage Month, Natl (Pres Proc) Sept 15
Hispanic Heritage Month, Natl 6592 .... Page 488
Historic Preservation Week, Natl .... May 14
Historic sites: Count Your Losses (Clinton, MD) .... Aug 31
Historical Assn Annual Meeting, Amer (Chicago, IL) .... Jan 5
Historical Festival, Fort Sisseton (Lake City, SD) .... June 3
History, American: After Revolution/Life (Athens, GA) .... Jan 28
History, Family, Day .... June 14
History Month, Afro-American .... Feb 1
History Month, American (MA) .... Feb 1
Hitchcock, Alfred: Birth Anniv .... Aug 13
Hite, Shere: Birth .... Nov 2
Hitler, Adolph: Birth Anniv .... Apr 20
Hmong New Year Celebration (Sheboygan, WI) .... Nov 25
Ho Chi Minh: Birth Anniv .... May 19
Ho, Don: Birth .... Aug 13
Hoban, James: Death Anniv .... Dec 8
Hoban, Russel: Birth .... Feb 4
Hobart, Garret A.: Birth Anniv .... June 3
Hobbit Day .... Sept 22
Hobbs, Lucy: First Woman Graduate Dental School: Anniv .... Feb 21
Hobby Industry Association Trade Show (Anaheim, CA) .... Jan 29
Hobby Show, Railroad and (Springfield, MA) Feb 4
Hobo Convention, Natl (Britt, IA) .... Aug 10
Hockey: First Black Pro Hockey Player: Anniv .... Nov 15
Hockey: Natl Coll Div I Men's Chmpshp (Providence, RI) .... Mar 30
Hockey Tourn, Co-ed Broomball (Sheboygan, WI) .... Feb 4
Hockney, David: Birth .... July 9
Hodag Muskie Challenge (Rhinelander, WI) .... Sept 16
Hodges, Craig: Birth .... June 27
Hoffa, James: Disappearance Anniv .... July 30
Hoffman, Dustin: Birth .... Aug 8
Hog Calling Contest (Baltimore, MD) .... July 12
Hog Day (Hillsborough, NC) .... June 17
Hogan, Ben: Birth .... Aug 13
Hogan, Hulk: Birth .... Aug 11
Hogarth, William: Birth Anniv .... Nov 10
Hogmanay (Scotland) .... Dec 31
Hoiby, Lee: Birth .... Feb 17
Holbrook, Hal: Birth .... Feb 17
Holiday Bowl Parade/Game (San Diego, CA) .... Dec 30
Holiday Day, Make Up Your Own .... Mar 26
Holiday, First US by Presidential Proclamation: Anniv .... Nov 26
Holladay, Wilhelmina: Birth .... Oct 10
Holland: Art and Antique Fair (Breda) .... Apr 8
Holland Fest, Cedar Grove (Cedar Grove, WI) .... July 28
Holland Tunnel: Anniv .... Nov 13
Holliday, Polly: Birth .... July 2
Holliger, Heinz: Birth .... May 21
Hollings, Ernest F.: Birth .... Jan 1
Holloway, Sterling: Birth Anniv .... Jan 4
Holly, Buddy: Birth Anniv .... Sept 7
Holly, Charles Hardin "Buddy": "Day the Music Died" .... Feb 3
Hollyhock Festival (Kyoto, Japan) .... May 15
Hollywood and Broadway, Move, to Lebanon, PA Day .... Feb 6
Holm, Celeste: Birth .... Apr 29
Holmes, Anna Marie: Birth .... Apr 17
Holmes, Oliver W.: Birth Anniv .... Aug 29
Holmes, Rupert: Birth .... Feb 24
Holmes, Sherlock: Baker St Irregulars' Dinner (New York, NY) .... Jan 6
Holocaust Day (Israel) .... Apr 27
Holocaust: Days of Remembrance .... Apr 23
Holocaust Museum, US: Anniv .... Apr 26
Holtz, Lou: Birth .... Jan 6
Holy Humor Month .... Apr 1
Holy Innocents Day .... Dec 28
Holy See: National Holiday .... Oct 22
Holy Thursday .... Apr 13
Holy Week .... Apr 9
Home Care Week, Natl .... Nov 26
Home Improvement Time .... Apr 1
Home: Natl Woodworking Month .... Apr 1
Home Run Record: Anniv .... Apr 8
Home Safety Week, Natl .... Apr 9
Home Shows and Tours
Bath: Natl Kitchen and Bath Month .... Oct 1
Candlelight Walk Tour/Chestertown (Chestertown, MD) .... Sept 16
Christmas Tour of Homes (Pella, IA) .... Dec 7
Festival of Houses and Gardens (Charleston, SC) .... Mar 16
Fredericksburg Day (Fredericksburg, VA) Apr 25
Great Northeast Home Show (Albany, NY) Feb 3
Home Improvement Show (West Allis, WI) . Feb 9
International Home Furnishings Market (High Point, NC) .... Apr 27
Maryland Home and Flower Show (Baltimore, MD) .... Mar 8
Miami/Ft. Lauderdale Home Show (Miami Beach, FL) .... May 27
Michigan Home & Garden Show (Pontiac, MI) .... Mar 2
Midsummer Night's Stroll (New Harmony, IN) .... July 20
Natchez Fall Pilgrimage (Natchez, MS) .... Oct 7
Natchez Spring Pilgrimage (Natchez, MS) . Mar 4
Original Massachusetts Home Show (West Springfield, MA) .... Mar 21
Plainfield Home and Product Show (Plainfield, CT) .... Mar 25
Showcase Trade and Home Show (Zion, IL) .... Apr 21
Snow Hill Heritage Weekend (Snow Hill, MD) .... Sept 22
Spring Pilgrimage to Antebellum Homes (Columbus, MS) .... Mar 28
Victorian Christmas Home Tour (Leadville, CO) .... Dec 2
WCUZ West Michigan Home/Garden Show (Grand Rapids, MI) .... Mar 2
Wright (Frank Lloyd) Plus (Oak Park, IL) . May 20
Wyandotte Heritage Days (Wyandotte, MI) .... Sept 8
Home-Based Business Week .... Oct 8
Homebrew Day, Natl .... May 6
Homeless Animals Day, Natl .... Aug 20
Homeless: Roof-Over-Head Day, Natl .... Dec 24
Homemakers of America FHA/Hero Week, Natl Future .... Feb 12
Homer, Louise: Birth .... Apr 28
Homer, Winslow: Birth Anniv .... Feb 24
Homer, Winslow: engravings exhibit (Austin, TX) .... Jan 23
Homesteader Harvest Festival (Brandon, SD) .... Sept 10
Hometown Reunion, Lakeside Historical Soc (Lakeside, CA) .... Oct 22
Honduras
Dia De Las Americas .... Apr 14
Francisco Morazan Holiday .... Oct 3
Independence Day .... Sept 15
Hone, William: Birth Anniv .... June 3
Honest Abe Awards: Natl Honesty Day .... Apr 30
Honesty Day, Natl .... Apr 30
Honey Month, Natl .... Sept 1
Honeymoon, First Balloon: Anniv .... June 20
Hong Kong
Birth of Lu Pan .... July 10
Birthday of Confucius (Observance) .... Sept 21
Birthday of Lord Buddha .... May 7
Chung Yeung Fest .... Nov 1
Hong Kong Food Fest .... Mar 4
International Dragon Boat Fest .... June 10
Lease Anniv .... June 9
Liberation Day .... Aug 28
Tin Hau Festival .... May 3
Hoodie-Hoo Day, Northern Hemisphere .... Feb 20
Hook, Sidney: Birth .... Dec 20
Hooker, John Lee: Birth .... Aug 22
Hooper, William: Birth Anniv .... June 17
Hoover, Herbert Clark: Birth Anniv .... Aug 10
Hoover, Herbert: Day (IA) .... Aug 13
Hoover, Lou H.: Birth Anniv .... Mar 29
Hope, Bob: Birth .... May 29
Hope, Bob: Chrysler Golf Classic (Bermuda Dunes, CA) .... Feb 15
Hopkins, Mark: Birth Anniv .... Feb 4
Hopkins, Stephen: Birth Anniv .... Mar 7
Hopkins, Telma: Birth .... Oct 28
Hoppe, Willie: Birth Anniv .... Oct 11
Hopper, Dennis: Birth .... May 17
Hopper, Grace: Birth Anniv .... Dec 9
Horn, Paul: Birth .... Mar 17
Hornacek, John: Birth .... May 3
Horne, Lena: Birth .... June 30
Horne, Marilyn: Birth .... Jan 16
Horowitz, Vladimir: Birth Anniv .... Oct 1
Horseless Carriage Day .... Oct 26
Horses
Augusta Futurity (Augusta, GA) .... Jan 6
Badminton Horse Trials (Badminton, England) .... May 4
Belmont Stakes (Belmont Park, NY) .... June 10
Belmont Stakes, First Running of .... June 19
Block House Steeplechase (Tryon, NC) .. Apr 22
Bonnie Blue Natl Horse Show (Lexington, VA) .... May 10
Breeders' Cup Chmpshp (Elmont, NY) ... Oct 28
Britt Draft Horse Show (Britt, IA) .... Sept 1
Butch Cassidy Outlaw Trail Ride (Vernal, UT) .... June 16
Central Montana Fair (Lewistown, MT) .. July 24
Cheltenham Gold Cup Mtg (Prestbury, England) .... Mar 14
Chincoteague Pony Penning (Chincoteague Island, VA) .... July 26
Church Point Buggy Festival (Church Point, LA) .... June 2
East Texas Walking/Racking Show (Marshall, TX) .... Mar 1
Epsom Derby (Surrey, England) .... June 7
Festival of the Horse (Oklahoma City, OK) .... Oct 20
Foxfield Races (Charlottesville, VA) .... Apr 29
Grand Natl Horseracing Meeting (Liverpool, England) .... Apr 6
Hoosier Horse Fair Expo (Indianapolis, IN) Apr 7
Horse and Carriage Weekend (Mystic, CT) .... July 14
Horse of the Year Show (London, England) .... Sept 26
Intl Gold Cup (The Plains, VA) .... Oct 21
Iron Horse Outraced by Horse: Anniv... Sept 18
Iroquois Steeplechase (Nashville, TN) .... May 13
Kentucky Derby (Louisville, KY) .... May 6
Kentucky Derby Fest (Louisville, KY) .... Apr 21
Kentucky State Fair (Louisville, KY) .... Aug 17
Masters (Calgary, AB, Canada) .... Sept 6
Middle of Nowhere Trail Ride, Ainsworth, NE) .... June 10
Miles City Jaycee Bucking Horse Sale (Miles City, MT) .... May 19
Mormon Trail Ride (Summit Valley, CA) .. May 19
Nationals (Calgary, AB, Canada) .... May 31
North American (Calgary, AB, Canada) .. July 5
Of Field and Forest (East Meredith, NY) .. June 4
Pony Express Festival (Hanover, KS) .... Aug 27
Pony Express, Inauguration of: Anniv .... Apr 3
Preakness Stakes (Baltimore, MD) .... May 20
Preakness Stakes: Anniv .... May 27
Provincial Ex (Brandon, Man, Canada) .. June 14
Racking Horse Spring Celebration (Decatur, AL) .... Apr 26
Racking World Celebration (Decatur, AL) .... Sept 22
Roanoke Valley Horse Show (Salem, VA) June 19
Rothmans July Handicap (Durban, South Africa) .... July 1
Royal Ascot (Ascot, England) .... June 13
Royal Red Arabian Horse Show (Regina, Canada) .... Aug 16
Royal Windsor Horse Show (Windsor, England) .... May 10
Steeplechase at Callaway Gardens (Pine Mountain, GA) .... Nov 4
Strawberry Hill Races (Richmond, VA) ... Apr 15
Sussex Farm and Horse Show (Augusta, NJ) .... Aug 4
Tennessee Walking Horse Natl Celeb (Shelbyville, TN) .... Aug 24
Upperville Colt/Horse Show (Warrenton, VA) .... June 6
Virginia Gold Cup (Warrenton, VA) .... May 6
Virginia State Horse Quarter House Div (Richmond, VA) .... Aug 4
Virginia State Horse Show All Breed Event (Richmond, VA) .... June 26
Winter Equestrian Fest (Tampa, FL) .... Mar 15
Winterthur Point-to-Point (Winterthur, DE) May 7
Horseshoe Tourn, Fall Classic (Fairmont, WV) .... Sept 23
Horseshoe Tourn, Head-of-the-Mon-River (Fairmont, WV) .... May 27
Horseshoe Tourn, Las Vegas Open (Las Vegas, NV) .... Jan 8
Horseshoe Tourn, World (Perry, GA) .... July 24
Hoskins, Bob: Birth .... Oct 26
Hospice Month, Natl .... Nov 1
Hospice Month, Natl 6630 .... Page 488
Hospital Week, Natl .... May 7
Hospitals Week, Children and .... Mar 19

Index ☆ Chase's 1995 Calendar of Events ☆

Hostage Released, Last American: Anniv	Dec 4
Hot and Spicy Food Intl Day	Jan 21
Hot Dog Month, Natl	July 1
Hot Dog: Natl Kraut and Frankfurter Week	Feb 9
Hot Dog Night (Luverne, MN)	July 20
Houdini, Harry: Birth Anniv	Mar 24
Houdini, Harry: Death Anniversary	Oct 31
Hough, Charles: Birth	Jan 5
House of Representatives, First Black Serves in: Anniv	Dec 12
House of Representatives, First Brawl: Anniv	Jan 30
House of Representatives, US: First Quorum Anniv	Apr 1
House Tours. See Home Shows and Tours	
Housekeepers Week, Natl	Sept 10
Housework Day, No	Apr 7
Housing Week, Natl 6702	Page 488
Houston, James: Birth	June 12
Houston, Whitney: Birth	Aug 9
Hovick, Rose L (Gypsy Rose Lee): Birth Anniv	Feb 8
Howard, Ken: Birth	Mar 28
Howard, Leslie: Death Anniv	June 1
Howard, Ron: Birth	Mar 1
Howard, Susan: Birth	Jan 28
Howe, Elias: Birth Anniv	July 9
Howe, Gordie: Birth	Mar 31
Howell, C. Thomas: Birth	Dec 7
Howes, Sally Ann: Birth	July 20
Hoyle, Edmund: Death Anniv	Aug 29
Hoyt, LaMarr: Birth	Jan 1
Hubbard, Elbert: Birth Anniv	June 19
Hubbard, L. Ron: Birth	Mar 13
Hubble, Edwin Powell: Birth Anniv	Nov 20
Hubble Space Telescope Deployed: Space Milestone	Apr 25
Huckleberry Finn Raft Race (New Preston, CT)	Sept 4
Huff 'n Puff Hot Air Balloon Rally (Topeka, KS)	Sept 8
Hug an Australian Day	Apr 26
Hug Holiday	June 15
Hug-a-Bear Sunday	Nov 5
Hugging Day, Natl	Jan 21
Hughes, Bernard: Birth	July 16
Hughes, Charles E.: Birth Anniv	Apr 11
Hughes, Howard: Birth Anniv	Dec 24
Hugo, Victor: Birth Anniv	Feb 26
Hula Bowl, Game (Honolulu, HI)	Jan 22
Hulce, Thomas: Birth	Dec 6
Hull, Bobby: Birth	Jan 3
Hull, Brett: Birth	Aug 9
Hull, Cordell: Birth Anniv	Oct 2
Hull, John: First Mint in America: Anniv	June 10
Human Beings Week, Universal	Mar 1
Human Relations. See also United Nations	
Afro-American History Month	Feb 1
Amateur Radio Month, Intl	Apr 1
Be an Angel Day	Aug 22
Be an Encourager® Day	Feb 1
Be Kind to Humankind Week	Aug 25
Black Solidarity Day (New Haven, CT)	Nov 6
Blame Someone Else Day	Jan 13
Brotherhood/Sisterhood Week	Feb 19
Canada–US Goodwill Week	Apr 23
Cave Man Never Days	July 25
Celebration of Love Week	Feb 12
Celebration of Peace Day (Concord, MA)	Aug 6
Children's Art Exhibit, Intl (San Francisco, CA)	May 15
Circle K Service Day, Intl	Nov 11
Coming Out Day, Natl	Oct 11
Communication Week, World	Nov 1
Courtesy Month, Natl	Sept 1
Cross Culture Day, Intl	Sept 14
Cultural Diversity Day	Feb 16
Day of Fellowship and Hope, Natl 6525	Page 488
Dream Day	Aug 28
Emancipation Proclamation: Anniv	Jan 1
First Joint Siberian-American Musical: Anniv	Dec 20
First National Convention for Blacks: Anniv	Sept 15
First US Breach of Promise Suit: Anniv	June 14
Forgiveness Day, Natl	June 24
Freedom Day, Natl (Pres Proc)	Feb 1
Freedom Forum Asian Center Opening: Anniv	Jan 17
Freedom of Information Day	Mar 16
Freedom Week	July 4
Friendship Day	Aug 6
Friendship Week, Intl	Feb 19
Good Neighbor Day, Natl	Sept 24
Good Samaritan Involvement Day	Mar 13
Greensboro Sit-in Anniv	Feb 1
Help Someone See Week Observance (Rockford, OH)	Mar 1
Holocaust: Days of Remembrance	Apr 23
Honesty Day, Natl	Apr 30
Hug an Australian Day	Apr 26
Hug Holiday	June 15
Human Relations Day (Flint, MI)	Jan 16
Human Relations Month (New Hanover County, NC)	Feb 1
Human Resources Month	Jan 1
Human Rights Day (Pres Proc)	Dec 10
Human Rights Day (UN)	Dec 10
Human Rights Day and Wk/Bill of Rights Day 6637	Page 488
Human Rights Month, Universal	Dec 1
Human Rights Week (Pres Proc)	Dec 10
Humanitarian Day	Jan 15
I Want You to Be Happy Day	Mar 3
Interfaith Fellowship Anniv Celeb (New York, NY)	Sept 10
International Week (College Station, TX)	Feb 27
Jefferson Awards (Washington, DC)	June 21
Jewish Heritage Week	Apr 30
Joygerm Day, Natl	Jan 8
Kiss-and-Make-Up-Day	Aug 25
Language Week, Intl	Dec 15
League of Nations: Anniv	Jan 10
Loving v Virginia: Anniv	June 12
Lumpy Rug Day	May 3
Man Watchers' Compliment Week	July 3
Meredith (James) Enrolls at Ole Miss: Anniv	Sept 30
Multicultural Communication Month	Apr 1
Nude Recreation Week	July 3
Organization Development Info Exchange (Wms Bay, WI)	May 23
Peace and Music Festival, Chicago (Chicago, IL)	Aug 5
Peace Corps: Birth	Oct 14
Peace Corps: Founding Anniv	Mar 1
Peace, Friendship and Good Will Week	Oct 25
Pen-Friends Week Intl	May 1
Pet Peeve Week, Natl	Oct 9
Poverty, Intl Day for Eradication	Oct 17
Race Relations Day	Feb 14
Race Unity Day	June 11
Ralph Bunche Awarded Nobel Peace Prize: Anniv	Dec 10
Red Cross Day, World	May 8
Red Cross: Founding Anniv	May 21
Red Kettle Kick Off Campaign for Salvation Army, Natl	Nov 24
Refugee Day 6499	Page 488
Relationship Renewal Day	May 21
Religion Day, World	Jan 15
Religious Freedom Week	Sept 24
Remembrance Day	Apr 15
Research/Study Team Nonviolent Change (Wms Bay, WI)	May 21
Respect Canada Day	July 15
Romance Awareness Month	Aug 1
Roof-Over-Your-Head Day, Natl	Dec 24
Rosa Parks Day	Dec 1
Salvation Army Advisory Organizations Sunday	May 21
Salvation Army in USA: Anniv	Mar 10
Service Day, Natl	Nov 4
Smile Week, Natl	Aug 7
St. Louis Race Riots: Anniv	July 2
Stop the Violence Day, Natl	Nov 22
Swap Ideas Day	Sept 10
Thank You Day, Natl	Jan 11
True Confessions Day	Mar 15
Universal Children's Week	Oct 1
Universal Family Week	May 7
Universal Human Beings Week	Mar 1
Victims of Violence Holy Day	Apr 4
Volunteer Week, Natl	Apr 23
Volunteers of America Week, Natl	Mar 5
Volunteers Week, Intl	June 1
World Community Day	Nov 3
World Day of Prayer	Mar 3
World Gratitude Day	Sept 21
World Hello Day	Nov 21
World Peace Meditation, Annual	Dec 31
Worldwide Kiwanis Week	Jan 15
Year for Tolerance, UN	
Humbard, Rex: Birth	Aug 13
Humbug Day	Dec 21
Hummel, Sister Maria Innocentia: Birth Anniv	May 21
Humor	
Alascattalo Day (Anchorage, AK)	Nov 19
Creativity, Positive Power (Saratoga Spgs, NY)	Apr 28
Humor Month, Holy	Apr 1
Humor Month, Natl	Apr 1
Humorists Are Artists Month (HAAM)	Mar 1
Kurtzman, Harvey (Mad magazine): Birth Anniv	Oct 3
Love May Make/Laughter Keeps	Feb 8
O. Henry Pun-Off (Austin, TX)	May 7
Rough River Humor Fest (Falls of Rough, KY)	Mar 31
Someday We'll Laugh About This Week	Jan 8
Humperdinck, Engelbert: Birth	May 3
Humphrey, Hubert: Birth Anniv	May 27
Humphrey, Muriel Fay Buck: Birth	Feb 20
Hundertwasser, Friedrich: Birth	Dec 15
Hungary	
Anniversary of 1956 Revolution	Oct 28
Hungary Declares Its Independence: Anniv	Oct 23
National Day	Aug 20
Hunger Awareness Month	Oct 1
Hungerford, Margaret Wolf: Duchess Who Wasn't Day	Aug 27
Hunt, Guy: Birth	June 17
Hunt, Helen: Birth	June 15
Hunt, James B.: Birth	May 16
Hunt, Linda: Birth	Apr 2
Hunt, Nelson Bunker: Birth	Feb 22
Hunt Weekend, Thanksgiving (Cismont, VA)	Nov 23
Hunter, Holly: Birth	Mar 20
Hunter, Kim: Birth	Nov 12
Hunter, Tab: Birth	July 11
Hunter-Gault, Charlayne	Feb 27
Hunter's Moon	Oct 8
Hunters' Moon, Feast of (Lafayette, IN)	Oct 7
Hunting and Fishing Day, Natl (Pres Proc)	Sept 23
Hunting: Cheltenham Gold Cup Mtg (Prestbury, England)	Mar 14
Hunting: Southeastern Sports-A-Rama (Chattanooga, TN)	Aug 11
Hurricane, Bayou: Anniv	Oct 1
Hurricane Hugo Hits American Coast: Anniv	Sept 21
Hurricane Season, Atlantic, Caribbean and Gulf	June 1
Hurricane Season, Central Pacific	June 1
Hurricane Season, Eastern Pacific	May 15
Hurricane Season, Western Pacific	Jan 1
Hurricane: Southern Cyclone: Anniv	Aug 24
Hurricane Supplication Day (Virgin Islands)	July 24
Hurricane Thanksgiving Day (Virgin Islands)	Oct 16
Hurston, Zora: Zora Neale Hurston Fest (Eatonville, FL)	Jan 26
Hurt, John: Birth	Jan 22
Hurt, William: Birth	Mar 20
Husky, Ferlin: Birth	Dec 3
Hussein: King of Jordan, Birth	Nov 14
Hussey, Olivia: Birth	Apr 17
Huston, Anjelica: Birth	July 8
HutchFest (Hutchinson, KS)	June 28
Hutchison, Kay Bailey: Birth	July 22
Hutin, Mme Francisquy: Ballet Introduced to the US	Feb 7
Hutton, Betty: Birth	Feb 26
Hutton, Lauren: Birth	Nov 17
Hutton, Timothy: Birth	Aug 16
Huxley, Aldous: Birth Anniv	July 26
Hynde, Chrissie: Birth	Sept 7

I

I Want You to Be Happy Day	Mar 3
Iacocca, Lee A: Birth	Oct 15
Ibsen, Henrik: Birth Anniv	Mar 20
Ice Break-up: Nenana Tripod Raising Fest (Nenana, AK)	Feb 25
Ice Cream Cone: Birth	Sept 22
Ice Cream Day, Natl	July 16
Ice Cream Fest, Old-Fashioned (Utica, OH)	May 27
Ice Cream Fest, Old-Fashioned (Wilmington, DE)	July 8
Ice Cream Flavor Day and Contest, Creative (Chicago, IL)	July 1
Ice Cream Month, Natl	July 1
Ice Cream Social (Indianapolis, IN)	July 4
Ice Cream Social (Rockford, OH)	Aug 26
Ice Cream Social, Old-Fashioned (Fond du Lac, WI)	Aug 6
Ice Fest, Tip-Up Town USA (Houghton Lake, MI)	Jan 21
Icebox Days (International Falls, MN)	Jan 20

☆ Chase's 1995 Calendar of Events ☆ Index

Iced Tea—Indiana

Iced Tea Month, Natl .... **June 1**
Iceland
Bun Day .... **Feb 27**
Bursting Day .... **Feb 28**
Independence Day .... **June 17**
Leif Erikson Day .... **Oct 9**
Shop and Office Workers' Holiday .... **Aug 7**
University Students' Celebration .... **Dec 1**
Icescape (Appleton, WI) .... **Feb 17**
Idaho
Admission Day .... **July 3**
Boise River Festival (Boise) .... **June 22**
Lionel Hampton Jazz Fest (Moscow) .... **Feb 22**
NAIA Men's Division II Basketball Chmpshp (Nampa) .... **Mar 9**
Secret Pal Day (Hayden Lake) .... **Jan 14**
Simplot Games (Pocatello) .... **Feb 17**
Ides Of March .... **Mar 15**
Idle, Eric: Birth .... **Mar 29**
Idol, Billy: Birth .... **Nov 30**
Iglesias, Julio: Birth .... **Sept 23**
Ilitch, Mike: Birth .... **July 20**
Illinois
Admission Day .... **Dec 3**
American Booksellers Assn Exhib/Conv (Chicago) .... **June 3**
American Brain Tumor Assn Symposium (Rosement) .... **June 9**
American Historical Assn An Mtg (Chicago) .... **Jan 5**
American Indian Celebration (Belleville, IL) .... **May 11**
American Soc of Assn Executives Mgmt Conf (Chicago) .... **Dec 3**
Animal's Messiah (Chicago) .... **Dec 10**
Anti-Cruelty Society Dogwash (Chicago) .... **July 15**
Arts/Quincy Fest (Quincy) .... **Sept 17**
Assn Jewish Libraries (AJL) Conv (Chicago) .... **June 18**
Assn of Amer Geographers Annual Mtg (Chicago) .... **Mar 14**
Bagelfest (Mattoon) .... **July 28**
Big Ten Men's Indoor Track/Field Chmpshp (Champaign) .... **Feb 24**
Bon Odori Festival of the Lanterns (Chicago) .... **July 8**
Brach's Holiday Christmas Parade (Chicago) .... **Nov 25**
Bud Billiken Parade (Chicago) .... **Aug 12**
Burgoo Fest (Utica) .... **Oct 8**
Cedarhurst Craft Fair (Mt Vernon) .... **Sept 9**
Celebration of Chocolate (Oregon) .... **Feb 1**
Celebration of Our Lady of the Snows (Belleville) .... **July 28**
Center City Church Conf (Aurora) .... **Oct 25**
Central States Threshermen's Reunion (Pontiac) .... **Aug 30**
Chicago Day at World's Fair: Anniv (Chicago) .... **May 1**
Chicago Golf Club: Anniv (Wheaton) .... **July 18**
Chicago Intl Art Expo (Chicago) .... **May 12**
Chicago Jazz Fest (Chicago) .... **Sept 1**
Chicago Marathon (Chicago) .... **Oct 29**
Chicago Peace and Music Festival (Chicago) .... **Aug 5**
Chicago's Freak Gas Explosion (Chicago) .... **Jan 17**
Chicago's Windy City Jitterbug Dance (Franklin Pk) .... **Jan 7**
Chicagoland Gems and Minerals Assn Show (Wheaton) .... **May 27**
Chocolate Harvest (Mattoon) .... **Oct 7**
Civil War Days (Rockford) .... **Aug 12**
Clarence Darrow Death Commemoration (Chicago) .... **Mar 13**
Clock Tower Jazz Fest (Rockford) .... **Jan 20**
Creative Ice Cream Flavor Day and Contest (Chicago) .... **July 1**
Decatur Celebration (Decatur) .... **Aug 4**
Dixon Petunia Fest (Dixon) .... **June 30**
Family Folklore (Skokie) .... **Apr 22**
Festival of Trees (Rock Island) .... **Nov 18**
Freedom Fest (Mahomet) .... **July 4**
Galesburg Chocolate Fest (Galesburg) . . **Feb 11**
Galesburg Railroad Days (Galesburg) . . **June 24**
Gay and Lesbian Pride Parade/Rally (Chicago) .... **June 25**
Ginza Holiday: Japanese Cultural Festival (Chicago) .... **Aug 18**
Gold Coast Art Fair (Chicago) .... **Aug 11**
Good, the Bad and the Cuddly (Chicago) **May 13**
Grand Detour Art Fair (Grand Detour) . . **Sept 17**
Great Cardboard Boat Regatta (Carbondale) .... **Apr 29**
Great Cardboard Boat Regatta (Rock Island) .... **July 1**
Great Chicago Flood: Anniv .... **Apr 13**
Handel Oratorio Soc Spring Performance (Rock Is) .... **Apr 8**
Handel's Messiah (Rock Island) .... **Dec 15**
Hard Rock Cafe 5K (Chicago) .... **June 18**
Harvard Milk Day Festival (Harvard) ..... **June 2**
Herb Fest (Mattoon) .... **Apr 29**
Historic Farm Days (Penfield) .... **July 7**
Intl Wizard of Oz Club Convention (Rosemont) .... **June 16**
Jubilee Days Fest (Zion) .... **Sept 2**
Knox County Scenic Drive (Galesburg) . . . **Oct 7**
Louis Armstrong:Cultural Legacy (Chicago) .... **Apr 15**
Manufacturing Week, Natl (Chicago) .... **Mar 13**
Marigold Fest (Pekin) .... **Sept 8**
Marriage Celebration (Belleville) .... **Sept 10**
Marriage Enrichment Weekend (Belleville) .... **Nov 10**
Midwest Doll Show and Sale (Zion) ...... **May 6**
Mobius Awards (Chicago) .... **Feb 2**
Morton Pumpkin Fest (Morton) .... **Sept 13**
Mud Volleyball Tourn (Rockford) .... **Aug 5**
Multi-Cultural Easter Egg Display (Belleville) .... **Apr 9**
Multimedia Schools Conference .... **Oct 31**
NJCAA Div II Natl Basketball Finals (Danville) .... **Mar 16**
Northern Illinois Farm Show (Rockford) . . . **Jan 3**
Old Town Art Fair Chicken Teriyaki Lunch (Chicago) .... **June 10**
Oldsmobile Balloon Classic (Danville) . . . **June 9**
On the Waterfront (Rockford) .... **Sept 1**
Pec Thing (Pecatonica) .... **May 20**
Pet Parade (LaGrange) .... **June 3**
Prime Beef Festival, Wyatt Earp Days (Monmouth) .... **Sept 6**
Printers Row Book Fair (Chicago) ...... **June 17**
Quincy Preserves Christmas Tour (Quincy) .... **Dec 10**
Rockford Storytelling Fest (Rockford) . . . **Sept 15**
Rocvale Balloon Rally (Rockford) .... **Sept 9**
Sa'Fest: Cartoon Art Compet (Chicago) . . **Oct 1**
St. Charles Scarecrow Fest (St. Charles) . **Oct 14**
Scientific Matls Mgrs Conf/Show, Natl Assn (Chicago) .... **July 24**
Senior Festival and Expo (Rockford) .... **May 19**
Sez Who? Fourplay March Madness (Chicago) .... **Mar 12**
Showcase Trade and Home Show (Zion) . **Apr 21**
Stearman Fly-In, Natl (Galesburg) .... **Sept 6**
Superman Celebration (Metropolis) .... **June 15**
Sweetcorn Fest, Natl (Hoopeston, IL) . . . **Aug 31**
Turtle Races (Danville) .... **June 3**
US Industrial Film/Video Awards (Chicago) .... **May 31**
Way of Lights (Belleville) .... **Nov 24**
Who's in Charge: Workers/Managers (Elmhurst) .... **July 22**
Wild West Days (Rockford) .... **May 20**
Windy City Classic Conv (Schaumburg) . **Sept 29**
World Chmpshp Old-Time Piano Playing Contest (Decatur) .... **May 26**
World Freefall Convention (Quincy) ...... **Aug 4**
Wright (Frank Lloyd) Plus (Oak Park) .... **May 20**
WROK/WZOK Annual Chili Shoot-Out (Rockford) .... **Oct 7**
WROK/WZOK Anything that Floats Raft Race (Rockford) .... **July 4**
Wyatt Earp Birthday Celebration (Monmouth) .... **Mar 18**
Wyatt Earp Christmas Walk (Monmouth) . **Sept 9**
Wyatt Earp Homecoming Western Days (Monmouth) .... **Aug 5**
Wyatt Earp OK Corral Anniv (Monmouth) **Oct 21**
Youth Sing Praise Performance (Belleville) .... **June 24**
Illustrators, Soc of: Annual Exhib (New York, NY) .... **Feb 11**
Imbolc .... **Feb 2**
Immaculate Conception, Feast Of .... **Dec 8**
Immigrant/Nonimmigrant Suspension of Status/ Nigeria 6636 .... **Page 488**
Immigration: Ellis Island Opened: Anniv ..... **Jan 1**
Impeachment: Vote to Impeach Andrew Johnson: Anniv .... **Feb 24**
Inauguration Day, Old .... **Mar 4**
Incandescent Lamp Demonstarted: Anniv . . **Oct 21**
Income Tax Birthday .... **Feb 3**
Income Tax Due Date, Quarterly Estimated Federal .... **Jan 15, Apr 15, June 15, Sept 15**
Income Tax Pay Day—But Not This Year . . . **Apr 15**
Income Tax Pay Day Extension .... **Apr 17**
Independence Day, (US)
Aalborg and Rebild Fest (Aalborg and Rebild, Denmark) .... **July 1**
Black Hills Roundup (Belle Fourche, SD) . . **July 2**
Bristol Civic, Military/Firemen's Parade (Bristol, RI) .... **July 4**
Cactus Pete's Carl Hayden Daze Celeb (Jackpot, NV) .... **July 2**
Ennis Rodeo & Parade (Ennis, MT) .... **July 3**
Fireworks Celebration (Demopolis, AL) . . . **July 4**
Fourth of July/Torchlight Procession (Winston-Salem, NC) .... **July 2**
Fredericksburg Heritage Fest (Fredericksburg, VA) .... **July 4**
Freedom Days (Farmington, NM) .... **July 1**
Freedom Fest (Mahomet, IL) .... **July 4**
Hood River Old-Fashioned 4th of July (Hood River, OR) .... **July 4**
Ice Cream Social (Indianapolis, IN) .... **July 4**
Independence Day .... **July 4**
Independence Day Celebration (Washington, DC) .... **July 4**
Independence Day Challenge Run (Boyne City, MI) .... **July 4**
KSNT Go 4th (Topeka, KS) .... **July 1**
Lewistown 4th Parade/Celeb (Lewistown, MT) .... **July 4**
Lizard Race, World's Greatest (Lovington, NM) .... **July 4**
Manistee Natl Forest Fest (Manistee, MI) **June 30**
Maryland Symphony at Antietam (Sharpsburg, MD) .... **July 1**
Mount Rushmore July 4 Celeb (Mt Rushmore, SD) .... **July 4**
Mystic Seaport Museum Independence Day Celeb (Mystic, CT) .... **July 4**
Old Vermont Fourth (Woodstock, VT) .... **July 4**
Old-Fashioned 4th of July (Shakopee, MN) .... **July 1**
Old-Fashioned Fourth (Dallas, TX) .... **July 4**
Old-Fashioned Picnic in Park (Pittsburgh, KS) .... **July 3**
Old-Time Farm Day/Grand Old Fourth (Shelburne, VT) .... **July 4**
Patriots Month, Natl .... **June 6**
Riverfest (Lacrosse, WI) .... **June 30**
Sandblast (Ft Lauderdale, FL) .... **July 4**
Spirit of Freedom Fest (Florence, AL) .... **July 4**
Stock Exchange Holiday (Independence Day) .... **July 4**
Sundown Salute (Junction City, KS) ...... **July 1**
Ten Thousand Crestonians (Creston, IA) . **July 4**
WOOD Radio/Smiths Ind Fireworks Gala (Grand Rapids, MI) .... **July 4**
Zoobalee (Garden City, KS) .... **July 4**
Independence Day, Pet Owners .... **Apr 18**
Independence Sunday (IA) .... **July 2**
India
Bhopal Poison Gas Disaster Anniv .... **Dec 3**
Children's Day .... **Nov 14**
Independence Day .... **Aug 15**
Rajiv Gandhi Assassinated: Anniv .... **May 21**
Republic Day .... **Jan 26**
Rohini I: First Satellite Launched .... **July 18**
Indiana
Admission Day .... **Dec 11**
Amateur Athletic Union Sullivan Award Dinner (Indpls) .... **Feb 27**
ARC (Assn Retarded Citizens) Convention (Indianapolis) .... **Oct 26**
Auburn-Cord-Duesenberg Fest (Auburn) **Aug 31**
Benjamin Harrison's Birthday Celeb (Indianapolis) .... **Aug 20**
Big Ten M/W Outdoor Track/Field (West Lafayette) .... **May 19**
Big Ten Men's Tennis Team Chmpshp (Bloomington) .... **Apr 27**
Big Ten Women's Swimming/Diving Chmpshp (Indianapolis) .... **Feb 16**
Big Ten Wrestling Chmpshp (Bloomington) .... **Mar 4**
Big Whopper Liar's Contest (New Harmony) .... **Sept 16**
Championship Cat Show (Indianapolis) .... **Jan 7, Oct 28**
Chautauqua of the Arts (Columbus) .... **Sept 16**
Checkerfest (Indianapolis) .... **May 1**
Christmas Gift and Hobby Show (Indianapolis) .... **Nov 4**
Circle City Classic (Indianapolis) .... **Oct 7**
Colt League Pony Baseball World Series (Lafayette) .... **Aug 10**
Elkhart Grand Prix/Motor Sports Weekend (Elkhart) .... **July 7**
Elks Natl Hoop Shoot (Indianapolis) ..... **Apr 28**

Index ☆ Chase's 1995 Calendar of Events ☆

Indiana (cont'd)——Jackson

Elwood Glass Fest (Elwood)............Aug 17
Ethnic Festival (South Bend)...........June 30
Fairmount Fest/Remembering James Dean (Fairmount)........................Sept 22
Feast of the Hunters' Moon (Lafayette)....Oct 7
Foxtail Drive-in Car Show (Auburn).....May 28
Germanfest (Fort Wayne).............June 11
Golden Raintree Festival (New Harmony)..........................June 16
Heritage Festival (New Harmony)........Apr 17
Historical Soc Christmas Candlelight Tour (Crawfordsv)..........................Dec 9
Holiday at Benjamin Harrison Home (Indianapolis).......................Nov 10
Hoosier Horse Fair Expo (Indianapolis)...Apr 7
Hoosier St Games Finals (Indianapolis)..July 14
Hoosier St Games Regionals (various)..June 24
Hot Rod Assn US Natls (Indianapolis)...Aug 31
Ice Cream Social (Indianapolis)..........July 4
Indiana Fiddlers Gathering (Battle Ground)...........................June 23
Indiana Flower and Patio Show (Indianapolis)........................Mar 11
Indianapolis Boat/Sport/Travel Show (Indianapolis)........................Feb 17
Indy 500-Mile Race (Indianapolis).......May 28
Indy Sr Olympics (Indianapolis).........June 7
Johnny Appleseed Fest (Fort Wayne)...Sept 16
Kunstfest (New Harmony).............Sept 16
Literally Haunted House (New Albany)....Oct 6
Little 500 (Anderson)..................May 27
Majic Bodacious Bunny Hunt (Ft Wayne).Apr 14
Maple Syrup Festival (Wakarusa)........Mar 24
Marion Easter Pageant (Marion).........Apr 14
Midsummer Nights' Stroll (New Harmony)..........................July 20
Military Air Transport crashes in Indiana: Anniv................................Feb 6
Mississinewa 1812 (Marion)............Oct 14
Natl Women's Music Fest (Bloomington) .June 1
NCAA Men's Swimming/Diving Chmpshps (Indianapolis).......................Mar 23
NCAA Men's/Women's Indoor Track/Field (Indianapolis).......................Mar 10
NCAA Women's Div I Swimming/Diving Chmpshp (Austin).....................Mar 16
Ohio River Fest for the Arts (Evansville) .May 13
Paradise Spring Treaties Commemoration....................Aug 26
Parke County Covered Bridge Fest (Rockville)..........................Oct 13
Parke County Maple Fair (Rockville).....Feb 25
Pigeons Return to City-County Bldg (Ft Wayne)............................Mar 20
Pin Ear on the Van Gogh (Ft Wayne).....Mar 30
Redbud Trail Rendezvous (Rochester)...Apr 29
Riley Fest (Greenfield)..................Oct 5
Round Barn Fest (Rochester)..........June 16
Santa Spectacular (Indianapolis)........Nov 24
Scarborough-Indianapolis Peace Games (Indianapolis).......................July 21
Songs of My People (Fort Wayne)........May 6
State Fair (Indianapolis)................Aug 9
Steamboat Days Fest (Jeffersonville)....Sept 8
Strawberry 100 Bike Tour/Fest (Crawfordsville).....................June 9
Sugar Creek Canoe Race and Tri Sport Fest (Crawford).........................Apr 23
Three Rivers Fest (Ft Wayne)............July 8
Trail of Courage Living-History Fest (Rochester)........................Sept 16
USA/Mobil Indoor M/W Chmpshps (Indianapolis)........................Mar 3
WCVL/WIMC Senior Cit Health Fair/Expo (Crawfordsville)......................Sept 26
World's Largest Garage Sale (South Bend)..............................June 3
Indiana, Robert: Birth..................Sept 13
Indonesia
Independence Day....................Aug 17
Kartini Day...........................Apr 21
Industrial Development Decade for Africa, Second................................Jan 1
Industry Day (Beatrice, NE)..............Apr 23
Indy 500-Mile Race (Indianapolis, IN).....May 28
Infant Death Syndrome, Sudden, Awareness Month, Natl............................Oct 1
Infertility Awareness Week, Natl...........Oct 1
Inge Fest, William (Independence, KS).....Apr 20
Ingram, James: Birth.....................Feb 16
Inouye, Daniel Ken: Birth.................Sept 7
Interfaith Fellowship Anniv Celeb (New York, NY).................................Sept 10
International Language Week............Dec 15
International Week (College Station, TX)...Feb 27
Interplanetary Confederation Days........Oct 14
Intertribal Indian Ceremonial (Gallup, NM) .Aug 8
Inventors Congress, Minnesota (Redwood Falls, MN).................................June 9
Investment: Own Your Own Share of America Day.................................June 15
Iowa
Admission Day........................Dec 28
Bald Eagle Appreciation Days (Keokuk)..Jan 21
Balloon Classic, Natl (Indianola)........July 28
Bentonsport Arts Festival (Keosauqua)..Oct 14
Bike Van Buren (Van Buren County).....Aug 19
Bird Day.............................Mar 21
Bix Beiderbecke Mem Jazz Fest (Davenport)........................July 27
Bridal Expo (Davenport, IA).............Jan 28
Britt Draft Horse Show (Britt)...........Sept 1
Burlington Cruise Nights (Burlington)....Apr 28
Burlington Steamboat Days/Amer Music Fest (Burlington)........................June 13
Christmas Tour of Homes (Pella).........Dec 7
Christmas Walk (Pella).................Nov 18
Civil War Muster and Mercantile Exposition (Davenport)........................Sept 15
Civil War Reenactment (Keokuk)........Apr 28
Donna Reed Fest for the Performing Arts (Denison)..........................June 10
Dragon Boat Races (Keokuk)...........Aug 19
Dubuquefest/Very Special Arts (Dubuque)..........................May 17
Dyersville Festival of the Arts (Dyersville).........................Sept 23
Eisenfest (Amana).....................Sept 15
Field of Dreams Extravaganza (Dyersville)Sept 1
Forest Craft/Scenic Drive Festival (Van Buren County)............................Oct 14
Fort Atkinson Rendezvous (Fort Atkinson)..........................Sept 23
Glenn Miller Birthplace Society Fest (Clarinda)...........................June 8
Grant Wood Art Festival (Stone City-Anamosa)..........................June 11
Great Cardboard Boat Regatta (Leon)....Aug 6
Healthy Exchanges Annual Potluck (DeWitt).............................July 8
Herbert Hoover Day...................Aug 13
Hobo Convention, Natl (Britt)...........Aug 10
Holzfest (Amana).......................Aug 18
Homefront: America in the 1940s (West Branch)............................Apr 15
Independence Sunday..................July 2
James T. Kirk's Future Birthday (Riverside, IA)..................................Mar 22
Jule Fest (Elk Horn)....................Nov 24
Kaleidoscope Intl Fair (Dubuque)........Apr 21
Marble Meet at Amana (Amana).........June 3
Melon City Criterium (Muscatine).......May 28
Oktoberfest (Amana)...................Sept 29
Pancake Day (Centerville).............Sept 23
Pella Tulip Time Fest (Pella)............May 11
Perry's "BRR" (Bike Ride to Rippey) (Perry)..............................Feb 4
Register's Bicycle Ride Across Iowa (Des Moines)............................July 23
Rivers Hall of Fame Inductions/Awards, Natl (Dubuque)..........................May 27
Southwest Iowa Prof Hot Air Balloon Races (Creston)...........................Sept 16
State Fair, Iowa (Des Moines)...........Aug 10
Stubb's Eddy River Rendezvous (Davenport)........................May 19
Ten Thousand Crestonians (Creston).....July 4
Trek Fest (Riverside)..................June 24
USA Intl Dragon Boat Fest (Dubuque)...Sept 8
Van and Connie Intl Leisure Suit Convention (Des Moines).........................Apr 1
Van Buren Co Literacy Day............Sept 28
Victorian Christmas Walk (Davenport)....Dec 1
World of Christmas (West Branch)......Nov 11
Youth Honor Day.......................Oct 31
Iran
National Day..........................Feb 11
New Year.............................Mar 21
Noruz (New Year)......................Mar 21
Seizure of US Embassy: Anniv...........Nov 4
Yalda.................................Dec 21
Iranian Embassy, Seizure: Anniv...........Apr 30
Iraq
Bush Orders Commencement Desert Shield: Anniv................................Aug 7
Congress Authorizes Force Against Iraq: Anniv...............................Jan 12
Death of Those Aboard Amer Helicopters 6671..............................Page 488
Desert Shield: UN Deadline Resolution: Anniv..............................Nov 28
Ground War Against Iraq Begins: Anniv..Feb 23
Iraq Invades Kuwait: Anniv..............Aug 2
Kuwait City Liberated/100-Hour War Ends: Anniv...............................Feb 27
National Day..........................July 17
War Against Iraq Begins: Anniv..........Jan 16
Ireland
Day of the Wren........................Dec 26
Ivy Day................................Oct 6
Joyce Papers Released: Anniv...........Apr 6
National Day..........................Mar 17
Orangemen's Day.....................July 12
Ireland, Patricia: Birth....................Oct 19
Irish Fest, Milwaukee (Milwaukee, WI).....Aug 18
Irish Heritage Season (Biloxi, MS)..........Mar 4
Irish-American Heritage Month 6533, 6656...............................Page 488
Irons, Jeremy: Birth......................Sept 19
Iroquois Indian Fest (Howes Cave, NY)....Sept 2
Iroquois Steeplechase (Nashville, TN).....May 13
Irving, Amy: Birth.......................Sept 10
Irving, John: Birth........................Mar 2
Irving, Washington: Birth Anniv.............Apr 3
Irwin, Bill: Birth..........................Apr 11
Isherwood, Christopher: Birth Anniv......Aug 26
Ising, Rudolf C.: Birth Anniv...............Aug 7
Isle of Eight Flags Shrimp Festival (Fernandina, FL).................................Apr 29
Isle of Wight: Intl Oboe Competition (Pondwell, Ryde)..................................May 4
Isozaki, Arata: Birth......................July 23
Israel
Hashoah/Holocaust Day................Apr 27
Israeli Olympiad Massacre: Anniv........Sept 5
Yom Ha'atzma'ut (Independence Day)....May 5
Italian Festival (McAlester, OK)...........May 27
Italian Heritage Fest, West Virginia (Clarksburg, WV)..................................Sept 1
Italian Street Fair (Brentwood, TN)........Sept 1
Italy
Carnival Week (Milan)..................Feb 26
Epiphany Fair..........................Jan 5
Explosion of the Cart (Florence)........Apr 16
Feast of San Pellegrino (Gualdo Tadino) .Apr 30
Feast of the Incappucciati (Gradoli).....Feb 23
Feast of the Redeemer................July 16
Fest of St. Efisio.......................May 1
Gioco Del Calcio.....................June 24
Gioco Del Ponte (Battle of Bridge)......June 4
Giostra della Quintana (Joust)..........Sept 10
Good Friday (Calitri)...................Apr 14
Historical Regatta......................Sept 3
Joust of the Quintana...................Aug 6
Joust of the Saracen....................Sept 3
La Befana...............................Jan 6
Palio Dei Balestrieri (crossbow).........May 28
Palio del Golfo.........................Aug 13
Procession of Addolorata and Mysteries (Taranto)............................Apr 13
Purgatory Banquet (Gradoli).............Mar 1
Republic Day..........................June 2
Stresa Musical Weeks (Stresa)..........Aug 23
Vesuvius Day..........................Aug 24
Ives, Burl: Birth........................June 14
Ivy Day (Ireland)..........................Oct 6

J

Jackson, Andrew: Birth Anniv.............Mar 15
Jackson, Anne: Birth.....................Sept 3
Jackson, Bo: Birth......................Nov 30
Jackson, Glenda: Birth....................May 9
Jackson, Jackie: Birth.....................May 4
Jackson, Janet: Birth....................May 16
Jackson, Jermaine: Birth.................Dec 11
Jackson, Jesse: Birth......................Oct 8
Jackson, Joe: Birth.......................Aug 11
Jackson, Kate: Birth......................Oct 29
Jackson, Marlon: Birth...................Mar 12
Jackson, Michael: Birth..................Aug 29
Jackson, Milt: Birth.......................Jan 1
Jackson, Phil: Birth.....................Sept 17
Jackson, Rachel D: Birth Anniv..........June 15
Jackson, Randy: Birth....................Oct 29
Jackson, Reggie: Birth...................May 18
Jackson, Stonewall: Historic Birthday Parties................................Jan 19
Jackson, Thomas J. "Stonewall": Birth Anniv................................Jan 21
Jackson, Tito: Birth......................Oct 15

_ ☆ _Chase's 1995 Calendar of Events_ ☆   Index

Jacobi, Derek: Birth.....Oct 22	Clock Tower Jazz Fest (Rockford, IL)....**Jan 20**	Jillian, Ann: Birth.....**Jan 29**
Jacobi, Lou: Birth.....**Dec 28**	French Quarter Fest (New Orleans, LA)...**Apr 7**	Jimmy Doolittle Quarter Midget Natl (Falls of
Jaeger, Andrea: Birth.....**June 4**	Gainesville Downtown Fest/Art Show	Rough, KY).....**Sept 8**
Jagger, Bianca: Birth.....**May 2**	(Gainesville, FL).....**Nov 18**	Jinnah, Mohammed Ali: Birth Anniv.....**Dec 25**
Jagger, Mick: Birth.....**July 26**	Guthrie Jazz Banjo Fest (Guthrie, OK)..**May 26**	Job. See Employment
Jahn, Helmut: Birth.....**Jan 1**	Hampton Jazz Festival (Hampton, VA)..**June 23**	Joe Cain Procession (Mobile, AL).....**Feb 26**
Jakes, John: Birth.....**Mar 31**	Jazz at the Philharmonic: Anniv.....**Mar 25**	Joel, Billy: Birth.....**May 9**
Jamaica	Jazz Fest (Lake Havasu, AZ).....**Jan 13**	John, Elton: Birth.....**Mar 25**
Abolition of Slavery.....**Aug 1**	Jazz Fest, Intl (Martinique, French West	John Gilpin's Ride: Publication Anniv.....**Nov 14**
Discovery by Columbus: Anniv.....**May 4**	Indies).....**Nov 30**	John Parker Day.....**Apr 19**
Independence Day.....**Aug 7**	Jazz Fest, Intl (Pori, Finland).....**July 15**	John Paul II: Assassination Attempt: Anniv **May 13**
Maroon Fest.....**Jan 6**	Jazz in June (Norman, OK).....**June 22**	John Paul II, Pope.....**May 18**
National Heroes Day.....**Oct 16**	King Biscuit Blues Fest (Helena, AR).....**Oct 13**	John, Tommy: Birth.....**May 22**
James, John: Birth.....**Apr 18**	Lionel Hampton Jazz Fest (Moscow, ID) .**Feb 22**	Johnny Appleseed Day.....**Mar 11**
James, Rick: Birth.....**Feb 1**	Louis Armstrong: Cultural Legacy	Johnny Appleseed Days (Lake City, MN)....**Oct 7**
James, Sonny: Birth.....**May 1**	(Charleston, SC).....**July 29**	Johnny Appleseed Fest (Fort Wayne, IN)..**Sept 16**
Jamestown Christmas (Williamsburg, VA)...**Dec 3**	Louis Armstrong: Cultural Legacy (Chicago,	Johns, Glynis: Birth.....**Oct 5**
Jamestown Landing Day (Williamsburg,	IL).....**Apr 15**	Johns, Jasper: Birth.....**May 15**
VA).....**May 13**	Louis Armstrong: Cultural Legacy (Dallas,	Johnson, Amy: Flight Anniv.....**May 5**
Jamestown Weekend Celeb (Jamestown Island,	TX).....**Jan 28**	Johnson, Andrew: Birth Anniv.....**Dec 29**
VA).....**May 13**	Louis Armstrong: Cultural Legacy (New	Johnson, Andrew: Impeachment Proceedings:
Jamestown, VA: Founding Anniv.....**May 14**	Orleans, LA).....**Oct 28**	Anniv.....**Mar 5**
Jamieson, Bob: Birth.....**Feb 1**	Montana Traditional Jazz Fest (Helena,	Johnson, Andrew: Reward Offered for Jefferson
Jamison, Judith: Birth.....**May 10**	MT).....**June 28**	Davis.....**May 2**
Janeway, Elliot: Birth.....**Jan 1**	Monterey Bay Blues Fest (Monterey,	Johnson, Arte: Birth.....**Jan 20**
Janis, Byron: Birth.....**Mar 24**	CA).....**June 24**	Johnson, Ben: Birth.....**Dec 30**
January Diet Month.....**Jan 1**	Monterey Jazz Fest (Monterey, CA).....**Sept 15**	Johnson, Betsy: Birth.....**Aug 10**
Japan	Montreux Detroit Jazz Fest (Detroit, MI)..**Sept 1**	Johnson, Claudia Alta "Lady Bird": Birth . **Dec 22**
Adults' Day.....**Jan 15**	New Orleans Jazz/Heritage Fest (New Orleans,	Johnson, Don: Birth.....**Dec 15**
Birthday of the Emperor.....**Dec 23**	LA).....**Apr 28**	Johnson, Earvin "Magic": Birth.....**Aug 14**
Bon Festival (Feast of Lanterns).....**July 13**	North Sea Jazz Fest (The Hague,	Johnson, Eliza M.: Birth Anniv.....**Oct 4**
Children's Day.....**May 5**	Netherlands).....**July 14**	Johnson, Kevin: Birth.....**Mar 4**
Constitution Memorial Day.....**May 3**	Parker, Charlie: Birth Anniv.....**Aug 29**	Johnson, Lyndon B.: Birth Anniv.....**Aug 27**
Cormorant Fishing Festival.....**May 11**	Parker, Charlie: Death Anniv.....**Mar 12**	Johnson, Lyndon B: Monday Holiday Law:
Culture Day.....**Nov 3**	Russian River Jazz Fest (Guerneville,	Anniv.....**June 28**
Day of the Rice God (Chiyoda).....**June 4**	CA).....**Sept 16**	Johnson, Marv: Death Anniv.....**May 16**
Doll Fest (Hinamatsuri).....**Mar 3**	Saskatchewan Jazz Fest (Saskatoon, Sask,	Johnson, Pres Andrew, Vote to Impeach:
Flower Fest (Hana Matsuri).....**Apr 8**	Canada).....**June 23**	Anniv.....**Feb 24**
Foundation Day, Natl.....**Feb 11**	SunFest (W Palm Beach, FL).....**May 3**	Johnson, Richard M: Birth Anniv.....**Oct 17**
Ha-Ri-Ku-Yo (Needle Mass).....**Feb 8**	Telluride Jazz Celebration (Telluride, CO) .**Aug 4**	Johnson, Samuel: Birth Anniv.....**Sept 18**
Haru-No-Yabuiri.....**Jan 16**	Town Point Jazz Fest (Norfolk, VA).....**Aug 25**	Johnson, Vance: Birth.....**Mar 13**
Health-Sports Day.....**Oct 10**	W.C. Handy Fest (Florence, AL).....**Aug 6**	Johnson, Vinnie: Birth.....**Sept 1**
Hollyhock Festival (Kyoto).....**May 15**	Wayne State Univ: Funeral for Winter (Detroit,	Johnson, Virginia: Birth.....**Feb 11**
Japan Agrees to End Use of Drift Nets:	MI).....**Apr 13**	Johnston, J. Bennett: Birth.....**June 10**
Anniv.....**Nov 26**	Jeffers, Robinson: Birth Anniv.....**Jan 10**	Johnston, Joseph: Surrender at Durham Station:
Japanese Era New Year.....**Jan 1**	Jefferson, Martha: Birth Anniv.....**Oct 19**	Anniv.....**Apr 18**
Kakizome.....**Jan 2**	**Jefferson, Thomas, 251st Anniv of Birth**	Johnstown Flood Anniv.....**May 31**
Kanto Earthquake Memorial Day.....**Sept 1**	**6669**.....**Page 488**	Joke Day, Joe Miller's.....**Aug 16**
Labor Thanksgiving Day.....**Nov 23**	Jefferson, Thomas: Birth Anniv.....**Apr 13**	Jolson, Al: Birth Anniv.....**May 26**
Moment of Silence (Nagasaki).....**Aug 9**	Jeffords, James M.: Birth.....**May 11**	Jones, Bobby: Birth Anniv.....**Mar 17**
Namahage.....**Dec 31**	Jeffries, John: Weatherman's Day.....**Feb 5**	Jones, Brereton C.: Birth.....**June 27**
Nanakusa.....**Jan 7**	Jenner, Bruce: Birth.....**Oct 28**	Jones, Casey: Birth Anniv.....**Mar 14**
Newspaper Week.....**Oct 1**	Jennings, Peter: Birth.....**July 29**	Jones, Davy: Birth.....**Dec 30**
Old People's Day.....**Sept 15**	Jennings, Waylon: Birth.....**June 15**	Jones, Dean: Birth.....**Jan 25**
Peace Festival (Hiroshima).....**Aug 6**	**Jesse James Days, Defeat of (Northfield,**	Jones, Edward "Too Tall": Birth.....**Feb 23**
Rice Planting Festival.....**June 14**	**MN)**.....**Sept 6**	Jones, George: Birth.....**Sept 12**
Roughouse Fest.....**Oct 14**	Jett, Joan: Birth.....**Sept 22**	Jones, Grace: Birth.....**May 19**
Shichi-Go-San.....**Nov 15**	**Jewish Holidays**	Jones, Howard: Birth.....**Feb 23**
Snow Festival.....**Feb 8**	AJL (Assn Jewish Libraries Conv) (Chicago,	Jones, James Earl: Birth.....**Jan 17**
Soma No Umaoi (Wild Horse Chasing) . **July 23**	IL).....**June 18**	Jones, Jennifer: Birth.....**Mar 2**
Tanabata (Star Festival).....**July 7**	Chanukah.....**Dec 18**	Jones, K.C.: Birth.....**May 25**
Usokae (Bullfinch Exchange Festival).....**Jan 7**	Fast of Gedalya.....**Sept 27**	Jones, Leroi: Birth.....**Oct 7**
Water-Drawing Fest.....**Mar 1**	Fast of Tammuz.....**July 16**	Jones, Mary H.: Birth Anniv.....**May 1**
Japanese: Bon Odori Festival of Lanterns	Hadassah: Anniv.....**Feb 24**	Jones, Quincy: Birth.....**Mar 14**
(Chicago, IL).....**July 8**	Holocaust: Days of Remembrance.....**Apr 23**	Jones, Ricki Lee: Birth.....**Nov 8**
Japanese: Cherry Blossom Fest (San Francisco,	Israel Yom Ha'atzma'ut (Independence	Jones, Shirley: Birth.....**Mar 31**
CA).....**Apr 14**	Day).....**May 5**	Jones, Terry: Birth.....**Feb 1**
Japanese Festival: Ginza Holiday (Chicago,	Jewish Book Month.....**Nov 18**	Jones, Tom: Birth.....**June 7**
IL).....**Aug 18**	Jewish Heritage Week.....**Apr 30**	Jones, Tommy Lee: Birth.....**Sept 15**
Japanese Festival: Hatsume Fair (Delray Beach,	Jewish Heritage Week 6550, 6665.....**Page 488**	Jong, Erica: Birth.....**Mar 26**
FL).....**Feb 25**	Jewish Renaissance Fair (Morristown, NJ)**Sept 3**	Jonson, Ben: Birth Anniv.....**June 11**
Japanese Internment (WWII): Anniv.....**Feb 19**	Kristallnacht Anniv.....**Nov 9**	Joplin, Janis: Birth Anniv.....**Jan 19**
Japanese: Northern California Cherry Blossom	Lag B'Omer.....**May 18**	Joplin, Scott: Birth Anniv.....**Nov 24**
Fest (San Francisco, CA).....**Apr 15**	Liberation of Buchenwald: Anniv.....**Apr 10**	Joplin, Scott Ragtime Fest (Sedalia, MO) ..**June 1**
Japanese: O-Bon Festival (Delray Beach,	Passover Begins.....**Apr 14**	**Jordan**
FL).....**Aug 12**	Pesach (Passover).....**Apr 15**	Arab League Day.....**Mar 22**
Jarreau, Al: Birth.....**Mar 12**	Purim.....**Mar 16**	Great Arab Revolt and Army Day.....**June 10**
Jarriel, Thomas Edwin: Birth.....**Dec 29**	Rosh Hashanah (New Year).....**Sept 25**	Independence Day.....**May 25**
Jaruzelski, Wojciech: Birth.....**July 6**	Rosh Hashanah Begins.....**Sept 24**	King Hussein's Birthday.....**Nov 14**
Jarvik, Robert: Birth.....**May 11**	Shavuot.....**June 4**	Jordan, Barbara: Birth.....**Feb 21**
Jarvis, Gregory B.: Birth Anniv.....**Aug 24**	Shavuot Begins.....**June 3**	Jordan, Michael: Birth.....**Feb 17**
Jay, John: Birth Anniv.....**Dec 12**	Shemini Atzeret.....**Oct 16**	Jordan, Vernon Eulion, Jr: Birth.....**Aug 15**
Jazz and Blues	Simchat Torah.....**Oct 17**	Joseph, Chief, Surrender: Anniv.....**Oct 5**
America's Intl Dixieland Jazz Fest (Sacramento,	Sukkot Begins.....**Oct 8**	Jouett, Jack: Ride Anniv Obsv.....**June 3**
CA).....**May 26**	Sukkot/Succoth/Feast of Tabernacles....**Oct 9**	Joule, James: Birth Anniv.....**Dec 24**
Bessie Smith Traditional Jazz Fest	Ta'anit Esther (Fast of Esther).....**Mar 15**	**Journalism/Media**
(Chattanooga, TN).....**May 5**	Tisha B'Av (Fast of Ab).....**Aug 6**	All the News That's Fit to Print: Anniv....**Feb 10**
Beyond Category: Genius Ellington (Los	Tu B'Shvat.....**Jan 16**	American Sportscasters Hall of Fame (New
Angeles, CA).....**Jan 7**	US Holocaust Museum: Anniv.....**Apr 26**	York, NY).....**Dec 7**
Beyond Category: Genius Ellington	Yom Hashoah/Holocaust Day (Israel).....**Apr 27**	Around the World in 72 Days Anniv.....**Nov 14**
(Rochester, NY).....**Apr 8**	Yom Kippur.....**Oct 4**	Beginning of the Penny Press: Anniv.....**Sept 3**
Bix Beiderbecke Mem Jazz Fest (Davenport,	Yom Kippur Begins.....**Oct 3**	Bly, Nellie: Birth Anniv.....**May 5**
IA).....**July 27**	Jewison, Norman: Birth.....**July 21**	Cartoon Art Appreciation Week.....**May 2**
Chicago Jazz Fest (Chicago, IL).....**Sept 1**	Jillette, Penn: Birth.....**Mar 5**	Columnist's Day, Natl.....**June 27**

Index ☆ Chase's 1995 Calendar of Events ☆

Journalism/Media (cont'd)—Kentucky

First American Daily Newspaper Published: Anniv....May 30
First American Newspaper: Anniv....Sept 25
First Magazine Published in America: Anniv....Feb 13
First TV Presidential News Conf: Anniv..Jan 25
Free Paper Week, Natl....Mar 19
Freedom Forum Asian Center Opening: Anniv....Jan 17
Health Information Management Week, Natl 6620....Page 488
Japan: Newspaper Week....Oct 1
Most Dubious News Stories of the Year..Dec 29
Multicultural Communication Month....Apr 1
New York Weekly Journal Anniv First Issue....Nov 5
New Yorker magazine: Birth....Feb 21
Newspaper Assn America Conv (New Orleans, LA)....Apr 23
Newspaper Carrier Day....Sept 4
Newspaper in Education & Literacy Conf (Seattle, WA)....June 7
Newspaper Week, Natl....Oct 9
Newspapers in Education Week....Mar 6
No News Is Good News Day....Sept 11
People Magazine: Anniv....Mar 4
Pulitzer, Joseph: Birth Anniv....Apr 10
Pyle, Ernest: Birth Anniv....Aug 3
Technical Expo and Conf (Atlanta, GA).June 24
Ten Best Censored Stories 1994 (Rohnert Park, CA)....Apr 1
Thomas, Lowell: Birth Anniv....Apr 6
Top Junk-Food News Stories....Jan 15
UN: World Press Freedom Day....May 3
Winchell, Walter: Birth Anniv....Apr 7
Zenger, John P.: Arrest Anniv....Nov 17
Joust of the Quintana (Italy)....Aug 6
Joust. See also Italy
Jousting Tourn, Natl Jousting Hall Fame (Mt Solon, VA)....June 17
Jousting Tourn, Natural Chimneys (Solon, VA)....Aug 19
Joyce, James: Birth Anniv....Feb 2
Joyce, James, Papers Released: Anniv....Apr 6
Joyce, James: Ulysses, Marathon Reading (Washington, DC)....June 15
Joyful Summer Celebration....June 10
Joygerm Day, Natl....Jan 8
Joyner-Kersee, Jackie: Birth....Mar 3
Juarez, Benito Pablo: Birth Anniv....Mar 21
Jubilee Days Fest (Zion, IL)....Sept 2
Jubilee—Grand Celeb (Chattanooga, TN)..Apr 28
Judo Chmpshps, European (Birmingham, England)....May 10
Judson, Emily: Birth Anniv....Aug 22
Juggling Day, Natl....June 17
Jule Fest (Elk Horn, IA)....Nov 24
Julia, Raul: Birth....Mar 9
Julian Calendar New Year's Day....Jan 14
Julian, Percy: Birth Anniv....Apr 11
Jumping Frog Jubilee (Angel Camp, CA)..May 18
June Is Turkey Lovers' Month....June 1
Juneteenth....June 19
Junk-Food News Stories, Top....Jan 15
Junkanoo (Bahamas)....Dec 26
Jupiter, Comet Crashes into: Anniv....July 16
Jupiter Effect: Anniv....Mar 10
Jury, First All-Woman: Anniv....Sept 22
Just For Laughs Festival (Montreal, Que, Canada)....July 20
Just Pray No: Worldwide Day of Prayer....Apr 22
Justice, David: Birth....Apr 14
Justice, Peace with, Week....Oct 16
Justice, US Dept of: Anniv....June 22

K

Kael, Pauline: Birth....June 19
Kahn, Madeline: Birth....Sept 29
Kainen, Jacob: Birth....Dec 7
Kalb, Marvin: Birth....June 9
Kamali, Norma: Birth....June 27
Kamehameha Day (HI)....June 11
Kander, George: Birth....Mar 18
Kane, Carol: Birth....June 18
Kanin, Garson: Birth....Nov 24
Kansas
Admission Day....Jan 29
American Rowing Natl Chmpshps (Topeka)....July 20
Annual Opolis Reunion Picnic (Opolis)..June 4
Antique Airplane Fly-In (Atchison)....May 26
Apple Fest (Topeka)....Oct 1
Atwood Early Rod Run (Atwood)....May 20
Beef Empire Days (Garden City)....June 2
Buffalo Bill Days (Leanvenworth)....Sept 10
Bullwhacker Days (Olathe)....June 24
Christmas in Old Dodge City (Dodge City)....Nov 25
Columbus Day Fest/Hot Air Balloon Regatta (Columbus)....Oct 6
Dalton Defenders Day (Coffeyville)....Oct 7
Dodge City Days (Dodge City)....July 28
Eagle Days (Junction City)....Jan 21
Festival of Carols (Topeka)....Nov 24
Festival of Trees (Topeka)....Nov 30
Five State Free Fair (Liberal)....Aug 13
Flatlanders Fall Fest (Goodland)....Sept 22
Grant County Home Products Dinner (Ulysses)....Sept 19
Great Plains Rowing Chmpshp (Topeka).Apr 14
Guitar Flat-Pickin Chmps/Walnut Valley Fest (Winfield)....Sept 14
Halloween Frolic (Hiawatha)....Oct 31
Heritage Christmas (Lindsborg)....Dec 2
Heritage Day Celebration (Leavenworth)....June 10
Hoisington Celebration (Hoisington)....Sept 1
Huff 'n Puff Hot Air Balloon Rally (Topeka)....Sept 8
HutchFest (Hutchinson)....June 28
Inter-State Fair/Rodeo (Coffeyville)....Aug 12
Intl Pancake Day (Liberal)....Feb 28
JCBC Century Ride (Junction City and Fort Riley)....Sept 10
Kansas City Renaissance Fest (Bonner Springs)....Sept 2
Kansas River Valley Art Fair (Topeka)....July 15
Kansas St Barbeque Champshp (Lenexa)....June 24
KSNT Go 4th (Topeka)....July 1
Lansing Daze (Leavenworth)....May 6
Lenexa Spinach Festival (Lenexa)....Sept 9
Little Balkans Days/Folklife Fest (Pittsburg)....Sept 1
Mennonite Relief Sale (Hutchinson)....Apr 7
Mid America All-Indian Center PowWow (Wichita)....July 28
Millfest (Lindsborg)....May 6
Neewollah (Independence)....Oct 20
New Beginning Fest (Coffeyville)....Apr 27
Old Sedgwick County Fair (Wichita)....Oct 7
Old-Fashioned Picnic in Park (Pittsburgh) July 3
Pony Express Festival (Hanover)....Aug 27
Produce for Victory: Posters/Home Front (Logan)....Jan 14
Ranch Rodeo (Liberal)....May 20
Salter Elected First Woman Mayor in US: Anniv....Apr 4
Sand Creek Folk Life Fest (Newton)....Apr 29
Sippin' Cider Days (Leavenworth)....Oct 7
Spring Prairie Fest (Olathe)....Apr 15
State Fair (Hutchinson)....Sept 8
Strength/Diversity: Japanese-Ameri Women (Logan)....Apr 15
Sundown Salute (Junction City)....July 1
Topeka Railroad Days Fest (Topeka)....Sept 1
Twelve Villages of Christmas (Washington County)....Dec 2
US Senior Amateur (Golf) Chmpshp (Hutchinson)....Sept 26
Wichita Winterfest (Wichita)....Dec 2
William Inge Fest (Independence)....Apr 20
Zoobalee (Garden City)....July 4
Kansas City Hotel Disaster: Anniv....July 17
Kaplan, Gabe: Birth....Mar 31
Karras, Alex: Birth....July 15
Kart Races, World Enduro Chmpshp (Daytona Beach, FL)....Dec 27
Karting: Elkhart Grand Prix (Elkhart, IN)...July 7
Kasdan, Lawrence: Birth....Jan 14
Kassebaum, Nancy Landon: Birth....July 29
Kauffmann, Stanley J.: Birth....Apr 24
Kaufman, Irving: Birth Anniv....June 24
Kavner, Julie: Birth....Sept 7
Kazakhstan: Day of the Republic....Dec 16
Kazan, Elia: Birth....Sept 7
Kazan, Lainie: Birth....May 16
Kazoo Day, Natl....Jan 28
Kazurinsky, Tim: Birth....Mar 3
KCCR Farm, Home and Sport Show (Pierre, SD)....Feb 25
Keach, Stacy, Jr: Birth....June 2
Keaton, Diane: Birth....Jan 5
Keaton, Michael: Birth....Sept 9
Keats, John: Birth Anniv....Oct 31
Keel, Howard: Birth....Apr 13
Keene, Donald: Birth....June 18
Keep America Beautiful Annual Meeting (Washington, DC)....Dec 6
Keep America Beautiful Month, Natl....Apr 1
Keeshan, Bob (Captain Kangaroo): Birth.June 27
Keillor, Garrison: Birth....Aug 7
Keith, Brian: Birth....Nov 14
Kellems, Vivien: Memorial Day....June 7
Keller, Helen: Birth Anniv....June 27
Keller, Helen, Festival (Tuscumbia, AL)....June 23
Kellerman, Sally: Birth....June 2
Kelley, Kitty: Birth....Apr 4
Kelly, Gene: Birth....Aug 23
Kelly, Grace P.: Birth Anniv....Nov 12
Kelly: Princess Grace: Death Anniv....Sept 14
Kelly Tire Blue-Gray All Star Classic (Montgomery, AL)....Dec 25
Kelly, Walt: Birth Anniv....Aug 25
Kelsey, Linda: Birth....July 28
Kemble, Fanny: Birth Anniv....Nov 27
Kemmler, William: First Electrocution for Death Penalty....Aug 6
Kemp, Jack: Birth....July 13
Kempthorne, Dirk: Birth....Oct 29
Keneally, Thomas: Birth....Oct 7
Kennan, George Frost: Birth....Feb 16
Kennedy, Anthony M.: Birth....July 23
Kennedy Assassination: Warren Report Anniv....Sept 27
Kennedy, Edward Moore: Birth....Feb 22
Kennedy, Ethel: Birth....Apr 11
Kennedy, George: Birth....Feb 18
Kennedy Intl Airport Dedication: Anniv....July 31
Kennedy, John F: Assassination Anniv....Nov 22
Kennedy, John F: Birth Anniv....May 29
Kennedy, John F: Committee on Assassinations Report....Mar 29
Kennedy, John F.: Day (MA)....Nov 26
Kennedy, John: First TV Presidential News Conf: Anniv....Jan 25
Kennedy, Robert F.: Assassination Anniv...June 5
Kennedy, Robert F: Birth Anniv....Nov 20
Kennedy, Robert: Committee on Assassinations Report....Mar 29
Kennedy, Rose: Birth....July 22
Kennedy-Cheng Summit Anniv....July 31
Kennerly, David Hume: Birth....Mar 9
Kenneth (Kenneth Everette Battelle): Birth.Apr 19
Kent State Students' Memorial Day....May 4
Kentucky
A. Lincoln Birthplace Founder's Day Wknd (Hodgenvi)....July 15
Admission Day....June 1
American Quilter's Soc Show (Paducah).Apr 27
Bar-B-Q Festival, Intl (Owensboro)....May 12
Battle of Blue Licks Celebration (Mount Olivet)....Aug 19
Berea Craft Fest (Berea)....July 14
Boone Day....June 7
Carter Caves Crawlathon (Olive Hill)....Jan 27
Corn Island Storytelling Fest (Louisville) Sept 15
Dulcibrrr (Falls of Rough)....Feb 3
Everly Brothers Homecoming (Central City)....Aug 28
Family Folklore (Lexington)....Jan 14
Festival of the Bluegrass (Lexington)....June 8
Forgotten Past Steam and Gas Show (Benton)....Aug 5
450-Mile Outdoor Fest....Aug 17
Fraley Family Music Weekend (Olive Hill).Sept 5
Great American Brass Band Fest (Danville, KY)....June 16
Great American Dulcimer Convention (Pineville)....Sept 22
Great Houseboat/Cruiser Parade (Falls of Rough)....July 1
Halloween Harvest of Horrors (Louisville) Oct 20
Internatl Strange Music Weekend (Olive Hill)....June 9
Jimmy Doolittle Quarter Midget Natl Race (Falls of Rough)....Sept 8
Kentucky Apple Fest (Paintsville)....Oct 1
Kentucky Derby (Louisville)....May 6
Kentucky Derby Fest (Louisville)....Apr 21
Kentucky Hills Weekend (Corbin)....Mar 3
Kentucky St Chmp Fiddler's Contest (Falls of Rough)....July 21
Lincoln's Birthplace Cabin Wreath Laying (Hodgenville)....Feb 12
Maifest (Covington)....May 19
Marion County Country Ham Days (Lebanon)....Sept 23
Memory Days (Grayson)....May 25
Mint Julep Scale Meet (Falls of Rough)..May 19
Musical Tribute to Dr. M. L. King (Hodgenville)....Jan 15
Oktoberfest (Covington)....Sept 8
Owensboro Show (Owensboro)....Aug 5

☆ Chase's 1995 Calendar of Events ☆ Index

Kentucky (cont'd)—Largent

Quartet Convention, Natl (Louisville) .... Sept 18
Rough River Humor Fest (Falls of Rough) Mar 31
Running of the Rodents (Louisville) ...... Apr 26
Scottish Weekend (Carrollton) .... May 12
Send a Kid to Kamp Radiothon (Lexington) .... May 20
Spring Ghost Tales (Louisville) .... May 12
Spring Stroll (Covington) .... Mar 18
State Fair (Louisville) .... Aug 17
Sternwheel Days (Augusta) .... June 24
Summer Sunfest (Covington) .... July 14
Swappin' Meetin' (Cumberland) .... Oct 6
Tater Days (Benton) .... Apr 1
Williamsburg Old-Fashioned Trading Days (Williamsburg) .... Sept 7
World's Largest Halloween Party (Louisville) .... Oct 19
Kentucky Derby, First: Anniv .... May 17
Kenya
Jamhuri Day .... Dec 12
Kenyatta Day .... Oct 20
Madaraka Day .... June 1
Kepler, Johannes: Birth Anniv .... Dec 27
Kercheval, Ken: Birth .... July 15
Kern, Jerome: Birth Anniv .... Jan 27
Kerns, Joanna: Birth .... Feb 12
Kerr, Deborah: Birth .... Sept 30
Kerr, Jean: Birth .... July 10
Kerr, Steve: Birth .... Sept 27
Kerr, Walter: Birth .... July 8
Kerrey, J. Robert: Birth .... Aug 27
Kerry, John F.: Birth .... Dec 11
Kersey, Jerome: Birth .... June 26
Kerwin, Brian: Birth .... Oct 25
Kerwin, Lance: Birth .... Nov 6
Kesey, Ken: Birth .... Sept 17
Ketcham, Hank: Birth .... Mar 14
Key Club Intl Convention (Anaheim, CA) ... July 1
Key Club Intl Week .... Nov 5
Key, Francis Scott: Star-Spangled Banner Inspired: Anniv .... Sept 13
Khachaturian, Aram (Ilich): Birth Anniv .... June 6
Khan, Chaka: Birth .... Mar 23
Kiam, Victor Kermit, II: Birth .... Dec 7
Kidd, Michael: Birth .... Aug 12
Kidder, Margot: Birth .... Oct 17
Kidnapping, Lindbergh: Anniv .... Mar 1
Kid'rific (Hartford, CT) .... Sept 9
Kids' Day (Baltimore, MD) .... Aug 2
Kids' Day, Kiwanis, Natl .... Sept 23
Kid's Swap Shop (Baltimore, MD) .... Aug 7
Kienzle, William Xavier: Birth .... Sept 11
Kiker, Douglas: Birth .... Jan 7
Kiley, Richard: Birth .... Mar 31
Killebrew, Harmon: Birth .... June 29
Killy, Jean-Claude: Birth .... Aug 30
Kilmer, Joyce: Birth Anniv .... Dec 6
Kilpatrick, James Jackson: Birth .... Nov 1
Kindergarten Day .... Apr 21
Kindergarten: Peabody, Elizabeth P.: Birth Anniv .... May 16
Kinetic Sculpture Race (Omaha, NE) .... Sept 9
King, Alan: Birth .... Dec 26
King, Albert: Birth .... Dec 17
King, B.B.: Birth .... Sept 16
King, Ben E.: Birth .... Sept 28
King, Billy Jean: Birth .... Nov 22
King, Bruce: Birth .... Apr 6
King, Carole: Birth .... Feb 9
King, Coretta Scott: Birth .... Apr 27
King, Don: Birth .... Aug 20
King, Martin Luther, Jr.
Assassination Anniv .... Apr 4
Birth Anniv .... Jan 15
Birthday Observed .... Jan 16
Committee on Assassinations Report .... Mar 29
Dream Day .... Aug 28
Human Relations Day (Flint, MI) .... Jan 16
Humanitarian Day .... Jan 15
King Awarded Nobel: Anniv .... Oct 14
Martin Luther King, Jr, Fed Holiday 6524, 6645 .... Page 488
Musical Tribute to Dr. M. L. King (Hodgenville, KY) .... Jan 15
Victims of Violence Holy Day .... Apr 4
King, Rodney: Los Angeles Riots: Anniv ... Apr 29
King, Stacey: Birth .... Jan 29
King, Stephen: Birth .... Sept 21
King, Wayne: Birth Anniv .... Feb 16
Kingsley, Ben: Birth .... Dec 31
Kipling, Rudyard: Birth Anniv .... Dec 30
Kiribati: Independence Day .... July 12
Kirk, James T: Future Birthday (Riverside, IA) .... Mar 22
Kirkpatrick, Jeane: Birth .... Nov 19
Kirshner, Don: Birth .... Apr 17
Kirtland, Jared: Birth Anniv .... Nov 10
Kiss-and-Make-Up-Day .... Aug 25
Kiss-Your-Mate Day .... Apr 28
Kissinger, Henry: Birth .... May 27
Kitchen and Bath Month, Natl .... Oct 1
Kites and Kite-Flying
Bucks County Kite Day (Langhorne, PA) .. May 7
Buffalo Beano Kite Fly/Frisbee Fling (Lubbock, TX) .... Apr 22
Frankenmuth Skyfest (Frankenmuth, MI) .. May 7
Kite Fest (Lorton, VA) .... Mar 19
Kite Fest of Santiago Sacatepequez (Guatemala) .... Nov 1
Kite Fly In, Kwin (Stockton, CA) .... May 29
Kitefest (Kalamazoo, MI) .... Apr 29
Mid-American Sport Kite Classic .... Sept 9
Mid-Atlantic Stunt Kite Chmpshp (Ocean City, MD) .... May 6
Sno'Fly (Kalamazoo, MI) .... Jan 1
Kite, Tom: Birth .... Dec 9
Kitt, Eartha: Birth .... Jan 16
Kittle, Ron: Birth .... Jan 5
Kiwanis: Circle K Intl Conv (Phoenix, AZ) ... Aug 5
Kiwanis: Circle K Intl Week .... Feb 5
Kiwanis Intl: Anniv .... Jan 21
Kiwanis Intl Convention (New Orleans, LA) .... June 23
Kiwanis: Key Club Intl Week .... Nov 5
Kiwanis Kids' Day, Natl .... Sept 23
Kiwanis Prayer Week .... May 14
Kiwanis Week, Worldwide .... Jan 15
Kiwifruit Day, California .... Dec 22
Klein, Calvin: Birth .... Nov 19
Klein, Robert: Birth .... Feb 8
Klemperer, Werner: Birth .... Mar 22
Kline, Kevin: Birth .... Oct 24
Klondike Days (Edmonton, Alta, Canada) .. July 20
Klugman, Jack: Birth .... Apr 27
Kneivel, Evel: Birth .... Oct 17
Knight, Bob: Birth .... Oct 25
Knight, Gladys: Birth .... May 28
Knight, O. Raymond: Birth Anniv .... Apr 8
Knights of Columbus Founders' Day .... Mar 29
Knights of Pythias: Founding Anniv .... Feb 19
Knights of St. Yago-Illuminated Night Parade(Tampa, FL) .... Feb 11
KNIM Radio's Big Fish 15 (Maryville, MO) .. June 9
Knitting, Needlecraft/Design Exhib (Esher, England) .... Jan 19
Knitting/Needlecraft Exhibit, NW (Manchester, England) .... Feb 23
Knitting/Needlecraft Exhibit, Southern (Bristol, England) .... Mar 17
Knotts, Don: Birth .... July 21
Knowles, John: Birth .... Sept 16
Knuckles Down Month, Natl .... Apr 1
Koch, Edward Irwin: Birth .... Dec 12
Koestler, Arthur: Birth .... Sept 5
Kohl, Helmut: Birth .... Apr 3
Kohl, Herb: Birth .... Feb 7
Koop, Charles Everett .... Oct 14
Kopell, Bernie: Birth .... June 21
Koppel, Ted: Birth .... Feb 8
Korbut, Olga: Birth .... May 16
Korea
Alphabet Day (Hangul) .... Oct 9
Children's Day .... May 5
Chusok .... Sept 9
Constitution Day .... July 17
Korea, North and South, End War: Anniv Dec 13
Korean Air Lines Flight 007 Disaster: Anniv .... Sept 1
Korean War Armistice: Anniv .... July 27
Korean War: Hayes, Ira: Birth Anniv ..... Jan 12
Korean War: US Invades Korea: Anniv .. June 25
Liberation Day .... Aug 15
Memorial Day .... June 6
National Day .... Sept 9
National Foundation .... Oct 3
Samiljol (Independence Day) .... Mar 1
Tano Day .... June 2
Korean Armistice, 40th Anniv 6582 ...... Page 488
Korman, Harvey: Birth .... Feb 15
Kosar, Bernie: Birth .... Nov 25
Koufax, Sandy: Birth .... Dec 30
Krakatoa: Eruption Anniv .... Aug 26
Krantz, Judith: Birth .... Jan 9
Krassner, Paul: Birth .... Apr 9
Kraus, Alfredo: Birth .... Nov 24
Kraut and Frankfurter Week, Natl .... Feb 9
Kreps, Juanita: Birth .... Jan 11
Kreutzmann, Bill: Birth .... June 7
Krim, Mathilde: Birth .... July 9
Kristallnacht: Anniv .... Nov 9
Kristofferson, Kris: Birth .... June 22
Krone, Julie: Birth .... July 24
Kruger, Paul: Birth Anniv .... Oct 10
Kruk, John: Birth .... Feb 9
Krupp, Alfried: Birth Anniv .... Aug 13
Ku Klux Klan Members Convicted of Conspiracy: Anniv .... Oct 20
Kubek, Tony: Birth .... Oct 12
Kuhio Day (HI) .... Mar 26
Kuhn, Bowie: Birth .... Oct 28
Kuhn, Maggie: Birth .... Aug 3
Kunin, Madeleine M.: Birth .... Sept 28
Kunstler, William M.: Birth .... July 7
Kupcinet, Irv: Birth .... July 31
Kuralt, Charles Bishop: Birth .... Sept 10
Kurowawa, Akira: Birth .... Mar 23
Kurtis, Bill: Birth .... Sept 21
Kurtz, Swoosie: Birth .... Sept 6
Kurtzman, Harvey: Birth Anniv .... Oct 3
Kutner, Luis: Birth Anniv .... June 9
Kuwait
Iraq Invades Kuwait: Anniv .... Aug 2
Kuwait City Liberated/100-Hour War Ends: Anniv .... Feb 27
National Day .... Feb 25
Kwanzaa .... Dec 26
Kyrgyzstan: Independence Day .... Aug 31
Kyser, Kay: Birth Anniv .... June 18

L

L'Enfant, Pierre C: Birth Anniv .... Aug 2
L'Engle, Madeleine: Birth .... Nov 29
La Befana (Italy) .... Jan 6
LaBelle, Patti: Birth .... Oct 4
Labor. See Employment
Labor: AFL-CIO Founding Anniv .... Dec 5
Labor: AFL-CIO Merger: Anniv .... Feb 9
Labor Day .... Sept 4
Labor Day (Bahamas) .... June 2
Labor Day (Zambia) .... May 1
Labor Day Bridge Walk, Horton Bay (Horton Bay, MI) .... Sept 4
Labor Day in the Pines (Angel Fire, MN) ... Sept 2
Labor: Five-Dollar-a-Day Minimum Wage: Anniv .... Jan 5
Labor: Full Employment Week .... Sept 4
Labor History Month 6688 .... Page 488
Lacoste, Rene: Birth .... July 2
Lacroix, Christian: Birth .... May 17
Lacrosse Classic, Lee-Jackson (Lexington, VA) .... May 6
Ladd, Cheryl: Birth .... July 2
Ladd, Diane: Birth .... Nov 29
Ladies' Day Initiated in Baseball: Anniv ... June 16
Lafayette Day (MA) .... May 20
Laffer, Arthur: Birth .... Aug 14
LaFleur, Guy: Birth .... Sept 20
Lag B'Omer .... May 18
Lagerlof, Selma: Birth Anniv .... Nov 20
Lahti, Christine: Birth .... Apr 4
Laimbeer, Bill: Birth .... May 19
Laine, Frankie: Birth .... Mar 30
Laing, R.D.: Birth .... Oct 7
Lake Festival, Virginia (Clarksville, VA) .... July 14
Lakefront Festival of Arts (Milwaukee, WI) June 16
Laker, Freddie: Birth .... Aug 6
Lamas, Lorenzo: Birth .... Jan 20
Lamb, Charles: Birth Anniv .... Feb 10
Lamour, Dorothy: Birth .... Dec 10
Lancaster, Burt: Birth .... Nov 2
Land, Edwin H.: Birth .... May 7
Landers, Ann: Birth .... July 4
Landesberg, Steve: Birth .... Nov 3
Landon, Michael: Birth Anniv .... Oct 31
Landry, Thomas Wade: Birth .... Sept 11
Lands Day, Public .... Sept 9
Lane, Mark: Birth .... Feb 24
Lange, Hope: Birth .... Nov 28
Lange, Jessica: Birth .... Apr 20
Langella, Frank: Birth .... Jan 1
Langer, Susanne: Birth Anniv .... Dec 20
Langley, Samuel Pierpont: Birth Anniv ..... Aug 22
Langston, Mark: Birth .... Aug 20
Language Week, Intl .... Dec 15
Lansbury, Angela: Birth .... Oct 16
Lansing Daze (Leavenworth, KS) .... May 6
Lansing, Robert: Birth .... June 5
Lanvin, Bernard: Birth .... Dec 27
Lao People's Dem Repub: National Holiday . Dec 2
Lappe, Frances Moore: Birth .... Feb 10
Larbaud, Valery N.: Birth Anniv .... Aug 29
Lardner, Ring, Jr: Birth .... Aug 19
Laredo, Ruth: Birth .... Nov 20
Largent, Steve: Birth .... Sept 28

Index ☆ Chase's 1995 Calendar of Events ☆

Larkin—Literature

Larkin, Barry: Birth....Apr 28
Larouche, Lyndon H., Jr: Birth....Sept 8
Larroquette, John: Birth....Nov 25
Larson, Gary: Birth....Aug 14
Las Vegas Night (Minocqua, WI)....July 20
Lasershow (Stone Mountain, GA)....May 5
Lasorda, Tom: Birth....Sept 22
Lasser, Louise: Birth....Apr 11
Late for Something Day, Be....Sept 5
Lathrop, Julia C.: Birth Anniv....June 29
Latvia: Baltic States' Independence Recognized: Anniv....Sept 6
Latvia: Independence Day....Nov 18
Lauda, Niki: Birth....Feb 22
Lauder, Estee: Birth....July 1
Laundry Day, Natl....Sept 22
Laundry Week, Natl....Sept 17
Lauper, Cyndi: Birth....June 20
Laurel and Hardy: Cuckoo Dancing Week..Jan 11
Lauren, Ralph: Birth....Oct 14
Laurie, Piper: Birth....Jan 22
Laurier, Sir Wilfred: Birth Anniv....Nov 20
Lautenberg, Frank R.: Birth....Jan 23
Laver, Rod: Birth....Aug 9
Lavin, Linda: Birth....Oct 15
Lavoisier, Antoine: Execution Anniv....May 8
Law Day (Pres Proc)....May 1
Law Day USA....May 1
Law Day, USA 6555, 6679,....Page 488
Law Enforcement Training Week, Natl 6523....Page 488
Law, Intl Decade of (UN)....Jan 1
Law, John Philip: Birth....Sept 7
Law: Mansfield, Arabella: Birth Anniv....May 23
Lawrence, Carol: Birth....Sept 5
Lawrence, David H.: Birth Anniv....Sept 11
Lawrence, James: Birth Anniv....Oct 1
Lawrence, Steve: Birth....July 8
Lawrence, Vicki: Birth....Mar 26
Lawson, Nigel: Birth....Mar 11
Leachman, Cloris: Birth....Apr 30
Leacock, Stephen: Birth Anniv....Dec 30
League of Nations: Anniv....Jan 10
League of Women Voters Council (Wash, DC)....June 10
League of Women Voters Formed: Anniv...Feb 14
Leahy, Patrick J.: Birth....Mar 31
Leakey, Richard E.: Birth....Dec 19
Lean, David: Birth Anniv....Mar 25
Leap Second Adjustment Time...June 30, Dec 31
Lear, Edward: Birth Anniv....May 12
Lear, Evelyn: Birth....Jan 18
Lear, Frances: Birth....July 14
Lear, Norman: Birth....July 27
Learned, Michael: Birth....Apr 9
Learning Disabilities Conf, Child/Adult (Orlando, FL)....Mar 1
Leary, Timothy Francis: Birth....Oct 22
Leavitt, Mike: Birth....Feb 11
Lebanon
  Independence Day....Nov 22
  Last American Hostage Released: Anniv..Dec 4
  Palestinian Massacre: Anniv....Sept 16
Lebanon Bologna Fest (Lebanon, PA)....Sept 15
LeBon, Simon: Birth....Oct 27
LeCarre, John: Birth....Oct 19
Lee, Brenda: Birth....Dec 11
Lee, Christopher: Birth....May 27
Lee, Harper: Birth....Apr 28
Lee, Michele: Birth....June 24
Lee, Peggy: Birth....May 26
Lee, Pinky (Pincus Leff): Birth Anniv....May 2
Lee, Robert E.
  Battle of Gettysburg: Anniv....July 1
  Birth Anniv....Jan 19
  Birthday Parties....Jan 19
  Defeat at Five Forks: Anniversary....Apr 1
  Fall of Richmond: Anniversary....Apr 3
  Lee Birth Celebration (Alexandria, VA)....Jan 22
  Peace Overtures in Civil War: Anniv....Mar 2
Lee-Jackson-King Day....Jan 16
Lee-Jackson Lacrosse Classic (Lexington, VA)....May 6
Left-Handed Golfers Am Chmpshp (Palm Beach Garden, FL)....June 27
Lefthanders Day, Intl....Aug 13
Legrand, Michel: Birth....Feb 24
LeGuin, Ursula: Birth....Oct 21
Lehrer, James: Birth....May 19
Lehrer, Tom: Birth....Apr 9
Lei Day (Hawaii)....May 1
Leigh, Janet: Birth....July 6
Leigh, Jennifer Jason: Birth....Feb 5
Leisure Suit Convention, Van/Connie Intl (Des Moines, IA)....Apr 1
Lemmon, Jack: Birth....Feb 8
Lemon, Meadowlark: Birth....Apr 25
LeMond, Greg: Birth....June 26
Lendl, Ivan: Birth....Mar 7
Lenin Assassination Attempt: Anniv....Aug 30
Lenin, Nikolai: Birth Anniv....Apr 22
Lennon, John: Birth Anniv....Oct 9
Lennon, John: Lennon-Ono Album Confiscation: Anniv....Jan 3
Lennon, Julian: Birth....Apr 8
Lennox, Annie: Birth....Dec 25
Leno, Jay: Birth....Apr 28
Lent Begins....Mar 1
Lent, Orthodox....Mar 6
Lenz, Kay: Birth....Mar 4
Leo Begins....July 23
Leo Sowerby Centennial and JubiLEO....May 1
Leonard, Roy: Birth....Jan 19
Leonard, Sugar Ray: Birth....May 17
Leroux, Charles: Last Jump: Anniv....Sept 12
Lesotho
  Army Day....Jan 20
  Moshoeshoe's Day....Mar 12
  National Day....Oct 4
  National Tree Planting Day (South Africa) Mar 21
Letterman, David: Birth....Apr 12
Letter-Writing Week, Universal....Jan 1
Leutze, Emanuel: Birth Anniv....May 24
Lever, Fat: Birth....Aug 18
Levertov, Denise: Birth....Oct 24
Levin, Carl: Birth....June 28
Levine, Charles A.: Death Anniv....Dec 6
Levine, David: Birth....Dec 20
Levine, Irving R.: Birth....Aug 26
Levine, James: Birth....June 23
Levingston, Cliff: Birth....Jan 4
Levy, Eugene: Birth....Dec 17
Lewis and Clark Rendezvous (St. Charles, MO)....May 20
Lewis, Anthony: Birth....Mar 27
Lewis, Carl: Birth....July 1
Lewis, Emmanuel: Birth....Mar 9
Lewis, Francis: Birth Anniv....Mar 21
Lewis, Huey: Birth....July 5
Lewis, Jerry Lee: Birth....Sept 29
Lewis, Jerry: Birth....Mar 16
Lewis, Jerry: Stars Across Amer MDA Labor Day Wknd....Sept 3
Lewis, John Llewellyn: Birth Anniv....Feb 12
Lewis, Meriwether: Birth Anniv....Aug 18
Lewis, Shari: Birth....Jan 17
Lewis, Sinclair: Birth Anniv....Feb 7
Lewis, Sinclair: Sinclair Lewis Days (Sauk Center, MN)....July 13
Liar's Contest, Big Whopper (New Harmony, IN)....Sept 16
Libby, Willard F: Birth Anniv....Dec 17
Liberation Day (Poland)....Jan 17
Liberia: National Day....July 26
Liberman, Alexander: Birth....Sept 4
Liberty Day....Mar 23
Liberty Tree Day (MA)....Aug 14
Libra Begins....Sept 23
Library, Librarian
  AJL (Assn Jewish Libraries Conv) (Chicago, IL)....June 18
  Librarians Day, Intl Special....Apr 13
  Library Assn, American: Birth....Oct 6
  Library Assn Conf, Church/Synagogue (Houghton, NY)....June 24
  Library Card Sign-up Month....Sept 1
  Library, First Presidential: Anniv....Nov 19
  Library Legislative Day (Washington, DC) Apr 11
  Library of Congress: Anniv....Apr 24
  Library Week, Natl....Apr 9
  New York Public Library: Anniv....May 23
  New York Public Library Centennial Celeb....May 20
  New York Public Library Centennial Gala May 22
  Night of a Thousand Stars (library read-alouds)....Apr 12
  Special Libraries Assn Annual Conf (Montreal, Canada)....June 10
  Thank You, School Librarian Day....Apr 12
  TV Turn-Off (Plymouth, MA)....Mar 5
Libyan Arab Jamahiriya: Revolution Day...Sept 1
Lichtenstein, Harvey: Birth....Apr 9
Lichtenstein, Roy: Birth....Oct 27
Liddy, G. Gordon: Birth....Nov 30
Lieberman, Joseph: Birth....Feb 24
Liebling, A.J.: Birth Anniv....Oct 18
Liechtenstein: National Day....Aug 15
Life Day, Evaluate Your....Oct 19
Life Is What You Make It, Sr Expo (Halifax, NS, Canada)....July 14
Light, Edison Fest of (Fort Myers, FL)....Feb 3
Light, Judith: Birth....Feb 9
Lightfoot, Gordon: Birth....Nov 17
Liliuokalani Deposed, Queen: Anniv....Jan 17
Lilly, William: Birth Anniv....Apr 30
Lily Fest, Easter in July (Smith River, CA)...July 8
Limerick Day....May 12
Limousine Scavenger Hunt (Tacoma, WA)..Oct 26
Lincoln, Abraham
  Assassination Anniv....Apr 14
  Assassination Conspirators Hanging....July 7
  Birth Anniv....Feb 12
  Birthday Observance (DE, OR)....Feb 6
  Enduring Mr. Lincoln, The (Clinton, MD)..Mar 2
  Gettysburg Address: Anniv....Nov 19
  John Wilkes Booth Escape Route Tour (Clinton, MD)....Apr 15, Sept 9
  Lincoln Birthplace Founder's Day Wknd (Hodgenvi, KY)....July 15
  Lincoln's Birthplace Cabin Wreath Laying (Hodgenville, KY)....Feb 12
  Signs Income Tax: Anniv....July 1
Lincoln, Mary Todd: Birth Anniv....Dec 13
Lind, Jenny: Birth Anniv....Oct 6
Lind, Jenny: US Premiere....Sept 11
Lindbergh, Anne M: Birth....June 22
Lindbergh, Charles Augustus: Birth Anniv..Feb 4
Lindbergh Flight: Anniv....May 20
Lindbergh Kidnapping: Anniv....Mar 1
Linden, Hal: Birth....Mar 20
Lingerie Week, Natl....Apr 23
Linn, Bambi: Birth....Apr 26
Linn-Baker, Mark: Birth....June 17
Linville, Larry: Birth....Sept 29
Lippman, Walter: Birth Anniv....Sept 23
Lister, Joseph: Birth Anniv....Apr 5
Liszt, Franz: Birth Anniv....Oct 22
Literacy Day, Intl (UN)....Sept 8
Literacy Day, Natl 6578....Page 488
Literacy Month, Natl....Sept 1
Literacy Volunteers America Conf (Buffalo, NY)....Oct 19
Literature
  A(ugusta) Baker's Dozen (Columbia, SC) Apr 28
  Aggiecon (College Station, TX)....Mar 23
  American Poet Laureate Establishment: Anniv....Dec 20
  Asimov, Isaac: Death Anniv....Apr 6
  Authors' Day, Natl....Nov 1
  Bad Poetry Day....Aug 18
  Be Kind to Editors and Writers Month....Sept 1
  Becky Thatcher Day (MO)....Dec 1
  Black Poetry Day....Oct 17
  Bollingen Prize Award: Anniv....Feb 19
  Boskone 32 (Framingham, MA)....Feb 17
  Cactus Pete's Carl Hayden Daze Celeb (Jackpot, NV)....July 2
  Cheltenham Festival of Literature (Cheltenham, England)....Oct 6
  Cowboy Christmas (Wickenburg, AZ)....Dec 1
  Dakota Cowboy Poetry Gathering (Medora, ND)....May 27
  Dickens Universe (Santa Cruz, CA)....July 30
  Edgar Allan Poe Fest (Cornwall, PA)....Oct 26
  Eliot, T.S.: Birth Anniv....Sept 26
  Eliza Doolittle Day....May 20
  First Magazine Published in America: Anniv....Feb 13
  Great Pikes Peak Cowboy Poetry Gath (Colo Sprngs, CO)....Aug 18
  Great Poetry Reading Day (Clio, MI)....Apr 28
  Hay-on-Wye Fest of Literature (Hay-on-Wye, Wales)....May 26
  Hemingway Days Fest (Key West, FL)...July 17
  Highlights Fdtn Writers Workshop (Chautauqua, NY)....July 15
  Hobbit Day....Sept 22
  John Steinbeck's Birthday (Salinas, CA) .Feb 27
  Joyce Papers Released: Anniv....Apr 6
  Laura Ingalls Wilder Gingerbread Sociable (Pomona, CA)....Feb 4
  Merriam, Eve: Death Anniv....Apr 11
  Montana Cowboy Poet Gathering (Lewistown, MT)....Aug 18
  Poetry Break....Jan 13
  Rochester Dickens Fest (Rochester, England)....June 1
  Salman Rushdie's Death Sentence: Anniv Feb 14
  Silent Spring Publication: Anniv....Apr 13
  Smith, Thorne: Birth Anniv....Mar 27
  Steinbeck Fest (Salinas, CA)....Aug 3
  Texas Cowboy Poetry Gathering (Alpine, TX)....Mar 3
  The Duchess Who Wasn't Day....Aug 27
  Tolkien Week....Sept 17

☆ Chase's 1995 Calendar of Events ☆ Index

Literature (cont'd)—Madonna

UK Year of Literature/Writing (Swansea, England) ... Jan 1
Ulysses, Marathon Reading of (Washington, DC) ... June 15
Underdog Day ... Dec 15
Vercors, Jean: Birth Anniv ... Feb 26
Ways With Words Literature Festival (Dartington, England) ... Aug 28
Wheatley, Phillis: Death Anniv ... Dec 5
Willa Cather Spring Conference (Red Cloud, NE) ... May 5
Zora Neale Hurston Fest (Eatonville, FL) ... Jan 26
Lithgow, John: Birth ... Oct 19
Lithuania: Baltic States' Independence Recognized: Anniv ... Sept 6
Lithuania: Independence Day ... Feb 16
Little Big Horn Days (Hardin, MT) ... June 22
Little, Cleavon: Birth Anniv ... June 1
Little 500 (Anderson, IN) ... May 27
Little League: See Baseball
Little League Baseball Week, Natl (Pres Proc) ... June 12
Little League Baseball World Series (Williamsport, PA) ... Aug 20
Little Richard: Birth ... Dec 25
Little, Richard Caruthers: Birth ... Nov 26
Live Aid Concerts: Anniv ... July 13
Liver Awareness Month, Natl ... Oct 1
Livermore, Mary: Birth Anniv ... Dec 19
Livestock: Acton Fair (Acton, ME) ... Aug 24
Livestock Show, Rio Grande Valley (Mercedes, TX) ... Mar 15
Livestock Show/Rodeo, Southwestern Expo (Ft Worth, TX) ... Jan 20
Livingston, Philip: Birth Anniv ... Jan 15
Livingston, Robert: Birth Anniv ... Nov 27
Livingston, Stanley: Birth ... Nov 24
Livingstone, David: Birth Anniv ... Mar 19
Lizard Race, World's Greatest (Lovington, NM) ... July 4
Lloyd, Christopher: Birth ... Oct 22
Lloyd Webber, Andrew: Birth ... Mar 22
Loan, US Takes Out Its First Loan: Anniv ... Sept 18
Lobster Fest, Maine (Rockland, ME) ... Aug 3
Lobster Race and Oyster Parade (Aiken, SC) ... May 5
Locke, Sondra: Birth ... May 28
Lockhart, June: Birth ... June 25
Locklear, Heather: Birth ... Sept 25
Lockwood, Belva A. Bennett: Birth Anniv ... Oct 24
Lodi Grape Festival (Lodi, CA) ... Sept 14
Loewy, Raymond: Birth Anniv ... Nov 5
Lofgren, Nils: Birth ... June 21
Lofton, James: Birth ... July 5
Log Cabin Day (Michigan) ... June 25
Log Show by the Sea (Brookings, OR) ... Sept 16
Logger Days (Libby, MT) ... July 13
Loggers/Sawdust Fest: State Chmpshp (Payson, AZ) ... July 15
Loggia, Robert: Birth ... Jan 3
Logging Museum Festival Days (Rangeley, ME) ... July 28
Loggins, Kenny: Birth ... Jan 7
Lolich, Michael Stephen: Birth ... Sept 12
Lollobrigida, Gina: Birth ... July 4
Loloma, Charles: Death Anniv ... June 9
Lombroso, Cesare: Birth Anniv ... Nov 18
London Bridge Days (Lake Havasu City, AZ) Oct 2
London, Jack: Birth Anniv ... Jan 12
London, Julie: Birth ... Sept 26
London: Iranian Embassy, Seizure: Anniv ... Apr 30
London Parade (England) ... Jan 1
Long Count Day ... Sept 22
Long, Crawford: First Use Anesthetic (Doctor's Day) ... Mar 30
Long, Howie: Birth ... Jan 6
Long, Huey P.: Assassination Anniv ... Sept 8
Long, Huey P.: Day ... Aug 30
Long, Shelley: Birth ... Aug 23
Longest War in History: Ending Anniv ... Feb 5
Longet, Claudine: Birth ... Jan 29
Longfellow, Henry Wadsworth: Birth Anniv. Feb 27
Longwood Gardens Autumn's Colors (Kennett Square, PA) ... Oct 1
Longwood Gardens Welcome Spring (Kennett Square, PA) ... Jan 21
Longyear, John M.: Birth Anniv ... Apr 15
Looff Carousel, Santa Cruz Beach: Anniv ... Aug 3
Look-Alike Day ... Apr 18
Loomis Day ... May 30
Loon Lake Harvest Festival (Aurora, MN) ... Sept 9
Loos, Anita: Birth Anniv ... Apr 26
Lopez, Nancy: Birth ... Jan 6
Lopez, Trini: Birth ... May 15
Lord, Jack: Birth ... Dec 30
Loren, Sophia: Birth ... Sept 20
Loring, Gloria: Birth ... Dec 10
Los Angeles (CA): Birth ... Sept 4
Lost Dutchman Days (Apache Junction, AZ) ... Feb 24
Lott, Ronnie: Birth ... May 8
Lott, Trent: Birth ... Oct 9
Loudon, Dorothy: Birth ... Sept 17
Louganis, Greg: Birth ... Jan 29
Louis, Joe: Birth Anniv ... May 13
Louis/Braddock/Schmeling Fight Annivs ... June 22
Louise, Tina: Birth ... Feb 11
Louisiana
Admissions Day ... Apr 30
Allons Manger (Belle Rose) ... Apr 23
American Tinnitus Assn Advisors Mtg (New Orleans) ... Sept 18
Bayou Hurricane: Anniv ... Oct 1
Black Heritage Festival of Louisiana (Lake Charles) ... Mar 3
Black History Parade (Monroe) ... Feb 25
Breaux Bridge Crawfish Fest (Breaux Bridge) ... May 5
Christmas Festival (Natchitoches) ... Dec 2
Christmas in Roseland (Shreveport) ... Nov 25
Church Point Buggy Festival (Church Point) ... June 2
Contraband Days (Lake Charles) ... May 2
Creole Christmas in French Quarter (New Orleans) ... Dec 1
Dubach Chuck Wagon Races (Dubach) ... June 29
Earth Day Regional Fest (Baton Rouge) ... Apr 23
Fest International de Louisiane (Lafayette) ... Apr 25
First Bloom Fest (Shreveport) ... Apr 29
French Quarter Fest (New Orleans) ... Apr 7
Gumbo Fest (Bridge City) ... Oct 13
Kiwanis Intl Convention (New Orleans) ... June 23
Louis Armstong: Cultural Legacy (New Orleans) ... Oct 28
Louisiana Passion Play (Ruston) ... May 5
Louisiana Purchase Day ... Dec 20
Louisiana Yambilee (Opelousas) ... Oct 25
Mamou Cajun Music Festival (Mamou) ... June 2
Mamou Mardi Gras Celebration (Mamou) Feb 28
Mardi Gras (Lafayette) ... Feb 25
Marksville Easter Egg Knocking Contest (Marksville) ... Apr 16
Melrose Plantation Arts/Crafts Fest (Natchitoches) ... June 10
Miss Louisiana Pageant (Monroe) ... June 15
Natchitoches/NW State U Folk Fest (Natchitoches) ... July 14
New Orleans Boat & Sportfishing Show (New Orleans) ... Feb 1
New Orleans Jazz/Heritage Fest (New Orleans) ... Apr 28
New Orleans Wine and Food Experience (New Orleans) ... July 13
Newspaper Assn America Conv (New Orleans) ... Apr 23
Office Olympics (Shreveport) ... Sept 15
Peach Fest, Louisiana (Ruston) ... June 9
Poke Salad Fest (Blanchard) ... May 11
St. Patrick's Day Parade (Baton Rouge) ... Mar 18
Shrimp and Petroleum Fest, Louisiana (Morgan City) ... Aug 31
Songs of My People (Shreveport) ... July 8
Southwest Louisiana State Fair/Expo (Lake Charles) ... Oct 26
State Fair of Louisiana (Shreveport) ... Oct 20
Taste of LA Food Festival (Monroe) ... Apr 23
Trash Fest (Anacoco) ... May 27
Twin Cities Krew Mardi Gras Parade (Monroe) ... Feb 4
Twin Cities Model Train Show (Monroe) ... Sept 2
USF&G Sugar Bowl (New Orleans) ... Jan 2
Veterans Day Celebration (Mamou) ... Nov 11
Lousma, Jack: Birth ... Feb 29
Love, Edmund George: Birth ... Feb 14
Love May Make World Go/Laughter Keeps Us ... Feb 8
Love, Mike: Birth ... Mar 15
Love Week, Celebration of ... Feb 12
Lovecraft, H.P.: Birth Anniv ... Aug 20
Lovejoy, Elijah P: Birth Anniv ... Nov 9
Lover's Day, Book Day and (Spain) ... Apr 23
Lover's Fair (Belgium) ... Dec 7
Lovera, Juan: Birth Anniv ... Dec 26
Lovitz, Jon: Birth ... July 21
Low, Juliette: Birth Anniv ... Oct 31
Lowe, Rob: Birth ... Mar 17
Lowell, Amy: Birth Anniv ... Feb 9
Lowell, James R.: Birth Anniv ... Feb 22
Lowell, Percival: Birth Anniv ... Mar 13
Lowry, Mike: Birth ... Mar 8
Loy, Myrna: Birth Anniv ... Aug 2
Loyalty Day (Pres Proc) ... May 1
Loyalty Day 6556, 6680 ... Page 488
Loyalty Days/Seafair Fest (Newport, OR) ... May 4
Lu Pan: Birth Anniv ... July 10
Lucas, George: Birth ... May 14
Lucci, Susan: Birth ... Dec 23
Luce, Clare Boothe: Birth Anniv ... Mar 10
Luck: Make Your Own Luck Day ... Aug 26
Luft, Lorna: Birth ... Nov 21
Lugar, Richard G.: Birth ... Apr 4
Lughnasadh ... Aug 1
Lukas, Anthony: Birth ... Apr 25
Lully, Jean Baptiste: Birth Anniv ... Nov 28
Lumet, Sidney: Birth ... June 25
Lumpy Rug Day ... May 3
Lunden, Joan: Birth ... Sept 19
Lung Assn, Amer: Christmas Seal Campaign Oct 1
Lupercalia ... Feb 15
Lupino, Ida: Birth ... Feb 4
LuPone, Patti: Birth ... Apr 21
Lupus Awareness Month ... Oct 1
Lusitania Sinking: Anniv ... May 7
Lutali, A.P.: Birth ... Dec 24
Luther, Martin: Birth Anniv ... Nov 10
Lutheran School Week, Natl ... Mar 5
Luxembourg
Beer Fest ... July 16
Blessing of the Wine (Greiveldange) ... Dec 26
Bretzelsonndeg (Pretzel Sunday) ... Mar 26
Burgsonndeg ... Feb 12
Candlemas ... Feb 2
Emaishen ... Apr 17
Ettelbruck Remembrance Day ... July 6
Liberation Ceremony ... Sept 9
National Holiday ... June 23
Osweiler ... Mar 27
Our Lady of Girsterklaus Procession ... Aug 20
Schuebermess (Shepherd's Fair) ... Aug 20
Luyts, Jan: Birth Anniv ... Sept 19
Lyman, Dorothy: Birth ... Apr 18
Lynch, David: Birth ... Jan 20
Lynch, Thomas: Birth Anniv ... Aug 5
Lynley, Carol: Birth ... Feb 13
Lynn, Loretta: Birth ... Apr 14
Lyon, Mary: Birth Anniv ... Feb 28

M

Ma, Yo-Yo: Birth ... Oct 7
Maass, Clara: Day (New Jersey) ... June 28
Mabon ... Sept 23
MacArthur, Douglas: Birth Anniv ... Jan 26
MacArthur, James: Birth ... Dec 8
Macau
Macau Day ... June 24
Portuguese Revolution Anniv ... Apr 25
Procession Our Lady Fatima ... May 13
Restoration of Independence Day ... Dec 1
MacaVulay, Thomas B: Birth Anniv ... Oct 25
Macchio, Ralph: Birth ... Nov 4
MacDonald, John A.: Birth Anniv ... Jan 11
Macedonian Uprising: St. Elias Day ... Aug 2
MacFadden, Bernarr: Birth Anniv ... Aug 16
MacGraw, Ali: Birth ... Apr 1
Mack, Connie: Birth ... Oct 29
Mackenzie, Gisele: Birth ... Jan 10
Mackie, Bob: Birth ... Mar 24
Mackinac Bridge Walk (St. Ignace, MI) ... Sept 4
MacLaine, Shirley: Birth ... Apr 24
MacLeish, Rod: Birth ... Jan 15
MacLeod, Gavin: Birth ... Feb 28
MacMurray, Fred: Birth Anniv ... Aug 30
MacNeil, Robert: Birth ... Jan 19
MacNelly, Jeff: Birth ... Sept 17
MacRae, Sheila: Birth ... Sept 24
Mad magazine: Kurtzman, Harvey: Birth Anniv ... Oct 3
Madagascar
Commemoration Day ... Mar 29
Independence Day ... June 26
National Holiday ... Dec 30
MADD Leadership Development Conf (Dallas, TX) ... May 11
MADD's Natl Candlelight Vigil (Washington, D.C.) ... Dec 2
MADD's Project Red Ribbon ... Nov 1
Madden, John: Birth ... Apr 10
Maddox, Lester Garfield: Birth ... Sept 30
Maddux, Greg: Birth ... Apr 14
Madigan, Edward: Birth ... Jan 13
Madison, Dolly: Birth Anniv ... May 20
Madison, James: Birth Anniv ... Mar 16
Madonna: Birth ... Aug 16

Index ☆ Chase's 1995 Calendar of Events ☆

Magazine—Maryland

Magazine, First Published in America: Anniv....Feb 13
Magellan, Ferdinand: Death Anniv....Apr 27
Magic Day, Natl....Oct 31
Magic Get-Together, Abbott's (Colon, MI)....Aug 2
Magic Week, Natl....Oct 25
Magna Carta Day....June 15
Maifest (Covington, KY)....May 19
Maifest (Hermann, MO)....May 20
Mail: Pony Express, Inauguration of: Anniv....Apr 3
Mail: V-Mail Delivery: Anniv....June 22
Mail: World Post Day (UN)....Oct 9
Mailer, Norman: Birth....Jan 31
Maiman, Theodore: Birth....July 11
Maine
  Acton Fair (Acton)....Aug 24
  Admission Day....Mar 15
  Bangor Labor Day Road Race (Bangor)....Sept 4
  Blue Hill Fair (Blue Hill)....Aug 31
  Central Maine Egg Fest (Pittsfield)....July 22
  Common Ground Country Fair (Windsor)....Sept 22
  Fryeburg Fair (Fryeburg)....Oct 1
  Intl Fest (Calais, ME/St. Stephen, NB, Canada)....Aug 5
  Kenduskeag Stream Canoe Race (Bangor)....Apr 22
  Logging Museum Festival Days (Rangeley)....July 28
  Maine Fest of the Arts (Brunswick)....Aug 3
  Maine Highland Games (Brunswick)....Aug 19
  Maine Law: Anniv....June 2
  Maine Lobster Fest (Rockland)....Aug 3
  Maine Potato Blossom Festival (Fort Fairfield)....July 16
  New England Sled Dog Races (Rangeley)....Mar 11
  Patriot's Day....Apr 17
  Thomas Pt Beach Bluegrass Fest (Brunswick)....Sept 1
  Windjammer Days (Boothbay Harbor)....June 28
  Yarmouth Clam Fest (Yarmouth)....July 14
Maintenance Day....Jan 18
Maintenance Personnel Day....Oct 16
Majerlee, Don: Birth....Sept 9
Majors, Lee: Birth....Apr 23
Majowski, Don: Birth....Feb 24
Makarova, Natalia: Birth....Nov 21
Make A Difference Day....Oct 21
Make Up Your Mind Day....Dec 31
Make Up Your Own Holiday Day....Mar 26
Make Your Own Luck Day....Aug 26
Makeba, Miriam: Birth....Mar 4
Makepeace, Chris: Birth....Apr 22
Malawi
  Independence Day....July 6
  Kamuzu Day....May 14
  Martyr's Day....Mar 3
Malaysia
  National Day....Aug 31
  Tioman Intl Regatta (Pahang)....Sept 3
Malcolm X: Assassination Anniv....Feb 21
Malcolm X: Birth Anniv....May 19
Malden, Karl: Birth....Mar 22
Maldives: National Day....July 26
Mali: Proclamation of the Republic Anniv....Sept 22
Malinowski, Bronislaw: Birth Anniv....Apr 7
Malkovich, John: Birth....Dec 9
Malle, Louis: Birth....Oct 30
Malone, Karl: Birth....July 24
Malone, Moses: Birth....Mar 23
Malpighi, Marcello: Birth Anniv....Mar 10
Malta
  Carnival....May 6
  Independence Day....Sept 21
  Republic Day....Dec 13
  Siege Broken: Anniv....Sept 8
Maltin, Leonard: Birth....Dec 18
Mamet, David: Birth....Nov 30
Mammography Day, Natl 6615....Page 488
Man Watchers' Compliment Week....July 3
Man Watchers Week....Jan 16
Man Will Never Fly Society Meeting (Kitty Hawk, NC)....Dec 16
Man-Powered Flight, First: Anniv....Aug 23
Manchester, Melissa: Birth....Feb 15
Manchester, William: Birth....Apr 4
Mandarich, Tony: Birth....Sept 23
Mandela Inuaguration: Anniv....May 10
Mandela, Nelson: Birthday....July 18
Mandela, Nelson: Capture Anniv....Aug 4
Mandela, Nelson: Prison Release Anniv....Feb 11
Mandlikova, Hana: Birth....Feb 19
Mandrell, Barbara Ann: Birth....Dec 25
Mangione, Chuck: Birth....Nov 29
Manilow, Barry: Birth....June 17
Mankiewicz, Joseph L.: Birth Anniv....Feb 11
Manley, Dexter: Birth....Feb 2
Mann, Horace: Birth Anniv....May 4
Mann, James: Birth Anniv....Oct 20
Mann, Marty: Birth Anniv....Oct 15
Manning, Danny: Birth....May 17
Mansfield, Arabella: Birth Anniv....May 23
Manson, Patrick: Birth Anniv....Oct 3
Manstein, Erich von: Birth Anniv....Nov 24
Mantle, Mickey: Birth....Oct 20
Mantooth, Randolph: Birth....Sept 19
Manufacturing Week, Natl....Mar 13
Mao Tse-Tung: Birth Anniv....Dec 26
Maple Fair, Parke County (Rockville, IN)....Feb 25
Maple Fest, Highland County (Highland County, VA)....Mar 11
Maple Fest of Nova Scotia (Cumberland Co, NS, Canada)....Mar 18
Maple Fest, Pennsylvania (Meyersdale, PA)....Apr 22
Maple Fest, Vermont (St. Albans, VT)....Apr 21
Maple Leaf Fest (Carthage, MO)....Oct 15
Maple Sugaring Fest (New Preston, CT)....Mar 13
Maple Syrup Fest (Beaver, PA)....Apr 8
Maple Syrup Festival (Wakarusa, IN)....Mar 24
Maple Syrup Saturday (Appleton, WI)....Mar 25
Maps: Natl Reading a Road Map Week....Apr 4
Maps: Oatmeal Fest (Bertram/Oatmeal, TX)....Sept 1
Maradona, Diego: Birth....Oct 30
Marathon. See Running Events
Marathon, Battle of: Anniv....Sept 9
Marathon, Days of: Anniv....Sept 2
Marble Meet at Amana (Amana, IA)....June 3
Marble Meet, Northeast (Marlborough, MA)....Oct 7
Marbles: Natl Knuckles Down Month....Apr 1
Marburg University Founding: Anniv (Germany)....May 30
Marceau, Marcel: Birth....Mar 22
March of Dimes Birth Defects Prevention Month....Jan 1
March of Dimes' WalkAmerica....Apr 29
Marchand, Nancy: Birth....June 19
Marciulionis, Sarunas: Birth....June 13
Marconi, Guglielmo: Birth Anniv....Apr 25
Mardi Gras....Feb 28
Mardi Gras (Lafayette, LA)....Feb 25
Mardi Gras Fest (Prince George, BC, Canada)....Feb 9
Mardi Gras, Snowmass (Snowmass Village, CO)....Feb 28
Marfan Syndrome Awareness Month....Feb 1
Margaret, Princess: Birth....Aug 21
Marie Antoinette: Execution Anniv....Oct 16
Marin, Cheech: Birth....July 13
Marinaro, Ed: Birth....Mar 31
Marine Corps Birthday....Nov 10
Marine War Memorial: Hayes, Ira: Birth Anniv....Jan 12
Marino, Dan: Birth....Sept 15
Marion Popcorn Fest (Marion, OH)....Sept 7
Maritime Day, Natl....May 22
Maritime Day, Natl 6564, 6692....Page 488
Maritime Fest (Charleston, SC)....Sept 15
Market Ability Month....Jan 1
Market Square Days (Portsmouth, NH)....June 9
Marley, Bob: Day....May 11
Maroon Fest (Jamaica)....Jan 6
Marquette, Jacques: Death Anniv....May 18
Marriage Celebration (Belleville, IL)....Sept 10
Marriage Day, World....Feb 12
Marriage Enrichment Weekend (Belleville, IL)....Nov 10
Marriage: Loving v Virginia: Anniv....June 12
Marriage of Elizabeth and Philip: Anniv....Nov 20
Marriage: Pleasure Your Mate Month....Sept 1
Marriage: Proposal Day....Mar 20, Sept 23
Marsalis, Branford: Birth....Aug 26
Marsalis, Wynton: Birth....Oct 18
Marsh, Jean: Birth....July 1
Marshall, Garry: Birth....Nov 13
Marshall, George: Birth Anniv....Dec 31
Marshall Islands: National Day....Oct 21
Marshall, John: Birth Anniv....Sept 24
Marshall, Penny: Birth....Oct 15
Marshall, Peter: Birth....Mar 30
Marshall, Thurgood: Birth Anniv....July 2
Marshall, Thurgood, Death of 6526....Page 488
Marshall, Thurgood: Sworn in to Supreme Court: Anniv....Oct 2
Marshall, Wilber: Birth....Apr 18
Martens, Wilfried: Birth....Apr 19
Martin, Billy: Birth Anniv....May 16
Martin, Dean: Birth....June 17
Martin, Dick: Birth....Jan 30
Martin, Harvey: Birth....Nov 16
Martin, Judith: Birth....Sept 13
Martin, Lynn: Birth....Dec 26
Martin, Pamela Sue: Birth....Jan 5
Martin, Steve: Birth....Aug 14
Martin Z. Mollusk Day (Ocean City, NJ)....May 4
Martinez, Jose: Birth....May 14
Martinmas....Nov 11
Martinmas Goose (Switzerland)....Nov 11
Martino, Al: Birth....Oct 7
Martyrs Day (Bangladesh)....Feb 21
Martyrs' Day (Panama)....Jan 9
Marvell, Andrew: Birth Anniv....Mar 31
Marx, Harpo: Birth Anniv....Nov 23
Marx, Richard: Birth....Sept 16
Mary Tyler Moore Show: Final Episode Anniv....Mar 19
Maryland
  African-American Heritage Fest (Snow Hill)....Sept 10
  All Dressed Up (Clinton, MD)....June 1
  Annapolis Run (Annapolis)....Aug 27
  Antietam Battlefield Meml Illumination (Sharpsburg)....Dec 2
  Antique Valentine Exhibit (Clinton)....Feb 9
  Autumn Glory Fest (Oakland)....Oct 12
  Bel Air Fest for the Arts (Bel Air)....Sept 17
  Boonesborough Days (Boonsboro)....Sept 9
  Booth, John Wilkes: Escape Route Tour (Clinton)....Sept 9
  Candlelight Walking Tour/Chestertown (Chestertown)....Sept 16
  Canoe House (Snow Hill)....Aug 6
  Catoctin Colorfest Arts and Crafts Show (Thurmont)....Oct 14
  Celebrity Read a Book Week (Williamsport)....Jan 22
  Celtic Fest, Southern Maryland (St. Leonard)....Apr 29
  Chesapeake Antique Fire Muster (Westminster)....May 13
  Chesapeake Club CFA All-Breed Cat Show (Baltimore)....Mar 11
  Chesapeake Turtle Derby (Baltimore)....July 5
  Chestertown Tea Party Fest (Chestertown)....May 27
  Chicken Clucking Contest (Baltimore)....June 21
  Chinese Lunar New Year Fest (Baltimore)....Feb 5
  City of Baltimore's Easter Celebration (Baltimore)....Apr 12
  Clover Leaf Planting (Annapolis)....Apr 29
  Count Your Losses (Clinton)....Aug 31
  Crab and Clam Bake (Crisfield)....July 19
  Defenders Day....Sept 12
  Delmarva Chicken Festival (Federalsburg)....June 9
  Doll Show (Baltimore)....July 19
  Elvis Presley Salute (Baltimore)....Aug 16
  Enduring Mr. Lincoln, The (Clinton)....Mar 2
  Fells Pt Fun Fest (Baltimore)....Oct 7
  French and Indian Rendezvous (Big Pool)....May 27
  Friendsville Fiddle/Banjo Contest (Friendsville)....July 15
  Governor's Bay Bridge Run (Annapolis)....May 7
  Governor's Invitl Firelock Match (Big Pool)....Sept 23
  Halfway Park Days (Hagerstown)....May 27
  Hard Crab Derby and Fair, Natl (Crisfield)....Sept 1
  Hog Calling Contest (Baltimore)....July 12
  Intergenerational Week (Williamsport)....Apr 23
  Intl Auto Show (Baltimore)....Feb 4
  John Wilkes Booth Escape Route Tour (Clinton)....Apr 15
  Jonathan Hager Frontier Craft Days (Hagerstown)....Aug 5
  Kids' Day (Baltimore)....Aug 2
  Kid's Swap Shop (Baltimore)....Aug 7
  Leitersburg Peach Festival (Leitersburg)....Aug 12
  Margaret Brent Demands a Political Voice: Anniv....June 24
  Maryland Day....Mar 25
  Maryland Home and Flower Show (Baltimore)....Mar 8
  Maryland Recreational Vehicle Show (Timonium)....Sept 9
  Maryland Renaissance Fest (Annapolis)....Aug 26
  Maryland Symphony at Antietam (Sharpsburg)....July 1
  McHenry Highland Fest (McHenry)....June 3
  Mid-Atlantic Stunt Kite Chmpshp (Ocean City)....May 6
  Military Field Days (Big Pool)....July 29
  Mitten Tree Kick-Off Ceremony (Baltimore)....Dec 1
  Museum Open House (Clinton)....Nov 11
  National Anthem Day....Sept 14

☆ Chase's 1995 Calendar of Events ☆ Index

Maryland (cont'd)—Meredith

Nautical/Wildlife Art Fest/Craft Show (Ocean City) . . . Jan 14
Pocomoke River Canoe Challenge (Snow Hill) . . . June 17
Postcard Show (Hagerstown) . . . May 26
Preakness Frog Hop (Baltimore) . . . May 17
Preakness Stakes (Baltimore) . . . May 20
Recreational Vehicle Show (Timonium) . . Feb 17
Rites of Spring (Patapsco St Pk) . . . Apr 22
St. Mary's County Oyster Fest (Leonardtown) . . . Oct 21
St. Patrick's Day Celebration (Baltimore) . Mar 17
Snow Hill Heritage Weekend (Snow Hill) Sept 22
Spring Day Celebration (Baltimore) . . . Mar 20
State Fair (Timonium) . . . Aug 26
Sugarloaf Craft Fest (Gaithersburg) . . . Apr 7, Nov 17, Dec 8
Sugarloaf Craft Festival (Timonium) . . . Apr 28, Oct 13
Sugarloaf's Crafts Fest (Somerset) . . . May 19, Sept 29
Thanksgiving Canned Goods Drive (Baltimore) . . . Nov 1
Towsontown Spring Festival (Towson) . . . May 6
Tree-Mendous Month (Annapolis) . . . Apr 6
US Mid-Amateur (Golf) Chmpshp (Owings Mill/ Pikesville) . . . Sept 16
US Senior Open (Golf) Championship (Bethesda) . . . June 29
Vegetarian Resource Group's Essay Contest for Kids . . . May 1
Victorian Christmas by Candlelight (Clinton) . . . Dec 9
Waterfowl Festival (Easton) . . . Nov 10
M*A*S*H: The Final Episode: Anniv . . . Feb 28
Mason, Bobbie Ann: Birth . . . May 1
Mason, Dave: Birth . . . May 10
Mason, Jackie: Birth . . . June 9
Mason, Marsha: Birth . . . Apr 3
Mason, Pamela: Birth . . . Mar 10
Masquerade Adventure, Mill Avenue (Tempe, AZ) . . . Oct 31
Massachusetts
American Heroine Rewarded: Anniv . . . June 8
American History Month . . . Feb 1
Amherst's Teddy Bear Rally (Amherst) . . . Aug 5
Antec '95 (Boston) . . . May 7
Army and Navy Union Day . . . Dec 9
Basketball Hall of Fame Dinner (Springfield) . . . May 6
Basketball Hall of Fame Enshrinement (Springfield) . . . May 8
Bastille Day Celebration (Boston) . . . July 14
Big E (West Springfield) . . . Sept 15
Boskone 32 (Framingham) . . . Feb 17
Boston Marathon (Boston) . . . Apr 17
Bridge over the Neponset: Anniv . . . Apr 1
Bunker Hill Day (Suffolk County) . . . June 17
Celebration of Peace Day (Concord) . . . Aug 6
Children's Day . . . June 11
Christmas Stroll (Nantucket) . . . Dec 1
Codman House Antique Vehicle Meet (Lincoln) . . . July 16
Codman House Artisans' Fair (Lincoln) . Sept 10
Cranberry Harvest Weekend (Nantucket Island) . . . Oct 13
Daffodil Fest (Nantucket Is) . . . Apr 28
Dietary Managers Assn Mtg/Expo (Boston) . . . July 30
Evacuation Day (Boston) . . . Mar 17
Expanded New England Kindergarten Conf (Randolph) . . . Nov 16
First Night (Boston) . . . Dec 31
First Women's Collegiate Basketball: Anniv . . . Mar 22
French Library Anniv Conf (Boston) . . . Oct 2
French Library Anniv Gala (Boston) . . . Nov 9
Grandparents Day . . . Oct 1
Harborfest (Nantucket Island) . . . June 16
Indian Day . . . Aug 12
John F. Kennedy Day . . . Nov 26
Lafayette Day . . . May 20
Liberty Tree Day . . . Aug 14
NBA Day in Basketball City, USA (Springfield) . . . Oct 27
Northeast Marble Meet (Marlborough) . . . Oct 7
Original Massachusetts Home Show (West Springfield) . . . Mar 21
Patriot's Day . . . Apr 17
Pet Cemeteries Fall Mtg, Intl Assn . . . Oct 6
Railroad and Hobby Show (Springfield) . . . Feb 4
Ratification Day . . . Feb 6
Reenactment of Boston Tea Party (Boston) . . . Dec 17
Sheep Shearing Fest (North Andover) . . . May 14
South Shore Singles Founding Anniv (Norwell) . . . May 26
Spalding Enshrinement Celebr Golf Tourn (Agawam, MA) . . . May 8
Starter Hall of Fame Tip-Off Classic (Springfield) . . . Nov 24
Student Government Day . . . Apr 7
Teacher's Day . . . June 4
34th St Express (Boston) . . . Dec 9
Try This On: Clothing/Gender/Power (New Bedford) . . . May 13
TV Turn-Off (Plymouth) . . . Mar 5
US Women's Amateur (Golf) Chmpshp (Brookline) . . . Aug 7
US Women's Mid-Amateur (Golf) Chmpshp (Manchester) . . . Sept 18
Youth Honor Day . . . Oct 31
Massacre, Valentine's Day: Anniv . . . Feb 14
Massey, Anna: Birth . . . Aug 11
Master Gardener Day . . . Mar 21
Masters, Edgar L.: Birth Anniv . . . Aug 23
Masters, William Howell: Birth . . . Dec 27
Mastroianni, Marcello: Birth . . . Sept 28
Mata Hari: Anniv of Execution . . . Oct 15
Matanzas Mule Day . . . Apr 27
Matchcover Society Conv, Rathkamp (Newark, NJ) . . . Aug 13
Mathematics Education Month . . . Apr 1
Mathers, Jerry: Birth . . . June 2
Matheson, Tim: Birth . . . Dec 31
Mathews, Harlan: Birth . . . Jan 17
Mathewson, Christy: Birth Anniv . . . Aug 12
Mathias, Bob: Birth . . . Nov 17
Mathis, Johnny: Birth . . . Sept 30
Matlin, Marlee: Birth . . . Aug 24
Matthau, Walter: Birth . . . Oct 1
Mattingley, Don: Birth . . . Apr 20
Maugham, W. Somerset: Birth Anniv . . . Jan 25
Mauldin, Bill: Birth . . . Oct 29
Maundy Thursday (Holy Thursday) . . . Apr 13
Mauritania: Independence Day . . . Nov 28
Mauritius: Independence Day . . . Mar 12
Max, Peter: Birth . . . Oct 19
Maxwell, James Clerk: Birth Anniv . . . Nov 13
May Day . . . May 1
May Day (Lumpkin, GA) . . . May 1
May Day Bank Holiday (United Kingdom) . . May 1
May, Elaine: Birth . . . Apr 21
May Fellowship Day . . . May 5
May in Montclair (Montclair Twshp, NJ) . . . May 1
May Ray Day . . . May 19
Mayall, John: Birth . . . Nov 29
Mayer, Maria Goeppert:Birth Anniv . . . June 28
Mayes, Ruben: Birth . . . June 6
Mayfield, Curtis: Birth . . . June 3
Mayflower Day . . . Sept 16
Mayo, Charles: Birth Anniv . . . July 19
Mayo, Virginia: Birth . . . Nov 30
Mayo, William J: Birth Anniv . . . June 29
Mayor: Stokes Becomes First Black Mayor in US: Anniv . . . Nov 13
Mays, Willie: Birth . . . May 6
Mazowiecki, Tadeusz: see Solidarity Founding . . . Aug 31
Mazursky, Paul: Birth . . . Apr 25
McAfee, Mildred: Birth . . . May 12
McAliskey, Bernadette Devlin: Birth . . . Apr 23
McArdle, Andrea: Birth . . . Nov 4
McAuliffe, Christa: Birth Anniv . . . Sept 2
McBride, Patricia: Birth . . . Aug 23
McCain, John Sidney, III: Birth . . . Aug 29
McCambridge, Mercedes: Birth . . . Mar 17
McCann, Les: Birth . . . Sept 23
McCarthy, Eugene: Birth . . . Mar 29
McCartney, Linda: Birth . . . Sept 24
McCartney, Paul: Birth . . . June 18
McClanahan, Rue: Birth . . . Feb 21
McClure, Doug: Birth . . . May 11
McConnell, Mitch: Birth . . . Feb 20
McCoo, Marilyn: Birth . . . Sept 30
McDaniel, Hattie: Birth Anniv . . . June 10
McDaniel, Xavier: Birth . . . June 4
McDowall, Roddy: Birth . . . Sept 17
McDowell, Jack: Birth . . . Jan 16
McDowell, Malcolm: Birth . . . June 13
McEnroe, John Patrick, Jr: Birth . . . Feb 16
McEntire, Reba: Birth . . . Mar 28
McEwen, Mark: Birth . . . Sept 16
McFarland, George (Spanky): Birth Anniv . . . Oct 2
McFerrin, Bobby: Birth . . . Mar 11
McGavin, Darren: Birth . . . May 7
McGinley, Phyllis: Birth . . . Mar 21
McGoohan, Patrick: Birth . . . Mar 19
McGovern, Elizabeth: Birth . . . July 18
McGovern, George Stanley: Birth . . . July 19
McGovern, Maureen: Birth . . . July 27
McGowan, William: Death Anniv . . . June 8
McGriff, Fred: Birth . . . Oct 31
McGuffey, William H.: Birth Anniv . . . Sept 23
McGuinn, Roger: Birth . . . July 13
McGuire, Dorothy: Birth . . . June 14
McGuire's St. Patrick's Day Run (Pensacola, FL) . . . Mar 11
McHale, Kevin: Birth . . . Dec 19
McInerney, Jay: Birth . . . Jan 13
McKay, Jim: Birth . . . Sept 24
McKean, Michael: Birth . . . Oct 17
McKeon, Doug: Birth . . . June 10
McKeon, Nancy: Birth . . . Apr 4
McKeon, Philip: Birth . . . Nov 11
McKernan, John R., Jr: Birth . . . May 20
McKinley, Ida Saxton: Birth Anniv . . . June 8
McKinley, William: Birth Anniv . . . Jan 29
McKinley, William: Death Anniv . . . Sept 14
McKuen, Rod: Birth . . . Apr 29
McLarty, Thomas F. (Mack): Birth . . . June 14
McLaughlin, John Joseph: Birth . . . Mar 29
McLean, Don: Birth . . . Oct 2
McLuhan, Marshall: Birth Anniv . . . July 21
McMahon, Ed: Birth . . . Mar 6
McMahon, Jim: Birth . . . Aug 21
McMichael, Steve: Birth . . . Oct 17
McNair, Barbara: Birth . . . Mar 4
McNair, Ronald E: Birth Anniv . . . Oct 12
McNamara, Robert S.: Birth . . . June 9
McNichol, Jimmy: Birth . . . July 2
McNichol, Kristy: Birth . . . Sept 9
McRae, Carmen: Birth . . . Apr 8
McVie, Christine: Birth . . . July 12
McWherter, Ned Ray: Birth . . . Oct 15
Mead, Margaret: Birth Anniv . . . Dec 16
Meadows, Audrey: Birth . . . Feb 8
Means, Russell Charles: Birth . . . Nov 10
Meara, Anne: Birth . . . Sept 20
Mears, Rick: Birth . . . Dec 3
Meat Loaf: Birth . . . Sept 27
Meatout, Great American . . . Mar 20
Mecklenburg Day (NC) . . . May 20
Medal of Honor, First World War II: Anniv . . Feb 10
Medical Lab Week, Natl . . . Apr 9
Medical School for Women Opened: Anniv . . Nov 1
Medieval Fair (Norman, OK) . . . Apr 7
Medieval Faire, Hoggetowne (Gainesville, FL) . . . Feb 17
Meditation, World Peace, Annual . . . Dec 31
Meese, Edwin, III: Birth . . . Dec 2
Mehta, Zubin: Birth . . . Apr 29
Meier, Garry . . . Dec 2
Mekka, Eddie: Birth . . . June 14
Mellencamp, John Cougar: Birth . . . Oct 7
Mellon, Andrew W.: Birth Anniv . . . Mar 24
Melon Days (Green River, UT) . . . Sept 15
Melville, Herman: Birth Anniv . . . Aug 1
Melville: Moby Dick Marathon (Mystic, CT) July 31
Memo Day, Natl . . . May 19
Memorial Day. See also Confederate
Memorial Day (Observed) . . . May 29
Memorial Day (Pres Proc) . . . May 29
Memorial Day (Traditional) . . . May 30
Memorial Day (Virgin Islands) . . . May 29
Memorial Day, Confederate (AL, MS) . . . Apr 24
Memorial Day, Confederate (FL, GA) . . . Apr 26
Memorial Day Parade and Services (Gettysburg, PA) . . . May 29
Memorial Day, Workers . . . Apr 28
Memory Day . . . Mar 21
Memory Days (Grayson, KY) . . . May 25
Men's Health Week, Natl 6700 . . . Page 488
Mencken, Henry Louis: Birth Anniv . . . Sept 12
Mendes, Sergio: Birth . . . Feb 11
Menendez de Aviles, Pedro: Birth Anniv . . . Feb 15
Mennonite: Pioneer Days (Steinback, Man, Canada) . . . Aug 4
Mennonite Relief Sale (Hutchinson, KS) . . . Apr 7
Menotti, Gian Carlo: Birth . . . July 7
Mensa Annual Meeting, American (St. Louis, MO) . . . July 4
Mental Health Month . . . May 1
Mental Illness Awareness Week . . . Oct 1
Mental Illness Awareness Week 6603 . . . Page 488
Mental Retardation Awareness Month . . . Mar 1
Menuhin, Yehudi: Birth . . . Apr 22
Mercer, John Herndon (Johnny): Birth Anniv . . . Nov 18
Merchant Sailing Ship Preservation Day . . . Nov 8
Mercouri, Melina: Birth Anniv . . . Oct 18
Meredith, Burgess: Birth . . . Nov 16
Meredith, Don: Birth . . . Apr 10
Meredith (James) Enrolls at Ole Miss: Anniv . . . Sept 30

Index ☆ Chase's 1995 Calendar of Events ☆

Meriwether, Lee: Birth	May 27
Merman, Ethel: Birth Anniv	Jan 16
Merriam, Eve: Death Anniv	Apr 11
Merrick, David: Birth	Nov 27
Merrill, Robert: Birth	June 4
Merrill, Stephen: Birth	June 21
Messina Earthquake Anniv	Dec 28
Messina, Jim: Birth	Dec 5
Metal craft: Eisenfest (Amana, IA)	Sept 15
Meteor Showers, Perseid	July 23
Meteorological Day, World (UN)	Mar 23
Metric Conversion Act: Anniv	Dec 23
Metric Week, Natl	Oct 8
Metropolitan Opera House: Opening Anniv	Oct 22
Metzenbaum, Howard M.: Birth	June 4
Mexican Fiesta (Milwaukee, WI)	Aug 25
Mexican-American: Day of the Teacher (El Dia Del Maestro)	May 10

Mexico
Anniv of Constitution	Feb 5
Blessing of Animals at Cathdral	Jan 17
Cinco de Mayo	May 5
Day of the Dead	Nov 1
Day of the Holy Cross	May 3
Dia de la Candelaria	Feb 2
Dia de la Raza	Oct 12
Feast of Our Lady of Solitude	Dec 18
Feast of the Radishes (Oaxaca)	Dec 23
Guadalupe Day	Dec 12
Independence Day	Sept 16
Mexico City Earthquake: Anniv	Sept 19
Posadas	Dec 16
Revolution Anniv	Nov 20
San Isidro Day	May 15
Sonora Showcase (Yuma, AZ)	Jan 28
Treaty of Guadalupe Hidalgo (with US)	Feb 2

Meyer, Nicholas: Birth	Dec 24
Micajah (Autryville, NC)	July 13
Micajah Autry Day (Autryville, NC)	July 15
Michael, George: Birth	June 25
Michaelmas	Sept 29
Michelangelo: Birth Anniv	Mar 6
Michelangelo: Pieta Attacked: Anniv	May 21
Michelson, Albert: First US Scientist Receives Nobel: Anniv	Dec 20
Michener, James: Birth	Feb 3

Michigan
Abbott's Magic Get-Together (Colon)	Aug 2
Admission Day	Jan 26
Africa's Legacy in Mexico/Photos (Ann Arbor)	Apr 15
Alma Highland Festival and Games (Alma)	May 26
Ann Arbor Film Festival (Ann Arbor)	Mar 14
Ann Arbor Summer Art/Street Fair (Ann Arbor)	July 19
Ann Arbor Winter Art Fair (Ann Arbor)	Oct 28
Antique Auto Show and Flea Market (Boyne City)	Aug 12
Antique Auto Show/Collector Car Fest (St. Ignace)	June 22
Arcadia Daze (Arcadia)	July 21
Arts and Crafts Show (East Tawas)	June 24
Asparagus Fest, Natl (Hart and Shelby)	June 9
Baby Food Fest, Natl (Fremont)	July 18
Bavarian Fest (Frankenmuth)	June 10
Big 10 Women's Golf Chmpshp (Ann Arbor)	May 5
Big 10 Women's Indoor Track/Field Chmpshp (Ann Arbor)	Feb 24
Blissfest (Cross Village)	July 14
Blossomtime Fest of SW Michigan (Benton Harbor)	Apr 30
Carry Nation Fest (Holly)	Sept 8
Cherry Fest, Natl (Traverse City)	July 2
Cherry Pit Spitting Contest, Intl (Eau Claire)	July 1
Children's Day (Detroit)	June 28
Chris Stratton Dart Tournament (Battle Creek)	Jan 22
Christmas in Bear Lake (Bear Lake)	Dec 2
Chrysler Opens Detroit Factory	Mar 31
Crim Festival of Races (Flint)	Aug 24
Curwood Fest (Owosso)	June 2
Danish Fest (Greenville)	Aug 18
Detroit Festival of the Arts (Detroit)	Sept 15
Dickens Olde Fashioned Christmas Fest (Holly)	Nov 24
Do Dah Parade (Kalamazoo)	June 3
Easter Monday Celebration (Portage)	Apr 17
Fall Harvest Fest (Boyne City)	Oct 7
First Ladies Gowns (Grand Rapids)	Jan 23
Frankenmuth Skyfest (Frankenmuth)	May 7
Freedom Fest, Intl (Detroit)	June 17
Fremont Car Show (Fremont)	July 22
From George to George: Monceaux (Grand Rapids)	Jan 1
Gillikins of Oz Convention (Escanaba)	May 6
Good, the Bad and the Cuddly (Royal Oak)	Feb 18
Grand Center Boat Show (Grand Rapids)	Feb 21
Grand Center Sport, Fishing & RV Show (Grand Rapids)	Mar 16
Grandeur/US Capitol 1793-1993 (Grand Rapids)	Jan 30
Great Fire of 1881: Anniv	Sept 5
Great Lakes Invit Barbershop Show (Grand Rapids)	Nov 4
Great Poetry Reading Day (Clio)	Apr 28
Grosse Pte Santa Claus Parade (Grosse Pte)	Nov 24
Holland Tulip Time Fest (Holland)	May 10
Horton Bay Labor Day Bridge Walk (Horton Bay)	Sept 4
Horton Bay Winter Olympics (Horton Bay)	Feb 26
Human Relations Day (Flint)	Jan 16
Independence Day Challenge Run (Boyne City)	July 4
International Festival of Lights (Battle Creek)	Nov 20
Julia A. Moore Poetry Festival (Flint)	May 24
Kalamazoo County Flowerfest (Kalamazoo/Portage)	July 28
Kalamazoo Holiday Parade (Kalamazoo)	Nov 11
Kaleva Days (Kaleva)	Aug 26
KIA Art Fair (Kalamazoo)	June 3
Kitefest (Kalamazoo)	Apr 29
Lakestride Half-Marathon (Ludington)	June 17
Log Cabin Day	June 25
Longhorn Chmpshp Rodeo (Auburn Hills)	Feb 24
Mackinac Bridge Walk (St. Ignace)	Sept 4
Mackinaw Mush Sled Dog Race (Mackinaw City)	Feb 4
Manistee Natl Forest Fest (Manistee)	June 30
Marinerfest (Tawas City)	July 12
Mary Stratton's Birthday (Detroit)	Mar 15
Michigan Fest (East Lansing)	Aug 4
Michigan Home & Garden Show (Pontiac)	Mar 2
Michigan Renaissance Fest (Holly)	Aug 12
Michigan Science Teachers Assn Conf (Lansing)	Feb 3
Michigan Sugar Festival (Sebewaing)	June 23
Michigan Thanksgiving Parade (Detroit)	Nov 23
Mid-American Sport Kite Classic	Sept 9
Mid-Winter's Day Celebration (Ann Arbor)	Feb 6
Month of the Young Child	Apr 1
Montreux Detroit Jazz Fest (Detroit)	Sept 1
Mushroom Hunting Chmpshps, Natl (Boyne City)	May 12
New Year's Fest (Kalamazoo)	Dec 31
Noel Night (Detroit)	Dec 13
On the Waterfront Swap Meet/Car Corral (St. Ignace)	Sept 15
Ozcanabans of Oz Convention (Escanaba)	Dec 9
Paczki Day (Hamtramck)	Feb 28
Perchville USA (Tawas Bay)	Feb 3
Pigeon Farmers' Festival (Pigeon)	July 26
Pontiac Boat, Sport & Fishing Show (Pontiac)	Feb 15
Pontiac Camper, Travel & RV Show (Pontiac)	Jan 25
Ribfest (Kalamazoo)	Aug 10
St. Johns Mint Fest (St. Johns)	Aug 12
St. Patrick's Day Dart Tourn (Battle Creek)	Mar 16
Salmon Tournament Week (Manistee)	June 23
Silver Bells in the City (Lansing)	Nov 17
Sno'Fly: First Kite Fly of the Year (Kalamazoo)	Jan 1
Snowman Sacrifice (Sault Ste Marie)	Mar 20
Soap Opera Fan Fair (Mackinaw City)	June 1
Spiders! (Bloomfield)	July 1
Spring Art Fair (Ann Arbor)	May 6
State Fair (Detroit)	Aug 22
Street Rod/Muscle Car Show (Boyne City)	Sept 2
Tawas Bay Waterfront Fine Art Show (Tawas City)	Aug 4
Tip-Up Town USA Ice Festival (Houghton Lake)	Jan 21
Trillium Fest (Muskegon)	May 13
Try This On: Clothing/Gender/Power (Flint)	Sept 16
US Women's Open (Golf) Chmpshp (Colorado Springs)	July 13
Victorian Christmas Sleighbell Parade (Manistee)	Nov 30
Wayne State Univ: Funeral for Winter (Detroit)	Apr 13
WCUZ West Michigan Home and Garden Show (Grand Rapids)	Mar 2
West Michigan Fair (Ludington)	Aug 21
Who's in Charge: Workers/Managers (Grand Rapids)	Jan 14
Wine/Harvest Fest, Michigan (Kalamazoo and Paw Paw)	Sept 6
Winkies of Oz Convention (Escanaba)	July 7
WOOD Radio St. Patrick's Day Parade (Grand Rapids)	Mar 11
WOOD Radio/Smiths Ind Fireworks Gala (Grand Rapids)	July 4
Wyandotte Heritage Days (Wyandotte)	Sept 8
Wyandotte Waterfest (Wyandotte)	Aug 11

Mickey Mouse's Birthday	Nov 18
Microbrewers Conf and Trade Show, Natl (Austin, TX)	Apr 23
Micronesia, Federated States of: Independence Day	Nov 3
Middle Children's Day	Aug 12
Middlemark, Marvin: Birth Anniv	Sept 16
Middleton, Arthur: Birth Anniv	June 26
Midler, Bette: Birth	Dec 1
Midnight Madness (Inuvik, NWT, Canada)	June 23
Midnight Madness Fun Bike Ride, HI-AYH (San Diego, CA)	Aug 19
Midnight Sun Baseball Game (Fairbanks, AK)	June 21
Midnight Sun Festival (Nome, AK)	June 17
Midnight Sun Festival (Yellowknife, NWT, Canada)	July 14
Midsummer (Wiccan)	June 21
Midsummer Day	June 23
Midsummer Night's Fair (Norman, OK)	July 14
Midsummer Nights' Stroll (New Harmony, IN)	July 20
Midwife's Day (Greece)	Jan 8
Mid-Winter's Day Celebration (Ann Arbor, MI)	Feb 6
Mikulski, Barbara Ann: Birth	July 20
Milano, Alyssa: Birth	Dec 19
Miles, Vera: Birth	Aug 23
Military Dictatorship Ended in Chile: Anniv	Dec 15
Military Families Recognition Week 6505 Page 488	
Military: Soldiers' Reunion Celebration (Newton, NC)	Aug 17
Military Through the Ages (Williamsburg, VA)	Mar 18
Milk Day Festival, Harvard (Harvard, IL)	June 2
Miller, Ann: Birth	Apr 12
Miller, Arthur: Birth	Oct 17
Miller, Bob: Birth	Mar 30
Miller, Glenn, Birthplace Society Fest (Clarinda, IA)	June 8
Miller, Glenn: Birth Anniv	Mar 1
Miller, Henry: Birth Anniv	Dec 26
Miller, Joe: Joke Day	Aug 16
Miller, Reggie: Birth	Aug 24
Miller, Roger: Birth Anniv	Jan 2
Miller, Steve: Birth	Oct 5
Miller, Walter Dale: Birth	Oct 5
Miller, Zell: Birth	Feb 24
Millett, Kate: Birth	Sept 14
Millfest (Lindsborg, KS)	May 6
Mills, Donna: Birth	Dec 11
Mills, Hayley: Birth	Apr 18
Milne, A.A.: Birth Anniv (Pooh Day)	Jan 18
Milsap, Ronnie: Birth	Jan 16
Milton, John: Birth Anniv	Dec 9
Milwaukee Journal: Al's Run (Milwaukee, WI)	Sept 23
Mimieux, Yvette: Birth	Jan 8
Mims, Marilyn: Birth	Sept 8
Mind, Make Up Your	Dec 31
Mind Mapping and Brainstorming Week, Natl	Sept 18
Mind Mapping for Problem Solving Week, Natl	Sept 25
Mind Mapping for Project Management Week, Natl	Sept 11
Mind Mapping in Schools Week, Natl	Sept 4
Mind Mapping Month, Natl	Sept 1
Minelli, Liza: Birth	Mar 12
Mining: Boom Days Celebration (Leadville, CO)	Aug 4
Mining: Oro City (Leadville, CO)	July 1
Mining: Yukon Gold Panning (Dawson City, Yukon, Canada)	July 1

Minnesota
Admission Day	May 11
Aebleskiver Days (Tyler)	June 24
Beargrease Sled Dog Marathon (Duluth)	Jan 15

✶ Chase's 1995 Calendar of Events ✶   Index

Big Island Rendezvous (Albert Lea).......Oct 7
Birdwatching Weekend (Madison).......Apr 28
Blackpowder Trade Fair/Gun Show (Albert Lea)..................Feb 11
Box Car Days (Tracy)..................Sept 2
Buffalo Days Celebration (Luverne).....June 3
Catfish Days (East Grand Forks).......July 14
Christmas at Pioneer Village (Worthington)...................Dec 7
Collegiate/Future Secretaries Conf (Bloomington).................Mar 23
Defeat of Jesse James Days (Northfield)..Sept 6
Edmund Fitzgerald Meml Beacon Lighting (Two Harbors)..................Nov 10
Festival of Adventures (Aitkin)......Sept 22
Festival of Nations (St. Paul)........Apr 27
Fifties Revival (Marshall)............June 2
Fish House Parade (Aitkin)...........Nov 24
Fishing Has No Boundaries (Bemidji)..June 24
Folkways of Christmas (Shakopee)....Nov 24
Giants Ridge Intl Classic Marathon (Biwabik)......................Mar 5
Giants Ridge Mountain Bike Fest (Biwabik).....................Sept 2
Halloween Celebration (Shakopee)...Oct 28
Hot Dog Night (Luverne)..............July 20
Icebox Days (International Falls).......Jan 20
Inventors Congress, Minnesota (Redwood Falls).......................June 9
Johnny Appleseed Days (Lake City).....Oct 7
King Turkey Days (Worthington)......Sept 9
Little Falls Sidewalk Arts/Crafts (Little Falls).....................Sept 9
Loon Lake Harvest Festival (Aurora)...Sept 9
Minneapolis Aquatennial (Minneapolis)..July 14
Minnesota Renaissance Fest (Shakopee) Aug 12
Muskies Inc Muskie Tourn, Intl (Northcentral)..................Sept 8
Nobel Conference (St. Peter)..........Oct 3
Norsefest (Madison)..................Nov 9
Norwegian Christmas (Brooklyn Park)....Dec 2
Old-Fashioned 4th of July (Shakopee)....July 1
Red Wing Sheep Dog Trial (Red Wing)...Oct 6
St. Olaf Christmas Fest (Northfield).....Nov 30
St. Paul Winter Carnival (St. Paul).......Jan 27
St. Urho's Day (Finland)..............Mar 18
Santaland (Madison)..................Nov 24
Scandinavian Hjemkomst Fest (Moorhead)....................June 21
Sinclair Lewis Days (Sauk Centre).......July 13
Song of Hiawatha Pageant (Pipestone)..July 21
Spring Breaking-Up at Giant Ridge (Biwabik)....................May 20
State Fair (St. Paul)..................Aug 24
Stiftungsfest (Young America).........Aug 25
Take a Kid Fishing Weekend (St. Paul)..June 10
Traverse des Sioux Encampment (St. Peter)........................Sept 8
Tri-State Band Fest (Luverne)........Sept 30
Ugly Truck Contest (Pelican Rapids).....July 8
US Senior Women's Amateur Golf Chmpshp (St. Paul)....................Sept 13
Water Ski Days (Lake City)...........June 23
White Oak Rendezvous (Deer River)....Aug 4
Winter Fest (Lake City)...............Jan 28
Wood Carvers Fest (Blackduck).......July 29
Woodcraft Fest (Grand Rapids).......July 15
Wrong Days (Wright).................July 14
Minority Enterprise Development Week (Pres Proc)........................Oct 1
Minority Enterprise Development Week 6591........................Page 488
Minow, Newton: Birth................Jan 17
Mint Fest, St. Johns (St. Johns, MI)....Aug 12
Mint Julep Scale Meet (Falls of Rough, KY) May 19
Mint, US: Anniv......................Apr 2
Mirror of the World Translation: Anniv, Caxton's........................Mar 8
Mischief Night......................Nov 4
Miss America Pageant Finale (Atlantic City, NJ)........................Sept 16
Miss America Pageant, First: Anniv...Sept 8
Missile, First Surface-to-Surface Missile: Anniv........................Dec 24
Missing Children's Day, Natl........May 25
Mission Delores: Founding Anniv......Oct 9
Mission San Carlos Borromeo de Carmelo Founding: Anniv..............June 3
Mission San Diego de Alcala: Founding Anniv........................July 16
Mission San Gabriel Archangel: Founding Anniv........................Sept 8
Mission San Juan Capistrano: Founding Anniv........................Nov 1

Mission San Luis Obispo de Tolosa: Founding Anniv........................Sept 1
Mission San Luis Rey de Francia Founding: Anniv........................June 13
Mission Santa Clara de Asis: Anniv....Jan 12
Mississinewa 1812 (Marion, IN).......Oct 14
Mississippi
Admission Day....................Dec 10
Battle West Point/Prairie (Columbus/West Pt).........................Feb 11
Broiler Fest, Mississippi (Forest).....June 1
Children's Victorian Christmas Open House (Washington)....................Dec 12
Choctaw Indian Fair (Philadelphia)....July 12
Civil Rights Workers Found Slain: Anniv..Aug 4
Confederate Memorial Day..........Apr 24
Copper Magnolia Fest (Washington)...Sept 23
Crop Day (Greenwood)..............Aug 5
Dixie Natl Livestock Show/Rodeo (Jackson)......................Jan 30
Farish Street Festival (Jackson).....Sept 1
Gautier Mullet Fest (Gautier).......Oct 14
Gem and Mineral Show (Jackson)....Feb 25
Ghost Tales Around Campfire (Washington)...................Oct 27
Grand Prix Catfish Races (Greenville)..May 13
Great Chunky River Raft Race (Chunky).June 3
Gum Tree Fest (Tupelo)............May 13
Irish Heritage Season (Biloxi).......Mar 4
Lighted Boat Parade on the Tenn-Tom Waterway (Columbus)............Dec 2
Meredith (James) Enrolls at Ole Miss: Anniv.........................Sept 30
Mississippi Deep Sea Fishing Rodeo (Gulfport)....................June 30
Natchez Fall Pilgrimage (Natchez)....Oct 7
Natchez Powwow (Natchez)........Mar 25
Natchez Spring Pilgrimage (Natchez)..Mar 4
Neshoba County Fair (Philadelphia)..July 28
Oleput Festival (Tupelo)............June 2
Oyster Festival (Biloxi).............Mar 16
Paint-a-Can-Contest (Biloxi)........May 27
Pioneer and Indian Festival (Ridgeland)..Oct 28
Pioneer Day (Washington)..........Oct 19
Pioneer Week (Washington)........June 26
Prairie Arts Festival (Westpoint).....Sept 2
Spring Pilgrimage to Antebellum Homes (Columbus)....................Mar 28
State Fair (Jackson)................Oct 4
Sun Herald Sand Sculpture Contest (Biloxi)........................Sept 16
Mississippi Riv Arts/Crafts Fest, Great (Hannibal, MO)..................May 27
Missouri
Admission Day....................Aug 10
American Mensa Annual Meeting (St. Louis)........................July 4
April in Trenchtown (St. Charles).....Apr 29
Autumn Historic Folklife Fest (Hannibal)..Oct 21
Becky Thatcher Day................Dec 1
Big Muddy Folk Music Fest (Boonville)..Apr 14
Blueberry Hill Open Dart Tourn (St. Louis)........................Feb 24
Cow Milked While Flying: Anniv......Feb 18
Dollhouse/Miniatures Show, Natl (St. Louis)........................Aug 12
Earth Day Community Fest (St. Louis)..Apr 22
Elvis Presley Birthday Party (St. Louis)..Jan 8
Elvis Presley Remembered (St. Louis)..Aug 13
Festival of the Little Hills (St. Charles)..Aug 18
Great Mississippi Riv Arts/Crafts Fest (Hannibal)....................May 27
Groundhog Run (Kansas City)......Feb 5
Harry's Hay Days (Grandview).....May 19
Jour de Fete (St. Genevieve).......Aug 12
Kahoka Fest of Bluegrass Music (Kahoka) Aug 9
KGBX Typewriter Toss (Springfield)..Apr 26
KGBX Valentine's Day Carnation Give Away (Springfield)..................Feb 14
KNIM Radio's Big Fish 15 (Maryville)..June 9
Kristkindl Markt (Hermann).........Dec 2
KRXL Car Cruise (Kirksville).......Aug 12
Lewis and Clark Rendezvous (St. Charles)......................May 20
Longhorn World Chmpshp Rodeo (Cape Girardeau)....................Feb 17
Maifest (Hermann).................May 20
Maple Leaf Fest (Carthage).........Oct 15
Michael Forbes Trolley Run (Kansas City) Apr 23
Midwinter Bluegrass Fest (Hannibal)...Feb 17
Missouri Day.......................Oct 18
Missouri River Fest of Arts (Boonville)..Aug 17
Natl FFA Convention (Kansas City)...Nov 9

Northeast Missouri Triathlon Chmpshp (Kirksville)..................Sept 10
Oktoberfest (Hermann)..............Oct 7
Oktoberfest (St. Charles)............Oct 7
Ozark Empire Fair (Springfield)......July 28
Pony Express-Jesse James Days (St. Joseph)........................Apr 1
Retailers Bakery-Deli Assn: Conv/Exhib (St. Louis)........................Mar 11
Riverfest (Cape Girardeau)...........June 9
Robidoux Rendezvous (St. Joseph)...May 13
St. Francois River Rendezvous (Farmington)...................Sept 22
St. Jo Heritage Tour (St. Joseph)......May 5
St. Louis Race Riots: Anniv..........July 2
St. Louis Storytelling Fest (St. Louis)..May 3
St. Louis Variety Club Telethon (St. Louis).Mar 4
St. Louis Walk of Fame Induction (St. Louis)........................May 21
Santa-Cali-Gon-Days (Independence)...Sept 1
Scott Joplin Ragtime Fest (Sedalia)....June 1
Show Me State Barbeque Cook-Off (St. Joseph)......................June 9
Show Me State Games (Columbia)...July 21
State Fair (Sedalia).................Aug 17
Tom Sawyer Days, Natl (Hannibal)...June 30
Trails West! (St. Joseph)............Aug 18
Valley of Flowers Festival (Florissant)...May 5
World Sheep and Crafts Festival (Bethel) Sept 2
World's Shortest St. Patrick's Day Parade (Maryville)...................Mar 17
Missouri River Expo (Bismarck, ND)...June 1
Mitchell, Cameron: Birth............Nov 4
Mitchell, Chad: Birth................Dec 5
Mitchell, George John: Birth.........Aug 20
Mitchell, Joni: Birth.................Nov 7
Mitchell, Margaret: Birth Anniv......Nov 8
Mitchell, Maria: Birth Anniv.........Aug 1
Mitchum, Robert: Birth.............Aug 6
Mitten Tree Kick-Off Ceremony (Baltimore, MD)..........................Dec 1
Mitterand, Francois: Birth...........Oct 26
Mix, Tom: Autumn Magic/Tom Mix Festival (Gutrie, OK)..................Sept 8
Mix, Tom: Birth Anniv..............Jan 6
Mize, Larry: Birth..................Sept 23
Mobius, August: Birth Anniv........Nov 17
Mobius Awards (Chicago, IL)........Feb 2
Model Airplane: Mint Julep Scale Meet (Falls of Rough, KY).............May 19
Model Airplane Show (Danville, VA)..June 24
Model Engineer Exhibition (London, England)......................Dec 30
Model Railroad Show (Wolfeboro, NH)..Aug 12
Moffo, Anna: Birth..................June 27
Mofford, Rose: Birth................June 10
Mohamed, Mahathir bin: Birth......Dec 20
Moldova, Republic of: National Day...Aug 27
Mole Day, Natl.....................Oct 23
Moliere Day........................Jan 15
Molineaux, Tom: First US Heavyweight Defeated: Anniv..................Dec 10
Molitor, Paul: Birth.................Aug 22
Mollusk Day, Martin Z. (Ocean City, NJ)..May 4
Molson, John: Birth Anniv..........Dec 28
Moment of Frustration Scream Day, Intl..Oct 12
Moment of Silence (Nagasaki, Japan)..Aug 9
Monaco
National Holiday....................Nov 19
Princess Grace: Death Anniv........Sept 14
Printemps des Arts de Monte-Carlo (Monte-Carlo)....................Apr 15
Monaghan, Tom: Birth..............Mar 25
Moncrief, Sidney: Birth..............Sept 21
Mondale, Joan A: Birth..............Aug 8
Mondale, Joan Adams: Birth........Aug 8
Monday Holiday Law: Anniv........June 28
Monet, Claude: Birth Anniv.........Nov 14
Money, Eddie: Birth.................Mar 2
Money Month, Soc/Prevention Cruely to Your Money........................Jan 1
Money, Paper, Issued: Anniv........Mar 10
Mongolian People's Republic: National Holiday........................July 11
Monitor Sinking: Anniv..............Dec 30
Monk, Art: Birth....................Dec 5
Monkey Trial: John T. Scopes Birth Anniv..Aug 3
Monroe Doctrine: Anniv.............Dec 2
Monroe, Elizabeth K.: Birth Anniv...June 30
Monroe, Harriet: Birth Anniv........Dec 23
Monroe, James: Birth Anniv.........Apr 28
Monroe, Marilyn: Birth Anniv.......June 1
Montagnier, Luc: Birth..............Aug 18
Montagu, Mary Wortley: Baptism Anniv..May 26

Minnesota (cont'd)—Montagu

☆ Chase's 1995 Calendar of Events ☆

Index

Montana
- Admission Day........Nov 8
- Arts in the Park (Kalispell)........July 21
- Bald Eagle Migration (Helena)........Nov 1
- Battle of Little Big Horn: Anniv........June 25
- Beartooth Run (Red Lodge)........June 24
- Big Sky State Games (Billings)........July 14
- C.M. Russell Auction Orig Western Art (Great Falls)........Mar 15
- Central Montana Fair (Lewistown)........July 24
- Christmas Stroll (Lewistown)........Dec 1
- Christmas to Remember (Laurel)........Dec 3
- Custer's Last Stand Reenactment (Hardin)........June 23
- Depot Fest of the Arts (Livingston)........July 1
- Ennis Rodeo & Parade (Ennis)........July 3
- Fest of Nations (Red Lodge)........Aug 5
- Forest Grove Arts and Crafts Show (Forest Grove)........Oct 22
- Fred E. Miller: Photographer of the Crows (Missoula)........Jan 14
- Frontier Days (Culbertson)........June 3
- Governor's Cup Marathon (Helena)........June 2
- Helena Railroad Fair (Helena)........Apr 23
- Last Chance Stampede and Fair (Helena)........July 27
- Laurel Herbstfest (Laurel)........Sept 21
- Lewistown 4th Parade/Celeb (Lewistown)........July 4
- Little Big Horn Days (Hardin)........June 22
- Livingston Railroad Swap Meet (Livingston)........Apr 22
- Logger Days (Libby)........July 13
- Miles City Jaycee Bucking Horse Sale (Miles City)........May 19
- Montana Cowboy Poet Gathering (Lewistown)........Aug 18
- Montana Governor's Cup Walleye Tourn (Ft Peck)........July 6
- Montana State Chokecherry Fest (Lewistown)........Sept 9
- Montana Traditional Jazz Fest (Helena)........June 28
- Montana Trappers Assn State Convention (Lewistown)........Sept 8
- MontanaFair (Billings)........Aug 12
- Mountain Man Rendezvous (Red Lodge)........July 21
- Nemont/Agribition (Wolf Point)........Oct 13
- Nordicfest (Libby)........Sept 8
- Nordicfest Parade (Libby)........Sept 9
- Northeast MT Threshing Bee/Antique Show (Culbertson)........Sept 23
- Northern Intl Livestock Expo (Billings)........Oct 16
- NRA Rodeo Finals (Billings)........Feb 3
- Oktoberfest/Thanks Canada (Helena)........Oct 13
- Pioneer Power Day Threshing Bee (Lewistown)........Sept 30
- Race to the Sky Dog Sled Race (Helena)........Feb 11
- Red Lodge Home of Champions Rodeo/Parade (Red Lodge)........July 2
- Running of the Sheep Sheep Drive (Reedpoint)........Sept 3
- Ski-Joring Finals, Natl (Red Lodge)........Mar 11
- State Fair (Great Falls)........July 29
- SummerFair (Billings)........July 15
- Tamarack Time (Bigfork)........Oct 14
- US Amateur Public Links (Golf) Chmpshp (Stow)........July 17
- US Girls' Junior (Golf) Chmpshp (Longmeadow)........July 31
- Western Heritage Days (Deer Lodge)........July 8
- Wild Horse Stampede (Wolf Point)........July 7
- Winter Carnival (Libby)........Mar 3
- Wolf Point's Hottest Chili Weekend (Wolf Point)........June 1
- Yellowstone Art Center's Auction (Billings)........Mar 4
- Montana, Joe: Birth........June 11
- Montgolfier, Jacques Etienne: Birth Anniv........Jan 7
- Montgolfier, Joseph M: Death Anniv........June 26
- Montgomery, Belinda: Birth........July 23
- Montgomery, Bernard Law: Birth Anniv........Nov 17
- Montgomery Boycott Arrests: Anniv........Feb 22
- Montgomery Bus Boycott Begins: Anniv........Dec 5
- Montgomery Bus Boycott Ends: Anniv........Dec 20
- Montgomery, Elizabeth: Birth........Apr 15
- Montgomery Ward Seized: Anniv........Apr 26
- Montoya, Carlos: Birth Anniv........Dec 3
- Moog, Robert: Birth........May 23
- Moon Day (First Moon Landing)........July 20
- Moon Fest........Sept 9
- Moon, Harvest........Sept 8
- Moon, Hunter's........Oct 8
- Moon Phases. See Contents for Astronomical Phenomena
- Moon, Warren: Birth........Nov 18
- Moore, Clayton: Birth........Sept 14
- Moore, Clement: Birth Anniversary........July 15
- Moore, Demi: Birth........Nov 11
- Moore, Dudley: Birth........Apr 19
- Moore, Julia A Davis: Birth Anniv........Dec 1
- Moore, Mary Tyler: Birth........Dec 29
- Moore, Melba: Birth........Oct 29
- Moore, Roger: Birth........Oct 14
- Moose Dropping Festival (Talkeetna, AK)........July 7
- Moran, Erin: Birth........Oct 18
- Moravian Candle Tea (Winston-Salem, NC)........Nov 30
- Moravian Easter Sunrise Service (Winston-Salem, NC)........Apr 16
- Morazan, Francisco: Holiday (Honduras)........Oct 3
- More Herbs, Less Salt Day........Aug 29
- More, Sir Thomas: Birth Anniv........Feb 7
- Moreau, Jeanne: Birth........Jan 23
- Moreno, Rita: Birth........Dec 11
- Morgan, Dennis: Birth........Dec 10
- Morgan, Harry: Birth........Apr 10
- Morgan, Jaye P.: Birth........Dec 3
- Morgan, John Hunt, Captured: Anniv........July 26
- Morgan, John P.: Birth Anniv........Apr 17
- Morgan, Michelle: Birth........Feb 29
- Moriarty, Cathy: Birth........Nov 29
- Moriarty, Michael: Birth........Apr 5
- Morione's Fest (Marinduque Island, Philippines)........Apr 13
- Mormon Battalion Recruitment Day: Anniv........July 16
- Mormon Church: Founding Anniv........Apr 6
- Mormon: Nauvoo Legion Chartered: Anniv........Feb 3
- Mormon Trail Ride (Summit Valley, CA)........May 19
- Mormons: Days of '47 Celebration (Salt Lake City, UT)........July 1
- Morocco: Anniv of the Throne........Mar 3
- Morris, Esther Hobart McQuigg: Birth Anniv........Aug 8
- Morris, Garrett: Birth........Feb 1
- Morris, Greg: Birth........Sept 27
- Morris, Jack: Birth........May 16
- Morris, Joe: Birth........Sept 15
- Morris, Lewis: Birth Anniv........Apr 8
- Morris, Mark: Birth........Aug 29
- Morris, William: Birth Anniv........Mar 24
- Morrison, Jim: Birth Anniv........Dec 8
- Morrison, Toni: Birth........Feb 18
- Morrison, Van: Birth........Aug 31
- Morro Bay Harbor Fest (Morro Bay, CA)........Oct 7
- Morrow, Rob: Birth........Sept 21
- Morse, Samuel F.: Birth Anniv........Apr 27
- Morse, Samuel: Opens First US Telegraph Line: Anniv........May 24
- Morton, Ferdinand: Birth Anniv........Sept 20
- Morton, Levi P.: Birth Anniv........May 16
- Mosbacher, Robert Adam: Birth........Mar 11
- Moscow Soccer Tragedy: Anniv........Oct 20
- Moseby, Lloyd: Birth........Nov 5
- Moseley-Braun, Carol: Birth........Aug 16
- Moses, Anna: Grandma Moses Day........Sept 7
- Moses, Edwin: Birth........Aug 31
- Moshoeshoe's Day (Lesotho)........Mar 12
- Mosquito Awareness Weekend (Walcott, AR)........Aug 20
- Mosquito Festival, Great Texas (Clute, TX)........July 27
- Most Dubious News Stories of the Year........Dec 29
- Moth-er Day........Mar 14
- Mother Goose Day........May 1
- Mother Goose Parade (El Cajon, CA)........Nov 19
- Mother Ocean Day........May 14
- Mother Teresa: Birth........Aug 27
- Mother's Day........May 14
- Mother's Day (Pres Proc)........May 14
- Mother's Day 6559, 6683........Page 488
- Mother's Day, Gold Star 6590........Page 488
- Mother's Whistler: Birthday........May 18
- Mother-in-Law-Day........Oct 22
- Mothering Sunday (England)........Mar 26
- Motley, Constance Baker: Birth........Sept 14
- Motor Voter Bill Signed: Anniv........May 20
- Motorcycle
  - Daytona Fall Motorcycle Races (Daytona Beach, FL)........Oct 20
  - Daytona Motorcycle Week (Daytona Beach, FL)........Mar 3
  - Daytona Supercross by Honda (Daytona Beach, FL)........Mar 11
  - Daytona 200 Motorcycle Road Race (Daytona Beach, FL)........Mar 12
  - Gold Wing Getaway (Daytona Beach, FL)........Mar 3
  - Golden Aspen Motorcycle Rally (Ruidoso, NM)........Sept 20
  - Intl TT Motorcycle Race (Isle of Man, England)........June 3
  - Sturgis Rally & Races (Sturgis, SD)........Aug 7
- Mott, Lucretia (Coffin): Birth Anniv........Jan 3
- Mott, Stewart Rawlings: Birth........Dec 4
- Mount Rushmore Completion: Anniv........Oct 31
- Mount Rushmore July 4 Celeb (Mt Rushmore, SD)........July 4
- Mount St. Helens Eruption: Anniv........May 18
- Mount Washington Tavern Candle Tours (Farmington, PA)........Dec 9
- Mountain Dance and Folk Fest (Asheville, NC)........Aug 3
- Mountain Eagle Indian Fest (Hunter, NY)........Sept 2
- Mountain Fair (Hiawassee, GA)........Aug 2
- Mountain Glory Fest (Marion, NC)........Oct 14
- Mountain Man Rendezvous (Felton, CA)........Nov 24
- Mountain Man Rendezvous (Red Lodge, MT)........July 21
- Mountaineer Folk Fest (Pikeville, TN)........Sept 9
- Mountbatten, Louis: Assassination Anniv........Aug 27
- Move Hollywood and Broadway to Lebanon, PA Day........Feb 6
- Moyers, Bill: Birth........June 5
- Moynihan, Daniel Patrick: Birth........Mar 16
- Mozambique: National Day........June 25
- Mozart Festival (San Luis Obispo, CA)........July 24
- Mozart Intl Fest (Bartlesville, OK)........June 8
- Mozart, Wolfgang Amadeus: Birth Anniv........Jan 27
- Mozart, Wolfgang Amadeus: Death Anniv........Dec 5
- MSC Student Conf Natl Affairs (College Station, TX)........Feb 17
- Mud Volleyball Tourn (Rockford, IL)........Aug 5
- Mudd Day........Dec 20
- Mudd, Roger: Birth........Feb 9
- Muhammad: Birth of Prophet Muhammad (Muslim Festival)........Aug 9
- Muir, John: Birth Anniv........Apr 21
- Muldaur, Maria: Birth........Sept 12
- Mule Day........Oct 26
- Mule: Matanzas Mule Day........Apr 27
- Mulgrew, Kate: Birth........Apr 29
- Mull, Martin: Birth........Aug 18
- Mullen, Larry: Birth........Oct 31
- Mullet Fest, Gautier (Gautier, MS)........Oct 14
- Mulligan, Richard: Birth........Nov 13
- Mullin, Chris: Birth........July 30
- Multicultural Communication Month........Apr 1
- Multicultural Substance Abuse Awareness Week........May 1
- Multimedia Schools Conference (Chicago, IL)........Oct 31
- Munoz-Rivera, Luis: Birth Anniv........July 17
- Munsel, Patrice: Birth........May 14
- Muppets: Henson, Jim: Birth Anniv........Sept 24
- Murkowski, Frank Hughes: Birth........Mar 28
- Murphy, Audie: Spirit America Awards (Decatur, AL)........July 3
- Murphy, Dale: Birth........Mar 12
- Murphy, Eddie: Birth........Apr 3
- Murray, Anne: Birth........June 20
- Murray, Bill: Birth........Sept 21
- Murray, Don: Birth........July 31
- Murray, Eddie: Birth........Feb 24
- Murray, Jan: Birth........Oct 4
- Murray, Patty: Birth........Oct 11
- Murray, Philip: Birth Anniv........May 25
- Musburger, Brent: Birth........May 26
- Muscular Dystrophy, Shamrocks for........Mar 17
- Muscular Dystrophy: Stars/J. Lewis Labor Day Wknd........Sept 3
- Museum Comes to Life Day (Ainsworth, NE)........June 24
- Museum Day, Intl........May 18
- Museum Open House (Clinton, MD)........Nov 11
- Museums: St. Jo Heritage Tour (St. Joseph, MO)........May 5
- Mushroom Fest, Telluride (Telluride, CO)........Aug 24
- Mushroom Hunting Chmpshps, Natl (Boyne City, MI)........May 12
- Musial, Stan: Birth........Nov 21
- **Music.** See also Bluegrass; Fiddlers; Jazz
  - Aberdeen Intl Youth Fest (Grampian, Scotland)........Aug 2
  - Accordion Awareness Month, Natl........June 1
  - Accordion Fest, Sata-Hame (Ikaalinen, Finland)........June 26
  - Altamont Concert: Anniv........Dec 6
  - America the Beautiful Published........July 4
  - Anderson, Marian: Easter Concert: Anniv........Apr 9
  - Ash Lawn-Highland Opera Fest (Charlottesville, VA)........June 25
  - Aspen Music Festival (Aspen, CO)........June 22
  - Barbershop Ballad Contest (Forest Grove, OR)........Mar 10
  - Barbershop Quartet Day........Apr 11
  - Bath Intl Fest (Bath, England)........May 19
  - Bayreuther Festspiele (Bayreuth, Germany)........July 25
  - Beatles Fest (Liverpool, England)........Aug 25

☆ *Chase's 1995 Calendar of Events* ☆ **Index**

Music (cont'd)—Native American

Beatles Last Album Released . . . . **Sept 26**
Beatles Last Public Appearance: Anniv . . **Jan 30**
Bergen Intl Festival (Bergen, Norway) . . . **May 24**
Big Muddy Folk Music Fest (Boonville, MO) . . . . **Apr 14**
Big Valley Jamboree (Craven, Sask, Canada) . . . . **July 12**
Biggest All Night Gospel Singing (Bonifay, FL) . . . . **July 1**
Bournemouth Musicmakers Fest (Bournemouth, England) . . . . **June 24**
Burlington Steamboat Days/Music Fest (Burlington, IA) . . . . **June 13**
Calgary Folk Fest (Calgary, AB, Canada) **July 17**
Cape May Music Fest (Cape May, NJ) . . . **May 14**
Caramoor Intl Music Fest (Katonah, NY) **June 24**
Cardiff Festival of Music (Cardiff, Wales) . **Sept 9**
Cardiff Singer of the World (Cardiff, Wales) . . . . **June 11**
Carols by Candlelight (Lorton, VA) . . . . **Dec 8**
Celtic Fest, Southern Maryland (St. Leonard, MD) . . . . **Apr 29**
Cheltenham Intl Fest of Music (Cheltenham, England) . . . . **July 1**
Chicago Peace and Music Festival (Chicago, IL) . . . . **Aug 5**
City Stages Music Festival (Birmingham, AL) . . . . **June 16**
Concert, World's Largest . . . . **Mar 2**
Country Jam USA (Eau Claire, WI) . . . . . . **July 20**
Country Jamboree (Fargo, ND) . . . . **Aug 13**
Country Music Day, Natl . . . . **July 4**
Country Music Fan Fair, Intl (Nashville, TN) . . . . **June 5**
Country Music Month 6606 . . . . **Page 488**
Dawson City Music Fest (Dawson City, Yukon, Canada) . . . . **July 21**
Depot Lane Singers (Schoharie, NY) . . . . . **May 6**
Desert Foothills Music Fest (Carefree, AZ) . . . . **Feb 15**
Dove Awards Presentation . . . . **Apr 27**
Drum Month, Intl . . . . **Nov 1**
Dulcibrrr (Falls of Rough, KY) . . . . **Feb 3**
Dulcimer Days (Coshocton, OH) . . . . **May 20**
Dulcimer Fest (Huntsville, AL) . . . . **Sept 17**
Eastern Music Festival (Greensboro, NC) . . . . **June 24**
Edmonton Folk Music Fest (Edmonton, AB, Canada) . . . . **Aug 10**
Everly Brothers Homecoming (Central City, KY) . . . . **Aug 28**
Fall Fair (Hiawassee, GA) . . . . **Oct 6**
Fall Gospel Jubilee (Live Oak, FL) . . . . **Oct 19**
Farish Street Festival (Jackson, MS) . . . . . **Sept 1**
Fest International de Louisiane (Lafayette, LA) . . . . **Apr 25**
Festival of Carols (Topeka, KS) . . . . **Nov 24**
Fraley Family Music Weekend (Olive Hill, KY) . . . . **Sept 5**
Frostbite Music Fest (Whitehorse, Yukon, Canada) . . . . **Feb 17**
Glenn Miller Birthplace Society Fest (Clarinda, IA) . . . . **June 8**
God Bless America 1st Performed: Anniv **Nov 11**
Gospel Music (Nashville, TN) . . . . **Apr 23**
Grand Teton Music Fest (Teton Village, WY) . . . . **June 27**
Great American Brass Band Fest (Danville, KY) . . . . **June 16**
Great American Dulcimer Convention (Pineville, KY) . . . . **Sept 22**
Great Lakes Barbershop Show (Grand Rapids, MI) . . . . **Nov 4**
Great Northern Music Fest (Inuvik, NWT, Canada) . . . . **July 28**
Guelph Spring Fest (Guelph, Ont, Canada) . . . . **Apr 28**
Guitar Flat-Pick Chmps, Natl (Winfield, KS) . . . . **Sept 14**
Guitar Month, Intl . . . . **Apr 1**
Handel Oratorio Soc Spring Performance (Rock Is, IL) . . . . **Apr 8**
Handel's Messiah (Rock Island, IL) . . . . . . **Dec 15**
Harrison Fest of Arts (Harrison Hot Sprgs, BC, Canada) . . . . **July 8**
Harrogate Intl Youth Music Fest (Harrogate, England) . . . . **Apr 14**
Harvest of Harmony (Grand Island, NE) . . . **Oct 7**
International Festival of Music (Lucerne, Switzerland) . . . . **Aug 16**
International Music Camp (Dunseith, ND) . . . . **June 11**
International Strange Music Weekend (Olive Hill, KY) . . . . **June 9**
Joensu Festival (Joensu, Finland) . . . . . . **June 18**
Kaustinen Folk Music Fest (Kaustinen, Finland) . . . . **July 15**
Kazoo Day, Natl . . . . **Jan 28**
Kentucky Chmp Fiddler's Contest (Falls of Rough, KY) . . . . **July 21**
Lahti Organ Fest (Lahti, Finland) . . . . **Aug 1**
Lennon-Ono Album Confiscation: Anniv . . **Jan 3**
Live Aid Concerts: Anniv . . . . **July 13**
Llangollen Intl Musical Eisteddfod (Llangollen, Wales) . . . . **July 4**
Madrigal Dinner and Concert (Milwaukee, WI) . . . . **Dec 2**
Mamou Cajun Music Festival (Mamou, LA) . . . . **June 2**
Mamou Mardi Gras Celebration (Mamou, LA) . . . . **Feb 28**
Maryland Symphony at Antietam (Sharpsburg, MD) . . . . **July 1**
Melbourne Music Fest (Australia) . . . . **Feb 10**
Mozart Festival (San Luis Obispo, CA) . . **July 24**
Mozart Intl Fest (Bartlesville, OK) . . . . **June 8**
Mozart Week (Salzburg, Austria) . . . . **Jan 20**
Music Brazos Nights (Waco, TX) . . . . **June 9**
Music Day, Natl (England) . . . . **June 24**
Music Educ East Div Conf (Rochester, NY) . . . . **Mar 31**
Music Educ NW Div In-Serv Conf (Spokane, WA) . . . . **Feb 17**
Music for Life Week . . . . **July 2**
Music in Our Schools Month . . . . **Mar 1**
Musical Instrument Auction (Seattle, WA) . . . . **May 26**
Musical Tribute Dr. M. L. King (Hodgenville, KY) . . . . **Jan 15**
Natl Women's Music Fest (Bloomington, IN) . . . . **June 1**
New World Symphony Premiere: Anniv . . **Dec 16**
Newport Music Fest (Newport, RI) . . . . **July 8**
Northern Lights Fest (Tromso, Norway) . . **Jan 19**
Nova Scotia Bluegrass/ Oldtime (Ardoise, NS, Canada) . . . . **July 28**
Oboe Compet, Intl (Pondwell, Ryde, Isle of Wight) . . . . **May 4**
Oklahoma All Night Singing (Konawa, OK) **Aug 5**
Old Bedford Village Gospel Fest (Bedford, PA) . . . . **Aug 12**
One-Hit Wonder Day, Natl . . . . **Sept 25**
Opera Debuts in the Colonies . . . . **Feb 8**
Opera Festival of New Jersey (Lawrenceville, NJ) . . . . **June 17**
Oregon Bach Festival (Eugene, OR) . . . . **June 23**
Oregon Coast Music Fest (Coos Bay, OR) . . . . **July 15**
Oslo Chamber Music Fest (Oslo, Norway) **Aug 4**
Pan Is Beautiful (Port of Spain, Trinidad) . **Oct 14**
Piano Month, Natl . . . . **Sept 1**
Piano Playing Contest, World Chmp Old-Time (Decatur, IL) . . . . **May 26**
Porter, Cole: Birth Anniv . . . . **June 9**
Pre-Opera Lecture Series (Charlottesville, VA) . . . . **June 26**
Printemps des Arts de Monte-Carlo (Monte-Carlo, Monaco) . . . . **Apr 15**
Quartet Convention, Natl (Louisville, KY) **Sept 18**
Radio City Music Hall: Anniv . . . . **Dec 27**
Rhythm on the River (Little Rock, AR) . . **Sept 23**
Rounds Resounding Day . . . . **Aug 1**
Sacred Harp Singing (Huntsville, AL) . . . . . **May 6**
St. Olaf Christmas Fest (Northfield, MN) . **Nov 30**
St. Petersburg Fest/Natl Band Chmpshp (St. Petersburg, FL) . . . . **Mar 17**
Sand Creek Folk Life Fest (Newton, KS) . . **Apr 29**
Santa Fe Chamber Music Fest (Santa Fe, NM) . . . . **July 7**
Savonlinna Opera Fest (Savonlinna, Finland) . . . . **July 8**
Saxophone Day . . . . **Nov 6**
Scott Joplin Ragtime Fest (Sedalia, MO) . **June 1**
Sea Music Fest (Mystic, CT) . . . . **June 9**
Sewanee Concerts and Fest (Sewanee, TN) . . . . **June 24**
Shrewsbury Intl Music Fest (Shropshire, England) . . . . **June 23**
Silent Record Week . . . . **Jan 1**
Singing on the Mountain (Linville, NC) . . **June 25**
Sing-Out Day, Natl . . . . **Mar 22**
Society Barbershop Sing Intl Conv (Miami, FL) . . . . **July 2**
Southern Appalachian Dulcimer Fest (McCalla, AL) . . . . **May 7**
Southern Cathedrals Fest (Chichester, Sussex) . . . . **July 20**
Southern Gospel Music Month . . . . **Sept 1**
Space Oddity Song Release: Anniv . . . . . **June 11**
SPEBSQSA (Barbershop Quartet) Conv (Tucson, AZ) . . . . **Jan 22**
Spring Music Fest (Hiawassee, GA) . . . . . **May 19**
Springsteen's First Album . . . . **May 22**
Steamboat's Rainbow Weekend (Steamboat Springs, CO) . . . . **July 14**
Stresa Musical Weeks (Stresa, Italy) . . . . . **Aug 23**
Summer Music Fest (Sitka, AK) . . . . **June 2**
Suwannee River Gospel Jubilee (Live Oak, FL) . . . . **June 15**
The Day the Music Died . . . . **Feb 3**
Three Choirs Festival (Glouchestershire, England) . . . . **Aug 19**
Time of Music (Viitasaari, Finland) . . . . **July 6**
Tri-State Band Fest (Luverne, MN) . . . . . . **Sept 30**
Tuba Day, Intl . . . . **May 5**
Turku Music Fest (Turku, Finland) . . . . **Aug 11**
Vietnam Moratorium Concert: Anniv . . . . . **Mar 28**
Voss Jazz Fest (Voss, Norway) . . . . **Apr 7**
W.C. Handy Fest (Florence, AL) . . . . **Aug 6**
Willamette Valley Folk Fest (Eugene, OR) **May 19**
Wratislavia Cantans: Intl Oratorio-Cantata Fest . . . . **Sept 3**
Youth Sing Praise Performance (Belleville, IL) . . . . **June 24**
Muskie, Edmund S.: Birth . . . . Mar 28
Muskies Inc Muskie Tourn, Intl (Northcentral, MN) . . . . Sept 8
Muslim Holidays
Ashura . . . . **June 8**
Id al-Fitr . . . . **Mar 2**
Id al-Hajj . . . . **May 10**
Lailat al Miraj . . . . **Dec 19**
Lailat al-Qadr . . . . **Feb 21**
New Year . . . . **May 30**
Prophet Muhammad's Birth . . . . **Aug 9**
Ramadhan . . . . **Feb 1**
Mussolini, Benito: Birth Anniv . . . . July 29
Mustard Day, Natl . . . . Aug 5
Muybridge, Eadweard: Birth Anniv . . . . Apr 9
My Lai Massacre: Anniv . . . . Mar 16
Myanmar: Independence Day . . . . Jan 4
Myers, Russell: Birth . . . . Oct 9
Myerson, Bess: Birth . . . . July 16

N

Nabors, Jim: Birth . . . . June 12
Nader, Ralph: Birth . . . . Feb 27
Nafels Pilgrimage (Canton Glarus, Switzerland) . . . . Apr 6
NAIA M&W Indoor Track/Field Chmpshp . . . Mar 3
NAIA Men's Division II Basketball Chmpshp (Nampa, ID) . . . . Mar 9
NAIA Outdoor Track/Field Chmpshps . . . . . May 25
Naismith, James: Birth Anniv . . . . Nov 6
Nakasone, Yasuhiro: Birth . . . . May 27
Namath, Joe: Birth . . . . May 31
Name Your Car Day . . . . Oct 2
Names: Z Day . . . . Jan 1
Namibia: Independence Day . . . . Mar 21
Nanakusa (Japan) . . . . Jan 7
Nancy Thurmond Organ/Tissue Donor Awareness Wk 6548, 6672 . . . . Page 488
Nanticoke Indian Powwow (Millsboro, DE) . Sept 9
Napa Valley Wine Auction (Napa Valley, CA) . . . . June 8
Nash, Graham: Birth . . . . Feb 2
Nash, Ogden: Birth Anniv . . . . Aug 19
Nast, Thomas: Birth Anniv . . . . Sept 27
Nastase, Ilie: Birth . . . . July 19
Natchez Fall Pilgrimage (Natchez, MS) . . . . . . Oct 7
Natchez Spring Pilgrimage (Natchez, MS) . . . Mar 4
Nation, Carry: Birth Anniv . . . . Nov 25
Nation, Carry, Fest (Holly, MI) . . . . Sept 8
National Anthem Day . . . . Mar 3
National Anthem Day (MD) . . . . Sept 14
Nationality Days, Ambridge (Ambridge, PA) . . . . May 19
Nations, Festival of (Red Lodge, MT) . . . . Aug 5
Nations, Festival of (St. Paul, MN) . . . . Apr 27
Native American
American Indian Celebration (Belleville, IL) . . . . **May 11**
American Indian Ceremonial Dancing (Taos, NM) . . . . **Sept 29**
American Indian Heritage Month, Natl 6511 . . . . **Page 488**
American Indian Time of Thanksgiving (Allentown, PA) . . . . **Oct 8**
Americana Indian and Western Art Show/Sale (Yuma, AZ) . . . . **Feb 3**
Apache Wars Begin: Anniv . . . . **Feb 4**

Index ☆ Chase's 1995 Calendar of Events ☆

Native American (cont'd)——New Hampshire

Battle of Little Big Horn: Anniv . . . **June 25**
Chehaw Natl Indian Fest' (Albany, NY) . . **May 19**
Cherokee Days of Recognition (Cleveland, TN) . . . **Aug 5**
Cherokee Strip Celebration (Perry, OK) . **Sept 16**
Chickahominy Fall Fest Powwow (Roxbury, VA) . . . **Sept 23**
Chief Joseph Surrender: Anniv . . . **Oct 5**
Choctaw Indian Fair (Philadelphia, MS) . . **July 12**
Cochise: Death Anniv . . . **June 7**
Corn Planting Ceremony (Allentown, PA) . **May 7**
Crow Reservation Opened for Settlement: Anniv . . . **Oct 15**
Custer Battlefield Becomes Little Bighorn . . . **Nov 26**
Custer's Last Stand Reenactment (Hardin, MT) . . . **June 23**
Fall Powwow and Fest (Lebanon, TN) . . . **Oct 6**
Festival of Native Arts (Fairbanks, AK) . . . **Feb 23**
Fred E. Miller: Photographer of the Crows (Missoula, MT) . . . **Jan 14**
Fred E. Miller: Photographer of the Crows (Pittsburgh, PA . . . **May 20**
Hayes, Ira Hamilton: Birth Anniv . . . **Jan 12**
Heritage Fair/Spring Fest (Ashland, VA) . . . **Apr 8**
Indian Dance and Crafts Fest (Childersburg, AL) . . . **Apr 1**
Indian Day (Bridgeport, AL) . . . **Apr 15**
Indian Day (OK) . . . **Sept 9**
Indian Heritage Festival (Huntsville, AL) . . **Oct 21**
Indian Heritage Festival (Martinsville, VA) **Sept 8**
Indian Powwow, Natl Chmpshp (Grand Prairie, TX) . . . **Sept 8**
Indian Summer Festival . . . **Sept 8**
Indian Summer Rendezvous, Ogallala's (Ogallala, NE) . . . **Sept 14**
Intertribal Indian Ceremonial (Gallup, NM) **Aug 8**
Iroquois Arts Showcase I, II (Howes Cave, NY) . . . **May 27**
Iroquois Indian Fest (Howes Cave, NY) . . . **Sept 2**
Kiamichi-Owa-Chito Fest of Forest (Broken Arrow, OK) . . . **June 14**
Last Formal Surrender of Confederate Troops: Anniv . . . **June 23**
Loloma, Charles: Death Anniv . . . **June 9**
Mid America All-Indian Center PowWow (Wichita, KS) . . . **July 28**
Minority Enterprise Development Week (Pres Proc) . . . **Oct 1**
Mountain Eagle Indian Fest (Hunter, NY) . **Sept 2**
Nanticoke Indian Powwow (Millsboro, DE) **Sept 9**
Natchez Powwow (Natchez, MS) . . . **Mar 25**
Native American Heritage Fest (Martinsville, VA) . . . **Sept 9**
Native Americans' Day (South Dakota) . . . **Oct 9**
Objiway Keeshigunun (Thunder Bay, Ont, Canada) . . . **Aug 19**
Oconaluftee Indian Village (Cherokee, NC) . . . **May 15**
Outdoor Summer Theater (Anasazi) (Farmington, NM) . . . **June 21**
Owensboro Show (Owensboro, KY) . . . **Aug 5**
Pendleton Round-Up (Pendleton, OR) . . . **Sept 13**
Pioneer and Indian Festival (Ridgeland, MS) . . . **Oct 28**
Poarch Creek Indian Thanksgiving Day Powwow (Atmore, AL) . . . **Nov 23**
Pocahontas: Death Anniv . . . **Mar 21**
Red Cloud: Death Anniv . . . **Dec 10**
Red Cloud Indian Art Show (Pine Ridge, SD) . . . **June 4**
Red Earth Native Amer Cultural Fest (Oklahoma City, OK) . . . **June 9**
Rising Water Falling Water Powwow (Richmond, VA) . . . **July 29**
Roasting Ears of Corn Food Fest (Allentown, PA) . . . **Aug 13**
St. Francois River Rendezvous (Farmington, MO) . . . **Sept 22**
Sedona Arts Fest (West Sedona, AZ) . . . **Oct 14**
Settlers Days (Bedford, PA) . . . **July 8**
Sitting Bull: Death Anniv . . . **Dec 15**
Southern Utah Folklife Festival (Springdale, UT) . . . **Sept 7**
Tecumseh! Epic Outdoor Drama (Chillicothe, OH) . . . **June 9**
Totah Festival (Farmington, NM) . . . **Sept 1**
Traditional Indian Dances (Gallup, NM) . . **May 29**
Trail of Courage Living-History Fest (Rochester, IN) . . . **Sept 16**
Traverse des Sioux Encampment (St. Peter, MN) . . . **Sept 8**
United Tribes PowWow (Bismarck, ND) . . **Sept 9**
Unto These Hills Cherokee Indian Drama (Cherokee, NC) . . . **June 15**
Virginia Indian Heritage Fest (Williamsburg, VA) . . . **June 17**
White Woman Made Indian Chief: Anniv . **Sept 18**
Wild Horse Stampede (Wolf Point, MT) . . . **July 7**
Wounded Knee Massacre: Anniv . . . **Dec 29**
NATO Founded: Anniv . . . Apr 4
Natural Bridges National Monument: Anniv Apr 16
Naturalists Rally, Roan Mountain (Elizabethton, TN) . . . Sept 9
Nature Camp, Junior (Wheeling, WV) . . . July 30
Nature: Garden Month, Natl . . . Apr 1
Nature: Oglebay Inst Winter Lodge Program (Wheeling, WV) . . . Jan 2
Nature: Roan Mtn Wildflow Tours/Birdwalks (Roan Mtn, TN) . . . May 6
Nature Tour, Cranberry Mountain (Richwood, WV) . . . May 13
Natwick, Mildred: Birth . . . June 19
Nauru: National Day . . . Jan 31
Nautilus: First Nuclear-Powered Submarine Voyage: Anniv . . . Jan 17
Nauvoo Legion Chartered: Anniv . . . Feb 3
Navratilova, Martina: Birth . . . Oct 10
Navy Day . . . Oct 27
Navy: Sea Cadet Month . . . Apr 1
Navy, US: Racial Brawl on Kitty Hawk: Anniv . . . Oct 12
Near Miss Day . . . Mar 23
Nearing, Scott: Birth Anniv . . . Aug 6
Neas, Ralph: Birth . . . May 14
Nebraska
Admission Day . . . **Mar 1**
Applejack Fest (Nebraska City) . . . **Sept 16**
Arbor Day: Anniv . . . **Apr 22**
Arbor Day Fest (Nebraska City) . . . **Apr 28**
Barn Day (Filley) . . . **July 9**
Blue River Fest (Crete) . . . **July 21**
Central Nebraska Ethnic Fest (Grand Island) . . . **July 21**
Cherished Treasures: A Quilt Display (Omaha) . . . **Feb 14**
Christmas at Union Station (Omaha) . . . **Nov 26**
Christmas on the Prairie (Wahoo) . . . **Dec 1**
Clarkson Czech Fest (Clarkson) . . . **June 22**
Classic Car Show (Norfolk) . . . **July 29**
Clearwater Chamber of Commerce Rodeo (Clearwater) . . . **June 23**
Cobblestone Fest (Falls City) . . . **Aug 17**
Columbus Days (Columbus) . . . **Aug 12**
Cottonwood Prairie Fest (Hastings) . . . **June 2**
Countryside Village Art Fair (Omaha) . . . **June 3**
Curtis Easter Pageant (Curtis) . . . **Apr 9**
Curtis Fall Fest (Curtis) . . . **Sept 16**
Eastpark Czech Fest (Lincoln) . . . **May 6**
Exeter Road Rally (Exeter) . . . **Aug 20**
Fairfest (Hastings) . . . **July 27**
Garfield County Fair (Burwell) . . . **July 25**
Gingerbread Village (Omaha) . . . **Nov 18**
Grundlovsfest (Dannebrog) . . . **June 3**
Halloween (Arapahoe) . . . **Oct 31**
Harvest of Harmony (Grand Island) . . . **Oct 7**
Holiday Highlights (Arapahoe) . . . **Nov 10**
Holiday Rhapsody in Lights (Norfolk) . . . **Nov 18**
Homespun History Day (Omaha) . . . **May 7**
Homestead Days (Beatrice) . . . **June 22**
Hot Air Balloon Show/Aero Extravaganza (Falls City) . . . **June 17**
Husker Feed Grains and Soybean Conference (Lincoln) . . . **Jan 11**
Industry Day (Beatrice) . . . **Apr 23**
Kass Kounty King Korn Karnival (Plattsmouth) . . . **Sept 7**
Kinetic Sculpture Race (Omaha) . . . **Sept 9**
Klown Karnival (Plainview) . . . **July 7**
Korn Klub Mud Drags (Plattsmouth) . . . **June 11**
Lavitsef (Norfolk) . . . **Sept 21**
Light of the World Christmas Pageant (Minden) . . . **Nov 25**
Lisco Old-timers Day (Lisco) . . . **Sept 10**
Lone Tree Days (Central City) . . . **July 1**
Middle of Nowhere Trail Ride (Ainsworth) . . . **June 10**
Milligan Mini-Polka Day (Milligan) . . . **Sept 24**
Museum Comes to Life Day (Ainsworth) **June 24**
Nebraska's Big Rodeo (Burwell) . . . **July 27**
Nebraskaland Days/Buffalo Bill Rodeo (North Platte) . . . **June 13**
Oakland Swedish Fest (Oakland) . . . **June 2**
Ogallala's Indian Summer Rendezvous (Ogallala) . . . **Sept 14**
Old German Christmas (Papillion) . . . **Nov 24**
Old-Fashioned Danish Christmas (Dannebrog) . . . **Dec 2**
Omaha Cactus/Succulent Society Show/Sale (Bellevue) . . . **Sept 2**
Oregon Trail Days (Gering) . . . **July 13**
Oregon Trails Rodeo (Hastings) . . . **Sept 2**
Papillion Days (Papillion) . . . **June 15**
Prairie Pioneer Days (Arapahoe) . . . **July 2**
Pumpkin Corners Pumpkin Day (Stamford) **Oct 2**
Ranch Rodeo (Burwell) . . . **July 30**
River City Roundup (Omaha) . . . **Sept 20**
Sandhill Crane Migration (Grand Island) . . **Mar 5**
Scotts Bluff County Fair (Mitchell) . . . **Aug 13**
Shakespeare on the Green (Omaha) . . . **June 22**
Sugar Valley Rally/Wkend (Scottsbluff) . . **June 2**
Summer Solstice at Carhenge (Alliance) . **July 29**
Triumph Agriculture Expo Farm/Ranch Machinery Show (Omaha) . . . **Mar 14**
Ugly Pickup Parade and Contest (Chadron) . . . **Oct 27**
Victorian Garden Walk (Omaha) . . . **Aug 12**
Wagons/Wires/Rails along Oregon Trail (Omaha) . . . **Mar 12**
Wayne Chicken Show (Wayne) . . . **July 8**
Willa Cather Spring Conference (Red Cloud) . . . **May 5**
Willow Tree Festival (Gordon) . . . **Sept 9**
Wings Over the Platte (Grand Island) . . . **Mar 17**
Necker, Jacques: Birth Anniv . . . Sept 30
Needham, Theresa: Birth Anniv . . . Apr 17
Nehru, Jawaharlal: Birth Anniv . . . Nov 14
Neighbor Day, Natl Good . . . Sept 24
Neither Snow Nor Rain Day . . . Sept 7
Nelson, Cindy: Birth . . . Aug 19
Nelson, David: Birth . . . Oct 24
Nelson, E. Benjamin . . . May 17
Nelson, Horatio: Birth Anniv . . . Sept 29
Nelson: Ozzie and Harriet Show Debut: Anniv . . . Oct 8
Nelson, Thomas: Birth Anniv . . . Dec 26
Nelson, Willie: Birth . . . Apr 30
Nemerov, Howard: Birth Anniversary . . . Feb 28
Nemont/Agribition (Wolf Point, MT) . . . Oct 13
Nepal
King's Birthday National Holiday . . . **Dec 29**
National Holiday . . . **Dec 28**
National Unity Day . . . **Jan 11**
Neptune Discovery: Anniv . . . Sept 23
Neshoba County Fair (Philadelphia, MS) . . July 28
Nesmith, Michael: Birth . . . Dec 30
Netherlands
Liberation Day . . . **May 5**
Midwinter Horn Blowing . . . **Dec 3**
National Windmill Day . . . **May 13**
Netherlands-United States: Diplomatic Anniv . . . **Apr 19**
North Sea Jazz Fest (The Hague) . . . **July 14**
Prinsjesdag (Parliament opening) . . . **Sept 19**
Queen's Birthday . . . **Apr 30**
Relief of Leiden . . . **Oct 3**
Scilly Isles Peace Anniv . . . **Apr 17**
Nettles, Graig: Birth . . . Aug 20
Neuhas, Richard John: Birth . . . May 14
Neutrality Appeal, American: Anniv . . . Aug 18
Nevada
ABC Seniors Tourn (Reno) . . . **May 6**
Admission Day . . . **Oct 31**
American Bowling Congress Chmpshp Tourn (Reno) . . . **Feb 4**
American Bowling Congress Masters Tourn (Reno) . . . **May 1**
Assn of College Unions: Intl Bowling Chmpshp (Reno) . . . **May 5**
Cactus Pete's Carl Hayden Daze Celeb (Jackpot) . . . **July 2**
Las Vegas Open Horseshoe Tournament (Las Vegas, NV) . . . **Jan 8**
Pizza Expo (Las Vegas) . . . **Mar 14**
Reno Rodeo (Reno) . . . **June 16**
State Fair (Reno) . . . **Aug 23**
Nevis: National Day . . . Sept 19
New Beginning Fest (Coffeyville, KS) . . . Apr 27
New England, Dark Day in: Anniv . . . May 19
New England Speakers Building Professl Competence Month . . . Apr 1
New Hampshire
Artists in the Park (Wolfeboro) . . . **Aug 16**
Build a Scarecrow Days (Milton) . . . **July 1**
Children's Day (Milton) . . . **July 22**
Fast Day . . . **Apr 24**
GMC Truck Mt Wash Hillclimb (Gorham) **June 23**
Harvest Days (Milton) . . . **Oct 7**
Herb Day (Milton) . . . **June 10**
League of NH Craftsmen's Fair (Sunapee) **Aug 5**
Loudon Camel Classic (Loudon) . . . **June 16**
Market Square Days (Portsmouth) . . . **June 9**
Miss New Hampshire Finals (Manchester) . . . **May 12**
Model Railroad Show (Wolfeboro) . . . **Aug 12**

✯ Chase's 1995 Calendar of Events ✯ Index

Mt Washington Bicycle Hillclmb (Gorham)..................Sept 10
New Hampshire Highland Games (Lincoln)..................Sept 15
Old-Time Farm Day (Milton)..........Aug 12
Riverfest in Manchester (Manchester)....Sept 8
Rochester Fair (Rochester)..........Sept 14

New Jersey
All-Breed Dog Show, Natl (Atlantic City)..Dec 3
Antique/Classic Car Auction/Flea Mrkt (Atlantic City)..................Feb 17
Archery Classic, Natl Atlantic City (Atlantic City)..................Apr 21
Atlantuce City Antique/Collectibles Expo (Atlantic City)..................Mar 18
Baby Parade (Ocean City)............Aug 10
Barnegat Bay Crab Race/Seafood Fest (Seaside)..................Aug 27
Be Nice to New Jersey Week............July 2
Big "C" Attus Day (Toms River)..........Oct 1
Board and Care Recognition Month: Share the Care..................Sept 1
Cape May Music Fest (Cape May)......May 14
Christmas Inns of Cape May (Cape May) Nov 25
Clara Maass Day..................June 28
Clownfest (Seaside Heights)..........Sept 23
Crispus Attucks Day..................Mar 5
Crustacean Beauty Pageant/Ocean City Creep (Ocean City)..................Aug 2
East Coast Craft Trade Show (Elmwood Park)..................May 6
Easter Parade (Atlantic City)..........Apr 16
Eggsibit (Phillipsburg)..................Apr 1
Festival of the Sea (Pt Pleasant Beach)..Sept 16
Festival-in-the-Park (Nutley)..........Sept 10
First Black Pro Hockey Player: Anniv..Nov 15
Flemington Agricultural Fair (Flemington) Aug 29
Flemington Speedway Racing Season (Flemington)..................Apr 29
Fresh Seafood Festival (Atlantic City)..June 10
Halloween Parade (Toms River)..........Oct 31
Harlem: Photographs by Aaron Siskind (Montclair)..................May 20
Heritage Day Fest (Lavallette)..........Sept 9
Hey Rube Get A Tube (Pt Pleasant)....Sept 17
Jersey Shore Kayak/Canoeing Show (Bayville)..................Sept 16
Jewish Renaissance Fair (Morristown)...Sept 3
Long Beach Island Chowder Cook-Off (Beach Haven)..................Sept 30
Martin Z. Mollusk Day (Ocean City)......May 4
May in Montclair (Montclair Twshp)......May 1
Mid-Summer Ocean Swim & Beach Barbecue (Seaside Park)..................Aug 19
Miss America Pageant Finale (Atlantic City)..................Sept 16
New Jersey Restaurant/Hospitality Expo (Somerset)..................Mar 19
Ocean County Folk Arts Festival (Lakewood)..................Sept 9
Old-Time Barnegat Decoy/Gun Show (Tuckerton)..................Sept 23
Opera Festival of New Jersey (Lawrenceville)..................June 17
Palm Sunday Egg Hunt (Seaside Heights).Apr 9
Postcard Show (Mt Laurel)............Sept 8
Precanex (Wildwood)..................Aug 16
Rathkamp Matchcover Society Conv (Newark)..................Aug 13
Ratification Day..................Dec 18
Seafood in Seaside (Seaside Hts)......June 17
Sessex Farm and Horse Show (Augusta)..Aug 4
Songs of My People (Paterson)..........Mar 4
South Jersey Canoe/Kayak Classic (Lakewood)..................June 3
Suffragists' Voting Attempt: Anniv......Nov 19
Toms River Wildfowl Art & Decoy Show (Toms River)..................Feb 4
Tour of Somerville (Somerville)........May 29
US Junior Amateur (Golf) Chmpship (Fargo)..................July 25
US Women's Am Public Links (Golf) Chmpshp (Colts Neck)..................June 21
Victorian Week (Cape May)............Oct 6
Wedding of the Sea (Atlantic City)......Aug 15
Wings 'n' Water Festival (Stone Harbor).Sept 16

New Mexico
Admission Day..................Jan 6
American Indian Ceremonial Dancing (Taos, NM)..................Sept 29
America's Sexy Wives Contest (Albuquerque)..................Oct 13
Angel Fire Ski Area Opening (Angel Fire).Dec 7
Balloon Rally and Arts and Crafts Show (Angel Fire)..................July 14

Chile Challenge Off-Road Bike Race (Angel Fire)..................Sept 22
Cosmonauts' Day (Alamogordo)..........Apr 8
Electric Light Parade (Lovington)......Nov 25
Enchanted Circle Century Bike Tour (Red River)..................Sept 10
Farmington Invitational Balloon Fest (Farmington)..................May 27
Freedom Days (Farmington)............July 1
Fun Flight (Alamogordo)..............June 16
Golden Aspen Motorcycle Rally (Ruidoso)..................Sept 20
Great American Duck Race (Deming)...Aug 25
Intertribal Indian Ceremonial (Gallup)....Aug 8
Labor Day in the Pines (Angel Fire)....Sept 2
More Than Meets the Eye (Albuquerque)..Jan 7
Noon Year's Eve Party (Albuquerque)..Dec 31
Outdoor Summer Theater (Farmington) June 21
Santa Fe Chamber Music Fest (Santa Fe).July 7
Southeastern New Mexico Arts/Crafts Fest (Lovington)..................Nov 4
Space Hall of Fame Induction, Intl (Alamogordo)..................Oct 7
Totah Festival (Farmington)..........Sept 1
Traditional Indian Dances (Gallup)......May 29
Volksmarch (Angel Fire)..............Sept 30
Winterfest Balloon Rally (Angel Fire)....Feb 3
World Shovel Racing Chmpshp (Angel Fire)..................Feb 3
World's Greatest Lizard Race (Lovington).July 4

New Orleans, Battle of: Anniv..........Jan 8
New World Symphony Premiere: Anniv....Dec 16
New Year: Dates and Celebrations
Baha'i New Year's Day (Naw-Ruz)......Mar 21
Chinese Lunar New Year Festival (Baltimore, MD)..................Feb 5
Chinese New Year..................Jan 31
Chinese New Year Festival (San Francisco, CA)..................Jan 31
Chinese New Year Golden Dragon Parade (San Francisco)..................Feb 11
CN Tower Celebration (Toronto, Ont)....Jan 1
Cowbellion Herd New Year's Eve (Mobile, AL)..................Dec 31
First Night (Boston, MA)..............Dec 31
First Night Hartford (Hartford, CT)......Dec 31
First Night Roanoke (Roanoke, VA)......Dec 31
First Night State College (State College, PA)..................Dec 31
First Night Waynesboro (Waynesboro, VA)..................Dec 31
Hmong New Year Celebration (Sheboygan, WI)..................Nov 25
Iranian New Year (Persian)............Mar 21
Japanese Era New Year................Jan 1
Jewish New Year (Rosh Hashanah)....Sept 25
Julian Calendar New Year's Day........Jan 14
Muslim New Year..................May 30
New Year's Day (Gregorian)............Jan 1
New Year's Eve..................Dec 31
New Year's Eve Celebration (New York, NY)..................Dec 31
New Year's Eve Fireworks (Brooklyn, NY) Dec 31
New Year's Fest (Kalamazoo, MI)......Dec 31
No Resolution Day..................Dec 31
Noon Year's Eve Party (Albuquerque, NM)..................Dec 31
Russia: New Year's Day Observance......Jan 1
St. Sylvester Fest (Madeira Is, Portugal).Dec 31
Stock Exchange Holiday..............Jan 2
Tempe Fiesta Bowl Block Party (Tempe, AZ)..................Dec 31
United Kingdom/Republic of Ireland New Year's Holiday..................Jan 2
Waterfront New Year's Eve Celebration (Norfolk, VA)..................Dec 31

New Year's Dishonor List..................Jan 1
New York
Albany Tulip Fest (Albany)............May 12
Altamont Fair Fest of Lights (Altamont).Nov 21
American Bible Society Annual Meeting (New York)..................May 11
American Sportscasters Hall of Fame (New York)..................Dec 7
Baker Street Irregulars' Dinner/Cocktail Party (New York)..................Jan 6
Belmont Stakes (Belmont Park)........June 10
Belmont Stakes, First Running of:......June 19
Beyond Category: Genius of Ellington (Rochester)..................Apr 8
Boxing Day Annual Celebration (Ney York)..................Dec 26
Breeders' Cup Chmpshp (Elmont)......Oct 28
Capital District Scottish Games (Altamont)..................Sept 2

Caramoor Intl Music Fest (Katonah)....June 24
Child Injury Prevention Week (New Hyde Park)..................Sept 1
Church/Synagogue Library Assn Conf (Houghton)..................June 24
Colonial Heritage Assn Antique Show/Sale (Schoharie)..................Mar 4
Colton Country Day (Colton)..........July 15
Depot Lane Singers (Schoharie)........May 6
Dozenal Soc of America: Annual Mtg (Garden City)..................Oct 14
Fall Postage Stamp Mega Event (New York City)..................Nov 2
Farm Sanctuary Hoedown (Watkins Glen, NY)..................July 1
Finger Lakes Christmas in October Craft Show (Rochester)..................Oct 28
Finger Lakes Craftsmen Christmas Arts/Crafts (Rochester)..................Nov 24
Finger Lakes Craftsmen Spring Arts/Crafts (Rochester)..................Apr 8
Frog Jumping Contest (Old Forge)....June 18
Giant Christmas Tree/Rockefeller Center (New York)..................Dec 4
Girls Incorporated Natl Conference (New York)..................Apr 27
Great New York State Snow Expo (Albany) Nov 3
Great Northeast Home Show (Albany)....Feb 3
Halloween Festival and Haunted Walk (Brooklyn)..................Oct 28
Headache Fdtn Fundraiser, Natl (New York)..................Apr 22
Health Education Week..............Feb 19
Highlights Fdtn Writers Workshop (Chautauqua)..................July 15
Hot Prospects (Brooklyn)..............July 2
Interfaith Fellowship Anniv Celeb (New York)..................Sept 10
Iroquois Arts Showcase I (Howes Cave).May 27
Iroquois Arts Showcase II (Howes Cave).July 8
Iroquois Indian Fest (Howes Cave)....Sept 2
Lilac Festival (Rochester)............May 12
Literacy Volunteers America Conf (Buffalo)..................Oct 19
Macy's Thanksgiving Day Parade (New York)..................Nov 23
Mountain Eagle Indian Fest (Hunter)....Sept 2
Music Educ E Div In-Service Conf (Rochester)..................Mar 31
National Online Meeting (New York)....May 2
New Prospects (Brooklyn)............Mar 5
New Year's Eve Celebration (New York)..Dec 31
New Year's Eve Fireworks (Brooklyn)..Dec 31
New York City Marathon (New York)....Nov 12
New York City Subway: Anniv..........Oct 27
New York Dollhouse/Miniatures Trade Show (NYC)..................Feb 18
New York Flower Show (New York)......Mar 2
New York Natl Boat Show (New York)....Jan 6
New York Public Library: Anniv........May 23
New York Public Library Centennial Celeb..................May 20
New York Public Library Centennial Gala May 22
New York Subway Accident: Anniv......Nov 2
Northeast Great Outdoors Show (Albany)..................Mar 17
Of Field and Forest (East Meredith)....June 4
Orange Cnty Delaware River Fest (Port Jervis)..................May 19
Palatine Museum Wool Day (Schoharie).Sept 23
Parade for the Planet (NYC)..........Apr 23
Positive Power Humor/Creativity Conf (Saratoga Spgs)..................Apr 28
Prospect Park Fishing Contest (Brooklyn)..................July 18
Ratification Day..................July 26
Reopening of 1743 Palatine House Museum (Schoharie)..................May 1
Reopening of the Railroad Car Museum (Schoharie)..................June 3
Rose Show (Long Island)............June 11
St. George's Day Annual Celebration (New York)..................Apr 23
St. Patrick's Day Parade (New York)....Mar 17
Society of Illustrators Annual Exhib (New York)..................Feb 11
Spiders! (New York)..................Mar 11
Spring Postage Stamp Mega Event (New York)..................Mar 16
State Fair (Syracuse)................Aug 24
Strength/Diversity: Japanese-Amer Women (Stony Brook)..................Jan 7
Try This On/Clothing, Gender, Power (New York)..................Jan 7

New Hampshire (cont'd)—New York

Index ★ Chase's 1995 Calendar of Events ★

Tupper Lake Woodsmen's Days (Tupper Lake)......July 8
US Open (Golf) Chmpshp (Southampton)......June 15
WICC Greatest Bluefish Tourn (Long Island Sound)......Aug 26
Women of Achievement Month......Sept 1
Yo-Yo Day, Natl (Arcade)......June 10
New York Stock Exchange Established: Anniv......May 17
New York Weekly: First Issue Anniv......Nov 5
New Yorker magazine: Birth......Feb 21
New Zealand
　Anzac Day......Apr 25
　Otago/Southland Provincial Anniv......Mar 23
　Waitangi Day......Feb 6
Newfoundland
　Discovery Day......June 23
　St. George's Day......Apr 24
Newhart, Bob: Birth......Sept 5
Newhouse, Samuel: Birth Anniv......May 24
Newley, Anthony: Birth......Sept 24
Newman, Arnold: Birth......Mar 3
Newman, Edwin: Birth......Jan 25
Newman, Laraine: Birth......Mar 2
Newman, Paul: Birth......Jan 26
Newman, Randy: Birth......Nov 28
Newmar, Julie: Birth......Aug 16
Newport Maritime Teddy Bear Fest (Newport)......July 23
Newspaper. See also Journalism
Newspaper Assn America Conv (New Orleans, LA)......Apr 23
Newspaper, First American: Anniv......Sept 25
Newspaper Week (Japan)......Oct 1
Newspaper Week, Natl......Oct 9
Newton, Isaac: Birth Anniv......Dec 25
Newton, Juice: Birth......Feb 18
Newton, Wayne: Birth......Apr 3
Newton-John, Olivia: Birth......Sept 26
Nez Perce: Chief Joseph Surrender: Anniv......Oct 5
Niarchos, Stavros Spyros: Birth......July 3
Nicaragua
　Chamorro, Violeta: Election Anniv......Feb 25
　Independence Day......Sept 15
　National Liberation Day......July 19
Nichols, Mike: Birth......Nov 6
Nicholson, Jack: Birth......Apr 22
Nicklaus, Jack: Birth......Jan 21
Nickles, Don: Birth......Dec 6
Nicks, Stevie: Birth......May 26
Nidetch, Jean: Birth......Oct 12
Nielsen, Arthur Charles: Birth Anniv......Sept 5
Nielsen, Carl: Birth Anniv......June 9
Nielsen, Leslie: Birth......Feb 11
Nietzsche, Friedrich Wilhelm: Birth Anniv......Oct 15
Niger: Republic Day......Dec 18
Nigeria: Independence Day......Oct 1
Night in Old Luling (Luling, TX)......Oct 14
Night of a Thousand Stars (library read-alouds)......Apr 12
Night Out, Natl......Aug 1
Night Watch (France)......July 13
Nightingale, Florence: Birth Anniv......May 12
Nimoy, Leonard: Birth......Mar 26
Nininger, Alexander, Jr: First WWII Medal of Honor......Feb 10
Nixon Birthday Holiday (Yorba Linda, CA)......Jan 9
Nixon, John: Death Anniv......Dec 31
Nixon, Richard M.: Anniv of Trip to China......Feb 21
Nixon, Richard M.: Birth Anniv......Jan 9
Nixon, Richard M.: Pardon Day......Sept 8
Nixon, Richard M.: Presidential Resignation: Anniv......Aug 9
Nixon, Richard M.: Shanghai Communique: Anniv......Feb 27
Nixon, Richard Milhous, Announcing Death of 6677......Page 488
Nixon, Thelma: Birth Anniv......Mar 16
No Housework Day......Apr 7
No Man's Land Celebration (Breckenridge, CO)......Aug 11
No News Is Good News Day......Sept 11
No Resolution Day......Dec 31
No-Tobacco Day, World......May 31
Noah, Yannick: Birth......May 18
Nobel, Alfred: Death Anniv......Dec 10
Nobel Conference (St. Peter, MN)......Oct 3
Nobel Prize Ceremonies (Oslo, Norway/Stockholm, Sweden)......Dec 10
Nobel Prize, First US Scientist Receives: Anniv......Dec 20
Nolte, Nick: Birth......Feb 8
Noodle Month, Natl......Mar 1
Noon Year's Eve Party (Albuquerque, NM)......Dec 31

Noriega Conviction: Anniv......Apr 9
Noriega, Manuel: Surrender......Dec 19
Norman, Greg: Birth......Feb 10
Norris, Chuck: Birth......Mar 10
North Carolina
　Albermarle Craftman's Fair (Elizabeth City)......Oct 27
　Appalachian Potters Market (Marion)......Dec 2
　Bald Is Beautiful Convention (Morehead)......Sept 8
　Battle of Guilford Courthouse: Observance (Greensboro)......Mar 15
　Bele Chere (Asheville)......July 28
　Black American Arts Festival (Greensboro)......Jan 14
　Block House Steeplechase (Tryon, NC)......Apr 22
　Boo at the Zoo (Asheboro)......Oct 28
　Carolina Craftsmen's Christmas Classic (Greensboro)......Nov 24
　Carolina Craftsmen's Labor Day Classic (Winston-Salem)......Sept 2
　Carolina Craftsmen's Spring Classic (Greensboro)......Mar 3, Apr 7
　Carolina Model Railroader's Show/Swap Meet (Greensboro)......Apr 1, Nov 18
　Charlotte Observer Marathon/Runner's Expo (Charlotte)......Jan 6
　Charlotte Observer Moonride Criterium (Charlotte)......Aug 25
　Chimney Rock Hillclimb (Chimney Rock)......Apr 28
　Chimney Rock Hillfall (Chimney Rock)......Sept 17
　Chimney Rock Park Photo Contest (Chimney Rock)......Jan 1
　Christmas at Biltmore (Asheville)......Nov 11
　Christmas at the Zoo (Asheboro)......Dec 24
　City Stage (Greensboro)......Oct 7
　Confederate Memorial Day......May 10
　Constance S. Larrabee: WWII Photo Journal (Raleigh)......Apr 22
　Craft Fair of the Southern Highlands (Asheville)......July 20
　Dare Day (Manteo)......June 3
　Earth Day Celebration (Chimney Rock)......Apr 21
　Easter Sunrise Service (Chimney Rock)......Apr 16
　Eastern Music Festival (Greensboro)......June 24
　Everybody's Day Fest (Thomasville)......Sept 30
　Fest of Flowers (Asheville)......Apr 7
　Folkmoot USA: The NC Intl Folk Fest (Waynesville)......July 20
　Fourth of July/Torchlight Procession (Winston-Salem)......July 2
　Fun Fly and Model Aircraft Show, Natl (Chimney Rock)......Aug 19
　Greensboro Sit-in Anniv......Feb 1
　Halifax Independence Day......Apr 12
　Hog Day (Hillsborough)......June 17
　Human Relations Month (New Hanover County)......Feb 1
　International Home Furnishings Market (High Point)......Apr 27
　Longhorn World Chmpshp Rodeo (Winston-Salem)......Nov 3
　Major/Minor League: Photos/BB Stadiums (Hendersonvi)......Oct 14
　Man Will Never Fly Society Meeting (Kitty Hawk)......Dec 16
　Mecklenburg Day......May 20
　Micajah (Autryville)......July 13
　Micajah Autry Day (Autryville)......July 15
　Moravian Candle Tea (Winston-Salem)......Nov 30
　Moravian Easter Sunrise Service (Winston-Salem)......Apr 16
　Mountain Dance and Folk Fest (Asheville)......Aug 3
　Mountain Glory Fest (Marion)......Oct 14
　NC RV and Camping Show (Charlotte)......Feb 3
　NC RV and Camping Show (Greensboro)......Mar 24
　NC RV and Camping Show (Raleigh)......Mar 10
　NC RV and Travel Show (Fayetteville)......Jan 13
　NCAA Women's Golf Chmpshp (Wilmington)......May 24
　Oconaluftee Indian Village (Cherokee)......May 15
　Palomino League World Series (Greensboro)......Aug 9
　Ratification Day......Nov 21
　Riverspree Fest (Elizabeth City)......May 27
　Salem Christmas (Winston-Salem)......Dec 16
　Sampson County Expo (Clinton)......Nov 2
　Shindig-on-the-Green (Asheville)......July 1
　Singing on the Mountain (Linville)......June 25
　Soldiers' Reunion Celebration (Newton)......Aug 17
　Songs of My People (Winston-Salem)......Sept 9
　Sound/Light Spectacular/Showboat (Wilmington)......June 2
　Southern Christmas Show (Greensboro, NC)......Nov 9
　Southern Spring Show (Charlotte)......Feb 25
　State Fair (Raleigh)......Oct 13
　State St Station Cookbook Fair (Greensboro)......Mar 25
　Surrender at Durham Station: Anniv......Apr 18
　Today's Woman (Winston-Salem)......Feb 17
　Unto These Hills Cherokee Indian Drama (Cherokee)......June 15
　Wassail Tea (Greensboro, NC)......Nov 12
　Wheelchair Tennis Tourn (Greensboro)......Sept 8
　Whistlers Convention, Natl (Louisburg)......Apr 20
　World Whimmy Diddle Competition (Asheville)......Aug 12
　Zoo Fling (Asheboro)......Apr 1
　Zoofest (Asheboro)......Oct 1
North Dakota
　Admission Day......Nov 2
　Bazaarfest (Fargo)......Oct 18
　Bismarck Folkfest (Bismarck)......Sept 15
　Bonanzaville USA Pioneer Days (West Fargo)......Aug 19
　Bonanzaville's Christmas on Prairie (W Fargo)......Dec 1
　Christmas Bazaar (Fargo)......Aug 26
　Cookbook Festival (Fargo)......Feb 11
　Country Jamboree (Fargo)......Aug 13
　Dakota Cowboy Poetry Gathering (Medora)......May 27
　Fort Union Trading Post Rendezvous (Williston)......June 15
　Frontier Army Days (Mandon)......June 24
　International Music Camp (Dunseith)......June 11
　International Old-Time Fiddlers Contest (Dunseith)......June 9
　Makoti Threshing Show (Makoti)......Oct 7
　Mandan Rodeo Days (Mandan)......July 1
　Medora Musical (Medora)......June 9
　Merry Prairie Christmas (Fargo-Moorhead)......Nov 26
　Missouri River Expo (Bismarck)......June 1
　Norsk Hostfest (Minot)......Oct 10
　North American Regatta (Lake Sakajawea)......July 21
　Potato Bowl (Grand Forks)......Sept 9
　Scandinavian Hjemkomst Fest (Fargo)......June 21
　Sodbuster Days (Fort Ransom State Park)......July 8
　State Fair (Minot)......July 21
　Tatanka Fest (Jamestown)......June 29
　Uff Da Day (Reeder)......May 17
　United Tribes PowWow (Bismarck)......Sept 9
　Wrestling Chmpshps, NAIA (Jamestown)......Mar 10
North, Oliver Laurence: Birth......Oct 7
North Sea Oil Rig Disaster: Anniv......Mar 27
North, Sheree: Birth......Jan 17
Northern Hemisphere Hoodie-Hoo Day......Feb 20
Northern Ireland
　Belfast Festival at Queen's (Belfast)......Nov 6
　Bloody Sunday: Anniv......Jan 30
　Christmas Holiday......Dec 25
　Game and Country Fair (Antrim)......June 30
　Orangemen's Day......July 12
　Oul' Lammas Fair......Aug 28
　St. Patrick's Day......Mar 17
Northwest Folklife Fest (Seattle, WA)......May 26
Northwest Ordinance: Anniv......July 13
Norton, Ken: Birth......Aug 9
Norway
　Bergen Intl Festival (Bergen)......May 24
　Constitution or Independence Day......May 17
　Easter Fest (Kautokeino, Karasjok)......Apr 13
　Midnight Sun at North Cape......May 14
　Midsummer Night......June 23
　Nobel Prize Awards Ceremony (Oslo)......Dec 10
　Northern Lights Fest (Tromso)......Jan 19
　Olsok Eve......July 29
　Oslo Chamber Music Fest (Oslo)......Aug 4
　Pageantry in Oslo (Oslo)......Oct 3
　Voss Jazz Fest (Voss)......Apr 7
Norwegian: Christmas (Brooklyn Park, MN)......Dec 2
Norwegian: Little Norway Fest (Petersburg, AK)......May 18
Norwegian: Nordicfest (Libby, MT)......Sept 8
Norwegian: Norsefest (Madison, MN)......Nov 9
Norwegian: Norsk Hostfest (Minot, ND)......Oct 10
Norwegian: 17th of May Fest (Seattle, WA)......May 17
Norwegian: Uff Da Day (Reeder, ND)......May 17
Nostradamus: Birth Anniv......Dec 14
Notary Public Day, Natl......Nov 7
Notary Public Week, Natl......Nov 5
Nothing Day, National......Jan 16
Nova Scotia Bluegrass/Oldtime Music Fest (Ardoise, NS)......July 28
Novak, Kim: Birth......Feb 13
Novello, Antonio: Birth......Aug 23
Novello, Don: Birth......Jan 1

☆ Chase's 1995 Calendar of Events ☆ Index

Nuclear—Onion

Nuclear Chain Reaction, 1st Self-Sustaining: Anniv . . . . Dec 2
Nuclear Power Plant Accident, Three Mile Island: Anniv . . . . Mar 28
Nuclear-Free World, First Step Toward a . . . . Dec 8
Nuclear-Powered Submarine Voyage, First: Anniv . . . . Jan 17
Nude Recreation Week . . . . July 3
Nugent, Ted: Birth . . . . Dec 13
Nujoma, Samuel: Birth . . . . May 12
Nunn, Sam: Birth . . . . Sept 8
Nureyev, Rudolf: Birth Anniv . . . . Mar 17
Nurse Anesthetist Week, Natl Certified Registered . . . . Aug 5
Nurse Assistants Day, Career . . . . June 1
Nurse: Delano, Jane: Birth Anniv . . . . Mar 26
Nurses: Natl School Nurse Day . . . . Jan 25
Nurses Week, Natl . . . . May 6
Nurses Week, Operating Room . . . . Nov 12
Nursing: Clara Maass Day (New Jersey) . . . . June 28
Nursing Conf Pediatric Primary Care (Nashville, TN) . . . . Mar 14
Nursing Home Week, Natl . . . . May 14
Nutrition Month, Natl . . . . Mar 1
Nutt Day, Emma M. . . . . Sept 1
Nykvist, Sven Vilhem: Birth . . . . Dec 3
Nylon Stockings: Anniv . . . . May 15
Nyro, Laura: Birth . . . . Oct 18

O

O. Henry: Birth Anniv . . . . Sept 11
O-Bon Fest (Delray Beach, FL) . . . . Aug 12
O'Brian, Hugh: Birth . . . . Apr 19
O'Brien, Edna: Birth . . . . Dec 15
O'Brien, Margaret: Birth . . . . Jan 15
O'Casey, Sean: Birth Anniv . . . . Mar 30
O'Connor, Carroll: Birth . . . . Aug 2
O'Connor, Donald: Birth . . . . Aug 28
O'Connor, Sandra Day: Birth . . . . Mar 26
O'Connor, Sinead: Birth . . . . Dec 8
O'Dell, William "Spike": Birth . . . . May 21
O'Hair, Madalyn Murray: Birth . . . . Apr 13
O'Hara, Catherine: Birth . . . . Mar 4
O'Hara, Maureen: Birth . . . . Aug 17
O'Higgins, Bernardo: Birth Anniv . . . . Aug 20
O'Keeffe, Georgia: Birth Anniv . . . . Nov 15
O'Leary, Hazel: Birth . . . . May 17
O'Loughlin, Gerald: Birth . . . . Dec 23
O'Neal, Patrick: Birth . . . . Sept 26
O'Neal, Ryan: Birth . . . . Apr 20
O'Neal, Tatum: Birth . . . . Nov 5
O'Neill, Death of Thomas P. 6644 . . . . Page 488
O'Neill, Eugene: Birth Anniv . . . . Oct 16
O'Neill, Jennifer: Birth . . . . Feb 20
O'Neill, Rose Cecil: Birth Anniv . . . . June 25
O'Neill, Thomas (Tip): Birth Anniv . . . . Dec 9
O'Sullivan, Maureen: Birth . . . . May 17
O'Toole, Peter: Birth . . . . Aug 2
Oakley, Annie: Birth Anniv . . . . Aug 13
Oakley, Charles: Birth . . . . Dec 18
Oates, John: Birth . . . . Apr 7
Oates, Joyce Carol: Birth . . . . June 16
Oatmeal Fest (Bertram/Oatmeal, TX) . . . . Sept 1
Oatmeal Month . . . . Jan 1
Occupational Therapy Month, Natl . . . . Apr 1
Occupations. See Employment
Ocean, Billy: Birth . . . . Jan 21
Ocean Fest Sea-Son, Natl Week (Ft Lauderdale, FL) . . . . Apr 14
Ocean: Mother Ocean Day . . . . May 14
Ocean, Natl Week of the . . . . Apr 16
Oconaluftee Indian Village (Cherokee, NC) May 15
Odetta: Birth . . . . Dec 31
Odetts, Clifford: Birth Anniv . . . . July 18
Odie's Birthday . . . . Aug 8
Office Olympics (Shreveport, LA) . . . . Sept 15
Oglebay Institute Winter Lodge Program (Wheeling, WV) . . . . Jan 2
Oglethorpe Day . . . . Feb 12
Oglethorpe, James: Birth Anniv . . . . Dec 22
Ohio
- Admission Day . . . . **Mar 1**
- Africa's Legacy in Mexico: Photos (Toledo) . . . . **Feb 11**
- All-American Soap Box Derby (Akron) . . . . **Aug 12**
- American Speech-Lang-Hearing Assn Conv (Cincinnati) . . . . **Nov 16**
- Annie Oakley Days (Greenville) . . . . **July 27**
- Anti-Saloon League Founded: Anniv (Oberlin) . . . . **May 24**
- Antique Power Exhib (Burton) . . . . **July 28**
- Apple Butter Fest (Burton) . . . . **Oct 7**
- Apple Butter Stirrin' (Coshocton) . . . . **Oct 21**
- Bob Evans Farm Fest (Rio Grande) . . . . **Oct 13**
- Budweiser Oktoberfest-Zinzinnati (Cincinnati) . . . . **Sept 16**
- Carnation City Fest (Alliance) . . . . **Aug 4**
- Catholic Educatl Assn Conv/Expo, Natl (Cincinnati) . . . . **Apr 18**
- Celebration of Quilts (Chardon) . . . . **May 20**
- Cheetah Run (Cincinnati) . . . . **Sept 3**
- Circleville Pumpkin Show (Circleville) . . . . **Oct 18**
- Clay Week, Natl (Uhrichsville) . . . . **June 13**
- Cleveland Natl Air Show (Cleveland) . . . . **Sept 2**
- Coshocton Canal Fest (Coshocton) . . . . **Aug 18**
- Country Christmas (Coshocton) . . . . **Dec 1**
- Dandelion May Fest (Dover) . . . . **May 6**
- Dulcimer Days (Coshocton) . . . . **May 20**
- EARTHFest (Cleveland) . . . . **Apr 23**
- Easter Egg Hunt (Rockford) . . . . **Apr 15**
- Fair at New Boston (Springfield) . . . . **Sept 2**
- Fall Fest of Leaves (Ross County) . . . . **Oct 20**
- Family Frolic (Coshocton) . . . . **June 17**
- Festival of Lights (Cincinnati) . . . . **Nov 24**
- Fishing Has No Boundaries (Sandusky) . . . . **July 8**
- General Tire PBA Men's Tourn (Fairlawn) . **Apr 17**
- Geneva Area Grape Jamboree (Geneva) **Sept 23**
- German Village Oktoberfest (Columbus) . **Sept 8**
- Goldstar Chili Fest (Cincinnati) . . . . **Oct 7**
- Gus Macker Charity Basketball Tourn (Chillicothe) . . . . **June 10**
- Halloween Happening (Cincinnati) . . . . **Oct 28**
- Help Someone See Week (Rockford) . . . . **Mar 1**
- Honda Starlight Celebration (Cincinnati) . **Nov 18**
- Ice Cream Social (Rockford) . . . . **Aug 26**
- Jackson County Apple Fest (Jackson) . . **Sept 19**
- Kahn's Kids Fest (Cincinnati) . . . . **June 3**
- Longhorn World Chmpshp Rodeo (Cincinnati) . . . . **Feb 10**
- Marion Popcorn Fest (Marion) . . . . **Sept 7**
- NCAA Div I Men's Golf Chmpshp (Columbus) . . . . **May 31**
- Ohio River Sternwheel Fest (Marietta) . . . . **Sept 8**
- Operation Youth (Cincinnati) . . . . **June 10**
- Paul Bunyan Show (Nelsonville) . . . . **Oct 6**
- Picnic on the Point (Cincinnati) . . . . **July 1**
- Piqua Heritage Fest (Piqua) . . . . **Sept 2**
- Planned Parenthood Book Fair (Dayton) . **Nov 10**
- Pro Football Hall of Fame Fest (Canton) . **July 21**
- Riverfest (Cincinnati) . . . . **Sept 2**
- Square Fair (Lima) . . . . **Aug 4**
- Stokes Becomes First Black Mayor in US: Anniv . . . . **Nov 13**
- Sweet Corn Fest (Millersport) . . . . **Aug 30**
- Swiss Fest (Sugarcreek) . . . . **Sept 29**
- Taste of Cincinnati (Cincinnati) . . . . **May 27**
- Tecumseh! Epic Outdoor Drama (Chillicothe) . . . . **June 9**
- Twins Day Fest (Twinsburg) . . . . **Aug 4**
- Utica Old-Fashioned Ice Cream Fest (Utica) . . . . **May 27**
- Walsh Invitational Rifle Tourn (Cincinnati) . **Nov 3**
- Who's in Charge: Workers/Managers (Youngstown) . . . . **May 20**
- Zoo Babies (Cincinnati) . . . . **June 3**

Oil and Gas Fest, West Virginia (Sisterville, WV) . . . . Sept 14
Oil Embargo Lifted, Arab: Anniv . . . . Mar 13
Oil: 55 mph Speed Limit: Anniv . . . . Jan 2
Oil Well, First Commercial: Anniv . . . . Aug 27
Oklahoma
- Admission Day . . . . **Nov 16**
- Aerospace America (Oklahoma City) . . . . **July 14**
- All-College Basketball Tourn (Oklahoma City) . . . . **Jan 6, Dec 29**
- American Bicycle Assn Grand Nationals (Oklahoma City) . . . . **Nov 23**
- Autumn Magic/Tom Mix Festival (Guthrie) **Sept 8**
- Azalea Fest (Muskogee) . . . . **Apr 8**
- Big Eight Men/Women Tennis Chmpshp (Oklahoma City) . . . . **Apr 21**
- Big Eight Men's/Women's Track/Field Chmpshp (Oklahoma City) . . . . **Feb 25**
- Bullnanza (Guthrie, OK) . . . . **Feb 3**
- Cherokee Strip Celebration (Perry) . . . . **Sept 16**
- Cherokee Strip Day . . . . **Sept 16**
- Chili and Bluegrass Fest (Tulsa) . . . . **Sept 8**
- Chocolate Fest (Norman) . . . . **Feb 11**
- Christmas Fair and Gallery (Norman) . . . . **Nov 25**
- Cimarron Territory Celebration (Beaver) . **Apr 17**
- Eighty-Niner Day Celeb (Guthrie) . . . . **Apr 18**
- Elizabeth Bower Day . . . . **May 25**
- Festival of the Arts (Weatherford) . . . . **Sept 9**
- Festival of the Horse (Oklahoma City) . . . **Oct 20**
- Garden of Lights (Muskogee) . . . . **Jan 1**
- Guthrie Jazz Banjo Fest (Guthrie) . . . . **May 26**
- Historical Day . . . . **Oct 10**
- Holy City Easter Pageant (Lawton) . . . . **Apr 15**
- Indian Day . . . . **Sept 9**
- International Finals Rodeo (OK City) . . . . **Dec 14**
- Inter-State Fair/Rodeo (Coffeyville, KS) . . **Aug 12**
- Iris Fest (Ponca City) . . . . **May 20**
- Italian Festival (McAlester) . . . . **May 27**
- Jazz in June (Norman) . . . . **June 22**
- Kiamichi-Owa-Chito Fest of the Forest (Broken Arrow) . . . . **June 14**
- Land Rush Begins . . . . **Apr 22**
- Last Formal Surrender of Confederate Troops: Anniv . . . . **June 23**
- Longhorn World Chmpshp Rodeo (Tulsa) . **Feb 3**
- Medieval Fair (Norman) . . . . **Apr 7**
- Men's Big 8 Baseball Chmpshp (Oklahoma City) . . . . **May 18**
- Midsummer Night's Fair (Norman) . . . . **July 14**
- Mozart Intl Fest (Bartlesville) . . . . **June 8**
- NAIA Men's Div I Basketball Chmpshp (Tulsa) . . . . **Mar 14**
- NAIA Men's Golf Chmpshp (Tulsa) . . . . **May 23**
- NAIA Men's Tennis Chmpshps (Tulsa) . . . **May 22**
- NAIA Women's Tennis Chmpshps (Tulsa) **May 22**
- NCAA Women's Softball World Series (Oklahoma City) . . . . **May 25**
- Octoberfest (Ponca City) . . . . **Oct 7**
- Oklahoma All Night Singing (Konawa) . . . . **Aug 5**
- Oklahoma Day . . . . **Apr 22**
- Oklahoma State Prison Rodeo (McAlester) . . . . **Sept 1**
- Oktoberfest (Tulsa) . . . . **Oct 19**
- 101 Wild West Rodeo (Ponca City) . . . . **Aug 16**
- Ponca City Ranch Rodeo (Ponca City) . . **May 12**
- Quadlings of Oz Convention (Tulsa) . . . . **Apr 29**
- Rattlesnake Derby (Mangum) . . . . **Apr 28**
- Rattlesnake Hunt (Waurika) . . . . **Apr 7**
- Red Earth Native Amer Cultural Fest (Oklahoma City) . . . . **June 9**
- Sand Plum Fest (Guthrie) . . . . **June 23**
- Senior Citizens Day . . . . **June 9**
- Sooner State Summer Games (Oklahoma City) . . . . **June 9**
- Sorghum Day Festival (Wewoka) . . . . **Oct 28**
- Sportsfest (Oklahoma City) . . . . **Feb 4**
- Springfest (Weatherford) . . . . **May 6**
- State Fair (Oklahoma City) . . . . **Sept 15**
- Teddy Bear Conv (Jenks) . . . . **Nov 11**
- Watermelon Fest (Rush Springs) . . . . **Aug 12**
- Will Rogers Cattle Sale/Cookoffs (Oolagah) . . . . **Mar 18**
- Will Rogers Days (Claremore) . . . . **Nov 3**
- World Chmpshp Hog Calling Contest (Weatherford) . . . . **Feb 25**
- World Cow Chip Throwing Chmpshp (Beaver) . . . . **Apr 17**

Okoye, Christian: Birth . . . . Aug 16
Oktoberfest (Amana, IA) . . . . Sept 29
Oktoberfest (Covington, KY) . . . . Sept 8
Oktoberfest (Denver, CO) . . . . Sept 15
Oktoberfest (Grand Prairie, TX) . . . . Oct 14
Oktoberfest (Hermann, MO) . . . . Oct 7
Oktoberfest (Ponca City, OK) . . . . Oct 7
Oktoberfest (Tulsa, OK) . . . . Oct 19
Oktoberfest, German Village (Columbus, OH) . . . . Sept 8
Oktoberfest/Thanks Canada (Helena, MT) . Oct 13
Oktoberfest-Zinzinnati, Budweiser (Cincinnati, OH) . . . . Sept 16
Olajuwon, Hakeem: Birth . . . . Jan 21
Old Home Days, Rockingham (Rockingham, VT) . . . . Aug 3
Old Inauguration Day . . . . Mar 4
Old West. See Cowboys/Old West
Oldenburg, Claes: Birth . . . . Jan 28
Older Americans Month (Pres Proc) . . . . May 1
Older Americans Month 6565 . . . . Page 488
Oldhem, Jawann: Birth . . . . July 4
Oleput Festival (Tupelo, MS) . . . . June 2
Olin, Ken: Birth . . . . July 30
Oliphant, Pat: Birth . . . . July 24
Olive Branch Petition: Anniv . . . . July 8
Oliver, Sy: Birth . . . . Dec 17
Olivier, Laurence: Death Anniv . . . . July 11
Ollie, Elm Farm: Cow Milked While Flying: Anniv . . . . Feb 18
Olmos, Edward James: Birth . . . . Feb 24
Olmsted, Frederick L.: Birth Anniv . . . . Apr 26
Olsen, Kenneth: Birth . . . . Feb 20
Olsen, Merlin: Birth . . . . Sept 15
Olympics: Israeli Olympiad Massacre: Anniv . . . . Sept 5
Oman: National Holiday . . . . Nov 18
Onassis, Jacqueline Kennedy: Birth Anniv . July 28
One-Hit Wonder Day, Natl . . . . Sept 25
One-Tooth Rhee Landing Day . . . . Jan 23
"On-Hold" Month, Natl . . . . Mar 1
Onion Fest, Vacaville (Vacaville, CA) . . . . Sept 9

Index ☆ Chase's 1995 Calendar of Events ☆

Onizuka——Parades

Onizuka, Ellison S.: Birth Anniv .... **June 24**
Online Conference (Chicago, IL) .... **Oct 30**
Online Meeting, Natl (New York, NY) .... **May 2**
Ono, Yoko: Birth .... **Feb 18**
Opera Debuts in the Colonies .... **Feb 8**
Opera Fest, Ash Lawn Highland (Charlottesville, VA) .... **June 25**
Operating Room Nurse Week .... **Nov 12**
Operation Youth (Cincinnati, OH) .... **June 10**
Orange Bowl Football Game (Miami, FL) .... **Jan 2**
Orangemen's Day (Ireland) .... **July 12**
Orangemen's Day (Northern Ireland) ...... **July 12**
Oregon
Admission Day .... **Feb 14**
American Business Women's Assn, Natl Conv (Portland) .... **Oct 3**
Artquake (Portland) .... **Sept 2**
Barbershop Ballad Contest (Forest Grove) .... **Mar 10**
Betty Picnic (Grants Pass) .... **June 10**
Brookings-Harbor Azalea Fest (Brookings) .... **May 26**
Children's Festival (Jacksonville) .... **July 9**
Christmas Greens Show (Salem) .... **Dec 1**
Cinco de Mayo Fest (Portland) .... **May 4**
Columbia River Cross Channel Swim (Hood River) .... **Sept 4**
Happy Birthday, A.C.! (Salem) .... **Feb 18**
High Wind Classic (Hood River) .... **June 24**
Hood River Apple Jam (Hood River) ..... **Aug 19**
Hood River County Fair (Hood River) ... **July 26**
Hood River Old-Fashioned Fourth of July (Hood River) .... **July 4**
Hood River Valley Blossom Fest (Hood River) .... **Apr 15**
Hood River Valley Harvest Fest (Hood River) .... **Oct 13**
Lincoln, Abraham: Birthday Observance (also DE) .... **Feb 6**
Log Show by the Sea (Brookings) ...... **Sept 16**
Loyalty Days/Seafair Fest (Newport) ..... **May 4**
Marion County Fair (Salem) .... **July 6**
Mother's Day Annual Rhododendron Show (Portland) .... **May 13**
NAIA Wom Div II Basketball Chmpshp Tourn (Monmouth) .... **Mar 9**
Newport Seafood and Wine Fest (Newport) .... **Feb 24**
Northwest Cribbage Tourn (Baker City) .. **Mar 10**
Oregon Bach Festival (Eugene) .... **June 23**
Oregon Coast Music Fest (Coos Bay) ... **July 15**
Oregon Shakespeare Fest (Portland) ..... **Jan 1**
Oregon State Fair (Salem) .... **Aug 24**
Pendleton Round-Up (Pendleton) .... **Sept 13**
Pole Pedal Paddle (Bend) .... **May 20**
Portland Rose Fest (Portland) .... **June 1**
Rhododendron Fest (Florence) .... **May 19**
Rogue River Jet Boat Marathon (Gold Beach) .... **June 2**
Salem Art Fair and Fest (Salem) .... **July 21**
Sea Lion Suds Fest (Gold Beach) .... **Mar 18**
Shakespeare Fest (Ashland) .... **Feb 17**
Sherwood Robin Hood Festival (Sherwood) .... **July 14**
Snowflake Fest (Klamath Falls) .... **Dec 2**
St. Urho's Day (Hood River) .... **Mar 16**
State Games of Oregon (Portland) .... **July 7**
Sternwheeler Days (Cascade Locks) ... **June 23**
Tinnitus Seminar, Intl (Portland) .... **July 12**
Trail's End Marathon (Seaside) .... **Mar 4**
Turkey Rama (McMinnville) .... **July 6**
Whale of a Wine Fest (Gold Beach) ...... **Jan 14**
Willamette Valley Folk Fest (Eugene) .... **May 19**
Winter Games of Oregon (various) .... **Mar 11**
Oregon Trail: Homestead Days (Beatrice, NE) .... **June 22**
Oregon Trail: Pony Express Festival (Hanover, KS) .... **Aug 27**
Organic Act Day (Virgin Islands) .... **June 20**
Organize Your Home Office Day, Natl ...... **Mar 28**
Orlando, Tony: Birth .... **Apr 3**
Oro City (Leadville, CO) .... **July 1**
Orr, Bobby: Birth .... **Mar 20**
Orr, Kay A.: Birth .... **Jan 2**
Ortega, Daniel: Chamorro, Violeta: Election Anniv .... **Feb 25**
Ortega Saavedra, Daniel: Birth .... **Nov 11**
Orthodox Ascension Day .... **June 1**
Orthodox Christmas, Old Calendar .... **Jan 7**
Orthodox Easter Sunday .... **Apr 23**
Orthodox: Festival of All Saints .... **June 18**
Orthodox Lent .... **Mar 6**
Orthodox Palm Sunday .... **Apr 16**
Orthodox Pentecost .... **June 11**
Orwell, George: Birth Anniv .... **June 25**
Osborne, Jeffrey: Birth .... **Mar 9**
Osbourne, Ozzy: Birth .... **Dec 3**
Osler, Sir William: Birth Anniv .... **July 12**
Osmond, Donny: Birth .... **Dec 9**
Osmond, Marie: Birth .... **Oct 13**
Ostara .... **Mar 20**
Osteopathic Medicine Week, Natl .... **Nov 5**
Osterwald, Bibi: Birth .... **Feb 3**
Oul' Lammas Fair (Northern Ireland) ...... **Aug 28**
Outhouse Race, Great Klondike (Dawson City, Yukon) .... **Sept 3**
Ovar Carnival Festival (Costa De Prata, Portugal) .... **Feb 25**
Overseas Chinese Day (Taiwan) .... **Oct 21**
Owen, Robert: Birth Anniv .... **May 14**
Owens, "Buck": Birth .... **Aug 12**
Owens, Gary: Birth .... **May 10**
Owens, Jesse: Birth Anniv .... **Sept 12**
Own Your Own Share of America Day .... **June 15**
Own Your Share of America Month .... **June 1**
Oxenberg, Catherine: Birth .... **Sept 22**
Oyster Bowl Game, Shrine (Norfolk, VA) ... **Nov 18**
Oyster Fest (Biloxi, MS) .... **Mar 16**
Oyster Fest (Chincoteague Island, VA) ...... **Oct 7**
Oyster Fest, St. Mary's County (Leonardtown, MD) .... **Oct 21**
Oz Club Convention, Wizard of (Rosemont, IL) .... **June 16**
Oz Convention, Gillikins of (Escanaba, MI) . **May 6**
Oz Convention, Munchkins of (Wilmington, DE) .... **Aug 4**
Oz Convention, Ozcanabans of (Escanaba, MI) .... **Dec 9**
Oz Convention, Quadlings of (Tulsa, OK) .. **Apr 29**
Oz Convention, Winkies of (Escanaba, MI) .. **July 7**
Oz, Frank: Birth .... **May 24**
Ozark Empire Fair (Springfield, MO) ...... **July 28**
Ozark Trail Festival (Heber Springs, AR) .... **Oct 6**
Ozawa, Seiji: Birth .... **Sept 1**
Ozick, Cynthia: Birth .... **Apr 17**
Ozma's Birthday .... **Aug 21**
Ozzie and Harriet Show Debut: Anniv .... **Oct 8**

P

Paar, Jack: Birth .... **May 1**
Paar, Jack: Water Closet Incident: Anniv ... **Feb 11**
Pacific Ocean Discovered: Anniv .... **Sept 25**
Pacing the Bounds (Liestal, Switzerland) .. **May 22**
Pacino, Al: Birth .... **Apr 25**
Packwood, Bob: Birth .... **Sept 11**
Paczki Day in Hamtramck (Hamtramck, MI) **Feb 28**
Paderewski, Ignace J: Birth Anniv .... **Nov 6**
Paganini, Nicolo: Birth Anniv .... **Oct 27**
Page, LaWanda: Birth .... **Oct 19**
Page, Patti: Birth .... **Nov 8**
Paige, Janis: Birth .... **Sept 16**
Paine, Thomas: Birth Anniv .... **Jan 29**
Pakistan
Founder's Death Anniv .... **Sept 11**
Republic Day .... **Mar 23**
Palance, Jack: Birth .... **Feb 18**
Palatine House Museum, Reopening 1743 (Schoharie, NY) .... **May 1**
Palestinian Massacre: Anniv .... **Sept 16**
Palestinian People, Intl Day of Solidarity with (UN) .... **Nov 29**
Palin, Michael: Birth .... **May 5**
Pallette, Royal Palette (Windsor, England) ... **Jan 1**
Palm Harbor Arts/Crafts/Music Fest (Palm Harbor, FL) .... **Dec 2**
Palm Sunday .... **Apr 9**
Palm Sunday Egg Hunt (Seaside Heights, NJ) .... **Apr 9**
Palm Sunday, Orthodox .... **Apr 16**
Palmeiro, Rafael: Birth .... **Sept 24**
Palmer, Alice Freeman: Birth Anniv .... **Feb 21**
Palmer, Arnold Daniel: Birth .... **Sept 10**
Palmer, Betsy: Birth .... **Nov 1**
Palmer, Jim: Birth .... **Oct 15**
Palmer, Lilli: Birth Anniv .... **May 24**
Palmer, Robert: Birth .... **Jan 19**
Palolo, Swarm of (American Samoa) .... **Oct 1**
Palomares Hydrogen Bomb Accident: Anniv .... **Jan 17**
Pan American Aviation Day (Pres Proc) ... **Dec 17**
Pan American Day (Pres Proc) .... **Apr 14**
Pan American Day and Week 6545, 6666 . **Page 488**
Pan American Flight 103 Explosion: Anniv . **Dec 21**
Pan American Flight 103 Victims Remembrance 6642 .... **Page 488**
Pan American Health Day (Pres Proc) ...... **Dec 2**
Pan American Week (Pres Proc) .... **Apr 9**
Panama
First Crossing of Panama Canal: Anniv .... **Jan 7**
Flag Day .... **Nov 4**
Independence from Spain Day .... **Nov 28**
Martyrs' Day .... **Jan 9**
National Holiday .... **Nov 3**
Noriega Conviction: Anniv .... **Apr 9**
Panama City Foundation Day .... **Aug 15**
United States Invasion of Panama: Anniv **Dec 19**
Pancake Day (Centerville, IA) .... **Sept 23**
Pancake Day, Intl (Liberal, KS) .... **Feb 28**
Pancake Week, Natl .... **Feb 26**
Panetta, Leon: Birth .... **June 28**
Panic Day .... **Mar 9**
Panizzi, Anthony: Birth Anniv .... **Sept 16**
Panoply (Huntsville, AL) .... **May 12**
Papal "No" on Ordaining Women: Anniv ... **May 30**
Papaya Month, Natl .... **Sept 1**
Paper Money Issued: Anniv .... **Mar 10**
Papillion Days (Papillion, NE) .... **June 15**
Papp, Joseph: Birth Anniv .... **June 22**
Papua New Guinea: Independence Day ... **Sept 16**
Parades
Auburn-Cord-Duesenberg Fest (Auburn, IN) .... **Aug 31**
Azalea Fest (Muskogee, OK) .... **Apr 8**
Baby Parade (Ocean City, NJ) .... **Aug 10**
Black History Parade (Monroe, LA) ...... **Feb 25**
Boom Box Parade (Willimantic, CT) ...... **July 4**
Brach's Holiday Christmas Parade (Chicago, IL) .... **Nov 25**
Bristol Civic, Military/Fireman Parade (Bristol, RI) .... **July 4**
Bud Billiken Parade (Chicago, IL) .... **Aug 12**
Canadian Tulip Fest (Ottawa, Ont) ...... **May 17**
Carnaval '95 (San Francisco, CA) .... **May 27**
Carry Nation Fest (Holly, MI) .... **Sept 8**
Channel 6 Thanksgiving Day Parade (Philadelphia, PA) .... **Nov 23**
Chinese New Year Golden Dragon Parade (Los Angeles, CA) .... **Feb 18**
Chinese New Year Golden Dragon Parade (San Francisco) .... **Feb 11**
Christmas on the Coosa (Wetumpka, AL) . **Dec 9**
Christmas on the River (Demopolis, AL) . **Nov 30**
Clownfest (Seaside Heights, NJ) .... **Sept 23**
Daffodil Fest (Tacoma/ Puyallup/Sumner/ Orting) .... **Apr 22**
Do Dah Parade (Kalamazoo, MI) .... **June 3**
Dodge City Days (Dodge City, KS) ...... **July 28**
Eighty-Niner Day Celeb (Guthrie, OK) ... **Apr 18**
Electric Light Parade (Lovington, NM) ... **Nov 25**
Ennis Rodeo & Parade (Ennis, MT) .... **July 3**
Festival of Trees (Rock Island) .... **Nov 18**
Fredericksburg Heritage Fest (Fredericksburg, VA) .... **July 4**
Gasparilla Pirate Fest/Parade (Tampa, FL) **Feb 4**
Gatlinburg July 4th Midnight Parade (Gatlinburg, TN) .... **July 4**
George Washington Birthday (Alexandria, VA) .... **Feb 20**
Golden Leaf Festival (Mullins, SC) ...... **Sept 23**
Great Houseboat/Cruiser Parade (Falls of Rough, KY) .... **July 1**
Grosse Pte Santa Claus Parade (Grosse Pte, MI) .... **Nov 24**
Halloween Parade (Toms River, NJ) ...... **Oct 31**
Hobo Convention, Natl (Britt, IA) .... **Aug 10**
Holiday Bowl Parade (San Diego, CA) ... **Dec 30**
Home of the Hamburger Celeb (Seymour, WI) .... **Aug 5**
Joe Cain Procession (Mobile, AL) .... **Feb 26**
Kalamazoo Holiday Parade (Kalamazoo, MI) .... **Nov 11**
Kass Kounty King Korn Karnival (Plattsmouth, NE) .... **Sept 7**
King Orange Jamboree Parade (Miami, FL) .... **Dec 31**
Knights of St. Yago-Illuminated Night (Tampa, FL) .... **Feb 11**
Lewistown 4th Parade/Celeb (Lewistown, MT) .... **July 4**
Lighted Boat Parade (Columbus, MS) .... **Dec 2**
Lobster Race and Oyster Parade (Aiken, SC) .... **May 5**
London Parade (England) .... **Jan 1**
Macon Cherry Blossom Fest (Macon, GA) .... **Mar 17**
Macy's Thanksgiving Day Parade (New York, NY) .... **Nov 23**
Maple Leaf Fest (Carthage, MO) .... **Oct 15**
Marigold Fest (Pekin, IL) .... **Sept 8**
Memorial Day Parade and Services (Gettysburg, PA) .... **May 29**
Memory Days (Grayson, KY) .... **May 25**
Michigan Thanksgiving Parade (Detroit, MI) .... **Nov 23**

☆ *Chase's 1995 Calendar of Events* ☆ Index

Parades (cont'd)—Pennsylvania

Mother Goose Parade (El Cajon, CA) .... **Nov 19**
News 4 Parade of Lights (Denver, CO) .... **Dec 1**
Nordicfest Parade (Libby, MT) .... Sept 9
Northern CA Cherry Blossom Fest (San Francisco, CA) .... **Apr 15**
Oktoberfest (Kitchener/Waterloo, Ont, Canada) .... **Oct 6**
Parade for the Planet (NY, NY) .... **Apr 23**
Parade of Wheels (Rehoboth Beach, DE) **May 13**
Pasadena Doo Dah Parade (Pasadena, CA) .... **Nov 26**
Pet Parade (LaGrange, IL) .... **June 3**
Pro-Am Snipe Excursion and Parade (Moultrie, GA) .... **Apr 1**
Pulaski Day Parade (Philadelphia, PA) .... **Oct 1**
Red Earth Native Amer Cultural Fest (Oklahoma City) .... **June 9**
Red Lodge Home Champs Rodeo/Parade (Red Lodge, MT) .... **July 2**
Remembrance Day Service/Parade (London, England) .... **Nov 12**
Rhododendron Fest Grand Floral Parade (Florence, OR) .... **May 19**
Rivers Fest (Ft Wayne, IN) .... **July 8**
St. Pat's Day Dog Fun Fair/Parade (Alexandria, VA) .... **Mar 11**
St. Patrick's Day Parade (Baton Rouge, LA) .... **Mar 18**
St. Patrick's Day Parade (New York, NY) .... **Mar 17**
St. Patrick's Day Parade (Roanoke, VA) .... **Mar 17**
St. Urho's Day (Finland, MN) .... **Mar 18**
St. Urho's Day (Hood River, OR) .... **Mar 16**
Santa by Stage Coach Parade (El Centro, CA) .... **Dec 2**
Shanghai Parade (Lewisburg, WV) .... **Jan 1**
Soldiers' Reunion Celebration (Newton, NC) .... **Aug 17**
Tournament of Roses Parade (Pasadena, CA) .... **Jan 2**
Twin Cities Krew/Janus MG Parade (Monroe, LA) .... **Feb 4**
Ugly Pickup Parade/Contest (Chadron, NE) .... **Oct 27**
Victorian Christmas Sleighbell Parade (Manistee, MI) .... **Nov 30**
Wild Horse Stampede (Wolf Point, MT) .... **July 7**
WOOD Radio St. Patrick's Day Parade (Grand Rapids, MI) .... **Mar 11**
World's Shortest St. Patrick's Parade (Maryville, MO) .... **Mar 17**
WSB-TV Salute 2 America Parade (Atlanta, GA) .... **July 4**
Yuma Jaycee's Silver Spur Parade (Yuma, AZ) .... **Feb 4**
Paradise Spring Treaties Commemoration .... Aug 26
Paraguay: Independence Day .... May 14
Parcell, Bill: Birth .... Aug 22
Pardon Day .... Sept 8
Paretsky, Sara: Birth .... June 8
Paris, Treaty of: Signing Anniv .... Sept 3
Parish, Robert: Birth .... Aug 30
Park Days, Halfway (Hagerstown, MD) .... May 27
Park, Grand Canyon Natl: Anniv .... Feb 26
Park Service Anniv Observ, Natl (Jamestown/ Yorktown, VA) .... Aug 25
Park Week, Natl 6670 .... Page 488
Parker: Bird Recorded: Anniv .... Apr 30
Parker, Charlie: Birth Anniv .... Aug 29
Parker, Charlie: Death Anniv .... Mar 12
Parker, Dave: Birth .... June 9
Parker, Eleanor: Birth .... June 26
Parker, Fess: Birth .... Aug 16
Parker, George: Death Anniv .... Mar 17
Parker: Jazz at the Philharmonic: Anniv .... Mar 25
Parker, John: Day .... Apr 19
Parkman, Francis: Birth Anniv .... Sept 16
Parks, Bert: Birth Anniv .... Dec 30
Parks, Gordon: Birth .... Oct 30
Parks Month, Natl Recreation and .... July 1
Parks, Rosa Lee: Birth .... Feb 4
Parks, Rosa: Day .... Dec 1
Parnell, Charles S.: Birth Anniv .... June 27
Parry, William: Birth Anniv .... Dec 19
Parseghian, Ara: Birth .... May 10
Parton, Dolly: Birth .... Jan 19
Partridge, John: Death Hoax: Anniv .... Mar 29
Passion Week .... Apr 2
Passiontide .... Apr 2
Passover .... Apr 15
Passover Begins .... Apr 14
Passport Presentation (Russia) .... Jan 2
Pasta Month, Natl .... Oct 1
Paste-Up Day .... May 7
Pasternak, Boris: Birth Anniv .... Feb 10
Pasteur, Louis: Birth Anniv .... Dec 27
Pasteur, Louis: First Successful Antirabies Inoculation .... July 6
Patent Granted for Hat Blocking Machine: Anniv .... Apr 3
Patent Issued for First Adding Machine: Anniv .... Oct 11
Patent Office Opens, US .... July 31
Patinkin, Mandy: Birth .... Nov 30
Patriot's Day (MA, ME) .... Apr 17
Patriots Month, Natl .... June 6
Patterson, Floyd: Birth .... Jan 4
Patton, George S, Jr: Birth Anniv .... Nov 11
Paul, Alice: Birth Anniv .... Jan 11
Paul Bunyan Show (Nelsonville, OH) .... Oct 6
Paul, Les: Birth .... June 9
Paul, Ron: Birth .... Aug 20
Pauley, Jane: Birth .... Oct 31
Paulus, Friedrich: Birth Anniv .... Sept 23
Pause for Pledge (Natl Flag Day USA) .... June 14
Pavan, Marisa: Birth .... June 19
Pavarotti, Luciano: Birth .... Oct 12
Pavlova, Anna: Birth Anniv .... Jan 3
Paxson, John: Birth .... Sept 29
Pay Your Bills Week .... Feb 17
Paycheck, Johnny: Birth .... May 31
Payne, John H: Birth Anniv .... June 9
Payton, Gary: Birth .... July 23
Payton, Walter: Birth .... July 25
PBX Telecommunications Week, Intl .... June 4
Peabody, Elizabeth Palmer: Birth Anniv .... May 16
Peabody, George: Birth Anniv .... Feb 18
Peace Conference, Versailles (WWI): Anniv Jan 18
Peace Corps: Birth .... Oct 14
Peace Corps: Founding Anniv .... Mar 1
Peace Day, Celebration of (Concord, MA) .... Aug 6
Peace Festival (Hiroshima, Japan) .... Aug 6
Peace, Friendship and Good Will Week .... Oct 25
Peace Games, Scarborough-Indianapolis (Indianapolis, IN) .... July 21
Peace, International Day of (UN) .... Sept 19
Peace Meditation, Annual World .... Dec 31
Peace Officer Memorial Day (Pres Proc) .... May 15
Peace Officer Memorial Day, Natl .... May 15
Peace with Justice Week .... Oct 16
Peace: World Hello Day .... Nov 21
Peach Festival, Leitersburg (Leitersburg, MD) .... Aug 12
Peach Festival, Louisiana (Ruston, LA) .... June 9
Peachtree Junior (Atlanta, GA) .... June 3
Peachtree Road Race (Atlanta, GA) .... July 4
Peale, Anna Claypoole: Birth Anniv .... Mar 6
Peale, Charles W.: Birth Anniv .... Apr 15
Peale, James: Death Anniv .... May 24
Peale, Raphael: Birth Anniv .... Feb 17
Peale, Rembrandt: Birth Anniv .... Feb 22
Peale, Sarah M.: Birth Anniversary .... May 19
Peale, Titian R: Birth Anniv .... Nov 17
Peanut Butter Lover's Month .... Nov 1
Peanut Month, Natl .... Mar 1
Peanut Tour, Great (Skippers, VA) .... Sept 7
Pearl Harbor Day .... Dec 7
Pearse, Richard: Flight Anniv .... Mar 31
Peary, Robert E.: Birth Anniv .... May 6
Pecan Day .... Mar 25
Peck, Annie S.: Birth Anniv .... Oct 19
Peck, Gregory: Birth .... Apr 5
Peel, Robert: Birth Anniv .... Feb 5
Peete, Calvin: Birth .... July 18
Pele: Birth .... Oct 23
Pell, Claiborne: Birth .... Nov 22
Pen-Friends Week Intl .... May 1
Pena, Federico: Birth .... Mar 15
Penderecki, Krzysztof: Birth .... Nov 23
Pendergrass, Teddy: Birth .... Mar 26
Penguin Plunge (Jamestown, RI) .... Jan 1
Penn, Arthur Heller: Birth .... Sept 27
Penn, Irving: Birth .... June 16
Penn, John: Birth Anniv .... May 6
Penn, Sean: Birth .... Aug 17
Penn, William: Birth Anniv .... Oct 14
Penn, William: Pennsylvania Deeded to: Anniv .... Mar 4
Pennario, Leonard: Birth .... July 9
Penniman, Little Richard: Birth .... Dec 25
Pennsylvania
A Is for Apples Weekend (Chadds Ford) .... **Oct 14**
Ambridge Nationality Days (Ambridge) .... **May 19**
American Indian Time of Thanksgiving (Allentown) .... **Oct 8**
Anthracite Coal Miners Day (Shenandoah) .... **Sept 4**
Apple Harvest Fest (Gettysburg) .... **Oct 7**
Army-Navy Football Game (Philadelphia) .... **Dec 2**
Art Faire (Lahaska) .... **June 3**
Battle of Gettysburg: Anniv .... **July 1**
Belsnickel Craft Show (Boyertown) .... **Nov 24**
Big 10 Men's Golf Chmpshp (University Park) .... **May 12**
Blueberry Fest (Montrose) .... **Aug 4**
Bucks County Kite Day (Langhorne) .... **May 7**
Buhl Day (Sharon) .... **Sept 4**
Central PA Fest of Arts (State College) .... **July 12**
Chadds Ford Days (Chadds Ford) .... **Sept 9**
Channel 6 Thanksgiving Day Parade (Philadelphia) .... **Nov 23**
Charles Dickens Victorian Christmas (Cornwall) .... **Nov 24**
Chester Old Fiddlers' Picnic (Coatesville) **Aug 12**
Children's Festival (Erwinna) .... **Sept 17**
Christmas Craft Show (York) .... **Dec 3**
Civil War Heritage Days (Gettysburg) .... **June 30**
Civil War Reenactment (Bedford) .... **Sept 9**
Civil War Relic/Collector's Show (Gettysburg) .... **June 30**
Clan Donald USA General Meeting (Pittsburgh) .... **Sept 28**
Corn Fest (Shippensburg) .... **Aug 26**
Corn Planting Ceremony (Allentown) .... **May 7**
Country Sidewalk sale (Lahaska) .... **Aug 12**
Covered Bridge Fest (Washington Cnty) **Sept 16**
Dankfest (Harmony) .... **Aug 26**
Easter Craft Show (York) .... **Feb 12**
Edgar Allan Poe Fest (Cornwall) .... **Oct 26**
Fabulous 1890s Weekend (Mansfield) .... **Sept 22**
Family Days Celebration (Farmington) .... **Aug 12**
Festival of the Red Rose (Manheim) .... **June 11**
Festifall (Point Marion) .... **Sept 24**
First American Abolition Soc Founded: Anniv .... **Apr 14**
First Commercial Oil Well: Anniv .... **Aug 27**
First National Convention for Blacks: Anniv .... **Sept 15**
First Night Football Game: Anniv (Mansfield) .... **Sept 28**
First Night State College (State College) .... **Dec 31**
Flag Day Ceremonies (Philadelphia) .... **June 14**
Flax Scutching Fest (Stahlstown) .... **Sept 9**
Fort Necessity Memorial (Farmington) .... **July 3**
Fred E. Miller: Photographer of the Crows (Pittsburgh) .... **May 20**
Gettysburg Address Memorial Ceremony (Gettysburg) .... **Nov 19**
Gettysburg Outdoor Antique Show (Gettysburg) .... **May 20, Sept 23**
Gettysburg Yuletide Fest (Gettysburg) .... **Dec 1**
Gift Festival, Intl (Fairfield) .... **Nov 9**
Gingerbread House Competition (Lahaska) .... **Nov 15**
Gnawed, Clawed Bird Feeder Fest (New Holland) .... **Sept 11**
God's Country Marathon (Coudersport) .... **June 3**
Great Pumpkin Carve (Chadds Ford) .... **Oct 26**
Great Pumpkin Fest (Bedford) .... **Oct 21**
Great Pumpkin Weekend in Chadds Ford (Chadds Ford) .... **Oct 28**
Groundhog Day in Punxsutawney, PA (Punxsutawney) .... **Feb 2**
Halloween Fright Night (Cornwall) .... **Oct 13**
Hanover Dutch Fest (Hanover) .... **July 29**
Harvest Fest (Intercourse) .... **Sept 15**
Harvest Show (Philadelphia) .... **Sept 16**
Holiday Craft Show (York) .... **Oct 15**
Johnstown Flood Anniv .... **May 31**
Kids Bridge Exhibition (Pittsburgh) .... **July 8**
Lapic Winefest (New Brighton) .... **Oct 7**
Lebanon Bologna Fest (Lebanon) .... **Sept 15**
Little League Baseball World Series (Williamsport) .... **Aug 20**
Littlestown Good Old Days (Gettysburg) **Aug 17**
Longwood Gardens Acres of Spring (Kennett Square) .... **Apr 1**
Longwood Gardens Autumn's Colors (Kennett Square) .... **Oct 1**
Longwood Gardens Christmas Display (Kennett Square) .... **Nov 23**
Longwood Gardens Chrysanthemum Fest (Kennett Square) .... **Oct 28**
Longwood Gardens Fest of Fountains (Kennett Square) .... **May 27**
Longwood Gardens Welcome Spring (Kennett Square) .... **Jan 21**
MADD's Natl Candlelight Vigil (Washington) .... **Dec 2**
Major/Minor League: Photos .... **June 10**
Maple Syrup Fest (Beaver) .... **Apr 8**
Memorial Day Parade and Services (Gettysburg) .... **May 29**
Mifflin-Juniata Arts Fest (Lewistown) .... **May 20**
Mt Washington Tavern Candle Tours (Farmington) .... **Dec 9**

Index ☆ *Chase's 1995 Calendar of Events* ☆

Pennsylvania (cont'd)——Poet Laureate

Northern Appalachian Storytelling Fest (Mansfield)..........Sept 15
New Oxford Flea Market/Art/Craft Show (Gettysburg)..........June 17
Old Bedford Village Bluegrass Fest (Bedford)..........July 22
Old Bedford Village Gospel Fest (Bedford)..........Aug 12
Old Bedford Village Senior Days (Bedford)..........Sept 16
Old-Fashioned Christmas Celebration (Bedford)..........Dec 1
Olde Time Christmas/Mt Washington Tavern (Farmington)..........Dec 9
Peanut-Kids-Baseball Day (Conshohocken)..........Apr 8
Penn State's AG Progress Days (Rock Springs)..........Aug 15
Penn's Landing In-Water Boat Show (Philadelphia)..........Sept 14
Pennsylvania Deeded to William Penn Anniv..........Mar 4
Pennsylvania Maple Fest (Meyersdale)...Apr 22
Pennsylvania Renaissance Faire (Cornwall)..........Aug 5
Philadelphia Antiques Show (Philadelphia) Apr 8
Philadelphia Boat Show (Philadelphia)...Jan 21
Philadelphia Flower Show (Philadelphia)..Mar 5
Pike Fest, Natl (Washington)..........May 20
Pioneer Days Celebration (Bedford).....Sept 2
Polish Christmas Open House (Philadelphia)..........Dec 16
Polish-American Weekend (Philadelphia) Aug 12
Polishfest (Pittsburgh)..........Nov 12
Pony League World Series (Washington) Aug 12
Pulaski Day Parade (Philadelphia, PA).....Oct 1
Punxsutawney Groundhog Fest (Punxsutawney)..........July 1
Quilt Competition/Display (Lahaska)...Jan 30
Rain Day (Waynesburg)..........July 29
Rally 'Round Bucks County (Langhorne)..Oct 1
Ratification Day..........Dec 12
Rhubarb Festival (Intercourse)..........May 20
Rittenhouse Square Art Annual (Philadelphia)..........June 7
Roasting Ears of Corn Food Fest (Allentown)..........Aug 13
St. Ann's Solemn Novena (Scranton)....July 17
Scarecrow Contest (Lahaska)..........Sept 11
Scarecrow Fest (Lahaska)..........Sept 16
Secretaries Leadership Conf, Prof (Philadelphia)..........Feb 10
Seneca Falls Survivor Votes: Anniv (Philadelphia)..........Nov 2
Settlers Days (Bedford)..........July 8
Snake Hunt (Cross Fork)..........June 24
Snow Shovel Riding Contest (Ambridge).Jan 14
South Mountain Fair (Gettysburg)......Aug 30
Spring Craft Show (York)..........Apr 30
State Craft Festival (Richboro)..........Oct 13
Summer Craft Show (York)..........July 16
Teddy Bear's Picnic (Lahaska)..........July 15
Three Rivers Arts Fest (Pittsburgh).....June 2
US's First Zoo: Anniv (Philadelphia)....July 1
Washington Crossing Delaware Reenact (Wash Crossing)..........Dec 25
World's Greatest Yard Sale (York)......June 10
Penny Press, Beginning of the: Anniv.....Sept 3
Pentecost..........June 4
People Magazine: Anniv..........Mar 4
Pepitone, Joe: Birth..........Oct 9
Pepper, Claude Denson: Birth Anniv......Sept 8
Pepper Week, Natl Pickled..........Oct 6
Peppercorn Ceremony (Bermuda).......Apr 23
Perchville USA (Tawas Bay, MI)..........Feb 3
Perdue, William: Birth..........Aug 29
Perez de Cuellar, Javier: Birth..........Jan 19
Perigean Spring Tides.....Mar 19, May 15, June 12, July 11, Dec 21
Perihelion, Earth at..........Jan 4
Perkins, Carl: Birth..........Apr 9
Perkins, Frances: Appointed to Cabinet....Mar 4
Perkins, Frances: Birth Anniv..........Apr 10
Perkins, Sam: Birth..........June 14
Perkins, Walter: Birth..........Dec 6
Perlman, Itzhak: Birth..........Aug 31
Perlman, Rhea: Birth..........Mar 31
Perot, H. Ross, Answers Question: Anniv..Feb 20
Perot, H. Ross: Birth..........June 27
Perrine, Valerie: Birth..........Sept 3
Perry, Gaylord: Birth..........Sept 15
Perry, Matthew: Commodore Perry Day..Apr 10
Perry, Oliver H.: Birth Anniv..........Aug 23
Perry, Steve: Birth..........Jan 22
Perry, William: Birth..........Oct 11

Perry, William "Refrigerator": Birth........Dec 16
Perseid Meteor Showers..........July 23
Pershing, John J.: Birth Anniv..........Sept 13
Person, Chuck: Birth..........June 27
Personal History Awareness Month......May 1
Pertinax: Assassination Anniv..........Mar 28
Peru
Countryman's Day..........June 24
Day of the Navy..........Oct 8
Democracy Stifled in Peru: Anniv......Apr 6
National Independence Day..........July 28
Saint Rose of Lima's Day..........Aug 30
Pesach (Passover)..........Apr 15
Pesach Begins..........Apr 14
Peshtigo Forest Fire Anniv..........Oct 8
Pest Control Month, Natl..........June 1
Pet. See also Animals
Pet Memorial Day, Natl..........Sept 10
Pet Owner Month, Responsible..........Feb 1
Pet Owners Independence Day..........Apr 18
Pet Parade (LaGrange, IL)..........June 3
Pet Peeve Week, Natl..........Oct 9
Pet Sitters Week, Natl Prof..........Mar 6
Pet Week, Natl..........May 7
Peter and Paul Day..........June 29
Peters, Bernadette: Birth..........Feb 28
Peters, Roberta: Birth..........May 4
Petersen, Donald: Birth..........Sept 4
Peterson, Cassandra: Birth..........Sept 17
Peterson, David: Birth..........Dec 28
Petrocelli, Rico: Birth..........June 27
Petroleum Fest, Louisiana Shrimp and (Morgan City, LA)..........Aug 31
Petty, Tom: Birth..........Oct 20
Petunia Fest, Dixon (Dixon, IL)..........June 30
Pfeiffer, Michelle: Birth..........Apr 29
Pharmacists Day..........Jan 12
Philadelphia Police Bombing: Anniv....May 13
Philip, King: Assassination Anniv......Aug 12
Philip, Prince: Birth..........June 10
Philippines
Aquino, Benigno: Assassination Anniv...Aug 21
Araw Ng Kagitingan..........May 6
Ati-Atihan Fest..........Jan 21
Black Nazarene Fiesta..........Jan 1
Bonifacio Day..........Nov 30
Carabao Fest..........May 14
Christmas Observance..........Dec 16
Constitution Day..........Jan 17
Earthquake Jolts: Anniv..........July 16
Feast of Our Lady of Peace/Good Voyage May 1
Feast of the Black Nazarene..........Jan 9
Fertility Rites..........May 17
Fil-American Friendship Day..........July 4
Holy Week..........Apr 9
Independence Day..........June 12
Morione's Fest (Marinduque Island)......Apr 13
Mount Pinatubo Erupts in Philippines: Anniv..........June 11
Philippine Independence: Anniv........Mar 24
Rizal Day..........Dec 30
Santacruzan..........May 1
Simbang Gabi..........Dec 16
US Military Leaves..........Nov 24
Philips, Wendy: Birth..........Jan 2
Phillips, MacKenzie: Birth..........Nov 10
Phillips, Michelle: Birth..........June 4
Phillips, Wally: Birth..........July 7
Phillips, Wendell: Birth Anniv..........Nov 29
Phillpotts, Eden: Birth Anniv..........Nov 4
Photography Exhibits
Abbott, Berenice: Birth Anniv..........July 17
Africa's Legacy in Mexico/Gleaton (Ann Arbor, MI)..........Apr 15
Africa's Legacy in Mexico: Photos (San Antonio, TX)..........June 17
Africa's Legacy in Mexico: Photos/Gleaton (Roanoke, V..........Aug 19
Africa's Legacy in Mexico: Photos/Gleaton (Toledo, OH)..........Feb 11
Baseball Stadiums (Hendersonville).....Oct 14
Baseball Stadiums (Scranton)..........June 10
Bourke-White, Margaret: Birth Anniv....June 14
Chimney Rock Park Photo Contest (Chimney Rock, NC)..........Jun 1
Constance S. Larrabee: Photo Journal (Raleigh, NC)..........Apr 22
Day in the Warsaw Ghetto (Denver, CO)..Apr 8
Day in the Warsaw Ghetto (Houston, TX)..Feb 4
Fred E. Miller: Photographer of the Crows (Missoula, MT)..........Jan 14
Fred E. Miller: Photographer of the Crows (Pittsburgh, PA)..........May 20
Harlem: Photographs by Aaron Siskind (Los Angeles, CA)..........Mar 4

Harlem: Photographs by Aaron Siskind (Montclair, NJ)..........May 20
Major/Minor League: Baseball Stadiums (Hollywood, FL)..........Feb 2
Major/Minor League: Baseball Stadiums (Naples, FL)..........Apr 8
Mystic Seaport Photo Weekend (Mystic, CT)..........Sept 9
Songs of My People (Fort Wayne, IN).....May 6
Songs of My People (Paterson, NJ)......Mar 4
Songs of My People (Shreveport, LA).....July 8
Songs of My People (Winston-Salem, NC) Sept 9
Physical Fitness and Sports Month, Natl....May 1
Piano Month, Natl..........Sept 1
Piano Playing, World Chmpshp Old-Time (Decatur, IL)..........May 26
Picasso, Pablo: Birth Anniv..........Oct 25
Piccard, Jacques: Birth..........July 28
Piccard, Jeannette Ridlon: Birth Anniv....Jan 5
Pickens, Slim: Birth..........June 29
Pickett, George: Defeat at Five Forks: Anniversary..........Apr 1
Pickett, Wilson: Birth..........Mar 18
Pickett's Charge: Battle Gettysburg: Anniv..July 1
Pickle Week, Intl..........May 19
Pickled Pepper Week, Natl..........Oct 6
Pickup Parade and Contest, Ugly (Chadron, NE)..........Oct 27
Pie Day, Natl..........Jan 23
Pied Piper of Hamelin Anniv..........July 22
Pied Piper: Rat-Catcher's Day..........July 22
Pierce, Franklin: Birth Anniv..........Nov 23
Pierce, Jane: Birth Anniv..........Mar 12
Pies Month, Great American..........Feb 1
Pieta Attacked: Anniv..........May 21
Pig Day, Natl..........Mar 1
Pigeons Return to City-County Bldg (Ft Wayne, IN)..........Mar 20
Pike Fest, Natl (Washington, PA)..........May 20
Pilgrim Landing Anniv..........Dec 21
Pinchbeck, Christopher: Death Anniv....Nov 18
Pinchot, Bronson: Birth..........May 20
Pinkerton, Allan: Birth Anniv..........Aug 25
Pinochet, Augusto: Military Dictatorship Ended..........Dec 15
Pinter, Harold: Birth..........Oct 10
Pioneer Days Celebration (Bedford, PA)..Sept 2
Pioneer Week (Washington, MS)..........June 26
Pippen, Scottie: Birth..........Sept 25
Piquet, Nelson: Birth..........Aug 17
Pirate Jamboree, Blackbeard (Norfolk, VA) July 28
Pirates: Billy Bowlegs Fest (Fort Walton, FL)..........June 2
Pirates: Contraband Days (Lake Charles, LA)..........May 2
Pirates: Gasparilla Pirate Fest/Parade (Tampa, FL)..........Feb 4
Pisces Begins..........Feb 20
Piscopo, Joe: Birth..........June 17
Pitcher, Molly: Birth Anniv..........Oct 13
Pitt, William: Birth Anniv..........May 28
Pizza Bake, The Great American..........Feb 7
Pizza Month, Natl..........Oct 1
Place, Mary Kay: Birth..........Sept 23
Plan Your Epitaph Day..........Apr 6, Nov 1
Planned Parenthood Book Fair (Dayton, OH)..........Nov 10
Plant, Robert: Birth..........Aug 20
Plants: Natl Garden Week..........Apr 9
Plants: Omaha Cactus/Succulent Soc Show/Sale (Bellevue, NE)..........Sept 2
Plants: Take Houseplants for Walk Day....July 27
Plato, Dana: Birth..........Nov 7
Play Presented in N American Colonies, First..........Aug 27
Player, Gary: Birth..........Nov 1
Pleasence, Donald: Birth..........Oct 5
Pleasure Your Mate Month..........Sept 1
Pledge of Allegiance, Pause for (Natl Flag Day USA)..........June 14
Pleshette, Suzanne: Birth..........Jan 31
Plimpton, George: Birth..........Mar 18
Plimsoll Day..........Feb 10
Plough Monday (England)..........Jan 9
Plummer, Amanda: Birth..........Mar 23
Plummer, Christopher: Birth..........Dec 13
Plutonium First Weighed: Anniv..........Aug 20
Poarch Creek Indian Thanksgiving Day Powwow (Atmore, AL)..........Nov 23
Poe, Edgar Allan: Birth Anniv..........Jan 19
Poe, Edgar Allan, Fest (Cornwall, PA)...Oct 26
Poet Gathering, Montana Cowboy (Lewistown, MT)..........Aug 18
Poet Laureate Establishment, American: Anniv..........Dec 20

☆ Chase's 1995 Calendar of Events ☆ Index

Poetry—Quadrangle

Poetry, Bad, Day....Aug 18
Poetry Break....Jan 13
Poetry Festival, Julia A. Moore (Flint, MI)..May 24
Poetry Gathering, Dakota Cowboy (Medora, ND)....May 27
Poetry Gathering, Texas Cowboy (Alpine, TX)....Mar 3
Poetry Reading Day, Great (Clio, MI)....Apr 28
Poindexter, John: Birth....Aug 12
Poinsettia Day....Dec 12
Pointer, Bonnie: Birth....July 11
Poison Prevention Awareness Month....Mar 1
Poison Prevention Week, Natl....Mar 19
Poison Prevention Week, Natl 6536....Page 488
Poitier, Sidney: Birth....Feb 20
Poke Salad Fest (Blanchard, LA)....May 11
Poland
  Constitution Day....May 3
  Liberation Day....Jan 17
  Solidarity Founding Anniv....Aug 31
  Solidarity Granted Legal Status....Apr 17
  Wratislavia Cantans: Intl Oratorio-Cantata Fest....Sept 3
Polanski, Roman: Birth....Aug 18
Polar Bear Jump Fest, Seward (Seward, AK)....Jan 20
Polar Bear Swim (Sheboygan, WI)....Jan 1
Pole, Pedal Paddle (Bend, OR)....May 20
Police: Peace Officer Memorial Day, Natl..May 15
Police Week, Natl....May 14
Polish Fest (Milwaukee, WI)....June 24
Polish Fest (Wisconsin Dells, WI)....Sept 8
Polish-American Christmas Gala (St. Francis, WI)....Dec 3
Polish-American Heritage Month....Oct 1
Polish-American Heritage Month 6493...Page 488
Polish-American Heritage Month Celebration (Philadelphia, PA)....Oct 1
Polish-American Weekend (Philadelphia, PA)....Aug 12
Polishfest (Pittsburgh, PA)....Nov 12
Polk, James: Birth Anniv....Nov 2
Polk, Sarah Childress: Birth Anniv....Sept 4
Polka. See also Dance
Polka Fest (Wisconsin Dells, WI)....Nov 3
Pollack, Sydney: Birth....July 1
Pollard, Michael J.: Birth....May 30
Polo
  Intl Open/Handicap (Wellington, FL)....Mar 1
  USPA Rolex Gold Cup (Wellington)....Mar 17
  World Cup (Wellington, FL)....Apr 1
Pompano Beach Seafood Fest (Pompano Beach, FL)....Apr 29
Ponce de Leon, Juan....Apr 2
Ponti, Carlo: Birth....Dec 11
Pony Express, Inauguration of: Anniv....Apr 3
Pony Express-Jesse James Days (St. Joseph, MO)....Apr 1
Pony League World Series (Washington, PA)....Aug 12
Pony Penning, Chincoteague (Chincoteague Island, VA)....July 26
Pooh Day (A.A. Milne Birth Anniv)....Jan 18
Poole, Cecil: First Black US State's Attorney: Anniv....July 6
Poor Richard's Almanac: Anniv....Dec 28
Pop, Iggy: Birth....Apr 21
Popcorn Introduced to Colonists: Anniv...Feb 22
Popcorn Poppin' Month, Natl....Oct 1
Pope, Alexander: Birth Anniv....May 21
Pope Benedict XV: Birth Anniv....Nov 21
Pope John Paul I: Birth Anniv....Oct 17
Pope John Paul II: Assassination Attempt Anniv....May 13
Pope John Paul II: Birth....May 18
Pope John XXIII: Birth Anniv....Nov 25
Pope Leo XIII: Birth Anniv....May 2
Pope Paul VI: Birth Anniv....Sept 26
Pope Pius XI: Birth Anniv....May 31
Pope Pius XII: Birth Anniv....Mar 2
Population: Day of Five Billion: Anniv....July 11
Population Day, World (UN)....July 11
Population, World, Awareness Week 6501 Page 488
Pork Month, Natl....Oct 1
Porter, Cole: Birth Anniv....June 9
Porter, Sylvia: Birth....June 18
Porter, Terry: Birth....Apr 8
Portugal
  Day of Portugal....June 10
  Holy Week Festivities....Apr 9
  Independence Day....Dec 1
  Last Pilgrimage to Fatima....Oct 12
  Our Lady of the Monte Festival (Funchal, Madeira Is)....Aug 15
  Ovar Carnival Festival (Costa De Prata)..Feb 25
  Pilgrimage to Fatima....May 12
  Portugal's Day....Apr 25
  St. Anthony of Padua: Feast Day....June 13
  St. Anthony's Eve and Festival (Lisbon).June 12
  St. John's Festival (Lisbon)....June 12
  St. John's Festival (Porto)....June 23
  St. Sylvester's Fest (Funchal, Madeira Island)....Dec 31
Post Day, World (UN)....Oct 9
Post, Markie: Birth....Nov 4
Post, Wiley: Birth Anniv....Nov 22
Postcard Shows
  Fun-in-the-Sun Postcard Sale (Orlando, FL)....Jan 21
  Postcard Show (Ft. Washington, VA)....June 10
  Postcard Show (Hagerstown, MD)....May 26
  Postcard Show (Mt Laurel, NJ)....Sept 8
  Postcard Week, Natl....May 7
Postmaster General Established, US: Anniv....Sept 22
Posture Month, Correct....May 1
Potato Blossom Festival, Maine (Fort Fairfield, ME)....July 16
Potato Bowl (Grand Forks, ND)....Sept 9
Potomac River Fest (Colonial Beach, VA)..June 9
Potomac South Branch Weekend (Terra Alta, WV)....July 14
Potter, Beatrix: Birth Anniv....July 6
Potters Market, Appalachian (Marion, NC)..Dec 2
Pound, Ezra: Birth Anniv....Oct 30
Pound, Ezra: Bollingen Prize Award....Feb 19
Poverty, Intl Day for Eradication....Oct 17
Poverty, War on: Anniv....Jan 8
Powell, Colin: Birth....Apr 5
Powell, Jane: Birth....Apr 1
Powell, John W.: Birth Anniv....Mar 24
Powell, Lewis Franklin, Jr.: Birth....Sept 19
Power, Tyrone: Birth Anniv....May 5
Powers, Francis G.: Birth Anniv....Aug 17
Powers, Stefanie: Birth....Nov 12
POW/MIA Recognition Day, Natl 6587...Page 488
Prairie Pioneer Days (Arapahoe, NE)....July 2
Prater's Mill Country Fair (Dalton, GA)....May 13
Prayer for Peace, Memorial Day 6566, 6696....Page 488
Prayer: Just Pray No....Apr 22
Prayer, National Day of (Pres Proc)....May 4
Prayer, Natl Day of 6553, 6668....Page 488
Prayer, Supreme Court Bans Official: Anniv....June 25
Prayer Week, Kiwanis....May 14
Prayer, World Day of....Mar 3
Preakness Stakes (Baltimore, MD)....May 20
Preakness Stakes: Anniv....May 27
Prentiss, Paula: Birth....Mar 4
Prescott, William: Birth Anniv....Feb 20
Presentation of the Lord (Candlemas Day)..Feb 2
Preservation Week, Natl Historic....May 14
Presidential Inaugural Ball: Anniv....May 7
Presidential Inauguration Anniv, George Washington....Apr 30
Presidential Joke Day....Aug 11
Presidential Resignation: Anniv....Aug 9
Presidents' Day....Feb 20
Presley, Elvis, Salute (Baltimore, MD)....Aug 16
Presley, Elvis: Birth Anniv....Jan 8
Presley, Elvis: Birthday Party (St. Louis, MO) Jan 8
Presley, Elvis: Death Anniv....Aug 16
Presley, Elvis, Remembered (St. Louis, MO)....Aug 13
Presley, Elvis: Week (Memphis, TN)....Aug 11
Presley, Priscilla Beaulieu: Birth....May 24
Pressler, Larry: Birth....Mar 29
Preston, Billy: Birth....Sept 9
Prevention of Animal Cruelty Month....Apr 1
Previn, Andre: Birth....Apr 6
Price, Alan: Birth....Apr 19
Price, Leontyne: Birth....Feb 10
Price, Mark: Birth....Feb 16
Price, Ray: Birth....Jan 12
Price, Vincent: Birth Anniv....May 27
Pride, Charley: Birth....Mar 18
Priesand, Sally: First Woman Rabbi in US: Anniv....June 3
Priestly, Joseph: Birth Anniv....Mar 13
Priestly, Joseph: Rubber Eraser Day....Apr 15
Prime Beef Festival, Wyatt Earp Days (Monmouth, IL)....Sept 6
Prince: Birth....June 7
Prince William: Birth....June 21
Princeton, USS, Explosion: Anniv....Feb 28
Principal, Victoria: Birth....Jan 3
Printemps des Arts de Monte-Carlo (Monte-Carlo, Monaco)....Apr 15
Printers Row Book Fair (Chicago, IL)....June 17
Printing Ink Day, Natl....Jan 17
Printing Week, Intl....Jan 15
Priorities Week, Intl....Sept 17
Prisoner of War, Former, Natl Recognition Day 6541, 6663....Page 488
Procession of the Addolorata (Taranto, Italy)....Apr 13
Procession of the Holy Blood (Belgium)...May 25
Procession of the Mysteries (Taranto, Italy) Apr 13
Proclamation 6491 of Oct 14, 1992 Revoked 6534....Page 488
Procrastination: Be Late for Something Day Sept 5
Procrastination Week, Natl....Mar 6
Produce Charity Ball, Imperial Valley (Imp Valley, CA)....Feb 11
Professional Secretaries Week....Apr 23
Prohibition Amendment: Anniv....Jan 16
Prohibition: Maine Law: Anniv....June 2
Prohibition Repeal: Anniv....Dec 5
Project Red Ribbon, MADD's....Nov 1
Project Safe Baby Month....May 1
Proposal Day....Mar 20, Sept 23
Proposition 13: Anniv....June 6
Proust, Marcel: Birth Anniv....July 10
Prowse, Juliet: Birth....Sept 25
Prune Breakfast Month, Natl....Jan 1
Pryor, David H.: Birth....Aug 29
Pryor, Richard: Birth....Dec 1
Psychic Week....Aug 7
PTA Convention, Texas (Austin, TX)....Nov 10
PTA Founder's Day, Natl....Feb 17
PTA Membership Enrollment Month in Texas....Sept 1
PTA Teacher Appreciation Week....May 7
Public Lands Day....Sept 9
Public Radio, Natl: Anniv....May 3
Public Relations: Getting World to Beat Path to Door Wk....Oct 15
Public School, First in America: Anniv....Feb 13
Public Service Recognition Week....May 1
Public Service Recognition Week 6682..Page 488
Publication of First Esperanto Book: Anniv July 26
Publishers, Associated, Inc: Founding Anniv....Nov 18
Puccini, Giacomo: Birth Anniv....Dec 22
Puckett, Kirby: Birth....Mar 14
Puerto Rico
  Constitution Day....July 25
  Discovery Day....Nov 19
  Loiza Aldea Fiesta....July 25
  Munoz-Rivera Day....July 17
  Navidades....Dec 15
Puig, Manuel: Birth....Dec 28
Pulaski, Casimir: Birth Anniv....Mar 4
Pulaski, Casimir: Memorial Day (Pres Proc) Oct 11
Pulaski, General, Memorial Day 6486....Page 488
Pulaski Memorial Day, General 6610....Page 488
Pulitzer, Joseph: Birth Anniv....Apr 10
Pulliam, Keshia Knight: Birth....Apr 9
Pullman, George Mortimer: Birth Anniv....Mar 3
Pumpkin Corners Pumpkin Day (Stamford, NE)....Oct 2
Pumpkin Fest, Great (Bedford, PA)....Oct 21
Pumpkin Fest, Morton (Morton, IL)....Sept 13
Pumpkin Fest, West Virginia (Milton, WV)...Oct 6
Pumpkin Show, Circleville (Circleville, OH) Oct 18
Pumpkin Weekend in Chadds Ford, Great (Chadds Ford, PA)....Oct 28
Pun-Off, O. Henry (Austin, TX)....May 7
Puns, Announcement of Ten Best of Year....Jan 1
Punsters Day, Abet and Aid....Nov 8
Purcell, Henry: Death Anniv....Nov 25
Purcell, Sarah: Birth....Oct 8
Purcell Tercentenary Tudeley Fest (Tudeley, England)....Nov 17
Purgatory Banquet (Gradoli, Italy)....Mar 1
Purim....Mar 16
Puri, Linda: Birth....Sept 2
Purple Heart: Anniv....Aug 7
Purpose, Natl Be On, Month....Jan 1
Pushkin, Alexander: Birth Anniv....May 26
Puzo, Mario: Birth....Oct 15
Puzzle Day, Natl....Jan 29
Puzzle Week, Natl Game and....Nov 24
Pyle, Denver: Birth....May 11
Pyle, Ernest: Birth Anniv....Aug 3
Pyle, Ernest: Death Anniv....Apr 18
Pynchon, Thomas: Birth....May 8
Pyrotechnics Guild Fireworks Conv (Stevens Point, WI)....Aug 6

Q

Qatar: Independence Day....Sept 3
Quadrangle Festival (Texarkana, AR and TX)....Sept 9

Index ☆ Chase's 1995 Calendar of Events ☆

Quaid—Reston

Quaid, Dennis: Birth....................Apr 9
Quaid, Randy: Birth....................Oct 1
Quail-Egg Eating/Prairie Dog Chili Cook (Grand Prairie, TX)....................Apr 1
Quartet Convention, Natl (Louisville, KY)..Sept 18
Quayle, Dan: Birth....................Feb 4
Quayle, Marilyn Tucker: Birth....................July 29
Queen Elizabeth I: Ascension Anniv....................Nov 17
Queen Elizabeth I: Birth Anniv....................Sept 7
Queen Mary, RMS: Anniv....................May 27
Queen Mother Elizabeth: Birth....................Aug 4
Queen's Official Birthday/Trooping Colours (England)....................June 10
Quick, Mike: Birth....................May 14
Quilt Competition/Display (Lahaska, PA)...Jan 30
Quilt Show (Woodstock, VT)....................July 29
Quilter's Soc Show, American (Paducah, KY)....................Apr 27
Quilting in the Tetons (Jackson, WY)........Oct 1
Quilts, Celebration of (Chardon, OH).....May 20
Quinlan, Karen Ann: Birth Anniv....................Mar 29
Quinn, Anthony: Birth....................Apr 21
Quisenberry, Dan: Birth....................Feb 7

R

Ra, Sun: Birth Anniv....................May 22
Rabbi, First Woman Rabbi in US: Anniv....June 3
Rabbit Show (Wheeling, WV)....................Apr 15
Rabbitt, Eddie: Birth....................Nov 27
Rabe, David: Birth....................Mar 10
Rabinowitz, Solomon: Birth Anniv....................Feb 18
Race Relations Day....................Feb 14
Race Riots, St. Louis: Anniv....................July 2
Race Unity Day....................June 11
Racial Discrimination, Intl Day for Elimination of (UN)....................Mar 21
Racicot, Marc: Birth....................July 24
Racism/Racial Discrimination, 3rd Decade to Combat (UN)....................Jan 1
Racism/Racial Discrimination, Solidarity Against (UN)....................Mar 21
Racism/Racial Discrimination, UN: 3rd Decade to Combat....................Dec 10
Racking Horse Spring Celebration (Decatur, AL)....................Apr 26
Racking World Celebration (Decatur, AL) .Sept 22
Radcliffe, Ann: Birth Anniv....................July 9
Radio
  CB: Intl REACT Month....................May 1
  Commercial Radio Broadcasting: Anniv .Aug 28
  First Play-by-Play Football Game Broadcast....................Nov 23
  Intl Amateur Radio Month....................Apr 1
  King Biscuit Blues Fest (Helena, AR).....Oct 13
  Loomis Day....................May 30
  Ozzie and Harriet Show Debut: Anniv.....Oct 8
  Radio Broadcast by a President, First...June 14
  Radio Broadcasting: Anniv....................Jan 13
  Radio City Music Hall: Anniv....................Dec 27
  Radio, Natl Public: Anniv....................May 3
  Radio Week, Amateur....................June 18
  Send a Kid to Kamp Radiothon (Lexington, KY)....................May 20
Radiologic Technology Week, Natl....................Nov 5
Radishes, Feast of (Oaxaca, Mexico)......Dec 23
Radziwill, Princess: Birth....................Mar 3
Raffin, Deborah: Birth....................Mar 13
Raft Race, Great Chunky River (Chunky, MS)....................June 3
Raft Race, WROK/WZOK Rock River (Rockford, IL)....................July 4
Raft. See also Boats
Railroad
  Box Car Days (Tracy, MN)....................Sept 2
  Carolina Model Railroader's Show (Greensboro, NC)....................Apr 1, Nov 18
  Fordyce on the Cotton Belt Fest (Fordyce, AR)....................Apr 24
  Galesburg Railroad Days (Galesburg, IL)June 24
  Golden Spike Driving: Anniv....................May 10
  Golden Spike Reenactment (Promontory, UT)....................May 10
  Handcar Races and Steam Fest (Felton, CA)....................July 8
  Helena Railroad Fair (Helena, MT).......Apr 23
  Hometown Days (Strasburg, CO).......Aug 19
  Hood River Valley Blossom Fest (Hood River, OR)....................Apr 15
  Iron Horse Outraced by Horse: Anniv...Sept 18
  Livingston Railroad Swap Meet (Livingston, MT)....................Apr 22
  Mid-Cont RR Steam Snow Train (New Freedom, WI)....................Feb 17
  Mid-Cont RR Steam Train Color Tours (New Freedom)....................Oct 7
  Model Railroad Show (Wolfeboro, NH) ..Aug 12
  New York City Subway: Anniv....................Oct 27
  Railroad and Hobby Show (Springfield, MA)....................Feb 4
  Railroader's Festival (Promontory, UT) ..Aug 12
  Reopening of the Railroad Car Museum (Schoharie, NY)....................June 3
  Roaring Camp Railroad Ride (Felton, CA)....................June 18
  34th St Express (Boston, MA)....................Dec 9
  Topeka Railroad Days Fest (Topeka, KS)..Sept 1
  Transcontinental US Railway Completion: Anniv....................Aug 15
Rain Day (Waynesburg, PA)....................July 29
Rainbow Warrior Sinking: Anniv....................July 10
Rainer, Luise: Birth....................Jan 12
Raines, Tim: Birth....................Sept 16
Rainey, Gertrude "Ma": Birth Anniv.........Apr 3
Rainey, Joseph: 1st Black in US House of Reps: Anniv....................Dec 12
Raitt, Bonnie: Birth....................Nov 8
Rally 'Round Bucks County (Langhorne, PA) Oct 1
Ramadhan (Muslim Holiday)....................Feb 1
Rambis, Kurt: Birth....................Feb 25
Rambo, Dack: Birth....................Nov 13
Rameau, Jean P: Birth Anniv....................Sept 25
Ramp Tramp Fest, Polk County (Benton, TN)....................Apr 22
Rampal, Jean-Pierre: Birth....................Jan 7
Ramses II Unearthed, Statue of: Anniv.....Nov 30
Ramson, Feast of (Richwood, WV).........Apr 15
Ranch Rodeo (Burwell, NE)....................July 30
Ranch Roundup, Texas (Wichita Falls, TX) .Aug 18
Rand, Sally: Birth Anniv....................Apr 3
Randall, Tony: Birth....................Feb 26
Randolph, Peyton: Death Anniv....................Oct 22
Ranger's Day, Woodmen....................May 12
Rankin, Jeanette: Birth Anniv....................June 11
Raphael: Birth Anniv....................Apr 6
Raphael, Sally Jessy: Birth....................Feb 25
Rashad, Ahmad: Birth....................Nov 19
Rashad, Phylicia: Birth....................June 19
Rat and Mouse, Fancy, Show (Riverside, CA)....................Jan 28
Rat-Catchers Day....................July 22
Rather, Dan: Birth....................Oct 31
Ratification Day....................Jan 14
Rattle, Simon: Birth....................Jan 19
Ratzenberger, John: Birth....................Apr 6
Rauschenberg, Robert: Birth....................Oct 22
Raven Mad Daze (Yellowknife, NWT, Canada)....................June 16
Rawls, Lou: Birth....................Dec 1
Ray Day, May....................May 19
Rayburn, Gene: Birth....................Dec 17
Raye, Martha: Birth....................Aug 27
Razor, Electric, First Marketed: Anniv......Mar 18
REACT Month, Intl....................May 1
Read, George: Birth Anniv....................Sept 18
Reader's Day, Natl Young/Book It..........Oct 2
Reading: Banned Books Week....................Sept 23
Reading Is Fun Week....................Apr 23
Reading Week, Natl Bathroom....................June 5
Reagan, Nancy: Birth....................July 6
Reagan, Ronald: Assassination Attempt Anniv....................Mar 30
Reagan, Ronald Prescott: Birth....................May 20
Reagan, Ronald Wilson: Birth....................Feb 6
Real Jewelry Month....................Nov 1
Receptionist Day, Natl....................May 10
Reconciliation, Natl Day of 6661........Page 488
Recreation and Parks Month, Natl..........July 1
Recreation Day, Special....................July 9
Recreation Week, Special....................July 9
Recreational Scuba Diving Week, Natl.....Aug 13
Recreational Vehicle Show (Timonium, MD)Feb 17
Recreational Vehicle Show, Maryland (Timonium, MD)....................Sept 9
Red Cloud: Death Anniv....................Dec 10
Red Cloud Indian Art Show (Pine Ridge, SD)....................June 4
Red Cross Day, World....................May 8
Red Cross: Founding Anniv....................May 21
Red Cross Month....................Mar 1
Red Cross Month, American (Pres Proc)....Mar 1
Red Cross Month, American 6535, 6653 .Page 488
Red Earth Native Amer Cultural Fest (Oklahoma City, OK)....................June 9
Red Kettle Kick Off for Salvation Army Christmas Season....................Nov 24
Red Ribbon Week for Drug-Free American 6497....................Page 488
Reddy, Helen: Birth....................Oct 25
Redford, Robert: Birth....................Aug 18
Redgrave, Lynn: Birth....................Mar 8
Redgrave, Vanessa: Birth....................Jan 30
Redneck Day, American....................July 4
Reed, Jerry: Birth....................Mar 20
Reed, Oliver: Birth....................Feb 13
Reed, Rex: Birth....................Oct 2
Reed, Walter: Birth Anniv....................Sept 13
Reed, Willis: Birth....................June 25
Reelfoot Eagle Tours (Tiptonville, TN)......Jan 1
Reenactment of Cowtown's Last Gunfight (Fort Worth, TX)....................Feb 8
Rees, Roger: Birth....................May 5
Reese, Della: Birth....................July 6
Reeve, Christopher: Birth....................Sept 25
Reeves, Jim: Death Anniv....................July 31
Reeves, Martha: Birth....................July 18
Reformation Day....................Oct 31
Reformation Sunday....................Oct 29
Refrigerator Day, Natl Clean Out Your.....Nov 15
Refugee Day 6499....................Page 488
Rehabilitation Week, Natl....................Sept 17
Rehabilitation Week, Natl 6596..........Page 488
Rehnquist, William Hubbs: Birth............Oct 1
Reich, Robert: Birth....................June 24
Reid, Harry: Birth....................Dec 2
Reid, Tim: Birth....................Dec 19
Reilly, Charles Nelson: Birth....................Jan 13
Re-Imaging Conference Controversy: Anniv Nov 4
Reiner, Carl: Birth....................Mar 20
Reiner, Rob: Birth....................Mar 6
Reinking, Ann: Birth....................Nov 10
Reitman, Ivan: Birth....................Oct 26
Rejection: Sorry Charlie Day....................Apr 1
Relationship Renewal Day....................May 4
Relaxation Day, Natl....................Aug 15
Relics and Rods Run to the Sun (Lake Havasu City, AZ)....................Oct 11
Religion Day, World....................Jan 15
Religion: Re-Imaging Conf Controversy: Anniv....................Nov 4
Religious Freedom Day....................Jan 16
Religious Freedom Day 6646............Page 488
Religious Freedom Week....................Sept 24
Religious Reference Books Week, Natl.....Sept 3
Religious Software Week, Natl............Aug 20
Remember the Maine Day....................Feb 15
Remembrance Day....................Apr 15
Remembrance Day (Canada)....................Nov 11
Remembrance Day (England)....................Nov 12
Remembrance, Days of....................Apr 23
Remington, Frederic S.: Birth Anniv.........Oct 4
Renaissance Fairs
  Alabama Renaissance Faire (Florence, AL)....................Oct 28
  Arizona Renaissance Fest (Apache Junction, AZ)....................Feb 11
  Georgia Renaissance Fall Fest (Fairburn, GA)....................Oct 7
  Georgia Renaissance Spring Fest (Atlanta, GA)....................Apr 22
  Jewish Renaissance Fair (Morristown, NJ)Sept 3
  Kansas City Renaissance Fest (Bonner Springs, KS)....................Sept 2
  Maryland Renaissance Festival (Annapolis, MD)....................Aug 26
  Michigan Renaissance Fest (Holly, MI)...Aug 12
  Minnesota Renaissance Fest (Shakopee, MN)....................Aug 12
  Pennsylvania Renaissance Faire (Cornwall, PA)....................Aug 5
Rendezvous, French and Indian (Big Pool, MD)....................May 27
Rendezvous, Great (Thunder Bay, Ont, Canada)....................July 7
Rendezvous, Lewis and Clark (St. Charles MO)....................May 20
Rendezvous, Mountain Man (Red Lodge, MT)....................July 21
Rendezvous, Redbud Trail (Rochester, IN)..Apr 29
Reno, Janet: Birth....................July 21
Renoir, Pierre: Birth Anniv....................Feb 25
Repeat Day (vocabulary)....................June 3
Republican Party Formed: Anniv...........July 6
Republican Symbol: Anniv....................Nov 7
Research Council, Natl: First Meeting Anniv....................Sept 20
Resnik, Judith A.: Birth Anniv....................Apr 5
Respect Canada Day....................July 15
Respighi, Ottorino: Birth Anniv............July 9
Responsible Pet Owner Month....................Feb 1
Reston, James: Birth....................Nov 3

☆ Chase's 1995 Calendar of Events ☆ Index
Retarded—Rodriguez

Retarded Citizens: ARC Annual Conv (Indianapolis, IN) .... Oct 26
Retrocession Day (Taiwan) .... Oct 25
Retton, Mary Lou: Birth .... Jan 24
Return Day (Georgetown, DE) .... Nov 9
Return Shopping Carts to the Supermarket Month .... Feb 1
Return the Borrowed Books Week .... Mar 1
Reunification of Germany: Anniv .... Oct 3
Reunion Party, All Class (Roanoke, VA) .... May 6
Reunion Picnic, Annual Opolis (Opolis, KS) June 4
Reuther, Walter Philip: Birth Anniv .... Sept 1
Revere, Paul: Birth Anniv .... Jan 1
Revere, Paul: Ride Anniv .... Apr 18
Revill, Clive: Birth .... Apr 18
Revise Your Work Schedule Month .... May 1
Revolution, American
- Anthony Wayne Day .... July 11
- Battle of Blue Licks Celebration (Mount Olivet, KY) .... Aug 19
- Battle of Guilford Courthouse Observ (Greensboro, NC) .... Mar 15
- Bennington Battle Day .... Aug 16
- Boston Tea Party: Anniv .... Dec 16
- Camden, Battle of: Anniv .... Aug 16
- Cessation of Hostilities: Anniv .... Jan 20
- Chestertown Tea Party Fest (Chestertown, MD) .... May 27
- Evacuation Day (Boston, MA) .... Mar 17
- Great Britain-US: Articles of Peace Anniv Nov 30
- Independence Day (United States) .... July 4
- John Parker Day .... Apr 19
- Liberty Day .... Mar 23
- Olive Branch Petition: Anniv .... July 8
- Paris, Treaty of: Signing Anniv .... Sept 3
- Paul Revere's Ride: Anniv .... Apr 18
- Revolutionary War Encampment (Alexandria, VA) .... Feb 19
- Richardson, Tacy: Ride Anniv .... Sept 22
- Sampson, Deborah: Birth Anniv .... Dec 17
- Washington Crossing Delaware Reenact (Wash Crossing, PA) .... Dec 25
- Washington, George: Takes Command of Continental Army .... July 3
- Yorktown Day (Yorktown, VA) .... Oct 19
- Yorktown Day Weekend (Yorktown, VA) .... Oct 14

Revolution, Russian: Anniv .... Nov 7
Rex Allen Days (Willcox, AZ) .... Oct 6
Reynolds, Burt: Birth .... Feb 11
Reynolds, Debbie: Birth .... Apr 1
Rhino Day, Save the .... May 1
Rhino Exhib, Southwest (Tuscon, AZ) .... Nov 10
Rhode Island
- Bristol Civic, Military/Firemen's Parade (Bristol) .... July 4
- Champagne Dinner (Newport) .... Dec 1
- Children's Party at Green Animals (Newport) .... July 14
- ESA Mid-Winter Surfing Chmpshp (Narragansett) .... Feb 18
- Gaspee Days (Warwick/Cranston) .... May 27
- Independence Day .... May 4
- NCAA Div I Men's Ice Hockey Chmpshp (Providence, RI) .... Mar 30
- Newport Maritime Teddy Bear Fest (Newport) .... July 23
- Newport Music Fest (Newport) .... July 8
- Newport Teddy Bear Fest (Newport) .... Aug 20
- Newport Yacht Club Frostbite Fleet Race (Newport) .... Jan 1
- Penguin Plunge (Jamestown) .... Jan 1
- Ratification Day .... May 29
- Tuna Tournament (Narragansett) .... Sept 2
- US Amateur (Golf) Chmpshp (Newport and Middletown) .... Aug 22
- Voters Reject Constitution: Anniv .... Mar 24

Rhodes, Cecil: Birth Anniv .... July 1
Rhododendron Fest (Florence, OR) .... May 19
Rhubarb Festival (Intercourse, PA) .... May 20
Rhue, Madlyn: Birth .... Oct 3
Ribbon, Project Red, MADD's .... Nov 1
Rice Month, Natl .... Sept 1
Rice, Jerry Lee: Birth .... Oct 13
Rice, Jim: Birth .... Mar 8
Rice, Tim: Birth .... Nov 10
Rich, Adam: Birth .... Oct 12
Rich, Charlie: Birth .... Dec 14
Richard, Maurice: Birth .... Aug 4
Richards, Ann W.: Birth .... Sept 1
Richards, Ivor A.: Birth Anniv .... Feb 16
Richards, Keith: Birth .... Dec 18
Richardson, Tacy: Ride Anniv .... Sept 22
Richie, Lionel: Birth .... June 20
Richmond, Mitch: Birth .... June 30
Richter Scale Day .... Apr 26
Rickenbacker, Edward V.: Birth Anniv .... Oct 8
Rickles, Don: Birth .... May 8
Rickover, Hyman George: Birth Anniv .... Jan 27
Ridby, Genl Matthew B., Death of 6583 .. Page 488
Riddles, Libby: Birth .... Apr 1
Ride, Sally Kristen: Birthday .... May 26
Ridgeley, Andrew: Birth .... Jan 26
Ridgway, Matthew Bunker: Birth Anniv .... Mar 3
Riegle, Donald W., Jr: Birth .... Feb 4
Riel, Louis: Hanging Anniv .... Nov 16
Rifle Chmpshp, NCAA Men's/Women's .... Mar 2
Rifle Tourn, Walsh Invitational (Cincinnati, OH) .... Nov 3
Rigby, Cathy: Birth .... Dec 12
Rigg, Diana: Birth .... July 20
Riggs, Bobby: Billy Jean King Wins: Anniv Sept 20
Riggs, Bobby: Birth .... Feb 25
Riley Fest (Greenfield, IN) .... Oct 5
Riley, James Whitcomb: Birth Anniv .... Oct 7
Riley, James Whitcomb: Riley Fest (Greenfield, IN) .... Oct 5
Riley, Pat: Birth .... Mar 20
Riley, Richard: Birth .... Jan 2
Ringwald, Molly: Birth .... Feb 18
Riot Act: Anniv .... July 20
Riot, Watts: Anniv .... Aug 11
Ripken, Cal, Jr: Birth .... Aug 24
Ritter, John: Birth .... Sept 17
River City Roundup (Omaha, NE) .... Sept 20
River Festival, Boise (Boise, ID) .... June 22
River Festival, Petaluma (Petaluma, CA) .... Aug 19
River Rendezvous, Stubb's Eddy (Davenport, IA) .... May 19
River Roast (Chattanooga, TN) .... May 19
Rivera, Chita: Birth .... Jan 23
Rivera, Diego: Birth Anniv .... Dec 8
Rivera, Geraldo: Birth .... July 4
Riverbend Festival (Chattanooga, TN) .... June 16
Riverfest (Cape Girardeau, MO) .... June 9
Rivers, Glenn: Birth .... Oct 13
Rivers Hall of Fame Inductions/Awards, Natl (Dubuque, IA) .... May 27
Rivers, Joan: Birth .... June 8
Rivers, Larry: Birth .... Aug 17
Rivers Month, American .... June 1
Riverspree Fest (Elizabeth City, NC) .... May 27
Rivlin, Alice: Birth .... Mar 4
Road Map Week, Natl Reading a .... Apr 4
Robb, Charles: Birth .... June 26
Robbins, Harold: Birth .... May 21
Robbins, Jerome: Birth .... Oct 11
Robert the Hermit: Death Anniv .... Apr 1
Roberts, Barbara: Birth .... Dec 21
Roberts, Eric: Birth .... Apr 18
Roberts, Julia: Birth .... Oct 28
Roberts, Oral: Birth .... Jan 24
Roberts, Pernell: Birth .... May 18
Robert's Rules Day .... May 2
Robertson, Cliff: Birth .... Sept 9
Robertson, Pat: Birth .... Mar 22
Robertson, Robbie: Birth .... July 5
Robeson, Paul: Birth Anniv .... Apr 9
Robin Hood Festival, Sherwood (Sherwood, OR) .... July 14
Robinson, Bill "Bojangles": Birth Anniv .... May 25
Robinson, Cliff: Birth .... Dec 16
Robinson Crusoe Day .... Feb 1
Robinson, David: Birth .... Aug 6
Robinson, Eddie: Birth .... Feb 12
Robinson, Edwin Arlington: Birth Anniv .... Dec 22
Robinson, Frank: Birth .... Aug 31
Robinson, Jackie: Birth Anniv .... Jan 31
Robinson Named First Black Manager: Anniv Oct 3
Robinson, Roscoe, Jr.: Birth Anniv .... Oct 11
Robinson, Smokey: Birth .... Feb 19
Robot Homicide, First: Anniv .... July 21
Roche, Eugene: Birth .... Sept 22
Rochester Fair (Rochester, NH) .... Sept 14
Rockbridge Community Fest (Lexington, VA) .... Aug 26
Rockefeller, Abby Greene Aldrich: Birth Anniv .... Oct 26
Rockefeller, David: Birth .... June 12
Rockefeller, John D., IV: Birth .... June 18
Rockhound Gemboree (Bancroft, Ont, Canada) .... Aug 3
Rocks: Chicagoland Gems/Minerals Assn Show (Wheaton, IL) .... May 27
Rockwell, Norman: Birth Anniv .... Feb 3
Rockwell, Norman: Colonial Sign Painter Cover: Anniv .... Feb 6
Rockwell, Norman: Home-Coming Marine Cover: Anniv .... Oct 13
Rockwell, Norman: Home-Coming Soldier Cover: Anniv .... May 26
Rockwell, Norman: Swimmer Cover: Anniv Aug 11
Roddenberry, Gene: Birth Anniv .... Aug 19
Rodents, Running of the (Louisville, KY) .... Apr 26
Rodeo
- Beef Empire Days (Garden City, KS) .... June 2
- Black Hills Roundup (Belle Fourche, SD) .. July 2
- Bullnanza (Guthrie, OK) .... Feb 3
- Calgary Exhibition/Stampede (Calgary, AB, Canada) .... July 7
- Canadian Finals Rodeo (Edmonton, Alta, Canada) .... Nov 8
- Cardston Stampede (Carston, Alta, Canada) .... Aug 10
- Central Montana Fair (Lewistown, MT) .. July 24
- Cheyenne Frontier Days (Cheyenne, WY) July 21
- Cimarron Territory Celebration (Beaver, OK) .... Apr 17
- Clearwater Chamber of Commerce Rodeo (Clearwater, NE) .... June 23
- Dixie Natl Livestock Show/Rodeo (Jackson, MS) .... Jan 30
- Eighty-Niner Day Celeb (Guthrie, OK) .... Apr 18
- Ennis Rodeo & Parade (Ennis, MT) .... July 3
- Grand Prairie Western Days (Grand Prairie, TX) .... May 18
- High School Rodeo, Natl Finals (Gillette, WY) .... July 24
- Homestead Rodeo (Homestead, FL) .... Feb 3
- Inter-State Fair/Rodeo (Coffeyville, KS) .. Aug 12
- International Finals Rodeo (OK City, OK) Dec 14
- Kernville Stampede Rodeo (Kernville, CA) . Oct 7
- La Fiesta De Los Vaqueros (Tucson, AZ) . Feb 22
- Last Chance Stampede and Fair (Helena, MT) .... July 27
- Longhorn Chmpshp Rodeo (Auburn Hills, MI) .... Feb 24
- Longhorn World Chmpshp Rodeo (Cape Girardeau, MO) .... Feb 17
- Longhorn World Chmpshp Rodeo (Chattanooga, TN) .... Mar 10
- Longhorn World Chmpshp Rodeo (Cincinnati, OH) .... Feb 10
- Longhorn World Chmpshp Rodeo (Greenville, SC) .... Nov 10
- Longhorn World Chmpshp Rodeo (Huntsville, AL) .... Mar 3
- Longhorn World Chmpshp Rodeo (Nashville, TN) .... Nov 17
- Longhorn World Chmpshp Rodeo (Tulsa, OK) .... Feb 3
- Longhorn World Chmpshp Rodeo (Winston-Salem, NC) .... Nov 3
- Mandan Rodeo Days (Mandan, ND) .... July 1
- Nebraskaland Days/Buffalo Bill Rodeo (North Platte, NE) .... June 13
- Nebraska's Big Rodeo (Burwell, NE) .... July 27
- NRA Rodeo Finals (Billings, MT) .... Feb 3
- Oklahoma State Prison Rodeo (McAlester, OK) .... Sept 1
- 101 Wild West Rodeo (Ponca City, OK) .. Aug 16
- Oregon Trails Rodeo (Hastings, NE) .... Sept 2
- Pendleton Round-Up (Pendleton, OR) .... Sept 13
- Ponca City Ranch Rodeo (Ponca City, OK) .... May 12
- Provincial Ex (Brandon, Man, Canada) .. June 14
- Ranch Rodeo (Burwell, NE) .... July 30
- Ranch Rodeo (Liberal, KS) .... May 20
- Raymond Stampede (Raymond, Alta, Canada) .... July 1
- Red Lodge Home of Champions Rodeo (Red Lodge, MT) .... July 2
- Reno Rodeo (Reno, NV) .... June 16
- Scotts Bluff County Fair (Mitchell, NE) .. Aug 13
- Senior Pro Rodeo (Payson, AZ) .... May 27
- Southwestern Expo Livestock Show/Rodeo (Ft Worth, TX) .... Jan 20
- Texas Ranch Roundup (Wichita Falls, TX) Aug 18
- Welling Pioneer Day (Welling, AB) .... Aug 26
- Western Stock Show and Rodeo, Natl (Denver, CO) .... Jan 10
- Wild Horse Stampede (Wolf Point, MT) .... July 7
- Wonago World Chmpshp Rodeo (Milwaukee, WI) .... June 16
- World's Oldest Continuous PRCA Rodeo (Payson, AZ) .... Aug 18
- World's Oldest Rodeo (Prescott, AZ) .... June 29
- Yuma Jaycee's Silver Spur Rodeo (Yuma, AZ) .... Feb 10

Roderick, David: Birth .... May 3
Rodin, Auguste: Birth Anniv .... Nov 12
Rodman, Dennis: Birth .... May 13
Rodriguez, Juan "Chi-Chi": Birth .... Oct 23

Index ☆ Chase's 1995 Calendar of Events ☆

Roe v Wade—St. Anthony's

Roe v Wade Decision: Anniv . . . . Jan 22
Roemer, Buddy: Birth . . . . Oct 4
Roentgen, Wilhelm K.: Birth Anniv . . . . Mar 27
Rogation Sunday . . . . May 21
Rogers, Edith Nourse: Birth Anniv . . . . Mar 19
Rogers, Fred: Birth . . . . Mar 20
Rogers, Ginger: Birth . . . . July 16
Rogers, Kenny: Birth . . . . Aug 21
Rogers, Roy: Birth . . . . Nov 5
Rogers, Wayne: Birth . . . . Apr 7
Rogers, Will: Birth Anniv . . . . Nov 4
Roget, Peter Mark: Birth Anniv . . . . Jan 18
Rohatyn, Felix: Birth . . . . May 29
Roller Coaster: Santa Cruz Beach Giant Dipper: Anniv . . . . May 17
Rollins, Wayne: Birth . . . . June 16
Roman, Ruth: Birth . . . . Dec 23
Romance Awareness Month . . . . Aug 1
Romance: Date Your Mate Month . . . . May 1
Romance: Kiss-Your-Mate Day . . . . Apr 28
Romance Month, Creative . . . . Feb 1
Romance: Sweetest Day . . . . Oct 21
Romania: National Day . . . . Dec 1
Rome: Birthday (Italy) . . . . Apr 21
Romer, Roy: Birth . . . . Oct 31
Rommel, Erwin: Birth Anniv . . . . Nov 15
Romney, George Wilcken: Birth . . . . July 8
Ronstadt, Linda: Birth . . . . July 15
Roof-Over-Your-Head Day, Natl . . . . Dec 24
Room of One's Own Day . . . . Jan 25
Rooney, Andy: Birth . . . . Jan 14
Rooney, Mickey: Birth . . . . Sept 23
Roosevelt, Anna Eleanor: Birth Anniv . . . . Oct 11
Roosevelt: First Fireside Chat: Anniv . . . . Mar 12
Roosevelt, Franklin D.: Birth Anniv . . . . Jan 30
Roosevelt, Franklin D.: Christmas Fireside Warning . . . . Dec 25
Roosevelt, Franklin D.: Death Anniv . . . . Apr 12
Roosevelt, Franklin D.: Elected to Fourth Term: Anniv . . . . Nov 7
Roosevelt, Theodore: Birth Anniv . . . . Oct 27
Roosevelt, Theodore: Medora Musical (Medora, ND) . . . . June 9
Roots: Alex Palmer Haley: Birth Anniv . . . . Aug 11
Rosacea Awareness Month . . . . Mar 1
Rose, Billy: Birth Anniv . . . . Sept 6
Rose Bowl Game (Pasadena, CA) . . . . Jan 2
Rose Bowl Game, Tourn of Roses: Anniv . . . . Jan 1
Rose, Fest of the Red (Manheim, PA) . . . . . June 11
Rose Fest, Portland (Portland, OR) . . . . June 1
Rose Fest/Tours, Cherokee (Gilmer, TX) . . . May 20
Rose Month, Natl . . . . June 1
Rose, Pete: Birth . . . . Apr 14
Rosenberg, Execution: Anniv . . . . June 19
Roses, South Carolina Fest of (Orangeburg, SC) . . . . Apr 28
Rosh Hashanah . . . . Sept 25
Rosh Hashanah Begins . . . . Sept 24
Ross, Betsy: Birth Anniv . . . . Jan 1
Ross, Diana: Birth . . . . Mar 26
Ross, George: Birth Anniv . . . . May 10
Ross, Herbert: Birth . . . . May 13
Ross, Katharine: Birth . . . . Jan 29
Ross, Marion: Birth . . . . Oct 25
Ross, Mrs William B.: 1st US Woman Governor Inaugurated . . . . Jan 5
Ross, Nellie Tayloe: Birth Anniv . . . . Nov 29
Rossellini, Isabella: Birth . . . . June 18
Rossner, Judith: Birth . . . . Mar 1
Rostropovich, Mstislav Leopoldovich: Birth Aug 12
Rotary Tiller Race, World Championship (Emerson, AR) . . . . June 24
Roth, David Lee: Birth . . . . Oct 10
Roth, Philip: Birth . . . . Mar 19
Roth, William V., Jr: Birth . . . . July 22
Roughouse Fest (Japan) . . . . Oct 14
Rounds Resounding Day . . . . Aug 1
Rousseau, Henri: Birth Anniv . . . . May 20
Rousseau, Jean J: Birth Anniv . . . . June 28
Route 66 Days (Ash Fork, AZ) . . . . June 24
Rowan, Carl: Birth . . . . Aug 11
Rowlands, Gena: Birth . . . . June 19
Royal Palette (Windsor, England) . . . . Jan 1
Royko, Mike: Birth . . . . Sept 19
Rozelle, Pete: Birth . . . . Mar 1
Rubber Eraser Day . . . . Apr 15
Rubens, Peter P.: Birth Anniv . . . . June 28
Rubik, Erno: Birth . . . . July 13
Ruble Becomes Convertible: Anniv . . . . July 1
Rudman, Warren Bruce: Birth . . . . May 18
Rudolph, Wilma: Birth . . . . June 23
Ruffin, Edmund: Birth Anniv . . . . Jan 5
Rug Day, Lumpy . . . . May 3
Rundgren, Todd: Birth . . . . June 22
Running Events
Annapolis Run (Annapolis, MD) . . . . Aug 27
Anvil Mountain Run (Nome, AK) . . . . July 4
Atlanta Marathon and Half Marathon (Atlanta, GA) . . . . Nov 23
Bangor Labor Day Road Race (Bangor, ME) . . . . Sept 4
Beartooth Run (Red Lodge, MT) . . . . June 24
Bolder Boulder 10K (Boulder, CO) . . . . May 29
Boston Marathon (Boston, MA) . . . . Apr 17
Charleston Distance Run (Charleston, WV) . . . . Sept 2
Charlotte Observer Marathon/Runner's Expo (Charlotte, NC) . . . . Jan 6
Cheetah Run (Cincinnati, OH) . . . . Sept 3
Chicago Marathon (Chicago, IL) . . . . Oct 29
Chieftains Run (Rome, GA) . . . . Dec 2
Comrades Marathon (Natal, South Africa) . . . . May 31
Crim Festival of Races (Flint, MI) . . . . Aug 24
Easter Beach Run (Daytona Beach, FL) . . Apr 15
Egg Races (Switzerland) . . . . Apr 17
Examiner Bay to Breaker Race (San Francisco, CA) . . . . May 21
Exit Glacier Run (Seward, AK) . . . . May 13
Gasparilla Distance Classic (Tampa, FL) . Feb 18
God's Country Marathon (Coudersport, PA) . . . . June 3
Governor's Bay Bridge Run (Annapolis, MD) . . . . May 7
Governor's Cup Marathon (Helena, MT) . . June 2
Groundhog Run (Kansas City, MO) . . . . Feb 5
Hard Rock Cafe 5K (Chicago, IL) . . . . June 18
Houston-Tenneco Marathon (Houston, TX) . . . . Jan 15
Independence Day Challenge Run (Boyne City, MI) . . . . July 4
Lakestride Half-Marathon (Ludington, MI) . . . . June 17
Leadville 100-Mile Ultra Run (Leadville, CO) . . . . Aug 19
Los Angeles Marathon/Family Reunion (Los Angeles, CA) . . . . Mar 5
Mad-City Marathon (Madison, WI) . . . . May 28
Michael Forbes Trolley Run (Kansas City, MO) . . . . Apr 23
Milwaukee Journal: Al's Run (Milwaukee, WI) . . . . Sept 23
Mount Marathon Race (Seward, AK) . . . . July 4
Nanisivik Midnight Sun Marathon (Nanisivik, NWT) . . . . June 29
New York City Marathon (New York, NY) . Nov 12
Peachtree Junior (Atlanta, GA) . . . . June 3
Peachtree Road Race (Atlanta, GA) . . . . July 4
Poarch Creek Indian TG Day Powwow/Run/ Walk (Atmore, AL) . . . . Nov 23
St. Patrick's Day Run, McGuire's (Pensacola, FL) . . . . Mar 11
Seward Silver Salmon 10K Run, Annual (Seward, AK) . . . . Sept 2
Shamrock Marathon and Sportsfest (Virginia Beach, VA) . . . . Mar 17
Tour o' Hawaii (Oahu, HI) . . . . Nov 2
Trail's End Marathon (Seaside, OR) . . . . Mar 4
Turkey Trot (Parkersburg, WV) . . . . Nov 23
Valentine Heart Throb Biathlon (Nome, AK) . . . . Feb 11
Runyan, Damon: Birth Anniv . . . . Oct 4
Rural Life Sunday . . . . May 21
Rush, Benjamin: Birth Anniv . . . . Dec 24
Rush, William: Death Anniv . . . . Jan 17
Rushdie, Salman: Birth . . . . June 19
Rushdie, Salman: Death Sentence Anniv . . . Feb 14
Rusk, Dean: Birth . . . . Feb 9
Russell, Anna: Birth . . . . Dec 27
Russell, Bill: Birth . . . . Feb 12
Russell, Charles M.: Birth Anniv . . . . Mar 19
Russell, Jane: Birth . . . . June 21
Russell, Kurt: Birth . . . . Mar 17
Russell, Leon: Birth . . . . Apr 2
Russell, Lillian: Birth Anniv . . . . Dec 4
Russell, Mark: Birth . . . . Aug 23
Russell, Nipsey: Birth . . . . Oct 13
Russia
Coup Attempt in the Soviet Union: Anniv Aug 19
Hope Sled Dog Friendship Run (Chukotka) . . . . Apr 1
New Year's Day Observance . . . . Jan 1
Passport Presentation . . . . Jan 2
Ruble Becomes Convertible: Anniv . . . . July 1
Russian Election Surprise: Anniv . . . . Dec 12
Soviet Cosmonaut Returns to New World: Anniv . . . . Mar 26
Soviet Union Dissolved: Anniv . . . . Dec 8
Victory Day . . . . May 9
Yeltsin Foes Freed: Anniv . . . . Feb 26
Yeltsin Shells White House: Anniv . . . . Oct 4
Russian River Jazz Fest (Guerneville, CA) . Sept 16
Rustin, Bayard: Birth Anniv . . . . Mar 17
Ruth, George "Babe"
Babe Calls His Shots . . . . Oct 1
Babe Ruth Day: Anniv . . . . Apr 27
Babe Ruth Dies: Anniv . . . . Aug 16
Babe Ruth's First Pro Homer: Anniv . . . . . . Sept 5
Babe Sets Home Run Record: ANNIV . . . Sept 30
Babe's Debut in Majors: Anniv . . . . July 11
Babe's Last Pitching: Anniv . . . . Oct 1
Babe's Pitching Debut: Anniv . . . . Apr 22
Babe's 714th Big One: Anniv . . . . May 25
Birth Anniv . . . . Feb 6
Commemorative Stamp: Anniv . . . . July 6
Farewell to the Yankees: Anniv . . . . Sept 24
First Major League Home Run: Anniv . . . . . May 6
Hangover Awareness Day, Natl . . . . Feb 7
House That Ruth Built: Anniv . . . . Apr 18
Last game as Yankee: Anniv . . . . Sept 30
Ruth Retires: Anniv . . . . June 2
Voted into Hall of Fame: Anniv . . . . Feb 2
Yankee Stadium Jubilee: Anniv . . . . June 13
Rutherford, Ann: Birth . . . . Nov 2
Rutledge, Edward: Birth Anniv . . . . Nov 23
Rutledge, John: Death Anniv . . . . July 18
Rwanda: Tragedy in Rwanda: Anniv . . . . Apr 6
Ryan, Nolan: Birth . . . . Jan 31
Rydell, Bobby: Birth . . . . Apr 26

S

Saarinen, Eliel: Birth Anniv . . . . Aug 20
Sabatini, Gabriela: Birth . . . . May 16
Saberhagen, Bret: Birth . . . . Apr 13
Sabo, Chris: Birth . . . . Jan 19
Sacagawea: Death Anniv . . . . Dec 20
Sacco-Vanzetti Memorial Day . . . . Aug 23
Sadat, Anwar El: Assassination Anniv . . . . Oct 6
Sadie Hawkins Day . . . . Nov 4
Safe Toys and Gifts Month . . . . Dec 1
Safer, Morley: Birth . . . . Nov 8
Safety
Auto Battery Safety Month . . . . Oct 1
Automobile Speed Reduction: Anniv . . . . . Nov 25
Baby Safety Month . . . . Sept 1
Cartoonists Against Crime Day . . . . Oct 25
Check the Smoke Alarms Day . . . . Dec 31
Child Injury Prevention Week (New Hyde Park, NY) . . . . Sept 1
Child Safety and Protection Month . . . . Nov 1
Children's Eye Health and Safety Month . . Sept 1
Crime Prevention Week, Natl . . . . Feb 5
Death Busters Day, Natl . . . . May 26
Farm Safety and Health Week, Natl 6595 . . . . Page 488
Fire Prevention Week . . . . Oct 8
Fire Prevention Week 6601 . . . . Page 488
Fire Safety Council, Natl: Anniv . . . . Dec 7
Firepup's Birthday . . . . Oct 1
Fireworks Safety Month . . . . June 1
Missing Children's Day, Natl . . . . May 25
Natl Child Safety Council: Founding Anniv Nov 9
Night Out, Natl . . . . Aug 1
Poison Prevention Week, Natl (Pres Proc) Mar 19
Poison Prevention Week, Natl 6536, 6651 . . . . Page 488
Project Red Ribbon, MADD's . . . . Nov 1
Project Safe Baby Month . . . . May 1
Public Safety Telecommunicators Week, Natl 6667 . . . . Page 488
REACT Month, Intl . . . . May 1
Sa'Fest: Cartoon Art Comp (Chicago, IL) . Oct 1
Safe Boating Week, Natl . . . . May 20
Safe Boating Week, Natl 6570, 6695 . . . Page 488
Safe Toys and Gifts Month . . . . Dec 1
Safetypup's Birthday . . . . Feb 12
Sleep Safety Month . . . . Feb 1
Sports Eye Safety Month . . . . Apr 1
Stop Violence/Save Kinds Month . . . . Sept 1
White Cane Safety Day 6612 . . . . Page 488
Safire, William: Birth . . . . Dec 17
Sagan, Carl: Birth . . . . Nov 9
Sager, Carole Bayer: Birth . . . . Mar 8
Sagittarius Begins . . . . Nov 23
Sahl, Mort: Birth . . . . May 11
Sailing Ship Preservation Day, Merchant . . . Nov 8
St. Andrew's Day . . . . Nov 30
St. Ann's Solemn Novena (Scranton, PA) . . July 17
St. Anthony of Padua: Feast Day (Portugal) . . . . June 13
St. Anthony's Day . . . . Jan 17

☆ Chase's 1995 Calendar of Events ☆ Index

St. Anthony's Eve and Festival (Lisbon, Portugal) . . . June 12
St. Apollinaris: Feast Day . . . July 23
St. Aubin, Helen "Callaghan": Birth Anniv . . Mar 13
St. Augustine, Feast of . . . Aug 28
St. Augustine of Canterburg, Feast of . . . May 26
St. Barbara's Day . . . Dec 4
St. Barthelemy: Patron Saint Day . . . Aug 24
St. Bartholomew's Day Massacre . . . Aug 24
St. Basil's Cathedral: Christmas Bells Ring Again: Anniv . . . Dec 24
St. Basil's Day . . . Jan 1
St. Bernard of Montjoux Feast Day . . . May 28
St. Catherine of Siena: Feast Day . . . Apr 29
St. Catherine's Day . . . Nov 25
St. Cecilia: Feast Day . . . Nov 22
St. Christopher: National Day . . . Sept 19
St. Clare of Assisi: Feast Day . . . Aug 11
St. David's Day (Wales) . . . Mar 1
St. Edward, The Confessor: Feast Day . . . Oct 13
St. Elias Day . . . Aug 2
St. Eustatius, West Indies: Statia and America Day . . . Nov 16
Saint, Eva Marie: Birth . . . July 4
St. Frances of Rome: Feast Day . . . Mar 9
St. Frances Xavier Cabrini: Birth Anniversary . . . July 15
St. Francis of Assisi: Feast Day . . . Oct 4
St. Gabriel Feast Day . . . Mar 24
St. Gabriel Possenti Feast Day . . . Feb 27
St. George: Feast Day (England) . . . Apr 23
St. George's Day (Newfoundland) . . . Apr 24
St. George's Day Annual Celebration (New York, NY) . . . Apr 23
St. Gotthard Auto Tunnel: Opening Anniv . . Sept 5
St. Gudula's Feast Day . . . Jan 8
St. Ignatius of Loyola: Feast . . . July 31
Saint James, Susan: Birth . . . Aug 14
St. Januarius: Feast Day . . . Sept 19
St. Jerome, Feast of . . . Sept 30
St. Joan of Arc: Feast Day . . . May 30
St. John, Apostle-Evangelist: Feast Day . . . Dec 27
St. John, Jill: Birth . . . Aug 19
St. John Nepomucene Neumann: Birth Anniv . . . Mar 28
St. John of Capistrano: Death Anniv . . . Oct 23
St. John the Baptist Day . . . June 24
St. John the Baptist, Beheading of . . . Aug 29
St. John's Festival (Porto, Portugal) . . . June 23
St. Joseph's Day . . . Mar 19
St. Jude's Day . . . Oct 28
St. Kitts/Nevis: Independence Day . . . Sept 19
St. Lasarus' Day (Bulgaria) . . . Apr 1
Saint Laurent, Yves: Birth . . . Aug 1
St. Lawrence Seaway Act: Anniv . . . May 13
St. Lawrence Seaway: Dedication Anniv . June 26
St. Lazarus, Procession of Icon of (Cyprus) Apr 15
St. Lucia: Independence Day . . . Feb 22
St. Luke: Feast Day . . . Oct 18
St. Nicholas' Day . . . Dec 6
St. Olaf Christmas Fest (Northfield, MN) . . . Nov 30
St. Oswald of Worcester Feast Day . . . Feb 28
St. Patrick's Day . . . Mar 17
St. Patrick's Day (Northern Ireland) . . . Mar 17
St. Patrick's Day Beard Growing Contest (Shamrock, TX) . . . Mar 17
St. Patrick's Day Celebration (Baltimore, MD) . . . Mar 17
St. Patricks's Day Celebration (Huntington, WV) . . . Mar 17
St. Patrick's Day Dart Tourn (Battle Creek, MI) . . . Mar 16
St. Patricks's Day Dog Fun Fair, Parade (Alexandria, VA) . . . Mar 11
St. Patrick's Day Parade (New York, NY) . . . Mar 17
St. Patrick's Day Parade (Roanoke) . . . Mar 17
St. Patrick's Day Parade, World Shortest (Maryville, MO) . . . Mar 17
St. Paul's Feast Day . . . June 28
St. Peter's Day . . . June 29
St. Pius X: Birth Anniv . . . June 2
St. Stephen's Day . . . Dec 26
St. Swithin's Day . . . July 15
St. Swithun's Celebration (Richmond Hill, Ont, Canada) . . . July 15
St. Sylvester's Day . . . Dec 31
St. Tamenend's Day . . . May 1
St. Thomas of Canterbury: Feast Day . . . Dec 29
St. Urho's Day (Finland, MN) . . . Mar 18
St. Urho's Day (Hood River, OR) . . . Mar 16
St. Valentine's Day . . . Feb 14
St. Vincent and the Grenadines: Independence Day . . . Oct 27
St. Vincent De Paul: Feast Day . . . Sept 27
St. Vincent De Paul: Old Feast Day . . . July 19
St. Vincent's Feast Day . . . Jan 22
Sainte-Marie, Buffy: Birth . . . Feb 20
Sajak, Pat: Birth . . . Oct 26
Sakharov, Andrei Dmitriyevich: Birth Anniv May 21
Salad Month, Natl . . . May 1
Salazar, Alberto: Birth . . . Aug 7
Salem Christmas (Winston-Salem, NC) . . . Dec 16
Sales, Soupy: Birth . . . Jan 8
Salinger, J.D.: Birth . . . Jan 1
Salk, Jonas Edward: Birth . . . Oct 28
Salk, Lee: Birth . . . Dec 27
Salk Vaccine: Anniv . . . Apr 12
Salley, John: Birth . . . May 16
Salmon Bake, Steilacoom (Steilacoom, WA) . . . July 30
Salomon, Haym: Death Anniv . . . Jan 6
Salt, Jennifer: Birth . . . Sept 4
Salter, Susanna, Elected 1st Woman Mayor in US: Anniv . . . Apr 4
Salute 2 America Parade, WSB-TV (Atlanta, GA) . . . July 4
Salvation Army Advisory Organizations Sunday . . . May 21
Salvation Army: Booth, William: Birth Anniv Apr 10
Salvation Army: Donut Day (Chicago, IL) . . June 2
Salvation Army in USA: Anniv . . . Mar 10
Salvation Army: Natl Red Kettle Kick Off . . . Nov 24
Salvation Army Week, Natl . . . May 15
Samhain . . . Oct 31
Samms, Emma: Birth . . . Aug 28
Samoa: Independence Day . . . June 1
Sampson County Expo (Clinton, NC) . . . Nov 2
Sampson, Deborah: Birth Anniv . . . Dec 17
Sampson, Ralph Lee: Birth . . . July 7
San Francisco 1906 Earthquake: Anniv . . . Apr 18
San Fransico 1989 Earthquake: Anniv . . . Oct 17
San Isidro Day (Mexico) . . . May 15
San Jacinto Day (TX) . . . Apr 21
San Marino: National Day . . . Sept 3
San Sebastian's Day (Brazil) . . . Jan 20
Sanborn, David: Birth . . . July 30
Sanctity of Human Life Day, Natl 6521 . . . Page 488
Sand Sculpture Contest, Sun Herald (Biloxi, MS) . . . Sept 16
Sand, George: Birth Anniv . . . July 1
Sand, Paul: Birth . . . Mar 5
Sandberg, Ryne: Birth . . . Sept 18
Sandblast (Ft Lauderdale, FL) . . . July 4
Sandburg, Carl: Birth Anniv . . . Jan 6
Sanders, Barry: Birth . . . July 16
Sanders, Colonel Harland David: Birth Anniv . . . Sept 9
Sanders, Richard: Birth . . . Aug 23
Sands, Tommy: Birth . . . Aug 27
Sandwich Day . . . Nov 3
Sandwich Day: John Montague Birth Anniv . Nov 3
Sandy, Gary: Birth . . . Dec 25
Sandys, Edwin: Birth Anniv . . . Dec 9
Sang, Samantha: Birth . . . Aug 5
Sanger, Margaret (Higgins): Birth Anniv . . . Sept 14
Santa Fe Trail Day (Las Animas, CO) . . . Apr 28
Santa Lucia Day (Sweden) . . . Dec 13
Santa-Cali-Gon Days (Independence, MO) . Sept 1
Santana, Carlos: Birth . . . July 20
Santayana, George: Birth Anniv . . . Dec 16
Santiago, Benito: Birth . . . Mar 9
Santiago, Saundra: Birth . . . Apr 13
Sao Tome and Principe: National Day . . . July 12
Sarah, Duchess of York: Birth . . . Oct 15
Sarandon, Susan: Birth . . . Oct 4
Sarbanes, Paul S.: Birth . . . Feb 3
Sarcastics Month, Natl . . . Oct 1
Sarducci, Father Guido (Don Novello): Birth . Jan 1
Saroyan, William: Birth Anniv . . . Aug 31
Sarrazin, Michael: Birth . . . May 22
Sartre, Jean Paul: Birth Anniv . . . June 21
Sasser, James Ralph: Birth . . . Sept 30
Sassoon, Vidal: Birth . . . Jan 17
Satanic Verses: Salman Rushdie Death Sentence Anniv . . . Feb 14
Saturday Night Massacre . . . Oct 20
Saturnalia . . . Dec 17
Sauce Month, Natl . . . Mar 1
Saudi Arabia: Kingdom Unification . . . Sept 23
Sauerkraut Salad and Sandwich Season . . . July 1
Sauntering Day, World . . . June 19
Savard, Denis: Birth . . . Feb 4
Save the Rhino Day . . . May 1
Save Your Back Week, Natl . . . Oct 22
Save Your Vision Week . . . Mar 5
Savings and Loan: Home Owners Loan Act: Anniv . . . June 13
Sawyer, Diane K.: Birth . . . Dec 22
Sawyer: Tom Sawyer's Cat's Birthday . . . Jan 3
Sax, Adolphe: Birth Anniv (Saxophone Day) Nov 6
Sax, Steve: Birth . . . Jan 29
Saxophone Day . . . Nov 6
Sayer, Leo: Birth . . . May 21
Sayers, Gale: Birth . . . May 30
Scaggs, Boz: Birth . . . June 8
Scalia, Antonin: Birth . . . Mar 11
Scaliger, Joseph J: Birth Anniv . . . Aug 4
Scandinavian Hjemkomst Fest (Fargo, ND/ Moorhead, MN) . . . June 21
Scandinavian: Dalesburg Midsummer Fest (Vermillion, SD) . . . June 23
Scarecrow Contest (Lahaska, PA) . . . Sept 11
Scarecrow Fest (Lahaska, PA) . . . Sept 16
Scarecrow Fest, St. Charles (St. Charles, IL) . . . Oct 14
Scarlatti, Domenico: Birth Anniv . . . Oct 26
Scavullo, Francesco: Birth . . . Jan 16
Schackelford, Ted: Birth . . . June 23
Schaefer, William D.: Birth . . . Nov 2
Schafer, Edward T.: Birth . . . Aug 8
Scharansky, Anatoly: Birth . . . Jan 20
Scheider, Roy: Birth . . . Oct 11
Schell, Maximilian: Birth . . . Dec 8
Schembechler, Glen Edward "Bo": Birth . . . Apr 1
Schirra, Wally: Birth . . . Mar 12
Schlafly, Phyllis Stewart: Birth . . . Aug 15
Schlesinger, Arthur Meier, Jr: Birth . . . Oct 15
Schmeckfest (Freeman, SD) . . . Mar 30
Schmidt, Mike: Birth . . . Sept 27
Schneider, Maria: Birth . . . Mar 27
Schneiderman, Rose: Birth Anniv . . . Apr 6
Schnitzler, Arthur: Birth Anniv . . . May 15
Schoelcher Day (French West Indies) . . . July 21
School. See Education
School Breakfast Week, Natl . . . Mar 6
School Celebration, Natl . . . Oct 13
School Counseling Week, Natl . . . Feb 6
School Family Day . . . May 9
School Lunch Week, Natl . . . Oct 9
School Spirit Season, Intl . . . Apr 30
Schopenhauer, Arthur: Birth Anniv . . . Feb 22
Schorr, Daniel: Birth . . . Aug 31
Schrempf, Detlef: Birth . . . Jan 21
Schubert, Franz: Birth Anniv . . . Jan 31
Schulz, Charles M.: Birth . . . Nov 26
Schulz, Charles M.: Charlie Brown and Snoopy Birthday . . . Oct 2
Schuman Plan Anniv: European Communities . . . May 9
Schuman, William Howard: Birth Anniv . . . Aug 4
Schurz, Carl: Birth Anniv . . . Mar 2
Schutz, Heinrich: Birth Anniv . . . Oct 8
Schwarzenegger, Arnold: Birth . . . July 30
Schwarzkopf, Norman H.: Birth . . . Aug 22
Schweitzer, Albert: Birth Anniv . . . Jan 14
Schwenkfelder Thanksgiving . . . Sept 24
Schwerner, Michael: Civil Rights Workers Disappear . . . June 21
Schwerner, Michael: Klan Members Convicted . . . Oct 20
Scialfa, Patty: Birth . . . July 29
Science Teachers Assn Conf, Michigan (Lansing, MI) . . . Feb 3
Science, Technology
Biomedical Research Day, Natl 6616 . . . Page 488
Edinburgh Intl Science Fest (Edinburgh, Scotland) . . . Apr 6
First Self-Sustaining Nuclear Chain Reaction . . . Dec 2
First US Scientist Receives Nobel: Anniv . Dec 20
Industry Day (Beatrice, NE) . . . Apr 23
Mole Day, Natl . . . Oct 23
Natl Chemistry Week . . . Nov 5
Nobel Conference (St. Peter, MN) . . . Oct 3
Plutonium First Weighed: Anniv . . . Aug 20
Science and Peace, Intl Week of (UN) . . . Nov 5
Science and Technology, Intl Fest (Edinburgh, Scotland) . . . Mar 23
Science and Technology Week, Natl . . . Apr 21
Scientific Matl Mgrs Conf/Show, Natl Assn (Chicago, IL) . . . July 24
Sky Awareness Week . . . Apr 23
Science Fiction: Asimov, Isaac: Death Anniv Apr 6
Science Fiction: Boskone 32 (Framingham, MA) . . . Feb 17
Scleroderma Awareness Week, Natl . . . June 1
Scobee, Francis R.: Birth Anniv . . . May 19
Scolari, Peter: Birth . . . Sept 12
Scopes, John T.: Birth Anniv . . . Aug 3
Scorpio Begins . . . Oct 23
Scorsese, Martin: Birth . . . Nov 17

St. Anthony's (cont'd)—Scorsese

Index ☆ Chase's 1995 Calendar of Events ☆

Scotland
- Aberdeen Intl Youth Fest (Aberdeen, Grampian)..........Aug 2
- Bell's Scottish Open..........July 5
- Borders Festival (Galashiels)..........Oct 13
- Braemar Royal Highland Gathering (Braemar)..........Sept 2
- Christmas Holiday..........Dec 25
- Cutty Sark Tall Ships Race (Weymouth, Dorset)..........July 15
- Edinburgh Book Festival (Edinburgh)....Aug 12
- Edinburgh Intl Fest/Fest Fringe (Edinburgh)..........Aug 13
- Edinburgh Intl Film Fest (Edinburgh, Lothian)..........Aug 12
- Edinburgh Intl Science Fest (Edinburgh)..Apr 6
- Edinburgh Military Tattoo (Edinburgh)....Aug 4
- Golf Open Chmpshp (St. Andrews)......July 20
- Hogmanay..........Dec 31
- Intl Fest Science/Technology (Edinburgh)Mar 23
- Mayfest (Glasgow)..........Apr 28
- New Year's Bank Holiday..........Jan 3
- Royal Scottish Auto Club Rally..........June 2
- Scottish Open Badminton Championship (Glasgow)..........Nov 22
- Summer Bank Holiday..........Aug 7
- Up Hellya AA..........Jan 31

Scotland Yard: First Appearance Anniv...Sept 29
Scott, Byron: Birth..........Mar 28
Scott, George C.: Birth..........Oct 18
Scott, Sir Walter: Birth Anniv..........Aug 15
Scott, Willard Herman: Birth..........Mar 7
Scott, Winfield: Birth Anniv..........June 13

Scottish
- Alma Highland Festival and Games (Alma, MI)..........May 26
- Bagpipe and Highland Dance Fest (New Preston, CT)..........Aug 6
- Capital District Scottish Games (Altamont, NY)..........Sept 2
- Christmas Walk (Alexandria, VA)..........Dec 2
- Clan Donald USA General Meeting (Pittsburgh, PA)..........Sept 28
- Gatlinburg Scottish Fest (Gatlinburg, TN)..........May 18
- Maine Highland Games (Brunswick, ME)..Aug 19
- McHenry Highland Fest (McHenry, MD)..June 3
- New Hampshire Highland Games (Lincoln, NH)..........Sept 15
- Scottish Tattoo/Fest/Highland Games (Stone Mountain, GA)..........Oct 19
- Scottish Weekend (Carrollton, KY)......May 12
- Southeast Florida Festival and Games (Key Biscayne, FL)..........Mar 4
- Virginia Scottish Games (Alexandria, VA)July 22

Scotto, Renata: Birth..........Feb 24
Scout Week, Girl..........Mar 5
Scouts. See also type of Scout
Scowcroft, Brent: Birth..........Mar 19
Scrabble Chmpshp, World (England)......Oct 12
Scratch Ankle (Milton, FL)..........Mar 23
Scream Day, Intl Moment of Frustration....Oct 12
Scruggs, Earl: Birth..........Jan 6
Scuba Diving: Seaspace (Houston, TX)......June 2
Scuba Diving Week, Natl Recreational....Aug 13
Sculley, John: Birth..........Apr 6
Scully, Vin: Birth..........Nov 29
Sea Cadet Month..........Apr 1
Sea, Festival of the (Pt Pleasant Beach, NJ)..........Sept 16
Sea Lion "Haul Out": Anniv (San Francisco, CA)..........Jan 19
Seaborg, Glenn: Plutonium First Weighed: Anniv..........Aug 20
Seafair (Seattle, WA)..........July 14
Seafood and Wine Fest, Newport (Newport, OR)..........Feb 24
Seafood Fest, Grant (Grant, FL)..........Feb 18
Seafood Month, Natl..........Oct 1
Seafood Month, Surimi..........June 1
Seaspace (Houston, TX)..........June 2
Seattle: Stay Away from Seattle Day......Sept 16
Seaver, Tom: Birth..........Nov 17
Second Day of Christmas..........Dec 26
Secret Pal Day..........Jan 14
Secretaries Day, Professional..........Apr 26
Secretaries Intl Conv, Professional (Seattle, WA)..........July 23
Secretaries Week, Professional..........Apr 23
Sedaka, Neil: Birth..........Mar 13
Seeger, Pete: Birth..........May 3
Segal, Erich: Birth..........June 16
Segal, George: Birth..........Feb 13
Seger, Bob: Birth..........May 6
Seidelman, Susan: Birth..........Dec 11
Seikaly, Rony: Birth..........May 10
Seinfeld, Jerry: Birth..........Apr 29
Self-Esteem Month, Natl Boost Your........Feb 1
Self-Improvement Month..........Sept 1
Self-University Week..........Sept 1
Selfridge, Thomas E.: Death Anniv..........Sept 17
Sellecca, Connie: Birth..........May 25
Selleck, Tom: Birth..........Jan 29
Sellers, Peter: Birth Anniv..........Sept 8
Senate Quorum, First: Anniv..........Apr 6
Sendak, Maurice: Birth..........June 10
Senegal: Independence Day..........Apr 4

Senior Citizens
- Adult Day Care Center Week..........Sept 17
- All States Picnic (Yuma, AZ)..........Jan 18
- AMD Awareness Month..........Feb 1
- California Senior Games (Sacramento, CA)..........June 2
- Centenarians Day, Natl..........Sept 22
- Easter Egg Hunt (Rockford, OH)..........Apr 15
- Elderly, Intl Day for..........Oct 1
- Hillhaven Ho Ho Hotline (Seattle, WA)...Dec 11
- Ice Cream Social (Rockford, OH).....Aug 26
- Indy Sr Olympics (Indianapolis, IN)......June 7
- Intergenerational Week (Williamsport, MD)..........Apr 23
- Lisco Old-timers Day (Lisco, NE)..........Sept 10
- Old Bedford Village Senior Days (Bedford, PA)..........Sept 16
- Older Americans Month 6565, 6687....Page 488
- Seaport Seniors Month (Mystic, CT)......Apr 1
- Senior Citizens Day (OK)..........June 9
- Senior Citizens Month..........May 1
- Senior Day (Milwaukee, WI)..........Feb 15
- Senior Expo: Life Is What You Make It (Halifax, NS)..........July 14
- Senior Festival and Expo (Rockford, IL)....May 19
- Senior Health and Fitness Day, Natl.....May 31
- Senior Travel Month, Natl..........May 1
- Smile Week, Natl Senior..........May 14
- Southwest Senior Chmpshp (Yuma, AZ)..Jan 17
- US Natl Senior Olympics (San Antonio, TX)..........May 17
- Veterans Golden Age Games Week, Natl 6581..........Page 488
- WCVL/WIMC Senior Health Fair/Expo (Crawfordsville, IN)..........Sept 26
- Yorkton Threshermen's Show/Sr Fest (Yorkton, Sask)..........Aug 5

Senses, Celebration of the..........June 24
September Fest (Childersburg, AL)..........Sept 30
Serkin, Peter: Birth..........July 24
Servan-Schreiber, Jean-Claude: Birth.....Apr 11
Servetus, Michael: Execution Anniv..........Oct 27
Service Day, Natl..........Nov 4
Service, Robert William: Birth Anniversary.Jan 16
Sessions, William: Birth..........May 27
Seton, Elizabeth Ann Bayley: Feast Day....Jan 4
Seton, Elizabeth Ann: Birth Anniv..........Aug 28
Seton, Elizabeth Ann: Canonization Anniv.Sept 14
Seurat, Georges: Birth Anniv..........Dec 2
Seuss, Dr.: Geisel, Theodor: Birth Anniv....Mar 2
Seuss, Dr.: Geisel, Theodor: Death Anniv..Sept 24
Severinsen, Doc: Birth..........July 7
Seward, William H.: Birth Anniv..........May 16
Seward's Day (AK)..........Mar 30
Sex Discrimination, Marth Griffiths Speaks Out Against..........Feb 8
Sex Month, Natl Talk With Your Teen About..Mar 1
Sexuality Education Month, Natl Family....Oct 1
Sexually Transmitted Diseases (STDs) Awareness Month, Natl..........Apr 1
Seychelles: Constitution Day..........June 18
Seymour, Jane: Birth..........Feb 15
Sez Who? Fourplay March Madness (Chicago, IL)..........Mar 12
Shaffer, Paul: Birth..........Nov 28
Shakespeare Fest, Colorado (Boulder, CO)..........June 23
Shakespeare Fest, Oregon (Ashland, OR)..Feb 17
Shakespeare Fest, Oregon (Portland, OR)..Jan 1
Shakespeare Theatre Season (Warwickshire, England)..........Mar 17
Shakespeare, William: Birth..........Apr 23
Shakespeare, Wm: Rev F Gastrell's Ejectment..........Nov 27
Shalala, Donna: Birth..........Feb 14
Shalikashvili: Appointed Chair Joint Chiefs: Anniv..........Aug 11
Shamrocks for Dystrophy..........Mar 17
Shamu's Birthday..........Sept 26
Shandling, Garry: Birth..........Nov 29
Shange, Ntozake: Birth..........Oct 18
Shanty Days (Algoma, WI)..........Aug 11
Shareware, Intl..........Dec 17
Sharif, Omar: Birth..........Apr 10
Shatner, William: Birth..........Mar 22
Shavuot..........June 4
Shavuot Begins..........June 3
Shaw, Artie: Birth..........May 23
Shaw Fest (Niagara-on-the-Lake, Ont, Canada)..........Apr 18
Shaw, George Bernard: Birth Anniv......July 26
Shaw, Irwin: Birth..........Feb 27
Shaw, Patty Hearst: Birth..........Feb 20
Shawn, Wallace: Birth Anniv..........Aug 31
Sheedy, Ally: Birth..........June 13
Sheehy, Gail Henion: Birth..........Nov 27
Sheen, Charlie: Birth..........Sept 3
Sheen, Martin: Birth..........Aug 3
Sheep and Crafts Festival, World (Bethel, MO)..........Sept 2
Sheep Market, Ho (Denmark)..........Aug 26
Sheep Shearing Fest (North Andover, MA)May 14
Shelby, Richard C.: Birth..........May 6
Sheldon, Sidney: Birth..........Feb 11
Shemini Atzeret..........Oct 16
Shepard, Alan: Birth..........Nov 18
Shepard, Sam: Birth..........Nov 5
Shepherd, Cybill: Birth..........Feb 18
Shepherd's Fair/Schueberness (Luxembourg)..........Aug 20
Sheridan, Richard B: Birth Anniv..........Oct 30
Sherman, Bobby: Birth..........July 22
Sherman, James S: Birth Anniv..........Oct 24
Sherman, Roger: Birth Anniv..........Apr 19
Sherman, William Tecumseh: Birth Anniv..Feb 8
Sherman, Wm T: Carolina Campaign Begins: Anniv..........Feb 1
Sherman, Wm T: Columbia Surrenders to Sherman: Anniversary..........Feb 17
Sherman, Wm T.: Surrender at Durham Station: Anniv..........Apr 18
Sherwood, Madeline: Birth..........Nov 13
Shevardnadze, Eduard: Birth..........Jan 25
Shields, Brooke: Birth..........May 31
Shilts, Randy: Birth Anniv..........Aug 8
Ships. See Boats, Ships, Other Vessels
Shire, Talia: Birth..........Apr 25
Shoemaker, Willie: Birth..........Aug 19
Shooting. See Gun and Shooting
Shopping Carts to the Supermarket Month, Return..........Feb 1
Shopping Reminder Day..........Nov 25
Shore, Dinah: Birth Anniv..........Mar 1
Short, Bobby: Birth..........Sept 15
Short, Martin: Birth..........Mar 26
Shostakovich, Dmitri: Birth Anniv..........Sept 25
Shovel Racing Chmpshp, World (Angel Fire, NM)..........Feb 3
Show and Tell Day at Work..........Jan 9
Shrimp and Petroleum Fest, Louisiana (Morgan City, LA)..........Aug 31
Shrimp Fest, Isle of Eight Flags (Fernandina, FL)..........Apr 29
Shrimp Fest, Low Country (McClellanville, SC)..........May 6
Shriver, Maria: Birth..........Nov 6
Shriver, Pam: Birth..........July 4
Shrove Monday..........Feb 27
Shrove Tuesday..........Feb 28
Shrovetide..........Feb 26
Shrovetide Pancake Race..........Feb 28
Shula, Don: Birth..........Jan 4
Shut-Ins Day, Natl..........Oct 15
Siberian Explosion Anniversary..........June 30
Sidney, Philip: Birth Anniv..........Nov 30
Sierra Leone: Independence Day..........Apr 27
Sierra Leone: National Holiday..........Apr 19
Sierra, Ruben: Birth..........Oct 6
Sight. See Health and Welfare
Sight-Saving Sabbath..........Jan 22
Signing Federal Credit Union Act: Anniv..June 26
Sigourney, Lydia: Birth Anniv..........Sept 1
Sikma, Jack: Birth..........Nov 14
Sikorsky, Igor: Birth Anniv..........May 25
Silent Record Week..........Jan 1
Silent Spring Publication: Anniv..........Apr 13
Sills, Beverly: Birth..........May 25
Silver Bells in the City (Lansing, MI)......Nov 17
Silverman, Fred: Birth..........Sept 13
Simchat Torah..........Oct 17
Simmons, Richard: Birth..........July 12
Simms, Phil: Birth..........Nov 3
Simon, Carly: Birth..........June 25
Simon, Neil: Birth..........July 4
Simon, Paul (musician): Birth..........Oct 13

☆ *Chase's 1995 Calendar of Events* ☆ **Index**

Simon (cont'd)—South Dakota

Simon, Paul (senator): Birth ... **Nov 29**
Simone, Nina: Birth ... **Feb 21**
Simplon Tunnel Opening: Anniv ... **May 19**
Simplot Games (Pocatello, ID) ... **Feb 17**
Simpson, Alan K.: Birth ... **Sept 2**
Simpson, O.J.: Birth ... **July 9**
Simpsons Premiere: Anniv ... **Jan 14**
Sims, John Haley "Zoot": Birth ... **Oct 29**
Sinai Day (Egypt) ... **Apr 25**
Sinatra, Frank: Birth ... **Dec 12**
Sinatra, Nancy: Birth ... **June 8**
Sinclair Lewis Days (Sauk Centre, MN) ... **July 13**
Sinclair, Upton: Birth Anniv ... **Sept 20**
Sinden, Donald: Birth ... **Oct 9**
Sing-Out Day, Natl ... **Mar 22**
Singapore
- Independence Day ... **Aug 9**
- Songkran Festival ... **Apr 13**

Singing, Oklahoma All Night Singing (Konawa, OK) ... **Aug 5**
Singing Telegram Birthday ... **July 28**
Singles Founding Anniv, South Shore (Norwell, MA) ... **May 26**
Singles Week, Natl ... **Sept 17**
Singletary, Mike: Birth ... **Oct 9**
Singleton, Raynoma Gordy: Birth ... **Mar 8**
Sinkie Day (Eat over kitchen sink) ... **Nov 24**
Sirica, John: Birth Anniv ... **Mar 19**
Siskel, Gene: Birth ... **Jan 26**
Sisley, Alfred: Birth Anniv ... **Oct 30**
Sitting Bull: Death Anniv ... **Dec 15**
Skaggs, Ricky: Birth ... **July 18**
Skelton, Red: Birth ... **July 18**
Skiing
- American Birkebeiner (Cable to Hayward, WI) ... **Feb 23**
- Angel Fire Ski Area Opening (Angel Fire, NM) ... **Dec 7**
- Giants Ridge Intl Classic Marathon (Biwabik, MN) ... **Mar 5**
- Pole, Pedal, Paddle (Bend, OR) ... **May 20**
- Pole, Pedal, Paddle (Jackson, WY) ... **Apr 1**
- Ski-Joring Finals, Natl (Red Lodge, MT) .. **Mar 11**
- Spearfish Cross Country Ski Challenge (Spearfish, SD) ... **Feb 18**
- Yosemite Nordic Holiday Race (Yosemite Natl Park, CA) ... **Mar 4**
- Yosemite Spring Skifest (Yosemite Natl Park, CA) ... **Apr 8**

Skin: Fungal Infection Awareness Month ... **May 1**
Skin: Rosacea Awareness Month ... **Mar 1**
Skin: Scleroderma Awareness Week, Natl .. **June 1**
Skinner, Samuel Knox: Birth ... **June 10**
Sky Awareness Week ... **Apr 23**
Skylab Falls to Earth ... **July 11**
Slavery Abolished in District of Columbia: Anniv ... **Apr 16**
Slavery Abolished: 13th Amendment ... **Dec 18**
Slavery Abolition Day (Martinique, French West Indies) ... **May 22**
Slavery, Abolition of (Jamaica) ... **Aug 1**
Slavery: First American Abolition Soc Founded: Anniv ... **Apr 14**
Slavery: Schoelcher Day (French West Indies) ... **July 21**
Slayton, Donald "Deke" K.: Birth Anniv ... **Mar 1**
Sled Dog
- Anchorage Fur Rendezvous (Anchorage, AK) ... **Feb 10**
- Beargrease Sled Dog Marathon (Duluth, MN) ... **Jan 15**
- Canadian Chmpshp Dog Derby (Yellowknife, NT) ... **Mar 24**
- Dog Sled Chmpshp/Winter Carnival (Wisconsin Dells, WI) ... **Feb 11**
- Festival Du Voyageur (Winnipeg, Man, Canada) ... **Feb 10**
- Hope Sled Dog Friendship Run (Nome, Ak to Russia) ... **Apr 1**
- Mackinaw Mush Race (Mackinaw City, MI) **Feb 4**
- Minden Sled Dog Derby (Minden, Ont, Canada) ... **Jan 21**
- New England Sled Dog Races (Rangeley, ME) ... **Mar 11**
- Race to the Sky (Helena, MT) ... **Feb 11**
- Yukon Quest 1,000-Mile Race (Whitehorse, Yukon) ... **Feb 11**

Sleep Month, Better ... **May 1**
Sleep Safety Month ... **Feb 1**
Sleidanus, Johannes: Death Anniv ... **Oct 31**
Slezak, Erika: Birth ... **Aug 5**
Slick, Grace: Birth ... **Oct 30**
Sloane, Hans: Birth Anniv ... **Apr 16**
Slote, Alfred: Birth ... **Sept 11**
Slovakia: National Day ... **Sept 1**
Slovenia: National Day ... **June 25**
Slovenian: St. Cyril's Parish Fest (Sheboygan, WI) ... **July 16**
Slovik, Eddie: Execution Anniv ... **Jan 31**
Slow Pitch Softball Tourn (Williamsport) ... **July 7**
Slugs Return from Capistrano ... **May 28**
Small Business Week ... **May 7**
Small Business Week 6561 ... **Page 488**
Smeal, Eleanor Marie Cutri: Birth ... **July 30**
Smile Week, Natl ... **Aug 7**
Smiles, Samuel: Birth Anniv ... **Dec 23**
Smirnoff, Yakov: Birth ... **Jan 24**
Smith, Adam: Birth Anniv ... **June 5**
Smith, Bessie: Birth Anniv ... **Apr 15**
Smith, Bessie: Jazz Fest (Chattanooga, TN) May 5
Smith, Bob C.: Birth ... **Mar 30**
Smith, Bubba: Birth ... **Feb 28**
Smith, Holland: Birth Anniv ... **Apr 20**
Smith, Howard K.: Birth ... **May 12**
Smith, J.T.: Birth ... **Oct 29**
Smith, Jaclyn: Birth ... **Oct 26**
Smith, James: Death Anniv ... **July 11**
Smith, Joseph: Birth Anniv ... **Dec 23**
Smith, Kate: Birth Anniv ... **May 1**
Smith, Kate: God Bless America 1st Perf: Anniv ... **Nov 11**
Smith, Liz: Birth ... **Feb 2**
Smith, Maggie: Birth ... **Dec 28**
Smith, Margaret Chase: Birth ... **Dec 14**
Smith, Michael J.: Birth Anniv ... **Apr 30**
Smith, Ozzy: Birth ... **Dec 26**
Smith, Robert: Alcoholics Anonymous Founding Anniv ... **June 10**
Smith, Robyn: Birth ... **Aug 14**
Smith, Roger: Birth ... **July 12**
Smith, Samantha: Death Anniv ... **Aug 25**
Smith, Sammi: Birth ... **Aug 5**
Smith, Stan: Birth ... **Dec 14**
Smith, Thorne: Birth Anniv ... **Mar 27**
Smith, Walter W. (Red): Birth Anniv ... **Sept 25**
Smits, Jimmy: Birth ... **July 9**
Smokeout, Great American ... **Nov 16**
Smoky Mountain Lights/Winterfest (Gatlinburg, TN) ... **Nov 15**
Smothers Brothers' Firing: Anniv ... **Apr 4**
Smothers, Dick: Birth ... **Nov 20**
Smothers, Tom: Birth ... **Feb 2**
Snack Food Month, Natl ... **Feb 1**
Snake Hunt (Cross Fork, PA) ... **June 24**
Sneak Some Zucchini onto Your Neighbors' Porch Night ... **Aug 8**
Snider, Dee: Birth ... **Mar 15**
Snipe Excursion and Hunt, Pro-Am (Moultrie, GA) ... **Apr 1**
Snodgrass, Carrie: Birth ... **Oct 27**
Snodgrass, W.D.: Birth ... **Jan 5**
Snow Fest (Pensacola Beach, FL) ... **Jan 28**
Snow Festival (Japan) ... **Feb 8**
Snow Sculpting Chmpshp, Colorado St (Breckenridge, CO) ... **Jan 18**
Snow Sculpting Competition, US Intl (Milwaukee, WI) ... **Jan 24**
Snow, Phoebe: Birth ... **July 17**
Snow Shovel Riding Contest (Ambridge, PA) ... **Jan 14**
Snowbird Breakfast (El Centro, CA) ... **Jan 14**
Snowflake Fest (Klamath Falls, OR) ... **Dec 2**
Snowman Sacrifice (Sault Ste Marie, MI) .. **Mar 20**
Snowmobile Hillclimb, World Chmpshp (Jackson, WY) ... **Mar 24**
Snyder, James "Jimmy the Greek": Birth ... **Sept 9**
Snyder, Tom: Birth ... **May 12**
Soap Box Derby, All-American (Akron, OH) Aug 12
Soccer Tragedy (Belgium): Anniv ... **May 29**
Soccer Tragedy, Moscow: Anniv ... **Oct 20**
Social Security Act: Anniv ... **Aug 14**
Social Welfare. See Health and Welfare ... **Aug 14**
Social Work Month, Natl Professional ... **Mar 1**
Socks Day, No ... **May 8**
Sodbuster Fest (Webster, SD) ... **June 10**
Softball
- NAIA Softball Chmpshp ... **May 17**
- NCAA Women's Softball World Series (Oklahoma City, OK) ... **May 25**
- Slow Pitch Softball Tourn (Williamsport, PA) ... **July 7**
- 365-Inning Softball Game: Anniv ... **Aug 14**

Software Week, Natl Religious ... **Aug 20**
Soil Stewardship Sunday ... **May 21**
Soldiers' Reunion Celebration (Newton, NC) ... **Aug 17**
Solemnity of Mary ... **Jan 1**
Solidarity Granted Legal Status ... **Apr 17**
Solidarity with Palestinian People, Intl day of (UN) ... **Nov 29**
Solomon Islands: National Holiday ... **July 7**
Solstice, Summer ... **June 21**
Solstice, Winter ... **Dec 22**
Solti, Sir Georg: Birth ... **Oct 21**
Solzhenitsyn, Aleksandr Isayevich: Birth .. **Dec 11**
Somalia: National Day ... **Oct 21**
Someday We'll Laugh About This Week ... **Jan 8**
Somers, Suzanne: Birth ... **Oct 16**
Sommer, Elke: Birth ... **Nov 5**
Sondheim, Stephen: Birth ... **Mar 22**
Sonora Showcase (Yuma, AZ) ... **Jan 28**
Sontag, Susan: Birth ... **Jan 28**
Sorghum Day Festival (Wewoka, OK) ... **Oct 28**
Sorghum/Harvest Fest, Fall (Huntsville, AL) ... **Sept 16**
Sorry Charlie Day ... **Apr 1**
Sorvino, Paul: Birth ... **Apr 13**
Soul, David: Birth ... **Aug 28**
Sound/Light Spectacular/Showboat (Wilmington, NC) ... **June 2**
Sounds of Season: Holiday Concert (Charlottesville, VA) ... **Dec 27**
Soup Month, Natl ... **Jan 1**
Sour Herring Premiere (Sweden) ... **Aug 17**
Sourest Day ... **Oct 25**
Sousa, John P: Birth Anniv ... **Nov 6**
Sousa: Stars and Stripes Forever Day ... **May 14**
Souter, David H.: Birth ... **Sept 17**
South Africa, Republic of
- Comrades Marathon (Natal) ... **May 31**
- Hansa Dusi Marathon (Natal) ... **Jan 19**
- Repeals Last Apartheid Law: Anniv ... **June 17**
- Republic Day ... **May 31**
- Rothmans July Handicap (Durban) ... **July 1**
- Standard Bank Natl Arts Fest (Grahamstown) ... **June 29**
- Van Riebeeck Day ... **Apr 6**

South Carolina
- A(ugusta) Baker's Dozen (Columbia) ... **Apr 28**
- Attack on Fort Sumter: Anniv ... **Apr 12**
- Attack on Fort Wagner: Anniv ... **July 19**
- Canadian American Days Fest (Myrtle Beach) ... **Mar 11**
- Carolina Campaign Begins: Anniv ... **Feb 1**
- Carolina Craftsmen's Christmas Classic (Columbia) ... **Nov 24**
- Carolina Craftsmen's Summer Classic (Myrtle Beach) ... **Aug 11**
- Charleston Maritime Fest (Charleston) .. **Sept 15**
- Christmas Craft Show (Aiken) ... **Dec 2**
- Columbia Surrenders to Sherman: Anniversary ... **Feb 17**
- Confederate Memorial Day ... **May 10**
- Fall Fiesta of the Arts (Sumter) ... **Oct 21**
- Fest-I-Fun (Fort Mill) ... **May 5**
- Festival of Houses and Gardens (Charleston) ... **Mar 16**
- Fort Sumter Returned to Union Control: Anniv ... **Feb 17**
- Fort Sumter Shelled by North: Anniv ... **Aug 17**
- Golden Leaf Festival (Mullins) ... **Sept 23**
- Gopher Hill Fest (Ridgeland) ... **Oct 7**
- Grand American Coon Hunt (Orangeburg) **Jan 6**
- Grand Strand Fishing Rodeo (Myrtle Beach) ... **Apr 1**
- Hartscapades (Hartsville) ... **May 18**
- Ironclads Battle in Charleston Harbor: Anniv ... **Jan 31**
- Jubilee (Bennettsville) ... **May 13**
- Lobster Race and Oyster Parade (Aiken) .. **May 5**
- Longhorn World Chmpshp Rodeo (Greenville) ... **Nov 10**
- Louis Armstrong: Cultural Legacy (Charleston) ... **July 29**
- Low Country Shrimp Fest (McClellanville) **May 6**
- Maritime Fest (Charleston) ... **Sept 15**
- MCI Heritage Classic (Hilton Head Island) ... **Apr 10**
- Ratification Day ... **May 23**
- Really Rosie, a Musical (Columbia) ... **Mar 16**
- Secession Anniv ... **Dec 20**
- South Carolina Fest of Roses (Orangeburg) ... **Apr 28**
- Southeastern Wildlife Expo (Charleston) . **Feb 17**
- Southern Cyclone: Anniv ... **Aug 24**
- Sun Fun Fest (Myrtle Beach) ... **June 2**
- Treasures by the Sea (Myrtle Beach) ... **Nov 1**

South Dakota
- Admission Day ... **Nov 2**
- Bicycle Trek Classic (Spearfish) ... **May 5**
- Black Hills Balloon Rally (Sturgis) ... **May 13**

Index ☆ Chase's 1995 Calendar of Events ☆

South Dakota (cont'd)—Spanish-American War

Black Hills Jeep Jamboree (Sturgis/ Deadwood) . . . Oct 6
Black Hills Roundup (Belle Fourche) . . . July 2
Black Hills Threshing Bee (Sturgis) . . . Aug 18
Buffalo Round-Up (Custer) . . . Oct 2
Buffalo Round-Up Bicycle Rally (Custer) Sept 30
Buffalo Wallow Chili Cookoff (Custer) . . . Oct 1
Center of Nation All-Car Rally (Belle Fourche) . . . June 9
Christmas at the Capitol (Pierre) . . . Nov 23
Corn Palace Fest (Mitchell) . . . Sept 8
Custer State Park Buffalo Auction (Custer) . . . Nov 18
Custer State Park Volksmarch (Custer) . . July 29
Czech Days (Tabor) . . . June 16
D.C. Booth Day (Spearfish) . . . May 21
Dalesburg Midsummer Festival (Vermillion) . . . June 23
Fishing Has No Boundaries (Pierre) . . . June 17
Fort Sisseton Historical Festival (Lake City) . . . June 3
Homesteader Harvest Festival (Brandon) . . . Sept 10
KCCR Farm, Home and Sport Show (Pierre) . . . Feb 25
Mount Rushmore July 4 Celeb (Mt Rushmore) . . . July 4
Native Americans' Day . . . Oct 9
Red Cloud Indian Art Show (Pine Ridge) . June 4
Schmeckfest (Freeman) . . . Mar 30
Sodbuster Fest (Webster) . . . June 10
Spearfish Cross Country Ski Challenge (Spearfish) . . . Feb 18
State Fair (Huron) . . . Aug 26
State Wild Game Cookoff (Oakwood Lakes State Park) . . . Aug 12
Sturgis Rally & Races (Strugis) . . . Aug 7
Tetonkaha Rendezvous (Bruce) . . . Aug 11
Yankton Riverboat Days/Summer Arts Fest (Yankton) . . . Aug 18
South Pole Discovery: Anniv . . . Dec 14
Southeastern Wildlife Expo (Charleston, SC) . . . Feb 17
Southern Festival of Books (Nashville, TN) . Oct 13
Soviet Rocket Threat: Anniv . . . July 9
Sowerby Centennial JubiLEO, Leo . . . May 1
Sowerby, Leo: Birth Anniv . . . May 1
Soyinka, Wole: Birth . . . July 13
Space — Excluding Space Milestones
Apollo I: Spacecraft Fire Anniv . . . Jan 27
Astronomy Day . . . May 6
Astronomy Week . . . Apr 10
Challenger Space Shuttle Explosion: Anniv . . . Jan 28
Christmas Greetings from Space: Anniv . Dec 19
Comet Crashes into Jupiter: Anniv . . . July 16
Cosmonauts' Day (Alamogordo, NM) . . . Apr 8
Discovery Launches Satellite of Discovery: Anniv . . . Sept 15
First American Woman in Space: Anniv . June 18
First Man in Space: Anniv . . . Apr 12
First Woman in Space: Anniv . . . June 16
First Woman to Walk in Space . . . July 17
Interplanetary Confederation Days . . . Oct 14
Jupiter Effect: Anniv . . . Mar 10
Near Miss Day . . . Mar 23
Ozark UFO Conf (Eureka Springs, AR) . . . Apr 7
Purple for Peace Day: Intergalactic Holiday . . . May 16
Soviet Cosmonaut Returns to New World: Anniv . . . Mar 26
Space Hall of Fame Induction, Intl (Alamogordo, NM) . . . Oct 7
Space Oddity Song Release: Anniv . . . June 11
Space Week . . . July 16
Spaceweek . . . July 16
US Space Exploration: Anniv . . . Jan 31
Space Milestone: First Man in Space: Anniv Apr 12
Space Milestones
Year 1 (1957)
Sputnik 1 . . . Oct 4
Sputnik 2 . . . Nov 3
Year 2 (1958)
Explorer I . . . Jan 31
Vanguard I . . . Mar 17
Year 3 (1959)
Luna 1 . . . Jan 2
Luna 2 . . . Sept 12
Luna 3 . . . Oct 4
Year 4 (1960)
Discover 13 . . . Aug 10
Echo 1 . . . Aug 12
Sputnik 5 . . . Aug 19
Year 5 (1961)
Project Mercury Test . . . Jan 31
Venera 1 . . . Feb 12
Sputnik 9 . . . Mar 9
Vostok 1 . . . Apr 12
Freedom 7 . . . May 5
Vostok 2 . . . Aug 6
Year 6 (1962)
Friendship 7 . . . Feb 20
Aurora 7 . . . May 24
Telstar . . . July 10
Vostok 3 . . . Aug 11
Year 7 (1963)
Faith 7 . . . May 15
Vostok 5 . . . June 14
Vostok 6 . . . June 16
Year 8 (1964)
Saturn SA-5 . . . Jan 29
Ranger 7 . . . July 28
Mariner 4 . . . Nov 28
Year 9 (1965)
Voskhod 2 . . . Mar 18
Gemini 4 . . . June 3
Gemini 5 . . . Aug 21
Pegasus 1 . . . Sept 17
Venera 3 . . . Nov 16
Year 10 (1966)
Luna 9 . . . Jan 31
Gemini 8 . . . Mar 16
Gemini 12 . . . Nov 11
Year 11 (1967)
Surveyor 3 . . . Apr 17
Venera 4 . . . June 12
Mariner 5 . . . June 14
Year 12 (1968)
OGO 5 . . . Mar 4
Soyuz 3 . . . Oct 26
Apollo 8 . . . Dec 21
Year 13 (1969)
Soyuz 4 . . . Jan 14
Apollo 10 . . . May 18
Apollo 11 . . . July 16
Moon Day . . . July 20
Apollo 12 . . . Nov 14
Year 14 (1970)
Osumi . . . Feb 11
Apollo 13 . . . Apr 11
China 1 . . . Apr 24
Soyuz 9 . . . June 1
Luna 16 . . . Sept 12
US/USSR Space Rescue Cooperation . Oct 28
Luna 17 . . . Nov 10
Year 15 (1971)
Apollo 14 . . . Jan 31
Soyuz 10 . . . Apr 23
Mars 2 and Mars 3 . . . May 19
Mariner 9 . . . May 30
Soyuz 11 . . . June 6
Apollo 15 . . . July 26
Intelsat 4 F-3 . . . Dec 19
Year 16 (1972)
Luna 20 . . . Feb 14
Pioneer 10 . . . Mar 2
Venera 8 . . . Mar 27
Apollo 16 . . . Apr 16
Copernicus, OAO 4 . . . Apr 21
Apollo 17 . . . Dec 7
Year 17 (1973)
Luna 21 . . . Jan 8
Skylab 2 . . . May 25
Skylab 3 . . . July 28
Intelstat-4 F-7 . . . Aug 23
Soyuz 12 . . . Sept 27
Skylab 4 . . . Nov 16
Soyuz 13 . . . Dec 18
Year 18 (1974)
NASA ATS-6 . . . May 30
Soyuz 14 . . . July 3
Soyuz 15 . . . Aug 26
Soyuz 16 . . . Dec 2
Year 19 (1975)
Soyuz 17 . . . Jan 10
Venera 9 and 10 . . . June 8
Apollo-Soyuz Test Project . . . July 15
Viking 1 and 2 . . . Aug 20
Year 20 (1976)
Soyuz 21 . . . July 6
Soyuz 22 . . . Sept 15
Soyuz 23 . . . Oct 14
Year 21 (1977)
Cosmos 954 . . . Jan 24
Soyuz 24 . . . Feb 7
Enterprise . . . Aug 12
Voyager 2 . . . Aug 20
Voyager 1 . . . Sept 5
Salyut 6 . . . Sept 29
Soyuz 25 . . . Oct 9
Soyuz 26 . . . Dec 10
Year 22 (1978)
Soyuz 27 . . . Jan 10
Soyuz 28 . . . Mar 2
Pioneer Venus I . . . May 20
Soyuz 29 . . . June 15
Soyuz 30 . . . June 27
Pioneer Venus . . . Aug 8
Soyuz 31 . . . Aug 26
Year 23 (1979)
Soyuz 32 . . . Feb 25
Skylab Falls to Earth . . . July 11
Year 24 (1980)
SMM . . . Feb 14
Soyuz 35 . . . Apr 9
Soyuz 36 . . . May 26
Soyuz T-2 . . . June 5
Rohini I . . . July 18
Soyuz 37 . . . July 23
Soyuz 38 . . . Sept 18
Soyuz T-3 . . . Nov 27
Year 25 (1981)
Soyuz T-4 . . . Mar 12
Soyuz 39 . . . Mar 22
STS-1 . . . Apr 12
Soyuz 40 . . . May 14
Ariane . . . June 19
STS-2 . . . Nov 12
Year 26 (1982)
STS-3 . . . Mar 22
Salyut 7 . . . Apr 19
Soyuz T-6 . . . June 24
STS-4 . . . June 27
Kosmos 1383 . . . July 1
Soyuz T-7 . . . Aug 19
Solyut 6 . . . Sept 29
STS-5 . . . Nov 11
Year 27 (1983)
NOAA-8 . . . Mar 28
STS-6 . . . Apr 4
STS-7 . . . June 18
Soyuz T-9 . . . June 27
STS-8 . . . Aug 30
STS-9 . . . Nov 28
Year 28 (1984)
STS-10 . . . Feb 3
STS-11 . . . Apr 6
Soyuz T-12 . . . July 17
Discovery . . . Aug 30
Record Time . . . Oct 2
Challenger . . . Oct 5
US Natl Comm . . . Oct 13
Vega I . . . Dec 15
Cosmos 1614 . . . Dec 19
Year 29 (1985)
Discovery US . . . Jan 24
Arabsat-1 . . . Feb 8
Brasilsat-1 . . . Feb 8
Discovery . . . Apr 12
Challenger . . . Apr 29
Year 30 (1986)
Mir Space Station . . . Feb 20
Arianne-2 . . . May 30
Year 31 (1987)
Soyuz TM-3 . . . July 22
Ariane-3 . . . Sept 15
Year 32 (1988)
Phobos 2 . . . July 12
Discovery . . . Sept 27
Discovery . . . Sept 29
Buran . . . Nov 15
Atlantis . . . Dec 1
Year 33 (1989)
Discovery . . . Mar 17
Atlantis . . . May 4
Voyager 2 . . . Aug 24
Year 34 (1990)
1st USSR Comm Satellite Mission . . . Feb 11
Hubble Space Telescope . . . Apr 25
Spacek, Sissy: Birth . . . Dec 25
Spaghetti Bridge Building Contest (Kelowna, BC, Canada) . . . Mar 10
Spain
Book Day and Lover's Day . . . Apr 23
National Holiday . . . Oct 12
Spanish American War
Maine Memorial Day . . . Feb 15
Matanzas Mule Day . . . Apr 27
Remember the Maine Day . . . Feb 15

☆ Chase's 1995 Calendar of Events ☆ Index

SPAR—Succoth

Spanish-American War Surrender of Guam to US: Anniv....June 20
SPAR Anniversary Week 6504....Page 488
Sparrow, Last Dusky Seaside: Death Anniv....June 16
Sparrow, Rory: Birth....June 12
Spassky, Boris: Birth....Jan 30
Speakes, Larry: Birth....Sept 13
SPEBSQSA Mid-Winter Convention (Tucson, AZ)....Jan 22
Special Librarians Day, Intl....Apr 13
Specter, Arlen: Birth....Feb 12
Spector, Phil: Birth....Dec 26
Speech Month, May Is Better Hearing and..May 1
Speech-Lang-Hearing Assn Conv, American (Cincinnati, OH)....Nov 16
Spelling Bee Finals, Natl....May 31
Spelvin, George: Day....Nov 15
Spengler, Oswald: Birth Anniv....May 29
Spielberg, Steven: Birth....Dec 18
Spieling Day, Natl....Jan 28
Spillane, Mickey: Birth....Mar 9
Spinach Fest (Alma, AR)....Apr 14
Spinks, Leon: Birth....July 11
Spinks, Michael: Birth....July 29
Spinning and Weaving Week, Natl....Oct 2
Spinoza, Baruch: Birth Anniv....Nov 24
Spirit of America (Decatur, AL)....July 3
Spirit of Freedom Fest (Florence, AL)....July 4
Spitz, Mark: Birth....Feb 10
Split Pea Soup Week, Natl....Nov 7
Spock, Benjamin: Birth....May 2
Spoon River Valley: Knox Cnty Scenic Dr (Galesburg, IL)....Oct 7
Spooner, William: Birth Anniv....July 22
Spooner's Day....July 22
Sport, Boat and RV Show, San Antonio (San Antonio, TX)....Jan 24
Sporting Assn World Sports Expo, Natl (Chicago, IL)....July 16
Sports Eye Safety Month....Apr 1
Sports, Physical Fitness and, Month, Natl...May 1
Sports: Sham El-Nesim (Egypt)....Apr 24
Sports Show, Milwaukee Sentinel (Milwaukee, WI)....Mar 10
Sports, Women and Girls in, Day 6527...Page 488
Sportscasters Hall of Fame, American (New York, NY)....Dec 7
Sportsfest (Oklahoma City, OK)....Feb 4
Spring begins....Mar 20
Spring Breaking-Up at Giant Ridge (Biwabik, MN)....May 20
Spring Day Celebration (Baltimore, MD)...Mar 20
Spring Festival (Lumpkin, GA)....Apr 6
Spring Festival, Intl (Lodi, CA)....May 20
Spring Festival, Towsontown (Towson, MD).May 6
Spring Fling (Wichita Falls, TX)....Apr 22
Spring Fling, Okefenokee (Waycross, GA)...Apr 8
Spring, Halfway Point....May 6
Spring Market Fair, Eighteenth-Century (McLean, VA)....May 20
Spring Pilgrimage to Antebellum Homes (Columbus, MS)....Mar 28
Spring Wildflower Pilgrimage (Gatlinburg, TN)....Apr 27
Springfest (Heber Springs, AR)....Apr 21
Springfest (Weatherford, OK)....May 6
Springfield, Rick: Birth....Aug 23
Springsteen, Bruce: Birth....Sept 23
Springsteen's First Album....May 22
Springtime Tallahassee (Tallahassee, FL)..Mar 15
Sprinkel, Beryl: Birth....Nov 20
Spruce Goose Flight Anniversary....Nov 2
Square Dance. See Dance....Aug 4
Squeeze Play First Used: Anniv....June 16
Squid Fest, Great Monterey (Monterey, CA)....May 27
Sri Lanka: Independence Day....Feb 4
Stack, Robert: Birth....Jan 13
Stage Days, Butterfield Overland (Benson, AZ)....Oct 13
Stagecoach Days (Marshall, TX)....May 20
Stahl, Lesley: Birth....Dec 16
Stalin, Joseph: Birth Anniv....Dec 21
Stallone, Sylvester: Birth....July 6
Stamos, John: Birth....Aug 19
Stamps
Americover (Irvine, CA)....Aug 4
Babe Ruth Commemoratiave: Anniv....July 6
Fall Postage Stamp Mega Event (New York, NY)....Nov 2
Postage Stamps, First Adhesive US....July 1
Precanex (Wildwood, NJ)....Aug 16
Spring Postage Stamp Mega Event (New York, NY)....Mar 16
Stamp Expo (Anaheim, CA)....Mar 24, Sept 15
Stamp Expo (Long Beach, CA)....Jan 6
Stamp Expo (Los Angeles, CA)....Apr 7
Stamp Expo (Palm Springs, CA)....Feb 17
Stamp Expo (Pasadena, CA)....Feb 17, July 14, Nov 10
Stamp Expo America (Anaheim, CA)....Nov 24
Stamp Expo USA (Anaheim, CA)....Feb 10
Stamp Expo: Anaheim (Anaheim, CA)...Oct 13
Stamp Expo: California (Anaheim, CA)..Aug 11
Stamp Expo: Fall (Los Angeles, CA)....Nov 17
Stamp Expo: South (Anaheim, CA)....Apr 28
Stamp Expo: West (Anaheim, CA)....Jan 13
Summer Stamp Expo (Anaheim, CA)....July 14
Standard Time Act, US: Anniv....Mar 19
Stanhope, Philip D.: Birth Anniv....Sept 22
Stanley, Henry Morton: Birth Anniv....Jan 28
Stanton, Elizabeth: Birth Anniv....Nov 12
Stanwyck, Barbara: Birth Anniv....July 16
Stapleton, Jean: Birth....Jan 19
Stapleton, Maureen: Birth....June 21
Star Festival (Tanabata) (Japan)....July 7
Star of Wonder Planetarium Show (Newport News, VA)....Jan 2
Star Trek: Anniv....Sept 8
Star Trek Fest (Riverside, IA)....June 24
Star Trek: Future Birthday of James T. Kirk (Riverside, IA)....Mar 22
Star-athon Telethon for Cerebral Palsy....Jan 21
Star-Spangled Banner Inspired: Anniv....Sept 13
Stark, USS Attack: Anniv....May 17
Starker, Janos: Birth....July 5
Starr, Bart: Birth....Jan 9
Starr, Ringo: Birth....July 7
Stars and Stripes Forever Day....May 14
State Dept, US: Birth....July 27
State Fairs. See Agriculture
States, St. Petersburg Fest of (St. Petersburg, FL)....Mar 17
Statue of Liberty: Dedication Anniv....Oct 28
Staub, Rusty: Birth....Apr 1
Staubach, Roger: Birth....Feb 5
Stay Away from Seattle Day....Sept 16
Stay Home Because You're Well Day....Nov 30
Stay Out of the Sun Day....July 3
Steamboat Days Fest (Jeffersonville, IN)...Sept 8
Steel, Danielle: Birth....Aug 14
Steele, Tommy: Birth....Dec 17
Steelmark Month (Pres Proc)....May 1
Steeplechase at Callaway Gardens (Pine Mountain, GA)....Nov 4
Steichen, Edward: Birth Anniv....Mar 27
Steiger, Rod: Birth....Apr 14
Stein, Gertrude: Birth Anniv....Feb 3
Steinbeck Fest (Salinas, CA)....Aug 3
Steinberg, David: Birth....Aug 9
Steinberg, Saul: Birth....June 15
Steinbrenner, George: Birth....July 4
Steinem, Gloria: Birth....Mar 25
Stella, Frank: Birth....May 12
Stendhal: Birth Anniv....Jan 23
Stephens, James: Birth....May 18
Stephens, Robert: Birth....July 14
Stephenson, George: Birth Anniv....June 9
Stephenson, Jan: Birth....Dec 22
Stern, Isaac: Birth....July 21
Sterne, Laurence: Birth Anniv....Nov 24
Sternwheel. See Boats/Ships
Sternwheel Fest, Ohio River (Marietta, OH).Sept 8
Stevens, Cat: Birth....July 21
Stevens, Connie: Birth....Aug 8
Stevens, John Paul: Birth....Apr 20
Stevens, Stella: Birth....Oct 1
Stevens, Ted: Birth....Nov 18
Stevenson, Adlai: Birth....Oct 23
Stevenson, Adlai Ewing: Birth Anniv....Feb 5
Stevenson, McLean: Birth....Nov 14
Stevenson, Robert L: Birth Anniv....Nov 13
Stewart, Don: Birth....Nov 14
Stewart, Jackie: Birth....June 11
Stewart, James "Jimmy": Birth....May 20
Stewart, Potter: Birth Anniv....Jan 23
Stewart, Rod: Birth....Jan 10
Stiers, David Ogden: Birth....Oct 31
Stiftungsfest (Young America, MN)....Aug 25
Stiller, Jerry: Birth....June 8
Stills, Stephen: Birth....Jan 3
Sting: Birth....Oct 2
Stock Exchange Holiday..Jan 2, Feb 20, April 14, May 29, July 4, Sept 4, Nov 23, Dec 25
Stock Exchange, NY, Established: Anniv...May 17
Stock Market Crash (1929): Anniv....Oct 29
Stock Market Panic: Anniv....Oct 24
Stock: Own Your Own Share of America Day....June 15
Stock Show and Rodeo, Natl Western (Denver, CO)....Jan 10
Stockings, Nylon: Anniv....May 15
Stocks: Dow-Jones Tops 1,000: Anniv....Nov 14
Stocks: Own Your Share of America Month June 1
Stockton, David: Birth....Nov 2
Stockton, John: Birth....Mar 26
Stockton, Richard: Birth Anniv....Oct 1
Stockwell, Dean: Birth....Mar 5
Stokes, Carl: Becomes First Black Mayor in US: Anniv....Nov 13
Stone, Harland Fiske: Birth Anniv....Oct 11
Stone, Lucy, Married: Anniv....May 1
Stone, Oliver: Birth....Sept 15
Stone, Sly: Birth....Mar 15
Stone, Steve: Birth....July 14
Stone, Thomas: Death Anniv....Oct 5
Stookey, Paul: Birth....Nov 30
Stop the Violence Day, Natl....Nov 22
Stop Violence/Save Kids Month....Sept 1
Stoppard, Tom: Birth....July 3
Storey, David: Birth....July 13
Storm, Gale: Birth....Apr 5
Story, Joseph: Birth Anniv....Sept 18
Storytelling
A(ugusta) Baker's Dozen (Columbia, SC) Apr 28
Connecticut Storytelling Fest (New London, CT)....Apr 21
Corn Island Storytelling Fest (Louisville, KY)....Sept 15
Ghost Tales Around Campfire (Washington, MS)....Oct 27
Halloween Harvest of Horrors (Louisville, KY)....Oct 20
N Appalachian Storytelling Fest (Mansfield, PA)....Sept 15
Rockford Storytelling Fest (Rockford, IL) Sept 15
St. Louis Storytelling Fest (St. Louis, MO). May 3
Spring Ghost Tales (Louisville, KY)....May 12
Storytelling Fest, Natl (Jonesborough, TN) Oct 6
Summer Storyteller-in-Residence (Atlanta, GA)....June 13
Tellabration Storytelling/Grown-ups (New London, CT)....Nov 17
Welsh Intl Fest Storytelling (Wales)....June 30
Yukon Intl Storytelling Fest (Whitehorse, Yukon)....June 23
Stradivari, Antonio: Death Anniv....Dec 18
Straight, Beatrice: Birth....Aug 2
Strange, Curtis: Birth....Jan 30
Strasberg, Susan: Birth....May 22
Strassman, Marcia: Birth....Apr 28
Stratemeyer, Edward L.: Birth Anniv....Oct 4
Stratton, Dorothy C.: Birthday....Mar 24
Stratton, Mary's Birthday (Detroit, MI)....Mar 15
Strauss, Levi: Birth Anniv....Feb 26
Strauss, Richard G: Birth Anniv....June 11
Stravinsky, Igor F.: Birth Anniv....June 17
Straw Hat Week....Apr 2
Strawberry, Darryl: Birth....Mar 12
Strawberry Fest, California (Oxnard, CA)..May 20
Strawberry Fest, Poteet (Poteet, TX)....Apr 7
Strawberry Fest, West Virginia (Buckhannon, WV)....May 17
Strawberry Fest/Hillsborough County Fair (Plant City, FL)....Mar 2
Strawberry Hill Races (Richmond, VA)....Apr 15
Strawberry Month, Natl....May 1
Strawberry 100 Bike Tour/Fest (Crawfordsville, IN)....June 9
Streep, Meryl: Birth....June 22
Streeter, Ruth Cheney: Birth Anniv....Oct 2
Streisand, Barbra: Birth....Apr 24
Stress Awareness Day, Natl....Apr 16
Stress Awareness Month....Apr 1
Strike Ends, First Baseball: Anniv....Apr 13
Stritch, Elaine: Birth....Feb 2
Stroke Awareness Month....May 1
Struthers, Sally Anne: Birth....July 28
Stuart, Gilbert: Birth Anniv....Dec 3
Student: See Education
Student Volunteer Day....Feb 20
Students' Memorial Day, Kent State....May 4
Styron, William: Birth....June 11
Submarine: Anniv of First Nuclear-Powered Voyage....Jan 17
Subway Accident, New York: Anniv....Nov 2
Subway, New York City: Subway Anniv....Oct 27
Succoth....Oct 9

Index ☆ Chase's 1995 Calendar of Events ☆

Sudan—Television

Sudan: Independence Day ... Jan 1
Sudden Infant Death Syndrome Awareness Month, Natl ... Oct 1
Suez Canal Anniv ... Nov 17
Suez Canal: Evacuation Day (Egypt) ... June 18
Suffrage Parade Attached, Woman: Anniv ... Mar 3
Sugar Bowl, USF&G (New Orleans, LA) ... Jan 2
Sugar Festival, Michigan (Sebewaing, MI) . June 23
Suhey, Matt: Birth ... July 7
Sukkot ... Oct 9
Sukkot Begins ... Oct 8
Sullivan, Anne: Birth Anniv ... Apr 14
Sullivan, Arthur: Birth Anniv ... May 13
Sullivan, Barry: Birth ... Aug 29
Sullivan, Louis Wade: Birth ... Nov 3
Sullivan, Mike: Birth ... Sept 22
Sullivan, Susan: Birth ... Nov 18
Sultana Explosion Anniv ... Apr 27
Sumac, Yma: Birth ... Sept 10
Summer Arrival: Martin Z. Mollusk Day (Ocean City, NJ) ... May 4
Summer Begins ... June 21
Summer Daylight-Saving Time (Europe) ... Mar 26
Summer, Donna: Birth ... Dec 31
Summer Fest, Quebec Intl (Quebec, Que, Canada) ... July 6
Summer Fest, Swanton (Swanton, VT) ... July 27
Summer, Halfway Point ... Aug 7
Summer: Joyful Summer Celebration ... June 10
Summer Market Fair, Eighteenth-Century (McLean, VA) ... July 15
Summer Music Fest (Sitka, AK) ... June 2
Summer Solstice at Carhenge (Alliance, NE) ... July 29
Summer Sunfest (Covington, KY) ... July 14
Summer Time (England) ... Mar 26
Summerfest (Harrisonburg, VA) ... July 30
Summerfolk Fest, Owen Sound (Owen Sound, Ont, Canada) ... Aug 18
Sun, Cancer in the Sun, Month ... June 1
Sun Day, Stay Out of the ... July 3
Sun Fun Fest (Myrtle Beach, SC) ... June 2
Sun: May Ray Day ... May 19
Sun Yat-Sen: Birth Anniv ... Nov 12
Sun Yat-Sen: Death Anniv ... Mar 12
Sundlun, Bruce: Birth ... Jan 20
Sung, Kil Il: Death Anniv ... July 8
Sunrise Fest (Inuvik, NWT, Canada) ... Jan 5
Sununu, John H.: Birth ... July 2
Super Bowl XXIX (Miami, FL) ... Jan 29
Superman Celebration (Metropolis, IL) ... June 15
Supreme Court Bans Official Prayer: Anniv ... June 25
Supreme Court: Brown v Board of Education: Anniv ... May 17
Supreme Court: Roe v Wade Decision: Anniv ... Jan 22
Supreme Court Term Begins ... Oct 2
Supreme Court: Wm Brennan Resigns: Anniv ... July 20
Surfing Chmpshp, ESA Mid-Winter (Narragansett, RI) ... Feb 18
Surfing: East Coast Surfing Chmp/Fest (Virginia Bch, VA) ... Aug 23
Surfing: Triple Crown: Hawaiian Pro Men's (Oahu, HI) ... Nov 14
Surfside Salutes Canada Week (Surfside, FL) ... Mar 5
Surfside's Intl Fiesta (Surfside, FL) ... Aug 5
Suriname
  Independence Day ... Nov 25
  Revolution Day ... Feb 25
Surratt, Mary: Execution Anniv ... July 7
Sutcliffe, Rick: Birth ... June 21
Sutherland, Donald: Birth ... July 17
Sutherland, Joan: Birth ... Nov 7
Swallows Depart San Juan Capistano ... Oct 23
Swallows Return to San Juan Capistrano . Mar 19
Swan, Joseph W: Birth Anniv ... Oct 31
Swanson, Gloria: Birth Anniv ... Mar 27
Swap Ideas Day ... Sept 10
Swappin' Meetin' (Cumberland, KY) ... Oct 6
Swayze, Patrick: Birth ... Aug 18
Swaziland: Independence Day ... Sept 6
Sweden
  All Saint's Day ... Nov 4
  Crayfish Premiere ... Aug 9
  Feast of Valborg ... Apr 30
  Flag Day ... June 6
  Gustavus Adolphus Day ... Nov 6
  Linnaeus Day (Stenbrohult) ... May 23
  Midsummer ... June 23
  Nobel Prize Awards Ceremony (Stockholm) ... Dec 10
  Sour Herring Premiere ... Aug 17
  St. Martin's Day ... Nov 11
  Santa Lucia Day ... Dec 13
Swedenborg, Emanuel: Birth Anniv ... Jan 29
Swedish Fest, Oakland (Oakland, NE) ... June 2
Sweetcorn Fest, Natl (Hoopeston, IL) ... Aug 31
Sweetest Day ... Oct 21
Swift, Jonathan: Birth Anniv ... Nov 30
Swimming
  Big 10 Men's Swimming/Diving Chmpshp Feb 23
  Columbia River Cross Channel Swim (Hood River, OR) ... Sept 4
  Hawaiian Christmas Rough Water Swim (Honolulu, HI) ... Dec 17
  Mid-Summer Ocean Swim/Beach Barbecue (Seaside Pk, NJ) ... Aug 19
  NAIA Swimming and Diving Chmpshps ... Mar 1
  NCAA Men's Div I Swim/Dive Chmpshps (Indianapolis, IN) ... Mar 23
  NCAA Women's Swimming/Diving Chmpshp (Austin, TX) ... Mar 16
  Penguin Plunge (Jamestown, RI) ... Jan 1
  Polar Bear (Sheboyan, WI) ... Jan 1
  Polar Bear Jump Fest (Seward, AK) ... Jan 20
  Recreational Scuba Diving Week, Natl ... Aug 13
  Seaspace (Houston, TX) ... June 2
  Swimming School Opens, First US ... July 23
  Waikiki Roughwater Swim (Honolulu, HI) . Sept 4
  Women's Big 10 Swimming/Diving Chmpshp (Indpls, IN) ... Feb 16
Swine Time (Climax, GA) ... Nov 25
Swiss Fest, Ohio (Sugarcreek, OH) ... Sept 29
Swiss: Volksfest (New Glarus, WI) ... Aug 6
Swit, Loretta: Birth ... Nov 4
Switzerland
  Chalandra Marz ... Mar 1
  Dornach Battle Commemoration ... July 23
  Egg Races ... Apr 17
  Homstrom (Scuol) ... Feb 5
  International Festival of Music (Lucerne) Aug 16
  Martinmas Goose ... Nov 11
  Meitlisunntig ... Jan 8
  Nafels Pilgrimage (Canton Glarus) ... Apr 6
  National Day ... Aug 1
  Onion Market (Zibelemarit) ... Nov 27
  Pacing the Bounds (Liestal) ... May 22
  St. Gotthard Auto Tunnel Opened ... Sept 5
  Sempach Battle Commemoration ... July 10
Symington, Fife: Birth ... Aug 12
Synge, John M.: Birth Anniv ... Apr 16
Syrian Arab Republic
  Independence Day ... Apr 17
  Revolution Day ... Mar 8
Szold, Henrietta: Birth Anniv ... Dec 21
Szymanowski, Karol: Birth Anniv ... Oct 6

T

T, Mr.: Birth ... May 21
Ta'anit Esther (Fast of Esther) ... Mar 15
Tabernacles, Feast of: First Day ... Oct 9
Tabori, Kristofer: Birth ... Aug 4
Taft, Helen Herron: Birth Anniv ... Jan 2
Taft, Howard: Taft Opened Baseball Season: Anniv ... Apr 14
Taft, William H.: Birth Anniv ... Sept 15
Tagore, Rabindranath: Birth Anniv ... May 6
Tahiti: Tahiti and Her Islands Awareness Month ... July 1
Taiwan
  Birthday of Kuan Yin, Goddess of Mercy . Mar 19
  Cheng Cheng Kung Birth Anniv ... Aug 9
  Cheng Cheng Kung Landing Day ... Apr 29
  Chiang Kai-Shek Day ... Oct 31
  Confucius's Birthday and Teachers' Day Sept 28
  Lantern Fest and Tourism Day ... Feb 14
  Overseas Chinese Day ... Oct 21
  Retrocession Day ... Oct 25
  Tomb-Sweeping Day, Natl ... Apr 5
  Tourism Week ... Feb 11
  Youth Day ... Mar 29
Tajikistan: Independence Day ... Sept 9
Take Your Houseplants for a Walk Day ... July 27
Talese, Gay: Birth ... Feb 7
Talk: Natl Spieling Day ... Jan 28
Talk With Your Teen About Sex Month, Natl . Mar 1
Talking Book Week ... Apr 17
Tallchief, Maria: Birth ... Jan 24
Tambo, Oliver Reginald: Birth Anniv ... Oct 27
Tammuz, Fast of ... July 16
Tandy, Jessica: Birth ... June 7
Tannehill Village Christmas (McCalla, AL) ... Dec 4
Tanner, Henry Ossawa: Birth Anniv ... June 21
Tanner, Roscoe: Birth ... Oct 15
Tanzania
  CCM Day ... Feb 5
  Independence and Republic Day ... Dec 9
  Saba Saba Day ... July 7
  Union Day ... Apr 26
  Zanzibar Revolution Day ... Jan 12
Tap Dance Day, Natl ... May 25
Tarbell, Ida M: Birth Anniv ... Nov 5
Tarkenton, Fran: Birth ... Feb 3
Tarkington, Booth: Birth Anniv ... July 29
Tarpley, Roy: Birth ... Nov 28
Tartikoff, Brandon: Birth ... Jan 13
Tasso, Torquato: Birth Anniv ... Mar 11
Taste of Madison (Madison, WI) ... Sept 2
Tater Days (Benton, KY) ... Apr 1
Tattoo, Nova Scotia Intl (Halifax, NS, Canada) ... July 1
Taurus Begins ... Apr 20
Tax. See Income Tax Due Date
Tax Protest: Vivien Kellems Memorial Day . June 7
Taylor, Billy: Birth ... July 24
Taylor, Elizabeth: Birth ... Feb 27
Taylor, George: Death Anniv ... Feb 23
Taylor, James: Birth ... Mar 12
Taylor, Lawrence: Birth ... Feb 4
Taylor, Lucy Hobbs: Birth Anniv ... Mar 14
Taylor, Margaret S.: Birth Anniv ... Sept 21
Taylor, Paul: Birth ... July 29
Taylor, Rod: Birth ... Jan 11
Taylor, Zachary: Birth Anniv ... Nov 24
Tchaikovsky, Peter Ilich: Birth Anniv ... Nov 6
Tea Month, Natl Hot ... Jan 1
Tea Party Fest, Chestertown (Chestertown, MD) ... May 27
Teacher. See Education
Teacher Appreciation, Natl ... Apr 7
Teacher Appreciation Week, PTA ... May 7
Teacher Day, Elementary School ... Jan 16
Teacher Day, Natl ... May 9
Teacher, Day of the (El Dia Del Maestro) . May 10
Teacher "Thank You" Week ... June 4
Teachers' Day, Confucius's Birthday and (Taiwan) ... Sept 28
Tebaldi, Renata: Birth ... Jan 2
Tebbit, Norman: Birth ... Mar 29
Technology, Intl Fest Science/ (Edinburgh, Scotland) ... Mar 23
Tecumseh: Death Anniv ... Oct 5
Tecumseh! Epic Outdoor Drama (Chillicothe, OH) ... June 9
Teddy Bear Conv (Jenks, OK) ... Nov 11
Teddy Bear Fest, Newport (Newport, RI) ... Aug 20
Teddy Bear Jamboree Show and Sale, Great (Bristol, CT) ... Oct 21
Teddy Bear Rally, Amherst's (Amherst, MA) . Aug 5
Teddy Bear Tea (Barstow, CA) ... Feb 11
Teddy Bear's Picnic (Lahaska, PA) ... July 15
Teen Day, Good, Natl 6648 ... Page 488
Teicher, Louis: Birth ... Aug 24
Teilhard De Chardin, Pierre: Birth Anniv ... May 1
Telecommunication Day, World (UN) ... May 17
Telecommunications Week, Intl PBX ... June 4
Telegram, Singing: Birthday ... July 28
Telegraph, Atlantic Cable Laid: Anniv ... July 27
Telegraph Line, Morse Opens First US: Anniv ... May 24
Telephone Anniv ... Mar 10
Telephone Operator, First: Emma M. Nutt Day ... Sept 1
Television
  Ball, Lucille: Birth Anniv ... Aug 6
  Bullwinkle Anniv; First Broadcast Anniv . Sept 24
  Court TV Debut: Anniv ... July 1
  Dove Awards ... Apr 27
  Dr. Who Premiere: Anniv ... Nov 23
  Fearless Forecasts of TV's Fall Flops ... Sept 11
  First Baseball Games Televised: Anniv ... Aug 26
  First TV Presidential News Conf: Anniv . . Jan 25
  Guiding Light: Anniv ... June 30
  Magnum, P.I. Series Premiere Anniv ... Dec 11
  Mary Tyler Moore Show: Final Episode Anniv ... Mar 19
  M*A*S*H: The Final Episode: Anniv ... Feb 28
  Middlemark, Marvin: Birth Anniv ... Sept 16
  Newton Minow Birthday ... Jan 17
  Ozzie and Harriet Show Radio Debut: Anniv ... Oct 8
  Simpsons Premiere: Anniv ... Jan 14
  Smothers Brothers' Firing: Anniv ... Apr 4
  Soap Opera Fan Fair (Mackinaw City, MI) June 1
  Society for Eradication of TV Conv (San Diego, CA) ... Sept 22
  Star Trek: Anniv ... Sept 8
  Star-athon Telethon for Cerebral Palsy ... Jan 21

☆ Chase's 1995 Calendar of Events ☆ Index

Television Academy Hall of Fame List. See contents
Television Academy Hall of Fame First Inductees...........Mar 4
There Went Johnny Night...........May 22
TV Talk Show Host Day...........Oct 23
TV Turn-Off (Plymouth, MA)...........Mar 5
Twilight Zone, The: The First Episode...........Oct 2
Ukrainian Famine Film Broadcast:Anniv...........Nov 30
Water Closet Incident: Anniv...........Feb 11
Welk, Lawrence: Birth Anniv...........Mar 11
World's Largest Concert...........Mar 2
Tell Someone They're Doing a Good Job Week...........Dec 17
Tell, Wilhelm, Fest (New Glarus, WI)...........Sept 2
Teller, Edward: Birth...........Jan 15
Temperature, Canada's Coldest Recorded: Anniv...........Jan 31
Ten-Four Day...........Oct 4
Tennessee
Activity Professionals Conv, Natl Assn (Nashvil)...........Mar 29
Admission Day...........June 1
African Cultural Ball (Chattanooga)...........Feb 3
African-American Cultural Festival (Chattanooga)...........Aug 19
American Soc Assn Execs Mgmt/Mtgs Forum (Nashville)...........Mar 12
Battle of Chattanooga: Anniv...........Nov 24
Belle Meade Plantation Fall Fest (Nashville)...........Sept 16
Bessie Smith Traditional Jazz Fest (Chattanooga)...........May 5
Cherokee Days of Recognition (Cleveland)...........Aug 5
Chocolate Fest (Knoxville)...........Jan 14
Christmas on the River (Chattanooga)...........Dec 2
Christmas Past at Audubon Acres (Chattanooga)...........Dec 1
Country Music Fan Fair, Intl (Nashville)...........June 5
Covered Bridge Celebration (Elizabethton)...........June 5
Crafts Fair (Nashville)...........May 5
Destiny in Dayton—Scopes Play (Dayton)...........July 20
Dogwood Arts Fest (Knoxville)...........Apr 7
Dogwood Arts Fest (Maryville)...........Apr 6
Elvis Week (Memphis)...........Aug 11
Fall Color Cruise and Folk Fest (Chattanooga)...........Oct 21
Fall Powwow and Fest (Lebanon)...........Oct 6
Festival in the Schools (Nashville area)...........Oct 9
Fiddlers Jamboree and Crafts Fest (Smithville)...........June 30
Field Trial Chmp (bird dog), Natl (Gr Junction)...........Feb 13
450-Mile Outdoor Fest...........Aug 17
Gatlinburg Scottish Fest (Gatlinburg)...........May 18
Gatlinburg's July 4th Midnight Parade (Gatlinburg)...........July 4
Goat Days (Millington)...........Sept 15
Gospel Music (Nashville)...........Apr 23
Iroquois Steeplechase (Nashville)...........May 13
Italian Street Fair (Brentwood)...........Sept 1
Jonesborough Days (Jonesborough)...........July 1
Jubilee—Grand Celeb (Chattanooga)...........Apr 28
Longhorn World Chmpshp Rodeo (Chattanooga)...........Mar 10
Longhorn World Chmpshp Rodeo (Nashville)...........Nov 17
Mid-South Fair (Memphis)...........Sept 22
Mountaineer Folk Fest (Pikeville)...........Sept 9
Museum of Appalachia TN Fall Homecoming (Norris)...........Oct 11
NAIA Wom Div I Basketball Chmpshp (Jackson)...........Mar 15
Nashville's Country Holidays (Nashville)...........Nov 1
NCAA Div I Men's/Women's Track Chmpshp (Knoxville)...........May 31
Nursing Conf Pediatric Primary Care (Nashville)...........Mar 14
Oak Ridge Atomic Plant Begun: Anniv...........Aug 1
Pat Boone Celebrity Spectacular (Chattanooga)...........May 4
Polk County Ramp Tramp Fest (Benton)...........Apr 22
Read Me Day...........Apr 26
Reelfoot Eagle Tours (Tiptonville)...........Jan 1
River Bend (Chattanooga)...........May 19
Riverbend Festival (Chattanooga)...........June 16
Roan Mountain Naturalists Rally (Elizabethton)...........Sept 9
Roan Mountain State Park Fall Fest (Roan Mountain)...........Sept 16
Roan Mountain State Park's Fireworks Jamboree (Roan Mountain)...........July 4
Roan Mountain Wildflower Tours/Birdwalks (Roan Mountain)...........May 6
Rock City Fairy Tale Fest (Rock City)...........Aug 12
Sewanee Concerts and Fest (Sewanee)...........June 24
Smoky Mountain Lights/Winterfest (Gatlinburg)...........Nov 15
Southeastern Sports-A-Rama (Chattanooga)...........Aug 11
Southern Festival of Books (Nashville)...........Oct 13
Spring Wildflower Pilgrimage (Gatlinburg)...........Apr 27
State Fair (Nashville)...........Sept 8
Storytelling Fest, Natl (Jonesborough)...........Oct 6
Tennessee Walking Horse Natl Celeb (Shelbyville)...........Aug 24
Tennis: Indoor Chmpshp (Memphis)...........Feb 13
Try This On: Clothing/Gender/Power (Clarksville)...........Nov 11
Wild Foods Day (Pikeville)...........Oct 28
Tenniel, John: Birth Anniv...........Feb 28
Tennille, Toni: Birth...........May 8
Tennis
Big Eight Men/Women Chmpshp (Oklahoma City, OK)...........Apr 21
Big Ten Men's Tennis Team Chmpshp (Bloomington, IN)...........Apr 27
Lawn Tennis Chmpshps at Wimbledon (London, England)...........June 26
NAIA Men's Tennis Chmpshps (Tulsa, OK)...........May 22
NAIA Women's Tennis Chmpshps (Tulsa, OK)...........May 22
Tennis: Indoor Chmpshp (Memphis, TN)...........Feb 13
Tennis Month, Natl...........July 1
Wheelchair Tennis Tourn (Greensboro, NC)...........Sept 8
Tennyson, Alfred Lord: Birth Anniv...........Aug 6
Tennyson, Alfred Lord: Death Anniv...........Oct 6
Tereshkova-Nikolaeva, Valentina: Birth...........Mar 6
Terkel, Studs: Birth...........May 16
Territorial Days (Prescott, AZ)...........June 10
Test Tube Baby Birthday...........July 25
Testaverde, Vinny: Birth...........Nov 13
Tet Offensive Begins: Anniv...........Jan 30
Tetonkaha Rendezvous (Bruce, SD)...........Aug 11
Texas
Admission Day...........Dec 29
African-American History and Heritage Tour (Dallas)...........Feb 4
Africa's Legacy in Mexico: Photos (San Antonio)...........June 17
Aggiecon (College Station)...........Mar 23
Air Show and Fireworks Display (Rockport)...........July 4
Antique Auto Swap Meet (Grand Prairie)...........June 10
Brazos Nights (Waco)...........June 9
Buffalo Beano Kite Fly/Frisbee Fling (Lubbock)...........Apr 22
Candlelight (Dallas)...........Dec 7
Charro Days (Brownsville)...........Feb 23
Cherokee Rose Fest/Tours (Gilmer)...........May 20
Chili Cookoff/Quail-Egg Eat, Prairie Dog (Gr Prairie)...........Apr 1
Chisholm Trail Round-Up Fest (Ft Worth)...........June 9
Cinco de Mayo Fiesta (Grand Prairie)...........May 7
Comal County Fair (New Braunfels)...........Sept 26
Come and Take It Day (Gonzales)...........Oct 6
Conference on Student Govt Assns (College Station)...........Feb 18
Copperas Cove Crime Stoppers Dog Show (Copperas Cove)...........Apr 30
Cotton Bowl Classic, Mobil (Dallas)...........Jan 2
Country Fair on Square/All Car Show (Gainesville)...........Oct 7
Dairy Day (Dallas)...........June 3
Day in the Warsaw Ghetto (Houston)...........Feb 4
Depot Days (Gainesville)...........May 19
Dickens on the Strand (Galveston)...........Dec 2
East Texas Walking/Racking Horse Show (Marshall)...........Mar 1
East Texas Yamboree (Gilmer)...........Oct 19
Electra Goat BBQ Cook-Off (Electra)...........May 13
Emancipation Day...........June 19
Entomological Soc of America, Annual Mtg (Dallas)...........Dec 13
Festival of Flags (Killeen)...........May 27
Fiesta San Antonio (San Antonio)...........Apr 21
Fireant Fest (Marshall)...........Oct 6
First Thanksgiving Festival (El Paso)...........Apr 29
Fulton Oysterfest (Fulton)...........Mar 3
Funfest (Amarillo)...........May 27
Global Fest (Waco)...........Apr 7
Grand Prairie Western Days (Grand Prairie)...........May 18
Great Texas Mosquito Fest (Clute)...........July 27
Harvest Weekends (Bryan)...........July 15
Hotter 'n Hell Hundred Weekend (Wichita Falls)...........Aug 24
Houston-Tenneco Marathon (Houston)...........Jan 15
Houston-Tenneco Marathon Running World Expo (Houston)...........Jan 13
Hummer/Bird Celebration (Rockport, Fulton)...........Sept 7
Independence Day...........Mar 2
International Week (College Station)...........Feb 27
Jalapeno Fest (Laredo)...........Feb 17
Josey's World Champion Jr Barrel Race (Marshall)...........May 5
Kids Bridge Exhibition (Houston)...........Oct 28
La Posada de Kingsville/Celeb of Lights (Kingsville)...........Nov 18
Louis Armstrong: Cultural Legacy (Dallas)...........Jan 28
MADD Leadership Development Conf (Dallas)...........May 11
Marfa Lights Fest (Marfa)...........Sept 1
Microbrewers Conference and Trade Show, Natl (Austin)...........Apr 23
MSC Student Conf Natl Affairs (College Station)...........Feb 17
Mustang League World Series (Irving)...........Aug 9
Muster (College Station)...........Apr 21
Natl Chmpshp Indian Powwow (Grand Prairie)...........Sept 8
Night in Old Luling (Luling)...........Oct 14
O. Henry Pun-Off (Austin)...........May 7
Oatmeal Fest (Bertram/Oatmeal)...........Sept 1
Oktoberfest (Grand Prairie)...........Oct 14
Old-Fashioned Fourth (Dallas)...........July 4
Possum Pedal 100 Bicycle Ride/Race (Graham)...........Mar 25
Poteet Strawberry Festival (Poteet)...........Apr 7
PTA Membership Enrollment Month in Texas...........Sept 1
Quadrangle Fest (Texarkana)...........Sept 9
Ranch Hand Breakfast (Kingsville)...........Nov 18
Reenactment of Cowtown's Last Gunfight (Fort Worth)...........Feb 8
Rio Grande Valley Livestock Show (Mercedes)...........Mar 15
Rockport Art Festival (Rockport)...........July 1
Rockport Seafair (Rockport)...........Oct 7
St. Patrick's Day Beard Growing Contest (Shamrock)...........Mar 17
San Antonio Sport, Boat and RV Show (San Antonio)...........Jan 24
San Jacinto Day...........Apr 21
Seaspace (Houston)...........June 2
Southwestern Expo Livestock Show/Rodeo (Ft Worth)...........Jan 20
Spring Fling (Wichita Falls)...........Apr 22
Stagecoach Days (Marshall)...........May 20
Southwestern Black Student Leadership Conf (College Station)...........Jan 19
Texas Aggie Bonfire (College Station)...........Nov 22
Texas Citrus Fiesta (Mission)...........Jan 30
Texas Cowboy Poetry Gathering (Alpine)...........Mar 3
Texas Folklife Fest (San Antonio)...........Aug 3
Texas Heritage Day (Dallas)...........Sept 23
Texas Musical Drama (Canyon)...........June 7
Texas PTA Conv (Austin)...........Nov 10
Texas Ranch Roundup (Wichita Falls)...........Aug 18
Texas State Chmpshp Domino Tournament (Hallettsville)...........Jan 22
Texas State Chmpshp Fiddlers Frolics (Hallettsville)...........Apr 20
Texas, State Fair of (Dallas)...........Sept 28
US Natl Senior Olympics (San Antonio)...........May 17
Waco Wind Fest (Waco)...........Mar 18
Washington's Birthday Celeb (Laredo)...........Feb 9
Watermelon Thump (Luling)...........June 22
Westfest (West)...........Sept 2
Wine & Roses Fest (Bryan)...........May 6
Winefest (Bryan)...........Nov 11
Winslow Homer engravings exhibit (Austin)...........Jan 23
Wonderland of Lights (Marshall)...........Nov 22
World Chmpshp BBQ Goat Cookoff/Crafts Fair (Brady)...........Sept 2
Wurstfest (New Braunfels)...........Nov 3
Thai Heritage Month...........Apr 1
Thailand
Birth of the Queen...........Aug 12
Chakri Day...........Apr 6
Chulalongkorn Day...........Oct 23
Constitution Day...........Dec 10
Coronation Day: Anniv...........May 5

Index ☆ Chase's 1995 Calendar of Events ☆

Thailand (cont'd)—Trevor

Elephant Round-Up at Surin . . . . Nov 18
King's Birthday and National Day . . . . Dec 5
Thank You Day, Natl . . . . Jan 11
Thank You Santa Hotline, Hillhaven (Seattle, WA) . . . . Dec 26
Thank You, School Librarian Day . . . . Apr 12
Thanksgiving, American Indian Time of (Allentown, PA) . . . . Oct 8
Thanksgiving Canned Goods Drive (Baltimore, MD) . . . . Nov 1
Thanksgiving Day . . . . Nov 23
Thanksgiving Day (Canada) . . . . Oct 9
Thanksgiving Day (Pres Proc) . . . . Nov 23
Thanksgiving Day 6625 . . . . Page 488
Thanksgiving Festival, Virginia (Charles City County) . . . . Nov 5
Thanksgiving Hunt Weekend (Cismont, VA) Nov 23
Thanksgiving Parade, Michigan (Detroit, MI) . . . . Nov 23
Thanksgiving: Turkey Trot (Parkersburg, WV) . . . . Nov 23
Tharp, Twyla: Birth . . . . July 1
Thatcher, Margaret Hilda Roberts: Birth . . . Oct 13
Thayer, Ernest L.: Birth Anniv . . . . Aug 14
Theatre
American Theatrical Productions Suspended: Anniv . . . . Oct 20
Arundel Festival (Arundel, England) . . . . . Aug 25
Colorado Shakespeare Fest (Boulder, CO) . . . . June 23
Curtis Easter Pageant (Curtis, NE) . . . . Apr 9
Destiny in Dayton—Scopes Play (Dayton, TN) . . . . July 20
Donna Reed Fest for the Performing Arts (Denison, IA) . . . . June 10
First Joint Siberian-American Musical: Anniv . . . . Dec 20
First Play Presented in N American Colonies . . . . Aug 27
Fringe Theater Event (Edmonton, Alta, Canada) . . . . Aug 18
George Spelvin Day . . . . Nov 15
Hair Broadway Opening: Anniv . . . . Mar 28
Holy City Easter Pageant (Lawton, OK) . . Apr 15
Louisiana Passion Play (Ruston, LA) . . . . . May 5
Medora Musical (Medora, ND) . . . . June 9
Micajah: Drama (Autryville, NC) . . . . July 13
Odetts, Clifford: Birth Anniv . . . . July 18
Oregon Shakespeare Fest (Portland, OR) . Jan 1
Outdoor Summer Theater (Farmington, NM) . . . . June 21
Papp, Joseph: Birth Anniv . . . . June 22
Really Rosie, a Musical (Columbia, SC) . . Mar 16
Royal Shakespeare Season (Warwickshire, England) . . . . Mar 17
Shakespeare Fest (Ashland, OR) . . . . Feb 17
Shakespeare on Saskatchewan Fest (Saskatoon, Canada) . . . . July 6
Shakespeare on the Green (Omaha, NE) June 22
Song of Hiawatha Pageant (Pipestone, MN) . . . . July 21
Tecumseh! Epic Outdoor Drama (Chillicothe, OH) . . . . June 9
Texas Musical Drama (Canyon, TX) . . . . . . June 7
Theatre Fest, Tampere (Tampere, Finland) . . . . Aug 15
Trial of Louis Riel (Regina, Sask, Canada) . . . . July 26
Unto These Hills Cherokee Indian Drama (Cherokee, NC) . . . . June 15
Walter Plinge Day (England) . . . . Dec 2
William Inge Fest (Independence, KS) . . . . Apr 20
Williams, Tennessee: Birth Anniv . . . . Mar 26
Theisman, Joe: Birth . . . . Sept 9
Therapeutic Recreation Week, Natl . . . . July 9
There Went Johnny Night . . . . May 22
Theroux, Paul: Birth . . . . Apr 10
Theus, Reggie: Birth . . . . Oct 13
Thible, Marie: First Free Flight by a Woman: Anniv . . . . June 4
Thicke, Alan: Birth . . . . Mar 1
Thigpen: Robert: Birth . . . . July 17
Thinnes, Roy: Birth . . . . Apr 6
Third Shift Workers Day, Natl . . . . May 10
Third World Day . . . . Apr 18
Thomas, B.J.: Birth . . . . Aug 7
Thomas, Betty: Birth . . . . July 27
Thomas, Clarence: Birth . . . . June 23
Thomas Crapper Day . . . . Jan 27
Thomas, Dave: Birth . . . . July 2
Thomas, Debi: Birth . . . . Mar 25
Thomas, Dylan: Birth Anniv . . . . Oct 27
Thomas, Heather: Birth . . . . Sept 8
Thomas, Helen: Birth . . . . Aug 4
Thomas, Isaiah: Birth Anniv . . . . Jan 19
Thomas, Kurt: Birth . . . . Mar 29
Thomas, Lowell: Birth Anniv . . . . Apr 6
Thomas, Marlo: Birth . . . . Nov 21
Thomas, Martha Carey: Birth Anniv . . . . Jan 2
Thomas, Philip Michael: Birth . . . . May 26
Thomas, Richard: Birth . . . . June 13
Thomas, Robert B.: Birth Anniv . . . . Apr 24
Thompson, Hunter S.: Birth . . . . July 18
Thompson, James R.: Birth . . . . May 8
Thompson, John: Birth . . . . Sept 9
Thompson, Sada: Birth . . . . Sept 27
Thompson, Tommy G.: Birth . . . . Nov 19
Thomson, Charles: Birth Anniv . . . . Nov 29
Thoreau, Henry David: Birth Anniv . . . . July 12
Thornburgh, Richard L.: Birth . . . . July 16
Thorpe, James: Birth Anniv . . . . May 28
Three Kings Day . . . . Jan 6
Three Rivers Fest and River Regatta (Fairmont, WV) . . . . May 25
Threshermen's Reunion, Central States (Pontiac, IL) . . . . Aug 30
Threshing Bee, Black Hills (Sturgis, SD) . . . Aug 18
Threshing Day, 18th-Century (McLean, VA) Nov 19
Threshing Show, Makoti (Makoti, ND) . . . . . . Oct 7
Thumb, Tom: Birth Anniversary . . . . Jan 4
Thurber, James: Birth Anniv . . . . Dec 8
Thurmond, Nate: Birth . . . . July 25
Thurmond, Strom: Birth . . . . Dec 5
Thyroid Disease Awareness Month . . . . Jan 1
Tiananmen Sq Massacre: Anniv . . . . June 4
Tickner, Charlie: Birth . . . . Nov 13
Tides, Perigean Spring . . Mar 19, May 15, June 12, July 11, Dec 2, Dec 21
Tierney, Gene: Birth Anniv . . . . Nov 20
Tiffany, Charles L.: Birth Anniv . . . . Feb 15
Tiffany, Louis C.: Birth Anniv . . . . Feb 18
Tilbrook, Glenn: Birth . . . . Aug 31
Tillis, Mel: Birth . . . . Aug 8
Time: Clock Month, Natl . . . . Oct 1
Time: Daylight-Saving Time (US) . . . . Apr 2
Time: Leap Second Adjustment Time . . . . . . Dec 31
Time: Leap Second Adjustment Time . . . . . June 30
Time: Standard Time Resumes . . . . Oct 29
Time: Summer Daylight-Saving Time (Europe) . . . . Mar 26
Time: Summer Time (England) . . . . Mar 26
Time: US Standard Time Act: Anniv . . . . Mar 19
Time Zone Plan, US Uniform: Anniv . . . . Nov 18
Tinker, Grant: Birth . . . . Jan 11
Tinnitus Assn Advisors Mtg, American (New Orleans, LA) . . . . Sept 18
Tiny Tim: Birth . . . . Apr 12
Tippett, Michael: Birth . . . . Jan 2
Tip-Up Town USA Ice Festival (Houghton Lake, MI) . . . . Jan 21
Tisch, Laurence: Birth . . . . Mar 5
Tisdale, Wayman: Birth . . . . June 9
Tisha B'Av (Fast of Ab) . . . . Aug 6
Titan II Missle Explosion: Anniv . . . . Sept 19
Titanic, Sinking of the: Anniv . . . . Apr 15
Tito (Josip Broz): Birth Anniv . . . . May 25
Tittle, Y.A.: Birth . . . . Oct 24
Tivoli—A Bible of Scandinavia (Seattle, WA) July 8
Tivoli Gardens Season (Copenhagen, Denmark) . . . . May 1
Toad Suck Daze (Conway, AR) . . . . May 5
Tobacco Auctioneer Chmpshp/Harvest Jubilee (Danville, VA) . . . . Oct 13
Tobacco Harvest (McLean, VA) . . . . Aug 20
Tobacco: World No-Tobacco Day . . . . May 31
Tobacco: Youth Against Month, Natl . . . . Oct 1
Tobago
Emancipation Day . . . . Aug 1
Independence Day . . . . Aug 31
Tobago Heritage Fest . . . . July 15
Toffler, Alvin: Birth . . . . Oct 4
Togo
Economic Liberation Day . . . . Jan 24
Independence Day . . . . Apr 27
Tojo Hideki: Birth Anniv . . . . Dec 30
Tojo Hideki: Execution Anniv . . . . Dec 23
Tolkien: Hobbit Day . . . . Sept 22
Tolkien Week . . . . Sept 17
Tom Sawyer Days, Natl (Hannibal, MO) . . . June 30
Tom Sawyer's Cat's Birthday . . . . Jan 3
Tomato Fest, Bradley County Pink (Warren, AR) . . . . June 8
Tomb-Sweeping Day, Natl (Taiwan) . . . . Apr 5
Tomczak, Mike: Birth . . . . Oct 23
Tomjanovich, Rudy: Birth . . . . Apr 24
Tomlin, Lily: Birth . . . . Sept 1
Tompkins, Daniel D.: Birth Anniv . . . . June 21
Toney, Andrew: Birth . . . . Nov 23
Tonga: National Day . . . . June 4
Toon, Al: Birth . . . . Apr 30
Tork, Peter: Birth . . . . Feb 13
Torme, Mel: Birth . . . . Sept 13
Torn, Rip: Birth . . . . Feb 6
Torquemada, Tomas de: Death Anniv . . . . . . Sept 16
Torture Abolition Day . . . . Feb 4
Tosafest (Wauwatosa, WI) . . . . Sept 8
Toure, Kwame: Birth . . . . June 29
Tourism Week, Natl . . . . May 7
Tournament of Roses Parade (Pasadena, CA) . . . . Jan 2
Towers, Constance: Birth . . . . May 20
Town Meeting Day (Vermont) . . . . Mar 7
Town Meeting on Main St, Natl (Little Rock, AR) . . . . May 14
Town Watch: Natl Night Out . . . . Aug 1
Towne, Benjamin: First Amer Daily Newspaper: Anniv . . . . May 30
Townshend, Peter: Birth . . . . May 19
Toy Show, Natl Farm (Dyersville, IA) . . . . Nov 3
Toynbee, Arnold J.: Birth Anniv . . . . Apr 14
Track and Field
Big Eight Men's/Women's Chmpshp (Oklahoma City, OK) . . . . Feb 25
Big Ten Men's Track/Field Chmpshp (Champaign, IL) . . . . Feb 24
Big Ten Men's/Women's Outdoor Track/Field (West Lafayette, IN) . . . . May 19
Big Ten Women's Track/Field Chmpshp (Ann Arbor, MI) . . . . Feb 24
Braille Institute Track/Field Olympics (Los Angeles) . . . . May 13
NAIA Men's/Women's Indoor Track/Field . . Mar 3
NAIA Outdoor Track/Field Chmpshps . . . May 25
NCAA Div I Men's/Women's Track Chmpshp (Knoxville, TN) . . . . May 31
NCAA Men's/Women's Indoor Track/Field (Indpls, IN) . . . . Mar 10
Simplot Games (Pocatello, ID) . . . . Feb 17
USA/Mobil Indoor M/W Chmpshps (Indianapolis, IN) . . . . Mar 3
Tractor. See Truck and Tractor
Trade Fair, Canton Autumn (China) . . . . Oct 15
Trade Week, World (Pres Proc) . . . . May 14
Trade Week, World 6563, 6690 . . . . Page 488
Trade: Amend Preferences 6599, 6635, 6650, 6655, 6676 . . . . Page 488
Trade: Dairy Limitations 6640 . . . . Page 488
Trade: Modify Duty re: Andean Trade Preferences Act 6544 . . . . Page 488
Trade: Modify Duty-Free 6575 . . . . Page 488
Trade: NAFTA Implementation 6641 . . . . . Page 488
Trade: Peru and Andean Trade 6585 . . . . . Page 488
Trade: Relations US and Romania 6577 . . Page 488
Trade: Special Rules re: Canadian Textiles 6543 . . . . Page 488
Trade: US-Canada Tarrif Schedule 6579 . Page 488
Trading Post Rendezvous, Fort Union (Williston, ND) . . . . June 15
Trafalgar, Battle of: Anniv . . . . Oct 21
Trail of Courage Living-History Fest (Rochester, IN) . . . . Sept 16
Trail's End Marathon (Seaside, OR) . . . . Mar 4
Trammel, Alan: Birth . . . . Feb 21
Transatlantic Flight, First Nonstop: Anniv . June 14
Transfer Day (Virgin Islands) . . . . Mar 31
Transport/Communications Decade for Asia and Pacific (UN) . . . . Jan 1
Transport/Communications Decade in Africa, Second (UN) . . . . Jan 1
Transportation Week, Natl (Pres Proc) . . . . May 14
Trappers Assn State Convention, Montana (Lewistown, MT) . . . . Sept 8
Trash Fest (Anacoco, LA) . . . . May 27
Travalena, Fred: Birth . . . . Oct 6
Travanti, Daniel J.: Birth . . . . Mar 7
Travel Month, Natl Senior . . . . May 1
Travers, Mary: Birth . . . . Nov 7
Travis, Randy: Birth . . . . May 4
Travolta, John: Birth . . . . Feb 18
Treasures by the Sea (Myrtle Beach, SC) . . . Nov 1
Treasury Dept, US: Anniv . . . . Sept 2
Treaty of Guadalupe Hidalgo . . . . Feb 2
Tree Planting Day, Natl (Lesotho, South Africa) . . . . Mar 21
Trees. See Arbor Day
Trees, Festival of (Rock Island, IL) . . . . Nov 18
Trek Fest (Riverside, IA) . . . . June 24
Trelawney, Edward J: Birth Anniv . . . . Nov 13
Trevino, Lee: Birth . . . . Dec 1
Trevor, Claire: Birth . . . . Mar 8

☆ Chase's 1995 Calendar of Events ☆ Index

Triathlon
Big Lake Regatta Water Fest (Big Lake, AK)....July 8
Northeast Missouri Triathlon Chmpshp (Kirksville, MO)....Sept 10
Trick or Treat Night....Oct 31
Trillin, Calvin: Birth....Dec 5
Trillium Fest (Muskegon, MI)....May 13
Trinidad
Boat Racing....Aug 1
Carnival (Port of Spain)....Feb 27
Emancipation Day....Aug 1
Independence Day....Aug 31
Pan Is Beautiful (Port of Spain)....Oct 14
Republic Day (Port of Spain)....Sept 24
Trinity Sunday....June 11
Tripod Raising Festival, Nenana (Nenana, AK)....Feb 25
Tripucka, Kelly: Birth....Feb 16
Trivia Contest, World's Largest (Stevens Point, WI)....Apr 7
Trivia Day....Jan 4
Trollope, Anthony: Birth Anniv....Apr 24
Truancy Law: Anniv....Apr 12
Trucks/Tractors
Alabama Natl Truck/Tractor Pull (Lexington, AL)....Aug 18
Ugly Pickup Parade/Contest (Chadron, NE)....Oct 27
Ugly Truck Contest (Pelican Rapids, MN)....July 8
Trudeau, Pierre: Birth....Oct 18
True Confessions Day....Mar 15
Truman, Bess (Elizabeth): Birth Anniv....Feb 13
Truman, Harry S.: Birth Anniv....May 8
Truman, Harry S: Harry's Hay Days (Grandview, MO)....May 19
Trumball, Jonathan: Birth Anniv....Oct 12
Truth, Sojourner: Death Anniv....Nov 26
Tu B'Shvat....Jan 16
Tub Race: Chimney Rock Hillfall (Chimney Rock, NC)....Sept 17
Tuba Day, Intl....May 5
Tubb, Ernest: Birth Anniv....Feb 9
Tuberous Sclerosis Awareness Month, Natl....May 1
Tubman, Harriet: Death Anniv....Mar 10
Tuchman, Barbara W.: Birth Anniv....Jan 30
Tucker, Jim Guy: Birth....June 13
Tucker, Michael: Birth....Feb 6
Tucker, Tanya: Birth....Oct 10
Tulip Fest, Albany (Albany, NY)....May 12
Tulip Fest, Skagit Valley (Mt Vernon, WA)....Mar 31
Tulip Time Fest, Holland (Holland, MI)....May 10
Tulip Time Fest, Pella (Pella, IA)....May 11
Tuna Tournament (Narragansett, RI)....Sept 2
Tune, Tommy: Birth....Feb 28
Tunisia
Independence Day....Mar 20
Martyrs' Day....Apr 9
Revolution Day....Jan 18
Tree Fest....Nov 12
Women's Day....Aug 13
Tunney, Gene: Long Count Day....Sept 22
Tunney, James Joseph (Gene): Birth Anniv....May 25
Turkey
National Holiday....Oct 29
National Sovereignty/Child's Day....Apr 23
St. Peter's Day....June 29
Victory Day....Aug 20
Youth and Sports Day....May 19
Turkey Days, King (Worthington, MN)....Sept 9
Turkey Lovers' Month, June Is....June 1
Turkey Olimpiks, Live (New Preston, CT)....Nov 19
Turkey Rama (McMinnville, OR)....July 6
Turkmenistan: National Day....Oct 27
Turner, Daryl: Birth....Dec 15
Turner, Ike: Birth....Nov 5
Turner, Kathleen: Birth....June 19
Turner, Lana: Birth....Feb 8
Turner, Ted: Birth....Nov 19
Turner, Tina: Birth....Nov 26
Turow, Scott: Birth....Feb 12
Turtle Derby, Canadian (Boissevain, Man, Canada)....July 14
Turtle Derby, Chesapeake (Baltimore, MD)....July 5
Tushingham, Rita: Birth....Mar 14
Tussaud, Marie: Birth Anniv....Dec 7
Twain, Mark (Samuel Clemens): Birth Anniv....Nov 30
Tweed Day....Apr 3
Twelfth Day (Epiphany)....Jan 6
Twelfth Night....Jan 5
Twiggy: Birth....Sept 19
Twilight Zone: The First Episode....Oct 2
Twin-O-Rama (Cassville, WI)....July 21
Twins Day Fest (Twinsburg, OH)....Aug 4
Twit Award Month, Intl....Apr 1
Twitty, Conway: Birth Anniv....Sept 1
Two Thousand Days Before "2000"....July 11
Tylenol Deaths: Anniv....Sept 29
Tyler, Anne: Birth....Oct 25
Tyler, John: Birth Anniv....Mar 29
Tyler, Julia G.: Birth Anniv....May 4
Tyler, Letitia Christian: Birth Anniv....Nov 12
Tyler's Cabinet Resigns: Anniv....Sept 11
Tynwald Day (England)....July 5
Typewriter Toss, KGBX (Springfield, MO)....Apr 26
Typing Contest, Natl....Apr 24
Tyson, Cicely: Birth....Dec 19
Tyson, Mike: Birth....June 30

U

U-2 Incident: Anniv....May 1
Ueberroth, Peter: Birth....Sept 2
Uecker, Bob: Birth....Jan 26
UFO Days (Elmwood, WI)....July 28
Uganda
Independence Day....Oct 9
Liberation Day....Apr 11
Uggams, Leslie: Birth....May 25
Ukraine: Independence Day....Aug 24
Ukrainian Famine Film Broadcast: Anniv....Nov 30
Ullman, Tracey: Birth....Dec 30
Ullmann, Liv Johanne: Birth....Dec 16
Ullr Fest (Breckenridge, CO)....Jan 9
Ulysses, Marathon Reading of (Washington, DC)....June 15
Underdog Day....Dec 15
Underground America Day....May 14
UNICEF Anniv [UN]....Dec 11
Union: First Baseball Strike Ends: Anniv....Apr 13
Unions. See Employment
Unitas, Johnny: Birth....May 7
United Airlines Flight 232 Crashes: Anniv....July 19
United Arab Emirates: National Day....Dec 2
United Cerebral Palsy's Casual Day....June 16
United Kingdom
Boxing Day....Dec 26
Boxing Day Bank Holiday....Dec 26
Commonwealth Day....Mar 13
Easter Monday Bank Holiday....Apr 17
Good Friday Bank Holiday....Apr 14
May Day Bank Holiday....May 1
New Year's Holiday (UK and Republic of Ireland)....Jan 2
Spring Bank Holiday....May 29
Summer Bank Holiday....Aug 28
Westminster Antiques Fair (London)....Nov 30
United Nations
Africa Industrialization Day....Nov 20
Carnaval '95 (San Francisco, CA)....May 27
Children's Art Exhibit, Intl (San Francisco, CA)....May 15
Colonialism, Intl Decade for Eradication of....Jan 1
Cultural Development, World Decade for....Jan 1
Decade Against Drug Abuse....Jan 1
Desert Shield: Deadline Resolution: Anniv....Nov 28
Disabled Persons, Asian/Pacific Decade....Jan 1
Disabled Persons, Intl Day of....Dec 3
Disarmament Week....Oct 24
Drug Abuse/Illicit Trafficking, Intl Day Against....June 26
Elderly, Intl Day for....Oct 1
Fourth United Nations Development Decade....Jan 1
Global Youth Project (San Franciso, CA)....June 19
Heartland Train/Kansas City UN50 (San Francisco, CA)....June 21
Human Rights Day....Dec 10
Indigenous People, Intl Decade of World's....Jan 1
Industrial Development Decade for Africa, Second....Jan 1
International Day of Peace....Sept 19
International Law, Decade of....Jan 1
Intl Day of Families....May 15
Intl Decade of World's Indigenous People....Dec 10
Japan Agrees to End Use of Drift Nets: Anniv....Nov 26
Literacy Day, Intl....Sept 8
Model UN of the Far West (San Francisco, CA)....Apr 19
Moment in History (San Francisco, CA)....June 19
Natural Disaster Reduction, Intl Day for....Oct 11
Natural Disaster Reduction, Intl Decade for....Jan 1
Noah's Flood (San Francisco, CA)....June 23
Peace, Security Pacific Rim (San Francisco, CA)....Apr 28
Poverty, Intl Day for Eradication....Oct 17
Racial Discrimination, Intl Day for Elimination of....Mar 21
Racism/Racial Discrimination, 3rd Decade to Combat....Dec 10
Racism/Racial Discrimination, Solidarity Against....Mar 21
Racism/Racial Discrimination, Third Decade to Combat....Jan 1
Rediscovering Justice (San Francisco, CA)....June 22
Reflections on Future of UN (San Francisco, CA)....June 20
Revokes Resolution on Zionism: Anniv....Dec 16
San Francisco Revisited....Apr 26
Science and Peace, Intl Week of....Nov 5
Solidarity with Palestinian People, Intl day of....Nov 29
State of the World Forum (San Francissco, CA)....Apr 21
Telecommunication Day, World....May 17
Third Disarmament Decade....Jan 1
Torture Abolition Day....Feb 4
Transport/Communications Decade for Asia and Pacific....Jan 1
Transport/Communications Decade in Africa, Second....Jan 1
UN Assn USA Annual Conv (San Francisco, CA)....June 25
UN Charter Signed: Anniv....June 26
UN 50th Anniversary Celeb (San Francisco, CA)....June 25
UN Rocks the World (San Francisco, CA)....June 24
UN Teacher's Workshop (San Francisco, CA)....Apr 29
UNICEF Anniv....Dec 11
UNICEF Day, Natl (Pres Proc)....Oct 31
United Nations Day....Oct 24
United Nations Day 6618....Page 488
United Nations General Assembly Anniversary....Jan 10
United We Dance....May 9
Universal Children's Day....Oct 2
Visions of Peace (San Francisco, CA)....June 23
Volunteer Day for Economic/Social Dvmt, Intl....Dec 5
Water, World Day for....Mar 22
Women's Day, Intl....Mar 8
World AIDS Day....Dec 1
World Development Information Day....Oct 24
World Environment Day....June 5
World Food Day....Oct 16
World Habitat Day....Oct 2
World Health Day....Apr 7
World Meteorological Day....Mar 23
World Population Day....July 11
World Post Day....Oct 9
World Press Freedom Day....May 3
Year for Tolerance....Jan 1
United States
Air Force Established: Birth....Sept 18
Berlin Airlift: Anniv....June 24
Blacks Ruled Eligible to Vote: Anniv....Apr 3
Bush Orders Commencement Desert Shield: Anniv....Aug 7
Canada-US Goodwill Week....Apr 23
Civil Rights Act of 1964: Anniv....July 2
Coins Stamped "In God We Trust": Anniv....Apr 22
Congress Assembles....Jan 3
Congress Authorizes Force Against Iraq: Anniv....Jan 12
Congress: First Meeting Anniv....Mar 4
Constitution of the US: Anniv....Sept 17
Customs: Anniv....Aug 1
Department of Justice Anniv....June 22
Department of State Birthday....July 27
Distinguised Service Medal: Anniv....Mar 7
55 mph Speed Limit: Anniv....Jan 2
First Brawl in US House of Representatives: Anniv....Jan 30
First Foreign-Born Chair Joint Chiefs: Anniv....Aug 11
First Ladies Gowns (Grand Rapids, MI)....Jan 23
First Mint in America Opens: Anniv....June 10
First US Government Building....July 31
First Woman US Ambassador Appointed: Anniv....Oct 28
General Election Day....Nov 7
Gerald Ford: Assassination Attempts: Anniv....Sept 5
Grandeur/US Capitol 1793-1993 (Grand Rapids, MI)....Jan 30

Triathlon—United States

Index ☆ Chase's 1995 Calendar of Events ☆

United States (cont'd)—Virginia

Great Seal of the US: Design Recommended Anniv..............June 20
Great Seal of the US: First Use Anniv...Sept 16
Ground War Against Iraq Begins: Anniv..Feb 23
Home Owners Loan Act: Anniv..........June 13
Independence Day.....................July 4
Invasion of Panama: Anniv..............Dec 19
Japanese Internment: Anniv.............Feb 19
Johnson Impeachment Proceedings: Anniv............................Mar 5
Kuwait City Liberated/100-Hour War Ends: Anniv...........................Feb 27
Library of Congress: Anniv..............Apr 24
Lincoln Signs Income Tax: Anniv........July 1
Moscow Communique: Anniv.............May 29
Motor Voter Bill Signed: Anniv..........May 20
National Labor Relations Act: Anniv.....July 5
Nuclear-Free World, First Step..........Dec 8
Paper Money Issued: Anniv.............Mar 10
Peace Corps: Founding Anniv...........Mar 1
Philippine Independence Anniv..........Mar 24
Pony Express, Inauguration of: Anniv....Apr 3
Postmaster General Established: Anniv..Sept 22
Presidential Succession Act: Anniv......July 18
Racial Brawl on Kitty Hawk: Anniv......Oct 12
Ratification Day......................Jan 14
Senate Quorum: Anniv.................Apr 6
Shanghai Communique: Anniv...........Feb 27
Signing Federal Credit Union Act: Anniv June 26
Standard Time Act: Anniv...............Mar 19
Supreme Court Bans Official Prayer: Anniv..........................June 25
Treasury Department: Anniv............Sept 2
Treaty of Guadalupe Hidalgo (with Mexico)..........................Feb 2
Uniform Time Zone Plan: Anniv........Nov 18
United States Mint: Anniv..............Apr 2
US Capital Established at NYC: Anniv...Sept 13
US Takes Out Its First Loan: Anniv......Sept 18
Vietnam War Ended: Anniv.............Jan 27
Vote to Impeach Pres Andrew Johnson: Anniv............................Feb 24
WAAC: Anniv..........................May 14
War Against Iraq Begins: Anniv.........Jan 16
War Department: Establishment Anniv...Aug 7
War of 1812: Declaration Anniv.........June 18
War on Poverty: Anniv.................Jan 8
Water Pollution Control Act: Anniv......Oct 18
White House Easter Egg Roll: Anniv.....Apr 2
Woman Runs the House: Anniv.........June 20
UNIVAC Computer: Birth...............June 14
Universal Human Rights Month..........Dec 1
Universal Letter-Writing Week..........Jan 1
University of Virginia Founder's Day (Charlottesville, VA)................Apr 13
Unser, Al: Birth.......................May 29
Unser, Bobby: Birth...................Feb 20
Unto These Hills Cherokee Indian Drama (Cherokee, NC)....................June 15
Up Helly AA (Scotland)................Jan 31
Updike, John: Birth...................Mar 18
Upshaw, Gene: Birth..................Aug 15
Uranus (planet) Discovery: Anniv.......Mar 13
Urich, Robert: Birth...................Dec 19
Uris, Leon: Birth......................Aug 3
Uruguay: Independence Day...........Aug 25
US Navy: Authorization Anniv..........Oct 13
US-Mexico War Declaration: Anniv.....May 13
USO Birthday........................Feb 4
USS Iowa: Explosion..................Apr 19
USS Kitty Hawk, Racial Brawl on: Anniv..Oct 12
USS Liberty, Attack on: Anniv..........June 8
USS Princeton Explosion: Anniv........Feb 28
USS Stark: Attack Anniv...............May 17
Ustinov, Peter: Birth..................Apr 16
Utah
  Admission Day.....................Jan 4
  Butch Cassidy Outlaw Trail Ride (Vernal).........................June 16
  Days of '47 Celebration (Salt Lake City)..July 1
  Dickens Festival (Salt Lake City).....Nov 24
  Golden Spike Reenactment (Promontory)...................May 10
  Greek Festival (Salt Lake City).......Sept 8
  Melon Days (Green River)...........Sept 15
  Pioneer Day........................July 24
  Plowing Match (Woodstock).........May 7
  Railroader's Festival (Promontory)...Aug 12
  Southern Utah Folklife Festival (Springdale)......................Sept 7
  State Fair (Salt Lake City)...........Sept 8
  Summerfest Art Faire (Logan).......June 15
  Sundance Film Festival (Park City)...Jan 19

World Folkfest (Springville)............July 8
Utley, Garrick: Birth..................Nov 19
Uzbekistan: Independence Day.........Sept 1

V

V-E Day..............................May 8
V-Mail Delivery: Anniv................June 22
Vaccaro, Brenda: Birth................Nov 18
Vachon, Rogie: Birth..................Sept 8
Vadim, Roger: Birth...................Jan 26
Valentine Exhibit, Antique (Clinton, MD)..Feb 9
Valentine, Karen: Birth................May 25
Valentine, Scott: Birth.................June 3
Valentine's Day......................Feb 14
Valentine's Day Carnation Give Away (Springfield, MO)..................Feb 14
Valentine's Day Massacre: Anniv......Feb 14
Valentino, Rudolph: Birth Anniv.......May 6
Valenzuela, Fernando: Birth...........Nov 1
Valley of Flowers Festival (Florissant, MO)..May 5
Valli, Frankie: Birth...................May 3
Van Ark, Joan: Birth..................June 16
Van Buren, Abigail: Burthday..........July 4
Van Buren Co Literacy Day...........Sept 28
Van Buren, Hannah Hoes: Birth Anniv..Mar 8
Van Buren, Martin: Birth Anniv........Dec 5
Van Devere, Trish: Birth...............Mar 9
Van Dyke, Dick: Birth.................Dec 13
Van Fleet, Jo: Birth...................Dec 30
Van Gogh, Pin in the Ear on (Ft Wayne, IN)..Mar 30
Van Gogh, Vincent: Birth Anniv........Mar 30
Van Hemmel, Martine: Birth...........Nov 16
Van Patten, Dick: Birth................Dec 9
Van Patten, Joyce: Birth...............Mar 9
Van Slyke, Andy: Birth................Dec 21
Vancouver, George: Death Anniv......May 10
Vanderbilt, Gloria: Birth...............Feb 20
Vandross, Luther: Birth................Apr 20
Vanuatu: Independence Day..........July 30
Vatican City: Independence Anniv.....Feb 11
Vaughn, Robert: Birth.................Nov 22
Vaughn, Sarah: Birth Anniv...........Mar 27
Veal Ban Action Day.................May 14
Veblen, Thorstein: Birth Anniv.........July 30
Vegetables Month, Fresh Fruits and....June 1
Vegetarian Awareness Month.........Oct 1
Vegetarian Day, World................Oct 1
Vegetarian Resource Group's Essay Contest for Kids............................May 1
Venezuela: Independence Day........July 5
Ventura, Charlie: Birth................Dec 2
Venture Capital, French Lib Conf (Boston, MA)..............................Oct 2
Vercors, Jean: Birth Anniv.............Feb 26
Verdi, Giuseppi: Birth Anniv...........Oct 10
Verdon, Gwen: Birth..................Jan 13
Vereen, Ben: Birth...................Oct 10
Verity, C. William: Birth...............Jan 26
Vermont
  Admission Day....................Mar 4
  Antique/Classic Car Show (Bennington) Sept 15
  Arts and Crafts Fair, Annual (Jay)....Oct 7
  Arts and Crafts Fest (Bennington)...Sept 16
  Bennington Museum Antiques Show/Sale (Bennington)....................July 7
  Brookfield Ice Harvest (Brookfield)...Jan 28
  Champlain Valley Fair (Essex Junction)..Aug 26
  Children's Day (Woodstock).........Aug 26
  Fall Foliage Fest (Walden)..........Oct 2
  Green Mountain Nationals Car Show (Bennington)...................June 25
  Harvest Celeb (Woodstock).........Oct 8
  Lake Champlain Balloon Fest (Essex Jct) June 2
  Old Vermont Fourth (Woodstock)....July 4
  Quechee Balloon Festival/Crafts Fair (Quechee).......................June 16
  Quilt Show (Woodstock)............July 29
  Rockingham Old Home Days (Rockingham)...................Aug 3
  Shelburne Museum Presents Lilac Sunday (Shelburne).....................May 21
  Shelburne Old-Time Farm Day/Grand Fourth (Shelburne)....................July 4
  State Fair (Rutland)................Sept 1
  Stratton Arts Festival (Stratton Mt, VT)..Sept 17
  Summer Social (Woodstock).......July 16
  Swanton Summer Fest (Swanton)...July 27
  Town Meeting Day................Mar 7
  Vermont Apple Festival (Springfield)..Oct 1
  Vermont Forestry Expo (Rutland)...July 22
  Vermont Maple Fest (St. Albans)...Apr 21

Wassail Celebration (Woodstock).......Dec 7
Wool Day: Sheep to Shawl/Border Collies (Woodstock)......................Sept 17
Verne, Jules: Birth Anniv..............Feb 8
Verrazano Day.......................Apr 17
Versailles Peace Conference: Anniv....Jan 18
Vespucci, Amerigo: Birth Anniv.......Mar 9
Vesuvius Day........................Aug 24
Veterans Bonus Army Eviction: Anniv..July 28
Veterans Day........................Nov 11
Veterans Day (Observed).............Nov 10
Veterans Day 6621..................Page 488
Veterans Day Celebration (Mamou, LA)..Nov 11
Veterans Golden Age Games Week, Natl 6581.............................Page 488
Veterans, Women, Recognition Week, Natl 6622.............................Page 488
Victims of Violence Holy Day...........Apr 4
Victoria Day (Canada)................May 22
Victorian. See also Christmas
Victorian Christmas Home Tour (Leadville, CO)............................Dec 2
Victorian Week (Cape May, NJ)........Oct 6
Victory Day or V-J Day (Announcement)..Aug 14
Victory Day or V-J Day (Ratification)....Sept 2
Vidal, Gore: Birth....................Oct 3
Video Games Day....................July 12
Vieira da Silva, Maria-Helena: Birth Anniv. June 13
Vietnam
  Battle of Dien Bien Phu Ends: Anniv...May 7
  Independence Day.................Sept 2
Vietnam Veterans Memorial 10th Anniv Day 6506.............................Page 488
Vietnam Veterans Memorial Statue: Unveiling Anniv..........................Nov 9
Vietnam War Ended: Anniv............Jan 27
Vietnam War: Hair Opening: Anniv....Mar 28
Vietnam War: Tet Offensive Begins.....Jan 30
Vietnam War: Vietnam Moratorium Concert: Anniv.........................Mar 28
Vigoda, Abe: Birth...................Feb 24
Viking Festival (Denmark).............June 23
Viking Festival, Jorvik (York, England)..Feb 10
Viking: Up Helly AA (Scotland)........Jan 31
Vilas, Guillermo: Birth................Aug 17
Vincent, Fay: Birth...................May 29
Vincent, Jan-Michael: Birth...........July 15
Vinson, Fred M.: Birth Anniv.........Jan 22
Vintners' Holidays, Yosemite (Yosemite Natl Park, CA)............................Nov 6
Vinton, Bobby: Birth.................Apr 16
Viola, Frank: Birth...................Apr 19
Violence Day, Natl Stop the...........Nov 22
Virchow, Rudolf: Birth Anniv.........Oct 13
Virgin Islands
  Danish West Indies Emancipation Day....July 3
  Hurricane Supplication Day.........July 24
  Hurricane Thanksgiving Day........Oct 16
  Liberty Day.......................Nov 1
  Memorial Day.....................May 29
  Nicole Robin Day..................Aug 4
  Nurse of the Year Award...........May 12
  Organic Act Day...................June 20
  Puerto Rico Friendship Day.........Oct 9
  Transfer Day......................Mar 31
Virginia
  Africa's Legacy in Mexico: Photos/ (Roanoke).......................Aug 19
  All Class Reunion Party (Roanoke)...May 6
  Amer Heritage Fest (Yorktown).....May 20
  Ash Lawn-Highland Opera Fest (Charlottesville)..................June 25
  Battle of Waynesborough: Anniv....Mar 2
  Bayou Boogalou & Cajun Food Fest (Norfolk)........................June 16
  Blackbeard Pirate Jamboree (Norfolk)..July 28
  Bonnie Blue Natl Horse Show (Lexington).....................May 10
  Budweiser Fest and BBQ Rib Burn-Off (Norfolk)........................Sept 15
  Carols by Candlelight (Lorton)......Dec 8
  Celebration of America's Bounty (McLean).......................Sept 16
  Chautauqua Festival in the Park (Wytheville)....................June 17
  Chickahominy Fall Fest Powwow (Roxbury)......................Sept 23
  Children's Colonial Days Fair (Yorktown)..July 1
  Chincoteague Easter Decoy Fest (Chincoteague Is)................Apr 15
  Chincoteague Pony Penning (Chincoteague Island).........................July 26
  Christmas Candlelight Tour (Alexandria)..Dec 9

584

☆ Chase's 1995 Calendar of Events ☆ Index

Virginia (cont'd)—Ward

Christmas Candlelight Tour (Fredricksburg) . . . . Dec 3
Civil War Peace Talks: Anniversary . . . . Feb 3
Civil War Weekend (Yorktown) . . . . May 27
Confederate Memorial Day . . . . May 29
Conservation Festival (Roanoke) . . . . June 10
Cut Your Own Christmas Tree (Charlottesville) . . . . Dec 2
Defeat at Five Forks: Anniversary . . . . Apr 1
East Coast Surfing Chmp/Sports Fest (Virginia Bch) . . . . Aug 23
Eighteenth-Century Autumn Market Fair (McLean) . . . . Oct 21
Eighteenth-Century Christmas Wassail (McLean) . . . . Dec 17
Eighteenth-Century Spring Market Fair (McLean) . . . . May 20
Eighteenth-Century Summer Market Fair (McLean) . . . . July 15
Eighteenth-Century Threshing Day (McLean) . . . . Nov 19
Fall of Richmond: Anniversary . . . . Apr 3
Fiddlefest (Galax) . . . . Aug 10
First Assembly Day (Yorktown) . . . . July 29
First Night Roanoke (Roanoke) . . . . Dec 31
First Night Waynesboro (Waynesboro) . . . Dec 31
First US Breach of Promise Suit: Anniv. . June 14
Foods/Feasts in 17th-Century Virginia (Williamsburg) . . . . Nov 23
Foxfield Races (Charlottesville) . . . . Apr 29
Fredericksburg Day (Fredericksburg) . . . . Apr 25
Fredericksburg Heritage Fest (Fredericksburg) . . . . July 4
George Washington Birthday Parade (Alexandria) . . . . Feb 20
George Washington Birthnight Banquet and Ball (Alexandria) . . . . Feb 18
Gingerbread/Lace Christmas Celeb (Charlottesville) . . . . Dec 1
Great Peanut Tour (Skippers) . . . . Sept 7
Gunston Hall Car Show (Lorton) . . . . Sept 10
Hampton Jazz Festival (Hampton) . . . . . . June 23
Harborfest (Norfolk) . . . . June 2
Heritage Fair/Spring Fest (Ashland) . . . . Apr 8
Highland County Maple Fest (Highland County) . . . . Mar 11
Historic Birthday Parties in Lexington (Lexington) . . . . Jan 19
Historic Garden Week . . . . Apr 22
Indian Heritage Festival (Martinsville) . . . . Sept 8
International Gold Cup (The Plains) . . . . . Oct 21
Jamestown Christmas (Williamsburg) . . . . . Dec 3
Jamestown Landing Day (Williamsburg) . May 13
Jamestown Weekend Celeb (Jamestown Island) . . . . May 13
Jousting Hall Fame Jousting Tourn, Natl (Mt Solon) . . . . June 17
Kite Fest (Lorton) . . . . Mar 19
Lee Birthday Celebrations (Alexandria) . . Jan 22
Lee-Jackson Lacrosse Classic (Lexington) . . . . May 6
Lee-Jackson-King Day . . . . Jan 16
Market Square Fair (Fredericksburg) . . . . May 13
Merrie Olde England Xmas Fest (Charlottesville) . . . . Dec 23
Military Through the Ages (Williamsburg) Mar 18
Model Airplane Show (Danville) . . . . June 24
National Park Service Anniv (Jamestown/Yorktown) . . . . Aug 25
Native American Heritage Fest (Martinsville) . . . . Sept 9
Natural Chimneys Jousting Tourn (Solon) . . . . Aug 19
NCAA Div I Women's Basketball Chmpshp Final's (Richmond) . . . . Apr 1
Old Fiddlers' Convention (Galax) . . . . Aug 9
Oyster Fest (Chincoteague Island) . . . . Oct 7
Postcard Show (Ft. Washington) . . . . June 10
Potomac River Fest (Colonial Beach) . . . . June 9
Pre-Opera Lecture Series (Charlottesville) . . . . June 26
Produce for Victory: Posters/Home Front (Hampton) . . . . Mar 18
Rainbow of Arts Craft Fest (Chesterfield) . . . . Sept 23
Ratification Day . . . . June 25
Revolutionary War Encampment (Alexandria) . . . . Feb 19
Rising Water Falling Water Powwow (Richmond) . . . . July 29
Roanoke Beach Party (Roanoke) . . . . Aug 26
Roanoke Festival in the Park '95 (Roanoke) . . . . May 25
Roanoke Valley Horse Show (Salem) . . . June 19
Rockbridge Community Fest (Lexington) Aug 26
Rockbridge Regional Fair (Lexington) . . . July 25
St. Patrick's Day Dog Fun Fair, Parade (Alexandria) . . . . Mar 11
St. Patrick's Day Parade (Roanoke) . . . . . . Mar 17
Salem Fair and exposition (Salem) . . . . . . June 30
Scottish Christmas Walk (Alexandria) . . . . Dec 2
Shamrock Marathon and Sportsfest (Virginia Beach) . . . . Mar 17
Shenandoah Apple Blossom Fest (Winchester) . . . . May 4
Shrine Oyster Bowl Game (Norfolk) . . . . . Nov 18
Sounds of Season: Holiday Concert (Charlottesville) . . . . Dec 27
Star of Wonder Planetarium Show (Newport News) . . . . Jan 2
State Fair on Strawberry Hill (Richmond) Sept 21
Strawberry Hill Races (Richmond) . . . . Apr 15
Streetscene (Covington) . . . . Aug 5
Sugarloaf Craft Fest (Manassas) . . . . Sept 8
Summerfest (Harrisonburg) . . . . July 30
Thanksgiving Hunt Weekend (Cismont) . . Nov 23
Tobacco Harvest (McLean) . . . . Aug 20
Town Point Jazz Fest (Norfolk) . . . . Aug 25
Town Point Park Wine Fest (Norfolk) . . . . . Oct 21
University of VA Founder's Day (Charlottesville) . . . . Apr 13
Upperville Colt/Horse Show (Warrenton) June 6
Virginia Boat Show (Richmond) . . . . Jan 16
Virginia Children's Fest (Norfolk) . . . . Oct 7
Virginia Christmas Show (Manassas) . . . . Oct 13
Virginia Christmas Show (Richmond) . . . . . Nov 2
Virginia Fest of American Film (Charlottesville) . . . . Oct 26
Virginia Gold Cup (Warrenton) . . . . May 6
Virginia Highlands Fest (Abingdon) . . . . . July 29
Virginia Indian Heritage Fest (Williamsburg) . . . . June 17
Virginia Lake Festival (Clarksville) . . . . July 14
Virginia Scottish Games (Alexandria) . . . . July 22
Virginia Spring Show (Richmond) . . . . Mar 10
Virginia State Horse Quarter Horse Division (Richmond) . . . . Aug 4
Virginia State Horse Show All Breed Event (Richmond) . . . . June 26
Virginia Thanksgiving Festival (Charles City County) . . . . Nov 5
Virgina-Carolina Craftsmen Christmas Classic (Richmond) . . . . Nov 3
Virginia-Carolina Craftmen's Fall Classic (Roanoke) . . . . Sept 29
Waterfront New Year's Eve Celebration (Norfolk) . . . . Dec 31
Wheat Harvest, 18th Century (McLean) . June 18
World Tobacco Auctioneer Chmpshp/Harvest Jub (Danville) . . . . Oct 13
Yorktown Christmas (Yorktown) . . . . Dec 2
Yorktown Day (Yorktown) . . . . Oct 19
Yorktown Day Weekend (Yorktown) . . . . . . Oct 14
Yuletide Traditions (Charlottesville) . . . . Dec 1
Virginia Company Expedition to America: Anniv . . . . Dec 20
Virginia Plan Proposed: Anniv . . . . May 29
Virgo Begins . . . . Aug 23
Vision. See Health and Welfare
Visit Your Relatives Day . . . . May 18
Visitor Appreciation Day, Natl . . . . May 10
Viticulturist's Day (Bulgaria) . . . . Feb 14
Vitti, Monica: Birth . . . . Nov 3
Vocabulary: Repeat Day . . . . June 3
Voight, Jon: Birth . . . . Dec 29
Voinovich, George V.: Birth . . . . July 15
Volcanoes: Mount Pinatubo Erupts in Philippines: Anniv . . . . June 11
Volcanoes: Vesuvius Day . . . . Aug 24
Volksfest (New Glarus, WI) . . . . Aug 6
Volksmarch (Angel Fire, NM) . . . . Sept 30
Volksmarch, Custer State Park (Custer, SD) . . . . July 29
Volleyball, Mud, Tourn (Rockford, IL) . . . . Aug 5
Volleyball: NFL Pro Bowl Beach Challenge (Oahu, HI) . . . . Feb 3
Voltaire, Jean F: Birth Anniv . . . . Nov 21
Volunteer Day for Economic/Social Dvmt, Intl (UN) . . . . Dec 5
Volunteer Day, Student . . . . Feb 20
Volunteer Week, Natl . . . . Apr 23
Volunteer Week, Natl 6546, 6673 . . . . Page 488
Volunteers: Make A Difference Day . . . . Oct 21
Volunteers of America Week, Natl . . . . Mar 5
Volunteers Week, Intl . . . . June 1
Von Braun: First Surface-to-Surface Missile: Anniv . . . . Dec 24
Von Furstenberg, Diane Halfin: Birth . . . . . . Dec 31
Von Richtofen: Red Baron Shot Down: Anniv . . . . Apr 21
Von Sydow, Max: Birth . . . . Apr 10
Vonnegut, Kurt, Jr: Birth . . . . Nov 11
Vote: Blacks Ruled Eligible to Vote: Anniv . . . Apr 3
Vote: Motor Voter Bill Signed: Anniv . . . . May 20
Voting: Women's Suffrage Anniv Celeb . . . . Aug 26

W

Wade, Virginia: Birth . . . . July 10
Wadlow, Robert Pershing: Birth Anniversary . . . . Feb 22
Wagner, Honus: Birth Anniv . . . . Feb 24
Wagner, Lindsay: Birth . . . . June 22
Wagner, Robert: Birth . . . . Feb 10
Wagon trains: Santa-Cali-Gon Days (Independence, MO) . . . . Sept 1
Wagon Trek, Wood Mountain (Willow Bunch, Canada) . . . . July 2
Wagoner, Porter: Birth . . . . Aug 12
Waihee, John D., III: Birth . . . . Mar 12
Waitangi Day (New Zealand) . . . . Feb 6
Waite, Morrison R.: Birth Anniv . . . . Nov 29
Waite, Terry: Birth . . . . May 31
Waitresses Day, Natl . . . . May 21
Waits, Tom: Birth . . . . Dec 7
Waitz, Grete: Birth . . . . Oct 1
Wald, Lillian: Birth Anniv . . . . Mar 10
Walden, Robert: Birth . . . . Sept 25
Waldheim, Kurt: Birth . . . . Dec 21
Wales
Cardiff Festival of Music (Cardiff) . . . . Sept 9
Cardiff Singer of the World (Cardiff) . . . . June 11
Christmas Holiday . . . . Dec 25
Hay-on-Wye Fest of Literature (Hay-on-Wye) . . . . May 26
Llandrindod Wells Victorian Fest (Llandrindod Wells) . . . . Aug 19
Llangollen Intl Musical Eisteddfod (Llangollen, Clwyd) . . . . July 4
North Wales Music Festival (North Wales) . . . . Sept 16
Potters Fest, Intl (Aberystwyth) . . . . July 14
Royal Natl Eisteddfod (Colwyn Bay) . . . . . . Aug 5
Royal Welsh Show (Powys) . . . . July 24
St. David's Day . . . . Mar 1
Swansea Festival of Music and the Arts (Swansea) . . . . Sept 25
Walker Cup Golf Chmpshp (Portcawl) . . . . Sept 9
Welsh Intl Fest Storytelling (Llantwit Major) . . . . June 30
Walesa, Lech: Solidarity Founding Anniv . . Aug 31
Walk: Horton Bay Labor Day Bridge Walk (Horton Bay, MI) . . . . Sept 4
Walk of Fame Induction, St. Louis (St. Louis, MO) . . . . May 21
Walk: World Sauntering Day . . . . June 19
WalkAmerica, March of Dimes . . . . Apr 29
Walken, Christopher: Birth . . . . Mar 31
Walker, Alice: Birth . . . . Feb 9
Walker, Clint: Birth . . . . May 30
Walker, Herschel: Birth . . . . Mar 3
Walker, Jimmie: Birth . . . . June 25
Walker, Mary E.: Birth Anniv . . . . Nov 26
Walker, Mort: Birth . . . . Sept 3
Walker, Wesley: Birth . . . . May 26
Walking Week, Natl 6558, 6684 . . . . Page 488
Walktoberfest (Amer Diabetes Assn) . . . . . . Sept 30
Wallace, George, shot: Anniv . . . . May 15
Wallace, Henry A.: Birth Anniv . . . . Oct 7
Wallace, Mike: Birth . . . . May 9
Wallach, Eli: Birth . . . . Dec 7
Wallenberg, Raoul: Birth Anniv . . . . Aug 4
Wallet, Skeezix: Birth . . . . Feb 14
Wallop, Malcolm: Birth . . . . Feb 27
Walpurgis Night . . . . Apr 30
Walter Plinge Day (England) . . . . Dec 2
Walters, Barbara: Birth . . . . Sept 25
Walters, David: Birth . . . . Nov 20
Walters, Vernon A.: Birth . . . . Jan 3
Walton, George: Death Anniv . . . . Feb 2
Walton, Izaac: Birth Anniv . . . . Aug 9
Wambaugh, Joseph: Birth . . . . Jan 22
Wapner, Joseph: Birth . . . . Nov 15
War Against Iraq Begins: Anniv . . . . Jan 16
War Crimes Trial (Germany) Anniv . . . . Oct 18
War of 1812: Declaration Anniv . . . . June 18
War of the Worlds Broadcast: Anniv . . . . Oct 30
War on Poverty Anniv . . . . Jan 8
Ward, Benjamin: Birth . . . . Aug 10
Ward, Burt: Birth . . . . July 6

Index ☆ Chase's 1995 Calendar of Events ☆

Ward (cont'd)—Whistling

Ward, Montgomery, Seized: Anniv . . . . Apr 26
Warden, Jack: Birth . . . . Sept 18
Warner, Charles Dudley: Birth Anniv . . . . . . Sept 12
Warner, David: Birth . . . . July 29
Warner, John William: Birth . . . . Feb 18
Warner, Malcom Jamal: Birth . . . . Aug 18
Warren Commission Report: Anniv . . . . Sept 27
Warren, Earl: Birth Anniv . . . . Mar 19
Warren, Lesley Ann: Birth . . . . Aug 16
Warren, Michael: Birth . . . . Mar 5
Warren, Robert: Birth Anniv . . . . Apr 24
Warrick, Ruth: Birth . . . . June 29
Warwick, Dionne: Birth . . . . Dec 12
Washbourne, Mona: Birth . . . . Nov 27
Washington (State)
Admission Day . . . . Nov 11
Bumbershoot: The Seattle Arts Fest (Seattle) . . . . Sept 1
Corvette and High Performance Meet (Puyallup) . . . . Feb 11, June 24
Daffodil Fest Parade (Tacoma/Puyallup/Sumner/Orting) . . . . Apr 22
Everett Salty Sea Days (Everett) . . . . June 1
Hillhaven Ho Ho Hotline (Seattle) . . . . Dec 11
Hillhaven Thank You Santa Hotline (Seattle) . . . . Dec 26
Limousine Scavenger Hunt (Tacoma) . . . . Oct 26
Music Educ NW Div In-Service Conf (Spokane) . . . . Feb 17
Musical Instrument Auction (Seattle) . . . . May 26
NAIA Swimming and Diving Chmpshps (Federal Way) . . . . Mar 1
Newspaper in Education and Literacy Conf . . . . June 7
Nordic Yule Fest (Seattle) . . . . Nov 18
Northwest Flower and Garden Show (Seattle) . . . . Feb 22
Northwest Folklife Fest (Seattle) . . . . May 26
Norwegian 17th of May Fest (Seattle) . . . May 17
Pacific Rim Wildlife Art Show (Tacoma) . Sept 22
Play Tacoma Celebration (Tacoma) . . . . July 1
Professional Secretaries Intl Conv (Seattle) . . . . July 23
Puyallup Spring Fair (Puyallup) . . . . Apr 20
Seafair (Seattle) . . . . July 14
Seattle Intl Children's Fest (Seattle) . . . . . . May 8
Skagit Valley Tulip Fest (Mt Vernon) . . . . . Mar 31
Stay Away from Seattle Day . . . . Sept 16
Steilacoom Apple Squeeze (Steilacoom) . Oct 15
Steilacoom Salmon Bake (Steilacoom) . . July 30
Tivoli—A Bible of Scandinavia (Seattle) . . . July 8
Toppenish Mural Society's Mural-in-a-Day (Toppenish) . . . . June 3
Washington State Activity Professionals Conv (Walla Walla) . . . . Aug 17
Washington State Apple Blossom Fest (Wenatchee) . . . . Apr 27
Western Washington Fair (Puyallup) . . . . . . Sept 8
ZooLights (Tacoma) . . . . Dec 1
Washington, Booker T.: Birth Anniv . . . . Apr 5
Washington, DC
American Soc of Assn Executives Mtg/Expo . . . . Aug 13
Bastille Day Moonlight Golf . . . . July 14
Black History Month Kickoff . . . . Jan 27
District Establishing Legislation Anniv . . . July 16
FHA Natl Leadership Meeting (Washington) . . . . July 9
Frisbee Disc Fest, Natl (Washington) . . . . . Sept 2
Geography Bee Finals, Natl . . . . May 23
Independence Day Celebration . . . . July 4
Intl Day of the Seal . . . . Mar 5
Invasion Anniv . . . . Aug 24
Keep America Beautiful Annual Meeting . . Dec 6
League of Women Voters Council (Wash) . . . . June 10
Library Legislative Day . . . . Apr 11
Natl Farm Animals Awareness Week (Washington) . . . . Sept 17
Spelling Bee Finals, Natl . . . . May 31
The Jefferson Awards . . . . June 21
White House Easter Egg Roll . . . . Apr 17
Youth of the Year, Natl . . . . Sept 18
Washington, Denzel: Birth . . . . Dec 28
Washington, George
Birth Anniv . . . . Feb 22
Birthday Observance (Legal Holiday) . . . . Feb 20
Death Anniv . . . . Dec 14
George Washington Birthday Parade (Alexandria, VA) . . . . Feb 20
George Washington Birthnight Banquet/Ball (Alexandria, VA) . . . . Feb 18
Jalapeno Fest (Laredo, TX) . . . . Feb 17
Presidential Inauguration Anniv . . . . Apr 30
Takes Command of Continental Army: Anniv . . . . July 3
Washington Birthday Fest (Eustis, FL) . . . Feb 10
Washington Cross Delaware Reenact (Wash Crossing, PA) . . . . Dec 25
Washington Monument: Dedication Anniv Feb 21
Washington's Birthday Celeb (Laredo, TX) Feb 9
Washington, Grover: Birth . . . . Dec 12
Washington, Harold: Elected First Chicago Black Mayor . . . . Apr 11
Washington, Martha: Birth Anniversary . . . June 21
Wasserstein, Wendy: Birth . . . . Oct 18
Watchers Week, Man . . . . Jan 16
Water Closet Incident: Anniv . . . . Feb 11
Water: Conserve Water/Detect-a-Leak Week May 7
Water Quality Month, Natl . . . . Aug 1
Water Ski Days (Lake City, MN) . . . . June 23
Water Ski Race, Catalina (Long Beach, CA) Aug 13
Water-Drawing Fest (Japan) . . . . Mar 1
Waterfowl Festival (Easton, MD) . . . . Nov 10
Waterfront, On the (Rockford, IL) . . . . Sept 1
Watergate Day . . . . June 17
Watermelon Fest (Rush Springs, OK) . . . . . . Aug 12
Watermelon Fest, Hope (Hope, AR) . . . . Aug 17
Watermelon Thump (Luling, TX) . . . . June 22
Waters, Ethel: Birth Anniv . . . . Oct 31
Waterston, Sam: Birth . . . . Nov 15
Watley, Jody: Birth . . . . Jan 30
Watson, Doc: Birth . . . . Mar 2
Watson, Tom: Birth . . . . Sept 4
Watt, James: Birth Anniv . . . . Jan 19
Watticism Day . . . . Sept 21
Wattleton, Alyce Faye: Birth . . . . July 8
Watts, Andre: Birth . . . . June 20
Watts, Charlie: Birth . . . . June 2
Watts Riot: Anniv . . . . Aug 11
Wayne, Anthony: Day . . . . July 11
Wayne, David: Birth . . . . June 30
Wayne, John: Birth Anniv . . . . May 26
Wear Purple for Peace Day: Intergalactic Holiday . . . . May 16
Weather: Canada's Coldest Recorded Temperature: Anniv . . . . Jan 31
Weather: Meteorological Day, World (UN) . . Mar 23
Weather Observer's Day, Natl . . . . May 4
Weather Vane, World's Largest: Anniv . . . . Sept 15
Weather: Warmest US Winter on Record Declared . . . . Mar 8
Weatherman's Day . . . . Feb 5
Weaver, Dennis: Birth . . . . June 4
Weaver, Fritz: Birth . . . . Jan 19
Weaver, Robert C.: First Black US Cabinet Member: Anniv . . . . Jan 18
Weaver, Sigourney: Birth . . . . Oct 8
Weaving, Spinning and Week, Natl . . . . Oct 2
Webb, James: Birth . . . . Feb 9
Webb, Spud: Birth . . . . July 13
Weber, Carl Maria Von: Birth Anniv . . . . Dec 18
Webster, Daniel: Birth Anniv . . . . Jan 18
Webster, Noah: Birth Anniversary . . . . Oct 16
Webster-Ashburton Treaty Signed . . . . Aug 9
Wedding: Cocoanut Grove Bridal Expo (Santa Cruz, CA) . . . . Jan 22
Wedding of Giants (Belgium) . . . . Aug 27
Wedding of the Sea (Atlantic City, NJ) . . . . . Aug 15
Weddings: Bridal Expo (Davenport, IA) . . . . Jan 28
Weddings Month, Natl . . . . Feb 1
Wedgewood, Josiah: Birth Anniv . . . . July 12
Week of the Ocean, Natl . . . . Apr 16
Weems, Mason L (Parson): Birth Anniv . . . . . Oct 11
Wegman, William: Birth . . . . Dec 2
Weicker, Lowell P.: Birth . . . . May 16
Weights and Measures Day . . . . May 20
Weinberg, Max M.: Birth . . . . Apr 13
Weinberger, Caspar Willard: Birth . . . . Aug 18
Weird Contest Week . . . . Aug 14
Weiskopf, Tom: Birth . . . . Nov 9
Weitz, Bruce: Birth . . . . May 27
Weizmann, Chaim: Birth Anniv . . . . Nov 27
Welch, John Francis, Jr: Birth . . . . Nov 19
Welch, Raquel: Birth . . . . Sept 5
Welcomegiving Day, You're . . . . Nov 24
Weld, Tuesday: Birth . . . . Aug 27
Weld, William F.: Birth . . . . July 31
Welding Month, Natl . . . . Apr 1
Welfare. See Health and Welfare
Welk, Lawrence: Birth Anniv . . . . Mar 11
Wells, H.G.: Birth Anniv . . . . Sept 21
Wells, Ida B.: Birth Anniv . . . . July 16
Wells, Kitty: Birth . . . . Aug 30
Wells, Mary: Birth Anniv . . . . May 13
Wellstone, Paul D.: Birth . . . . July 21
Welty, Eudora: Birth . . . . Apr 13
Wendell, Bill: Birth . . . . Mar 22
Wendt, George: Birth . . . . Oct 17
Wenner, Jann: Birth . . . . Jan 7
Werfel, Franz: Birth Anniv . . . . Sept 10
West. See Cowboys/Old West
West, Adam: Birth . . . . Sept 19
West, Dottie: Birth . . . . Oct 11
West Michigan Fair (Ludington, MI) . . . . Aug 21
West, Rebecca: Birth Anniv . . . . Dec 25
West Virginia
Admission Day . . . . June 20
Boo at the Zoo (Wheeling) . . . . Oct 21
Boulevard Events/Toyota Times! (Charleston) . . . . Aug 27
Burlington Old-Fash Apple Harvest Fest (Burlington) . . . . Sept 30
Canaan Valley Holiday Weekend (Terra Alta) . . . . May 26
Cathedral State Park Weekend (Terra Alta) . . . . Sept 8
Charleston Distance Run (Charleston) . . . Sept 2
Charleston Sternwheel Regatta Fest (Charleston) . . . . Aug 25
Cherry River Festival (Richwood) . . . . Aug 6
Cranberry Mountain Spring Nature Tour (Richwood) . . . . May 13
Cranesville Weekend (Terra Alta) . . . . June 2
Dogwood Arts & Crafts Fest (Huntington) . . . . Apr 28
Dolly Sods Weekend (Terra Alta) . . . . Aug 4
Fall Classic Horseshoe Tourn (Fairmont) Sept 23
Fall Folk Dance Camp (Wheeling) . . . . Sept 1
Farm Days Weekend (Wheeling) . . . . June 17
Feast of the Ramson (Richwood) . . . . Apr 15
Good Egg Treasure Hunt (Wheeling) . . . . . Apr 14
Halloween Night Walks (Oglebay) . . . . Oct 27
Haunted Trail and Kiddie Trail (Huntington) . . . . Oct 13
Head-of-the-Mon-River Horseshoe Tourn (Fairmont) . . . . May 27
Junior Nature Camp (Wheeling) . . . . July 30
Mountain State Forest Fest (Elkins) . . . . . Sept 30
Nongame Wildlife Weekend (Davis) . . . . . . June 2
Oglebay Institute Winter Lodge Program (Wheeling) . . . . Jan 2
Potomac South Branch Weekend (Terra Alta) . . . . July 14
Preston County Buckwheat Fest (Kingwood) . . . . Sept 28
Rabbit Show (Wheeling) . . . . Apr 15
Richwood Winterfest (Richwood) . . . . Feb 9
St. Patrick's Day Celebration (Huntington) . . . . Mar 17
Shanghai Parade (Lewisburg) . . . . Jan 1
Spring Folk Dance Camp (Wheeling) . . . . May 26
Strawberry Fest (Buckhannon) . . . . May 17
Three Rivers Fest and River Regatta (Fairmont) . . . . May 25
Tri-State Anti-Drug Air Show (Kenova) . . May 26
Turkey Trot (Parkersburg) . . . . Nov 23
Webster County Woodchopping Fest (Webster Springs) . . . . May 20
West Virginia Day Celebration (Huntington) . . . . June 20
West Virginia Glass Fest (Wheeling) . . . . . . Aug 4
West Virginia Italian Heritage Fest (Clarksburg) . . . . Sept 1
West Virginia Oil and Gas Fest (Sisterville) . . . . Sept 14
West Virginia Pumpkin Fest (Milton) . . . . . . Oct 6
West Virginia Square/Round Dance Conv (Buckhannon) . . . . Aug 4
Winter Festival of Lights (Wheeling) . . . . . . Nov 1
Western Heritage Days (Deer Lodge, MT) . . July 8
Western Samoa
Anzac Day . . . . Apr 25
Arbor Day . . . . Nov 3
National Holiday . . . . Sept 5
Samoan Fire Dance . . . . Dec 31
White Sunday . . . . Oct 8
Westinghouse, George: Birth Anniv . . . . Oct 6
Whale, Humpback Awareness Month (Honolulu, HI) . . . . Feb 1
Whales: Shamu's Birthday . . . . Sept 26
Wharton, Edith: Birth Anniv . . . . Jan 24
Wheatley, Phillis: Death Anniv . . . . Dec 5
Wheeler, William A.: Birth . . . . June 30
Whimmy Diddle Competition, World (Asheville, NC) . . . . Aug 12
Whiner's Day, Natl . . . . Dec 26
Whipple, William: Birth Anniv . . . . Jan 14
Whiskey Flat Days (Kernville, CA) . . . . Feb 17
Whistler, James: Birth Anniv . . . . July 10
Whistlers Convention, Natl (Louisburg, NC) Apr 20
Whistling: Birthday of Mother's Whistler . . . May 18

☆ Chase's 1995 Calendar of Events ☆ Index

Whitaker, Louis: Birth .... May 12
White, Barry: Birth .... Sept 12
White, Betty: Birth .... Jan 17
White, Byron Raymond: Birth .... June 8
White Cane Safety Day (Pres Proc) .... Oct 15
White, Edward Douglass: Birth Anniv ...... Nov 3
White, Gilbert: Birth Anniversary .... July 18
White House Easter Egg Roll (Washington, DC) .... Apr 17
White House 200th Anniversary 6588 .... Page 488
White, Jo Jo: Birth .... Nov 16
White, Ryan: Death Anniv .... Apr 8
White Shirt Day .... Feb 11
White, Stanford: Birth Anniv .... Nov 9
White Sunday (American Samoa) .... Oct 8
White Sunday (Western Samoa) .... Oct 8
White, Vanna: Birth .... Feb 18
White, William Allen: Birth Anniv .... Feb 10
Whitehead, Robert: Birth .... Mar 3
Whitelaw, Billie: Birth .... June 6
Whiting, Margaret: Birth .... July 22
Whitman, Christine T.: Birth .... Sept 26
Whitman, Slim (Otis): Birth .... Jan 20
Whitman, Stuart: Birth .... Feb 1
Whitman, Walt: Birth Anniv .... May 31
Whitmire, Kathryn: Birth .... Aug 15
Whitmonday .... June 5
Whitsunday .... June 4
Whittier, John Greenleaf: Birth Anniv .... Dec 17
Who, Dr., Premiere: Anniv .... Nov 23
Wiccan Holidays
  Beltane .... Apr 30
  Imbolc .... Feb 2
  Lughnasadh .... Aug 1
  Mabon .... Sept 23
  Midsummer .... June 21
  Ostara .... Mar 20
  Samhain .... Oct 31
  Yule .... Dec 22
Wicker, Tom: Birth .... June 18
Widmark, Richard: Birth .... Dec 26
Wiest, Dianne: Birth .... Mar 28
Wiggin, Kate Douglas: Birth Anniv .... Sept 28
Wilbur, Richard: Birth .... Mar 1
Wild Foods Day (Pikeville, TN) .... Oct 28
Wild Game Cookoff, State (Oakwood Lakes State Park, SD) .... Aug 12
Wild West Days (Rockford, IL) .... May 20
Wilde, Oscar: Birth Anniv .... Oct 16
Wilder, Gene: Birth .... June 11
Wilder, L. Douglas: First Black Governor Elected: Anniv .... Nov 7
Wilder, Laura Ingalls Gingerbread Sociable (Pomona, CA) .... Feb 4
Wilder, Thornton: Birth Anniv .... Apr 17
Wildflower Fest of the Arts (Dahlonega, GA) .... May 20
Wildflower Tours/Birdwalks, Roan Mtn (Roan Mountain, TN) .... May 6
Wildfowl Art & Decoy Show, Toms River (Toms River, NJ) .... Feb 4
Wilhelm Tell Fest (New Glarus, WI) .... Sept 2
Wilkins, Dominique: Birth .... Jan 12
Wilkins, Roy: Birth Anniv .... Aug 30
Will, George F.: Birth .... May 4
Will Rogers Day .... Nov 4
Will Rogers Days (Claremore, OK) .... Nov 3
Willard, Archibald M.: Birth Anniv .... Aug 22
Willard, Frances E. C.: Birth Anniv ...... Sept 28
William, Prince: Birth .... June 21
William the Conqueror: Death Anniv .... Sept 9
Williams, Andy: Birth .... Dec 3
Williams, Archie: Birth Anniv .... May 1
Williams, Billy Dee: Birth .... Apr 6
Williams, Cindy: Birth .... Aug 22
Williams, Clarence, III: Birth .... Aug 21
Williams, Curtis: Birth .... Dec 11
Williams, Deniece: Birth .... June 3
Williams, Esther: Birth .... Aug 8
Williams, Hank, Jr: Birth .... May 26
Williams, Hank, Sr.: Birth Anniv .... Sept 17
Williams, John: Birth .... Feb 8
Williams, Mason: Birth .... Aug 24
Williams, Paul: Birth .... Sept 19
Williams, Robin: Birth .... July 21
Williams, Ted: Birth .... Aug 30
Williams, Tennessee: Birth Anniv .... Mar 26
Williams, William: Birth Anniv .... Apr 8
Williamson, Fred: Birth .... Mar 5
Willis, Bruce: Birth .... Mar 19
Willkie, Wendell L.: Birth Anniv .... Feb 18
Willow Tree Festival (Gordon, NE) .... Sept 9
Wilson, Ann: Birth .... June 19
Wilson, Brian: Birth .... June 20

Wilson, Edith: Birth Anniv .... Oct 15
Wilson, Ellen L.: Birth Anniv .... May 15
Wilson, Flip: Birth .... Dec 8
Wilson, Gahan: Birth .... Feb 18
Wilson, Harold: Birth .... Mar 11
Wilson, Henry: Birth Anniv .... Feb 16
Wilson, Mary: Birth .... Mar 4
Wilson, Nancy: Birth .... Feb 20
Wilson, Tom: Birth .... Aug 1
Wilson, William G.: Alcoholics Anonymous Founding Anniv .... June 10
Wilson, Woodrow: Birth Anniv .... Dec 28
Winchell, Paul: Birth .... Dec 21
Winchell, Walter: Birth Anniv .... Apr 7
Wind Fest, Waco (Waco, TX) .... Mar 18
Windmill Day, Natl (Netherlands) .... May 13
Windom, William: Birth .... Sept 28
Windsor, Duke of: Marriage Anniv .... June 3
Windstorms: Discovery Launches Satellite of Discovery Anniv .... Sept 15
Windsurfing: Aloha Classic Pro Chmshp (Maui, HI) .... Oct 29
Windsurfing: Tioman Intl Regatta (Pahang, Malaysia) .... Sept 3
Wine & Roses Fest (Bryan, TX) .... May 6
Wine, American, Appreciation Wk 6530 .. Page 488
Wine and Harvest Fest, Michigan (Kalamazoo/Paw Paw, MI) .... Sept 6
Wine, Blessing of the (Greiveldange, Luxembourg) .... Dec 26
Wine Fest, Town Pt Pk (Norfolk, VA) ...... Oct 21
Winefest, Lapic (New Brighton, PA) .... Oct 7
Winfield, David: Birth .... Oct 3
Winfrey, Oprah: Birth .... Jan 29
Winger, Debra: Birth .... May 16
Wings 'n' Water Festival (Stone Harbor, NJ) .... Sept 16
Wings Over the Platte (Grand Island, NE) . Mar 17
Winkerbean, Funky: Anniv .... Mar 27
Winkler, Henry: Birth .... Oct 30
Winston Cup, Daytona 500 NASCAR-FIA (Daytona Bch, FL) .... Feb 19
Winston Select 500 Race Week (Talladega, AL) .... May 4
Winter Begins .... Dec 22
Winter Edgar: Birth .... Dec 28
Winter Festivals
  Aspen/Snowmass Winterskol (Snowmass Village, CO) .... Jan 11
  Atlantic Winter Fair (Halifax, NS, Canada) . Oct 7
  Badger State Winter Games (Wausau, WI) . Feb 3
  Brookfield Ice Harvest (Brookfield, VT) . Jan 28
  Calgary Winter Fest (Calgary, AB, Canada) .... Feb 17
  Dickside Festival (Gravenhurst, Ont, Canada) .... Aug 11
  Dog Sled Chmpshp/Winter Carnival (Wisconsin Dells, WI) .... Feb 11
  Horton Bay Winter Olympics (Horton Bay, MI) .... Feb 26
  Icescape (Appleton, WI) .... Feb 17
  Jasper in January (Jasper, AB, Canada) . Jan 13
  Leadville's Crystal Carnival (Leadville, CO) Mar 3
  Mitten Tree Kick-Off Ceremony (Baltimore, MD) .... Dec 1
  Muskoka Winter Carnival (Gravenhurst, Ont, Canada) .... Feb 23
  Richwood Winterfest (Richwood, WV) ..... Feb 9
  Rossland Winter Carnival (Rossland, BC, Canada) .... Jan 27
  St. Paul Winter Carnival (St. Paul, MN) . Jan 27
  Snowfest (Tahoe City, CA) .... Mar 3
  Snowflake Fest (Klamath Falls, OR) ..... Dec 2
  Snowman Sacrifice (Sault Ste Marie, MI) . Mar 20
  Ullr Fest (Breckenridge, CO) .... Jan 9
  Warmest US Winter on Record Declared .. Mar 8
  Wayne State Univ: Funeral for Winter (Detroit, MI) .... Apr 13
  Winter Carnival (Bancroft, Ont, Canada) . Feb 10
  Winter Carnival (Red Lodge, MT) .... Mar 3
  Winter Carnival Bon Soo (Sault Ste Marie, Ont, Canada) .... Jan 27
  Winter Fest (Lake City, MN) .... Jan 28
  Winter Fest (New Glarus, WI) .... Jan 13
  Winter Fest of Lights (Wheeling, WV) ... Nov 1
  Winter Fun Fest (New Preston, CT) ...... Feb 12
  Winterfest Balloon Rally (Angel Fire, NM) . Feb 3
  Winterlude (Ottawa, Ont, Canada) .... Feb 3
  Wisconsin Dells Flake Out Fest (Wisconsin Dells, WI) .... Jan 20
Winter, Halfway Point .... Feb 4
Winter, Johnny: Birth .... Feb 23
Winter Solstice: Yalda (Iran) .... Dec 21
Winters, Jonathan: Birth .... Nov 11

Winters, Shelley: Birth .... Aug 18
Winwood, Steve: Birth .... May 12
Wisconsin
  Admission Day .... May 29
  Aerobatic Chmpshps, Intl (Fond du Lac) . Aug 6
  American Birkebeiner (Cable to Hayward) .... Feb 23
  Automotion (Wisconsin Dells) .... May 13
  Badger State Summer Games .... June 9
  Badger State Winter Games (Wausau) ..... Feb 3
  Bald Eagle Day (Cassville) .... Jan 28
  Bastille Days (Milwaukee) .... July 13
  Beef-a-Rama/Arts and Crafts Show (Minocqua) .... Sept 30
  Bratwurst Days (Sheboygan) .... Aug 4
  Buck Fever days and Chili Feed (Monocqua) .... Nov 17
  Cedar Grove Holland Fest (Cedar Grove) . July 28
  Celebrate (Appleton) .... May 13
  Christmas Craft Fair USA: Indoor Show (Milwaukee) .... Dec 9
  Co-Ed Broomball Hockey Tourn (Sheboygan) .... Feb 4
  Coho Family Festival (Sheboygan) .... July 21
  Country Affair (Menomonee Falls) .... Oct 21
  Country Jam USA (Eau Claire) .... July 20
  Cows on the Concourse (Madison) .... June 3
  Craft Fair USA: Indoor Show (Milwaukee) .... Apr 22, July 22, Sept 23, Oct 28
  Dog Sled Chmpshp/Winter Carnival (Wisconsin Dells) .... Feb 11
  EAA Intl Fly-In Conv/Sport Aviation Exhib (Oshkosh) .... July 27
  Elm Farm Ollie Day Celebration .... Feb 18
  Farm Woman of Year Month .... Feb 1
  Fashion Show (Milwaukee) .... May 12
  First Friday (Milwaukee) .... Jan 6
  Fishing Has No Boundaries (Eagle River) . June 2
  Fishing Has No Boundaries (Hayward) .... May 19
  Frog Days (Melvina) .... Aug 4
  Gallery Night (Milwaukee) .... Jan 20
  German Fest (Milwaukee) .... July 28
  Great Cardboard Boat Regatta (Sheboygan) .... July 4
  Great Wisc Dells Balloon Rally (Wisconsin Dells) .... June 3
  Groundhog Day in Sun Prairie, WI (Sun Prairie) .... Feb 2
  Heritage Day Celebration (WI Dells) .... June 10
  Hmong New Year Celebration (Sheboygan) .... Nov 25
  Hodag Muskie Challenge (Rhinelander) . Sept 16
  Holiday Craft and Gift Show (Milwaukee) Nov 24
  Holiday Folk Fair (Milwaukee) .... Nov 17
  Home Improvement Show (West Allis) .... Feb 9
  Home of the Hamburger Celeb (Seymour) . Aug 5
  Icescape (Appleton) .... Feb 17
  Indian Summer Festival .... Sept 8
  Kohler Outdoor Arts Festival (Sheboygan) .... July 15
  Lake Winnebago Sturgeon Season (Fond du Lac) .... Feb 12
  Lakefront Festival of Arts (Milwaukee) . June 16
  Las Vegas Night (Minocqua) .... July 20
  Mad-City Marathon (Madison) .... May 28
  Madrigal Dinner and Concert (Milwaukee) Dec 2
  Maple Syrup Saturday (Appleton) .... Mar 25
  Masters of the Grill (Kenosha) .... June 25
  Mexican Fiesta (Milwaukee) .... Aug 25
  Mid-Cont RR Steam Snow Train (New Freedom) .... Feb 17
  Mid-Cont RR Steam Train Color Tours (New Freedom) .... Oct 7
  Milk Jug Regatta (Minocqua) .... June 11
  Milwaukee Boat Show at MECCA (Milwaukee) .... Feb 15
  Milwaukee Irish Fest (Milwaukee) .... Aug 18
  Milwaukee Journal: Al's Run (Milwaukee) Sept 23
  Milwaukee Sentinel Sports Show (Milwaukee) .... Mar 10
  Mini Grand Prix (Fond du Lac) .... May 28
  Northeastern WI Antiq Power Thresheree (Sturgeon Bay) .... Aug 19
  Oktoberfest (LaCrosse) .... Sept 29
  Old Crafts Day (Fond du Lac) .... July 9
  Old-Fashioned Ice Cream Social (Fond du Lac) .... Aug 6
  Organization Development Info Exchange (Wms Bay) .... May 23
  Polar Bear Swim (Sheboygan) .... Jan 1
  Polish Fest (Milwaukee) .... June 24
  Polish Fest (Wisconsin Dells) .... Sept 8
  Polish-American Christmas Gala (St. Francis) .... Dec 3

Index ☆ Chase's 1995 Calendar of Events ☆

Wisconsin (cont'd)—World War II

Polka Fest (Wisconsin Dells)............. **Nov 3**
Pyrotechnics Guild Fireworks Conv (Stevens Point)............. **Aug 6**
Research/Study Team Nonviolent Change (Wms Bay)............. **May 21**
Riverfest (Lacrosse)............. **June 30**
St. Cyril's Parish Fest (Sheboygan)............. **July 16**
St. Spyridon Greek Orthodox Fest (Sheboygan)............. **July 29**
Senior Day (Milwaukee)............. **Feb 15**
Shanty Days (Algoma)............. **Aug 11**
Spring Craft/Gift Show (Milwaukee)............. **Mar 11**
State Fair (Milwaukee)............. **Aug 3**
State Parks Open House and Free Fishing Day............. **June 4**
Sun Prairie's Sweet Corn Fest (Sun Prairie)............. **Aug 17**
Taste of Madison (Madison)............. **Sept 2**
Timberfest (Woodruff)............. **June 16**
Tosafest (Wauwatosa)............. **Sept 8**
Try This On: Clothing, Gender, Power (Janesville)............. **Mar 11**
Twin-O-Rama (Cassville)............. **July 21**
UFO Days (Elmwood)............. **July 28**
US Intl Snow Sculpting Competition (Milwaukee)............. **Jan 24**
Volksfest (New Glarus)............. **Aug 6**
Wausau Air Show and Fly-in (Wausau)............. **June 18**
Who-Zha-Wa Fall Fest (Wisconsin Dells) **Sept 15**
Wilhelm Tell Fest (New Glarus)............. **Sept 2**
Winter Fest (New Glarus)............. **Jan 13**
Wisconsin Cow Chip Throw (Prairie du Sac)............. **Sept 1**
Wisconsin Dells Flake Out Fest (Wisconsin Dells)............. **Jan 20**
Wollersheim Winery Grape Stomp Fest (Prairie du Sac)............. **Oct 7**
Wonago World Chmpshp Rodeo (Milwaukee)............. **June 16**
World Beef Expo (Madison)............. **Apr 6**
World Dairy Expo (Madison)............. **Oct 4**
World's Largest Trivia Contest (Stevens Point)............. **Apr 7**

Wise, Robert: Birth............. Sept 10
Wise, Thomas James: Birth Anniv............. Oct 7
Witherspoon, John: Birth Anniv............. Feb 5
Witt, Paul Junger: Birth............. Mar 20
Wizard of Oz Club Convention, Intl (Rosemont, IL)............. June 16
Wodehouse, Pelham G: Birth Anniv............. Oct 15
Wofford, Harris: Birth............. Apr 9
Wojtyla: Pope John Paul II: Birth............. May 18
Wolcott, Oliver: Birth Anniv............. Nov 20
Wolf, Peter: Birth............. Mar 7
Wolfe, James: Birth Anniv............. Jan 2
Wolfe, Tom: Birth............. Mar 2
Wolfman Jack: Birth............. Jan 21
Wollersheim Winery Grape Stomp Fest (Prairie du Sac, WI)............. Oct 7
Women
American Business Women's Assn, Natl Conv of (Portland, OR)............. **Oct 3**
American Heroine Rewarded: Anniv............. **June 8**
Around the World in 72 Days: Anniv............. **Nov 14**
Aspinwall Crosses US on Horseback: Anniv............. **July 8**
Becky Thatcher Day (MO)............. **Dec 1**
Bloomer, Amelia Jenks: Birth Anniv............. **May 27**
Business Women's Day, American............. **Sept 22**
Business Women's Week, Natl............. **Oct 16**
Canada: Persons Day............. **Oct 18**
Domestic Violence Awareness Month 6619............. **Page 488**
Equal Rights Party Founding: Anniv............. **Sept 20**
Felton, Mrs W.H., First Woman US Senator **Oct 3**
Feminine Empowerment Month, Natl............. **Mar 1**
First All-Woman Jury: Anniv............. **Sept 22**
First Doctor of Science Degree Earned by a Woman............. **June 20**
First Female FBI Agent: Anniv............. **Oct 25**
First US Woman Governor Inaugurated............. **Jan 5**
First Woman Rabbi in US: Anniv............. **June 3**
First Woman to Graduate Dental School: Anniv............. **Feb 21**
First Woman to Walk in Space............. **July 17**
First Woman US Ambassador Appointed: Anniv............. **Oct 28**
First Women's Collegiate Basketball Game: Anniv............. **Mar 22**
Friedan, Betty N.: Birth............. **Feb 4**
Fuller, Margaret: Birth Anniv............. **May 23**
Healthy Weight, Healthy Look Day............. **Jan 19**
Healthy Weight Week............. **Jan 15**
King, Billie Jean, Wins Battle of Sexes: Anniv............. **Sept 20**
League of Women Voters Anniv Celeb............. **Feb 14**
League of Women Voters Council (Wash, DC)............. **June 10**
League of Women Voters Formed: Anniv **Feb 14**
Lingerie Week, Natl............. **Apr 23**
Lucy Stone Married: Anniv............. **May 1**
Mammography Day, Natl 6615............. **Page 488**
Margaret Brent Demands a Political Voice: Anniv............. **June 24**
Martha Griffiths Speech Against Sex Discrimination............. **Feb 8**
Medical School Opened: Anniv............. **Nov 1**
Meitlisunntig (Switzerland)............. **Jan 8**
Merriam, Eve: Death Anniv............. **Apr 11**
Natl Women's Music Fest (Bloomington, IN)............. **June 1**
Papal "No" on Ordaining Women: Anniv **May 30**
Perkins, Frances (1st Woman Appointed to US Cabinet)............. **Mar 4**
Pocahontas: Death Anniv............. **Mar 21**
Pony Girls Softball Natl Championships **Aug 5**
Queen Liliuokalani Deposed: Anniv............. **Jan 17**
Re-Imaging Conf Controversy: Anniv............. **Nov 4**
Salter Elected First Woman Mayor in US: Anniv............. **Apr 4**
Secretary's Day Tea (Wilmington, DE)............. **Apr 26**
Seneca Falls Survivor Votes: Anniv............. **Nov 2**
Strength/Diversity: Japanese American Women (Logan, KS)............. **Apr 15**
Strength/Diversity: Japanese-American Women (Stony Brook, NY)............. **Jan 7**
Suffrage Parade Attacked: Anniv............. **Mar 3**
Suffragists' Voting Attempt: Anniv............. **Nov 19**
Surratt, Mary: Execution Anniv............. **July 7**
Take Our Daughters to Work Day............. **Apr 27**
Today's Woman (Winston-Salem)............. **Feb 17**
Tubman, Harriet: Death Anniv............. **Mar 10**
Universal Women's Week............. **Mar 8**
WAAC: Anniv............. **May 14**
White Woman Made Indian Chief: Anniv **Sept 18**
WIBC Queen's Tourn (Tucson, AZ)............. **May 14**
Wisconsin Farm Woman of Year Month............. **Feb 1**
Woman Runs the House: Anniv............. **June 20**
Women and Girls in Sports Day, Natl 6527, 6649............. **Page 488**
Women in Agriculture Day, Natl 6698............. **Page 488**
Women of Achievement Month (NY)............. **Sept 1**
Women Veterans Recognition Week, Natl 6622............. **Page 488**
Women's Convention Held (Seneca Falls): Anniv............. **July 19**
Women's Day, Intl (UN)............. **Mar 8**
Women's Equality Day............. **Aug 26**
Women's Equality Day 6586............. **Page 488**
Women's Get-Away Weekend, Natl............. **Mar 10**
Women's Hall of Fame, Natl: Dedication Anniv............. **July 21**
Women's Heart Health Day............. **Feb 1**
Women's History Month 6537, 6654............. **Page 488**
Women's History Month 6654............. **Page 488**
Women's History Month, Natl............. **Mar 1**
Women's Rights Conv Called: Anniv............. **July 14**
Women's Suffrage Amendment Introduced: Anniv............. **Jan 10**
Women's Suffrage Anniv Celeb............. **Aug 26**
Working Womens' Day, Intl............. **Mar 8**

Wonder, Stevie: Birth............. May 13
Wood Carvers Fest (Blackduck, MN)............. July 29
Wood, Grant, Art Fest (Stone City-Anamosa, IA)............. June 11
Wood, Grant: Birth Anniv............. Feb 13
Wood, Ron: Birth............. June 1
Woodard, Alfre: Birth............. Nov 2
Woodchopping Fest, Webster County (Webster Springs, WV)............. May 20
Woodcock, Leonard F.: Birth............. Feb 16
Woodcraft Fest (Grand Rapids, MN)............. July 15
Woodcraft: Holzfest (Amana, IA)............. Aug 18
Woodhull, Victoria C.: Birth Anniv............. Sept 23
Woodmen of the World Founders Day............. June 6
Woodmen Ranger's Day............. May 12
Woodruff, Judy: Birth............. Nov 20
Woods, Granville T.: Birth Anniv............. Apr 23
Woods, James: Birth............. Apr 18
Woodsmen's Days, Tupper Lake (Tupper Lake, NY)............. July 8
Woodson, Carter G: Associated Publ Founded: Anniv Celeb............. Nov 18
Woodstock Anniv............. Aug 15
Woodward, Bob: Birth............. Mar 26
Woodward, Joanne: Birth............. Feb 27
Woodward, Robert B.: Birth Anniv............. Apr 10
Woodworking Month, Natl............. Apr 1
Wool Day, Palatine Museum (Schoharie, NY)............. Sept 23
Woolf, Virginia: Birth Anniv............. Jan 25
Woolley, Mary E.: Birth Anniv............. July 13
Word Processing Transcriptionist Week, Natl Jan 8
Wordsworth, William: Birth Anniv............. Apr 7
Work. See Employment-Related Events
Workaholic's Day............. July 5
Workers, Third Shift Day, Natl............. May 10
Working Women's Day, Intl............. Mar 8
World AIDS Day, UN............. Dec 1
World Campaign for the Biosphere............. June 5
World Day of Prayer............. Mar 3
World Sauntering Day............. June 19
World Trade Week (Pres Proc)............. May 14
World Trade Week 6563............. Page 488
World War I
Alvin C. York Day............. **Oct 8**
Anzac Day............. **Apr 25**
Baseball Declared Non-Essential: Anniv **July 20**
Beginning and Ending Anniv............. **June 28**
Campbell Becomes 1st Amer Air ACE: Anniv............. **Apr 14**
Mata Hari: Execution Anniv............. **Oct 15**
Neutrality Appeal, American: Anniv............. **Aug 18**
Red Baron Shot Down: Anniv............. **Apr 21**
US Food Rationing: Anniv............. **July 1**
Versailles Peace Conference: Anniv............. **Jan 18**
Veterans Bonus Army Eviction: Anniv............. **July 28**
World War II
Americans Land at Puerto Princesa: Anniv............. **Feb 28**
Arnold, Henry H.: Birth Anniv............. **June 25**
Atomic Bomb Delivered: Anniv............. **July 26**
Atomic Bomb Tested: Anniv............. **July 16**
Badoglio, Pietro: Birth Anniv............. **Sept 28**
Bataan Death March: Anniv............. **Apr 10**
Battle of Okinawa Begins: Anniv............. **Apr 1**
Battle of Okinawa Ended: Anniv............. **June 21**
Belsen Concentration Camp Liberated: Anniv............. **Apr 15**
Berlin Surrenders: Anniv............. **May 2**
Bormann, Martin: Birth Anniv............. **June 17**
Bourke-White, Margaret: Birth Anniv............. **June 14**
Braun, Eva: Birth Anniv............. **Apr 30**
Christmas Fireside Chat Warning: Anniv **Dec 25**
Churchill Enters Germany: Anniv............. **Mar 25**
Clark, Mark: Birth Anniv............. **May 1**
D-Day: Anniv............. **June 6**
Day in the Warsaw Ghetto (Denver, CO)............. **Apr 8**
Day in the Warsaw Ghetto (Houston, TX)............. **Feb 4**
De Gaulle, Charles: Birth Anniv............. **Nov 22**
Declaration Anniv............. **Sept 3**
Diary of Anne Frank: Last Entry: Anniv............. **Aug 1**
East Meets West: Anniv............. **Apr 25**
First Medal of Honor: Anniv............. **Feb 10**
First Surface-to-Surface Missile: Anniv............. **Dec 24**
German 16-Year-Olds Drafted: Anniv............. **Mar 5**
Germans Try to Deal: Anniv............. **Apr 22**
Germany's First Surrender: Anniv............. **May 7**
Germany's Second Surrender: Anniv............. **May 8**
Gillars, Mildred E.: Death Anniv............. **June 25**
Graham, Calvin ("Baby Vet") Birth Anniv............. **Apr 3**
Halsey, William "Bull": Birth Anniv............. **Oct 30**
Himmler, Heinrich: Birth Anniv............. **Oct 7**
Hirohito's Radio Address: Anniv............. **Aug 15**
Hitler Moves to Bunker: Anniv............. **Apr 1**
Hitler Takes Command of Berlin: Anniv............. **Apr 23**
Howard, Leslie: Death Anniv............. **June 1**
Imperial Palace Attacked: Anniv............. **Aug 14**
Indianapolis Sunk: Anniv............. **July 29**
Italy Declares War on Japan: Anniv............. **July 14**
Japan's Unconditional Surrender: Anniv **Aug 10**
Japanese Internment: Anniv............. **Feb 19**
Japanese Reject Surrender: Anniv............. **Aug 9**
Kristallnacht Anniv............. **Nov 9**
Krupp, Alfried: Birth Anniv............. **Aug 13**
Liberation of Buchenwald: Anniv............. **Apr 10**
Liberation of Dachau: Anniv............. **Apr 29**
Liberation of Milan: Anniv............. **Apr 25**
Manstein, Erich von: Birth Anniv............. **Nov 24**
Marshall, George C.: Birth Anniv............. **Dec 31**
McAfee, Mildred: Birth............. **May 12**
Montgomery, Bernard Law: Birth Anniv............. **Nov 17**
Mussolini Executed: Anniv............. **Apr 28**
Napalm Used: Anniv............. **July 11**
Nimitz, Chester: Birth Anniv............. **Feb 24**
Oak Ridge Atomic Plant Begun: Anniv............. **Aug 1**
Operation Floating Chrysanthemum: Anniv............. **Apr 6**
Operation Overcast............. **July 6**

✯ Chase's 1995 Calendar of Events ✯ Index

Paulus, Friedrich: Birth Anniv .... Sept 23
Potsdam Conference: Anniv .... July 17
Potsdam Declaration: Anniv .... July 24
Produce for Victory: Posters/Home Front (Hampton, VA) .... Mar 18
Produce for Victory: Posters/Home Front (Logan, KS) .... Jan 14
Pyle, Ernest: Death Anniv .... Apr 18
R. Prager Meml Day: German-Amer Remembrance .... Apr 5
Remagen Bridge Capture: Anniversary ... Mar 7
Ridgway, Matthew Bunker: Birth Anniv ... Mar 3
Rogers, Edith Nourse: Birth Anniv .... Mar 19
Rommel, Erwin: Birth Anniv .... Nov 15
Roosevelt, Franklin D.: Death Anniv ...... Apr 12
Russia: Victory Day .... May 9
Russians Fire on Berlin: Anniv .... Apr 20
Smith, Holland: Birth Anniv .... Apr 20
Soviet Advance in Manchuria: Anniv ..... Aug 14
Stratton, Dorothy C.: Birthday .... Mar 24
Suicide Weapon Introduced: Anniv ...... Mar 18
Time for Natl Observance of 50th Anniv 6568 .... Page 488
Tokyo Blanket Bombing: Anniv .... Mar 9
Train for Paris: Anniv .... Jan 15
UN Charter Signed: Anniv .... June 26
US Landing on Iwo Jima: Anniv .... Feb 19
US Landing on Luzon: Anniv .... Jan 9
V-E Day .... May 8
V-Mail Delivery: Anniv .... June 22
Vietnam Conflict Begins: Anniv .... Aug 22
WAAC: Anniv .... May 14
Warbirds in Action Air Show (Shafter, CA) .... Apr 22
Yamamoto, Isoroku: Birth Anniv .... Apr 4
Yamato Suicide: Anniv .... Apr 7
Yugoslavis/USSR Treaty: Anniv .... Apr 11
World's End Day .... Oct 22
Worley, Jo Anne: Birth .... Sept 6
Worthy, James: Birth .... Feb 27
Wouk, Herman: Birth .... May 27
Wounded Knee Massacre: Anniv .... Dec 29
Wren, Christopher: Birth Anniv .... Oct 20
Wren, Day of the (Ireland) .... Dec 26
Wrestling Chmpshp, Big Ten (Bloomington, IN) .... Mar 4
Wrestling Chmpshps, NAIA .... Mar 10
Wright Brothers Day (Pres Proc) .... Dec 17
Wright Brothers Day 6638 .... Page 488
Wright Brothers First Powered Flight: Anniv .... Dec 17
Wright, Frank Lloyd: Birth Anniv .... June 8
Wright (Frank Lloyd) Plus (Oak Park, IL) . May 20
Wright, Gary: Birth .... Apr 26
Wright, Orville: Birth Anniv .... Aug 19
Wright, Rick: Birth .... July 28
Wright, Wilbur: Birth Anniv .... Apr 16
Wristwrestling, World Chmpsh (Petaluma, CA) .... Oct 14
Wrong Days (Wright, MN) .... July 14
Wrong Way Corrigan Day .... July 17
Wurstfest (New Braunfels, TX) .... Nov 3
Wyatt Earp Christmas Walk (Monmouth, IL) Sept 9
Wyatt Earp Homecoming Western Days (Monmouth, IL) .... Aug 5
Wyatt Earp OK Corral Anniv (Monmouth, IL) .... Oct 21
Wyatt, Jane: Birth .... Aug 12
Wyman, Bill: Birth .... Oct 24
Wyman, Jane: Birth .... Jan 4
Wynette, Tammy: Birth .... May 4
Wyoming
Admission Day .... July 10
Cheyenne Frontier Days (Cheyenne) ... July 21
Cowboy State Games Winter Sports Fest (Casper) .... Feb 18
Cowboy State Summer Games (Casper) . June 8
Don Macleod Cookout (Jackson) .... May 20
Fall Arts Fest (Jackson) .... Sept 15
First US Woman Governor Inaugurated ... Jan 5
Fishing Has No Boundaries (Thermopolis) .... June 6
Grand Teton Music Fest (Teton Village) . June 27
National Finals High School Rodeo (Gillette) .... July 24
Old West Days (Jackson) .... May 26
Pole, Pedal, Paddle (Jackson) .... Apr 1
Quilting in the Tetons (Jackson) .... Oct 1
Shriner's Invitational Cutter Races (Jackson) .... Feb 18
State Fair (Douglas) .... Aug 21
Thermopolis Day (Thermopolis) .... Jan 6
World Chmpshp Snowmobile Hillclimb (Jackson) .... Mar 24
Wythe, George: Death Anniv .... June 8

X

X-Ray Discovery Day: Anniv .... Nov 8
Xerox 914 Donated to Smithsonian: Anniv . Aug 20

Y

Yacht Race, Intl: Anniv .... Aug 22
Yalda (Iran) .... Dec 21
Yale, Linus: Birth Anniv .... Apr 4
Yale University Founded: Anniv .... Oct 16
Yalow, Rosalyn: Birth .... July 19
YAM Programs for Youth Art Month (Huntsville, AL) .... Mar 12
Yamaguchi, Kristi: Birth .... July 12
Yamamoto, Isokoru: Birth Anniv .... Apr 4
Yambilee, Louisiana (Opelousas, LA) .... Oct 25
Yamboree, East Texas (Gilmer, TX) ..... Oct 19
Yankovic, Alfred Matthew "Weird Al": Birth . Oct 23
Yankton Riverboat Days/Summer Arts Fest (Yankton, SD) .... Aug 18
Yarborough, Cale: Birth .... Mar 27
Yard Sale, World's Greatest (York, PA) .. June 10
Yarrow, Peter: Birth .... May 31
Yastrzemski, Carl: Birth .... Aug 22
Yeager, Chuck: Birth .... Feb 13
Yeats, William B: Birth Anniv .... June 13
Yell "Fudge" at the Cobras in North America Day .... June 2
Yeltsin, Boris, Inaugurated Russian President: Anniv .... July 10
Yeltsin Shells White House: Anniv ...... Oct 4
Yemen Arab Republic: National Holiday ... Sept 26
Yemen: Natl Day .... May 22
Yevtushenko, Yevgeny Alelesandrovich: Birth .... July 18
Yo-Yo Day, Natl (Arcade, NY) .... June 10
Yom Hashoah (Israel) .... Apr 27
Yom Kippur .... Oct 4
Yom Kippur Begins .... Oct 3
York, Alvin C: Day .... Oct 8
York, Dick: Birth Anniv .... Sept 4
York, Michael: Birth .... Mar 27
Yorktown Day .... Oct 19
Yorktown Day (Yorktown, VA) .... Oct 19
Yosemite Chefs' Holidays (Yosemite Natl Park, CA) .... Jan 1
Yosemite Vintners' Holidays (Yosemite Natl Park, CA) .... Nov 6
Yothers, Tina: Birth .... May 5
You're All Done Day .... Dec 31
You're Welcomegiving Day .... Nov 24
Young, Andrew: Birth .... Mar 12
Young, Brigham: Birth Anniv .... June 1
Young, Coleman Alexander: Birth .... May 24
Young, Faron: Birth .... Feb 25
Young, Loretta: Birth .... Jan 6
Young, Neil: Birth .... Nov 12
Young, Paul: Birth .... Jan 17
Young, Robert: Birth .... Feb 22
Yount, Robin: Birth .... Sept 16
Yours, Mine and Ours Month, Natl ...... Jan 1
Youth Against Tobacco Month, Natl .... Oct 1
Youth Art Month .... Mar 1
Youth Day (Cameroon) .... Feb 11
Youth Day (People's Republic of China) . May 4
Youth Hostels, American: Christmas Trip (San Diego, CA) .... Dec 26
Youth of the Year, Natl (Washington, DC) . Sept 18
Youth, Operation (Cincinnati, OH) ...... June 10
Youth Service Day, Natl .... Apr 18
Youth Service Day, Natl 6674 .... Page 488
Youth Sing Praise Performance (Belleville, IL) .... June 24
Youth Sports Program Day, Natl 6576 . . Page 488
Yugoslavia
Civil War: Anniv .... June 25
Tito (Josip Broz): Birth Anniv .... May 25
Yukon Gold Panning (Dawson City, Yukon, Canada) .... July 1
Yule (Wiccan) .... Dec 22

Z

Z Day .... Jan 1
Zaharias, Mildred "Babe": Birth Anniv .... June 26
Zaire
Independence Day .... June 30
Regime Anniv .... Nov 24
Zaire Immigration 6574 .... Page 488
ZAM! Zoo and Aquarium Month .... June 1
Zambia
African Freedom Day .... May 25
Heroes Day .... July 3
Independence Day .... Oct 24
Labor Day .... May 1
Mutomboko Ceremony .... July 29
Unity Day .... July 4
Youth Day .... Aug 7
Zamenhof, Dr. L.L.: Birth Anniv .... Dec 15
Zamenhof: Publication of Esperanto: Anniv .... July 26
Zanuck, Darryl F.: Birth Anniv .... Sept 5
Zappa, Frank: Birth Anniv .... Dec 21
Zeffirelli, Franco: Birth .... Feb 12
Zenger, John P.: Arrest Anniv .... Nov 17
Zetkin, Clara: Birth Anniv .... July 5
Zimbabwe: Independence Day .... Apr 18
Zimbalist, Efrem, Jr: Birth .... Nov 30
Zimbalist, Stephanie: Birth .... Oct 6
Zimmer, Don: Birth .... Jan 17
Zindel, Paul: Birth .... May 15
Zionism, UN Revokes Resolution on: Anniv .... Dec 16
Zoeller, Frank Urban "Fuzzy": Birth ..... Nov 11
Zola, Emile: Birth Anniv .... Apr 2
Zolotow, Charlotte: Birth .... June 26
Zolotow, Maurice: Birth .... Nov 23
Zoo. See also Animals
Zoo, Boo at the (Asheboro, NC) .... Oct 28
Zoo Fling (Asheboro, NC) .... Apr 1
Zoo, US's First: Anniv (Philadelphia, PA) .. July 1
Zoobalee (Garden City, KS) .... July 4
Zoofest (Asheboro, NC) .... Oct 1
ZooLights (Tacoma, WA) .... Dec 1
Zucchini onto Your Neighbors' Porch Night, Sneak Some .... Aug 8
Zukerman, Pinchas: Birth .... July 16
Zwingli, Ulrich: Birth Anniv .... Jan 1

World War II (cont'd)—Zwingli

☆ Chase's 1995 Calendar of Events ☆

ANNIVERSARY GIFTS

1st	Paper, plastics, clocks		43rd	Trips
2nd	Calico, cotton, china		44th	Groceries
3rd	Leather, simulated leather, crystal, glass		45th	Sapphire
4th	Silk, synthetic material, electrical appliances, fruit, flowers		46th	Original poetry tributes
5th	Wood, silverware		47th	Books
6th	Iron, wood, candy		48th	Optical (spectacles, microscopes, telescopes)
7th	Copper, wool, desk sets		49th	Luxuries of any kind
8th	Electrical appliances, linen, lace, bronze, pottery		50th	Gold
9th	Pottery, willow, leather		55th	Emerald
10th	Tin, aluminum, diamond jewelry		60th	Diamond Jubilee
11th	Steel, fashion jewelry, accessories		80th	Diamond and pearl
12th	Linen, silk, pearls, colored gems		85th	Diamond and sapphire
13th	Lace, textiles, furs		90th	Diamond and emerald
14th	Ivory, gold jewelry		95th	Diamond and ruby
15th	Crystal, glass, watches		100th	10-carat diamond
16th	Silver hollowware			
17th	Furniture			
18th	Porcelain			
19th	Bronze			
20th	Platinum, china			
21st	Brass, nickel			
22nd	Copper			
23rd	Silver plate			
24th	Musical instruments			
25th	Silver			
26th	Original pictures			
27th	Sculpture			
28th	Orchids			
29th	New furniture			
30th	Pearl, diamond			
31st	Time pieces			
32nd	Conveyances (including automobiles)			
33rd	Amethyst			
34th	Opal			
35th	Coral, jade			
36th	Bone china			
37th	Alabaster			
38th	Beryl, tourmaline			
39th	Lace			
40th	Ruby			
41st	Land			
42nd	Improved real estate			

List compiled from those of Emily Post, Amy Vanderbilt, *The Hostess Book* and *Hallmark Date Book*.

Order additional copies of CHASE'S ANNUAL EVENTS 1995 with this form.

Ship to _____

Address _____

_____

City _____ State _____ Zip _____

Phone (___) _____

Please send me _____ copies of the 1995 edition of
CHASE'S ANNUAL EVENTS at $47.95 each $_____

Add applicable sales tax in AL, CA, FL, IL, NC, NY, OH, PA, TX $_____

Shipping & Handling: Add $3.75 for first copy,
$2.00 for each additional copy $_____

                                                   Total $_____

☐ Check or money order enclosed payable to: Contemporary Books

Charge my ☐ VISA ☐ MasterCard

Acct. # _____ Exp. Date ____/____

X _____
Signature (if charging to bankcard)

Name (please print) _____

STANDING ORDER AUTHORIZATION

To assure that I automatically receive each year's new edition of CHASE'S ANNUAL EVENTS as soon as it's available, please accept this Standing Order Authorization to ship me _____ copies of each annual edition, beginning with the 1996 edition, and to bill me at the address above. I understand I may cancel my Standing Order at any time.

X _____
         Signature   Date

_____
Name (please print)

GUARANTEE: Any book you order is unconditionally guaranteed and may be returned within 10 days of receipt for full refund.

BB95

mail to: **Contemporary Books, Dept. C**
Two Prudential Plaza, Suite 1200
Chicago, Illinois 60601-6790

HOW TO SUBMIT A NEW ENTRY

Submit your own suggestions for new entries for forthcoming editions of *Chase's s Calendar of Events* to the editors. Additional background information about the history and observance of each event will be appreciated for our reference files. Please be sure your dates are confirmed for 1996, as tentative dates cannot be included. Use a separate sheet for each event submitted. Information selected by the editors may be used and publicized through their books, syndicated services and/or other related products and services. The editors reserve the right to select and edit information received. There is no charge for being listed in *Chase's*. **Please submit all information to:**

Calendar Editor
Chase's Calendar of Events
Two Prudential Plaza, Suite 1200
Chicago, Illinois 60601-6790

☞ DEADLINE FOR 1996 EDITION: MAY 15, 1995

---

1. Exact name of event:
2. Exact INCLUSIVE DATES for 1996:
3. Location (site, city and state):
4. Brief description of event:

5. Formula (ONLY if used to set date each year):
   (Example: Annually, the first Friday in June.)
6. Name, title and address of person or agency to appear in your entry in *Chase's 1996*.

7. Phone number you would like to appear in *Chase's 1996*. _____
8. Sponsor (if any):
9. Print name and phone number of person whom we can call with questions regarding your entry:
   (for our records only) _____ ( ) _____
10. Signature of person furnishing above information: _____
11. Please print signature _____
12. Approximate number of people attending event. _____
13. **Please circle the exact inclusive dates for your 1996 event on the calendar below.**

Key dates in 1996:

M. L. King Birth	Jan 15
Chinese New Year	Feb 19
Presidents' Day	Feb 19
Ash Wednesday	Feb 21
Passover	Apr 4
Easter	Apr 7
Mother's Day	May 12
Memorial Day	May 27
Father's Day	June 16
Labor Day	Sept 2
Rosh Hashanah	Sept 14
Columbus Day	Oct 14
Thanksgiving	Nov 28
Chanukah	Dec 6

1996

JAN.
 S M T W T F S
        1 2 3 4 5 6
 7 8 9 10 11 12 13
14 15 16 17 18 19 20
21 22 23 24 25 26 27
28 29 30 31

FEB.
             1 2 3
 4 5 6 7 8 9 10
11 12 13 14 15 16 17
18 19 20 21 22 23 24
25 26 27 28 29

MAR.
                1 2
 3 4 5 6 7 8 9
10 11 12 13 14 15 16
17 18 19 20 21 22 23
24 25 26 27 28 29 30
31

APR.
    1 2 3 4 5 6
 7 8 9 10 11 12 13
14 15 16 17 18 19 20
21 22 23 24 25 26 27
28 29 30

MAY
          1 2 3 4
 5 6 7 8 9 10 11
12 13 14 15 16 17 18
19 20 21 22 23 24 25
26 27 28 29 30 31

JUNE
                   1
 2 3 4 5 6 7 8
 9 10 11 12 13 14 15
16 17 18 19 20 21 22
23 24 25 26 27 28 29
30

JULY
    1 2 3 4 5 6
 7 8 9 10 11 12 13
14 15 16 17 18 19 20
21 22 23 24 25 26 27
28 29 30 31

AUG.
             1 2 3
 4 5 6 7 8 9 10
11 12 13 14 15 16 17
18 19 20 21 22 23 24
25 26 27 28 29 30 31

SEPT.
 1 2 3 4 5 6 7
 8 9 10 11 12 13 14
15 16 17 18 19 20 21
22 23 24 25 26 27 28
29 30

OCT.
       1 2 3 4 5
 6 7 8 9 10 11 12
13 14 15 16 17 18 19
20 21 22 23 24 25 26
27 28 29 30 31

NOV.
                1 2
 3 4 5 6 7 8 9
10 11 12 13 14 15 16
17 18 19 20 21 22 23
24 25 26 27 28 29 30

DEC.
 1 2 3 4 5 6 7
 8 9 10 11 12 13 14
15 16 17 18 19 20 21
22 23 24 25 26 27 28
29 30 31

Note: This page may be photocopied in order to submit additional event entries to *Chase's 1996 Calendar of Events*.

Two Prudential Plaza, Suite 1200, Chicago, Illinois 60601-6790 • Phone (312) 540-4500 • Fax (312) 540-4687